BIOCHEMISTRY
The Chemical Reactions of Living Cells

ACADEMIC PRESS

New York   San Francisco   London

A SUBSIDIARY OF HARCOURT BRACE JOVANOVICH, PUBLISHERS

BIOCHEMISTRY

The Chemical Reactions of Living Cells

DAVID E. METZLER

Iowa State University

ACADEMIC PRESS, INC.
111 Fifth Avenue, New York, New York 10003

United Kingdom Edition published by
ACADEMIC PRESS, INC. (LONDON) LTD.
24/28 Oval Road, London NW1

Library of Congress Cataloging in Publication Data

Metzler, David E

    Biochemistry (TX)

    Includes bibliographical references and index.
      1.   Cytochemistry.   2.   Chemical reactions.   I.   Ti-
tle.   [DNLM: 1.   Cells.   2.   Biochemistry.   3.   Histo-
cytochemistry.   QH611 M596c]
QH611.M4   574.8'76   75–32031
ISBN 0–12–492550 2

Preface

Because the study of the chemical reactions of living cells is a vast and complex field that touches almost every branch of chemistry and biology, writing an introductory textbook in this field is a difficult task. I would never have undertaken such a project if my goal had been merely to try to improve on the textbooks currently available, but I am convinced that a new approach to biochemistry will be useful to a broad range of students and teachers. Therefore, rather than dividing biochemistry into segments centered around specific chemical compounds such as proteins, nucleic acids, lipids, and carbohydrates, I treat chemical reactions of cells as a primary theme. While stressing biological concerns, I try to trace all physiological phenomena back to the underlying chemistry.

For the student, the study of biochemistry in all of its complexities and nuances may seem overwhelming. I have tried in various ways to make this task easier for students with diverse backgrounds. For instance, the main body of the text focuses on a clear and uncluttered presentation of biochemical principles. Ancillary information on vitamins, metallic elements, poisons, research techniques, and metabolic diseases may be fascinating to some students but of lesser interest to others. Therefore, such information is set off from the main text in special boxes, where it will not interrupt the flow of the fundamental material. (These boxes are noted on the detailed table of contents, and are also listed under eight subject headings on p. xxix.) In addition to trying to provide a clear and highly readable introduction to the basic principles of biochemistry, I have also attempted to provide the student with a book that can be used in the future as a reference source. Literature citations of three types have been included: (1) those to standard reference books such as ''The Enzymes'' (edited by Paul Boyer; Academic Press, New York; a treatise in 13 volumes); (2) references to especially important papers, some old and some recent; and (3) current references that should help the student become acquainted with the research literature. Study questions and problems accompany each chapter.

In brief, I hope this book will provide a readable, student-oriented introduction to the chemical structures and reactions of living cells. The end result is not a radical departure from previous textbooks, but rather a distinctive perspective on the subject combined with some fresh approaches.

The first section of the book (Chapters 1–3) contains introductory material on cell structure, molecular architecture, and energetics. Chapter 1 provides quantitative information on sizes of cells and organelles and on the genetic complexity of organisms, and also describes characteristics of important species considered in later chapters. It is designed especially for students with a minimum of biological knowledge and will serve as a review for others. In this chapter, as well as throughout the book, the coverage includes bacteria, plants, and animals. In the second chapter a brief review of structural principles for molecules is followed by a systematic consideration of molecular structure and chemical properties of proteins, carbohydrates, nucleic acids, and lipids. A modern approach is followed whereby conformational properties as well as structures and reactivities are emphasized. A section on inorganic elements is followed by a review of methods of structure determination, including nmr techniques.

The third chapter, which starts with a review of thermodynamic equations, presents a practical approach to biochemical thermodynamics with a series of tabulated data. All of the free energy and enthalpy values cited later in the text are consistent with those in the tables. I chose to use the SI unit of kilojoules per mole throughout after I became convinced that this is the accepted norm in the international biochemical literature. The change from the more familiar kilocalories per mole is simple. An innovation is the use of free energies of combustion by NAD^+ which simplifies the arithmetic in evaluating free energy changes for metabolic processes.

Chapters 4 and 5 (the second major section of the book) deal with the ways in which biomolecules interact. From the quantitative treatment of binding to the structures of oligomeric enzymes, microtubules, viruses, and muscle, the treatment in Chapter 4 pursues a rigorous contemporary approach. Allosteric effects are considered quantitatively and systematically. Chapter 5 covers the structures and chemical properties of membranes and of the surrounding cell coats. A major aim of this and other chapters is to provide students with sufficient background so that they can begin to read the current research literature without immediately having to consult a variety of other background sources.

In the third major section of the book (Chapters 6–8) the general properties of enzymes, the kinetics of chemical reactions, and the various mechanisms employed in enzymatic catalysis are discussed. Chapter 6 presents enzyme kinetics in sufficient detail to satisfy the needs of most students and provides a description of the mechanisms by which enzymatic reactions are controlled within cells. In Chapter 7 a systematic and rational classification of enzymatic reactions is given, including information about a variety of enzymes and the techniques used to study them. Chapter 8 treats the chemistry and special functions of coenzymes, relating these functions to reaction types considered in preceding chapters. These chapters contain much reference material, and do not have to be read completely in sequence. However, it will be easy for the student or instructor to discern the principles emphasized and to apply them. Materials from these chapters may, if desired, be incorporated with material from Chapter 2 into units dealing with proteins, carbohydrates, nucleic acids, and lipids without difficulty.

The fourth major section (Chapters 9–14) of the book treats reaction sequences found in metabolism. Chapter 9 explores the "logic" of metabolic cycles and other pathways. Patterns are shown to arise as a natural result of the kind of chemistry needed to obtain a desired product from a given reactant. I hope that this presentation will be a distinct improvement over the conventional ones. Chapter 10 deals with electron transport and oxidative phosphorylation as well as with the energy-yielding reactions of chemiautotrophs. Many otherwise confusing aspects of metabolism are shown to be simple when it is understood (Chapter 11) that certain steps are needed to couple cleavage of ATP to biosynthesis. In the specialized chapter on carbohydrate and lipid metabolism (Chapter 12) most of the sequences are easy to understand in the light of preceding discussions.

A unique chapter, Light in Biology (Chapter 13), covers not only photosynthesis, vision, and other biological responses to light, but light absorption spectra, circular dichroism, and fluorescence. Chapter 14 provides details of the biosynthesis and catabolism of an enormous number of nitrogenous compounds. It can serve as a starting point for individual literature research projects by biochemistry students and could profitably be used in courses on natural products chemistry.

The last section (Chapters 15 and 16) deals with topics of genetic and hormonal control of metabolism, development, and brain function. Chapter 15 not only covers biosynthesis of nucleic acids and proteins but also includes a succinct summary of methods used in the study of biochemical genetics. Although this material has been placed near the end of the book, it is entirely possible to present part or all of it to a class at an earlier stage of the course. The final chapter (16) provides a brief introduction to the problems of cell communication, neurochemistry, differentiation, and ecological problems (metabolism in the environment).

I would like to acknowledge with thanks the many people who have helped me in the preparation of this text: among them my colleagues at Iowa State University, especially John Foss and Bernard White, and many scientists in other institutions. Special thanks are due the John Simon Guggenheim Foundation for a fellowship, Iowa State University for a faculty improvement leave, and the staff and students of the Department of Biochemistry of the University of California, Berkeley, for many helpful discussions and criticisms. Special thanks go to Peggy Johnston, Jeanne Peters, and Wilma Holdren for excellent and dedicated help in the preparation of the manuscript; to Carol Harris for computations, supervising the proofreading, and for checking every numerical value; and to the graduate students who checked every equation and reference. The members of the staff of Academic Press have my gratitude for their patience and encouragement. It has been a pleasure to work with them.

David E. Metzler

Acknowledgments

I wish to express my appreciation to the following reviewers, each of whom read and criticized part or all of the manuscript, for their generous assistance: Michael J. Chamberlin, University of California, Berkeley; Eric E. Conn, University of California, Davis; Charles H. Doering, Stanford University; Harrison Echols, University of California, Berkeley; Lloyd L. Ingraham, University of California, Davis; Martin D. Kamen, University of Southern California; Edward A. Khairallah, University of Connecticut; Daniel E. Koshland, Jr., University of California, Berkeley; Christopher K. Mathews, University of Arizona; Donald B. McCormick, Cornell University; J. B. Neilands, University of California, Berkeley; Lester Packer, University of California, Berkeley; Daniel L. Purich, University of California, Santa Barbara; P. K. Stumpf, University of California, Davis; R. G. Wolfe, University of Oregon; and W. A. Wood, Michigan State University.

Brief Table of Contents

Contents

2

The Molecules from Which We Are Made, 47

Molecules **5.** Detecting Products **6.** Determination of Molecular Weight **7.** Conformations of Macromolecules

References, 144

Study Questions, 148

3

Energetics of Biochemical Reactions, 151

4

How Molecules Stick Together, 182

5

Membranes and Cell Coats, 252

6

Enzymes: The Protein Catalysts of Cells, 301

7

The Kinds of Reactions Catalyzed by Enzymes, 353

8

Coenzymes—Nature's Special Reagents, 428

9

Organization of Metabolism: Catabolic Pathways, 517

10

How Oxygen Meets the Electrons with Generation of ATP, and Other Stories, 559

11

Biosynthesis: How New Molecules Are Put Together, 630

12

Some Specific Pathways of Metabolism of Carbohydrates and Lipids, 680

13

Light in Biology, 744

14

The Metabolism of Nitrogen-Containing Compounds, 805

15

Biochemical Genetics and the Synthesis of Nucleic Acids and Proteins, 891

16

Growth, Differentiation, and Chemical Communication between Cells, 1000

A Note about
the Boxed Information

Boxes containing ancillary information on various subjects are included throughout the text. These boxes fall under eight subject headings. These groups themselves can be combined in order to develop various themes. Thus, a *nutritional theme* is provided by the boxes on *Vitamins* together with those on *Essential Elements*. A complete list of boxes, in eight groups, follows:

1. One series describes the properties of *The Vitamins* and the history of their discovery. The corresponding coenzymatic functions are discussed in the main text, as is the biosynthesis of several of the vitamins.

2. Another series describes functions of most of the *Essential Elements*. In addition, such topics as complex formation, ion transport, cobalt in Vitamin B_{12} and others are dealt with in the main text.

Some of the other sections dealing with metal ions are:

3. The importance of mutations is introduced in Chapter 1 and many examples of known *Metabolic Defects* in humans are considered, both in the main text and in the following boxes:

Some related sections in the main text are:

4. Structures and properties of many of the best known *Antibiotics* are given, often in boxes:

Material on antibiotics in the main text includes:

Chapter 5, Section B,2

Chapter 12, Sections G; I,4

Chapter 15, Section C,2

5. The actions of a variety of *Poisons* including their uses in biochemical investigations are described:

See also Chapter 12, Section I,4

6. Several boxes deal with *Methods* and *Measurements:*

7. Topics in *Physiological Chemistry* characterize one set of boxes. Material on related topics is found throughout the text:

8. *Other Topics* include the following:

The following journals are cited with nonstandard abbreviations: ABB, *Arch. Biochem. Biophys.*; BBA, *Biochim. Biophys. Acta*; BJ, *Biochem. J.*; BBRC, *Biochem. Biophys. Res. Commun.*; EJB, *Eur. J. Biochem.*; JACS, *J. Amer. Chem. Soc.*; JBC, *J. Biol. Chem.*; JMB, *J. Mol. Biol.*; PNAS, *Proc. Nat. Acad. Sci. U.S.*

BIOCHEMISTRY
The Chemical Reactions of Living Cells

1

The Scene of Action

This book is about the unceasing, complex series of chemical reactions by which cells grow and reproduce, feed and excrete wastes, move and communicate with each other. Thousands of reactions, each one catalyzed by its own specific enzyme, are linked together in branching and interconnecting sequences to form a network of great complexity. The whole of these reactions, **metabolism,** is a main theme of this book. Equally important though are the **structures** of the remarkable molecules of which cells are composed. Proteins, nucleic acids, carbohydrates, lipids, and coenzymes are all needed for life. All have precise structures that fit them for their unique roles in living cells. All are continually being created and destroyed and at the same time are exerting control over each other's reactions in surprising ways.

Metabolism includes both the synthesis and degradation of the many chemical constituents of cells. In animals, foods are broken down to provide not only energy but also chemical intermediates for synthesis of molecules necessary for growth. Similarly, every cell of every organism either makes or gathers from its surroundings small molecules and combines these chemical building blocks to form larger molecules. At the same time, cells contain enzymes for tearing down

all of the molecules that have been made. Thus, we have a **steady state** in which complex substances are continuously being formed in one series of reactions and degraded in another. The result is a marvelous system of self-renewal of our tissues.

Human beings are not alone in the world but are surrounded by a multitude of other living things whose metabolism is vitally important to us. Photosynthetic organisms obtain energy from sunlight and synthesize compounds that the human body requires but cannot make. Microorganisms, using a variety of reactions to obtain their energy, produce decay and a return of organic matter to forms usable by plants. This book deals with chemical reactions occurring in the widest diversity of different organisms. Some strange and unusual reactions will be considered, along with those metabolic sequences common to most living things.

Biologists have described over a million different species of living things, many with very specialized ways of life. A surprising fact is that all of these organisms have a great deal of chemistry in common. Thus, formation of lactic acid both in bacteria and in human muscle requires the same enzymes. The same 20 amino acids can be isolated from proteins of plants, animals, and microorganisms. The "genetic code" of the DNA (**deoxyribonucleic acid**) molecules appears to be universal—the same for all organisms. We see that there is a unity of life and that we can study metabolism as the whole of the chemical transformations going on in all living things.

However, no matter how impressive the similarities among organisms, the differences are just as striking. As the sizes and shapes of living things vary, so do many aspects of their metabolism. These differences all reflect differences in the **genes,** in the coded information carried in the molecules of DNA. Such genetic differences lead to differences in the structures of the **protein molecules** made by cells. Among these proteins are the **enzymes,** *complex little machines, each one designed to catalyze a single specific chemical reaction.*

Consider a chemical reaction that takes place in virtually every living cell (there are many such reactions). If the enzyme that catalyzes the reaction is isolated from tissues of a number of different organisms it will usually be found to be similar in general properties and in the mechanism of catalysis. Nevertheless, in most cases the *exact* makeup of amino acids will be peculiar to the organism that produced it. Often such species differences will affect only the external shape of an enzyme molecule, the catalytic machinery being essentially the same. However, in other cases the structure of the **active site** of the enzyme may vary enough between two species to cause significant alterations in metabolism. Differences in metabolism can lead to differences in form and behavior. It is possible that the distinction between a cow and a horse is to be found in the summation of many subtle variations in the structures of enzymes and other proteins.

Variations in protein structures are not limited to differences between species. Individuals differ from one another. Serious genetic diseases such as **sickle cell anemia** sometimes result from the replacement of a single amino acid unit in a protein by a different amino acid.

Genetic deviations from the "normal" structure of a protein result from **mutations.** Most mutations, whether they occurred initially in our own cells or in those of our ancestors, are detrimental. However, such mutations also account for variation among individuals of a species and provide a basic driving force for evolution. For these reasons we will be much concerned in this book with the chemical nature and consequences of mutations.

A. THE SIMPLEST LIVING THINGS

The simplest organisms are the **bacteria** and **blue-green algae** which make up the kingdom Prokaryotae (Monera).[1,2] Their cells are called **prokaryotic** because no membrane-enclosed nucleus is present. Cells of all other organisms contain nuclei separated from the cytoplasm by membranes. They are called **eukaryotic.** While some biologists classify viruses as organisms, these amazing objects (Box 4-C) are less than alive and are usually devoid of any metabolism of their own.

1. Mycoplasmas

The simplest of the prokaryotes and simplest of all known living things are the **mycoplasmas.** These little bacteria do not have the rigid cell wall characteristic of most bacteria. For this reason they are easily deformed and often pass through filters designed to stop bacteria. They are nutritionally fussy and are usually, if not always, parasitic. Some live harmlessly in mucous membranes of people, but others cause diseases such as primary atypical pneumonia (caused by *Mycoplasma pneumoniae*).

The mycoplasma cell may be as small as a sphere 0.33 μm (0.00033 mm) in diameter.[3] It is bounded by a thin **cell membrane** about 8 nm (80 Å) thick. The membrane encloses the **cytoplasm,** a fluid material containing many dissolved substances as well as submicroscopic particles. At the center is a single, highly folded molecule of deoxyribonucleic acid (DNA), which constitutes the bacterial **chromosome.** Together with surrounding material it

BOX 1-A

ABOUT MEASUREMENTS

In 1960 the International General Conference on Weights and Measures adopted an improved form of the metric system, **The International System of Units** (SI). The units of mass, length, and time are the kilogram (kg), meter (m), and second (s). The following prefixes are used for fractions and multiples:

10^{-3}, *milli* (m) 10^{-6}, *micro* (μ) 10^{-9}, *nano* (n)
10^{-12}, *pico* (p) 10^{-15}, *femto* (f) 10^{-18}, *atto* (a)
10^{3}, *kilo* (k) 10^{6}, *mega* (M) 10^{9}, *giga* (G)
10^{12}, *tera* (T)

There is an inconsistency in that the prefixes are applied to the gram (g) rather than to the basic unit, the kilogram.

An attempt has been made to use SI units throughout the book whenever possible. There will be no feet, microns, miles, or tons. Molecular dimensions are given uniformly in nanometers rather than in angstrom units (Å). Remember that 1 Å = 0.1 nm. Likewise the calorie and kilocalorie have been replaced by the SI unit of energy, the **joule** (J).

Two changes, resulting from adoption of the SI, have already been made by most journals: **micrometer** (μm) has replaced the older term **micron** (μ) and **nanometer** (nm) has replaced **millimicron** (mμ). Do not confuse μm (new) with mμ (old).

Throughout the book frequent use is made of the following symbols:

$\approx$, "approximately equal to"
$\sim$, "approximately" or "about"

BOX 1-B

MOLECULAR WEIGHTS AND DALTONS

Atomic and molecular masses are all referred to the mass of the carbon isotope, ^{12}C, whose atomic weight is defined as exactly 12. The actual mass of a single atom of ^{12}C is 12 **daltons**, one dalton being 1.661×10^{-24} g. The masses of molecules can be given in daltons and are numerically equivalent to **molecular weights**. However, molecular weights (designated MW in this book) refer to molar masses (gram per mole), and it is not correct to use the dalton as a unit of molecular weight. The dalton is most useful in referring to structures such as chromosomes, ribosomes, mitochondria, viruses, and whole cells where the term molecular weight would be inappropriate.[a]

[a] J. T. Edsall, *Nature* (*London*) **228**, 888 (1970).

teins, but there is room for a total of only about 50,000 protein molecules. **Ribonucleic acids** (RNA) of several types and many smaller molecules are also present.

What is the minimum quantity of proteins, DNA, and other chemical machinery needed to make a living cell? It is not yet known, but it is clear that it must all fit into the tiny cell of the mycoplasma.[3]

2. *Escherichia coli*

The biochemist's best friend is *Escherichia coli*, an ordinarily harmless inhabitant of the intestinal tract of men and beasts. This bacterium is easy to grow in the laboratory and has become "the best understood organism at the molecular level."[4] It may also be regarded as a *typical* true bacterium. The cell of *E. coli* is a rod ~ 2 μm long and 0.8 μm in diameter with a volume of ~ 1 μm^3 and a density of ~ 1.1 g/cm^3. The mass is $\sim 1 \times 10^{-12}$ g (1 pg or $\sim 0.7 \times 10^{12}$ daltons).[5] It is about 100 times bigger than the smallest mycoplasma. The structure, as revealed by the electron microscope, is

may be called a **nucleus*** or nucleoid. Besides the DNA there are about 400 roughly spherical particles ~ 20 nm in diameter, the **ribosomes.**† These are the centers of protein synthesis. Included in the cytoplasm are many different kinds of pro-

---

* Many biologists restrict the use of the word *nucleus* to eukaryotic cells.

† The number is a rough estimate made by comparing sizes and RNA contents of mycoplasmas and *E. coli*.

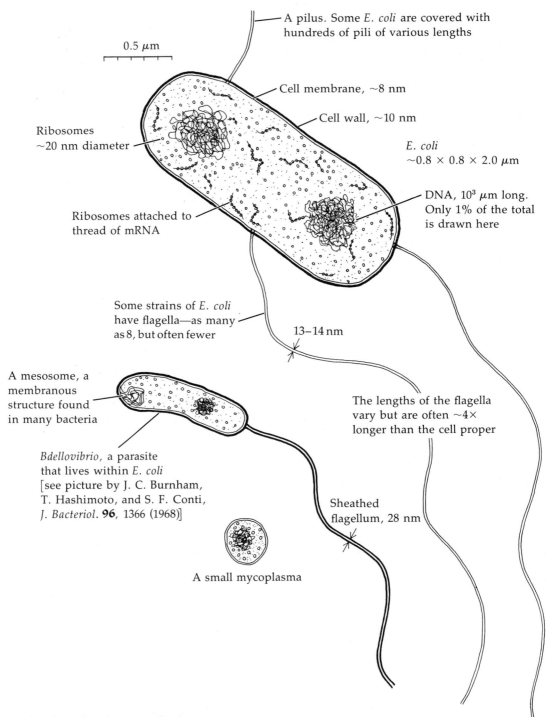

0.5 µm

A pilus. Some *E. coli* are covered with hundreds of pili of various lengths

Cell membrane, ~8 nm

Cell wall, ~10 nm

Ribosomes ~20 nm diameter

E. coli ~0.8 × 0.8 × 2.0 µm

DNA, 10^3 µm long. Only 1% of the total is drawn here

Ribosomes attached to thread of mRNA

Some strains of *E. coli* have flagella—as many as 8, but often fewer

13–14 nm

A mesosome, a membranous structure found in many bacteria

The lengths of the flagella vary but are often ~4× longer than the cell proper

Bdellovibrio, a parasite that lives within *E. coli* [see picture by J. C. Burnham, T. Hashimoto, and S. F. Conti, *J. Bacteriol.* **96**, 1366 (1968)]

Sheathed flagellum, 28 nm

A small mycoplasma

Fig. 1-1 *Escherichia coli* and some smaller bacteria.

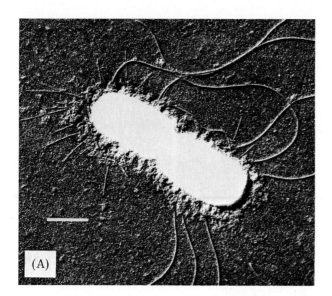

Fig. 1-2 Electron micrographs of bacteria. (A) An intact cell of *E. coli* shadowed with platinum and showing flagella (curved) and pili (straight). Scale, 0.5 μm. Note that these are not F pili (sex pili). Courtesy of F. D. Williams, Gail E. VanderMolen, and C. F. Amstein.

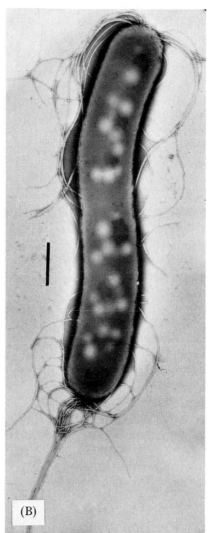

Fig. 1-2 (B) A cell of a *Spirillum* negatively stained with phosphotungstic acid. Note the tufts of flagella at the ends, the rough appearance of the outer surface, the dark inclusions of poly-β-hydroxybutyric acid and light-colored granules of unknown nature. Scale, 1 μm. Courtesy of F. D. Williams, Gail E. VanderMolen, and C. F. Amstein.

simple and internally closely resembles that of the mycoplasma (Figs. 1-1 and 1-2A).

Each cell of *E. coli* contains from one to four identical DNA molecules, the number depending on how fast the cell is growing, as well as ~ 15,000–30,000 ribosomes. Other particles are sometimes seen within bacteria. They include food stores, e.g., fat droplets and granules (Fig. 1-2B) of poly-β-hydroxybutyric acid (accounting for up to 25% of the weight of *Bacillus megaterium*) or, occasionally, glycogen. Granules of highly polymerized phosphoric acid (polymetaphosphate) known as metachromatic granules are common. **Vacuoles,** droplets of a separate aqueous phase, may also be present.

3. The Genome

The genetic instructions for a cell are found in the DNA molecules. Both DNA and proteins are chain molecules, the former made up of **nucleo-**

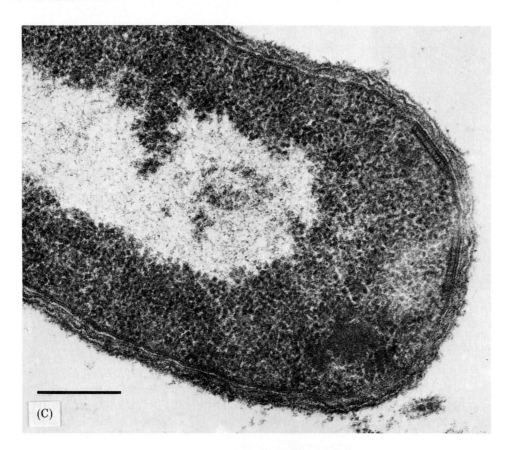

(C)

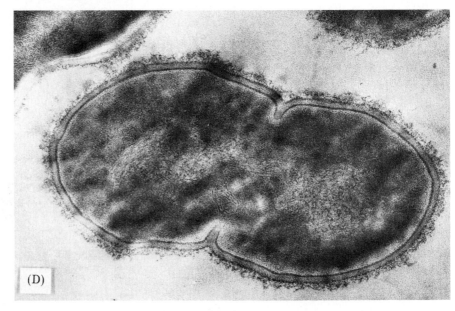

(D)

Fig. 1-2 (C) Thin (~60 nm) sections of an aquatic gram-negative bacterium, *Aquaspirillum fasciculus*. Note the light-colored DNA, the dark ribosomes, the double membrane characteristic of gram-negative bacteria (Chapter 5, Section D,1) and the cell wall. In addition, an internal "polar membrane" is seen at the end. It may be involved in some way in the action of the flagella. Scale, 0.25 μm. (D) A thin section of a dividing cell of *Streptococcus*, a gram-positive organism. Note the DNA (light-stranded material). A portion of a mesosome is seen in the center and the septum can be seen forming between the cells. Micrographs courtesy of F. D. Williams, Gail E. Vander-Molen, and C. F. Amstein.

tides and the latter of **amino acids.** The molecule of DNA is "double-stranded," i.e., there are actually two chains coiled together and consisting of **nucleotide pairs** (see Fig. 2-21). It is now firmly established that much of the genetic message in DNA is a series of code "words" or **codons.** Each codon is composed of three nucleotides (three nucleotide pairs in double-stranded DNA) and stands for one of the 20 different kinds of amino acids that make up proteins. The sequence of codons in the DNA tells a cell how to order the amino acids for construction of its many different proteins.

Assume that an "average" protein molecule consists of a folded chain of ~ 300 amino acids. The part of the DNA molecule (*one gene*) that contains the code for this protein will therefore be a sequence of about 900 nucleotide pairs. Allowing a few more nucleotides to form spacer regions between genes *we can take 1000 as the number of nucleotide pairs in an average gene.*

The **genome** is the quantity of DNA that carries a complete set of genetic instructions for an organism. In bacteria, the genome is a single chromosome consisting of one double-stranded DNA molecule. The DNA of *Mycoplasma arthritidis* has a mass of 0.5×10^9 daltons (molecular weight, MW = 0.5×10^9). The DNA of *E. coli* is about five times larger (MW = 2.5×10^9).

The average molecular weight of a nucleotide pair (as the disodium salt) is 664. It follows that the DNA of *M. arthritidis* contains $0.5 \times 10^9/664 = 8 \times 10^5$ nucleotide pairs, enough to code for 800 different average-sized proteins. The DNA of *E. coli* contains 3.8×10^6 nucleotide pairs, enough to code for ~ 4000 different proteins. Since some of the DNA is needed for other purposes, the number of genes specifying proteins is probably somewhat less.

The lengths of bacterial chromosomes are remarkable. Each nucleotide pair contributes 0.34 nm to the length of the molecule; thus, the total length of the DNA of an *E. coli* chromosome is 1.3 mm, more than 600 times that of the cell which contains it. Clearly, the DNA molecules are highly folded, and they appear in the electron microscope as dense "nuclei," occupying only about one-fifth of the cell volume (Figs. 1-2C,D). Each bacterial nucleus contains a complete set of genetic "blueprints" and functions more or less independently. Each nucleus is **haploid,** meaning that it contains only *one* complete set of genes.*

4. Ribonucleic Acids (RNA) and the Transcription and Translation of Genetic Information

The genetic messages in the DNA are not translated directly by the protein-synthesizing machinery of cells. Instead, molecules of ribonucleic acid (RNA) are synthesized according to the instructions encoded in the DNA, a process called **transcription.** It has been established that RNA molecules carry exactly the same coded information as is found in one or more of the segments (genes) of DNA. Thus, if DNA is the "master blueprint" of the cell, molecules of RNA are "secondary blueprints." This concept is embodied in the name **messenger RNA** (mRNA) which is applied to a small, rapidly metabolizing fraction of RNA that carries information specifying amino acid sequences of proteins. Messenger RNA carries the genetic messages from the genes to the ribosomes where the proteins are made.

Ribosomes are extraordinarily complex little protein-synthesizing machines. Each ribosome of *E. coli* has a mass of 2.7×10^6 daltons and contains $\sim 65\%$ of a special **ribosomal RNA** and $\sim 35\%$ protein. About 50 different kinds of protein molecules are present as parts of the ribosomal structure. Ribosomes are able to read the genetic messages from mRNA and to accurately assemble any kind of protein molecule that a gene may specify (**translation** of the genetic message).

---

* Haploid nuclei are usual in prokaryotes, but sexual conjugation among bacteria (as well as certain laboratory manipulations) can produce "partially diploid" organisms containing two sets of some of the genes.

BOX 1-C

**THE ELECTRON MICROSCOPE,
THIN SECTIONS AND REPLICAS**

Since it first became commercially available in 1939, the electron microscope has gradually become one of the most important tools of the cell biologist. With a practical resolution of about 0.4 nm, the electron microscope allows us to "see" protein and nucleic acid molecules as well as fine detail in organelles of cells. A major breakthrough in use of the electron microscope came in 1950 with the development of microtomes and knives capable of cutting thin (20–200 nm) sections of tissues embedded in plastic.

If a slice of fresh (frozen) tissue is examined directly, little is seen because most of the atoms found in cells are of low atomic weight and scatter electrons weakly and uniformly. Hence, thin sections must be "stained" with atoms of high atomic weight, e.g., by treatment with potassium permanganate. Tissues must also be "fixed" to prevent disruption of cell structures during the process of removal of water and embedding in plastic. Fixatives such as formaldehyde react with amino groups and other groups of proteins and nucleic acids. Some proteins are precipitated in place and digestive enzymes that otherwise would destroy much of the fine structure of the cell are inactivated. Glutaraldehyde (a 5-carbon dialdehyde) is a widely used and excellent fixative which cross-links protein molecules in the tissue. Osmium tetroxide is frequently used both as the fixing agent and as the staining material (Chapter 5, Section A,1). A continuing worry is that some structures seen in sections of fixed tissues may be "artifacts" of the fixing and staining procedures.

A bacterium such as *E. coli* can be sliced into as many as 10 thin longitudinal slices (see Fig. 1-2) and a eukaryotic cell of 10 μm diameter into as many as 100 slices. Sharply focused images can only be obtained with such thin sections. Serial sections can be examined to determine three-dimensional structures.[a]

Small particles, including macromolecules, may be "shadowed." A metal such as chromium or platinum is evaporated in a vacuum from an angle onto the surface of the specimen. Individual DNA molecules can be "seen" in this way, but the resolution is relatively low. Only the "shadows" are seen and they are 2–3 times wider than the DNA molecules proper. The **negative contrast**

5. Membranes and Cell Walls

Like the mycoplasma, the *E. coli* cell is bounded by a thin (8 nm) membrane which consists of ~50% protein and ~50% lipid.* When "stained" (e.g., with permanganate) for electron microscopy this membrane appears as two very thin (~2.0 nm) dark lines separated by an unstained center band (~3.5 nm) (Fig. 1-2C). *Single membranes of approximately the same thickness and staining behavior occur in all cells,* both of bacteria and of eukaryotes.

A cell membrane is much more than just a sack.

---

* Membranes of true bacteria contain no cholesterol, but oddly enough those of most mycoplasmas, like our own cell membranes, contain this sterol. Members of the genus *Mycoplasma* are parasitic and the cholesterol, essential to the mycoplasmas, is provided by the host organism.

It serves to control the passage of small molecules into and out of the cell. The oxidative metabolism of bacterial cells is catalyzed by enzymes found on the inside surface of the membrane. Bacterial cell membranes are often folded in to form **mesosomes,** which appear in sections as multilayered structures (Figs. 1-1 and 1-2D). Mesosomes are thought to be centers of specialized metabolism and of DNA replication. While mesosomes are not always seen in *E. coli* the replication of DNA in this organism seems to occur on certain parts of the membrane surface, probably under the control of membrane-bound enzymes. The formation of the new membrane, which divides multiplying cells, proceeds synchronously with the synthesis of DNA.

A characteristic of true bacteria is a rigid **cell wall** which surrounds the cell membrane. In

Box 1-C (*Continued*)

method is more often used for visualizing proteins. A thin layer of a solution containing the molecules to be examined together with an electron-dense material such as 1% sodium phosphotungstate is spread on a thin carbon support film. Upon drying, a uniform electron-dense layer is formed. Where the protein molecules lie, the phosphotungstate is excluded; hence, the term **negative contrast**.

Surfaces of cells, slices or intact bacteria can be coated with a deposit of platinum or carbon. The coating, when removed, provides a "negative" **replica** which can be examined in the microscope. Alternatively, a thin plastic replica can be made and can be shadowed to reveal topography. Among the most important replica methods are "freeze fracture" and "freeze etching." Fresh tissue (which may contain glycerol to prevent formation of large ice crystals) is frozen rapidly. Such frozen cells can often be revived; hence, may be regarded as still alive. The frozen tissue is placed in a vacuum chamber within which it is sliced or fractured with a cold knife. If desired, the sample can be kept in the vacuum chamber at about $-100°C$ for a short time during which some water molecules evaporate from the surface. The

resultant etching reveals fine structure of cell organelles and membranes in sharp relief. After etching, a suitable replica is made and examined (Fig. 1-11). Recent evidence suggests that fracturing tends to take place through lipid portions of cell membranes.

Every biochemist should examine some of the high quality electron micrographs that appear in books on the subject[b–g] and in such journals as the *Journal of Cell Biology*, *Journal of Ultrastructure Research*, and *Journal of Bacteriology*.

[a] For a recently published example, see H. P. Hoffmann and C. J. Avers, *Science* **181**, 749–751 (1973).

[b] M. C. Ledbetter and K. R. Porter, "Introduction to the Fine Structure of Plant Cells." Springer-Verlag, Berlin and New York, 1970.

[c] D. W. Fawcett, "An Atlas of Fine Structure, the Cell, Its Organelles, and Inclusions." Saunders, Philadelphia, Pennsylvania, 1966.

[d] G. H. Haggis, "The Electron Microscope in Molecular Biology." Wiley (Interscience), New York, 1967.

[e] K. R. Porter and M. A. Bonneville, "The Fine Structure of Cells and Tissues," 3rd ed. Lea & Febiger, Philadelphia, Pennsylvania, 1968.

[f] S. Wischnitzer, "Introduction to Electron Microscopy," 2nd ed. Pergamon, New York, 1970.

[g] M. A. Hayat, "Basic Electron Microscopy Techniques," 2 vols. Van Nostrand-Reinhold, Princeton, New Jersey, 1972.

E. coli, the wall is ~ 40 nm thick, a complex, layered structure whose chemical makeup is considered in Chapter 5. In some bacteria the wall may be as much as 80 nm thick, and it may be further surrounded by a thick capsule or slime layer. The main function of the wall seems to be to prevent osmotic swelling and bursting of the bacterial cell when the surrounding medium is a very dilute solution.

If the osmotic pressure of the medium is not too low, bacterial cell walls can sometimes be dissolved leaving living cells bounded only by their membranes. Such **protoplasts** can be produced by action of the enzyme lysozyme on gram-positive bacteria such as *Bacillus megaterium*. One reason that protoplasts are useful in biochemical studies is that many substances enter cells more readily when the cell wall is absent. The term **spheroplast**

is usually used for cells of gram-negative bacteria in which the wall has been incompletely disrupted. Penicillin (which interferes with synthesis of cell walls in many bacteria) can convert *E. coli* and other gram-negative bacteria into spheroplasts. Strains of bacteria lacking rigid walls are known as **L forms.**

6. Flagella and Pili

Many bacteria, as anyone with a good microscope can verify, swim. Very thin threadlike flagella of diameter 10–20 nm provide the motion. Sometimes there is a single flagellum, while in other cases pairs or groups of flagella, at one or both ends of the bacterium, move synchronously. The corkscrewlike *Spirillum* (Fig. 1-2) moves tufts of flagella at each end in such a way that they appear to rotate at 50 revolutions/s while the body of the cell rotates more slowly in the other direction. The bacterium moves ahead at a speed of up to 50 μm/s.[6] Some strains of *E. coli* have no flagella, but others contain up to ~8 per cell distributed more or less randomly over the surface. The flagella tend to stream out behind in a bundle when the bacterium swims.

Other appendages on bacteria were not even suspected until bacteria were studied with electron microscopes. These are the extremely thin, long filaments known as **pili** (or fimbriae), which range in thickness from 3 to 25 nm (Fig. 1-2A). Their functions are largely unknown, but cells with many pili tend to stick together readily. Some pili ("F pili") in *E. coli* and related organisms appear to have a specific role in sexual conjugation.

7. The Rapid Metabolism of Bacteria

One of the most impressive things about bacteria is their incredibly rapid rate of metabolic activity and growth. Under some conditions, it takes a bacterial cell only 20 min to double its size and to divide to form two cells. An animal cell may take 24 h for the same process. Equally impressive are the rates at which bacteria transform their foods into other materials. The high rate of bacterial metabolism has often been attributed to the large surface to volume ratio (see also Chapter 3, Section A,5). For a small spherical bacterium (coccus) of diameter 0.5 μm, the ratio of the surface area to the volume is 12×10^6 m^{-1}, while for an ameba of diameter 150 μm the ratio is only 4×10^4 m^{-1} (the ameba can increase this quite a bit by sticking out some pseudopods). Thimann estimated[7] that for a 90 kg man, the ratio is only 30 m^{-1}.

8. Spores

Under some conditions, certain bacteria (e.g., *Bacillus subtilis*) are able to form **spores** (endo-spores). These are compact little cells that form inside the vegetative cell, sometimes having only $\frac{1}{10}$ the volume of the latter. Their water content is very low, their metabolic rate is near zero, and they are extremely resistant to heat and desiccation. Under suitable conditions, the spores can "germinate" and renew their vegetative growth. Studies of bacterial sporulation have become popular because they may provide a basic understanding of the ways in which cells can change a developmental pattern in response to an alteration in the environment.

9. Classification of Bacteria

Bacteria vary enormously in their chemistry and metabolism, and it has been difficult to classify them in a rational way. In higher organisms species are often defined as forms that cannot interbreed. Such a criterion is meaningless for bacteria, and the classification into species and genera must therefore be somewhat arbitrary. A currently used scheme (Table 1-1) divides the kingdom of the prokaryotes into 19 parts. The grouping is based on many characteristics including shape, staining behavior, and chemical activities. The table includes genus names of all the bacteria mentioned throughout the text.

Bacteria may have the shape of spheres or straight or curved rods. Some (e.g., the **actinomycetes**) grow in a branching filamentous form. Words used to describe bacteria often refer to these shapes: a **coccus** is a sphere, a **bacillus** a rod, and a **vibrio** a curved rod with a flagellum at one end. A **spirillum** is screw-shaped with a tuft of flagella at one or both poles. The whole matter is confusing because these same words are frequently used to name particular genera or families. Many other names are derived from some chemical activity of the bacterium being described.

10. Nutrition of Bacteria

Autotrophic (self-nourishing) bacteria can synthesize all of their organic cell constituents from

TABLE 1-1
Current Classification Scheme for Bacteria[a,b]

Kingdom Procaryotae
    Division I: The Cyanobacteria (botanical name, Cyanophyta; blue-green algae)
    Division II: The bacteria

The bacteria are further divided into the following 19 parts. Within these some genera are grouped into families and orders while others are left with "uncertain affiliation." A few genera, most of which are mentioned elsewhere in this book, are listed by name. Members of a single family are placed together and are separated by semicolons from members of other families.

1. Photosynthetic (phototrophic) bacteria
Rhodospirillum, Rhodopseudomonas (purple nonsulfur bacteria); *Chromatium, Thiospirillum* (purple sulfur bacteria); *Chlorobium* (green sulfur bacteria)
2. The gliding bacteria
Beggiatoa (a filamentous bacterium containing sulfur granules)
3. Sheathed bacteria
4. Budding and appendaged bacteria
5. Spirochetes (long bacteria, up to 500 μm, that are apparently propelled by the action of filaments wrapped around the cell between the membrane and wall)
Treponema (*T. palladum*, syphilis), *Leptospira*
6. Spiral and curved bacteria
Spirillum, Bdellovibrio
7. Gram-negative, aerobic rods and cocci
Pseudomonas, Gluconobacter; Azotobacter; Rhizobium; Acetobacter; Brucella (*B. abortus*, brucellosis), *Halobacterium*
8. Gram-negative, facultatively anaerobic rods
Escherichia, Salmonella (*S. typhi*, typhoid fever), *Shigella* (*S. dysenteriae*, bacterial dysentery), *Klebsiella, Serratia; Proteus, Yersinia* (*Y. pestis*, plague) *Enterobacter*[c], *Vibrio* (*V. cholerae*, Asiatic cholera), *Zymomonas; Flavobacterium, Haemophilus*
9. Gram-negative, anaerobic bacteria
Desulfovibrio, Butyrivibrio
10. Gram-negative, cocci and coccobacilli (aerobes)
Neisseria (*N. gonorrhea*, gonorrhea)
11. Gram-negative cocci (anaerobes)
Veillonella
12. Gram-negative chemilithotrophic bacteria
Nitrobacter, Nitrosomonas; Thiobacillus
13. Methane-producing bacteria
Methanobacterium
14. Gram-positive cocci
Micrococcus, Staphylococcus (*S. aureus*, boils, infections); *Streptococcus* (*S. pyogenes*, scarlet fever, throat infections); *S. pneumoniae*, pneumonia), *Leuconostoc; Peptococcus*
15. Endospore-forming rods and cocci
Aerobic: *Bacillus* (*B. anthracis*, anthrax)
Anaerobic: *Clostridium* (*C. tetani*, tetanus; *C. botulinum*, botulism)
16. Gram-positive asporogenous rod-shaped bacteria
Lactobacillus
17. Actinomycetes and related organisms
Corynebacterium (*C. diphtheriae*, diphtheria); *Propionibacterium; Actinomyces, Bifidobacterium; Mycobacterium* (*M. tuberculosis*, tuberculosis); *Streptomyces*
18. The rickettsias (parasitic bacteria with exacting nutritional requirements and small genome sizes)
Rickettsia (*R. rickettsii*, Rocky Mountain spotted fever); *Chlamydia* (*C. trachomatis*, trachoma)
19. The mycoplasmas
Mycoplasma; Acholeplasma; Thermoplasma

[a] Data from "Bergey's Manual of Determinative Bacteriology," 8th ed. Williams & Wilkins, Baltimore, Maryland, 1973.
[b] The human diseases caused by some species are also listed.
[c] Formerly *Aerobacter*.

carbon dioxide, water, and inorganic forms of nitrogen and sulfur. The **photoautotrophs** extract their energy from sunlight, while the **chemoautotrophs** obtain energy from inorganic chemical reactions. For example, **hydrogen bacteria** oxidize H_2 to H_2O and **sulfur bacteria** (found in "sulfur springs") oxidize H_2S to H_2SO_4.

Most bacteria like the fungi and animals are **chemoheterotrophic,** i.e., they obtain energy from breakdown of organic compounds. Some heterotrophic bacteria are **anaerobes.** They metabolize complex organic substances such as sugars in the complete absence of oxygen, a process called **fermentation.** Some anaerobes oxidize organic compounds with an inorganic oxidant such as nitrate (**denitrifying bacteria**) or sulfate (**sulfate-reducing bacteria**). Some of the anaerobes, many of which are members of the genus *Clostridium,* are poisoned by oxygen and are known as **obligate anaerobes.** Others, including *E. coli,* are **facultative anaerobes,** able to grow either in the presence or absence of oxygen. **Obligate aerobes** depend upon combustion of organic compounds with oxygen for energy.

One of the largest groups of strictly aerobic heterotrophic bacteria, the pseudomonads (*Pseudomonas* and related genera), are of interest to biochemists because of their ability to oxidize organic compounds, such as alkanes, aromatic hydrocarbons, and steroids, which are not attacked by most other bacteria. Often, the number of oxidative reactions used by bacteria is limited. For example, some **acetic acid bacteria** obtain all of their energy by oxidation of ethanol to acetic acid:

$$CH_3CH_2OH + O_2 \longrightarrow CH_3COOH + H_2O \quad (1\text{-}1)$$

An important criterion of classification is the **gram stain** which distinguishes bacterial cells as **gram-positive** or **gram-negative** according to their ability to retain the basic dye, crystal violet, as an iodine complex. The specificity depends upon differences in the structure of the cell wall (see Chapter 5). Most actinomycetes, the spore-forming bacilli, and most cocci are gram-positive, while *E. coli,* other enterobacteria, and pseudomonads are gram-negative (Table 1-1).

11. Photosynthetic and Nitrogen-Fixing Prokaryotes

It is usually thought that the earth was once a completely anaerobic place containing methane, formaldehyde, and more complicated organic compounds. The first forms of life may have been similar to present-day anaerobic bacteria such as *Clostridium.*[8] The present-day **photosynthetic bacteria** (purple and green bacteria) may be related to organisms that developed at a second stage of evolution: those able to capture energy from sunlight. It is interesting that most of these (gram-negative) photosynthetic bacteria are strict anaerobes. None can make oxygen as do higher plants. Rather, the hydrogen needed to carry out the reduction of carbon dioxide in the photosynthetic process is obtained by the splitting of inorganic compounds, such as H_2S, thiosulfate, or H_2, or is taken from organic compounds.

In all photosynthetic organisms, including the higher plants, photosynthesis takes place in membranous structures. In the purple bacteria the light-absorbing pigments (bacterial chlorophylls and carotenes) are embedded in membranes that appear to be infoldings of the outer cell membrane. These infoldings form characteristic structures of interconnecting hollow vesicles, parallel tubes, or stacks of layers (**lamellae**) about 50–100 nm in diameter and known as **chromatophores.** In the green bacteria, the photosynthetic pigments line vesicles within the cell. Today, photosynthetic bacteria are found principally in sulfur springs and in deep lakes, but at one time they were probably far more abundant and the only photosynthetic organisms on earth.

Before organisms could produce oxygen a second complete photosynthetic system, which could cleave H_2O to O_2, had to be developed (see Chapter 13). The simplest oxygen-producing creatures existing today are the blue-green algae[9] (cyanobacteria). Some of these plants are unicellular and are structurally similar to colorless bacteria. Others, such as *Oscillatoria,* a slimy plant that often coats the inside walls of household aquaria, consist of long filaments about 6 μm in diameter (see Fig. 1-9). These filaments are capable of slow gliding movements by which they can creep out onto a clean surface of the fishbowl. All of the blue-green algae contain two pigments not found in other prokaryotes: **chlorophyll** *a* and **β-carotene.** These same pigments are found in the true algae and in higher plants.

Some blue-green algae, e.g., *Nostoc,* contain, in addition to pigmented cells, paler cells known as **heterocysts.** The heterocysts appear to have a specialized function of fixing molecular nitrogen. The development of the ability to convert N_2 into organic nitrogen compounds represents another important evolutionary step. Because of their ability both to fix nitrogen and to carry out photosynthesis, the blue-green algae have the simplest nutritional requirements of any organisms. They need only N_2, CO_2, water, light, and minerals for growth.

Evolution of the photosynthetic cleavage of water to oxygen was doubtless a major event with far-reaching consequences. Biologists generally believe that as oxygen accumulated in the earth's atmosphere, the obligate anaerobes (which are poisoned by oxygen) became limited to strictly anaerobic environments. Meanwhile, new classes of bacteria appeared with mechanisms for detoxifying oxygen and for using oxygen to oxidize complex organic compounds to obtain energy.

B. EUKARYOTIC CELLS

The second major subdivision of living things are the **eukaryotes,** organisms whose cells contain true nuclei. Eukaryotic cells are much larger and more complex internally than are prokaryotes. A membrane-enclosed nucleus contains most of the DNA, which is thus separated from the cytoplasm. Within the cytoplasm are various **organelles** with characteristic structures, e.g., the **mitochondria, lysosomes,** and **centrioles.** Eukaryotic cells come in so many sizes and shapes and with so many specialized features that it is impossible to describe the "typical" cell. Nevertheless, Fig.

TABLE 1-2
Approximate Sizes of Some Cells

Cell	Diameter (μm)	Approximate volume (μm^3)
E. coli	1	1.0
Small thymus cell	6	120
Liver cell	20	4,000
Human ovum (mature)	120	500,000
Hen's egg (white excluded)	20,000	4×10^{12}
Yeast cell	10	500
Onion root (meristematic cell)	17	2,600
Parenchyma cell of a fruit	1,000	1×10^8

1-3 is an attempt to portray some sort of "average" cell,* partly plant and partly animal.

1. The Size and Complexity of Eukaryotic Cells

As can be seen from Table 1-2, which lists the diameters and volumes of several roughly spherical cells, the variation in size is enormous. However, a diameter of 10 to 20 μm may be regarded as typical for both plants and animals. In a large cell like the ovum, many adjacent cells have assisted in synthesis of foodstuffs which have been transferred to the developing egg cell.

Plant cells are often large but usually the bulk (90% or more) of the cell is filled with a **vacuole** or **tonoplast**[9a] (unrealistically small in the drawing in Fig. 1-3). The metabolically active protoplasm of plant cells often lies in a thin layer at the periphery of the cell. Many cells are far from spherical; for example, the red blood cells of the human are discs $8 \times 8 \times 1$ to 2 μm with a volume of 80 μm^3. Plant fiber cells may be several millimeters in length. Nerve cells of animals have long exten-

* See the famous drawing of J. Brachet, *Sci. Am.* **205,** 55 (Sep 1961).

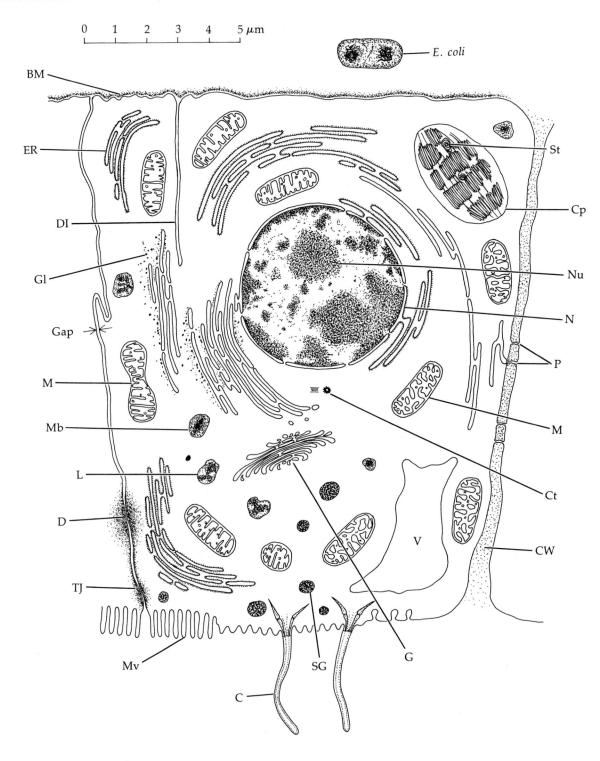

sions, the **axons,** which in the human sometimes attain a length of a meter.

The genome sizes of several eukaryotic cells are compared with those of *E. coli* in Table 1-3.

It can be seen that yeast contains about five times as much genetic matter as *E. coli* and that man (and the mouse) has about 600 times as much. However, genes may be duplicated in higher organisms and large amounts of **repetitive DNA** of unknown significance are often present (some amphibians have 25 times more DNA per cell than do humans). Watson suggested[10] that the true genetic complexity of a vertebrate cell must be at least 20–50 times that of *E. coli.* Thus, a human cell may contain on the order of 10^5 genes.

2. The Nucleus

In a typical animal cell the nucleus has a diameter of ~ 5 μm and a volume of ~ 65 μm^3. Except at the time of cell division it is densely and almost uniformly packed with DNA. Even with the electron microscope, it is difficult to see any distinct structures. Because of its acidic character, DNA is stained by basic dyes. Long before the days of modern biochemistry, the name **chromatin** was given to the material in the nucleus that was colored by basic dyes. At the time of cell division, the chromatin is consolidated into distinct **chromosomes** which contain, in addition to 15% DNA, about 10% **ribonucleic acid** (RNA) and 75% protein.

Fig. 1-3 The "average" eukaryotic cell. This composite drawing shows the principal organelles of both animal and plant cells approximately to the correct scale. Abbreviations: BM, basement membrane; ER, endoplasmic reticulum (rough, with ribosomes attached; smooth ER is depicted nearer the nucleus and on the right side of the cell); DI, deep indentation of plasma membrane; Gl, glycogen granules; Gap, space $\sim$10–20 nm thick between adjacent cells; M, mitochondrion; Mb, microbody; L, lysosome; D, desmosome; SG, secretion granule; TJ, tight junction; Mv, microvilli; C, cilium; G, Golgi apparatus; V, vacuole; CW, cell wall (of a plant); Ct, centrioles; P, plasmodesmata; N, nucleus; Nu, nucleolus; Cp, chloroplast; St, starch granule. Adapted from a drawing by Michael Metzler.

TABLE 1-3
Haploid Genome Sizes for Several Organisms

Organism	Weight in daltons[a,b]	No. of nucleotide pairs
E. coli	2.5×10^9	3.8×10^6
Saccharomyces cerevisiae (a yeast)	13×10^9	20×10^6
Sea urchin	600×10^9	900×10^6
Drosophila melanogaster (fruit fly)	60×10^9	90×10^6
Man and mouse	1500×10^9	2300×10^6

[a] C. A. Thomas, Jr., and L. A. MacHattie, *Annu. Rev. Biochem.* **36,** 485–518 (1967).

[b] H. R. Mahler and E. H. Cordes, "Biological Chemistry," 2nd ed., p. 187. Harper, New York, 1971.

Nearly all of the RNA of the cell is synthesized (transcribed) in the nucleus according to the instructions coded in the DNA. For the most part the RNA then moves out of the nucleus into the cytoplasm where it functions in protein synthesis (translation of the genetic information) and very likely in other ways not yet suspected.

Each cell nucleus contains one or more dense **nucleoli,** regions that are extremely rich in RNA (they contain 10–20% of the total RNA of cells). Nucleoli are sites of synthesis and of temporary storage or ribosomal RNA, which is needed in abundance for assembly of ribosomes.

The **nuclear envelope** is a pair of membranes, usually a few tens of nanometers apart, that surround the nucleus and separate off a **perinuclear space.** The membranes contain "pores" about 40–100 nm in diameter with sievelike structure. The pores contain tubular channels of diameter $\sim$4.5 nm, which provide a route for passage of RNA and other substances from the nucleus into the cytoplasm.[11]

3. The Plasma Membrane

The thin ($\sim$8 nm) outer cell membrane or **plasmalemma** (Fig. 1-4) controls the flow of materials into and out of cells, conducts impulses in nerve cells and along muscle fibrils, and participates in chemical communication with other cells. Deep infoldings of the outer membrane sometimes run into the cytoplasm, e.g., in striated muscle they form the "T system" of tubules which functions in excitation of muscle contraction (Chapter 4). Infoldings of the plasma membrane may sometimes connect with the nuclear envelope to provide one or more direct channels between the exterior of the cell and the perinuclear space.[12]

Membrane-enclosed vesicles sometimes bud inward from the plasma membrane and fuse with lysosomes. In this manner the cell may engulf particles **(phagocytosis)** or droplets of the external medium **(pinocytosis).** Surfaces of cells designed to secrete materials or to absorb substances from the surrounding fluid (e.g., cells lining kidney tubules and pancreatic secretory cells) are often covered with very fine projections or **microvilli** which greatly increase the surface area. In other cases projections from one cell interdigitate with those of an adjacent cell to give more intimate contact.

4. The Cytoplasmic Membranes

Although cytoplasm is fluid and in some organisms can undergo rapid streaming, the electron microscope has revealed that the liquid portion, the **cytosol,** is crossed by a series of membranes known as the **endoplasmic reticulum** (ER), a complex network of tubes, vesicles, and flattened sacs **(cisternae).** The **intracisternal space** of the ER appears to connect with the perinuclear space and to a series (usually 3–12) of flattened, slightly curved disk-shaped membranes, the **Golgi apparatus.**[13] This organelle (Fig. 1-3) was first reported by Camillo Golgi[13a] in 1898, but its existence was long doubted.

Parts of the ER **(the rough endoplasmic reticulum)** are lined with many ribosomes of ~21–25 nm diameter. While resembling those of bacteria, these eukaryotic ribosomes are about 50% heavier (4×10^6 daltons). The **smooth endoplasmic retic-**

ulum lacks ribosomes but proteins made in the rough ER may be modified, e.g., by addition of carbohydrate chains in the smooth ER.

The Golgi membranes lie close to the smooth ER on the side toward the center of the cell. At the outer edges the Golgi membranes pinch off to form vacuoles which are often densely packed with enzymes or other materials. These **secretion granules** move to the surface and are released from the cell. In the process **(exocytosis)** the membranes surrounding the granules fuse with the outer cell membrane.[14,15]

The Golgi apparatus is more than just a protein-packaging organelle—it is the site of various synthetic reactions. Like the smooth ER, Golgi membranes may add carbohydrates to proteins (to form glycoproteins) and sulfate groups to polysaccharides.[16,17] Golgi membranes of liver cells are involved in secretion of lipoproteins and possibly of fat-soluble vitamins (such as vitamin A) into the blood.[18] Thus the ER, the Golgi membranes, and secretion granules represent an organized system of synthetic units.

This system functions not only in the synthesis of enzymes to be secreted but also in the formation of new membranes. The rough ER appears to contribute membrane material to the smooth ER and Golgi apparatus, while material from Golgi membranes can become incorporated into the outer cell membrane. Outer mitochondrial membranes and membranes around vacuoles in plant cells may also be derived directly from the ER.[19] Outer membrane material is probably "recycled" by **endocytosis.**[20]

The term **microsome,** frequently met in the biochemical literature, refers to small particles of 50–150 nm diameter, mostly fragments of the ER but in part derived from the plasma membrane. Microsomes are formed when cells are ground or homogenized. Upon centrifugation of the disrupted cells, nuclei and other large fragments settle first, then the mitochondria. At very high speeds, e.g., at 100,000 times the force of gravity, the microsomes (whose mass is $\sim 10^8$–10^9 daltons) settle. In the electron microscope one sees that in the microsomes the membrane fragments have

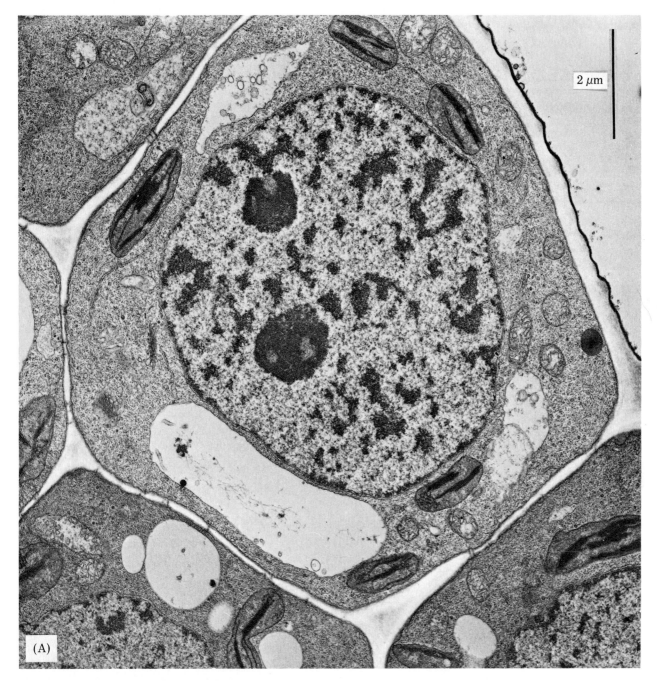

2 μm

Fig. 1-4 (A) Electron micrograph of a thin section of a young epidermal cell of a sunflower. The tissue was fixed and stained with uranyl acetate and lead citrate. Clearly visible are the nucleus, mitochondria, chloroplasts, a Golgi body (dictyosome, in the left center), endoplasmic reticulum, cell wall, plasmodesmata, and cuticle (upper right, thin dark layer). Micrograph courtesy of H. T. Horner.

Fig. 1-4 (B) Two dictyosomes (plant Golgi bodies) from a cell of the glandular trichome of *Psychotria punctata*. Note the netted appearance from which the term dictyosome is derived. (C) Higher magnification micrograph of stacked membranous **grana** (photosynthetic lamellae) of an alfalfa leaf chloroplast. Micrographs courtesy of H. T. Horner.

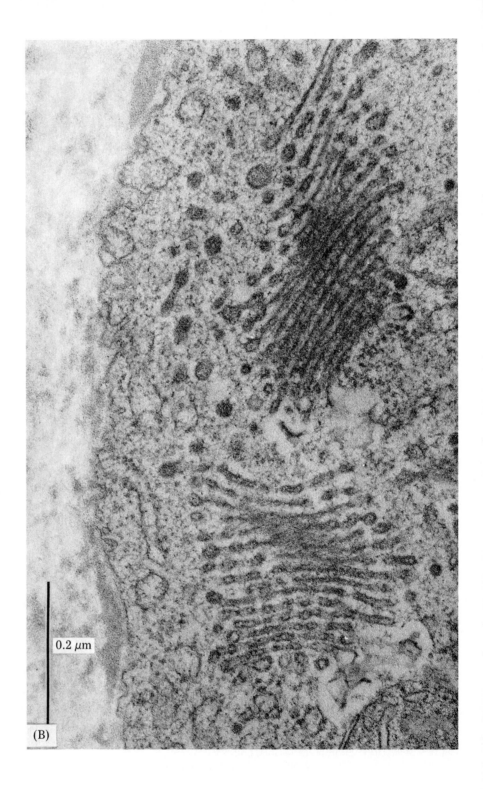

0.2 μm

(B)

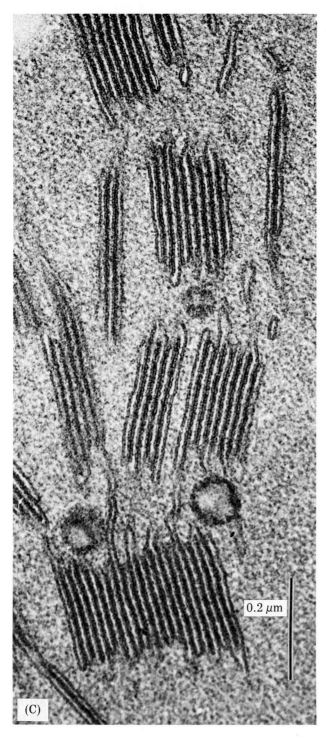

(C)

0.2 μm

closed to give small sacs to the outside of which the ribosomes still cling:

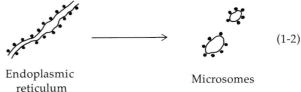

Endoplasmic
reticulum

Microsomes

(1-2)

Such regeneration of a closed vesicle structure appears to be a general property of membrane fragments. **Synaptosomes** are closed vesicles formed from the **synaptic** endings of nerve cells during homogenization. However, synaptosomes are derived mainly from the plasma membrane, not the ER, and often contain mitochondria.

5. Mitochondria and Plastids

The presence of **mitochondria,** complex bodies about the size of bacteria and bounded on the outside by a double membrane (Figs. 1-3 and 1-4), is characteristic of eukaryotic cells. The inner membrane of a mitochondrion is highly folded to form the so-called **cristae** (crests). The outer membrane is porous to small molecules but the passage of substances into and out of the inner space of the mitochondrion (the **matrix**) is tightly controlled by the inner membrane. Although some of the oxidative chemical activities of the cells are located in the ER, the principal energy-yielding reactions for aerobic organisms are found in the mitochondria; these organelles are the site of utilization of most oxygen. It was a surprise to most biochemists when it was found that each mitochondrion contains a small circular molecule of DNA. Furthermore, mitochondria contain ribosomes of a size similar to those of bacteria and smaller than those lining the rough ER.

Mitochondria are present in all eukaryotic cells that use oxygen for respiration. The numbers per cell vary from the *one* mitochondrion of certain tiny **trypanosomes** to as many as 3×10^5 in some oocytes. A typical liver cell contains more than 1000 mitochondria.[21] A new aspect of mitochondria of yeast cells has been reported.[22] Serial sectioning (Box 1-C) of a single cell showed that all of the mitochondria are interconnected. Thus, in yeast cells mitochondria are to be thought of as a single interconnected "compartment" of the cell rather than as individual organelles. It is not clear how broadly this concept will apply.

Plastids are organelles of plant cells that serve a variety of purposes. The most important are the **chloroplasts,** the chlorophyll-containing sites of photosynthesis. Like mitochondria they contain folded internal membranes and several small molecules of DNA.

6. Lysosomes and Microbodies

Lysosomes are vesicles bounded by a single membrane and containing a whole battery of enzymes powerful enough to digest almost anything in the cell. They appear to originate from Golgi membranes. In cells that engulf bits of food (e.g., the ameba) lysosomes apparently provide the digestive enzymes. Lysosomes may also digest "worn-out" or excess cell parts including mitochondria. Lysosomes are vital components of cells,[23,24] and several serious human diseases result from a lack of specific lysosomal enzymes.

Microbodies occur in many cells,[24,25] in some green leaves in numbers up to $\frac{1}{3}$ those of mitochondria. Microbodies are often about the size of mitochondria but are surrounded by a *single* membrane and sometimes contain an apparently crystalline "core." The membrane is porous to small molecules such as sucrose, permitting these organelles to be separated from mitochondria by centrifugation in a sucrose gradient where the microbodies assume a density of about 1.25 g cm^{-3} compared to 1.19 for the impervious mitochondria.

Microbodies are rich in enzymes that produce and decompose hydrogen peroxide. Two distinct types have been described: **peroxisomes** are found in liver and kidney cells and in green leaves, while **glyoxysomes** are present in germinating oilseeds. The glyoxysomes assume a special role in catalyzing reactions of the "glyoxylate pathway" of metabolism (Chapter 11, Section D,4).

7. Centrioles, Cilia, Flagella, and Microtubules

Many cells contain centrioles, little cylinders about 0.15 μm in diameter and 0.5 μm long, which are *not* bound by membranes. Each centriole contains a series of fine **microtubules** of 20 nm diameter. A pair of centrioles are present near the nucleus in most animal cells and play an important role in cell division. However, they have never been observed in plant cells.

Centrioles are related in structure to the long **flagella** and shorter **cilia** (the two words are virtually synonymous) commonly present as organelles of locomotion in eukaryotic cells. Stationary cells of our own bodies also often have cilia. For example, there are 10^9 cilia/cm^2 in bronchial epithelium.[26] Modified flagella form the receptors of light in our eyes and of taste in our tongues. Flagella and cilia are somewhat larger in diameter (about 0.2 μm) than are centrioles and have a characteristic internal structure: 11 hollow microtubular filaments $\sim$24 nm in diameter are arranged in a "9 + 2" pattern (Figs. 1-5 and 1-6). Each of the microtubules resembles a single bacterial flagellum in appearance, but there are distinct and significant chemical differences. The **basal body** or **kinetisome** (Fig. 1-5) resembles a centriole both in structure and dimensions and in its mode of replication.

Microtubules similar to those of flagella are also found within the cytoplasm of cells.[27] They look like little "pipes" but there is no evidence that they function as such. They more likely serve a structural function as a "cytoskeleton." In nerve axons the microtubules run parallel to the length of the axon and may be part of a mechanical transport system for cell constituents.

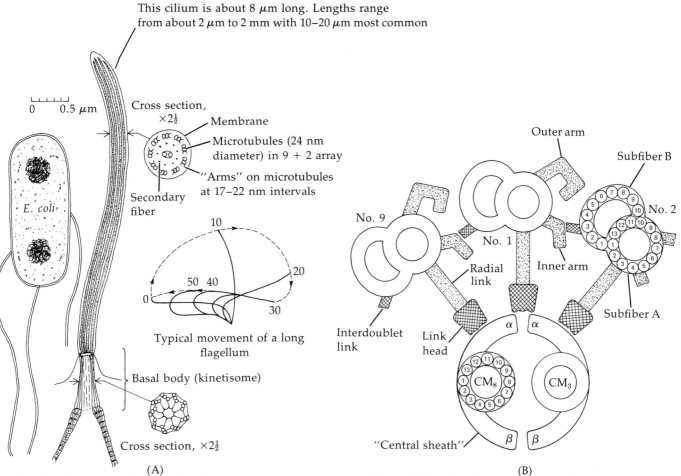

This cilium is about 8 μm long. Lengths range from about 2 μm to 2 mm with 10–20 μm most common

0   0.5 μm

Cross section, ×2½

E. coli

Membrane

Microtubules (24 nm diameter) in 9 + 2 array

"Arms" on microtubules at 17–22 nm intervals

Secondary fiber

Typical movement of a long flagellum

Basal body (kinetisome)

Cross section, ×2½

(A)

Outer arm

Subfiber B

No. 9

No. 1

No. 2

Radial link

Inner arm

Subfiber A

Interdoublet link

Link head

α   α

CM₈   CM₃

"Central sheath"

β   β

(B)

Fig. 1-5 (A) Structure of cilia and flagella of eukaryotes. After P. Satir, *Sci. Am.* **204**, 108–116 (Feb 1961). Copyright 1961 by Scientific American, Inc. All rights reserved. (B) Schematic diagram of a portion of a typical 9 + 2 axoneme (outer portion of a cilium) as viewed from base to tip. The outer doublet tubules (No. 1, 2, and 9) are joined to the central sheath or paired projections (α and β) attached to each of the two central microtubules (CM₃ and CM₈). Note the construction of the microtubules from 13 strands, each a chain of protein subunits (see also Chapter 4, Section D,2). Drawn approximately to scale. From M. A. Sleigh, ed., "Cilia and Flagella," p. 14. Academic Press, New York, 1974. Movement diagram in (A) from M. A. Sleigh, *Endeavour* **30**, 11 (1971).

8. Cell Coats, Walls, and Shells

Like bacteria, cells of higher plants and animals are frequently coated with extracellular materials. Plants have rigid walls rich in cellulose and other carbohydrate polymers. Cells on outside surfaces of plants are covered with a layer of wax. Surfaces of animal cells are usually coated with carbohydrate molecules which are attached to specific surface proteins to form **glycoproteins.** Spaces between cells are filled with such "cementing substances" as **pectins** in plants and **hyaluronic acid** in animals. Insoluble proteins such as **collagen** and **elastin** are secreted by connective tissue cells. Cells

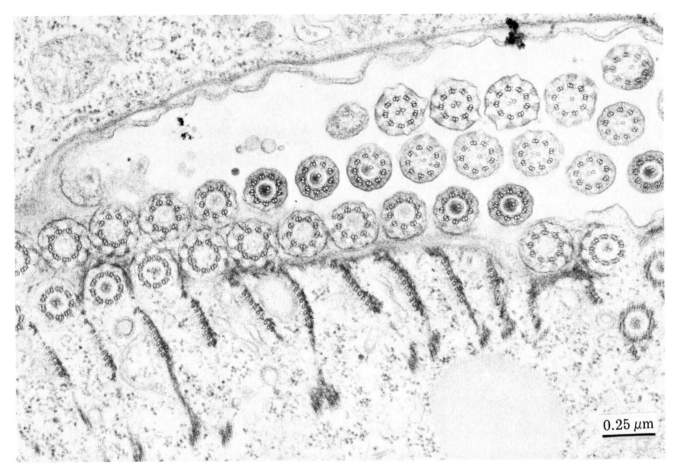

0.25 μm

Fig. 1-6 Cross section of cilia from the oral membranelle of *Tetrahymena*. These cilia are used to direct food into the mouth. The plane of the section cuts through, from left to right, the ciliary bases (nine sets of triplet microtubules) and the cilium proper (nine sets of doublet microtubules). The latter show zero, one, and two central microtubules from lower left to upper right center, respectively, corresponding to sections through progressively more distal portions of the cilia. Stained with uranyl acetate-lead citrate. From A. M. Elliott and D. E. Outka, "Zoology" 5th ed., p. 122. Copyright 1976. Reproduced by permission of Prentice-Hall, Inc., Englewood Cliffs, New Jersey.

that lie on a surface (epithelial and endothelial cells) are often lined on one side with a thin, collagen-containing **basement membrane** (Fig. 1-3). Inorganic deposits such as calcium phosphate (in bone), calcium carbonate (egg shells and spicules of sponges), and silicon dioxide (shells of diatoms) are laid down, often by cooperative action of several or many cells. Thus, a considerable amount of metabolism goes on outside of cells.

C. THE EVOLUTION OF COMPLEX ORGANISMS

The striking differences between eukaryotic and prokaryotic cells have led to many speculations about the evolutionary relationship of these two great classifications of living organisms. A popular theory is that mitochondria arose from aerobic bacteria. It is speculated that after blue-green algae had developed and oxygen had become abundant, a **symbiotic** relationship arose in which some small aerobes lived within cells of larger bacteria that had previously been obligate anaerobes. The aerobes used any oxygen present, protecting the surrounding anaerobic organism from its toxicity. The relationship became permanent and led eventually to the mitochondria-containing eukaryotic cell.[28,29] Further symbiosis with bluegreen algae could have led to the chloroplasts of the eukaryotic plants.

A fact that supports such ideas is the existence among present-day organisms of many symbiotic relationships. Thus, the green paramecium (*Paramecium bursaria*) contains, within its cytoplasm, an alga (*Chlorella*), a common green plant that is quite capable of living on its own. Perhaps by accident it took up residence within the paramecium.[28]

Biologists and biochemists have been quick to accept the symbiotic theory of origin of mitochondria. However, Raff and Mahler suggested that mitochondria are more likely to have arisen from mesosomal membranes and the DNA within them from extrachromosomal genetic material (**plasmids** or **episomes,** see Chapter 15, Section D,7) of a kind frequently found within prokaryotic organisms.[30] A lively debate is in progress.[30-32]

Fossils of bacteria and blue-green algae have been obtained from rocks whose age, as determined by geochemical dating, is more than three billion years. It seems likely that at about the same time the first eukaryotic cells appeared and began to evolve into the more than a million species now known.

1. Gene Duplication

By what means has it been possible for the genome size of an organism to increase as it evolved from a lower form to a higher one? Simple mutations that cause alterations in protein sequences could lead to changes in form and behavior of the organisms. However, they could not account for the increase in genetic material that accompanied evolution. A possible explanation is that sometimes a second copy of a gene, or of several genes, accidentally became incorporated into a cell nucleus.[32a] Now, with an extra copy of a gene, a cell would survive even when mutations rendered the protein for which one of these copies coded unusable. As long as one of the genes remained "good," the organism could grow and reproduce. The extra, mutated gene could be carried, unfunctioning, for many generations. As long as it produced only harmless, nonfunctioning proteins there might be no selection pressure to eliminate it and it might undergo repeated mutations. Finally, after many mutations, the protein for which it coded could prove useful to the cell in some new way.

An example of evolution via gene duplication is provided by the oxygen-carrying proteins of blood. It appears that about a billion years ago, the gene for an ancestral protein (**globin**) was doubled. One gene evolved into that of present-day **hemoglobin** and the other into that of the muscle protein **myoglobin.** At still a later date, the hemoglobin gene again doubled leading to the present-day α chain and β chain of hemoglobin (Chapter 4, Section D,8). These are two distinctly different but related protein subunits whose genes are not even on the same chromosome.

2. Sex, Chromosomes, and Cell Division

Bacteria usually reproduce by simple fission. The single DNA molecule of the chromosome is duplicated and the bacterium divides, each daughter cell receiving an identical chromosome. However, the beginnings of sexual reproduction are already present among some bacteria and **genetic recombination** (Chapter 15, Section G) provides a deliberate process for mixing of the bacterial genes.

Sexual reproduction is fully developed in eukaryotic organisms, in which the growth of a multicelled individual begins with the fusion of two haploid **gametes,** an egg and a spermatozoon. Each gamete carries a complete set of genetic instructions, and after the nuclei fuse, the fertilized egg or **zygote** is **diploid.** Each diploid cell contains *two* complete sets of genetic blueprints of quite different origin. This provides a very important advantage to the organism. If a gene from one parent is defective, the chances are that the gene from the other parent will be good. Furthermore, sexual reproduction provides a means for mixing of genes so that each of us not only receives half his genes from his mother and half from his father but also some genes from each grandparent on both sides of the family, some from each great-grandparent, etc.

When eukaryotic cells prepare to divide, the DNA molecules of the nucleus, which have been spread out through a large volume, coil and fold. Together with proteins and other molecules they form the dense bodies known as chromosomes. Some organisms, such as *Ascaris* (a roundworm), have only two chromosomes. One chromosome contains the genetic information that the baby *Ascaris* inherited from its father and the other the genes inherited from its mother. Thus, there is only one **homologous pair** of chromosomes, each one a complete set of genetic instructions for the cell. Both chromosomes of the pair divide with each cell division so that every cell of the organism has the homologous pair. Like the *Ascaris,* we too have diploid cells in our bodies but each cell contains 23 homologous pairs of chromosomes.* Human chromosomes vary in size but are usually 4–6 μm long and ~1 μm in diameter.

3. Mitosis and Meiosis

The process of cell division, **mitosis,** begins and ends the **cell cycle** by which a single diploid cell

divides. From a biochemical viewpoint mitosis is seen as the duplication of the genetic blueprint followed by its consolidation into compact packages, the chromosomes. The latter are then divided equally among the two forming cells as is described further in Chapter 15 (Section D,9).

What happens to the mitochondria during mitosis? These also divide, as do the chloroplasts in plant cells. Thus, replication of the DNA in these organelles must also occur at some stage in the cell cycle. In at least some cases, the division of the mitochondria is coordinated with cell division so that the average number of mitochondria per cell stays remarkably constant. The same thing seems to be true for cells of lower organisms that contain symbiotic algae inside, even though in this case the algae can live apart from their host in a free form. Here is a biochemical puzzle. What is it that induces the algae inside the host to divide in synchrony with the host's cell division?

By the successive divisions of mitosis, a single fertilized eukaryotic egg cell can grow to an adult. Only 40 to 50 successive mitotic divisions will produce a human. However, formation of gametes (germ cells), which are haploid, requires the special process of **meiosis,** by which the number of chromosomes is divided in half. During meiosis one chromosome of each of the homologous pairs of the diploid cell is passed to each of the gametes formed. In an organism like the *Ascaris,* which contains only a single pair of chromosomes, a gamete receives either the chromosome of maternal origin or that of paternal origin but not both. In organisms having several pairs of chromosomes, one chromosome of each pair is passed to the gamete in a random fashion during meiosis, each gamete receiving some chromosomes of maternal and some of paternal origin.

Meiosis is considered further in Chapter 15, where special attention is given to **crossing-over.** This essential feature of meiosis breaks the linkage between genes and provides a means of mixing genetic information within chromosomes. Crossing-over is closely analogous to genetic recombination in bacteria and at the molecular level may be the same.

---

* Some other chromosome numbers are as follows: mouse, 20; toad, 11; onion, 8; mosquito, 3; and *Drosophila,* 4.

BOX 1-D

INHERITED METABOLIC DISEASES

In 1908 Archibald Garrod[a] proposed that **cystinuria** (Chapter 5, Section B,2,b) and certain other defects in amino acid and sugar metabolism were inherited diseases. Since that time the number of recognized genetic defects of human metabolism has increased at an accelerating rate to over 1500.[b] For over 100 of these diseases the nature of the protein defect is recognized.[c–e] An example is **sickle cell anemia** (Box 4-D). However, most of the defects involve a loss of activity of some important enzyme.

While many genetic diseases are rare, affecting about one person in 10,000, others such as **cystic fibrosis** affect one in 2500. Since there are so many metabolic diseases the total number affected is thought to exceed 2% of persons born. Many of these die at an early age. A much greater number (>5%) are affected by such conditions as diabetes and mental illness which are, in part, of genetic origin. Since new mutations are always arising, genetic diseases present a problem of increasing significance.

At what rate are new mutations occurring? From the haploid DNA content (Table 1-2) we can estimate that the total coding capacity of the DNA in a human cell exceeds two million genes (actually two million *pairs* of genes in diploid cells). However, it is generally thought that only a fraction of the DNA codes for proteins. Various authors estimate from 20,000 to 100,000 pairs of structural genes for the human. Turning for some guidance to bacteria, the *easily detectable* rate of mutation in bacteria is about 10^{-6} per gene, or 10^{-9} per base pair per replication.[f] Thus, in the replication of the 2×10^{-9} base pairs in human chromosomes we might anticipate about two mistakes per cell division. Since there are several cell divisions of the germ cells between each human generation, the number of new mutations in one generation may be substantial. Another source of new mutations is damage to double-stranded DNA during the 20 years or more of a human generation. No doubt most damage is repaired by sophisticated repair systems (Chapter 15) but some remains. Fortunately, many mutations may be harmless or nearly so and a few may be beneficial. It is also possible that eukaryotic replication systems are more error-free than those of bacteria. On the other hand, the introduction by man of mutagenic chemicals into the environment is a cause for concern.

Many mutations are *lethal*. The homozygote does not survive and is lost in an early (and usually undetected) spontaneous abortion. Among healthy individuals it has been estimated that as many as ten lethal recessive mutations may be present, as well as at least 3–5 autosomal recessive mutations of a seriously harmful type. Harmful dominant mutations are also frequent in the population. These include an elevated lipoprotein content of the blood and an elevated cholesterol level which is linked to a high incidence of early heart disease.

Biochemical disorders are important because of the light they shed on metabolic processes, and frequent reference will be made to these diseases throughout the book. Of course, a principal reason for their study is the hope of finding cures. In some cases, for example, in the treatment of **phenylketonuria** (Chapter 14, Section H,5) or of **galactosemia** (Chapter 12, Section A,1), a change in the diet can prevent irreversible damage to the brain, the organ most frequently affected in many of these diseases. In many other cases no satisfactory therapy is presently available, but the possibilities of finding some way to supply missing enzymes or to carry out "genetic surgery" are among the most exciting developments of contemporary medical biochemistry (Chapter 15, Section H,4).

[a] A. E. Garrod, "Inborn Errors of Metabolism." Oxford, London, 1909.

[b] T. Friedmann and R. Roblin, *Science* **175,** 949–955 (1972).

[c] J. B. Stanbury, J. B. Wyngaarden, and D. S. Fredricksen, eds., "The Metabolic Basis of Inherited Disease," 3rd ed. McGraw-Hill, New York, 1972.

[d] D. J. H. Brock and O. Mayo, eds., "The Biochemical Genetics of Man." Academic Press, New York, 1972.

[e] R. H. S. Thompson and I. D. P. Wootton, "Biochemical Disorders in Human Disease," 3rd ed. Academic Press, New York, 1970.

[f] J. D. Watson, "Molecular Biology of the Gene," 3rd ed., p. 254. Benjamin, Menlo Park, California, 1976.

4. Sex Chromosomes and Autosomes

Diploid cells of most higher organisms have one pair of chromosomes which, among other things, determines the sex of the individual. Cells of the human female contain a pair of "X chromosomes," while cells of the male contain only one X chromosome, which is paired with a short stubby "Y chromosome."*

Genes occurring on the sex chromosomes are said to be **sex-linked.** Since most of these genes are present in the long X chromosome but not in the Y chromosome, genetic defects in the X chromosome show up in males, who have only one copy of the gene, much oftener than in females, who have two copies. Thus, hemophilia and color blindness are predominately afflictions of the human male. Females are mostly **heterozygous** for the defective gene, having a good copy in one of their X chromosomes.† When, as is usually the case, one good copy of the gene is sufficient for normal health, the genetic trait is referred to as **recessive.** For example, hemophilia and color blindness are both sex-linked and recessive traits.

Chromosomes other than the sex chromosomes are called **autosomes** and genes carried by them are autosomal.

5. Haploid and Diploid Phases

In the human being and other higher animals, meiosis leads directly to formation of the gametes, the egg and sperm cells. When these fuse, a diploid nucleus is created again and the adult develops by repeated mitosis of the diploid cells. While meiosis also occurs in the life cycle of all other eukaryotic creatures, it is not always at a point corresponding to that in the human life cycle. Thus, the cells of many protozoa and of fungi are ordinarily haploid. After two haploid nuclei fuse to form a diploid cell, meiosis quickly occurs again to produce haploid individuals. An alternation of haploid and diploid phases of the life cycle is common, both among lower plants and lower animals. Thus, gametes of ferns fall to the ground and germinate to form a low growing green mosslike haploid or **gametophyte** form. The latter produces motile haploid gametes which fuse to a diploid zygote that grows into the larger and more obvious **sporophyte** form of the fern.

It is presumably the ability to survive as a heterozygote, even with one or more highly deleterious mutations, that has led to the dominance of the diploid phase in higher plants and animals. However, to the biochemical geneticist organisms with a haploid phase offer enormous experimental advantages because recessive mutants can be detected readily.

D. THE KINGDOM PROTISTA

Unicellular eukaryotes are often grouped together with multicellular organisms in which all cells have similar functions (little or no differentiation into tissues) as the kingdom **Protista.** The fungi may also be included but are often regarded as a separate kingdom.

1. Protozoa

Among the best known of the animal-like protista is the **ameba** (subphylum **Sarcodina** or Rhizopoda). The most striking feature of the ameba (Fig. 1-7) is its method of locomotion which involves the transformation of cytoplasm from a liquid state to a semisolid gel. As the ameba moves, the cytoplasm at the rear liquifies and flows to the

---

* In mammals the presence or absence of the Y chromosome determines the sex, but in other organisms the Y chromosome may be completely absent. For example, among some insects (Orthoptera) the females are XX and the males XO. In other insects (Lepidoptera) as well as in birds and some other vertebrates the males are XX and females XY.

† Although human females have two X chromosomes per cell only one of the two becomes active in the synthesis of messenger RNA. In some cells the maternal X chromosome is active and in some the paternal X chromosome. Thus, in the body as a whole the genes from both parents are "expressed."

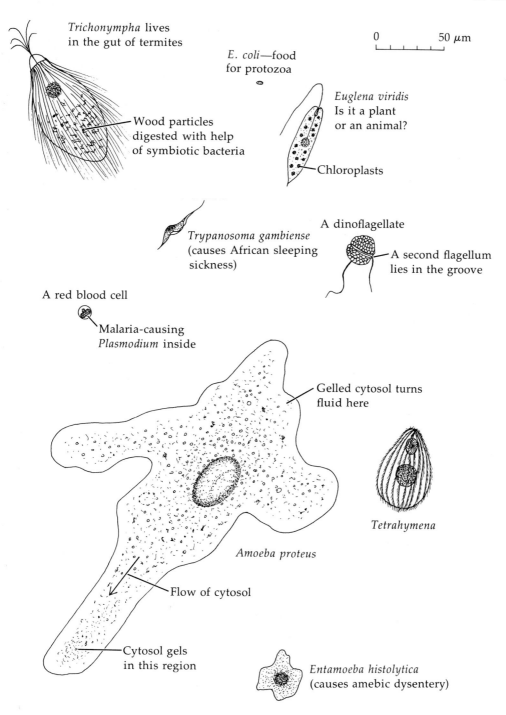

Trichonympha lives
in the gut of termites

E. coli—food
for protozoa

0 ‖‖‖‖‖‖‖ 50 μm

Wood particles
digested with help
of symbiotic bacteria

Euglena viridis
Is it a plant
or an animal?

Chloroplasts

Trypanosoma gambiense
(causes African sleeping
sickness)

A dinoflagellate

A second flagellum
lies in the groove

A red blood cell

Malaria-causing
Plasmodium inside

Gelled cytosol turns
fluid here

Tetrahymena

Amoeba proteus

Flow of cytosol

Cytosol gels
in this region

Entamoeba histolytica
(causes amebic dysentery)

Fig. 1-7 A few well-known Protista.

front and into the extending pseudopodia where it solidifies along the edges. The ameba poses several important biochemical questions. For example, what remarkable chemistry underlies the reversible change from liquid to solid cytoplasm? What chemical processes cause the **contractile vacuoles**[33] of the ameba periodically to eject excess fluid—acting as a rudimentary excretory system within a single cell? How can the cell membranes break and reform so quickly when an ameba engulfs food particles?

Relatives of the ameba include the **Radiolaria**, marine organisms of remarkable symmetry with complex internal skeletons containing the carbohydrate polymer **chitin** together with silica (SiO_2) or strontium sulfate. The **Foraminifera** are little ameba that deposit external shells of calcium carbonate or silicon dioxide. Over 20,000 species are known and now as in the distant past their minute shells fall to the bottom of the ocean and form limestone deposits.

Tiny ameboid parasites of the subphylum **Sporozoa** attack members of all other animal phyla. Several genera of **Coccidia** parasitize rabbits and poultry, doing enormous damage. Man is often the victim of species of the genus *Plasmodium* (Fig. 1-7) which invade red blood cells and other tissues to cause **malaria,** one of man's most important ailments on a worldwide basis. Throughout history malaria has probably killed more humans than any other disease.

Another subphylum of protozoa, the **Mastigophora,** are propelled by a small number of flagella and are the connecting link between animals and the algae. *Euglena viridis,* a small freshwater organism with a long flagellum in front, has a flexible tapered body, green chloroplasts, and a light-sensitive "eye spot" which it apparently uses to keep itself in the sunshine (Fig. 1-7). *Euglena* is also able to live as a typical animal if there is no light. Treatment with streptomycin (Box 12-A) causes the *Euglena* to lose its chloroplasts and to become an animal permanently. The **dinoflagellates** (Fig. 1-7), some colorless and some green, occur in great numbers among the plankton of the sea.

The **hemoflagellates** are responsible for some of man's most terrible diseases. Trypanosomes (genus *Trypanosoma*) invade the cells of the nervous system causing African sleeping sickness. Other flagellates live in a symbiotic relationship. The most complex flagellates (Fig. 1-7) known live in the alimentary canals of termites and roaches. Termites depend upon **bacteria** living within the cells of these symbiotic protozoans to provide the essential enzymes needed to digest the cellulose in wood.

Members of the subphylum **Ciliophora,** structurally the most complex of the protozoa, are covered with a large number of **cilia** which beat together in an organized pattern. A question that immediately comes to mind is: How are the cilia able to communicate with each other to provide this organized pattern? Two of the ciliates most commonly studied by biochemists are *Tetrahymena*[34] (Fig. 1-7), one of the simplest, and *Paramecium,* one of the more complex.

The **Myxomycetes** or "slime molds" are probably more closely related to protozoa than to fungi. Members of the family Acrasieae, the best studied member of which is *Dictyostelium discoideum*, start life as small amebas. After a time, when the food supply runs low, some amebas begin to secrete pulses of a chemical attractant **cyclic AMP** (Chapter 6, Section F,5). Neighboring amebas respond to the pulses of cyclic AMP by emitting their own pulses about 15 s later, then moving toward the original source.[35] The ultimate effect is to cause the amebas to stream to centers where they aggregate and form funguslike fruiting bodies. Asexual spores are formed and the cycle begins again. Other Myxomycetes grow as a multinucleate (diploid) **plasmodium** containing thousands of nuclei but no individual cell membranes.

2. Fungi

Lacking photosynthetic ability, living mostly in soil but sometimes in water, the fungi are represented by almost half as many species ($\sim 10^5$) as are the vascular plants. The distinguishing characteristics of fungi are the lack of chlorophyll and growth as a series of many branched tubules (usually 6–8 μm diameter), the hyphae, which constitute the **mycelium.** The organisms are mostly coenocytic, i.e., the hyphae are not made up of separate cells but contain a mass of protoplasm with many nuclei. Only occasional septa divide the tubules. Most fungi are saprophytic but some are parasites producing serious and difficult to treat infections in man. An important medical problem is the lack of adequate antibiotics for treating fungal infections (mycoses). On the other hand, fungi produce important antibiotics, e.g., **penicillin.** Still others form some of the most powerful toxins known!

The lower fungi **(Phycomycetes)** include simple aquatic molds and mildew organisms. Higher fungi are classified as **Ascomycetes** or **Basidiomycetes** according to the manner in which the sexual spores are born. In the Ascomycetes these spores are produced in a small sac called an **ascus** (Fig. 1-8). Each ascus contains four or eight spores in a row, a set of four representing the results of a single pair of meiotic divisions (a subsequent mitotic division will give eight spores). This is one of the features that has made *Neurospora crassa* (Fig. 1-8) a favorite subject for genetic studies.[36] The ascospores can be dissected out in order from the ascus and cultivated separately to observe the results of crossing-over during meiosis.

Neurospora also reproduces via haploid spores **conidia.** The haploid mycelia exist as two mating types and conidia or mycelia from one type can fertilize cells in a special body (the protoperithecium) of the other type to form zygotes. The latter immediately undergo meiosis and mitosis to form the eight ascospores.

Among the Ascomycetes are the highly prized edible truffles and morels. However, most mushrooms and puffballs are fruiting bodies of Basidiomycetes. Among other Basidiomycetes are the **rusts** which do enormous damage to wheat and other grain crops.

Yeasts are fungi adapted to life in an environment of high sugar content and which usually remain unicellular and reproduce by budding (Fig. 1-8). Occasionally the haploid cells fuse in pairs to form diploid cells and sexual spores. Some yeasts are related to the Ascomycetes, others to Basidiomycetes. *Saccharomyces cerevisiae,* the organism of both baker's and brewer's yeast, is an Ascomycete. It can grow indefinitely in either the haploid or diploid phase, the latter having slightly larger cells than the former.[37]

3. Algae

Algae are a very diverse group of chlorophyll-containing eukaryotic organisms which may be either unicellular or colonial. The colonial forms are usually organized as long filaments, either straight or branched, but in some cases, as blades resembling leaves. However, there is little differentiation among cells. The gold-brown, brown, and red algae contain special pigments in addition to the chlorophylls.

The euglenids **(Euglenophyta)** and dinoflagellates **(Pyrrophyta),** discussed in the protozoa section, can equally well be regarded as algae. The bright green **Chlorophyta,** unicellular or filamentous algae, are definitely plants, however. Important to the biochemist is *Chlamydomonas,* still a rather animal-like creature with two flagella and a carotenoid-containing eyespot or **stigma** (Fig. 1-9). *Chlamydomonas* contains a single chloroplast. A **pyrenoid,** a center for the synthesis of starch, lies, along with the eyespot, within the chloroplast. The organism is haploid with "plus" and "minus" strains and motile gametes. Zygotes immediately undergo meiosis to form haploid spores. Chromosome mapping has progressed to the point where *Chlamydomonas* is an important organism for biochemical genetic studies.

Among the filamentous algae, *Ulothrix* shows its relationship to the animals through formation of asexual spores with four flagella and biflagellate gametes. Only the zygote is diploid. On the other hand, the incomparably beautiful *Spirogyra* (Fig. 1-9) has no motile cells and the ameboid male gamete flows through a tube formed between the two mating cells. This behavior suggests a relationship of *Spirogyra* to the higher green plants.

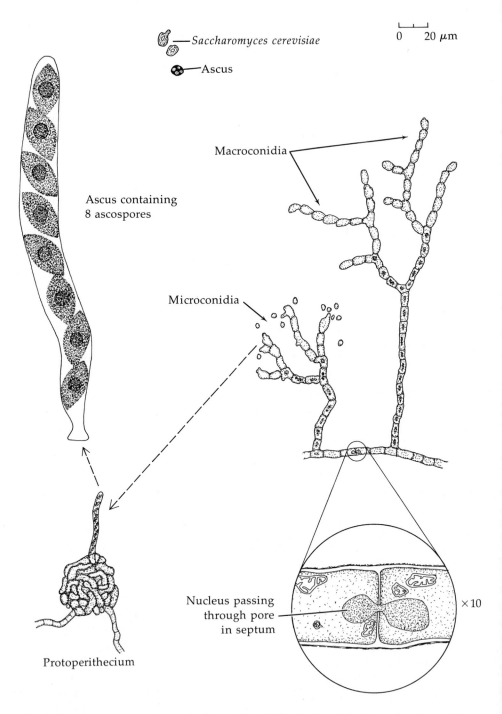

Saccharomyces cerevisiae

Ascus

0   20 μm

Ascus containing
8 ascospores

Macroconidia

Microconidia

Protoperithecium

Nucleus passing
through pore
in septum

×10

Fig. 1-8 Two frequently studied fungi, *Neurospora crassa* and the yeast *Saccharomyces cerevisiae*. After I. Webster, "Introduction to Fungi." Cambridge Univ. Press, London and New York, 1970.

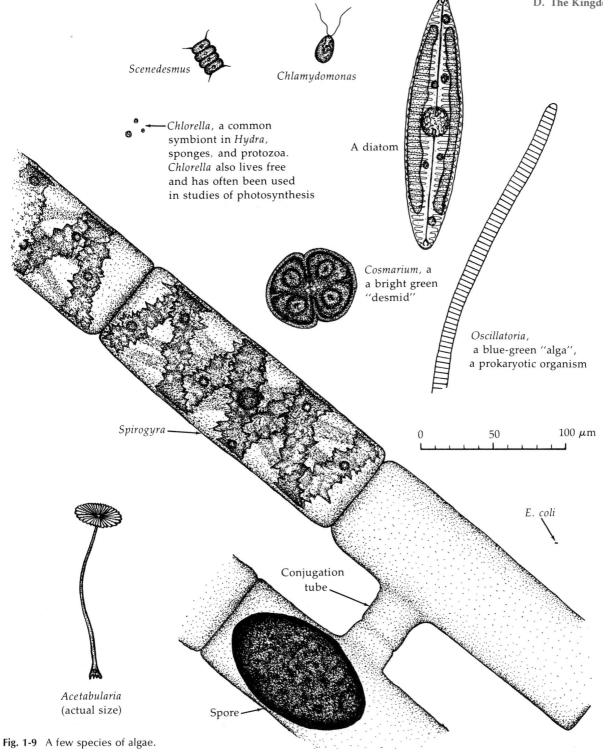

Scenedesmus

Chlamydomonas

Chlorella, a common
symbiont in *Hydra*,
sponges, and protozoa.
Chlorella also lives free
and has often been used
in studies of photosynthesis

A diatom

Cosmarium, a
a bright green
"desmid"

Oscillatoria,
a blue-green "alga",
a prokaryotic organism

Spirogyra

| | | |
0   50   100 μm

E. coli

Conjugation
tube

Acetabularia
(actual size)

Spore

Fig. 1-9 A few species of algae.

Some unicellular algae grow to a remarkable size. One of these is *Acetabularia* (Fig. 1-9), which lives in the warm waters of the Mediterranean and other tropical seas. The cell contains a single nucleus which lies in the base or rhizoid portion. In the mature alga, whose life cycle is 6 months (laboratory) to 1 year (nature), a cap of characteristic form develops. When cap development is complete, the nucleus divides into about 10^4 secondary nuclei which migrate up the stalk and out into the rays of the cap where they form cysts. After the cap decays and the cysts are released, meiosis occurs and the flagellated gametes fuse in pairs to form zygotes which again grow into diploid algae.

Because of its large size and the location of the nucleus in the base, *Acetabularia* has been used for important studies of **morphogenesis.**[38,39] The growing alga can be cut in several places to give rhizoid, basal, and apical sections. Furthermore, the nucleus can be removed and transplanted from one piece into another. Pieces can be grafted together to form algae with more than one nucleus; the apical portion from one species can be grafted onto the rhizoid portion of another, and so forth. Such investigations have shown that the morphology (shape) of the cap depends upon information sent from the nucleus. However, protein synthesis and growth occur even if the nucleus is removed. Thus, it appears that early in growth information-containing molecules (thought to be messenger RNA) migrate from the nucleus up into the growing tip. Continuing activity of the nucleus is not needed for protein synthesis in the rest of the cell.

Look through the microscope at almost any sample of algae from a pond or aquarium and you will see little boatlike **diatoms** slowly gliding through the water. The most prominent members of the division **Chrysophyta,** diatoms are noted for their external "shells" of silicon dioxide. Large and ancient deposits of "diatomaceous earth" contain these extremely durable silica skeletons which are finely marked, often with beautiful patterns (Fig. 1-9). The slow motion of diatoms is accomplished in a most unusual way by streaming of protoplasm through a groove on the surface of the cell. Diatoms are an important part of marine plankton, and it is estimated that three-quarters of the organic material of the world is produced by diatoms and dinoflagellates. Like the brown algae, Chrysophyta contain the pigment **fucoxanthin.**

Other groups of algae are the brown and red marine algae or seaweed. The former **(Phaeophyta)** include the giant kelps from which the polysaccharide **algin** is obtained. The **Rhodophyta** are delicately branched plants containing the red pigment, **phycoerythrin.** The polysaccharides, **agar** and **carrageenin,** a popular additive to chocolate drinks and other foods, come from red algae.

4. Lichens

Over 15,000 varieties of lichens grow on rocks and in other dry and often cold places. A lichen is an association of a fungus and either a true alga or a prokaryotic blue-green alga. Algae appear to benefit little from formation of lichens, but the fungi penetrate the algae cells and derive nutrients from them.[40] Although either of the two partners in a lichen can be cultured separately, the combination of the two is capable of producing special pigments and phenolic substances known as **depsides** which are not formed by either partner alone.

E. MULTICELLULAR ORGANISMS

No attempt will be made to describe the many multicellular animals (Metazoa) and plants (Metaphyta), but the following sections will attempt to pinpoint a few matters of biochemical significance.

1. The Variety of Animal Forms

In the past work with the higher vertebrates has dominated biochemical studies. However, a host of lower forms are now receiving increased attention. The simplest are tiny symbiotic* worms of the phylum **Mesozoa,** which live in the kidneys of deep sea-dwelling cephalopods.[41] They are made up of only 25 cells in a single layer enclosing one or a small number of elongated axial cells (Fig. 1-10). Study of the chemical communication between this small number of cells could be very revealing.

The **Porifera** or sponges are the most primitive of multicelled animals. They lack distinct tissues but contain several specialized types of cells. The body is formed by stationary cells that pump water through the pores to bring food to the sponge. Within the body **amebocytes** work in groups to form the **spicules** of calcium carbonate, silicon dioxide, or the protein **spongin** (Fig. 1-10). Sponges appear to lack any nervous system.

The next most complex major phylum **Cnidaria** (formerly Coelenterata) consists of radially symmetrical individuals with two distinct cell layers, the **endoderm** and **ectoderm.** Many species exist both as a polyp or **hydra** form (Fig. 1-10), and as a **medusa** or jellyfish. As far as we can tell, the jellyfish has no brain but the ways in which the neurons interconnect in a primitive radial net are of interest.

The Cnidaria have an extremely simple body form with remarkable regenerative powers. The freshwater hydra, a creature about a centimeter long (Fig. 1-10), has become a very important object of biological studies. Containing a total of about 10^5 cells, a complete hydra can be regenerated from a small piece of tissue if the latter contains some of both the inner and outer cell layers.[42]

The body of flatworms (phylum **Platyhelminthes**) consists of two external cell layers (endoderm and ectoderm) with a third layer between. A distinct excretory system is present. In addition to a nerve net resembling that of the Cnidaria there are a cerebral ganglion and distinct eyes. One large group of flatworms, the **planarians** (typically about 15 mm in length), inhabit freshwater streams. They are said to be the simplest creatures in which *behavior* can be studied.

Many parasitic flatworms (tapeworms and flukes) attack higher organisms. Among them are the **Schistosoma,** tiny worms that are transmitted to man through snails and attack the blood vessels. The resulting **schistosomiasis** is one of the most widespread debilitating diseases on earth today, affecting 200 million people or more.

The roundworms (**Nematoda**) and rotifers (**Rotifera**) are sometimes placed in separate phyla and sometimes together as the **Aschelminthes.** In these organisms, in addition to the **enteron** (alimentary tract) there is a separate body cavity. Free living nematodes abound in water and soil but many species are parasitic. They do enormous damage to plants and some species, e.g., trichina, hookworms, and filaria worms, attack man.

Rotifers with their whirling "wheels" of cilia on their heads and transparent bodies are a delight to the microscopist. To the cell biologist, they are interesting because they are "cell constant" organisms. The total number of cells in the body is said to be constant as is that in every part of every organ.

The **Annelida** (segmented worms) are believed to be evolutionary antecedents of the arthropods. Present-day members include earthworms, leeches, and $\sim 10^5$ species of marine **polychaetes.** Annelids have a true body cavity separate from the alimentary canal and lined by a peritoneum. They have a well-developed circulatory system and their blood usually contains hemoglobin or the closely related **erythrocruorin.**

About 10^6 species of **arthropods** (80% of all known animals) have been described. These creatures with a segmented exoskeleton of **chitin** and other materials include the horseshoe crabs, the Arachnida (scorpions, spiders, and mites), the Crustacea, Myriopoda (centipedes and millipedes), and the Insecta. Important biochemical problems are associated with the development and use of insecticides and with our under-

---

* Mesozoa have been regarded as parasitic, but they appear to facilitate excretion of NH_3 by the host by acidifying the urine.[40a] Note that Mesozoa are regarded by some biologists as representing a subkingdom rather than a phylum.

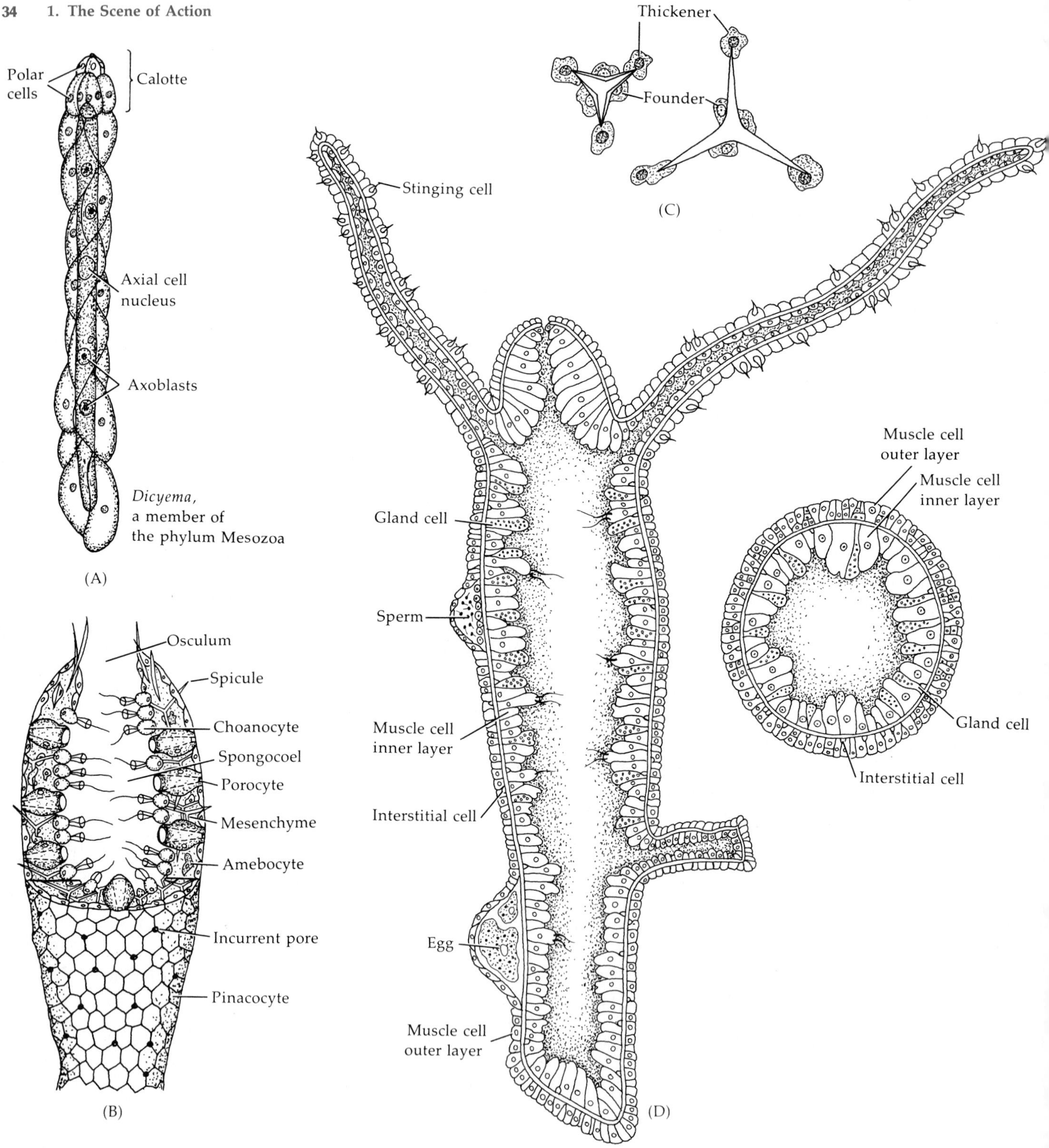

Polar cells

Calotte

Axial cell nucleus

Axoblasts

Dicyema,
a member of
the phylum Mesozoa

(A)

Osculum

Spicule

Choanocyte

Spongocoel

Porocyte

Mesenchyme

Amebocyte

Incurrent pore

Pinacocyte

(B)

Thickener

Founder

Stinging cell

(C)

Gland cell

Sperm

Muscle cell
inner layer

Interstitial cell

Egg

Muscle cell
outer layer

Muscle cell
outer layer

Muscle cell
inner layer

Gland cell

Interstitial cell

(D)

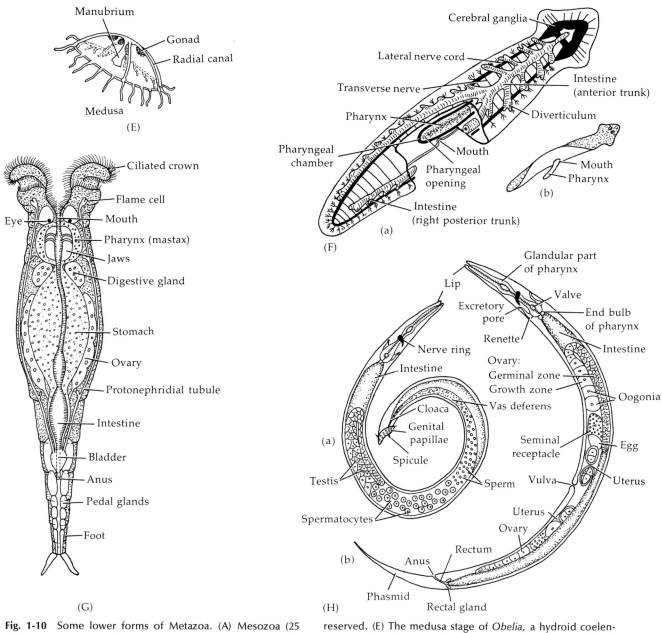

Fig. 1-10 Some lower forms of Metazoa. (A) Mesozoa (25 cells). After C. P. Hickman, "Biology of the Invertebrates." Mosby, St. Louis, Missouri, 1973. (B) A small asconoid sponge. After C. A. Villee, W. F. Walker, Jr., and R. D. Barnes, "General Zoology." Saunders, Philadelphia, Pennsylvania, 1973. (C) Ameboid cells of a sponge forming spicules. After Hickman, 1967. (D) Hydra. After W. F. Loomis, *Sci. Am.* **200**, 150 (Apr 1959). Copyright (1959) by Scientific American, Inc. All rights reserved. (E) The medusa stage of *Obelia,* a hydroid coelenterate. (F) A planarian, length 15 mm. After Hickman, 1973. (a) Diagram of digestive and nervous systems; cutaway section shows ventral mouth. (b) Pharynx extended through ventral mouth. (G) A rotifer, *Philodina* ($\sim 10^3$ cells). After C. A. Villee *et al.,* (1973). "General Zoology." (H) A nematode *Rhabditis. Ascaris* is very similar in appearance. After Hickman, (1973).

standing of the metamorphosis that occurs during the growth of many arthropods.[43]

Among the molluscs (phylum **Mollusca**) the squids and octopuses have generated the most interest among biochemists. Nerve cells **(neurons)** of squid contain giant axons, the study of which has led to much of our knowledge of nerve conduction. Octopuses show signs of intelligence not observed in other invertebrates whose nervous reactions seem to be entirely "preprogramed." The brains of some snails contain only 10^4 neurons, some of which are unusually large. Molluscan brains are under active study in an attempt to understand their organization and biochemistry.

The **Echinodermata** or spiny-skinned animals (starfish, sea urchins, and sea cucumbers) are regarded as a highly advanced phylum.* Their embryological development has been studied intensively.

The phylum **Chordata,** to which we ourselves belong, includes not only the vertebrates but also more primitive marine animals. Among these primitive species, which may be related to early ancestral forms, are the **tunicates** or sea squirts. They have a very high concentration of vanadium in their blood.

2. Animal Cell Types and Tissues

Isolated cells in tissue culture, no matter how highly differentiated, tend to revert quickly to three basic types known as **epitheliocytes, mechanocytes,** and **amebocytes.** Epitheliocytes are closely adherent cells derived from epithelial tissues and thought to be related in their origins to the two surface layers of the embryonic blastula. Mechanocytes, often called **fibroblasts** or **fibrocytes,** are derived from muscle, supporting or connective

---

* The apparent fivefold radial symmetry of adult echinoderms might suggest a relationship to the primitive Cnidaria. However, the larvae have bilateral symmetry and the body organization of the adults is misleading.

tissue. Like the amebocytes, they arise from embryonic mesenchymal tissue cells that have migrated inward from the lower side of the blastula (Chapter 16, Section C,2). **Neurons, neuroglia,** and **lymphocytes** are also regarded as distinct cell types.

a. Tissues

Cells aggregate to form four major kinds of tissue. **Epithelial tissues** line the primary surfaces of the body: the skin, the digestive tract, urogenital tract, and glands. Two common types are flat platelike cells **(squamous epithelium)** and **columnar epithelial cells.** Glands (sweat, oil, mammary, and internal secretory) as well as the sensory organs of the tongue, nose, and ear are all composed of epithelial cells. Epithelial cells are among the most highly polarized of cells. One side of each cell faces the outside, either air or water, while the other side is frequently directly against a basement membrane.

Supporting and connective tissue include fatty **(adipose)** tissue as well as **cartilage** and **bone.** Both of the latter contain large amounts of nonliving, intercellular material or **ground substance,** mostly complex polysaccharides. Embryonic **fibroblasts** differentiate into two types of fibers: white fibers produce the protein collagen and yellow fibers form elastin. Both of these proteins are assembled in the intercellular space and are embedded in the ground substance. **Osteoblasts** form bone by deposition (in layers 3–7 μm thick) of calcium phosphate, calcium carbonate, and organic cement.

A third tissue is **muscle** which is classified into three types: **striated** (voluntary skeletal muscle), **cardiac** (involuntary striated muscle), and **smooth** (involuntary) muscle. There are two major kinds of cells in **nervous tissue,** the fourth tissue type. Neurons are the actual conducting cells whose cell membranes carry nerve impulses. Glial cells lie between and around the neurons.

b. Blood Cells

Blood and the linings of blood vessels are sometimes regarded as a fifth tissue type. The human body contains 5×10^9 erythrocytes or red blood cells per milliliter, a total of 2.5×10^{13} cells in the 5 l of blood present in the body. Erythrocytes are rapidly synthesized in the bone marrow. The nucleus is squeezed out to leave a cell almost completely filled with hemoglobin. With an average lifetime of 125 days, human red blood cells are destroyed by leukocytes in the spleen and liver.

The white blood cells or **leukocytes** are nearly a thousandfold less numerous than red cells. About 7×10^6 are present per ml of blood. There are three types of leukocytes: **lymphocytes** ($\sim 26\%$ of the total), **monocytes** ($\sim 7\%$ of the total), and **polymorphonuclear leukocytes** or **granulocytes** (~ 70–75% of the total). Lymphocytes are about the same size as erythrocytes and are made in lymphatic tissue. Individual lymphocytes may survive for as long as 10 years. They function in antibody formation and are responsible for maintenance of long-term immunity. Monocytes, two times larger, are active in ingesting bacteria. Granulocytes of diameter 9–12 μm and formed in the red bone marrow serve a variety of functions. Three kinds of granulocytes are distinguished by staining: **neutrophils, eosinophils,** and **basophils.** Neutrophils work like monocytes in ingesting bacteria, while the function of the last two groups is not as clear. The number of eosinophils increases in hay fever and asthma and under the influence of some body parasites, while the basophil count is greatly increased in leukemia and is also increased by inflammatory diseases. Blood **platelets (thrombocytes),** tiny (2–3 μm diameter) cell-like bodies essential for rapid coagulation of blood, are formed by fragmentation of the cytoplasm of bone marrow **megakaryocytes.** One mature megakaryocyte may contribute 3000 platelets to the 1–3×10^8 per ml present in whole blood.

c. Cell Culture

The laboratory growth of isolated animal cells is becoming increasingly important in biochemistry. Sometimes it is important to have many cells with as nearly as possible identical genetic makeup. With bacteria such cells are obtained by **plating out** the bacteria and selecting a small colony that has

grown from a single cell to propagate a "pure strain." Similarly, single eukaryotic cells may be selected for tissue culture and give rise to a **clone** of cells which remain genetically identical until altered by mutations.

An important development is the culture of embryonic fibroblasts to obtain enough cells to perform prenatal diagnosis of inherited metabolic diseases (Box 1-D). Tissue culture is easiest with embryonic or cancer cells, but many other tissues can be propagated. Bear in mind that the cells that grow best are not entirely normal; for example, the well-known **HeLa** strain of human cancer cells which has been grown for many years throughout the world contains 70–80 chromosomes per cell compared with the normal 46.

3. Cell Contacts and Communication

Two basic questions about multicelled creatures are: How do cells stick together? and How do cells communicate one with another?

To answer the first question note that plant cells are surrounded by thick walls that cement the cells together and hold them in place. Animal cells lack rigid walls, but do associate in highly specific ways. Otherwise our bodies would collapse.

a. Cell Contacts and Junctions

Many epithelial cells, e.g., those lining the border of kidney tubules and secretory glands form **tight junctions** with adjacent cells. Electron microscopy shows that in these junctions the outer portions of the membranes actually fuse in some places (Fig. 1-3).[44,45] Freeze-etched surfaces (Box 1-C) of fractured tissues sometimes pass through tight junctions. Study of electron micrographs of such surfaces shows that some cells are completely surrounded by belts of tight junctions, sometimes referred to as **occlusion zones** or **terminal bars** (Fig. 1-11). Tight junctions between endothelial cells of blood capillaries in the brain prevent free diffusion of compounds from the bloodstream into brain cells and form the **blood–brain**

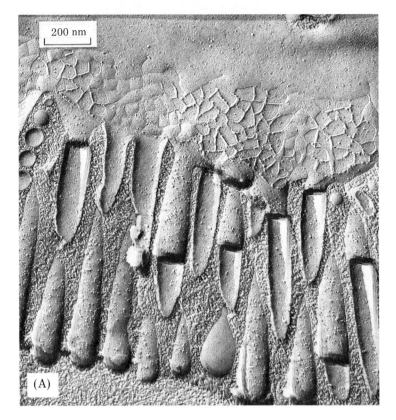

(A)

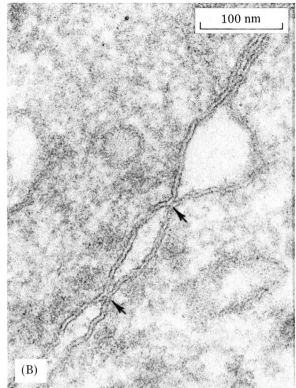

(B)

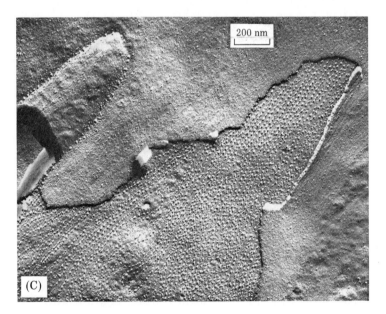

(C)

Fig. 1-11 Electron micrographs of cell junctions of three types. (A) Freeze-fractured zona occludens (occlusion zone) between epithelial cells of the rat small intestine. The tight junctions are represented as a meshwork of ridges (in the P or protoplasmic fracture face) or grooves (in the E fracture face which looks toward the extracellular space). These represent the actual sites of membrane fusion. Microvilli are seen in the lower part of the photograph. Scale, 200 nm. From D. S. Friend and N. B. Gilula, *J. Cell. Biol.* **53,** 771 (1972). (B) Thin cross section of tight junction between mouse hepatocytes. The arrows indicate points of membrane fusion. Scale, 100 nm. From N. B. Gilula, *in* "Cellular Membranes and Tumor Cells," p. 221. Copyright 1975 by The Williams & Wilkins Co., Baltimore. (C) A freeze-fractured septate junction from ciliated epithelium of a mollusc. This type of junction forms a belt around the cells. Fracture face P (central depressed area) contains parallel rows of membrane particles that correspond to the arrangement of the intracellular septa seen in thin sections. The surrounding fracture face E contains a complementary set of grooves. Particles in nonjunctional membrane regions (upper right corner) are randomly arranged. Scale, 200 nm. (D) Thin section of a septate junction of the type shown in (C). The plasma membranes of the two cells are joined by a periodic arrangement of electron-dense bars or septa, which are present within the intercellular space. Note the Golgi membranes in the lower part of the photograph. Scale, 100 nm.

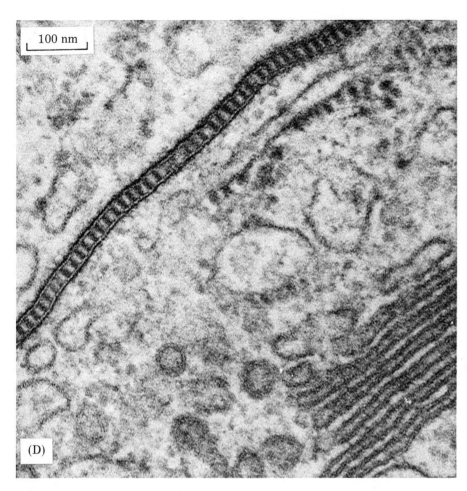

100 nm

(D)

barrier.[46] Tight junctions between neurons and adjacent cells surround the **nodes of Ranvier** (Chapter 5, Section B,3).

Contacts of another type, known as **septate desmosomes** or **adhesion discs,** form a belt around the cells of invertebrate epithelia. In these contacts a space of ~18 nm between adjacent cell membranes is bridged in a number of places by thin walls. Behind the desmosomes the membrane often is backed up at these points by an electron-dense region to which are attached many fine fibrils **(microfilaments)** of ~6–10 nm diameter.[47]

In addition to tight junctions cell interfaces often contain an extensive region with a 10–20 nm gap between adjacent cells and in which 6.0 nm fibrils are present within the cytoplasm. In freeze-etched preparations, surfaces within the gap junctions show a polygonal lattice of structures with 10 nm repeat distances. These **gap junctions** are probably regions of strong cohesion and may be very significant in cell communication.

b. Communicating Junctions

In addition to cohesion there must be communication among cells; without this, harmonious growth and differentiation of cells would be impossible. One method of communication is by passage of chemical substances through special junctions.[48] One of many observations suggesting the importance of such communication is "electrical coupling" of cells. Whereas cell membranes usually have a very high electrical resistance,

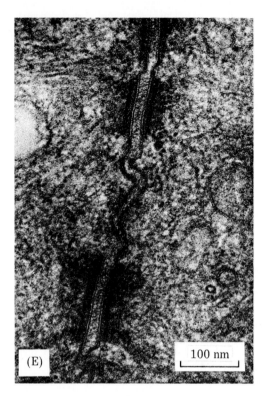

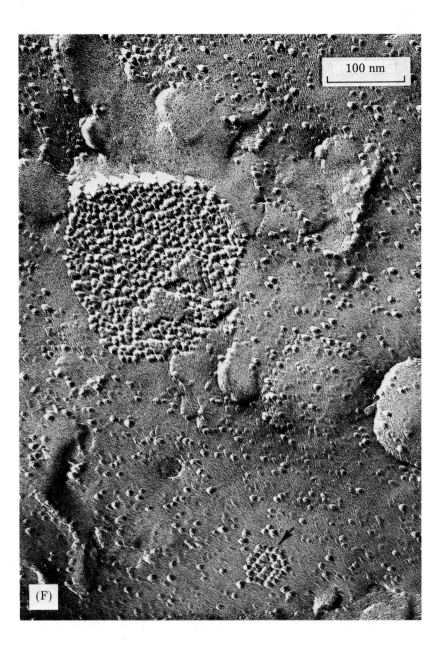

Fig. 1-11 (E) Desmosomes (macula adherans) in rat intestinal epithelium. Features include a wide (25–35 nm) intercellular space containing dense material, two parallel cell membranes, a dense plaque associated with the cytoplasmic surface and cytoplasmic tonofilaments that converge on the dense plaque. Scale, 100 nm. (C), (D), and (E) are from N.B. Gilula[44] (pp. 18, 19, and 21 respectively). (F) Freeze-fractured surface through gap junctions between communicating cells in culture. Both a large junction and a smaller one below (arrow) can be seen. Scale, 100 nm. (G) Gap junctions in thin section. Scale, 100 nm. (F) and (G) are from N. B. Gilula, *Nature* (*London*) **235**, 262 (1972).

membranes of coupled cells possess regions of low resistance which have been associated with gap junctions.[49] One of the most highly developed communicating junctions is the **synapse,** a specialized contact between neurons. A nerve impulse transmitted through the membrane of one neuron triggers the release of a chemical substance (**trans-mitter)** which passes across the gap between cells of the synapse and initiates a nerve impulse in the second neuron.

It has been theorized that cancer cells have fewer communicating junctions than normal cells,[50] but experiments in tissue culture have raised doubts.[51]

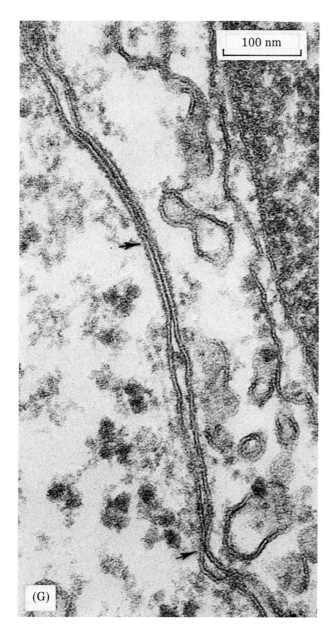

(G)

ment with the enzyme **trypsin** (which digests protein "cement" between the cells). When dissociated cells of orange sponges were mixed with those from yellow sponges, the cells clumped together to reform small sponges.[52,53] Furthermore, orange cells stuck to orange cells and yellow to yellow cells. Similar results have been obtained using liver, kidney, and embryonic brain cells in culture. Again liver cells clumped together with liver cells and kidney cells with other kidney cells. How do the cells recognize each other? Answers are coming from much biochemical work now in progress.

When a wound heals, epithelial cells grow and move across the wound surface but they stop when they meet. Likewise, cells and tissue culture growing on a glass surface experience this same **contact inhibition** and spread to form a unicellular layer. Cancer cells in culture do not stop but climb one on top of the other, apparently lacking proper recognition and communication. The biochemical understanding of this phenomenon will be of tremendous medical significance.

4. Higher Plants and Plant Tissues

Botanists recognize two divisions of higher plants. The **Bryophyta** or moss plants consist of the **Musci** (mosses) and **Hepaticae** (liverworts). The plants grow predominantly on land, are characterized by swimming sperm cells, and a dominant gametophyte (haploid) phase. The second division are the **Tracheophyta** or vascular plants, which contain conducting tissues. About 2×10^5 species are known. The ferns (class Filicineae, formerly Pteridophyta) are characterized by a dominant diploid plant and alternation with a haploid phase (Section C,5).

The seed plants are represented by two classes: the **Gymnosperms** (cone-bearing trees) and the **Angiosperms,** the true flowering plants.

Several kinds of plant tissues are recognized. Undifferentiated, embryonic cells found in rap-

c. Cell Recognition

It is important for cells of higher organisms to be able to recognize other cells as being identical, as belonging to another tissue, or as being "foreign."

A striking experiment was performed with sponges, whose cells can be separated by treat-

idly growing regions of shoots and roots form the **meristematic tissue.** By differentiation, the latter yields the simple tissues, the parenchyma, collenchyma, and sclerenchyma. **Parenchyma** cells are among the most abundant and least specialized in plants. They give rise through further differentiation to the **cambium layer,** the growing layer of roots and stems. They also make up the pith or pulp in the center of stems and roots, where they serve as food storage cells.

The **collenchyma,** present in herbs, is composed of elongated supporting cells and the **sclerenchyma** of woody plants is made up of supporting cells with hard lignified cell walls and a low water content. This tissue includes **fiber cells,** which may be extremely long; e.g., pine stems contain fiber cells of 40 μm diameter and 4 mm long.

Two complex tissues, the **xylem** and **phloem,** provide the conducting network or "circulatory system" of plants. In the xylem or woody tissue, most of the cells are dead and the thick-walled tubes **(tracheids)** serve to transport water and dissolved minerals from the roots to the stems and leaves. The phloem cells provide the principal means of downward conduction of foods from the leaves. Phloem cells are joined end to end by **sieve plates,** so-called because they are perforated by numerous minute pores through which cytoplasm of adjoining sieve cells appears to be connected by strands 5–9 μm in diameter.[54] Mature sieve cells have no nuclei, but each sieve cell is paired with a nucleated "companion" cell.

Epidermal tissue of plants consists of flat cells, usually containing no chloroplasts, with a thick outer wall covered by a heavy waxy **cuticle** about 2 μm thick. Only a few specialized cells are found in the epidermis. Among them are the paired **guard cells** that surround the small openings known as **stomata** on the undersurfaces of leaves and control transpiration of water. Specialized cells in the root epidermis form **root hairs,** long extensions (~ 1 mm) of diameter 5–17 μm. Each hair is a single cell with the nucleus located near the tip.

Figure 1-12 shows a section from a stem of a typical angiosperm. Note the thin cambium layer between the phloem and the xylem. Its cells continuously undergo differentiation to form new layers of xylem increasing the woody part of the stem. New phloem cells are also formed and as the stem expands, all of the tissues external to the cambium are renewed and the older cells are converted into bark.

Plant **seeds** consist of three distinct portions. The **embryo** develops from a zygote formed by fusion of a sperm nucleus originating from the pollen and an egg cell. The fertilized egg is surrounded in the gymnosperms by a nutritive layer or **endosperm** which is **haploid** and is derived from the same gametophyte tissue that produced the egg. In angiosperms *two* sperm nuclei form; one of these fertilizes the egg, while the other fuses with *two* haploid **polar nuclei** derived from the female gametophyte. (The polar nuclei are formed by the same mitotic divisions that formed the egg.) From this develops a $3n$ **triploid** endosperm.

F. THE REGULATION OF CELL CHEMISTRY

Biologists have long been fascinated by the ability of organisms to maintain a nearly constant internal environment despite drastic changes outside the organism. For example, the pH of our blood is constant at 7.40 ± 0.05. The concentration of free glucose (5 mM) in blood may rise briefly after a meal but otherwise it is remarkably constant. The same could be said about most of the other constituents of body fluids and of the cells themselves. This phenomenon, **homeostasis,** depends upon a complex set of regulatory mechanisms.

Consider the entrance of a nutrient compound A into a cell. If the compound simply came into the cell and reacted in various ways, and if none of the products could leave the cell, an **equilibrium** would be reached. Such a system would have none of the properties of a living cell. Living cells take in nutrients and excrete products. Within the cell, reactions take place in sequences that often

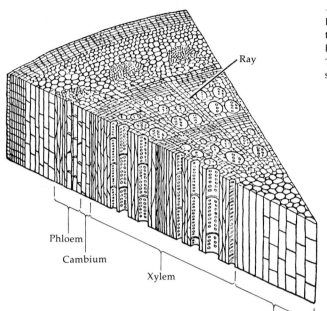

Ray

Phloem

Cambium

Xylem

Pith

Fig. 1-12   Section of the stem of an angiosperm. Enlarged sections showing tubes of the phloem (left) and xylem (right). From S. Biddulph and O. Biddulph, *Sci. Am.* **200**, 44–49 (Feb 1959). Copyright 1959 by Scientific American, Inc. All rights reserved.

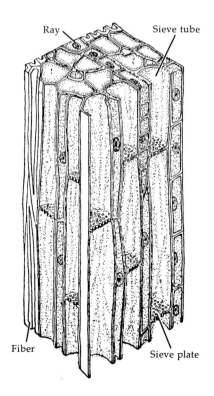

Ray

Sieve tube

Fiber

Sieve plate

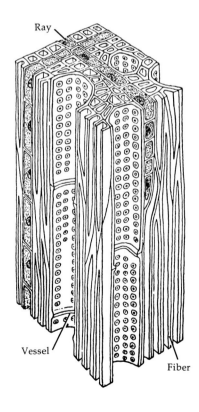

Ray

Vessel

Fiber

branch and converge. A simple hypothetical pathway with one branch is indicated in Eq. 1-3.

$$A \xleftarrow{} A \longrightarrow B \xrightarrow{\begin{array}{c} C \rightleftharpoons D \\ \nearrow \qquad \searrow \\ \searrow \qquad \nearrow \end{array}} E \xrightarrow{} \begin{array}{c} \text{excretion} \\ \text{or} \\ \text{immobilization} \end{array} \qquad (1\text{-}3)$$

cell membrane $C' \longrightarrow D'$

Some of the properties of homeostasis are already seen in such a simple **open system.**[55] A **steady state** (not an equilibrium) is set up in which the rates of production and destruction of the intermediate compounds exactly balance. The change of the rate of transport of A into the cell or of a product out of the cell, or a change in the catalytic activity of an enzyme promoting an irreversible step in the chain will significantly alter the steady state concentrations of material within the system.

Living cells are more complex in that they have sensing devices which detect and counteract changes in steady state concentrations. In most instances these are **feedback** devices, closely analogous in their action to that of a thermostat in a heating system.

The **active site** or catalytic center of an enzyme constitutes a relatively small part of the entire molecule. Much of the rest of the molecule, particularly the outer surface, may have its amino acid sequence altered substantially by mutations without altering the catalytic activity. Nevertheless, *interactions* of various portions of the enzyme surface with other molecules can indirectly affect catalysis. Molecules in a concentrated solution, as in the cytoplasm, tend to stick together. The binding of a second molecule to a particular site on an enzyme surface may distort the enzyme structure and bring about an increase or a decrease in the catalytic efficiency. The interaction of enzymes with **feedback inhibitors** that "turn off" enzymes when the end products of metabolic sequences accumulate is one facet of regulatory biochemistry that depends upon such interactions.

A similar mechanism underlies the binding of **hormone** molecules to specific **receptors** on cell membranes or within cells. Hormones are regulatory compounds that are released by cells of one tissue and which act to regulate the metabolism of **target cells** in another tissue. The binding of the hormones to the receptor presumably distorts the latter in such a way as to trigger a response.

REFERENCES

1. R. Y. Stanier, M. Doudoroff, and E. A. Adelberg, "The Microbial World," 3rd ed. Prentice-Hall, Englewood Cliffs, New Jersey, 1970.
2. J. Mandelstam and K. McQuillen, "Biochemistry of Bacterial Growth." Wiley, New York, 1968.
3. J. Maniloff and H. J. Morowitz, *Bacteriol. Rev.* **36**, 263–290 (1972)
4. J. D. Watson, "Molecular Biology of the Gene," 3rd ed., p. 61. Benjamin, Menlo Park, California, 1976.
5. S. E. Luria, *in* "The Bacteria" (I. C. Gunsalus and R. Y. Stanier, eds), Vol. 1, p. 1. Academic Press, New York, 1960.
6. H. C. Berg, *Nature (London)* **254**, 389–392 (1975).
7. K. V. Thimann, "The Life of Bacteria," 2nd ed. Macmillan, New York, 1963.
8. K. Decker, K. Jungermann, and R. R. Thauer, *Angew. Chem., Int. Ed. Engl.* **9**, 138–158 (1970).
9. R. Y. Stanier, R. Kunisawa, M. Mandel, and G. Cohen-Bazire, *Bacteriol. Rev.* **35**, 171–205 (1971).

9a. G. J. Wagner and H. W. Siegelman, *Science* **190**, 1298–1299 (1975).
10. J. D. Watson, "Molecular Biology of the Gene," 3rd ed., p. 499. Benjamin, Menlo Park, California, 1976.
11. P. L. Paine, L. C. Moore, and S. B. Horowitz, *Nature (London)* **254**, 109–114 (1975).
12. Z. B. Carothers, *Science* **175**, 652–654 (1972).
13. D. H. Northcote, *Endeavour* **30**, 26–33 (1971).
13a. C. Golgi, *Arch. Ital. Biol.* **30**, 60 and 278 (1898).
14. G. Palade, *Science* **189**, 347–358 (1975).
15. B. Satir, *Sci. Am.* **233**, 29–37 (Oct. 1975).
16. M. Neutra and C. P. Leblond, *Sci. Am.* **220**, 100–107 (Feb 1969).
17. W. G. Whaley, M. Dauwalder, and J. E. Kephart, *Science* **175**, 596–599 (1972).
18. S. E. Nyquist, F. L. Crane, and D. J. Morré, *Science* **173**, 939–941 (1971).
19. J. Reinert and H. Ursprung, eds., "Origin and Continuity of Cell Organelles," Springer-Verlag, Berlin and New York, 1971.
20. L. Orci, F. Malaisse-Lagae, M. Ravazzola, M. Amherdt, and A. E. Renold, *Science* **181**, 561–562 (1973).

21. A. V. Loud, *J. Cell Biol.* **37,** 27–46 (1968).
22. H. P. Hoffmann and C. J. Avers, *Science* **181,** 749–751 (1973).
23. J. T. Dingle, "Lysosomes, A Laboratory Handbook," North-Holland Publ., Amsterdam, 1972.
24. C. de Duve, *Science* **189,** 186–194 (1975).
25. N. E. Tolbert, *Annu. Rev. Plant Physiol.* **22,** 45–74 (1971).
26. M. A. Sleigh, *Endeavour* **30,** 11–17 (1971).
27. J. B. Olmsted and G. G. Borisy, *Annu. Rev. Biochem.* **42,** 507–540 (1973).
28. L. Margulis, *Sci. Am.* **225,** 49–57 (Aug 1971).
29. P. John and F. R. Whatley, *Nature (London)* **254,** 495–498 (1975).
30. R. A. Raff and H. R. Mahler, *Science* **177,** 575–582 (1972).
30a. T. Cavalier-Smith, *Nature (London)* **256,** 463–468 (1975).
31. T. Uzzell and C. Spolsky, *Science* **180,** 516–517 (1973).
32. R. A. Raff and H. R. Mahler, *Science* **180,** 517 (1973).
32a. C. L. Markert, J. B. Shaklee, and G. S. Whitt, *Science* **189,** 102–114 (1975).
33. J. A. McKanna, *Science* **179,** 88–90 (1973).
34. D. L. Hill, "Biochemistry and Physiology of Tetrahymena," Academic Press, New York, 1972.
35. A. Robertson, D. J. Drage, and M. H. Cohen, *Science* **175,** 333–335 (1972).
36. R. H. Davis and F. J. deSerres, *in* "Methods in Enzymology" (H. Tabor and C. W. Tabor, eds.), Vol. 17A, p. 79. Academic Press, New York, 1970; a summary of techniques.
37. G. R. Fink, *in* "Methods in Enzymology" (H. Tabor and C. W. Tabor, eds.), Vol. 17A, p. 59. Academic Press, New York, 1970.
38. J. L. A. Brachet, *Endeavour* **24,** 155–161 (1965).
39. G. Rickter, *in* "Physiology and Biochemistry of Algae" (R. A. Lewin, ed.), Vol. 2, pp. 633–652. Academic Press, New York, 1962.
40. V. Ahmadjian, *in* "Physiology and Biochemistry of Algae" (R. A. Lewin, ed.), Vol. 2, p. 817. Academic Press, New York, 1962.
40a. E. A. Lapan, *Comp. Biochem. Physiol.* **52A,** 651–657 (1975).
41. E. A. Lapan and H. Morowitz, *Sci. Am.* **227,** 94–101 (Dec. 1972).
42. W. F. Loomis, *Sci. Am.* **200,** 145–146 (Apr 1959).
43. See M. Florkin and B. T. Scheer, eds., "Chemical Zoology," Vols. 5 and 6. Academic Press, New York, 1970, 1971.
44. N. B. Gilula, *in* "Cell Communication" (R. P. Cox, ed.) Wiley, New York, 1974, pp. 1–29.
45. N. S. McNutt and R. S. Weinstein, *Prog. Biophys. Mol. Biol.* **26,** 47–101 (1973).
46. T. S. Reese and M. J. Karnovsky, *J. Cell Biol.* **34,** 207–217 (1970).
47. A. J. Hudspeth and J. P. Revel, *J. Cell Biol.* **50,** 92–101 (1971).
48. W. R. Loewenstein, *Sci. Am.* **222,** 79–86 (May 1970).
49. B. W. Payton, M. V. L. Bennett, and G. D. Pappas, *Science* **166,** 1641–1643 (1969).
50. W. R. Loewenstein and Y. Kanno, *J. Cell Biol.* **33,** 225–234 (1967).
51. R. G. Johnson and J. D. Sheridan, *Science* **174,** 717–719 (1971).
52. T. Humphreys, *Dev. Biol.* **8,** 27–47 (1963).
53. A. A. Moscona, *Sci. Am.* **205,** 143–162 (Sep 1961).
54. P. Jarvis and R. Thaine, *Nature (London), New Biol.* **232,** 236 (1971).
55. L. von Bertalanffy, *Science* **113,** 23–29 (1950).

STUDY QUESTIONS

1. Describe the principal structural or organizational differences between prokaryotic and eukaryotic cells.

2. Describe two principal functions of proteins within cells, one function of DNA, two functions of RNA, one function of lipids.

3. Compare the chemical makeup of ribosomes, of cell membranes, and of bacterial flagella.

4. Assume the following dimensions: Mycoplasma, sphere, 0.33 μm diameter; *E. coli*, cylinder, 0.8 μm diameter $\times$ 2 μm; liver cell, sphere 20 μm; root hair, cylinder, 10 μm diameter $\times$ 1 mm.

   a. Calculate for each cell the total volume, the mass in grams and in daltons (assume a specific gravity of 1.0).

   b. Assume that bacterial ribosomes are approximately spherical with a diameter of 23 nm. What is their volume? If the mass of a bacterial ribosome is 2.7×10^6 daltons (Chapter 15, Section C,1), what is its apparent density (divide mass by volume)? Experimentally the buoyant density of bacterial ribosomes in a cesium chloride gradient (Chapter 2) is about 1.6 g/cm^3. How can this difference be explained?

   If eukaryotic ribosomes are 1.17 times larger than bacterial ribosomes in linear dimensions, what is the volume of a eukaryotic ribosome?

   c. What fraction by volume of *E. coli* consists of cell wall, plasma membrane, ribosomes (assume 15,000). Assuming that a cell of *E. coli* is 80% water, what fraction by weight of the total solids consists of ribosomes? of DNA (assuming 2 chromosomes per cell)?

   d. What fraction by volume of a liver cell is composed of ribosomes, nucleus, plasma membrane, mitochondria (assume 1000 mitochondria)?

5. a. What is the molar concentration of an enzyme of which only one molecule is present in an *E. coli* cell?

   b. Assume that the concentration of K^+ within an *E. coli* cell is 150 mM. Calculate the number of K^+ ions in a single cell.

   c. If the pH inside the cell is 7.0 how many H^+ ions are present?

6. If chromosomes (and chromatin) are 15% DNA, what will be the mass of the 23 pairs of chromosomes in a human diploid cell? If the nucleus has a diameter of 5 μm and a density of 1.1 g/cm³, what fraction by weight of the nucleus is chromatin?

7. Compare the surface to volume ratios for an *E. coli* cell, a liver cell, a nucleus of an eukaryotic cell, a root hair. If a cell of 20 μm diameter is 20% covered with microvilli of 0.1 μm diameter and 1 μm length centered on a 0.2 μm spacing, how much will the surface/volume ratio be increased?

8. It has been shown that the code for specifying a particular amino acid in a protein is determined by a sequence of three nucleotides in a DNA chain. Each such sequence is known as a codon. There are four different kinds of nucleotide units in DNA. How many different codons exist? Note that this is larger than the number of different amino acids (20) incorporated into proteins plus the three stop (termination) codons (see Tables 15-2 and 15-3 for a list of codons).

9. State two similarities and two differences between blue-green algae and green algae.

10. Compare the sizes and structures of bacterial and eukaryotic flagella.

11. Compare the chemical makeup of the extracellar "coat" or "matrix" materials secreted by the following cells: bacteria, fibroblasts, osteoblasts, plant cells, fungi.

12. Define the "steady state" of living cells and contrast with a state of chemical equilibrium.

2

The Molecules from Which We Are Made

From the biochemist's viewpoint life begins with the small chemical "building blocks" which cells either make or gather from another source. The 20-odd amino acids, a few sugars, acetic acid, some longer chain fatty acids, glycerol, 2 purine and 3 pyrimidine "bases," and phosphoric acid serve as the raw materials for construction of thousands of more elaborate compounds. A dozen compounds of miscellaneous structure, the vitamins, together with several inorganic ions, are also present universally in all living things.

The small building blocks or **monomers** are joined together by cells into giant **macromolecules** or **polymers** in which the monomer units are linked by strong covalent chemical bonds. A polymer may consist of just a few monomer units (an **oligomer**) or it may contain hundreds, thousands, or even millions of units. A typical protein contains 100 to several hundred amino acids, the DNA molecule of the genome of *E. coli* contains ~4 × 10^6 nucleotide pairs, and a highly branched starch molecule may contain over 10^6 sugar units. Biopolymers may be straight chainlike molecules or they may be branched. The chains are sometimes coiled into rigid rods with a helical configuration maintained by many weaker secondary bonds. More often they are folded into irregular

structures of great complexity. Frequently they are stacked together to form networks, fibers, and membranes. In some cases (e.g., in collagen of connective tissue), the protein molecules are "cross-linked" by strong covalent chemical bonds. However, in most instances, the macromolecules of cells are tied together by weaker electrostatic and van der Waals forces.

The shapes and dimensions of molecules are of special importance in biochemistry. The following section provides a review of basic principles.

A. STRUCTURAL PRINCIPLES FOR SMALL MOLECULES

Stable organic molecules are held together with covalent bonds which are usually very strong, the standard free energies of formation (ΔG_f^0) being of the order of −400 kJ/mol. The bonds have definite directions (measured by bond angles) and lengths.

1. Bond Angles

There is usually a tetrahedral arrangement of bonds around single-bonded carbon atoms and most phosphorus atoms. All of the bond angles (there are six of them) about this carbon atom have nearly the same tetrahedral angle of 109.5°. Re-

109.5°

member this angle. Bond angles connecting chains of carbon atoms in organic compounds vary only slightly from it, and even atoms that are attached to fewer than four groups usually have similar angles; for example, the H—O—H angle in the water molecule is 105°, and the H—N—H angles of ammonia are 107°. In ethers the C—O—C angle is 111°. However, angles of nearly 90° are seen in some simple compounds; e.g., 92° in H_2S and PH_3, and 101° in H_2O_2.

The presence of **double bonds** leads to **planarity** and to compounds with bond angles of $\sim 120°$ (the internal angle in a hexagon). In many cases what at first glance appears to be nonplanar is actually planar because of **resonance;** for example, the peptide linkage (see Fig. 2-4) is nearly planar and the angles all fall within four degrees of 120°.

2. Bond Lengths

Chemists describe bond lengths as the distances between the nuclei of bonded atoms. Remember that the C—C single bond has a length of 0.154 nm (1.54 Å). The C—O bond is ~ 0.01 nm shorter (0.143 nm), and the typical C—H bond has a length of ~ 0.109 nm. The C—N bond distance is halfway between that for C—C and C—O (0.149 nm). Other lengths, such as that of O—H, can be estimated from the covalent radii given in Table 2-1.

The length of a double bond between any two atoms (e.g., C=C) is almost exactly 0.020 nm less than that for a single bond between the same atoms. If there is resonance, hence only partial double bond character, the shortening will be somewhere in between. For example, the length of the C—C bond in benzene is 0.140 nm; the C—O distances in the carboxylate anion are 0.126 nm.

Carboxylate ion

Using simple geometry, it is easy to calculate overall lengths of molecules; two distances worth remembering are

Distance between alternate C atoms in fully extended hydrocarbon chain

Distance across a benzene ring*

3. Contact Distances

The covalent bond distances and angles tell us how the atomic nuclei are arranged in space but they do not tell us anything about the outside surfaces of molecules. The distance from the center of an atom to the point at which it contacts an adjacent atom in a packed structure such as a crystal is known as the **van der Waals radius.** The ways in which biological molecules fit together are determined largely by the van der Waals contact radii. These, too, are listed in Table 2-1. In every case they are approximately equal to the *covalent radius plus 0.080 nm.* Van der Waals radii are not as con-

* Chemists often use simplified formulas, such as the one for benzene. The six hydrogen atoms have been omitted. Resonance between the two possible arrangements of the three double bonds is indicated by the circle. Other chemical shorthand of the following type is used throughout the book. (Oxygen and nitrogen atoms will always be shown.)

TABLE 2-1
The Sizes of Some Atoms[a,b]

Element	Covalent radius (nm)	van der Waals radius (nm)
H	0.030	0.12
F	0.064	0.135
C	0.077	0.16
N	0.070	0.15
O	0.066	0.145
Cl	0.099	0.180
Si	0.117	—
P	0.110	0.19
S	0.104	0.185
Br	0.114	0.195
I	0.133	0.215
"Radius of methyl group"		0.20
Half-thickness of aromatic molecules		0.170

[a] L. Pauling, "The Nature of the Chemical Bond," 3rd ed., pp. 224–227 and 260. Cornell Univ. Press, Ithaca, New York, 1960.
[b] Covalent radii for two atoms can be summed to give the interatomic distance. The van der Waals radii determine how closely molecules can pack. Thus, the closest observed contacts between atoms in macromolecules are approximately 0.02 nm less than the sum of the van der Waals radii; V. Sasisekharian, A. V. Lakshminarayanan, and G. N. Ramachandran, *in* "Conformation of Biopolymers" (G. N. Ramachandran, ed.), Vol. 2, p. 641. Academic Press, New York, 1967.

stant as covalent radii because atoms can be "squeezed" a little, but only enough to decrease the contact radii by 0.005–0.01 nm.

4. Asymmetry: Right-Handed and Left-Handed Molecules

The left hand looks much like the right hand, but they are different. One is the **mirror image** of the other. A more practical difference is that your right hand will not go into a left-handed glove. Despite our daily acquaintance with "handedness" it isn't easy to explain in words how a right and a left hand differ. However, since most biochemicals, monomers and polymers alike, are asymmetric, it is important to understand molecular asymmetry. It is essential to be able to visualize asymmetric molecules in three dimensions and to write their structures on paper. One of the best

ways of learning to do this is to study molecular models. You may learn even more by making your own models (see Appendix).

Whenever four different groups are bonded to a central carbon atom, the molecule is asymmetric and the four groups can be arranged in two different ways, i.e., in two different **configurations.** Consider an α-amino acid:

$$R—\overset{\displaystyle NH_2}{\underset{\displaystyle H}{\overset{|}{\underset{|}{C}}}}—COOH$$

α-Amino acid

To indicate its three-dimensional structure on a flat piece of paper, the bonds that project out of the plane of the paper and up toward the reader are often drawn as elongated triangles, while bonds that lie behind the plane of the paper are shown either as thin lines or as dotted lines.

L-Amino acid

The family of amino acids having the configuration about the α-carbon atom shown in the preceding structural formulas are known as L-amino acids. Those amino acids which are mirror images of the L-amino acids are D-amino acids.* Pairs of D and L compounds are known as **enantiomorphic** forms or **enantiomers.**

* The D and L designations give the absolute configuration of a molecule. An experimentally measurable quantity related to the asymmetry of molecules is the optical rotation (Chapter 13, Section B,5). The sign of the optical rotation (+ or −) is sometimes given together with the name of a compound, e.g., D(+)-glucose. The older designations *d* (dextro) and *l* (levo) indicate + and −, respectively. However, compounds with D configuration may have either + or − optical rotation.

The L-amino acid structure above is shown in four different ways. To recognize them all as the same structure, they can be turned in space to an orientation in which the carboxyl group is up, the side chain (or R group) is down (and projecting behind the paper) while the amino group and hydrogen atom project upward from the paper at the sides. According to a convention introduced by Emil Fischer, an amino acid is L if, when oriented in this manner, the amino group lies to the left and D if it lies to the right. Fischer further pro-

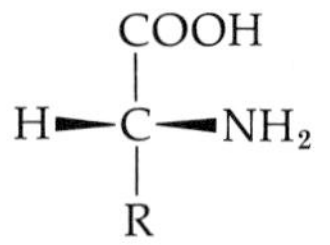

$$
\begin{array}{c}
\text{COOH} \\
| \\
\text{H}\!\!-\!\!\text{C}\!\!-\!\!\text{NH}_2 \\
| \\
\text{R}
\end{array}
$$

D-Amino acid

posed that the amino acid in this orientation could be projected onto the paper and drawn with ordinary lines for all the bonds as shown in the **Fischer projection formula** of L-alanine. The configuration is (S) in the Cahn–Ingold–Prelog terminology (Section 5).

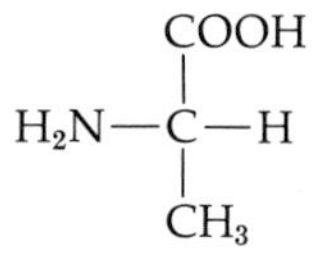

$$
\begin{array}{c}
\text{COOH} \\
| \\
\text{H}_2\text{N}\!-\!\text{C}\!-\!\text{H} \\
| \\
\text{CH}_3
\end{array}
$$

Fischer projection formula of L-alanine

Biochemical reactions in general are highly stereospecific and a given enzyme catalyst will act upon molecules of only one of the two configurations. A closely related fact is that proteins ordinarily consist entirely of amino acids of the L series.

In the older literature **optical isomerism** of the type represented by D and L pairs was usually discussed in terms of "asymmetric carbon atoms" or **asymmetric centers.** Now the terms **chiral** (pronounced *ki-ral*) **molecules,** chiral centers, and **chirality** (Greek: "handedness") are widely used. Whereas compounds with one chiral center exist as an enantiomorphic pair, molecules with two or more chiral centers also exist as **diastereomers.** These are pairs of isomers with an opposite configuration at one or more of the chiral centers, but which are not complete mirror images of each

other. An example is L-threonine* which has the 2(S), 3(R) configuration. The diastereomer with the 2(S), 3(S) configuration is known as L-*allo*-threonine.

5. The *RS* Notation for Configuration

In 1956 Cahn, Ingold, and Prelog proposed an unambiguous way of specifying configuration at any chiral center.[1] It is especially useful for those classes of compounds where no well-established DL system is available. The four groups surrounding the central carbon atom (or other central atom) are ranked according to a **priority sequence.** The priority of a group is determined by a number of sequence rules, the most important of which is *"higher atomic number precedes lower."* In the illustration below, the priorities of the groups in the D-alanine molecule are indicated by the letters $a > b > c > d$. The highest priority is assigned to the NH$_2$ groups which contain nitrogen bonded to the central atom. To establish the configuration, the observer views the molecule down the axis connecting the central atom to the group having the lowest priority, i.e., to group d. When viewed in this way, the sequence of groups a, b, and c can either be that of a right-handed turn (clockwise) as shown in the drawing below or that of a left-handed turn (counterclockwise).

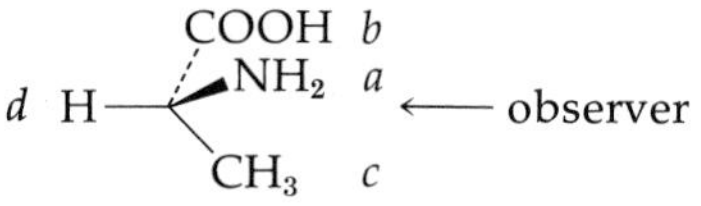

$$
\begin{array}{c}
\text{COOH} \;\; b \\
\nearrow \text{NH}_2 \;\; a \\
d \;\; \text{H}\!\!-\!\!\diagdown \qquad \longleftarrow \text{observer} \\
\text{CH}_3 \;\; c
\end{array}
$$

D-Alanine

* Sometimes the subscript s or g is added to the D or L prefix to indicate whether a compound is being related to serine or to glyceraldehyde (sugar convention, Section C,1) as a configurational standard. Ordinary threonine is L$_s$- or D$_g$-threonine. The configuration of dextrorotatory (+)-tartaric acid can be given as 2(R),3(R), or as D$_s$, or as L$_g$.

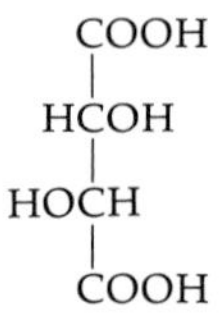

$$
\begin{array}{c}
\text{COOH} \\
| \\
\text{HCOH} \\
| \\
\text{HOCH} \\
| \\
\text{COOH}
\end{array}
$$

2(R),3(R)-Tartaric acid

The view down the axis toward the group of lowest priority (d), which lies behind the page. The right-handed turn indicates the configuration R (rectus = (right); the opposite configuration is S (sinister = left)

To establish the priority sequence of groups one first looks at the atoms that are bonded directly to the central atom, arranging them in order of decreasing atomic number. Then if necessary, one moves outward to the next set of atoms, again comparing atomic numbers. In the case of alanine, groups b and c are ordered in this way because they both contain carbon directly bonded to the central atom. When double bonds are present at one of the atoms being examined, e.g., the carboxyl group in alanine, "phantom atoms" replicating the real ones are imagined at the ends of the bonds, e.g.,

$$-C\overset{O}{\underset{OH}{\diagdown}} \quad \rightarrow \quad -C\overset{(O)}{\underset{OH}{\diagdown}}O-(C) \quad \leftarrow \text{phantom atoms}$$

The complete rules are not given here but the following functional groups are ordered in terms of decreasing priority[2]: $SH > OR > OH > NH\text{-}COCH_3 > NH_2 > COOR > COOH > CHO > CH_2OH > C_6H_5 > CH_3 > {}^3H > {}^2H > H$. Although the RS system is unambiguous, often two very closely related compounds that belong to the same configurational family in the DL system have opposite configurations in the RS system. For this reason, the DL system is customarily used in talking about the configurations of amino acids and sugars.

6. Conformations: The Shapes That Molecules Can Assume

Equally as important to the biochemist as the stable arrangements of bonded atoms (configurations) are **conformations,** the various orientations of groups that are caused by rotation about single bonds.[3] In many molecules such rotation occurs rapidly and freely at ordinary temperatures and we can think of a —CH_3 group as a kind of erratic windmill, turning in one direction, then another. However, even the simplest molecules have **preferred conformations** and in more complex structures, rotation is usually very restricted.

Consider a molecule in which two groups, A and B, are joined by two CH_2 (methylene) groups. If A and B are pulled as far apart as possible, the molecule is in its fully extended or *anti* conformation:

View down the axis joining the C atom as indicated by arrow (Newman projection)

In this molecule, not only are A and B as far apart as possible but also all of the hydrogen atoms are at their maximum distance one from the other. This can be seen readily by viewing the molecule down the axis joining the carbon atoms (Newman projection). Rotation of the second carbon atom 180° around the single bond yields the **eclipsed** conformation. If A and B are large bulky groups they will bump together, attainment of the eclipsed conformation will be almost impossible, and rotation will be severely restricted. Even if A and B are hydrogen atoms (ethane), there will be a **rotational barrier** in the eclipsed conformation which amounts to ~13 kJ/mol, because of the crowding of the hydrogen atoms as they pass each other.[3] (This can be appreciated readily by examination of space-filling molecular models.)

Eclipsed conformation

If groups A and B are methyl groups (in butane), the steric hindrance between A and B leads to a rotational barrier of about 25 kJ (~6 kcal) per mol. The consequence of this simple fact is that in many biological molecules, chains of CH_2 groups tend to assume fully extended conformations. The same zigzag structure is found in polyethylene, which consists of long chains of CH_2 groups.

In addition to the fully extended conformation there are two **gauche** or **skewed** conformations which are only slightly less stable than the *anti* conformation and in which A and B interfere only if they are very bulky. In one of the two **gauche** conformations B lies to the right of A when viewed down the axis while in the other B lies to

One of the two gauche conformations

ϕ, the *torsion angle*

the left of A. The two are related to right-handed and left-handed screws, respectively (the threads on an ordinary right-handed household screw when viewed down the axis from either end move backward from left to right in the same fashion as the groups A and B in the illustration). The angle ϕ is the **torsion angle** and is positive for right-handed conformations (which may also be designated P or + in contrast to M or − for left-handed). Gauche conformations are important in many biological molecules; for example, the sugar alcohol **ribitol** stacks in crystals in the "sickle" conformation,[4] where the chain starts out (at the left) in the zigzag arrangement but shifts to a gauche confor-

Ribitol in sickle conformation

mation around the fourth carbon atom, minimizing steric interference between the OH groups on the second and fourth carbons.

Many molecules contain rings of carbon atoms with 3, 4, 5, 6, or more atoms. While most rings containing two or more double bonds are almost flat, five- and six-membered rings with single bonded atoms are not. Single-bonded six-membered rings, such as those in cyclohexane and in sugars, most often assume a "chair" conformation, as shown for glucose.

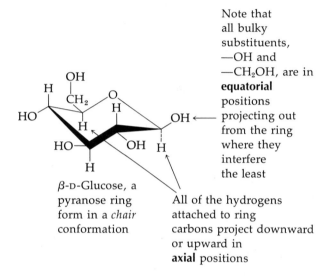

Note that all bulky substituents, —OH and —CH_2OH, are in **equatorial** positions projecting out from the ring where they interfere the least

β-D-Glucose, a pyranose ring form in a *chair* conformation

All of the hydrogens attached to ring carbons project downward or upward in **axial** positions

In addition to the chair conformations of six-membered rings less stable "boat" conformations are also possible.[3,5,6] The six boat forms are all freely interconvertible through six intermediate "skew" forms.

Boat

Skew

Since the internal angle in a pentagon is 108° (close to the tetrahedral angle) we might anticipate a nearly planar structure for a five-membered ring. However, because of the eclipsing of hydrogen atoms on adjacent carbons in such a flat structure, one of the carbons may be buckled out of the plane of the other four about 0.05 nm, into an "envelope conformation" (see Fig. 2-17 for an example of this and related conformations).

7. Hydrogen Bonds and Hydrophobic Interactions

While covalent bonds hold molecules together weaker forces acting between molecules are responsible for many of the most important properties of biochemical substances. Among those noncovalent interactions two are of special importance: **hydrogen bonds** and **hydrophobic interactions.** Hydrogen bonds represent electrostatic attractions resulting from the uneven distribution of electrons in bonds; for example, the bonding electron pairs of the H—O bonds of water molecules are attracted more tightly to the oxygen atoms than to the hydrogen atoms. A small net positive charge is left on the hydrogen and a small net negative charge on the oxygen. Such **polarization** is sometimes indicated by putting arrowheads on the bonds or by the use of the symbols δ^+ and δ^-.

$$\overset{\delta^+}{H} \longrightarrow \overset{\delta^-}{O}$$
$$\overset{\delta^+}{H}$$

Molecules with strongly polarized bonds are referred to as **polar molecules** and functional groups with such bonds as **polar groups.** They are to be contrasted with such nonpolar groups as the —CH_3 group in which the electrons in the bonds are nearly equally shared by carbon and hydrogen.

A hydrogen bond is formed when the positively charged end of one of the dipoles (polarized bonds) is attracted to the negative end of another dipole. Water molecules tend to hydrogen bond strongly one to another and each oxygen atom can be hydrogen bonded to two other molecules.

A water molecule whose oxygen atom is hydrogen bonded to two other water molecules; note the tetrahedral arrangement of bonds around the oxygen

Hydrogen bonds have a strongly directional character and are strongest when the line connecting all three atoms in the bond is straight. The enthalpy of formation ΔH^0 of a linear hydrogen bond may be as much as -20 kJ mol^{-1} (-5 kcal mol^{-1}).

Hydrogen bonds are longer than covalent bonds but are distinctly shorter (by ~ 0.06–0.08 nm) than the contact distances given by van der Waals radii. Remember that hydrogen bonds are always formed between pairs of groups, one containing the negative end of a dipole and the other providing the proton. The proton acceptor group can be thought of as donating an unshared pair of electrons. In discussing organic reaction mechanisms, the direction of motion of **electrons** is often designated with arrows. In this book arrows from electron donor groups to hydrogen atoms (as in the preceding drawing of the hydrogen bonds between water molecules) are also sometimes used to show the directionality of hydrogen bonds.

Fats, hydrocarbons, and other materials whose molecules consist largely of nonpolar groups have a low solubility in water and a high solubility in nonpolar solvents. In an aqueous solution nonpolar groups tend to stick together, a phenomenon that is often called **hydrophobic bonding.** In contrast, sugars and other substances containing many polar groups are very soluble in water and do not tend to associate in this solvent.

The solubility of sugar in water can be understood readily in terms of the ability of the numerous hydroxyl groups in the sugar to form hydrogen bonds with water molecules. Hydrophobic bonding, discussed further in Chapter 4, Section B,4 is a little harder to comprehend, but results principally from the strong internal cohesion of the hydrogen-bonded water structure.[7]

Many molecules with flat ring structures, such as purines and pyrimidines, are only slightly soluble in either water or organic solvents. Molecules of these substances, which contain both polar and nonpolar regions, prefer neither type of solvent but adhere tightly to each other in solid crystals.

8. Tautomerism

Many simple organic compounds exist as mixtures of two or more rapidly interconvertible isomers or **tautomeric forms.** Tautomers, at least in principle, can be separated one from the other at low temperatures where the rate of interconversion is low.

The classic example is the **keto–enol** equilibrium (Eq. 2-1). Although usually less stable than

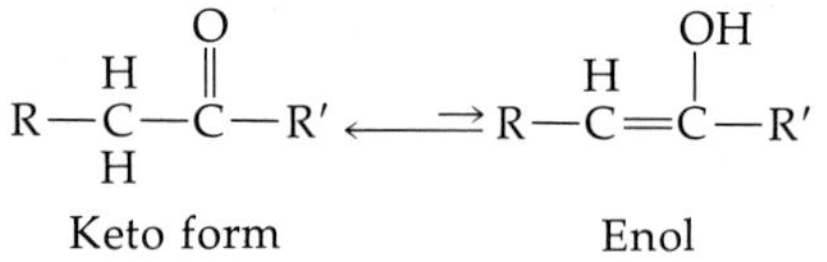

$$R-\overset{H}{\underset{H}{C}}-\overset{O}{\overset{\|}{C}}-R' \rightleftharpoons R-\overset{H}{C}=\overset{OH}{\underset{}{C}}-R' \quad (2\text{-}1)$$

Keto form   Enol

the keto form, the enol is present in small amount. It is readily formed from the keto tautomer by virtue of the fact that hydrogen atoms attached to carbon atoms that are immediately adjacent to carbonyl (C=O) groups are remarkably acidic. Easy dissociation of a proton is a prerequisite for tautomerism. Since most hydrogen atoms bound to carbon atoms do not dissociate readily, tautomerism is unusual unless a carbonyl or other "activating group" is present.

Since protons bound to oxygen and nitrogen atoms are usually readily dissociable tautomerism is possible in amides and in ring systems containing O and N (Eqs. 2-2 to 2-5).

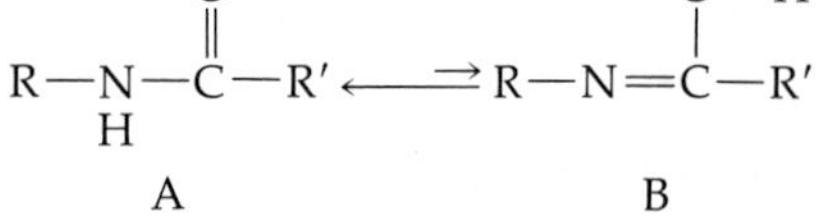

$$R-\overset{}{\underset{H}{N}}-\overset{O}{\overset{\|}{C}}-R' \rightleftharpoons R-N=\overset{O-H}{\underset{}{C}}-R' \quad (2\text{-}2)$$

A   B

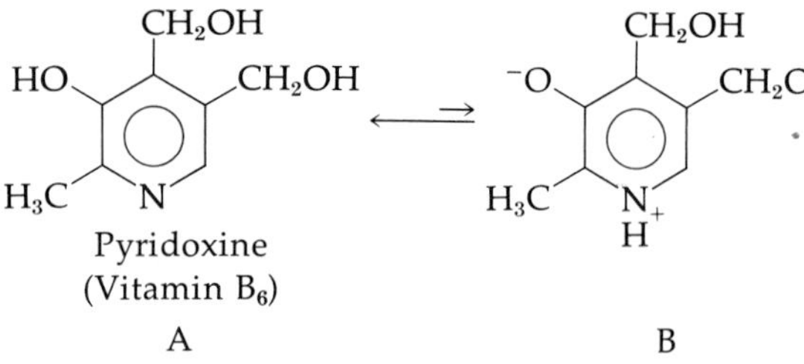

Pyridoxine
(Vitamin B_6)
A   B   (2-3)

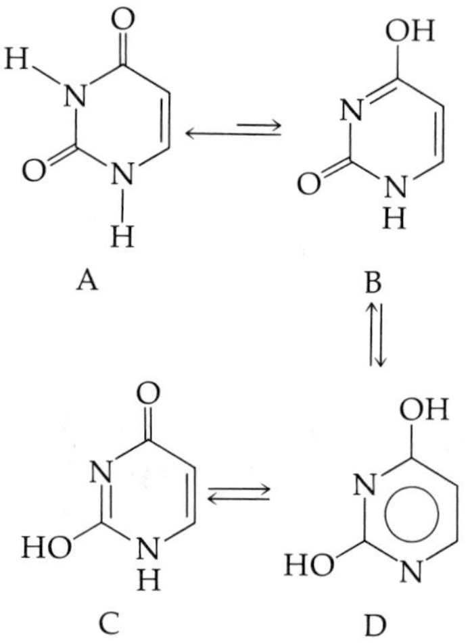

A   B   (2-4)

C   D

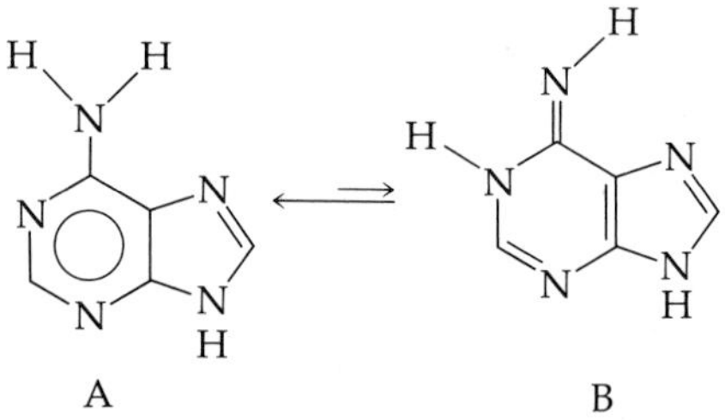

A   B   (2-5)

The tautomerism in Eq. 2-2 is the counterpart of that in the keto–enol transformation; form B is occasionally present in a peptide. Pyridoxine (Eq. 2-3) dissolved in water exists primarily as the **dipolar ionic** tautomer B but in methanol as the uncharged tautomer A. Pyrimidines (Eq. 2-4) and purines (Eq. 2-5) can form a variety of tautomers. The existence of form D of Eq. 2-4 is the basis for referring to uracil as dihydroxypyrimidine. However, the diketo tautomer A predominates. In a pair of tautomers, a hydrogen atom always moves from one position to another and the lengths and bond character of other bonds also change.

The equilibrium constant for a tautomeric interconversion is simply the ratio of the mole fractions of the two forms; for example, the ratio of enol to keto forms of acetone in water[8] is $\sim 2 \times 10^{-6}$; the ratio of dipolar ion to uncharged pyridoxine (Eq. 2-3) is ~ 4 at 25°C.[9] The ratios of tautomers B, C, and D to the tautomer A of uracil (Eq. 2-4) are thought to be small, but it is difficult to measure them quantitatively.[10,11] These tautomeric ratios (when defined for given overall states of protonation) are *not* influenced by pH changes but may be altered by changes in temperature or solvent or by binding to a protein or other molecule.

9. Resonance

It is important to differentiate tautomerism from **resonance,** a term used to indicate that the properties of a given molecule cannot be represented by a *single* valence structure but can be represented as a hybrid of two or more structures in which *all the nuclei remain in the same places.* Only the bonding electrons move to convert one resonance form into another. An example is the **enolate anion** which

$$R-\underset{\underset{\text{A}}{H}}{\overset{\overset{O}{\|}}{C}}-C-R' \longleftrightarrow R-\underset{\underset{\text{B}}{H}}{\overset{\overset{O^-}{|}}{C}}=C-R'$$

Enolate anion

can be thought of as a hybrid of structures A and B. A double-headed arrow is often used to indicate that two structures drawn are resonance structures rather than tautomers or other separable isomers.

Although they must be distinguished carefully, tautomerism and resonance are closely related. Thus, the acidity of carbon-bound hydrogens in ketones, which allows formation of enol tautomers, is a direct result of the fact that the enolate anion produced by dissociation of one of these hydrogens is stabilized by resonance.

Similarly, tautomerism in the imidazole group, present in most proteins, is related to resonance in the imidazolium cation. Because of this resonance, if a proton approaches structure A of Eq. 2-6 and becomes attached to the left-hand nitrogen atom (N-1), the positive charge in the resulting intermediate is distributed over both nitrogen atoms. This makes the proton on N-3 acidic, permitting it to dissociate to tautomer B. This tautomerism of the imidazole group is believed important to the function of many enzymes and other proteins; for example, if N-3 of structure A (Eq. 2-6) is embedded in a protein, a proton approaching from the outside can induce the tautomerism shown with the release of a proton in the interior of the protein, perhaps at the active site of an enzyme Chapter 7, Section J,2).

B. THE PROTEINS

The monomer units from which proteins are derived are the 20 α-**amino acids.** These small molecules share a property essential to their role in polymer formation: *they contain at least two different chemical groups able to react with each other to form a covalent linkage.* In the amino acids these are the amino ($-NH_2$) and carboxyl ($-COOH$) groups, and the characteristic linkage in the protein polymer is the **peptide** (amide) linkage. Formation of a peptide linkage can be imagined to occur by the splitting out of water between the linking carboxyl and amino groups (see Eq. 2-7).

$$H_2N-\underset{\underset{R}{|}}{CH}-\underset{\underset{}{\overset{\overset{O}{\|}}{C}}}-\boxed{OH + H}-\underset{\underset{H}{|}}{N}-\underset{\underset{}{|}}{CH}-COOH \longrightarrow$$

α-Amino acids

$$H_2N-\underset{\underset{R}{|}}{CH}-\underset{\overset{O}{\|}}{C}-\underset{\underset{H}{|}}{N}-\underset{}{CH}-COOH + H_2O \quad (2\text{-}7)$$

A peptide linkage
in a dipeptide

The equilibrium in a reaction of this type, in an aqueous solution, favors the free amino acids rather than the peptide. Therefore, both biochemical and laboratory syntheses of peptides are indirect and do not involve a simple splitting out of water.

Since the dipeptide of Eq. 2-7 still contains reactive carboxyl and amino groups other amino acid units can be joined by additional peptide linkages to form **polypeptides** (proteins).

1. The α-Amino Acids

Three simple facts are basic to an understanding of the behavior of amino acids as free metabolic intermediates and as components of proteins.

a. Amino acids exist mostly as dipolar ions (zwitterions).

$$H_3{}^+N-\underset{\underset{R}{|}}{\overset{\overset{H}{|}}{C}}-C\overset{\overset{\overset{\displaystyle O}{\diagup\!\!\diagup}}{}}{\diagdown_{O^-}} \quad \text{instead of} \quad H_2N-\underset{\underset{R}{|}}{\overset{\overset{H}{|}}{C}}-C\overset{\overset{\displaystyle O}{\diagup\!\!\diagup}}{-OH}$$

b. Amino acids are asymmetric and form two families, D and L (Section A,4).

c. Amino acids are differentiated by the structures of their **side chain groups,** designated R in the foregoing formulas. These groups are of varying chemical structure. Because they stick out and provide the bulk of the external surface of the polymers, they determine many of the chemical and physical properties of proteins.

Table 2-2 shows the structures of the side chains only (except for proline where the complete structure is given) of the amino acids commonly found in proteins. Three letter abbreviations, commonly used in describing sequences of amino acids in proteins as well as one letter abbreviations, used by evolutionists and computers, are also given.

In discussions of protein structure the amino acids of groups a, b, and c of Table 2-2 plus phenylalanine and methionine are sometimes grouped together as **nonpolar.** They tend to seek a hydrophobic environment on the "inside" of a protein molecule. The opposite group are the **polar,** charged molecules (groups f and i) which usually protrude into the water surrounding the protein. The rest are classified as polar but noncharged.

To get acquainted with amino acid structures, the reader might first learn those of **glycine, alanine, serine, aspartic acid,** and **glutamic acid.** Note that structures of many other amino acids can be related to that of alanine by replacement of a hydrogen by another group. Thus, replacement of a β hydrogen of alanine gives the following:

by phenyl	$\longrightarrow$ phenylalanine
by —OH	$\longrightarrow$ serine
by —SH	$\longrightarrow$ cysteine
by —COOH	$\longrightarrow$ aspartic acid

Metabolic interrelationships will help you learn structures of the rest of the amino acids later.

Table 2-2 includes information on acid dissociation constants (as pK_a values) for amino acid side chains. The terminal amino and carboxyl groups of protein chains when free, can also participate in acid-base reactions with approximately the following pK_a values (see Chapter 4, Section B for a discussion of the meaning of pK_a).

terminal —COOH, $pK_a = 3.6\text{–}3.7$
terminal —NH$_3{}^+$, $pK_a = 7.5\text{–}7.9$

Since the –COOH groups of glutamic and aspartic acids are completely dissociated at neutral pH, it is customary in the biochemical literature to refer to these amino acids as **glutamate** and **aspartate** without reference to the nature of the cation or cations present. Likewise "-ate" endings are used

TABLE 2-2
Structure and Chemical Properties of Side Chain Groups of Amino Acids

a. Glycine "side chain" = —H (Gly, G)[a]

Strictly a link in the peptide chain, glycine provides a minimum of steric hindrance to rotation and to placement of adjacent groups

b. Amino acids with alkyl groups as side chains

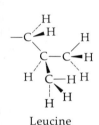

These bulky groups of distinctive shape are important in hydrophobic bonding and in forming binding sites of specific shapes in enzymes

Alanine   Valine
(Ala, A)   (Val, V)

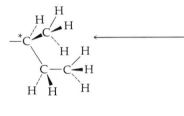

* Note that isoleucine has a *second* chiral center

Leucine   Isoleucine
(Leu, L)   (Ile, I)

c. Proline (Because the side chain is fused back to the amino group, the entire structure not just the side chain is shown)

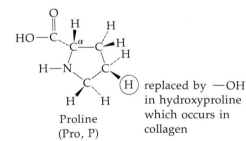

Note the *secondary* amino group (proline is sometimes called an **imino acid**) and the rigid conformation. The presence of proline strongly influences the folding of protein chains

(H) replaced by —OH in hydroxyproline which occurs in collagen

Proline
(Pro, P)

d. Aromatic amino acids

Phenylalanine   Tyrosine
(Phe, F)   (Tyr, Y)

Acidic proton, pK_a ~9.5–10.9. H donor in hydrogen bonds and functional group in enzymatic catalysis

(continued)

TABLE 2-2 (*Continued*)

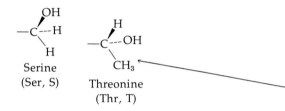

Tryptophan
(Trp, W)

Tyrosine, phenylalanine, and tryptophan can form hydrophobic bonds and may be especially effective in bonding to other flat molecules

e. Amino acid alcohols

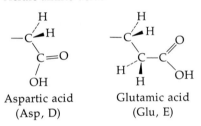

Serine
(Ser, S)

Threonine
(Thr, T)

The —OH group has very weakly acidic properties (pK_a ~13.6). It can form esters with phosphoric acid or organic acids and is a site of attachment of sugar rings in glycoproteins. Hydroxyl groups of serine are found at the active centers of some enzymes

Note the second chiral center. The side chain of L-threonine has the configuration shown here, that of D-threonine the opposite. The L-amino acid with the opposite configuration in the side chain only is L-*allo*-threonine

f. Acidic amino acids

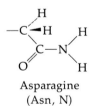

Aspartic acid
(Asp, D)

Glutamic acid
(Glu, E)

Carboxyl groups of these side chains are dissociated at neutral pH (pK_a values are 4.3–4.7) and provide anionic (−) groups on the surfaces of proteins

g. Amides of aspartic acid and glutamic acid

Asparagine
(Asn, N)

The amide group is not acidic but is polar and able to participate in hydrogen bonding

Glutamine
(Gln, Q)

If it is uncertain whether a position in a protein is occupied by aspartic acid or asparagine, it may be designated Asx or B. If glutamic or glutamine, as Glx or Z

(*continued*)

TABLE 2-2 (*Continued*)

h. Sulfur-containing amino acids

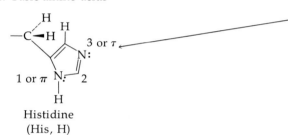

Cysteine
(Cys, C)

S—H ← a weak acid.
pK_a ~8.3–8.6

Methionine
(Met, M)

Cysteine is most noted for its ready spontaneous oxidation by O_2 to the "double-headed" amino acid cystine; a process that often forms disulfide cross-links between peptide chains:

2 —CH_2—SH

—H_2C
S—S
CH_2—

i. Basic amino acids

Histidine
(His, H)

3 or τ

1 or π

This basic site accepts a proton to form a conjugate acid of pK_a ~6.4–7.0 and carrying a positive charge

The **imidazole** groups in histidine side chains are parts of the active sites of many enzymes. Like other basic groups in proteins they also may bind metal ions

Lysine
(Lys, K)

A flexible side arm with a potentially reactive amino group at the end. The high pK_a of ~10.5 means that lysine side chains are ordinarily protonated in neutral solutions

Arginine
(Arg, R)

The **guanidino** group has a high pK_a of over 12 and remains protonated under most circumstances. It is stabilized by resonance as indicated by the curved arrows. Guanidino groups are sometimes important sites for binding of phosphate groups

a The three-letter and one-letter abbreviations used for the amino acid residues in peptides and proteins are given in parentheses.

for most other acids, e.g., malate, oxaloacetate, phosphate, adenylate, and in names of *enzymes*, e.g., lactate dehydrogenase.

2. Polypeptides

Chains formed by polymerization of 100–1000 or more amino acid molecules provide the **primary structure** of proteins. The monomer units in the chain are known as amino acid **residues,** each amino acid unit having lost one molecule of H_2O during the polymerization.* A polypeptide chain usually has one free terminal amino group and a terminal carboxyl group at the other end. However, these groups are sometimes linked to form a **cyclic** peptide. Peptides are named according to the amino acid residues present and beginning with the one bearing the terminal amino group. Thus, L-alanyl-L-valyl-L-methionine has the structure:

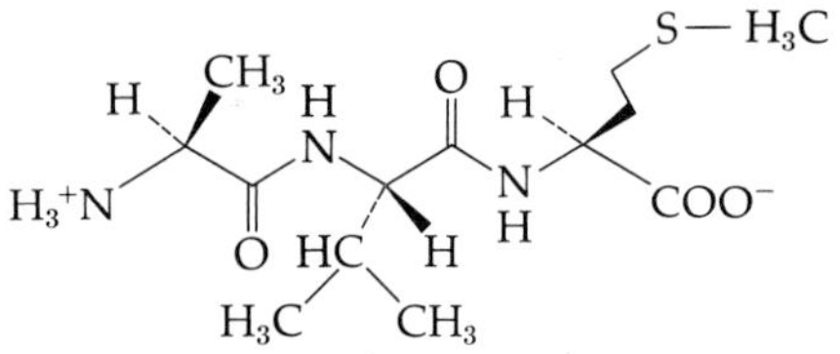

Note that this tripeptide, like amino acids, is a dipolar ion. The same structure can be abbreviated Ala·Val·Met. In such abbreviations the **N-terminal** amino acid residue is always placed at the left and the **C-terminal** amino acid residue at the right end.

Recall that the **sequences** of amino acids in protein chains are always precisely specified by the genes. Knowledge of amino acid sequences is of major importance in understanding the behavior of specific proteins. For this reason a major fraction of recent biochemical effort has gone into "sequencing" hundreds of proteins. One protein for which this task has been accomplished is a human

* Strictly speaking, one water molecule per peptide linkage is lost, i.e., *one less* than the number of amino acid residues.

γ-immunoglobulin with 446 residues in one chain and 214 in another. An example of a complete sequence for another protein is given in Fig. 2-1. Sequences of some small peptide hormones and antibiotics are shown in Fig. 2-2.

3. Conformations of Peptide Chains

Consider a chain of —CH₂— (methylene) groups, such as those in the synthetic polymer polyethylene. The hydrogen atoms on alternate C atoms of the fully extended chain barely touch, but larger atoms cannot be accommodated.

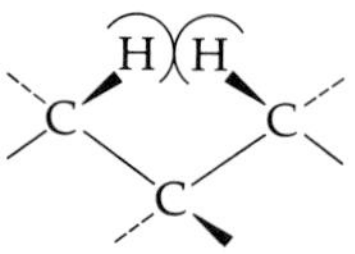

Even when fluorine atoms (van der Waals radius, 0.135 nm) replace the hydrogen atoms (radius, 0.12 nm) of polyethylene, the extended chain is no longer possible. The torsion angle in polyfluoroethylene is changed from the 180° of polyethylene to 166°, enough to relieve the congestion but not enough to cause severe eclipsing of the fluorines on adjacent carbons. The resulting **helical structure** is reminiscent of those occurring in proteins and other biopolymers. Thus, helix formation is a natural result of steric repulsion between groups of atoms.[12,13]

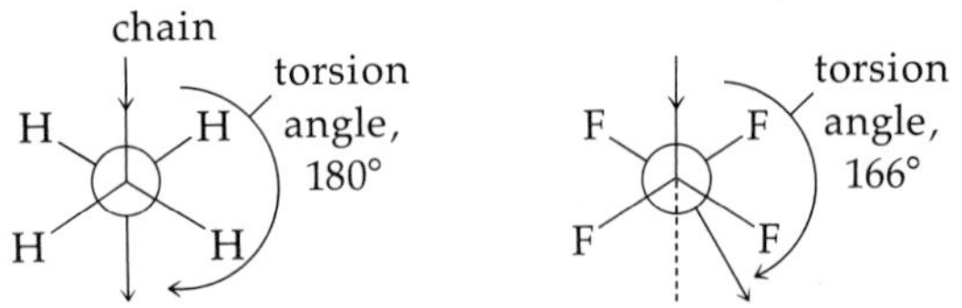

Polyethylene   Polyfluoroethylene (Teflon)

The conformational properties of biopolymers are complex but are considerably simplified by the fact that most of the monomer units are quite rigid.

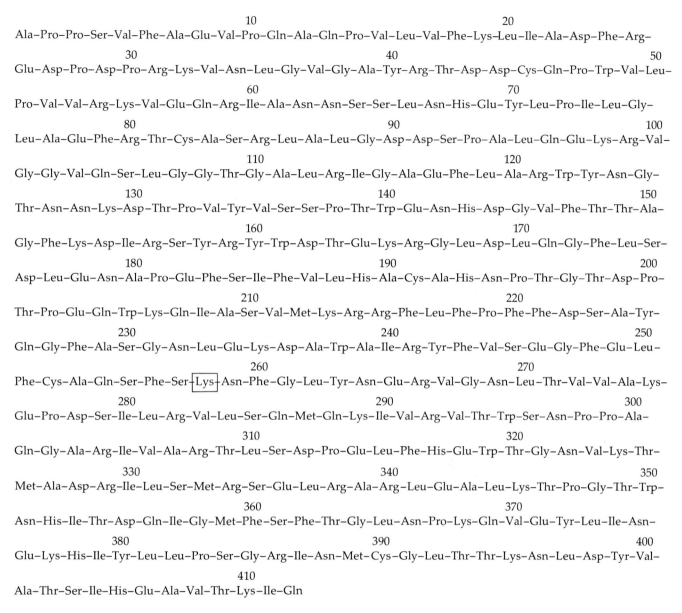

10 20

Ala–Pro–Pro–Ser–Val–Phe–Ala–Glu–Val–Pro–Gln–Ala–Gln–Pro–Val–Leu–Val–Phe–Lys–Leu–Ile–Ala–Asp–Phe–Arg–

30 40 50

Glu–Asp–Pro–Asp–Pro–Arg–Lys–Val–Asn–Leu–Gly–Val–Gly–Ala–Tyr–Arg–Thr–Asp–Asp–Cys–Gln–Pro–Trp–Val–Leu–

60 70

Pro–Val–Val–Arg–Lys–Val–Glu–Gln–Arg–Ile–Ala–Asn–Asn–Ser–Ser–Leu–Asn–His–Glu–Tyr–Leu–Pro–Ile–Leu–Gly–

80 90 100

Leu–Ala–Glu–Phe–Arg–Thr–Cys–Ala–Ser–Arg–Leu–Ala–Leu–Gly–Asp–Asp–Ser–Pro–Ala–Leu–Gln–Glu–Lys–Arg–Val–

110 120

Gly–Gly–Val–Gln–Ser–Leu–Gly–Gly–Thr–Gly–Ala–Leu–Arg–Ile–Gly–Ala–Glu–Phe–Leu–Ala–Arg–Trp–Tyr–Asn–Gly–

130 140 150

Thr–Asn–Asn–Lys–Asp–Thr–Pro–Val–Tyr–Val–Ser–Ser–Pro–Thr–Trp–Glu–Asn–His–Asp–Gly–Val–Phe–Thr–Thr–Ala–

160 170

Gly–Phe–Lys–Asp–Ile–Arg–Ser–Tyr–Arg–Tyr–Trp–Asp–Thr–Glu–Lys–Arg–Gly–Leu–Asp–Leu–Gln–Gly–Phe–Leu–Ser–

180 190 200

Asp–Leu–Glu–Asn–Ala–Pro–Glu–Phe–Ser–Ile–Phe–Val–Leu–His–Ala–Cys–Ala–His–Asn–Pro–Thr–Gly–Thr–Asp–Pro–

210 220

Thr–Pro–Glu–Gln–Trp–Lys–Gln–Ile–Ala–Ser–Val–Met–Lys–Arg–Arg–Phe–Leu–Phe–Pro–Phe–Phe–Asp–Ser–Ala–Tyr–

230 240 250

Gln–Gly–Phe–Ala–Ser–Gly–Asn–Leu–Glu–Lys–Asp–Ala–Trp–Ala–Ile–Arg–Tyr–Phe–Val–Ser–Glu–Gly–Phe–Glu–Leu–

260 270

Phe–Cys–Ala–Gln–Ser–Phe–Ser–⎡Lys⎤–Asn–Phe–Gly–Leu–Tyr–Asn–Glu–Arg–Val–Gly–Asn–Leu–Thr–Val–Val–Ala–Lys–

280 290 300

Glu–Pro–Asp–Ser–Ile–Leu–Arg–Val–Leu–Ser–Gln–Met–Gln–Lys–Ile–Val–Arg–Val–Thr–Trp–Ser–Asn–Pro–Pro–Ala–

310 320

Gln–Gly–Ala–Arg–Ile–Val–Ala–Arg–Thr–Leu–Ser–Asp–Pro–Glu–Leu–Phe–His–Glu–Trp–Thr–Gly–Asn–Val–Lys–Thr–

330 340 350

Met–Ala–Asp–Arg–Ile–Leu–Ser–Met–Arg–Ser–Glu–Leu–Arg–Ala–Arg–Leu–Glu–Ala–Leu–Lys–Thr–Pro–Gly–Thr–Trp–

360 370

Asn–His–Ile–Thr–Asp–Gln–Ile–Gly–Met–Phe–Ser–Phe–Thr–Gly–Leu–Asn–Pro–Lys–Gln–Val–Glu–Tyr–Leu–Ile–Asn–

380 390 400

Glu–Lys–His–Ile–Tyr–Leu–Leu–Pro–Ser–Gly–Arg–Ile–Asn–Met–Cys–Gly–Leu–Thr–Thr–Lys–Asn–Leu–Asp–Tyr–Val–

410

Ala–Thr–Ser–Ile–His–Glu–Ala–Val–Thr–Lys–Ile–Gln

Fig. 2-1 The complete amino acid sequence of cytoplasmic aspartate aminotransferase from pig heart. The peptide has the composition: Lys_{19}, His_8, Arg_{26}, $(Asp + Asn)_{42}$, Ser_{26}, Thr_{26} $(Glu + Gln)_{41}$, Pro_{24}, Gly_{28}, Ala_{32}, Cys_5, Val_{29}, Met_6, Ile_{19}, Leu_{38}, Tyr_{12}, Phe_{23}, Trp_9. The molecular weight is 46,344 and the complete enzyme is a dimer of MW = 93,147 containing 2 mol of the bound coenzyme pyridoxal phosphate (Chapter 8, Section E) attached to lysine-258 (enclosed in box). [From Yu. A. Ovchinnikov et al., FEBS Lett. **29**, 31–33 (1973) and S. Doonan et al., FEBS Lett. **38**, 229–233 (1974)].

$$\text{Cys—Tyr—Ile—Gln—Asn—Cys—Pro—Leu—Gly NH}_2$$

with S—S bridge connecting the two Cys residues

N-terminal   Glycine amide, C-terminal

Human oxytocin

$$\text{Cys—Tyr—Phe—Gln—Asn—Cys—Pro—Arg—GlyNH}_2$$

with S—S bridge connecting the two Cys residues

Human vasopressin

1 Ser—Tyr—Ser—Met—Glu—His—Phe—Arg—Trp—Gly—Lys—Pro—Val—Gly—Lys 15

30 Glu—Asp—Glu—Ala—Gly—Asn—Pro—Tyr—Val—Lys—Val—Pro—Arg—Arg—Lys

Ser—Ala—Glu—Ala—Phe—Pro—Leu—Glu—Phe   39

Human adrenocorticotropin (ACTH)

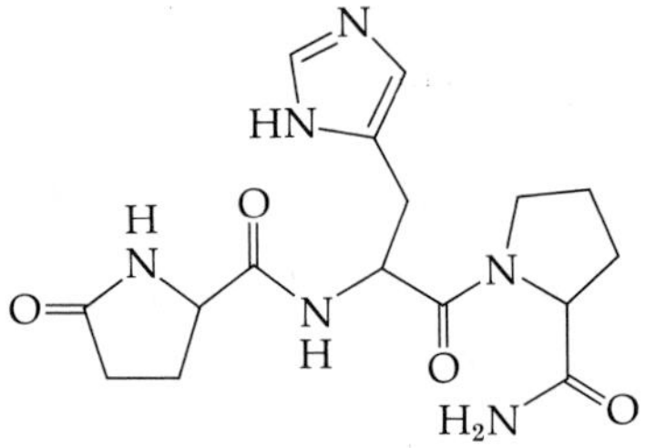

pyroglutamyl—His—Trp—Ser—Tyr

Pro—Arg—Leu—Gly

GlyNH$_2$

Hypothalamic thyrotropic
hormone releasing factor
(TRF); pyroglutamyl-histidyl-
prolinamide

Luteinizing hormone
releasing factor (LRF)

$$\text{D-Phe}\rightarrow\text{L-Leu}\rightarrow\text{L-Orn}\rightarrow\text{L-Val}\rightarrow\text{L-Pro}$$
$$\uparrow\qquad\qquad\qquad\qquad\qquad\downarrow$$
$$\text{L-Pro}\leftarrow\text{L-Val}\leftarrow\text{L-Orn}\leftarrow\text{L-Leu}\leftarrow\text{D-Phe}$$

Gramicidin S

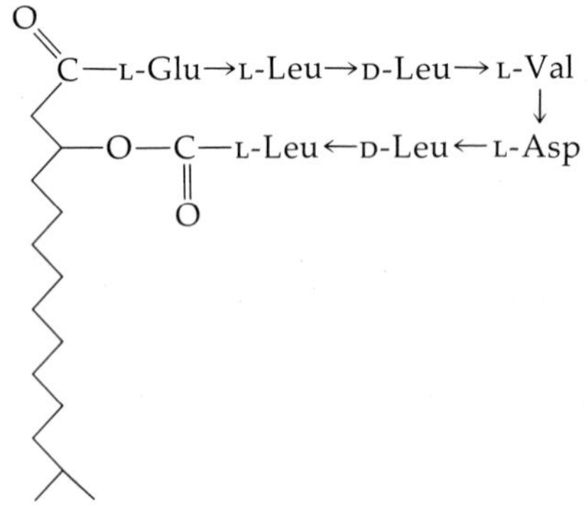

$$\text{C—L-Glu}\rightarrow\text{L-Leu}\rightarrow\text{D-Leu}\rightarrow\text{L-Val}$$
$$\downarrow$$
$$\text{O—C—L-Leu}\leftarrow\text{D-Leu}\leftarrow\text{L-Asp}$$

Surfactin

a. The Peptide Unit

In polypeptides, rigidity is provided by the planarity of the peptide linkages which is favored by resonance of the following type:

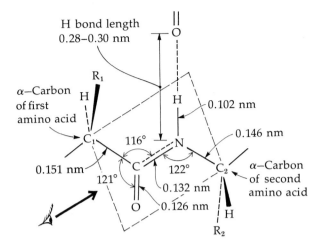

The predicted shortening of the C—N bond and the lengthening of the C=O bond (as a result of this resonance) have been verified experimentally by X-ray diffraction measurements. Each peptide unit can be thought of as a rigid sheet with the dimensions given in Fig. 2-3 (top). However, the bonds around the nitrogen retain some pyramidal character[14,15] (Fig. 2-3, bottom).

While the trans peptide linkage shown in Fig. 2-3 is usual, the following *cis*-peptide linkage (which is about 8 kJ mol^{-1} less stable than the trans linkage) also occurs occasionally, especially when the nitrogen atom is from proline.[16,17]

cis-Peptide linkages

Fig. 2-2 Structures of a few naturally occurring peptides. Oxytocin and vasopressin are hormones of the neurohypophysis (posterior lobe of the pituitary gland). Adrenocorticotropin is a hormone of the adenohypophysis (anterior pituitary). The hormones of the adenohypophysis are released under the influence of releasing factors (regulatory factors) produced in the neighboring hypothalamus (a portion of the brain) in response to neural stimulation. Structures of two releasing factors are shown. Note that the γ-carboxyl groups of the N-terminal glutamate residues have reacted with the neighboring terminal —NH$_2$ groups to form cyclic amide (pyroglutamyl) groups [see R. Guillemin and R. Burgus, *Sci. Am.* **227**, 24–33 (Nov 1972)]. Gramicidin S is an antibiotic made by *Bacillus brevis*, surfactin is a depsipeptide (containing an ester linkage), a surface active antibiotic of *Bacillus subtilus*. See R. M. Hochster, M. Kates, and J. H. Quastel, eds., "Metabolic Inhibitors," Vol. 3, p. 312. Academic Press, New York, 1973.

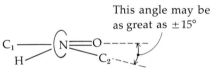

View down C—N axis
(as indicated by heavy arrow above)

Fig. 2-3 Dimensions of the peptide linkage. Interatomic distances, including the hydrogen bond length to an adjacent peptide linkage, are indicated. The atoms enclosed by the dotted lines all lie *approximately* in a plane. However, as indicated in the lower drawing, the nitrogen atom tends to retain some pyramidal character.

A polypeptide can be thought of as a chain of flat peptide units fastened together as in Fig. 2-4. Each peptide unit is connected by the α-carbon of an amino acid. This carbon provides *two* single bonds to the chain and rotation can occur about both of them (except in the cyclic amino acid proline). To specify the conformation of an amino acid unit in a protein chain, it is necessary to specify torsion angles about both of these single bonds. These torsion angles are indicated by the symbols ϕ and ψ and are assigned the value 180° for the fully extended chain* as shown in Fig. 2-4.[18] Since both ϕ and ψ can vary, a large number of conformations are possible. However, many are excluded because they bring certain atoms into

* Note that a different convention, according to which $\psi = \phi = 0$ for an *extended* chain, has been used widely.[13]

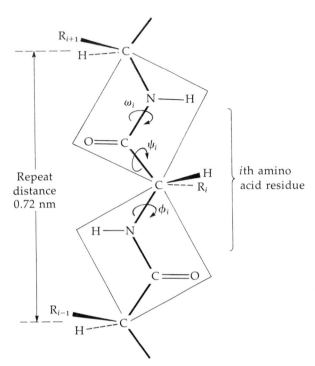

Fig. 2-4 Two peptide units in the completely extended β conformation. The torsion angles ϕ_i, ψ_i, and ω_i are defined as 0° when the main chain atoms assume the cis or eclipsed conformation. The angles in the completely extended chain are all 180°.

collision, as can be quickly established by study of molecular models.

Nowadays it is customary to examine the whole range of possible combinations of ϕ and ψ using computers. This has been done for the peptide linkage by Ramachandran and the results are often presented as plots of ϕ vs. ψ (Ramachandran plots or **conformational maps**) in which possible combinations of the two angles are indicated by blocked out areas. The conformational map for peptides (Fig. 2-5) indicates that a large number of combinations of the two torsion angles are possible. Within real proteins most of the possible ones exist. However, most pairs of angles are truly impossible except for glycine-containing peptides, which have more flexibility (Fig. 2-5B). Approximate torsion angles for some regular peptide structures are given in Table 2-3.

b. The Extended Chain β Structures

As first pointed out by Linus Pauling, an important structural principle for proteins is that the maximum number of possible hydrogen bonds involving the C=O and N—H groups of the peptide chain must be formed. One simple way to do this is to line up fully extended chains ($\phi = \psi = 180°$) and to form hydrogen bonds between them. Such a structure exists for polyglycine and resembles that in Fig. 2-6 (upper left). Note that adjacent chains run in opposite directions; hence, the term **antiparallel β structure.** The antiparallel arrangement not only gives the best hydrogen bond formation between chains but also permits a single chain to fold back on itself, a fact that may be of great importance in formation of cell structures.

While a β structure of fully extended polyglycine chains is possible, side chains (R groups) of other amino acids cannot be accommodated without some distortion of the structure. Thus, silk fibroin has a repeat distance of 0.70 nm compared with 0.72 nm for the fully extended chain (Fig. 2-4). Pauling and Corey showed that this shortening of the chain could result from rotation of angle ϕ by $\sim 40°$ (to $-140°$) and rotation of ψ in the opposite direction by $\sim 55°$ (to $+135°$) to give a slightly puckered chain. The resulting multichain structure is known as a **pleated sheet** (Fig. 2-6). A parallel chain β structure is also found very frequently within proteins.

c. Polyglycine II and Collagen

A second form of polyglycine is known in which each amino acid residue is rotated 120° from the preceding one about a 3-fold screw axis (see end view in Fig. 2-7A). The angle ψ is about 150° while ϕ is $\sim -80°$ for each residue. The distance along the axis is 0.31 nm/residue and the repeat distance is 0.93 nm. The molecules can coil either into a right-handed or left-handed helix. In this structure, the N—H and C=O groups protrude perpendicular to the axis of the helix. As in the β structure, they can form hydrogen bonds with adjacent chains.

Fig. 2-5 (A) Calculated Ramachandran diagram or conformational map showing allowed values of ϕ and ψ. The two irregularly shaped boxes on the left side of the figure represent the regions in which no steric hindrance exists. Note that the β pleated sheet, collagen, and α helix structures fall within these regions. The stippled regions are those in which some hindrance exists, but for which real molecular structures are possible.

(B) Ramachandran plot for the B chains of insulin (Chapter 4, Section D,7): (●) molecule 1 and (□) molecule 2. The angles ϕ and ψ were determined experimentally from X-ray diffraction at 0.19 nm resolution and model building. The majority of conformations are those of a distorted α helix or approximately those of a β structure. Residues 8, 20, and 23 are glycine. From T. Blundell, G. Dodson, D. Hodgkin, and D. Mercola, *Adv. Protein Chem.* **26**, 279–402 (1972).

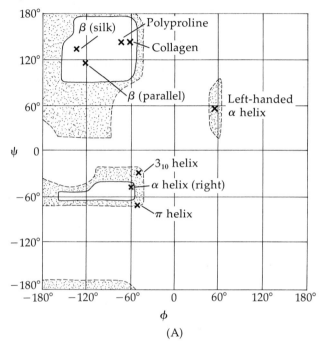

(A)

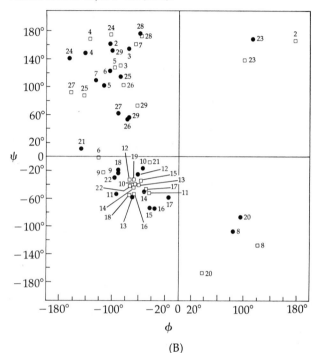

(B)

TABLE 2-3
Approximate Torsion Angles for Some Regular Peptide Structures[a]

Structure	ϕ	ψ
Hypothetical fully extended polyglycine chain	$-180°$	$+180°$[b]
β-Poly(L-alanine) in antiparallel-chain pleated sheet	$-139°$	$+135°$
Parallel-chain pleated sheet	$-119°$	$+113°$
Polyglycine II	$-80°$	$+150°$
Poly(L-proline) II	$-78°$	$+149°$
Collagen[c]	$-51°, -76°, -45°$	$+153°, +127°, +148°$
Right-handed α helix	$-57°$	$-47°$

[a] From IUPAC-IUB Commission on Biochemical Nomenclature, *Biochemistry* **9**, 3471–3479 (1970).
[b] The torsion angles for the fully extended chain can be designated either $+180°$ or $-180°$, the two being equivalent. They are given as $\phi = -180°$ and $\psi = +180°$ to facilitate comparison with the other structures.
[c] Three sets of angles were found for the three residues in the repeating sequence $(Pro-Gly-Pro)_n$ which predominates in collagen by A. Yonath and W. Traub, *JMB* **43**, 461–477 (1969).

Poly(L-proline) assumes a helical structure very similar to that of polyglycine II. Because of the presence of the bulky side chain groups a *left-handed* helix is favored. It is generally true that for a given helical arrangement of a peptide chain containing all L-amino acids, the right-handed and left-handed helices will be of different stability.

Collagen, the principal protein of connective tissue, basement membranes, and other structures, is one of the most abundant proteins in the animal body. The fundamental unit of collagen structure is referred to as **tropocollagen** and is believed to be a triple helix of overall dimensions 1.5×280 nm. It resembles the structure of polyglycine II (Fig. 2-7A) but contains only three chains. However, the left-handed helices of the individual chains are further wound into a **right-handed superhelix.**

Whereas most proteins do not contain a high percentage of any one amino acid, collagen contains 33% glycine, 21% proline + hydroxyprolines, and 11% alanine. The reason is apparently that bulky side chain groups cannot fit on the inside of the triple helix. Similar considerations apply to silk fibroin which consists in large part of chains composed of the repeating unit.

$$(Gly—Ser—Gly—Ala—Gly—Ala)_n$$

In the pleated sheet all the alanine and serine side chains protrude on one side of the sheet while the other side has only the hydrogen atoms of the glycine. This permits an efficient stacking of the sheets with interdigitation of the side chains of alanine and serine in adjacent sheets.[19]

d. The α Helix

An important series of conformations are obtained if ϕ and ψ are both near $-60°$. These are the hydrogen-bonded helices, the most important of which is the α helix (Fig. 2-8). In this case, the right-handed helix is far more stable for L-amino acids than is the left-handed helix. Only the right-handed α helix has been found in nature. Note that all of the C=O and N—H groups of the pep-

Fig. 2-6 The extended chain β pleated sheet structures. At the left is the antiparallel structure found in silk fibroin. The 0.70 nm spacing is slightly decreased from the fully extended length. The amino acid side chains (R) extend alternately above and below the plane of the "accordion pleated" sheet. The pairs of linear hydrogen bonds between the chains impart great strength to the structure. The chain can fold back on itself using a "β turn" in which the peptide unit in the center of the turn is oriented almost perpendicular to the plane of the pleated sheet. The parallel chain structure (right) is similar but with a less favorable hydrogen bonding arrangement. At the bottom is a stereoscopic view[26] of the peptide backbone of the enzyme carboxypeptidase A. While the hydrogen bonds have not been drawn and only the positions of the α-carbon atoms are shown, it is possible to trace the course of eight peptide strands running from the upper left to lower right through the center of the structure. These strands, some in parallel and some in antiparallel arrangement, are all linked by hydrogen bonds. The drawing may be observed with a special viewer (a suitable viewer is available from Abrams Instrument, Corp., Lansing, Michigan) but, with a little practice, it is possible to obtain a stereoscopic view unaided. (Since many stereo views of molecular structures are being published in journals, this is a useful thing to be able to do.) Hold the book with good illumination about 20–30 cm from your eyes. Allow your eyes to relax as if viewing a distant object. Of the four images that are visible, the two in the center can be fused to form the stereoscopic picture. For a good schematic view of this backbone of β structure, see Dickerson and Geis,[13] p. 89.

tide linkages lie roughly parallel to the axis of the helix, each carbonyl group being hydrogen bonded to the fourth N—H group on up the chain. The number of amino acid units per turn of the helix is 3.61. The **pitch** (repeat distance) of the helix (which can be determined experimentally from X-ray diffraction data) is 0.541 nm.

The horny layers of skin, hair, fingernails, and porcupine quills are made up of a mixture of proteins known as **keratin.** One of the major components,* the "low-sulfur protein," consists of α helices but with an observed pitch of 0.51 rather than 0.54 nm. This shortening may result from the winding of two or three right-handed α helices around each other in a left-handed superhelix to

---

* The other proteins of keratin appear to form an amorphous "matrix" in which the microfibrils are embedded. One of the matrix proteins is a "high-sulfur protein," rich in cystine and highly cross-linked. Human hair is one of the richest known sources of cystine.

N-terminus   C-terminus

Antiparallel

Parallel

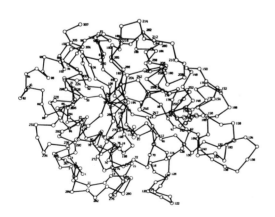

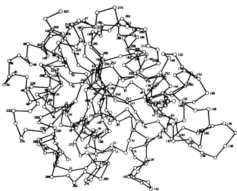

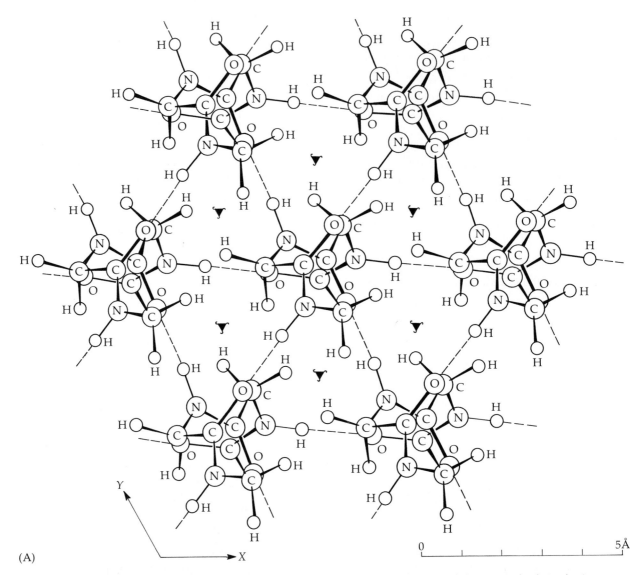

Fig. 2-7 Structures of polyglycine II and collagen. (A) The structure of polyglycine II viewed down the axis of a helix chain. Note the threefold symmetry of the structure. Polyproline forms a similar structure and collagen is thought to be a triple-stranded "rope" of three peptide chains having essentially the same structure but in addition containing a right-handed supercoil.

form a **microfibril** of about 2.0 nm diameter.[19-21] The same kind of "coiled-coil" structure is thought to exist in the muscle proteins **myosin**[22] and **tropomysin**[23] (Chapter 4, Section F,1). Both are thought to be two-stranded "ropes" in which hydrophobic side chains from one strand fit into gaps in the surface of the adjacent strand—a knobs-into-holes bonding arrangement.*

Other helices of smaller and larger diameter

* Helical strands tend to coil into ropes because a favorable interstrand contact can be repeated along the length of the strands only if the strands coil about each other (Crick[23a]).

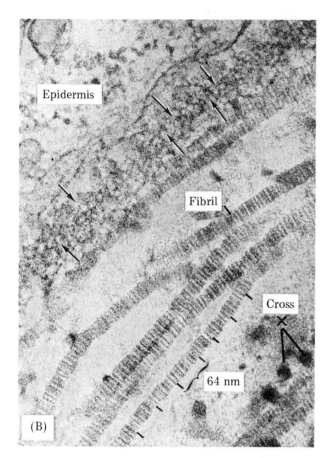

(B)

Fig. 2-7 (B) Electron micrograph of the basement membrane (between the arrows) and underlying collagen fibrils in a section of amphibian skin. In the basement membrane, the filaments 1.5 nm in diameter (arrows) are believed to be individual tropocollagen molecules arranged in a meshwork. In the fibrils, the tropocollagen molecules are arranged parallel to one another in a "quarter stagger" pattern which results in a striation with a 64 nm repeat period. Several fibrils are cross sectioned (cross X). Courtesy of Elizabeth D. Hay.

than the α helix are possible and have some importance in protein structure.[13] The **3₁₀ helix** has exactly three residues per turn; each carbonyl hydrogen bonds to the third N—H on up the chain. Thus, the helix is tighter than the α helix. The **π helix** with 4.4 residues per turn is of larger diameter than the α helix. While the 3_{10} and π helices are not major protein structures, they do occur, often as single turns at the ends of helices.

4. Globular Proteins

The majority of proteins found in living cells are folded in a much more complex way than are the chains of the fibrous proteins.* The first protein for which the complete three-dimensional structure was worked out by X-ray crystallography is **myoglobin,** a small (MW = 17,500) oxygen-carrying protein of muscle. The 153 amino acid residues in the chain are for the most part arranged in eight different α-helical segments containing from 7 to 26 residues each. These rodlike helices are stacked together in an irregular fashion as shown in Fig. 2-8. The figure gives an incomplete picture because the space between the rods and inside the molecule is tightly packed with amino acid side chains almost all of which are hydrophobic. On the other hand, nearly all of the polar side chains project from the outsides of the helices into the surrounding water.

As the structures of more and more proteins have been determined in recent years, it has become clear that, as in the case of myoglobin, globular proteins are usually held together largely by interactions between hydrophobic residues. Within the interior of proteins the side chain groups are packed together remarkably well. While occasional holes are present, they usually are filled by water molecules.[24,25] For example, the packing density (volume enclosed by the van der Waals envelope divided by the total volume) is ~0.75 for the interior of the lysozyme and ribonuclease molecules compared with the theoretical value of 0.74 for close-packed spheres. Polar groups are usually on the surface but are sometimes "buried" and hydrogen bonded to other groups in the interior of the protein. The other surfaces of proteins also contain nonpolar side chains which are sometimes

---

* The pattern of folding of a peptide chain into a helix or a β structure is often referred to as the **secondary structure** of the protein. The further folding, which involves interactions between groups that are distant in the primary sequence, is called **tertiary structure.** The aggregation of monomeric protein subunits into oligomers (Chapter 4) provides the **quaternary structure.**

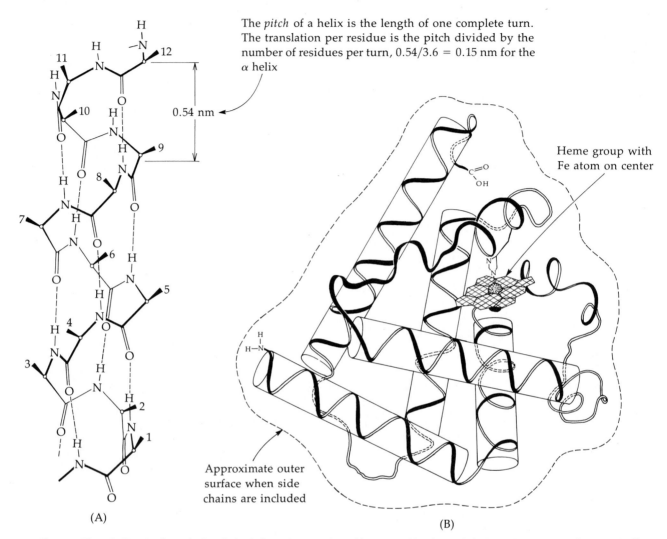

The *pitch* of a helix is the length of one complete turn. The translation per residue is the pitch divided by the number of residues per turn, $0.54/3.6 = 0.15$ nm for the α helix

0.54 nm

Heme group with Fe atom on center

Approximate outer surface when side chains are included

(A)

(B)

Fig. 2-8 The α helix. (A) The right-handed α helix with vertical hydrogen bonds indicated by dotted lines. The positions of the amino acid side chains are indicated by the numbers. (B) The conformation of the peptide backbone of myoglobin.

[See J. C. Kendrew, *Sci. Am.* **205**, 96–110 (Dec 1961). Five long α helices are indicated as rods. Several shorter helices can also be seen. The overall size of the molecule is approximately $4.4 \times 4.4 \times 2.5$ nm.

clustered into **hydrophobic regions.** The latter may be sites of interaction with other proteins or with lipid portions of membranes.

Myoglobin is exceptional in that most globular proteins have a relatively small amount of α helix. For example, the 129-residue chain of **lysozyme** (Fig. 2-9), one of the smallest enzymes (MW = 14,600), contains only a few short helices. For the

most part, the folding pattern is complex and irregular. Note that one region contains an antiparallel β pleated sheet. The sheet begins between residues 42 and 45. The chain then turns in a hairpin loop and residues 51–54 are hydrogen bonded to residues 42–45 as indicated in Fig. 2-9. The pleated sheet structure continues to a certain extent to other parts of the chain as well.

In larger protein molecules an extensive β structure is often present in the center of the molecule. This can be seen in the structure of the 307-residue enzyme **carboxypeptidase A** (Fig. 2-6, bottom).[26] The β structure, which is twisted into a kind of left-handed propeller, appears to provide a rigid "wall" upon which other parts of the protein molecule are hung. Many proteins contain sections of parallel as well as antiparallel β structure. In many of the larger proteins, such as **glyceraldehyde-phosphate dehydrogenase** (334 residues) the peptide chain is folded into two or more distinct **domains** with a "hinge" region between.[27] The folding patterns of the two domains of glyceraldehydephosphate dehydrogenase (Fig. 2-10) are both predominantly β pleated sheet. Note that the NAD^+-binding domain has almost entirely parallel β structure while the "catalytic" domain contains both parallel and antiparallel strands. Both domains contain α-helical segments, lying on both sides of the central sheet.

While the three-dimensional structure of a protein can best be displayed with stereo drawings, the internal hydrogen bonding pattern is often summarized most clearly in a two-dimensional diagram of the type shown in Fig. 2-11.

Studies of synthetic polypeptides as well as examination of known structures of proteins obtained by X-ray crystallography reveal that some amino acids, *e.g., glutamic acid, alanine, and leucine tend to promote α-helix formation.* Others, like *methionine, valine,* and *isoleucine are more often present in β structure,* while *glycine, proline,* and *asparagine tend to be present in bends.* It appears that the folding pattern of a peptide is, in effect, encoded in the sequence itself. Thus, when several residues that favor helix formation are clustered together, a helix may form. It may then be elongated in both directions until a region containing "helix breakers" such as proline is reached. Similarly, regions of β structure may form when appropriate residues are clustered. Random folding of other regions leads to additional hydrogen bonding and to the hydrophobic interactions that are found in the interior of globular proteins.

Now that the three-dimensional structures of many proteins are known, serious attempts are being made to predict the folding patterns of other proteins from their amino acid sequences.[28-30] In one method numerical "conformational parameters" are computed to give the probability of a particular residue being found in helix, β structure, or β turn (Table 2-4). According to Chou and Fasman[29] when four helix formers out of six residues are clustered, nucleation of a helix can be assumed. Propagation can then occur in both directions until it is terminated by a tetrapeptide of helix breakers. Likewise, if three out of five β formers are clustered, nucleation of a β sheet may occur. In this case, bonding to other parts of the chain separated by means of β turns or regions of **random coil** is essential. Protein prediction contests are being held in which various groups are attempting, with considerable success, to predict the folding pattern of a protein before its crystal structure is established.[31]

Recently Levitt and Chothia have classified 31 globular proteins of known structure into four clearly separated structural groups: (I) all -α proteins, such as myoglobin and hemoglobin; (II) proteins built from β sheets stacked to form a layered structure. These include the small rubredoxin (Fig. 10-4) and concanavalin A (Fig. 5-7); (III) the α + β proteins which contain a mixture of all -α and all -β (usually antiparallel) regions within the same polypeptide chain. Papain and thermolysin (Chapter 7) are examples; (IV) the α/β proteins, which have α helices and β strands, usually occurring one after the other in alternating fashion. There is usually one main β sheet in the center with helices packed on both sides, as in Fig. 2-6, bottom (carboxypeptidase), in Fig. 2-10 (glyceraldehyde-phosphate dehydrogenase) and in Fig. 7-5 (hexokinase and phosphoglycerate kinase).

5. Disulfide Bridges

Another structural feature of the lysozyme molecule (Fig. 2-9) is the presence of four disulfide cross-links or bridges between different parts of

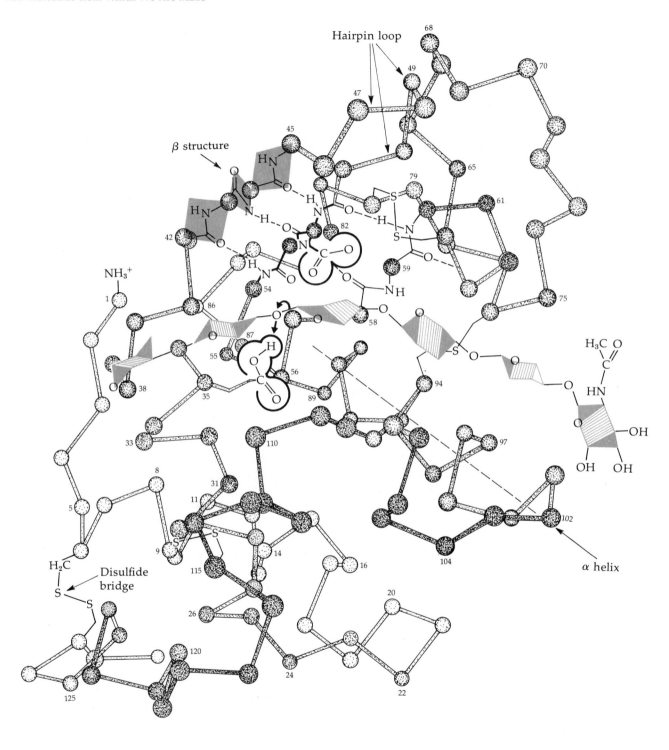

Hairpin loop

β structure

Disulfide
bridge

α helix

Fig. 2-9 Folding of the peptide chain of the enzyme lysozyme. (From R. E. Dickerson and I. Geis, "The Structure and Action of Proteins," p. 71. Benjamin, New York, 1969. Copyright 1969 by Dickerson and Geis.) For most of the molecule only the positions of attachment of the amino acid side chains (the α-carbon atoms) are shown. However, in one region the peptide linkages are drawn in to show the hydrogen bonding pattern of the antiparallel β structure present in the center of the molecule. One of the α-helical sections is indicated by an arrow. The shaded hexagons represent the backbone of a polysaccharide substrate bound in the "active site." The carboxyl group at position 35 is believed to assist in cleaving the O—C bond as indicated by the arrows (Chapter 7, Section C,4,a). The arrow at the left indicates the location of one of the four disulfide bridges. Others are visible in the upper right and lower left toward the center.

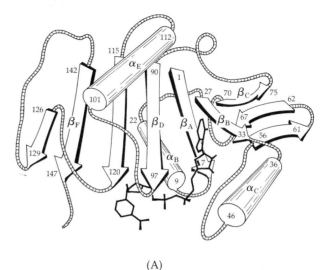

(A)

the chain. These form spontaneously when two —SH groups of cysteine side chains are close together and are oxidized by O_2 or some other reagent (see Eq. 2-8). Disulfide linkages are fre-

$$2 \ —CH_2—SH \longrightarrow —H_2C{\diagdown}S—S{\diagup}CH_2— \qquad (2\text{-}8)$$

quently present in proteins that are secreted from cells but are less common in enzymes that stay within cells. Perhaps because the latter are in a protected environment, the additional stabilization provided by disulfide bridges is not needed.

One of the most highly cross-linked proteins known occurs in the keratin "matrix." Breakage of the —S—S— linkages of this protein in hair is an essential step in the chemical "permanent wave" process. A thiol compound is used to reductively cleave the cross-links and after the hair is reset new cross-links are formed by air oxidation.

The geometry of —S—S— bridges is of some concern to protein chemists. The preferred torsion angle about the S—S bond appears to be 90° but considerable variation may exist.

Two S atoms viewed
down S—S axis

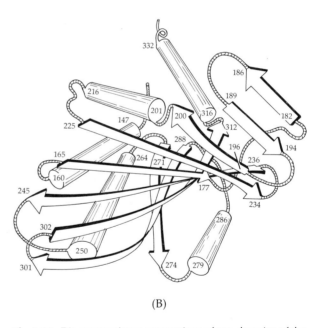

(B)

Fig. 2-10 Diagrammatic representations of two domains of the glyceraldehyde-phosphate dehydrogenase molecule (from lobster). (A) The NAD^+-binding domain. The numbers designate the sequence of amino acids and the symbols αB, αC, etc., and βA, βB, etc., designate α-helical segments and strands in the β pleated sheet. (B) The catalytic domain viewed in order to see clearly the large pleated sheet forming the subunit interface. From Buehner et al.[27]

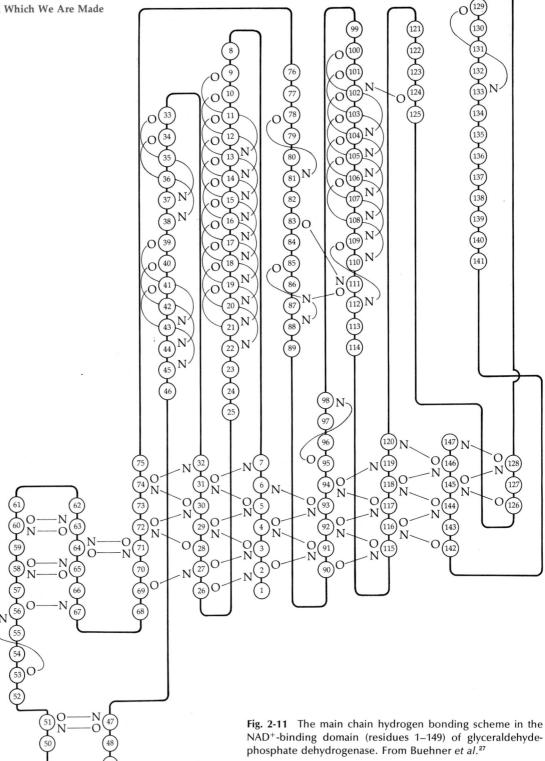

Fig. 2-11 The main chain hydrogen bonding scheme in the NAD⁺-binding domain (residues 1–149) of glyceraldehyde-phosphate dehydrogenase. From Buehner *et al.*[27]

TABLE 2-4
Classification of Protein Residues according to Their Tendencies to Form α Helix, β Structure, and β Turns[a]

Amino acid	P_α	Helix-forming tendency	P_β	β-structure-forming tendency	P_t
Glu$^-$	1.53	++	0.26	br+	0.44
Ala	1.45	++	0.97	w	0.57
Leu	1.34	++	1.22	+	0.53
His$^+$	1.24	+	.71	br	0.69
Met	1.20	+	1.67	++	0.67
Gln	1.17	+	1.23	+	0.56
Trp	1.14	+	1.19	+	1.11
Val	1.14	+	1.65	++	0.30
Phe	1.12	+	1.28	+	0.71
Lys$^+$	1.07	w	0.74	br	1.01
Ile	1.00	w	1.60	++	0.58
Asp$^-$	0.98	i	0.80	++	1.26
Thr	0.82	i	1.20	+	1.00
Ser	0.79	i	0.72	br	1.56
Arg$^+$	0.79	i	0.90	i	1.00
Cys	0.77	i	1.30	+	1.17
Asn	0.73	br	0.65	br	1.68
Tyr	0.61	br	1.29	+	1.25
Pro	0.59	br+	0.62	br	1.54
Gly	0.53	br+	0.81	i	1.68

[a] The conformational parameters P_α, P_β, and P_t (β turn) are the frequencies of finding a particular amino acid in an α helix, β structure, or β turn (in 15 proteins of known structure) divided by the average frequency of residues in those regions. The residues in the table are arranged in order of decreasing tendency toward helix formation. The symbols ++, +, w, i, br, and br+ designate "strong former," "former," "weak former," "indifferent," "breaker," and "strong breaker," respectively. From data of Chou and Fasman.[29]

6. Sizes and Shapes of Protein Molecules

Both internal structure and overall size and shape of proteins vary enormously. To give some idea of the possibilities Fig. 2-12 shows several different ways in which a polypeptide chain 300 amino acid residues in length could be folded. The chain could stretch in fully extended form for ~100 nm. If the chain were folded back on itself about 13 times it could form a 7 nm square pleated sheet about 0.5 nm thick. The same polypeptide could form a thin α-helical rod 45 nm long and ~1.1 nm thick. Together with two other similar chains it could form a collagen-type triple helix (if it had the right amino acid composition) of 87 nm length and about 1.5 nm diameter. (Note that this is about $\frac{1}{3}$ the length of tropocollagen.)

Globular proteins vary considerably in the tightness of packing and the amount of internal water of hydration.[24,25] However, a density of ~1.4 g cm^{-3} is typical. With an average mass per residue of 115 daltons our 300 residue polypeptide would have a mass of 34,500 daltons or 5.74×10^{-20} g and would occupy a volume of 41 nm^3. This might be a cube 3.45 nm in width, a "brick" of dimensions $1.8 \times 3.6 \times 6.3$ nm, a sphere of diameter 4.3 nm, or an object of very irregular shape. For purposes of calculation idealized ellipsoid and rod shapes are often assumed.

It is informative to compare these dimensions with those of the smallest structures visible in cells; for example, a bacterial flagellum is ~13 nm in diameter and a cell membrane ~8–10 nm in thickness. Bricks of the size of the 300-residue polypeptide could be used to assemble a bacterial flagellum or a eukaryotic microtubule. An α-helical polypeptide could extend through a cell membrane and project on both sides while a globular protein of the same chain length might be completely embedded in the membrane.

7. Conformational Changes in Polypeptides

Peptide chains can assume extended conformations in which hydrogen bonds are formed between folds of the same chain (β structure) and can also exist in one of several helical conformations. In some cases, the same protein can do both. An elegant example is provided by the α-helical low-sulfur protein of hair (Section B,3,d). Hair can be stretched greatly and when this is done, the α helices uncoil and the chains assume a β configuration with hydrogen bonds *between* chains instead of *within* a single chain. Thus, a given polymer may have more than one conforma-

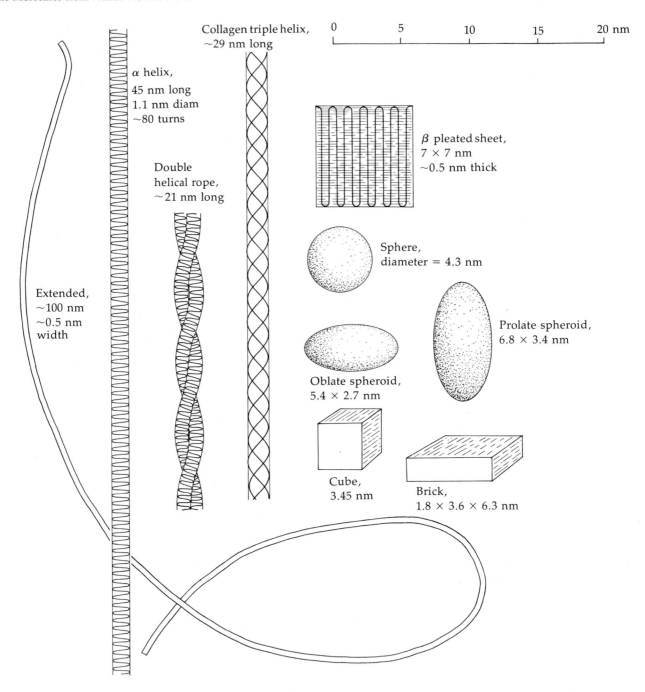

Collagen triple helix,
~29 nm long

α helix,
45 nm long
1.1 nm diam
~80 turns

Double
helical rope,
~21 nm long

β pleated sheet,
7 × 7 nm
~0.5 nm thick

Sphere,
diameter = 4.3 nm

Prolate spheroid,
6.8 × 3.4 nm

Oblate spheroid,
5.4 × 2.7 nm

Extended,
~100 nm
~0.5 nm
width

Cube,
3.45 nm

Brick,
1.8 × 3.6 × 6.3 nm

Fig. 2-12   Some shapes a protein molecule of 300 amino acids (MW ~34,500) might assume.

BOX 2-A

THE PROTEINS OF BLOOD PLASMA

Among the most studied of all proteins are those present in blood plasma.[a,b,c] Their ready availability together with the clinical significance of their study led to the early development of electrophoretic separations of these proteins. Electrophoresis at a pH of 8.6 (in barbital buffer) indicates six main components. The major, and one of the fastest moving proteins, is **serum albumin.** Trailing behind it in the electrophoretic pattern are the **α_1-, α_2-,** and **β-globulins, fibrinogen,** and **γ-globulins.** Upon starch gel electrophoresis, each of these bands breaks up into a larger number, a total of about 20 zones being separated. Most zones contain more than one protein and over 50 well-characterized plasma proteins have been isolated. Sixty or more enzymes are present, some in very small quantities which may have leaked from body cells.

The normal concentration range of total serum protein is from 5.7–8.0 g per 100 ml ($\sim$1 millimolar). The amounts of the individual proteins vary from that of albumin (3.5–4.5 g per 100 ml) to a few milligrams or less. The next most abundant proteins are the **immunoglobulins** (Box 5-F) one of them (IgG) being present to the extent of 1.2–1.8 g per 100 ml. Also present in amounts greater than 200 mg per 100 ml are the **α-** and **β-lipoproteins,** the **α_1 anti-trypsin, α_2-macroglobulin, haptoglobin, transferrin,** and fibrinogen.

Plasma proteins have a variety of functions. One, fullfilled principally by serum albumin, is to impart a high enough osmotic pressure to plasma to match that of the cytoplasm of cells. Human serum albumin consists of a single chain of 584 amino acid residues of MW = 69,000. There are three homologous repeat units or domains, each containing six disulfide bridges, a fact that suggests that gene duplication occurred twice during the evolution of serum albumins.[d] The relatively low molecular weight and high density of negative charges on the surface make serum albumin well adapted for the role of maintaining osmotic pressure.

A second major function of serum proteins is transport. Thus, serum albumin binds to and carries many sparingly soluble metabolic products. Transferrin transports iron and **ceruloplasmin** (an $\alpha2$ protein, Box 10-H) transports copper. **Transcortin** carries corticosteroids and progesterone, the **retinol-binding protein** carries vitamin A and **cobalamin-binding proteins** vitamin B_{12}. The lipoproteins, of which there are three principal classes, carry phospholipids, neutral lipids and cholesterol esters.[e] The majority of the mass of these substances is lipid. The $\alpha1$ fraction of serum contains the **high density lipoprotein** (HDL).[f] The pre-β fraction contains **very low density lipoprotein** (VLDL) and the β fraction, the **low density lipoprotein** (LDL). All these proteins have been under intensive investigation as a result of interest in artery disease and the fact that cholesterol and other lipids carried by these proteins may be deposited in arteriosclerotic plaques.

The immunoglobulins, the α_1 trypsin inhibitor (Box 7-C), the ten or more blood clotting factors (Fig. 6-16), and the proteins of the complement system (Box 5-G) all have protective functions that are discussed elsewhere in this book. Hormones, many of them proteins (Table 16-1), are present in the blood as they are carried to their target tissues. Many serum proteins have unknown or poorly understood functions. Among these are a group of glycoproteins present in substantial quantities. The concentrations of some of them, such as haptoglobin (or α_2-macroglobulin) tend to be increased in a variety of pathological conditions.

[a] F. W. Putnam, ed., "The Plasma Proteins," 2nd ed., Vols. 1 and 2. Academic Press, New York, 1975.

[b] A. White, P. Handler, and E. L. Smith, "Principles of Biochemistry," 5th ed., Chapter 30. McGraw-Hill, New York, 1973.

[c] M. W. Turner and B. Hulme, "The Plasma Proteins: An Introduction." Pitman, London, 1971.

[d] P. Q. Behrens, A. M. Spiekerman, and J. R. Brown, *Fed. Proc.* **34**, 591 (1975).

[e] J. D. Morrisett, R. L. Jackson, and A. M. Gotto, Jr., *Annu. Rev. Biochem.* **44**, 183–207 (1975).

[f] G. Schonfeld, B. Pfleger, and R. Roy, *JBC* **250**, 7943–7950 (1975).

tion in which the folding and hydrogen bonding are different.

The situation with soluble proteins is more complex, but again, more than one folded conformation is possible with different sets of hydrogen bonds and of hydrophobic interactions between side chains. Some of the conformations of a globular protein are more stable than others, and a protein will ordinarily assume one of the energetically most favorable conformations. Nevertheless, there are probably other conformations of almost equal energy.* It appears that *there is often an easy interconversion from one conformation to another,* a fact of great biological significance.

Conformational alterations in proteins are probably facilitated by the fact that some hydrogen-bonded groups are found in the hydrophobic interior of protein molecules. All of the "buried" hydrogen atoms suitable for H bond formation are ordinarily bonded to a suitable electron donor group. However, there are, in general, more of these electron donor groups than there are hydrogen atoms to which to bind. This sets the stage for a competition between electronegative centers for a particular proton suitable for hydrogen bonding and provides a molecular basis for the easy triggering of conformational changes.[32]

An extreme conformational alteration is the **denaturation** of proteins, which may be caused by heating or by treatment with various reagents such as strong acids and bases, **urea, guanidine hydrochloride,** and **sodium dodecyl sulfate** (SDS). Denaturation leads to unfolding of the protein to a more or less random coil conformation (in which the chain is folded randomly rather than into a helix, β sheet, or other regular pattern). In the denatured state the amide groups of the peptide chain form hydrogen bonds with surrounding water molecules rather than with each other. Characteristic biochemical activities are lost and physical properties such as sedimentation constant, viscosity, and light absorption are altered. The ease of denaturation of proteins and the fact

that denaturation is sometimes reversible show that the energy differences between the folded conformations and the open random coil conformation are not great.

Complete denaturation of a protein had generally been regarded as an irreversible process prior to 1956 when Anfinsen showed that denatured **ribonuclease** (Chapter 7, Section E,2) could refold spontaneously.[33] The 124-residue ribonuclease molecule contains four disulfide (—S—S—) bridges and thus is firmly tied together. When these bridges are reductively broken (Section H,3,a), the enzyme is readily denatured and becomes inactive. However, Anfinsen found that upon reoxidation under appropriate conditions, complete activity reappeared. The molecule was clearly able to fold spontaneously into the correct conformation—one in which the correct *one* of 105 possible pairings of the eight —SH groups present needed to reform the four disulfide bridges correctly took place. This observation has had an important influence on our thinking about protein synthesis and the folding of polypeptide chains into biologically active molecules.[34,34a]

It is generally conceded that folding of the peptide chain in a specific way probably commences while the chain is still being synthesized on a ribosome. The growing chain probably begins to fold in a random way until it finds a stable conformation. As the synthesis of the peptide chain is completed, the molecule spontaneously folds further into a "native" conformation. However, the final biological activity of a protein often does not appear until it has undergone alteration by other enzymes, sometimes accompanied by a further change in conformation.

C. SUGARS AND POLYSACCHARIDES

The simple sugars (**monosaccharides**) and their oligomers and polymers (**polysaccharides**) constitute a major class of cell components, known collectively as **carbohydrates.** A typical monomer unit is **α-D-glucose** (α-D-glucopyranose), a cyclic **ring form** of the sugar glucose:

---

* No computer is large enough to systematically examine the energy of formation of all the possible conformations of the simplest protein. Therefore, this statement cannot be made quantitative.

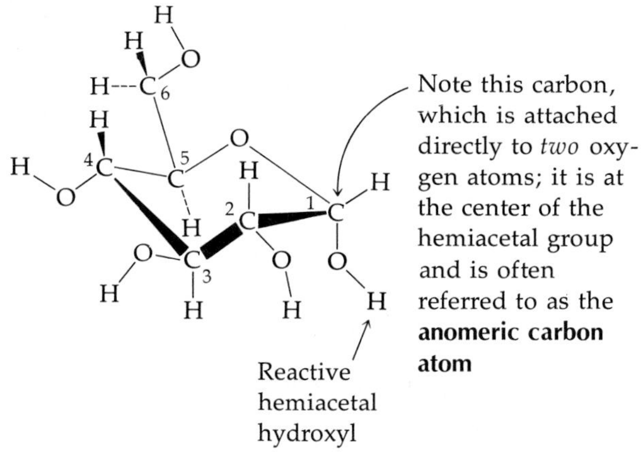

Note this carbon, which is attached directly to *two* oxygen atoms; it is at the center of the hemiacetal group and is often referred to as the **anomeric carbon atom**

Reactive hemiacetal hydroxyl

The reactive linking groups are two types of —OH: *the hemiacetal hydroxyl group* plus *one other hydroxyl* that can be located on any carbon atom. The variety of monomer units and positions of linkage make possible a bewildering variety of polysaccharide structures.

The sugar units in polysaccharides are joined by **glycosidic** (acetal) linkages (Eq. 2-9). Equation 2-9

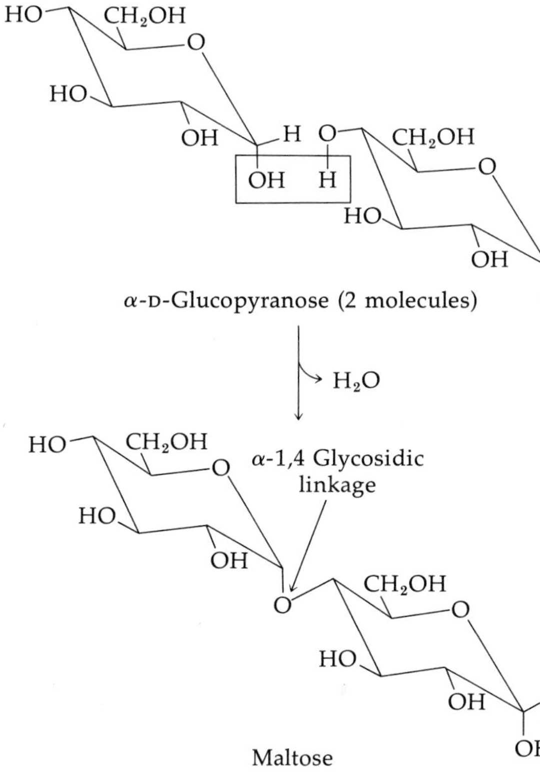

α-D-Glucopyranose (2 molecules)

↓ H_2O

α-1,4 Glycosidic linkage

(2-9)

Maltose

shows the formation of the disaccharide maltose from two molecules of glucose. The joining of additional glucose units in a similar manner would lead to a polymer. The symbol α-1,4 refers to the fact that in maltose the glycosidic linkage connects carbon atom 1 (the **anomeric** carbon atom) of one ring with C-4 of the other and that the configuration about the anomeric carbon atom is α (see next section).

Functional groups present in the polymers are the numerous free hydroxyl groups, some of which are used to form additional glycosidic linkages forming **branched chains.** The hydroxyl groups also permit structural modifications of the basic sugar ring prior to polymerization. Thus, polysaccharides may contain —COOH, —NH_2, sulfate, —$NHCOCH_3$, and other functional groups in addition to hydroxyl groups.

1. Structures and Properties of Sugars

Sugars are polyhydroxyaldehydes **(aldoses)** or polyhydroxyketones **(ketoses).** The carbonyl group is a highly reactive group, and a characteristic reaction is **addition** of electron-rich groups such as the —OH group (Chapter 7, Section H). If a sugar chain is long enough (4–6 carbon atoms) one of the hydroxyl groups of the same molecule can add to the carbonyl group to form a cyclic **hemiacetal** or ring form, which reaches an equilibrium with the free aldehyde or ketone form (Eq. 2-10). The six-

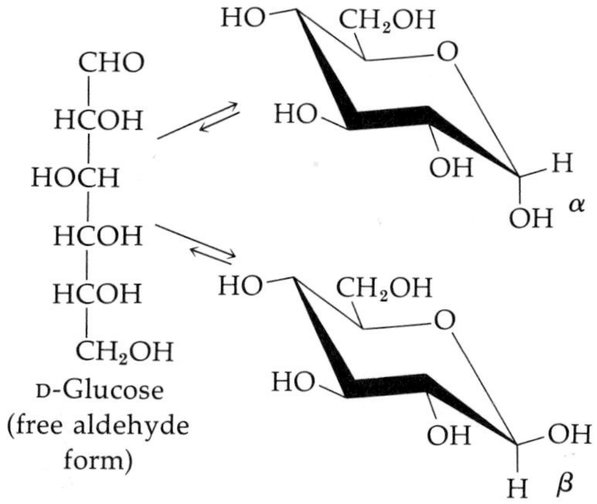

CHO
HCOH
HOCH
HCOH
HCOH
CH_2OH
D-Glucose (free aldehyde form)

(2-10)

membered rings formed in this way (**pyranose rings**) are especially stable, but five-membered **furanose** rings also exist in some carbohydrates. It is the natural tendency of 5- and 6-carbon sugars (**pentoses** and **hexoses**) to cyclize that permits formation of stable sugar polymers from the highly reactive and unstable monomers.

When a sugar cyclizes a new chiral center is formed at the carbon atom (anomeric carbon atom) derived from the original carbonyl group. The two configurations about this carbon atom are designated α and β as indicated in Eq. 2-10. In an equilibrium mixture, ring forms of most sugars predominate over open chains. Thus, at 25°C in water glucose reaches an equilibrium containing ~0.02% free aldehyde, 38% α-pyranose form, 62% β-pyranose form, and <0.5% each of the much less stable furanose forms. Although polysaccharides are composed almost exclusively of sugar residues in ring form, the open chain forms are sometimes metabolic intermediates.

Sugars contain several chiral centers and the various diastereomers are given different names. Thus, **glucose, mannose,** and **galactose** are just three of the eight diastereomeric aldohexoses (the others are allose, altrose, gulose, idose, and talose).[6] Each of these sugars exists as a pair of D and L forms (enantiomers), complete mirror images one of the other. Fischer projection formulas (Section A,4) are often used to show relationships between sugars, as in Fig. 2-13. These projection formulas are convenient in relating sugar structures one to another, but they give an unrealistic picture of three-dimensional structure. According to the Fischer convention each carbon atom must be viewed with both vertical bonds projecting behind the atom viewed. In fact, the molecule cannot assume such a conformation; for example, compare the three-dimensional structure of the sugar alcohol ribitol (Section A,6), which is formed by reduction of the carbonyl of ribose with its Fischer formula.

Monosaccharides are classified as D or L according to the configuration at the chiral center farthest from the carbonyl group (Fig. 2-13). If the —OH group attached to this carbon atom lies to the right when the sugar is oriented according to the Fischer convention, the sugar belongs to the D family. The simplest of all the chiral sugars is glyceraldehyde.

$$
\begin{array}{c}
\text{O} \\
\parallel \\
\text{HC} \\
\mid \\
\text{H—C—OH} \\
\mid \\
\text{CH}_2\text{OH}
\end{array}
$$

D-Glyceraldehyde

Ring forms of sugars also are often drawn according to the Fischer convention, e.g., α-D-glucopyranose.

$$
\begin{array}{c}
\quad\quad \overset{\alpha}{} \\
\text{H—C—OH} \\
\mid \\
\text{H—C—OH} \\
\mid \\
\text{HO—C—H} \\
\mid \\
\text{H—C—OH} \quad \text{O} \\
\mid \\
\text{H—C} \\
\mid \\
\text{CH}_2\text{OH}
\end{array}
$$

α-D-Glucopyranose

When drawn in this way the hydroxyl at the anomeric carbon atom is on the right side of the molecule for the α form in the D series of sugars and at the left side for the L sugars. Thus, α-D-glucopyranose and α-L-glucopyranose are enantiomers.

To simplify the drawing of sugar rings **Haworth structural formulas** are often used.

$$
\text{CH}_2\text{OH}
$$

α-D-Glucopyranose
(Haworth structural
formula)

The lower edge of the ring (heavy line) is to be thought of as projecting out toward the reader and

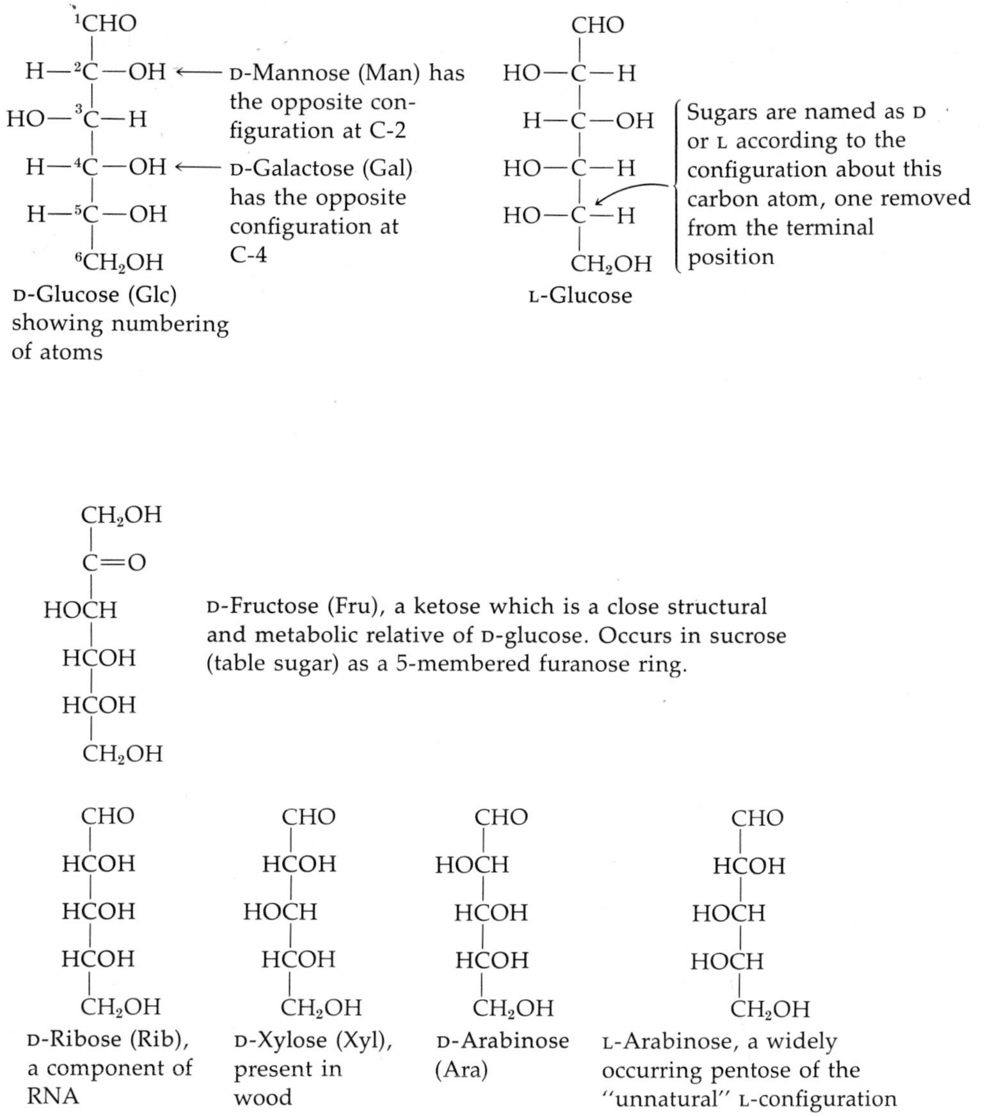

$$
\begin{array}{c}
^1CHO \\
| \\
H-{}^2C-OH \\
| \\
HO-{}^3C-H \\
| \\
H-{}^4C-OH \\
| \\
H-{}^5C-OH \\
| \\
^6CH_2OH
\end{array}
$$

⟵ D-Mannose (Man) has the opposite configuration at C-2

⟵ D-Galactose (Gal) has the opposite configuration at C-4

D-Glucose (Glc) showing numbering of atoms

$$
\begin{array}{c}
CHO \\
| \\
HO-C-H \\
| \\
H-C-OH \\
| \\
HO-C-H \\
| \\
HO-C-H \\
| \\
CH_2OH
\end{array}
$$

Sugars are named as D or L according to the configuration about this carbon atom, one removed from the terminal position

L-Glucose

$$
\begin{array}{c}
CH_2OH \\
| \\
C=O \\
| \\
HOCH \\
| \\
HCOH \\
| \\
HCOH \\
| \\
CH_2OH
\end{array}
$$

D-Fructose (Fru), a ketose which is a close structural and metabolic relative of D-glucose. Occurs in sucrose (table sugar) as a 5-membered furanose ring.

$$
\begin{array}{c}
CHO \\
| \\
HCOH \\
| \\
HCOH \\
| \\
HCOH \\
| \\
CH_2OH
\end{array}
\qquad
\begin{array}{c}
CHO \\
| \\
HCOH \\
| \\
HOCH \\
| \\
HCOH \\
| \\
CH_2OH
\end{array}
\qquad
\begin{array}{c}
CHO \\
| \\
HOCH \\
| \\
HCOH \\
| \\
HCOH \\
| \\
CH_2OH
\end{array}
\qquad
\begin{array}{c}
CHO \\
| \\
HCOH \\
| \\
HOCH \\
| \\
HOCH \\
| \\
CH_2OH
\end{array}
$$

D-Ribose (Rib), a component of RNA

D-Xylose (Xyl), present in wood

D-Arabinose (Ara)

L-Arabinose, a widely occurring pentose of the "unnatural" L-configuration

Fig. 2-13   Some sugar (monosaccharide) structures drawn according to the Fischer convention.

the other edge as projecting behind the plane of the paper. The Haworth structures are easy to draw and unambiguous in depicting configurations but they do not show the spatial relationships of groups attached to the rings correctly. For this reason conformational formulas of the type shown in Fig. 2-14 are often employed.

Most sugars occur in the chair conformation that places most of the substituents in equatorial positions. For D-aldoses this is usually the "C1" conformation shown on p. 79, whereas for L-aldoses it is the other ("1C") chair form. There are exceptions to this rule. Thus, α-D-idopyranose also assumes the 1C conformation. It is also noteworthy

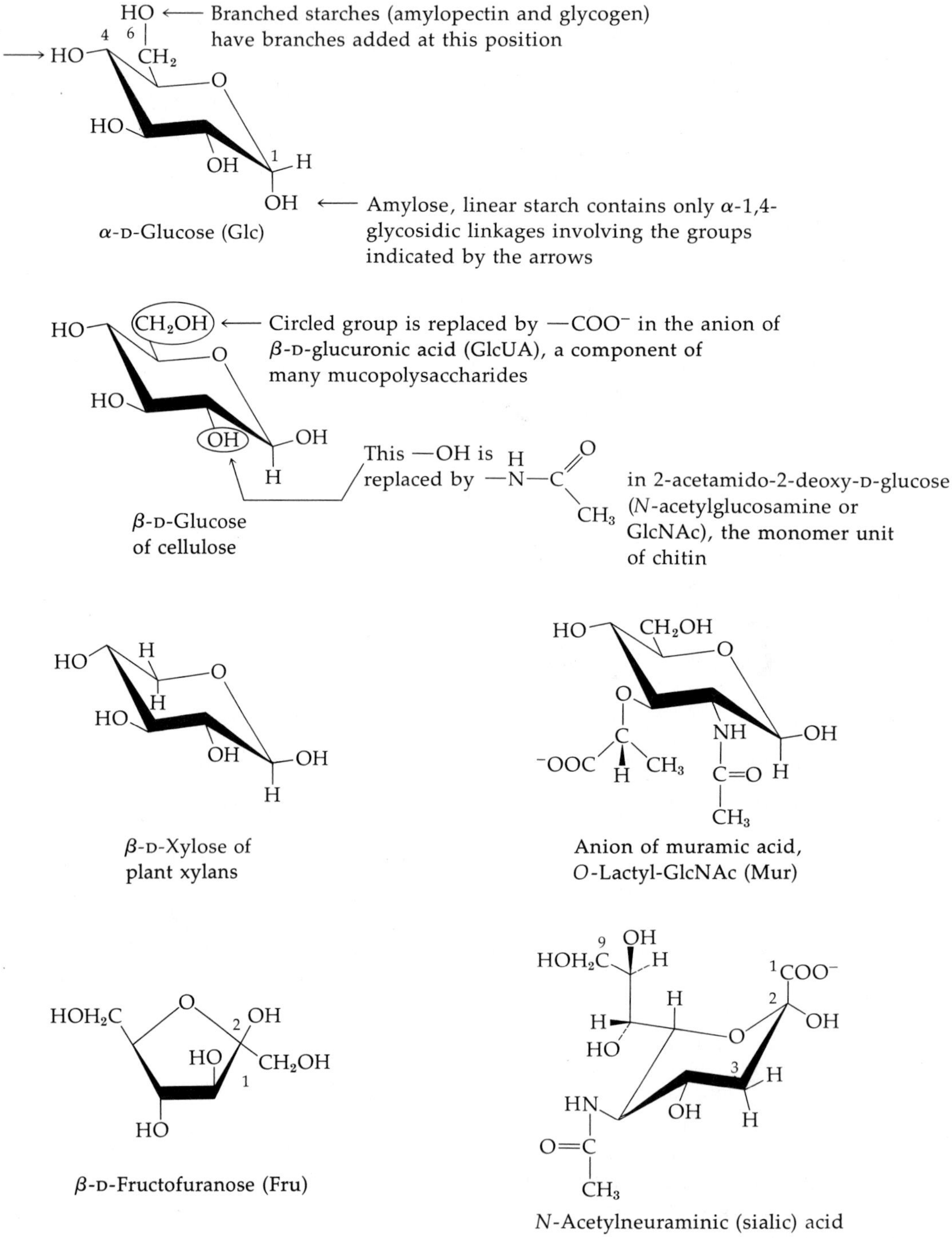

HO ⟵ Branched starches (amylopectin and glycogen)
have branches added at this position

α-D-Glucose (Glc)

OH ⟵ Amylose, linear starch contains only α-1,4-
glycosidic linkages involving the groups
indicated by the arrows

(CH₂OH) ⟵ Circled group is replaced by —COO⁻ in the anion of
β-D-glucuronic acid (GlcUA), a component of
many mucopolysaccharides

This —OH is
replaced by —N—C   in 2-acetamido-2-deoxy-D-glucose
(N-acetylglucosamine or
GlcNAc), the monomer unit
of chitin

β-D-Glucose
of cellulose

β-D-Xylose of
plant xylans

Anion of muramic acid,
O-Lactyl-GlcNAc (Mur)

β-D-Fructofuranose (Fru)

N-Acetylneuraminic (sialic) acid

Fig. 2-14   Some sugar rings present in polysaccharides.

that the substituent on the anomeric carbon atom of a sugar ring often prefers an axial orientation.[35]

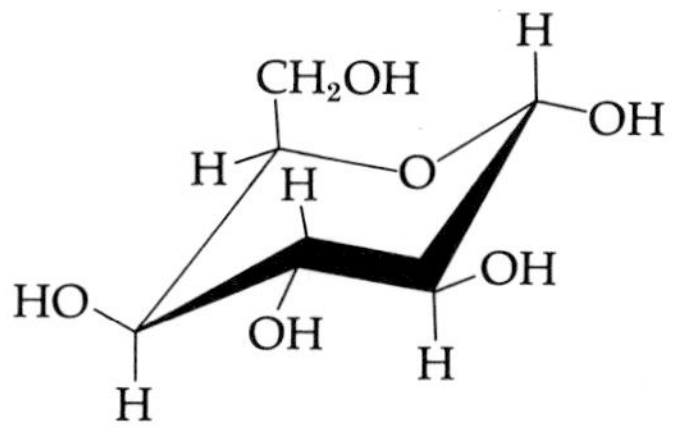

α-Pyranose form of
D-idose in 1C conformation

Reduction of the carbonyl group of a sugar (e.g., with sodium borohydride) yields a **sugar alcohol** such as ribitol (Section A,6) or **sorbitol** (from reduction of glucose, Eq. 11-8). The aldehyde groups of aldoses can also be oxidized by a variety of agents to form carboxylic acids known as **aldonic acids.** This fact accounts for the **reducing properties** of aldoses. For example, in alkaline solution aldoses reduce cupric ions to cuprous oxide, silver ions to the free metal or ferricyanide to ferrocyanide. The latter reaction provides the basis for sensitive analytical procedures. Even though the aldoses tend to exist largely as hemiacetals (Eq. 2-10) the reducing property is fully evident. While reduction by metal-containing reagents is usually via the free aldehyde, oxidation by hypobromite (Br_2 in alkaline solution) yields the lactone as in the enzymatic reaction of Eq. a, Table 8-4.

Many sugar derivatives occur in nature. Among these are aldonic acids such as 6-phosphogluconic acid (Eq. 9-12) and **uronic acids,** which have —COOH in the terminal position (position 6 in **glucuronic** acid, GlcUA, Fig. 2-14).* The —OH group in the 2 position of glucose may be replaced by —NH_2 to form 2-amino-2-deoxyglucose, commonly called **glucosamine** (GlcN) or by —NH—CO—CH_3 to form **N-acetylglucosamine** (GlcNAc). Similar derivatives of other sugars exist. In many polysaccharides sulfate groups are attached in ester linkage to the sugar units.

* The abbreviation GlcA is also used, but many chemists prefer GlcUA to avoid confusion with gluconic acid.

Two common *6-deoxy* sugars (lacking the oxygen atom at C-6) are **rhamnose** and **fucose**. Both are of the "unnatural" L-configuration but are derived metabolically from D-glucose and D-mannose, respectively.

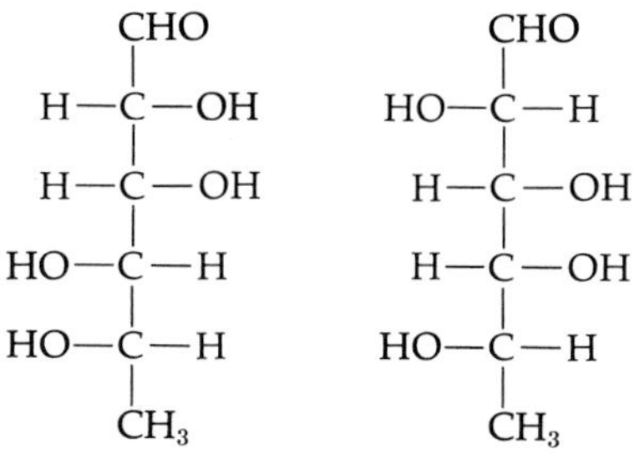

L-Rhamnose (Rha)   L-Fucose (Fuc)

2. Oligosaccharides

The **disaccharides** and other oligosaccharides composed of a small number of sugar units are important to the metabolism of plants and animals. Examples are **lactose** (milk), **sucrose** (green plants), and **trehalose** (fungi).

While the α and β ring forms of free sugars can usually undergo ready interconversion the configuration at the anomeric carbon atom is "frozen" when a glycosidic linkage is formed. To describe such a linkage, we must state this configuration together with the positions joined in the two rings (see Eq. 2-9). Thus, lactose can be described as a disaccharide containing one galactose unit in a β-pyranose ring form and whose anomeric carbon atom (C-1) is joined to the 4 position of glucose (a β-1,4 linkage):

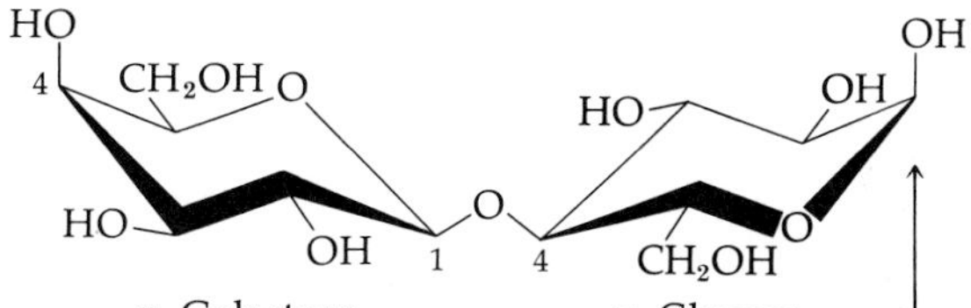

D-Galactose   D-Glucose

α-Lactose

Note α configuration at this "reducing end" of the molecule of α-lactose

The systematic name for α-lactose, O-β-D-galactopyranosyl-(1$\rightarrow$4)-α-D-glucopyranose, provides a complete description of the stereochemistry, ring sizes, and mode of linkage. It can be abbreviated β-D-Gal$p$-(1$\rightarrow$4)-α-D-Glc$p$. Maltose is α-D-Glc$p$-(1$\rightarrow$4)-α-D-Glc$p$. Since pyranose rings are so common and since most natural sugars belong to the D family, the designations D and p are often omitted. It may be assumed that in this book sugars are always D unless they are specifically designated as L. Sometimes, after the sequence of sugar residues in a polysaccharide has been worked out, the linkage remains uncertain and incomplete abbreviated formulas are given.

Note that in the drawing of the lactose structure the glucose ring has been "flipped over" with respect to the orientation of the galactose ring, a consequence of the presence of the β-1,4 linkage. For maltose, where the linkage is α-1,4, the two rings are usually drawn with the same orientation (Eq. 2-9).

Because the glucose ring in lactose is free to open to an aldehyde and to equilibrate (in solution) with other ring forms, the name lactose does not imply a fixed ring structure for the glucose half. Thus, lactose can be abbreviated as βGal-(1$\rightarrow$4)-Glc. (Nevertheless, in crystalline form it exists either as α-lactose or β-lactose.*) On the other hand, in sucrose and trehalose the reducing groups of two rings are joined. The molecules exist in only one form.

Sucrose [αGlc$p$-(1$\rightarrow$2)-βFru$f$]

---

*β-Lactose is more soluble and sweeter than α-lactose. The latter may crystallize in ice cream upon prolonged storage to give a "sandy" texture.

α,α-Trehalose [αGlc$p$-(1$\rightarrow$1)αGlc$p$]

3. Polysaccharides

Polymers of sugar are present in all cells and serve a variety of functions. Thus, **cellulose** and **chitin** provide strength to green plants and to skeletons of arthropods, respectively. **Hyaluronic acids** and other mucopolysaccharides provide a protective material between animal cells and **pectins** and related polysaccharides play a similar role in plants. Cell surfaces are usually heavily coated with polysaccharide materials of widely varying structures. Differences in these external polysaccharides are important in giving immunological identity to organisms. **Starch, glycogen,** and other storage polysaccharides serve as rapidly metabolizable food reserves for cells.[35a]

One component of plant starch, **amylose,** is a linear polymer of many α-D-glucopyranose units in 1,4 linkage. Starch granules always contain a second kind of starch **amylopectin.*** Both amylopectin and **glycogen** (animal starch) consist of bushlike molecules. Branches are attached to α-1,4-linked chains through α-1,6 linkages (Figs. 2-14 and 2-15). Some bacteria, e.g., *Leuconostoc mesenteroides,* make a linear 1,6-linked poly-D-glucose or **dextran** and some species tack on branches with α-1,4 or α-1,3 linkages. Dextrans formed by bacteria growing on the surfaces of teeth are an important component of **dental plaque.** Bacterial dextrans are also produced commercially and are chemically "cross-linked" to form gels (trade name, Sephadex) which are widely used in biochemical separation procedures (Section H,1,d).

Cellulose, probably the most abundant carbohydrate on earth (plants produce $\sim 10^{14}$ kg/year), is a

---

* Some starches, such as that of "waxy" maize contain only amylopectin and lack amylose.

β-1,4-linked polyglucose. **Chitin,** another abundant polysaccharide, is found in cell walls of fungi and in exoskeletons of arthropods. It is a linear β-1,4-linked polymer of *N*-acetylglucosamine, structurally similar to cellulose. Pectins are β-1,4-linked polygalacturonic acids in which some of the carboxyl groups are methylated.

Cell walls of yeasts contain mannose polymers **(mannans)** in which the main α-1,6-linked chain carries short branches (1–3 mannose units) joined in α-1,2 and α-1,3 linkage. In addition to cellulose, plants all contain **xylans** which consist mainly of chains of β-1,4-linked xylopyranose units. Note that xylose, a 5-carbon sugar, exists in the six-membered pyranose ring form in this polymer. On the other hand, fructose, a 6-carbon sugar, is present as five-membered furanose rings in **inulin,** the storage polysaccharide of the Jerusalem artichoke. The difference has to do with biosynthetic pathways. Pyranose rings are always more stable thermodynamically than are furanose rings. The latter arise in inulin and in sucrose because the biosynthesis occurs via the 6-phosphate ester of fructose.

A number of polysaccharides, including the starches, have one "reducing end" in which the ring can open to form a free aldehyde group with reducing properties. In other cases the reducing end is "blocked." One way in which that may happen (suggested for inulin) is for the chain to terminate in a sucrose structure. A terminal trehalose structure is possible for other polysaccharides. Sometimes sugar chains end in cyclic polysaccharides with no free reducing end and in many cases they are attached to protein or lipid molecules.

Numerous polysaccharides are made up of repeating units of more than one monomer. Some of these **heteropolysaccharides** are composed of two sugars in a simple alternating sequence.[36,37] Examples are **hyaluronic acid,** the **chondroitin sulfates,** and **dermatan sulfate,** important components of the "ground substance" or intracellular cement of connective tissue. Hyaluronic acid is a repeating polymer of glucuronic acid and *N*-acetylglucosamine with the structure shown in

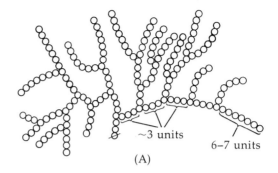

~3 units

6–7 units

(A)

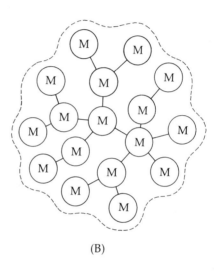

(B)

Fig. 2-15 Schematic diagrams of the glycogen molecule. (A) The structure of a segment of glycogen as proposed by K. H. Meyer, *Adv. Enzymol.* **3,** 109–136 (1943). The circles represent glucose residues which are connected by α-1,4 linkages and, at the branch points, by α-1,6 linkages. The symbol φ designates the reducing group. From D. French, *in* "Symposium on Foods: Carbohydrates and Their Roles" (H. W. Schultz, ed.), pp. 26–54. Avi Publ., Westport, Connecticut, 1969. (B) Schematic two-dimensional representation of a glycogen particle of MW = 760,000 and containing ~4700 glucose units (outlined by dashed lines). Digestion by β-amylase (Chapter 7, Section C,6) yields a "limit dextrin" of MW ≈ 410,000 which is represented by the circles and connecting lines. The regions labeled M are densely branched and are held together by regions of low branch density. Upon digestion with α-amylase (Chapter 7, Section C,6) the M regions are incompletely digested leaving limit dextrins, each consisting of about 34 glucose units. From G. A. Brammer, M. A. Rougvie, and D. French, *Carbohyd. Res.* **24,** 343–354 (1972).

→

Fig. 2-16 (A) The repeating disaccharide units of hyaluronic acid and related mucopolysaccharides. (B) Tentative model of the molecular architecture of proteoglycan aggregates of cartilage showing the arrangement of hyaluronic acid and chondroitin and keratan sulfates. The latter are bound covalently to a protein core to form a proteoglycan subunit which aggregates with hyaluronic acid. The drawing is approximately to scale except that the lengths of the chondroitin sulfate chains have been reduced about one-half. From L. Rosenberg, *in* "Dynamics of Connective Tissue Macromolecules" (P. M. C. Burleigh and A. R. Poole, eds.), p. 107. North-Holland, Amsterdam, 1975. The mucopolysaccharide chains may interact electrostatically with collagen fibrils in cartilage. (C) Dark field electron micrograph of a proteoglycan aggregate from bovine articular cartilage (from bearing surfaces of joints). The filamentous backbone consists of hyaluronic acid (as in B) and the proteoglycan subunits (individual polysaccharide chains cannot be seen) extend from the backbone. From Rosenberg *et al.*[38b] Photograph courtesy of Lawrence Rosenberg.

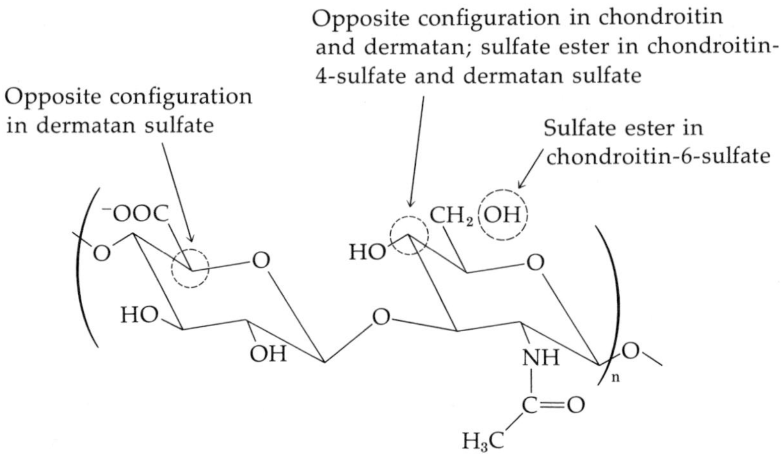

Opposite configuration in dermatan sulfate

Opposite configuration in chondroitin and dermatan; sulfate ester in chondroitin-4-sulfate and dermatan sulfate

Sulfate ester in chondroitin-6-sulfate

D-Glucuronic acid   N-Acetyl-D-glucosamine

Hyaluronic acid $[-\beta GlcUA-(1\rightarrow3)-\beta GlcNAc-(1\rightarrow4)-]_n$

Chondroitin $[-\beta GlcUA-(1\rightarrow3)-\beta GalNAc-(1\rightarrow4)-]_n$

Dermatan sulfate $[-\alpha\text{-L-iduronic acid-}(1\rightarrow3)\text{-GalNAc-4-sulfate-}(1\rightarrow4)-]_n$

(A)

Fig. 2-16. The chondroitin sulfates and dermatan sulfate are closely similar but with substitution by N-acetylgalactosamine and α-L-iduronic acid, respectively, and with sulfate ester groups in the positions indicated in Fig. 2-16A.

Heparin is a mucopolysaccharide with anticoagulant properties that is secreted into the bloodstream by mast cells present in the lungs, liver, and various tissues. The linear polysaccharide appears to contain disaccharide units of the following type: [-uronic acid-$(1\rightarrow4)$-GlcN-2,6 disulfate-$(1\rightarrow4)$-]$_n$. Both the amino groups and the 6-hydroxyls of the glucosamine residues carry sulfate groups. In some units D-glucuronic acid is present in α-1,4 linkage, but more often L-iduronic acid-2-sulfate is the first unit in the disaccharide.[38]

The **keratans** of skin contain the repeating unit $[-\beta Gal-(1\rightarrow4)-\beta GlcNAc-6\text{-sulfate-}(1\rightarrow3)-]_n$ or the corresponding structure containing galactosamine sulfate. Because of the carboxylate and sulfate groups all of these polymers are bristling with negative charges.

4. Peptidoglycans and Glycoproteins

Many simple heteropolysaccharides are, in fact, **proteoglycans** in which a **terminal unit** of the polysaccharide is covalently attached through an O-glycosidic linkage to a serine residue of a protein.[39–43] Thus, the complete chondroitin chain is as shown below.

Repeating unit   Terminal unit

$$\left[-\beta GlcUA-(1\rightarrow3)\text{-GalNAc-}(1\rightarrow4)-\right]_n-\beta GlcUA-(1\rightarrow3)-\beta Gal-(1\rightarrow3)-\beta Gal-(1\rightarrow4)-\beta Xylp$$

Serine
(in protein)

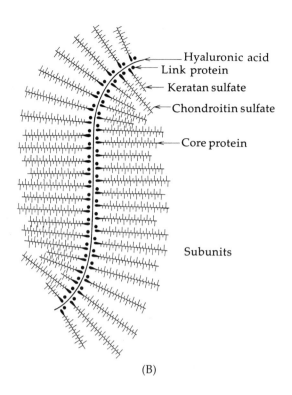

Hyaluronic acid
Link protein
Keratan sulfate
Chondroitin sulfate

Core protein

Subunits

(B)

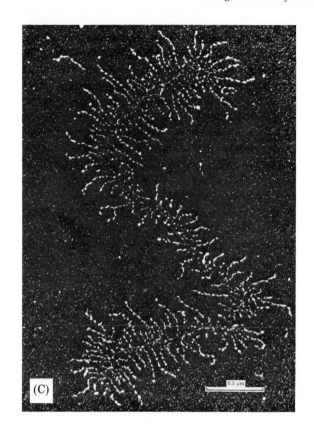

0.5 μm

(C)

Dermatan sulfate and heparin are attached to proteins by the identical linkage.[43]

Short oligosaccharide groups are often attached to proteins of cell surfaces and to secreted proteins. The sugar chains are attached in *O*-glycosidic linkage to —OH groups of serine, threonine, or hydroxylysine (in collagen only) or in *N*-glycosidic linkage to the amide nitrogen of asparagine.[43a] These **glycoproteins** include enzymes, hormones, and structural proteins. In most cases, the function of the carbohydrate portion is unknown but in the **mucins,** which serve as lubricants, the presence of many negatively charged sialic acid groups is thought to cause expansion and rigidity which increases the viscosity of the protein.[41]

Some mucins are over 50% carbohydrate; for example, that of the submaxillary (salivary) gland of sheep contains up to 800 chains per protein molecule of α-D-*N*-acetylneuraminic acid-(2$\rightarrow$6)-α-D-GlcNAc- attached to serine and threonine residues.[40] Other glycoproteins contain only one or a few carbohydrate chains. Molecules of ovalbumin (MW $\sim$44,500) from hen eggs each carry a single carbohydrate chain. At least five different oligosaccharides of the following type are present[39]:

$$\alpha Man\text{-}(1\rightarrow4)\text{-}\beta GlcNAc\text{-}(1\rightarrow4)\text{-}\beta GlcNAc\text{-}N\text{-}Asp\ (protein)$$

$$(GlcNAc)_{\overline{0,1\ or\ 2}}\quad (Man)_{4\ or\ 5}$$

$$(GlcNAc)_{0\ or\ 1}$$

Similar carbohydrate groups are present in the enzymes ribonuclease (bovine and porcine), α-amylase (*Aspergillus oryzae*), and bromelain (pineapple).[44] Glycoproteins in the serum of Antarctic fishes have an unusually effective "antifreeze" activity.[45-47]

5. Conformations of Polysaccharide Chains

Despite the variety of different monomer units and kinds of linkage present in carbohydrate chains, the conformational possibilities are limited. The sugar ring is a rigid unit and the connection of one unit to the next can be specified by means of two torsion angles ϕ and ψ just as with peptides.[48,49] However, no clear-cut consensus exists on the conformation for which ϕ and ψ should both be 0° in the case of polysaccharides. While a tentative general rule for all polymers has been suggested[18] it is inconvenient in its application to polysaccharides. One convenient convention is to take ϕ and ψ as 0° when the two *midplanes* of the sugar rings are coplanar:

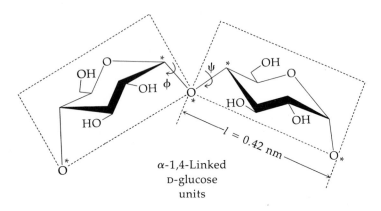

α-1,4-Linked
D-glucose
units

These planes are defined by the atoms marked by asterisks and are, in general, not exactly perpendicular to the planes of the rings. A systematic examination of the possible values for ϕ and ψ for cellulose (poly-β-D-glucose) (see structure at top of p. 89) shows that these angles are constrained to an extremely narrow range placing the monomer units in an almost completely extended conformation.

Each glucose unit is flipped over 180° from the previous one. The polymer has a twofold screw axis and there is a slight zigzag in the plane of the rings.[50,51] Remember that in the chair form of glucose all of the OH groups lie in equatorial positions and are able to form hydrogen bonds with neighboring chains. This feature, together with the rigidity of conformation imposed by the β configuration of the monomer units themselves, doubtless accounts for the ability of cellulose to form strong fibers.

In starch and glycogen, glucose units are again used to form a chain but this time with α-1,4 linkages. An extended conformation is no longer possible and the chains tend to undergo helical coiling. One of the first helical structures of a biopolymer to be discovered (in 1943)[52] is the left-handed helix of **amylose** wound around molecules of iodine (I_2) in the well-known blue starch–iodine complex (Fig. 2-17). The helix contains 6 residues per turn with a pitch of 0.8 nm and a diameter of nearly 1.4 nm.[53,54]

Another more tightly coiled double helical form of amylose has also been proposed.[55] Each chain would contain 6 glucose units per turn and the two chains could be arranged in either parallel or antiparallel directions. The average amylose molecule contains 1000 glucose units and could be stretched to a slender chain 500 nm long, longer than the crystalline regions observed in starch granules. Thus, the chains within the granules must fold back on themselves, possibly in hairpin fashion:

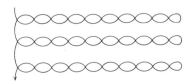

Crystalline array of
hairpin folds
proposed for linear
starch molecules[56]

Cellulose

Agarose (MW ~ 120,000) is an alternating carbohydrate polymer.

Agarose:

$$\text{---}\!\!\left[\!\text{-}\beta\text{-}\text{D-Gal-}(1\rightarrow 4)\text{-}3,6\text{-anhydro-}\alpha\text{-L-Gal-}(1\rightarrow 3)\text{-}\!\right]_n\text{---}$$

It is the principal component of agar and the compound that accounts for most of the gelling properties of that remarkable substance. A solid agar gel containing 99.5% water can be formed. The molecular structure of agarose is that of a left-handed double helix with a threefold screw axis, a pitch of 1.90 nm, and a central cavity containing water molecules.[57] A similar structure has been established for the gel-forming **carrageenans** from red seaweed. X-Ray data suggest that three of the disaccharide units form one turn of a right-handed

---

BOX 2-B

SILICON: AN ESSENTIAL TRACE ELEMENT

No one can doubt that diatoms, which make their skeletons from SiO_2, have an active metabolism of silicon. Likewise, the radiolaria, some higher plants, and sponges often accumulate silicon. Limpets make opal base plates for their teeth.[a] Even so, the metabolism of silicon has received little attention and it has only recently been demonstrated that this element is essential for growth and development of higher animals.[b–e] In the chick, silicon is found in active calcification sites of young bone.[d] It is present in low amounts in the internal organs of mammals but makes up ~0.01% of the skin, cartilage, and ligaments. Schwarz[a] has traced this silicon to the mucopolysaccharides such as chondroitin-4-sulfate, dermatan sulfate, and heparan sulfate. All of these contain ~0.04% silicon or one atom of silicon per 130–280 repeating units of the polysaccharides. In the plant kingdom pectins contain about five times this amount. The silicon present is tightly bound, apparently in ether linkage. Schwarz suggested that orthosilicic acid, $Si(OH)_4$,

reacts with hydroxyl groups of the carbohydrates to form ether linkages which may bridge between two chains as follows:

$$R_1\text{---O---}\underset{\underset{OH}{|}}{\overset{\overset{OH}{|}}{Si}}\text{---O---}R_2 \qquad R_1\text{---O---}\underset{\underset{OH}{|}}{\overset{\overset{OH}{|}}{Si}}\text{---O---}\underset{\underset{OH}{|}}{\overset{\overset{OH}{|}}{Si}}\text{---O---}R_2$$

Note that in each of these formulas additional free OH groups are available on the silicon so that it is possible to cross-link to more than two polysaccharide chains. These results, while preliminary, suggest that silicon may function as a biological cross-linking agent in connective tissue (see also Chapter 11, Section E,3).

[a] K. Schwarz, *PNAS* **70**, 1608–1612 (1973).
[b] K. Schwarz, *in* "Trace Element Metabolism in Animals" (F. Mills, ed.), pp. 25–38. Livingstone, Edinburgh, 1970.
[c] K. Schwarz and D. B. Milne, *Nature* (*London*) **239**, 333–334 (1972).
[d] E. M. Carlisle, *Science* **178**, 619–621 (1972).
[e] W. H. Hoekstra, J. W. Suttie, H. E. Ganther, and W. Mertz, eds., "Trace Element Metabolism in Animals—2." Univ. Park Press, Baltimore, Maryland, 1974.

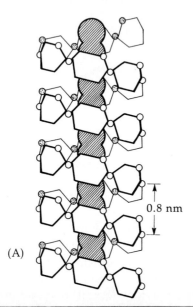

(A)

0.8 nm

(B)

Fig. 2-17 (A) Structure of the helical complex of amylose with iodine (I₂). The iodine is located in the interior of a helix having 6 glucose residues per turn. (B) Model of a parallel double helix. There are 6 glucose units per turn of each strand. The repeat period measured from the model is 2.1 nm per 6 glucose units. Courtesy of Dexter French.

ι-Carrageenan: $\dashv\!-\beta$-D-Gal-4-sulfate-

(1→4)-3,6-anhydro-α-D-Gal-2-sulfate-(1→3) $\dashv_n$

helix with a pitch of 2.6 nm. A second chain with a parallel orientation, but displaced by half a turn, wraps around the first helix.[58] Such double helical regions provide "tie-points" for the formation of gels[59] (Fig. 2-18). Sulfate groups protrude from the structure in pairs and provide binding sites for calcium ions which stabilize the gel.

The presence of occasional extra sulfate groups in these polymers causes kinks in the chains (because the derivatized pyranose rings reverse their conformation to the other chair form) and prevent the entire polysaccharide chain from assuming a regular helical structure.[59a]

A three-stranded right-handed triple helix has been proposed for **xylan** of higher plants, a β-1,3-linked polymer of D-xylose.[60] A double helical structure has been proposed for hyaluronic acid,[61] and various single helices have been discovered for chondroitin sulfates.[62]

D. NUCLEIC ACIDS

The nucleic acids DNA and RNA are **polynucleotides.** The monomer units (nucleotides) are themselves made up of three parts:

1. A **purine** or **pyrimidine** "base" (structures in Fig. 2-19)

2. A sugar, either D-ribose or D-2-deoxy-ribose

3. Phosphoric acid

These parts are joined by elimination of two molecules of water as indicated in Fig. 2-20. The nucleotides are combined through **phosphodiester** linkages between the 5'-hydroxyl of the sugar in one nucleotide and the 3'-hydroxyl of another (see Eq. 2-11). The structures of a pair of short polynucleotide strands are shown in Fig. 2-21.

$$
\underset{3'}{\overset{5'}{HO_3^-P-O-sugar\text{-}base}}
$$

(2-11)

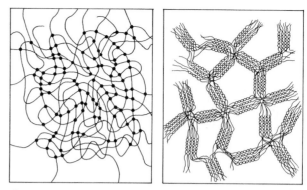

Fig. 2-18 A schematic representation of the agarose gel network (right), in comparison with a network such as that of Sephadex which is formed from random chains. Note that the aggregates in agarose gels may actually contain $10–10^4$ helices rather than the smaller numbers shown here. From Arnott *et al.*[58]

Fig. 2-19 Structures of pyrimidine and purine bases. Most text and reference books show these rings turned over with C-5 of the pyrimidine ring and C-8 of purines to the right. The presentation used here was chosen to be consistent with that in the drawings of nucleotides in Fig. 2-20.

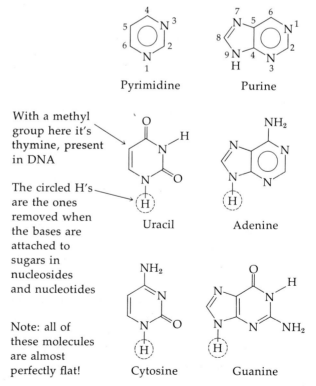

1. Names and Abbreviations

The purine and pyrimidine ring compounds formed in nucleic acids are known simply as bases, even though some of them have almost no basic character. **Nucleosides** are the *N*-glycosides (or *N*-glycosyl derivatives) of the bases with ribose or deoxyribose. The **nucleotides** are phosphate esters of nucleosides. Similar names are applied to related biochemicals which are not present in DNA or RNA.

The naming of nucleosides and nucleotides is confusing. Table 2-5 gives names of the principal nucleotides from which the nucleic acids are formed. Even worse than the names in the table are **hypoxanthine** (derived by replacement of the —NH₂ group of adenine with —OH and tauto-

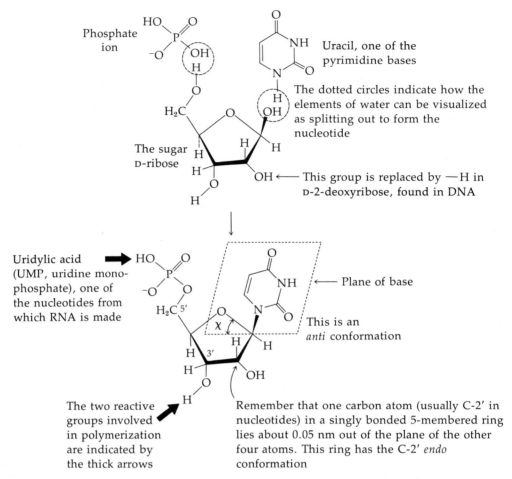

Phosphate ion

Uracil, one of the pyrimidine bases

The dotted circles indicate how the elements of water can be visualized as splitting out to form the nucleotide

The sugar D-ribose

This group is replaced by —H in D-2-deoxyribose, found in DNA

Uridylic acid (UMP, uridine monophosphate), one of the nucleotides from which RNA is made

Plane of base

This is an *anti* conformation

The two reactive groups involved in polymerization are indicated by the thick arrows

Remember that one carbon atom (usually C-2' in nucleotides) in a singly bonded 5-membered ring lies about 0.05 nm out of the plane of the other four atoms. This ring has the C-2' *endo* conformation

Fig. 2-20 The formation of a nucleotide from a pyrimidine base, D-ribose and phosphate ion. The reactions are imagined to occur by splitting out water molecules. Note that the positions in the base are numbered from 1–6 or 1–9, as shown in Fig. 2-19, while those in the sugar are numbered 1'–5'.

merization of the

$$-\overset{\overset{\displaystyle OH}{|}}{C}=N-$$

group of the resulting hydroxypurine to

$$-\overset{\overset{\displaystyle O}{\|}}{C}-\overset{\overset{\displaystyle H}{|}}{N}-$$

The nucleoside formed from hypoxanthine and ribose is known as **inosine** (I) and the corresponding nucleotide as **inosinic acid.**

A commonly used abbreviation for nucleic acid structures portrays the sugar rings as vertical lines. The abbreviations A, C, U, T, and G for the individual bases or Pu (purine) or Py (pyrimidine) are placed at the upper ends of the lines, and slanted lines with P in the centers represent the 3'–5' phosphodiester linkages:

A   U   G   Pu   Py

Terminal 5'-phosphate   terminal 3'—OH

The same polynucleotide structure can be further abbreviated

pApUpGpPupPy

5' end   *or*   3' end

A·U·G·Pu·Py

By convention the 5' end of a polynucleotide is ordinarily placed to the left in formulas like the above.

2. Functional Groups in Nucleic Acids

The ionized phosphate groups of the polymer "backbone" give nucleic acid molecules a high negative charge. For this reason DNA in cells is usually associated with (1) basic proteins such as the **histones** or **protamines** (in spermatozoa of fish), with (2) polycations of amines such as **spermidine** (H_3^+N—$CH_2CH_2CH_2CH_2$—NH_2^+—$CH_2CH_2CH_2$—NH_3^+), or with (3) alkaline earth cations such as Mg^{2+}.

Free hydroxyl groups are present in the 2 positions of the sugars in RNA. Perhaps they function as identifying markers for the RNA monomers. They affect the conformation of the polymer. The lack of the 2-OH in deoxyribose confers a greater stability toward degradation by base to DNA.

The purine and pyrimidine bases are the "side chains" of the nucleic acids. The polar groups

$\diagdown$C=O, $\diagdown$NH, and $\diagdown\!\!\!\!\rangle$—NH_2

present in the bases are able to form hydrogen bonds, both to other nucleic acid chains (e.g., in the DNA double helix) and to proteins. The flat hydrophobic bases can interact with each other **(stacking)** as well as with aromatic amino acid side chains, drugs, and mutagenic chemicals.

3. Chemical Reactions of Polynucleotides[63]

Both DNA and RNA are easily broken down by acid-catalyzed hydrolysis (Section H,2). The gly-

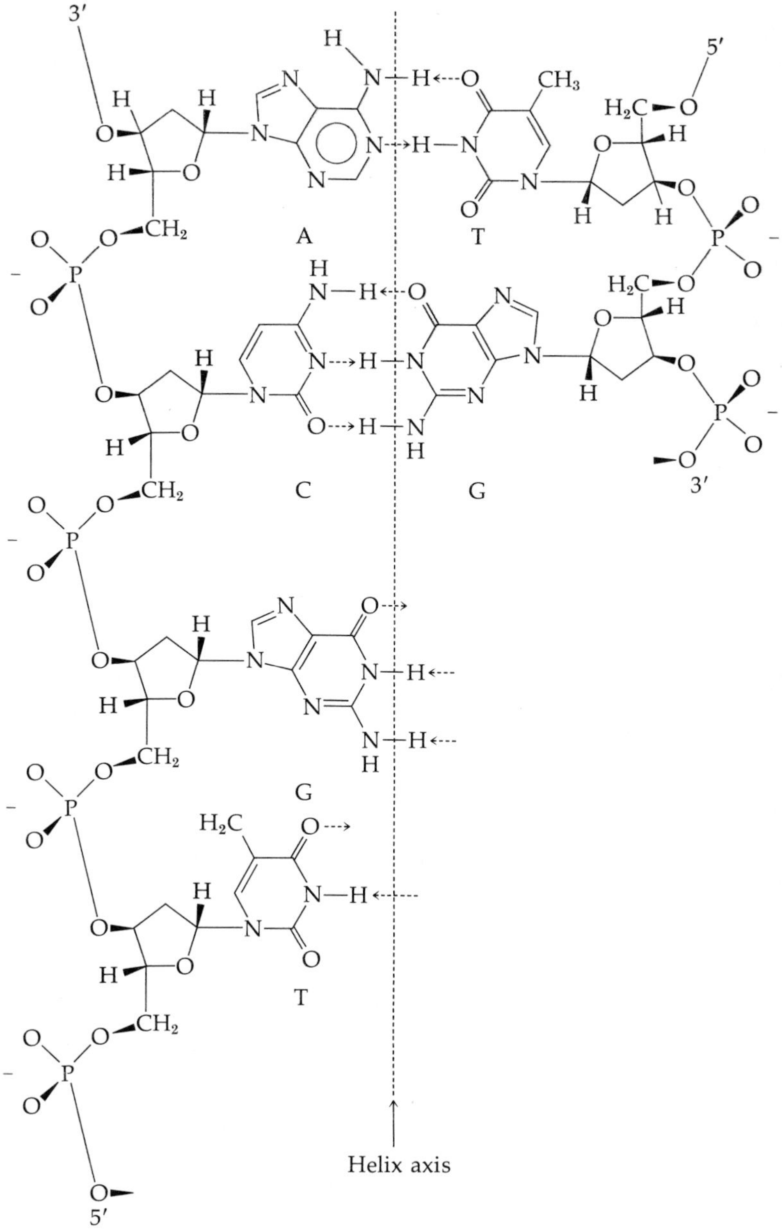

Fig. 2-21 A distorted (flattened) view of the Watson–Crick structure of DNA.

TABLE 2-5
Names of Pyrimidine and Purine Bases, Nucleosides, and 5'-Nucleotides

	A. Nucleotide units of RNA[a]			
Base:	Uracil	Cytosine	Adenine	Guanine
Abbrev:	Ura (U)	Cyt (C)	Ade (A)	Gua (G)
Nucleoside:	Uridine	Cytidine	Adenosine	Guanosine
Abbrev:	Urd	Cyd	Ado	Guo
5'-Nucleotide:	Uridine 5'-phosphate	Cytidine 5'-phosphate	Adenosine 5'-phosphate	Guanosine 5'-phosphate
	or	or	or	or
	5'-uridylic acid	5'-cytidylic acid	5'-adenylic acid	5'-guanylic acid
Abbrev:	Urd-5'-P	Cyd-5'-P	Ado-5'-P	Guo-5'-P
	or	or	or	or
	UMP	CMP	AMP	GMP

B. Nucleotide units of DNA

These contain 2-deoxyribose; hence, the nucleosides and nucleotides are deoxyadenosine (dAdo), deoxyadenosine 5'-phosphate (dAMP), etc.

DNA contains thymine (Thy) rather than uracil. Thymidine (dThd) and thymidine 5' phosphate are deoxyribose derivatives. The ribose derivatives of thymine are the nucleoside ribosylthymidine (Thd) and ribosylthymidine 5'-phosphate (Thd-5'P).

[a] Isomers of the 5'-nucleotides, in which the phosphate is attached to the oxygen on C-3', are the 3'-nucleotides. These are also known and care must be taken to avoid ambiguity. The simple abbreviations UMP, CMP, AMP, and GMP always refer to the 5'-nucleotides.

cosidic linkages of DNA are more labile than those of RNA. Purine nucleotides are more easily hydrolyzed than are pyrimidine nucleotides. Aside from the ease of degradation by acids and bases, the covalent structure of nucleic acids is relatively stable, a fact that makes these molecules suitable as genetic materials. Even so, a variety of chemical reactions can take place.

Nitrous acid reacts with amino groups of the bases converting them to —OH groups. The resulting

$$-N\!\!=\!\!\overset{|}{C}\!\!-\!\!OH$$

groups are tautomerized to

$$-NH\!\!-\!\!\overset{\overset{\displaystyle O}{\|}}{C}\!\!-$$

Thus, nitrous acid converts cytosine to uracil in nucleic acids and acts as an efficient **mutagenic** chemical (Chapter 15, Section H,1). Similarly, adenine is converted to hypoxanthine and guanine to xanthine. Another mutagenic substance **hydroxylamine** (H_2N—OH) reacts with carbonyl groups, especially those of pyrimidines, even though these carbonyl groups are part of cyclic amide structures and are relatively unreactive.

Pyrimidines are readily reduced by sodium borohydride, $NaBH_4$ (reaction 2-12). Uridine, thy-

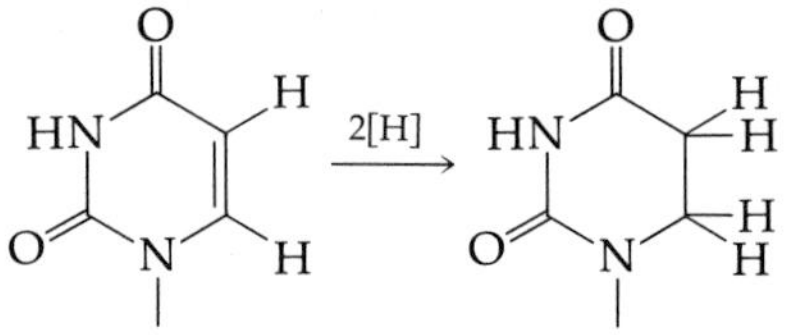

$$(2\text{-}12)$$

midine, and to a lesser extent cytosine, in nucleic acids undergo photochemical dimerization and hydration reactions (Eqs. 13-22 and 13-23).

Formaldehyde reacts with amino groups of purine and pyrimidine bases (Chapter 7, Section H,3).[64]

4. Conformations of Nucleotides

The furanose ring of ribose or deoxyribose can exist in several "envelope" and skew conforma-

Fig. 2-22 Conformational properties of nucleosides. (A) Representations of several conformations of a ribose or deoxyribose ring in a nucleoside (from Saenger[66]). (B) View down the N—C axis joining pyrimidine base to the sugar in a nucleoside. The angle χ' is measured from 0° to 360°. *Syn* conformations are those for which χ falls in the region of the heavy semicircle (from W. Saenger[66]). (C) A second convention for measuring the torsion angle of the glycoside C—N linkage. The angle χ is measured from 0° to ±180° as indicated. An anti conformation is shown. (D) Labeling of the conformational angles of the main chain in a polynucleotide.[68] (At least four other conventions for naming these angles have also been used.[70])

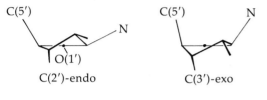

C(2')-endo   C(3')-exo   C(2')-endo-C(3')-exo   C(3')-endo

(A)

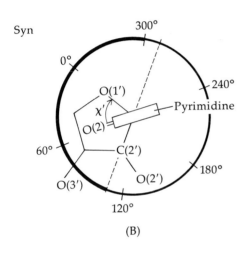

(B)

(C)

(D)

tions. It is ordinarily C-2' or C-3' that is out of the plane of the other four ring atoms. If this carbon atom lies above the ring, i.e., toward the base, the ring conformation is known as **endo;** when below the ring, as **exo** (Fig. 2-22A). The C(2')-*endo* and C(3')-*endo* conformations are most common in individual nucleotides but the C(3')-*exo* conformation is apparently present in the B form of DNA.[65] Skew conformations such as C(2')-*endo*-C(3')-*exo* (Fig. 2-22A) are readily interconvertible with the envelope conformations.[66,67]

The orientation of the base with respect to the sugar is specified by the torsion angle χ (Figs. 2-22B,C). There is no uniformity yet concerning symbols for torsion angles and reference 0° angles for nucleotides. Thus, χ has been specified in at least two ways, as indicated in Fig. 2-22. The zero angle is often assigned to the conformation in which the C(2)-N(1) bond of a pyrimidine or the C(4)-N(9) bond of a purine is cis to the C(1')-O(1') bond of the sugar[66] with angles (designated χ' in Fig. 2-22B) running from 0° to 360°. Alternatively, the zero angle may be that in which the bond in the base is cis to the C(1')-C(2') bond with χ varying from $-180°$ to $+180°$ (Fig. 2-22C).[65]

Measured values of χ vary among different nucleotides, a typical value being $\sim 127°$. In this **anti** conformation the CO and NH groups in the 2 and 3 positions of the pyrimidine ring (or in positions 1, 2, and 6 of the purine ring) are *away* from the sugar ring, while in the **syn** conformation (180° different) they lie over the ring. The anti conformation is the one commonly present in nucleic acids and in most free nucleotides.

An additional five torsion angles are needed to specify the backbone conformation of a polynucleotide[65–70] (Fig. 2-22D). Angles ω, ξ, and θ are within the nucleotide, whereas ϕ and ψ have the same meaning as in polypeptide and polysaccharide chains. The possible values of ω, ξ, and θ are quite limited. Another torsion angle σ that lies within the sugar ring (Fig. 2-22D) is indicative of the position of C-3' in the ring. In the C(3')-*endo* conformations $\sigma = 80 \pm 10°$, while in the C(3')-*exo* conformation of the B form of DNA it is $\sim 156°$.

5. Double Helices

One of the most exciting biological discoveries of this century came in 1953 when James Watson and Francis Crick recognized DNA as a double helix of two antiparallel polynucleotide chains.* A most significant feature of the proposed structure was the pairing of bases in the center through hydrogen bonding. The pairs and triplets of hydrogen bonds (indicated as --> in Fig. 2-21) could form only if adenine (A) was paired with thymine (T) and cytosine (C) with guanine (G) at every point in the entire DNA structure. Thus, the nucleotide sequence in one chain is complementary to but not identical to that in the other chain. It was apparent almost immediately that the sequence of bases in a DNA chain must convey the encoded genetic information. The complementarity of the two strands suggested a simple mechanism for replication of genes during all divisions. The two strands could separate and a complementary strand could be synthesized along each strand to give two molecules of the DNA, one for each of the two cells. The essential correctness of this concept has now been proved.

The geometry of the DNA double helix is depicted in Fig. 2-23. In one type of DNA fiber (A form)† the base pairs are inclined to the helix axis by about 20°, while in another (B form), which is stable under conditions of high humidity, the base pairs lie are almost normal to the axis (the backbone torsion angles are $\omega = 155°$, $\xi = 36°$, $\theta = -146°$, $\phi = -96°$, and $\psi = -46°$).[65] The B form is thought to approximate the conformation of most DNA in cells.

If we look directly at the axis of the double helix and perpendicular to one of the base pairs, and ignoring the fact that the base pair itself is asymmetric, we see that the nucleotide unit in one chain is related to the nucleotide unit lying across from it

---

* The structure was deduced from model building together with X-ray diffraction data of M. F. Wilkins and R. Franklin on artificial DNA fibers. Watson, Crick, and Wilkins received a Nobel Prize for their discovery in 1962.

† Actually, a series of closely related forms.

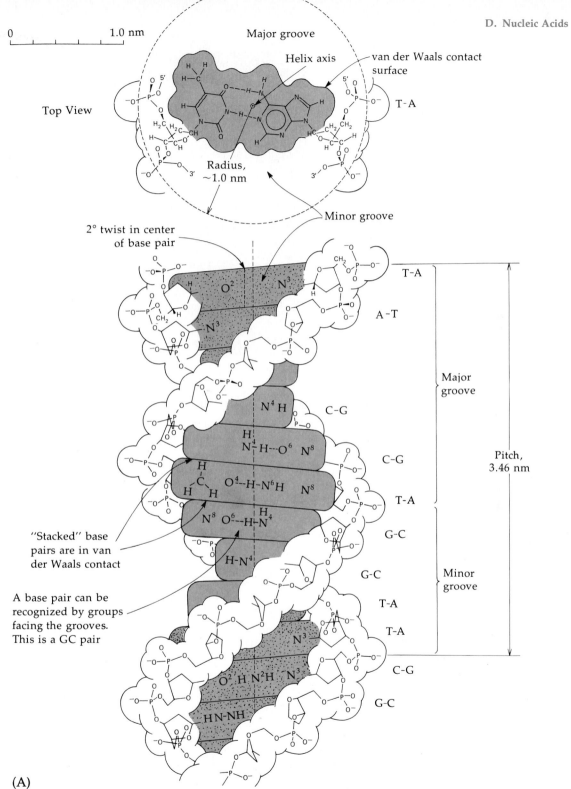

Fig. 2-23 (A) The double helix of DNA. The structure shown is that of the B form. Based on data of S. Arnott and D. W. L. Hukins, *JMB* **81**, 93–105 (1975)

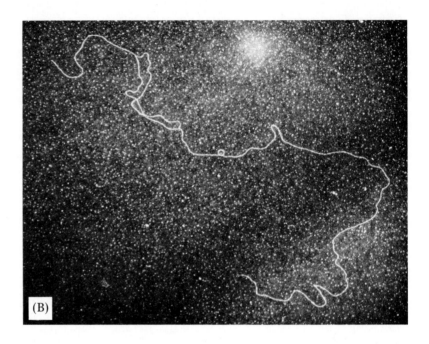

Fig. 2-23 (B) Electron micrograph of a DNA molecule (from a bacterial virus (bacteriophage T7) undergoing replication. The viral DNA is a long (∼14 μm) duplex rod containing about 40,000 base pairs. In this view of a replicating molecule an internal "eye" in which DNA has been duplicated is present. The DNA synthesis was initiated at a special site (origin) about 17% of the total length from one end of the duplex. The DNA was stained with uranyl acetate and viewed by dark field electron microscopy. Micrograph courtesy J. Wolfson and D. Dressler.

(B)

in the opposite plane by a *twofold axis of rotation* **(dyad axis).** This symmetry element, which arises from the antiparallel arrangement of the chains, makes the DNA molecule from the outside look identical whether viewed from one end or the other—and whether viewed as a model by the human eye or through contact with an enzyme which might act on the molecule. Actually, the two chains are not identical, and the genetic information can be read off from the surfaces exposed in the major groove (see Fig. 2-23).

Note the dimensions of the double helix (Fig. 2-23). The diameter, measured between phosphorus atoms, is just 2.0 nm. The pitch is 3.4 nm. There are 10 base pairs per turn. Thus, the rise per base pair is 0.34 nm, just the van der Waals radius of an aromatic ring (Table 2-1). It is clear that the bases are "stacked" in the center of the helix. A typical gene of 1000 bases would be a segment of DNA rod about 340 nm long (Fig. 2-23B).

As is appropriate for the cell's master blueprint, DNA in the double helix is extremely stable. Despite its long length, it does not often break in nature. Several factors contribute to this stability. (1) The pairs and triplets of hydrogen bonds between the bases. (2) The van der Waals attraction between the flat bases which stack together. (3) The fact that on the outside of the molecule are many oxygen atoms, some negatively charged, which are able to form strong hydrogen bonds with water or with special proteins surrounding the molecule. (4) The ability to form various superhelices (see below).

The fact that DNA can also exist as a paracrystalline A form (in which the bases are tilted and there are 11 base pairs per turn) suggests the possibility that both conformations may be important in nature. While **RNA** molecules usually exist as single chains, they often form **hairpin loops** consisting of double helices in the A conformation.[71] The B conformation is impossible because of the presence of the 2'-hydroxyl groups on the ribose rings in RNA. Transient "hybrid" DNA-RNA double helices are also thought to exist within cells and they too may be constrained to the A structure. It is noteworthy that the A structures differ from the B structure of DNA in having a large (∼0.8 nm) hole along the axis and a very deep major groove.[71a] The base pairs do not lie on the axis as in Fig. 2-23.

The best known forms of RNA are the low

molecular weight tRNA molecules. In all of them the bases can be paired to form a "cloverleaf" structure with three hairpin loops and sometimes a fourth. The structure of one of these molecules has been determined by X-ray diffraction[72-74] (Fig. 2-24). The irregular body is nearly as complex in its conformation as is a globular protein. Note at the bottom of the structure the **anticodon,** a triplet of bases having the correct structures to permit pairing with the three bases of the codon specifying a particular amino acid, in this case phenylalanine.

6. Base Pairs

The base pairs proposed by Watson and Crick are shown in Fig. 2-21. While diffraction studies indicate that it is these pairs that actually exist in DNA other possibilities have been considered. Figure 2-25 shows the shapes and the hydrogen bonding groups available in the bases. The number of electron donor and acceptor groups available for hydrogen bonding is large and more than one mode of base pairing is possible; for example, Hoogsteen proposed the accompanying pairing (illustrated for A-T). Note that the distance

Hoogsteen base pair

spanned by the base pair (between the C-1' sugar carbons) is 0.88 nm, less than the 1.08 nm of the Watson–Crick pairs. While this type of pairing has not been detected in double helices, a **triple helix** structure can be formed by adding a third base to a base pair using the Hoogsteen pairing arrangement.[75] Indeed, base triplets of exactly this

type have now been identified in certain locations in tRNA molecules.[74]

Purine and pyrimidine bases interact in living things not only with each other but also with proteins. They bind to the enzymes of nucleic acid and nucleotide metabolism and they serve as "handles" by which proteins can hold on to many metabolic intermediates and coenzymes (Chapter 8).

How strong are the bonds formed between pairs of bases in DNA? The question is hard to answer, because of the strong interactions of the molecules with polar solvents. Some insight comes from studies of the association of bases in nonpolar solvents. Thus, 1-cyclohexyluracil was shown to form a dimer involving either hydrogen bonding or stacking, but the association was weak with ΔG^0 of formation of ~ 5 kJ mol^{-1}. When the same compound was mixed with 9-ethyladenine, a base-paired complex formed between the two compounds with a formation constant (see Eq. 4-2) over 10-fold greater than that for the dimer[68] (Eq.

2-13). When the circled hydrogen atom in Eq. 2-13 was replaced by $-CH_3$, blocking pairing, K_f fell below 1 ($\Delta G^0 > 0$). The difference in the free energy of formation in the two cases was only 7 kJ mol^{-1}, but such small energies summed over the many base pairs present in the DNA molecule help provide stability to the structure.

Recent measurements of formation of base-paired complexes of all of the nucleotides permit a quantitative estimation of free energies of formation of helical regions of RNA molecules.[76] Table 2-6 shows the increment in the free energy of formation of such a helix from a single-stranded molecule for the *addition of one base pair* at the end of an existing helix. Addition of an AU pair supplies only -5 to -7 kJ to ΔG_f^0. The exact amount depends upon whether an A or a U is at the 5' end in the existing helix. If an AU pair is added to the helix terminating in CG or GC, -9 kJ/mol is added. Larger increases result from addition of GC pairs, which contain three hydrogen bonds between the bases versus the two in AU pairs.

A UG pair also provides a very small amount of stabilization to an RNA double helix, while the presence of unpaired bases has a destabilizing effect. The most stable hairpin loops contain 4–5 bases. Depending upon whether the loop is "closed" by CG or AU, the helix is destabilized by 20–30 kJ/mol. "Bulge loops," which protrude from one side of a helix, have a smaller destabilizing effect. An example of the way in which Table 2-6 can be used to estimate the energies of formation of a loop in a straight-chain RNA is illustrated in Fig. 2-26.

7. Tautomerism and Base Pairing

Tautomerism has an interesting relationship to the formation of the pairs and triplets of hydrogen bonds in DNA. Each base exists predominately as one preferred tautomer, but at any moment a few molecules of a given base are present as less stable tautomers. This fact may be responsible for the occurrence of some mutations. Thus, tautomer B of Eq. 2-5 would not be able to pair with thymine, its proper pairing partner, but could pair with cytosine. If this happened during gene replication an incorrect copy of the gene, differing in a single "code letter" would be formed.

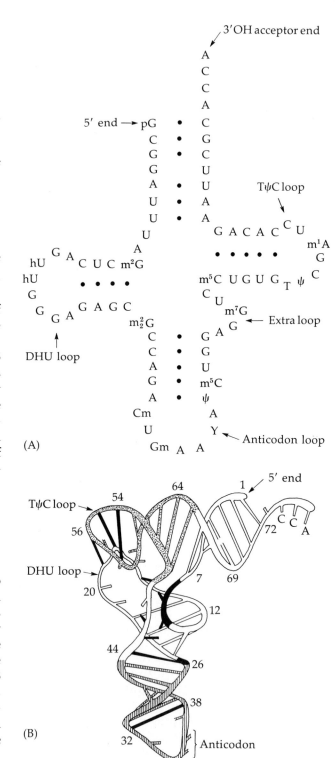

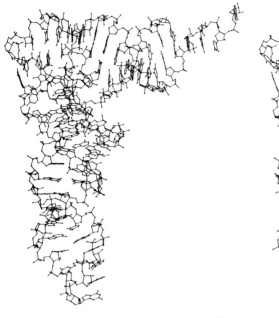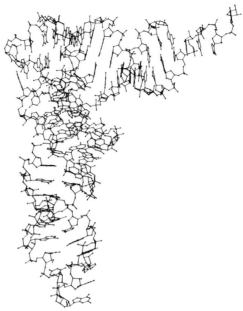

(C)

Fig. 2-24 The structure of phenylalanine transfer RNA of yeast. (A) The sequence of nucleotides shown in conventional cloverleaf diagram. (B) Perspective diagram of folding of polynucleotide chain. The ribose phosphate backbone is drawn as a continuous cylinder with bars to indicate hydrogen-bonded base pairs. The positions of single bases are indicated by rods which are intentionally shortened. The TψC arm is heavily stippled, and the anticodon arm is marked by vertical lines. Tertiary structure interactions are illustrated by black rods. Redrawn from G. J. Quigley, and A. Rich, *Science* **194**, 796–806 (1976). Copyright 1976 by the American Association for the Advancement of Science. (C). Stereo view of the structure of yeast phenylalanyl-tRNA as revealed by X-ray crystallography. The acceptor stem with the protruding ACCA sequence at the 3′ end is to the right. The anticodon GAA is at the bottom right side of the drawing. The guanine ring is clearly visible at the very bottom. The middle adenine is seen exactly edge-on as is the "hypermodified" base Y (see Fig. 15-10) which lies just above the anticodon. Its side chain is visible on the back side of the drawing. Preceding the anticodon on the left (5′ side) are two unpaired bases (C and U). The 2′-hydroxyl groups of the cytosine and of the guanosine in the anticodon are methylated. Moving up the anticodon stem one can see two groups of base triplets which utilize both Watson–Crick and Hoogsteen types of base pairing. For suggestions on viewing this figure, see the legend to Fig. 2-6. Drawing courtesy of Alexander Rich.

8. The Base Composition of DNA and RNA

The nucleotide composition of DNA is surprisingly variable. The sum of the percent cytosine plus the percent guanine (C + G) for bacteria varies all the way from 22 to 74%. That for *E. coli* is 51.7%. Among eukaryotic organisms, the range is somewhat narrower (28 to 58%; for man it is 39.7%). The fact that bacterial DNA molecules are more varied than those of higher organisms is not surprising. The prokaryotes have evolved for just as many million years as we. Perhaps because of their simpler structure and rapid rate of division, nature has done more experimentation with genetic changes than she has done with ours.

Comparisons of the C + G content of organisms is sometimes used as a basis for establishing genetic relatedness. However, thymine is especially susceptible to photochemical alteration by ultraviolet light. Thus, bacteria with a high (C + G) content may have evolved in environments subject to strong sunlight, whereas those with a low (C + G) content have developed in more protected locations.[77]

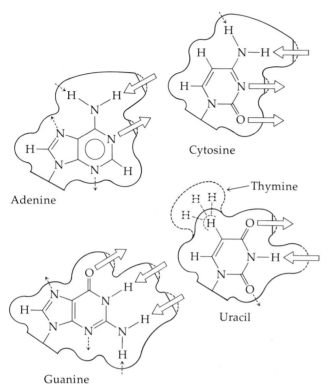

Cytosine

Adenine

Thymine

Uracil

Guanine

Fig. 2-25 Outlines of the purine and pyrimidine bases of nucleic acids showing van der Waals contact surfaces and some of the possible directions in which hydrogen bonds may be formed. Open arrows are used to indicate the hydrogen bonds present in the Watson–Crick base pairs.

The picture of DNA as a chain consisting of only four kinds of nucleotides must be modified somewhat. DNA from many sources contains a significant number of methylated bases. Among these are 5-methylcytosine and 6-methyladenine. It is likely that these methylated bases mark special points in the genetic blueprint. Methylation is accomplished after the synthesis of the polynucleotide. Another function of methylation appears to be to protect DNA from attack by enzymes formed at the command of invading viruses (Chapter 15, Section F). Some viruses, notably the bacteriophage of the T-even series that attack *E. coli* (Box 4-E), have developed their own protective devices. They contain **hydroxymethylcytosine** in place of cytosine. The extra hydroxyl groups provided in this fashion often carry one to two glu-

cose units in glycosidic linkage.[78] A bacteriophage attacking *Bacillus subtilus* substitutes hydroxymethyluracil for uracil and 5-dihydroxypentyluracil for thymine.[79]

The modifications carried out on RNA molecules are more varied and more extensive than those of DNA and are dealt with in Chapter 15.

9. Rings, Supercoils, and Intercalation

While DNA molecules may exist as straight rods, the two ends are often covalently joined. Thus, the chromosome of *E. coli* is a single closed circle. Circular DNA molecules are also found in mitochondria and in some viruses.[80]

As with proteins, simple helices can be distorted to form **supercoils** (superhelices). To accomplish this a torque must be applied at one end. Thus, when a loosely twisted heavy rubberband is twisted far enough (as when a model airplane is being prepared for flight) positive supercoiling takes place. In a similar fashion DNA can undergo positive (or negative) superhelix formation **(tertiary coiling)**. Circular DNA molecules are frequently supercoiled. When an uncoiled duplex is "wound," the number of turns made (the **winding number**), α, equals the sum of the secondary turns, β, expected for an unconstrained helical duplex (e.g., the Watson–Crick structure) plus the number of superhelical turns τ:

$$\alpha = \tau + \beta \qquad (2\text{-}14)$$

The value of β is always positive but that of τ can be negative, the secondary structure (Watson–Crick helix) being fully formed but with left-handed superhelical turns present.[81]

The **superhelix density** of a DNA molecule is usually expressed as σ = number of superhelical turns per 10 base pairs.[80,82] In naturally occurring circular DNA molecules σ is often negative, a typical value being -0.05 ($\sim$5 superhelical turns per 1000 base pairs). The presence of superhelices in circular DNA molecules can be recognized readily by its effect upon the sedimentation constant of

TABLE 2-6
Free Energies of Formation ΔG_f^0 at 25°C for Addition of One Base Pair to an Existing RNA Helix[a,b]

Base pair at end of existing helix	Base pair added	ΔG_f^0 $\left(\dfrac{\text{kcal mol}^{-1}}{\pm 10\%}\right)$	ΔG_f^0 $\left(\dfrac{\text{kJ mol}^{-1}}{\pm 10\%}\right)$
$\overset{A}{\underset{\cdot}{U}}$	$\overset{A}{\underset{\cdot}{U}}$	−1.2	−5.0
$\overset{U^c}{\underset{\cdot}{A}}$	$\overset{A}{\underset{\cdot}{U}}$	−1.8	−7.5
$\overset{C}{\underset{\cdot}{G}}$ or $\overset{G^d}{\underset{\cdot}{C}}$	$\overset{A}{\underset{\cdot}{U}}$	−2.2	−9
$\overset{C}{\underset{\cdot}{G}}$	$\overset{G}{\underset{\cdot}{C}}$	−3.2	−13
$\overset{G}{\underset{\cdot}{C}}$	$\overset{C}{\underset{\cdot}{G}}$ or $\overset{G}{\underset{\cdot}{C}}$	−5.0	−21
$\overset{G}{\underset{\cdot}{U}}$	$\overset{U}{\underset{\cdot}{G}}$	−0.3	−1
Hairpin loops of 4–5 bases			
Closed by GC		+5	+21
Closed by AU		+7	+29
Bulge loops			
1 base		+3	+12
4–7 bases		+5	+21

[a] Table modified from that of Tinoco *et al.*[76]

[b] All base pairs on the page are oriented as follows:

$$5'\longrightarrow A \longrightarrow 3'$$
$$3'\longleftarrow U \longleftarrow 5'$$

[c] ΔG_f^0 is the same for $\overset{U}{\underset{\cdot}{A}}$ added to an $\overset{A}{\underset{\cdot}{U}}$ end.

[d] ΔG_f^0 is the same for $\overset{C}{\underset{\cdot}{G}}$ or $\overset{G}{\underset{\cdot}{C}}$ added to an $\overset{A}{\underset{\cdot}{U}}$ end.

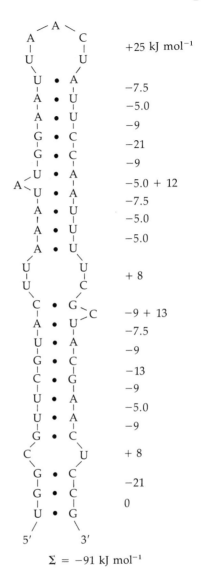

$$\Sigma = -91 \text{ kJ mol}^{-1}$$

Fig. 2-26 The contribution of base-paired regions and loops to the free energy of a possible secondary structure for a 55 base fragment from R17 virus. The loop shown here is part of a larger one considered by Tinoco *et al.*[76]

the DNA.[81] Thus, naturally occurring supercoiled DNA from polyoma virus sediments rapidly. After "nicking" of one of the strands of the double helix by brief exposure to a DNA-hydrolyzing enzyme, the resulting "relaxed" form of the molecule sediments more slowly. Supercoiling also affects viscosity of solutions of DNA as well as electrophoretic mobility (Fig. 2-27A)[82a] and may sometimes be recognized by electron microscopy.

What causes the formation of supercoils in nature? A clue comes from studies of **intercalation**,[83–86] the insertion of flat aromatic rings between base pairs in a DNA duplex. Many antibiotics, drugs, dyes, and other substances are capable of inserting themselves into DNA molecules in this way. Among them are daunomycin, proflavine, ethidium bromide, and hycanthone (Fig. 2-27B). Hycanthone, one of the most widely used drugs in the world, is employed in the treatment

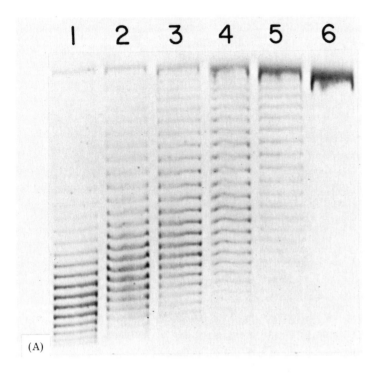

(A)

Fig. 2-27 (A) Electrophoresis of DNA from virus SV40 (Box 4-C) with varying numbers of superhelical turns. Molecules of the native DNA (lane 1) move rapidly toward the anode as a series of bands, each differing from its neighbors by one superhelical turn. The average number of superhelical turns is about 25. Incubation with a "DNA-relaxing enzyme" from human cells (Chapter 15, Section E,1) causes a stepwise removal of the superhelical turns by a cutting and resealing of one DNA strand. The DNA incubated with this enzyme at 0°C for periods of 1, 3, 6, 10, and 30 min (lanes 2–6) is gradually converted to a form with an average of zero supercoils. For details see Keller.[82a] Electrophoresis was carried out in an 0.5% agarose −1.9% polyacrylamide slab gel (17 × 18 × 0.3 cm). After electrophoresis the bands of DNA were visualized by staining with the fluorescent intercalating dye ethidium bromide. Photograph courtesy of Walter Keller. (B) Structures of some substances that tend to "intercalate" into DNA structures. See also the structure of the actinomycin–DNA complex (Box 15-B).

Daunomycin
an antitumor antibiotic

Hycanthone
a widely used
drug for
schistosomiasis

Ethidium

Acridine yellow

Proflavin

(B)

of schistosomiasis (Chapter 1, Section E,1). Since intercalating agents can be mutagenic such drugs are not without their hazards.

Intercalation is often used to estimate the amount of negative supercoiling of DNA molecules. Varying amounts of the intercalating agent are added, and the sedimentation constant (or other property) of the DNA is observed. As increasing intercalation occurs, the secondary turns of DNA are unwound (the value of β in Eq. 2-14 decreases). Each intercalated ring causes an unwinding of the helix of $\sim 26°$. Since for a closed covalent duplex α (Eq. 2-14) is constant, the decrease in β caused by increased intercalation leads to an increase in the value of τ. When sufficient intercalation has occurred to raise τ to zero, a minimum sedimentation rate is observed. Addition of further intercalating agent then causes positive supercoiling.

The "replicative form" of DNA of the virus ϕX174 (Box 4-C), a small circular molecule containing ~ 5000 base pairs, was treated with proflavine.[82] The binding of 0.06 mol of proflavine per mole of nucleotides reduced τ to zero. From this it could be estimated* that $\sigma = -0.055$, corresponding to -27 superhelical turns at 25°, pH 6.8, ionic strength ≈ 0.2. Changes in temperature, pH, and ionic environment strongly influence supercoiling. In general σ becomes less negative by $\sim 3.3 \times 10^{-4}$ per degree of temperature increase.[86c] For example, the observed value [82] of σ for ϕX174 DNA was -0.059 at 15°C and -0.040 (-20 superhelical turns) at 75°C at an ionic strength of ~ 0.2.

Recent studies have shown that naturally occurring or artificially prepared supercoiled DNA molecules can often be separated by electrophoresis into about 10 forms, each differing from the other by one supercoiled turn. (Fig. 2-27A, lane 1). The relative amounts of these topological isomers form an approximately Gaussian distribution. The isomers are believed to arise as a result of thermal fluctuations in the degree of supercoiling at the time that the circles were enzymatically closed.

Does intercalation of flat molecules into nucleic acid chains have a biochemical function? It seems likely that the answer is yes; for example, aromatic rings of amino acid side chains in proteins designed to interact with nucleic acids might intercalate into nucleic acid helices serving a kind of "bookmark" function.[85,86] Changes in superhelix density caused by intercalation or by changes in the ionic environment may be important in the orderly handling of DNA by enzyme systems within the cell.

10. Denaturation and Hybridization

Like proteins, nucleic acids can undergo denaturation with separation of the double helix of DNA and of the double-stranded regions of RNA molecules (e.g., of tRNA, Fig. 2-24). Denaturation can be accomplished by addition of acids, bases, alcohols, or by removal of stabilizing counterions such as Mg^{2+}. The product is a random coil and denaturation can be described as a helix $\rightarrow$ coil transition. Denaturation of nucleic acids by heat, like that of proteins, is "cooperative" (Chapter 4, Section C,7) and sets in rather suddenly at a characteristic **melting temperature.**

A plot of the optical absorbance at 260 nm (the wavelength of maximum light absorption by nucleic acids) versus temperature is known as a **melting curve** (Fig. 2-28). The absorbance is lower, by up to 34%, for native than for denatured nucleic acids. This **hypochromic effect** (Chapter 13, Section B,4,e) is a result of the interaction between the closely stacked bases in the helices of the native molecules. The melting temperature T_m is taken as the midpoint of the increase in absorbance (Fig. 2-28). As the percent G + C increases, the nucleic acid becomes more stable toward denaturation and T_m is found to increase almost linearly with increases in the G + C content. In 0.15 M NaCl + 0.015 M sodium citrate buffer pH 7.0,

---

* The estimate given here is based on an unwinding angle of $\sim 16°$ for intercalation of a proflavin molecule as compared with 26° for intercalation of ethidium chloride. In the older literature the unwinding angle for ethidium was usually taken as 12° and the resulting values of the superhelix density ($^{12}\sigma$) were correspondingly lower than current values ($^{26}\sigma$) which assume an unwinding angle of $\sim 26°$ for ethidium.[84,86a,86b]

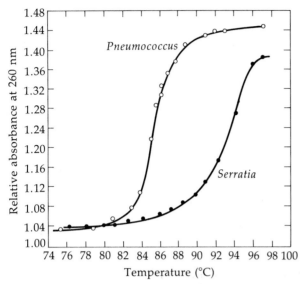

Fig. 2-28 A melting curve for DNA molecules from two different sources. From J. N. Davidson, "The Biochemistry of Nucleic Acids," 7th ed., p. 148. Academic Press, New York, 1972.

Eq. 2-15 holds. The exact numerical relationship depends strongly upon the ionic composition and pH of the medium.[87,88]

$$\% \ (G + C) = 2.44 \ (T_m - 69.3) \ \text{in} \ °C \qquad (2\text{-}15)$$

Complete denaturation of a molecule of DNA leads to separation of the two complementary strands. If a solution of denatured DNA is cooled quickly, the denatured strands remain separated. However, if the temperature is held for some time just below T_m (a process known as **annealing**), the native double-stranded structure can be reformed. This fact has provided an important tool for the study of nucleic acids by permitting the formation of hybrid duplexes.[89,90]

One kind of artificial double helix that can be formed is a **DNA-RNA hybrid.** Thus, it has been observed that messenger RNA (mRNA) molecules hybridize with only one of the two separated strands of the DNA carrying the gene for the mRNA. The hybridization technique can also be used to prepare **heteroduplex** DNA molecules in which one strand comes from one strain of an organism and the other strand from a genetic

variant of the same organism. Some mutations consist of **deletions** or **additions** of one or a substantial number of bases to a DNA chain. Heteroduplexes prepared from DNA of such mutants hybridized with that from a nonmutant strain have a normal hydrogen-bonded Watson–Crick structure for the most part. However, they have single-stranded loops in regions where deletions or additions prevent complementary base pairing (see Fig. 15-24).

Hybridization measurements have permitted many studies of **homology** of nucleic acids from different species. A nucleic acid is cut (e.g., by sonic oscillation) into pieces of moderate length (~1000 nucleotides) and is denatured. The denatured DNA fragments are mixed with denatured DNA of another species. Nucleotide sequences that are shared with little change between species tend to hybridize, whereas sequences that are drastically different between two species do not. One way to do such an experiment is to immobilize the long-chain denatured DNA from the one organism by embedding it in an agar gel[90] or by absorbing it onto a nitrocellulose filter.[91] The DNA fragments from the second organism are passed through a column containing "beads" of the DNA-containing agar or through the filter with adsorbed DNA. Pairing of fragments with complementary sequences occurs and such paired fragments are retained while strands that do not pair pass on through the column (or filter).

A very important technique is the study of the kinetics of reassociation of denatured DNA fragments.[92–95] Reassociation is found to obey second-order kinetics (Chapter 6, Section A,3) and Eq. 2-16, which is readily derived by integrating Eq. 6-8 for $[A] = [B] = C$ from time $= 0$ to t:

$$C/C_0 = 1/(1 + k \ C_0 t) \qquad (2\text{-}16)$$

The initial concentration of denatured DNA, C_0, is related in this way to the concentration C of DNA remaining undissociated at time t. A plot of the fraction of molecules reassociated versus the logarithm of $C_0 t$ (Fig. 2-29A) is a convenient way of displaying data. As indicated in Fig. 2-29B, the value of $C_0 t$ increases in direct proportion to the

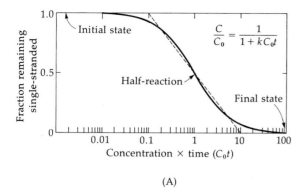

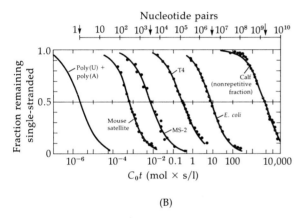

length of the DNA chain in the genome, but it is very much decreased if the sequence of bases is highly repetitive [poly(U) and poly(A)]. The slope of the plot at the midpoint gives an indication of the heterogeneity of the DNA fragments in a solution (Chapter 15, Section I,1).

11. Conformational Changes and Palindromes in Nucleic Acids

The possibility that both the B- and A-helical conformations of DNA are biologically important has already been suggested. Another type of conformational change that may be of biological significance is associated with **palindromes,** sequences of DNA that read the same in the forward and reverse directions.[95] Such sequences have been found at many places in the DNA of viruses and bacteria. For example, consider the gene that specifies the sequence of nucleotides for the tRNA molecule of Fig. 2-24. It is a double-stranded DNA segment in which one strand has a sequence identical to that of Fig. 2-24 except for the substitution of T for U and for ψ (pseudouridine, Fig. 15-10) and for the lack of methylation and other base modifications. The second strand is the exact complement. Figure 2-30 shows part of this gene, that corresponding to the 3' end of the tRNA molecule down to the arrow marked "extra loop" in Fig. 2-24. This DNA molecule could exist in a second conformation having a loop on each side of the molecule (Fig. 2-30). Note that the "stems" of the two loops in this "cruciform" conformation are identical and symmetrically disposed around the center of the molecule.

There is no evidence yet that DNA actually assumes cruciform conformations in cells, and many biochemists doubt their importance. On the other hand, there is little doubt that palindromic sequences in DNA, whether in a linear or looped conformation, are of great importance in the interaction of nucleic acids with symmetrical dimeric and tetrameric protein molecules of a kind considered in Chapter 4.[95a]

Fig. 2-29 Reassociation curves for DNA. [from R. J. Britten and D. E. Kohne, *Science* **161,** (1968). Copyright 1968 by the American Association for the Advancement of Science, and R. J. Britten and J. Smith.[94]] (A) Time course of an ideal, second-order reaction to illustrate the features of the log C_0t plot. The equation represents the fraction of DNA which remains single-stranded at any time after the initiation of the reaction. For this example, K is taken to be 1.0, and the fraction remaining single-stranded is plotted against the product of total concentration and time on a logarithmic scale. (B) Reassociation of double-stranded nucleic acids from various sources. The genome size is indicated by the arrows near the upper nomographic scale. Over a factor of 10^9, this value is proportional to the C_0t required for half-reaction. The DNA was sheared, and the other nucleic acids are reported to have approximately the same fragment size (about 400 nucleotides, single-stranded). Correction has been made to give the rate that would be observed at 0.18 M sodium-ion concentration. The temperature in each case was optimal, i.e., ~30° below the melting temperature T_m. The extent of reassociation was established by measuring optical rotation (calf thymus DNA), ribonuclease resistance (MS-2), or hypochromicity.

* Some well-known palindromes in the English language are "Madam, I'm Adam" and "A man, a plan, a canal. Panama." M. Gardner, *Sci. Am.* **223,** 110–112 (Aug 1970).

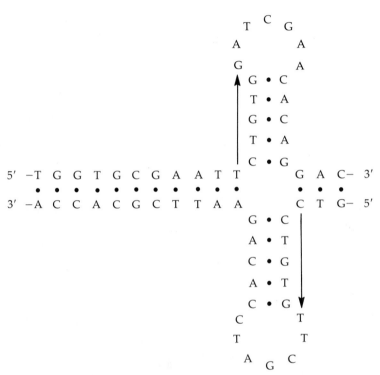

5′ –T G G T G C G A A T T C T G T G G A T C G A A C A C A G G A C– 3′

3′ –A C C A C G C T T A A G A C A C C T A G C T T G T G T C C T G– 5′

CψT

TψC loop

Fig. 2-30 Two conformations of a segment of the yeast phenylalanine tRNA gene. The segment shown codes for the 3′-end of the tRNA molecule shown in Fig. 2-24, including the TψC loop.

Those readers anxious to know more about the biological role of nucleic acids at this point may wish to read all or part of Chapter 15 now.

E. THE LIPIDS[96−99]

The name **lipid** refers to a large and heterogeneous group of substances. They are classified together by their property of high solubility in apolar solvents or by their relatedness to compounds with such solubility properties. Most lipids are not high polymers but are made by linking together smaller molecules. Some of these "building blocks" are derived from acetic acid units by complex polymerization reactions. The resulting molecules, such as the **fatty acids,** are mostly hydrophobic in character. However, they usually contain at least one polar group which may serve as a linkage point to other components. Ionic groups (e.g., phosphate, —NH$_3^+$) and polar carbohydrate units are frequently present. Lipids that contain both nonpolar and polar groups are usually found in membranes and at other surfaces between water and hydrophobic regions of cells.

1. Lipid Building Blocks

See Fig. 2-31 for examples of some structures.

a. Derived from Acetic Acid, CH_3COOH

Common (even carbon number) fatty acids and fatty alcohols, some containing double bonds, almost always of cis configuration (Table 2-7).

Odd carbon number fatty acids

Branched, cyclopropane-containing, and other specialized fatty acids

Polyprenyl compounds: terpenes, carotenoids, sterols and steroids which are formed from acetate by reductive condensation and decarboxylation. The structures are such that they might have been formed by polymerization of isoprene

$$CH_2{=}\underset{\underset{\displaystyle CH_3}{|}}{C}{-}CH{=}CH_2$$

hence, the older name *isoprenoid*.

TABLE 2-7
Some Important Fatty Acids

No. carbon atoms	Systematic name	Common name	Abbreviation[a]	Common name of acyl group
		Saturated fatty acids		
1	Methanoic	Formic		Formyl
2	Ethanoic	Acetic		Acetyl
3	Propanoic	Propionic		Propionyl
4	Butanoic	Butyric	4:0	Butyryl
12	Dodecanoic	Lauric	12:0	Lauroyl[b]
14	Tetradecanoic	Myristic	14:0	Myristoyl
16	Hexadecanoic	Palmitic	16:0	Palmitoyl
18	Octadecanoic	Stearic	18:0	Stearoyl
20	Eicosanoic	Arachidic	20:0	
22	Docosanoic	Behenic	22:0	
24	Tetracosanoic	Lignoceric	24:0	
		Unsaturated fatty acids[c]		
4		Crotonic	4:1(2t)	Crotonoyl
16		Palmitoleic	16:1(9c)	
18		Oleic	18:1(9c)	Oleoyl
		Vaccenic	18:1(11c)	
18		Linoleic	18:2(9c,12c)	
18		Linolenic	18:3(9c,12c,15c)	
20		Arachidonic	20:4(5c,8c,11c,14c)	

[a] The number of carbon atoms is given first, then the number of double bonds. The positions of the lowest numbered carbon of each double bond and whether the configuration is cis (c) or trans (t) is indicated in parentheses.

[b] Official IUPAC names of these and other acyl groups have been designated by the Commission of the Nomenclature of Organic Chemistry in *Pure and Applied Chemistry* **10**, 111–125 (1965). In a number of cases IUPAC inserted an *o* in the traditional name, e.g., palmityl became palmitoyl and crotonyl became crotonoyl. However, acetyl was not changed. The logic is not clear and many authors continue to use the older names. In many cases the systematic names, e.g., hexadecanoyl (from hexadecanoic acid), are preferable and IUPAC-IUB recommends that alkyl radicals always be designated by systematic names, e.g., hexadecyl, *not* palmityl alcohol. Indeed, the older use of palmityl for both acyl and alkyl radicals was one reason for IUPAC's adoption of new names for acyl radicals.

[c] Systematic names are not often used because of their complexity, e.g., linolenic acid is *cis,cis,cis*-9,12,15-octadecatrienoic acid.

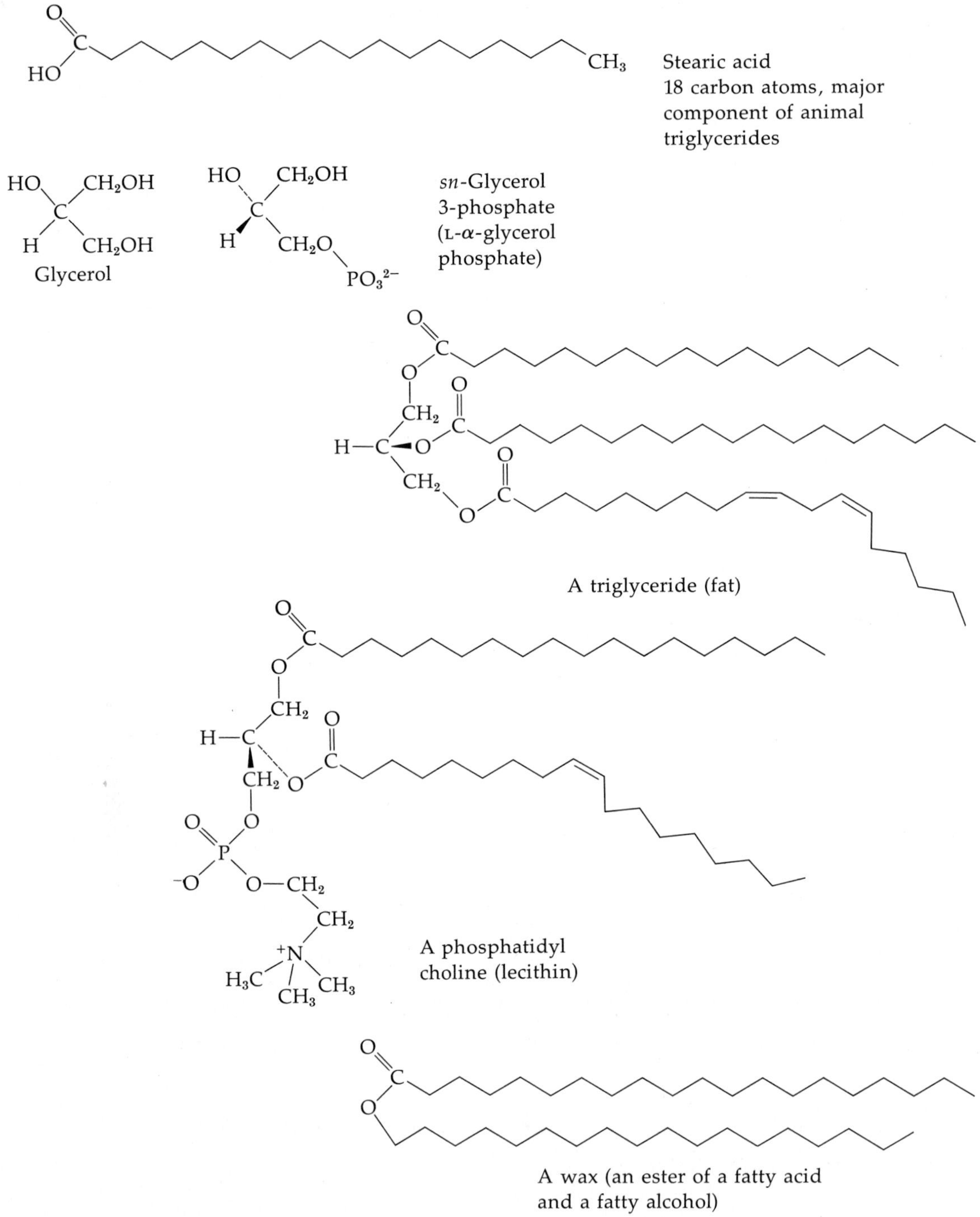

Fig. 2-31 Structures of some representative lipid molecules.

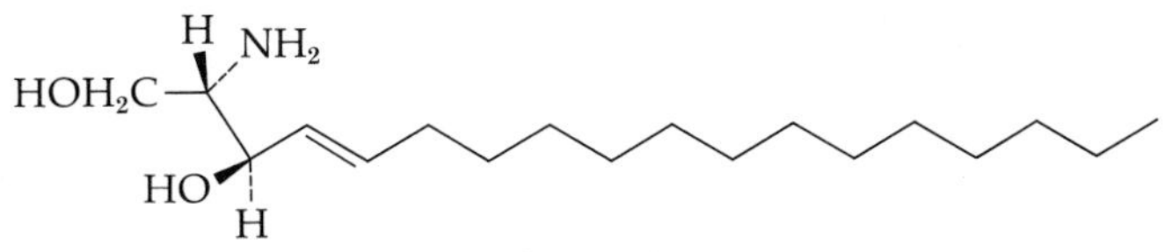

Sphingosine

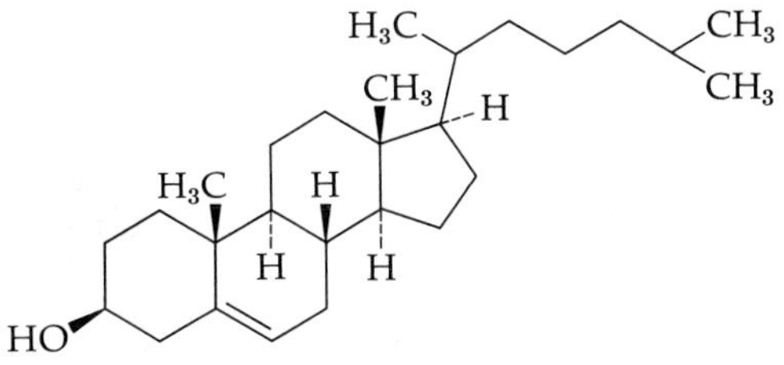

Cholesterol

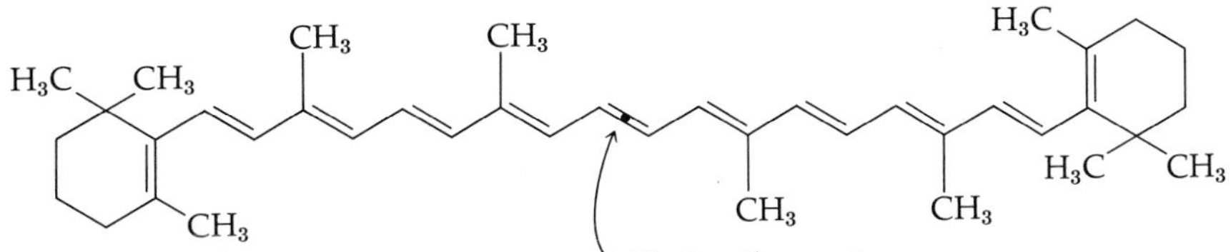

β-Carotene
a yellow pigment of
carrots and other plants
converted in the human
body to vitamin A

Notice the center
of symmetry in
this 40-carbon
molecule derived
from 8 prenyl
units

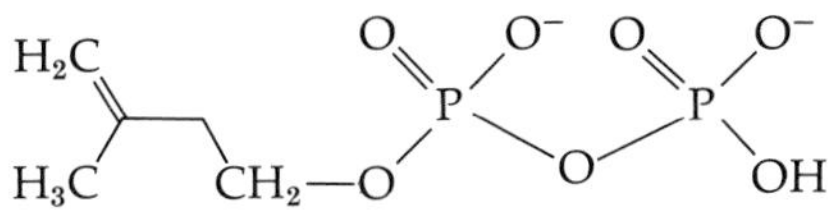

Prenyl pyrophosphate
(isopentenyl pyrophosphate),
precursor of steroid,
carotenoid and other
isoprenoid substances

Fig. 2-31 (*Continued*)

b. **Polar Constituents,**
   Not Derived from Acetate

Glycerol

Serine, ethanolamine, choline
   (Note the structural relationships of these
   three; ethanolamine and choline are synthesized from serine)

Sphingosine (partially derived from a fatty
   acid)

Inorganic phosphate and sulfate as well as
   organic phosphonates and sulfonates

Inositol, sugars

2. Linkage

The components of complex lipids are linked in
a variety of ways, some of which are illustrated in
Figs. 2-31 and 2-32. Often glycerol acts as the central unit, e.g. combining in ester linkage with
three fatty acids to form **triglycerides** (the common
fats of adipose tissues and plant oils). The hydrocarbon chains of the fatty acids tend to assume the
extended conformation, but double bonds are
often present and induce kinks and bends. The
phosphatides (Table 2-8) are derivatives of *sn*-

TABLE 2-8
Structures of Some Phosphatides and Sphingolipids

The ester linked fatty acid in this position is replaced in plasmalogens by a vinyl ether group:

and in ether phosphatides by an alkoxy group:

Phosphatides (glycerol phospholipids)   Abbrev.

Y = —OH, phosphatidic acid

—O—CH₂CH₂—N⁺(CH₃)₃   Phosphatidylcholine (lecithin)   PC

—O—CH₂—CH—NH₃⁺
              |
            COO⁻   Phosphatidylserine   PS

—O—CH₂CH₂—NH₃⁺   Phosphatidylethanolamine   PE

Phosphatidylinositol   PI

—O—CH₂—CH—CH₂OH
              |
            OH   Phosphatidylglycerol

glycerol 3-phosphate* (L-α-glycerol phosphate), a typical member being phosphatidylcholine or **lecithin.** The phosphate and choline groups are both electrically charged providing a polar "head" for the molecule. A second group of phospholipids, the **sphingomyelins,** contain a special long-chain base **sphingosine** as the central unit. Unlike the triglycerides, which are all liquid at body temperature, phospholipids are solids, a property that is doubtless related to their suitability for construction of biological membranes (Chapter 5).

* The stereochemical numbering system assigns the 1 position to the group (of two equivalent groups) occupying the *pro-S* position (see Chapter 6, Section D,2).

The **cerebrosides** and **gangliosides** are sphingosine-containing lipids **(sphingolipids)** in which the polar constituent is a sugar unit rather than phosphate. Other **glycolipids,** found in bacteria and in green plants, contain glycerol and fatty acids as well as α-D-galactose, glucose, and mannose. Chloroplasts contain a large amount of a special **sulfolipid** (Fig. 2-32).

3. Fatty Acids of Lipids

Except for the first few members of the series which are soluble in water, fatty acids have strongly hydrophobic characteristics. However, they are all acids with pK_a values of ~4.8. To the

TABLE 2-8 (*Continued*)

$-O-CH_2-CH-CH_2-O$ Diphosphatidylglycerol (cardiolipin), a
           | special component of bacteria and mitochondria
            OH

An —OH group occurs here in some brain cerebrosides

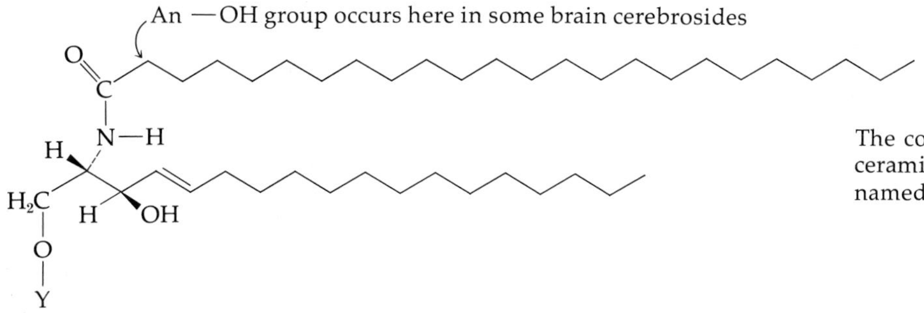

The compounds in which Y = H are ceramides. Sphingolipids are often named as ceramide derivatives

Sphingolipids

$Y = -\boxed{P}-$ choline ($-O-\overset{\overset{\displaystyle O}{\|}}{\underset{\underset{\displaystyle O^-}{|}}{P}}-O-CH_2CH_2-N^+(CH_3)_3$), sphingomyelins

$-\overset{\overset{\displaystyle O}{\|}}{\underset{\underset{\displaystyle O^-}{|}}{P}}-CH_2CH_2NH_3^+$, ceramide aminoethyl phosphonates

$-\beta$-D-Gal or β-D-Gal(1→4)Glc—;   cerebrosides or ceramide mono- and oligosaccharides

The galactose bears a 3-sulfate group in cerebroside sulfatides, e.g., in lactosyl ceramide sulfate

GalNAc(1→3)Gal(1→4)Gal(1→4)Glc — is present in sphingolipid of red blood cell membranes

---

extent that free fatty acids occur in nature, they are likely to be found in interfaces between lipid and water with the carboxyl groups dissociated and protruding into the water. However, most naturally occurring fatty acids are esterified or combined in amide linkage in the complex lipids.

Although a seemingly endless variety of structures of fatty acids exist within a given organism, a small number usually predominate. Higher plants contain mostly the C_{16} **palmitic** acid and the C_{18} unsaturated **oleic** and **linoleic** acids. The C_{18} saturated **stearic** acid is almost completely absent and C_{20} to C_{24} acids are rarely present except in the outer cuticle of leaves. At the same time, certain taxonomic groups contain unusual fatty acids which may be characteristic of a group; e.g., the Compositae (daisy family) contain acetylenic fatty acids and the castor bean a special hydroxy fatty acid (Fig. 2-32).

Like plants, animals contain palmitic and oleic acids but in addition, a large amount of stearic acid. Furthermore, the C_{20}, C_{22}, and C_{24} acids are present and the variety of fatty acids found is generally larger than in a given plant species. A striking fact is that in most organisms the unsaturated fatty acids contain strictly cis double bonds. Table 2-9 shows the fatty acid composition of some typical triglyceride mixtures.

Within bacteria, polyunsaturated fatty acids are largely absent but branched fatty acids occur fre-

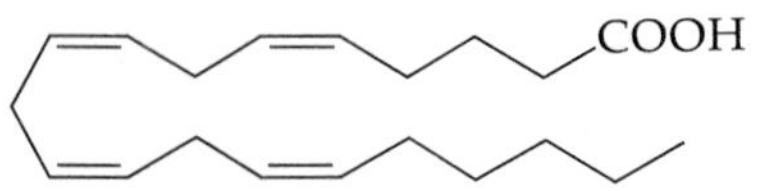

Arachidonic acid, a nutritionally essential fatty acid

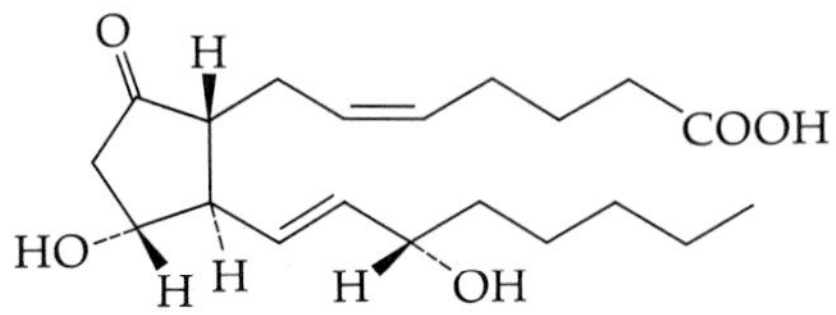

Prostaglandin PGE$_2$, one of a family of hormones, this one synthesized in tissues from arachidonic acid

A second galactose may be added here

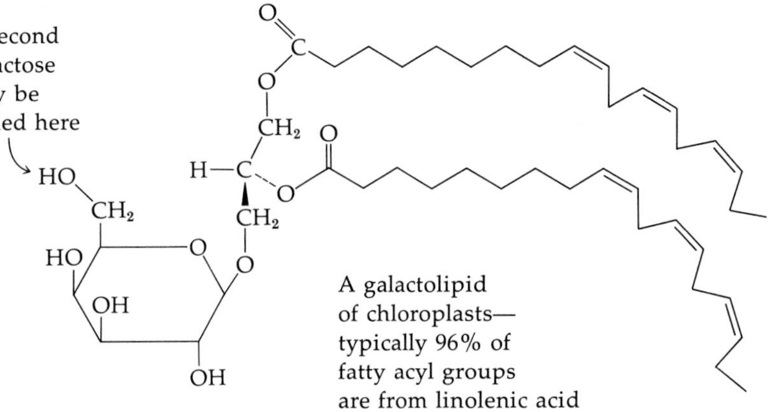

A galactolipid of chloroplasts—typically 96% of fatty acyl groups are from linolenic acid

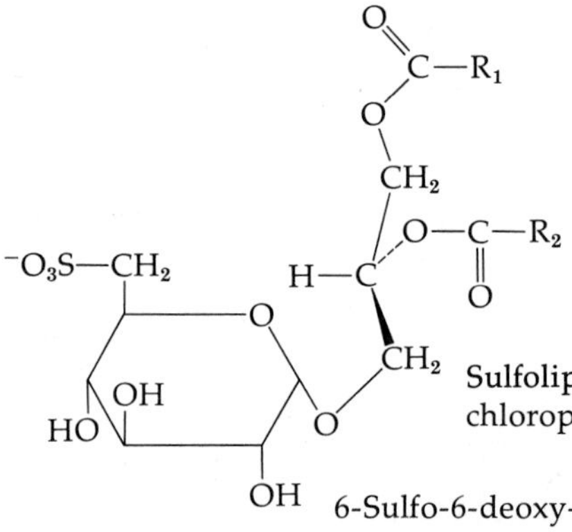

Sulfolipid of chloroplasts

6-Sulfo-6-deoxy-α-D-glucopyranosyl diglyceride

Branched fatty acids of the *anti-iso* series have a branch here and a 5-carbon "starter piece" derived from isoleucine

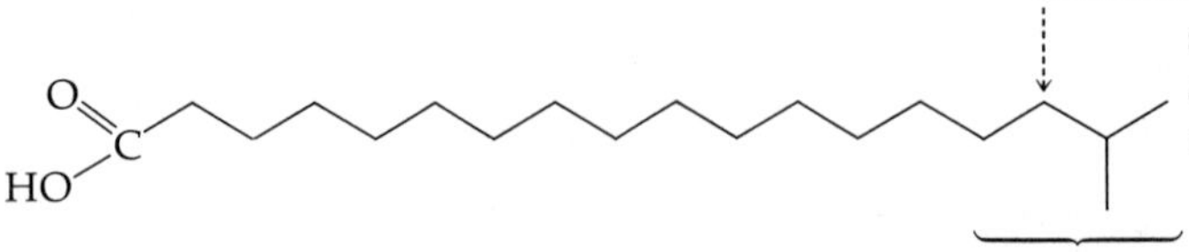

Branched fatty acids of the *iso* series contain a 5-carbon "starter piece" derived from leucine or a 4-carbon piece derived from valine

Fig. 2-32 More lipid structures.

Fig. 2-32 *(Continued)*

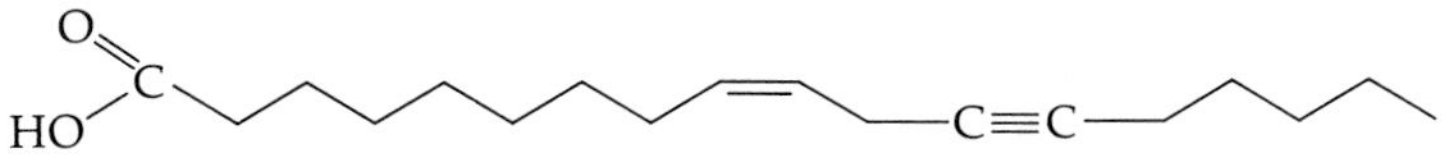

Crepenynic acid 18:2 (9c, 12a), accounts for 60% of the fatty acids in seeds of *Crepis foetida,* a member of the compositae family

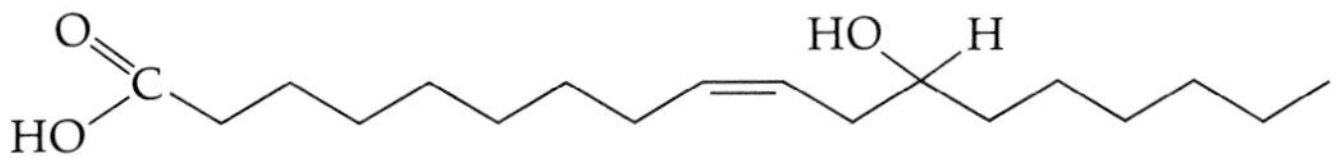

Ricinoleic acid (12-hydroxyoleic acid), accounts for up to 90% of the fatty acids of *Ricinus communis* (castor bean)

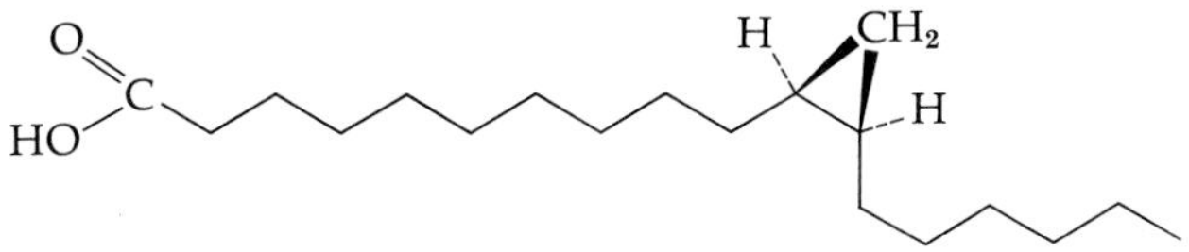

Lactobacillic acid, a major fatty acid of lactobacilli

quently as do cyclopropane-containing acids, hydroxy acids, and free unesterified fatty acids. Distribution of fatty acids within a given organism is not uniform; for example, the lipid portion of biological membranes may be 90% phospholipid. Phospholipids in turn tend to contain a higher proportion of unsaturated fatty acids than do the triglyceride stores.

4. Poly-β-hydroxybutyrate

While lipids are not true polymers, the important bacterial storage material polyhydroxybutyric

acid is related metabolically and structurally to the lipids. This highly reduced polymer is made up exclusively of D-β-hydroxybutyric acid units in ester linkage. About 1500 residues are present per chain. The structure is that of a compact right-handed coil with a 2-fold screw axis and a pitch of 0.60 nm.[100] Within bacteria, it occurs in thin lamellae ~5.0 nm thick. Since a chain of 1500 residues stretches to 440 nm, there must be ~88 folds in a single chain.

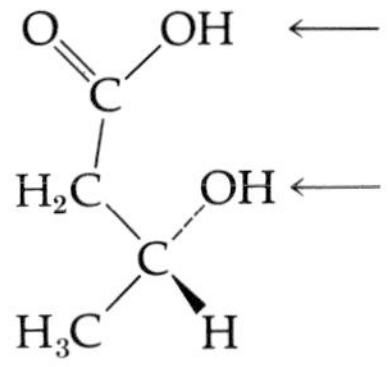

Linking groups in polymer

D-β-Hydroxy-butyric acid

TABLE 2-9
Fatty Acid Composition of Some Typical Fats and Oils[a]

Fats and oils	No. of carbon atoms and (following colon) the No. of double bonds					
	14	16	18	16:1	18:1	18:2
Man: depot fat	3	23	4	8	45	10
Beef tallow	4	30	25	5	36	1
Corn oil		13	2		31	54
Lard	1	28	15	3	42	9

[a] F. D. Gunstone, "An Introduction to the Chemistry and Biochemistry of Fatty Acids and Their Glycerides," 2nd ed., Chapter 6. Chapman and Hall, London, 1967.

F. SOME IMPORTANT SMALL MOLECULES AND IONS

Cells contain a large number of small organic molecules and metal ions. Among these the **co-**

enzymes (Chapter 8) serve as "special reagents" that assist enzymes in carrying out what would otherwise be difficult chemical reactions. They often serve as oxidizing agents, carrying electrons from reduced organic materials to oxygen during biological oxidation. They also carry electrons needed for biological reduction, e.g., for light-driven photosynthetic processes. **Adenosine 5'-triphosphate (ATP)** and related substances (Box 3-A) play extraordinarily active roles within cells in energy metabolism. Hormones and other small regulatory molecules as well as a host of metabolic intermediates **(intermediary metabolites)** all have important functions within cells.

1. Inorganic Components

The solid matter of cells consists principally of C, H, O, N, S, and P. However, if a tissue or an entire organism is burned, the remaining ash (typically 3–5% of the total solids, but much higher for mineralized tissues) is found to contain a large number of chemical elements. Among the cations, Na^+, K^+, Ca^{2+}, and Mg^{2+} are present in relatively large amounts. Thus, the body of a 70 kg man contains 1050 g Ca (mostly in the bones), 245 g K, 105 g Na, and 35 g Mg. Iron (3 g), zinc (2.3 g), and rubidium (1.2 g) are the next most abundant. Of these iron and zinc are essential to life but rubidium is probably not.

The other metallic elements in the human body amount to less than 1 g each, but at least seven of them play essential roles. They include copper (100 mg), manganese (20 mg), and cobalt (~5 mg). Others, such as chromium (<6 mg), tin, and vanadium have only recently been shown essential for higher animals.[101–103] It is likely that nickel, lead, and others may also be needed.[103]

Nonmetallic elements predominating in the ash are phosphorus (700 g in the human body), sulfur (175 g), and chlorine (105 g). Not only are these three elements essential to all living cells but also selenium, fluorine, silicon (Box 2-B), and iodine are needed by higher animals, and boron is needed by plants (Fig. 2-33).

What is the likelihood that other elements will be found essential? Consider a human red blood cell, a disc-shaped object of volume ~80 μm^3 and containing about 3×10^8 protein molecules (mostly hemoglobin). About 7×10^5 atoms of the "trace metal" copper and 10^5 atoms of the nutritionally essential tin are present in a single red cell. Also present are 2×10^4 atoms of silver, a toxic metal. Its concentration, over 10^{-7} M, is sufficient that it could have an essential catalytic function. However, we know of none and it may simply have gotten into our bodies from handling money, jewelry, and other silver objects. The red blood cell also contains boron and aluminum (3×10^5 atoms each), arsenic (7×10^5 atoms), lead (7×10^4 atoms), and nickel (2×10^4 atoms). Of the

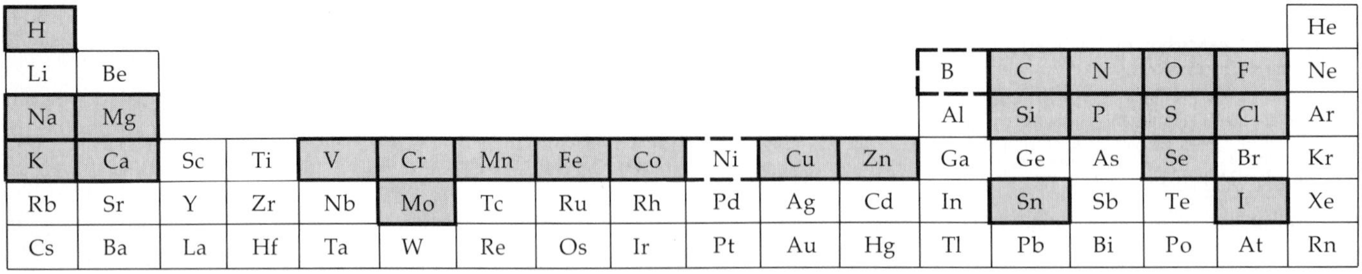

H																	He
Li	Be											B	C	N	O	F	Ne
Na	Mg											Al	Si	P	S	Cl	Ar
K	Ca	Sc	Ti	V	Cr	Mn	Fe	Co	Ni	Cu	Zn	Ga	Ge	As	Se	Br	Kr
Rb	Sr	Y	Zr	Nb	Mo	Tc	Ru	Rh	Pd	Ag	Cd	In	Sn	Sb	Te	I	Xe
Cs	Ba	La	Hf	Ta	W	Re	Os	Ir	Pt	Au	Hg	Tl	Pb	Bi	Po	At	Rn

Fig. 2-33 Elements known to be essential to living things (after Frieden[102]). Essential elements are enclosed within shaded boxes. The 11 elements, C, H, O, N, S, P, Na, K, Mg, Ca, and Cl make up 99.9% of the mass of man. An additional 13 are known to be essential for higher animals in trace amounts. Boron is essential to higher plants but apparently not to animals, microorganisms, or algae.

elements (uranium and below) in the periodic table, only four (Ac, Po, Pa, and Ra) are present, on the average, in quantities less than 1 atom per cell.[101]

Of the apparently nonessential elements, several, e.g., Cs, Rb, Sr, and Ni (possibly essential); are *relatively* nontoxic. Others are highly toxic, e.g., Sb, As, Ba, Be, Cd, Pb, Hg, Ag, Tl, and Th.

The ionic compositions of tissues and of body fluids vary substantially. Blood of marine organisms is similar to that of seawater in its content of Na^+, Cl^-, Ca^{2+}, and Mg^{2+}. Blood of freshwater and terrestrial organisms contains about 10 times less Na and Cl and several times less Ca and Mg than is present in seawater, but it is nevertheless relatively rich in these ions.

In general, cells are rich in K^+ and Mg^{2+}, the K^+ predominating by far, and are poor in Na^+ and Ca^{2+}. Chloride is the principal inorganic anion, but organic carboxylate and phosphate groups contribute most of the negative charges (Table 2-10), many of which are fixed to proteins or other macromolecules. Ling estimated that cells typically contain about 1.66 M of amino acid residues in their proteins. Of these residues, 10% have negatively charged side chains and 8% positively charged. The difference is a net negative charge amounting to 33 mM within cells.[104]

2. Selective Accumulation of Ions

Living organisms concentrate many elements from water, often by a factor of 10^4 or more. Even marine species concentrate nearly all of the elements except sodium and chlorine. Frequently, a single species or a group of organisms has a special affinity for a particular element and acts as an **accumulator organism.**[101] Thus, Al is accumulated by club mosses. As by brown algae and coelenterates; B by brown algae and sponges; F by vertebrates (in the skeleton); and Fe by some bacteria, plankton, and horsetails. Silicon is accumulated by horsetails, diatoms, some protozoa, and

sponges; Sr in preference to Ca by brown algae; V by some ascidians; Y by ferns; and Zn by coelenterates. Specific accumulator species are also known for Ba, Co, Li, Ni, Ti, Se, La, and Nd.

Mammalian tissues also often tend to concentrate particular ions. Thus, Al, As, I, and V accumulate in hair and nails; Cd, Hg, and Mn in kidneys; Sn in intestinal tissues; Zn and Sr in the prostate; Cu in the brain; and Ba in the choroid of the eye.[101]

Within cells the inorganic ion content varies greatly between different organelles and subcellular regions. Thus, copper is 10–15 times as concentrated in mitochondria as in whole beef heart tissues.[105]

3. Functions of Inorganic Ions

Numerous cases are known in which a specific metal ion or an anion such as Cl^- is required for the activity of an enzyme. In a number of instances it has been shown that a metal ion binds at a particular point on or within an enzyme molecule. There it may affect the catalyzed reaction by virtue of its intensely concentrated electrical charge. Some metal ions are able to undergo reversible oxidation and reduction. This property permits iron, copper, and cobalt to become parts of the active centers of enzymes catalyzing many oxidation–reduction processes. Also important is the ability of metal ions to act as centers for orientation of different parts of a protein or other macromolecule. The binding of a metal ion may cause enormous changes in the conformation of a molecule (Chapter 4, Section C,8,c).

G. THE CHEMICAL COMPOSITION OF CELLS

How much protein, nucleic acid, carbohydrate, lipid, and other materials are typically present in living cells? It is hard to give a simple answer because cells vary so much. Some specialized cells store large amounts of triglycerides or of carbohydrates. Glandular cells often have an unusually high content of both protein and RNA. The composition of organisms varies with age. For example, the pig embryo is 97% water; at birth it is only 89% water. A lean 45 kg pig may contain 67% water, but a very fat 135 kg animal only 40% water.

A simple approximate analysis of tissue composition can be made in the following way. The tissue is thoroughly dried at low temperature in vacuum and the solid material is extracted with suitable solvents to remove lipid. After evaporation of the solvent the lipid residue is weighed. By this procedure a young leafy vegetable might be found to contain 2–5% lipid on a dry weight basis. Even very lean meats would contain 10–30% lipid.

Protein content is frequently estimated by determining the content of nitrogen and multiplying by 6.25. In a young green plant, 20–30% of the dry matter may be protein, while in very lean meat it may reach 50–70%.

A sample of tissue is burned at a high temperature, and the resulting ash is weighed and provides a measure of the inorganic constituents of tissues. The "ash" commonly amounts to 3–10% and is higher in specialized tissues such as bone.

In this approximate method of analysis carbohydrate is estimated by the difference from 100%. It amounts to 50–60% in young green plants and only 2–10% in typical animal tissues. In exceptional cases the carbohydrate content of animal tissues may be higher. Thus, the glycogen content of oysters is 28%.

The amount of nucleic acid in cells varies from 0.1% in yeast and 0.5–1% in muscle and in bacteria to 15–40% in thymus gland and sperm cells. In these latter materials of high nucleic acid content it is clear that multiplication of % N by 6.25 is not a valid measure of protein content.

Table 2-10 compares the composition of a bacterium, of a green plant, and of an active animal tissue (rat liver).

TABLE 2-10
Approximate Composition of Metabolically Active Cells and Tissues

Component	E. coli[a] (%)	Green plant (spinach, Spinacia oleracea)[b] (%)	Rat liver[c] (%)
H_2O	70	93	69
Protein	15	2.3	21
Amino acids	0.4		
DNA	1		0.2
RNA	6		1.0
Nucleotides	0.4		
Carbohydrates	3	3.2	
Cellulose		0.6	
Glycogen			3.8
Lipids	2	0.3	6
Phospholipids			3.1
Neutral lipids			1.6
Sterols			0.3
Other small molecules	0.2		
Inorganic ions	1	1.5	
K^+			0.4

Component	Concentration in rat liver (M)
Amino acid residues	2.1
Nucleotide units	0.03
Glycogen	0.22
K^+	0.1

[a] From J. D. Watson, "Molecular Biology of the Gene," 3rd ed., p. 69. Benjamin, New York, 1976. The amounts of amino acids, nucleotides, carbohydrates, and lipids include precursors present in the cell.
[b] From B. T. Burton, "Human Nutrition," 3rd ed., p. 505. McGraw-Hill, New York, 1976.
[c] From C. Long, ed., "Biochemists' Handbook," pp. 677–679. Van Nostrand-Reinhold, Princeton, New Jersey, 1961.

H. HOW WE HAVE LEARNED THE STRUCTURES

How have chemists deduced the thousands of structural formulas that we write for the substances found in nature? The answer is far too complex to give here in detail. However, the separation of compounds, the analysis of mixtures, and the unraveling of structures remain essential parts of modern biochemistry. Therefore a "mini-review" follows with a few references to help the reader to get started with the literature.

1. Separations

Before structural work can begin, pure substances must be separated from the complex mixtures in which they occur in cells and tissues. Often, a substance must be isolated from a tissue in which it is present in a very low concentration. The molecules of the pure compounds must then be cut up and the pieces separated, purified, and identified. Accurate quantitative analysis is required to determine the ratios of component fragments. Then considerable ingenuity must usually be exercised in putting the pieces of the jigsaw puzzle back together to determine the structure of the original molecules.

a. Fractionation of Cells[105-110]

A fresh tissue or a paste of packed cells of a microorganism (usually collected by centrifugation) may be the starting material.

Tissue is often ground in a kitchen-type blender or, for gentler treatment, in a special **homogenizer.*** Microbial cells are frequently broken with supersonic oscillation **(sonication)** or in special pressure cells. It is important to pay proper attention to the pH, buffer composition, and, if subcellular organelles are to be separated, to the osmotic pressure. To preserve the integrity of organelles, 0.25 M sucrose is frequently used as the suspending medium, and $MgCl_2$ as well as a metal complexing agent such as ethylenediaminetetraacetate (EDTA) (Table 4-2) is added. Soluble enzymes are usually extracted without addition of sucrose, but reducing compounds such as glutathione (Box 7-B), mercaptoethanol, or dithiothreitol (Section H,3,a) may be added.

The crude **homogenate** may be strained and is

* Most popular is the Potter–Elvehjem homogenizer, a small apparatus in which a glass or plastic pestle rotates inside a tight-fitting mortar tube (see standard laboratory equipment catalogs for pictures).

usually centrifuged briefly to remove cell fragments and other "debris." Cell organelles are usually separated by centrifugation.[106,107] In one procedure a homogenate in 0.25 M sucrose (**isotonic** with most cells) is centrifuged for 10 min at a field of 600–1000 times the force of gravity (600–1000 g) to sediment nuclei and whole cells. The supernatant is then centrifuged another 10 min at ~ 10,000 g to sediment mitochondria and lysosomes. Finally, centrifugation at ~ 100,000 g for about an hour yields a pellet of microsomes. Each of the separated components can be resuspended and recentrifuged to obtain cleaner preparations of the organelles. The sedimented particles can often be solubilized by chemical treatment, for example, by the addition of detergents. The soluble **supernatant** remaining after the highest speed centrifugation provides the starting material for isolation of soluble enzymes and many small molecules.

b. Separations Based on Solubility

Some fibrous proteins are almost insoluble in water and everything else can be dissolved away. More often soluble proteins are precipitated from aqueous solutions by **salting out,** addition of large amounts of salts such as ammonium sulfate. Different proteins precipitate at different salt concentrations. Hence, a fraction of proteins precipitating between two different concentrations of salt can be selected for further purification. Precipitation methods are popular first steps in purification schemes because they can be carried out on a large "batch" scale.

RNA is often prepared by treating a solution with aqueous phenol to precipitate the protein. Depending upon the conditions, DNA may also be removed. Various precipitation and extraction procedures may be used to separate the RNA in the aqueous layer from polysaccharides, from DNA (if present), and from other components.[111-113] Nucleic acids may also be separated from proteins by dissolving the latter away with a protein-hydrolyzing enzyme.

Some carbohydrates (e.g., cellulose and glycogen) are sufficiently stable that other materials can be removed by boiling in basic solution. Lipids are usually extracted from tissues using nonpolar solvents,[114] e.g., a 2:1 mixture of $CHCl_3$ and CH_3OH.

c. Separation by Partition

Many of the most important separation methods are based on repeated equilibration of a material between two separate phases, at least one of which is usually liquid. Small molecules may be separated by **countercurrent distribution** in which a material is repeatedly partitioned between two liquid phases, one more polar than the other. New portions of both liquids are moved in a special machine in a "countercurrent manner" between the equilibration steps.[115]

A similar result is accomplished by using as one phase a solid powder or fine "beads" packed in a vertical column or spread in a thin layer on a plate of glass. The methods are usually referred to as **chromatography,** a term proposed by M. Tswett, who first described isolation of plant leaf pigments (chlorophylls and carotenes) by this procedure in 1903. Tswett passed solutions of the pigments in nonpolar solvents such as hexane through columns of **alumina** and of various other adsorbents and observed separation of colored bands which moved down the column as more solvent was passed through. Individual pure pigments could be **eluted** from the column by continued passage of solvent. The method is still much used and is often referred to as **adsorption chromatography.** It is assumed that the pigments are absorbed on the surface rather than being dissolved in the solid material.

Other column packing materials such as **silica gel** contain a large amount of water, and the separation may be thought of as partition between an immobilized aqueous phase in the silica gel and the one flowing through the column. Lipids may be separated on **reverse phase** columns in which silica gel, alumina, or other inert material is impregnated with a nonpolar liquid. The mobile phase is a more polar solvent.

A sheet of high quality filter paper containing adsorbed water serves as the immobile phase in **paper chromatography.** Recently, **thin layer chromatography,** which employs a layer of silica gel spread on a glass plate, has often supplanted paper chromatography because of its rapidity and sharp separations (Fig. 2-34). For volatile materials **vapor phase chromatography** permits equilibration between the gas phase and immobilized liquids at relatively high temperatures. The formation of volatile derivatives, e.g., methyl esters or trimethylsilyl derivatives of sugars, extends the usefulness of the method.

An important concept is that of **affinity chromatography.**[116,117] This term is applied to the use of specially designed adsorbents which depend upon specific biochemical interactions to selectively hold a macromolecule. An example already discussed in Section D,10 is the adsorption of complementary nucleic acid fragments to immobi-

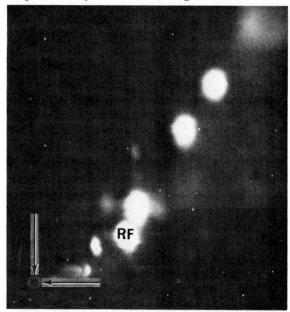

Fig. 2-34 Photograph of a two-dimensional thin layer (silica gel) chromatogram of a mixture of flavins formed by irradiation of the vitamin riboflavin. The photograph was made by the fluorescence of the compounds under ultraviolet light. Some riboflavin (the spot labeled RF) remains. The arrows indicate the location of the sample spot before chromatography. Chromatography solvents: a mixture of acetic acid, 2-butanone, methanol, and benzene in one direction and n-butanol, acetic acid, and water in the other.[147]

lized DNA strands. Affinity chromatography can also be used to purify enzymes, antibodies, and other proteins designed to bind tightly to specific small molecules.

Because of their open gel structure (Fig. 2-18) agarose derivatives in bead form provide a good solid support matrix for preparation of adsorbents. The hydroxyl groups of agarose are often linked to amino compounds. The agarose is first treated with cyanogen bromide (Br—$C \equiv N$) in base to "activate" the carbohydrate (the chemistry is discussed by Porath[118]). Then the amino compound is added. The overall reaction is given by Eq. 2-17.

$$\text{Agarose—OH} + \text{Br—CN} + \text{R—NH}_2 \xrightarrow{\text{Br}^-}$$

$$\underset{\substack{\|\\ \text{NH}}}{\text{agarose—O—C—NH—R}}$$

$$+ \underset{\substack{\|\\ \text{O}}}{\text{agarose—O—C—NH—R}}$$

$$+ \text{other products} \quad (2\text{-}17)$$

Adsorbents containing a large variety of R groups of specific shapes can be made in this way. Furthermore, if the coupling is done with a diamine [$R = (CH_2)_n$—NH_2], the resulting ω-aminoalkyl agarose can be coupled with other compounds by reaction with a water-soluble **carbodiimide** (Eq. 2-18).

$$\underset{\substack{\|\\ \text{O}}}{\text{Agarose—O—C—NH—CH}_2\text{—CH}_2\text{—NH}_2}$$

$$+ \text{R—COOH} \xrightarrow[\text{carbodiimide}]{\text{water-soluble}}$$

$$\underset{\substack{\|\\ \text{O}}}{\text{agarose—O—C—NH—CH}_2\text{—CH}_2\text{—}}\underset{\substack{\text{H}\\|}}{\text{N}}\underset{\substack{\|\\ \text{O}}}{\text{—C—R}}$$

$$(2\text{-}18)$$

Carbodiimides are widely used for forming amide or phosphodiester linkages in the laboratory. The formation of an amide as in Eq. 2-18 can be pictured as in Eq. 2-19.

$$\underset{\substack{\text{Carbodiimide}}}{\text{R''—N=C=N—R''}}$$

$$R—C\underset{O^-}{\overset{O}{<}} \longrightarrow R—\underset{\substack{\|\\ \text{O}}}{C}—O—C\underset{\substack{\\ N—R''}}{\overset{\substack{HN—R''\\ }}{<}}$$

$$R'—NH_2$$

$$O=C\underset{\substack{N—R''\\ H}}{\overset{\substack{H\\ N—R''}}{<}}$$

$$\underset{\substack{\text{H}}}{R—\overset{\substack{O\\ \|}}{C}—N—R'} \quad (2\text{-}19)$$

For reaction in nonaqueous medium dicyclohexylcarbodiimide (R'' = cyclohexyl) is often used, but for linking groups to agarose a water-soluble reagent such as 1-ethyl-3-(3-dimethylaminopropyl)carbodiimide is recommended.[119]

Many other means of preparation of adsorbents for affinity chromatography are also available.

An example of the use of affinity chromatography in purification of an enzyme is the isolation of a staphylococcal nuclease (DNA-hydrolyzing enzyme) by chromatography on an agarose column containing the following group.[120]

This diphosphonucleoside group fits to the active site of the enzyme

d. Separations Based on Molecular Size

In **dialysis**[121] and **ultrafiltration**,[122] a thin membrane, e.g., made of cellulose acetate (cellophane)

and containing holes 1–10 nm in diameter (typically 5 nm), is used as a semipermeable barrier. Small molecules pass through but large ones are retained. Dialysis depends upon diffusion (and can be hastened by adequate stirring) while ultrafiltration requires a pressure difference across the membrane. A more sophisticated procedure is **gel filtration.**[123] A column is packed with material such as a **cross-linked dextran** (Sephadex*) in the form of soft beads. The interior of the beads is a three-dimensional network of carbohydrate strands (Fig. 2-18). The interstices between strands (whose size depends upon the degree of cross-linking introduced chemically into the gel) are small enough to exclude large molecules but to admit smaller ones. If a mixture of materials of different molecular size is passed through such a column the smaller molecules are retarded because of diffusion into the gel, while the larger molecules pass through unretarded (Fig. 2-35). Sephadex G-25 excludes all but low molecular weight salts and compounds no larger than a simple sugar ring. Sephadex G-200, which is much less cross-linked, permits separation of macromolecules in the molecular weight range of 5000–200,000.

e. Centrifugation

Some of the most versatile separation methods make use of the preparative ultracentrifuge.[124,125] The sizes and shapes of molecules determine their **rates of sedimentation** in a given centrifugal field. Each molecule has its characteristic **sedimentation constant,** usually expressed in Svedberg units S. On the other hand, the *density* of a particle is a major factor in determining *where* it will come to rest if centrifugation is carried to equilibrium.

While cell fragments can be separated in a crude way by centrifuging at a series of different speeds (centrifugal fields) for fixed lengths of time (Section H,1,a), the cleanest results, both for organelles and molecules, are obtained by centrifugation in **density gradients.** For example, RNA can be separated into several fractions of differing sedimentation constants by first preparing a cen-

* Trade name for product of Pharmacia Fine Chemicals, Uppsala, Sweden.

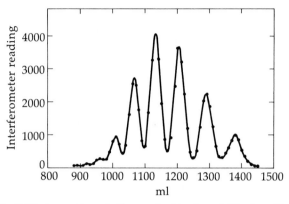

Fig. 2-35 Separation of oligosaccharides by gel filtration. The sugars dissolved in distilled water were passed through a column of Sephadex G-25. The peaks contain (right to left) glucose, cellobiose, cellotriose, etc. From P. Flodin and K. Aspberg, *Biol. Struct. Funct., Proc. IUB/IUBS Int. Symp., 1st., 1960* **1,** 345 (1961).

trifuge tube that contains sucrose ranging in concentration from 25% at the bottom to 5% at the top. If the solution of RNA is carefully layered on the top and the tube is centrifuged at a high speed for several hours, the different RNA fractions separate into slowly sedimenting sharp bands which are stabilized by the sucrose gradient. If the tube is punctured at the bottom and the contents are allowed to run dropwise into a series of test tubes in a **fraction collector** the positions and amounts of each RNA fraction can be determined.

If centrifugation is continued long enough, particles reach an equilibrium position in the density gradient. Cell organelles can be separated according to their density in a sucrose gradient at equilibrium (Chapter 1, Section B,6). Molecules, such as DNA are often separated in **CsCl gradients.**[126] The gradients in the dense salt solution are stable, and the sharpness of banding of particles is insured by use of a high centrifugal field. Single-stranded DNA may be separated from double-stranded DNA, and DNA's of differing G + C content can be separated. The latter separation is based on differences in buoyant densities ρ in CsCl which are approximately shown in Eq. 2-20.

$$\rho = 1.660 + 0.098 \ (\text{mole fraction C + G}) \qquad (2\text{-}20)$$

f. Electrical Charge

Several methods of separation depend directly upon the net charge which is carried by a molecule at the pH of the medium. The student can easily estimate this net charge for compounds containing various combinations of acidic and basic groups. Consider the pK_a of each group and the extent to which that group is dissociated at the selected pH. At some pH, the **isoelectric point,** a molecule will carry no net charge and will be immobile in an electric field. At any other pH it will move toward the anode (+) or cathode (−).

Electrophoresis, the process of separating molecules by migration in an electrical field, is conducted in many ways. A tiny sample of protein solution, perhaps of blood serum, is placed in a thin line on a piece of paper or cellulose acetate. The sheet is moistened with a buffer and electric current is passed through it. An applied voltage of a few hundred volts suffices to separate serum proteins in about an hour. To hasten the process and prevent diffusion of low molecular weight materials, high voltage electrophoresis has become popular. Two or three thousand volts are applied and the sample is cooled by water-chilled plates or by immersion in a tank of kerosene. Large-scale electrophoretic separations are conducted in beds of starch or of other gels. One of the most popular and sensitive methods for separation of proteins is electrophoresis in a column filled with **polyacrylamide gel.** The method, now highly developed, depends both upon electrical charge and molecular size and has been referred to as electrophoretic molecular sieving.[127,128]

In **isoelectric focusing,** a pH gradient is developed electrochemically in a vertical column between an anode and a cathode. The pH gradient is stabilized by the presence of a density gradient, often formed with sucrose, and the apparatus is thermostated at a very constant temperature. Proteins within the column migrate in one direction or the other until they reach the pH of their isoelectric point where they are "focused" into a narrow band. As little as 0.01 pH unit may separate two adjacent protein bands. The isoelectric points of the proteins under the usual experimental conditions are said to be close to those of their **isoionic points,** the pH values at which the proteins are isoelectric in the complete absence of added electrolyte.[129]

An extremely important separation method based upon electrical charge is **ion exchange chromatography.** Aqueous solutions are usually employed and the columns are packed with beads of **ion exchange resins,** porous materials containing bound ionic groups such as $-SO_3^-$, $-COO^-$, $-NH_3^+$, or quaternary nitrogen atoms. Synthetic resins based on a cross-linked polystyrene are usually employed for separation of small molecules. For larger molecules chemical derivatives of cellulose or of cross-linked dextrans (Sephadex) are more appropriate. Positively charged ions, such as amino acids in acidic solution, are placed on a cation exchange resin such as Dowex 50, which contains dissociated sulfonic acid groups. The adsorbed amino acids are eluted with HCl or with buffers of increasing pH. The procedure has been adapted for automatic quantitative analysis of amino acid mixtures obtained by hydrolysis of a protein or peptide (Fig. 2-36). Purine and pyrimidine bases can also be separated on sulfonated polystyrene. The negatively charged nucleotides are usually separated on a quaternary base-type resin.[130] The procedure has been developed on the ultramicro scale; for example, the nucleotides from a sample of rat liver mitochondria containing from 3 to 8 mg of protein can be determined quantitatively.[131]

2. Hydrolysis

Almost all of the biopolymers are inherently unstable with respect to cleavage to the monomer units by reaction with water. **Hydrolysis** can be catalyzed by protons, by hydroxyl ions, or by enzymes. Mechanisms are discussed in Chapter 7. Hydrolysis may be complete or partial and may be random or specifically directed at certain linkages in a polymer.

Complete hydrolysis of proteins is usually carried out by heating under N_2 with 6 N HCl at

BOX 2-C

ISOTOPES IN BIOCHEMICAL INVESTIGATIONS

Both stable[a] and radioactive[b–d] isotopes are widely used in chemical and biological investigations. The study of metabolism was revolutionized by the introduction of isotopic tracers. In one of the first biological experiments with the stable isotope [15]N (detected by mass spectrometry), Schoenheimer and associates in 1937 established the previously unsuspected turnover of protein in living tissues (Chapter 14, Section B). Calvin and associates first traced the pathway of carbon in photosynthesis using the radioactive [14]CO_2 (Box 11-A). In a similar way, ^{32}P and ^{35}S have served to elucidate the metabolism of phosphorus and sulfur; and tritium (^{3}H) has been used to label a variety of organic substances such as thymine. Radioactive isotopes provide the basis for sensitive analytical procedures such as **radioimmunoassays** of minute quantities of hormones (Box 16-A). Through **radioautography** they facilitate numerous analytical procedures (see accompanying photo) and provide the basis for important end-group methods used in sequence determination on polynucleotides (Fig. 2-37).

A change in isotopic mass, especially from ^{1}H to ^{2}H or ^{3}H, often produces a strong effect on reaction rates and the study of **kinetic isotope effects** has provided many insights into the mechanisms of enzymatically catalyzed reactions. Isotopes have permitted a detailed understanding of the stereochemistry of enzymatic reactions, an impressive example being the synthesis and use of chiral acetate (Chapter 7, Section J,2,g). These studies led to a Nobel Prize for J. W. Cornforth in 1975.[f] Specific isotopic properties provide the basis for nmr studies (Section I).

Several isotopes that have found important and widespread use in biochemistry are listed in the following table. For the radioactive isotopes, the half-life is given, as is the type of particle emitted, and the energy of the particle. Gamma rays, such as those given off in decay of ^{125}I or ^{131}I are very penetrating and easy to count precisely as is the energetic β radiation from [32]P. On the other hand, [3]H (tritium) is relatively difficult to detect[g]

but its weak β particle, which can travel only a short distance through a sample, makes it uniquely suitable for radioautography on a microscopic scale. The half-life (Eq. 6-4) determines the isotopic abundance needed to achieve a given radiation rate, a practical matter in providing a sufficient number of decays per minute to permit counting with an acceptably low statistical error. The amount of an isotope giving 3.7×10^{10} disintegrations per second (this is 1 g of pure radium, 0.3 mg of ^{3}H or 0.22 g of ^{14}C is known as the **curie** (Ci). One millicurie (mCi) provides 2.22×10^9 disintegrations/min (dpm). Radio-labeled substances ordinarily contain a relatively small fraction of the unstable isotope together with a larger number of unlabeled molecules. Compounds are usually sold in millicurie or microcurie quantities and with a stated specific activity as mCi mmol^{-1}. For example, a compound labeled at a single position with ^{3}H and having a specific activity of 50 mCi mmol^{-1} would contain about 0.17% [3]H at that position.

Isotope	Half-life	Energy of radiation (MeV)	
		β	γ
[2]H (deuterium)	Stable		
[3]H (tritium)	12.26 y	0.018	
[13]C	Stable		
[14]C	5760 y	0.159	
[15]N	Stable		
[18]O	Stable		
[22]Na	2.6 y	0.54	1.28
[32]P	14.3 d	1.17	
[35]S	87.2 d	0.167	
[36]Cl	4×10^5 y	0.716	
[40]K	1.4×10^9 y	1.4	1.5
[45]Ca	152 d	0.255	
[59]Fe	45 d	0.467	7.70
[65]Zn	250 d	0.32	1.14
[90]Sr	25 y	0.54	
[125]I	60 d	0.035	0.030
[131]I	8.07 d	0.606	0.365

Box 2-C (*Continued*)

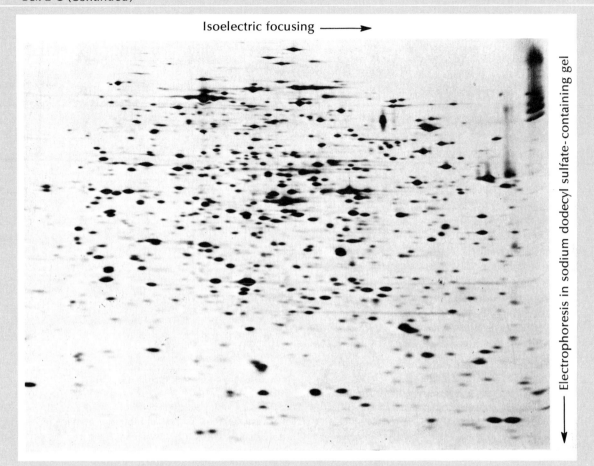

Isoelectric focusing ⟶

Electrophoresis in sodium dodecyl sulfate-containing gel

Radioautogram showing the separation of proteins of *Escherichia coli* labeled with ¹⁴C-amino acids. From O'Farrell.[e] Twenty-five μl of sample containing 180,000 cpm and ~10 μg of protein were subjected to isoelectric focusing (Section H,1,f) in a 2.5 × 130 mm tube containing polyacrylamide gel to separate proteins according to isoelectric point. The gel was then extruded from the column and was placed on one edge of a slab of polyacrylamide gel. Then SDS electrophoresis in the second dimension separated the proteins according to size. Over 1000 spots could be seen in the original radioautogram, which was obtained by placing a piece of photographic film over the gel slab and exposing it to the radiation for 875 hours. For details see the paper by O'Farrell.[e]

[a] N. A. Matwiyoff and D. G. Ott, *Science* **181**, 1125–1132 (1973).

[b] C. H. Wang, D. L. Willis, and W. D. Loveland, "Radiotracer Methodology in the Biological, Environmental, and Physical Sciences." Prentice-Hall, Englewood Cliffs, New Jersey, 1975.

[c] Y. Wang, ed., "Handbook of Radioactive Nuclides." The Chemical Rubber Co., Cleveland, Ohio, 1969.

[d] C. C. Thornburn, "Isotopes and Radiation in Biology." Butterworth, London, 1972.

[e] P. H. O'Farrell, *JBC* **250**, 4007–4021 (1075).

[f] J. W. Cornforth, *Science* **193**, 121–125 (1976).

[g] E. D. Bransome, Jr., ed., "Liquid Scintillation Counting." Grune and Stratton, New York, 1970.

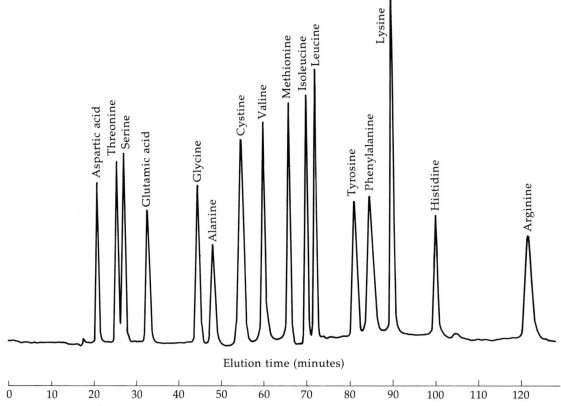

Fig. 2-36 Ion exchange chromatogram of a mixture of amino acids produced by a commercial amino acid analyzer. A mixture of amino acids (0.5 nmol each) was placed on an 0.9 cm (diameter) ×58 cm (length) column of micro beads of sulfonated polystyrene (Durrum type DC-4A) and was eluted with citrate buffers whose pH was varied in three steps from 3.5 to 6.4. The effluent solution was reacted continuously with fluorescamine (see p. 134) and the fluorescence of the product was recorded continuously. Courtesy of Durrum Chemical Corp.

about 110°C for 12–96 h. Some amino acids, especially tryptophan are destroyed. In fact, no procedure has been found which gives the ideal complete hydrolysis.* Complete base-catalyzed hydrolysis of proteins is possible, but extensive racemization of amino acids results.

Hydrolysis of nucleic acids is also catalyzed by strong acid; for example, heating at 100° C for an hour in 12 normal perchloric acid is sufficient to break nucleic acids down to their constituent bases. The N-glycosidic linkages of DNA are more labile than those of RNA and those to purines

more so than those to pyrimidines. A useful procedure based on these differences is to leave DNA overnight in the cold at pH 2 to cleave off all of the purine bases. The resulting polymer is known as an **apurinic** acid (Table 2-11).

In alkaline solutions RNA is hydrolyzed to a mixture of 2'- and 3'-nucleotides.

* Recently complete hydrolysis with 4-N-methanesulfonic acid containing 3-(2-aminoethyl)indole has been reported to give better results.[131a]

The mechanism involves participation of the free 2'-OH of the ribose groups and formation of cyclic 2',3'-phosphates and is similar to that of pancreatic ribonuclease (Chapter 7, Section E,2). Because deoxyribose lacks the free 2'-OH, DNA is stable in base.

Enzymatic methods of hydrolysis are especially useful because of the specificity that they sometimes provide. **Trypsin,** a so-called **endopeptidase,** cleaves peptide chains at a rapid rate only if the carbonyl group of the amide linkage cleaved is contributed by one of the basic amino acids, lysine or arginine. Thus, trypsin converts a protein into a relatively small number of **tryptic peptides** which may be separated and characterized. Trypsin acts only on denatured proteins, and to obtain good results the disulfide bridges must be broken first.

Other enzymes such as **chymotrypsin** and **pepsin** (Chapter 7, Section D,2) are less specific but may nevertheless be used to cut a peptide chain into smaller fragments whose structures can be determined. To establish the complete amino acid sequence for a protein, "overlapping" peptide fragments must be found that contain sequences from ends of two different tryptic fragments. In this way the tryptic peptides can be placed in the order in which they occurred in the native protein.

TABLE 2-11
Some Hydrolytic Cleavage Reactions of Polynucleotides[a]

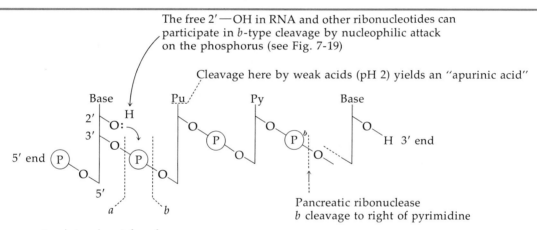

The free 2'—OH in RNA and other ribonucleotides can participate in b-type cleavage by nucleophilic attack on the phosphorus (see Fig. 7-19)

Cleavage here by weak acids (pH 2) yields an "apurinic acid"

Pancreatic ribonuclease
b cleavage to right of pyrimidine

A. Cleavage at points *a* is catalyzed
   1. Throughout the molecule by endonucleases
       Pancreatic deoxyribonuclease I
   2. Only at the 3' end by exonucleases
       Venom diesterase, nonspecific, attacks DNA and RNA. A free 3'-OH is essential

B. Cleavage at point *b* is catalyzed
   1. Randomly throughout the molecule by endonucleases and by bases (nonenzymatically)
       Pancreatic ribonuclease cleaves only to the right of a pyrimidine-containing nucleotide
       Ribonuclease T1 of *Aspergillus oryzae* cleaves to the right of a guanine-containing residue (3'-guanylate)
       Ribonuclease T2 of *Aspergillus oryzae* cleaves to the right of an adenine-containing residue (3'-adenylate)
       Pancreatic deoxyribonuclease (DNase) II
       Micrococcal DNase
   2. Only at the 5' end by exonucleases
       Bovine spleen phosphodiesterase hydrolyzes both polyribo- and polydeoxyribonucleotides

   [a] Reactions of both RNA and DNA are included.
   [b] The symbol Ⓟ in these locations stands for —PO_2^-—.

Peptide mapping or "fingerprinting" is a procedure that begins with cleavage of the disulfide linkages, denaturation, and digestion with an enzyme such as trypsin or pepsin. The sizes and amino acid compositions of the resulting series of peptides are characteristic of the protein under study. The mixture of peptides is placed on a sheet of paper and subjected to chromatography in one direction, then to electrophoresis in the other direction, the peptides separating into a characteristic pattern or fingerprint. Fingerprinting has been especially useful in looking for small differences in protein structure, for example, between genetic variants of the same protein (Fig. 4-20).

Several enzymes catalyze stepwise removal of amino acids from one or the other end of a peptide chain. **Carboxypeptidases** remove amino acids from the carboxyl terminal end, while **aminopeptidases** attack the opposite end. Using chromatographic methods, the amino acids released by these enzymes may be examined at various times and some idea of the sequence of amino acids at the chain ends may be obtained.

Complete enzymatic digestion of proteins can be accomplished with a mixture of enzymes including proteases produced by fungi (Pronase). However, the enzymes themselves attack each other, and enzymatic hydrolysis has in general not proved satisfactory for quantitative conversion of proteins into their constituent amino acids. A new procedure employs hydrolytic enzymes immobilized in a column of agarose gel. The protein to be hydrolyzed is passed through the gel and the constituent amino acids emerge from the bottom of the column.[132]

Exonucleases cleave nucleotides from the ends of polynucleotide chains while **endonucleases** cleave within chains. Some attack only single-stranded nucleic acids, while others attack double-stranded molecules. Some nucleases cut both strands of a DNA helix while others "nick" the molecule by cleaving just one strand. The specificities of a few nucleases are indicated in Table 2-11. Partial enzymatic hydrolysis of RNA provides short nucleotide sequences suitable for sequence determination. (The first RNA molecule sequenced was that of alanine tRNA. Cleavage was accomplished with pancreatic ribonuclease and with ribonuclease T.[133]) Two-dimensional polyacrylamide gel electrophoresis provides an excellent method of fingerprinting.[134]

While methods for determining nucleotide sequences of DNA developed more slowly they are now well established.[134a] The cleavage of DNA molecules at specific sequences (usually palindromic, Section D,11) is usually accomplished by **restriction endonucleases** (Chapter 15, Section F,1). The fragments may then be degraded by the acid depurination reaction and by exonucleases.[135] Electrophoresis on cellulose acetate at pH 3.5 followed by a special type of "homochromatography" on thin layer plates of DEAE-cellulose[136] yields excellent fingerprints.* Oligonucleotides carrying a radioactive label at one end may be sequenced simply by conducting a partial hydrolysis with an exonuclease (see subsequent paragraphs) followed by electrophoresis and homochromatography (Fig. 2-37). The smallest oligonucleotide moves the farthest during the homochromatography; each spot below it represents a compound one nucleotide residue longer than the preceding one. From the movement of each spot relative to the one above it during electrophoresis a tentative sequence can be deduced.[137,137a] Thus, in Fig. 2-37 the uppermost spot is an oligonucleotide of uncertain composition but the one below it contains an additional thymidylate (T) residue, resulting in a distinctly more rapid migration during electrophoresis at pH 3.5. Addition of a G unit has a lesser effect and addition of an A a very small effect; addition of C causes a decrease in electrophoretic mobility. The sequence can be read off directly from the fingerprint as (5')–TTATTAGCCAGCCGT–(3'). This deduction was confirmed by further studies utilizing

---

* Radioactive oligonucleotides are separated in a solvent containing a random mixture of unlabeled oligonucleotides (an RNA digest). During development of the plate, the unlabeled nucleotides displace the labeled nucleotides under study and migrate, approximately according to their size, to different positions on the plate.

Complex lipids may be cleaved by **lipases.** Pancreatic lipase removes the 1- and 3-acyl groups from triglycerides to leave 2-monoglycerides. The **phospholipases A,** present in tissues, in venom, and in bacteria, specifically remove an acyl group from either the 1 or 2 position of phospholipids. The **phosphodiesterases** known as phospholipases C and D, occurring in bacteria and plant tissues, respectively, cleave on the two sides of the phosphodiester linkage as indicated below.

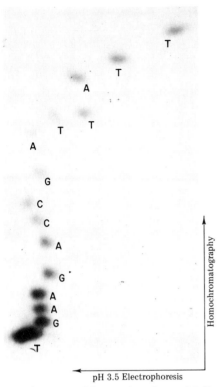

A mild base-catalyzed hydrolysis can also be used to specifically cleave the carboxylate esters (points A_1 and A_2 in diagram). The remaining phosphodiesters, amides (sphingolipids), and ethers can be separated and identified.

3. Some Special Chemical Cleavages

Numerous reagents have been discovered or have been designed specifically to assist the chemist in tackling structural problems. Only a few will be mentioned here.

a. Cleavage of Disulfide Bridges in Proteins

Before a polypeptide chain can be degraded it is usually necessary to break the disulfide bridges.[138–141] Three reactions for accomplishing this are shown in Eqs. 2-20 to 2-22. Oxidation with performic acid (Eq. 2-20) has been used on ri-

$$P_1\text{—}CH_2\text{—}S\text{—}S\text{—}CH_2\text{—}P_2 \xrightarrow{\overset{\overset{\textstyle O}{\|}}{H\text{—}C\text{—}O\text{—}OH}}$$

$$P_1\text{—}CH_2\text{—}SO_3^- + P_2\text{—}CH_2\text{—}SO_3^- \qquad (2\text{-}20)$$

bonuclease but is not often employed because the performic acid also oxidizes tryptophan residues. Cleavage by sulfite (Eq. 2-21) also has disadvan-

Fig. 2-37 Two-dimensional fingerprint of a partial digest of a ^{32}P-labeled oligonucleotide cut from a cDNA copy (Chapter 15, Section G,4) of a globin mRNA strand. The oligonucleotide was treated with snake venom exonuclease (phosphodiesterase, Table 2-11) which removes nucleotides one at a time from the 3' end. The lowest spot is an oligonucleotide of unspecified composition while each successive spot below it contains one additional nucleotide unit. The sequence (5')–TTATTAGCCAGAAGT–(3') was deduced from the relative mobilities of the intermediates (see text). From Proudfoot and Brownlee.[137]

depurination and nearest-neighbor analysis. Another recent approach to rapid sequencing of DNA is described by Sanger and Coulson.[137b]

A battery of special enzymes are available for attacking polysaccharides. In general, these enzymes are specific for a particular sugar that is joined in a given type of glycosidic linkage. Examples are the starch digesting enzymes. The α-amylases of saliva and of the pancreas cleave starch chains randomly, while the β-amylases of plants cleave maltose units from the ends of straight starch chains (Chapter 7, Section C,6).

$$P_1—CH_2—S—S—CH_2—P_2 + SO_3^{2-} + H^+ \longrightarrow$$

$$P_1—CH_2—S—SO_3^- + HS—CH_2—P_2 \qquad (2\text{-}21)$$

tages and the best of the three methods is probably the reduction by **dithiothreitol** or **dithioerythritol** (Eq. 2-22). These two dithiols cyclize to

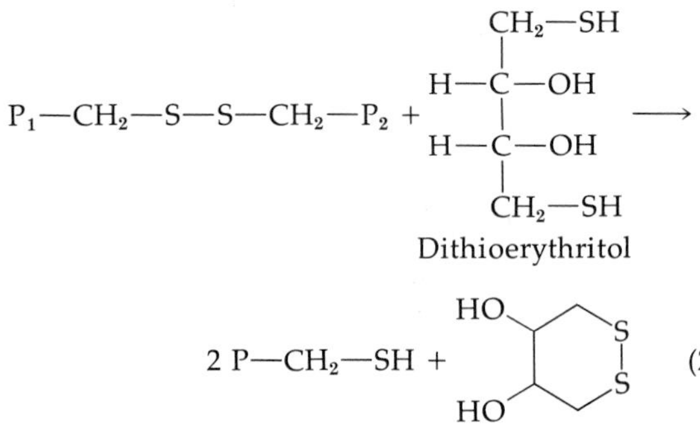

$$P_1—CH_2—S—S—CH_2—P_2 \; + \;
\begin{array}{c}
CH_2—SH \\
| \\
H—C—OH \\
| \\
H—C—OH \\
| \\
CH_2—SH
\end{array}
\longrightarrow$$

Dithioerythritol

$$2\,P—CH_2—SH \; + \; \text{(ring structure)} \qquad (2\text{-}22)$$

stable disulfides upon oxidation. Not only are the reagents useful in cleaving disulfide bridges in proteins but also they are widely used to protect SH groups in enzymes against accidental oxidation by oxygen. Mercaptoethanol, HS—CH$_2$—CH$_2$—OH, is used for the same purpose but is usually not as effective.

Once cleaved, disulfide bridges may be prevented from reforming by conversion of the resulting thiol groups to stable derivatives. Iodoacetate and iodoacetamide are commonly used for this purpose (Eq. 2-23).

$$P—CH_2—SH \; + \; ICH_2COO^- \longrightarrow$$
$$(ICH_2CONH_2)$$

$$P—CH_2—S—CH_2—COO^- + H^+ + I^- \qquad (2\text{-}23)$$
$$(P—CH_2—S—CH_2—CONH_2)$$

A better reagent is acrylonitrile (Eq. 2-24).

$$P—CH_2—SH \; + \; CH_2{=}CH—CN \longrightarrow$$

$$P—CH_2—S—CH_2CH_2—CN \qquad (2\text{-}24)$$

Acrylonitrile

b. Special Cleavage Reagents for Peptide Chains

Among the most useful protein reagents is **phenylisothiocyanate** whose use was developed by P. Edman. The reagent reacts with the N-terminal amino group of peptides (Eq. 2-25). The resulting adduct, because of the favorable geometry, undergoes a cyclization with cleavage of the peptide linkage (Eq. 2-25) under acid conditions. The re-

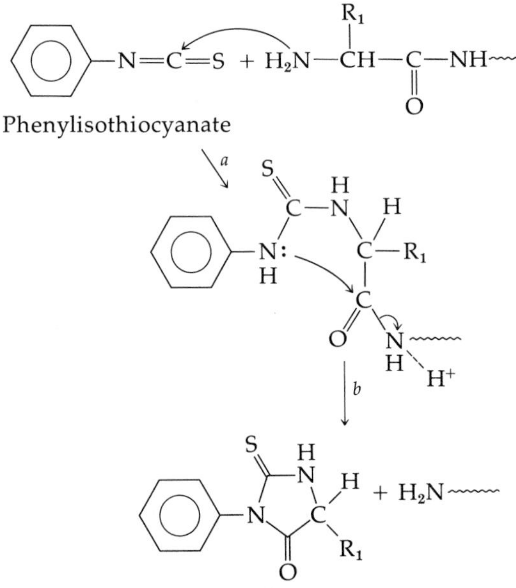

Phenylisothiocyanate

$$(2\text{-}25)$$

Phenylthiohydantoin (PTH)
of N-terminal amino acid

sulting **phenylthiohydantoin** of the N-terminal amino acid can be identified. The procedure can then be repeated on the shortened peptide chain to identify the amino acid residue in the 2 position. With careful work this Edman degradation can be carried on down the chain for several tens of residues. Indeed, a special machine, the "protein sequenator" does the whole thing automatically.

Many other reagents have been proposed for use in cleaving peptides at specific residues within the chain.[138] One has been outstandingly useful; cyanogen bromide, N≡C—Br cleaves adjacent to methionine residues. As indicated in Eq. 2-26, the sulfur of methionine displaces a bromide ion. Because of a favorable spatial relationship, the resulting sulfonium compound undergoes C—S bond cleavage through participation of the adjacent peptide group (Eq. 2-26). The C=N of the product is hydrolyzed with cleavage of the peptide chain (Eq. 2-26).

Peptidyl
homoserine lactone

The linkage Asp-Gly can often be cleaved specifically by treatment with hydroxylamine at high pH. The possible chemistry involved is discussed by Bornstein.[142]

4. Labeling Ends of Molecules

The Nobel Prize in chemistry was awarded in 1957 to F. Sanger for determination of the primary structure of insulin. Sanger had devoted 10 years of his life to this first sequence determination of a protein. His approach made use of partial hydrol-

ysis of peptide chains that had been labeled by reaction of free amino groups with **fluorodinitrobenzene** (Eq. 2-27). The linkage to the dinitro-

Dinitrophenyl-labeled
peptide (yellow)

phenyl group is stable to acid, and complete acid hydrolysis of the labeled peptide gave a yellow dinitrophenylated amino acid which originated from the N-terminal position. In addition ϵ-amino groups of lysine residues were labeled. Partial acid hydrolysis of the labeled peptides led to smaller yellow fragments whose amino acid composition was determined. Eventually, Sanger put the pieces of the jigsaw puzzle together to learn the sequences of the 21- and 30-residue chains which are linked by disulfide bridges to form the insulin molecule (Fig. 4-13). In recent years **dansyl chloride** has largely replaced fluorodinitroben-

Dansyl chloride
(5-Dimethylaminonaphthylsulfonyl chloride)

zene. The resulting peptide derivatives are brilliantly fluorescent, permitting "end group analysis" on much smaller amounts of peptide.

The SH groups in protein side chains may also be labeled in a number of ways. One of the most popular employs Ellman's reagent 5,5'-dithiobis(2-nitrobenzoic acid) DTNB) (Eq. 2-28).

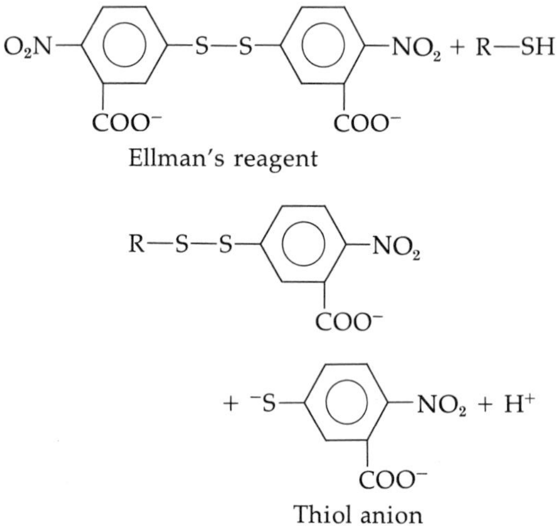

Ellman's reagent

Thiol anion

$$+ \, ^-S-\!\!\!\langle\!\!\langle \, \rangle\!\!\rangle\!-NO_2 + H^+ \qquad (2\text{-}28)$$

This reagent reacts quantitatively with —SH groups to form mixed disulfides with release of the thiol anion which absorbs light at 412 nm. This permits quantitative determination of the content of —SH groups in the protein. Additional disulfide exchange reactions can occur leading to reaction of an additional molecule of DTNB with cleavage of a protein disulfide linkage.[143] A variety of reagents are available for modification of other protein side chains.[138,143a]

Turning to carbohydrates, the classical end group method is **exhaustive methylation.** Repeated treatment with a methylating agent such as dimethylsulfate converts all free OH groups to OCH_3 groups. Complete acid hydrolysis, followed by separation of the methylated sugars and their quantitative determination, reveals the relative amounts of **end units** (containing four methoxyl groups), straight **chain units** (containing three methoxyl groups), and **branch points** (containing two methoxyl groups). Furthermore, the structures of the methylated derivatives provide information on the positions of the linkages in the sugar rings.

A relatively new method of labeling the reducing ends of carbohydrate chains is reduction of

the aldehyde groups with sodium borotritide (NaB^3H_4). The radioactive label so introduced can be used through radioautography of chromatograms to visualize fragments derived from the reducing end.

One of the most important reagents in investigations of carbohydrate structure is periodic acid (or sodium periodate). This reagent oxidatively cleaves C—C bonds bearing adjacent OH groups to form dialdehydes (Eq. 2-29). The method is

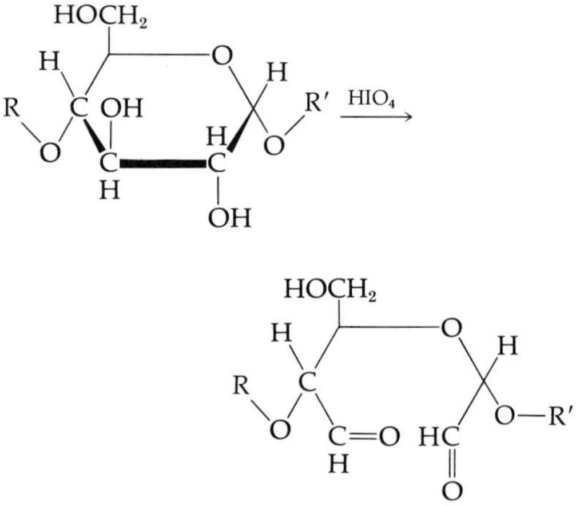

$$(2\text{-}29)$$

quantitative. After some hours of reaction, excess periodate can be destroyed with ethylene glycol, or the amount of periodate consumed in the oxidation can be determined. If three consecutive carbon atoms bear hydroxyl groups, formic acid is liberated from the central atom and can also be measured quantitatively. Furthermore, after destruction of excess periodate the dialdehyde can be reduced by addition of solid sodium borohydride to form stable CH_2OH groups. Following mild acid hydrolysis to split the acyclic acetal linkages, the fragments can be separated and identified. The sequence of reactions is known as the Smith degradation.[144]

An example of the combined use of several techniques in determining the structure of a complex polysaccharide is shown in Table 2-12.

End groups of **polyribonucleotides** can also be cleaved by periodate after any phosphate groups on the 3'-hydroxyls are removed by phospho-

TABLE 2-12
Determination of Structure of the 6-O-Methylglucose-Containing Lipopolysaccharide of *Mycobacterium phlei*[a,b]

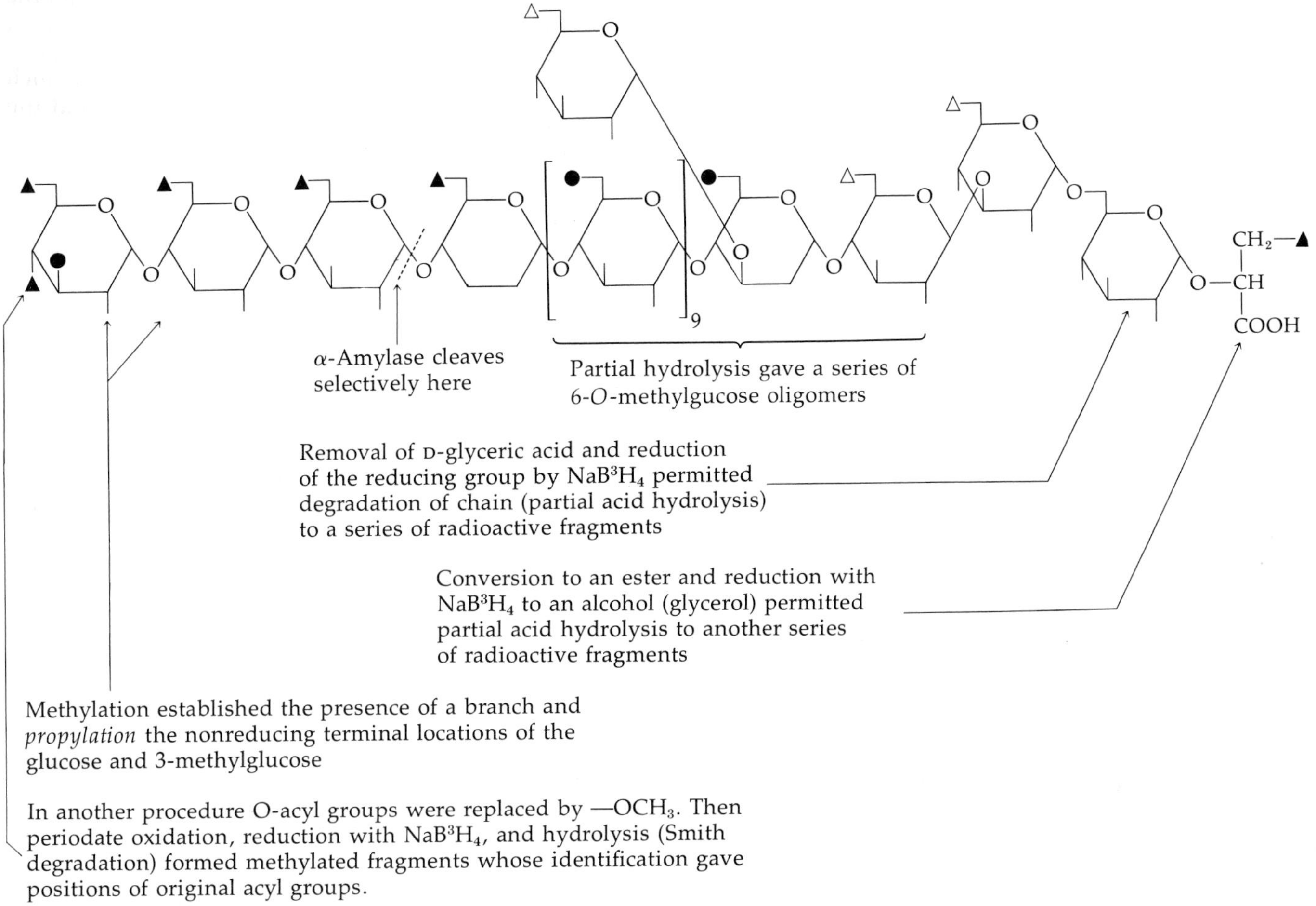

α-Amylase cleaves
selectively here

Partial hydrolysis gave a series of
6-O-methylgucose oligomers

Removal of D-glyceric acid and reduction
of the reducing group by NaB^3H_4 permitted
degradation of chain (partial acid hydrolysis)
to a series of radioactive fragments

Conversion to an ester and reduction with
NaB^3H_4 to an alcohol (glycerol) permitted
partial acid hydrolysis to another series
of radioactive fragments

Methylation established the presence of a branch and
propylation the nonreducing terminal locations of the
glucose and 3-methylglucose

In another procedure O-acyl groups were replaced by —OCH₃. Then
periodate oxidation, reduction with NaB^3H_4, and hydrolysis (Smith
degradation) formed methylated fragments whose identification gave
positions of original acyl groups.

---

[a] From M. H. Saier, Jr., and C. E. Ballou, *J. Biol. Chem.* **243,** 4332 (1968); G. R. Gray and C. E. Ballou, *J. Biol. Chem.* **247,** 8129–8135 (1972); W. L. Smith and C. E. Ballou, *J. Biol. Chem.* **248,** 7118–7125 (1973).
[b] Fatty acids are attached at positions designated ▲: 3 acetyl, 1 propionyl, 1 isobutyryl, and 1 octanoyl are present as well as up to 3 succinyl groups at positions marked △. Methyl groups are designated ●.

---

monoesterase. The dialdehydes formed can be reduced with radioactive NaB^3H_4. Another method of introducing a radioactive end group on polynucleotides involves the transfer of a radioactive phosphoryl group from ATP by the action of a nonspecific **polynucleotide kinase** to the free 5′-hydroxyl group of either a polyribo- or polydeoxyribonucleotide.[145]

An important technique in the study of DNA structure (developed by A. Kornberg and associates) is **nearest neighbor sequence analysis.**[146] Using a single radioactive ^{32}P-containing nucleoside triphosphate (deoxyadenosine triphosphate in the example in Eq. 2-30) together with the three other unlabeled nucleoside triphosphate precursors of DNA, a primer chain of DNA is elongated from the 3' end using DNA polymerase as a catalyst (Chapter 15, Section A,3,a). A single-stranded **template chain** of DNA (not shown in Eq. 2-30) directs placement of the proper nucleotide in

guanine gives the nearest neighbor frequencies for the adjacent pairs, TA, CA, and GA. Using the other radioactive nucleoside triphosphates one at a time in separate experiments, all of the nearest neighbor frequencies can be obtained. From such an experiment it was possible to deduce that the strands in the double helix were oriented in an antiparallel fashion, as predicted by Watson and Crick (see Davidson[146] for details). If the strands had been parallel, different nearest neighbor frequencies would have been observed.

In a similar way but using alkaline b-type cleavage, nearest neighbor analysis of RNA synthesized on a DNA primer has been made.[145]

$$(2\text{-}30)$$

the next position at the growing end in such a way that the final product is a Watson–Crick duplex. The chemistry of this synthetic reaction is explained later (Chapter 15, Section A,3,a), but it is not important to understand it here. The important fact is that the incorporation of ^{32}P occurs in the bridge phosphates which connect the nucleotide originally carrying the ^{32}P to the 3' position of the neighboring nucleotide.

The ^{32}P-containing product of the reaction is next cleaved with a mixture of micrococcal DNase and spleen phosphodiesterase (Table 2-11) to give fragments as indicated in Eq. 2-30. Since these hydrolytic enzymes promote only b-type cleavage (Table 2-11), the ^{32}P will now be attached to what was the nearest neighbor to adenosine in the DNA. Measurement of the radioactivity in each of the 3'-nucleotides of thymine, cytosine, and

5. Detecting Products

Important to almost all biochemical activity is the ability to detect, and to measure quantitatively, tiny amounts of specific compounds. "Color reagents," which develop characteristic colors with specific compounds are most popular. For example, **ninhydrin** (Box 8-F) can be used as a "spray reagent" to detect a small fraction of a micromole of an amino acid or peptide in a spot on a chromatogram. It can also be used for a quantitative determination, the color being developed in a solution. Similarly, phenols and concentrated sulfuric acid convert sugars (in solution or on chromatograms) to red pigments. This is the basis for a general colorimetric method for carbohydrates. Reducing sugars are specifically located on chromatograms by spraying with silver nitrate.

Nucleotides as well as many other light absorbing materials can be quantitatively determined by their characteristic light absorption (Figs. 13-11 and 13-12).[146] Even more sensitive than ultraviolet absorption is fluorescence. For example, 3 pmol (1 ng) of riboflavin can be detected in a spot on a thin layer chromatogram (Fig. 2-34).[147] A newly developed reagent, "fluorescamine," reacts with any primary amine to form a highly fluorescent product. As little as 50 pmol of amino acid can be determined quantitatively (Fig. 2-36).[148]

"Fluorescamine"

$$+ RNH_2 \longrightarrow$$

(2-31)

(fluorescent)

Extremely sensitive methods are available for detecting compounds leaving vapor-phase chromatographic columns. Flame ionization detectors can measure as little as a few picomoles of almost any substance. The importance of developing new, more sensitive analytical methods by which the quantity of material investigated can be scaled down can hardly be overemphasized. Increasingly sensitive methods of detection, including mass spectrometry, now permit measurement of fmol quantities in some cases. Analysis of the output of neurotransmitters from single neurons in the brain may be critically important in understanding the functions of nerve cells, and the ability to analyze the contents of single cells may be crucial in elucidating many aspects of the biochemistry of higher organisms.

A useful variation on the usual thin layer or paper chromatographic technique is **diagonal chromatography.** A sample is chromatographed in one direction. Then some chemical reaction is carried out on the thin layer plate or paper and chromatography is carried out in the second direction. Compounds unmodified by the chemical treatment all lie on the diagonal across the plate, but those modified are usually found at other places.

The technique has been used for studying photochemical[149] and other[147] reactions of flavins. A similar diagonal electrophoresis has been used to identify the pairs of —SH groups forming disulfide linkages in proteins.[150] Peptide fragments containing S—S bridges are subjected to paper electrophoresis. Then the paper is exposed to performic acid vapor to cleave the bridges according to Eq. 2-20 and electrophoresis is conducted in the second direction and the paper is sprayed with ninhydrin. The spots falling off the diagonal are those that participated in S—S bridge formation. They can be associated in pairs from their positions on the paper and can be identified with peptides characterized during standard sequencing procedures on performic acid oxidized protein.

6. Determination of Molecular Weight

The evaluation of the molecular weight of a biopolymer is often of critical importance. Minimum molecular weights can frequently be computed from the content of a minor constituent (e.g., the tryptophan content of a protein, the iron content of hemoglobin). However, physicochemical techniques provide the basis for most molecular weight measurements.[151,152] Measurements of **osmotic pressure** or **light scattering** provide determinations of molecular weight that are simple in principle, but which have their pitfalls. The most reliable methods depend upon the **ultracentrifuge.** A straightforward determination of molecular weight is obtained by centrifuging until an equilibrium distribution of the polymer molecules is obtained.[125,151] Using short cells, this **sedimentation equilibrium** can be attained in a few hours. The molecular weight is computed from the observed variation in the concentration from the center to the periphery of the centrifuge cell.

The ultracentrifuge is more frequently used to measure the sedimentation constant s (Section H,1,e). The value of s depends not only upon the molecular weight but also upon the density and the shape of the molecule. However, to the extent that protein molecules can be regarded as spheres, s will increase with molecular weight approximately as $MW^{2/3}$. A plot of log s against log M should be a straight line. Figure 2-38 shows such a plot for a number of proteins. Note that the plots for nucleic acids (which in many instances can be approximated as rods rather than spheres) fall on a different line from that of proteins. Furthermore, the sedimentation constant falls off more rapidly with increasing molecular weight than it should for spheres.

Several new methods of molecular weight determination rival ultracentrifugation. One is a simple gel filtration. A column of gel such as Sephadex is prepared carefully and is calibrated by passing a series of protein solutions through it. The volume V_e at which a protein peak emerges from the column is divided by the elution volume V_0 which is observed for a very large particle that is completely excluded from the gel. The ratio V_e/V_0 is plotted against log MW (Fig. 2-39A) for a series of reference proteins of known molecular weights. As with estimation of molecular weights from sedimentation constants, it is assumed that all the molecules are roughly spherical and that the molecular weight of an unknown protein can be estimated from its position on the graph.[153,154] Another modification of the method depends upon chromatography in a high concentration of the denaturing salt **guanidinium chloride.** The assumption is made that proteins generally are denatured into a random coil conformation in such a solvent.[154]

NH₂ structure

H₂N—C   Cl⁻
       \
        NH₂
Guanidinium chloride

Probably the most important of the new methods of molecular weight determination is gel electrophoresis in the presence of the denaturing detergent **sodium dodecyl sulfate** (SDS). In an SDS solution, the protein molecules are not only denatured, but also all appear to be more or less evenly

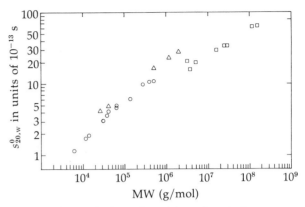

Fig. 2-38 Plots of the logarithm of the sedimentation constant s against the logarithm of the molecular weight for a series of proteins and nucleic acids: (○) globular proteins, (△) RNA, and (□) DNA. Proteins include a lipase (milk), cytochrome c, ribonuclease (pancreatic), lysozyme (egg white), follicle-stimulating hormone, bacterial proteases, human hemoglobin, prothrombin (bovine), malate dehydrogenase, γ-globulin (horse), tryptophanase (*E. coli*), glutamate dehydrogenase (chicken), and cytochrome a. Double-stranded DNA molecules are those of bacteriophage φX174 (replicative form), T7, λ_b2, T₂, and T₄, and that of a papilloma virus. The RNA molecules are tRNA, rRNA, and mRNA of *E. coli*, and that of turnip yellow mosaic virus. Sources: "CRC Handbook of Biochemistry." Chem. Rubber Publ. Co., Cleveland, Ohio, 1968; S. P. Colowick and N. O. Kaplan, eds., "Methods in Enzymology," Vol. 12B, pp. 388–389. Academic Press, New York, 1968.

coated with detergent.[155] The resulting rodlike molecules generally show a uniform dependence of electrophoretic mobility on molecular weight (plotted as log MW). An example is shown in Fig. 2-39B. Again, the molecular weight of the protein under investigation is determined by comparison of its rate of migration with that of a series of "marker proteins."[153,156]

7. Conformations of Macromolecules

Many of the physical methods mentioned in the preceding section measure quantities that depend upon both the molecular weight and shape of molecules. Thus, the same methods may provide some information about conformation. However, two other methods stand out as especially important: **X-ray diffraction** studies of crystalline compounds and **nuclear magnetic resonance** (nmr) measurements on compounds in solution.

a. X-Ray Diffraction

Soon after the discovery of X-rays, the diffraction patterns obtained by passing an X-ray beam through crystals were used to measure interatomic distances and to determine crystal structure. The approach has been refined to the point that structures of relatively complex organic compounds can often be obtained from diffraction patterns within a few days. Indeed, the biggest problem in using X-ray diffraction is often the preparation of suitable nearly perfect crystals of adequate size.

In recent years the use of X-ray diffraction has been extended to determination of structures of proteins and nucleic acids. Precise measurements of bond distances and angles with small reference compounds have provided standard bond lengths and angles that can be assumed to hold in the more complex polymeric structures. Determination of protein and nucleic acid structures has depended upon the use of these standard bond distances and angles in model building together with the diffraction data (Fig. 4-19B) on the polymers.[71]

b. Nuclear Magnetic Resonance (nmr)[157-162]

The absorption of electromagnetic radiation of radiofrequencies by magnetic atomic nuclei provides the basis for nmr spectroscopy. All nuclei with odd mass numbers (e.g., ^1H, ^{13}C, ^{15}N, ^{17}O,

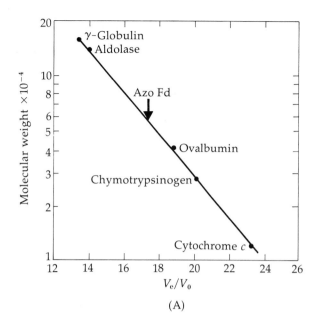

(A)

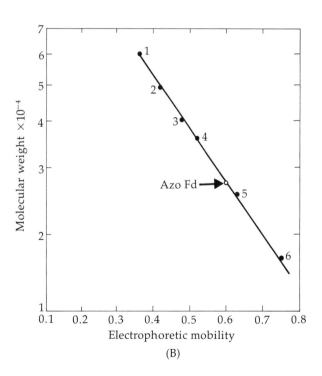

(B)

Fig. 2-39 (A) Molecular weight estimation of "native" azoferredoxin by Sephadex G-200 filtration. The solid circles indicate the position of peak midpoints for proteins of known molecular weights. The arrow indicates the position of azoferredoxin. Each protein peak was found in 5–7 tubes. The column flow rate was 3.2 ml/h. Fractions of 1.6 ml were collected by use of a fraction collector. (B) Estimation of the molecular weight of polypeptide chain of azoferredoxin using SDS-polyacrylamide electrophoresis; from a set of four standard curves. The marker proteins are (1) catalase, (2) fumarase, (3) aldolase, (4) glyceraldehyde-phosphate dehydrogenase, (5) α-chymotrypsinogen A, and (6) myoglobin. (○) indicates position of azoferredoxin. (A) and (B) reprinted with permission from G. Nakos and L. Mortenson, *Biochemistry* **10**, 457 (1971). Copyright by the American Chemical Society.

^{19}F, and ^{31}P), as well as those with an even mass number but an odd atomic number, have magnetic properties; however, ^{2}H, ^{12}C, and ^{16}O are nonmagnetic. Absorption of a quantum of energy $E = h\nu$ occurs only when the nuclei are in the strong magnetic field of the nmr spectrometer. For a 100 megahertz (100 MHz) spectrometer (for which $\nu = 100$ MHz) the energy of a quantum is only $E = 6.625 \times 10^{-34} \times 10^8$ J $= 0.04$ J mol^{-1}, five orders of magnitude less than the average energy of thermal motion of molecules (3.7 kJ mol^{-1}). Thus, the spin transitions induced in the nmr spectrometer have no significant effect on the chemical properties of molecules.

The resonance frequency ν at which absorption occurs in the spectrometer is given by Eq. (2-32)

$$\nu = \mu H_0/h \qquad (2\text{-}32)$$

where H_0 is the strength of the external magnetic field, μ is the magnetic moment of the nucleus being investigated, and h is the Planck's constant.

In principle, an nmr spectrum can be obtained by holding the magnetic field H_0 constant and observing the values of the frequency at which absorption occurs, just as is done for ultraviolet visible, and infrared spectra (Chapter 13). In practice, nmr spectrometers operate at a fixed radiofrequency, most often 60, 100, and 220 MHz. The most powerful nmr spectrometers built operate at over 300 MHz. To scan a spectrum, the magnetic field established by the powerful electromagnet of the spectrometer is changed at a uniform rate. Nevertheless, nmr spectra are always reported in terms of **frequency shifts** in hertz (cycles per second), as if it were the frequency of the oscillator that had been changed.

The **proton nmr (pmr)** spectrum of the coenzyme pyridoxal phosphate in ^{2}H$_2$O (deuterium oxide) is shown in Fig. 2-40 as obtained with a 60 MHz spectrometer (see also Korytnyk and Ahrens[163]). Four things can be measured from such a spectrum: (1) The **intensity** (area under the band). In a proton nmr spectrum, areas are usually proportional to the numbers of equivalent protons giving rise to absorption bands. (2) The **chemical shift,** the difference in frequency between the peak observed for a given proton and a peak of some standard reference compound. In the case of Fig. 2-40 the reference peak at the right of the spectrum is that of Tier's salt (sodium 3-trimethylsilyl 1-propane sulfonate). (3) The **width** at half-height (in hertz). For molecules in rapid motion in solution the width for protons is ~ 0.1 Hz. (4) **Coupling constants** which measure interactions between nearby magnetic nuclei.

With a magnetic field of $H_0 = 14,000$ G and a 60 MHz oscillator, the position of proton bands in organic compounds is spread over a range of approximately 700 Hz. The positions are always measured in terms of a shift from some standard whose peak appears at the high energy end of the spectrum. For protons this is most often **tetramethylsilane (TMS)** an inert substance that can be added directly to the sample in its glass tube. Since biochemists most often use the nonmagnetic ^{2}H$_2$O as solvent, the water-soluble Tier's salt is frequently used as a standard. Its position is insignificantly different from that of TMS.

When nmr spectra are measured in ^{2}H$_2$O, it is often desirable to give the "pD" of the medium. This is usually taken as the pH meter reading plus 0.40.

In a molecule like TMS, the electrons surrounding the nuclei "shield" the nucleus so that it does not experience the full external magnetic field. For this reason, absorption occurs at a high frequency (high energy). Protons that are bound to a carbon atom or other atom deficient in electrons (because of attachment to electron withdrawing atoms or groups) are **deshielded.** The greater the deshielding, the further **downfield** from the TMS position is the nmr peak. In the spectrum shown in Fig. 2-40, the 2-methyl protons appear 146 Hz below the Tier's salt peak but still at a relatively high field. Aromatic rings lead to strong deshielding of attached protons because of a **ring current** induced in the circulating π electrons. Thus, the peaks of the 5-methylene protons which are adjacent to the aromatic ring occur 303 and 310 Hz downfield. The 6-H, which is bound directly to the ring, is more strongly deshielded and appears 463 Hz downfield. The 4'-hydrogen

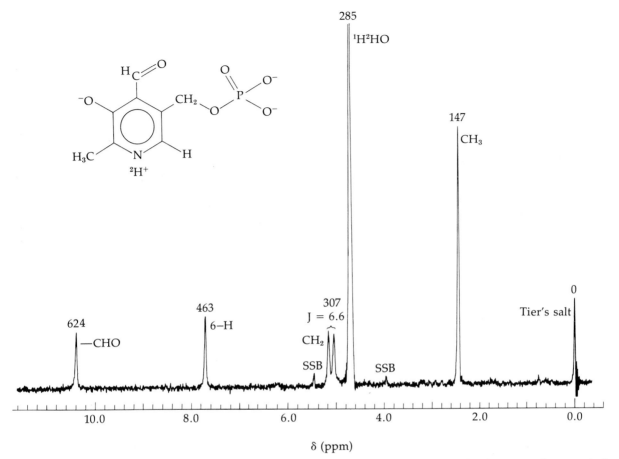

Fig. 2-40 The 60 MHz proton magnetic resonance spectrum of pyridoxal 5'-phosphate at neutral pH (pD = 7.05). The internal standard is Tier's salt. Chemical shifts in hertz are indicated beside the peaks, and the δ scale is shown below the spectrum. SSB marks the two *spinning side bands* that arise from inhomo- geneities in the magnetic field and are spaced symmetrically about the H²HO peak. The sample tube is spun rapidly about its axis during the measurements to average inhomogeneities in the magnetic field. The side bands arise because such averaging is incomplete. Spectrum courtesy of John Likos.

of the aldehyde group is deshielded as a result of a similar "diamagnetic electronic circulation" in the carbonyl group. Its peak is even further downfield.

The magnitude of a chemical shift may be expressed directly in hertz, as a shift from the position of the reference compound. When expressed in this way the chemical shift increases as the magnetic field of the spectrometer increases. Thus, the shift is greater for a 100 MHz than for a 60 MHz spectrometer. To make the chemical shift independent of the instrument used, the frequency independent unit δ is often cited (Eq. 2-33).

$$\delta \text{ (ppm)} = \frac{\Delta\nu \text{ (Hz)} \times 10^6}{\nu \text{ (Hz) of oscillator}} \qquad (2\text{-}33)$$

The value of δ is the shift in frequency relative to frequency of the oscillator in parts per million and is independent of the field strength. However, it still depends upon use of a particular reference standard which must be stated when a δ value is given. Somewhat less frequently used is the τ scale (Eq. 2-34).

$$\tau = 10 - \delta \text{ (relative to TMS)} \qquad (2\text{-}34)$$

The energy of the spin transition of a hydrogen nucleus is strongly influenced by the local presence of other magnetic nuclei, e.g., other protons. This **spin–spin interaction (coupling)** leads to a splitting of nmr bands of protons into two or more closely spaced bands. Thus, the ethyl group often appears in nmr spectra as a "quartet" of four peaks at evenly spaced intervals arising from the CH_2 group and a triplet of peaks arising from the CH_3 protons. While the protons attached to the same carbon do not ordinarily split each other's peaks, the protons on the neighboring carbon do. These neighboring protons can be in either of the two spin states, a fact that results in easily measured differences in the energy of the nmr transition under consideration.[157–161]

The **coupling constant** J is the difference in hertz between the successive peaks in a multiplet. It is a field independent quantity and the same no matter what the frequency of the spectrometer. In Fig. 2-40 the peak of the 5'-methylene protons is split by ^1H-^{31}P coupling, with a value of $J \approx 7$ Hz. While spin–spin coupling is most pronounced when magnetic nuclei are close together in a structure, the effect can sometimes be transmitted through up to five covalent bonds.

When spin coupling is suspected, the technique of **double irradiation** or **spin decoupling** is frequently employed. The sample is irradiated at the resonance frequency of one of the nuclei involved in the coupling while the spectrum is observed in the frequency region of the other nucleus of the coupled pair. Under these conditions the multiplet collapses into a singlet and the mutual coupling of the two nuclei is established.

The coupling constant between two protons attached to adjacent carbon atoms (or other atoms) depends upon the torsion angle ϕ (Eq. 2-35).

$$J_{H,H'} \approx A \cos^2 \phi + B \cos \phi + C \qquad (2\text{-}35)$$

The Karplus equation (Eq. 2-35)* was predicted on theoretical grounds, but the constants A, B, and C are empirical.[164] In the following approximate equations (Eqs. 2-36 and 2-37)[164] the middle term of Eq. 2-35 has been omitted:

$$H—C—C—H' \quad J = 17 \cos^2 \phi + 1.1 \qquad (2\text{-}36)$$

$$H—C—N—H' \quad J = 12 \cos^2 \phi + 0.2 \qquad (2\text{-}37)$$

$$H—C—O—H' \quad J = 10 \cos^2 \phi - 1.0 \qquad (2\text{-}38)$$

Since the second term of these equations amounts to 1 Hz or less, single-term Karplus equations are frequently used.

The torsion angle between the hydrogen atoms on the α carbon of an amino acid residue and on the adjacent nitrogen of a peptide linkage (ϕ', not ϕ)† can also be observed by nmr and can be used to establish whether the peptide linkages are cis or trans:

$$J = 8.5 \cos^2 \phi' \text{ for } \phi' \text{ between } 0° \text{ and } 90°$$
$$= 9.5 \cos^2 \phi' \text{ for } \phi' \text{ between } 90° \text{ and } 180°$$

Nuclear magnetic resonance measurements have been used for determination of the conformation of small peptides and polyesters[165–167] and in determination of the structure of carbohydrates[168] and nucleotides.

A useful technique is to measure an nmr spectrum in the presence of paramagnetic **lanthanide** ions such as those of **europium** (Eu) and **praseodymium** (Pr).[169,169a] These ions induce strong shifts in the positions of many nmr peaks, and it is possible to develop empirical relationships helpful in establishing structures.

Proteins give extraordinarily complex pmr spectra, but a considerable amount of success has been achieved in understanding them.[170–175] Figure 2-41 shows the pmr spectrum of the enzyme **ribonuclease** at 60 and 220 MHz. Note the increased resolution obtained with the higher frequency instrument and the increased sharpness of many

---

* The equation can also be expressed as $J \approx A' + B' \cos \phi + C' \cos 2\phi$ [M. Karplus, *JACS* **85**, 2870–2871 (1963)]. Since $\cos 2\phi = 2 \cos^2 \phi - 1$, this is equivalent to Eq. 2-35 but the constants A', B', and C' are quite different from A, B, and C. Forms containing $\sin^2 \phi$ are also seen.

† If ϕ (Section B,3) = 180°, $\phi' \approx +120°$.

bands in the 220 MHz spectrum of the enzyme denatured by heating (at 72.5°C). The latter observation can be interpreted to mean that the envi-

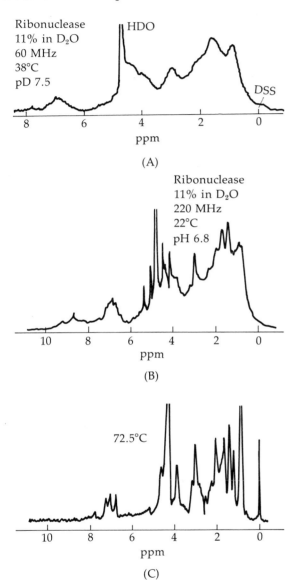

Fig. 2-41 Proton magnetic resonance spectra of ribonuclease: (A) at 60 MHz, (B) at 220 MHz, and (C) at 220 MHz on heat denatured enzyme. Concentration: 11% in ^2H_2O, pD in (A) was 7.5 and in (B) was 6.8. The standard is Tier's salt. Reprinted with permission from C. C. MacDonald and W. D. Phillips, *J. Am. Chem. Soc.* **89**, 6333 (1967). Copyright by the American Chemical Society.

ronments of all the side chain groups of one type within a protein become essentially equivalent as a result of denaturation. In line with this interpretation, it has been shown that pmr spectra of "random coil" proteins can be predicted well from tables of standard chemical shifts for the individual amino acids.[171]

As a rule, pmr spectra of proteins are run in D_2O because the H_2O absorption band obscures much of the spectrum. However, at the far downfield end ($\delta > 10$) there are some weak peaks arising from NH protons of imidazole side chains that can be observed in H_2O (Fig. 2-42). The spectra at 3 pH values show a change in position of the lowest energy peak from 12.9 to 11.1 ppm as the pH is raised from 4.4 to 8.9. At the intermediate pH of 5.31 the peak is in an intermediate position, closer to that at pH 4.4 than that at pH 5.8. When the chemical shift is plotted as a function of pH an S-shaped titration curve (Chapter 4, Section C) is obtained from which it is possible to estimate pK_a values for dissociable groups. In this case the histidine (which is believed to function at the active center of the enzyme, see Chapter 7, Section E,2) was estimated to have a pK_a of 5.8.

Proton nmr spectra (up to 360 MHz) of tRNA molecules have permitted identification of as many as 26 resonances which may be assigned to the 20 N—H—N hydrogen-bonded protons in the base pairs (see Fig. 2-24) and to protons involved in tertiary hydrogen bonding interactions (e.g., in Hoogsteen base triplets, Section D,6).[176]

c. Carbon-13 Nuclear Magnetic Resonance (cmr)[177-180]

Use of the magnetic nucleus ^{13}C in nmr developed slowly because the low natural abundance of this isotope made it difficult to obtain spectra. Another complication was the occurrence of ^{13}C—^1H coupling involving the many protons normally present in organic compounds. A striking improvement in cmr spectra was obtained when "wide band proton decoupling" (**noise decoupling**) was developed. With a natural abundance of only 1.1%, ^{13}C is rarely present in a molecule at adjacent positions. Thus, ^{13}C—^{13}C coupling does not intro-

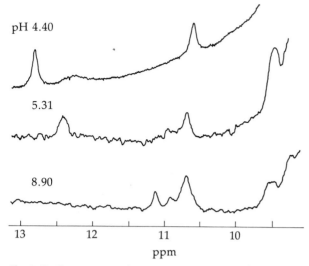

pH 4.40

5.31

8.90

13   12   11   10

ppm

Fig. 2-42 Proton magnetic resonance spectra of ribonuclease A in H_2O–0.1 M NaCl, 22°, at pH 4.4, 5.3, and 8.9. Different numbers of scans were averaged to obtain each spectrum using the Varian C1024 computer. Reprinted with permission from J. H. Griffen et al., Biochemistry **12**, 2097 (1973). Copyright by the American Chemical Society.

duce any complexities and in a noise-decoupled natural abundance cmr spectrum each carbon atom gives rise to a single peak. Even so, ^{13}C nmr spectroscopy did not become practical until the development of pulsed **Fourier transform** (FT) spectrometers.[181] In such instruments a strong pulse of radio frequency radiation is delivered over a period of a few microseconds and absorption is observed at all frequencies simultaneously. Although 1–2 s must be allowed before the next pulse is delivered, the equivalent of one complete ordinary nmr spectrum is obtained each 1–2 s. By attaching the instrument to a computer and allowing repeated spectra to accumulate over a period of minutes, hours, or a day or more, it is possible to obtain precise cmr spectra.

Chemical shifts in cmr spectra are often 100 ppm or more relative to TMS. The effects of substituents attached to a carbon atom are often additive when two or more substituents are attached to the same atom.[177]

The cmr spectrum in Fig. 2-43 is that of the high affinity iron-complexing agent of E. coli, **entero-**

bactin, a cyclic trimer of 2,3-dihydroxy-N-benzoyl-L-serine, whose structure is shown in Fig. 2-44. Besides illustrating use of nmr spectra in structure determination, the report from which Fig. 2-43 is taken[167] presents results of considerable biochemical interest. The complexing of a trivalent ion with enterobactin induces a large change in the overall conformation of the molecule. While the angle ψ is almost the same in the free enterobactin and in the gallium chelate, ϕ changes from 60° to −150°, and the sidechain angle χ changes from 162° to 60°. The amide linkage in free enterobactin is almost planar, but in the chelate the torsion angle ω has changed to −133°.

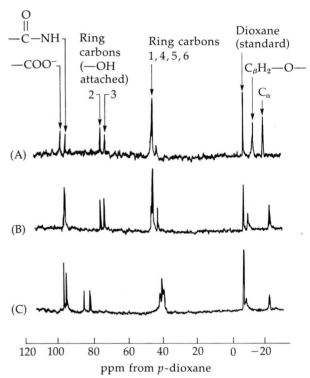

Fig. 2-43 Carbon-13 nmr proton noise-decoupled spectra of (A) enterobactin monomer (2,3-dihydroxy-N-benzoyl-L-serine), (B) free enterobactin, and (C) Ga^{3+}-enterobactin. Compounds were dissolved in $(CD_3)_2SO$ at ~50°. Chemical shifts are referred to internal p-dioxane. From 17,000 to 49,000 spectra were accumulated. Reprinted with permission from M. Llinas et al., Biochemistry **12**, 3840 (1973). Copyright by the American Chemical Society.

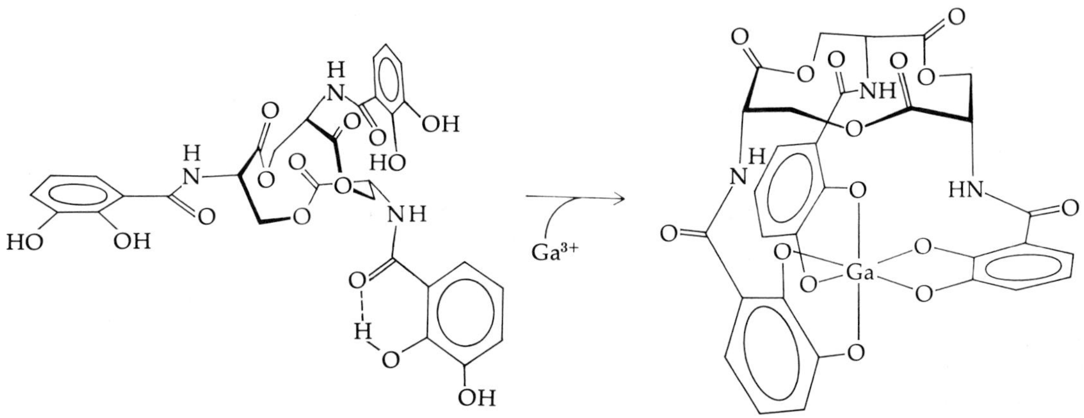

$\omega \approx 180°$
$\phi \approx 60°$
$\psi \approx 100°$
$\chi \approx 162°$

$\omega \approx 133°$
$\phi \approx -150°$
$\psi \approx 100°$
$\chi \approx 60°$

Fig. 2-44 The structure of the iron-complexing agent entero-bactin of *E. coli*, together with the structure of its complex with the gallium ion Ga^{3+}.[167]

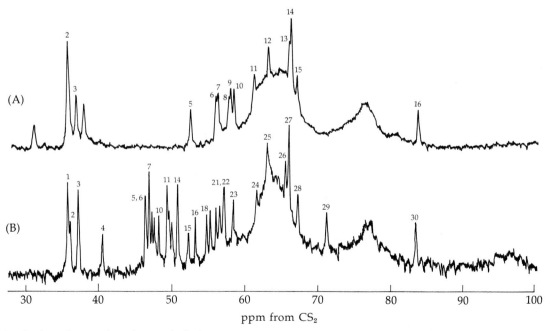

Fig. 2-45 Region of aromatic carbon and C^ζ of arginine residues in the proton decoupled natural abundance ^{13}C nmr spectrum of horse heart cytochrome *c* at pH 6.7, 41°C recorded at 15.18 MHz. (A) Ferricytochrome *c* (14.4 m*M*) after 46,000 accumulations (14 h). (B) Ferrocytochrome *c* (11.5 m*M*) after 16,384 accumulations. From Oldfield and Allerhand.[178]

Carbon-13 nmr is also being applied to proteins. Figure 2-45 shows just a portion of the cmr spectrum of **cytochrome c** (Chapter 10, Section B,5). Many of the resonances can be assigned to individual carbon atoms, either in aromatic amino acids or in the porphyrin ring. Note the striking changes in the spectrum accompanying reduction of the iron from the ferric to the ferrous state.

d. Phosphorus-31 Nuclear Magnetic Resonance

Phosphorus nmr has only 1/15 the sensitivity of proton nmr, but its utility in biochemistry is considerable. Of special interest is the ability to measure intracellular pH by observing the chemical shift of the orthophosphate (P_i) signal.[182]

Phosphorus-31 nmr can also be used to follow the course of chemical reactions of phosphorylated substrates within cells.[183-185] Recently natural abundance nitrogen-15 nmr has also been applied to proteins and other biochemical substances.[186]

e. Optical Methods for Investigating Conformations

Changes in the absorption of ultraviolet light by aromatic amino acid side chains and by purine and pyrimidine bases (Fig. 2-28) are often interpreted in terms of conformational changes of polymers. Other important techniques are infrared spectroscopy, Raman spectroscopy, fluorescence, and circular dichroism spectroscopy, which are discussed in Chapter 13.

REFERENCES

1. R. S. Cahn, *J. Chem. Educ.* **41**, 116–125 (1964).
2. R. Bentley, "Molecular Asymmetry in Biology," Vol. 1, pp. 49–56. Academic Press, New York, 1969.
3. J. B. Lambert, *Sci. Am.* **222**, 58–70 (Jan 1970).
4. G. Strahs, *Adv. Carbohydr. Chem. Biochem.* **25**, 53–107 (1970).
5. S. J. Angyal, *Angew. Chem., Int. Ed. Engl.* **8**, 157–166 (1969).
6. W. W. Pigman and D. Horton, eds., "The Carbohydrates," 2nd ed., Vol. 1A, pp. 56–57. Academic Press, New York, 1972.
7. W. P. Jencks, "Catalysis in Chemistry and Enzymology," pp. 4 and 7. McGraw-Hill, New York, 1969.
8. D. J. Cram and G. S. Hammond, "Organic Chemistry," 2nd ed., p. 221. McGraw-Hill, New York, 1964.
9. D. E. Metzler, C. M. Harris, R. J. Johnson, D. B. Siano, and J. A. Thomson, *Biochemistry* **12**, 5377–5392 (1973).
10. N. K. Kochetkov and E. I. Budovskii, "Organic Chemistry of Nucleic Acids," Part A, pp. 137–147. Plenum, New York, 1971.
11. M. Daniels, *PNAS* **69**, 2488–2491 (1972).
12. A. Rich, *in* "Biophysical Science, a Study Program" (J. L. Oncley, ed.), pp. 50–60. Wiley, New York, 1959. This article also appeared in *Rev. Mod. Phys.* **31**, 50–60 (1959).
13. R. E. Dickerson and I. Geis, "The Structure and Action of Proteins." Benjamin, New York, 1969.
14. G. N. Ramachandran, A. V. Lakshminarayanan, and A. S. Kolaskar, *BBA* **303**, 8–13 (1973).
15. G. N. Ramachandran and A. S. Kolaskar, *BBA* **303**, 385–388 (1973).
16. R. Huber and W. Steigemann, *FEBS Lett.* **48**, 235–237 (1974).
17. I. L. Karle, *Biochemistry* **13**, 2155–2162 (1974).
18. IUPAC-IUB Commission on Biochemical Nomenclature, *Biochemistry* **9**, 3471–3479 (1970).
19. R. E. Dickerson and I. Geis, "The Structure and Action of Proteins," pp. 33–43. Benjamin, New York, 1969.
20. R. D. B. Fraser, and J. M. Gillespie, *Nature (London)* **261**, 650–654 (1976).
21. E. Suzuki, W. G. Crewther, R. D. B. Fraser, T. P. MacRae, and N. M. McKern, *JMB* **73**, 275–278 (1973).
22. C. Cohen and K. C. Holmes, *JMB* **6**, 423–432 (1963).
23. J. Sodek, R. S. Hodges, L. B. Smillie, and L. Jurasek, *PNAS* **69**, 3800–3804 (1972).
23a. F. H. C. Crick, *Acta Crystallogr.* **6**, 689–697 (1953).
24. F. M. Richards, *JMB* **82**, 1–14 (1974).
25. C. Chothia, *Nature (London)* **254**, 304–308 (1975).
26. F. A. Quiocho, P. H. Bethge, W. N. Lipscomb, J. F. Studebaker, R. D. Brown, and S. H. Koenig, *Cold Spring Harbor Symp. Quant. Biol.* **36**, 561–567 (1971).
27. M. Buehner, G. C. Ford, D. Moras, K. W. Olsen, and M. G. Rossmann, *JMB* **90**, 25–49 (1974).
28. K. Nagano, *JMB* **75**, 401–420 (1973).
29. P. Y. Chou and G. D. Fasman, *Biochemistry* **13**, 222–245 (1974).
30. T. T. Wu and E. A. Kabat, *JMB* **75**, 13–31 (1973).
31. G. E. Schulz, *et al. Nature (London)* **250**, 140–141 (1974).
31a. M. Levitt and C. Chothia, *Nature (London)* **261**, 552–558 (1976).
32. R. H. Kretsinger and C. E. Nockolds, *JBC* **248**, 3313–3326 (1973).
33. C. B. Anfinsen, *Science* **181**, 223–230 (1973).
34. H. R. Mahler and E. H. Cordes, "Biological Chemistry," 2nd ed., pp. 235–245. Harper, New York, 1971.
34a. C. B. Anfinsen and H. A. Scheraga, *Adv. Protein Chem.* **29**, 205–300 (1975).

35. R. D. Guthrie, "Guthrie and Honeyman's Introduction to Carbohydrate Chemistry," 4th ed., p. 17. Oxford Univ. Press (Clarendon), London and New York, 1974.

35a. D. French, "Chemistry and Biochemistry of Starch," MTP Int. Ser. Sci. Vol. 5, pp. 267–335, Butterworths, London, 1975.

36. A. White, P. Handler, and E. L. Smith, "Principles of Biochemistry," 5th ed., pp. 52–58. McGraw-Hill, New York, 1973.

37. R. W. Jeanloz, *in* "The Carbohydrates" (W. Pigman and D. Horton, eds.), 2nd ed., Vol. 2B, pp. 589–625. Academic Press, New York, 1970.

38. M. E. Silva and C. P. Dietrich, *JBC* **250**, 6841–6846 (1975).

38a. J. Woodhead-Galloway and D. W. L. Hukins, *Endeavour* **35**, 73–78 (1976).

38b. L. Rosenberg, W. Hellmann and A. K. Kleinschmidt, *JBC* **250**, 1877–1883 (1975).

39. R. Montgomery, *in* "The Carbohydrates" (W. Pigman and D. Horton, eds.), 2nd ed., Vol. 2B, pp. 627–709. Academic Press, New York, 1970.

40. E. C. Heath, *Annu. Rev. Biochem.* **40**, 29–56 (1971).

41. R. G. Spiro, *Annu. Rev. Biochem.* **39**, 599–638 (1970).

42. R. D. Marshall, *Annu. Rev. Biochem.* **41**, 673–702 (1972).

43. E. L. Stern, B. Lindahl, and L. Rodén, *JBC* **246**, 5707–5715 (1971).

43a. P. Hallgren, A. Lundblad, and S. Svensson, *JBC* **250**, 5312–5314 (1975).

44. Y. C. Lee and J. R. Scocca, *JBC* **247**, 5753–5758 (1972).

45. A. L. DeVries, *Science* **172**, 1152–1155 (1971).

46. J. R. Vandenheede, A. I. Ahmed, and R. E. Feeney, *JBC* **247**, 7885–7889 (1972).

47. B. Pessac and V. Defendi, *Science* **175**, 898–900 (1972).

48. V. S. R. Rao, N. Yathindra and P. R. Sundararajan, *Biopolymers* **8**, 325–333 (1969).

49. B. K. Sathyanarayana and V. S. R. Rao, *Biopolymers* **11**, 1379–1394 (1972).

50. D. A. Rees, *Adv. Carbohydr. Chem. Biochem.* **24**, 267–332 (1969).

51. B. J. Poppleton and A. McL. Mathieson, *Nature (London)* **219**, 1046–1048 (1968).

52. R. E. Rundle and D. French, *JACS* **65**, 1707–1710 (1943).

53. D. French, *J. Anim. Sci.* **37**, 1048–1061 (1973).

54. A. Hybl, R. E. Rundle, and D. E. Williams, *JACS* **87**, 2779–2788 (1965).

55. K. Kainuma and D. French, *Biopolymers* **11**, 2241–2250 (1972).

56. D. French, *Dempun Kagaku* **19**, 8–25 (1972).

57. S. Arnott, A. Fulmer, W. E. Scott, I. C. M. Dea, R. Moorehouse and D. A. Rees, *JMB* **90**, 269–284 (1974).

58. S. Arnott, W. E. Scott, D. A. Rees, and C. G. A. McNab, *JMB* **90**, 253–267 (1974).

59. D. A. Rees, *BJ* **126**, 257–273 (1972).

59a. S. Kirkwood, *Ann. Rev. Biochem.* **43**, 401–417 (1974).

60. R. D. Preston, *Sci. Am.* **218**, 102–108 (June 1968).

61. E. D. T. Atkins and J. K. Sheehan, *Science* **179**, 562–564 (1973).

62. S. Arnott, J. M. Guss, D. W. L. Hukins, and M. B. Mathews, *Science* **180**, 743–745 (1973).

63. D. M. Brown, *in* "Basic Principles in Nucleic Acid Chemistry" (P. O. P. Ts'o, ed.), Vol. 2, pp. 1–90. Academic Press, New York, 1974.

64. M. Ya. Feldman, *Prog. Nucleic Acid Res. Mol. Biol.* **13**, 1–49 (1973).

65. S. Arnott and D. W. L. Hukins, *JMB* **81**, 93–105 (1973).

66. W. Saenger, *Angew Chem., Int. Ed. Engl.* **12**, 591–601 (1973).

67. C. Altona and M. Sundaralingam, *JACS* **94**, 8205–8212 (1972).

68. D. Voet and A. Rich, *Prog. Nucleic Acid Res. Mol. Biol.* **10**, 183–265 (1970).

69. S. Arnott, *Prog. Biophys. Mol. Biol.* **21**, 265–319 (1970).

70. P. O. P. Ts'o, ed., "Basic Principles in Nucleic Acid Chemistry," Vol. 2, pp. 305–469. Academic Press, New York, 1974.

71. S. Arnott, D. W. L. Hukins, S. D. Dover, W. Fuller, and A. R. Hodgson, *JMB* **81**, 107–122 (1973).

71a. J. M. Rosenberg, N. C. Seeman, and R. O. Day, *BBRC* **69**, 979–987 (1976).

72. S. H. Kim, F. L. Suddath, G. J. Quigley, A. McPherson, J. L. Sussman, A. H. J. Wang, N. C. Seeman, and A. Rich, *Science* **185**, 435–440 (1974).

73. J. D. Robertus, J. E. Ladner, J. T. Finch, D. Rhodes, R. S. Brown, B. F. C. Clark, and A. Klug, *Nature (London)* **250**, 546–551 (1974).

73a. A. Rich and U. L. Raj Bhandary, *Annu. Rev. Biochem.* **45**, 805–860 (1976).

74. S. H. Kim, J. L. Sussman, F. L. Suddath, G. J. Quigley, A. McPherson, A. H. J. Wang, N. C. Seeman, and A. Rich, *PNAS* **71**, 4970–4974 (1974).

75. S. Arnott and P. J. Bond, *Nature (London), New Biol.* **244**, 99–101 (1973).

75a. E. De Clercq, P. F. Torrence, P. De Somer, and B. Witkop, *JBC* **250**, 2521–2531 (1975).

76. I. Tinoco, Jr., P. N. Borer, B. Dengler, M. D. Levine, O. C. Uhlenbeck, D. M. Crothers, and J. Gralla, *Nature (London), New Biol.* **246**, 40–41 (1973).

77. C. E. Singer and B. N. Ames, *Science* **170**, 822–826 (1970); **175**, 1393 (1972).

78. J. N. Davidson, "The Biochemistry of the Nucleic Acids," 7th ed., pp. 204–205. Academic Press, New York, 1972.

79. C. Krasuski, H. Hayashi, and K. Nakanishi, *Fed. Proc.* **31**, 444 (abstr.) (1972).

80. D. R. Helinski and D. B. Clewell, *Annu. Rev. Biochem.* **40**, 899–942 (1971).

81. J. Vinograd, J. Lebowitz, and R. Watson, *JMB* **33**, 173–197 (1968).

82. A. M. Campbell and D. J. Jolly, *BJ* **133**, 209–226 (1973).

82a. W. Keller, *PNAS* **72**, 2550–2554 (1975).

83. W. J. Pigram, W. Fuller, and M. E. Davies, *JMB* **80**, 361–365 (1973).

84. L. F. Liu and J. C. Wang, *BBA* **395**, 405–412 (1975).

85. E. J. Gabbay, K. Sanford, C. S. Baxter, and L. Kapicak,

Biochemistry **12,** 4021–4029 (1973).

86. E. J. Gabbay, R. E. Scofield, and C. S. Baxter, *JACS* **95,** 7850–7857 (1973).

86a. W. Keller, *PNAS* **72,** 4876–4880 (1975).

86b. M. Shure and J. Vinograd, *Cell* **8,** 215–226 (1976).

86c. R. E. Depew and J. C. Wang, *PNAS* **72,** 4275–4279 (1975).

87. G. Felsenfeld, *Proced. Nucleic Acid Res.* **2,** 233–244 (1971).

88. D. E. Kennell, *Prog. Nucleic Acid Res. Mol. Biol.* **11,** 259–301 (1971).

89. K. Borre and N. Szybalski, *in* "Methods in Enzymology" (L. Grossman and K. Moldave, eds.), Vol. 21D, pp. 350–383. Academic Press, New York, 1971.

90. A. J. Bendich and E. T. Bolton, *in* "Methods in Enzymology" (L. Grossman and K. Moldave, eds.), Vol. 12B, pp. 635–640. Academic Press, New York, 1968.

91. D. Gillespie, *in* "Methods in Enzymology" (L. Grossman and K. Moldave, eds.), Vol. 12B, pp. 641–668. Academic Press, New York, 1968.

92. R. J. Britten and D. E. Kohne, *Carnegie Inst. Washington, Yearb.* **65,** 78–106 (1967).

93. R. J. Britten and D. E. Kohne, *Science* **161,** 529–540 (1968).

94. R. J. Britten and J. Smith, *Carnegie Inst. Washington, Yearb.* **68,** 378–391 (1970).

94a. J. G. Wetmur, *Ann. Rev. Biophys. Bioeng.* **5,** 337–361 (1976).

95. D. A. Wilson and C. A. Thomas, Jr., *JMB* **84,** 115–144 (1974).

95a. H. M. Sobell, *Ann. Rev. Biophys. Bioeng.* **5,** 307–335 (1976).

96. W. W. Christie, "Lipid Analysis; Isolation, Preparation, Identification and Structural Analysis of Lipids," Pergamon, Oxford, 1973.

97. C. Hitchcock and B. W. Nichols, "Plant Lipid Biochemistry," Academic Press, New York, 1971.

98. M. Kates, *in* "Laboratory Techniques in Biochemistry and Molecular Biology" (T. S. Work and E. Work, eds.), Vol. 3, Part II. North-Holland Publ., Amsterdam, 1972.

99. M. I. Gurr and A. T. James, "Lipid Biochemistry, an Introduction," Cornell Univ. Press, Ithaca, New York, 1971.

100. K. Okamura and R. H. Marchessault, *in* "Conformation of Biopolymers" (G. N. Ramachandran, ed.), Vol. 2, p. 709–720. Academic Press, New York, 1967.

101. H. J. M. Bowen, "Trace Elements in Biochemistry," Academic Press, New York, 1966.

102. E. Frieden, *Sci. Am.* **227,** 52–60 (Jul 1972).

103. W. G. Hoekstra, J. W. Suttie, H. E. Ganther, and W. Mertz, eds., "Trace Element Metabolism in Animals—2," Univ. Park Press, Baltimore, Maryland, 1974.

104. G. N. Ling, "A Physical Theory of the Living State," Ginn (Blaisdell), Boston, Massachusetts, 1962.

105. P. O. Wester, *BBA* **109,** 268–283 (1965).

106. G. D. Birnie and S. M. Fox, eds., "Subcellular Components," Butterworth, London, 1969.

107. H. R. Mahler and E. H. Cordes, "Biological Chemistry," 2nd ed., pp. 441–457. Harper, New York, 1971.

108. For several methods of preparation of mitochondria, see R. W. Estabrook and M. E. Pullman, eds., "Methods in Enzymology," Vol. 10. Academic Press, New York, 1967.

109. For preparation of plasma membranes and the Golgi apparatus, see W. B. Jakoby, ed., "Methods in Enzymology," Vol. 22, pp. 99–148. Academic Press, New York, 1971.

110. For isolation of lysosomes, see J. T. Dingle, "Lysosomes, A Laboratory Handbook," North-Holland Publ., Amsterdam, 1972.

111. J. N. Davidson, "The Biochemistry of Nucleic Acids," 7th ed., pp. 73–82. Academic Press, New York, 1972.

112. L. Grossman and K. Moldave, eds., "Methods in Enzymology," Vol. 12A, pp. 531–708. Academic Press, New York, 1967.

113. K. S. Kirby, *in* "Methods in Enzymology" (L. Grossman and K. Moldave, eds.), Vol. 12B, pp. 87–99. Academic Press, New York, 1968.

114. M. I. Gurr and A. T. James, "Lipid Biochemistry, an Introduction." Cornell Univ. Press, Ithaca, New York, 1971.

115. C. J. O. R. Morris and P. Morris, "Separation Methods in Biochemistry," Wiley (Interscience), New York, 1963.

116. P. Cuatrecasas and C. B. Anfinsen, *in* "Methods in Enzymology" (W. B. Jakoby, ed.), Vol. 22, pp. 345–378. Academic Press, New York, 1971.

117. W. B. Jakoby and M. Wilchek, eds., "Methods in Enzymology," Vol. 34. Academic Press, New York, 1974.

118. J. Porath, *Nature (London)* **218,** 834–838 (1968).

119. J. K. Inman, *in* "Methods in Enzymology" (W. B. Jakoby and M. Wilchek, eds.), Vol. 34, pp. 30–58. Academic Press, New York, 1974.

120. P. Cuatrecasas, M. Wilchek, and C. B. Anfinsen, *PNAS* **61,** 636–643 (1968).

121. P. McPhie, *in* "Methods in Enzymology" (W. B. Jakoby, ed.), Vol. 22, pp. 23–32. Academic Press, New York, 1971.

122. W. F. Blatt, *in* "Methods in Enzymology" (W. B. Jakoby, ed.), Vol. 22, pp. 39–49. Academic Press, New York, 1971.

123. J. Reiland, *in* "Methods in Enzymology" (W. B. Jakoby, ed.), Vol. 22, pp. 287–321. Academic Press, New York, 1971.

124. G. B. Cline and R. B. Ryel, *in* "Methods in Enzymology" (W. B. Jakoby, ed.), Vol. 22, pp. 168–204. Academic Press, New York, 1971.

125. T. J. Bowen, "An Introduction to Ultracentrifugation," Wiley (Interscience), New York, 1970.

126. J. N. Davidson, "The Biochemistry of Nucleic Acids," 7th ed., p. 132. Academic Press, New York, 1972.

127. L. Shuster, *in* "Methods in Enzymology" (W. B. Jakoby, ed.), Vol. 22, pp. 412–433. Academic Press, New York, 1971.

128. A. Chrambach and D. Rodbard, *Science* **172,** 440–451 (1971).

129. O. Vesterberg, *in* "Methods in Enzymology" (W. B. Jakoby, ed.), Vol. 22, pp. 389–412. Academic Press, New York, 1971.

130. W. E. Cohn, *in* "Methods in Enzymology" (S. P. Colowick and N. O. Kaplan, eds.), Vol. 3, pp. 724–743. Academic Press, New York, 1957.

131. H. W. Heldt and M. Klingenberg, *in* "Methods in Enzymology" (R. W. Estabrook and M. E. Pullman, eds.), Vol. 10, pp. 482–487. Academic Press, New York, 1967.

131a. R. J. Simpson, M. R. Neuberger, and T.-Y Liu, *JBC* **251**, 1936–1940 (1976).

132. C. C. Q. Chin and F. Wold, *Anal. Biochem.* **61**, 379–391 (1974).

133. R. W. Holley *et al., Science* **147**, 1462–1465 (1965).

134. C. Weissman, M. A. Billeter, H. M. Goodman, J. Hindley, and H. Weber, *Annu. Rev. Biochem.* **42**, 303–328 (1973).

134a. R. Wu, R. Bambara, and E. Jay, *CRC Crit. Rev. Biochem.* **2**, 455–512 (1975).

134b. S. Mandeles, "Nucleic Acid Sequence Analysis." Columbia Univ. Press, New York, 1972.

135. E. B. Ziff, J. W. Sedat, and F. Galibert, *Nature (London), New Biol.* **241**, 34–37 (1973).

136. G. G. Brownlee and F. Sanger, *EJB* **11**, 395–399 (1969).

137. N. J. Proudfoot and G. G. Brownlee, *Nature (London)* **252**, 359–362 (1974).

137a. V. Ling, *JMB* **64**, 87–102 (1972).

137b. F. Sanger and A. R. Coulson, *JMB* **94**, 441–448 (1975).

138. R. Barker, "Organic Chemistry of Biological Compounds," Prentice-Hall, Englewood Cliffs, New Jersey, 1971.

139. Yu. M. Torchinskii, "Sulfhydryl and Disulfide Groups of Proteins," Nauka, Moscow, 1971 (in Russian); Plenum, New York, 1973 (in English).

140. M. Friedman, "The Chemistry and Biochemistry of the Sulfhydryl Group in Amino Acids, Peptides and Proteins," Pergamon, Oxford, 1973.

141. P. C. Jocelyn, "Biochemistry of the SH Group," Academic Press, New York, 1972.

142. P. Bornstein, *Biochemistry* **9**, 2408–2421 (1970).

143. R. J. Ackerman and J. F. Robyt, *Anal. Biochem.* **50**, 656–659 (1972).

143a. G. E. Means and R. E. Feeney, "Chemical Modification of Proteins" Holden-Day, San Francisco, California, 1971.

144. J. H. Sloneker, D. G. Orentas, C. A. Knutson, P. R. Watson, and A. Jeanes, *Can. J. Chem.* **46**, 3353–3361 (1968).

145. M. Székely, *Proced. Nucleic Acid Res.* **2**, 780–795 (1971).

146. J. N. Davidson, "The Biochemistry of the Nucleic Acids," 7th ed., pp. 242–245, 294, and 295. Academic Press, New York, 1972.

147. G. E. Treadwell, W. L. Cairns, and D. E. Metzler, *J. Chromatogr.* **35**, 376–388 (1968).

148. S. Udenfriend, S. Stein, P. Böhlen, W. Dairman, W. Leimgruber, and M. Weigele, *Science* **178**, 871–872 (1972); see also J. R. Benson and P. E. Hare [*PNAS* **72**, 619–622 (1975)] for an even more sensitive method.

149. S. Svobodová, I. M. Hais, and J. V. Koštíř, *Chem. Listy* **47**, 205–212 (1953).

150. J. R. Brown and B. S. Hartley, *BJ* **101**, 214–228 (1966).

151. K. E. Van Holde, "Physical Biochemistry," Prentice-Hall, Engelwood Cliffs, New Jersey 1971.

152. H. R. Mahler and E. H. Cordes, "Biological Chemistry," 2nd ed., pp. 75–91. Harper, New York, 1971.

153. G. Nakos and L. Mortenson, *Biochemistry* **10**, 455–458 (1971).

154. K. G. Mann and W. W. Fish, *in* "Methods in Enzymology" (C. H. W. Hirs and S. N. Timasheff, eds.), Vol. 26C, pp. 28–42. Academic Press, New York, 1972.

155. D. K. Igou, J.-T. Lo, and D. S. Clark, W. L. Mattice, and E. S. Younathan, *BBRC* **60**, 140–145 (1974).

156. K. Weber, J. R. Pringle, and M. Osborn, *in* "Methods in Enzymology" (C. H. W. Hirs and S. N. Timasheff, eds.), Vol. 26C, pp. 1–27. Academic Press, New York, 1972.

156a. D. M. Collins, F. A. Cotton, E. E. Hazen, Jr., E. F. Meyer, Jr., and C. N. Morimoto, *Science* **190**, 1047–1053 (1975).

157. J. D. Roberts, "Nuclear Magnetic Resonance," McGraw-Hill, New York, 1959.

158. G. C. K. Roberts and O. Jardetzky, *Adv. Protein Chem.* **24**, 447–545 (1970).

159. O. Jardetzky and N. G. Wade-Jardetzky, *Annu. Rev. Biochem.* **40**, 605–634 (1971).

160. F. A. Bovey, "Nuclear Magnetic Resonance Spectroscopy," Academic Press, New York, 1969.

161. J. W. Emsley, J. Feeney, and L. H. Sutcliffe, "High Resolution Nuclear Magnetic Resonance Spectroscopy," Vols. 1 and 2. Pergamon, Oxford, 1965, 1966.

162. T. L. James, "Nuclear Magnetic Resonance in Biochemistry: Principles and Applications," Academic Press, New York, 1975.

163. W. Korytnyk and H. Ahrens *in* "Methods in Enzymology" (D. B. McCormick and L. D. Wright, eds.), Vol. 18A, pp. 475–483. Academic Press, New York, 1970.

164. M. Barfield and M. Karplus, *JACS* **91**, 1–10 (1969).

165. A. E. Tonelli, *Biochemistry* **12**, 689–692 (1973).

166. F. A. Bovey, *in* "Chemistry and Biology of Peptides" (J. Meienhofer, ed.), pp. 3–28. Arbor Sci. Publ., Ann Arbor, Michigan, 1972.

167. M. Llinás, D. M. Wilson, and J. B. Neilands, *Biochemistry* **12**, 3836–3843 (1973).

168. P. L. Durette and D. Horton, *Adv. Carbohydr. Chem. Biochem.* **26**, 49–125 (1971).

169. W. DeW. Horrocks, Jr. and J. P. Sipe, III, *JACS* **93**, 6800–6804 (1971).

169a. M. R. Willcott, III, and R. E. Davis, *Science* **190**, 850–857 (1975).

170. C. C. McDonald and W. D. Phillips, *JACS* **89**, 6332–6341 (1967).

171. C. C. McDonald and W. D. Phillips, *JACS* **91**, 1513–1521 (1969).

172. J. S. Cohen, W. R. Fisher, and A. N. Schechter, *JBC* **249**, 1113–1118 (1974).

173. J. H. Griffin, J. S. Cohen, and A. N. Schechter, *Biochemistry* **12**, 2096–2099 (1973).

174. J. L. Markley, *Biochemistry* **14,** 3546–3554 and 3554–3561 (1975).

174a. K. Würthrich, "NMR in Biological Research: Peptides and Proteins", North-Holland, Amsterdam, 1976.

174b. H. Shindo, M. B. Hayes and J. S. Cohen, *JBC* **251,** 2644–2647 (1976).

175. M. B. Hayes, H. Hagenmaier, and J. S. Cohen, *JBC* **250,** 7461–7472 (1975).

176. B. R. Reid and G. T. Robillard, *Nature* (*London*) **257,** 287–291 (1975).

177. F. A. L. Anet and G. C. Levy, *Science* **180,** 141–148 (1973).

178. E. Oldfield and A. Allerhand, *PNAS* **70,** 3531–3535 (1973).

179. E. Oldfield, R. S. Norton, and A. Allerhand, *JBC* **250,** 6381–6402 (1975).

180. E. Oldfield and A. Allerhand, *JBC* **250,** 6403–6407 (1975).

181. E. D. Becker and T. C. Farrar, *Science* **178,** 361–368 (1972).

182. R. B. Moon and J. H. Richards, *JBC* **248,** 7276–7278 (1973).

183. D. I. Hoult, S. J. W. Busby, D. G. Gadian, G. K. Radda, R. E. Richards, and P. J. Seeley. *Nature* (*London*) **252,** 285–287 (1974).

184. J. M. Salhany, T. Yamane, R. G. Shulman and S. Ogawa, *PNAS* **72,** 4966–4970 (1975).

185. G. T. Burt, T. Glonek and M. Bárány, *JBC* **251,** 2584–2591 (1976).

186. D. Gust, R. B. Moon and J. D. Roberts, PNAS **72,** 4696–4700 (1975).

STUDY QUESTIONS

1. Draw the following hydrogen-bonded structures:
   a. A dimer of acetic acid.
   b. A tyrosine–carboxylate bond in the interior of a protein.
   c. A phosphate–guanidinium ion pair in an enzyme–substrate complex.
   d. The base pairs GC and GU. (Note that the GU pair is not a Watson–Crick pair).

2. Draw the tautomeric structures possible for the cation formed by protonation of 9-methyladenine.

3. Complete the following table:

Name	Monomer	Linkage	Range of molecular weights
Protein			
Polysac-charide			
Nucleic acid			
Teichoic acid			
Poly-β-hydroxy-butyrate			

4. Predict whether the following peptide segments will be likely to exist as an α helix or as part of a β structure within a protein:
   a. Poly-L-leucine
   b. Poly-L-valine
   c. Pro · Glu · Met · Val · Phe · Asp · Ile
   d. Pro · Glu · Ala · Leu · Phe · Ala · Ala

5. Contrast the properties of the amino acids with those of the saturated fatty acids with respect to solubility in water and in ether and to physical state. How are these differences explained in terms of structure?

6. What functional groups are found in protein side chains? Of what importance to protein structure and function are (a) hydrophobic groups, (b) acidic and basic groups, (c) sulfhydryl groups?

7. Make a table of characteristic pK_a values for acidic and basic groups in proteins. Which of these groups contribute most significantly to the titration curves of proteins?

8. The tripeptide L-Ala · L-His · L-Gln had the following pK_a values: 3.0 (α-COOH), 9.1 (α-NH$_3^+$), 6.7 (imidazolium).
   a. What is the isoelectric pH (pI) of the peptide, i.e., the pH at which it will carry no net charge? Hint, the pI for amino acids is usually given approximately as the arithmetic mean of two pK_a values. (See E. J. Cohn and J. T. Edsall, "Proteins, Amino Acids and Peptides," pp. 90–93. Reinhold, New York, 1943.)
   b. Draw the structures of the ionic forms of the peptide that occur at pH 5 and at pH 9. At each pH compute the fraction of the peptide in each ionic form. NOTE: For definitions and for problems dealing with pH, pK_a's and buffers, see Chapter 4.

9. Name all of the isomeric tripeptides which could be formed from one molecule each of tyrosine, alanine, and valine.

10. a. Write the structure for glycyl-L-tryptophanyl-L-prolyl-L-seryl-L-lysine.
    b. What amino acids could be isolated from it following acid hydrolysis?
    c. Following alkaline hydrolysis?

d. After nitrous acid treatment followed by acid hydrolysis?

e. In an electrolytic cell at pH 7.0 would the peptide migrate toward the cathode or toward the anode? What is the approximate isoelectric point of the peptide?

f. If a solution of this peptide were adjusted to pH 7, then titrated with sodium hydroxide in the presence of 10% formaldehyde, how many equivalents of base would be required per mole of peptide to raise the pH to 10?

11. Compare structural features and properties of the following proteins: silk fibroin, α-keratin, collagen, bovine serum albumin.

12. In what way do the solubilities of proteins usually vary with pH? Why?

13. A peptide is shown to contain only L-lysine and L-methionine. Titration of the peptide shows 3 free amino groups for each free carboxyl group present, and each amino group liberates 1 mol of N_2 when the peptide is treated with HNO_2 in the Van Slyke apparatus. When the deaminated peptide is hydrolyzed completely in acid and the hydrolyzate again treated with HNO_2, the same amount of N_2 is liberated as that derived from the intact peptide. A sample of the original peptide is treated with excess dinitrofluorobenzene to give a dinitrophenyl (DNP)-peptide, which is shown spectrophotometrically to contain 3 DNP groups per free carboxyl group. When this DNP-peptide is completely hydrolyzed, the following products are found: a colorless compound containing S (A_1); a yellow compound containing S (A_2); and a yellow compound *not* containing S (A_3). Partial hydrolysis of the DNP-peptide yields A_1, A_2, A_3, plus 4 additional yellow compounds, B_1, B_2, B_3, and B_4. On complete hydrolysis B_1 yields A_1, A_2, and A_3; B_2 yields A_1 and A_2; B_3 yields A_1 and A_3; and B_4 yields A_3 only.

What is the most probable structure of the original peptide?

14. A nonreducing disaccharide gives an octamethyl derivative with dimethyl sulfate and alkali. On acid hydrolysis, this derivative yields 1 mol of 2,3,4,6-tetramethyl-D-glucose and 1 mol of 2,3,4,6-tetramethyl-D-galactose. The disaccharide is hydrolyzed rapidly by either maltase or lactase (a β-galactosidase).

Give an adequately descriptive name of the disaccharide, and draw its Haworth projection formula.

15. An aldopentose (A) of the D-configuration on oxida-

tion with concentrated nitric acid gives a 2,3,4-trihydroxypentanedioic acid (a trihydroxyglutaric acid) (B) which is optically inactive. (A) on addition of HCN, hydrolysis, lactonization, and reduction gives two stereoisomeric aldohexoses (C) and (D). (D) on oxidation affords a 2,3,4,5-tetrahydroxyhexanedioic acid (a saccharic acid) (E) which is optically inactive. Give structures of compounds (A)–(E).

16. What products are formed when periodic acid reacts with sorbitol?

17. A 10.0 g sample of glycogen gave 6.0 millimol of 2,3-di-O-methylglucose on methylation and acid hydrolysis.

a. What percent of the glucose residues in glycogen have chains substituted at the α-1→6 position?

b. What is the average number of glucose residues per chain?

c. How many millimols of 2,3,6-tri-O-methylglucose were formed?

d. If the molecular weight of the polysaccharide is 2×10^6, how many glucose residues does it contain?

e. How many nonreducing ends are there per molecule or equivalently how many chains are there per molecule?

18. What is meant by the T_m of a DNA sample? How does T_m vary with base composition and what is the explanation of this?

19. Draw a schematic representation of the polynucleotide portion of a DNA molecule and of an RNA molecule and indicate positions of cleavage by the following treatments:

a. Mild HCl
b. More vigorous HCl
c. Mild NaOH
d. More vigorous NaOH
e. Pancreatic RNase
f. Pancreatic DNase
g. Splenic DNase
h. Splenic phosphodiesterase
i. Snake venom phosphodiesterase
j. DNase from *Micrococcus*

20. A sample of DNA was hydrolyzed by acid and was found to have the following composition (in mol %): adenine, 24.0; thymine, 33.0; guanine, 23.0; cytosine, 20.0. This composition differs from that of most DNA preparations. Indicate two such differences and offer a possible explanation in terms of DNA structure.

21. The iodine number of a compound is defined as the number of grams of I_2 absorbed (through addition to $C = C$ bonds to give a diiodo derivative) per 100 g of fat. NOTE: Iodine monochloride or iodine monobromide are the usual halogenating reagents but the iodine number is expressed in terms of grams of I_2. The saponification number is the number of milligrams of KOH needed to completely saponify (hydrolyze and neutralize the resulting fatty acids) 1 g of fat. A pure triglyceride has a saponification number of 198 and an iodine number of 59.7. What is (a) the molecular weight? (b) The average chain length of the fatty acids? (c) The number of double bonds in the molecule?

22. Spermaceti (a wax from the head of the sperm whale) resembles high molecular weight hydrocarbons in physical properties and inertness toward $Br_2/CHCl_3$ and $KMnO_4$; on qualitative analysis, it gives positive tests only for carbon and hydrogen. However, its IR spectrum shows the presence of an ester linkage, and quantitative analysis gives the empirical formula $C_{16}H_{32}O$. A solution of the wax in alcoholic KOH is refluxed for a long time. Titration of an aliquot shows that one equivalent of base is consumed for every 475 g of wax. Water and ether are added to the cooled reflux mixture, and the aqueous and etheral layers are separated. Acidification of the aqueous layer yields a solid A with a neutralization equivalent of 260 ± 5. Evaporation of the ether layer gives solid B, which could not be titrated. Reduction of either spermaceti or A by lithium aluminum hydride gave B as the only product. What is the likely structure of spermaceti?

23. (a) Explain *briefly* two advantages of the isotope dilution method of analysis. (b) From the following data, calculate the amount of cAMP (cyclic AMP) present per ml of human gluteus maximus muscle cells. Cells were treated with ^{32}P-cAMP, S.A. = 50 μCi/μmol for 0.2 h, (all cAMP was taken up by cells), cells were homogenized, and the soluble cAMP was isolated and purified. The specific activity of the isolated cAMP = 10 μCi/μmol. The total amount of cAMP added was 1.0×10^{-7} mol per ml of cells.

24. The figure in Box 2-C shows a high resolution separation of the soluble proteins of *E. coli*. The author labeled the proteins with ^{14}C-containing amino acids.
    a. How would you carry out the labeling experiment?
    b. What other isotope(s) could be used to label proteins? What chemical form(s) would you use? What limitations might there be?
    c. What soluble components of an *E. coli* cell sonicate might interfere with the 2-D separation, and how could they be removed?
    d. What technique(s) other than radioactive labeling could be used for locating proteins?
    e. If *all* the soluble proteins of *E. coli* were detected, about how many separate proteins would you expect to see?
    f. Indicate two or more properties of the resolution technique which are most significant in making it applicable to a system containing a very large number of proteins.

25. ^{35}S is a beta emitter, with no gamma or other type of radiation. It has the following properties: $t_{1/2} = $ 86.7 days, $\epsilon_{max} = 0.168$ MeV
    a. Write the equation for the radiochemical decomposition of ^{35}S.
    b. Discuss the advantages and limitation of the use of ^{35}S as an isotopic tracer.

26. Given a quantity of agarose, cyanogen bromide, 6-aminohexanoic acid, and any other necessary reagents show the chemical reactions for making an affinity column bearing the amino acid tryptophan bound through its amino group.

3
Energetics of Biochemical Reactions

We all know from experience the importance of energy to life. We know that we must eat and that hard work not only tires us but makes us hungry. Our bodies generate heat, an observation that led Lavoisier around 1780 to the conclusion that respiration represented slow combustion of foods within the body. Later the discovery of the first and second laws of thermodynamics permitted the development of precise, quantitative relationships between heat, energy, and work. Modern biochemical literature abounds with references to the thermodynamic quantities **energy** E, **enthalpy** H, **entropy** S, and **Gibbs free energy** G.

The purposes of this chapter are threefold: (1) to provide a short review of thermodynamic equations, (2) to provide tables of thermodynamic quantities for biochemical substances and to explain the use of these data in the consideration of equilibria in biochemical systems, and (3) to introduce the **adenylate system** (consisting of adenosine triphosphate, adenosine diphosphate, adenosine monophosphate, and inorganic phosphate) and its central role in energy metabolism.

Some of the many books on thermodynamics available for further study are mentioned in references 1–5.

A. THERMODYNAMICS

Thermodynamics[1–5] is concerned with the quantitative description of heat and energy changes and of chemical equilibria. Knowledge of *changes* in thermodynamic quantities, such as ΔH and ΔS, enables us to predict the equilibrium positions in reactions and whether or not under given circumstances a reaction will or will not take place. Furthermore, the consideration of thermodynamic quantities may provide insight into the nature of forces responsible for bonding between molecules.

An important restriction in the use of classical thermodynamic information is that it deals only with equilibria and says nothing about kinetics. This has led to the occasional assertion that thermodynamics is not relevant to biochemistry. This is certainly not true; it is important to understand energy relationships in biochemical reactions. At the same time, it is essential to avoid the trap of assuming that thermodynamic calculations appropriate for **equilibrium** situations can be applied directly to the **steady state** found in a living cell (Chapter 1).

1. The System and Its Surroundings

In using thermodynamic equations, we must specify precisely the **system** under consideration; for example, the system might be the solution in a flask resting in a thermostated bath. The flask, the bath, and everything else would be **surroundings** or **environment.** The system plus surroundings is sometimes referred to as **the universe** in discussions of thermodynamics.

2. The First Law

The first law of thermodynamics asserts the conservation of energy and also the equivalence of **work** and **heat.** Work and heat are both regarded as energy in transit. Heat may be absorbed by a system from the surroundings or evolved by a system and absorbed in the surroundings. Work can be done by a system on the surroundings. The first law postulates that there is an **internal energy function** E (also designated U), which is dependent only on the present **state** of the system and in no way is dependent upon the history of the system. The first law states that E *can be changed only by the flow of energy as heat or work.* In other words, the law states that energy can neither be created nor destroyed.

In mathematical form, the first law is given by Eq. 3-1

$$\Delta E = E \text{ (products)} - E \text{ (reactants)} = Q - W \quad (3\text{-}1)$$

where Q is the heat absorbed by the system from the surroundings and W is the work done by the system on the surroundings. Energy, heat, and work are all measured in the same units. Chemists have traditionally used the **calorie** (cal) or **kilocalorie** (kcal) but are switching to the SI unit, the **joule** (see Table 3-1). Work done by the system may be **mechanical** (e.g., a change in the volume of the surroundings), **electrical** (e.g., the charging of a battery), or **chemical** (e.g., effecting the synthesis of a polypeptide from amino acids).

3. Enthalpy Changes and Thermochemistry

We are most often interested in the *changes* in the thermodynamic functions when a chemical reaction takes place; for example, the heat absorbed by the system within a **bomb calorimeter** where the volume stays constant (Q_V) is a direct measure of the change in E (Eq. 3-2).

$$Q_V = \Delta E \quad (3\text{-}2)$$

To measure ΔE for combustion of a biochemical compound, the substance might be placed in a bomb together with gaseous oxygen and the mixture ignited within the calorimeter by an electric spark. In this case, heat would be evolved from the bomb and would pass into the surroundings. Thus, Q_V and ΔE would be negative. The bomb calorimeter is designed to measure Q_V and thereby to give us a way of determining ΔE for reactions.

4. Enthalpy and Processes at Constant Pressure

Chemical and biochemical reactions are much more likely to be conducted at constant pressure (usually 1 atm) than they are at constant volume. For this reason, chemists tend to use the enthalpy function H more often than the internal energy E.

$$H = E + PV \quad (3\text{-}3)$$

It follows from Eq. 3-3 that *if the pressure is constant,* $\Delta H)_P = \Delta E)_P + P\,\Delta V$. Since in a process at constant pressure, $P\,\Delta V$ is exactly the pressure–volume work done on the surroundings, the heat absorbed at constant pressure (Q_P) is a measure of $\Delta H)_P$ (Eq. 3-4).

$$Q_P = \Delta E)_P + P\,\Delta V = \Delta H)_P \quad (3\text{-}4)$$

The term enthalpy was coined to distinguish H from E, but chemists sometimes tend to be careless about language and many discussions of energy in the literature are, in fact, about enthalpy.

TABLE 3-1
Units of Energy and Work and the Values of Some Physical Constants

The joule, SI unit of energy
$1 \text{ J} = 1 \text{ kg m}^2 \text{ s}^{-2}$
$= 1 \text{ N m}$ (newton meter)
$= 1 \text{ W s}$ (watt second)
$= 1 \text{ C V}$ (coulomb volt)
Thermochemical calorie
$1 \text{ cal} = 4.184 \text{ J}$
Large calorie
$1 \text{ Cal} = 1 \text{ kcal} = 4.184 \text{ kJ}$
Work required to raise 1 kg 1 m on earth
(at sea level) $= 9.807 \text{ J}$
Free energy of hydrolysis of 1 mol of ATP
at pH 7, millimolar concentrations $= -12.48 \text{ kcal} = -52.2 \text{ kJ}$
Work required to concentrate 1 mol of a substance
1000-fold, e.g., from 10^{-6} to 10^{-3} M
$= 4.09 \text{ kcal} = 17.1 \text{ kJ}$
Avogadro's number, the number of particles in a mole
$N = 6.0220 \times 10^{23}$
Faraday $1 \text{ F} = 96,485 \text{ C mol}^{-1}$ (coulombs per mole)
Coulomb $1 \text{ C} = 1 \text{ A s}$ (ampere second)
$= 6.241 \times 10^{18}$ electronic charges
The Boltzmann constant
$k_B = 1.3807 \times 10^{-23} \text{ J deg}^{-1}$
The gas constant, $R = N\,k_B$
$R = 8.3144 \text{ J deg}^{-1} \text{ mol}^{-1}$
$= 1.9872 \text{ cal deg}^{-1} \text{ mol}^{-1}$
$= 0.08206 \text{ l atm deg}^{-1} \text{ mol}^{-1}$
and at 25° $RT = 2.479 \text{ kJ mol}^{-1}$
The unit of temperature is °K (or simply K); 0°C = 273.16°K
$\ln x = 2.3026 \log x$

Often the difference is insignificant because if the pressure–volume work is negligible, E and H are the same.

5. Enthalpies of Combustion and Physiological Fuel Values

Enthalpy changes can be obtained directly from measurement of heat absorption at constant pressure. Hence, enthalpy changes accompanying oxidation of foods long ago attracted the attention of chemists and physiologists alike. The **heat of combustion** $(-\Delta H_c)$ is usually determined from ΔE_c which is measured in a bomb calorimeter. Since

$\Delta E)_V$ and $\Delta E)_P$ are nearly identical, it follows that $\Delta H)_P = \Delta E)_V + P\,\Delta V$ where ΔV is the volume change which would have occurred if the reaction were carried out at constant pressure P and can be estimated by calculation. Since ΔH is desired for combustion to carbon dioxide, water, elemental nitrogen (N_2), and sulfur, correction must be made for amounts of the latter elements converted into oxides. By these procedures, it is possible to obtain highly accurate values of ΔH_c both for biochemicals and for mixed foodstuffs. In recent years direct determination of even small values of ΔH for chemical and biochemical reactions using a microcalorimeter has become routine.[6,7]

In nutrition, the quantity $-\Delta H_c$ is sometimes referred to as the gross energy. Values are usually expressed in kilocalories which is abbreviated *kcal* by biochemists but *Cal* (with a capital C) in the nutritional literature.

Physiological fuel values or caloric values of foods are corrected enthalpies of combustion (ΔH_c), but with an opposite sign. While enthalpies of combustion of foods are all negative, the caloric values are given as positive numbers. Typical values are shown in the following tabulation.

	Caloric values per gram	
Carbohydrates	4.1 kcal	17 kJ
Pure glucose	3.75 kcal	15.7 kJ
Lipids	9.3 kcal	39 kJ
Proteins[a]	4.1 kcal	17 kJ

[a] Nitrogen excreted as urea.

Note that the value for proteins has been calculated for the conversion of the nitrogen to urea (the major nitrogenous excretion product in mammals) rather than to elemental nitrogen.

From a thermochemical viewpoint, can a human or animal be regarded as just a catalyst for the combustion of foodstuffs? To answer this question, large calorimeters have been constructed into which an animal or a human can be placed. If, while in the calorimeter, the subject neither gains nor loses weight, the heat evolved should be just equal to $-\Delta H$ for combustion of the food consumed to CO_2, water, and urea. That this prediction has been verified experimentally does not seem surprising, but at the time that the experiments were first done in the early years of the century, there were probably those who doubted that the first law of thermodynamics could be applied to animals.

In practice, animal calorimetry is quite complicated because of the inherent difficulty of accurate heat measurements, uncertainties about the amount of food stored, and the necessity of corrections for ΔH_c of the waste products. However, the measurement of energy metabolism has been of considerable importance in nutrition and medicine. Indirect methods of calorimetry have been developed for use in measuring the **basal metabolic rate** of humans. For a good discussion see White *et al.*[8]

The basal metabolic rate is the rate of heat evolution in the resting, postabsorptive state (postabsorptive means that the subject has not eaten recently). In this condition, the subject is using stored foods to provide energy and is oxidizing them at a relatively constant rate. The basal metabolic rate tends to be *proportional to the surface area*; for a young female adult, it is typically about 154 kJ h^{-1} m^{-2}, and for a young adult male ~172 kJ h^{-1} m^{-2}. This comes to ~320–360 kJ h^{-1} for a 70 kg person. Note that 360 kJ h^{-1} is the same as the power output of a 100 W light bulb. While there is considerable variation among individuals, basal metabolic rates far below or above normal may indicate a pathological condition such as an insufficiency or oversupply of the thyroid hormone thyroxine. Metabolic rates fall somewhat below the basal value during sleep and are much higher than basal during hard exercise. A human may attain rates as high as 2500 kJ (600 kcal) per hour. At a basal rate of 320 kJ (76 kcal) per hour, a person requires 7680 kJ (1835 kcal) each 24 h to supply his basal needs, plus additional energy for muscular exercise. Routine light exercise as in office work or housework increases metabolism to about double the basal rate.

6. Entropy and the Flow of Heat

Why does heat flow from a warm body into a cold one? Why doesn't it ever flow in the reverse direction? We can see that differences in temperature control the direction of flow of heat, but that observation raises still another question: What *is* temperature? Reflection on these questions, and on the interconversion of heat and work led to discovery of the second law of thermodynamics and to definition of a new thermodynamic function, the entropy S.

Consider a phase transition at constant temperature and pressure, e.g., the melting of ice. At a temperature just above 0°C ice melts completely, but at a temperature just below 0°C it does not melt at all. At 0°C we have an equilibrium; in the language of thermodynamics, the melting of ice at 0°C is a **reversible reaction.** What criterion could be used to predict this behavior for water? For many familiar phenomena, e.g., combustion, a spontaneous reaction is accompanied by the evolution of a large amount of heat, i.e., ΔH is negative. However, when ice melts it absorbs heat. The ΔH of fusion amounts to 6.008 kJ mol^{-1} at 0°C and is nearly the same just below 0°C, where the ice does not melt, and just above 0°C where the ice melts completely. In the latter case the melting of ice is a spontaneous reaction for which ΔH is positive. Consideration of such facts made it clear that the sign of the enthalpy change does *not* serve as a criterion of spontaneity.

A correct understanding of the ice–water transition came when it was recognized that when ice melts not only does H increase by 6.008 kJ mol^{-1} (with the molecules acquiring additional internal energy of translation, vibration, and rotation), but *the molecules become more disordered.* Although historically entropy was introduced in a different context, it is now recognized to be a measure of "microscopic disorder." When ice melts, the entropy S increases because the structure becomes less ordered.

7. The Second Law of Thermodynamics

The second law of thermodynamics is stated in many different ways, but the usual mathematical formulation asserts that for the "universe" (or for an isolated system)

ΔS (system + surroundings) = 0
for reversible processes
$\Delta S > 0$ for real (nonreversible) processes

The second law is sometimes stated in another way: *the entropy of the universe always increases.*

The second law also defines both S and the thermodynamic temperature scale (Eq. 3-5).

$$dS_{\text{reversible}} = q/T \qquad (3\text{-}5)$$

where q is an infinitesimal quantity of heat absorbed. For a *reversible phase transition* such as the melting of ice at constant pressure and temperature, the change in entropy of the H_2O is just $\Delta H/T$ (Eq. 3-6). Entropy is measured in units of

$$\Delta S)_{P,T,\text{reversible}} = Q/T = \Delta H/T \qquad (3\text{-}6)$$

joules per °K or calories per °K, the latter often being abbreviated as *eu* (**entropy units**). Since the melting of ice is a reversible process, the second law asserts that the entropy of the surroundings decreases by the same amount that the entropy of the water increases. Note that $T \Delta S$ is numerically equal to the heat of fusion, 6.008 kJ mol^{-1} in the case of water at 0°C. Thus, the entropy increase in the ice as it is melted at 0°C is 6.008 $\times$ 10^3 J/273.16°K = 22.0 J/deg.

The definition of thermodynamic temperature (Eq. 3-7; treated further in textbooks of thermodynamics) also follows from Eq. 3-5.

$$T \ (°K) = (\partial E/\partial S)_V = (\partial H/\partial S)_P \qquad (3\text{-}7)$$

The entropy of a substance can be given a precise mathematical formulation involving the degree of molecular disorder as in Eq. 3-8.

$$S = k_B \ln \Omega \qquad (3\text{-}8)$$

Here k_B is the **Boltzmann constant** (see Table 3-1) and Ω is given precisely as the number of microscopic states (different arrangements of the particles) of the system corresponding to a given macroscopic state, i.e., to a given temperature, pressure, and quantity. It increases as volume or temperature is increased and in going from solid to liquid to gaseous states. Equation 3-8 is not part of classical thermodynamics (which deals only with macroscopic systems, i.e., with large collections of molecules). However, using the methods of statistical thermodynamics,[9] Eq. 3-8 can be used successfully to predict the entropies of gases.

A biochemical example which can be related directly to Eq. 3-8 is the *racemization of an amino acid.* A solution of an L-amino acid will be efficiently changed into the racemic mixture of 50% D and 50% L by the action of a special enzyme (a **ra-**

cemase) with no uptake or evolution of heat. Thus, $\Delta H = 0$ and the only change is an entropy change. Let us designate Ω for the pure isomer as Ω'. Noting that there are just two choices of configuration for each of the N molecules in 1 mol of the racemate we see that for the racemate.

$$\Omega = 2^N \Omega' \qquad (3\text{-}9)$$

Applying Eq. 3-8 we calculate (Eq. 3-10):

$$\begin{aligned} \Delta S &= k_B (\ln 2^N + \ln \Omega') - k_B \ln \Omega' \\ &= N k_B \ln 2 = R \ln 2 = 5.76 \text{ J }^\circ\text{K}^{-1} \text{ mol}^{-1} \end{aligned} \qquad (3\text{-}10)$$

8. Entropies from Measurement of Heat Capacities

It follows from Eq. 3-8 that $S = 0$ when $T = 0$ for a perfect crystalline substance in which no molecular disorder exists. The *third law of thermodynamics* asserts that as the thermodynamic temperature T approaches $0°\text{K}$ the entropy S also approaches zero for perfect crystalline substances. From this it follows that at any temperature above $0°\text{K}$, the entropy is given by Eq. 3-11.

$$S = \int_0^T C_p \, d \ln T \qquad (3\text{-}11)$$

In this equation C_P is the heat capacity at constant pressure (Eq. 3-12).

$$C_P = (\partial H / \partial T)_P \qquad (3\text{-}12)$$

If C_P is measured at a series of low temperatures down to near $0°\text{K}$, Eq. 3-11 can be used to evaluate the absolute entropy S. If phase transitions occur as the temperature is raised, entropy increments given by Eq. 3-6 must be added to the value of S given by Eq. 3-11. For a few compounds (such as water, Chapter 4, Section B,4) molecular disorder is present in the crystalline state even at $0°\text{K}$. For these substances a term representing the entropy at $0°\text{K}$ must be added to Eq. 3-11.

The entropies of a few substances are given in Table 3-2. Note how the entropy increases with increasing complexity of structure, with transitions from solid to liquid to gas, and with decreasing hardness of solid substances.

TABLE 3-2
Entropies of Selected Substances[a]

Substance	State[b]	Entropy S	
		eu	(J $°\text{K}^{-1}$ mol^{-1})
C (diamond)	s	0.55	2.3
C (graphite)	s	1.36	5.7
Cu	s	8.0	33
Na	s	12.2	51
H_2O (ice)	s	9.8	41
H_2O	l	16.7	70
H_2O	g (1 atm)	45.1	189
He	g	30.1	126
H_2	g	31.2	131
N_2	g	45.8	192
CO_2	g	51.1	214
Benzene	g	64.3	269
Cyclohexane	g	71.3	298

[a] All values are given in entropy units (eu) of calories per degree Kelvin per mole and in joules per degree per mole at 25°C (298.16°K).
[b] Here s stands for solid, l for liquid, and g for gaseous.

9. A Criterion of Spontaneity

We have seen that while many spontaneous processes, e.g., combustion of organic compounds, are accompanied by liberation of heat (negative ΔH), others are accompanied by absorption of heat from the surroundings (positive ΔH). An example of the latter is the melting of ice at a temperature just above 0°C. We have seen that in this case a large change in entropy of the water occurs upon melting and that at equilibrium at 0°C the numerical value of $T \Delta S$ is just equal to $-\Delta H$ (Eq. 3-6).

The recognition that $\Delta H - T \Delta S = 0$ at equilibrium led J. W. Gibbs to realize that the proper thermodynamic function for determining the spontaneity of a reaction is what is now known as the Gibbs free energy G (Eq. 3-13).

$$G = H - TS \qquad (3\text{-}13)$$

For a process at constant temperature and pressure the change in G is given by Eq. 3-14.

$$\Delta G_{T,P} = \Delta H - T \Delta S \qquad (3\text{-}14)$$

Furthermore, for a reversible (equilibrium) process doing only pressure–volume work:

$$\Delta G)_{T,\text{reversible}} = \Delta H - T\,\Delta S = 0 \qquad (3\text{-}15)$$

It can also be shown readily that ΔG *is negative for any spontaneous* (irreversible) *process.* Such a process is called **exergonic.** Likewise, if ΔG is *positive*, a given reaction will *not* proceed spontaneously and is called **endergonic.** The magnitude of a free energy decrease $(-\Delta G)$ is a direct measure of the maximum work which could be obtained from a given chemical reaction if that reaction could be coupled in some fashion reversibily to a system able to do work. It represents the maximum amount of electrical work that could be extracted or the maximum amount of muscular work or osmotic work obtainable from a reaction in a biological system. In any real system, the amount of work obtainable is necessarily less than $-\Delta G$ because real processes are irreversible, i.e., entropy is created.

Returning to the older assumption that the magnitude of ΔH might be an index of work obtainable, we note that $T\,\Delta S$ amounts to only a few kilocalories for most reactions. Therefore, if ΔH is large, as in the combustion of foodstuffs, it is not greatly different from ΔG for the same process. Hence, the use of the caloric value of a food as an approximate measure of the work obtainable from its utilization in the body is justified.

10. Standard States

For thermodynamic data to be useful in chemical calculations, we must agree upon standard states for elements and compounds. For example, if we wish to talk about free energy changes occurring when one or more pure compounds are converted to other pure substances, we must agree upon a state (crystalline, liquid, gaseous, or in solution) and upon a pressure (especially when gases are involved) at which the data apply. The standard pressure is usually 1 atm. Standard states of the elements are the pure *crystalline, solid,* or *gaseous materials,* e.g., C (graphite), S (crystalline,

rhombic), P (crystalline, white), and N_2, O_2, and H_2 (gaseous). It is also essential to specify the temperature. Thermodynamic data are most often given for 25°C, but there is a standard state for each substance at each temperature.

It is usually impractical to measure the values of G or H, but ΔG and ΔH for a chemical reaction can be evaluated. To compile useful tables of free energy data, it is customary to arbitrarily assume that G is zero for all of the elements in their standard states. Then the values of ΔG for formation of any compound by combination of the elements in the correct ratio can be obtained (e.g., by measuring ΔH of combustion and by obtaining entropies from heat capacity measurements). The resulting **standard free energies of formation** are given the symbol ΔG_f^0.

11. Summing Free Energy Changes

A convenient feature of thermodynamic calculations is that if two or more chemical equations are summed, ΔG for the resulting overall equation is just the sum of the ΔG's for the individual equations as illustrated in Eqs. 3-16 to 3-19. The same applies for ΔH and ΔS.

$$CH_3COOH\ (l) \longrightarrow 2\ C + 2\ H_2 + O_2 \qquad (3\text{-}16)$$
$$-\Delta G_f°\ \text{for acetic acid} = +396.4\ \text{kJ mol}^{-1}$$

$$2\ O_2 + 2\ C \longrightarrow 2\ CO_2 \qquad (3\text{-}17)$$
$$2\ \Delta G_f°\ \text{for } CO_2 = -788.8\ \text{kJ mol}^{-1}$$

$$O_2 + 2\ H_2 \longrightarrow 2\ H_2O \qquad (3\text{-}18)$$
$$2\ \Delta G_f°\ \text{for } H_2O = -474.4\ \text{kJ mol}^{-1}$$

$$CH_3COOH\ (l) + 2\ O_2 \longrightarrow 2\ CO_2 + 2\ H_2O\ (l) \qquad (3\text{-}19)$$
$$\Delta G°\ \text{for combustion of acetic acid} =$$
$$-866.8\ \text{kJ mol}^{-1}$$

In this example an equation for the decomposition of acetic acid into its elements (Eq. 3-16) has been summed with Eqs. 3-17 and 3-18, representing the formation of the proper number of molecules of CO_2 and H_2O. The sum of the three equations gives the equation for the combustion of acetic acid to CO_2 and water, and the sum of the ΔG values for the three equations gives ΔG for combustion of acetic acid. Note that the resulting value of ΔG is that for combustion of pure liquid acetic acid by oxygen at 1 atm to give CO_2 at 1 atm and pure liquid water, all reactants and products being in their standard states.

The process described in the preceding paragraph is represented by Eq. 3-20, which is a general equation for calculation of ΔG^0 for any reaction from ΔG_f^0 of products and reactants.

$$\Delta G^0 = \Sigma \Delta G_f^0 \text{ (products)} - \Sigma \Delta G_f^0 \text{ (reactants)} \quad (3\text{-}20)$$

How does the free energy change vary if we go from the standard state of a compound to some other state? Consider a change of pressure in a gas. It is easy to show (see any thermodynamics text) that

$$(\partial G / \partial P)_T = V \quad (3\text{-}21)$$

Using Eq. 3-21 together with the perfect gas law, we obtain the relationship (Eq. 3-22) between the free energy $\overline{G}$ of 1 mol of a substance* at pressure P and the standard free energy $\overline{G}^0$, at pressure P^0.

$$\overline{G} - \overline{G}^0 = RT \ln \frac{P}{P^0} = RT \ln P \quad (3\text{-}22)$$

Since P^0 is by definition 1 atm, the free energy change per mole upon changing the pressure from P^0 to P is just $RT \ln P$. It is customary in books on thermodynamics to develop most of the important thermodynamic equations as applied to a perfect gas, but we will move at this point to a consideration of biochemical substances in solution.

12. Reactions in Solution

Biochemists are usually interested in the behavior of substances dissolved in relatively dilute aqueous solutions (as in cytoplasm), although in some instances, the interest may be in nonaqueous solutions. In either case, it is necessary to establish a standard state for the solute. The stan-

dard state of a substance in aqueous solution is customarily taken as a strictly hypothetical one **molal** solution (1 mol of solute per kilogram of water) *whose properties are those of a solute at infinite dilution*. An equation exactly analogous to Eq. 3-22 can be written relating the free energy of 1 mol of dissolved solute, $\overline{G}_i$ to the free energy $\overline{G}_i^0$ in the hypothetical standard state of unit activity and to the **activity** a_i of the solute* (Eq. 3-23).

$$\overline{G}_i = \overline{G}_i^0 + RT \ln a_i \quad (3\text{-}23)$$

Equation 3-23 and the equations that follow from it apply to molal activities. However, concentration can be substituted for activity in very dilute solution where the behavior of the dissolved molecules approximates that of the hypothetical ideal solution for which the standard state is defined. For any real solution, the activity can be expressed as the product of an activity coefficient and the concentration (Eq. 3-24)

$$a = \gamma c \quad (3\text{-}24)$$

where a = activity, γ = activity coefficient, and c = molal concentration. Thus, to use the tabulations of thermodynamic functions for substances in solution to predict behavior in other than very dilute solution, we must multiply the concentration of every component by the appropriate activity coefficient. For the very approximate calculations which are often of interest to biochemists, it is customary to equate concentration with activity. Furthermore, in dilute solutions, the more usual **molar** concentrations (moles per liter) are nearly equal to molal concentrations.†

From Eq. 3-23 it follows that the *free energy change for dilution* from one activity a_1 to another a_2 is

$$\Delta \overline{G} \text{ (dilution from } a_1 \text{ to } a_2) = RT \ln(a_2/a_1) \quad (3\text{-}25)$$

---

* A bar over the symbol G or ΔG always indicates that the free ene   free energy change is for 1 mol of substance.

* Here the subscript i designates a particular component in a solution which also contains solvent and, perhaps, other components. To be precise, $\overline{G}_i$ is a partial molar free energy, i.e., the change in total free energy of a very large volume of solution when 1 mol of the component is added.

† The same equations are also used with *mole fractions* rather than molal concentrations.

13. ΔG^0 and the Equilibrium Constant

Consider the following generalized chemical equation (Eq. 3-26) for reaction of a mol of A with b mol of B to give products C and D, etc.

$$a\,A + b\,B + \cdots = c\,C + d\,D + \cdots \quad (3\text{-}26)$$

The standard free energy change ΔG^0 for the process is given by Eq. 3-27:

$$\Delta G^0 = c\,\overline{G}^0(C) + d\,\overline{G}^0(D) + \cdots \\ - a\,\overline{G}^0(A) - b\,\overline{G}^0(B) \cdots \quad (3\text{-}27)$$

The symbol $\overline{G}^0(A)$ designates the free energy of A, etc. The value of ΔG for any desired concentrations of reactants of products can be related to this ΔG^0 by applying to each component Eq. 3-23 with the following result (Eq. 3-28):

$$\Delta G = \Delta G^0 + RT \ln \frac{a_C{}^c\, a_D{}^d \cdots}{a_A{}^a\, a_B{}^b \cdots} \quad (3\text{-}28)$$

Here a_C represents the activity of component C, etc. This useful equation permits us to calculate ΔG for the low concentrations usually found in biochemical systems (more often in the millimolar range rather than approaching the hypothetical $1\,M$ of the standard state). Often concentrations are substituted in Eq. 3-28 for activities:

$$\Delta G \approx \Delta G^0 + RT \ln \frac{[C]^c[D]^d}{[A]^a[B]^b} \quad (3\text{-}29)$$

Equation 3-28 is used in another way by noting that $\Delta G = 0$ when a system is at equilibrium. Furthermore, at equilibrium, the product $a_C{}^c a_D{}^d \ldots / a_A{}^a a_B{}^b \ldots$ is just the equilibrium constant K. Hence, it follows that

$$\Delta G^0 = -RT \ln K = -2.303 RT \log K \\ = -19.145T \log K \text{ J mol}^{-1} \\ = -5.708 \log K \text{ kJ mol}^{-1} \text{ at } 25°C \\ = -1.364 \log K \text{ kcal mol}^{-1} \text{ at } 25°C \quad (3\text{-}30)$$

Note that although the units of ΔG° are kJ mol^{-1}, the free energy change in Eq. 3-30 is that for the reaction of a mol of A, b mol of B etc., as in the equation used to define K (Eq. 3-26 in this instance).

14. Activity Coefficients and Apparent Equilibrium Constants

Strictly speaking, Eq. 3-30 applies only to thermodynamic equilibrium constants, that is, to constants that employ activities rather than concentrations. The experimental determination of such constants requires measurements of the apparent equilibrium constant K' at a series of different concentrations and extrapolation to infinite dilution (Eq. 3-31).

$$K' = \text{apparent equilibrium constant} \\ = \frac{[C]^c[D]^d}{[A]^a[B]^b} \text{ at equilibrium} \quad (3\text{-}31)$$

Extrapolation of K' to infinite dilution to give K is usually easy because the activity coefficients of most ionic substances vary in a regular manner with **ionic strength** and follow the **Debye–Hückel** equation (Eq. 3-32) in very dilute solutions (ionic strength <0.01).

$$\log \gamma = -0.509 Z_1 Z_2 \sqrt{\mu} \quad (3\text{-}32)$$

The integers Z_1 and Z_2 are the numbers of charges (valences) for the cation and anion of the salt. The ionic strength (μ) is evaluated by Eq. 3-33.

$$\mu = \tfrac{1}{2} \Sigma_i C_i Z_i{}^2 \quad (3\text{-}33)$$

Here C_i are the molar concentrations of the ions. The summation is carried out over all the ions present. The activity coefficient γ (Eq. 3-32) is the mean activity coefficient for both the cation and anion.

Equation 3-32 suggests that extrapolation of equilibrium constants to infinite dilution is done appropriately by plotting $\log K'$ vs. $\sqrt{\mu}$. Such a plot[10] is shown in Fig. 3-1. Here values of $pK_a{}'$ for dissociation of $H_2PO_4{}^-$, AMP$^-$, and ADP^{2-}, and ATP^{3-} are plotted vs. $\sqrt{\mu}$. The variation of $pK_a{}'$ with $\sqrt{\mu}$ at low concentrations (Eq. 3-34) is derived by application of the Debye–Hückel equation (Eq. 3-32):

$$pK_a{}' = pK_a - 0.509\,(Z_A{}^2 - Z_{HA}{}^2)\sqrt{\mu} \quad (3\text{-}34)$$

Straight lines of slope $-0.509\,(Z_A{}^2 - Z_{HA}{}^2)$ are expected. The observed (negative) slopes (Fig. 3-1) are ~ 1.5 for $H_2PO_4^-$ and AMP^-, ~ 2.5 for ADP^{2-}, and ~ 3.5 for ATP^{3-}. The data over the entire range of ionic strength are fitted by empirical relationships of the type of Eq. 3-35

$$pK_a' = pK_a - a\sqrt{\mu} + b\mu \quad \text{for } \mu < 0.2 \quad (3-35)$$

in which a and b are empirically determined constants. For example, for $H_2PO_4^-$ $a = 1.52$ and $b = 1.96$. The value of pK_a found was 7.18, about 0.22 greater than the value at $\mu = 0.2$, an ionic strength more commonly used in the laboratory and close to that found in tissues. Note that the difference between the extrapolated pK_a for ATP^{3-} of 7.68 and the observed value of ~ 7.04 at $\mu = 0.2$ is even greater. Serious errors can be introduced into calculations by using extrapolated values for K for solutions of appreciable ionic strength. The errors will be maximal for ions of high charge type such as ATP^{3-} and ATP^{4-}.

Another problem with equilibrium constants for reactions that use or produce hydrogen ions is

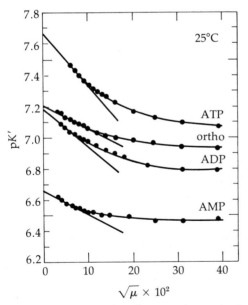

Fig. 3-1 The apparent pK values for secondary ionizations of AMP, ADP, ATP, and H_3PO_4 (abbreviated ortho.) plotted against $\sqrt{\mu}$. Temperature: 25°C. Reprinted from R. C. Phillips et al., *Biochemistry* **2**, 503 (1963). Copyright by the American Chemical Society.

that there is no rigorous relationship between pH and a_{H^+} or $[H^+]$. Indeed, the concept of an activity of a single ion has little meaning in thermodynamics. Nevertheless, in the pH range of interest to biochemists results that are very close to those obtained by more rigorous methods are achieved by assuming that the pH meter responds to hydrogen ion activity. The almost universal practice of biochemists is to assume the pH meter reading obtained with a glass electrode equal to $-\log_{H^+}$ and to substitute the value of a_{H^+} so obtained for $[H^+]$ in defining apparent equilibrium constants.

Often equilibrium data are not extrapolated to $\mu = 0$ but K' is used to define an apparent free energy change $\Delta G'$.

15. Changes in Equilibria with Temperature

At constant pressure ΔG varies with absolute temperature according to Eq. 3-36:

$$\frac{d(\Delta G/T)}{dT} = -\Delta H/T^2 \quad (3-36)$$

The corresponding variation in K is described by the **van't Hoff equation** (Eq. 3-37)

$$\frac{d \ln K}{dT} = \frac{\Delta H^0}{RT^2}$$

or

$$\frac{d \ln K}{d(1/T)} = \frac{-\Delta H^0}{R} \quad (3-37)$$

or

$$\Delta H^0 (\text{kJ mol}^{-1}) = \frac{-0.01914 \, d \log_{10} K}{d(1/T)}$$

If ΔH^0 can be assumed constant over the temperature range of an experiment, a plot of $\ln K$ vs. $1/T$ provides a convenient estimate of ΔH^0 (or $\Delta H'$ if $\ln K'$ is plotted). The slope of the line will be $-\Delta H^0/R$. Since ΔG^0 can be calculated from K, the method also permits evaluation of ΔS^0 using Eq. 3-14. However, the method is of low accuracy and it is preferable to establish ΔH by direct calorimetry. Also, especially for proteins, the assumption that ΔH^0 is constant over a significant temperature range may be erroneous.

B. TABLES OF ΔG^0 VALUES FOR BIOCHEMICAL COMPOUNDS

1. Free Energies of Formation

Table 3-3 gives, in the first column, standard values of free energies of formation from the elements ΔG_f^0 for a variety of pure solids, gases, and liquids, as well as values for substances in solution at the hypothetical 1 M activity. A few comments are in order concerning the latter.

Consider the value of ΔG_f^0 for pure liquid acetic acid, -389.1 kJ mol^{-1}. The equation for the reaction of formation from the elements is (3-38):

2 C (s) + 2 H$_2$ (g, 1 atm)
$$+ \text{O}_2 \text{ (g, 1 atm)} \longrightarrow \text{C}_2\text{H}_4\text{O}_2 \text{ (l)} \quad (3\text{-}38)$$
$\Delta G_f^0 = -389.1$ kJ mol^{-1}

To obtain the free energy of formation in aqueous solution, we must have solubility data as well as activity coefficients of acetic acid at various concentrations. From these data the free energy change for solution of the liquid acetic acid in water to give aqueous acetic acid in the hypothetical 1 molal standard state (Eq. 3-39) can be obtained.

$$\text{Acetic acid (l)} \longrightarrow \text{acetic acid (aq)} \quad (3\text{-}39)$$
$$\Delta G = -7.3 \text{ kJ mol}^{-1}$$

From Eqs. 3-38 and 3-39 we obtain

$$2 \text{ C} + 2 \text{ H}_2 + \text{O}_2 \longrightarrow \text{acetic acid (aq)} \quad (3\text{-}40)$$
$$\Delta G_f^0 = -396.4 \text{ kJ mol}^{-1}$$

In many computations it is convenient to have ΔG values for single ions, e.g., for acetate$^-$. We can obtain ΔG_f^0 of acetate$^-$ (aq) from that of acetic acid (aq) by making use of ΔG^0 of dissociation (Eq. 3-41).

$$\text{Acetic acid} \longrightarrow \text{H}^+ + \text{acetate}^- \quad (3\text{-}41)$$
$$\Delta G^0 = -5.708 \log K = 5.708 \text{ p}K$$
$$= +27.2 \text{ kJ mol}^{-1} \text{ at } 25°C$$

By convention *we define the free energy of formation of H$^+$ as zero.* Then by summing Eqs. 3-40 and 3-41 we obtain ΔG_f^0 of acetate$^-$ = -369.2 kJ mol^{-1}.

2. Free Energies of Dissociation of Protons

Table 3-4 gives thermodynamic dissociation constants and values of ΔG^0 and ΔH^0 for a number of acids of interest in biochemistry. Some of these values were used in obtaining the values of ΔG_f^0 for the ions of Table 3-3. The data of Table 3-4 can also be used in evaluation of free energy changes for reactions of ionic forms not given in Table 3-3.

3. Group Transfer Potentials

Recall that the equilibria for reactions by which monomers are linked to form biopolymers (whether amides, esters, phosphodiesters, or glycosides) usually favor *hydrolysis* rather than formation. The equilibrium positions depend on the exact structures. Some linkages are formed easily if monomer concentrations are high enough, but others are never formed in significant concentrations. Likewise, hydrolysis may be partial at equilibrium or it may be 99.9% or more complete.

Consider hydrolysis (Eqs. 3-42 and 3-43) of two organic phosphates, **adenosine triphosphate** (ATP) and **glucose 6-phosphate** (see Fig. 7-1 for structure).

$$\text{HATP}^{3-} + \text{H}_2\text{O} \longrightarrow \text{HADP}^{2-} + \text{H}_2\text{PO}_4^- \quad (3\text{-}42)$$
$$\Delta G^0 \approx -32.9 \text{ kJ mol}^{-1}$$
$$\Delta H^0 \approx -22.6 \text{ kJ mol}^{-1} \text{ at } 25°C, \mu = 0.25$$

$$\alpha\text{-D-Glucose 6-phosphate}^- + \text{H}_2\text{O} \longrightarrow$$
$$\text{glucose} + \text{H}_2\text{PO}_4^- \quad (3\text{-}43)$$
$$\Delta G^0 = -16.4 \text{ kJ mol}^{-1} \text{ at } 25°C$$

Note that the free energy decrease upon hydrolysis is twice as large for ATP as it is for glucose 6-phosphate. Glucose phosphate is thermodynamically more stable than ATP. It would be easier to form than would ATP by a reversal of the hydrolysis reaction and also easier to form biosynthetically. Furthermore, from the free energies of

TABLE 3-3
Free Energies of Formation and of Oxidation at 25°C for Compounds of Biochemical Interest[a,b]

Compound	Formula	ΔG_f^0 (kJ mol^{-1})	ΔG_c^0 (kJ mol^{-1})	For oxidation by NAD$^+$		No. of electrons
				ΔG_{ox}^0 (kJ mol^{-1})	$\Delta G_{ox}'$ (pH 7) (kJ mol^{-1}	
Acetaldehyde	C_2H_4O	−139.7	−1123.5	171.5	−28.3	10
Acetic acid	$C_2H_4O_2$	−396.4	−866.8	169.2	9.3	8
Acetate$^-$		−369.2	−894.0	142.0	22.1	8
Acetyl-CoA		−374.1*	−889.1*	146.9*	−13.0*	8
Acetyl-P		−1218.4	−901.7	134.3	−25.6	8
Acetylene^c	C_2H_2	209.2	−1235.2	59.8	−140.0	10
Acetoacetate$^-$	$C_4H_5O_3{}^-$	−493.7	−1795.4	276.5	−3.2	16
Acetone	C_3H_6O	−161.2	−1733.6	338.4	18.7	16
cis-Aconitate^{3-}	$C_6H_3O_6{}^{3-}$	−920.9	−2157.0	173.9	−65.8	18
L-Alanine	$C_3H_7O_2N$	−371.3	−1642.0	300.4	0.8	15
L-Asparagine	$C_4H_8O_3N_2$	−526.6	−1999.7	331.2	−28.4	18
L-Aspartate$^-$	$C_4H_6O_4N^-$	−700.7	−1707.0	235.4	−24.3	15
n-Butanol	$C_4H_{10}O$	−171.8	−2591.7	516.2	36.7	24
n-Butyric acid	$C_4H_8O_2$	−380.2	−2146.1	443.8	44.2	20
n-Butyrate$^-$	$C_4H_7O_2{}^-$	−352.6	−2173.7	416.2	56.6	20
Butyryl-CoA		−357.5*	−2168.8*	421.1*	21.5*	20
Caproate$^-$	$C_6H_{11}O_2{}^-$	−329.7	−3459.7	684.1	84.7	32
CO_2 (g)		−394.4	0.0	0.0	0.0	0
CO_2 (aq)		−386.2	−8.2	−8.2	−8.2	0
$HCO_3{}^-$		−587.1	−44.5	−44.5	−4.6	0
CO (g)		−137.3	−257.1	1.9	−38.1	2
Citrate^{3-}	$C_6H_5O_7{}^{3-}$	−1166.6	−2148.4	182.5	−57.3	18
Creatine	$C_4H_9O_2N_3$	−264.3	−2380.6	338.8	−80.8	21
Creatinine	$C_4H_7ON_3$	−28.9	−2378.8	340.6	−79.0	21
Crotonate$^-$	$C_4H_5O_2{}^-$	−275.7	−2013.4	317.5	−2.1	18
Cysteine	$C_3H_7O_2NS$	−339.8	−2178.3	541.1	121.5	21
Cystine	$C_6H_{12}O_4N_2S_2$	−665.3	−4133.8	1046.0	246.9	40
Dihydroxyacetone-P^d		−1293.2	−1458.4	95.5	−144.2	12
Erythrose 4-P^d		−1439.1	−1944.1	127.8	−191.9	16
Ethanol	C_2H_6O	−181.5	−1318.8	235.1	−4.6	12
Ethylene (g)^c	C_2H_4O	68.1	−1331.3	222.7	−17.1	12
Formaldehyde	CH_2O	−130.5	−501.0	16.9	−63.0	4
Formic acid	CH_2O_2	−356.1	−275.5	−16.5	−56.5	2
Formate$^-$	$CHO_2{}^-$	−350.6	−281.0	−22.0	−22.0	2
Fructose	$C_6H_{12}O_6$	−915.4	−2874.1	233.8	−245.7	24
Fructose 6-P^d		−1758.3	−2888.1	219.8	−259.7	24
Fructose di-P^d		−2600.8	−2902.5	205.4	−274.1	24
Fumaric acid	$C_4H_4O_4$	−647.1	−1404.8	149.2	−90.6	12
Fumarate$^-$	$C_4H_3O_4{}^-$	−604.2	−1447.7	106.2	−93.6	12
α-D-Galactose	$C_6H_{12}O_6$	−923.5	−2865.9	242.0	−237.5	24
α-D-Glucose	$C_6H_{12}O_6$	−917.2	−2872.5	235.6	−243.8	24
Glucose 6-P		−1760.3	−2886.0	221.8	−257.6	24
L-Glutamate$^-$	$C_5H_8O_4N^-$	−696.8	−2342.5	376.9	−2.7	21
L-Glutamine	$C_5H_{10}O_3N_2$	−524.8	−2633.1	474.8	−4.7	24

(continued)

TABLE 3-3 (*Continued*)

Compound	Formula	ΔG_f^0 (kJ mol^{-1})	ΔG_c^0 (kJ mol^{-1})	For oxidation by NAD$^+$		No. of electrons
				ΔG_{ox}^0 (kJ mol^{-1})	$\Delta G_{ox}'$ (pH 7) (kJ mol^{-1})	
Glycerol	$C_3H_8O_3$	−488.5	−1643.4	169.5	−110.2	14
Glycerol-P		−1336.2	−1652.6	160.3	−119.4	14
Glycine	$C_2H_5O_2N$	−373.5	−1008.3	157.2	−22.6	9
Glycogen	$C_6H_{10}O_5$	−665.3	−2887.0	220.9	−258.6	24
Glycolate$^-$	$C_2H_3O_3^-$	−523.4	−739.7	37.2	−42.7	6
Glyoxylate$^-$	$C_2HO_3^-$	−461.1	−564.9	−46.9	−86.9	4
3-P glycerate^{-d}		−1515.7	−1235.9	59.0	−100.8	10
2-P glycerate^{-d}		−1509.9	−1241.8	53.2	−106.6	10
Glyceraldehyde 3-P^d		−1285.6	−1466.0	87.9	−151.8	12
H_2O (l)		−237.2	0.0	0.0	0.0	0
OH$^-$		−157.3	−79.9	−79.9	−39.9	0
H$^+$		0.0	0.0	0.0	0.0	0
H_2 (g)		0.0	−237.2	21.8	−18.2	2
H_2O_2		−136.8	−100.4	−359.4	−319.4	−2
H_2S		−27.4	−714.6	321.3	161.5	8
HS$^-$		12.6	−754.5	281.4	161.5	8
β-Hydroxybutyric acid	$C_4H_8O_3$	−531.4	−1994.9	336.0	−23.6	18
β-Hydroxybutyrate$^-$	$C_4H_7O_3^-$	−506.3	−2020.0	310.9	−8.8	18
Hydroxypyruvate	$C_3H_4O_4$	−615.9	−1041.6	−5.7	−165.5	8
Hypoxanthine	C_5H_6O	89.5	−2773.0	334.8	−144.6	24
Isocitrate^{3-}	$C_6H_5O_7^{3-}$	−1160.0	−2155.1	175.8	−63.9	18
α-Ketoglutarate^{2-}	$C_5H_4O_5^{2-}$	−798.0	−1885.5	186.4	−53.3	16
Lactate$^-$	$C_3H_5O_3^-$	−516.6	−1378.1	175.9	−23.9	12
α-Lactose	$C_{12}H_{22}O_{11}$	−1515.2	−5826.5	389.3	−569.7	48
L-Leucine	$C_6H_{13}O_2N$	−356.3	−3551.7	721.6	62.3	33
Mannitol	$C_6H_{14}O_6$	−942.6	−3084.0	282.8	−236.6	26
Malate^{2-}	$C_4H_4O_5^{2-}$	−845.1	−1444.0	109.9	−49.9	12
Methane (g)	CH_4	−50.8	−818.0	218.0	58.2	8
Methanol	CH_4O	−175.2	−693.5	83.4	−36.4	6
NH$_4^+$		−79.5	−276.3	112.2	12.3	3
NO$_2^-$		−34.5	−84.1	−472.6	−372.7	−3
NO (g)		86.7	−86.7	−345.7	−305.7	−2
NO$_3^-$		−110.5	−8.1	−655.6	−515.7	−5
Oxalate^{2-}	$C_2O_4^{2-}$	−674.9	−351.1	−92.1	−52.1	2
Oxaloacetate^{2-}	$C_4H_2O_5^{2-}$	−797.2	−1254.7	40.2	−79.7	10
H_3PO_4 (aq)^e		−1147.3	0.0	0.0	0.0	0
$H_2PO_4^-$ (aq)^e		−1135.1	−12.1	−12.1	27.8	0
HPO$_4^{2-}$ (aq)^e		−1094.1	−53.1	−53.1	26.8	0
n-Propanol	C_3H_8O	−175.8	−1956.1	374.8	15.2	18
Isopropanol	C_3H_8O	−185.9	−1946.0	384.9	25.3	18
Propionate$^-$	$C_3H_5O_2^-$	−360.0	−1534.2	278.2	38.5	14
Pyruvate$^-$	$C_3H_3O_3^-$	−474.5	−1183.1	111.9	−47.9	10
Phosphoenolpyruvate$^-$		−1269.5	−1245.0	50.0	−109.8	10
Ribose 5-P^d		−1599.9	−2414.9	175.0	−224.6	20
Ribulose 5-P^d		−1597.6	−2417.1	172.8	−226.8	20

(*continued*)

TABLE 3-3   *(Continued)*

Compound	Formula	ΔG_f^0 (kJ mol^{-1})	ΔG_c^0 (kJ mol^{-1})	ΔG_{ox}^0 (kJ mol^{-1})	$\Delta G_{ox}'$ (pH 7) (kJ mol^{-1})	No. of electrons
				For oxidation by NAD$^+$		
Sedoheptulose 7-P[d]		−1913.3	−3364.6	261.2	−298.2	28
Sedoheptulose di-P[d]		−2755.8	−3379.0	246.9	−312.5	28
Sorbitol	$C_6H_{14}O_6$	−942.7	−3083.9	282.9	−236.5	26
Succinate^{2-}	$C_4H_4O_4^{2-}$	−690.2	−1598.9	214.1	14.3	14
Succinyl-CoA		−686.7*	−1602.4*	210.6*	−29.2*	14
Sucrose	$C_{12}H_{22}O_{11}$	−1551.8	−5789.9	425.9	−533.1	48
SO$_4^{2-}$		−742.0	0.0	0.0	79.9	0
SO$_3^{2-}$		−497.1	−244.9	14.1	54.0	2
S$_2$O$_3^{2-}$		−513.4	−733.4	302.5	222.6	8
L-Threonine	$C_4H_9O_3N$	−514.6	−2130.3	330.1	−49.5	19
L-Tyrosine	$C_9H_{11}O_3N$	−387.2	−4466.8	842.5	23.4	41
Urea	CH_4ON_2	−203.8	−664.9	112.0	−7.8	6
Uric acid	$C_5H_4O_3N_4$	−356.9	−2089.4	241.5	−118.1	18
L-Valine	$C_5H_{11}O_2N$	−360.0	−2916.5	579.9	40.5	27
Xanthine	$C_5H_5O_2N_4$	−139.3	−2425.6	293.8	−125.7	21
D-Xylulose	$C_5H_{10}O_5$	−748.1	−2409.8	180.1	−219.5	20

[a] The quantities tabulated are ΔG_f^0, the standard free energy of formation from the elements; ΔG_c^0, the standard free energy of combustion: ΔG_{ox}^0, the standard free energy of oxidation by NAD$^+$ to products NADH + H$^+$, CO$_2$, H$_2$O, N$_2$, HPO$_4^{2-}$, and SO$_4^{2-}$; $\Delta G_{ox}'$ (pH 7), the apparent standard free energy change at pH 7. All values are in kJ mol^{-1} at 25°C in aqueous solution unless indicated otherwise. If a compound is designated (g) the values are for the gaseous phase at 1 atm pressure. The number of electrons involved in complete oxidation to CO$_2$, H$_2$O, N$_2$, and H$_2$SO$_4$ is given in the final column. If this number is negative, the compound must be reduced to obtain the products, e.g., 2 NO$_3^-$ + 10 e$^-$ + 12 H$^+$ → N$_2$ + 6 H$_2$O. The data for phosphate esters refers to the compounds with completely dissociated phosphate groups (−O−PO$_3^{2-}$). The values of ΔG_f^0 for many of these compounds were calculated as ΔG_f^0 (nonphosphorylated compound) −ΔG^0 for hydrolysis (to HPO$_4^{2-}$, Table 3-5) −ΔG_f^0 for H$_2$O (one molecule for each phosphate ester formed) + ΔG_f^0 for HPO$_4^{2-}$ (from this table). Data from Bassman and Krause[d] were used directly. For acyl-CoA derivatives CoA(−SH) is treated as an "element," i.e., the values of ΔG_f^0 given and designated with an asterisk (*) are for formation from the elements plus free CoA. The values of ΔG_c^0 and ΔG_{ox} are for oxidation to the usual products plus CoA. Values of ΔG^0 of hydrolysis (Table 3-5) were used in computing ΔG_f^0 for each of these compounds from that of the corresponding alcohol or carboxylate anion. Another source containing an extensive table of free energy values is R. C. Wilhoit, *in* "Biochemical Microcalorimetry" (H. D. Brown, ed.), pp. 305–317. Academic Press, New York, 1969.

[b] The major source is C. Long, ed., "Biochemists Handbook," pp. 90–92. Van Nostrand, Reinhold, Princeton, New Jersey, 1961. Most of the values in this collection are from K. Burton, *Ergeb. Physiol., Biol. Chem. Exp. Pharmakol.* **49,** 275–298 (1957).

[c] From D. R. Stull, E. F. Westrum, Jr., and G. C. Sinke, "The Chemical Thermodynamics of Organic Compounds." Wiley, New York, 1969.

[d] J. A. Bassham and G. H. Krause, *BBA* **189,** 207–221 (1969).

[e] J. R. Van Wazer, "Phosphorus and Its Compounds," Vol. I, p. 889. Wiley (Interscience), New York, 1958.

hydrolysis it follows that a phosphoryl group could be transferred spontaneously from ATP to glucose in the presence of a suitable catalyst but not vice versa.

Because it reflects quantitatively the thermodynamic tendency for a group to be transferred to another nucleophile (see Chapter 7), *the free energy decrease upon hydrolysis* is sometimes called the **group transfer potential.** Thus, the phosphoryl group

$$-\overset{\displaystyle O}{\underset{\displaystyle OH}{\overset{\|}{P}}}-O^-$$

of ATP is transferred during hydrolysis to a hydroxyl ion from water with a group transfer potential of 32.9 kJ mol^{-1}. While the choice of water as the reference nucleophile for expression of the group transfer potential is somewhat arbitrary, it is customary. Transfer of groups is important in energy metabolism and in biosynthesis of polymers. Hence, biochemists often refer to group transfer potentials (negative free energies of hy-

drolysis) of biochemical compounds. Some of these are given in Table 3-5.

4. "Constants" That Vary with pH

Equation 3-42 is written for hydrolysis of HATP^{3-} to HADP^{2-} + H$_2$PO$_4^-$, a stoichiometry

TABLE 3-4
Values of pK_a, ΔG^0, and ΔH^0 for Ionization of Acids at 25°C^{a,b}

Acid	pK_a	ΔG^0 (kJ mol^{-1})	ΔH^0 (kJ mol^{-1})	Acid	pK_a	ΔG^0 (kJ mol^{-1})	ΔH^0 (kJ mol^{-1})
Formic acid	3.75	21.4	0.04	Glucose 6-phosphate	6.50	37.1	−1.8
Acetic acid	4.76	27.2	−0.1	Pyrophosphoric			
Propionic acid	4.87	27.8	−0.6	acid, H$_4$P$_2$O$_7$			
Lactic acid	3.97 (35°C)	23.4	2.2	pK_3	6.7	38.1	−1.3
Pyruvic acid	2.49	14.2	12.1	(apparent)	6.12^e	34.9	0.5
NH$_4^+$	9.25	52.8	52.2	pK_4	9.4	53.6	−7.1
CH$_3$NH$_4^+$	10.59	60.4	55.4	(apparent)	8.95^e	51.2	1.7
Alanine				Adenosine	3.5	20.1	13.0
—COOH	2.35	13.4	3.1	AMP			
—NH$_3^+$	9.83	56.1	45.4	pK_1 (ring,			
β-Alanine				apparent)	3.74^e	21.3	4.2
—COOH	3.55	20.3	4.5	pK_2 (phosphate)	6.67^d	38.1	3.6
—NH$_3^+$ (apparent)	10.19	58.2		(apparent)	6.45^e	36.8	3.6
L-Alanyl-L-alanine	3.34	19.1	−0.5	ADP			
Aspartic acid				pK_2 (ring,			
—COOH	2.05	11.7	7.7	apparent)	3.93^e	22.4	4.2
—COOH	3.87	22.1	4.0	pK_3 (diphosphate)	7.20^d	41.1	−5.7
—NH$_3^+$	10.60	60.5	38.8	(apparent)	6.83^f	39.0	−5.7
H$_2$CO$_3$, pK_1	6.35^c	36.2	9.4	ATP			
pK_2	10.33	59.0	15.1	pK_3 (ring,			
H$_3$PO$_4$, pK_1	2.12	12.1	−7.9	apparent)	4.06	23.2	0
pK_2	7.18^d	41.0	3.8	pK_4 (triphosphate)	7.68^d	43.8	−7.0
(apparent)	6.78^e	38.7	3.3	(apparent)	7.06^d	40.2	−7.0
pK_3	12.40	70.8	17.6	Pyridine	5.17	29.5	20.1
Glycerol				Phenol	9.98	56.9	23.6
1-phosphate, pK_2	6.66	38.0	−3.1				

^{a} These are thermodynamic values (infinite dilution) except for those labeled apparent. The latter apply at an ionic strength of 0.2–0.25.

^{b} Most data are from W. P. Jencks, and J. Regenstein, *in* "Handbook of Biochemistry and Molecular Biology," 3rd ed., Vol. I (G. D. Fasman, ed.), pp. 305–351. © CRC Press, Inc., Cleveland, 1976.

^{c} Here, pK_1 is for K_1 = [H$^+$][HCO$_3^-$]/[CO$_2$] + [H$_2$CO$_3$]. From R. E. Forster, J. T. Edsall, A. B. Otis, and F. J. W. Roughton, eds., *NASA Spec. Publ.* **188** (1969).

^{d} From R. C. Phillips, P. George, and R. J. Rutman, *Biochemistry* **2**, 501–508 (1963).

^{e} From R. A. Alberty, *JBC* **244**, 3290–3302 (1969).

^{f} Values used by R. A. Alberty ("Horizons of Bioenergetics," pp. 135–147. Academic Press, New York, 1972) calculated for 0.2 ionic strength from equations of R. C. Phillips, P. George, and R. J. Rutman [*JACS* **88**, 2631–2640 (1966)].

BOX 3-A

THE ADENYLATE SYSTEM

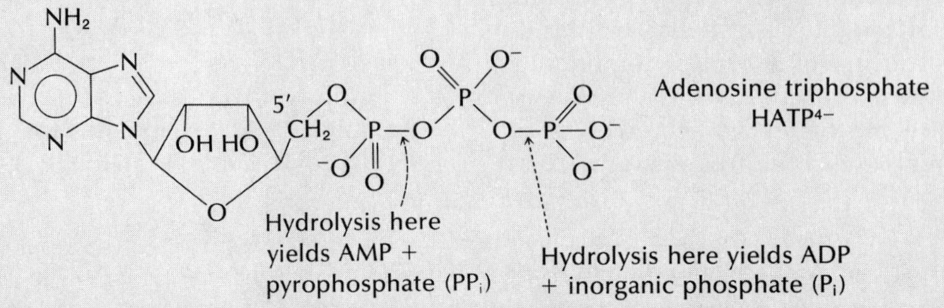

Hydrolysis here
yields AMP +
pyrophosphate (PP$_i$)

Hydrolysis here yields ADP
+ inorganic phosphate (P$_i$)

Adenosine triphosphate
HATP^{4-}

Of central importance to the energy metabolism of all cells is the **adenylate system** which consists of the triphosphate (ATP), diphosphate (ADP), and 5′-monophosphate (AMP) of adenosine together with **inorganic phosphate** (P$_i$) and **magnesium** ions. Adenosine triphosphate is a thermodynamically unstable molecule with respect to hydrolysis to ADP or AMP as indicated in the diagram above. It is this instability toward hydrolysis (high group transfer potential) that permits ATP to serve as a carrier of chemical energy for use in most energy-requiring processes in cells, including the contraction of muscles.

The **phosphate anhydride** (pyrophosphate) linkages of ATP are generated by the joining of ADP and inorganic phosphate by means of special **phosphorylation** reactions. The most important of the latter occur in the photosynthetic membranes

of chloroplasts and in oxygen-utilizing membranes of bacteria and of mitochondria. Conversion of AMP to ADP is accomplished by transfer of the terminal phosphoryl group from an ATP molecule to AMP (Chapter 7) in a reaction catalyzed by an extremely active enzyme (**adenylate kinase**) that is found in all cells. The following equations indicate how one of the special phos-

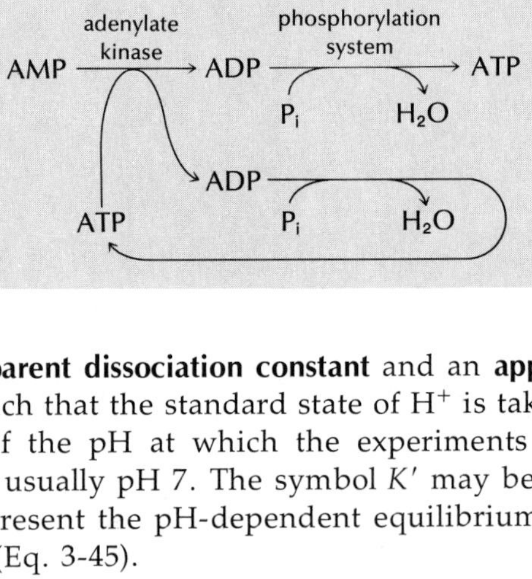

that applies well in the pH range around 6. However, at a pH above ~7 most of the ATP is in the form ATP^{4-} and is cleaved to HPO$_4^{2-}$ according to Eq. 3-44.

$$\text{ATP}^{4-} + \text{H}_2\text{O} \longrightarrow \text{ADP}^{3-} + \text{HPO}_4^{2-} + \text{H}^+ \qquad (3\text{-}44)$$
$$\Delta G^0 \approx +5.4 \text{ kJ mol}^{-1} \text{ at } 25°\text{C}, \quad \mu = 0.25$$
$$\Delta H^0 \approx -19.7 \text{ kJ mol}^{-1}$$

The value of $\Delta G^0 = +5.4$ kJ mol^{-1} for this reaction is hardly the large negative number expected for a highly spontaneous reaction. What is the matter? The problem is that H$^+$ is produced and that the standard state of H$^+$ is $1M$, not 10^{-7} M. Because of this biochemists often prefer to use another kind

of **apparent dissociation constant** and an **apparent ΔG** such that the standard state of H$^+$ is taken as that of the pH at which the experiments were done, usually pH 7. The symbol K' may be used to represent the pH-dependent equilibrium constant (Eq. 3-45).

$$K' = \frac{[\text{ADP}^{3-}][\text{HPO}_4^{2-}]}{[\text{ATP}^{4-}]} \qquad (3\text{-}45)$$

If a single proton is produced in the reaction as in Eq. 3-44, the following relationship will hold (Eq. 3-46).

$$\Delta G' = \Delta G^0 - 5.708 \times \text{pH kJ mol}^{-1} \text{ at } 25°\text{C} \quad (3\text{-}46)$$

Box 3-A (*Continued*)

phorylation reactions must be used twice for the conversion of one molecule of AMP into one molecule of ATP.

Various measures of the phosphorylating potential of the adenylate system within cells have been proposed. One measure is the product [ATP]/[ADP] [P_i] (which will be called the **phosphorylation state ratio** or R_p). This ratio directly affects the free energy of hydrolysis of ATP (Eq. 3-28). The value of R_p may be as high as 10^5 M^{-1} within cells,[a] adding -22.8 kJ mol^{-1} to ΔG of hydrolysis of ATP. Another quantity, proposed by Atkinson and associates,[b–d] is the "energy charge," the mole fraction of adenylic acid "charged" by conversion to ATP. The ADP is regarded as "half-charged."

$$\text{Energy charge} = \frac{[\text{ATP}] + \frac{1}{2}[\text{ADP}]}{[\text{ATP}] + [\text{ADP}] + [\text{AMP}]}$$

The energy charge varies from 0 if only AMP is present to 1.0 if all of the AMP is converted to ATP. Measurements on a variety of cells and tissues[d] show that the energy charge is usually between 0.75 and 0.90. Although it is easy to calculate its numerical value, the energy charge cannot be used in chemical equations. The idea that the energy charge of a cell may play a key role in regulation of metabolism has been challenged.[e]

Abbreviations: Remember that P_i refers to the mixture of ionic forms of phosphoric acid present under experimental conditions. Between pH 4 and pH 10 this will be $H_2PO_4^-$ ($pK_a' = 6.8$) and HPO_4^{2-}. Likewise the symbols AMP, ADP, and ATP refer to mixtures of ionic forms and PP_i refers to a mixture of the ions of pyrophosphoric acid. Above pH 4.4 only $H_2P_2O_7^{2-}$ ($pK_a' = 6.1$), $HP_2O_7^{3-}$ ($pK_a' = 9.0$), and $P_2O_7^{4-}$ contribute appreciably.

As far as is known pyrophosphate is usually hydrolyzed rapidly to two molecules of P_i by pyrophosphatases in cells. This process, too, serves an essential function in the adenylate system (Chapter 7, Section E,1).

Both the positions of equilibria in reactions of the adenylate system and the rates of reaction depend upon the concentrations of metal ions present. Magnesium ion is especially important and complexes such as $MgATP^{2-}$ are regarded as the true substrates for many ATP-utilizing enzymes.

[a] L. J. Eilerman and E. C. Slater, *BBA* **216**, 226–228 (1970).

[b] J. S. Swedes, R. J. Sedo, and D. E. Atkinson, *JBC* **250**, 6930–6938 (1975).

[c] L. C. Shen, L. Fall, G. M. Walton, and D. E. Atkinson, *Biochemistry* **7**, 4041–4045 (1968).

[d] A. G. Chapman, L. Fall, and D. E. Atkinson, *J. Bacteriol.* **108**, 1072–1086 (1971).

[e] D. L. Purich and H. J. Fromm, *JBC* **248**, 461–466 (1973).

Note that $\Delta G' = -RT \ln K'$ and that [H^+] does not appear in the expression for K' given by Eq. 3-45. From the value $\Delta G^0 = +5.4$ kJ mol^{-1} we obtain for the hydrolysis of ATP at 25°C, $\mu = 0.25$, according to Eq. 3-44,

$$\Delta G' \text{ (pH 7)} = -34.5 \text{ kJ mol}^{-1} \qquad (3\text{-}47)$$
$$= -8.26 \text{ kcal mol}^{-1}$$

An additional set of standard states is frequently met in the biochemical literature. An equilibrium constant, designated here as $K^\dagger$ (and often in the literature as K_{obs}), is used to relate the *total concentrations of all ionic forms* of the components present at the pH of the experiment. Thus,

$$K^\dagger = \frac{[\text{ADP, all forms}][\text{phosphate, all forms}]}{[\text{ATP, all forms}]} \qquad (3\text{-}48)$$

and

$$\Delta G^\dagger = -RT \ln K^\dagger \qquad (3\text{-}49)$$

The free energy change $\Delta G^\dagger$ can be related to $\Delta G'$ by considering the relationship of $K^\dagger$ to K'. For ATP hydrolysis in the pH range of 2–10 $K^\dagger$ is given by Eq. 3-50.

$$K^\dagger = \frac{K' \left(1 + \dfrac{[\text{H}^+]}{K_{\text{HADP}^{2-}}} + \dfrac{[\text{H}^+]^2}{K_{\text{HADP}^{2-}} K_{\text{H}_2\text{ADP}^-}}\right)\left(1 + \dfrac{[\text{H}^+]}{K_{\text{HPO}_4^{2-}}}\right)}{\left(1 + \dfrac{[\text{H}^+]}{K_{\text{HATP}^{3-}}} + \dfrac{[\text{H}^+]^2}{K_{\text{HATP}^{3-}} K_{\text{H}_2\text{ATP}^{2-}}}\right)}$$

$$(3\text{-}50)$$

TABLE 3-5
Free Energies of Hydrolysis at 25°C (in kJ mol^{-1})^a

Compound	Products	ΔG^0	$\Delta G'$ (pH 7)	ΔH^0
ATP^{4-}	ADP^{3-} + HPO$_4^{2-}$ + H$^+$	5.41^b	−34.54	−19.71
ATP^{4-}	AMP^{2-} + HP$_2$O$_7^{3-}$ + H$^+$	2.54^c	−37.4	−19.0
MgATP^{2-}	MgADP$^-$ + HPO$_4^{2-}$ + H$^+$	16.0^d	−24.0	−14.2
ADP^{3-}	AMP^{2-} + HPO$_4^{2-}$ + H$^+$	3.67^c	−36.3	−13.5
AMP^{2-}	Adenosine + HPO$_4^{2-}$	−9.6^e	−9.6	0
ATP^{4-}	Adenosine + HP$_3$O$_{10}^{4-}$	−36.0^e	−36.0	−7.9
HP$_2$O$_7^{3-}$	2 HPO$_4^{2-}$ + H$^+$	6.54^c	−33.4	−12.6
Acetyl phosphate^{2-}	Acetate$^-$ + HPO$_4^{2-}$ + H$^+$	−7.7^f	−47.7	
1,3-Diphosphoglycerate^{4-}	3-Phosphoglycerate^{3-} + HPO$_4^{2-}$ + H$^+$	−14.5^g	−54.5	
Phosphoenolpyruvate^{3-}	Pyruvate$^-$ + HPO$_4^{2-}$	−61.9	−61.9	−25.1
Carbamoyl phosphate^{2-}	H$_2$N—$\overset{\overset{\displaystyle O}{\|}}{C}$—O$^-$ + HPO$_4^{2-}$ + H$^+$	−11.5	−51.5	
Creatine phosphate$^-$	Creatine$^+$ + HPO$_4^{2-}$	−43.1	−43.1	
Phosphoarginine$^-$	Arginine + HPO$_4^{2-}$		−38.1^h	(Mg^{2+} present)
Glycerol phosphate^{2-}	Glycerol + HPO$_4^{2-}$	−9.2	−9.2	
α-D-Glucose 6-phosphate^{2-}	α-D-Glucose + HPO$_4^{2-}$	−13.8	−13.8	−2.5
Glucose 1-phosphate^{2-}	Glucose + HPO$_4^{2-}$	−20.9	−20.9	
Maltose (or glycogen)	2-Glucose	−16.7	−16.7	
Sucrose	Glucose + fructose	−29.3	−29.3	
UDPglucose^{2-}	Glucose + UDP^{3-} + H$^+$	9.4	−30.5	
N^{10}-Formyltetrahydrofolic acid	Formate$^-$ + H$^+$ + tetrahydrofolic acid	14.1	−25.9^h	
Acetic anhydride	2-Acetate$^-$ + 2H$^+$	31.1	−48.9	
Acetyl-CoA	Acetate$^-$ + H$^+$ + CoA	4.9	−35.1^i	
Succinyl-CoA$^-$	Succinate^{2-} + H$^+$ + CoA	−3.5	−43.5^j	
Ethyl acetate	Ethanol + acetate$^-$ + H$^+$	20.2	−19.7	
Asparagine	Aspartate$^-$ + NH$_4^+$	−15.1	−15.1	
Glycine ethyl ester$^+$ (39°C)	Glycine + ethanol + H$^+$	4.9	−35.1	
Valyl-tRNA$^+$	Valine + tRNA + H$^+$	4.9	−35.1	

^{a} Unless indicated otherwise, the values below are based on tables from W. P. Jencks, *in* "Handbook of Biochemistry and Molecular Biology," 3rd ed., Vol I (G. D. Fasman, ed.), pp. 296–304. © CRC Press, Inc., Cleveland, 1976. The value of $\Delta G'$ at pH 7 for a reaction producing one proton equals $\Delta G^0 − 39.96$ kJ mol^{-1}.

^{b} R. W. Guynn and R. L. Veech, *JBC* **248**, 6966–6972 (1973).

^{c} Based on the value of +11.80 kcal mol^{-1} for hydrolysis to P$_2$O$_7^{4-}$ plus ΔG^0 of dissociation of HP$_2$O$_7^{3-}$ as quoted by R. A. Alberty [*JBC* **244**, 3290–3324 (1969)]. However, 1.017 kcal mol^{-1} was added to the value 11.80 to bring it into line with that for hydrolysis of ATP to ADP.

^{d} R. A. Alberty, "Horizons of Bioenergetics," Academic Press, New York, 1972, pp. 135–147.

^{e} P. George, R. J. Witonsky, M. Trachtman, C. Wu, W. Dorwart, L. Richman, W. Richman, F. Shurayh, and B. Lentz, *BBA* **223**, 1–15 (1970).

^{f} Based on $\Delta G†$ = 3.0 kcal mol^{-1} for acetyl phosphate + CoA → acetyl-CoA + P$_i$, E. R. Stadtman [*in* "The Enzymes" (P. D. Boyer, ed.), 3rd ed., Vol. 8, pp. 1–49. Academic Press, New York, 1973], together with $\Delta G'$ (pH 7) for hydrolysis of acetyl-CoA.

^{g} Estimated from ΔG^0 for ATP hydrolysis + $\Delta G' = −19.9$ kJ mol^{-1} for the 3-phosphoglycerate kinase reaction: K. Burton and H. A. Krebs, *BJ* **54**, 94–107 (1953); **59**, 44–46 (1955).

^{h} Estimated from data at pH 7.7 or 8.0 (tables of Jencks).

^{i} R. W. Guynn, H. J. Gelberg, and R. L. Veech [*JBC* **248**, 6957–6965 (1973)] found $\Delta G° = −35.75$ kJ mol^{-1} at 38°C. Without data on ΔH, correction to 25°C is difficult. K. Burton [*BJ* **59**, 44–46 (1955)] gave $\Delta G'$ (pH 7) for ATP^{4-} + acetate$^-$ + CoA → ADP^{3-} + acetyl-CoA + HPO$_4^{2-}$ as approximately zero at 25°C. Guynn *et al.* found −0.56 kJ mol^{-1} at 38°C. This same value (−0.56 kJ) at 25°C was assumed to obtain the figure given here. This is equivalent to assuming ΔH^0 of hydrolysis as almost the same for ATP and acetyl-CoA, an unsupported assumption.

^{j} Assumed 2 kcal mol^{-1} more negative than that of acetyl-CoA as in tables of Jencks.

In this equation $K_{HADP^{2-}}$, etc., are stepwise dissociation constants as given in Table 3-4. The expressions in parentheses are the **Michaelis pH functions** (see also Chapter 6, Section E,5). In this instance they serve to relate the total concentration of a particular component to the concentration of the most highly dissociated form. Thus, for the pH range 2–10 (Eq. 3-51)

$$[P_i] \text{ total} = [HPO_4^{2-}](1 + [H^+]/K_{H_2PO_4^-}) \quad (3\text{-}51)$$

Using apparent pK_a values ($\mu = 0.2$) for $H_2PO_4^-$, $HADP^{2-}$, and $HATP^{3-}$ of 6.78, 6.83 and 7.06 (Table 3-4), we compute $\Delta G^{\dagger} = -35.0$ kJ mol^{-1} at pH 7 (if we take $\Delta G'$ at pH 7 as -34.5 kJ mol^{-1}). The difference between $\Delta G'$ and $\Delta G^{\dagger}$ in this case is small, but it would be larger if the ionic forms in Eq. 3-44 were not the ones predominating at pH 7.

To obtain the free energy change for a reaction under *other than standard conditions*, Eq. 3-28 must be applied. Thus, at pH 7, and 0.01 M activities of ADP^{3-}, ATP^{4-}, and HPO_4^{2-}, ΔG for hydrolysis of ATP according to Eq. 3-44 is $-34.5 - 2 \times 5.71 = -45.9$ kJ mol^{-1} = -11.0 kcal mol^{-1}. Thus, at concentrations existing in cells (usually in the millimolar range) ATP has a substantially higher group transfer potential than under standard conditions.

To obtain ΔG at a temperature other than 25°C we must know ΔH for the reaction. Using Eq. 3-37 it is easy to show that ΔG at temperature T_2 is related to that at T_1 as in Eq. 3-52.

$$\Delta G_2 \approx \frac{T_2 \Delta G_1 - (T_2 - T_1) \Delta H}{T_1} \quad (3\text{-}52)$$

The enthalpy of hydrolysis of ATP according to Eq. 3-44 is ~ -19.7 kJ mol^{-1}. Using this value and applying Eq. 3-52 we find* that $\Delta G'$ (pH 7) for hydrolysis of ATP at 38°C is -35.2 kJ mol^{-1}. Bear in mind that all of the foregoing free energy changes are *apparent* values applying to solutions of ionic strength ~ 0.25.

---

* The exact value of $\Delta G'$ (pH 7) at 25°C used in obtaining this answer is -34.54 kJ mol^{-1}. In fact, the present author obtained this latter value from the value -35.19 kJ mol^{-1} at 38°C reported by Guynn and Veech.[11]

5. Metal Ions Have Their Say

Both ADP and ATP as well as inorganic pyrophosphate form complexes with metal ions. Since the magnesium complexes are often the predominant forms of ADP and ATP under physiological conditions, we must consider the free energy changes in Eqs. 3-53 and 3-54[12] (apparent values for $\mu = 0.2$ in kJ mol^{-1} at 25°C).

$$ATP^{4-} + Mg^{2+} \longrightarrow Mg\,ATP^{2-} \quad (3\text{-}53)$$
$$\Delta G° = -26.3; \Delta H° = 13.8$$

$$ADP^{3-} + Mg^{2+} \longrightarrow Mg\,ADP^- \quad (3\text{-}54)$$
$$\Delta G° = -19.8; \Delta H° = 15.1$$

Using the apparent $\Delta G°$ values for Eqs. 3-53 and 3-47, we obtain

$$MgATP^{2-} + H_2O \longrightarrow MgADP^- + HPO_4^{2-} + H^+ \quad (3\text{-}55)$$
$$\Delta G' \text{ (pH 7)} = -28.0 \text{ kJ mol}^{-1}$$

The stoichiometry of Eq. 3-55 never holds exactly; some of the Mg^{2+} dissociates from the $MgADP^-$; both protons and Mg^{2+} bind to $H_2PO_4^-$; $HATP^{3-}$ and $HADP^{2-}$ are present and bind Mg^{2+} weakly.[13] Thus, the observed value of $\Delta G^{\dagger}$ varies with both pH and magnesium concentration, as well as with changes in ionic strength. Tables and graphs showing the apparent value of $\Delta G^{\dagger}$ under various conditions have been prepared by Alberty[13] and by Phillips *et al.*[14] An example plotted using the constants and equations of Alberty* is shown in Fig. 3-2. From this graph we find that $\Delta G^{\dagger}$ for hydrolysis of ATP at pH 7, 25°C, $\mu = 0.2$ and 1 mM Mg^{2+} (a relatively high intracellular concentration[15]) is -30.35 kJ mol^{-1} (-7.25 kcal mol^{-1}).

We see that complexing with Mg^{2+} somewhat decreases the group transfer potential of the phosphoryl group of ATP. Furthermore, changes in the concentration of free Mg^{2+} with time and between

---

* However, the value for $\Delta G'$ of hydrolysis of ATP = 34.54 kJ mol^{-1} at $[Mg^{2+}] = 0$ based on results of Guynn and Veech[11] was used. The pK_a values and formation constants of Mg^{2+} complexes were those of Alberty.[13] Note that George *et al.*[15] have provided formation constants of these complexes at infinite dilution where the values of $\Delta G°$ of formation are considerably more negative than those given in Eqs. 3-53 to 3-55.

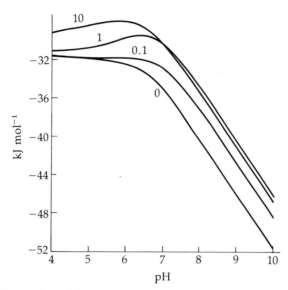

Fig. 3-2 Plots of the apparent free energy $\Delta G^\dagger$ for hydrolysis of ATP as a function of pH at a series of different concentrations of free magnesium ions. Millimolar [Mg^{2+}] is indicated by the numbers by the curves. Computer-drawn graphs courtesy of Carol M. Harris.

different regions of a cell may have significant effects.[16] While Mg^{2+} is a principal cation in tissues, it is by no means the only one. Thus, Ca^{2+}, Mn^{2+}, and even K^+ will affect equilibria involving polyphosphates such as ATP.

6. Uncertainties

It is not easy to measure the group transfer potential of ATP and published values vary greatly. George *et al.*[15] reported ΔG^0 for Eq. 3-42 as -39.9 kJ mol^{-1} at 25°C, $\mu = 0.2$ and -41.25 kJ mol^{-1} at infinite dilution. They regarded these values as good to ± 4 kJ mol^{-1}. However, most other estimates,[17-21] including the recent one of Guynn and Veech,[11] have been at least 4 kJ mol^{-1} less negative. The self-consistent set of thermodynamic data used throughout this book are based in part on the value of ΔG for hydrolysis of ATP obtained by Guynn and Veech.[11]

7. Bond Energies and Approximate Methods for Estimation of Thermodynamic Data

For approximate estimation of enthalpy changes during reactions, use can be made of empirical bond energies (Table 3-6) which represent the approximate enthalpy changes for formation of compounds in a gaseous state from atoms in the gas phase.

8. Free Energies of Combustion by O_2 and by NAD^+

Since oxidation processes are so important in the metabolism of aerobic organisms, it is often convenient to discuss free energies of combustion. These are easily derived from the free energies of formation. For example, ΔG_c for acetate ion (aqueous) may be obtained as follows:

$$\text{Acetate}^- + H^+ \longrightarrow 2\,C + 2\,H_2 + O_2$$
$$\Delta G° = +369.2 \text{ kJ mol}^{-1}$$
$$= -\Delta G_f° \text{ of acetate (from Table 3-3)}$$

$$2\,H_2 + O_2 \longrightarrow 2\,H_2O$$
$$\Delta G° = 2 \times -237.2 = -474.4 \text{ kJ mol}^{-1}$$
$$= 2 \times \Delta G_f° \text{ of } H_2O$$

$$2\,C + 2\,O_2 \longrightarrow 2\,CO_2$$
$$\Delta G° = 2 \times -394.4 = -788.8 \text{ kJ mol}^{-1}$$
$$= 2 \times \Delta G_f° \text{ of } CO_2$$

$$\text{Acetate}^- (1\,M) + H^+ (1\,M) + 2\,O_2 (1 \text{ atm}) \longrightarrow$$
$$2\,CO_2 (1 \text{ atm}) + H_2O \text{ (l)} \quad (3\text{-}56)$$
$$\Delta G° = -894.0 \text{ kJ mol}^{-1}$$
$$= \Delta G_c° \text{ of acetate}^-$$

Table 3-3 lists values of ΔG_c^0 as well as those of ΔG_f^0.

Much of the oxidation occurring in cells is carried out by a special oxidizing agent **nicotinamide adenine dinucleotide (NAD$^+$)** or by the closely related NADP$^+$ (Chapter 8). It is convenient to tabulate values of ΔG^0 for complete oxidation of compounds to CO_2 using NAD$^+$ rather than O_2 as the oxidizing agent.* These values, designated ΔG_{ox}^0

* A quite similar approach has been used by Decker *et al.*[22]

TABLE 3-6
Empirical Bond Energies and Resonance Energies[a]

Energy Values for Single Bonds (kJ mol^{-1})

H—H	436	Si—H	295	C—O	351
C—C	346	N—H	391	C—S	259
Si—Si	177	P—H	320	C—F	441
N—N	161	O—H	463	C—Cl	328
O—O	139	S—H	339	C—Br	276
S—S	213	C—Si	290	C—I	240
C—H	413	C—N	292	Si—O	369

Energy Values for Multiple Bonds (kJ mol^{-1})

C=C	615	C=S	477
N=N	418	C≡C	812
O=O	402 ($^1\Delta$ state)	N≡N	946 (N_2)
C=N	615	C≡N	866 (HCN)
C=O	686 (formaldehyde)		891 (nitriles)
	715 (aldehydes)		
	728 (ketones)		

Empirical Resonance Energy Values (kJ mol^{-1})

Benzene	155	—C(=O)—OH	117
Naphthalene	314		
Styrene	155 + 21	—C(=O)—OR	100
Phenol	155 + 29		
Benzaldehyde	155 + 17	—C(=O)—NH$_2$	88
Pyridine	180	Urea	155
Pyrrole	130	CO	439
Indole	226	CO$_2$	151

[a] From L. Pauling, "The Nature of the Chemical Bond," 3rd ed., pp. 85, 189, and 195–198. Cornell Univ. Press, Ithaca, New York, 1960.

and $\Delta G'_{ox}$ (pH 7), are given in Table 3-3. Note that these values are relatively small, corresponding to the fact that little energy is made available to cells by oxidation with NAD^+; for example (Eq. 3-57):

$$\text{Acetate}^- + 2\, H_2O + 4\, NAD^+ \longrightarrow$$
$$2\, CO_2 + 4\, NADH + 3\, H^+ (10^{-7}\, M) \quad (3\text{-}57)$$
$$\Delta G'_{ox}\ (\text{pH 7}) = +22.1\ \text{kJ mol}^{-1}$$

When the reduced NAD^+ (designated NADH) so formed is reoxidized in mitochondria (Eq. 3-58) a large amount of energy is made available to cells

$$4\, NADH + 4\, H^+ + 2\, O_2 \longrightarrow 4\, H_2O + 4\, NAD^+$$
$$\Delta G'_{ox}\ (\text{pH 7}) = -876.1\ \text{kJ mol}^{-1} \quad (3\text{-}58)$$

The sum of Eqs. 3-57 and 3-58 is the equation for combustion of acetate by O_2 (Eq. 3-59) and the two ΔG values sum to ΔG_c for acetate$^-$.

$$\text{Acetate}^- + H^+ + 2\, O_2 \longrightarrow 2\, CO_2 + 2\, H_2O \quad (3\text{-}59)$$
$$\Delta G_c^0 = -894.0\ \text{kJ mol}^{-1}$$
$$\Delta G_c'\ (\text{pH 7}) = -854.0\ \text{kJ mol}^{-1}$$

The values of ΔG_{ox} (Table 3-3) not only give an immediate indication of the relative amounts of energy available from oxidation of substrates with NAD^+ but also they are very convenient in evaluating ΔG for fermentation reactions. For example, consider the fermentation of glucose to ethanol (Eq. 3-60):

$$\alpha\text{-D-Glucose} \longrightarrow 2\, CO_2 + 2\ \text{ethanol} \quad (3\text{-}60)$$

The free energy change $\Delta G'$ (pH 7) for fermentation of glucose to ethanol and CO_2 can be written immediately from the data of Table 3-3 (Eq. 3-61).

$$\Delta G'\ (\text{pH 7}) = -243.8 - 2(-4.6)$$
$$= -234.6\ \text{kJ mol}^{-1} \quad (3\text{-}61)$$

The values of ΔG_{ox} for H_2O, CO_2, and H^+ are always zero and need not be considered. The same computation can be made using the awkwardly large values of ΔG_c^0 or using values of ΔG_f^0. The latter are also large, and CO_2 and water must be considered in the equations. Table 3-3 can be used to obtain values of ΔG^0 for many metabolic reactions considered later in the book. Data from any column in the table may be used for this purpose, but for simplicity try the values in the $\Delta G'_{ox}$ column.

Note, however, that for reactions involving oxidation by O_2 or by any oxidant, other than NAD^+, not appearing in Table 3-3, the following two-step procedure is necessary if the ΔG_{ox} values are used. From the ΔG_{ox} values compute ΔG^0 or $\Delta G'$ for the reaction under consideration using NAD^+ as the oxidant. Then add to this the free energy of oxidation by O_2 (or other oxidant) of the appropriate number of moles of NADH formed. The latter is given for O_2 in Table 3-7 and can also be evaluated for a number of other oxidants, such as Fe^{3+}, and cytochrome c, from the data in Table 3-7.

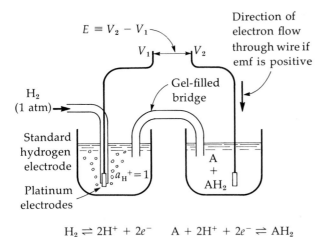

$$H_2 \rightleftharpoons 2H^+ + 2e^- \qquad A + 2H^+ + 2e^- \rightleftharpoons AH_2$$

Fig. 3-3 Device for measurement of electrode potentials. The electrode reactions are indicated below each half-cell. The maximum electrical work that can be done by such a cell on its surroundings is $-\Delta G = nEF$ where $E = V_2 - V_1$ as measured by a potentiometer. If A is reduced to AH_2 by H_2, electrons will flow through an external circuit as indicated. A will be reduced in the right-hand cell. H_2 will be oxidized to H^+ in the left-hand cell. Protons will flow through the gel bridge from left to right as one of the current carriers in the internal circuit.

C. ELECTRODE POTENTIALS AND FREE ENERGY CHANGES FOR OXIDATION–REDUCTION REACTIONS

We live under a blanket of the powerful oxidant O_2. By cell respiration oxygen is reduced to H_2O which is a very poor reductant. Toward the other end of the scale of oxidizing strength lies the very weak oxidant H^+, which some bacteria are able to convert to the strong reductant H_2. The O_2–H_2O and H^+–H_2 couples define two biologically important oxidation reduction ("redox") systems. Lying between these two systems are a host of other pairs of metabolically important substances engaged in oxidation–reduction reactions within cells.

There are two common methods for expressing the oxidizing or reducing powers of redox couples in a quantitative way. On the one hand, we can list values of ΔG for oxidation of the reduced form of a couple to the oxidized form by O_2. A compound with a large value of $-\Delta G$ for this oxidation will be a good reductant. An example is H_2 for which ΔG of combustion at pH 7 (Table 3-3) is -237 kJ/mol. Poor reductants such as Fe^{2+} are characterized by small values of ΔG of oxidation (-8.5 kJ mol^{-1} for 2 $Fe^{2+} \rightarrow$ 2 Fe^{3+}). The free energies of oxidation of biological hydrogen carriers, discussed in Chapter 8, for the most part fall between those of H_2 and Fe^{2+}.

A second way of expressing the same information is to give **electrode potentials** (Table 3-7). Electrode potentials are important for another reason in that their direct measurement sometimes provides an experimental approach to the study of oxidation–reduction reactions within cells.

1. Measurement of Electrode Potentials

To measure an electrode potential it must be possible to reduce the oxidant of the couple by flow of electrons (Eq. 3-62) from an electrode surface, often of specially prepared platinum.

$$A + 2\,H^+ + 2\,e^- \rightleftharpoons AH_2 \qquad (3\text{-}62)$$

Equation 3-62 represents a reaction taking place at a single electrode. A complete electrochemical cell has two electrodes and the reaction occurring is the sum of two half-reactions. The electrode potential of a given half-reaction is obtained from the measured electromotive force of a complete cell in which one half-reaction is that of a standard **reference electrode** of known potential. Figure 3-3 indicates schematically an experimental setup for measurement of an electrode potential. The standard hydrogen electrode consists of platinum over which is bubbled hydrogen gas at one atmosphere pressure. The electrode is immersed in a solution containing hydrogen ions at unit activity ($a_{H^+} = 1$). The potential of such an electrode is conventionally taken as zero. In practice it is more likely that the reference electrode is a calomel electrode or some other electrode that has been estab-

TABLE 3-7
Reduction Potentials of Some Biologically Important Systems[a,b]

Half-reaction	E^0 (V)	$E^{0'}$ (pH 7) (V)	$-\Delta G'$ (pH 7) (kJ mol^{-1}) for oxidation by O_2 (per 2 electrons)
$O_2 + 4\,H^+ + 4\,e^- \longrightarrow 2\,H_2O$	+1.229	+0.815	0.0
$Fe^{3+} + e^- \longrightarrow Fe^{2+}$	0.771	0.771	8.5
$NO_3^- + 2\,H^+ + 2\,e^- \longrightarrow NO_2^- + H_2O$		0.421	76.0
Cytochrome f (Fe^{3+}) $+ e^- \longrightarrow$ cytochrome f (Fe^{2+})		0.365	86.8
Fe (CN)$_6^{3-}$ (ferricyanide) $+ e^- \longrightarrow$ Fe (CN)$_6^{4-}$		0.36	87.8
$O_2 + 2\,H^+ + 2\,e^- \longrightarrow H_2O_2$	0.709	0.295	100.3
Cytochrome a (Fe^{3+}) $+ e^- \longrightarrow$ Cytochrome a (Fe^{2+})		0.29	101.3
p-Quinone $+ 2\,H^+ + 2\,e^- \longrightarrow$ hydroquinone	0.699	0.285	102.3
Cytochrome c (Fe^{3+}) $+ e^- \longrightarrow$ cytochrome c (Fe^{2+})		0.254	108.3
Adrenodoxin (Fe^{3+}) $+ e^- \longrightarrow$ adrenodoxin (Fe^{2+})		0.15	128.3
Cytochrome b_2 (Fe^{3+}) $+ e^- \longrightarrow$ cytochrome b_2 (Fe^{2+})		0.12	134.1
Ubiquinone $+ 2\,H^+ + 2\,e^- \longrightarrow$ ubiquinone H_2		0.10	138.0
Cytochrome b (Fe^{3+}) $+ e^- \longrightarrow$ cytochrome b (Fe^{2+})		0.075	142.8
Dehydroascorbic acid $+ 2\,H^+ + 2\,e^- \longrightarrow$ ascorbic acid		0.058	146.1
Fumarate^{2-} $+ 2\,H^+ + 2\,e^- \longrightarrow$ succinate^{2-}		0.031	151.3
Methylene blue $+ 2\,H^+ + 2\,e^- \longrightarrow$ leucomethylene blue (colorless)		0.011	155.2
Crotonyl-CoA $+ 2\,H^+ + 2\,e^- \longrightarrow$ butyryl-CoA		−0.015	160.2
Glutathione $+ 2\,H^+ + 2\,e^- \longrightarrow$ 2-reduced glutathione		~−0.10	176.6
Oxaloacetate^{2-} $+ 2\,H^+ + 2\,e^- \longrightarrow$ malate^{2-}		−0.166	189.3
Pyruvate$^-$ $+ 2\,H^+ + 2\,e^- \longrightarrow$ lactate^{1-}		−0.185	193.0
Acetaldehyde $+ 2\,H^+ + 2\,e^- \longrightarrow$ ethanol		−0.197	195.3
Riboflavin $+ 2\,H^+ + 2\,e^- \longrightarrow$ dihydroriboflavin		−0.208	197.4
Acetoacetyl-CoA $+ 2\,H^+ + 2\,e^- \longrightarrow$ β-hydroxybutyryl-CoA		−0.238 (38°C)	203.2
$S + 2\,H^+ + 2\,e^- \longrightarrow H_2S$	0.14	−0.274	210.2
Lipoic acid $+ 2\,H^+ + 2\,e^- \longrightarrow$ dihydrolipoic acid		−0.29	213.2
$NAD^+ + H^+ + 2\,e^- \longrightarrow$ NADH	−0.113	−0.32	219.0
$NADP^+ + H^+ + 2\,e^- \longrightarrow$ NADPH		−0.324	219.8
Ferredoxin (Fe^{3+}) $+ e^- \longrightarrow$ ferredoxin (Fe^{2+}) (*Clostridia*)		−0.413	237.0
$2\,H^+ + 2\,e^- \longrightarrow H_2$	0	−0.414	237.2
$CO_2 + H^+ + 2\,e^- \longrightarrow$ formate$^-$		−0.42 (30°C)	238.3
Ferredoxin (Fe^{3+}) $+ e^- \longrightarrow$ ferredoxin (Fe^{2+}) (spinach)		−0.432	240.6

[a] A compound with a more positive potential will oxidize the reduced form of a substance of lower potential with a standard free energy change $\Delta G^0 = -nF\,\Delta E^0 = -n\,\Delta E^0 \times 96.49$ kJ mol^{-1} where n is the number of electrons transferred from reductant to oxidant. The temperature is 25°C unless otherwise indicated. E^0 refers to a standard state in which the hydrogen ion activity = 1; $E^{0'}$ refers to a standard state of pH 7, but in which all other activities are unity.

[b] The major source is P. A. Loach, *in* "Handbook of Biochemistry and Molecular Biology" 3rd ed. Vol. I (G. D. Fasman, ed.), pp. 122–130. © CRC Press, Inc., Cleveland, 1976.

lished experimentally as reliable and whose potential is accurately known.

The standard electrode is connected to the experimental electrode compartment by an electrolyte-filled bridge. In the experimental compartment the reaction represented by Eq. 3-62 occurs at the surface of another electrode (often platinum). The voltage difference between the two electrodes is measured with a potentiometer. The difference between the observed voltage and that of the reference electrode gives the electrode potential of the couple under investigation. It is important that the electrode reaction under study be strictly reversible. When the electromotive force (emf) of the experimental cell is balanced with the potentiometer against an external voltage source no current flows through the cell. However, for a reversible reaction a slight change in the applied voltage will lead to current flow. The flow will be in either of the two directions, depending upon whether the applied voltage is raised or lowered.

Not all redox couples are reversible. This is especially true of organic compounds; for example, it is not possible to readily determine the electrode potential for an aldehyde–alcohol couple. In some cases, e.g., with enzymes, a readily reducible dye of potential similar to that of the couple being measured can be added. (A list of suitable dyes has been described by Dutton.[23]) If the dye is able to rapidly exchange electrons with the couple being studied, it is still possible to measure the electrode potential directly. In many cases electrode potentials appearing in tables have been calculated from free energy data. (The student should be able to calculate many of the potentials in Table 3-7 from free energy data from Table 3-3.) If A, H^+, and AH_2 are all present at unit activity in the experimental cell, the observed potential for the half-reaction will be the **standard electrode potential** E^0. If the emf of the hypothetical cell with the standard hydrogen electrode is positive when electron flow is in the direction indicated by the arrow in Fig. 3-3, the potential of the couple A/AH_2 is also taken as positive (and is often called a **reduction potential**). This is the convention used in establishing Table 3-7, but the student should note that potentials of exactly the opposite sign (oxidation potentials) are used by some chemists. To avoid confusion in reading it is best to be familiar with one or two potentials such as those of the O_2–H_2O and NAD^+–NADH couples.

When electrons flow in the external circuit the maximum amount of work that can be done per mole of electrochemical reaction ($-\Delta G$) is given by Eq. 3-63

$$-\Delta G = nEF = nE \times 96.487 \text{ kJ mol}^{-1} \text{ V}^{-1}$$
$$= nE \times 23.061 \text{ kcal mol}^{-1} \text{ V}^{-1} \quad (3\text{-}63)$$

where F equals the number of coulombs per mole of electrons (Avogadro's number times the charge

on the electron = 96,487 coulombs), and E is measured in volts and represents the difference of the electrode potentials of the two half-cells. In the case of a cell using the standard hydrogen electrode, E is the electrode potential of the experimental couple. The number of moles of electrons transferred in the reaction equation (n) is usually 1 or 2 in biochemical reactions (2 for Eq. 3-62).

Since the reactants and products need not be at unit activity, we must define the observed electrode potential E as a function of E^0 and the activities (concentrations) of A, AH_2, and H^+ (Eq. 3-64).

$$E = E^0 + \frac{RT}{2F} \ln \frac{[A][H^+]^2}{[AH_2]} \quad \text{where } n = 2$$
$$= E^0 + 0.0296 \log \frac{[A][H^+]^2}{[AH_2]} \text{ volts at } 25°C \quad (3\text{-}64)$$

In the biochemical literature values of the apparent standard electrode potential at pH 7 ($E^{0'}$) are usually tabulated instead of values of E^0 (Table 3-7, second column). Note that $E^{0'}$ (pH 7) for the hydrogen electrode is not 0 but -0.414 V. Values of E are related to $E^{0'}$ by Eq. 3-64 with $[H^+]^2$ deleted from the numerator (since the term in log $[H^+]$ is contained in $E^{0'}$). On the scale of $E^{0'}$ (pH 7) the potential of the oxygen–water couple is 0.815 V, while that of the NAD^+–NADH couple is -0.32 V.

D. THERMODYNAMICS AND LIFE PROCESSES

Can thermodynamics be applied to living organisms? Classical thermodynamics deals with equilibria, but living beings are never in equilibrium. The laws of thermodynamics are usually described as statistical laws. How can such laws apply to living things, some of which contain all of their genetic information in a single molecule of DNA? The ideal, reversible reactions of classical thermodynamics occur at infinitesimal speeds. How can thermodynamics be applied to the very rapid chemical reactions that take place in organisms? One answer is that thermodynamics can be used to decide *whether or not a reaction is possible* under given conditions, even if it cannot be used for more. Thus, if we know the steady state concentrations of reactants and products within a cell, we can state whether a reaction will or will not tend to go in a given direction.

Nevertheless, it has seemed to many persons that there ought to be generalities comparable to the laws of thermodynamics that would apply to the kind of steady state or "dynamic equilibrium" existing in organisms. In fact, **nonequilibrium** or **irreversible thermodynamics** constitutes a well-developed branch of modern physical chemistry.

A characteristic of irreversible thermodynamics is that *time* is explicitly introduced. Furthermore, **open systems** are considered. These are systems in which materials flow into and out of the system. Clearly, a living organism is an open system not a closed one of classical thermodynamics. In any open system, the flow of materials is significant. Concentration gradients are set up and transport phenomena may become of primary importance. A serious problem limiting the application of irreversible thermodynamics in biology is that most of the relationships derived apply only to *near equilibrium* situations. However, living things appear often to operate very far from equilibrium. Because biochemical reactions can be very fast, it is not clear that irreversible thermodynamics, as now developed, can be applied fruitfully to most biochemical problems. Even so, the approach is an important one that should not be overlooked by the serious student.

For an introduction to nonequilibrium thermodynamics and to the literature in the field, see the chapter by Caplan[24] and books by Prigogine[25] and Katchalsky and Curran.[26]

Taking a somewhat different approach, Mc-Clare[27-29] maintains that the second law of thermodynamics is not, as usually stated, a statistical law. Rather, when expressed in proper form, it also applies to the bacterium with its single molecule of DNA. Thus, classical thermodynamics can be applied to living cells. The author pointed out that a characteristic of living things that differentiates them from the systems usually considered by thermodynamicists is the rapidity of the reactions. He asserted that a process such as that by which molecular energy is transformed into mechanical energy in the muscle must depend upon the fact that the energy transfer occurs sufficiently rapidly that it does not become dissipated as heat. McClare regarded enthalpy changes as more significant than free energy or entropy changes in reactions of metabolism.

Another important idea, expressed in detail by Morowitz,[30] is that *the flow of energy through any open system is sufficient to "organize" that system.* Thus, if a flask of water is placed on a hot plate, a cycle is established. The water moves via cyclic convection currents that develop as a result of the flow of energy through the system. Morowitz reckoned that the $6 \pm 3 \times 10^{18}$ kJ/year of solar energy that fall on the earth supplies the organizing principle for life. Just as it drives the great cycles within the atmosphere and within the seas, it gives rise to the branching and interconnecting cycles of metabolism. Perhaps this idea may even make the spontaneous development of the organized systems that we call life from inanimate precursors through evolution seem a little more understandable.

1. Growth Yields of Bacteria

While attempts to find generalizations from thermodynamics applicable to living systems have not been very fruitful, strictly empirical observations about growth and energy consumption have revealed some interesting facts. For a large number of anaerobic fermentation processes the energy yielding reactions by which cells generate ATP from fermentation are quite well understood (Chapter 9). The stoichiometry is often remarkably exact, and it is possible to estimate with considerable accuracy the quantity of ATP generated by the fermentation of a given quantity of substrate. Also easy to measure is the amount of living material generated during the course of a fermentation; for example, bacteria grow rapidly and the cells can be harvested, washed, dried, and weighed.

When this is done, it is found that no matter what the substrate fermented, with a few exceptions, the quantity Y_{ATP}, the *grams of dry cells per mole of ATP generated is nearly constant*[22,31] and approximately *equal to 10.5*. Another generalization is that for bacteria growing and producing only cells, CO_2, and water (cells growing aerobically), $40 \pm 5\%$ of the carbon and hydrogen available is oxidized to CO_2 and water while $60 \pm 5\%$ is assimilated into the cells. Note that the percent assimilated is very much greater than in anaerobic fermentations where the vast majority of the substrate is fermented and *not* assimilated. As we shall see later, this difference results from the very much greater yield of ATP from oxidation than from fermentation.

Of the free energy of oxidation of substrates $\sim 62\%$ is conserved as the free energy of combustion of the components of the dried bacteria. Thus, $-\Delta G_c \simeq -\Delta H_c \simeq 22$ kJ g^{-1} of dry weight of bacteria.[31]

E. LINEAR FREE ENERGY RELATIONSHIPS

Organic functional groups exert characteristic electronic effects upon other groups to which they are attached. The quantitative expression of such effects by means of "linear free energy relationships" provides some very useful correlations of chemical results. The best known linear free energy relationship is the **Hammett equation** which deals with the transmission of electronic effects across a benzene or other aromatic ring. Consider the acid dissociation constants of three classes of compounds:

Benzoic acids   Phenylacetic acids   Phenols

The values of pK_a given in Table 3-8 have been observed for the parent compounds and for the

TABLE 3-8
Values of pK_a for Unsubstituted and Substituted Benzoic Acids, Phenylacetic Acids, and Phenols

Meta substituent	Benzoic[a] acids	Phenylacetic acids	Phenols
—H (parent compound)	4.202	4.31	9.92
—Cl	3.837	4.13	9.02
—NO₂	3.460	3.97	8.39

[a] The pK_a values for substituted benzoic acids were recently reported by P. D. Bolton, K. A. Fleming, and F. M. Hall [*JACS* **94**, 1033–1034 (1972)]. The values of σ calculated from them are slightly different from those appearing in Table 3-9.

meta-chloro- and the *meta*-nitro-substituted compounds.

Recall that pK_a is the negative logarithm of a dissociation constant. It follows, from Eq. 3-30, that values of pK_a are directly proportional to values of ΔG^0 for dissociation of protons. In the Hammett treatment differences in pK_a values, rather than differences in ΔG^0, are considered. When a hydrogen atom in the meta position of benzoic acid is replaced by the electron-withdrawing Cl or NO_2, pK_a is lowered, i.e., the basicity of the conjugate base of the acid is decreased. The decrease in pK_a amounts to -0.365 for *m*-chlorobenzoic acid and -0.742 for *m*-nitrobenzoic acid. The Hammett treatment asserts that these changes in pK_a are a measure of the electron-withdrawing power of the meta substituent.[32] Thus, the nitro group is about twice as strong as the chloro group in this respect. The numerical values of these changes in the dissociation constant of benzoic acid define the **substituent constants** σ, which are used in the Hammett equation (Table 3-9).

$$pK_0 - pK = \sigma \text{ for substituted benzoic acids} \quad (3\text{-}65)$$

The decreases in the pK_a of phenylacetic acid occasioned by replacement of the *meta*-hydrogen with Cl or NO_2 are -0.18 and -0.34, respectively, substantially less than for the benzoic acids. On the other hand, for the phenols the differences amount to -0.90 and -1.53, much *greater* than for the benzoic acids. The Hammett equation (Eq.

TABLE 3-9
Substituent Constants for Use in the Hammett and Taft Equations[a]

Substituent	σ_m	σ_p	σ_p^-	σ_p^+	σ^* (Taft)
—O⁻	−0.71	−0.52			
—NH₂	−0.16	−0.66		−1.11	0.62
—OH	0.121	−0.37		−0.85	1.34
—OCH₃	0.115	−0.27		−0.78	1.81
—CH₃	−0.069	−0.17		−0.31	0.00
—NH—COCH₃	0.21	−0.01		−0.25	
—H	0	0		0	0.49
—CH₂OH	0.08	0.08			0.56
—COO⁻	−0.10	0.00		0.11	−1.06
—SO₃⁻	0.05	0.09		0.12	
—SH	0.25	0.15		0.019	1.68
—CH₂Cl		0.18			1.05
—CONH₂	0.28	0.36			
—F	0.337	0.06	0.02	−0.07	
—I	0.352	0.18		0.135	
—Cl	0.373	0.227		0.114	
—CHO	0.36	0.22	0.37		
—COCH₃	0.376	0.502	0.85		
—COOH	0.37	0.45		0.42	2.08
—COOCH₃	0.39	0.31		0.49	
—SO₂NH₂	0.55	0.62		0.61	
—CN	0.56	0.66	0.89	0.66	
—C≡CH					2.18
—CF₃	0.43	0.54		0.61	2.61
—CCl₃					2.65
—NO₂	0.710	0.778	1.25	0.790	4.0
—NH₃⁺	1.13	1.70			3.76
⟩N (pyridine)	0.73	0.83			
⟩NH⁺ (pyridine)	2.18	2.42		4.0	

[a] Sources are as follows: L. P. Hammett, "Physical Organic Chemistry," 2nd ed., p. 356. McGraw-Hill, New York, 1970; C. G. Swain and E. C. Lupton, Jr., *JACS* **90**, 4328–4337 (1968); H. H. Jaffé, *ibid.* **77**, 4445–4448 (1955); G. B. Barlin and D. D. Perrin, *Q. Rev., Chem. Soc.* **20**, 75–101 (1966).

3-66) asserts that for reactions such as the dissociation of protons from phenylacetic acids or from phenols, the changes in ΔG occasioned by meta substitutions are proportional to the σ values, i.e., to the changes in ΔG for the standard reaction—dissociation of a proton from benzoic acid.[32–34]

$$\log (K/K_0) = pK_0 - pK$$
$$= \rho\sigma \quad \text{the Hammett equation* (3-66)}$$

The proportionality constant ρ, which also appears in the equation, is a measure of the "sensitivity" of the reaction to the presence of electron-withdrawing or electron-donating substituents in the ring. For benzoic acid, ρ is taken as 1.00. Using the data of Table 3-8, together with many other data, an average value of $\rho = 0.49$ has been found for phenylacetic acids. Likewise, $\rho = 2.23$ for phenols, and $\rho = 5.7$ for dissociation of protons from substituted pyridinium ions. The sensitivity to substituent changes is highest (highest ρ) in the latter case where proton dissociation is directly from an atom in the ring and is lowest when the basic center is removed farthest from the ring (phenylacetic acid).

Through knowledge of the value of ρ for a given reaction, it is possible to predict the effect of a substituent on pK_a through the use of tabulated values of σ. In many cases the effects are additive for multisubstituted compounds (Eq. 3-67).

$$pK = pK_0 - \rho \Sigma\sigma \quad (3\text{-}67)$$

While the examples chosen here concern only dissociation of protons, the Hammett equation has a much broader application. Equilibria for other types of reaction can be treated. Furthermore, since rates of reactions are related to free energies of activation, many rate constants can be correlated (Chapter 6, Section E,5,e).

Since substituents in o, m, and p positions have quantitatively different influences, different substituent constants σ are defined for each position. Moreover, since special complications arise from ortho interactions, it is customary to tabulate σ values only for meta and para positions.† These are designated σ_m and σ_p. An additional complication is that certain reactions are unusually sensi-

---

* The equation can be written[32] in the more general form $\log k_{ij} - \log k_{0j} = \rho_j\sigma_i$ in which k may be either an equilibrium constant or a rate constant. The subscript j denotes the reaction under consideration and i the substituent influencing the reaction.

† However, apparent σ values for *ortho* substituents are available.[34]

tive to para substituents that are able to interact by resonance directly across the ring. An example is the acid dissociation of phenols. While σ_p for the nitro group is ordinarily 0.778, a correct prediction of the effect of the p-nitro group on dissociation of phenol is given only if σ_p is taken as 1.25. This higher σ value is designated σ_p^-. The resonance in the phenolate anion giving rise to this enhanced effect of the nitro group may be indicated as follows:

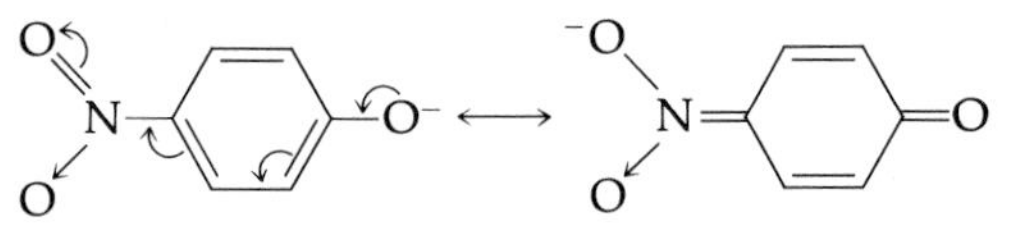

(Here the curved arrows represent the direction of flow of electrons needed to convert from the one resonance structure to the other.) For similar reasons, some reactions require the use of special σ_p^+ constants for strongly electron-donating substituents such as OH, which are able to interact across the ring by resonance. Thus σ_p for the OH group is -0.37, while σ_p^+ is $-0.85.$[35] For the methoxyl group ($-OCH_3$) $\sigma_p = -0.27$, $\sigma_p^+ = -0.78$, and $\sigma_m = +0.12$.

The use of different kinds of substituent constants complicates the application of the Hammett equation. Indeed, there are over 20 different sets of σ values in use. A simplification is the representation of substituent constants as linear combinations of two terms, one representing "field" or "inductive" effects and the other resonance effects.[35,36] For a given para substituent, the appropriate substituent constant for a given reaction series is described by Eq. 3-68:

$$\sigma = fF + rR \qquad (3\text{-}68)$$

In this equation the coefficients f and r are characteristic of the reaction series under consideration

and F and R of the particular substituent. Thus, values of σ_p of the ordinary Hammett equation become (Eq. 3-69)

$$\sigma_p = 0.56F + R \qquad (3\text{-}69)$$

Values of R and F for some substituents are shown in the following tabulation.

	R	F
$-NH_2$	-0.68	0.04
$-OH$	-0.64	0.49
$-OCH_3$	-0.50	0.41
$-CH_3$	-0.14	-0.05
$-COOH$	$+0.14$	0.55
$-NO_2$	$+0.20$	1.11

The substituents are listed in the order of decreasing resonance electron donation. The substituents carboxyl and nitro, having positive values of R, are electron acceptors. All substituents, except the methyl group, are seen to exert electron-withdrawing inductive effects.

The values of σ_m tend to emphasize the nonresonance influence of a group and hence to parallel the F component, whereas ρ_p^+ values are more influenced by R.

Many other linear free energy relationships have been proposed; for example, the acid strengths of aliphatic compounds can be correlated using the "Taft polar substituent constants" σ^* and Eq. 3-70.

$$\log (K/K_0) = \rho^* \sigma^* \qquad (3\text{-}70)$$

For example, Eqs. 3-71 and 3-72 give good approximations of pK_a values.[34]

for R—COOH $\qquad pK_a = 4.66 - 1.62\ \sigma^* \qquad (3\text{-}71)$

for R—CH$_2$—COOH $\quad pK_a = 5.16 - 0.73\ \sigma^* \qquad (3\text{-}72)$

REFERENCES

1. B. H. Mahan, "Elementary Chemical Thermodynamics." Benjamin, New York, 1963.
2. G. N. Lewis and M. Randall, "Thermodynamics and the Free Energy of Chemical Substances" (rev. by K. S. Pitzer and L. Brewer), 2nd ed. McGraw-Hill, New York, 1961.
3. M. H. Everdell, "Introduction to Chemical Thermodynamics." Norton, New York, 1965.
4. K. G. Denbigh, "The Principles of Chemical Equilibrium." Cambridge Univ. Press, London and New York. 1955.
5. R. E. Dickerson, "Molecular Thermodynamics." Benjamin, New York, 1969.
6. J. M. Sturtevant in "Techniques of Chemistry" (A. Weiss-

berger, ed.), Vol. I, Part V, pp. 347–425. Wiley (Interscience), New York, 1971.

7. H. D. Brown, ed., "Biochemical Microcalorimetry." Academic Press, New York, 1969.

8. A. White, P. Handler, and E. L. Smith, "Principles of Biochemistry," 4th ed., pp. 291–301. McGraw-Hill, New York, 1968.

9. R. W. Gurney, "Introduction to Statistical Mechanics," Chapter 1. McGraw-Hill, New York, 1949.

10. R. C. Phillips, P. George, and R. J. Rutman, *Biochemistry* **2**, 501–508 (1963).

11. R. W. Guynn and R. L. Veech, *JBC* **248**, 6966–6972 (1973).

12. R. A. Alberty, "Horizons of Bioenergetics" (A. San Pietro and H. Gest, eds.), pp. 135–147. Academic Press, New York, 1972.

13. R. A. Alberty, *JBC* **244**, 3290–3302 (1969).

14. R. C. Phillips, P. George, and R. J. Rutman, *JBC* **244**, 3330–3342 (1969). Note that the thermodynamic data in this paper are all based on extrapolation to infinite dilution.

15. P. George, R. C. Phillips, and R. J. Rutman, *Biochemistry* **2**, 508–512 (1963).

16. D. L. Purich and H. J. Fromm, *Curr. Top. Cell. Regul.* **6**, 131–167 (1972).

17. W. P. Jencks, *in* "Handbook of Biochemistry" (H. A. Sober, ed.), p. J-148. Chem. Rubber Publ. Co., Cleveland, Ohio, 1968.

18. M. R. Atkinson and R. K. Morton, *Comp. Biochem.* **2**, 1–95 (1960).

19. J. A. Bassham and G. H. Krause, *BBA* **189**, 207–221 (1969).

20. G. E. Vladimirov, V. G. Vlassova, A. Y. Kolotilova, S. N. Lyzlova, and N. S. Panteleyeva, *Nature (London)* **179**, 1350–1351 (1957).

21. J. Rosing and E. C. Slater, *BBA* **267**, 275–290 (1972).

22. K. Decker, K. Jungermann, and R. K. Thauer, *Angew. Chem., Int. Ed. Engl.* **9**, 138–158 (1970).

23. P. L. Dutton, *BBA* **226**, 63–80 (1971).

24. S. R. Caplan, *Curr. Top. Bioenerg.* **4**, 1–79 (1971).

25. I. Prigogine, "Introduction to Thermodynamics of Irreversible Processes," 3rd ed. Wiley, New York, 1967.

26. A. Katchalsky and P. F. Curran, "Nonequilibrium Thermodynamics in Biophysics." Harvard Univ. Press, Cambridge, Massachusetts, 1965.

27. C. W. F. McClare, *J. Theor. Biol.* **30**, 1–34 (1971).

28. C. W. F. McClare, *J. Theor. Biol.* **35**, 233–246 (1972).

29. C. W. F. McClare, *J. Theor. Biol.* **35**, 569–595 (1972).

30. H. J. Morowitz, "Energy Flow in Biology." Academic Press, New York, 1968.

31. W. J. Payne, *Annu. Rev. Microbiol.* **24**, 17–52 (1970).

32. L. P. Hammett, "Physical Organic Chemistry," 2nd ed., pp. 347–390. McGraw-Hill, New York, 1970.

33. P. R. Wells, *Chem. Rev.* **63**, 171–219 (1963).

34. G. B. Barlin and D. D. Perrin *Q. Rev., Chem. Soc.* **20**, 75–101 (1966).

35. C. G. Swain and E. C. Lupton, Jr., *JACS* **90**, 4328–4337 (1968).

36. L. D. Hansen and L. G. Hepler, *Can. J. Chem.* **50**, 1030–1035 (1972).

STUDY QUESTIONS

1. (a) From ΔG^0 for hydrolysis of sucrose (Table 3-5) calculate the equilibrium constant

$$K = \frac{[\text{glucose}][\text{fructose}]}{[\text{sucrose}]}$$

at 25°C. Call this hydrolysis reaction 1. (b) Is the sucrose in a 1 M solution stable? Explain. (c) If acid is added to a 1 M sucrose solution to catalyze its hydrolysis, what will be the final sucrose concentration at equilibrium? (Assume that concentrations equal activities for the purpose of these calculations.) (d) Reaction 2 is the hydrolysis of α-D-glucose 1-phosphate to glucose and inorganic phosphate (P_i). Using ΔG^0 for this reaction (Table 3-5) calculate the equilibrium constant. (e) Sucrose phosphorylase from the bacterium *Pseudomonas saccharophila* catalyzes the following reaction (reaction 3):

$$\text{Sucrose} + P_i \longrightarrow$$
$$\alpha\text{-D-glucose 1-phosphate} + \text{fructose}$$

Calculate the equilibrium constant and the standard free energy change at 25°C for reaction 3 from the equilibrium constants obtained above for reactions 1 and 2. Show that ΔG^0 for reaction 3 = ΔG^0 of reaction 1 − ΔG^0 of reaction 2. (f) Could the bacterium carry out reaction 3 in the following two consecutive steps? Explain.

$$\text{Sucrose} \longrightarrow \text{glucose} + \text{fructose}$$
$$\text{Glucose} + \text{phosphate} \longrightarrow \text{glucose 1-phosphate}$$

2. For each of the following reactions, state whether the equilibrium constant will be between 0.1 and 10 (i.e. about one), greater than 100, or less than 0.01. Assume that the pH is constant at 7.0.

a. $2\ ADP^{3-} \longrightarrow ATP^{4-} + AMP^{2-}$

b. $ATP^{3-} + glucose \longrightarrow$
   glucose 6-phosphate$^{2-} + ADP^{2-} + H^+$

c. $ADP^{2-} + HPO_4^{2-} + H^+ \longrightarrow ATP^{3-}$

d. Glucose 6-phosphate$^{2-} \longrightarrow$
   fructose 6-phosphate^{2-}

e. Phosphoenolpyruvate$^{3-} + glucose \longrightarrow$
   glucose 1-phosphate$^{2-} + pyruvate^-$

3. The combustion of 1 mol of solid urea to liquid water and gaseous carbon dioxide and nitrogen (N_2) in a bomb colorimeter at 25°C (constant volume) liberated 666 kJ of heat energy. Calculate ΔH, the change in heat content (enthalpy) for this reaction.

4. Using data of Table 3-3 calculate $\Delta G'$ (pH 7) for the following reactions:

   a. Glucose $\longrightarrow$ 2 lactate$^- + 2H^+$

   b. $2\ NH_4^+ + HCO_3^- \longrightarrow$
      urea $+ 2H_2O + H^+$

   c. α-Ketoglutarate$^{2-} + 1/2\ O_2 + CoA + H^+ \longrightarrow$
      succinyl-CoA $+ H_2O + CO_2$

5. What is the ionic strength of a 0.2 M solution of NaCl? of 0.2 M Na$_2$SO$_4$?

6. The [ATP]/[ADP] ratio in an actively respiring yeast cell is about 10. What would be the intracellular [3-phosphoglycerate]/[1,3-diphosphoglycerate] ratio have to be to make the phosphoglycerate kinase reaction (Fig. 9-7, reaction 7) proceed toward 1,3-diphosphoglycerate synthesis at 25°C, pH 7?

7. (a) Using data from Table 3-7 determine the equilibrium constant for the reaction between malate and methylene blue, assuming all reactants present initially at the same concentration. Indicate clearly the direction of the reaction for which the free energy change is written. (b) Calculate the percentage of the reduced (leuco) form of methylene blue present at pH 7 and 25°C in a system for which the measured electrode potential is 0.065 V.

8. NAD$^+$ is a coenzyme for both pyruvate dehydrogenase and ethanol dehydrogenase. Using the values of E_0' from Table 3-7 calculate the free energy change and the equilibrium constant for the reaction.

   Lactate$^- +$ acetaldehyde $\longrightarrow$
   pyruvate$^- +$ ethanol

9. Consider the oxidation of acetate at 25°C:

$CH_3COO^-(0.1\ M) + 2O_2(g, 0.2\ atm)$
$\qquad + H^+(10^{-7}\ M) \longrightarrow 2H_2O(l) + 2CO_2(g)$

a. What is the equilibrium pressure of CO_2 if the reaction is not coupled to any other reaction?

b. What is the equilibrium pressure of CO_2 if the reaction is coupled to the formation of 0.01 M ATP from 0.01 M ADP and 0.01 M HPO$_4^{2-}$ in the citric acid cycle?

c. What do the above calculations tell you about the prospects of gaining 100% efficiency of energy storage in ATP in the citric acid cycle?

d. If the actual pressure of CO_2 is 0.01 atm, what is the efficiency of energy storage under the conditions in (b)?

10. The equilibrium constant for the following reaction, which is catalyzed by ATP-creatine transphosphorylase, has been determined by chemical analysis. The data are given below. [S. A. Kuby and E. A. Noltman, in "The Enzymes," 2nd ed. (P. D. Boyer, H. Lardy, and K. Myrbäck, eds), Vol. VI, pp. 515–602. Academic Press, New York, 1962].

$ATP^{4-} + creatine \longrightarrow$
$\qquad ADP^{3-} + creatine\ phosphate^{2-} + H^+$

t (°C)	K
20	6.30×10^{-9}
30	5.71×10^{-9}
38	5.47×10^{-9}

a. What are ΔG^0, ΔH^0, and ΔS^0 for the reaction at 25°C?

b. What are $\Delta G'$, $\Delta H'$, and $\Delta S'$ (pH 7) for the reaction at 25°C?

c. What are $\Delta G'$, $\Delta H'$, and $\Delta S'$ (pH 7) for the hydrolysis of creatine phosphate at 25°C?

11. The following reaction was carried out in a calorimeter at 25°C in 0.1 M phosphate buffer at pH 7.4 in the presence of a particulate suspension containing the mitochondrial electron transport system [M. Poe, H. Gutfreund, and R. W. Estabrook, ABB **122**, 204–211 (1967)]:

$NADH + H^+ + 1/2\ O_2$ (aq, sat.) $\longrightarrow$
$\qquad NAD^+ + H_2O$

The oxygen consumption was monitored continuously with an oxygen electrode. The temperature was monitored simultaneously with a thermocouple immersed in the solution. At the start of the reaction 96 μmol NADH was added to 29.0 ml buffer containing O_2. A nearly zero-order reaction was observed with the rate of O_2 consumption of 6.87 μmol/min and the rate of temperature rise of 0.01171°K/min. The heat capacity of the calorimeter and contents was 254.6 J/°K. What is ΔH for the above reaction? NOTE: The H^+ is supplied by the phosphate buffer, which has a ΔH of dissociation of 5.4 kJ mol^{-1}.

12. Enthalpy and free energy changes for the following reaction at 25°C are given in Table 3-5 (p. 168).

$$ATP^{4-} \ (1 \ M) + H_2O \longrightarrow ADP^{3-} \ (1 \ M)$$
$$+ \ H^+ \ (10^{-7} \ M) + HPO_4{}^{2-} \ (1 \ M)$$

a. How much heat is evolved at constant temperature and pressure if the reaction takes place in a test tube without doing any work other than $p \ \Delta V$ work?

b. How much heat is evolved or absorbed by the foregoing reaction if it is coupled with 100% efficiency to an endergonic reaction?

c. What efficiency of coupling to an endergonic reaction is required in order that the foregoing reaction neither evolve nor absorb heat?

4

How Molecules Stick Together

Since cells are made up of molecules, the ways in which these molecules cling together to provide adequate strength and rigidity is vitally important. In addition, it is clear that the binding of small molecules to large ones is basic to many biological phenomena such as metabolism of foodstuffs and action of hormones. Interactions between macromolecules are necessary for the motion of flagella, the contraction of muscle, the action of antibodies, the transmission of nerve impulses, and numerous other phenomena.

A. PRINCIPLE OF COMPLEMENTARITY

Since the forces acting between them are weak, two molecules will cling together tightly only if there is a close fit between their surfaces. Many atoms must be in contact. To obtain a strong bond, there must be an *exact* fit, i.e., the two molecular surfaces must be complementary one to the other. If a "knob" (e.g., a —CH₃ group) is present on one surface, a hollow must exist in the complementary surface. A positive charge in one surface must be opposite a negative charge in the

other. A proton donor group can form a hydrogen bond only if it is opposite a group with unshared electrons; nonpolar (hydrophobic) groups must be opposite each other if hydrophobic bonding is to occur. One of the most important principles of biochemistry is that *two molecules with complementary surfaces tend to join together and interact, whereas molecules without complementary surfaces do not interact.* Watson called this selective stickiness.[1] Selective stickiness permits the "self-assembly" of biological macromolecules having surfaces of complementary shape into fibers, tubes, membranes, and polyhedra. It provides the means for specific pairing of bases during the replication of DNA.

Complementarity of surfaces is equally important to the chemical reactions of cells. These transformations are catalyzed or "directed" by enzymes, each one of which contains reactive chemical groupings in exactly the right place and the right orientation to interact with and promote a chemical change in another molecule, the **substrate.** Specific catalysis is characteristic of living things, and much of biochemistry is devoted to its study. Furthermore, changes in cell structure are linked closely with chemical reactions. The movements of muscle fibers and the flowing of the cytoplasm in the ameba are striking examples.

B. FORCES ACTING BETWEEN MOLECULES

Biochemists usually explain the quantitative aspects of interactions between molecules in terms of *van der Waals forces, electrostatic interactions, hydrogen bonding,* and *hydrophobic forces.* The net effect is most frequently given quantitative expression in an equilibrium constant and in the corresponding enthalpy and entropy changes.

1. Formation Constants and Dissociation Constants

The strength of bonding between two particles can be expressed as a **formation constant** K_f. Consider the binding of a molecule X to a second molecule P, which might be a protein, a nucleic acid, a metal ion, or other particle. If there is on the surface of P only a single binding site for X, the process can be described by Eq. 4-1 and the equilibrium constant K_f by Eq. 4-2.

$$X + P \rightleftharpoons PX \qquad (4\text{-}1)$$
$$K_f = [PX]/[P][X] \qquad (4\text{-}2)$$

The units for Eq. 4-2 are liters per mole (or M^{-1}). The formation constant K_f is a direct measure of the strength of the binding: the higher the constant, the stronger the interaction. This fact can be expressed in an alternative way by giving the standard free energy change ($\Delta G°$) for the reaction (Eq. 4-3). The more negative $\Delta G_f°$, the stronger the binding.

$$\Delta G_f° = -RT \ln K_f = -2.303RT \log K_f$$
$$= -5.708 \log K_f \quad \text{kJ mol}^{-1} \text{ at } 25°C \quad (4\text{-}3)$$

To avoid confusion, it is important to realize that **dissociation constants** (Eq. 4-4) are frequently used (especially in enzymology and to describe the strengths of acids).

$$K_d = 1/K_f \qquad (4\text{-}4)$$

Unfortunately, the use of association or formation constants and of dissociation constants are both firmly entrenched in different parts of the chemical literature; be sure to keep them straight.

It is often convenient to tabulate logarithms of the formation constants since these are proportional to the free energies. Note the following relationship (Eq. 4-5).

$$\log K_f = -\log K_d = pK_d \qquad (4\text{-}5)$$

The logarithms of formation constants and pK values of dissociation constants are identical and are a measure of the standard free energy decrease in the association reaction. Note further, that the difference in $\Delta G°$ corresponding to a change of one unit in $\log K_f$ or pK_d is -5.7 kJ mol^{-1}, a handy number to remember.

The average kinetic energy of motion of a molecule in solution is about $\frac{3}{2}k_B T$, where k_B is Boltzmann's constant; for 1 mol the kinetic energy is $\frac{3}{2}RT$ or 3.7 kJ (0.89 kcal) per mol at 25°C. Thus, if $K_f = 10$ ($\Delta G° = -5.7$ kJ, -1.36 kcal mol^{-1}), the binding energy is only slightly in excess of the thermal energy of the molecules and the complex is weakly bound. In this instance, if X and P are both present in 10^{-4} molar concentrations (typical enough for biochemical systems), only 0.1% of the molecules will exist as the complex ([complex] = $K_f[X][P]$). If the formation constant is higher by a factor of 1000 i.e., $K_f = 10^4$ ($\Delta G° = -22.8$ kJ mol^{-1}), 38% of the molecules will exist as the complex; while if $K_f = 10^7$ (extremely strong binding, $\Delta G° = -40$ kJ mol^{-1} or -9.55 kcal), 97% of the molecules will be complexed.

2. van der Waals Forces

All atoms have a weak tendency to stick together, and because of this even helium liquifies at a low enough temperature. This **van der Waals attraction** acts only at a very short distance. It results from the electrostatic force between the positively charged nucleus of one atom and the negatively charged electrons of the other. Because nuclei are screened by the electron clouds surrounding them, the force is weak and fluctuating; for example, a small decrease in enthalpy ($\Delta H° = -2.0$ kJ mol^{-1}) accompanies the transfer of methane (CH_4) from the gas phase into solution in an "inert" hydrocarbon solvent. This small energy change reflects the van der Waals attraction between the four hydrogen atoms of the methane and the atoms of the solvent.[2]

3. Attraction between Charged Groups (Salt Linkages)

Fixed positive and negative charges attract each other strongly; for example, consider a carboxylate ion in contact with $-NH_3^+$ or with an ion of calcium:

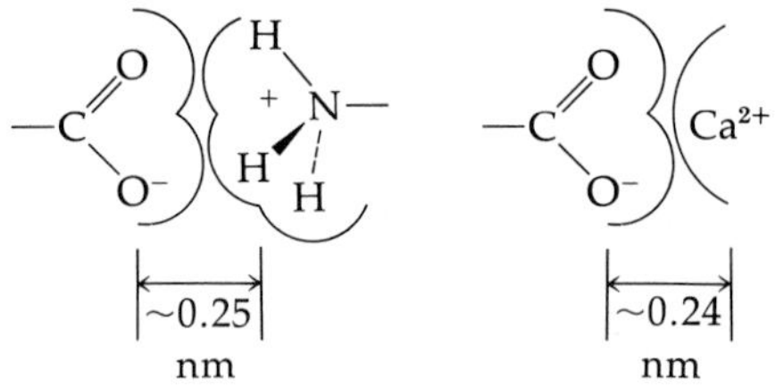

From the van der Waals radii of Table 2-1 and the ionic crystal radius of Ca^{2+} of 0.10 nm, we can estimate an approximate distance between the centers of positive and negative charge of 0.25 nm in both cases. (However, the distance may be considerably greater if the charged groups are surrounded by hydration "shells" of oriented water molecules.) It is of interest to apply Coulomb's law to compute the force F between two charged particles which are almost in contact, e.g., at a distance of 0.30 nm (Eq. 4-6).

$$F = 8.9875 \times 10^9 \times \frac{qq'}{r^2\epsilon} \quad \text{newtons} \quad (4\text{-}6)$$

In this equation r is the distance in meters, q and q' are the charges in coulombs (one electronic charge $= 1.6021 \times 10^{-19}$ coulombs), ϵ is the dielectric constant, and F is the force in newtons. The force per mole is NF where N is Avogadro's number.

An uncertainty in this kind of calculation is in the dielectric constant ϵ which is 1.0 for a vacuum, about 2 for hydrocarbons, and 78.5 for water at 25°C. If ϵ is taken as 2 the force, for $r = 0.30$ nm, is 7.7×10^{14} newtons mol^{-1} (it would be twice as great for the Ca^{2+} —COO^- case). This means that to move the charges further apart by just 0.01 nm would require 7.7 kJ mol^{-1}, a substantial amount of energy. However, if the dielectric constant were that of water, this would be reduced almost 40-fold and the electrostatic force would not be highly significant in binding.

A calculation that is often made is the work required to completely remove two charges from a given distance, e.g., 0.30 nm, to an infinite distance (Eq. 4-7).

$$W \text{ (kJ mol}^{-1}） = 8.9875 \times 10^6 \times \frac{qq'}{r\epsilon} N$$
$$= \frac{138.94}{\epsilon\, r \text{ (in nm)}} \quad (4\text{-}7)$$

If $\epsilon = 2$ this amounts to 232 kJ mol^{-1} for single charges at a distance of 0.30 nm; 69 kJ mol^{-1} at 1 nm, and only 6.9 kJ mol^{-1} at 10 nm, the distance across a cell membrane. Use will be made of these estimates later. At this point it is sufficient to point out that enormous forces exist between closely spaced charges.

Electrostatic forces are of great significance in interactions between molecules and in the induction of changes in conformations of molecules; for example, attraction between $-COO^-$ and $-NH_3^+$ groups must be important in interactions between proteins. Calcium ions often interact with carboxylate groups of proteins and carbohydrates, sometimes converting solutions of these molecules to rigid "gels" (e.g., agarose, Chapter 2, Section C,5). The doubly charged Ca^{2+} ion can bridge between two carboxylate or other polar groups.

An important aspect of all electrostatic interactions in aqueous solutions is the **hydration** of ions. Each ion is surrounded by a shell of oriented water molecules held by the attraction of the water dipoles to the charged ion. The hydration of ions has a strong influence on all aspects of electrostatic interactions and plays a dominant role in determining such matters as the strength of acids and bases, the free energy of hydrolysis of ATP, and the strength of bonding of metal ions to negatively charged groups.

Effects of hydration are difficult to assess quantitatively. Note, for example that $\Delta G°$ for dissociation of acetic acid is 27.2 kJ mol^{-1} (Table 3-4), much less than the 232 kJ mol^{-1} calculated as the work required to separate two charges in a medium of dielectric constant 2. Note that $\Delta H°$ for dissociation of acetic acid is almost zero (-0.1 kJ mol^{-1}) and that $\Delta S°$ is consequently -91.6 J °K^{-1}. This large entropy decrease reflects the increased amount of water that is immobilized in the hydration spheres of the H$^+$ and acetate$^-$ ions formed in the reaction. In contrast (Table 3-4), $\Delta H°$ and $\Delta S°$ for dissociation of NH$_4^+$ to NH$_3$ and H$^+$ converts one positive ion to another with $\Delta H° = +52.5$ kJ mol^{-1} and a small entropy change: $\Delta S° = -2.0$ J °K^{-1}.

4. Hydrogen Bonds and the Structure of Water

One of the most important weak interactions between biologically important molecules is the hydrogen bond (Chapter 2, Section A,7). We have seen the importance of this type of dipole–dipole interaction in the structure of proteins, carbohydrates, and nucleic acids. We must also consider its significance to the properties of the biological solvent *water*.

Biochemists often talk about "water structure," by which they mean the tendency for clusters of water molecules to be hydrogen bonded in an ice-like array. In ordinary ice all of the water molecules are connected by hydrogen bonds, six molecules forming a hexagonal ring resembling that of cyclohexane. The structure is extended in all directions by formation of additional hydrogen bonds to adjacent molecules. Each oxygen atom forms covalent bonds with two hydrogen atoms and also forms hydrogen bonds to two hydrogen atoms in other molecules. (Some of these hydrogen bonds are indicated in the drawing by dashed arrows.) The molecules in ice assume various orientations in the hexagonal array (see drawing). Many orientations are possible and the molecules frequently rotate and form hydrogen bonds in different ways. In fact, this randomness remains as the temperature is lowered, and ice is one of the relatively few substances with a residual entropy at absolute zero. Ice is also unusual in that the molecules do

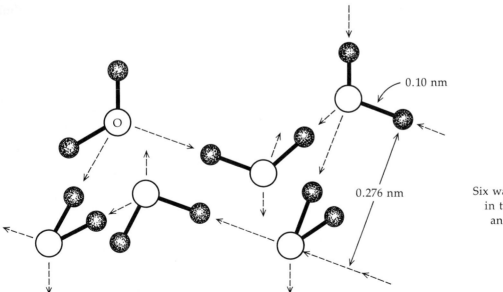

0.10 nm

0.276 nm

Six water molecules
in the lattice of
an ice crystal

not assume closest packing in the crystal but form an open structure. The hole through the middle of the hexagon and on through the hexagons lying below it is ∼0.06 nm in diameter.

The nature of liquid water is still incompletely understood, but we know that the icelike clusters of molecules are continually breaking up and re-forming in what has been called a flickering cluster structure.

Water molecules are able to hydrogen bond not only to each other but also to polar groups of dissolved compounds. Thus, every group that is capable of forming a hydrogen bond to another organic group is also able to form hydrogen bonds of a somewhat similar strength with water. For this reason, hydrogen bonding is not always a significant force in holding small molecules together in aqueous solutions. While two polar molecules may have a strong tendency to stick together through hydrogen bonding when dissolved in a nonpolar solvent, they often do not associate in water. How then can biochemists assert that hydrogen bonding is of extreme importance in the structure of macromolecules and in interactions between biochemical substances? The answer lies in the delicately poised equilibria according to which two molecules can either be held together by hydrogen bonding or can be dissociated and dissolved in water. Proteins and nucleic acids can be either properly folded with internal hydrogen bonds formed or they can be denatured with hydrogen bonds from those same groups to water. The free energy change between these two states is relatively small.

Perhaps the most significant thing about hydrogen bonding in biochemistry is that it often provides the specificity necessary to bring surfaces together in a complementary way. In other words, the location of hydrogen bond forming groups in surfaces between molecules is one of the important forces insuring an exact alignment of the surfaces. However, hydrogen bonding does not offer an explanation of the force which causes so many organic molecules to aggregate and bind to each other in an aqueous solution.

5. Hydrophobic Bonding[2,3,3a]

How can we explain the tendency for nonpolar groups to aggregate in micelles, in the centers of protein molecules, and in cell membranes? The following attempt is adapted from Jencks.[2]

Consider the changes occurring in an aqueous solution when a hydrophobic molecule is transferred into the water from an inert solvent, such as carbon tetrachloride, in two steps: (1) A cavity of about the right size to accommodate the molecule is formed in the water. Since many hydrogen bonds are broken, the free energy of cavity formation is high. This is principally an enthalpy (ΔH) effect. (2) The water molecules in the solvent now make changes in their orientations to accommodate the nonpolar molecule that has been placed in the cavity. The water molecules move to give good van der Waals contacts and also reorient themselves to give the maximum number of hydrogen bonds. Since hydrogen bonds can be formed in many different ways in water, there may be as many or even more hydrogen bonds after the reorientation of water molecules has taken place than before. This is true especially at low temperature where a substantial amount of ice structure exists within liquid water. In many cases, it appears that the restriction of the mobility of the water molecules surrounding the hydrophobic groups, i.e., the increase in "structure" of the water, is the most important aspect of hydrophobic bonding. In the case of dissolved hydrocarbons, the enthalpy of formation of the new hydrogen bonds often almost exactly balances the enthalpy of creation of the cavity initially so that ΔH for the overall process (transfer from inert solvent into water) is almost zero (it may be either a small negative or a small positive quantity). However, the restriction of the water mobility results in a very large *decrease* in the entropy, i.e., ΔS is negative. Since $\Delta G = \Delta H - T \Delta S$ and the term $-T \Delta S$ is positive, the free energy for the transfer from inert solvent to water is also positive; i.e., the transfer is unfavorable. This accounts for the low solubility of hydrocarbons in water.

The same explanation accounts for the tendency of hydrocarbon molecules to stick together in water. Think of the formation of a hydrophobic bond as a transfer of nonpolar portions of molecules from water into a close association with one another and in which these hydrophobic regions are in an environment similar to that in an inert solvent. The number of water molecules immobilized around the surfaces of the hydrophobic regions is decreased. As a consequence water molecules are freed from the structured region around the hydrophobic surfaces and the entropy increases. The entropy change ΔS is usually positive for formation of a "hydrophobic bond" between two hydrocarbons or alkyl groups.

Since the entropy term $T \Delta S$ is often the predominant one in the free energy expression for K_f, it is often stated that hydrophobic bonding is strictly an entropy effect. However, as Jencks has pointed out, the ultimate explanation lies in the great cohesiveness of water molecules and both entropy and enthalpy effects may be important. Depending on the interactions of the solute with water, ΔS_f for hydrophobic bonds may sometimes be zero or even negative. This is the case with heterocyclic rings that contain both hydrophobic regions and polar groups able to form hydrogen bonds with water. Although they participate in hydrogen bonding, polar groups tend to cause a *decrease* of structure in the surrounding water. This decrease may equal or exceed the increased structure that surrounds hydrophobic regions. Consequently, ΔS for transfer of heterocyclic rings into water may be positive. Conversely, ΔS may be negative for association of heterocyclic molecules in water.

Because the structuring of the water around heterocyclic bases is less than that around completely nonpolar molecules, the enthalpy change for hydrophobic association of these rings may be negative enough to make association favorable, despite the decrease in entropy (see Eq. 2-13, Table 2-6). Consequently, the principal force providing stability in DNA is the "stacking" of bases, again a hydrophobic interaction, but this time

with a negative entropy change ($\Delta S \approx -30$ J $°K^{-1}$ per base pair) and with an enthalpy decrease of ~ -14 to -30 kJ mol^{-1} (Table 2-6).[4]

It is often observed that the formation constant K_f for hydrophobic associations increases with increasing temperature. This is in contrast to the behavior of K_f for many association reactions involving polar molecules. Since $R \ln K_f = -\Delta G°/T = -\Delta H°/T + \Delta S°$, it is clear that for strong association to occur, either $\Delta H°$ must have a rather large negative value or $\Delta S°$ a substantial positive value. If ΔH is negative, as in the majority of exergonic reactions (e.g., for protonation of NH_3, $\Delta H° = -52$ kJ mol^{-1}), K_f will decrease with increasing temperature. However, if $\Delta S°$ has a large positive value, $\Delta H°$ may be positive, a situation often observed for hydrophobic associations. In this case K_f will increase with increasing temperature.

An increase in stability at higher temperatures is sometimes used as a criterion for hydrophobic bonding. However, note that it is not observed for base stacking interactions in polynucleotides for which "melting" occurs at high temperatures (Chapter 2, Section D,10).

C. THE QUANTITATIVE MEASUREMENT OF BINDING

Measurement of the strength of association of molecules is an everyday aspect of modern biochemical research. For example, it may be important to know how strongly a hormone binds to a cell membrane or a feedback inhibitor (Chapter 1, Section F) to an enzyme to determine whether the interaction is significant physiologically. It is important to understand how binding is measured and to appreciate some of the complexities encountered in the measurement of binding.

1. Analyzing Data

The extent of binding of a molecule X to another molecule P (Eq. 4-1) is measured by varying the concentrations of X and P and observing changes in the concentration of the complex [PX]. The first prerequisite is to find some measurable property which is different for the complex than for either of the free components. For example, the complex may be colored and the components colorless. More commonly, the complex simply has a different light absorbance at a certain wavelength than do the components. Likewise, the circular dichroism or the chemical shift of a peak in the nmr spectrum may change. If P is an enzyme, only the complex PX will undergo decomposition to products. Sometimes (but not always) the rate of breakdown of PX (the enzyme–substrate complex) to form products is relatively slow compared to the rate at which the equilibrium between X, P, and PX is established. If this is so, the concentration of the complex PX will be proportional to the observed rate of formation of product.

Whatever change of property is measured experimentally, its value will increase with increasing concentrations of X if the concentration of the macromolecule P is kept constant. In the usual experimental design, the molar concentration of P is small and it is possible to increase the concentration of X to quite large values. When this is done, it is usually observed that at high enough values of [X] most of the P is converted to PX, and the change being measured (say ΔA) no longer increases. This effect is known as **saturation** and is observed in most binding studies and also in many physiological phenomena.

The property or change in property being measured (ΔA) reaches a maximum value ΔA_{max} at saturation, i.e., when all of compound P is converted to PX. The ratio of [PX] to the total concentration of all forms of P present [P]$_t$ is known as the **saturation fraction** and is given the symbol $\bar{y}$. If P has more than one binding site for X, y is defined as the fraction of the total binding sites occupied. If n is the number of sites per molecule, the total number of sites is n[P]. The value of $\bar{y}$ is often taken as equal to $\Delta A/\Delta A_{max}$, an equality that holds for multisite macromolecules only if the change in A is the same for each successive molecule of X

added. This is not always true, but when it is Eq. 4-8 is followed.

$$\sum_i i \frac{[PX_i]}{n[P]_t} = \bar{y} = \frac{\Delta A}{\Delta A_{max}} \tag{4-8}$$

Here i represents the number of ligands X bound to P and may vary from 0 to n. When $n = 1$ the saturation fraction $\bar{y}$ and ΔA are related to the concentration of free unbound X and the formation constant by Eqs. 4-9.

$$\bar{y} = \frac{K_f[X]}{1 + K_f[X]} \qquad \Delta A = \frac{\Delta A_{max} K_f[X]}{1 + K_f[X]} \tag{4-9}$$

A plot of $\bar{y}$ or ΔA against [X] (sometimes referred to as an **adsorption isotherm**—because it had better be done at constant temperature if good results are to be obtained) for a hypothetical experiment is shown in Fig. 4-1. Note (from both Fig. 4-1 and Eq. 4-9) that $\bar{y}$ reaches a value of 0.5 when [X] is just equal to $1/K_f$ (or to K_d). Note also that saturation is reached slowly and that even at the point representing the highest concentration of X ($8/K_f$) saturation is less than 90%. Since in the usual experimental situation, we do not know $\bar{y}$ but only ΔA, it is difficult (unless K_f is very high) to estimate the limiting value ΔA_{max} from a plot of this

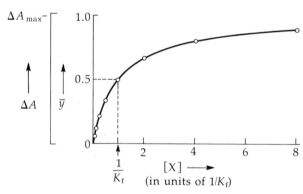

Fig. 4-1 An adsorption isotherm, a plot of the saturation fraction $\bar{y}$ or of some change in a measured property ΔA vs. [X], the concentration of a substance that binds reversibly to a macromolecule. The curve is hyperbolic and [X] = $1/K_f$ when $\bar{y}$ = 0.5.

type. However, we need to know $\Delta A_{\max}$ to evaluate K_f. For this reason, plots like Fig. 4-1 are little used, this one being included mainly to illustrate a point of nomenclature. The curve shown in Fig. 4-1 is a rectangular hyperbola, and the type of saturation curve shown is frequently referred to as **hyperbolic.** This is in contrast to certain other binding curves (Section C,7) which, when plotted in this way are **sigmoidal** (S-shaped).

A better type of plot is that of $\bar{y}$ against $\log[X]$ (Fig. 4-2). Note the following features. (1) The curve is symmetrical about the midpoint at $\log[X] = -\log K_f$. (2) No matter how high or low the concentration range used in the experiments, it is easy to choose a scale that puts all the points on the same sheet of paper. (3) Spacing between points tends to be more uniform than in a plot against $[X]$ (compare Figs. 4-1 and 4-2 for which the experimental points represent the same data and for which values of $[X]$ for successive points are each twofold greater than the preceding one). (4) The same logarithmic scale can be used for all compounds, no matter how strong or weak the binding, and the same shape curve is obtained for all 1:1 complexes. The midpoint slope, $d\bar{y}/d \log[X]$, is 0.576; the change in $\log[X]$ in going from 10 to 90% saturation is 1.81. The curve is familiar to most chemists because it is frequently used for pH titration curves in which pH substitutes for $-\log[X]$. To represent a complex with tighter binding, the curve is simply moved to the left, and for weaker binding, to the right. Mathematical description of curves of this type with hyperbolic functions has proved useful.[4a]

Saturation data are often plotted in yet another form known as the **Scatchard plot*** (Fig. 4-3). The value of $\Delta A/[X]$ (or of $\bar{y}/[X]$) is plotted against ΔA (or $\bar{y}$) and a straight line is fitted to the points, preferably using the "method of least squares." The intercept on the x axis and the slope of the fitted line give values of $\Delta A_{\max}$ and K_f, respec-

tively, as indicated by Eq. 4-10 (which follows directly from Eq. 4-9).

$$\bar{y}/[X] = K_f - \bar{y}K_f$$
$$\Delta A/[X] = \Delta A_{\max} K_f - \Delta A K_f \qquad (4\text{-}10)$$

Before measuring saturation curves the student should read the excellent paper of Deranleau[5] and, for further discussion, an article by Dowd and Riggs.[6]

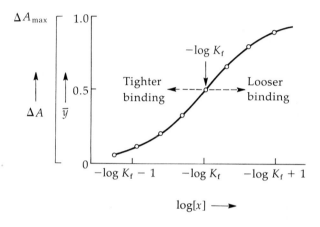

Fig. 4-2 A saturation curve plotted on a logarithmic scale for $[X]$. The data points are the same as those used in Fig. 4-1.

Fig. 4-3 A Scatchard plot of the same data shown in Figs. 4-1 and 4-2. This is the best of the linear plots for studying binding.

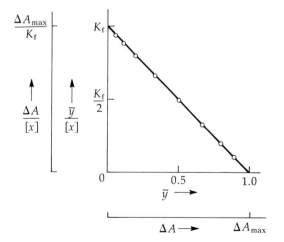

---

* The Scatchard plot is the best of the various linear transformations of the saturation equation and is much to be preferred to "reciprocal plots" (see Fig. 6-3).

2. Multiple Binding Sites on a Single Molecule

From the heading this may appear an unexciting topic, but molecules in living things interact with more than one other molecule at a given instant, and it is just this that keeps us glued together and functioning. First, consider the simplest situation in which a macromolecule P binds successively one molecule of X, then a second, and a third, up to a total of n. We define the stepwise formation constants, $K_1, K_2, \ldots, K_n$, as in Eq. 4-11.

$$P + X \underset{}{\overset{K_1}{\rightleftharpoons}} PX$$

$$PX + X \underset{}{\overset{K_2}{\rightleftharpoons}} PX_2 \cdots \qquad (4\text{-}11)$$

$$PX_{n-1} + X \underset{}{\overset{K_n}{\rightleftharpoons}} PX_n$$

The general expression for the ith stepwise formation constant is given by Eq. 4-12.

$$K_i = \frac{[PX_i]}{[PX_{i-1}][X]} \qquad (4\text{-}12)$$

Remember that $\bar{y}$ is the fraction of total binding sites saturated. The number of moles of X bound per mole of P is $n\bar{y}$ and is obtained by summing the concentrations $[PX] + 2[PX_2] + \cdots$ and dividing by the sum of all the forms of P (Eqs. 4-13 and 4-14).

For two binding sites ($n = 2$)

$$2\bar{y} = \frac{[PX] + 2[PX_2]}{[P] + [PX] + [PX_2]} \qquad (4\text{-}13)$$

For the general case

$$n\bar{y} = \left(\sum_{i=1}^{n} i[PX_i] \right) \Big/ \left([P] + \sum_{i=1}^{n} [PX_i] \right) \qquad (4\text{-}14)$$

Here the summations are over all the integral values of i from 1 to n. Now by expressing each concentration, $[PX_i]$, in terms of the concentrations $[X]$ and $[P]$ of *free* X and P, together with the stepwise formation constants, we obtain Eq. 4-15.

For $n = 2$

$$2\bar{y} = \frac{K_1[X] + 2K_1K_2[X]^2}{1 + K_1[X] + K_1K_2[X]^2} \qquad (4\text{-}15)$$

A similar equation can be written for the general case. Note that the concentration of P does not appear in Eq. 4-15 and that $\bar{y}$ is a function only of $[X]$ and the stepwise formation constants. Such equations define the isotherms for the binding of two or more molecules of X to P. From an experimental plot of $\bar{y}$ (or of ΔA) vs. $[X]$ or log $[X]$, it is possible in favorable cases to determine the stepwise constants $K_1, K_2, \ldots, K_n$. However, this becomes quite complicated. To simplify Eq. 4-15 and the corresponding equation for the general case, we can group the constants together and designate the products of constants (K_1, K_1K_2, $K_1K_2K_3$, etc.) as $\psi_1, \psi_2, \ldots, \psi_n$. Our equations are now as follows (Eqs. 4-16 and 4-17).

For $n = 2$

$$2\bar{y} = \frac{\psi_1[X] + 2\psi_2[X]^2}{1 + \psi_1[X] + \psi_2[X]^2} \qquad (4\text{-}16)$$

For the general case

$$n\bar{y} = \sum_{i=1}^{n} i\psi_i[X]^i \Big/ \left(1 + \sum_{i=1}^{n} \psi_i[X]^i \right) \qquad (4\text{-}17)$$

From experimental data, it is usually easiest to first determine the ψ's (there are n of them), then to calculate from the ψ's the stepwise constants.

For example (Eq. 4-18),

$$K_1 = \psi_1 \qquad K_2 = \psi_2/K_1, \quad \text{etc.} \qquad (4\text{-}18)$$

While Eq. 4-17 (the **Adair equation**) might seem to provide a complete description of the binding process, it usually does not. In many cases, there is more than one kind of binding site on a macromolecule and Eq. 4-17 tells us nothing about the distribution of the ligand X among different sites in complex PX. Furthermore, if n is large, it is impossible experimentally to determine as many as n different constants. We would like to find some simplification. To handle both these aspects of the problem, we must consider the **microscopic binding constants.**

3. Microscopic Binding Constants[†]

A microscopic constant is that which applies to a single binding site; for example, consider the association of a proton with a carboxylate ion (Eq. 4-19).

$$R—COO^- + H^+ \rightleftarrows R—COOH \qquad (4-19)$$

The association constant[‡] is $K_f \approx 6 \times 10^4$ and log $K_f = 4.8$. Since there is only one binding site on the carboxylate ion, the observed K_f *is* the microscopic binding constant. Now consider the pyridoxine anion in which there are *two* basic centers, the —O^- and the N.

CH$_2$OH

HOH$_2$C

O^- ⟵ log $K_f^* = 8.20^{§} = pK_b^*$
$K_f^* = 1.6 \times 10^8$

N   CH$_3$

log $K_f^* = 8.79 = pK_a^*$
$K_f^* = 6.2 \times 10^8$

Pyridoxine
(vitamin B$_6$)

A proton can attach to either of them. However, the microscopic binding constants K_f^* for the two groups are distinctly different. The nitrogen atom has the higher affinity for a proton (higher basicity). Thus, at 25°C in the neutral (monoprotonated) form 80% of the molecules carry a proton on the N, while the other 20% are protonated on the less basic —O^-.

[†] The microscopic constants are sometimes referred to as *intrinsic* constants, but the latter term is used in a more restricted way in this book. Microscopic constants are designated by asterisks here.

[‡] Chemists more customarily talk about dissociation constants for protons: K_d (often designated K_a for acids) = 1.7 $\times$ 10^{-5}; $pK_d = 4.8$ for R—COOH. However, to make this discussion uniform only binding constants will be used.

[§] The subscripts a and b used in this discussion (and also often in the literature) do *not* refer to acidic and basic but to the individual association steps shown in Eq. 4-21.

4. Tautomerism and Proton Binding

The two monoprotonated forms of pyridoxine are the tautomeric pair shown in Eq. 2-3. The **tautomeric ratio** R = [dipolar ion]/[neutral form] is a pH-independent equilibrium constant with a value[7] of 0.796/0.204 = 3.9 at 25°C. It should be noted that evaluation of microscopic constants for proton binding to compounds containing non-identical groups depends upon measurement of the tautomeric ratio (or ratios if more than two binding sites are present). In the case of pyridoxine, a spectrophotometric method was used to obtain R.

To calculate microscopic constants from stepwise constants and tautomeric ratios, consider Eq. 4-20 in which [PX]$_A$ and [PX]$_B$ are the concentrations of the two tautomers.

$$K_1 = \frac{[PX]_A + [PX]_B}{[P][X]} = K_a^* + K_b^* \qquad (4-20)$$

It is clear that K_1 is just the sum of the two microscopic constants for protonation to form tautomers PX(A) and PX(B). In a similar fashion it can be shown that the second stepwise constant K_2 is related to the microscopic constants for binding the second molecule of X as in Eqs. 4-21 and 4-22.[†]

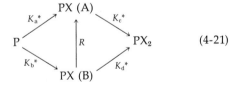

$$\qquad (4-21)$$

$$1/K_2 = 1/K_c^* + 1/K_d^* \qquad (4-22)$$

Since the tautomeric ratio R equals [PX]$_A$/[PX]$_B$, Eqs. 4-20 and 4-22 can be rewritten as Eqs. 4-23 and 4-24.

[†] This discussion concerns binding (association) constants, but the chemical literature deals largely with dissociation constants. Thus, in the literature the scheme of Eq. 4-21 is usually written in the opposite way when protons are involved, PH$_2$ dissociating with microscopic dissociation constants K_a and K_b, and K_1 being the first stepwise dissociation constant.

$$K_1 = K_a{}^*(1 + 1/R) \quad \text{and} \quad K_a{}^* = RK_b{}^* \quad (4\text{-}23)$$

$$1/K_2 = (1 + 1/R)/K_c{}^* \quad \text{and} \quad K_d{}^* = RK_c{}^* \quad (4\text{-}24)$$

In the case of pyridoxine, values of $\log K_1 = 8.89$ and $\log K_2 = 4.94$ were determined spectrophotometrically. These, together with the experimental value of R, were used to estimate microscopic constants.

Because it is often difficult to measure tautomeric ratios, microscopic constants are frequently assumed identical to binding constants for compounds in which one of the basic groups is methylated, esterified, or otherwise blocked. Consider the protonation of the two tautomers of the very weakly basic 1-methyluracil (Eq. 4-25).

(4-25)

The apparent pK values ($\log K_f$) were measured for protonation of the following two dimethylated derivatives.[8] In both cases cations of structure analogous to that in Eq. 4-25 are obtained.

$$pK_a = -3.25 \approx \log K_c{}^*$$

$$pK_a = +0.65 \approx \log K_d{}^*$$

It can reasonably be assumed that these pK_a values approximate $\log K_c{}^*$ and $\log K_d{}^*$ as indicated.

Thus, applying Eq. 4-24 $\log R \approx 0.65 + 3.25 = 3.9$. This result indicates that tautomer A of Eq. 4-25 is overwhelmingly predominant.[†]

Tautomerism among monoprotonated forms of cysteine, glutathione, and other amino acids has also received considerable attention by biochemists[9,9a] as has tautomerism in binding of protons and other small ligands to proteins.[10,11]

5. Statistical Effects

Consider a straight-chain dicarboxylic acid which has two *identical* binding sites. If the chain connecting the two carboxylate anions is long enough, the carboxylate groups will be far enough

$$K^* = 5 \times 10^4 \qquad K^* = 5 \times 10^4$$

apart that they do not influence each other through electrostatic interaction. Each group has a microscopic binding constant (K^*) of 5×10^4 (the latter can also be called an **intrinsic binding constant** because it is characteristic of a carboxylate group that is free of interactions with other groups). Intuition tells us (correctly) that, in its binding of protons, a solution of this dicarboxylic acid dianion will behave exactly like a solution of the monovalent anion R—COO^- at twice the concentration. A single intrinsic binding constant suffices to describe both binding sites. It may seem surprising then that the *stepwise* formation constants K_1 and K_2 differ: $K_1 = 10 \times 10^4$ and $K_2 = 2.5 \times 10^4$. This fact reflects the so-called statistical effect. *Either* of the two carboxylate groups in the molecule can bind a proton in the first step to give two indistinguishable molecules, PH (Eq. 4-26).

[†] This result also predicts that, within experimental error, the pK_a ($\log K_2$) for 1-methyluracil will equal $\log K_c{}^*$, namely, -3.25. In fact, it is close to this (-3.40), but not close enough to allow much confidence in the quantitative accuracy of this estimate of R. The authors of the study have estimated R by another method also.[8]

$$(4\text{-}26)$$

If we label the two forms of PH as A and B (Eq. 4-26) and consider that each one of them is independently in equilibrium with P through formation constant K^* it is clear (Eq. 4-27, see also Eqs. 4-23 and 4-24) that

$$K_1 = \frac{[PH]_A + [PH]_B}{[P][H^+]} = 2K^* \quad \text{and} \quad K_2 = K^*/2 \quad (4\text{-}27)$$

If this reasoning bothers you, it may help to recognize that this result is related to probability and arises for the same reason that if you reach into a barrel containing 50% white balls and 50% black balls, you will pull out one of each just twice as often as you will pull out a pair of white or a pair of black. In the general case of n equivalent binding sites, the microscopic formation constants K_i^* are related[12,13] to the stepwise constants K_i by Eq. 4-28.

$$K_i = \frac{(n + 1 - i)}{i} K_i^* \quad (4\text{-}28)$$

It is also easy to show,[14] using Eqs. 4-17 and 4-28, that for n completely equivalent and independent binding sites Eqs. 4-29 and 4-30 hold:

$$\bar{y} = \frac{K^*[X](1 + K^*[X])^{n-1}}{(1 + K^*[X])^n} \quad (4\text{-}29)$$

or

$$\bar{y} = \frac{K^*[X]}{1 + K^*[X]} \quad (4\text{-}30)$$

In this case the microscopic association constants are all identical and represent a single **intrinsic** constant applicable to all of the sites. In fact, Eq. 4-30 is identical to that for association of a single

proton (or other ligand) with a single binding site, satisfying our intuitive notion that a set of n completely independent binding sites ought to behave just like a solution of an n-fold more concentrated compound with a single binding site. Thus, our arithmetic has led us only to a conclusion that was already obvious. However, it is rarely true that binding sites on a single macromolecule are completely independent; there is almost always *interaction* between them, and the equations that we have derived for evaluation of stepwise and intrinsic constants will be applicable to such cases.

6. Electrostatic Repulsion; Anticooperativity

As we have seen, a hypothetical acid with an infinite distance between the carboxylate groups and $\log K^* = 4.8$ would have two binding constants separated by the statistical distance ($\log 4 = 0.6$) as shown in the first line of Table 4-1. The observed binding constants for protons with the dianions of acids containing 7, 4, 2, and 1 CH_2 groups are also shown in the table. For the longest chain (that of azelaic acid) the $\log K$ values are not far different from those of the hypothetical long chain acid. However, as the groups come closer together, the deviation from the predicted behav-

TABLE 4-1
Binding Constants of Protons to Dicarboxylic Acids[a]

Acid dianion	No. of CH₂ groups	log K_1 (pK_2)	log K_2 (association) (pK_1) (dissociation)
Hypothetical dianion with log $K^* = 4.8$	∞	5.1	4.5
Azelaic	7	5.41	4.55
Adipic	4	5.41	4.42
Succinic	2	5.48	4.19
Malonic	1	5.69	2.83

[a] From R. P. Bell, "The Proton in Chemistry," 2nd ed., p. 96. Cornell Univ. Press, Ithaca, New York, 1973.

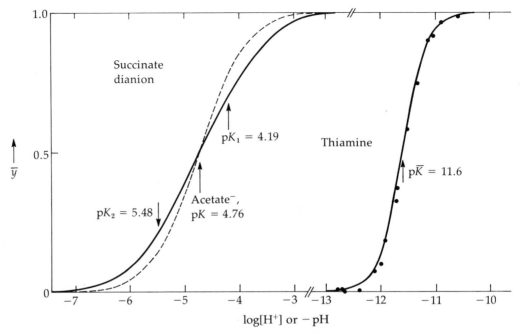

Fig. 4-4 Binding of protons to acetate ion, succinate dianion, and thiamine anion. Acetate (dashed line) binds a single proton with a normal width binding curve. Succinate dianion binds two protons with anticooperativity; hence, a broad-ening of the curve. Thiamine anion (yellow form, see Eq. 4-31) binds two protons with complete cooperativity and a steep binding curve.

ior becomes very large. The first binding constant is increased markedly because of the additional electrostatic attraction and the second is decreased.[†] The spread between the two $\log K$ values increases from 0.6 to as much as 2.9 as a result of the interaction between the binding sites. Thus, the binding of the first proton makes it harder to bind a second proton. Such negative interaction or **anticooperativity** between binding sites is most common and always leads to a spread of the formation constants and a broadening of the curve of $\bar{y}$ vs. $\log [X]$. For example, Fig. 4-4 compares the binding curve for protons with acetate ion and with succinic acid dianion.

Simple electrostatic theory predicts anticooperativity in the binding of protons. Equation 4-7 can

be applied if an appropriate dielectric constant can be chosen. The difference between the two successive $\log K$ values reduced by 0.6 (the statistical factor) is a measure of the electrostatic effect. For malonic acid this ΔpK is 2.25 and for succinic acid it is 0.69 (Table 4-1). In 1923, N. Bjerrum proposed that this value of ΔpK could be equated directly with the work needed to bring two negative charges together to a distance representing the charge separation in the malonate dianion. Thus,

applying Eq. 3-30, $\Delta G = 5.708\ \Delta pK$ kJ mol^{-1} = 12.84 kJ mol^{-1} for malonate. Equating this with W in Eq. 4-7 and assuming a dielectric constant of 78.5 (that of water) the distance of charge separation r is calculated to be 0.138 nm. This is clearly

[†] Note that in malonic and succinic acids the first proton bound can be shared by both carboxyl groups through formation of a hydrogen bond. Additional factors operate in oxalic acid where the carboxyl groups are connected directly and for which pK values are 4.19 and 1.23.

too small. The computation was improved by Westheimer and Kirkwood who assumed a dielectric constant of two *within* the molecule. By approximating the molecule as an ellipsoid of revolution, they were able to make reasonably accurate calculations of electrostatic effects on pK_a values.[15] Thus, for malonic acid Westheimer and Shookhoff [16] predicted $r = 0.41$ nm for malonic acid dianion.

Electrostatic theory has also been successfully used to interpret titration curves of proteins in which the net negative or positive charge distributed over the surface of the protein varies continuously from high pH to low as more and more protons are added.[17]

Electrostatic effects can be transmitted extremely effectively through aromatic ring systems, a fact that doubtless explains much of the significance of heterocyclic aromatic systems in biochemical molecules. Consider the microscopic binding constant of the phenolate anion of pyridoxine as influenced by the state of protonation of the ring nitrogen:

We see that $\Delta pK = 3.26$, even greater than that of malonic acid.

In such aromatic systems the electrostatic effects are usually satisfactorily predicted by the Hammett equation. An exercise for the student is first to calculate the four microscopic dissociation constants for 3-hydroxypyridine from the stepwise pK_a values of 4.91 and 9.62, together with the tautomeric ratio[7] $R =$ [dipolar ion]/[uncharged tautomer] = 1.05. Then use the Hammett equation (Eq. 3-66) to predict the same constants and compare the results.[18]

7. Cooperative Processes

Can it ever happen that interaction between groups leads to a *decrease* from the statistical separation between values of the stepwise constants instead of to an increase? At first glance, the answer seems to be no. A decreased separation would imply that the intrinsic binding constant for the second proton bound is higher than that for the first, but common sense tells us that the first proton ought to bind at the site with the highest binding constant. However, look at the experimental binding curve of protons with the anion of thiamine shown in Fig. 4-4. Instead of being broadened from the curve for acetate, it is just half as wide. The explanation depends upon some rather amusing chemistry of thiamine. Under suitable conditions, this vitamin can be crystallized as a yellow sodium salt, the structure of whose anion is shown in Eq. 4-31. Weak binding of a proton to one of the nitrogens as shown in Eq. 4-31 creates electron deficiency at the adjacent

(4-31)

carbon and the S^- anion adds closing the ring to an unstable tricyclic form of thiamine.[†] This tricyclic form can be observed in methanol and can be crystallized. It is very unstable in water because the central ring can open, the electrons flowing as indicated by the small arrows to create a strongly basic site on the same nitrogen. Then, a second proton combines at this basic nitrogen with a very high binding constant to form the cation. *The key to the reversed order of strength of the binding constants lies in the molecular rearrangements intervening between the two binding steps.* In this particular case, we cannot even measure the successive binding constants K_1 and K_2 because K_2 is very much larger than K_1 (probably two or more orders of magnitude). Consequently, the binding curve shown in Fig. 4-4 is (within experimental error) exactly twice as steep at the center (the slope is 2×0.576) as that for acetate ion and is accurately represented in Eq. 4-32, where $\bar{K} = \sqrt{K_1 K_2}$. Comparison of Eq. 4-32 with Eq. 4-15 shows how the latter has been simplified because no significant concentration of the form PX is present in the cooperative case.

$$\bar{y} = \frac{\bar{K}^2[X]^2}{1 + \bar{K}^2[X]^2} \qquad (4\text{-}32)$$

The binding of protons by the thiamine anion is an example of a **cooperative process,** so named because binding of the first proton makes binding of the second easier. Although relatively rare among small molecules, cooperative processes are extremely common and important in biochemistry. A cooperative binding curve is sometimes referred to as sigmoidal because the plot of $\bar{y}$ against [X] (the binding isotherm) is S-shaped. A binding process can be described as *completely cooperative* when the maximum possible cooperativity is observed. This implies that the nth binding site for ligand X must have essentially no affinity for X until all of the $n-1$ other sites have been filled. However, after its conversion to a strong binding site, the affinity of the nth site for X must be so

---

[†] This is only part of the story about the acid-base chemistry of thiamine. For the rest, see Chapter 8, Section D and Metzler[19] and Hopmann and Brugnoni.[20]

strong that in any equilibrium mixture significant amounts of only P and PX_n occur.

It is easy to show that for n binding sites with completely cooperative binding the saturation fraction is (Eq. 4-33)

$$\bar{y} = \frac{\bar{K}^n[X]^n}{1 + \bar{K}^n[X]^n} \qquad (4\text{-}33)$$

where $\bar{K} = (K_1 \cdots K_n)^{1/n}$. The midpoint slope in the binding curve (y vs. $\log[X]$) is $0.576n$ and the change, $\Delta \log[X]$, between $\bar{y} = 0.1$ and $\bar{y} = 0.9$ is $1.81/n$.

Equation 4-33 can be rewritten as

$$\bar{y}/(1 - \bar{y}) = \bar{K}^n [X]^n \qquad (4\text{-}34)$$

Taking logarithms (Eq. 4-34),

$$\log [\bar{y}/(1 - \bar{y})] = n \log \bar{K} + n \log [X] \quad (4\text{-}35)$$

A plot of $\log [\bar{y}/(1 - \bar{y})]$ vs. $\log[X]$ is known as a **Hill plot.** It is, according to Eq. 4-34, linear with a slope of n. Remember that this equation was derived for an ideal case of completely cooperative binding at n sites. However, Hill plots are often used by biochemists to plot experimental data for systems in which cooperativity is *incomplete.* Thus, the experimentally measured slope of a Hill plot (n_{Hill}) is usually less than n, the number of binding sites.

A comparison of n_{Hill} with n is often used as a measure of the degree of cooperativity. Thus, $n_{Hill}/n = 1.00$ for complete cooperativity, but is less than one if cooperativity is incomplete. It is not necessary to make a Hill plot to get n_{Hill}. From the usual binding curve of $\bar{y}$ (or ΔA) vs. $\log [X]$ the midpoint slope can be measured with satisfactory precision. Alternatively, the difference, $\Delta \log [X]$ between 0.1 and 0.9 saturation can be evaluated and n_{Hill} calculated from Eq. 4-36:

$$n_{Hill} = \frac{\text{midpoint slope}}{0.576} = \frac{1.81}{\Delta \log[X]} \qquad (4\text{-}36)$$

Now a little warning: Binding curves sometimes have "wiggles" in them; in such cases Hill plots are not linear and no simple measure of cooperativity can be defined.

A second example of cooperativity is provided

by the reversible denaturation of coiled peptide chains. Some proteins can be brought to a pH of ~4 by addition of acid but without protonation of buried groups with intrinsic pK_a values greater than four. Then when a little more acid is added, some less basic group is protonated, permitting the protein to unfold and to expose the more basic hidden groups. Thus, cooperative proton binding is observed. As in the case of thiamine the cooperativity depends upon occurrence of a conformational change in the molecule linked to protonation of a particular group.

Another type of cooperative phenomenon is found in the reversible transformation between an α helix and a random coil conformation. In this case, once a helix is started, additional turns form rapidly and the molecule is completely converted into the helix. Likewise, once it unfolds it tends to unfold completely. Melting of DNA (Chapter 2, Section D,10) or indeed, of any crystal, is cooperative.[21] The stacking of nucleotides alongside a template polynucleotide can also be cooperative. For example, the binding of one adenosine molecule to two strands of polyuridylic acid leads to cooperative formation of a triple helical complex (Chapter 2, Section D,6). Here the stacking interactions make helix growth energetically easier than initiation of new helical regions.[22] A substantial literature on cooperative binding exists.[23-25]

8. Binding of Metal Ions

Equilibria in the formation of complex ions with metals are treated exactly as is the binding of small molecules and ions to macromolecules.[26-28] Stepwise constants are defined for the formation of complexes containing one, two, or more ligands X bound to a central metal ion. The symbol P in the preceding equations is usually replaced by M (for metal), the symbol L (ligand) is often used in place of X, and the ψ's are usually referred to as β's. Thus

$$\beta_1 = K_1 \qquad \beta_n = K_1 K_2 \cdots K_n \qquad (4\text{-}37)$$

Many important questions can be asked about the binding of metal ions within living cells. For example: What fraction of a given metal ion is free and what fraction is bound to organic molecules? To what ligands is a metal bound? Since many metal ions are highly toxic in excess, it is clear that homeostatic mechanisms must exist. How do such mechanisms sense the free metal ion activity within cells? How does the body get rid of unwanted metal ions? Answers to all these questions depend upon the quantitative understanding of the binding of metal ions to the variety of potential binding sites found within a cell.

Table 4-2 gives formation constants for 1:1 complexes of several metal ions and a number of inorganic as well as organic ligands.[28,28a] Only the values of $\log K_1$ are given when a series of stepwise constants have been established. However, in many cases two or more ligands can bind to the same metal ion. Thus for cupric ion and ammonia there are four constants.

$$Cu^{2+} + NH_3 \qquad \log K_i: 4.0, 3.3, 2.7, 2.0$$

Note that they are all separated by more than the statistical distance (which in this case is less than the 0.6 logarithmic units for two equivalent binding sites). Thus, anticooperativity in binding of successive ligands is observed here and in most other cases.

Most metal ions will bind two or three successive amino acids. In the case of copper, whose preferred coordination number is four, two ligands may be bound. In this case, a distinct anticooperativity is evident in the spread of the two constants.

$$Cu^{2+} + \text{alanine} \qquad \log K_i: 8.1, 6.8$$

a. Factors Affecting the Strength of Binding of a Metal in a Complex

An obvious factor in metal binding is the basicity of the ligand. The more basic ligands tend to bind metal ions more tightly just as they do protons. However, the strength of bonding to metal ions is more nearly proportional to the **nucleophilic character** (Chapter 7, Section C,1, only partly determined by basicity to protons) of a group than to basicity itself.

TABLE 4-2
Logarithms of Binding Constants for Some Metal Complexes at 25°C[a]

Ligand	H⁺	Mg²⁺	Ca²⁺	Mn²⁺	Cu²⁺	Zn²⁺
Hydroxide, OH^-	14.0	2.5	1.4		6.5	4.4
Acetate⁻	4.7	~0.65	0.5	~1.0	2.0	1.5
Lactate⁻	3.8	~1.0	~1.2	1.3	3.0	2.2
Succinate²⁻	5.2	1.2	1.2		3.3	2
NH_3	9.3	~0	−0.2	0.8	4.0	2.4
Ethylenediamine	10.2	0.4		2.8	10.8	6.0
Glycine⁻	9.6	2.2	1.4*	2.8	8.2	5.0
Glycine amide	8.1			~1.5	5.3	3.3
Alanine⁻	9.7	2.0*	1.2*	3.0*	8.1	4.6
Aspartate²⁻	9.6	2.4	1.6		8.6	5.8
Glycylglycine⁻	8.1		1.2*	2.2*	5.5	3.4
Pyridine	5.2			0.1	2.5	1
Imidazole	7.5			1.6	4.6	2.6
Histidine	9.1			3.3	10.2	6.6
Adenine	9.8				8.9	6.4
Citrate³⁻	5.6	3.2	4.8	3.5	~4	4.7
EDTA⁴⁻ [b]	10.2	8.8	10.6	13.8	18.7	16.4
EGTA⁴⁻ [c]	9.4	5.3	10.9	12.2	17.6	12.6
ATP⁴⁻	6.5	4.2	3.8	4.8	6.1	4.9

[a] All values are for log K_1. Included is the highest pK_a for protons. Data for amino acids are from Martell and Smith.[28] Others are from Sillén and Martell.[28a] Most constants for amino acids are for ionic strength 0.1. Some (designated by asterisks) are for zero ionic strength. The values shown for other ligands have been selected from a large number reported without examination of the original literature.

[b] Ethylenediaminetetraacetic acid, a chelating agent widely

$$\begin{array}{ccc} ^-OOC-H_2C & & CH_2-COO^- \\ & N-CH_2-CH_2-N & \\ ^-OOC-H_2C & & CH_2-COO^- \end{array}$$

used by biochemists for preventing unwanted reactions of metal ions. The high formation constants insure that most metal ions remain bound to the EDTA.

[c] EGTA is similar to EDTA but has the group $-CH_2-CH_2-O-CH_2-CH_2-O-CH_2-CH_2-$ joining the two nitrogen atoms in place of $-CH_2-CH_2-$ of EDTA. Note that EGTA has a higher selectivity for Ca^{2+} compared to Mg^{2+} than does EDTA.

The pH of the medium always has a strong effect on metal binding. Competition with protons means that metal complexes tend to be of weak stability at low pH. Thus, anions of carboxylic acids are completely protonated below a pH of ~4 and a metal can combine only by displacing a proton. However, at pH 7 or higher, there is no competition from protons. On the other hand, protons are still very strong competitors, even with a strongly complexing metal ion such as Cu^{2+}, in the case of ethylenediamine, whose pK_a values are 10.7 and 7.5 (Table 4-2). At high pH another factor is competition between the ligand and hydroxyl ion. Thus, at pH 7 about one-half of Cu^{2+} dissolved in water is complexed as $CuOH^+$.

One of the biggest single factors in determining the affinity of organic molecules for metal ions is the **chelate effect.**[†] This term refers to the greatly enhanced binding of metal ions resulting from the presence of two or more complexing groups in the same organic molecule. The chelate effect has been exploited to good advantage by nature in the design of such important metal binding molecules as the porphyrins (Fig. 10-1), chlorophyll (Fig. 13-19), enterobactin (Fig. 2-44), calcium-binding proteins (Section C,8,c), and many others. Also note, from Table 4-2, that many relatively simple compounds such as the α-amino acids and citric acid form strong **chelate complexes** with metal ions.

b. Effects of the Metal Ion on Chelation

The **charge,** the **ionic radius** (Table 4-3), the **degree of hydration,** and the **geometry of orbitals** used in covalent bonding between metal and chelating groups all affect the formation constants of a complex. Generally speaking, multicharged ions form stronger complexes than do monovalent ions which have a lower charge density.

Among ions of a given charge type (e.g., Na^+ vs. K^+; Mg^{2+} vs. Ca^{2+}), the smaller ions are more strongly hydrated than are the larger ions in which the charge is dispersed over a greater surface area. Most cations, except for the largest ones, have a primary hydration sphere containing about six molecules of water. Four molecules of water can be placed around the ion in one plane as shown on p. 199 for the coordination of water molecules to Mg^{2+}.[29]

[†] Chelate (pronounced "keel-ate") is from a Greek word meaning crab's claw.

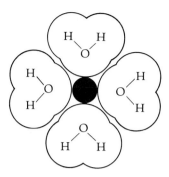

One additional water molecule can coordinate above and another below to provide six molecules in an array of *octahedral* geometry. Other solvent molecules are held in a looser secondary sphere. Thus, electrochemical transference experiments indicate a total of ~16 molecules of water around Na^+ and ~10 around K^+.

To form a chelate complex a metal ion must usually lose most of its hydration sphere. For this reason, the larger, less hydrated metal ions often bind more strongly than do the smaller more hydrated ones. Thus, Ca^{2+} binds more tightly than does Mg^{2+} to EDTA (Table 4-2). However, the reverse order is also observed, especially with negatively charged ligands (e.g., OH^-) in which the

charge is highly concentrated. Note that Mg^{2+} binds more tightly than does Ca^{2+} to ATP^{4-} (Table 4-2; Chapter 3, Section B,5). Differences of this type lie at the basis of highly important differences in biological effects of ions.

The following is a well-known sequence of the stabilities of complexes of metals of the first transition series.

$$Mn^{2+} < Fe^{2+} < Co^{2+} < Ni^{2+} < Cu^{2+} > Zn^{2+}$$

Simple electrostatic theory based upon differences in the ionization potentials or electronegativity of the ions would predict a gradual monotonic increase in chelate stability from manganese to zinc. In fact, with nitrogen-containing ligands Cu^{2+} usually forms by far the strongest complexes (Table 4-2). Cobalt, nickel, and iron ions also show an enhanced tendency to form complexes with nitrogen-containing ligands. The explanation is thought to lie in the ability of the transition metals to supply d orbitals which can participate in covalent bond formation by accepting electrons from the ligands. It is significant that iron, copper, and cobalt are often located in the centers of specially designed nitrogen-containing structures such as the heme of our blood (iron, Fig. 10-1) and vitamin B_{12} (Box 8-K).

To what ligands will specific ions tend to bind in cells? The alkali metal ions Na^+ and K^+ are mostly free but are in part bound to specific sites in proteins. Likewise, Ca^{2+} and Mg^{2+} will tend to remain partially free and complexed with the numerous phosphate and carboxylate ions present in cells. On the other hand, the heavier metal ions including those of zinc, copper, iron, and other transition metals will tend to bind to nitrogen or sulfur atoms, often in special molecules such as the porphyrins.

c. Metal Binding Sites in Proteins

Many **metalloproteins** contain special metal-binding **prosthetic groups** (e.g., the porphyrin group in hemoglobin, Fig. 10-1). In other instances clusters of carboxylate, imidazole, or other groups are used to create a specific binding site. In

TABLE 4-3
Ionic Radii for Some Metallic and Nonmetallic Ions[a]

		Mn^{2+}	0.080		
Li^+	0.060	Fe^{2+}	0.076	H^-	0.21
Na^+	0.095	Co^{2+}	0.074	F^-	0.136
K^+	0.133	Ni^{2+}	0.069	Br^-	0.195
Rb^+	0.148	Cu^{2+}	0.072^b	I^-	0.216
		Zn^{2+}	0.074		
		Cd^{2+}	0.097		
Be^{2+}	0.031				
Mg^{2+}	0.065	Al^{3+}	0.050		
Ca^{2+}	0.099	Fe^{3+}	0.064		
Sr^{2+}	0.113	Mo^{4+}	0.070		
Ba^{2+}	0.135	Mo^{6+}	0.062		

[a] Radii are calculated according to the method of Pauling and are taken from F. A. Cotton and G. Wilkinson, "Advanced Inorganic Chemistry," 3rd ed. Wiley (Interscience), New York, 1972.

[b] From L. H. Ahrens as given by M. J. Sienko and R. A. Plane, "Physical Inorganic Chemistry," pp. 68–69. Benjamin, New York, 1963.

some instances an NH group of a peptide linkage can serve as one of the ligands by losing a proton. Small peptides react with Cu^{2+} to form complexes,[30,31] some of which contain a bond to the amide nitrogen (Eq. 4-38, step b).

Glycylglycine
anion

$$(4\text{-}38)$$

A linkage of Ca^{2+} to an amide as well as to a cluster of carboxylate ions has been found in a special **calcium-binding protein** or "parvalbumin." The three-dimensional structure of the protein from carp muscle has been established (Fig. 4-5).[32] There are two calcium-binding sites. In one (to the left in Fig. 4-5A) the Ca^{2+} ion is bound by four carboxylate groups from aspartate and glutamate side chains, a hydroxyl group of serine, and the residue 57 carbonyl oxygen of the peptide backbone. Note that the same peptide group is hydrogen bonded to a carbonyl group of another segment of peptide chain near the second Ca^{2+} site (to the right in Fig. 4-5A). This site contains four carboxylate ions (one of which coordinates the Ca^{2+} with both oxygen atoms) and another peptide carbonyl group. The significance of this structure is not clear, but inspection of Fig. 4-5 should emphasize an important fact: By pulling several basic groups together, chelation can lead to a major conformational change in a peptide. Chelation also imparts a more rigid geometry to molecules, a fact that may be important to its biological activity.

It is well known that calcium ions are released into the cytoplasm in response to stimulation by nerves and that they trigger various responses such as the contraction of muscle. It is quite likely that the binding of Ca^{2+} at sites such as those in the carp calcium-binding protein induce conformational changes that initiate biological responses. The calcium-binding protein contains an interesting network of internal polar groups that are hydrogen bonded in a specific way (Fig. 4-5B). A rearrangement of these internal bonds (Chapter 2, Section B,7) could be induced by the binding of calcium ions and could alter the way in which this protein (whose function is obscure) reacts with some other protein (compare with the action of troponin C, Section F,1).

Another series of Ca^{2+}-binding sites in proteins contain the special chelating amino acid α-carboxyglutamate (Box 10-D) at their active sites.

D. HOW MACROMOLECULES PACK TOGETHER

Just as amino acids, sugars, and nucleotides are the building blocks for formation of proteins, polysaccharides, and nucleic acids, so these macromolecules are the units from which larger structures are assembled. Fibers, microtubules, virus "coats," and small symmetrical groups of **subunits** in **oligomeric enzymes** all result from the packing of macromolecules in well-defined ways (sometimes referred to as **quaternary structure**). Consider first the aggregation of *identical* protein subunits. We know that many protein molecules are nearly spherical, but they are nevertheless distinctly asymmetric. In the drawings that follow the asymmetry is exaggerated, but the principles illustrated are valid.

1. Rings with Cyclic Symmetry

Consider a subunit **(protomer)** of the shape shown in Fig. 4-6 and containing a region a that is

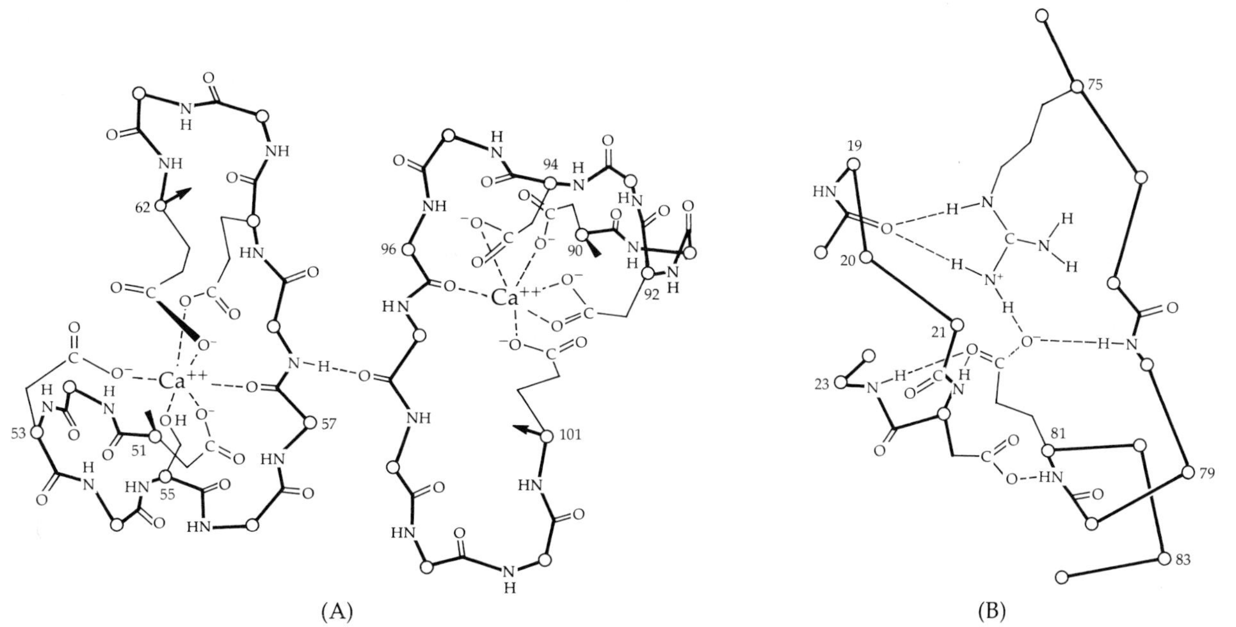

Fig. 4-5 (A) Part of the 108-residue peptide chain of the calcium-binding protein of carp muscle. The two calcium-binding loops are shown together with a hydrogen bond between them. (B) A view of the intricate network of hydrogen bonds linking two segments of the peptide chain in the interior of the molecule. Note especially the bonding of the guanidinium group from arginine-75 to the carboxylate of glutamic acid 81 and to the peptide carbonyl of residue 18. Note that the carboxylate group also interacts with two different peptide NH groups. From Kretsinger and Nockolds.[32] See also Kretsinger.[32a]

complementary to the surface j on another part of the same molecule. Two such protomers will tend to stick together to form a dimer, region a of one protomer sticking to region j of the other. The dimer will still contain a free region a at one end and a region j at the other which are not involved in bonding. Other protomers can stick to these free ends. In some instances long chains can be formed. However, if the geometry is just right, as in Fig. 4-6A, a third subunit can fit in to form a closed ring "trimer." Depending on the geometry of the subunits the ring can be even smaller (a dimer) or it can be larger (a tetramer, pentamer, etc.). The bonding involved is between two dif-

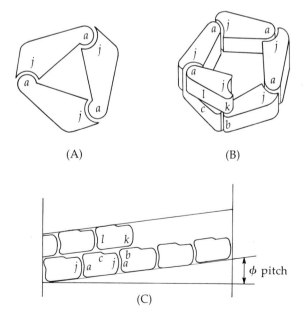

Fig. 4-6 Heterologous bonding of subunits: (A) in a ring; (B) in a helix, and (C) "radial projection" of subunits arranged as in helix B. Different bonding regions of the subunit are designated a, b, c, j, k, and l.

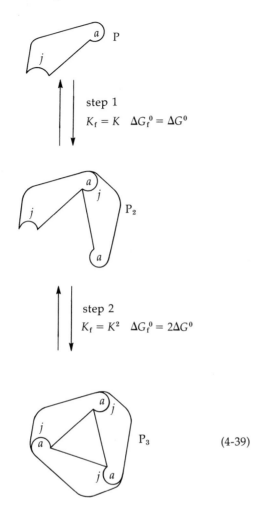

step 1
$K_f = K$ $\Delta G_f^0 = \Delta G^0$

step 2
$K_f = K^2$ $\Delta G_f^0 = 2\Delta G^0$

(4-39)

ferent regions (a and j) of a subunit and is sometimes described as **heterologous.**[33] To obtain a closed ring of subunits, the angle between the bonding groups a and j must be correct or the ring cannot be completed.

A ring formed using exclusively heterologous interactions possesses **cyclic symmetry,** e.g., the trimer in Fig. 4-6A has a **3-fold axis:** Each subunit can be superimposed on the next by rotation through 360°/3. The oligomer is said to have C_3 symmetry. Many real proteins, including all of those with 3, 5, or other uneven number of identical protomers, appear to be formed of subunits arranged with cyclic symmetry.

Now consider the quantitative aspects of heterologous interactions with ring formation. Let K be the formation constant and ΔG^0 the free energy change for the reaction of the j end of protomer P with the a end of a second protomer to form the dimer P_2 (Eq. 4-39). In the second step (Eq. 4-39) a third protomer combines. Note that it forms *two* new aj interactions. Therefore, for this step ΔG_f^0 is $2 \Delta G^0$ and K_f is equal to K^2. The overall formation constant for formation of a trimer from three protomers is given[†] by Eq. 4-40.

$$3 P \longrightarrow P_3 \qquad (4\text{-}40)$$
$$K_f = K^3$$
$$\Delta G_f^0 = 3 \Delta G^0$$

Now, consider a hypothetical example: Protomer P is continuously synthesized by a cell and at the same time is degraded to some nonaggregating form in a second metabolic reaction. The two reactions are balanced so that [P] is always present at a steady state value of $10^{-5} M$. Suppose that a value for a single aj interaction of $K = 10^4$ (or $\Delta G^0 = -22.8$ kJ mol^{-1}) governs aggregation to form dimers and trimers (rings). What concentration of dimers and of trimer rings will be present in the cell in equilibrium with the 10^{-5} concentration of P? Using Eq. 4-39 we see that the concentration of dimers $[P_2]$ is $10^4 \times (10^{-5})^2 = 10^{-6} M$. (Note that the amount of material in this concentration of dimer is equivalent to $2 \times 10^{-6} M$ of the monomer units.) The concentration of rings $[P_3]$ is $(10^4)^3 \times (10^{-5})^3 = 10^{-3} M$ (equivalent to $3 \times 10^{-3} M$ of the monomer units). Thus, of the *total* P present in the cell ($10^{-5} + 0.2 \times 10^{-5} + 300 \times 10^{-5} M$), 99.6% is associated to trimers, 0.33% is still monomers, and only 0.07% exists as dimers. Thus, the formation of two heterologous bonds simultaneously to complete a ring imparts a

---

[†] This treatment assumes that ΔG° for formation of both new aj bonds in the trimer is exactly the same as that for formation of the aj bond in the dimer. The reader may wish to criticize this assumption and to suggest conditions that might lead to overestimation or underestimation of K_f for the trimer as calculated above.

high degree of cooperativity to the association reaction of Eq. 4-39. We will find in a cell mostly either rings or monomer, but little dimer.

Now consider what will happen to the little rings within the cell if the process that removes P to a nonassociating form suddenly becomes more active so that [P] falls to 10^{-6} M. If K is still 10^4, what will be the percentages of P, P_2, and P_3 at equilibrium? Here we note a characteristic of cooperative processes: a higher than first power dependence on a concentration.

2. Helices

If the angle at the interface aj is slightly different, we obtain instead of a closed ring, a helix as shown in Fig. 4-6B. The helix may have an integral number of subunits per turn or a nonintegral number, as in the figure. The same type of heterologous interaction aj is involved in joining each subunit to the preceding one, but in addition other interactions occur. If their surfaces are complementary and the geometry is correct, groups from two different parts of the molecule (e.g., b and k) may fit together to form another heterologous bond. Still a third heterologous interaction cl may be formed between two other parts of the subunit surfaces. If interactions aj, bk, and cl are strong (i.e., if the surfaces are highly complementary over large areas), extremely strong microtubular structures may be formed, such as those in the flagella of eukaryotic organisms (Fig. 1-5). If the interactions are weaker, **labile microtubules,** such as are often observed to form and dissociate within cells, may arise.

The geometry of subunits within a helix is often advantageously displayed by imagining that the surface of the structure can be unfolded to give a radial projection (Fig. 4-6C). Here subunits corresponding to those in the helix in Fig. 4-6B are laid out on a plane obtained by slitting the cylinder representing the surface of the helix and laying it out flat. In the example shown, the number of subunits per turn is about 4.3 but it can be an integral number. The interactions bk between

subunits along the direction of the fiber axis may sometimes be stronger than those (aj) between adjacent subunits around the spiral. In such cases the microtubule becomes frayed at the ends through breaking of the aj interactions. This phenomenon can be observed under the electron microscope for the microtubules from flagella of eukaryotic organisms.

Figure 4-7 shows the artist's conception of four helical structures from the molecular domain: a pilus from *E. coli,* an **actin** microfibril ("F-actin") from muscle, a bacterial flagellum (*E. coli*), and **tobacco mosaic virus.** Each is thought to be composed of a single kind of protomer. The virus has been studied most. The amino acid sequence of the 158 amino acids in the protein subunits (MW = 17,500) is known; about 2200 units combine to form a rod ~300 nm long. A single strand of RNA containing ~6600 nucleotides (~3 per protein subunit) lies coiled in a groove.[34-36a] It would appear that only heterologous bonds hold the protein subunits together in this helix as in the simpler bacterial pilus.[37]

Another structure, which has recently been worked out in detail, is the protein "coat" of a filamentous bacterial virus.[38,39] The protein subunits of bacteriophage Pf1 (Box 4-C) have a molecular weight of 5000. They exist as α-helical rods of 7 nm length which are arranged in the virus in a left-handed helical pattern with a pitch of 1.5 nm and with 4.4 subunits per turn (Fig. 4-8). The protein rods are inclined to the helix axis and extend inward. A given rod makes close contact with rods 5 units (−5) and 9 units (−9) back down the helix (Figs. 4-8A,B). This arrangement permits a "knobs-in-holes" (Chapter 2, Section B,3,d) type of hydrophobic bonding between subunits. It is quite likely that bacterial pili and flagella have somewhat similar structures rather than the bricklike ones shown in Fig. 4-7.

Both the helical viruses and bacterial pili can be regarded as *single* helical coils of subunits, sometimes called **one-start helices.** The actin microfilament (Fig. 4-7) has the geometry of *two chains* of subunits coiled around one another,[40] i.e., it is a two-start helix. The bacterial flagella of *E. coli* and

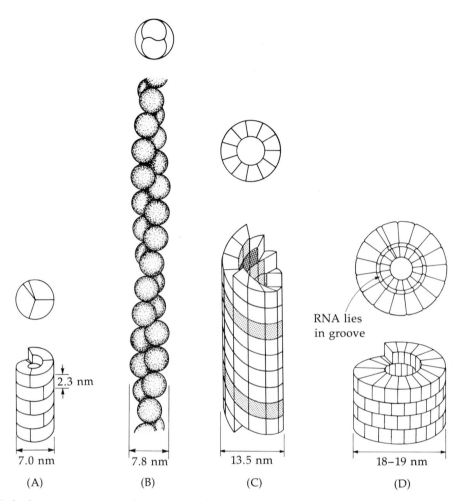

2.3 nm

7.0 nm   7.8 nm   13.5 nm   18–19 nm

RNA lies
in groove

(A)   (B)   (C)   (D)

Fig. 4-7 Some helical structures composed of protein sub-units. (A) Pilus from *E. coli*: pitch, 2.3 nm; hole, 2.0–2.5 nm; MW = 17,000; $3\frac{1}{8}$ units/turn. (B) F-Actin from muscle: pitch, 70 nm; 2 strands; MW = 60,000; 13 units/turn. (C) Bacterial flagellum: pitch, 25 nm; hole, 6 nm; 5 strands; MW = 40,000; 11 units/turn. (D) Tobacco mosaic virus: pitch, 2.5 nm; MW = 17,500; length; 300 nm (~2200 units); 16.3 sub-units/turn.

Salmonella can be thought of as *five chains* coiled together about the same axis (one of the chains is shaded in the drawing in Fig. 4-7). Alternatively, the same flagellum can be viewed as composed of 11 parallel strands forming a helix of much longer pitch.[41] Bacterial flagella have many fascinating properties (see Box 4-B). For example, they usually appear under the electron microscope to be super-coiled with a "wavelength" (pitch) of ~2.5 μm. What does this fact imply about the molecular packing? It is hoped that the reader will consider this question but will not try to answer it too hastily.

3. Isologous Bonds: Paired Interactions

If two subunits are held together with interactions *aj* and are related by a twofold axis of rota-

Fig. 4-8 (A) A sketch showing the proposed arrangement of α-helical protein rods of 1.0×7.0 nm in the protein coat of the filamentous bacteriophage. Pf1.[38] Dotted line at the top traces the left-handed α helix on which the N-termini of the helical rods are thought to lie. (B) Ball-and-stick representation of α-carbon atoms in neighboring α helices of the current Pf1 model. Atoms are projected on to a plane parallel to the z axis of the virion. Numbers indicate residues. The three subunits illustrate the contacts between any subunit (0) and the subunits five units (-5) and nine units (-9) down the virus helix. From Nakashima *et al.*[39]

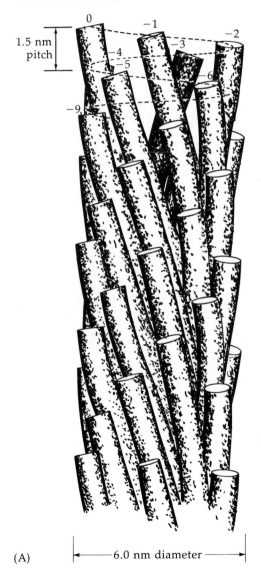

(A)

(B)

BOX 4-A

MICROTUBULES AND THE ACTION OF COLCHICINE

A prominent component of cytoplasm consists of microtubules of 24 ± 2 nm diameter with a 13–15 nm hollow core. They are present in the most striking form in the flagella and cilia of eu-

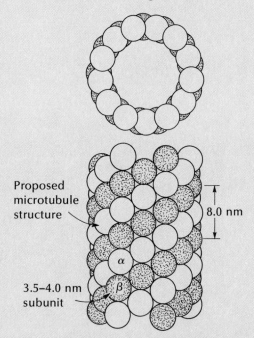

Proposed microtubule structure

8.0 nm

3.5–4.0 nm subunit

α

β

karyotic cells (Fig. 1-5). The stable microtubules of cilia are thought to be an integral part of the machinery causing the motion of the flagella.[a,b] (Figure shown above reprinted from J. Bryan.[b]) Labile microtubules, which form and then disappear, are often found in cytoplasm in which motion is taking place, for example, in the pseudopodia of the ameba. The mitotic spindle (Chapter 15, Section D,9) consists of a series of microtubules which appear to function in the movement of chromosomes in a dividing cell. Microtubules are also found in the cleavage planes of plant cells during division.

Many microtubules are found in the long axons of nerve cells and are thought to function in the "fast transport" of proteins and other materials from the cell body down the axons. Microtubules

of unknown function are found in many sensory cells. Recently, microtubules have been demonstrated throughout the cytoplasm of a variety of cells. The accompanying micrograph of a mouse embryo fibroblast was obtained by Weber et al. using indirect immunofluorescence techniques.[c] The cells were fixed with formaldehyde, dehydrated, and treated with antibodies (formed in a rabbit) to microtubule protein. The cells were then treated with fluorescent goat antibodies to rabbit γ-globulins (see Box 5-F) and the photograph was taken by fluorescent light emission.

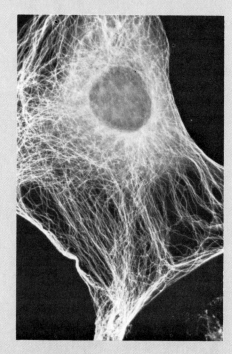

(Micrograph courtesy of Klaus Weber.) It is thought that motion in microtubular systems depends upon cooperation with other proteins. Thus, the arms on the microtubules of cilia (Fig. 1-5) catalyze the hydrolysis of ATP, and in this respect resemble the muscle protein myosin. Motion of cilia probably results from the sliding of the microtubules in a manner related to that in skeletal muscle (Section F).

Microtubules are all made of **tubulins**, mixed dimers (αβ) of two closely related subunits of

Box 4-A (*Continued*)

MW = 60,000 together with smaller amounts of higher molecular weight protein.[d] The microtubule itself can be thought of as a series of parallel filaments formed by the end-to-end aggregation of tubulin molecules. Each tubulin dimer binds one molecule of guanosine triphosphate (GTP) strongly, and a second molecule more loosely. In this respect tubulin resembles actin whose subunits are of about the same size. However, there is little similarity between the two proteins in amino acid sequence.

It is commonly believed that the labile microtubules of cytoplasm are in a "dynamic equilibrium" with monomer or dimer units. Thus, the tubules can grow or be disassembled, depending upon metabolic conditions. GTP appears to be required for assembly of microtubules, and it is possible that the hydrolysis of this nucleotide triphosphate provides an essential step.[e] Recent reports of phosphorylation of microtubular protein suggest that the picture is even more complex.

Of special interest is the reaction of microtubules with the alkaloid **colchicine** which is pro-

duced by various members of the family Liliaciae. This compound with its tropolone ring system is specifically and tightly bound to tubulin. The striking result of this binding within living cells is the disassembly of labile microtubules including the mitotic spindle. Dividing cells treated with colchicine appear to be blocked at metaphase (Chapter 15, Section D,9) and daughter cells with a high degree of polyploidy are formed. This has led to the widespread use of colchicine in inducing formation of tetraploid varieties of flowering plants. Similar effects upon microtubules are produced by the antitumor agents **vincristine** and **vinblastine**, alkaloids formed by the common *Vinca* (periwinkle).[g]

The microtubules of eukaryotic cilia (and flagella) represent a special case, for in the main part of the cilium, they exist as fused pairs. The A tubule carries the arm and the B tubule is fused to it with common subunits in the center. As with labile microtubules two kinds of tubulin can be isolated, but the relationship to the structure of the paired tubules is uncertain.

[a] J. A. Snyder and J. R. McIntosh, *Annu. Rev. Biochem.* **45**, 699–720 (1976).

[b] J. Bryan, *Fed. Proc., Fed. Am. Soc. Exp. Biol.* **33**, 152–157 (1974).

[c] K. Weber, R. Pollack, and T. Bibring, *PNAS* **72**, 459–463 (1975).

[d] D. B. Murphy and G. G. Borisy, *PNAS* **72**, 2696–2700 (1975).

[e] M. Jacobs, H. Smith, and E. W. Taylor, *JMB* **89**, 455–468 (1974).

[f] T. N. Margulis, *JACS* **96**, 899–902 (1974).

[g] L. Wilson, J. R. Bamberg, S. B. Mizel, L. M. Grisham, and K. M. Creswell, *Fed. Proc., Fed. Am. Soc. Exp. Biol.* **33**, 158–166 (1974).

Colchicine[f]

tion as shown in Fig. 4-9A, we obtain an **isologous dimer.** Each point such as *a* in one subunit is related to the same point in the other subunit by reflection through the axis of rotation. In the center, along the twofold axis, points *c* and *c'* are *directly opposite the same points* in the other subunit. Figure 4-9A is drawn with a hole in the center so that groups *c* and *c'* do not actually touch, and it is the paired interactions such as *aj* of groups not adjacent to the axis that contribute most to the bonding. However, a real protein dimer may or

may not have such a hole. The pair of identical interactions in an isologous dimer is usually referred to as a single **isologous bond.** Such a bond always contains the paired interactions between complementary groups (*aj*) and has pairs of identical groups along the axis.[†]

Isologous bonding is extremely important in oligomeric enzymes, and it has been suggested

† The paired interactions do not have to be highly specific ones involving complementary surfaces to meet the definition of isologous bonding.

BOX 4-B

BACTERIAL FLAGELLA

One of the mysteries of biology is the mechanism of conversion of chemical energy into mechanical work. The smallest organs of propulsion are the bacterial flagella and we might hope to unravel some of the mystery by looking at them. Prokaryotic flagella are composed of only one kind of protein, **flagellin**. Flagellin molecules contain no cysteine or tryptophan and usually little phenylalanine, proline, or histidine. They have a high content of hydrophobic amino acids and contain one residue of the unusual **ε-N-methyllysine**. The subunits are arranged in a helix (Fig. 4-7) in which they also form 11 nearly longitudinal rows.[a-e] A structural feature that cannot be explained by a simple helical stacking of subunits is the **supercoil** of pitch ∼2.3 μm. This feature is essential for function, and mutant bacteria with straight flagella are nonmotile.

The supercoiled structure is presumed to arise from a conformational difference, with altered dimensions, in the subunits of one longitudinal row.[d] Here, as with the icosahedral viruses, quasiequivalence permits formation of a structure that would be impossible with full equivalence of subunits.

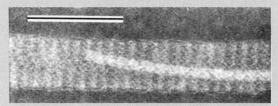

Terminal portion of a bacterial flagellum within a sheath. This micrograph was of an unidentified bacterium found in pond water and negatively stained with phosphotungstic acid. Bar marker = 100 nm. The function of the loose-fitting sheath is uncertain. Courtesy of F. D. Williams.

How does a bacterial flagellum move? The individual flagella are too small to be seen with a light microscope, and no one has yet obtained electron micrographs of a functioning flagellum. However, the supercoiled structure suggested a possibility. Since all of the flagellum subunits are

identical, it might be possible to induce a cyclic contraction of one longitudinal row after another around the tubule leading to the propagation of a helical wave.[f] An energy-dependent conformational change induced within the bacterial cell at the base of the flagellum could provide the needed energy.

Despite the attractiveness of this idea, various experiments suggest that the flagellum is probably a rigid "propeller" that is rotated by a "motor" at the base.[g] Some of the evidence comes from the observation that a bacterium linked artificially (by means of antibodies) to a short stub of a flagellum of another bacterium can be rotated by the second bacterium.

A remarkable proposal comes from study of a mutant of *Salmonella* with "curly" flagella which have a superhelix of one-half the normal pitch. The presence of *p*-fluorophenylalanine in the medium also produces curly flagella, and normal flagella can be transformed to curly ones by a suitable change of pH.

Although it is impossible to see individual flagella, on live bacteria, bundles of flagella can be viewed by dark-field light microscopy. Normal flagella appear to have a left-handed helical form, but curly flagella form a right-handed helix.[h] Normal bacteria swim in straight lines but periodically "tumble" before swimming in a new random direction. (This behavior is part of the system of **chemotaxis** by which the organism moves toward a food supply, Chapter 16, Section B,7.) Curly mutants tumble continuously. It is suggested that when bacteria tumble the flagella change from normal to curly. The pitch is reversed and shortened. A proposed mechanism for the change of pitch involves propagation of conformational changes down additional rows of flagellin subunits.[d]

What kind of "motor" powers bacterial flagella? Electron microscopy reveals a "hook" attached to a rod that passes through the cell wall and is in turn attached to a thin disc (the **M ring**) embedded in the cytoplasmic membrane, as indicated by the accompanying drawing[g] for a gram-positive bacterium. The torque is thought to be generated between the M ring and the **S ring**,

Box 4-B (*Continued*)

which is mounted on the cell wall. There is (as yet) no evidence for muscle-type proteins in the motor, and it has been suggested that translocation of ions from the interior of the bacterium to the surface of the M ring (in a cyclic pattern) may be involved. Those translocated charges could interact with fixed charges in the S ring to cause the motion.

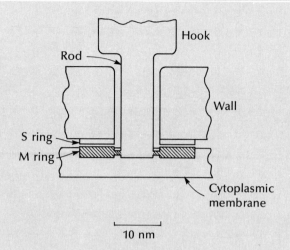

10 nm

How do flagella grow? Iino added *p*-fluorophenylalanine to a suspension of bacteria whose flagella had been broken off (by rapid stirring) at various lengths from the body. Curly ends appeared as the flagella grew out.[i] Unlike the growth of hairs on our body, the flagella grew from the outer ends. Since no free flagellin is found in the surrounding medium, it appears that the flagellin monomers of diameter 4.0–4.5 nm are synthesized within the bacterium, then pass out through the ~6 nm hole in the flagella and bind at the ends. Flagella grow at the rate of 1 μm in 2–3 min initially, then more slowly until they attain a length of ~15 μm (*Salmonella*). Now some questions for the reader: Why don't the flagellin molecules moving down the "pipe" simply pass out into the medium? Why does the flagellum stop growing at ~15 μm?

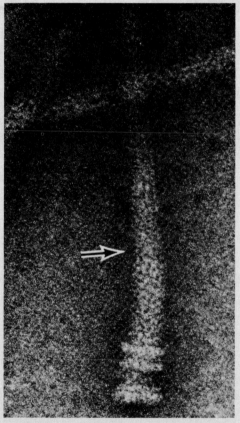

Two additional rings are present at the basal ends of flagella from gram-negative bacteria. This is an electron micrograph[i] of a flagellum from *E. coli* stained with uranyl acetate. The N and S rings are seen at the end. Above them are the P ring, thought to connect to the peptidoglycan layer, and the L ring, thought to connect to the outer membrane or lipopolysaccharide layer (see Fig. 5-8). An arrow marks the junction between hook and thinner filament. The hook is often bent to form an elbow.

[a] T. Iino, *Bacteriol. Rev.* **33**, 454–475 (1969).
[b] W. Bode, *Angew. Chem., Int. Ed. Engl.* **12**, 683–693 (1973).
[c] E. J. O'Brien and D. M. Bennett, *JMB* **70**, 133–152 (1972).
[d] C. R. Calladine, *Nature* (*London*) **255**, 121–124 (1975).
[e] C. Gonzalez-Beltran and R. E. Burge, *JMB* **88**, 711–716 (1974).
[f] W. F. Harris, *J. Theor. Biol.* **47**, 295–308 (1974).
[g] H. C. Berg, *Nature* (*London*) **254**, 389–392 (1975).
[h] K. Shimada, R. Kamiya, and S. Asakura, *Nature* (*London*) **254**, 332–334 (1975).
[i] M. L. DePamphilis and J. Adler, *J. Bacteriol.* **105**, 384–395 (1971).

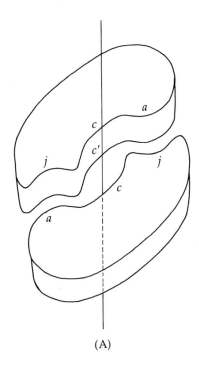

(A)

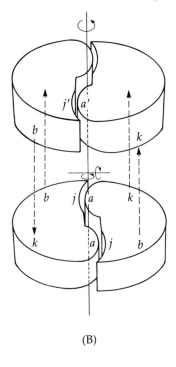

(B)

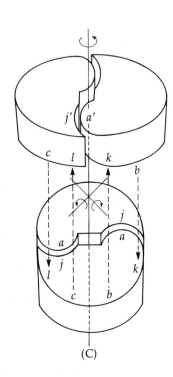

(C)

Fig. 4-9 (A) Isologous bonding between pairs of subunits; (B) an "isologous square" arrangement of subunits; (C) a "tetrahedral" arrangement of subunits. Note the three 2-fold axes.

that isologous interactions evolved very early. Initially there may not have been much complementarity in the bonding but two "hydrophobic spots" on the surface of the subunits came together in a nonspecific association.[42] Later in evolution the more specific paired interactions could have been added.

Isologous dimers can serve as subunits in the formation of larger closed oligomers and helices; for example, an isologous pair of the sort shown in Fig. 4-9A can be flipped over onto the top of another similar pair as shown in Figs. 4-9B and C. Again, if the proper complementary surfaces exist, bonds can form as shown (*bk* in Fig. 4-9B and *bk* and *cl* in Fig. 4-9C). Both the structures in Fig. 4-9B and C possess **dihedral (D$_2$) symmetry.**[43] In addition to the twofold axis of rotation lying perpen-

dicular to the two rings, there are two other two-fold axes of rotation as indicated in the drawings. Note that again the interactions are paired, i.e., there are two *bk* interactions and two *cl* interactions for each pair of subunits. Thus, the new bonds are also isologous. There are a total of six pairs of interactions, one between each combination of two subunits. This may be a little more difficult to see in Fig. 4-9B than in Fig. 4-9C because in the former the subunits are arranged in a more or less square configuration. Nevertheless, an isologous interaction between the left-hand subunit in the top ring and the subunit in the lower ring at the right does exist, even if it is only electrostatic and at a distance.

An example of a tetrameric enzyme with perfect dihedral symmetry of the type shown in Fig. 4-9B is **lactate dehydrogenase** (Chapter 8, Section H,2). The plant agglutinin **concanavalin A** (Fig. 5-7) resembles the structure in Fig. 4-9C.[44,45]

4. Oligomers Containing Both Heterologous and Isologous Interactions

Square arrays of four subunits can be formed using either heterologous or isologous interactions. Both types of bonding can occur in larger aggregates. For example, two heterologous trimers such as that shown in Fig. 4-6 can associate to a hexamer having dihedral (D_3) symmetry; a heterologous "square tetramer" can dimerize to give a dihedral (D_4) octamer.[43] The enzyme **glutamine synthetase** is a double ring of 12 subunits. Probably the upper ring is flipped over onto the lower giving dihedral symmetry (D_6) with one 6-fold axis and six 2-fold axes at right angles to it.[46]

Isologous bonds may appear in helices, too. Actin monomers form dimers as well as chains. Thus, the subunits in the structure shown in Fig. 4-7 may be composed of isologous dimers formed by pairs lying across the axis from each other. Now the reader may wish to consider whether or not the bacterial flagellum shown in Fig. 4-7 could be made from isologous dimers.

5. Oligomers with Cubic Symmetry (Polyhedra)

Symmetrical arrangements containing more than one axis of rotation of order higher than 2-fold are said to have cubic symmetry. The **tetrahedron** is the simplest example. It contains four 3-fold axes which pass through the vertices and the centers of the faces and three 2-fold axes which pass through the midpoints of the six edges. *Since protein subunits are always asymmetric, a tetrameric protein cannot possess cubic symmetry.*[†] However, a

[†] As we have already seen, tetrameric enzymes have dihedral symmetry. Thus, the many theoretical treatments of the behavior of "tetrahedral" arrangements of four subunits found in recent literature are not very useful.

heterologous trimer with 3-fold symmetry can form a face of a tetrahedron containing a total of 12 asymmetric subunits (Fig. 4-10). We see that around each vertex a second kind of heterologous trimer is formed involving interactions *bk*, while at the 2-fold axis, pairs of subunits possess an isologous interaction *cl*.[43]

Twenty-four subunits can interact to form a **cube.** Three 4-fold axes pass through the centers of the faces, four 3-fold axes pass through the vertices, and six 2-fold axes pass through the edges (see Fig. 8-17). The largest structure of cubic symmetry that can be made is the **icosahedron,** a regular solid with 20 triangular faces. Sixty subunits

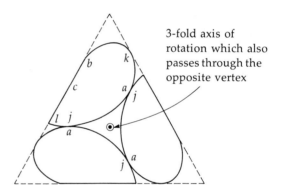

3-fold axis of rotation which also passes through the opposite vertex

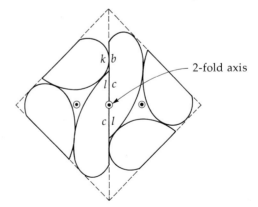

2-fold axis

Fig. 4-10 An arrangement of subunits with cubic symmetry. A tetrahedron composed of 12 identical, asymmetric subunits. Viewed down 3-fold and 2-fold axes. The letters represent bonding regions as in Fig. 4-9. See also Fig. 8-17.

BOX 4-C

THE VIRUSES[a-c]

Attacking every living thing from the smallest mycoplasma to man the nucleoprotein particles known as viruses have no metabolism of their own but "come alive" when the nucleic acid that they contain enters a living cell. Viruses are significant to us not only because of the serious disease problems that result from their activities but also as tools in the study of molecular biology. A mature virus particle or **virion** consists of one or more nucleic acid molecules in a protein coat or **capsid**, usually of helical or icosahedral form. The capsid is made up of "morphological subunits" or **capsomers**. The latter can sometimes be seen clearly with the electron microscope. The capsomers in turn are usually composed of a number of smaller protein subunits. Some of the larger viruses are surrounded by membranous envelopes. Others, such as the T-even **bacteriophage** which attack *E. coli*, have an extraordinarily complex structure (Box 4-E).

Most viruses contain a genome of either double-stranded DNA or single-stranded RNA, but some small viruses have single-stranded DNA and others have double-stranded RNA. The number of nucleotides in a virus genome may vary from a few thousand to several hundred thousand, and the number of genes from 3 to over 200. Sometimes the nucleic acid molecules within the virion are circular, but in other cases they are linear.

On p. 213 list of a few of the known virus types and individual viruses. Shapes are indicated as: I, icosahedral; H, helical; and C, complex. The second dimension given for some helical and complex viruses is the length (nm). The length of the nucleic acid in thousands of bases (single-stranded DNA or RNA) or base pairs (double-stranded nucleic acids) is shown. The number of genes present may exceed this number somewhat.

Among the very small viruses are the helical bacteriophages such as fd, f1, and M13 (Fig. 4-8) which resemble thin bacterial pili. Containing *single-stranded* circular DNA molecules of MW $\approx 2 \times 10^6$, these viruses attach themselves to the sex pili of male bacteria (e.g., *E. coli*). They may introduce their DNA into the bacteria through the pili (Chapter 15, Section A,1). Bacteriophage ϕX174, an icosahedral DNA-virus 25 nm in diameter, is only three times as thick as the thinnest cell membrane.[d] Its single-stranded DNA contains only $\sim$5000 nucleotides and appears to contain 9 genes.[d] It is remarkable that such a tiny virus is able to seize control of the metabolic machinery of the cell and turn it all in the direction of synthesis of more virus particles. Obviously, the virus makes use of many genes and other components of the cell that it infects.

A large group of animal viruses, the **parvoviruses**, are similar in size and architecture to the bacteriophage ϕX174. Some of them are unable to reproduce unless the cell is also infected by a larger adenovirus.

Among the icosahedral viruses containing double-stranded DNA, are the **papovaviruses**, some of which cause warts and others malignant tumors. Much studied by biochemists is **simian virus 40** (SV40), a monkey virus capable of inducing tumors in other species. Another tumor virus is the **polyomavirus** of the mouse. Slightly larger are the **papillomaviruses**, including the one causing human warts. Still larger (70 nm diameter) are the **adenoviruses**, of which some 32 types infect humans. The **herpesviruses** are very large and are enveloped by a lipid-containing membrane. The largest of the icosahedral viruses are those causing polyhedroses in insects. One, infecting the fly, *Tipula*, measures 130 nm in diameter. Another group of large DNA-containing viruses are the tailed bacteriophage (such as the T-even phage, Box 4-E).

The smallest of the RNA-containing viruses are the bacteriophage R17, MS2, and Qβ, whose nucleic acids contain 3500–4500 nucleotides and only three genes. Essentially the complete nucleotide sequence of the RNA of these viruses has been determined (Chapter 15, Section C,2,i). Only a little larger are the **picornaviruses** (from "picoRNA," meaning very little RNA). Among these small icosahedral viruses of 15–30 nm diameter are many that attack man. Among them are the **enteroviruses** including the **polioviruses**, the

Box 4-C (*Continued*)

Type of genome and group or individual virus name	Shape	Diameter (nm)	Masses in daltons × 10⁻⁶		Thousands of bases or base pairs (kilobases)
			Total	DNA or RNA	
DNA, single-stranded					
Bacteriophage fd, f1, M13	H	60 × 1–2 μm		2.0	6.0
Bacteriophage ϕX174	I	25	6.2	1.8	5.5
Parvoviruses	I	18–25		1.8	5.5
DNA, double-stranded					
Papovaviruses	I	35–55			
SV40 (monkey)	I		17.3	3.4	5.2
Polyoma (mouse)	I		23.6	3.0	4.5
Papilloma (human wart)	I			5.3	8.0
Adenoviruses	I	70		20–25	30–38
Polyhedrosis viruses of insects	I	70–130			
Herpesviruses	I				
core		78	~1000	50–90	76–136
envelope		150–200			
Pox viruses, e.g., Smallpox, vaccinia (cowpox)	C	160 × 250	~4000	160–240	240–360
T-even bacteriophage	C		215	130	197
RNA, single-stranded					
Small bacteriophage, R17, MS2, Qβ	I	23–26	3.6–4.0	1.2–1.5	3.5–4.5
Picornaviruses	I		8.4	2.6	7.9
Polioviruses	I	27	6.8	2	6.1
Rhinoviruses	I	27–30	7–8	2.2–2.8	6.7–8.5
Turnip yellow mosaic virus	I	28	5.0–6.0	2.0	6.1
Tobacco mosaic virus	H	18 × 300	40	2.2	6.7
Influenza virus	I	80–100	200	2.0	6.1
Bullet-shaped viruses					
Rhabdoviruses	C	20 × 130			
RNA, double-stranded					
Reoviruses	I	55–60			
RNA, unsheathed					
Potato spindle tuber viroid				0.1	0.30

coxsackieviruses and some of the **echoviruses**. A second class of picornaviruses include **rhinoviruses** responsible for the common cold. Already ~200 different types are known. Many icosahedral RNA viruses attack plants, e.g., the turnip yellow mosaic virus, of diameter 28–30 nm.

Large viruses of 80–100 nm diameter bearing 8–10 spikes at the vertices of the icosahedra cause influenza, mumps, and related diseases. The internal structure must be complex. Only 1% of the virus is RNA and that appears to consist of several pieces of relatively low molecular weight (~0.5 × 10⁶).

Best known of the helical RNA viruses is the tobacco mosaic virus (Section D,2).[e] Of more complex structure are the "bullet-shaped viruses" of which the rabies virus is an example. The diameter of these viruses is 65–90 nm and the length 120–500 nm. The internal structure includes a helical arrangement of nucleoprotein.

Double-stranded RNA is unusual in nature but constitutes the genome of some viruses, e.g., the

Box 4-C (*Continued*)

reoviruses. The RNA of reoviruses fragments into about 10 segments upon infection.

Several plant diseases including the **potato spindle tuber disease** are caused by very small molecules of RNA of MW = 120,000 or less with a folded structure.[f] Since such an RNA could code for a protein containing only about 100 amino acids, it may be unlikely that the virus (or **viroid** as it is often called) carries a gene for a protein. Whatever its genetic message, the viroid causes the plant cell to replicate many copies of the molecule, which may be then transmitted to other plants by aphids, or on the surface of tools, by man.

[a] R. W. Horne, *Sci. Am.* **208**, 48–56 (Jan 1963).
[b] K. Maramorosch and E. Kurstak, eds., "Comparative Virology." Academic Press, New York, 1971.
[c] C. A. Knight, "Chemistry of Viruses," 2nd ed. Springer Publ., New York, 1975.
[d] R. M. Benbow, R. F. Mayol, J. C. Picchi, and R. L. Sinsheimer, *J. Virol.* **10**, 99–114 (1972).
[e] H. Fraenkel-Conrat, *Sci. Am.* **211**, 47–54 (Oct 1964).
[f] J. L. Marx, *Science* **178**, 734 (1972).

are required and at each vertex they form a heterologous **pentamer.** As with the tetrahedron, each face contains a heterologous trimer and isologous dimers form across the edges.

6. Icosahedral Viruses[47]

Many viruses consist of roughly spherical protein shells (coats) containing DNA or RNA inside (Box 4-C). The coat protein is usually made of many identical subunits, a fact that can be rationalized in terms of economy from the genetic viewpoint. Thus, only one gene is needed to specify the structure of a large number of subunits.[48] Under the electron microscope the viruses often have an icosahedral appearance (Fig. 4-11), and chemical studies show that the number of subunits is a multiple of 60. For example, the coat of the RNA-containing **cowpea chlorotic mottle virus** (diameter ~25 nm) contains 180 subunits of MW = 19,600 and containing 183 amino acid residues.[49] Small RNA-containing bacteriophage such as f2 contain 180 subunits[50] of MW $\approx$ 13,750 enclosing an RNA molecule of MW $\approx 1.1 \times 10^6$. The coat of the **bushy stunt virus** (diameter ~33 nm) also contains 180 subunits, while that of the human **wart virus** (diameter ~56 nm) contains 420 subunits, seven times the number in a regular icosahedron. According to a theory of **quasi-equivalence** of subunits[47] (see next section), the distances between the centers of subunits are preserved in a family of **icosadeltahedra** containing subunits in multiples of 20. However, the angles must vary somewhat from those in a regular icosahedron (compare with geodesic shells in which the angles are constant but the distances are not all the same). Icosadeltahedra contain hexamers as well as pentamers at vertices; for example, the shells of the 180-subunit viruses contain clusters of subunits forming 12 pentamers, 20 hexamers, 60 trimers (on the faces), and 90 dimers (across the edges).

Coats of some spherical viruses contain more than one kind of protein. Thus, the polioviruses (diameter 25 nm) contain three major coat proteins (α, β, and γ) formed by cleavage of one large precursor protein into at least four pieces.[51] The three largest pieces of MW ~ 33,000, 30,000, and 25,000 aggregate probably as $(\alpha\beta\gamma)_{60}$. The small (diameter 25 $\pm$ 1 nm) single-stranded DNA bacteriophage such as ϕX174 contain small hollow spikes at the vertices of the icosohedral shell (Fig. 4-11).[52]

7. Quasi-equivalence of Subunit Conformations

It is rather natural to think about association of subunits in symmetrical ways and the observation of square, pentagonal, and hexagonal arrangements of subunits directly with the electron microscope led to a ready acceptance of the idea that protomers tend to associate symmetrically.

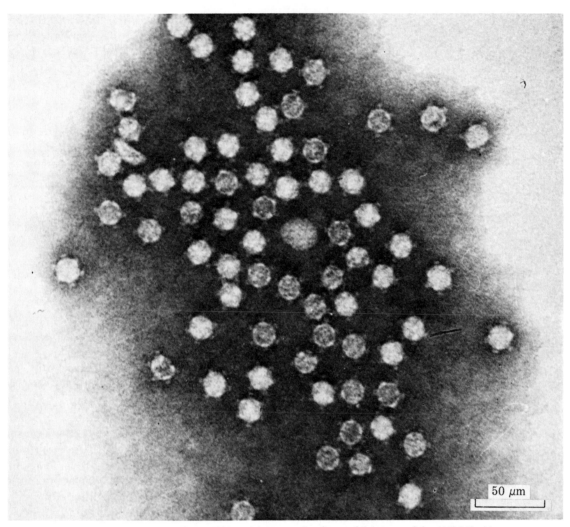

50 µm

Fig. 4-11 Bacteriophage Lφ7 of *E. coli* (similar to φX174). Courtesy of Dr. A. S. Tikhonenko, Institute of Molecular Biology, Academy of Sciences of the U.S.S.R., Moscow.

However, consider the predicament of the two molecules shown in Fig. 4-12. They might get together to form an isologous dimer if it were not for the fact that their "noses" are in the way. Despite the obvious steric hindrance, an isologous dimer can be formed in this case if one subunit is able to undergo a small change in conformation (Fig. 4-12). In the resulting dimer the two subunits are only quasi-equivalent.

Unsymmetrical dimerization of proteins appears to be a very common phenomenon; for example, the enzymes **malic dehydrogenase**[53] and

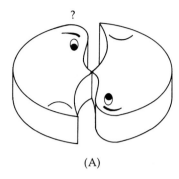

(A)

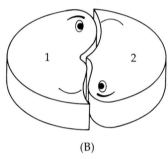

(B)

Fig. 4-12 Nonsymmetrical bonding in a dimer. (A) Two molecules which cannot dimerize because of a bad fit at the center. (B) A solution: Molecule 1 has refolded its peptide chain a little, changing shape enough to fit to molecule 2.

glyceraldehyde-phosphate dehydrogenase (Fig. 2-10) are tetramers of approximate dihedral symmetry. Crystal structure determinations have been made by assuming all subunits equivalent. Nevertheless, the existence of asymmetry shows up in a striking way, namely, that *only one of the two identical peptide chains of lactic dehydrogenase is able to bind the coenzyme NAD⁺.* The other binding site is apparently distorted so much that it can no longer bind this small, essential cocatalyst. A similar phenomenon has been reported for glyceraldehyde-phosphate dehydrogenase (see Chapter 8, Sections H,3 and 5).

The polypeptide hormone **insulin** is a small protein made up of two chains (designated A and B) which are held together by disulfide bridges (Fig. 4-13A). Figure 4-13B is a sketch of the structure as revealed by X-ray crystallography,[54,55] only the backbone of the peptide chains and a few side

chains being shown. In the drawing, the B chain lies behind the A chain. Beginning with the N-terminal phenylalanine-1, the peptide backbone makes a broad curve, then falls into an α helix of three turns lying more or less in the center of the molecule. Finally, after a sharp turn, it continues upward on the left side of the drawing in a nearly completely extended β structure. The A chain has an overall U shape with two roughly helical portions. The U shape is partly maintained by a disulfide bridge running between two parts of the A chain. Two disulfide bridges (one at the top of the drawing and one near the bottom) hold the A and B chains together, and hydrophobic bonding of internal side chain groups helps to stabilize the molecule.

Insulin in solution dimerizes readily, the subunits occupying quasi-equivalent positions. Under proper conditions, three dimers associate to form a hexamer of approximate dihedral (D_3) symmetry that is stabilized by the presence of two zinc ions. Figure 4-13C is a crude sketch of the hexamer showing the three dimers, the 3-fold axis of symmetry, and the two pseudo-2-fold axes, one passing between the two subunits of the dimer and the other between two adjacent dimers.

Note the positions of the two zinc atoms. Each lies on the 3-fold axis and is bound by three imidazole rings from histidines B-10. The significance of the zinc binding is uncertain but these hexamers readily form rhombohedral crystals, even within the pancreatic cells that synthesize insulin. The structure illustrates a feature common to many oligomeric enzymes of circular or dihedral symmetry. As in the insulin hexamer, the central parts of the molecule are often quite open and protruding side chain groups, such as the imidazole groups in insulin, form handy nests

Fig. 4-13 The structure of insulin of the pig. (A) The amino acid sequence of the A and B chains linked by disulfide bridges. (B) Sketch showing the backbone structure of the insulin molecule as revealed by X-ray analysis. The A and B chains have been labeled. Positions and orientations of aromatic side chains are also shown and may be related to those in Figs. 2-14 and 2-15. (C) Schematic drawing showing packing of six insulin molecules in the zinc-stabilized hexamer.

1
A. Gly·Ile·Val·Glu·Gln·Cys~S——S~Cys·Ser·Leu·Tyr·Gln·Leu

Cys—Tyr—Ser—Ile

1
B. Phe·Val·Asn·Gln·His·Leu·Cys

Gly·Ser·His

Ala·Leu
Gln   Tyr
Val   Leu
Leu   Val~S—S~Cys
      Cys
      Gly   Asn 21
      Gln
      Arg
      Gly
      Phe
      Phe
      Tyr
      Thr
      Pro
      Lys
      Ala 30

Gln
Asn
Tyr

(A)

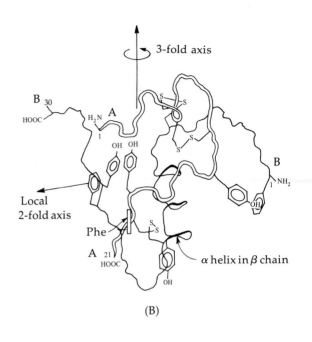

3-fold axis

B 30
HOOC   H₂N A
         OH OH

Local
2-fold axis

Phe

A 21
HOOC

α helix in β chain

B

NH₂

OH

OH

(B)

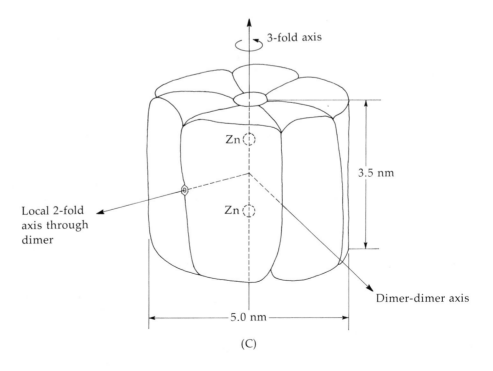

3-fold axis

Zn

3.5 nm

Zn

Local 2-fold
axis through
dimer

Dimer-dimer axis

5.0 nm

(C)

into which ions or molecules regulating activity of proteins can fit. However, the function of the zinc in insulin is not known.

Figure 4-14 shows some details of the bonding between the subunits in the insulin dimer with a view from the outside of the molecule down the 2-fold axis (marked by the X in the center of the phenylalanine 25 ring in the right-hand chain) through the dimer. The C-terminal ends of the B chains are seen in an extended conformation. The two antiparallel chains form a β structure with two pairs of hydrogen bonds. If there were perfect isologous bonding, these two pairs would be entirely equivalent and symmetrically related one to the other. A straight line drawn from a position in one chain and passing through the 2-fold axis (X) would also pass approximately through the corresponding position in the other chain. However, careful examination of the drawing will reveal many deviations from perfect symmetry.

Perhaps the most striking asymmetry in Fig. 4-14 is at the center where the Phe-25 from the right-hand chain projects upward and to the left. If the symmetry were perfect the corresponding side chain from the left-hand chain would project upward and to the right and the two phenylalanines would collide, exactly as do the "noses" in Fig. 4-12. In insulin one phenylalanine side chain has been flipped back out of the way.

Figure 4-15 shows one more view of insulin, a section through the molecule in a plane that passes approximately through both of the 2-fold axes (Fig. 4-13C). Note the packing of the side chains along both axes. These groups correspond to groups c and c' of Fig. 4-9. Note that they are identical, as demanded in isologous bonding, and they are, for the most part, hydrophobic, aromatic or alkyl (principally Leu). The symmetry is not perfect and the methyl groups of the leucines tend to interdigitate with each other to form a tighter interaction than would be possible with a completely symmetrical disposition. In summary, examination of the structure of the insulin dimer and hexamer illustrates principles of isologous bonding of protomers but also emphasizes that

two protomers often associate in an asymmetric way.

The existence of quasi-equivalence of subunits in icosahedral virus coats and in certain enzymes has been mentioned. The same phenomenon provides the supercoil in bacterial flagella and accounts for some interesting aspects of the structure of tobacco mosaic virus. The protein subunits of the virus can exist either as a helix with 16.3 subunits per turn (Fig. 4-7) or as a flat ring of 17 subunits.[36a] A very small conformational difference is involved. These rings dimerize but do not form larger aggregates. What is surprising is that the dimeric rings do not have dihedral symmetry. Rather, all the subunits in the dimeric disk are oriented in one direction, but have two different conformations. It has been proposed that the disk serves as an intermediate in virus assembly. X-Ray diffraction studies show that the inner portions of the quasi-equivalent disk subunits assume a jawlike appearance as if awaiting the incorporation of RNA. It is postulated that as the RNA becomes bound, the disks dislocate to a "lockwasher" conformation to initiate and propagate growth of the helical virus particle.[36a] These and many other curious observations suggest that both quasi-equivalence and conformational change lie at the heart of many biological phenomena.

8. Oligomers with More than One Kind of Subunit

Many enzymes, virus coats, and more complex molecular structures are made up of two or more different protomers. Probably the most studied example is **hemoglobin,** a tetramer ($\alpha_2\beta_2$) composed of two similar but different subunits designated α and β (both of MW = 16,100).

Fig. 4-14   View of the paired N-terminal ends of the B chains in the insulin dimer. View is approximately down the pseudo-2-fold axis toward the center of the hexamer.

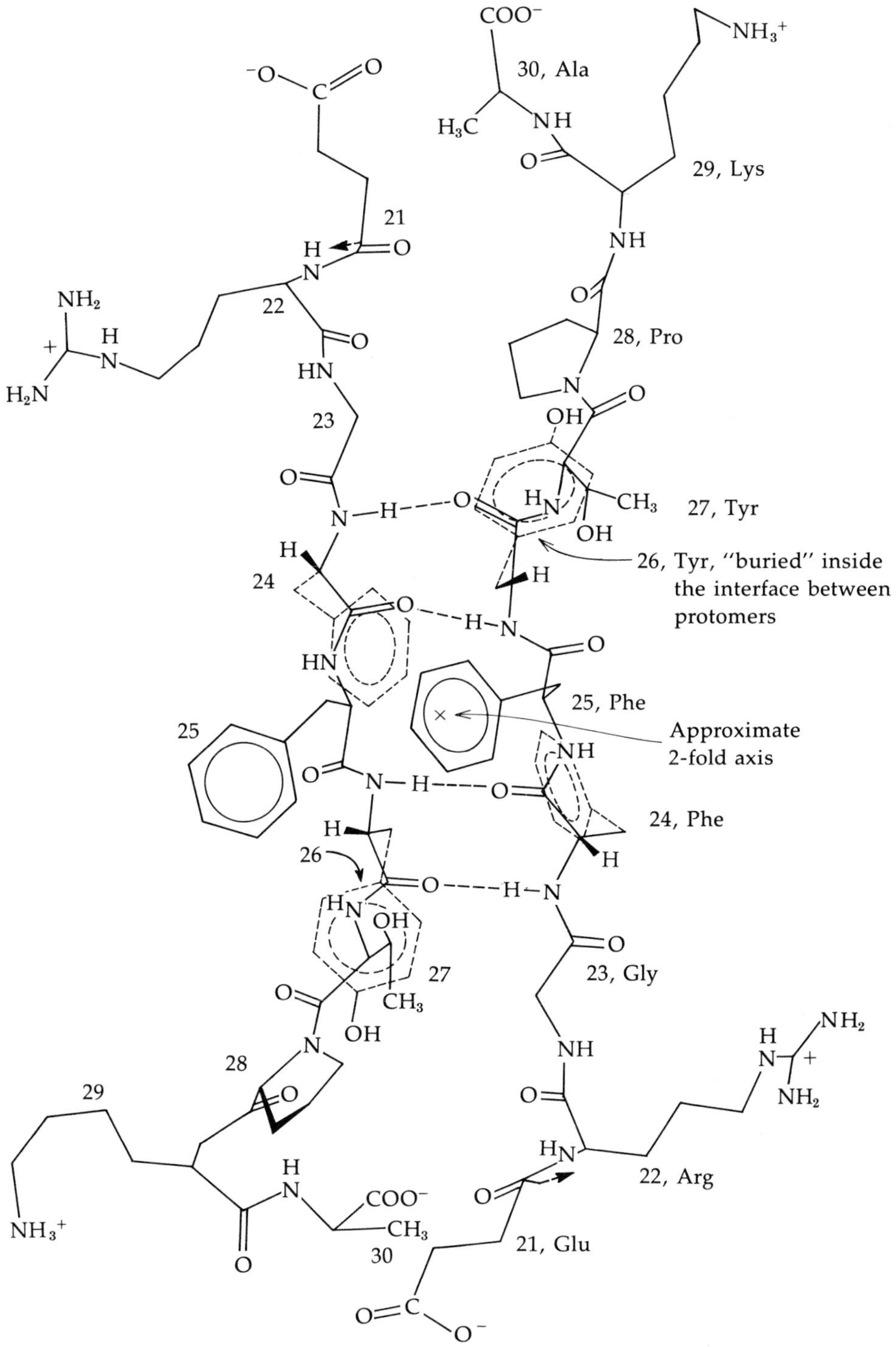
30, Ala
29, Lys
21
22
23
24
28, Pro
27, Tyr
26, Tyr, "buried" inside the interface between protomers
25, Phe
25
Approximate 2-fold axis
24, Phe
26
27
23, Gly
28
29
30
22, Arg
21, Glu

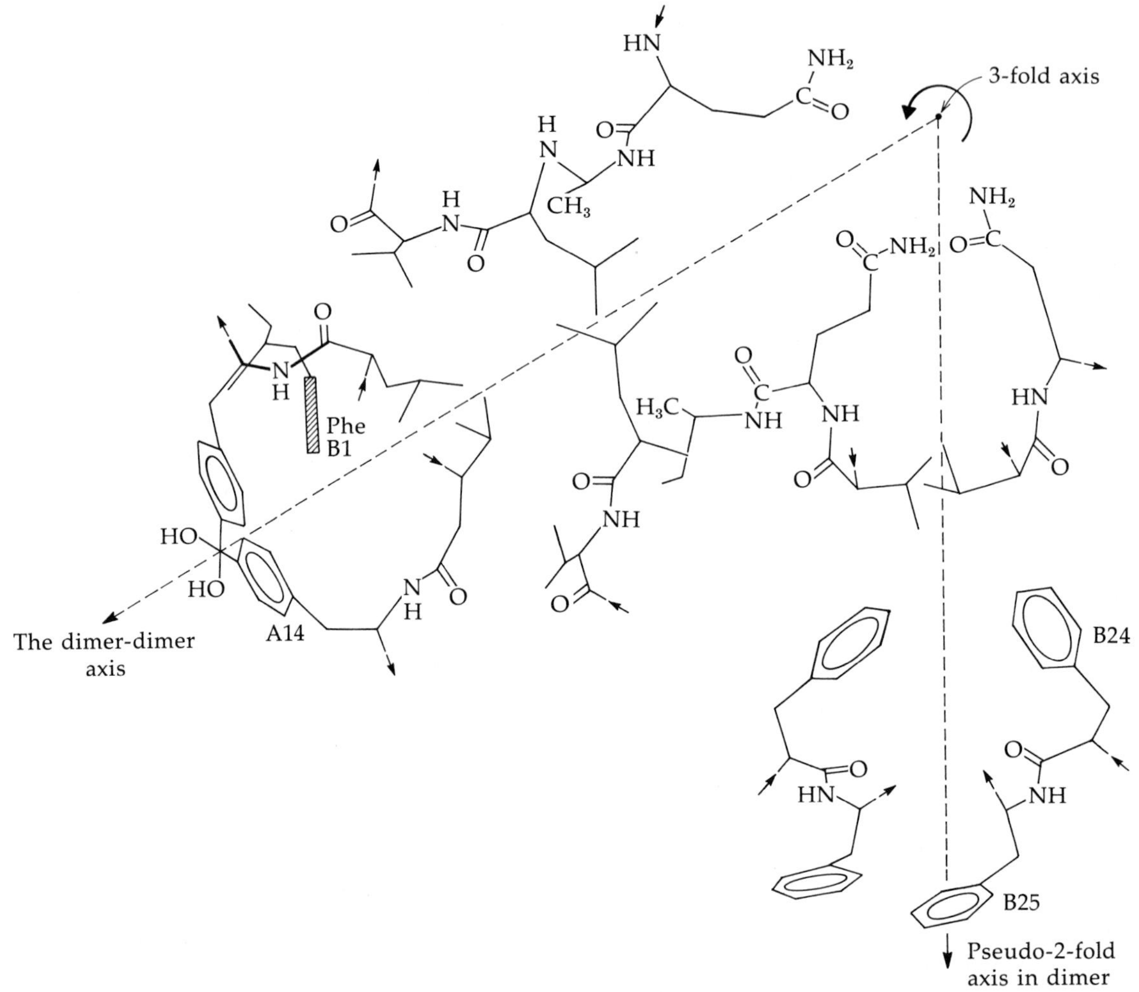

Fig. 4-15 Another view of the insulin molecule. Some groups lying near a plane passing through both the true and pseudo-2-fold axes. After Blundell et al.[54]

Although the folding of the peptide chain is almost the same in both subunits (and almost identical to that of the monomeric **myoglobin**),[56] there are numerous differences in the amino acid sequence. If it were not for these differences, hemoglobin would be a highly symmetric molecule with the bonding pattern indicated in Fig. 4-9C with three 2-fold axes of rotation. Hemoglobin is often said to have one true axis of rotation and two *pseudo*-2-fold axes. There are two sets of true iso-

logous interactions (those between the two α subunits and between the two β subunits) and two pairs of unsymmetrical interactions (between α and β subunits). The nearly symmetric orientation of different portions of the peptide backbone is clearly seen in the beautiful drawings of Dickerson and Geis.[57]

The contact region involved in one pair of αβ interactions in hemoglobin ($\alpha_1\beta_1$) is more extensive than the other. There is close contact between 34 different amino acid side chains and 110 atoms lie within 0.4 nm of each other.[55] Hydrophobic bonding is the principal force holding the two subunits together, and only a few reciprocal contacts of the type found in a true isologous bond remain. The second contact designated $\alpha_1\beta_2$ involves only 19 residues and a total of 80 atoms. Because this interaction is weaker, hemoglobin dissociates relatively easily into $\alpha\beta$ dimers held together by the $\alpha_1\beta_1$ contacts and motion occurs along the $\alpha_1\beta_2$ contacts during oxygenation (Section E,5,a). The truly isologous interactions (i.e., $\alpha\alpha$ and $\beta\beta$) are weak because the identical protomers hardly touch each other.

Aspartate carbamoyltransferase (MW = 310,000), a highly regulated enzyme obtained from *E. coli*, can be dissociated into two trimers referred to as **catalytic subunits** (MW = 100,000 for the trimer) and into three dimers, the **regulatory subunits** (MW = 34,000 for the dimer). The molecule is roughly triangular in shape[58,59] with a thickness of 9.2 ± 1.0 nm and a length of the triangular side of 10.5 ± 1.0 nm. The symmetry is 3:2, i.e., it is dihedral with one 3-fold axis of rotation and three 2-fold axes. The two trimers of catalytic subunits apparently lie back-to-back with the dimeric regulatory subunits lying between them and fitting into the grooves around the edges of the trimers. The dimers are not aligned exactly parallel with the 3-fold axis, but to avoid eclipsing, the upper half of the array is rotated around the 3-fold axis with respect to the lower half. In the center is an aqueous cavity of dimensions $\sim 2.5 \times 5.0 \times 5.0$ nm. The active sites of the enzyme appear to be inside this cavity which is reached through six openings of 1.5 nm diameter around the sides.

Many other oligomeric enzymes and other complex assemblies of more than one kind of protein subunit are known. For example, the **α-ketoacid dehydrogenases** are huge complexes of MW $\approx$ 2–4×10^6 and containing three different proteins in a cubic array (Fig. 8-18). The filaments of striated muscle (Section F), antibodies (Box 5-F), complement of blood (Box 5-G), and the amazing tailed bacteriophage (Box 4-E) all have fascinating and complex molecular architectures.

E. COOPERATIVE CHANGES IN CONFORMATION

A substrate will bind better to some conformations of a protein than it will to others. This simple fact, together with the tendency for protein monomers to associate into clusters, leads to remarkable behavior which puzzled biochemists for many years. We now recognize that cooperative changes in conformation within oligomeric proteins provide the basis for important aspects of the regulation of enzyme action and of metabolism. They impart cooperativity to the binding of small molecules such as that of oxygen to hemoglobin and of substrates and regulating molecules to enzymes. It is probable that many of the most fundamental and mysterious properties of living things are linked directly to cooperative changes within the fibrils, membranes, and other internal structures of the cell. For these reasons it will be appropriate to consider the matter quantitatively and in some detail.

While cooperative binding of oxygen to hemoglobin has been known for years, the widespread significance was not generally recognized. In 1965 Monod, Wyman, and Changeux[33] (MWC) provided a simple, appealing mathematical description that has served to focus new attention on the phenomenon. Since the MWC model is too restrictive for many situations, a more general treatment that uses the terminology of Koshland[60-62] is described here.

Consider an equilibrium (Eqs. 4-41 and 4-42) between protein molecules in two different conformations A and B (T and R in the MWC terminology) and containing a single binding site for molecule X.

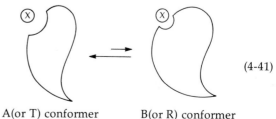

$$\text{A(or T) conformer} \qquad \text{B(or R) conformer} \tag{4-41}$$

$$K_t = [B]/[A] \tag{4-42}$$

If the equilibrium constant K_t is approximately 1, the two **conformers** have equal energies, but if $K_t < 1$, A is more stable than B.

1. Unequal Binding of Substrate and "Induced Fit"

Assume that conformer B binds X more strongly than does conformer A (as is suggested by the shapes of the binding sites in Eq. 4-41). The intrinsic binding constants to the A and B conformers K_{AX} and K_{BX} (or K_T and K_R) are (Eq. 4-43)

$$K_{AX} = [AX]/[A][X]$$
$$K_{BX} = [BX]/[B][X] \qquad (4\text{-}43)$$

The entire set of equilibria for this system are shown in Eq. 4-44. Note that the constant relating

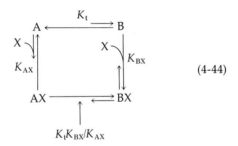

$$(4\text{-}44)$$

BX to AX is not independent of the other three constants but is given by the expression $K_t K_{BX}/K_{AX}$. Now consider the following situation. Suppose that A predominates in the absence of X but that X binds more tightly to B than to A. There will tend to be largely either free A or BX in the equilibrium mixture (with smaller amounts of AX and B present also). An interesting kinetic question arises. By which of the two possible pathways from A to BX (Eq. 4-44) will the reaction take place? The first possibility, assumed in the MWC model, is that X binds only to preformed B, which is present in a small amount in equilibrium with A. The second possibility is that X can bind to A but that AX is then rapidly converted to BX. We could say that X *induces a conformational* change that leads to a better fit. This is the basis for the **in-**

duced fit theory of Koshland. Bear in mind that the equilibrium constants can give us the equilibrium concentrations of all four forms in Eq. 4-44. However, rates of reaction are often important in metabolism and we cannot say *a priori* which of the two pathways will be followed.

Note that if K_{BX}/K_{AX} is very large, an insignificant amount of AX will be present at equilibrium. Experimentally, in such a case there is no way to determine K_{AX}, and the two constants K_t and K_{BX} are sufficient to describe the *equilibria*. Nevertheless, an induced fit mechanism may still hold.

2. Formation of Oligomers

Now consider the association of A and B to form oligomers in which the intrinsic binding constants K_{AX} and K_{BX} have the same values as in the monomers. Since more enzymes apparently exist as isologous dimers than as any other oligomeric form,[63] it is appropriate to consider the behavior of dimers in some detail. Monod *et al.* have emphasized that both conformers A and B (T and R) can associate to form isologous dimers in which symmetry is preserved (Eq. 4-45).

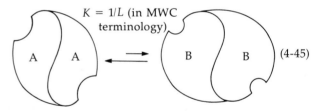

$$(4\text{-}45)$$

On the other hand, association of B and A would lead to an unsymmetric dimer in which bonding between subunits might be poor:

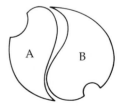

Mixed AB dimer which
associates weakly

In the MWC treatment, the assumption is made that the mixed dimer AB can be neglected entirely. However, a general treatment requires that we consider all dimeric forms.[64] The formation constants K_{AA}, K_{BB}, and K_{AB} are defined as follows[60,61] (Eqs. 4-46 to 4-48; note the statistical factor of 2 relating K_{AB} to the association constant K_f):

$$2A \rightleftharpoons A_2 \quad K_{AA} = \frac{[A_2]}{[A]^2} \tag{4-46}$$

$$2B \rightleftharpoons B_2 \quad K_{BB} = \frac{[B_2]}{[B]^2} = \frac{[B_2]}{K_t^2[A]^2} \tag{4-47}$$

$$A + B \rightleftharpoons AB \quad K_f = 2K_{AB} = \frac{[AB]}{[A][B]} = \frac{[AB]}{K_t[A]^2} \tag{4-48}$$

3. Binding Equilibria for a Dimerizing Protein

All of the equilibria of Eqs. 4-42 through 4-48 involved in formation of dimers A_2, AB, and B_2 and in the binding of one or two molecules of X per dimer are depicted in Fig. 4-16. Above each arrow the microscopic constant associated with that step is shown multiplied by an appropriate statistical factor. The fractional saturation $\bar{y}$ is given by Eq. 4-49.

$2\bar{y}$ (based on dimer) =

$$\frac{\begin{array}{c}[AX] + [BX] + [A_2X] + 2[A_2X_2] + [ABX] \\ + [AXB] + 2[ABX_2] + [B_2X] + 2[B_2X_2]\end{array}}{\begin{array}{c}\frac{1}{2}([A] + [AX] + [B] + [BX]) + [A_2] + [A_2X] \\ + [A_2X_2] + [AB] + [ABX] + [AXB] \\ + [ABX_2] + [B_2] + [B_2X] + [B_2X_2]\end{array}} \tag{4-49}$$

Each of the nine terms in the numerator gives the concentration of bound X represented by one of the nine forms containing X in Fig. 4-16. The 14 terms in the denominator represent the concentration of protein in each form including those containing no bound X. Protein concentrations are given in terms of the molecular weight of the dimer; hence, some of the terms in the denominator are multiplied by ½.

All of the terms in both the numerator and the

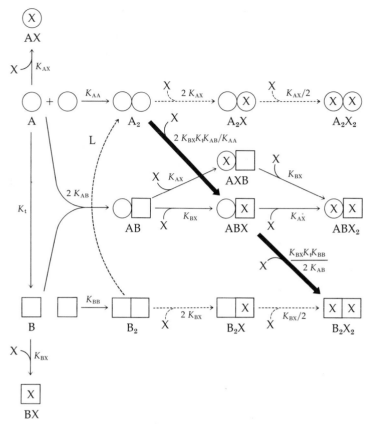

Fig. 4-16 Possible forms of a dimerizing protein existing in two conformations with a single binding site per protomer for X. Dashed arrows indicate equilibria considered by MWC. Solid arrows indicate equilibria considered by Koshland *et al.*[61,62] Heavy arrows are for the simplest induced fit model with no dissociation of the dimer. Note that all equilibria are regarded as reversible (despite the unidirectional arrows). K_{AX} and K_{BX} are assumed the same for subunits in monomeric and dimeric forms.

denominator of Eq. 4-49 can be related back to [X], using the microscopic constants from Fig. 4-16 to give an equation (comparable to Eq. 4-15) which presents $\bar{y}$ in terms of [X], K_{AX} and K_{BX}, K_t and the interaction constants K_{AA}, K_{AB}, and K_{BB}. Since the equation is too complex to grasp immediately, let us consider several specific cases in which it can be simplified.

a. No Dissociation of Oligomers

If both K_{AA} and K_{BB} are large enough no dissociation into monomers takes place. The transition between conformation A and conformation B can still occur within the dimer or higher oligomer, and the mathematical relationships shown in Fig. 4-16 are still appropriate. Further restrictions are needed to describe the most popular models of oligomeric proteins.

i. The Monod–Wyman–Changeux Model.[33] If only symmetric dimers are allowed, i.e., K_{AA} and $K_{BB} \gg K_{AB}$ (see Eq. 4-45), only those equilibria indicated with dotted arrows in Fig. 4-16 need be considered. In the absence of ligand X the ratio $[B_2]/[A_2]$ is a constant, $1/L$ in the MWC terminology (Eq. 4-50; see also Eq. 4-45).

$$\frac{[B_2]}{[A_2]} = \frac{1}{L} = \frac{K_{BB}}{K_{AA}} K_t^2 \qquad (4\text{-}50)$$

Both of the association constants K_{AA} and K_{BB} and the transformation constant K_t affect the position of the equilibrium. Thus, a low ratio of $[B_2]$ to $[A_2]$ could result if K_{BB} and K_{AA} were similar but K_t was small. If K_t were ~ 1 a low ratio could still arise because $K_{AA} \gg K_{BB}$, i.e., because the subunits are associated more tightly in A_2 than in B_2. For this case Eq. 4-49 simplifies to Eq. 4-51.

$$2\bar{y} = \frac{[A_2X] + 2[A_2X_2] + [B_2X] + 2[B_2X_2]}{[A_2] + [A_2X] + [A_2X_2] + [B_2] + [B_2X] + [B_2X_2]}$$

$$= \frac{\begin{array}{c} 2K_{AA}K_{AX}[X] + 2K_{AA}K_{AX}^2[X]^2 \\ + 2K_{BB}K_{BX}K_t^2[X] + 2K_{BB}K_{BX}K_t^2[X]^2 \end{array}}{\begin{array}{c} K_{AA} + 2K_{AA}K_{AX}[X] + K_{AA}K_{AX}^2[X]^2 + K_{BB}K_t^2 \\ + 2K_{BB}K_{BX}K_t^2[X] + K_{BB}K_{BX}^2K_t^2[X]^2 \end{array}} \qquad (4\text{-}51)$$

Substituting from Eq. 4-50 into Eq. 4-51 we obtain (Eq. 4-52):

$\bar{y}$ (for dimer)

$$= \frac{L \cdot K_{AX}[X](1 + K_{AX}[X]) + K_{BX}[X](1 + K_{BX}[X])}{L(1 + K_{AX}[X])^2 + (1 + K_{BX}[X])^2} \qquad (4\text{-}52)$$

For an oligomer with n subunits Monod *et al.* assumed that all sites in either conformer are independent and equivalent. The equation for $\bar{y}$ (based on Eq. 4-29) is

$$\bar{y} = \frac{L \cdot K_{AX}[X](1 + K_{AX}[X])^{n-1} + K_{BX}[X](1 + K_{BX}[X])^{n-1}}{L(1 + K_{AX}[X])^n + (1 + K_{BX}[X])^n} \qquad (4\text{-}53)$$

We assume initially that B_2 binds X more strongly than does A_2. Hence, if the equilibrium in Eq. 4-50 favors B_2 strongly (L is small), the addition of X to the system will not shift the equilibrium between the two conformations, and binding will be noncooperative (Eq. 4-53 will reduce to Eq. 4-30). However, if the equilibrium favors A_2 (L is large), addition of X will shift the equilibrium in favor of B_2 (which binds X more tightly). Furthermore, since the expression for $\bar{y}$ (Eq. 4-52) contains a term in $K_{BX}^2[X]^2$ in the numerator, binding will tend to be cooperative. In the extreme case that L is large and $K_{AX} \sim 0$, most of the terms in Eq. 4-52 drop out and it approaches the equation previously given for completely cooperative binding (Eq. 4-33) with $\bar{K} = K_{BX}^2 L$. With other values of K_{AX}, K_{BX}, and L incomplete cooperativity is observed.[60]

ii. Koshland's Induced Fit Model.[61,62] In this model only A_2, ABX, and B_2X_2 are considered (heavy arrows in Fig. 4-16). The expression for $2\bar{y}$ is[†]

$$2\bar{y} = \frac{[ABX] + 2[B_2X_2]}{[A_2] + [ABX] + [B_2X_2]}$$

$$= \frac{2K_{BX}K_t \dfrac{K_{AB}}{K_{AA}}[X] + 2(K_{BX}K_t)^2 \dfrac{K_{BB}}{K_{AA}}[X]^2}{1 + 2K_{BX}K_t \dfrac{K_{AB}}{K_{AA}}[X] + (K_{BX}K_t)^2 \dfrac{K_{BB}}{K_{AA}}[X]^2} \qquad (4\text{-}54)$$

if K_{AB} is small (no "mixed" dimer) this expression also simplifies to Eq. 4-33 for completely cooperative binding with the value $\bar{K}$ given by Eq. 4-55.

$$\bar{K} = K_{BX}^2 K_t^2 \frac{K_{BB}}{K_{AA}} = \frac{K_{BX}^2}{L} \qquad (4\text{-}55)$$

On the other hand, if K_{AB} is large compared to K_{AA} and K_{BB}, anticooperativity (negative cooperativity) will be observed. The saturation curve will contain two separate steps just as in the binding of protons by succinate dianion (Fig. 4-4).

[†] Koshland sometimes arbitrarily set $K_{AA} = 1$, redefining K_{BB} as an *interaction constant* equal to K_{BB}/K_{AA}. Although this simplifies the algebra it is appropriate only for completed associated systems and may prove confusing. Therefore, the constants will be used in the book as defined in Eqs. 4-46 through 4-48.

b. One Conformational State Dissociated

It may happen that A_2 is a dimer but that B_2 dissociates into monomers because K_{BB} is very small. In such a case binding of X leads to dissociation of the protein. A well-known example is provided by hemoglobin of the lamprey which is a dimer and which dissociates to a monomer upon binding of oxygen.[64] Equation 4-49 simplifies to Eq. 4-56.

$$2\bar{y} = \frac{[BX]}{[A_2] + \frac{1}{2}[BX]} \qquad (4\text{-}56)$$

The reader may wish to consider whether the weakly cooperative binding of oxygen by lamprey hemoglobin is predicted by this equation.

Look again at the expression for L, the constant determining the relative amounts of a protein in conformations A and B in the absence of ligand. From Eq. 4-50 we see that a large value of L (conformer A favored) can result either because K_t is very small or because $K_{BB} \ll K_{AA}$. Thus, if $K_t \sim 1$ and L is large, the subunits must associate much more weakly in B_2 than in A_2 and the chances are that binding of X will dissociate the molecule as in the case of lamprey hemoglobin. On the other hand, if K_t is very small, implying that the molecule is held in conformation A because of some intrinsically more stable folding pattern in that conformation, K_{BB} might exceed K_{AA} very much; if K_{AA} were low enough A_2 could be completely dissociated. Binding of ligand would lead to association and to cooperative binding. This can be easily verified by writing down the appropriate terms from Eq. 4-49.

4. Higher Oligomers

Mathematical treatment of binding curves for oligomers containing more than two subunits is more complex, but the algebra is straightforward and it is easy to program a computer to do necessary calculations. The student should beware of picking an equation from the literature and assuming that it is likely to be satisfactory. Consider the following two tetrameric structures:

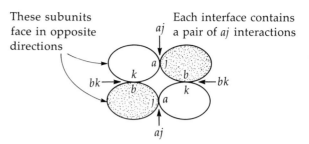

Isologous square

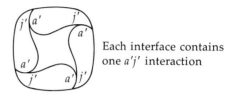

Heterologous square

In the isologous square (also shown in Fig. 4-9B) separate contributions to the free energy of binding can be assigned to the individual pairs of interactions aj and bk.

Thus, following Cornish–Bowden and Koshland[62] for assembly of the tetramer (Eq. 4-57):

$$\Delta G_f = 2\,\Delta G_{ajAA} + 2\,\Delta G_{bkAA}$$
$$K_f = K_{ajAA}^2 K_{bkAA}^2 \qquad (4\text{-}57)$$

Since free energies are additive, the formation constant will be the product of formation constants representing the individual interactions; thus, K_{ajAA} represents the formation constant of a dimer in which only the aj pair of bonds is formed.

In the isologous tetrahedron (Fig. 4-9C) the third set of paired interactions cl must be taken into account. (However, the third interaction constant will not be an association constant of the type represented by K_{ajAA} and K_{bkAA} but a dimensionless number.) On the other hand, the heterologous square has only a single interaction constant.

Now consider the binding of one molecule of X to the isologous tetramer with a conformational change in one subunit (Eq. 4-58).

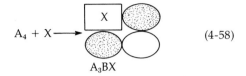

$$A_4 + X \longrightarrow \qquad (4\text{-}58)$$

$$A_3BX$$

We see that one pair of aj interactions and one pair of bk interactions have been altered. The equilibrium constant for the binding of X to the tetramer will be (Eq. 4-59)

$$K = 4\, \frac{K_{aj\,AB}K_{bk\,AB}}{K_{aj\,AA}K_{bk\,AA}}\, K_{BX}\, K_t \qquad (4\text{-}59)$$

in which the 4 is a statistical factor arising from the fact that there are four different ways in which to form A_3BX. When a second molecule of X is added three geometrical arrangements are possible:

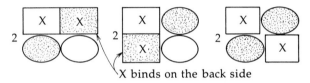

X binds on the back side

Each one can be formed in two ways. It is a simple matter to write down the microscopic constants for addition of the second molecule of X as the sum of three terms. Because values of the constants for aj and bk differ, it will be clear that the three ways of adding the second molecule of X are not equally probable. Thus, the oligomer will show preferred orders of "loading" with ligand X.

Two different geometries for the heterologous tetramer are possible in form $A_2B_2X_2$. Again, the different arrangements need not be equally probable and the relative distribution of each will be determined by the specific values of the interaction constants.[†]

While the foregoing may seem like an unnecessarily long exercise, it should provide a basic approach which can be applied to specific problems.

[†] Note that in the heterologous tetramer A_4 only one type of a interaction is present between subunits. However, as soon as a single molecule of X is bound and one subunit of conformation B is present, two kinds of aj interactions exist. (One in which group a is present in conformation A and the other in which it is present in conformation B.) Since there interactions always occur in equal numbers they can be lumped together.

However, remember that mathematical models require simplification. Real proteins may have more than two stable conformations.[11] The entire outer surface of a protein is made up of potential binding sites for a number of different molecules, both small and large. Filling of almost any of these sites can affect the functioning of a protein.

5. Hemoglobin

Now let us look at the nature of the cooperative binding of oxygen to the tetrameric ($\alpha_2\beta_2$) hemoglobin molecule (Section D,8) and at its physiological significance.[65] Within each subunit of hemoglobin the peptide chain folds around the single large flat iron-containing **heme** ring in a characteristic pattern (Fig. 4-17). The folding is essentially the same in all hemoglobins, both in the α and β subunits and in the monomeric muscle oxygen storage protein, myoglobin. The imidazole group of histidine F-8 is coordinated with the iron in the center of the heme. The other side of the iron atom is the site of binding of a single molecule of O_2.

The cooperativity in the binding of O_2 is illustrated in Fig. 4-18. Depending upon conditions, values of n_{Hill} (Eq. 4-35) may be as high as 3. The physiological significance is clear. In the capillaries of the lungs at a partial pressure of oxygen of ~ 100 mm of mercury, the hemoglobin is nearly saturated with oxygen; however, when the red cells pass through the capillaries of tissues in which oxygen is utilized the partial pressure of oxygen falls to about 5 mm of mercury. The cooperativity means that the oxygen is more completely "unloaded" than it would be if all four heme groups acted independently.

Deoxyhemoglobin has a low affinity for O_2, but after one or more of the subunits have become oxygenated the affinity of the remaining subunits increases about 500-fold.[†] The monomeric myoglobin also has a high affinity for oxygen, as does the abnormal **hemoglobin H,** which is made up of

[†] This applies at physiological concentrations of the effector 2,3-diphosphoglycerate.

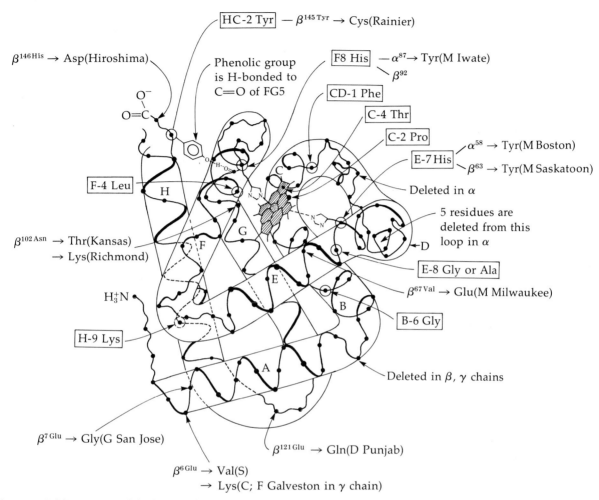

Fig. 4-17 Folding pattern of the hemoglobin monomers. The pattern shown is for the β chain of human hemoglobin. Some of the differences between this and the α chain and myoglobin are indicated. Evolutionarily invariant residues are indicated by boxes. Other markings show substitutions observed in some abnormal human hemoglobins. Invariant residues are numbered according to their location in one of the helices A–H while mutant hemoglobins are indicated by the position of the substitution in the entire α or β chain.

four β subunits. The latter also completely lacks cooperative binding.[66] These results can be interpreted to indicate that deoxyhemoglobin exists in the A (T) conformation, whereas oxyhemoglobin is in the B (R) conformation. The separated subunits and myoglobulin stay in the B conformation in *both* states of oxygenation. The subunits of hemoglobin H also appear to be frozen in the B conformation, even though the quaternary structure is similar to that of deoxyhemoglobin.[66] Fitting data for oxygenation of hemoglobin at pH 7 with parameters of the MWC model it is estimated[67] that $L = 9 \times 10^3$ and $c = K_{fA}/K_{fB}$ (see Eq. 6-52) = 0.014.

While no detectable dissociation of the tetramer into subunits takes place in deoxyhemoglobin, oxyhemoglobin does dissociate to some extent ($K = 2 \times 10^{-6}$) into αβ dimers. There is an exten-

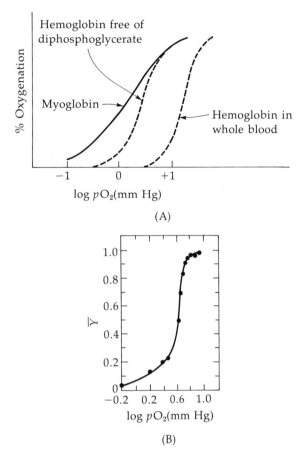

Fig. 4-18 Cooperative binding of oxygen by hemoglobins. (A) Binding curve for myoglobin: noncooperative, and for hemoglobin in whole blood in the presence of 2,3-diphosphoglycerate. Note the decrease in oxygen affinity caused by diphosphoglycerate. (B) Saturation curve for hemoglobin (erythrocruorin) of *Arenicola*, a spiny annelid worm. The molecule contains 192 subunits and 96 hemes. It shows very strong cooperativity with $n_{Hill} \approx 6$. From (A) Benesch and Benesch,[75] and (B) L. Waxman, *JBC* **246**, 7318–7327 (1971).

sive literature on oxygen-binding equilibria of hemoglobins. A number of recent papers are cited by Edelstein,[67] Herzfield and Stanley,[68] and Baldwin.[68a]

a. Structural Changes Accompanying Oxygen Binding

Perutz and associates, using X-ray crystallography, found surprisingly *small* but real dif-ferences in the conformation of the subunits of deoxy- and oxyhemoglobin.[69,70] More striking is the fact that upon oxygenation, both α and β sub-units undergo substantial amounts of *rotation*, the net result being that the hemes of the two β subunits move about 0.07 nm closer together in the oxy form than in the deoxy form. Within the $\alpha_1\beta_1$ contacts (Fig. 4-19) little change is seen. On the other hand contact $\alpha_1\beta_2$ is altered drasti-cally. As Perutz expressed it, there is a "jump in the dovetailing" of the CD region of the α sub-unit relative to the FG region of the β subunit. The hydrogen bonding pattern is changed.

Another obvious difference is seen in the pres-ence of salt bridges at the ends of the molecules of deoxyhemoglobin. The $—NH_3^+$ group of Lys H-10 in each α subunit is hydrogen bonded to the car-boxyl group of the C-terminal arginine of the op-posite α chain. The guanidinium group of each C-terminal arginine is H-bonded to the carboxyl group of aspartate H-9 in the opposite α chain. It is also hydrogen bonded to an inorganic anion (phosphate or Cl^-) which in turn is H-bonded to the α amino group of valine-1 of the opposite α chain (a pair of isologous interactions). At the other end of the molecule, the C-terminal group of histidine-146 of each β chain binds to the amino group of lysine C-6 of the α chain while the imida-zole side chain binds to aspartate FG-1 of the same β chain. These salt bridges appear to provide extra stability to deoxyhemoglobin and account for the high value of the constant L.

Tyrosine HC-2, the second amino acid residue from the C-terminal end, is one of the few **in-variant residues** in hemoglobin. The location of this amino acid has been conserved throughout evolution in hemoglobins and myoglobins of all species studied. In deoxyhemoglobin, tyrosine HC-2 lies tucked into a pocket between the H and F helices and is hydrogen bonded to the main chain carbonyl of residue FG-5 (Figs. 4-17 and 4-19). Perutz and associates have observed that upon oxygenation this tyrosine is released from its pocket; the salt bridges at the ends of the mole-cules are broken and the subunits shift into the new bonding pattern characteristic of oxyhemo-

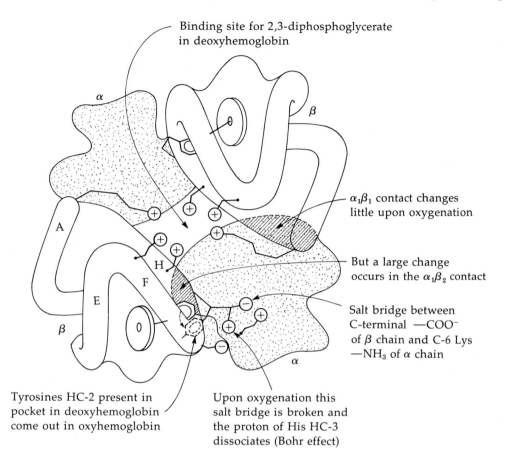

Binding site for 2,3-diphosphoglycerate in deoxyhemoglobin

$\alpha_1\beta_1$ contact changes little upon oxygenation

But a large change occurs in the $\alpha_1\beta_2$ contact

Salt bridge between C-terminal —COO⁻ of β chain and C-6 Lys —NH₃ of α chain

Upon oxygenation this salt bridge is broken and the proton of His HC-3 dissociates (Bohr effect)

Tyrosines HC-2 present in pocket in deoxyhemoglobin come out in oxyhemoglobin

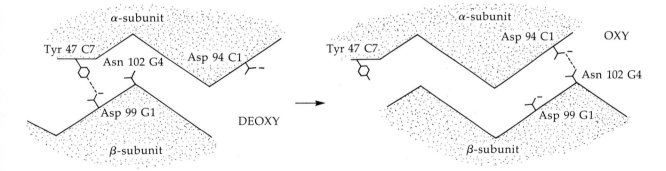

Fig. 4-19 (A) Structural changes occurring upon oxygenation of hemoglobin. After Dickerson[72] and Perutz.[71] "Rotation at the contact $\alpha_1\beta_2$ causes a jump in the dovetailing of the CD region of α relative to the FG region of β and a switch of hydrogen bonds as shown" (Perutz).

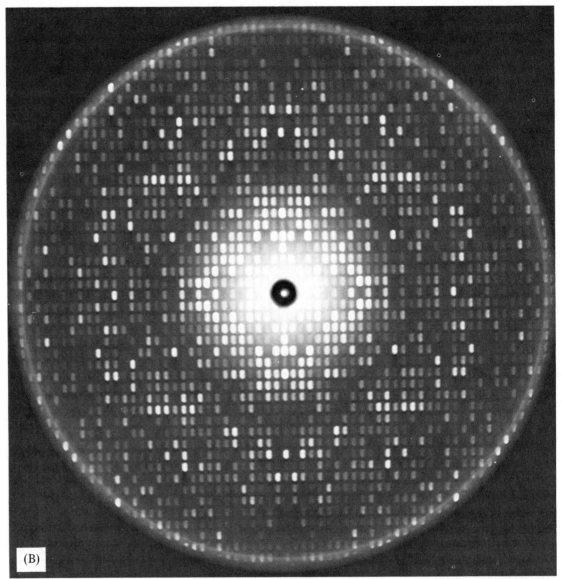

(B)

Fig. 4-19 (B) An X-ray diffraction photograph such as was used to determine the structure of hemoglobin. This precession photograph was obtained from a crystal of human deoxyhemoglobin by rotating the crystal along two different axes in a defined manner before a narrow X-ray beam. The film also was moved synchronously. The periodicity observed in the photograph is a result of a diffraction phenomenon arising from the periodic arrangement of atoms in the crystal. The distances of the spots from the origin (center) are inversely related to the distances between planes of atoms in the crystal. In this photograph (which shows only two dimensions of the three-dimensional diffraction pattern) the spots at the periphery represent a spacing of 0.28 nm. If the intensities of the spots are measured, and if the phases of the harmonic functions required for the Fourier synthesis can be assigned correctly, the structure can be deduced to a resolution of 0.28 nm from a set of patterns of this type. In the case of human deoxyhemoglobin, a complete set of data would consist of about 27,000 spots. Photograph courtesy of Arthur Arnone.

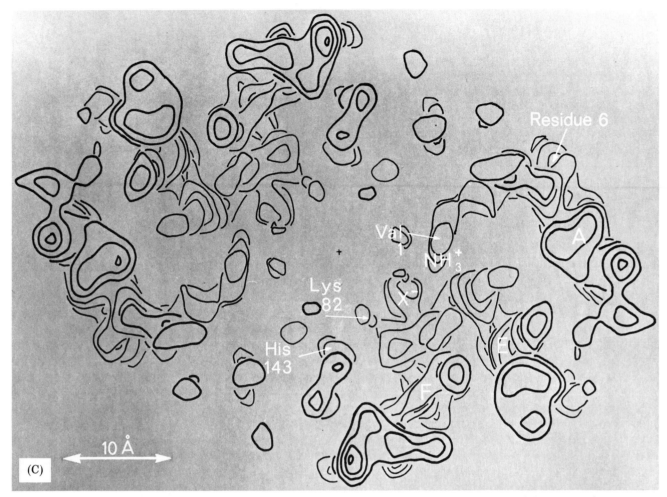

Fig. 4-19 (C) Electron density map obtained from X-ray diffraction data for human deoxyhemoglobin at 0.35 nm resolution. The contour lines indicate regions of high electron density corresponding to parts of the hemoglobin molecule. This map represents a section, largely through the β subunits and perpendicular to the 2-fold axis at the level of glutamate-6, the position of the sickle cell substitution. The section includes parts of helices A, E, and F and Val 1, Glu 6, Lys 82, and His 143. The peak marked X⁻ is an inorganic anion, probably sulfate or phosphate.

globin. Oxygenation of two of the hemes (Perutz thought those of the α chains) leads to a cooperative conformational change for all four subunits.[71,72]

How does the binding of O_2 to the iron of heme trigger the conformational change in hemoglobin? As pointed out in Chapter 10 (Section B,4), the iron atom in the heme is probably moved into the plane of the heme group about 0.06 nm when oxygen binds.[73] This displacement is transmitted through histidine F-8 pulling helix F toward the heme and setting in motion changes in tertiary structure which loosen the hydrogen bonds at the $a_1\beta_2$ contacts and the salt bridges between the subunits. Despite the careful X-ray studies the details of the triggering mechanism are not yet obvious. Bear in mind that the resolution obtainable in X-ray analysis of crystalline proteins

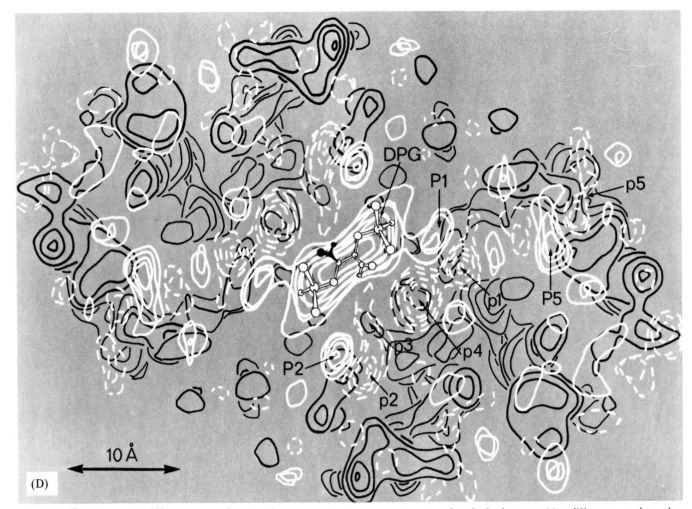

Fig. 4-19 (D) A difference map showing changes in electron density upon binding of 2,3-diphosphoglycerate (DPG) is superimposed (in white contours) on the map shown in (C). Positive difference electron density is marked by solid white contours and upper case letters. Negative difference electron density is marked by dashed white contours and lower case letters. A "symmetry-averaged" molecule of DPG has been superimposed to fit the large positive difference peak on the dyad axis. The pairs of positive and negative peaks labeled P1, p1 and P2, p2 show that the N-terminal α-amino groups and histines H-21 move inward when DPG binds. The large negative peak p4 indicates that the anion is displaced by the binding of DPG at an adjacent location.

does not usually allow positions of light atoms to be determined to an accuracy better than 0.03 nm. As a consequence, the mechanism of transmission of cooperative effects cannot be observed directly, but must be guessed from changes in the tertiary structure of subunits seen on loss of ligand from the R (e.g., oxy-) or addition of ligand to the T (e.g., deoxy-) structure. It is clear that it is very difficult to understand something as subtle as the molecular machinery underlying the cooperative binding of oxygen by hemoglobin.

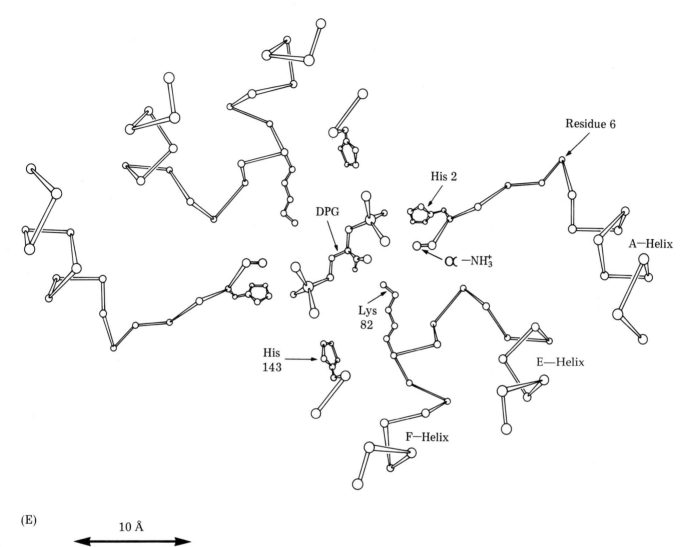

Residue 6

His 2

DPG

α —NH$_3^+$

A—Helix

Lys
82

His
143

E—Helix

F—Helix

(E)

10 Å

Fig. 4-19 (E) An interpretation of the electron density maps of (C) and (D). The phosphate groups of 2,3-diphosphoglycerate (DPG) form salt bridges with valines 1 and histidines 2 and 143 of both β chains and with lysine 82 of one chain. This binding pulls the A helix and residue 6 toward the E helix accounting for the pairs of positive and negative electron density peaks P1, p1 and P2, p2 in (D). From Arnone.[74]

---

b. The Bohr Effect

The breaking of the salt bridges at the ends of the hemoglobin molecule upon oxygenation has another interesting result. The pK_a values of the N-terminal valines of the α subunits and of histidines HC-3 of the β subunits are abnormally high in the deoxy form because they are tied up in the salt bridges. In the oxy form where the groups are freed, the pK values are lower. If hemoglobin is held at a constant pH of 7, these protons are observed to dissociate upon oxygenation. This **Bohr effect** is important because acidification of hemoglobin tends to stabilize the deoxy form. In capillaries in which oxygen pressure is low and in which carbon dioxide and lactic acid may have accumulated, the lowering of the pH causes oxyhemoglobin to release oxygen more efficiently.

6. Allosteric Regulators

Just as the conformational equilibria in hemoglobin can be shifted by attachment of oxygen to the heme groups, so the binding of certain other molecules at different sites can also affect the conformation. Such compounds are called **allosteric effectors** or **regulators** because they bind at a site other than the "active site." They are considered in more detail in Chapter 6 (Section B,6). An important allosteric effector for hemoglobin is **2,3-diphosphoglycerate,** a compound found in human red blood cells in an unusually high concentration (in approximately equimolar amount with hemoglobin).

2,3-Diphosphoglycerate

One molecule of diphosphoglycerate binds to a single hemoglobin tetramer in the deoxy form with $K_f = 1.4 \times 10^5$. It has about half this affinity for oxyhemoglobin.[74] X-Ray crystallography shows that diphosphoglycerate binds between the two β chains of deoxyhemoglobin directly on the two-fold axis (Fig. 4-19).[71] It has long been known that the affinity of oxygen for hemoglobin in whole blood is less than that for isolated hemoglobin[75,76] (Fig. 4-18). We now see that this change is a result of the presence of 2,3-diphosphoglycerate in erythrocytes. The shift is important for it allows a larger fraction of the oxygen carried to be un-loaded from red corpuscles in body tissues. The diphosphoglycerate level of red cells varies with physiological condition, e.g., people living at higher elevations have a higher concentration.[76] It has been suggested that artificial manipulation of the level of this regulatory substance in erythrocytes may be of clinical usefulness for disorders of oxygen transport. Not all species contain 2,3-diphosphoglycerate in erythrocytes, and in birds and turtles its function appears to be served by inositol pentaphosphate.

Another allosteric effect on hemoglobin has already been considered in the preceding section. The Bohr effect can be viewed as resulting from the action of **protons** as allosteric effectors that bind to the amino and imidazole groups involved in forming the salt linkages. **Carbon dioxide** also acts as a physiological effector by combining reversibly with NH_2-terminal groups of the α and β subunits to form **carbamino** (carbamate, $-NH-COO^-$) groups.[77,78] The affinity for CO_2 is highest in deoxygenated hemoglobin. Consequently, unloading of O_2 is facilitated in the CO_2-rich respiring tissues. Hemoglobin carries a significant fraction of CO_2 to the lungs, and there the oxygenation of hemoglobin facilitates the dissociation of CO_2 from the carbamino groups.

7. Comparative Biochemistry of Hemoglobin; Abnormal Hemoglobins

Even within the human, the hemoglobin family is quite large. In addition to myoglobin and adult hemoglobin A $(\alpha_2\beta_2)$ there is a minor hemoglobin A_2 $(\alpha_2\delta_2)$. Prior to birth the blood contains **fetal hemoglobin,** also called hemoglobin F $(\alpha_2\gamma_2)$. In the presence of 2,3-diphosphoglycerate it has a higher oxygen affinity than hemoglobin A as befits its role in supplying oxygen to the fetus. Hemoglobin F disappears a few months after birth and is replaced by hemoglobin A. Each of the hemoglobins differs from the others in amino acid sequence. When we turn to other species we find that the amino acid composition of hemoglobins

BOX 4-D

SICKLE CELL ANEMIA

A gene that is found frequently in persons of African descent produces the crippling and highly lethal disease sickle cell anemia in homozygotes.[a] In 1949 Pauling and Itano and associates discovered that hemoglobin from sickle cell anemia patients migrated unusually rapidly upon electrophoresis.[b] Later (1957) Ingram[c] devised the method of protein "fingerprinting" (Chapter 2, Section H,2, Fig. 4-20) and applied it to hemoglobin. He split the hemoglobin molecule into 15 tryptic peptides which he separated by electrophoresis and chromatography. From these experiments he was able to locate the abnormality in sickle cell hemoglobin (hemoglobin S) at position 6 in the β chain (see Fig. 4-17). The glutamic acid normally present in this position was replaced by valine in hemoglobin S. This was the first instance in which a genetic disease was traced directly to the presence of a single amino acid substitution in a specific protein.

When hemoglobin S is deoxygenated it tends to "crystallize" in the red blood cells, which contain 33% by weight of hemoglobin. The crystallization leads to distortion of the red cells into a sickle shape and these distorted corpuscles are easily destroyed, leading to anemia. Apparently the introduction of the hydrophobic valine residue at position 6 near the end of the molecule helps form a new bonding domain by which the hemoglobin tetramers can associate to form long microtubular arrays.[d] These crystallize within the blood cells.

Why is there such a high incidence of the sickle cell gene, estimated to be present in three million Americans? The survival and spread of the gene in Africa was apparently the result of a balance between its harmful effects and a beneficial effect under circumstances formerly existing. The malaria parasite, the greatest killer of all time, lives in red blood cells during part of its life cycle (Fig. 1-7). For some reason red cells that contain hemoglobin S as well as hemoglobin A are less suitable then cells containing only hemoglobin A for growth of the malaria organism. Thus, heterozygotic carriers of the sickle cell gene survived epidemics of malaria but at the price of seeing one-fourth of their offspring die of sickle cell anemia.

What is the outlook for a sufferer of sickle cell anemia today? Careful medical care, including blood transfusion, can prolong life greatly but there is a need for a cure. Recent experiments have shown that cyanate reacts with the terminal amino group (of valine) in the β chain of hemoglobin S and tends to prevent "sickling." The reaction is one of carbamoylation.

$$\overset{\overset{\displaystyle H_3C \quad CH_3}{\diagdown \diagup}}{\underset{\underset{\displaystyle H}{|}}{\overset{\displaystyle C}{\underset{\displaystyle H_3\overset{+}{N}-C-R}{}}}} + N\equiv C-O^- \longrightarrow \overset{\overset{\displaystyle H_3C \quad CH_3}{\diagdown \diagup}}{\underset{\underset{\displaystyle H \quad H}{|\quad|}}{\overset{\overset{\displaystyle O \qquad C}{\| \qquad}}{H_2N-C-N-C-R}}}$$

Cyanate has been cautiously tested in humans, but appears too toxic for use. However, there is hope that new drugs can be designed that prevent hemoglobin S from crystallizing.[e]

[a] The much more numerous heterozygotes carry one hemoglobin S gene but have, at most, minor problems.

[b] L. Pauling, H. A. Itano, S. J. Singer, and I. C. Wells, *Science* **110**, 543–548 (1949).

[c] V. M. Ingram, *Nature* (*London*) **180**, 326–328 (1957).

[d] B. C. Wishner, K. B. Ward, E. E. Lattman, and W. E. Love, *JMB* **98**, 179–194 (1975).

[e] D. R. Harkness, *Trends Biochem. Sci.* **1**, 73–76 (1976).

---

varies even more. The interactions between subunits also vary and in one class of hemoglobins, the **erythrocruorins,** found in certain invertebrates, as many as 192 subunits are present.[79]

What is common to all of the hemoglobins? The same folding pattern of the peptide chain is always present. The protein is always wrapped around an identical, or a very similar, heme group. What is most amazing is that in spite of the striking conservation of overall structure, when all hemoglobins are compared, *there are only nine invariant residues.* A tenth is *almost* invariant. These

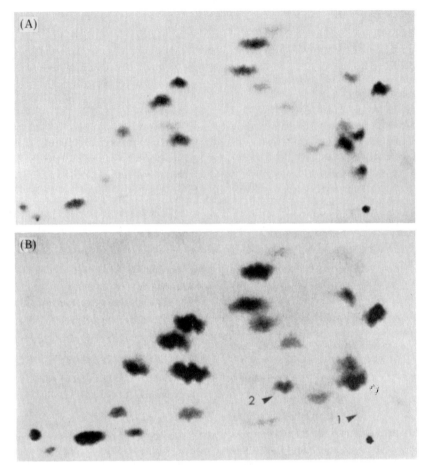

Fig. 4-20 "Fingerprints" of human hemoglobins. The denatured hemoglobin was digested with trypsin and the 28 resulting peptides were separated on a sheet of paper by electrophoresis in one direction (horizontal in the figures; anode to the left) and by chromatography in the other direction (vertical in the figure). The peptides were visualized by spraying with ninhydrin or with specific reagents for histidine or tyrosine residues. (A) The fingerprint of normal adult hemoglobin A. (B) Fingerprint of hemoglobin S (sickle cell hemoglobin). Note that one histidine-containing peptide (1) is missing and a new one (2) is present. This altered peptide contains the first 8 residues of the N-terminal chain of the subunit of the protein. From H. Lehmann and R. G. Huntsman, "Man's Haemoglobin." North-Holland, Amsterdam, 1974.

ten residues are indicated in Fig. 4-17 by the boxes. The two glycines (or alanine) at B-6 and E-8 are invariant because the close contact between the B and E helices does not permit a larger side chain. Proline C-2 helps the molecule turn a corner. Four of the other invariant residues are directly associated with the heme group. Two of them, His E-7 and His F-8, are the "heme-linked" histidines. The ninth residue, tyrosine HC-2, previously mentioned (Section 5,a) plays a role in the cooperativity of oxygen binding. Only lysine H-9 is on the outside of the molecule. The reasons for its conservation are unclear.[80]

Many alterations in the hemoglobin structure have arisen by mutations in the human population. It is estimated that the blood of one person in

600 contains a mutant hemoglobin, usually with a harmless substitution of one amino acid for another (Fig. 4-20). However, substitutions near the heme group often adversely affect the binding of oxygen and substitutions in one of the interfaces between subunits may decrease the cooperative interaction between subunits.[81] One of the most common and serious abnormal hemoglobins is **hemoglobin S,** which is present in individuals suffering from **sickle cell anemia** (see Box 4-D). In hemoglobin S, glutamic acid 6 of the β chain is replaced by valine. Replacement of the same amino acid by lysine leads to hemoglobin C and is associated with a mild disease condition quite unlike sickle cell anemia. A few of the many other substitutions that have been studied are indicated in Fig. 4-17.

Of interest are the family of **hemoglobins M.** Only heterozygotic individuals survive. Their blood is dark because in hemoglobin M the iron in half of the subunits is oxidized irreversibly to the ferric state. The resulting **methemoglobin** is present in normal blood to the extent of about 1%. While normal methemoglobin is reduced by a special **methemoglobin reductase** system (Box 10-A), the methemoglobins M cannot be. All of the five hemoglobin M's result from substitution, near the heme group. In four of them one of the heme-linked histidines (either F-8 or E-7) of either the α or β subunits is substituted by tyrosine. In the fifth, valine 67 of the β chains is substituted by glutamate. The two hemoglobins M that carry substitutions in the α subunits (M_{Boston} and M_{Iwate}) are frozen in the T (deoxy) conformation and therefore have low oxygen affinities and lack cooperative effects.

In hemoglobin Rainier the usually invariant tyrosine H-22 is substituted by cysteine. Oxygen affinity is very high and again cooperativity is lacking. Hemoglobin Kansas, in which the β^{102} asparagine is substituted by threonine, also lacks cooperativity and has a very low oxygen affinity, while hemoglobin Richmond, in which the same amino acid is substituted by lysine, functions normally. In hemoglobin Hiroshima the C-terminal histidine in the β chain is replaced by aspartic acid. This C-terminal histidine is one that donates a Bohr proton and in the mutant hemoglobin the oxygen affinity is increased threefold and the Bohr effect is halved.[82]

F. MUSCLE

Among the more complex aggregates of protein subunits, none has attracted more attention than the contractile fibers of muscles. Several types of muscle exist within our bodies. **Striated skeletal muscles** act under voluntary control. Closely related is the **involuntary striated heart muscle. Smooth involuntary muscles** constitute a third type. Within other organisms special types of muscle occur. Thus, the asynchronous flight muscles of certain insects are able to cause the wings to beat at rates of 100–1000 cycles/s. In these muscles nerve impulses are used only to start and to stop the action; otherwise the cycle of contraction and relaxation continues automatically.

1. The Structural Organization of Striated Muscle

Skeletal muscles consist of bundles of long fibers of diameter 10–100 μm which are usually formed by the fusion of many embryonic cells. The lengths are typically 2–3 cm in mammals but may sometimes be as great as 50 cm. Each fiber can be considered a single cell with up to 100–200 nuclei. Typical cell organelles are present but are often given special names. Thus, the plasma membrane (plasmalemma) of muscle fibers is called the **sarcolemma.** The cytoplasm is **sarcoplasm** and mitochondria may be called **sarcosomes.** The major characteristic of muscle cells is the presence of the contractile **myofibrils,** organized bundles of proteins which are *not* separated by membranes from the cytoplasm.

Under the light microscope cross striations can be seen in the myofibrils with a repeating distance of ~2.5 μm (Figs. 4-21 and 4-22). The space

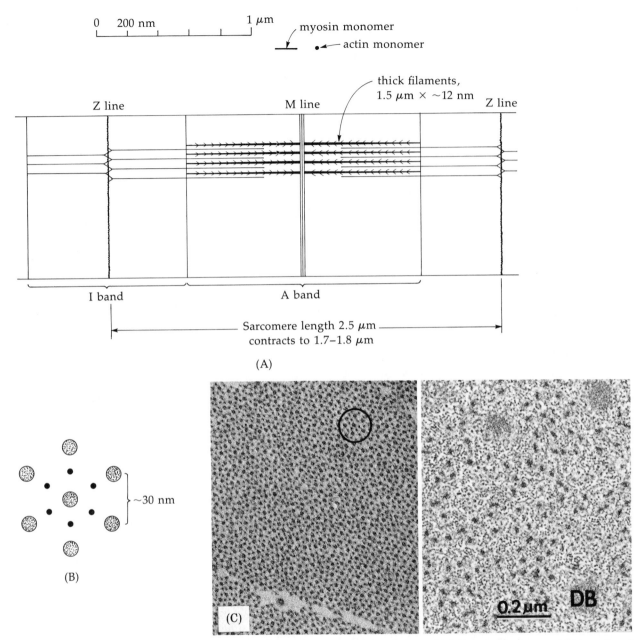

(A)

(B)

(C)

Fig. 4-21 (A) Diagram of the structure of a typical sarcomere of skeletal muscle. The longitudinal section depicted corresponds to that of the electron micrograph, Fig. 4-22A. (B) A sketch showing the arrangement of thick and thin filaments as seen in a transverse section of a striated muscle fiber. (C) Left: electron micrograph of a transverse section of a glycerated rabbit psoas muscle. The hexagonal arrangement of six thin fil- aments around one thick filament can be seen in the center of the circle. Six other thick filaments form a larger concentric circle as in (B). Right: transverse section of a smooth muscle fiber. Notice the irregular arrangement of thick and thin fila- ments. Filaments of intermediate diameter are also present, as are dense bodies (DB). The latter are characteristically present in smooth muscle.

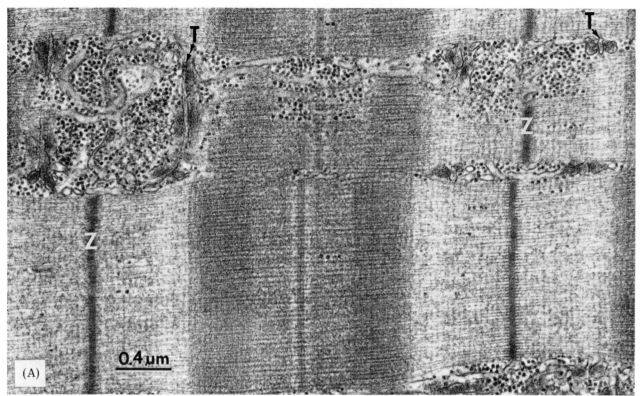

Fig. 4-22 (A) Electron micrograph of a longitudinal section of a mammalian skeletal muscle (pig biceps muscle). The tissue was doubly fixed, first with formaldehyde and glutaraldehyde, then with osmium tetroxide. It was then stained with uranyl acetate and lead citrate. The section shows a white muscle fiber containing few mitochondria and narrow Z lines. The Z lines are marked and the M line, the A and I bands, thick and thin filaments are all clearly seen. Notice the periodicity of ~40 nm along the thin filaments and M line. This spacing along the thin filaments corresponds to the length of the tropomyosin molecule and the cross striation is thought to repre-sent bound tropomyosin and troponin. The numerous dense particles in the upper part of the micrograph are glycogen granules while the horizontal membranous structures are transverse tubules of the sarcoplasmic reticulum (endoplasmic reticulum). These come into close apposition to the T tubules leading from the surface of the muscle fiber. The T tubules (marked T) are visible in longitudinal section at the upper left of the micrograph on both sides of the Z line and in cross section in the upper right-hand corner. There a T tubule is seen lying between two lateral cisternae of the sarcoplasmic reticulum.

between two of the dense **Z lines** defines the **sarcomere,** the basic contractile unit. In the center of the sarcomere is a dense **A band** (anisotropic band). The name refers to the intense birefringence of the band. Straddling the Z lines are less dense **I bands** (the abbreviation stands for *isotropic,* a misnomer; the bands lack birefringence, but are not isotropic). Weakly staining **M lines** (usually visible only with an electron microscope) mark the centers of the A bands and of the sarcomeres.

The fine structure of the sarcomere was a mys-tery until 1953, when H. E. Huxley, examining thin slices of muscle with the electron microscope, discovered a remarkably regular array of protein filaments.[83] Thick filaments, 12–16 nm in diameter and ~1.5 μm long, are packed in a hexagonal array on 40–50 nm centers throughout the A bands (Fig. 4-21B). In between these **thick filaments** are **thin filaments** only 8 nm in diameter and extending from the Z line for a length of ~1.0 μm. When contracted muscle was examined it was found that the I bands had nearly disappeared and that the

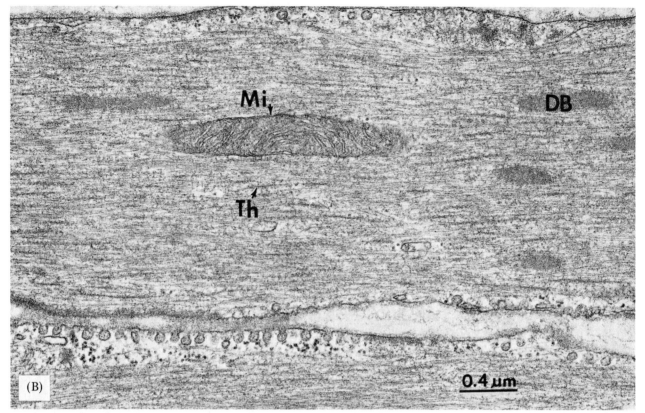

Fig. 4-22 (B) Longitudinal section of smooth muscle (chicken gizzard) fixed as in A. Thick filaments (Th), which are considerably thicker than those in striated muscle and less regular, can be seen throughout the section. They are surrounded by many thin filaments which are often joined to dense bodies (DB). A mitochondrion (Mi) is seen in the center of the micrograph and at the lower edge is a boundary between two adjacent cells. Notice the pinocytotic vesicles which are present in large numbers in the plasma membrane and which are extremely active in smooth muscle. Micrographs courtesy of Marvin Stromer, Iowa State University.

amount of overlap between the thick and the thin filaments had increased in such a way as to indicate that contraction had consisted of the sliding movement of the thick and thin filaments with respect to each other. In skeletal muscle the sarcomere shortens to a length of ~1.7–1.8 μm. In insect flight muscle a much smaller shortening occurs repetitively at a very high rate.

The myofibrillar proteins make up 50–60% of the total protein of muscle cells. Insoluble at low ionic strengths, they dissolve when the ionic strength exceeds ~0.3 and can be extracted with salt solutions. The major protein myosin constitutes the bulk of the thick filaments. The second protein actin makes up much of the thin filaments as indicated in Fig. 4-7. Tropomysin and **troponin** are "regulatory proteins" associated with the actin filaments,[84] while **α-actinin** is found in the Z lines. A recently discovered **M protein** resides in the M lines.[85]

Myosin is a remarkably long slender molecule of dimensions ~160 × 2 nm (Fig. 4-23). Most of the molecule is made up of two apparently identical chains in the form of α helices.[86] These are coiled around each other to form a double-stranded rope (Chapter 2, Section B,3,d). While the C-terminal ends of the two parallel chains appear to form a simple rod, four other smaller subunits of

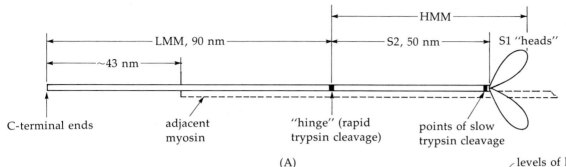

(A)

Fig. 4-23 (A) An approximate scale drawing of the myosin molecule. The "hinge" is a region that is rapidly attacked by trypsin to yield the light and heavy meromyosins (LMM and HMM). Total length ~160 nm, MW ~ 470,000; two heavy chains, MW ~ 200,000; two pairs of light chains, MW = 16,000–21,000; heads: ~15 × 4 × 3 nm. (B) Radial projection illustrating structure suggested by Squire[87] for thick filaments of vertebrate skeletal muscle. The region of the bare zone at the M line is shown. The filled circles represent the head ends of the myosin molecules and the arrowheads represent the other end of the rod (i.e., the end of the LMM portion of the molecule). Antiparallel molecules interacting with overlaps of 43 and 130 nm are shown joined by single and triple cross-lines, respectively. Positions where two arrowheads meet are positions of end-to-end butting. O is an "up" molecule (thin lines) and A a "down" molecule (thick lines). The molecules move from the core at the C-terminal end to the filament surface at the head end. The levels marked B may be the levels of attachment of M-bridge material to the myosin filament. The level M-M is the center of the M line and of the whole filament. Note that the lateral scale is exaggerated more than 3-fold.

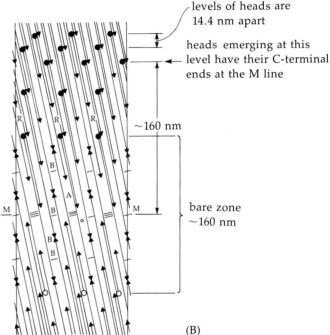

(B)

MW ~ 16,000–21,000 join with the two heavy chains at the N-terminal end to form a pair of "heads." A short treatment with trypsin cuts the myosin molecule into two pieces. From the tail end a **light meromyosin** molecule (LMM) 90 nm in length is formed. The remainder of the molecule including the heads is designated **heavy meromysin** (HMM). The latter can be cleaved by a longer trypsin treatment into one S2 fragment of MW ~ 62,000 and 40 nm length, and two S1 fragments of MW ~ 110,000, each representing one of the two heads (Fig. 4-23).

Dissociated myosin molecules can be induced to aggregate into rods similar to the thick filaments of muscle. Since the filaments have a thickness of ~ 14 nm, a large number of the thin 2 nm myosin molecules must be packed together. Electron microscopy reveals the presence of the heads projecting from the thick filaments at intervals of ~ 43 nm. However, there is a bare zone at the M line, a fact that suggests tail-to-tail aggregation of the myosin monomers at the M line in the centers of the thick filaments. A packing arrangement involving 300 myosin molecules (up to 30 rods in a single cross section) in close packing with a small central open core has been proposed for skeletal muscle myosin fibers[87] and is illustrated in Fig. 4-23. A somewhat different arrangement with four heads (rather than three) per helical repeat distance of 14.3 nm may represent the true structure.[88] Another protein, the C protein, also occurs in smaller amounts within the myosin filament.[88] A different packing of the myosin rods is found in insect flight muscle.

The protein actin has its own special properties. The native fibrous F-actin (Fig. 4-7) is made up of monomers of MW ~43,000 that contain ~374 amino acids per chain. The presence of one molecule of N^7-methylhistidine at position 73 is of interest. Actin filaments can be dissolved in a low ionic strength medium containing ATP to give a soluble monomeric **G-actin.** Each G-actin monomer usually contains one molecule of bound ATP and a calcium ion. Addition of 1 mM Mg^{2+} or 0.1 M KCl leads to spontaneous transformation into filaments similar to the thin filaments of muscle, each of which contains 340–380 actin monomers. The ATP is hydrolyzed in the process, leaving bound ADP in the F-actin filament. We see a striking similarity to the binding of nucleotides to microtubule subunits (Box 4-A) and in the contractile phage tail (Box 4-E). However, no common functional role is evident.

That actin and myosin are jointly responsible for contraction was demonstrated long before the fine structure of the myofibril became known. In about 1929 ATP was recognized as the energy source for muscle contraction, but it was not until 10 years later that Englehardt and Ljubimowa showed that isolated myosin preparations catalyzed the hydrolysis of ATP.[88a] Thus, the enzymatic machinery for harnessing the free energy of hydrolysis of ATP was suggested to reside in the major fibrous protein of muscle. Later A. Szent-Györgi[88b,c] showed that a combination of the two proteins actin[†] and myosin (**actomyosin**) was required for Mg^{2+}-stimulated ATP hydrolysis ("ATPase" activity). He also demonstrated and popularized the phenomenon of **superprecipitation** by which actomyosin could be induced to contract rapidly in a test tube upon addition of ATP. Although the significance of superprecipitation to

---

[†] Actin was discovered by F. Straub.[88d]

muscle contraction is still debated, it served to focus attention upon the cooperative interaction of actin and myosin in the action of muscle.

Chemical investigations on myosin have revealed that the ATPase activity induced by actin is all located in the head part of the myosin molecule. Under the electron microscope the myosin heads can sometimes be seen to be attached to the nearby thin actin filaments as **cross-bridges.** When skeletal muscle is relaxed (not activated by a nerve impulse) the cross-bridges are not attached and the muscle can readily be stretched. However, when the muscle is activated and under tension the cross-bridges form more frequently. When ATP is exhausted (e.g., after death) muscle enters the state of **rigor.** The cross-bridges can be seen by the electron microscope to be almost all attached to thin filaments, accounting for the complete immobility of muscle in rigor. These observations led naturally to the hypothesis that during contraction the myosin heads attach themselves to the thin actin filaments. The hydrolysis of ATP is then coupled in some way to the generation of a tension that causes the thick and thin filaments to be pulled past each other. The heads then release themselves and become attached at new locations. Repetition of this process leads to the sliding motion of the filaments. The chemical mechanisms, which are still poorly understood,[89,90] will be considered further in Box 10-F.

Equally as remarkable as the contractile process are the controls required to regulate voluntary muscles. In the muscle cells the endoplasmic reticulum (**sarcoplasmic reticulum**) is organized in a striking regular manner.[91,92] Interconnecting tubules run longitudinally through the fibers among the bundles of contractile elements. At regular intervals they come in close contact with infoldings of the outer cell membrane (the **T system** of membranes, Fig. 4-22A). A nerve impulse enters the muscle fiber traveling along the plasmalemma and into the T tubules. At the points of close contact the signal is somehow transmitted to the longitudinal tubules of the sarcoplasmic reticulum which contain a high concentration of calcium ions. The arriving nerve signal causes a sudden release of the calcium ions into the cytoplasm and the myofibrils. There the calcium binds to the **C subunit** of troponin, an oligomeric protein that together with tropomyosin (Chapter 2, Section B,3,d) forms a regulatory complex. This complex is attached to the actin fibrils (Fig. 4-24).

When the regulatory proteins are completely removed from the actin fibrils contraction can occur freely until the ATP is exhausted. However, in the presence of the regulatory proteins and in the absence of calcium, both contraction and hydrolysis of ATP are blocked. A working hypothesis for the functioning of this system[93,94] postulates that an elongated tropomyosin rod fits into the grooves between the actin and the myosin heads.[92] Figure 4-24 shows an end view looking down the actin helix. The head (S1) of a myosin molecule is shown attached to one of the actin subunits. In resting muscle the tropomyosin is bound to actin near the site to which the S1 portion of the myosin binds. As a consequence, the tropomyosin rod blocks attachment of the S1 cross-bridges of myosin to actin and prevents actin-stimulated hydrolysis of ATP. Tropomyosin is about 41 nm long and contacts about seven actin subunits at once.[95] Thus, one tropomyosin–troponin complex synchronously controls seven actin subunits.

Troponin consists of three polypeptides that range in mass from 18,000 to 37,000 daltons. One peptide (T) binds troponin tightly to tropomyosin at a site located about one-third of way from the C-terminus to the N-terminus of the tropomyosin molecule. The second troponin peptide (I) interacts with actin in the absence of Ca^{2+} and works with the other two peptides to keep the tropomyosin in the proper position to inhibit ATP hydrolysis. When the C subunit binds calcium ions the inhibition is removed and contraction can occur. What is still missing is a reasonable picture of how all this "machinery" functions. Evidence from X-ray diffraction and electron microscopy suggests[93,94] that when calcium binds to troponin the tropomyosin moves through an angle of $\sim 20°$ away from S1, permitting uncovering of the active site for the myosin–ATP–actin interactions (Fig. 4-24). Can it be that the tropomyosin rolls like a

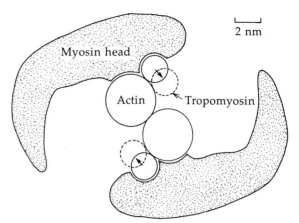

Fig. 4-24 Sketch of hypothetical structure of the actomyosin–tropomysin complex viewed down the axis of an actin filament.[92,94] Interaction of actin with the myosin head S1 and ATP is blocked by tropomyosin (small circles). However, in the presence of Ca^{2+} it is proposed that the tropomyosin moves to the positions indicated by the dashed circles to allow reaction to occur. After Cohen[93] and Wakabayashi et al.[94]

log along the surface of the actin, uncovering sites on seven actin molecules at once? If so, what kind of "motor" is used to roll the log and what keeps the log from falling off of the actin? We can only speculate. Perhaps side chain knobs, protruding from the tropomyosin like teeth on a submicroscopic gear, engage complementary holes in the actin.[†] If so, how does the binding of calcium ion to the troponin cause the tropomyosin log to roll? We know that binding of metals to proteins can cause enormous changes in conformation (Section C,8,c). Perhaps a conformational change induced in the C subunit of troponin can somehow be translated into the necessary force for causing the motion.

2. Contraction in Nonmuscle Cells

The motion of eukaryotic flagella is thought to involve a sliding of the microtubular filaments (Box 4-A) somewhat analogous to the sliding of

† A set of magnesium ion bridges between zones of negative charge on tropomyosin and actin has been suggested.[95a]

muscle filaments.[96] The presence of a protein with ATPase activity also suggests a relationship to muscle. Actomyosin-like proteins have now been identified in many other cells. In the brain such a contractile protein may be responsible for the rapid release of vesicles of neurotransmitter compounds at synaptic endings.[97] Actin has been isolated from the primitive *Dictyostelium* and appears to be a universal constituent of eukaryotic cells.[98] Microfilaments in cytoplasm appear often to be actin. In fibroblasts these filaments form a three-dimensional matrix about 100 nm thick just under the cell membrane. Other microfilaments aggregate to form "stress fibers" running along the bottom and out into the edges of cells. Preparation of a fluorescent antibody specific for actin has permitted direct visualization of the latter by light microscopy[99] (see also Box 4-A). Myosin may be just as widespread.[98,100] A possible way in which a nonmuscle contractile apparatus might function[98] is indicated in Fig. 4-25.

G. SELF-ASSEMBLY OF MACROMOLECULAR STRUCTURES

While it is easy to visualize the growth of bacterial flagella (Box 4-B) by passage of subunits out through the holes in the center, it is not as easy to imagine how more complex objects such as the sarcomere of muscle are formed. Recently, a great deal of insight has been obtained, much of it through observation of the development of bacteriophage. For example, the filamentous phage (Fig. 4-8) are assembled from hydrophobic protein subunits and DNA. The protein subunits are synthesized and stored within the cytoplasmic membrane of the bacteria. The small α helices can easily fit within the membrane, and they remain there until a DNA molecule enters the membrane. We do not know how the process is initiated, but it is likely that each subunit contains a nucleotide binding site that interacts with the DNA. Adja-

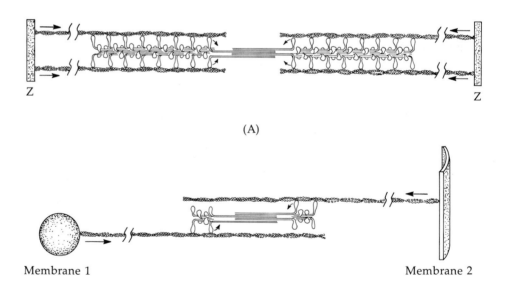

(A)

Membrane 1   Membrane 2

(B)

Fig. 4-25 (A) A schematic drawing of a sarcomere from striated muscle. (B) A representation of the interaction of membrane-associated actin with myosin to produce oriented movement in nonmuscle cells. The drawing shows a membrane-bound vesicle being drawn toward another membrane, e.g., the plasma membrane. From Spudich and Lord.[98] The bipolar nature of the myosin aggregates are an essential feature of the model.

BOX 4-E

THE T-EVEN BACTERIOPHAGE

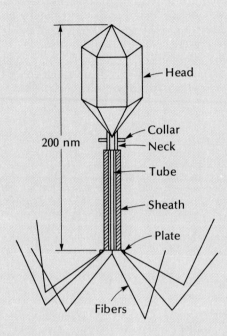

200 nm

Head

Collar

Neck

Tube

Sheath

Plate

Fibers

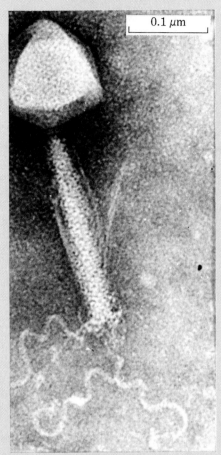

0.1 μm

A tailed phage from *Bacillus subtilis*. Courtesy of A. S. Tikhonenko.

Among the most remarkable objects made visible by the electron microscope are the T-even bacteriophage (T2, T4, and T6) which attack *E. coli*. [a–c] While it is often not evident how a virus gains access to a cell, these "molecular syringes" literally inject their DNA through a hole dissolved in the cell wall of the host bacterium. The virus, of length ~200 nm and mass ~225×10^6 daltons, contains 130×10^6 daltons of DNA in a 100×70 nm head with an elongated icosahedral shape. The head surface appears to be formed from ~840 copies of a protein of MW = 45,000 (specified by gene 23) arranged as 140 hexamers together with ~55 copies of a different protein arranged as 11 pentamers.[d] The head contains at least nine other proteins including three internal, basic proteins that enter the bacterium along with the DNA.

The phage **tail** contains an **internal tube** with a 2.5 nm hole, barely wide enough to accommo-

date the flow of the DNA molecule into the bacterium. An 8×10^6 dalton **sheath** surrounding the tail tube is made up of 144 subunits of MW ~ 55,000, arranged in the form of 24 rings of six subunits each. The sheath has contractile properties and shortens from ~80 to ~30 nm, forcing the inner tube through the wall of the bacterium. At the end of the tail is a baseplate, a hexagonal structure bearing a series of short **pins** as well as six elongated **fibers**, each consisting of four subunits of MW ~ 100,000. A special lysozyme[e] is among the 10 proteins known to make up the baseplate. The virus contains smaller mol-

Box 4-E (*Continued*)

ecules as well. For example, about 30% of the basic groups of the DNA are neutralized by the polyamines **putrescine** and **spermidine** (Chapter 14, Section C,4). The tailplate contains six molecules of the coenzyme 7,8-dihydropteroylhexaglutamate (Chapter 8, Section K).

How is infection by a T-even virus initiated? The tail fibers bind to specific receptor sites on the bacterial surface. This triggers a sequence of conformational changes in the fibers, baseplate, and sheath. The lysozyme is released from the baseplate and etches a hole in the bacterial cell wall. Contraction of the sheath is initiated at the baseplate and continues to the upper end of the sheath. The tail tube is forced into the bacterium and the DNA rapidly flows through the narrow hole into the host cell.

During contraction the subunits of the sheath undergo a remarkable rearrangement into a structure containing 12 larger rings of 12 subunits each.[f] Thus, a kind of mutual "intercalation" of subunits occur. Its unidirectional and irreversible nature indicates that the shortening of the phage tail differs from the contraction of muscle. It appears that the protein subunits may be in an unstable high energy state when the tail sheath of the phage is assembled. The stored energy remains available for later contraction.

[a] W. B. Wood and R. S. Edgar, *Sci. Am.* **217**, 60–72 (Jul 1967).
[b] C. K. Mathews, "Bacteriophage Biochemistry." Van Nostrand-Reinhold, Princeton, New Jersey, 1971.
[c] D. J. Cummings, N. C. Couse, and C. L. Forrest, *Adv. Virus Res.* **16**, 1–41 (1970).
[d] D. Branton and A. Klug, *JMB* **92**, 559–565 (1975).
[e] Not to be confused in its catalytic properties with the better known lysozyme of egg white (Fig. 2-9 and Chapter 7, Section C,4,a).
[f] M. F. Moody, *JMB* **80**, 613–635 (1973).

cent sides of the subunits are hydrophobic and interact with other subunits to spontaneously coat the DNA. As the rod is assembled, the hydrophobic groups become "buried." It is postulated that the remaining groups on the outer surface of the virus are hydrophilic and that the formation of this hydrophilic rod provides a driving force for automatic extrusion of the phage from the membrane.[39] Bacterial pili may be extruded in a similar manner. They arise very rapidly and may possibly be retracted again into the bacterial membrane.

Perhaps the most remarkable known example of self-assembly is that of the T-even phage (Box 4-E).[101–103] From careful genetic analysis (Chapter 15, Section D,2) it is known that at least 18 genes are required for formation of the heads, 21 genes for the tails, and 7 genes for the tail fibers. Most of these genes code for proteins that are actually incorporated into the mature virus, but a few specify enzymes needed in the assembly process. Mutant strains of the virus have been obtained that synthesize all but one of the structural proteins and in which the remaining proteins accu-

mulate within the bacterial host. Remarkably, these proteins have no tendency to aggregate spontaneously. However, when the missing protein (synthesized by bacteria infected with another strain of virus) is added complete virus particles are formed rapidly. This and other observations have led to the conclusion that during assembly each different protein is added to the growing aggregate in a strictly specified sequence. The addition of each protein in some manner creates a binding site for the next protein.

The assembly of the phage tail has been studied in detail. Six copies of each of three different proteins are assembled in one sequence to form a "hub" with hexagonal symmetry (Fig. 4-26). In another assembly sequence, seven different proteins join together to form a wedge-shaped piece. Six of these pieces then assemble around the hub to form the hexagonal baseplate. Only after this, two more proteins add to the surface of the base plate and activate it for the polymerization of the tail tube. Only after the polymerization of the internal tube begins do the sheath units polymerize,

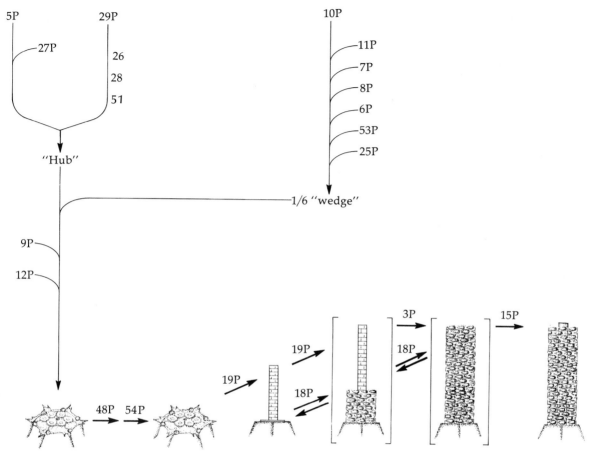

Fig. 4-26 Assembly sequence for the bacteriophage T4 tail. The numbers refer to the genes in the T4 chromosome map (Fig. 15-19). A "P" after the number indicates that the protein gene product is incorporated into the phage tail. Other numbers indicate gene products that are thought to have essential catalytic functions in the assembly process. Adapted from King and Mykolajewycz[101] and Kikachi and King.[102]

and only when these tubular structures have reached the correct length, is a cap protein placed on top. The head is then attached and only then do the tail fibers, which have been assembled separately, join at the opposite end.

It is difficult to imagine how each step in this complex assembly process can set the stage for the next step. Yet the evidence is compelling. It appears that the structure of newly synthesized protein is stable only until a specific interaction with another protein takes place. The binding energy of this interaction is sufficient to trigger a conforma-

tional alteration that affects some other part of the protein surface and generates a complementarity toward a binding site on the next protein that is to be added. That every one of the baseplate proteins should have such self-activating properties is remarkable.

The triggering of a change in one protein by interaction with another protein is not limited to such bizarre objects as the tailed phages. It is a principle that doubtless applies to the construction of microtubules, the myofibrillar assemblies of muscle, and to many more labile but equally real cascade systems of protein–protein interactions such as that involved in the clotting of blood (Fig. 6-16). Membranes also grow by self-assembly and have incorporated in them delicately poised receptors ready to register through suitable and, as yet, largely unknown means responses to chemical signals from the environment. When viewed in this way, it becomes clear that there is no line between the protein–protein interactions that lead to formation of small aggregates of proteins and the much more complex phenomena that are displayed by intact cells in their responses to hormones and other external stimuli.

REFERENCES

1. J. D. Watson, "Molecular Biology of the Gene," 3rd ed., p. 98. Benjamin, New York, 1976.
2. W. P. Jencks, "Catalysis in Chemistry and Enzymology," pp. 393–436. McGraw-Hill, New York, 1969.
3. C. Tanford, "The Hydrophobic Effect: Formation of Micelles and Biological Membranes." Wiley, New York, 1973.
3a. C. Chothia and J. Janin, Nature (London) 256, 705–708 (1975).
4. D. D. F. Shiao and J. M. Sturtevant, Biopolymers 12, 1829–1836 (1973).
4a. H. B. F. Dixon, BJ 137, 443–447 (1974).
5. D. A. Deranleau, JACS 91, 4044–4049 (1969).
6. J. E. Dowd and D. S. Riggs, JBC 240, 863–869 (1965).
7. D. E. Metzler, C. M. Harris, R. J. Johnson, D. B. Siano, and J. A. Thomson, Biochemistry 12, 5377–5392 (1973).
8. A. R. Katritzky and A. Waring, J. Chem. Soc. pp. 1540–1548 (1962).
9. D. L. Vander Jagt, L. D. Hansen, E. A. Lewis, and L. B. Han, ABB 153, 55–61 (1972).
9a. H. B. F. Dixon and K. F. Tipton, BJ 133, 837–842 (1973).
10. I. M. Klotz and D. L. Hunston, JBC 250, 3001–3009 (1975).
11. G. Weber, Adv. Protein Chem. 29, 1–83 (1975).
12. J. Steinhardt and J. A. Reynolds, "Multiple Equilibria in Proteins." Academic Press, New York, 1969.
13. D. E. Koshland, Jr., in "The Enzymes" (P. D. Boyer, ed.), 3rd ed., Vol. 1, pp. 341–396. Academic Press, New York, 1970.
14. C. Tanford, "Physical Chemistry of Macromolecules," pp. 533–534. Wiley, New York, 1961.
15. J. T. Edsall and J. Wyman, "Biophysical Chemistry," Vol. I, pp. 457–476. Academic Press, New York, 1958.
16. F. H. Westheimer and M. W. Shookhoff, JACS 61, 555–560 (1939).
17. C. Tanford, "Physical Chemistry of Macromolecules," Chapter 8. Wiley, New York, 1961.
18. See also H. H. Jaffe, JACS 77, 4445–4448 (1955).
19. D. E. Metzler, in "The Enzymes" (P. D. Boyer, H. Lardy, and K. Myrback, eds.), 2nd ed., Vol. 2, pp. 295–337. Academic Press, New York, 1960.
20. R. F. W. Hopmann and G. P. Brugnoni, Nature (London), New Biol. 246, 157–158 (1973).
21. D. M. Crothers, Biopolymers 10, 2147–2160 (1971).
22. V. N. Damle, Biopolymers 11, 1789–1816 (1972).
23. T. Hill, "Introduction to Statistical Thermodynamics," pp. 235–241. Addison-Wesley, Reading, Massachusetts, 1962.
24. G. Schwarz, EJB 12, 442–453 (1970).
25. V. Bloomfield, D. M. Crothers, and I. Tinoco, "Physical Chemistry of Nucleic Acids," Harper, New York.
26. J. Bjerrum, "Metal Amine Formation in Aqueous Solution." Haase & Son, Copenhagen, 1941.
27. G. L. Eichhorn, ed., "Inorganic Biochemistry," 2 vols. Elsevier, Amsterdam, 1973.
28. A. E. Martell and R. M. Smith, "Critical Stability Constants," 3 vols. Plenum, New York, 1974, 1975.
28a. L. G. Sillén and A. E. Martell, "Stability Constants of Metal-Ion Complexes," Spec. Publ. No. 17. Chemical Society, London, 1964.
29. J. T. Edward, in "Intestinal Absorption of Metal Ions, Trace Elements and Radionuclides" (S. C. Skoryna and D. Waldron-Edward, eds.) pp. 3–19. Pergamon, Oxford, 1971.
30. H. C. Freeman, in "Inorganic Biochemistry" (G. L. Eichhorn, ed.), Vol. 1, pp. 121–166. Elsevier, Amsterdam, 1973.
31. R. J. Angelici, in "Inorganic Biochemistry" (G. L. Eichhorn, ed.), Vol. 1, pp. 63–101. Elsevier, Amsterdam, 1973.
32. R. H. Kretsinger and C. E. Nockolds, JBC 248, 3313–3326 (1973).
32a. R. H. Kretsinger, Annu. Rev. Biochem. 45, 239–266 (1976).
33. J. Monod, J. Wyman, and J.-P. Changeux, J. Mol. Biol. 12, 88–118 (1965).
34. M. A. Lauffer and C. L. Stevens, Adv. Virus Res. 13, 1–63 (1968).
35. K. C. Holmes, G. J. Stubbs, E. Mandelkow, and U. Gallwitz, Nature (London) 254, 192–196 (1975).
36. R. Sperling, L. A. Amos, and A. Klug, JMB 92, 541–558 (1975).
36a. J. N. Champness, A. C. Bloomer, G. Bricogne, P. J. G. Butler, and A. Klug, Nature (London) 259, 20–24 (1976).
37. C. C. Brinton, Jr., in "The Specificity of Cell Surfaces" (B. D. Davis and L. Warren, eds.), p. 37. Prentice-Hall, Englewood Cliffs, New Jersey, 1967.

38. D. A. Marvin and E. J. Wachtel, *Nature (London)* **253,** 19–23 (1975).

39. Y. Nakashima, R. L. Wiseman, W. Konigsberg, and D. A. Marvin, *Nature (London)* **253,** 68–70 (1975).

40. J. Hanson, *Nature (London)* **213,** 353–356 (1967).

41. W. Bode, *Angew. Chem.* **12,** 683–693 (1973).

42. R. C. Valentine, *in* "Symmetry and Function of the Biological Systems at the Macromolecular Level" (A. Engström and B. Strandberg, eds.), 11th Nobel Symp., pp. 165–180. Wiley, New York, 1969.

43. I. M. Klotz, D. W. Darnall, and N. R. Langerman, *in* "The Proteins," 3rd ed. (H. Neurath and R. L. Hill, eds.), Vol. 1, pp. 293–411. Academic Press, New York, 1975.

44. G. M. Edelman, B. A. Cunningham, G. N. Reeke, Jr., J. W. Becker, M. J. Waxdal, and J. L. Wang, *PNAS* **69,** 2580–2584 (1972).

45. K. D. Hardman and C. F. Ainsworth, *Biochemistry* **11,** 4910–4919 (1972).

46. D. Eisenberg, E. G. Heidner, P. Goodkin, M. V. Dastoor, B. H. Weber, F. Wedler, and J. O. Bell, *Cold Spring Harbor Symp. Quant. Biol.* **36,** 291–294 (1971).

47. R. A. Crowther, *Endeavour* **30,** 124–129 (1971).

48. J. D. Watson, "Molecular Biology of the Gene," 3rd ed., p. 107. Benjamin, New York, 1976.

49. J. B. Bancroft, *Adv. Virus Res.* **16,** 99–134 (1970).

50. J. Hohn and B. Hohn, *Adv. Virus Res.* **16,** 43–98 (1970).

51. R. R. Rueckert, *in* "Comparative Virology" (K. Maramorosch and E. Kurstak, eds.), pp. 255–306. Academic Press, New York, 1971.

52. A. S. Tikhonenko, "Ultrastructure of Bacterial Viruses." Plenum, New York, 1970.

53. D. Tsernoglou, E. Hill, and L. J. Banaszak, *Cold Spring Harbor Symp. Quant. Biol.* **36,** 171–178 (1971).

54. T. L. Blundell, J. F. Cutfield, E. J. Dodson, G. G. Dodson, D. C. Hodgkin, and D. A. Mercola, *Cold Spring Harbor Symp. Quant. Biol.* **36,** 233–241 (1971).

55. D. C. Hodgkin, *Nature (London)* **255,** 103 (1975).

56. M. F. Perutz, H. Muirhead, J. M. Cox, and L. C. G. Goaman, *Nature (London)* **219,** 131–139 (1968).

57. R. E. Dickerson and I. Geis, "The Structure and Action of Proteins," p. 56. Harper, New York, 1969.

58. D. C. Wiley, D. R. Evans, S. G. Warren, C. H. McMurray, B. F. P. Edwards, W. A. Franks, and W. N. Lipscomb, *Cold Spring Harbor Symp. Quant. Biol.* **36,** 285–290 (1971).

59. D. R. Evans, S. G. Warren, B. F. P. Edwards, C. H. McMurray, P. H. Bethge, D. C. Wiley, and W. N. Lipscomb, *Science* **179,** 683–685 (1973).

60. See G. G. Hammes and C.-W. Wu, *Science* **172,** 1205–1211 (1971).

61. D. E. Koshland, Jr., *in* "The Enzymes" (P. D. Boyer, ed.), Vol. I, p. 341. Academic Press, New York, 1970.

62. A. J. Cornish-Bowden and D. E. Koshland, Jr., *JBC* **245,** 6241–6250 (1970); **246,** 3092–3102 (1971).

63. D. W. Darnall and I. M. Klotz, *ABB* **149,** 1–14 (1972).

64. Y. Dochi, Y. Sugita, and Y. Yoneyama, *JBC* **248,** 2354–2363 (1973).

65. E. Antonini and M. Brunori, "Hemoglobin and Myoglobin in Their Reactions with Ligands." North-Holland Publ., Amsterdam, 1971.

66. R. Benesch and R. E. Benesch, *Science* **185,** 905–908 (1974).

67. S. J. Edelstein, *Annu. Rev. Biochem.* **44,** 209–232 (1975).

68. J. Herzfield and H. E. Stanley, *JMB* **82,** 231–265 (1974).

68a. J. N. Baldwin, *Prog. Biophys. Mol. Biol.* **29,** 225–320 (1975).

69. M. F. Perutz and L. F. Ten Eyck, *Cold Spring Harbor Symp. Quant. Biol.* **36,** 295–310 (1971).

70. M. F. Perutz, H. Muirhead, J. M. Cox, L. C. G. Goaman, F. S. Mathews, E. L. McGandy, and L. E. Webb, *Nature (London)* **219,** 29–32 (1968).

71. M. F. Perutz, *Nature (London)* **228,** 726–734 (1970).

72. R. E. Dickerson, *Annu. Rev. Biochem.* **41,** 815–841 (1972).

73. J. L. Hoard and W. R. Scheidt, *PNAS* **70,** 3919–3922 (1973).

74. A. Arnone, *Nature (London)* **237,** 146–149 (1972).

75. R. Benesch and R. E. Benesch, *Nature (London)* **221,** 618–622 (1969).

76. G. J. Brewer and J. W. Eaton, *Science* **171,** 1205–1211 (1971).

77. C. Bauer, R. Baumann, U. Engels, and B. Pacyna, *JBC* **250,** 2173–2176 (1975).

78. M. Perrella, D. Bresciani, and L. Rossi-Bernardi, *JBC* **250,** 5413–5418 (1975).

79. L. Waxman, *JBC* **250,** 3790–3795 (1975).

80. E. Zuckerkandl, *Sci. Am.* **212,** 110–118 (May 1965).

81. M. F. Perutz and H. Lehmann, *Nature (London)* **219,** 902–909 (1968).

82. J. S. Olson, Q. H. Gibson, R. L. Nagel, and H. B. Hamilton, *JBC* **247,** 7485–7493 (1972).

83. H. E. Huxley, *Sci. Am.* **199,** 67–82 (Nov 1958).

84. J. M. Murray and A. Weber, *Sci. Am.* **230,** 59–71 (Feb 1974).

85. M. F. Landon and C. Oriol, *BBRC* **62,** 241–245 (1975).

86. A. E. Huxley, *Cold Spring Harbor Symp. Quant. Biol.* **37,** 689–693 (1972). The entire volume is concerned with the mechanism of muscle contraction.

87. J. M. Squire, *JMB* **77,** 291–323 (1973).

88. K. Morimoto and W. F. Harrington, *JMB* **83,** 83–97 (1974).

88a. W. A. Engelhardt and M. N. Ljubimowa, *Nature (London),* **144,** 668–669 (1939).

88b. A. Szent-Györgyi, *Studies Inst. Med. Chem., Univ. Szeged* **1,** 17–26 (1941) [reprinted in H. M. Kalckar, "Biological Phosphorylations," pp. 465–472. Prentice-Hall, Englewood Cliffs, New Jersey, 1969].

88c. A. Szent-Györgyi, "Chemistry of Muscular Contraction," Academic Press, New York, 1947.

88d. F. B. Straub, *Studies Inst. Med. Chem. Univ. Szeged* **2,** 3–15 (1942) [reprinted in H. M. Kalckar, "Biological Phosphorylation," pp. 474–483. Prentice-Hall, Englewood Cliffs, New Jersey, 1969].

89. M. Elzinga and J. H. Collins, *Cold Spring Harbor Symp. Quant. Biol.* **37,** 1–7 (1973).

90. Y. Tonomura and F. Oosawa, *Annu. Rev. Biophys. Bioeng.* **1,** 159–190 (1972).

91. K. R. Porter and C. Franzini-Armstrong, *Sci. Am.* **212**, 73–80 (Mar 1965).
92. G. Hoyle, *Sci. Am.* **222**, 85–93 (Apr 1970).
93. C. Cohen, *Sci. Am.* **233**, 36–45 (Nov 1975).
94. T. Wakabayashi, H. E. Huxley, L. A. Amos, and A. Klug, *JMB* **93**, 477–497 (1975).
95. M. Stewart and A. D. McLachlan, *Nature* (*London*) **257**, 331–333 (1975).
95a. M. Stewart and A. D. McLachlan, *Nature* (*London*) **257**, 331–333 (1975).
96. C. J. Brokaw, *Science* **178**, 455–562 (1972).
97. S. Berl, S. Puszkin, and W. J. Nicklas, *Science* **179**, 441–446 (1973).
98. J. D. Spudich and K. Lord, *JBC* **249**, 6013–6020 (1974).
99. R. Pollack, M. Osborn, and K. Weber, *PNAS* **72**, 994–998 (1975).
100. R. G. Painter, M. Sheetz, and J. J. Singer, *PNAS* **72**, 1359–1363 (1975).
101. J. King and N. Mykolajewycz, *JMB* **75**, 338–358 (1973).
102. Y. Kikuchi and J. King, *JMB* **99**, 645–672 (1975).
103. S. Casjens and T. King, *Annu. Rev. Biochem.* **44**, 555–611 (1975).

STUDY QUESTIONS

1. NOTE: This problem makes use of a dissociation constant rather than a formation constant, as is used in most places in the text. The apparent dissociation constant K_a for the $H_2PO_4^-$ ion at 25°C and 0.5 M total phosphate concentration is 1.380×10^{-7}. (This is based on a National Bureau of Standards buffer consisting of a solution 0.025 M in KH_2PO_4 and 0.025 M in Na_2HPO_4). (See R. G. Bates, "Determination of pH." Wiley, New York, 1964.)

   a. Calculate the negative logarithm of K_a. This is known as pK_a. Plot this number against the square root of ionic strength on Fig. 3-1.

   b. From the expression for the dissociation constant of an acid, HA, derive the logarithmic form

   $$pH - pK_a = \log [A^-]/[HA] = \log_{10} \alpha/(1 - \alpha)$$

   where α is the fraction of the acid in the ionized form.

   c. Suppose that you wanted to prepare a buffer of pH = 7.00 at 25°C from anhydrous KH_2PO_4 (MW = 136.09) and Na_2HPO_4 (MW = 141.98). If you placed 3.40 g of KH_2PO_4 in a 1-l volumetric flask, how much anhydrous Na_2HPO_4 would you have to weigh and add before making to volume to obtain the desired pH? If you wanted to have the correct pH to ± 0.01 unit, how accurately would you have to weigh your salts? NOTE: It is quicker to prepare a buffer of precise pH this way than it is to titrate a portion of buffer acid to the desired pH with sodium hydroxide.

2. The apparent pK_a for 0.1 M formic acid is 3.7 at 25°C.

   a. Concentrated HCl was added to a liter of 0.1 M sodium formate until a pH of 1.9 was attained. Calculate the concentration of formate ion and that of unionized formic acid.

   b. Calculate the hydrogen ion concentration.

   c. How many equivalents of HCl had to be added in part (a) to bring the pH to 1.9?

3. Exactly 0.01 mol portions of glycine were placed in several 100-ml volumetric flasks. The following exact amounts of HCl or NaOH were added to the flasks, the solutions were made up to volume with water, mixed, and the pH measured. Calculate the pK_a values for the carboxyl and amino groups from the following, making as many independent calculations of each pK as the data permit. Note that at low pH values you must correct for the free hydrogen ion concentration (see question 2c).

Flask No.	Mol HCl	Mol NaOH	pH
1	0.010		1.71
2	0.009		1.85
3	0.006		2.25
4	0.002		2.94
5		0.002	9.00
6		0.004	9.37
7		0.005	9.60

4. Using the pK_a values from problem 3, construct the theoretical titration curve showing the equivalents of H^+ or OH^- reacting with 1 mol of glycine as a function of pH. Note that the shape of this curve is independent of the pK_a. Sketch similar curves for glutamic acid (pK_a's equal 2.19, 4.25, and 9.67), histidine (pK_a's equal 1.82, 6.00, and 9.17) and lysine (pK_a's equal 2.18, 8.95 and 10.53).

   Compare your plot for glycine with a plot of 1 N acid or base *added* to 0.01 mol of glycine in 100 ml of water. Also compare your curves with those for glycine published in other textbooks.

5. Aliphatic amines react with formaldehyde according to the following reversible reactions:

$$R—NH_2 + HCHO \rightleftharpoons R—NH—CH_2OH$$

$$R—NH—CH_2OH + HCHO \rightleftharpoons R—N\begin{smallmatrix}CH_2OH\\ \\CH_2OH\end{smallmatrix}$$

$$R—N\begin{smallmatrix}CH_2OH\\ \\CH_2OH\end{smallmatrix} + HCOH \rightleftharpoons$$

$$R—N\begin{smallmatrix}CH_2—O\\ \\CH_2—O\end{smallmatrix}CH_2 + H_2O$$

Indicate qualitatively how the presence of 9% formaldehyde in the solvent will affect the titration curves sketched in question 4.

6. The pK_a values of side chain groups in proteins sometimes depart from those given in Table 2-2. This is often true for groups that are "buried" in the protein or which are very close to other ionic groups. Suggest plausible explanations for the following:
   a. An unusually high pK_a for an aspartate carboxyl (See Chapter 5, Section C,4,a).
   b. An unusually low pK_a for an aspartate carboxyl.
   c. An unusually high pK_a for a phenolic group of tyrosine.

7. Rewrite Equations 4-20 through 4-25 in terms of dissociation constants. These may be labeled K_1, K_2, K_a, etc., as is conventional, but you may prefer to use K_{1d}, K_{2d}, K_{ad}, etc., to avoid confusion.

8. (a) Describe two ways in which the tautomeric ratio R in Eq. 4-21 could be determined or estimated. (b) Calculate for pyridoxine pK_c^* for Eq. 4-21 and compare it with pK_b^* and with the corresponding pK for phenol. Comment on the differences.

9. A molecule has two identical binding sites for a ligand X. The free energy of interaction between ligands bound to the same molecule, ϵ, is defined as the change in free energy of binding of the ligand to the molecule that results from the prior binding of a ligand at the adjacent site. If the saturation fraction is $\bar{y}$, show from the equation for the binding isotherm that the following equation holds when $\bar{y} = \frac{1}{2}$:

$$d\bar{y}/d \ln [X] = \frac{1}{2(1 + e^{\epsilon/2RT})}$$

10. The hydrogen ion binding curve for succinate is shown in Fig. 4-4. From the curve estimate ϵ and the microscopic association constants.

11. A linear chain molecule has a very large number of identical binding sites for a ligand X. The free energy of interaction between ligands bound to adjacent sites is ϵ. Interactions between non-nearest neighbors are considered negligible. If the binding constant for a site adjacent to unoccupied sites is K_r, the binding isotherm is given by

$$\bar{y} = \frac{1}{2} + \frac{K[X]e^{-\epsilon/RT} - 1}{2\{(K[X]e^{-\epsilon/RT} - 1)^2 + 4K[X]\}^{1/2}}$$

[J. Applequist, *J. Chem. Ed.* 54, 417 (1977)]. Show from the equation for the binding isotherm that the following equation holds at $\bar{y} = \frac{1}{2}$:

$$d\bar{y}/d \ln [X] = \frac{1}{4e^{\epsilon/2RT}}$$

12. The binding of adenosine to polyribouridylic acid [poly(U)] has been studied by the method of equilibrium dialysis [Huang and Ts'o, *JMB* **16**, 523 (1966)]. The table below gives the fraction of poly(U) sites occupied, $\bar{y}$ at various molar concentrations of free adenosine [A] at 5°C. Assuming that the nearest-neighbor interaction model is correct, determine the intrinsic association constant for the binding of adenosine to poly U and the free energy of interaction of adjacent bound adenosines. Do the bound molecules attract or repel each other?

(A) × 10³	$\bar{y}$	(A) × 10³	$\bar{y}$
0.51	0	3.07	0.72
2.10	0	4.00	0.92
2.70	0.15	6.50	0.93
2.96	0.36	8.50	0.93
3.01	0.52	10.00	1.00

5

Membranes and Cell Coats

We have considered ways in which protein subunits can be stacked to form closed oligomers and long helices. Another exceedingly important arrangement of cell constituents is that of flat sheets or membranes[1-10] which, from a molecular viewpoint, can be regarded as almost infinite in extent in two dimensions. This chapter will deal with the function, composition, and chemical properties of biological membranes and also of the cell walls of bacteria, fungi, and plants.

Membranes serve many purposes within cells. The most obvious is to divide space into "compartments." Thus, the plasma membrane bounds cells and mitochondrial membranes separate the enzymes and metabolites of mitochondria from those of the cytosol. Membranes are semipermeable and regulate the penetration into cells and organelles of both ionic and nonionic substances. Many of these materials are brought into the cell against a concentration gradient. Thus, osmotic work is done in a process known as **active transport.** In bacteria and in mitochondria, oxidative phosphorylation occurs in the membranes to provide energy for the organism. Within the chloroplasts of green leaves, highly folded membranes containing chlorophyll absorb energy from the sunlight. Within our eyes, thin membranes contain the photoreceptor proteins that function in vision. Electrical impulses are transmitted along the membranes of nerve cells.

Several membranes have been subjects of intensive investigation over a period of many years and have provided much of our knowledge of the subject. (1) **Myelin,** consisting of a few to many membranous layers derived from the plasma membrane of the **Schwann cells** which lie adjacent to many neurons. The Schwann cells literally wrap themselves around the neuronal axons; the cytoplasm is squeezed out leaving little but tightly packed membrane layers which provide an effective "insulation" around the axons. Myelin membranes are the most stable known and also have the highest lipid content (80%). (2) The plasma membrane of the human red blood cell, which can be prepared for study by osmotic rupture of the cells. The remaining **ghosts** contain ~1% of the dry matter of the cell and have probably been studied more than any other membrane. (3) Membranes of bacteria, including *E. coli.* (4) The outer segment of the visual receptor cells known as **rods** (Chapter 13), which consists of a closely packed and regular array of flat discs, each one consisting of a pair of membranes. (5) The double membrane system of beef heart mitochondria. (6) The regularly packed membranes **(grana)** of plant chloroplasts (Chapter 13).

A. THE STRUCTURE OF MEMBRANES

Membranes are made up largely of protein and lipid,[*,10] the ratio (by weight) of protein to lipid varying from 0.25 in myelin to ~3.0 in bacterial membranes. A 1:1 ratio by weight may be regarded as typical. In addition, small amounts of carbohydrates (<5%) are present and traces of RNA (<0.1%) may also be found. The lipid content of membranes is undoubtedly responsible for such properties as high electrical resistance and impermeability to ions and to other polar substances. Because of the lipid content nonpolar materials often pass through membranes readily. For example, anesthetics usually have a high solubility in lipids enabling them to penetrate nerve membranes.

Under the electron microscope the typical membrane appears to be ~7 nm thick, but low angle X-ray diffraction suggests a thickness closer to 11 nm. Thus, membranes are exceedingly thin, the thickness approaching molecular dimensions.

1. The Lipid Bilayer Model of Membrane Structure

In 1926, Gorter and Grendel calculated that the erythrocyte ghost contained just enough lipid to form a layer 3.0–4.0 nm thick around the cell.[†] This kind of information, together with the known propensity of lipids to aggregate in "mi-

celles" in which the hydrocarbon "tails" clustered together and the polar "heads" protruded into the surrounding water,[12a] led J. F. Danielli in the 1930's to propose the well-known **lipid-bilayer** model for membrane structure. The essential features of this model are indicated in Fig. 5-1. Hydrophobic bonding holds the fully extended hydrocarbon chains together while the polar groups of the phospholipid molecules interact with proteins that line both sides of the lipid bilayer.

While many alternative models of membrane structure have been proposed, the lipid bilayer concept continues to provide the starting point for most discussions.[1] Direct and strong support seems to come from electron microscopy. When stained with osmium tetroxide or potassium permanganate, most membranes show a characteristic three-layered structure (Fig. 5-1) consisting of two darkly stained lines ~2–2.5 nm thick with a clear space ~2.5–3.5 nm wide in the center. Both myelin and the retinal rod outer segments show closely spaced pairs of such membranes with a combined width of 18 nm. Phospholipid solutions in the **lamellar** (smectic) "liquid crystalline" phase when stained, embedded, and sectioned for electron microscopy show a similar structure. Nevertheless, many questions must be raised about the interpretation of these results. Why does OsO_4 stain only the outer protein layer when it is known to react with double bonds of hydrocarbon side chains of lipids to form osmate esters which are readily reduced to a diol and osmium? Why do membranes from which most

Osmate ester

of the lipid has been extracted still stain with OsO_4 to give three-layered electron micrographs? Perhaps little can be concluded from the three-layered appearance. What we are learning is that it is extremely difficult to determine even the

---

[*] An exception is provided by membranes of **gas vacuoles** present in some bacteria and blue-green algae. These thin (~2 nm) membranes are apparently entirely protein.[10a]

[†] Gorter and Grendel reached a correct conclusion, but only because their measurements of pressures of surface films contained compensating errors.[11]

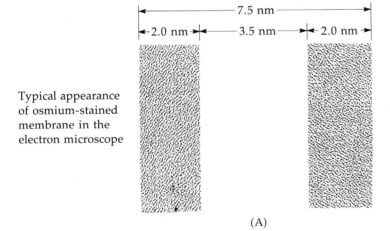

O

‖

O   C

‖ |

O   O   O   CH₃

‖ | |

⁻O—P   C

| ‖

O   O

| H

CH₂   CH₃

|

⁺N

H₃C   CH₃

|

CH₃

|← 2.0 nm →|

Cross-sectional area
of hydrocarbon chain
is 0.2 nm²

Distearoyl phosphatidylcholine

Schematic diagram
of phospholipid bilayer

|← 4.0 nm →|

Dimension for fully
extended C₁₈ chains

Polar
"heads"

|← 7.5 nm →|

|←2.0 nm→|← 3.5 nm →|← 2.0 nm →|

Typical appearance
of osmium-stained
membrane in the
electron microscope

(A)

Fig. 5-1   (A) Bimolecular lipid layers and membranes.

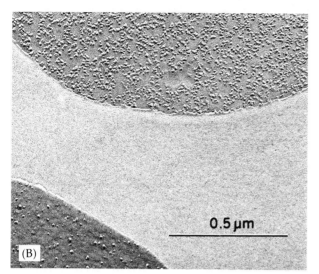

0.5 μm

(B)

Fig. 5-1 (B) Freeze-fractured membranes of two erythrocyte "ghosts." The upper fracture face (PF) shows the interior of the membrane "half" closest to the cytoplasm. The smooth region is lipid and contains numerous particles. The lower face, the extracellular half (EF) possesses fewer particles. The space between the two is non-etched ice. See Figs. 1-2, 1-4, and Box 1-C for electron micrographs of biological membranes. Courtesy of Knute A. Fisher.

thickness, let alone the complete structure of an object that is only 6–10 nm thick.

2. Artificial Membranes

Probably the strongest support for the lipid bilayer model comes from the preparation of very thin artificial membranes, usually made from a solution of phospholipid (e.g., phosphatidylcholine or a mixture of phospholipids plus cholesterol) in a hydrocarbon solvent. A droplet of solution is placed on a small orifice in a plastic sheet, separating two compartments filled with an aqueous medium. The solution in the orifice quickly "drains," just as does a soap bubble, and the resulting film eventually becomes so thin that the bright colors disappear and a "black membrane" is formed.[12] Similar membranes, but without a residual content of hydrocarbon solvent, have been formed by apposition of two lipid mono-

layers formed at an air–water interface.[13] The thickness of such artificial membranes is thought to be only 6–9 nm. Resilient and self-sealing, the membranes can be stained with OsO_4 to give a typical three-layered pattern.

Biological membranes serve as permeability barriers to the passage of polar ions and molecules, a fact that is reflected in their high electrical resistance and capacitance. The electrical resistance of membranes is usually $\sim 10^3$ ohms cm^{-2}, while the capacitance is ~ 0.5–1.5 microfarad (μF) cm^{-2}. The corresponding values for artificial membranes are $\sim 10^7$ ohms cm^{-2} and 0.6–0.9 μF cm^{-2}. The lower resistance of the biological membranes must result from the presence of proteins and other ion-carrying substances or of "pores" in the membranes, but the capacitance values for the two types of membrane are impressively close to those expected according to the bilayer hypothesis.[14–16]

3. Lipids of Membranes

Approximately 1500 different lipids have been identified in myelin of the central nervous system of humans. About 30 are present in substantial amounts.[12] The distribution of the different kinds of lipids varies markedly between membranes from different sources, making it difficult to state generalizations about membrane composition. However, **phospholipids** are apparently always present and make up from 40% to over 90% of the total lipid (Table 5-1).

Five types of phospholipids predominate: phosphatidylcholine, phosphatidylethanolamine, phosphatidylserine, cardiolipin (diphosphatidylglycerol), and sphingomyelin. Small amounts of phosphatidylinositol also occur. Significant amounts of cardiolipin are found only in bacteria and in the *inner* membrane of mitochondria. Sphingomyelin is absent from mitochondria. Ceramide aminoethylphosphonate is a common sphingolipid of invertebrates. Chloroplast membranes contain a special sulfolipid with negatively charged "head" groups (see Chapter 2, Section E for structures).

TABLE 5-1
Estimated Chemical Compositions of Some Membranes

	Percentage of total dry weight of membrane[a]					
Compound	Myelin (bovine)	Retinal rod	Plasma membrane (human erythrocyte)	Mitochondrial membranes	*E. coli*[b,c,d] (inner and outer membranes)	Chloroplasts[e]
Protein	22	59	60	76	75	48
Total lipid	78	41	40	24	25	52[f]
Phosphatidylcholine	7.5	13	6.9	8.8		
Phosphatidylethanolamine	11.7	6.5	6.5	8.4	18	
Phosphatidylserine	7.1	2.5	3.1			
Phosphatidylinositol	0.6	0.4	0.3	0.75		
Phosphatidylglycerol					4	
Cardiolipin[g]		0.4		4.3	3	
Sphingomyelin	6.4	0.5	6.5			
Glycolipid	22.0	9.5	Trace	Trace		23
Cholesterol	17.0	2.0	9.2	0.24		
Total phospholipid	33	27	24	22.5	25	4.7
Phospholipid as a percent of total lipid	42	66	60	94	>90%	9

[a] M. M. Dewey and L. Barr, *Curr. Top. Membr. Transp.* **1,** 6 (1970).
[b] H. R. Kaback, *Curr. Top. Membr. Transp.* **1,** 35–99 (1970).
[c] S. Mizushima and H. Yamada, *BBA* **375,** 44–53 (1975).
[d] I. Yamato, Y. Anraku, and H. Hirosawa, *J. Biochem.* (*Tokyo*) **77,** 705–718 (1975). These investigators found 67% protein, 21% lipids, 10% carbohydrate, and 2% RNA.
[e] H. K. Lichtenthaler and R. B. Park, *Nature* (*London*) **198,** 1070–1072 (1963).
[f] About 14% is accounted for by chlorophyll, carotenoids, and quinones.[e]
[g] Diphosphatidylglycerol (Table 2-8).

Glycolipids are important constituents of the plasma membranes, of myelin (which is derived from plasma membranes), of the endoplasmic reticulum, and of chloroplasts. Of these the cerebrosides are especially abundant in myelin. Myelin also contains sulfate esters of cerebrosides (**sulfatides**). In plant membranes, the predominant lipids (galactosyl diglycerides) are based on glycerol rather than sphingosine.

Cholesterol makes up 17% of myelin and is present in plasma membranes. However, it usually does not occur in bacteria and is present in trace amounts in mitochondria. Esters of sterols, which occur as transport forms, do not exist in membranes. Membranes, likewise, contain no triglycerides, the latter being found as droplets in the cytoplasm.

Quantitatively minor membrane components that are known to have important biological functions include **ubiquinone** (Chapter 10, Section D, present in the inner mitochondrial membrane), the **tocopherols** (Chapter 10, Section D), and **polyprenyl alcohols** (Chapter 12, Section H). Plant chloroplast membranes carry chlorophyll, carotenes, and other lipid-soluble pigments.

4. Physical Properties of Membranes in Relation to Lipid Composition

A completely extended C_{18} fatty acid chain as shown in Fig. 5-1 has a length of ~2.0 nm and occupies when viewed "end-on," an area of ~0.20 nm². The hydrocarbon layer in a lipid bilayer containing such chains would have a thickness of about 4.0 nm. Experimentally, the thickness in myelin is ~3.5 nm, but that in the membranes of the rod outer segment is only 1.8 nm and that in artificial black films can be as little as 3.1 nm when all solvent is removed.[15] These and many other results suggest that the hydrocarbon chains are to some extent folded and that the membrane is expanded over that expected according to the simplest model. The prevalence of unsaturated fatty acids with cis double bonds in membranes encourages folding as does the inclusion of other molecules of irregular shape such as sterols and proteins. Membrane expansion is also suggested from measurements of the area of monolayers of plasma membrane lipids. Cross-sectional areas for phospholipids average 0.52 nm² compared to ~0.40 nm² (2 × the value for a single extended chain) expected for closest packing.[16]

At a low enough temperature, lipid bilayers behave as solids. X-Ray diffraction patterns indicate an 0.42 nm spacing corresponding to close packing of the fatty acid chains in a hexagonal array. As the temperature is raised above a **transition temperature** (T_t), the spacing increases to 0.46 nm. The bilayer continues to hold together, but the fatty acid chains have melted and are now free to rotate and undergo twisting movements more freely than at lower temperatures.[16a] Melting of membrane lipids has been demonstrated strikingly by several techniques. For example, sharp bands in both the proton- and ^{13}C-nmr spectra of membranes above the transition temperature are identifiable with the methylene and methyl hydrogens of the fatty acid side chains. Below T_t the same bands become extremely broad because of the lack of mobility of the hydrocarbon chains (see Box 5-A). Similar conclusions come from study of **spin-labeled** bilayers containing covalently bound stable free radicals whose unpaired electrons can be observed with **electron paramagnetic resonance** (epr, see Box 5-B).

The transition temperature or "melting point" of lipid bilayers depends on the fatty acid composition. Saturated, long-chain fatty acids have high transition temperatures. Myelin is especially rich in long-chain sphingolipids and cholesterol, both of which tend to stabilize artificial bilayers. Consequently, within our bodies, the bilayers of myelin tend to be almost solid.

The transition temperature of a bilayer often appears quite sharp but more careful studies show that, as with impure crystals, melting begins considerably below T_t. Thus, the paramagnetic nitroxide 2,2,6,6-tetramethylpiperidine-1-oxyl is

2,2,6,6-Tetramethylpiperidine-1-oxyl

more soluble in liquid regions of bilayers than it is in solid. As bilayers are warmed in the epr spectrometer, the solubility of this spin-labeled compound in the lipid can be followed (see Box 5-B). For dipalmitoylphosphatidylcholine (a lecithin) T_t is 40.5°C, but the first detectable melting occurred at a **pretransition temperature** of 29.5°C.[17] Similar conclusions have been reached by following the fluorescence of a "polarity-dependent fluorescence probe" such as N-phenylnaphthylamine in 1 μM concentration (see Chapter 13, Section C,1). The compound is incorporated into the membrane and becomes more strongly fluorescent. It is thought that between the pretransition temperature and T_t solid and liquid regions coexist within a bilayer.[17a] The term **lateral phase separation** has been applied to the phenomenon.[17] Since changes in the equilibrium between solid and liquid can be induced readily, e.g., by changes in the ionic environment surrounding the bilayer, it is thought that lateral phase separation may be of fundamental significance in such phenomena as nerve conduction.[18]

It is believed that in all organisms the lipids of most membranes are partially liquid at those temperatures suitable for life processes.[19] At least three distinct means have evolved for keeping membrane lipids liquid. (1) In our bodies (as well

BOX 5-A

RELAXATION TIMES IN MAGNETIC RESONANCE SPECTROSCOPY

Much of the current literature on membranes, enzymes, and conformational analysis of small molecules refers to **longitudinal** and **transverse relaxation times** and **correlation times**. To appreciate the significance of these terms, the student will have to read further,[a–e] but these paragraphs will serve as a brief introduction.

An initial aim in the development of nmr spectroscopy was to obtain a high resolution spectrum in which each magnetic nucleus gave rise to a sharp absorption band or bands. The narrowness of such bands is limited by the **Heisenberg uncertainty principle** which states that $\Delta E \times \Delta t = h/2\pi$, where h is Planck's constant, ΔE is the uncertainty in the energy, and Δt is the lifetime of the magnetically excited state. Since $E = h\nu$ for electromagnetic radiation, ΔE is directly proportional to the bandwidth. The magnetic nucleus is well shielded from external influences and the lifetime of its excited state tends to be long. Hence, $\Delta\nu$ is small, often amounting to less than 0.2 Hz. This fact is very favorable for the success of high resolution proton magnetic resonance (e.g., see Figs. 2-40 and 2-41).

If a very strong pulse of electromagnetic radiation is applied in the nmr spectrometer, virtually all of the nuclei can be placed in the magnetically excited state. If another pulse is applied immediately, little energy will be absorbed because the system is **saturated**. In the most common "continuous wave" nmr spectrometers, the energy is always kept small so that little saturation occurs. However, in pulsed nmr instruments such as those used for Fourier transform methods, the strong pulses lead to a high degree of saturation. Application of repeated pulses would produce no useful information were it not for the fact that the excited nuclei **relax** back to their equilibrium energy distribution at a reasonably rapid rate. Relaxation occurs through interactions of the nuclei with fluctuating magnetic fields in the environment. For organic molecules in solution the fluctuations that are most often effective in bringing about relaxation are the result of moving electrical dipoles in the immediate vicinity. Even so, relaxation times for protons amount to seconds in water.

The relaxation of nuclear magnetic states is characterized by two relaxation times. The longitudinal or **spin–lattice relaxation time** T_1 measures the rate of relaxation of the net magnetic vector of the nuclei in the direction of H_0, the field of the spectrometer magnet. The transverse or **spin–spin relaxation time** T_2 measures the relaxation in a plane perpendicular to the direction of H_0. Under suitable circumstances, the two relaxation times can be measured independently. In general, $T_2 \leq T_1$. For solids, T_2 is quite short, $\sim 10^{-5}$ s, whereas, as previously noted, relaxation times of seconds are observed in solutions. This lengthening of the lifetime of the excited state in going from solid to liquid leads to a narrowing of absorption lines, an important aspect of nmr spectroscopy. This explains why nmr bands in liquids are often narrow but an increase in viscosity or a loss of fluidity in a membrane leads to broadening.

How can T_2 and T_1 be measured? To a first approximation, T_2 for fluids is often estimated from the width of the band $\Delta\nu$ at half-height:

$$T_2 \approx 1/\pi\,\Delta\nu$$

However, special pulsed nmr methods are usually employed. The measurement of T_1 is especially easy using Fourier transform spectrometers and ^{13}C. An attempt is often made to relate T_1 and T_2 to the molecular dynamics of a system. For this purpose a relationship is sought between T_1 or T_2 and the correlation time τ_c of the nuclei under investigation, where τ_c is the time constant for exponential decay of the fluctuations in the medium that are responsible for relaxation of the magnetism of the nuclei. In general, $1/\tau_c$ can be thought of as a rate constant made up of the sum of all the rate constants for various independent processes that lead to relaxation. One of the most important of these is molecular tumbling, for which $1/\tau_r = (3k_BT)/(4\pi\eta r^3)$. Note the close relationship of this equation to that of rotational diffusion (Eq. 6-32).

Box 5-A (*Continued*)

Another term is the reciprocal of the **residence time** τ_m, which represents the mean time that a pair of dipoles are close enough together to lead to relaxation.

In the usual solvents at room temperature, τ_c is of the order of 10^{-12} to 10^{-10} s. Thus, relaxation rates in solutions are considerably faster than the frequencies of radiation absorbed in the nmr spectrometer ($\sim 10^8$ s^{-1}). Under these circumstances, relaxation is relatively ineffective and T_1 and T_2 are large and equal one to the other under most circumstances. Bands remain sharp. As the correlation time increases (as happens, for example, if the viscosity is increased), T_1 and T_2 fall, T_1 reaching a minimum when $\tau_c \approx \nu$, the frequency of the absorbed radiation. Under these circumstances, lines are broadened and hyperfine lines (from coupling between nuclei) cannot be resolved. As τ_c is increased further, T_2 reaches a constant low value, while T_1 rises again. Nmr measurements can be made in the region where τ_c exceeds ν, a circumstance that is favored by the use of high frequency spectrometers. On the other hand, in fluids it is more customary to work in the range of "extreme motional narrowing" at low values of τ_c. Both T_1 and T_2 rise as the mobility of the molecules increases.

A serious limitation on use of nmr measurements in the study of proteins comes from the increased tumbling time of large molecules. Since $1/\tau_r$ is often the most important term in the relaxation rate constant, only small proteins (MW $\sim$ 20,000 or less)[b] give sharp bands.

A practical problem in ^{13}C-nmr arising from slow relaxation (long T_1) is that partial saturation is attained and reduced signal intensities are observed for those carbon atoms for which relaxation is especially ineffective. Relaxation times

can be measured separately for each carbon atom in a molecule. Proper interpretation can yield a wealth of information about the "segmental motion" of groups within a molecule. Although the relationships between relaxation times and molecular motion are complex, the matter is often simplified in the case of ^{13}C-nmr.[f] Since the carbon atoms are usually surrounded by attached hydrogen atoms, it is dipole–dipole interactions with these hydrogen atoms that cause most of the nuclear relaxation. For a carbon atom attached to N equivalent protons in a molecule undergoing very rapid tumbling

$$1/T_1 \approx \frac{N\hbar^2\gamma_c^2\gamma_H^2\tau_{eff}}{r^6}$$

where $\hbar = h/2\pi$ and γ_c and γ_H are the magnetogyric ratios of carbon and hydrogen nuclei. This equation permits a calculation of an effective correlation time τ_{eff} for each carbon atom.

An example of the kinds of interpretation that can be made is provided by Goodman *et al.*[g] Some problems of interpretation of linewidths in membranes are considered by Seiter and Chan.[h]

[a] F. A. Bovey, "Nuclear Magnetic Resonance Spectroscopy." Academic Press, New York, 1969.

[b] T. C. Farrar and E. D. Becker, "Pulse and Fourier Transform NMR." Academic Press, New York, 1971.

[c] J. T. Swift, *Tech. Chem.* **6**, Part II, 521–563 (1974).

[d] J. R. Lyerla, Jr. and D. M. Grant, *in* "Magnetic Resonance" (C. A. McDowell, ed.), MTP Int. Rev. Sci. No. 4, pp. 155–200. Butterworth, London, and Univ. Park Press, Baltimore, 1972.

[e] A. G. Lee, N. J. M. Birdsall, and C. Metcalfe, *in* "Methods in Membrane Biology" (E. D. Korn, ed.), Vol. 2, pp. 1–156. Plenum, New York, 1974.

[f] G. A. Gray, *CRC Crit. Rev. Biochem.* **1**, 247–364 (1973).

[g] R. A. Goodman, E. Oldfield, and A. Allerhand, *JACS* **95**, 7553–7558 (1973).

[h] C. H. A. Seiter and S. I. Chan, *JACS* **95**, 7541–7553 (1973).

as in *E. coli*) *unsaturated fatty acids* are present and lower the melting point. Mutants of *E. coli* that are unable to synthesize unsaturated fatty acids cannot live unless these materials are supplied in the medium.[20] (2) In *Bacillus subtilis*, which contains no unsaturated fatty acids when grown at

37°C, and in other gram-positive bacteria, more than 70% of membrane fatty acids contain *methyl branches* (Chapter 12, Section E).[21] Like cis double bonds, methyl side chains can decrease the melting point and increase the monolayer surface area by a factor of as much as 1.5. (3) Yet another mechanism for lowering the melting point of fats is the incorporation of *cyclopropane-containing fatty acids* (Chapter 12, Section E,1).

On the other hand, the presence of cholesterol tends to reduce the mobility of molecules in membranes and causes phospholipid molecules to occupy a smaller area than they would otherwise.[22]

Why must membrane lipids be mobile? One reason is probably to be found in the participation of membranes in vital transport processes. Biological membranes have a relatively high permeability to neutral molecules (including H_2O) and it has been suggested that above T_t fatty acid chains are free to rotate by 120° around single bonds from trans to gauche conformations. When such rotation occurs about adjacent, or nearly adjacent

---

BOX 5-B

ELECTRON PARAMAGNETIC RESONANCE SPECTRA AND "SPIN LABELS"

Unpaired electrons have magnetic moments and are therefore suitable objects for magnetic resonance spectroscopy. The technique is similar to nmr spectroscopy, but microwave frequencies of $\sim 10^{10}$ Hz are employed, the energies being ~ 100 times greater than those used in nmr.[a–c] Unpaired electrons are found in organic free radicals and in certain transition metal ions, two classes of intermediates that are important in some types of enzymatic processes. Furthermore, **spin labels** in the form of stable organic free radicals, can be attached to macromolecules at many different points. Coupling of such artificially introduced unpaired electrons with the magnetic moments of other unpaired electrons or of magnetic nuclei can often be observed using either epr or nmr techniques.

The conditions for absorption of energy in the epr spectrometer are given by the equation

$$h\nu = g\beta H_0$$

which is identical in form to that for nmr spectroscopy. Here β is a constant called the Bohr magneton. The value of g, the **spectroscopic splitting factor**, is one of the major characteristics needed to describe an epr spectrum. The value of g is exactly 2.000 for a free electron but may be somewhat different in radicals and substantially different in transition metals. One factor that causes g values to vary with environment is "**spin–orbit coupling**" which arises because the p and d orbitals of atoms have directional character. For the same reason g sometimes has three discrete values for the three different directions (g value

anisotropy). In other instances the g value parallel to the direction of H_0 ($g_{||}$) differs from that in the perpendicular direction ($g_\perp$). Both values can be ascertained experimentally.

A second feature of an epr spectrum is **hyperfine structure** which results from coupling of the magnetic moment of the unpaired electron with nuclear spins. The coupling is analogous to the spin–spin coupling of nmr (Chapter 2, Section H). The hyperfine splitting constant A, like the coupling constant J of nmr spectroscopy, is given in Hertz. Splitting may be caused by a magnetic atomic nucleus about which the electron is moving, or by some adjacent nucleus or other unpaired electron. Sometimes important chemical conclusions can be drawn from the presence or absence of splitting. Thus, the epr spectrum of a metal ion in a complex will be split by nuclei in the ligand only if covalent bonding takes place.

It is customary in epr spectroscopy to plot the first derivative of the absorption, rather than the absorption itself. Thus, for a nitroxide spin label the epr spectrum consists of three equally spaced bands whose peaks are marked at the points where the steep lines in the middle of the first derivative plots cross the horizontal axis.

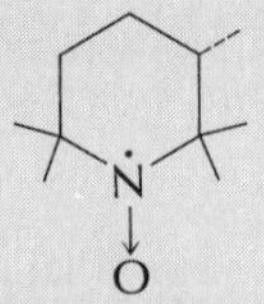

A nitroxide spin label

Coupling with the ^{14}N nuclear spin causes splitting into three lines as shown in the following figure.

Box 5-B (*Continued*)

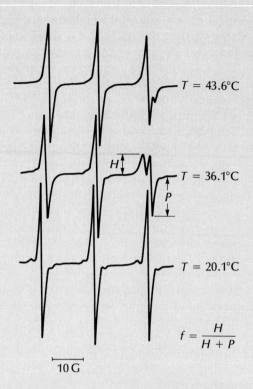

$$f = \frac{H}{H + P}$$

├─ 10 G ─┤

Epr spectrum of tetramethylpiperidine-1-oxyl (Section A,4) dissolved in an aqueous dispersion of phospholipids. Top: above T_t; center between T_t and pretransition temperature; bottom: below pretansion temperature. From Shimshick and McConnell.[d] Reprinted with permission from Biochemistry **12**, 2353. Copyright 1973 by the American Chemical Society.

The lower of the three spectra approximates that of the spin label in water alone, while the others are composite spectra for which part of the spin label has dissolved in the phospholipid bilayers.

Since frequencies for epr spectroscopy are ~100 times higher than those for nmr spectroscopy, correlation times (Box 5-A) must be less than ~10^{-9} s to obtain sharp spectra. While sharp bands may be obtained in solutions, samples are often frozen to eliminate molecular motion and spectra are taken at very low temperatures. In the case of investigation of spin labels in lipid bilayers, the bandwidth and shape is a sensitive function of molecular motion which may be either random or of a restricted type. Computer simulations are often used to match observed band shapes under varying conditions with those predicted for particular theories of motional broadening of lines.[e,f] Among the spin-labeled compounds that have been incorporated into lipid bilayers are the following:

Much of the interpretation of the observed changes in epr spectra of spin labels is strictly empirical. For example, the spectra in the foregoing figure can be interpreted to indicate that the spin label dissolves in the lipid to a greater extent at higher temperatures. The ratio f (defined in the figure) is an empirical quantity whose change can be monitored as a function of temperature. Plots of f vs. T have been used to identify transition and pretransition temperatures in bilayers.[d]

[a] P. F. Knowles, *Essays Biochem.* **8**, 79–106 (1972).

[b] K. S. Chen and N. Hirota, *Tech. Chem.* **6**, Part II, 565–636. (1974).

[c] O. H. Griffith and A. S. Waggoner, *Acc. Chem. Res.* **2**, 17–24 (1969).

[d] E. J. Shimshick and H. M. McConnell, *Biochemistry* **12**, 2351–2360 (1973).

[e] P. Deraux and H. M. McConnell, *JACS* **94**, 4475–4481 (1972).

[f] E. Sackmann and H. Trauble, *JACS* **94**, 4482–4491, 4492–4498, and 4499–4510 (1972).

single bonds, *kinks* are formed. If a kink originates near the bilayer surface (as will usually be the case), a small molecule may jump into the void created. Since the kink can easily migrate through the bilayer, a small molecule may be carried through with it.[23] Perhaps the same factors assist transport of larger molecules which function as carriers in membrane transport.

Not only can molecules diffuse through membranes but also membrane lipids and proteins can move with respect to neighboring molecules. The rate of **lateral diffusion** of lipids in bilayers and of antigenic groups (proteins) on cell surfaces are rapid. If diffusion of phospholipids is assumed to occur by a pairwise exchange of neighboring molecules, the frequency of such exchanges can be estimated[24] as $\sim 10^7 \text{ s}^{-1}$.

Experimental data (nmr and esr) show that the outer portions of bilayers are more solid than the inside. It may be no coincidence that the first double bond in polyunsaturated fatty acids usually comes between C-9 and C-10. Thus, the double bonds in galactosyl diglycerides of the lamellae of chloroplasts appear at a distance from the surface that matches that of the methyl branches of the phytyl side chains (Fig. 13-19) of chlorophyll.[25] A molten center of bilayer may be able to accommodate irregularities introduced by the methyl groups in the phytyl chains which are thought to be embedded in the chloroplast membranes and to anchor the chlorophyll molecules.*

5. The Two Sides of a Membrane

Many observations indicate that there are gross differences between the inside and outside of the membranes that bound cells. For example, Bretscher[3,26] observed that, among the phospholipids of the erythrocyte membrane, phosphatidylcholine predominates in many mammals, but is replaced by sphingomyelin in ruminants. Sheep erythrocytes are resistant to phospholipase A of

cobra venom (Chapter 2, Section H,2), which is known to remove the fatty acid from the central position on the glycerol of phosphatidylcholine, causing lysis of the cells. The resistance of sheep erythrocytes suggests that the sphingomyelin is on the outside of the membrane while the phosphatidylethanolamine and other phospholipids are inside. By inference, phosphatidylcholine may be on the outside of those membranes containing it. Supporting this idea is the observation that most of the reactive amino groups are found on the inside surfaces. Since the total content of phosphatidylcholine and sphingomyelin often exceeds that of phosphatidylethanolamine and phosphatidylserine, the bilayer would be incomplete on the inside of the membrane were it not for the presence of proteins. The latter may reasonably be expected to contribute more to the inside than to the outside surface of a membrane.

Glycolipids are presumably on the outside of the membrane with the attached sugar chains projecting into the surrounding water. A generalization that is often made is that sugar groups attached either to lipids or proteins tend to be on outer cell surfaces or on materials that are being exported from cells.

6. Membrane Proteins

It has been difficult to isolate and study proteins from membranes because of the tendency of these proteins to become denatured. However, new approaches have resulted in rapid progress. Some proteins can be solubilized by treatment with detergents. For example, **rhodopsin,** the light-sensitive pigment and major protein component of the outer segments of retinal rods, can be obtained in a solubilized form which is bleached by light in an apparently normal way (Chapter 13, Section F). A few enzymes from membranes have been purified by fractionation in organic solvents such as methanol. Membrane proteins usually do not dissolve in water, but red cell membranes have been almost completely solubilized in water using a 5×10^{-3} M solution of the chelating agent

---

* According to one theory; however, see Chapter 3, Section E,3 for other possibilities.

EDTA (Table 4-2) or in 0.1 M tetramethyl ammonium bromide.[27] These observations suggest that ionic linkages between proteins (or between proteins and phospholipids) are important in membrane stability.

Gel electrophoresis of plasma membrane proteins in sodium dodecyl sulfate (SDS, Chapter 2, Section H,6) yields 1–6 prominent bands and at least 35 less intense bands ranging in molecular weight from 10,000 to $\sim$360,000.[28] However, some very important proteins known to be present in membrane, such as **(Na$^+$ + K$^+$)-activated ATPase** (Section B,2,c), are present in such low quantities (e.g., a few hundred molecules in a single red blood cell)[3,3a] that they do not show up on the electropherogram. Mitochondrial membranes may be even more complex than plasma membranes. Myelin appears somewhat simpler.

About one-third of the protein of the red blood cell membrane is accounted for by a pair of hydrophobic peptides which have been called **spectrin** (or tektin). With a molecular weight of 220,000–250,000 they, together with a smaller peptide of MW = 43,000, appear to form rods $\sim$3 $\times$ 200 nm in dimension. There are $\sim$2.2 $\times$ 10^5 of these rods per red blood cell, and if laid side by side they would just cover the inner surface (130 μm^2) of the erythrocyte. Since various chemical treatments[28a] that covalently label groups on proteins exposed on the outer surface of erythrocytes (e.g., iodination with lactoperoxidase, Chapter 10, Section B,6) do not label these peptides, it is believed that they lie on the inner surface of the membrane.[3,29] It has been suggested that they resemble myosin, but this relationship is disputed and their function remains unknown.

A second major component of the red blood cell membrane is a glycoprotein of MW $\sim$100,000 that is apparently firmly seated within the membrane but whose carbohydrate chains are thought to project into the surrounding medium.[29] Electron micrographs of freeze-fractured surfaces through the membrane bilayer (Fig. 5-1B) show $\sim$4200 particles of 8 nm diameter per square micrometer, randomly distributed and apparently embedded in the membrane. These may represent the glyco-

protein. The function of the protein is unknown, but it has been suggested that its molecules contain "channels" needed for transport of water[29a] and of anions[29b] (exchange of bicarbonate and chloride across the red cell membranes occurs very actively in association with the functioning of hemoglobin in oxygen transport). That the glycoprotein penetrates the membrane is shown by the fact that trypsin attacks and partially degrades the protein of intact erythrocytes without disruption of the membrane. Furthermore, a series of chemical reactions covalently labels this protein but not most membrane proteins.[3,3a]

Another glycoprotein of erythrocytes, **glycophorin** (also designated PAS-1),[30,31] is of special interest. This protein of MW $\sim$31,000 also appears to be embedded in the membrane but is remarkable in containing $\sim$60% by weight of carbohydrate. It is a single peptide chain of $\sim$131 amino acids of known sequence[31] with an extremely high threonine and serine content in the first 50 residues at the N-terminal end. This end, which is believed to protrude from the cell membrane, carries $\sim$16 oligosaccharide branches bearing a total of $\sim$160 molecules of sugar per peptide chain (Fig. 5-2). Because of a high content of sialic acid (Fig. 2-14), the glycoprotein carries a high negative charge. The glycophorin molecule consists of three "domains." In addition to the carbohydrate-bearing N-terminal portion, there are $\sim$32 residues of a completely hydrophobic nature that are thought to form an α-helical segment that penetrates the bilayer. The third domain consisting of $\sim$35 residues at the C-terminal end is hydrophilic and rich in proline, glutamate, and aspartate. It may extend into the cytoplasm (Fig. 5-2) and could bind calcium ions or interact with —NH$_3^+$ groups on phospholipid heads.

If all the carbohydrate residues of the glycophorin molecules were spread over the surface of the cell they could cover approximately one-fifth of the surface in a loose network. However, it is more likely that they form bushy projections of a more localized sort. That the projecting carbohydrates are not mere decoration is shown by the fact that they carry **blood group antigens** of the MN

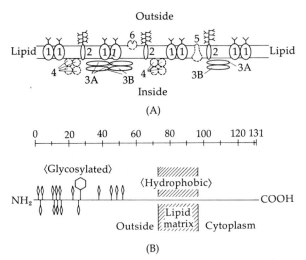

Fig. 5-2 (A) Schematic model of the erythrocyte membrane. (1) The MW = 100,000 glycoprotein, (2) sialoglycoprotein (glycophorin), (3) spectrin, (4) glyceraldehyde-3-phosphate dehydrogenase, (5) (Na⁺ + K⁺)-ATPase, and (6) acetylcholinesterase. There is no correlation intended between the number of copies depicted here and the number of polypeptide chains present in the membrane. The stalklike structures include carbohydrate (Y). From R. J. Juliano, *BBA* **300**, 341–378 (1973). (B) Schematic diagram of the glycophorin molecule. Diamonds, *O*-glycosidic linkage; hexagon, *N*-glycosidic linkage. From Tomita and Marchesi.[31]

type (Section C,1) as well as other immunological determinants. They also serve as receptors for influenza viruses and are sites of attachment of plant agglutinins (Section C,3).

Since the 20 or more glycoproteins of the red blood cell[28b] are embedded in the membrane, they are sometimes referred to as **intrinsic** or **integral proteins.** Those proteins that are more loosely associated with the membrane, principally at the inner surface, may be called **peripheral.**[32]

Myelin contains a simple and yet remarkable assortment of proteins. A very basic **A-1 protein** accounts for 30% of the total protein.[33] It has a molecular weight of ~18,000, and the amino acid sequence of the human protein is known. It is thought that the basic protein may exist in an extended β configuration and that a series of four closely spaced proline residues near the center permit the peptide chain to make a hairpin loop.

This protein has attracted a great deal of interest because it can induce in animals an **autoimmune disease,** experimental **allergic encephalomyelitis,** which mimics such human demyelinating diseases as **multiple sclerosis.** This implies that the protein in the intact myelin is subject to attack by circulating antibodies and suggests that the protein is exposed on the surface of the myelin membrane. However, its function is unknown.

The second major component of myelin is a **proteolipid** exceedingly rich in hydrophobic amino acid residues and containing fatty acids probably bound by ester linkages.[9] Similar proteolipids are of widespread occurrence.[33a] A protein of MW ≈ 12,000 and soluble in a 2:1 chloroform–methanol mixture has been extracted from the ER of muscle cells. Subunits (sex pilins) of the F pili of *E. coli* (Chapter 1, Section A,6) of similar size can remain dissolved in the outer membrane of the bacterial cell wall.[33b]

Many of the proteins of membranes carry out enzymatic functions. Thus, the entire electron transport system of mitochondria is embedded in membranes (Chapter 10) and a number of highly lipid-soluble enzymes have already been isolated. An example is **phosphatidylserine decarboxylase,** which converts phosphatidylserine to phosphatidylethanolamine in the biosynthesis of the latter (Chapter 12, Section F,2). **Glycosyltransferases** required in the biosynthesis of **lipopolysaccharides** of the outer cell coats of bacteria also appear to be embedded in the plasma membrane. These enzymes are inactive when removed from the lipid environment but are active in the presence of phosphatidylethanolamine.[34] A protein of MW = 26,000 present in the "purple membranes" of *Halobacterium halobium* (Chapter 13, Section F,6) is of particular interest. Not only does it appear to constitute a light-dependent proton pump, but also its structure to 0.7 nm resolution has been determined by electron microscopy.[35] Each molecule is folded into seven closely packed α-helical segments which extend roughly perpendicular to the membrane. The protein molecules form an extremely regular array with phospholipid molecules filling the spaces between them.

The preceding paragraphs give the briefest of introductions to what is already known about the content and composition of cell membrane proteins. Investigations of membrane proteins will doubtless provide many interesting and surprising developments in the near future.

B. MEMBRANE METABOLISM AND FUNCTION

Why are membranes so important to cells? The most obvious answer is that membranes enclose and define the limits of living cells. However, we should also note that a membrane represents a natural kind of aggregation of **amphipathic** molecules, i.e., molecules containing both hydrophobic and hydrophilic ends. The packing of such molecules in a bilayer also represents a natural arrangement for a boundary between two different aqueous phases. In addition, membranes are the natural "habitat" for many relatively nonpolar molecules formed by metabolism. Included are many proteins with hydrophobic surfaces. Indeed, some proteins such as **cytochrome b₅** (Chapter 10, Section B,5) appear to contain special hydrophobic appendages that anchor the proteins to membrane surfaces. The semiliquid interior of the membrane permits distortion of the bilayer and the addition or subtraction of proteins and low molecular weight materials in response to metabolic processes in the adjacent cytoplasm.

It has been suggested[36] that the principal factor providing stability to macromolecules and membranes is the hydrophobic nature of reduced organic compounds. This leads to the separation of lipids, proteins, and other molecules from the aqueous cytoplasm into oligomeric aggregates and membranes. However, the best catalysts (e.g., most of the enzymes) are soluble in water. Thus, membranes represent thin regions of relative stability adjacent to aqueous regions in which chemical reactions occur readily and which tend to contain the more polar, the smaller, and the more water-soluble materials.

It is also notable that the stability of a membrane surface provides the means of bringing together reactants and of promoting complex sequences of biochemical reactions. Thus, it is the membrane–cytoplasmic interfaces that often appear to be the metabolically most active regions of cells.

Although relatively stable, membrane components are not inert chemically. They have a metabolism of their own which is related to the high concentrations of oxidizing enzymes located in or on membranes. They also contain dissolved quinones and other low molecular weight catalysts. Oxidative reactions provide an important means of modification of hydrophobic membrane constituents. Thus, the sterols, prostaglandins, and other regulatory molecules are initially synthesized as hydrophobic chains attached to water-soluble carriers (Chapter 12). The hydrophobic synthetic products tend to be deposited in membranes (the prostaglandin precursors are polyunsaturated fatty acids of phospholipids). However, attack by oxygen leads to introduction of hydroxyl groups and to a gradual increase in water solubility. As the hydrophilic nature of the compound is increased through successive hydroxylation reactions, the hydrophobic membrane constituents eventually redissolve in the water and are completely metabolized.

Another process that actively degrades membrane lipids is attack by hydrolytic enzymes such as the phospholipases.

1. The Flow of Membrane Constituents

Observation of cells under the microscope with time lapse photography reveals that the plasma membrane as well as the mitochondria and other organelles are in a constant state of motion. Mitochondria twist and turn and the surface membrane undulates continuously. Vesicles empty their contents to the outside of the cells while materials are taken into cells through endocytosis

(Chapter 1, Section B,4). In addition, chemical evidence indicates a directed flow of materials of which membranes are constructed from the endoplasmic reticulum (ER) to the Golgi vesicles, excretion granules, and plasma membrane. An active aspect of the biosynthetic processes within cells is the addition of carbohydrate (glycosyl) residues to glycoproteins and glycolipids. The glycosyltransferases responsible for this biosynthetic activity (Chapter 12) are found in the ER and in the Golgi vesicles. These enzymes catalyze the addition of carbohydrate units, one at a time, to specific sites on proteins, lipids, and other materials that are being excreted from cells. Other enzymes bring about other modifications, e.g., the addition of sulfate or acetyl groups to carbohydrate units of glycoproteins.

The glycoproteins and the glycolipids of the external surface of the plasma membrane are presumably also constructed within the ER and Golgi vesicles. These membrane materials must then be transported from the interior of the cell to the plasma membrane. All along the route new materials may be inserted from the cytoplasmic side of the membrane, while enzymes within the vesicles add glycosyl units and make other modifications. If this picture is correct, new material is being added to the plasma membrane surface quite rapidly. Obviously there must be a counterbalancing process that *removes* membrane from the surface. The very active endocytosis of fluids and of solid materials from outside the cells presumably accomplishes this "recycling" of membrane components. A rapid, oriented flow of lipid molecules in eukaryotic plasma membranes with uptake through a molecular filter has been postulated.[36a] As we shall see later, many lysosomal enzymes are needed by cells. One of the functions of these enzymes appears to be to degrade excess membrane constituents, including the complex polysaccharides brought back into the cell from its external surfaces.

An important property of membranes is the tendency for any small piece of membrane to fold up and seal itself into a more or less spherical vesicle. Examination of dispersions of phospholipids in water with the electron microscope shows that concentric layered structures (**liposomes**) have been formed. Sonic oscillation breaks these into smaller vesicles surrounded by phospholipid bilayers similar to those in cell membranes. Under appropriate conditions these small vesicles may fuse to form larger ones. Cells can also be induced to fuse giving rise to multinucleate cells, for example, by increasing the membrane fluidity and possibly also by modifying the orientation of polar groups in the phospholipids.[37] The phenomenon is of considerable practical biological significance in plant breeding and in studies of human chromosomes (Chapter 15, pp. 962–963).

2. The Transport of Molecules through Membranes[38,39]

It is of utmost importance that cells be able to take in essential nutrients and to secrete other materials. Small neutral molecules can penetrate membranes by **simple diffusion** as mentioned in Section A,4. The rate of simple diffusion of a substance is determined by its solubility in the membrane, by its diffusion coefficient (Chapter 6, Section A,7) in the membrane, and by the difference in its concentration between the outside and the inside of the cell. This concentration difference is commonly referred to as the **concentration gradient** across the membrane. When electrically charged ions are involved, any electrical potential difference across the membrane, resulting from accumulation of excess negative ions within the cell, will also affect the diffusion process.

While simple diffusion may account for the entrance of water, carbon dioxide, oxygen, and anesthetic molecules into cells, **facilitated diffusion**[40] is a much more common process. Like simple diffusion, facilitated diffusion depends upon a concentration gradient and molecules always flow from the higher concentration to the lower. A distinguishing feature of facilitated diffusion is the presence of a **saturation** effect, i.e., a tendency to reach a maximum rate of flow through the membrane as the concentration of the diffusing substance (on the high concentration side) is increased. Saturation is also observed in enzymatic action (Chapter 6, Section A,2), and its observation in facilitated diffusion suggests a common mechanism in the two processes. It appears that a **mobile carrier**, often a protein, combines with the material to be transported. The carrier then diffuses the short distance to the other side of the membrane and discharges the bound molecule or ion. If the rates of binding to the carrier and of release from the carrier are greater than those of the diffusion process, so-called Michaelis–Menten kinetics are observed (Chapter 6, Section A,5). The "kinetic parameters" V_{max} (maximum velocity) and K_m (Michaelis constant) can be defined as in Eq. 6-15.

The most intriguing transport process of membranes is **active transport** by which a material is carried across a membrane *against* a concentration gradient, i.e., from a lower concentration to a higher concentration. This process necessarily has a positive free energy change as given by Eq. 3-25 of approximately 5.71 log c_2/c_1 kJ mol^{-1} where c_2 and c_1 are the higher and lower concentrations, respectively. Thus, it is necessary that the transport process be coupled in some manner with a spontaneous exergonic reaction. Such coupling can be accomplished in at least two ways. **Primary active transport** depends upon a direct coupling to a reaction such as the hydrolysis of ATP to "pump" the solute across the membrane. **Secondary active transport** utilizes the energy of an electrochemical gradient established for a second solute; that is, a second solute is pumped against a concentration gradient and the first solute is then allowed to cross the membrane through some kind of exchange process with the second solute. Another modification of active transport has been described as **group translocation**.[41] This is a process in which the substance to be transported undergoes covalent modification and the modified product enters the cell.

Transport processes, whether facilitated or active, appear to be quite complex and to require the participation of more than one membrane protein. Sometimes the name **permease** is used to describe the transport system. Since very small amounts of proteins are involved, a principal tool in study of transport systems has been genetic analysis. It is hoped that this approach will allow us to ascertain the number of genes involved in transporting a given substance.

a. Binding Proteins

Recently, considerable progress has been made in the isolation of **binding proteins** that are thought to be parts of permease systems. Most of these molecules can be dissociated from the surfaces of bacteria, for example, by sudden changes in the osmotic pressure of the medium **(osmotic shock)**.[42–44] Thus, cells of *E. coli* suspended in 0.5 *M* sucrose, treated with 10^{-4} *M* EDTA for 10 min and then diluted with cold water release a variety of proteins that bind sugars, amino acids, metal ions, and other substances. One has a molecular weight of $\sim 35,000$ and binds specifically the sugar **galactose.** It is not known exactly where the binding proteins are located in the bacteria. Binding proteins are usually referred to as "periplasmic" (Section D), but they may be bound loosely to the plasma membrane.

While a binding protein could itself be the mobile carrier for facilitated diffusion, most of those that have been isolated appear to be component parts of *active* transport systems and it is not yet clear how they function. One idea is that a binding protein with a high affinity for the substance to be transported binds its substrate tightly at the outer surface of a cell. It then diffuses to the inside of the membrane. There, in a process that is coupled to a spontaneous exergonic reaction such as the hydrolysis of ATP, it is changed to a different conformation with a lower affinity for the substrate. As a result, the compound being transported is released on the inside of the membrane and the carrier diffuses back to the outside. There its conformation again reverts to the high affinity one, probably in response to some chemical alteration.

What classes of molecules and ions are transported across membranes? Inorganic ions needed by cells can all be concentrated, often very highly, from surrounding media (Chapter 2, Section F,2). Thus, green plants extract their essential nutrients from the extremely dilute solutions in contact with their roots. Microorganisms such as yeast and bacteria have the same ability, and specific concentrating systems for many ions such as K^+, Ca^{2+}, sulfate, and phosphate have been identified. A frog skin can take up Na^+ from a $10^{-5} M$ solution of NaCl and extrude it into the internal fluids at greater than $0.1 M$. The lining of the stomach is able to concentrate hydrogen ions in gastric juice to $\sim 0.16 M$.

Amino acids are actively transported by bacteria and by animal cells.[38,39] In *E. coli* there appears to be a specific transport system for almost every individual amino acid. For several amino acids there is more than one system. Often there is a **high affinity system** able to pump a material into the cell from an extremely low concentration, whereas other parallel systems have lower affinity receptors. Systems of transport of amino acids as well as of sugars have been studied in detail in bacteria.[38,45,46] One transport system of bacteria that has been subjected to detailed chemical and genetic study couples the cleavage of **phosphoenolpyruvate** (Table 3-5) to the entrance of a variety of sugars (including the common aldohexoses) into cells. The sugars apparently cross the inner membrane as phosphate esters (group translocation).[46a,46b] Another couples electron transport (Chapter 10) through a membrane-bound redox system to transport of amino acids or of lactose. The system does not appear to depend upon generation of ATP.[46c]

Organelles within cells have their own ion-concentrating mechanisms. Thus, mitochondria can concentrate K^+, Ca^{2+}, Mg^{2+}, and other divalent metal ions, as well as dicarboxylic acids (Chapter 10). It appears that the entrance and exit of many substances from mitochondria occur by **exchange diffusion,** i.e., by secondary active transport.

b. Genetic Aspects of Transport

Innumerable mutations that affect uptake of nutrients by microorganisms have been observed.[38]

It will suffice here to mention only the potassium transport system of *E. coli*.[45,47] One *E. coli* mutant lives normally in $0.1 M$ K^+ but cannot survive at the much lower concentrations that can be tolerated by most bacteria. In *E. coli* strain K12 at least six genes have been identified that are required for three distinct potassium uptake systems. Two of these systems transport potassium into the cell (against a concentration gradient) from relatively high concentrations of K^+ in the surrounding medium. The third system is able to pump K^+ from a very low concentration, the half-saturation value (K_m) being $\sim 10^{-6} M$. Interestingly, if the bacteria are grown at a higher concentration of K^+, this high affinity system is inactive, i.e., the gene is "turned off" (**repressed**). However, if the bacteria are cultured in a medium of very low potassium concentration, the gene is **expressed** and the transport system appears in the membrane.

Not only are numerous transport defects known in bacteria, but also a substantial number of human diseases affecting membrane transport have been identified.[48] Several of these involve faulty reabsorption of materials in kidney tubules and in the small intestine. In **cystinuria**, stones of cystine develop in the kidneys and bladder. Patients may excrete over 1 g of cystine in 24 h compared to a normal of ~ 0.05 g. Excessive amounts of lysine, arginine, and ornithine are also excreted. The existence of such hereditary diseases makes it clear that, like bacteria, human cells can concentrate a variety of amino acids (see also Chapter 14, Section B,3) and other substances. In the case of a kidney tubule cell, the substances are taken up on one side of the cell. (e.g., at the bottom of the cell in Fig. 1-3) and discharged into the bloodstream from the other side of the cell. A second well known but very rare human defect of absorption is **renal glycosuria**. Again, the proximal tubule is involved. This autosomal dominant trait can lead to misdiagnosis as diabetes mellitus. In fact, persons with the defect are usually quite healthy and the condition is not considered a disease.

c. "Pumps" for Sodium, Potassium, and Calcium

Within virtually all cells the sodium concentration is relatively low, while that of potassium is high (Table 5-2, Box 5-C). One theory[49] accounts for this by regarding the cytoplasm as analogous to an ion exchange resin with fixed charges in a lattice. Highly cross-linked ion exchange resins exhibit specificity toward binding of certain ions; e.g., sulfonic acid resins tend to bind potassium preferentially and phosphonic acid resins tend to bind sodium. Support for the theory has been offered in the form of altered nmr relaxation times of protons in cell water and of $^{39}K^+$ and $^{23}Na^+$ within cells.[50] However, experimental measurements are difficult and the evidence has been contested.[51]

In contrast to the ion exchange theory, it is now almost universally accepted that cells have an active **ion pump** that removes Na^+ from cells and introduces K^+. Various experimental approaches have been used to study this process. For example, the cytoplasm of the giant axons of nerves of squid can be squeezed out and replaced by various ionic solutions. Similarly, erythrocyte ghosts can be allowed to reseal with various materials inside. Ion transport into or out of cells has been demonstrated with such preparations and with intact cells of many types. Such transport is found to be blocked by such inhibitors as cyanide ion, which prevents nearly all oxidative metabolism. However, the cyanide block can be relieved by introduction of ATP and other phosphate compounds of high group transfer potential into the cells.

Since, for eukaryotic cells, ATP is the most effective of such compounds, it is suggested as the natural source of energy for ion accumulation. However, as pointed out in Section a, bacteria (and mitochondria) may use energy from other sources.

Uptake of K^+ by cells and extrusion of Na^+ from cells are specifically blocked by certain of the "cardiac glycosides." Among them **ouabain** (Fig. 12-18) is most commonly employed. Ouabain labeled with ^3H binds to the outer surface of cells, and from this binding an estimate of the number of ion pumping sites per cell can be obtained.[52] For erythrocytes there are 100–200 sites/cell (or ~ 1 site/μm^2), while for the HeLa cell (a widely studied strain of human cancer cells that has been grown in culture around the world for many years) 10^5 to 10^6 sites/cell (~ $10^3/\mu m^2$) are found. Further experiments show that in the presence of Na^+ within the cell and K^+ on the outside of the cell, ATP is hydrolyzed. The rate of hydrolysis is directly related to the concentrations of the two alkali metal ions and to the number of ouabain-binding sites and also requires the presence of ions of magnesium.

The foregoing observations have led to the concept of a **(sodium + potassium)-activated ATPase** as synonymous with the membrane-bound ion pump. Within the cell Na^+ must be located on one side of the membrane and K^+ on the other to activate this enzyme system. However, the purified enzyme would be expected to hydrolyze ATP in the test tube in the presence of $Na^+ + K^+ + Mg^{2+}$.

TABLE 5-2
Principal Ionic Constituents (in mmol/kg H_2O) of Human Blood Plasma and of the Intracellular Fluid of Skeletal Muscle[a]

Ion	Blood plasma	Skeletal muscle (intracellular)
Na^+	150	14
K^+	5	150
Mg^{2+}	0.9	8
Ca^{2+}	2.5	1
Cl^-	105	16
HCO_3^-	27	10
Protein$^-$	17[b]	50[b]
Other anions[c]	6	146

[a] Composite of data from E. Muntwyler, "Water and Electrolyte Metabolism and Acid-Base Balance," p. 14. Mosby, St. Louis, Missouri, 1968; A White, P. Handler, and E. L. Smith, "Principles of Biochemistry," 5th ed., p. 802. McGraw-Hill, New York, 1973; C. Long "Biochemist's Handbook," p. 670. Van Nostrand, Princeton, New Jersey, 1961. The reader should be aware that reported ranges for some constituents are very wide.

[b] Milliequivalents/kg H_2O.

[c] Phosphates and other nonprotein anions.

BOX 5-C

THE ALKALI METAL IONS

Although sodium and potassium occur in similar amounts in the crust of the earth, living cells all accumulate potassium ions almost to the exclusion of sodium.[a–c] Indeed, it appears that sodium ions may be required only by certain marine organisms and by multicellular animals that regulate their internal body fluids. Most non-marine plants have no demonstrable need for sodium.

The tendency to accumulate potassium ions is the more remarkable since seawater is $\sim 0.46\ M$ in Na^+ and only $0.01\ M$ in K^+. The other alkali metals occur in even smaller amounts, e.g., $0.026\ mM\ Li^+$, $0.001\ mM\ Rb^+$, and a trace of Cs^+. Soil water is $\sim 0.1\ mM$ in K^+ and $0.65\ mM$ in Na^+. Again, strong discrimination in favor of potassium is observed in uptake by plants.

Intracellular concentrations of K^+ range from 200 mM in *E. coli* and 150 mM in mammalian muscle to ~ 30 mM in freshwater invertebrates such as clams, hydra, and some protozoa. The latter is about the lowest concentration found. While K^+ cannot be replaced by Na^+, a partial replacement by Rb^+ and to a lesser extent by Cs^+ is usually possible. In many microorganisms rubidium can almost completely replace potassium, and even a rat can survive for a *short* time with almost complete substitution of Rb^+ for K^+. Protons replace most K^+ in brown algae.[d] The human nutritional requirement for potassium is quite high, amounting to ~ 2 g/day. It has been suggested that present populations may suffer a chronic deficiency of potassium as a result of food processing and boiling of vegetables.[e]

Sodium is also essential to higher animals, and rats die on a sodium-free diet. Within cells, the sodium content varies among species, but is usually no more than 0.1–0.2 times the K^+ content. In blood, the relationship is reversed, human plasma being $0.15\ M$ in Na^+ and $0.005\ M$ in K^+. Curiously, the taste for salt in the diet appears to be largely an acquired one.[f]

It is not immediately obvious why K^+ is the preferred counterion within tissues, but a fundamental reason may lie in the differences in hydration (Chapter 4, Section C,8,b) between Na^+ and K^+. On the other hand, the relationship of these ions to the excitability of membranes (Section B,3) may be of paramount importance, even in bacteria. The concentration differences in the two ions across membranes represent a readily available source of free energy for a variety of membrane-associated activities.

Many intracellular enzymes require K^+ for activity.[b,c] These include enzymes promoting phosphorylation of carboxyl groups or enolate anions, and elimination reactions yielding enols.

[a] R. P. Kernan, "Cell K." Butterworth, London, 1965.
[b] C. H. Suelter, *in* "Metal Ions in Biological Systems" (H. Sigel, ed.), Vol. 3, pp. 201–251. Dekker, New York, 1974.
[c] C. H. Suelter, *Science* **168**, 789–795 (1970).
[d] H. B. Steinbach, *Comp. Biochem.* **4**, pp. 677–720 (1962).
[e] C. E. Weber, *J. Theor. Biol.* **29**, 327–328 (1970).
[f] H. Kaunitz, *Nature* (*London*) **178**, 1141–1144 (1956).

In fact, the isolation of such a protein has been successful.[53–56] Isolated $(Na^+ + K^+)$-activated ATPase subjected to SDS gel electrophoresis separates into two peptide chains, a large one of MW $\sim 95,000$–$100,000$, and a smaller glycoprotein of MW $\sim 50,000$. Antibodies to the isolated large subunit bind to membrane fragments on what is believed to correspond to the inner surface of the plasma membrane.[57] It is reasonable to suppose that the glycoprotein is on the outer surface.

The sodium-potassium pump displays a curious stoichiometry. *Three sodium ions are pumped from the inside and two potassium ions from the outside of a cell for each molecule of ATP cleaved.* Thus, an excess of positive ions is pumped out with the result that a negative charge develops inside the cell. The presence of such a charge has long been recognized by the electrical membrane potential that it produces (Section B,3). Since the cell membrane is somewhat permeable to K^+, the membrane potential induces a diffusion of K^+ through the somewhat "leaky" membrane with partial neutralization of the negative charge on the membrane. A steady state is reached at which the rate of passive diffusion of ions just balances the membrane potential set up by the active transport.

The energy for transport of Na^+ and K^+ by the ion pump is clearly supplied by ATP.* The function of the ($Na^+ + K^+$)-activated ATPase is not merely to catalyze the hydrolysis of ATP, and the name ATPase is a misnomer. The splitting of ATP must be coupled by suitable "machinery," which remains entirely unknown, to the pumping of ions. A strictly hypothetical model of the way in which this could happen is shown in Fig. 5-3. It is assumed that there are two conformations of the ion pump proteins. In one conformation (A) the protein binds three sodium ions tightly, while in the other conformation (B) it binds two potassium ions. The ATP operates the "motor" that carries out the conformational changes and may also participate directly in the formation of ion-binding sites. In Fig. 5-3 the ion pump, in conformation A, is shown embedded in a membrane with the large peptide on the inside and the glycoprotein facing the outside. In the center (which could be at an interface between two subunits of a dimer as in Fig. 4-9) there is a narrow cavity into which chelating groups (e.g., C=O groups of the peptide chain) protrude. These groups form the three binding sites for the 0.095 nm diameter Na^+ ion. The spontaneous binding of the sodium ions triggers a phosphorylation reaction by which a phosphoryl group from the $MgATP^{2-}$ complex is transferred to a group Y. This phosphorylation in turn triggers a change to conformation B in which the channel to the outside is open and that to the inside is closed. At the same time the affinity for Na^+ is decreased and the sodium ions dissociate on the outside.

The next step is loading with two K^+ ions. The affinity for K^+ in conformation B is high. The return to conformation A with release of K^+ to the inside is triggered by hydrolytic removal of the phosphoryl group as inorganic phosphate (P_i). It may seem unreasonable that a channel could be opened and closed so readily with synchronous changes in the number and specificity of ion-binding sites. However, recall the type of structural alteration occurring upon oxygenation of hemoglobin (Fig. 4-19). Rotation of the subunits with respect to one another causes small changes in the geometrical relationships of groups protruding into the central channel. This affects strongly the binding of 2,3-diphosphoglycerate. Very small movements could open up the Na^+ binding groups and create new binding sites for the larger K^+ ion, using in part the same chelating groups.[57]

The question of how phosphorylation could induce conformational changes is dealt with in Chapter 7, Section F,5. Since transfer between different side chain groups of the protein could be involved, a shift of the phosphoryl group from group Y to a second group X is indicated to accompany the A → B conformational transition in Fig. 5-3.

The chemical nature of the ion-binding sites for Na^+ and K^+ in the ion pump is unknown. However, some ideas can be obtained from study of the peptide antibiotics, many of which bind metal ions and catalyze their diffusion through membranes.[58] An example is the cyclic **depsipeptide** (a peptide containing ester linkages as well as amide linkages) **valinomycin.** The antibiotic is made up of D- and L-valine, L-lactic acid, and D-hydroxyisovaleric acid. This **ionophore,** when

L-Val   D-Val

D-hydroxy-   L-lactic acid
isovaleric acid

Valinomycin

---

* The pumping of sodium and potassium ions is one of the most important energy-requiring activities of cells. It is said to account for half of the ATP utilization in resting muscles and an even higher fraction of that in nerve cells. Thus, it constitutes an important fraction of the basal metabolic activity (Chapter 3, Section A,5).

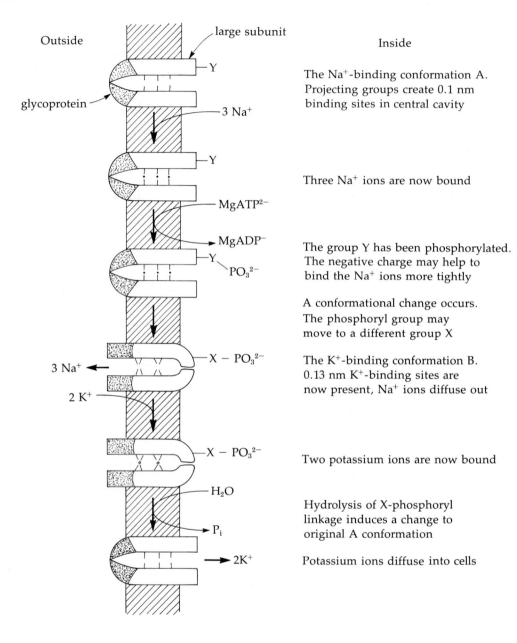

Fig. 5-3 A strictly hypothetical model of a Na$^+$ + K$^+$ pump which operates by ATP-driven opening and closing of a channel at opposite ends and with alternate tight binding of Na$^+$ and K$^+$.

BOX 5-D

ANTIBIOTICS

Many organisms produce chemical substances that are toxic to other organisms. Some plants secrete from roots or leaves compounds that block the growth of other plants. More familiar to us are the medicinal antibiotics produced by fungi and bacteria. Although the growth inhibition of one kind of organism upon another was well known in the last century, modern interest in the phenomenon began in 1929 when Alexander Fleming noticed the inhibition of growth of staphylococci by *Penicillium notatum*. His observation led directly to the isolation of penicillin, first used on a human patient in 1941. A few years later **actinomycin** (Box 15-B) and **streptomycin** (Box 12-A) were isolated from soil actinomycetes (streptomyces) by S. A. Waksman, who coined the name **antibiotic** for these compounds. Streptomycin was effective against tuberculosis, a finding that helped to stimulate an intensive search for additional antibiotic substances. Since that time new antibiotics have been discovered at the rate of more than 50 a year. Over 60 are in commercial production.

Major classes of antibiotics are (1) more than 200 peptides such as the **gramicidins** and **tyrocidins**; (2) **penicillin** and the **cephalosporins** (Box 7-D); (3) the **tetracyclines** (Fig. 12-10); (4) the **macrolides**, large ring lactones such as the **erythromycins** (Fig. 12-10); and (5), the **polyene** antibiotics (Fig. 12-10).

How do antibiotics act? Some, like penicillin, block specific enzymes (Box 7-D). The peptide antibiotics (Section B,2,c) often form complexes with metal ions and apparently disrupt the control of ion permeability in bacterial membranes. The polyene antibiotics interfere with proton and ion transport in fungal membranes. Tetracyclines and many other antibiotics interfere directly with protein synthesis (Chapter 15, Section C,2,h). Others intercalate into DNA molecules (Chapter 2, Section D,9; Box 15-B). Thus, there is no single mode of action. The search for a suitable antibiotic for human use consists in finding a compound highly toxic to infective organisms but with low toxicity to human cells.

Do we still need new antibiotics? Yes. Better antibiotics are needed against gram-negative organisms such as *Salmonella* which sometimes causes severe infections. Present antifungal agents are inadequate, and better antibiotics against protozoa are badly needed. While searches for antibiotics will doubtless continue among the mixed populations of microbes from the soil, swamps, and lakes of the world, Perlman believed it likely that chemical modification will play an increasing role in the design of better antibiotics.[a] Already the semisynthetic penicillins hold an important place in medicine, and chemical modification of other antibiotics such as rifamycin (Box 15-A) has led to a new series of effective drugs.

[a] D. Perlman, *in* "Medicinal Chemistry" (A. Burger, ed.), 3rd ed., Part I, pp. 305–370. Wiley, New York, 1970.

---

incorporated into an artificial membrane bathed in a K^+-containing medium, enormously increases the conductance of the membrane.* The molecule has been studied intensively from a structural viewpoint.[58,60–63a] The uncomplexed molecule has a more extended conformation than it does in the potassium complex. The conformational change results in the breaking of a pair of hydrogen bonds and formation of new hydrogen bonds as the molecule folds around the potassium ion (Fig. 5-4). Valinomycin facilitates potassium transport in a passive manner, but we see that transport involves cyclic changes between two conformations as the carrier complexes with the ion, diffuses across the membrane, and releases the ion on the other side. The rate of transport is remarkably rapid, each valinomycin molecule being able to carry $\sim 10^4$ potassium ions across a membrane per second. Thus, a very small amount of an ionophore is sufficient to alter the perme-

---

* However, there is no proof that these antibiotics have a natural function of this type in the sporulating cells in which they are synthesized. A role for **bacitracin** in uptake of Mn^{2+} has been proposed.[59]

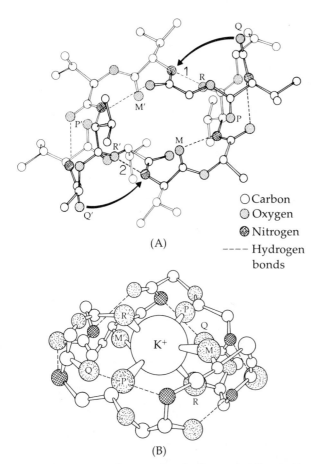

○ Carbon
◎ Oxygen
◉ Nitrogen
---- Hydrogen bonds

(A)

(B)

Fig. 5-4 Structures of uncomplexed valinomycin (A) and of its complex with K^+ (B). It is proposed that the most exposed carbonyl oxygen atoms P, P', M, and M' form the initial complex with K^+. Then the hydrogen bonds marked 1 and 2 break, allowing atoms R and R' to provide the remaining ligands for six-fold coordination of the K^+ ion. The rounding out of the molecule as the ion enters is accomplished by minor conformational changes which bring atoms Q and Q' into position to complete the intramolecular hydrogen bonding of the K^+-complexed ion. From Duax *et al. Science* **176**, 911 (1972). Copyright 1972 by the American Association for the Advancement of Science.

ability and the conductance of a membrane enormously.

A remarkable feature of valinomycin is that the stability constant for potassium is very much greater than that of sodium. Thus, valinomycin acts as a relatively specific potassium transport ionophore. Indeed, if added to a suspension of

mitochondria, it becomes incorporated into the membranes and specifically promotes uptake of potassium ions.* In contrast to valinomycin, the peptide **antamanide** from mushrooms (Box 15-C) has a binding cavity of a different geometry and shows a strong preference for *sodium* over *potassium*. The structure of the Na^+-antamanide complex is shown in Fig. 5-5. Another ionophore that has been mentioned previously is the iron transport agent enterobactin (Fig. 2-44). Note that in this instance, too, binding to a metal ion is accompanied by a large conformational change in the peptide. A possible role for certain phospolipids as ionophores has been suggested.[64a]

Calcium ions are also actively pumped out of most cells by a system that appears to have properties similar to those of the $(Na^+ + K^+)$-ATPase.[65,66] An intestinal calcium-binding protein whose synthesis appears to depend upon vitamin D has been isolated[44,67] (see Box 5-E). The protein is of low molecular weight and may be similar to calcium-binding proteins of muscle (Chapter 4, Section C,8,c).

Transport systems for many other ions probably exist. One ion that does *not* appear to be transported actively is chloride. Membranes are relatively permeable to chloride ion (see Section A,6), the principal anion of plasma (Table 5-2). In most instances chloride ions appear to be distributed passively according to the **Donnan equilibrium** as given by Eq. 5-1.[68,69]

$$[K^+]_i[Cl^-]_i = [K^+]_o[Cl^-]_o \qquad (5-1)$$

Here the subscripts i and o refer to the inside and the outside of the cell, respectively. According to this equation, since the potassium ion concentration within a cell is maintained at a high value by the operation of the $Na^+ + K^+$ pump and by the presence of nondiffusible anions within the cell, the internal chloride concentration must be low. Thus, the product of the two equals that of the low exterior $[K^+]$ and high exterior $[Cl^-]$.

---

* When valinomycin is added to a suspension of cells of *Streptococcus faecalis* the high ratio of $[K^+]_{inside}/[K^+]_{outside}$ rapidly falls.[64] Loss of K^+ (or Rb^+) from cells is presumed to explain the antibiotic activity.

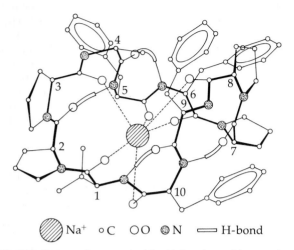

Na⁺ ○ C ○○ O ◉ N ⊂⊃ H-bond

Fig. 5-5 Proposed structure of the Na⁺–antamanide complex. From Ovchinnikov.[60]

3. The Conduction of Nervous Impulses

A major practical consequence of the maintenance of concentration differences of ions between cytoplasm and external environment is to be found in the excitability of membranes. Thus, the concentration differences of the ions provide a "storage battery" of readily available energy that can be used for propagation of electrical signals across the surfaces of cells. While the phenomenon is most highly developed in the axons of nerves, it is not limited to nerve cells and is found even in protozoa such as *Paramecium*[70,70a] and probably in bacteria.

While the chemical basis of nerve conduction is still little known, the electrical events have been described with precision. If a microelectrode is inserted through a cell membrane and the potential difference is measured between the inside and outside of the cell, a **resting potential** which, in nerve cells, is as high as 90 mV, is observed. The origin of the potential appears to lie in the concentration differences of ions. Thus, from the value of ΔG for dilution of an ion (Eq. 3-25) and the relationship between ΔG and electrode potential (Eq. 3-63), it is easy to derive the **Nernst equation** (Eq.

5-2). According to this equation, which applies to a single ion only, a 10-fold concentration difference across the membrane for a monovalent ion ($n = 1$) would lead to a 59 mV potential. Since membranes are relatively impermeable to sodium ions, it is generally conceded that the origin of the membrane potential lies mainly with the potassium ion concentration difference. However, a complete equation takes account of K⁺, Na⁺, and Cl⁻ together with their respective permeabilities.[69,71,72]

$$E = \frac{RT}{nF} \, ln\left(\frac{c_1}{c_2}\right) = \frac{0.059}{n} \log\left(\frac{c_1}{c_2}\right) \qquad \text{at } 25°C \qquad (5\text{-}2)$$

If the permeability of a membrane toward sodium ions is increased in a local region, sodium ions flow through the membrane into the cell neutralizing the negative charge inside and **depolarizing** the membrane. Such depolarization leads to propagation of an electrical signal of diminishing intensity over the surface of the membrane in a manner analogous to the flow of electrical current along a coaxial cable. It is thought that local increases in Na⁺ permeability are often involved in triggering nerve impulses. Other ions such as Ca²⁺ may also play a role. While the kind of passive transmission of electrical signals that results from a local depolarization of the membrane is suitable for very short nerve cells, it cannot be used to send signals long distances. Most nerve axons employ a more efficient means of transmission through development of an **action potential.** The action potential is an impulse that passes along the axon and for a short fraction of a second (~0.5 ms in mammalian nerves) changes the membrane potential in a characteristic way (Fig. 5-6). Initially, the negative potential of 50–70 mV drops rapidly to zero and then becomes positive by as much as 40–50 mV, after which it returns to the resting potential. The remarkable thing about the action potential is that it is propagated down the axons at velocities of 1–100 m/s without loss of intensity.

To establish the chemical basis of the action potential, A. L. Hodgkin and A. F. Huxley in the 1950's devised the **voltage clamp,** a sophisticated

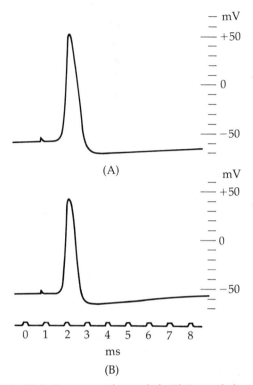

— mV
— +50
— 0
— −50

(A)

mV
+50
0
−50

0 1 2 3 4 5 6 7 8

ms

(B)

Fig. 5-6 (A) Action potential recorded with internal electrode from extruded axon filled with potassium sulfate (16°C). (B) Action potential of an intact axon, with same amplification and time scale (18°C). The voltage scale gives the potential of the internal electrode relative to its potential in the external solution—with no correction for junction potential. From A. Hodgkin, "Conduction of Nervous Impulses," 1964. Courtesy of Charles C Thomas, Publisher, Springfield, Illinois.

device by which the transmembrane current can be measured while using a feedback mechanism to fix the membrane potential at a preselected value.[69,71,73] Using the voltage clamp the membrane conductance could be measured as a function of the membrane potential and of time. It was found that immediately after a decrease in membrane potential was imposed with the voltage clamp the permeability of the membrane toward sodium ions rose rapidly. Since an increased sodium ion permeability automatically leads to depolarization in an adjacent region of the membrane, a self-propagating wave is established and moves down the axon. Unclear is the nature of the chemical changes that alter permeability. The volt-

age clamp studies also revealed that after a fraction of a millisecond the permeability to potassium ions also increases. At the same time, the sodium ion permeability decreases again and the normal membrane potential is soon reestablished.* The sequence of events can be described as the opening of sodium "channels" (not the same as the pores in the Na^+ pump) followed by the opening of potassium channels, and then by a closing of the channels in the same sequence. The results of these investigations led Hodgkin and Huxley to propose equations that quantitatively describe the action potential and that predict the observed conduction velocities and other features of nerve impulses.

A special feature of nerves that are designed to transmit impulses very rapidly is the presence of the wrapping of myelin. The axon is effectively insulated from the surrounding medium by the myelin sheets except for special regions, the **nodes of Ranvier**, which lie at 1–2 mm intervals along the nerve. The nerve impulse in effect jumps from one nerve to the next. This **saltatory conduction** occurs much more rapidly (up to 100 m/s) than conduction in unmyelinated axons.

What is known about the channels through which Na^+ and K^+ flow during nerve excitation? That the channels for the two ions are separate has been shown conclusively by the fact that **tetrodotoxin** of the puffer fish (Fig. 16-7) exerts its toxic action by blocking the Na^+ channels while having no effect upon conductance for K^+. At the same time the K^+ channels can be blocked by various quaternary ammonium salts. Since the binding constant for tetrodotoxin is high ($K_f \sim 3 \times 10^8 \ M^{-1}$), it is possible to titrate the sodium channels. While different investigators have obtained different numbers, the maximum appears to be quite small—of the order of 40–75 Na^+ channels/μm^2 of surface[74] (the same surface area contains 2×10^6 phospholipid molecules). The number of conduction channels for Na^+ appears to be 10 times less than the number of pumping channels, i.e., of ($Na^+ + K^+$)-ATPase.[75] The number of K^+ channels is not known.

* However, during an **absolute refractory period** of ~ 0.5 ms no other nerve impulse can be passed.

Since the number of ion-conducting channels is small, the rate of sodium passage through the open channels must be extremely rapid and has been estimated as $\sim 10^8$ ions/s. It has also been argued on this basis that the channels cannot act by means of ionophoric carriers but must be "pores" that can be opened and closed in response to changes in the membrane potential. It has been suggested that the sodium and potassium ions must be stripped of their hydration spheres before they pass through the channels which are likely to be formed by protein molecules embedded in the membrane.

How do the "gates" to the ion channels open? The rate of increase of conductance with change in membrane potential is such as to indicate cooperativity.[74,76] Little more can be said at present, and the understanding of the chemical basis for ion conductance in excitable membranes remains at the frontier of biochemical knowledge.

C. CELL SURFACE ANTIGENS AND RECEPTORS

The attention of many biochemists is now focused on ways in which cell surfaces interact with other biological objects. For example, membrane surfaces contain groups capable of acting as **antigens.** Antigens are specific chemical structures that can elicit the production of antibodies that will bind specifically to them. Over 250 different antigenic groups ("determinants") have already been described for the surface of the red blood cell. They determine the type of the blood, while similar groups on the surfaces of other cells determine whether a transplanted tissue will be rejected. Various proteins from plant and other sources act as **agglutinins** by binding to surface groups much as do antibodies. Viruses that attack cells become adsorbed at specific **surface receptors** that may be identical to particular antigenic determinants. There is special interest in trying to understand how cells recognize other cells as "foreign." The prevention of tissue rejection and the treatment of serious **autoimmune diseases**

(Chapter 16, Section C,7) are among the reasons for this interest.

1. The Blood Group Substances

A striking effect of genetic variations on the surfaces of red blood cells is evident in the human blood types.[77] This is most clearly seen in the ABO system first described by Landsteiner in 1900. Individuals are classified into four types: A, B, AB, and O. Blood of individuals of the same type can be mixed without clumping of cells, but serum from a type O individual contains antibodies that agglutinate erythrocytes of persons of types A and B.* Serum of persons of type B causes type A cells to clump and vice versa. The phenomenon depends upon the presence on the cell surface of specific antigens consisting of highly branched oligosaccharides containing **L-fucose.** The antigens are found not only on the surface proteins of erythrocytes (e.g., attached to glycophorin) but also attached to both proteins and lipids in other parts of the body. For example, in $\sim 80\%$ of the population they are present in glycoproteins of saliva.

Although many complexities remain, recent work has both established the structures of the antigenic determinants and has elucidated the genetic basis for the ABO blood groups. There are two types of terminal nonreducing ends in each molecule of blood group substance (types 1 and 2). They are distinguished by whether the linkage at the left side of the following oligosaccharide structure is 1–3 or 1–4 (see diagram on p. 278). Both types of chains terminate in α-linked N-acetylgalactosamine unit in A type individuals, but in galactose in B type individuals. In blood group substance H, found in type O individuals, this terminal residue is completely lacking, and the chains are one unit shorter than in the A and B substances. Type AB individuals are heterozygous and contain both A and B substances.

---

* The popular but inaccurate description of type O individuals as "universal donors" arises from the absence of antibodies against the type O blood cells in serum of persons of type A, B, or AB.

BOX 5-E

CALCIUM

The essentiality of calcium ions to living things was recognized in the last century by S. Ringer, who showed that ~1 μM Ca²⁺ was needed to maintain the beat in a perfused frog heart. Later, calcium was shown essential for repair to ruptures in the cell membrane of the protozoan *Stentor* and for the motion of amebas. The animals quickly died in its absence. The role in the frog heart was traced to transmission of the nerve impulse from the nerve to the heart muscle.

Like Na⁺, the calcium ion is actively excluded from cells.[a,b] Indeed, the vast bulk of the calcium in the human body is present in the bones.[c-e] In the human the blood serum concentration of Ca²⁺ is 2.5 μM of which ~1.5 μM is free and the rest is chelated with proteins, carbohydrates, and other materials. Within cells the concentration is much lower. For example, in red blood cells a total [Ca²⁺] of 3 μM has been found for the cyto-

plasm but less than 1 μM is free. A gradient in [Ca²⁺] of 10^2 to 10^5 is maintained across membranes by the calcium ion pump. This is counteracted by a very slow diffusion back into cells.

A characteristic function of Ca²⁺ in living things is **activation** of a variety of metabolic processes. This occurs when a sudden change in permeability of the plasma membrane or in the membranes of the endoplasmic reticulum allows Ca²⁺ to diffuse into the cytoplasm. Thus, during the contraction of muscle, the Ca²⁺ concentration rises from ~0.1 to ~10 μM[f] as a result of release from storage in the endoplasmic reticulum. The calcium ions bind to troponin C initiating contraction (Chapter 4, Section F,1).[g] The ER membranes of muscle are rich in the Ca²⁺ pump protein and in addition, contain a series of calcium-binding proteins (see Chapter 4, Section C,8c).[h] One of the Ca²⁺-binding proteins of rabbit muscle **calsequestrin** has a molecular weight of 46,500 and can bind up to 43 mol of Ca²⁺ per mol of protein.[i,j]

This terminal unit is αGal in individuals of blood type B and is completely lacking in persons of type O

Linkage is to position 3 in type 1 chains and to position 4 in type 2 chains

$$\alpha GalNAc(1\longrightarrow3)\beta Gal(1\longrightarrow3,4)\beta GlcNAc-$$

An A blood group antigen

α-L-Fucose

A fucose residue in α-1,4 linkage is present here in individuals of types Leᵇ and Leᵃ

This fucose is absent in persons of type Leᵃ and no fucose groups are present in type I individuals

The genetic basis for the ABO blood groups is simple. There are three **alleles,** variants of a gene, that appears to be the code for synthesis of a **glycosyltransferase.** In A type individuals, this enzyme transfers N-acetylgalactosamine onto the terminal

positions of the blood group substances. The enzyme specified by the *B* allele transfers galactose. The structural difference between the two enzymes could be a very small one resulting in an altered substrate specificity. The *O* gene apparently produces inactive enzyme. Another gene, the **H gene,** has been identified as that for a **fucosyltransferase** that places α-L-fucose on the galactose unit in the foregoing structure. Persons with an inactive *H* gene either have the rare blood type I or, more often, contain another active gene **Le** which codes for a transferase adding fucose in α-1,4 linkage to the N-acetylglucosamine. Such individuals have blood type Leᵃ, whereas individuals with both active *H* and *Le* genes are of type Leᵇ.

Individuals with active gene *Se* (for secretion) secrete glycoproteins bearing the blood group substance into saliva and other body fluids. Large amounts of these soluble blood group substances have been obtained from ovarian cysts and structures have been determined. A proposed complete structure for the Leᵇ substance from such a glycoprotein follows.[78,79]

Box 5-E *(Continued)*

Another triggering function for calcium is the release of neurotransmitters at nerve endings (synapses). In most instances an inflow of Ca^{2+} precedes the initiation of an impulse in a "post-synaptic" neuron (Chapter 16, Section B,4). It has been proposed that release of Ca^{2+} is one of two basic elements in a common mechanism by which cells react to external stimuli. The other is the release of cyclic AMP (Chapter 6, Section F,5).[f,k,l]

In the human body bone plays a dominant role in calcium metabolism. Bone consists of a dense matrix of the protein collagen together with a crystalline calcium phosphate mineral whose approximate composition is that of **hydroxylapatite**, $Ca_{10}(PO_4)_6(OH)_2$ (see structure at the right). A small amount of Mg^{2+} is present in place of Ca^{2+} and a very small fraction of the OH^- is replaced by fluoride ion which has a bone strengthening effect. Embedded in the bone matrix are spidery cells, the **osteocytes**. Among these are osteoblasts

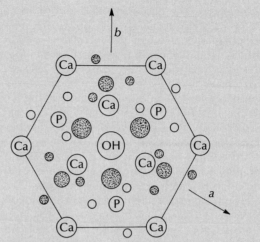

An end-on view of a crystallite of hydroxylapatite. The shaded atoms of Ca, P, and O represent an underlying layer. The OH^- groups form a longitudinal H-bonded array in the center. From J. A. Weatherell and C. Robinson in Zipkin,[e] p. 66.

$$\begin{array}{ccc}
\alpha\text{-L-Fuc} & & \alpha\text{-LFuc} \\
\quad|1 & & \quad|1 \\
\text{(Type 2 chain)} \quad\downarrow 2 & & \downarrow 3 \\
\beta\text{Gal-}(1\rightarrow4)\text{-}\beta\text{GlcNAc} & & \\
& \searrow 6 & \\
\end{array}$$

βGal-$(1\rightarrow3)$-βGlcNAc-$(1\rightarrow4)\beta$Gal-$(1\rightarrow3)$-GlcNAc-βGal-$(1\rightarrow3)$-GalNAc-Ser or Thr

$$\begin{array}{ccc}
& \nearrow 3 & \\
& 1 & \\
\beta\text{Gal-}(1\rightarrow3)\text{-}\beta\text{GlcNAc} & & \beta\text{GlcNAc} \\
\uparrow 2 \qquad \uparrow 4 & & \uparrow 3 \\
1 \qquad 1 & & 1 \\
\alpha\text{-L-Fuc} \quad \alpha\text{-L-Fuc} & & \beta\text{Gal}
\end{array}$$

(Type 1 chain)

Similar substances from hog gastric mucin contain sulfate groups on some carbohydrate units.[80]

While it is thought that the A and B antigens responsible for clumping of mature erythrocytes by specific antibodies are primarily those attached to glycoproteins,[81,82] the antigens are also found attached to **glycosphingolipids** (Table 2-8) of the outer cell membrane.[83-85] A B-active glycolipid is among those accumulating in excess in Fabry's disease (Chapter 12, Section D,1).[84]

$$\alpha\text{Gal-}(1\rightarrow3)\text{-}\beta\text{Gal-}(1\rightarrow3,4)\text{-}\beta\text{GlcNAc-}(1\rightarrow3)\text{-}\beta\text{Gal-}(1\rightarrow4)\text{-Glc-ceramide}$$
$$\uparrow 2$$
$$1$$
$$\text{Fuc} \qquad \text{B-Active glycolipid}$$

Box 5-E (*continued*)

which secrete the fibrous material and promote the laying down of calcium phosphate. The mineral phase of bone is essentially in chemical equilibrium with the calcium and phosphate ions present in the blood serum. Thus, bone cells can easily promote either the deposition or dissolution of the mineral phase by localized change in pH in concentrations of Ca^{2+} or HPO_4^{2-} or of chelating compounds. The large multinucleate **osteoclasts** reabsorb calcium.

Homeostasis of calcium ion is regulated in a complex manner. The **parathyroid hormone** (PTH) and the thyroid hormone **calcitonin** play key roles. If the concentration of Ca^{2+} falls, secretion of the 83 amino acid PTH is increased. This hormone acts directly on the osteoclasts to increase dissolution of bone minerals. PTH also increases the reabsorption of Ca^{2+} in kidney tubules, the overall effect being to elevate the serum level of calcium. Calcitonin, on the other hand, is secreted when the level of Ca^{2+} rises. It acts to lower the level of Ca^{2+} by promoting deposition of calcium by osteoblasts. The two hormones function as a "push–pull feedback system" (Chapter 6, Section F,4). Vitamin D also functions in the regulation of calcium ion concentrations (Box 12-D). The vitamin appears to be required for synthesis of calcium-binding proteins needed for intestinal uptake, kidney reabsorption, and dissolution of bone.[m] An adequate supply of vitamin D is essential for normal calcification.

Aside from the specific calcium-binding proteins, many other proteins bind this metal ion.

Among them are the α-amylases (Chapter 7, Section C,6), thermolysin (Chapter 7, Section D,4), staphylococcal nuclease (Chapter 7, Section E,5), concanavalin A (Fig. 5-7), and several of the blood clotting proteins (Fig. 6-16). The latter are of interest because special binding sites dependent upon vitamin K are present (Box 10-D). Calcium ions also bind to various carbohydrates, e.g., carageenin gels (Chapter 2, Section C,5).

The free calcium ion concentration in tissues is remarkably low, and until recently no reliable estimate of the quantity was available. One technique now in use is to measure light emitted by the calcium-dependent fluorescent protein aequorin (Chapter 13, Section H).

[a] C. P. Bianchi, "Cell Ca++." Appleton, New York, 1968.

[b] A. W. Cuthbert, "Calcium and Cellular Function." St. Martins, New York, 1970.

[c] I. Zipkin, ed., "Biological Mineralization." Wiley, New York, 1973.

[d] G. H. Bourne, ed., "The Biochemistry and Physiology of Bone," 2nd ed., Vol. I. Academic Press, New York, 1972.

[e] T. S. Teo and J. H. Wang, *JBC* **248**, 5950–5955 (1973).

[f] R. S. Mani, W. D. McCubbin, and C. M. Kay, *Biochemistry* **13**, 5003–5007 (1974).

[g] R. P. Rubin, "Calcium and the Secretory Process." Plenum, New York, 1974.

[h] F. L. Siegel, *Struct. Bonding* (Berlin) **17**, 221–268 (1973).

[i] T. J. Ostwald and D. H. MacLennan, *JBC* **249**, 974–979 (1974).

[j] T. J. Ostwald, D. H. MacLennan, and K. J. Dorrington, *JBC* **249**, 5867–5871 (1974).

[k] H. Rasmussen, *Science* **170**, 404–412 (1970).

[l] D. McMahon, *Science* **185**, 1012–1021 (1974).

[m] K. J. Dorrington, A. Hui, T. Hofmann, A. J. W. Hitchman, and J. E. Harrison, *JBC* **249**, 199–204 (1974).

Another series of blood cell antigens determine the MN specificities. Again, the antigens appear to be oligosaccharides attached to serine, threonine, and asparagine residues of glycophorin molecules (Fig. 5-2). Two identified structures are as follows[86]:

αNeurNAc-(2→6)-βGal-(1→3,4)-GlcNAc

Gal-βGlcNAc-(βMan)$_3$-GlcNAc-Asn

αFuc-(1→2,6) αGal——βGlcNAc

αNeurNAc-(2→3)-βGal-(1→3)-GalNAc-Ser (Thr)

$\uparrow$6
$|$2

αNeurNAc

Note that *N*-acetylneuraminic acid (Fig. 2-14) terminates the chains. This negatively charged sugar derivative is a major component of the surface polysaccharides of cells. It is a member of a group of **sialic acids** which include *N*-glycolylneuraminic acid and methylated neuraminic acid derivatives.

2. Histocompatibility Antigens

Transplantation of tissue (e.g., skin grafts) can be carried out successfully between members of a single inbred line of mice. Such **isografts** are not rejected, but grafts between members of two different inbred lines **(allografts)** are rejected within about two weeks. Even tumors transplanted from one strain to another are destroyed. A single group of genes, the **major** histocompatibility complex (MHC or H-2 complex), is the primary determinant of histocompatibility of mouse tissues,[87] but differences in other genes may also lead to a slow rejection of transplanted tissue. Since there are many different MHC genes, transplantation is successful only within inbred lines. With an understanding of the mouse MHC system, it has been possible to turn attention to the corresponding gene complex in humans, the **leukocyte locus A** (HLA). Like the mouse system, it contains many genes and each gene may have many alleles. In fact, it is extremely unlikely that two individuals will have identical histocompatibility genes.

Tissue rejection is an immunological response involving the action of **lymphocytes** (Chapter 2, Section E,2). There are two kinds of lymphocytes: **B cells** and **T cells.** Embryonic bone marrow cells migrate to other parts of the lymphatic system where they proliferate. Those in the thymus gland multiply to form the T cells and those in certain other locations into B cells. One class of T cells (killer T cells) cause tissue rejection by a mechanism that is not yet clear.

What is the nature of the surface antigens responsible for the rejection of cells by T lymphocytes? They appear to be glycoproteins.[30,88−89a] However, it is the protein portion rather than the carbohydrate that reacts with the T cells. The HLA antigens appear to contain two heavy peptide chains (MW $\sim$ 46,000) and two light chains of MW 12,000.[89b] The latter are identical to β_2-**microglobulin,** a protein normally found in small amounts in serum and in urine. The amino acid sequence of β_2-microglobulin is closely homologous with the constant domains of human immunoglobulin IgG (Box 5-F). The obvious sugges-

tion is that a structural similarity exists between the histocompatibility antigens and antibodies. However, sequencing of the mouse H-2 and human HLA antigens has just begun[89c−d] and it is not yet possible to draw firm conclusions about the structures.

3. Protein Agglutinins (Lectins)[90,91]

A series of plant proteins with a remarkable ability to agglutinate erythrocytes have been the subject of intense study during recent years. Not only are these **lectins** of value as sugar specific protein reagents for study of cell surfaces, but some of them, such as **concanavalin A** (which makes up 2–3% of the protein of the jack bean), preferentially agglutinate cancer cells. Concanavalin A and some other lectins are also **mitogenic,** i.e., they stimulate resting lymphocytes to undergo mitosis and to multiply. Lectins appear to have binding sites with a specificity for particular sugars. Thus, concanavalin A binds residues of α-D-mannopyranose and α-D-glucopyranose with unmodified hydroxyl groups at C-3, C-4, and C-6.[92] The protein also has specific binding sites for Ca^{2+} and for a transition metal ion such as Mn^{2+}. Soybean lectin binds D-N-acetylgalactosamine and D-galactose units, while wheat lectin is specific for D-N-acetylglucosamine.

The complete structure of concanavalin A has recently been obtained by X-ray crystallography[93,94] (Fig. 5-7). The protein is a tetramer with a subunit molecular weight of 27,500. It dissociates to a dimer at pH 5.8 or lower. Concanavalin A is unusual in having no α helix. Of the 237 residues in the monomer, 57% exist in three sheets of β structure (Fig. 5-7C). Carbohydrate does not bind until the two metal-binding sites are filled. A conformational change in the protein accompanying the ion binding may be essential to the creation of the sugar-binding site.

What is the role of lectins in plants and other organisms that form them? Are they the plant equivalent of antibodies designed to protect against attack by bacteria and fungi? In view of their mitogenic properties, do they have some function in controlling cell division and germination in plants? At present we do not know. Formation of lectinlike proteins by slime molds[95] and of proteins with agglutinating properties by embryonic chick fibroblasts[96] suggests another function. Lectins may be components of cell recognition systems and may serve to impart cohesiveness to cell surfaces by linking polysaccharide groups from two adjacent cells.

4. Binding of Antibodies to Cell Surfaces

Immunoglobulins (Box 5-F) are best known as circulating antibodies able to agglutinate cells of foreign organisms and to "fix" **complement** (Box 5-G). Another function of antibodies is to cause the "triggering" of activity within specialized cells. Many cell surfaces contain receptors (**Fc receptors**) that bind to the C-terminal ends of immunoglobulin molecules. For example, most molecules of IgE are bound to the blood cells known as basophils (Chapter 1, Section E,2,b) and to the **mast cells** of tissues. The binding of an antigen (**allergen**) with these bound IgE molecules stimulates the release of granules containing histamine and may be the cause of allergic reactions.

Immunoglobulins are also present on the surfaces of B lymphocytes. The binding of specific antigens to these surface antibodies causes the B cells to multiply and to produce large quantities of antibodies in response to infection. The T cells, of which there are several types, are also necessary for formation of IgG and in their absence only IgM is formed. It is thought that some T cells also recognize antigens and then stimulate the B cells into synthesis. Before the lymphocytes start to divide interesting surface phenomena occur. If fluorescent antigens are allowed to bind to a lymphocyte, it can be seen that the cell surface is relatively evenly covered with the antibody–antigen complexes but that after a short time the antibodies aggregate to form "patches" and then begin to migrate to one side of the cell, where they

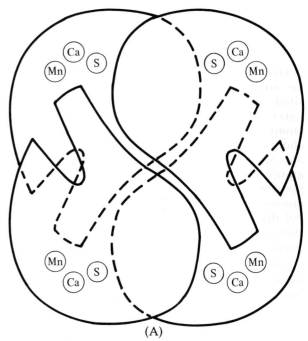

(A)

Fig. 5-7 The structure of concanavalin A, a lectin from jack beans. (A) Schematic representation of the tetrameric structure of the protein.[93] The binding sites for Ca²⁺ and Mn²⁺ and carbohydrate (S) are marked. (G. M. Edelman, J. W. Becker, G. N. Reeke, Jr., and B. A. Cunningham, private communication.) (B) Stereogram of the α-carbon positions of the 237 amino acid residues of the monomer.[94a] The positions of the metal ion binding sites are marked CA and MN. The carbohydrate binding site is marked CHO, a nonpolar binding site is labeled NP, and the N- and C-termini as N and C. (C) Stereogram of the binding sites for Ca²⁺ and Mn²⁺. The two ions are both apparently hexacoordinated. Note that the carboxyl groups of both residues 10 and 19 donate oxygen ligands to both ions. Among other ligands are a peptide carbonyl group and an imidazole side chain. The latter donates N-3 to the Mn²⁺ and is apparently hydrogen bonded to a peptide carbonyl group. The small circles bonded to the Mn and Ca ions are water molecules. The carbohydrate binding position is also marked (CHO). Drawing courtesy of K. D. Hardman.

eventually form a "cap." At still longer times the cap material is engulfed by the lymphocytes.

There is no evidence that "patching" and "capping" has anything to do with initiating antibody synthesis. However, it is being studied intensively because of the possibility that it may shed light on the mobility of the bound immunoglobulins and other receptors in cell membranes. It

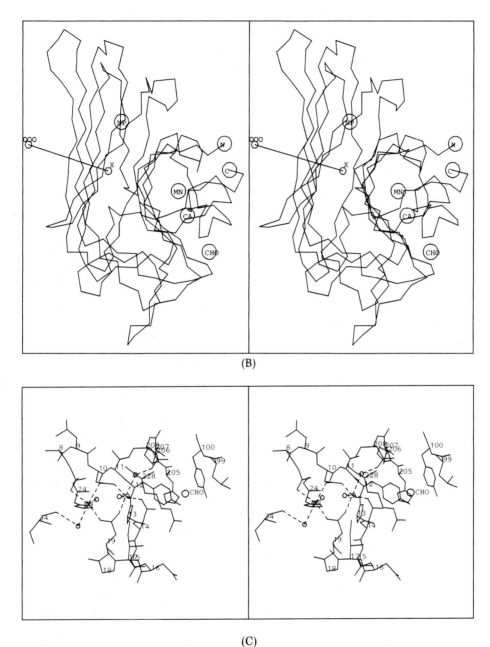

(B)

(C)

has been suggested that receptor molecules, like glycophorin, may pass through the membrane and be attached on the inside to a cytoskeleton of microfilaments and microtubules.[97] A receptor in one state might be free to diffuse laterally and form patches, a process that apparently requires no metabolic energy. Alternatively it could be attached to the cytoskeleton. Movements in the latter would be essential for the energy-dependent capping phenomenon. The binding of lectins can sometimes trigger lymphocytes into antibody synthesis. Since the structure of concanavalin A and the nature of its binding to carbohydrate groups are quite well understood (Section C,3), it is hoped that study of the interaction of lectins with cell surfaces will shed some light on the complex processes of cellular responses to antigens.[98,99]

An alternative explanation of capping is based on the postulated oriented flow of liquid within membranes (Section B,1).[36a]

5. Receptors for Hormones

Many hormones, e.g., **insulin, glucagon, ACTH** (adrenocorticotropic hormone), and **epinephrine**

BOX 5-F

THE ANTIBODIES

Among the most interesting of the soluble serum proteins are the **immunoglobulins** (γ-globulins) which serve as antibodies to combat invasion of the body by foreign materials.[a,b] From the molecular weights and chemical properties five classes of immunoglobulins are recognized. The first three, IgG, IgM, and IgA, are quantitatively the most significant, but IgD and IgE are also important. For example, the content of IgE is elevated in allergic persons.

The basic structure of all of the immunoglobulins is that of a quasi-symmetrical dimer composed of a pair of light chains and a pair of heavy chains whose lengths vary among the different classes of immunoglobulins. Two classes of **light chains** κ and λ are found in human antibodies. The **heavy chains** are designated γ, μ, α, δ, and ϵ (see accompanying tabulation). Both IgM and IgA contain an additional J chain.[c]

Symbol	MW	Formula
IgG	150,000	$\kappa_2\gamma_2$ or $\lambda_2\gamma_2$
IgM	950,000	$(\kappa_2\mu_2)_5$ or $(\lambda_2\mu_2)_5$
IgA	300,000 or more	$(\kappa_2\alpha_2)_n$ or $(\lambda_2\alpha_2)_n$
IgD	160,000	$\kappa_2\delta_2$ or $\lambda_2\delta_2$
IgE	190,000	$\kappa_2\epsilon_2$ or $\lambda_2\epsilon_2$

Treatment with mercaptoethanol splits the disulfide linkages holding the chains together permitting preparation of monomeric light and heavy chains. When the peptide chains of the immunoglobulins were enzymatically hydrolyzed it was found that the resulting peptide fragments were extremely heterogeneous. They appeared to be mixtures of many different kinds of pep-

tides. The result was not unexpected, for it had long been recognized that the body contains literally thousands of different antibodies, each with a specific binding site for a different antigenic determinant. While it had been unclear previously how different binding sites could be formed, the marked heterogeneity in amino acid sequence suggested the correct answer: Each antibody has its own sequence.

Progress toward understanding of the detailed structure of antibodies came when it was recognized that patients with tumors of the lymphatic system (e.g., the bone marrow tumors **multiple myeloma**) produced tremendous quantities of *homogeneous* immunoglobulins or parts thereof. Similar tumors were soon discovered in mice, providing a ready source of experimental material. The **Bence-Jones proteins** that are secreted in the urine of myeloma patients turned out to be light chains of immunoglobulins. Sequence determinations showed each Bence-Jones protein was homogeneous, even though no two patients secreted the same protein. Later, intact myeloma globulins and macroglobulins (IgM) of a homogeneous kind were also obtained.

The first complete amino acid sequence of an IgG molecule was announced in 1969.[d] The protein contained 446 amino acids in each heavy chain and 214 in each light chain, for a total of 1320. The longer heavy chains of IgM molecules contain 576 amino acids.[e] In all of the immunoglobulins the heavy and light chains are linked through disulfide linkages and the chains are constrained to fold into loops by the presence of intrachain disulfide linkages. The IgM molecule is polymerized through additional disulfide linkages to form a pentamer readily visible with the electron microscope.

(adrenaline) appear not to enter cells but to bind to specific protein receptors on the surfaces of cell membranes.[99a] For example, it has been estimated that a **fat cell** (of adipose tissue) of ~50 μM diameter contains 160,000 insulin receptor sites or 21 receptors/μm^2 of cell surface.[100,100a]

How can the combining of a hormone with a receptor on the outside of a membrane affect chemical events within the cell? It is likely that the receptor extends through the membrane. In some cases it may be in contact with an enzyme attached to the inner surface of the membrane. A

Box 5-F (*Continued*)

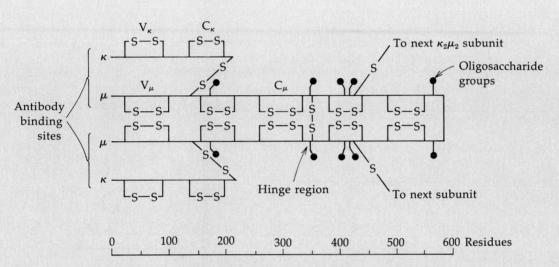

Symmetrical structure of one-fifth of an IgM macroglobulin

The heavy chains also carry oligosaccharide units. In IgM there are five of these, as indicated in the accompanying drawing. They contain mannose and *N*-acetylglucosamine units linked to asparagine. Other immunoglobulins (IgA, IgE, and IgG) contain fucose, galactose, and *N*-acetylneuraminic acid as well.[f]

Digestion of an intact molecule of IgG with papain cleaves both heavy chains in the so-called "hinge" region near the interchain disulfide bridge. This splits the molecule into three parts, two "Fab" antibody-binding fragments, each containing the *N*-terminal end of a heavy chain together with a linked light chain, and an "Fc" fragment. Even before it was known that IgG could be split to two Fab fragments, the antibody was known to be divalent, i.e., capable of binding with two different antigens (drawing at right).

The shape and overall structure of IgG mole-

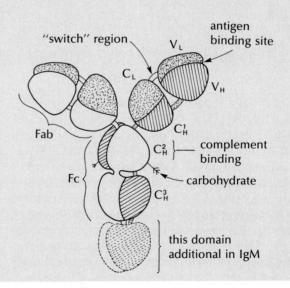

Box 5-F (*Continued*)

cules has been verified by electron microscopy and X-ray diffraction studies.[g]

Sequence determinations on immunoglobulins led to a surprise. In some regions of the molecules there is extreme variation in the amino acid sequence between one homogeneous antibody and the next, but other regions have a constant sequence. The molecule can be divided into domains. The **variable regions**, which occupy the N-terminal ends of the chains, are designated V_L and V_H for the light and heavy chains respectively. The constant regions are C_L and C_H. Examination of the C_H region showed that much of the sequence is repeated after ~ 110 residues. In the IgG molecule the constant region of the heavy chains is made up of three such domains ($C_H{}^1$,

$C_H{}^2$, and $C_H{}^3$) with a great deal of sequence homology. A fourth C_H domain is present in IgM. These facts suggest that duplication of a smaller gene coding for about 110 amino acids took place in the evolution of the immunoglobulins.

Within the variable regions of immunoglobulin chains are **hypervariable regions** that are thought to form the antigen binding sites. These regions are located at the ends of the Fab fragments and involve both the light and heavy chains.

Crystallographic studies show that within all of the domains each of the two peptide chains is folded in a similar way. Seven extended lengths of chain form two mostly antiparallel β sheets between which hydrophobic side chains are packed. The overall size of the unit is $\sim 4.0 \times$

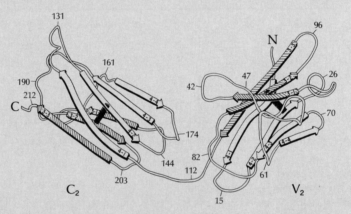

Folding patterns of one chain in a constant and a variable domain of a Bence-Jones protein.[j]

conformational change in the receptor protein induced by binding of the hormone may be transmitted directly to the enzyme causing it to become active. An example is provided by the enzyme that generates **cyclic AMP** (Chapter 6, Section F,5) in response to the binding of a variety of hormones to surface receptors. Cyclic AMP spreads by diffusion through the cell and affects many aspects of metabolism. In other cases, hormones (e.g., the steroid hormones) are known to penetrate the membrane and to become attached to receptors within the cytoplasm. Even in these cases

it is possible that there are special surface sites to which the free hormone or a hormone–protein complex binds before being taken into the cell.

Cell membranes may also harbor enzymes that destroy hormones. Thus, the polypeptide hormone glucagon (formed by special cells in the pancreas) enters the bloodstream and is destroyed largely by the liver cells that are the target cells which the hormone influences. The hormone is present in blood in an extraordinarily low concentration (10^{-10} to 10^{-9} M) and in the human being has a half-time within the body of only 10 min.[101]

Box 5-F (*Continued*)

2.5 × 2.5 nm. An S—S bridge links the two sheets in the center of each domain. The folding patterns in the variable domains are somewhat more complex. Different domains are linked by segments of extended peptide chain in what is called the **switch region.**[h–n]

Recently, it has been possible to establish the exact mode of binding to Fab fragments of specific **haptens.** Haptens are small molecules having the binding properties of antigenic determinants but unable by themselves to induce formation of antibodies when injected into animals. Binding of the hapten phosphorylcholine to one Fab fragment[h] and of vitamin K to another[i] has been shown to involve the hypervariable regions of both the heavy and light chains.

How do immunoglobulins function? One effect, **agglutination**, results when a multivalent antibody combines with two different cells. However, antibody–antigen interactions more often trigger other responses. One of these is the binding of protein **C1q**, a component of complement (Box 5-G). It has been established that it is the C_H2 domain of the Fc region of IgG that binds to C1q.[o] The binding occurs only after antigen (but not hapten) binds to the immunoglobulin. Furthermore, it appears than an aggregate of two or more molecules of IgG (or a single molecule of the pentameric IgM) is required to activate complement. It has been suggested that a change of conformation within the immunoglobulin molecule may accompany antibody binding and be responsible for generation of a binding site for C1q. However, it is strange that haptens cannot cause complement binding (they also cause no detectable conformational alterations in Fab) and that only multivalent antigens (able to bind to more than one antibody) can cause complement binding. Thus, there are some important mysteries remaining about the function of these amazing molecules.

[a] G. M. Edelman, *Science* **180**, 830–840 (1973).

[b] R. R. Porter, *Science* **180**, 713–716 (1973).

[c] S. P. Hauptman and T. B. Tomasi, *JBC* **250**, 3891–3896 (1975).

[d] G. E. Edelman, B. A. Cunningham, W. E. Gall, P. D. Gottlieb, U. Rutishauser, and M. J. Waxdal, *PNAS* **63**, 78–85 (1969).

[e] F. W. Putnam, G. Florent, C. Paul, T. Shinoda, and A. Shimizu, *Science* **182**, 287–291 (1973).

[f] J. Baenziger, S. Kornfeld, and S. Kochwa, *JBC* **249**, 1889–1896 (1974).

[g] V. R. Sarma, E. W. Silverton, D. R. Davies, and W. D. Terry, *JBC* **246**, 3753–3759 (1971).

[h] D. M. Segal, E. A. Padlan, G. H. Cohen, S. Rudikoff, M. Potter, and D. R. Davies, *PNAS* **71**, 4298–4302 (1974).

[i] L. M. Amzel, R. J. Poljak, F. Saul, J. M. Varga, and F. F. Richards, *PNAS* **71**, 1427–1430 (1974).

[j] M. Schiffer, R. L. Girling, K. R. Ely, and A. B. Edmundson, *Biochemistry* **12**, 4620–4631 (1973).

[k] O. Epp, P. Colman, H. Fehlhammer, W. Bode, M. Schiffer, R. Huber, and W. Palm, *EJB* **45**, 513–524 (1974).

[l] C. C. F. Blake, *Nature* (*London*) **253**, 158 (1975).

[m] A. Nisonoff, J. E. Hopper, and S. B. Spring, "The Antibody Molecule." Academic Press, New York, 1975.

[n] D. R. Davies, E. A. Padlan, and D. M. Segal, *Annu. Rev. Biochem.* **44**, 639–667 (1975).

[o] T. L. K. Low, Y.-S. V. Liu, and F. W. Putnam, *Science* **191**, 390–392 (1976).

D. STRUCTURE OF BACTERIAL CELL WALLS

The plasma membrane of bacterial cells (except for mycoplasmas) is surrounded by a multilayered wall which may be separated from the membrane by a thin **periplasmic space.** The thickness of this space depends upon the osmotic pressure of the medium and normally is very small. The innermost layer of the wall (Fig. 5-8) is a highly crosslinked material known as **peptidoglycan** or **murein.** The backbone of the peptidoglycan is a β-1,4-

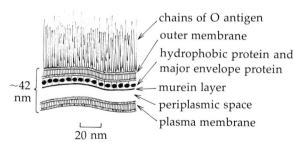

chains of O antigen
outer membrane
hydrophobic protein and
major envelope protein
murein layer
periplasmic space
plasma membrane

~42 nm

20 nm

Fig. 5-8 Structure of cell walls of bacteria. Schematic drawing showing plasma membrane and wall of a gram-negative bacterium. See C. A. Schnaitman, *J. Bacteriol.* **108**, 553–563 (1971). See also Figs. 1-2C and D.

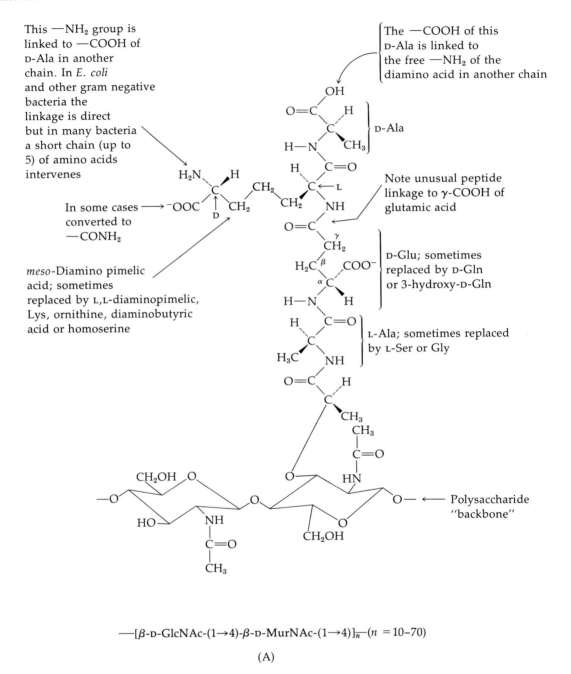

This —NH₂ group is linked to —COOH of D-Ala in another chain. In *E. coli* and other gram negative bacteria the linkage is direct but in many bacteria a short chain (up to 5) of amino acids intervenes

In some cases → converted to —CONH₂

meso-Diamino pimelic acid; sometimes replaced by L,L-diaminopimelic, Lys, ornithine, diaminobutyric acid or homoserine

The —COOH of this D-Ala is linked to the free —NH₂ of the diamino acid in another chain

D-Ala

Note unusual peptide linkage to γ-COOH of glutamic acid

D-Glu; sometimes replaced by D-Gln or 3-hydroxy-D-Gln

L-Ala; sometimes replaced by L-Ser or Gly

Polysaccharide "backbone"

$$—[\beta\text{-D-GlcNAc-}(1{\rightarrow}4)\text{-}\beta\text{-D-MurNAc-}(1{\rightarrow}4)]_{\overline{n}}—(n = 10\text{–}70)$$

(A)

Fig. 5-9 (A) Repeating unit of structure of a bacterial peptidoglycan (murein). Some connecting bridges are pentaglycine (*Staphylococcus aureus*), trialanylthreonine (*Micrococcus ro-* *seum*), and polyserine (*Staphylococcus epidermis*). (B) Schematic drawing of the peptidoglycan of *Staphylococcus aureus*. From Osborn.[107]

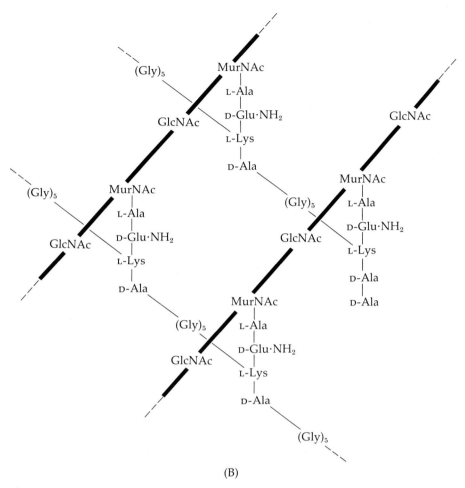

(B)

Fig. 5-9 (Continued)

linked polymer of alternating *N*-acetylgluco-samine and *N*-acetylmuramic acid residues. Alternate units of the resulting chitinlike molecule carry an unusual peptide attached to the lactyl groups of the *N*-acetylmuramic acid units (Fig. 5-9). The peptide side chains are cross-linked as indicated in the figure. In *E. coli* and other gram-negative bacteria the peptidoglycan forms a thin (2 nm) continuous network around the cell. In gram-positive bacteria the highly cross-linked peptidoglycan forms a layer as much as 10 nm thick.[102–104]

1. The Wall of Gram-Negative Bacteria

Outside the murein layer of *E. coli* and other gram-negative organisms is a phospholipid-containing layer or **outer membrane** which has the thickness and possibly the structure of a typical biological membrane (Fig. 5-8). This membrane appears to be attached to the murein layer in the following way. A small **hydrophobic protein** (containing 57 amino acids) is attached through an amide linkage between the side chain amino

BOX 5-G

COMPLEMENT

Complement is a group of about a dozen proteins found in blood serum which are activated in a cascade mechanism when antibody and antigen combine.[a-e]

Antibody + antigen

$$\downarrow$$

complex $\xrightarrow{\text{complement}}$ cell lysis, phagocytosis, etc.

The ultimate effects of the action of complement include destruction of cells by lysis and by the activation of leukocytes which engulf foreign cells by phagocytosis. Complement also induces the release of **chemotactic factors** that attract polymorphonuclear leukocytes (Chapter 1, Section E,2) to the site involved. The attention of biochemists has been attracted especially to the recognition component C1 of complement—three proteins designated C1q, C1r, and C1s. Protein C1q binds to the C_H2 domain of antibodies that have combined with antigens. However (Box 5-F), it takes at least a dimer of IgG to activate C1q. On the other hand, a single molecule of the naturally pentameric IgM suffices. Factor C1q, whose molecular weight is ~400,000, has a most surprising structure.[d] There is a central portion of diameter 3–6 nm and length 10–12 nm to which are attached six very thin connecting strands 10–13 nm long and ~1.5 nm in diameter which

terminate in globular ends of ~6 nm diameter. These ends are thought to be the sites of combination with the immunoglobulins.

A chemical surprise came when the thin connecting strands were found to have a high content of hydroxyproline and hydroxylysine. They appear to be made up of three chains each and apparently have the collagen structure (Chapter 2, Section B,3c). However, the globular ends probably have a more typical protein structure. Thus, protein C1q is another molecular machine whose secrets are far from understood. Antigens bind to antibodies activating their complement-binding regions. The activated antibodies bind to C1q and this binding in some manner activates C1q, which in turn activates C1r. The latter is thought to bind at the center of C1q, while the antibodies bind at the outer ends of the delicate molecule. How then is the activation "message" carried from the outer arms to the center?

Activated C1r is known to convert C1s, a proenzyme, into an active esterase that attacks complement components C2 and C4 and initiates a branching cascade of reactions causing lysis, chemotaxis, and other observed responses.

[a] A. White, P. Handler, and E. Smith, "Principles of Biochemistry," 5th ed., pp. 830–832. McGraw-Hill, New York, 1973.

[b] M. M. Mayer, *Sci. Am.* **229**, 54–66 (Nov 1973).

[c] D. R. Schultz, "The Complement System." Karger, Basel, 1971.

[d] E. Shelton, K. Yonemasu, and R. M. Stroud, *PNAS* **69**, 65–68 (1972).

[e] H. J. Müller-Eberhard, *Annu. Rev. Biochem.* **44**, 697–724 (1975).

group of the C-terminal lysine of the protein and a diaminopimelic acid residue of the murein.[105] Thus, the protein replaces one of the terminal D-alanine residues of about one in ten of the murein peptides. The N-terminal serine of the protein is in turn covalently attached to a lipid that is presumed to be buried in the outer membrane. There are ~10^5 molecules of the protein per cell spread over a surface area of peptidoglycan of ~3 μm^2. Consequently, the small protein molecules might be spaced about 5 nm apart on the average.

About 10^5 copies/cell of a larger protein are also present in the space between the membranes in *E. coli*. Molecules of this **envelope protein** of MW = 36,500 are arranged on the outer surface of the peptidoglycan in a hexagonal array with a 7.5 nm spacing.[106] The envelope protein may provide rigidity to the rod-shaped bacteria. Larger (MW 65,000–150,000) proteins form regular arrays on the surface of many bacteria.[*, 106a]

* In addition to the two proteins mentioned, cell envelopes of *E. coli* contain one other protein in amounts comparable to those of the envelope protein mentioned above. See Garten *et al.* for a review.[106b]

The outer surface of the outer membrane of gram-negative bacteria is covered by a **lipopolysaccharide** of marvelous complexity.[107,108] The outermost layer of the lipopolysaccharide consists of long projecting polysaccharide chains, with specific repeating units, that have antigenic properties and are called **O antigens.** Specific antibodies can be prepared against these polysaccharides, and so varied are the structures that 1000 different "serotypes" of *Salmonella* are known. These are classified into 17 principal groups. For example, group E3 contains the repeating unit

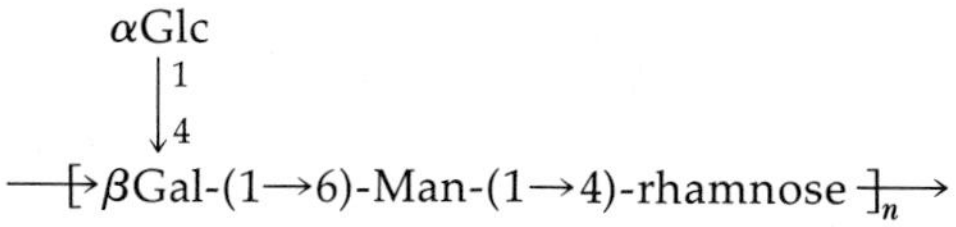

$$\alpha Glc$$
$$\downarrow{1}$$
$$\downarrow{4}$$
$$-\!\!\rightarrow\!\!\beta Gal\text{-}(1\!\rightarrow\!6)\text{-}Man\text{-}(1\!\rightarrow\!4)\text{-}rhamnose\!\rightarrow_{n}\!\!\rightarrow$$

where n may be ~ 50 on the average. Groups A, B, and D contain the repeating unit

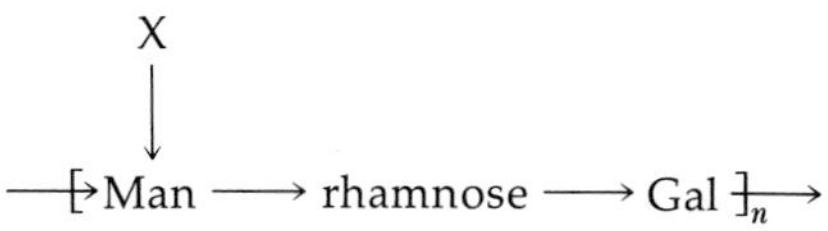

$$X$$
$$\downarrow$$
$$-\!\!\rightarrow\!Man\longrightarrow rhamnose \longrightarrow Gal\!\rightarrow_{n}\!\!\rightarrow$$

where X is a 3,6-dideoxyhexose: **paratose** in type A, **abequose** in type B, and **tyvelose** in type D (Fig. 5-10). The existence of the many serotypes depends both on the variety of components and types of linkage in the repeating units and on further structural variations at the chain ends.

At the inner end of the O antigen is a shorter polysaccharide chain whose structure is less

Fig. 5-10 Some "rare" sugars, components of the outer cell wall "antigens" of gram-negative bacteria.

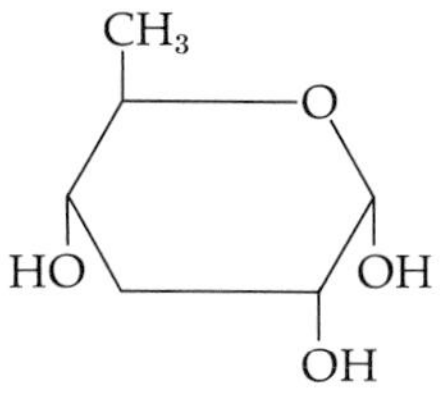

Paratose (3,6-dideoxy-D-ribohexose)

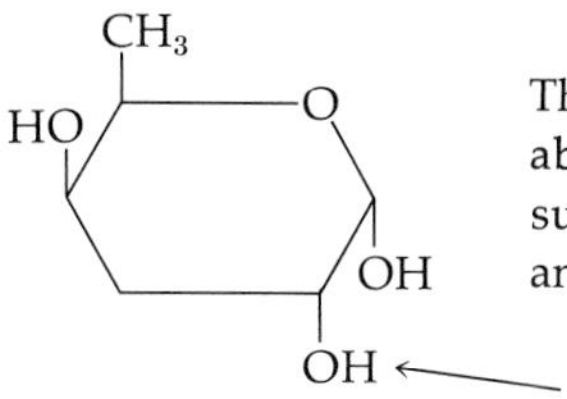

Abequose (3,6-dideoxy-D-xylohexose)

The enantiomorph of abequose is colitose, a sugar found in surface antigens of *E. coli*

Abequose is acetylated here in some antigens

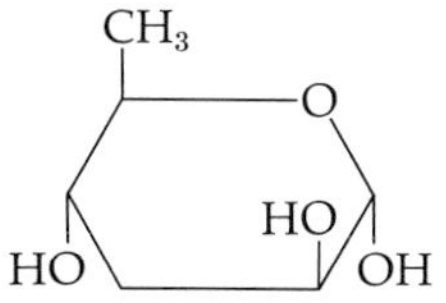

Tyvelose (3,6-dideoxy-D-arabohexose)

The enantiomorph of tyvelose is ascarylose, found not only in certain bacterial antigens but also in the eggs of *Ascaris*

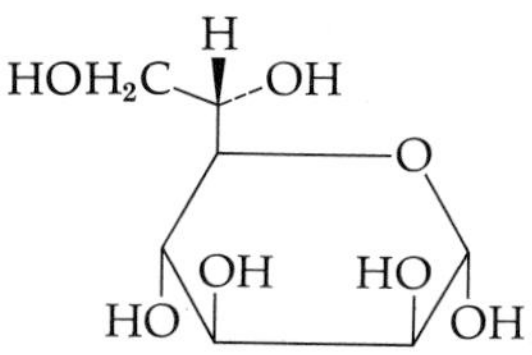

L-*Glycero*-D-*Manno*heptose

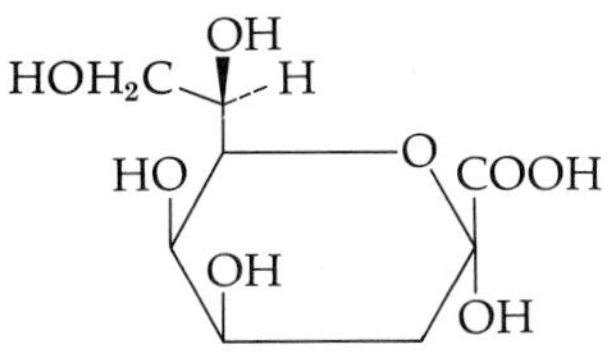

3-Deoxy-D-mannooctulosonic acid, also called 2-keto-3-deoxyoctonate or KDO

varied than that of the outer ends but which is nevertheless remarkable in containing two special sugars found only in bacterial cell walls: a seven carbon **heptose** and an eight carbon α-keto sugar acid, **ketodeoxyoctonate** (KDO). The structures are given in Fig. 5-10 and the arrangement of these sugars in the *Salmonella* lipopolysaccharide is shown in Fig. 5-11. The figure also shows the manner in which the oligosaccharide bearing the O antigen is attached to a repeating unit of two molecules of *N*-acetylglucosamine in β-1,6 linkage. These disaccharide units are not linked together in a conventional way but by **pyrophosphate bridges.** Furthermore, the free positions in the disaccharide are esterified with fatty acyl residues that presumably penetrate the outer membrane and provide an anchor for the entire lipopolysaccharide. The fatty acids, **lauric, myristic,** and **palmitic** are present in a 1:1:1 ratio. In addition, three molecules of **3-D-hydroxymyristic acid** are combined, two of them in amide linkage, one with each of the amino groups of the GlcN units. The third residue of hydroxymyristic acid is esterified through its hydroxyl group to one of the other fatty acid molecules, and it is thought that this particular structure may be responsible for the unusually strong fever-inducing properties of the lipopolysaccharide.

Why do so many different surface antigens exist in *Salmonella*? A clue comes from the fact that the ends of these carbohydrate chains are the groups to which the antibodies in animals clamp themselves should the bacteria enter the bloodstream. A number of mutants known as R forms (because of the growth as rough colonies on agar plates) completely lack the outer O antigen. The R mutants of *Salmonella* are nonpathogenic, whereas the "smooth" strains with intact O antigen often cause illness. Perhaps, if the O antigen has the right cluster of sugar rings at the ends, the host organism does not recognize it as dangerous. Whatever the explanation, it appears that carbohydrate groups on cell surfaces play an important part in the mechanism by which cells recognize one another.

A remarkable fact comes from the study of certain **temperate bacteriophage,** bacterial viruses whose genome can become incorporated into that of bacteria (Chapter 15, Section D,8). Sometimes the incorporation of the virus genes into an enterobacterium causes a change in the O antigen structure. Infection by one virus causes the loss of *O*-acetyl groups on some sugar residues; other viruses cause addition of substituents. Still others cause changes from α to β linkage or from 1,4 to 1,6 linkage at a particular point in the oligosaccharide. Apparently virus genes can both code for new enzymes, glycosyltransferases that are involved in assembly of O antigens and can act to repress synthesis of the transferases of the host. Another interesting fact is that the dangerous pathogenic organism of typhoid fever *Salmonella typhi* is only dangerous if it is infected with a certain temperate bacteriophage. An interesting parallel is suggested by the fact that cancer-causing **(oncogenic)** viruses cause transformations of mammalian cells in culture in such a way that surface antigens are altered.

Studies of temperate phage of *Salmonella* have given us some insight into the ways in which these bacterial viruses attach themselves to their host cell walls. It appears to be the O antigens themselves that are the initial sites of attachment. The slender tail fibers of the phage (Box 4-E), acting much like antibodies, bind to specific groupings of the polysaccharide. However, when the phage genome becomes incorporated and the O antigen structure is altered, further attachment of viruses is blocked. At the same time the cell becomes susceptible to attack by *another* strain of virus.[109]

2. Teichoic Acids

Found in association with the peptidoglycan layer and cell wall of gram-positive bacteria are **teichoic acids** which, in some species, account for 50% of the dry weight of the cell walls.[110,111] These

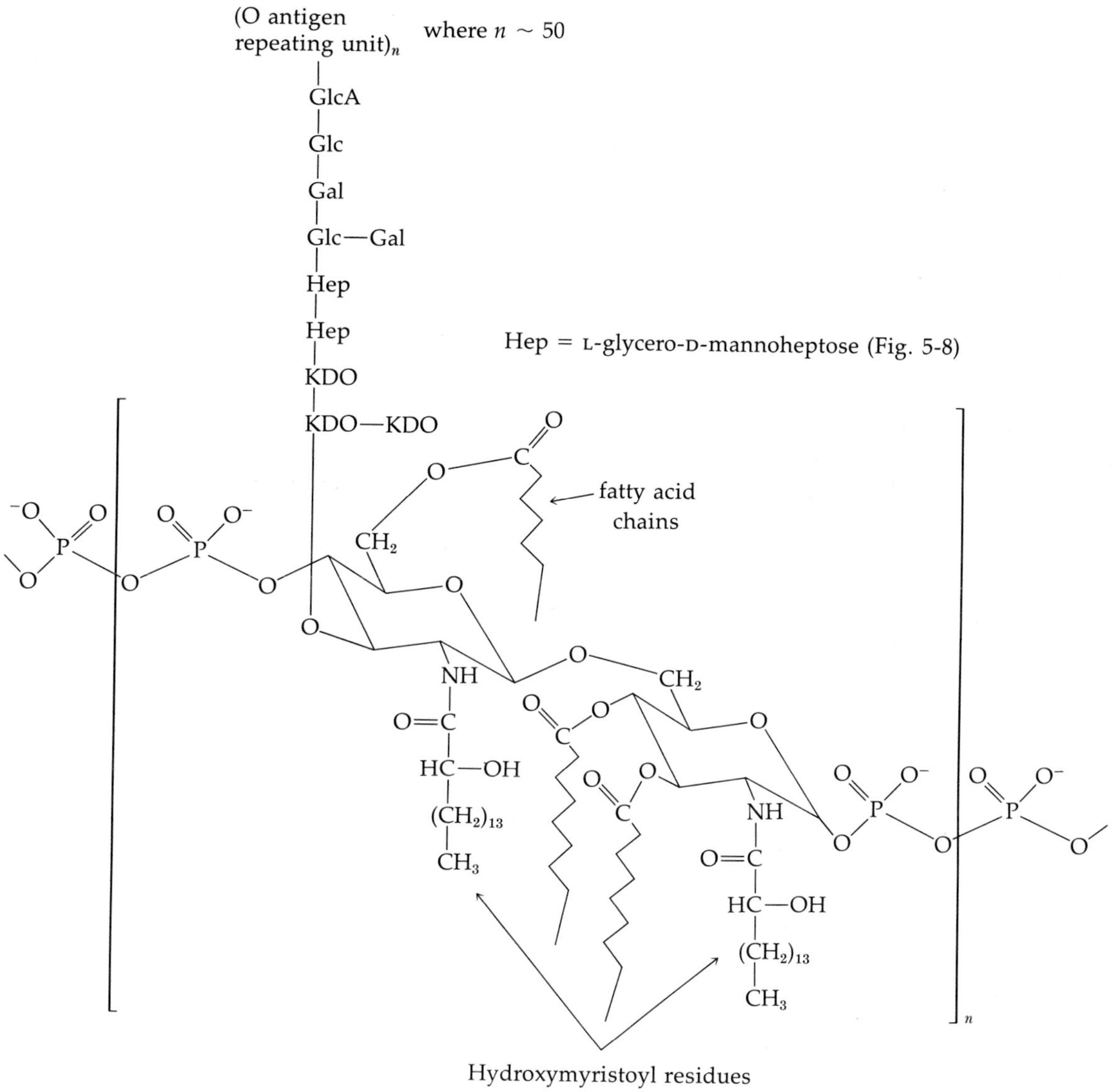

(O antigen repeating unit)$_n$ where $n \sim 50$

GlcA
|
Glc
|
Gal
|
Glc—Gal
|
Hep
|
Hep   Hep = L-glycero-D-mannoheptose (Fig. 5-8)
|
KDO
|
KDO—KDO

fatty acid chains

CH$_2$

O=C
|
HC—OH
|
(CH$_2$)$_{13}$
|
CH$_3$

NH

O=C
|
HC—OH
|
(CH$_2$)$_{13}$
|
CH$_3$

Hydroxymyristoyl residues

Of the 3 fatty acid residues indicated one is palmitoyl, one is lauroyl and one is myristoyl hydroxymyristoyl:

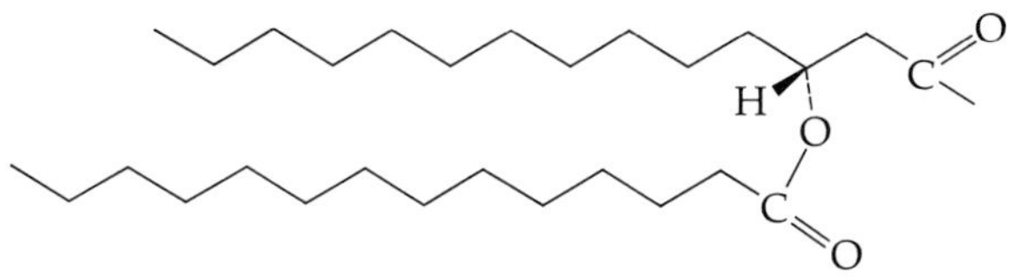

Fig. 5-11 Repeating unit of the outer glycolipid layer of the cell wall of *Salmonella*.[108]

are high polymers of the general types shown below.

1. Ribitolteichoic acids

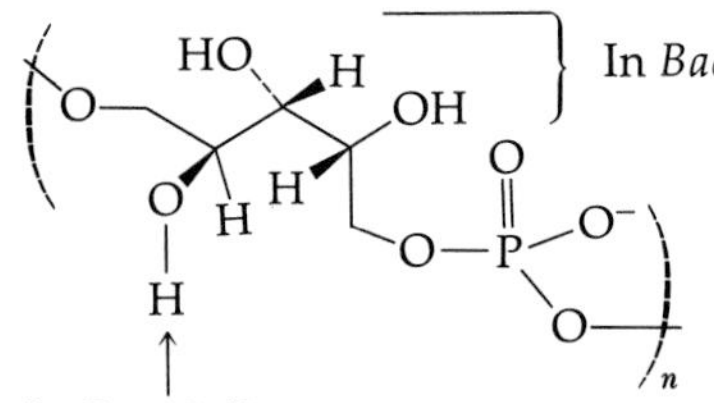

In *Bacillus subtilus* D-alanine is attached to one of these —OH groups in at least half of the units

In *B. subtilus* β-D-glucose is attached here

2. Glycerolteichoic acids

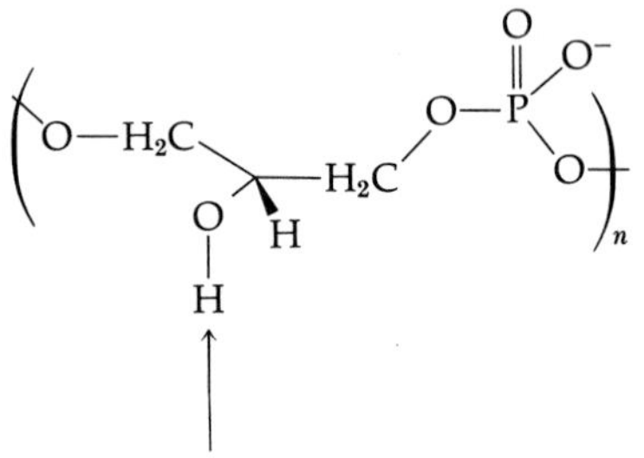

In *Lactobacillus arabinosus* D-alanine in ester linkage occupies this position but is replaced by β-D-glucose on about one unit in nine

The teichoic acids of cell walls are covalently attached through phosphodiester linkages to muramic acid residues of the peptidoglycan. Two possible arrangements have been suggested.[112] In one, the peptidoglycan strands are oriented perpendicular to the outer ends. According to this picture the wall structure would consist of a 10 nm layer of peptidoglycan with a 12 nm layer of teichoic acid on the outside. Another possibility is that the peptidoglycan strands are oriented parallel to the surface of the bacterium. Since the teichoic acid is uniformly distributed in its attachment to the peptidoglycan, such an arrangement would mean that the teichoic acid is intimately associated with peptidoglycan throughout the wall.

Other teichoic acids are covalently attached to glycolipids (e.g., to an oligosaccharide joined in glycosidic linkage to a diglyceride) which are part of the plasma membrane.[111]

E. PLANT CELL WALLS

The thick plant cell wall (Fig. 1-3) has a structure of remarkable complexity[113–116] which provides strength and rigidity to plants and is at the same time able to elongate rapidly during periods of growth. Northcote[113] likened the wall structure to the glass fiber-reinforced plastic fiber glass. Thus, the cell wall contains **microfibrils** of cellulose and other polysaccharides embedded in a **matrix,** also largely polysaccharide. The **primary cell wall** laid down in green plants during early stages of growth contains loosely interwoven cellulose fibrils ~10 nm in diameter and with an ~4 nm crystalline center. The cellulose in these fibrils has a degree of polymerization of 8000–12,000 glucose units. As the plant cell matures a secondary cell wall is laid down on the inside of the primary wall. This contains many layers of closely packed microfibrils, alternate layers often being laid down at different angles to one another. The microfibrils in green plants are most often cellulose, but may contain other polysaccharides as well. Some algae are rich in fibrils of xylan and mannan.

The materials present in the matrix phase vary with the growth period of the plant. During initial phases **pectic substances** (polygalacturonic acid derivatives) predominate but later xylans and a variety of neutral polysaccharides ("hemicelluloses") appear. Primary cell wall constituents of dicotyledons include **xyloglucans** (containing a linear glucan chain with xylose, galactose, and fucose units in branches), **arabinogalactans,** and **rhamnogalactouronans** (consisting of straight chains of galacturonic acid units with rhamnose rings inserted at intervals). All three of these complex polysaccharides are covalently linked to each other and to the cellulose microfibrils[114] (Fig. 5-12).

Although the principal cell wall components of plants are carbohydrates, a small amount of a glycoprotein named **extensin** is also present.[117] This material, like collagen of the intercellular matrix of animal tissues, is rich in 4-hydroxyproline. It contains arabinose and galactose attached as an oligosaccharide of ~9 units to the hydroxyl groups of

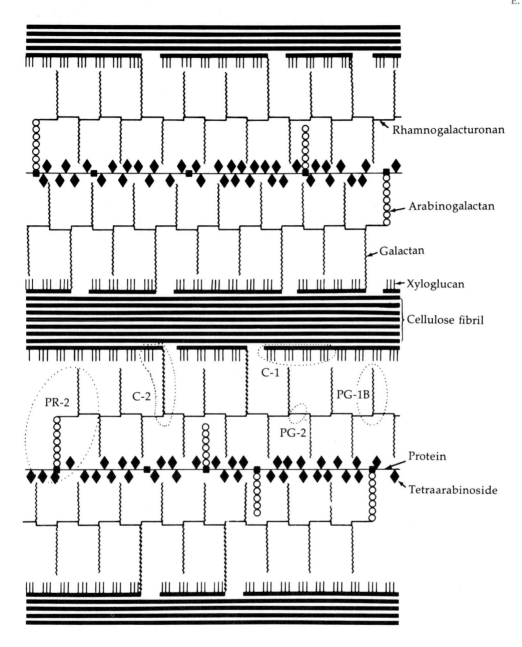

Fig. 5-12 Tentative structure of the walls of suspension-cultured sycamore cells. This model is not intended to be quantitative, but an effort has been made to present the wall components in approximately proper proportions. The distance between cellulose elementary fibrils is expanded to allow room to present the interconnecting structure. There are probably between 10 and 100 cellulose elementary fibrils across a single primary cell wall. From Albersheim *et al.*[115]

the protein. During advanced stages, as the walls harden into wood, large amounts of **lignins** (Chapter 14, Section H,6) are laid down in plant cells. These chemically resistant polymers contain many aromatic rings.

The site of biosynthesis of pectins and hemicelluloses is probably Golgi vesicles which pass to the outside via exocytosis. On the other hand, cellulose fibrils may be extruded through the plasma membrane. A remarkable aspect of primary plant cell walls is their ability to elongate extremely rapidly during cell growth. While the driving force for cell expansion is thought to be the development of pressure within the cell, the manner in which the wall expands is closely regulated. After a certain point in development elongation occurs in one direction only and under the influence of a series of plant hormones. Most striking is the effect of the **gibberellins** (Chapter 12, Section H,1)

which cause very rapid elongation.

Elongation of plant cell walls doubtless depends upon chemical cleavage and reforming of cross-linking polysaccharides. However, the cellulose fibrils may remain intact and slide past each other.[114]

Yeasts and other fungi have cell walls made up of **glucans, chitin,** and a **mannan-protein complex.** Some of the highly branched mannan chains serve as characteristic species-specific antigens.[118] Like those of the bacterial and animal cell surfaces, the antigens are extremely varied in structure and an understanding of them has important medical implications. The yeast system is a favorable one for studying the genetics of the biosynthetic enzymes producing the mannans. Yeasts can be grown either as haploid or hybrid diploid forms, facilitating genetic analysis.

REFERENCES

1. G. Rouser, G. J. Nelson, S. Fleischer, and G. Simon, *in* "Biological Membranes" (D. Chapman, ed.), p. 5. Academic Press, New York, 1968.
1a. P. J. Quinn, "Molecular Biology of Cell Membranes." University Park Press, Baltimore, Maryland, 1976.
2. C. F. Fox and A. D. Keith, eds., "Membrane Molecular Biology." Sinauer Assoc., Stamford, Connecticut, 1972.
3. M. S. Bretscher, *Science* **181**, 622–629 (1973).
3a. M. S. Bretscher and M. C. Raff, *Nature (London)* **258**, 43–49 (1975).
4. J. D. Robertson, *in* "Cell Biology in Medicine" (E. E. Bittar, ed.), pp. 3–48. Wiley, New York, 1973.
5. R. A. Capaldi, *Sci. Am.* **230**, 27–33 (Mar 1974).
6. N. A. Machtiger and C. F. Fox, *Annu. Rev. Biochem.* **26**, 575–600 (1973).
7. M. K. Jain, "The Bimolecular Lipid Membrane." Van Nostrand-Reinhold, Princeton, New Jersey, 1972.
8. A. R. Oseroff, P. W. Robbins, and M. M. Burger, *Annu. Rev. Biochem.* **26**, 647–682 (1973).
9. G. Guidotti, *Annu. Rev. Biochem.* **41**, 731–752 (1972).
9a. H. T. Tien, "Bilayer Lipid Membranes." Dekker, New York, 1974.
10. G. L. Nicolson, *BBA* **457**, 57–108 (1976).
10a. D. D. Jones and M. Jost, *Arch. Mikrobiol.* **70**, 43–64 (1970).
11. E. D. Korn, *Science* **153**, 1491–1498 (1966).
12. P. Mueller, D. O. Rudin, H. T. Tien, and W. C. Wescott, *Nature (London)* **194**, 979–980 (1962).

12a. C. Tanford, "The Hydrophobic Effect." Wiley, New York, 1973.
13. M. Montal and P. Mueller, *PNAS* **69**, 3561–3566 (1972).
14. J. S. O'Brien, *FEBS Symp.* **20**, 33–38 (1970).
15. R. Fettiplace, D. M. Andrews, and D. A. Haydon, *J. Membr. Biol.* **5**, 277–296 (1971).
16. D. Wolff, M. Canessa-Fischer, F. Vargas, and G. Díaz, *J. Membr. Biol.* **6**, 304–314 (1971).
16a. D. L. Melchior and J. M. Steim, *Annu. Rev. Biophys. Bioeng.* **5**, 205–238 (1976).
17. E. J. Shimshick and H. M. McConnell, *Biochemistry* **12**, 2351–2360 (1973).
17a. S. W. Hui and D. F. Parsons, *Science* **190**, 383–384 (1975).
18. H. Träuble and H. Eibl, *PNAS* **71**, 214–219 (1974).
19. D. M. Engelman, *JMB* **47**, 115–117 (1970).
20. P. Overath, H. U. Schairer, and W. Stoffel, *PNAS* **67**, 606–612 (1970).
21. K. Willecke and A. B. Pardee, *JBC* **246**, 5264–5272 (1971).
22. M. K. Jain, *Curr. Top. Membr. Transp.* **6**, 1–57 (1975).
23. H. Träuble, *J. Membr. Biol.* **4**, 193–208 (1971).
24. P. Devaux and H. M. McConnell, *JACS* **94**, 4475–4481 (1972).
25. A. Rosenberg, *Science* **157**, 1191–1196 (1967).
26. M. S. Bretscher, *Nature (London), New Biol.* **236**, 11–12 (1972).
27. J. A. Reynolds and H. Trayer, *JBC* **246**, 7337–7342 (1971).
28. D. M. Neville, Jr. and H. Glossmann, *JBC* **246**, 6335–6338 (1971).
28a. K. L. Carraway, *BBA* **415**, 379–410 (1975).
28b. C. G. Gahmberg, *JBC* **251**, 510–515 (1976).

29. E. Reichstein and R. Blostein, *JBC* **250**, 6256–6263 (1975).

29a. P. A. Brown, M. B. Feinstein, and R. I. Sha'afi, *Nature (London)* **254**, 523–525 (1975).

29b. A. Rothstein, Z. I. Cabantchik, and P. Knauf, *Fed. Proc., Fed. Am. Soc. Exp. Biol.* **35**, 3–10 (1976).

30. T. L. Steck and G. Dawson, *JBC* **249**, 2135–2142 (1974).

31. V. T. Marchesi, H. Furthmayr and M. Tomita, *Annu. Rev. Biochem.* **45**, 667–698 (1976).

32. S. J. Singer and G. L. Nicolson, *Science* **175**, 720–731 (1972).

33. E. H. Eylar, S. Brostoff, G. Hashim, J. Caccam, and P. Burnett, *JBC* **246**, 5770–5784 (1971).

33a. P. Laggner, *Nature (London)* **255**, 427–428 (1975).

33b. J. P. Beard and J. C. Connally, *J. Bacteriol.* **122**, 59–65 (1975).

34. A. Endo and L. Rothfield, *Biochemistry* **8**, 3508–3515 (1969).

35. R. Henderson and P. N. T. Unwin, *Nature (London)* **257**, 28–32 (1975).

36. S. Black, *Adv. Enzymol.* **38**, 193–234 (1973).

36a. M. S. Bretscher, *Nature (London)* **260**, 21–23 (1976).

37. Q. F. Ahkong, F. C. Cramp, D. Fisher, J. I. Howell, W. Tampion, M. Veninder, and J. A. Lucy, *Nature (London), New Biol.* **242**, 215–217 (1973).

38. L. E. Hokin, ed., "Metabolic Pathways," 3rd ed., Vol. 6. Academic Press, New York, 1972.

39. H. N. Christensen, "Biological Transport," 2nd ed. Benjamin, New York, 1975.

40. J. Wyman, *in* "Structure and Function of Oxidation-Reduction Enzymes" (A. Åkeson and A. Ehrenberg, eds.), pp. 133–138. Pergamon, Oxford, 1972.

41. S. Roseman, *in* "Metabolic Pathways" (L. E. Hoken, ed.), 3rd ed., Vol. 6, pp, 41–89. Academic Press, New York, 1972.

42. H. M. Kalckar, *Science* **174**, 557–565 (1971).

43. R. R. Aksamit and D. E. Koshland, Jr., *Biochemistry* **13**, 4473–4478 (1974).

44. D. L. Oxender, *in* "Biomembranes" (L. A. Manson, ed.), Vol. 5, pp. 25–79. Plenum, New York, 1974.

45. D. L. Oxender, *Annu. Rev. Biochem.* **41**, 777–809 (1972).

46. G. Kaczorowski, L. Shaw, M. Fuentes, and C. Walsh, *JBC* **250**, 2855–2865 (1975).

46a. B. Anderson, N. Weigel, W. Kundig, and S. Roseman, *JBC* **246** 7023–7033 (1971).

46b. M. H. Saier, Jr., B. U. Feucht, and L. J. Hofstadter, *JBC* **251**, 883–892 (1976).

46c. J. Boonstra, M. T. Huttunen, W. N. Konings, and H. R. Kaback, *JBC* **250**, 6792–6798 (1975).

47. F. M. Harrold and K. Altendorf, *Curr. Top. Membr. Transp.* **5**, 1–50 (1974).

48. L. E. Rosenberg, *in* "Biological Membranes" (R. M. Dowben, ed.), pp. 255–295. Little, Brown, Boston, Massachusetts, 1969.

49. G. N. Ling, "A Physical Theory of the Living State." Ginn (Blaisdell), Boston, Massachusetts, 1962.

50. R. Damadian, *Biophys. J.* **11**, 773–785 (1971).

51. R. Cooke and I. D. Kuntz, *Annu. Rev. Biophys. Bioeng.* **3**, 95–126 (1974).

52. P. F. Baker and J. S. Willis, *Nature (London)* **226**, 521–527 (1970).

53. J. R. Perrone, J. F. Hackney, J. F. Dixon, and L. E. Hokin, *JBC* **250**, 4178–4184 (1975).

54. K. Taniguchi and R. L. Post, *JBC* **250**, 3010–3018 (1975).

55. L. K. Lane, J. H. Copenhaver, Jr., G. E. Lindenmayer, and A. Schwartz, *JBC* **248**, 7197–7200 (1973).

56. R. Blostein, *JBC* **250**, 6118–6124 (1975).

57. J. Kyte, *JBC* **249**, 3652–3660 (1974).

58. B. C. Pressman, *Annu. Rev. Biochem.* **45**, 501–530 (1976).

59. H. I. Haavik and Ø. Froyshov, *Nature (London)* **254**, 79–81 (1975).

60. Y. A. Ovchinnikov, *Mitochondria Biomembr., Proc. Fed. Eur. Biochem. Soc., Meet., 8th, 1972* Vol. 28, pp. 279–306 (1972).

61. P. Läuger, *Science* **178**, 24–30 (1972).

62. W. Simon, W. E. Morf, and P. C. Meier, *Struct. Bonding (Berlin)* **16**, 113–160 (1973).

63. W. L. Duax, M. Hauptman, C. M. Weeks, and D. A. Norton, *Science* **176**, 911–915 (1972).

63a. G. D. Smith, W. L. Duax, D. A. Langs, G. T. DeTitta, J. W. Edmonds, D. C. Rohrer, and C. M. Weeks, *JACS* **97**, 7242–7247 (1975).

64. F. M. Harrold and J. R. Baarda, *J. Bacteriol.* **94**, 53–60 (1967).

64a. C. A. Tyson, H. Vande Zande, and D. E. Green, *JBC* **251**, 1326–1332 (1976).

65. F. Bastide, G. Meissner, S. Fleischer, and R. L. Post, *JBC* **248**, 8385–8391 (1973).

66. H. J. Schatzmann, *Curr. Top. Membr. Transp.* **6**, 125–168 (1975).

67. K. J. Dorrington, A. Hui, T. Hofmann, A. J. W. Hitchman, and J. E. Harrison, *JBC* **249**, 199–204 (1974).

68. H. B. Bull, "An Introduction to Physical Biochemistry," 2nd ed., pp. 174–178. Davis Co., Philadelphia, Pennsylvania, 1971.

69. D. J. Aidley, "The Physiology of Excitable Cells." Cambridge Univ. Press, London and New York, 1971.

70. R. Eckert, *Science* **176**, 473–480 (1972).

70a. J. L. Browning, D. L. Nelson, and H. G. Hansma, *Nature (London)* **259**, 491–494 (1976).

71. W. J. Adelman, Jr., ed., "Biophysics and Physiology of Excitable Membranes." Van Nostrand-Reinhold, Princeton, New Jersey, 1971.

72. R. A. Nystrom, "Membrane Physiology." Prentice-Hall, Englewood Cliffs, New Jersey, 1973.

73. A. L. Hodgkin, "The Conduction of the Nervous Impulse." Thomas, Springfield, Illinois, 1964.

74. R. D. Keynes, *Nature (London)* **239**, 29–32 (1972).

75. J. M. Ritchie, *Prog. Biophys. Mol. Biol.* **26**, 149–187 (1973).

76. R. Lefever and J. L. Deneuborg, *Adv. Chem. Phys.* **29**, 349–374 (1975).

77. V. Ginsburg, *Adv. Enzymol.* **36**, 131–149 (1972).

78. K. O. Lloyd and E. A. Kabat, *PNAS* **61**, 1470–1477 (1968).

79. E. A. Kabat, *in* "Blood and Tissue Antigens" (D. Aminoff, ed.), pp. 187–198. Academic Press, New York, 1970.

80. B. L. Slomiany and K. Meyer, *JBC* **248**, 2290–2295 (1973).

81. S. Takasaki and A. Kobata, *JBC* **251**, 3610–3615 (1976).

82. H. S. Slayter, A. G. Cooper, and M. C. Brown, *Biochemistry* **13**, 3365–3371 (1974).

83. J. Kościelak, A. Piasek, H. Gorniak, A. Gardas, and A. Gregor, *EJB* **37**, 214–225 (1973).

84. J. R. Wherret and S. Hakomori, *JBC* **248**, 3046–3051 (1973).

85. A. Slomiany and M. I. Horowitz, *JBC* **248**, 6232–6238 (1973).

86. R. C. Hughes, *Prog. Biophys. Mol. Biol.* **26**, 189–268 (1973).

87. J. Klein, "Biology of the Mouse Histocompatibility-2 Complex." Springer-Verlag, Berlin and New York, 1975.

88. R. A. Reisfeld and B. D. Kahan, *Sci. Am.* **226**, 28–37 (Jun 1972).

89. D. E. Isenman, R. H. Painter, and K. J. Dorrington, *PNAS* **72**, 548–552 (1975).

89a. M. C. Raff, *Nature* (*London*) **254**, 287–288 (1975).

89b. R. Henning, R. J. Milner, K. Reske, B. A. Cunningham, and G. M. Edelman, *PNAS* **73**, 118–122 (1976).

89c. E. S. Vitetta, J. D. Capra, D. G. Klapper, J. Klein, and J. W. Uhr, *PNAS* **73**, 905–909 (1976).

89d. J. Bridgen, D. Snary, M. J. Crumpton, C. Barnstable, P. Goodfellow, and W. F. Bodmer, *Nature* (*London*) **261**, 200–205 (1976).

89e. J. L. Strominger, R. E. Humphreys, J. M. McCune, P. Parham, R. Robb, T. Springer, and C. Terhorst, *Fed. Proc.* **35**, 1177–1182 (1976).

90. N. Sharon and H. Lis, *Science* **177**, 949–958 (1972).

91. H. Lis and N. Sharon, *Annu. Rev. Biochem.* **42**, 541–574 (1973).

92. R. Kornfeld and C. Ferris, *JBC* **250**, 2614–2619 (1975).

93. J. W. Becker, G. N. Reeke, Jr., B. A. Cunningham, and G. M. Edelman, *Nature* (*London*) **259**, 406–409 (1976).

94. K. D. Hardman and C. F. Ainsworth, *Biochemistry* **15**, 1120–1128 (1976).

94a. K. D. Hardman and I. J. Goldstein *in* "Immunochemistry of Proteins" (M. Z. Atassi, ed.) Plenum, New York, 1977.

95. W. A. Frazier, S. D. Rosen, R. W. Reitherman, and S. H. Barondes, *JBC* **250**, 7714–7721 (1975).

96. K. M. Yamada, S. S. Yamada, and I. Pastan, *PNAS* **72**, 3158–3162 (1975).

97. I. Yahara and G. M. Edelman, *PNAS* **72**, 1579–1583 (1975).

98. G. M. Edelman, *Science* **192**, 218–226 (1976).

99. H. R. Bourne, L. M. Lichtenstein, K. L. Melmon, C. S. Henney, Y. Weinstein, and G. M. Shearer, *Science* **184**, 19–28 (1974).

99a. P. Cuatrecasas and M. D. Hollenberg, *Adv. Prot. Chem.* **30**, 251–451 (1976).

100. T. Kono and F. W. Barham, *JBC* **246**, 6210–6216 (1971).

100a. L. Jarett and R. M. Smith, *PNAS* **72**, 3526–3530 (1975).

101. S. L. Pohl, H. Michiel, J. Krans, L. Birnbaumer, and M. Rodbell, *JBC* **247**, 2295–2301 (1972).

102. D. A. Reaveley and R. E. Burge, *Adv. Microb. Physiol.* **7**, 1–81 (1972).

103. W. J. Lennarz, *Acc. Chem. Res.* **5**, 361–367 (1972).

104. J. W. Costerton, J. M. Ingram, and K.-J. Cheng, *Bacteriol. Rev.* **38**, 87–110 (1974).

105. V. Braun, *BBA* **415**, 335–377 (1975).

106. J. P. Rosenbusch, *JBC* **249**, 8019–8029 (1974).

106a. V. B. Sleytr, *Nature* (*London*) **257**, 400–402 (1975).

106b. W. Garten, I. Hindennach, and U. Henning, *EJB* **59**, 215–221 (1975).

107. M. J. Osborn, *Annu. Rev. Biochem.* **38**, 501–538 (1969).

108. E. T. Rietschel, H. Gottert, O. Lüderitz, and O. Westphal, *EJB* **28**, 166–173 (1972).

109. R. Losick and P. W. Robbins, *Sci. Am.* **221**, 121–124 (Nov 1969).

110. J. Baddiley, *Essays Biochem.* **8**, 35–77 (1972).

111. A. J. Wicken and K. W. Knox, *Science* **187**, 1161–1167 (1975).

112. A. R. Archibald, J. Baddiley, and J. E. Heckels, *Nature* (*London*), *New Biol.* **241**, 29–31 (1973).

113. D. H. Northcote, *Annu. Rev. Plant Physiol.* **23**, 113–132 (1972).

114. P. Albersheim, *Sci. Am.* **232**, 80–95 (Apr 1975).

115. P. Albersheim, W. D. Bauer, K. Keestra, and K. W. Talmadge, *in* "Biogenesis of Plant Cell Wall Polysaccharides (F. Loewus, ed.), pp. 117–147. Academic Press, New York, 1973.

115a. R. D. Preston, *in* "Dynamic Aspects of Plant Ultrastructure," (A. W. Robards, ed.), pp 256–309. McGraw-Hill, New York, 1974.

115b. R. D. Preston, "The Physical Biology of Plant Cell Walls." Halsted Press, New York, 1974.

116. R. D. Preston, "The Physical Biology of Plant Cell Walls." Chapman & Hall, London, 1974.

117. M. F. Heath and D. H. Northcote, *BJ* **125**, 953–961 (1971).

118. C. E. Ballou and W. C. Raschke, *Science* **184**, 127–134 (1974).

STUDY QUESTIONS

1. Stearic acid (1.16 g) was dissolved in 100 ml of ethanol. A 10 μl portion of the resulting solution was pipetted onto a clean surface of a dilute HCl solution (in a shallow tray) where it spread to form a monolayer of stearic acid. The layer was compressed (by moving a Teflon barrier across the tray) until the surface pressure π started to rise sharply and reached $\sim$ 20 dyn/cm. Note that $\pi = \gamma_0 - \gamma$ where γ is the measured surface tension with the film present and γ_0 is the higher surface tension of water alone. The compressed film occupied a 20 $\times$ 24 cm area. Calculate the cross-sectional area of an alkyl chain in stearic acid. [See J. B. Davenport, in "Biochemistry and Methodology of Lipids" (A. R. Johnson and J. B. Davenport, eds.), pp. 47–83. Wiley-Interscience, New York, 1971; and M. C. Phillips, in "Progress in Surface and Membrane Science" (J. F. Danielli, D. M. Rosenberg, and D. A. Cadenhead, eds.), Vol. 5, pp. 139–221. Academic Press, New York, 1972.

2. In 1925 E. Gorter and F. Grendel (*J. Exp. Med.* **41,** 439) reported measurements in which they extracted lipid from red blood cell membranes with acetone, spread the lipids as a monolayer, and measured the area of the compressed monolayer. They then estimated the surface area of an erythrocyte and calculated that the ratio of the lipids (as a monolayer) to the surface area of the red blood cell was 1.9–2.0. More modern experiments[11] gave the following: each erythrocyte membrane contains 4.5×10^{-16} mol of phospholipid and 3.1×10^{-16} mol of cholesterol.

   a. If the cross-sectional areas of phospholipid and cholesterol molecules in a membrane are taken as 0.70 and 0.38 nm², respectively, what surface area would be occupied in a monolayer?

   b. If the measured surface area of an erythrocyte is 167 μm², what is the ratio of the area calculated in (a) to the area of the cell surface?

   c. How might you explain the difference between this answer and that of Gorter and Grendel? See Korn.[11]

3. The following experimental observations are related to biological membrane structure and function. Discuss the implications of each observation with respect to membrane structure.

   a. Many macrocyclic antibiotics (nonactin, valinomycin, and others) form 1:1 complexes with al-

kali metal cations in a highly selective manner. The complexes are readily soluble in nonpolar organic solvents. These antibiotics increase the electrical conductance and permeability to alkali metal cations of synthetic phospholipid membranes. Valinomycin increases the electrical conductance of thylakoid membranes of chloroplasts in the presence of K^+ but not in the presence of Na^+; it also uncouples oxidative phosphorylation in mitochondria. (See Chapter 10).

   b. Treatment of intact chloroplasts with a galactolipase releases galactose from galactosyl diglycerides. Treatment of red blood cell ghosts with phopholipase C releases about 75% of the lipid P in water-soluble form. In neither case is the structural integrity of the membrane destroyed.

   c. Using sphingomyelin as a hapten, antibodies specific for this lipid can be produced. When red blood cell ghosts are exposed to these antibodies, it can be shown that the antibodies react, but only on one side of the membrane.

4. Describe the structure of biological membranes and the characteristic functions of lipid-, protein-, and carbohydrate-containing components. Describe the differences between inner and outer membrane surfaces.

5. Compare the distribution of triglycerides, phosphatidylcholine, phosphatidylethanolamine, sphingomyelins, glycolipids, and cholesterol within cells. Consider differences between the two sides of membranes.

6. Consider the chemistry underlying the labeling of cell surfaces with each of the following: (a) Lactoperoxidase, (b) galactose oxidase, (c) formylmethionylsulfone methyl phosphate, (d) the diazonium salt of diiodosulfanilic acid, (e) fluorescent antibodies, (f) antibodies conjugated with ferritin. Write the equations for the chemical reactions involved. State what surface groups will be labeled by each reagent. List special advantages of each of these reagents.

7. Which would be the more effective detergent in the pH range 2 to 3, sodium lauryl sulfonate or sodium laurate? Why?

8. Suppose that a cell contains 10 mM Na^+ and 100 mM K^+ and that it is bathed in extracellular fluid containing 100 mM Na^+ and 5 mM K^+. How much energy will be required to transport 3 equivalents of

Na^+ out and 2 equivalents of K^+ in? Compare this with $\Delta G'$ for hydrolysis of ATP at pH 7. Assume that the membrane is permeable to Cl^-.

9. An *E. coli* cell is said to contain about 10^5 molecules of an envelope protein of MW = 36,500. If the latter is spherical and the spheres are closely packed in a hexogonal lattice, how much of the surface area of the bacterium would be covered? What would the diameter of the protein be? What spacing would be required if 10^5 molecules covered the surface completely? Suggest a shape for the protein molecule that is consistent with the requirement.

10. Starting at the point where a nerve impulse has reached a muscle cell (mammalian skeletal) and ending with cessation of contraction, trace or indicate the movement within the cell of the Ca^{2+} ions that trigger contraction of the myofibrils (include names of specific proteins and/or structures).

6

Enzymes: The Protein Catalysts of Cells

Most of the machinery of cells is made of enzymes. Hundreds of them have been extracted from living cells, purified and crystallized. Many others are recognized only by their catalytic action and have not yet been isolated in pure form. While most of the known enzymes are soluble globular proteins, the structural proteins of the cell may also exhibit catalysis. Thus, actin and myosin together catalyze the hydrolysis of ATP (Chapter 4, Section F). (However, we do not understand how this enzymatic reaction is coupled to movement of the muscle filaments.)

A vast literature documents the properties of enzymes as catalysts and as proteins. Because of this the beginner is apt to lose sight of some simple and fundamental questions: "How did we learn that the cell is crammed with enzymes?" "How do we recognize that a protein *is* an enzyme?" Part of the answer to both questions is that enzymes are recognized *only* by their *ability to catalyze chemical reactions*. Thus, an everyday operation for most biochemists is the measurement of catalytic activity of enzymes. Only by measuring rates of catalysis carefully and quantitatively has it been possible to isolate and purify these remarkable molecules.

The quantitative study of catalysis by enzymes (enzyme kinetics) is a highly developed mathematical branch of science that is of utmost practical importance to the biochemist. Study of kinetics is our most important means of learning about the mechanisms of catalysis at the active sites of enzymes. Kinetic studies are used to measure the affinity and the specificity of binding of substrates and of inhibitors to enzymes, to establish maximum rates of catalysis by specific enzymes, and in many other ways.

The following section contains a brief summary of basic concepts together with some practical tips. Note that a major goal in kinetic studies is to establish a **rate equation** which attempts to describe the velocity of a reaction in terms of experimentally measurable parameters. The serious student should read some of the excellent articles and books on the subject.[1-5]

A. PRIMER ON ENZYME KINETICS

To measure the rate of a chemical reaction, one has to start the reaction at a definite time, e.g., by rapidly mixing together two reactants. Then, while keeping the mixture at an accurately constant temperature and pH (very important!), the concentration of a reactant or product is measured after a fixed time interval, or at various times. No end of ingenuity has gone into devising the enormous variety of ways of doing this for particular enzymes. Whatever the procedure, the information sought is the **rate** at which some concentration changes with time. Very often a **progress curve** showing the decrease in the concentration [S] of a reactant (substrate) or the increase in the concentration [P] of a product with time is constructed (Fig. 6-1).

The velocity of an enzymatic reaction is given by Eq. 6-1. The units of velocity are moles per liter

$$v = -d[S]/dt$$
(and usually also)† $$v = d[P]/dt$$ (6-1)

per second (M s^{-1})‡ or more traditionally in enzymology moles per liter per minute. We are interested in the **instantaneous velocity,** which at any time is given by the slope of the progress curve (Fig. 6-1). Usually we want to measure the velocity immediately after the reaction is started. In some cases, as in the example given in Fig. 6-1, this is hard to do with accuracy.

In some chemical reactions, e.g., in first order processes, a logarithmic plot of the progress curve (log[S] vs. t) gives a straight line so that the initial slope need not be determined. However, in most experiments with enzymes, the **assay** is set up in such a way that the progress curve is nearly linear for at least short periods of time. A very sensitive method of detecting a product may be needed. The use of radioactive substrates is popular for this reason.

It is important to note that if the progress curve is not a straight line at the beginning and if the amount of compound that reacts in a fixed interval of time is taken as the rate, an erroneous answer will be obtained. Sometimes an **integrated rate ex-**

---

† The equality of rates of substrate decrease and product increase holds only under steady state conditions. See Section A,5.

‡ Velocities should always be expressed *intensively* as M s^{-1}, rather than *extensively*, e.g., as μmoles s^{-1} without specification of the volume of solution.

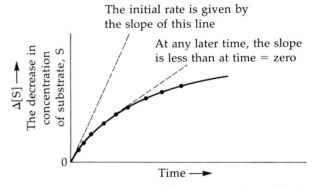

Fig. 6-1 The progress curve for an enzymatic reaction in which the substrate S is converted to products.

---

pression (describing the time-course of product formation) can be used, and there are other tricks for comparing relative rates even when the progress curve is of the sort shown in Fig. 6-1.

1. First Order Reactions

In many chemical reactions the rate of decrease of the concentration [A] of a given reactant is found experimentally to be directly proportional to the concentration of that reactant (Eq. 6-2).

$$v = -d[A]/dt = k[A]$$ (6-2)

The proportionality constant k is known as the rate constant. First order kinetics is observed for **unimolecular processes** in which a molecule of A is converted to product P in a given time interval with a probability that does not depend on interaction with another molecule. A good example is radioactive decay. Enzyme–substrate complexes often react by unimolecular processes. In many other cases, a first order reaction is **pseudounimolecular;** compound A, in actuality, reacts with a second molecule, e.g., water. However, the second molecule is present in great excess so that its concentration does not change during the experiment. Consequently, the velocity is apparently proportional only to [A].

A first order rate constant k has units of s^{-1}. Note that when [A] = 1, $v = k$. Thus, the first order rate constant k is a measure of the speed in $M\ s^{-1}$ of the reaction of a substance at unit activity. As a first order reaction proceeds, [A] decreases and at time t is given by any one of the following three equivalent expressions (Eq. 6-3; obtained by integration of Eq. 6-2) in which t_0 is the time at which the reaction was started.

$$[A] = A_0 e^{-kt}$$
$$\ln([A_0]/[A]) = kt \qquad (6\text{-}3)$$
$$\log[A_0] - \log[A] = kt/2.303$$

Equation 6-3 is the equation of exponential decay, a characteristic of which is that [A] is halved in a time that is independent of concentration. The **half-life** is $t_{1/2}$ (Eq. 6-4):

$$t_{1/2} = \ln 2/k = 0.693/k \qquad (6\text{-}4)$$

The **relaxation time** τ for A is defined by Eq. 6-5

$$\tau = 1/k = t_{1/2}/\ln 2 \qquad (6\text{-}5)$$

and represents the time required for the concentration [A] to fall to $1/e$ (or ~ 0.37) of its initial value.

2. Turnover Numbers

When an enzyme is catalyzing product formation at the maximum possible rate V_{max}, we can often assume that the intermediate ES is being converted to products according to Eq. 6-6:

$$d[P]/dt = V_{max} = k[ES] = k[E]_t \qquad (6\text{-}6)$$

Here E_t is the total enzyme, namely, the free enzyme E plus enzyme–substrate complex ES. The equation holds only at **substrate saturation**, i.e., when the substrate concentration is high enough that essentially all of the enzyme has been converted to the intermediate ES. The process is not first order, because ES is continually being regenerated from free enzyme. However, the rate constant k is comparable to the first order rate constants that we considered in the preceding section and gives a measure of the speed at which the enzyme operates. When the concentration $[E]_t$ is given in moles per liter of *active sites* (actual molar

concentration multiplied by the number of active sites per mole) the constant k is known as the **turnover number** or **molecular activity.**[6]

Turnover numbers can be measured only with pure enzymes. Partly for this reason the activity of an enzyme is more often given as *units of activity per milligram of protein* (**specific activity**). One **international unit** is the amount of enzyme that produces, under standard (usually optimal) conditions, 1 μmol of product per minute. The International Union of Biochemistry[7] now recommends a new unit, the **katal** (kat), the amount of enzyme that converts one mol s^{-1} of substrate to product.

$$1\ kat = 6 \times 10^7\ \text{international units}$$
$$1\ \text{international unit} = 16.67\ nkat\ \text{(nanokatals)}$$

If, under the standard assay conditions, the enzyme is pure and is saturated with substrate, the turnover number is

Turnover No. = katals/mol of active sites
= [nkat/mg] $\times$ MW $\times 10^{-6}/n$
= [international units/mg]
$\times$ MW $\times 10^{-3}/60n$

where MW is the molecular weight of the enzyme and n is the number of active sites per molecule. Since the activity of an enzyme is temperature and pH dependent, these variables must be specified.

Turnover numbers of enzymes vary from ~ 1 to $\sim 10^6\ s^{-1}$. Trypsin, chymotrypsin, and many intracellular enzymes have turnover numbers of $\sim 10^2\ s^{-1}$. Among the fastest enzymes are **catalase,** which converts H_2O_2 to H_2O and O_2 (Chapter 10 Section B,6); **carbonic anhydrase,**[8] which equilibrates carbonic acid and carbon dioxide ($H_2CO_3 \rightleftarrows H_2O + CO_2$); and **$\Delta^5$-3-ketosteroid isomerase.**[9] These enzymes have maximum turnover numbers of $2 \times 10^5\ s^{-1}$ or more. Compare these reaction rates with those of the typical organic synthesis in the laboratory. A reaction mixture must often be heated for hours ($k < 10^{-3}\ s^{-1}$). In fact, enzymes often accelerate rates by factors of 10^6 or more over those observed in the absence of enzyme at a comparable temperature and pH. Since enzymes often bring two or more substrates together at specific binding sites (active sites) rapid reaction can be catalyzed even when the reactants are present in low concentrations.

3. Second Order Reactions

In most cases, two molecules, A and B, must meet and collide for a chemical reaction to occur (Eq. 6-7):

$$A + B \xrightarrow{k_2} P \qquad (6\text{-}7)$$

The velocity of such a process is characterized by a **bimolecular rate constant** k_2 and is proportional to the product of the concentrations of A and B (Eq. 6-8):

$$v = k_2[A][B] \qquad (6\text{-}8)$$

The units of k_2 are $M^{-1}\,s^{-1}$. Note that if [B] is present at unit activity, the rate is $k_2[A]$, a quantity with units of s^{-1}. Thus, the bimolecular or second order rate constant for reaction of A with B may be compared with first order constants when the second reactant B is present at unit activity. In many real situations, reactant B is present in large excess and in a virtually constant concentration. In such a case, the experimentally observed rate constant is $k_2[B]$, the *apparent first order rate constant* for a pseudo-unimolecular reaction. The bimolecular rate constant k_2 can be obtained by dividing the apparent constant by [B].

4. Reversible Chemical Reactions

In any reversible process, we must consider rate constants for both the forward and reverse reactions. At equilibrium a reaction proceeds in the forward direction at exactly the same velocity as in the reverse reaction so that no change occurs. There is always a relationship between the equilibrium constant and the rate constants. For example, Eq. 6-9.

$$A + B \underset{k_2}{\overset{k_1}{\rightleftharpoons}} P \qquad (6\text{-}9)$$

$$\text{Equilibrium constant} = K = \frac{[P]}{[A][B]} = \frac{k_1}{k_2}$$

Here k_1 is a bimolecular rate constant for the forward reaction and k_2 a unimolecular rate constant for the reverse reaction.† The equilibrium constant K can easily be shown (from Eqs. 6-2 and 6-8) to equal k_1/k_2 for the reaction of Eq. 6-9.

What relationships exist between experimentally observable rates and k_1 and k_2 for a reversible reaction? Consider first the simplest case (Eq. 6-10):

$$A \underset{k_2}{\overset{k_1}{\rightleftharpoons}} B \qquad (6\text{-}10)$$

If pure A is placed in a solution, its concentration will drop until it reaches an equilibrium with the B which has formed. In this case [A] does not decay exponentially but $[A] - [A]_{equil}$ does (the student should be able to demonstrate this fact readily). Therefore, if $\log([A] - [A]_{equil})$ is plotted against time a first order rate constant k, characteristic of *the rate of approach to equilibrium*, will be obtained. Its relationship to k_1 and k_2 is given by Eq. 6-11.

$$k = k_1 + k_2 \qquad (6\text{-}11)$$

Furthermore, the relaxation time τ for approach to equilibrium equals $1/k$. τ is also related to the relaxation times for the forward (τ_1) and reverse (τ_2) reactions alone (Eq. 6-12).

$$\tau^{-1} = \tau_1^{-1} + \tau_2^{-1} \qquad (6\text{-}12)$$

For the more complex case[10] of Eq. 6-9, Eq. 6-13 holds

$$k = \tau^{-1} = k_1([A]_e + [B]_e) + k_2 \qquad (6\text{-}13)$$

where $[A]_e$ and $[B]_e$ are the equilibrium concentrations of A and B.

† In kinetic equations rate constants are usually numbered consecutively and the subscripts do not imply anything about the molecularity. Odd numbered constants usually refer to steps in the forward direction and even numbered constants to steps in the reverse direction.

5. Formation and Reaction of ES Complexes

Enzymes cannot act at a distance.[†] Hence, the first step in enzymatic catalysis is the combining of the enzyme and substrate to form a complex, ES (Eq. 6-14).

$$E + S \underset{k_2}{\overset{k_1}{\rightleftharpoons}} ES \overset{k_3}{\rightleftharpoons} E + P \qquad (6\text{-}14)$$

The complex formation is reversible and the complex undergoes conversion, usually in a unimolecular reaction, to a product or products. Three rate constants are needed to describe this system for a reaction that is irreversible overall. A complete description of the kinetic behavior under all circumstances is fairly involved, but in the vast majority of cases the enzyme is present in an extremely low concentration (e.g., 10^{-8} M) while the substrate is present in large excess.[‡] Under these circumstances and in the majority of situations considered by biochemists, the **steady state approximation** can be used. It is assumed that the rate of formation of ES from free enzyme and substrate is exactly balanced by the rate of conversion of ES on to P; that is, for a relatively short time during the duration of the experimental measurement of velocity, the concentration of ES remains essentially constant. In fact, this is an oversimplified statement. As Jencks has pointed out, the steady state criterion is met if the *absolute* rate of change of a concentration of a transient intermediate is very small compared to that of the reactants and products.[12]

The **Michaelis–Menten equation** for the initial reaction rate of a single substrate with an enzyme is (Eq. 6-15)

$$v = \frac{V_{max}}{1 + K_m/[S]} = \frac{V_{max}[S]}{K_m + [S]} \qquad (6\text{-}15)$$

[†] There were times in the recent past when chemists thought that substrate and enzyme might remain 10 nm or more apart. Can you imagine what types of experiments were done to disprove this possibility?

[‡] This relationship may be reversed within cells, however.[11]

where $K_m = (k_2 + k_3)/k_1$. The equation can be derived by setting the rate of formation of the ES complex (k_1 [E][S]) equal to its rate of breakdown, ($[k_2 + k_3][ES]$). It follows (Eq. 6-16) that

$$[E][S] = \frac{(k_2 + k_3)}{k_1} [ES] = K_m [ES] \qquad (6\text{-}16)$$

Using this equation together with a mass balance (conservation of enzyme) relationship ([E] = $[E]_t - [ES]$) we can solve for the ratio $[ES]/[E]_t$, the fraction of enzyme combined as enzyme–substrate complex (Eq. 6-17).

$$\frac{[ES]}{[E]_t} = \frac{[S]}{K_m + [S]} \qquad (6\text{-}17)$$

Maximum velocity $V_{max} = k_3 [E]_t$ is attained only when all of the enzyme is converted into ES. Under other conditions $v = k_3 [ES]$ and Eq. 6-18 holds.

$$[ES]/[E]_t = v/V_{max} \qquad (6\text{-}18)$$

Substituting from Eq. 6-18 into Eq. 6-17 gives the Michaelis–Menten equation (Eq. 6-15).

Equation 6-15 provides a relationship between the velocity observed at any particular substrate concentration and the maximum velocity that would be achieved at infinite substrate concentration. The two quantities V_{max} and K_m are often referred to as the **kinetic parameters** of an enzyme and their determination is an important part of the characterization of any enzyme.

In many cases, the rate at which ES is converted back to free E and S is much greater than the rate of conversion of ES to products ($k_2 \gg k_3$). In such cases K_m equals k_2/k_1, the dissociation constant for breakdown of ES to free enzyme and substrate. Thus, K_m *sometimes* has a close inverse relationship to the strength of binding of substrate to enzyme. In such cases, $1/K_m$ is a measure of the affinity of the substrate for its binding site on the enzyme and for a series of different substrates acted on by the same enzyme. The more tightly bound substrates have the lower values of K_m. But beware! The condition that k_3 is negligible compared to k_2 may be met with some (poorer) substrates, but may not always be met by others.

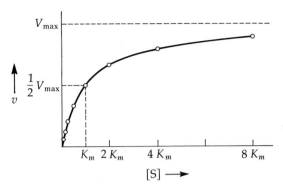

Fig. 6-2 Plot of observed velocity v vs. substrate concentration [S] for a simple enzymatic process.

Figure 6-2 shows a plot of velocity against substrate concentration as given by Eq. 6-15. The position of V_{max} on the ordinate is marked, but it clear that the experimental velocity (v) can never attain V_{max} unless K_m is extremely low. Rather, the value of v approaches V_{max} slowly and asymptotically. The value of K_m can be estimated from Fig. 6-2; it is equal to [S] when $v = V_{max}/2$. However, because of the difficulty of establishing the value of V_{max} a plot of the type shown in Fig. 6-2 is rarely used. It does have the advantage of being easy to comprehend and with the advent of computer-assisted evaluation of V_{max} and K_m from experimental data, we may see a return to plots of this type. Figure 6-2 is identical in form to the saturation curves for reversible binding shown in Fig. 4-1.

6. Linear Forms for Rate Equations

To obtain K_m and V_{max} from experimental rate data, Eq. 6-15 can be transformed to one of several linear forms, for example

$$\frac{1}{v} = \frac{1}{V_{max}} + \frac{K_m}{V_{max}} \frac{1}{[S]} \qquad (6\text{-}19)$$

From a **double reciprocal** or **Lineweaver–Burk** plot of $1/v$ against $1/[S]$ (Fig. 6-3A), Eq. 6-19 can be used to evaluate K_m/V_{max} and $1/V_{max}$ from the slope and intercept, respectively. A better linear plot (**Eadie–Hofstee plot**) is that of $v/[S]$ vs. v (Fig. 6-3B) which is related to the Scatchard plot (Fig. 4-3) and is fitted by Eq. 6-20.

$$\frac{v}{[S]} = \frac{V_{max}}{K_m} - v \frac{1}{K_m} \qquad (6\text{-}20)$$

While the point for [S] = 0 and v = 0 cannot be plotted, the ratio $v/[S]$ approaches V_{max}/K_m as v approaches zero. Note the distribution of the points in Fig. 6-3. Substrate concentrations were chosen such that the increase in velocity from point to point is more or less constant, a desirable experimental situation. The points on the Eadie–Hofstee plot are also nearly evenly distributed, but those of the Lineweaver–Burk plot are compressed at one end. A second advantage of the Eadie–Hofstee plot is that the entire range of possible substrate concentrations from near zero to infinity can be fitted onto a single plot.

In discussions of the *control* of metabolism through regulation of enzymatic activity, it is often preferable to plot v against log [S] as in Fig. 6-4. This plot also has the virtue that the entire range of attainable substrate concentrations can be plotted on one piece of paper if the point for [S] = 0 at minus infinity is omitted. The same scale can be used for all enzymes. The plot is S-shaped, both for simple cases that are represented by Eq. 6-15 and for enzymes that bind substrate cooperatively (Section B,5). Thus, the classification of "hyperbolic" vs. "sigmoidal" is lost. However, the degree of cooperativity can be directly measured from the midpoint slopes of curves of this type (Chapter 4, Section C,7). Since it is awkward to measure V_{max} from a plot of this type, it may be preferable to obtain these parameters from a linear plot (Fig. 6-3B). Alternatively, computer-assisted methods can be used to obtain both K_m and V_{max} and to fit a curve to the experimental points as in Fig. 6-4. For cases involving cooperative binding of substrates (Section B,5) computer fitting is almost essential.

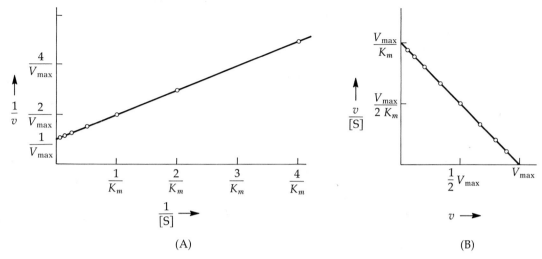

(A)   (B)

Fig. 6-3 (A) Double reciprocal or Lineweaver–Burk plot of $1/v$ vs. $1/[S]$. The intercept on the vertical axis gives $1/V_{max}$ and the slope gives K_m/V_{max}. (B) The Eadie–Hofstee plot of $v/[S]$ against v. The slope is $-1/K_m$; the intercept on the vertical axis is V_{max}/K_m and that on the horizontal axis is V_{max}.

7. Diffusion and the Rate of "Encounter" of Enzyme with Substrate

What determines the value of k_1 in Eq. 6-14? This rate constant represents the process by which the substrate and enzyme find each other, become mutually oriented, and bind to form the ES complex. If orientation and binding are rapid enough, the rate will be determined by the speed with which the molecules can come together by diffu-

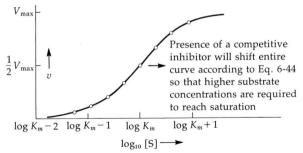

Fig. 6-4 Plot of v against log [S] for an enzyme-catalyzed reaction.

sion. Molecules in solution are not free to travel for more than a tiny fraction of their diameter as a result of their frequent collision with solvent molecules. During diffusion the molecules of a liquid twist and turn and squeeze past each other. The result is visible in the **Brownian movement** of microscopic particles suspended in a fluid. If an individual particle is followed, it is seen to undergo a "random walk," moving in first one direction then another. Einstein showed that if the distances transversed by such particles in a given time Δt are measured, the mean square of these Δx values Δ^2 can be related to the diffusion constant D (which is usually given in units of cm^2 s^{-1}) by Eq. 6-21.

$$\Delta^2 = 2D \, \Delta t \qquad (6\text{-}21)$$

For molecules, the Brownian movement cannot be observed directly but the diffusion constant can be measured, for example, by observing the rate of spreading of a boundary between two different concentrations of the substance.[13] The diffusion constant for $^1H^2HO$ (HDO) in H_2O at 25°C is 2.27×10^{-5} cm^2 s^{-1}, and the values for the ions

K^+ and Cl^- are about the same.[14] For many small molecules D is approximately 10^{-5} cm^2 s^{-1}, and the value decreases as the size of the molecule increases. Thus, for ribonuclease (MW = 13,683) $D = 1.1 \times 10^{-6}$ cm^2 s^{-1} and for myosin (MW = 5×10^5) D is $\sim 1 \times 10^{-7}$ cm^2 s^{-1}. The **Stokes–Einstein equation** (Eq. 6-22) can be used to relate the diffusion coefficient to the radius of a spherical particle, the coefficient of viscosity η and the Boltzmann constant k_B.

$$D = \frac{k_B T}{6\pi\eta r} \qquad (6\text{-}22)$$

To estimate the rate constant for a reaction that is controlled strictly by the frequency of collisions of particles, we must ask how many times per second will one of a number n of particles be hit by another of the particles as a result of Brownian movement. The problem was first approached in 1917 by Smoluchowski[15,16] who considered the rate at which a particle B diffuses toward a second particle A and disappears when the two collide. Using Fick's law of diffusion, he concluded that the number of **encounters** per milliliter per second was

Number of encounters/ml s^{-1}
$$= 4\pi(D_A + D_B)(r_A + r_B)n_A n_B \qquad (6\text{-}23)$$

Here D_A and D_B are the two diffusion coefficients, r_A and r_B are the radii of the two particles, and n_A and n_B the numbers of particles per milliliter. The number of encounters per liter per second is $(4\pi/1000)(D_A + D_B)(r_A + r_B)N^2[A][B]$ where N is Avogadro's number. Dividing this frequency by N gives the rate of collision v in M s^{-1}, a velocity that can be equated with $k_D[A][B]$, where k_D is a second order rate constant (Eq. 6-24).

$$v = \frac{4\pi}{1000}(D_A + D_B)(r_A + r_B)[A][B]N = k_D[A][B] \qquad (6\text{-}24)$$

Equation 6-25 follows:

$$k_D = \frac{4\pi N}{1000}(D_A + D_B)(r_A + r_B)M^{-1}\,s^{-1} \qquad (6\text{-}25)$$

While Eq. 6-25 is thought to overestimate the diffusion-limited rate constant a little, it is a good approximation.

If it is assumed that the diffusing particles are spherical, diffusion constants D_A and D_B can be calculated from Eq. 6-22, and Eq. 6-25 becomes (Eq. 6-26)

$$k_D = \frac{2RT}{3000\eta}\left(2 + \frac{r_A}{r_B} + \frac{r_B}{r_A}\right) \qquad (6\text{-}26)$$

Note that the value of k_D does not vary rapidly as the ratio of radii r_A/r_B is changed. Thus, in most cases it is possible to assume that $r_A/r_B \approx 1$, in which case the equation simplifies (Eq. 6-27).

$$k_D \approx \frac{8RT}{3000\eta} M^{-1}\,s^{-1} \qquad (6\text{-}27)$$

For water at 25°C, the coefficient of viscosity η is ~ 0.01 poise (1 poise = 10^{-5} newton cm^{-2}) which leads, according to Eq. 6-27, to $k_D \approx 0.7 \times 10^{10}$ M^{-1} s^{-1}. It was shown by Debye[17] that this rate constant must be multiplied by a correction factor when charged particles rather than uncharged spheres diffuse together. This factor may be of the order of 5 to 10 for a substrate and enzyme carrying two or three charges and may act to either increase or decrease reaction rates.

A simple alternative derivation of Eq. 6-27 has been given by D. French.† Consider a small element of volume ΔV swept out by a particle as it moves through the solution for a distance equal to its own radius. This element of volume will equal πr^3 (Eq. 6-28).

$$\Delta V = \pi r^2 \times r = \pi r^3 \qquad (6\text{-}28)$$

It will be swept out in a time Δt which can be calculated from Eq. 6-21 as $r^2/2D$. Substituting the value of D given by Eq. 6-22, we obtain for Δt (Eq. 6-29)

$$\Delta t = \frac{3\pi\eta r^3}{k_B T} \qquad (6\text{-}29)$$

Division of Eq. 6-28 by Eq. 6-29 gives the volume swept out per second by one particle (Eq. 6-30):

$$\Delta V/\Delta t = k_B T/3\eta \text{ cm}^3\,s^{-1} \qquad (6\text{-}30)$$

---

† Private communication from D. French, to whom the author is indebted for most of this discussion on encounter theory. Part of the discussion has been published.[18]

Since the collision radius for two particles of equal size is two times the particle radius, the effective volume swept out will be four times that given by Eq. 6-30. Since both particles are diffusing, the effective diffusion constant will be twice that used in obtaining Eq. 6-25. Thus, the effective volume swept out by the particle in a second will be eight times that given by Eq. 6-30. The volume swept out by one mole of particles is equal to k_D (recall that the second order rate constant has dimensions of $l \ mol^{-1} \ s^{-1}$). Thus, when converted to a moles per liter basis and multiplied by 8, Eq. 6-30 should (and does) become identical with the Smoluchowski equation (Eq. 6-27).

The volume given by Eq. 6-30 is about $1.4 \times 10^{-11} \ cm^2$, which could be represented approximately by a cube 2.4 μm on a side. If we compare this volume with that of a cell (Table 1-2) or of an organelle, we see that in one second an enzyme molecule will sweep out a large fraction of the volume of a small cell, mitochondrion, chloroplast, etc.

8. The "Cage Effect" and Rotation of Molecules

It is of interest to compare the bimolecular rate constant for encounters calculated by the Smoluchowski theory (Eq. 6-26) with the corresponding bimolecular rate constant for molecular collisions given by the kinetic theory of gases (Eq. 6-31).

k (collision)

$$= \frac{N(r_A + r_B)^2}{1000} \left[8\pi k_B T \left(\frac{1}{m_A} + \frac{1}{m_B} \right) \right]^{1/2} \quad (6\text{-}31)$$

Here m_A and m_B are the masses of the two particles. This rate constant is also relatively independent of molecular size and for spheres varies from 4 to $11 \times 10^{11} \ M^{-1} \ s^{-1}$, something over an order of magnitude greater than the rate constant for encounters. In a gas, molecules collide, then bounce far away at the frequency of collision. In a solution, they still collide at about the same rate so that *100 to 200 collisions occur between two particles for each encounter*. During the time of the single en-

counter, the particles are together in a solvent "cage," a fact of great significance for enzymatic reactions.

During the time of an encounter between substrate and enzyme both molecules undergo random rotational motions. Therefore successive collisions bring them together in different orientations, one of which is likely to lead to a sufficiently close match of complementary surfaces (of substrate and binding site) that formation of a "productive" ES complex takes place.

Molecular rotation in a solution is described quantitatively by diffusion laws (analogous to Fick's laws) for which a **rotary diffusion constant** θ is defined.[19,20] Consider a group of molecules all oriented the same way initially, then undergoing rotary diffusion until their orientations became random. If we measure the orientation of each molecule by an angle α we see that initially the value of $\cos \alpha$ is 1 but that when the angles become random the mean value of $\cos \alpha$ averaged over all molecules is zero. The rotary relaxation

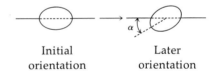

| Initial orientation | Later orientation |

time τ is the time required for the mean value of $\cos \alpha$ to fall to $1/e$ (which occurs when $\alpha = 68\frac{1}{2}°$). For a sphere θ is given by Eq. 6-32.

$$\theta = \tfrac{1}{2}\tau = \frac{k_B T}{8\pi \eta r^3} \quad (6\text{-}32)$$

Ellipsoidal or rod-shaped molecules have two different rotary diffusion constants while, if the dimensions of the molecule are different along all three axes, three constants must be specified.

From Eq. 6-32 we can calculate (if $\eta = 0.01$ poise) the following values for a small spherical molecule (substrate) of ~1 nm length and for a spherical enzyme of 5 nm diameter:

$r = 0.5$ nm	$\theta \approx 1.3 \times 10^9 \ s^{-1}$
$r = 2.5$ nm	$\theta \approx 1.0 \times 10^7 \ s^{-1}$

We see that smaller molecules rotate much faster than large ones and that rotational relaxation times for small proteins are of the same order of magnitude as k_D for diffusion-limited encounter. However, for very large molecules, especially long rods, the rotary relaxation time about a short axis may be a large fraction of a second.

9. Reversible Enzymatic Reactions

Although in many enzyme-catalyzed reactions, the equilibrium lies far to one side, other reactions are freely reversible. Since a catalyst promotes reactions in both directions, we must consider the action of an enzyme on the reverse reaction. Let us designate the maximum velocity in the forward direction as V_f and that in the reverse direction as V_r. There will be a Michaelis constant for reaction of enzyme with product K_{mP}, while K_{mS} will refer to the reaction with substrate.

Just as in any other chemical reaction, there is a relationship between the rate constants for forward and reverse reactions and the equilibrium constant. This is simply derived by setting $v_f = v_r$ for the condition that product and substrate concentrations are those at equilibrium. For a single substrate–single product system, the resulting **Haldane relationship*** is given by Eq. 6-33.

$$K_{eq} = V_f K_{mP}/V_r K_{mS} \qquad (6\text{-}33)$$

Because the Haldane relationship imposes constraints on the values of the velocity constants and Michaelis constants, it is of some importance in considering regulation of metabolism. Consider the case that the maximum forward velocity V_f is high and that K_{mS} has a moderately low value (fairly strong binding of substrate). If the reaction is freely reversible ($K_{eq} \sim 1$) and the velocity of the reverse reaction V_r is about the same as that of the forward reaction, it will necessarily be true (from Eq. 6-33) that the product P will also be fairly tightly bound. On the other hand, if $V_r \ll V_f$, the value of K_{mP} will have to be much *lower* than that of K_{mA}. In such a situation P will remain tightly

*Named for J. B. S. Haldane, British kineticist and proposed in his book, ''Enzymes.''[21]

bound to the enzyme and since V_r is low, it will tend to clog the enzyme. Such **product inhibition** may sometimes slow down a whole pathway of metabolism. In such a case, the only way that an enzymatic sequence can keep going in the forward direction is for product P to be rapidly removed by a subsequent reaction with a second enzyme. The enzyme may be thought of as operating as a kind of ''one-way valve'' that turns off the flow in a metabolic pathway when the concentration of its product rises.

10. Reactions of Two or More Substrates

Enzymes frequently catalyze the reaction of two, three, or even more different molecules to give one, two, three, or more products. Sometimes all of the molecules must be bound to active sites at the same time and are presumably lined up on the molecule in such a way that they can react in proper sequence. In other cases, the enzyme may transform molecule A to a product, then cause the product to react with molecule B. The number of variations is enormous.[1-5]

The order in which two molecules A and B bind to an enzyme to form a complex EAB may be completely *random* or it may be strictly *ordered*. Both situations occur with real enzymes. Cleland has introduced a widely used method of depicting the various possibilities. The following example (Eq. 6-34) shows the reaction of A and B in an ordered sequence to form the complex EAB which is then isomerized to complex EPQ:

$$(6\text{-}34)$$

The latter can be thought of as the complex formed by binding the two products P and Q to the enzyme. The rate constant to the left of each vertical arrow or above each horizontal arrow refers to the reaction in the forward direction as indicated by the arrow while the other constants (to the right or below the arrows) refer to the reverse reactions. The velocity in the forward direction for an enzyme with ordered binding is given by Eq. 6-35a,

$$v_f = \frac{V_f[A][B]}{K_{eqA}K_{mB} + K_{mB}[A] + K_{mA}[B] + [A][B]} \quad (6\text{-}35a)$$

which may also be written† in the reciprocal form (Eq. 6-35b):

$$\frac{1}{v_f} = \frac{1}{V_f}\left(1 + \frac{K_{mA}}{[A]} + \frac{K_{mB}}{[B]} + \frac{K_{eqA}K_{mB}}{[A][B]}\right) \quad (6\text{-}35b)$$

The kinetic parameters are V_f, the maximum velocity in the forward direction, the two Michaelis constants, K_{mB} and K_{mA}, and an equilibrium constant K_{eqA}, which is the constant for reversible dissociation of the complex EA and is equal to k_2/k_1. The relationship between the parameters of Eq. 6-35a (K_m's, V's, K_{eqA}'s) and the kinetic constants (k_1–k_{10}) is not obvious. However, remember that the parameters are experimental quantities determined by measurements on the enzyme. Sometimes, but not always, it is possible to ascertain some of the values of individual kinetic constants from the experimental parameters.

An equation entirely similar to Eq. 6-35a can be written for the velocity v_r of the reaction involving molecules P and Q. Also an equation can be written in similar form for $v_f - v_r$, i.e., the instantaneous velocity of the reaction in any mixture of all four components A, B, P, and Q.

The kinetic parameters of Eq. 6-35b are often obtained from experimental data through the use of reciprocal plots (Fig. 6-5). Note that Eq. 6-35b is linear only if the concentration of one or the other of the substrates A and B is kept constant. Thus a series of experiments is usually performed in which [A] is varied while [B] is held constant. Then [A] is held constant and [B] is varied. Each of these experiments leads to a family of lines (Fig.

6-5A) whose slopes and intercepts are measured. Then, the slopes and intercepts of this family of curves are plotted against the reciprocal of the second concentration (i.e., the one that was held fixed in obtaining a particular line).

From these **secondary plots,** V_f and one of the Michaelis constants can be determined (Figs. 6-5B and C). Using two sets of secondary plots, all of the constants of Eq. 6-35b may be established (except that the product $K_{eqA}K_{mB}$ is treated as a single constant K_{AB}). Alternatively, a computer can be used to examine all of the data at once and to obtain the best values of the parameters. The latter approach is desirable because reliable estimates of the standard deviations of the parameters can be obtained.[1]

The meaning of the kinetic parameters may be a little hard to grasp. V_f is the velocity that would be obtained if both [A] and [B] were infinite. Each K_m corresponds to that for a simple system in which the concentration of the second substrate is high enough to saturate the enzyme.

For the bimolecular reaction that we have considered, there are two Haldane relationships (Eq. 6-36):

$$K_{eq} = \frac{V_f K_{mP} K_{dQ}}{V_r K_{dA} K_{mB}} = \left(\frac{V_f}{V_r}\right)^2 \frac{K_{dP} K_{mQ}}{K_{mA} K_{dB}} \quad (6\text{-}36)$$

Of these only the first is ordinarily used.

a. "Ping-Pong" Mechanisms

A common type of mechanism that is especially prevalent with coenzyme-requiring reactions has been dubbed **ping-pong** because the enzyme alternates between two forms E and E′ (Eq. 6-37). Sub-

$$
\begin{array}{ccccccccc}
 & A & & & P & B & & & Q \\
 & k_1 \downarrow k_2 & & & k_5 \downarrow k_6 & k_7 \downarrow k_8 & & & k_{11} \downarrow k_{12} \\
E & \longrightarrow EA & \xrightarrow[k_4]{k_3} & E'P & \longrightarrow E' & \longrightarrow E'B & \xrightarrow{k_9}_{k_{10}} & EQ & \longrightarrow E
\end{array}
$$
(6-37)

Ping-pong mechanism for reaction of A and B to form P and Q

strate A reacts via complex EA to form E′, a modified enzyme—often one in which the coenzyme has been changed (e.g., via transamination, Chapter 8, Section E,1). At the same time A has

---

† Another much used form of this equation was proposed by Dalziel[22]:

$$\frac{[E]_t}{v_f} = \phi_0 + \frac{\phi_1}{[A]} + \frac{\phi_2}{[B]} + \frac{\phi_{12}}{[A][B]}$$

The total enzyme concentration $[E]_t$ usually equals v_f/k_f (Eq. 6-6). It follows from Eq. 6-36 that $\phi_0 = 1/k_f$ in the present case where k_f can sometimes be equated with k_5 of Eq. 6-34.

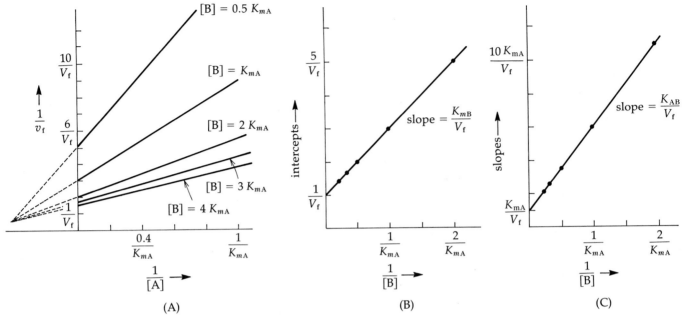

Fig. 6-5 Reciprocal plots used to analyze kinetics of two-substrate enzymes. (A) Plot of $1/v_f$ against $1/[A]$ for a series of different concentrations of the second substrate B. (B) A secondary plot in which the intercepts from graph A are plotted against $1/[B]$. (C) Secondary plot in which the slopes from graph A have been plotted against $1/[B]$. The figures have been drawn for the case that $K_{mA} = 10^{-3} \, M$, $K_{mB} = 2K_{mA}$, and $K_{AB} = K_{eqA}K_{mB}$ (Eq. 6-36) $= K_{mA}/200$ and [A] and [B] are in units of moles per liter. Eadie–Hofstee plots of $v_f/[A]$ vs. v_f at constant [B] can also be used as the primary plots. The reader can easily convert Eq. 6-35 to the proper form analogous to Eq. 6-20.

been changed to product P still bound to the enzyme. Product P dissociates leaving E′ which is then able to react with the second substrate B and to go through the second half of the cycle during which E′ is converted back to E.

The rate equations for the ping-pong mechanism resemble those considered for the ordered bimolecular reaction (Eqs. 6-35 and 6-36), but each has one less term† (Eqs. 6-38a and 6-38b):

or

$$v_f = \frac{V_f[A][B]}{K_{mB}[A] + K_{mA}[B] + [A][B]} \tag{6-38a}$$

$$\frac{1}{v_f} = \frac{1}{V_f}\left(1 + \frac{K_{mA}}{[A]} + \frac{K_{mB}}{[B]}\right) \tag{6-38b}$$

† Note that if $K_{AB}/[A][B]$ is small compared to the other terms in Eq. 6-36 the reaction will appear to be ping-pong even though it is sequential and goes through ternary complex EAB.

One less kinetic parameter can be obtained from an analysis of the data than can be obtained for ordered reactions. Nevertheless, in the scheme for the ping-pong mechanism (Eq. 6-37), twelve kinetic constants are indicated. At least this many steps must be considered to describe the behavior of the enzyme. Clearly, not all of these constants can be determined from a study of steady state kinetics but must be obtained in other ways.

An interesting feature of the ping-pong mechanism is that the families of lines (in double reciprocal plots) obtained when one substrate is held constant while the other is varied, no longer intersect as they do in Fig. 6-5A but are parallel as in Fig. 6-7 (for noncompetitive inhibition).

b. Nonproductive Complexes

In the steady state of action of an enzyme with ping-pong kinetics, part of the enzyme is in form

E and part in form E'. In the idealized case, E has affinity only for A and Q while E' has affinity only for B and P. However, in many real situations, P and B have *some* affinity also for E (and A and Q for E'). In many cases this is intuitively understandable because the products and reactants often all have some features of structure in common. Thus, we might expect some affinity for all four forms at the binding sites of both E and E'. In special cases it appears that the ability of enzymes with ping-pong kinetics to form **nonproductive** (sometimes called abortive) **complexes** has been exploited for regulatory purposes. The scheme shown in Eq. 6-39 is Eq. 6-37 rearranged to depict a situation in which product P normally undergoes a sequence of further reactions. However, if P accumulates to

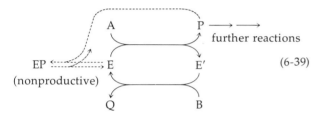

(6-39)

a high enough concentration it can react with form E to give a nonproductive complex EP. This represents an effective form of product inhibition which can be relieved only by a lowering of the concentration of P by its further metabolism. Concrete examples of this type of inhibition and its importance in metabolic regulation have been described.

c. Methods of Handling Rate Equations for Complex Mechanisms

While steady state rate equations can be derived easily for the simple cases discussed in the preceding sections, enzymes are often considerably more complex and the derivation of the correct rate equations can be extremely tedious. The topological theory of graphs, widely used in analysis of electrical networks, has been applied to both steady state and nonsteady state enzyme kinetics.[23-25] The methods employ diagrams of the type shown in Eq. 6-40. Here the reaction of an enzyme with two substrates A and B with a *random* order

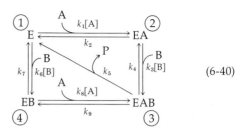

(6-40)

of binding is depicted. (In contrast, Eq. 6-34 applies to the case of *ordered* binding of two substrates.) When complex EAB is formed, it can decompose to free enzyme and to the single product P. Each one of the **nodes,** which are numbered 1–4 in the diagram, corresponds to a single form of the enzyme. The appropriate first order rate constant or apparent first order constant is placed by each arrow. The methods provide easy to follow rules for deriving from such a scheme the steady state rate equation.[1]

The importance of the simplified schematic methods is apparent when one considers the more complex case in which EAB breaks down to two products P and Q with a random order of release. The rate equation, in reciprocal form, contains 672 terms in the denominator. A simple method is needed. For that matter, in such complex cases it is worthwhile to enlist the help of a computer in deriving the equation.[26]

d. The Rapid Equilibrium Assumption

Rate equations for enzymes are often simplified if a single step, e.g., that of reaction of complex EAB to product in Eq. 6-40, is **rate limiting.** If it is assumed that all reaction steps preceding or following the rate limiting step are at equilibrium, the equation for random binding with a two-substrate and two-product reaction simplifies to one whose form is similar to that obtained for ordered binding (Eq. 6-35). This simplification is often valid but it is not always so, especially for very rapidly acting enzymes.

e. Kinetics with High Enzyme Concentrations

Laboratory studies of the kinetics of purified enzymes are usually conducted with enzyme concentrations of 10^{-7} to 10^{-10} M, but within cells enzyme concentrations are probably often in the range 10^{-6} to 10^{-5} M.[27] Thus, within cells enzyme concentrations may be higher than those of the substrates upon which the enzymes act. Great caution must be observed in drawing conclusions about the kinetics under such circumstances. Methods have been devised for handling kinetic data when the concentration of enzyme is greater than K_m, a condition that leads to intolerably high errors if the usual equations are applied.[28]

11. Kinetics of Rapid Reactions

The fastest steps in an enzymatic process cannot be observed by conventional steady state kinetic methods because the latter cannnot be applied to reactions with half-times of less than about 10 s. Consequently, a whole variety of newer methods have been developed[10,28-30] to measure rates in the range of 1 to 10^{-13} s.

a. Flow Methods

One of the first rapid kinetic methods to be devised consists in rapidly mixing reactants by flowing two solutions together in a special mixing device and allowing the mixture to flow at a rate of several meters per second down a straight tube. At a flow velocity of 10 m s^{-1} a solution will move 1 cm in 10^{-3} s. Observations of the mixture can be made at a suitable distance, e.g., 1 cm, and with various flow rates. Using spectrophotometry or any number of other observation techniques, the formation or disappearance of a product or reactant can be followed. The special advantage of this technique is that observation can be made slowly. However, it is uneconomical of the often precious reactant solutions, e.g., those of purified enzymes.

The widely used **stopped flow** technique involves a rapid mixing of two solutions by the flow technique during a period of only 1–2 (or a few) milliseconds. Flow is initiated and stopped by means of a plunger which drives two hypodermic

syringes. The two solutions are injected into the mixing chamber and pass on into an observation cell where (after flow has stopped) light absorption emf or conductivity, etc., can be measured. The method demands a means of rapid observation. For example, if light absorption is measured, a photomultiplier is attached to an oscilloscope and the changes in absorbance over a period of a small fraction of a second are displayed and the trace is photographed. Relaxation times of down to a few milliseconds can be measured in this way.

b. Relaxation Methods

Kinetic measurements over periods of tens of microseconds or less can be made by rapidly inducing a small displacement from the equilibrium position of a reaction (or series of reactions) and observing the rate of return (relaxation) of the system to equilibrium. Best known is the **temperature jump** method devised by Eigen and associates. Over a period of about 10^{-6} s a potential difference of ~100 kV is applied across the experimental solution. A rapid electrical discharge from a bank of condensers passes through the solution (without any sparking) raising the temperature 2–10 degrees. All the chemical equilibria for which $\Delta H \neq 0$ are perturbed. If some property such as the absorbance at a particular wavelength or the conductivity of the solution is measured, very small relaxation times can be determined.

While it may not be intuitively obvious, if the displacement from equilibrium is small, the rate of return to equilibrium can always be expressed as a first order process (e.g., see Eq. 6-13). In the event that there is more than one chemical reaction required to reequilibrate the system, each reaction has its own characteristic relaxation time. If these relaxation times are close together, it is difficult to unravel them; however, they often differ, one from the other, by an order of magnitude or more. Thus, two or more relaxation times can often be evaluated for a given solution. In favorable circumstances these relaxation times can be related directly to rate constants for particular steps. For example, Eigen measured the conductivity of water following a temperature jump[10] and observed the rate of combination of H^+ and OH^- for which τ at 23°C equals 37×10^{-6} s. From this, the rate constant for

$$H^+ + OH^- \longrightarrow H_2O \qquad (6\text{-}41)$$

combination of OH^- and H^+ (Eq. 6-41) was calculated as follows (Eq. 6-42):

$$k = 1/\{\tau([OH^-] + [H^+])\} = 1.3 \times 10^{11}\ M^{-1}\ s^{-1} \quad (6\text{-}42)$$

Pressure jump and electric field jump methods have also been used as have methods depending upon periodic changes in some property. For example, absorption of ultrasonic sound causes a periodic change in the pressure of the system.

c. Flash Methods

Another widely useful method is to discharge a condenser through a flash tube over a period of 10^{-12} to 10^{-4} s causing a rapid light absorption in a sample in an adjacent parallel tube. Following the flash, changes in absorption spectrum or fluorescence of the sample are followed. The availability of extremely intense lasers as light sources now makes it possible to follow the results of light flashes of nanosecond duration and to measure extremely short relaxation times.[31]

d. Some Results

Rapid kinetic methods have revealed that enzymes often combine with substrates extremely rapidly,[32] and values of k_1 in Eq. 6-14 frequently fall in the range of 10^6 to $10^8\ M^{-1}\ s^{-1}$. Even so, combination of substrate and enzyme is slower than the diffusion-controlled limit of 10^9 to $10^{10}\ M^{-1}\ s^{-1}$ indicating that a certain time is required for a substrate molecule to become oriented and seated in the active sites of the enzyme. Helix-coil transitions of polypeptides have relaxation times of about 10^{-8} s, but renaturation of a denatured protein may be much slower. Interconversion between chair and boat forms of cyclohexane derivatives may have $\tau \sim 10^{-5}$ s at room temperature while rotation about a C—N bond in an amide linkage may be very slow with $\tau \sim 0.1$ s. The nonenzymatic hydration of the aldehyde pyridoxal phosphate (Chapter 8, Section E) occurs

with $\tau = 0.01 - 0.1$ s, depending upon the pH. The T jump study upon which this result is based[33] is a careful one that would merit the reader's attention.

B. INHIBITION AND ACTIVATION OF ENZYMES

The action of most enzymes is inhibited by a variety of substances. Inhibition is often highly specific, and studies of the relationship between inhibitor structure and activity have been important to the development of our concepts of active sites and of complementarity of surfaces of biomolecules. Inhibition of enzymes also lies at the basis of the action of a very large fraction of drugs.

Inhibition may be **reversible** or **irreversible.** The latter refers to reactions that lead to permanent inactivation of an enzyme.[33a] An example is the action of organophosphorus nerve poisons (Chapter 7, Section D,1) on the enzyme **acetylcholinesterase.** Often, but not always, irreversible inhibition is preceded by reversible binding of the inhibitor at a complementary site on the enzyme surface. The mathematical treatment of irreversible inhibition will not be considered, but quantitative aspects of the action of reservible inhibitors will be examined.

1. Competitive Inhibitors

Inhibitors with close structural similarities to a substrate tend to be bound to the substrate-binding site. For truly competitive inhibition, substrate and inhibitor must not only compete for the same site but their binding must be mutually exclusive. The affinity of the inhibitor for the enzyme is expressed quantitatively through the **inhibition constant** K_i which is the *dissociation constant of the enzyme inhibitor-complex EI* (Eq. 6-43).

$$K_i = [E][I]/[EI] \qquad (6\text{-}43)$$

Using the steady state assumption and writing a mass balance equation that includes not only free enzyme and ES but EI as well, we obtain an equation relating rate to substrate concentration that is entirely analogous to Eq. 6-15. However, K_m is replaced by an apparent K_m' defined as

$$K_m' = K_m \left(1 + \frac{[\text{I}]}{K_i}\right) \qquad (6\text{-}44)$$

The relationship between [S] and v is given by Eq. 6-45:

$$\frac{v}{[\text{S}]} = \frac{V_{\max}}{K_m'} - \frac{v}{K_m'} \qquad (6\text{-}45)$$

and between $1/v$ and $1/[\text{S}]$ by Eq. 6-46:

$$\frac{1}{v} = \frac{1}{V_{\max}} + \frac{K_m}{V_{\max}[\text{S}]}\left(1 + \frac{[\text{I}]}{K_i}\right) \qquad (6\text{-}46)$$

A commonly used test for competitive inhibition is to plot either $v/[\text{S}]$ vs v (Eq. 6-45) or $1/v$ vs $1/[\text{S}]$ (Eq. 6-46), both in the absence of inhibitor and in the presence of one or more fixed concentrations of I. The result is a family of lines (Fig. 6-6) that converge on one of the axes at the value $1/V_{\max}$. We see that the maximum velocity is unchanged by the presence of inhibitor. If sufficient substrate is added the enzyme can always be saturated with substrate and the inhibitor can be completely excluded. From the change in slope caused by addition of inhibitor the value of K_i can be calculated by using Eq. 6-45 or Eq. 6-46.

The effect of a fixed concentration of a competitive inhibitor on a plot of v against log[S] (Fig. 6-7) is to shift the curve to the right, i.e., toward higher values of [S], but without any change in shape (or in the value of $V_{\max}$).

2. Noncompetitive Inhibition and Activation

If an inhibitor binds not only to free enzyme but also to the enzyme substrate complex ES, inhibition is **noncompetitive**. In this case, S and I do not mutually exclude each other and both can be bound to the enzyme at the same time. Why does such an inhibitor slow an enzymatic reaction? In most instances, the structure of the inhibitor does

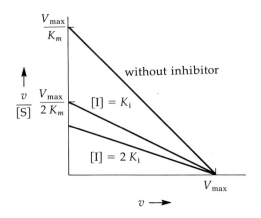

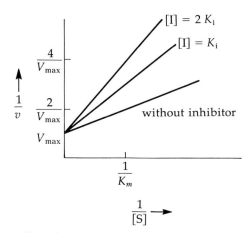

Fig. 6-6 Effect of a competitive inhibitor on the Eadie–Hofstee plot (top) and double reciprocal plot (bottom). The apparent K_m (Eq. 6-44) is increased by increasing [I], but $V_{\max}$ is unchanged.

Fig. 6-7 Plots of v vs. log [S] for competitive and noncompetitive inhibition.

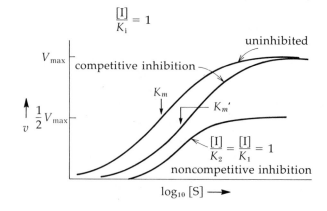

not show a close similarity to that of substrate, a fact which suggests that the binding of inhibitor is at an allosteric site, i.e., a site other than that of the substrate. The inhibition of the enzyme may result from a distortion of the three-dimensional structure of the enzyme which is caused by the binding of the inhibitor. This distortion may be transmitted to the active site even though the inhibitor binds far from that site. On the other hand, in some cases the bound inhibitor may interfere with the catalytic action by partially overlapping the active site. In either case the ES complex reacts to give product in a normal way, but the ESI complex reacts more slowly or not at all.

Binding of a substance to an allosteric site sometimes has the effect of *increasing* the activity of an enzyme rather than of inhibiting it. The quantitative treatment of such activation is similar to that of inhibition; allosteric inhibitors and activators are often considered together and are referred to as **modifiers** or **effectors**. A general scheme[34] is given by Eq. 6-47.

$$
\begin{array}{ccccc}
E + S & \underset{K_{ds}}{\rightleftarrows} & ES & \xrightarrow{k_3} & product \\
+ & & + & & \\
M \text{ (modifier)} & & M & & \\
K_1 \updownarrow & & K_2 \updownarrow & & \\
EM & \underset{K'_{ds}}{\rightleftarrows} & ESM & \xrightarrow{k_4} & product
\end{array}
\qquad (6\text{-}47)
$$

Here K_1 and K_2 are equilibrium constants for dissociation of M from EM and ESM, respectively, while K_{ds} and K'_{ds} are the dissociation constant of ES and of ESM (to S and EM), respectively. Note that the latter equilibrium (indicated by dashed arrows) is not independent of the others (Eq. 6-48):

$$
K'_{ds} = K_{ds}K_2/K_1 \qquad (6\text{-}48)
$$

Note also that we are making the simplifying assumption of a rapid equilibrium (Section A,10d).

We have already considered competitive inhibition which is obtained when $K_2 = 0$ (and therefore $K'_{ds} = 0$). For this case M is always an inhibitor and no activation is possible. Noncompetitive inhibition is obtained if ESM does not react at all,

i.e., if $k_4 = 0$. The rate constant in reciprocal form is given by Eq. 6-49.

$$
\frac{1}{v} = \frac{1}{V_{max}}\left(1 + \frac{[I]}{K_2}\right) + \frac{K_m}{V_{max}[S]}\left(1 + \frac{[I]}{K_1}\right) \qquad (6\text{-}49)
$$

We see that $1/V_{max}$ is multiplied by a term containing [I] and K_2. Thus, a characteristic of noncompetitive inhibition is that the maximum velocity is decreased from that observed in the absence of the inhibitor. No matter how high the substrate concentration, inhibition cannot be reversed completely.

Figure 6-8 shows a plot of $1/v$ against $1/[S]$ at a series of fixed values of [I]. For the case that $K_1 = K_2$ (**purely noncompetitive inhibition**), a family of reciprocal plots that intersect on the horizontal axis at a value of $-1/K_m$ is obtained. On the other hand, if K_1 and K_2 differ (**partially competitive inhibition**), the family of curves intersect at some other point to the left of the vertical axis and, depending upon the relative values of K_1 and K_2, either above or below the horizontal axis. The example illustrated is for $K_2 = 0.5K_1$; i.e., binding of M to ES is twice as strong as to E.

Figure 6-7 shows inhibition data for both the noncompetitive and competitive cases plotted vs log[S]. The shift of the midpoint to the right in each case reflects the tendency of the inhibitor to

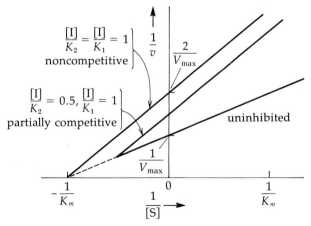

Fig. 6-8 Double reciprocal plots for noncompetitive and partially competitive inhibition.

exclude the substrate from binding while the lowered value of the maximum velocity in the case of noncompetitive inhibition results from the failure of the substrate to completely displace inhibitor from the enzyme even as [S] becomes very high.

If inhibitor binds only to ES and not to E, i.e., $K_1 = 0$, a family of parallel double reciprocal plots of $1/v$ vs $1/[S]$ will be obtained. This case, usually referred to as **uncompetitive inhibition** is rarely observed with one-substrate systems. However, multisubstrate enzymes with ping-pong mechanisms often give parallel line plots with inhibitors.

If rate constant k_4 of Eq. 6-47 is not zero we may have either inhibition or activation. If $k_3 = k_4$ the effect of a modifier will be to affect the apparent K_m, either increasing it (inhibition) or decreasing it (activation). The maximum velocity will be unchanged. Monod *et al.*[35] referred to enzymes showing such behavior as **K systems.** On the other hand, if $K_1 = K_2$ and k_3 differs from k_4 we have a purely **V system.** In the general case a modifier affects both the apparent K_m and V_{max}.

Activation by specific metallic ions is observed with a large variety of enzymes. In many instances the metallic ion is properly regarded as a *second substrate* which must bind along with the first substrate before reaction can occur. Alternatively, the complex of the organic substrate with the metal ion can be considered the "true substrate." Thus, many enzymes act upon the magnesium complex of ATP (Chapter 3, Section B,5). The enzymes can either be regarded as two-substrate enzymes requiring $Mg^{2+} + ATP^{4-}$ or as one-substrate enzymes acting upon $ATPMg^{2-}$.

3. Inhibitors in the Study of Enzyme Mechanisms[1,36]

A substrate analogue will frequently inhibit only one of the two forms of a multisubstrate enzyme with a ping-pong mechanism. Reciprocal plots made for various inhibitor concentrations consist of a family of parallel lines as in uncompetitive in-

hibition. Observation of such parallel line plots for an enzyme can support a ping-pong mechanism. (However, it cannot prove it for in some cases parallel lines are observed for inhibition of enzymes acting by an ordered sequential mechanism.) In the case of an ordered bimolecular reaction (Eq. 6-34), a question that naturally arises is: Of the two substrates required by the enzyme, which one binds to the enzyme first? If the concentration of one substrate is kept constant but varying concentrations of an inhibitory analogue of that substrate are added and $1/v$ is plotted against the reciprocal of the concentration of the other substrate, parallel lines are obtained if, and only if, the substrate of fixed concentration is B (the substrate adding second in the binding sequence) and if I is its analogue. The substrate binding first (A) is the one whose concentration was varied in the experiment.

Product inhibition (Section A,9) is also of value in providing information about mechanisms. For example, if $1/v$ is plotted against $1/[A]$ in the presence and absence of the product Q, the product will be found to compete with A and to give a typical family of lines for competitive inhibition. On the other hand, a plot of $1/v$ vs. $1/[B]$ in the presence and absence of Q will indicate noncompetitive inhibition if the binding of substrates is ordered (Eq. 6-34). In other words, only the A-Q pair of substrates are competitive. Product inhibition is also observed with enzymes having ping-pong kinetics as a result of formation of nonproductive complexes (Section A,10,b).

4. Agonists vs. Antagonists

Inhibition of enzymes provides the basis for the effects of antibiotics and other chemotherapeutic substances (e.g., see Box 6-A). However, many drugs act on cell surface receptors which are not ordinarily regarded as enzymes. According to **receptor theory,** developed around 1937, structually similar drugs often elicit similar responses because they bind to the same receptor. The receptor might normally bind a hormone, neurotrans-

BOX 6-A

THE SULFONAMIDES AS ANTIMETABOLITES

The development of the "sulfa drugs,"[a-c] derivatives of **sulfanilamide,** originated with studies by Paul Erhlich of the staining of protozoal parasites by synthetic dyes. In 1932 it was shown that the red dye 2,4-diaminoazobenzene-4'-sulfonamide (Prontosil) dramatically cured systemic infections by gram-positive bacteria. Subsequent

Prontosil

studies revealed that bacteria convert the azo dye to sulfanilamide, a compound with marked bacteriostatic activity (i.e., it inhibits growth without killing the bacteria). Although sulfonilamide had been used in large quantities since 1908 as a dye intermediate, its usefulness as an antibacterial agent had not been recognized.

In 1935 D. D. Woods found that the growth inhibition of sulfanilamide was reversed by yeast extract.[d] From this source, in 1940, he isolated **p-aminobenzoic acid** and demonstrated that the inhibitory effect of $3 \times 10^{-4}\,M$ sulfanilamide was overcome by $6 \times 10^{-8}\,M$ p-aminobenzoate. The relationship between the two compounds was shown to be strictly competitive. If the sulfonilamide concentration was doubled, twice as much p-aminobenzoate was required to reverse the inhibition as before. These facts led to the formula-

Sulfonamides
R=H, in sulfonilamide

p-Aminobenzoate

tion by Woods and by P. Fildes (in 1940) of the **antimetabolite theory.**[d,e] It was proposed that p-aminobenzoate was needed by bacteria and that sulfanilamide competed for a site on an enzyme designed to act on p-aminobenzoate. We now know that the idea was correct and that the enzyme on which the competition occurs catalyzes the synthesis of dihydropteroic acid (Fig. 14-34), a precursor to folic acid.

While sulfanilamide itself is somewhat toxic, a variety of related drugs of outstanding value have been developed. Over 10,000 sulfonamides and related compounds have been tested for antibacterial action.

[a] E. F. Gale, E. Cundliffe, P. E. Reynolds, M. H. Richmond, and M. J. Waring, "The Molecular Basis of Antibiotic Action." Wiley, New York, 1972.

[b] T. J. Bardos, *Top. Curr. Chem.* **52,** (1974).

[c] R. G. Shepherd, in "Medicinal Chemistry" (A. Burger, ed.), 3rd ed., pp. 255–304. Wiley (Interscience), New York, 1970.

[d] D. D. Woods, *Brit. J. Exp. Pathol.* **21,** 74–90 (1940).

[e] P. Fildes, *Lancet* **I** (1940), 955–957 (1940).

mitter, or other metabolite whose geometry is partially shared by the drug. Binding of a drug of one class, termed agonists in the pharmacological literature, to an appropriate receptor triggers a response in a cell, just as does a hormone. On the other hand, compounds of related structure often act as **antagonists,** binding to receptor but failing to elicit a response. Agonist and antagonist often act in a strictly competitive fashion as in competitive inhibition of enzyme action.

5. Isotope Exchange at Equilibrium

Consider the reaction of substrates A and B to form P and Q (Eq. 6-50). If both reactants and both products are present with the enzyme and in the ratio found at equilibrium no net reaction will take place. However, the reactants and products will be continually interconverted under the action of the enzyme. Now if a small amount of

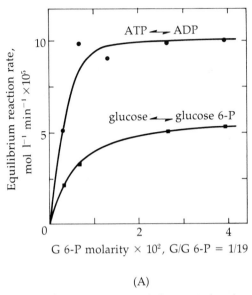

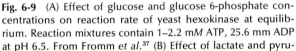

(A)

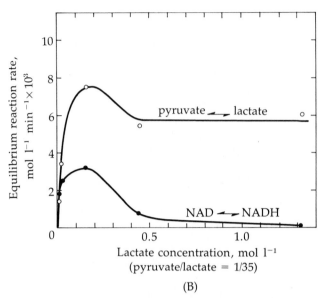

(B)

Fig. 6-9 (A) Effect of glucose and glucose 6-phosphate concentrations on reaction rate of yeast hexokinase at equilibrium. Reaction mixtures contain 1–2.2 mM ATP, 25.6 mm ADP at pH 6.5. From Fromm *et al.*[37] (B) Effect of lactate and pyruvate concentrations on equilibrium reaction rates of rabbit muscle lactate dehydrogenase. Reaction mixtures contained 1.7 mM NAD$^+$, 30–46 μM NADH in Tris-nitrate buffer, pH 7.9, 25°C. From Silverstein and Boyer.[38]

highly labeled reactant (A* or B*) is added, the rate at which isotope is transferred from the labeled reactant into one or the other of the products can be measured. In many cases, a label in molecule A will appear in only one of the products (e.g., Q) and a label in B will appear in P where P is that product whose structure most closely resembles that of B.

$$A + B \rightleftharpoons P + Q \qquad (6\text{-}50)$$

Figure 6-9A shows the rate of the glucose*–glucose 6-phosphate exchange catalyzed by the enzyme hexokinase (Chapter 7, Section E,6). The exchange rate is plotted against the concentration of glucose 6-phosphate with the ratio of [glucose]/[glucose 6-phosphate] constant at 1/19, such that an equilibrium ratio of reactants and products is always maintained. As can be seen from the graph, this exchange rate increases monotonically as substrate concentrations are increased and so does the rate of ATP–ADP exchange. The fact that both exchange rates increase continuously

suggests random binding of substrates.[37] The inequality of the two maximal exchange rates suggests that release of glucose 6-phosphate may be slower than that of ADP.

Figure 6-9B shows similar plots for lactate dehydrogenase.[38] In this case the pyruvate*-lactate exchange reaches a high constant value† as the amount of pyruvate is increased (with a constant [pyruvate]/[lactate] ratio of 1/35). However, the NAD*–NADH exchange increases rapidly at first but then drops abruptly as the pyruvate and lactate concentrations continue to increase. This suggests an ordered mechanism (Eq. 6-34) in which NAD$^+$ and NADH represent A and Q, respectively, and pyruvate and lactate represent B and P. As the concentrations of B and P become very high, the enzyme shuttles back and forth between EA and EQ, but these two complexes rarely dissociate to give free enzyme and A or Q. Hence, the A*–Q exchange rate drops.

---

† The initial rise before the constant value is attained is not regarded as significant.

In other cases a label may be transferred from A into P or from B into Q. Information on such exchanges has provided a valuable criterion of mechanistic type, as we shall see in later chapters.

6. Allosteric Effectors and Changes in Enzyme Conformation

The binding of a substance at an allosteric site with the induction of a conformational change appears to form the basis for many aspects of regulation. The term **allostery** (or allosterism) generally has reference to the effects of allosteric inhibitors or activators on *oligomeric* enzymes. However, monomeric enzymes may also be subject to allosteric regulation and we should start our discussion with a monomer that contains binding sites for substrate, inhibitor, and activator and which exists in conformations A and B (Fig. 6-10). Let us assume (see Eq. 4-42) that conformer B binds both substrate and activator well but that it binds inhibitor poorly or not at all. On the other hand, A binds inhibitor well but binds substrate and activator poorly. This simple combination of two conformers with different binding properties provides a means by which enzymes can be turned "on" or "off" in response to changing conditions.

If an inhibitory substance builds up to a high concentration within a cell, it binds to conformer A; if the inhibitor concentration is high enough, virtually all of the enzyme will be locked in the inactive conformation A. The enzyme will be turned off or at least reduced to a low activity. On the other hand, in the presence of a high concentration of activator the enzyme will be turned on because it is locked in the B conformation. The relative concentrations of inhibitor, activator, and substrate within a cell at any given time will determine what fraction of the enzyme is in active conformation B. It is this interplay of inhibitory and activating effects that provides the basis for much of the regulation of cell chemistry (Chapter 1, Section F).

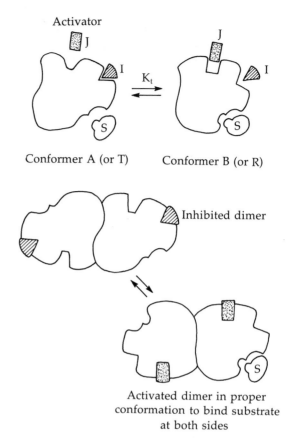

Fig. 6-10 Top: An enzyme with binding sites for allosteric inhibitor I and activator J. Conformer A binds inhibitor I strongly but has little affinity for activator J or for substrate S. Conformer B binds S and catalyzes its reaction. It also binds activator J whose presence tends to lock the enzyme in the "on" conformation B. Conformers A and B are designated T and R in the popular model of Monod *et al.*[35] Center and bottom: Inhibited and activated dimeric enzymes.

The effects of inhibitors or activators on the monomeric enzyme of Fig. 6-10 can be described by Eqs. 6-47 to 6-49. Separate terms for inhibition and activation can be included. The equilibrium between the two conformers can also be explicitly indicated according to Eq. 4-44. For monomeric enzymes it is usually not profitable to try to separate the constants K_t and K_{BX} (Eq. 4-44) for conformational change and substrate or activator binding.

Most intracellular enzymes are oligomeric, and binding of allosteric effectors leads to interesting effects. For mathematical treatment binding constants must be defined for both inhibitor and activator to both conformers A and B. Since all species must be taken into account in the mass balance, the equations are complex. The Monod–Wyman–Changeux (MWC) model (Chapter 4, Section E) gives an especially simple picture. From Eq. 4-53 the saturation curve for an oligomeric enzyme according to this model is (Eq. 6-51)

$$\bar{y} = \frac{Lc\alpha(1 + c\alpha)^{n-1} + \alpha(1 + \alpha)^{n-1}}{L(1 + c\alpha)^n + (1 + \alpha)^n} \qquad (6\text{-}51)$$

In this equation L is the **allosteric constant** and is given (for a dimer) by Eq. 4-50. The constant c (Eq. 6-52) is the ratio of dissociation constants K_{BS} and K_{AS} for the two conformers[†]

$$c = K_{BS}/K_{AS} \qquad (6\text{-}52)$$

and α is defined (Eq. 6-53) as[‡]

$$\alpha = [S]/K_{BS} \qquad (6\text{-}53)$$

To take account of effects of inhibitor and activator the *ratios* of dissociation constants of I from BI and AI and of activator J from BJ and AJ are defined as in Eq. 6-54. Likewise "normalized concentrations" of I and J are defined (Eq. 6-55) as β and γ, respectively.

$$d = \frac{K_{BI}}{K_{AI}} > 1 \qquad e = \frac{K_{BJ}}{K_{AJ}} < 1 \qquad (6\text{-}54)$$

$$\beta = [I]/K_{AI} \qquad \gamma = [J]K_{BJ} \qquad (6\text{-}55)$$

According to the MWC model, in the presence of inhibitor and activator at normalized concentrations β and γ an enzyme model will still follow Eq. 6-51, but the allosteric constant L will be replaced by an apparent allosteric constant L' (Eq. 6-56).[39]

$$L' = L\left[\frac{(1 + \beta d)(1 + \gamma e)}{(1 + \beta)(1 + \gamma)}\right]^n \qquad (6\text{-}56)$$

† Note that in this chapter *dissociation* constants of ES complexes are being used, whereas all of Chapter 4, including Eq. 4-53, is written in terms of association constants.

‡ Using the notation of Chapter 4, $\alpha = [S]K_{BX}$.

Figure 6-11 shows plots of $\bar{y}$ vs. log α for two different values of L' for a tetramer with a specific value assumed for c. In both cases, $\bar{y}$ approaches 1 as log α increases, but since we are dealing with noncompetitive inhibition at high values of L', much of the enzyme will be in the T (A) conformation at saturation.

As we have seen previously, noncompetitive inhibition cannot be completely reversed by very high substrate concentrations. Monod *et al.* defined a **function of state R** (Eq. 6-57) which is the fraction of total enzyme in the R (B) conformation.

$$\bar{R} = \frac{(1 + \alpha)^n}{L(1 + c\alpha)^n + (1 + \alpha)^n}$$
$$= \text{function of state} \qquad (6\text{-}57)$$

In a K system (Section B,2) it is the value of $\bar{R}$ that determines the velocity with which an enzyme reacts. Figure 6-11 also shows $\bar{R}$ as a function of $\log[\alpha]$. Note that when L' is low $\bar{R}$ does not approach zero even when $[S] \to 0$. In other words, the enzyme is never completely turned off, just as when L' is high the enzyme is never completely turned on.

Figure 6-11 may be compared with Fig. 6-7 which shows similar curves for noncompetitive

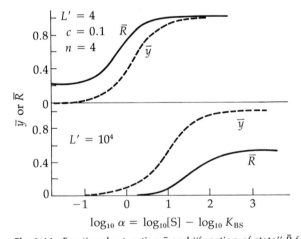

Fig. 6-11 Fractional saturation $\bar{y}$ and "function of state" $\bar{R}$ for hypothetical tetrameric enzymes following the MWC model. Curves are calculated for two different values of the apparent allosteric constant L' (Eq. 6-56) and for $c = 0.1$ (Eq. 6-52). After Rubin and Changeux.[39]

inhibition of a monomeric enzyme. The significant difference between the two figures is that saturation of the oligomeric enzyme occurs over a narrower concentration range than does that of the monomer, i.e., saturation of the oligomeric enzyme, especially in the presence of inhibitor, is *cooperative*. Note that cooperative binding of substrate requires that the free enzyme be largely in conformation T (A), as it is in the presence of an inhibitor.

Allosteric interactions between two identical molecules, whether of substrate or of effector, are described by Monod *et al.*[35] as **homotropic interactions.** Such interactions lead to cooperativity or anticooperativity in binding. Allosteric interaction between two different molecules, e.g., a substrate and an activator is designated **heterotropic.**

For many enzymes the MWC model is unrealistically simple. The more general treatment of binding equilibria given in Chapter 4 may be applicable. However, remember that in addition to K systems there are V systems in which a conformational change alters the maximum velocity (see Eq. 6-47) and sometimes both substrate affinity and maximum velocity.

The fact that data can be fitted to an equation is no proof that a mathematical model is correct; other models may predict the results just as well. For example, a simple alternative explanation of cooperative response to increasing substrate concentration has been proposed by Rabin.[40] A monomeric enzyme with a single substrate-binding site is all that is required. In its active conformation E the enzyme reacts with substrate rapidly and the ES complex rapidly yields product as in the upper loop of Eq. 6-58. However, a slow conformational change can convert E to E', a less active form with much lower affinity for substrate. At the same time, the complex ES', if formed, can

$$
\begin{array}{c}
S \longrightarrow ES \longrightarrow E + P \\
\quad\quad \updownarrow \quad\quad \vdots \text{ slow} \quad\quad (6\text{-}58) \\
S \longrightarrow ES' \\
\quad\quad\quad\quad \cdots E' \text{ (inactive)}
\end{array}
$$

equilibrate with ES and thereby alter the conformational state of the protein. At low substrate concentrations E' will predominate and the enzymatic activity will be correspondingly low. At high substrate concentrations E will tend to be in the active conformation long enough to bind another substrate molecule and to stay in the active conformation. Details of this and other kinetic models predicting a sigmoidal dependence of velocity on substrate concentration are discussed by Newsholme and Start.[41]

The physiological significance of cooperative binding of substrates to enzymes may sometimes be analogous to that of cooperative binding of oxygen by hemoglobin which provides for more efficient release of oxygen to tissues (Chapter 4, Section E,5). However, in the presence of excess activator an enzyme is locked in the R (B) conformation and no cooperativity is seen in the binding of substrate. In this case, each binding site behaves independently. On the other hand, *there will be strong cooperativity in the binding of the activator*. The result is that control of the enzyme is sensitive to a higher power than the first of the activator concentration. Likewise, the turning off of the enzyme is more sensitive to inhibitor concentration as a result of cooperative binding of the latter. It seems likely that the evolution of oligomeric enzymes is at least partly a result of the greater efficiency of control mechanisms based on cooperative binding of effectors.

7. Regulatory Subunits

A growing number of enzymes are known that contain in addition to the "catalytic subunits" which bear the active sites, **regulatory subunits** which bind either loosely or strongly to the catalytic subunits and serve as allosteric modifiers. The regulatory subunits in turn are subject to conformational changes induced by binding of inhibitors or activators. The best known example is aspartate carbamoyltransferase (Chapter 4, Section D,8). The regulatory subunits carry binding sites for cytidine triphosphate (CTP), which acts as a specific inhibitor of the enzyme. The significance of this fact is that aspartate carbamoyltransferase catalyzes the first reaction specific to the pathway of synthesis of pyrimidine nucleotides (Chapter 14, Section K,1). CTP is an end product of that pathway and exerts feedback inhibition on the enzyme that initiates its synthesis.

C. GROWTH RATES OF CELLS[42]

How do we correctly describe the rate at which cells grow? Consider bacteria in the rapidly growing "log phase" of growth. Each cell divides after a fixed length of time, the **doubling time,** which may be as small as 20 min for *E. coli.*† If a given volume of culture contains N_0 bacteria initially, the number N_n after n cell divisions will be (Eq. 6-59)

$$N_n = 2^n N_0 \qquad (6\text{-}59)$$

Thus, a single bacterium with a generation time of 20 min can produce 2^{144} cells in 48 h of exponential growth.

As **exponential growth rate constant** k is equal to *the number of doublings per unit time.* Thus, k is the reciprocal of the doubling time. It is easy to show that the number of bacteria present at time t will be given by Eq. 6-60.

$$N_t = 2^{kt} N_0 \qquad (6\text{-}60)$$

From this it follows that

$$kt = \log_2(N_t/N_0) = \log_{10}(N_t/N_0)/0.301 \qquad (6\text{-}61)$$

Equation 6-61 can be used to determine k by counting the number of bacteria at zero time and at time t.

Another way of expressing the growth is to equate the rate of increase of the number of bacteria with a growth rate constant μ multiplied by the

† The doubling time may be as little as 17 min in the early stages of growth but slows somewhat as time goes on. A mean value of about 26 min for *E. coli* at 37°C is typical. Doubling times for mammalian cells in tissue culture are typically about 1 day.

number of bacteria present at that time (Eq. 6-62).

$$dN/dt = \mu N \qquad (6\text{-}62)$$

This is a general equation for an autocatalytic reaction and N could be replaced with a concentration, for example, the total content of cellular matter per liter in the medium. From Eqs. 6-61 and 6-62 it can readily be shown[42] that μ is given by Eq. 6-63.

$$\mu = k \ln 2 = 0.69k \qquad (6\text{-}63)$$

When bacteria are transferred to new medium there is usually a **lag** before exponential growth begins, and eventually exponential growth stops and the culture enters **stationary phase.** The latter is usually followed by relatively rapid death of cells in the culture.

It is often desirable to study cell growth under conditions where a constant generation time is maintained but in which the density of cells in the medium does not increase. This is done with a simple device known as the **chemostat.** A culture vessel containing a certain number of bacteria is stirred to ensure homogeneity. Fresh culture medium continuously enters the vessel from a reservoir and part of the content of the vessel, suspended bacteria included, is continuously removed through another tube. The bacterial population in the vessel builds up to a constant level and can be maintained at the same level for relatively long periods of time.

D. SPECIFICITY OF ENZYMATIC ACTION

One of the most impressive facts about enzymes is that they are usually highly specific in their action. In some cases, the specificity toward substrate is almost absolute. For many years urea was believed to be the *only* substrate for the enzyme **urease** and succinate the only substrate for **succinate dehydrogenase.** Even after much searching for other substrates, only one or two closely related compounds could be found that were acted on at all. In other cases enzymes have specificity

Urea

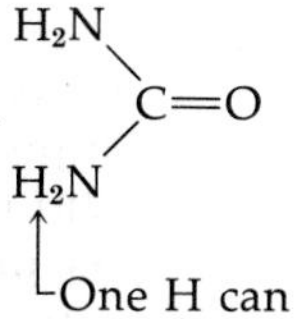

$$H_2N$$
$$\quad\quad C{=}O$$
$$H_2N$$

↑
└One H can be replaced by —OH and some substrate activity remains

Succinate

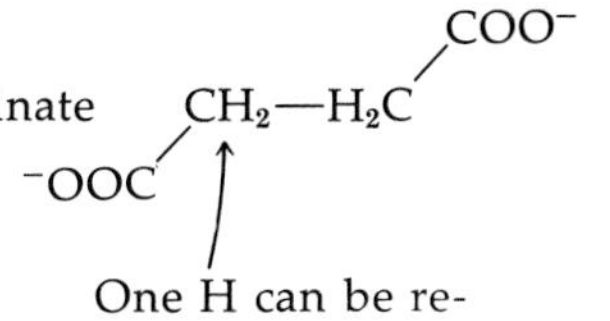

$$COO^-$$
$$CH_2{-}H_2C$$
$$^-OOC$$

↑
One H can be replaced by Cl with retention of substrate activity and by —CH₃ with partial retention of activity

for a class of compounds. For example, the **D-amino-acid oxidase** of kidney oxidizes a whole variety of D-amino acids but does not touch L-amino acids.

Almost as impressive as the substrate specificity of enzymes is the specificity for a given reaction. Often a substrate is capable of reacting in a variety of ways and the enzyme chooses one and catalyzes only that reaction. Side reactions do occur to a small extent, but the most impressive thing to a

BOX 6-B

UREASE AND THE TRACE METAL NICKEL

In addition to being noted for the near complete specificity toward its substrate urea, the enzyme urease of the jack bean has a special place in biochemical history. It was the first enzyme to be crystallized (by J. B. Sumner in 1926). Although Sumner eventually obtained the Nobel Prize for his accomplishment, his first reports were greeted with skepticism and outright disbelief.

Urease catalyzes the hydrolytic cleavage of urea to two molecules of ammonia and one of bicarbonate and is useful in the analytical determination of urea. It is not generally present in organisms. Recently it was discovered that each molecule of urease (MW = 105,000) contains two atoms of bound nickel.[a,b] The presence of the metal ion had been overlooked previously, despite the fact that the absorption spectrum of the purified enzyme contains an absorption "tail" extending into the visible region with a shoulder at 425 nm and weak maxima at 725 and 1060 nm.

Nickel has recently been shown a dietary essential for animals.[c] Thus, nickel-deficient chicks grow poorly, have thickened legs, and have dermatitis. Tissues of deficient animals contain swollen mitochondria and swollen perinuclear space suggesting a role in membranes. The toxicity of Ni is very low and homeostatic mechanisms exist in the animal body for regulating its concentration. Within tissues the nickel content ranges from 1 to 5 μg/l. Some of the nickel in serum is present as low molecular weight complexes and some is bound to serum albumin. In addition, there is a specific nickel-containing

protein of the macroglobulin class known as **nickeloplasmin**.[d] Nickel is present in plants; in some it accumulates to high concentrations.[e] Because of its ubiquitous occurrence it is difficult to prepare a totally Ni-free diet.

Although nickel can occur in a variety of oxidation states, only Ni(II) is common. This ion contains eight 3d electrons a situation that leads naturally to square-planar coordination of four ligands. However, the ion is "ambivalent" in that it is also able to form a complex with six ligands and an octahedral geometry. It has been suggested that this ambivalence may be of biochemical significance. While the role of Ni^{2+} in urease is not precisely understood it is possible that it participates in the catalysis just as does Zn^{2+} in carboxypeptidase (Fig. 7-3). Alternatively, it may coordinate a molecule of NH₃ as it is split from the substrate. It has been suggested that a variety of other enzymes that generate ammonia at active sites through the hydrolysis of glutamine (Chapter 14, Section C,2) may also contain nickel or some other transition metal.[b]

[a] N. E. Dixon, C. Gazzola, R. L. Blakeley, and B. Zerner, *JACS* **97**, 4131–4133 (1975).

[b] N. E. Dixon, C. Gazzola, R. L. Blakeley, and B. Zerner, *Science* **191**, 1144–1150 (1976).

[c] F. H. Nielson, *in* "Trace Element Metabolism in Animals-2" (W. G. Hoekstra, J. W. Suttie, H. E. Ganther and W. Mertz, eds.), pp. 381–395. University Park Press, Baltimore, Maryland, 1974; see also A. Schnegg and M. Kirchgessner, *Int. J. Vitamins Nutr. Res.* **46**, 96–99 (1976).

[d] S. Nomoto, M. D. McNeely and F. W. Sunderman, Jr., *Biochemistry* **10**, 1647–1651 (1971).

[e] B. C. Severne, *Nature (London)* **248**, 807–808 (1974).

chemist comparing an enzyme-catalyzed reaction with an uncatalyzed organic reaction is that the latter is accompanied by large amounts of side reaction products, but the enzymatic reaction is extraordinarily clean.

1. Complementarity of Substrate and Enzyme Surfaces

Impressed by the specificity of enzymatic action, biochemists early adopted a **lock and key theory** which stated that the substrate must fit to an active site precisely for reaction to occur. Modern experiments have amply verified this idea, but with one important modification. If the enzyme is the "lock" and the substrate the "key," it appears that the entrance of the key into the lock often induces conformational changes in the protein. Numerous instances can be cited in which an enzyme appears to fold around the substrate to create a tighter fit. Measurement of changes in circular dichroism, ultraviolet spectra, and sedimentation constants as well as X-ray diffraction experiments with enzyme–inhibitor complexes all support this idea. As we have already seen (Chapter 4, Section E,1), "induced fit" is also an important concept in consideration of subunit interactions.

What are active sites like in terms of structure? From crystallographic studies we are "seeing" more and more of them directly. However, X-ray diffraction studies usually give us no certain information about the conformational changes involved in induced fit. Furthermore, high-resolution crystallographic studies have been made with relatively few enzymes. Thus, more traditional chemical methods of "mapping" active sites still occupy much of the effort of enzymologists. By measuring binding constants of inhibitors in which structure is systematically varied and by altering the structure of substrates, the effect of changes in size and shape on both binding and reaction velocity can be learned. A good example is provided by the work of Meister and associates on glutamine synthetase of sheep brain.

Both D- and L-glutamic acid as well as α-aminoadipic acid are substrates but of 10 monomethyl derivatives of D- and L-glutamic acid only three are substrates. If the assumption is made that the substrates bind in a completely extended conformation, the hydrogens that can be substituted with a retention of substrate activity all lie on one side of the molecule, the one that lies behind the paper in the two drawings that follow.

CH_3 can be substituted for one of the circled H's with retention of substrate activity. This suggests

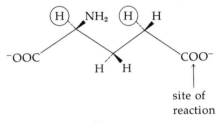

that the side of the molecules that is toward the reader faces the enzyme. On the basis of this hypothesis, Gass and Meister[43] prepared the dicarboxycyclohexane analogue (L-cis-1-amino-1,3-dicarboxycyclohexane) of L-glutamate in which the cyclohexane ring was connected at the points of the two replaceable hydrogens. This compound also was found to be a good substrate verifying

L-cis-1-Amino-1,3-dicarboxycyclohexane

the hypothesis and establishing something concrete about the geometry of the binding site of the enzyme.

2. Prochiral Centers

Most enzymes possess an infallible ability to recognize the difference between the right side and the left side of an organic substrate even when the latter has perfect bilateral symmetry. Before considering this fact further we must be acquainted with a system for the naming of two identical groups attached to a carbon atom that bears two other nonidentical groups.[44] Consider

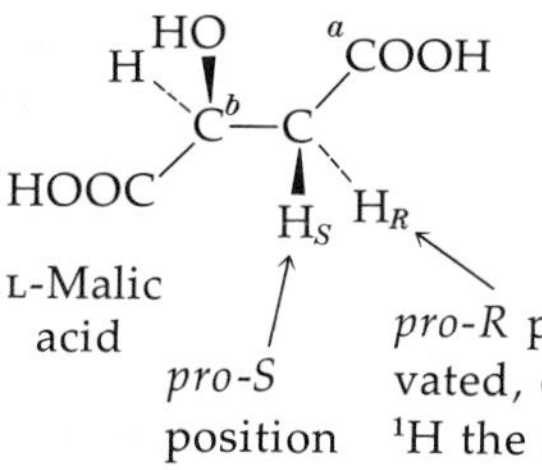

L-Malic
acid

pro-S
position

pro-R position. If priority is elevated, e.g., by substitution of ^2H for ^1H the configuration of the molecule becomes R. In this case the priority of this position would be c and that of the other hydrogen d

carbon atom 2 of malic acid which bears two H atoms as well as two other groups. Priorities of the groups attached to this carbon are assigned according to the R,S system (Chapter 2, Section A,5). Now, we ask what will be the configuration R or S when the priority of one of the two identical groups is raised, e.g., by substitution of one of the hydrogen atoms by deuterium? If the configuration becomes R that group occupies a *pro-R* position; if the configuration becomes S it occupies a *pro-S* position. Referring to the diagram above, it is easy to see (by viewing down the bond to the group of lowest priority and applying the usual rule for determining configuration) that if the *pro-R* hydrogen (H_R) is replaced by deuterium, the configuration will be R. Conversely, replacement of H_S by deuterium will lead to the S configuration.

It is a striking fact that when L-malic acid (above) is dehydrated by the enzyme **fumarase**

(Chapter 7, Section H,6), the hydrogen in the *pro-R* position is removed but that in the *pro-S* position is not touched. This can be demonstrated nicely by allowing the dehydration product, fumarate, to be hydrated to malate in ^2H_2O (Eq. 6-64). The malate formed contains deuterium in

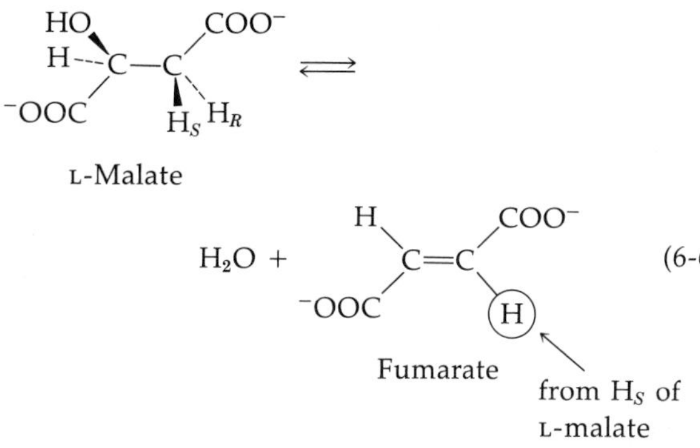

$$\text{(6-64)}$$

the *pro-R* position. If this malate is now isolated and placed with another portion of enzyme in H_2O, the deuterium is cleanly removed. The fumarate produced contains no deuterium, which it would if the enzyme were not completely stereospecific in its action.

Another striking example is the dehydrogenation (oxidation) of ethanol catalyzed by **alcohol dehydrogenase** (Eq. 6-65).

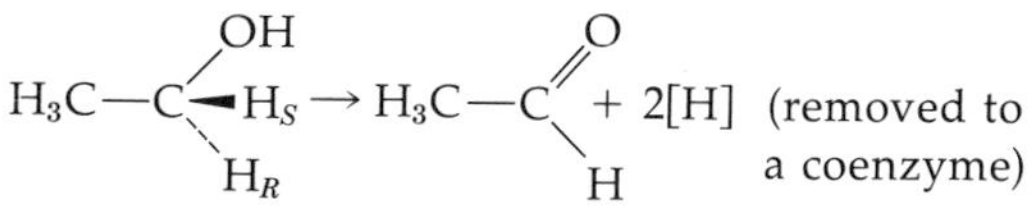

$$\text{(6-65)}$$

It is the *pro-R* hydrogen which is removed. If the reaction is reversed in such a way that deuterium is introduced into ethanol from the reduced coenzyme optically active R-2-deuterioethanol is formed.

The ability of an enzyme to recognize a single hydrogen of a pair of hydrogens on a CH_2 group was at first a surprise to biochemists, but it is now realized that it is the rule rather than the exception. It is no more surprising than the fact that a right foot fits only a right shoe and not a left. The stereospecificity of enzymes is a natural result of the complementarity of enzyme and substrate surfaces, just as the fit of a shoe is determined by the complementarity of surfaces of foot and shoe.

3. Stereochemical Numbering

In some molecules, such as citrate, the two ends of the main carbon chain form identical groups. Since the two ends are distinguishable to an enzyme, it is important to decide which should be labeled C-1 and which C-5. Hirschmann has proposed a stereochemical numbering system[45] in which the carbons are numbered beginning with the end of the chain that occupies the pro-S position.

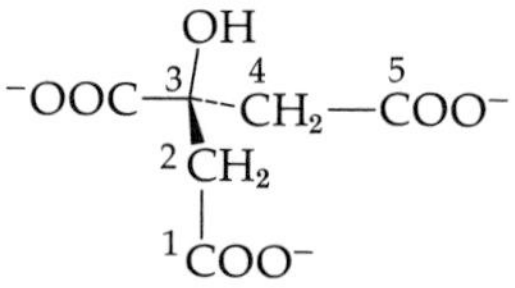

$$^-OOC \overset{3}{\underset{\overset{|}{\underset{1}{\text{COO}^-}}}{\overset{OH}{\underset{2}{\overset{|}{\text{CH}_2}}}} \overset{4}{CH_2} - \overset{5}{COO}^-$$

Stereochemical numbering for a symmetrical molecule. Carbons are numbered beginning with the end of the chain that occupies the pro-S position

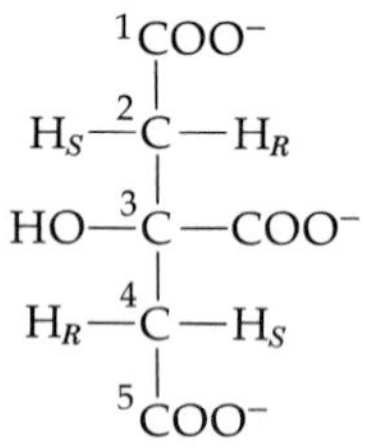

Citrate

$$\begin{array}{c} ^1COO^- \\ | \\ H_S - ^2C - H_R \\ | \\ HO - ^3C - COO^- \\ | \\ H_R - ^4C - H_S \\ | \\ ^5COO^- \end{array}$$

Citrate ion according to the stereochemical numbering systems and the Fischer convention

Citrate

Citrate containing ^{14}C in the 1 position as shown in the diagrams is referred to as sn-citrate[1-^{14}C]. This is the form of labeled citrate that is synthesized in living cells from oxaloacetate and [1-^{14}C]acetyl coenzyme A (Chapter 9, Section B,1). The first step in the further metabolism of citrate is the elimination of the OH group from C-3 together with the H_R proton from C-4 through the action of the enzyme **aconitase**. Note that in this case the proton at C-4 is selected rather than that of C-2. Glycerol phosphate, used by cells to construct phospholipids, always bears a phosphate group on the —CH_2OH in the pro-R position and is therefore sn-3-glycerol phosphate.

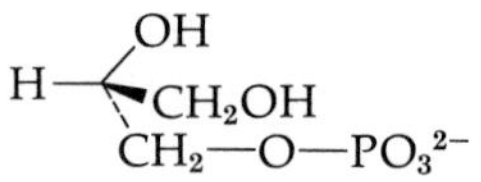

$$H - \overset{OH}{\underset{\underset{CH_2-O-PO_3^{2-}}{\overset{|}{CH_2OH}}}{<}}$$

sn-3-Glycerol phosphate

E. MECHANISMS OF CATALYSIS

Kinetic studies tell how fast enzymes act but by themselves they say nothing about *how* enzymes catalyze reactions; i.e., they do not give the **mechanism** of catalysis. By mechanism, we imply a complete description of the steps by which a reaction takes place. Most of the steps involve the simultaneous breaking of a chemical bond and formation of a new bond. For example, consider a simple displacement reaction, that of a hydroxyl ion reacting with methyl iodide to give the products methanol and iodide ion (Eq. 6-66). The reac-

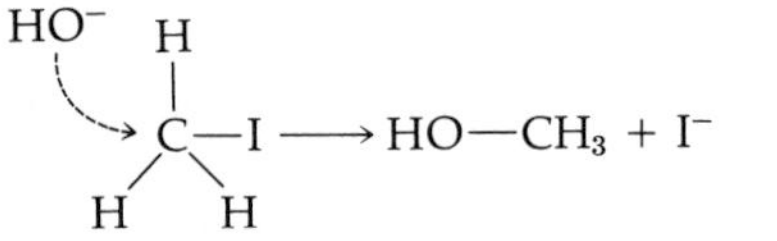

$$\begin{array}{c} HO^- \quad H \\ \diagdown \quad | \\ \rightarrow C - I \longrightarrow HO - CH_3 + I^- \qquad (6\text{-}66) \\ \diagup \quad \diagdown \\ H \quad H \end{array}$$

tion can be thought of as an "attack" by the OH^- ion on the carbon atom of the methyl group with a simultaneous pushing off of the I^-.

1. The "Transition State"

A reaction such as that of Eq. 6-66 is by no means instantaneous, and at some point in time between that at which the reactants exist and that at which products have been formed the C—I bond will be stretched and partially broken and the new C—O bond will be partially formed. The structure at this point is not that of an ordinary compound and is energetically unstable with respect to both the reactants and the products. The intermediate structure of the least stability is known as the **transition state.** For the displacement reaction of Eq. 6-66 it might be represented as follows:

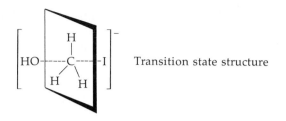

Transition state structure

The negative charge is distributed between the attacking OH group and the departing iodide. Note that the bonds to the central carbon atom are no longer tetrahedrally arranged but that the C—H bonds lie in a single plane and the partial bonds to the OH and I lie at right angles to that plane.

It is useful in discussing a reaction mechanism to construct a **transition state diagram** in which free energy† G is plotted against **reaction coordinate** (Fig. 6-12A). The reaction coordinate is usually not assigned an exact physical meaning but represents the progress from reactants toward products. It is directly related to the extent to which an existing bond has been stretched and broken or a new one formed. The high energy point is the transition state. A somewhat more detailed idea of a transition state is obtained from a contour diagram such as that of Fig. 6-12B. Here energy is plotted as a function of two bond distances, e.g., the lengthening C—I bond distance and the shortening C—O bond distance for Eq. 6-66. The path of minimum energy across the "saddle point" representing the transition state is indicated by the dashed line.

In a reaction coordinate diagram the difference in value of G between *reactants and products* is the overall free energy change ΔG for the reaction, while the difference in G between the *transition state and reactants* is $\Delta G^{\ddagger}$, the **free energy of activation**. The magnitude of $\Delta G^{\ddagger}$ represents the "energy barrier" to a reaction and largely determines reaction rate.

The diagrams in Fig. 6-12 are too simple for enzymatic reactions which usually occur in several intermediate steps. There will be transition states for each step and valleys in between. The valleys

† Energy E or enthalpy H may be plotted in the same way. Authors frequently do not state whether G, E, or H is meant.

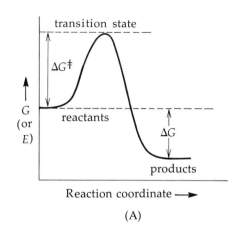

(A)

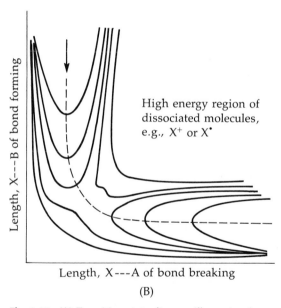

High energy region of dissociated molecules, e.g., X^+ or $X^{\bullet}$

Length, X---A of bond breaking

(B)

Fig. 6-12 (A) Transition state diagram illustrating free energy vs. reaction coordinate for conversion of reactants to products in a chemical reaction. (B) Contour map of free energy vs. interatomic bond distances for reaction $B + X—A \rightarrow BX + A$.

correspond to the sometimes very unstable intermediates. The passing from reactants to products in an enzymatic reaction can be likened to wandering through a series of mountain ranges of various heights and finally reaching the other side.

2. Quantitative Transition State Theory[46,47]

In the 1880's Arrhenius observed that the rate of chemical reactions varies with temperature according to Eq. 6-67 in a manner similar to the variation of an equilibrium constant with temperature

$$k = Ae^{-E_a/RT} \qquad (6\text{-}67)$$

(integrate Eq. 3-37 and compare). The quantity E_a is known as the **Arrhenius activation energy** and the constant A as the "preexponential factor." These observations, together with studies of the effects of salts on reaction rates and observation of quantitative correlations between rates and equilibrium constants, all suggested that a rate constant for a reaction might be a product of a constant term which is nearly independent of temperature and a constant $K^{\ddagger}$ which has the properties of an equilibrium constant for formation of the transition state.[46] Eyring made this quantitative as Eq. 6-68, where k_B is the Boltzmann con-

$$k = [(k_B T)/h]K^{\ddagger} \qquad (6\text{-}68)$$

stant and h is Planck's constant. This equation is based on Eyring's proposal that all transition states break down with the same rate constant $k_B T/h$. The equation is sometimes modified by multiplying the right side by a **transmission coefficient κ** which is thought to be nearly 1 for most reactions. Equation 6-68 can be rewritten as Eq. 6-69, in which $\Delta G^{\ddagger}$ is the free energy of acti-

$$k \approx [(k_B T)/h]e^{-\Delta G^{\ddagger}/RT} \qquad (6\text{-}69)$$

vation. (At 25°C $\Delta G^{\ddagger} = 5.71 \log k + 73.0$ kJ mol^{-1} when k is in units of s^{-1}.) The relation is approximate and a more correct statistical mechanical treatment is available.[46] It follows from Eq. 6-68 and Eq. 3-14 that Eq. 6-70 holds.

$$k \approx \left(\frac{k_B T}{h} e^{\Delta S^{\ddagger}/R}\right)e^{-\Delta H^{\ddagger}/RT} \qquad (6\text{-}70)$$

From this it appears that $\Delta H^{\ddagger} \approx E_a$, the Arrhenius activation energy. A more correct treatment gives $\Delta H^{\ddagger} = E_a - RT$ for reactions in solution. Since RT

at 25°C is only 2.5 kJ mol^{-1} the approximation that $\Delta H^{\ddagger} = E_a$ is often used. The preexponential term (in parentheses in Eq. 6-70) is seen to depend principally on $\Delta S^{\ddagger}$, the entropy change accompanying formation of the transition state. The quantities $\Delta G^{\ddagger}$, $\Delta H^{\ddagger}$, and $\Delta S^{\ddagger}$ are sometimes determined for enzymatic reactions but useful interpretations have proved difficult.

Anything that tends to stabilize the transition state (decrease $\Delta G^{\ddagger}$) increases the rate of reaction. In general, the role of a catalyst is to permit the formation of a transition state of lower energy (higher stability) than that for the uncatalyzed reaction. Stabilization of the transition state of a reaction by an enzyme suggests that the enzyme has a higher affinity for the transition state than it does for substrate or products, an idea that appears to have been expressed first by Pauling.[48]

> I think that enzymes are molecules that are complementary in structure to the activated complexes of the reactions that they catalyze. The attraction of the enzyme molecule for the activated complex would thus lead to a decrease in its energy, and hence to a decrease in the energy of activation of the reaction and to an increase in the rate of the reaction.

3. Transition State Inhibitors[49,49a]

Suppose that a chemical reaction of compound S takes place with rate constant k_N through transition state T. Let the equilibrium constant for formation of T be $K_N^{\ddagger}$. Assume that an enzyme E combines with S with dissociation constant K_{dS} and with T with dissociation constant K_{dT} (Eq. 6-71).

$$
\begin{array}{ccc}
\text{S} & \xrightarrow{K_N^{\ddagger}} & \text{T} \longrightarrow \text{product} \\
+ & & + \\
\text{E} & & \text{E} \\
\Big\updownarrow K_{dS} & & \Big\updownarrow K_{dT} \\
\text{ES} & \underset{K_E^{\ddagger}}{\rightleftharpoons} & \text{ET} \longrightarrow \text{product}
\end{array}
\qquad (6\text{-}71)
$$

If equilibrium is assumed for all four sets of double arrows it is easy to show that Eq. 6-72 holds.

$$K_E{}^\ddagger/K_N{}^\ddagger = K_{dS}/K_{dT} \qquad (6\text{-}72)$$

According to transition state theory, if the transmission coefficient $\kappa = 1$, T and ET will be transformed to products at the same rate. Thus, if the mechanisms of the nonenzymatic and enzymatic reactions are assumed the same, the ratio of maximum velocities for first order transformation of ES and S will be given by Eq. 6-73.

$$k_E/k_N = K_E{}^\ddagger/K_N{}^\ddagger = K_{dS}/K_{dT} \qquad (6\text{-}73)$$

For some enzymes the ratio k_E/k_N may be 10^8 or more. Thus, if $K_{dS} \approx 10^{-3}$ the constant K_{dT} would be $\approx 10^{-11}$. The enzyme would be expected to bind the transition state structure (T) 10^8 times more tightly than it binds S.

The foregoing reasoning suggests that if structural analogues of T could be found for a particular reaction, they too might be very tightly bound—more so than ordinary substrate analogues. Wolfenden has listed a series of proposed **transition state inhibitors** that are very tightly bound to enzymes.[49]

4. Microscopic Reversibility

The important statistical mechanical **principle of microscopic reversibility** asserts that *the mechanism of any chemical reaction considered in the reverse direction must be exactly the inverse of the mechanism of the forward reaction.* A consequence of this principle is that if the mechanism of a reaction is known, that of the reverse reaction is also known. Furthermore, it follows that the *forward and reverse reactions catalyzed by any enzyme must occur at the same active site* on the enzyme. The principle of microscopic reversibility is especially useful when the likelihood of a given mechanism is being considered. If a mechanism is proposed for a reversible reaction the principle of microscopic reversibility will give an unambiguous mechanism for the reverse reaction. Sometimes this reverse mechanism will look ridiculous and, recognizing this, the chemist can search for a better one.

5. Acid and Base Catalysis

Many reactions that are promoted by enzymes can also be catalyzed by acids or bases and often by both. A much-studied example is **mutarotation** the reversible interconversion of the α- and β-anomeric forms of sugars such as glucose (Eq. 6-75). This reaction is catalyzed by a specific **mutarotase** and also by inorganic acids and bases. Such facts suggest that there is something in common between simple acids and bases and enzymes as catalysts. Since many side chains of amino acids contain acidic and basic groups, we are led to a natural conclusion that these groups must participate in catalysis as acids and bases. However, to understand *how* they participate we must look quantitatively at some equilibrium and rate constants.

a. Strengths of Acids

Acids are proton donors and bases are proton acceptors. The strength of an acid is measured by the dissociation constant or by its negative logarithm, the pK_a (Eq. 6-74, see also Chapter 4, Section C).

$$pK_a \equiv pK = -\log_{10} K_a \qquad (6\text{-}74)$$

where $K_a = [H^+][B^-]/[HB]$. Remember that $[H^+]$ represents the concentration of the H_3O^+ ion (or more properly, its activity) rather than that of a free proton. For this discussion acids will be given the symbol HB or H^+B and the **conjugate bases** formed by their dissociation as B^- or B.

Remember that *strong acids have low pK_a values* and that the *conjugate bases formed from them are weak bases.* Likewise, *very weak acids have high pK_a values and their conjugate bases are strong.* It will be worthwhile in considering enzyme mechanisms to remember the following pK_a values.

−1.74 H_3O^+ in 55.5 M water

4.7   —COOH in glutamic and aspartic acid side chains

6.8   imidazole (histidine) and $-O-\overset{\overset{O}{\|}}{\underset{\underset{O^-}{|}}{P}}-OH$

7.8   —NH_3^+, terminal in peptide

10.0   —OH

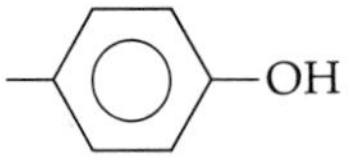

10.2   —NH_3^+ in lysine side chain

13.6   —CH_2OH of serine side chain

15.74   free ^-OH in 55.5 M H_2O, 25°C

These values are not constant but vary by about ±0.5 unit depending upon the exact structures and the environment of the particular side chain in a protein. In some cases the variations are even greater.

b. Acid-Base Catalysis of Mutarotation

The mutarotation of glucose proceeds through the free aldehyde form as an intermediate (Eq. 6-75). Looking at the first half of the reaction, we

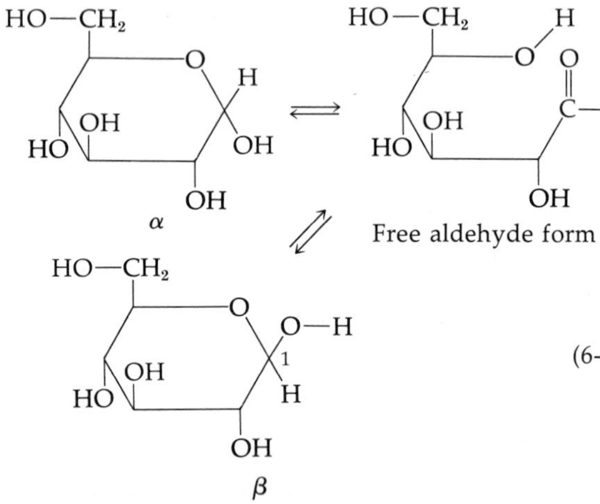

(6-75)

see that a hydrogen atom is removed from the oxygen at carbon 1 and that a proton (probably a different one) is transferred to the oxygen of the ring with cleavage of the C—O bond. A similar process

in reverse is required for the second half of the reaction.

Transfers of hydrogen, most often between atoms of oxygen, nitrogen, and sulfur atoms, are a common feature of biochemical reactions. The bonds between hydrogen and oxygen, nitrogen and sulfur atoms tend to be polarized strongly, leaving a partial positive charge on the hydrogen atoms. Thus, the groups are weakly acidic and protons can be transferred from them relatively easily. It is reasonable to suppose that acid and base catalysis are related to these hydrogen transfers.

c. General Base and General Acid Catalysis

Base-catalyzed mutarotation might be formulated as follows: A base such as an OH^- ion attacks the proton on the hydroxyl group, removing it to form an anion and the conjugate acid BH^+ (Eq. 6-76, step a). The anion is then isomerized to

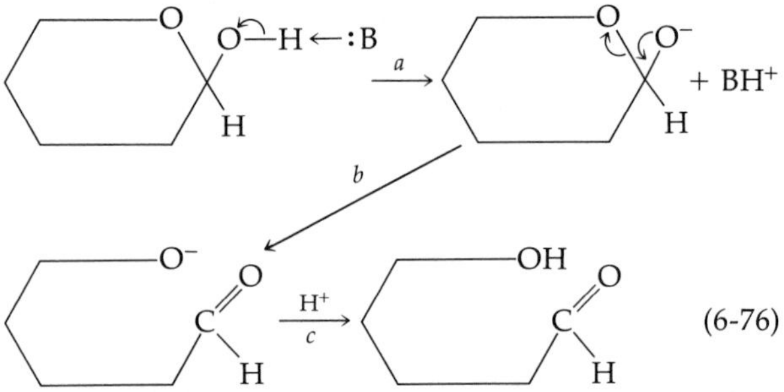

(6-76)

form an anion with the ring opened (Eq. 6-76, step b). Addition of a proton (transfer of a proton from H_3O^+) gives the free aldehyde form of the sugar (Eq. 6-76, step c).

The catalytic base in the foregoing equation might be OH^-, but it could equally well be some other weaker base such as ammonia or even water. In some cases it is found that the rate of a catalyzed reaction is proportional only to the concentration of OH^- and the presence of other weaker bases has no effect on the rate. Such catalysis is referred to as **specific hydroxyl ion catalysis.** More commonly the rate is found to depend both on $[OH^-]$ and on the concentration of other weaker bases. In such cases the apparent first order rate constant (k_{obs}) for the process is represented by a sum of terms (Eq. 6-77):

$$k_{obs} = k_{H_2O} + K_{OH^-}[OH^-] + k_B[B] \qquad (6\text{-}77)$$

The term k_{H_2O} is the rate in pure water and represents catalysis occurring by the action of water alone as either an acid or a base. The last two terms represent the contributions to the catalysis by OH^- and by the other base, respectively. The term $k_B[B]$ represents **general base catalysis,** an effect that is believed to be very important to the mechanisms of action of many enzymes. In enzymology it refers to the ability of some basic group in the enzyme to act as a proton acceptor.

Catalysis of mutarotation by acids occurs if an acid donates a proton to the oxygen in the sugar ring as shown in Eq. 6-78. Again, either **specific**

$$(6\text{-}78)$$

acid catalysis (by H_3O^+) or **general acid catalysis** is possible.

Both general base and general acid catalyses are thought to be used by enzymes. Enzymes are not able to concentrate protons or hydroxyl ions to the point that *specific* base or acid catalysis would be effective. However, general acid and general base catalysis can be accomplished by groups in their normal state of protonation at the pH of the cell.

d. Concerted Mechanisms

A third possible type of catalysis requires that a base and an acid act *synchronously* to effect the breaking and formation of bonds in a single step. Thus, tetramethylglucose mutarotates very slowly in benzene containing either pyridine (a base) or phenol (an acid), but when both pyridine and phenol are present, mutarotation is rapid. This suggested to Swain and Brown[50] a **concerted** mechanism in which *both* an acid and a base participate (Eq. 6-79).

Transition state structure   (6-79)

Note that during the reaction the acid BH^+ is converted to its conjugate base B and the base B' to its conjugate acid HB'. It might seem that these agents, having been altered by the reaction, are not serving as true catalysts. However, a simple proton exchange will restore the original forms and complete the catalytic cycle. In aqueous solutions, water itself might act as the acid or the base or even both in concerted catalysis.

The experimental evidence for concerted acid–base catalysis of the mutarotation of tetramethyl-glucose by phenol and pyridine in benzene is now considered unsound.[51,52] Concerted acid–base catalysis is difficult to prove for nonenzymatic reactions in aqueous solution.† However, it may be very important in enzymatic action. Amino acid side chains could supply two properly placed acid–base groups.

† Recently evidence for concerted acid–base catalysis of the enolization of acetone by acetic acid and acetate has been reported.[52a]

e. The Brönsted Relationships

The effectiveness of a given base as a general base catalyst can usually be related to its basicity (pK_a) through the **Brönsted equation** (Eq. 6-80).[53]

$$\log k_B = \log G_B + \beta(pK_a) \qquad (6\text{-}80)$$

Here k_B is defined for Eq. 6-77 and G_B is a constant for a particular reaction. A similar equation (6-81)

$$\log k_{HB} = \log G_A - \alpha(pK_a) \qquad (6\text{-}81)$$

relates the constant k_A for general acid catalysis to pK_a. These equations are linear free energy relationships (Chapter 3, Section E). However, notice that they apply to *rates* rather than to equilibria. Indeed, the Hammett equation (Eq. 3-66) is also very often applied to rates as well as to equilibria. The condition that must be met if the Brönsted equations are to hold is that the free energy of activation for the reaction be directly related to the basicity (or acidity) of the catalyst.

The exponents β and α of Eqs. 6-80 and 6-81 measure the sensitivity of a reaction toward the basicity or acidity of the catalyst. It is easy to show that as β or α approach 1.0 general base or general acid catalysis is usually lost and that the rate is exactly that of specific hydroxyl ion or specific hydrogen ion catalysis.[53] As β or α approach zero basic or acidic catalysis is indetectable. Thus, general base or general acid catalysis is most significant when β or α is in the neighborhood of 0.5. Under these circumstances it is easy to see how a moderately weak base like imidazole (from histidine side chains) can be an unusually effective catalyst at pH 7.

To determine α or β experimentally a plot of log k_B or log k_{HB} vs. pK_a (a **Brönsted plot**) is made and the slope is measured. Statistical corrections (Chapter 4, Section C,5) should be applied for dicarboxylic acids and for ammonium ions from which one of three protons may be lost from the nitrogen atom.

General base or general acid catalysis implies an important feature of any mechanism for which it is observed, namely, that removal of a proton or addition of a proton is involved in the rate-determining step of a reaction.

f. Tautomeric Catalysis

In an impressive experiment Swain and Brown showed that a much more effective catalyst for the mutarotation of sugars than a mixture of an acid and a base can be designed by incorporating the acid and base groups into *the same molecule*.[50] Thus, with 0.1 M tetramethylglucose in benzene solution, 0.001 M α-hydroxypyridine is 7000 times as effective a catalyst as a mixture of 0.001 M pyridine + 0.001 M phenol. Swain and Brown proposed the following completely concerted reaction mechanism for the **polyfunctional catalyst α-hydroxypyridine.** They proposed that the hydrogen bonded complex formed before reaction (Eq. 6-82)

2-Hydroxy-pyridine

Complex

2-Pyridone   (6-82)

is analogous to an enzyme–substrate complex. The product of the catalyst is 2-pyridone, a tautomer of 2-hydroxypyridine, the two forms being in a rapid reversible equilibrium with one another.

Rony preferred to call catalysis by α-hydroxypyridine **tautomeric catalysis** and believed that its efficiency lies not simply in the close proximity of acidic and basic groups in the same molecule but also in the ability of the catalyst to repeatedly cycle between two tautomeric states.[51]

g. Rates of Proton Transfer

The diffusion-controlled limit for second order rate constants (Section A,7) is $\sim 10^{10} \, M^{-1} \, s^{-1}$. In 1956, Eigen, who had developed new methods for studying very fast reactions, made the remarkable discovery that protons and hydroxyl ions react much more rapidly when present in a lattice of ice than when in solution.[54] He observed second order rate constants of 10^{13} to $10^{14} \, M^{-1} \, s^{-1}$. These reaction rates are almost as fast as the rates of molecular vibration; e.g., the frequency of vibration of the OH bond in water† is about $10^{14} \, s^{-1}$. The explanation seems to be as follows: The OH^- ion and the proton, which is combined with a water molecule to form H_3O^+, are both hydrogen bonded to adjacent water molecules. Since the water molecules in ice are all hydrogen bonded, a chain of water molecules links the hydroxyl and the hydrogen ions (Eq. 6-83). By a synchronous

$$(6-83)$$

movement of the electrons from the OH^- ion and through each of the water molecules in the chain (as indicated by the little arrows) the neutralization can take place during the time of one

† The OH stretching frequency is the same as that of the infrared light absorbed in exciting this vibration. The frequency ν is equal to the wave number (3710 cm^{-1} for —OH stretching) times c, the velocity of light (3×10^{10} cm s^{-1}). Thus, for the OH stretch in water, $\nu = 3 \times 10^{10} \times 3710 = 1.14 \times 10^{14} \, s^{-1}$.

molecular vibration. Note that the positions of the oxygen atoms are unchanged at the end of the reaction but that the protons that were engaged in hydrogen bond formation have moved a little toward the left. This kind of reaction is not only a remarkable demonstration of the mobility of hydrogen ions but also could be closely related to tautomeric catalysis in enzymes. Thus, Wang has suggested that proton transfer along rigidly and accurately held hydrogen bonds in the ES complex may be an essential feature of enzymatic catalysis.[55] A synchronous shift of protons in carboxylic acid, imidazolium, and phosphate groups can readily be envisioned (Eq. 6-84). The net effect of

$$(6-84)$$

the process is to transfer a proton from one end of the chain to the other (as in Eq. 6-83) with facile tautomerization reactions providing the pathway. It is easy to imagine that such a pathway might be constructed by protein side chains to join the two sides of an active center promoting a concerted acid–base catalyzed reaction such as that of Eq. 6-79.

Other tautomerization processes are possible within proteins if the existence of less stable "minor tautomers" is allowed. Thus, the shifts shown in Eq. 6-85, involving peptide bonds in an

$$(6-85)$$

α helix or a β sheet, could be imagined. The electron flow is shown as occurring toward a guanidine group of an arginine side chain. Many other possibilities can be envisioned, especially if coenzymes or purine or pyrimidine bases are also involved in the reaction. All could occur with extraordinary rapidity and might be hard to detect.

h. The Effect of pH on Enzymatic Action

The preceding arguments for participation of acidic and basic groups in enzymatic catalysis are based largely on studies of nonenzymatic **model reactions.** Is there any evidence that enzymes *really* contain such groups? The answer is "yes," the most obvious being the variation of enzymatic activity with pH. It is very often found that a plot of V_{max} against pH gives a bell-shaped curve (Fig. 6-13). An **optimum** rate is observed at some intermediate pH, often (but not always) in the range of pH 6–9. This type of curve can be interpreted most simply by assuming the presence in the active site of two ionizable groups a and b. As a result, there are three forms of the enzyme with different degrees of protonation: E, EH, and EH_2 (Eq. 6-86).

$$
\begin{array}{ccc}
\text{E} & & \text{ES} \\
K_{bE} \Updownarrow & \xrightarrow{\ k_1\ } & K_{bES} \Updownarrow \\
\text{EH} & \underset{k_2}{\overset{}{\rightleftharpoons}} \text{EHS} & \xrightarrow{\ k_3\ } \text{EH + products} \\
K_{aE} \Updownarrow & & K_{aES} \Updownarrow \\
\text{EH}_2 & & \text{EH}_2\text{S}
\end{array}
\tag{6-86}
$$

Let us designate the acid dissociation constants for the two groups in the enzyme as K_{aE} and K_{bE} and those of the ES complex as K_{aES} and K_{bES}. The rate constants k_1, k_2, and k_3 define the rates of formation and breakdown of the ES complex.

If it is assumed that only form EHS reacts to give products, the bell-shaped curves of Fig. 6-13 are obtained. The frequent observation of such curves supports the model of Eq. 6-86 and also suggests a mechanism: If group a of the enzyme must be dissociated to its conjugate base and group b must be protonated for reaction to take place, it is reasonable to suppose that these two groups participate in acid–base catalysis.

For the simple case illustrated in Eq. 6-86, the pH dependence of the initial maximum velocity and of the Michaelis constant are given by Eqs. 6-87 and 6-88, respectively.

$$
V_{max} = \frac{k_3[\text{E}]_t}{1 + \dfrac{[\text{H}^+]}{K_{aES}} + \dfrac{K_{bES}}{[\text{H}^+]}}
\tag{6-87}
$$

$$
K_m = \frac{(k_2 + k_3)}{k_1} \cdot \frac{\left(1 + \dfrac{[\text{H}^+]}{K_{aE}} + \dfrac{K_{bE}}{[\text{H}^+]}\right)}{\left(1 + \dfrac{[\text{H}^+]}{K_{aES}} + \dfrac{K_{bES}}{[\text{H}^+]}\right)}
\tag{6-88}
$$

If the enzyme is completely saturated with substrate only forms EH_2S, EHS, and ES are present. From the definitions of K_{aES} and K_{bES} it follows (Eq. 6-89) that under these circumstances

$$
[\text{E}]_t = [\text{EHS}]\left(1 + \frac{[\text{H}^+]}{K_{aES}} + \frac{K_{bES}}{[\text{H}^+]}\right)
\tag{6-89}
$$

The quantity in parentheses is the same as that in the denominators of Eqs. 6-88 and 6-89 and is sometimes called a **Michaelis pH function.**[56] A similar pH function for *free* enzyme appears in the numerator of Eq. 6-88. The pH dependence of enzymatic action is often more complex than that shown in Fig. 6-13 and given by the foregoing equations. However, it is easy to write Michaelis pH functions for enzymes with any number of dissociable groups in both E and ES and to write appropriate equations of the type of Eqs. 6-86 to 6-89 easily. Bear in mind that if the *free substrate* contains groups dissociating in the pH range of interest, a Michaelis pH function for the free substrate will also appear in the numerator of Eq.

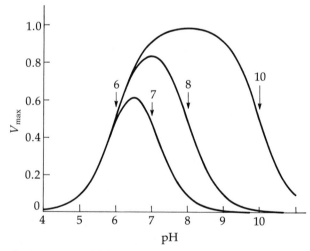

Fig. 6-13 Expected dependence of V_{max} on the pH according to Eq. 6-87 with k_3 [E_t] = 1, pK_{aES} = 6, and pK_{bES} as given on the graph. After Alberty.[58] Computer-drawn graph courtesy C. Harris.

6-88. If the pH dependence of the enzyme is regulated by a conformational change in the protein, there may be a cooperative gain or loss of more than one proton and the Michaelis pH function must reflect this fact. This can sometimes be accomplished by addition of a term related to Eq. 4-32.

Dixon has developed the subject in detail[56] and has suggested that log V_{max} (or log specific activity) and $-$log K_m be plotted vs. pH. This leads to graphs of the type shown in Fig. 6-14. The curved segments of the graphs that extend for about $1\frac{1}{2}$ pH units on either side of each pK_a are assymptotic to straight lines of slope 1 when a single proton is involved or of a higher slope for multiple cooperative proton dissociation. The straight lines can be extrapolated and intersect at the pK_a values. (However, it is better to fit a complete curve of theoretically correct shape to the points.) Note that the curved line always passes below or above the intersection point at the value of log 2 = 0.30 except in the case of cooperative proton dissociation when it is closer.

Upward turns in the curve of log K_m vs. pH correspond to pK_a values in the free enzyme or substrate and downward turns to pK_a values in ES. This approach to analysis of pH dependence has been adopted widely but often incorrectly. For example, many published curves have very sharp bends in which the curved portion covers less than 3 pH units and the curve is much closer than 0.30 to the extrapolated point.† This suggests cooperative proton binding and an **apparent pK_a** that is related to $\overline{K}$ of Eq. 4-33. The reader may wish to pursue the matter further by considering pK_a values in the transition state.[57] Interesting discussions of certain causes of unusually sharp optima in pH-dependence curves have been published.[57a,57b]

Fumarase, which catalyzes the reversible hydration of fumaric acid to malic acid (Eq. 6-64), has

† Strangely, a large fraction of the illustrative curves used by Dixon and Webb[56] show this same deviation from theoretical behavior. Many of the pK_a values may be a result of multiple proton dissociations which yield sharp bends or of experimental errors (buffer effects for example).

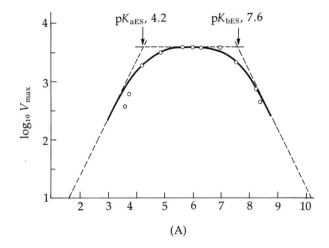

(A)

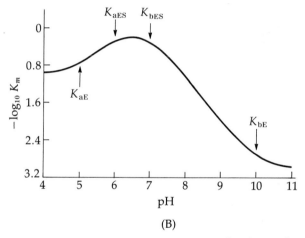

(B)

Fig. 6-14 (A) Plot of log V_{max} vs. pH for a crystalline bacterial α-amylase. From S. Ono, K. Hiromi, and Y. Yashikawa, *Bull. Chem. Soc. Jpn.* **31,** 957–962 (1958). (B) Theoretical curve of log K_m vs. pH for Eq. 6-88 with $K_{aE} = 5$, $K_{bE} = 10$, $K_{aES} = 6$, and $K_{bES} = 7$. Courtesy of C. Harris.

been intensively studied. A bell-shaped pH dependence for both forward and reverse reactions was observed by Alberty and co-workers in an important early study.[58] Using Eqs. 6-88 and 6-89, apparent pK_a values for groups *a* and *b* were measured for the enzyme in 0.01 ionic strength buffers (Table 6-1). It is important to realize that in this reversible reaction, the kinetics are more complex than is indicated by the preceding equations and the apparent pK_a values are not

TABLE 6-1
Apparent pK_a Values for Fumarase and for Its Complexes with Fumarate and with Malate[a]

	Enzyme	Enzyme–fumarate complex	Enzyme–malate complex (reverse reaction)
pK_a	6.2	5.3	6.6
pK_b	6.8	7.3	8.5

[a] From Alberty.[58]

necessarily directly related to the true pK_a values. However, it is tempting to assume[58] that the two pK_a values of 6.2 and 6.8 in the free enzyme result from identical groups, probably imidazole, with intrinsic pK_a values of ~6.5. Fumarase is discussed further in Chapter 7, Section H,6.

Mutarotation of glucose in *E. coli* is catalyzed by a specific mutarotase[59] which has a turnover number of 10^4 s^{-1}. The plot of $-\log K_m$ vs. pH indicates two pK_a values in the free enzyme at 5.5 and 7.6 while the plot of $\log V_{max}$ yields a single pK_a of 4.75 for the ES complex.[59] The latter might represent a catalytic group (group B' of Eq. 6-79), possibly imidazole in its conjugate base form. Why doesn't the group having pK_a 7.6 in the free enzyme also show a pK_a in the ES complex? Either the group has no catalytic function or its pK_a is so strongly shifted by substrate binding that it is not detected in the $\log V_{max}$ plot. Thus, the experimental data do not say conclusively whether or not a group comparable to $-BH^+$ of Eq. 6-79 functions in the enzyme.

6. Covalent Catalysis

In addition to participating in acid–base catalysis certain amino acid side chains may enter into covalent bond formation with substrate molecules. This phenomenon has been referred to as **covalent catalysis.** Since basic groups participate most often, it is also called **nucleophilic catalysis.** Covalent catalysis is very common in enzymes catalyzing nucleophilic displacement reactions and

specific examples will be considered in Chapter 7. Coenzymes frequently participate in covalent catalysis as detailed in Chapter 8.

7. Proximity and Orientation Effects

Despite the high state of development of enzymology and the fact that we have the detailed structures of several enzymes from X-ray diffraction experiments, there is still something mysterious about these remarkable catalysts. Neither acid–base catalysis nor covalent catalysis seems able to explain the enormous rate enhancements produced by enzymes. What other factors are involved?

One of the earliest ideas about enzymes was that they simply brought reactants together and bound them side by side for a long enough time that the reactive groups might bump together and finally react. Intuitively this **proximity factor** seems important, but early attempts to estimate its effect quantitatively led to the conclusion that it is a relatively minor factor. However, Page and Jencks argued that earlier reasoning was in error and that rate enhancements by factors of 10^3 or more may be expected solely from the loss in the entropy of two reactants when they are brought together and anchored on an enzyme surface.[60,61] In view of the entropy decrease involved, it is clear that the enthalpy of binding must be high, and if this explanation is correct that the *binding of the substrates* to the enzyme itself provides much of the driving force for catalysis. The latter idea was originally proposed by Westheimer who stated that enzymes use the substrate-binding force as an **entropy trap.**[62] The losses of translational and rotational entropy, which Page and Jencks estimated as up to -160 to -210 kJ/deg/mol, overcome the unfavorable entropy of activation usual in bimolecular reactions.

Much recent discussion has centered on the question of how precise the orientation of substrates must be for rapid reaction.[63,63a] Several model reactions have been studied. For example, compounds such as the acid shown in Eq. 6-90 spontaneously form an internal ester (lactone) with elimination of water.

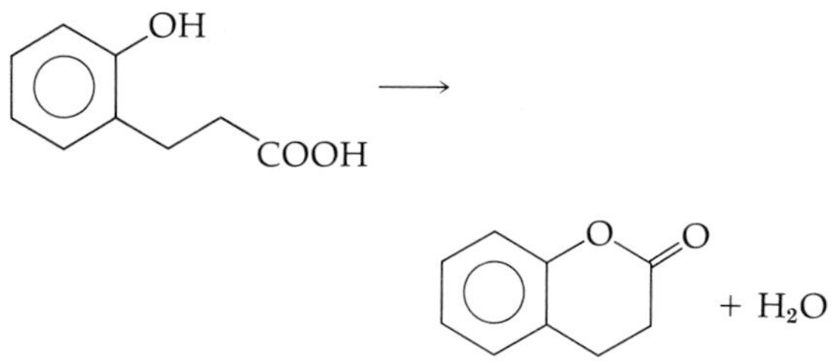

(6-90)

The compound shown below reacts much more rapidly because its conformation is highly restricted.[64,65] The reaction rate is over 10^{11} times that of the compound shown in Eq. 6-90, presumably because the —COOH is constrained to collide with the —OH group very frequently. The three methyl groups interdigitate and form a trialkyl lock. The results suggest that orientation effects may play a large role in enzymatic catalysis.

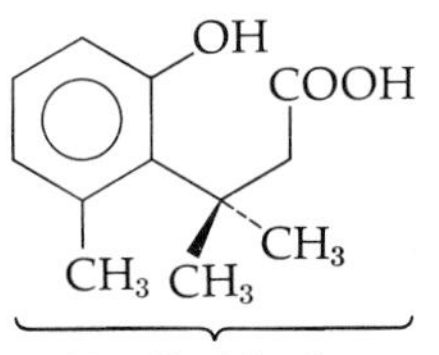

Trialkyl lock

Another important concept is that enzymes are able to induce **strain** or "distortion" in the substrates leading to weakening of the specific bonds (see *lysozyme*, Chapter 7, Section C,4). Such strain may be accompanied by or may be a result of a conformational change in the protein itself. Still another factor that must be considered is that some reactions proceed more rapidly in a low dielectric medium than they do in water. Thus, the polar groups of a substrate adsorbed to an enzyme active site may become dehydrated and more reactive.

Substantial volume changes ($\Delta V^{\ddagger}$) occur during formation of the transition state for many enzyme-catalyzed reactions. It has been suggested that these may result largely from changes in hydration

of groups on the enzyme surface and that these changes may play an important role in catalysis.[65a]

8. Conclusions

In summary, it appears that enzymes exert their catalytic powers by **first** of all bringing together substrates and binding them in proper orientations at the active site. **Second,** they provide acidic and basic groups in the proper orientations to promote proton transfers within the substrate. **Third,** groups within the enzyme (especially nucleophilic groups) may enter into covalent interaction with the substrates to form structures that are more reactive than those originally present in the substrate. **Fourth,** the enzyme may be able to induce strain or distortion in the substrate perhaps accompanying a conformational change within the protein. The question is often asked: "Why are enzymes such large molecules?" At least part of the answer is immediately evident when we consider that the formation of a surface complementary to that of the substrates and possessing reasonable rigidity requires a complex geometry in the peptide backbone. In addition, the enzyme must provide functional groups at the proper places to enter into catalysis. It may require a certain bulk to provide a low dielectric medium. Finally, if conformational changes occur during the course of the catalysis, we can only be surprised that nature has succeeded in packing so much machinery into such a small volume.

F. THE REGULATION OF ENZYMATIC ACTIVITY

The control of metabolism is accomplished largely through mechanism that regulate *the locations, the amounts, and the catalytic activities of enzymes.*[41,66] The purpose of this section is to summarize these control mechanisms and to introduce terminology and shorthand notations that will be used throughout the book. Many of the control elements considered are summarized in Fig. 6-15.

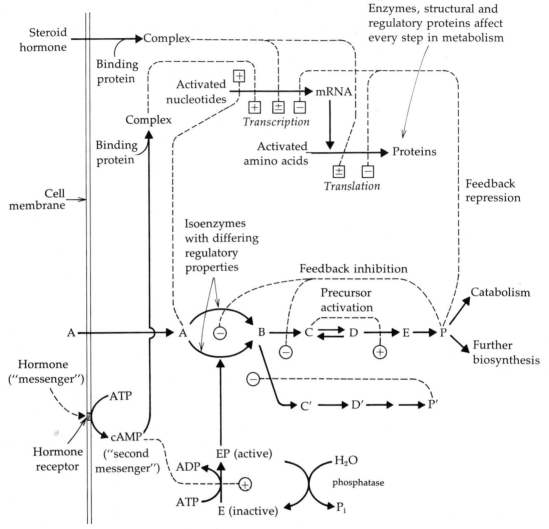

Fig. 6-15 Some control elements for metabolic reactions. Throughout the book modulation of the activity of an enzyme by allosteric effectors or of transcription and translation of genes is indicated by dotted lines from the appropriate metabolite. The lines terminate in a minus sign for inhibition or repression and in a plus sign for activation or derepression. Circles indicate direct effects on enzymes, while boxes indicate repression or induction of enzyme synthesis. See G. G. Hammes and C.-W. Wu, *Science* **172**, 1205–1211 (1971) for a similar scheme and discussion.

1. The Pacemaker Concept

Understanding metabolic control becomes easier if we focus our attention on those enzymes that under a given set of conditions, catalyze rate-limiting steps. Such **pacemaker** enzymes[67] often are involved in (a) reactions that determine the overall respiration rate of a cell, (b) reactions that initiate attack on a substrate when subsequent intermediates in a metabolic sequence do not accumulate, and (c) reactions that initiate branch pathways in metabolism. Usually the first step in a unique biosynthetic pathway for a compound (often referred to as the **committed step**) acts as the pacemaker. Such reactions often proceed with a large free energy decrease and tend to be highly controlled by cells. On the other hand, enzymes catalyzing in-between reactions may not be regulated and may operate at or near equilibrium.

If conditions within a cell change, a pacemaker reaction may cease to be rate limiting. For example, a reactant may be depleted to the point where the rate of its formation from a preceding reaction determines the overall rate. Thus, metabolism of glucose in our bodies occurs through the rapidly interconvertible phosphate esters **glucose 6-phosphate** and **fructose 6-phosphate.** The pacemaker enzyme in utilization of glucose is often **phosphofructokinase** (Chapter 11, Section F,4; Eq. 6-91, step *b*), which catalyzes further metabolism of fructose 6-phosphate. However, if metabolism by this route is sufficiently rapid, the rate of formation of glucose 6-phosphate from glucose (the **hexokinase** reaction, Eq. 6-91, step *a*) may become rate-limiting.

$$\text{Glucose} \xrightarrow[\substack{a \\ \text{hexokinase}}]{\text{ATP} \quad \text{ADP}} \text{Glucose 6-phosphate}$$

$$\Updownarrow$$

$$\text{Fructose 6-phosphate}$$

$$\underset{\substack{\text{phosphofructo-} \quad b \\ \text{kinase}}}{} \begin{array}{c} \text{ATP} \\ \text{ADP} \end{array}$$

$$\text{Fructose 1,6-diphosphate}$$

$$\downarrow$$

$$\text{Further metabolism} \qquad (6\text{-}91)$$

Some catabolic reactions depend upon ADP, but under conditions of high metabolic rate, the ADP concentration may become very low because it is all phosphorylated to ATP. Under these conditions reactions utilizing ADP may become rate-

limiting pacemakers in reaction sequences. Depletion of a reactant often has an additional effect of changing the whole pattern of metabolism. Thus, if oxygen is unavailable to a yeast, the reduced coenzyme NADPH accumulates and reduces pyruvic acid to lactic acid (Chapter 7, Section A,6). The result is a shift from oxidative metabolism to fermentation.

The conclusion is that the pacemaker concept is a useful one but that, depending upon conditions, different enzymes may serve in this capacity. It is also important to realize that reaction rates may be determined by the rate of diffusion of a compound through a membrane. Thus, membrane transport processes can also serve as pacemakers.

2. Genetic Control of Enzyme Synthesis

Ultimately all cellular regulatory mechanisms depend upon genetic information and its expression. Within a given cell many genes are transcribed continuously but others may remain unexpressed. Both the rates of transcription and of degradation of mRNA molecules are among the factors that affect the rates of synthesis of enzymes on the ribosome of the cytoplasm.

a. Repression and Induction

The production of some enzymes is referred to as **constitutive,** implying that the enzyme is formed no matter what the environmental conditions of the cell. For example, bacteria synthesize the enzymes required to catabolize glucose under all conditions of growth. Another group of enzymes, known as **inducible,** are often produced only in trace amounts. However, if cells are grown in the presence of a substrate for these enzymes, they are synthesized in much larger amounts. Thus, when *E. coli* is cultivated in the presence of lactose, a series of enzymes required for the catabolism of that disaccharide are formed. Investigation of the phenomenon has shown that the enzymes required for lactose degradation are normally repressed. The genes coding for these proteins are kept turned off through the action of a specific **repressor** protein with allosteric properties. The repressor binds to a specific site on the DNA and blocks transcription of the genes that it controls (Fig. 6-15). However, an **inducer** such as lactose† binds to an allosteric site of the repressor and causes a decrease in the affinity of the repressor for DNA. The appropriate genes are then **derepressed.**

Synthesis of most enzymes is apparently repressed most of the time. The appearance of specific enzymes at a particular time in the life of an organism or in a certain differentiated tissue results from derepression brought about by the accumulation of specific metabolites or by other, as yet unknown, mechanisms. In eukaryotic cells control may be exerted both at the transcriptional level and at the translational level.

Not only can repression be relieved by the presence of an inducer, but also it can be enhanced by an end product of a metabolic sequence. Such **feedback repression** is, in some cases, also mediated by allosteric alteration of the repressor protein. In eukaryotes feedback control may presumably occur at either the transcriptional or translational levels, as indicated in Fig. 6-15.

In addition to the rate of initial synthesis, other factors affect the amounts of active enzymes present in cells. Some enzymes are synthesized as enzymatically inactive **proenzymes** which are later modified to the active forms, usually by hydrolytic degradation. Sometimes there is more than one active species, depending upon the extent of the degradation. Finally, active enzymes are destroyed, both by accident and by deliberate mechanisms that involve hydrolytic digestion. Thus, as with other cell constituents, the synthesis of enzymes is balanced by degradation. The overall process is usually described as **protein turnover.**[68]

A characteristic of control mechanisms acting at the transcriptional and translational levels is that they are slow, often with response times of hours or even days.

† The actual inducer is an isomer of lactose, **allolactose** (Chapter 15, Section B,1).

b. Isoenzymes (Isozymes)

An important aspect of metabolic control is apparent from the existence of multiple forms of many enzymes. As a rule, these **isoenzymes** are *not* isomers. Rather, they are *similar but chemically distinct protein molecules.* The existence of isoenzymes has been known for a long time but interest was focused on them a few years ago when it was discovered that **lactate dehydrogenase** in the human and in most animals consists of five forms easily separable by electrophoresis and evenly spaced on electropherograms. The enzyme is a tetramer, and it was soon established that there are two kinds of subunits. The tetramer with the highest electrophoretic mobility (isoenzyme 1) has four identical subunits of type B (or H) and is often designated B_4. The slowest moving tetramer (isoenzyme 5) consists of four type A (or M) subunits. The other three forms AB_3, A_2B_2 and A_3B contain both subunits in different proportions. The two subunits are encoded by entirely separate genes, and the two genes are expressed to different extents in different tissues. Thus, heart muscle and liver produce mainly subunit B(H), while skeletal muscle produces principally subunit A(M).

Why do cells produce isoenzymes? The first reason appears to be that enzymes with differing kinetic parameters are needed to fulfill varying functions.[69] Thus, substrate concentrations may vary greatly between different tissues; between mitochondria, nucleus, and cytoplasm; and at different developmental stages of an organism. In the case of lactate dehydrogenase, isoenzyme 1 is inhibited by pyruvate, a product of the action of the

L-Lactate   Pyruvate

enzyme when oxidizing lactate. While the reason for this product inhibition is not obvious,† it

† The special chemical basis for the inhibition is discussed in Chapter 8, Section H,7,a.

seems at least somewhat appropriate for an aerobic organ such as heart in which pyruvate is removed by oxidation and in which overactivity of the lactate dehydrogenase would lead to accumulation of pyruvate and inhibition of the enzyme. On the other hand, isoenzyme 5 of skeletal muscle is less inhibited by pyruvate and is appropriate for an enzyme designed to reduce pyruvate to lactate during a burst of muscular activity.

A clearer example is provided by the multiple forms of hexokinase (Eq. 6-91).[70] The brain enzyme has a low value of $K_m = 0.05$ mM for glucose. It is thus able to phosphorylate glucose and to make that substrate available to the brain for metabolism, even when the glucose concentration in the tissues falls to a low value. On the other hand, **glucokinase,** the isozyme of liver, has a much higher K_m of ~ 10 mM and reaches its maximal activity in removing glucose from blood when the latter rises to high concentrations (blood normally contains ~ 5.5 mM glucose).

Other reasons for the appearance of isoenzymes are the cleavage of proenzymes to multiple forms, partial hydrolytic breakdown of enzymes, and the reversible modifications discussed in Section F,4.

A third and important origin of isoenzymes is genetic variation in heterozygotes. Thus, if a genetic variant of a particular protein carries one more or one less positive or negative charge than the "standard" enzyme, electrophoresis will reveal a new isoenzyme in the heterozygotic individual. Note that while electrophoresis is most often used to detect isoenzymes, the many genetic variants in which amino acid substitutions have occurred without change in the charge are not detected in this way. Many such variants probably exist but remain unreported.

While isoenzymes have been designated in various ways, it has now been agreed that *they should be numbered in the order of decreasing electrophoretic mobility.* Electrophoresis is usually conducted at pH 7 to 9; most enzymes are negatively charged in this range; the one that migrates most rapidly toward the cathode is designated number 1. This is the same convention that has been used in the electrophoresis of blood proteins for many years. Thus, α_1-, α_2-, etc., globulins are numbered in order of decreasing mobility[71] (see Box 2-A).

3. Spatial Separation: Compartments and Organized Assemblies

The geometry of cell construction provides important aspects of metabolic control. In a bacterium, the periplasmic space (Chapter 5, Section D) provides a separate compartment from the cytosol. Some enzymes are localized in this space and do not mix with those within the cell. Other enzymes are fixed within or attached to the membrane. Eukaryotic cells have more compartments: nuclei, mitochondria (containing both matrix space and intermembrane spaces), lysosomes, microbodies, and vacuoles. Within the cytosol tubules and vesicles of the endoplasmic reticulum separate off another membrane-bound compartment.

The rates of transport of metabolites through the membranes between compartments is limited and may be controlled tightly. Indeed, membrane-bound receptors are the targets for most hormones, for action of neurotransmitters, and for various chemical modification reactions.

While many enzymes appear to be dissolved in the cytosol and without any long-term association with other proteins, in other instances enzymes that catalyze a series of consecutive reactions may be attached to a membrane and held close together as an organized assembly. This appears to be the case for oxidative enzymes of mitochondria (Chapter 10, Section E). In other instances several enzymes associate to form a high molecular weight complex. The ketoacid dehydrogenases (Chapter 8, Section J) and the cytoplasmic fatty acid synthetases (Chapter 11, Section D,6) are examples. In both these instances it appears that the product of the first enzyme is covalently attached to a "carrier" and, while so attached, is subjected to the action of a series of other enzymes.

4. Regulation of the Activity of Enzymes

Rapid regulatory mechanisms often act directly on enzymes. In some cases an enzyme is normally inactive or almost so but is converted to an active form by a **covalent modification**.[72] Alternatively, covalent modification may inactivate an enzyme. Thus, the activities of two enzymes of glycogen metabolism, **glycogen phosphorylase** and **glycogen synthetase,** are altered by phosphorylation (transfer of the terminal phosphoryl group of ATP to a specific serine residue, Chapter 11, Section F,3). The effect of the two modifications is to convert the glycogen-degrading enzyme **phosphorylase b** into a more active form **phosphorylase a**, and to inactivate the glycogen-synthesizing enzyme. This switches the cellular metabolism from that designed to deposit the storage polysaccharide glycogen to one that degrades glycogen to provide the cell with energy. The modification of each enzyme is reversed by another enzyme, a **phosphatase,** which restores the enzyme to the original state (Fig. 6-15, bottom). Both the modification enzyme (a **kinase,** Chapter 7, Section E,6) and the phosphatase are under allosteric control. These rather complex regulatory mechanisms act to provide the cell with modified enzyme for what may be a very short period of time.

Glutamine synthetase (Chapter 14, Section B,2) is modified (using ATP) by **adenylylation** of a specific tyrosine side chain.[73] Relaxation to the unmodified enzyme is catalyzed by a deadenylylating enzyme.

Probably the most common and widespread control mechanism in cells is **allosteric activation or inhibition,** which has previously been discussed (Section B,6). Allosteric control mechanisms are incorporated into metabolic pathways in many different ways, but two patterns are met most frequently. The first may be referred to as **precursor activation.** A metabolite acting as an allosteric effector turns on an enzyme that either acts directly on that metabolite or acts on a product that lies a little further ahead in the sequence. For example, in Fig. 6-15 metabolite C activates the enzyme that

catalyzes an essentially irreversible reaction of compound D. In other cases, activation is less direct. The enzyme turned on may be needed for formation of a second substrate. The latter may be required to react with the activating metabolite to give a needed metabolic product.

More common than precursor activation is **feedback inhibition.** An end product of a metabolic sequence accumulates and turns off enzymes needed for its own formation. Most often *the first enzyme unique to the specific biosynthetic pathway for the product* is inhibited. On the other hand, a product often inhibits more than one of the enzymes in a biosynthetic sequence, as illustrated in Fig. 6-15. Often, when a cell makes two or more isoenzymes, only one is inhibited by a particular product. This is illustrated in Fig. 6-15: Product P inhibits only one of the two isoenzymes that catalyze conversion of A to B, while the other is controlled by an enzyme modification reaction.

A real example is illustrated in Fig. 14-6. Three isoenzymes convert aspartate to β-aspartyl phosphate, the precursor to the end products threonine, isoleucine, methionine, and lysine. Each product inhibits only one of the isoenzymes, as indicated in Fig. 14-6.

5. Hormones

The binding of hormones to receptors on the cell surface is a major control element. In some cases, e.g., in the binding of glucagon and ACTH to their receptors, it appears that the entire effect can be explained on the basis of activation of the enzyme **adenylate cyclase** (Eq. 6-92, step a). This enzyme converts ATP to **cyclic AMP (cAMP,** cyclic

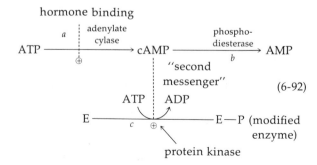

adenosine monophosphate). The chemistry of the reaction is discussed in Chapter 7, Section E,8. Cyclic AMP is sometimes referred to as a "second messenger," for it carries the "message" brought by the "first messenger" (the hormone) into the cell. Cyclic AMP is rapidly hydrolyzed by a **phosphodiesterase** (Eq. 6-92, step b, see also Chapter 7, Section E,8) to AMP. However, while it exists, cAMP acts as an allosteric effector for **protein kinases** (Eq. 6-92, step c) which catalyze the modification reactions of proteins such as the phosphorylation of glycogen synthetase, discussed in the preceding section and in Chapter 11, Section F,3. Representative of this group of enzymes is a soluble cAMP-dependent kinase of broad specificity, which has been isolated from muscle as an $\alpha_2\beta_2$ dimer. The two catalytic subunits are inactive until cAMP binds to the regulatory subunits. The complex then dissociates into active catalytic subunits and a cAMP-containing regulatory subunit dimer.[73a,73b]

In 1956, Earl Sutherland first recognized the existence of cAMP as a compound mediating the action of the hormones **epinephrine** (adrenaline) and glucagon on glycogen phosphorylase. For many years most biochemists regarded cAMP as a curiosity and the strange regulatory chemistry of phosphorylase as an unusual specialization. Recently, that view has been altered drastically as cAMP has been reported to function as a second messenger in the action of over 20 different hormones. Cyclic AMP also appears to mediate the action of neurotransmitters released at synapses between nerve cells. Even in *E. coli* cAMP is generated and acts as a positive effector in initiating transcription of certain genes (Chapter 15, Section B,2). In 1971, Sutherland was awarded a Nobel Prize for his almost single-handed development of this field.[74,75]

While in higher organisms cAMP appears to act primarily within the cells, it serves as a signal *between* cells of the cellular slime mold *Dictyostelium discoideum* (Chapter 1, Section D,1). It is noteworthy that the same mechanism of inactivation with phosphodiesterase is used.[76] The oscillatory production of cAMP has been attributed to

autocatalytic control: activation by 5'-AMP and a separate cAMP-activated hydrolase that cleaves ATP directly to 5'-AMP.[77]

Adenylate cyclase is apparently attached to the internal surfaces of membranes. However, its activity is strongly influenced by the binding of hormones to receptors on the *external* membrane surface (Chapter 5, Section C,5). The mechanism of transmission of the chemical signal through the membrane remains uncertain. A very interesting discovery is that the protein toxin of *Vibrio cholerae* induces the profuse diarrhea and loss of salt from the body characteristic of Asiatic cholera by stimulating adenylate cyclase in the cells of the wall of the small intestine.[78,79]

The effects of cAMP within cells are numerous. Allosteric activation of protein kinases affects a whole series of enzymes concerned with energy metabolism. Proteins of membranes,[79a] microtubules, and ribosomes[79b] are also phosphorylated as are the nuclear proteins known as histones (Chapter 15, Section I,2). Since the same mechanism can lead to quite different responses in different specialized cells, specific examples of these effects will be discussed later. Many enzymes that are modified by protein kinases require calcium ion for activity. Thus, for the effects of adenylate cyclase activity to be fully felt, a release of Ca^{2+} into the cytoplasm is also necessary. The latter is usually triggered by nervous stimulation. However, the binding of concanavalin A to T lymphocytes (Chapter 5, Section C,3) increases uptake of Ca^{2+} by these cells and may play a role in inducing mitosis.[80] Control of the entry of Ca^{2+} into cells may be an important facet of communication between cells.

The possibility that **guanosine triphosphate** (GTP) may be converted to **cyclic GMP (cGMP)** and that the latter may antagonize cAMP in many aspects of metabolic control is under active consideration.[81–83] However, cyclic GMP and other cyclic nucleotides are present in much smaller amounts than is cAMP. The picture is further complicated by the apparent need for GTP as an additional allosteric activator for efficient response of adenylate cyclase of liver cells to glucagon.[84]

While specific receptors for insulin have been identified (Chapter 5, Section C,5), the mechanism by which the hormone affects metabolism is obscure. The major effect on carbohydrate metabolism appears to be one of controlling the rate of entry of glucose into cells.[85] The possibility of mediation by cyclic GMP has been suggested.

The **steroid hormones** act in quite a different way. These molecules enter cells and bind to specific receptor proteins that are dissolved in the cytosol.[86-88] The hormone-protein complexes then move into the nucleus where they apparently induce changes in gene expression through regulation of transcription or translation (Fig. 6-15, top).

6. Amplification of Regulatory Signals

The effect of a regulated change in the activity of an enzyme is often amplified through a **cascade mechanism.** The first enzyme acts on a second enzyme, the second on a third, etc. The effect is to create very rapidly a large amount of the active form of the last enzyme in the series. An example is the blood clotting mechanism[89] presented in simplified form in Fig. 6-16. Here we see a series of five enzymes (beginning with factor XII), each of which activates the next, usually by cutting off a small piece of the peptide chain by proteolytic action. Finally, thrombin attacks fibrinogen removing a small peptide and converting it to **fibrin,** a specialized protein that spontaneously coagulates. What is to keep a cascade mechanism from getting out of control? Why doesn't the slightest bruise convert all the prothrombin in our bodies to thrombin and coagulate our blood? Just as there is a special enzyme to remove cAMP quickly once it is produced, there are doubtless mechanisms for removing the activated enzymes from the cascade sequence of Fig. 6-16. There is also a special enzyme system that dissolves the clot as healing occurs.[89]

In another case the details of the mechanism by which a cascade effect is reversed are known. Glycogen phosphorylase of muscle is activated in a cascade sequence that begins with release of epinephrine, the latter being under the control of the autonomic nervous system (Chapter 16, Section B,3). Binding of epinephrine to the cell membrane releases cAMP which activates a protein kinase. The kinase phosphorylates another enzyme, **phosphorylase kinase.** At this point the muscles are prepared for the rapid breakdown of glycogen. However, the final initiating signal is the release of Ca^{2+} into the cytoplasm in response to impulses to specific muscles via the motor neurons (Chapter 4, Section F,1). Phosphorylase kinase is activated by calcium ions, and in their presence it converts inactive phosphorylase b to the active phosphorylase a. Spontaneous reversion to the resting state occurs through the action of phosphatases that cut off the phosphoryl groups placed on the protein by the kinases. Also essential are the phosphodiesterase, which destroys the cAMP, and the calcium ion pump, which reduces the concentration of the activating calcium ion to a low level.

Another amplification mechanism, whose details remain to be worked out, is found in the retina of the eye. A single quantum of light falling on a receptor cell can, under appropriate conditions, initiate a nerve impulse (Chapter 16, Section B,3). The latter requires the flow of a large number of Na^+ ions across the membrane, and it is hard to believe that absorption of one quantum could initiate a photochemical reaction leading to that much sodium transport unless some intermediate amplification stages exist.

Other amplification mechanisms are of a quite different type. One that depends upon **substrate cycling** is considered in Chapter 11, Section F,6.

7. Cross Regulation

A metabolite may not only regulate its own production by feedback inhibition but also may control the production of something else that is formed by quite an independent pathway.[66] A particularly striking example is the synthesis of ATP and GTP from their common precursor **inosine 5'-phosphate** (IMP, Fig. 6-17). A major need

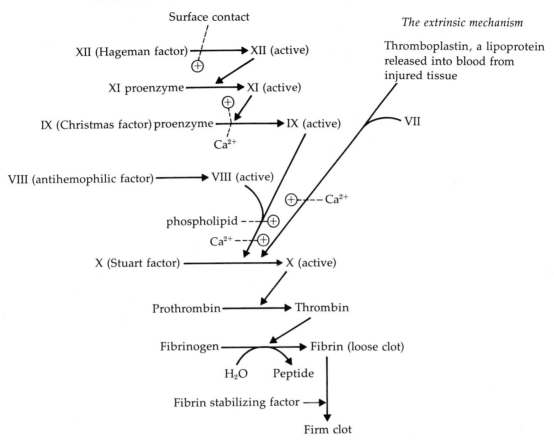

Fig. 6-16 Cascade mechanism by which blood clotting is initiated. There are two ways to initiate the cascade: the intrinsic mechanism is triggered by surface contact; the extrinsic mechanism by release of thromboplastin from injured tissues. See O. D. Ratnoff and B. Bennett, *Science* **179**, 1291–1298 (1973), and Montgomery *et al.*[89]

for both ATP and GTP in cells is for synthesis of RNA and DNA so it is not surprising that the production of the two is kept in balance by special regulatory mechanisms. We see from Fig. 6-17 that the synthesis of AMP from IMP requires the actual participation of GTP. Furthermore, ATP participates in the synthesis of GMP. Production of both AMP and GMP is inhibited by feedback inhibition. In addition, deliberate mechanisms exist for hydrolysis of excesses of both of these compounds (the outer loops in Fig. 6-17). However, the hydrolysis of AMP is inhibited by GTP and the reductive deamination of GMP is inhibited by ATP.

8. Species Differences In Regulation

Catalytic mechanisms of particular enzymes have usually been conserved throughout evolution, and certain residues in a protein may be invariant. However, species differences in amino acid sequence affecting residues on the surface of

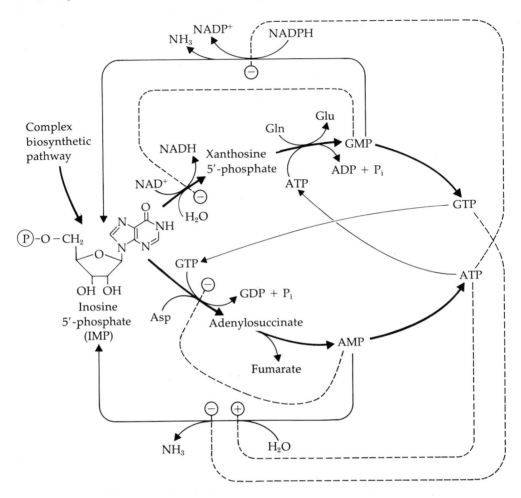

Fig. 6-17 Cross regulation of synthesis of the purine nucleoside triphosphates ATP and GTP. After Stadtman,[66] p. 431.

proteins are numerous. Since these changes in the surface shape of a protein molecule may radically affect sensitivity to a potential allosteric effector, it should not be surprising that when a given enzyme is studied from more than one species, very different regulatory properties are often found. Thus, after having arduously worked out regulatory mechanisms in one species, the biochemist is apt to find that many of them do not apply at all to another organism. This fact not only makes life maddeningly complex for the person trying to understand metabolism, but it has many practical consequences. Metabolism *is* different in different

species and so is the sensitivity of key enzymes in the body to poisons, drugs, and environmental pollutants. Those concerned about human life should be very, very cautious in extrapolating to humans from data obtained with other animals.

G. CLASSIFICATION OF ENZYMES

An official commission of the International Union of Biochemistry (IUB) has classified enzymes in the following six categories[90]:

1. *Oxidoreductases*
2. *Transferases.* These catalyze group transfer reactions.
3. *Hydrolases.* Enzymes catalyzing transfer of groups to H_2O.
4. *Lyases.* Enzymes promoting addition to double bonds or the reverse.
5. *Isomerases*
6. *Ligases* (synthetases). Enzymes that catalyze condensation with simultaneous cleavage of ATP and related reactions.

As an example, chymotrypsin is classified EC 3.4.4.5, according to the IUB system.[90] In this book a much more mechanistically based classification is used. Because the official names are quite long, in most cases the traditional trivial names for enzymes will be retained.

Remember that to be precise it is always necessary to mention the species from which an enzyme was isolated and if possible, the strain. Also remember that every genetic difference is reflected in some change in some protein. It just might be the enzyme that you are working with is slightly different from the same enzyme prepared in a different laboratory.

REFERENCES

1. H. Fromm, "Initial Rate Enzyme Kinetics." Springer-Verlag, Berlin and New York, 1975.
2. I. H. Segal, "Enzyme Kinetics." Wiley, New York, 1975.
3. M. Dixon and E. C. Webb, eds., "Enzymes," 2nd ed. Academic Press, New York, 1964.
4. W. W. Cleland, *in* "The Enzymes," 3rd ed. (P. D. Boyer, ed.) Vol. 2, pp. 1–65. Academic Press, New York, 1970.
4a. H. Gutfreund, "Enzymes: Physical Principles." Wiley, New York, 1975.
5. J. T.-F. Wong, "Kinetics of Enzyme Mechanisms." Academic Press, New York, 1975.
6. T. E. Barnum, "Handbook of Enzymes," Vol. I. Springer-Verlag, Berlin and New York, 1969.
7. Commission on Biochemical Nomenclature of the International Union of Biochemistry, "Enzyme Nomenclature." Elsevier, Amsterdam, 1972.
8. R. G. Khalifah, *JBC* **246**, 2561–2573 (1971).
9. A. M. Benson, R. Jarabah, and P. Talalay, *JBC* **246**, 7514–7525 (1971).
10. E. F. Caldin, "Fast Reactions in Solution." Wiley, New York, 1964.
11. A. Marco and R. Sols, *Curr. Top. Cell. Regul.* **2**, 227–273 (1970).
12. W. P. Jencks, "Catalysis in Chemistry and Enzymology," p. 570. McGraw-Hill, New York, 1969.
13. H. B. Bull, "An Introduction to Physical Biochemistry," 2nd ed. Davis, Philadelphia, Pennsylvania, 1971.
14. L. G. Longsworth, *Phys. Tech. Biol. Res., 2nd Ed.* **2A**, 85–120 (1968).
15. M. Smoluchowski, *Z. Phys. Chem.* **92**, 129–168 (1917).
16. E. F. Caldin, "Fast Reactions in Solution," pp. 10 and 279. Wiley, New York, 1964.
17. P. Debye, *Trans. Electrochem. Soc.* **82**, 265–272 (1942).
18. D. French, *Brew. Dig.* **32**, 50–56 (1957).
19. H. B. Bull, "An Introduction to Physical Biochemistry," 2nd ed., p. 299. Davis, Philadelphia, Pennsylvania, 1971.
20. E. J. Cohn and J. T. Edsall, "Proteins, Amino Acids and Peptides as Ions and Dipolar Ions," pp. 506ff. Van Nostrand-Reinhold, Princeton, New Jersey, 1943.
21. J. B. S. Haldane, "Enzymes." Longmans, Green, New York, 1930.
22. K. Dalziel, *Acta Chem. Scand.* **11**, 1706–1723 (1957).
23. M. V. Volkenstein and B. N. Goldstein, *Biokhimiya* **31**, 541–547 (1966); *BBA* **115**, 471–477 (1966).
24. H. J. Fromm, *BBRC* **40**, 692–697 (1970).
25. N. Seshagiri, *J. Theor. Biol.* **34**, 469–486 (1972).
26. F. B. Rudolph and H. J. Fromm, *JBC* **246**, 6611–6619 (1971).
27. P. A. Srere, *Science* **158**, 936–937 (1967).
28. S. Cha, *JBC* **245**, 4814–4818 (1970).
29. A. N. Schechter, *Science* **170**, 273–280 (1970).
30. G. G. Hammes, ed., *Tech. Chem.* **6**, Part II (1974).
31. G. Porter and M. A. West, *Tech. Chem.* **6**, Part II, 367–462 (1974).
32. G. G. Hammes and P. R. Schimmel, *in* "The Enzymes" (P. D. Boyer, ed.), 3rd ed., Vol. 2, pp. 67–114. Academic Press, New York, 1970.
33. M.-L. Ahrens, G. Maass, P. Schuster, and H. Winkler, *JACS* **92**, 6134–6139 (1970).
33a. R. R. Rando, *Science* **185**, 320–324 (1974).
34. C. Frieden, *JBC* **239**, 3522–3531 (1964).
35. J. Monod, J. Wyman, and J. D. Changeux, *JMB* **12**, 88–118 (1965).
36. D. L. Purich and H. J. Fromm, *BBA* **268**, 1–3 (1972).
37. H. J. Fromm, E. Silverstein, and P. D. Boyer, *JBC* **239**, 3645–3652 (1964).
38. E. Silverstein and P. D. Boyer, *JBC* **239**, 3901–3907 (1964).
39. M. M. Rubin and J. D. Changeux, *JMB* **21**, 265–274 (1966).
40. B. R. Rabin, *BJ* **102**, 22c (1967).
41. E. A. Newsholme and C. Start, "Regulation in Metabolism." Wiley, New York, 1973.
42. See R. Y. Stanier, M. Douderoff, and E. A. Adelberg, "The Microbial World," p. 298ff. Prentice-Hall, Englewood Cliffs, New Jersey, 1970, for a more detailed treatment.
43. J. D. Gass and A. Meister, *Biochemistry* **9**, 842–846 (1970).

44. K. R. Hanson, *JACS* **88,** 2731–2742 (1966).

45. H. Hirschmann, *JBC* **235,** 2762–2767 (1960).

46. L. P. Hammett, "Physical Organic Chemistry," 2nd ed., pp. 101ff. McGraw-Hill, New York, 1970.

47. W. P. Jencks, "Catalysis in Chemistry and Enzymology," p. 599ff. McGraw-Hill, New York, 1969.

48. L. Pauling, *Nature (London)* **161,** 707–713 (1948).

49. R. Wolfenden, *Annu. Rev. Biophys. Bioeng.* **5,** 271–306 (1976).

49a. G. E. Lienhard, *Science* **180,** 149–154 (1973).

50. C. G. Swain and J. F. Brown, Jr., *JACS* **74,** 2534–2537 and 2538–2543 (1952).

51. P. R. Rony, *JACS* **91,** 6090–6096 (1969).

52. W. P. Jencks, "Catalysis in Chemistry and Enzymology." p. 199. McGraw-Hill, New York, 1969.

52a. A. F. Hegarty and W. P. Jencks, *JACS* **97,** 7188–7189 (1975).

53. W. P. Jencks, "Catalysis in Chemistry and Enzymology," pp. 170–199. McGraw-Hill, New York, 1969.

54. M. Eigen, *Angew. Chem., Int. Ed. Engl.* **3,** 1–28 (1964).

55. J. H. Wang, *Science* **161,** 328–334 (1968).

56. M. Dixon and E. C. Webb, "Enzymes," 2nd ed., pp. 118–145. Academic Press, New York, 1964.

57. J. E. Critchlow and H. B. Dunford, *J. Theor. Biol.* **37,** 307–320 (1972).

57a. H. B. F. Dixon, *BJ* **131,** 149–154 (1973).

57b. J. L. Wood, *BJ* **143,** 775–777 (1974).

58. R. A. Alberty, *Adv. Enzymology* **17,** 1–64 (1956).

59. F. Hucho and K. Wallenfels, *EJB* **23,** 489–496 (1971).

60. M. I. Page and W. P. Jencks, *PNAS* **68,** 1678–1683 (1971).

61. J. F. Kirsch, *Annu. Rev. Biochem.* **42,** 205–234 (1973).

62. F. H. Westheimer, *Adv. Enzymol.* **24,** 441–482 (1962).

63. D. E. Koshland, Jr., K. W. Carraway, G. A. Dafforn, J. D. Goss, and D. R. Storm, *Cold Spring Harbor Symp. Quant. Biol.* **36,** 13–20 (1971).

63a. W. P. Jencks and M. I. Page, *BBRC* **57,** 887–892 (1974).

64. S. Milstien and L. A. Cohen, *JACS* **94,** 9158–9165 (1972).

65. J. M. Karle and I. L. Karle, *JACS* **94,** 9182–9189 (1972).

65a. P. S. Low and G. N. Somero, *PNAS* **72,** 3305–3309 (1975).

66. E. R. Stadtman, *in* "The Enzymes," 3rd ed. (P. D. Boyer, ed.), Vol. 1, pp. 397–459. Academic Press, New York, 1970.

67. H. A. Krebs, *Endeavour* **16,** 125–132 (1957).

68. G. Palade, *Science* **189,** 347–358 (1975).

69. D. L. Purich and H. J. Fromm, *Curr. Top. Cell. Regul.* **6,** 131–167 (1972).

70. D. L. Purich, H. F. Fromm, and F. R. Rudolph, *Adv. Enzymol.* **39,** 249–326 (1973).

71. J. H. Wilkinson, "Isoenzymes," 2nd ed. Lippincott, Philadelphia, Pennsylvania, 1970.

72. H. L. Segal, *Science* **180,** 25–32 (1973).

73. S. P. Adler, D. Purich, and E. R. Stadtman, *JBC* **250,** 6264–6272 (1975).

73a. J. Erlichman, C. S. Rubin, and O. M. Rosen, *JBC* **248,** 7607–7609 (1973).

73b. J. A. Beavo, P. J. Bechtel, and E. G. Krebs, *in* "Methods in Enzymology," (G. Hardman and B. W. O'Malley, eds.) Vol. 38, pp. 299–308. Academic Press, New York, 1975.

74. E. W. Sutherland, *Science* **177,** 401–408 (1972).

75. G. A. Robinson, R. W. Butcher, and E. W. Sutherland, "Cyclic AMP." Academic Press, New York, 1971.

76. J. D. Gross, *Nature (London)* **255,** 522–523 (1975).

77. A. Goldbeter, *Nature (London)* **253,** 540–542 (1975).

78. N. Hirschhorn and W. B. Greenough, III, *Sci. Am.* **225,** 15–21 (Aug 1971).

79. N. Sahyoun and P. Cuatrecasas, *PNAS* **72,** 3438–3442 (1975).

79a. C. S. Rubin, *JBC* **250,** 9044–9052 (1975).

79b. A. M. Gressner and I. W. Wool, *JBC* **251,** 1500–1504 (1976).

80. M. H. Freedman and M. C. Raff, *Nature (London)* **255,** 378–382 (1975).

81. N. D. Goldberg, R. F. O'Dea, and M. K. Haddox, *Adv. Cyclic Nucleotide Res.* **3,** 1–48 (1973).

81a. Y. Takai, S. Nakaya, M. Inoue, A. Kishimoto, K. Nishiyama, H. Yamamura, and Y. Nishizuka, *JBC* **251,** 1481–1487 (1976).

82. G. B. Kolata, *Science* **182,** 149–151 (1973).

83. E. J. Neer and E. A. Sukiennik, *JBC* **250,** 7905–7909 (1975).

84. Y. Salomon, M. C. Lin, C. Londos, M. Rendell, and M. Rodbell, *JBC* **250,** 4239–4245 (1975).

85. J. Elbrink and I. Bihler, *Science* **188,** 1177–1184 (1975).

86. W. T. Schrader, D. O. Toft, and B. W. O'Malley, *JBC* **247,** 2401–2417 (1972).

87. E. V. Jensen and E. R. DeSombre, *Science* **182,** 126–134 (1973).

88. R. W. Kuhn, W. T. Schrader, R. G. Smith, and B. W. O'Malley, *JBC* **250,** 4220–4228 (1975).

89. R. Montgomery, R. L. Dryer, T. W. Conway, and A. A. Spector, "Human Biochemistry." Mosby, St. Louis, Missouri, 1974.

90. Commission on Biochemical Nomenclature, International Union of Pure and Applied Chemistry and the International Union of Biochemistry, "Enzyme Nomenclature." Elsevier, Amsterdam, 1972.

STUDY QUESTIONS

1. If an enzyme catalyzes the reaction $A \rightarrow B + C$ will it also catalyze the reaction $B + C \rightarrow A$?
2. Give two reasons why enzymes are important to living organisms.
3. How is the instantaneous initial velocity of an enzymatic reaction measured? What precautions must be taken to ensure that a true instantaneous velocity is being obtained?
4. What is meant by the statement that biochemical reactions are stereochemically specific? Why is such stereospecificity to be expected in organisms (which are constructed from asymmetric units)?
5. How is the rate of an enzyme action influenced by changes in (a) temperature, (b) enzyme concentration, (c) substrate concentration, (d) pH?
6. Can an enzyme use a different site to catalyze a reverse reaction than it does to catalyze the reaction in the forward direction? Explain your answer.
7. In what ways are enzymatic reactions typical of ordinary catalytic reactions in organic or inorganic chemistry, and in what ways are they distinct?
8. In general, for a reaction that can take place with or without catalysis by an enzyme, what is the effect of the enzyme on (a) standard free energy change of the reaction, (b) energy of activation of the reaction, (c) initial velocity of the reaction, (d) temperature coefficient of the rate constant?
9. What are the dimensions of the rate constant for zero-, first-, and second-order reactions? If a first-order reaction is half completed in $\frac{1}{2}$ min, what is its rate constant?
10. A sample of hemin containing radioactive iron, ^{59}Fe, was assayed with a Geiger–Müller counter at intervals of time with the following results:

Time (days)	Radioactive counts per min
0	3981
2	3864
6	3648
10	3437
14	3238
20	2965

Determine the half-life of ^{59}Fe and the value for the decay constant.

11. The kinetics of the aerobic oxidation of enzymatically reduced nicotinamide adenine dinucleotide (NADH) have been investigated at pH 7.38 at 30°C. The reaction rate was followed spectrophotometrically by measuring the decrease in absorbance at 340 nm over a period of 30 min. The reaction may be represented as

$$NADH + H^+ + riboflavin \longrightarrow$$
$$NAD^+ + riboflavin \cdot H_2$$

Time (min)	O.D. (at 340 mn)
1	0.347
2	0.339
5	0.327
9	0.302
16.5	0.275
23	0.254
27	0.239
30	0.229

Determine the rate constant and the order of the reaction.

12. Using the method of graphs,[1,24] write the initial rate equation for the following system with A, B, P, and Q present.

a. Satisfy yourself that

$$\frac{-d[A]}{dt} = k_1[A][E] - k_2[EA]$$

$$= \frac{-d[B]}{dt} = k_3[B][EA] - k_4[EXY]$$

$$= \frac{d[P]}{dt} = k_5[EXY] - k_6[P][EQ]$$

$$= \frac{d[Q]}{dt} = k_7[EQ] - k_8[E][Q]$$

The determinants (given by the method of graphs) which provide the steady-state concentrations of [E] and of the various enzyme-substrate and enzyme–product complexes are

$$[E] = k_2k_7[k_4 + k_5] + k_3k_5k_7[B] + k_2k_4k_6[P]$$

$$[EA] = k_1k_7[k_4 + k_5][A] + k_1k_4k_6[A][P]$$
$$+ k_4k_6k_8[P][Q]$$

$$[EXY] = k_1 k_3 k_7 [A][B] + k_2 k_6 k_8 [P][Q]$$
$$+ k_1 k_3 k_6 [A][B][P] + k_3 k_6 k_8 [B][P][Q]$$

$$[EQ] = k_2 k_8 [k_4 + k_5][Q] + k_1 k_3 k_5 [A][B]$$
$$+ k_3 k_5 k_8 [B][Q]$$

b. To obtain an expression for v_f the expression for $-d[A]/dt$ may be multiplied by $[E]_t$ and divided by $[E] + [EA] + [EXY] + [EQ] = [E]_t$. Then the above determinants may be substituted.

c. What is $V_{max,forward}$ and $V_{max,reverse}$? HINT: $V_{max,forward}$ is obtained when $[P] = [Q] = 0$ and $[A] = [B] = \infty$. What are K_{mA} and K_{mB}? HINT: K_{mA} is obtained when $[P] = [Q] = 0$, $[B] = \infty$, and $v = \frac{1}{2} V_{max,forward}$.

d. With a knowledge of the kinetic parameters, indicate how the eight rate constants may be obtained if the total concentration of enzyme, $[E]_t$ is known.

13. Anticooperativity was observed in the plot of velocity vs. substrate concentration for an enzyme. Can this observation be explained by the Monod–Wyman–Changeux model for oligomeric enzymes? By the model of Koshland? Explain.

14. Illustrated is a generalized metabolic pathway in which capital letters indicate major metabolites in the pathway, small letters indicate cofactors and numbers indicate enzymes catalyzing the reactions.

$$A \xrightarrow[\substack{m \quad n}]{1} B \rightleftharpoons[\;]{2} C \xrightarrow[\substack{o \quad p}]{3} D \rightleftharpoons[\;]{4} E \xrightarrow{5} F$$

List and describe four different ways in which the pathway might be regulated, referring to the specific enzymes, reactants and cofactors indicated in the diagram. NOTE: Do not just refer to four different reactions as possible sites of regulation, but give four different general methods for regulation.

15. Interpret, for each of the following cases, the curve showing measured initial velocity at constant substrate concentration (*not* maximum velocity) against pH for an enzyme-catalyzed reaction.

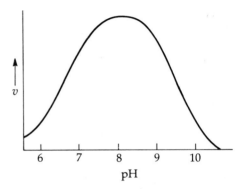

a. The substrate is neutral and contains no acidic or basic groups. The Michaelis constant is found to be independent of pH over the range studied.

b. The substrate is neutral and contains no acidic or basic groups, and the maximum velocity is found to be independent of pH.

c. The substrate is an α-amino acid, and the maximum velocity is found to be independent of pH.

7

The Kinds of Reactions Catalyzed by Enzymes

The remainder of this book is largely about the reactions of metabolism. **Metabolic pathways** consist of sequences of reactions that form a complex, branched, and interconnected network. It would be pointless to try to memorize all of these pathways. However, it is quite important to learn some of the major metabolic sequences and to understand their significance. The attainment of both of these goals will be easier if we look first at the individual chemical reactions catalyzed by enzymes and *then* at the pathways. Understanding the logic of the combining of the individual reactions into metabolic sequences will then be simpler. Nevertheless, before pursuing this approach it will be worthwhile to have a quick preview of the central pathways of metabolism. This will make it easier to view the individual chemical reactions with some biological perspective.

A. A PREVIEW OF SOME METABOLIC PATHWAYS

The sole function of many of the most active metabolic pathways is to provide the cell with ATP and other "high-energy" intermediates that can be used in biosynthesis and in other energy-requiring activities. Thus, the first pathways to be considered are those of the catabolism of foodstuffs and its coupling to synthesis of ATP.

1. Priming or Activation of Metabolites

After a polymeric nutrient is digested (hydrolyzed) and the monomeric products are absorbed into a cell, an energy-requiring "priming" reaction is usually required. For example, the hydrolysis of fats (whether in the gut or within cells) produces free fatty acids. Before further metabolism, the fatty acids are combined with a special coenzyme, **coenzyme A,** to form a **fatty acyl-CoA** derivative. This reaction requires the "expenditure" of ATP, i.e., hydrolysis to AMP and PP_i (Box 3-A). Likewise, glucose, when taken into cells, is converted to the phosphate ester glucose 6-phosphate, again at the expense of utilization of ATP. The major metabolic pathways often start with one of these two substances: a *fatty acyl-CoA derivative* or *glucose 6-phosphate*. The structures of both are given at the top of Fig. 7-1.

2. Beta Oxidation

Cells may use fatty acids to obtain energy by oxidizing them to CO_2 and water. Fatty acids may also enter biosynthetic pathways and be used to create other compounds. In either case, cells usually cut off two-carbon units from the fatty acyl-CoA chain to form acetyl-CoA (solid vertical arrow in Fig. 7-1). This process, known as **beta oxidation,** takes place within the mitochondria of eukaryotic cells. Each time that a 2-carbon unit is removed, two oxidation steps are required. The hydrogen atoms removed in these oxidations are transferred to special **hydrogen carriers NAD$^+$** and **FAD** to form **NADH** and **FADH$_2$** (see Figs. 8-10 and 8-14 for structures). *Transfer of hydrogen atoms from substrates to these hydrogen carriers is typical of biological oxidation processes.*

3. The Electron Transport Chain, Oxidative Phosphorylation

In aerobic organisms the reduced hydrogen carriers are reoxidized by molecular oxygen in the **electron transport chain,** shown below the circle near the center of Fig. 7-1. The oxidation of reduced NADH by O_2 is a highly exergonic process ($\Delta G'$ at pH 7 = -219 kJ mol^{-1}) and is accompanied by the generation (from ADP and inorganic phosphate) of three molecules of ATP. This process of **oxidative phosphorylation** (Chapter 10) is the principal source of usable energy (in the form of ATP) provided by breakdown of fats in the human body.

4. The Citric Acid Cycle

The 2-carbon acetyl units removed from fatty acid chains must be completely oxidized to carbon dioxide to provide cells with a maximum of energy. Chemically the oxidation of an acetyl group is not easy, and probably for this reason nature has devised an elegant catalytic cycle, the citric acid cycle, which is shown at the lower right in

Fig. 7-1. The 4-carbon compound **oxaloacetate** is condensed with an acetyl group from acetyl-CoA to form the 6-carbon **citrate;** then in the remaining reactions of the cycle, two carbon atoms are removed as CO_2 and oxaloacetate is regenerated. Several oxidation steps are involved, each one of which feeds additional **reducing equivalents** (i.e., hydrogen atoms removed from substrates) into the pool of hydrogen carriers and allows for more synthesis of ATP via the electron transport chain.

While oxaloacetate is regenerated and, therefore, not consumed during the operation of the citric acid cycle, it actively enters other metabolic pathways. To replace losses, oxaloacetate can be synthesized from **pyruvate** and CO_2 in a reaction that uses ATP as an energy source. This is indicated by the heavy dashed line leading downward to the right from pyruvate in Fig. 7-1. Pyruvate itself is formed from breakdown of carbohydrates such as glucose, and the need for oxaloacetate in the citric acid cycle makes the oxidation of fats in the human body absolutely dependent on the concurrent metabolism of carbohydrates.

5. Glycolysis

A major pathway for breakdown of carbohydrates is the "Embden–Meyerhof–Parnas" or **glycolysis** pathway (left side of Fig. 7-1). The strategy employed by most cells in the catabolism of 6-carbon sugars is to convert them to glucose 6-phosphate and (in several steps) to cleave that substance to two equivalent molecules of the triose phosphate **glyceraldehyde 3-phosphate.** The latter is converted via 3-phosphoglycerate to pyruvate. One oxidation step is involved, providing additional NADH to the electron transport chain. Pyruvate is then oxidized further to acetyl units of acetyl-CoA, which may be oxidized completely in

Fig. 7-1 A preview of some metabolic pathways. Several major pathways of catabolism are indicated by heavy lines. Biosynthetic routes are shown with dashed lines. A few of the points of synthesis and utilization of ATP are indicated, as are a few oxidation–reduction reactions that produce or utilize the reduced hydrogen carriers NADH, NADPH, and FADH$_2$.

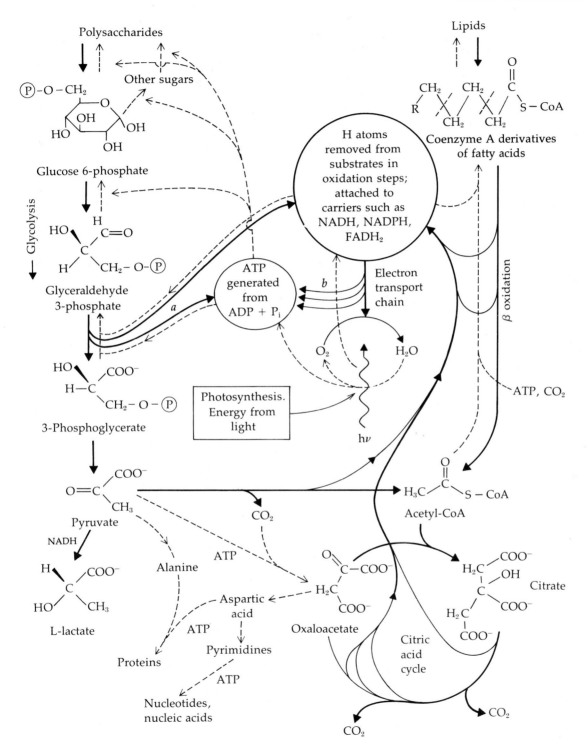

the citric acid cycle. We see that the citric acid cycle and associated reactions such as the interconversion of pyruvate and oxaloacetate, and of pyruvate and acetyl-CoA, lie at the very center of both fat and carbohydrate metabolism. They are also closely related to metabolism of amino acids and many other compounds. For this reason, it will be a good idea now to become thoroughly familiar with the names and structures of the substances shown in Fig. 7-1.

6. Fermentation

While the conversion of pyruvate to acetyl-CoA and the oxidation of the latter accomplishes the complete combustion of glucose to carbon dioxide and water, another variation of the glycolysis pathway permits fermentation of sugars in the absence of oxygen. For example, lactic acid bacteria can reduce pyruvate to **lactate** with NADH (lower left side of Fig. 7-1). Note that this reduction exactly balances the oxidation step preceding it, i.e. the oxidation of glyceraldehyde 3-phosphate to 3-phosphoglycerate. With a balanced sequence of an oxidation reaction, followed by a reduction reaction, the conversion of glucose to lactate, a **fermentation reaction,** can occur in the absence of oxygen and *without* passage of electrons through the electron transport chain.

Organisms cannot live without energy so the most essential part of a fermentation sequence is the reaction by which ATP is generated. In the case of the lactic acid fermentation and most others, this reaction is the oxidation of glyceraldehyde 3-phosphate to 3-phosphoglycerate. The oxidation of an aldehyde to a carboxylic acid is a strongly exergonic process, and the synthesis of ATP is coupled to this reaction. Since each glucose molecule yields two triose phosphates, two molecules of ATP are generated by the fermentation of one glucose molecule. This is enough for a bacterium to live on if it ferments enough sugar. Anaerobic conversion of glucose to lactate is only one example of a large number of different fermentation processes which are discussed in Chapter 9.

7. Biosynthesis

At the same time that cells tear down foodstuffs to obtain energy, they are continuously creating new materials. The dashed lines in Fig. 7-1 indicate pathways by which such biosynthesis takes place. Starting again at the right side, we see that fatty acid synthesis proceeds from acetyl-CoA and reverses fatty acid breakdown. However, both carbon dioxide and ATP, a source of energy, are needed for the synthesis. Furthermore, while oxidation of fatty acids requires NAD^+ as one of the oxidants, and generates NADH, the biosynthetic process often requires a related hydrogen carrier NADPH. The pattern seen in biosynthesis of fatty acids is typical. Synthetic reactions resemble the catabolic sequences in reverse, but distinct differences are evident. These can usually be related to the requirement for energy and to control mechanisms.

8. Photosynthesis

A principal means of formation of glucose in nature is through photosynthesis in green plants (Fig. 7-1, center). Light energy is captured by chlorophyll and is used to transfer electrons from chlorophyll to other electron carriers. For example, $NADP^+$ is reduced to NADPH and the latter is used to reduce carbon dioxide to sugar phosphates in a complex series of reactions known as the **reductive pentose phosphate pathway.** ATP is also required for photosynthetic reduction of CO_2 and is generated by allowing some of the electrons to flow back through an electron transport chain in the membranes of the chloroplasts. The generation of ATP in this **photosynthetic phosphorylation** doubtless occurs in a manner analogous to that in the electron transport chain of mitochondria. The electrons removed from chlorophyll in the light-requiring reaction are replaced (in green plants) by the cleavage of water with release of oxygen, O_2, and generation of hydrogen ions (H^+) plus electrons.

A small number of other biosynthetic pathways

are indicated in Fig. 7-1. For example, pyruvate is readily converted to the amino acid **alanine** and oxaloacetate to **aspartic acid;** the latter may in turn be converted to pyrimidines. Other amino acids, purines, and additional compounds needed for construction of cells are formed in pathways, most of which branch from some compound shown in Fig. 7-1 or from a point on one of the pathways shown in the figure. In virtually every instance biosynthesis is dependent upon a supply of energy furnished by the cleavage of ATP. In many cases it also requires one of the hydrogen carriers in a reduced form.

While Fig. 7-1 outlines in briefest form a minute fraction of the metabolic pathways known, the ones shown are of central importance and will become our starting points in exploring metabolism.

B. CLASSIFYING ENZYMATIC REACTIONS

The majority of enzymes appear to contain, in their active centers, only the side chains of amino acids. Most of the reactions catalyzed can be classified into three types (Table 7-1): (1) **displacement** or **substitution** reactions in which one base or nucleophile replaces another, (2) **addition** reactions in which a reagent adds to a double bond, and (3) **elimination** reactions by which groups are removed from a substrate to create double bonds. Note that the latter are simply the reverse of addition reactions.

Two other groups of reactions depend upon formation of transient **enolate anions** or related intermediates. These include (4) **isomerases** and (5) a large and important group of reactions that **form or cleave carbon-carbon bonds**. Finally, there is a group (6) of **isomerization** and **rearrangement** reactions that do not appear to fit any of the foregoing categories.

Biochemical reactions of the first type (displacement) include all of the hydrolytic reactions by which biopolymers are broken down to monomers as well as most of the reactions by which the monomers are linked together to form

polymers. Addition reactions are used to introduce oxygen, nitrogen, and sulfur atoms into biochemical compounds and elimination reactions often provide the driving force for biosynthetic sequences. Complex enzymatic processes are often combinations of several steps involving displacement, addition, or elimination. The reactions involving formation or cleavage of C—C bonds are essential to biosynthesis and degradation of the various carbon skeletons found in biomolecules, while the isomerization reactions provide connecting links between the other kinds of reactions in the establishment of metabolic pathways.

The remainder of this chapter contains a discussion of the reaction types and of experimental investigations of specific enzymes. While it is hoped that the reader will find all of this material of interest it is not necessary to read it in sequence. In fact, after studying Table 7-1, the reader may move directly to Chapter 8 for a brief look at coenzyme functions and then go on to the chapters dealing with metabolism. In those chapters, cross-references back to Chapters 7 and 8 point out discussions of individual enzymes.

Nevertheless, the novice will probably want to read at least the sections on the following topics immediately: chymotrypsin (Section D,1), carboxypeptidase (Section D,4), ribonuclease (Section E,2), multiple displacement reactions (Section F), imines (Section H,3), addition to C=C bonds (Section H,5), and beta cleavage and condensation (Section J).

C. NUCLEOPHILIC DISPLACEMENT REACTIONS (REACTION TYPE 1)

Nucleophilic displacement is a simple substitution process in which a **nucleophilic reagent** (base) approaches an electron-deficient carbon or phosphorus atom **(electrophilic center)** and forms a bond to it, at the same time displacing some other atom, usually O, N, or S. The displaced atom leaves with its bonding electron pair and with

TABLE 7-1
Types of Biochemical Reaction

Reaction	Examples
1. Nucleophilic Displacement	

A. $B:^- + \overset{\mid}{\underset{\mid}{C}}-Y + H^+ \longrightarrow B-\overset{\mid}{\underset{\mid}{C}} + YH$ Transmethylases, glycosidases

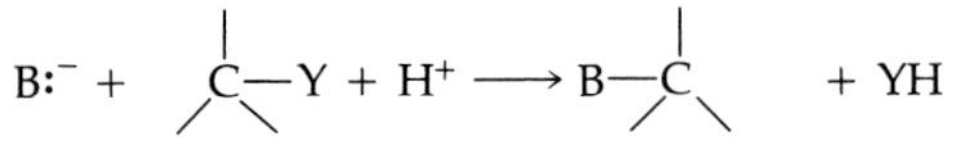

B. $B:^- + \overset{O}{\overset{\parallel}{\underset{\mid}{C}}}-Y + H^+ \longrightarrow B-\overset{O}{\overset{\parallel}{\underset{\mid}{C}}} + YH$ Esterases, peptidases

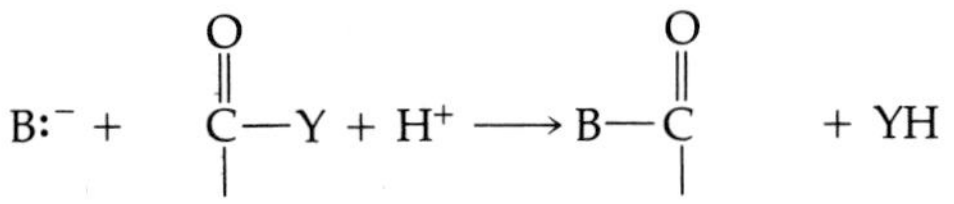

C. $B:^- + \overset{O}{\overset{\parallel}{\underset{O^-}{\underset{O}{P}}}}-Y + H^+ \longrightarrow B-\overset{O}{\overset{\parallel}{\underset{O^-}{\underset{O}{P}}}} + YH$ Phosphatases, phosphokinases, phosphomutases

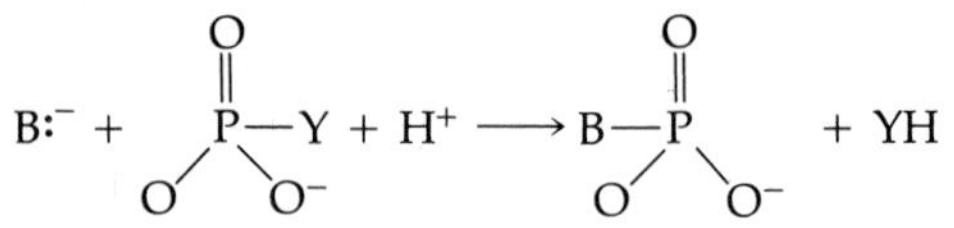

D. $B:^- + \overset{O}{\overset{\parallel}{\underset{O^-}{\overset{}{S}}}}-Y + H^+ \longrightarrow B-\overset{O}{\overset{\parallel}{\underset{O}{\underset{\parallel}{S}}}}-O^- + YH$ Sulfatases, sulfotransferases

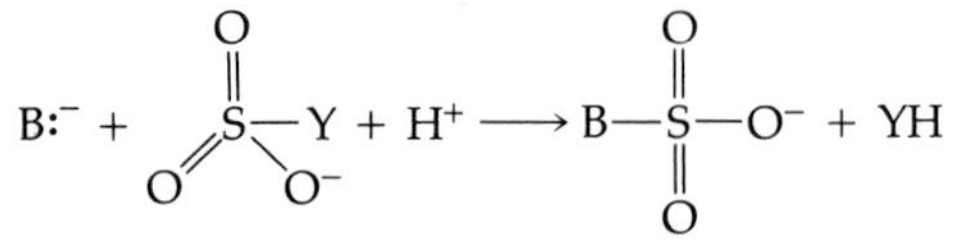

2. Addition

A. To polarized double bonds such as $\overset{\diagdown}{\underset{\diagup}{C}}{=}O$

    and $\overset{\diagdown}{\underset{\diagup}{C}}{=}N{-}$

This reaction most often occurs as a step in an enzymatic process formation of hemiacetals, hemiketals, hemimercaptals, carbinolamines

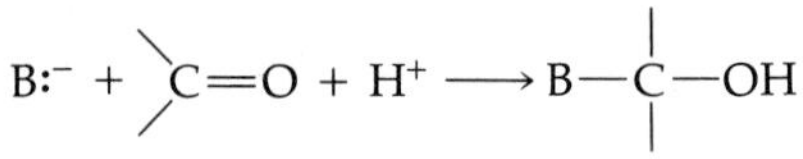

    $B:^- + \overset{\diagdown}{\underset{\diagup}{C}}{=}O + H^+ \longrightarrow B-\overset{\mid}{\underset{\mid}{C}}-OH$ Mutarotase

B. To a double bond conjugated with $\overset{\diagdown}{\underset{\diagup}{C}}{=}O$ or

    $\overset{\diagdown}{\underset{\diagup}{C}}{=}N{-}$

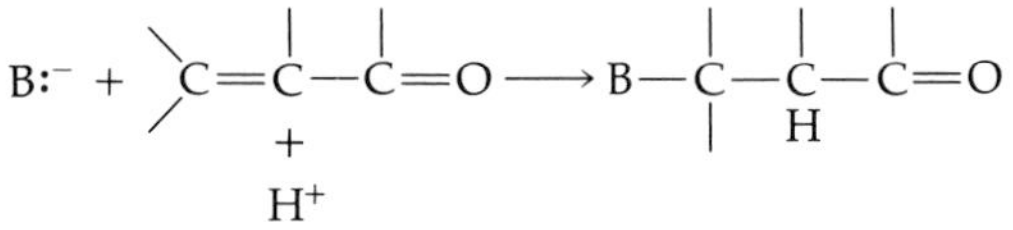

    $B:^- + \overset{\diagdown}{\underset{\diagup}{C}}{=}\overset{\mid}{\underset{\mid}{C}}-\overset{\mid}{C}{=}O \longrightarrow B-\overset{\mid}{\underset{\mid}{C}}-\overset{\mid}{\underset{H}{C}}-\overset{\mid}{C}{=}O$ Acyl-CoA hydratases, fumarase, aspartase
    $ +$
    $ H^+$

C. To isolated double bonds

      Oleate hydratases

3. Elimination

  A, B. Precisely the opposite of *addition*

Eliminations that are the reverse of type 2A are frequent steps in more complex enzyme mechanisms

  C. Decarboxylative elimination

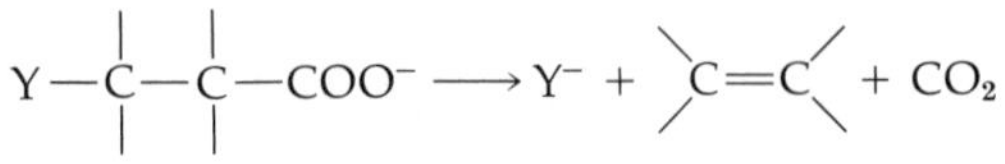

    $Y-\overset{\mid}{\underset{\mid}{C}}-\overset{\mid}{\underset{\mid}{C}}-COO^- \longrightarrow Y^- + \overset{\diagdown}{\underset{\diagup}{C}}{=}\overset{\diagup}{\underset{\diagdown}{C}} + CO_2$

(continued)

TABLE 7-1 (*Continued*)

Reaction	Examples

4. Formation of Enolate Anions and Eneamines

A. $$-\overset{O}{\underset{}{\overset{\|}{C}}}-\overset{}{\underset{H}{\overset{|}{C}}}- \longrightarrow -\overset{O}{\underset{}{\overset{\|}{C}}}-\overset{\ominus}{\underset{}{\overset{|}{C}}}- \longleftrightarrow -\overset{O^-}{\underset{}{\overset{|}{C}}}=\overset{}{\underset{}{\overset{|}{C}}}-$$

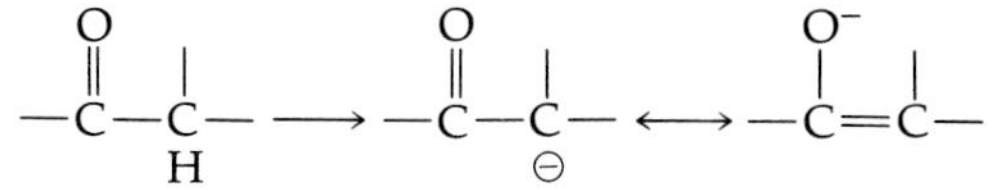

B. $$-\overset{H^+N\diagup}{\underset{H}{\overset{|}{C}}}-\overset{}{\underset{}{\overset{}{C}}}- \longrightarrow -\overset{H^+N\diagup}{\underset{}{\overset{\|}{C}}}-\overset{\ominus}{\underset{}{\overset{|}{C}}}- \longleftrightarrow -\overset{HN\diagup}{\underset{}{\overset{|}{C}}}=\overset{}{\underset{}{\overset{|}{C}}}-$$

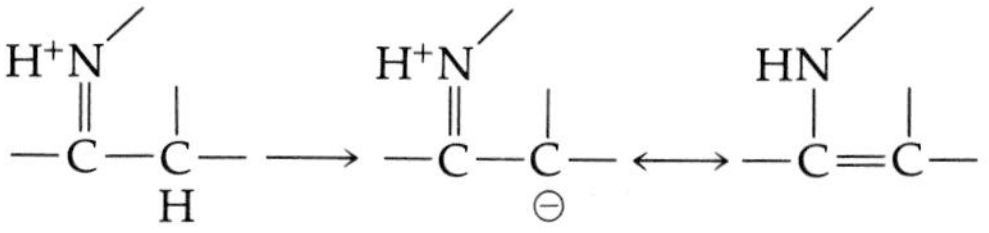

Isomerization reactions

C. $$-\overset{O}{\underset{}{\overset{\|}{C}}}-\overset{OH}{\underset{H}{\overset{|}{C}}}- \longrightarrow -\overset{HO}{\underset{}{\overset{|}{C}}}=\overset{OH}{\underset{enediol}{\overset{|}{C}}}- \longrightarrow -\overset{OH}{\underset{H}{\overset{|}{C}}}-\overset{O}{\underset{}{\overset{\|}{C}}}-$$ Sugar isomerases

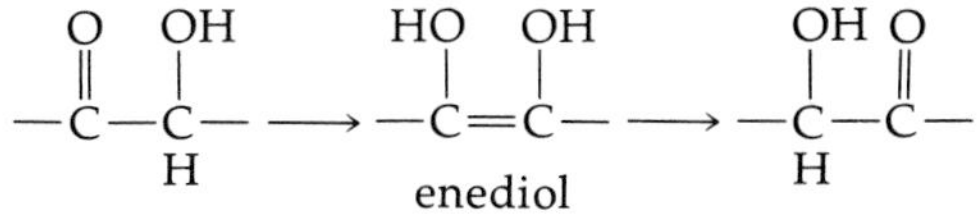

D. $$-\overset{O}{\underset{}{\overset{\|}{C}}}-\overset{H}{\underset{}{\overset{|}{C}}}-\overset{}{\underset{}{\overset{|}{C}}}=\overset{}{\underset{}{\overset{|}{C}}}-$$ Ketosteroid isomerases

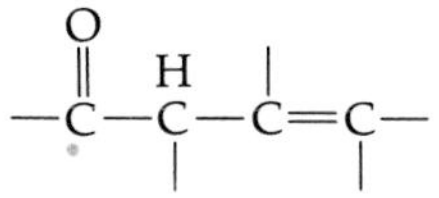

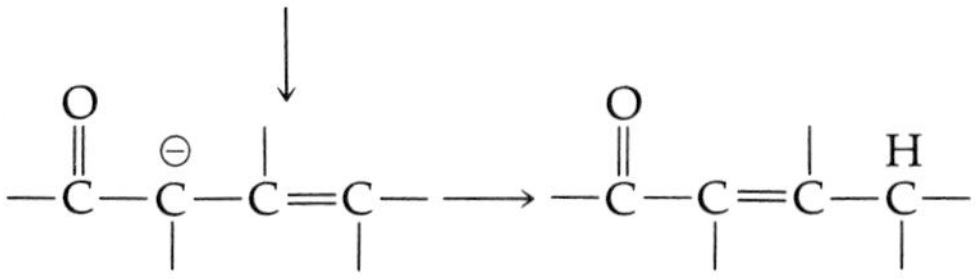

$$\downarrow$$

$$-\overset{O}{\underset{}{\overset{\|}{C}}}-\overset{\ominus}{\underset{}{\overset{|}{C}}}-\overset{}{\underset{}{\overset{|}{C}}}=\overset{}{\underset{}{\overset{|}{C}}}- \longrightarrow -\overset{O}{\underset{}{\overset{\|}{C}}}-\overset{}{\underset{}{\overset{|}{C}}}=\overset{}{\underset{}{\overset{|}{C}}}-\overset{H}{\underset{}{\overset{|}{C}}}-$$ *cis,trans*-Aconitate isomerases

5. Enolate Anions as Nucleophiles: Formation of Carbon-Carbon Bonds (β Condensation)

A. Displacement on a carbonyl group

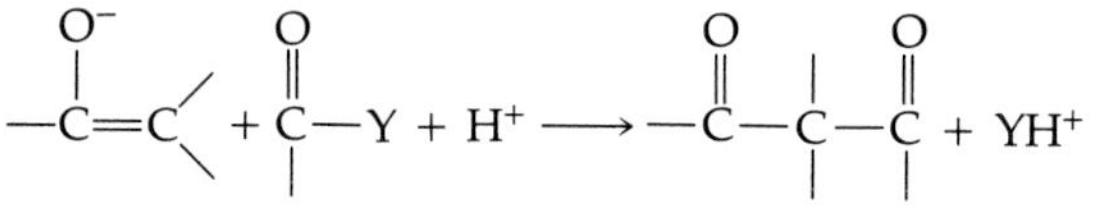

$$-\overset{O^-}{\underset{}{\overset{|}{C}}}=\overset{}{\underset{}{\overset{\diagup}{C}}}\diagdown + \overset{O}{\underset{}{\overset{\|}{C}}}-Y + H^+ \longrightarrow -\overset{O}{\underset{}{\overset{\|}{C}}}-\overset{}{\underset{}{\overset{|}{C}}}-\overset{O}{\underset{}{\overset{\|}{C}}} + YH^+$$ β-Ketothiolases (Y = —S—CoA)

B. Addition to a carbonyl group: aldol condensations

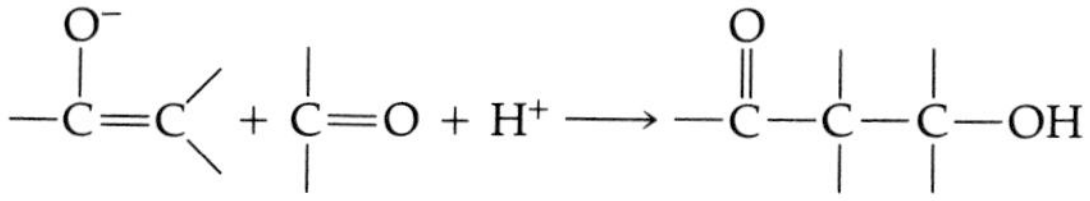

$$-\overset{O^-}{\underset{}{\overset{|}{C}}}=\overset{}{\underset{}{\overset{\diagup}{C}}}\diagdown + \overset{}{\underset{}{\overset{|}{C}}}=O + H^+ \longrightarrow -\overset{O}{\underset{}{\overset{\|}{C}}}-\overset{}{\underset{}{\overset{|}{C}}}-\overset{}{\underset{}{\overset{|}{C}}}-OH$$ Aldolases, citrate synthetases

C. Addition to carbon dioxide (β carboxylation)

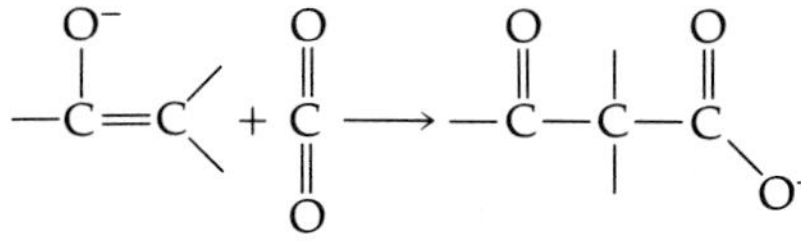

$$-\overset{O^-}{\underset{}{\overset{|}{C}}}=\overset{}{\underset{}{\overset{\diagup}{C}}}\diagdown + \overset{O}{\underset{O}{\overset{\|}{C}}} \longrightarrow -\overset{O}{\underset{}{\overset{\|}{C}}}-\overset{}{\underset{}{\overset{|}{C}}}-\overset{O}{\underset{O^-}{\overset{\|}{C}}}$$ Oxaloacetate decarboxylases, phosphoenolpyruvate carboxylase

(*continued*)

TABLE 7-1 (*Continued*)

Reaction	Examples

6. Some Isomerization and Rearrangement Reactions

  A. Allylic rearrangement (1,3-proton shift)

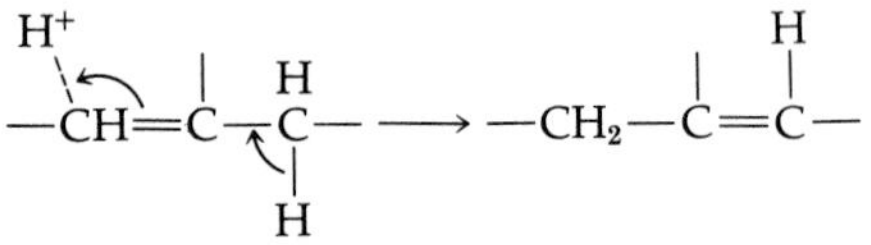

  B. Allylic rearrangement with condensation

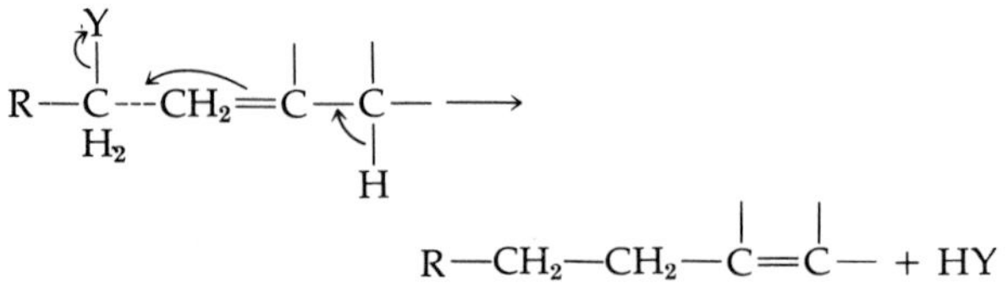

Condensation of dimethylallyl pyrophosphate with isopentenyl pyrophosphate

  C. Rearrangements with alkyl or hydride ion shift

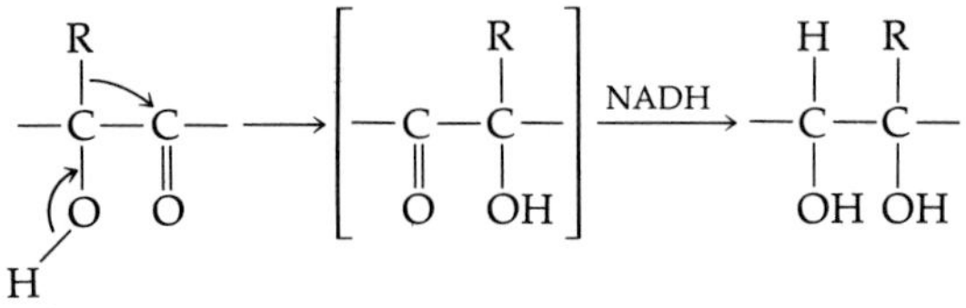

Biosynthesis of leucine and valine

Glyoxalase

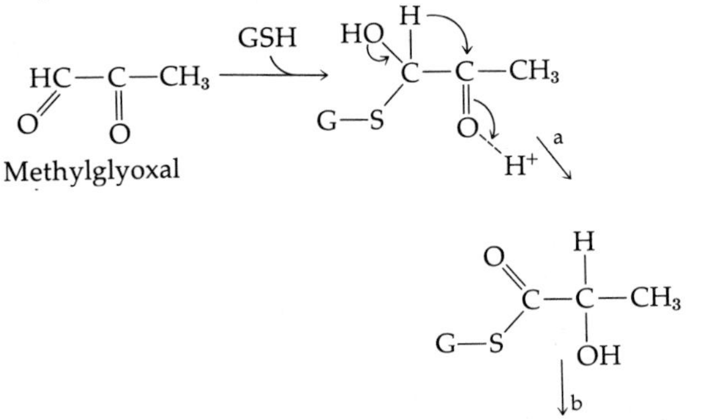

Methylglyoxal

D-Lactate + GSH

(continued)

whatever other chemical group is attached, the whole thing being called the **leaving group.** Simultaneous or subsequent donation of a proton from an acidic group of the enzyme, or from water, to the O, N, or S atom of the leaving group is usually required to complete the reaction. Note that the base B (which may or may not carry a negative charge) must often be generated by enzyme-catalyzed removal of a proton from the conjugate acid BH.

For purposes of classifying the reactions of metabolism, the nucleophilic displacements may be divided into four subtypes (Table 7-1). These are displacements on (A) a saturated carbon atom

TABLE 7-1 (*Continued*)

Reaction	Examples

D. A complex rearrangement in the synthesis of lanosterol

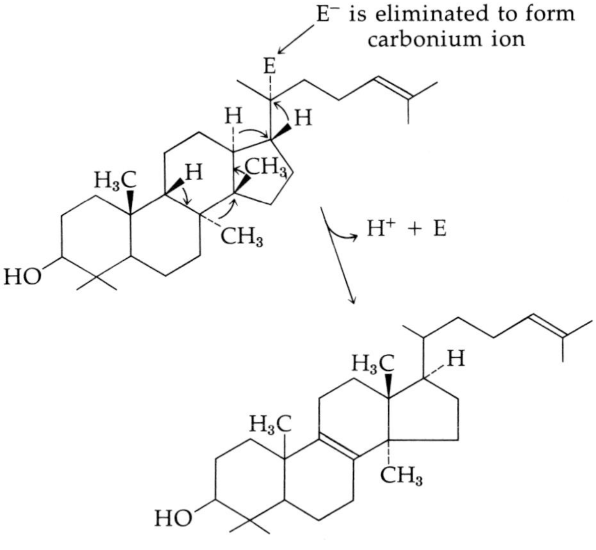

Lanosterol

(very often a methyl group), (B) a carbonyl group of an ester or amide, (C) a phosphate group, or (D) a sulfate group. In addition, many enzymes employ in sequence a displacement on a carbon atom followed by a second displacement on a phosphorus atom (or vice versa).

1. Factors Affecting Rates of Displacement Reactions[1-3]

In Chapter 6, Section E, the displacement of an iodide ion from methyl iodide by a hydroxide ion was considered. Can we similarly displace a methyl group from ethane, $CH_3—CH_3$, thereby breaking the C—C bond and forming CH_3OH? The answer is no. Ethane is perfectly stable in sodium hydroxide, nor is it cleaved by a simple displacement process within our bodies. Likewise, long hydrocarbon chains such as those in the fatty acids cannot be cut by a corresponding process during metabolism of fatty acids. Thus, nucleophilic displacement reactions take place when displacement is easy, but when it is very difficult they do not occur. Likewise, not every anion or base B can act to displace another group.

At least four factors affect the likelihood that a displacement reaction will take place: (1) the position of the equilibrium in the overall reaction, (2) the reactivity (**nucleophilicity**) of the entering nucleophile, (3) the chemical nature of the leaving group that is displaced, and (4) special structural features present in the substrate.

The first factor is a thermodynamic one. An example is provided by **hydrolases** which catalyze cleavage of amide, ester, and phosphodiester linkages using water as the entering nucleophile. Because enzymes usually act in an environment of high water content, the equilibrium virtually always favors hydrolysis rather than the reverse reactions of synthesis.

Nucleophilic reactivity (nucleophilicity) is partly determined by basicity, i.e., compounds that are strong bases tend to react more rapidly in nonenzymatic reactions than do weaker bases. The hydroxyl ion (OH^-) is a better nucleophile than $-COO^-$. However, enzymes usually act optimally near a neutral pH. The $-COO^-$ group may be more reactive at pH 7 than strongly basic groups like $-NH_2$ or OH^- because those groups will be almost completely protonated and the active nucleophile will be present in low concentrations.

A second factor affecting nucleophilic reactivity is **polarizability.** By this we mean the ease with which the electron distribution around an atom or within a chemical group can be distorted. A large atomic radius and the presence of double bonds in a group both tend to increase polarizability. In general, the more highly polarizable a group, the more rapidly it will react in a nucleophilic displacement (apparently because a polarizable group is able to form a partial bond at a greater distance than can a nonpolarizable group). Thus, I^- is more reactive than Br^-, which is more reactive than Cl^-. Polarizable bases such as imidazole are often much more reactive than nonpolarizable ones like $-NH_2$. Sulfur compounds tend to have a high nucleophilic reactivity, a fact that may contribute to the importance of $-SH$ groups in biochemistry. Interestingly, in displacement reactions on carbonyl groups (reaction type 1B), the less polarizable (hard) nucleophiles are more reactive than polarizable ones such as I^-.

Attempts have often been made to relate nucleophilic reactivity to basicity plus some other single quantity such as polarizability.[1]

Another effect, which may not be very important in enzymatic catalysis but which helps explain the action of certain poisons, is called the **α effect.** This refers to the observed fact that certain combinations of atoms (in which an atom with unpaired electrons is directly bonded to the nucleophilic center undergoing reaction) are more nucleophilic than other compounds of similar basicity. The high reactivity of the poisons **hydroxyl-amine** (NH_2OH) and **cyanide ion** can be explained in this way.[4]

The chemistry of the leaving group greatly affects both rate and equilibrium position in nucleophilic displacements. The leaving group must accommodate a pair of electrons and often must bear a negative charge. A methyl group pushed off from ethane or methane as CH_3^- would be an extremely poor leaving group (the pK_a of methane as an acid[5] has been estimated as 47). Iodide ion is a good leaving group, but F^- is over 10^4 times poorer.[5] In an aqueous medium, phosphate is a much better leaving group than OH^- and pyrophosphate and tripolyphosphate are even better.

2. Nucleophilic Displacements on Singly Bonded Carbon Atoms

Nucleophilic attack on a methyl group or other alkyl group occurs readily only if the carbon is attached to some atom bearing a positive charge.

$$\text{(7-1)}$$

e.g., $H_3C—\overset{\displaystyle |}{\underset{\displaystyle |}{S}}{}^{+}$

The displacement reaction on such a sulfonium ion results in the transfer of the methyl group from the sulfur to the attacking nucleophile and is referred to as **transmethylation.** Transmethylation is an important metabolic process in which the methyl group donor is usually **S-adenosylmethionine.** A typical example is the methylation of an amine such as nicotinamide (Eq. 7-1). Related reactions include the nonenzymatic cleavage of thiamine by bisulfite (Box 8-D), transmethylation from folic acid derivatives (Eq. 8-85), and the reaction by which S-adenosylmethionine is synthesized (Eq. 11-3).

Another simple displacement on a saturated carbon is catalyzed by **haloacetate halidohydrolases** of soil pseudomonads (Eq. 7-2).[6] Even the poor leaving group F^- can be displaced by OH^-.

$$HO^- + F—CH_2—COO^- \longrightarrow$$
$$HO—CH_2—COO^- + F^-   (7\text{-}2)$$

a. Inversion as a Criterion of Mechanism

An interesting result was obtained when one of the halidohydrolases was tested with a substrate containing a chiral center at the carbon on which displacement occurs.[6] Reaction of L-2-chloropropionate with hydroxyl ion gave only D-lactate, a compound of configuration opposite to that of the reactant. A plausible explanation (Eq. 7-3) is that the hydroxyl ion attacks the central

carbon from behind the chlorine atom. The resulting five-bonded transition state (center) loses chloride ion to form the product D-lactate in which **inversion of configuration** has occurred. Inversion always accompanies single displacement reactions in which bond breaking and bond formation occur synchronously, as in Eq. 7-3. However, the occurrence of inversion does not rule out more complex mechanisms.

b. Displacement Reactions of Glycosides

The polarization of a single C—O bond is quantitatively much less than that of the $C—S^+$ bond of S-adenosylmethionine, and simple ethers are not readily cleaved by displacement reactions. However, glycosides which contain a carbon atom attached to *two* oxygen atoms do undergo displacement reactions leading to **phosphorolysis, hydrolysis,** or **glycosyl exchange.** Equation 7-4

$$+ —B: + HOR   (7\text{-}4)$$

pictures the reaction of a glycoside (such as a glucose unit at the end of a starch chain) with phosphate as the displacing agent. An enzyme-bound acidic group $—BH^+$ is shown assisting the process. Again, an inversion of configuration with formation of a product of the β configuration at the anomeric carbon atom would be predicted. In fact, phosphorylases usually do *not* cause inversion. Thus, **sucrose phosphorylase** from *Pseudomonas saccharophilia* catalyzes the reaction shown in Eq. 7-5. Two possible explanations for these reactions are that there might either be a **double displacement** reaction or a stabilized **carbonium ion** intermediate.

Sucrose (an α-glucoside of fructose) + phosphate
$$\longrightarrow \alpha\text{-}D\text{-glucose 1-phosphate} + \text{fructose} \quad (7\text{-}5)$$

3. Double Displacement Mechanisms

Sucrose phosphorylase could catalyze two consecutive single displacements, each with inversion. The first would require displacement by a nucleophilic group of the enzyme (Eq. 7-6, step a)

$(7\text{-}6)$

to form a **glycosyl enzyme.** A second displacement (Eq. 7-6, step b), in which phosphate attacks, would regenerate the enzyme with its free nucleophilic group B^-.

What predictions can be made and tested exper-

imentally to prove a double displacement mechanism? The four types of experiments discussed in Section C,3,a–c have been performed with sucrose phosphorylase; they are also appropriate for many other enzymes.

a. Exchange Reactions

A double displacement mechanism predicts that an enzyme will catalyze partial exchange reactions between one of the two substrates and a labeled product. For example, sucrose containing ^{14}C in the fructose portion of the molecule should react with sucrose phosphorylase to form glucosyl enzyme and free radioactive fructose as the displaced product (Eq. 7-7). Very low molar concentrations of enzymes are usually used in experiments, and the easiest way to detect reaction (7-7) is to add a large excess of nonradioactive fructose.

$$\begin{array}{l} D\text{-Glucose 1-fructoside*} + E \rightleftharpoons \text{glucosyl—E} \\ \quad \text{(sucrose)} \qquad\qquad\qquad\quad + \text{fructose*} \quad (7\text{-}7) \end{array}$$

Here the asterisks designate the ^{14}C-containing group. Under these conditions, the enzyme catalyzes no overall chemical reaction but changes back and forth repeatedly between the free enzyme and glucosyl enzyme. Each time, in the reverse reaction, it makes use primarily of unlabeled fructose. The net effect is catalysis by the enzyme of the *exchange reaction* of fructose with sucrose (Eq. 7-8).

$$\text{Sucrose*} + \text{fructose} \rightleftharpoons \text{sucrose} + \text{fructose*} \quad (7\text{-}8)$$

This fructose–sucrose exchange has been observed experimentally[7] as has a second predicted exchange: that of glucose 1-phosphate with radioactive phosphate (Eq. 7-9).

$$\begin{array}{l} \text{Glucose 1—PO}_4^{2-} + H^{32}PO_4^{2-} \longrightarrow \\ \qquad\qquad \text{glucose 1—}^{32}PO_4^{2-} + HPO_4^{2-} \quad (7\text{-}9) \end{array}$$

Related to these exchange reactions are transfer reactions in which a ketose sugar other than fructose, e.g., D-ketoxylose, reacts with the glucosyl enzyme to form a new disaccharide containing glucose and the added ketose.[7]

The exchange criterion of mechanism is often

applied to complex enzymatic processes in all areas of metabolism and it is important to understand it. It is equally as important to be aware of the limitations. A double displacement mechanism predicts certain exchange reactions, but observation of these exchanges does not *prove* the existence of a covalently enzyme-bound intermediate. Furthermore, enzymes using double displacement mechanisms may not always catalyze the expected exchange reactions (see Section C,5).

b. Arsenolysis

Sucrose phosphorylase also catalyzes the cleavage of sucrose by arsenate, and in addition promotes a rapid cleavage (**arsenolysis**) of glucose 1-phosphate to free glucose. This reaction is most readily understandable as a displacement by arsenate on a glucosyl enzyme intermediate. The product is the unstable glucose 1-arsenate (see Box 7-A). Arsenolysis is a general way of trapping reactive enzyme-bound intermediates that normally react with phosphate groups. Arsenate is one of many substrate analogues that can also be used to siphon off reactive enzyme-bound intermediates into nonproductive side paths.

c. Kinetics

A double displacement mechanism requires that the enzyme shuttle back and forth between free enzyme and the intermediate carrying the substrate fragment (e.g., a glycosyl enzyme). Kinetic measurements with sucrose phosphorylase show that the rate varies with the concentrations of sucrose and HPO_4^{2-} in the characteristic fashion expected for a ping-pong mechanism (Chapter 6, Section A,10,a).[8]

d. Direct Isolation
of Intermediates

Techniques for isolation of pure enzymes and for working with very small quantities of material have now developed to the point where it is often possible to confirm directly the existence of postulated enzyme-bound intermediates. Isolation of a glucosyl enzyme in denatured form after reaction

of highly purified sucrose phosphorylase with radioactive sucrose has been accomplished.[8] The glucosyl enzyme is too labile to easily permit degradation of the peptide chain and isolation of a glucosyl-labeled fragment; however, indirect evidence[9] strongly suggests that —B: is a carboxylate group —COO⁻.

4. Carbonium Ions

In a second well-known mechanism of nucleophilic displacement the leaving group departs first (often in a protonated form) leaving a carbonium ion (Eq. 7-10). In the common terminology of

$$(7\text{-}10)$$

Nearby charged group attached to enzyme

Stabilized carbonium ion

physical organic chemistry this is an S_N1 reaction* rather than an S_N2 reaction of the kind shown in Eqs. 7-3 and 7-6. Note that the carbonium ion is depicted as a resonance hybrid of two forms, one of which (oxonium ion) contains a double bond between carbon and oxygen. This double-bonded structure can be visualized as arising from the original structure by a kind of internal displacement by the unshared electrons on oxygen, as indicated by the small arrows. Such an internal displacement can also be described as an **elimination.** Thus, as is often the case for enzymatic reactions,

---

* However, the terminology is inappropriate for enzymes because although the reactions are nucleophilic substitutions (S_N), the breakdown of enzyme substrate complexes to product is usually a zero order process and the numbers 1 and 2 in the symbols S_N1 and S_N2 refer to the order or molecularity of the reaction.

BOX 7-A

ARSENIC

Arsenate, AsO_4^{3-}, is chemically similar to phosphate, in size, geometry, and in its ability to enter into biochemical reactions. However, arsenate esters are far less stable than phosphate esters. If formed on an enzyme surface, they are immediately hydrolyzed upon dissociation from the enzyme. This fact accounts for much of the toxic nature of arsenic compounds.

Arsenate will replace phosphate in all phosphorolytic reactions, e.g., in the cleavage of glycogen by glycogen phosphorylase and sucrose by sucrose phosphorylase (Section 3,b). In both cases glucose 1-arsenate is presumably a transient intermediate but is hydrolyzed quickly to glucose. The overall process is called **arsenolysis**. Another reaction in which arsenate can replace phosphate is the oxidation of glyceraldehyde 3-phosphate in the presence of P_i to form 1,3-diphosphoglycerate:

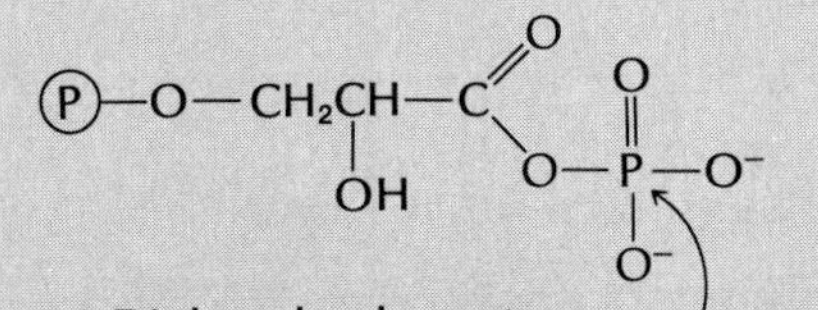

1,3-Diphosphoglycerate

This phosphoryl group is normally transferred to ADP to form ATP

The subsequent transfer of the 1-phosphoryl group to ADP is an important energy-yielding step in metabolism (Chapter 8, Section H,5). When arsenate substitutes for phosphate the acyl arsenate (1-arseno-3-phosphoglycerate) is hydrolyzed to 3-phosphoglycerate. Thus, in the presence of arsenate oxidation of 3-phosphoglyceraldehyde continues but ATP synthesis ceases. Arsenate is said to **uncouple** phosphorylation from oxidation. Arsenate can partially replace phosphate in stimulating the respiration of mitochondria and is an uncoupler of oxidative phosphorylation (Chapter 10, Section E,5).

Enzymes that normally act on a phosphorylated substrate will usually catalyze a slow reaction of the corresponding unphosphorylated substrate if arsenate is present. Apparently the arsenate ester of the substrate forms transiently on the enzyme surface, permitting the reaction to occur.

Arsenite is noted for its tendency to react rapidly with thiol groups, especially with dithiols such as lipoic acid.

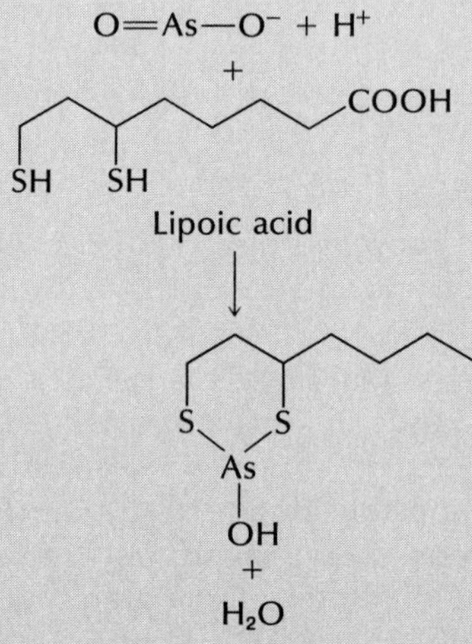

Lipoic acid

By blocking oxidative enzymes requiring lipoic acid (Chapter 8, Section J), arsenite causes the accumulation of pyruvate and of other α-keto acids.

Compounds of arsenic have been used in medicine for over 2000 years, but only in this century have specific arsenicals been created as drugs. In 1905 it was discovered that sodium arsanilate is toxic to trypanosomes. The development by P. Ehrlich of **arsenicals** for the treatment of syphilis (in 1909) first focused attention on the possibility of effective chemotherapy against bacteria.

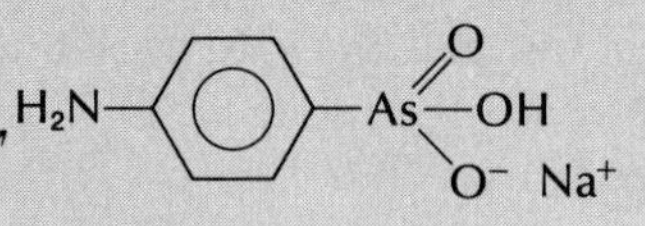

Substitution by —CH_2CONH_2 yields "tryparsamide," a more effective anti-trypanosome drug

Sodium arsanilate

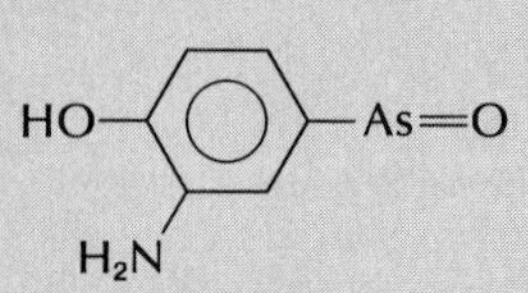

"Oxophenarsine," one of Ehrlich's anti-syphilis drugs

we have a semantic problem. Is this an elimination or just half of a displacement reaction? For purposes of classifying metabolic reactions it is simplest to look at the overall process which is displacement.

In the stabilized carbonium ion, the ring atoms C-2, C-1, and C-5 and the oxygen atom are almost coplanar. The ring conformation is known as **half-chair.**

a. Lysozyme

A carbonium ion mechanism has been proposed for lysozyme, an enzyme specifically designed to attack and cleave polysaccharide chains in the peptidoglycan layer of the cell walls of bacteria.[10] Lysozyme occurs as a protective agent in tears and other body secretions and in very large amounts in egg white. Egg white lysozyme was the first enzyme for which a complete three-dimensional structure was determined by X-ray diffraction.[11]

It appears that six N-acetylglucosamine or N-acetylmuramic acid rings of the polysaccharide substrate fit precisely into a groove on the side of the molecule. The bond between the fourth and fifth rings is then cleaved (Fig. 2-9). At the presumed active site, a glutamic acid residue (No. 35) is in just the right position to serve as the proton donor (i.e., BH in Eq. 7-10) while an aspartic acid residue (No. 52) lies on the opposite side of the groove. Both Glu 35 and Asp 52 have abnormally high pK_a values (microscopic pK_a's are ~5.3 and 4.6 respectively,* in the fully protonated active site[12]) as a result of the hydrophobic environment and hydrogen bonding to other groups. Aspartic acid 52 usually dissociates first and the resulting electrostatic interaction tends to keep Glu 35 protonated until ~pH 6. Positively charged basic groups nearby affect the pK_a values; hence, the behavior of the enzyme is sensitive to the ionic strength of the medium.[12] The Asp 52 anion lies close (~0.3 nm)[13] to the center of positive charge expected in the carbonium ion and would be expected to stabilize the carbonium ion (see Eq. 7-10).

* At an ionic strength of ~0.2.

The lysozyme reaction is completed by stereospecific addition of a hydroxyl ion to the carbonium ion. The original β configuration is retained in the product. Such stereospecificity for reactions of enzyme-bound carbonium ions is not surprising for the enzyme probably assists in generation, at the appropriate location, of the attacking hydroxyl ion.*

A study of models shows that for all six sugar rings of the substrate to bind tightly, the ring containing the carbon atom on which the displacement occurs must be distorted from its normal "chair" conformation into the "half-chair" required by the carbonium ion mechanism.[15,16] Thus, by binding the substrate chain at six different sites the enzyme provides leverage to distort a particular ring into a conformation similar to that of the transition state. This may be a generally important aspect of enzymatic catalysis.

b. Kinetic Isotope Effects

A carbon–hydrogen bond breaks more easily than does a carbon-deuterium bond and much more easily than a carbon-tritium bond. Thus, when breaking of a C—H bond is suspected to be part of the rate-determining step, a comparison of the rates for the C—^{1}H and C—^{2}H bond is often made. In the case of lysozyme, there is no carbon-hydrogen bond broken in the slow step, but a **secondary kinetic isotope effect** has been observed.[17] Small differences in vibrational energies of molecules containing ^{1}H and ^{2}H result from the difference in mass of the isotopes. Consequently, a molecule with ^{1}H in the number 1 position can be converted to the corresponding carbonium ion (Eq. 7-11) somewhat more easily than the mole-

$$\tag{7-11}$$

cule with ^{2}H in the same position. For example, in the nonenzymatic acid-catalyzed hydrolysis of a

* A lucid account of lysozyme action is provided by Dickerson and Geis.[14]

phenyl glucoside, a reaction believed to proceed through a carbonium ion intermediate, the ratio k_{1H}/k_{2H} is 1.14, whereas in the base-catalyzed hydrolysis of the same compound (believed to occur by a double displacement reaction involving participation of the neighboring OH group on C-2) the ratio k_{1H}/k_{2H} is 1.03. The corresponding ratio measured for the action of lysozyme is 1.11, much closer to that of the carbonium ion mechanism than to that of the double displacement mechanism.[17]

As attractive as the arguments for the carbonium ion mechanism in lysozyme are, it might still be possible to explain the experimental observations by a double displacement in which Asp 52 serves as the nucleophile in forming an unstable intermediate glycosyl enzyme. It is interesting that both sucrose phosphorylase and lysozyme apparently contain a carboxylate ion at the active site. In the former enzyme the carboxylate ion forms a covalent glucosyl enzyme, whereas in the latter it apparently stabilizes a carbonium ion. Is there really a difference in mechanism? Is it possible that the glucosyl enzyme is formed only upon denaturation of sucrose phosphorylase? Nature hides her secrets well. The difficulty in pinning down the fine mechanistic details of enzymatic action makes it essential to be skeptical, to examine data critically, to dream up all possible alternatives—even when things seem to be proved beautifully.

5. Glycogen Phosphorylase

Many phosphorylases behave in a puzzling way. For example, **glycogen phosphorylase** of muscle, which catalyzes the conversion of glycogen (α-glycosidic linkages) to α-D-glucose 1-phosphate, shows neither partial exchange reactions nor inversion of configuration, as might be expected for a single nucleophilic displacement. Similarly, no arsenolysis occurs when glucose 1-phosphate and arsenate are incubated with the enzyme.[18] A possible explanation is that the enzyme does nothing until both substrates are

bound. This might happen if the active site of the enzyme is operative only when the enzyme has properly folded itself around the two substrates, i.e., the active conformation of the protein is stabilized by the presence of the substrates.

A carbonium ion mechanism for glycogen phosphorylase has been proposed, partly on the basis of a strong inhibition by **5-gluconolactone**,[18,19] a compound having a half-chair conformation.

5-Gluconolactone

Glycogen phosphorylase is an unusual glycolytic enzyme. It contains a bound coenzyme pyridoxal 5'-phosphate (Chapter 8, Section E,3) whose function in this enzyme is unknown.[18,20]

6. Other Enzymes Acting on Glycosides

The widely distributed **α-amylases**[21,22] are *endo*-glycosidases which hydrolyze starch chains by random attack at points far from chain ends to form short polysaccharide chains (**dextrins**) as well as simpler sugars. The reaction proceeds with retention of configuration, the reducing groups created being in the α-anomeric forms. α-Amylases are found in both plants and animals. For example, a powerful enzyme of this class is found in human saliva* and another is formed by the pancreas. All α-amylases appear to have an absolute requirement for **chloride ion** as an activator. However, any significance to the mechanism of the action is not yet clear.

* However, some perfectly normal individuals are completely lacking in this enzyme.

Studies of the **action patterns** of α-amylases have revealed many details of the probable mode of binding as well as possible mechanisms.[22,23] The substrate binding region of the porcine* pancreatic enzyme appears to have five subsites, each capable of holding a single glucose residue. The chain is cleaved between the residues bound at the second and third subsites (numbered from the reducing end of the substrate). Using various O-substituted substrates it was found that the presence of hydroxyethyl groups at many locations does not interfere with enzymatic action. However, substitution in either the 2, 3, or 6 position of the glycosyl unit upon which displacement occurs does block enzymatic attack. The lactone of maltobionic acid (the 4-glucosyl derivative of 5-gluconolactone) is a powerful competitive inhibitor, suggesting that the substrate of this enzyme also assumes a half-chair conformation in or near the transition state. French *et al.*[23] suggested that the enzyme holds the glucosyl unit firmly through hydrogen bonding to the C-3 and C-6 hydroxyl groups and then exerts torsion about the C-2, C-3 bond by literally grasping and moving the C-2 hydroxyl (Eq. 7-12, step *a*). Thus, a conformational motion within the enzyme could be coordinated with conversion of the substrate to the transition state structure. As with lysozyme, protonation of the bridge oxygen (possibly with an imidazole group) is postulated as is stabilization of the carbonium ion by an adjacent carboxylate ion. A further possibility is that the carboxylate ion forms an actual bond to C-1, as in Eq. 7-6, and that this is accomplished through further conformational change in the sugar ring, leading to a transient high-energy boat conformation[22] (Eq. 7-12, step *b*).

The study of amylases has been important to our understanding of enzymatic action in another way. While endolytic enzymes are generally thought to carry out random attack on biopolymers, it has been shown that after the initial catalytic reaction one of the polysaccharide prod-

Glucosyl unit of starch held to amylase by H-bonding

(7-12)

ucts of amylase action often does not leave the enzyme. The polysaccharide simply "slides over" until it fully occupies all of the subsites of the substrate binding site and a second "attack" occurs. Porcine pancreatic α-amylase produces an average of seven catalytic events each time it forms a complex with amylose.[24] An important question is whether or not there is a mechanism by which the enzyme deliberately promotes the "sliding" of the substrate or whether the dissociated product simply diffuses a short distance while enclosed in a solvent cage. The latter explanation is apparently adequate.[24] The concept of **multiple attack** arising from these studies may be very important to our understanding of the biosynthetic mechanisms such as DNA replication and transcription (Chapter 15).

---

* Some words that biochemists use instead of genus names for denoting the source of a material: *porcine*, pig; *bovine*, cow; *ovine*, sheep; and *murine*, mouse or rat.

Beta amylases, characteristic plant enzymes, have an *exo* action, cutting off chain ends two sugar units at a time as maltose. The original α linkage is inverted, the product being β-maltose. Thus, either a single direct displacement by water or a carbonium ion mechanism might be possible.[25]

Many other glycosidases are known. Some are hydrolytic enzymes designed to remove specific sugar rings from substrates. Many biosynthetic **glycosyltransferases** transfer glycosyl units from one molecule to another (Chapter 5, Sections B,1, C,1, and D,1; Chapter 12, Section C,1).

D. DISPLACEMENT REACTIONS ON CARBONYL GROUPS

Reaction type 1B (Table 7-1) consists of nucleophilic displacement of a group Y attached to a carbonyl carbon. If Y is OR, the substrate is an **ester,** if SR a **thioester,** and if NHR an **amide.** When B^- is OH^- (from H_2O), the reaction is hydrolysis; while if B^- is the anion of an alcohol, thiol, or amine, the reaction is **transacylation.** Transacylation processes are extremely important in biosynthesis, but it is the enzymes catalyzing hydrolysis that have been studied most intensively.

The carbonyl group

$$\diagdown C = O$$

is highly polarized, the resonance form

$$\diagdown C^+ - O^-$$

contributing substantially to its structure. An attack by a base will take place readily on the electrophilic carbon atom. While the reactivity of the carbonyl group in esters and amides is greatly decreased because of the resonance stabilization of these groups, the carbon atom still maintains an electrophilic character and may combine with basic groups. Thus, in the base-catalyzed hydrolysis of esters a hydroxyl ion may add to the carbonyl group to form a transient single-bonded "tetrahedral" intermediate (Eq. 7-13). Similar

$$R - \overset{\overset{\textstyle O}{\|}}{\underset{\underset{\textstyle OH^-}{|}}{C}} - OR' \longrightarrow R - \overset{\overset{\textstyle O^-}{|}}{\underset{\underset{\textstyle OH}{|}}{C}} - OR' \xrightarrow{H^+}$$

$$R - \overset{\overset{\textstyle O}{\|}}{\underset{\underset{\textstyle OH}{}}{C}} \quad + \quad HOR' \quad (7\text{-}13)$$

"Tetrahedral" intermediate

intermediates may form during the action of many enzymes catalyzing reactions of type 1B. However, for purposes of classification these reactions can all be regarded as simple displacement reactions on a carbon atom with the understanding that transient single-bonded intermediates may sometimes exist.

Among the hydrolases are acetylcholinesterase of nerve cells (Box 7-B) and a large number of digestive enzymes. Of the latter, the **proteases** and **peptidases** have received the most attention. Pepsin, trypsin, chymotrypsin, and carboxypeptidases are all powerful catalysts for the breakdown of the proteins. They are all secreted as inactive proenzymes (Chapter 6, Section G,2) or **zymogens.**[26] After synthesis on the ribosomes of the endoplasmic reticulum of special secretory cells, the proenzymes are "packaged" as **zymogen granules** which travel to the surface of the cell and are secreted into the surrounding medium. Pepsinogen is a component of the gastric juice, while chymotrypsinogen, trypsinogen, and the other pancreatic proenzymes are discharged via the pancreatic duct into the small intestine. At their site of action, the zymogens are converted to active enzymes by attack of another enzyme molecule which cuts out a piece (sometimes a rather large piece) of the precursor.[25]

1. Chymotrypsin and Trypsin

The most thoroughly studied protease, **chymotrypsin,** exists in several slightly different forms obtained by the breaking of certain peptide linkages in chymotrypsinogen. The latter is a single peptide chain of 245 amino acids; the amino acids in the active enzyme are usually numbered according to their position in the original zymogen. An important clue to the mechanism of action of chymotrypsin came originally from investigation of acetylcholinesterase. This key enzyme of the nervous system was shown to be inactivated irreversibly by a class of powerful phosphorus-containing poisons that had been developed as insecticides and as war gases (nerve gases).

a. Serine at the Active Center

Around 1949, one of the nerve gases, diisopropylphosphofluoridate (DFP), was shown to inactivate chymotrypsin. When radioactive ^{32}P-containing DFP was allowed to react with chymotrypsin, the ^{32}P became firmly attached to the enzyme in covalent linkage. Subsequent experi-

$$CH_3-CH(CH_3)-O-\overset{\overset{\displaystyle O}{\|}}{\underset{\underset{\displaystyle CH_3-CH-CH_3}{|}}{P}}-F$$

Diisopropylphosphofluoridate
[also called diisopropylfluoro-
phosphate (DFP)]

ments showed that when the labeled enzyme was denatured and subjected to acid hydrolysis the phosphorus stuck firmly. The radioactive fragment was identified as O-phosphoserine.

$$^-O-\overset{\overset{\displaystyle O}{\|}}{\underset{\underset{\displaystyle HO}{|}}{^{32}P}}-O-CH_2-\overset{\overset{\displaystyle COO^-}{|}}{\underset{\underset{\displaystyle H}{|}}{C}}-NH_3^+$$

O-Phosphoserine

The chemistry of the reaction of DFP and chymotrypsin is straightforward. The hydroxyl group of the serine side chain attacks the phosphorus displacing the fluoride ion. Note that this is a nucleophilic displacement on phosphorus (reaction type 1C), but in this instance it occurs on an enzyme that normally catalyzes displacements on C=O.

The DFP molecule can be regarded as a **pseudosubstrate** (quasi-substrate) which reacts with the enzyme in a manner analogous to that of a true substrate, but which does not complete the reaction sequence in the normal way.

From study of peptides formed by partial hydrolysis of the ^{32}P-labeled chymotrypsin, the sequence of amino acids surrounding the reactive serine was established and eventually the position of that serine in the sequence of the entire chain was identified as position 195. Preceding it at positions 193 and 194 are a glycine and an aspartic acid, while at position 196 there is another glycine.

$$\overset{\overset{\displaystyle O-\textcircled{P}}{|}}{\underset{\underset{\displaystyle 195}{}}{Gly-Asp-Ser-Gly}}$$

The same sequence was soon discovered around reactive serine residues in **trypsin, thrombin, elastase,** and in the trypsinlike **cocoonase** used by silkmoths to escape from their cocoons.[27] A closely related sequence containing Glu-Ser-Ala was found in acetylcholinesterase. Thus, a family of **serine peptidases** and esterases exists, all of which have some similarities in the sequence of amino acids around the reactive serine and all of which are inhibited by DFP.[28]

b. An Acyl-Enzyme Intermediate

Another pseudosubstrate **p-nitrophenyl acetate** reacts with chymotrypsin at pH 4 (far below the optimum pH for hydrolysis) with rapid release of

$$CH_3-\overset{\overset{\displaystyle O}{\|}}{C}-O-\underset{}{\bigcirc}-NO_2$$

p-Nitrophenyl acetate

p-nitrophenol and formation of an **acetyl derivative** of the enzyme. This acetyl enzyme hydrolyzes very slowly at pH 4 but rapidly at higher pH. These experiments suggested that chymotrypsin, like sucrose phosphorylase, acts by a double displacement mechanism (Eq. 7-14):

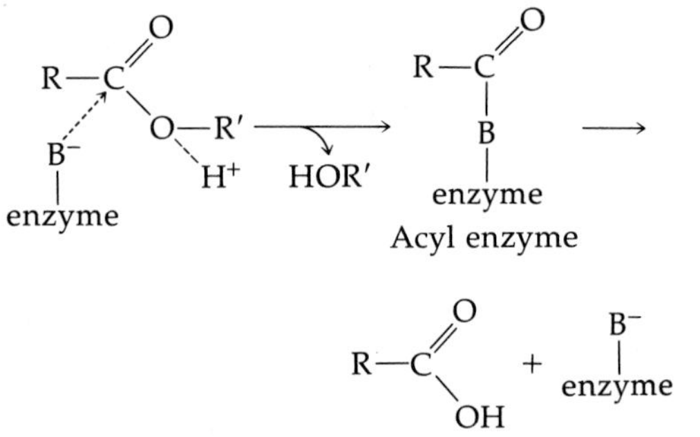

What is group —B⁻ in the foregoing equation? The experiments with DFP immediately suggested the —O⁻ group from Ser 195. However, there was reluctance to accept this deduction because of the very weak acidity of the —CH$_2$OH group. Furthermore, other data had suggested an imidazole group of a histidine side chain. For example, the activity of chymotrypsin in catalyzing the hydrolysis of esters varies with pH. While the Michaelis constant changes very little, V_{max} follows a typical bell-shaped curve (Fig. 6-13) with an optimum between pH 8 and 9. The shape of the curves for a series of different esterases suggested one pK_a of an enzyme group between 6.1 and 6.8 and a second between 9 and 9.6. The pK_a between 6.1 and 6.8 could most logically be identified with an imidazole group which would be unprotonated in the catalytically active form of the enzyme. Other experiments also implicated a histidine residue. For example, chymotrypsin reacts with 2,4-dinitrofluorobenzene and inactivation accompanies the attack of this reagent on one of the two histidine residues in the enzyme.

BOX 7-B

INSECTICIDES

Over 200 organic insecticides, designed to kill insects without excessive danger to humans and animals, are presently in use.[a–d] Many of these compounds act by inhibiting cell respiration; others "uncouple" ATP synthesis from electron transport. The chlorinated hydrocarbons such as DDT act on nerves in a manner that is still not fully understood. One of the largest classes of organic insecticides act on the specific nerve enzyme **acetylcholinesterase.** Acetylcholine, a neurotransmitter, is released at many nerve synapses (Chapter 16). The acetylcholine (which is very toxic in excess) must be destroyed rapidly to prepare the synapse for transmission of another impulse:

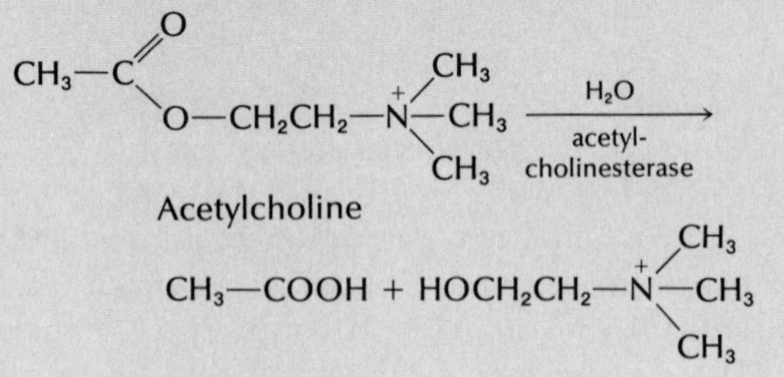

Like chymotrypsin, acetylcholinesterase contains an active site serine residue that reacts with organophosphorus compounds such as DFP (Section D,1,a).

The extreme toxicity of esters of pyrophosphate and of dialkylphosphorofluoridates appears to have become known in the 1930's and led to their development in Germany and in England as insecticides and as nerve gases. Among the most notorious is DFP itself, for which the LD$_{50}$ (dose lethal to 50% of the animals tested) is only 0.5 mg kg^{-1} intravenously. This exceedingly dangerous compound can cause rapid death by absorption through the skin. Many organophosphorus compounds and other acetylcholinesterase inhibitors that are selectively toxic to insects have been found, e.g.,

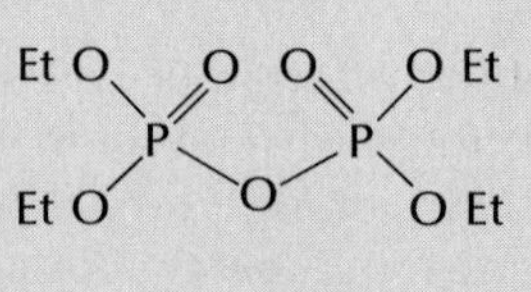

Tetraethyl pyrophosphate (TEPP)

Box 7-B (*Continued*)

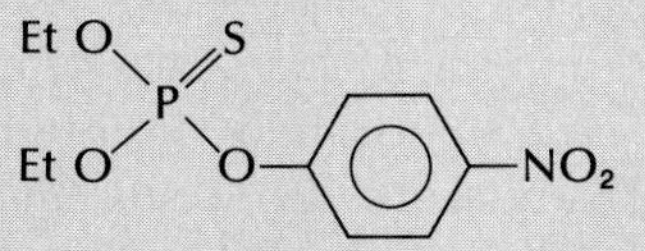

$$\text{EtO} - \overset{\displaystyle S}{\underset{\displaystyle |}{P}} - \text{O} - \text{C}_6\text{H}_4 - \text{NO}_2$$

Parathion

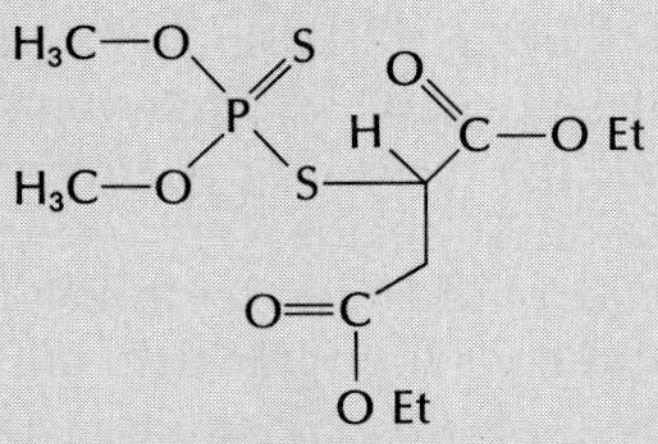

Malathion

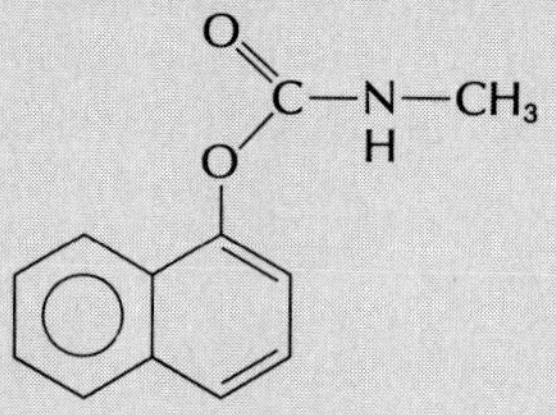

Carbaryl

The characteristic high group transfer potential of a phosphoryl group in pyrophosphate linkage, which makes ATP so useful in cells, also permits tetraethyl pyrophosphate to phosphorylate active sites of acetylcholinesterases. While TEPP is exceedingly toxic, it is rapidly hydrolyzed; even a few hours after use all harmful residues are gone.

Two of the most widely used insecticides at present are **parathion** and **malathion**. These compounds are much less toxic than DFP or TEPP. They do not become effective insecticides until they undergo bioactivation during which conversion from a $P{=}S$ to a $P{=}O$ compound occurs:

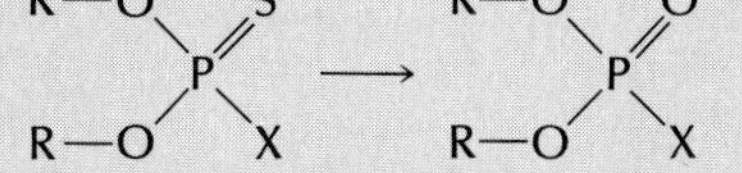

$$\begin{array}{ccc} \text{R}-\text{O} & & \text{R}-\text{O} \\ & \overset{S}{\underset{|}{P}} & \longrightarrow \quad \overset{O}{\underset{|}{P}} \\ \text{R}-\text{O} \quad \text{X} & & \text{R}-\text{O} \quad \text{X} \end{array}$$

Highly toxic

The desulfuration reaction involves microsomal oxidases of the liver, the sulfur being oxidized ultimately to sulfate.[e]

The reactivity of parathion with cholinesterases depends upon the high group transfer potential imparted by the presence of the excellent leaving group, the *p*-nitrophenolate anion (see also Section D,1,b). If the P—O linkage to this group is hydrolyzed before the desulfuration reaction takes place, the phosphorus compound is rendered harmless. Thus, the design of an effective insecticide involves finding a compound which insects activate rapidly but which is quickly degraded by higher animals. Other factors, such as rate of penetration of the insect cuticle and rate of excretion from the organism, are also important.

The phosphorylated esterases formed by the action of organophosphorus inhibitors are very stable, but antidotes have been found that can reverse the inhibition. The oxime of 2-formyl-1-methylpyridinium ion (pralidoxime) is very effective.[f] Its positive charge is thought to permit it to bind to the site normally occupied by the quaternary nitrogen of acetylcholine and to displace the dialkylphosphoryl group:

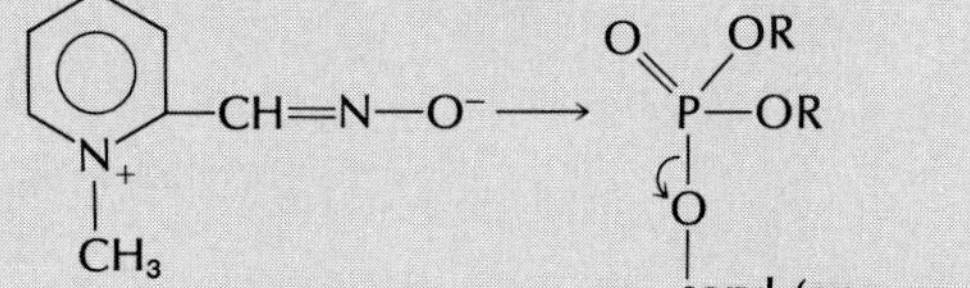

Pralidoxime   seryl (enzyme)

Carbaryl, a widely used methyl carbamate, is a pseudosubstrate of acetylcholinesterase that reacts 10^5 to 10^6 times more slowly than do normal substrates. The carbamoylated enzyme formed is not as stable as the phosphorylated enzymes and the inhibition is reversible.

[a] J. E. Casida, *Annu. Rev. Biochem.* **42**, 259–278 (1973).
[b] D. F. Heath, "Organophosphorus Poisons." Pergamon, Oxford, 1961.
[c] G. Schrader, "Die Entwicklung Neuer Insektizider Phosphorsäure-Ester." Verlag Chemie, Weinheim, 1963.
[d] *Bull. W. H. O.* **44**, 1–470 (1971).
[e] T. Nakatsugawa, N. M. Tolman, and P. A. Dahm, *Biochem. Pharmacol.* **18**, 1103–1114 (1969).
[f] I. B. Wilson and S. Ginsburg, *BBA* **18**, 168–170 (1955).

c. Nonenzymatic Models

Imidazoles themselves have been found to catalyze the nonenzymatic hydrolysis of *p*-nitrophenyl acetate with formation of unstable acetyl imidazoles as intermediates.

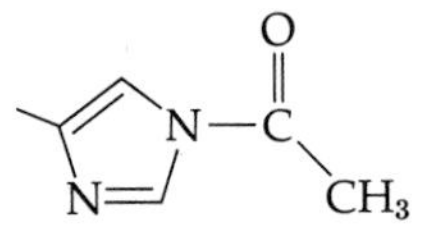

It was attractive to think that the acyl-enzyme intermediate of Eq. 7-14 might have such a structure. Thus, there were two candidates for group —B:, *serine* and *histidine*. While the stable end products of reactions with pseudosubstrates were unquestionably derivatives of serine, the possibility remained that these were side products and that histidine was involved in transient, rapidly forming and reacting intermediates.

d. The Three-Dimensional Structure

The structures of chymotrypsin and trypsin have now been revealed by X-ray diffraction studies,[29-32] which have confirmed what the chemical experiments had suggested. Both Ser 195 and His 57 are at the active site (Fig. 7-2). Bear in mind that the X-ray diffraction results do not show where the hydrogen atoms are and that these have been added in the figure. The short contact distance (0.30 nm) between the nitrogen of His 57 and the oxygen of Ser 195 suggests a hydrogen bond. The same logic was used to deduce the presence of the other hydrogen bonds shown. If the histidine is unprotonated and a hydrogen is present on the serine OH group, we see that histidine could act as a proton acceptor (general base catalyst) to assist in removing the proton from the —CH_2OH of a serine making the hydroxyl group more nucleophilic than it would be otherwise.

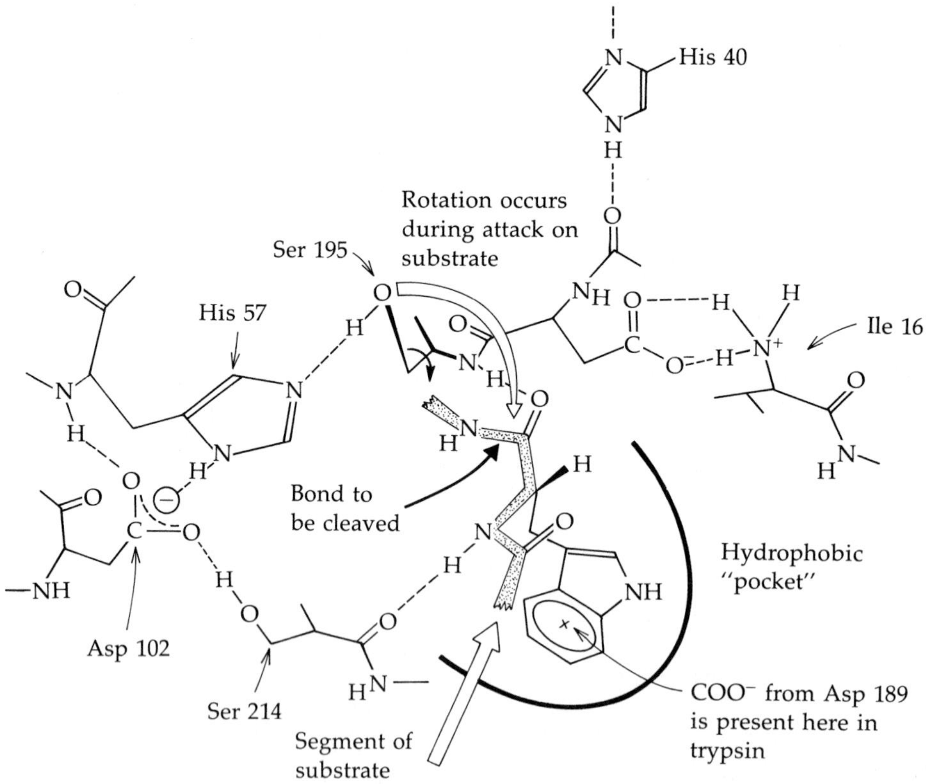

Fig. 7-2 The active site of chymotrypsin with a segment of substrate present. After Blow[29] and Henderson and Wang.[13]

The X-ray studies revealed that His 57 is also hydrogen bonded on the other side to the carboxylate group of Asp 102, which is in turn hydrogen bonded to two other groups. Aspartate 102 has one of the few carboxylate side chains that is buried inside the protein. To Blow[29] the structure suggested a **charge-relay system** by which protons might move synchronously, from Ser 195 to the imidazole and from the imidazole to Asp 102. The significance of this system in the serine proteases is still not fully understood, but it should be considered in the light of the concept of "tautomeric catalysis" (Chapter 6, Section E,5,f). It may be that electronic displacements running out into the peptide backbone are also coordinated with operation of the charge relay. The independent evolution of identically the same charge-relay system in an otherwise quite different serine protease **subtilisin** (*Bacillus subtilis*) suggests that the system may be truly essential to the catalytic mechanism.[33,34]

Direct nmr observations of the hydrogen-bonded proton between His 57 and Asp 102 are interpreted as supporting the charge-relay concept.[35] However, attempts to design an effective nonenzymatic model incorporating the charge-relay system have not been successful.[36]

Making the serine hydroxyl more nucleophilic is not the only possible function of His 57. The group $R'—NH^-$ displaced from the substrate is a poor leaving group unless protonated first. A protonated histidine could donate a proton to that leaving group (general acid catalysis, Eq. 7-15). As

$$\text{(7-15)}$$

Jencks has pointed out,[37] it is often experimentally difficult to distinguish general base catalysis from general acid catalysis. Perhaps His 57 performs both functions, first removing a proton from

serine, then, a moment later, donating it to the leaving group.[13,38]

Another obvious mechanism by which an enzyme could assist in a displacement reaction on a carbonyl group is through protonation of the carbonyl oxygen atom by an acidic group of the enzyme (Eq. 7-16). This would greatly increase the

$$\text{(7-16)}$$

positive charge on the carbon atom and would make attack by a nucleophile easier. It would also stabilize a tetrahedrally bonded intermediate (Eq. 7-13). The carbonyl oxygen is very weakly basic, but it could be protonated by a suitably oriented acidic group of the enzyme (HB in Eq. 7-16). In the serine proteases this function is apparently fulfilled by two NH groups of amide linkages, one being the NH group of Ser 195 in chymotrypsin (Fig. 7-2). It appears that the fit of substrate into the **oxyanion hole** between the two NH groups is good only for the tetrahedral intermediate.[33]

It has been suggested[38] that formation of the tetrahedral oxyanion (with charge flowing from the buried Asp 102 carboxylate into the substrate) triggers a conformational change that permits the His 57 imidazole to accept a proton back from Asp 102 and to protonate the leaving group as the tetrahedral intermediate breaks down (Eq. 7-17). Why would a conformational change occur? The enzyme contains an intricate network of hydrogen bonds. The buildup of negative charge on the oxygen of the tetrahedral intermediate will certainly affect the electron distribution within some of these hydrogen bonds at a distance from the active center. If there are several energetically similar conformations of the protein, it is possible that an

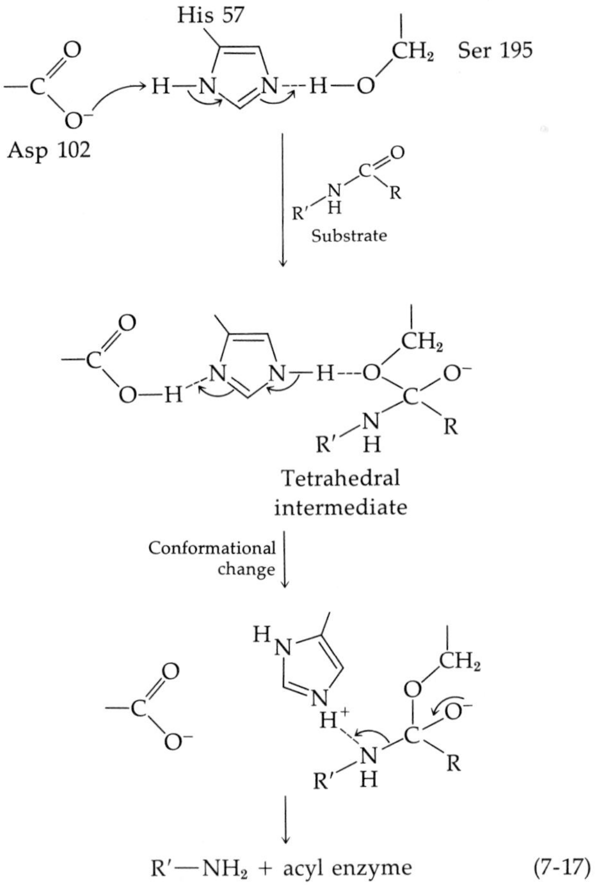

Substrate

Tetrahedral
intermediate

Conformational
change

$$R'—NH_2 + \text{acyl enzyme} \qquad (7\text{-}17)$$

altered charge distribution could cause a sudden jump from one conformation to another and that this could be an essential part of the mechanism of catalysis.

At high pH, the velocity of chymotrypsin action falls off, indicating a pK_a of ~8 to 9. This pK_a could represent the N-terminal amino group of isoleucine 16. One of the bonds cleaved in conversion to the active enzyme is connected to the amino group Ile 16. This amino group forms a salt linkage (ion pair) with Asp 194 (Fig. 7-2) which is next to the serine at the active center. Perhaps the salt linkage helps to hold the enzyme in the required conformation for reaction. Deprotonation above pH 8–9 would cause inactivation.[39]

An alternative is that the high pK_a belongs to the second histidine in the enzyme, His 40. How-

ever, this residue is lacking in a bacterial serine protease which has the same type of pH profile. Histidine 40 is located close to Asp 194 and is hydrogen bonded to a peptide carbonyl. It is thought that it may play some role in the zymogen-active enzyme interconversion.

e. Substrate Specificity

Like most enzymes, trypsin and chymotrypsin display distinct specificities for certain substrates. For rapid action by chymotrypsin the C=O group of the peptide linkage being cleaved must be donated by one of the aromatic amino acids. Thus, there must be a part of the substrate binding site that binds preferentially to large, flat aromatic groups. The crystal structure shows this site (Fig. 7-2) to be composed of hydrophobic side chain groups of amino acids. On the other hand, in trypsin the specificity portion of the binding site contains a fixed negative charge provided by the carboxylate side chain group of Asp 189. This explains the specificity of trypsin in acting only upon peptide linkages containing adjacent arginine or lysine residues, both of which contain positive charges in a neutral medium.[32]

2. Pepsin

Several different families of proteases are known, not all of which contain serine at the active center. One family consists of pepsin, of the stomach and related enzymes such as **rennin,** an enzyme obtained from the fourth stomach of the calf. Rennin causes a rapid clotting of milk and is widely used in manufacture of cheese. Some intracellular **cathepsins** and various proteases of fungi are also related. The pepsin family of proteases is unusual in being most active in the pH range 1–5, a characteristic that probably explains why neither serine nor histidine is a component of the active site. It is thought that a carboxylate ion is the nucleophile in a double displacement mechanism and that a second carboxyl is the proton donor to the leaving group. Thus, the mechanism may be similar to that of lysozyme.

BOX 7-C

PROTEASE INHIBITORS OF ANIMALS AND PLANTS

Premature conversion of proenzymes such as trypsinogen into active proteases in the pancreas would be disastrous. To prevent this, the pancreas also produces specific inhibitors. The pancreatic **trypsin inhibitor** is a small protein of MW = 6500 which specifically binds at the active site of trypsin with $K_f = 10^{12}$ M (at alkaline pH values).[a] Crystal structure determinations have been made with both the trypsin inhibitor and trypsin itself, and it is found that the two molecules fit snugly together.[b] The inhibitor is bound as if it were a peptide substrate, one edge of the inhibitor molecule forming an antiparallel β structure with a peptide chain in the enzyme. Lysine 15, which forms part of this β structure, enters the specific binding site for a basic amino acid in a substrate. Thus, the protease inhibitor is a modified substrate which may actually undergo attack at the active site. However, the fit between the two molecules is so tight that there is no possibility of a water molecule entering to complete the catalytic act and the complex remains inactive. (There is not enough inhibitor to interfere with the action of the large amounts of trypsin formed from chymotrypsinogen in the small intestine.)

Protease inhibitors which block the action of trypsin are also found in many plants. The antiprotease activity is usually highest in seeds and tubers, but synthesis of protease inhibitors can be induced in other parts of plants by wounding. Perhaps these inhibitors function to protect plants against attack by insects.[c] The structure of the soybean trypsin inhibitor and of its complex with trypsin has been determined. The complex resembles that with the pancreatic trypsin inhibitor. However, the soybean inhibitor is slowly cleaved and the diffraction studies[d] show that the complex exists as a tetrahedrally bonded adduct as in Eq. 7-13.

Some other purpose must be served by the α_1-**antitrypsin** found in the α-globulin fraction of blood serum. There is no trypsin in tissues, but this inhibitor blocks the action of a variety of serine proteases. A hereditary absence of α_1-antitrypsin often leads to severe **pulmonary emphysema** at an early age.[e–g] It is known that **elastase** and a neutral protease are released by lysosomal granules of blood granulocytes at sites of inflammation. Without the protease inhibitor in the environment, the neutral protease and the elastase (which causes collagen breakdown in a localized area) may react in an uncontrolled fashion. The hereditary defect may be in the slow release of the protease inhibitor from its site of synthesis in liver cells.[f] The matter is complicated by the fact that patients deficient in α_1-antitrypsin also tend to be deficient in a **chemotactic factor inactivator** (see Box 5-G).

[a] R. Huber, D. Kukla, A. Rühlmann, and W. Steigemann, *Cold Spring Harbor Symp. Quant. Biol.* **36**, 141–150 (1971).
[b] R. M. Stroud, L. M. Kay, and R. E. Dickerson, *Cold Spring Harbor Symp. Quant. Biol.* **36**, 125–140 (1971).
[c] T. R. Green and C. A. Ryan, *Science* **175**, 776–777 (1972).
[d] D. M. Blow, J. Janin, and R. M. Sweet, *Nature* (*London*) **249**, 54–57 (1974).
[e] P. A. Ward and R. C. Talamo, *J. Clin. Invest.* **52**, 516–519 (1973).
[f] J. Lieberman, C. Mittman, and H. W. Gordon, *Science* **175**, 63–65 (1972).
[g] A. B. Cohen, *JBC* **248**, 7055–7059 (1973).

3. Papain

Papain from the latex of the papaya is one of a family of plant enzymes that includes **bromelin** (pineapple), **ficin** (fig), and several enzymes of bacterial origin. Another bacterial protease **clostripain** (*Clostridium histolyticum*) appears to have a catalytic site like that of papain but a specificity similar to that of trypsin.[40]

The participating nucleophile in papain appears to be an SH group.[41] From a structure determination by X-ray diffraction we know that there is an imidazole group near the SH group where it could act as a proton donor to the leaving group.[42] Thus, despite the differences, there are similarities in the action of all of these families of protein-hydrolyzing enzymes.

An inactive form of papain, thought to be a

proenzyme, is activated by treatment with thiols. Propapain seems to be a structural isomer of papain, which must undergo intramolecular thiol-disulfide exchange to generate the active site —SH group.[43]

4. Metal Ions in Protease Action: Carboxypeptidase A

In addition to trypsinogen and chymotrypsinogen, the pancreatic secretion contains other zymogens which are converted to enzymes that cleave amino acids from the ends of peptide chains (not at positions *within* the chain as do trypsin and chymotrypsin). The **carboxypeptidases** attack only at the C-terminal ends removing one amino acid at a time, a property that has made them valuable reagents for determination of amino acid sequences in peptides. Carboxypeptidases have also been used for modification of proteins by removal of one or a few amino acids from the ends. The most striking feature of carboxypeptidase A is the presence of one atom of firmly bound **zinc ion.** It can be removed from the protein and can be substituted by other metal ions, in some cases with reconstitution of catalytic activity.

The presence of a zinc ion in carboxypeptidase immediately suggested a role in catalysis. Unlike protons, which have a weak affinity for the oxygen of an amide carbonyl group, a metal ion can form a strong complex. If held in position by other ligands from the protein, a properly placed zinc ion might be expected to greatly enhance the electrophilic nature of the carbon atom of the C=O group. The three-dimensional structure of carboxypeptidase A is now known, and the X-ray diffraction studies have confirmed the hunch about the zinc ion and have suggested a great deal more about the mechanism.[44-46]

The zinc ion is chelated by two imidazole groups and a glutamate side chain (Fig. 7-3). Arginine 145 and Tyr 248 form hydrogen bonds to the substrate, the latter in such a way that it can serve as a general acid catalyst in protonation of

the leaving group. Either a carboxylate group of Glu 270 or a water molecule appears to be the attacking nucleophile. If the former, an anhydride intermediate must be formed. The presence of a hydrophobic "pocket" accounts for the preference of the enzyme for C-terminal amino acids with bulky, hydrophobic side chains.

Another zinc-containing protease is **thermolysin,** a notably heat-stable enzyme produced by *Bacillus thermoproteolyticus.* It also contains four bound calcium ions.[47,48]

5. Acyltransferases

Acyl groups are frequently transferred from amides or esters to various acceptors in biosynthetic reactions. For example, the final step in formation of peptide linkages during protein synthesis on ribosomes is the transfer of a peptidyl group joined in ester linkage on a tRNA molecule onto an amino group of an "activated" amino acid (Chapter 11, Section E,1).

The last step in synthesis of peptidoglycans of bacterial cell walls is also a **transacylation** (transpeptidation). The amino group of the diamino acid (see Fig. 5-9) in one peptide chain attacks an amide linkage in an adjacent chain. A molecule of D-alanine is displaced and a cross-link is formed (Box 7-D). Acyl groups, especially the acetyl group, are often transferred onto nucleophilic sites from thioesters of coenzyme A (Chapter 8, Section B). An example is the formation of acetylcholine (Box 7-B) from choline and acetyl-CoA by transacylation. Note the high group transfer potential of thioesters insures that the reactions proceed to completion.

E. DISPLACEMENT ON A PHOSPHORUS ATOM

Nucleophilic displacement on a phosphorus atom (Table 7-1, reaction type 1C) is a very important reaction type in terms of the number of enzymes involved and their central functions in nucleic acid biochemistry and in energy metabolism.

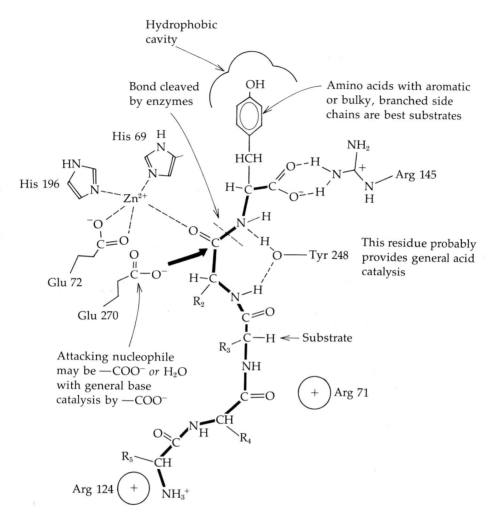

Fig. 7-3 Structure of the active site of carboxypeptidase A with a peptide substrate present. After Lipscomb *et al.*[44-46]

1. Phosphatases

The phosphatases catalyze hydrolysis of phosphate esters to produce inorganic phosphate (Eq. 7-18). Two groups, the **acid phosphatases** and **alkaline phosphatases,** are nonspecific and cleave many different phosphate esters, whereas other phosphatases, e.g., **glucose-6-phosphatase** and **fructose-1,6-diphosphatase,** are specific for single substrates. The function of the nonspecific phosphatases is somewhat obscure. The alkaline phosphatases are found in bacteria, fungi, and higher animals but not in higher plants. In *E. coli* alkaline phosphatase appears to be concentrated in the periplasmic space. It is found in the brush

$$R-O-\overset{\displaystyle O}{\underset{\displaystyle O^-}{\overset{\displaystyle \|}{P}}}-O^- + H_2O \longrightarrow R-OH + HPO_4^{2-}$$

$$\text{(7-18)}$$

or

$$R-O-\textcircled{P} + H_2O \longrightarrow R-OH + P_i$$

BOX 7-D

PENICILLINS AND RELATED ANTIBIOTICS[a]

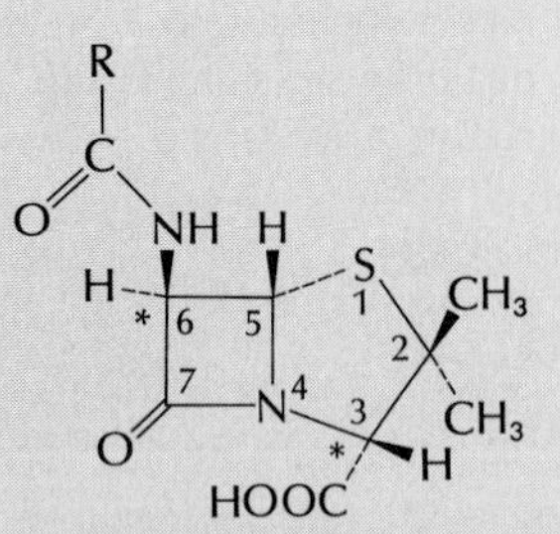

Penicillins

Penicillin (from the fungus *Penicillium*) was the first antibiotic to find practical use in medicine. Commercial production began in the early 1940's. Benzypenicillin (penicillin G), one of several natural penicillins that differ in the R group (see structure above), has become one of the most important of all drugs. Most effective against gram-positive bacteria, it also attacks gram-negative bacteria including *E. coli* at higher concentrations.

A popular semisynthetic penicillin **ampicillin** (R = D-α-aminobenzyl) is a broad range antibiotic that attacks both gram-negative and gram-positive organisms. It shares with penicillin an extremely low toxicity but some danger of allergic reactions. Other semisynthetic penicillins are resistant to **penicillinases**. These enzymes, produced by penicillin-resistant bacteria (Chapter 15, Section D,7), cleave the four-membered β-lactam ring of natural penicillins and inactivate them.

Penicillin kills only *growing* bacteria by preventing proper cross-linking of the peptidoglycan layer of their cell walls (Chapter 5, Section D). An amino group from a diamino acid in one peptide chain of the peptidoglycan displaces a D-alanine group in a transpeptidation (transacylation) reaction. It has been suggested that the penicillins are structural analogues of D-alanyl-D-alanine and bind to the active site of the transacylase.[b,c] The β-lactam ring of penicillins is unstable, making penicillins powerful acylating agents. If the transacylase acts by a double displacement mechanism the initial attack of a nucleophilic group of the enzyme on penicillin bound at the active site of the transacylase would lead to formation of an inactivated, penicillinoylated enzyme. Experimental evidence supports this mode of action but reveals that more than one protein is derivatized by penicillin; hence, there may be other sites of action, too.[b,d,e]

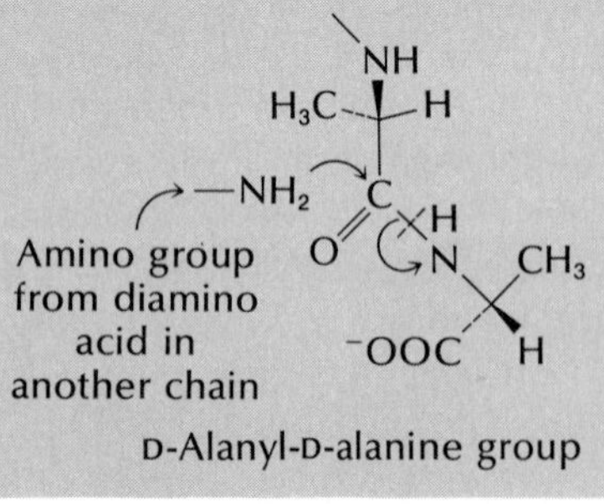

Closely related to penicillin is another antibiotic **cephalosporin C**. It contains a D-α-aminoadipoyl side chain which can be replaced to form semisynthetic cephalosporins.

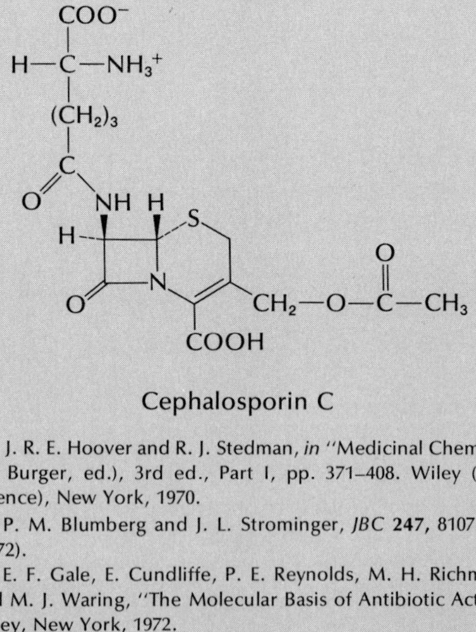

Cephalosporin C

Amino group from diamino acid in another chain

D-Alanyl-D-alanine group

[a] J. R. E. Hoover and R. J. Stedman, *in* "Medicinal Chemistry" (A. Burger, ed.), 3rd ed., Part I, pp. 371–408. Wiley (Interscience), New York, 1970.

[b] P. M. Blumberg and J. L. Strominger, *JBC* **247**, 8107–8113 (1972).

[c] E. F. Gale, E. Cundliffe, P. E. Reynolds, M. H. Richmond, and M. J. Waring, "The Molecular Basis of Antibiotic Action." Wiley, New York, 1972.

[d] P. M. Blumberg and J. L. Strominger, *Bacteriol. Rev.* **38**, 291–375 (1974).

[e] B. G. Spratt and A. B. Pardee, *Nature (London)* **254**, 516–517 (1975).

border of kidney cells, in cells of the intestinal mucosa, and in the osteocytes and osteoblasts of bone. It is almost absent from red blood cells, muscle, and other tissues which are not involved extensively in transport of nutrients. One theory is that the enzyme is responsible for providing inorganic phosphate at places where it is needed.

The alkaline phosphatase of *E. coli* is a dimer of MW ~89,000 which requires Zn^{2+}, is allosterically activated by Mg^{2+}, and has a pH optimum above 8.[49] At a pH of ~4, incubation of the enzyme with inorganic phosphate leads to formation of a phosphoenzyme. Using ^{32}P-labeled phosphate, it was established that the phosphate becomes attached to a reactive serine residue in the sequence:

$$Asp \cdot Ser \cdot Ala$$
$$\underset{\textcircled{P}}{|}$$

The same sequence is also found in mammalian alkaline phosphatases (which also have similar metal ion requirements[50]). Aside from the replacement of Ala for Gly, it is also the same as the sequence at the active centers of serine proteases. Perhaps the two groups of enzymes evolved from a common ancestral protein.[51]

There is a tendency for only one of the active sites of *E. coli* alkaline phosphatase to react at a time. This behavior has been interpreted[52] as indicating a catalytic cooperativity between subunits and a "flip-flop" mechanism (see Chapter 8, Section H,3). However, the proposal has been challenged.[53]

Acid phosphatases have a pH optimum of ~5 and are inhibited by fluoride ion. They occur in plants as well as animals. In bone, the acid phosphatase content is high in the osteoclasts which function in the resorption of calcium from bone. A phosphoryl enzyme has been trapped from both plant and animal acid phosphatases.[54,55] Brief incubation with ^{32}P-labeled *p*-nitrophenyl phosphate followed by quenching in an alkaline denaturing medium gave a covalently labeled protein. After alkaline hydrolysis ^{32}P-containing N^1-phosphohistidine was isolated.

N^1-Phosphohistidine

Pyrophosphatases, present in all cells, catalyze hydrolysis of inorganic pyrophosphate to 2 molecules of P_i (Box 3-A). The very active pyrophosphatase of *E. coli* has a turnover number of over 2×10^4 s^{-1} at 37°C. The 1000 molecules per cell are sufficient to immediately hydrolyze any pyrophosphate produced by bacterial metabolism.[56]

2. Ribonuclease

The 1972 Nobel Prize for chemistry was awarded to Stanford Moore, William H. Stein, and Christian B. Anfinsen for studies on ribonuclease, the pancreatic digestive enzyme responsible for breakdown of RNA. Stein and Moore, who had earlier developed ion exchange methods for separating amino acids and peptides, determined the amino acid sequence of the 124-residue enzyme from the bovine pancreas.[57] Before the X-ray structural studies were done, Lys 41 was identified as being unusually reactive with dinitrofluorobenzene. Furthermore, Stein and Moore had concluded (because their photooxidation inactivates the enzyme) that His 12 and His 119, almost at opposite ends of the chain, were both at the active site. The newer crystallographic results show that Lys 41 and the two histidines do all lie close together and on a cleft that contains the active site.[58]

Ribonuclease catalyzes cleavage of the phosphodiester linkages of RNA in two steps. The adjacent hydroxyl group on the 2' position of the ribose ring is presumably deprotonated by attack of a base B (Eq. 7-19, step *a*). The alkoxide ion generated then attacks the phosphorus, displacing

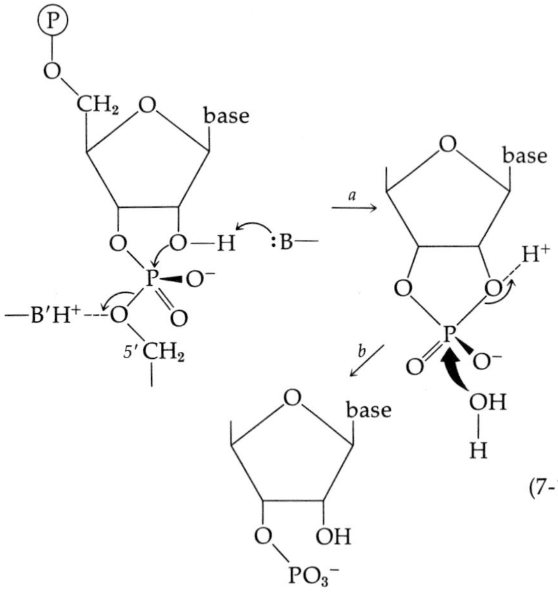

(7-19)

the oxygen attached to the 5'-carbon of the next nucleotide unit. The cleavage may be further aided by donation of a proton from an acid group B'H$^+$. The intermediate is a cyclic 2',3'-diphosphate which then undergoes hydrolysis by attack of a water molecule in step b to give the free 3'-nucleotide. Thus, the overall reaction is a two-step double displacement, analogous to that of chymotrypsin, except that a neighboring group in the substrate rather than an amino acid side chain is the nucleophilic catalyst.

The best guess from X-ray evidence is that the basic group that removes a proton from the 2'-hydroxyl is His 12, while the acidic group donating a proton to the departing 5'-oxygen is protonated His 119.[59] (However, an artificial N^τ-carboxymethyl–His 12 derivative of ribonuclease has some catalytic activity, a fact that raises some questions.[60]) The pH dependence of the enzyme is in line with such a mechanism for there are two pK_a values of 5.4 and 6.4 which regulate the activity of the enzyme. (A pK_a of 5.8 was deduced from the nmr spectra of Fig. 2-42.) Lysine 41 lies immediately above the two histidines. Perhaps its positive charge is used to neutralize some of the negative charge on the oxygens of the phosphate group making attack by the entering

nucleophile easier. An interesting facet of ribonuclease chemistry comes from the finding that a bacterial peptidase splits a 20-residue fragment from the N-terminal end. This "S-peptide" could be recombined with the rest of the molecules to give a functional enzyme called ribonuclease S. Its structure was determined by X-ray diffraction and found to be essentially the same as that of native ribonuclease.

Techniques for protein synthesis have been developed to the point that small enzymes can be synthesized in the laboratory. This permits the creation of new modified enzymes and a critical examination of functions of groups at the active center. For example, a 70-residue peptide analogue of ribonuclease S with a number of deletions and a total lack of disulfide bridges still retained measurable catalytic activity.[61]

3. Pentacovalent Intermediates and Permutational Rearrangements during Displacements on Phosphorus

Whereas a carbon atom can form only four stable covalent bonds, phosphorus is able to form five. Nucleophilic attack on carbon leads to a *transient* five-bonded transition state, but attack on phosphorus produces a pentacovalent intermediate that may be relatively stable and long-lived. Note that our conventional way of drawing phosphate esters with a double bond from phosphorus to one of the oxygens is misleading. All of the P—O bonds share some of the double bond character and the phosphate group has many characteristics of a completely single-bonded structure:

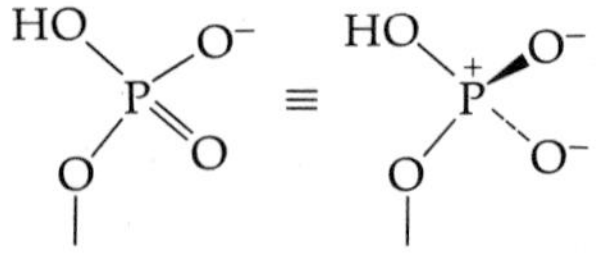

Equation 7-20 shows a nucleophile X$^-$ attacking the tetrahedral phosphate group on the face opposite the group, O—R, to form a pentacovalent intermediate whose geometry is that of a **trigonal**

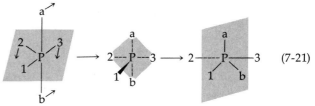

$$\text{(7-20)}$$

Trigonal bipyramid

bipyramid. In this structure, not all of the bond angles are equal. Those in the **equatorial** plane are 120°, whereas all of the angles between any of those bonds and the two that attach to the groups in the **apical** positions are 90°. This disparity arises naturally from the fact that it is impossible to place five points on the surface of a sphere all equidistant one from the other (whereas it *is* possible to group *four* points equidistant one from the other on a sphere to form a tetrahedron). The attack of X^- from the side opposite O—R leads to a trigonal bipyramid in which —O—R and —X occupy the two apical positions. However, if X^- attacks a face opposite one of the other oxygens, —O—R takes an equatorial position.

The chemical reactivity of groups in the apical and equatorial positions of pentacovalent intermediates is quite different.[62] In particular, elimination of a nucleophilic group to form a tetrahedral phosphate is easier from an apical position than from an equatorial position. Consider ribonuclease again. In Eq. 7-19 the 2′-hydroxyl group is shown attacking directly behind the 5′-oxygen of the second nucleotide, placing both the attacking and leaving groups in apical positions in the intermediate pentacovalent compound. This provides for easy elimination of the leaving group. In step *b* of Eq. 7-19, the water molecule is shown (arbitrarily) attacking the face opposite the 3′-oxygen. This would lead to an intermediate in which the group to be eliminated (the 2′-oxygen) lies in an equatorial position. Before such a group could be eliminated, it would probably have to undergo a **permutational rearrangement** by which it would be transferred from an equatorial to an apical position.

One type of permutational rearrangement, known as **pseudorotation,** can be visualized as shown in Eq. 7-21.[63,64] The axial groups a and b move back by means of a "vibratory motion," while equatorial groups 2 and 3 move forward, still in the same equatorial plane. Equatorial group 1 does not move. This has the effect of decreasing the 120° bond angles between the equatorial groups and increasing the bond angles between group 1 and the axial groups until all four bond angles to group 1 are equivalent. The re-

$$\text{(7-21)}$$

Tetragonal bipyramid

sulting **tetragonal bipyramid** is a "high-energy" transition state structure in the pseudorotation process and can either revert to the original structure or, by continued motion of the groups in the same directions as before, to the structures shown at the right of Eq. 7-21. In the final structure, groups 2 and 3 are axial and the original axial groups a and b are equatorial. A second mechanism of permutational rearrangement named turnstile rearrangement has the same end result as pseudorotation.[65] Pseudorotation is slow enough that it could be a rate-limiting process in some enzymatic reactions.

The attack of a displacing group in a position opposite the leaving group may be referred to as **in-line** while attack from another face to give an intermediate requiring pseudorotation may be called an **adjacent process.**

If two of the atoms attached to phosphorus are part of a five-membered ring (as is the case with the cyclic ribonuclease intermediate of Eq. 7-19), one must occupy an apical position and the other an equatorial position. This means that in step *b* of Eq. 7-19, the water molecule must necessarily attack a face behind *either* the 2′- or 3′-oxygen atoms rather than behind either of the other two oxygens.

On the basis of the stereochemistry of the action of ribonuclease on diastereoisomers of uridine 2',3'-cyclic phosphorothionate[66] and other evidence,[67] an in-line mechanism is favored.

HOCH₂   O   Uracil

Uridine 2',3'-cyclic phosphorothionate

← Two configurations are possible about this chiral center

4. Formation of Metaphosphate

Displacement reactions on phosphorus might also take place via an internal displacement (elimination) to form a **metaphosphate** ion (Eq. 7-22, step *a*). A nucleophilic reagent would then add to the eliminated metaphosphate (Eq. 7-22, step *b*). The formation of metaphosphate is analogous to for-

$$R-O-\overset{O}{\underset{H-O}{\overset{\|}{P}}}-O^- \xrightarrow{a} R-OH + \overset{O}{\underset{^-O}{\overset{\|}{P}}}=O$$

Metaphosphate

$$Y-OH + \overset{O}{\underset{^-O}{\overset{\|}{P}}}=O \xrightarrow{b} Y-O-\overset{O}{\underset{O}{\overset{\|}{P}}}-OH$$

(7-22)

mation of carbonium ions during the action of lysozyme (Eq. 7-10). Just as it has been difficult to establish a carbonium ion mechanism conclusively, the role of metaphosphate in enzymatic catalysis remains elusive.[68]

5. Micrococcal (Staphylococcal) Nuclease

A bacterial polynucleotide-hydrolyzing enzyme widely used in nearest neighbor analysis (Chapter 2, Section H,4) can hydrolyze either DNA or RNA to 3'-nucleotides. The three-dimensional structure of the 149-residue **staphylococcal nuclease** is known.[69–71] As with pancreatic ribonuclease, the molecule can be split into two peptides (residues 6–48 and 49–149) which combine to form a complex with active enzymatic properties.[71,72] The complex is formed even when residues 43–48 are removed from the smaller of the two peptides. However, glutamine 43, which binds to an essential Ca^{2+} ion, is needed for enzymatic activity,[72] as is the peptide bond to the adjacent threonine 44.

The crystal structure determination on the nuclease shows that the majority of the acidic and basic side chains of the protein interact with each other through clusters of hydrogen bonds. Equally interesting is the arrangement of hydrogen bonds between the enzyme and the 5'-phosphate group of the specific inhibitor deoxythymidine 3',5'-diphosphate. Two guanidinium ions from arginine side chains interact with the 5'-phosphate group (Fig. 7-4). It is clear that these positively charged groups are almost tailor-made for interaction with phosphate ions and that their positive charges strongly polarize the phosphoryl group and facilitate nucleophilic attack. Further-

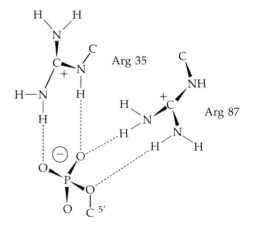

Fig. 7-4 Drawing showing the hydrogen bonding interactions between the guanidinium ions of arginines 35 and 87 of the staphylococcal nuclease with the 5'-phosphate of the inhibitor thymidine 3',5'-diphosphate in the complex of E + I + Ca^{2+}. After Cotton *et al.*[73]

more, arginine 87 is in a position to protonate the leaving group —O⁻.

As Cotton *et al.* pointed out,[73] the binding of phosphate-containing molecules to proteins is a "pervasive feature of the chemistry of living systems." The nature of the binding is of special interest. Arginine side chains can form up to five hydrogen bonds each and are able to interact simultaneously with phosphate groups and with other groups in a protein. For example, in the staphylococcal nuclease Arg 35 bonds both to the phosphate and to a carbonyl group of the peptide backbone.

6. Kinases

A large class of enzymes catalyze transfer of phosphoryl groups from one atom to another.* Among these, the kinases transfer phosphoryl groups from polyphosphates such as ATP to oxygen, nitrogen, or sulfur atoms of the second substrates. An example is **hexokinase** (Eq. 6-91; Fig. 9-7), the enzyme responsible for synthesis of glucose 6-phosphate from free glucose (Eq. 7-23). Hexokinase, like most enzymes catalyzing transfer of phosphoryl groups, has an absolute requirement for a divalent cation, usually Mg^{2+}. While the magnesium complex of ATP is regarded as the true substrate for hexokinase, the exact method of complexing of the metal to the polyphosphate on the enzyme surface is not known. The metal probably binds to groups on the enzyme as well as to the substrate.

Equation 7-23 provides a hypothetical picture and illustrates a possible function of the metal ion in these reactions. By complexing with the two phosphate groups in the portion of the ATP molecule that becomes the leaving group in the displacement reaction, the metal ion attracts electrons toward it and makes the bond cleavage easier. Another possible location for the metal in the ES complex would be on the terminal phosphoryl

---

* Phosphatases are also phosphotransferases that transfer phosphoryl groups onto hydroxyl ions. However, we usually classify such enzymes as hydrolases rather than transferases.

$$(7\text{-}23)$$

α-D-Glucose   Glucose 6-phosphate

group (the one being transferred) where it could serve to neutralize the negative charge and make approach of the attacking nucleophile easier. However, if the metal bridged across between the first and second phosphates, it would form a strong chelate ring and might prevent rather than assist the reaction. Thus, the position indicated in Eq. 7-23 is probably more likely. Perhaps, as in ribonuclease, some basic group from the enzyme neutralizes the negative charge on the terminal phosphate.

Equation 7-23 is drawn as an in-line attack of the hydroxyl group of the glucose. This direct single displacement mechanism seems likely because no partial exchange reactions have been observed nor do the kinetics indicate a ping-pong mechanism that might be indicative of the transient formation of a phosphoryl enzyme. Nevertheless, the latter possibility cannot be excluded.[74] It may be that both substrates must be bound at the active site before the enzyme assumes a conformation suitable for initiation of catalysis. The phosphoryl enzyme intermediate could be extremely reactive and difficult to trap. From some phosphotransferases rather stable phosphoenzymes have been prepared[75] by reaction with γ-^{32}P-labeled ATP in the absence of the second substrate. Mildvan[67] suggested that these are side reactions, possible indications of metaphosphate intermediates.

Three-dimensional structures have been described recently for a yeast hexokinase (Fig. 7-5A),[76] the 355-residue **phosphoglycerate kinase** (Fig. 7-5B),[77] and **adenylate kinase** (Box 3-A).[78] The latter, with MW $\approx 22,000$, is the smallest of the known kinases. In all three cases the protein is organized into two lobes or domains. For phosphoglycerate kinase and adenylate kinase the ATP-binding site has been located and the two phosphoryl groups closest to the adenosine (P_α and P_β) are bound to a single domain (the lower domain A in Fig. 7-5B). The terminal phosphoryl group of ATP (P_γ) fits into the cleft between the two domains. It appears that the magnesium ion is bound to P_α and P_β as in Eq. 7-23.

Of great current interest are protein kinases (Chapter 6, Section F,2) which transfer phosphoryl groups from ATP to hydroxyl groups of specific serine or theonine residues in their substrates. A common feature of the sites of phosphorylation in the substrate proteins is the presence of a lysine or arginine residue, separated from the serine or threonine by only one residue.[78a]

Because of its tightness of binding inhibitory action and other properties, the planar nitrate ion may serve as an analogue of the transferable phosphoryl group in the trigonal bipyramidal configuration of the transition states (Eq. 7-20) for kinases.[78b] Another ion of interest as a highly reactive analogue of the phosphate ion is ferrate, FeO_4^{2-}. A potent oxidizing agent, the ferrate ion is nevertheless similar in geometry and acid–base properties to phosphate and may be a useful reagent for modifying phosphate-binding sites.[78c]

7. Nuclear Relaxation by Paramagnetic Ions

The use of paramagnetic ions, e.g., Mn^{2+}, Cu^{2+}, or Cr^{3+} to induce nuclear relaxation in substrate and coenzyme molecules at active sites of enzymes is a powerful method that is being increasingly employed.[79-83] Flavin radicals and specifically introduced nitroxide spin labels can serve as well.

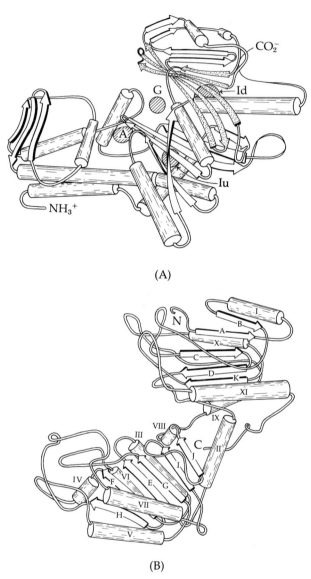

(A)

(B)

Fig. 7-5 (A) Drawing showing the polypeptide backbone of a subunit of hexokinase B of yeast. The position at which glucose binds is indicated by a G and an AMP binding site by A. A binding site for ATP between two subunits in a different type of crystal is indicated by Iu and Id. From Fletterick *et al.*[76] (B) Drawing of the phosphoglycerate kinase molecule. Cylinders representing α-helical segments are numbered with Roman numerals from the N-terminus. Arrows represent the individual strands in the β sheets and their direction, and are labeled alphabetically from the N-terminus. The ATP-binding domain is at the bottom of the drawing. From Blake and Evans.[77]

BOX 7-E

MAGNESIUM

It has been estimated that the average adult ingests 10–12 mmol of magnesium ion daily (~1/4 g). Of this, approximately one-third is absorbed from the digestive tract. An equivalent amount is excreted in the urine to maintain homeostasis. Sixty percent of the magnesium in the body is found in the bones. The Mg^{2+} content of serum is ~0.85 mM, while in tissues the total magnesium content is 5–8 mM. However, only ~1 mM Mg^{2+} is free. The remainder is bound to proteins and to soluble compounds such as ATP, ADP, and other phosphate- and carboxylate-containing substances.

It has been suggested that [Mg^{2+}], like [H^+], remains relatively constant within cells and that these two ions are in free equilibrium with the blood serum.[a] Nevertheless, there are probably instances in which at least temporary alterations in concentrations of both free Mg^{2+} and H^+ occur.[b] During rapid catabolism of carbohydrates glycolysis may lead to acidification of muscle cells, the pH falling from 7.3 to as low as 6.3. A drop in pH causes a remarkably large decrease in the extent of binding of Mg^{2+} to molecules such as ATP and to a transient increase in [Mg^{2+}]. Similarly, the release of diphosphoglycerate from hemoglobin upon oxygenation leads to a decreased concentration of free Mg^{2+} as the latter coordinates with diphosphoglycerate.[c] Such changes in the free Mg^{2+} concentrations may be of great significance in metabolic regulation.[d]

The magnesium ion has a smaller radius than Ca^{2+}, a fact that may account for its freer entry into cells. Mg^{2+} can often be replaced by Mn^{2+} with full activity for enzymes that require it. On the other hand, high concentrations of Ca^{2+} are often antagonistic to Mg^{2+}. This[e] antagonism is clearly seen in the effect of the two ions on irritability of cells. A deficiency of magnesium or an excess of calcium in the surrounding medium leads to increased irritability. On the other hand, excess magnesium leads to anesthesia. It is of interest that the Mg^{2+} concentration is high in hibernating animals.

Many enzymes are dependent upon Mg^{2+}, the largest single group being the phosphotransferases, for which MgATP (Chapter 3, Section B,5) may be regarded as the substrate. Among the Mg^{2+}-dependent enzymes are phosphatases and other enzymes catalyzing transfer of phosphoryl groups. A special role for magnesium is participation in photosynthesis as a component of chlorophyll.

One of the most toxic metals is beryllium. It appears that Be^{2+} competes with Mg^{2+} at many enzyme sites, among them those of phosphoglucomutase and of phosphatases.

[a] D. Veloso, R. W. Guynn, M. Oskarrson, and R. L. Veech, *JBC* **248**, 4811–4819 (1973).

[b] D. L. Purich and H. J. Fromm, *Curr. Top. Cell. Regul.* **6**, 131–167 (1972).

[c] H. F. Bunn, B. J. Ransil, and A. Chao, *JBC* **246**, 5273–5279 (1971).

[d] H. Rubin, *PNAS* **72**, 3551–3555 (1975).

[e] J. Meli and F. L. Bygrave, *BJ* **128**, 415–420 (1972).

Paramagnetic ions are well known to affect the magnetic relaxation of nearby nuclei (Box 5-A). Thus, small amounts of Mn^{2+} in a sample lead to broadening of lines in ordinary ^1H, ^{13}C, and ^{31}P nmr spectra.

Useful information about enzymes can sometimes be obtained by looking at the nmr signal of protons in the solvent water. The relaxation time of solvent protons is usually greater than 1 s. However, in the ion $Mn(H_2O)_6^{2+}$ the protons of the coordinated water molecules have greatly increased relaxation rates, both T_1 and T_2 being of the order of 10^{-5} s. Since the coordinated water molecules usually exchange very rapidly with the bulk solvent, a small number of manganese ions can cause a significant increase in relaxation rate for all of the water protons. Measurable broadening of the proton band is observed and differences in T_1 and T_2 can be measured by appropriate methods. It is known that paramagnetic relaxation effects usually increase as the inverse sixth power of the internuclear distance. Knowing the Mn^{2+}–H distance to be 0.287 ± 0.005 nm for hydrated Mn^{2+}, it is possible to relate the effects on T_1 and T_2 to the number of H_2O molecules coordinated at any one time to a protein-bound metal ion and to their rate of exchange with the bulk solvent.

While relaxation effects on ^1H, ^{13}C, and ^{31}P are more difficult to observe, they are capable of giving an enormous amount of information about geometry at active sites. The basis is the inverse sixth power relationship between the efficiency of nuclear relaxation and the internuclear distance. While the theory is complex, under some conditions the equations governing the phenomena simplify to (Eq. 7-24). In this equation, r is the internuclear distance, C is a combination of physical constants, and T_{1M} is the longitudinal relaxation time for the paramagnetically induced relaxation. The complex function $f(\tau_c)$ depends upon the correlation time τ_c, the resonance frequency of the nucleus being observed, and the frequency of precession of the electron spins at the paramagnetic centers. Various methods are available for estimating τ_c and in turn the distance r according to Eq. 7-24.

$$r = C[T_{1M}f(\tau_c)]^{1/6} \qquad (7-24)$$

An example of the results of such studies[84] is shown in Fig. 7-6. Both a bound Mn^{2+} ion and a nitroxide spin label were used with a 220 MHz Fourier transform nmr spectrometer. In addition, epr measurements gave the Mn^{2+}-nitroxide distance. The manganese ion is seen to be bonded to the α and β phosphates of ADP, as in Eq. 7-23.

Complexes of ADP and ATP with Cr^{3+} have been prepared and have been tested as substrates for various enzymes. These complexes are totally inactive with kinases, possibly because the chromium ion is coordinated in the wrong way or too tightly. Perhaps some movement of the ion or of the groups around it occurs during action of the enzyme and this is blocked in the chromium complexes.[67]

8. Adenylate Cyclase and Cyclic AMP

The formation of 3',5'-cyclic AMP (cAMP) consists of a displacement on the internal phosphorus

atom (P_α) of ATP by the 3'-hydroxyl group of the ribose ring (Eq. 7-25, step *a*).

$$(7-25)$$

The unique properties of adenylate cyclase[85] and the related guanylate cyclase[86] are discussed in Chapter 5, Section C,5 and Chapter 6, Section F,5. Kinetically cyclic AMP is an extremely stable molecule, but thermodynamically it is unstable with respect to hydrolysis. A specific phosphodiesterase cleaves cAMP to 5'-adenylic acid (AMP, Eq. 7-25, step *b*). The two consecutive reactions of Eq. 7-25 provide for the production and destruction of cAMP. Note the similarity of this sequence to the two steps of the ribonuclease reaction (Eq. 7-19).

9. Mutases

Another group of phosphotransferases shift phosphoryl groups from one position to another

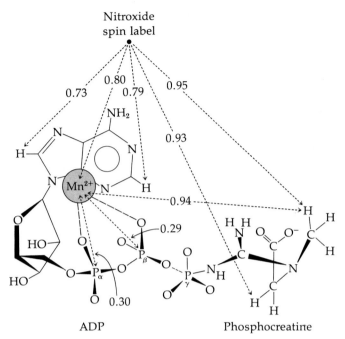

Fig. 7-6 Topography of substrates at the active site of creatine kinase based on distances measured by magnetic resonance methods. After McLaughlin *et al.*[84] The structure is based on that of the enzyme–Mn^{2+}–ADP–phosphocreatine complex.

within a substrate. An example is **phosphoglucomutase**, an enzyme that operates through formation of an intermediate **phosphoenzyme**. The phosphoryl group in the phosphoenzyme is attached to a reactive serine residue and can be transferred either to the 6 or the 1 position of a glucose phosphate (step *a* and reverse of step *b* in Eq. 7-26).[87] The two steps, Eq. 7-26, steps *a* and *b*, accomplish the reversible isomerization of glucose 1-phosphate and glucose 6-phosphate via **glucose 1,6-diphosphate**. Since the phosphoryl enzyme is relatively unstable and subject to hydrolysis to E and P_i, a separate reaction (catalyzed by a kinase, Eq. 7-26, step *c*) is available for generation of glu-

(7-26)

cose 1,6-diphosphate, which rephosphorylates any free enzyme formed by hydrolysis of the phosphoryl enzyme.

Phosphoglycerate mutase functions by a similar mechanism, but the phosphoryl group of the phosphoenzyme is attached to the imidazole group.[87a]

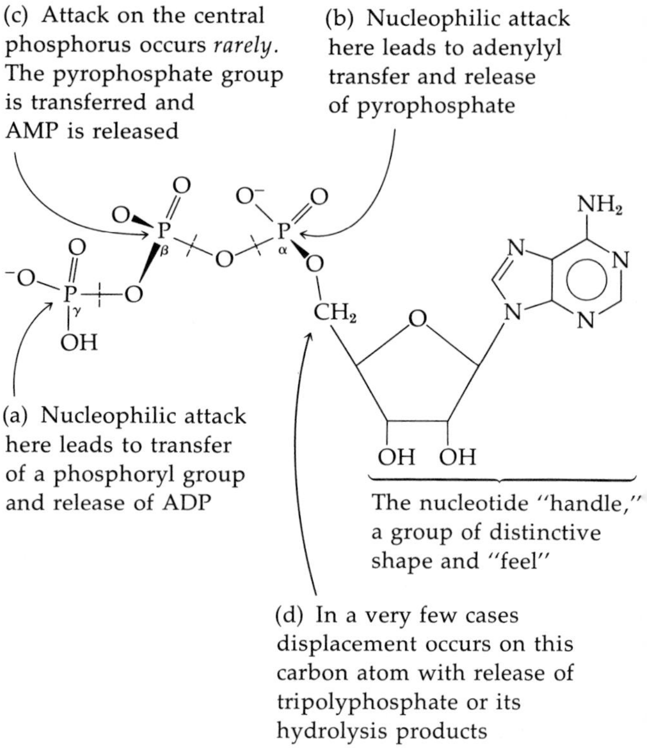

F. MULTIPLE DISPLACEMENT REACTIONS AND THE COUPLING OF ATP CLEAVAGE TO ENDERGONIC PROCESSES

A combination of two types of displacement reactions, *one on phosphorus followed by one on carbon*, is required in many enzymatic reactions including most of those by which the cleavage of ATP is coupled to biosynthesis. However, to harness the group transfer potential of ATP to drive an endergonic metabolic process there must be a mechanism of **coupling.** Otherwise hydrolysis of ATP within a cell would simply generate heat. *An essential part of the coupling mechanism usually consists of a nucleophilic displacement on phosphorus followed by displacement on carbon.*

1. Transfer of Phosphoryl, Pyrophosphoryl, and Adenylyl Groups from ATP

The first step in coupling ATP cleavage to any process is transfer of part of the ATP molecule itself to a nucleophile Y, usually by displacement on one of the three phosphorus atoms (Fig. 7-7). The nucleophilic attack may be on the terminal phosphorus (P_γ) with displacement of ADP or on the internal phosphorus (P_α) with displacement of inorganic pyrophosphate. In the first case

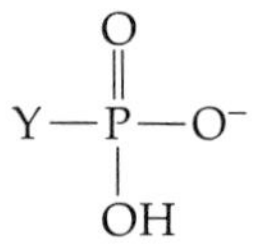

$$Y-\overset{\overset{\displaystyle O}{\|}}{\underset{\underset{\displaystyle OH}{|}}{P}}-O^-$$

is formed, in the latter, **Y-adenylyl** (often shortened to Y-adenyl). More rarely displacement occurs on

(c) Attack on the central phosphorus occurs *rarely.* The pyrophosphate group is transferred and AMP is released

(b) Nucleophilic attack here leads to adenylyl transfer and release of pyrophosphate

(a) Nucleophilic attack here leads to transfer of a phosphoryl group and release of ADP

The nucleotide "handle," a group of distinctive shape and "feel"

(d) In a very few cases displacement occurs on this carbon atom with release of tripolyphosphate or its hydrolysis products

Fig. 7-7 Four ways in which ATP is cleaved.

the central phosphorus (P_β) with transfer of a pyrophosphoryl group to the nucleophile. Still less frequent is a displacement on C-5' (Fig. 7-7) as described in Eq. 11-3. If the nucleophile Y in one of these displacement reactions is H_2O, the resulting hydrolysis tends to go to completion, i.e., the phosphoryl, adenylyl, and pyrophosphoryl groups of ATP all have high group transfer potentials (Table 3-5). If Y is an —OH group in an ordinary alcohol the transfer reaction also tends to go to completion because the group transfer potential of a simple phosphate ester is relatively low. Consequently, phosphorylation by ATP is often used as a means of introducing an essentially irreversible step in a metabolic pathway.

2. Acyl Phosphates

Transfer of a phosphoryl or adenylyl group from ATP to the oxygen atom of a carboxylate group yields an **acyl phosphate,** a type of metabolic intermediate of special significance. In acyl phosphates

$$R-\overset{O}{\overset{\|}{C}}-O-\overset{O}{\underset{\underset{O^-}{|}}{\overset{\|}{P}}}-O-H(R)$$

Acyl phosphate

both the acyl group and the phosphoryl group have high group transfer potentials. Consequently, acyl phosphates can serve as the metabolic intermediates through which the group transfer potential of ATP is transferred into other molecules and is harnessed to do chemical work.

A typical example of the coupling of hydrolysis of ATP to a synthetic reaction through formation of an acyl phosphate is the synthesis of acetyl coenzyme A shown in Eq. 7-27 (see also Chapter

$$CoA-SH + CH_3-COO^- + H^+ \longrightarrow$$
Coenzyme A

$$CH_3-\overset{O}{\overset{\|}{C}}-S-CoA + H_2O \quad (7\text{-}27)$$
Acetyl coenzyme A

$$\Delta G'(\text{pH } 7) = +35.1 \text{ kJ mol}^{-1} (+8.4 \text{ kcal mol}^{-1})$$

8, Section B). Because the acetyl group in the product also has a high group transfer potential, $\Delta G'$ for this reaction is highly positive and the formation of acetyl-CoA will not occur spontaneously.

However, the sum of $\Delta G'$ for reaction 7-27 plus that for the hydrolysis of ATP (Eq. 7-28) is nearly zero ($+0.6 \text{ kJ mol}^{-1}$).

$$ATP^{4-} + H_2O \longrightarrow ADP^{3-} + HPO_4^{2-} + H^+ \quad (7\text{-}28)$$
$$\Delta G' = -34.5 \text{ kJ mol}^{-1}(-8.3 \text{ kcal mol}^{-1})$$

Coupling of the two reactions can be accomplished by first letting an oxygen atom of the nucleophilic carboxylate group attack P_γ of ATP to form acetyl phosphate (Eq. 7-29, step a). In the second step (Eq. 7-29, step b) the sulfur atom of the —SH group of CoA attacks the carbon atom of the acetyl phosphate with displacement of the good leaving group P_i. While $\Delta G'$ for Eq. 7-29, step a is moderately positive (meaning that a relatively low concentration of acetyl phosphate will accumulate unless the [ATP]/[ADP] ratio is high),

the equilibrium in Eq. 7-29, step b favors the products.

The two reactions of Eq. 7-29 are catalyzed by **acetate kinase** and an **S-acetyltransferase,** respectively. The sequence represents an essential first step in bacterial utilization of acetate for growth. It is also used in a few bacteria in reverse as a way of *generating* ATP in fermentation reactions. On the other hand, most eukaryotic cells make acetyl-CoA from acetate by coupling the synthesis to cleavage of ATP to AMP and P_i. A single enzyme **acetate thiokinase** catalyzes both steps in the reaction (Eq. 7-30). The sequence is entirely parallel to

$$CH_3-\overset{O}{\overset{\|}{C}}-O^- \xrightarrow[\Delta G'(\text{pH } 7) = +13.2 \text{ kJ mol}^{-1}]{ATP^{4-} \quad ADP^{3-}} H_3C-\overset{O}{\overset{\|}{C}}-\overset{O}{\overset{\|}{P}}-O^-$$

CoA—SH   Acetyl phosphate

$$\Delta G'(\text{pH } 7) = -12.6 \text{ kJ mol}^{-1} \downarrow^b P_i$$

$$CoA-S-\overset{O}{\overset{\|}{C}}-CH_3 \quad (7\text{-}29)$$
Acetyl-CoA

$$CH_3-\overset{O}{\overset{}{C}}\overset{}{\underset{O^{*-}}{}}$$

$$\text{adenosine}-O-\overset{O}{\overset{\|}{P}}-O-\overset{O}{\overset{\|}{P}}-O-\overset{O}{\overset{\|}{P}}-O^-$$

$$a \searrow HP_2O_7{}^{3-}$$

$$\text{adenosine}-O-\overset{O}{\overset{\|}{P}}-O^*-\overset{CH_3}{\underset{O}{\overset{\|}{C}}} \dashleftarrow \overset{CH_3}{\underset{H}{}}S-CoA$$

Acyl adenylate

$$b \searrow$$

$$\text{adenosine}-O-\overset{O}{\overset{\|}{P}}-O^* \qquad H_3C-\overset{O}{\overset{\|}{C}}-S-CoA \quad (7\text{-}30)$$

AMP   Acetyl-CoA

that of Eq. 7-29, but the initial displacement is on P_α of ATP to form **acetyl adenylate**. This intermediate remains tightly bound to the enzyme until the second step in the sequence takes place. It has been shown that ^{18}O present in acetate (designated by the asterisks in Eq. 7-30) appears in the phosphoryl group of AMP as expected for the indicated mechanism.

3. General Mechanism of Formation of Thioesters, Esters, and Amides

The sequences of Eqs. 7-29 and 7-30 are general ones used by cells for linking carboxylic acids to —OH, —SH, and —NH_2 groups. For example, that of Eq. 7-30 is used in the formation of **aminoacyl-tRNA** molecules required for the synthesis of proteins. For reference purposes the reactions are summarized in Table 7-2. The sequences are designated S1A, S1B, or S1C, depending on whether a thioester, oxygen ester or amide is formed. The symbols α or γ indicate whether ATP is cleaved at P_α or P_γ. For example, formation of acetyl-CoA in eukaryotes is via sequence S1A (α). The sequence is understood to include the hydrolysis of inorganic pyrophosphate (PP_i) to P_i, an aspect of the coupling of ATP cleavage to biosynthesis that is discussed further in Chapter 11, Section B,2.

While it is easy to generalize, each specific enzyme has its peculiarities. **Succinyl-CoA synthetase** (succinate thiokinase) of *E. coli* probably acts by sequence S1A (γ) with succinyl phosphate

$$^-OOC-CH_2-CH_2-C\overset{\displaystyle O}{\underset{}{\diagdown}}O-\textcircled{P}$$

as an intermediate. However, the first step may be the formation of a phosphoenzyme in which the phosphoryl group is attached to N^3 of a histidine side chain on the α subunit of the $\alpha_2\beta_2$ tetramer.[88,88a,b]

Synthesis of glutamine by glutamine synthetase probably occurs via sequence S1C (γ). However, it has not been possible to prove directly the existence of the assumed intermediate γ-glutamyl

phosphate. The partial isotope exchange reactions which might be expected are not observed.[89] Apparently, the intermediate acyl phosphate has only a transient existence and all three reactants must be bound to the enzyme concurrently before the active site becomes functional. Evidence for the acyl phosphate [90,91] includes isolation of the internal amide of glutamic acid **5-oxoproline** (pyrrolidonecarboxylic acid, Eq. 7-31) and reduction by sodium borohydride to an alcohol (Eq. 7-32).

$$(7-31)$$

Because of the difficulty in demonstrating directly the acyl phosphate intermediate, it has been suggested that glutamine synthetase and some other enzymes catalyzing multiple displacements may promote a "completely concerted" reaction. Both displacement steps would occur simultaneously via a single "key transition state" as shown in Eq. 7-32 for glutamine synthetase.

$$(7-32)$$

TABLE 7-2
Synthesis of Thioesters, Esters, and Amides[a]

At P_α

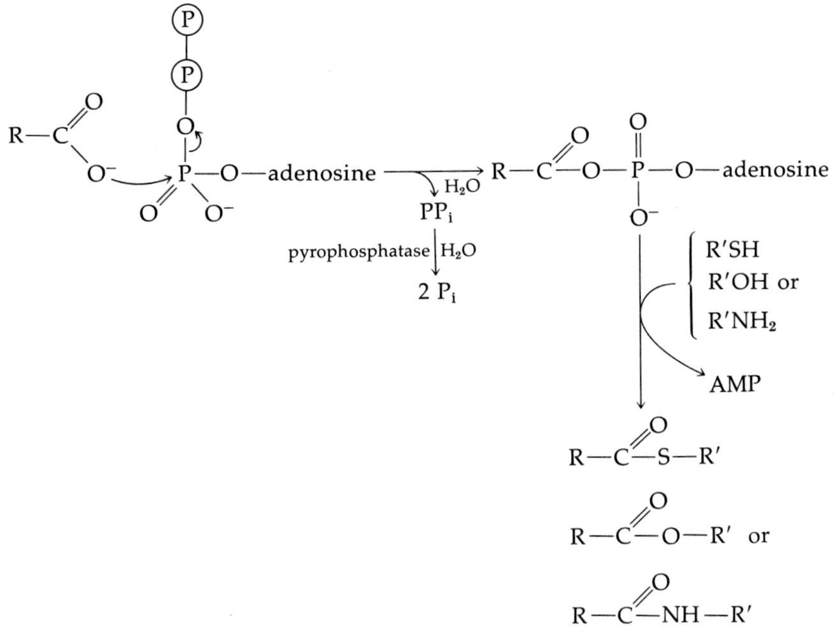

At P_γ

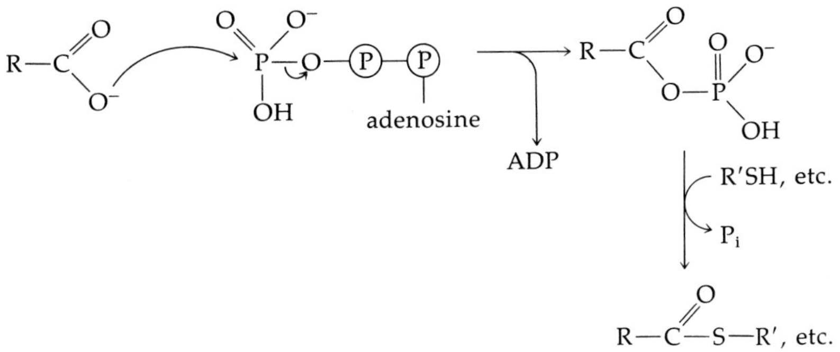

Designation of Reaction Sequences Used in This Book

S1A (α or γ), formation of thioesters $R-\overset{O}{\overset{\|}{C}}-SR'$,   e.g., an acyl-CoA

S1B (α), formation of esters $R-\overset{O}{\overset{\|}{C}}-OR'$,   e.g., aminoacyl-tRNA

S1C (γ), formation of amides (direct) $R-\overset{O}{\overset{\|}{C}}-NHR$, e.g., glutamine

[a] Consecutive displacements on P, then C are used to link small molecules in a process coupled to cleavage of ATP or a related compound. Cleavage of ATP may be at P_α or P_γ.

However, no convincing evidence for such a mechanism of any chemical reaction has ever been presented. It seems much more likely that enzymatic reactions occur in a series of discrete steps.

4. Coenzyme A Transferases

The following problem in energy transfer arises frequently: A thioester, for example, succinyl-CoA, is available to a cell and the energy available in its unstable linkage is needed for synthesis of a different thioester. It would be possible for a cell to first form ATP or GTP, using the thiokinase reaction in reverse; then the ATP or GTP formed could be used to make the new linkage, again through the action of a thiokinase. However, special enzymes, the **CoA transferases,** function more directly (Eq. 7-33). The mechanism is not obvious.

$$Succinyl\text{-}CoA + R\text{---}COOH \longrightarrow$$

$$R\text{---}\overset{\overset{\displaystyle O}{\|}}{C}\text{---}CoA + succinite \qquad (7\text{-}33)$$

$$\Delta G \sim 0$$

How can the CoA be transferred from one acyl group to another while still retaining the high group transfer potential of the acyl group?

The following experiments shed some light. Kinetic studies of **succinyl-CoA–acetoacetate CoA transferase** indicate a ping-pong mechanism (Chapter 6). Thus, the enzyme alternates between two distinct forms, one of which has been shown to contain bound CoA.[92] The E-CoA intermediate formed from enzyme plus acetoacetyl-CoA was reduced with ^{3}H-containing sodium borohydride and the protein was completely hydrolyzed with HCl. Tritium-containing α-amino-δ-hydroxy-

$$HOH_2C\text{---}CH_2\text{---}CH_2\text{---}\overset{\overset{\displaystyle NH_2}{|}}{HC}\text{---}COOH$$

α-Amino-δ-hydroxyvaleric acid

valeric acid was isolated. Since thioesters (as well as oxygen esters) are cleaved (in a two-step

process) to alcohols by reduction with borohydride, it was concluded that the intermediate E-CoA is a thioester of a glutamic acid side chain.[93] Studies of exchange reactions using ^{18}O-labeled succinate show that [18]O from succinate enters both the E-CoA intermediate and the carboxyl group of acetoacetate.[94]

At least two mechanisms can be visualized. The first involves formation of a transient anhydride and is mechanistically similar to reactions discussed in preceding sections.[94a] The student should write out the step-by-step details of this mechanism. Does it explain the [18]O exchange data? The second possibility[92] is a **4-center reaction**

(Eq. 7-34). While mechanisms of this type have not been proved for enzymatic reactions, they are possible in principle.

5. Phosphorylation-Induced Conformational Changes

A possible result of transfer of a phosphoryl group onto a functional group of a protein is the induction of a conformational change in the protein. Indeed, there is strong suggestive evidence that such changes occur during the action of ATP-dependent ion pumps (Chapter 5, Section B,2,c) and during the action of muscle (Box 10-F). Conformational changes may also result from phosphorylation of regulatory sites on proteins. Phosphorylation of an imidazole group hydrogen bonded to a C=O group of an amide backbone group in a protein could be a way of causing a tautomerization similar to that shown in Eq. 6-84. This might facilitate a conformational change or might convert the protein into a "high-energy" form able to spontaneously alter its shape as in muscular contraction.

G. TRANSFER OF SULFATE

While esters of sulfuric acid (organic sulfates) do not play as central a role in metabolism as do phosphate esters, they occur widely. Both oxygen esters ($R—O—SO_3^-$, often referred to as O-sulfates*) and derivatives of sulfamic acid ($R—NH—SO_3^-$, N-sulfates*) are found, the latter occurring in mucopolysaccharides such as heparin. Sulfate esters of mucopolysaccharides and of steroids are ubiquitous. Choline sulfate and ascorbic acid 2-sulfate are found in cells. Sulfate esters of phenols and many other organic sulfates are found as urinary excretion products.

Sulfotransferases, including sulfokinases, transfer sulfuryl groups to —O and —N of suitable acceptors (reaction type 1D, Table 7-1). Often transfer is from the special "active sulfate," **3'-phosphoadenosine 5'-phosphosulfate** (Eq. 11-4). However, transfer from ascorbic acid 2-sulfate may also be important.[95,96] **Sulfatases** catalyze hydrolysis of sulfate esters. The importance of such enzymes is demonstrated by the genetic disease **mucopolysaccharidosis III** (Sanfilippo syndrome), in which excessive secretion of heparan sulfate is observed, apparently because of the absence of a specific sulfatase.[97] The disease causes very severe mental retardation.

A large literature on sulfate-transferring enzymes is available,[98] but few studies of pure enzymes have been made.

H. ADDITION AND ELIMINATION REACTIONS

In addition to direct nucleophilic displacement, the most common mechanistic processes in enzymatic catalysis are addition to double bonds and its reverse, elimination.

---

* Although these names are widely used, they are not strictly correct; it is not a sulfate but a *sulfuryl* radical which is attached to O and N.

1. Addition to Polarized Double Bonds (Reaction Type 2A, Table 7-1)

Alcohols, thiols, and amines add readily to the electrophilic carbon of the carbonyl group to form the **hemiacetals, hemiketals, hemimercaptals,** and **carbinolamines.** The formation of ring structures of sugars (Eq. 6-75) is a well-known example. Water can also add to carbonyl groups. Thus, acetaldehyde in aqueous solution exists as a mixture of about 50% free aldehyde and 50% hydrate in rapidly reversible equilibrium[99] (Eq. 7-35) and formaldehyde is over 99.9% hydrated.[100]

$$CH_3—C{\overset{O}{\underset{H}{}}} + H_2O \rightleftharpoons$$

$$CH_3—C{\overset{OH}{\underset{H}{}}}—OH \qquad (7\text{-}35)$$

Addition reactions often occur as parts of more complex reactions catalyzed by enzymes. For example, a thiol group of glyceraldehyde-3-phosphate dehydrogenase reacts with the aldehyde substrate to form a hemimercaptal, which is subsequently oxidized to a thioester (Eq. 7-36) (see also Fig. 8-13).

$$E—S—H + R—C{\overset{O}{\underset{H}{}}} \longrightarrow E—S—\overset{H}{\underset{R}{C}}—OH \overset{ox}{\longrightarrow}$$

$$E—S—\underset{R}{C}=O \qquad (7\text{-}36)$$

2. Carbonic Anhydrase

One of the simplest addition reactions is the hydration of CO_2 to bicarbonate ion (Eq. 7-37).

$$O=C=O + H_2O \longrightarrow H^+ + HO-C{\overset{O}{\underset{O^-}{\Big\backslash}}} \qquad (7-37)$$

Without catalysis the reaction may require several seconds,[101] the apparent first order rate constant being ~ 0.03 s^{-1} at 25°C. Cells must often hasten the process. The specific catalyst **carbonic anhydrase** is widespread in its distribution and is especially active in tissues (e.g., red blood cells) involved in respiration. One liter of mammalian blood contains 1–2 g of this enzyme, a monomeric protein of MW $\sim 30,000$ containing ~ 260 amino acids and one tightly bound ion of Zn^{2+}. The three-dimensional structures of the two major isoenzymes from human blood have been determined by X-ray diffraction.[102,103] Both molecules have an ellipsoidal shape of dimensions $\sim 4.1 \times 4.1 \times 4.7$ nm. The zinc atom in each molecule lies in a deep pocket ~ 1.2 nm from the surface and is surrounded by three histidine side chains and one H$_2$O or OH$^-$ ion. The four ligands form a distorted tetrahedron (Fig. 7-8).

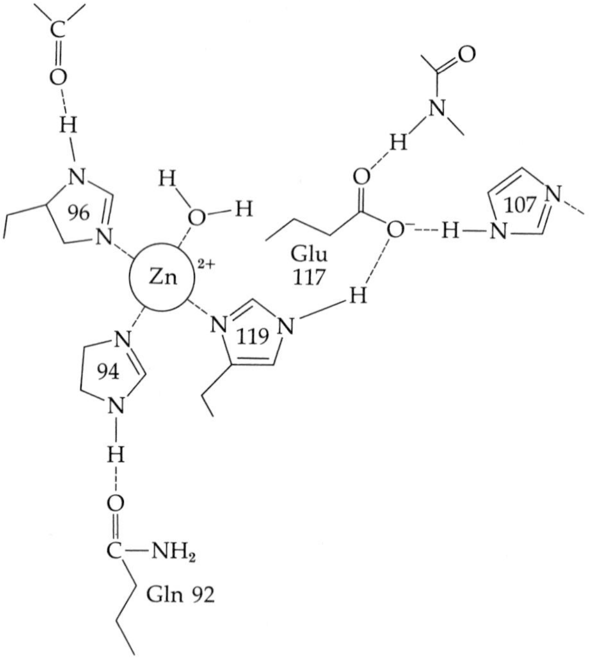

Fig. 7-8 The chelated zinc ion at the active site of human carbonic anhydrase B. From Kannan et al.[103]

One of the coordinating imidazole groups (No. 119) is hydrogen bonded to a carboxylate group of a Glu side chain and the carboxylate group is also bound into a more extended H-bonded network, part of which is indicated in Fig. 7-8. This feature of the structure, reminiscent of the charge-relay system of chymotrypsin, may enhance the affinity of imidazole 119 for the metal and may also function during catalysis. The other imidazole groups also form H bonds to protein groups.

Carbonic anhydrase is one of the most active enzymes known, the turnover number at 25°C for hydration of CO$_2$ being $\sim 10^6$ s^{-1}. The same enzyme catalyzes hydration of acetaldehyde (Eq. 7-35) but at a 1000-fold slower rate. A pK_a of ~ 7 regulates the activity. The most popular theory assumes that the zinc ion binds a water molecule and that in the Zn—OH$_2$ complex a proton is lost (with a pK_a of ~ 7—unusually low perhaps because of the hydrophobic surroundings[104,104a] to give Zn$^+$—OH. The latter is in effect a stabilized hydroxyl ion existing at a pH where OH$^-$ is normally not present in quantity. It is this hydroxyl ion that adds to the CO$_2$ or to the aldehyde substrate. Thus, the Zn^{2+} in this enzyme functions in generation of the attacking base rather than in polarization of a carbonyl group, the role attributed to Zn^{2+} in carboxypeptidase (Chapter 7, Section D,4).

Related to the reaction catalyzed by carbonic anhydrase is the addition of amino groups of hemoglobin to CO$_2$ to form carbamino groups (—NH—COO$^-$) as discussed in Chapter 4, Section E,6.

3. Imines

Many enzymatic reactions depend upon formation of **imines (Schiff bases)**. The two-step reaction consists of *addition* of an amino group to a carbonyl group to form a carbinolamine followed by *elimination* of water (Eq. 7-38). Many **aldolases** (Section J,2) have lysine side chains at their active centers. The amino groups of these side chains form Schiff bases with the ketone substrates prior

BOX 7-F

ZINC

The average human ingests 10–15 mg of zinc a day.[a] Although it is poorly absorbed, tissue concentrations of zinc are relatively high and the metal plays an essential role in a multitude of enzymes. The total zinc content of a 70 kg person is 1.4–2.3 g. A typical tissue concentration of Zn^{2+} is 0.3–0.5 mM; an unusually high content of ~15 mM is found in the prostate gland.

Zinc ion is much more tightly bound to most organic ligands than is Mg^{2+} (Table 4-2). It has a filled $3d$ shell and tends to form four covalent bonds with tetrahedral symmetry, often with nitrogen- or sulfur-containing ligands. Unlike Mg^{2+}, which interacts rapidly and reversibly with enzymes, Zn^{2+} tends to be tightly bound within **metalloenzymes.** The three-dimensional structures of several of these enzymes are known. A common feature is the surrounding of the Zn^{2+} at the active center by three imidazole groups, the fourth coordination position being free for interaction with substrate. It is also of considerable interest that the second nitrogen of the imidazole ring in many instances is hydrogen bonded to a main chain carbonyl group of the peptide.[b] This feature is also shared by the iron atoms of heme proteins (Fig. 10-1).

The chemical function of Zn^{2+} in enzymes is doubtless that of a Lewis acid providing a concentrated center of positive charge at a nucleophilic site on the substrate.[c] This role is discussed in Section D,4 for carboxypeptidase (Fig. 7-3). Zinc ions are also required for thermolysin (Section D,4), dipeptidases, alkaline phosphatase (Section E,1), RNA polymerases, DNA polymerases,[d] carbonic anhydrase (Fig. 7-8), Class II aldolases (Section J,2c), some alcohol dehydrogenases (Chapter 8, Section H,2), and superoxide dismutases (Box 10-H). Zinc binds to insulin hexamers (Fig. 4-13C).

An unusual protein **metallothionein,** present in all animal tissues, binds large amounts of many metal ions and especially Zn^{2+}. The small protein of MW ~6600 contains 33% cysteine and binds as many as six metal ions per molecule.[e] The function of metallothionein is unknown. Perhaps it is a metal ion buffer; it could serve to remove unwanted metals; in its uncomplexed form it could serve as a redox buffer like glutathione (Box 7-G).

In carnivores, the **tapetum,** the reflecting layer behind the retina of the eye of many animals, contains crystals of the Zn^{2+}-cysteine complex. Since zinc ions have no color their presence has often been overlooked. Zinc ions will doubtless be found in many more places within cells.

Zinc ions in enzymes can often be replaced by Mn^{2+}, Co^{2+}, and other ions with substantial retention of catalytic activity.[c,e,f] From a nutritional viewpoint, Cu^{2+} competes with zinc ion as does the very toxic Cd^{2+}. The latter accumulates in the cortex of the kidney. Dietary cadmium in concentrations less than those found in human kidneys shortens the lives of rats and mice.

[a] B. L. O'Dell and B. J. Campbell, *Comp. Biochem.* **21**, 179–216 (1971).

[b] A. Liljas and M. G. Rossman, *Annu. Rev. Biochem.* **43**, 475–507 (1974).

[c] A. S. Mildvan, *Annu. Rev. Biochem.* **43**, 357–399 (1974).

[d] B. J. Polesz, G. Seal, and L. A. Loeb, *PNAS* **71**, 4892–4896 (1974).

[e] J. H. R. Kägi, S. R. Himmelhoch, P. D. Whanger, J. L. Bethune, and B. L. Vallee, *JBC* **249**, 3537–3542 (1974).

[f] W. L. Bigbee and F. W. Dahlquist, *Biochemistry* **13**, 3542–3549 (1974).

$$R-\overset{\displaystyle O}{\underset{\displaystyle H}{C}} + H_2N-R' \longrightarrow R-\overset{\displaystyle OH}{\underset{\displaystyle H}{\underset{|}{C}}}-\overset{}{\underset{\displaystyle H}{N}}-R'$$

(carbinolamine)

$$\Big\downarrow {-H_2O}$$

$$R-\underset{\displaystyle H}{C}=N-R' \quad (7\text{-}38)$$

(imine or Schiff Base)

to the principal reactions of breaking or forming C—C bonds. The initial reaction of the coenzyme **pyridoxal phosphate** with amino acid substrates is the formation of Schiff bases (Chapter 8, Section E). Indeed, the groups

$$\overset{}{\underset{}{C}}=O \quad \text{and} \quad \overset{\displaystyle H}{\underset{\displaystyle H}{N}}-$$

are inherently complementary and their interaction through imine formation is an extremely common biochemical process.

Schiff bases form rapidly (often in a fraction of a second) but not instantaneously, and one or both steps of Eq. 7-38 may require catalysis to achieve enzymatic velocities.[105] The reaction is usually completely reversible and formation constants are often low so that a carbonyl compound present in small amounts may not react extensively with an amine unless the two are brought together on an enzyme surface.

Most Schiff bases readily undergo reduction by sodium borohydride $NaBH_4$ to form secondary amines in which the original aldehydes are bound covalently to the original amino groups (Eq. 7-39).

$$\diagdown C{=}N\diagup \xrightarrow{\text{NaBH}_4^*} -\underset{|}{\overset{\text{H}^*}{C}}-N\diagup_{\text{H}} \qquad (7\text{-}39)$$

Secondary amine

This provides a widely used method of deducing sites of Schiff base formation in proteins and of introducing isotopic labels. For example, an isotopically substituted aldehyde or amine can be employed, or a radioactive label can be introduced, by using ^3H-containing sodium borohydride in the reduction (Eq. 7-39). Subsequent hydrolysis of the protein (either acid-catalyzed or enzymatic) can establish to what amino acid side chain the substrate has been bound, and partial hydrolysis can lead to identification of the specific site in the peptide chain.

4. Stereochemistry and the Naming of the Faces of Trigonal Carbon Atoms

Adducts formed by addition of nucleophiles to carbonyl groups are often asymmetric. Thus, addition of an amino group to the "front" side of the carbonyl group in Eq. 7-40 creates a carbinolamine of the *S* configuration.

$$\qquad (7\text{-}40)$$

Addition from "front"   *S* configuration
(*re*-face)

Addition of the amino group from behind the plane of the paper would lead to a carbinolamine of the *R* configuration. To describe these facts without reference to the plane of the paper Hanson proposed a systematic way of labeling the faces of "trigonal atoms" such as the carbon of the C=O group.[106,107] The trigonal atom is viewed from one side as in the drawing below. The three groups surrounding the carbon atom are given priorities a, b, and c, as in the *RS* system (Chapter

2, Section A,5). If the sequence a, b, c of priorities of the groups is clockwise, the face toward the reader is *re* (*rectus*), if counterclockwise, *si* (*sinister*). Priorities can usually be assigned on the basis of the atomic numbers of the first atoms in the three groups (as in the example shown) but, if necessary, "phantom atoms" attached to C and O must be added as described in Chapter 2. For a trigonal carbon the phantom atom on the carbon is ignored completely, whereas that on oxygen is used to establish the priorities of

$$={O}, \quad ={C}\diagup_{\diagdown}, \quad ={N}{-}, \quad \text{etc.}$$

5. Addition to a Double Bond Conjugated with a Carbonyl Group (Reaction Type 2B)

Conjugation with a double bond transmits the polarization of the carbonyl group to a position lo-

$$R-\overset{|}{C}=\overset{|}{C}-\overset{|}{C}=O \longleftrightarrow R-\overset{|}{C}=\overset{|}{C}-\overset{|}{\underset{+}{C}}-O^- \longleftrightarrow$$

$$R-\overset{|}{\underset{+}{C}}-\overset{|}{C}=\overset{|}{C}-O^-$$

cated two carbon atoms further along the chain.

Because of this effect bases can add to a carbon-carbon double bond in a position β to the carbonyl group as in Eq. 7-41, step a. The product of addition of OH^- is the anion of an **enol** in which the negative charge is distributed by resonance between the oxygen of the carbonyl group and the carbon adjacent to the carbonyl. A stable product results if a proton adds to the latter position (Eq. 7-41, step b):

$$R-\overset{|}{\underset{-OH}{C}}=\overset{|}{C}-\overset{}{C}=\overset{|}{O} \overset{a}{\longrightarrow} R-\overset{|}{\underset{HO}{C}}-\overset{|}{C}=\overset{|}{C}-O^-$$

$$\downarrow$$

$$R-\overset{|}{\underset{OH}{C}}-\overset{|}{C}-\overset{|}{\underset{\ominus}{C}}=O$$

Enolate
anion

$$b \Big\downarrow H^+$$

$$R-\overset{|}{\underset{OH}{C}}-\overset{|}{\underset{H}{C}}-\overset{|}{C}=O \qquad (7\text{-}41)$$

The reverse of an addition reaction of this type is also known as an **elimination reaction.**

Enzymatic reactions involving addition to a $C=C$ bond adjacent to a carbonyl group (or in which elimination occurs α, β to a carbonyl) are numerous. The fact that the nucleophilic group always adds at the β position suggests that the mechanism portrayed by Eq. 7-41 is an important one. It is noteworthy that *frequently in a metabolic sequence a carbonyl group is deliberately introduced to facilitate elimination or addition at adjacent carbon atoms.* The carbonyl may be formed by oxidation of a hydroxyl group or it may be provided by a

thioester formed with CoA or with an **acyl carrier protein** (Chapter 8, Section B,3).

The addition of water to trans-α,β-unsaturated CoA derivatives according to Eq. 7-42 is catalyzed

$$H_2O + \quad \underset{R}{\overset{H}{\diagdown}}C=C\underset{H}{\overset{}{\diagup}}\overset{O}{\overset{\|}{C}}-S-CoA \xrightarrow[\text{hydratase}]{\text{enoyl}}$$

$$HO\diagdown \overset{H}{\underset{R}{\overset{}{C}}}\overset{}{\underset{C}{\overset{}{C}}}\overset{O}{\overset{\|}{C}}-S-CoA \qquad (7\text{-}42)$$

$$\underset{H_2}{}$$

L-β-Hydroxyacyl-CoA

by **enoyl hydratase** (crotonase) from mitochondria and represents a step in the oxidation of fatty acids.[108,109] A similar enzyme in the bacterium *Rhodospirillum rubrum* dehydrates L-β-hydroxybutyryl-CoA (reverse of Eq. 7-42) to the trans unsaturated compound **crotonyl-CoA:** A second enoyl hydratase with the opposite stereospecificity catalyzes hydration of crotonyl-CoA to D-β-hydroxybutyryl-CoA.[110] Together they serve to convert L-β-hydroxybutyryl-CoA synthesized from acetate units into D-β-hydroxybutyryl-CoA, which may be further converted to the storage polymer poly-D-β-hydroxybutyrate (Chapter 2, Section E,4).

6. Addition to Double Bonds Adjacent to Carboxylate Groups

Often in biochemical reactions addition is to $C=C$ bonds that are not conjugated with a true carbonyl group but with the far poorer electron acceptor $-COO^-$. While held on an enzyme a carboxylate group may be protonated, making it a better electron acceptor. Nevertheless, we must ask whether the mechanism of Eq. 7-41 holds for these enzymes. Indeed, we are forced by various experimental data to consider a quite different mechanism, one that has been established for the simple, nonenzymatic hydration of alkenes. An example is the hydration of ethylene by hot water with dilute sulfuric acid as a catalyst (Eq. 7-43), an

$$CH_2{=}CH_2 \xrightarrow{\;H^+\;} \underset{\substack{Carbonium \\ ion}}{CH_3{-}CH_2{}^+} \xrightarrow{\;H_2O \quad H^+\;}$$

$$CH_3{-}CH_2OH \qquad (7\text{-}43)$$

industrial method of preparation of ethanol. The π electrons of the double bond form the point of attack by a proton, and the resulting carbonium ion readily abstracts a hydroxyl ion from water. On the other hand, direct addition of OH^- to form a carbanion is not favored, for there is no adjacent carbonyl group to stabilize the negative charge.

a. Fumarase[111]

The most studied enzyme of this group is probably fumarase (fumarate hydratase; Chapter 6, Sections D,2 and E,5), a tetramer of MW $\sim 200,000$ with a turnover number of $\sim 2 \times 10^3$ s^{-1}. The product is L-malate (R-malate). If the reaction is carried out in ^2H_2O an atom of ^2H is incorporated in the pro-R position, i.e., the proton is added from the re-face of the trigonal carbon (Eq. 7-44).

$$(7\text{-}44)$$

To obtain L-malate the hydroxyl must be added from the opposite side of the double bond. Such anti (trans) addition is much more common in both nonenzymatic and enzymatic reactions than is addition of both H and OH (or $-Y$) from the same side (syn, cis, or adjacent addition). With two exceptions, in which unusual structures are involved, all enzymatic addition and elimination reactions studied to date are anti with the proton entering from the re-face. This suggested to Rose the conservation throughout evolution of a single basically similar mechanism.[112]

The pH dependence of the action of fumarase suggests participation of both an acidic and a basic group at the active site (Chapter 6, Section E,5). However, this information does not help in deciding between anion and carbonium ion mechanisms. That the cleavage of the C—H bond is not rate limiting is suggested by the observation that malate containing ^2H in the pro-R position is dehydrated at the same rate as ordinary malate. If the anion mechanism (Eq. 7-41) is correct, this might mean that the ^2H from the pro-R position of specifically labeled malate is removed in a fast step and that loss of OH^- is slower. If so, the ^2H should be "washed out" of L-malate faster than could happen by conversion to fumarate followed by rehydration to malate. In fact, the opposite was observed. The hydroxyl group was lost rapidly and the ^2H more slowly. This result suggests a carbonium ion mechanism (Eq. 7-45):

$$(7\text{-}45)$$

While no primary isotope rate effect has been observed, when the hydrogen at C-2 or the pro-S hydrogen at C-3 of malate was replaced by ^2H or ^3H distinct secondary isotope effects were seen. Thus, $k(^1H)/k(^2H) = 1.09$ for both the pro-S and C-3 hydrogen atoms.[113] These findings appear to support the carbonium ion mechanism and suggest that step b of Eq. 7-45 is rate limiting (see also Section C,4,b). The relative values of V_{max} for hydration of fumarate, fluorofumarate, and difluorofumarate (104, 410, and 86 μmol ml^{-1} min^{-1} mg^{-1}) also seem to support the carbonium ion mechanism (the student should try to interpret the foregoing data before consulting the original paper).[114]

Recently, it has become clear that protons removed from a substrate to a basic group in a protein need not exchange rapidly with solvent (see Section I,2). Indeed, it has been shown that the proton removed by fumarase from malate is held by the enzyme for relatively long periods of time. Its rate of exchange between malate and solvent is slower than the exchange of a bound fumarate ion on the enzyme surface with another substrate molecule from the medium.[115] Thus, it appears that the overall rate is determined by the speed of dissociation of products from the enzyme and that we still do not know whether removal of a proton precedes or follows loss of OH$^-$. There is still a third possibility: The proton and hydroxyl groups may be added *simultaneously* in a concerted reaction.[115]

b. Enolase

A key reaction in the metabolism of sugars is the dehydration of 2-phosphoglycerate (Eq. 7-46)

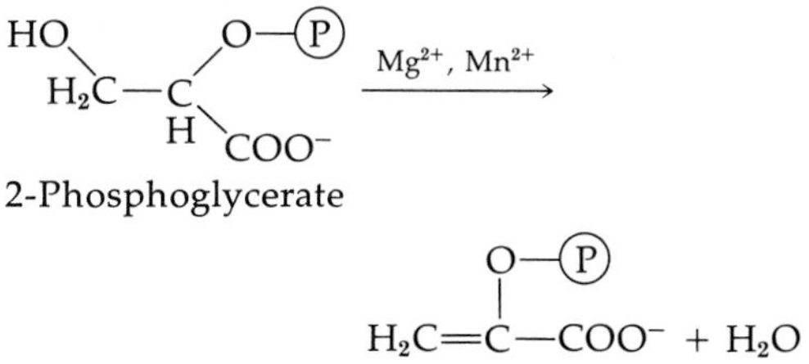

2-Phosphoglycerate

$$H_2C{=}C-COO^- + H_2O \qquad (7\text{-}46)$$

Phosphoenolpyruvate

to form **phosphoenolpyruvate,** the phosphoryl derivative of the enolic form of pyruvic acid. Isotopic exchange studies with this enzyme suggest that the proton is rapidly removed to form a carbanion intermediate whose breakdown is rate limiting.[116] The enzyme has a complex metal ion requirement,[117] usually met by Mg^{2+} and Mn^{2+}. From nmr studies of the relaxation of water pro-

tons, it is concluded that the Mn^{2+} ion coordinates two rapidly exchangeable water molecules in the free enzyme.[118] When substrate binds, one of these water molecules may be immobilized and undergo the addition reaction. The phosphate group of the substrate may serve as a general base catalyst in removing a proton from the water[118]:

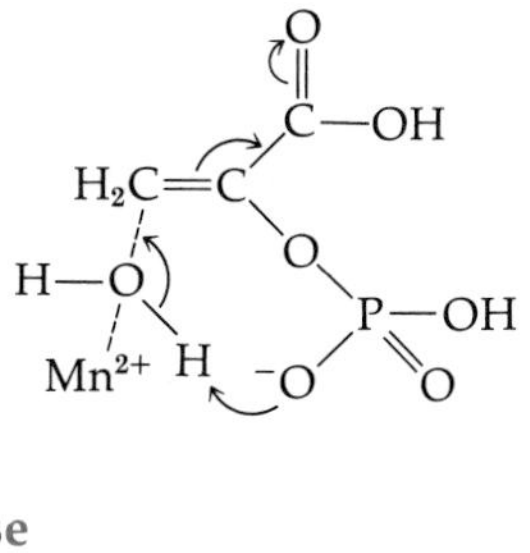

c. Aconitase

Another enzyme of this group, aconitase (aconitate hydratase), catalyzes two reactions: (a) the dehydration of citrate to form *cis*-aconitate and (b) the rehydration in a different way to form isocitrate (Eq. 7-47). Both reactions are completely stereospecific. In the first (Eq. 7-47, step a), the

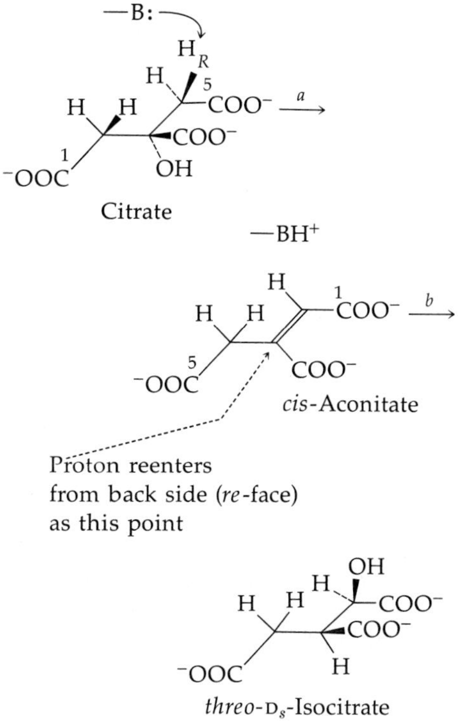

Citrate

cis-Aconitate

Proton reenters from back side (*re*-face) as this point

threo-D$_s$-Isocitrate   (7-47)

pro-R proton from C-4 (stereochemical numbering) of citrate is removed and in Eq. 7-47, step *b* *threo*-D_s-isocitrate is formed. Proton addition is to the *re*-face in both cases.

As with fumarase, the enzyme holds the abstracted proton (for up to 7×10^{-5} s), long enough that a *cis*-aconitate molecule sometimes diffuses from the enzyme and (if excess *cis*-aconitate is present) is replaced by another. The result is that the new *cis*-aconitate molecule sometimes receives the proton (intermolecular proton transfer). The proton removed from citrate is often returned to the molecule in Eq. 7-47, step *b*, but (as indicated by the dashed arrow in the central structure of Eq. 7-47) the position of reentry is different from that of removal. It seems likely that after the initial proton removal, the *cis*-aconitate formed "flips over" so that it can be rehydrated with participation of the same groups involved in dehydration but with formation of the new product.[119] Perhaps this complex mechanism accounts for the low turnover number of 15 s^{-1}.

Aconitase contains and requires ferrous iron (Fe^{2+}). Although it has been suggested that the ferrous ion may engage in an oxidation–reduction process (implying a mechanism quite different from any discussed here), it seems more likely that it aids in attachment of the substrate to the enzyme and in generation of a hydroxyl ion in a manner similar to that suggested for Zn^{2+} in carbonic anhydrase (Section H,2).[120]

d. Addition and Elimination of Other Nucleophiles

Groups other than OH$^-$ can be added or eliminated from a position β to —COO$^-$. Thus, **aspartase,** a bacterial enzyme, catalyzes the addition of ammonia to fumarate to form L-aspartate, a reaction analogous to that catalyzed by fumarase. Trans addition is observed and isotope rate effects are consistent with a carbonium ion mechanism.[121] On the other hand, β-**methylaspartase** catalyzes a similar addition but causes a rapid exchange of deuterium from water into the substrate.[122]

Polymers of uronic acids with 1,4 linkage such as hyaluronic acid, dermatan sulfate (Fig. 2-16), and the pectins (Chapter 2, Section C,3) are susceptible to cleavage by a group of bacterial enzymes that employ an elimination mechanism.[123] The geometry of the β-linked galacturonic acid units of pectin is favorable for a trans elimination of the 5-H and the O-glycosyl group in the 4 position (Eq. 7-48). However, the corresponding hyal-

Galacturonic acid unit
of pectin

$$R—OH + \qquad\qquad\qquad\qquad (7\text{-}48)$$

uronidase acting on glucuronic acid residues causes a cis elimination. This result suggests that the activating effect of the carboxyl group in facilitating removal of the 5-hydrogen as a proton is important[124] and that an anionic intermediate is formed. Elimination of the β substituent would then be possible from either the equatorial or axial position.

An interesting use is made of addition to a double bond by certain cis-trans isomerases.[125] These are sulfhydryl-containing enzymes. One isomerase converts **maleate** to fumarate with a turnover number of 300 s^{-1} and similar enzymes isomerize **maleylacetoacetate** and **maleylpyruvate** to the corresponding fumaryl derivatives (Eq. 7-49). The —SH group of the enzyme is thought to add to the double bond. Rotation can then occur in the enolic intermediate. Thiocyanate ion catalyzes the isomerization of maleic acid nonenzymatically, presumably by a similar mechanism.

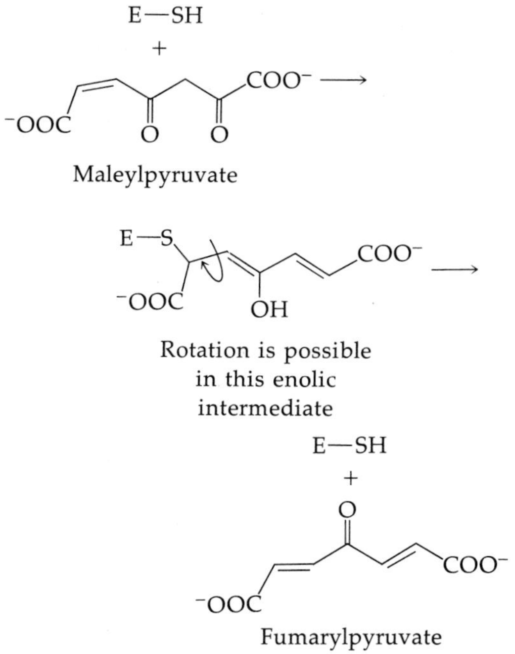

Maleylpyruvate

Rotation is possible
in this enolic
intermediate

(7-49)

Fumarylpyruvate

7. Addition to Isolated Double Bonds

Only a few examples are known in which an enzyme induces addition to a double bond that is *not conjugated* with a carbonyl or carboxyl group. Bacterial cells (of a pseudomonad) have been observed to catalyze stereospecific hydration of oleic acid to D-10-hydroxystearate.[126] As in preceding examples, addition is anti and the proton enters from the *re*-face.

8. Conjugative Elimination

Elimination can occur if the electrophilic and nucleophilic groups to be removed are located not on adjacent carbon atoms but are separated from each other by a pair of atoms joined by a double bond. Such a conjugative elimination of phosphate is the last step in the biosynthesis of **chorismate** (Eq. 7-50, step *a*). The latter compound, in two enzymatic steps (Eq. 7-50, step *b*), is converted to **prephenate** which undergoes another conjugative

elimination (Eq. 7-50, step *c*) with loss of both water and CO_2 to form **phenylpyruvate**, the immediate biosynthetic precursor of phenylalanine. These reactions provide a good example of how *elimination reactions can be used to generate aromatic groups*. In fact, this is the usual method of synthesis of such groups in nature.

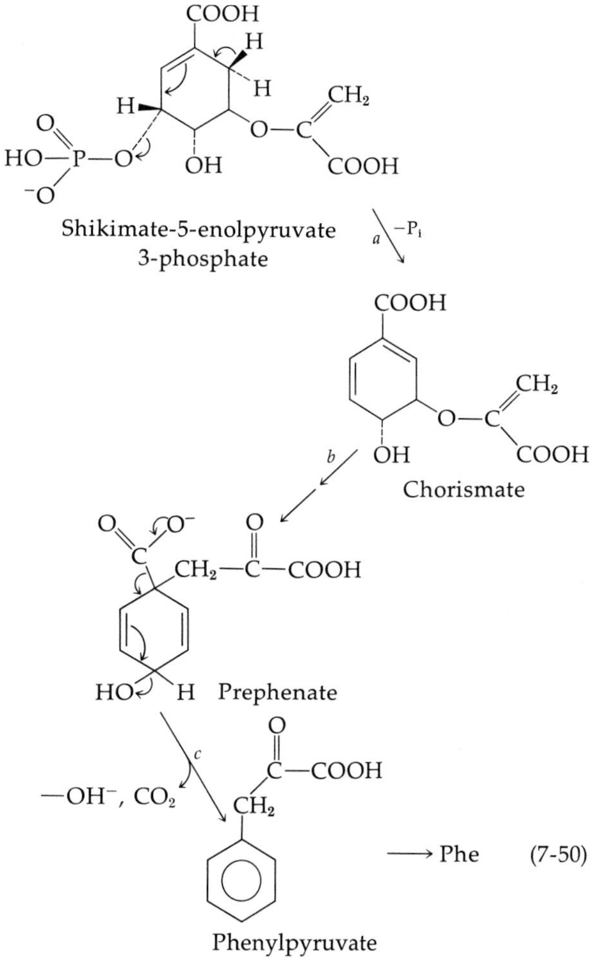

Shikimate-5-enolpyruvate
3-phosphate

Chorismate

Prephenate

Phenylpyruvate

$\longrightarrow$ Phe   (7-50)

Note the stereochemistry of Eq. 7-50, step *a*. Orbital interaction rules predict that if the elimination is a concerted process it should be syn. The observed anti elimination suggests a more complex mechanism involving participation of a nucleophilic group of the enzyme.[127] Step *b* occurs nonenzymatically,[128] but the enzyme enhances the rate by a factor of 2×10^6.

9. Decarboxylative Elimination

While elimination usually involves a proton together with a nucleophilic group such as —OH, —NH$_3^+$, phosphate or pyrophosphate, another electrophilic group such as —COO$^-$ can replace the proton. Thus, a C—C bond can be cleaved with loss of CO$_2$ together with a nucleophilic group from the β position. One example is shown by Eq. 7-50, step c. Another is the conversion of **mevalonic acid-5-pyrophosphate** to **isopentenyl pyrophosphate** (Eq. 7-51): This is a key reaction

Mevalonic acid-5-
pyrophosphate

Isopentenyl pyrophosphate (7-51)

in the biosynthesis of isoprenoid compounds. The phosphate ester shown in brackets is a probable intermediate.

10. Reversibility of Addition and Elimination Reactions

Many addition and elimination reactions are strictly reversible. This is usually true when water is added to a double bond as in the action of fumarase or in nonenzymatic hydration of aldehydes and ketones. However, in biosynthetic sequences, the reactions are often nearly irreversible because of the fact that the group eliminated is phosphate or pyrophosphate. Both of these ions occur in low concentrations within cells so that the reverse reaction does not tend to take place. In

the decarboxylative eliminations, carbon dioxide is produced and reversal becomes extremely unlikely because of the high stability of CO$_2$. Further irreversibility is introduced when the major product is an aromatic ring, as in the formation of phenylpyruvate.

I. ENOLIC INTERMEDIATES IN ISOMERIZATION REACTIONS

Enolate anions (Chapter 2, Section A,9) formed from carbonyl compounds (reaction type 4A, Table 7-1) play a central part in metabolism. They are intermediates in isomerization processes and they are instrumental in creating reactive nucleophilic centers on carbon atoms. Carbonyl groups are often converted to N-protonated imines (Schiff bases) from which a proton or carbon can be dissociated to form an **eneamine,** electronically analogous to an enolate anion (reaction type 4B, Table 7-1).

A group of tautomerases catalyze the keto–enol transformation itself. The activity of the widely distributed **oxaloacetate keto–enol-tautomerase** (Eq. 7-52) is especially high in animal tissues.[129]

$$^-OOC—CH_2—\overset{\overset{O}{\|}}{C}—COO^- \xrightleftharpoons{\text{tautomerase}}$$

$$^-OOC—\overset{H}{C}=\overset{OH}{C}—COO^- \quad (7\text{-}52)$$

Oxaloacetate is a metabolically important dicarboxylic acid which exists to a substantial extent in the enolic form. Presumably the slow uncatalyzed rate of interconversion accounts for the existence of this enzyme.

1. Aldose-Ketose Isomerases

The oxidation of one functional group of a molecule by an adjacent group in the same molecule is a feature of many metabolic sequences. In most cases an enol is formed, as an intermediate, either from a ketone as in Eq. 7-53, or by a dehydration

reaction (Eq. 7-59). One group of enzymes catalyze the interconversion of aldose sugars with the
corresponding 2-ketoses (Table 7-1, reaction 4C).
Glucose-6-phosphate isomerase appears to function
in all cells with a high efficiency.[130] The enzyme
from rabbit muscle (MW ~ 132,000, a dimer) converts glucose 6-phosphate to fructose 6-phosphate
with a turnover number of ~ 10^3 s^{-1}. The nonenzymatic counterpart of this and other similar reactions is the base-catalyzed **Lobry de Bruyn–Alberda
van Ekenstein transformation**[131] which was investigated in 1895 by the two Dutch chemists for
whom the reaction is named. In the same year
Emil Fisher proposed for the reaction an **enediol**
intermediate whose existence has been verified by
modern mechanistic studies.

When glucose 6-phosphate is isomerized by
glucose-6-phosphate isomerase in ^2H_2O, deuterium enters the molecule of fructose 6-phos-
phate at C-1, a result expected according to
the enediol mechanism (Eq. 7-53). In the reverse

1 position →

Glucose 6-phosphate   cis-Enediol

(7-53)

Fructose 6-phosphate

reaction ^2H-containing fructose 6-phosphate was
found to react at only 45% of the rate of the ^1H-
containing compound. Thus, the primary deuterium isotope effect expected for a rate-limiting
cleavage of the C—H bond was observed.

A surprising result was noted when fructose 6-
phosphate containing both ^2H and ^{14}C in the 1 position was isomerized in the presence of a large
amount of nonlabeled fructose 6-phosphate. The
product glucose 6-phosphate contained not only

^{14}C but also ^2H, and the distribution indicated that
the ^2H had entered (at the C-2 position) by direct
transfer from the C-1 position.[130] This indicates
that in over half the turnovers of the enzyme, ^2H
removed from C-1 is put back on the same molecule at C-2. This intramolecular transfer of a proton is taken as evidence of a syn transfer, i.e., the
proton is removed and put back on the same side
of the molecule.* Together with the known configuration of glucose at C-2, the occurrence of syn
transfer suggests that the intermediate is the
cis-enediol and that addition of a proton at either
C-1 or C-2 of the endiol is to the re-face. In every
case investigated, the other aldose-ketose isomerases also make use of the cis-enediol and in
most cases, addition is to the re-face. However,
mannose-6-phosphate isomerase catalyzes addition to the si-face.

2. Diffusion-Controlled Dissociation of Protons

The direct proton transfer between C-1 and C-2
during the action of sugar isomerases may seem
puzzling. How can a highly mobile proton remain
attached to a group in the enzyme for a millisecond or more instead of being transferred out to a
solvent molecule? Could this mean that the enzyme promotes the transfer of a hydride ion or of a
hydrogen atom rather than a proton? If so, the observed proton exchange with solvent would be an
unimportant side reaction. On the other hand,
could the group in the enzyme that removes the
proton be out of contact with the aqueous medium
and thus able to hold onto the proton more tightly?
In recent years, it has been recognized that neither
of these explanations is necessary. An imidazole
group is the most likely proton-carrying group at
the active sites. It is now believed that a proton
cannot be expected to transfer out from an imidazolium group with a rate constant much greater

---

* However, this argument should be accepted with caution.
There are nonenzymatic reactions in which a proton is removed and returned to the opposite side of a molecule without
exchange with solvent.[132]

than 10^3 s^{-1}. The same is true for the conjugate acids of other moderately strong basic groups such as phosphate.

The argument is as follows.[133] The rate of donation of a proton from H_3O^+ to imidazole (reverse of Eq. 7-54) is known to be diffusion controlled with a rate constant of 1.5×10^{10} M^{-1} s^{-1}.

$$ImH^+ + H_2O \rightleftharpoons \left[ImH^+\text{---}OH_2\right] \rightleftharpoons Im + OH_3^+ \qquad (7\text{-}54)$$

Very fast diffusion-controlled
$k \sim 1.5 \times 10^{10}$ M^{-1} s^{-1}

The equilibrium constant for reaction 7-54, calculated from the pK_a of 7.0 for imidazole, is 10^{-7} M. Since K_{eq} is also the *ratio* of the overall rate constants for the forward and reverse reactions, we see that for the forward reaction $k_f = 10^{-7} \times 1.5 \times 10^{10} = 1.5 \times 10^3$ s^{-1}. This slow rate results from the fact that in the intermediate complex (in brackets in Eq. 7-54) the proton is on the imidazole group most of the time. For a small fraction of the time it is on the coordinated molecule H_2O, but reverts to being on the imidazole many times before the imidazole and OH_3^+ separate. Because of this unfavorable equilibrium within the complex, the diffusion-controlled rate of proton transfer from a protonated imidazole to water is far less than for proton transfer in the reverse direction.

3. Ring Opening Promoted by Isomerases

Glucose-6-phosphate isomerase catalyzes a second reaction, namely, the opening of the ring of the α-anomer of glucose 6-phosphate (one-half of the mutarotation reaction, Eq. 6-75). Noltmann suggested that a protonated base BH^+ (perhaps an ϵ-amino group of a lysine side chain) and another basic group B (perhaps imidazole) participate in the ring opening as indicated in Fig. 7-9. On the other hand, a mechanism can be written in which ring opening and formation of the *cis*-enediol occur simultaneously through an internal dis-

placement reaction. For such a mechanism Schray and Rose suggested the skew-boat conformation of glucose 6-phosphate as a "high-energy" intermediate.[134]

4. Other Sugar Phosphate Isomerases

The enzyme **mannose-6-phosphate isomerase,** that converts mannose 6-phosphate to fructose 6-phosphate, is a Zn^{2+}-containing monomer of MW $\approx 45,000$. The dimeric **triosephosphate isomerase** of MW $\sim 52,000$ interconverts glyceraldehyde and dihydroxyacetone phosphates and is said to be the fastest enzyme participating in glycolysis (Section A,5). Its molecular activity is ~ 2800 s^{-1} in the direction shown in Eq. 7-55 and ~ 250 s^{-1} in the

$$\qquad (7\text{-}55)$$

Glyceraldehyde 3-phosphate $\qquad$ Dihydroxyacetone phosphate

reverse direction (the predominant direction in metabolism). Although its high catalytic activity might favor intramolecular transfer of the proton removed by the enzyme, very little such transfer has been observed. This suggests that a relatively weak base such as a carboxylate group may serve as the proton acceptor at the C-3 position of dihydroxyacetone phosphate.[135] Various covalent labeling experiments also implicate a carboxylate group (of Glu 165).[130]

The structure of triosephosphate isomerase of chicken muscle has recently been determined by X-ray diffraction.[136] About 22% of the 247 residues of each monomer form an eight-stranded parallel β-pleated sheet. This is twisted into a barrel structure (see Chapter 2, Section B,4). Alternating with the strands of the β structure are four α-helical

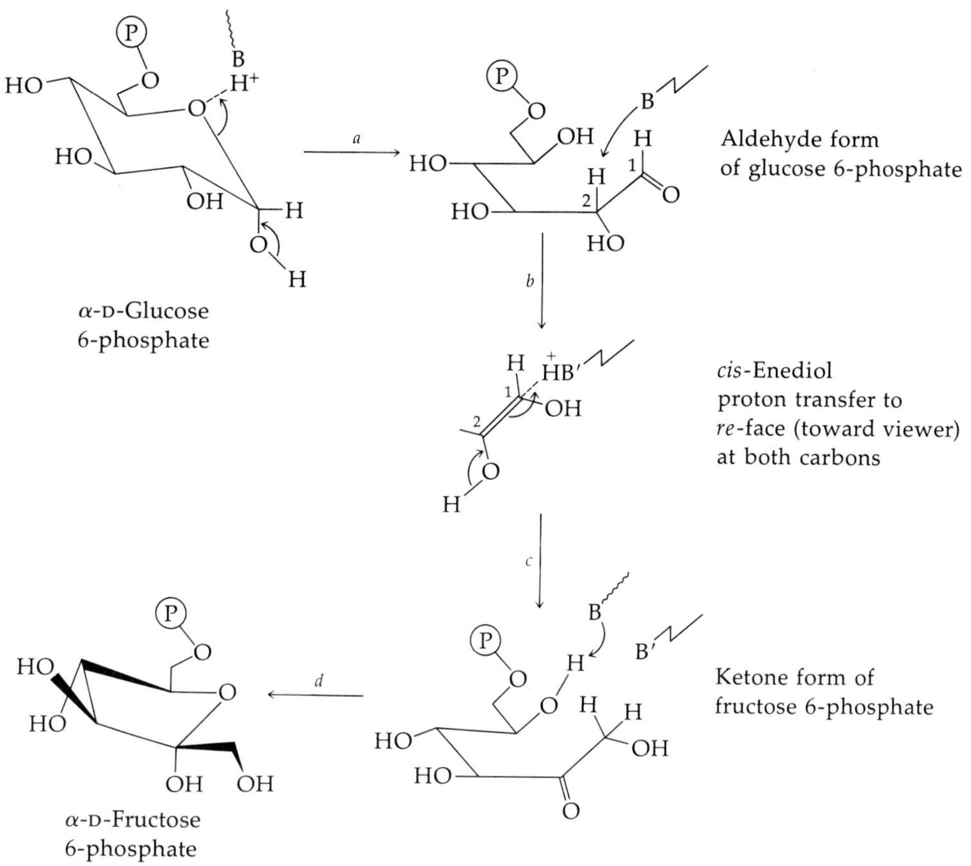

Fig. 7-9 Ring opening and isomerization catalyzed by glucose-6-phosphate isomerase. After Noltmann.[130]

regions comprising 55% of all residues. The structure somewhat resembles that of glyceraldehyde-phosphate dehydrogenase (Fig. 2-10).

An enzyme that interconverts **ribose 5-phosphate** and **ribulose 5-phosphate** is another important isomerase.

5. Δ⁵-3-Ketosteroid Isomerase and Other Enzymes Catalyzing 1,3-Proton Shifts

Cholesterol serves in the animal body as a precursor of all of the steroid hormones including the 3-ketosteroids **progesterone** and **testosterone** (Chapter 12, Section I,3). While cholesterol con-

tains a double bond in the 5,6 position, as in the structure at the left in Eq. 7-56, in the ketosteroids,

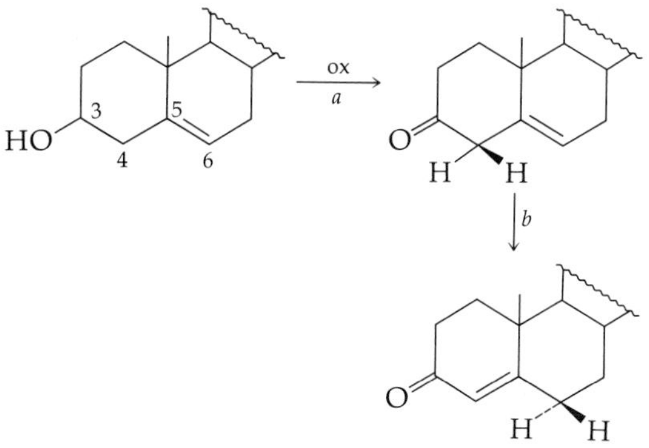

$$(7\text{-}56)$$

the bond has moved into conjugation with the carbonyl group. The hydroxyl group of cholesterol is first oxidized (Eq. 7-56, step a) to a keto group. This is followed by an essentially irreversible migration of the double bond (Eq. 7-56, step b) catalyzed by a **Δ^5-3-ketosteroid isomerase.**[137] The enzyme has been completely purified from *Pseudomonas testosteroni* and is an oligomeric protein of MW $\sim 40,000$. It has a remarkably high molecular activity ($\sim 10^5$ s^{-1}), mostly accounted for by an unusually low value of $\Delta H^{\ddagger}$. The protein is very hydrophobic and is soluble in high concentrations of ethanol. This property, if it is shared by the animal enzyme, is compatible with the location of the latter in the endoplasmic reticulum.

Substrates containing ^2H in the 4 position react only one-fourth as rapidly as normal substrates. The large isotope effect indicates that cleavage of the C—H bond to form an enolate anion (presumably stabilized by the enzyme) is rate limiting (Eq. 7-57, step a). The proton in the axial position at C-4 is removed[138] and must be carried by some group on the enzyme over to the 6 position, where it is returned to the substrate, again in an axial position (Eq. 7-57, step b), a syn transfer. No exchange of the proton with solvent is observed, presumably because of the extreme rapidity of the isomerase action. That the reaction does involve transfer of H$^+$ and does not occur by some other mechanism is demonstrated by the fact that poor substrates and competitive inhibitors such as nortestosterone (which has a structure with a double bond arrangement of the product in Eq. 7-57) do undergo complete exchange of one of the hydrogens at C-4 with the medium. Furthermore, the ultraviolet absorption band of the inhibitor at 248 nm is shifted to 258 nm upon combination with the enzyme, a fact that suggests formation of the stabilized enolate anion of Eq. 7-57.

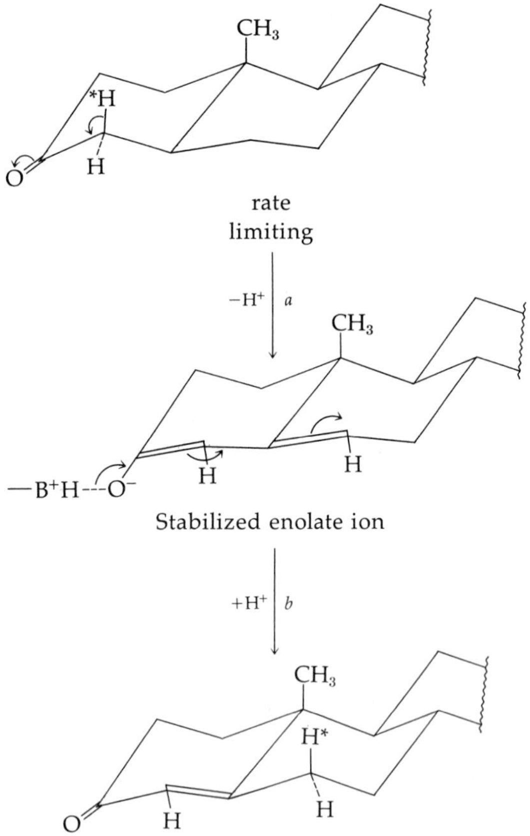

The enzymatic isomerization of *cis*-aconitate to *trans*-aconitate (Eq. 7-58) apparently also involved

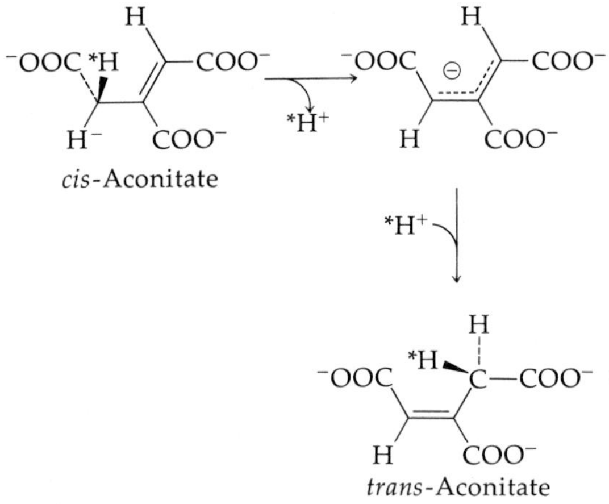

rate limiting

Stabilized enolate ion

(7-57)

cis-Aconitate

(7-58)

trans-Aconitate

proton transfer.[139] While it is not clear whether resonance in the anion extends into the carboxylate groups, it seems likely that it does and that the mechanism is directly related to that of the ketosteroid isomerase. However, there are other allylic rearrangements in which neither a carbonyl nor carboxyl group is present in the substrate. It is not known whether these occur via anion or carbonium ion intermediates.

Allylic rearrangements involving proton shifts

can also occur as part of the mechanism of enzymes catalyzing other types of reaction.

6. Internal Oxidation–Reduction by Dehydration to an Enol or Enol Phosphate

When a carboxylic acid contains hydroxyl groups in both the α and β positions, dehydration leads to formation of an enol that can tautomerize to a 2-keto-3-deoxy derivative of the original acid (Eq. 7-59). Thus, phosphogluconate dehydratase

$$
\begin{array}{c}
\text{COO}^- \\
| \\
\text{HC}-\text{OH} \\
| \\
\text{HO}-\text{CH} \\
| \\
\text{R}
\end{array}
\xrightarrow[\text{H}_2\text{O}]{}
\left[
\begin{array}{c}
\text{COO}^- \\
| \\
\text{C}-\text{OH} \\
|| \\
\text{CH} \\
| \\
\text{R}
\end{array}
\right]
\longrightarrow
\begin{array}{c}
\text{COO}^- \\
| \\
\text{C}=\text{O} \\
| \\
\text{CH}_2 \\
| \\
\text{R}
\end{array}
\quad (7\text{-}59)
$$

yields 2-keto-3-deoxyphosphogluconate as the product.* This reaction initiates a unique pathway of sugar breakdown (the Entner–Douderoff pathway, Chapter 9, Section E,4) in certain organisms. The 6-phosphogluconate is formed by oxidation of the aldehyde group of glucose 6-phosphate. This pattern of oxidation of a sugar to an aldonic acid followed by dehydration according to Eq. 7-59 is met frequently in metabolism. A related reaction is the dehydration of 2-phosphoglyceric acid by enolase (Eq. 7-46). The product is phosphoenolpyruvic acid, a stabilized form of the enolic intermediate of Eq. 7-59.

A more complex internal oxidation reduction (Eq. 7-60) is catalyzed by **methylglyoxal synthetase.**[140] Free glyceraldehyde 3-phosphate is not acted on by the enzyme, but it is likely that an enzyme-bound form of the compound (indicated by brackets in Eq. 7-60) undergoes elimination to

---

* When the reaction is carried out in $^2\text{H}_2\text{O}$ the ^2H is found to be incorporated with a random configuration at C-3 indicating that the enzyme catalyzes only the dehydration and that the tautomerization of the enol to the ketone is nonenzymatic.

$$
\begin{array}{c}
\text{CH}_2\text{OH} \\
| \\
\text{C}=\text{O} \\
| \\
\text{CH}_2\text{O}\,\textcircled{P}
\end{array}
\longrightarrow
\left[
\begin{array}{c}
\text{CHO} \\
| \\
\text{H}-\text{C}-\text{OH} \\
| \\
\text{CH}_2\text{O}\,\textcircled{P}
\end{array}
\right]
\xrightarrow{\;\text{P}_i\;}
$$

Dihydroxyacetone phosphate

$$
\left[
\begin{array}{c}
\text{CHO} \\
| \\
\text{C}-\text{OH} \\
|| \\
\text{CH}_2
\end{array}
\right]
\longrightarrow
\begin{array}{c}
\text{CHO} \\
| \\
\text{C}=\text{O} \\
| \\
\text{CH}_3
\end{array}
\quad (7\text{-}60)
$$

Methylglyoxal

form an enol. The latter gives rise to methylglyoxal.

J. BETA CLEAVAGE AND CONDENSATION

In the preceding sections of this chapter we examined displacement and addition reactions that usually involved a nucleophile containing O, N, or S. Bonds from carbon to these atoms can usually be broken easily by acidic or basic catalysis. The breaking and making of C—C bonds does not occur as readily and the "carbon skeletons" or organic molecules often stick together tenaciously. Yet living cells both form and destroy a variety of complex, branched carbon compounds.

A major mechanistic problem in cleavage or formation of carbon–carbon bonds is the creation of a nucleophilic center on a carbon atom. The problem is most often solved by using the *activating influence of a carbonyl group* to generate a resonance-stabilized enolate anion.

Just as the presence of a carbonyl group facilitates cleavage of an adjacent C—H bond, so it can also assist the cleavage of a C—C bond. The best known reactions of this type are the **aldol cleavage**

Cleavage of C—H bond facilitated by a carbonyl group

β Cleavage with release of CO_2

and the **decarboxylation of β-keto acids.** The latter has been referred to as β decarboxylation and its reverse as β carboxylation. The term will be extended to include other reactions by which bonds between the α and β carbon atoms of a carbonyl compound are broken or formed, and these will be referred to as **β cleavage and β condensation.**

The β *condensation* reactions consist of displacement or addition reactions in which an enzyme-bound enolate anion acts as the nucleophile. We can group these condensation reactions into three categories as indicated by reaction type 5A, 5B, and 5C of Table 7-1.

1. Displacement on a Carbonyl Group (Reaction Type 5A)

A β-keto acid, with proper catalysis, is susceptible to hydrolysis by attack of water on the carbonyl group. An **oxaloacetate acetylhydrolase** has been isolated from *Aspergillus niger*[141] (Eq. 7-61).

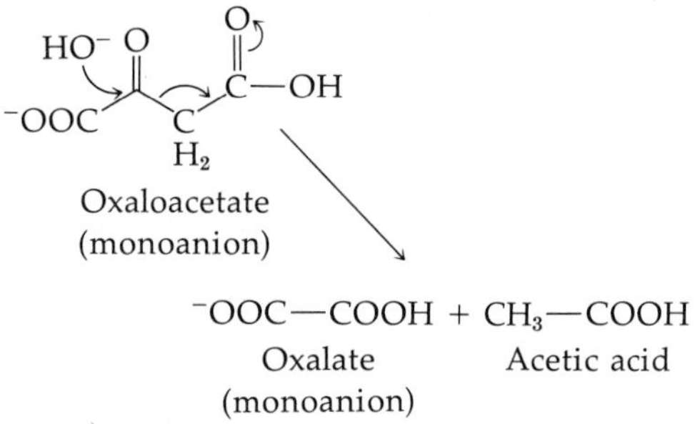

$$^-OOC—COOH + CH_3—COOH \quad (7\text{-}61)$$

Oxaloacetate (monoanion)

$$^-OOC—COOH + CH_3—COOH \quad (7\text{-}61)$$

Oxalate   Acetic acid
(monoanion)

Better known are the **thiolases.**[142,143] One function of these enzymes is to cleave β-ketoacyl derivatives of CoA by displacement with a thiol group of another CoA molecule (Eq. 7-62). This is the chain cleavage step in the β oxidation sequence by which fatty acid chains are degraded (Fig. 7-1). Since the thiolases are inhibited by —SH reagents it has been suggested that a thiol group in the enzyme reacts initially with the β-carbonyl group as in Eq. 7-62 to give an enzyme-bound S-acyl intermediate. The acyl group is then transferred to CoA in a second step. Some thiolases may form Schiff bases with their substrates.[143]

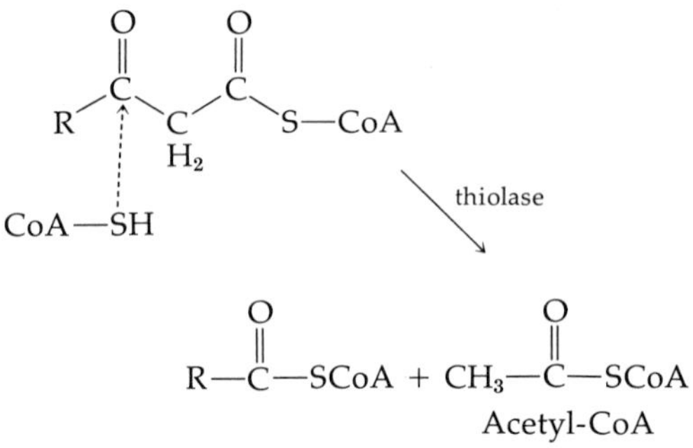

$$R—C—SCoA + CH_3—C—SCoA \quad (7\text{-}62)$$

Acetyl-CoA

2. Addition of an Enolate Anion to a Carbonyl Group or an Imine (Reaction Type 5B)

The **aldol condensation** (Eq. 7-63) is one of the commonest reactions by which C—C bonds are

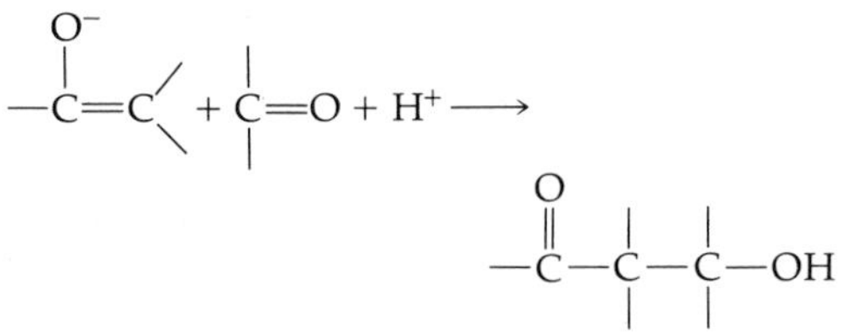

$$—C=C \Big< + C=O + H^+ \longrightarrow \quad —C—C—C—OH \quad (7\text{-}63)$$

formed and (in the reverse reaction) cleaved in metabolism.

a. Fructose-Diphosphate Aldolase

The best known aldolase cleaves (reversibly) fructose diphosphate into 2 molecules of triose phosphate (Eq. 7-64) during glycolysis. The en-

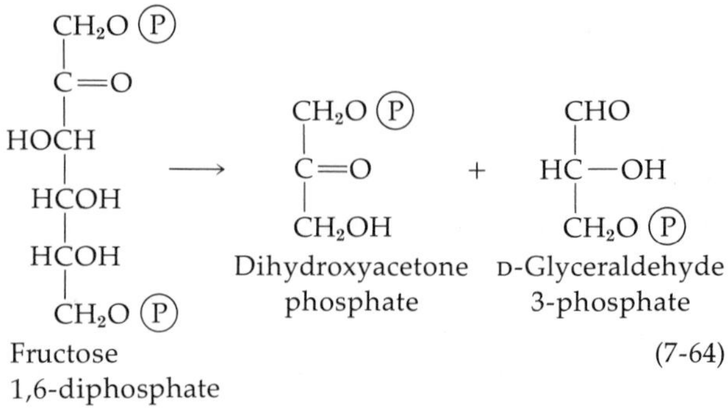

Fructose
1,6-diphosphate

Dihydroxyacetone   D-Glyceraldehyde
phosphate   3-phosphate

$$(7\text{-}64)$$

zyme is found in all plant and animal tissues but is absent from a few specialized bacteria. The much-studied rabbit muscle aldolase is a protein of MW $\sim$ 160,000; a tetramer of four nearly identical peptide chains.[144–146]

Treatment of the enzyme–substrate complex of aldolase and dihydroxyacetone phosphate with sodium borohydride at pH 6, 0°C, leads to formation of a covalent linkage between the protein and substrate. This and other evidence suggested a Schiff base intermediate (Fig. 7-10). Using ^{14}C-containing substrate and borohydride reduction (Eq. 7-41) an active site lysine was labeled. The radioactive label was followed through the sequence determination and the position of the lysine was identified as 227 in a chain of 361 amino acids.

The presence of an essential histidine residue at the active center is suggested by photochemical oxidation experiments. For example, irradiation with visible light in the presence of the halogenated fluorescein dye **rose bengal** catalyzes photoinactivation of aldolase with destruction of a large number of histidines. A more specific reagent is pyridoxal phosphate (Chapter 8, Section E) which binds to the active site, apparently by forming a Schiff base with a second lysine residue. When bound pyridoxal phosphate was irradiated, a single histidine (probably His 359) was destroyed in each subunit.

Reaction of 5,5'-dithiobis-2-nitrobenzoate (DTNB, Eq. 2-28) with aldolase caused rapid blocking of one SH group in each subunit. The presence of substrate prevented the reaction, suggesting that the SH groups (probably Cys 72 or Cys 336) are also at the active sites. Other evidence suggests that the carboxyl terminal Tyr 361 may be at or near the active site.[146]

This information has been used to develop the hypothetical active site chemistry[145] indicated in Fig. 7-10. Note that the enzyme is shown as catalyzing the opening of a ring form of fructose diphosphate, the phosphate group of the substrate possibly participating.[147] The β cleavage may be initiated by a —S$^-$ group. The active site imidazole is thought to donate a proton, and it has

been shown to enter the same position, stereochemically speaking, as was occupied by the bond between the carbon atoms in fructose 1,6-diphosphate. It has been suggested that the proton from imidazole is donated within a hydrophobic region of the enzyme as described in Chapter 2, Section A,9.

Aldolase is one of the enzymes that has shown a special propensity for reaction with **tetranitromethane.** This reagent reacts relatively slowly with tyrosyl groups to form 3-nitrotyrosyl groups (Eq. 7-65). The by-product **nitroform** is intensely

$$O_2N-\underset{\underset{NO_2}{|}}{\overset{\overset{NO_2}{|}}{C}}-NO_2 \; + \; \left\langle \bigcirc \right\rangle -O^- \longrightarrow$$

$$O_2N-\underset{\underset{H}{|}}{\overset{\overset{NO_2}{|}}{C}}-NO_2 \; + \; \left\langle \overset{NO_2}{\bigcirc} \right\rangle -O^- \qquad (7\text{-}65)$$

Nitroform

yellow[148] with $\epsilon_{350} = 14,400$. The reagent also oxidizes SH groups and reacts with other anionic groups. Aldolase converts tetranitromethane to nitroform much more rapidly in the presence of substrate than in its absence. This suggests that the reagent, which can be thought of as a producer of the electrophilic nitronium ion NO_2^+, competes with the imidazole in reacting with the enamine intermediate of Fig. 7-10.

b. Isoenzymes of Aldolase

Most mammalian tissues contain three aldolase isoenzymes: A, B, and C. Isoenzymes A and C appear to be predominantly embryonic and in adult, differentiated tissues isoenzyme B predominates.

In certain tumors isoenzyme B is replaced by the embryonic form A. This is especially true in poorly differentiated (highly dedifferentiated) rapidly growing cancers, e.g., certain hepatomas.

c. Two Classes of Aldolases

Like fructose-1,6-diphosphate aldolase of animal tissues, a large number of other aldolases are

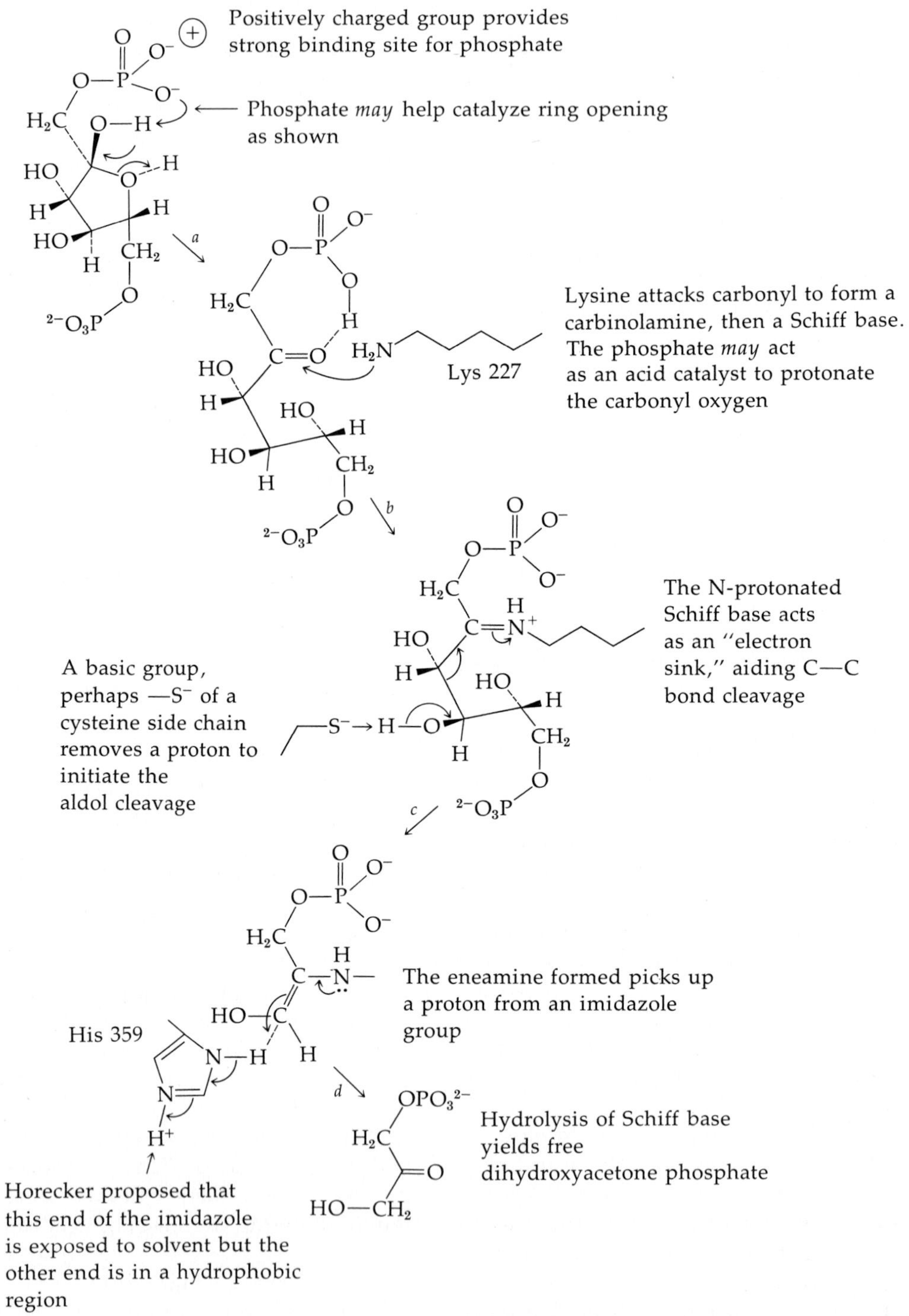

Positively charged group provides strong binding site for phosphate

← Phosphate *may* help catalyze ring opening as shown

Lysine attacks carbonyl to form a carbinolamine, then a Schiff base. The phosphate *may* act as an acid catalyst to protonate the carbonyl oxygen

Lys 227

The N-protonated Schiff base acts as an "electron sink," aiding C—C bond cleavage

A basic group, perhaps —S⁻ of a cysteine side chain removes a proton to initiate the aldol cleavage

The eneamine formed picks up a proton from an imidazole group

His 359

Hydrolysis of Schiff base yields free dihydroxyacetone phosphate

Horecker proposed that this end of the imidazole is exposed to solvent but the other end is in a hydrophobic region

Fig. 7-10 Possible mechanism of action of fructose-1,6-diphosphate aldolase.

also inactivated by sodium borohydride in the presence of substrate. These "Class I" aldolases have no metal requirement and are not inhibited by EDTA. On the other hand, aldolases of bacteria and fungi are inhibited by EDTA and contain a metal (usually Zn^{2+}) at their active centers. These "Class II" aldolases are not inactivated by sodium borohydride in the presence of substrate. The reader will be able to propose a reasonable mechanism and a function for the metal in these enzymes.

Some blue-green algae contain both types of aldolase, as do the flagellates *Euglena* and *Chlamydomonas*.

d. Transaldolase

An important enzyme in the pentose phosphate pathways of metabolism (Chapter 9, Section E,3) catalyzes aldol cleavage of one substrate, such as fructose 1,6-diphosphate, but instead of releasing free dihydroxyacetone phosphate transfers this 3-carbon unit to another aldose.[149] The mechanism of transaldolase (Eq. 9-15) appears to be basically the same as that of fructose-1,6-diphosphate aldolase.

e. Other Aldolases

Special aldolases too numerous to mention cleave and form C—C bonds throughout metabolism. A substantial number of them act on 2-keto-3-deoxy substrates[150] forming pyruvate as one product (Eq. 7-66). The aldehyde product varies.

$$
\begin{array}{c}
COO^- \\
| \\
C{=}O \\
| \\
CH_2 \\
\text{-----}|\text{-----} \\
HC{-}OH \\
| \\
R
\end{array}
\longrightarrow
\begin{array}{c}
COO^- \\
| \\
C{=}O \\
| \\
CH_3
\end{array}
+
\begin{array}{c}
H{\diagdown} \\
C{=}O \\
| \\
R
\end{array}
\qquad (7\text{-}66)
$$

Pyruvate

In the Entner–Doudoroff pathway of carbohydrate metabolism (Chapter 9, Section E,4) 2-keto-3-deoxy-6-phosphogluconate is cleaved to pyruvate and 3-phosphoglyceraldehyde. The 8-carbon sugar acid "KDO" of bacterial cell walls (Fig. 5-10)

is cleaved by another aldolase. The catabolism of hydroxyproline leads to 4-hydroxy-2-ketoglutarate, which is cleaved to pyruvate and glyoxylate. An aldolase, functioning in the catabolism of deoxynucleotides, cleaves 2-deoxyribose 5-phosphate to acetaldehyde and glyceraldehyde 3-phosphate.

f. Polycarboxylic Acid Synthetases*

A number of enzymes, including the key one which catalyzes the formation of citric acid in the first step of the citric acid cycle, promote aldol condensations of acetyl-CoA with ketones (Eq. 7-67). An α-keto acid is most often the second substrate, and the bracketed structure of Eq. 7-67

$$
\begin{array}{c}
O \\
\| \\
H_3C{-}C{-}SCoA \\
+ \\
O{=}C{-}COO^- \\
| \\
R
\end{array}
\longrightarrow
\left[
\begin{array}{c}
O \\
\| \\
H_2C{-}C{+}SCoA \\
| \\
HO{-}C{-}COO^- \\
| \\
R
\end{array}
\right]
\xrightarrow[H_2O]{CoA{-}SH}
$$

$$
\begin{array}{c}
H_2C{-}COO^- \\
| \\
HO{-}C{-}COO^- \\
| \\
R
\end{array}
\qquad (7\text{-}67)
$$

often appears to be an intermediate.[151] In some cases the acetyl group may be transferred from acetyl-CoA onto an SH group of the enzyme prior to the condensation, however.[151a] The same enzyme catalyzes the second step, hydrolysis of the CoA ester. These enzymes are important in biosynthesis. They carry out the initial steps in a *general chain elongation process* (Chapter 11, Section D,7). One function of the thioester group in acetyl-CoA is to activate the methyl hydrogens toward the aldol condensation. In addition the

* According to the IUB classification (Chapter 6, Section G) these enzymes are lyases and the name synthetase should not be used. The latter is reserved for ligases. According to the IUB recommendations, the name **synthase** can be used to emphasize the synthetic aspect of a lyase reaction. However, the use of the two similar words synthase and synthetase is confusing and seems quite unnecessary to many biochemists. Only the term synthetase is used in this book.

subsequent hydrolysis of the thioester linkage provides for overall irreversibility and "drives" the synthetic reaction.

Figure 7-11 shows the stereochemistry of the reaction, and lists ketone reactants and products. These enzymes may be classified by designating the face of the carbonyl group that is attacked by the enolate anion. The common citrate synthetase of animal tissues condenses with the *si*-face and is designated *si*-citrate synthetase. On the other hand, a few anaerobic bacteria possess citrate (*re*)-synthetase having the opposite stereochemistry.[152,153]

g. Chiral Acetates and Their Use in Stereochemical Studies

Consider the series of enzyme-catalyzed reactions (Eq. 7-68). Fumarate is hydrated in ^{3}H-

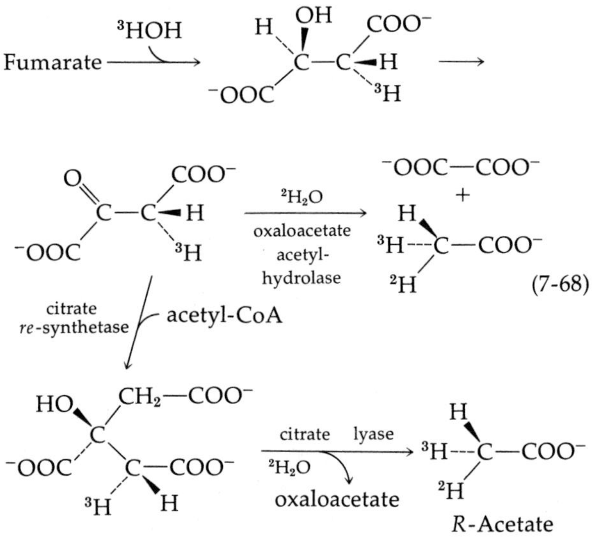

(7-68)

containing water to malate which is oxidized to oxaloacetate. Hydrolysis of the latter with oxaloacetate acetylhydrolase (Eq. 7-61) in ^{2}H$_2$O gives oxalate and *chiral* (R) acetate. The identical product can be obtained by condensing oxaloacetate with acetyl-CoA using citrate (*re*)-synthetase. The resulting citrate is cleaved in ^{2}H$_2$O using a citrate lyase having the *si* specificity.[154,155] Acetate of the opposite chirality can be formed enzymatically beginning with [2,3-^{3}H]fumarate hydrated

by fumarase in ordinary water.* Chiral acetates have also been prepared nonenzymatically,[154] and their configuration has been established unequivocally.

An interesting fact about the stereochemistry of both the oxaloacetate acetylhydrolase and the citrate lyase (Eq. 7-68) is that inversion of configuration occurs about the carbon atom that carries the negative charge in the departing enolate anion. That inversion also occurs during the action of citrate (*re*)-synthetase and other related enzymes has been demonstrated through the use of chiral acetates.[154,155] The findings with malate synthetase[151] are illustrated in Eq. (7-69). Presumably a

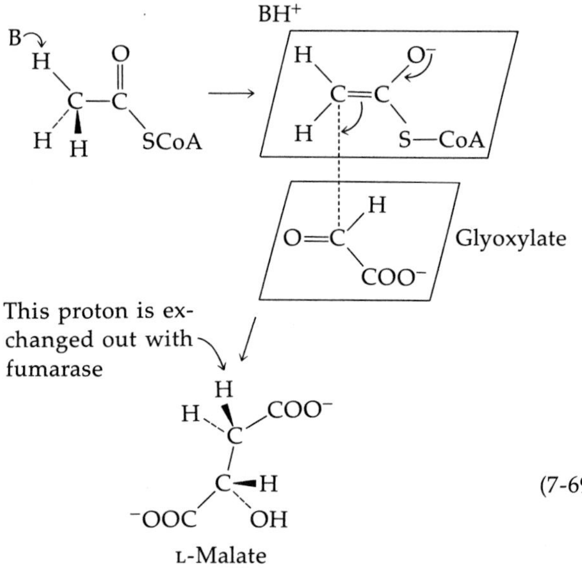

(7-69)

basic group B on the enzyme removes a proton to form the planar enolate anion. The second substrate glyoxylate approaches from the other side of the molecule and condenses as shown. The reader will be able to imagine how inversion was proved. Since any one of the three protons in either *R* or *S* chiral acetyl-CoA might have been abstracted by base B, several possible combinations of isotopes would be possible in the L-malate formed. However, making use of the fact that the *pro-R* hy-

* John W. Cornforth was awarded a Nobel prize in 1975 in part for this work [see *Science* **173**, 121–125 (1976) for his Nobel Lecture].

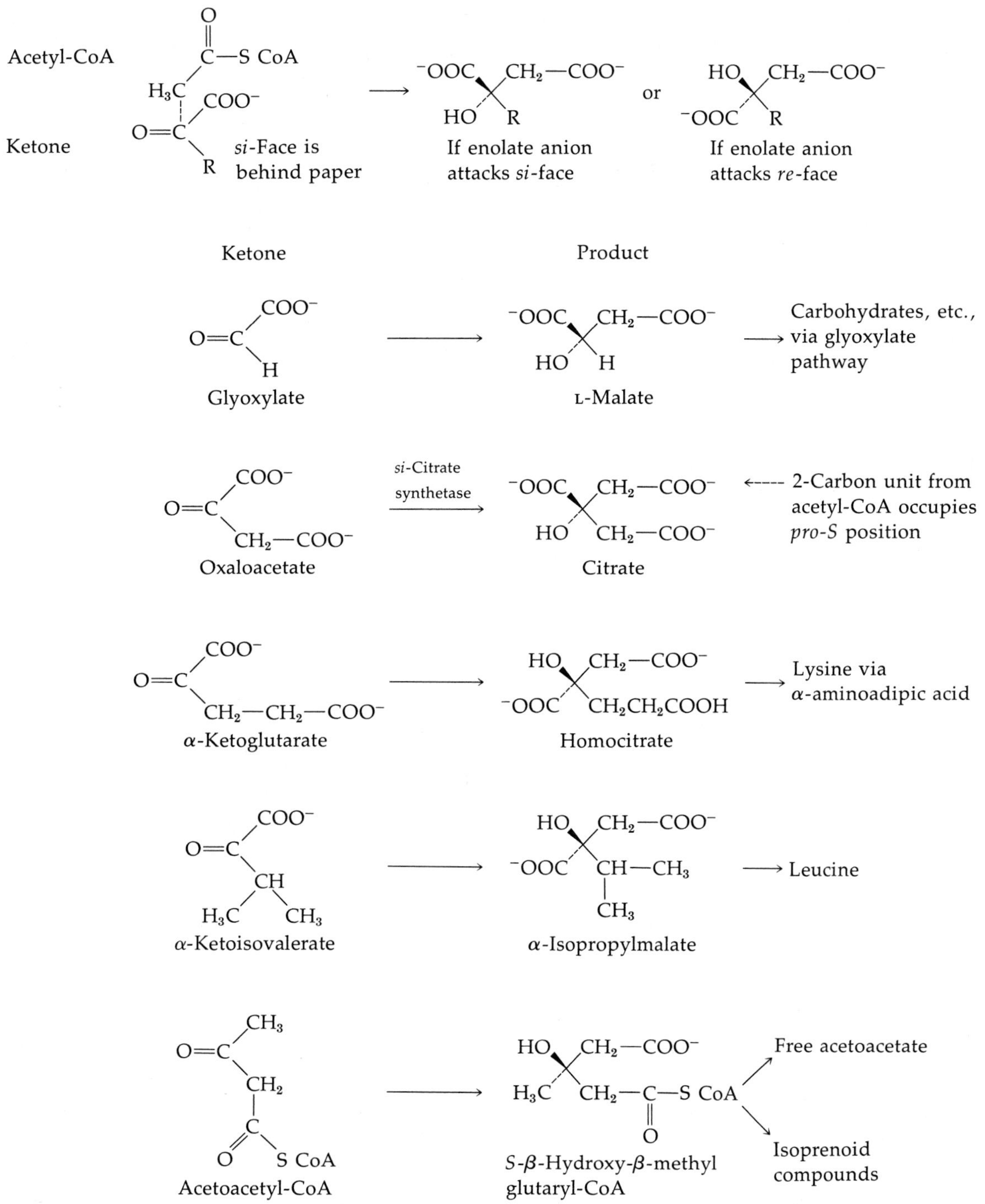

Fig. 7-11 Reactants and products of some polycarboxylic acid synthetases.

drogen at C-3 in malate is specifically exchanged out into water by the action of fumarase the problem was solved.

h. Citrate Cleaving Enzymes

In eukaryotic organisms the synthesis of citrate takes place within the mitochondria, but under some circumstances citrate is exported into the cytoplasm. There it is cleaved by a **citrate lyase.** To ensure that the reaction goes to completion, this cleavage is coupled to the hydrolysis of ATP to ADP and inorganic phosphate (Eq. 7-70). The

$$\text{Citrate} + \text{ATP} + \text{CoA} \rightleftharpoons \text{oxaloacetate}$$
$$+ \text{ acetyl-CoA} + \text{ADP} + P_i \qquad (7\text{-}70)$$
$$\Delta G' \text{ (pH 7)}^* = +0.9 \text{ kJ mol}^{-1}$$

reaction sequence (Eq. 7-71) is quite complex but can be understood in terms of an initial synthesis of citryl-CoA using a mechanism similar to that in Eq. 7-29. However, there is evidence for both phosphoenzyme and citryl enzyme intermediates (Eq. 7-71).[152] The ATP-dependent cleavage of mal-

ate to acetyl-CoA and glyoxylate uses the same sequence but requires two enzymes, malyl-CoA being an intermediate.[157]

A substrate-induced citrate lyase has been found in a few bacteria that promote the anaerobic dissimilation of citrate. Citrate is split to oxaloacetate and acetate. Evidence has been presented

* The position of the equilibrium is extremely dependent on the concentration of Mg^{2+} as a result of the strong chelation of Mg^{2+} by citrate.[156]

that the enzyme is initially in an *acetylated form* able to undergo acyl exchange to form a citryl enzyme before the aldol cleavage takes place (Eq. 7-72). These large proteins of MW $\approx 585,000$ are

oligomers $(\alpha\beta\gamma)_6$ whose small subunit (γ) of MW $\approx 10,000$ carries an unusual covalently bound derivative of coenzyme A.[158,158a]

The first unique enzyme of the important glyoxylate pathway (Chapter 11, Section D,4), **isocitrate lyase,** cleaves isocitrate to succinate and glyoxylate (Eq. 7-73). This enzyme may exist in a succinyl form[152] with a mechanism of cleavage analogous to that of Eq. 7-72.

$$+ \ ^-\text{OOC}-\text{CH}_2-\text{CH}_2-\text{COO}^- \qquad (7\text{-}73)$$
$$\text{Succinate}$$

$$\Delta G' \text{ (pH 7)} = 8.7 \text{ kJ mol}^{-1}$$

3. Addition of an Enolate Ion to Carbon Dioxide (Reaction Type 5C and Its Reverse)

The addition of an enolate anion to CO_2 to form a β-keto acid represents one of the commonest means of incorporation of CO_2 into organic compounds. The reverse, **decarboxylation,** is a major mechanism of biochemical formation of CO_2. The equilibrium constant usually favors decarboxylation but the cleavage of ATP can be coupled to drive carboxylation when it is needed, e.g., in photosynthesis.

a. Spontaneous Decarboxylation of β-Keto Acids

β-Keto acids such as oxaloacetic acid and acetoacetic acid are notoriously unstable. Their decarboxylation is catalyzed by many substances including amines and metal ions such as Zn^{2+}, Cu^{2+}, and Fe^{2+}. The amine catalysis[159] depends, in part, upon Schiff base formation. The metal ions form a chelate in which they can promote electron withdrawal from the labile C—C bond to form an enolate anion.[160]

Can we apply any of this information from non-enzymatic catalysis to decarboxylating enzymes? Some decarboxylases do form Schiff bases with their substrates, and an important group of carboxylases (those containing the vitamin biotin) also contain bound Mn^{2+}.

b. Acetoacetate Decarboxylase

In the acetone-forming fermentation of *Clostridium acetobutylicum* (see Chapter 9, Section F,4) large amounts of an enzyme catalyzing the decarboxylation of acetoacetate (Eq. 7-74) are required.

$$CH_3-\overset{O}{\overset{\|}{C}}-CH_2-COO^- + H^+ \longrightarrow$$

Acetoacetate

$$CH_3-\overset{O}{\overset{\|}{C}}-CH_3 + CO_2 \quad (7\text{-}74)$$

Acetone

The enzyme (MW ~340,000) contains 12 subunits and has been crystallized.[161] It is inactivated by borohydride in the presence of substrate, and acid hydrolysis of the inactivated enzyme yielded ε-N-isopropyllysine. Decarboxylation probably occurs from a Schiff base by a mechanism analogous to that of the aldol cleavage shown in Fig. 7-10.

c. Linked Oxidation and Decarboxylation

A common metabolic pathway is *oxidation of a β-hydroxy acid to a β-keto acid followed by decar-*

boxylation. The two steps are usually catalyzed by the same enzyme. While the *free β-keto acid* is not an intermediate the *bound β-keto acid* may be. (A concerted decarboxylation and dehydrogenation is also a possibility.) The **malic enzyme,** which catalyzes conversion of L-malate to pyruvate (Eq. 7-75), is found in most organisms. The initial de-

L-Malate

$$\left[^-OOC-CH_2-\overset{O}{\overset{\|}{C}}-COO^- \right] \xrightarrow[H^+]{CO_2}$$

Hypothetical bound oxaloacetate

$$CH_3-\overset{O}{\overset{\|}{C}}-COO^- \quad (7\text{-}75)$$

Pyruvate

hydrogenation involves the transfer of hydrogen to NAD^+ (see Fig. 8-10).

Other reactions of this type are the oxidation of isocitrate to α-ketoglutarate in the citric acid cycle (Fig. 9-2, steps *b* and *c*) and oxidation of 6-phosphogluconate to ribulose 5-phosphate (Eq. 9-12).

d. Phosphoenolpyruvate, a Key Metabolic Intermediate

A compound of central importance in metabolism is the phosphate ester of the enol form of pyruvate, commonly known simply as phosphoenolpyruvate (PEP). (According to good biochemical etiquette, this is pronounced "P-E-P," not "pep") PEP is formed in the glycolysis pathway by dehydration of 2-phosphoglycerate

$$\begin{array}{c} COO^- \\ | \\ C-O-PO_3^{2-} \\ \| \\ CH_2 \end{array}$$

Phosphoenolpyruvate (PEP)

(Eq. 7-46) or by decarboxylation of oxaloacetate. It is utilized in metabolism in a variety of ways, serving as a preformed enol from which a reactive enolate anion can be released for condensation reactions. Its place in metabolism will be considered more fully in later chapters, but let us consider here the way in which it is generated by decarboxylation of oxaloacetate. The latter reaction, which plays a crucial role in the metabolism of animals, other higher organisms, and bacteria, is catalyzed by **PEP carboxykinase.** A molecule of guanosine triphosphate or of inosine triphosphate captures and phosphorylates the enolate anion generated by the decarboxylation (Eq. 7-76). The stereochemistry is such that CO_2 de-

$$(7\text{-}76)$$

si-Face
is toward reader

parts from the *si*-face of the forming enol. Another enzyme **PEP carboxytransphosphorylase** catalyzes the following decarboxylation (Eq. 7-77). The enzyme has been found only in propionic acid bacteria and in *Entamoeba*. It seems likely that this enzyme normally functions in a reverse direction, providing a means of synthesis of oxaloacetate from PEP in these organisms. The pyrophosphate formed is presumably hydrolyzed to "pull" the reaction in the desired direction.[162]

$$\text{Oxaloacetate}^- + PP_i \longrightarrow PEP + CO_2 + P_i \qquad (7\text{-}77)$$

Oxaloacetate can also be enzymatically decarboxylated to pyruvate (see Eq. 9-8), presumably via the enolate anion but without the phosphorylation shown in Eq. 7-76. The enzyme **pyruvate kinase** (Chapter 9, Section E,1), which normally produces the enolate anion of pyruvate as a product, appears to be identical to oxaloacetate decarboxylase.[162a]

e. Carbon Dioxide or Bicarbonate Ion?

A question that is bound to arise in the consideration of carboxylation and decarboxylation reactions is whether the reactant or product is CO_2 or HCO_3^-. An approach to answering the question was first suggested by Krebs and Roughton,[163] who pointed out that the equilibrium between free CO_2 and HCO_3^- (Eq. 7-78) is not attained

$$CO_2 + H_2O \underset{}{\overset{a}{\rightleftharpoons}} H_2CO_3 \underset{}{\overset{b}{\rightleftharpoons}} H^+ + HCO_3^- \qquad (7\text{-}78)$$

instantaneously but may require several seconds. If an enzyme produces CO_2 as a product and progress of the reaction is followed manometrically, the pressure will rise higher than the equilibrium value as the CO_2 is evolved. Later when substrate is exhausted and the CO_2 equilibrates with bicarbonate, the pressure will fall again. The addition of carbonic anhydrase, which catalyzes Eq. 7-78a abolishes the "overshoot." If bicarbonate is the primary product of a decarboxylation, there is a lag in the appearance of free CO_2.

A second approach is to use [^{18}O]bicarbonate and to follow the incorporation of ^{18}O into a carboxylated substrate. If CO_2 is the primary substrate only two labeled oxygen atoms enter the compound, whereas, if HCO_3^- is the reactant three are incorporated.[164] A third technique is measurement of the *rate* of incorporation of CO_2 or bicarbonate into the carboxylated product. Over a short interval of time, e.g., 1 min, different kinetics will be observed for the incorporation of CO_2 and of bicarbonate.[165]

Using these methods, it has been established that the product formed in Eqs. 7-76 and 7-77 is CO_2. However, the carboxylation enzymes considered in the next section use bicarbonate as the substrate.

f. Incorporation of Bicarbonate into Carboxyl Groups

An important enzyme with a biosynthetic function in many bacteria and plants is **PEP carboxylase,** which catalyzes the reaction of Eq. 7-79. This enzyme, in effect, accomplishes the reverse of Eq.

$HCO_3^- + PEP \longrightarrow$

$$^-OOC-\overset{\overset{O}{\|}}{C}-CH_2-COO^- + P_i \qquad (7\text{-}79)$$

7-76 by converting the 3-carbon PEP into the 4-carbon oxaloacetate. The latter is needed for "priming" of the citric acid cycle and for biosynthesis of such amino acids as aspartate and glutamate. That the enzyme functions in this way is indicated by the fact that mutants of *Salmonella* defective in this enzyme do not grow unless oxaloacetate or some other intermediate in the citric acid cycle is added to the medium. The enzyme from *S. typhimurium* is a tetramer of MW ~ 400,000 with complex regulatory properties. The corresponding enzyme from spinach has 12 subunits and 12 bound Mn^{2+} ions. The enzyme has a special function in the so-called C_4 plants where it functions in a carbon dioxide concentrating system (Chapter 13, Section E,9,c).

When [^{18}O]bicarbonate is a substrate, two of the labeled oxygen atoms enter the oxaloacetate while the third appears in P_i. A possible explanation is illustrated in Eq. 7-80; the ^{18}O atoms are desig-

nated with asterisks. The carboxyl group enters on the *si*-face of PEP.

PEP carboxylase is lacking from animal tissues and fungi. In these creatures PEP is converted to pyruvate which is then carboxylated to oxaloacetate with coupled cleavage of ATP. The enzyme that forms oxaloacetate, **pyruvate carboxylase,** not only utilizes bicarbonate ion but also contains the vitamin **biotin** bound at the active site (Chapter 8, Section C). Nevertheless, there may be mechanistic similarities to the action of PEP carboxylase (compare Eqs. 7-80 and Eq. 8-6).

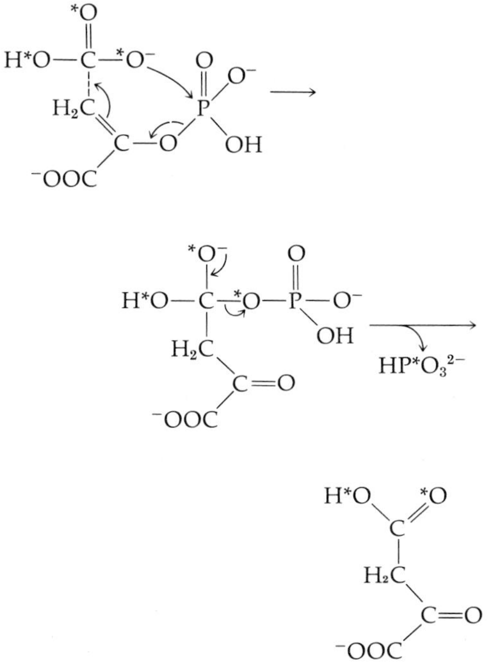

g. Ribulose Diphosphate Carboxylase

When ^{14}C-labeled CO_2 enters chloroplasts of green plants, the first organic ^{14}C-containing compound detected is 3-phosphoglycerate. Two molecules of this compound are formed through the action of **ribulose-1,5-diphosphate carboxylase,** an enzyme present in chloroplasts and making up 16% of the protein of spinach leaves. The enzyme is found throughout green plants and in purple and green bacteria. The reaction differs from other carboxylations in that the carboxylated product is split by the same enzyme. The

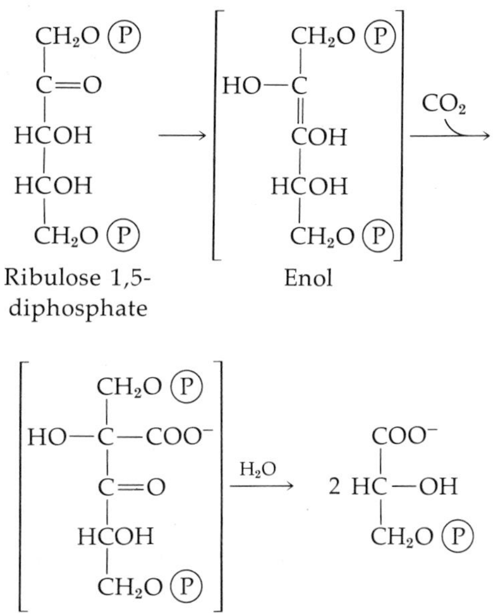

(7-80)

(7-81)

3-Phosphoglycerate

structure of the substrate, for which the enzyme is absolutely specific, does not permit a direct β carboxylation to form the observed product. Indirect evidence suggests the mechanism shown in Eq. 7-81. Ribulose diphosphate is converted to the enol (or possibly on to a 3-keto compound). Loss of a proton from the 3-OH group forms the enolate anion needed for the carboxylation. The first carboxylated product (the second bracketed intermediate in Eq. 7-81) is a β-keto acid. It has been shown[166] to undergo enzyme-catalyzed hydrolytic cleavage, as in Eq. 7-61. Support for this mechanism comes from the observation that the following carboxylic acid is a potent inhibitor, possibly a transition state analogue.[167]

$$
\begin{array}{c}
CH_2O\,\textcircled{P} \\
| \\
HO-C-COO^- \\
| \\
HCOH \\
| \\
HCOH \\
| \\
CH_2O\,\textcircled{P}
\end{array}
$$

Ribulosediphosphate carboxylase from spinach has a molecular weight of ~560,000, contains eight pairs of nonidentical subunits and eight active sites. It requires a divalent metal such as Mg^{2+}. The value of K_m for total CO_2 ($CO_2 + HCO_3^-$) is high, 11–30 mM. For the true substrate CO_2, K_m is only 0.45 mM. In intact chloroplasts the affinity for substrate is distinctly higher, K_m for total CO_2 dropping to ~0.6 mM. Part of the difference appears to result from allosteric activation by fructose 6-phosphate and ribose 5-phosphate in the chloroplasts.[168,169]

It is well known that O_2 inhibits photosynthesis in most plants. Part of the explanation is probably that O_2 competes directly for CO_2 at the active site of ribulose diphosphate carboxylase.[169,170] Chloroplasts inhibited by oxygen produce **glycolic acid** in relatively large amounts.[171] One origin seems to be

the reaction of the enolate ion derived from the intermediate in Eq. 7-81 with O_2 (Eq. 7-82). The

Enolate anion   Peroxide

Phosphoglycolate   3-Phospho-
                      glycerate   (7-82)

resulting peroxide would break up under the hydrolytic action of the enzyme to form phosphoglycolate and 3-phosphoglycerate. Although this conclusion is generally accepted it is surprising. Molecular oxygen usually does not react rapidly with organic substrates (with a few exceptions, e.g., dihydroflavins, Chapter 8, Section I,7) except under the influence of a transition metal ion (Chapter 10, Section B,3). Further investigation of this unexpected reaction of ribulosediphosphate carboxylase should be of great theoretical and practical interest, the latter because of its significance in lowering the yield in photosynthesis (Chapter 13, Section E,9).

K. SOME ISOMERIZATION AND REARRANGEMENT REACTIONS

There are a few metabolic reactions that do not fit into any of the categories discussed so far and which apparently do not depend upon a coenzyme. The reactions involve transfer of alkyl groups or of hydrogen atoms from one carbon to

BOX 7-G

GLUTATHIONE, INTRACELLULAR TRIPEPTIDE[a,b]

γ-L-Glutamyl-L-cysteinylglycine

In 1929, F. G. Hopkins discovered the tripeptide glutathione (GSH) and recognized it as a constituent of most if not all cells. Within animal cells the compound is typically present in a concentration of ~1–5 mM. Lower levels are found in bacteria. Glutathione also occurs in green plants and fungi, the usual source of isolation being yeast. A closely related peptide of unknown function, **ophthalmic acid** (originally isolated from the lens of the eye), is identical to glutathione in structure except for the replacement of the SH group by CH_3.

The most interesting chemical characteristics of glutathione are the γ-glutamyl linkage and the presence of a free SH group. The latter can be oxidized to form a disulfide bridge linking two glutathione molecules.

$$2\ GSH - 2\ [H] \longrightarrow G\!-\!S\!-\!S\!-\!G$$
$$E^{\circ\prime}\ (pH\ 7) = -0.25\ V\ ^c$$

It is this chemistry that has focused the attention of biochemists on glutathione as an intracellular reducing agent whose primary function may be to protect the SH groups of proteins by keeping them reduced. Glutathione also has a specific role in the reduction of **hydrogen peroxide** (Box 10-A) and of the oxidized form of ascorbic acid (Box 10-G).

Glutathione serves as a coenzyme for a select

number of enzymes including **glyoxalase** (Section K), **maleylacetoacetate isomerase** (Eq. 7-49), and **DDT dehydrochlorinase** (an enzyme that catalyzes the elimination of HCl from molecules of the insecticide and which is especially active in DDT-resistant flies[a]). Glutathione is also a coenzyme for oxidation of formaldehyde to formate, presumably via the hemimercaptal.[b,d]

Meister has proposed a function for glutathione in transport of amino acids across membranes. The details are given in Chapter 14, Section B,3 as is the pathway for biosynthesis of glutathione.

Glutathione is said to be the specific factor eliciting the feeding reaction of *Hydra*; that is, the release of glutathione from injured cells causes the *Hydra* to engulf food.

A bit of special chemistry of glutathione arises from the overlapping pK_a values of the SH group and of the NH_3^+ group. The two higher stepwise pK_a values are 8.74 and 9.62. Various experiments have been designed to establish the microscopic constants (Chapter 4, Section C,3). In one experiment the ratio of the two forms

was found to be ~1.8. The student should be able to use this value together with the stepwise pK_a values to establish the microscopic constants K_a, K_b, K_c, and K_d.

[a] S. G. Whaley, *Adv. Protein Chem.* **21**, 1–34 (1966).

[b] A. Meister, *Metab. Pathways*, 3rd ed. **7**, 101–188 (1975).

[c] G. Gorin, A. Esfandi, and G. B. Guthrie, Jr., *ABB* **168**, 450–454 (1975).

[d] J. I. Goodman and T. R. Tephly, *BBA* **252**, 489–505 (1971).

another. The hydrogen atoms move by direct transfer without exchange with the medium. All of the reactions could involve carbonium ions but there is often more than one mechanistic possibility.

A simple 1,3-proton shift is shown in Table 7-1

as reaction type 6A. An example is the isomerization of oleic acid to *trans*-Δ^{10}-octadecenoic acid (Eq. 7-83) catalyzed by a soluble enzyme from a pseudomonad.[172] A second example is isomerization of **isopentenyl pyrophosphate** to **dimethylallyl**

$$\text{Oleic acid (cis-}\Delta^9\text{-octadecenoic acid)} \longrightarrow \textit{trans-}\Delta^{10}\text{-Octadecenoic acid} \tag{7-83}$$

Pinacol

$$\text{Pinacolone} \tag{7-85}$$

pyrophosphate (Eq. 7-84).[173] The stereochemistry has been investigated using the ^{3}H-labeled compound shown in Eq. 7-84. The *pro-R* proton is lost from C-2 and a proton is added to the *re*-face at C-4. When the reaction was carried out in ^{2}H$_2$O a chiral methyl group was produced[173] as shown in Eq. 7-84. Carbonium ion, anion, and addition–elimination mechanisms must all be considered for this reaction. A concerted proton addition and abstraction is also possible, the observed trans stereochemistry being expected for such a mechanism.

Isopentenyl pyrophosphate

Dimethylallyl pyrophosphate

$$\tag{7-84}$$

Reaction type 6B is allylic rearrangement with simultaneous condensation with another molecule. The reaction illustrated in Table 7-1 occurs during the polymerization of polyprenyl compounds (Chapter 12, Section H). It could be initiated by formation of a carbonium ion through dissociation of Y$^-$ (a pyrophosphate group) or it could begin with the creation of an anion through dissociation of the proton or through addition of a nucleophilic group from the enzyme. Evidence favoring the carbonium ion mechanism has been reported.[173a]

Reaction type 6C (Table 7-1) occurs in the biosynthesis of leucine and valine (Fig. 14-10). The rearrangement is often compared with the nonenzymatic acid-catalyzed pinacol-pinacolone re-

arrangement in which a similar shift of an alkyl group takes place (Eq. 7-85). The enzyme-catalyzed rearrangement presumably gives the structure drawn in brackets in Table 7-1, but the same enzyme always catalyzes reduction with NADH to the diol (the Mg^{2+}-dependent enzyme is called **acetohydroxy acid isomeroreductase**). In fact, a rearrangement has never been observed without the accompanying reduction,[174] and mechanisms other than the one shown in Table 7-1 are possible.

A related reaction is catalyzed by **glyoxalase I**. An SH group from the tripeptide **glutathione** (Box 7-G) adds to the aldehyde carbonyl of the substrate **methylglyoxal**. The adduct undergoes rearrangement with a hydrogen atom shift (Table 7-1). The product, a thioester of D-lactic acid and glutathione, is hydrolyzed by a second enzyme **glyoxalase II**.[175,176] Franzen has designed an elegant model using such compounds as dimethylthioethanolamine.[175] [*Note added in proof.* However, more recent high resolution nmr experiments with glyoxalase I have shown that when the reaction occurs in ^{2}H$_2$O some deuterium is incorporated into the product lactate. With the model system, one atom of ^{2}H was found in the product.[176a] These results suggest that the mechanism of action of glyoxalase I involves an enediol generated from the adduct of methylglyoxal and glutathione. The reaction should therefore be classified with those catalyzed by the sugar isomerases (Section I,1).]

A complex rearrangement initiated by a carbonium ion follows the introduction of a hydroxyl group into sterols (reaction 6D, Table 7-1; see also Eq. 12-31).

REFERENCES

1. W. P. Jencks, "Catalysis in Chemistry and Enzymology," pp. 78–110. McGraw-Hill, New York, 1969.

2. J. N. Lowe and L. L. Ingraham, "An Introduction to Biochemical Reaction Mechanisms." Prentice-Hall, Englewood Cliffs, New Jersey, 1974.

3. T. C. Bruice and S. J. Benkovic, "Bioorganic Mechanisms," 2 vols. Benjamin, New York, 1966.

4. J. E. Dixon and T. C. Bruice, *JACS* **93**, 6592–6597 (1971).

5. E. M. Kosower, "An Introduction to Physical Organic Chemistry," pp. 28 and 81. Wiley (Interscience), New York, 1968.

6. P. Goldman, G. W. A. Milne, and D. B. Keister, *JBC* **243**, 428–434 (1968).

7. M. Doudoroff, H. A. Barker, and W. Z. Hassid, *JBC* **168**, 725–732 and 733–746 (1947).

8. R. Silverstein, J. Voet, D. Reed, and R. H. Abeles, *JBC* **242**, 1338–1346 (1967).

9. F. DeToma and R. H. Abeles, *Fed. Proc., Fed, Am. Soc. Exp. Biol.* **29**, 461 (1970).

10. T. Imoto, L. N. Johnson, A. C. T. North, D. C. Phillips, and J. A. Rupley, *in* "The Enzymes," 3rd ed. (P. D. Boyer, ed.), Vol. 7, pp. 665–868. Academic Press, New York, 1972.

11. D. C. Phillips, *Sci. Am.* **215**, 78–90 (Nov 1966); *PNAS* **57**, 484–495 (1967).

12. S. M. Parsons and M. A. Raftery, *Biochemistry* **11**, 1623–1628 (1972).

13. R. Henderson and J. H. Wang, *Annu. Rev. Biophys. Bioeng.* **1**, 1–26 (1972).

14. R. E. Dickerson and I. Geis, "The Structure and Action of Proteins," pp. 69–78. Harper, New York, 1969.

15. S. K. Banerjee, E. Holler, G. P. Hess, and J. A. Rupley, *JBC* **250**, 4355–4367 (1975).

16. S. K. Banerjee and J. A. Rupley, *JBC* **250**, 8267–8274 (1975).

17. F. W. Dahlquist, T. Rand-Meir, and M. A. Raftery, *PNAS* **61**, 1194–1198 (1968).

18. J.-I. Tu, G. R. Jacobson, and D. J. Graves, *Biochemistry* **10**, 1229–1236 (1971).

19. A. M. Gold, E. Legrand, and G. R. Sánchez, *JBC* **246**, 5700–5706 (1971).

20. E. H. Fischer, A. Pocker, and J. C. Saari, *Essays Biochem.* **6**, 23–68 (1970).

21. J. F. Robyt, C. G. Chittenden, and C. T. Lee, *ABB* **144**, 160–167 (1971).

22. J. F. Robyt and D. French, *JBC* **245**, 3917–3927 (1970).

23. D. French, Y. Chan, and B. England, *Fed. Proc., Fed. Am. Soc. Exp. Biol.* **33**, 1313 (1974).

24. J. F. Robyt and D. French, *ABB* **138**, 662–670 (1970).

25. J. A. Thoma, J. E. Spradlin, and S. Dygert, *in* "The Enzymes," 3rd ed. (P. D. Boyer, ed.), Vol. 5, pp. 115–189. Academic Press, New York, 1971.

26. B. Kassel and J. Kay, *Science* **180**, 1022–1027 (1973).

27. K. J. Kramer, R. L. Felsted and J. H. Law, *JBC* **248**, 3021–3028 (1973).

28. H. Neurath, K. A. Walsh, and W. P. Winter, *Science* **158**, 1638–1644 (1967).

29. D. M. Blow, *in* "The Enzymes," 3rd ed. (P. D. Boyer, ed.), Vol. 3, pp. 185–212. Academic Press, New York, 1971.

30. R. M. Stroud, L. M. Kay, and R. E. Dickerson, *JMB* **83**, 185–208 (1974).

31. M. Krieger, L. M. Kay, and R. M. Stroud, *JMB* **83**, 209–230 (1974).

32. R. M. Stroud, L. M. Kay, and R. E. Dickerson, *Cold Spring Harbor Symp. Quant. Biol.* **36**, 125–140 (1971).

33. J. D. Robertus, J. Kraut, R. A. Alden, and J. J. Birktoft, *Biochemistry* **11**, 4293–4303 (1972).

34. L. T. J. Delbaere, W. L. B. Hutcheon, M. N. G. James, and W. E. Thiessen, *Nature (London)* **257**, 758–763 (1975).

35. G. Robillard and R. G. Shulman, *JMB* **86**, 541–558 (1974).

36. G. A. Rogers and T. C. Bruice, *JACS* **96**, 2473–2481 (1974).

37. W. P. Jencks, "Catalysis in Chemistry and Enzymology," pp. 218–226. McGraw-Hill, New York, 1969.

38. W. H. Cruickshank and H. Kaplan, *JMB* **83**, 267–274 (1974).

39. A. R. Fersht, *Cold Spring Harbor Symp. Quant. Biol.* **36**, 71–73 (1971).

40. W. H. Porter, L. W. Cunningham, and W. M. Mitchell, *JBC* **246**, 7675–7682 (1971).

41. A. N. Glazer and E. L. Smith, *in* "The Enzymes," 3rd ed. (P. D. Boyer, ed.), Vol. 3, pp. 501–546. Academic Press, New York, 1971.

42. J. Drenth, J. N. Jansonius, R. Koekoek, and B. G. Wolthers, *Adv. Protein Chem.* **25**, 79–115 (1971); *in* "The Enzymes," 3rd ed. (P. D. Boyer, ed.), Vol. 3, pp. 485–499. Academic Press, New York, 1971.

43. K. Brocklehurst and M. P. J. Kierstan, *Nature (London), New Biol.* **242**, 167–170 (1973).

44. F. A. Quiocho and W. N. Lipscomb, *Adv. Protein Chem.* **25**, 1–78 (1971).

45. J. A. Hartsuck and W. N. Lipscomb, *in* "The Enzymes," 3rd ed. (P. D. Boyer, ed.), Vol. 3, pp. 1–56. Academic Press, New York, 1971.

46. W. N. Lipscomb, *Chem. Soc. Rev.* **1**, 319–336 (1972).

47. W. L. Bigbee and F. W. Dahlquist, *Biochemistry* **13**, 3542–3549 (1974).

48. B. W. Matthews, L. H. Weaver, and W. R. Kester, *JBC* **249**, 8030–8044 (1974).

49. R. A. Anderson, W. F. Bosron, F. S. Kennedy, and B. L. Vallee, *PNAS* **72**, 2989–2993 (1975).

50. G. Cathala, C. Brunel, D. Chappelet-Tordo, and M. Lazdunski, *JBC* **250**, 6046–6053 (1975).

51. T. W. Reid and I. B. Wilson, *in* "The Enzymes," 3rd ed. (P. D. Boyer, ed.), Vol. 4, pp. 373–415. Academic Press, New York, 1971.

52. D. Chappelet-Tordo, M. Iwatsubo, and M. Lazdunski, *Biochemistry* **13**, 3754–3762 (1974).

53. W. Bloch and M. J. Schlesinger, *JBC* **249**, 1760–1768 (1974).

54. R. L. VanEtten and M. E. Hickey, *Fed. Proc., Fed. Am. Soc. Exp. Biol.* **31**, 451 (abstr.) (1972).

55. R. L. VanEtten, P. P. Waymack, and D. M. Rehkop, *JACS*

96, 6783–6785 (1974).

56. J. Josse and S. C. K. Wong, *in* "The Enzymes," 3rd ed. (P. D. Boyer, ed.), Vol. 4, pp. 499–541. Academic Press, New York, 1971.

57. S. Moore and W. H. Stein, *Science* **180,** 458–464 (1973).

58. F. M. Richards and H. W. Wyckoff, *in* "The Enzymes," 3rd ed. (P. D. Boyer, ed.), Vol. 4, pp. 647–806. Academic Press, New York, 1971.

59. G. C. K. Roberts, E. A. Dennis, D. H. Meadows, J. S. Cohen, and O. Jardetsky, *PNAS* **62,** 1151–1158 (1969).

60. E. Machuga and M. H. Klapper, *JBC* **250,** 2319–2323 (1975).

61. B. Gutte, *JBC* **250,** 889–904 (1975).

62. F. H. Westheimer, *Acc. Chem. Res.* **1,** 70–78 (1968).

63. K. Mislow, *Acc. Chem. Res.* **3,** 321–331 (1970).

64. C. A. Bunton, *Acc. Chem. Res.* **3,** 257–265 (1970).

65. F. Ramirez and I. Ugi, *Adv. Phys. Org. Chem.* **9,** 25–126 (1971).

66. F. Eckstein, *Angew Chem.* **14,** 160–166 (1975).

67. A. S. Mildvan, *Annu. Rev. Biochem.* **43,** 357–399 (1974).

68. S. J. Benkovic and K. J. Schray, *in* "The Enzymes," 3rd ed. (P. D. Boyer, ed.), Vol. 8, pp. 201–238. Academic Press, New York, 1972.

69. A. Arnone, C. J. Bier, F. A. Cotton, V. W. Day, E. E. Hazen, Jr., D. C. Richardson, J. S. Richardson, and A. Yonath, *JBC* **246,** 2302–2316 (1971).

70. F. A. Cotton, C. J. Bier, V. W. Day, E. E. Hazen, Jr., and S. Larsen, *Cold Spring Harbor Symp. Quant. Biol.* **36,** 243–249 (1971).

71. C. B. Anfinsen, P. Cuatrecasas, and H. Taniuchi, *in* "The Enzymes," 3rd ed. (P. D. Boyer, ed.), Vol. 4, pp. 177–204. Academic Press, New York, 1971.

72. G. R. Sánchez, I. M. Chaiken, and C. B. Anfinsen, *JBC* **248,** 3653–3659 (1973).

73. F. A. Cotton, V. W. Day, E. E. Hazen, Jr., and S. Larsen, *JACS* **95,** 4834–4840 (1973).

74. A. J. Brake and B. H. Weber, *JBC* **249,** 5452–5457 (1974).

75. R. S. Anthony and L. B. Spector, *JBC* **246,** 6129–6135 (1971).

76. R. J. Fletterick, D. J. Bates, and T. A. Steitz, *PNAS* **72,** 38–42 (1975).

77. C. C. F. Blake and P. R. Evans, *JMB* **84,** 585–601 (1974).

78. G. E. Schulz, M. Elzinga, F. Marx, and R. H. Schirmer, *Nature (London)* **250,** 120–123 (1974).

78a. R. E. Williams, *Science* **192,** 473–474 (1976).

78b. D. H. Buttlaire and M. Cohn, *JBC* **249,** 5733–5740 (1974).

78c. Y. M. Lee and W. F. Benisek, *JBC* **251,** 1553–1560 (1976).

79. A. S. Mildvan and M. Cohn, *Adv. Enzymol.* **33,** 1–70 (1970).

80. A. S. Mildvan and J. L. Engle, *in* "Methods in Enzymology" (C. H. W. Hirs and S. N. Timasheff, eds.), Vol. 26, Part C, pp. 654–682. Academic Press, New York, 1972.

81. W. J. O'Sullivan, G. H. Reed, K. H. Marsden, G. R. Gough, and C. S. Lee, *JBC* **247,** 7839–7843 (1972).

82. C. H. Fung, A. S. Mildvan, A. Allerhand, R. Komoroski,

and M. C. Scrutton, *Biochemistry* **12,** 620–629 (1973).

83. A large book of papers on magnetic resonance and biochemical systems is Vol. **222** of *Ann. N.Y. Acad. Sci.* (1973).

84. A. C. McLaughlin, J. S. Leigh, Jr., and M. Cohn, *JBC* **251,** 2777–2787 (1976).

85. K. Takai, Y. Kurashina, C. Suzuki-Hori, H. Okamoto, and O. Hayaishi, *JBC* **249,** 1965–1972 (1974).

86. V. Macchia, S. Varrone, H. Weissbach, D. L. Miller, and I. Pastan, *JBC* **250,** 6214–6217 (1975).

87. P. P. Layne and V. A. Najjar, *JBC* **250,** 966–972 (1975).

87a. Z. B. Rose, N. Hamasaki, and S. Dube, *JBC* **250,** 7939–7942 (1975).

88. J. S. Nishimura and F. Grinnell, *Adv. Enzymol.* **36,** 183–202 (1972).

88a. C. M. Bowman and J. S. Nishimura, *JBC* **250,** 5609–5613 (1975).

88b. P. H. Pierson and W. A. Bridger, *JBC* **250,** 8524–8529 (1975).

89. F. C. Wedler and P. D. Boyer, *JBC* **247,** 984–992 (1971).

90. Y. Tsuda, R. A. Stephani, and A. Meister, *Biochemistry* **10,** 3186–3189 (1971).

91. J. A. Todhunter and D. L. Purich, *JBC* **250,** 3505–3509 (1975).

92. L. B. Hersh and W. P. Jencks, *JBC* **242,** 3468–3480 and 3481–3486 (1967).

93. F. Solomon and W. P. Jencks, *JBC* **244,** 1079–1083 (1969).

94. R. W. Benson and P. D. Boyer, *JBC* **244,** 2366–2371 (1969).

94a. H. White and W. P. Jencks, *JBC* **251,** 1688–1699 (1976).

95. C. G. Mead and F. J. Finamore, *Biochemistry* **8,** 2652–2655 (1969).

96. A. D. Bond, B. W. McClelland, J. R. Einstein, and F. J. Finamore, *ABB* **153,** 207–214 (1972).

97. K. Kresse and E. F. Neufeld, *JBC* **247,** 2164–2170 (1972).

98. A. B. Roy and P. A. Trudinger, "The Biochemistry of Inorganic Compounds of Sulphur." Cambridge Univ. Press, London and New York, 1970.

99. W. P. Jencks, "Catalysis in Chemistry and Enzymology," p. 465. McGraw-Hill, New York, 1969.

100. R. G. Kallen and W. P. Jencks, *JBC* **241,** 5851–5863 (1966).

101. J. T. Edsall and J. Wyman, "Biophysical Chemistry," Vol. 1, p. 550ff. Academic Press, New York, 1958.

102. S. Lindskog, L. E. Henderson, K. K. Kannan, A. Liljas, P. O. Nyman, and B. Strandberg, *in* "The Enzymes," 3rd ed. (P. D. Boyer, ed.), Vol. 5, pp. 587–665. Academic Press, New York, 1971.

103. K. K. Kannan, B. Notstrand, K. Fridborg, S. Lövgren, A. Ohlsson, and M. Petef, *PNAS* **72,** 51–55 (1975).

104. J. H. Coates, G. J. Gentle, and S. F. Lincoln, *Nature (London)* **249,** 773–775 (1974).

104a. P. Woolley, *Nature (London)* **258,** 677–682 (1975).

105. For detailed discussion, see W. P. Jencks' "Catalysis in Chemistry and Enzymology." McGraw-Hill, New York, 1969.

106. K. R. Hanson, *JACS* **88,** 2731–2742 (1966).

107. R. Bentley, "Molecular Asymmetry in Biology," Vol. 1, p. 183. Academic Press, New York, 1969.

108. R. M. Waterson and R. L. Hill, *JBC* **247**, 5258–5265 (1972).

109. H. Schulz, *JBC* **249**, 2704–2709 (1974).

110. G. J. Moskowitz and J. M. Merrick, *Biochemistry* **8**, 2748–2755 (1969).

111. R. L. Hill and J. W. Teipel, *in* "The Enzymes," 3rd ed. (P. D. Boyer, ed.), Vol. 5, pp. 539–571. Academic Press, New York, 1971.

112. I. A. Rose, *CRC Crit. Rev. Biochem.* **1**, 33–57 (1972).

113. D. E. Schmidt, Jr., W. G. Nigh, C. Tanzer, and J. H. Richards, *JACS* **91**, 5849–5854 (1969).

114. W. G. Nigh and J. H. Richards, *JACS* **91**, 5847–5848 (1969).

115. J. N. Hansen, E. C. Dinovo, and P. D. Boyer, *JBC* **244**, 6270–6279 (1969).

116. E. C. Dinovo and P. D. Boyer, *JBC* **246**, 4586–4593 (1971).

117. D. P. Hanlon and E. W. Westhead, *Biochemistry* **8**, 4247–4260 (1969).

118. T. Nowak, A. S. Mildvan, and G. L. Kenyon, *Biochemistry* **12**, 1690–1701 (1973).

119. I. A. Rose and E. L. O'Connell, *JBC* **242**, 1870–1879 (1967).

120. J. P. Glusker, *in* "The Enzymes," 3rd ed. (P. D. Boyer, ed.), Vol. 5, pp. 413–439. Academic Press, New York, 1971.

121. T. B. Dougherty, V. R. Williams, and E. S. Younathan, *Biochemistry* **11**, 2493–2498 (1972).

122. L. L. Ingraham, "Biochemical Mechanisms," pp. 48–58. Wiley, New York, 1962.

123. J. Kiss, *Adv. Carbohydr. Chem. Biochem.* **29**, 229–303 (1974).

124. F. G. Bordwell, J. Weinstock, and T. F. Sullivan, *JACS* **93**, 4728–4735 (1971).

125. S. Seltzer, *in* "The Enzymes," 3rd ed. (P. D. Boyer, ed.), Vol. 6, pp. 381–406. Academic Press, New York, 1972.

126. G. J. Schroepfer, Jr., *JBC* **241**, 5441–5447 (1966).

127. R. K. Hill and G. R. Newkome, *JACS* **91**, 5893–5894 (1969).

128. P. R. Andrews, G. D. Smith, and I. G. Young, *Biochemistry* **12**, 3492–3498 (1973).

129. R. G. Annett and G. W. Kosicki, *JBC* **244**, 2059–2067 (1969).

130. E. A. Noltmann, *in* "The Enzymes," 3rd ed. (P. D. Boyer, ed.), Vol. 6, pp. 271–354. Academic Press, New York, 1972.

131. J. C. Speck, Jr., *Adv. Carbohydr. Chem.* **13**, 63–103 (1958).

132. D. J. Cram and L. Gosser, *JACS* **86**, 2950–2952 (1964).

133. W. P. Jencks, "Catalysis in Chemistry and Enzymology," pp. 207–213. McGraw-Hill, New York, 1969.

134. K. J. Schray and I. A. Rose, *Biochemistry* **10**, 1058–1062 (1971).

135. J. R. Knowles, P. L. Leadlay, and S. G. Maister, *Cold Spring Harbor Symp. Quant. Biol.* **36**, 157–164 (1971).

136. D. W. Banner, *et al.*, *Nature (London)* **255**, 609–614 (1975).

137. P. Talalay and A. M. Benson, *in* "The Enzymes," 3rd ed. (P. D. Boyer, ed.), Vol. 6, pp. 591–618. Academic Press, New York, 1972.

138. S. K. Malhotra and H. J. Ringold, *JACS* **87**, 3228–3236 (1965).

139. J. P. Klinman and I. A. Rose, *Biochemistry* **10**, 2259–2266, 2267–2272 (1971).

140. D. J. Hopper and R. A. Cooper, *BJ* **128**, 321–329 (1972).

141. H. Lenz, P. Wunderwald, V. Buschmeier, and H. Eggerer, *Hoppe-Seyler's Z. Physiol. Chem.* **352**, 517–519 (1971).

142. U. Gehring and F. Lynen, *in* "The Enzymes," 3rd ed. (P. D. Boyer, ed.), Vol. 7, pp. 391–405. Academic Press, New York, 1972.

143. J. A. Kornblatt and H. Rudney, *JBC* **246**, 4417–4423 (1971).

144. B. L. Horecker, O. Tsolas, and C. Y. Lai, *in* "The Enzymes," 3rd ed. (P. D. Boyer, ed.), Vol. 7, pp. 213–258. Academic Press, New York, 1972.

145. C. Y. Lai and B. L. Horecker, *Essays Biochem.* **8**, 149–178 (1972).

146. C. Y. Lai, N. Nakai, and D. Chang, *Science* **183**, 1204–1205 (1974).

147. P. Model, L. Ponticorvo, and D. Rittenberg, *Biochemistry* **7**, 1339–1347 (1968).

148. P. Christen and J. F. Riordan, *Biochemistry* **7**, 1531–1538 (1968).

149. O. Tsolas and B. L. Horecker, *in* "The Enzymes," 3rd ed. (P. D. Boyer, ed.), Vol. 7, pp. 259–280. Academic Press, New York, 1972.

150. W. A. Wood, *in* "The Enzymes," 3rd ed. (P. D. Boyer, ed.), Vol. 7, pp. 281–302. Academic Press, New York, 1972.

151. M. J. P. Higgins, J. A. Kornblatt, and H. Rudney, *in* "The Enzymes," 3rd ed. (P. D. Boyer, ed.), Vol. 7, pp. 407–434. Academic Press, New York, 1972.

151a. H. M. Miziorko, K. C. Clinkenbeard, W. D. Reed, and M. D. Lane, *JBC* **250**, 5768–5773 (1975).

152. L. B. Spector, *in* "The Enzymes," 3rd ed. (P. D. Boyer, ed.), Vol. 7, pp. 357–389. Academic Press, New York, 1972.

153. G. Gottschalk, S. Dittbrenner, H. Lenz, and H. Eggerer, *EJB* **26**, 455–461 (1972).

154. H. Lenz, W. Buckel, P. Wunderwald, G. Biedermann, V. Buschmeier, H. Eggerer, J. W. Cornforth, J. W. Redmond, and R. Mallaby, *EJB* **24**, 207–215 (1971).

155. J. Rétey, J. Lüthy, and D. Arigoni, *Nature (London)* **226**, 519–521 (1970).

156. R. W. Guynn and R. L. Veech, *JBC* **248**, 6966–6972 (1973).

157. L. B. Hersch, *JBC* **248**, 7295–7303 (1973).

158. J. B. Robinson, Jr., M. Singh and P. A. Srere, *PNAS* **73**, 1872–1876 (1976).

158a. M. Singh and P. A. Srere, *JBC* **250**, 5818–5825 (1975).

159. S. P. Bessman and E. C. Layne, Jr., *Arch. Biochem.* **26**, 25–32 (1950).

160. R. Steinberger and F. H. Westheimer, *JACS* **73**, 429–435 (1951).

161. I. Fridovich, *in* "The Enzymes," 3rd ed. (P. D. Boyer, ed.), Vol. 6, pp. 255–270. Academic Press, New York, 1972.

162. M. F. Utter and H. M. Kolenbrander, *in* "The Enzymes,"

3rd ed. (P. D. Boyer, ed.), Vol. 6, pp. 117–168. Academic Press, New York, 1972.

162a. P. S. Noce and M. F. Utter, *JBC* **250**, 9099–9105 (1975).

163. H. A. Krebs and F. J. W. Roughton, *BJ* **43**, 550–555 (1948).

164. Y. Kaziro, L. F. Hass, P. D. Boyer, and S. Ochoa, *JBC* **237**, 1460–1468 (1962).

165. T. G. Cooper, T. T. Tchen, H. G. Wood, and C. R. Benedict, *JBC* **243**, 3857–3863 (1968).

166. M. I. Siegel and M. D. Lane, *JBC* **248**, 5486–5498 (1973).

167. M. I. Siegel, M. Wishnik, and M. D. Lane, *in* "The Enzymes," 3rd ed. (P. D. Boyer, ed.), Vol. 6, pp. 169–192. Academic Press New York, 1972.

168. B. B. Buchanan and P. Schürmann, *Curr. Top. Cell. Regul.* **7**, 1–20 (1973).

169. F. J. Ryan and N. E. Tolbert, *JBC* **250**, 4229–4233 (1975).

170. G. Bowes and W. L. Ogren, *JBC* **247**, 2171–2176 (1972).

171. J. A. Bassham and M. Kirk, *Plant Physiol.* **52**, 407–411 (1973).

172. C. E. Mortimer and W. G. Niehaus, Jr., *JBC* **249**, 2833–2842 (1974).

173. K. Clifford, J. W. Cornforth, R. Mallaby, and G. T. Phillips, *J. Chem. Soc., Chem. Commun.* pp. 1599–1600 (1971).

173a. C. D. Poulter, D. M. Satterwhite, and H. C. Rilling, *JACS* **98**, 3376–3377 (1976).

174. S. Dagley and D. E. Nicholson, "An Introduction to Metabolic Pathways," pp. 48–52. Wiley, New York, 1970.

175. V. Franzen, *Chem. Ber.* **90**, 623–633 (1957).

176. I. A. Rose, *BBA* **25**, 214–215 (1957).

176a. S. Hall, A. M. Doweyko, and F. Jordan, *JACS* **98**, 7460–7461 (1976).

STUDY QUESTIONS

1. Glycogen synthetase catalyzes the addition of glucosyl units to the nonreducing ends of glycogen chains from uridine diphosphate-glucose according to Eq. 11-24c. Discuss possible mechanisms for this reaction. Does inversion occur? What experiments would you suggest to clarify the mechanism? [See W. Stalmans and H. G. Hers, *in* "The Enzymes," 3rd ed. (P. D. Boyer, ed.), Vol. IX, pp. 309–326. Academic Press, New York, 1973].

2. Pancreatic lipase catalyzes hydrolysis of triglycerides to diglycerides and monoglycerides in the intestinal tract. Diethyl *p*-nitrophenyl phosphate inactivates the enzyme with stoichiometric release of *p*-nitrophenol. The phosphate becomes bound to a serine residue in the sequence Leu·Ser· Gly·His. The enzyme is inactivated by photooxidation and has a pH–activity profile suggesting an essential group with a $pK_a = 5.8$. [P. Desnuelle, *in* "The Enzymes," 3rd ed. (P. D. Boyer, ed.), Vol. VII, pp. 575–616. Academic Press, New York, 1972]. Suggest a reasonable mechanism for the action of this enzyme.

3. Distinguish between (a) hydrolases and hydratases, (b) phosphatases and phosphorylases, (c) exopeptidases and endopeptidases, (d) pepsin and cathepsins, (e) trypsin and chymotrypsin, (f) trypsin and trypsinogen.

4. Compare the properties and mechanisms of action of trypsin, pepsin, and carboxypeptidase; carboxypeptidase and carbonic anhydrase; fructosediphosphate aldolase and acetoacetate decarboxylase.

5. One of the steps in the biosynthesis of purines is the following reaction that is catalyzed by a single enzyme, glycineamide ribotide synthetase. (Fig. 14-31, reaction *b*).

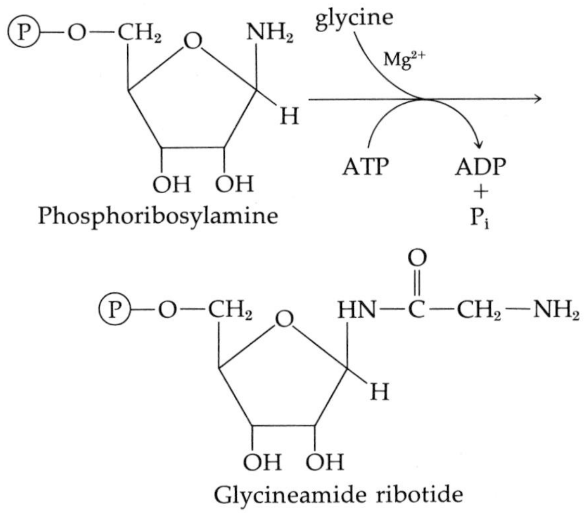

Phosphoribosylamine

Glycineamide ribotide

The reaction is reversible, and in the reverse reaction arsenate can replace phosphate (but without synthesis of ATP). When ^{18}O-containing orthophosphate was added, one atom of ^{18}O from this phosphate was found in the glycine produced by the reverse reaction.

Propose a reasonable mechanism.

6. The following results were obtained with the enzyme ethanol kinase from a pseudomonad:

Plots of the reciprocal of the initial velocity against 1/[ethanol] at different fixed concentrations of ATP gave converging lines. The same was true when the reciprocal of initial reaction velocity was plotted against 1/[ATP] at different fixed levels of ethanol. There was a very appreciable rate for the ATP $\rightleftarrows$ ADP exchange reaction, but no ethanol $\rightleftarrows$ ethanol phosphate exchange with the kinase preparation. When the enzyme is incubated with [γ-^{3}P]ATP it is found to contain ^{32}P after passage through Sephadex. The Sephadex-treated enzyme will not transfer the ^{32}P to ethanol; however, [^{32}P]-phosphate is slowly liberated into the medium.

a. Suggest a mechanism for the action of this enzyme.

b. Explain the various experimental observations in light of your answer in (a).

7. A 2-keto-3-deoxy-L-arabonate dehydratase from *Pseudomonas saccharophila* catalyzes the following reaction:

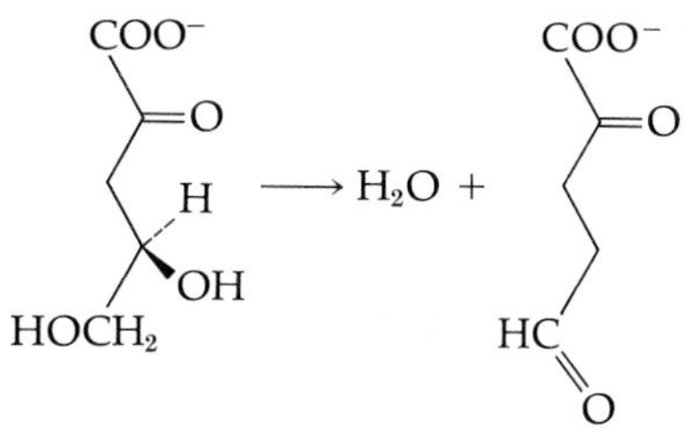

Borohydride reduction in the presence of substrate fixed the substrate to the enzyme. Propose a reasonable mechanism.

8

Coenzymes— Nature's Special Reagents

Most of the reactions discussed in Chapter 7 are promoted by enzymes that contain only those functional groups found in the side chains of the amino acids. *Coenzymes often serve as additional reagents needed for reactions that would be difficult or impossible using only simple acid–base catalysis.* In many instances *coenzymes also serve as carriers,* alternating catalysts that accept and donate chemical groups, hydrogen atoms, or electrons. Coenzymes can be considered in three groups:

1. Compounds of high group transfer potential such as ATP and GTP that function in energy coupling within cells. Because ATP is cleaved and then dissociates from the enzyme to which it is bound, ATP is most often regarded as a substrate rather than a coenzyme. However, it is appropriate to view it as a phosphorylated form of AMP or ADP serving as a carrier of "high-energy" phosphoryl groups.

2. Compounds, often derivatives of vitamins (Box 8-A) which, while at the active site of an enzyme, react to alter the structure of a substrate in a way that permits it to react more readily. The majority of the coenzymes including coenzyme A, pyridoxal phosphate, thiamine diphosphate, and vitamin B_{12} coenzymes fall into this group.

3. Oxidative coenzymes with special structures of precisely determined oxidation–reduction potential which serve as carriers of hydrogen atoms or electrons, e.g., NAD^+, $NADP^+$, FAD, and lipoic acid.

Some of the coenzymes such as NAD^+ and $NADP^+$ are easily dissociable from proteins and function by carrying hydrogen atoms from one enzyme to another. On the other hand, many coenzymes, including FAD, are more tightly bound and rarely if ever dissociate from the protein catalyst. Heme groups are covalently linked to proteins such as cytochrome *c* and cannot be dissociated without destroying the enzyme. Very tightly bound coenzyme groups are often called **prosthetic groups,** but there is no sharp line dividing prosthetic groups from the loosely bound coenzymes.

A. ATP AND THE NUCLEOTIDE "HANDLES"

The role of ATP in "driving" biosynthetic reactions has been considered in Chapter 7, Section F. In that discussion attention was focused entirely on the polyphosphate group which undergoes cleavage. What about the adenosine end? Here is a shapely structure borrowed from the nucleic acids. What is it doing as a carrier of the phosphoryl groups of ATP? At least part of the answer seems to be that the adenosine monophosphate (AMP) portion of the molecule is a "handle" which can be "grasped" by catalytic proteins. In

The adenosine "handle"

the case of acetate thiokinase (Eq. 7-30), the handle is important because the intermediate acyl adenylate must remain tightly bound to the protein. Without the large adenosine group, there would be little for the protein to hold on to.

AMP is only one of several handles to which nature attaches phosphoryl groups to form di- and triphosphate derivatives. Like AMP, the other handles too are nucleotides, the monomer units of nucleic acids. Thus, one enzyme requiring a polyphosphate as an energy source selects ATP, another CTP or GTP. Furthermore, the nucleotide handles not only carry polyphosphate groups but also are present in other coenzymes, such as CoA, NAD^+, $NADP^+$, and FAD. In addition, they often serve as carriers for various small organic molecules. The latter become active metabolic intermediates such as **uridine diphosphate glucose** (UDP-glucose or UDPG), important in sugar metabolism (Chapter 11, Section E,1,b) and **cytidine diphos-**

phate choline, an intermediate in synthesis of phospholipids (Eq. 11-26).

Recalling that acetyl adenylate (acetyl AMP) is an intermediate in synthesis of acetyl-CoA and comparing the biosynthesis of sugars, of phospholipids and of acetyl-CoA, we see that in each case the enzyme involved is specific for a different nucleotide handle. The handle provides a means of recognition by which an enzyme can pick the right bit of raw material out of the sea of molecules surrounding it. Of course, that is not the whole story because the molecule selected by an enzyme must be correct in its entirety, not just in the handle.

Figure 2-25 shows the shapes of the four purine and pyrimidine bases forming the most common nucleotide handles. The outlines represent the surfaces as given by the van der Waals radii and the arrows show *some* of the directions in which hydrogen bonds can be formed to adjacent groups. The distinctive differences between the four bases, both in shape and hydrogen bond patterns, are immediately obvious. In binding to proteins, the hydrogen bond-forming groups in the purine and pyrimidine bases may interact with precisely positioned groups in the protein. The ribose or deoxyribose ring contains additional groups that can hydrogen bond to a protein and the negatively charged oxygen atoms of the 5'-phosphate can interact with positively charged protein side chains.

Are the nucleotide handles inert groups of constant shape or do they enter into some active chemistry? The question is rarely asked by biochemists, perhaps because nothing to date has suggested direct participation in catalysis. However, in biochemistry we are always prepared for surprises and it may turn out that the "handles" participate in catalysis just as do the traditional "business ends" of the coenzymes.

B. COENZYME A AND PHOSPHOPANTETHEINE

The existence of a special coenzyme required in biological acetylation was recognized by Fritz Lip-

THE DISCOVERY OF THE VITAMINS

Up until the present century, several mysterious and often fatal diseases which resulted from vitamin deficiencies were common. Sailors on long sea voyages were often the victims, and centuries ago some perceptive individuals recognized that diet was at fault. In the Orient, the disease **beriberi** was rampant and millions died of its strange paralytic effects ("polyneuritis"). A Chinese friend tells me that generations of Chinese knew that a tea made from rice bran cured beriberi, but this knowledge was either not widely available or was not believed.

In 1893, a Dutch physician named C. Eijkman working in Java produced paralysis in chicks by feeding them the white rice consumed by the local populace. Eijkman also showed that the paralysis could be relieved promptly by feeding an extract of rice polishings. At first he thought there was something toxic in the white rice which was neutralized by a material from the bran. Later he concluded correctly that the rice bran contained an essential nutrient.

In 1912, the Polish biochemist Casimir Funk formulated the "vitamine theory" according to which the diseases **beriberi, pellagra, rickets,** and **scurvy** resulted from lack in the diet of four different vital nutrients. Funk imagined them all to be amines, hence, the name **vitamine.** In the same year in England, F. G. Hopkins announced that he had fed rats on purified diets and discovered that amazingly small amounts of "accessory growth factors," which could be obtained from milk, were necessary for normal growth.[a]

By 1915 McCollum and Davis at the University of Wisconsin had recognized that rat growth depended on not one but two accessory factors. The first, soluble in fatty solvents, they called "A" and the other, soluble in water, they designated "B." Factor B cured beriberi in chicks. After some years when it was shown that vitamine A was not an amine, the "e" was dropped and **vitamin** became the general term for the accessory growth factors.

Progress in isolation of the vitamins was slow, principally because of a lack of interest. According to R. R. Williams, when he started his work on isolation of the anti-beriberi factor in 1910 most people were convinced that his efforts

mann in 1945.[1] The joining of acetyl groups to other molecules is a commonplace reaction within living cells, one example being the formation of **acetylcholine,** an essential chemical "transmitter" of nervous impulses (Chapter 16, Section B,4). In the organic chemical laboratory acetylation is carried out with reactive compounds such as acetic anhydride or acetyl chloride. Lipmann wondered what nature used in their place. His approach in seeking the biological "active acetate" is one that has been used successfully in solving many biochemical problems.

1. Isolating the New Coenzyme

Lipmann first set up a test system: He examined the ability of extracts prepared from fresh liver tissue to catalyze the acetylation of sulfanilamide

(Eq. 8-1). A specific color test was available for quantitative determination of very small amounts of the product. Thus, the rate of acetylation of sulfanilamide under standard conditions was a measure of the activity of the biochemical acetylation system. Lipmann soon discovered that the reaction required ATP and that the ATP was cleaved to

$$\text{Sulfanilamide} \xrightarrow{\text{acetylation}}$$

(8-1)

Box 8-A (*Continued*)

were doomed to failure, so ingrained was Pasteur's idea that diseases were caused only by bacteria. In 1926, Jansen isolated a small amount of thiamine, but it was not until 1933 that R. R. Williams, working almost without financial support, succeeded in preparing a large amount of the crystalline compound from rice polishings. Characterization and synthesis followed rapidly.

It soon became apparent that the new vitamin alone would not satisfy the dietary need of rats for the "B" factor. A second thermostable factor (B_2) was required in addition to thiamine (B_1), which was very labile and easily destroyed by heating. Within a few years, it was clear that this second factor also contained more than one component and the mixture was designated **vitamin B_2 complex.** It was not until relatively specific animal tests for each one of the members had been devised that the confusion was resolved. Then it became clear that **riboflavin** was most responsible for the stimulation of rat growth, while **vitamin B_6** was needed to prevent a facial dermatitis or rat "pellagra." **Pantothenic acid** was especially effective in curing a chick dermatitis, while **nicotinamide** was required to cure human pellagra. **Biotin**

was required for growth of yeast.

The anti-scurvy ("antiscorbutic") activity was called **vitamin C,** and when its structure became known, **ascorbic acid.** The fat-soluble factor preventing rickets was designated **vitamin D.** By 1922 the need for another fat-soluble factor, **vitamin E,** essential for full-term pregnancy in the rat, was recognized, and by the early 1930's **vitamin K** and the **essential fatty acids** were added to the list. Study of blood disorders of man, "tropical macrocytic anemia" and "pernicious anemia," led to recognition of two more water-soluble vitamins **folic acid** and **vitamin B_{12}.** The latter is required in minute amounts and was not isolated until 1948.

Have all the vitamins been discovered? Most investigators think so. Rats can be reared on an almost completely synthetic diet. However, there is always the possibility that for good health humans require some as yet undiscovered compounds in our diet.

[a] Funk and Hopkins received the Nobel prize in medicine in 1929 for these discoveries.

ADP concurrently with the formation of acetyl sulfanilamide.

A second discovery was that dialysis or ultrafiltration rendered the liver extract almost inactive in acetylation. Apparently some essential material passed out through the semipermeable dialysis membrane. When the dialysate or ultrafiltrate was concentrated and added back, acetylation activity was restored. The unknown material was not destroyed by boiling, and Lipmann postulated that it was a new coenzyme which he called **coenzyme A** (for acetylation). Now the test system was used to estimate the amount of the coenzyme in a given volume of dialysate or in any other sample. When small amounts of CoA were supplied to the test system, only partial restoration of the acetylation activity was observed and the amount of restoration was proportional to the amount of CoA. With test system in hand to monitor various frac-

tionation methods, Lipmann soon isolated the acetylation coenzyme in pure form from yeast and liver.

2. Structure and Functions of Coenzyme A

The structure of CoA (Fig. 8-1) is surprisingly complex. The handle is AMP with an extra phosphoryl group on the 3'-hydroxyl. The phosphate of the 5'-carbon is linked in anhydride (pyrophosphate) linkage to another phosphoric acid, which is in turn esterified with **pantoic acid.** Pantoic acid is linked to **β-alanine** and the latter to **β-mercaptoethylamine** through amide linkages, the reactive SH group being attached to a long (1.9 nm) semiflexible chain.

Coenzyme A can be cleaved by hydrolysis to **pantetheine, pantetheine 4′-phosphate,** and **pantothenic acid** (Fig. 8-1). These three compounds are all **growth factors.** Pantetheine is required by *Lactobacillus bulgaricus,* an organism that lives in milk (and converts milk to yogurt). The bacterium, which finds a ready supply of pantetheine in milk, has lost the ability to synthesize this compound. However, it can convert pantetheine to CoA. Pantetheine 4′-phosphate is required by *Acetobacter suboxydans* and pantothenic acid is a vitamin (Box 8-B).

While CoA was discovered as the "acetylation coenzyme," we know now that it has a far more general function. It is required (as acetyl-CoA) in the citric acid cycle, in the β oxidation of fatty acids, and in innumerable other ways. Because of its importance throughout biochemistry and the stimulation of other investigations which the discovery of CoA produced, Lipmann was awarded a Nobel Prize in 1953.

The two distinct chemical functions of CoA, which have already been considered in Chapter 7 (Sections D,5, H,5, and J,1), are summarized in Table 8-1.

3. Acyl Carrier Proteins and Phosphopantetheine

Synthesis of fatty acids requires a small (MW = 8700 in *E. coli*) acyl carrier protein (ACP) whose functions are similar to those of CoA. However, in acyl carrier proteins the nucleotide handle of the coenzyme is absent and pantetheine 4′-phosphate is covalently bonded through phosphoester linkage directly to a serine residue of the ACP (in the *E. coli* at position 36 in the 77-residue protein[2,3]). We see that the nucleotide handle of CoA has been replaced with a much larger and more complex protein which doubtless interacts in specific ways with the multiprotein complex required for fatty acid synthesis (Chapter 11, Section B,4).

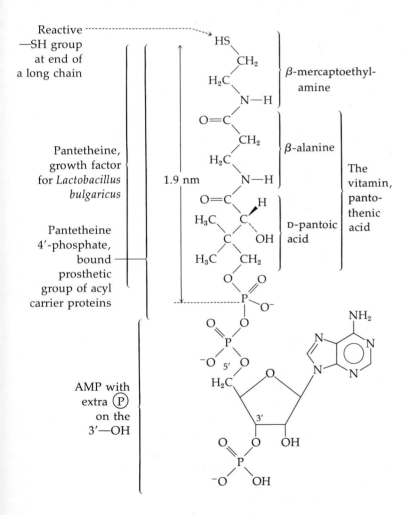

Fig. 8-1 Coenzyme A, an acyl-activating coenzyme.

BOX 8-B

PANTOTHENIC ACID

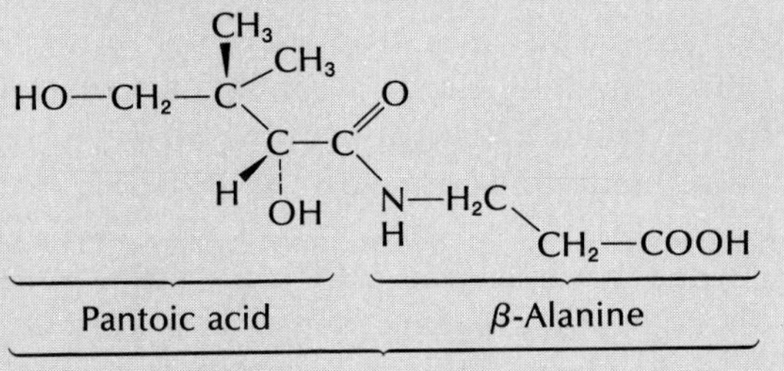

Pantoic acid — β-Alanine

Pantothenic acid MW = 219.2

The recognition of pantothenic acid as a vitamin stems from studies of the role of the "vitamin B_2 complex" in preventing dermatitis of chicks. Investigation of the nutrition of yeast and of lactic acid bacteria also suggested new growth factors. By 1938 it was realized that a single acidic substance present in most natural materials was responsible for all three activities. The new vitamin was soon isolated and characterized. It was given the name pantothenic acid in recognition of its universal occurrence.

Pantothenic acid becomes part of the CoA structure within cells. The reactive center of the latter (Fig. 8-1) is the SH group and the β-alanine portion of pantothenic acid is part of a flexible arm on which the SH group is carried. The mystery is why pantoic acid, a small odd-shaped molecule that the human body cannot make, is so essential for life. Some enzymes will act on simple derivatives of CoA lacking both the nucleotide and pantoic acid portions. However, there must be in our bodies some enzymes that depend upon the unique structure of pantoic acid. Possibly the hydroxyl group reacts in some way. Possibly the two methyl groups enter into formation of a "trialkyl lock" (Chapter 6, Section E,7) part of a sophisticated "elbow" or shoulder for the SH-bearing arm.

Physical properties: A white material that crystallizes poorly. The calcium salt is the usual commercial form.

Nutritional requirement: 10–15 mg/day. Deficiency causes apathy, depression, impaired adrenal function, and muscular weakness. ω-Methylpantothenic acid is a specific antagonist.

Bound phosphopantetheine is also found in a citrate cleaving enzyme (Eq. 7-72) and in enzymes involved in synthesis of peptide antibiotics (Chapter 11, Section E,1,a).

C. BOUND BIOTIN AS A PROSTHETIC GROUP

The vitamin biotin is found in tissues tightly bonded through a covalent linkage to proteins. The first clue to this fact was obtained from isolation of a biotin-containing material **biocytin,** ϵ-*N*-biotinyl-L-lysine, from autolysates of rapidly growing yeast.[4] We know now that the lysine residue of the biocytin was originally present in a protein chain at the active site of biotin-containing enzymes. Thus, in its functional forms the double-ring system of biotin is attached to a protein by a flexible arm.

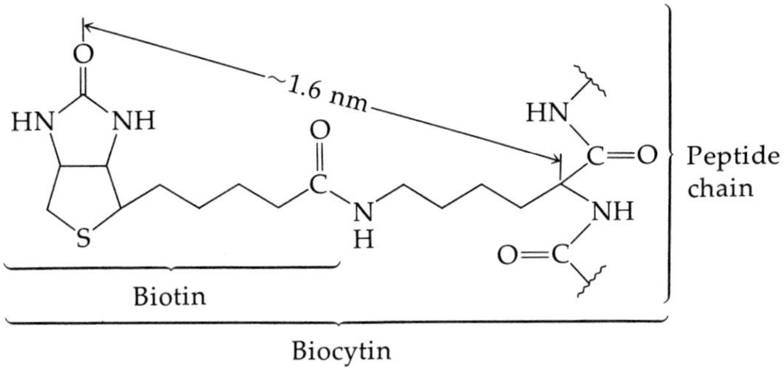

Biotin — Biocytin

Biotin serves as a **carboxyl group carrier** in a series of β carboxylation reactions of type 5C (Table 7-1). This function was originally suggested by the fact that aspartate partially replaces biotin in promoting the growth of the yeast *Torula cremonis.* Aspartate was known to arise by transamination from oxaloacetate which in turn could be formed by β carboxylation of pyruvate. Subsequent studies showed that biotin was needed for an ATP-dependent reaction of pyruvate with bicarbonate ion (Eq. 8-2). Other biotin-requiring

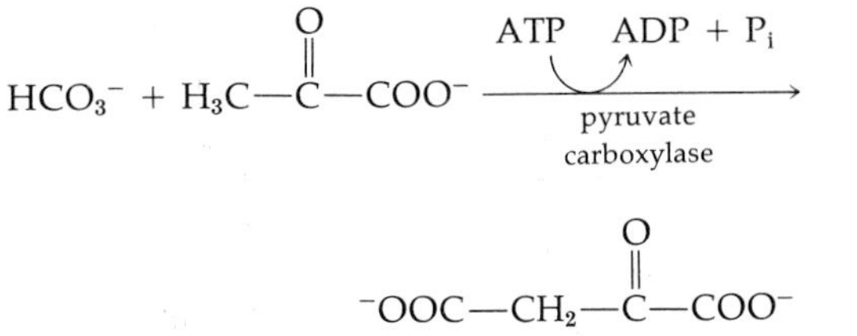

$$HCO_3^- + H_3C-\overset{\overset{\displaystyle O}{\|}}{C}-COO^- \xrightarrow[\substack{\text{pyruvate}\\ \text{carboxylase}}]{ATP \quad ADP + P_i}$$

$$^-OOC-CH_2-\overset{\overset{\displaystyle O}{\|}}{C}-COO^- \qquad (8\text{-}2)$$

carboxylases act on acetyl-CoA, propionyl-CoA, and β-methylcrotonyl-CoA, using HCO_3^- to add carboxyl groups at the sites indicated by the arrows in the accompanying structures. Note that

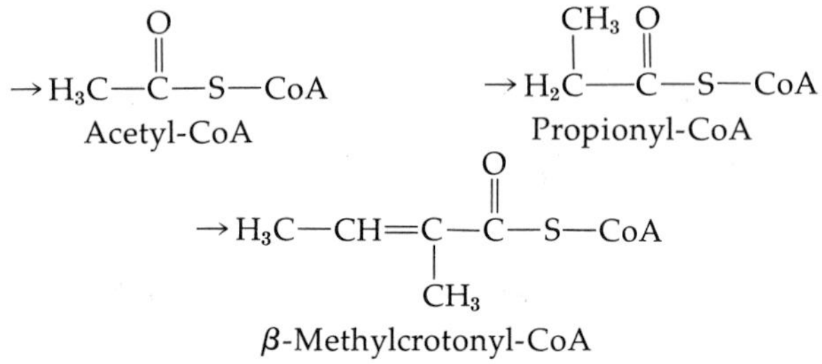

$$\rightarrow H_3C-\overset{\overset{\displaystyle O}{\|}}{C}-S-CoA$$
Acetyl-CoA

$$\rightarrow H_2\overset{\overset{\displaystyle CH_3}{|}}{C}-\overset{\overset{\displaystyle O}{\|}}{C}-S-CoA$$
Propionyl-CoA

$$\rightarrow H_3C-CH=\overset{\underset{\displaystyle CH_3}{|}}{C}-\overset{\overset{\displaystyle O}{\|}}{C}-S-CoA$$
β-Methylcrotonyl-CoA

while carboxylation of β-methylcrotonyl-CoA is not, strictly speaking, β carboxylation (Chapter 7, Section J,3), the presence of the C=C double bond conjugated with the carbonyl group makes it electronically analogous to β carboxylation.

1. The Mechanism of Biotin Action

It may seem surprising that a coenzyme is needed for these carboxylation reactions. However, unless the cleavage of ATP were coupled to the reactions, the equilibria would lie far in the direction of decarboxylation. As a result of coupling with the cleavage of ATP, the carboxylation becomes spontaneous. For example, the measured apparent equilibrium constant K' for carboxylation of propionyl-CoA to S-methylmalonyl-CoA at pH 8.1 and 28°C is[5] given by Eq. 8-3.

$$K' = \frac{[ADP][P_i][\text{methylmalonyl-CoA}]}{[ATP][HCO_3^-][\text{propionyl-CoA}]} = 5.7 \quad (8\text{-}3)$$

$$\Delta G' = -4.36 \text{ kJ mol}^{-1}$$

The function of biotin is to mediate the coupling of ATP cleavage to carboxylation. This is accomplished by a two-stage process in which a **carboxybiotin** intermediate is formed. There is one biotin-containing enzyme that does not utilize ATP. Propionic acid bacteria contain a **carboxyltransferase** which transfers a carboxyl group reversibly from methylmalonyl-CoA to pyruvate to form oxaloacetate and propionyl-CoA. No ATP is needed because free HCO_3^- is not a substrate. However, biotin again plays the role of a carboxyl group carrier. Table 8-2 summarizes the functions of biotin in ATP-dependent carboxylation and transcarboxylation reactions.

The structure of biotin (Box 8-C) suggested that bicarbonate might be incorporated reversibly into position 2' of the vitamin. However, this proved not to be true and it remained for F. Lynen and associates to obtain a clue from a "model reaction." They showed that purified β-methylcrotonyl-CoA carboxylase from rats[6] promotes the carboxylation of *free* biotin with bicarbonate ($H^{14}CO_3^-$) and ATP. While the carboxylated biotin was very labile, treatment with diazomethane (Eq. 8-4) gave a more stable dimethyl ester that was characterized as a derivative of N-1'-carboxybiotin. The impli-

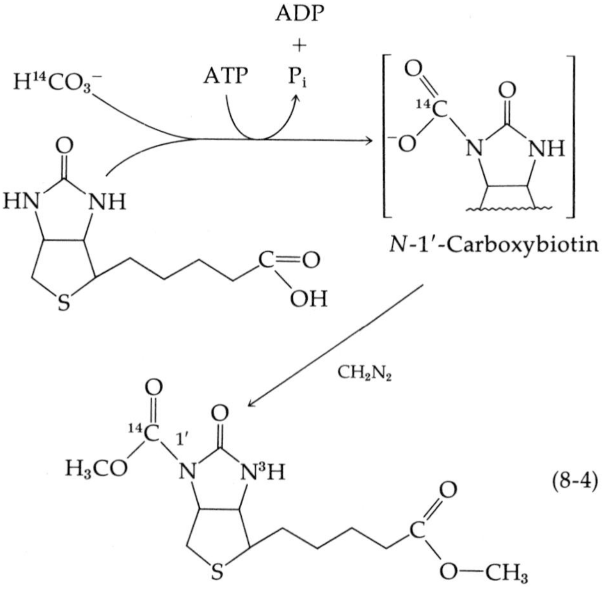

N-1'-Carboxybiotin

(8-4)

TABLE 8-2
Biotin-Assisted β Carboxylation (Reaction Sequence S5A)[a]

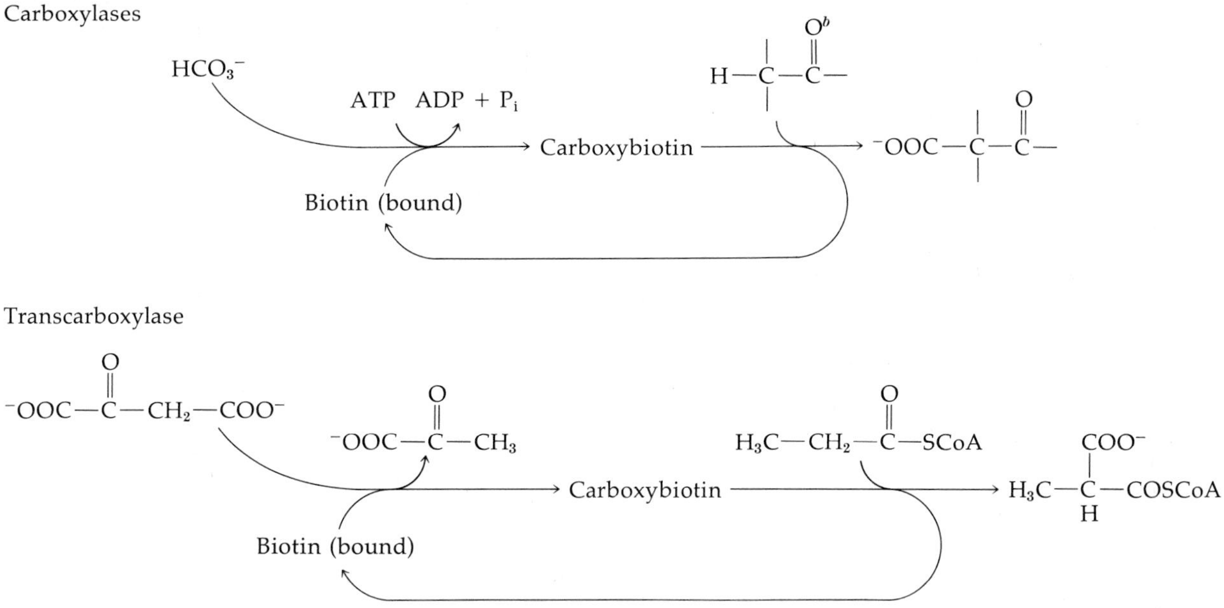

Carboxylases

Transcarboxylase

[a] See Table 7-2.

[b] $HC\overset{*}{-}C=C-\overset{O}{\underset{\|}{C}}-$ can also be carboxylated at the asterisked position.

cation was clear: The carboxylated form of biotin is N-1'-carboxybiotin. The cleavage of ATP is required in the initial step to couple the CO_2 from HCO_3^- to the biotin. The covalently bound biotin at active sites of enzymes was also successfully labeled with $^{14}CO_2$ and treatment with diazomethane followed by hydrolysis with trypsin and pepsin gave authentic N-1'-carboxybiocytin. To complete the catalytic action of an enzyme the carboxyl group need only be transferred from carboxybiotin to the substrate to be carboxylated. Enzymatic transfer of a carboxyl group from chemically synthesized carboxybiotin onto specific substrates has now been demonstrated, in confirmation of the proposed mechanism.[7]

In at least some cases biotin-dependent enzymes can be dissociated into subunits. Thus, **acetyl-CoA carboxylase** of *E. coli* is dissociable into three parts: (1) a **biotin carboxyl carrier protein** (MW = 22,000) containing the covalently bound biotin, (2) **biotin carboxylase** (MW = 100,000, a dimer), and (3) **transcarboxylase** (MW = 90,000, also a dimer). Biotin carboxylase is needed to catalyze the initial ATP-dependent carboxylation of bound biotin attached to the biotin carboxyl carrier protein. The transcarboxylase contains the binding site for acetyl-CoA and catalyzes the transfer of the carboxyl group from the bound carboxybiotin to acetyl-CoA to form pyruvate. It is probably significant that biotin is attached to the protein at the end of a flexible 1.6 nm "arm" which may permit the biotin to move from a site on the carboxylase to a site on the transcarboxylase.

There are important mechanistic questions about both the formation of carboxybiotin by the biotin carboxylase subunit and the transcarboxylase reaction. The fact that ^{18}O from labeled bicarbonate enters the P_i split from ATP in the first

BOX 8-C

BIOTIN

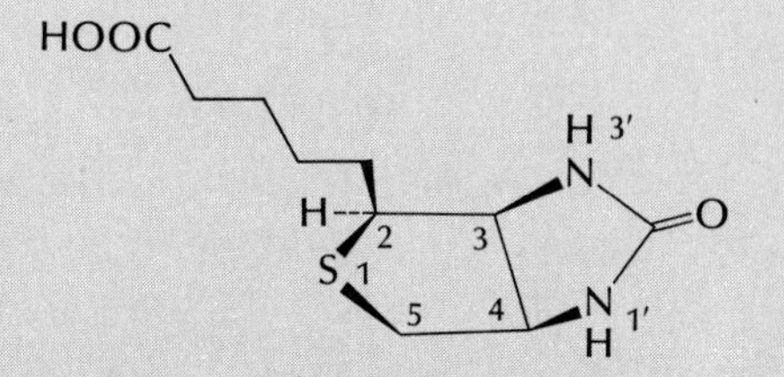

HOOC

H 3'
N
O
H---
S 1 2 3
5 4
N 1'
H

(+)-Biotin
MW = 244.3

By 1901 it was recognized that yeast required for its growth an unknown material which was referred to as **bios.** Bios was eventually shown to be a mixture of pantothenic acid, inositol, and a third component which was given the name biotin. By the 1930's the vitamin had been recognized in two other ways: as a factor promoting growth and respiration of the clover root nodule organism *Rhizobium trifolii* and as "vitamin H." This vitamin was required by rats to prevent a dermatitis and paralysis which developed when the rats were fed large amounts of uncooked egg white. Isolation of the pure vitamin was a heroic task accomplished by Kögl in 1935. In one preparation 250 kg of dried egg yolk yielded 1.1 mg of crystalline biotin (1.4% of the total biotin present in the starting material).

Isomers and derivatives[a]: Biotin contains three chiral centers and therefore has eight stereoisomers. Of these, only one, the dextrorotatory

(+)-biotin is biologically active.[b] The vitamin is readily oxidized to the sulfoxide

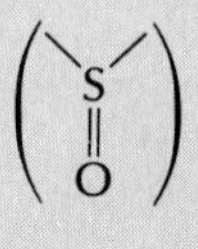

and to the sulfone

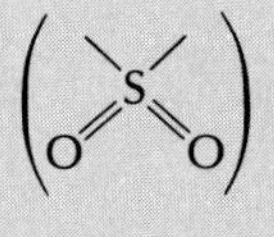

Desthiobiotin, in which the sulfur has been removed and replaced by two hydrogen atoms, can replace biotin in some organisms and appears to lie on one pathway of biosynthesis.[c] **Oxybiotin,** in which the sulfur has been replaced by oxygen, is active for many organisms and partially active for others. No evidence for conversion to biotin itself has been reported, and oxybiotin may function satisfactorily in at least some enzymes.

Daily requirement: For an adult human an intake of 0.15–0.3 mg/day of biotin is regarded as adequate.

[a] B. W. Langer, Jr. and P. György, in "The Vitamins" (W. H. Sebrell, Jr. and R. S. Harris, eds.), 2nd ed., Vol. 2, pp. 294–322. Academic Press, New York, 1968.
[b] G. T. DeTitta, J. W. Edmonds, W. Stallings, and J. Donohue, *JACS* **98**, 1920–1926 (1976).
[c] R. J. Parry and M. G. Kunitani, *JACS* **98**, 4024–4025 (1976).

reaction suggests transient formation, by nucleophilic attack of HCO_3^- on ATP, of **carbonyl phosphate.** The carboxyl group of this reactive anhy-

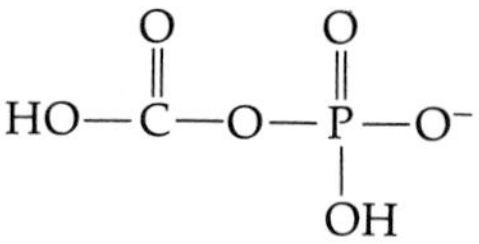

$$HO-\overset{O}{\underset{}{C}}-O-\overset{O}{\underset{OH}{P}}-O^-$$

Carbonyl phosphate

dride would then be transferred to biotin by attack of N-1' of biotin. The two step sequence is essentially sequence S1C(γ) (Table 7-2). Suggestive evi-

dence that carbonyl phosphate *is* an intermediate comes from the observation that biotin carboxylase promotes the transfer of a phosphoryl group to ADP from **carbamoyl phosphate,** an analogue of carbonyl phosphate (Eq. 8-5). The reaction is analogous to the reverse of the reaction step by which

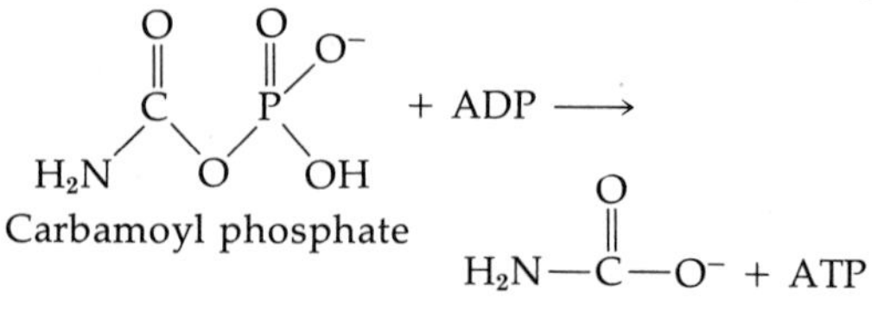

$$H_2N-\overset{O}{\underset{O}{C}}-\overset{O}{\underset{OH}{P}}-O^- + ADP \longrightarrow$$

Carbamoyl phosphate

$$H_2N-\overset{O}{\underset{}{C}}-O^- + ATP \quad (8\text{-}5)$$

carbonyl phosphate is postulated to be formed from ATP and bicarbonate.

Another possibility is that the terminal phosphoryl group of ATP is transferred to biotin to form an enol phosphate (Eq. 8-6, step *a*) and that

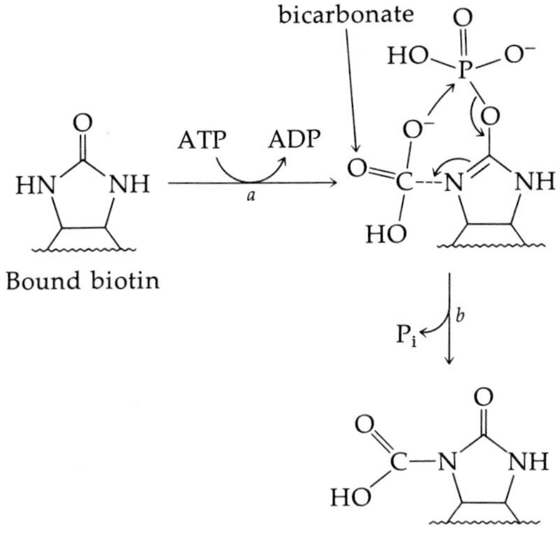

Bound biotin

$$\text{(8-6)}$$

Carboxybiotin

the latter reacts with bicarbonate in a sequence (Eq. 8-6, step *b*) analogous to that of PEP carboxylase (Eq. 7-80). Cleavage of the enol phosphate in step *b* by attack of HCO_3^- would create a highly nucleophilic center at N-1'. At the same time it would form a transient reactive carbonyl phosphate ready to react with N-1'.

The carboxyl group of carboxybiotin is presumably transferred into the final products by nucleophilic attack of an enolate anion on the carbonyl carbon (Eq. 8-7). In the carboxyltransferase reaction (Table 8-2) transfer of tritium from

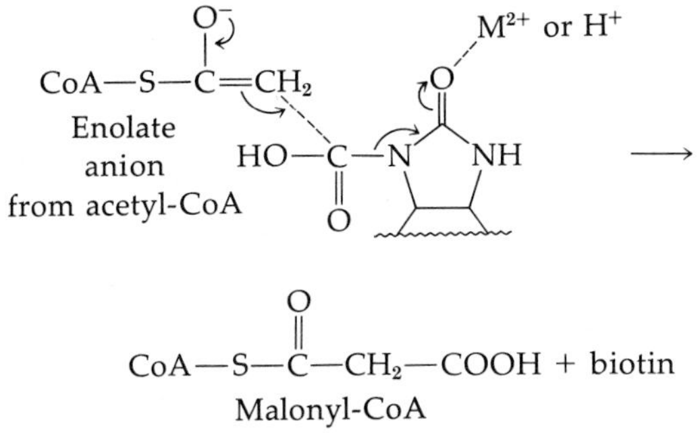

$$\text{(8-7)}$$

Malonyl-CoA

[3-[3H]pyruvate into propionyl-CoA has been observed. This suggests the possibility that the 2'-carbonyl of biotin is the proton-accepting group for the formation of the transient enolate ion.[7a]

A bound divalent metal ion, usually Mn^{2+}, is specifically required in the transcarboxylation step. This fact has permitted the study of the geometry of the binding of substrates relative to Mn^{2+} using epr and nmr relaxation techniques.[8–10] One possible function of the metal would be to assist in the enolization of the carboxyl acceptor. However, in the case of pyruvate carboxylase measurement of the effect of the bound Mn^{2+} on ^{13}C relaxation times in the substrate indicates a distance of ~ 0.7 nm between the carbonyl carbon and the Mn^{2+}, too great for direct coordination of the metal to the carbonyl oxygen. An attractive alternative possibility is that the metal binds to the carbonyl of biotin as indicated in Eq. 8-7. The effect, which could also be brought about by a hydrogen-bonded proton, is to make the biotin a better leaving group in the displacement.[11]

2. Other Properties of Biotin-Dependent Enzymes[5]

Most biotin-dependent carboxylases are large proteins. For example, the protomer of pyruvate carboxylase has a molecular weight of 410,000 per biotin molecule. In the presence of the allosteric activator, citrate, it polymerizes to a "large form" of MW = 4–8×10^6. Ten to twenty protomers associate into filaments 7–10 nm wide by 400 nm long. Only this large form is enzymatically active. Acetyl-CoA carboxylase from rat and chicken liver behaves in a similar way, a monomer (MW = 410,000) associating to a polymer of MW = 8×10^6.

Like the acetyl-CoA carboxylase from *E. coli*, the transcarboxylase of propionic acid bacteria is a large (MW $\sim 792,000$) and complex enzyme. It consists of 18 subunits and contains 6 biotin molecules and 6 metal ions, both Co^{2+} and Zn^{2+} being present.[12]

Recently, three more biotin-dependent enzymes have been discovered. **Carbamoylphosphate synthetase** of *E. coli*, [13,13a] essential in synthesis of pyrimidines, catalyzes Eq. 8-8. Presumably one ATP is used to form a carboxybiotin intermediate while the second phosphorylates the carboxyl group and perhaps assists in the cleavage of the amide bond of glutamine as well.

$$\text{Glutamine} + 2\ \text{ATP} + \text{HCO}_3^- \longrightarrow$$

$$\text{glutamate} + 2\ \text{ADP} + P_i$$

$$+\ \ \underset{\text{Carbamoyl phosphate}}{H_2N-\overset{\overset{\displaystyle O}{\|}}{C}-O-\text{\textcircled{P}}}\qquad (8\text{-}8)$$

Yeast (*Saccharomyces cerevisiae*) cannot make biotin and requires an unusually large amount of the vitamin when urea, allantoin, allantoic acid, and certain other compounds are supplied as the sole source of nitrogen for growth. The reason is that urea must first be carboxylated by the biotin-containing **urea carboxylase**[13] (Eq. 8-9) before it can be hydrolyzed to NH_3 and CO_2. In *Candida utilis* the carboxylase and hydrolase activities reside in a single enzyme **urea amidolyase**.[15]

$$\underset{\text{Urea}}{H_2N}\overset{\overset{\displaystyle O}{\underset{\displaystyle \|}{C}}}{}\ NH_2 + HCO_3^- \xrightarrow[\text{carboxylase}]{\text{ATP ADP} + P_i \atop \text{urea}}$$

$$\underset{\substack{N\text{-Carboxyurea}\\ \text{(allophanate)}}}{H_2N}\overset{\overset{\displaystyle O}{\underset{\displaystyle \|}{C}}}{}\ NH-COO^-$$

$$\Big\downarrow\ \overset{\text{allophanate}}{\underset{\text{amidolyase}}{}}\ H_2O$$

$$2\ HCO_3^- + 2\ NH_4^+\qquad (8\text{-}9)$$

3. Avidin

A biochemical mystery is the presence in egg white of the glycoprotein **avidin**.[16] A tetramer of MW = 68,000, each subunit contains a specific binding site for biotin. Avidin binds biotin extraordinarily tenaciously with $K_f \approx 10^{15}\ M$. While nature's purpose in placing this unusual protein in egg white is unknown, avidin has provided an important tool to enzymologists interested in biotin-containing enzymes. Addition of avidin invariably inhibits these enzymes and inhibition by avidin is diagnostic of a biotin-containing protein. Because of its specific production in the hen oviduct, formation of avidin is a process of interest to the endocrinologist. Formation of the protein is regulated by hormones, both by estrogens and by progesterone.

Historically, avidin was important to the discovery of biotin. The bond between avidin and biotin is so tight that inclusion of raw egg white in the diet of animals is sufficient to cause a severe biotin deficiency. Prevention of this "egg white damage" was one of the early tests of the presence of biotin in a natural material.

4. Attachment of Biotin to Apoenzymes

A special **holoenzyme synthetase** attaches biotin to the proper ϵ-amino group at the active centers of biotin enzymes.[17] It may activate all of the biotin enzymes and appears to use reaction sequence S1C (α) of Table 7-2 through an intermediate biotinyl AMP.

D. THIAMINE DIPHOSPHATE[18–20]

In Chapter 7 we considered the breaking of a bond between two carbon atoms, one of which is also bonded to a carbonyl group. These β cleavages are catalyzed by simple acidic and basic groups of the protein side chains. On the other hand, the decarboxylation of α-keto acids (Eq. 8-10) and the cleavage and formation of α-hydroxyketones (Eq. 8-11) depend upon thiamine

$$R-\overset{\overset{O}{\|}}{C}-COO^- + H^+ \longrightarrow R-\overset{\overset{O}{\|}}{C}H + CO_2 \qquad (8\text{-}10)$$

$$R-\overset{OH}{\underset{H}{\overset{|}{C}}}-\overset{\overset{O}{\|}}{C}-R' \longrightarrow R-\overset{\overset{O}{\|}}{C}H + H\overset{\overset{O}{\|}}{C}-R' \qquad (8\text{-}11)$$

diphosphate (Fig. 8-2). These reactions represent a second important method of making and breaking carbon-carbon bonds (**α condensation** and **α cleavage**). The common feature of all thiamine-catalyzed reactions is that the bond broken (or formed) is *immediately adjacent to the carbonyl group*, not one carbon removed, as in β cleavage reactions. No simple acid-base catalyzed mechanisms can be written.

The first real clue to the mechanism of thiamine action came in about 1950 when S. Mizuhara (drawing on previous experiments of T. Ugai) showed that thiamine at pH 8.4 catalyzes the non-enzymatic conversion of pyruvate into acetoin (Eq. 8-12). This reaction can be thought of as a combination of two consecutive steps: decarboxylation of pyruvate according to Eq. 8-10 and con-

densation of the two resulting acetaldehydes according to Eq. 8-11 in reverse.

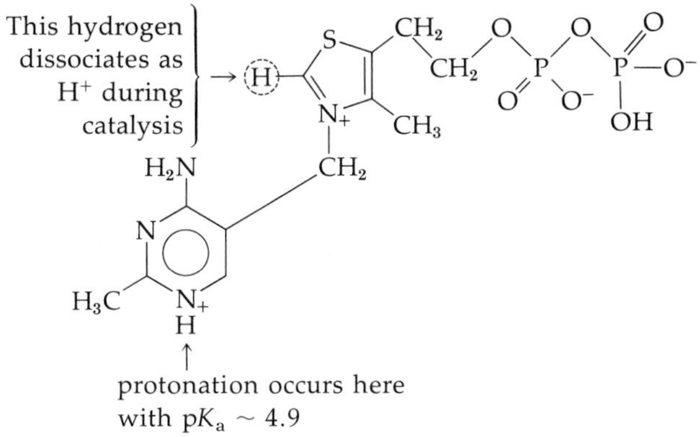

$$2\ CH_3-\overset{\overset{O}{\|}}{C}-COO^- + 2\ H^+ \longrightarrow$$

$$CH_3-\overset{}{\underset{O}{\overset{\|}{C}}}-\overset{\overset{H}{|}}{\underset{OH}{\overset{|}{C}}}-CH_3 + 2\ CO_2 \qquad (8\text{-}12)$$

1. The Mechanism of α Cleavage

Following Mizuhara's lead, R. Breslow investigated the same reaction using the then new nmr method. He made the surprising discovery that the hydrogen atom in the 2 position of the thiazolium ring (between the sulfur and the nitrogen) exchanged easily with deuterium of ^2H_2O. (The pK_a of this proton is now estimated[21] to be about 12.6.) Breslow proposed that the **thiazolium dipolar ion** (or **ylid**) formed by this dissociation (Eq. 8-13, step *a*) is the key intermediate in reactions of thiamine-dependent enzymes, a proposal that is now generally accepted. The anionic center of the dipolar ion, which is stabilized by the adjacent positive charge on the nitrogen atom, can react with a substrate such as an α-keto acid or α-keto alcohol (α-ketol) by addition to the carbonyl group (Eq. 8-13, step *b* or 8-13, step *b'*).[22,23]

The adducts formed in Eqs. 8-13 step *b* and 8-13, step *b'* are able to undergo β cleavage readily, as indicated by the arrows showing the electron flow toward the $=N^+-$ group. Below the structures of the adducts are those of a β-keto acid and a β-ketol with arrows indicating the electron flow in decarboxylation and in the aldol cleavage. The similarities to the thiamine-dependent cleavage reaction are especially striking if one remembers that in some aldolases and decarboxylases the substrate carbonyl group is first converted to an N-protonated Schiff base before the bond cleavage.

This hydrogen dissociates as H$^+$ during catalysis

protonation occurs here with $pK_a \sim 4.9$

Fig. 8-2 Thiamine diphosphate (thiamine pyrophosphate), the principal coenzyme form of thiamine.

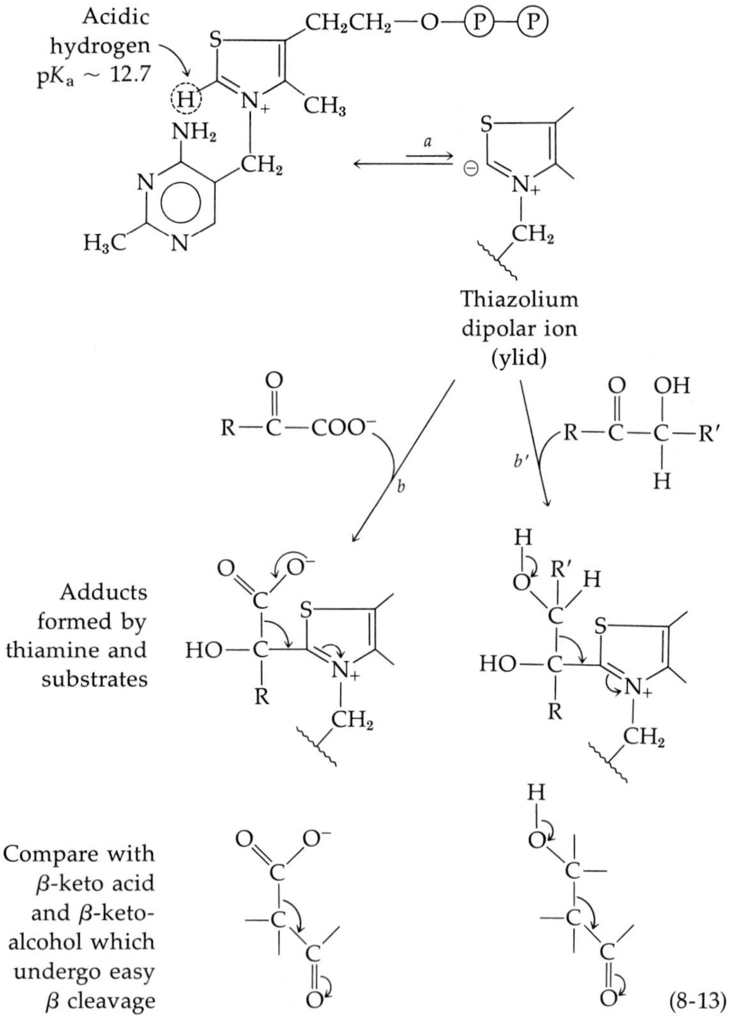

Acidic hydrogen $pK_a \sim 12.7$

CH_2CH_2—O—(P)—(P)

S

N_+ CH_3

NH_2 CH_2

H_3C N

a

Thiazolium dipolar ion (ylid)

R—C—COO$^-$ (with O double bond)

b

R—C—C—R' (with O, OH)

b'

Adducts formed by thiamine and substrates

Cleavage product

Thiazolium dipolar ion

$$(8\text{-}14)$$

Compare with β-keto acid and β-keto-alcohol which undergo easy β cleavage

$$(8\text{-}13)$$

We see that *the essence of the action of thiamine diphosphate as a coenzyme is to convert the substrate into a form in which electron flow can occur from the bond to be broken into the structure of the coenzyme.* Because of this alteration in structure, a bond breaking reaction that would not otherwise have been possible occurs readily. To complete the catalytic cycle, the electron flow has to be reversed again. The thiamine-bound cleavage product from either of the adducts in Eq. 8-13 can be reconverted to the thiazolium dipolar ion and an aldehyde as shown in Eq. 8-14.

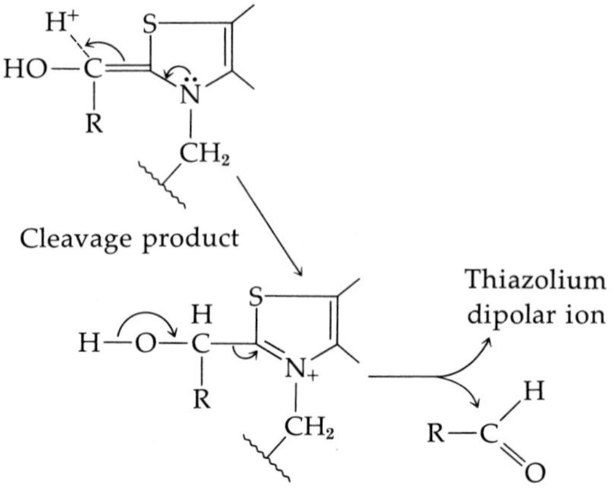

What is the role of the pyrimidine portion of the coenzyme in these reactions? It has often been suggested that the —NH₂ group of the pyrimidine is properly placed to function as a basic catalyst in the generation of the thiazolium dipolar ion. X-Ray diffraction studies show that in crystalline thiamine compounds the two rings are oriented in a manner* that would favor this function.[25] However, the amino group of thiamine is not very basic; indeed, the site of protonation at low pH ($pK_a \approx 4.8$) is largely N-1 of the pyrimidine ring (Fig. 8-2). In the protonated form the —NH₂ group is even less basic because of electron withdrawal into the ring. Thus, an enzyme could alter the basicity of the —NH₂ group at various stages of a reaction by controlling the protonation of the ring. One very speculative idea is that an enzyme might stabilize a *minor tautomer* of the aminopyrimidine ring. Together with a nucleophile from the protein it would constitute a "charge-relay system" analogous to that in chymotrypsin (Fig. 7-2). At the top of p. 441 is a drawing of a hypothetical "charge relay" system by which the aminopyrimidine ring of thiamine could assist in removal of the 2'-proton of the thiazolium ring. The pyrimidine ring also has a major effect on the basicity of the thiazolium nitrogen and may affect the ease of dissociation of the C-2 proton.[26]

* However, the orientation is somewhat different in crystals of 2-α-hydroxyethyl thiamine, an intermediate in thiamine-catalyzed decomposition of pyruvate to acetaldehyde.[24]

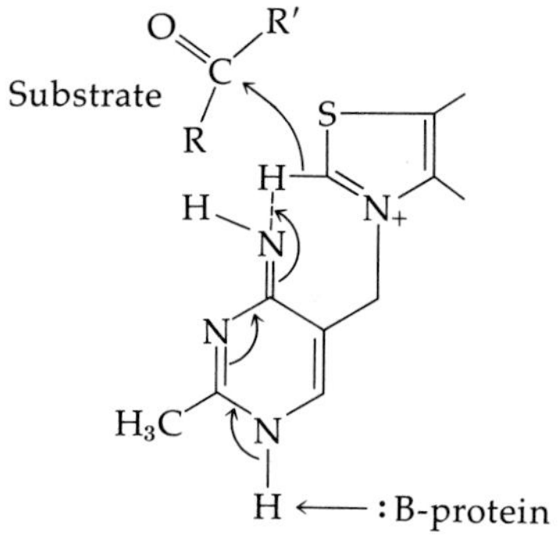

Thiamine-dependent enzymes usually require a divalent metal such as Mg^{2+} or Mn^{2+}, and it is logical to suppose that the principal site of coordination is with the pyrophosphate group.[27]

2. Enzymatic Reactions Involving Thiamine

All of the known thiamine diphosphate-dependent reactions can be derived from the five half-reactions, a through e, shown in Fig. 8-3. Each of these half-reactions is an α cleavage which leads to a thiamine-bound "active aldehyde" (center, Fig. 8-3) identical to the structure at the left hand side of Eq. 8-14. The decarboxylation of an α-keto acid to an aldehyde is represented by step b followed by a in reverse. The most studied enzyme catalyzing a reaction of this type is **pyruvate decarboxylase** of yeast, a Mg^{2+}-requiring dimeric enzyme of MW $\sim$ 200,000.[28] Thiamine diphosphate and Mg^{2+} dissociate from the enzyme at pH 8 but active enzyme can be reconstituted below pH 7.

Formation of α-ketols from α-keto acids also starts with step b, but is followed by condensation with another carbonyl compound in step c, in reverse. A well-known example involves decarboxylation of pyruvate and condensation of the resulting active acetaldehyde with another pyruvate molecule to give **α-acetolactate.** The reaction is catalyzed by **acetolactate synthetase,** sometimes referred to as a "carbolyase." Acetolactate is the precursor to valine and leucine (Fig. 14-10). A

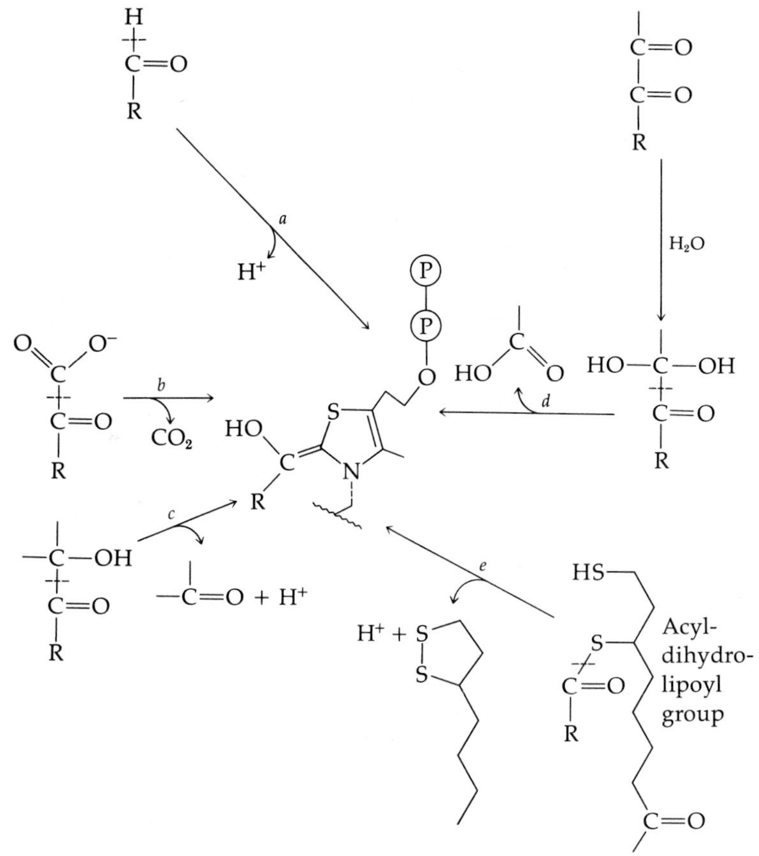

Fig. 8-3 Half-reactions making up the thiamine-dependent α-cleavage and α-condensation reactions (reaction type 7, see Table 9-1).

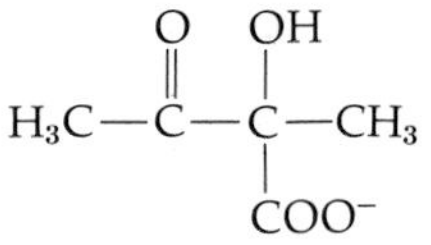

α-Acetolactate

similar ketol condensation is required in the biosynthesis of isoleucine (Fig. 14-10). Acetolactate is a β-keto acid and is readily decarboxylated to acetoin, a reaction of importance in certain bacterial fermentations (Eq. 9-28). The ketol condensation of two molecules of glyoxylate with decarboxylation is catalyzed by **glyoxylic carboligase** (Chapter 9, Section C,2).

BOX 8-D

THIAMINE (VITAMIN B₁)

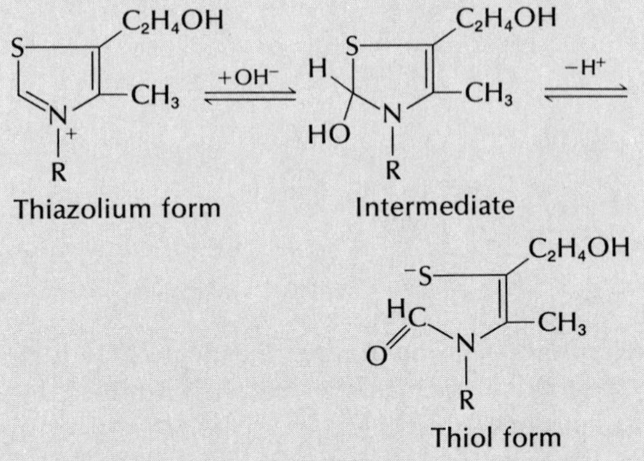

The history of the discovery and isolation of thiamine is described in Box 8-A. In 1937, K. Lohman and P. Schuster isolated pure "cocarboxylase," a dialyzable coenzyme required for decarboxylation of pyruvate by an enzyme from yeast. It was shown to be thiamine diphosphate (Fig. 8-2). Mono-, tri-, and tetraphosphates also occur naturally in smaller amounts.

Physical properties: Thiamine and its coenzyme forms are white solids, very soluble in water. Thiamine is usually sold as the chloride hydrochloride (MW = 337.3) and thiamine diphosphate as the monochloride (MW = 460.8).

Acid–base chemistry: In basic solution, thiamine reacts in two steps with an opening of the thiazole ring to give the anion of a thiol form which may be crystallized as the sodium salt. This reaction, like the competing reaction that has been described in Chapter 4 (Eq. 4-31) and which leads to a yellow unstable form of the thiamine anion, is an example of almost completely cooperative 2-proton dissociations with linked structural changes. No significant concentration of the intermediate is present during the titration. This property is unusual among small molecules and was instrumental in leading Williams *et al.* to the correct structure for the vitamin. Do these reac-

Thiazolium form   Intermediate

Thiol form

tions have any biological significance? The thiol form depicted in the accompanying structure or the "yellow form" (Eq. 4-31) could become attached to active sites of proteins through disulfide linkages, but if such enzyme chemistry of thiamine occurs, we are not yet aware of it.

Basic conditions also promote degradation of thiamine. Thus, the vitamin is destroyed by the cooking of foods under mildly basic conditions. The thiol form undergoes hydrolysis and oxidation by air to a disulfide. The tricyclic form (Eq. 4-31) is oxidized to **thiochrome**, a fluorescent compound whose formation from thiamine by treatment with alkaline ferricyanide is the basis of a much used fluorimetric assay.

Ketols can also be formed enzymatically by cleavage of an aldehyde (step *a*) followed by condensation with a second aldehyde (step *c*, in reverse). α-Diketones can be cleaved (step *d*) to a carboxylic acid plus active aldehyde, which can react either via *a* or *c* in reverse. These and other combinations of steps are often observed as side reactions of such enzymes as pyruvate decarboxylase. A related thiamine-dependent reaction is that of pyruvate and acetyl-CoA to give the α-diketone, diacetyl $CH_3COCOCH_3$.[29] The reaction can be viewed as a displacement of the CoA anion from acetyl-CoA by attack of thiamine-bound active acetaldehyde derived from pyruvate (reverse of step *d*, Fig. 8-3 with release of CoA).

Step *e* of Fig. 8-3 represents the cleavage of an

acyl-dihydrolipoic acid derivative. It usually operates in the reverse direction and is part of the process of *oxidative decarboxylation of an α-keto acid* which begins with step *b* (see Section J).

An important cleavage reaction of α-ketols utilizes step *c* of Fig. 8-3 followed by the same step in reverse but with a different aldehyde acceptor. The enzyme is **transketolase** (Eq. 9-15), essential in the pentose phosphate pathways of metabolism and in photosynthesis. A related reaction (Fig. 8-4) that is mechanistically more complex is catalyzed by the enzyme **phosphoketolase** whose action is essential to the energy metabolism of some bacteria. A product of phosphoketolase is acetyl phosphate, whose cleavage can be coupled to synthesis of ATP (Fig. 8-4).

Box 8-D (*Continued*)

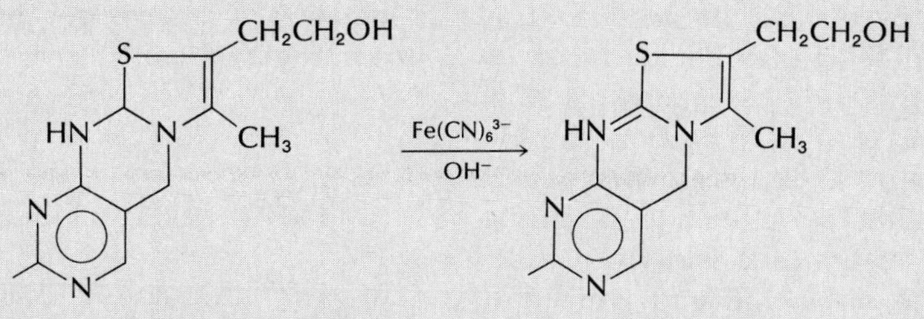

Tricyclic form of thiamine   Thiochrome

Sulfite cleavage: In a sodium sulfite solution at pH 5, thiamine is cleaved by a nucleophilic displacement reaction on the methylene group to give the free thiazole and a sulfonic acid. A similar cleavage is catalyzed by thiamine-degrading enzymes (thiaminases).

$$Thiamine + HSO_3^- \longrightarrow$$

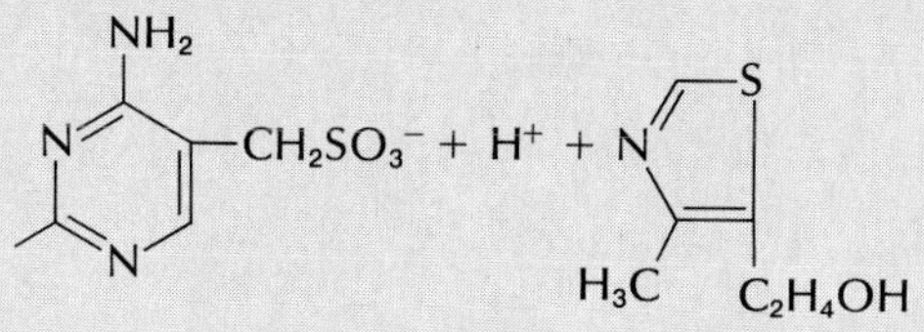

Analogues: Replacement of the methyl group on the pyrimidine ring by ethyl, propyl, or iso-

propyl gives compounds with some vitamin activity, but replacement by hydrogen cuts activity to 5% of the original. The butyl analogue is a competitive inhibitor. Treatment of thiamine with boiling $5N$ HCl deaminates it to the hydroxy analogue **oxythiamine,** a potent antagonist. **Pyrithiamine,** another competitor containing

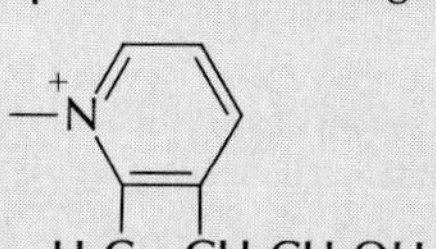

in place of the thiazolium ring, is very toxic especially to the nervous system.

Daily requirement: 0.23 mg or more per 1000 kcal of food consumed and a minimum total of 0.8 mg/day.

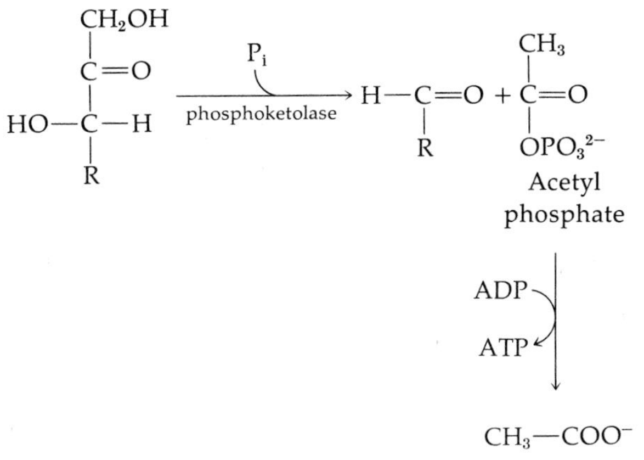

Fig. 8-4 Cleavage of an α-ketol to an aldehyde and a carboxylic acid, a reaction sequence that is coupled to phosphorylation of ADP (reaction sequence S7A).

We expect the required α cleavage to yield a thiamine-containing structure (first structure in

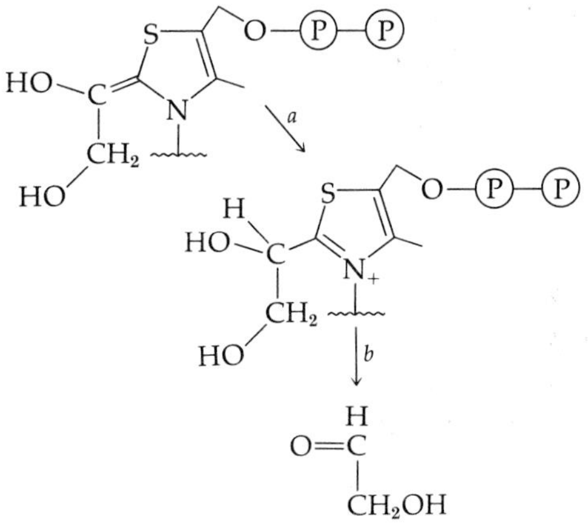

Glycolaldehyde

(8-15)

Eq. 8-15). Protonation of this intermediate, as shown in Fig. 8-15, step a, could be followed (step b) by breakdown to glycolaldehyde. In fact, it is thought that glycolaldehyde may be produced in chloroplasts in just this way as a side reaction from transketolase (Chapter 13, Section E,9,b). However, phosphoketolase catalyzes what is, in effect, an internal oxidation–reduction reaction. The aldehyde group of the glycolaldehyde fragment is oxidized to a carboxylic acid while the CH_2OH group is reduced to CH_3. One possible mechanism involves 2-acetylthiamine diphosphate as an intermediate while another is based on addition of P_i to a double bond.

3. Thiamine Coenzymes in Nerve Action

The striking paralysis caused by thiamine deficiency together with studies of thiamine analogues as metabolites suggest a special action for thiamine in nerves.[18a] It has been suggested that thiamine diphosphate or possibly thiamine triphosphate plays an essential role in the sodium transport system of nerve membranes.[20,30] Support for this idea comes from the observation that the analogue **pyrithiamine** displaces thiamine from nerve preparations. The potent nerve poison **tetrodotoxin** (Fig. 16-7) not only blocks nerve conduction by inhibiting inward diffusion of sodium but also promotes release of thiamine from nerve membranes. Support for a metabolic significance of thiamine triphosphate comes from identification of both soluble and membrane-associated thiamine triphosphatases in rat brain.[31,32]

E. PYRIDOXAL PHOSPHATE

The phosphate ester of the aldehyde form of vitamin B_6, **pyridoxal phosphate** (pyridoxal-P or PLP), is required by many enzymes catalyzing reactions of amino acids and of amines. The reactions are numerous, and pyridoxal phosphate is

surely one of nature's most versatile catalysts. The story begins with the biochemical reaction of **transamination,** a process of central importance in nitrogen metabolism.

1. Transamination[33]

In 1937 Alexander Braunstein and M. G. Kritsman, in Moscow, described a new reaction by which amino groups could be transferred from one carbon skeleton to another. For example, the amino group of glutamic acid was transferred to the carbon skeleton of oxaloacetic acid to form aspartic acid and α-ketoglutarate (Eq. 8-16). Braun-

$$\overset{\hspace{2.2cm}NH_3^+}{{}^-OOC-CH_2-CH_2-\underset{H}{C}-COO^-}$$

Glutamate

$$+\ {}^-OOC-CH_2-\overset{\overset{\displaystyle O}{\|}}{C}-COO^- \rightleftharpoons$$

Oxaloacetate

$$\overset{\hspace{3.2cm}\overset{\displaystyle O}{\|}}{{}^-OOC-CH_2-CH_2-C-COO^-}$$

α-Ketoglutarate

$$+\ {}^-OOC-CH_2-\overset{\overset{\displaystyle NH_3^+}{|}}{\underset{H}{C}}-COO^- \quad (8\text{-}16)$$

Aspartate

stein recognized the widespread and general significance of this **transamination** process in the nitrogen metabolism of organisms. A series of **aminotransferases** (transaminases), for which glutamate was usually one of the reactants, were shown to catalyze the transformations.

A few years later Esmond Snell reported the nonenzymatic conversion of pyridoxal into pyridoxamine (Box 8-E) by heating with glutamate and noted that this was also a transamination. He

BOX 8-E

THE VITAMIN B₆ FAMILY: PYRIDOXINE, PYRIDOXAL, and PYRIDOXAMINE

This group is replaced by —CHO in pyridoxal CH₂NH₂ in pyridoxamine

Site of esterification with phosphate to form coenzymes

Pyridoxine (pyridoxol)
MW = 169.2 (for free base)

The existence of a new vitamin was recognized in 1934 when it was observed that rats on a vitamin-free diet supplemented with thiamine and riboflavin developed a facial dermatitis, "rat pellagra." In 1938, **pyridoxine** (the usual commercial form of vitamin B₆) was isolated and synthesized. Studies of the nutritional requirements of lactic acid bacteria soon indicated other naturally occurring forms of the new vitamin which were more active than pyridoxine in promoting growth of some of these bacteria. In 1944, Snell identified the unknown substances as an amine **pyridoxamine,** and an aldehyde **pyridoxal.** Pyridoxal could be formed from pyridoxine by mild oxidation, and pyridoxamine could be formed (by transamination) from pyridoxal by heating with glutamic acid in solution. In fact, it was these simple experiments which suggested to Snell the correct structures for the new forms of vitamin B₆.

The acid-base chemistry and tautomerism of pyridoxine are discussed in Chapter 4, Sections C,3 and C,4.

Nutritional requirements: Vitamin B₆ is widely distributed in foods, and symptoms of severe deficiency are seldom observed in humans. An in-

take of 1.5–2 mg/day appears adequate for most adults, and 0.4 mg/day for infants. However, the latter figure may be close to the border line, and a number of cases of convulsions have been attributed to partial destruction of vitamin B₆ in liquid milk formulas. Convulsions were observed when the vitamin B₆ content was reduced to about one-half that normally present in human milk.

Vitamin B₆ in animal tissues and products is largely in the form of pyridoxal, pyridoxamine, and their phosphate esters. The lability of the aldehyde explains the ease of destruction of the vitamin by excessive heat or by light. On the other hand, plant tissues contain mostly pyridoxine, which is more stable. The phosphorylated forms are interconvertible within cells.[a,b] Pyridoxine 5'-phosphate can be oxidized to PLP and the latter may undergo transamination to PMP.

Several cases of children with abnormally high vitamin B₆ requirement (2–10 mg/day) have been reported, and a number of rare metabolic diseases are known[c–e] in which specific enzymes, e.g., cystathionine synthetase, have a reduced affinity for PLP. Patients with these diseases also benefit from a higher than normal intake of the vitamin. Cases of excessive excretion of the vitamin are known, the most clear-cut example being provided by a strain of laboratory mice that require twice the normal amount of vitamin B₆ and which die in convulsions after a brief period of vitamin B₆ depletion.[c]

[a] E. E. Snell and B. E. Haskell, *Compr. Biochem.* **21**, 47–71 (1970).

[b] S. Johansson, S. Lindstedt, and H.-G. Tiselius, *JBC* **249**, 6040–6046 (1974).

[c] S. H. Mudd, *Fed. Proc., Fed. Am. Soc. Exp. Biol.* **30**, 970–976 (1971).

[d] R. R. Bell and B. E. Haskell, *ABB* **147**, 588–601 (1971).

[e] T. A. Pascal, G. E. Gaull, N. G. Beratis, B. M. Gillam, H. H. Tallan, and K. Hirschhorn, *Science* **190**, 1209–1211 (1975).

proposed that pyridoxal might be a part of a coenzyme needed for aminotransferases. The hypothesis was soon verified and the coenzyme was identified as pyridoxal 5'-phosphate (Fig. 8-5). At about the same time, I. C. Gunsalus and co-

workers had noticed that the activity of **tyrosine decarboxylase** produced by lactic acid bacteria was unusually low when the medium was deficient in pyridoxine. Addition of pyridoxal plus ATP increased the decarboxylase activity. PLP was syn-

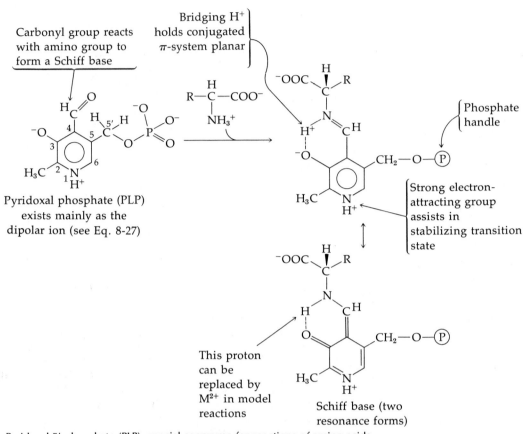

Fig. 8-5 Pyridoxal 5'-phosphate (PLP), special coenzyme for reactions of amino acids.

thesized and was found to be the essential co-enzyme. It quickly became apparent that PLP also functions with many other enzymes.

2. Pyridoxal Phosphate as a Catalyst

Both pyridoxal and PLP, in the complete absence of enzymes, are able not only to undergo transamination with amino acids but also to serve as catalysts for a variety of other reactions of amino acids identical to those catalyzed by PLP-dependent enzymes. Thus, the coenzyme itself can be regarded as the "active site" of the enzymes and can be studied in nonenzymatic model reactions. From early model studies the following conclusions about PLP were reached[34]:

a. The aldehyde group of PLP reacts readily and reversibly with amino acids to form Schiff bases (Fig. 8-5). The Schiff bases react further to give products.

b. For an aldehyde to be a catalyst a strong electron-attracting group, e.g., the ring nitrogen of pyridine (as in PLP) must be ortho or para to the —CHO group. A nitro group, also strongly electron-attracting, can replace the pyridine nitrogen in model reactions.

BOX 8-F

NINHYDRIN

An important analytical reagent in the biochemistry laboratory, ninhydrin (1,2,3-indantrione monohydrate) forms reactive Schiff bases (ketimines) with amino acids just as does pyridoxal phosphate. Decarboxylation of the ketimines followed by hydrolysis of the resulting aldimines yields an intermediate amine that can couple with a second molecule of ninhydrin to form a characteristic purple color.[a] The reaction

is widely used in chromatography and in quantitative amino acid analysis (Chapter 2, Section H,5). While α-amino acids react most readily, primary amines and peptides also form Ruhemann's purple. In these cases a proton rather than CO_2 is lost from the ketimine. When pyridoxamine (on chromatograms) reacts, a bright orange product, presumably the aldimine, appears. Secondary amines, such as proline, give a yellow color.

The intermediate amine can be hydrolyzed to free ammonia. To ensure maximum color yield

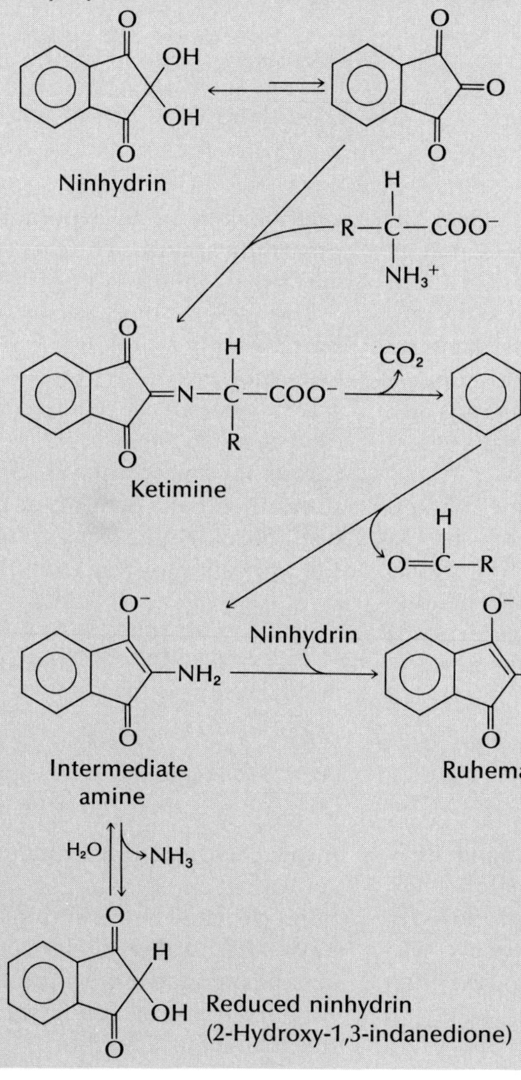

Ninhydrin

R—C—COO⁻
with NH₃⁺

Ketimine

Aldimine

Intermediate amine

Ruhemann's purple

Reduced ninhydrin
(2-Hydroxy-1,3-indanedione)

Box 8-F (*Continued*)

ninhydrin solutions for quantitative analysis usually contain some form of reduced ninhydrin, which can react with free NH_3 and ninhydrin to form Ruhemann's purple (see scheme on p. 447).

Both the ninhydrin reaction and pyridoxal phosphate-catalyzed decarboxylation of amino acids are examples of the Strecker degradation.

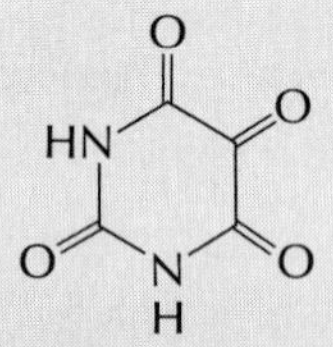

Alloxan

Strecker reported in 1862 that alloxan causes the decarboxylation of alanine to acetaldehyde, CO_2 and ammonia.[b] Many other carboxyl compounds, e.g., those of the general structure

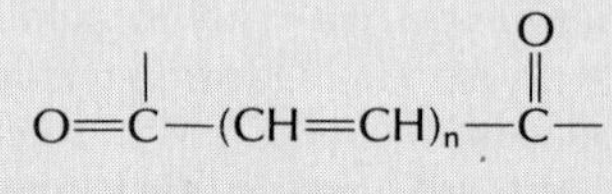

and *p*-nitrosalicylaldehyde also cause the Strecker degradation.[c]

[a] M. Friedman and L. D. Williams, *Bioorganic Chem.* **3**, 267–280 (1974).
[b] A. Strecker, *Annalen* **123**, 363–365 (1862).
[c] A. Schönberg and R. Moubacher, *Chem. Rev.* **50**, 261–277 (1952).

c. The presence of an —OH group adjacent to the —CHO group greatly enhances the catalytic activity. Since certain metal ions, such as Cu^{2+} and Al^{3+} increase the rates in model systems and are known to chelate with Schiff bases of the type formed with PLP, it was concluded that either a metal ion or a proton formed a chelate ring and helped to hold the Schiff base in a planar conformation (Fig. 8-5).*

d. In model systems the 5-hydroxymethyl and 2-methyl groups are not needed for catalysis. On the other hand, in enzymes the 5-CH_2OH group is essential for attachment of the phosphate handle. We now know that the 2-CH_3 group is *unnecessary* for coenzymatic activity with many enzymes.

3. The General Function of PLP

From a consideration of the various known PLP-dependent enzymes of amino acid metabolism, Braunstein and Shemyakin in 1953 proposed a general mechanism of action.[35] The theory was completely in accord with the results of the then

* However, such a function for metal ions has *not* been found in PLP-dependent enzymes.

unpublished model experiments mentioned in the preceding section. The theory has since been fully verified by studies of a variety of enzymes.

The general mechanism of PLP action can be stated simply as follows: *Pyridoxal phosphate reacts to replace the —NH_2 (or —NH_3^+) group of amino acid substrates with a group that is electronically the equivalent of an adjacent carbonyl.* This is done through formation of a Schiff base (Fig. 8-5). However, note that a Schiff base of an amino acid with a simple aldehyde (for example, acetaldehyde) has the opposite polarity from that of C=O. Such an

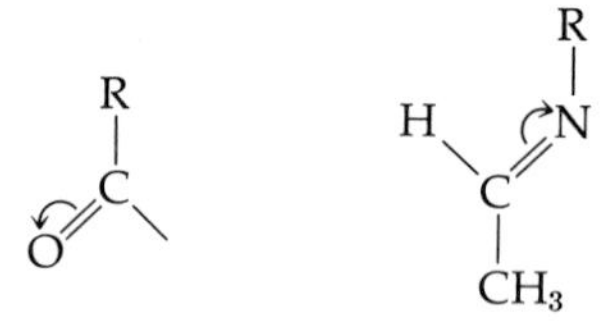

Carbonyl group

Schiff base with acetaldehyde

imine could not substitute for a carbonyl group in activating an α-hydrogen or in facilitating C—C bond cleavage in the amino acid. It is necessary to have the strongly electron-attracting pyridine group conjugated with the C=N group in such a way that electrons can flow from the substrate into the coenzyme (Fig. 8-6).

The reactions of PLP-amino acid Schiff bases may be compared with those of a β-keto acid as shown in Fig. 8-6. The reactions fall naturally into three groups (a,b,c), depending upon whether the bond cleaved is from the α-carbon of the substrate to the hydrogen atom, to the carboxyl group, or to the side chain. A fourth group of reactions (d) also involve removal of the α-hydrogen but are mechanistically more complex than those in group (a).

a. Loss of the α-Hydrogen

Dissociation of the α-hydrogen from a Schiff base of PLP leads to a **quinonoid intermediate** which, like an enolate anion, can react in a variety of ways (Fig. 8-6A):

i. Elimination. When a good leaving group is present in the β position of the amino acid it can be eliminated (Fig. 8-6B).[36] A large number of enzymes catalyze such reactions. Among them are **serine** and **threonine dehydratases** (elimination of

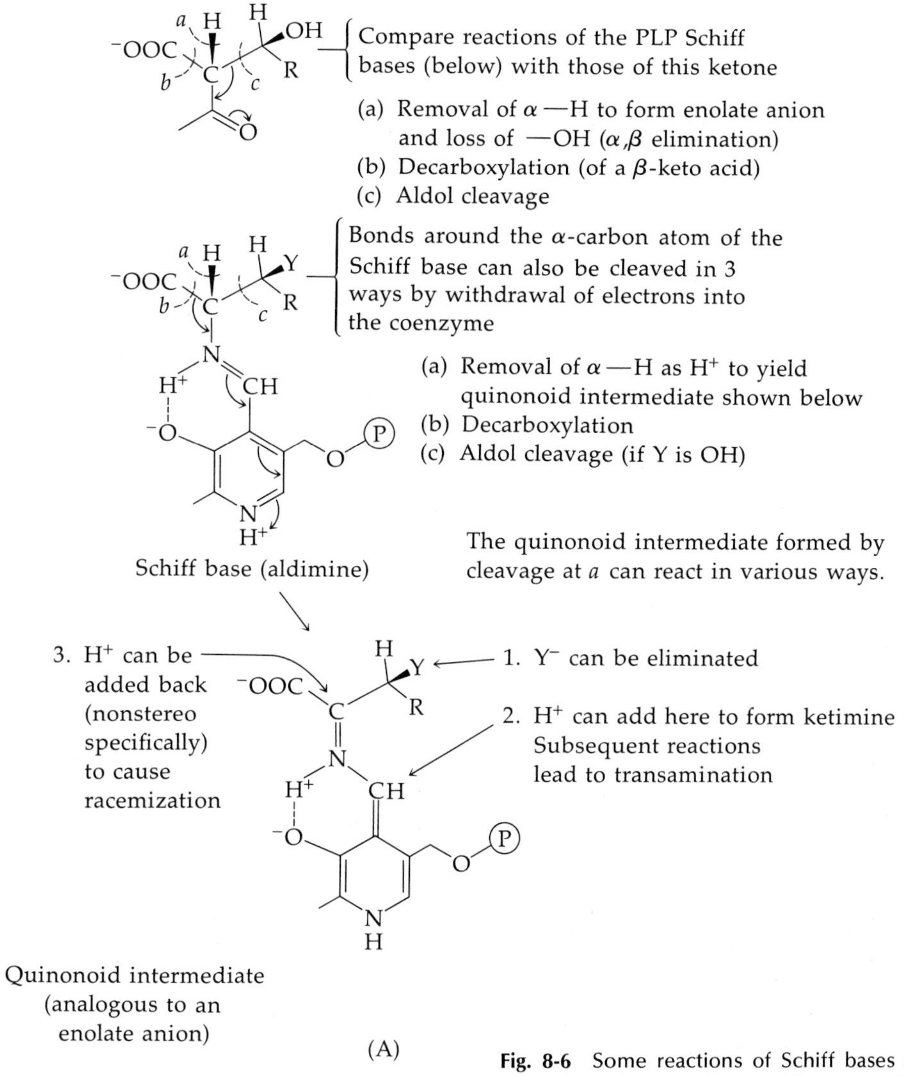

Compare reactions of the PLP Schiff bases (below) with those of this ketone

(a) Removal of α —H to form enolate anion and loss of —OH (α,β elimination)
(b) Decarboxylation (of a β-keto acid)
(c) Aldol cleavage

Bonds around the α-carbon atom of the Schiff base can also be cleaved in 3 ways by withdrawal of electrons into the coenzyme

(a) Removal of α —H as H^+ to yield quinonoid intermediate shown below
(b) Decarboxylation
(c) Aldol cleavage (if Y is OH)

Schiff base (aldimine)

The quinonoid intermediate formed by cleavage at a can react in various ways.

3. H^+ can be added back (nonstereo specifically) to cause racemization

1. Y^- can be eliminated

2. H^+ can add here to form ketimine Subsequent reactions lead to transamination

Quinonoid intermediate (analogous to an enolate anion)

(A)

Fig. 8-6 Some reactions of Schiff bases of pyridoxal phosphate (reaction type 8, see Table 9-1). (A) Formation of the quinonoid intermediate.

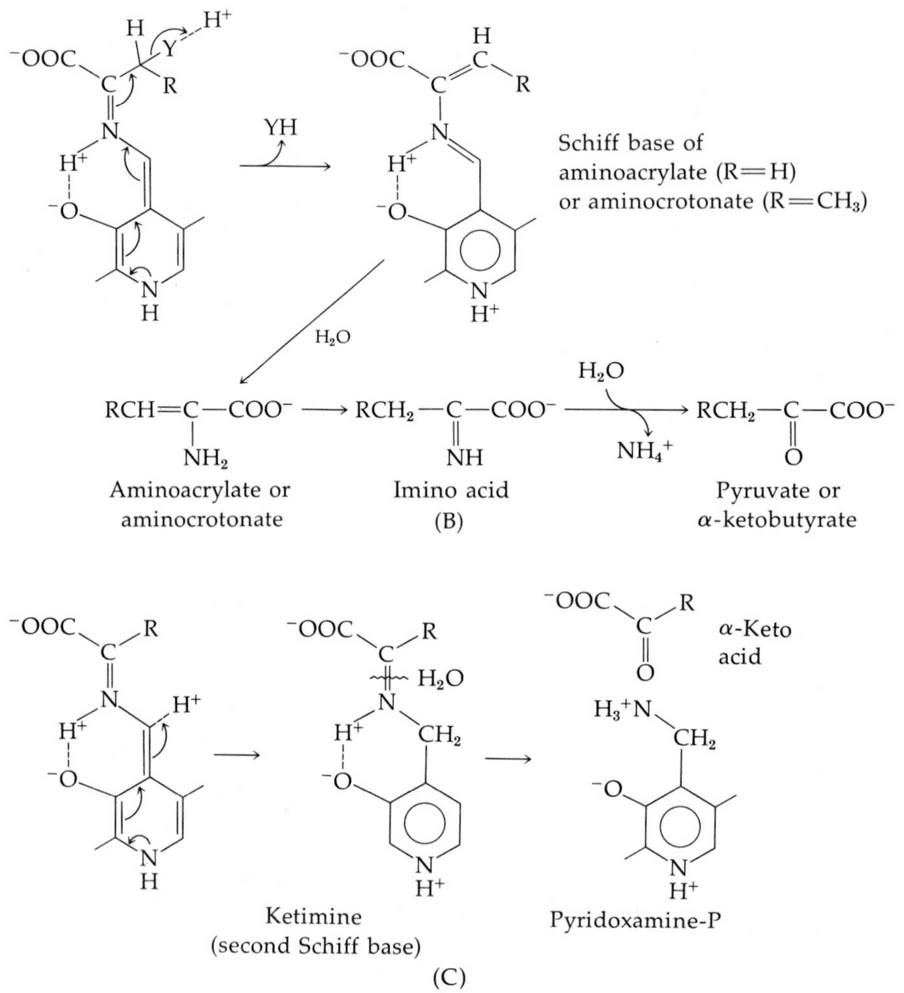

Fig. 8-6 (B) Elimination of a β substituent, and (C) transamination.

OH$^-$, as H_2O), **tryptophanase** of bacteria (elimination of indole), and **alliinase** of onions and garlic (elimination of 1-propenylsulfenic acid, the lachrymator formed on crushing these herbs).

Beta replacement is catalyzed by such enzymes as **tryptophan synthetase** (Chapter 14, Section I,3) and **cysteine synthetase** (Chapter 14, Section G), essential in amino acid biosynthesis. In both elimination and β replacement an unsaturated Schiff base, usually of aminoacrylate or aminocrotonate is a probable intermediate. (Fig. 8-6B). Conversion to the final products is usually assumed to be via hydrolysis to free aminoacrylate, tautomerization to an imino acid, and hydrolysis of the latter to pyruvate and ammonium ion (Fig. 8-6B). However, stereospecific addition of a proton at the β-C atom of α-ketobutyrate has been reported.[37] Thus, the student may wish to consider alternative ways in which the unsaturated Schiff base of Fig. 8-6B might break up (see also Eq. 8-28).

ii. Transamination. A proton can add to the carbon attached to the 4 position of the PLP ring (Fig. 8-6C) to form a second Schiff base, often referred to as a **ketimine.** The latter can readily undergo hydrolysis to **pyridoxamine phosphate** (PMP) and an α-keto acid. This sequence represents one of two half-reactions (Eq. 8-17 and 8-18) required for enzymatic transamination:

$$\text{Amino acid 1} + \text{PLP} \rightleftharpoons \text{keto acid 1} + \text{PMP} \quad (8\text{-}17)$$

$$\text{PMP} + \text{keto acid 2} \rightleftharpoons \text{PLP} + \text{amino acid 2} \quad (8\text{-}18)$$

Sum: Amino acid 1 + keto acid 2 $\rightleftharpoons$ keto acid 1 + amino acid 2

Transaminases are extraordinarily important in metabolism of amino acids, over 50 different enzymes being known.[33] Best studied is **aspartate aminotransferase** of pig heart cytoplasm, a dimer of subunit weight MW = 46,344 (Fig. 2-1). A mitochondrial isoenzyme has somewhat different properties.

iii. Racemization. A proton can be added back at the original alpha position but without stereospecificity. Racemases which do this are important to bacteria which must synthesize D-alanine and D-glutamic acid from the corresponding L-isomers for use in peptidoglycan synthesis.

b. Decarboxylation

The bond to the carboxyl group can be broken (Fig. 8-6A) rather than that to the α-hydrogen. This reaction, catalyzed by **amino-acid decarboxylases**, also leads to a quinonoid intermediate. The sequence is completed by addition of a proton at the original site of decarboxylation followed by breakup of the Schiff base. Decarboxylation of amino acids is essentially irreversible and frequently appears as a final step in synthesis of amino compounds. For example, glutamic acid is decarboxylated to **γ-aminobutyric acid** and 3,4-dihydroxyphenylalanine (dopa) to **dopamine** in the brain. Histidine yields **histamine**. Lysine is formed in bacteria by decarboxylation of *meso*-diaminopimelic acid (Chapter 14, Section D,2) and phosphatidylethanolamine by decarboxylation of phosphatidylserine (Chapter 12, Section F,2).

c. Side Chain Cleavage

In a third type of reaction the side chain of the Schiff base of Fig. 8-6A can undergo aldol cleav-

age. Conversely, a side chain can be added by β condensation. The best known enzyme catalyzing side chain cleavage is **serine transhydroxymethylase** which converts serine to glycine and formaldehyde.[38] The latter is not released in a free form but is transferred (Eq. 8-19) by the same enzyme specifically to **tetrahydrofolic acid** with which it forms a cyclic adduct (Eq. 8-69). Threonine is cleaved to acetaldehyde by the same enzyme. Another related reaction is indicated in Fig. 14-27 (top).

L-Serine + tetrahydrofolic acid $\longrightarrow$ glycine + 5,10-methylenetetrahydrofolic acid   (8-19)

Two ester condensation reactions join acyl groups from CoA derivatives to Schiff bases derived from glycine or serine. Succinyl-CoA is the acyl donor in the biosynthesis (Eq. 8-20) of δ-**aminolevulinic acid**, an intermediate in synthesis

$$^-\text{OOC}-\text{CH}_2\text{CH}_2-\overset{\displaystyle O}{\overset{\|}{\text{C}}}-\text{S}-\text{CoA}$$

Glycine + PLP

Quinonoid intermediate

CoASH

Schiff base

decarboxylation $\searrow$ CO_2

PLP $\nwarrow$ H_2O

$$^-\text{OOC}-\text{CH}_2\text{CH}_2-\overset{\displaystyle O}{\overset{\|}{\text{C}}}-\text{CH}_2-\text{NH}_2 \quad (8\text{-}20)$$

δ-Aminolevulinate

of hemes (Chapter 14, Section F,4). Since the enzyme does not catalyze decarboxylation of glycine in the absence of succinyl-CoA, the decarboxylation probably follows the condensation as indicated in Eq. (8-20).[39]

In the biosynthesis of **sphingosine** (Eq. 12-24) serine is condensed, again with decarboxylation, with palmitoyl-CoA to form an aminoketone intermediate (Eq. 8-21).[40] The hydrogen marked by the asterisk in the equation has been shown to come from the solvent. Thus, decarboxylation follows the condensation in this case as well.

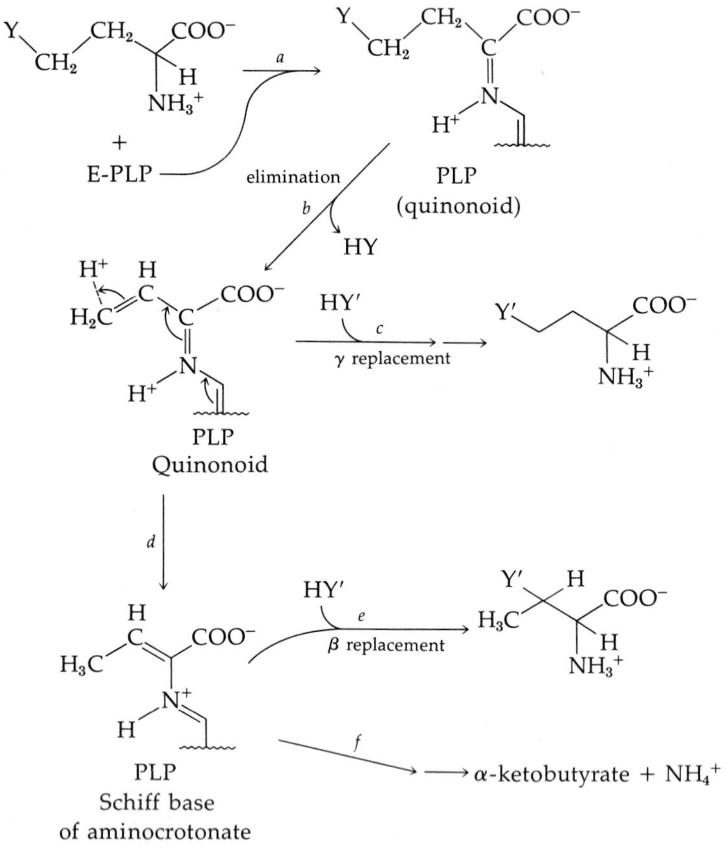

$$\text{CH}_3\text{—(CH}_2)_{14}\text{—C—C—CH}_2\text{OH}$$

Sphingosine   (8-21)

d. The Quinonoid Intermediate as Electron Acceptor

The fourth group of PLP-dependent reactions are thought to depend upon either the quinonoid intermediate or the ketimine of Fig. 8-6C. In both of these forms the original α-hydrogen of the amino acid has been removed and the $C=NH^+$ bond is polarized, again in a direction that favors electron withdrawal from the amino acid into the coenzyme. This permits another series of reactions analogous to those of β-keto acids to occur. Both elimination and C—C bond cleavage α, β to the $C=N$ group of the quinonoid (or ketimine) intermediate occur.

Among this type of enzyme is a group that catalyzes elimination of a γ-substituent from an amino acid (Fig. 8-7). The eliminated group is sometimes replaced by another substituent, either in the γ or the β position. The quinonoid intermediate formed initially undergoes elimination of the γ substituent (β with respect to the $C=N$ group) along with a proton from the β position of the original amino acid[36] (Fig. 8-7, steps a and b). The resulting intermediate can react in any of three ways, depending upon the enzyme. Addition of HY' leads to γ replacement (step c) while addition of a proton at the γ position (indicated by arrows in the structure) leads, via reaction step d, to an α,β-unsaturated Schiff base. The latter can react by addition of HY' (β replacement, step e) or it can break down to an α-keto acid and ammonium ion (step f), just as in the β-elimination reactions of Fig. 8-6B. An important γ-replacement reaction is conversion of **O-succinylhomoserine** to *cystathionine* (Eq. 8-22). This reaction lies on the pathway

Fig. 8-7 Some PLP-dependent reactions involving elimination of a γ substituent. Replacement by another γ substituent or by a substituent in the β position is possible as is deamination to an α-keto acid.

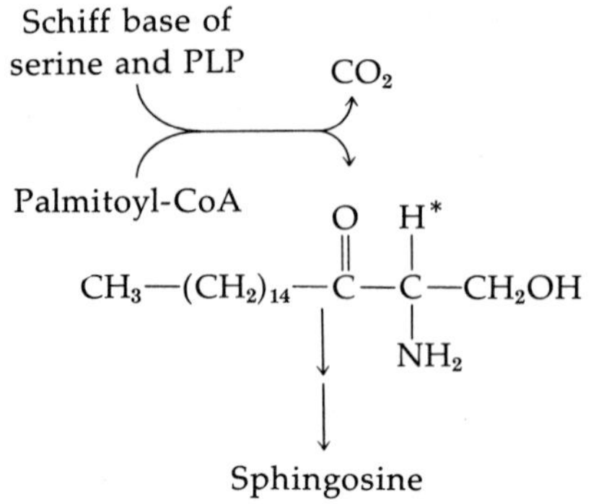

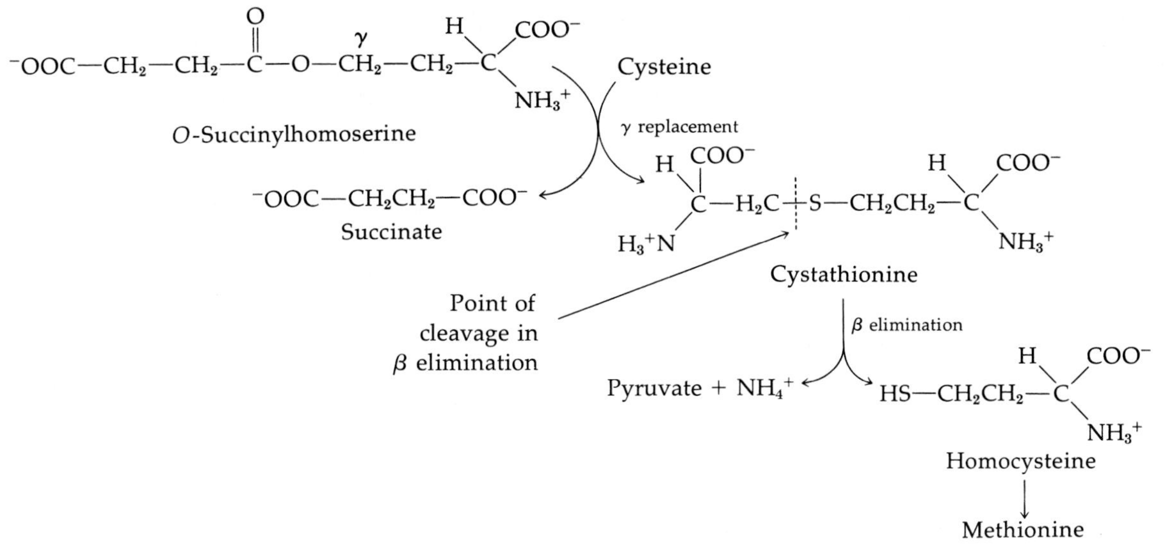

$$-OOC-CH_2-CH_2-\overset{\overset{O}{\|}}{C}-O-CH_2-CH_2-\overset{\overset{H}{\underset{NH_3^+}{|}}}{C}\diagup^{COO^-}$$

O-Succinylhomoserine

Cysteine

γ replacement

$$-OOC-CH_2CH_2-COO^-$$
Succinate

$$\overset{H}{\underset{H_3^+N}{}}\diagdown\overset{COO^-}{\underset{}{}}C-H_2C\vdots S-CH_2CH_2-\overset{H}{\underset{NH_3^+}{}}C\diagup^{COO^-}$$

Cystathionine

β elimination

Point of
cleavage in
β elimination

Pyruvate + NH$_4^+$ ← HS—CH$_2$CH$_2$—$\overset{H}{\underset{NH_3^+}{}}C\diagup^{COO^-}$

Homocysteine

↓

Methionine   (8-22)

of biosynthesis of methionine in *Salmonella*. Subsequent reactions are β elimination from cystathionine of **homocysteine** which is then converted to methionine (Eq. 8-85). A related pathway in fungi and higher plants uses O-acetylhomoserine. Threonine is formed from **O-phosphorylhomoserine** via γ elimination followed by β replacement with —OH. The reaction is catalyzed by **threonine synthetase** (Fig. 14-6).

The loss of a β-carboxyl group as CO_2 can also occur through a quinonoid intermediate. The best known example is provided by the bacterial **aspartate β-decarboxylase**[41] which converts aspartate to alanine and CO_2. A related reaction, the conversion of the amino acid **kynurenine** to alanine and anthranilic acid (Fig. 14–26) presumably depends upon hydration of the carbonyl group prior to β cleavage (Eq. 8-23). An analogous thiolytic cleavage is the reaction of CoA with 2-amino-4-ketopentanoate to form acetyl-CoA and alanine.[42]

e. The Mystery of Glycogen Phosphorylase

While pyridoxal phosphate is ideally designed to catalyze reactions of amino compounds it is surprising to find it as an essential cofactor for glycogen phosphorylase (Chapter 7, Section C,5). The PLP coenzyme is linked to phosphorylase in much the same way as in transaminase (see Section E,6), but there is no obvious function for the coenzyme.[43] Also surprising is the estimate that 50% of the vitamin B$_6$ in our body is present as PLP in muscle phosphorylase.[44] From studies of vitamin B$_6$-deficient rats, it appears that PLP in phosphorylase may serve as a reserve supply much of which can be taken for other purposes during times of deficiency.

Apart from its function as a coenzyme, PLP is a specific inhibitor, possibly an allosteric effector, for a number of different enzymes, including aldolase, glutamate dehydrogenase, and hexo-

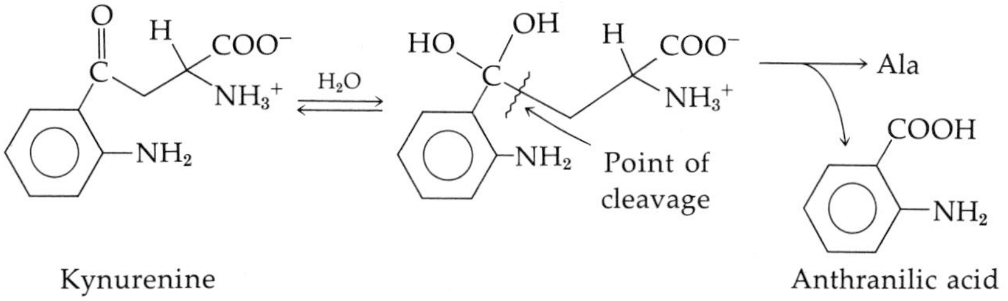

$$\overset{\overset{O}{\|}}{C}\underset{}{}\overset{H}{\underset{NH_3^+}{}}\diagup^{COO^-} \xrightarrow{H_2O} \overset{HO}{\underset{}{}}\overset{OH}{\underset{}{}}C\underset{}{}\overset{H}{\underset{NH_3^+}{}}\diagup^{COO^-} \longrightarrow \text{Ala}$$

Kynurenine   Point of cleavage   Anthranilic acid

COOH
NH$_2$ (8-23)

kinase. It also binds specifically to a variety of other proteins.[45]

4. Pyridoxamine Phosphate as a Coenzyme

If PLP is a specific cofactor designed to react with amino groups in substrates, might not pyridoxamine phosphate (PMP) act as a coenzyme for reactions of carbonyl compounds? The first example of this kind of function has been found[46] in the formation of **3,6-dideoxyhexoses** needed for bacterial cell surface antigens (Fig. 5-8). Glucose (as cytidine diphosphate glucose, CDPglucose) is first converted to 4-keto-6-deoxy-CDPglucose. The conversion of the latter to 3,6-dideoxy-CDPglucose (Eq. 8-24) requires PMP as well as a reducing agent (NADH or NADPH).

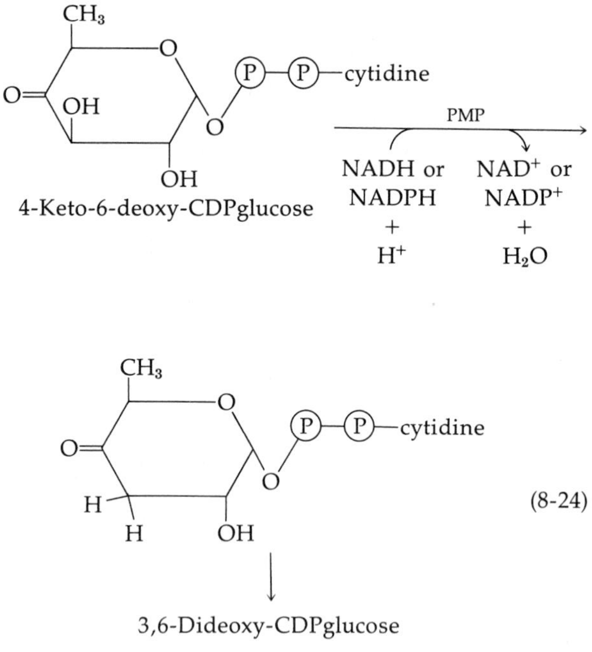

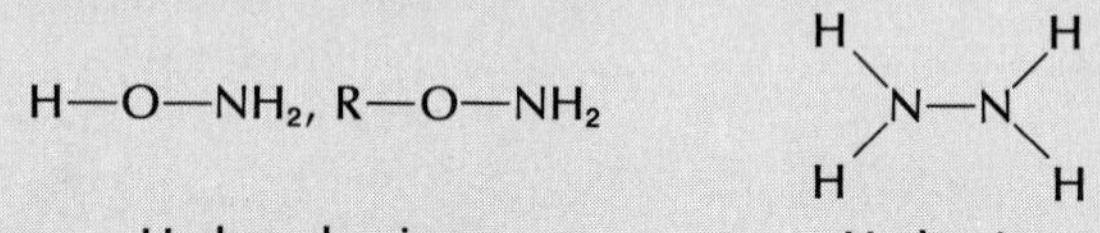

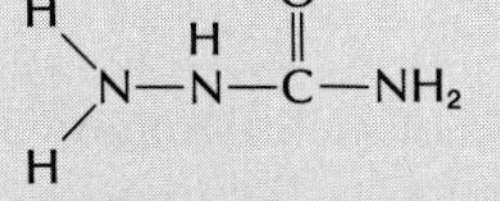

4-Keto-6-deoxy-CDPglucose

$$\xrightarrow[\substack{\text{NADH or} \\ \text{NADPH} \\ + \\ H^+}]{\text{PMP}} \substack{\text{NAD}^+ \text{ or} \\ \text{NADP}^+ \\ + \\ H_2O}$$

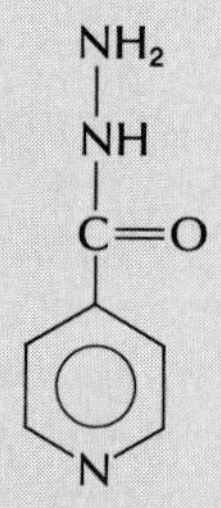

3,6-Dideoxy-CDPglucose

(8-24)

The student should find it of interest to propose a mechanism for this reaction, taking into account the expected direct transfer of a hydrogen from NADH as described in Section G.

Box 8-G (*Continued*)

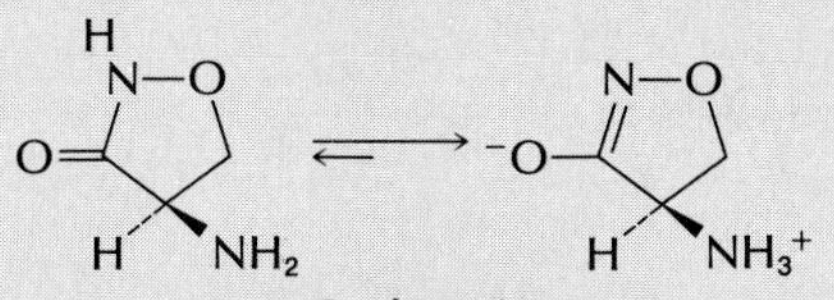

D-Cycloserine

alanine racemase. It also inhibits the ATP-dependent **D-alanyl-D-alanine synthetase** needed in biosynthesis of the peptidoglycan of bacterial cell walls (Box 7-D),[b] a non-PLP enzyme.

The synthetic isomer L-cycloserine is toxic to humans and inhibits many PLP enzymes. This observation led Khomutov *et al.* to synthesize a series of substituted cycloserines[c] including two "cycloglutamates":

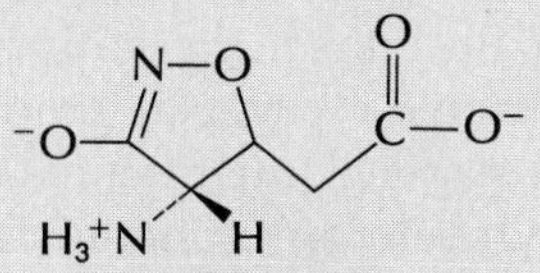

A "cycloglutamate" that
inhibits aspartate
aminotransferase

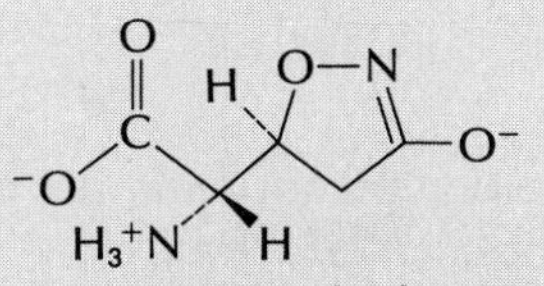

An isomeric "cycloglutamate"
tricholomic acid found in
certain mushrooms

One of these reacts strongly with aspartate aminotransferase while the other specifically inhibits glutamate decarboxylase of *E. coli.*[d] The compounds can be regarded as structural analogues of glutamic acid with fixed conformations (see also Chapter 6, Section D,1). It seems that nature anticipated the synthetic chemist, for it has been reported that the mushroom *Tricholoma muscarium* contains the same compound, "tricholomic acid." It is said to impart two interesting properties to the mushroom[e]: a delicious flavor and a lethal action on flies that alight on the mushroom's surface.

The substituted cysteine derivative L-peni-

cillamine causes convulsions and low glutamate decarboxylase levels in the brain, presumably because the Schiff base formed with PLP can then undergo cyclization, the SH group adding to the C=N to form a stable thiazolidine ring:

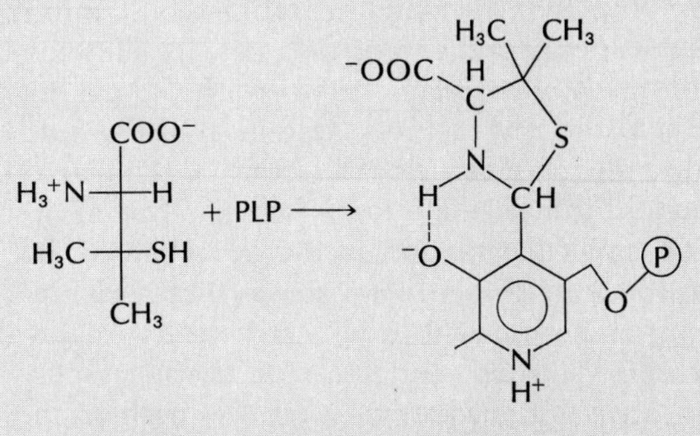

Toxopyrimidine, the alcohol derived from the pyrimidine portion of the thiamine molecule, is a structural analogue of pyridoxal. When fed to rats or mice it induces running fits which can be stopped by administration of vitamin B_6. Phosphorylation of toxopyrimidine by pyridoxal kinase is thought to produce an antagonistic analogue of PLP. In a similar fashion, 4-deoxypyridoxine induces death in chicks and causes convulsions and other symptoms of vitamin B_6 deficiency in man (tests were done on humans in the hope that the compound would be a successful anticancer drug). A host of synthetic PLP derivatives have been made, some of which are effective in blocking PLP enzymes.[f,g]

[a] E. E. Snell and B. E. Haskell, *Compr. Biochem.* **21**, 47–71 (1970).

[b] E. F. Gale, E. Cundliffe, P. E. Reynolds, M. H. Richmond, and M. J. Waring, "The Molecular Basis of Antibiotic Action," pp. 61–71. Wiley, New York, 1972.

[c] R. M. Khomutov, G. K. Kovaleva, E. S. Severin, and L. V. Vdovina, *Biokhimiya* **32**, 900–907 (1967).

[d] L. P. Sastchenko, E. S. Severin, D. E. Metzler, and R. M. Khomutov, *Biochemistry* **10**, 4888–4894 (1971).

[e] H. Iwasaki, T. Kamiya, O. Oka, and J. Veyanagi, *Chem. Pharm. Bull.* **17**, 866–872 (1969).

[f] E. E. Snell, *Vitam. Horm. (N.Y.)* **28**, 265–290 (1970).

[g] W. Korytnyk and M. Ikawa, *in* "Methods in Enzymology" (D. B. McCormick and L. D. Wright, eds.), Vol. 18A, pp. 524–566. Academic Press, New York, 1970.

5. Stereochemistry of PLP-Requiring Enzymes

In 1966, Dunathan postulated that the bond in the substrate amino acid that is to be broken by a PLP-dependent enzyme should lie in a plane perpendicular to the plane of the cofactor-imine π system (Fig. 8-8). Such an orientation would minimize the energy of the transition state by allowing maximum σ-π overlap between the breaking bond and the ring-imine π system. It also would provide the geometry closest to that of the planar quinonoid intermediate to be formed, thus minimizing molecular motion in the approach to the transition state.[47] Figure 8-8 shows three orientations of an amino acid in which the α-hydrogen, the carboxyl group, and the side chain, respectively, are positioned for cleavage. For each orientation shown another geometry suitable for cleavage of the same bond is obtained by rotating the amino acid through 180°.

Dunathan's proposal explains certain side reactions observed with PLP-requiring enzymes and has received support from experiments of Bailey and associates with an **α-dialkylaminotransferase** isolated from soil bacteria.[48] The enzyme ordinarily catalyzes, as one half-reaction, the combination decarboxylation-transamination reaction shown in Eq. 8-25. The enzyme also acts on both

$$\text{PMP} + CO_2 + \overset{H_3C \quad CH_3}{\underset{O}{\overset{|}{C}}} \qquad (8\text{-}25)$$

D- and L-alanine, decarboxylating the former but catalyzing only removal of the α-H from L-alanine. The results can be rationalized by assuming that the enzyme possesses a definite site for one alkyl group but that the position of the second alkyl group can be occupied by —H or —COO$^-$ and

that the group labilized lies perpendicular to the π system:

Alkyl binding site   L-Alanine   D-Alanine

Glycine is completely unreactive, suggesting that occupation of the alkyl binding site is required for catalysis to occur.

If Dunathan's postulate is accepted for aminotransferases, there are only two possible orientations of the amino acid substrate. One is shown in Fig. 8-8. In the other, the amino acid is rotated 180° so that the α-hydrogen protrudes *behind* the plane of the paper. Dunathan chose to study **pyridoxamine:pyruvate** aminotransferase, an enzyme closely related to PLP-requiring aminotransferases but for which pyridoxamine and pyridoxal are substrates (Eq. 8-26):*

Pyridoxamine + pyruvate $\rightleftharpoons$ pyridoxal

$$\qquad\qquad + \text{ L-alanine} \quad (8\text{-}26)$$

Using pyridoxal and L-alanine containing ^{2}H in the α position as substrates, he demonstrated a direct transfer of the α proton of the amino acid to the 4' position of the product pyridoxamine (indicated by the asterisks in Fig. 8-8). It was further shown that the ^{2}H was incorporated into pyridoxamine specifically in the *pro-S* position at C-4'. The results suggest that some group on the protein abstracts a proton from the α position and transfers it on the same side of the π system (syn transfer), adding it to the *si*-face of the C=N group. Dunathan concluded that the orientation of the amino acid substrate is that shown in Fig. 8-8.

Other experiments have shown that reduction with ^{3}H-containing $NaBH_4$ of either the aldimine

* The reaction of Eq. 8-26 is also catalyzed by the apoenzyme of aspartate aminotransferase and the same direct, stereospecific proton transfer has been observed as for pyridoxamine-pyruvate aminotransferase.[47] It is assumed that the same results would hold for PLP.

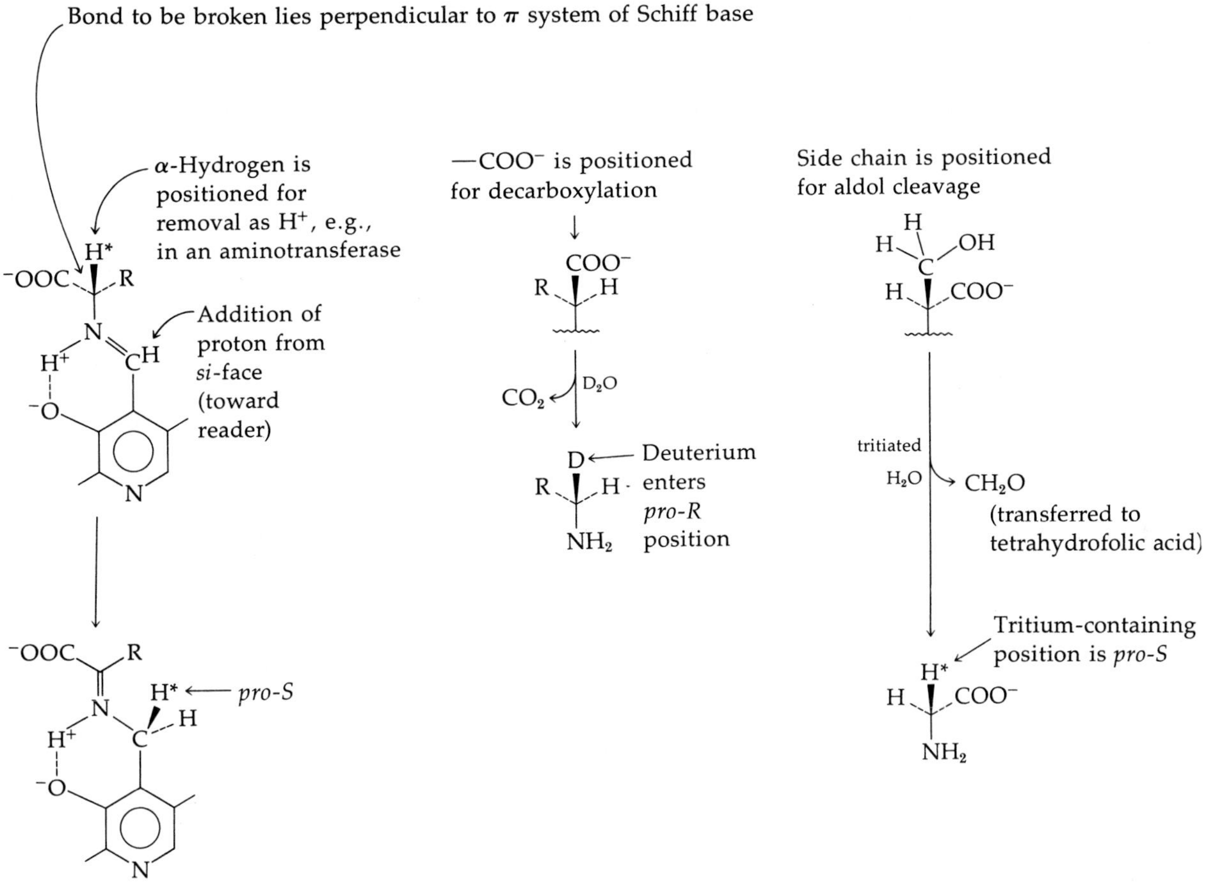
Bond to be broken lies perpendicular to π system of Schiff base
α-Hydrogen is positioned for removal as H+, e.g., in an aminotransferase
H*
-OOC
R
N
H+
CH
Addition of proton from si-face (toward reader)
-O
N
-OOC
R
H* ← pro-S
N
H
H+
C
-O
N
—COO− is positioned for decarboxylation
COO−
R
H
CO2
D2O
D ← Deuterium enters pro-R position
R
H
NH2
Side chain is positioned for aldol cleavage
H
H
OH
C
H
COO−
tritiated H2O
CH2O (transferred to tetrahydrofolic acid)
Tritium-containing position is pro-S
H*
H
COO−
NH2

Fig. 8-8 Stereochemical aspects of catalysis by PLP-requiring enzymes.

or the ketimine formed by the substrates of this enzyme also occurs from the *si*-face, suggesting that this face lies toward the outside of the protein.[47]

Decarboxylation of an amino acid in ^2H_2O leads to incorporation of ^2H in the *pro-R* position, the position originally occupied by the carboxyl group (Fig. 8-8). Cleavage of serine by serine hydroxymethylase in ^3H-containing water leads to incorporation of ^3H in the *pro-S* position. Stereospecific introduction of ^2H or ^3H has also been observed in the β position of α-ketobutyrate formed in β or γ elimination reactions. Conversion of serine to tryptophan by tryptophan synthetase occurs without inversion at C-3.[49]

6. Seeing Changes in the Optical Properties of the Coenzyme

A striking characteristic of many coenzymes is light absorption (see Chapter 13). This property is measured by the **absorption spectrum** and may also give rise to **circular dichroism** and to **fluorescence.** The optical properties of the vitamin B_6 coenzymes are especially sensitive to changes in environment and in the state of protonation of the various groups in the molecule. For example, PMP in the neutral dipolar ionic form, which exists at pH 7, has three strong light absorption bands centered at 327, 253, and 217 nm.[50]

Most of the various ionic forms of vitamin B_6 derivatives have three absorption bands spaced at roughly similar intervals, but the positions and intensities vary. For example, the minor tautomer of PMP containing an uncharged ring (Eq. 8-27)

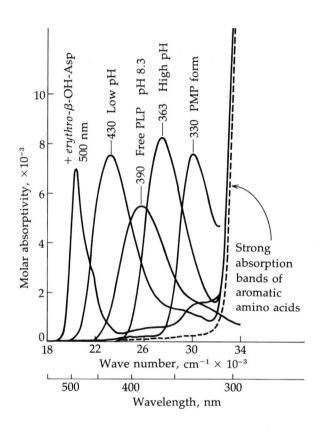

Dipolar ionic form

327 nm $(30.6 \times 10^3$ cm$^{-1})$*
253 nm $(39.6 \times 10^3$ cm$^{-1})$
217 nm $(46.2 \times 10^3$ cm$^{-1})$

(8-27)

Minor tautomer

283 nm $(35.3 \times 10^3$ cm$^{-1})$

has its low energy (long wavelength) band at 283 nm. When both the ring nitrogen and phenolic oxygen are protonated the band shifts again to 294 nm $(34.0 \times 10^3$ cm^{-1} and if both groups are deprotonated the resulting anion absorbs at 312 nm $(32.1 \times 10^3$ cm$^{-1})$. Thus, observation of the absorption spectrum of the coenzyme bound to an enzyme surface can tell us whether particular groups are protonated or unprotonated. For example, the peak of bound PMP at 330 nm in aspartate aminotransferase (Fig. 8-9) is indicative of the dipolar ionic form (Eq. 8-27).†

Pyridoxal phosphate exists in an equilibrium between the aldehyde and its covalent hydrate (as in Eq. 7-35). The aldehyde has a yellow color and absorbs at 390 nm $(25.6 \times 10^3$ cm^{-1}, Fig. 8-9), whereas the hydrate absorbs at nearly the same

Fig. 8-9 Absorption spectra of various forms of aspartate aminotransferase compared with that of free pyridoxal phosphate. Absorbance (Eq. 13-5) is plotted against increasing energy of light quanta in units of wave number. The more commonly used wavelength scale is also given. Spectra are most often presented with the low wavelength side to the left. However, the convention adopted here is becoming more common. The low pH form of the enzyme observed at pH <5 is converted to the high pH form with pK_a ~6.3. Addition of *erythro-β*-hydroxyaspartate produces a quinonoid form whose spectrum here is shown only $\frac{1}{3}$ its true height. The spectrum of free PLP is shown at pH 8.3. The spectrum of the apoenzyme (---) contains a small amount of residual absorption in the 300–400 nm region, whose origin is uncertain.

---

* Positions of absorption bands are given both in nanometers and in wave numbers. The wave number $(\bar{\nu})$ in cm^{-1} is equal to 10^7 divided by the wavelength in nanometers (see Chapter 13).

† However, the 5 nm shift from the position of free PMP suggests a distinct change in environment.

position as does PMP. The absorption bands of Schiff bases of PLP are shifted even further to longer wavelengths, N-protonated forms absorbing at 415–430 nm (23–24×10^3 cm^{-1}), whereas forms with an unprotonated C=N group absorb at shorter wavelengths.

a. Imine Groups in Free Enzymes

When he examined the first highly purified aspartate aminotransferase preparation (in 1957) W. T. Jenkins noted a surprising fact: The bound coenzyme (at pH ~5) absorbed not at 390 nm, as does PLP, but at 430 nm, like a Schiff base (Fig. 8-9). Furthermore, when the pH was raised, the absorption band shifted to 363 nm. The result suggests dissociation of a proton (with a pK_a of ~6.3) from the hydrogen-bonded position in a Schiff base of the type shown in Fig. 8-5. It was quickly demonstrated for this enzyme and for a series of others that (as would be expected for a Schiff base) reduction with sodium borohydride caused the spectrum to revert to one similar to that of PMP and fixed the coenzyme to the protein. After complete HCl digestion of such borohydride-reduced proteins a fluorescent amino acid containing the reduced pyridoxyl group was obtained and in every case was identified as **ε-pyridoxyllysine.** Thus, PLP-

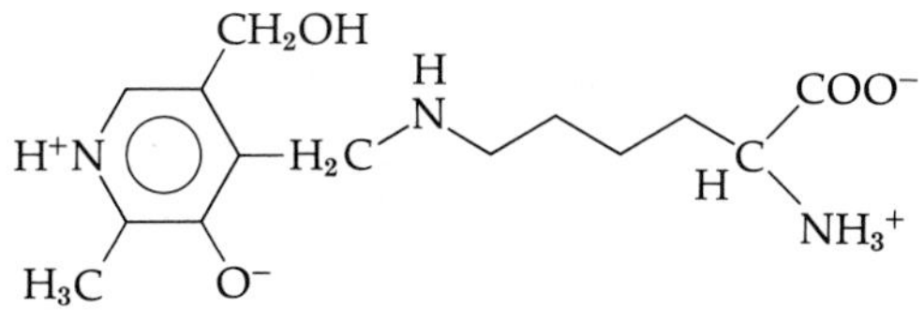

ε-Pyridoxyllysine

containing enzymes in the absence of substrates exist as Schiff bases with lysine side chains of the proteins.* An exception may be δ-aminolevulinate synthetase (Eq. 8-20) for which an adduct of PLP and an active site SH group has been proposed.[51]

* Reduction of glycogen phosphorylase by sodium borohydride yields the same coupling of the PLP to a lysine side chain, but surprisingly the enzyme remains active.

b. Transimination

Formation of a Schiff base of PLP with a substrate probably occurs not by addition of the amino group to C=O but to C=N. Addition is followed by elimination, not of —OH but of the ε-NH$_2$ group (Eq. 8-28). This mechanism may be employed by the enzyme because addition to the

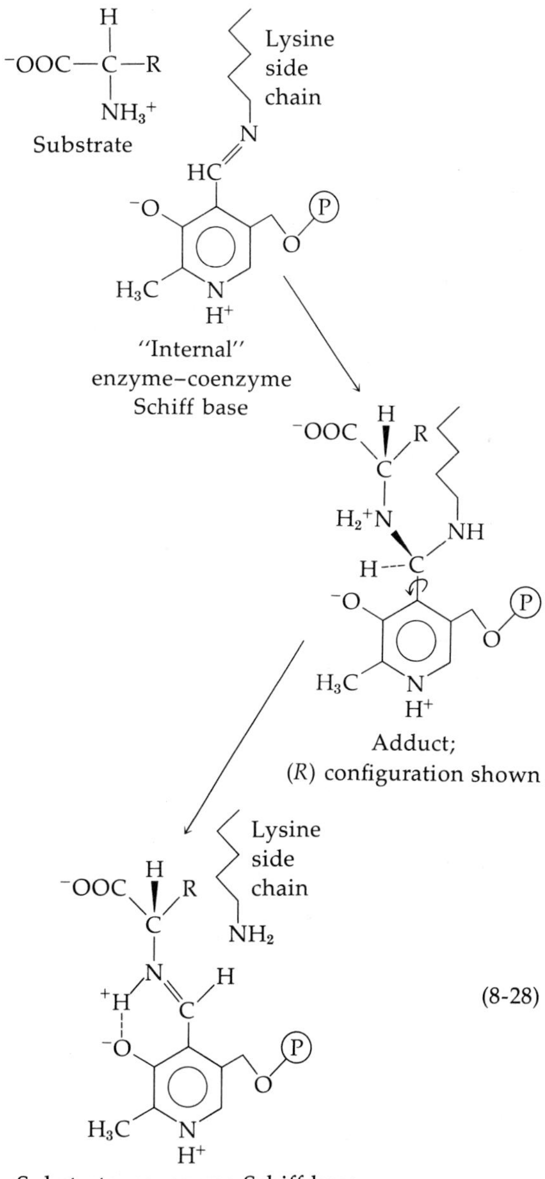

"Internal" enzyme–coenzyme Schiff base

Adduct; (R) configuration shown

Substrate–coenzyme Schiff base

(8-28)

—HC=NH$^+$ is faster[52] than is addition to —HC=O. Transimination is confirmed by the transient loss of the positive circular dichroism (CD, Chapter 13, Section B,5) of the enzyme absorption bands when substrate is added. After the substrate is used, the circular dichroism returns.*

c. Absorption Bands at 500 nm

The most striking spectra, obtained with many PLP enzymes under special conditions, contain intense and unusually narrow bands at ~500 nm (20×10^3 cm^{-1}). Such a band is observed with aspartate aminotransferases acting on erythro-β-hydroxyaspartate (Fig. 8-9). This pseudosubstrate undergoes transamination very slowly, and the 500 nm absorbing form which accumulates could be an intermediate in the normal reaction sequence. A similar spectrum is produced by tryptophanase acting on the competitive inhibitor L-alanine. Under the same conditions the enzyme promotes a rapid exchange of the α-hydrogen of the alanine with ^{2}H of ^{2}H$_2$O. Serine hydroxymethylase gives a 495–500 nm band with both D-alanine and with the normal product glycine.[53] A similar spectrum has been produced in a model reaction.[54] It is generally believed that these bands represent the postulated quinonoid intermediates.

7. Constructing a Detailed Picture of the Action of a PLP Enzyme

Consider the number of different steps that must occur in ~10^{-3} s during the action of an aminotransferase. First, the substrate binds to form the "Michaelis complex." Then the transimination (Eq. 8-28) takes place in two steps and is followed by the removal of the α-hydrogen to form

the quinonoid intermediate. An additional four steps are then needed to form the ketimine, to hydrolyze it, and to release the keto acid product to give the PMP form of the enzyme. The reaction sequences in some of the other enzymes are even more complex.

In the case of aspartate aminotransferase, it is reasonable to assume that the positive charge on the amino group of the substrate aids in guiding the substrate toward a correct fit through attraction of the negative charge on the —O$^-$ of the coenzyme (Eq. 8-28). Likewise, a positively charged group on the protein would attract the α-COO$^-$ of the substrate, and, in the ES complex, would neutralize it. This would in turn cause a drop in the pK_a of the —NH$_3^+$ group of the substrate permitting it to lose a proton.[55] It is attractive to think that this proton might then be transferred (perhaps with the aid of a group in the protein) to the nitrogen of the C=N group of the internal Schiff base.[55] Thus, the nucleophilic —NH$_2$ group would be generated by a process that would at the same time increase the electrophilic properties of the carbon atom of the imine group. This would favor the immediate addition of —NH$_2$ to —C=N$^+$H— to give the adduct of Eq. 8-28.

As Ivanov and Karpeisky pointed out,[55] each step in the overall sequence appears to change the electronic or steric characteristics of the whole complex in a way that facilitates the next step. While as yet unproved, this may be an important principle applicable throughout enzymology: *An enzyme will be an efficient catalyst if each step leads to a change that sets the stage for the next step.*

Many steps in the action of PLP-dependent enzymes require proton transfers, and each such transfer will influence the subsequent step in the sequence. Certain steps require alterations in conformation both of the substrate, coenzyme, and enzyme. For example, conversion of the adduct of Eq. 8-28 to the substrate–coenzyme Schiff base involves a spatial rearrangement. This may be accomplished by rotation about a single bond as shown in Eq. 8-28 or by rotation of the coenzyme.[33,35] Notice that the ε-amino group, which is eliminated (Eq. 8-28) is strongly basic. It has often

---

* It is not clear why the PLP absorption bands of enzymes almost always have a positive CD. The CD is probably induced in the symmetric *chromophore* (light-absorbing group) by an electrically asymmetric environment of the protein.

been suggested that this basic group functions in the next step, in the abstraction of the α-H and in its transfer to the 4'-carbon.

The PLP can be removed from aspartate aminotransferase, and coenzyme analogues of various kinds can be added back. Some of these bind well, but act very slowly. For example, N-methylated PLP does not "turn over" at all when glutamate is added. This fact suggests that at some stage in the reaction sequence the proton on the pyridine nitrogen may have to be removed by deliberate action of the enzymatic machinery.

The student should consider what steric problems might be encountered in the reaction sequence and the possibility of formation in Eq. 8-28 of an adduct of the (S) configuration instead of the (R) configuration.

F. KETO ACIDS AND OTHER UNUSUAL ELECTROPHILIC CENTERS

A few enzymes that might be expected to have pyridoxal phosphate at their active sites have bound α-keto acids instead. These and several apparently related enzymes of unusual function are the subject of this section.

1. Decarboxylases

In some species decarboxylases for histidine[56] and for S-adenosylmethionine[57] contain a covalently bound keto acid at the active site. The enzymes are inhibited by carbonyl reagents and by borohydride. When ^{3}H-containing borohydride was used to reduce the histidine decarboxylase of *Lactobacillus*, ^{3}H was incorporated and was recovered in lactic acid following hydrolysis.[56] This suggested the presence of pyruvic acid attached by an amide linkage and undergoing the chemical reactions that are shown in Eq. 8-29. Reduction in the presence of histidine fixed this substrate co-

Pyruvate linked to enzyme through amide group

$$^3H-\overset{\overset{\displaystyle CH_3}{|}}{\underset{\underset{\displaystyle COOH}{|}}{C}}-OH \qquad (8\text{-}29)$$

Lactic acid

valently to the bound pyruvate. Thus, as with the PLP-containing decarboxylases, a Schiff base is formed with the substrate. Decarboxylation is presumably accomplished by using the electron-attracting properties of the carbonyl group of the amide (Eq. 8-30).

Bound pyruvate   Histidine combined as a Schiff base

When ^{14}C-labeled serine was fed to organisms producing histidine decarboxylase, ^{14}C was incorporated into the bound pyruvoyl group. Thus, serine is the precursor of the bound pyruvate. The enzyme is manufactured in the cell as a longer proenzyme of MW $\sim$ 37,000 which is cleaved to two peptide chains. One of MW $\sim$ 28,000 contains the N-terminal pyruvoyl residue and the other of MW $\sim$ 9000 contains a carboxyl-terminal serine.[58] The activation occurs spontaneously by incubation of the proenzyme for 24–48 h at pH $\sim$ 7 and 37°C. Substituted serine residues under mildly alkaline conditions readily undergo α,β elimination to form **dehydroalanine** residues. Conversion of proenzyme to active histidine decarboxylase could occur by such an elimination followed by a specific hydrolysis (Eq. 8-31).[58]

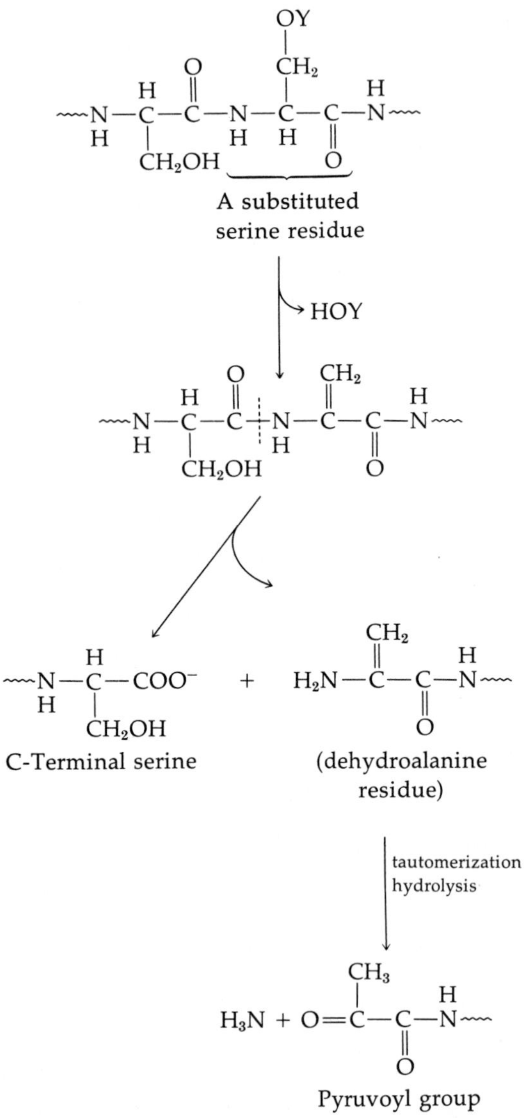

A substituted
serine residue

↓ HOY

C-Terminal serine   +   (dehydroalanine
                           residue)

tautomerization and
hydrolysis

$$H_3N + O=\overset{CH_3}{\underset{O}{\overset{|}{C}}}-\overset{H}{\underset{}{C}}-N\sim \quad (8\text{-}31)$$

Pyruvoyl group

This histidine decarboxylase is also unusual in having five subunits of each of the MW = 9000 and MW = 28,000 chains with a 5-fold rotational symmetry. The proenzyme also is a pentamer.

2. Threonine Dehydratase

While threonine dehydratases from most bacteria require PLP that from *Pseudomonas putida*

does not[59] nor does that from sheep liver. The cofactor for the sheep liver enzyme is an α-ketobutyryl group[60] (the student will be able to conclude from which amino acid this prosthetic group has been derived).*

3. Urocanase

Catabolism of histidine in most organisms proceeds via an initial elimination of NH_3 (Eq. 8-32,

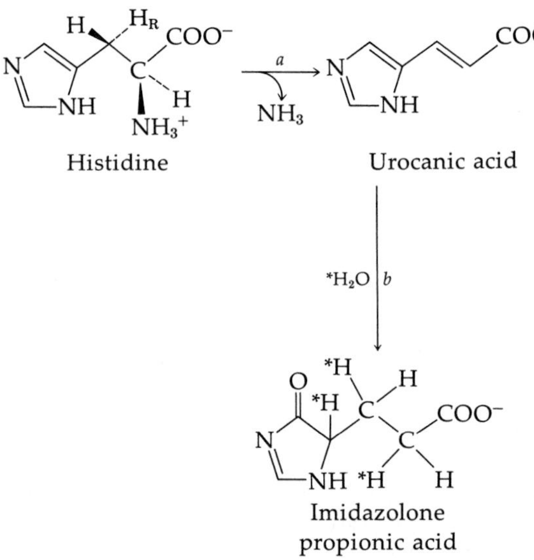

Histidine   Urocanic acid

*H_2O | b

Imidazolone   (8-32)
propionic acid

step *a*). The reaction is unusual because the nucleophilic substituent eliminated is on the α-carbon atom rather than the β (compare with Eq. 7-41, reverse). In the second step, another unusual enzyme urocanase converts urocanic acid to imidazolonepropionic acid (Eq. 8-32, step *b*). By reduction of urocanase from *Pseudomonas putida* with ³H-containing $NaBH_4$ the protein has been shown to contain an N-terminal α-ketobutyryl residue.[61]

The reaction catalyzed (Eq. 8-32, step *b*) can be visualized as a 1,4 addition of water to urocanic acid followed by rearrangement (Eq. 8-33). George

---

* However, a recent report claims PLP as the coenzyme.[60a] Perhaps there are isoenzymes of two types.

and Phillips[61] suggested that the α-ketobutyryl group forms a positively charged imine (Eq. 8-33, step b) which can then rearrange as indicated. When the reaction is run in labeled H_2O, three protons from solvent are incorporated into the product[62] as shown by the asterisks in Eq. 8-32.

4. Proline Reductase

An enzyme needed in the anaerobic breakdown of proline[62a] utilizes a special dithiol-containing protein to reductively open the ring (Eq. 8-34).

$$(8-34)$$

Like histidine decarboxylase, the enzyme contains an N-terminal pyruvoyl residue. Hodgin and Abeles[63] suggested formation of a positively charged Schiff base with the substrate (Eq. 8-35). The bond breaking mechanism is not obvious, but that in Eq. 8-35 has been proposed.[64]

$$(8-35)$$

5. Histidine and Phenylalanine Deaminases (Ammonia-Lyases)

Returning to Eq. 8-32, step a, we note that there is nothing in the structure of the substrate that would permit an easy elimination of the α-amino group. Thus, it should not be surprising that the enzyme **L-histidine ammonia-lyase** contains a special active center, as does the related **L-phenylalanine ammonia-lyase.** The latter eliminates

—NH_3^+ along with the *pro-S* hydrogen in the β position to form *trans*-cinnamic acid (Eq. 8-36). The

$$(8\text{-}36)$$

trans-Cinnamic acid

latter is formed in higher plants and is converted into a bewildering array of derivatives (Box 12-B; Chapter 14, Section H,6). When ^2H is introduced into the β position, no isotope effect on the rate is observed. Rather, the rate-limiting step appears to be release of ammonia from the coenzyme group. It appears that the enzyme must in some way make the amino group a much better leaving group than it would be otherwise.

Both enzymes are inhibited by sodium borohydride and also by nitromethane. It has been argued that both enzymes contain N-terminal dehydroalanine residues, the amino groups of which exist as Schiff bases, possibly with aldehyde groups derived by the oxidation of serine side chains.[65,66] A possible mechanism, based on this assumption, has been proposed. The possible participation of a transition metal has been suggested.[66a]

All of the mechanisms proposed in this section must be regarded as strictly tentative. Since they imply unusual reactivities of bound keto acid residues, there is probably more to the story.

G. HYDROGEN TRANSFER COENZYMES

One of the most common of biological oxidation reactions is **dehydrogenation** of an alcohol to a ketone or aldehyde (Eq. 8-37). It was recognized

$$H-\overset{|}{\underset{|}{C}}-OH \longrightarrow \overset{|}{\underset{|}{C}}{=}O + 2[H] \qquad (8\text{-}37)$$

early in this century that the two hydrogens removed are transferred to hydrogen-carrying coenzymes such as **nicotinamide-adenine dinucleotide** (NAD$^+$),* **nicotinamide adenine dinucleotide phosphate** (NADP$^+$), **flavin adenine dinucleotide** (FAD), and **riboflavin 5'-phosphate** (also known as flavin mononucleotide, FMN).

When NAD$^+$ becomes reduced, only one of the hydrogen atoms removed from the alcohol becomes firmly attached to the NAD$^+$, converting it to NADH, while the other one becomes a free proton (Eq. 8-38). From the study of ^2H-labeled al-

$$NAD^+ + 2[H] \longrightarrow NADH + H^+ \qquad (8\text{-}38)$$

cohols and their oxidation by NAD$^+$, it has been shown that *dehydrogenases catalyze direct transfer of the hydrogen that is attached to carbon in the alcohol to NAD$^+$*. Furthermore, there is never any exchange with protons of the medium. At the same time, the hydrogen attached to the oxygen of the alcohol is released into the medium as H$^+$:

$$H-\overset{|}{\underset{|}{C}}-OH$$

This hydrogen is released as a proton

This H is transferred directly to NAD$^+$

The foregoing observations suggested that it might be profitable to view these biological oxidations (dehydrogenations) as removal of a hydride ion H$^-$ together with a proton H$^+$ rather than as removal of two hydrogen atoms. Thus, NAD$^+$ and NADP$^+$ are usually regarded as *hydride ion accepting coenzymes*. Nevertheless, it is important to realize that it has been impossible to conclusively establish whether the transfer of a hydrogen atom to these coenzymes is followed by or preceded by transfer of an electron or whether the two are transferred simultaneously, i.e., as H$^-$.

The situation is even less clear for the flavin

---

* NAD$^+$ is often called diphosphopyridine nucleotide (DPN$^+$) and NADP$^+$ is called TPN$^+$. Many biochemists prefer these older names and they still appear frequently in current literature.

coenzymes FAD and riboflavin phosphate. However, regardless of the actual mechanism, it is convenient to classify most hydrogen transfer reactions of metabolism as if they occurred by transfer of a hypothetical hydride ion. *The hydride ion can be regarded as a nucleophile* which can add to double bonds or can be eliminated from substrates in reactions of types that we have already considered and which are listed in Table 8-3. While the reactions are all shown as reductions, many are reversible and some normally proceed in the direction opposite that in the table.

Why are there *four* major hydrogen transfer coenzymes, NAD$^+$, NADP$^+$, FAD, and riboflavin phosphate, instead of just one? Part of the answer is that the reduced **pyridine nucleotides** NADPH

TABLE 8-3
Some Hydrogen Transfer Reactions[a]

Reaction	Example
A. Reduction of carbonyl group	
$H^- + \;>\!\!C\!=\!\!O + H^+ \longrightarrow H\overset{\textstyle |}{\underset{\textstyle |}{C}}\!-\!OH$	Alcohol dehydrogenase
B. Reduction of a Schiff base	
$>\!\!C\!=\!\!O + NH_3 \rightleftharpoons \;>\!\!C\!=\!\!NH \xrightarrow{+H^-,\,H^+} H\overset{\textstyle |}{\underset{\textstyle |}{C}}\!-\!NH_2$	Amino-acid dehydrogenases, amine oxidases
C. Reduction of thioesters	
$-\overset{\textstyle O}{\overset{\|}{C}}\!-\!SR \xrightarrow{H^-,\,H^+} -\overset{\textstyle OH}{\underset{\textstyle H}{\overset{\textstyle |}{C}}}\!-\!SR \rightleftharpoons -\overset{\textstyle O}{\overset{\|}{C}}H + R\!-\!SH$	Aldehyde dehydrogenases
D. Reduction of α,β-unsaturated acyl groups	
$R\!-\!C\!\!\begin{smallmatrix}H^-\\\searrow H\\ \end{smallmatrix}\!\!\underset{H}{\overset{}{C}}\!-\!\overset{O}{\overset{\|}{C}}\!-\!Y \longrightarrow R\!-\!CH_2\!-\!CH_2\!-\!\overset{O}{\overset{\|}{C}}\!-\!Y$ (Y = —S—CoA or —OH)	Acyl-CoA dehydrogenase, succinic dehydrogenase
E. Reduction of an isolated double bond	
$R\!-\!\overset{H}{\overset{}{C}}\!=\!C\!\!<\xrightarrow{H^-,\,H^+} R\!-\!CH_2\!-\!HC\!\!<$	Reduction of desmosterol to cholesterol

[a] Reaction type 9 (see Table 9-1): A hypothetical hydride ion H^- is transferred onto the substrate from a coenzyme of suitable reduction potential such as NADH (DPNH), NADPH (TPNH), FADH$_2$ or reduced riboflavin 5′-phosphate. Enzymes are usually named for the reverse (dehydrogenation) reactions.

and NADH are more powerful reducing agents than are reduced flavins (Table 3-7). Conversely, flavin coenzymes are more powerful oxidizing agents than are NAD$^+$ and NADP$^+$. Flavin coenzymes are based on the vitamin **riboflavin** and pyridine nucleotides on **nicotinamide.** The structures of these two vitamins have no doubt been selected by nature to give just the right redox potentials (Chapter 3, Section C) to the coenzymes. However, it is not this simple. NAD$^+$ and NADP$^+$ tend to be present in free forms within cells, diffusing from a site on one enzyme to a site on another. Flavin coenzymes are usually tightly bound to proteins, fixed and unable to move. Thus, they tend to accept hydrogen atoms from one substrate and to pass them to a second substrate while attached to a single enzyme.

The redox potential of a pyridine nucleotide coenzyme system is determined by the standard redox potential for the free coenzyme together with the ratio of concentrations of oxidized to reduced coenzyme ([NAD$^+$]/[NADH], Eq. 3-64). Thus, a precise potential can be defined for the NAD$^+$ system within a cell. The redox potential may vary in different parts of the cell because of differences in the [NAD$^+$]/[NADH] ratio, but within a given region of the cell it is constant. On the other hand, the redox potentials of flavoproteins vary. Since the flavin coenzymes are not dissociable, two flavoproteins may operate at very different potentials even when they are physically close together.

Why are there two pyridine nucleotides, NAD$^+$ and NADP$^+$ differing only in the presence or absence of an extra phosphate on the nucleotide handle? Part of the answer seems to be that they are members of two different oxidation–reduction systems both based on nicotinamide but independent one of the other. The experimentally measured ratios [NAD$^+$]/[NADH] and [NADP$^+$]/[NADPH] differ markedly. Thus, the two systems appear to operate at different redox potentials in the same region of the cell. A generalization that holds much of the time (but not always) is that *NAD$^+$ serves as an oxidizing agent for removal of hydrogen from substrates, while NADP$^+$ is* *more often reduced to NADPH and used as a reducing agent in biosynthetic processes.*

H. PYRIDINE NUCLEOTIDE COENZYMES AND DEHYDROGENASES

In 1904 A. Harden and W. Young showed that cell-free yeast "juice"* lost its ability to ferment glucose to alcohol and carbon dioxide when it was dialyzed. Apparently fermentation depended upon a low molecular weight substance able to pass through the pores of the dialysis membrane. Fermentation could be restored by adding back to the yeast juice either the concentrated dialysate or boiled yeast juice (in which the enzyme proteins had been destroyed). The heat-stable material, which Harden and Young called **cozymase,** was eventually found to be a mixture of inorganic phosphate ion, thiamine diphosphate, and NAD$^+$. However, characterization of NAD$^+$ was not accomplished until 1935.

Pure NADP$^+$ was isolated from red blood cells in 1934 by Otto Warburg and W. Christian who had been studying the oxidation of glucose 6-phosphate by erythrocytes.[67] They demonstrated a requirement for a dialyzable coenzyme which they characterized and named **triphosphopyridine nucleotide** (TPN$^+$, now officially NADP$^+$, Fig. 8-10). It is interesting that even before its recognition as an important vitamin in human nutrition, nicotinamide was identified as a component of NAD$^+$.

Warburg and Christian recognized the relationship of NADP$^+$ to NAD$^+$ and proposed that both of these compounds act as hydrogen carriers through alternate reduction and oxidation of the pyridine ring. They showed that the coenzymes could be reduced both enzymatically and nonenzymatically with sodium dithionite Na$_2$S$_2$O$_4$ (Eq. 8-39).

---

* Formed by grinding yeast with sand and filtering. The discovery (by Bücher in 1899) that such yeast juice fermented sugar represented one of the starting points for the development of modern biochemistry.

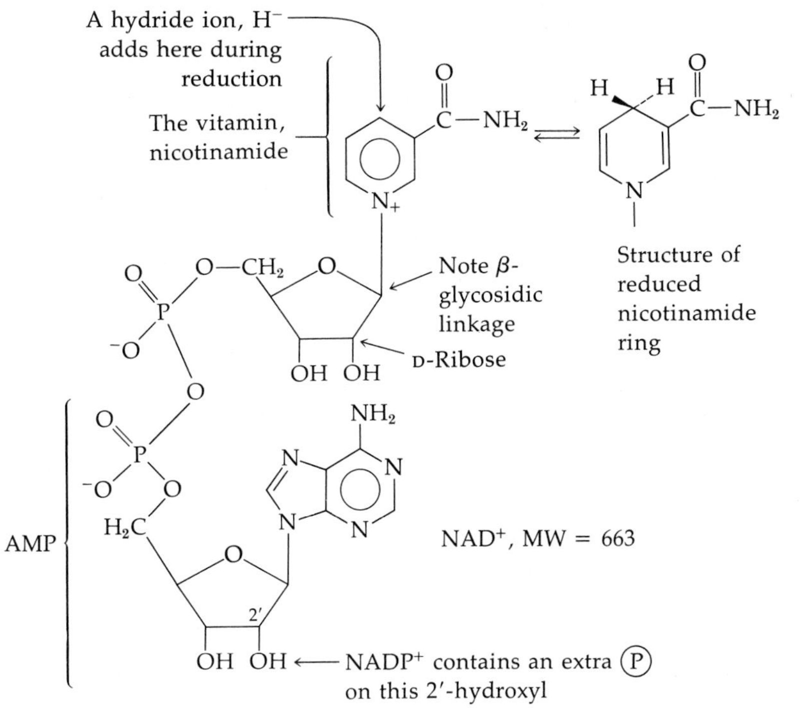

Fig. 8-10 The hydrogen-carrying coenzymes NAD⁺ (DPN⁺) and NADP⁺ (TPN⁺). Note that we use the abbreviations NAD⁺ and NADP⁺, even though the net charge on the entire molecule at pH 7 is negative.

$$S_2O_4^{2-} + NAD^+ + H^+ \longrightarrow NADH + 2\ SO_2 \quad (8\text{-}39)$$

The reduced coenzymes NADH and NADPH were characterized by a new light-absorption band at 340 nm not present in the oxidized forms, which absorb maximally at 260 nm (Fig. 8-11). The reduced forms are stable in air, but their reoxidation was found to be catalyzed by certain "yellow enzymes."

1. Direct Transfer of H Atoms

Although it was clear that the pyridine ring was reduced when NAD⁺ was converted to NADH the position at which the hydrogen was added was not established until 1944 when the use of isotopic labels in biochemical studies had begun. When NAD⁺ was reduced in ^2H_2O by dithionite (Eq. 8-39) an atom of ^2H was introduced into the reduced pyridine ring and was shown by chemical degradation of the ring to be present at the 4 position para to the ring nitrogen[68] (see Fig. 8-10). Later, Westheimer and associates showed that during enzymatic reduction of NAD⁺ by $CH_3—C^2H_2OH$, one of the ^2H atoms (see Eq. 6-65) was transferred into the NADH formed, thus establishing the direct transfer of a hydrogen atom.[69] Furthermore, when the NAD^2H so formed was reoxidized enzymatically with acetaldehyde, regenerating NAD⁺ and ethanol, the ^2H was completely removed.

This was one of the first examples of the ability of an enzyme to choose between two identical atoms at a *pro*-chiral center (Chapter 6, Section D,2). The two hydrogen atoms at the 4 position in NADH were designated as H_A (now known as *pro-R*) and H_B (*pro-S*) and the two sides of the ni-

BOX 8-H

NICOTINIC ACID AND NICOTINAMIDE

Nicotinic acid   MW = 123.1   Nicotinamide

Nicotinic acid was prepared in 1867 by oxidation of nicotine. While it was isolated by Funk and independently by Suzuki (1911–1912) from yeast and rice polishings, it was not recognized as a vitamin. Its biological significance was first established in 1935 when nicotinamide was found to be a component of NAD⁺ (H. von Euler and associates) and of NADP⁺ (Warburg and Christian).

In 1937, C. Elvehjem and co-workers demonstrated the cure of canine "blacktongue" with nicotinic acid, and later in the same year cure of human **pellagra** was demonstrated by several groups. At that time pellagra was a very important disease in the United States, especially in the south. In 1912–1916, the U.S. Public Health Service estimated that there were 100,000 victims and 10,000 deaths a year.[a,b] (Pellagra is characterized by weakness, indigestion, and loss of appetite followed by dermatitis, diarrhea, mental disorders, and eventually death.)

Physical properties: Both forms of the vitamin are stable, colorless, and highly soluble in water.

Daily requirement: About 7.5 mg for an adult. The amount is decreased by the presence in the diet of tryptophan, which can be converted partially to nicotinic acid (Chapter 14, Section I). Tryptophan is about 1/60 as active as nicotinic acid itself. The one-time prevalence of pellagra in the southern United States was a direct consequence of a diet high in corn whose proteins have a very low tryptophan content.

[a] A. F. Wagner and K. Folkers, "Vitamins and Coenzymes," p. 73. Wiley (Interscience), New York, 1964.

[b] For reviews of two interesting books dealing with the history of pellagra, see B. G. Rosenkrantz, *Science* **183**, 949–950 (1974).

---

cotinamide ring as A and B. Alcohol dehydrogenase always removes the H_A (*pro-R*) hydrogen. Malate, isocitrate, lactate, and D-glycerate dehydrogenases select the same hydrogen.[70] However, dehydrogenases acting on glucose 6-phosphate, glutamate, 6-phosphogluconate, and 3-phosphoglyceraldehyde remove the *pro-S* hydrogen.

$pro\text{-}R \rightarrow H_A \qquad H_B \leftarrow pro\text{-}S$

$CONH_2$

When a hydrogen atom is transferred by an enzyme from the 4 position of NADH or NADPH to an aldehyde or ketone to form an alcohol, the placement of the hydrogen atom on the alcohol is also stereospecific. Thus, alcohol dehydrogenase acting on NAD²H converts acetaldehyde to (R)-mono-[²H]ethanol (Eq. 6-65). Likewise, pyruvate is reduced by lactate dehydrogenase to L-lactate, and so on.

Many attempts have been made to establish whether a hydride ion or a hydrogen atom is transferred in the rate-determining step for dehydrogenases. In one study para-substituted benzaldehydes were reduced with NADH and NAD²H using yeast alcohol dehydrogenase as a catalyst.[71] This permitted the application of the Hammett equation (Eq. 3-66) to rate data. For a series of benzaldehydes for which σ^+ varied widely, a value of $\rho = +2.2$ was observed for the rate constant with both NADH and NAD²H. Thus, electron-accepting substituents in the para position hasten the reaction. While the significance of this observation is not immediately obvious,[71] the relatively low value of ρ is probably incompatible with a mechanism that requires complete transfer of a single electron from NADH to acetaldehyde in the first step. A primary isotope effect on the rates was $k_H/k_{2H} = 3.6$, indicating that the C—H bond

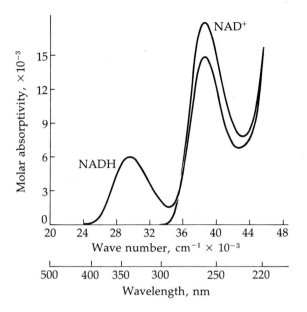

Fig. 8-11 Absorption spectra of NAD+ and NADH. Spectra of NADP+ and NADPH are nearly the same as these. The difference in absorbance between oxidized and reduced forms at 340 nm is the basis for what is probably the single most often used spectral measurement in biochemistry. Reduction of NAD+ or NADP+ or oxidation of NADH or NADPH is measured in dozens of methods of enzyme assay. If a pyridine nucleotide is not a reactant for the enzyme being studied, a "coupled assay" is often possible. For example, the rate of enzymatic formation of ATP can be measured by adding to the reaction mixture glucose + hexokinase + glucose-6-phosphate dehydrogenase + NADP+. ATP, as it is formed, phosphorylates glucose (hexokinase). Then NADP+ oxidizes the glucose 6-phosphate formed with production of NADPH, whose rate of appearance is monitored at 340 nm.

plished for several dehydrogenases. The structure of lactate dehydrogenase from the dogfish is known to 0.2 nm resolution.[72,73] The enzyme is a tetramer of MW = 140,000. The coenzyme curls around one end of the protein in a sort of "C" conformation with the nicotinamide ring lying in a cleft (Fig. 8-12). While the enzyme–substrate complex is too short-lived to examine directly, its structure can be inferred from those of various enzyme–inhibitor complexes, e.g., that of the dehydrogenase with an NAD+–pyruvate adduct (Eq. 8-47).[73] From these the active site structure of Fig. 8-12 has been deduced. The L-lactate is seen in position for transfer of a hydride ion to the A side of the NAD+ ring. The lactate carboxylate ion is held and neutralized by the guanidinium group of Arg 171, and His 195 is in position to serve as a general base catalyst to abstract a proton from the hydroxyl group of the substrate.

Many interesting interactions of protein side chains appear to occur. Thus, the carboxylate side chain of glutamate 140 neutralizes the positive charge on the NAD+. That this may have a functional significance is suggested by study of a model reaction. Reduction of the N-methylacridinium ion by the following substituted dihydronicotinamide derivative in acrylonitrile proceeds much more rapidly than for the corresponding compound lacking the carboxylate ion.[74]

in NADH is broken in the rate-limiting step. The fact that the isotope effect is the same for all of the substituted benzaldehydes argues in favor of hydride ion transfer.[71]

2. Alcohol Dehydrogenase Proteins

Most dehydrogenases are dimers or tetramers formed from subunits of MW = 20,000–40,000, but some are larger. Complete structure determinations by X-ray diffraction have been accom-

Even before the crystal structure of lactate dehydrogenase was known, the lack of pH dependence of coenzyme binding from pH 5 to 10 together with observed inactivation by butanedione suggested that the pyrophosphate group of NAD+ binds to a guanidinium group of an arginine side chain.[75] The X-ray diffraction studies indicate that

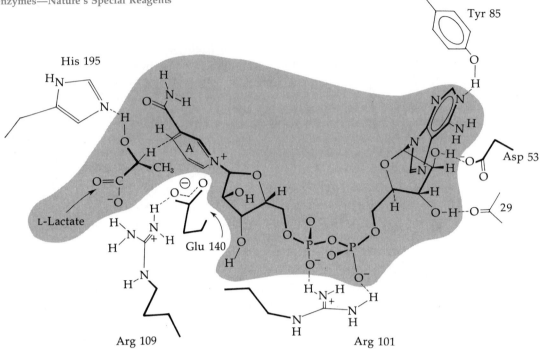

Fig. 8-12 Schematic diagram of the binding of NAD$^+$ and of L-lactate at the active site of lactate dehydrogenase. After Adams *et al.*[73,74]

Arg 101 does function in this way (Fig. 8-12). Somewhat surprisingly the amino group of the adenine ring does not appear to be hydrogen bonded to the protein. Rather, the adenine is bound in a hydrophobic crevice and the amino group is pointed out into the solvent.

Other dehydrogenases whose structures are known include malate dehydrogenase,[76] the nonspecific liver alcohol dehydrogenase,[77] and glyceraldehyde-3-phosphate dehydrogenase (Section H,5). It is noteworthy that in all three of these dehydrogenases and in lactate dehydrogenase there is a nearly constant structural feature consisting of a six-stranded parallel β sheet together with several α-helical coils[77,78,78a] (Fig. 2-10A). This coenzyme-binding substructure may be specially adapted for interaction with NAD$^+$ and, in modified form, may be present in a variety of other enzymes.[78a] Some dehydrogenases, including liver alcohol dehydrogenase,[77] contain an essential atom of Zn^{2+}. Coordination with the hy-

droxyl group of the alcohol or the carbonyl group of the aldehyde substrate is probable.[78b]

3. Conformational Changes during Dehydrogenase Action

Crystals of pig heart malate dehydrogenase bind only 1 NAD$^+$ per dimer, a second NAD$^+$ going on with great difficulty.[79] Similar anticooperativity has been reported for glyceraldehyde-3-phosphate dehydrogenase and alkaline phosphatase (Chapter 7, Section E,1). However, there is controversy about experimental observations[80,80a] and the student should look carefully at the data. An intriguing idea is that the anticooperativity in coenzyme binding reflects a cooperative action between subunits during catalysis. Suppose that only conformation (A) binds reduced substrate and NAD$^+$, while conformation (B) binds NADH and oxidized substrate. If reduced substrate and NAD^+ were present in excess and if oxidized substrate were efficiently removed from the scene by further oxidation, the following events could occur in the mixed AB dimer. The subunit of conformation A would bind substrates, react, and be converted to conformation B. At the same time, because of the strong AB interaction, the subunit that was originally in conformation B would be converted back to A and would be ready to initiate a new round of catalysis. Since conformation A has a low affinity for NADH, dissociation of the reduced coenzyme would be facilitated.[81] It is known that dissociation of NADH is often the slow step in dehydrogenase action. Such a **reciprocating** or **flip-flop** mechanism was first suggested by Harada and Wolfe.[81a] The idea is attractive in that it provides a natural basis for the existence of the many known dimeric enzymes that do not exhibit evident allosteric properties. Lazdunski[82] suggested that flip-flop mechanisms occur widely among dimeric enzymes, but whether they actually exist remains uncertain.

4. Glutamate Dehydrogenase

The reduction of a Schiff base (Table 8-3, reaction B) is catalyzed by glutamate dehydrogenase (Chapter 14, Section B,1). The bovine enzyme is a large hexameric protein with a subunit weight of ~56,000. Each subunit contains 500 residues whose sequence is now known.[83] Borohydride reduction of the complex of the enzyme with the substrate α-ketoglutarate indicates formation of a Schiff base with Lys 126. This seems strange, for to be reduced to glutamate, ketoglutarate must react with ammonia rather than with a lysine side chain. Perhaps ammonia adds to the Schiff base of ketoglutarate with lysine and transimination (Section E,6,b) takes place to form a Schiff base with ammonia. Alternatively, the NH_3 adduct might be reduced by displacement with a hydride ion from NADH or NADPH (Eq. 8-40). An imidazole group may also be present at the active center of the enzyme.[84]

Schiff base
with lysine

NH_3 adduct

Glutamic acid

(8-40)

5. Glyceraldehyde-3-Phosphate Dehydrogenase and the Generation of ATP in Fermentation Reactions

A standard biochemical method for reduction of carboxyl groups to aldehyde groups is conversion, in an ATP-requiring process, to a thioester followed by reduction of the thioester (reaction type 9C of Table 8-3). Conversely, the oxidation of an aldehyde to a carboxylic acid, a highly exergonic process, usually proceeds through a thioester intermediate whose cleavage can then be coupled to synthesis of ATP. This sequence (S9C) which is extremely important to the energy metabolism of cells, is set off for future reference in Fig. 8-13.

The most noteworthy example is provided by glyceraldehyde-3-phosphate dehydrogenase, a tetramer of identical subunits of MW = 45,000 whose three-dimensional structure has been determined (Fig. 2-10).[78] Recall that aldehydes are in equilibrium with their covalent hydrates (Eq. 8-41). Dehydrogenation of the latter would yield an acid (Eq. 8-41, step b). However, such a mecha-

Covalent
hydrate

(8-41)

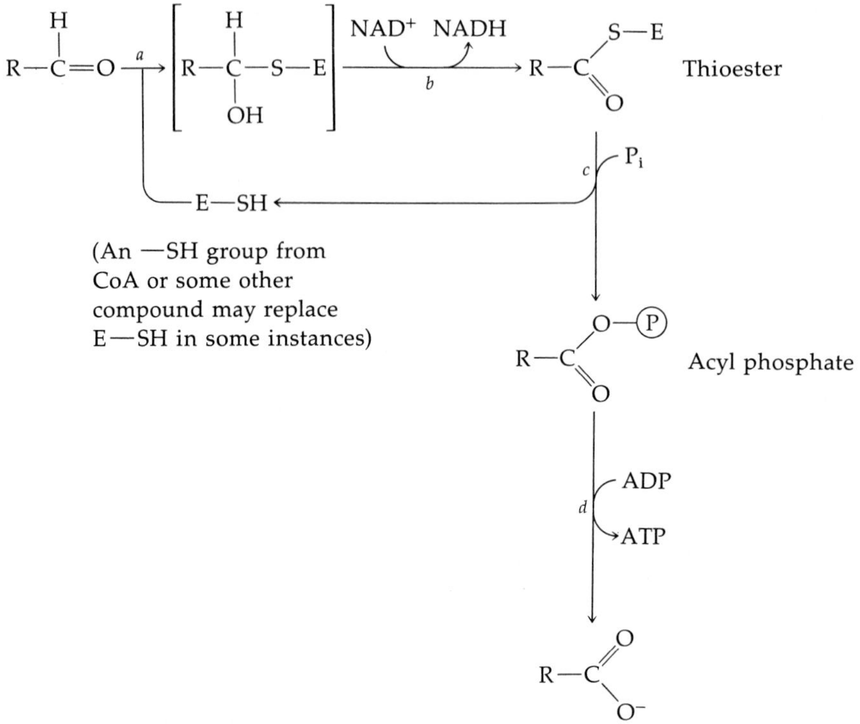

Fig. 8-13 Generation of ATP coupled to oxidation of an aldehyde to a carboxylic acid (reaction sequence S9C). The most important known example of this sequence is the oxidation of glyceraldehyde 3-phosphate to 3-phosphoglycerate (Fig. 9-7). Other important sequences for "substrate level" phosphorylation are shown in Figs. 8-4, 8-19, and 8-21.

nism would offer no possibility of conserving the energy available from the reaction. In the case of glyceraldehyde-phosphate dehydrogenase the enzyme possesses a special sulfhydryl group that adds to the carbonyl in the first step (step a, Fig. 8-13) to form an adduct (thiohemiacetal). The adduct is then oxidized by NAD^+ to a thioester, an S-acyl enzyme (step b).[85] The latter is cleaved by the same enzyme through displacement on carbon by an oxygen atom of P_i (phosphorolysis, step c). The sulfhydryl group of the enzyme is released simultaneously and the product, an acyl phosphate (1,3-diphosphoglycerate) is formed. A separate enzyme then transfers the phosphoryl group from the 1 position of 1,3-diphosphoglycerate to ADP to form ATP and 3-phosphoglycerate. The overall sequence of Fig. 8-13 is the synthesis of one mole of ATP coupled to the oxidation of an aldehyde to a carboxylic acid.

Note that the last two reaction steps in Fig. 8-13, steps c and d, are just the reverse of sequence S1A (α) of Table 7-2. Thus, the chemistry by which ATP is generated during glycolysis and that by which it is utilized in biosynthesis is nearly the same. However, in the former case, the "high-energy" thioester has been created by oxidation (steps a and b of Fig. 8-13). Arsenate uncouples ATP synthesis from the oxidation in this sequence (Box 7-A).

A reaction resembling that of glyceraldehyde-3-phosphate dehydrogenase is catalyzed by **glucose-6-phosphate dehydrogenase**. This "ferment" was the enzyme that originally attracted Warburg's attention and led to the discovery of $NADP^+$. The substrate is not the free aldehyde but the hemiacetal ring form and it is oxidized to a lactone. The lactone is then hydrolyzed to 6-phosphogluconate (Eq. 8-42). Apparently cells are not able to con-

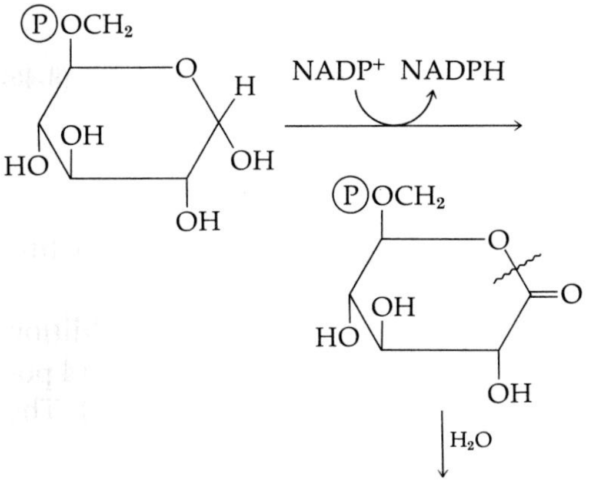

6-Phosphogluconate   (8-42)

serve any energy from this oxidation of an aldehyde to a carboxylic acid but the ring opening step ensures that the reaction goes to completion. It is noteworthy that this reaction is a major supplier of reduced NADP (NADPH) for reductive biosynthesis and that the large free energy decrease for the overall reaction ensures that the ratio [NADPH]/[NADP$^+$] is kept high within cells.

6. Other Dehydrogenases

Some pyridine nucleotide-dependent dehydrogenases are able to catalyze the reduction of isolated double bonds (reaction type 9E). An example is the hydrogenation of **desmosterol** by NADPH (Eq. 8-43), the final step in one of the

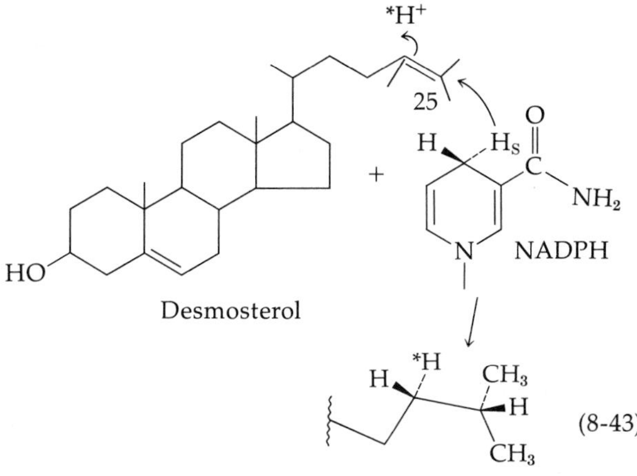

Desmosterol

Cholesterol   (8-43)

pathways of biosynthesis of cholesterol. In this and in two other reactions of the same type hydrogen transfer has been shown to be from the *pro-S* position in NADPH directly to C-25 of the sterol. The proton introduced from the medium (designated by the asterisks in Eq. 8-43) enters trans to the H$^-$ ion from NADPH. The proton always adds to the more electron-rich terminus of the double bond, i.e., it follows the Markovnikov rule. This result suggests that protonation of the double bond may precede H$^-$ transfer.[86]

Dehydrogenation of alcohols is often used to generate carbonyl groups needed for some special chemical purpose. For example, a carbonyl compound may be a symmetric intermediate in a reaction that inverts the configuration about a chiral center. Consider **UDPglucose 4-epimerase,** an enzyme that converts UDPgalactose to UDPglucose (Eq. 8-44; Chapter 11, Section E,1b) and is essen-

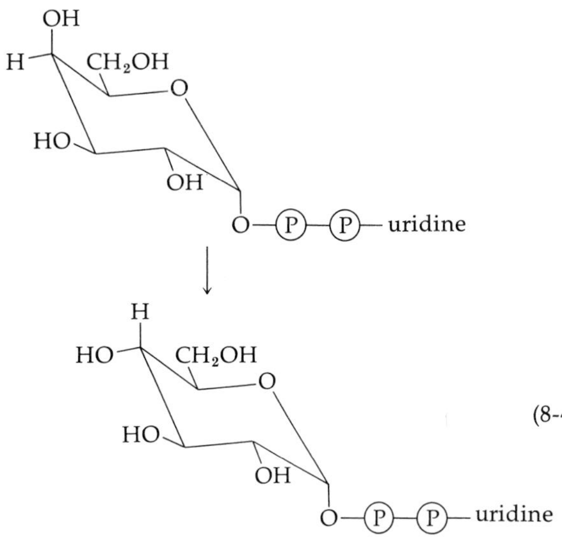

(8-44)

tial in the metabolism of galactose in our bodies. The enzyme contains bound NADP$^+$ and evidence has been provided for a 4-keto intermediate.[87] Another way that formation of a keto group can assist in epimerization of a sugar is through enolization with nonstereospecific return of a proton to the intermediate enediol.[88] Still another possibility is through aldol cleavage followed by aldol condensation, again with inversion of configuration. In each of these cases the initial creation of a keto group by dehydrogenation is essential to permit further metabolic reactions to occur.

7. Some Surprising Chemistry of the Pyridine Nucleotides

Despite the apparent simplicity of their structures, the chemistry of the nicotinamide ring in NAD^+ and $NADP^+$ is amazingly varied. For example, NAD^+ is extremely unstable in basic solutions, whereas NADH is just as unstable in slightly acidic media. These facts, together with the ability of NAD^+ to undergo condensation reactions with other compounds, have sometimes caused serious errors in interpretation of experiments. They may also be of significance to biological function.

a. Addition to NAD^+ and $NADP^+$

Many nucleophilic reagents add reversibly at the para position (Eq. 8-45) to form adducts

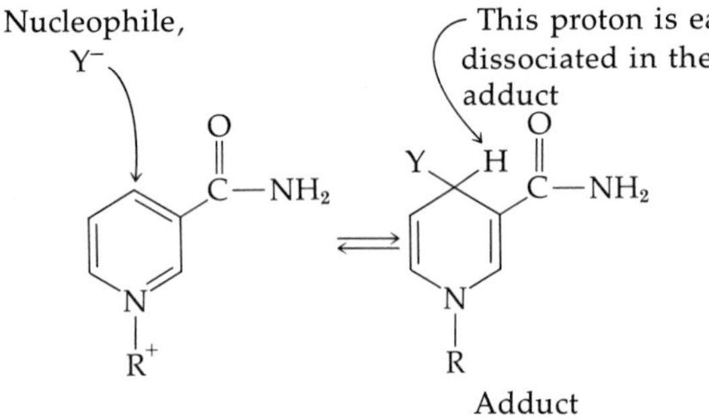

(8-45)

Adduct

having structures reminiscent of those of the reduced coenzymes. Formation of the cyanide adduct, whose absorption maximum is at 327 nm, has been used to introduce deuterium into the para position of the pyridine nucleotides. In the adduct, the proton adjacent to the highly polarized $C\equiv N$ is easily dissociable as a proton. Other anions such as $-S^-$, bisulfite, and dithionite also add. Addition can also occur at the two ortho positions.

The adducts of OH^- in the 2 or 4 positions of NAD^+ undergo ring opening (Eq. 8-46) in base-

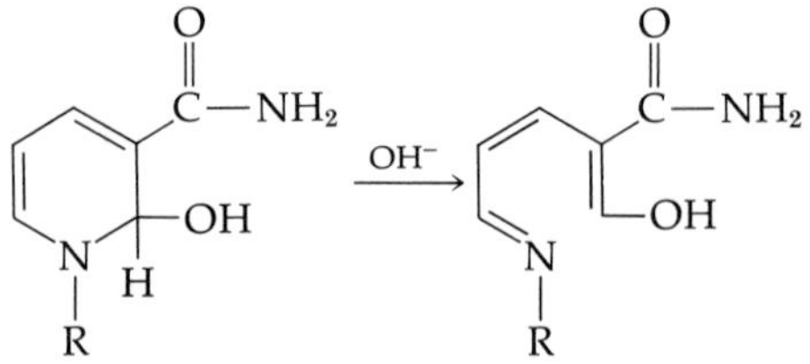

(8-46)

catalyzed reactions which can be followed by further degradation.[87-89]

Another base-catalyzed reaction is the addition of enolate anions derived from ketones to the 4 position of the pyridine nucleotides (Eq. 8-47). The

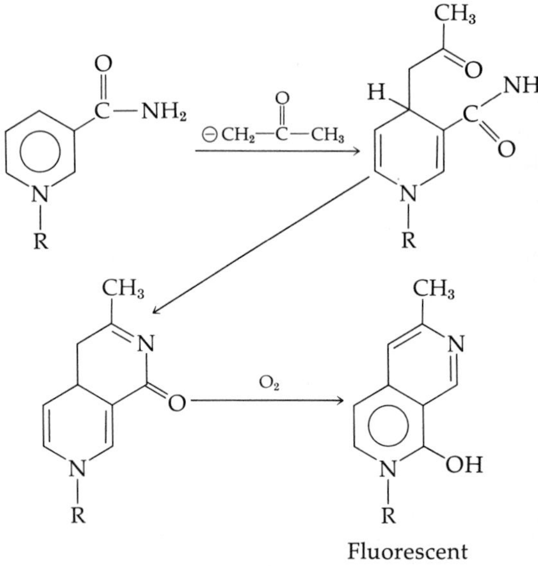

(8-47)

Fluorescent

adducts undergo ring closure and in the presence of oxygen are slowly converted to fluorescent materials. While forming the basis for a useful analytical method for determination of NAD^+ (using 2-butanone), these reactions also have created a troublesome enzyme inhibitor. Traces of acetone present in commercial NADH cause the problem.[90]

While the reactions of Eq. 8-47 occur nonenzymatically only under the influence of strong base, dehydrogenases often catalyze similar condensations in a relatively rapid and reversible fashion. The reactions are specific for those ketones that are products of action of the dehydrogenases: pyruvate inhibits only lactate dehydrogenase, α-

ketoglutarate inhibits glutamate dehydrogenase, etc.[91] We have seen (Chapter 6, Section A,9) that product inhibition is one standard control element in regulation of metabolism; the phenomena which we are discussing here may be part of such control mechanisms.

b. Modification of NADH in Acid

Reduced pyridine nucleotides are destroyed rapidly in dilute HCl and more slowly at pH 7 in reactions catalyzed by buffer acids.[92–95] Apparently the reduced nicotinamide ring is first protonated at C-5 and then a nucleophile Y^- adds at the 6 position (Eq. 8-48). The nucleophile may

$$(8-48)$$

adduct

be OH^-, and the adduct may undergo further reactions. For example, water can also add to the other double bond and the compound can undergo ring opening on either side of the nitrogen. The early steps in the modification reaction are partially reversible, but the overall sequence is irreversible. A product of surprising structure formed by the acid modification of NADH has been characterized by crystal structure determination.[96] The group Y of Eq. 8-48 is the C-2' hydroxyl

An acid modification
product from NADH

of the ribose ring. Before this hydroxyl can react, the configuration of the glycosidic linkage must change from β to α. The reader can suggest a mechanism for acid-catalyzed epimerization at C-1' preceding addition.

The foregoing reactions have attracted considerable interest because glyceraldehyde-3-phosphate dehydrogenase, in a side reaction, converts NADH to a substance referred to as NADH-X whose properties are similar to those of the acid modification product. In this case, reaction of NADH-X with ATP and an enzyme from yeast reconverts it to NADH. The possibility that chemistry of this type might play a role in the process of oxidative phosphorylation (Eq. 10-14) has often been considered.

c. Other Reactions of Pyridine Nucleotides

Alkaline ferricyanide oxidizes NAD^+ and $NADP^+$ to 2-, 4-, and 6-pyridones. Such pyridones, especially the 6-pyridone of N-methylnicotinamide, are well-known excretion products

A 6-pyridone formed from a
pyridine nucleotide

of nicotinic acid in mammals. Reoxidation of NADH and NADPH to NAD^+ and $NADP^+$ can be accomplished with ferricyanide, quinones, and riboflavin but not by H_2O_2 or O_2. When heated in 0.1 N alkali at 100°C for 5 min, NAD^+ is hydrolyzed to nicotinamide and adenosinediphosphate-ribose.

8. Analogues of Pyridine Nucleotides[97]

Treatment of NAD^+ with nitrous acid deaminates the adenine ring. The resulting deamino NAD^+ as well as synthetic analogues containing

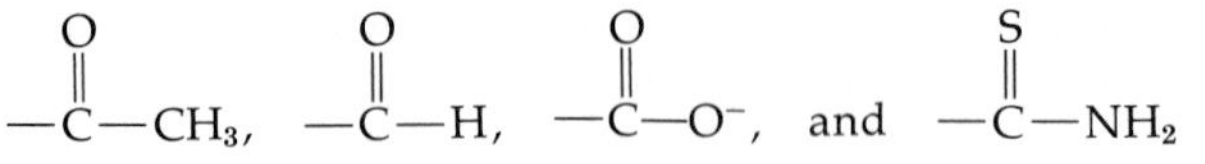

$$-\overset{\overset{\displaystyle O}{\|}}{C}-CH_3, \quad -\overset{\overset{\displaystyle O}{\|}}{C}-H, \quad -\overset{\overset{\displaystyle O}{\|}}{C}-O^-, \quad \text{and} \quad -\overset{\overset{\displaystyle S}{\|}}{C}-NH_2$$

in place of the carboxamide group have been used widely in enzyme studies.

I. THE FLAVIN COENZYMES

Flavin adenine dinucleotide (FAD) and **riboflavin 5'-phosphate** (FMN, Fig. 8-14) are perhaps the most versatile of all the oxidation coenzymes. The name flavin-adenine dinucleotide is not really appropriate because the D-ribityl group is not linked to the riboflavin in a glycosidic linkage; hence, the molecule is not a dinucleotide. However, the name has stuck. Flavin mononucleotide (FMN) is an even less appropriate designation for riboflavin 5'-phosphate.

Discovery of the role of riboflavin in biological oxidation was an outgrowth of biochemists' interest in respiration of cells. In the 1920's Warburg

had provided evidence that oxygen reacted with an iron-containing respiration catalyst. Later it was shown that the dye **methylene blue** could often substitute for oxygen as an oxidant. Oxidation of glucose 6-phosphate by methylene blue within red blood cells required both a "ferment" (enzyme) and a "coferment" later identified as NADP+. A yellow protein, isolated from yeast, was found to have the remarkable property of being decolorized by the reducing system of glucose 6-phosphate plus the protein and "coferment" from red blood cells.

Warburg and Christian showed that the yellow color of this **old yellow enzyme** came from a pigment of the flavin class and proposed that its cyclic reduction and reoxidation played a role in cellular oxidation. When NADP+ was isolated the proposal was extended to encompass a **respiratory chain.** The two hydrogen carriers NADP+ and flavin would work in sequence to link dehydrogenation of glucose to the iron-containing catalyst that interacted with oxygen. While it is now

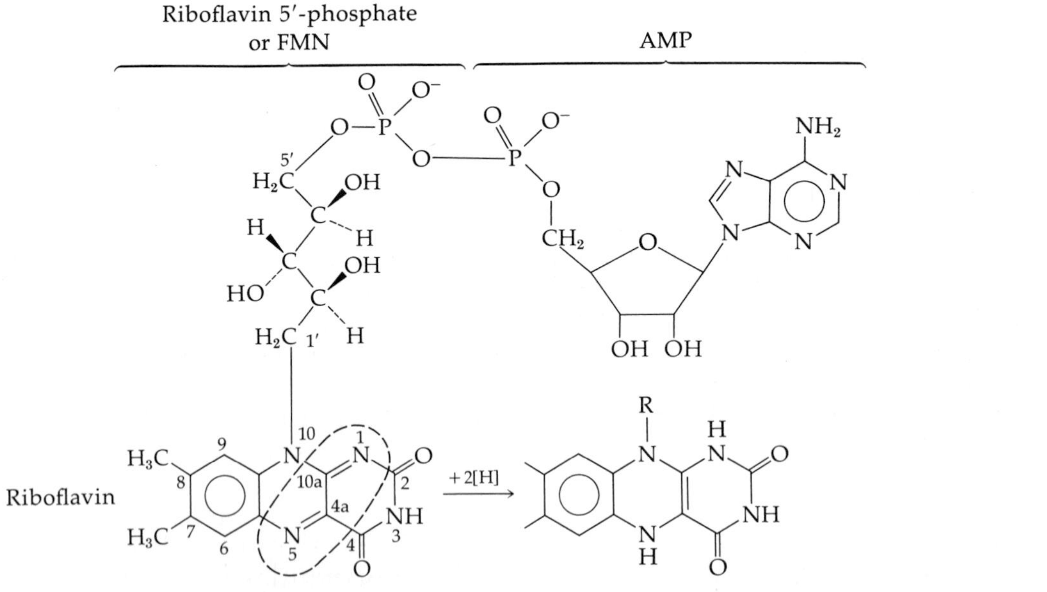

Fig. 8-14 The flavin coenzymes FAD and riboflavin 5'-phosphate. Dotted lines enclose the region that is altered upon reduction. Note that in the older literature a different numbering of the isoalloxazine ring is used. Riboflavin is now known as 7,8-dimethyl-10-(1'-D-ribityl) isoalloxazine; the old name was 6,7-dimethyl-9-(1'-D-ribityl) isoalloxazine. The present position 5 was numbered 10.

RIBOFLAVIN

MW = 376.4

The bright orange-yellow color and brilliant greenish fluorescence of riboflavin first attracted the attention of chemists. Blyth isolated the vitamin from whey in 1879 and others later obtained the same fluorescent, yellow compound from eggs, muscle, and urine. All of these substances, referred to as **flavins** because of their yellow color, were eventually recognized as identical.

The structure of riboflavin was established in 1933 by R. Kuhn and associates, who had isolated 30 mg of the pure material from 30 kg of dried egg albumin (10,000 eggs). The intense fluores-

cence assisted in the final stages of purification. The vitamin was synthesized in 1935 by P. Karrer.

Riboflavin, a yellow solid, has a low solubility of ~100 mg/l at 25°C. Three crystalline forms are known. One of these ("readily soluble form") is 10 times more soluble than the others and can be used to prepare metastable solutions of higher concentration. One crystalline form is platelike and occurs naturally in the tapetum (Box 7-F) of the nocturnal lemur.

Nutritional requirement: Approximately 2 mg/day. Because of its wide distribution in food, deficiency is rarely seen in humans. The skin and eyes are affected first. Riboflavin is produced commercially in large quantities by fungi such as *Eremothecium asbyii*, which, apparently because of some metabolic anomaly, produce the vitamin in such copious amounts that it crystallizes in the culture medium.

Riboflavin is stable to heat but is extremely sensitive to light, a fact of some nutritional significance. Do not leave milk in bright sunshine (see Fig. 2-34).

thought that the "old yellow enzyme" was a form of dihydrolipoyl dehydrogenase (Section J,2) and does not accept electrons from NADPH, the concept of a respiratory chain was correct.

The old yellow enzyme was shown by Hugo Theorell in Stockholm to contain riboflavin 5′-phosphate. By 1938, FAD was recognized as the coenzyme of another yellow protein, **D-amino-acid oxidase** of kidney tissue. Like the pyridine nucleotides, the new flavin coenzymes were reduced by dithionite to a nearly colorless dihydro form (Figs. 8-14 and 8-15). Thus, the chemical basis for their function as hydrogen carriers was evident.

Three facts account for the need of cells for both the flavin and pyridine nucleotide coenzymes: (a) Flavins are usually stronger oxidizing agents than is NAD^+. This property fits them for a function in the electron transport chain of mitochondria where a sequence of more and more powerful oxidants is needed and makes them ideal oxidants in

a variety of other dehydrogenations. (b) Flavins can be reduced either by one- or two-electron processes. This enables them to participate in oxidation reactions involving free radicals and in reactions with metal ions. (c) Reduced flavins can be reoxidized directly and rapidly by O_2, a property shared with relatively few other organic substances. For example, NADH and NADPH are not spontaneously reoxidized by oxygen. Their "autoxidizability" allows flavins of some enzymes to pass electrons directly to O_2 and also provides a basis for the functioning of flavins in hydroxylation reactions.

1. Flavoproteins and Their Reduction Potentials

Flavin coenzymes are usually tightly bound to proteins and cycle between reduced and oxidized

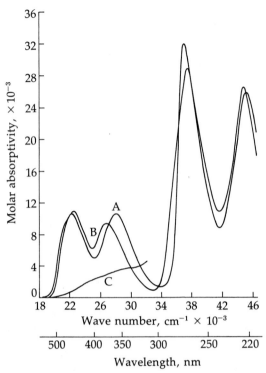

Fig. 8-15 Absorption spectrum of neutral riboflavin (A), the riboflavin anion (B), and reduced to the dihydro form (Fig. 8-14) by the action of light in the presence of EDTA (C). A solution of 1.1×10^{-4} M riboflavin containing 0.01 M EDTA was placed 11.5 cm from a 40 W incandescent lamp for 30 min.

and reduced flavin coenzymes to a protein have a distinct and strong effect on the reduction potential. If the oxidized form is bound weakly, but the reduced form is bound tightly, the bound flavin will have a greater tendency to stay in the reduced form than it did when free. The reduction potential $E^{0'}$ will be less negative than it is for the free flavin–dihydroflavin couple. On the other hand, if the oxidized form of the flavin is bound more tightly by the protein than is the reduced form, $E^{0'}$ will be more negative and the flavoenzyme will be a less powerful oxidizing agent than free riboflavin. In fact, the values of $E^{0'}$ (pH 7) for flavoproteins span a remarkably wide range from -0.49 to $+0.19$ V.

Note that each flavoprotein accepts electrons from a substrate which it is oxidizing and passes the electrons on to another substrate, an oxidant. This is one reason that the chemistry of flavoproteins is so varied. To describe a flavoprotein we must be able to say how the electrons get into the flavin from the oxidizable substrate, how they flow out of the flavin, and what is the final electron acceptor.

2. Typical Dehydrogenation Reactions Catalyzed by Flavoproteins

The functions of flavoproteins are so numerous (a survey in 1970 listed 45 different flavoproteins)[98] and so diversified that books have been written about them.[99] We can consider only a few of the enzymes here. Flavin-containing enzymes catalyze oxidation of hemiacetals to lactones (Eq. a, Table 8-4), of alcohols to aldehydes (Eq. b), and of amines to imines (Eq. c). They oxidize carbonyl compounds or carboxylic acids to α,β-unsaturated carbonyl compounds (Eq. d) and flavoproteins in electron transport chains oxidize NADH and NADPH (Eq. c). A group of flavoproteins oxidize dithiols to disulfides (Eq. f).

The first three of these reactions could equally well be catalyzed by pyridine nucleotide-requiring dehydrogenases. Recall that D-glucose-

states while attached to the same protein molecule. What determines the reduction potential of the flavin in such a **flavoprotein?** In a free unbound coenzyme the redox potential depends upon the structures of the oxidized and reduced forms of the couple. Both riboflavin and the pyridine nucleotides contain aromatic ring systems that are stabilized by resonance. Part, but not all, of this resonance is lost upon reduction. The value of $E^{0'}$ depends upon the relative amounts of resonance in the oxidized and reduced forms and upon any other factors that stabilize one form more than the other. The structures of the coenzymes reflect a design that results in values of $E^{0'}$ appropriate for biological functions.

The relative strengths of binding of oxidized

TABLE 8-4
Some Typical Flavoprotein-Mediated Dehydrogenation Reactions

D-Glucose   Lactone of D-gluconic acid   D-Gluconic acid   (a)

$$HOOC-CH_2OH \xrightarrow[\text{glycolate oxidase}]{-2[H]} HOOC-CH=O \quad (b)$$

Amino-acid oxidases   (c)

$$R-CH_2-CH_2-\overset{O}{\overset{\|}{C}}-SCoA \xrightarrow{-2[H]} R-CH=CH-\overset{O}{\overset{\|}{C}}-S-CoA \quad (d)$$

$$NADH \text{ (or NADPH)} + H^+ \xrightarrow{-2[H]} NAD^+ (NADP^+) \quad (e)$$

Dihydrolipoic acid amide   (f)

6-phosphate dehydrogenase uses $NADP^+$ as the oxidant (Eq. 8-42). The first product is the lactone which is hydrolyzed to 6-phosphogluconic acid. A similar reaction of *free* glucose (Eq. a, Table 8-4) is catalyzed by **glucose oxidase,** a flavoprotein of MW ~ 154,000 containing two molecules of FAD and produced by *Penicillium notatum* and other fungi. The plant enzyme **glycolate oxidase,** a dimer containing riboflavin 5'-phosphate, catalyzes the reaction of Eq. b, Table 8-4. Two amino-acid oxidases (Eq. c) are well known. One, specific for D-amino acids and obtained from kidney, was the original source from which Warburg isolated FAD. Its molecular weight is ~38,000.[100] Many snake venoms contain a very active L-amino-acid oxidase of MW = 140,000, which contains two molecules of FAD. **Amine oxidases** catalyze the related reaction in which the nitrogen is substituted and in which the carboxyl group need not be present. In all of these cases, the reduced flavin produced is reoxidized with molecular oxygen and hydrogen peroxide is the product. Nature has chosen to forego the use of an electron transport chain, giving up the possible gain of ATP in favor of simplicity and more direct reaction with oxygen. In some cases there may be a specific value to the organism in forming H_2O_2.

By contrast the reaction of Eq. d, Table 8-4, could not be accomplished by a pyridine nucleotide system because the reduction potential is inappropriate. The more powerfully oxidizing flavin system is needed. (However, the reverse reaction, hydrogenation of a $C{=}C$ bond, is often carried out biologically with a reduced pyridine nucleotide.) Equation d represents a type of reaction important in the energy metabolism of aerobic cells. For example, the first oxidative step in the β oxidation of fatty acids (Chapter 9, Section A,1) is the α,β dehydrogenation of fatty acyl-CoA derivatives. A related reaction occurring in the citric acid cycle is dehydrogenation of succinate to fumarate (Eq. 8-49). The dehydrogenation in-

$$(8\text{-}49)$$

volves trans removal of one of the pro-S hydrogens and one of the pro-R hydrogens.[101] Neither succinate dehydrogenase nor fatty acyl-CoA dehydrogenases react with O_2. Rather, the reduced flavins pass their electrons to the electron transport chain of mitochondria.

The NADH and NADPH dehydrogenases of mitochondria (Eq. e, Table 8-4) are also linked to oxygen through a series of iron catalysts discussed in Chapter 10.

A group of three enzymes **dihydrolipoyl dehy-**

drogenase (lipoamide dehydrogenase), **glutathione reductase,** and **thioredoxin reductase** make up a distinct subclass of flavoproteins. The reaction catalyzed by the first of these is illustrated in Eq. f, Table 8-4. The other two enzymes usually promote the reverse type of reaction, the reduction of a disulfide to two SH groups by NADPH (Eq. 8-50).

G—S—S—G + NADPH + H⁺ $\xrightarrow[\text{reductase}]{\text{glutathione}}$

Oxidized
glutathione

2 G—SH + NADP⁺   (8-50)

Glutathione
(see Box 7-G for structure)

Glutathione reductase splits the disulfide into two halves, but reduction of the small protein thioredoxin (MW = 12,000) simply opens a "loop" in the peptide chain, as does reduction of lipoic acid. In all three of these flavoproteins, there is a disulfide group present, apparently adjacent to the flavin and also participating in the chemical reactions.[102,102a] It is attractive to think that when dihydrolipoic acid approaches its dehydrogenase, it is oxidized to the disulfide by the internal disulfide group in the enzyme which is thereby opened to form two SH groups. Disulfides and thiols are well known to participate in such disulfide exchange processes nonenzymatically. On the surface of the enzyme such reactions may be extremely rapid. The flavin coenzyme may then oxidize the —SH groups of the enzyme back to the internal disulfide, the reduced flavin being reoxidized by external NAD⁺. There is evidence for transient formation of a half-reduced flavin which may be accompanied by a radical on a sulfur atom.

3. Covalently Linked and Other Modified Flavin Coenzymes

An extremely active flavin-containing enzyme of animal mitochondria is **succinate dehydrogenase** (Eq. 8-49). Not only is the enzyme tightly embedded in the membranes of the cristae of the mitochondria but also the flavin was found to be attached to the protein through a covalent linkage. Recently, the chemistry has become clear and a modified FAD containing **8α-(N-3-histidyl)riboflavin** has been isolated.[103-105]

Covalently bound modified FAD
of succinate dehydrogenase

The same prosthetic group has been identified in a 6-hydroxynicotine oxidase of *Athrobacter oxidans*[106] and is probably present in sarcosine (*N*-methylglycine) dehydrogenase.[107] In liver monoamine oxidase FAD is attached through the same 8α position to a sulfur atom of a cysteine residue[108,109] and in the "flavocytochrome" cytochrome c_{552} of *Chromatium* still another covalent linkage, a thiohemiacetal, is found.[110] Cytochrome b reductase also contains covalently linked flavin.[111] It will be interesting to see how many variations on this theme nature has devised.

While the significance is still unknown, an NADH dehydrogenase of *Peptostreptococcus elsdenii* contains large amounts of 6-hydroxy-FAD and an 8-hydroxy-FAD (in which the 8-methyl is replaced by OH) as well as FAD itself.[112] Another modified flavin **roseoflavin** is not a coenzyme but an antibiotic from *Streptomyces davawensis*. Its probable structure is deduced from X-ray analysis of a derivative.[113]

Probable structure
of roseoflavin

4. Mechanisms of Flavin Dehydrogenase Action

The argument was made at the beginning of Section G that all hydrogenation and dehydrogenation reactions can be rationalized in terms of a hypothetical hydride ion H⁻ acting as a nucleophile and either being donated by a reduced coenzyme or accepted by an oxidized coenzyme. A reasonable hydride-transfer mechanism for flavoprotein dehydrogenases can be written as shown in Eq. 8-51. The hydride ion is donated at

$$(8\text{-}51)$$

N-5 and a proton is accepted at N-1. Oxidation of alcohols, amines, ketones, and reduced pyridine nucleotides can all be visualized in this way.

While very few of the reactions catalyzed by flavoproteins have been successfully carried out in model systems the nonenzymatic oxidation of NADH by flavins occurs at moderate speed in water at room temperature. A variety of flavins and dihydropyridine derivatives have been studied, and the electronic effects observed for the reaction are compatible with the hydride ion transfer mechanism.[114]

According to the mechanism of Eq. 8-51, a hydride ion would be transferred directly from a carbon atom in a substrate to the flavin, as occurs in reduction of NAD⁺ or NADP⁺ by dehydrogenases. However, if a labeled hydrogen were transferred to N-5 it would immediately exchange with the medium, rapid exchange being characteristic of hydrogens attached to nitrogen. To avoid this problem, Brüstlein and Bruice used a **5-deazaflavin** to oxidize NADH nonenzymatically[115]

(Eq. 8-52). When the reaction was carried out in ^2H_2O no ^2H entered the product at C-5, indicating that a hydrogen atom (circled in Eq. 8-52) had

$$\text{(8-52)}$$

been transferred directly from the NADH to the C-5 position. A similar direct transfer to C-5 of 5-deazariboflavin 5'-phosphate by the flavoprotein N-methylglutamate synthetase has been demonstrated.[116] (This enzyme catalyzes a two-step reaction of type c of Table 8-4 by which glutamate reacts with methylamine to give N-methylglutamate and ammonia.) Direct stereospecific transfer of ^3H from [α-^3H]alanine to deaza-FAD has also been shown.[116a] While these experiments demonstrate direct hydrogen transfer they do not *prove* a hydride ion mechanism.

A second possible mechanism of flavin reduction is suggested by studies of addition reactions involving the isoalloxazine ring of flavins. Sulfite adds by forming an N—S bond at the 5 position. However, Hamilton[117] suggested that a more usual position for addition of nucleophiles is to carbon 4a, which together with N-5 forms a cyclic Schiff base. Hamilton argued that other electrophilic centers in the molecule, such as carbons 2, 4, and 10a, would be unreactive because of their involvement in amide or amidine type of resonance. An alcohol (or other substrate*) could add to a flavin

* Hamilton suggested that adducts with acyl-CoA derivatives might form by reaction of the carbonyl oxygen of the substrate with C-4a of the flavin. The student may wish to write detailed mechanisms for dehydrogenation reactions through such intermediates and also for the oxidation of a dithiol to a disulfide (Eq. f, Table 8-4).

as is shown in Eq. 8-53 (step a). The cleavage of the newly formed C—O bond could then occur by movement of electrons from the alcohol part of the adduct into the flavin, Eq. 8-53, step b. The prod-

$$\text{(8-53)}$$

ucts are the reduced flavin and an aldehyde. The same thing is accomplished as by a hydride ion transfer from the carbon atom of the alcohol, but in this case the hydrogen has been released from carbon as a proton. In fact, both the hydrogens in the original substrate (that on oxygen and that on carbon) have dissociated as protons, the electrons having moved as a pair during the cleavage of the adduct. Hamilton argued that hydride ion transfers are rare in biochemistry, one reason being that an isolated hydride ion has a large diameter while a proton is small and highly mobile. Hamilton suggested that for this reason dehydrogenation may most often take place by proton transfer mechanisms.

Experimental support for a proton transfer mechanism analogous to that of Eq. 8-53 has been obtained by using D-chloroalanine as a substrate for D-amino-acid oxidase.[118] Chloropyruvate is the expected product, but under anaerobic conditions pyruvate is formed (Eq. 8-54).

$$\underset{\underset{NH_2}{|}}{Cl-CH_2-\overset{\overset{H}{|}}{C}-COOH} \xrightarrow{-2H} \left[\underset{\underset{NH}{\|}}{Cl-CH_2-C-COOH}\right]$$

$$\Big\downarrow \text{anaerobic} \qquad\qquad \Big\downarrow H_2O$$

$$\underset{\underset{O}{\|}}{Cl-CH_2-C-COOH} + NH_3$$

$$\underset{\underset{O}{\|}}{CH_3-C-COOH} + NH_4^+ + Cl^- \qquad (8\text{-}54)$$

Kinetic data obtained with $\alpha\text{-}^2H$ and $\alpha\text{-}^3H$ substrates suggest a common intermediate for both reactions of Eq. 8-54. This intermediate could be an anion formed by loss of H^+ from an adduct

This H removed as H^+
in rate-limiting step

analogous to that of Eq. 8-53. The anion could eliminate chloride ion as indicated by the dashed arrows in the foregoing structure. This would lead to formation of pyruvate without reduction of the flavin. Alternatively, the electrons from the carbanion could flow into the flavin, reducing it as in Eq. 8-53. The fact that this occurs only in the presence of O_2 tells us that the tale is more complex. Oxygen must be bound *before* the flavin is reduced. But where and how it is bound remains to be determined.

An obvious question is whether the direct transfer of hydrogen mentioned in the preceding section is compatible with mechanisms of the type shown in Eq. 8-53.

A third possible mechanism of flavin dehydrogenation consists of consecutive transfer of a hydrogen atom and of an electron with intermediate radicals both on the flavin and on the substrate being oxidized. Such a mechanism takes full advantage of the tendency for flavins to form stable radicals as described in the next section. It has often been assumed that the intermediate radicals formed from alcohols, amines, etc., are so unstable that their formation is thermodynamically impossible. However, Bruice and Yano[119] and others[119a] have argued that radical mechanisms are plausible.

5. Half-Reduced Flavins

One of the most characteristic properties of flavins is their ability to accept a single electron to form a radical or **semiquinone**. If the oxidized form F of a flavin is mixed with the reduced form FH_2, a single hydrogen atom is transferred from FH_2 to F to form two FH radicals (Eq. 8-55). The equilib-

$$F + FH_2 \underset{}{\overset{K_f}{\rightleftharpoons}} 2\,FH^{\cdot} \qquad (8\text{-}55)$$

rium represented by this equation is independent of pH, but because all three forms of the flavin have different pK_a values (Fig. 8-16), the apparent equilibrium constants relating total concentrations of oxidized, reduced, and radical forms vary with pH.[120,121] The fraction of radicals present is greater at low pH and at high pH than at neutrality. For a 3-alkylated flavin[120] the formation constant K_f has been estimated as 2.3×10^{-2} and for riboflavin[121] as 1.5×10^{-2}. From these values and the pK_a values in Fig. 8-16, it is possible to estimate the amount of radical present at any pH.

Neutral flavin radicals have a blue color (the wavelength of the maximum, λ_{max} is ~ 580 nm) but either protonation at N-1 or dissociation of a proton from N-5 leads to red cation or anion radicals with λ_{max} at ~ 470 nm. (For complete spectra, see Mahler and Cordes.[122]) Both blue and red radicals are observed in enzymes, some enzymes favoring one and some the other. Hemmerich has suggested that enzymes forming red radicals have a strong hydrogen bonding group binding to the

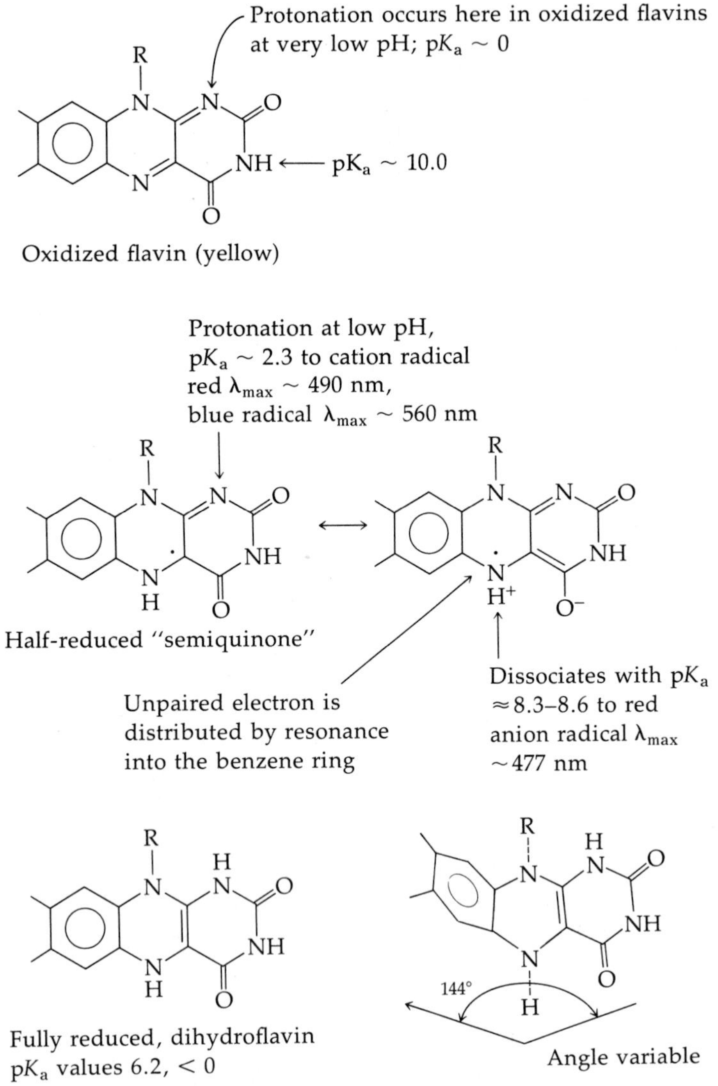

Protonation occurs here in oxidized flavins at very low pH; $pK_a \sim 0$

$pK_a \sim 10.0$

Oxidized flavin (yellow)

Protonation at low pH, $pK_a \sim 2.3$ to cation radical
red $\lambda_{max} \sim 490$ nm,
blue radical $\lambda_{max} \sim 560$ nm

Half-reduced "semiquinone"

Unpaired electron is distributed by resonance into the benzene ring

Dissociates with pK_a ≈ 8.3–8.6 to red anion radical λ_{max} ~ 477 nm

Fully reduced, dihydroflavin
pK_a values 6.2, < 0

$144°$

Angle variable

Fig. 8-16 Properties of oxidized, half-reduced, and fully reduced flavins. See Müller et al.[120]

proton in the 5 position. This increases the basicity of N-1 leading to its protonation and formation of the red cation radical.

If an enzyme binds a flavin radical much more tightly than the fully oxidized or reduced forms, the flavoprotein is reduced in two one-electron steps. In such proteins the values of $E^{\circ\prime}$ for the two steps may be widely separated. For example, a small electron-carrying **flavodoxin** (azotoflavin)

from *Azotobacter* is believed to function in fixation of N_2 (Chapter 14, Section A,2). Azotoflavin forms a blue semiquinone and has values of $E^{\circ\prime}$ (pH 7.7) of -0.270 and -0.464 V.[123] The latter is the lowest potential known for a flavoprotein.

Similar flavodoxins have been obtained from anaerobic bacteria and from blue-green algae (phytoflavin). Determination of the three-dimensional structures of two flavodoxins has revealed that the tightly bound riboflavin 5'-phosphate[124,125] is partially "buried" near the surface of the 138-residue protein. An aromatic side chain, from tryptophan or tyrosine, lies against the flavin on the outside of the molecule. A surprising discovery (with clostridial flavodoxin) is that the phosphate group of the coenzyme is *not* bound to a basic arginine or lysine side chain but to a cluster of neutral polar groups—four hydroxyl groups from serine and threonine residues and four backbone NH groups.[124] Flavodoxins can be crystallized in all three forms: oxidized, semiquinone, and fully reduced. In the crystals the flavin semiquinone, like the oxidized flavin, is nearly planar. In all the flavodoxins the two reduction steps are well separated. For example, that of *Peptococcus elsdenii* shows values of $E^{0\prime}$ (pH 7) of -0.115 and -0.373 V.

6. Metal Complexes of Flavins and Metalloflavoproteins

The presence of metal ions in some flavoproteins[125a] suggests a direct association of metal ions and flavins. Although oxidized flavins do not readily bind most metal ions, they do form red complexes with Ag^+ and Cu^+ with a loss of a proton from N-3.[126,127] Flavin semiquinones (radicals)

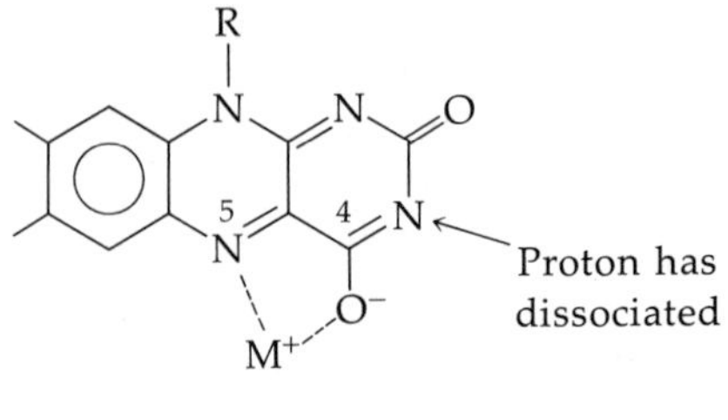

Proton has dissociated

form stronger red complexes with many metals.[128] Again, the metal ion is thought to interact with N-5 and with the oxygen atom at C-4. If the same site is occupied by a metal ion with more than one oxidation state, electron transfer between the flavin and a substrate could take place through the metal atom.

An unusual flavoprotein **xanthine oxidase** catalyzes both Eqs. 8-56 and 8-57. The closely related

$$-C\overset{O}{\underset{H}{\big<}} + H_2O \xrightarrow{-2[H]} -C\overset{O}{\underset{OH}{\big<}} \qquad (8\text{-}56)$$

Hypoxanthine   Xanthine

aldehyde oxidase carries out only the reaction of Eq. 8-56. Xanthine oxidase can also oxidize xanthine further by a repetition of the same type of oxidation process at positions 8 and 9 (see Eq. 8-57) to form **uric acid.** Perhaps the most interesting thing about xanthine oxidase is that for each molecule of bound FAD it contains a tightly bound atom of molybdenum (see Box 14-A) as well as four atoms of iron. (The molecule is a dimer of MW ~ 275,000 containing 2 FAD, 2 Mo, and 8 Fe.) In fact, the first hint of the role of molybdenum in metabolism came from the discovery that a deficiency of molybdenum in the diet of animals led to a decrease in the activity of xanthine oxidase in the liver (the enzyme is also found in milk).

Equations 8-56 and 8-57 have been written as if xanthine oxidase were a dehydrogenase. This would imply that water first adds to the C=O or C=N bond at the site of oxidation after which the adduct is dehydrogenated. Such a mechanism does not explain the need for Mo and Fe.

The observation of characteristic epr signals from the molybdenum during the action of the enzyme suggests that Mo does participate.[129,129a] The first event in the enzymatic reaction may be the

reduction of Mo(VI) to Mo(V). Hamilton suggested the mechanism[117] shown in Eq. 8-58.

The adduct formed initially by addition of a metal-coordinated hydroxyl group would be oxidized by electron flow through the oxygen and into the two molybdenum atoms. Each of the latter would receive one electron to form two atoms of Mo(V). The electrons could then be passed to the flavins, to the iron atoms of the protein, and to O_2 through a miniature electron transport chain. Xanthine oxidase is one of a group of flavin enzymes that contain tightly bound iron that is not associated with a heme ring. It is a member of the **iron–sulfur proteins** discussed in Chapter 10, Section C. Xanthine oxidase and aldehyde oxidase also contain a persulfide group (—S—S⁻) which is essential for activity.[130]

Other flavin-containing iron-sulfur proteins are NADH dehydrogenase (Eq. e, Table 8-4) and succinate dehydrogenase (Eq. 8-49). Both of these mitochondrial enzymes pass electrons on, presumably through the bound iron atoms, to the cytochrome system of the electron transport chain (Chapter 10).[130a]

Some metalloflavoproteins contain **heme** groups (Chapter 10, Section B). An example is **L-lactate dehydrogenase** of yeast, an enzyme that is also called **cytochrome b_2.** The protein is a tetramer of MW ~ 235,000, each unit of which contains two different peptide chains, one molecule of riboflavin phosphate, and one of heme.[131] In all probability the flavin and heme are attached to different peptide chains and the enzyme acts as a typical dehydrogenase, the electrons being passed out from the reduced flavin through the iron atom of the bound heme. The external acceptor is probably cytochrome c.

7. Reactions of Reduced Flavins with Oxygen

Free dihydroriboflavin reacts nonenzymatically in seconds and reduced flavin dehydrogenases even faster with molecular oxygen to form hydrogen peroxide. An intermediate, possibly an adduct at position 4a (Eq. 8-59) can be detected

$$\text{(8-59)}$$

spectrophotometrically in the nonenzymatic case.[132,133] (Other positions for the addition may also be possible.) Formation of a cyclic peroxide by addition of O_2 at both carbons 4a and 10a, has also been suggested[134]:

Cyclic peroxide

The products observed for many different flavoprotein reactions could be explained on the basis of peroxide intermediates. For example, breakdown of the adduct of Eq. 8-59 by protonation on the inner oxygen would yield H_2O_2 and oxidized flavin. On the other hand, the C—O bond could cleave homolytically to give two radicals: a flavin

radical and a perhydroxy radical $\cdot O_2H$. The latter could in turn dissociate to form the **superoxide anion radical** (Eq. 8-60):

$$\cdot O_2H \rightarrow H^+ + \cdot O_2^- \text{ (superoxide anion radical)} \quad \text{(8-60)}$$

Various metal-containing **superoxide dismutases** catalyze the reaction shown in Eq. 8-61. When a

$$2 \cdot O_2^- + 2 H^+ \rightleftharpoons O_2 + H_2O_2 \quad \text{(8-61)}$$

reduced flavin is allowed to be reoxidized nonenzymatically in the presence of superoxide dismutase, the rate is much reduced over that in its absence. This result suggests that the nonenzymatic reoxidation of reduced flavins occurs by a radical mechanism in which superoxide anion is formed. Further conversion of the flavin radical to the fully oxidized form would require a second step as shown in Eq. 8-62. However, superoxide dismu-

$$FH\cdot + \cdot O_2^- + H^+ \longrightarrow F + H_2O_2 \quad \text{(8-62)}$$

tases usually have little effect on the reoxidation of flavin in dehydrogenases. An exception is xanthine oxidase, which is a very strong producer of superoxide anions. Fried and associates suggested that an important function of xanthine oxidase is to provide hydrogen peroxide and superoxide radicals for use in coupled biological oxidation.[134a]

Another class of flavoproteins, discussed in Chapter 10, Section G,2 are the **monooxygenases** or **hydroxylases**. In these enzymes the reduced flavin helps to introduce one atom of oxygen from O_2 into the substrate while converting the other atom to H_2O. Formally the reaction could be depicted as a cleavage between the two oxygens in the adduct of Eq. 8-59 with generation of OH^+ from the terminal OH group.

8. Absorption of Light

Flavins are among the natural light receptors (Fig. 8-15)[135] and display an interesting and much studied photochemistry.[136,136a] Flavins may function in various photoresponses of plants (Chapter 13, Section G,2), and serve as light emitters in some bioluminescent species (Chapter 13, Section H).

J. LIPOIC ACID AND THE OXIDATIVE DECARBOXYLATION OF α-KETO ACIDS

The isolation of lipoic acid in 1951 followed the discovery only 10 years earlier that the ciliate protozoan *Tetrahymena geleii* required an unknown factor for growth. Studies with bacteria also pointed toward a new coenzyme. Acetic acid was observed to promote rapid growth of *Lactobacillus casei* but could be replaced by an unknown "acetate replacing factor." Another lactic acid bacterium *Streptococcus faecalis* was unable to oxidize pyruvate without addition of "pyruvate oxidation factor." By 1949, all three unknown substances were recognized as identical. After a heroic effort in which they worked up the equivalent of 10 tons of water-soluble residue of liver, Lester Reed at the University of Texas and his collaborators at Eli Lilly and Co. isolated 30 mg of a fat-soluble acidic material which was named **lipoic acid** (the compound is also known as 6-thioctic acid).

While *Tetrahymena* must have lipoic acid in its diet, we humans apparently can make our own, and it is not considered a vitamin. Lipoic acid is present in tissues in extraordinarily small amounts. Its only certain function is to participate in the oxidative decarboxylation of α-keto acids.[137,137a] The structure is simple, and the functional group is clearly the cyclic disulfide which swings on the end of a long arm. Lipoic acid is not found free but, like biotin, is combined with enzymes in covalent linkage through an amide group involving a lysine side chain:

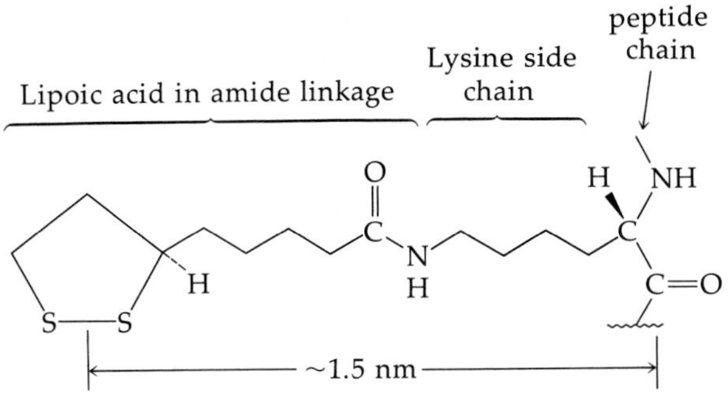

Lipoic acid in amide linkage   Lysine side chain   peptide chain

~1.5 nm

A special enzyme catalyzes the attachment of lipoic acid to the enzymes by sequence S1C (Table 7-2) which involves cleavage of ATP to AMP and pyrophosphate.

1. Chemical Reactions of Lipoic Acid

The most striking chemical property of lipoic acid is the presence of ring strain of $\sim 17\text{--}25$ kJ mol^{-1} in the cyclic disulfide. Because of this, thiol groups and cyanide ions react readily with oxidized lipoic acid to give mixed disulfides (Eq. 8-63) and isothiocyanates (Eq. 8-64), respectively.

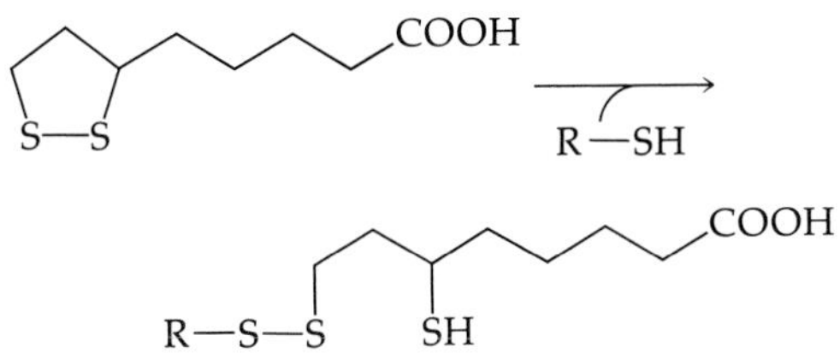

(8-63)

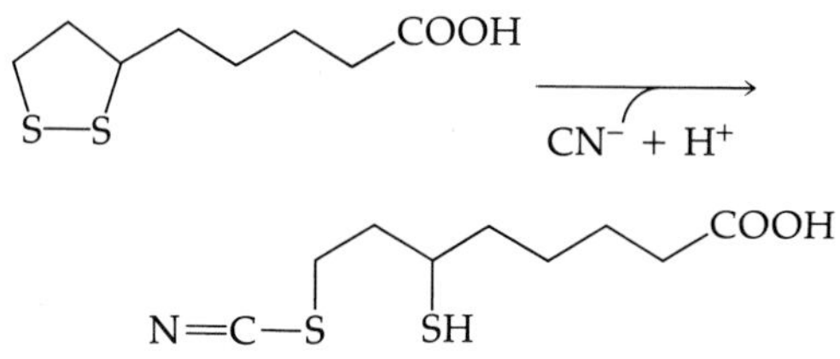

(8-64)

Another result of the ring strain is that the reduction potential $E^{0'}$ (pH 7, 25°C), is -0.30 V, almost the same as that of reduced NAD (-0.32 V). Thus, reoxidation of reduced lipoic acid amide by NAD^+ is thermodynamically feasible.[137] Yet another property attributed to the ring strain in lipoic acid is the presence of an absorption maximum at 333 nm.

2. Enzymatic Function

The oxidative decarboxylation of α-keto acids involves cleavage of the keto acid with production of CO_2 and combination of the remaining acyl group with CoA (Eq. 8-65). NAD^+ serves as an ox-

$$R-\overset{\overset{\displaystyle O}{\|}}{C}-COO^- \quad \xrightarrow[\text{thiamine diphosphate}]{\underset{\text{CoASH}}{\overset{CO_2}{\longrightarrow}}} \xrightarrow[\quad NAD^+ \quad NADH \quad]{\overset{\text{lipoic acid, FAD}}{\downarrow}}$$

$$R-\overset{\overset{\displaystyle O}{\|}}{C}-S-CoA \qquad (8\text{-}65)$$

idant. The reaction is catalyzed by a complex of enzymes of MW ~ 1–9×10^6, depending on the species and exact substrate.[137–139] Separate ketoacid dehydrogenase systems are known for pyruvic acid, α-ketoglutaric acid, and for α-keto acids with branched side chains derived metabolically from leucine, isoleucine, and valine. The **pyruvate** and **α-ketoglutarate dehydrogenases** of *E. coli* have been studied most. In both cases, the complex can be separated into three components. The first is a **decarboxylase** (also referred to as a dehydrogenase) for which the **thiamine diphosphate** is the dissociable cofactor. The second is the flavoprotein **dihydrolipoyl dehydrogenase,** and the third is a lipoic acid containing "core" enzyme.

Electron microscopy of the core enzyme reveals a striking octahedral symmetry which has been confirmed by X-ray diffraction. The core enzyme **dihydrolipoyl transacetylase** from pyruvate dehydrogenase has a molecular weight of $\sim 1.7 \times 10^6$ and contains 24 subunits of MW = 70,000 apparently all identical. Each subunit carries a bound lipoyl group.[138,139] The structure is shown schematically in Fig. 8-17. To the core are bound about 12 of the dimeric decarboxylase-dehydrogenase units of MW = 192,000 and 6 molecules of the dimeric flavoprotein of MW = 112,000. The 12 decarboxylase–dehydrogenase dimers appear to be distributed symmetrically on the 12 edges of the transacetylase cube and the 6 flavoprotein dimers on the 6 faces of the cube. It is likely that the active centers of all

of the subunits come close together in the regions where the subunits touch. In such a region, the sequence of catalytic reactions indicated in Fig. 8-18 may take place.

The unique function of lipoic acid is in the oxidation of the thiamine-bound active aldehyde in such a way that when the complex with thiamine breaks up, the acyl group formed by the oxidative decarboxylation of the keto acid is attached to dihydrolipoic acid. Since the lipoic acid is known to be attached to the subunits of the core enzyme, it appears that using its 1.5 nm long arm, lipoic acid reaches over to the thiamine diphosphate site on one of the decarboxylase subunits. Bearing the acyl group, it now swings back to a site on the core enzyme where CoA is bound. The acyl group is transferred to CoA producing dihydrolipoic acid which then swings to the third subunit where there is an FAD and disulfide waiting to reoxidize the lipoic acid. The reduced flavin-disulfide enzyme is then oxidized by NAD^+ (Fig. 8-18).

An alternative to the mechanism presented in Fig. 8-18 has been proposed. Ferricyanide, which can replace NAD^+ as an oxidant for pyruvate dehydrogenase, is also able nonenzymatically to oxidize thiamine-bound active acetaldehyde to 2'-acetylthiamine. The acetyl group in this com-

$$\underset{H_3C}{\overset{\overset{\displaystyle O}{\|}}{\underset{}{}}}\overset{}{C}\text{—} \begin{array}{c} S \\[-2pt] \end{array} \text{—CH}_2\text{CH}_2\text{OH}$$

2'-Acetylthiamine

pound has a high group transfer potential. Thus, the possibility existed that lipoic acid first oxidizes the active aldehyde to a thiamine-bound acyl derivative, then in the second step accepts the acyl group by a nucleophilic displacement reaction. However, this mechanism fails to explain the unique role of lipoic acid in oxidative decarboxylation.

The activity of the pyruvate dehydrogenase complex of many mammalian tissues is controlled

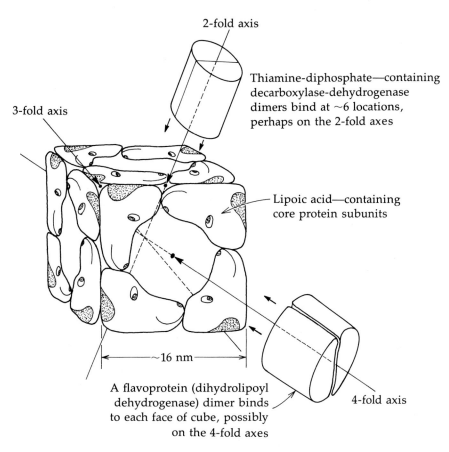

2-fold axis

Thiamine-diphosphate—containing
decarboxylase-dehydrogenase
dimers bind at ~6 locations,
perhaps on the 2-fold axes

3-fold axis

Lipoic acid—containing
core protein subunits

~16 nm

A flavoprotein (dihydrolipoyl
dehydrogenase) dimer binds
to each face of cube, possibly
on the 4-fold axes

4-fold axis

Fig. 8-17 Schematic arrangement of 24 subunits in transsuccinylase "core" of α-ketoglutarate dehydrogenase of *E. coli*. To obtain the observed cubic symmetry the core subunits must be grouped in trimers at the corners of the cube giving a structure with three 4-fold axes, four 3-fold axes, and six 2-fold axes of symmetry (4:3:2 symmetry). However (see text), symmetry is not perfect; only half of the subunits carry bound lipoic acid and only 6 decarboxylase dimers are present. In the case of pyruvate dehydrogenase from *E. coli*, each transacetylase unit carries lipoic acid and there are 12 decarboxylase dimers.

in part by a phosphorylation–dephosphorylation mechanism.[138,139a] Phosphorylation of the decarboxylase (dehydrogenase) subunit by an ATP-dependent kinase produces an inactive phosphoenzyme. A phosphatase reactivates the dehydrogenase (see Fig. 6-15).

In addition to the lipoic acid-dependent pyruvate dehydrogenase, cells of *E. coli* contain a **pyruvate oxidase,** a soluble flavoprotein that acts together with a membrane-bound electron transport system to convert pyruvate to acetate and CO_2. The mechanism is obscure.[138b]

3. Pyruvate-Ferredoxin Oxidoreductase[140]

An enzyme similar to lipoic acid-dependent ketoacid dehydrogenases is found in clostridia and other strict anaerobes. This **pyruvate-ferredoxin oxidoreductase** catalyzes *reversible* decarboxylation of pyruvate (Eq. 8-66). The oxidant **ferredoxin** is a low redox potential iron-sulfur protein discussed in Chapter 10, Section C. Ferredoxins from clostridia are two-electron oxidants. Comparison with Eq. 8-65 shows that oxidized ferredoxin substi-

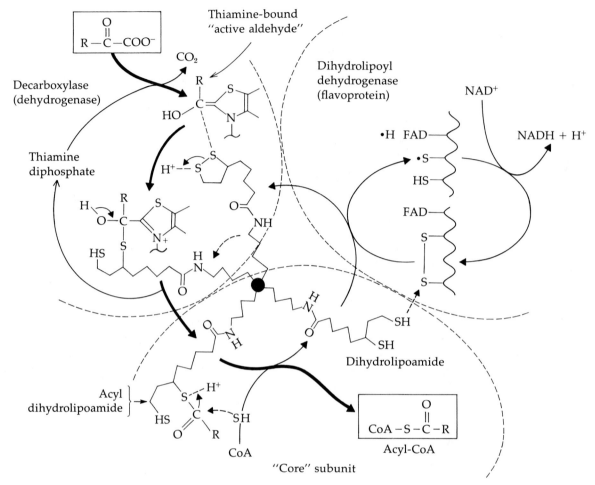

Fig. 8-18 Sequence of reactions catalyzed by α-ketoacid dehydrogenases. The substrate and product are shown in boxes, and the path of the oxidized keto acid is traced by the heaviest arrows. The lipoic acid "head" is shown rotating about the point of attachment to a "core" subunit.

$$\text{Pyruvate}^- + \text{CoA} \xrightarrow[\text{ferredoxin (ox)} \quad \text{ferredoxin (red)}]{} \text{acetyl-CoA} + \text{CO}_2 \tag{8-66}$$

$$\Delta G' \ (\text{pH } 7) = -16.9 \ \text{kJ mol}^{-1}$$

tutes for NAD^+. However, the reaction does not require lipoic acid. It seems likely that a thiamine-bound active acetaldehyde is oxidized by an iron-sulfide center in the oxidoreductase to an acetyl-thiamine intermediate (Section J,2) which then reacts with CoA.

A similar enzyme that apparently is involved in synthesis of α-ketoglutarate from succinyl-CoA and CO_2 (α-**ketoglutarate synthetase**) has been purified from photosynthetic bacteria.[140a]

The cleavage of pyruvate to CO_2 and acetyl-CoA is thought to serve in some instances as a means of generating a very low potential reductant for other biochemical processes. An example is biological fixation of N_2 (Chapter 14, Section A). It is possible that either reduced ferredoxin or, in *Azotobacter*, the very low potential flavodoxin (Section I,5) may be generated by cleavage of pyruvate and be used in the N_2 fixation process.[141] On the other hand, Eq. 8-66 may more often operate in reverse during biosynthesis (Chapter 11, Section C,1).

4. Oxidative Decarboxylation by Hydrogen Peroxide

Nonenzymatic oxidative decarboxylation of α-keto acids by hydrogen peroxide is a well-known reaction. The first step is presumably the formation of an adduct, an organic peroxide, which breaks up as indicated in Eq. 8-67. What may be an enzyme-catalyzed version of this reaction is promoted by **lactate oxygenase,** a hexameric flavoprotein of MW ~350,000 obtained from *Mycobacterium.* Under anaerobic conditions, the enzyme produces pyruvate by a simple dehydrogenation.[142] However, in the presence of oxygen it forms acetic acid and one of the oxygen atoms of the carboxyl group comes from O_2. Since hydrogen peroxide is the usual product formed from oxygen by flavoproteins, it is possible that with lactate oxygenase the hydrogen peroxide formed immediately oxidizes the pyruvate* according to Eq. 8-67.

$$R-C\overset{O}{\underset{COO^-}{\Big\langle}} + H_2O_2 \longrightarrow \left[R-\overset{OH}{\underset{}{\overset{|}{C}}}-C\overset{O}{\underset{O}{\Big\langle}} \right] \longrightarrow$$

$$R-C\overset{O}{\underset{O^-}{\Big\langle}} + CO_2 + H_2O \qquad (8\text{-}67)$$

Nevertheless, the mechanism of action of this oxygenase is not entirely clear. The enzyme does produce pyruvate in the absence of O_2, but neither trapping reagents for pyruvate nor for H_2O_2 inhibit the reaction.[143] Possibly H_2O_2 and pyruvate are generated in close proximity one to the other and react so rapidly that neither can be trapped. On

* The α-keto acids formed by the action of amino acid oxidases are also oxidized by the accumulating hydrogen peroxide. Addition of catalase (Chapter 10, Section B,6) destroys the H_2O_2 and allows the keto acid to accumulate.

the other hand, superoxide anion radical rather than H_2O_2 could be formed (Eq. 8-60). Other mechanisms have also been proposed.[144,145] Some other flavoprotein hydroxylases may function in similar ways (Chapter 10, Section G,2).

5. The Pyruvate-Formate-Lyase Reaction

An anaerobic α cleavage of pyruvate to formate and acetyl-CoA (Eq. 8-68) is essential to the en-

$$CH_3-C\overset{O}{\underset{COO^-}{\Big\langle}} + CoA-SH \longrightarrow$$

$$CH_3-\overset{O}{\overset{||}{C}}-S-CoA + H-C\overset{O}{\underset{O^-}{\Big\langle}} \qquad (8\text{-}68)$$

$$\Delta G' \text{ (pH 7)} = -12.9 \text{ kJ mol}^{-1}$$

ergy economy of many cells. No external oxidant is needed, and the reaction does not appear to require lipoic acid. The reaction is sometimes referred to as the "phosphoroclastic reaction" from the fact that the product acetyl-CoA usually reacts further with phosphate to form acetyl phosphate (Fig. 8-19). The latter can then transfer its phosphoryl group to ADP to form ATP.

The mechanism of the cleavage of pyruvate in Eq. 8-68 is still obscure. Thiamine diphosphate is not involved,[146,147] and free CO_2 is not an intermediate.[148] It is possible that an intermediate electron carrier first reduces the carboxyl group of pyruvate via a thioester intermediate. Later, after C—C bond cleavage, it could oxidize an intermediate active aldehyde, similar to that formed with thiamine, returning the carrier to its reduced state. The unstable enzyme system required consists of several proteins, some of which seem to be "auxiliary" in that they are needed to "activate" the lyase proper. Both the *E. coli* and clostridial enzymes are activated by reducing agents including reduced ferredoxin or reduced flavodoxin and by *S*-adenosylmethionine[148] (structure in Eq. 7-1),

Reaction Sequence S7B. Oxidative decarboxylation of an α-keto acid with thiamine diphosphate.
   (1) with lipoic acid and NAD^+ as oxidants
   (2) with ferredoxin as oxidant

$$(1) \; \Delta G' \; (pH \; 7, \; for \; pyruvate) = -35.5 \; kJ \; mol^{-1} \; overall$$
$$(2) \; \Delta G' \; (pH \; 7, \; for \; pyruvate) = -13.9 \; kJ \; mol^{-1} \; overall$$

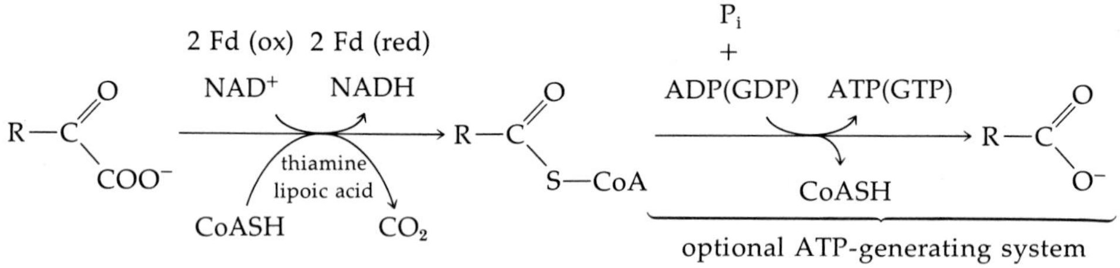

Reaction Sequence S7C. The pyruvate-formate-lyase reaction

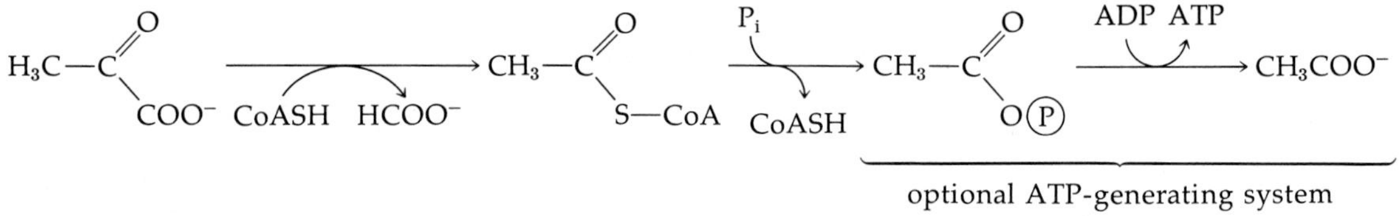

$$\Delta G' \; (pH \; 7) = -13.5 \; kJ \; mol^{-1} \; overall$$

Fig. 8-19 Two systems for oxidative decarboxylation of α-keto acids and for "substrate level" phosphorylation. The value of $\Delta G' = +34.5 \; kJ \; mol^{-1}$ (Table 3-5) was used for the synthesis of ATP^{4-} from ADP^{3-} and HPO^{2-} in computing the values of $\Delta G'$ given.

which may be an allosteric effector. Inactive and active forms may be interconvertible by alkylation and reduction.[149]

Bear in mind that while *E. coli* grown anaerobically produces the pyruvate-formate-lyase, the same bacterium grown in the presence of air forms the lipoic acid-containing pyruvate dehydrogenase complex (Fig. 8-17).

In many bacteria, including *E. coli*, the formic acid produced by Eq. 8-67 can be interconverted with CO_2 and hydrogen has through the action of the two-enzyme **formic-hydrogen-lyase** system (Chapter 9, Section F,2).

6. Cleavage of α-Keto Acids and Substrate Level Phosphorylation

The α-ketoacid dehydrogenases yield CoA derivatives which may enter biosynthetic reactions: Alternatively, the acyl-CoA compounds may be cleaved with generation of ATP. The pyruvate-formate-lyase system also operates as part of an ATP-generating system for anaerobic organisms. [e.g., in the "mixed acid fermentation" of enterobacteria such as *E. coli* (Chapter 9, Section F,2)]. These two reactions constitute an important pair of processes providing for "substrate-level" phosphorylation. For this reason they are set off in Fig. 8-19. They should be compared with two previously considered examples of substrate level phosphorylation (Figs. 8-4 and 8-3).

K. TETRAHYDROFOLIC ACID AND OTHER PTERIN COENZYMES[150]

Reduced forms of the vitamin **folic acid** are **carriers for one-carbon units** at all oxidation levels, except that of CO_2 (for which biotin is the carrier). Moreover, folic acid is just one of the derivatives of the **pteridine** ring system that enjoy a widespread distribution throughout nature. Pteridines provide coloring to insect wings and eyes and to the skins of amphibians and fish. Pteridines appear to act as protective filters in insect eyes; the isolation of several photosensitive pteridines suggest that some may function as light receptors. In addition to the reduced folic acid coenzymes, the mammalian liver contains at least one other pteridine with a function in hydroxylation of aromatic amino acids.

Because of the prevalence of derivatives of 2-amino-4-hydroxypteridine, this compound has been given the trivial name **pterin.** If the structure of guanine is learned it will be easy to remember that of pterin, and it will also be easy to remember that pterins are derived biosynthetically from

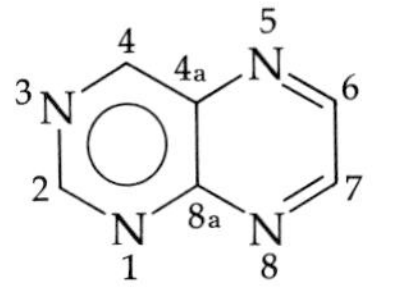

Pteridine

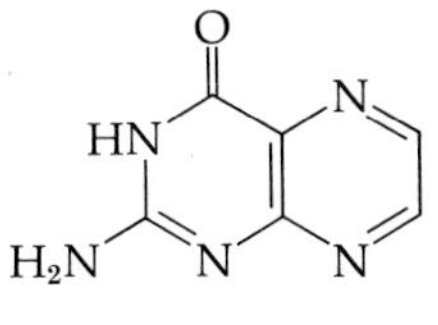

Pterin: 2-amino-4-hydroxypteridine 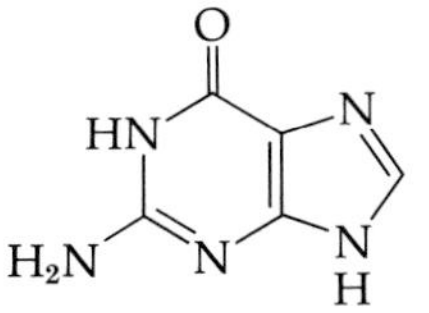   Guanine

guanine. Also note the similarity of the two-ring system of pterin to that of riboflavin (Box 8-I).

The first biochemist to become interested in pteridines was apparently F. G. Hopkins, who in 1891 began the investigation of the yellow and white pigments of the common sulfur and cabbage butterflies. Almost 50 years and a million butterflies later, the structures of the two pigments **xanthopterin** and **leucopterin** were established.

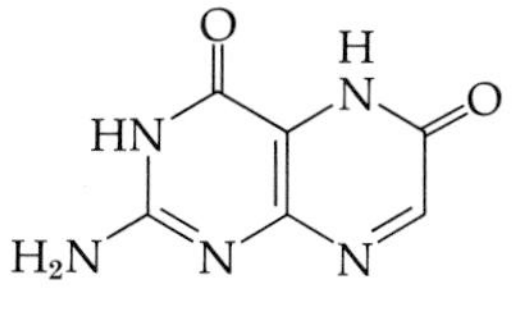

Xanthopterin (yellow) 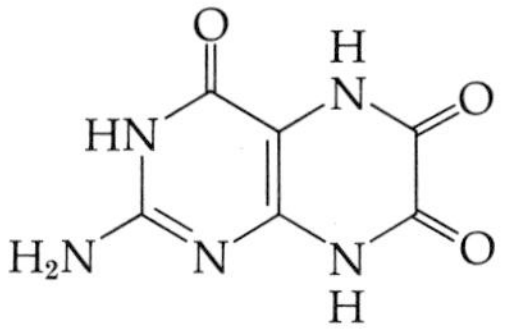   Leucopterin (white)

While the function of pterins in insect wings may appear to be solely one of providing color, these pigments are produced in such quantities as to suggest that their synthesis may be a means of deposition of nitrogenous wastes in dry form.

Among the pterins isolated from the eyes of *Drosophila*[151] is **sepiapterin,** in which the pyrazine

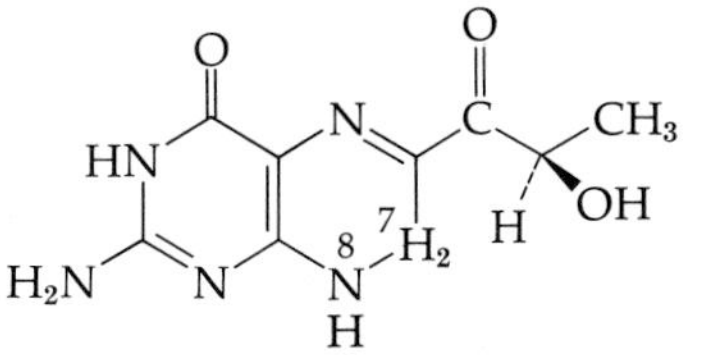

Sepiapterin

ring has been reduced in the 7,8 position. Reduction of the carbonyl group of sepiapterin with $NaBH_4$ followed by air oxidation produces **biopterin,** a compound of wide distribution which was

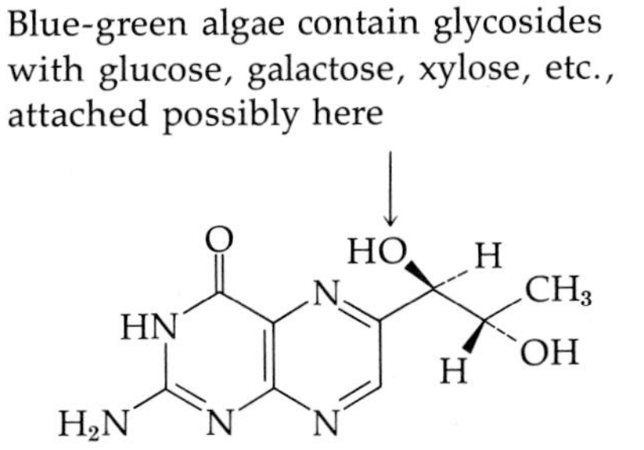

Blue-green algae contain glycosides with glucose, galactose, xylose, etc., attached possibly here

L-*erythro*-Biopterin

BOX 8-J

**FOLIC ACID
(PTEROYLGLUTAMIC ACID)**

para-Aminobenzoic   L-Glutamic
acid   acid

In 1931, L. Wills, working in India, observed that patients with **tropical macrocytic anemia**, an anemia in which the erythrocytes are enlarged but reduced in numbers, were cured by administration of yeast or liver extracts. The disease could be mimicked in monkeys fed the local diet, and a similar anemia could be induced in chicks. By 1938 it had also been shown that a factor present in yeast, alfalfa, and other materials was required for the growth of chicks. Isolation of the new vitamin came rapidly after it was recognized that it was also an essential nutrient for *Lactobacillus casei* and *Streptococcus faecalis* R.[a] Spinach was a rich source of the new compound, and it was named folic acid (from the same root as the word foliage).

Metabolic functions for folic acid were suggested by the observations that the requirement for *Streptococcus faecalis* could be replaced by thymine plus serine plus a purine base. Folic acid is required for the biosynthesis of all of these substances. A function in the interconversion of serine and glycine was suggested by the observation that certain mutants of *E. coli* required either serine or glycine for growth. Isotopic labeling experiments established that in the rat as well as in the yeast *Torulopsis* serine and glycine could be interconverted. It was also shown that the amount of interconversion decreased in the folate-deficient rat.

[a] The microbiological activities were actually those of di- and triglutamyl derivatives, one of the facts that has led to the description of the history of folic acid as "the most complicated chapter in the story of the vitamin B complex" [A. F. Wagner and K. Folkers, "Vitamins and Coenzymes," p. 114. Wiley (Interscience), New York, 1964].

first isolated from human urine.[152,153] Biopterin is present in liver (as well as other tissues) and is thought to function in a reduced form as a **hydroxylation coenzyme**. A variety of other functions in oxidative reactions, in regulation of electron transport and in photosynthesis have been proposed.[154] **Neopterin,** found in honey bee larvae resembles biopterin but has a D-*erythro* configuration in the side chain. The content of biopterin and related pterins in natural materials can be assayed on the basis of cofactor activity for a *Pseudomonas* phenylalanine hydroxylase. A less specific but widely used assay for pterins is measurement of growth response of *Crithidia fasciculata*, a trypanosomid parasite of mosquitos. Folate derivatives are not active in this test, and the concentrations of nonfolate pterins found in blood by the *Crithidia* assay (of the order of 0.1 mg/l) exceed those of the folate derivatives.

1. Folate Coenzymes

The coenzymes responsible for carrying single carbon units are derivatives of **5,6,7,8-tetrahydrofolic acid** abbreviated H_4PteGlu or THF. The folate coenzymes are present in very low concentrations and are readily oxidized by air, making studies difficult. In addition to the single L-glutamate unit present in tetrahydrofolic acid itself, the coenzymes occur as **conjugates** in which additional molecules of L-glutamic acid have been combined in amide linkage, always through the atypical linkage of the γ (side chain) carboxyl rather than the α-carboxyl. Coenzyme forms may contain *from one to seven glutamate residues* and with few exceptions, the distribution of the various forms and the significance of their existence is unknown. The bacterium *Clostridium acidi-urici* contains exclusively the triglutamate

Tetrahydrofolic acid (THF) or
tetrahydropteroylglutamic acid (H₄PteGlu)

Tetrahydrofolyltriglutamic acid
(THF-triglutamic acid or H₄PteGlu₃)

derivatives[156] and contains at least one enzyme with an absolute specificity for THF-triglutamic acid. On the other hand, the serum of many species contains only derivatives of folic acid itself. Pteroylpentaglutamate is the major folate derivative present in rat liver.[157]

Reduction creates a new chiral center at C-6. The enzymatically active form of THF exhibits negative optical rotation at 589 nm and is designated (−)THF. However, because some forms have positive rotation, it has been suggested that the designation (l) be used.[156]

2. Dihydrofolate Reductase

Folic acid and its polyglutamyl derivatives can be reduced to the THF coenzymes in two stages: the first step is a slow reduction with NADPH to 7,8-dihydrofolate. The reaction is apparently catalyzed by the same enzyme that rapidly reduces the dihydrofolates to tetrahydrofolates. Again, NADPH is the reducing agent and the enzyme has been given the name **dihydrofolate reductase.** Aside from its role in providing reduced folate coenzymes for cells, this enzyme has attracted enormous attention because it appears to be the site of action of the two important anticancer drugs **aminopterin** and amethopterin (**methotrexate**). These compounds inhibit dihydrofolate reductase in concentrations as low as 10^{-8} to

10^{-9} M. Because folate coenzymes are required in the biosynthesis of both purines and thymine, rapidly growing cancer cells have a high requirement for activity of this enzyme. An enormous number of synthetic compounds have been prepared in the hope of finding still more effective inhibitors. Unfortunately, all cells require dihydrofolate reductase; hence, the compounds are toxic and cannot be used for prolonged therapy. Even so, treatment with these drugs has returned many leukemic patients to months of normal life.*

Aminopterin (4-amino-4-deoxyfolic acid)

Complete cures of the relatively rare choriocarcinoma have been achieved and new methods of chemotherapy using the antifolates in combination with other drugs offer hope for the future.

* Some patients have lived for 5 years or more, whereas before 1960 persons with acute leukemia lived no more than 3–6 months.

3. Single Carbon Compounds and Groups in Metabolism

Some single carbon compounds and groups important in metabolism are shown in Table 8-5. Groups at three different oxidation levels, corresponding to **formic acid, formaldehyde,** and the **methyl group,** are carried by tetrahydrofolic acid coenzymes. While the most completely reduced compound, methane, cannot exist in a combined form, its biosynthesis depends on THF, as does that of carbon monoxide. Figure 8-20 summarizes the known metabolic interrelationships of the compounds and groups of Table 8-5.

For many organisms **serine** is the major precursor of C-1 units. The β-carbon of serine is removed as formaldehyde through direct transfer to tetrahydrofolate with formation of methylene-

THF and glycine (Eq. 8-19). The latter, in turn can yield another single carbon unit by loss of CO_2 under the influence of the THF and the PLP-requiring glycine decarboxylase system (Eq. 14-32). Free formaldehyde can also combine with THF to form methylene-THF (Eq. 8-69).[155] In some

$$(8\text{-}69)$$

methylene-THF

TABLE 8-5
Single Carbon Compounds in Order of Oxidation State

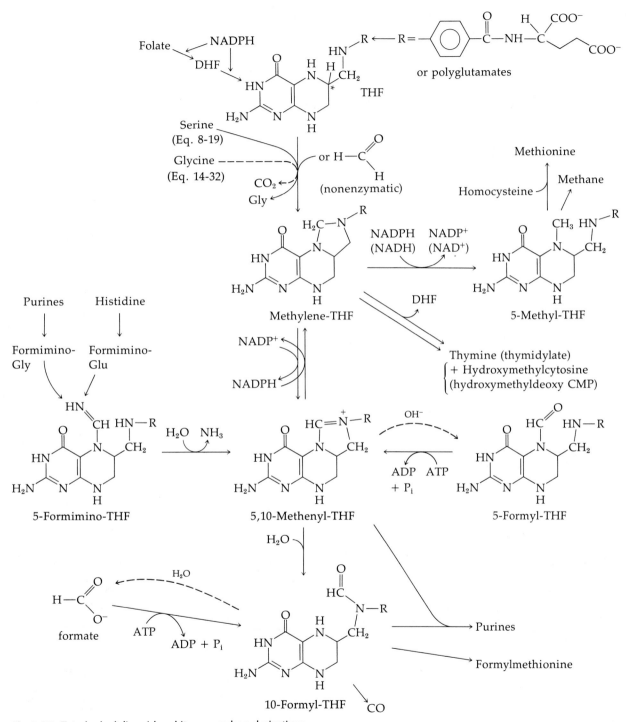

Fig. 8-20 Tetrahydrofolic acid and its one-carbon derivatives.

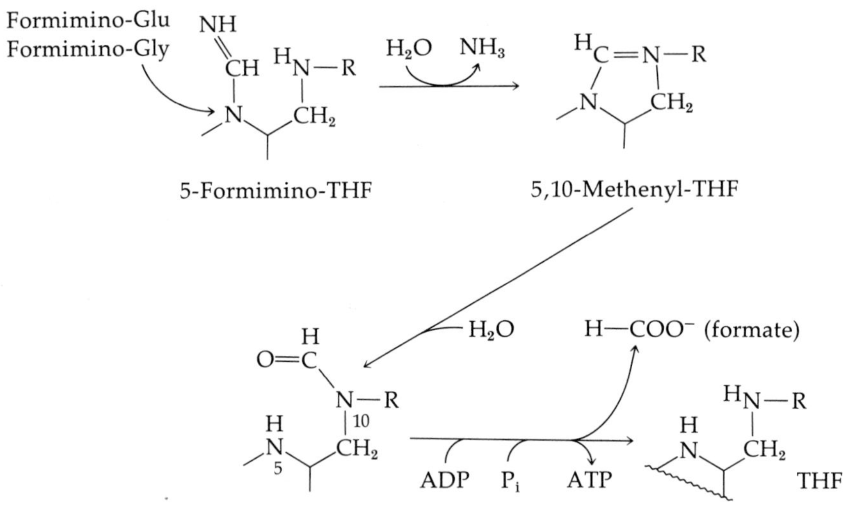

Fig. 8-21   A reaction providing for substrate-level phosphorylation via 10-formyl-THF in clostridia.

organisms, notably bacteria that are able to subsist on formate as a sole carbon source, the initial combination of formate with THF yields 10-formyl-THF (lower left corner, Fig. 8-21).

While attached to THF, the single carbon unit can either be oxidized from the formaldehyde level to 5,10-methenyl-THF and 10-formyl-THF or it can be reduced to 5-methyl-THF. Likewise, organisms utilizing formate can reduce 10-formyl-THF to methylene-THF and transfer the single carbon unit to glycine to form serine.

The significance of 5-formyl-THF (leucovorin or citrovorum factor")* in metabolism is less clear. It can arise by transfer of a formyl group from formylglutamate and an enzyme exists that converts 5-formyl-THF to 5,10-methenyl-THF with concurrent cleavage of ATP. The compound is being used in a remarkable way in treatment of certain highly malignant cancers. Following surgical removal of the tumor the patient is periodically given what would normally be a lethal dose of methotrexate. Then, about 36 h later, the patient is rescued by injection of 5-formyl-THF. The mechanism of rescue is not fully understood.

*N^5-Formyl-THF is a growth factor for *Leuconostoc citrovorum* and following this discovery in 1949 was referred to as the citrovorum factor.

Another source of single carbon units in metabolism is the degradation of histidine which occurs both in bacteria and in animals via **formiminoglutamic acid.** The latter transfers the —CH=NH group to THF forming 5-formimino-THF which is in turn converted to 5,10-methenyl-THF and ammonia. In bacteria that ferment purines, **formiminoglycine** is an intermediate. Again, the formimino group is transferred to THF and deaminated, the eventual product being 10-formyl-THF. In these organisms, an enzyme capable of converting formate to 10-formyl-THF has a very high activity[158] and it is thought that it operates in reverse as a mechanism for synthesis of ATP in this type of fermentation (Fig. 8-21). Formyl phosphate is a probable intermediate.[158a]

Both 5,10-methenyl-THF and 10-formyl-THF serve as biological formylating agents and *both of them are needed in the synthesis of purines* (Chapter 14, Section K,3). A remarkable reaction occurs in the gas gland of the Portugese man-of-war, which produces large amounts of carbon monoxide, apparently from the 10-formyl-THF. Carbon monoxide also arises naturally from the metabolism of bacteria. This aspect of ecological biochemistry is of current interest because of the large amount of man-made CO entering the biosphere. The bacteria make much more of it than does man.

Methylene-THF serves as the direct precursor of the 5-methyl group of thymine as well as of the hydroxymethyl group of hydroxymethylcytosine. During thymine formation the coenzyme is oxidized to dihydrofolic acid (Eq. 14-51), which must be reduced by dihydrofolate reductase to complete the catalytic cycle. Reduction of methylene-THF to 5-methyl-THF is one source of methane in bacterial metabolism (Section L,8) and within all living organisms provides the many methyl groups required in formation of methionine and in modification of proteins, nucleic acids, and other biochemicals through methylation of specific groups.

L. VITAMIN B_{12} COENZYMES

The nutritional studies that led to the isolation of vitamin B_{12} (**cyanocobalamin**) in 1948 gave few clues about the coenzymatic functions, and no concrete information was obtained for several years. A major reason for the delay was the extreme sensitivity of the coenzymes to decomposition by light. Progress came from an unexpected source. The initial step in the anaerobic fermentation of glutamate by *Clostridium tetanomorphum* was found by H. A. Barker to be a remarkable rearrangement to β-methylaspartate[159] (Eq. 8-70).

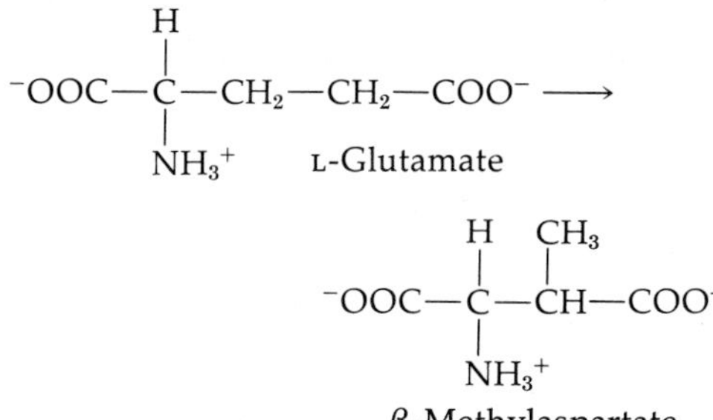

$$^-OOC-\underset{\underset{NH_3^+}{|}}{\overset{\overset{H}{|}}{C}}-CH_2-CH_2-COO^- \longrightarrow$$

L-Glutamate

$$^-OOC-\underset{\underset{NH_3^+}{|}}{\overset{\overset{H}{|}}{C}}-\overset{\overset{CH_3}{|}}{CH}-COO^- \qquad (8\text{-}70)$$

β-Methylaspartate

The latter compound can be catabolized in more or less standard ways by reactions which cannot be used on glutamate itself. Thus, this initial rearrangement is an indispensable step in the energy metabolism of the bacterium.

A new coenzyme required for this reaction was isolated in 1958 after it was discovered that protection from light during the preparation was necessary. The structure of this **5'-deoxyadenosyl coenzyme B_{12}** (5'-deoxyadenosylcobalamin) was determined in 1961 by X-ray diffraction.[160]

The vitamin B_{12} coenzyme is related to cyanocobalamin (Box 8-K) by replacement of the CN group by a 5'-deoxyadenosyl group as indicated in the abbreviated formulas shown below.[161–166]

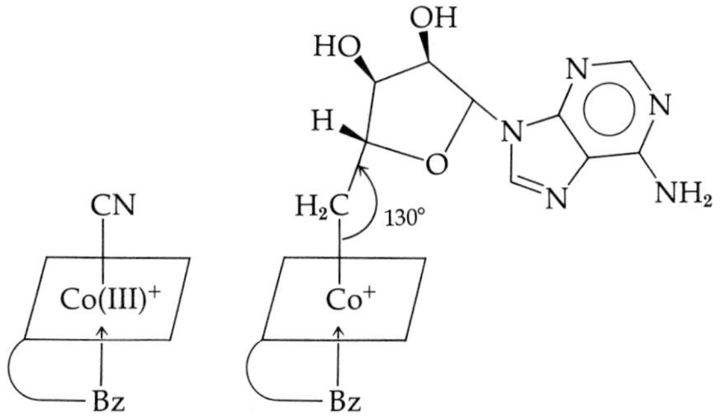

Cyanocobalamin   5'-Deoxyadenosylcobalamin

Here the squares represent the corrin ring system and Bz the dimethylbenzimidazole coordinated with the cobalt from below the ring.

The most surprising structural feature is the Co—C σ bond of length 2.05 Å. Thus, the coenzyme is an alkyl cobalt, the first such compound found in nature. Prior to 1961 all alkyl cobalts had been thought to be unstable. While the bond of 5'-deoxyadenosylcobalamin is covalent, the Co—C—C angle of 130° suggests some ionic character.[160] The oxidation state of the cobalt is $3+$, and cyanocobalamin can be imagined as arising by replacement of a single hydrogen on the inside of the corrin ring by Co^{3+} plus CN^-. However, bear in mind that three other nitrogens of the corrin ring and a nitrogen of dimethylbenzimidazole also bond to the cobalt. Each nitrogen atom donates an electron pair to form coordinate covalent linkages. The latter are drawn in Box 8-K with arrowheads to indicate a formal difference between them and the other Co—N bond. However because of resonance in the conjugated double bond system of the corrin, all four of the Co—N bonds in the ring are nearly equivalent and the positive charge is distributed over the nitrogen atoms surrounding the cobalt.

BOX 8-K

COBALAMIN (VITAMIN B$_{12}$)

Cyanocobalamin
MW = 1355
($C_{63}H_{88}N_{14}PO_{14}Co$)

The story[a] of vitamin B$_{12}$ begins with pernicious anemia, largely a disease of persons of age 60 or more, but which occasionally strikes children. Before 1926 the disease was incurable and usually fatal. Abnormally large, immature, and fragile red blood cells are produced but the total number of erythrocytes is much reduced ($1–3 \times 10^6$ mm^{-3} vs. $4.5–6 \times 10^6$ mm^{-3}). Within the bone marrow mitosis appears to be blocked and DNA synthesis is suppressed. The disease also affects other rapidly growing tissues such as the gastric mucous membranes (which stop secreting HCl) and nervous tissues. Demyelination of the central nervous system is often observed with loss of muscular coordination (ataxia) and psychotic symptoms.

In 1926, G. Minot and W. Murphy discovered that pernicious anemia could be controlled by eating one-half pound of raw or lightly cooked liver per day, a treatment which not all patients accepted with enthusiasm. It was not until 22

years later that vitamin B$_{12}$ was isolated (as the crystalline derivative cyanocobalamin) and was shown to be the curative agent. It is present in liver to the extent of ~ 1 mg kg^{-1} or $\sim 10^{-6}$ M. Although great amounts of effort were expended in preparation of concentrated liver extracts for the treatment of pernicious anemia, the lack of an assay other than treatment of human patients made progress slow.

In the early 1940's nutritional studies of young animals raised on diets lacking animal proteins and maintained out of contact with their own excreta (which contained vitamin B$_{12}$) demonstrated the need for "animal protein factor" which was soon shown to be the same as vitamin B$_{12}$. The animal feeding experiments also demonstrated that waste liquors from streptomyces fermentations such as those used in production of streptomycin and other antibiotics were extremely rich in vitamin B$_{12}$. A later development which greatly aided the isolation efforts was the recognition of

Box 8-K (*Continued*)

vitamin B$_{12}$ as a growth factor for a strain of *Lactobacillus lactis*. This bacterium responded with half-maximum growth to as little as 0.013 μg/l (10^{-11} *M*).

Another line of investigation that also converged in demonstrating the importance of vitamin B$_{12}$ was study of the unusually high nutritional requirement of ruminant animals for cobalt. It seems likely that this need arises from the requirement of vitamin B$_{12}$ for the microorganisms of the rumen. In regions of the earth with a low soil cobalt content such as Australia, cobalt deficiency in sheep and cattle is a serious problem.

In 1948, red cobalt-containing crystals of vitamin B$_{12}$ were obtained almost simultaneously by two large pharmaceutical firms. Charcoal adsorption from liver extracts was followed by elution with alcohol and numerous other separation steps. Later fermentation broths provided a richer source.

Structure: Chemical studies revealed that the new vitamin had an enormous molecular weight, that it contained one atom of phosphorus which could be released as P$_i$, a molecule of aminopropanol and a ribofuranoside of dimethyl benzimidazole with the unusual α configuration:

α-Ribofuranoside of dimethylbenzimidazole

D$_g$-1-Amino-2-propanol

Note the relationship of the dimethylbenzimidazole to the structure of the ring system of riboflavin (Box 8-I). Several molecules of ammonia could also be released from amide linkages by hydrolysis. However, all attempts to reversibly remove the cobalt from the ring system were unsuccessful. The structure was determined in 1956 by Dorothy C. Hodgkin and co-workers using X-ray diffraction methods.[b] At that time, it was the largest organic structure determined by X-ray diffraction. The complete laboratory synthesis was accomplished in 1972.[c]

The ring system of vitamin B$_{12}$, like that of porphyrins (Fig. 10-1), is made up of four pyrrole rings whose biosynthetic relationship to the corresponding rings in porphyrins is obvious from the structures. In addition, a number of "extra" methyl groups are present. A less extensive conjugated system of double bonds is present in the **corrin** ring of vitamin B$_{12}$ than in porphyrins, and as a result, a large number of chiral centers are found around the periphery of the somewhat nonplanar rings.

The form of vitamin B$_{12}$ isolated initially, cyanocobalamin, contained cyanide attached to one of the coordination positions of the cobalt. It occurs in minor amounts if at all in nature, but is generated through the addition of cyanide during the isolation procedures. **Hydroxocobalamin** (B$_{12a}$) containing OH$^-$ in place of CN$^-$ does occur in nature. However, the predominant forms are the B$_{12}$ coenzymes in which an alkyl group replaces the CN$^-$ of cyanocobalamin.

Nutritional requirements: Intramuscular injection of as little as 3–6 μg of crystalline vitamin B$_{12}$ is sufficient to bring about a remission of pernicious anemia and 1 μg daily provides a suitable maintenance dose (often administered as hydroxycobalamin injected once every 2 weeks). The required oral dose is probably 2 to 50 times higher, but there is rarely difficulty in meeting the requirement from ordinary diets. Vitamin B$_{12}$ has the distinction of being formed principally by bacteria, and most plants contain little or none. Thus, pernicious anemia symptoms are sometimes observed among strict vegetarians.

Pernicious anemia is usually caused not by lack

Box 8-K (*Continued*)

of vitamin B_{12} but by poor absorption of the vitamin. Absorption depends upon the so-called **intrinsic factor,** a mucoprotein (or mucoproteins) synthesized by the stomach lining. Pernicious anemia patients often have a genetic predilection toward decreased synthesis of the intrinsic factor. Gastrectomy (which decreases synthesis of the intrinsic factor) and infection with fish tapeworms (which compete for available vitamin B_{12} and interfere with absorption) can also induce the disease.

Normal blood levels of vitamin B_{12} are $\sim 2 \times 10^{-10}$ M or a little more, but in vegetarians the level may drop to less than one-half this value. A deficiency of folic acid can also cause megaloblastic anemia, and a large excess of folic

acid can, to some extent, reverse the anemia of pernicious anemia. Fear that ingestion of folic acid in large amounts might interfere with a diagnosis of pernicious anemia and at the same time permit severe neurological damage led the U.S. Food and Drug Agency in 1960 to set a maximum allowed folic acid content of 0.25 mg in multivitamin tablets. However, this amount of folic acid is probably too low in high demand situations, e.g., during the third trimester of pregnancy. Some authorities feel that the danger of folic acid masking B_{12} deficiency has been greatly exaggerated.

[a] See Smith[161] for a historical account.
[b] D. C. Hodgkin, *Science* **150,** 979–988 (1965).
[c] See report in *Science* **179,** 266–267 (1973).

In both bacteria and in liver, the 5'-deoxyadenosyl coenzyme is the most abundant form of vitamin B_{12}. However, lesser amounts of **methylcobalamin** also exist. A variety of naturally

Methylcobalamin

occurring analogues of the coenzymes have also been isolated. For example, **pseudo vitamin B_{12}** contains, in place of the dimethylbenzimidazole, adenine, which like dimethylbenzimidazole is combined with ribose in the unusual α linkage. A compound called factor A is the vitamin B_{12} analogue with 2-methyladenine. A whole series of other factors have been isolated from such sources as sewage sludge which abounds in anaerobic bacteria. It has been suggested that plants may contain vitamin B_{12}-like materials which do not support growth of bacteria. Thus, we may not have discovered all of the alkyl cobalt coenzymes.

1. Reduction of Cyanocobalamin and Synthesis of Alkyl Cobalamines

Cyanocobalamin can be reduced in two one-electron steps (Eq. 8-71).[163,167] The cyanide ion is

Cyanocobalamin ⟶ B_{12r} ⟶

$$\qquad (8\text{-}71)$$

B_{12s}

lost in the first step (Eq. 8-71, step a), which may be accomplished with chromous acetate at pH 5 or by catalytic hydrogenation. The product is the brown paramagnetic compound B_{12r}, a tetragonal low-spin cobalt(II) complex. In the second step (Eq. 8-71, step b), an additional electron is added, e.g., from sodium borohydride or from chromous acetate at pH 9.5, to give the gray-green exceedingly reactive B_{12s}. The latter is thought to be in equilibrium with cobalt(III) hydride, as shown in Eq. 8-72, step a. The hydride is unstable and

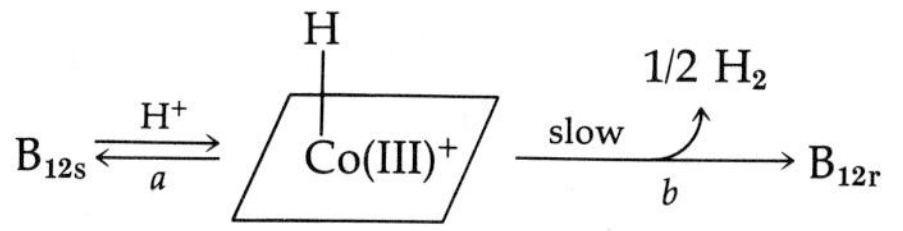

$$B_{12s} \underset{a}{\overset{H^+}{\rightleftharpoons}} \overset{H}{Co(III)^+} \overset{slow}{\underset{b}{\longrightarrow}} B_{12r} \qquad (8\text{-}72)$$

breaks down slowly to H$_2$ and B$_{12r}$ (Eq. 8-72, step b).[168]

Vitamin B$_{12}$ reacts rapidly with alkyl iodides (e.g., methyl iodide or a 5'-chloro derivative of adenosine) via nucleophilic displacement to form the alkyl cobalt forms of vitamin B$_{12}$ (Eq. 8-73).

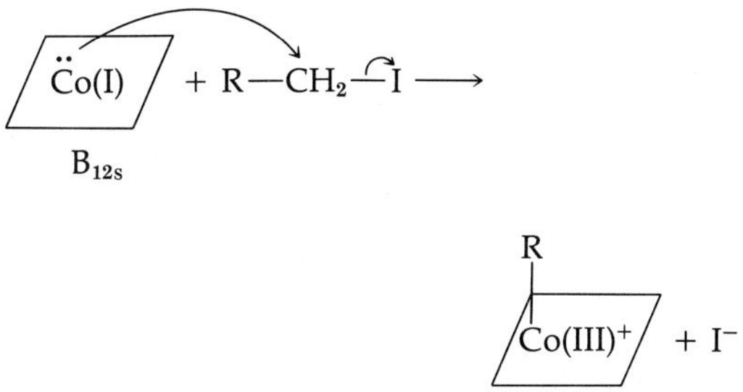

These reactions provide a convenient way of preparing isotopically labeled alkyl cobalamins, including those selectively enriched in ^{13}C for use in nmr studies.[169] The biosynthesis of 5'-deoxyadenosylcobalamin utilizes exactly the same type of reaction with ATP as a substrate.[170] A **B$_{12s}$ adenosyltransferase** catalyzes nucleophilic displacement on the 5'-carbon of ATP with formation of the coenzyme and displacement of inorganic tripolyphosphate PPP.

2. Three Nonenzymatic Cleavage Reactions of Vitamin B$_{12}$ Coenzymes

The 5'-deoxyadenosyl coenzyme is easily decomposed by a variety of agents. Anaerobic irradiation yields principally vitamin B$_{12r}$ and a cyclic 5',8-deoxyadenosine (in the presence of air a variety of compounds are formed). The latter is probably formed through an intermediate radical (Eq. 8-74):

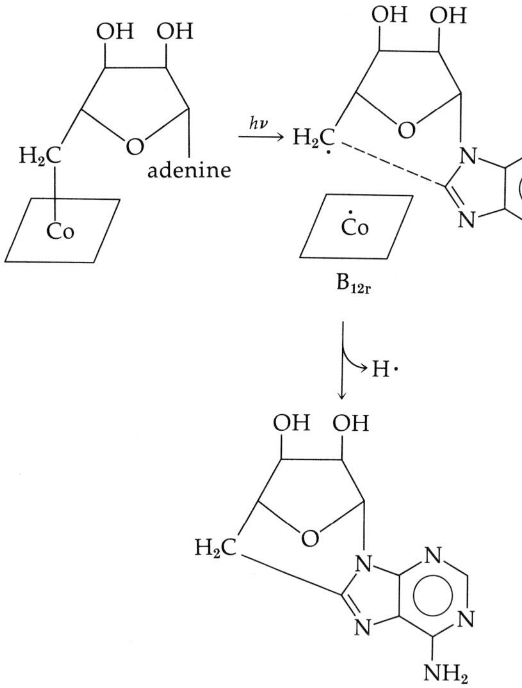

Hydrolysis of deoxyadenosylcobalamin by acid (1 N HCl, 100°C, 90 min) yields (Eq. 8-75) hydroxycobalamin, adenine, and an unsaturated sugar.

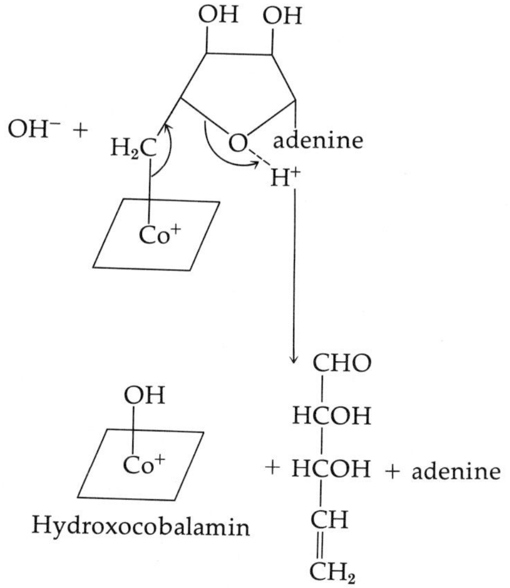

The initiating reaction step is thought to be protonation of the oxygen of the ribose ring. A related cleavage by alkaline cyanide can be viewed as a nucleophilic displacement of the deoxyadenosyl anion by cyanide. The end product is **dicyanocobalamin,** in which the loosely bound dimethyl benzimidazole-containing nucleotide is replaced by a second cyanide ion. Methyl and other simple alkyl cobalamins are stable to alkaline cyanide.

3. Enzymatic Functions of B$_{12}$ Coenzymes

Reactions dependent upon alkyl corrin coenzymes are of two types. Those of the first type depend upon 5'-deoxyadenosylcobalamin and can be depicted formally as shown in Eq. 8-76. Some

$$-\overset{|}{\underset{\boxed{X}}{C}}-\overset{|}{\underset{H}{C}}- \longrightarrow -\overset{|}{\underset{H}{C}}-\overset{|}{\underset{\boxed{X}}{C}}- \qquad (8\text{-}76)$$

group X, which may be attached by a C—C, C—O, or C—N bond, is transferred to an adjacent carbon atom bearing a hydrogen. At the same time, the hydrogen is transferred to the carbon to which X was originally attached. In one exceptional case (that of **ribonucleotide reductase,** Table 8-6) the hydrogen transfer is *intermolecular.* The second group of reactions involves transfer of methyl groups via methylcobalamin and certain related reactions.[163]

Conclusive evidence has now been obtained to show that in reactions of the first type the hydrogen is transferred via the B$_{12}$ coenzyme. In no case, except that of ribonucleotide reductase, does any exchange with the medium take place. Since X may be an electronegative group like OH, the reactions may all be treated formally as hydride ion transfers. However, it is more likely that they occur via homolytic cleavages with intermediate

radical formation.

Abeles and associates showed that when **dioldehydratase** (Table 8-6) catalyzes the conversion of 1,2-[1-^{3}H]propanediol to propionaldehyde, tritium appears in the coenzyme as well as in the final product. When ^{3}H containing coenzyme is incubated with unlabeled propanediol, the product also contains ^{3}H. Chemical degradation of the labeled coenzyme showed that the radioactivity was exclusively at C-5'. Furthermore, when synthetic 5'-deoxyadenosyl coenzyme containing ^{3}H in the 5' position was synthesized, it, too, transferred ^{3}H to product. Most important, using a mixture of propanediol and ethylene glycol, a small amount of *intramolecular* transfer was demonstrated; that is, ^{3}H was transferred into acetaldehyde, the product of dehydration of ethylene glycol.

Another important experiment[171] showed that ^{18}O from [2-^{18}O]propanediol was transferred to the 1 position without exchange with solvent. Furthermore, ^{18}O from (*S*)-[1-^{18}O]propanediol was retained in the product while that from the (*R*) isomer was not. Thus, it appears that the enzyme stereospecifically dehydrates the final intermediate. From these and other experiments, the partial mechanism shown in Eq. 8-77 was proposed. In Eq. 8-77, step *a* the hydrogen marked with an asterisk is transferred to the coenzyme with cleavage of the cobalt-5'-deoxyadenosyl bond. The 5'-deoxyadenosine now contains hydrogen from the substrate and that hydrogen becomes essentially equivalent to the two already present in the coenzyme. The nature of the substrate after this hydrogen transfer is not established, but it seems reasonable to suppose that the hydrogen removed is replaced by the cobalt. This substrate-cobalamin compound must then undergo an isomerization reaction, which, in the case of dioldehydratase, leads to intramolecular transfer of the OH group (Eq. 8-77, step *b*). In Eq. 8-77, step *c* the hydrogen atom is transferred back from the 5'-deoxyadenosine to its new location in the product, and in step *d* the resulting *gem*-diol is dehydrated to form the aldehyde product.

$$(8\text{-}77)$$

According to the mechanism of Eq. 8-77, 5'-deoxyadenosine is freed from its bond to cobalt during the action of the enzyme. Why then does the deoxyadenosine not escape from the coenzyme entirely, leading to its inactivation? Substrate-induced inactivation is not ordinarily observed with coenzyme B_{12}-dependent reactions, but it has been shown for several enzymes with pseudo substrates. For example, glycolaldehyde converts the coenzyme of dioldehydratase to 5'-deoxyadenosine while ethylene glycol does the same with ethanolamine deaminase. When 5'-deoxyinosine (Chapter 2, Section D,1) replaces the 5'-deoxyadenosine of the normal coenzyme in dioldehydratase, 5'-deoxyinosine is quantitatively released by substrate. This suggests that normally the protein may hold the adenine group of 5'-deoxyadenosine through hydrogen bonding to the amino group. Because the OH group of inosine tautomerizes to C=O, inosine may not be held as tightly.

Evidence favoring a mechanism analogous to that of Eq. 8-77 for the ethanolamine ammonia-lyase is the transfer of ^3H from [5'-^3H]deoxyadenosylcobalamin into both reactant L-2-aminopropanol and product at kinetically significant rates.[172,173] Transfer of ^2H from the 1-position of deuterated 2-aminopropanol into the methyl group of [5',5'-di-^2H]adenosylcobalamin to form 5'-deoxyadenosine containing a —C^2H_3 group provides further support for the mechanism.[173a]

Despite the evidence in its favor, there has been some reluctance to accept 5'-deoxyadenosine as an intermediate in vitamin B_{12}-dependent isomerization reactions. It is hard to believe that a methyl group could exchange hydrogen atoms so rapidly. Protonation of the oxygen of the ribose ring as in Eq. 8-75 might facilitate release of a hydrogen atom. However, substitution of the ring oxygen by CH_2 in a synthetic analogue does not destroy the coenzymatic activity.[163] Another possibility is that the methyl group has an unusual reactivity if the cobalt is reduced to Co(II) (see the next section).

4. Ribonucleotide Reductase

Ribonucleotides are reduced to deoxyribonucleotides needed for DNA synthesis by special enzymes acting on the di- or triphosphates of the purine and pyrimidine nucleosides (Chapter 14, Section K,1,a). In some organisms the reduction requires vitamin B_{12}. For example, in *Lactobacillus leichmanni* the triphosphates are reduced by transfer of hydrogen to a dithiol such as dihydroli-

TABLE 8-6
B$_{12}$ Coenzyme-Dependent Reactions Involving H Transfer Reaction Type 10

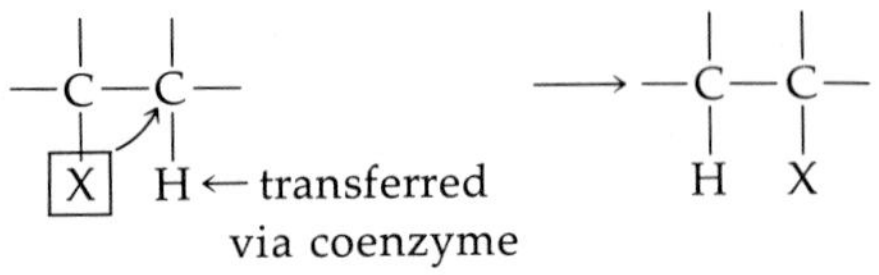

General reaction. Migrating group is enclosed in a box

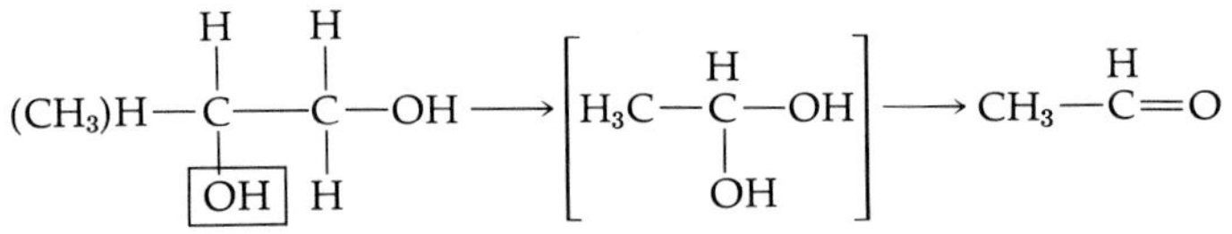

Diol dehydratase. Glycerol dehydratase catalyzes the same type reaction

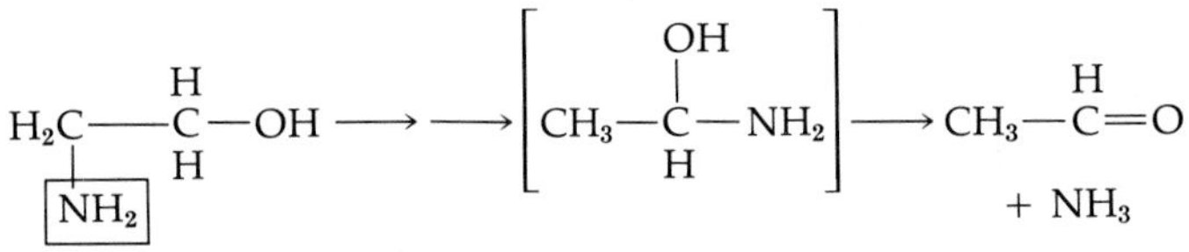

Ethanolamine ammonia-lyase

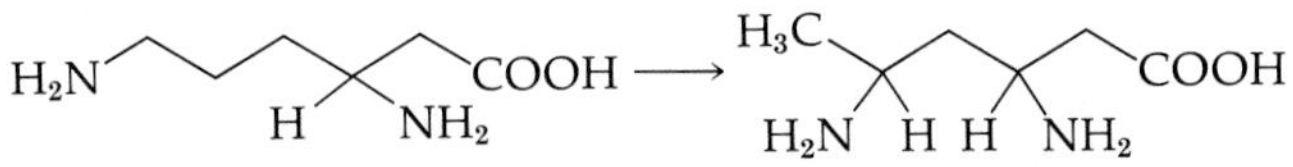

L-β-Lysine mutase. D-α-Lysine mutase and ornithine mutase catalyze the same type of reaction

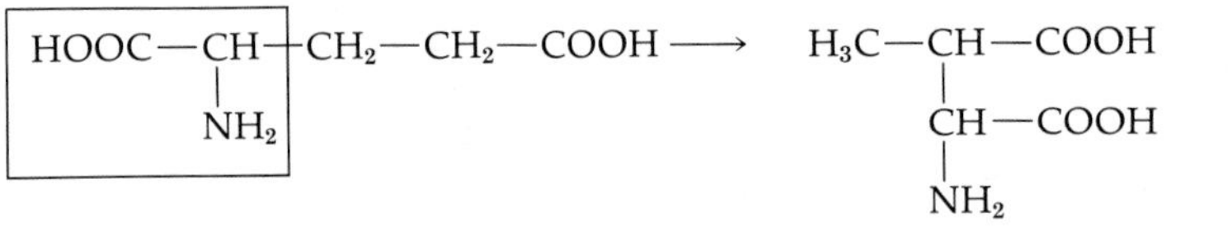

Glutamate mutase

threo-β-Methyl-L-aspartate

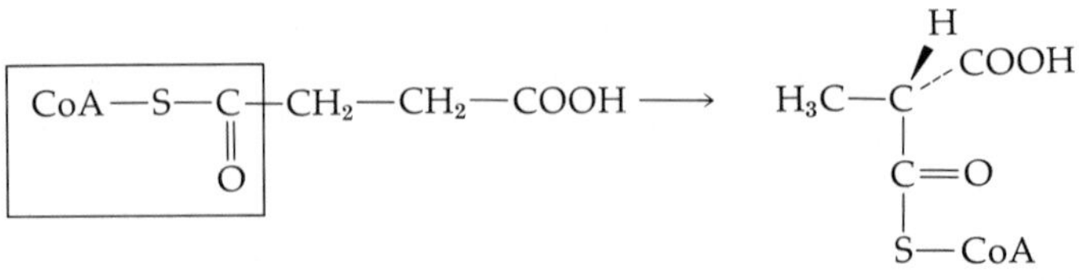

Methylmalonyl-CoA mutase

(R)-Methylmalonyl-CoA

(continued)

TABLE 8.6 (*Continued*)

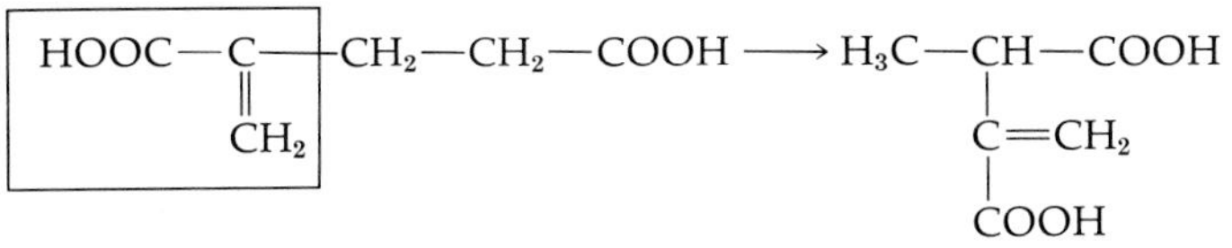

α-Methyleneglutarate mutase

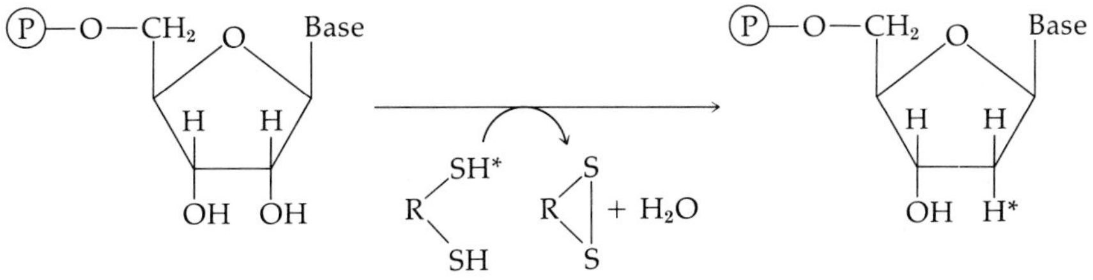

Ribonucleotide reductase

poic acid or the dithiol protein thioredoxin (Table 8-6). Protons from water are reversibly incorporated at C-2′ of the reduced nucleotide with retention of configuration. This result suggests the possibility that the C-2′ hydrogen is transferred via 5′-deoxycobalamin into an SH group of the thioredoxin and that the latter exchanges with the medium.[174,175]

Reaction of the reductase with dihydrolipoic acid in the presence of deoxy-GTP (which apparently serves as an allosteric activator) leads to rapid (a few milliseconds) formation of a radical with a characteristic epr spectrum that can be studied when the reaction mixture is rapidly cooled to 130°K. When GTP (a true substrate) is used instead of dGTP, the radical signal reaches a maximum in about 20 ms, then decays.

Of the various oxidation states of cobalt (3+, 2+, and 1+) only the 2+ state of vitamin B$_{12r}$ is paramagnetic and gives rise to an epr signal. The electronic absorption spectrum of the coenzyme of ribonucleotide reductase also changes in a way suggesting formation of B$_{12r}$. Thus, it has been proposed that a *homolytic* cleavage to B$_{12r}$ and a stabilized 5′-deoxyadenosyl radical may occur (Eq. 8-78).[175a] The epr signal of the 5′-deoxyadenosyl radical is not seen. Therefore, it is not

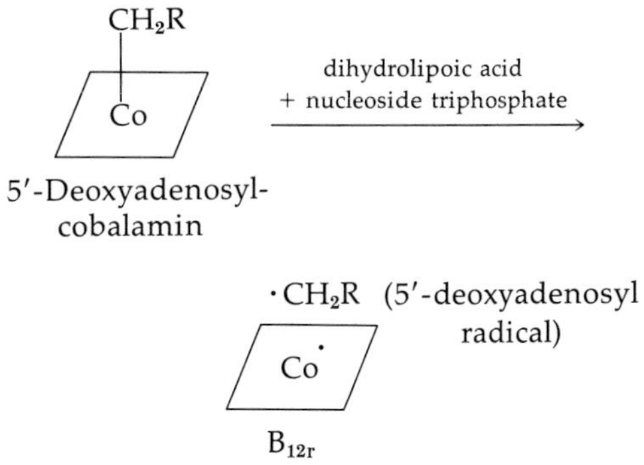

(8-78)

clear whether it has accepted a hydrogen atom from the dihydrolipoate or not. Nor is it clear by what mechanism the subsequent stereospecific replacement of the 3′-OH group of the ribonucleotide substrate by a hydrogen atom takes place.

Since the initial observation of an epr signal with ribonucleotide reductase similar signals have been observed with other enzymes.[176-178] Thus, it is possible that reaction 8-78 occurs as an initial step in every case. If the mechanism of Eq. 8-77 holds for the rearrangement reactions, step *a* would be formulated as an attack of either the reduced cobalt or the 5′-deoxyadenosyl radical of Eq. 8-78 on the substrate.

5. Isomerization of Substrates

How does the substrate isomerize in step b of Eq. 8-77? Radical,[164] carbonium ion, and carbanion intermediates are all possible.[164a] For dioldehydratase cleavage of the C—Co bond to form B_{12s} and a carbonium ion is possible (Eq. 8-79). The carbonium ion would presumably be cyclized to an epoxide which could react with the B_{12s} (Eq. 8-79, step b) to complete the isomerization. Alternatively, elimination of OH^- could yield a cobalt-bound carbonium ion in equilibrium with a cobalt-olefin π complex.[178a]

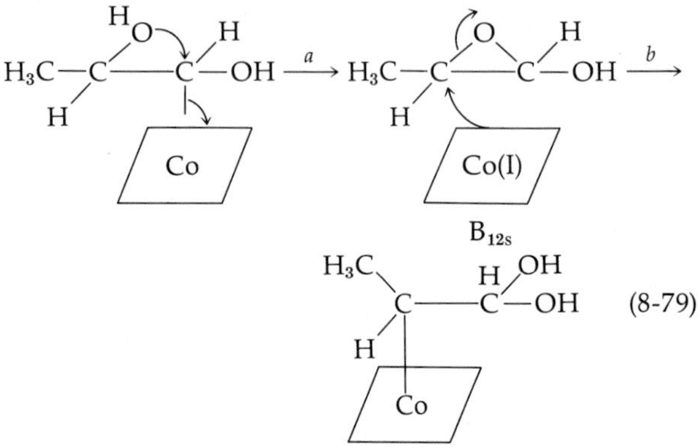

$$(8-79)$$

For methylmalonyl mutase Ingraham favored cleavage of the Co—C bond to form a carbanion, the substrate-cobalamin compound serving as a sort of "biological Grignard reagent." The carbanion would be stabilized by the carbonyl group of the thioester forming a "homoenolate anion"

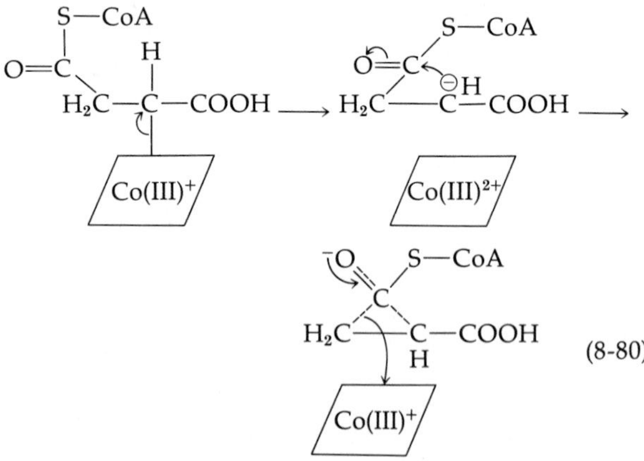

$$(8-80)$$

Homoenolate anion

(Eq. 8-80). The latter could break up in either of two ways reforming a C—Co bond and causing the isomerization.[179,180]

In the case of ribonucleotide reductase we must ask whether the Co(II) intermediate reacts with C-3' of the ribonucleotide substrate to initiate the reduction, whether a 5'-deoxyadenosyl radical abstracts a proton from the reduced thioredoxin to give a radical on sulfur ($-S^{\cdot}$) or both. Much remains to be learned.

Interesting nonenzymatic model systems are being studied. Nonenzymatic rearrangement of a carbon-cobalt bonded B_{12} derivative has been observed.[181] Other investigations are based upon complexes known as **cobaloximes.**[182] A cobaloxime

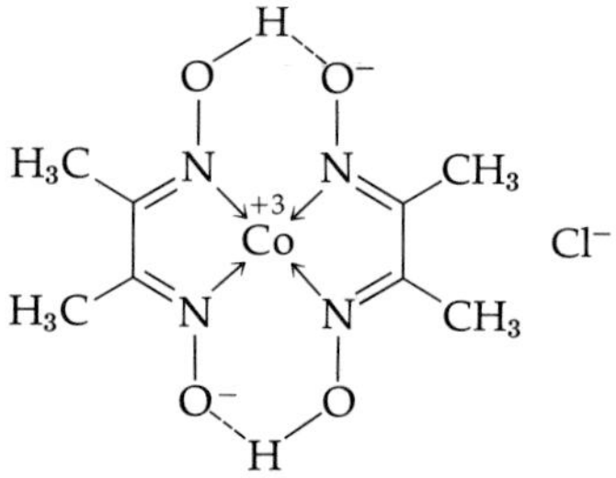

Cobaloxime

is a complex of Co^{3+} and two molecules of dimethylglyoxime. If such a compound is coordinated with a suitable basic group in one of the axial positions, it behaves in many ways like vitamin B_{12}. It can be reduced to the Co(II) state and reacted with alkyl halides to form alkyl cobaloximes analogous in many properties and chemical reactivity to the vitamin B_{12} coenzymes.

6. Stereochemistry of Vitamin B_{12}-Dependent Enzymes

Dioldehydratase acts on either the (R) or (S) isomers of 1,2-propanediol (Eqs. 8-81 and 8-82;

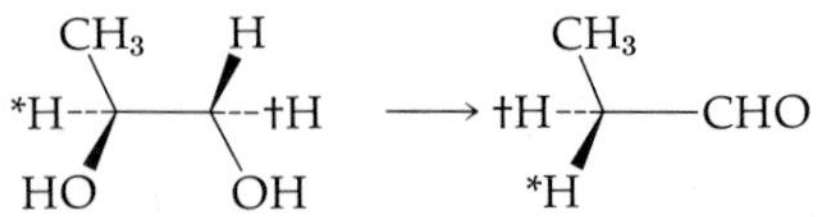

$$(8-81)$$

(S)-1,2-Propanediol

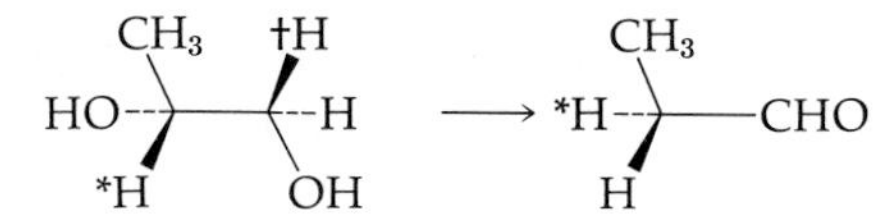

(8-82)

(R)-1,2-Propanediol

asterisks and daggers mark positions of labeled atoms in specific experiments). In both cases the reaction proceeds with retention of configuration at C-2 and with stereochemical specificity[183] for one of the two hydrogens at C-1. The reaction catalyzed by methylmalonyl-CoA mutase likewise proceeds with retention of configuration at C-2 (Eq. 8-83)[184] while the glutamate mutase reaction is accompanied by inversion (Eq. 8-84).

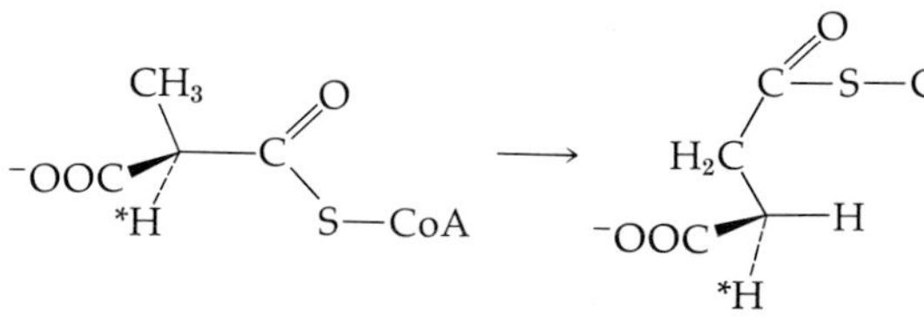

(8-83)

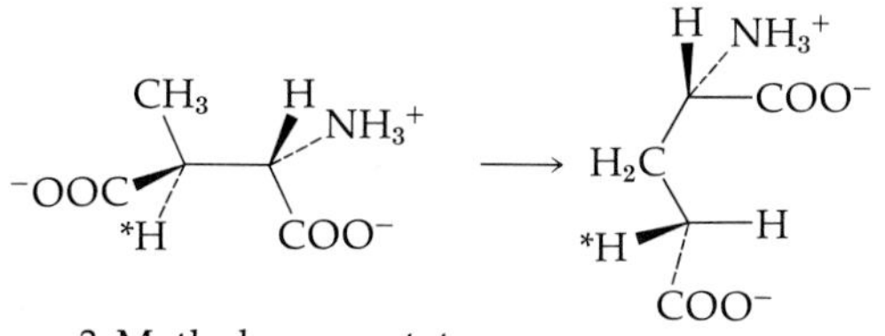

(8-84)

threo-3-Methyl-L-aspartate

During the action of D-α-lysine mutase (Table 8-6) a hydrogen atom is transferred from C-5 to C-6.[185] Two proteins are needed for the reaction; pyridoxal phosphate is required and is apparently directly involved in the amino group migration. The closely related L-β-lysine mutase requires pyruvate as a cofactor.[163]

7. Transfer Reactions of Methyl Groups

The generation and utilization of methyl groups is a quantitatively important aspect of the metabolism of all cells. As we have seen (Section K,3) methyl groups can be created by the reduction of one-carbon compounds attached to tetrahydrofolic acid. Methyl groups of methyltetrahydrofolic acid (N^5-CH_3THF) can then be transferred to the sulfur atom of homocysteine to form methionine (Eq. 8-85). The latter is converted to *S*-

$$N^5\text{-}CH_3\text{-THF} + HS-CH_2-CH_2-\underset{NH_3^+}{\overset{H}{\underset{|}{C}}}-COO^- \longrightarrow$$

Homocysteine

$$H_3C-S-CH_2CH_2-\underset{NH_3^+}{\overset{H}{\underset{|}{C}}}-COO^- \quad (8\text{-}85)$$

Methionine + THF

adenyosylmethionine (SAM), the nearly universal methyl group donor in transmethylation reactions (Eq. 7-1). In some bacteria, in fungi, and in higher plants, the methyl-THF-homocysteine transmethylase does not depend upon vitamin B_{12}. However, man shares with certain strains of *E. coli* and some other bacteria the need for a methylcobalamin-requiring enzyme. It presumably functions in a cyclic process. The cobalt is thought to alternate between the 1+ and 3+ oxidation states (Eq. 8-86). The first indication of this

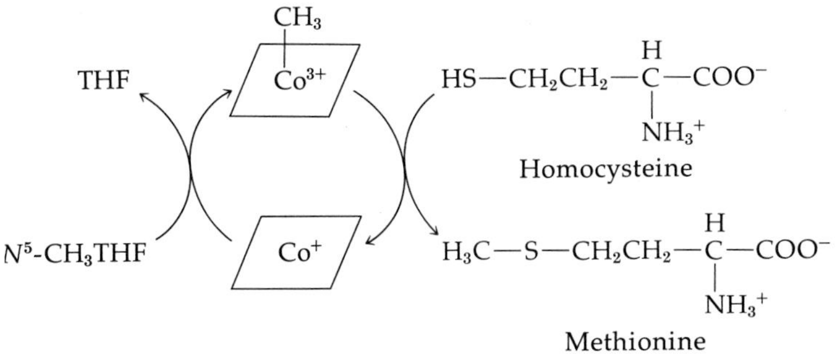

(8-86)

was the report by Weissbach that ^{14}C-labeled methylcobalamin could be isolated following treatment of the enzyme with such methyl donors as SAM and methyl iodide after reduction (e.g., with reduced riboflavin phosphate). The sequence parallels that of Eq. 8-73 for the laboratory synthesis of methylcobalamin. Nevertheless, the trans-methylase demonstrates many complexities. Initially, it must be "activated" by SAM or methyl iodide, after which it cycles according to Eq. 8-86 but is gradually inactivated. It is thought that there is another form of the enzyme that must be methylated with SAM to generate the active form.[186]

8. Methane Formation

A second type of methyl transfer from methylcobalamin occurs in the generation of methane by anaerobic bacteria, a quantitatively important reaction in the biosphere. Methanogenic bacteria can convert methyl groups of methanol, acetate or N^5-CH_3THF to methane and can also reduce CO_2, formaldehyde, or formate to methane. It is noteworthy that it is not vitamin B_{12} as such but a factor A corrinoid containing 5-hydroxybenzimidazole (p. 502) that serves as the coenzyme. The reactions by which the methyl corrinoid is formed are uncertain, but when vitamin B_{12s} (Co^+) is added the methyl groups can be trapped as methylcobalamin suggesting the Co^+ state in the natural corrinoid carrier. For some time it was thought that the methyl corrinoid was reduced directly by hydrogen to methane, but recently it has been shown that a low molecular weight coenzyme designated **coenzyme M** is necessary.[187] This coenzyme is the simple sulfonic acid shown below in Eq. 8-87. In its reduced form, the thiol group of the coenzyme reacts with the methyl corrinoid to form methylated coenzyme M from which methane is released by reduction with hydrogen in an ATP- and Mg^{2+}-requiring process.[188] The natural reductant may be either H_2 together with hydrogenase (Chapter 10, Section F,1) or pyruvate possibly via reduced ferredoxin generated by the pyruvate-ferredoxin oxidoreductase reaction (Section J,3).

A third important methyl transfer reaction is that from methyl corrinoids to mercury, arsenic, selenium, and tellurium. These reactions have special interest because of the generation of toxic methyl and dimethyl mercury and dimethylarsine.[189] We have previously considered transfer reactions in which a methyl group is transferred from a methyl corrinoid to nucleophilic groups such as the SH group of homocysteine. In such transfers, the methyl group is in effect transferred as a positive CH_3^+ ion, i.e., by a nucleophilic displacement on a carbon. This is the same type of methyl transfer that can occur from SAM or methyl-THF. By contrast, the transfer to Hg^{2+} is that of a carbanion, CH_3^-, with no valence change occurring in the cobalt (Eq. 8-88). Methyl corrinoids are able to undergo this type of reaction nonenzymatically, and the ability to transfer a methyl anion is a property of methyl corrinoids not shared by other transmethylating agents such as SAM. Note that at the conclusion of reaction 8-88 the cobalt is in the 3+ state. To be remethylated, it must presumably be reduced to Co(II). [A second methyl group can be transferred by the same type of reaction to form $(CH_3)_2Hg$.]

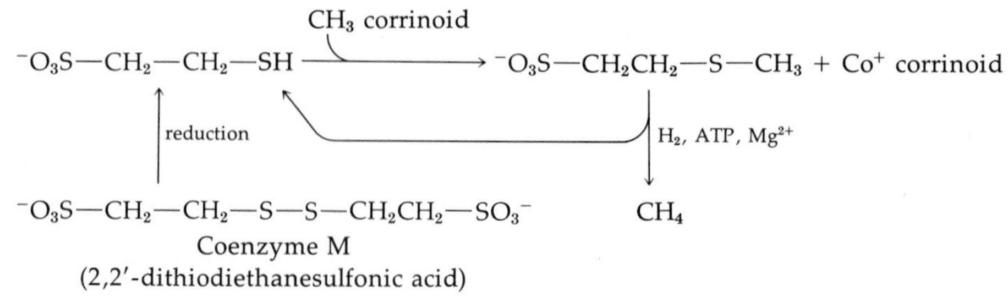

$$^-O_3S-CH_2-CH_2-SH \xrightarrow{\text{CH}_3\text{ corrinoid}} {}^-O_3S-CH_2CH_2-S-CH_3 + Co^+ \text{ corrinoid}$$

reduction ↑   ↘   ↓ H_2, ATP, Mg^{2+}

$$^-O_3S-CH_2-CH_2-S-S-CH_2CH_2-SO_3^- \qquad CH_4$$

Coenzyme M
(2,2'-dithiodiethanesulfonic acid)

(8-87)

$$H_2O + \underset{\underset{\text{Co}^{3+}}{|}}{\overset{\overset{CH_3}{|}}{\boxed{}}} + Hg^{2+} \longrightarrow$$

$$\underset{\underset{\text{Co}^{3+}}{|}}{\overset{\overset{OH_2}{|}}{\boxed{}}} + CH_3Hg^+ \qquad (8\text{-}88)$$

Methylation of arsenic is an important pollution problem because of the widespread use of arsenic compounds in insecticides and because of the presence of arsenate in the phosphate used in household detergents.[189,190] After reduction to arsenite, methylation occurs in two steps (Eq. 8-89).

Arsenate:

$$\overset{\overset{OH}{|}}{HO-\underset{\underset{O}{\|}}{As}-O^-} \xrightarrow{2e^-} \overset{\overset{OH}{|}}{HO-\underset{\;}{As}-O^-}$$

$$O=\underset{\;}{As}-O^-$$

Arsenite

$$CH_3^+ \swarrow$$

$$\underset{\underset{O}{\|}}{HO-\underset{\underset{CH_3}{|}}{As}}-O^- \quad \text{Methylarsonate}$$

$$CH_3^+ \swarrow 2e^-$$

$$\underset{\underset{O}{\|}}{H_3C-\underset{\underset{CH_3}{|}}{As}}-O^- \quad \text{Dimethylarsinate}$$

$$\swarrow 4e^-$$

$$\underset{\underset{H}{|}}{H_3C-\underset{\overset{CH_3}{|}}{As}} \quad \text{Dimethylarsine}$$

$$(8\text{-}89)$$

Together with additional reduction steps, this produces **dimethylarsine,** one of the principal products of action of methanogenic bacteria on arsenate. While the methyl transfer is shown as occurring through CH_3^+, with an accompanying loss of a proton from the substrate, the details of the reaction are still under investigation. A CH_3 radical may be transferred with formation of a cobalt(II) corrinoid.[191]

The anaerobic bacterium *Clostridium aceticum* can obtain its energy for growth by reduction of CO_2 with hydrogen (Eq. 8-90).[163,191] Presumably

$$4 H_2 + 2 CO_2 \longrightarrow CH_3COO^- + H^+ + 2 H_2O \qquad (8\text{-}90)$$

$$\Delta G' \text{ (pH 7)} = -94.9 \text{ kJ mol}^{-1}$$

one of the CO_2 molecules is reduced to formate and thence to a methyl corrinoid. The latter, in a process that is not understood, combines with another CO_2 to form acetate. The existence of a cobalt carboxymethyl intermediate has been suggested.

$$\underset{\text{Co}^{3+}}{\overset{\overset{CH_2COO^-}{|}}{\boxed{}}}$$

There is evidence that the CO_2 does not enter directly but is transferred from pyruvate.[192]

As we have seen, vitamin B_{12} and its relatives are remarkable catalysts of versatile function. How many more reactions dependent upon them remain to be discovered? How are the corresponding reactions catalyzed in green plants and other organisms lacking the cobalt-containing catalysts? The answers to these questions may hold many new surprises.

REFERENCES

1. F. Lipmann, *JBC* **160,** 173–190 (1945).
2. P. R. Vagelos and A. R. Larrabee, *JBC* **242,** 1776–1781 (1967).
3. T. C. Vanaman, S. J. Wakil, and R. L. Hill, *JBC* **243,** 6420–6431 (1968).
4. L. D. Wright, E. L. Cresson, H. R. Skeggs, R. L. Peck,

D. E. Wolf, T. R. Wood, J. Valent, and K. Folkers, *Science* **114,** 635–636 (1951).
5. A. W. Alberts and P. R. Vagelos, *in* "The Enzymes," 3rd ed. (P. D. Boyer, ed.), Vol. 6, pp. 37–82. Academic Press, New York, 1972.
6. F. Lynen, J. Knappe, E. Lorch, G. Jütting, and E. Ringelmann, *Angew. Chem.* **71,** 481–486 (1959).
7. S. E. Polakis, R. B. Guchhait, and M. D. Lane, *JBC* **247,**

1235–1337 (1972).

7a. I. A. Rose, E. L. O'Connell, and F. Solomon, *JBC* **251**, 902–904 (1976).

8. M. C. Scrutton and M. R. Young, *in* "The Enzymes," 3rd ed. (P. D. Boyer, ed.), Vol. 6, pp. 1–35. Academic Press, New York, 1972.

9. C. H. Fung, A. S. Mildvan, A. Allerhand, R. Komoroski, and M. C. Scrutton, *Biochemistry* **12**, 620–629 (1973).

10. G. H. Reed and M. C. Scrutton, *JBC* **249**, 6156–6162 (1974).

11. R. Griesser, H. Sigel, L. D. Wright, and D. B. McCormick, *Biochemistry* **12**, 1917–1922 (1973).

12. N. M. Green, R. C. Valentine, N. G. Wrigley, F. Ahmad, B. Jacobson, and H. G. Wood, *JBC* **247**, 6284–6298 (1972).

13. F. Ahmad, D. G. Lygre, B. E. Jacobson, and H. G. Wood, *JBC* **247**, 6299–6305 (1972).

13a. S. G. Powers and J. F. Riordan, *PNAS* **72**, 2616–2620 (1975).

14. P. A. Whitney and T. Cooper, *JBC* **248**, 325–330 (1973).

15. R. J. Ronn, J. Hampshire, and B. Levenberg, *JBC* **247**, 7539–7545 (1972).

16. N. M. Green, *in* "Methods in Enzymology" (D. B. McCormick and L. D. Wright, eds.), Vol. 18A, pp. 414–424. Academic Press, New York, 1970.

17. L. Siegel, J. L. Foote, and M. J. Coon, *JBC* **240**, 1025–1031 (1965).

18. L. O. Krampitz, "Thiamin Diphosphate and Its Catalytic Functions," Dekker, New York, 1970.

18a. C. J. Gubler, ed. "Thiamine." Wiley, New York, 1976.

19. D. E. Metzler, *in* "The Enzymes," 2nd ed. (P. D. Boyer, H. Lardy, and K. Myrbäck, eds.), Vol. 2, pp. 295–337. Academic Press, New York, 1960.

20. I. G. Leder, *Metab. Pathways, 3rd Ed.* **7**, 57–85 (1975).

21. R. F. W. Hopmann and G. P. Brugnoni, *Nature* (*London*), *New Biol.* **246**, 157–158 (1973).

22. R. Breslow, *JACS* **80**, 3719–3726 (1958).

23. R. Breslow and E. McNelis, *JACS* **81**, 3080–3082 (1959).

24. M. Sax, P. Pulsinelli, and J. Pletcher, *JACS* **96**, 155–165 (1974).

25. J. Pletcher and M. Sax, *JACS* **94**, 3998–4005 (1972).

26. A. A. Gallo and H. Z. Sable, *JBC* **249**, 1382–1389 (1974).

27. H. J. Grande, R. L. Houghton, and C. Veeger, *EJB* **37**, 563–569 (1973).

28. A. D. Gounaris, I. Turkenkopf, S. Buckwald, and A. Young, *JBC* **246**, 1302–1309 (1971).

29. L. F. Chuang and E. B. Collins, *J. Bacteriol.* **95**, 2083–2089 (1968).

30. Y. Itokawa and J. R. Cooper, *Science* **166**, 759–761 (1969).

31. Y. Hashitani and J. R. Cooper, *JBC* **247**, 2117–2119 (1972).

32. R. L. Barchi and R. O. Viale, *JBC* **251**, 193–197 (1976).

33. A. E. Braunstein, *in* "The Enzymes," 3rd ed. (P. D. Boyer, ed.), Vol. 9, pp. 379–481. Academic Press, New York, 1973.

34. D. E. Metzler, M. Ikawa, and E. E. Snell, *JACS* **76**, 648–652 (1954).

35. A. E. Braunstein and M. M. Shemyakin, *Biokhimiya* **18**, 393–411 (1953).

36. L. Davis and D. E. Metzler, *in* "The Enzymes," 3rd ed. (P. D. Boyer, ed.), Vol. 7, pp. 33–74. Academic Press, New York, 1972.

37. S. Guggenheim and M. Flavin, *JBC* **244**, 6217–6227 (1969).

38. L. Schirch and T. Gross, *JBC* **243**, 5651–5655 (1968).

39. Z. Zaman, P. M. Jordan, and M. Akhtar, *BJ* **135**, 257–263 (1973).

40. K. Krisnangkura and C. C. Sweeley, *JBC* **251**, 1597–1602 (1976).

41. S. E. Tate and A. Meister, *Adv. Enzymol.* **35**, 503–543 (1971).

42. I. M. Jeng, R. Somack, and H. A. Barker, *Biochemistry* **13**, 2898–2903 (1974).

43. D. J. Graves and J. H. Wang, *in* "The Enzymes," 3rd ed. (P. D. Boyer, ed.), Vol. 7, pp. 435–482. Academic Press, New York, 1972.

44. E. G. Krebs and E. H. Fischer, *Vitam. Horm.* (*N.Y.*) **22**, 399–410 (1964).

45. K. D. Schnackertz and E. A. Noltmann, *Biochemistry* **10**, 4837–4843 (1971).

46. P. A. Rubenstein and J. L. Strominger, *JBC* **249**, 3776–3781 (1974).

47. H. C. Dunathan, *Adv. Enzymol.* **35**, 79–134 (1971).

48. G. B. Bailey, O. Chotamangsa, and K. Vuttivej, *Biochemistry* **9**, 3243–3248 (1970).

49. G. E. Skye, R. Potts, and H. G. Floss, *JACS* **96**, 1593–1595 (1974).

50. D. E. Metzler, C. M. Harris, R. J. Johnson, D. B. Siano, and J. A. Thomson, *Biochemistry* **12**, 5377–5392 (1973).

51. P. L. Scholnick, L. E. Hammaker, and H. S. Marver, *JBC* **247**, 4132–4137 (1972).

52. E. H. Cordes and W. P. Jencks, *Biochemistry* **1**, 773–778 (1962).

53. L. Schirch and W. T. Jenkins, *JBC* **239**, 3801–3807 (1964).

54. S. Matsumoto and Y. Matsushima, *JACS* **94**, 7211–7213 399–410 (1964).

55. V. I. Ivanov and M. Ya. Karpeisky, *Adv. Enzymol.* **32**, 21–53 (1969).

56. W. D. Riley and E. E. Snell, *Biochemistry* **7**, 3520–3528 (1968).

57. R. B. Wickner, C. W. Tabor, and H. Tabor, *JBC* **245**, 2132–2139 (1970).

58. P. A. Recsei and E. E. Snell, *Biochemistry* **12**, 365–371 (1973).

59. M. S. Cohn and A. T. Phillips, *Biochemistry* **13**, 1208–1212

60. G. Kapke and L. Davis, *Biochemistry* **14**, 4273–4276 (1975).

60a. R. S. Greenfield and D. Wellner, *Fed. Proc., Fed. Am. Soc. Exp. Biol.* **35**, 1749 (1976).

61. D. J. George and A. T. Phillips, *JBC* **245**, 528–537 (1970).

62. R. M. Egan and A. T. Phillips, *Fed. Proc., Fed. Am. Soc. Exp. Biol.* **35**, 1749 (abstr.) (1976).

62a. B. Seto and T. C. Stadtman, *JBC* **251**, 2435–2439 (1976).

63. D. S. Hodgins and R. H. Abeles, *ABB* **130**, 274–285 (1969).

64. G. A. Hamilton, *Prog. Bioor. Chem.*, **1**, 83–157 (1971).

65. I. L. Givot, T. A. Smith, and R. H. Abeles, *JBC* **244**,

6341–6353 (1969).

66. K. R. Hanson and E. A. Havir, *in* "The Enzymes," 3rd ed. (P. D. Boyer, ed.), Vol. 7, pp. 75–166. Academic Press, New York, 1972.

66a. N. E. Dixon, C. Gazzola, R. L. Blakeley, and B. Zerner, *Science* **191**, 1144–1150 (1976).

67. For an English translation of part of one of Warburg and Christian's early papers, see H. M. Kalckar, "Biological Phosphorylations," pp. 86–97, Prentice-Hall, Englewood Cliffs, New Jersey, 1969.

68. M. E. Pullman and S. P. Colowick, *JBC* **206**, 129–141 (1954).

69. H. F. Fisher, E. E. Conn, B. Vennesland, and F. H. Westheimer, *JBC* **202**, 687–697 (1953).

70. D. J. Walton, *Biochemistry* **12**, 3472–3478 (1973).

71. J. P. Klinman, *JBC* **247**, 7977–7987 (1972).

72. M. G. Rossman, M. J. Adams, M. Buehner, G. C. Ford, M. L. Hackert, P. J. Lentz, Jr., A. McPherson, Jr., R. W. Schevitz, and I. E. Smiley, *Cold Spring Harbor Symp. Quant. Biol.* **36**, 179–191 (1972).

73. M. J. Adams, M. Buehner, K. Chandrasekhar, G. C. Ford, M. L. Hackert, A. Liljas, M. G. Rossman, I. E. Smiley, W. S. Allison, J. Everse, N. O. Kaplan, and S. S. Taylor, *PNAS* **70**, 1968–1972 (1973).

74. J. Hajdu and D. S. Sigman, *JACS* **97**, 3524–3526 (1975).

75. P. C. Yang and G. W. Schwert, *Biochemistry* **11**, 2218–2224 (1972).

76. D. Tsernoglou, E. Hill, and L. J. Banaszak, *JMB* **69**, 75–87 (1972).

77. C.-I. Brändén, H. Jörnvall, H. Eklund, and B. Furugren, *in* "The Enzymes," 3rd ed. (P. D. Boyer, ed.), Vol. 11, pp. 103–190. Academic Press, New York, 1975.

78. M. G. Rossman, D. Moras, and K. W. Olsen, *Nature (London)* **250**, 194–199 (1974).

78a. M. G. Rossman, A. Liljas, C.-I. Brändén, and L. J. Banaszak, *in* "The Enzymes," 3rd ed. (P. D. Boyer, ed.), Vol. 11, pp. 61–102. Academic Press, New York, 1975.

78b. D. J. Creighton, J. Hajdu, and D. S. Sigman, *JACS* **98**, 4619–4625 (1976).

79. L. A. Banaszk, and R. A. Bradshaw, *in* "The Enzymes," 3rd ed. (P. D. Boyer, ed.), Vol. 11, pp. 369–396. Academic Press, New York, 1975.

80. B. D. Peczon and H. O. Spivey, *Biochemistry* **11**, 2209–2217 (1972).

80a. M. Hadorn, V. A. John, F. K. Meier, and H. Dutler, *EJB* **54**, 65–73 (1975).

81. M. Lazdunski, C. Petitclerc, D. Chappelet, and C. Lazdunski, *EJB* **20**, 124–139 (1971).

81a. K. Harada and R. G. Wolfe, *JBC* **243**, 4131–4137 (1968).

82. M. Lazdunski, *Curr. Top. Cell. Regul.* **6**, 267–310 (1972).

83. K. Moon, D. Piszkiewicz, and E. L. Smith, *PNAS* **69**, 1380–1383 (1972).

84. F. Hucho, U. Markau, and H. Sund, *EJB* **32**, 69–75 (1973).

85. J. I. Harris and M. Waters, *in* "The Enzymes," 3rd ed. (P. D. Boyer, ed.) Vol. 13, pp. 1–49. Academic Press, New York, 1976.

86. I. A. Watkinson, D. C. Wilton, A. D. Rahimtula, and M. M. Akhtar, *EJB* **23**, 1–6 (1971).

87. G. L. Nelsestuen and S. Kirkwood, *JBC* **246**, 7533–7543 (1971).

88. L. Glaser, *in* "The Enzymes," 3rd ed. (P. D. Boyer, ed.), Vol. 6, pp. 355–380. Academic Press, New York, 1972.

89. C. C. Guilbert and S. L. Johnson, *Biochemistry* **10**, 2313–2316 (1971).

90. S. Chaykin, *Annu. Rev. Biochem.* **36**, Part I, 149–170 (1967).

91. J. Everse and N. O. Kaplan, *Adv. Enzymol.* **37**, 61–133 (1973).

92. A. Stock, E. Sann, and G. Pfleiderer, *Justus Liebig's Ann. Chem.* **647**, 188–219 (1961).

93. C. C. Johnston, J. L. Gardner, C. H. Suelter, and D. E. Metzler, *Biochemistry* **2**, 689–696 (1963).

94. S. G. A. Alivisatos, F. Ungar, and G. J. Abraham, *Biochemistry* **4**, 2616–2630 (1965).

95. B. M. Anderson, M. L. Reynolds, and C. D. Anderson, *ABB* **110**, 577–582 (1965).

96. N. J. Oppenheimer, *BBRC* **50**, 683–690 (1973).

97. N. O. Kaplan, M. M. Ciotti, M. Hamolsky, and R. E. Bieber, *Science* **131**, 392–397 (1960).

98. P. Hemmerich, G. Nagelschneider, and C. Veeger, *FEBS Lett.* **8**, 69–83 (1970).

99. See, e.g., H. Kamin, ed., "Flavins and Flavoproteins." Univ. Park Press, Baltimore, Maryland, 1971.

100. S. C. Tu and D. B. McCormick, *JBC* **248**, 6339–6347 (1973).

101. J. Rétey, J. Seibl, D. Arigoni, J. W. Cornforth, G. Ryback, W. P. Zeylemaker, and C. Veeger, *EJB* **14**, 232–242 (1970).

102. L. Thelander, *JBC* **245**, 6026–6029 (1970).

102a. E. T. Jones and C. H. Williams, Jr., *JBC* **250**, 3779–3784 (1975).

103. J. Salach, W. H. Walker, T. P. Singer, A. Ehrenberg, P. Hemmerich, S. Ghisla, and U. Hartmann, *EJB* **26**, 267–278 (1972).

104. W. H. Walker, T. P. Singer, S. Ghisla, P. Hemmerich, U. Hartmann, and E. Zeszotek, *EJB* **26**, 279–289 (1972).

105. W. C. Kenney, W. H. Walker, and T. P. Singer, *JBC* **247**, 4510–4513 (1972).

106. H. Möhler, M. Brühmüller, and K. Decker, *EJB* **29**, 152–155 (1972).

107. D. R. Patek and W. R. Frisell, *ABB* **150**, 347–354 (1972).

108. E. B. Kearney, J. I. Salach, W. H. Walker, R. L. Seng, W. Kenney, E. Zeszotek, and T. P. Singer, *EJB* **24**, 321–327 (1971).

109. W. H. Walker, E. B. Kearney, R. L. Seng, and T. P. Singer, *EJB* **24**, 328–331 (1971).

110. W. H. Walker, W. C. Kenney, D. E. Edmonson, T. P. Singer, J. R. Cronins, and R. Hendriks, *EJB* **48**, 439–448 (1974).

111. P. Strittmatter, *JBC* **246**, 1017–1024 (1971).

112. S. Ghisla and S. G. Mayhew, *EJB* **63**, 373–390 (1976).

113. R. Miura, K. Matsui, K. Hirotsu, A. Shimada, M. Takatsu,

and S. Otani, *J. Chem. Soc., Chem. Commun.* pp. 703–704 (1973).

114. J. L. Fox and G. Tolin, *Biochemistry* **5**, 3865–3872 (1966).
115. M. Brüstlein and T. C. Bruice, *JACS* **94**, 6548–6549 (1972).
116. M. S. Jorns and L. B. Hersch, *JBC* **250**, 3620–3628 (1975).
116a. L. B. Hersh and M. S. Jorns, *JBC* **250**, 8728–8734 (1975).
117. G. H. Hamilton, *Prog. Bioorg. Chem.* **1**, 83–157 (1971).
118. C. T. Walsh, A. Schonbrunn, and R. H. Abeles, *JBC* **246**, 6855–6866 (1971).
119. T. C. Bruice and Y. Yano, *JACS* **97**, 5263–5271 (1975).
119a. J. Fisher, R. Spencer, and C. Walsh, *Biochemistry* **15**, 1054–1064 (1976).
120. F. Müller, P. Hemmerich, and A. Ehrenberg, *in* "Flavins and Flavoproteins" (H. Kamin, ed.), pp. 107–120. Univ. Park Press, Baltimore, Maryland, 1971.
121. B. G. Barman and G. Tollin, *Biochemistry* **11**, 4760–4765 (1972).
122. H. R. Mahler and E. H. Cordes, "Biological Chemistry," 2nd ed., p. 650. Harper, New York, 1971.
123. D. C. Yoch, *BBRC* **49**, 335–342 (1972).
124. K. D. Watenpaugh, L. C. Sieker, L. H. Jensen, J. Legall, and M. Dubourdieu, *PNAS* **69**, 3185–3188 (1972).
125. R. M. Burnett, G. D. Darling, D. S. Kendall, M. E. Le-Quesne, S. G. Mayhew, W. W. Smith, and M. L. Ludwig, *JBC* **249**, 4383–4392 (1974).
125a. U. Hatefi and D. L. Stiggall, *in* "The Enzymes," 3rd ed. (P. D. Boyer, ed.), Vol. 13, pp. 175–297, Academic Press, New York, 1976.
126. C. J. Fritchie, Jr., *JBC* **248**, 7516–7521 (1973).
127. M. W. Yu and C. J. Fritchie, Jr., *JBC* **250**, 946–951 (1975).
128. F. Müller, L. E. G. Eriksson, and A. Ehrenberg, *EJB* **12**, 93–103 (1970).
129. R. C. Bray and J. C. Swann, *Struct. Bonding (Berlin)* **11**, 107–144 (1972).
129a. R. C. Bray, D. J. Lowe, and M. J. Barber, *BJ* **141**, 309–311 (1974).
130. U. Branzoli and V. Massey, *JBC* **249**, 4346–4349 (1974).
130a. T. Ohnishi, J. Linn, D. B. Winter, and T. E. King, *JBC* **251**, 2105–2109 (1976).
131. J. P. Forestier and A. Baudras, *in* "Flavins and Flavoproteins" (H. Kamin, ed.), pp. 599–605 Univ. Park Press, Baltimore, Maryland, 1971.
132. V. Massey, G. Palmer, and D. Ballou, *in* "Flavins and Flavoproteins" (H. Kamin, ed.), pp. 349–361. Univ. Park Press, Baltimore, Maryland, 1971.
133. S. B. Smith, M. Brüstlein, and T. C. Bruice, *JACS* **96**, 3696–3697 (1974).
134. M. Yamasaki and T. Yamano, *BBRC* **51**, 612–619 (1973).
134a. R. Fried, L. W. Fried, and D. R. Babin, *EJB* **33**, 439–445 (1973).
135. F. Müller, S. G. Mayhew, and V. Massey, *Biochemistry* **12**, 4654–4662 (1973).
136. B. Holmström, *Ark. Kemi* **22**, 329–346 (1964).
136a. W. L. Cairns and D. E. Metzler, *JACS* **93**, 2772–2777 (1971).

137. U. Schmidt, P. Grafen, K. Altland, and H. W. Goedde, *Adv. Enzymol.* **32**, 423–469 (1969).
137a. M. Koike and K. Koike, *Metab. Pathways, 3rd Ed.* **7**, 87–99 (1975).
138. L. J. Reed, *Acc. Chem. Res.* **7**, 40–46 (1974).
138a. L. J. Reed, F. H. Pettit, M. H. Eley, L. Hamilton, J. H. Collins, and R. M. Oliver, *PNAS* **72**, 3068–3072 (1975).
138b. C. C. Cunningham and L. P. Hager, *JBC* **250**, 7139–7146 (1975).
139. N. Tanaka, K. Koike, K. Otsuka, M. Hamada, K. Ogasahara, and M. Koike, *JBC* **249**, 191–198 (1974).
139a. P. K. Chiang and B. Sacktor, *JBC* **250**, 3399–3408 (1975).
140. K. Uyeda and J. C. Rabinowitz, *JBC* **246**, 3120–3125 (1971).
140a. U. Gehring and D. I. Arnon, *JBC* **247**, 6963–6969 (1972).
141. H. Bothe and B. Falkenberg, *Z. Naturforsch., Teil B* **27**, 1090–1094 (1972).
142. M. Katagiri and S. Takemori, *in* "Flavins and Flavoproteins" (H. Kamin, ed.), pp. 447–461. Univ. Park Press, Baltimore, Maryland, 1971.
143. T. Yamauchi, S. Yamamoto, and O. Hayaishi, *JBC* **248**, 3750–3752 (1973).
144. H. W. Orf and D. Dolphin, *PNAS* **71**, 2646–2650 (1974).
145. C. Walsh, O. Lockridge, V. Massey, and R. Abeles, *JBC* **248**, 7049–7054 (1973).
146. J. Knappe, J. Schacht, W. Möckel, T. Höpner, H. Vetter, Jr., and R. Edenharder, *EJB* **11**, 316–327 (1969).
147. J. Knappe, H. P. Blaschkowski, P. Gröbner, and T. Schmitt, *EJB* **50**, 253–263 (1974).
148. R. J. Thauer, F. H. Kirchniawy, and K. A. Jungermann, *EJB* **27**, 282–290 (1972).
149. N. P. Wood and K. Jungermann, *FEBS Lett.* **27**, 49–52 (1972).
150. R. L. Blakley, "The Biochemistry of Folic Acid and Related Pteridines." North-Holland Publ., Amsterdam, 1969.
151. E. Hadorn, *Sci. Am.* **206**, 101–110 (1962).
152. T. Fukushima and T. Shiota, *JBC* **247**, 4549–4556 (1972).
153. E. C. Taylor and P. A. Jacobi, *JACS* **96**, 6781–6782 (1974).
154. H. Rembold and W. L. Gyure, *Angew. Chem., Int. Ed. Engl.* **11**, 1061–1072 (1972).
155. R. G. Kallen and W. P. Jencks, *JBC* **241**, 5845–5850 and 5851–5863 (1966).
156. N. P. Curthoys, J. M. Scott, and J. C. Rabinowitz, *JBC* **247**, 1959–1964 (1972).
157. C. M. Houlihan and J. M. Scott, *BBRC* **48**, 1675–1681 (1972).
158. J. A. K. Harmony, P. J. Shaffer, and R. H. Himes, *JBC* **249**, 394–401 (1974).
158a. D. H. Buttlaire, R. H. Himes, and G. H. Reed, *JBC* **251**, 4159–4161 (1975).
159. H. A. Barker, *in* "The Enzymes," 3rd ed. (P. D. Boyer, ed.), Vol. 6, pp. 509–537. Academic Press, New York, 1972.
160. P. G. Lenhert and D. C. Hodgkin, *Nature (London)* **192**, 937–938 (1961).
161. E. L. Smith, "Vitamin B$_{12}$," 3rd ed. Methuen, London, 1965.

162. J. M. Pratt, "Inorganic Chemistry of Vitamin B_{12}." Academic Press, New York, 1972.

163. T. C. Stadtman, *Science* **171**, 859–867 (1971).

164. J. M. Wood and D. G. Brown, *Struc. Bonding (Berlin)* **11**, 47–105 (1972).

164a. R. H. Abeles and D. Dolphin, *Accts. Chem. Res.* **9**, 114–120 (1976).

165. B. M. Babior, ed. "Cobalamin." Wiley, New York, 1975.

166. D. G. Brown, *Prog. Inorg. Chem.* **18**, 177–286 (1974).

167. F. Wagner, *Annu. Rev. Biochem.* **35**, Part 1, 405–428 (1966).

168. G. N. Schrauzer, E. Deutsch, and R. J. Windgassen, *JACS* **90**, 2441–2442 (1968).

169. T. E. Needham, N. A. Matwiyoff, T. E. Walker, and H. P. C. Hogenkamp, *JACS* **95**, 5019–5024 (1973).

170. S. H. Mudd, *in* "The Enzymes," 3rd ed. (P. D. Boyer, ed.), Vol. 8, pp. 121–154. Academic Press, New York, 1973.

171. J. Rétey, A. Umani-Ronchi, J. Seibl, and D. Arigoni, *Experientia* **22**, 502–503 (1966).

172. T. J. Carty, B. M. Babior, and R. H. Abeles, *JBC* **249**, 1683–1688 (1974).

173. B. Baboir, T. J. Carty, and R. H. Abeles, *JBC* **249**, 1689–1695 (1974).

173a. K. Sato, J. C. Orr, B. M. Babior, and R. H. Abeles, *JBC* **251**, 3734–3737 (1976).

174. H. P. C. Hogenkamp, R. K. Ghambeer, C. Brownson, R. L. Blakley, and E. Vitols, *JBC* **243**, 799–808 (1968).

175. W. H. Orme-Johnson, H. Beinert, and R. L. Blakley, *JBC* **249**, 2338–2343 (1974).

175a. G. N. Sando, R. L. Blakely, H. P. C. Hogenkamp, and P. J. Hoffman, *JBC* **250**, 8774–8779 (1975).

176. S. A. Cockle, H. A. O. Hill, R. J. P. Williams, S. P. Davies, and M. A. Foster, *JACS* **94**, 275–277 (1972).

177. T. H. Finlay, J. Valinsky, K. Sato, and R. H. Abeles, *JBC* **247**, 4197–4207 (1972).

178. B. M. Babior, T. H. Moss, and D. C. Gould, *JBC* **247**, 4389–4392 (1972).

178a. R. B. Silverman and D. Dolphin, *JACS* **98**, 4626–4633 (1976).

179. L L. Ingraham, *Ann. N. Y. Acad. Sci.* **112**, 713–720 (1964).

180. J. N. Lowe and L. L. Ingraham, *JACS* **93**, 3801–3802 (1971).

181. P. Dowd and M. Shapiro, *JACS* **98**, 3724–3725 (1976).

182. G. N. Schrauzer and R. J. Windgassen, *JACS* **88**, 3738–3743 (1966).

183. B. Zagalak, P. A. Frey, G. L. Karabatsos, and R. H. Abeles, *JBC* **241**, 3028–3035 (1966).

184. M. Sprecher, M. J. Clark, and D. B. Sprinson, *JBC* **241**, 872–877 (1966).

185. C. G. D. Morley and T. C. Stadtman, *Biochemistry* **10**, 2325–2329 (1971); **11**, 600–605 (1972).

186. G. T. Burke, J. H. Mangum, and J. D. Brodie, *Biochemistry* **10**, 3079–3085 (1971).

187. B. C. McBride and R. S. Wolfe, *Biochemistry* **10**, 2317–2324 (1971).

188. C. D. Taylor and R. S. Wolfe, *JBC* **249**, 4879–4885 (1974).

189. J. M. Wood, *Science* **183**, 1049–1052 (1974).

190. B. C. McBride and R. S. Wolfe, *Biochemistry* **10**, 4312–4317 (1971).

191. D. J. Parker, H. G. Wood, R. K. Ghambeer, and L. G. Ljungdahl, *Biochemistry* **11**, 3074–3080 (1972).

192. M. Schulman, R. K. Ghambeer, L. G. Ljungdahl, and H. G. Wood, *Fed. Proc., Fed. Am. Soc. Exp. Biol.* **32**, 627 (abstr.) (1973).

STUDY QUESTIONS

1. The following diseases result from deficiencies of particular vitamins. What vitamin do you associate with each disease? (a) Scurvy, (b) rickets, (c) beriberi, (d) pellagra.

2. Complete the following table.

Reaction type or enzyme	Coenzyme or prosthetic group
Decarboxylation of amino acid	
Decarboxylation of α-keto acid	
Oxidative decarboxylation of α-keto acid	
Transketolase	
Pyruvate carboxylase	
β-ketothiolase	
Transamination	
Formyl group transfer	
Formation of methane	

3. What coenzymes have ADP as part of their structure?

4. Discuss the cleavage of carbon–carbon bonds in metabolic systems, with examples of specific enzymatic reactions. Summarize what appear to you to be general principles for this type of reaction.

5. What is a typical substrate for pyridoxal phosphate-requiring enzymes? Draw the structure of a coenzyme–substrate intermediate. Explain the mechanisms of transamination, decarboxylation, and deamination (of β-hydroxyamino acids) as catalyzed by enzymes of this group.

6. What is thiamine diphosphate (thiamine pyrophosphate)? How does it react with α-keto acid substrates? Show in detail the mechanisms of decarboxylation of α-keto acids (a) to aldehydes and (b) to acyl coenzyme A derivatives. (c) Outline the mechanism action of the enzyme transketolase.

7. Give the structure of lipoic acid and explain the binding of this compound to proteins. Show exactly how this compound is involved in the oxidative decarboxylation of α-keto acids.

8. What is the structure of biotin and how is it bound to enzymes? Write equations for the two steps involved in the action of biotin enzymes.

9. Illustrate the conversion of folic acid to tetrahydrofolic acid. Write equations for the reaction of the latter with (a) formaldehyde, (b) formic acid, (c) serine. Show how the single-carbon fragment is transferred into (d) purines, (e) methionine, (f) the methyl group of thymine.

10. Illustrate typical oxidative reactions of NAD^+ and $NADP^+$. What is the principal source of NADH in cells? Compare the relative ratios of [NADH]/[NAD^+] and [NADPH]/[$NADP^+$] in tissues. What is the usual metabolic fate of NADH?

11. List some typical substrates of flavoproteins. Illustrate the reversible reduction of the prosthetic group. Give the chemical reactions involving flavoproteins in (a) fatty acid oxidation, (b) citric acid cycle, (c) the electron transport chain.

12. Write the equation for the reaction in propionate metabolism which depends on a vitamin B_{12} coenzyme.

13. Thymidine 5'-phosphate (thymidylic acid, dTMP) is formed by reactions of deoxyuridine 5'-phosphate (dUMP) involving transfer of a 1-carbon unit from methylenetetrahydrofolic acid.

   a. Draw the structures for deoxyuridine 5'-phosphate and for methylenetetrahydrofolic acid. Propose a step-by-step mechanism for the reaction of these two compounds to form dTMP.

   b. T-even bacteriophage contain 5-hydroxymethylcytidylic acid. Propose a mechanism for synthesis of this nucleotide. NOTE: These reactions are discussed in Chapter 14.

Organization of Metabolism: Catabolic Pathways

Metabolism involves a bewildering array of chemical reactions, many of them organized as complex cycles which sometimes appear difficult to understand. Yet there is a logic and orderliness. With few exceptions, metabolic pathways can be regarded as sequences of the reactions considered in Chapters 7, 8, and 10 (and summarized in Table 9-1) organized to accomplish specific chemical goals.

In this chapter we will examine some of the major pathways of **catabolism** of foods and of cell constituents. Reactions of **anabolism** or biosynthesis will be dealt with in later chapters.

A. THE OXIDATION OF FATTY ACIDS

Hydrocarbons yield more energy upon combustion than do most other organic compounds, and it is not surprising that one important type of food reserve, the fats, is essentially hydrocarbon in nature. From an energy viewpoint, the component fatty acids are the most important part. Most aerobic cells can oxidize fatty acids completely to CO_2 and water, a process that takes place within the matrix space of the mitochondria of eukaryotic cells.

The oxidized end of a fatty acid provides a point for chemical attack, the first step of which is a "priming reaction" in which the fatty acid is converted via sequence S1A (α) to a water-soluble acyl-CoA derivative in which the α-hydrogens of the fatty acyl radicals are "activated" (Eq. 9-1).

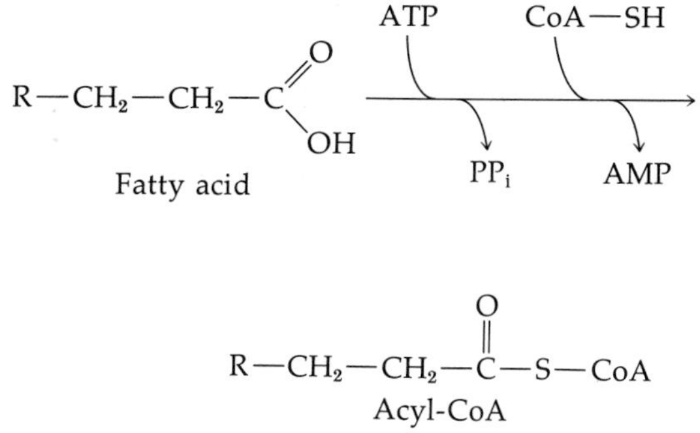

$$R-CH_2-CH_2-C\underset{OH}{\overset{O}{\lessgtr}} \qquad \xrightarrow[\text{PP}_i \quad \text{AMP}]{\text{ATP} \quad \text{CoA}-\text{SH}}$$

Fatty acid

$$R-CH_2-CH_2-\overset{O}{\overset{\|}{C}}-S-CoA \qquad (9\text{-}1)$$

Acyl-CoA

The reaction is catalyzed by **acetate thiokinase** (Eq. 7-30) and other acyl-CoA synthetases (fatty acid activating enzymes). There are at least two types, one specific for medium length carbon chains (4 C to 12 C) and the other for longer chains. Mitochondria also contain **acyl-CoA synthetases**[1] that

TABLE 9-1
Short Summary of Metabolic Reaction Types

Reaction	Table pages	Figures	Text page
1. Nucleophilic displacement	358		357
A. On —CH$_2$Y			362
B. On —CO—Y	432		370
C. On phosphorus			378
D. On sulfur			395
S1 Consecutive displacement on P and C often with cleavage of ATP. Used in synthesis of esters, amides, thio-esters, and in substrate level phosphorylation	393		390
2. Addition			
A. To C=O or C=N	358		395
B. To C=C			398
3. Elimination	358		
A. To form C=O or C=N			395
B. To form C=C			398
C. Decarboxylative			404
4. Formation of enolate anions and eneamines and their participation in isomerization reactions	359	7-9	404
5. Enolate anions as nucleophiles	359		409
A. Displacement on C=O			410
S5A Biotin-dependent carboxylation	435		433
B. Addition to C=O (aldol condensation)		7-10, 7-11	410
C. Addition to CO$_2$ (β carboxylation)			416
6. Some rearrangement reactions	360		420
7. Thiamine-dependent α cleavage		8-3	438
S7A Phosphoketolase in ATP synthesis		8-4	442
S7B Oxidative decarboxylation of α-keto acids		8-19	487
S7C Pyruvate-formate-lyase		8-19	491

Reaction	Table pages	Figures	Text page
8. Reactions of Schiff bases of pyridoxal phosphate		8-6, 8-7	444
9. Hydrogen and electron transfer reactions	465, 479		464
A. NAD$^+$- and NADP$^+$-dependent	465	8-10	466
B. Flavin-dependent	479	8-14	476
C. Lipoic acid-dependent		8-18	487
D. Reactions of iron–sulfur proteins		10-4, 10-5, 10-6	573
E. Reactions of quinones		10-8	576
F. Reactions of cytochromes		10-3	568
G. Selenium-dependent dehydrogenation			536
S9C Generation of ATP coupled to oxidation of an aldehyde		8-13	471
10. B$_{12}$-coenzyme-dependent reactions	506		499
Isomerization reactions			504
Ribonucleotide reductase			505
Methyl transfer reactions			509
11. Hydroxylation			
Dioxygenases			615
Monooxygenases (hydroxylases)			616
Flavin-containing			617
Pteridine-dependent			618
Ketoglutarate-dependent			619
Ascorbate-dependent			620
Cytochrome P-450-dependent			620
12. Reactions of peroxides			
Oxidative decarboxylation			491
Glutathione peroxidase			564
13. Folic acid-dependent reactions	496	8-20, 8-21	493

cleave GTP to GDP and P_i; i.e., they use sequence S1A (γ).

1. Beta Oxidation

Inspection of the acyl-CoA molecule and consideration of the biochemical reaction types available shows that the only reasonable mode of further attack is oxidation by a flavoprotein to remove hydrogens from the α and β positions to give an unsaturated acyl-CoA derivative (Fig. 9-1, Eq. *a*). One of the few possible reactions of the unsaturated compound so formed is nucleophilic addition at the β position. Water is added (Eq. *b*) and the resulting alcohol is oxidized to a ketone by NAD^+ (Eq. *c*). This series of three reactions is the well-known β-oxidation sequence. Figure 9-1 also shows another β-oxidation sequence that converts succinic acid to oxaloacetic acid within the citric acid cycle.

Just as there are several different fatty acid activating enzymes with specificities for different chain lengths, there are several **acyl-CoA dehydrogenases** that catalyze Eq. *a* of Fig. 9-1. All contain FAD. In each case, the reduced $FADH_2$ on the enzyme is reoxidized by a special **electron transferring flavoprotein**[2,3] that also contains FAD. This protein is thought to carry the electrons abstracted in the oxidation process to the inner membrane of the mitochondrion where they enter the mitochondrial electron transport system.

At the end of the β-oxidation sequence, the β-ketoacyl-CoA derivative is cleaved (Eq. 9-2, step *b*) by a thiolase (see also Eq. 7-62). Again, there is a family of enzymes with differing chain length specificities. One of the products (Eq. 9-2) is acetyl-CoA which can be catabolized to CO_2 through the citric acid cycle. The second product of the thiolytic cleavage is an acyl-CoA derivative two carbon atoms shorter than the original. It is recycled through the β-oxidation process, a two-carbon acetyl unit being removed as acetyl-CoA during each turn of the cycle (Eq. 9-2). The process continues until the fatty acid chain is completely

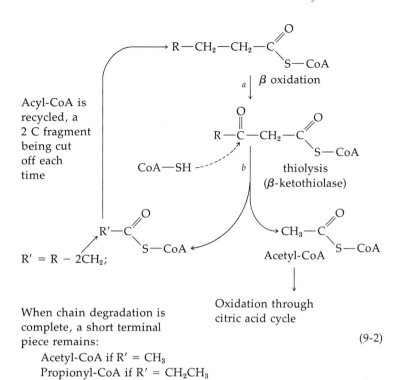

Acyl-CoA is recycled, a 2 C fragment being cut off each time

When chain degradation is complete, a short terminal piece remains:
   Acetyl-CoA if $R' = CH_3$
   Propionyl-CoA if $R' = CH_2CH_3$

(9-2)

degraded. If the original fatty acid contained an even number of carbon atoms in a straight chain, acetyl-CoA *is the only product*. If the fatty acid contained an odd number of carbon atoms, **propionyl-CoA** is formed at the end.

2. Branched Chain Fatty Acids

Most of the fatty acids in our bodies as well as in plant fats have straight unbranched chains. However, branches, usually consisting of methyl groups, are present in lipids of certain microorganisms and in waxes of plant surfaces. As long as there are not too many branches and if they occur only in the even numbered positions, i.e., on carbons 2, 4, etc., β oxidation proceeds normally. Propionyl-CoA is formed in addition to acetyl-CoA as a product of the chain degradation. On the other hand, if methyl groups occur in positions 3,5, etc., β oxidation is blocked at Eq. *c* of Fig. 9-1.

AN EARLY LABELING EXPERIMENT

Long before the advent of radioactive tracers, Knoop in 1904 synthesized fatty acids labeled by chemical attachment of a benzene ring at the end opposite the carboxyl group. He prepared these labeled fatty acids with both odd and even numbers of carbon atoms in the straight chain and fed them to dogs. Then from the dogs' urine, he isolated two compounds: **hippuric acid** and **phenylaceturic acid,** amides of glycine with benzoic acid and phenylacetic acid, respectively. Knoop showed that phenylacetic acid was pro-

duced from those fatty acids with an even number of carbon atoms, while benzoic acid was produced from those with an odd number. From these experimental results, Knoop deduced that fatty acid degradation occurs two carbon atoms at a time and proposed his famous β oxidation theory.

Later experiments with isotopic labeling verified Knoop's proposals, but study of isolated enzymes was not possible until after the discovery of CoA in 1950. Then studies of fatty acid oxidation by extracts from isolated mitochondria quickly established details of the pathway.

A striking example of the effect of such blockage was provided by the synthetic detergents in common use until about 1966. These detergents contained a hydrocarbon chain with methyl groups distributed more or less at random along the chain. β Oxidation was blocked at many points and the result was a foamy pollution crisis in sewage plants in the United States and in some other countries. Since 1966, only biodegradable detergents having straight hydrocarbon chains have been sold.

3. Oxidation of Saturated Hydrocarbons

While the oxidized end of a fatty acid chain provides the point of attack for β oxidation, oxidation of a saturated hydrocarbon chain cannot be initiated so easily. Even so, tissues of our bodies can

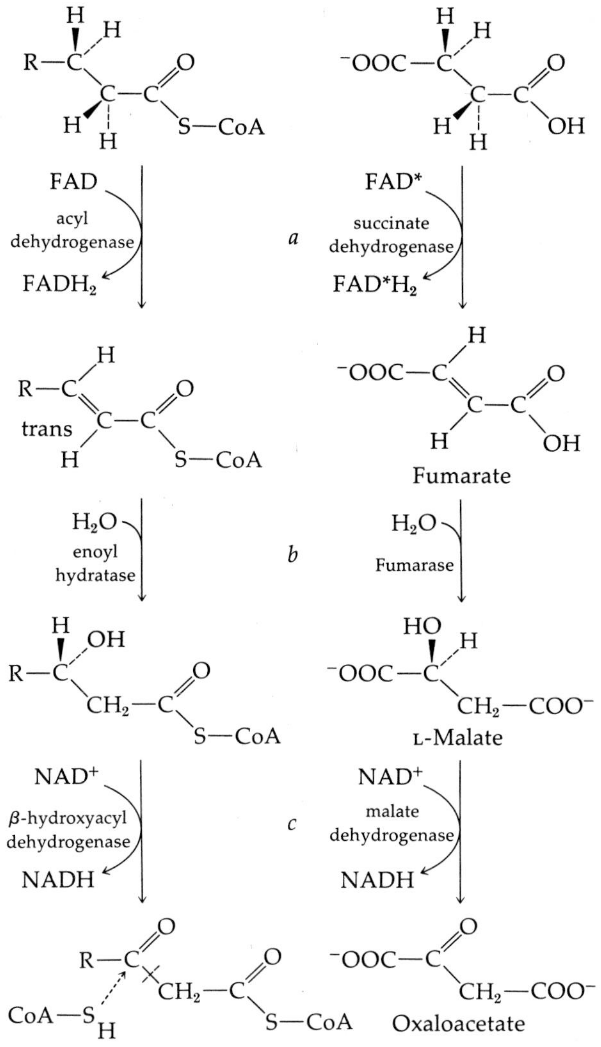

Fig. 9-1 Reactions of the β-oxidation sequence. Left: oxidation of CoA derivatives of fatty acids; right: oxidation of succinate. Both reaction sequences occur within mitochondria of eukaryotic cells and are catalyzed by specific enzymes, probably (with one exception) dissolved within the matrix space. FAD* refers to a special FAD derivative found in succinic dehydrogenase (Chapter 8, Section I,3). This enzyme is tightly bound to the inner mitochondrial membrane.

metabolize hydrocarbons such as n-heptane very slowly. Some microorganisms can oxidize straight chain hydrocarbons rapidly, and strains of *Pseudomonas* and of the yeast *Candida* have been studied

as possible agents for conversion of petroleum into edible proteins.[4]

The first step in oxidation of n-octane by *Pseudomonas* is a **hydroxylation** using molecular oxygen (Chapter 10, Section G,2f) to form n-octanol (Eq. 9-3). Further oxidation of octanol to an acyl-CoA derivative, presumably via the aldehyde (Eq. 9-3), is a standard biochemical oxidation sequence.

$$n\text{-Octane} \xrightarrow{\text{O}_2 \text{ hydroxylation}} C_7H_{15}CH_2\text{—OH}$$

$$\text{Octanol}$$

$$\downarrow$$

$$C_7H_{15}CHO$$

$$\downarrow$$

$$C_7H_{15}\overset{\overset{\displaystyle O}{\|}}{C}\text{—S—CoA} \qquad (9\text{-}3)$$

4. Alpha Oxidation and Omega Oxidation

In animal tissues such straight chain fatty acids as palmitic acid, stearic acid, and oleic acid are degraded almost entirely via β oxidation.[5] However, plant cells often oxidize fatty acids one carbon at a time. While detailed pathways have not been established, the initial attack is usually hydroxylation on the α-carbon atom (Eq. 9-4) to either the D-

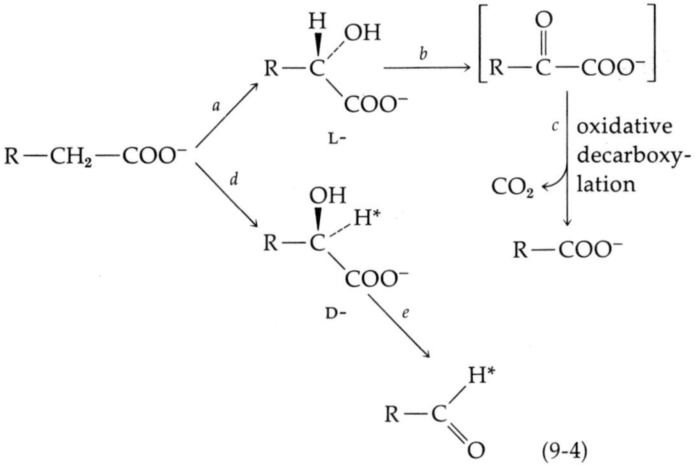

or the L-2-hydroxy acid.[6–8] The L-hydroxy acids are oxidized rapidly, perhaps by dehydrogenation to the α-keto acid (Eq. 9-4, step b) and oxidative decarboxylation (possibly by H_2O_2, Eq. 8-67). The D-hydroxy acids tend to accumulate and are normally present in green leaves. However, they too are oxidized further, with retention of the α-hydrogen as indicated by the asterisks in Eq. 9-4, step e. This suggests a new type of dehydrogenation with concurrent decarboxylation. (Compare with Eq. 7-75.)

α Oxidation also occurs to some extent in animal tissues, especially brain.[9] When β oxidation is blocked by the presence of a methyl side chain, the body sometimes uses α oxidation to get past the block (see **Refsum's disease**, Box 9-B). On other occasions, hydroxylation occurs at the opposite end of the chain (**ω oxidation**) to yield a dicarboxylic acid. Thus, 3,6-dimethyloctanoic acid is degraded in the human body largely via ω oxidation.

5. Oxidation of Unsaturated Fatty Acids

β Oxidation of such unsaturated acids as the Δ^9-oleic acid leads to a Δ^3-*cis*-enoyl-CoA intermediate. A special Δ^3-*cis*-Δ^2-*trans*-enoyl-CoA isomerase must act on this intermediate (Eq. 9-5) before β oxidation can proceed[1]:

$$R \overset{3}{\diagup} \overset{CH_2}{\diagdown} \overset{O}{\underset{\|}{C}} - S - CoA \xrightarrow{\text{isomerase}}$$

$$R \diagdown \diagup \overset{2}{\diagdown} \overset{O}{\underset{\|}{C}} - S - CoA \qquad (9\text{-}5)$$

Note the similarity to Eqs. 7-57 and 7-58.

6. Carnitine and Mitochondrial Permeability

One of the principal factors controlling the rate of oxidation of fatty acids is the rate of entrance into the mitochondria.[10,11] While some long chain fatty acids (perhaps 30% of the total) enter mitochondria as such and are converted to CoA derivatives inside the matrix the majority are "activated" to acyl CoA derivatives outside the mitochondria. Penetration of these acyl-CoA derivatives through the mitochondrial inner membrane is greatly facilitated by the compound **carnitine**, γ-trimethylamino-β-hydroxybutyrate:

$$\begin{array}{c} H_3C \\ \overset{+}{N} COO^- \\ H_3C | H OH \\ CH_3 \end{array}$$

Carnitine

Carnitine is present in nearly all organisms and in all animal tissues. The highest concentration is found in muscle where carnitine accounts for almost 0.1% of the dry matter. In fact it was first isolated from meat extracts in 1905. No clue to its biological action was obtained until 1948 when G. Fraenkel and associates described a new dietary factor required by the mealworm, *Tenebrio molitor*. At first designated simply **vitamin B_t**, the active compound was identified in 1952 as carnitine. Some bacteria also require the compound preformed, but most organisms are able to synthesize their own. The inner membrane of mitochondria contains a long-chain acyltransferase that catalyzes transfer of the fatty acyl group from CoA to the hydroxyl group of carnitine (Eq. 9-6). Perhaps,

$$R - \overset{O}{\underset{\|}{C}} - S - CoA$$

Acyl-CoA

$$\xrightarrow[\text{CoA—SH}]{\text{carnitine}} \qquad (9\text{-}6)$$

$$R - \overset{O}{\underset{\|}{C}} - O \diagdown \overset{H}{\diagup} \diagdown \overset{N^+(CH_3)_3}{\diagdown} \overset{O^-}{\underset{C}{\diagdown}} \overset{}{\underset{\|}{O}}$$

Acyl carnitine

BOX 9-B

GENETIC DISEASES: REFSUM'S DISEASE

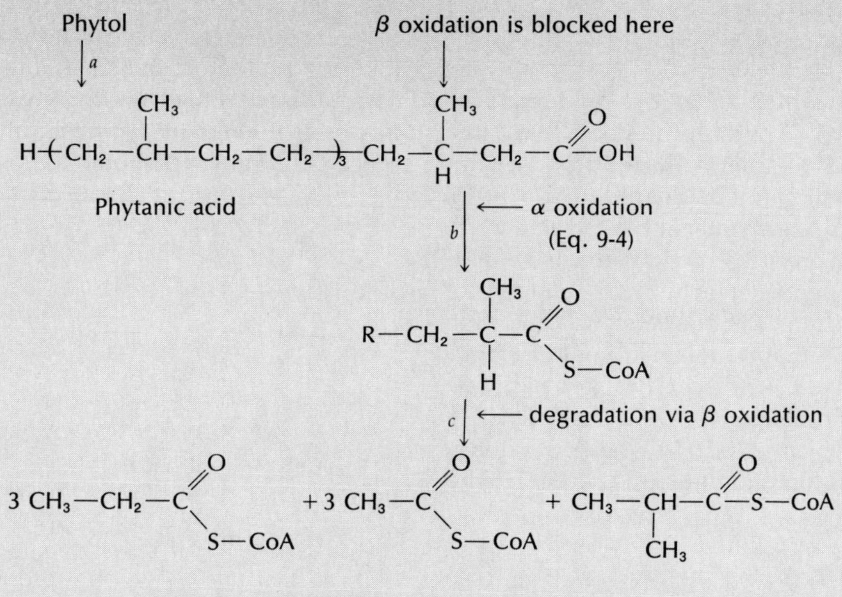

Phytol

$\downarrow a$

β oxidation is blocked here

$$\overset{\displaystyle CH_3}{\underset{\displaystyle H}{H\text{---}(CH_2\text{---}CH\text{---}CH_2\text{---}CH_2)_3\text{---}CH_2\text{---}C\text{---}CH_2\text{---}C}}\overset{\displaystyle O}{\underset{}{}}\text{---}OH$$

Phytanic acid

b $\leftarrow \alpha$ oxidation (Eq. 9-4)

$$R\text{---}CH_2\overset{\displaystyle CH_3}{\underset{\displaystyle H}{\text{---}C}}\text{---}C\overset{\displaystyle O}{\underset{\displaystyle S\text{---}CoA}{}}$$

c $\leftarrow$ degradation via β oxidation

$$3\ CH_3\text{---}CH_2\text{---}C\overset{\displaystyle O}{\underset{\displaystyle S\text{---}CoA}{}} + 3\ CH_3\text{---}C\overset{\displaystyle O}{\underset{\displaystyle S\text{---}CoA}{}} + CH_3\text{---}\underset{\displaystyle CH_3}{CH}\text{---}C\overset{\displaystyle O}{\underset{}{}}\text{---}S\text{---}CoA$$

In this inherited (autosomal recessive) disorder of lipid metabolism the 20-carbon branched chain fatty acid **phytanic acid** accumulates in tissues. Phytanic acid is normally formed in the body from the plant alcohol **phytol** present as an ester in chlorophyll (Fig. 13-19). Because β oxidation is blocked, the first step in the degradation of phytanic acid is α oxidation, following which the complete molecule undergoes β oxidation to three molecules of propionyl-CoA, three of acetyl-CoA, and one of isobutyryl-CoA.

Patients with Refsum's disease are unable to oxidize phytanic acid because of a defect in the α-oxidation system.[a] Whether the serious neurological symptoms of Refsum's disease result from the accumulation of phytanic acid in the brain or because the brain is unable to produce some essential odd chain fatty acids is not known with certainty.

A question for the reader is "What would be a likely effect of incorporation of phytanic acid into cell membranes of the brain?"

[a] D. Steinberg, J. H. Herndon, Jr., B. W. Uhlendorf, C. E. Mize, J. Avigan, and G. W. A. Milne, *Science* **156**, 1740–1742 (1967).

as proposed by Severin, acyl carnitine derivatives pass through the membrane more easily than acyl-CoA derivatives because the positive and negative charges can swing together and neutralize each other as indicated in Eq. 9-6. Once inside the mitochondrion the acyl group is transferred back from carnitine onto CoA prior to initiation of the β oxidation sequence.

Recently, a case of an inborn carnitine deficiency in a human has been reported. The patient's muscles were weak and lipid-filled vacuoles accumulated in the cytoplasm.[12]

7. Ketone Bodies

When a fatty acid with an even number of carbon atoms is broken down through β oxidation the last intermediate before complete conversion to acetyl-CoA is the four-carbon **acetoacetyl-CoA:**

$$CH_3-\overset{\overset{O}{\|}}{C}-CH_2-\overset{\overset{O}{\|}}{C}-S-CoA$$

Acetoacetyl-CoA

Acetoacetyl-CoA, which appears to be in equilibrium with acetyl-CoA within the body, is an important metabolic intermediate for several reasons.[12a] Not only can it be cleaved to two molecules of acetyl-CoA and enter the citric acid cycle but also it is a precursor for synthesis of isoprenoid compounds including cholesterol (Chapter 12, Section I). Equally significant, free **acetoacetate** is an important constituent of blood.

Acetoacetate, is a β-keto acid that can readily be decarboxylated to acetone (Eq. 7-74), and can also be reduced by an NADH-dependent dehydrogenase to D-3-hydroxybutyrate. Note that the

$$CH_3-\overset{\overset{H}{|}}{\underset{CH_2-COO^-}{C}}\diagdown^{OH}$$

D-3-Hydroxybutyrate

configuration of this compound is opposite to that of L-3-hydroxybutyryl-CoA which is formed during β oxidation of fatty acids (Fig. 9-1). D-3-Hydroxybutyrate is stored as a polymer in bacteria (Chapter 2, Section E,4).

The three compounds, *acetoacetate, acetone,* and *3-hydroxybutyrate* are known as **ketone bodies.** Their concentrations increase abnormally in various pathological conditions, the commonest of which is **diabetes mellitus** (Box 11-C), and also during starvation. The resulting condition of **ketosis** is dangerous if severe because formation of ketone bodies produces hydrogen ions (see Eq. 9-7) and acidifies the blood.

Normal rat blood contains about 0.07 mM acetoacetate, 0.18 mM hydroxybutyrate,* and a variable amount of acetone.[13] Of these, acetoacetate

* These amounts increase to 0.5 and 1.6 mM after 48 h of starvation. On the other hand, the blood glucose concentration (6 mM) falls to 4 mM after 48 h starvation.[13]

and hydroxybutyrate are important alternate energy supplies for muscle and other tissues when carbohydrates are in short supply. Acetoacetate can be thought of as a transport form of acetyl units which can be reconverted to acetyl-CoA and oxidized in the citric acid cycle.

While some free acetoacetate is formed by hydrolysis of acetoacetyl-CoA, most arises in the liver indirectly in a two-step process (Eq. 9-7) that is closely linked to synthesis of cholesterol and other polyprenyl compounds.

2 Acetyl-CoA

$\searrow$ CoASH

$$CH_3-\overset{\overset{O}{\|}}{C}-CH_2-\overset{\overset{O}{\|}}{C}-SCoA$$
Acetoacetyl-CoA

$$CoA-S-\overset{\overset{|}{\underset{\|}{C}}}{C}-CH_3$$
$\overset{O}{}$
Acetyl-CoA $\quad a \quad$ H$_2$O

$\searrow$ CoASH

$$CH_3-\overset{\overset{CH_2-COO^-}{|}}{\underset{OH}{C}}-CH_2-\overset{O}{C}\diagup SCoA$$

β-Hydroxy-β-methylglutaryl CoA (HMG-CoA)

Polyprenyl synthesis

b $\searrow CH_3-\overset{\overset{O}{\|}}{C}-S-CoA$

$$CH_3-\overset{\overset{O}{\|}}{C}-CH_2-COO^-+H^+ \qquad (9\text{-}7)$$
Acetoacetate

Equation 9-7, step a is an aldol condensation followed by hydrolysis of one thioester linkage (Chapter 7, Section J,2,f) while Eq. 9-7, step b is a simple aldol cleavage. The overall reaction has the stoichiometry of a direct hydrolysis of acetoacetyl-CoA to acetoacetate. Perhaps the more complex mechanism is required for proper control, but the reason is not obvious.

Utilization of ketone bodies for energy requires reconversion to acetyl-CoA (Eq. 9-8). Krebs *et al.*[13] suggested that all of the reactions of Eq. 9-8 may be nearly at equilibrium in tissues that use ketone bodies for energy.

$$\text{3-Hydroxy butyrate} \xrightarrow[\text{NAD}^+ \quad \text{NADH}]{} \text{Acetoacetate}$$

(see Eq. 7-33)

Succinyl-CoA

Succinate

Acetoacetyl-CoA

thiolase ⟍ CoA—SH

2 Acetyl-CoA

Oxidation via citric acid cycle   (9-8)

B. THE CITRIC ACID CYCLE

To complete the oxidation of fatty acids the acetyl units of acetyl-CoA generated in the β oxidation sequence must be oxidized to carbon dioxide and water.[14] The **citric acid cycle** by which this oxidation is accomplished is a vital part of the metabolism of almost all aerobic creatures. It occupies a central position in metabolism because of the fact that acetyl-CoA is also generated by the catabolism of carbohydrates and of certain amino acids.

1. A Clever Way to Cleave a Reluctant Bond

Oxidation of the chemically resistant two-carbon acetyl group presents a chemical problem.

As we have seen, cleavage of a C—C bond most frequently occurs between atoms that are α and β to a carbonyl group. Such β cleavage (Chapter 7, Section J) is clearly impossible within the acetyl group. The only other common type of cleavage is that of a C—C bond adjacent to a carbonyl group (α cleavage), a thiamine-dependent process (Chapter 8, Section D). However, α cleavage would require the prior oxidation (hydroxylation) of the methyl group of acetate. While many such biological hydroxylation reactions are known (Chapter 10, Section G) they appear to be used rarely in the major pathways of catabolism.*

The solution to the chemical problem of oxidizing acetyl groups efficiently is one very commonly found in nature; a catalytic cycle. Although direct cleavage is impossible, the two-carbon acetyl group of acetyl-CoA *can* undergo an aldol condensation with a second carbonyl compound. The condensation product has more than two carbon atoms and a β cleavage to yield CO_2 is possible. Since the cycle is designed to oxidize acetyl units we will regard acetyl-CoA as the **primary substrate** for the cycle. The second carbonyl compound with which it condenses can be called the **regenerating substrate.** To complete the catalytic cycle, it is necessary that two carbons be removed from the compound formed by condensation of the two substrates and that the remaining molecule be reconvertible to the original regenerating substrate. The reader may wish to play a game by devising suitable sequences of reactions for an acetyl-oxidizing cycle and finding the simplest possible regenerating substrate. Ask yourself whether nature could have used anything simpler than **oxaloacetate,** the molecule actually employed in the citric acid cycle.

The reactions of the citric acid cycle are shown in Fig. 9-2. The first step (*a*) is the condensation of

---

* It is tempting to speculate that hydroxylation is a metabolically difficult reaction—too slow to serve in a major catabolic pathway. On the other hand, there does not appear to be any obvious reason why fast and efficient hydroxylases could not have evolved. Perhaps hydroxylation is avoided because the overall yield of energy obtainable is less than that from dehydrogenation and use of an electron transport chain.

BOX 9-C

DISCOVERY OF THE CITRIC ACID CYCLE

One of the first persons to study the oxidation of organic compounds by animal tissues was T. Thunberg, who between 1911 and 1920 discovered about 40 organic compounds that could be oxidized by animal tissues. Succinate, fumarate, malate, and citrate were oxidized the fastest. Well aware of Knoop's β oxidation theory, Thunberg proposed a cyclic mechanism for oxidation of acetate. Two molecules of this two-carbon compound were supposed to condense (with reduction) to succinate, which was then oxidized as in the citric acid cycle to oxaloacetate. The latter was decarboxylated to pyruvate, which was oxidatively decarboxylated to acetate to complete the cycle. One of the reactions essential for this cycle could not be verified experimentally (it is left to the reader to recognize which one).

In 1935, A. Szent-Györgyi discovered that all of the carboxylic acids that we now recognize as members of the citric acid cycle stimulated respiration of animal tissues that were oxidizing another substrate such as glucose. Drawing on this knowledge, H. A. Krebs[a] and W. A. Johnson in 1937 proposed the citric acid cycle. Krebs provided further confirmation in 1940 by the observation that malonate, a close structural analogue and competitive inhibitor of succinate, in concentrations as low as $0.01\ M$ blocked the respiration of tissues by stopping the oxidation of succinate to fumarate. In muscle, 90% of all respiration was inhibited and succinate was shown to accumulate, powerful proof of the importance of the citric acid cycle in the respiration of animal tissues.

$$^-OOC-CH_2-COO^-$$
Malonate

$$^-OOC-CH_2-CH_2-COO^-$$
Succinate

[a] Nobel Laureate, 1953.

acetyl-CoA with oxaloacetate. Note that the same enzyme citrate synthetase which catalyzes the condensation also removes the CoA by hydrolysis after it has served its function of activating a methyl hydrogen (Eq. 7-67). Before the condensation product citrate can be degraded through a β cleavage, the hydroxyl group must be moved from its tertiary position to an adjacent carbon where, as a secondary alcohol, it can be oxidized to a carbonyl group. This is accomplished through steps b and c, both catalyzed by the enzyme aconitase (Eq. 7-47). Isocitrate can be oxidized to the β-keto acid **oxalosuccinate,** which does not leave the enzyme surface but readily decarboxylates (steps c and d; also see Eq. 7-75).

The second carbon to be removed from citrate is also released as CO_2 through **oxidative decarboxylation** of the α-keto acid **ketoglutarate** (α-oxoglutarate, Chapter 8, Section J,2). All that remains to complete the cycle is to convert the four-carbon succinyl unit of succinyl-CoA back to oxaloacetate through two oxidation steps. This involves the conversion of succinyl-CoA to free succinate (step f) followed by a β-oxidation sequence (steps g–i, Fig. 9-2; see also Fig. 9-1). Steps e and f accomplish a **substrate level phosphorylation** (sequence S7B, Fig. 8-19).[15] Succinyl-CoA is a "high energy" unstable thioester, and whereas step f could also be accomplished by a simple hydrolysis of the thioester this would be energetically wasteful. Hence, the cleavage is coupled to synthesis of ATP (*E. coli* and higher plants) or GTP

Fig. 9-2 Reactions of the citric acid cycle (the tricarboxylic acid cycle). Asterisk designates positions of label from entrance of carboxyl-labeled acetate into the cycle. Note that it is *not* the two carbon atoms from acetyl-CoA that are immediately removed as CO_2 but two atoms from oxaloacetate. Only after several turns of the cycle are the carbon atoms of the acetyl-CoA completely converted to CO_2. Nevertheless, the cycle can properly be regarded as a mechanism of oxidation of acetyl groups to CO_2. Dagger designates position of ^2H introduced into malate as $^2H^+$ from the medium. FAD§ designates covalently bound 8-histidyl-FAD (Chapter 8, Section I,3).

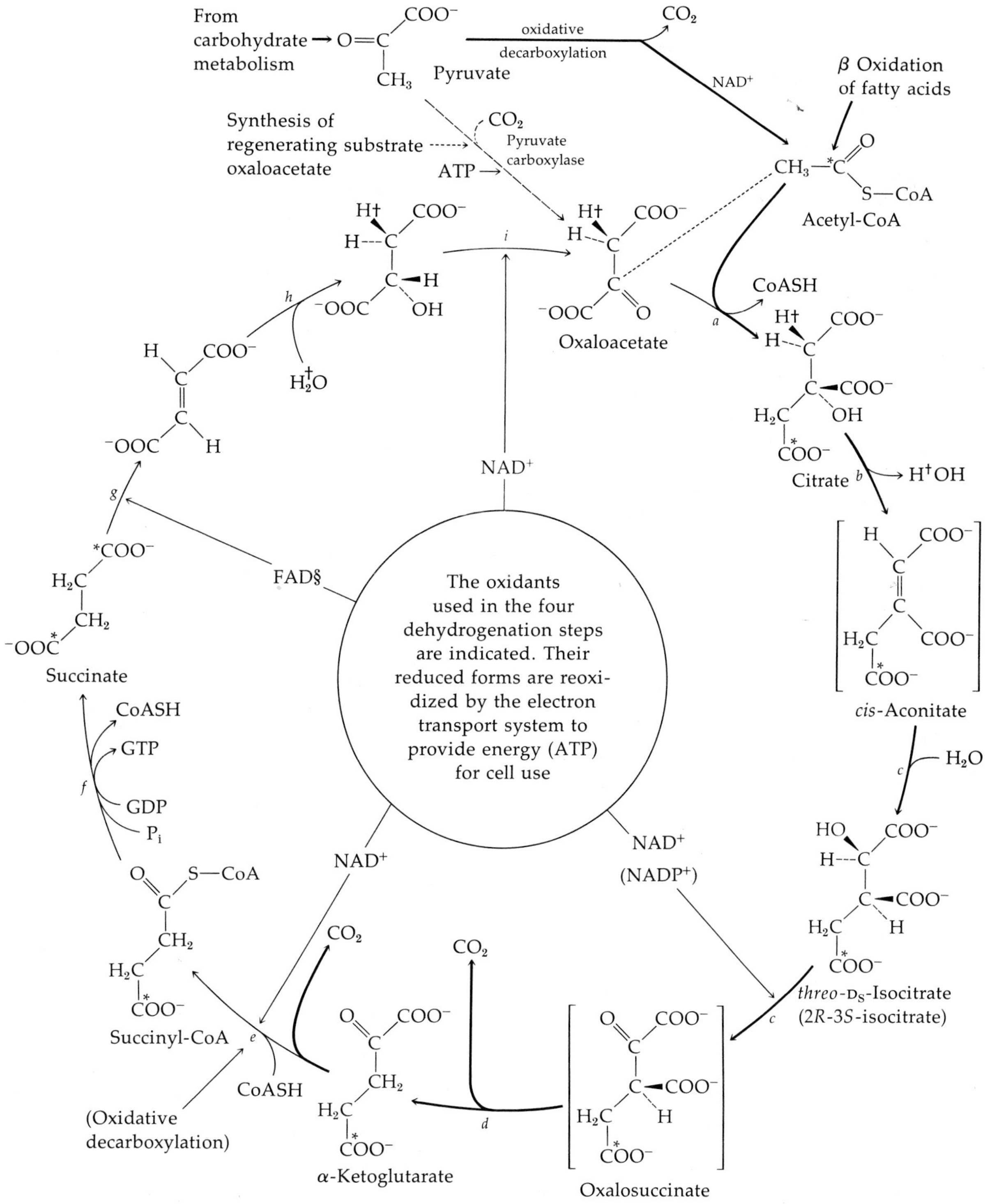

From carbohydrate metabolism → Pyruvate

oxidative decarboxylation → CO_2

NAD^+

β Oxidation of fatty acids

Synthesis of regenerating substrate oxaloacetate

CO_2
Pyruvate carboxylase
$ATP →$

Acetyl-CoA

Oxaloacetate

CoASH

Citrate

H^+OH

cis-Aconitate

H_2O

H_2^+O

Succinate

CoASH
GTP
GDP
P_i

Succinyl-CoA

(Oxidative decarboxylation)

The oxidants used in the four dehydrogenation steps are indicated. Their reduced forms are reoxidized by the electron transport system to provide energy (ATP) for cell use

FAD§

NAD^+

NAD^+

NAD^+
($NADP^+$)

CO_2

CO_2

CoASH

α-Ketoglutarate

Oxalosuccinate

threo-D_S-Isocitrate
(2R-3S-isocitrate)

BOX 9-D

POISONS: FLUOROACETATE AND "LETHAL SYNTHESIS"

Among the most deadly of simple compounds is sodium fluoroacetate, the LD_{50} (the dose lethal for 50% of animals receiving it) is only 0.2 mg/kg for rats, over tenfold less than that of the nerve poison diisopropylphosphofluoridate (Chapter 7, Section D,1).[a,b] Popular (but controversial) as the rodent poison "1080," fluoroacetate is also found in the leaves of a number of different poisonous plants found in Africa, Australia, and South America. Remarkably, difluoroacetate HCF_2-COO^- is nontoxic.

Biochemical studies reveal that fluoroacetate has no toxic effect on cells until it is converted metabolically to fluorocitrate and that the latter serves as a highly specific inhibitor of aconitase (Chapter 7, Section H,6).[b,c] This fact is remarkable since citrate formed by the reaction of fluoroxaloacetate and acetyl-CoA has only weak inhibitory activity toward the same enzyme. Yet, it is the fluorocitrate formed in the latter fashion that contains a fluorine atom at a site that is attacked in the action of aconitase.

The small van der Waals radius of fluorine (0.135 nm), comparable to that of hydrogen (0.12 nm) is often cited as the basis for the ability of fluoro compounds to "deceive" enzymes. However, it is more likely that the high electronegativity and ability to enter into hydrogen bonds make F more comparable to —OH in metabolic effects. In the case of fluorocitrate it has been proposed that the inhibitory isomer

Fluorocitrate bound at active site in proper way

Fluorocitrate bound in inhibitory manner with F coordinated with Fe^{2+}

binds in the "wrong way" to aconitase in such a manner that the fluorine atom is coordinated with the ferric ion at the catalytic center.[c]
Many naturally occurring fluorine compounds are known.[c] Further studies of the biochemistry of these substances and of fluorine-containing drugs are needed.

[a] G. W. Gibble, *J. Chem. Educ.* **50**, 460–462 (1973).
[b] K. Elliot and J. Birch, eds. Ciba Foundation, "Carbon-Fluorine Compounds." Elsevier, Amsterdam, 1972.
[c] J. P. Glusker, *in* "The Enzymes" (P. D. Boyer, ed.), 3rd ed., Vol. 5, pp. 413–439. Academic Press, New York, 1971.

(mammals). Some of the succinyl-CoA formed in mitochondria is used in other ways, e.g., as in Eq. 9-8.

2. Synthesis of the Regenerating Substrate Oxaloacetate

The primary substrate of the citric acid cycle is acetyl-CoA. Despite many references in the biochemical literature to substrates "entering" the cycle as oxaloacetate (or as one of the immediate precursors succinate, fumarate, or malate), these compounds *are not consumed* by the citric acid cycle. Oxaloacetate is completely regenerated; hence, the term *regenerating substrate*. A prerequisite for the operation of a catalytic cycle is that the regenerating substrate be readily available and that its concentration be increased if necessary to accommodate a more rapid rate of reaction of the cycle. Oxaloacetate can normally be formed in any amount needed for operation of the citric acid cycle from **PEP** or from **pyruvate** (Eq. 8-2), both of these compounds being readily available from metabolism of sugars.

In bacteria and green plants **PEP carboxylase** (Eq. 7-79), a highly regulated enzyme, is responsible for synthesizing oxaloacetate. In animal tissues **pyruvate carboxylase** (Eq. 8-2) plays the same role. The latter enzyme is almost inactive in the absence of the allosteric effector acetyl-CoA.

For this reason, it went undetected for many years. In the presence of high concentrations of acetyl-CoA the enzyme is fully activated and provides for synthesis of a high enough concentration of oxaloacetate to permit the cycle to function. Even so, the oxaloacetate concentration in mitochondria is remarkably low—only 2–4×10^{-7} M (20–40 molecules per mitochondrion).[16]

BOX 9-E

USE OF ISOTOPIC TRACERS IN STUDY OF THE TRICARBOXYLIC ACID CYCLE

The first use of isotopic labeling in the study of the citric acid cycle and one of the first in the history of biochemistry was carried out by H. G. Wood and C. H. Werkman at Iowa State University.[a] The aim was to study the fermentation of glycerol by propionic acid bacteria, a process that was not obviously related to the citric acid cycle.

$$\text{Glycerol} \longrightarrow \text{propionate}^- + H^+ + H_2O$$

$$\Delta G^0 = -69 \text{ kJ/mol}$$

Some succinate was also formed in the fermentation, and on the basis of simple measurements of the fermentation balance reported in 1938, it was suggested that CO_2 was incorporated into oxaloacetate which was then reduced to succinate. As we now know, this is indeed an essential step in the propionic acid fermentation (Section F,3). At the time ^{14}C was not available but the mass spectrometer, newly developed by A. O. Nier, provided a means of using the stable isotope ^{13}C as a tracer. Wood and Werkman constructed both a thermal diffusion column for the purpose of preparing bicarbonate enriched in ^{13}C and a mass spectrometer. By 1941 it was established unequivocally that carbon dioxide was incorporated into succinate by the bacteria.[b] To test the idea that animal tissues could also incorporate CO_2 into succinate Wood examined the metabolism of a pigeon liver preparation to which malonate had been added to block succinate dehydrogenase (Box 9-C). Surprisingly, the accumulating succinate contained no ^{13}C. Soon,

however, it was shown that CO_2 was incorporated into the carboxyl group of α-ketoglutarate that is adjacent to the carbonyl group. That carboxyl is lost in conversion to succinate (Fig. 9-2) explaining the lack of ^{13}C in succinate. It is of historical interest that these observations were incorrectly interpreted by a majority of the biochemists of the time. They agreed that citrate could not be a member of the tricarboxylic acid cycle.

Since citrate is a symmetric compound it was assumed that any ^{13}C incorporated into citrate would be present in equal amounts in both terminal carboxyl groups. This would necessarily result in incorporation of ^{13}C into succinate. It was not until 1948 that Ogston popularized the concept that by binding with substrates at at least three points, enzymes were capable of asymmetric attack upon symmetric substrates.[c] In other words, an enzyme could synthesize citrate with the carbon atoms from acetyl-CoA occupying a particular one of the two $-CH_2COOH$ groups surrounding the prochiral center (Chapter 6, Section D,2).

In recent years the complete stereochemistry of the citric acid cycle has been elucidated through the use of a variety of isotopic labels (Chapter 7, Section J,2,g). Some of the results are indicated by the asterisks and daggers in the structures in Fig. 9-2.

[a] H. G. Wood, in "The Molecular Basis of Biological Transport" (J. F. Woessner, Jr. and F. Huijing, eds.), pp. 1–54. Academic Press, New York, 1972.

[b] H. G. Wood, C. H. Werkman, A. Hemingway, and A. O. Nier, *JBC* **139**, 377–381 (1941).

[c] A. G. Ogston, *Nature* (*London*) **162**, 963 (1948).

3. Common Features of Catalytic Cycles

The citric acid cycle is not only one of the most important metabolic cycles in aerobic organisms (bacteria, protozoa, fungi, higher plants, and man), but also *it is a typical catalytic cycle.* Other cycles also have one or more primary substrates and at least one regenerating substrate. Thus, there is, associated with every catalytic cycle, a metabolic pathway that provides for synthesis of the regenerating substrate. Although it usually need operate only slowly to replenish regenerating substrate lost in side reactions, the pathway also provides *a mechanism for the net biosynthesis of any desired quantity of any intermediate in the cycle.* Thus, cells draw off from the citric acid cycle considerable amounts of oxaloacetate, α-ketoglutarate, and succinyl-CoA for synthesis of other cell constituents. For example, aspartate and glutamate are formed directly from oxaloacetate and α-ketoglutarate by transamination (Eq. 8-16). It is often stated that the citric acid cycle functions in biosynthesis, but note that when intermediates in the cycle are drawn off for synthesis, the complete cycle does not operate. Rather the metabolic pathway for synthesis of the regenerating substrate, together with some of the enzymes of the cycle, are used to construct a biosynthetic pathway.

Sometimes the word **amphibolic** is applied to those metabolic sequences that are part of a catabolic cycle and at the same time are involved in a biosynthetic (anabolic) pathway. Another term that is applied to pathways used for the synthesis of regenerating substrates is **anaplerotic.** This word suggested by H. L. Kornberg comes from a Greek root meaning "filling up."[17]

4. Control of the Cycle

What factors determine the rate of oxidation by the citric acid cycle? As with most other important pathways of metabolism, a number of different control mechanisms operate and different steps become rate limiting under different conditions.[18] The following appear to be major factors: (1) the rate of generation of acetyl units (which may in turn depend upon the availability of free unacylated CoA), (2) the availability of oxaloacetate, and (3) the rate of reoxidation of NADH to NAD$^+$ in the electron transport chain (Chapter 10). Note (Fig. 9-3) that acetyl-CoA is a positive effector for conversion of pyruvate to oxaloacetate. Thus, acetyl-CoA "turns on" the formation of a substance required for its own further metabolism. However, when no pyruvate is available, operation of the cycle may be impaired by lack of oxaloacetate. This appears to be the case when liver metabolizes high concentrations of ethanol. The latter is oxidized to acetate but it cannot provide oxaloacetate. The accumulating acetyl units are converted to ketone bodies but are only slowly oxidized in the cycle. A similar problem may arise during metabolism of fatty acids with inadequate sugar metabolism, e.g., in diabetes (Box 11-C).

The rate of oxidative steps in the cycle is determined by the rate of reoxidation of NADH in the electron transport chain. This rate may, under some conditions, be limited by availability of O_2. However, in aerobic organisms it is usually determined by the concentration of ADP and/or P_i available for conversion to ATP in the "oxidative phosphorylation" process (Chapter 10). If catabolism supplies more than enough ATP to meet the cell's energy needs the concentration of ADP falls to a low level, cutting off phosphorylation. At the same time ATP is present in high concentration and acts as a feedback inhibitor for the catabolism of carbohydrates and fats. This inhibition is exerted at many points, a few of which are indicated in Fig. 9-3. An important site of inhibition is the **pyruvate dehydrogenase** complex (Chapter 8, Section J,2).[19] Another is **citrate synthetase,** the enzyme catalyzing the first reaction of the citric acid cycle.[20] However, there is some doubt about the physiological significance of this inhibition.[16] Still another way in which the level of phosphorylation of the adenylate system can regulate the cycle depends upon the need for GDP in step f of the cycle (Fig. 9-2). Within mitochondria GTP is largely used to reconvert AMP to ADP. Consequently, formation of GDP is promoted by AMP, a compound that arises in mitochondria from the utilization of ATP for fatty acid activation (Eq. 9-1).

In some bacteria (including *E. coli*) ATP does not inhibit citrate synthetase but NADH does; the control is via the redox potential of the NAD⁺ system rather than by the level of phosphorylation of the adenine nucleotide system.[21] Regulation of succinic dehydrogenase activity may depend upon the redox state of ubiquinone (Chapter 10, Section D).

A second signal to the cell that catabolism is proceeding more rapidly than necessary is pro-

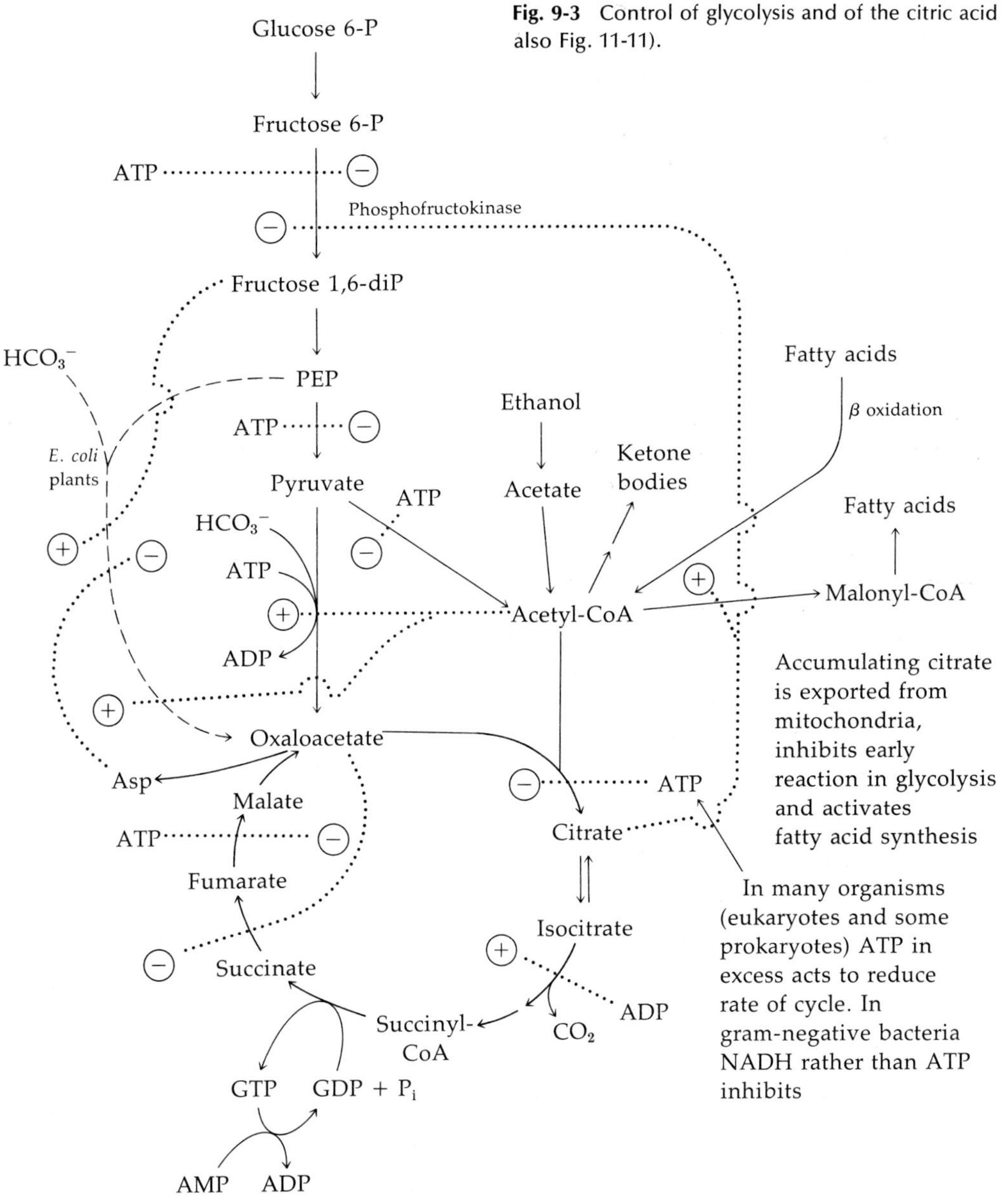

Fig. 9-3 Control of glycolysis and of the citric acid cycle (see also Fig. 11-11).

Accumulating citrate is exported from mitochondria, inhibits early reaction in glycolysis and activates fatty acid synthesis

In many organisms (eukaryotes and some prokaryotes) ATP in excess acts to reduce rate of cycle. In gram-negative bacteria NADH rather than ATP inhibits

vided by the export of accumulated citrate from the mitochondria. While such export is essential for fatty acid synthesis, citrate also acts as a negative effector for **phosphofructokinase** (Fig. 9-3) and thus decreases the overall rate of glycolysis. On the other hand, if the ATP level drops and ADP accumulates, the latter acts as a *positive* effector for oxidation of isocitrate and causes a rapid drop in citrate concentration.

The student will be able to think of still other possible control features. Consider, for example, the levels of P_i and CoA. It has been suggested that *every* enzyme of the cycle may have control mechanisms, each of which comes into action under appropriate circumstances.[21]

5. Catabolism of Intermediates of the Citric Acid Cycle

It sometimes happens that a cell must oxidize large amounts of one of the compounds found in the citric acid cycle.[22] Thus, bacteria subsisting on succinate as a carbon source must oxidize it for energy as well as convert some of it to carbohydrates, lipids and other materials. Complete combustion of any citric acid cycle intermediate can be accomplished by conversion to malate followed by (Eq. 9-9, step *a*) oxidation of malate to oxaloacetate and decarboxylation (β cleavage) to pyruvate, or (Eq. 7-9, step *b*) oxidation and decarboxylation of malate by a single enzyme, the "malic enzyme" (Eq. 7-75) without free oxaloacetate as an intermediate. Pathway *b* is probably the most important. There are actually two malic enzymes in animal mitochondria, one specific for $NADP^+$ and one reacting with NAD^+ as well.[23] They both have complex regulatory properties. For example, the $NADP^+$-dependent enzyme is activated by a high concentration of free CoA and is inhibited by NADH. It has been suggested that when glycolysis becomes slow the free CoA level rises and "turns on" malate oxidation.[24] On the other hand, rapid glycolysis increases the NADH concentration which inhibits the malic enzyme. The result

Acetyl-CoA   (9-9)

is a buildup of the oxaloacetate concentration and an increase in activity of the citric acid cycle. The malic enzyme is also thought to function as part of an NADPH-generating cycle (Eq. 11-13).

C. OXIDATIVE PATHWAYS RELATED TO THE CITRIC ACID CYCLE

In this section we will consider some other catalytic cycles as well as some noncyclic pathways of

* One oxaloacetate decarboxylase activity has been identified as a side reaction of pyruvate kinase.[22a] An enolate ion is a common intermediate for the phosphotransferase and decarboxylase activities.

oxidation of one- and two-carbon substrates that are utilized by microorganisms.

1. The γ-Aminobutyrate Shunt

An interesting variation of the citric acid cycle occurs in brain tissue (Fig. 9-4). Acetyl-CoA and oxaloacetate are condensed (step a) in the usual way to form citrate which is then converted to α-ketoglutarate. The latter is transformed to L-glutamate either by direct amination (b) or by transamination (c), the amino donor being γ-aminobutyrate. Both glutamate and γ-aminobutyrate occur in high concentrations in brain (10 and 0.8 mM, respectively), and both have been suggested to be important neurotransmitters. γ-Aminobutyrate is of special interest because it is not found in significant amounts in other mammalian tissues, and much evidence points to its role as a principal neuronal inhibitory substance (Chapter 16, Section B,4,g).[25]

γ-Aminobutyrate is formed by decarboxylation of glutamate (step d) and is catabolized via transamination (e) to succinic semialdehyde followed by oxidation to succinate and oxaloacetate. If the two transamination steps in the pathway are linked (as indicated in Fig. 9-4), a complete cycle is formed which parallels the citric acid cycle but in which α-ketoglutarate is oxidized to succinate via glutamate and γ-aminobutyrate. Note that no thiamine pyrophosphate is required but that a reductive amination of α-ketoglutarate to glutamate is necessary. The cycle may be described as "the γ-aminobutyrate shunt," and it has been suggested that it may play a significant role in the overall oxidative processes of brain tissue.

This γ-aminobutyrate shunt is also prominent in green plants. For example, under anaerobic conditions the radish *Raphanus sativus* accumulates large amounts of γ-aminobutyrate.[26]

2. The Dicarboxylic Acid Cycle

Some bacteria can subsist solely on glycolate, glycine, or oxalate, all of which are oxidized to glyoxylate (Eq. 9-10). Glyoxylate is both oxidized

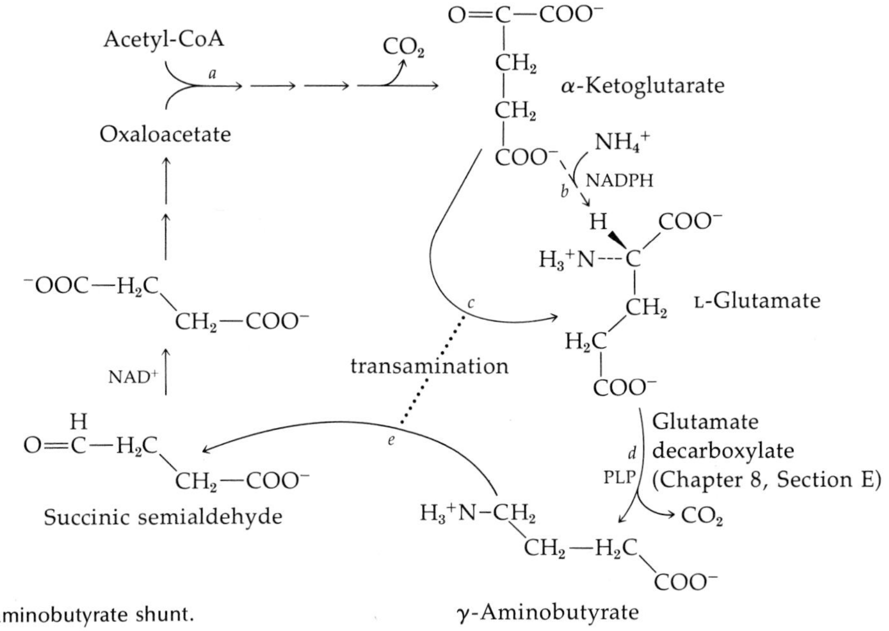

Fig. 9-4 The γ-aminobutyrate shunt.

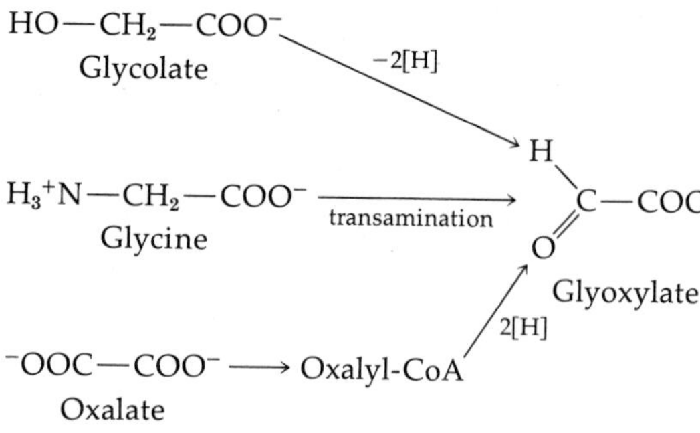

$$HO-CH_2-COO^-$$
Glycolate

$$H_3^+N-CH_2-COO^- \xrightarrow{\text{transamination}}$$
Glycine

$$^-OOC-COO^- \longrightarrow \text{Oxalyl-CoA}$$
Oxalate

$$\xrightarrow{-2[H]}$$

$$\begin{array}{c} H \\ C-COO^- \\ O \end{array}$$
Glyoxylate

$$2[H]$$

(9-10)

to CO_2 and water to provide energy to the bacteria and is utilized for biosynthetic purposes. The energy-yielding process is found in the **dicarboxylic acid cycle** (Fig. 9-5), which catalyzes the complete oxidation of glyoxylate to CO_2. Four hydrogen atoms are removed with generation of two molecules of NADH which can be oxidized by the respiratory chain to provide energy.[27] Note that in the dicarboxylic acid cycle glyoxylate is the principal substrate and acetyl-CoA is the regenerating substrate rather than the principal substrate as it is for the citric acid cycle.

The logic of the dicarboxylic acid cycle is simple. Acetyl-CoA contains a potentially free carboxyl group. After the acetyl-CoA has condensed with glyoxylate and the resulting hydroxyl group has been oxidized this carboxyl group appears in oxaloacetate in a position β to the carbonyl group. The carboxyl donated by the glyoxylate is still in the α position. A consecutive β cleavage and an oxidative α cleavage release the two carboxyl groups as carbon dioxide to reform the regenerating substrate. The cycle is simple and efficient. Like the citric acid cycle, it depends upon thiamine diphosphate without which the α cleavage would be impossible. Comparing the citric acid cycle (Fig. 9-2) with the simpler dicarboxylic acid cycle we see that in the former the initial condensation product citrate contains a hydroxyl group attached to a tertiary carbon atom. With no adjacent hydrogen, it is impossible to oxidize it directly to the carbonyl group which is essential for subsequent chain cleavage. Hence, the dependence on aconitase to shift the OH to an adjacent carbon. Both cycles involve oxidation of a hydroxy

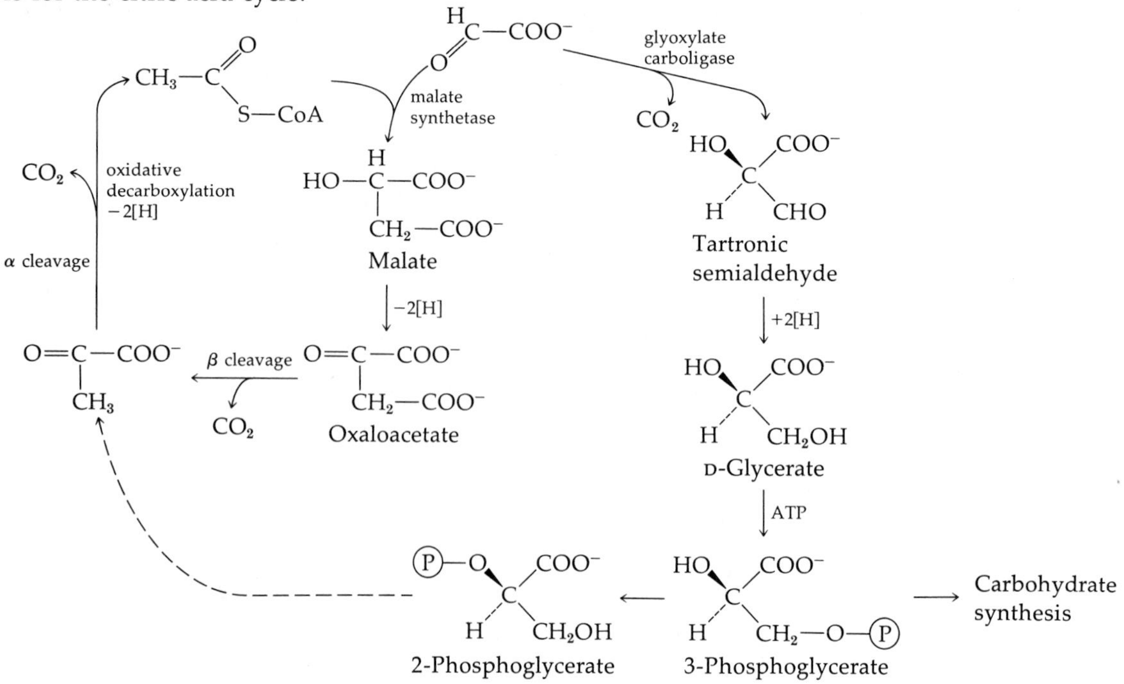

Fig. 9-5 The dicarboxylic acid cycle for oxidation of glyoxylate to carbon dioxide.

acid to a ketone followed by β cleavage and oxidative α cleavage. In the citric acid cycle additional oxidation steps are needed to convert succinate back to oxaloacetate, corresponding to the fact that the citric acid cycle deals with a more reduced substrate than does the dicarboxylic acid cycle.

The synthetic pathway for the regenerating substrate of the dicarboxylic acid cycle is quite complex. Two molecules of glyoxylate undergo α condensation with decarboxylation by glyoxylate carboligase (Chapter 8, Section D,2) to form **tartronic semialdehyde.** The latter is reduced to D-glycerate which is phosphorylated to 3-phosphoglycerate and 2-phosphoglycerate. Since the phosphoglycerates are carbohydrate precursors (Chapter 11, Section A), this **glycerate pathway** provides the organisms with a means for synthesis of carbohydrates and other complex materials from glyoxylate alone. At the same time, 2-phosphoglycerate is readily converted to pyruvate (see Fig. 9-7, reactions 9 and 10) and pyruvate, by oxidative decarboxylation, to the regenerating substrate acetyl-CoA. An alternate route for synthesis of the regenerating substrate (used by *Micrococcus denitrificans*) is described by Eq. 11-17.

3. Oxidation of Oxalate to CO₂ via Formate

Formic acid is a suitable energy fuel for some organisms, e.g., certain pseudomonads. Utilization of oxalate by bacteria (Eq. 9-11) also involves formate as well as oxyalyl- and formyl-CoA intermediates. Note that a thiamine diphosphate-dependent α cleavage is required and that the energy of the formyl-CoA is preserved by use of a CoA transferase (Chapter 7, Section F,4) to transfer the CoA to oxalate.[28]

The oxidation of formate to CO_2 is catalyzed by **formate dehydrogenase,** an NAD^+-dependent enzyme that contains **selenium** as well as **molybdenum** and **iron.**[29,30] The membrane-bound protein from *E. coli* contains one atom each of selenium and molybdenum, together with one heme group and iron–sulfur centers (Chapter 10, Section C).

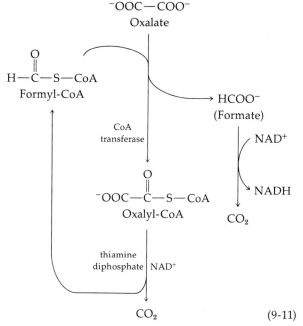

(9-11)

There are three kinds of peptide chain, and it appears that four of each associate to form an aggregate of MW $\sim 590,000$ (see also Box 9-F). Oxidation of formate also occurs in animals and plants.[31] However, free formate is not a usual metabolic intermediate. Thus, glyoxylate is most often degraded via the dicarboxylic acid cycle (Fig. 9-5) rather than through oxidative decarboxylation to formyl-CoA and formate. In most instances both formate and formaldehyde enter metabolic pathways only while firmly attached to tetrahydrofolic acid (Fig. 8-20).

D. CATABOLISM OF PROPIONYL COENZYME A AND PROPIONATE

β Oxidation of fatty acids with an odd number of carbon atoms leads to the formation of propionyl-CoA in addition to acetyl-CoA. The three-carbon propionyl unit is also produced by degradation of isoprenoid compounds, of isoleucine, threonine, and methionine. Humans ingest small amounts of free propionic acid, e.g., from

BOX 9-F

SELENIUM, A DEADLY POISON AND A NUTRITIONAL ESSENTIAL

It was a surprise when K. Schwartz and associates in 1957 showed that the extremely toxic element selenium was a "nutritional factor" essential for prevention of the death of liver cells in rats.[a] Liver necrosis could be prevented by as little as 0.1 part per million (ppm) of selenium in the diet. Similar amounts of selenium were shown to prevent muscular dystrophy ("white muscle disease") in cattle and sheep grazing on selenium-deficient soil. Sodium selenite and other inorganic selenium compounds were more effective than organic compounds in which Se had replaced S.

Recently, four proteins have been shown to contain selenium.[b,c] Using radioactive ^{75}Se as a tracer it has been possible to isolate small amounts of these **selenoproteins** in pure form. The selenium appears to be covalently attached to the proteins, possibly through an aromatic or heterocyclic prosthetic group.

One selenoprotein of unknown function has been isolated from muscle. It is lacking in animals with white muscle disease. A small protein of MW $\sim 10,000$ it also contains a heme prosthetic group.

A second selenoprotein is **glutathione peroxidase** which catalyzes the peroxidation of glutathione (GSH, Box 7-G)[c,d]:

$$H_2O_2 + 2\ GSH \longrightarrow 2\ H_2O + G{-}S{-}S{-}G$$

or

$$2\ ROOH + 2GSH \longrightarrow$$
$$ROH + H_2O + G{-}S{-}S{-}G$$

Unlike most peroxidases the enzyme is not a heme protein but contains one atom of selenium attached to a peptide chain of MW $\sim 22,000$. The native enzyme is a tetramer consisting of four of these chains. Glutathione peroxidase is a principal protective agent against accumulation of H_2O_2 and organic peroxides within cells (Box 10-A).

A third selenoprotein **formate dehydrogenase** (described in text) contains both selenium and molybdenum.

The fourth selenoprotein functions in the reductive deamination of glycine by *Clostridium sticklandii*.

$$\begin{array}{c} CH_2{-}COOH \\ | \\ NH_2 \end{array} + R\ (SH)_2 \xrightarrow{\quad\quad} $$
$$ADP + P_i \qquad ATP$$

$$CH_3COOH + NH_3 + R{\Big\langle}\begin{array}{c}S\\|\\S\end{array}$$

The mechanism of this remarkable reduction of an amino acid by a dithiol compound is a mystery, but one of the three to four proteins of the reductase system is a heat-stable acidic protein of MW $\sim 12,000$ and containing one atom of selenium per mole.[b]

Other biochemical functions for selenium have been proposed. A selenium-containing protein may mediate the transfer of electrons from reduced glutathione into the cytochrome system.[e] Selenium has been reported to protect animals against the toxicity of mercury, a fact that may be related to the ease with which selenium undergoes methylation (Chapter 8, Section L,7). Much evidence supports a relationship between the nutritional need for selenium and that for vitamin E. Lack of either causes muscular dystrophy in many animals as well as severe edema ("exudative diathesis") in chicks. Since vitamin E-deficient rats also have a low selenide (Se^{2-}) content, it has been suggested that vitamin E protects reduced selenium from oxidation.[f] These observations have also led to the suggestion that there may be Fe–Se proteins analogous to the known Fe–S proteins (formate dehydrogenase from *Clostridium thermoaceticum* contains nonheme iron[b] as well as selenium).

Relatively little is known about the metabolism of selenium.[b] Through the use of ^{75}Se as a tracer, normal rat liver has been shown to contain Se^{2-}, SeO_3^{2-}, and selenium in a higher oxidation state.[f] It has been suggested that glutathione may be involved in reduction of selenite to selenide.[g] The nonenzymatic reduction of selenite by glutathione yields a selenotrisulfide derivative:

$$4\ GSH + SeO_3^{2-} + 2\ H^+ \xrightarrow{\text{pH 7}} G-S-Se-S-G + G-S-S-G + 3\ H_2O$$

(selenotrisulfide)

spontaneous

glutathione reductase

$$G-S-S-G + Se$$

$$G-S-Se-H$$

$$G-SH + Se$$

The latter is spontaneously decomposed to oxidize glutathione and elemental selenium or by the action of glutathione reductase to glutathione and selenium. It is conceivable that selenium could be inserted into organic groups from a selenopersulfide intermediate of the kind postulated in the latter reaction.

[a] E. Frieden, *Sci. Am.* **227**, 52–60 (Jul 1972).

[b] T. C. Stadtman, *Science* **183**, 915–922 (1974).
[c] H. E. Ganther, in "Selenium" (R. A. Zingaro and W. C. Cooper eds.), pp. 546–614. Van Nostrand-Reinhold, Princeton, New Jersey, 1974.
[d] J. T. Rotruck, A. L. Pope, H. E. Ganther, A. B. Swanson, D. G. Hafeman, and W. G. Hoekstra, *Science* **179**, 588–590 (1973).
[e] O. A. Levander, V. C. Morris, and D. J. Higgs, *Fed. Proc., Fed. Am. Soc. Exp. Biol.* **32**, 886 (abstr.) (1973).
[f] A. T. Diplock and J. A. Lucy, *FEBS Lett.* **29**, 205–210 (1973).
[g] M. Sandholm and P. Sipponen, *ABB* **155**, 120–124 (1973).

Swiss cheese (which is cultivated with propionic acid-producing bacteria) and from propionate added to bread as a fungicide. In **ruminant** animals, such as the cow,* propionate is a major source of energy.

1. The Malonic Semialdehyde Pathways

The most obvious route of metabolism of propionyl-CoA would be further β oxidation to the CoA derivative of malonic semialdehyde. The latter could in turn be oxidized to malonyl-CoA, a β-keto acid which decarboxylates readily to acetyl-CoA (pathway a, Fig. 9-6). While the necessary enzymes have been found in *Clostridium kluyveri*,[32] the pathway appears little used. However, a closely related route, found in green plants and in many microorganisms, is hydrolysis of β-hydroxypropionyl-CoA to *free* β-hydroxypropionate which is then oxidized to the semialde-

hyde of malonic acid (pathway b, Fig. 9-6).

A possible advantage of pathway b over a could exist if β-hydroxypropionyl-CoA is not cleaved by simple hydrolysis as indicated in Fig. 9-6. If the energy of the thioester linkage is conserved by generation of ATP or GTP (as occurs during the conversion of succinyl-CoA to succinate in the citric acid cycle) an additional substrate level phosphorylation step would be present.*

It has often been suggested that the three-carbon compound propionate can be converted directly to pyruvate, also a three-carbon compound. The simplest way to do this would be through α oxidation to lactate, but there seems to be little evidence for such a pathway. Another possible route to lactate would be via addition of water to acrylyl-CoA, an intermediate in pathway a of Fig. 9-6. The water molecule would have to add in the "wrong way," the OH^- ion going to the α-carbon instead of the β.† The resulting lactyl-

---

* In cattle, sheep, and other ruminants the ingested food undergoes extensive fermentation in the **rumen,** a large digestive organ containing cellulose-digesting bacteria and protozoa. Major products of the rumen fermentations include acetate, propionate, and butyrate.

* Can the reader suggest a reason why the coupling of synthesis of another molecule of ATP to pathway a might be awkward?

† What kind of active center would the reader propose for an enzyme catalyzing such a reaction? See Eq. 8-36 for a similar reaction.

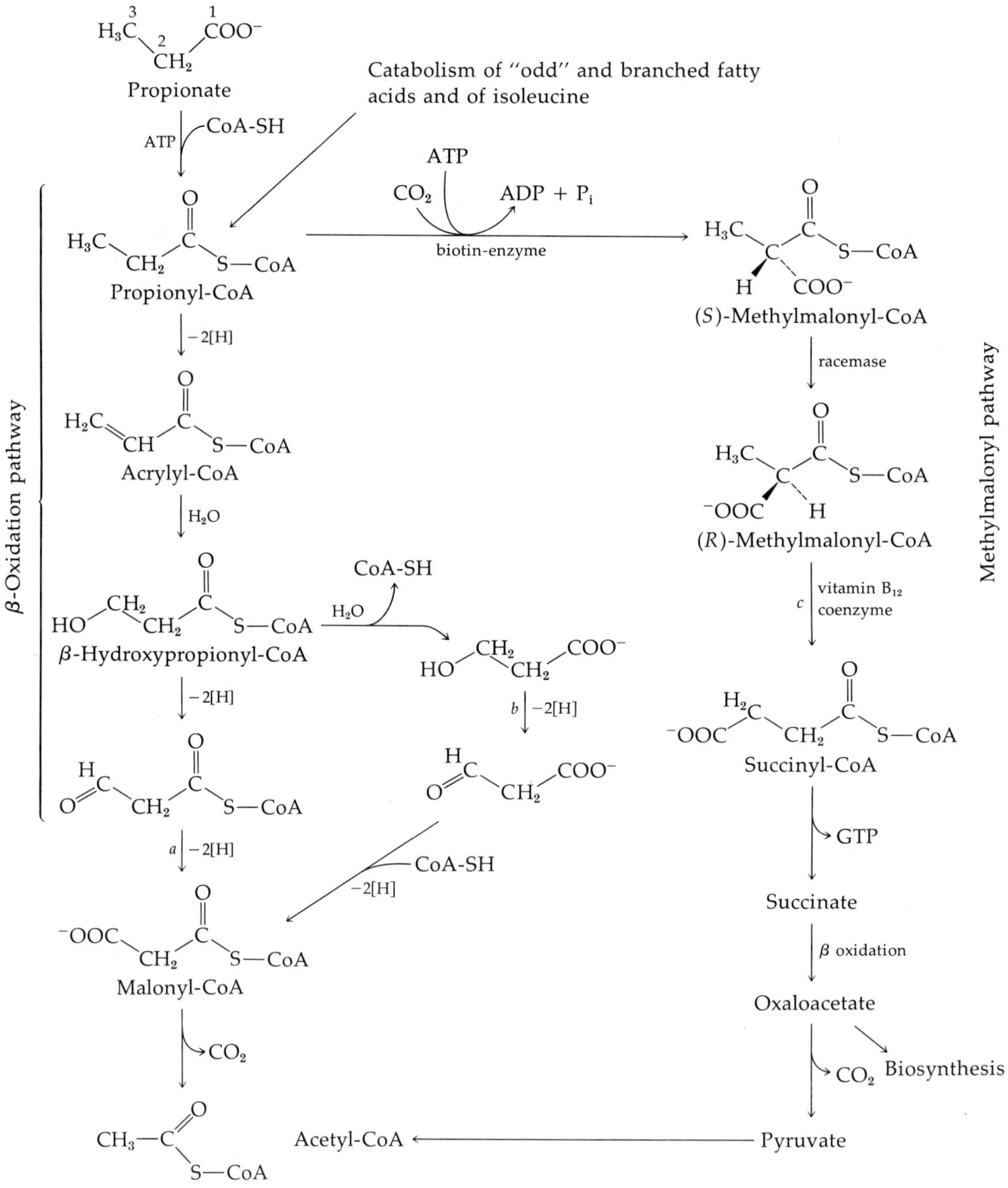

Fig. 9-6 Catabolism of propionate and propionyl-CoA. Note that in pathway *b* the malonyl-CoA produced will have the thioester linkage to C-3 of the original propionate instead of to C-1 as shown.

CoA could be converted to pyruvate readily. While evidence has been provided that *Clostridium propionicum* interconverts propionate, lactate, and pyruvate, attempts to demonstrate the necessary enzymes have failed.[33] In the same organism the chemically more reasonable addition of ammonium ion to acrylyl-CoA to form β-alanyl-CoA does occur.

2. The Methylmalonyl Pathway of Propionate Utilization

Despite the simplicity and logic of the β-oxidation pathway of propionate metabolism, higher animals use the more complex methylmalonyl pathway (Fig. 9-6, step *c*). This is one of the two processes in higher animals presently known to depend upon vitamin B_{12}. Recall that this vitamin has never been found in higher plants, nor does the methylmalonyl pathway occur in plants.

The pathway (Fig. 9-6) begins with a biotin and ATP-dependent carboxylation of propionate. The (*S*)-methylmalonyl-CoA so formed is isomerized to (*R*)-methylmalonyl-CoA (the reader can suggest a simple mechanism for this reaction) after which the methylmalonyl-CoA is converted to succinyl-CoA in a vitamin B_{12} coenzyme-requiring step (Table 8-6). The succinyl-CoA is converted to free succinate (with formation of GTP compensating for the ATP used initially). The succinate, by β oxidation, is converted to oxaloacetate which is decarboxylated to pyruvate (in effect removing the carbon dioxide that was put on at the beginning of the sequence). Pyruvate is converted by oxidative decarboxylation to acetyl-CoA. A natural question is "Why did nature go to all this trouble to do something that could have been done much more directly?" The answer is not clear but here are some ideas.

Malonyl-CoA, the product of the β-oxidation pathway of propionate metabolism (Fig. 9-6, step *a*), is also an intermediate in biosynthesis of fatty acids. Perhaps the presence of too much malonyl-CoA would interfere with lipid metabo-

---

BOX 9-G

GENETIC DISEASES: METHYLMALONIC ACIDURIA

In this severe and often fatal disease up to 1–2 g of methylmalonic acid per day (compared to a normal of < 5 mg/day) are excreted in the urine. Two forms of the disease are known. One is responsive to vitamin B_{12} and cultured fibroblasts from patients contain a very low level of the vitamin B_{12} coenzyme (Chapter 8, Section L). Addition of excess vitamin B_{12} restores coenzyme synthesis to normal.[a] The second form of the disease does not respond to vitamin B_{12} suggesting a defect in the mutase protein. A closely related disease is caused by a deficiency of propionyl-CoA carboxylase.

Methylmalonic aciduria and propionyl-CoA decarboxylase deficiency are both accompanied by severe ketosis, hypoglycemia, and hyperglycinemia. The cause of these conditions is not entirely clear. However, methylmalonyl-CoA, which accumulates in methylmalonic aciduria, is a known inhibitor of pyruvate carboxylase. Therefore, ketosis may develop because of impaired conversion of pyruvate to oxaloacetate.

[a] L. E. Rosenberg, *in* "The Metabolic Basis of Inherited Disease" (J. B. Stanbury, J. B. Wyngaarden, and D. S. Fredrickson, eds.), 3rd ed., pp. 440–456. McGraw-Hill, New York, 1972.

---

lism. On the other hand, the tacking on of an extra CO_2 and the use of ATP at the beginning makes the methylmalonyl pathway look suspiciously like a biosynthetic rather than a catabolic route (see Chapter 11). The methylmalonyl pathway provides a means for converting propionate to oxaloacetate, a transformation that is chemically difficult.

E. CATABOLISM OF SUGARS

1. Glycolysis

Because in most sugars each carbon atom bears an oxygen atom, chemical attack by oxidation is possible at any point in the molecule. Every sugar contains a potentially free aldehyde or ketone group and the carbonyl function can readily be moved to adjacent positions by isomerases. Consequently, aldol cleavage is also possible at many points. For these reasons, the metabolism of carbohydrates is complex and varied. However, in the energy economy of most organisms including man the **glycolysis pathway*** by which hexoses are converted to pyruvate (Chapter 7, Section A,5, Fig. 9-7) stands out above all others.

The discovery of glycolysis followed directly the early observations of Buchner and of Harden and Young on fermentation of sugar by yeast juice (Chapter 8, Section H). Another line of research, the study of muscle, soon converged with the investigations of alcoholic fermentation. Physiologists were interested in the process by which an isolated muscle could obtain energy for contraction in the absence of oxygen. It was shown by A. V. Hill that glycogen was converted to lactate to supply the energy, and Meyerhof later demonstrated that the chemical reactions were related to those of alcoholic fermentation. The establishment of the structures and functions of the pyridine nucleotides in 1934 (Chapter 8, Section H) coincided with important studies of glycolysis by G. Embden in Frankfurt and of J. K. Parnas in Poland. The sequence of reactions in glycolysis (the **Embden–Meyerhof–Parnas pathway**) soon became clear. All of the enzymes catalyzing the individual steps in the sequence have now been isolated and crystallized and are being studied in detail.

a. Formation of Pyruvate

The conversion of glucose to pyruvate requires ten enzymes (Fig. 9-7) and the sequence can be divided into four stages: preparation for chain cleavage (reactions 1–3), cleavage and equilibration of triose phosphates (reactions 4 and 5), oxidative

---

* The word glycolysis means cleavage of sugars or of glycogen. Originally, it referred only to anaerobic fermentations by which lactic acid or ethanol and CO_2 were formed but now it has a more general meaning and is used to describe the breakdown of sugars via glucose 6-phosphate, fructose diphosphate, and pyruvate both in the presence or absence of oxygen.

generation of ATP (reactions 6 and 7), and conversion of 3-phosphoglycerate to pyruvate (reactions 8–10).

In preparation for chain cleavage, free glucose is phosphorylated by ATP under the action of hexokinase (reaction 1). There are several hexokinases in tissues, some of which are relatively nonspecific and will phosphorylate other sugars such as mannose and fructose (Chapter 6, Section F,2; Chapter 7, Section E,6). The phosphorylated product glucose 6-phosphate can also arise without the expenditure of ATP by cleavage of a glucosyl unit from glycogen by the consecutive action of glycogen phosphorylase (reaction 1a) and phosphoglucomutase (reaction 1b, see also Eq. 7-26). The latter enzyme transfers a phosphoryl group from the oxygen at C-1 to that at C-6.

Why do cells attach a phosphoryl group to a sugar to initiate metabolism of the sugar? For one thing, the phosphoryl group constitutes an electrically charged handle for binding the sugar phosphate to enzymes. For another, there is a kinetic advantage in initiating a reaction sequence with a highly irreversible reaction such as the phosphorylation of glucose. Beyond that, there is a possibility that the phosphoryl group may function in catalysis (Chapter 7, Section J,2,a).

Reaction 2 of Fig. 9-7 is a simple isomerization to move the carbonyl group to C-2 so that β cleavage to two three-carbon fragments can occur. Before cleavage, a second phosphorylation (reaction 3) takes place to form fructose 1,6-diphosphate. When fructose diphosphate is cleaved by aldolase, each of the two halves has a phosphate handle. This second "priming reaction" (reaction 3) is the first step in the series that is unique to glycolysis. Therefore, it should not be surprising to learn that the enzyme catalyzing it, **phosphofructokinase,** is apparently one of the most highly regulated catalysts of the glycolytic sequence (Eq. 6-91; Chapter 11, Section F,4).

Cleavage of fructose diphosphate (reaction 4) is catalyzed by an aldolase (Eq. 7-64) with formation of glyceraldehyde 3-phosphate and dihydroxyacetone phosphate. These two triose phosphates are then equilibrated by an isomerase (reaction 5, see

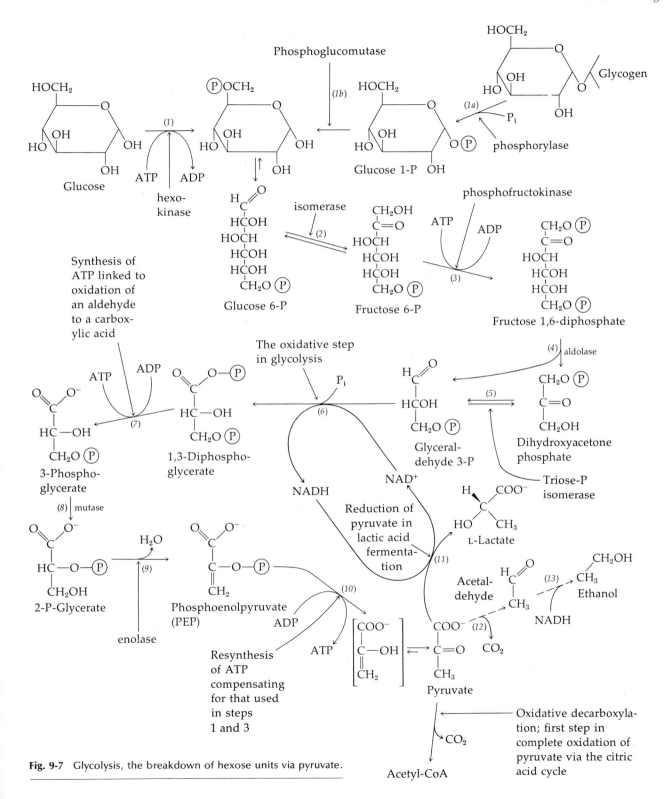

Fig. 9-7 Glycolysis, the breakdown of hexose units via pyruvate.

also Chapter 7, Section I,4). As a result, both halves of the hexose can be further metabolized via glyceraldehyde 3-P to pyruvate. On the other hand, an alternate pathway exists for metabolism of dihydroxyacetone-P. It can be reduced to **glycerol phosphate,** a precursor of lipids and an intermediate in some fermentations.

The oxidation of glyceraldehyde 3-P to the corresponding carboxylic acid, 3-phosphoglyceric acid (Fig. 9-7, reactions 6 and 7), is coupled to synthesis of a molecule of ATP from ADP and P_i (Fig. 8-13). Note that two molecules of ATP are formed per hexose cleaved and that two molecules of NAD^+ are converted to NADH in the process.

The conversion of 3-phosphoglycerate to pyruvate begins with transfer of a phosphoryl group from the C-3 to the C-2 oxygen (reaction 8) and is followed by dehydration through a standard α,β elimination catalyzed by **enolase** (reaction 9). The product, phosphoenolpyruvate (PEP, Chapter 7, Section J,3,d), is a "high energy" compound whose phosphoryl group can be transferred easily to ADP (by the action of the enzyme **pyruvate kinase**) to leave the enol of pyruvic acid (shown in brackets in Fig. 9-7) which spontaneously converts to the much more stable pyruvate ion (cf. Eq. 7-59). Because two molecules of PEP are formed from each glucose molecule, this process provides for the recovery of the two molecules of ATP that were expended in the initial formation of fructose 1,6-diphosphate from glucose.

b. The Further Metabolism of Pyruvate

In the **aerobic metabolism,** utilized by most tissues of our bodies, pyruvate is oxidatively decarboxylated to acetyl-CoA which can then be completely oxidized in the citric acid cycle (Fig. 9-2). The NADH produced in reaction 6 as well as in the oxidative decarboxylation of pyruvate and in subsequent reactions of the citric acid cycle is reoxidized in the electron transport chain of the mitochondria. An important alternative fate of pyruvate is to enter into fermentation reactions. For example, the enzyme lactate dehydrogenase (reaction 11) catalyzes reduction of pyruvate with

NADH to L-lactate (for many bacteria D-lactate is formed). As indicated in Fig. 9-7, this reaction can be coupled to the NADH-producing reaction 6 to give a balanced process by which glucose can be fermented to lactate in the absence of oxygen (Chapter 7, Section A,6). In a very similar process, yeast cells decarboxylate pyruvate (α cleavage) to acetaldehyde which is reduced to ethanol using the NADH produced in reaction 6 (Fig. 9-7, reactions 12 and 13). These fermentation reactions and many others are discussed further in Section F.

2. Generation of ATP through Substrate Oxidation

The formation of ATP from ADP and P_i is a vital process for all cells. It is frequently referred to as "phosphorylation," and the process is subdivided as (1) **oxidative phosphorylation** associated with the passage of electrons through an electron transport chain—usually in mitochondria; (2) **photosynthetic phosphorylation,** a similar process occurring in chloroplasts under the influence of light; and (3) **substrate level phosphorylations.** Only the latter are understood chemically and the oxidation of glyceraldehyde-3-P and the accompanying ATP formation (reactions 6 and 7, Fig. 9-7, Fig. 8-13) provides the best known example. The dehydrogenation of glyceraldehyde 3-P is tremendously important for yeasts and many other microorganisms that live anaerobically and depend entirely on this single reaction for their supply of energy. Since only two molecules of ATP are produced for each molecule of hexose degraded in this way,* it is not surprising that yeast

* The conversion of glucose to lactate or to ethanol and CO_2 is accompanied by a net synthesis of two molecules of ATP. It seems most logical to view these as arising from oxidation of glyceraldehyde-3-P. The formation of ATP from PEP and ADP in step 10 of Fig. 9-7 can be regarded as recapturing of ATP "spent" in the priming reactions. Note that a glucose unit of glycogen can be converted to pyruvate with an apparent net gain of *three* molecules of ATP. However, two molecules of ATP were needed for the initial synthesis of each hexose unit of glycogen (Eq. 11-24). Therefore, the overall net yield for fermentation via stored polysaccharide is only one ATP per hexose.

must ferment enormous quantities of sugar.

Within aerobic tissues, reoxidation of NADH occurs via the electron transport chain of mitochondria as described in Chapter 10.

3. The Pentose Phosphate Pathways

Important metabolic sequences involving the five-carbon pentose sugars are referred to as the **pentose phosphate pathways** or the phosphogluconate pathway or the hexose monophosphate shunt. Historically, the evidence for such routes dates from the experiments of Warburg on the oxidation of glucose 6-P to 6-phosphogluconate. Recall that it was study of this reaction which led to discovery of NADP$^+$ (Chapter 8, Section H). For many years the oxidation remained an enzymatic reaction without a defined pathway. However, it was assumed to be part of an alternative method of degradation of glucose. Supporting evidence was found in the observation that tissues continue to respire in the presence of a high concentration of fluoride ion, a known inhibitor of the enolase reaction and capable of almost completely blocking glycolysis. Some tissues (for example, liver) are especially active in respiration through this alternate pathway. We now know that the pentose phosphate pathways are multiple as well as multipurpose. They not only function in catabolism but, when operating in the reverse direction (*reductive* pentose phosphate pathway), lie at the heart of the sugar-forming reactions of photosynthesis.

The oxidative pentose pathways provide a means for *cutting the carbon chain of a sugar molecule one carbon at a time*, the carbon removed appearing as CO_2. The enzymes required constitute three distinct systems, all of which are found in the cytosol of animal cells: (i) a dehydrogenation—decarboxylation system, (ii) an isomerizing system, and (iii) a sugar rearrangement system. The dehydrogenation—decarboxylation system cleaves glucose 6-P to CO_2 and the pentose phosphate, ribulose 5-P (Eq. 9-12). Three enzymes are

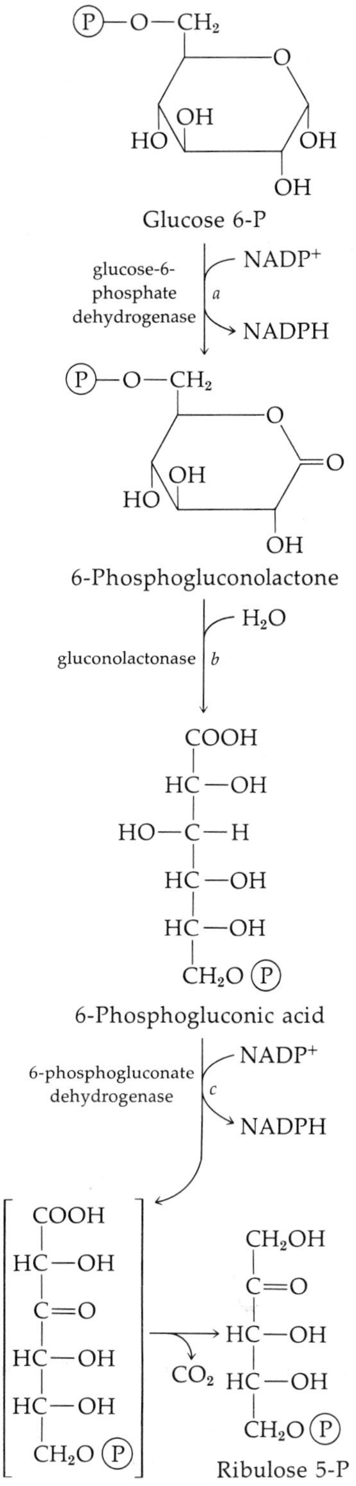

Glucose 6-P

glucose-6-phosphate dehydrogenase — *a*   NADP$^+$ → NADPH

6-Phosphogluconolactone

gluconolactonase — *b* H_2O

6-Phosphogluconic acid

6-phosphogluconate dehydrogenase — *c*   NADP$^+$ → NADPH

(9-12)

Ribulose 5-P

required, the first being glucose-6-P dehydrogenase (Eq. 9-12, step a, see also Eq. 8-42). The immediate product, a lactone, is spontaneously hydrolyzed, but the action of **gluconolactonase** (Eq. 9-12, step b) causes a more rapid ring opening. A second dehydrogenation is catalyzed by **6-phosphogluconate dehydrogenase** (Eq. 9-12, step c) and this reaction is immediately followed by a β decarboxylation catalyzed by the same enzyme (as in Eq. 7-75). The value of ΔG^0 for oxidation of glucose 6-P to ribulose 5-P by $NADP^+$ according to Eq. 9-12 is -30.8 kJ mol^{-1}, a negative enough value to drive the $[NADPH]/[NADP^+]$ ratio to an equilibrium value of over 2000 at a CO_2 tension of 0.05 atm.

The isomerizing system, consisting of two enzymes, interconverts three pentose phosphates (Eq. 9-13). As a consequence the three pentose phosphates exist as an equilibrium mixture. Both xylulose 5-P and ribose 5-P are needed for further reactions in the pathways.

$$(9\text{-}13)$$

The ingenious sugar rearrangement system uses two enzymes **transketolase** and **transaldolase**. Both enzymes catalyze chain cleavage and transfer reactions (Eqs. 9-14 and 9-15) involving the same group of substrates. These enzymes use the two basic types of C—C bond cleavage, adjacent to a carbonyl group (α) and one carbon removed from

a carbonyl group (β). Both types are needed in the pentose phosphate pathways as in the citric acid cycle.

$$(9\text{-}14)$$

$$(9\text{-}15)$$

a. An Oxidative Pentose Phosphate Cycle

Putting the three systems together, we can form a cycle that oxidizes hexose phosphates. Three carbon atoms are chopped off one at a time (Fig. 9-8A), leaving a three-carbon triose phosphate as the product. Since the dehydrogenation system works only on glucose 6-P, a part of the sugar rearrangement system must be utilized between each of the three oxidation steps. Note that a C_5 unit (ribose 5-P) is used in the first reaction with transketolase but is regenerated at the end of the sequence. Thus, this C_5 unit is the regenerating substrate for the cycle. As indicated by the dotted arrow, it is readily formed in any quantity needed by oxidation of glucose 6-P. Before the C_5 unit formed in each oxidation step can be processed by the sugar rearrangement reactions, it must be isomerized from ribulose 5-P to xylulose 5-P; before the C_5 unit, produced at the end of the sequence in Fig. 9-8, can be reutilized as a regenerating substrate, it must be isomerized to ribose 5-P. Thus, the pathway is quite complex. Note too, that the same C_5 substrates appear at several points in Fig.

9-8A. Thus, substrates from different parts of the cycle become scrambled and the pathway does not degrade all the hexose molecules in a uniform manner.

The pentose phosphate cycle is often presented as a means for complete oxidation of hexoses to CO_2. For this to happen the C_3 unit indicated as the product in Fig. 9-8A must be converted (through the action of aldolase, a phosphatase and hexose phosphate isomerase) back to one-half molecule of glucose-6-P which can enter the cycle at the beginning. On the other hand, alternative ways of degrading the C_3 product glyceraldehyde-P are available. For example, using glycolytic enzymes, it can be oxidized to pyruvate and to CO_2 via the citric acid cycle.

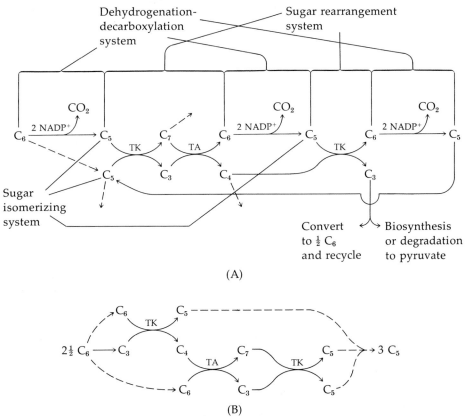

Fig. 9-8 The pentose phosphate pathways. (A) Oxidation of a hexose (C_6) to three molecules of CO_2 and a three-carbon fragment with the option of removing C_3, C_4, C_5, and C_7 units for biosynthesis (dashed arrows). (B) Nonoxidative pentose pathways:

$$2\tfrac{1}{2}\,C_6 \longrightarrow 3\,C_5 \quad \text{or} \quad 2\,C_6 \longrightarrow 3\,C_4 \quad \text{or} \quad 3\tfrac{1}{2}\,C_6 \longrightarrow 3\,C_7.$$

As a general rule, NAD$^+$ is associated with catabolic reactions and it is somewhat unusual to find NADP$^+$ acting as an oxidant. However, in mammals the enzymes of the pentose phosphate pathway are specific for NADP$^+$. The reason is thought to lie in the need of NADPH for biosynthesis (Chapter 11, Section C). On this basis the occurrence of the pentose phosphate pathway in tissues having an unusually active biosynthetic function (liver and mammary gland) is understandable. It is possible that in these tissues the cycle operates as indicated in Fig. 9-8A with the C$_3$ product also being used in biosynthesis. Furthermore, as the reader should be able to demonstrate, any of the products from C$_4$ to C$_7$ may be withdrawn in any desired amounts without disrupting the smooth operation of the cycle. For example, we know that the C$_4$ intermediate **erythrose 4-P** is required in synthesis of aromatic amino acids by bacteria and plants (but not in animals). Likewise, **ribose 5-P** is needed for formation of nucleic acids and of several amino acids.

b. Nonoxidative Pentose Phosphate Pathways

The sugar rearrangement system together with the glycolytic enzymes (which convert glucose 6-P to glyceraldehyde 3-P) can function to transform hexose phosphates into pentose phosphates (Fig. 9-8B).[34] The overall equation is given by Eq. 9-16.

$$2\tfrac{1}{2}\ C_6 \longrightarrow 3\ C_5 \qquad (9\text{-}16)$$

The reader can easily show that the same enzymes will catalyze the net conversion of hexose phosphate to erythrose 4-phosphate and to sedoheptulose 7-phosphate (Eq. 9-17):

$$2\ C_6 \longrightarrow 3\ C_4 \qquad 3\tfrac{1}{2}\ C_6 \longrightarrow 3\ C_7 \qquad (9\text{-}17)$$

An investigation of metabolism of the red lipidforming yeast *Rhodotorula gracilis* (which lacks phosphofructokinase and is thus unable to break down sugars through the glycolytic pathway) indicates that 20% of the glucose is *oxidized* through the pentose phosphate pathway while 80% is altered by the nonoxidative pentose phosphate

pathway[*,35,36] of Eq. 9-16. A number of fermentations are also based on the pentose phosphate pathways (Section F,6).

Evaluation of the relative contributions of the glycolysis pathway, the pentose phosphate pathway, and of other metabolic routes within living cells is difficult and is the subject of continuing efforts.[35,36]

4. The Entner–Doudoroff Pathway

There are other ways in which a six-carbon sugar chain can be cut. One of these provides the basis for the **Entner–Doudoroff pathway,** used by *Zymomonas lindneri* and many other species of bacteria. Glucose-6-P is oxidized first to 6-phosphogluconate which is converted by dehydration to a 2-keto-3-deoxy derivative (Eq. 7-59; Eq. 9-18, step *a*). The resulting 2-keto-3-deoxy sugar is cleaved by an aldolase (Eq. 7-66; Eq. 9-18, step *b*) to pyruvate and triose-P which are then metabolized in standard ways.

$$(9\text{-}18)$$

* There is a mystery, though. If the glycolysis pathway is blocked, how can the C$_3$ unit be formed for operation of the nonoxidative pathway (Fig. 9-8B)? The answer may lie in the further breakdown of a pentose phosphate to a three-carbon unit and a two-carbon unit.

F. FERMENTATION: "LIFE WITHOUT OXYGEN"

Louis Pasteur recognized in 1860 that fermentation was not a spontaneous process but a result of life in the absence of air.* He realized that yeasts decompose much more sugar under anaerobic conditions than they do aerobically, and that the anaerobic fermentation was essential to the life of these organisms. There are many kinds of fermentation and because in every case the amounts of substrates decomposed are large, they have provided attractive subjects for biochemical study. Furthermore, if we believe current theories about the history of the earth, life evolved at a time when no oxygen was available. Therefore, the most primitive organisms must have used fermentations which may thus be the oldest as well as the simplest ways in which cells obtain energy.

Fermentation is also a vital process in the human body. While in most of our everyday activities our muscles receive enough oxygen to oxidize pyruvate and obtain ATP through aerobic metabolism, there are circumstances in which the oxygen supply is inadequate. For example, during extreme exertion, after all oxygen is consumed, muscle cells produce lactate by fermentation. In fact, white muscle of fish and fowl has little aerobic metabolism and normally yields L-lactate as a principal end product. Likewise, a variety of tissues within the human body are poorly supplied with blood, e.g., the transparent lens and cornea of our eyes. Cells of these tissues also have little oxidative metabolism and depend for their energy upon fermentation of glucose to lactate.

Some of the lactate formed in muscle and other tissues enters the bloodstream and is carried to the liver where it is reoxidized to pyruvate. Part of the pyruvate is then oxidized via the citric acid cycle while a larger part is reconverted to glucose (Chapter 11, Section D,5). The latter may be released into the bloodstream and returned to the muscles. The overall process is known as the **Cori cycle.***

1. Fermentations Based on the Embden–Meyerhof Pathway

a. Homolactic and Alcoholic Fermentations

We have already considered briefly (Section E,1b) the processes by which glucose can be converted to lactate and, by yeast cells, to ethanol and CO_2. The overall features of these two fermentations are summarized in Eq. 9-19. Solid lines show

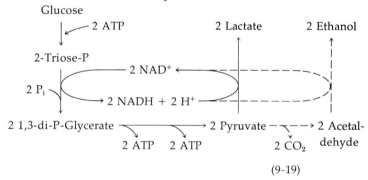

$$(9\text{-}19)$$

the conversion of glucose to lactate, while the dashed lines indicate the alternative pathway by which pyruvate is converted to ethanol in the alcoholic fermentation. Several features common to all fermentations are illustrated by Eq. 9-19. The NADH produced in the oxidation step (to the left of the equation) is reoxidized in a reaction by which substrate is reduced to the final end product. The NAD alternates between oxidized and reduced forms. This coupling of oxidation steps with reduction steps in exact equivalence is characteristic of all true (anaerobic) fermentations. The formation of ATP from ADP and P_i by substrate level phosphorylation is also common to all fermentations. The stoichiometry is usually exact

---

* In fact, we now know that Brewer's yeast *Saccharomyces cerevisiae* cannot grow in the complete absence of O_2 unless it is supplied with preformed sterols and unsaturated fatty acids.[37]

* Named for Nobel Prize winning investigators of glycogen metabolism, Carl and Gerty Cori. Recent evidence indicates that in humans the Cori cycle functions very little in utilization of lactate from muscle tissue. Most lactate is reconverted to glycogen within muscle after exercise. The cycle is more important in other less aerobic tissues.

and simple. For example, according to Eq. 9-19 a net total of two moles of ATP are formed per mole of glucose fermented.

b. Energy Relationships

If we disregard the synthesis of ATP, the equations for the overall reactions of Eq. 9-19 are given by Eqs. 9-20 and 9-21. Note that the free energy changes are negative and of sufficient magnitude that the reactions will unquestionably go to completion. However, the synthesis of two molecules of ATP from inorganic phosphate and ADP, a reaction (Eq. 9-22) for which $\Delta G'$ is substantially positive, is coupled to the fermentation:

Glucose $\longrightarrow$ 2 lactate$^-$ + 2 H$^+$ (9-20)
$\Delta G'$ (pH7) = -196 kJ mol^{-1} (-46.8 kcal mol^{-1})

Glucose $\longrightarrow$ 2 CO$_2$ + 2 ethanol (9-21)
ΔG^0 = -235 kJ mol^{-1}

ADP^{3-} + HPO$_4{}^{2-}$ + H$^+$ $\longrightarrow$ ATP^{4-} + H$_2$O (9-22)
$\Delta G'$ (pH 7) = $+34.5$ kJ mol^{-1} (Table 3-5)

To obtain the net free energy change for the complete reaction we must add $2 \times 34.5 = +69.0$ kJ to the values of $\Delta G'$ for Eqs. 9-20 and 9-21. When this is done we see that the net free energy changes are still highly negative, that the reactions will proceed to completion, and that these fermentations can serve as an usable energy source for organisms.

Biochemists often divide ΔG for the ATP synthesis in a coupled reaction sequence (in this case $+69$ kJ) by the overall ₊ree energy decrease for the coupled process (196 or 235 kJ mol^{-1}) to obtain an "efficiency." In the present case the efficiency would be 35 and 29% for coupling of Eq. 9-22 (for 2 mol of ATP) to Eqs. 9-20 and 9-21, respectively. According to this calculation, nature is approximately one-third efficient in the utilization of available metabolic free energy for ATP synthesis. However, it is important to realize that this kind of calculation of "efficiency" has no exact thermodynamic significance. Furthermore, the utilization (for various purposes) by a cell of the ATP formed is far from 100% efficient.

Why are the free energy decreases for Eqs. 9-20 and 9-21 so large? No overall oxidation takes place; there is only a rearrangement of the existing bonds between atoms of the substrate. Why does this rearrangement of bonds lead to a substantial negative ΔG? An answer is suggested by an examination of the numbers of each type of bond in the substrate and in the products. During the conversion from glucose to two molecules of lactate one C—C bond, one C—O bond, and one O—H bond are lost and one C—H bond and one C=O are gained. If we add up the bond energies for these bonds (Table 3-6) we find that the difference (ΔH) between substrate and products amounts to only about 20 kJ/mol. However, lactic acid contains a carboxyl group, and carboxyl groups have a special stability as a result of resonance. The extra resonance energy of a carboxyl group (Table 3-6) is ~ 117 kJ (28 kcal) per mol or 234 kJ/mol for two carboxyl groups. This is approximately the same as the free energy change (Eq. 9-20) for fermentation of glucose to lactate. Thus, the energy available results largely from the rearrangement of bonds by which the carboxyl groups of lactate are formed. Likewise, the resonance stabilization of CO$_2$ is given by Pauling as 151 kJ/mol, again of just the right magnitude to explain ΔG in alcoholic fermentation (Eq. 9-21).

As a general rule, we can say that fermentations can occur when substrates consisting of largely singly bonded atoms and groups, such as the carbonyl group that are not highly stabilized by resonance, are converted to products containing carboxyl groups or to CO$_2$. If we assume an efficiency of $\sim 30\%$, the energy available will be about sufficient for synthesis of one ATP molecule for each carboxyl group or CO$_2$ created. Bear in mind that generation of ATP also depends upon availability of a mechanism. It is of interest that most synthesis of ATP is linked directly to the chemical processes by which carboxyl groups or CO$_2$ molecules are created in a fermentation process. The most important single reaction is the oxidation of the aldehyde group of glyceraldehyde 3-P to the carboxyl group of 3-phosphoglycerate (Fig. 8-13).

Compare the fermentation of glucose with the complete oxidation of the sugar to carbon dioxide and water (Eq. 9-23), a process which yeast cells (as well as our own cells) carry out in the presence of air. The overall free energy change is over 10 times greater than that for fermentation, a fact that permits the cell to form an enormously greater quantity of ATP. The net gain in ATP synthesis, accompanying Eq. 9-23 is usually about 38 mol of ATP—19 times more than is available from fermentation of glucose. Thus, the explanation of Pasteur's observation that yeast decomposes much less sugar in the presence of air than in its absence is clear. Also, we can understand why a cell, living anaerobically, must metabolize an enormous amount of substrate to grow. (Recall from Chapter 3, Section D,1 that ~1 mol of ATP energy is needed to produce 10 g of cells.)

$$\text{Glucose} + 6\,O_2 \longrightarrow 6\,CO_2 + 6\,H_2O \qquad (9\text{-}23)$$
$$\Delta G' = -2872 \text{ kJ } (-686.5 \text{ kcal) mol}^{-1}$$

c. Variations of the Alcoholic and Homolactic Fermentations

The course of a fermentation is often affected drastically by changes in conditions. Many variations can be visualized by reference to Fig. 9-9 which shows a number of available metabolic sequences. We have already discussed the conversion of glucose to triose phosphate and via reaction pathway a to pyruvate and through reaction c to lactate or reactions d to ethanol.

If bisulfite is added to a fermenting culture of yeast, the acetaldehyde formed through reaction d is trapped as the bisulfite adduct blocking the reduction of acetaldehyde to ethanol, an essential part of fermentation, according to Eq. 9-21. Yeast cells accommodate to this change by using the accumulating NADH to reduce half of the triose-P to glycerol through pathway b. Two enzymes are needed, a dehydrogenase and a phosphatase to hydrolytically cleave off the phosphate. The balanced reaction is given by Eq. 9-24:

$$\text{Glucose} \longrightarrow \text{glycerol} + \text{acetaldehyde (trapped)} + CO_2$$
$$\Delta G' \text{ (pH 7)} = -105 \text{ kJ mol}^{-1} \qquad (9\text{-}24)$$

Note that one molecule of CO_2 is produced and that the overall free energy change is still adequate to make the reaction highly spontaneous. However (referring to Fig. 9-9), note that the net synthesis of ATP is now apparently zero. The fermentation apparently does not permit cell growth. Nevertheless, it has been used industrially for production of glycerol.

A related variation of alcoholic fermentation occurs with yeast grown in an alkaline medium. Under these circumstances acetaldehyde is oxidized to acetate by an NAD^+-dependent dehydrogenase. The NADH formed in this step is used to reduce an equivalent amount of acetaldehyde to ethanol. At the same time, NADH produced in the initial oxidation of triose-P is used to reduce half of the triose-P formed to glycerol-P. The overall reaction is given by Eq. 9-25:

$$2 \text{ Glucose} + H_2O \longrightarrow 2 \text{ glycerol} + \text{ethanol}$$
$$+ \text{ acetate} + H^+ + 2\,CO_2 \quad (9\text{-}25)$$
$$\Delta G' \text{ (pH 7)} = -285 \text{ kJ mol}^{-1}$$

The reaction in this form is probably beneficial to the cell because the acetic acid produced tends to neutralize the alkali. The pH of the medium returns to neutrality, at which point the standard alcoholic fermentation resumes again. The overall value of $\Delta G'$ is such that we might expect synthesis of two or three molecules of ATP, but it is not clear where or how this takes place. The obvious logical spot is in the oxidation of acetaldehyde to acetate which could occur via acetyl-CoA and acetyl phosphate (see Fig. 9-9), a pathway utilized in many bacterial fermentations. However, yeast contains an enzyme that oxidizes acetaldehyde directly to acetate and no associated ATP synthesis has been demonstrated.

Reduction of dihydroxyacetone phosphate to glycerol phosphate also occurs in insect flight muscle and apparently operates as an alternative to lactic acid formation in that tissue. While there is no net gain of ATP in the conversion of free glucose to glycerol phosphate and pyruvate, in muscle stored glycogen is the starting material

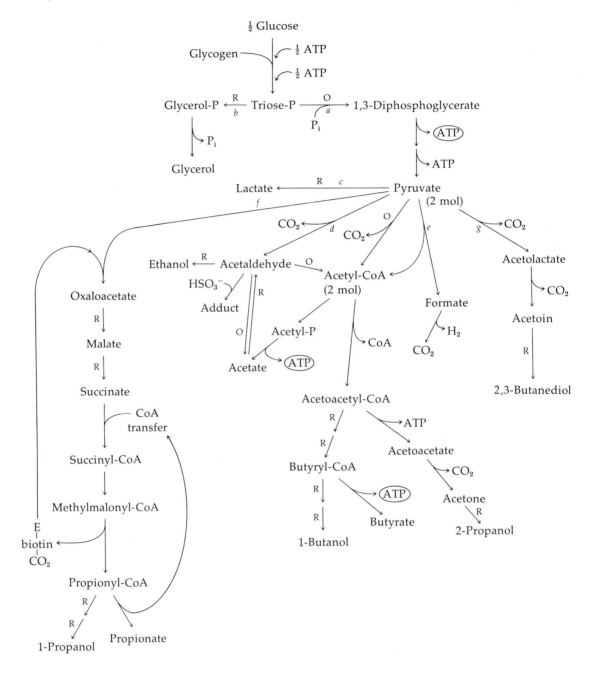

Fig. 9-9 Reaction sequences in fermentations based on the Embden–Meyerhof pathway. Oxidation steps (producing NADH) are marked "O"; reduction steps (using NADH) are marked "R".

used. Only half as much ATP is required in the priming reactions as is required for use of free glucose. Thus, the dismutation of triose-P to glycerol-P and pyruvate does provide a way of making ATP rapidly available during the vigorous contraction of the powerful insect flight muscle. During the slower recovery phase, glycerol-P is thought to be reoxidized, apparently after entering the mitochondria of these highly aerobic cells. Thus, the transport of glycerol-P into mitochondria serves as a means for transporting reducing equivalents derived from reoxidation of NADH into the mitochondria. It is possible that its significance to the muscle metabolism is more related to this function than to the rapid formation of ATP.

2. The Mixed Acid Fermentation

The enterobacteria, including *E. coli*, convert glucose to ethanol and acetic acid and either formic acid (or CO_2 and H_2 derived from it). The stoichiometry is variable but Eq. 9-26 can be used to represent an idealized fermentation:

$$\text{Glucose} + H_2O \longrightarrow \text{ethanol} + \text{acetate}^-$$
$$+ H^+ + 2 H_2 + 2 CO_2 \quad (9\text{-}26)$$
$$\Delta G' \text{ (pH 7)} = -225 \text{ kJ mol}^{-1}$$

The details of the process and the oxidation–reduction balance can be pictured as in Eq. 9-27.

Glucose is initially converted to pyruvate with formation of two molecules of ATP and two of NADH. Pyruvate is cleaved by the pyruvate formate-lyase reaction (Chapter 8, Section J,5). The products are acetyl-CoA and formic acid (Eq. 9-27, pathway *e* in Fig. 9-9).

Half of the acetyl-CoA is cleaved to acetate via acetyl-P with generation of ATP while the other half is reduced (two steps) to ethanol using the two molecules of NADH produced in the initial oxidation of triose phosphate (Eq. 9-27). The overall energy yield is three molecules of ATP per glucose. The "efficiency" is thus $3 \times 34.5 \div 225 = 46\%$. Some glucose is also converted to D-lactic and succinic acids (pathway *f*, Fig. 9-9); hence, the name **mixed acid fermentation.**

In some mixed acid fermentations (e.g., that of *Shigella*) formic acid accumulates, but in other cases (e.g., with *E. coli*) it is converted to CO_2 and H_2 (Eq. 9-27). The equilibration of formic acid with CO_2 and hydrogen is catalyzed by the **formic hydrogen-lyase** system which consists of two enzymes. The selenium-containing formate dehydrogenase (Section C,3) catalyzes oxidation of formate to CO_2 by NAD^+ while the membrane-bound **hydrogenase,** probably an iron–sulfur protein (Chapter 10, Section F,1,a), equilibrates $NADH + H^+$ with $NAD^+ + H_2$. Hydrogenase also serves to release H_2 from excess NADH. Krebs has pointed out that an excess of NADH may arise because growth of cells requires biosynthesis of many components such as amino acids. When glucose is the sole source of carbon the biosynthetic reactions involve an excess of oxidation steps over reduction steps.[38] The excess of reducing equivalents may be released as H_2 or may be used to form highly reduced products such as succinate.

Among such genera as *Aerobacter* and *Serratia* part of the pyruvate formed is condensed with decarboxylation to form **acetolactate** (also an intermediate in the biosynthesis of valine, Fig. 14-10), which is in turn decarboxylated to acetoin (Eq. 9-28; pathway *g* of Fig. 9-9). The acetoin is reduced with NADH to **2,3-butanediol,** while a third

$1\frac{1}{2}$ Glucose

↘ 3 ATP

↘ 3 NADH

3 Pyruvate

CO_2 ← (two molecules) → $H_2 + CO_2$

Acetyl-CoA

$$H_3C-\underset{O}{\overset{O}{C}}-\underset{\underset{COO^-}{|}}{\overset{OH}{C}}-CH_3$$

Acetolactate

Ethanol

↙ 2 NADH

CO_2 ↰

$$H_3C-\underset{O}{\overset{O}{C}}-\underset{\underset{H}{|}}{\overset{OH}{C}}-CH_3 \xrightarrow{\text{NADH}} \text{2,3-Butanediol} \qquad (9\text{-}28)$$

Acetoin

molecule of pyruvate is converted to ethanol, hydrogen, and CO_2 (Eq. 9-28). The following equation (Eq. 9-29) can be written:

$$1\tfrac{1}{2}\text{ Glucose} \longrightarrow \text{butanediol} + 3\ CO_2$$
$$+ H_2 + \text{ethanol} \qquad (9\text{-}29)$$

However, the organisms also produce acid by the reactions previously considered. The reaction of Eq. 9-29 provides the basis for industrial production of butanediol (which can be dehydrated nonenzymatically to butadiene).

The mixed acid fermentations are not limited to bacteria. Thus, trichomonads, parasitic flagellate protozoa, also have an anaerobic metabolism by which pyruvate is converted to acetate, succinate, CO_2, and H_2. These organisms contain no mitochondria but have microbodylike particles that have been called **hydrogenosomes** and which can convert pyruvate to acetate, CO_2, and H_2.[39] The enzyme catalyzing pyruvate cleavage apparently does not contain lipoate and may be related to the pyruvate-ferredoxin oxidoreductase of clostridia (Eq. 8-66). The hydrogenosomes also contain an active hydrogenase.

Many invertebrate animals are true facultative anaerobes, able to survive for long periods, sometimes indefinitely, without oxygen.[39a,b] Among these are *Ascaris* (Fig. 1-10), oysters, and other molluscs. Succinate and alanine are among the main end products of anaerobic metabolism. The former may arise by a mixed acid fermentation that also produces pyruvate. The pyruvate is converted to acetate to balance the fermentation in *Ascaris lumbricoides,* which is in effect an obligate anaerobe. However, in molluscs the pyruvate may undergo transamination with glutamate to form alanine and α-ketoglutarate, and the ketoglutarate may be oxidatively decarboxylated to succinate. These reactions obviously depend upon the availability of a store of glutamate or of other amino acids, such as arginine, that can give rise to glutamate.

3. The Propionic Acid Fermentation

Propionic acid producing bacteria are especially numerous in the digestive tract of ruminants. Within the rumen some bacteria digest cellulose to form glucose which is then converted to lactate and other products. The propionic acid bacteria can convert either glucose or lactate into propionic and acetic acids which are absorbed into the bloodstream of the host. Some succinic acid is usually formed also.

The basis for the propionic acid fermentation is conversion of pyruvate to oxaloacetate by carboxylation and the further conversion through succinate and succinyl-CoA to methylmalonyl-CoA and propionyl-CoA, reactions which are almost the exact reverse of those discussed in Section D,2 for the oxidation of propionate in the animal body. However, whereas the carboxylation of pyruvate to oxaloacetate in the animal body requires ATP, the propionic acid bacteria save one equivalent of ATP by using carboxyltransferase (Chapter 8, Section C,1). This enzyme donates a carboxyl group from a preformed carboxybiotin compound generated in the decarboxylation of methylmalonyl-CoA in the next to final step of the reaction sequence (Fig. 9-10). A second molecule

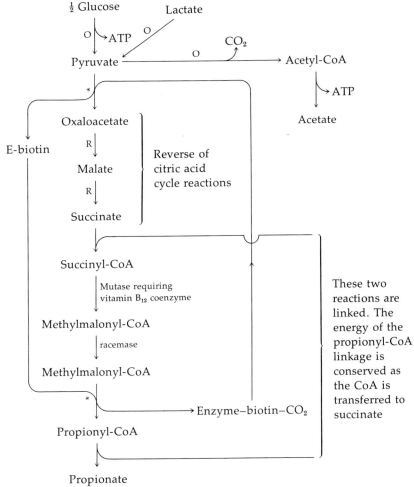

Fig. 9-10 Propionic acid fermentation of *Propionobacteria* and *Veillonella*. Oxidation steps are designated by the symbol "O" and reduction steps by "R". The two coupled reactions marked by asterisks are catalyzed by carboxyltransferase.

of ATP is saved by linking directly the conversion of succinate to succinyl-CoA to the cleavage of propionyl-CoA to propionate through the use of a CoA transferase (Chapter 7, Section F,4). To provide for oxidation–reduction balance, two-thirds of the glucose goes to propionate and one-third to acetate (Eq. 9-30):

$$1\tfrac{1}{2} \text{ Glucose} \longrightarrow 2 \text{ propionate}^- + \text{acetate}^-$$
$$+ 3 \text{ H}^+ + \text{CO}_2 + \text{H}_2\text{O} \quad (9\text{-}30)$$
$$\Delta G' \text{ (pH 7)} = -465 \text{ kJ per } 1\tfrac{1}{2} \text{ mol of glucose}$$

Note that more carboxyl groups and CO_2 molecules are formed in this fermentation ($2\tfrac{2}{3}$ per glucose molecule) than in the regular lactic acid fermentation. $\Delta G'$ is correspondingly more negative as is the yield of ATP (also $2\tfrac{2}{3}$ mol/mol of glucose fermented).

Using the same mechanism (Fig. 9-10), propionic acid bacteria are also able to take lactate, the product of fermentation by other bacteria, and ferment it further to propionate and acetate (Eq. 9-31).

3 Lactate$^-$ $\longrightarrow$ 2 propionate$^-$ + acetate$^-$

$$+ H_2O + CO_2 \quad (9\text{-}31)$$

$\Delta G'$ (pH 7) $= -171$ kJ per 3 mol of lactate

The net gain is one molecule of ATP. This reaction probably accounts for the particular niche in the ecology of the animal rumen which is occupied by propionic acid bacteria.

4. Butyric Acid and Butanol-Forming Fermentations

A variety of fermentations are carried out by bacteria of the genus *Clostridium* and by the rumen organisms *Eubacterium* (*Butyribacterium*) and *Butyrivibrio*. For example, glucose may be converted to butyric and acetic acids together with CO_2 and H_2 (Eqs. 9-32 and 9-33).

2 Glucose + 2 H_2O $\longrightarrow$ butyrate$^-$ + 2 acetate$^-$

$$+ 4 CO_2 + 6 H_2 + 3 H^+ \quad (9\text{-}32)$$

$\Delta G'$ (pH 7) $= -479$ kJ per 2 mol of glucose

2 Glucose

$\rightarrow$ 4 NADH $\longrightarrow$ 2 H_2

$\searrow$ 4 ATP   two used below

4 Pyruvate

$\rightarrow$ 4 CO_2 + 4 H_2

4 Acetyl-CoA

$\rightarrow$ 2 Acetate$^-$

2 ATP

1 Acetoacetyl-CoA

$\swarrow$ NADH

β-Hydroxybutyryl-CoA $\longrightarrow$ Crotonyl-CoA

$\swarrow$ NADH

Butyryl-CoA $\longrightarrow$ Butyrate$^-$

ATP   (9-33)

The yield of ATP ($3\frac{1}{2}$ mol/mol of glucose) is the highest we have met giving an efficiency of 50%.

Another fermentation yields butanol and isopropanol as well as ethanol and acetone. Formation of the first two of these products yields $2\frac{1}{2}$ mol of ATP formed per mol of glucose fermented (Eq. 9-34). Note that in both Eqs. 9-33 and 9-34 the in-

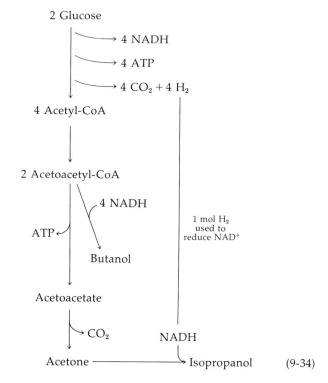

Sum:

2 glucose $\longrightarrow$ butanol + isopropanol + 5 CO_2 + 3 H_2

$\Delta G^0 = -495$ kJ per 2 mol of glucose

terconversion between hydrogen and NADH plays an essential role. Bacteria of the genus Clostridium contain the ferredoxin-dependent hydrogenase that is needed for this interconversion (Chapter 10, Section F,1,a).

Closely related and interesting fermentations (Eq. 9-35) are catalyzed by *Clostridium kluyveri*.

2 $CH_3CH\!=\!CH\!-\!COO^-$ + 2 H_2O $\longrightarrow$

   (Crotonate)

$$\text{butyrate}^- + 2 \text{ acetate}^- + H^+ \quad (9\text{-}35)$$

$\Delta G'$ (pH 7) $= -105$ kJ mol^{-1}

Note that the value of $-\Delta G'$ is one of the lowest that we have considered in any fermentation but is still enough to easily provide for the synthesis of one molecule of ATP. The fermentation is accomplished according to the scheme of Eq. 9-36. The energy of the butyryl-CoA linkage and of one of the acetyl-CoA linkages is conserved and utilized in the initial formation of crotonyl-CoA. That leaves one acetyl-CoA which can be converted via acetyl-P to acetate with formation of ATP.

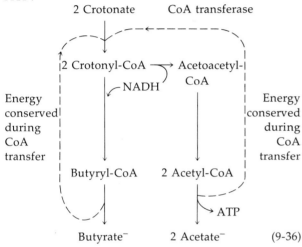

$$\text{Butyrate}^- \qquad 2 \text{ Acetate}^- \qquad (9\text{-}36)$$

5. Fermentation of Ethanol and Acetate to Butyrate and Caproate

One of the most unusual fermentations is the conversion of ethanol and acetate to butyrate and caproate by *Clostridium kluyveri* (Eqs. 9-37 and 9-38):

$$\text{Ethanol} + \text{acetate}^- \longrightarrow \text{butyrate}^- + H_2O \qquad (9\text{-}37)$$
$$\Delta G^0 = -39 \text{ kJ mol}^{-1}$$

$$2 \text{ Ethanol} + \text{acetate}^- \longrightarrow \text{caproate}^- + 2 H_2O \qquad (9\text{-}38)$$
$$\Delta G^0 = -72 \text{ kJ mol}^{-1}$$

The free energy decreases are very small, and the fact that the organism grows while fermenting substrates in this way is remarkable. As the reader can easily verify by examining the reaction pathways (the same as those in Eq. 9-33), there is no obvious mechanism for obtaining any ATP. A

plausible explanation, based on "nonstoichiometric coupling" of ATP synthesis has been advanced.*

6. Fermentations Based on the Phosphogluconate and Pentose Phosphate Pathways

Some lactic acid bacteria of the genus *Lactobacillus* as well as *Leuconostoc mesenteroides* carry out the **heterolactic** fermentation which is based on the reactions of the pentose phosphate pathway. The reason is doubtless that these organisms lack the key enzyme aldolase necessary to cleave fructose 6-P to two molecules of triose-P. The pathway (Eq. 9-39) involves reactions with which the

$$\text{Lactate} \qquad (9\text{-}39)$$

reader is already familiar. Glucose is converted to ribulose-5-P using reactions of the pentose phosphate pathway. The ribulose-P is cleaved by phosphoketolase (Fig. 8-4) to acetyl-P and glyceraldehyde-P which are converted to ethanol

* The explanation is somewhat lengthy and the reader is referred to the excellent review entitled "Energy Production in Anaerobic Organisms" by Decker *et al.*[37]

and lactate, respectively. The overall yield is only one ATP per glucose fermented.*

A simple modification of Eq. 9-39 permits fermentation of pentoses to acetate and lactate with the generation of one ATP via acetyl-P. A second ATP is generated from the oxidation of glyceraldehyde-3-P. Since one ATP is needed initially to "prime" the pentose, the net ATP yield is two.

Still another variation (Eq. 9-40)[40] is found in the genus *Bifidobacterium*. Phosphoketolase and a **phosphohexoketolase** (that cleaves fructose-6-P to erythrose-4-P and acetyl-P) are required, as are the enzymes of the sugar rearrangement system (Section E,3). The net yield of ATP is $2\frac{1}{2}$ per glucose molecule.

Another fermentation based on oxidation to 6-phosphogluconate utilizes the Entner–Doudoroff

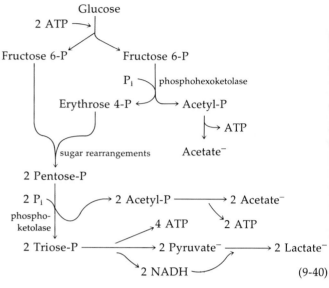

(9-40)

pathway (Eq. 9-18). Using this route the reader can easily construct a balanced fermentation process by which glucose yields ethanol and CO_2 as in a yeast fermentation. What yield of ATP is expected?

---

* The phosphoketolase cleavage can be regarded as an oxidative phosphorylation process. The aldehyde group of the "active glycolaldehyde" cleaved from ribulose-P is oxidized to an acyl phosphate at the expense of reduction of $-CH_2OH$ to $-CH_3$.

REFERENCES

1. P. K. Stumpf, *Annu. Rev. Biochem.* **38**, 159–212 (1969).
2. C. L. Hall and H. Kamin, *JBC* **250**, 3476–3486 (1975).
3. C. D. Whitfield and S. G. Mayhew, *JBC* **249**, 2801–2810 and 2811–2815 (1974); isolation of an electron-transferring flavoprotein from *Peptostreptococcus elsdenii*.
4. M. J. Johnson, *Science* **155**, 1515–1519 (1967).
5. G. J. Antony and B. R. Landau, *J. Lipid Res.* **9**, 267–269 (1968).
6. C. Hitchcock and B. W. Nichols, "Plant Lipid Biochemistry," pp. 213–221. Academic Press, New York, 1971.
7. A. J. Markovetz, P. K. Stumpf, and S. Hammarström, *Lipids* **7**, 159–164 (1972).
8. C. Hitchcock and A. Rose, *BJ* **125**, 1155–1156 (1971).
9. M. Hoshi and Y. Kishimoto, *JBC* **248**, 4123–4130 (1973).
10. S. Friedman and G. S. Fraenkel, *in* "The Vitamins" (W. H. Sebrell, Jr. and R. S. Harris, eds.), Vol. 5, pp. 329–355. Academic Press, New York, 1972.
11. R. Bressler, *Compr. Biochem.* **18**, 331–359 (1970).
12. A. G. Engel and C. Angelini, *Science* **179**, 899–901 (1973).
12a. W. Huth, R. Jonas, I. Wunderlich, and W. Seubert, *EJB* **59**, 475–489 (1975).
13. H. A. Krebs, D. H. Williamson, M. W. Bates, M. A. Page, and R. A. Hawkins, *Adv. Enzyme Regul.* **9**, 387–409 (1971).
14. J. M. Lowenstein, *Metab. Pathways, 3rd Ed.* **1**, 146–270 (1967).
15. J. S. Nishimura and F. Grinnell, *Adv. Enzymol.* **36**, 183–202 (1972).
16. F. S. Rolleston, *Curr. Top. Cell. Regul.* **5**, 47–75 (1972).
17. H. L. Kornberg, *Essays Biochem.* **2**, 1–31 (1966).
18. E. A. Newsholme and C. Start, "Regulation in Metabolism," pp. 124–145. Wiley, New York, 1973.
19. P. K. Chiang and B. Sacktor, *JBC* **250**, 3399–3408 (1975).
20. H. A. Krebs, *Adv. Enzyme Regul.* **8**, 335–353 (1970).
21. P. A. Srere, *Adv. Enzyme Regul.* **9**, 221–233 (1971).
22. M. Lopes-Cardozo and S. G. Van den Bergh, *BBA* **357**, 193–203 (1974).
22a. D. J. Creighton and I. A. Rose, *JBC* **251**, 61–68 (1976).
23. R. D. Mandella and L. A. Sauer, *JBC* **250**, 5877–5884 (1975).
24. A. R. Macrae, *BJ* **122**, 495–501 (1971).
25. R. Balázs, Y. Machiyama, B. J. Hammond, T. Julian, and D. Richter, *BJ* **116**, 445–467 (1970).
26. J. G. Streeter and J. F. Thompson, *Plant Physiol.* **49**, 579–584 (1972).
27. W. Bartley, H. L. Kornberg, and J. R. Quayle, "Essays in Cell Metabolism", p. 125. Wiley (Interscience), New York, 1970.
28. J. R. Quayle, *BJ* **89**, 492–503 (1963).
29. T. C. Stadtman, *Science* **183**, 915–922 (1974).

30. H. G. Enoch and R. L. Lester, *JBC* **250**, 6693–6705 (1975).
31. D. Peacock and D. Boulter, *BJ* **120**, 763–769 (1970).
32. P. R. Vagelos, *JBC* **235**, 346–350 (1960).
33. P. R. Vagelos, J. M. Earl, and E. R. Stadtman, *JBC* **234**, 490–497 (1959).
34. B. L. Horecker, *J. Chem. Educ.* **42**, 244–253 (1965).
35. M. Höfer, K. Brand, K. Deckner, and J.-U. Becker, *BJ* **123**, 855–863 (1971).
36. J. F. Williams, K. G. Rienits, P. J. Schofield, and M. G. Clark, *BJ* **123**, 923–943 (1971).
37. K. Decker, K. Jungermann, and R. K. Thauer, *Angew.*

Chem., Int. Ed. Engl. **9**, 138–158 (1970).
38. H. A. Krebs, *Essays Biochem.* **8**, 1–34 (1972).
39. D. G. Lindmark and M. Müller, *JBC* **248**, 7724–7728 (1973).
39a. P. W. Hochachka and T. Mustafa, *Science* **178**, 1056–1060 (1972).
39b. P. W. Hochachka and G. N. Somero, "Strategies of Biochemical Adaptation," pp. 46–61. Saunders, Philadelphia, 1973.
40. R. Y. Stanier, M. Doudoroff, and E. A. Adelberg, "The Microbial World," 3rd ed., p. 191. Prentice-Hall, Englewood Cliffs, New Jersey, 1970.

STUDY QUESTIONS

1. How many molecules of ATP can be formed per mole of palmitic acid oxidized to CO_2 and water by a cell? Assume that reoxidation of NADH by the electron transport chain of mitochondria produces 3 molecules of ATP and that reoxidation of $FADH_2$ produces 2.

2. If 1 mol of ATP can be used to synthesize 10.5 g of dry matter in a cell (See Chapter 3, Section D,1), what will be the yield of cells formed (in grams) per gram of palmitic acid metabolized, assuming that all of the ATP formed is used for growth?

3. How many molecules of ATP can be formed by complete oxidation of 1 mol of acetic acid? By complete oxidation of 1 mol of glucose?

4. Compare the values of $\Delta G°$ and the number of moles of ATP formed for complete oxidation of each of the following:
   a. Acetate (pH 7)
   b. A 2-carbon unit from a fatty acid
   c. Lactate (pH 7)
   d. A 3-carbon unit from glucose ($\frac{1}{2}$ of a molecule)

5. The branched chain amino acids valine, leucine, and isoleucine are often degraded by organisms as follows: Transamination yields an α-keto acid that is oxidatively decarboxylated to an acyl-CoA derivative. The latter undergoes β oxidation. What products are formed in this way from isoleucine? How are they metabolized to CO_2? What problems with this route of catabolism can you see for valine and leucine? Can you suggest reasonable detailed catabolic pathways? Compare your proposals with the established routes shown in Figure 14-11.

6. The citric acid cycle must be "primed" with the regenerating substrate oxaloacetate. Name four metabolic sources of this compound (not necessarily

all available to animals). Note that two of these are amino acids.

7. Metabolism of glucose is initiated by the action of hexokinase:

$$\text{Glucose} + \text{ATP}^{4-} \xrightarrow{\text{hexokinase}}$$

$$\text{glucose 6-phosphate}^{2-} + \text{ADP}^{3-} + \text{H}^+$$

$\Delta G° = -8.4$ kJ/mol at 25°C

The reaction is not readily reversible under physiological conditions. Why?

8. Many compounds found in cells are synthesized from substances found in or closely related to the glycolysis pathway and the citric acid cycle. Starting with pyruvate outline the reactions leading to each of the following:
   a. aspartic acid   f. alanine
   b. lactic acid   g. acetic acid
   c. ethanol   h. glutamic acid
   d. acetoacetic   i. δ-aminolevulinic
      acid   acid
   e. acetone   j. propionic acid

9. If radioactive glucose, labeled in carbons 3 and 4 with ^{14}C, were incubated with a cell-free liver homogenate under anaerobic conditions, what positions in the lactate produced would be labeled with ^{14}C?

10. Consider the metabolism of the following ^{14}C-labeled compounds via the citric acid cycle. [1-^{14}C]pyruvate, [2-^{14}C]pyruvate, [2-^{14}C]acetate, [1-^{14}C]succinate. Show positions of any label appearing in oxaloacetate after one complete turn of the cycle. State which substrates will yield $^{14}CO_2$ during one turn of the cycle.

11. Assume that within a yeast cell the reactions be-

tween NAD$^+$, NADH, glyceraldehyde 3-phosphate, 1,3-diphosphoglycerate, ethanol, and acetaldehyde are all at equilibrium.

a. What is the standard free energy change for the coupled pair of redox reactions linking these compounds, i.e., for

$$\text{Glyceraldehyde-3-phosphate}$$
$$+ \text{ acetaldehyde} \longrightarrow$$
$$\text{3-phosphoglycerate + ethanol}$$

b. If the ethanol concentration is 1 M, and glyceraldehyde 3-phosphate and 3-phosphoglycerate 10 mM each, what will be the concentration of acetaldehyde at equilibrium?

c. What will $\Delta G°$ be if the sequence is coupled to the synthesis of 1 mol of ATP?

d. What will the acetaldehyde concentration be at equilibrium if the concentrations given in (b) are assumed, if the reaction is coupled to synthesis of 1 molecule of ATP and if the phosphorylation state ratio R_p (Box 3-A) is 10^4?

12. The nematode *Ascaris lumbricoides* (as well as various other invertebrates) can live under totally anaerobic conditions by fermenting glucose to succinate plus pyruvate [See W. J. Landsperger and B. G. Harris, *JBC 251*, 3599–3602 (1976)].

a. What is $\Delta G'$ (pH 7) for this fermentation?

b. Although pyruvate is a product, the organism is almost totally lacking in pyruvate kinase. It apparently converts PEP to malate (Eq. 7-76) which enters the mitochondria. Write a reaction scheme for the conversion of malate to the observed products within mitochondria.

c. How many moles of ATP per mole of glucose are formed within the cytosol by this fermentation? It is thought that 1 mol of ATP is also generated in the mitochondria. Suggest a way in which this could happen (See Fig. 10-11).

d. Acetate also accumulates in *Ascaris*. Explain how this could happen under anaerobic conditions.

13. In oysters the pyruvate produced from the fermentation discussed in question 12 reacts with glutamate to form alanine and α-ketoglutarate. The latter can be oxidatively decarboxylated and converted to propionate. Write a balanced fermentation by which glutamate and glucose are converted to alanine, propionate and succinate. What is $\Delta G'$ (pH 7)? The yield of ATP per mole of glucose fermented?

10

How Oxygen Meets the Electrons with Generation of ATP, and Other Stories

In this chapter we will look at the processes by which reduced carriers such as NADH and $FADH_2$ are oxidized. Most familiar to us, because we use it in our own bodies, is **aerobic respiration.** Hydrogen atoms of NADH, $FADH_2$, and other reduced carriers are transferred through a chain of additional carriers of increasingly positive reduction potentials and are finally combined with O_2 to form H_2O. In fact, it is not the hydrogen atoms themselves but *electrons* that are deliberately transferred. The hydrogen nuclei appear to move freely as protons, mixing with the surrounding solvent. For this reason, the chain of carriers is often called the **electron transport chain** or **respiratory chain.** Since far more energy is available to cells from oxidation of NADH and $FADH_2$ than can be obtained by fermentation, the chemistry of the electron transport chain and of the associated reactions of ATP synthesis assumes great importance.

In some organisms, especially bacteria, energy may be obtained through oxidation of H_2, H_2S, or Fe^{2+} rather than of hydrogen removed from organic substrates. Furthermore, some specialized bacteria use **anaerobic respiration** in which NO_3^-, SO_4^{2-}, or CO_2 act as oxidants of either reduced carriers or reduced inorganic substances. In the present chapter, we will consider these energy yielding processes as well as the chemistry of reactions of oxygen that lead to incorporation of atoms from O_2 into organic compounds.

The oxidative processes of cells have been hard to study, largely because the enzymes responsible are located in or on cell membranes. In bacteria the enzymes lie on the inside of the plasma membrane or on membranes of mesosomes. In eukaryotes, they are found in the inner membranes of the mitochondria, and to a lesser extent in the endoplasmic reticulum. The study of oxidative phosphorylation (p. 582) has been especially frustrating. It has been difficult to isolate the components that participate in the process and even harder to put them back together and reconstitute an active system.

A. HISTORICAL NOTES

Animal respiration has been of serious interest to chemists since 1777, when Lavoisier recognized that foods undergo slow combustion within the body—in the blood according to Lavoisier. Perhaps it was L. Spallanzani whose observations of 1803–1807 established for the first time that the tissues were the actual site of respiration. However, these observations were largely ignored. C. A. MacMunn, in 1884, discovered that cells contain the heme pigments which are now known as **cytochromes.** However, the leading biochemists of the day were content to dismiss MacMunn's observations as experimental error and it was not until the present century that serious study of the chemistry of biological oxidations began.[1]

Recognition that substrates are oxidized by **dehydrogenation** is usually attributed to H. Wieland. During the years 1912–1922 he showed that various synthetic dyes such as methylene blue could be substituted for oxygen and would support the respiration of cells. Subsequent experiments (Chapter 8, Section H) led to isolation of the soluble pyridine nucleotides and flavoproteins and to development of the concept of an electron transport chain. Looking at the other end of the respiratory chain, Warburg noted (1908) that all aerobic cells contain iron. Moreover, iron-containing charcoal prepared from blood catalyzed nonenzymatic oxidation of many substances, but iron-free charcoal prepared from cane sugar did not. Cyanide was found to inhibit tissue respiration at low concentrations similar to those needed to inhibit nonenzymatic catalysis by iron salts. On the basis of these investigations, Warburg proposed in 1925 that aerobic cells contain an iron-based *Atmungsferment* (respiration enzyme), later called **cytochrome oxidase.** The *Atmungsferment* was shown to be inhibited by carbon monoxide.

Knowing that carbon monoxide complexes of hemes are dissociated by light, Warburg and Negelein (in 1928) determined the photochemical **action spectrum** (see footnote Chapter 13, Section C) for reversal of the carbon monoxide inhibition of respiration of the yeast *Torula utilis.* The spectrum closely resembled the absorption spectrum of other heme derivatives. Thus, it was proposed that O_2, as well as CO, combines with the iron of the heme group in the *Atmungsferment.*

Meanwhile, during the years 1919–1925, D. Keilin was peering through a microscope, equipped with a spectroscope ocular, at thoracic muscles of flies and other insects. He observed a pigment with four distinct absorption bands which at first he thought was derived by some modification of hemoglobin. However, when he found the same pigment in fresh baker's yeast, he recognized that he was observing a new substance. Its importance became apparent in the following way (quoted from Keilin's own account[2]):

> One day while I was examining a suspension of yeast freshly prepared from a few bits of compressed yeast shaken vigorously with a little water in a test-tube, I failed to find the characteristic four-banded absorption spectrum, but before I had time to remove the suspension from the field of vision of the microspectroscope the four absorption bands suddenly reappeared. This experiment was repeated time after time and always with the same result: the absorption bands disappeared on shaking the suspension with air and reappeared within a few seconds on standing.
>
> I must admit that this first visual perception of an intracellular respiratory process was one of the most impressive spectacles I have witnessed in the course of my work. Now I had no doubt that cytochrome is not only widely distributed in nature and completely independent of haemoglobin, but that it is an intracellular respiratory pigment which is much more important than haemoglobin.

Keilin soon realized that three of the absorption bands, those at 604, 564, and 550 nm (*a, b,* and *c*) represented three different pigments, while that at 521 nm was common to all three. Keilin proposed the names cytochromes *a, b,* and *c.* The idea of an electron transport or respiratory chain quickly followed[2] as the flavin and pyridine nucleotide coenzymes were recognized to play their role at the dehydrogenase level. Hydrogen removed from substrates by these carriers could be used to oxidize reduced cytochromes. The latter would be oxidized by oxygen under the influence of cytochrome oxidase.

In 1929 Fiske and Subbarow,[2a] curious about the occurrence of purine compounds in muscle extracts, discovered and characterized ATP. It was soon shown (largely through the work of

E. Lundsgaard and K. Lohman) that hydrolysis of ATP provided energy for muscular contraction. At about the same time, it was learned that synthesis of ATP accompanied glycolysis. That ATP could also be formed as a result of electron transport became clear following an observation of V. A. Engelhardt in 1930 that methylene blue stimulates ATP synthesis by tissues.

The study of electron transport chains and of oxidative phosphorylation began in earnest after E. D. Kennedy and A. L. Lehninger in 1949 showed that mitochondria were the site not only of ATP synthesis but also of the operation of the citric acid cycle and fatty acid oxidation pathways. By 1959, B. Chance had introduced elegant new techniques of spectrophotometry that led to formulation of the electron transport chain as follows:

$$\text{Substrate} \longrightarrow \text{pyridine nucleotides} \longrightarrow$$
$$\text{flavoprotein} \longrightarrow \text{cyt } b \longrightarrow \text{cyt } c \longrightarrow$$
$$\text{cyt } a \longrightarrow \text{cyt } a_3 \longrightarrow O_2 \qquad (10\text{-}1)$$

Since that time, some new components have been added, notably **ubiquinones** and **nonheme iron proteins**. Some of the details beginning with properties of the heme proteins and of oxygen itself are given below.

B. HEME PROTEINS

In 1879, two of the most striking colors of nature were shown to be related by the German physiological chemist, Hoppe-Seyler. The red iron-containing heme from blood and the green magnesium complex chlorophyll *a* of leaves were found to have similar ring structures. However, it remained for H. Fischer in Munich during the years 1910–1940 to prove the structures and to provide the names and numbering systems that are used today.

1. Some Names to Remember

Porphins are large rings each made up of four smaller pyrrole rings connected by four methene bridges. **Chlorins,** found in the chlorophylls, have one ring (D) reduced. A specific class of porphins, the **porphyrins,** have eight substituents around the periphery of the large ring. Like the chlorins and the corrins of vitamin B_{12} (Box 8-K), the porphyrins are all formed biosynthetically from **porphobilinogen.** This compound is polymerized in

Porphobilinogen

two ways (Fig. 14-13) to give porphyrins of types I and III (Fig. 10-1). In type I porphyrins, polymerization of porphobilinogen has taken place in a regular way so that the sequence of the carboxymethyl and carboxyethyl side chains (often referred to as acetic acid and propionic acid side chains) are the same all the way around the outside of the molecule. However, most biologically important porphyrins belong to type III in which the first three rings A, B, and C have the same sequence of carboxymethyl and carboxyethyl side chains, but in which ring D has been incorporated in a reverse fashion. Thus, the carboxyethyl side chains of rings C and D are adjacent one to another (see Fig. 10-1). Porphyrins containing all four carboxymethyl and four carboxyethyl side chains intact are known as **uroporphyrins.** Uroporphyrins I and III are both excreted in small amounts in the urine. Another excretion product is **coproporphyrin** III in which all of the carboxymethyl side chains have been decarboxylated to methyl groups. The feathers of the tropical touraco are colored with a copper(II) complex of coproporphyrin III and this porphyrin as well as others is commonly found in birds' eggs.

The heme proteins are all derived from protoporphyrin IX, which is formed by decarboxylation of two of the carboxyethyl side chains of uroporphyrin III to vinyl groups (Fig. 10-1).

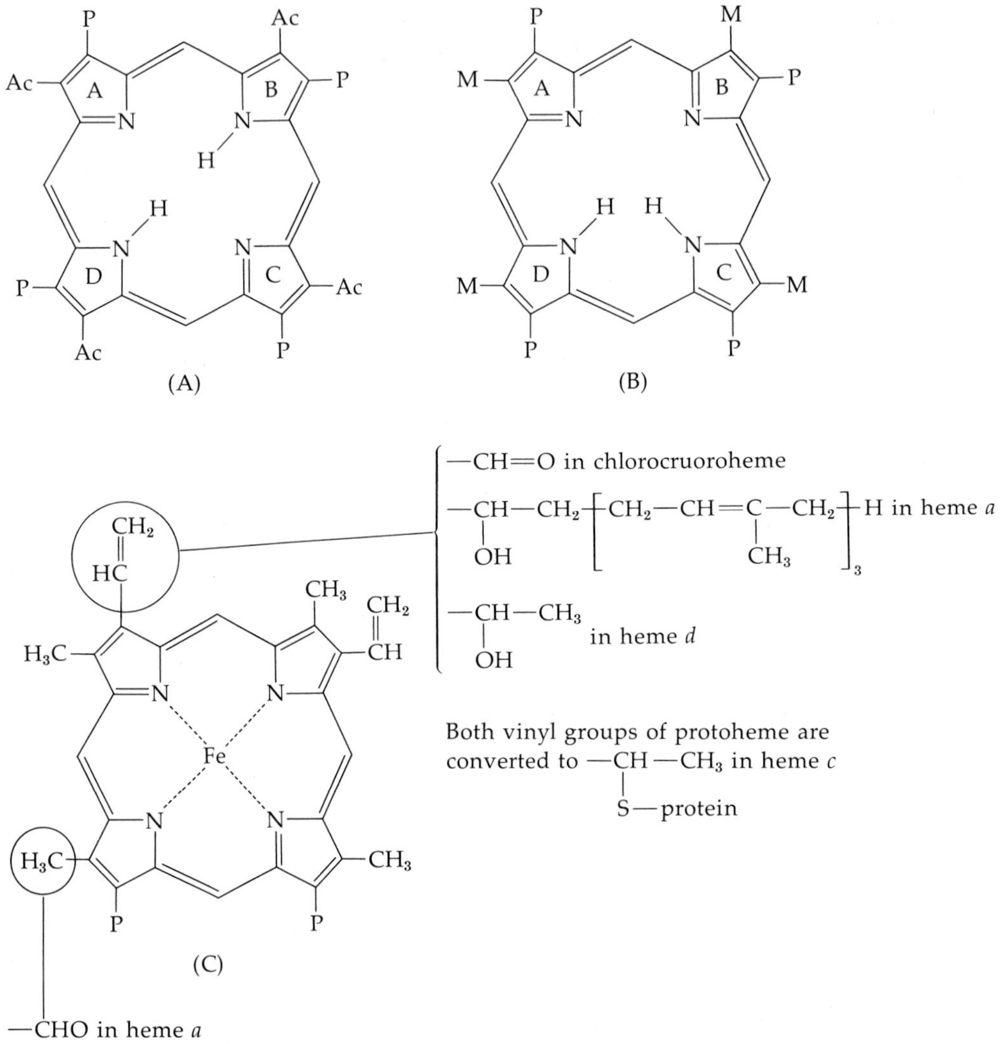

−CH=O in chlorocruoroheme

$-\underset{\underset{OH}{|}}{CH}-CH_2\left[CH_2-CH=\underset{\underset{CH_3}{|}}{C}-CH_2\right]_3H$ in heme a

$-\underset{\underset{OH}{|}}{CH}-CH_3$ in heme d

Both vinyl groups of protoheme are converted to $-\underset{\underset{S-protein}{|}}{CH}-CH_3$ in heme c

−CHO in heme a

Fig. 10-1 Structures of some biologically important porphyrins. (A) Uroporphyrin I, Ac = —CH₂COOH and P = —CH₂CH₂COOH. (B) Coproporphyrin III. (Note that a different tautomeric modification of the ring is pictured here than in (A). This kind of tautomerism occurs within all of the porphyrins.) (C) Protoheme: The Fe^{2+} complex of protoporphyrin IX, present in hemoglobin, cytochromes b, and various other proteins.

2. Hemes

Note that porphyrins contain a completely conjugated system of double bonds and that in the center there are two hydrogen atoms attached to two of the nitrogens. In fact, the two hydrogens are free to move to other nitrogens in the center with rearrangement of the double bonds. Thus, we have tautomerism as well as resonance in the heme ring. The two central hydrogens can be replaced by many metal ions to form highly stable chelates. The complexes with Fe(II) are known as **hemes** and the complex of protoporphyrin IX with Fe(II) as **protoheme.** These compounds may also be designated as **ferrohemes** and the Fe(III) compounds formed by oxidation as **ferrihemes.** Since iron tends to have a coordination number of six, other ligands can attach to the iron from the two free sides of the heme. When these are nitrogen ligands, such as pyridine or imidazole, the resulting compounds have characteristic absorption spectra and are referred to as **hemochromes.** Some of the heme-containing proteins such as cytochrome b_5, which contains two imidazole groups in these positions, can also be regarded as hemochromes and have the characteristic hemochrome spectra.

Several modifications of protoheme are known as indicated in Fig. 10-1. To determine which heme exists in a particular protein, it is customary to split off the heme by treatment with acetone and hydrochloric acid and to convert it by addition of pyridine to the pyridine hemochrome for spectral analysis. By this means, protoheme has been shown to occur in hemoglobin, myoglobin, cytochromes of the b, o, and P-450 types and in catalases and peroxidases. Cytochromes a and a_3 contain **heme a**,[3] while the terminal oxidase system of many bacteria contains **heme d** (formerly a_2). **Heme c** (present in cytochromes c, b_4, and f) is a variation in which SH groups of the protein have added to the vinyl groups of protoheme. Recently discovered hemes include siroheme (Fig. 14-16) and a formyl group containing heme from human erythrocytes.[3a]

3. Functions of Heme Proteins

Heme is found in all organisms except the anaerobic clostridia and lactic acid bacteria. The lactic acid bacteria are remarkable in that apparently they contain no iron whatsoever, whereas clostridia are rich in nonheme iron compounds. Heme proteins of blood carry oxygen reversibly, whereas those of terminal oxidase systems and hydroxylases and oxygenases "activate" oxygen, allowing it to combine with carbon compounds or with hydrogen. Other heme compounds catalyze reactions not with O_2 but with H_2O_2. These are the **peroxidases** and **catalases.** Still another group of heme enzymes are purely electron transferring compounds.

As Ingraham has remarked,[4] "living in a bath of 20% oxygen, we tend to forget how reactive it is." From a thermodynamic viewpoint, all living matter is extremely unstable with respect to combustion by oxygen. Ordinarily a high temperature is required and if we are careful with fire, we can expect to escape a catastrophe. However, 1 mol of copper properly chelated can catalyze consumption of all the air in an average room in a second.[4] Thus, as biochemists we are interested both in the fact that O_2 is kinetically stable and unreactive but that oxidative enzymes are able to promote rapid reactions.

Two oxygen atoms, each with six valence electrons might reasonably be expected to form a double-bonded structure with one σ and one π bond as follows:

$$:\overset{..}{O}=\overset{..}{O}:$$

In fact, it has long been known that O_2 is paramagnetic and contains *two* unpaired electrons. From this evidence we might assign O_2 the structure shown below at the left. However, Pauling suggests that the structure to the right containing two three-electron bonds (each involving one of these unpaired electrons) also contributes in an important way:

$$:\overset{..}{O}-\overset{.}{O}: \qquad :\overset{.}{O}-\overset{...}{O}:$$

The oxygen molecule is very stable, and it is difficult to add an electron to form the reactive **superoxide anion radical** O_2^-:

$$:\overset{..}{O}-\overset{..}{O}:^-$$

For this reason, oxidative attack by O_2 tends to be slow. However, once an electron has been acquired, it is easy for additional electrons to add to the structure and further reduction occurs easily. The biochemical question suggested is, "How can some heme proteins carry O_2 reversibly without any oxidation of the iron contained in them while others *activate* oxygen toward reaction with substrates?"

4. Oxygen Carrying Proteins

In Chapter 4, Section E,5, we examined the remarkable behavior of hemoglobin of red blood cells in cooperative binding of four molecules of O_2. We also looked at its structural relationship to the monomeric myoglobin of muscle, whose role is thought to be one of assisting the diffusion of O_2 into tissues[5] and possibly of providing a store of oxygen. The iron in hemoglobin and myoglobin is always ferrous. Red cells contain a special system for reducing any iron that accidentally is converted to the ferric state (see p. 237; Box 10-A). The majority opinion is that the binding of O_2 to

---

BOX 10-A

GLUTATHIONE PEROXIDASE AND ABNORMALITIES OF RED BLOOD CELLS

The processes by which hemoglobin is kept in the Fe(II) state and functioning normally within intact erythrocytes is vital to our health. Numerous hereditary defects leading to a tendency toward anemia have helped to unravel the biochemistry indicated in the accompanying scheme.[a]

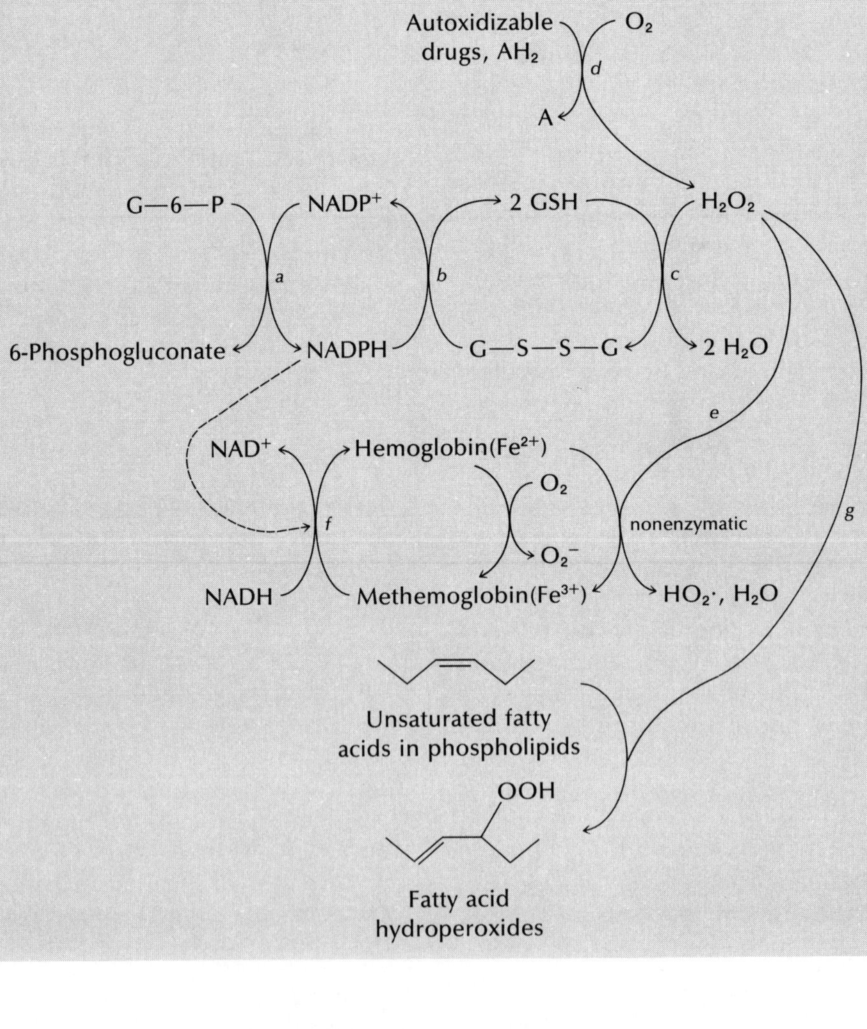

the iron in the heme does *not* cause a temporary change in the oxidation state of the metal. However, there is some controversy on this point.[6] According to Ingraham, oxidation of the metal does not occur because of the previously mentioned difficulty of adding one electron to the oxygen molecule. At the same time, it is difficult to

transfer two electrons from the metal to oxygen because Fe(IV) is unstable.

Bonding of the heme iron to oxygen is thought to occur by donation of a pair of electrons by the oxygen to the metal. In the unoxygenated hemoglobin the ferrous ion is in the "high-spin" state; four of the five $3d$ orbitals in the valence shell of

---

Box 10-A (*Continued*)

About 90% of the glucose utilized by erythrocytes is converted by glycolysis to lactate but ~10% is oxidized (via glucose 6-phosphate) to 6-phosphogluconate. The oxidation (reaction *a*) is catalyzed by glucose-6-phosphate dehydrogenase (Eq. 8-42) using $NADP^+$. This is the principal reaction providing the red cell with NADPH for reduction of glutathione (Box 7-G) according to reaction *b*. Despite the important function of glucose-6-P dehydrogenase, over 100 million persons, principally in tropical and Mediterranean areas, have a hereditary deficiency of this enzyme. Furthermore, genetic variations are numerous, at least 22 types having been identified. Nevertheless, the lack of the enzyme is truly detrimental and leads to excessive destruction of red cells and anemia during some sicknesses and in response to administration of certain drugs. The survival of the defective genes, like that for sickle cell hemoglobin (Box 4-D) is thought to result from increased resistance to malaria parasites.

Other erythrocyte defects that lead to drug sensitivity include a deficiency of glutathione (resulting from a decrease in its synthesis) and a deficiency of glutathione reductase (reaction *b*). The effects of drugs have been traced to the production of H_2O_2 from oxygen (reaction *d*). Current thinking is that the function of glutathione and of the enzymes catalyzing reactions *a*, *b*, and *c* is to destroy hydrogen peroxide arising naturally or from autoxidation of drugs. The selenium-containing peroxidase (Box 9-F) is the principal enzyme destroying H_2O_2 (reaction *c*) in red blood cells; catalase is thought to function in a similar way (Eq. 10-5) and both enzymes are probably necessary for optimal health.

An excess of H_2O_2 can damage erythrocytes in

two ways. One is to cause excessive oxidation of functioning hemoglobin to the Fe(III)-containing methemoglobin. (Methemoglobin is also formed spontaneously during the course of the oxygen-carrying function of hemoglobin. It is estimated that normally as much as 3% of the hemoglobin may be oxidized to methemoglobin daily.) The methemoglobin formed is reduced back to hemoglobin through the action of **NADH-methemoglobin reductase** (reaction *f*). A smaller fraction of the methemoglobin is reduced by a similar enzyme requiring NADPH (as indicated by the dashed arrow). A hereditary lack of the NADH-methemoglobin reductase is also known.

A second destructive function of H_2O_2 is attack on double bonds of unsaturated fatty acids of the phospholipids in cell membranes. The resulting fatty acid hydroperoxides can react further with C—C chain cleavage and disruption of the membrane. This is thought to be the principal cause of the hemolytic anemia induced by drugs in susceptible individuals. Glutathione peroxidase is thought to decompose these fatty acid hydroperoxides. Vitamin E, acting as an antioxidant, is also needed for good health of erythrocytes (Box 10-C).

Some cases of granulomatous disease (Section B,6) are accompanied by a decreased glutathione peroxidase activity and a decreased microbicidal activity of phagocytes. It has been suggested that hydroperoxides of fatty acids interfere with normal phagocytosis by inhibiting required enzymes.[b]

[a] I. Chanarin, *in* "Biochemical Disorders in Human Disease," 3rd ed. (R. H. S. Thompson and I. D. P. Wooton, eds.), pp. 163–173. Academic Press, New York, 1970.

[b] R. E. Serfass and H. E. Ganther, *Nature* (*London*) **255**, 640–641 (1975).

the iron ion each contain one unpaired electron. The addition of oxygen causes the iron to revert to the "low-spin" state in which all of the electrons are paired. In fact, the loss of paramagnetism upon oxygenation of hemoglobin attracted the interest of chemists many years ago. It is generally thought that the stability of heme-oxygen complexes is enhanced by "back double bonding," i.e., the donation of an electron pair from one of the filled d orbitals of the iron atom to form a π bond with the adjacent oxygen.[7] This can be indicated symbolically as follows:

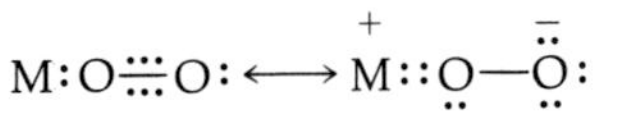

$$M{:}O{\vdots\vdots}O{:} \longleftrightarrow M{::}\overset{+}{O}{-}\overset{\overline{}}{O}{:}$$

These structures, which have been formulated by assuming that one of the unshared electron pairs on O_2 forms the initial bond to the metal, are expected to lead to an angular geometry which has recently been observed unequivocally in a model complex.[8]

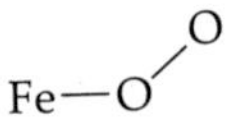

$$Fe{-}O\overset{\displaystyle O}{\diagup}$$

The possibility that one of the pairs of π electrons of oxygen forms the initial bond has also been proposed[4]:

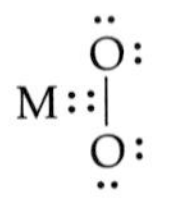

$$M{::}\overset{\overset{\displaystyle \ddot{O}:}{\big|}}{\underset{\displaystyle \ddot{O}:}{}}$$

However, the observed geometry appears to exclude this structure.

All oxygen-carrying heme proteins have an imidazole group of a histidine side chain in the coordination position of iron on the side opposite the oxygen binding site. Without such a group, heme does not combine with oxygen, and it is noteworthy that coordination with heterocyclic nitrogen compounds is known to favor formation of low-spin iron complexes. Some very simple compounds that closely mimic the behavior of myoglobin have been prepared by attaching an imidazole group by a chain of appropriate length to the edge of a heme ring.[7,8] Similar compounds bearing a pyridine ring in the fifth coordination

position have also been prepared, but they have a low affinity for oxygen. Thus, the polarizable imidazole ring itself seems to play an essential role in promoting oxygen binding. The participation of the π electrons of the imidazole ring in bonding to the iron has been suggested (Eq. 10-2).[7] The π bonding to the iron would allow the iron to back bond more strongly with an O_2 atom entering the sixth coordination position. The diagrams in Eq.

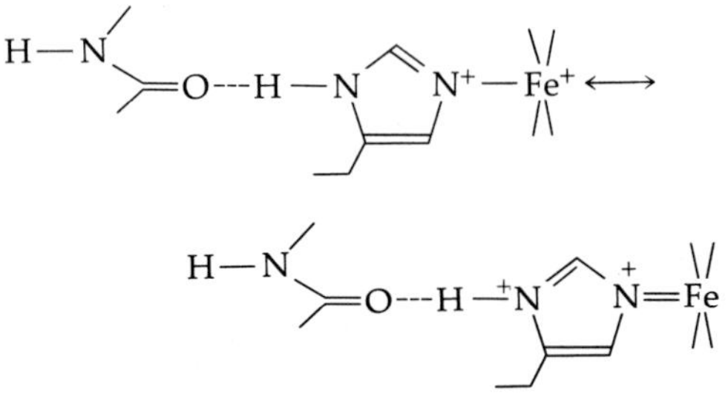

$$\tag{10-2}$$

10-2 illustrate another feature found frequently in heme proteins. The N—H group of the imidazole is hydrogen bonded to a peptide backbone C=O group. It is easy to see how oxygenation could induce a change in the position of the H-bonded proton and of the charge distribution in the H-bonded network of the protein.*

An extraordinarily interesting discovery is that the coordination of the heme iron with histidine appears to provide the basis for the cooperativity in binding of oxygen by hemoglobin.[9,10] The radius of high spin iron, whether ferric or ferrous, is so large that the iron cannot fit into the center of the porphyrin ring but is displaced toward the coordinated imidazole group by a distance of 0.06 nm for Fe(II). Thus, in deoxyhemoglobin both iron and the imidazole group lie further from the center of the ring than they do in oxyhemoglobin. In the latter, the iron lies in the center of the por-

---

* The carbonyl group shown in Eq. 10-2 is embedded in the F helix and is hydrogen-bonded to other amide groups. Electron withdrawal into the heme-oxygen complex would tend to strengthen the H bond shown in Eq. 10-2 and to weaken a competing H bond in the helix. This would affect the charge distribution in the upper end of the F helix and could conceivably help to induce a momentary conformational change that could trigger the rearrangement of structure shown in Fig. 4-19. The $\beta_1\alpha_2$ contact in which a change of H bonding takes place is located nearby behind the F and G helices.

phyrin ring because the change to the low-spin state is accompanied by a decrease in ionic radius.[9,11] The change in protein conformation induced by this small shift in the position of the iron ion has already been described (Chapter 4, Section E,5). However, the exact nature of the "triggering" of the conformational changes is not yet clear. To some extent it can be viewed as a mechanical reaction to the shortening of the distance between the porphyrin ring and helix F of hemoglobin (Fig. 4-17). It has been proposed that this shortening induces a tilting of the heme which in turn loosens part of the H-bond network of the protein permitting the other conformational changes previously described (Chapter 4, Section E,5).[10] On the other hand, perhaps sufficient consideration has not been given to changes in electron distribution mentioned in the preceding paragraph. In any event, it is remarkable that nature has so effectively made use of the subtle differences in the properties of iron induced by changes in the electron distribution within the *d* orbitals of this transition metal. The possibility

that transition metal ions in other biochemical structures induce conformational changes in a similar manner should be kept in mind.

The oxygen carrier found in a few groups of invertebrates, such as the sipunculid worms, is a nonheme iron-containing protein, **hemerythrin.**[11a] The subunits, containing about 113 amino acid residues each, are often associated as octamers of C_4 symmetry. Each monomer has an active site that contains two atoms of Fe(II) 0.34 nm apart. An oxygen molecule is thought to fit between an Fe pair as indicated in the following drawing (after Klotz *et al*).[11a]

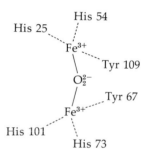

BOX 10-B

VANADIUM

A possible biochemical function of vanadium is suggested by the presence of the **vanadocytes,** green blood cells containing 4% of V(III) and 1.5–2 N H_2SO_4, present in the tunicates (sea squirts, Chapter 1, Section E,1).[a] It has been proposed that the V-containing protein **vanadochrome** is an oxygen carrier. However, this is uncertain and the function remains unknown. Vanadium is accumulated by a number of other marine organisms and is present in animal tissues to the extent of ~0.1 part per million.

Recently vanadium has been shown a dietary essential for rats.[b,c] It is presumably also needed by humans, who typically consume ~2 mg/day. The adult body contains a total of ~30 mg of V. A possible function is in lipid metabolism. Vanadium can assume a variety of oxidation states from +2 to +5. The vanadate ion VO_4^{3-} is the predominant form of V(V) in basic solution. The

VO^{2+} ion is an especially stable double-bonded unit in compounds of V(IV). The chemistry of the element suggests a possible redox function.

Attention has been drawn to the nutritional significance of vanadium by the observation that in high doses it inhibits cholesterol synthesis and lowers the phospholipid and cholesterol content of blood. Vanadium is reported to inhibit development of caries by stimulating mineralization of teeth. Unlike tungsten (Box 14-A) vanadium does not compete with molybdenum in the animal body.[d] However, an inactive vanadium-containing nitrogenase has been identified in *Azotobacter.*

[a] D. B. Carlisle, *Proc. Roy. Soc. B* **171**, 31–42 (1968).

[b] K. Schwarz and D. B. Milne, *Science* **174**, 426–428 (1971).

[c] L. L. Hopkins, Jr., in "Trace Element Metabolism in Animals," (W. A. Hoekstra, J. W. Suttie, H. E. Ganther and W. Mertz, eds.), Vol. 2, pp. 397–406. University Park Press, Baltimore, 1974.

[d] J. L. Johnson, K. V. Rajagopalan and H. J. Cohen, *JBC* **249**, 859–866 (1974).

The O_2 is thought to accept two electrons, oxidizing the two iron atoms to Fe(III) and itself becoming a peroxide dianion O_2^{2-}. The process is completely reversible. The conversion of the oxygen to a bound peroxide ion is supported by studies of resonance Raman spectra (see Chapter 13, Section B,3,b).

Similarly, the blue copper-containing **hemocyanin** of many invertebrates binds one O_2 for two Cu(I) atoms. The oxygen is thought to bridge between the two copper atoms. Since CO does not

$$Cu^+ \text{---} O_2$$
$$| \atop Cu^+$$

bind (as it does to the iron in hemoglobin) Ingraham[4] suggested a nonlinear arrangement. The oxygenated compound is extremely blue with an intensity of light absorption 5–10 times stronger than that of known cupric complexes. Because of this fact it is thought that some of the copper may undergo a change in oxidation state upon binding of oxygen. Further support for the idea comes from the observation that treatment of oxygenated hemocyanin by glacial acetic acid leads to formation of cupric and cuprous ions in equivalent amounts:

$$(CuO_2Cu)^{2+} + H^+ \longrightarrow Cu^{2+} + Cu^+ + HO_2 \qquad (10\text{-}3)$$

A hydroperoxide radical is assumed to be the other product. The hemocyanins are large oligomeric molecules with a striking appearance under the electron microscope.[11b,11c]

Many iron and copper proteins bind O_2 in a manner similar to that of hemoglobin, myoglobin, or hemocyanin but the oxygen is then "activated" and undergoes further reaction. We will look at such enzymes shortly, but first we will go to the other extreme and examine a group of heme enzymes that function strictly as *electron* carriers.

5. The Cytochromes

The changes in the absorption bands seen by Keilin within cells result from oxidation of the ferrous iron of cytochromes to ferric iron. Thus,

these proteins serve as one-electron carriers whose active centers are heme groups. Numerous cytochromes from various sources have been described.[12–14] The classification into groups *a*, *b*, and *c* according to the position of the α band (the longest wavelength band, Fig. 10-2) in the absorption spectrum follows the practice introduced by Keilin. However, nowadays it is more common to designate a new cytochrome by giving the wavelength of the α band, e.g., cytochrome 552 or cyt *b*-557.5. Furthermore, present practice ties the designations *a*, *b*, *c*, and *d* to the heme type (Fig. 10-1).

Cytochromes of the *b* type contain protoheme, which is also found[13–16] in the bacterial **cytochromes *o***. Typical cytochromes *b* do not react with O_2 but the cytochromes *o* serve as terminal electron acceptors (cytochrome oxidases) and are autoxidizable by O_2. Another protoheme-

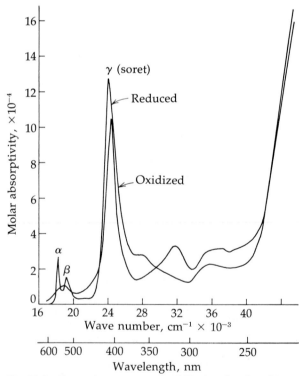

Fig. 10-2 Absorption spectra of oxidized and reduced horse heart cytochrome c at pH 6.8. From data of E. Margoliash and N. Frohwirt, *BJ* **71,** 570–572 (1959).

containing cytochrome, involved in hydroxylation, is called **cytochrome P-450.** Here the 450 refers to the position of the intense "Soret band" (also called the γ band) of the spectrum (Fig. 10-2) rather than to that of the α band. Other properties are also used in arriving at designations for cytochromes. For example, cytochrome a_3 has a spectrum similar to that of cytochrome a but it reacts readily with both CO and O_2.

Cytochrome c is one of the few intracellular heme pigments that is soluble in water and that can be removed easily from membranes. Mitochondrial cytochrome c was the first of the cytochromes to be purified and crystallized. Its structure in both the ferric and ferrous forms has been determined by X-ray diffraction. A small protein of MW ~13,000 containing 104 amino acid residues, cytochrome c has been isolated from plants, animals, and eukaryotic microorganisms.[12,17,18] Complete amino acid sequences have been determined for over 50 species. Within the peptide chain 28 positions are invariant. A number of other positions contain only "conservative" substitutions. Cytochrome c is one of the proteins that is being used in attempting to trace evolutionary relationships between species by observing differences in sequences. For example, man and the chimpanzee have identical cytochrome c, but 12 differences in amino acid sequence occur between man and the horse and 44 between man and *Neurospora*.[17] The related cytochrome c_2 of the photosynthetic bacterium *Rhodospirillum rubrum* is thought to have diverged in evolution 2×10^9 years ago from the precursor of mammalian cytochrome c. Even so, 15 residues remain invariant and functional equivalence is found at other positions.[19]

Structural studies[12,17-19] on cytochrome c and on cytochrome c_2 show that the heme group provides the backbone around which the peptide chain is wound. The 104 residues of cytochrome c are barely enough to envelope the heme. In both the oxidized and reduced forms of the protein, methionine 80 (to the left in Fig. 10-3) and histidine 18 (to the right) fill the fifth and sixth coordination positions of the iron. The heme is thus tightly wrapped by the polypeptide chain and nearly inaccessible to the surrounding solvent. This fact poses a problem.

How does an electron enter ferricytochrome c to reduce it and how does an electron exit from the tightly closed ferrocytochrome c to reduce cytochrome oxidase? One possibility is that sulfur (of methionine 80) donates an electron to Fe^{3+} leaving an electron deficient radical. Conceivably the "hole" so created could be filled by an electron jumping from the —OH group of the adjacent tyrosine 67. This tyrosine could then swing to the surface and make contact with tyrosine 74, accepting an electron from it.[17] Thus, the reductase acting on ferricytochrome c would interact at the surface with tyrosine 74. While such a mechanism involving transient swinging tyrosine radicals is unprecedented, it merits consideration. Related suggestions,[20,21] are that electrons flow singly or as pairs from the surface through hydrogen-bonded paths to reach tyrosine 67 and methionine 80. Electron exchange through the exposed edge of the heme has also been suggested.[22] With regard to the reoxidation of cytochrome c, Dickerson speculated that a group from cytochrome oxidase, the next enzyme in the electron transport chain, may protrude into the crevice that opens in ferrocytochrome c to accept an electron.

The structure of cytochrome b_5 solubilized from liver microsomes has also been determined. Although its exact function is not known, it presumably plays a role similar to that of cytochrome c in delivering electrons to a fatty acid desaturating system located in the endoplasmic reticulum. The protein contains 93 amino acids plus another 44 (largely hydrophobic) which are cleaved from the N-terminus during solubilization of the protein. Presumably this N-terminal portion provides a hydrophobic tail buried in the ER membranes. The heme in cytochrome b_5 is not covalently bonded to the protein but is held tightly between two histidine side chains. The folding pattern of the protein is completely unlike that of either cytochrome c or myoglobin. Again, no pathway for transport of an electron between

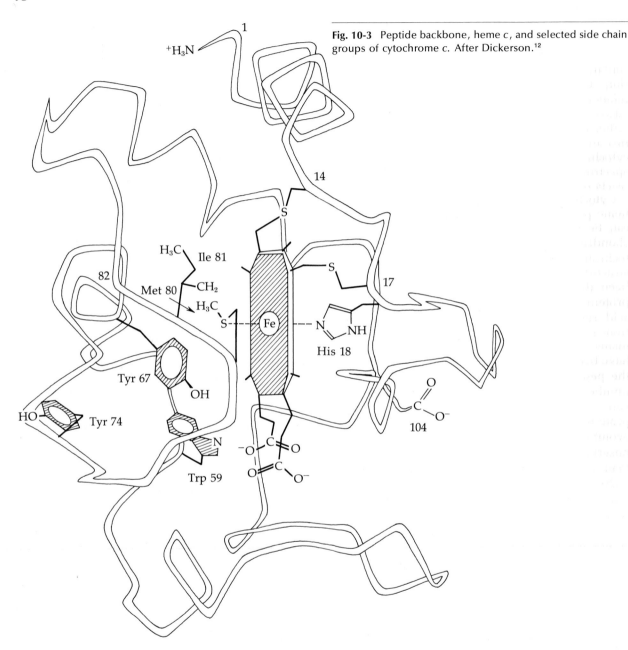

Fig. 10-3   Peptide backbone, heme *c*, and selected side chain groups of cytochrome *c*. After Dickerson.[12]

the iron atom and the surface of the molecule is immediately evident.[23]

Structures of other cytochromes functioning in the electron transport chain are not yet known. Cytochrome c_1 (α band of reduced compound at 554 nm compared with 550 nm in cyt *c*) contains heme bound much as in cytochrome *c* but possibly to a higher molecular weight protein. Cytochrome *b* may exist in several forms (Table 10-5). The molecular weight is higher than that of cytochrome *c*, and cytochrome *b* may exist as a tight complex with ubiquinone.[24]

Cytochrome oxidases perform a unique function in aerobic organisms by combining with O_2 in much the same way as does hemoglobin and then rapidly reducing the O_2 to two molecules of H_2O.[24a] Not only is the O—O bond broken but also four electrons are required for the reduction. Clearly the process is complex and we know little about it. It is significant that the cytochrome oxidase of mammalian mitochondria contains two hemes (cytochrome *a*) plus two atoms of Cu(I) per functional unit. Thus, by reducing both molecules of cytochrome *a* and the two copper atoms, four electrons can be stored for the reduction of one molecule of O_2. The chemistry of cytochrome oxidase is poorly understood. As originally observed by Keilin, only half of the cytochrome *a* combines with CO. It is designated cytochrome a_3. From results of SDS-polyacrylamide gel electrophoresis six to seven subunits of MW = 5000 to 42,000 have been identified in the cytochrome oxidase of yeast.[24b,c] It is of interest that the three larger subunits appear to be encoded by genes in the mitochondrial DNA. The heme groups are bound to some of the smaller peptides. It has been proposed that in the intact enzyme O_2 is bound initially between the Fe atom of cytochrome a_3 and a cuprous ion: $a_3{}^{2+}$—O_2—Cu^+. In the next step the O_2 would be reduced in a two-electron process to a bridged peroxide and subsequently to two water molecules.

Several other cytochromes are discussed in Sections F,1 and F,2. Detailed reviews of bacterial cytochromes[13,15] as well as more general discussions[12,24] have been published.

6. Catalase and Peroxidases

Another group of heme enzymes catalyze reactions not of O_2 but of hydrogen peroxide H_2O_2. The peroxidases are widespread in plant tissues where they are found especially in the peroxisomes. Also occurring to some extent in animal tissues, they catalyze the following reaction (Eq. 10-4):

$$H_2O_2 + AH_2 \longrightarrow 2\,H_2O + A \qquad (10\text{-}4)$$

Catalase, an enzyme of general occurrence in aerobic cells, may sometimes amount to as much as 1% of the dry weight of bacteria. The enzyme catalyzes the breakdown of H_2O_2 to water and oxygen. It is thought to do this by a mechanism fundamentally the same as that employed by peroxidases. If Eq. 10-4 is rewritten with H_2O_2 for AH_2 and O_2 for A, we have the following (Eq. 10-5):

$$2\,H_2O_2 \xrightarrow{\text{catalase}} 2\,H_2O + O_2 \qquad (10\text{-}5)$$

The action of catalase is very fast, almost 10^4 times faster than that of peroxidases. The molecular activity per catalytic center is about 2×10^5 s^{-1}.

Catalase is thought to exert a protective function preventing the accumulation of H_2O_2 which might be harmful to cell constituents. The complete intolerance of obligate anaerobes to oxygen may result from their lack of this enzyme. Some support for this idea comes from the existence of a hereditary condition of **acatalasemia**.[25] Persons with extremely low catalase activity are found worldwide but are especially numerous in Korea. In Japan it is estimated that there are 1800 persons lacking catalase. Since about half of these persons have no symptoms, catalase might be judged unessential. However, many of the individuals affected develop ulcers around their teeth which may lead to serious problems. Apparently hydrogen peroxide produced by bacteria accumulates and oxidizes hemoglobin to methemoglobin (Box 10-A) depriving the affected tissues of oxygen.

The most widely studied peroxidase is one isolated from horseradish and containing one heme in a protein of MW $\sim$40,000. Catalases are usually tetramers of MW $\sim$250,000. Both peroxidases and catalase contain high-spin Fe(III) and resemble metmyoglobin in properties. The enzymes are reducible to the Fe(II) state in which form they are able to combine (irreversibly) with O_2. We see that the same active center found in hemoglobin is present in peroxidases and catalase but its chemistry has been modified by the proteins. The affinity for O_2 has been altered drastically and a new group of catalytic activities has emerged for the ferriheme-containing proteins.

The mechanisms of action of catalases and peroxidases are incompletely understood but attention has focused on a series of striking colored intermediates formed in the presence of substrates.[25a] When a slight excess of H_2O_2 is added to a solution of peroxidase, the dark brown enzyme first turns olive green (compound I), then pale red as it forms compound II. Compound II slowly reacts with substrate AH_2 (or with another H_2O_2 molecule) to regenerate the original enzyme. This sequence of reactions is indicated by the top horizontal row of arrows in Eq. 10-6.

$$
\begin{array}{c}
\text{from } AH_2 \text{ or } H_2O_2 \\
H_2O_2 \quad \overbrace{1\,e^- \quad 1\,e^-} \\
E\!-\!Fe(III) \xrightarrow{} I \xrightarrow{} II \xrightarrow{} E\!-\!Fe(III) \\
\downarrow \qquad\qquad\qquad H_2O_2 \\
E\!-\!Fe(II) \text{---} \underset{H_2O_2}{\text{-----}} \\
\qquad\qquad\qquad \longrightarrow E\!-\!Fe(II)O_2 \\
O_2 \qquad\qquad \text{Oxyperoxidase} \\
\text{(compound III)}
\end{array}
\qquad (10\text{-}6)
$$

Titrations have established that compound I is converted to compound II by a one-electron reduction and compound II to free peroxidase by another one-electron reduction (e.g., using ferrocyanide). Thus, the iron in compound I could formally be designated Fe(V) and that in II as Fe(IV). However, this tells us little about the fate of the H_2O_2 in its reaction with the enzyme to form compounds I and II. Are the oxygen atoms of H_2O_2 present in compounds I and II or not? The answer has been the topic of hot debate and is still not settled. However, consider the following information. The enzyme in the Fe(III) form can be reduced to Fe(II), as previously mentioned. When the Fe(II) enzyme reacts with H_2O_2, it is apparently converted to compound II, suggesting that the latter may be an Fe(II) complex of the peroxide anion[26–28]:

$$E\!-\!Fe^+(II)\!-\!OOH$$

It is clear that the iron is in a low-spin state. Nevertheless, workers prefer to regard compound II as

a **ferryl iron** complex which could be derived from the preceding structure by addition of a proton and loss of water.

$$E\!-\!Fe^{2+}(IV)O$$

Addition of H_2O_2 appears to convert compound II to compound III which is thought to be the same as "oxyperoxidase." The latter is formed by addition of O_2 to the ferrous form of the free enzyme (Eq. 10-6). The H_2O_2 used in conversion of II to oxyperoxidase is presumably reduced to two molecules of water.

If the structure of compound II is uncertain, that of compound I is even more so. Initially, it was thought to be a complex of H_2O_2 or its anion with Fe(III). However, the magnetic and spectral properties are thought inappropriate for this structure. Much speculation about the location of the attached peroxide oxygens has ensued. Here are a few more details for the reader to consider in thinking about the problem. The H_2O_2 in Eq. 10-5 can be replaced by ROOH, an organic peroxide, and the interconversions from I to II and on to free enzyme can be carried out by a one-electron donor such as K_2IrBr_6. If an electron is donated by substrate AH_2, a free radical $AH\cdot$ will be the product of the conversion of compound I to compound II. That free radical could then donate a second electron to form free enzyme or a second molecule of AH_2 could be used to form two radicals $AH\cdot$. The latter could be disproportionate to A and AH_2.

Chloroperoxidase,[29] isolated from the mold *Caldariomyces fumago*, carries out organic chlorination reactions of the following type (Eq. 10-7) using H_2O_2 plus Cl^-.

$$
\begin{array}{c}
\text{[cyclopentanedione]} + Cl^- + H_2O_2 \longrightarrow \\
\\
\text{[chloro cyclopentanedione]} + H_2O + OH^- \qquad (10\text{-}7)
\end{array}
$$

Chlorination proceeds through a transient iron-bound intermediate spectrally indistinguishable from compound I. It has been suggested[29] to have the structure Fe—OCl. The protein has a molecular weight of ~42,000 and is isolated in a low-spin ferric state. The reduced ferrous enzyme is a high-spin form with spectroscopic properties almost identical to those of cytochrome P-450 (Section G,2f).[29,30] **Lactoperoxidase** of milk with $I^- + H_2O_2$, promotes an entirely analogous *iodination* of tyrosine and histidine residues of proteins. With radioactive $^{125}I^-$ or $^{131}I^-$ it provides a convenient and much used method for labeling of proteins in exposed surfaces of membranes.[31]

Myeloperoxidase, present in polymorphonuclear leukocytes (Chapter 1, Section E,2,b), utilizes H_2O_2 and a halide ion to kill ingested bacteria.[32,33] Phagocytosis induces increased respiration by the leukocyte and generation of H_2O_2, partly in a membrane-bound NADPH oxidase. Some of the H_2O_2 is used by myeloperoxidase to attack the bacteria, apparently through generation of HOCl by peroxidation of Cl^-.[33a] Other oxygen-dependent killing mechanisms also may be used.[34] Hereditary lack of NADPH oxidase (an X-linked trait) may be a cause of the serious **granulomatous disease** in which resistance to infection by many common bacteria is lacking.

C. NONHEME IRON PROTEINS

By no means is all of the iron within cells chelated by porphyrin groups. While hemerythrin (Section B,4) has been known for many years, the general significance of nonheme iron proteins was not seen until large-scale preparation of mitochondria was developed by D. Crane in about 1945. It was noted that the iron content of mitochondria far exceeded that of the heme proteins present. An important discovery was made in 1960 by H. Beinert, who was studying the mitochondrial dehydrogenase systems for succinate and for NADH. He observed that when the electron transport chain was partially reduced by these substrates and the solutions were frozen at low temperature and examined, a strong epr signal was

observed at $g = 1.94$. The signal was obtained only upon reduction by substrate, and fractionation pointed to the nonheme iron proteins. While their functions are still something of a mystery, it has been suggested that at least six proteins of this type may be present in the electron transport chain.[35-36a] Three others are present in association with the flavin-containing succinate dehydrogenase (Chapter 8, Section I,3).[36b]

The presence of nonheme iron proteins is even more evident in the clostridia which contain no heme at all. It was from these bacteria that the first nonheme iron protein was isolated and named **ferredoxin.** This protein, which has a remarkably low reduction potential of $E^0 = -0.41$ V, participates in the pyruvate-ferredoxin oxidoreductase reaction (Chapter 8, Section J,3), in nitrogen fixation in some species, and in formation of H_2. A small green-brown protein, it contains only 54 amino acids but complexes eight atoms of iron. If the pH is lowered to ~1, eight molecules of H_2S are released. Thus, the protein contains eight "labile sulfur" atoms in some kind of iron sulfide linkage. Ferredoxins were only the first members to be discovered of a large family of **iron–sulfur proteins.**[37-39] Most contain iron and "labile sulfur" in a 1:1 ratio, but the number of iron atoms per protein varies. In addition, one group contains no labile sulfur, the iron being held by four cysteine side chains. The simpler iron-sulfur proteins may be classified as shown in the accompanying tabulation. In addition, there are more complex iron-sulfur proteins such as **nitrogenase,** (Chapter 14,

Iron and labile sulfur content	Names of proteins
1 Fe	Rubredoxins
2 Fe, 2 S^{2-}	Chloroplast ferredoxin, adreno-redoxin, putidaredoxin, *E. coli* ferredoxin[40]
4 Fe, 4 S^{2-}	High potential iron protein (*Chromatium*), some bacterial ferredoxins, hydrogenase
n Fe, n S^{2-}	Bacterial ferredoxins (usually $n = 8$)

Section A,2) which also contains molybdenum. The standard reduction potentials of iron–sulfur proteins cover a remarkable span all the way from the -0.42 V of spinach ferredoxin to $+0.35$ V of the so-called high potential iron protein of *Chromatium*.

The structures of several of these proteins have been determined by X-ray diffraction.[38,41] The simplest is the **rubredoxin** of *Clostridium pasteurianum,* a small peptide of MW ~6000 (Fig. 10-4). Among the 54 amino acids are four cysteine residues whose side chains form a distorted tetrahedron about a single iron atom (Fig. 10-4).[38] Three of the Fe—S bond distances are "normal"—about 0.23 nm; but the fourth, that to cysteine 41, is unusually short (0.205 nm). The exact function of the clostridial rubredoxin is not known, but it is thought to participate in electron transport and can substitute for ferredoxin in

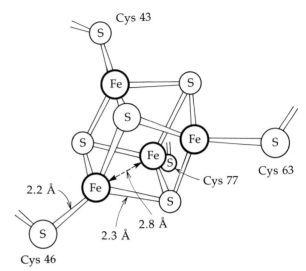

Fig. 10-5 The Fe_2S_4 cluster of the *Chromatium* high potential iron protein.

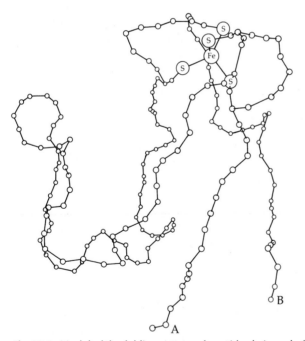

Fig. 10-4 Model of the folding pattern of peptide chain and of iron coordination in rubredoxin of *Clostridium pasteurianum* A and B are the N- and C-terminals. From J. R. Herriott, L. C. Sieker, L. H. Jensen, and W. Lovenberg, *JMB* **50,** 391–406 (1970).

some reactions. A larger rubredoxin of MW ~19,000 and able to bind two iron ions is involved in electron transport in a hydroxylase system of *Pseudomonas* (Section G,2,f).[42]

X-Ray diffraction studies have shown that the 86 amino acid polypeptide chain of the high potential iron protein of *Chromatium* is wrapped around an **iron–sulfur cluster** containing the side chains of four cysteine residues plus four iron atoms and four sulfur atoms (Fig. 10-5).[41] Each cysteine sulfur attaches to one atom of Fe, the four Fe atoms forming an irregular tetrahedron with an Fe—Fe distance of ~0.28 nm. The four labile sulfur atoms (S^{2-}) form an interpenetrating tetrahedron 0.35 nm on a side with each of the sulfur atoms bonded to three iron atoms. The cluster is ordinarily able to accept only a single electron. The iron–sulfur cluster structure was a surprise, but after its discovery it was found that ions such as $[Fe_4S_4(S—CH_2CH_2COO^-)_4]^{6-}$ assemble spontaneously from their components and have a similar cluster structure.[43,44] Thus, as is usual, living things have simply improved upon a bonding arrangement that arose naturally.

The structure of a bacterial ferredoxin[38,45] from *Peptococcus aerogenes* is shown in Fig. 10-6. Note

Fig. 10-6 Plot of the α-carbon, iron, and sulfur positions for ferredoxin of *Peptococcus aerogenes*[45]: ($\odot$) Fe, ($\bigcirc$) S^{2-}, ($\otimes$) S (cysteine), and ($\bullet$) C_α.

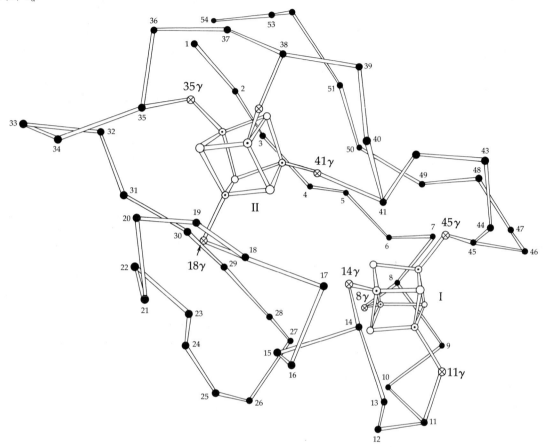

that this protein with eight Fe and eight labile sulfur atoms contains two Fe_4S_4 clusters with essentially the same structure as that of the *Chromatium* protein. Each cluster appears to accept one electron. Much of the amino acid sequence in the first half of the ferredoxin chain is repeated in the second half, suggesting that the present chain may have originated as a result of gene duplication. Many invariant positions are present in the sequence, including those of the cysteine residues forming the Fe—S cluster.

A surprising fact is that the reduction potentials are very different for bacterial ferredoxins (Fd) and for the high potential protein of *Chromatium*

although structures of their active centers appear virtually identical. It is thought that the Fe_4S_4 clusters may exist in three oxidation states (Eq. 10-8),

$$[Fe_4S_4(SR)_4]^{3-} \xrightarrow[e^-]{} [Fe_4S_4(SR)_4]^{2-} \xrightarrow[e^-]{} [Fe_4S_4(SR)_4]^- \quad (10\text{-}8)$$

| Reduced Fd | Oxidized Fd reduced high potential iron protein | Oxidized high potential iron protein; superoxidized Fd |

each differing by one electron.[46-49] The *Chromatium* protein and the ferredoxins would have the

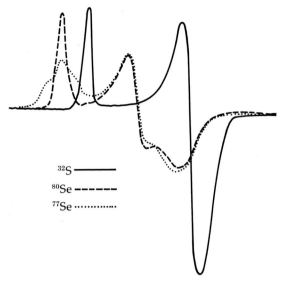

^{32}S ———
^{80}Se - - - - -
^{77}Se ·············

Fig. 10-7 Electron paramagnetic resonance spectrum of the Fe-S protein putidaredoxin in the natural form (^{32}S) and with labile sulfur replaced by selenium isotopes. Well-developed shoulders are seen in the low-field end of the spectrum of the ^{77}Se (spin = ½)-containing protein. From W. H. Orme-Johnson, R. E. Hansen, H. Beinert, J. C. M. Tsibris, R. C. Bartholomaus, and I. C. Gunsalus, *PNAS* **60,** 369–372 (1968).

middle oxidation state in common.* The size of the cluster is a little less in the more oxidized states (in the *Chromatium* protein the Fe—Fe distance changes from 0.281 to 0.272 nm upon oxidation). Synthetic iron–sulfur clusters have weakly basic properties[47] and accept protons with a pK_a of from 3.9 to 7.4.

It is possible to remove iron and labile sulfur from some iron-sulfur proteins and to again reconstitute active enzyme by adding sulfide and iron atoms in an appropriate manner. Likewise, it has been possible to exchange the natural isotope ^{56}Fe (nuclear spin zero) with ^{57}Fe, which has a

* However, an 8-iron ferredoxin from *Azotobacter* contains one cluster with $E^0 = -0.42$ V and a second with $E^0 = +0.34$ V; from epr measurements it appears that *both* clusters function between the -2 and -1 (oxidized and superoxidized) states of Eq. 10–8, despite the widely differing potentials.[49a] Soluble mammalian succinate dehydrogenase contains three iron–sulfur centers of $E^0 = -0.40$, -0.005, and $+0.06$ V, respectively. The highest potential center in this enzyme also appears to operate between the -2 and -1 states of Eq. 10–8.[49b]

magnetic nucleus.[50] In a similar manner ^{32}S can be replaced by ^{77}Se. The resulting proteins appear to function naturally and give epr spectra containing hyperfine lines that result from interaction of these nuclei with unpaired electrons in the Fe_4S_4 clusters (Fig. 10-7). These observations suggest that electrons accepted by Fe_4S_4 clusters are not localized on a single type of atom but interact with nuclei of both Fe and S.

Ferredoxins from chloroplasts contain two iron atoms and two labile sulfur atoms, presumably with the following structure:

$$\text{Protein}\left|\begin{array}{c} -S \\ \\ -S \end{array}\begin{array}{c} S \\ Fe \quad Fe \\ S \end{array}\begin{array}{c} S- \\ \\ S- \end{array}\right|\text{protein}$$

Although the three-dimensional structure of chloroplast ferredoxin has not been established, structures of related model compounds have been determined.[51]

D. QUINONES AS HYDROGEN CARRIERS

In 1955, R. A. Morton and associates in Liverpool announced the isolation of a widely distributed quinone of unknown structure.[52] Named **ubiquinone** for its ubiquitous occurrence, it was characterized in 1958 as a benzoquinone attached to an unsaturated isoprenoid side chain (Fig. 10-8). In fact, there is a family of ubiquinones: that from bacteria typically contains six isoprenoid units in the side chain, while that from mammalian mitochondria contains 10. Ubiquinone was isolated independently by F. L. Crane and associates using isooctane extraction of mitochondria. These workers proposed that the new quinone, which they called **coenzyme Q,** might function in electron transport. The name ubiquinone and the abbreviation Q have now been accepted. A subscript indicates the number of isoprenoid units, e.g., Q_{10}. Ubiquinone can be reversibly reduced to the hydroquinone (Fig. 10-8) providing a basis for its function in electron transport.[52–56]

These two methoxyl groups are replaced by CH_3 groups in plastoquinones

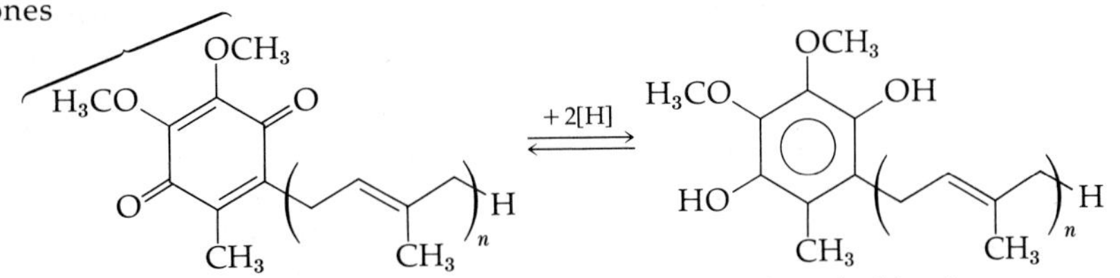

Fig. 10-8 Structures of the isoprenoid quinones and vitamin E.

Ubiquinones (coenzyme Q)
$n = 6-10$

$+2[H]$

Reduced ubiquinone

Ubichromanol

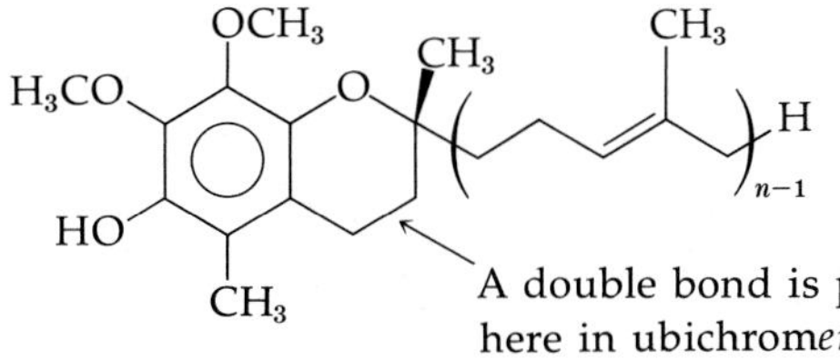

A double bond is present here in ubichrom*en*ol

Tocotrienols contain double bonds here

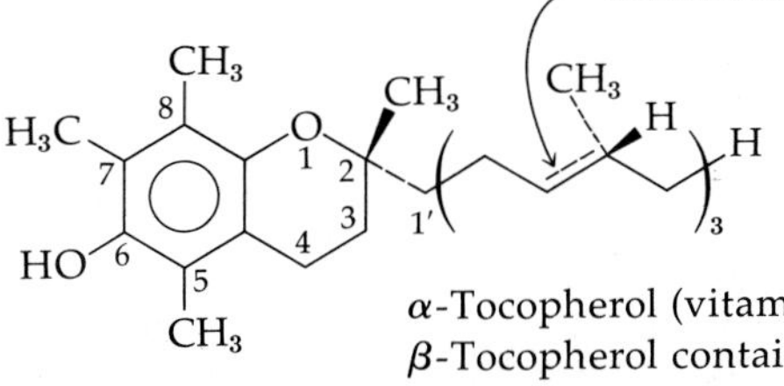

α-Tocopherol (vitamin E)
β-Tocopherol contains H in position 7
γ-Tocopherol contains H in position 5
δ-Tocopherol contains H in positions 5 and 7

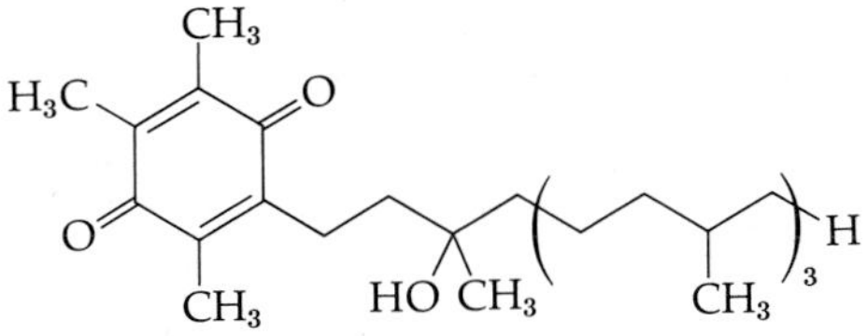

Tocopherolquinone of chloroplasts

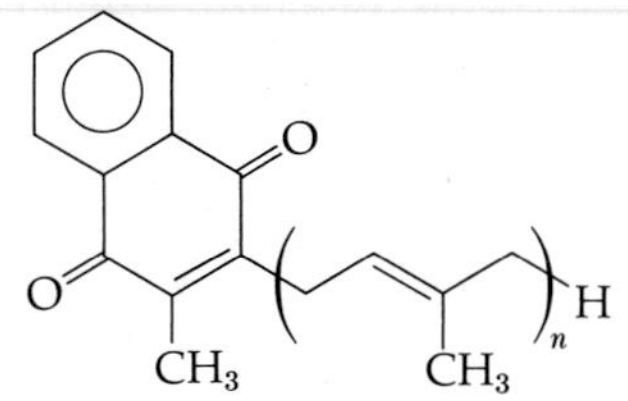

Vitamins K

$n = 4,5$. In the phylloquinone (vitamin K_1) series only one double bond is present in the innermost unit of the isoprenoid side chain

A closely related series of **plastoquinones** occur in chloroplasts (Chapter 13, Section E,6). In these compounds the two methoxyl groups of ubiquinone are replaced by methyl groups. The most abundant member, plastoquinone A, contains nine isoprenoid units.[56]

Addition of the hydroxyl group of the reduced ubiquinone or plastoquinone to the adjacent double bond leads to a chroman-6-ol structure. The compounds derived from ubiquinone in this way are called **ubichromanols** (Fig. 10-8). The corresponding ubichromenol (Fig. 10-8) has been isolated from human kidney. **Plastochromanols** are derived from plastoquinone. That from plastoquinone A, first isolated from tobacco, is also known as solanochromene. A closely related and important family of chromanols are the **tocopherols**, vitamin E (Fig. 10-8). Tocopherols are plant products found primarily in plant oils and are essential for the nutrition of humans and other animals. α-Tocopherol is the most important form of the vitamin E family; smaller amounts of the β, γ, and δ forms occur. In addition, a series of **tocotrienols** contain unsaturated isoprenoid units. The configuration of α-tocopherol is 2R,4'R,8'R as indicated in Fig. 10-8. When α-tocopherol is oxidized, e.g., with ferric chloride, the ring can be opened by hydrolysis to give **tocopherolquinones** (Fig. 10-8), which can in turn be reduced to tocopherolhydroquinones. Large amounts of the tocopherolquinones have been found in chloroplasts.

Another important family of quinones, quite similar in structure to those already discussed, are the **vitamins K** (Fig. 10-8). These occur naturally as two families. The vitamins K_1 **(phylloquinones)** have only one double bond in the side chain and that is in the isoprenoid unit closest to the ring. This fact suggests again the possibility of chromanol formation. In the vitamin K_2 **(menaquinone)** series, a double bond is present in each of the isoprenoid units. A synthetic compound **menadione** completely lacks the isoprenoid side chain and bears a hydrogen in the corresponding position on the ring. Nevertheless, menadione serves as a synthetic vitamin K and can apparently be con-

---

BOX 10-C

THE VITAMIN E FAMILY:
THE TOCOPHEROLS

Vitamin E was recognized in 1926 as a factor preventing sterility in rats that had been fed rancid lipids.[a] The curative factor, present in high concentration in wheat germ and lettuce seed oils, is a family of vitamin E compounds, the tocopherols (Fig. 10-8). The first of these was isolated by H. M. Evans and associates in 1936.

Vitamin E deficiency in the rabbit or rat is accompanied by muscular degeneration (**nutritional muscular dystrophy;** see also Box 9-F). A variety of other symptoms, varying from one species to another, are observed. Animals deficient in vitamin E display obvious physical deterioration followed by sudden death. Muscles of deficient rats show abnormally high rates of oxygen uptake, and abnormalities appear in the membranes of the endoplasmic reticulum as viewed with the electron microscope. It is thought that deterioration of lysosomal membranes may be the immediate cause of death.

The tocopherol requirement for humans is not known with certainty but about 5 mg/day plus an additional 0.6 mg for each gram of polyunsaturated fatty acid consumed may be adequate. It is estimated that the average daily intake is about 14 mg, but the increasing use of highly refined foods may lead to dangerously low consumption. Plant oils are usually the richest sources of tocopherols while animal products contain lower quantities.

The function of the tocopherols remains elusive.[a-c] A commonly accepted theory is that the principal function is to serve as an antioxidant for unsaturated lipids. In this role, tocopherols may protect lipid membranes from attack by free radicals.[d] Radicals, which may be generated by enzymes or in nonenzymatic reactions catalyzed by traces of transition metals, can initiate an autocatalytic chain reaction of the following type[e]:

Box 10-C (*Continued*)

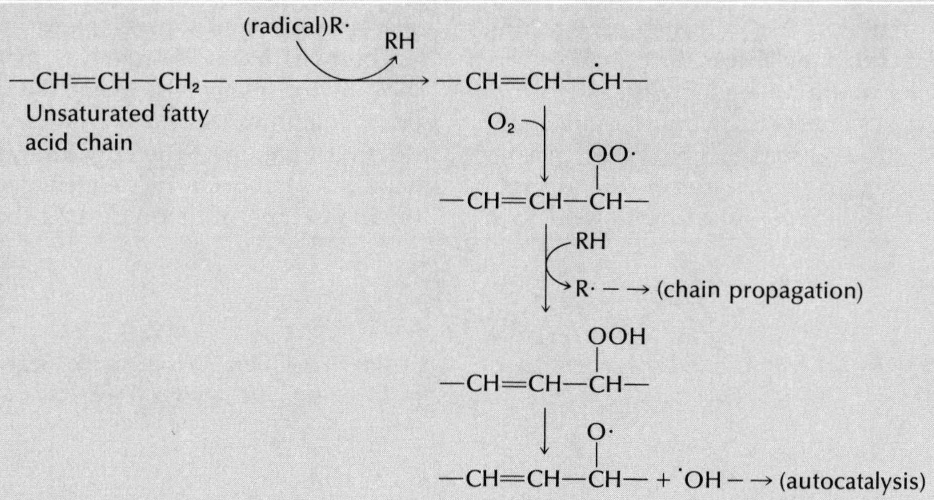
(radical)R·
RH
—CH=CH—CH₂—
Unsaturated fatty acid chain
—CH=CH—ĊH—
O₂
OO·
—CH=CH—CH—
RH
R· – → (chain propagation)
OOH
—CH=CH—CH—
O·
—CH=CH—CH— + ˙OH – → (autocatalysis)

The reaction sequence is pictured as regenerating the radical R· in a chain propagation sequence and at the same time producing an organic peroxide that can be cleaved to *two* radicals that can react further. Thus, an autocatalytic process is set up, leading to rapid development of rancidity in fats. A variety of dimerization and chain cleavage reactions ensue.

The presence of a small amount of tocopherol inhibits this peroxidation of fats presumably by trapping radicals to form more stable tocopherol radicals (Eq. 10-9)[f] which may themselves dimerize or react with other radicals to terminate the chain. That vitamin E does function in this way seems to be supported by the observation that much of the tocopherol requirement of some species can be replaced by *N,N'*-diphenyl-*p*-phenylenediamine, a synthetic antioxidant (see Table 10-3 for the structure of a related substance).

One of the deleterious consequences of lipid peroxidation is thought to be the formation of malonaldehyde, which arises by radical cleavage reactions of polyenoic acids.[g] This bifunctional aldehyde forms Schiff bases with protein amino

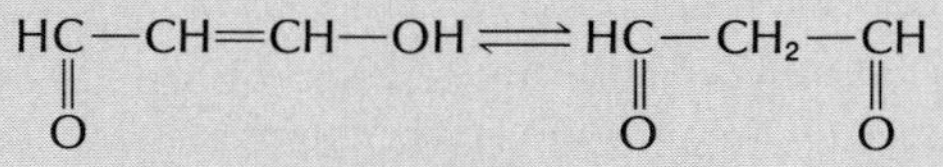
HC—CH=CH—OH ⇌ HC—CH₂—CH
O
O
O

Malonaldehyde

groups and acts as a cross-linking agent. The **age pigments** (also called **lipofuscin**) are thought to represent precipitated lipid-protein complexes resulting from such reactions.[h]

While three generations of rats have been raised on a tocopherol-free diet containing *N,N'*-diphenyl-*p*-phenylenediamine, not all of the deficiency symptoms are prevented. Thus, tocopherols may play a more specific function, possibly within cell membranes. Whether it is a protective function for some members of the electron transport system or a role in a redox process remains uncertain.

[a] W. H. Sebrell, Jr. and R. S. Harris, eds., "The Vitamins," Vol. 5. Academic Press, New York, 1972.

[b] R. H. Wasserman and A. N. Taylor, *Annu. Rev. Biochem.* **41,** 179–202 (1972).

[c] H. F. DeLuca and J. W. Suttie, eds., "The Fat-Soluble Vitamins." Univ. of Wisconsin Press, Madison, Wisconsin, 1970.

[d] I. Molenaar, J. Vos, and F. A. Hommes, *Vitam. Horm.* (*N.Y.*) **30,** 45–82 (1972).

[e] N. R. Artman, *Adv. Lipid Res.* **7,** 245–330 (1969).

[f] J. L. G. Nilsson, G. D. Daves, Jr., and K. Folkers, *Acta Chem. Scand.* **22,** 207–218 (1968).

[g] K. S. Chio and A. L. Tappel, *Biochemistry* **8,** 2827–2832 (1969).

[h] Several forms of a neuronal storage disease *uroid lipofuscinosis* have been found. Lipofuscin pigments accumulate in nerve cells leading to progressive cerebral degeneration. Leukocytes of patients have been found deficient in peroxidase [D. Armstrong, S. Dimmitt, D. H. Boehme, S. C. Leonberg, Jr., and W. Vogel, *Science* **186,** 155–156 (1974)].

verted in the body to forms containing isoprenoid side chains.

What are the functions of these interesting quinones and chromanols? It is now generally accepted that the ubiquinones function as electron transport agents soluble in the lipid of mitochondrial membranes. Presumably, the plastoquinones play an analogous function in electron transport systems of chloroplast membranes. On the other hand, the functions of vitamins E and K are uncertain. There is evidence that in some mycobacteria vitamin K participates in the electron transport chain in the same way that ubiquinone functions in mammals. Some bacteria contain both menaquinones and ubiquinones. However, in higher organisms the only known function of vitamin K at present is related to synthesis of proteins needed in blood clotting (Box 10-D).

BOX 10-D

THE VITAMIN K FAMILY

The existence of an "antihemorrhagic factor" required in the diet of chicks to ensure rapid clotting of blood was reported in 1929 by H. Dam.[a] The fat-soluble material, later designated vitamin K, causes a prompt (2–6 h) decrease in the clotting time when administered to deficient animals and birds. Thus, the clotting time for a vitamin K-deficient chick may be greater than 240 s, but 6 h after injection of 2 μg of vitamin K_3 it falls to 75 $\pm$ 27 s.[b]

The pure vitamin (Fig. 10-8), a 1,4-naphthoquinone, was isolated from alfalfa in 1939. Within a short time two series, the phylloquinones (vitamin K_1) and the menaquinones (vitamin K_2), were recognized. The phylloquinones contain four isoprenoid units in the side chain and occur in plants, while animals and bacteria contain the menaquinones with five isoprenoid units (Fig. 10-8). The synthetic menadione (vitamin K_3, Table 10-3) is also active as a vitamin.

The only established function of vitamin K is related to blood clotting. The effect of vitamin K deficiency has been traced directly to a decrease in the **prothrombin** activity (Fig. 6-16), as well as in that of clotting factors VII, IX, and X and another plasma protein of unknown function. In 1972, it was discovered that abnormal prothrombin formed by the liver in the absence of vitamin K lacked the ability to bind calcium ions essential for the binding of prothrombin to phospholipids and to its activation to thrombin. With this knowledge it was possible to pinpoint the structural differences between the normal and abnormal protein hear the amino terminal end of the ~560

residue glycoprotein.[c] From tryptic digests of normal prothrombin peptides were isolated that differed from the corresponding peptides from the abnormal prothrombin in their electrophoretic mobility. Careful chemical analysis, together with nmr measurements, established that in normal prothrombin residues 7, 8, 15, 17, 20, 21, 26, 27, 30, and 33, all identified as glutamic acid in the sequence analysis, are actually **γ-carboxyglutamate** residues:

$$
\begin{array}{c}
\text{COO}^- \\
|\\
^+\text{H}_3\text{N}-\text{CH} \\
|\\
\text{CH}_2 \\
|\\
\text{CH} \\
^-\text{OOC}\diagup\ \ \diagdown\text{COO}^-
\end{array}
$$

γ-Carboxyglutamate

The fact that γ-carboxyglutamate had never been identified as a protein substituent before is explained by the easy decarboxylation of this malonic acid derivative to normal glutamic acid. The function of vitamin K must be to assist in the incorporation of the additional carboxyl group into the glutamate residues of preformed prothrombin. Similar post-transcriptional modifications of the other blood clotting factors are thought to occur.[d] The increased calcium ion binding resulting from these modifications is clearly the result of the addition of the extra carboxylate anion groups to the metal chelating centers of the proteins.

Box 10-D (*Continued*)

An interesting facet of vitamin K metabolism was revealed by the observation that cattle fed on spoiled sweet clover develop a fatal hemorrhagic disease. The causative agent was identified as a **dicoumarol,** a compound arising from coumarin, a

OH   OH

CH₂

Dicoumarol

natural constituent of clover. Dicoumarol and the closely related synthetic **warfarin** are both potent vitamin K antagonists. Interestingly, the latter has been employed both as a rat poison and in the treatment of thrombo-embolic disease. As rodenticides hydroxycoumarin derivatives are usually safe because a single accidental ingestion by a child or pet animal does little harm, whereas regular ingestion by rodents is fatal. Apparently, vitamin K in the body undergoes a regular oxidative conversion to phylloquinone 2,3-epoxide. The lat-

O

CH₃

O

O

CH₃

CH₃   CH₃   CH₃

Phylloquinone 2,3-epoxide

ter is converted back to phylloquinone in the liver and the coumarin anticoagulants inhibit this reconversion. The matter is of considerable practical importance because of the spread of warfarin-resistant rats in Europe and the United States. It is thought that the resistance mutation has altered the enzyme that converts the epoxide back to phylloquinone so that the enzyme is no longer susceptible to inhibition by warfarin.[e]

Although the mechanism of participation of vitamin K in carboxylation of glutamate residues is obscure, some clues have been obtained.[f] The

hydroquinone form of the vitamin is apparently required as is O_2. The same microsomal preparation that catalyzes the carboxylation forms phylloquinone 2,3-epoxide and the carboxylation is inhibited by warfarin. This suggests a mechanism related to that of hydroxylases in which the epoxide of vitamin K would be alternately formed and reduced. A possibility is that during formation of the epoxide from the hydroquinone (perhaps via a peroxide) the second oxygen atom of O_2 could be released as OH^- at a location adjacent to the γ-methylene group of the glutamic acid residue to be carboxylated. This hydroxyl ion could abstract a proton from the methylene group and carboxylation could then be effected by a reagent such as carbonyl phosphate (Chapter 8, Section C).

Whether vitamin K has other functions in the human body is uncertain. The naphthoquinone ring may be an oxidant in electron transport in some bacteria. In addition, a specific coenzyme role for menaquinone in bacteria in the conversion of dihydroorotate to orotate (Chapter 14, Section K,1), with fumarate as the ultimate oxidant, has been postulated.[f]

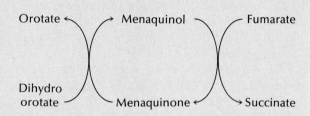

[a] R. H. Wasserman and A. N. Taylor, *Annu. Rev. Biochem.* **41,** 179–202 (1972).

[b] R. E. Olson, *Science* **145,** 926–928 (1964).

[c] J. Stenflo, *JBC* **251,** 355–363 (1976).

[d] J. B. Howard and G. L. Nelsestuen, *PNAS* **72,** 1281–1285 (1975).

[e] R. G. Bell and P. T. Caldwell, *Biochemistry* **12,** 1759–1762 (1973).

[f] J. A. Sadowski, C. T. Esmon, and J. W. Suttie, *JBC* **251,** 2770–2776 (1976).

[g] N. A. Newton, G. B. Cox, and F. Gibson, *BBA* **244,** 155–166 (1971).

The most widely accepted function of the readily oxidizable tocopherols is to serve as **antioxidants,** "scavengers" for free radicals arising in cells (Box 10-C). When a single hydrogen atom is removed from a hydroquinone or from a chromanol such as a tocopherol, a free radical is formed (Eq. 10-9). Phenols substituted in the 2, 4, and 6 positions give especially stable radicals. Thus, the

$$\text{OH} \quad \xrightarrow{-[\text{H}]} \quad :\overset{..}{\text{O}}\cdot \tag{10-9}$$

ability to form a relatively stable free radical may account for both the methyl substituents in the tocopherols and their chromanol structures.

Whether the ability to form stable radicals is also important in the function of ubiquinones and plastoquinones remains uncertain. A further question about this entire family of compounds is "Why the long isoprenoid side chain?" A simple answer is that it serves to anchor the compound in the lipid portion of the cell membranes where it is to function. In fact, in the case of ubiquinones it is usually thought that the quinone moves freely through the lipid phase in both the oxidized and reduced forms shuttling electrons between carriers.

Is ubiquinone to be regarded as a new vitamin? Evidence to date suggests that it is not and that animals are able to make these compounds in adequate quantity.[52] However, it has been claimed that ubiquinones may have beneficial effects for patients with muscular dystrophies. It has also been claimed that a reduced level of ubiquinone is found in gum tissues of patients with peridontal disease.[57]

E. THE ELECTRON TRANSPORT CHAIN AND OXIDATIVE PHOSPHORYLATION

One of the central questions in modern biochemistry is "How is ATP generated by flow of electrons through a series of carriers?" The question is important because most of the ATP formed in aerobic and in some anaerobic organisms is made by this process of oxidative phosphorylation. Moreover, the energy captured during photosynthesis is used to form ATP in a similar manner. The mechanism of ATP generation may also be intimately tied to the function of membranes in transport of ions. It is quite possible that the mechanism of oxidative phosphorylation is related in a reverse fashion to the utilization of ATP in providing energy for the contraction of muscles.

During the 1940's when it had become clear that formation of ATP from ADP and inorganic phosphate was coupled to electron transport in the mitochondria, biochemists began their first attempts to pick apart the system and to understand the molecular mechanisms. However, nature sometimes strongly resists attempts to pry out her secrets and the situation today has been aptly summarized by Ephraim Racker: "Anyone who is not confused about oxidative phosphorylation just doesn't understand the situation."[58] The confusion has not arisen through any lack of effort or imagination. Numerous speculations on the mechanism of oxidative phosphorylation have been published, but no completely convincing explanation has appeared. Furthermore, the failure to explain oxidative phosphorylation in straightforward chemical terms has led to sometimes vague proposals with fancy sounding names. The "chemiosmotic hypothesis" and "mechanochemical coupling" have been placed in opposition to what has been dubbed the "chemical hypothesis." By the latter term are meant proposals involving formation of discrete but unidentified "high energy" chemical intermediates.

Within bacteria the site of the electron transport and oxidative phosphorylation systems appears to be in the cytoplasmic membranes. In eukaryotic cells the major site is within the mitochondria. For this reason we should probably start with a closer look at mitochondria, the "power plants of the cell."

1. The Architecture of the Mitochondrion[59–61]

A typical mitochondrion is about as large as a cell of *E. coli* but both the form and size of the organelle vary enormously. In every case a mitochondrion is made up of two concentric membranes, an *outer* and an *inner* membrane, each ~ 5–7 nm thick (Fig. 10-9). However, in liver there is little inner membrane and a large matrix space, while in heart mitochondria there are more folds and a higher rate of oxidative phosphorylation. The enzymes catalyzing the tricarboxylic acid cycle are also more active in heart mitochondria. Furthermore, because of its high metabolic activity almost one-third of the bulk of heart muscle consists of mitochondria. A typical heart mitochondrion has a volume of 0.55 μm^3; for every cubic micrometer of mitochondrial volume there are 89 μm^2 of inner mitochondrial membranes.[62]

Mitochondria can swell and contract, and forms other than the *orthodox* form usually seen in osmium-fixed electron micrographs have been described. In *condensed forms* of mitochondria the cristae are swollen, the matrix volume is much reduced, and the inner membrane space is increased. Rapidly respiring mitochondria fixed for electron microscopy exhibit forms that have been referred to as "energized" and "energized-twisted."[63]

The outer membranes of mitochondria can be removed from the inner membranes by osmotic rupture.[64] Analyses on the separated outer and inner membrane fractions show that the outer membrane is less dense (density ~ 1.1 g/cm^3) than the inner. It is highly permeable to most substances of MW = 10,000 or less. The ratio of phospholipid to protein is high (~ 0.82 on a weight basis) and extraction of the phospholipids by acetone destroys the membrane. Of the lipids present, there is a low content of cardiolipin, a high content of phosphoinositol and cholesterol, and no ubiquinone. The inner membrane, whose density is ~ 1.2 g/cm^3, is impermeable to many substances. In fact, with the exception of neutral molecules of MW < 150, permeability for all materials appears to be tightly controlled. The ratio of phospholipid to protein in the inner membrane is low (~ 0.27), and cardiolipin makes up $\sim 20\%$ of the total phospholipid. Ubiquinone and other components of the respiratory chain are present in the inner membrane.

Another characteristic of the inner mitochondrial membrane is the presence, under certain conditions, of projections on the inside surface (the surface facing the mitochondrial matrix). Fernandez-Moran, who discovered these particles in 1962, suggested that they might contain the enzymes of the electron transport system, but subsequent investigation has shown this not to be the case. Rather, the spherical particles of MW $\sim 85,000$ attached to the membrane through a "stalk" display "ATPase" activity. The latter is a clue that they may participate in *synthesis* of ATP during oxidative phosphorylation. In fact, the protein present is now known as **coupling factor 1** (F_1 or ATP synthetase). Since the knobs have not been seen except in negatively stained phosphotungstate preparations there is doubt that they exist in intact mitochondria. They may be developed during the staining procedure and represent some new aggregation of enzymes originally present within the inner membrane.

In addition to the bacterialike ribosomes (mitoribosomes) and the small circular molecules of DNA, mitochondria contain variable numbers of dense granules[65] of calcium phosphate, either $Ca_3(PO_4)_2$ or hydroxylapatite (Box 5-E).

2. The Chemical Activities of Mitochondria

Mention of mitochondria usually brings to the mind of the biochemist the *citric acid cycle,* the β-*oxidation pathway* of fatty acid metabolism, and *oxidative phosphorylation.* In addition to these major processes, many other chemical events occur in the mitochondria. Perhaps most notable is the ability of mitochondria to concentrate ions

~7 nm

outer
membrane

inner
membrane

intermembrane space

5-7 nm membrane thickness

8.5 nm

Outer membrane is
freely permeable to
ATP, ADP, sucrose,
salts, etc.

Matrix

When negatively stained
with phosphotungstate
the matrix face of the
inner membrane appears
covered with knobs of
coupling factor F_1, sometimes attached by stalks

Inner membrane
is not freely
permeable

Cristae

Cytochromes, ubiquinone,
and nonheme iron carriers
are embedded in inner
membrane

Submitochondrial particles
usually have the matrix
side of the inner membrane
on the outside

(A)

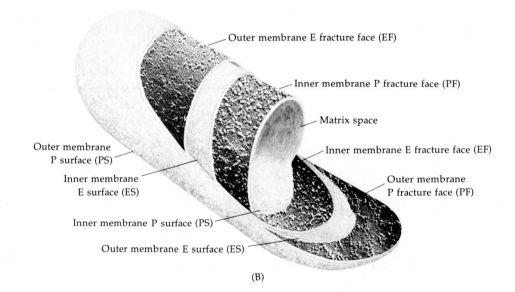

Outer membrane E fracture face (EF)

Inner membrane P fracture face (PF)

Matrix space

Inner membrane E fracture face (EF)

Outer membrane
P fracture face (PF)

Outer membrane
P surface (PS)

Inner membrane
E surface (ES)

Inner membrane P surface (PS)

Outer membrane E surface (ES)

(B)

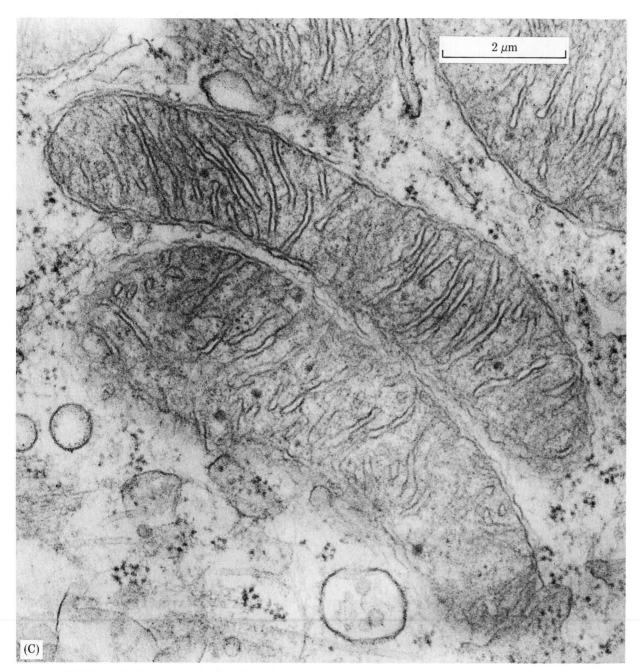

2 μm

(C)

Fig. 10-9 (A) Schematic diagram of mitochondrial structure. (B) Model showing organization of particles in mitochondrial membranes revealed by freeze-fracture electron microscopy. The characteristic structural features seen in the four half-membrane faces (EF and PF) that arise as a result of fracturing of the outer and inner membranes are shown. The four smooth membrane surfaces (ES and PS) are revealed by etching (See Box 1-A). From L. Packer and L. Worthington.[59a] (C) Thin section of mitochondria of cultured kidney cells from a chicken embryo. Courtesy of Judie Walton.

such as Ca^{2+}. Mitochondria also control the entrance and exit of many substances including ATP. Thus they exert important regulatory functions both on catabolic and biosynthetic sequences. Mitochondria make some of their own proteins and take in many others from the cytoplasm as they grow and multiply.

Where within the mitochondria are specific enzymes localized? One approach to this question is to see how easily the enzymes can be dissociated from mitochondria. Some enzymes come out readily under hypotonic conditions. Some enzymes are released only upon sonic oscillation, suggesting that they are inside the matrix space. Others, including the cytochromes and the flavoproteins that act upon succinate and NADH, are so tightly bound to the mitochondrial membranes that they can be dissociated only through the use of detergents. It is now generally agreed that these tightly bound components are embedded in the inner membrane.

If, as is also usually agreed, the enzymes of the citric acid cycle and β oxidation are present in the matrix, it would seem that the reduced carriers must approach the inner membrane from the matrix side (the M side). Thus, the embedded enzymes designed to oxidize NADH, succinate, and other reduced substrates must be accessible from the matrix side. However, an α-glycerolphosphate dehydrogenase, a flavoprotein, is accessible from the "outside" (the C side) of the inner membrane.

Specific fluorescent antibodies to cytochrome c bind only to the C side of the inner membrane but antibodies to cytochrome oxidase label both sides, suggesting that this protein spans the membrane.[66,66a] However, oxidation of cytochrome c (mediated by cytochrome a) occurs only on the C surface and reduction of O_2 (cytochrome a_3) only on the M side.[66] Furthermore, antibodies to the "coupling factor" that makes up the "knobs" bind strictly to the matrix side. These results are among those that have led to the picture in Fig. 10-10B.

The outer mitochondrial membrane contains monoamine oxidase and cytochrome b_5 as well as other proteins. Its composition may be similar to

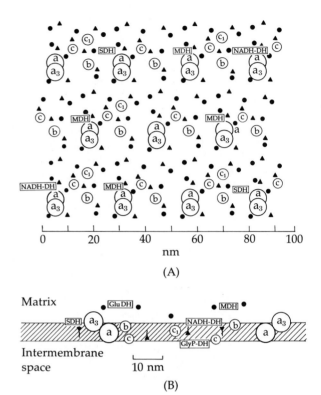

Fig. 10-10 (A) Proposed distribution of enzymes in mitochondrial inner membrane based on analytical data for heart mitochondria. Abbreviations: a, a_3, b, c, c_1, cytochromes; NADH-DH, GluDH, GlyP-DH, MDH, SDH, dehydrogenases for NADH, glutamate, glycerolphosphate, malate, and succinate; (▲) ubiquinone and (●) NAD. (B) Proposed transmembrane arrangement of some of the respiratory carriers and dehydrogenases. From Kröger and Klingenberg.[67a]

that of the membranes of the endoplasmic reticulum. **Adenylate kinase** (myokinase),[67] a key enzyme involved in equilibrating ATP and AMP with ADP (Box 3-A; Chapter 7, Section E,6), is one of the enzymes thought to be characteristically present in the **intermembrane space** (between inner and outer membranes).*

What are the concentrations of the electron carriers in mitochondrial membranes? In one experiment, cytochrome b was found in rat liver mitochondria to the extent of $0.28 \ \mu mol/g$ of protein. If

---

* Smaller amounts of other isoenzymes of adenylate kinase are present in the nucleus and in the cytoplasm.[66]

TABLE 10-1
Ratios of Components in the Electron Transport Chain of Mitochondria[a,b]

Electron carrier	Rat liver mitochondria	Beef heart mitochondria
Cytochrome a_3	1.0	1.1
Cytochrome a	1.0	1.1
Cytochrome b	1.0	1.0
Cytochrome c_1	0.63	0.33–0.51
Cytochrome c	0.78	0.66–0.85
Pyridine nucleotides	24	
Flavins	3	1
Ubiquinone	3–6	7
Copper		2.2
Nonheme iron		5.5

[a] From Wainio,[59] and references cited therein.
[b] Molecular ratios are given. Those for the cytochromes refer to the relative numbers of heme groups.

we take a total mitochondrion as about 22% protein, the average concentration of the cytochrome would be ~ 0.06 mM. Since all the cytochromes are concentrated in the inner membranes which may account for 10% or less of the volume of the mitochondrion, the concentration of cytochromes may approach 1 mM in these membranes. This is sufficient to ensure rapid reactions with substrates. Many analytical data are available on the *ratios* of different components of the respiratory chain (Table 10-1).

Kröger and Klingenberg[67a] have portrayed the distribution of enzymes in the mitochondrial inner membrane in a way that gives a better feeling for what the membrane surface may actually be like (Fig. 10-10). Of course, we do not know that the proteins are associated in regular arrays or whether they float at random in the membrane and make contact one with another by diffusion.

3. The Sequence of Electron Carriers

What is the sequence of carriers through which electron flow takes place? Figure 10-11 will serve as a basis for discussion even though several details are uncertain. Several lines of evidence have been used to deduce the scheme shown. In the first place it seems reasonable to suppose that the carriers should lie in order of increasing reduction potential going from left to right of Fig. 10-11. This need not be strictly true, however; especially since the reduction potentials existing in the mitochondria may be somewhat different from those in isolated enzyme preparations.

The development by Chance of a dual wavelength spectrophotometer permits easy observation of the state of oxidation or reduction of a given carrier within mitochondria.[68] This technique together with the study of specific inhibitors (some of which are indicated in Fig. 10-11 and Table 10-2) allowed some electron transport sequences to be assigned. For example, blockage with **rotenone** or **amytal** prevents reduction of the cytochrome system by NADH but allows reduction by succinate and by other substrates having their own flavoprotein components in the chain. Another approach to ascertaining the sequence of carriers has been the use of artificial electron acceptors some of which are shown in Table 10-3. These serve to bypass parts of the chain as indicated in Fig. 10-11.

An important method of studying the electron transport chain is to break mitochondrial membranes into fragments that retain an ability to catalyze some of the reactions of the chain. Many methods have been employed to obtain **submitochondrial particles.** The well-known Keilin–Hartree preparation of heart muscle is obtained by homogenizing mitochondria and precipitating at low pH. While the resulting particles have a low cytochrome c content and do not carry out oxidative phosphorylation, they respire actively. Other electron transport particles have been prepared by sonic oscillation. Under the electron microscope such particles appear to be small membranous vesicles resembling mitochondrial cristae.

While many detergents are strong denaturants of proteins, some detergents disrupt mitochondrial membranes without destroying enzymatic activity. A favorite is **digitonin** (Fig. 12-18) which causes disintegration of the outer membrane. The remaining fragments of inner membrane still

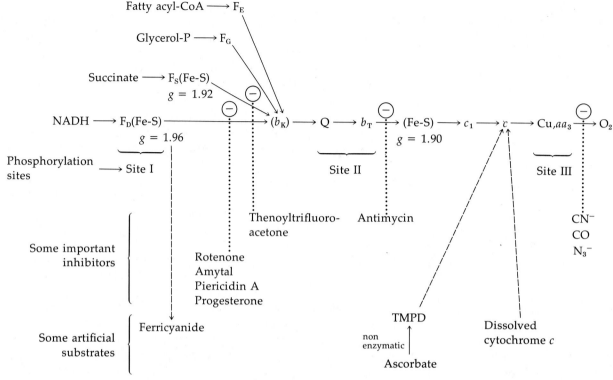

Fig. 10-11 A current concept of the electron transport chain of mitochondria.

retain activity for oxidative phosphorylation. Submitochondrial particles have been further fractionated by chemical treatments. In one procedure, separate "complexes" are obtained that catalyze reactions of four different portions of the electron transport process.[69] The complexes I, II, III, and IV catalyze the reactions indicated in Eq. 10-10:

$$\text{Succinate} \xrightarrow{\text{II}}$$
$$\text{NADH} \xrightarrow{\text{I}} Q \xrightarrow{\text{III}} \text{cyt } c \xrightarrow{\text{IV}} O_2 \qquad (10\text{-}10)$$

Chemical analysis of the complexes has helped to verify the probable location of some components in the intact chain. For example, a high iron content is found in both complexes I and II and copper is found in complex IV.

A word is in order about the special role of ubiquinone in the electron transport chain. There is much evidence that it is a true and essential elec-tron carrier. For example, mutants of *E. coli* lacking the ability to synthesize ubiquinone can live by fermentation of glucose but cannot subsist on substrates such as succinate that require oxida-tion to supply the cell with energy.[70] However, the lipid-soluble ubiquinone is present in bacterial and mitochondrial membranes in relatively large amounts compared to other electron carriers (Table 10-1). It seems to be located at a point of convergence of the NADH and succinate branches of the chain and is usually regarded as a "soluble" carrier in contrast to "fixed" carriers such as the flavoproteins and cytochromes. In this respect, ubiquinone plays a role somewhat like that of NAD^+ which carries electrons from various solu-ble dehydrogenases in the aqueous matrix solu-tion to the flavoprotein NADH dehydrogenase in the membrane. A third mitochondrial component that could function in a similar way is cytochrome *c*. Unlike the other cytochromes, it is water soluble

TABLE 10-2
Some Well-Known Respiratory Inhibitors[a]

Rotenone, an insecticide and fish poison from a plant root	
Piericidin A, a structural analogue of ubiquinone	
Amytal	
Progesterone	(see Fig. 12-17)
Antimycin A, a *Streptomyces* antibiotic	R = *n*-butyl or *n*-hexyl
Thenoyltrifluoroacetone (TTB, 4,4,4-trifluoro-1-(2-thienyl)-1,3-butadione	
Cyanide	$^-C{\equiv}N$
Azide	$N{\equiv}N{=}N^-$
Carbon monoxide	CO

[a] See Fig. 10-11 for sites of inhibition.

TABLE 10-3
Some Artificial Electron Acceptors[a,b]

Compound	Structure	$E^{0\prime}$ (pH 7), 30°C
Ferricyanide	$Fe(CN)_6{}^{3-}$	+0.36 V (25°C)
Oxidized form of tetramethyl-*p*-phenylenediamine (TMPD)		+0.260 V
2,6-Dichlorophenol-indophenol (DCIP)		+0.217 V
Phenazine methosulfate (PMS)		+0.080 V
Ascorbate	(See Box 10-G)	+0.058 V
Methylene blue		+0.011 V
Menadione		+0.008 V (25°C)
Tetrazolium salts, e.g., "neotetrazolium chloride"		−0.125 V

[a] From Wainio,[59] pp. 106–111.
[b] See Fig. 10-11 for sites of action.

and easily leached from mitochondrial membranes. Nevertheless, it is usually present in a roughly 1:1 ratio with the fixed cytochromes. It seems unlikely that it is as free to diffuse as are ubiquinone and NAD^+.

4. The Stoichiometry and Sites of Oxidative Phosphorylation

Synthesis of ATP *in vitro* by tissue homogenates was first demonstrated in 1937 by H. M. Kalckar who has written an interesting historical account of the subject.[71] An important accomplishment came in 1941 when S. Ochoa obtained the first reliable measurement of the **P/O ratio.** The P/O ratio is the *number of moles of ATP generated per atom of oxygen utilized* in respiration. It is also equal to the number of moles of ATP formed for each pair of electrons passing through the electron transport chain. Ochoa established that for the oxidation of pyruvate to acetyl-CoA and CO_2 (a process that should send two electrons down the chain) the P/O ratio was approximately three. This value has since been confirmed many times. However, the student should be aware that experimental difficulties in measuring the P/O ratio are numerous and many errors have been made, even in recent years. One method for measuring the P/O ratio is based on the ATP assay described in the legend to Fig. 8-11.

The experimental observation of a P/O ratio of ~3 for oxidation of pyruvate and of a variety of other substrates that donate NADH to the electron transport chain led to the concept that there are three sites for generation of ATP. It was soon shown that the P/O ratio was only two for oxidation of succinate. This suggested that one of the sites (site I) is located between NADH and ubiquinone and precedes the joining of the succinate pathway.

In 1949, Lehninger used ascorbate plus tetramethylphenylenediamine (TMPD, Table 10-3) to introduce electrons into the chain at cytochrome c. The sequence ascorbate → TMPD → cytochrome c occurs nonenzymatically. Later it was possible to use cytochrome c as an electron donor directly. In either case only one ATP was generated, as would be anticipated if only site III were found to the right of cytochrome c. Site I was further localized by H. Lardy who used ferricyanide as an artificial oxidant to oxidize NADH in the presence of antimycin a. Again a P/O ratio of one was observed. Finally, in 1955, Slater showed that passage of electrons from succinate to cytochrome c also gave only one ATP, that generated at site II.

5. Respiratory Control, Uncoupling, and Exchange Reactions

With proper care, relatively undamaged mitochondria can be obtained. Such mitochondria are **tightly coupled.** By this we mean that electrons cannot pass through the electron transport chain without generation of ATP. Furthermore, if the concentration of ADP or of P_i becomes too low, both phosphorylation and respiration cease. This **respiratory control** is a property of undamaged mitochondria. On the other hand, damaged mitochondria or submitochondrial particles are often able to transfer electrons at a rapid rate *without* synthesis of ATP and with no inhibition at low ADP concentrations. A related phenomenon is **uncoupling** which is brought about by various compounds, the best known of which is **2,4-dinitrophenol.** Even before the phenomenon of uncoupling was discovered, it had been known that dinitrophenol substantially increased the respiration rates of animals. The compound had even been used (with some fatal results) in weight control pills. The chemical basis of uncoupling is uncertain, but a possibility is indicated in Eq. 10-11 which shows one strictly hypothetical model for the chemistry of oxidative phosphorylation. Here three carriers A, B, and C operate in the electron transport chain. Carrier C is a better oxidizing agent than B or A and carrier B has some special chemistry that permits it, in the reduced state, to react with a protein Y to form Y—BH_2. The latter, an unidentified adduct, is converted by oxidation

Y

A ⟵ ⟶ BH_2 ⟶ Y—BH_2 ⟶ C

AH_2 ⟶ B ⟵ ⟵ Y~B ⟵ ⟶ CH_2

Spontaneous hydrolysis
of intermediate ⟶ X~Y
X ~ Y would cause
the ATPase reaction

X ⟵ Uncouplers could
substitute for X
in this reaction

P_i

⟶ Y

X~(P)   Reactions causing P_i—ATP
and other exchanges and
ATPase reactions

ADP ⟶

⟶ X

ATP   (10-11)

Hypothetical coupling
site for ATP synthesis

with carrier C to a "high energy" oxidized form indicated as Y ~ B. Once the possibility of generating such an intermediate is conceded, it is easy to imagine plausible ways in which the energy of this intermediate could be transferred into forms with which we are already familiar. For example, another protein X could react to form X ~ Y in which the X ~ Y linkage could be a thioester, an acyl phosphate, or other high energy form. Furthermore, it might not be necessary to have two proteins; X and Y could be different functional groups of the same protein. They might be nonprotein components, e.g., Y might be a phospholipid.

Generation of ATP by the remaining reactions of Eq. 10-11 is straightforward. For example, if X ~ Y were a thioester the reactions would be the reverse of sequence S1A of Table 7-2. These reaction steps would also be responsible for various observed exchange reactions. For example, mitochondria catalyze exchange of inorganic phosphate ($H^{32}PO_4^{2-}$) into the terminal position of ATP. Mitochondria and submitochondrial par-

ticles also contain ATP-hydrolyzing (*ATPase*) activity. While ATP hydrolysis is presumably a nonphysiological reaction, there is a strong correlation between the presence of ATPase and the ability of submitochondrial particles to catalyze oxidative phosphorylation. Both the ATPase and the exchange reactions appear to utilize the same machinery that synthesizes ATP in tightly coupled mitochondria. In the scheme of Eq. 10-11 ATPase activity would be observed if some spontaneous hydrolysis of X ~ Y were to occur. Partial disruption of the system would lead to increased ATPase reactivity, as is observed. Furthermore, it would be easy to imagine that uncouplers such as the dinitrophenolate ion or arsenate ion, acting as nucleophilic displacing groups, could substitute for a group such as X. Spontaneous breakdown of labile intermediates would permit oxidation to proceed unimpaired. This would be true even when X was entirely tied up as X ~ Y and unable to react further because of a lack of inorganic phosphate. Since there are three different sites of phosphorylation we might expect to have three different enzymes of the type Y in the foregoing scheme, but it would be necessary to have only one X.

Another observation consistent with the scheme of Eq. 10-10 is the inhibition of the oxidation of NADH by the antibiotic **oligomycin** (rutamycin). This compound also inhibits mitochondrial ATPase activity. However, the inhibition is released by dinitrophenol, a fact that suggests that oligomycin binds to the enzymes catalyzing the exchange reactions and not to the electron transport chain itself. An important experimental discovery is that the knobs that are visible in negatively stained mitochondrial fragments catalyze the ATPase and exchange reactions. The knob protein F_1 (Section E,8) is one of several "coupling factors" needed to reconstitute phosphorylation by disrupted mitochondria.

We will return to the coupling scheme of Eq. 10-11 again, but the reader should note at this point that all attempts to identify an intermediate such as X ~ Y have failed. Furthermore, most claims to have seen Y ~ B by any means have been disproved.

BOX 10-E

USING METABOLISM TO GENERATE HEAT: THERMOGENIC TISSUES

A secondary but important role of metabolism in warm-blooded animals is to generate heat. In most instances, the heat evolved from ordinary metabolism is quite adequate and the organism can control temperature by regulating the heat exchange with the environment. An intriguing biochemical finding is the presence in warm-blooded animals of a **brown fat** tissue that contrasts strikingly with the normally white adipose tissue. Brown fat is found in small quantities in the newborn human, and in a newborn rabbit it accounts for 5–6% of the body weight.[a–c] It is especially abundant in the newborn of those species that are born without fur and in hibernating animals. It is now believed that the function of the tissue is to produce heat. The unusually high concentration of blood vessels and mitochondria, and the large number of sympathetic nerve connections in brown fat are all related to efficient heat generation. An interesting biochemical question remains. Is the energy available from electron transport in the mitochondria dissipated as heat because ATP synthesis is uncoupled from electron transport? Or does ATP synthesis take place but the resulting ATP is hydrolyzed wastefully through the action of an active ATPase?

Some plant tissues are thermogenic. For example, the spadix (sheathed floral spike) of the skunk cabbage *Symplocarpus foetidus* can maintain a 10–25° higher temperature than that of the surrounding air.[c,d]

The bombardier beetle generates a hot, quinone-containing defensive discharge which is sprayed from a special reaction chamber at a temperature of 100°C. The reaction mixture of 25% hydrogen peroxide and 10% hydroquinone plus methylhydroquinone is stored in a reservoir [see

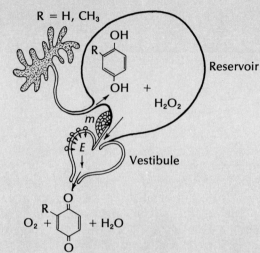

accompanying figure where E stands for enzyme and m, muscle that controls valve (from D. J. Aneshansley et al.,[e] copyright 1969 by the American Association for the Advancement of Science)] and reacts with explosive force when it comes into contact with catalase and peroxidases in the reaction chamber.[c,e,f] The synthesis and storage of 25% H_2O_2 is remarkable and poses some interesting biochemical questions.

[a] O. Lindberg, ed., "Brown Adipose Tissue." Am. Elsevier, New York, 1970.
[b] M. J. R. Dawkins and D. Hull, *Sci. Am.* **213**, 62–67 (1965).
[c] P. W. Hochachka, *Fed. Proc., Fed. Am. Soc. Exp. Biol.* **33**, 2162–2169 (1974).
[d] J. T. Bahr and W. D. Bonner, Jr., *JBC* **248**, 3441–3445 (1973).
[e] D. J. Aneshansley, T. Eisner, J. M. Widom, and B. Widom, *Science* **165**, 61–63 (1969).
[f] T. Eisner and J. Meinwald, *Science* **153**, 1341–1350 (1966).

6. "States" of Mitochondria and Spectrophotometric Observation

Chance and Williams defined five **states** of mitochondria[68]; of these, states 3 and 4 are often mentioned in current literature. The definitions of all five states are given in Table 10-4. Note that the definitions apply only to tightly coupled mitochondria. Such mitochondria, when in the actively phosphorylating state 3, go spontaneously into state 4 as they convert ADP to ATP.

Chance and associates have employed spectrophotometry on intact mitochondria or sub-

TABLE 10-4
The States of Tightly Coupled Mitochondria

Variable conditions	State 1 no added substrate	State 2 from starved animals	State 3 actively phosphorylating	State 4 ADP depleted	State 5 anaerobic
O_2	+	+	+	+	Absent
[ADP]	Low	High	High	Low	High
Oxidizable substrate	Low	~0	High	High	High
Respiratory rate	Slow	Slow	Fast	Slow	0
Rate-limiting component	ADP	Substrate	Electron transport chain	ADP	O_2

mitochondrial particles to investigate both the sequence of carriers and the sites of phosphorylation. Using a special dual wavelength spectrophotometer the light absorption at the absorption maximum (λ_{max}) of a particular component is followed relative to the light absorption of some other reference wavelength (λ_{ref}). The principal wavelengths used are given in Table 10-5.

From these measurements the state of oxidation or reduction of each one of the carriers can be observed in the various "states" of Table 10-4 and in the presence of inhibitors. The experiments served to establish that electrons passing down the chain do indeed reside for a certain length of time on particular carriers. That is, in a given state, each carrier exists in a defined ratio of oxidized to reduced forms ([ox]/[red]). Such a result would presumably not be seen if the entire chain functioned in some cooperative manner with electrons passing from the beginning to the end in a single reaction. Furthermore, by observing changes in the ratio [ox]/[red] under different conditions some localization of the three phosphorylation sites could be made. In one experiment antimycin a was added to block the chain ahead of cytochrome c_1. Then tightly coupled mitochondria were allowed to go into state 4 by depletion of ADP. Since the concentration of oxygen was high and cytochrome a_3 has a low K_m for O_2 (~3 μM) cytochrome a_3 was kept in a

highly oxidized state. Cytochrome a was also observed to be oxidized but cytochrome c_1 and c remained reduced. The presence of this **crossover point** suggested at the time that cytochrome c might be at or near one of the "energy conserva-

TABLE 10-5
Wavelengths of Light Used to Measure States of Oxidation of Carriers in the Electron Transport Chain of Mitochondria[a]

Carrier	λ_{max} (nm)[b]	λ_{ref} (nm)
NADH	340	374
Flavins	465	510
Cytochromes		
b^{2+}	564(α)	575
	530(β)	
	430(γ)	
c_1^{2+}	534(α)	
	523(β)	
	418(γ)	
c^{2+}	550(α)	540
	521(β)	
	416(γ)	
a^{2+}	605(α)	630(590)
	450(γ)[a]	
a_3^{2+}	600(α)[a]	
	445(γ)	455

[a] After Chance and Williams.[68]

[b] The wavelengths used for each carrier in dual wavelength spectroscopy appear opposite each other in the two columns. Some positions of other absorption bands of cytochromes are also given.

tion sites." Accounts of more recent experiments using the same approach are given by Wilson *et al.*[72-75]

7. Thermodynamics and "Reverse Electron Flow"

From Table 3-7 the value of $\Delta G'$ for oxidation of one mole of NADH by oxygen (1 atm) is -219 kJ. At a pressure of $\sim 10^{-2}$ atm O_2 in tissues the value is ~ -213 kJ. However, when the reaction is coupled to the synthesis of three molecules of ATP ($\Delta G' = +34.5$ kJ mol^{-1}) the net free energy change for the overall reaction becomes -110 kJ mol^{-1}. This is still overwhelmingly negative. However, we must remember that the concentrations of ATP, ADP, and P_i can depart very far from the 1:1:1 ratio implied by the standard free energy changes. An interesting experiment is to allow oxidative phosphorylation to proceed until the mitochondria reach state 4 and to measure the "mass action ratio" $[ATP]/[ADP][P_i]$ that is attained. This **phosphorylation state ratio* R_p** (see Box 3-A) may attain values of 10^4 M^{-1} or more.[73,76] As a result, the overall value of ΔG for oxidation of NADH in the coupled electron transport chain is less negative than is ΔG^0. Indeed, if synthesis of three molecules of ATP is coupled to electron transport the system would reach an equilibrium if $R_p = 10^{6.4}$ at 25°C, the difference in ΔG and ΔG^0 being $3RT \ln R_p = 3 \times 5.708 \times 6.4 = 110$ kJ mol^{-1}.

While it is probably impossible to raise R_p high enough to reach a true equilibrium between NADH, O_2 and the adenylate system, within more restricted parts of the chain, equilibrium can be attained. It is even possible to have reversed electron flow. Consider the passage of electrons from NADH part way through the chain

---

* This ratio has also been called the phosphate potential or the phosphorylation potential. However, others define phosphate potential as the free energy of formation of ATP under existing conditions, i.e., as $+34.5$ kJ mol^{-1} + RT ln $([ATP][ADP][P_i])$. Logically, a potential should be expressed in volts as in Eq. 3-64. In view of the possibility of confusion, it seems better to avoid use of these terms.

and back out to fumarate, the oxidized form of the succinate-fumarate couple. The free energy change $\Delta G'$ (pH 7) for oxidation of NADH by fumarate is -67.7 kJ mol^{-1}. In uncoupled mitochondria electron flow would always be from NADH to fumarate. However, in tightly coupled mitochondria, in which ATP is being generated at site I, the overall value of $\Delta G'$ becomes much less negative. If $R_p = 10^4$ M^{-1}, $\Delta G'$ for the coupled process becomes approximately zero ($-67.7 + 68$ kJ mol^{-1}). Electron flow can easily be reversed so that succinate reduces NAD$^+$. In fact, such ATP-driven reverse flow may occur under some physiological conditions within mitochondria of living cells. As we shall see later, some anaerobic bacteria generate all of their NADH by reversed electron flow.

Another experiment involving equilibration with the electron transport chain is to measure the "observed potential" of a carrier in the chain as a function of the concentrations of ATP, ADP, and P_i. The observed potential E is obtained by measuring log([ox]/[red]) and applying Eq. 10-12 in which $E^{0\prime}$ is the known midpoint potential of the couple (Table 3-7) and n is the number of electrons required to reduce one molecule of the carrier.

$$E = \frac{-\Delta G}{nF} = E^{0\prime} + \frac{0.0592}{n} \log \frac{[\text{ox}]}{[\text{red}]}$$

$$= \text{observed potential of carrier} \qquad (10\text{-}12)$$

If the system is equilibrated with a "redox buffer" consisting of a mixture of some pair that equilibrates readily with the chain (Chapter 3, Section C,1),[73] E can be fixed at a preselected value. For example, a 1:1 mixture of succinate and fumarate would fix E at $+0.03$ V while the couple β-hydroxybutyrate–acetoacetate in a 1:1 ratio would fix it at $E^{0\prime} = -0.266$ V. Consider the potential of one of the cytochromes of the b type, the one designated b_K by Wilson and associates. Cytochrome b_K has an $E^{0\prime}$ value of 0.030 V. Substituting this in Eq. 10-12 and using $E = -0.266$ V (as obtained by equilibration with β-hydroxybutyrate–acetoacetate) the reader can easily calculate that at equilibrium the ratio [ox]/[red] for cytochrome b_K

is about 10^{-5}. In other words, in the absence of O_2 this cytochrome will be kept almost completely in the reduced form in an uncoupled mitochondrion.

However, if the electron transport between β-hydroxybutyrate and cytochrome b_K is tightly coupled to the synthesis of one molecule of ATP the observed potential of the carrier will be determined not only by the imposed potential E_i of the equilibrating system but also by the phosphorylation state ratio of the adenylate system (Eq. 10-13):

$$
\begin{aligned}
E(\text{observed}) &= E^{\circ\prime} + \frac{0.0592}{n} \log_{10} \frac{[\text{ox}]}{[\text{red}]} \\
&= E_i + \frac{\Delta G'_{\text{ATP}}}{96.5n'} + \frac{RT}{n'F} \ln \frac{[\text{ATP}]}{[\text{ADP}][\text{P}_i]} \\
&= E_i + \frac{0.358}{n'} + \frac{0.0592}{n'} \log_{10} R_p \quad (10\text{-}13)
\end{aligned}
$$

Here $\Delta G'_{\text{ATP}}$ is the group transfer potential ($-\Delta G'$ of hydrolysis) of ATP at pH 7 (Table 3-5) and n' is the number of electrons passing through the chain required to synthesize one ATP. Note, however, that in the upper part of the equation n is the number of electrons required to reduce the carrier, namely, one in the case of cytochrome b_K.

From Eq. 10-13 it is clear that in the presence of a high phosphorylation state ratio a significant fraction of cytochrome b_K may remain in the reduced form at equilibrium. Thus, if $R_p = 10^4$, if $E^{0\prime}$ for cytochrome b_K is 0.030 V, if $n' = 2$ and the potential E is fixed at -0.25 V using the hydroxybutyrate–acetoacetate couple, we calculate, from Eq. 10-13, that the ratio [ox]/[red] for cytochrome b_K will be 1.75. Now, if R_p is varied the observed potential of the carrier should change as predicted by Eq. 10-13. This variation has been observed.[73] Furthermore, for a tenfold change in R_p the observed potential of cytochrome b_K changed by 0.030 V, just that predicted if $n' = 2$. On the other hand, the observed potential of cytochrome c varied by 0.059 V for every tenfold change in the ratio. This is just as expected if $n' = 2$ and if synthesis of two molecules of ATP are coupled to the electron transport to cytochrome c. Thus, we have experimental verification of an interesting fact: Even when one-electron carriers such as the cytochromes are involved the

passage of *two electrons* is required to synthesize one molecule of ATP. Furthermore, from experiments of this type it can be deduced that the sites of phosphorylation are localized approximately as indicated in Fig. 10-11.

Another kind of experiment can be done by equilibrating the electron transport chain with an external redox pair of known potential using *uncoupled* mitochondria. The value of $E^{0\prime}$ of a particular carrier can then be measured by observation of the ratio [ox]/[red] and applying Eq. 10-12. While changes in the equilibrating potential E will be reflected by changes in [ox]/[red] the value of $E^{0\prime}$ will remain constant. The $E^{0\prime}$ values of Fe—S proteins and copper atoms in the electron transport chain can be obtained by equilibrating mitochondria and then rapidly freezing them in liquid nitrogen. The ratios [ox]/[red] are obtained from esr measurements at 77°K. The results of such measurements as reported by Wilson *et al.*[72-75] are shown in Table 10-6.

The values of $E^{0\prime}$ of the mitochondrial carriers fall into four **isopotential groups** at ~ -0.30, ~ 0, ~ 0.22, and ~ 0.39 V (Table 10-6). When tightly coupled mitochondria are allowed to go into state 4 (low ADP, high ATP, O_2 present but low respiration rate) the observed potentials change. That of the lowest isopotential group (which includes $NAD^+/NADH$) falls to ~ -0.38 V, corresponding to a high state of reduction of the carriers to the left of the first phosphorylation site in Fig. 10-11. Groups 2 and 3 remain close to their midpoint potentials at ~ -0.05 and $+0.26$ V. In this condition the potential difference between each successive group of carriers amounts to ~ 0.32 V, just enough to balance the formation of one molecule of ATP for each two electrons passed at a ratio $R_p \approx 10^4 M^{-1}$ (Eq. 10-13).

Two cytochromes show exceptional behavior and appear twice in Table 10-6. The midpoint potential $E^{0\prime}$ of cytochrome b_T changes from -0.030 V in the absence of ATP to $+0.245$ V in the presence of a high concentration of ATP. On the other hand, $E^{0\prime}$ for cytochrome a_3 drops from $+0.385$ to 0.155 V in the presence of ATP. This shift in potential suggests that a high energy *reduced* form of

TABLE 10-6
Electrode Potentials of Mitochondrial Electron Carriers and Free Energy Changes Associated with Passage of Electrons[a]

	Electron carrier	$E^{0\prime}$ (pH 7) isolated	$E^{0\prime}$ (pH 7.2) in mitochondria	ΔG (kJ mol^{-1}) for 2 e^- flow to O_2 at 10^{-2} atm, carriers at pH 7
	NADH/NAD$^+$	-0.320		-213
Group I ~-0.30 V	Flavoprotein		~-0.30	
	Fe—S protein		~-0.305	
	β-Hydroxybutyrate-acetoacetate	-0.266		-203
	Lactate-pyruvate	-0.185		-187
	Succinate-fumarate	0.031		-146
Group II ~0 V	Flavoprotein		~-0.045	
	Cytochrome b_T		-0.030	
	Cu		0.001	
	Fe—S protein		0.030	
	Cytochrome b_K		0.030	
	Ubiquinone	0.10	0.045	-132
	Cytochrome a_3 + ATP		0.155	
Group III ~0.22 V	Cytochrome c_1		0.215	
	Cytochrome c	0.254	0.235	-102
	Cytochrome b_T + ATP		0.245	
	Cytochrome a	0.29	0.210	
	Cu		0.245	
	Fe—S protein		0.28	
Group IV	Cytochrome a_3		0.385	-77
	O_2 (10^{-2} atm)	0.785		0.0
	1 atm	0.815		

[a] Data from Wilson et al.[72,73]

cytochrome a_3 is coupled to synthesis of ATP. The presence of a high concentration of ATP tends to make the formation of this intermediate by reduction more difficult (Section E,9,a). On the other hand, the change of $E^{0\prime}$ for cytochrome b_T in the opposite direction suggests a high energy oxidized form (Eq. 10-11). Such conclusions depend upon the precision and certainty with which the ratio [ox]/[red] can be measured by spectroscopic means. These results have also been interpreted to mean that cytochromes b_T and a_3 are *directly* involved in the oxidative phosphorylation process.[72-75] However, not everyone agrees.[77]

8. Reconstitution of Phosphorylating Particles

Many attempts have been made to take apart and to reconstitute phosphorylating submitochondrial particles. The first accomplishment has been the removal of several "coupling factors" of which the best known is coupling factor F_1 (p. 583). Removal of F_1 from submitochondrial particles always leads to loss of ATP synthesis, but electron transport may proceed unimpaired. It has been possible to regenerate phosphorylating ability by adding F_1 back to membrane preparations. Thus, F_1 is apparently intimately involved in ATP synthesis. This important protein has been isolated both from mitochondria and from chloroplasts as homogeneous particles of MW ~285,000. It seems to be composed of five different kinds of polypeptide chains with molecular weights ~60,000, 56,000, 36,000, 17,000, and 13,000.[78,79]

Solubilization of the inner mitochondrial membrane itself can be accomplished with detergents. For example, Ragan and Racker[80] homogenized mitochondria in sucrose solution and added the steroidal detergent **sodium cholate** (Fig. 12-16) as well as other reagents. After stirring the mixture gently and centrifuging, the supernatant was subjected to ammonium sulfate fractionation. A **hydrophobic protein fraction** depleted in phospholipids, cytochrome oxidase, and soluble components such as cytochrome c was obtained. When an appropriate mixture of phospholipids (including phosphatidylethanolamine and phosphatidylcholine) was added to the hydrophobic protein complex, the ability to phosphorylate ADP during oxidation of NADH by Q_1 was restored. The reconstituted system was sensitive to inhibition by rotenone (Table 10-2) uncoupling agents and oligomycin. As much as 0.5 mol of ATP was generated per mole of NADH oxidized. Similar success in regenerating sites 2 and 3 has also been reported.[81]

The efforts to resolve and reconstitute oxidative phosphorylation are extremely important in pointing the way to future experiments. Nevertheless, there are difficulties in interpreting the results. Few of the components have been isolated in a completely homogeneous state. It will be necessary to further purify and characterize them and to learn how to handle each protein without denaturation. However, we can probably look forward confidently to the time when it will be possible to mix the many components of the mitochondrial membranes, each one highly purified, and to reconstitute a functioning electron transport and phosphorylation system. Such experiments may also answer the question "Is a membrane needed for oxidative phosphorylation?" While some workers feel that current experiments have demonstrated that an intact membrane is *not* essential, no one has obtained phosphorylation with truly soluble enzyme preparations. ATP synthesis occurs only when the appropriate proteins are embedded in a phospholipid "vesicle."

9. Theories of Oxidative Phosphorylation and Model Experiments

Many suggestions have been offered for formation of high energy intermediates by electron flow. The idea is a natural one that comes from studies of substrate level phosphorylation in which high energy intermediates are generated by the passage of electrons through the substrates. As we have already seen (Chapter 8, Section H,5) the aldehyde group of glyceraldehyde 3-phosphate is converted to an acyl phosphate, which after transfer of the phosphoryl group of ADP becomes a carboxylate group. In this case the free energy of oxidation of the aldehyde to the carboxylate group provides the energy for the synthesis of ATP. The reaction differs from mitochondrial electron transport in that the product, 3-phosphoglyceric acid, is not reconverted to glyceraldehyde 3-phosphate. Electron carriers of the respiratory chain must be regenerated in some cyclic process. The latter requirement makes it more different to imagine a practical mechanism for oxidative phosphorylation.

a. Theories Based on Special Chemistry of an Electron-Carrying Coenzyme or Prosthetic Group

Equation 10-11 represents a suitable starting point for examining theories[82,83] of oxidative phosphorylation.* Lipmann[84] proposed a general scheme related to this equation. Group Y'—OH (group Y of Eq. 10-11) is visualized as adding to a suitable carbon–carbon double bond in carrier BH_2 to initiate the sequence. Although the isotopic exchange reactions (Section E,5) rule out the possibil-

---

* For a list of proposed mechanisms of oxidative phosphorylation see Lardy and Ferguson.[82]

ity that either ADP or P_i serve as Y, it is attractive to think that a bound phosphate ion, e.g., in a phospholipid or coenzyme, could be involved. The low energy adduct Y—BH_2 of Eq. 10-11 would be converted by oxidation to Y ~ B which would be similar in reactivity to an acyl phosphate or thioester.

It is important to realize that whatever the nature of Y ~ B, part of group Y is left attached to B after the transfer of Y to X. Thus, Eq. 10-11 is more complete if Y is replaced by Y'OH. In this case compound X ~ Y' is formed and the carrier is left in the form of B—OH. Elimination of a hydroxyl group is required to regenerate B. Part of the problem in identifying suitable chemistry for these steps in oxidative phosphorylation is the necessity for easy occurrence of such an elimination.

While Eq. 10-11 shows the addition of compound Y to a reduced form of carrier B with subsequent oxidation to a high energy form of the oxidized carrier, it is possible that Y could add to B, the oxidized form of the carrier. Then B—Y could be reduced to BH_2 ~ Y, a high energy form of the reduced carrier. After reaction with group X (as in Eq. 10-11) a modified reduced carrier B'H_2 would be left and could be converted by elimination to BH_2 to complete the cycle. Thus, we should be alert to the possibility of either an oxidized high energy form of the carrier or a reduced high energy form.

One variation of Lipmann's general scheme postulates nucleophilic addition to a double bond of NADH (see Eq. 8-48). The oxidative phosphorylation cycle would operate as shown in Eq. 10-14.[85,86] Note that the elimination of water to regenerate NADH is the reverse of Eq. 8-48 and would have an unfavorable equilibrium position, a weakness of this scheme. Mechanisms based on addition of Y—O^- to the 6 position of NAD^+ can also be visualized.*

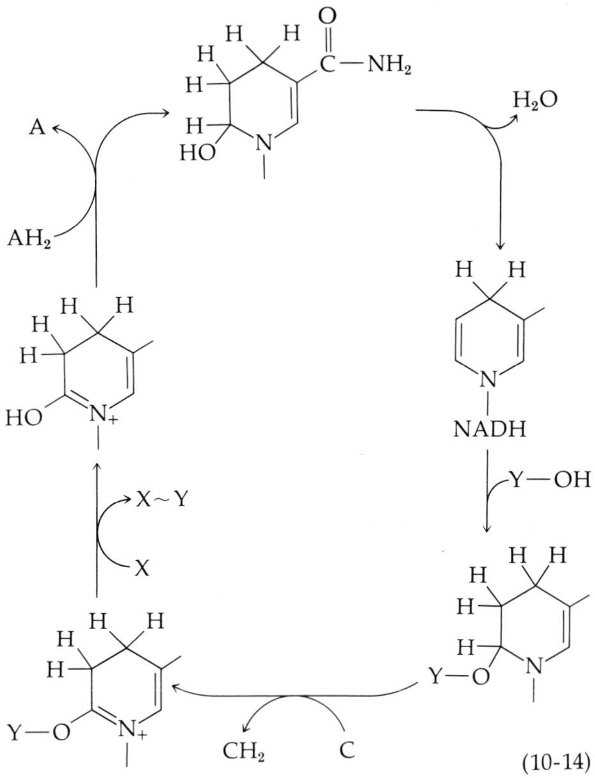

$$(10\text{-}14)$$

---

* The reader may wish to consider the following scheme: Y—O^- adds to the 6-position of NAD^+. The adduct is then oxidized (removal of H^- from the 6 position). The resulting derivative transfers group Y to generate X~Y, leaving 6-hydroxy NAD^+. The latter is reduced by carrier AH_2 to 6-hydroxy NADH which, after a tautomerization step, loses OH^- to regenerate NAD^+.

Many investigators have wondered whether some special chemistry of ubiquinone or other quinones may be involved in oxidative phosphorylation. Harrison showed that if a phosphate ester of a hydroquinone (a quinol phosphate) is oxidized in a suitable model system a high energy oxidized intermediate is formed.[87] If P_i is present (X in Eq. 10-15) pyrophosphate (X ~ Y) is generated. (Alternatively Y may be released as a metaphosphate group able to interact with X.) An uncertainty in the scheme of Eq. 10-15 is the means by which the quinol phosphate could be formed. As shown in the equation, it could result from addition of YOH to the carbonyl group of the oxidized quinone followed by loss of hydroxyl ion during the reduction by AH_2. A modification of this scheme[88] postulates loss of a proton from the methyl group of ubiquinone, the ring being closed at the same time to form a chromanol with

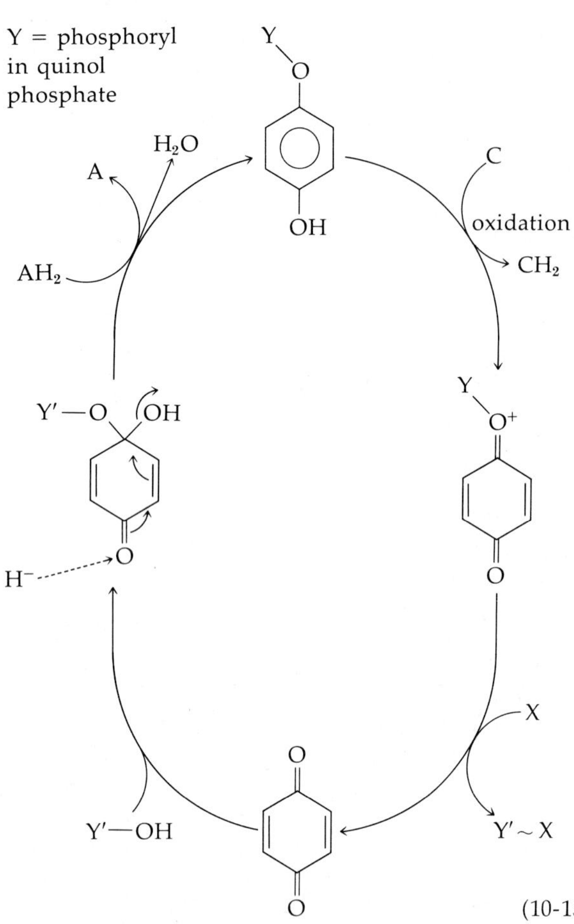

Y = phosphoryl
in quinol
phosphate

(10-15)

an *ortho*-quinonoid (quinone methide) structure (Eq. 10-16). Addition of phosphate and reduction

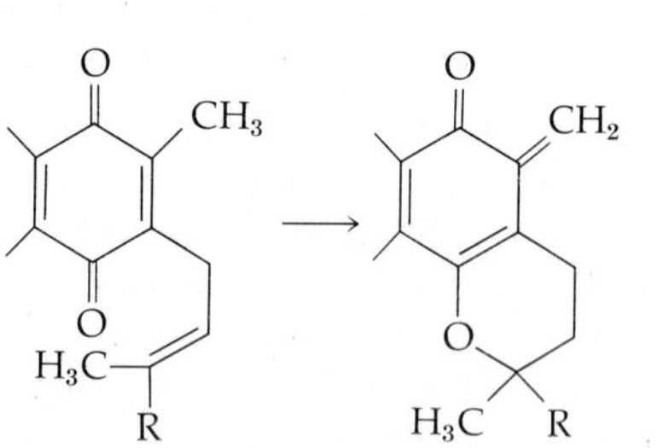

(10-16)

could occur as in Eq. 10-15 (lower center and left side). Another theory postulates that the oxidized

form of a quinone acts as the reduced carrier.[89]

A proposal of Wang links some special chemistry of cytochromes to phosphorylation (Eq. 10-17).[90,91] A reduced form of a cytochrome with iron in the Fe(II) state (upper right of Eq. 10-17) is

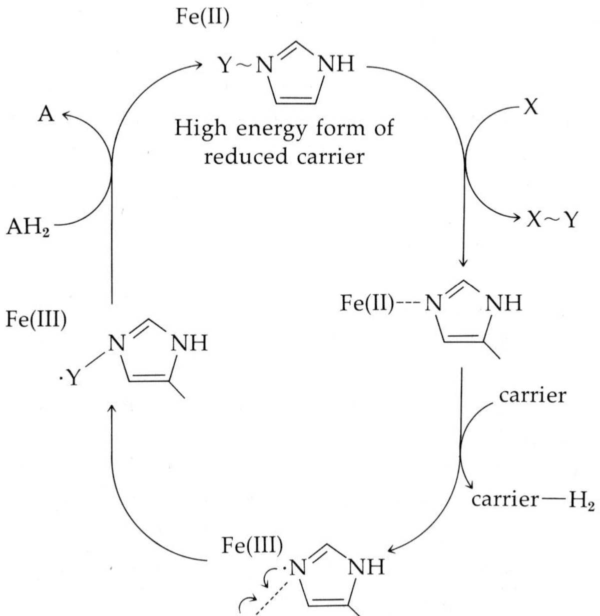

(10-17)

coordinated with an imidazole group of the protein. Now a two-electron oxidation by carrier removes one electron from the iron and the other from the imidazole group to create a radical. The latter undergoes a coupling reaction with Y. Then in a two-electron reduction by AH_2 both the intermediate radical and the Fe(III) accept electrons to form a high energy derivative of the reduced carrier. The latter transfers Y to X.

Another possibility is suggested by results of oxidation of thioethers by iodine.[91,92] The cyclic participation of a methionyl residue of an electron carrier would be possible. It is also entirely reasonable to suppose that some as yet unknown aspect of the chemistry of the iron–sulfur proteins of mitochondria may provide the basis for oxidative phosphorylation.

b. Theories Based on Conformational Changes of Proteins

Passage of electrons through the electron transport chain may cause changes in protein conformation that could lead to synthesis of high energy intermediates. The fact that cytochrome c is known to undergo small but distinct conformational changes upon oxidation or reduction supports this notion.[12] The tight association of one protein with another in the mitochondrial inner membrane suggests that any conformational change induced at a site of electron transfer might be transmitted through one or more proteins to a distant site (e.g., in coupling factor F_1) where ATP might be formed.

If a conformational change in a protein were large enough it might cause a carboxyl group and an SH group to be squeezed together causing spontaneous formation of a thioester linkage. Then, when the protein relaxed to its other conformation, the high energy character of the thioester would become apparent and it could enter into the exchange reactions necessary for synthesis of ATP. Another possibility is that a protein might bind ADP and inorganic phosphate at adjacent sites. Then, following a conformational change the two components might be literally squeezed together leading to spontaneous displacement of an OH ion with formation of ATP and H_2O.

A most interesting idea, backed by some experimental evidence,[93] is that ADP and P_i bind at adjacent sites and, in the hydrophobic environment of the active site of F_1, spontaneously eliminate water to form tightly bound ATP. The driving force would reside in the strong binding to one conformer of the protein. Electron transport would induce a linked conformational change that would cause the F_1 molecule to release the synthesized ATP.

Ideas of conformational coupling of ATP synthesis and electron transport are especially attractive when we recall that ATP is used in muscle to carry out mechanical work. Here we have the hydrolysis of ATP coupled to motion in the protein components of the muscle (Box 10-F). Would it not be reasonable to form ATP as a result of some mo-

---

BOX 10-F

THE CHEMISTRY OF MUSCULAR CONTRACTION

While the properties of the protein assemblies found in muscle are being described in elegant detail (Chapter 4, Section F,1), the most important question remains unanswered.[a,b] How can the muscle machinery use the free energy of hydrolysis of ATP to do mechanical work? From electron microscopy and X-ray diffraction studies, it has been shown that in rigor the crossbridges formed by the myosin heads are all firmly attached to the thin actin filaments. However, the addition of ATP causes an instantaneous release of the crossbridges from the thin filaments. In relaxed muscle, the thin filaments are free to move past the thick filaments and the muscle has some of the properties of a weak rubber band. However, activation by a nerve impulse, with associated release of calcium ions (Chapter 4, Section F,1), causes the thin filaments to slide between the thick filaments with shortening of the muscle.

An activated muscle shortens if a low tension is applied to the muscle, but at a higher tension it maintains a constant length. Since the maximum tension developed is proportional to the length of overlap between the thick and thin filaments, it is natural to identify the individual crossbridges as the active centers for generation of the force needed for contraction.

The widely accepted "rowing hypothesis" of H. E. Huxley postulates a cyclic series of changes in the crossbridges. As shown in the accompanying figure,[c] the upper end of the myosin rod above

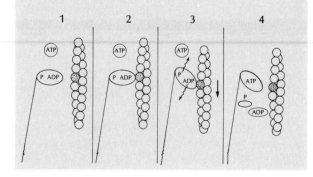

Box 10-F (*Continued*)

the "hinge region" that extends 70–110 nm from the rod tail, swings freely in the space between the thick and thin filaments. When ATP is added it binds to the myosin head detaching the latter from the thin filament. The bound ATP is immediately hydrolyzed, but the ADP and P_i formed remain bound at the active site. Furthermore, some of the free energy of hydrolysis of the ATP is stored, perhaps through a conformational change in the myosin head. As long as calcium ions are absent, only a slow release of the bound ADP and P_i and replacement with fresh ATP takes place. Thus, myosin alone shows a weak "ATPase" activity. On the other hand, in activated muscle it is proposed that the head with the split ATP products combines with an actin subunit. The crossbridges appear to attach themselves at right angles to the thin filaments. Then the stored energy in the myosin head (or in the actin) is used to bring about a conformational change that alters the angle of attachment of the myosin head to the thin filament from 90° to about 45°. Such a change can be observed directly by microscopy. In the absence of ATP (rigor) the heads make a chevronlike structure binding at an angle of about 45° to the actin filaments. The change in this angle is believed sufficient to move the actin filament by ~10 nm (the length of two actin subunits). The P_i and ADP can now be dissociated from the head and replaced by ATP causing a release of the actomyosin complex. In the meantime, other crossbridges are attached so that the thin filaments do not slip back to their original positions.

The Huxley model is appealing and enjoys considerable experimental support, but it is by no means clear how the energy of ATP hydrolysis is linked to the conformational change. It is possible that quite different processes account for the elementary cycle of muscle contraction. One idea is that the conformational change does not occur in the myosin head or in the actin but in the "hinge" region of myosin. There is a stretch of 200–300 residues containing a large number of positively charged side chains and in which the α helix is apparently inherently unstable. Perhaps the interaction with ATP triggers a cooperative

conformational change by which this region is converted from α helix to a "random coil" in which the two peptide chains of the myosin rod are more extended.[d,e] Perhaps the incoming flow of calcium ions is sufficient to destabilize the structure and permit such a conformational change. Reaction with actin would have to trigger a reverse conformational change leading to a shortening of the distance and pulling the thin filament. The reader may be skeptical about the possibility of a cycle of reversible expansion and contraction of this type. However, using a similar helix–random coil transition of collagen, an interesting machine has been constructed as shown in the accompanying figure.[f] A belt of collagen

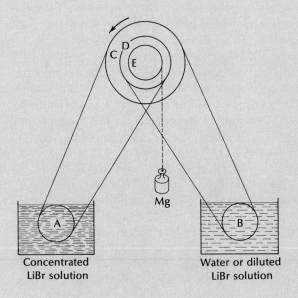

passes first through a concentrated lithium bromide solution where it undergoes contraction and later through a water bath where the concentrated salt is diluted and the helical collagen chains revert to random coil conformations. The machine really runs using the free energy of dilution of the LiBr as a source of energy.

Some authors wondered whether the most important aspect of the muscle machinery is present in the actin.[g] For example, it has been proposed

Box 10-F (*Continued*)

that ATP hydrolysis induces a few percent shortening in a series of 15–20 actin molecules, sufficient to account for the 10 nm movement required in contraction. Others suggested that the crossbridges are not part of the contractile machinery but serve as "catches." Since it is known that muscle contracts nearly *isovolumetrically*, anything that expands the sarcomere will cause a contraction. It is proposed that the hydrolysis of ATP deposits negatively charged phosphoryl groups on the actin filaments and that the electrostatic repulsion is responsible for lateral expansion of the sarcomere.[h] Still other authors have emphasized the possibility that cleavage of ATP may lead to a resonant transfer of energy into a vibrational mode of the amide linkages in the α-helical sections of myosin.[i,j] This vibrational energy could move a long distance through the hydrogen-bonded lattice of the proteins and in some way be harnessed in a contractile process. While the idea may seem farfetched, it emphasizes that we should not think of the myosin rods nor of the thin filaments as rigid inert materials. We do not know in what part of the apparatus the contractile machinery is concentrated and somebody's speculative scheme may turn out to be right.

The reader might try to imagine how a molecular machine could split ATP and contract. It may help to look at the structure of ATP itself. Note the triphosphate group with the many negative charges repelling each other. Think of what must happen when this molecule displaces ADP and P_i from the attached myosin head. The process breaks a protein–protein bond, most likely by inducing an electrostatic repulsion at the right point in the interface between the proteins. Think about the problem of generating ATP in oxidative phosphorylation and of the possible role of protons in inducing ATP synthesis (Section E,9c). What could protons do to a protein surrounding ATP in the inverse process? Think about the effect of Mg^{2+} complexed with the polyphosphate group of ATP and about what might happen if Ca^{2+} binds to a nearby site on a protein. Consider the fact that evidence of transient phosphorylation of protein side chains exists. What

could happen if a histidine side chain hydrogen-bonded to a peptide backbone position at the end of a helix were phosphorylated? The author has not been able to imagine how all these factors can be put together to make a muscle machine, but maybe some clever reader will.

While ATP provides the immediate source of energy for operating the muscle, its concentration is only about 5 mM. However, muscle contains in addition, a **phosphagen,** an N-phosphoryl derivative of a guanidinium compound. In mammalian muscle, the phosphagen is **creatine phosphate. Arginine phosphate** and related compounds

Creatine phosphate

can serve in invertebrates. The group transfer potential for creatine phosphate is -43.1 kJ mol^{-1}. Thus, the transfer to ATP is spontaneous with $\Delta G' = -8.6$ kJ mol^{-1}. Creatine phosphate is present at a concentration of 20 mM providing a reserve of high energy phosphoryl groups and keeping the adenylate system of muscle buffered at a high "energy charge." (See Chapter 14, Section C,3 for the biosynthesis of creatine.)

[a] Y. Tonomura and F. Oosawa, *Annu. Rev. Biophysics Bioeng.* **1,** 159–190 (1972).

[b] E. W. Taylor, *Annu. Rev. Biochem.* **41,** 577–616 (1972).

[c] Drawing from H. G. Mannherz, J. B. Leigh, K. C. Holmes, and G. Rosenbaum, *Nature (London), New Biol.* **241,** 226–229 (1973).

[d] W. F. Harrington, *PNAS* **68,** 685–689 (1971).

[e] M. Burke, S. Himmelfarb, and W. F. Harrington, *Biochemistry* **12,** 701–710 (1973).

[f] I. Z. Steinberg, A. Oplatka, and A. Katchalsky, *Nature (London)* **210,** 568–571 (1966).

[g] K. Laki, *J. Theor. Biol.* **44,** 117–130 (1974).

[h] R. Ashley, *J. Theor. Biol.* **36,** 339–354 (1972).

[i] C. W. F. McClare, *J. Theor. Biol.* **35,** 569–595 (1972).

[j] A. S. Davydov, *J. Theor. Biol.* **38,** 559–569 (1973).

tion induced in the protein components of the mitochondrial membrane? The dramatic alteration in the shape of mitochondria between states 4 (depleted in ADP) and 3 (actively respiring) has suggested to some that there are underlying conformational changes in membrane proteins that are involved in phosphorylation.[94] Similar considerations may apply to phosphorylation in chloroplasts.[95]

A **phonon theory** does not propose a conformational change but assumes that a particular lattice vibration (e.g., an N—H stretching frequency) in the frequency region 2.7–5.4 kK, i.e., 32–65 kJ mol^{-1} (see Chapter 13, Section B,3a) might be excited in a membrane protein by electron transport.[96] As a result of this specific molecular motion within the protein suitably placed phosphates (probably hydrogen bonded to amide backbone groups) would be snapped together to form pyrophosphate linkages.

c. Theories Based on Proton Gradients

To account for the inability to identify high energy intermediates as well as the apparent necessity for an intact membrane, P. Mitchell, in 1961, offered the **chemiosmotic theory** of oxidative phosphorylation.[97,98] The theory also takes into account the existence of **energy-linked** processes such as the accumulation of cations by mitochondria. The principal features of the Mitchell theory are illustrated in Fig. 10-12. The inner membrane of the mitochondrion is supposed to contain a **proton pump** which is operated by electron flow and which causes protons to be expelled through the membrane from the matrix space. The idea of pumping protons by electron transport was not a new one and had previously been suggested as a means of accumulating hydrochloric acid in the stomach. As indicated in Fig. 10-12, an oxidized carrier B, upon reduction to BH_2, acquires two protons. These protons are not necessarily acquired from reduced carrier AH_2, and Mitchell supposed that they are picked up from the solvent on the matrix side of the membrane. Then, when BH_2 is reoxidized by carrier C, protons are released

on the outside of the membrane. Mitchell offered evidence for this stoichiometry of two protons expelled for each ATP synthesized. It follows that there should be three different proton pumps in the electron transport chain corresponding to the three phosphorylation sites.

The postulated proton pumps can either lead to bulk accumulation of protons in the intermembrane space, with a corresponding drop in pH, or to an accumulation of protons on the membrane itself. The latter will be expected if counterions X^- do not pass through the membrane with the protons. The result in such a case will be the development of a **membrane potential,** a phenomenon already well documented for nerve membranes (Chapter 5, Section B,3).

A fundamental postulate of the chemiosmotic theory is the presence of an oriented ATPase or ATP synthetase that utilizes the free energy difference of the proton gradient to drive the synthesis of ATP (Fig. 10-12). Since $\Delta G'$ (pH 7) for ATP synthesis is $+ 34.5$ kJ mol^{-1} and it is assumed that the passage of two protons through the ATPase is required to form one ATP, the necessary pH gradient (given by Eq. 3-25) would be $34.5/(2 \times 5.708) = 3.0$ pH units at 25°C. On the other hand, if the phosphorylation state ratio is $\sim 10^4 M^{-1}$, the pH difference would have to be ~ 5 units. Experiments can be cited that show that passage of electrons does induce a pH difference and that an artificially induced pH difference across mitochondrial membranes leads to ATP synthesis. However, pH gradients of the required order of the magnitude have not been observed. Nevertheless, if the membrane were charged as indicated in Fig. 10-12, without accumulation of protons in the bulk medium, a membrane potential would be developed and could drive the ATP synthetase, just as would a proton gradient. A membrane potential $\Delta\psi$ of ~ 0.177 V at 25°C would be (by Eq. 3-63) equivalent to a 3.0 unit change in pH. The free energy change for passage of one ion across the membrane is (Eq. 10-18):

$$\Delta G = 5.708 \, \Delta pH = 96.5 \, \Delta\psi \text{ kJ} \qquad (10\text{-}18)$$

The term **proton motive force** has been used to

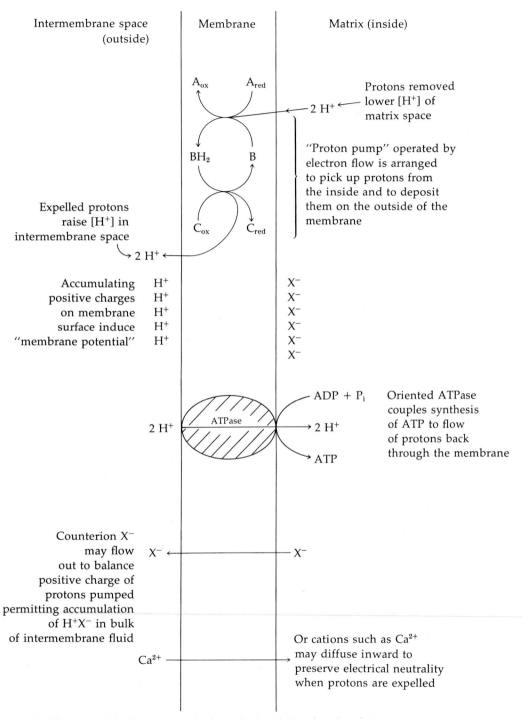

Intermembrane space (outside)	Membrane	Matrix (inside)

A_{ox} A_{red}

$2\ H^+ \longleftarrow$ Protons removed lower $[H^+]$ of matrix space

BH_2 B

"Proton pump" operated by electron flow is arranged to pick up protons from the inside and to deposit them on the outside of the membrane

Expelled protons raise $[H^+]$ in intermembrane space

C_{ox} C_{red}

$2\ H^+ \leftarrow$

Accumulating	H^+	X^-
positive charges	H^+	X^-
on membrane	H^+	X^-
surface induce	H^+	X^-
"membrane potential"	H^+	X^-
		X^-

$ADP + P_i$

Oriented ATPase couples synthesis of ATP to flow of protons back through the membrane

$2\ H^+$ ATPase $\rightarrow 2\ H^+$

$\searrow$ ATP

Counterion X^- may flow out to balance positive charge of protons pumped permitting accumulation of H^+X^- in bulk of intermembrane fluid

$X^- \longleftarrow \qquad X^-$

$Ca^{2+} \longrightarrow$ Or cations such as Ca^{2+} may diffuse inward to preserve electrical neutrality when protons are expelled

Fig. 10-12 Principal features of the "chemiosmotic theory" of oxidative phosphorylation.

describe the sum of the membrane potential and pH gradient, both in volts (Eq. 10-19),

$$\text{Proton motive force} = \Delta\psi - (RT/F)\,\Delta pH \quad (10\text{-}19)$$

where $RT/F = 0.0592$ at 25°C. Since it is very difficult to measure the membrane potential of mitochondrial or chloroplast membranes directly, the postulate has not been tested adequately.* However, the Mitchell hypothesis has stimulated intensive experimental efforts and is usually regarded as one of the currently most important concepts in membrane biology.

How can the accumulation of ions by mitochondria be explained? As indicated in the lower part of Fig. 10-12 there are two possibilities for preservation of electrical neutrality. The counterions X^- may flow out to balance the protons discharged on the outside. On the other hand, if a cation such as Ca^{2+} flows inward to balance the two protons flowing outward, neutrality will be preserved; the mitochondrion will be observed to accumulate calcium ions. Experimentally such accumulations are observed to accompany electron transport. Not only is calcium accumulated but also in the presence of a suitable ionophore (Chapter 5, Section B,2,c) "energy-dependent" accumulation of potassium ions also takes place.[99]

Can one imagine a suitable proton pump and an oriented ATPase driven by proton flow? Here is one strictly hypothetical model. For a nucleophile Y to be able to generate a high energy linkage Y ~ P directly by attack on the phosphorus atom of P_i an OH^- ion must be eliminated. This is not a very probable reaction at pH 7 but it would be reasonable at low pH. Thus, we might imagine that the function of the oriented ATPase is to deliver a proton specifically and to hold it against the oxygen to be eliminated (Eq. 10-20). How could a proton be directed to the proper spot to assist in this way? Perhaps it could move down a channel

Nucleophile Y attacks P to form high energy Y ~ P directly

(10-20)

through the membrane to the correct position. On the other hand, it might be easier to imagine that the proton is specifically generated at the right site through the electron transport process.

Think about the effect of oxidizing an iron atom from Fe^{2+} to Fe^{3+} in a cytochrome. Perhaps it is not farfetched to imagine that removal of an electron could be accompanied by electron displacements within the coordinated histidine, peptide, and other polar groups (Eq. 10-21) to generate a proton

(10-21)

next to the phosphate. It is possible to imagine that such displacements are even transmitted into adjacent proteins to reach the F_1 active sites. At this point we have come full circle and are again considering something akin to "high-energy chemical intermediates." Many biochemists feel that when oxidative phosphorylation is understood there will be something of truth from all of the currently competing views.

10. Energy-Linked Processes in Mitochondria

The synthesis of ATP by mitochondria is strongly inhibited by oligomycin. On the other hand, there are processes that require energy from electron transport and that are not inhibited by oligomycin. These include the transport of ions across the mitochondrial membrane. Another

---

* The membrane potential has been estimated indirectly, e.g., from the distribution of K^+ ions across mitochondrial membranes in the presence of valinomycin[98a] or by using dyes that alter their fluorescence in response to membrane potential.[98b] Membrane vesicles from *E. coli* have also been studied,[98c] but interpretations remain controversial.[98d]

energy-linked process is "reverse electron flow" from succinate to NAD^+ (Section E,7). In both of these instances oligomycin has no effect but dinitrophenol and other uncouplers block the reactions. This fact can be rationalized if we assume that the high energy intermediate $X \sim Y$ is formed in the presence of oligomycin and that the reverse electron flow and ion pumping can be driven by the free energy of hydrolysis of this intermediate without formation of ATP. Dinitrophenol uncouples all of the reactions by promoting the hydrolysis of $X \sim Y$ while oligomycin affects only ATP synthesis. The observations are also explained by the Mitchell hypothesis which proposes that ion transport precedes ATP synthesis.

Another energy-linked process is the transhydrogenase reaction (Eq. 11-12).

11. Transport across Mitochondrial Membranes

Like the external plasma membrane of the cells the inner mitochondrial membrane is selective. Some nonionized materials pass through readily but the transport of ionic substances, including the anions of the dicarboxylic and tricarboxylic acids, is tightly controlled. In some cases energy-dependent "active transport" is involved. In other cases one anion may pass inward in exchange for another anion passing outward. In either case specific translocating carrier proteins are needed (Chapter 5, Section B,2).

One translocation system exchanges ADP for ATP. This adenine nucleotide carrier takes ADP into the matrix for phosphorylation in a 1:1 ratio with ATP exported into the cytoplasm.[60,100–102] A separate carrier allows P_i to enter, probably as $H_2PO_4^-$. It is usually assumed that the phosphorylation state ratio $R_p = [ATP]/[ADP][P_i]$ is the same both inside and outside the mitochondria. However, Klingenberg finds R_p 10 times higher outside than inside.[102] This suggests that newly synthesized ATP may be released largely on the outside rather than the inside of the mitochondrial inner membrane. A smaller amount would be released

inside for use in activating fatty acids, for protein synthesis, etc. Pyruvate also appears to enter mitochondria on its own carrier, probably together with a proton. On the other hand, dicarboxylic acid anions such as malate and α-ketoglutarate can be exchanged in a 1:1 ratio, as can aspartate and glutamate.

The membranes of mitochondria *are not permeable to NADH*. Thus, the transfer of reducing equivalents from NADH produced in the cytoplasm into the mitochondria poses an important problem. In fungi and in green plants the problem is met by providing *two* NADH dehydrogenases (flavoproteins) embedded in the inner mitochondrial membranes.[61,103] One faces the matrix space and oxidizes the **endogenous** NADH produced in the matrix while the second faces outward to the intermembrane space and is able to oxidize the **exogenous** NADH formed in the cytoplasm. Both feed electrons into the chain via ubiquinone, but the exogenous NADH dehydrogenase is *not* inhibited by rotenone (see Fig. 10-11).

In animals the reducing equivalents from NADH enter the mitochondria indirectly. A variety of mechanisms have been suggested and it is likely that several may operate at once. In insect flight muscle NADH reduces dihydroxyacetone phosphate. The resulting α-glycerol phosphate (*sn*-3-glycerol phosphate) passes through the permeable outer membrane of the mitochondria where it is reoxidized to dihydroxyacetone phosphate by means of the FAD-containing glycerol-phosphate dehydrogenase embedded in the outer surface of the inner membrane. The dihydroxyacetone can then be returned to the cytoplasm. The overall effect of this **glycerol phosphate shuttle** (Fig. 10-13) is to provide for mitochondrial oxidation of NADH produced in the cytoplasm. In mammals the same function appears to be served by a more complex **malate-aspartate shuttle** (Fig. 10-13). Reduction of oxaloacetate to malate by NADH, transfer of malate into mitochondria and reoxidation with NAD^+ accomplishes the transfer of reducing equivalents into the mitochondria. However, mitochondrial membranes are relatively impermeable to oxaloacetate. Therefore, return of

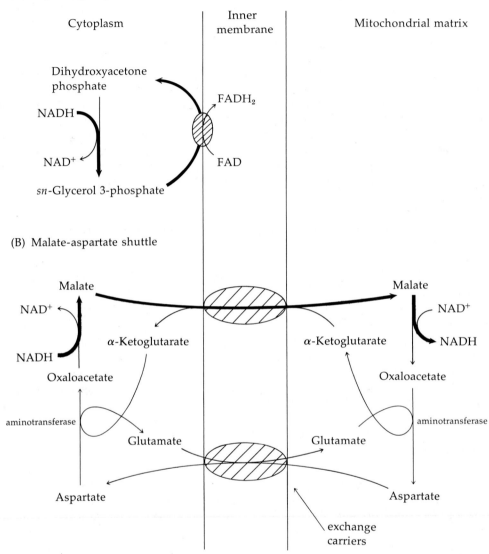

(A) Glycerol phosphate shuttle

Cytoplasm   Inner membrane   Mitochondrial matrix

Dihydroxyacetone phosphate

NADH

NAD⁺

FADH₂

FAD

sn-Glycerol 3-phosphate

(B) Malate-aspartate shuttle

Malate

NAD⁺

NADH

Oxaloacetate

α-Ketoglutarate

aminotransferase

Glutamate

Aspartate

Malate

NAD⁺

NADH

Oxaloacetate

α-Ketoglutarate

aminotransferase

Glutamate

Aspartate

exchange carriers

Fig. 10-13   (A) The glycerol phosphate shuttle and (B) the malate-aspartate shuttle for transport of electrons from cytoplasmic NADH into mitochondria. The heavy arrows trace the pathway of the electrons transported.

the latter to the cytoplasm is via transamination to aspartate which leaves the mitochondria together with α-ketoglutarate. At the same time glutamate enters the mitochondria in exchange for aspartate. The α-ketoglutarate presumably exchanges with the entering malate as is indicated in Fig. 10-13.

It has been suggested that the export of aspartate from mitochondria is "energy driven" like the Na⁺ pump of the cytoplasmic membrane.[104] The mechanism of such transport may be related to that of amino acid *uptake* by cytoplasmic membrane vesicles from bacteria.[105–107] It appears that accumulation of amino acids by such vesicles does not depend upon ATP but does require electron transport from specific membrane-bound flavin-containing dehydrogenases. Dehydrogenation of **D-lactate** is especially effective in supporting uptake of amino acids by vesicles from *E. coli.* Dehydrogenation of L-lactate or of glycerol phosphate specifically promotes uptake by vesicles of *Staphylococcus aureus.* The affinity of an amino acid for the membrane "pumps" of these vesicles appears to vary with the redox state of electron carriers in the membrane. The results suggest a "pump" in which the redox state of a carrier embedded in the pump protein controls the affinity for a substrate and also induces a conformational change. For example, changes analogous to those induced by phosphorylation and dephosphorylation in the hypothetical Na^+ pump of Fig. 5-3 could occur. However, the interpretations of experimental data upon which these proposals are based have been questioned.[107,108]

The role of carnitine in facilitating transport of fatty acids into mitochondria has already been considered in Chapter 9, Section A,6.

F. ENERGY FROM INORGANIC REACTIONS

Some bacteria obtain all of their energy from inorganic reactions (Chapter 1, Section A,10). These **chemolithotrophic** organisms usually have a metabolism that is similar to that of heterotrophic organisms, but they have an additional capacity for obtaining energy from an inorganic process. They also have the ability to fix CO_2 in the same manner as do green plants. The chloroplasts of green plants, using energy from sunlight, supply the organism with both ATP and the reducing agent NADPH. In a similar way the lithotrophic bacteria must obtain both energy and reducing materials from the inorganic reactions.

Lithotrophic organisms often grow slowly, and study of their metabolism has been difficult.[109] Nevertheless, enough is now known to make it clear that in most cases these bacteria possess an electron transport chain similar to that present in mitochondria. Passage of electrons through this

chain leads to ATP formation. The amount of ATP formed depends upon the number of coupling sites which in turn depends upon the electrode potentials of the reactions involved. Thus, H_2, when oxidized by O_2, leads to passage of electrons through all three coupling sites with synthesis of three molecules of ATP. On the other hand, oxidation by O_2 of nitrite, for which $E^{0'}$ (pH 7) = $+0.42$ V, leads to ATP synthesis only at site III. In the case of nitrite oxidation not only is the yield of ATP less than in the oxidation of H_2 but also another problem arises. Whereas reduced pyridine nucleotides can be generated readily from H_2, nitrite is not a strong enough reducing agent to reduce NAD^+ to NADH. The only way that reducing agents can be generated in cells utilizing this reaction is via reverse electron flow driven by hydrolysis of ATP. Such reverse electron flow is a common process for many chemolithotrophic organisms.

Let us consider the inorganic reactions in two groups: (1) oxidation of reduced inorganic compounds by O_2 and (2) oxidation reactions in which an inorganic oxidant such as nitrate or sulfate substitutes for O_2. The latter reactions are often referred to as **anaerobic respiration.**

1. Reduced Inorganic Compounds as Substrates for Respiration

a. The Hydrogen Oxidizing Bacteria

Species from several genera including *Hydrogemonas, Pseudomonas,* and *Alcaligenes* can oxidize H_2 with oxygen (Eq. 10-22):

$$H_2 + \tfrac{1}{2} O_2 \longrightarrow H_2O \qquad (10\text{-}22)$$
$$\Delta G^0 = -237.2 \text{ kJ mol}^{-1}$$

Some can oxidize carbon monoxide (Eq. 10-23):

$$CO + \tfrac{1}{2} O_2 \longrightarrow CO_2 \qquad (10\text{-}23)$$
$$\Delta G^0 = -257.1 \text{ kJ mol}^{-1}$$

The hydrogen bacteria can also oxidize organic compounds. Their metabolism appears to be straightforward: Electrons are passed down the electron transport chain with generation of three molecules of ATP. The key enzyme is a membrane-bound hydrogenase which delivers the electrons into the chain. A separate soluble hydrogenase (sometimes called **hydrogen dehydrogenase**) transfers electrons to $NADP^+$ to form NADPH for use in the reductive pentose phosphate cycle and for other biosynthetic purposes.[110]

It may be appropriate at this point to comment on the widespread occurrence and varied nature of hydrogenases.[111] These enzymes are found in many organisms, including some plants and animals (Chapter 9, Section F,2), and often serve to release "excess" hydrogen from cells (p. 551). In the latter function hydrogenases operate in the opposite direction than in hydrogen-oxidizing bacteria. Some hydrogenases are membrane bound and are often linked through unidentified carriers to formate dehydrogenase (Chapter 9, Section C,3). In the strict anaerobes such as clostridia the hydrogenases are linked to ferredoxins. The purified hydrogenases are iron–sulfur proteins.[50,111,112] The enzyme from *Clostridium pasteurianum* has a molecular weight of 60,000 and contains four atoms of iron and four of labile sulfur. It has been suggested that hydrogenases contain an Fe_4S_4 cluster able to accept or donate either one or two electrons and that the cluster is the site of binding of or formation of H_2.[111]

b. Nitrifying Bacteria

Two genera of soil bacteria convert ammonium ion to nitrite and nitrate (Eqs. 10-24 and 10-25).[*,113]

$$NH_4^+ + \tfrac{3}{2}O_2 \longrightarrow NO_2^- + 2\,H^+ + H_2O \quad (Nitrosomonas)$$
(10-24)

$$\Delta G'\ (\text{pH 7}) = -272\ \text{kJ mol}^{-1}$$

* These free energy changes can be evaluated from Table 3-3. Using the first column (ΔG^0_f), ΔG^0 for Eq. 10-27 is calculated to be $-34.5 - 237.2 + 79.5$. To obtain $\Delta G'$ (pH 7) we must subtract $2 \times 7 \times 5.708$ kJ for dilution of the two hydrogen ions produced from unit activity to $10^{-7}\ M$. The $\Delta G'_{ox}$ (pH 7) values for oxidation by NAD^+ can also be used. In this case $\Delta G'$ (pH 7) for Eq. 10-27 is evaluated as $12.3 + 372.7 - 219.0 \times 3$ kJ. The third term represents the free energy of oxidation of three moles of NADH by $\tfrac{3}{2}O_2$.

$$NO_2^- + \tfrac{1}{2}O_2 \longrightarrow NO_3^- \quad (Nitrobacter)$$
(10-25)

$$\Delta G'\ (\text{pH 7}) = -76\ \text{kJ mol}^{-1}$$

The importance of these reactions to the energy metabolism of the bacteria was recognized in 1895 by S. Winogradsky, who first proposed the concept of chemiautotrophy. Because the nitrifying bacteria grow slowly (generation time ≈ 10–12 h) it has been hard to get enough cells for biochemical studies and progress has been slow. The reaction catalyzed by *Nitrosomonas* (Eq. 10-26) is the more complex and presumably

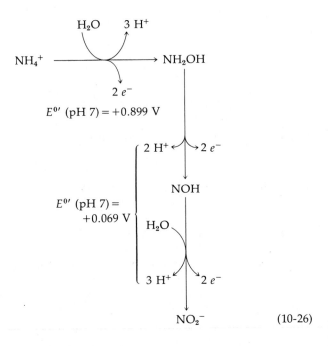

(10-26)

occurs in three stages. The presence of hydrazine blocks oxidation of hydroxylamine (NH_2OH) and permits that intermediate (Eq. 10-26) to accumulate. The oxidation of ammonium ion by O_2 to hydroxylamine is endergonic with $\Delta G'$ (pH 7) = 16 kJ mol^{-1} and is incapable of providing energy to the cell. On the other hand, the oxidation of hydroxylamine to nitrate by O_2 is highly exergonic with $\Delta G'$ (pH 7) = -228 kJ mol^{-1}. The corresponding electrode potentials for the two-

and four-electron oxidation steps are indicated in Eq. 10-26. It is apparent that the second step could feed four electrons into the electron transport system at about the potential of a flavoprotein. The synthesis of two molecules of ATP for each two electrons or a total of four ATP's for the overall reaction would be anticipated.

While attempts to prepare soluble NH_4^+-oxidizing systems have not been successful, a hydroxylamine-oxidizing particle containing b and c type cytochromes and possibly flavins has been obtained.[114,115] It presumably has broken away from the terminal oxidase (cyt a) system.

The other genus of nitrifying bacteria *Nitrobacter* depends on a simpler energy-yielding reaction (Eq. 10-25) with a relatively small free energy decrease. The two-electron oxidation delivers electrons to the electron transport chain at $E^{0'} = +0.42$ V. It is reasonable to anticipate that a single molecule of ATP should be formed for each pair of electrons. However, *Nitrobacter* contains a confusing array of different cytochromes in its membranes.[116,117] What is clear is that some of the ATP generated by passage of electrons from nitrate to oxygen is used to drive a reverse flow of electrons to generate reduced pyridine nucleotides required for biosynthetic reactions (Eq. 10-27).

An interesting feature of the structure of *Nitrobacter* is the presence of several double-layered membranes which completely envelop the interior of the cell. Nitrate entering the cell is oxidized on these membranes and cannot penetrate to the interior where it might have toxic effects.

c. The Sulfur Oxidizing Bacteria

Bacteria of the genus *Thiobacillus* are able to oxidize sulfide, elemental sulfur, thiosulfate, and sulfite to sulfate.[12,118-120] Most of these small gram-negative organisms, found in water and soil, are able to grow in a simple salt medium containing an oxidizable sulfur compound and CO_2. One complexity in understanding the energy-yielding reactions is the tendency of sulfur to form chain molecules. Thus, when sulfide is oxidized it is not clear whether it is necessarily converted to elemental sulfur as indicated in Eq. 10-28, step a (lower left). It might be oxi-

$$(10\text{-}28)$$

dized through some organically bound intermediate to sulfite. However, elemental sulfur (S_8^0) tends to be precipitated.* Thiosulfate is oxidized readily

$$(10\text{-}27)$$

---

* Sulfur is often precipitated as small globules within the cells of *Beggiatoa*, another sulfide-oxidizing bacterium.

by all species, the major pathway beginning with cleavage to S^0 and SO_3^{2-} (Eq. 10-28, step b). If a high thiosulfate concentration is present some may be oxidized to tetrathionate (Eq. 10-28, step c).

While details remain obscure, glutathione may be involved in the initial reaction (Eq. 10-28, step d) of oxidation of elemental sulfur (Eq. 10-29).

$$GSH + S_8^0 \longrightarrow G\text{—}S\text{—}S_8\text{—}H \qquad (10\text{-}29)$$

The linear polysulfide so obtained is oxidized, sulfur atoms being removed either one at a time to form sulfite or two at a time to form thiosulfate.[118,119]

Oxidation of sulfite to sulfate occurs by a **sulfite-cytochrome c reductase** (sulfite oxidase, Eq. 10-28, step e) or by a pathway through **adenylyl sulfate** (also called adenosine 5'-phosphosulfate and commonly abbreviated **APS**). The oxidation of sul-

Adenylyl sulfate (APS)

fite via adenylyl sulfate (Eq. 10-28, steps f and g) is of interest in that it provides a means of substrate level phosphorylation, the only one known in the chemolithotrophic bacteria. No matter which of the two pathways of sulfite oxidation is used the thiobacilli also obtain energy from electron transport. With a value of $E^{0'}$ (pH 7) of -0.454 V [$E^{0'}$ (pH 2) = -0.158 V] for the sulfate-sulfite couple an abundance of energy may be obtained from electron transport. Note that oxidation of sulfite to sulfate produces hydrogen ions. Indeed, pH 2 is optimum for growth of *Thiobacillus thiooxidans* and the bacterium withstands 5% sulfuric acid.[119]

The "iron bacterium" *Thiobacillus ferrooxidans* obtains energy from the oxidation of ferrous ion to ferric ion with subsequent precipitation of ferric hydroxide (Eq. 10-30).

$$ (10\text{-}30) $$

Since the reduction potential for the Fe(II)/Fe(III) couple is $+0.77$ V at pH 7, the energy obtainable in this reaction is small. It is significant that these bacteria always oxidize reduced sulfur compounds, too. Especially interesting is their oxidation of the mineral **pyrite**, ferrous sulfide (Eq. 10-31).*

$$2\ FeS + 4\tfrac{1}{2}\ O_2 + 5\ H_2O \longrightarrow$$
$$2\ Fe(OH)_3 + 2\ SO_4^{2-} + 4\ H^+ \qquad (10\text{-}31)$$

$$\Delta G'\ (\text{pH } 2) = -1340\ kJ$$

Because sulfuric acid is generated in this reaction a serious water pollution problem has been created by the bacteria living in abandoned mines. Water running out of the mines often has a pH of 2.3 or less.[121]

Bacteria are not the only organisms that can oxidize reduced sulfur compounds. For example, a molybdenum-containing **sulfite oxidase** (Chapter 14, Section G) whose function may be to detoxify sulfur dioxide by oxidation of sulfite to sulfate is present in animal livers.[122]

2. Anaerobic Respiration

a. Nitrate as an Electron Acceptor

The use of nitrate as an alternative oxidant to O_2 is widespread among bacteria. For example, *E. coli* can subsist anaerobically using nitrate, which is reduced to nitrite in the process (Eq. 10-32)[123, 124]:

$$NO_3^- + 2\ H^+ + 2\ e^- \longrightarrow NO_2^- + H_2O \qquad (10\text{-}32)$$
$$E^{0'}\ (\text{pH } 7) = +0.421\ V$$

The **nitrate reductase** of bacteria such as *E. coli* is tightly bound to the membrane. The solubilized enzyme from *E. coli* has a molecular weight of 720,000 and contains four large and four small

---

* The free energy data are from "CRC Handbook of Chemistry and Physics," 48th edition, 1967–1968. The solubility product of Fe(OH)$_3$ was used in estimating ΔG^0_f of Fe(OH) as -687.7 kJ mol^{-1}.

polypeptide chains and four atoms of molybdenum.[125, 126] It is thought that it is the latter metal that interacts directly with the nitrate. Electron flow is from the electron transport chain through a cytochrome of the b type (cyt b_{555}) and into the molybdenum atom.[126] Formate dehydrogenase (Chapter 9, Section C,3) appears to be closely associated with the nitrate reductase, and formate is a preferred electron donor for the reduction.

Nitrate is also reduced by fungi and higher plants prior to incorporation into amino acids and other cell constituents. Reduction to nitrite is a first step in this process. A much studied assimilatory nitrate reductase is that of *Neurospora crassa*. The enzyme is a large complex (MW $\sim 228,000$) containing not only molybdenum but also cytochrome b_{557}. The following electron transport scheme has been proposed[127]:

$$NADH \longrightarrow FAD \longrightarrow cyt\ b_{557} \longrightarrow Mo \longrightarrow NO_3^-$$
$$\searrow cyt\ c \qquad (10\text{-}33)$$

Several types of **denitrifying bacteria** reduce either nitrate or nitrite ions to N_2. *Micrococcus denitrificans* uses H_2 to reduce nitrate (Eq. 10-34):

$$5\ H_2 + 2\ NO_3^- + 2\ H^+ \longrightarrow N_2 + 6\ H_2O \qquad (10\text{-}34)$$
$\Delta G'(\text{pH } 7) = -561$ kJ mol^{-1} of nitrate reduced or -224 kJ mol^{-1} of H_2 oxidized

Thiobacillus denitrificans, like other thiobacilli, can oxidize sulfur (as well as H_2S or thiosulfate). Nitrate is the oxidant (Eq. 10-35):

$$5\ S + 6\ NO_3^- + 2\ H_2O \longrightarrow 5\ SO_4^{2-} + 3\ N_2 + 4\ H^+ \qquad (10\text{-}35)$$
$\Delta G'\ (\text{pH } 7) = -455$ kJ mol^{-1} of nitrate reduced or -546 kJ mol^{-1} of S oxidized

The enzymology of the denitrifying bacteria is not well known, but a stepwise reduction of nitrate (Eq. 10-36) is possible.[12, 124] All of these bacteria can reduce nitrite as well as nitrate and some can reduce N_2O. Others, under some conditions, *form* N_2O. Equation 10-36 can also portray the assimilatory reduction of nitrate and nitrite to NH_3 that is almost universal among bacteria, fungi, and green plants. The reduction of nitrite can proceed in

$$(10\text{-}36)$$

two-electron steps (Eq. 10-36, steps *a–c*) via the hypothetical nitroxyl (NOH) and hydroxylamine (NH_2OH) to ammonia. Formation of N_2 could occur by dimerization of NOH (Eq. 10-36, step *d*) followed by another two-electron reduction (Eq. 10-36, step *e*). Dehydration of the intermediate dimer would yield N_2O (Eq. 10-36, step *f*).

The electron transport processes associated with reduction of nitrate to nitrite appear to be straightforward. In view of the high value of $E^{0'}$ of $+0.42$ V (Eq. 10-32), we may conclude that phosphorylation site III is bypassed and that the yield of ATP will be one less than when O_2 is the oxidant. Similar considerations may apply to the other reduction steps of Eq. 10-36. The nitrite, hyponitrite, and hydroxylamine reductases may all tie into the electron transport chain at the level of the cytochromes b or c. However, nitrite reductases from *Neurospora* and from spinach contain **siroheme** (Fig. 14-16), previously found in sulfite reductases.[127a] Fungi supply electrons for reduction of nitrite from NADPH via FAD and siroheme, while green plants use reduced ferredoxin as the reductant.

b. Sulfate-Reducing Bacteria

A few obligate anaerobes obtain energy by using sulfate ion as an oxidant. For example, *De-*

sulfovibrio desulfuricans catalyzes a rapid reduction of sulfate by H_2 (Eq. 10-37):

$$4 H_2 + SO_4^{2-} + 2 H^+ \longrightarrow H_2S + 4 H_2O \qquad (10\text{-}37)$$
$$\Delta G' \text{ (pH 7)} = -154 \text{ kJ mol}^{-1} \text{ of sulfate reduced}$$

While this may seem like an esoteric biological process the reaction is quantitatively significant. For example, it has been estimated that within the Great Salt Lake basin bacteria release sulfur as H_2S in an amount of 10^4 metric tons (10^7 kg) per year.[128]

Perhaps because the reduction potential for sulfate is extremely low ($E^{0'}$, pH 7 $= -0.454$ V), organisms are not known to reduce it directly to sulfite. Rather a molecule of ATP is utilized to form **adenylyl sulfate** (p. 612) through the action of **ATP sulfurylase** (Eq. 11-4). Then the adenylyl sulfate is reduced by cytochrome c_3 as in Eq. 10-38. Cy-

tochrome c_3, a low potential carrier ($E^{0'}$, pH 7 $= 0.21$ V) of MW $\sim 13,000$, contains four heme groups and is found in high concentration in sulfate-reducing bacteria.[12,129] Its reduction of adenylyl sulfate (Eq. 10-38b) is mediated by **adenylyl sulfate reductase**, an FAD—Fe—S protein. It has been suggested[130] that an intermediate in the reaction is the adduct of sulfite with FAD (Eq. 10-39). The initial step in this hypothetical mechanism is displacement on sulfur by a strong nucleophile generated by reduction of the flavin.

Bisulfite produced according to Eq. 10-38 is reduced further by a **sulfite reductase** which is thought to receive electrons from flavodoxin, cyt c_3, and a hydrogenase (Eq. 10-40):

$$H_2 \longrightarrow \text{hydrogenase} \longrightarrow$$
$$\text{(Fe—S)}$$
$$\text{cyt } c_3 \longrightarrow \text{flavodoxin} \longrightarrow$$
$$\text{sulfite reductase} \longrightarrow SO_3^{2-} \quad (10\text{-}40)$$

In fact, there may be *three* reductases operating on bisulfite, trithionate, and thiosulfate in the cyclic scheme[131] of Eq. 10-41. A possible role for menaquinone (vitamin K_2), present in large amounts in *Desulfovibrio*, has been suggested.[132]

$$HSO_3^- \xrightarrow[X3]{2e^-} S_3O_3^{2-} \xrightarrow{2e^-} S_2O_3^{2-} \xrightarrow{2e^-} H_2S$$

$$HSO_3^- \qquad HSO_3^-$$

$$(10\text{-}41)$$

Sulfate is reduced in smaller amounts by many organisms including bacteria, fungi, and green plants to supply sulfur for incorporation into organic constitutents of a cell. This assimilatory process (Chapter 14, Section G) is quite similar to that employed by the sulfate-reducing bacteria.

c. Methane Bacteria

The methane-producing bacteria, (Chapter 8, Section L,7) are also classified as chemiautotrophic organisms. While they can utilize substances such as methanol and acetic acid, they can also reduce CO_2 to methane and water using H_2 (Eq. 10-42):

$$4\,H_2 + CO_2 \longrightarrow CH_4 + 2\,H_2O \qquad (10\text{-}42)$$
$$\Delta G^0 = -131 \text{ KJ mol}^{-1}$$

The electron transport is from hydrogenase through ferredoxin to formate dehydrogenase and to the tetrahydrofolic acid-dependent dehydrogenases that carry out the stepwise reduction of formate to methyl groups (Fig. 8-21). However, the mechanism of generation of ATP in the sequence is obscure.[133]

G. OXYGENASES AND HYDROXYLASES

For many years the idea of dehydrogenation dominated thinking about biological oxidation. It was generally assumed that the oxygen found in organic substances always came from water. A water molecule could be added to a double bond and the resulting alcohol dehydrogenated. Nevertheless, there were indications that small amounts of O_2 itself were essential, even to anaerobically growing cells.[134] In 1955, O. Hayaishi and H. Mason independently demonstrated that ^{18}O

was sometimes incorporated into organic compounds directly from $^{18}O_2$, as in Eq. 10-43. Today a bewildering variety of **oxygenases** are known to function in forming such essential metabolites as sterols, prostaglandins, and active derivatives of vitamin D. Oxygenases are also needed in the catabolism of many substances, acting most often on

$$\text{(benzene diol)} + {}^{18}O_2 \xrightarrow{\text{pyrocatechase}} \begin{array}{c} C^{18}OOH \\ C^{18}OOH \end{array} \quad (10\text{-}43)$$

nonpolar groups that cannot be readily attacked by other types of enzyme.[134–136]

Oxygenases are classified either as **dioxygenases** or as **monooxygenases** ("mixed function oxidases"). The monooxygenases are also called **hydroxylases.** While the dioxygenases catalyze incorporation of two atoms of oxygen, as in Eq. 10-43, the monooxygenases incorporate only one atom. The other oxygen atom from the O_2 is converted to water. A typical monooxygenase reaction is the hydroxylation of an alkane to an alcohol:

$$CH_3-(CH_2)_n-CH_3 + O_2 + BH_2 \longrightarrow$$
$$CH_3-(CH_2)_n-CH_2OH + H_2O + B \quad (10\text{-}44)$$

A characteristic of the monooxygenases is that some additional reduced substrate, a **cosubstrate** (BH_2 in Eq. 10-44), is required to reduce the second atom of the O_2 molecule to H_2O.

1. Dioxygenases

Among the best known of the oxygenases that incorporate both atoms of O_2 into the product are those that cleave double bonds of aromatic compounds at positions adjacent to OH groups or between OH groups as in Eq. 10-43. Other dioxygenases cleave aliphatic compounds. A well-known example is the cleavage of β-carotene to vitamin A (Box 12-C). Several dioxygenases crystallized from bacteria are nonheme iron proteins of MW = 50,000 or more for a single subunit. The proteins usually contain Fe(II) but have no labile sulfur.[1] On the other hand, **tryptophan dioxygenase** (tryptophan pyrrolase, Chapter 14, Section I) is a heme enzyme. It catalyzes the reaction of Eq. 10-45. The oxygen atoms designated by the asterisks are derived from O_2.

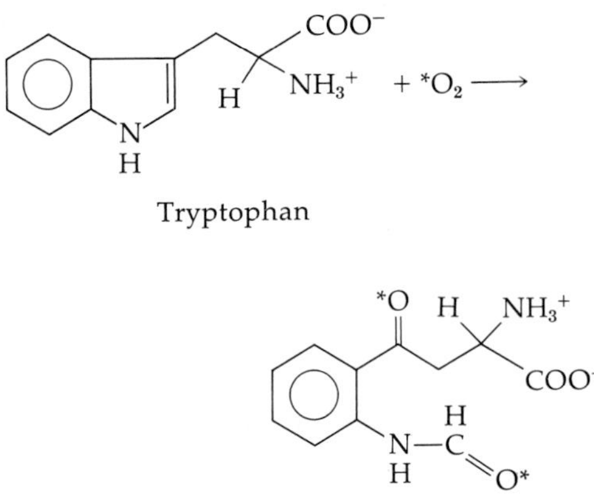

Tryptophan

$$(10\text{-}45)$$

Formylkynurenine

As with the nonheme oxygenases, tryptophan dioxygenase is active when the iron is the ferrous form. It seems reasonable to suppose that formation of a complex between Fe(II) and O_2 is an essential first step. However, tryptophan must also be present before combination with O_2 can take place. At 5°C the enzyme, tryptophan, and O_2 combine to give an altered spectrum reminiscent of that of compound III of peroxidase (Eq. 10-6). This oxygenated complex may, perhaps, then be converted to an Fe(III) complex with superoxide ion (Eq. 10-46). There is some evidence (based upon

$$\begin{array}{ccc}
\text{Fe(II)}\!-\!O_2 & \longrightarrow \text{Fe(III)}^+\!-\!O_2^- & \longrightarrow \text{Fe(III) (ferriheme)} \\
\text{Oxygenated} & & \downarrow \\
\text{complex} & & O_2^- \text{ (attacking reagent?)}
\end{array}$$

$$(10\text{-}46)$$

inhibition by superoxide dismutase of the enzyme from some organisms)[137] that superoxide anion radical may be the species that attacks the substrate to initiate the hydroxylation reaction (Eq. 10-47). Note that in the first step one electron is returned to the Fe(III) form of the enzyme to regenerate the original Fe(II) form.

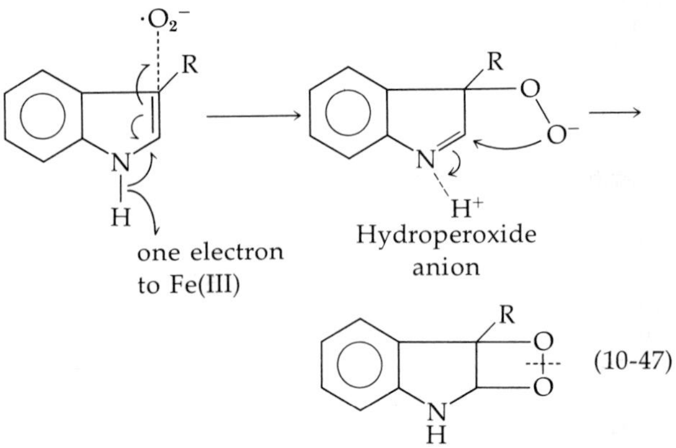

one electron
to Fe(III)

Hydroperoxide
anion

$$(10\text{-}47)$$

A plant enzyme usually classified as a dioxygenase is **lipoxygenase** (lipoxidase). Containing one atom of iron(II) per molecule, lipoxygenase catalyzes oxidation of polyunsaturated fatty acids in lipids (Eq. 10-48).[138] The formation of the hy-

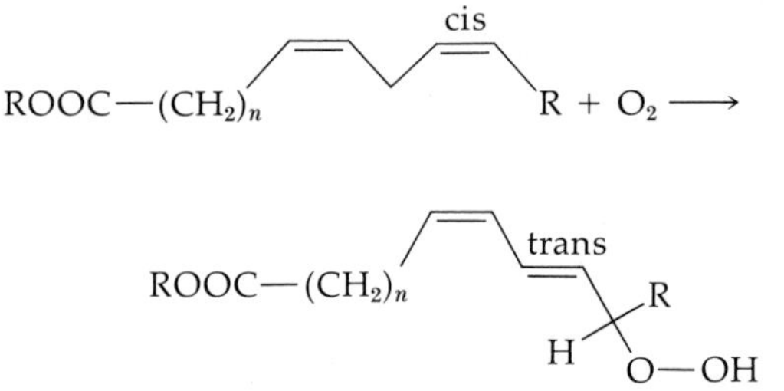

$$(10\text{-}48)$$

droperoxide product is accompanied by a shift of the double bond and conversion from cis to trans configuration. The enzyme appears to generate free radicals, but the mechanism of their formation is not clear. In forming a hydroperoxide as a product lipoxygenase has something in common with several of the monooxygenases discussed in the next section.

2. Monooxygenases

Two classes of monooxygenases are known. Those requiring a cosubstrate (BH_2 of Eq. 10-44) in addition to the substrate to be hydroxylated are known as **external monooxygenases.** In the other group, the **internal monooxygenases,** some portion of the substrate being hydroxylated also serves as the cosubstrate. Many internal monooxygenases contain flavin cofactors and are devoid of metal.

a. Flavin-Containing Monooxygenases

Recall that dihydroflavins react with O_2 to form H_2O_2 (Chapter 8, Section I,7). In view of this fact, it is attractive to think that monooxygenases catalyze dehydrogenation of their substrates by the flavin cofactors and that the reduced flavins react with O_2 to form H_2O_2. The latter would serve as the hydroxylating agent. An example has already been described in Chapter 8. **Lactate oxygenase** may dehydrogenate lactate to pyruvate and oxidatively decarboxylate pyruvate to acetate with H_2O_2 (Eq. 8-67). One atom of oxygen from O_2 is incorporated into the acetate formed.[139]

A second internal monooxygenase is **lysine oxygenase,** a tetrameric FAD-containing protein of subunit weight MW $\sim 61{,}000$.[140] Monooxygenases of similar type produced by bacteria attack arginine and other basic amino acids. Again, a dehydrogenation followed by oxidative decarboxylation with H_2O_2 as in Eq. 8-67 could account for the observed products (Eq. 10-49). When native lysine

$$(10\text{-}49)$$

monooxygenase is treated with various sulfhydryl-blocking reagents the modified enzyme

formed produces an α-keto acid, ammonia, and H_2O_2. These are just the products predicted if the bracketed intermediate of Eq. 10-49 were to break up hydrolytically.

One manner in which NADPH can serve as a cosubstrate is to reduce a flavoprotein enzyme. The reduced flavin can then react with O_2 to generate the hydroxylating reagent. An example is **4-hydroxybenzoate** hydroxylase, an enzyme that forms 3,4-dihydroxybenzoate as a product. NADH reacts only after the flavoprotein forms a complex with the substrate, and evidence has been presented for an oxygenated intermediate, possibly a hydroperoxide as indicated in Eq. 10-50

E·flavin (10-50)

(see also Chapter 8, Section I,7).[141] According to this mechanism, one of the two oxygen atoms in the hydroperoxide reacts with the aromatic substrate, effectively as OH^+ or as a superoxide radical. Another suggestion[142] is that the hydroperoxide of Eq. 10-50 may eliminate H_2O to form an **oxaziridine** which would be the active hydroxylating reagent (Eq. 10-51). Attack of the oxaziridine

$$(10\text{-}51)$$

on a nucleophilic center in the substrate would lead to the adduct shown. Elimination and tautomerization would give the product. An epoxide (see Eq. 10-54) could also be formed as an alternative initial product.

b. Reduced Pteridines as Cosubstrates

A dihydro form of biopterin (p. 493) serves as a cosubstrate that is reduced by NADPH (Eq.

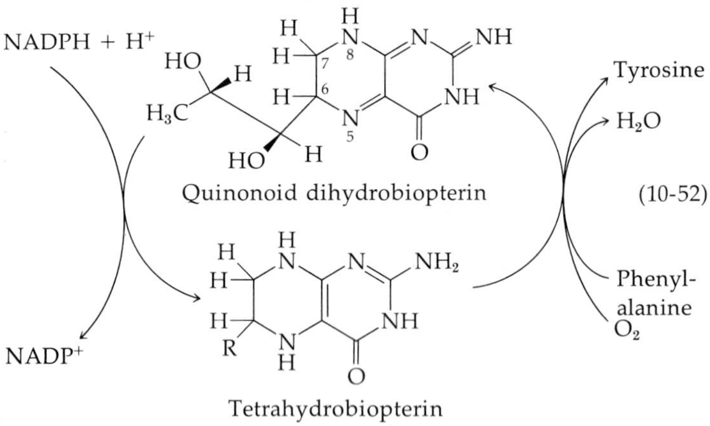

$$\text{(10-52)}$$

Quinonoid dihydrobiopterin

Tetrahydrobiopterin

10-52) with one group of hydroxylases including **phenylalanine hydroxylase** of human liver. The tetrahydrobiopterin formed is similar in structure to a reduced flavin. The mechanism of its interaction with O_2 and of the hydroxylation of phenylalanine to tyrosine could be essentially the same as that of 4-hydroxybenzoate hydroxylase. Dihydrobiopterin can exist as a number of isomers. The quinonoid form shown in Eq. 10-52 is a tautomer of 7,8-dihydrobiopterin, the form generated by dihydrofolate reductase (Chapter 8, Section K,2). The reader may wish to guess what arguments have been used[143, 144] in favor of the structure indicated in Eq. 10-52. A pyridine nucleotide-dependent **dihydropteridine reductase**[145] catalyzes the left-hand reaction of Eq. 10-52.

Phenylalanine hydroxylase is of special interest because one of the best known hereditary biochemical defects **phenylketonuria** (Chapter 14, Section H,5) is caused by its absence. Other pteridine-dependent hydroxylases are also known.

For example, **tryptophan hydroxylase** of the brain forms 5-hydroxytryptophan, the first step in synthesis of the neurotransmitter 5-hydroxytryptamine (Chapter 16, Section B,4).[146]

c. Hydroxylation-Induced Migration

A general result of enzymatic hydroxylation of aromatic compounds is the intramolecular migration of a hydrogen atom or of a substituent atom or group (Eq. 10-53).[147] Dubbed the NIH shift

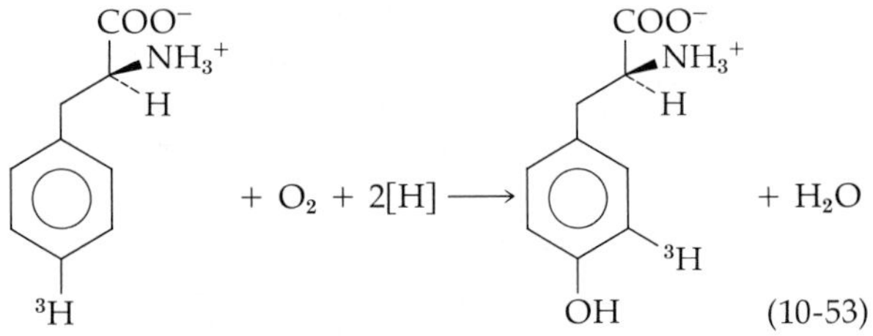

$$\text{(10-53)}$$

(because the workers discovering it were in a National Institutes of Health laboratory), the migration tells us something about possible mechanisms of hydroxylation. In Eq. 10-53 a tritium atom has shifted in response to the entering of the hydroxyl group. The migration can be visualized as resulting from electrophilic attack on the aromatic system, e.g., by the ion OH^+ (Eq. 10-54):

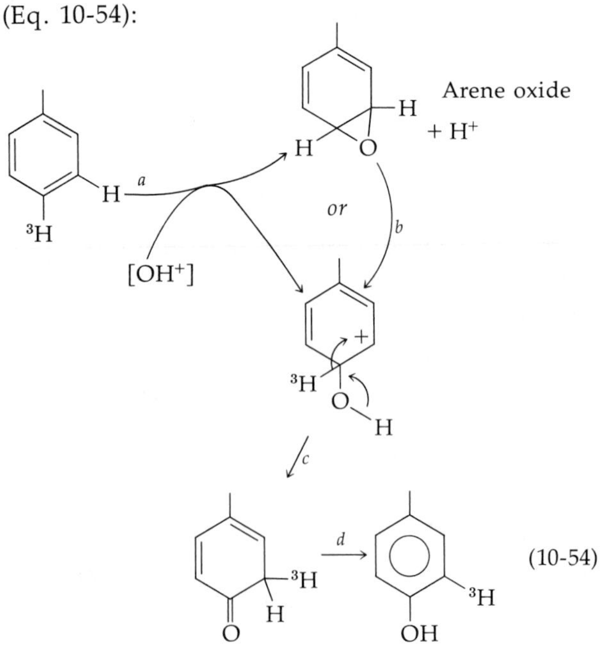

$$\text{(10-54)}$$

Such an attack could lead on the one hand to an **epoxide** (**arene oxide,** step *a*) or directly to a carbonium ion of the structure shown in Eq. 10-54. Arene oxides have been prepared and it has been shown that they can be converted (presumably via the carbonium ion, step *b*) to end products in which the NIH shift has occurred.[148, 149] Thus, either carbonium ions or arene oxides are logical initial products of attack on aromatic substrates. These results suggest that the hydroxylating reagent is of electrophilic character, a property expected for the previously discussed hydroperoxides, superoxide anion, oxaziridine intermediates, or O_2 coordinated to a metal ion.

An example of the operation of the NIH shift in causing migration of a large substituent is illustrated by the hydroxylation of *para*-hydroxyphenylpyruvic acid (Eq. 10-55), a key step in the

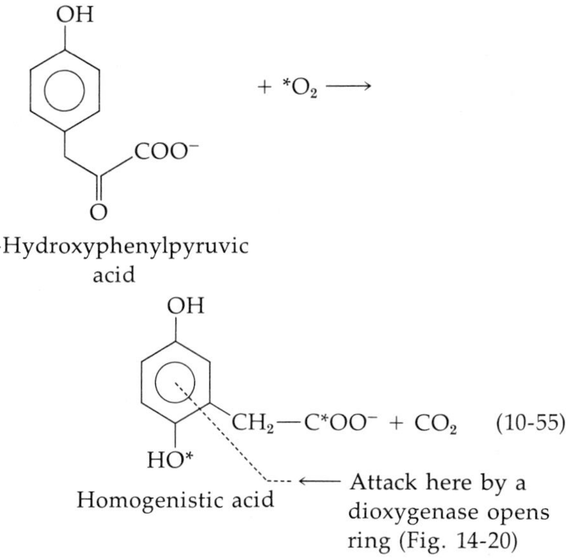

p-Hydroxyphenylpyruvic
acid

Homogenistic acid   ⟵ Attack here by a
dioxygenase opens
ring (Fig. 14-20)

catabolism of tyrosine. This reaction should be compared with that of Eq. 8-67. No doubt a similar mechanism is involved. Activated oxygen may in the first step add to the carbonyl group and then, in an electrophilic attack at the other end of the O_2, to the aromatic ring. Note that the electron donating effect of the *p*-hydroxyl group would assist in this attack. Subsequent decarboxylation would cleave the O—O bond as in Eq. 8-67. At the same time a hydroxylated carbonium ion of the

type shown in Eq. 10-54 would be generated and would undergo the NIH shift. While this enzyme is a dioxygenase it is probably related in its mechanism of action to the monooxygenases.

d. α-Ketoglutarate as a Decarboxylating Cosubstrate

A series of monooxygenases accept hydrogen atoms from α-ketoglutarate which is decarboxylated in the process to form succinate. Two of the enzymes are involved in hydroxylation of residues of lysine and proline, respectively, in the collagen precursor **procollagen** (Chapter 11, Section E,3). Another hydroxylates γ-butyrobetaine to carnitine (Chapter 9, Section A,6). The enzymes all contain Fe(II). The mechanism may be similar to that suggested for Eq. 10-55.[150, 151] That is, activated oxygen may add to the carbonyl group of α-ketobutyrate on the one hand (a nucleophilic addition) and carry out an electrophilic attack on the substrate to be hydroxylated (Eq. 10-56):

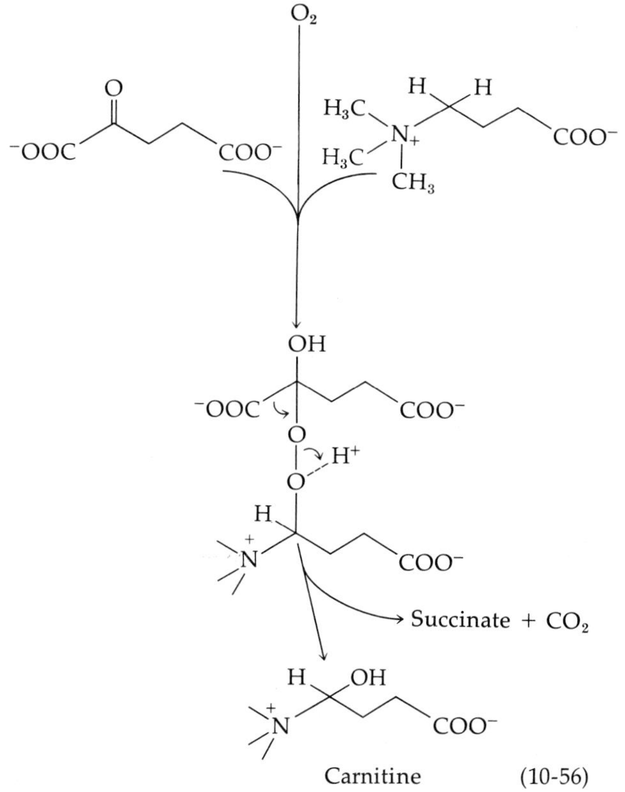

Carnitine   (10-56)

Alternatively, the metal-bound oxygen may react with ketoglutarate with decarboxylation and formation of persuccinic acid which might be the hydroxylating reagent.[151a]

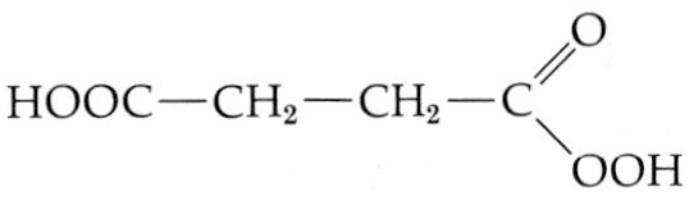

$$HOOC-CH_2-CH_2-C\overset{O}{\underset{OOH}{\big\backslash}}$$

Persuccinic acid

e. Ascorbic Acid as a Cosubstrate

Many hydroxylases require the presence of a reducing agent. Often **ascorbic acid** (vitamin C, Box 10-G) is especially effective. This is true for all of the α-ketoglutarate-requiring hydroxylases discussed in the preceding section. Ascorbic acid apparently serves to keep the metal-containing catalyst reduced. However, the enzyme **dopamine β-hydroxylase** utilizes ascorbic acid as a true cosubstrate in the synthesis of norepinephrine (**noradrenaline**, Eq. 10-57).[152] This reaction, important

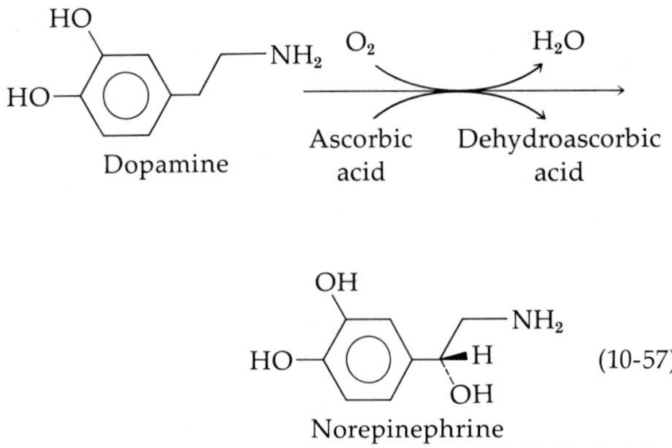

Dopamine → Norepinephrine (10-57)

in neurons of the brain, also occurs actively in the adrenal gland. The latter has long been known as especially rich in ascorbic acid. The structure of the oxidized form of vitamin C, **dehydroascorbic acid**, is also shown in Box 10-G. Dopamine β-hydroxylase contains several atoms of copper, and it is believed that ascorbic acid reduces two of the copper atoms from the $+2$ to the $+1$ state. Then O_2 combines, perhaps in much the same way as it does with hemocyanin (p. 568). Something resembling a metal-coordinated O_2^{2-} ion is then available for the hydroxylation reaction. It is

noteworthy that another copper-containing enzyme (ascorbic acid oxidase, Box 10-H) also acts on the vitamin C molecule.

f. Hydroxylation with Cytochrome P-450

A final class of hydroxylases require hemoproteins classified as cytochrome P-450.* These proteins are involved in the hydroxylation of steroids[153] and alkanes,[154, 155] in the hydroxylation of the methylene group of camphor,[156] and in the metabolism of various drugs.[135, 157, 158] The reactions show the following features. The hydroxyl group is introduced without inversion of configuration (Eq. 10-58). This equation represents the

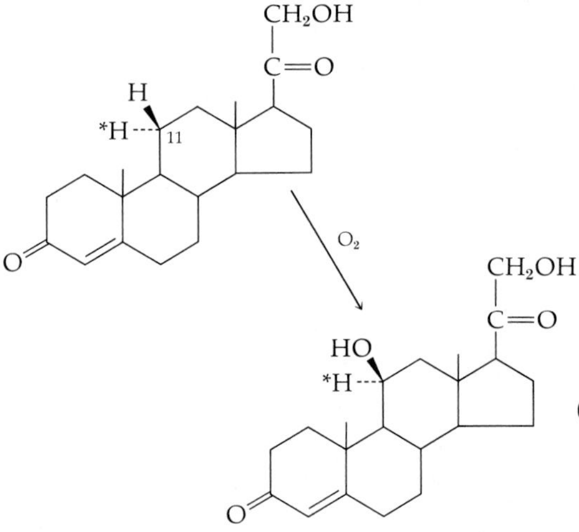

Corticosterone (10-58)

11β-hydroxylation of a steroid, an essential step in the biosynthesis of steroid hormones. Another feature of the 11β-hydroxylase system is that the same enzyme that catalyzes Eq. 10-58 converts an unsaturated derivative to an epoxide (Eq. 10-59). In all of the cytochrome P-450 hydroxylase

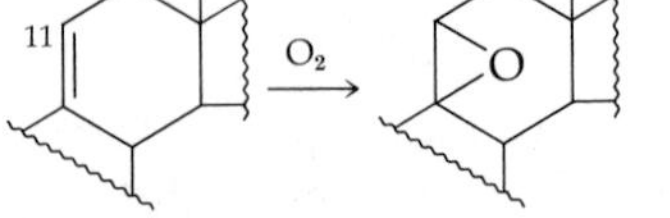

(10-59)

---

* Cytochrome P-450 is so named because, in its reduced form, it forms a pigment absorbing at 450 nm with carbon monoxide.

BOX 10-G

VITAMIN C: ASCORBIC ACID

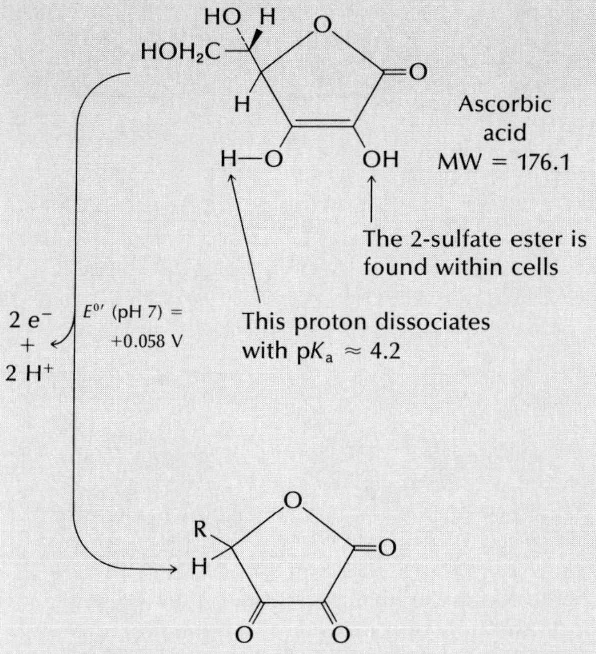

HO H
HOH$_2$C

O

=O

Ascorbic
acid
MW = 176.1

H

H—O   OH

The 2-sulfate ester is
found within cells

$2 e^-$
$+$
$2 H^+$

$E^{0'}$ (pH 7) =
+0.058 V

This proton dissociates
with p$K_a \approx 4.2$

R

O

=O

H

O   O

Dehydroascorbic acid

Hemorrhages of skin, gums, and joints were warnings that death was near for ancient sea voyagers stricken with **scurvy**. Even though it was recognized by the year 1700 that the disease could be prevented by eating citrus fruit, it was 200 years before efforts to isolate vitamin C were made. Ascorbic acid was first obtained in crystalline form in about 1930. The lively account of those researches by A. Szent-Györgyi makes interesting reading.[a]

Among mammals only man and the guinea pig require ascorbic acid in the diet, other species being able to make it themselves. Compared to that of other vitamins, the requirement is large. While 10 mg/day prevents scurvy, subclinical deficiency as judged by fragility of small capillaries in the skin is present at that level of intake. Various "official" recommendations for vitamin C intake range from 30 to 70 mg/day, and one of the most vigorous nutritional controversies of recent times has surrounded Linus Pauling's recom-

mendation of 0.25–10 g/day.[b] Pauling and others maintain that ascorbic acid has a specific beneficial effect in preventing or ameliorating symptoms of the common cold.[c] On the other hand, critics maintain that unrecognized hazards may exist in high doses of this seemingly innocuous compound. Although ascorbic acid has antioxidant properties, it also promotes the generation of free radicals in the presence of ferric ions, and it is conceivable that too much may be a bad thing.

The biochemical functions of ascorbic acid are not clearly defined. In addition to its well-established reducing properties and easy oxidation to **dehydroascorbic acid** (structure at left), it is a weak acid with metal complexing properties. Ascorbic acid is present in extremely high concentrations in adrenal glands where one function may be to act as a cosubstrate for dopamine β-hydroxylase (Eq. 10-57). A general role in the hydroxylation of procollagen is likely[c] (Section G,2e), and it is possible that any effect of ascorbic acid in preventing colds may be a result of increased hydroxylation of collagen. It has been shown that high levels of ascorbic acid in guinea pigs lead to more rapid healing of wounds.[d,e]

Ascorbic acid together with Fe(II) and O$_2$ is a powerful nonenzymatic hydroxylating reagent for aromatic compounds.[f,g] Like hydroxylases, the reagent attacks nucleophilic sites, e.g., converting phenylalanine to tyrosine. Oxygen atoms from ^{18}O$_2$ are incorporated into the hydroxylated products. While H$_2$O$_2$ is formed in the reaction mixture, it cannot replace ascorbate. However, the relationship of this system to biochemical functions of ascorbate is not clear.

[a] A. Szent-Györgyi, *Annu. Rev. Biochem,* **32,** 1–14 (1963).

[b] L. Pauling, "Vitamin C and the Common Cold." Freeman, San Francisco, California, 1970.

[c] M. J. Barnes and E. Kodicek, *Vitam. Horm.* (N.Y.) **30,** 1–43 (1972).

[d] M.-L.S. Yew, *PNAS* **70,** 969–972 (1973).

[e] R. Harwood, M. E. Grant, and D. S. Jackson, *BJ* **142,** 641–651 (1974).

[f] V. Ullrich and Hj. Standinger, K. Block, and O. Hayaishi, eds., "Biological and Chemical Aspects of Oxygenase," pp. 235–249. Maruzen, Tokyo, 1966.

[g] G. A. Hamilton, R. J. Workman, and L. Woo *JACS* **86,** 3391–3392 (1964).

systems an electron transport chain delivers electrons from NADH or NADPH through a flavoprotein to ferredoxin (Fd) or rubredoxin. Apparently one of these latter nonheme iron proteins then reduces the iron in a complex of substrate plus cytochrome P-450 from the Fe(III) to the Fe(II) state (Eq. 10-60, step a). Then oxygen combines with the ferrous iron, oxidizing it to the ferric state. At some point another electron is donated from the electron transport chain. The coordinated oxygen, as O_2^- or O_2^{2-}, attacks the substrate and cytochrome P-450 is released in the Fe(III) state. While the details are not clear, a sequence something like this must be operative in all the cases.

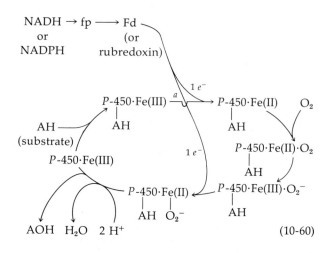

$$(10-60)$$

COPPER-CONTAINING PROTEINS

Despite the variety of functions of copper in all living things,[a] it was not until 1924 that this element was recognized as a nutritional essential. Copper is so broadly distributed in foods that a deficiency has never been observed in humans. Copper deficiency does occur occasionally in animals, sometimes because absorption of Cu^{2+} is antagonized by Zn^{2+} and sometimes because copper is tied up by molybdate as an inert complex. There are copper-deficient desert areas of Australia where neither plants nor animals survive. Copper-deficient animals have bone defects. Hair color is lacking and hemoglobin synthesis is impaired. Cytochrome oxidase activity is low. The protein elastin of arterial walls is poorly crosslinked and the arteries are weak.

An adult human ingests ~2–5 mg of copper per day, about 30% of which is absorbed. The total body content of copper is ~100 mg, and both its uptake and excretion (via the bile) are highly regulated. Since an excess of copper is toxic, regulation is important. A disease (Wilson's disease), in which copper accumulates in liver and brain, is known. The normal content of copper in those tissues amounts to ~10^{-4} gram atoms per liter. Cupric ion is the most tightly bound metal ion in most chelating centers (Table 4-2), which accounts for the fact that the copper present in living cells is almost entirely complexed with proteins.

Copper ions lie at the active centers of a versatile group of catalysts.[a–d] Like iron, the copper ion is a site for reaction with O_2. Its ability to undergo reversible reduction permits it to function in a variety of redox processes. The simplest function for copper proteins is to serve (like cytochrome c) as a single-electron carrier. Bacterial **azurins** are brilliant blue proteins of low molecular weight that are thought to function in electron transport chains. The "blue protein" of *Pseudomonas auruginosa* has a molecular weight of ~16,300 and contains one Cu^{2+} per mole. The absorption spectrum of this bright blue protein is shown in Fig. 13-8. The blue color reflects a property of Cu^{2+} that is observed in the hydrated ion $Cu(H_2O)_4^{2+}$ and more intensely in $Cu(NH_3)_4^{2+}$. The color, which arises from a transition of an electron from one d orbital to another within the copper atom, is still more intense in copper peptide chelates of the type shown in Eq. 4-38. In the beautiful blue proteins the intensity of the d-d absorption bands are an order of magnitude more intense than in the simpler compounds, a fact that is not completely understood. Coordination with one or more sulfur atoms of methionine side chains is a possibility.[e]

Other low molecular weight blue proteins include **stellacyanin,** a copper mucoprotein from

Box 10-H (*Continued*)

the Japanese lac tree.[f] The peptide of 108 residues contains 20% carbohydrate and a single copper atom. **Plastocyanin,** first isolated from the alga *Chlorella,* has since been found to be ubiquitous in green plants. It is thought to function in the electron transport chain between two light absorbing centers in photosynthesis (Chapter 13, Section E,6).

Most copper proteins react with O_2, sometimes reversibly as in the oxygen-carrying hemocyanin (p. 568). More often oxygen is "activated" to undergo a chemical reaction. In one group of enzymes the organic substrates are dehydrogenated and oxygen is reduced to H_2O_2. Thus, **galactose oxidase** (of *Polyporus*) catalyzes the conversion of the 6-hydroxymethyl group of galactose to an aldehyde group[g,h]:

$$CH_2OH \xrightarrow[]{} CHO$$
$$O_2 \quad H_2O_2$$

The large polypeptide chain of MW = 68,000 contains one copper ion. Unlike the blue proteins, galactose oxidase is a light murky green. Whereas neither oxygen nor galactose affects the spectrum of the enzyme, the combination of the two does. It is thought that both galactose and O_2 may bind to the copper and that the latter may cycle between Cu(I) and Cu(III). A similar mechanism may hold for tyrosinase (see below).[h] Galactose oxidase has been used to label glycoproteins of external cell membrane surfaces. Exposed terminal galactosyl or *N*-acetylgalactosaminyl residues are oxidized to the corresponding C-6 aldehydes and the latter are reduced under mild conditions with tritiated sodium borohydride.[i]

Amine oxidases that contain both Cu^{2+} and flavin coenzymes are similar in their action to the amino acid oxidases (Table 8-4). One of these amine oxidases is responsible for conversion of ϵ-amino groups of side chains of lysine to aldehyde groups in collagen and elastin (p. 665). Another copper enzyme is **urate oxidase,** which causes a decarboxylation of the substrate (Fig. 14-33).

Some copper enzymes, e.g., dopamine β-hydroxylase (Eq. 10-57) are typical hydroxylases. A copper hydroxylase of widespread occurrence,

tyrosinase (also known as **polyphenol oxidase**), catalyzes a two-step reaction of hydroxylation followed by dehydrogenation. First identified in

Tyrosine

3,4-Dihydroxyphenylalanine (dopa)

mushrooms, the enzyme has also been obtained from the Japanese lac tree. It is present in large amounts in plant tissues and is responsible for the darkening of cut fruits. In animals tyrosinase participates in the synthesis of **dihydroxyphenylalanine** ("dopa") and in the formation of the black **melanin** pigment of skin and hair. Either a lack of or blockage of this enzyme in the melanocytes (melanin-producing cells) causes **albinism.** Mushroom (*Polyporus*) tyrosinase has a molecular weight of ~120,000 and contains four subunits. Each subunit contains one atom of Cu(I).

An unusual reaction is catalyzed by the copper-containing **quercitinase** of *Aspergillus flavus.* The

$+ CO$

Box 10-H (*Continued*)

enzyme should be classified as an oxygenase. A reasonable mechanism for production of the carbon monoxide can be written via a bridged peroxy intermediate.[j,k]

The **blue copper oxidases**[l] are an unusual group of catalysts that are able to reduce both atoms of molecular oxygen to H_2O. In this respect they resemble cytochrome oxidase (which also contains copper, Section B,5), but they do not contain iron. **Ascorbic acid oxidase** of plant tissues converts ascorbate to dehydroascorbate (Box 10-G). When substrate is added to the enzyme, the blue color fades and it can be shown that the copper is reduced to the 1+ state. **Laccase,** from the latex of the Japanese lac tree, or from the mushroom *Polyporus*, catalyzes the same conversion of substrates as does tyrosinase, but with H_2O as the product instead of hydrogen peroxide. The *Polyporus* enzyme, which unlike tyrosinase is insensitive to CO, has a molar absorptivity of >1000 at 610 nm. It has been shown that laccase contains at least three kinds of copper. One has the blue color of the azurin type of molecule and binds oxygen. Another is a "nonblue" copper ion that may serve as an anion binding site, possibly for stabilization of an intermediate peroxide. Two other Cu^{2+} ions form a diamagnetic pair that acts as a two-electron acceptor receiving electrons from a substrate, then passing them on to oxygen, perhaps via an intermediate peroxide.

Ceruloplasmin is a blue protein of MW = 150,000 and contains 8 Cu^+ and 8 Cu^{2+} ions. The principal copper-containing protein of blood, it contains 3% of the total body copper. Ceruloplasmin appears to have something to do with regulation of the copper content of the body for in the copper storage disease (Wilson's disease) the ceruloplasmin content is low. In addition, ceruloplasmin has enzymatic properties resembling those of laccase and is also able to catalyze the oxidation of Fe^{2+} to Fe^{3+}. The latter reaction is important because only Fe^{3+} can combine with the iron transport protein **transferrin** (Box 14-D). For this reason, ceruloplasmin is sometimes referred to as **ferroxidase.**

A widely distributed group of proteins known variously as erythrocuprein, cerebrocuprein,

etc., were formerly regarded as copper storage proteins.[m] In recent years, however, it has been recognized that these proteins are **superoxide dismutases** (Eq. 8-61). The eukaryotic cytoplasmic forms of the enzyme are dimers of MW = 31,300, each subunit containing an atom of copper and an atom of zinc. The copper is thought to be bound to three imidazole nitrogen atoms. It probably undergoes alternate reduction and oxidation upon reaction with superoxide radicals. The crystal structure is being determined, and it should soon be possible to have a more definite picture of the environment of copper in this enzyme.[n]

In addition to these copper-zinc superoxide dismutases, mitochondria and many bacteria possess manganese-containing enzymes with the same function (Box 13-A). Cells of *E. coli* have both manganese-containing and iron-containing superoxide dismutases.[o]

A cytoplasmic **copper-binding protein** resembling metallothionein (Box 7-F) in its high cysteine content and other properties has been described.[p]

[a] E. Frieden, *Sci. Am.* **218**, 103–114 (1968).

[b] B. L. O'Dell and B. J. Campbell, *Compr. Biochem.* **21**, 191–203 (1971).

[c] J. Peisach, P. Aisen, and W. E. Blumberg, eds., "Biochemistry of Copper." Academic Press, New York, 1966.

[d] R. Malkin and B. G. Malmström, *Adv. Enzymol.* **33**, 177–244 (1970).

[e] T. E. Jones, D. B. Rorabacher, L. A. Ochrymowycz, *JACS* **97**, 7485–7486 (1975).

[f] J. Peisach, W. G. Levine, and W. E. Blumberg, *JBC* **242**, 2847–2858 (1967).

[g] M. J. Ettinger and D. J. Kosman, *Biochemistry* **13**, 1247–1251 (1974).

[h] G. R. Dyrkacz, R. D. Libby, and G. H. Hamilton, *JACS* **98**, 626–628 (1976).

[i] C. G. Gahmberg, *JBC* **251**, 510–515 (1976).

[j] H. G. Krishnamurty and F. J. Simpson, *JBC* **245**, 1467–1471 (1970).

[k] W. H. Vanneste and A. Zuberbühler, *in* "Molecular Mechanisms of Oxygen Activation" (O. Hayaishi, ed.), pp. 371–404. Academic Press, New York, 1974.

[l] J. A. Fee, *Struct. Bonding* (*Berlin*) **23**, 1–60 (1975).

[m] U. Weser, *Struct. Bonding* (*Berlin*) **17**, 1–65 (1973).

[n] K. A. Thomas, B. H. Rubin, C. J. Bier, J. S. Richardson, and D. C. Richardson, *JBC* **249**, 5677–5683 (1974).

[o] F. J. Yost, Jr., and I. Fridovich, *JBC* **248**, 4905–4908 (1973).

[p] D. R. Winge, R. Premakumar, R. D. Wiley, and K. V. Rajagopalan, *ABB* **170**, 253–266 (1975).

Several forms of mammalian liver cytochrome P-450 are known.[135] All are tightly bound to membranes of the endoplasmic reticulum and are difficult to solubilize. These enzymes may not require a protein of the rubredoxin type as an intermediate electron carrier but may react directly with a flavoprotein. An interesting property is their inducibility. Administration of various drugs, e.g., phenobarbital, and many other compounds causes as much as a 20-fold increase in cytochrome P-450 activity. Aromatic hydrocarbons induce a different form of the hydroxylase than do barbiturates. The role of cytochromes P-450 is often to convert a drug or other foreign compound to a form that is more readily excreted. However, the result is not always beneficial. Thus 3-methylcholanthrene, a strong inducer of cytochrome P-450, is converted to a powerful carcinogen by the hydroxylation reaction.[159]

3. Desaturation of Fatty Acids

Another enzymatic activity associated with the endoplasmic reticulum that requires O_2 is conversion of saturated fatty acids to cis-unsaturated acids (e.g., stearoyl-CoA and oleoyl-CoA, Table 2-7)[160] The mechanism of the reactions is uncertain. Cytochrome b_5 appears to be a specific electron carrier for the desaturase systems in animals; ferredoxins are involved in plants.

REFERENCES

1. H. M. Kalckar, "Biological Phosphorylations." Prentice-Hall, Englewood Cliffs, New Jersey, 1969.
2. D. Keilin, "The History of Cell Respiration and Cytochrome." Cambridge Univ. Press, London and New York, 1966.
2a. C. H. Fiske and Y. Subbarow, *Science* **70**, 381–382 (1929).
3. W. S. Caughey, G. A. Smythe, D. H. O'Keeffe, J. E. Maskasky, and M. L. Smith, *JBC* **250**, 7602–7622 (1975).
3a. D. E. Hultquist, R. T. Dean, and D. W. Reed, *JBC*, **251**, 3927–3932 (1976).
4. L. L. Ingraham, *Compr. Biochem.* **14**, 424–446 (1966).
5. B. A. Wittenberg, J. B. Wittenberg, and P.R.B. Caldwell, *JBC* **250**, 9038–9043 (1975).
6. R. J. P. Williams, *Biochem. Soc. Trans.* **1**, 1–26 (1973).
7. J. Geibel, C. K. Chang, and T. G. Traylor, *JACS* **97**, 5924–5926 (1975).
8. J. P. Collman, R. R. Gagne, C. A. Reed, T. R. Halbert, G. Lang, and W. T. Robinson, *JACS* **97**, 1427–1439 (1975).
9. J. L. Hoard, *Science* **174**, 1295–1302 (1971).
10. L. Anderson, *JMB* **79**, 495–506 (1973).
11. J. L. Hoard and W. R. Scheidt, *PNAS* **70**, 3919–3922 (1973).
11a. I. M. Klotz, G. L. Klippenstein, and W. A. Hendrickson, *Science* **192**, 335–344 (1976).
11b. L. Waxman, *JBC* **250**, 3796–3806 (1975).
11c. R. J. Siezen and E. F. J. van Bruggen, *JMB* **90**, 77–89 (1974).
12. R. E. Dickerson and R. Timkovich, in "The Enzymes," 3rd ed. (P. D. Boyer, ed.), Vol. 11, pp. 397–547. Academic Press, New York, 1975.
13. M. D. Kamen and T. Horio, *Annu. Rev. Biochem.* **39**, 673–700 (1970).
14. R. Lemberg and J. Barrett, "The Cytochromes." Academic Press, New York, 1972.
15. N. S. Gel'man, M. A. Lukoyanova, and D. N. Ostrovskii, "Bacterial Membranes and the Respiratory Chain," Biomembranes, Vol. 6. Plenum, New York, 1975.
16. B. Hagihara, N. Sato, and T. Yamanaka, in "The Enzymes," 3rd ed. (P. D. Boyer, ed.), Vol. 11, pp. 549–593. Academic Press, New York, 1975.
17. R. E. Dickerson, *Sci. Am.* **226**, 58–72 (Apr 1972).
18. T. Takano, O. B. Kallai, R. Swanson, and R. E. Dickerson, *JBC* **248**, 5234–5255 (1973).
19. F. R. Salemme, S. T. Freer, Ng. H. Xuong, R. A. Alden, and J. Kraut, *JBC* **248**, 3910–3921 (1973).
20. F. R. Salemme, J. Kraut, and M. D. Kamen, *JBC* **248**, 7701–7716 (1973).
21. J. E. Harrison, *PNAS* **71**, 2332–2334 (1974).
22. E. Stellwagen and R. D. Cass, *JBC* **250**, 2095–2098 (1975).
23. F. S. Mathews, M. Levine, and P. Argos, *JMB* **64**, 449–464 (1972); *Nature (London), New Biol.* **233**, 15–16 (1971).
24. P. Nicholls and W. B. Elliott, in "Iron in Biochemistry and Medicine" (A. Jacobs and M. Worwood, eds.). Academic Press, New York, 1974.
24a. B. Chance, C. Saronia, and J. S. Leigh, Jr., *JBC* **250**, 9226–9237 (1975).
24b. S. H. Phan and H. R. Mahler, *JBC* **251**, 257–269 (1976).
24c. G. Schatz and T. L. Mason, *Annu. Rev. Biochem.* **43**, 51–87 (1974).
25. H. Aebi and H. Suter, in "The Metabolic Basis of Inherited Disease" (J. B. Stanbury, J. B. Wyngaarden, and D. S. Fredrickson, eds.), McGraw-Hill, New York, 1972.
25a. G. R. Shonbaum and B. Chance, in "The Enzymes," 3rd ed. (P. D. Boyer, ed.), Vol. 13, pp. 363–408. Academic Press, New York, 1976.

26. A. S. Brill, *Compr. Biochem.* **14,** 447–479 (1972).

27. J. Peisach, W. E. Blumberg, B. A. Wittenberg, and J. B. Wittenberg, *JBC* **243,** 1871–1880 (1968).

28. R. W. Noble and Q. H. Gibson, *JBC* **245,** 2409–2413 (1970).

29. P. M. Champion, E. Münck, P. G. Debrunner, P. F. Hollenberg, and L. P. Hager, *Biochemistry* **12,** 426–435 (1973).

30. P. M. Champion, R. Chiang, E. Münck, P. Debrunner, and L. P. Hager, *Biochemistry* **14,** 4159–4166 (1975).

31. T. J. Mueller and M. Morrison, *JBC* **249,** 7568–7573 (1974).

32. M. Morrison and G. R. Schonbaum, *Annu. Rev. Biochem.* **45,** 861–888 (1976).

33. D. C. Hohn and R. I. Lehrer, *J. Clin. Invest.* **55,** 707–713 (1975).

33a. J. E. Harrison and J. Schultz, *JBC* **251,** 1371–1374 (1976).

34. R. I. Lehrer, *J. Clin. Invest.* **55,** 338–346 (1975).

35. D. F. Wilson, P. L. Dutton, M. Erecińska, J. G. Lindsay, and N. Sato, *Acc. Chem. Res.* **5,** 234–241 (1972).

36. P. D. Bragg, *in* "Microbial Iron Metabolism" (J. B. Neilands, ed.), pp. 303–348. Academic Press, New York, 1974.

36a. T. Ohnishi, *EJB* **64,** 91–103 (1976).

36b. T. Ohnishi, J. Lim, D. B. Winter, and T. E. King, *JBC* **251,** 2105–2109 (1976).

37. W. Lovenberg, ed., "Iron–Sulfur Proteins," Vols. I and II. Academic Press, New York, 1973.

37a. D. O. Hall, K. K. Rao, and R. Mullinger, *Biochem. Soc. Trans.* **3,** 472–479 (1975).

38. L. H. Jensen, *Annu. Rev. Biochem.* **43,** 461–474 (1974).

39. J. B. Neilands, ed., "Microbial Iron Metabolism." Academic Press, New York, 1974.

40. H.-E. Knoell and J. Knappe, *EJB* **50,** 245–252 (1974).

41. C. W. Carter, Jr., J. Kraut, S. T. Freer, Ng. H. Xuong, R. A. Alden, and R. G. Bartsch, *JBC* **249,** 4212–4225 (1974).

42. T. Ueda, E. T. Lode, and M. J. Coon, *JBC* **247,** 2109–2116 (1972).

43. T. Herskovitz, B. A. Averill, R. H. Holm, J. A. Ibers, W. D. Phillips, and J. F. Weiher, *PNAS* **69,** 2437–2441 (1972).

44. R. C. Job and T. C. Bruice, *PNAS* **72,** 2478–2482 (1975).

45. E. T. Adman, L. C. Sieker, and L. H. Jensen, *JBC* **248,** 3987–3996 (1973).

46. C. W. Carter, Jr., J. Kraut, S. T. Freer, R. A. Alden, L. C. Sieker, E. Adman, and L. H. Jensen, *PNAS* **69,** 3526–3529 (1972).

47. T. C. Bruice, R. Maskiewicz, and R. Job, *PNAS* **72,** 231–234 (1975).

48. R. H. Holm, *Endeavour* **121,** 38–43 (1975).

49. C. W. Carter, Jr., J. Kraut, S. T. Freer, and R. A. Alden, *JBC* **249,** 6339–6346 (1974).

49a. W. V. Sweeney, J. C. Rabinowitz, and D. C. Yoch, *JBC* **250,** 7842–7847 (1975).

49b. T. Ohnishi, J. Lim, D. B. Winter, and T. E. King, *JBC* **251,** 2105–2109 (1976).

50. W. H. Orme-Johnson, *Annu. Rev. Biochem.* **42,** 159–204 (1973).

51. R. W. Lane, J. A. Ibers, R. B. Frankel, and R. H. Holm, *PNAS* **72,** 2868–2872 (1975).

52. R. A. Morton, *Biol. Rev. Cambridge Philos. Soc.* **46,** 47–96 (1971).

53. R. A. Morton, *Vitamins* **5,** 355–391 (1972).

54. R. A. Morton, ed., "Biochemistry of Quinones." Academic Press, New York, 1965.

55. R. Bentley, *in* "Lipid Metabolism" (S. J. Wakil, ed.), p. 481. Academic Press, New York, 1970.

56. M. Gibbs, "Structure and Function of Chloroplasts." Springer-Verlag, Berlin and New York, 1971.

57. R. A. Morton, *in* "Current Trends in the Biochemistry of Lipids," *Biochem. Soc. Symp. No. 35* (J. Gunyuly and M. S. Smellie, eds.), pp. 203–217. Academic Press, New York, 1972.

58. Quoted by A. L. Lehninger, *in* "Biochemistry and Biophysics of Mitochondrial Membranes" (G. F. Azzone, ed.), p. 1. Academic Press, New York, 1972.

59. W. W. Wainio, "The Mammalian Mitochondrial Respiratory Chain." Academic Press, New York, 1970.

59a. L. Packer and L. Worthington, *in* "The Biogenesis of Mitochondria" (A. M. Kroon and C. Saccone, eds.), pp. 537–540. Academic Press, New York, 1974.

60. E. A. Munn, "The Structure of Mitochondria." Academic Press, New York, 1974.

61. D. Lloyd, "The Mitochondria of Microorganisms." Academic Press, New York, 1974.

62. L. D. Smith and R. K. Gholson, *JBC* **244,** 68–71 (1969).

63. R. A. Harris, C. H. Williams, M. Caldwell, D. E. Green and E. Valdivia, *Science* **165,** 700–703 (1969).

64. Ed. L. Ernster and Z. Drahota, *Mitochondria: Struct. Funct. Fed. Eur. Biochem. Soc., Meet., 5th, 1968* FEBS Symp., Vol. 17, pp. 5–31 (1969).

65. L. V. Sutfin, M. E. Holtrop, and R. E. Ogilvie, *Science* **174,** 947–949 (1971).

66. C. R. Hackenbrock and K. M. Hammon, *JBC* **250,** 9185–9197 (1975).

66a. G. D. Eytan, R. C. Carroll, G. Schatz, and E. Racker, *JBC* **250,** 8598–8603 (1975).

67. L. Noda, *in* "The Enzymes," 3rd ed. (P. D. Boyer, ed.), Vol. 8, pp. 279–305. Academic Press, New York, 1973.

67a. A. Kröger and M. Klingenberg, *in* "The Structure of Mitochondria" (E. A. Munn, ed.), p. 2820. Academic Press, New York, 1974.

68. B. Chance and G. R. Williams, *JBC* **217,** 409–427 (1955); *Adv. Enzymol.* **17,** 65–134 (1956).

69. Y. Hafeti, *Compr. Biochem.* **14,** 199–231 (1966).

70. G. B. Cox and F. Gibson, *BBA* **346,** 1–25 (1974).

71. H. M. Kalckar, "Biological Phosphorylations." Prentice-Hall, Englewood Cliffs, New Jersey, 1969.

72. D. F. Wilson, P. L. Dutton, M. Erecińska, J. G. Lindsay, and N. Soto, *Acc. Chem. Res.* **5,** 234–241 (1972).

73. D. F. Wilson, M. Erecińska, and L. P. Dutton, *Annu. Rev. Biophys. Bioeng.* **3,** 203–230 (1974).

74. D. F. Wilson, M. Stubbs, N. Oshino, and M. Erecińska, *Biochemistry* **13**, 5305–5311 (1974).

75. J. G. Lindsay, C. S. Owen, and D. F. Wilson, *ABB* **169**, 492–505 (1975).

76. L. J. M. Eilermann and E. C. Slater, *BBA* **216**, 226–228 (1970).

77. A. M. Lambowitz, W. D. Bonner, Jr., and M. K. F. Wikström, *PNAS* **71**, 1183–1187 (1974).

78. W. A. Catterall and P. L. Pederson, *JBC* **246**, 4987–4994 (1971).

79. N. Nelson, D. W. Deters, H. Nelson, and E. Racker, *JBC* **248**, 2049–2055 (1973).

80. C. I. Ragan and E. Racker, *JBC* **248**, 2563–2569 (1973).

81. K. H. Leung and P. C. Hinkle, *JBC* **250**, 8467–8471 (1975).

82. H. A. Lardy and S. M. Ferguson, *Annu. Rev. Biochem.* **38**, 991–1034 (1969).

83. D. E. Green, ed., *Ann. N. Y. Acad. Sci.* **227** (1974). A special volume on oxidative phosphorylation and related matters.

84. F. Lipmann, *in* "Currents in Biochemical Research" (D. E. Green, ed.), (Interscience), Wiley New York, 1946. Reprinted in F. Lipmann, "Wanderings of a Biochemist, " Wiley (Interscience), New York, 1971.

85. J. A. Barltrop, P. W. Grubb, and B. Hesp, *Nature (London)*, **199**, 759–761 (1963).

86. E. J. H. Bechara and C. Cilento, *Biochemistry* **11**, 2606–2610 (1972).

87. K. Harrison, *Nature (London)* **181**, 1131 (1958).

88. M. Vilkas and E. Lederer, *Experientia* **18**, 546–549 (1962).

89. H. I. Hadler, *Experientia* **17**, 268–269 (1961).

90. J. H. Wang, *Science* **167**, 25–30 (1970).

91. T. Higuchi and K.–H. Gensch, *JACS* **88**, 3874–3875 (1966).

92. T. Higuchi and K.–H. Gensch, *JACS* **88**, 5486–5491 (1966).

93. P. D. Boyer and C. Degani, *Fed. Proc., Fed. Am. Soc. Exp. Biol.* **33**, 1292 (1974).

94. J. T. Penniston, R. A. Harris, J. Asai, and D. E. Green, *PNAS* **59**, 624–631 (1968).

95. D. A. Harris and E. C. Slater, *BBA* **387**, 335–348 (1975).

96. K. D. Straub, *J. Theor. Biol.* **44**, 191–206 (1974).

97. P. Mitchell, *Biol. Rev. Cambridge Philos. Soc.* **41**, 445–502 (1966).

98. P. Mitchell, "Chemiosmotic Coupling and Energy Transduction," Glynn Res., Bodmin., Cornwall, England, 1968.

98a. P. Mitchell and J. Moyle, *EJB* **7**, 471–484 (1969).

98b. P. C. Laris, D. P. Bahr, and R. R. J. Chaffee, *BBA* **376**, 415–425 (1975).

98c. S. Ramos, S. Schuediner, and H. R. Koback, *PNAS* **73**, 1892–1896 (1976).

98d. H. Tedeschi, "Mitochondria: Structure, Biogenesis and Transducing Function," pp. 93–100. Springer, Vienna, 1976.

99. B. C. Pressman, *Fed. Proc., Fed. Am. Soc. Exp. Biol.* **27**, 1283–1288 (1968).

100. H. G. Shertzer and E. Racker, *JBC* **249**, 1320–1321 (1974).

101. F. E. Sluse, M. M. Ranson, and C. Liebecq, *EJB* **25**, 207–217 (1972).

102. M. Klingenberg, *FEBS Lett.* **6**, 145–154 (1970).

103. J. O. D. Coleman and J. M. Palmer, *EJB* **26**, 499–509 (1972).

104. K. F. LaNoue, J. Bryla, and D. J. P. Bassett, *JBC* **249**, 7514–7521 (1974).

105. S. A. Short, H. R. Kaback, and L. D. Kohn, *JBC* **250**, 4291–4296 (1975).

106. G. Kaczorowski, L. Shaw, M. Fuentes, and C. Walsh, *JBC* **250**, 2855–2865 (1975).

107. R. D. Simoni and D. W. Postma, *Annu. Rev. Biochem.* **44**, 523–554 (1975).

108. J. F. Hare, K. Olden, and E. P. Kennedy, *PNAS* **71**, 4843–4846 (1974).

109. H. G. Schlegel and V. Eberhardt, *Adv. Microb. Physiol.* **7**, 205–242 (1972).

110. C. T. Gray and H. Gest, *Science* **148**, 186–192 (1965).

111. D. L. Erbes, R. H. Burris, and W. H. Orme-Johnson, *PNAS* **72**, 4795–4799 (1975).

112. P. H. Gitlitz and A. I. Krasna, *Biochemistry* **14**, 2561–2568 (1975).

113. W. Wallace and D. J. D. Nicholas, *Biol. Rev. Cambridge Philos. Soc.* **44**, 359–391 (1969).

114. M. K. Rees, *Biochemistry* **7**, 353–366 and 366–372 (1968).

115. A. B. Hooper and A. Nason, *JBC* **240**, 4044–4057 (1965).

116. L. A. Kiesow, J. B. Shelton, and J. W. Bless, *ABB* **151**, 414–419 (1972).

117. W. J. Ingledew, J. G. Cobley, and J. B. Chappell, *Biochem. Soc. Trans.* **2**, 149–151 (1974).

118. P. A. Trudinger, *Adv. Microb. Physiol.* **3**, 111–158 (1969).

119. A. B. Roy and P. A. Trudinger, "The Biochemistry of Inorganic Compounds of Sulfur." Cambridge Univ. Press, London and New York, 1970.

120. R. M. Lyric and I. Suzuki, *Can J. Biochem.* **48**, 334–343, 344–354, and 355–363 (1970).

121. P. R. Dugan, "Biochemical Ecology of Water Pollution." Plenum, New York, 1972.

122. H. J. Cohen, S. Betcher-Lange, D. L. Kessler, and K. V. Rajagopalan, *JBC* **247**, 7759–7766 (1972).

123. H. D. Peck, Jr., *Annu. Rev. Microbiol.* **22**, 489–518 (1968).

124. H. W. Doelle, "Bacterial Metabolism." Academic Press, New York, 1969.

125. C. H. MacGregor, C. A. Schnaitman, and D. E. Normansell, *JBC* **249**, 5321–5327 (1974).

126. R. L. Lester and J. A. DeMoss, *J. Bacteriol.* **105**, 1006–1014 (1971).

127. R. H. Garrett and A. Nason, *JBC* **244**, 2870–2882 (1969).

127a. J. M. Vega, R. H. Garrett, and L. M. Siegel, *JBC* **250**, 7980–7989 (1975).

128. D. G. Grey and M. L. Jensen, *Science* **177**, 1099–1100 (1972).

129. D. V. DerVartanian and J. LeGall, *BBA* **346**, 79–99 (1974).

130. H. D. Peck, Jr., R. Bramlett and D. V. DerVartanian, *Z. Naturforsch. Teil B* **27**, 1084–1086 (1972).

131. K. Irie, K. Kobayashi, M. Kobayashi, and M. Ishimoto, *J. Biochem. (Tokyo)* **73**, 353–366 (1973).

132. G. C. Wagner, R. J. Kassner, and M. D. Kamen, *PNAS* **71**, 253–256 (1974).
133. H. A. Barker, *in* ''Horizons of Bioenergetics'' (A. San Pietro and H. Gest, eds.), pp. 7–31. Academic Press, New York, 1972.
134. O. Hayaishi and M. Nozaki, *Science* **164**, 389–396 (1969).
134a. O. Hayaishi, ed., ''Molecular Mechanisms of Oxygen Activation'' Academic Press, New York, 1974.
135. I. C. Gunsalus, T. C. Pederson, and S. G. Sligar, *Annu. Rev. Biochem.* **44**, 377–407 (1975).
136. P. D. Boyer, ed., ''The Enzymes,'' 3rd ed., Vol. 12. Academic Press, New York, 1975. This book contains several chapters on hydroxylases.
137. F. Hirata and O. Hayaishi, *JBC* **250**, 5960–5966 (1975).
138. E. K. Pistorius and B. Axelrod *JBC* **251**, 7144–7148 (1976).
139. O. Lockridge, V. Massey, and P. A. Sullivan, *JBC* **247**, 8097–8106 (1972).
140. M. I. S. Flashner and V. Massey, *JBC* **249**, 2579–2586 (1974).
141. B. Entsch, D. P. Ballou and V. Massey, *JBC* **251**, 2550–2563 (1976).
142. H. W. Orf and D. Dolphin, *PNAS* **71**, 2646–2650 (1974).
143. S. Kaufman, *JBC* **239**, 332–338 (1964).
144. R. L. Blakley, ''The Biochemistry of Folic Acid and Related Pteridines,'' pp. 293–314. North-Holland Publ., Amsterdam, 1969.
145. J. E. Craine, E. S. Hall, and S. Kaufman, *JBC* **247**, 6082–6091 (1972).
146. J. H. Tong and S. Kaufman, *JBC* **250**, 4152–4158 (1975).
147. G. Guroff, J. W. Daly, D. M. Jerina, J. Renson, B. Witkop, and S. Udenfriend, *Science* **157**, 1524–1530 (1967).
148. D. R. Boyd, J. W. Daly, and D. M. Jerina, *Biochemistry* **11**, 1961–1966 (1972).
149. G. J. Kasperek, T. C. Bruice, H. Yagi, N. Kaubisch, and D. M. Jerina, *JACS* **94**, 7876–7882 (1972).
150. B. Lindblad, G. Lindstedt, M. Tofft, and S. Lindstedt, *JACS* **91**, 4604–4606 (1969).
151. K. I. Kivirikko, K. Shudo, S. Sakakibara, and D. J. Prockop, *Biochemistry* **11**, 122–129 (1972).
151a. C. W. Jefford, A. F. Boschung, T. A. B. M. Bolsman, R. M. Moriarty, and B. Melnick, *JACS* **98**, 1017–1018 (1976).
152. M. Goldstein, T. H. Joh, and T. Q. Garvey, III, *Biochemistry* **7**, 2724–2730 (1968).
153. C. J. Sih, *Science* **163**, 1297–1300 (1969).
154. T. Ueda, E. T. Lode, and M. J. Coon, *JBC* **247**, 2109–2116 (1972).
155. A. M. Chakrabarty, C. Chou, and I. C. Gunsalus, *PNAS* **70**, 1137–1140 (1973).
156. C-A. Yu, I. C. Gunsalus, M. Katagiri, K. Suhara, and S. Takemori, *JBC* **249**, 94–101 (1974).
157. J. R. Gillette, D. C. Davis, and H. A. Sasame, *Annu. Rev. Pharmacol.* **12**, 57–84 (1972).
158. A. Y. H. Lu and W. Levin, *BBA* **344**, 205–240 (1974).
159. C. Heidelberger, *Annu. Rev. Biochem.* **44**, 79–121 (1975).
160. W.-H. Kunau, *Angew Chem. Intern. Ed.* **15**, 61–74 (1976).

STUDY QUESTIONS

1. Given 50 mg each of pure type I (rubredoxin) and type II (ferredoxin) iron–sulfur proteins, describe a simple chemical test to distinguish between the two substances.

2. What would be the difference in net charge at pH 7 between clostridial apoferredoxin (i.e., the iron-free protein) and reconstituted ferredoxin prepared from 4 ferrous ions, 4 ferric ions and 8 mol of H_2S plus apoenzyme? *Note:* The pK_a of a cysteinyl SH is ~ 9.

3. Two reactions of the citric acid cycle are

   L-Malate + NAD^+ $\rightleftharpoons$

   oxaloacetate + NADH + H^+

   Succinate + FAD (bound) $\rightleftharpoons$

   fumarate + $FADH_2$ (bound)

   a. How much ATP can be formed in a cell as a result of each of these oxidation reactions?
   b. How much ATP can be formed in association with oxidation of glyceraldehyde 3-phosphate

   by NAD^+ in the cytoplasm?
   Consider the role of various shuttle mechanisms for entrance of reducing equivalents into the mitochondria in your answer to (b).

4. (a) Calculate $\Delta G'$ (pH 7) for oxidation of pyruvate to CO_2 and water by NAD^+. (b) Compare this with the value of $\Delta G'$ for oxidation by O_2. (c) How much would this value be changed if the oxygen tension were only 0.01 atm? (d) Would mitochondrial oxidation of pyruvate be slowed significantly if the O_2 pressure were this low within a cell?

5. Suppose that a membrane-bound hydrogenase equilibrated H_2 with H^+ and with other components of the respiratory chain of an organism. (a) At pH 7 and an H_2 pressure of 0.1 atm, what would be the following ratios? [NADH]/[NAD^+]; [Fd (reduced)]/[Fd (oxidized)]; calculated for clostridial ferredoxin; [cyt c (reduced)]/[cyt c (oxidized)]? (b) Would these ratios be changed if the synthesis of 1

mol of ATP were tightly coupled to the passage of electrons through the respiratory chain at site I? If so, how much would the ratio change for a phosphorylation state ratio $R_p = 10^5 \ M^{-1}$?

6. The $NAD^+/NADH$ electrode potential in intact mitochondria has been determined by equilibrating the mitochondria with acetoacetate/β-hydroxybutyrate, for which the electrode potential is $E^{\circ\prime} = -0.266$ V. When the ratio $[NAD^+]/[NADH]$, determined spectrophotometrically, is unity, the ratio $[acetoacetate]/[\beta$-hydroxybutyrate$]$ is 0.16. [Erecinska *et al.*, *ABB* **160,** 412 (1974)].

    a. What is $E^{\circ\prime}$ for the $NAD^+/NADH$ couple under the conditions of these experiments?

    b. The value of $E^{\circ\prime}$ for free $NAD^+/NADH$ is -0.32 V. Suppose the discrepancy between this figure and the result in (a) is to be accounted for by the proposition that NAD^+ and NADH bind to a mitochondrial protein P with binding constants K_1 and K_2, respectively, and that the spectrophotometric method does not distinguish between the free and bound forms. Show that the $E^{\circ\prime}$ determined in (a) is greater than that in the free state by the amount

$$\frac{RT}{nF} \ln \left(\frac{1 + K_2[P]}{1 + K_1[P]} \right)$$

    Which of the constants K_1 and K_2 is the larger by this interpretation?

7. Consider the reaction by which methane is generated from H_2 and CO_2 (Eq. 10-42). Write out a detailed step-by-step sequence for this process. Can you offer suggestions concerning the generation of ATP coupled to this reaction? Are there any ATP-requiring steps in the process?

8. (a) Estimate the total amount in grams of nicotinamide, of pantothenic acid, and of thiamine in the body of a 70 kg person. (b) Calculate the amount of ATP (in millimoles; in grams) needed for such a person to climb from the basement to the third floor of a building (10 m) assuming a 50% efficiency in utilization and a constant value of the phosphorylation state ratio R_p of $10^3 \ M^{-1}$. (c) How much of this ATP could be provided by transfer of a phosphoryl group from stored creatine phosphate in muscle? Assume that muscle constitutes 40% of the total body mass. (d) Calculate how many times, on the average, each molecule of nicotinamide would turn over, i.e., undergo reduction and reoxidation, during the climb mentioned in (b). Do the same for pantothenic acid (cycling between acyl-CoA and free CoA forms), and for thiamine.

9. Dehydrated firefly tails have become available commercially from at least one supplier of biochemical reagents. For what purpose are they used in biochemical analysis?

10. Assume that the creatine phosphate content of muscle is 20 mM and that it might fall to 10 mM during muscular action. What would be the value of the phosphorylation state rates R_p if the creatine kinase reaction operates at equilibrium?

11

Biosynthesis: How New Molecules Are Put Together

In this chapter the general principles and strategy of synthesis of the many carbon compounds found in living things will be considered. Since green plants and autotrophic bacteria are able to assemble all of their needed carbon compounds from CO_2, we will want to examine the mechanisms by which this is accomplished. We will also need to ask how other organisms are able to subsist on simple compounds such as formate or acetate.

A. METABOLIC LOOPS AND BIOSYNTHETIC FAMILIES

The routes of biosynthesis (anabolism) often closely parallel pathways of biodegradation (catabolism) (Fig. 7-1). Thus, catabolism begins with hydrolytic breakdown of polymeric molecules; the resulting monomers are then cleaved into smaller two- and three-carbon fragments. Biosynthesis begins with formation of monomeric units from smaller pieces followed by assembly of the monomers into polymers. The mechanisms of the individual reactions of biosynthesis and biodegradation are also often closely parallel. Reactions

of carbon-carbon bond formation in biosynthesis are related to those of carbon-carbon bond breaking in catabolism, and there is a similarity between mechanisms of formation of polymers and of hydrolysis. Nevertheless, in most instances, there are clear-cut and distinct differences. Thus, a first principle of biosynthesis is: *Biosynthetic pathways, although related to catabolic pathways, differ from them in distinct ways and are often catalyzed by completely different sets of enzymes.*

1. Metabolic Loops

The sum of the pathways of biosynthesis and biodegradation form a continuous loop—a series of reactions that take place concurrently and often within the same part of a cell. Metabolic loops often begin in the central pathways of carbohydrate metabolism with a three- or four-carbon compound (e.g., phosphoglycerate, pyruvate, or oxaloacetate). After loss of some atoms as CO_2 the remainder of the compound rejoins the "mainstream" of metabolism by entering a major catabolic pathway leading to acetyl-CoA and oxidation in the citric acid cycle. However, many other variations occur. Not all of the loops are closed within a given species. Humans are unable to synthesize the vitamins and **essential amino acids.** We depend upon other organisms to make these compounds, but we do degrade them. Some metabolites (e.g., uric acid) are excreted by man and are further catabolized by bacteria. From a chemical viewpoint the whole of nature can be regarded as an enormously complex set of branching and interconnecting metabolic cycles. Thus, the synthetic pathways used by autotrophs are all parts of metabolic loops terminating in oxidation back to CO_2.

An important feature of metabolic loops is that it is often not possible to state at what point in the loop biosynthesis has been completed and biodegradation begins. An end product X that serves one need of a cell may be a precursor to another cell component Y which is then degraded to complete the loop. The reactions that convert X to Y can be regarded as either biosynthetic (for Y) or catabolic (for X).

2. Key Intermediates and Biosynthetic Families

In examining routes of biosynthesis it is helpful to identify some key intermediates. One of these is **3-phosphoglycerate.** A primary product of photosynthesis, 3-phosphoglycerate may reasonably be regarded as the starting material from which all other carbon compounds in nature are formed. Phosphoglycerate, in most organisms, is readily interconvertible with both **glucose** and **phosphoenolpyruvate.** Any of these three compounds can serve as the precursor for synthesis of other organic materials. A first stage in biosynthesis consists of those reactions by which 3-phosphoglycerate or phosphoenolpyruvate arise, whether it be from CO_2, formate, acetate, lipids, or polysaccharides.

The further biosynthetic pathways from 3-phosphoglycerate to the myriad amino acids, nu-

cleotides, lipids, and miscellaneous compounds found in cells are complex and numerous. However, the basic features are relatively simple. Figure 11-1 indicates the origins of many substances including the 20 amino acids present in proteins, nucleotides, and lipids. Among the additional key biosynthetic precursors that can be identified from this chart are **glucose 6-phosphate, pyruvate, oxaloacetate, acetyl-CoA, α-ketoglutarate,** and **succinyl-CoA.**

The amino acid **serine** originates almost directly from 3-phosphoglycerate, as does **aspartate** from oxaloacetate and **glutamate** from α-ketoglutarate. These three amino acids each give rise to "families" of other compounds.[1] A little attention paid to establishing correct family relationships will make the study of biochemistry easier. Besides the serine, aspartate, and glutamate-ketoglutarate families, a fourth large family originates directly from pyruvate and a fifth (mostly lipids) from acetyl-CoA. The aromatic amino acids are formed from erythrose 4-P and phosphoenolpyruvate, **chorismic acid** (Eq. 7-50) being the key intermediate. Other families of compounds arise from glucose 6-P and from the **pentose phosphates.** These groups have been set off roughly by the thick arcs in Fig. 11-1.

B. HARNESSING THE ENERGY OF ATP

Throughout the history of biochemistry it has often been proposed that a particular biosynthetic pathway was the exact reverse of a catabolic pathway. For example, protein-hydrolyzing enzymes under suitable conditions of amino acid concentration and pH catalyze formation of amino acid polymers resembling proteins. The enzymes that catalyze β oxidation of fatty acid derivatives, when isolated from mitochondria, catalyze formation of fatty acyl-CoA derivatives from acetyl-CoA and a reducing agent such as NADH. However, reactant concentrations within cells are rarely appropriate for reversal of a catabolic sequence.

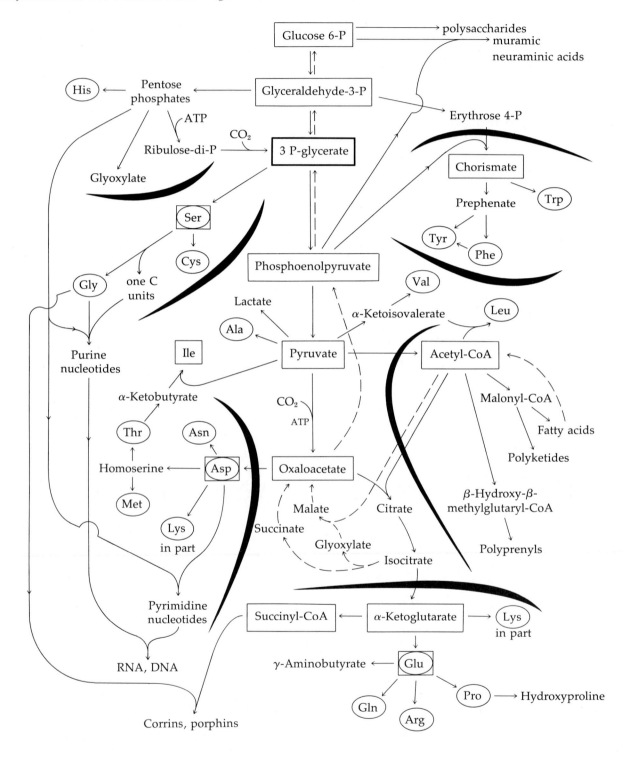

For catabolic sequences ΔG^0 is usually distinctly negative and reversal requires high concentrations of end products. However, the latter are often promptly removed from the cells. For example, NADH produced in degradation of fatty acids is oxidized to NAD^+ and is therefore never available in sufficient concentrations to reverse the β-oxidation sequence.

Nature's answer to the problem of reversing a catabolic pathway lies in the coupling of cleavage of ATP to the biosynthetic reaction. The concept has already been considered in Chapter 7, Section F, where one sequence for linking hydrolysis of ATP to biosynthesis is discussed. However, living cells employ several different methods of harnessing the free energy of hydrolysis of ATP to drive biosynthetic processes. Many otherwise strange aspects of metabolism become clear if it is recognized that they are simply a means for coupling ATP cleavage to biosynthesis. A few of the most important of these coupling mechanisms will be summarized in this section.

1. Group Activation

Consider the formation of an ester (or of an amide) from a free carboxylic acid and an alcohol (or amine) by the "splitting out" of a molecule of water (Eq. 11-1). The reaction is thermody-

$$R-C\overset{O}{\underset{OH}{\big\langle}} \;+\; HO-R' \;\longrightarrow$$
$$(H_2N-R')$$

$$R-C\overset{O}{\underset{O-R'}{\big\langle}} \;+\; H_2O \quad (11\text{-}1)$$
$$\left(\begin{array}{c} \\ N-R' \\ | \\ H \end{array}\right)$$

Fig. 11-1 Some major biosynthetic pathways. Key intermediates are enclosed in boxes and the 20 common amino acids of proteins are circled. Dotted lines indicate reactions of the glyoxylate pathway and of glucogenesis.

namically unfavorable with values of $\Delta G'$ (pH 7) ranging from $\sim +10$ to 30 kJ mol^{-1} depending on conditions and structures of the specific compounds. Long ago, organic chemists learned that reactions can be made to proceed by careful removal of the water that is generated from the reaction mixture. However, it is often better to "activate" the carboxylic acid by conversion to an acid chloride or an anhydride:

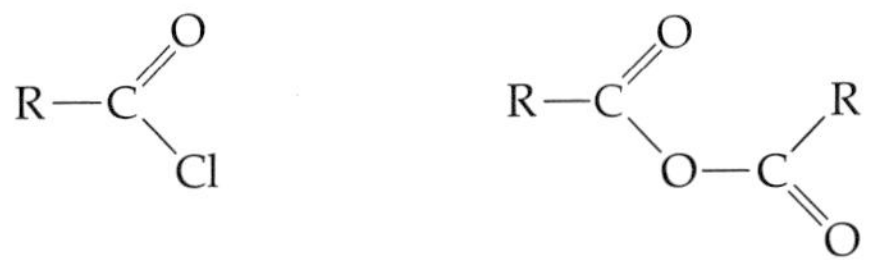

$$R-C\overset{O}{\underset{Cl}{\big\langle}} \qquad\qquad R-C\overset{O}{\underset{O-C}{\big\langle}}\overset{R}{\underset{O}{\big\langle}}$$

Nucleophilic attack on the carbonyl group of such a compound results in displacement of a good leaving group, Cl^- or $R-COO^-$. Nature has followed the same approach in forming from the carboxylic acid an **acyl phosphate** or an **acyl-CoA** derivative (see Chapter 3, Section B; Chapter 7, Section F; and Table 11-1).

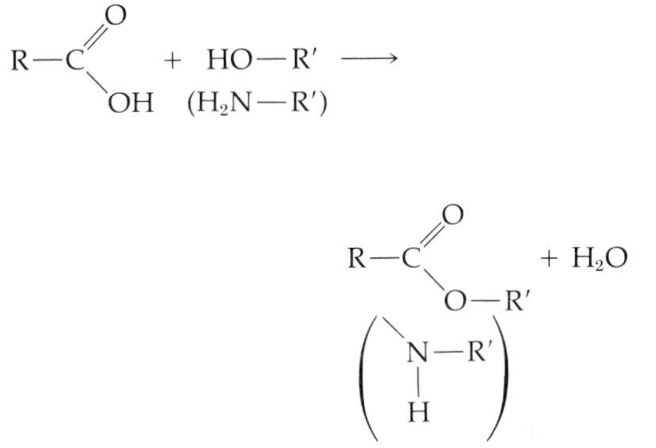

The virtue of these "activated" acyl compounds in biosynthetic reactions has been considered (Chapter 7, Section F; Table 8-1). Just as a carboxylic acid can be converted to an "active acyl" derivative, so other groups can be activated. The "high energy phosphate" compounds themselves can be thought of as **active phosphoryl** compounds. Sulfate is converted to a phosphosulfate anhydride, an **active sulfuryl** derivative. Sugars are converted to compounds like glucose 1-P or sucrose that contain **active glycosyl** groups. The group transfer potentials of the latter, though not as great as that of the phosphoryl groups of ATP, are still high enough to make glucose 1-phosphate and sucrose effective glycosylating reagents. Table 11-1 lists several of the more important activated groups.

Group activation usually takes place at the expense of ATP cleavage as discussed in Chapter 7, Section F. As pointed out in that discussion, acyl phosphates play a central role in metabolism by virtue of the fact that they contain both an activated acyl group and an activated phosphoryl

TABLE 11-1
"Activated" Groups Used in Biosynthesis

Group	Typical activated forms		

Phosphoryl,

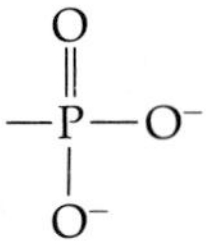

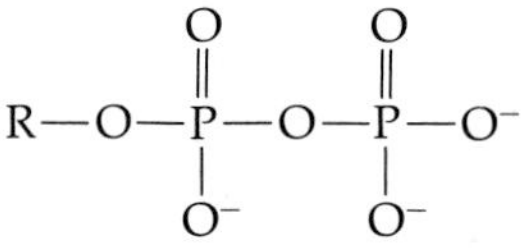

$$-\overset{\overset{\displaystyle O}{\|}}{\underset{\underset{\displaystyle O^-}{|}}{P}}-O^-$$

Pyrophosphate

$$R-O-\overset{\overset{\displaystyle O}{\|}}{\underset{\underset{\displaystyle O^-}{|}}{P}}-O-\overset{\overset{\displaystyle O}{\|}}{\underset{\underset{\displaystyle O^-}{|}}{P}}-O^-$$

Guanidine phosphate

$$R-\overset{\displaystyle H}{N}-\overset{\overset{\displaystyle}{}}{\underset{\underset{\displaystyle NH_2^+}{\|}}{C}}-\overset{\displaystyle H}{N}-\overset{\overset{\displaystyle O}{\|}}{\underset{\underset{\displaystyle O^-}{|}}{P}}-O^-$$

Enol phosphate

$$R-\overset{\overset{\displaystyle CH_2}{\|}}{C}-O-\overset{\overset{\displaystyle O}{\|}}{\underset{\underset{\displaystyle O^-}{|}}{P}}-O^-$$

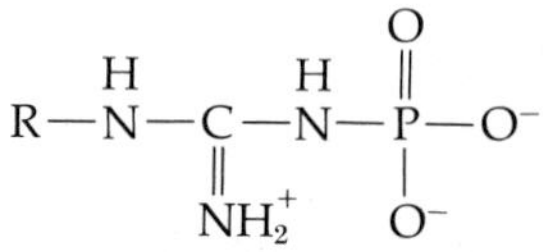

Acyl phosphate

$$R-\overset{\overset{\displaystyle O}{\|}}{C}-O-\overset{\overset{\displaystyle O}{\|}}{\underset{\underset{\displaystyle O^-}{|}}{P}}-O^-$$

Sulfuryl,

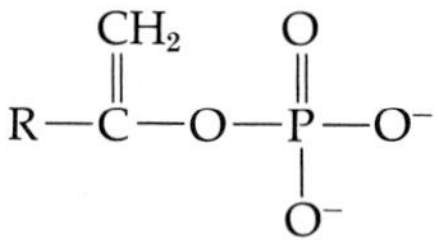

$$-\overset{\overset{\displaystyle O}{\|}}{\underset{\underset{\displaystyle O}{\|}}{S}}-O^-$$

$$R-O-\overset{\overset{\displaystyle O}{\|}}{\underset{\underset{\displaystyle O^-}{|}}{P}}-O-\overset{\overset{\displaystyle O}{\|}}{\underset{\underset{\displaystyle O}{\|}}{S}}-O^-$$

in PAPS

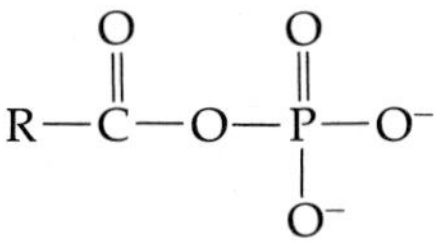

Acyl,

$$-\overset{\overset{\displaystyle O}{\|}}{C}-R$$

Thioester

$$R-\overset{\overset{\displaystyle O}{\|}}{C}-S-R'$$

Acyl phosphate

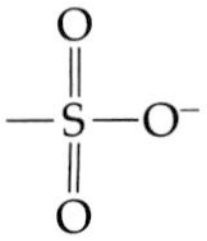

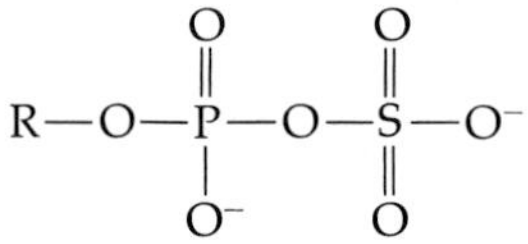

$$R-\overset{\overset{\displaystyle O}{\|}}{C}-O-\overset{\overset{\displaystyle O}{\|}}{\underset{\underset{\displaystyle O^-}{|}}{P}}-O-R'$$

Glycosyl,

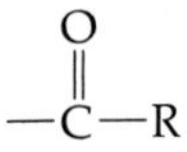

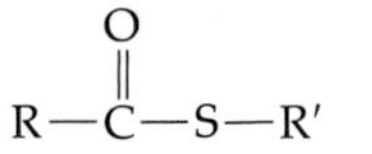

(pyranose ring with O)

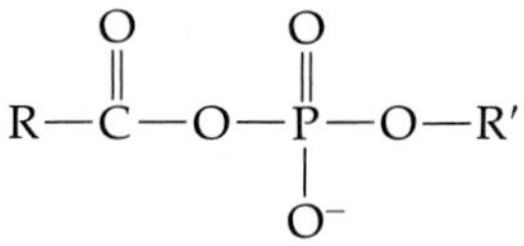

(pyranose ring with O)

$$O-\overset{\overset{\displaystyle O}{\|}}{\underset{\underset{\displaystyle O^-}{|}}{P}}-O^-$$

Sucrose

(pyranose ring–O–furanose ring)

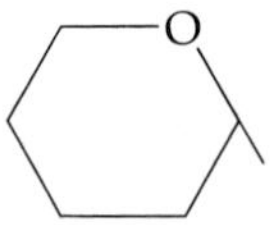

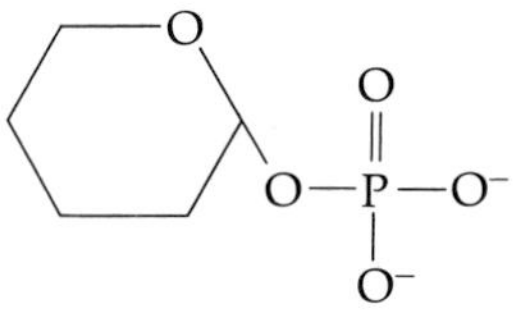

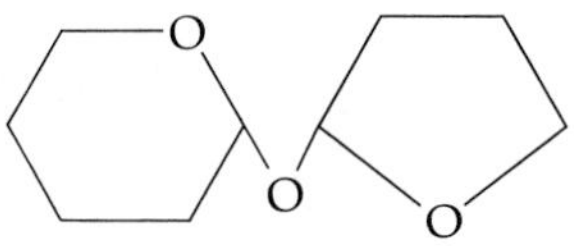

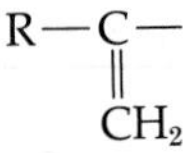

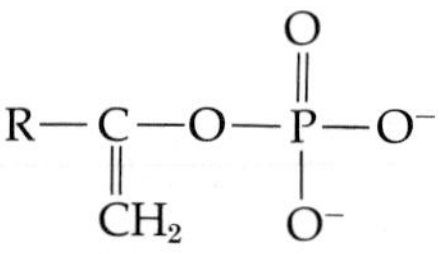

Enoyl,

$$R-\overset{\overset{}{}}{\underset{\underset{\displaystyle CH_2}{\|}}{C}}-$$

Enol phosphate

$$R-\overset{\overset{}{}}{\underset{\underset{\displaystyle CH_2}{\|}}{C}}-O-\overset{\overset{\displaystyle O}{\|}}{\underset{\underset{\displaystyle O^-}{|}}{P}}-O^-$$

Carbamoyl,

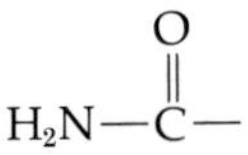

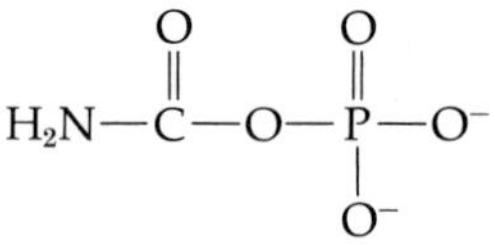

$$H_2N-\overset{\overset{\displaystyle O}{\|}}{C}-$$

Carbamoyl phosphate

$$H_2N-\overset{\overset{\displaystyle O}{\|}}{C}-O-\overset{\overset{\displaystyle O}{\|}}{\underset{\underset{\displaystyle O^-}{|}}{P}}-O^-$$

Alkyl, R—

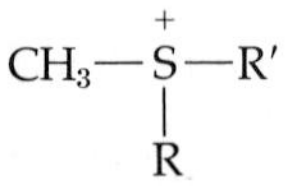

S-Adenosylmethionine

$$CH_3-\overset{\overset{\displaystyle +}{}}{\underset{\underset{\displaystyle R}{|}}{S}}-R'$$

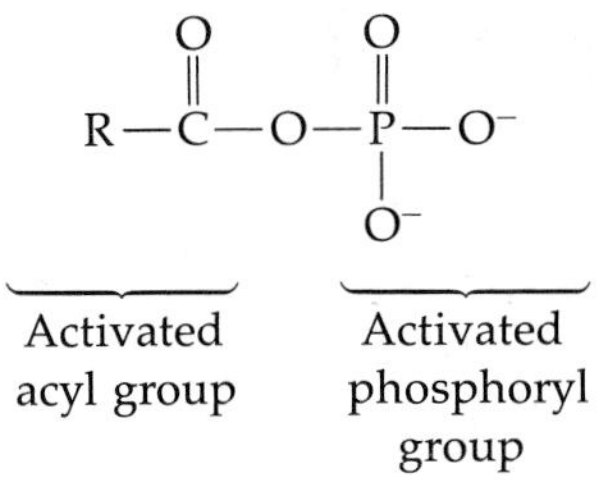

Acyl phosphate

Activated acyl group   Activated phosphoryl group

group. The high group transfer potential can be conserved in subsequent reactions *in either one group or the other* (but not in both). Thus, displacement on P by an oxygen of ADP will regenerate ATP and attack on C by an —SH will give a thioester. Several of the other compounds in Table 11-1 also can be split in two ways to yield different activated groups, e.g., the phosphosulfate anhydride, enoyl phosphate, and carbamoyl phosphate. It is probably only through intermediates of this type that cleavage of ATP can be coupled to synthesis of activated groups. The importance of such **common intermediates** in the synthesis of ATP by substrate level phosphorylation has already been stressed (Chapter 8, Section H,5).

2. Hydrolysis of Pyrophosphate

The splitting of inorganic pyrophosphate (PP_i) to two phosphate (P_i) groups (Chapter 7, Section E,1) is catalyzed by **pyrophosphatases** that apparently occur universally. Their function appears to be simply to remove the product PP_i from reactions that produce it (Table 7-2), shifting the equilibrium toward formation of a desired compound. An example is the formation of **activated amino acids** (aminoacyl-tRNA molecules) needed for protein synthesis. As shown in Eq. 11-2, the process requires the use of two ATP molecules to activate one amino acid.[2] While the "spending" of two ATP's for the addition of one monomer unit to a polymer does not appear necessary from a thermodynamic viewpoint, it is frequently observed, and there is no doubt that hydrolysis of PP_i assures that the reaction will go virtually to completion. Transfer RNA's tend to become saturated with amino acids according to Eq. 11-2 even if the concentration of free amino acid in the cytoplasm is low.

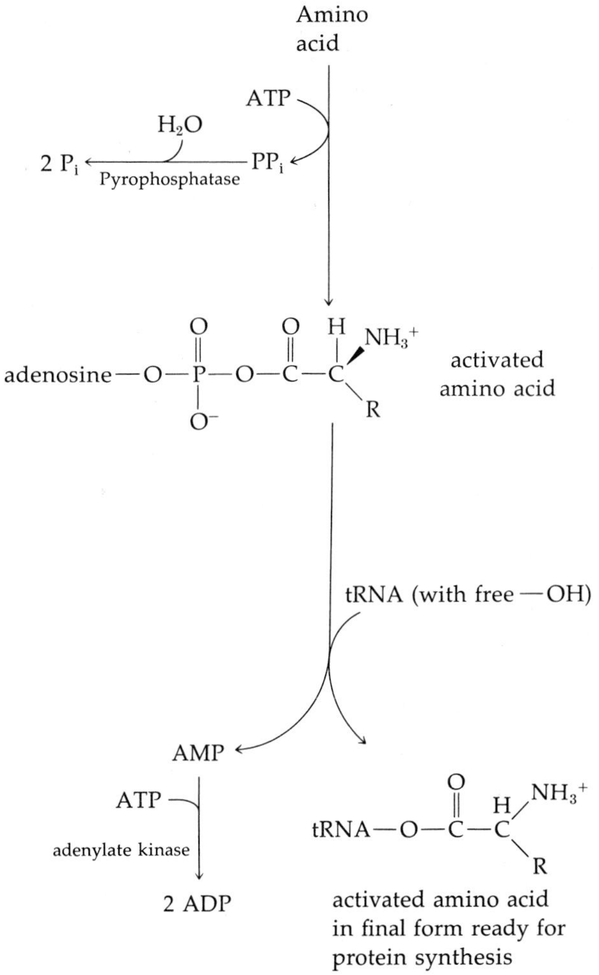

Net: 2 ATP + amino acid + tRNA $\longrightarrow$

aminoacyl tRNA + 2 ADP + 2 P_i

(11-2)

On the other hand, kinetic considerations may be involved. Perhaps the biosynthetic sequence would move too slowly if it were not for the extra boost given by the removal of PP_i. Furthermore, biosynthetic pathways are highly controlled and part of the explanation for the complexity may depend on control mechanisms which are only incompletely understood. Although pyrophosphatases are ubiquitous, it is still not absolutely certain that PP_i is *always* hydrolyzed. Under some circumstances, the energy of the remaining phosphoanhydride linkage may be conserved by a cell (Chapter 13, Section E,6).

In some metabolic reactions pyrophosphate esters are formed by consecutive transfer of the terminal phosphoryl groups of two ATP molecules to a hydroxyl group. Such esters often react with elimination of PP_i; e.g., in polymerization of prenyl units (reaction type 6B, Table 7-1; Fig. 12-11). Again, hydrolysis to P_i follows. Thus, *cleavage of pyrophosphate is a second very general method for coupling ATP cleavage to synthetic reactions.*

In a few instances group activation is coupled to cleavage of ATP at C-5' (Fig. 7-7) with formation of bound **tripolyphosphate** (PPP_i), which in turn is hydrolyzed to P_i and PP_i and ultimately to *three* molecules of P_i. An example is the formation of S-adenosylmethionine (SAM)[3] shown in Eq. 11-3.

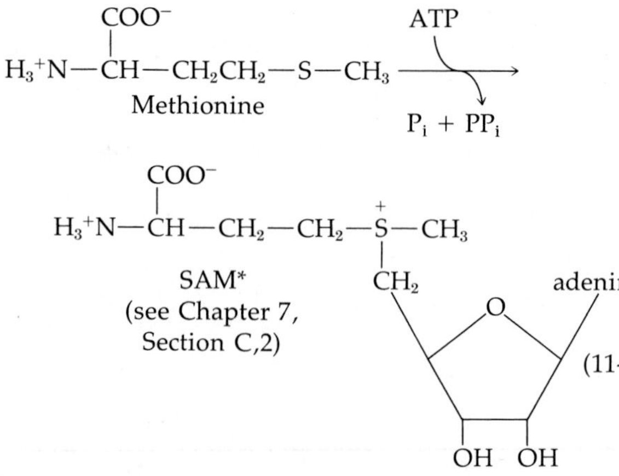

(11-3)

The reaction is a displacement on the 5'-methylene group of ATP by the sulfur atom of methionine. The initial product may be enzyme-bound PPP_i, but it is P_i and PP_i that are released from the enzyme.[3] The P_i has been established to arise from the terminal phosphorus (P_γ) of ATP.

* The bonding arrangement around the sulfur atom in sulfonium salts such as SAM is not planar. Furthermore, there is not rapid inversion as there is around a nitrogen atom. Therefore, stable enantiomorphic forms exist. Biologically formed SAM is the (−) isomer.

3. Coupling by Phosphorylation and Subsequent Cleavage by a Phosphatase

A third general method for coupling the hydrolysis of ATP to drive a synthetic sequence is to transfer the terminal phosphoryl group from ATP to a hydroxyl group *somewhere* on a substrate. Then, after the substrate has undergone a synthetic reaction, the phosphate is removed by action of a phosphatase. A good example is provided by the activation of sulfate (Eq. 11-4).[4] The overall standard free energy change for Eq. 11-4, steps *a* and *b*, is distinctly positive (+ 12 kJ mol⁻¹), and the equilibrium concentration of adenylyl sulfate formed in this group activation process is extremely low.[4] Nature's solution to this problem

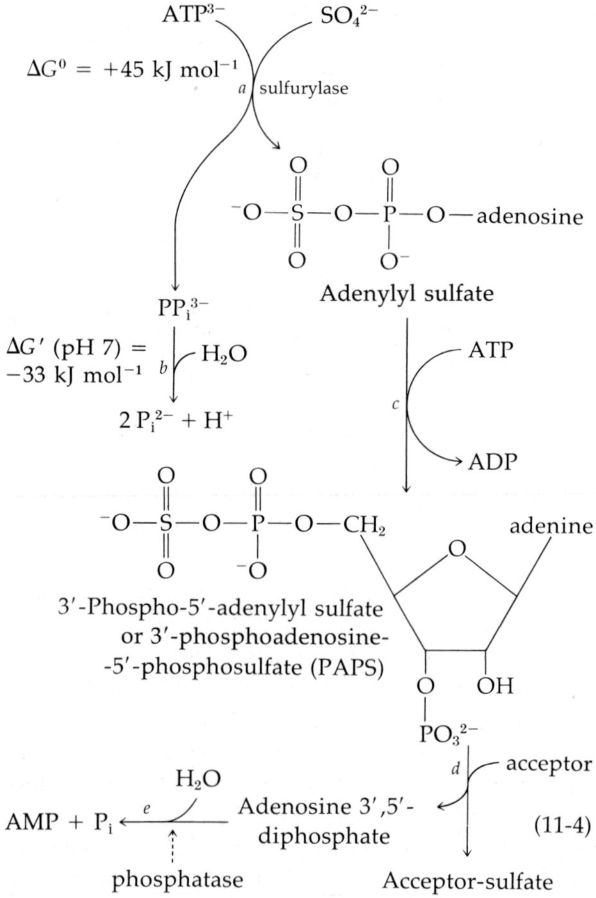

(11-4)

is to spend another molecule of ATP to phosphorylate the 3'—OH of adenosine phosphosulfate. Thus, as the latter is formed, it is promptly converted to 3'-phospho-5'-adenylyl sulfate (Eq. 11-4, step c). Since the equilibrium in Eq. 11-4, step c, lies far toward the right, the product accumulates in a substantial concentration (up to 1 mM in cell-free systems) and serves as the active sulfuryl donor in formation of sulfate esters. The reaction cycle is completed by two more reactions. In Eq. 11-4, step d, the sulfuryl group is transferred to an acceptor and in Eq. 11-4, step e, the extra phosphate group is removed from adenosine 3',5'-diphosphate by a specific phosphatase. Since the reconversion of AMP to ADP requires expenditure of still a third high energy linkage of ATP, the overall process makes use of three high energy phosphate linkages for formation of one sulfate ester.

An altogether analogous use of ATP is found in photosynthetic reduction of carbon dioxide in which ATP phosphorylates ribulose 5-P to ribulose diphosphate and the phosphate groups are removed later by phosphatase action on fructose diphosphate and sedoheptulose diphosphate (Section D,2). Phosphatases involved in synthetic pathways usually have a high substrate specificity and are to be distinguished from nonspecific phosphatases which are essentially digestive enzymes (Chapter 7, Section E,1).

4. Carboxylation and Decarboxylation; Synthesis of Fatty Acids

A fourth way in which cleavage of ATP can be coupled to biosynthesis was not recognized until a few years ago when S. J. Wakil and co-workers discovered that synthesis of fatty acids in animal cytoplasm is stimulated by carbon dioxide. However, when $^{14}CO_2$ was used in the experiment no radioactivity appeared in the fatty acids formed. Rather, it was found that acetyl-CoA was carboxylated to **malonyl-CoA** in an ATP- and biotin-requiring process (Eq. 11-5, see also Chapter 8, Section C). The carboxyl group formed in this

$$CH_3-\overset{\overset{\displaystyle O}{\|}}{C}-S-CoA \xrightarrow{\quad HCO_3^- \quad ATP \quad ADP + P_i \quad}$$

$$^-OOC-CH_2-\overset{\overset{\displaystyle O}{\|}}{C}-S-CoA \qquad (11\text{-}5)$$

activated hydrogens

group to be removed later
by decarboxylation

reaction is later converted back to CO_2 in a decarboxylation.

We know now that both an acetyl group of acetyl-CoA and a malonyl group of malonyl-CoA are transferred to the sulfur atoms of the phosphopantetheine groups of a low molecular weight **acyl carrier protein** (Chapter 8, Section B,3). The malonyl group of the malonyl acyl carrier protein (malonyl ACP) is then condensed with the acetyl ACP. The unstable β-keto acid formed (bracketed in Eq. 11-6) is presumably decarboxylated either

$$R-\overset{\overset{\displaystyle O}{\|}}{C}\diagdown_{S-ACP} + H_2C-\overset{\overset{\displaystyle COO^-}{|}}{\underset{}{C}}\diagup^{O}_{S-ACP}$$

$$\downarrow ACP-SH$$

$$\left[R-\overset{\overset{\displaystyle O}{\|}}{C}-\overset{\overset{\displaystyle COOH}{|}}{\underset{\underset{\displaystyle H}{|}}{C}}-C\diagup^{O}_{S-ACP} \right]$$

$$\downarrow CO_2$$

$$R-\overset{\overset{\displaystyle O}{\|}}{C}-CH_2-\overset{\overset{\displaystyle O}{\|}}{C}-S-ACP \qquad (11\text{-}6)$$

β-Ketoacyl ACP

by the condensing enzyme or later in the sequence. The net result of the decarboxylation is to drive the reaction to completion and in effect to link C—C bond formation to the cleavage of the ATP required for the carboxylation step.

A second function of the carboxylation–decarboxylation cycle can also be visualized. The added carboxyl in malonyl-CoA helps activate the methylene hydrogens toward dissociation as H^+ making condensation easier. Perhaps both of these reasons for synthesis of fatty acids via malonyl-CoA are important. As we have seen before, it is often difficult to state which is the most important feature that led to the evolution of a particular metabolic pathway.

Carboxylation followed by a later decarboxylation is an important pattern in other biosynthetic pathways, too. Sometimes the decarboxylation follows the carboxylation by many steps. For example, pyruvate (or PEP) is converted to uridylic acid (Eq. 11-7, details in Fig. 14-9):

$$\text{Pyruvate}$$

ATP, CO$_2$ / ADP + P$_i$

$$\text{Oxaloacetate} \longrightarrow \text{Asp} \longrightarrow \longrightarrow \longrightarrow$$

$$\text{CO}_2$$
$$\text{Orotidine 5'-P} \longrightarrow \text{5'-Uridylic acid (UMP)} \quad (11\text{-}7)$$

C. REDUCING AGENTS FOR BIOSYNTHESIS

Still another striking difference between biosynthesis of fatty acids and β oxidation (in mammals) is that the former has an absolute requirement for NADPH while the latter requires NAD^+ and flavoproteins (Fig. 11-2). This fact, together with many other observations, has led to the generalization that *biosynthetic reduction reactions usually require NADPH rather than NADH* (see Chapter 8, Section G). Many measurements have shown that in the cytosol of eukaryotic cells the ratio [NADPH]/[NADP$^+$] is high, while the ratio [NADH]/[NAD$^+$] is low. Thus, the NAD$^+$/NADH system is kept highly oxidized, in line with the role of NAD$^+$ as a principal biochemical oxidant,

while the NADP$^+$/NADPH system is kept reduced.

The use of NADPH in step g of Fig. 11-2 insures that significant amounts of the β-ketoacyl ACP derivative are reduced to the alcohol. At this point, note still another difference between β oxidation and biosynthesis. The alcohol formed in this reduction step in the biosynthetic process has the D configuration while the corresponding alcohol in β oxidation has the L configuration (Fig. 11-2).

1. Reversing an Oxidative Step with a Strong Reducing Agent

The second reduction step in biosynthesis of fatty acids in the rat liver (step i) also requires NADPH.* The corresponding step in β oxidation utilizes FAD, but NADPH is a stronger reducing agent than FADH$_2$. Therefore, use of a reduced pyridine nucleotide again provides a thermodynamic advantage in pushing the reaction in the biosynthetic direction. Interesting variations have been observed among different species. For example, fatty acid synthesis in the rat requires only NADPH, but the multienzyme complexes from *Mycobacterium phlei, Euglena gracilis,* and the yeast *Saccharomyces cerevisiae* all give much better synthesis with a mixture of NADPH and NADH than with NADPH alone.[5] Apparently, NADPH is required in step g and NADH in step i. This seems reasonable for the equilibrium in step i lies far toward the product formation, and NADH at a very low concentration could carry out the reduction.

2. Fructose for Sperm Cells

A striking example of the way in which the high [NADPH]/[NADP$^+$] ratio in cells can be used to

---

* While no bound cofactor has been identified in the *E. coli* enzyme, riboflavin 5'-phosphate is thought to be present in the yeast enzyme and is required for a maximum rate of reduction.

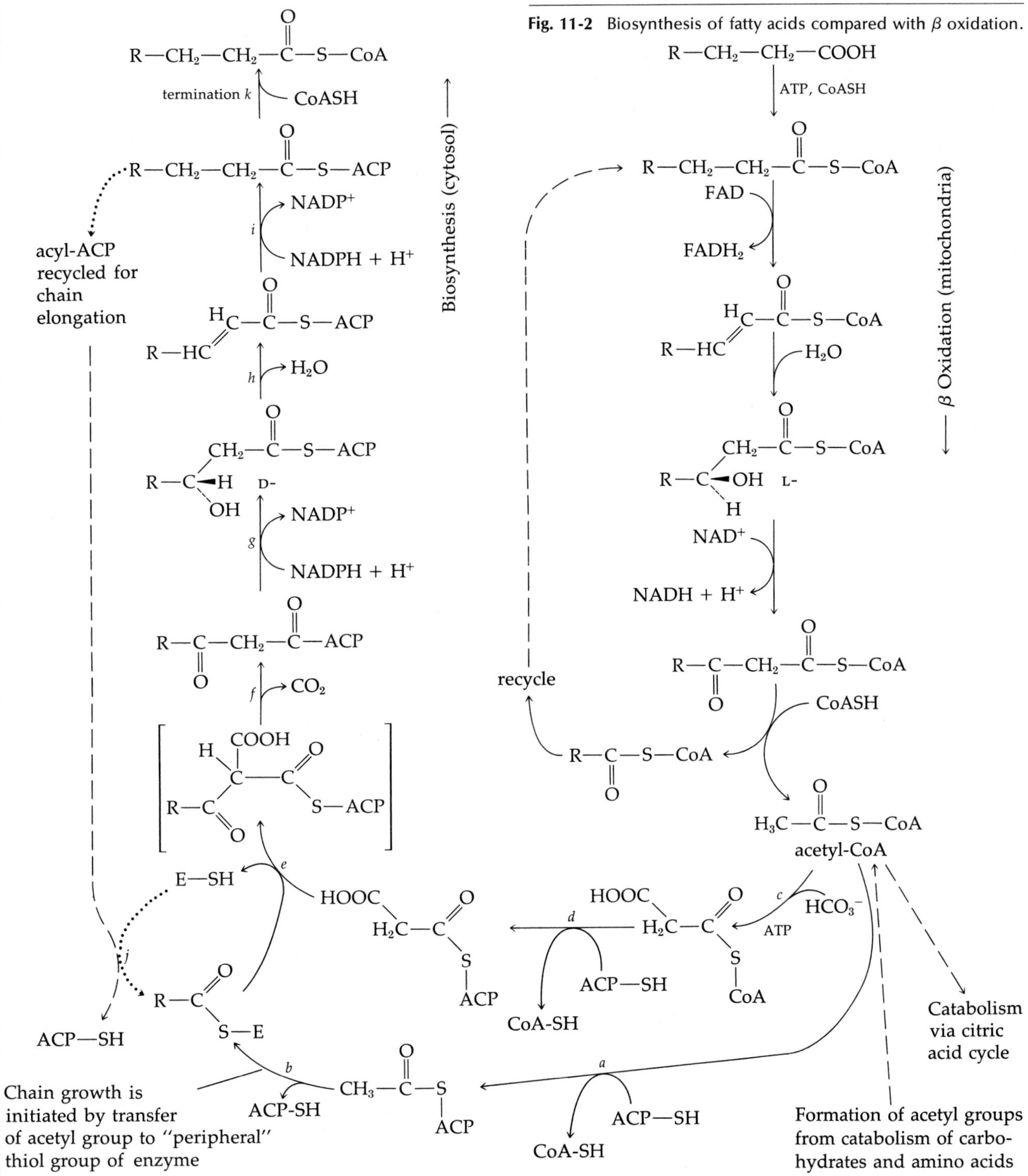

$R-CH_2-CH_2-\overset{\displaystyle O}{\overset{\|}{C}}-S-CoA$

termination k CoASH

$R-CH_2-CH_2-\overset{\displaystyle O}{\overset{\|}{C}}-S-ACP$

i NADP$^+$ / NADPH + H$^+$

acyl-ACP recycled for chain elongation

$\overset{H}{R-HC}{\displaystyle \diagup}C-\overset{\displaystyle O}{\overset{\|}{C}}-S-ACP$

h H$_2$O

$\underset{R-\overset{|}{C}\text{—H}}{\underset{\overset{|}{OH}}{}}\quad CH_2-\overset{\displaystyle O}{\overset{\|}{C}}-S-ACP$ D-

g NADP$^+$ / NADPH + H$^+$

$R-\overset{\displaystyle \|}{\underset{O}{C}}-CH_2-\overset{\displaystyle O}{\overset{\|}{C}}-ACP$

f CO$_2$

$\left[\begin{array}{c} H \quad COOH \\ R-\overset{\|}{\underset{O}{C}}-\overset{|}{C}-\overset{\displaystyle O}{\overset{\|}{C}}-S-ACP \end{array}\right]$

E—SH e

HOOC, $H_2C-\overset{\displaystyle O}{\overset{\|}{C}}-\underset{ACP}{\overset{|}{S}}$

d ACP—SH / CoA-SH

HOOC, $H_2C-\overset{\displaystyle O}{\overset{\|}{C}}-\underset{CoA}{\overset{|}{S}}$ c HCO$_3^-$ / ATP

$R-\overset{\displaystyle \|}{\underset{O}{C}}-\underset{S-E}{}$

j

ACP—SH

b ACP-SH

$CH_3-\overset{\displaystyle O}{\overset{\|}{C}}-\underset{ACP}{\overset{|}{S}}$

a ACP—SH / CoA-SH

Chain growth is initiated by transfer of acetyl group to "peripheral" thiol group of enzyme

← Biosynthesis (cytosol) →

Fig. 11-2 Biosynthesis of fatty acids compared with β oxidation.

$R-CH_2-CH_2-COOH$

ATP, CoASH

$R-CH_2-CH_2-\overset{\displaystyle O}{\overset{\|}{C}}-S-CoA$

FAD / FADH$_2$

$\overset{H}{R-HC}{\displaystyle \diagup}C-\overset{\displaystyle O}{\overset{\|}{C}}-S-CoA$

H$_2$O

$\underset{R-\overset{|}{C}\text{—OH}}{\underset{\overset{|}{H}}{}}\quad CH_2-\overset{\displaystyle O}{\overset{\|}{C}}-S-CoA$ L-

NAD$^+$ / NADH + H$^+$

$R-\overset{\displaystyle \|}{\underset{O}{C}}-CH_2-\overset{\displaystyle O}{\overset{\|}{C}}-S-CoA$

CoASH

recycle

$R-\overset{\displaystyle \|}{\underset{O}{C}}-S-CoA$

$H_3C-\overset{\displaystyle O}{\overset{\|}{C}}-S-CoA$
acetyl-CoA

→ β Oxidation (mitochondria) →

Catabolism via citric acid cycle

Formation of acetyl groups from catabolism of carbohydrates and amino acids

advantage by nature is supplied by the metabolism of sperm cells. Whereas D-glucose is the commonest sugar used as an energy source by mammalian cells, spermatozoa use principally D-fructose, a sugar that is not readily metabolized by cells of surrounding tissues.[6] Fructose, which is present in human semen at a concentration of ~12 mM, is made from glucose by cells of the seminal vesicle. The synthetic process is simple. Glucose is reduced by NADPH to the sugar alcohol D-**sorbitol,** which is in turn oxidized in the 2 position by NAD$^+$ (Eq. 11-8). The combination

CHO
|
HCOH   NADPH   NADP$^+$
|
HOCH
|
HCOH
|
HCOH
|
CH$_2$OH
D-Glucose

CH$_2$OH
|
HCOH   NAD$^+$ NADH
|
HOCH
|
HCOH
|
HCOH
|
CH$_2$OH
D-Sorbitol

CH$_2$OH
|
C=O
|
HOCH
|
HCOH   (11-8)
|
HCOH
|
CH$_2$OH
D-Fructose

of the high [NADPH]/[NADP$^+$] and high [NAD$^+$]/[NADH] ratios is sufficient to shift the equilibrium far toward fructose formation.[6a]

3. Regulation of the State of Reduction of the NAD and NADP Systems

The ratio [NAD$^+$]/[NADH] appears to be maintained at a relatively constant value and in equilibrium with a series of different reduced and oxidized substrate pairs. Thus, in the cytoplasm of rat liver cells, the dehydrogenations catalyzed by lactate dehydrogenase, α-glycerolphosphate dehydrogenase, and malate dehydrogenase are all at equilibrium with the same ratio of [NAD$^+$]/[NADH].[7] When rat liver was removed and rapidly frozen (less than 8 s)* and the concentrations of different components of the cytoplasm were determined,[8] the ratio [NAD$^+$]/[NADH] was found to be 634, while the ratio of [lactate]/[pyruvate] was 14.2. From these values an apparent equilibrium constant for reaction c of Eq. 11-9 was calculated as $K_c' = 9.0 \times 10^3$. The known equilibrium constant for

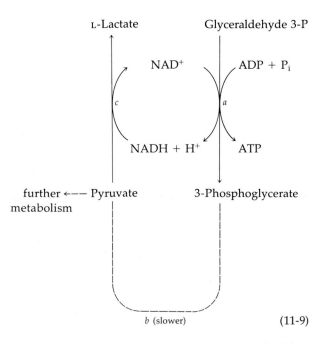

further ←–– Pyruvate   3-Phosphoglycerate
metabolism

b (slower)   (11-9)

the reaction (from *in vitro* experiments) is 8.8×10^3 (Eq. 11-10). In a similar way it can be

* The popular technique of "freeze-clamping" was used. See Section F,4.

shown that several other dehydrogenation reactions appear to be nearly at equilibrium.

$$K_c' \text{ (pH 7, 38°C)} = \frac{[\text{lactate}]}{[\text{pyruvate}]} \frac{[\text{NAD}^+]}{[\text{NADH}]}$$

$$= 8.8 \times 10^3 \qquad (11\text{-}10)$$

Now consider Eq. 11-9, step a, the ADP- and P_i-requiring oxidation of glyceraldehyde 3-phosphate (Fig. 8-13). Experimental measurements indicate that this reaction is also at equilibrium in the cytoplasm. In one series of experiments the measured phosphorylation state ratio $[\text{ATP}]/[\text{ADP}][P_i]$ was 709, while the ratio [3-phosphoglycerate]/[glyceraldehyde 3-phosphate] was 55.5. The overall equilibrium constant for Eq. 11-9a was given by Eq. 11-11. That calculated from known equilibrium constants is 60.

$$K_a' \text{ (pH 7, 38°C)} = \frac{[\text{ATP}]}{[\text{ADP}][P_1]}$$

$$\times \frac{[\text{3-phosphoglycerate}]}{[\text{glyceraldehyde phosphate}]} \frac{[\text{NADH}]}{[\text{NAD}^+]}$$

$$= 709 \times 55.5 \times 1/634 = 62 \quad (11\text{-}11)$$

From these data Krebs and Veech concluded that the oxidation state of the NAD system is determined largely by the phosphorylation state ratio of the adenine nucleotide system.[9] If the ATP level is high the equilibrium in Eq. 11-9a will tend to be reached at a higher $[\text{NAD}^+]/[\text{NADH}]$ ratio and lactate may be oxidized to pyruvate to adjust the [lactate]/[pyruvate] ratio.

It is important not to confuse the reactions of Eq. 11-9 as they occur in an aerobic cell with the tightly coupled pair of redox reactions in the homolactate fermentation (Chapter 9, Section F,1a). What Krebs and Veech stated is that the reactions of steps a and c of Eq. 11-9 are essentially at equilibrium under most conditions. On the other hand, the reaction of step b may be relatively slow. Furthermore, pyruvate is utilized in many other metabolic pathways and ATP is hydrolyzed and converted to ADP through innumerable processes taking place within the cell. Reduced NAD does not cycle between the two enzymes in a stoichiometric way and the "reducing equivalents" of NADH formed are, in large measure, transferred

to the mitochondria. The proper view of the reactions of Eq. 11-9 is that the redox pairs represent a kind of **redox buffer system** that poises the NAD^+/NADH couple at a ratio appropriate for its metabolic function.

Somewhat surprisingly, within the mitochondria the ratio $[\text{NAD}^+]/[\text{NADH}]$ is ~100 times lower than in the cytoplasm. Even though mitochondria are the site of oxidation of NADH to NAD^+ the intense catabolic activity occurring in the β-oxidation pathway and the citric acid cycle ensure extremely rapid production of NADH. Furthermore, the reduction state of NAD is apparently buffered by the low potential (Chapter 10, Section E,7) β-hydroxybutyrate-acetoacetate couple. Mitochondrial pyridine nucleotides also appear to be at equilibrium with glutamate dehydrogenase.[9]

How is the cytoplasmic $[\text{NADPH}]/[\text{NADP}^+]$ ratio set at a value higher than that of $[\text{NADH}]/[\text{NAD}^+]$? One reaction that has been considered is **transhydrogenation** (Eq. 11-12):

$$\text{NADH} + \text{NADP}^+ \longrightarrow \text{NAD}^+ + \text{NADPH} \quad (11\text{-}12)$$

There are soluble enzymes that catalyze this reaction for which K equals ≈ 1. In mitochondria an energy-linked system (Chapter 10, Section E,10) involving the membrane promotes the reaction and shifts the equilibrium to favor NADPH. However, this system is not believed to be of significance in the cytoplasm.

Another obvious way to raise the $[\text{NADPH}]/[\text{NADP}^+]$ ratio would be to convert NADH to NADPH by phosphoryl transfer from ATP. After utilization of NADPH the resulting NADP^+ could be hydrolyzed by a phosphatase to NAD^+ and reconverted to NADH by any of a large group of catabolic reactions. The process would almost exactly parallel the phosphorylation–dephosphorylation sequence used in coupling ATP cleavage to activation of sulfate (Eq. 11-4). Although the idea is attractive, there does not appear to be any evidence that it is actually used by organisms.

Within the human body the primary method of coupling ATP cleavage to reduction of NADP^+ is via carboxylation followed by eventual decarboxylation. One cycle that accomplishes this is given in Eq. 11-13. The first step (Eq. 11-13, step *a*) is

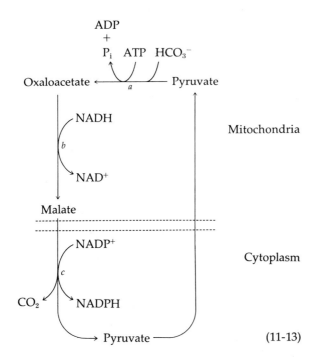

(11-13)

ATP-dependent carboxylation of pyruvate to oxaloacetate, a reaction that occurs actively within mitochondria (Eq. 8-2). Oxaloacetate can be reduced by malate dehydrogenase using NADH (Eq. 11-13, step *b*), and the resulting malate can be exported from the mitochondria. In the cytoplasm the malate is oxidized to pyruvate, with decarboxylation, by the malic enzyme (Chapter 9, Section B,5). The malic enzyme (Eq. 11-13, step *c*) is specific for $NADP^+$, is very active, and also appears to operate at or near equilibrium within the cytoplasm. On this basis, using known equilibrium constants, it is easy to show that the ratio $[NADP^+]/[NADPH]$ will be $\sim 10^5$ times lower at equilibrium than the ratio $[NAD^+]/[NADH]$.[9,10]

Note that since NADPH is continuously used in biosynthetic reactions, and is thereby reconverted to $NADP^+$, the cycle of Eq. 11-13 must operate continuously. As in Eq. 11-9, a true equilibrium does not exist but steps *b* and *c* are essentially at equilibrium. These equilibria, together with those

of Eq. 11-9 for the NAD system, ensure the correct redox potential of both pyridine nucleotide coenzymes in the cytoplasm.

One more aspect of the reaction of Eq. 11-13 should be mentioned. Malate is not the only form in which C_4 compounds are exported from mitochondria. Much oxaloacetate is combined with acetyl-CoA to form citrate; the latter leaves the mitochondria and is cleaved by the citrate-cleaving enzyme (Eq. 7-70). This, in effect, exports both acetyl-CoA (needed for lipid synthesis) and oxaloacetate which is reduced to malate within the cytoplasm. Alternatively, oxaloacetate may be transaminated to aspartate. The aspartate, after leaving the mitochondria, may be converted in another transamination reaction back to oxaloacetate. All of these are part of the nonequilibrium process by which C_4 compounds diffuse out of the mitochondria before completing the reaction sequence of Eq. 11-13 and entering into other metabolic processes. Note that the reaction of Eq. 11-13 leads to the *export* of reducing equivalents from mitochondria, the opposite of the process catalyzed by the malate-aspartate shuttle (Fig. 10-13). The two processes are presumably active under different conditions.

While the difference in the redox potential of the two pyridine nucleotide systems is clear cut in mammalian tissues, in *E. coli* the apparent potentials of the two systems are much more nearly the same.[11]

4. Reduced Ferredoxin in Reductive Biosynthesis

Both the NAD and NADP systems have standard electrode potentials $E^{0'}$ (pH 7) of -0.32 V. As we have seen, the NAD system operates at a somewhat less negative potential (-0.24 V) and the NADP system at a more negative potential (-0.38 V). In green plants and in some bacteria a more powerful reducing agent is available in the form of reduced ferredoxin. The value of $E^{0'}$ (pH 7) for clostridial ferredoxin is -0.41 V, corresponding to a free energy change for the two-electron reduction of a substrate ~ 18 kJ mol^{-1} more negative than the corresponding value of $\Delta G'$ for reduction by NADPH. Using reduced ferredoxin (Fd) some photosynthetic bacteria and anaerobic bacteria are able to carry out reductions that are virtually impossible with the pyridine nucleotide system. For example, pyruvate and α-ketoglutarate can be formed from acetyl-CoA (Chapter 8, Section J,3) and succinyl-CoA, respectively (Eq. 11-14).[12,13] In our bodies the reaction of

$$\text{Acetyl-CoA} \xrightarrow[\text{Fd}_{red} \quad \text{Fd}_{ox}]{\text{CO}_2} \text{Pyruvate}^- + \text{CoA} \qquad (11\text{-}14)$$

$$\Delta G' \text{ (pH 7)} = +17 \text{ kJ mol}^{-1}$$

Eq. 11-14, with NAD$^+$ as the oxidant goes only in the opposite direction and is essentially irreversible.

D. ASSEMBLING THE MONOMER UNITS

We will now look at the general problem of constructing the monomers from which biopolymers are made.

1. Carbonyl Groups in Chain Formation and Cleavage

Except for some vitamin B$_{12}$-dependent reactions, carbon–carbon bonds are rarely broken without the participation of a carbonyl group. Likewise, carbonyl groups have a central mechanistic role in biosynthesis. The activation of hydrogen atoms β to carbonyl groups permits β condensations to occur during biosynthesis. Condensations of carbonyl compounds with thiamine diphosphate make possible β condensations. Aldol condensations require the participation of two carbonyl compounds.

Because of the importance of carbonyl groups to the mechanisms of condensation reactions, we find that much of the assembly of either straight

chain or branched carbon skeletons takes place between compounds in which the average oxidation state of the carbon atoms is similar to that in carbohydrates (or in formaldehyde, H$_2$CO). The diversity of chemical reactions possible with compounds at this state of oxidation is a maximum, a fact that may explain why carbohydrates and closely related substances are major biosynthetic precursors and why the average state of oxidation of the carbon in most living things is close to that in carbohydrates.[14,*]

In Fig. 11-3 several biochemicals have been arranged according to the oxidation state of carbon. We see that most of the important biosynthetic intermediates lie within ± 2 electrons per carbon atom of the oxidation state of carbohydrates. As the chain length grows, they tend to fall even closer. It is extremely difficult to move through enzymatic processes between 2C, 3C, and 4C compounds (i.e., vertically in Fig. 11-3) except at the oxidation level of carbohydrates or somewhat to its right, at a slightly higher oxidation level. On the other hand, it is often possible to move horizontally with ease using redox reactions. Thus, fatty acids are assembled from acetate units, which lie at the same oxidation state as carbohydrates, and after assembly, are reduced.

Note another fact from Fig. 11-3. Among compounds of the same overall oxidation state, e.g., acetic acid and sugars, the oxidation states of individual carbon atoms can be quite different. Thus, in a sugar every carbon atom can be regarded as immediately derived from formaldehyde, but in acetic acid one end has been oxidized to a carboxyl group and the other has been reduced to a methyl group. Such internal oxidation–reduction reactions (Chapter 7, Section I,6) play an important role in the chemical manipulations necessary to assemble the carbon skeletons needed by a cell. It may be valuable for the student to arrange other compounds in the same manner as in Fig. 11-3 and to observe how the overall oxidation state of

---

* This fact may also be related to the presumed occurrence of formaldehyde as a principal component of the earth's atmosphere in the past and to the ability of formaldehyde to condense spontaneously to form carbohydrates.

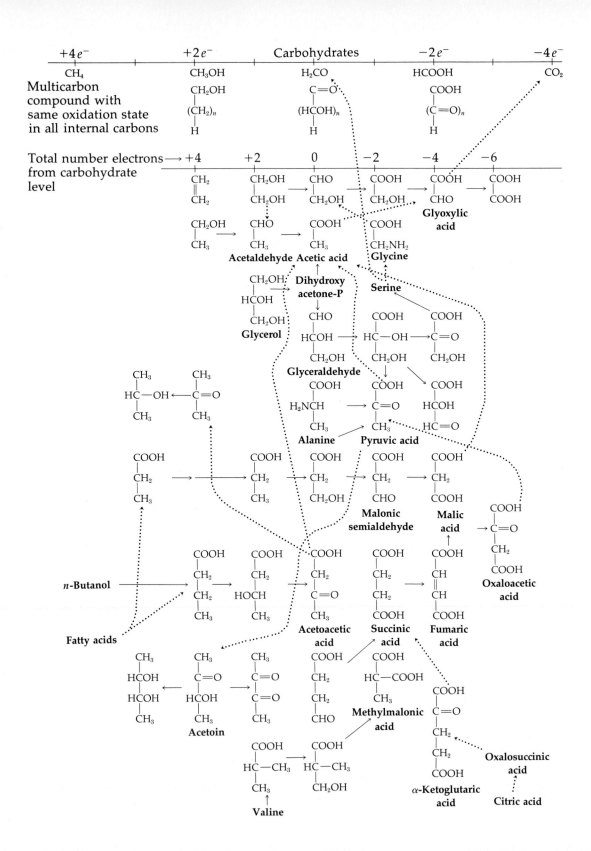

the compound changes as it moves along a metabolic pathway.

As previously mentioned, the loss of carboxyl groups as CO_2 (decarboxylation) is a feature of many biosynthetic routes. Referring again to Fig. 11-3, it may be of some interest to note that many of the biosynthetic intermediates such as pyruvate, ketoglutarate, and oxaloacetate are more oxidized than the carbohydrate level. However, their decarboxylation products, which become incorporated into the compounds being synthesized, are closer to the oxidation level of carbohydrates.

The next sections will deal with specific routes of biosynthesis of 3-phosphoglycerate, a three-carbon compound from which all other biochemical substances can be synthesized.

2. Starting with CO_2

Cells equipped with the proper enzymes and an adequate ratio of reduced to oxidized ferredoxin can use the reaction of Eq. 11-14 to incorporate CO_2 into pyruvate. Succinyl-CoA can react with CO_2 in the same type of reaction to form α-ketoglutarate (Chapter 8, Section J,3). This provides for reversal of the only irreversible step in the citric acid cycle. Using these reactions photosynthetic bacteria and some anaerobes carry out a **reductive tricarboxylic acid cycle.** Together with Eq. 11-14 the cycle provides for the complete synthesis of pyruvate from CO_2.

A quantitatively much more important pathway of CO_2 fixation is the **reductive pentose phosphate pathway,** often known as the **Calvin cycle** (Box 11-A). This sequence of reactions takes place in the chloroplasts of green plants as well as in chemiautotrophic bacteria. The Calvin cycle is essentially a way of reversing the oxidative pentose phosphate pathway (Fig. 9-8). The latter accomplishes the complete oxidation of glucose by $NADP^+$ (with one ATP required for initial conversion of glucose to glucose 6-phosphate, Eq. 11-15):

Fig. 11-3 Several biochemicals arranged according to the average oxidation state of the carbon atoms.

$$Glucose + ATP^{4-} + 7\ H_2O + 12\ NADP^+ \longrightarrow$$
$$6\ CO_2 + 12\ NADPH + ADP^{3-}$$
$$+ HPO_4^{2-} + 13\ H^+ \quad (11\text{-}15)$$

$$\Delta G'\ (pH\ 7) = -278\ kJ\ mol^{-1}$$

It would be almost impossible for a green plant to fix CO_2 using photochemically generated NADPH by an exact reversal of Eq. 11-15, because of the high positive free energy change. To meet this thermodynamic problem the reductive pentose phosphate pathway has been modified in a way that couples ATP cleavage to the synthesis.

The **reductive carboxylation** system is shown within the dashed box of Fig. 11-4. Ribulose 5-phosphate is the starting compound and in the first step one molecule of ATP is expended to form **ribulose 1,5-diphosphate.** The latter is carboxylated and cleaved to two molecules of 3-phosphoglycerate (the reaction has been discussed in Chapter 7, Section J,3,g). The reductive step (step c) of the system employs NADPH together with ATP. Except for the use of the NADP system instead of the NAD system it is exactly the reverse of the phosphoglyceraldehyde dehydrogenase of glycolysis (Chapter 8, Section H,5). Looking at the first three steps of Fig. 11-4 it is clear that in the reductive pentose phosphate pathway three molecules of ATP are utilized for each CO_2 incorporated. On the other hand, in the oxidative direction *no* ATP is generated by the operation of the pathway. Thus, the difference between catabolic and biosynthetic pathways is again one of coupling the hydrolysis of ATP to reverse an otherwise irreversible sequence.

The reactions enclosed within the dashed box of Fig. 11-4 do not give the whole story about the coupling mechanism. A phosphoryl group was transferred from ATP in step a and to complete the hydrolysis it must be removed in some future step. This is indicated in a general way in Fig. 11-4 by the reaction steps d, e, and f. Step f represents the action of specific phosphatases that remove phosphoryl groups from the seven-carbon sedoheptulose diphosphate and from fructose diphosphate. In either case the resulting ketose monophosphate reacts with an aldose (via transketo-

BOX 11-A

¹⁴C AND THE CALVIN CYCLE

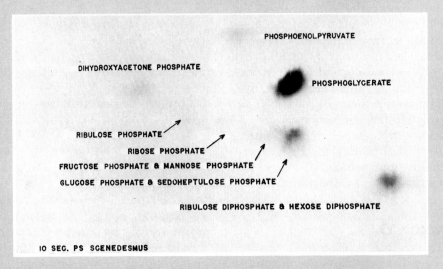

Chromatogram of extract of the alga *Scenedesmus* after photosynthesis in the presence of $^{14}CO_2$ for 10 s. Courtesy of J. A. Bassham.

The chemical nature of photosynthesis has intrigued chemists for centuries but little was learned about the details until radioactive ^{14}C became available. Discovered in 1940 by Samuel Rubin and Martin Kamen, the isotope was available in quantity by 1945 as a product of nuclear reactors. In that year, Melvin Calvin and associates began their studies that elucidated the mechanism of incorporation of CO_2 into organic materials, which led to a Nobel Prize for Calvin in 1961.

A key development was two-dimensional paper chromatography with radioautography (Box 2-C). A suspension of the alga *Chlorella* (Fig. 1-9) was allowed to photosynthesize in a stream of ordinary CO_2 in the light. At a certain time, a portion of $H^{14}CO_3$ was injected into the system, and after a few seconds of photosynthesis with ^{14}C present the suspension of algae was run into hot methanol to denature proteins and to stop the reaction. The soluble materials extracted from the algal cells were concentrated and chromatographed; radioautographs were then prepared. It was found that after 10 s of photosynthesis in the presence of $^{14}CO_2$, the algae contained a dozen or more ^{14}C labeled compounds. These included malic acid, aspartic acid, PEP, alanine, triose phosphates, and other sugar phosphates and diphosphates. However, during the first five seconds a single compound, 3-phosphoglycerate, contained most of the radioactivity.[a] This finding suggested that a two-carbon regenerating substrate might be carboxylated by $^{14}CO_2$ to phosphoglycerate. Search for this two-carbon compound was unsuccessful, but Andrew Benson, in Calvin's laboratory, soon identified ribulose diphosphate,[b] which kinetic studies proved to be the true regenerating substrate.[c,d] Its carboxylation and cleavage[e] represent the first step in what has come to be known as the Calvin cycle (Fig. 11-4).

[a] A. A. Benson, J. A. Bassham, M. Calvin, T. C. Goodale, V. A. Haas and W. Stepka, *JACS* **72**, 1710–1718 (1950).

[b] A. A. Benson, *JACS* **73**, 2971–2972 (1951).

[c] M. Calvin and P. Massini, *Experientia* **8**, 445–484 (1952).

[d] J. A. Bassham, A. A. Benson, L. D. Kay, A. Z. Harris, A. T. Wilson, and M. Calvin, *JACS* **76**, 1760–1770 (1954).

[e] M. Calvin and J. A. Bassham, "The Photosynthesis of Carbon Compounds." Benjamin, New York, 1962.

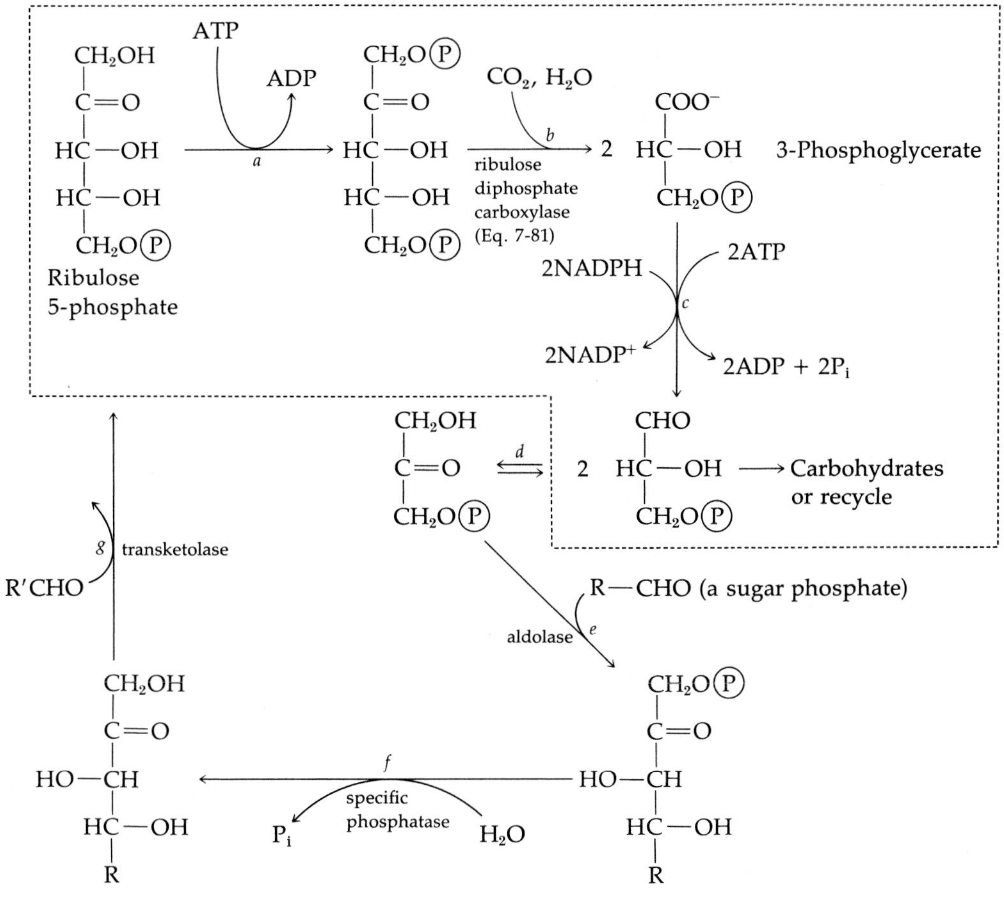

Fig. 11-4 (A) The reductive carboxylation system used in reductive pentose phosphate pathway. The essential reactions of this system are enclosed within the dashed box. Typical subsequent reactions follow. The phosphatase action completes the phosphorylation–dephosphorylation cycle. (B) The reductive pentose phosphate cycle arranged to show the combining of 3 CO_2 molecules to form one molecule of triose phosphate. Abbreviations are RCS, reductive carboxylation system (from above); A, aldolase, Pase, specific phosphatase; and TK, transketolase.

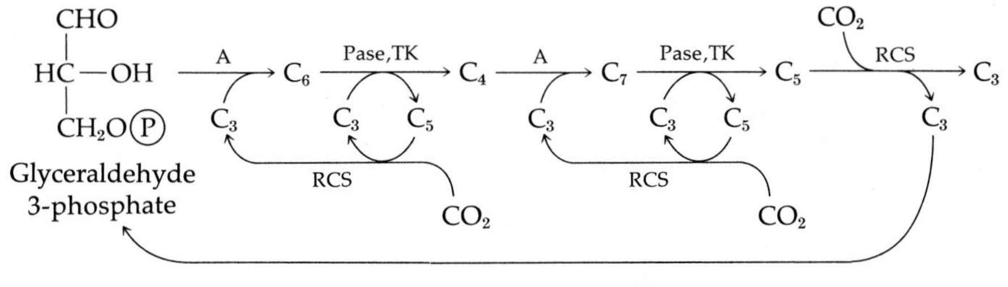

lase, step *g*) to regenerate ribulose 5-phosphate, the CO_2 acceptor. The overall reductive pentose phosphate cycle (Fig. 11-4B) is easy to understand as a reversal of the oxidative pentose phosphate pathway in which the oxidative decarboxylation system (Eq. 9-15) is replaced by the reductive carboxylation system of Fig. 11-4A. The scheme as written in Fig. 11-4B shows the incorporation of three molecules of CO_2. The reductive carboxylation system (RCS in the figure) operates three times with a net production of one molecule of triose phosphate. Of course, as with other biosynthetic cycles (p. 530) any amount of any of the intermediate metabolites may be withdrawn into various metabolic pathways without disruption of the flow through the cycle.

The overall reaction of carbon dioxide reduction in the Calvin cycle (Eq. 11-16) becomes

$$6\ CO_2 + 12\ NADPH + 18\ ATP^{4-} + 12\ H_2O \longrightarrow$$
$$glucose + 12\ NADP^+ + 18\ ADP^{3-}$$
$$+ 18\ HPO_4^{2-} + 6\ H^+ \quad (11\text{-}16)$$

The free energy change $\Delta G'$ (pH 7) is now -378 kJ mol^{-1} instead of the $+278$ kJ mol^{-1} required to reverse the reaction of Eq. 11-15.

3. Biosynthesis from Formate, Formaldehyde, or Methanol

Bacteria subsisting on one-carbon compounds are able to oxidize them to CO_2 to obtain energy, using formate dehydrogenase (Chapter 9, Section C,3) in the final step. They may incorporate CO_2 for biosynthetic purposes via the Calvin cycle. However, other routes of assimilation of the one-carbon compound are found in some species. Thus, a pseudomonad investigated by Quayle and associates[15] converts C_1 compounds to acetate via tetrahydrofolic acid-bound intermediates and CO_2 via the "serine pathway" shown in Fig. 11-5. This is a cyclic process for converting one molecule of formaldehyde (THF-bound) plus one of CO_2 into acetate. The regenerating substrate is **glyoxylate.**

H
|
O=C—COO⁻

Glyoxylate

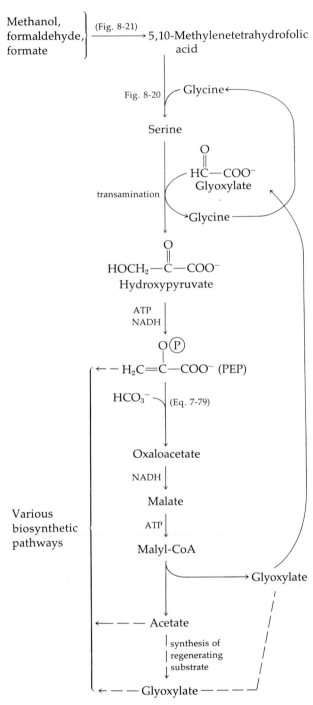

Fig. 11-5 The serine pathway for assimilation of one-carbon compounds.

Before condensation with the "active formalde-hyde" from methylene THF (Fig. 11-5) the glyoxylate undergoes transamination to glycine. The glycine plus formaldehyde forms serine which is then transaminated to hydroxypyruvate. Note that glyoxylate plus formaldehyde could have been joined in a thiamine-dependent α condensation but that, as in the γ-aminobutyrate shunt (Fig. 9-4), the two coupled transamination steps permits use of a pyridoxalphosphate-dependent C—C bond formation.

The reduction of hydroxypyruvate to phosphoenolpyruvate (Fig. 11-5) depends upon ATP and may be accomplished by reduction to 3-phosphoglycerate, isomerization to 2-phosphoglycerate, and elimination to form PEP exactly as during glycolysis (Fig. 9-7). The conversion of malate to acetate and glyoxylate via malyl-CoA (Chapter 7, Section J,2,h) forms the product acetate and regenerates glyoxylate. As with other metabolic cycles, various intermediates, e.g., PEP, can be withdrawn for biosynthesis. However, it is essential to have an independent method of synthesizing the regenerating substrate glyoxylate. This is accomplished from acetate as indicated in Fig. 11-5 by using the cyclic process discussed in the following paragraph.

4. The Glyoxylate Pathway

Recall that the reductive carboxylation of acetyl-CoA to pyruvate (Eq. 11-14) occurs only in a few bacteria. For most species, from microorganisms to animals, the irreversibility of the oxidative decarboxylation of pyruvate to acetyl-CoA is a fact with many important consequences. Thus, carbohydrate is readily converted to fat; because of the irreversibility of this process excess calories lead to the deposition of fats. However, fat cannot be used to generate most of the biosynthetic intermediates needed for formation of carbohydrates and proteins because those intermediates originate largely from C_3 units.

This limitation on the conversion of C_2 acetyl units to C_3 metabolites is overcome in many organisms, including E. coli, by the **glyoxylate pathway.** This reaction sequence converts *two* acetyl units into *one* C_3 unit with decarboxylation of the fourth carbon atom. The pathway provides a way for many organisms (including E. coli and Tetrahymena) to subsist on acetate as a sole or major carbon source. The glyoxylate pathway is especially prominent in plants that store large amounts of fat in their seeds (**oil seeds**). In the germinating oil seed it permits fat to be converted rapidly to sucrose, cellulose, and other carbohydrates needed for growth.

The first part of the glyoxylate pathway can be thought of as a modified citric acid cycle designed to oxidize acetate to glyoxylate (Fig. 11-6, upper left). This **acetyl-CoA–glyoxylate** cycle also serves to generate glyoxylate for the serine cycle shown in Fig. 11-5. The regenerating substrate of the acetyl-CoA–glyoxylate cycle is oxaloacetate which, in effect, serves as the oxidizing agent for conversion of the methyl group of acetyl-CoA to an aldehyde group. This four-electron oxidation occurs via the dehydration–rehydration mechanism of aconitase. Isocitrate is cleaved by **isocitrate lyase** (Eq. 7-73) to form glyoxylate. The other product, succinate, is reoxidized to regenerate oxaloacetate in a four-electron β oxidation which completes the cycle.

The independent pathway for synthesis of the regenerating substrate is condensation of glyoxylate with acetyl-CoA (malate synthetase, Fig. 7-11) to form malate and oxidation of the latter to oxaloacetate (Fig. 11-6). The same sequence completes the glyoxylate pathway by providing oxaloacetate for synthesis of carbohydrates and other substances.

The glyoxylate pathway is often drawn as a kind of internal cycle within the citric acid cycle. However, the scheme of Fig. 11-6 makes the stoichiometry clearer. It is likely that in bacteria there is no spatial separation of the citric acid cycle and glyoxylate pathways.* However, in some plants some of the enzymes needed for the glyoxylate

* It is interesting that the genes coding for the special enzymes of the glyoxylate pathway are found together in a cluster in the bacterial chromosome.

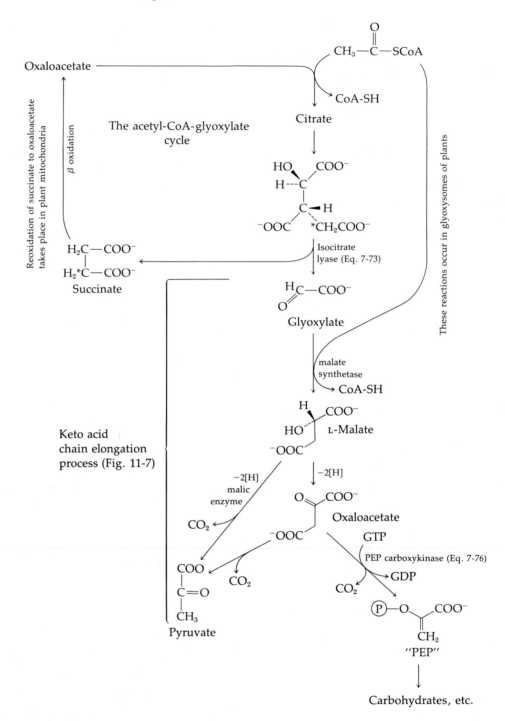

Fig. 11-6 The glyoxylate pathway. Asterisks trace pathway of labeled carbon from acetate.

pathway are localized in the glyoxysomes and the rest in the mitochondria (Fig. 11-6).

Malate and oxaloacetate, the immediate products of the glyoxylate pathway can be converted to pyruvate and phosphoenolpyruvate (PEP) as indicated in Fig. 11-6. Pyruvate is needed for synthesis of the pyruvate family of compounds while PEP can be converted to all the other intermediates required for biosynthesis (Fig. 11-1).

An interesting modification of the glyoxylate pathway is used by *Micrococcus denitrificans*.[16] One molecule of glyoxylate is transaminated to glycine (Eq. 11-17). Then, using a PLP-dependent reaction this glycine is condensed with a second glyoxylate to form β-hydroxyaspartate; elimination of ammonia and tautomerization of the resulting enol produces oxaloacetate (Eq. 11-17).

$$OHC—COO^-$$

transamination

$$H_3{}^+N—H_2C—COO^-$$

OHC—COO⁻

PLP

$$H_3{}^+N \quad H$$
$$C—COO^-$$
$$^-OOC—C \quad OH$$
$$H$$

erythro-β-Hydroxyaspartate

→ NH₃

$$\left[\begin{array}{c} H \\ C—COO^- \\ ^-OOC—C \\ OH \end{array} \right]$$

Acetyl-CoA ⟵ ⟵ Oxaloacetate

Carbohydrates   (11-17)

5. Biosynthesis of Glucose (Glucogenesis)

In preceding sections we have examined routes of biosynthesis of the three-carbon precursors of carbohydrates. Triose phosphates arise from the reductive pentose phosphate pathway (Fig. 11-4B). The reductive tricarboxylic acid cycle and glyoxylate pathways yield oxaloacetate which is readily converted to phosphoenolpyruvate. Now let us consider the further conversion of PEP and the triose phosphates to **glucose 1-phosphate,** the key intermediate in biosynthesis of the entire family of sugars and polysaccharides.

The conversion of PEP to glucose 1-P represents a reversal of part of the glycolysis sequence. It will be convenient to discuss this along with the reversal of the complete glycolysis sequence from lactic acid. The latter, which is called **gluconeogenesis,** is an essential part of the Cori cycle (Chapter 9, Section F). The same process may be used by the body to convert pyruvate derived from deamination of alanine or serine (Chapter 14) into carbohydrates.

Just as with the pentose phosphate cycle, an exact reversal of the glycolysis sequence (Eq. 11-18) is precluded on thermodynamic grounds.

$$2 \text{ Lactate}^- + 3 \text{ ATP}^{4-} + 2 \text{ H}_2\text{O} \longrightarrow \text{C}_6\text{H}_{10}\text{O}_5(\text{glycogen})$$
$$+ 3 \text{ ADP}^{3-} + 3 \text{ HPO}_4{}^{2-} + \text{H}^+ \quad (11\text{-}18)$$

$$\Delta G' \text{ (pH 7)} = + 107 \text{ kJ}$$

Even at very high values of R_p the reaction would be unlikely to go to completion. The actual pathways used for gluconeogenesis (Eq. 11-19, dashed lines) appear to differ from those of glycolysis (Eq. 11-19, solid lines) in three significant ways: (1) While glycogen breakdown is initiated by reaction of inorganic phosphate catalyzed by phosphorylase (Eq. 11-19, step a) the biosynthetic sequence from glucose 1-phosphate, via uridine diphosphate glucose (Eq. 11-19, step b, see also p. 660), is coupled to cleavage of a molecule of ATP. (2) In the catabolic process (glycolysis) fructose 6-P is converted to fructose 1,6-diphosphate through

Glycogen ← UDP

P_i ——— Glycogen ←————— → UDP   ATP

Phosphorylase a b UDPglucose

→ Glucose 1-P b → PP_i → 2 P_i

ADP

Glucose 6-P   ——— UTP

Fructose 6-P ←————————— P_i

ATP c ╲kinase d ╲phosphatase

ADP ←   → Fructose 1,6-diphosphate H_2O

↓

further degradation   (11-19)

the action of a special kinase (Eq. 11-19, step c). Fructose-diP is then cleaved by aldolase and the resulting triose phosphate is degraded further. In glucogenesis a phosphatase is used to form fructose-P from fructose-diP (Eq. 11-19, step d). (3) During glycolysis phosphoenolpyruvate is converted to pyruvate by a kinase with generation of ATP (Fig. 9-7, reaction 10). During glucogenesis

$$CH_3-\overset{O}{\overset{\|}{C}}-COO^- \xrightarrow{\quad HCO_3^- \quad}$$

ATP   ADP + P_i

pyruvate
carboxylase (Eq. 8-2)

$$^-OOC-CH_2-\overset{O}{\overset{\|}{C}}-COO^- \xrightarrow{\quad GTP \quad GDP \quad}$$

Oxaloacetate CO_2

PEP carboxykinase (Eq. 7-7b)

$$CH_2=\overset{O-\textcircled{P}}{\overset{\|}{C}}-COO^- \quad (11-20)$$

Phosphoenolpyruvate

* However, the usually accepted notion that no reversal of pyruvate kinase action occurs during gluconeogenesis has been challenged.[17]

pyruvate is converted to phosphoenolpyruvate indirectly via oxaloacetate (Eq. 11-20).* This is another example of the coupling of ATP cleavage through a carboxylation–decarboxylation sequence. The net effect is to use *two* molecules of ATP (actually one ATP and one GTP) rather than *one* to convert the pyruvate to PEP.

The overall reaction for reversal of glycolysis now has a comfortably negative standard free energy change (Eq. 11-21).

$$2 \text{ Lactate}^- + 7 \text{ ATP}^{4-} + 6 \text{ H}_2\text{O} \longrightarrow \text{glycogen} \\ + 7 \text{ ADP}^{3-} + 7 \text{ HPO}_4^{2-} + 5 \text{ H}^+ \quad (11\text{-}21)$$

$\Delta G'$ (pH 7) $= -31$ kJ mol^{-1}

Two enzymes that are able to convert pyruvate directly to PEP are found in bacteria and plants. In each case, as in the animal enzyme system discussed in the preceding paragraph, the conversion involves expenditure of two "high energy linkages" of ATP. The **PEP synthetase** of *E. coli* apparently acts by first transferring a pyrophosphoryl group from ATP to some group Y in the enzyme (Eq. 11-22). A phosphate group is hydrolyzed from this intermediate (dashed line in Eq. 11-22, step b) ensuring that intermediate E—Y—P is present in high amount. The latter reacts with

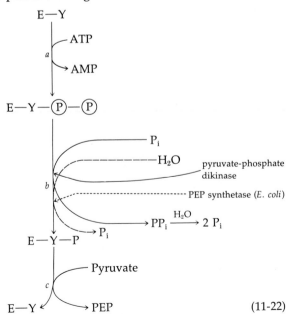

(11-22)

pyruvate to form PEP.[18] A similar enzyme is **pyruvate-phosphate dikinase,** first identified in tropical grasses and now known to play an important role in the CO_2 concentrating system of the so-called "C_4 plants" (Chapter 13, Section E,9).[19-22] The same enzyme participates in gluconeogenesis in *Acetobacter*.[20] The proposed mechanism for this enzyme is also portrayed in Eq. 11-22. The only difference from the PEP synthetase of *E. coli* is that inorganic phosphate (rather than water) is the attacking nucleophile in Eq. 11-22*b* and inorganic pyrophosphate is a product. The latter is probably hydrolyzed by phosphatase action, the end result being that the overall reaction is the same as with PEP synthetase.

Here is a question for the student. The same overall result would also be obtained if intermediate E—Y—P of Eq. 11-22 were formed by transfer of a phosphoryl group from ATP to Y. The product would be ADP. To insure that the process went to completion this ADP would have to be hydrolyzed to AMP and P_i by a special phosphatase. Why is this process never used in nature?

6. Building Hydrocarbon Chains with Two-Carbon Units

Fatty acid chains are taken apart two-carbon atoms at a time by the β-oxidation process. Biosynthesis of fatty acids reverses this process by using the two-carbon acetyl unit of acetyl-CoA as a starting material. The coupling of ATP cleavage to this process by a carboxylation–decarboxylation sequence, the role of acyl carrier protein (Section B,4), and the use of NADPH as a reductant (Section C) have already been discussed.

The complete sequence of reactions for fatty acid biosynthesis is given in Fig. 11-2, where it is compared with the β-oxidation sequence. Why does the latter require CoA derivatives while biosynthesis requires the more complex acyl carrier protein (ACP)? The reason is probably one of control. The ACP is a complex handle able to hold the growing fatty acid chain and to guide it from one

enzyme to the next. In *E. coli* the various enzymes catalyzing the reactions of Fig. 11-2 are found in the cytosol and behave as if they were independent proteins but in *Mycobacterium*, a yeast *Euglena*, pigeon, and rat tissues, a multienzyme complex is formed. This **fatty acid synthetase** from yeast has a molecular weight of 2.3 million and its appearance under the electron microscope is reminiscent of that of the ketoacid dehydrogenase complexes (Chapter 8, Section J).

It is thought that the ACP molecule lies at the center of the complex and that the growing fatty acid chain on the end of the phosphopantetheine prosthetic group moves from one subunit to the other.[23] A "primer" (usually acetyl-CoA in *E. coli*) transfers its acyl group first to the central molecule of ACP (step *a*, Fig. 11-2) and then to a "peripheral" thiol group, probably of a cysteine side chain on a separate protein subunit (step *b*, Fig. 11-2). Next a malonyl group is transferred (step *d*) from malonyl-CoA to the free thiol group on the ACP. The condensation (step *e*) occurs with the freeing of the peripheral thiol group. However, the latter does not come into use again until the β-ketoacyl group formed has undergone the complete sequence of reduction reactions (steps *g–i*). Then the growing chain is transferred to the peripheral —SH (step *j*) and a new malonyl unit is introduced on the central ACP.

After the chain reaches a length of 12 carbon atoms, there is a tendency for the acyl group to be transferred off to a CoA molecule (step *k*) rather than to pass around the cycle again. Thus, chain growth is terminated. This tendency systematically increases as the chain grows longer. In yeast, the principal chain lengths of the fatty acids released are 14, 16, and 18 carbon atoms.[24]

Although acetyl-CoA is most often the primer (or **starter piece**), butyryl-CoA is better with enzymes from the rabbit. Butyryl-CoA arises from acetyl-CoA by a reversal of β oxidation, the necessary enzymes occurring in significant amounts in the cytosol.[25] Another group of starter pieces are used in synthesis of branched-chain fatty acids (Section 10) and in the formation of flavonol pigments (Box 12-B).

7. The Keto Acid Chain Elongation Process

Note that the conversion of glyoxylate to pyruvate in Fig. 11-6 occurs in a series of reactions that is reminiscent of the conversion of oxaloacetate to α-ketoglutarate in the citric acid cycle (Fig. 9-2). Both of these sequences are examples of a general chain elongation process for α-keto acids that is used quite frequently in biosynthesis. For example, it occurs in the formation of leucine and of lysine. The generalized reaction sequence (Fig. 11-7) takes place in four steps: (1) condensation of the α-keto acid with an acetyl group, (2) isomerization by dehydration and rehydration (aconitase in the case of the citric acid cycle), (3) dehydrogenation, and (4) β decarboxylation. In many cases steps 3 and 4 are combined as a single enzymatic reaction. Note that isomerization of the intermediate hydroxy acid (Fig. 11-7) is required because the hydroxyl group is attached to a tertiary carbon bearing no hydrogen. The hydroxyl group must be moved to the adjacent carbon atom before oxidation to a ketone can take place. However, in the case of glyoxylate, isomerization is not necessary because R = H.

It may be protested that the reaction of the citric acid cycle by which oxaloacetate is converted to ketoglutarate does not follow exactly the pattern of Fig. 11-7. The carbon dioxide removed in the decarboxylation step does not come from the part of the molecule donated by the acetyl group but from that formed from oxaloacetate. However, the end result is the same. Furthermore, there are now known to be two citrate-forming enzymes with different stereospecificities (Chapter 7, Section J,2,f), and it is possible that originally the biosynthetic pathway was strictly according to the sequence of Fig. 11-7.

At the bottom of Fig. 11-7 several stages of the α-keto acid elongation process are arranged in tandem. It is seen that glyoxylate (a product of the acetyl-CoA–glyoxylate cycle) can be built up systematically to pyruvate, oxaloacetate, α-ketoglutarate, and a precursor of lysine using this one reaction sequence.

8. Decarboxylation as a Driving Force in Biosynthesis

Consider the relationship of the following prominent biosynthetic intermediates one to another:

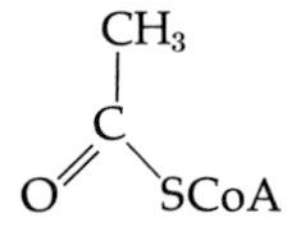

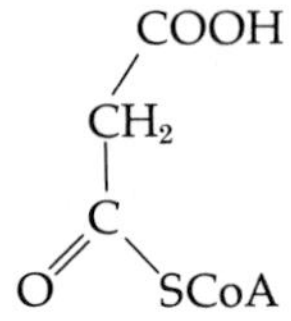

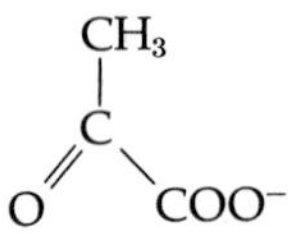

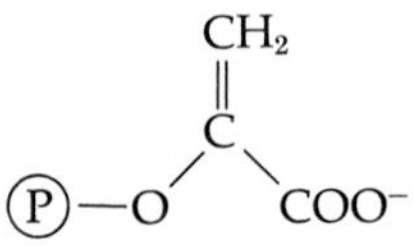

We have seen that acetyl-CoA can be utilized for the synthesis of long chain fatty acids and that this is done by carboxylation to malonyl-CoA. We can think of the malonyl group as a *β-carboxylated acetyl group*. When synthesis of a fatty acid occurs the carboxyl group is lost, and it is only the acetyl group that is ultimately incorporated into the fatty acid. In a similar way pyruvate can be thought of as an *α-carboxylated acetaldehyde* and oxaloacetate as an *α-* and *β*-dicarboxylated acetaldehyde. During the biosynthetic reactions these three- and four-carbon compounds very often undergo decarboxylation. Thus, they both can be regarded as "activated acetaldehyde units." Phosphoenolpyruvate is an *α*-carboxylated phosphoenol form of acetaldehyde and undergoes both decarboxylation and dephosphorylation before contributing a two-carbon unit to the final product.

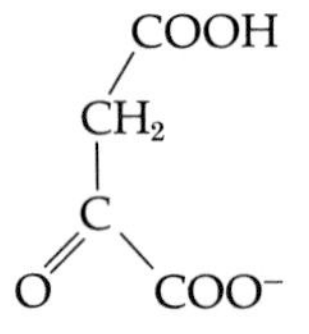

It is of interest to compare two chain elongation processes by which two-carbon units are combined. In fatty acid biosynthesis the acetyl units are reduced, after condensation, to form a straight hydrocarbon chain. In the keto acid chain elongation mechanism, the acetyl unit is introduced but is later decarboxylated. Thus, the chain is increased in length by one carbon atom at a time.

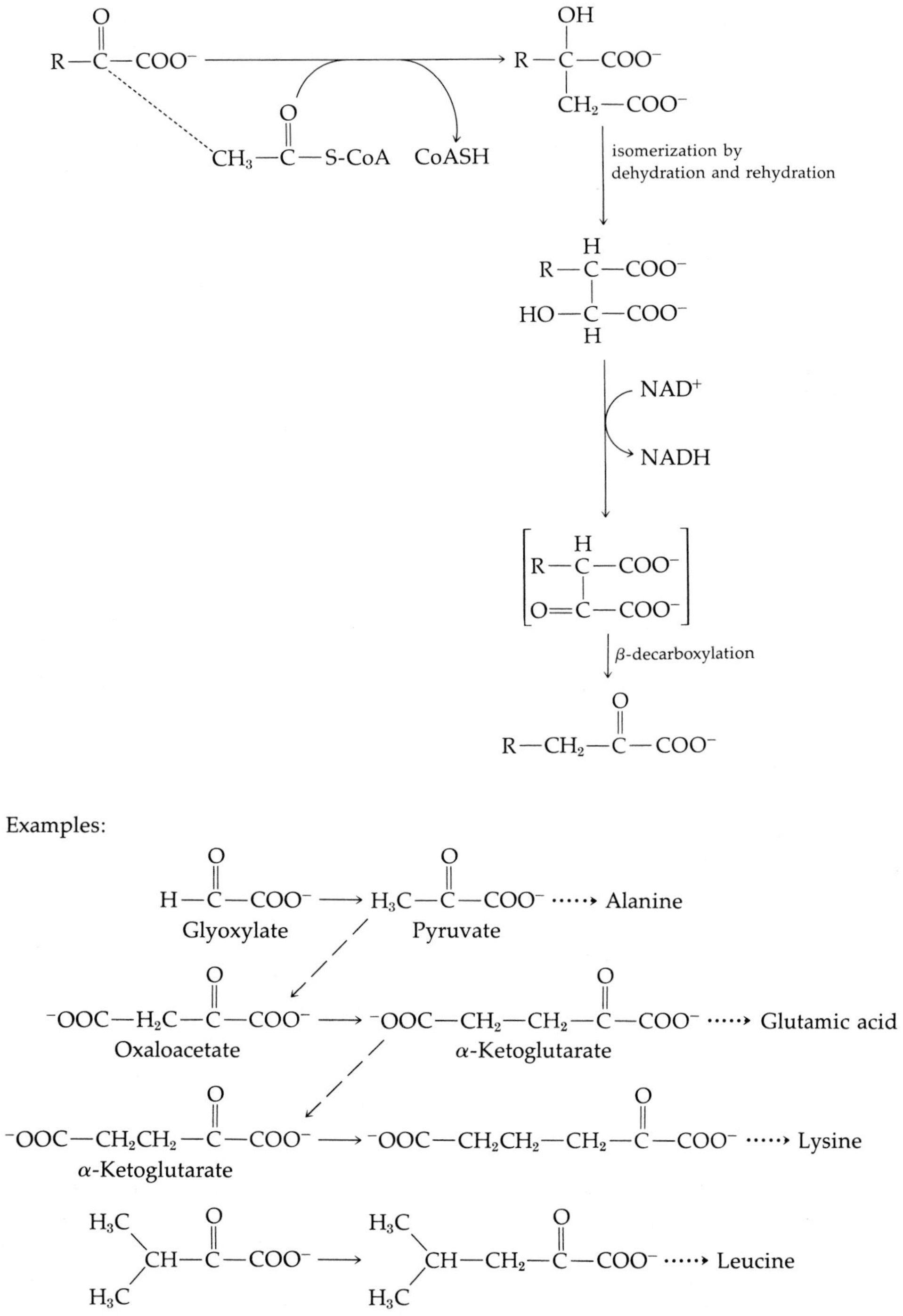

Fig. 11-7 The keto acid chain elongation process.

These two mechanisms account for a great deal of the biosynthesis of chain extension. However, other variations occur. For example, glycine (a carboxylated methylamine), under the influence of pyridoxal phosphate, is able to condense, with accompanying decarboxylation, with molecules such as succinyl-CoA (Eq. 8-20) to extend the carbon chain and at the same time to introduce an amino group. Likewise serine (a carboxylated ethanolamine) condenses with palmitoyl-CoA in biosynthesis of sphingosine (Eq. 8-21). Phosphatidylserine is decarboxylated to phosphatidylethanolamine in the final synthetic step for that phospholipid (Fig. 12-8).

9. Stabilization and Termination of Chain Growth by Ring Formation

Another synthetic principle is related to the stability of ring molecules containing five and six atoms. *Biochemical substances frequently undergo cyclization to stable ring structures.* For example, the three-carbon glyceraldehyde phosphate exists in solution primarily as the free aldehyde (in equilibrium with the covalent hydrate, Eq. 7-35). On the other hand, glucose 6-phosphate exists largely as the cyclic hemiacetal (Eq. 11-23). In this ring form

Glucose 6-P   Hemiacetal (pyranose)   Glucose 1-P

(11-23)

no carbonyl group is present and further chain elongation is inhibited.* When the hemiacetal is enzymatically isomerized to glucose 1-phosphate (Eq. 11-23, via glucose 1,6-diphosphate, Eq. 7-26)

* However, six-carbon chains *are* sometimes elongated as in formation of *N*-acetylneuraminic acid (Eq. 12-6).

the ring is firmly locked. Glucose 1-phosphate, in turn, serves as the biosynthetic precursor of polysaccharides and related compounds in all of which the sugar rings are stable. Ring formation can occur in lipid biosynthesis, too. Among the **polyketides** (Chapter 12, Section G) are many compounds in which ring formation has occurred by ester or aldol condensations followed by reduction and elimination processes. This is a typical sequence for biosynthesis of highly stable aromatic rings.

10. Formation of Branched Chains

Branched carbon skeletons are formed by standard reaction types but sometimes with addition of rearrangement steps. Compare the biosynthetic routes to three different branched five-carbon units. The first is the condensation of an *acetyl* group with a *propionyl* group (Fig. 11-8). Presumably propionyl-CoA is first carboxylated to methylmalonyl-CoA before the condensation. Decarboxylation and reduction yields an acyl-CoA derivative with a methyl group in the 2 position. The resulting branched compound (presumably bound to an acyl carrier protein) is thought to be an intermediate in the formation of branched chain fatty acids (Chapter 12, Section E).

A second five-carbon branched unit (Fig. 11-8) in which the branch is one carbon further down the chain, in the 3 position, is of extraordinary importance as an intermediate in the biosynthesis of **polyprenyl** (isoprenoid) compounds and steroids. In this case three two-carbon units are used as the starting material with decarboxylation of one unit. Two acetyl units are first condensed to form acetoacetyl-CoA. Then a third acetyl unit, which has been transferred from acetyl-CoA onto an SH group of the enzyme,[25a] is combined with the acetoacetyl-CoA through an ester condensation. The thioester linkage to the enzyme is hydrolyzed to free the product **3-hydroxy-3-methylglutaryl-CoA.** The thioester group in this compound is then reduced to an alcohol, a standard reaction process.

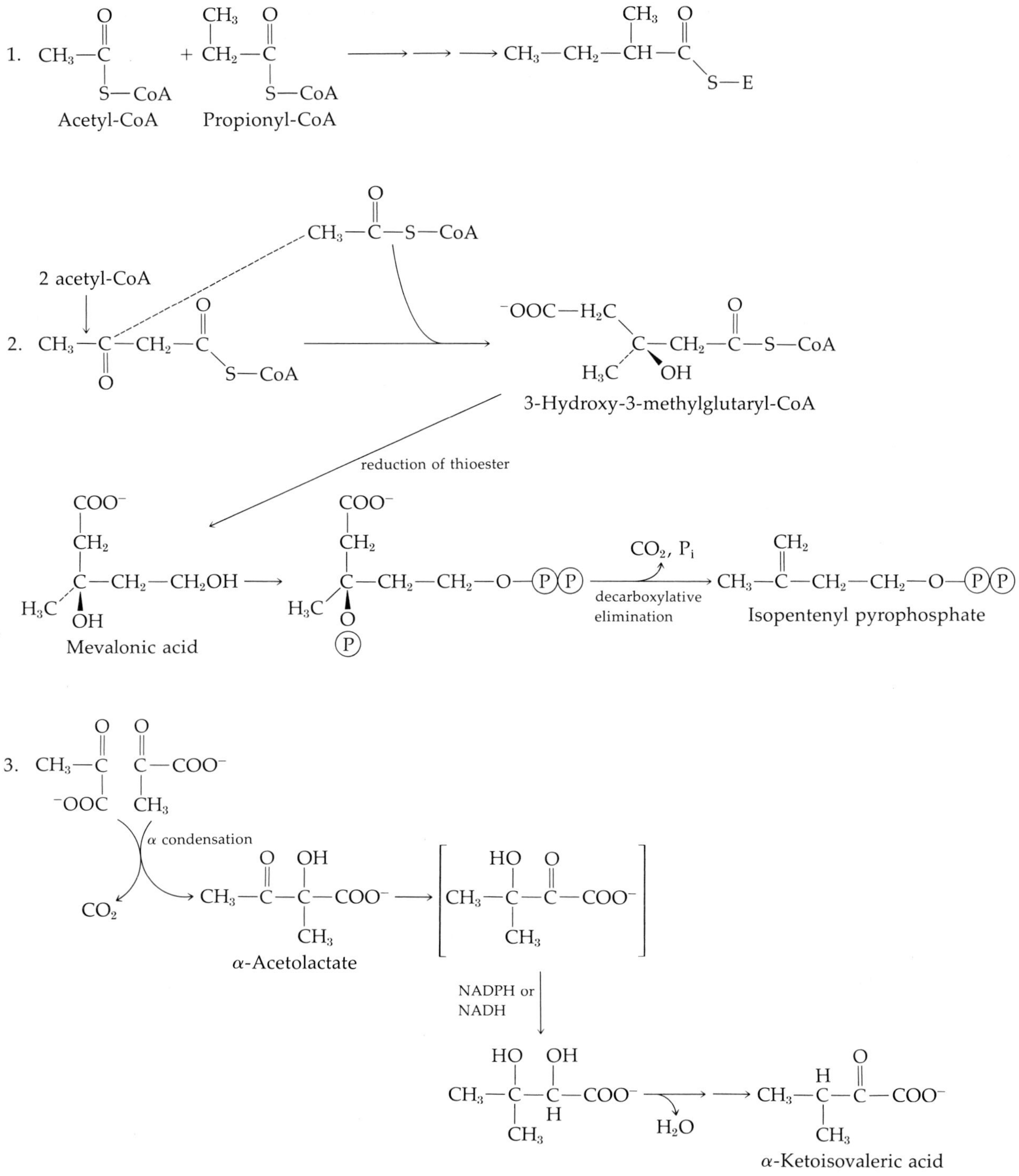

Fig. 11-8 Formation of three five-carbon branched units.

The resulting **mevalonic acid** was originally identified as a growth factor for microorganisms and was later shown to be an effective precursor for the biosynthesis of cholesterol in rat liver. In three successive phosphoryl transfers from ATP, followed by a decarboxylative elimination, mevalonic acid is converted to **isopentenyl pyrophosphate** (prenyl pyrophosphate).

The third type of carbon branched unit (Fig. 11-8) is α-ketoisovaleric acid, the immediate precursor from which valine is formed by transamination. The starting units are two molecules of pyruvate which combine in α condensation (thiamine pyrophosphate-dependent) with decarboxylation. The resulting α-acetolactate contains a branched chain but is quite unsuitable for formation of an α-amino acid. A rearrangement is required to move the methyl group to the β position (Chapter 7, Section K). Elimination of water from the diol forms the enol of the desired α-keto acid (Fig. 11-8). The precursor of isoleucine is formed in an analogous way. One molecule of pyruvate condenses (with decarboxylation) with a molecule of α-ketobutyrate. On the other hand, the keto acid precursor of leucine is formed by chain elongation of the five-carbon branched precursor of valine (Fig. 11-7).

E. BIOSYNTHESIS AND MODIFICATION OF POLYMERS

The biopolymers are the most characteristic components of living cells. These giant molecules are formed, chemically modified, and catabolized through irreversible reaction sequences, often all within a single cell. In addition, polymers are often reversibly altered as a means of metabolic control.

1. Patterns of Biosynthesis

At least three chemical problems may be defined for the assembly of a polymer. The first is *to overcome thermodynamic barriers*. The second is *to control the rate of synthesis*, and the third is *to establish the pattern or sequence in which the monomer units are linked together*. Let us look briefly at the principal routes of group activation and polymerization and consider how these problems are met.

a. Amino Acids and Peptides

Activation of amino acids for incorporation into oligopeptides and proteins can occur via the two routes of acyl activation shown in Table 7-2. In the first of these, S1C(γ), an acyl phosphate is formed and reacts with an amino group to form a peptide linkage. The tripeptide **glutathione** is formed in two steps of this type (Box 7-G).

In the second method of activation [S1C(α), Table 7-2] aminoacyl adenylates (the "activated" amino acid of Eq. 11-3) are formed and may transfer their aminoacyl groups onto specific tRNA molecules (Eq. 11-3).

In other cases activated aminoacyl groups are transferred onto —SH groups to form intermediate thioesters [Table 7-2, S1A(α)]. An example is found in the synthesis of the antibiotic **gramicidin S** formed by *Bacillus brevis*. The antibiotic is a cyclic decapeptide with the following five amino acid sequence repeated twice in the ringlike molecule[26]:

(—D-Phe-L-Pro-L-Val-L-Orn-L-Leu—)$_2$

The soluble enzyme system responsible for its synthesis contains a large protein of MW = 280,000 that activates the amino acids as aminoacyl adenylates and transfers them to thiol groups of 4'-phosphopantetheine molecules covalently attached to the enzyme.[26,27] Four amino acids Pro, Val, Orn (ornithine, Fig. 14-2), and Leu are all bound. A second enzyme (of MW = 100,000) is needed for activation of phenylalanine. It is apparently the activated phenylalanine* that initiates polymer formation in a manner analogous to that of fatty acid elongation (Section D,6). Initiation occurs when the amino group of the activated phenylalanine (on the second enzyme) attacks the acyl group of the amin-

---

* At some point in the process L-phenylalanine is isomerized to D-phenylalanine.

oacyl thioester by which the activated proline is held. Next the freed imino group of proline attacks the activated valine, etc., to form the pentapeptide. Then two pentapeptides are joined and cyclized to give the antibiotic. The sequence is absolutely specific, and it is remarkable that this relatively small enzyme system is able to carry out each step in the proper sequence. Some other peptide antibiotics such as the **tyrocidines** and **polymyxins** are synthesized in a similar way.

In contrast to the enzyme-controlled polymerization pattern of the biosynthesis of gramicidin S, proteins (as well as nucleic acids) are synthesized by **template** mechanisms. The sequence of monomer units follows the genetic code and sequences of an almost infinite variety are found in nature. A key reaction in the formation of proteins is the transfer of activated aminoacyl groups to molecules of tRNA (Eq. 11-2). The tRNA's act as carriers or "adapters" as explained in detail in Chapter 15. Each **amino acid activating enzyme** (aminoacyl-tRNA synthetase) must recognize the correct tRNA and attach to it the correct amino acid. The tRNA then carries the activated amino acid to the ribosome where it is placed, at the correct moment, in the active site. A **transacylation** reaction then joins the amino acid with the growing peptide chain.

b. Polysaccharides

The cleavage of two high energy phosphate linkages of ATP is usually required for the incorporation of one sugar monomer into a polysaccharide. However, the activation process for carbohydrates has its own distinctive pattern (Eq. 11-24). Usually a sugar is first phosphorylated by a kinase (Eq. 11-24, step a). Then a nucleoside triphosphate (NuTP) reacts under the influence of a second enzyme with elimination of pyrophosphate and formation of a **glycopyranosyl ester** of the nucleoside diphosphate, more often known as a "sugar nucleotide" (Eq. 11-24, step b). The inorganic pyrophosphate is hydrolyzed by pyrophosphatase while the sugar nucleotide serves as the immediate donor of the activated glycosyl group for polymerization (Eq. 11-24, step c). In the

Glycopyranosyl ester of nucleoside diphosphate

(11-24)

polymerization step the glycosyl group is transferred with displacement of the nucleoside diphosphate. Thus, the overall process involves first the cleavage of ATP to ADP and P_i, then the cleavage of a nucleoside triphosphate to a nucleoside diphosphate plus P_i. The nucleoside triphosphate in Eq. 11-24, step b is sometimes ATP itself, in which case the overall result is the splitting of two molecules of ATP to ADP. However, as shown in Chapter 12, the whole series of nucleotide "handles" can serve to carry activated glycosyl units.

UDPG

The first of the sugar nucleotides to be discovered was **uridine diphosphate glucose** (UDPG). The compound was discovered by Luis F. Leloir (around 1950) during the course of an investigation of the conversion of galactose 1-phosphate into glucose 1-phosphate.[28] This and other interconversions of hexoses take place at the sugar nucleotide level, a fact that was unknown at the time. Leloir's studies led to the discovery of both UDPglucose and UDPgalactose and to a Nobel Prize in 1970.

What determines the pattern of incorporation of sugar units into polysaccharides? Some homopolysaccharides, like cellulose and the linear form of starch (amylose), contain only one monosaccharide component in only one type of linkage. To form such chains a single enzyme can add unit after unit of activated sugar to the growing end. In contrast, at least two enzymes are needed to assemble the glycogen molecule. One is the synthetase that transfers activated glucosyl units from UDPglucose to the growing ends of the polymer. The second is a **branching enzyme,** a transglycosylase. After the chain ends attain a length of about ten glucose units the branching enzyme attacks a glycosidic linkage somewhere in the chain. Acting much like a hydrolase, it probably forms a glycosyl enzyme or a stabilized carbonium ion intermediate. In any event, the enzyme does not release the severed chain fragment (as does an α-amylase, Chapter 7, Section C,6) but transfers it to another nearby site on the glycogen molecule. There the enzyme rejoins the chain that it carries

to a free 6-hydroxyl group of the glycogen creating a new branch attached by an α-1,6-linkage.

Other carbohydrate polymers consist of repeating oligosaccharide units. Thus, in hyaluronic acid units of glucuronic acid and N-acetyl-D-glucosamine alternate. The "O antigens" of bacterial cell coats (Chapter 5, Section D,1) contain repeating subunits of about five different sugars. In these cases it appears that the pattern of polymerization is established by the specificities of the individual enzymes involved. An enzyme capable of joining an activated glucosyl unit to a growing polysaccharide will do so only if the proper structure has been built up to that point (see Chapter 12, Section C,1 for details).

c. The Biosynthesis of Nucleic Acids

The activated nucleotides are the nucleoside 5'-triphosphates. The activated ribonucleotides ATP, GTP, UTP, and CTP are needed for RNA synthesis and the 2'-deoxyribonucleotide triphosphates, dATP, dTTP, dGTP, and dCTP for DNA synthesis. In every case the addition of activated monomer units is directed by a protein that combines with the template nucleic acid. The choice of the proper nucleotide unit to place next in the growing strand is determined by the nucleotide already in place in the complementary strand, a matter that is dealt with in Chapter 15. The chemistry is a simple displacement of pyrophosphate (Eq. 11-25). Note that

$$\text{HO}-\overset{3'}{\text{Nu}}-\text{O}-\overset{5'}{\textcircled{P}}-\textcircled{P}-\textcircled{P} + \text{HO}-\overset{3'}{\text{polynucleotide}} \longrightarrow \text{PP}_i + \text{HO}-\overset{3'}{\text{Nu}}-\text{O}-\overset{\overset{\displaystyle O}{\|}}{\underset{\displaystyle}{\text{P}}}-\text{O}-\text{polynucleotide} \qquad (11\text{-}25)$$

it is always the 3'-hydroxyl of the polynucleotide that attacks the phosphorus atom of the activated nucleoside triphosphate. Thus, nucleotide chains grow *from the 5' end,* new units always being added *at the 3' end.*

d. Assembly Patterns of Phospholipids

One characteristic process of lipid biosynthesis deserves comment here. Choline and ethanolamine are activated in much the same way as are sugars (Eq. 11-26). For example, choline can be

$$H_3C-\overset{\overset{\displaystyle H_3C}{|}}{\underset{\underset{\displaystyle H_3C}{|}}{N^+}}-CH_2-CH_2-OH$$

Choline

$$a \Big\downarrow \quad \overset{ATP}{\underset{ADP}{}}$$

$$-\overset{|}{\underset{|}{N^+}}-CH_2CH_2-O-\boxed{P}$$

$$b \Big\downarrow \quad \overset{CTP}{\underset{PP_i \xrightarrow{H_2O} 2\,P_i}{}}$$

$$-\overset{|}{\underset{|}{N^+}}-CH_2CH_2-O-\boxed{P}\boxed{P}-O-cytidine$$

Cytidine diphosphate choline

Y—OH $\overset{c}{\Big\downarrow}$

Phosphorylcholine acceptor such as the hydroxyl group of a diglyceride

$\longrightarrow$ CMP

$$-\overset{|}{\underset{|}{N^+}}-CH_2-CH_2-O-\overset{\overset{\displaystyle O}{\|}}{\underset{\underset{\displaystyle O^-}{|}}{P}}-O-Y \quad (11\text{-}26)$$

Phospholipid (phosphatidylcholine if Y is a diglyceride)

phosphorylated using ATP (Eq. 11-26, step *a*) and the phosphoryl choline formed can be further converted (Eq. 11-26, step *b*) to **cytidine diphosphate**

choline. Phosphorylcholine is transferred from the latter onto a suitable acceptor to form the final product (Eq. 11-26, step *c*). Note the following difference from the polymerization pattern for polysaccharide synthesis. When the sugar nucleotides react, the entire nucleoside diphosphate is eliminated, but CDPcholine and CDPethanolamine react with elimination of CMP, leaving one phosphoryl group in the final product. The same thing is true in the synthesis of the bacterial teichoic acids (Chapter 5, Section D,2). Either CDPglycerol or CDPribitol is formed first and polymerization takes place with elimination of CMP to form the alternating phosphate-sugar alcohol polymer.[28a]

2. Irreversible Modification and Catabolism of Polymers

While polymers are being synthesized continuously by cells, they are also being torn down. For example, within the livers of rats or of rabbits the majority of proteins have half-lives of only 1–8 days.[29,30] However, at least one enzyme has a half-life of only 11 min, while others may last for weeks.[31] Among various species, half-lives of homologous proteins are roughly proportional to the life-span of the organism.[32]

Similar observations can be made for many of the lipid components of cells, for the ribonucleic acids, and for the polysaccharides of all surfaces. Indeed, it is probably the continuous flow of matter within cells that leads to the building up of membranes and organelles as well as to the "turnover" of the constituents from which they are formed. *This flow of matter is driven by irreversible alterations of the polymers,* including their eventual hydrolysis. It would hardly be possible to list all of the known modification reactions of biopolymers, but the following paragraphs describe a few of them.

a. Specific Hydrolysis Reactions

Enzymes are often secreted as proenzymes later to be activated (Chapter 6, Section F,2; Chapter 7,

Section D). The same is true for rat serum albumin[33] and for some peptide hormones (Chapter 16, Section A,1) including insulin. Recall that the latter is made up of one A and one B chain joined by disulfide linkages (Fig. 4-13). Insulin is synthesized in the β cells of the pancreatic islets of Langerhans as an 84 amino acid **proinsulin.** It is apparently after the proinsulin has been packaged in granules that the chain cleavage takes place (Fig. 11-9).[34-36] In a similar way the transfer RNA's are generated from longer chain RNA molecules transcribed from the DNA templates. After transcription a whole series of modifications of specific bases in the tRNA molecules occurs and hydrolytic cleavage cuts the molecules to the final size. Messenger RNA molecules in eukaryotic cells are believed to be transcribed as huge, long molecules that are cut into shorter pieces before they function (Chapter 15, Section B,5).

b. Modification by Group Transfer Reactions

Proteins, nucleic acids, and other cell components are modified by methyltransferases that transfer methyl groups from S-adenosylmethionine to special sites in the polymers. This phenomenon was first recognized in 1959 when ϵ-N-**methyllysine** was found in the flagellar protein of *Salmonella*. Since then both the ϵ-N-di- and trimethyllysines have been isolated, as have ω-N-methylarginine, 3-methylhistidine, and others.[37] These methylated amino acids are found in histones, muscle proteins, brain proteins, and in cytochrome c of some species. Enzymes are also known that transfer methyl groups to side chain carboxyl groups. Since the resulting methyl esters are unstable to hydrolysis, the extent and importance of this type of modification of proteins is uncertain.

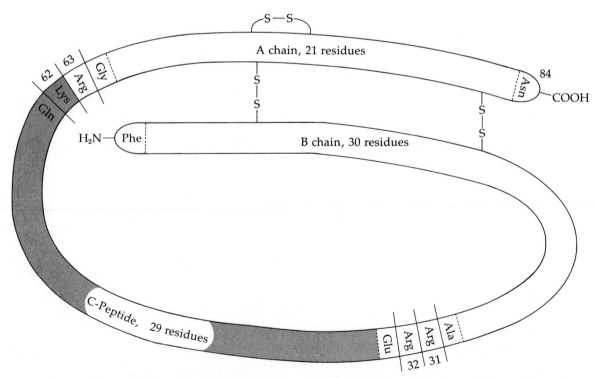

Fig. 11-9 Schematic diagram of the structure of porcine proinsulin. The 29-residue C-peptide as well as basic residues 31, 32, 62, and 63 are cut from the peptide upon conversion to insulin. See Fig. 4-13 for details of insulin structure.

Another modification reaction of proteins is the transfer of leucyl and/or phenylalanyl residues from tRNA molecules directly to certain proteins, altering them from the structures originally produced on the ribosomes.[38]

More impressive than the methylation of proteins is that of nucleic acids. Molecules of DNA are methylated in particular spots (Chapter 2, Section D,8) while tRNA undergoes extensive methylation and other modifications after its initial synthesis. Adenine rings may be methylated at N-1 or on the 6-NH$_2$ group. Uracil, cytosine, and guanine bases are also methylated, as are some $2'$ — OH groups of the ribose rings of RNA. N-Acetylation and N-isopentenylation of adenine rings are among a host of other known modification reactions (Chapter 15, Section B,4).

c. Modification by Attachment of Prosthetic Groups

In several instances a terminal stage in biosynthesis of a functional protein consists in the covalent attachment of a prosthetic group forming part of the active site of the enzyme. For example, biotin and lipoic acid are attached through an enzymatic process to enzymes requiring them. Riboflavin becomes covalently bound to a few proteins and a heme group is covalently joined to cytochrome c. It is also possible that noncovalently bonded coenzymes are attached to peptide chains at specific times, perhaps before the entire protein chain has been synthesized.

3. Hydroxylation and Other Modification Reactions of Connective Tissue Proteins

Collagen, the most abundant protein in the body, makes up much of the organic mass of skin, tendon, blood vessels, bone, the cornea, vitreous humor of the eye, and basement membranes. The closely related elastin is found in the elastic yellow connective tissue fibers which are found in abundance in ligaments and in walls of blood vessels. Collagen is synthesized by the fibroblasts and is

excreted into the extracellular space where it is polymerized into a durable long-lived material.[38a] The intracellular precursor procollagen, like mature collagen (Chapter 2, Section B,3,c), contains three chains. The major form of collagen in most tissues of most species (collagen I) contains two α 1(I) chains and one α 2 chain and may be designated $[\alpha\ 1(I)]_2\alpha\ 2$. Cartilage collagen (collagen II) contains three α 1(II) chains: $[\alpha\ 1(II)]_3$. Collagen III, found in various tissues, especially of the embryo is $[\alpha\ 1(III)]_3$.[38b]

Each procollagen chain, of MW = 140,000, contains over 1000 amino acid residues. Special hydroxylases (Chapter 10, Section G,2,d) convert some of the proline and lysine residues of the procollagen chains to **4-hydroxyproline** and **hydroxylysine** (Eqs. 11-27 and 11-28). Lesser amounts of 3-

Proline residue   4-Hydroxyproline residue   (11-27)

Lysine side chain

erythro-δ-Hydroxylysine residue   (11-28)

hydroxyproline are formed.[38c] Hydroxylation begins while the growing peptide chains are still attached to ribosomes along the rough endoplasmic reticulum. The hydroxylases are apparently localized in the membranous vesicles of the endoplasmic reticulum (ER).[39]

Galactosyl units are transferred onto some of the hydroxyl groups of the hydroxylysine side chains, and glucosyl groups are then transferred onto some of the galactosyl groups. The three procollagen chains are thought to be assembled as a triple helix before the protein is secreted. In the

BOX 11-B

GENETIC DEFECTS OF COLLAGEN STRUCTURE

Any major protein of the body is likely to be associated with a number of genetic problems. In the case of collagen, the possibility for harmful mutations is enhanced by the fact that there is more than one set of collagen genes.[a,b] At least four distinct molecular types of collagen have been identified, and it has been shown that the expression of the various collagen genes is different in different tissues. Thus, cartilage collagen contains predominantly molecules with three identical α chains different in amino acid sequence from the α1 and α2 chains of tendons and bone. The skin collagen of infants and the collagen of heart valves and large arteries represent two other types.

A clear-cut hereditary abnormality of collagen is **dermatosparaxis** of cattle, a disease in which the skin is extremely brittle. The collagen chains are disorganized and have poor fiber-forming properties. The procollagen peptidase that cleaves a peptide from the N-termini of the chains of procollagen is apparently defective.[c]

A human collagen disease is the **Ehlers–Danlos syndrome,** which in some instances is accompanied with recurrent joint dislocations and curvature of the spine. In one form of the disorder procollagen peptidase is lacking.[d] In another type III collagen is lacking.[e] The hydroxylysine content of collagen is low, preventing effective cross-linking.[f] The condition of **lathyrism** is induced in animals ingesting seeds of *Lathyris odoratus,* the common sweetpea. Since lathyris peas form part of the diet of some peoples, the condition is also well known in humans. Curvature of the spine and rupture of the aorta are common consequences. The biochemical problem has been traced to the presence in the seeds of **β-** cyanoalanine and of its decarboxylation product **β-aminopropionitrile.**[f]

$$N \equiv C - CH_2CH_2 - NH_3^+$$

Although the mode of action is not known, this compound appears to be an inhibitor of lysyl oxidase essential to the cross-linking of both collagen and elastin.[g] A hereditary defect with a similar effect has been described in the mouse. Either a defect in lysyloxidase or in copper metabolism is suspected.[e] One form of the human Ehlers–Danlos syndrome is thought to result from a deficiency of lysyl oxidase.[d]

Possible abnormalities in collagen structure in human arthritic conditions are being investigated. For example, in the very common **osteoarthritis** it has been reported that cartilage, which normally contains the $[\alpha1(II)]_3$ type of collagen, is replaced by collagen containing some α2 chains with less glycosylation.[h] On the other hand, the brittle bones of persons with one type of inherited **osteogenesis imperfecta** may contain type III collagen along with the type I collagen normally found in bones.[d]

Fibrous atherosclerotic plaques in human arteries have been reported to contain excessive amounts of type I collagen compared to collagen of normal arterial walls, which is largely type III[i]

[a] A. J. Bailey, *Compr. Biochem.* **26B,** 297–423 (1968).
[b] E. J. Miller and V. J. Matukas, *Fed. Proc., Fed. Am. Soc. Exp. Biol.* **33,** 1197–1204 (1974).
[c] K. von der Mark and P. Bornstein, *JBC* **248,** 2285–2289 (1973).
[d] P. K. Müller, C. Lemmen, S. Gay, and W. N. Meigel, *EJB* **59,** 97–104 (1975).
[e] F. M. Pope et al., *PNAS* **72,** 1314–1316 (1975).
[f] P. Bornstein, *Annu. Rev. Biochem.* **43,** 567–603 (1974).
[g] A. S. Narayanan, R. C. Siegel, and G. R. Martin, *BBRC* **46,** 745–751 (1972).
[h] M. Nimni and K. Deshmukh, *Science* **181,** 751–752 (1973).
[i] K. A. McCullagh and G. Balian, *Nature* (London) **258,** 73–75 (1975).

extracellular space two special **procollagen peptidases** act to cleave a peptide of MW ~35,000 from the C-terminus and a peptide of MW ~20,000 from the N-terminal end of each of the three chains.[40–41] The amino acid composition of the peptides removed is quite unlike that of the remaining **collagen monomer** (also called tropocollagen) which contains one-third glycine and much proline. These terminal peptides are linked in the procollagen molecule by disulfide bridges introduced prior to secretion.[41a]

The collagen monomers are three-stranded

"ropes" of dimensions ~1.5 × 300 nm. When they reach their final location they are crosslinked to form collagen.[40,42–44] This process is initiated by oxidation of some of the lysyl and hydroxylysyl side chains from amino groups to aldehyde groups under the action of a copper-containing oxidase (Eq. 11-29, Box 10-H). The aldehyde

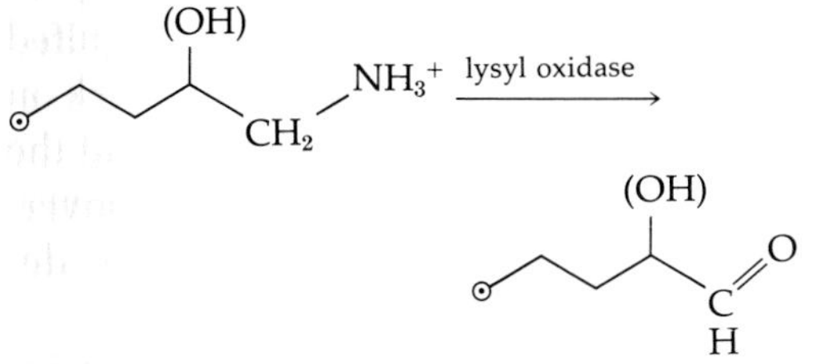

$$(11\text{-}29)$$

groups enter into a variety of reactions that lead to the crosslinking of the tropocollagen units and to the formation of insoluble fibers. One reaction is an aldol condensation followed by elimination of water (Eq. 11-30, step a). If one of the two alde-

hydes involved in the aldol condensation is derived from hydroxylysine, two isomeric condensation products are formed. The aldol condensation product can react further: An imidazole group from a histidine side chain can add to the carbon-carbon double bond and another lysine side chain can form a Schiff base with the free aldehyde. The results of these two processes are summarized in Eq. 11-30, step b. The final product **histidinohydroxymerodesmosine** links four different side chain groups. In other instances, simple Schiff bases are formed between aldehyde and ε-amino groups or only two of the three reactions depicted in Eq. 11-30 take place.

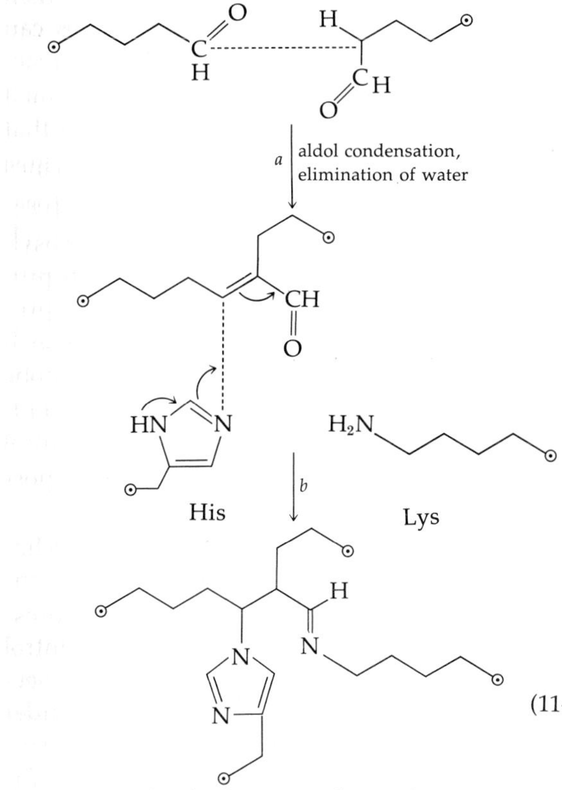

The cross-linkages are apparently not located at random but are found in certain positions, often toward the ends of the collagen monomers. Thus, histidine residues are found only at positions 89, 929, and 1034 in the peptide chains. The variety and number of cross-linkages vary among different species.

A remarkable kind of cross-linkage is found in elastin. Three aldehyde groups combine with one lysine amino group through aldol condensations, dehydration, and oxidation to form **desmosine** and **isodesmosine**. In the native elastin molecule all four amino groups and carboxyl groups are pre-

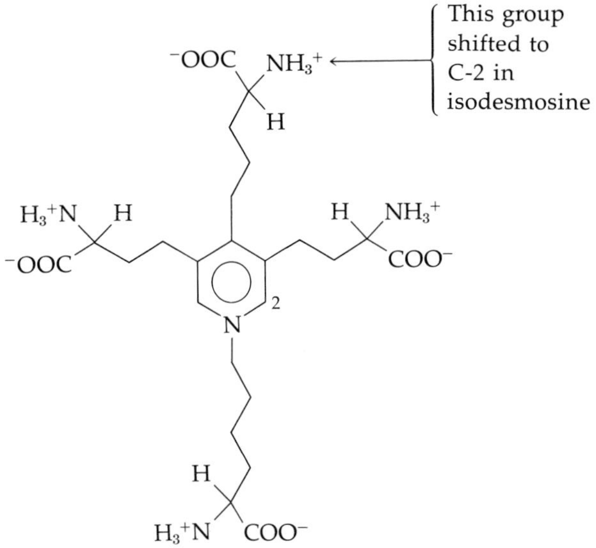

$$(11\text{-}30)$$

Histidinohydroxymerodesmosine

Desmosine

sumably present in peptide linkage. The reader should be able to propose biosynthetic mechanisms for both desmosine and isodesmosine.[43a]

F. REGULATION OF BIOSYNTHETIC PATHWAYS

Before examining some of the control mechanisms influencing biosynthesis, let us put together a brief summary picture of metabolism.

1. An Overall View of Cellular Metabolism

At one time it was popular to view a cell as a "bag of enzymes." Indeed, much of metabolism can be explained by the action of several thousand enzymes promoting specific reactions of their substrates. While the reactions catalyzed are based on the natural chemical reaction possibilities present in the substrates, the enzymes channel the reactions into a selected series of metabolic pathways most often organized as cycles.

Some of these pathways catalyze energy-yielding reactions that provide ATP. Much of the latter serves to drive reductive biosynthetic processes. It is these reductive processes that produce the less reactive hydrophobic lipid groupings and amino acid side chains so essential to the assembly of insoluble intracellular structure. Whether oligomeric proteins, membranes, microtubules, or filaments, these structures are the natural result of aggregation caused by hydrophobic forces together with electrostatic and hydrogen bonding interactions. A major part of metabolism is the creation of complex molecules that aggregate spontaneously in highly specific ways to generate structure. This includes the lipid-rich cytoplasmic membranes which, together with embedded carrier proteins, control the entry of substances into cells.

Nothing within a cell is static. All the structure that is being built up is also being torn down. Everything turns over at a slower or faster rate. Hydrolases attack all of the polymers of which cells are composed, and active catabolic reactions degrade the monomers formed. Membrane surfaces are altered by reactions of hydroxylation and by glycosylation. These reactions provide a driving force that doubtless assists in moving membrane materials into the outer surface of cells. At the same time other processes, including the breakdown by lysosomal enzymes, make it possible for membranous materials to be reengulfed into the cell. Oxidative processes lead to attack on hydrophobic materials such as the sterols and the fatty acids of membrane lipids and to their conversion into more soluble substances that are degraded and completely oxidized.

Anything that affects the rate of a reaction involved in either biosynthesis or degradation of any component of the cell will affect the overall picture in some way. Thus, we can correctly say that every chemical reaction that contributes to a quantitatively significant extent to metabolism has a controlling influence. Since molecules can interact with each other in so many ways, reactions of metabolic control are innumerable. Small molecules act on macromolecules as effectors that influence conformation and reactivity. Enzymes act on each other, cutting, oxidizing, and cross-linking. Transferases add phosphoryl, glycosyl, methyl, and other groups to various sites on proteins. The catalytic activities of such modified proteins may be altered greatly. The number of such interactions that may be significant to metabolic control within an organism could amount to millions. Small wonder that current biochemical journals are filled with a confusing number of postulated control mechanisms.

Despite this complexity some regulatory mechanisms do stand out in a clear-cut fashion. The control of enzyme synthesis through feedback repression (Chapter 6, Section F,2) and the rapid control of activity by feedback inhibition (Chapter 6, Section F,4) have been considered previously. Under some circumstances, in which there is a constant growth rate, these feedback controls may be sufficient to insure the harmonious and proportional increase of all constituents of a cell. Such may be the case for bacteria during logarithmic growth (Chapter 6, Section C) or for a mammalian embryo growing rapidly and drawing all its nutrients from the relatively constant supply in the maternal blood.

Contrast the situation in an adult human. Little growth takes place. The metabolism of many body constituents can vary greatly with time and physiological state. For example, the body makes drastic readjustments from normal feeding to a starvation situation and from resting to heavy exercise. The metabolism needed for rapid exertion is different from that needed for sustained work. A fatty diet requires different metabolism than a high carbohydrate diet. The required control mechanisms must be rapid and highly sensitive. In the successive sections we will examine some of the controlling features of the breakdown and biosynthesis of both carbohydrates and lipids in the animal body.

2. Glycogen and Blood Glucose

Two special features of glucose metabolism in animals are dominant.[44] The first is the storage of glycogen for use in providing muscular energy rapidly. This is a relatively short term matter but the rate of glycolysis can be intense: The entire glycogen content of muscle could be exhausted in only 20 s of anaerobic fermentation or in 3.5 min of oxidative metabolism.[45] Thus, there must be a way to turn on glycolysis quickly and to turn it off when it is no longer needed. At the same time, it must be possible to reconvert lactate to glucose or glycogen (gluconeogenesis). The glycogen stores of the muscle must be repleted from glucose of the blood. If insufficient glucose is available from the diet or from the glycogen stores of the liver it must be synthesized from amino acids.

A second special feature of glucose metabolism is that certain tissues, including brain, blood cells, kidney medulla, and testis, ordinarily obtain virtually all of their energy through oxidation of glucose. For this reason, the glucose level of blood cannot be allowed to drop much below the normal 5 mM. The mechanism of regulation of the blood glucose level is quite complex and incompletely understood. A series of hormones are involved.

Insulin (Chapter 4, Section D,7; Chapter 5, Section C,5; Box 11-C) possibly in cooperation with chromium (Box 11-D), promotes an increased rate of uptake of glucose by muscle and other tissues. **Glucagon** (Chapter 6, Section F,5), a 29-amino

---

BOX 11-C

DIABETES MELLITUS

The best known and one of the most prevalent metabolic defects in human beings is diabetes. Out of a million people about 400 develop juvenile diabetes between the ages of 8 and 12. Approximately 33,000 out of a million (over 3%) develop the disease by age 40–50, and by the late 70's over 7% are affected.[a] A propensity toward diabetes is partially hereditary, and at least two recessive "defective" genes are present in a high proportion of the population. The severity of the disease varies greatly. About half of the juvenile patients can be treated by diet alone, while the other half must receive regular insulin injections because of the atrophy of the insulin producing β cells of the pancreas. Among adult diabetics the problem is often a decrease in the sensitivity to insulin.[b] This may result from a lowered number of insulin receptors or from a defective structure in the receptor. A decrease in the number of receptors has also been observed in insulin-resistant obese individuals. For many diabetics treatment with sulfonylurea drugs such as the following somehow induces an increase in the number of insulin receptors formed. Dietary restriction has a similar effect.

$$H_3C \overline{\bigcirc} \overset{\overset{O}{\|}}{\underset{\underset{O}{\|}}{S}} - \overset{H}{\underset{}{N}} - \overset{\overset{O}{\|}}{\underset{}{C}} - \overset{H}{\underset{}{N}} - (CH_2)_3CH_3$$

1-Butyl-p-tolylsulfonylurea

Box 11-C (*Continued*)

In other patients, enzymatic reactions causing destruction of insulin are overactive. Glucagon production usually increases in diabetes, and this may contribute in an important way to the problems associated with the disease.[b]

Recent interest is centered on another peptide hormone **somatostatin** (Chapter 16, Section A,1) which inhibits the release of both glucagon and insulin from the pancreatic islet cells. Some diabetics are apparently benefited by treatment with this material.

A striking symptom of diabetes is the high blood glucose level which may range from 8 to 60 mM.[c] Defective utilization of glucose seems to be tied to a failure of glucose to exert proper feedback control. The result is that gluconeogenesis is increased with corresponding breakdown of proteins and amino acids. The liver glycogen is depleted and excess nitrogen from protein degradation appears in the urine. Products of fatty acid degradation accumulate, leading to ketosis (p. 676). The volume of urine is excessive and tissues are dehydrated.

Despite intensive study the chemical basis for the action of insulin remains unknown.[d,e] It is generally agreed that the hormone acts on the plasma membrane of all tissues, causing substantial permeability changes that increase the uptake of glucose, various ions, and other materials. These permeability changes may in turn lead to the major overall anabolic effect of insulin: Synthesis of glycogen, lipids, and proteins is en-

hanced. At the same time, catabolism is inhibited and the activities of catabolic enzymes such as glucose-6-phosphatase decrease. A key to the understanding of insulin action may be the identification of a "second messenger" comparable to cAMP. While cGMP has been suggested as the second messenger for insulin, an ion such as K^+ may be more likely.[f]

The completely chemical synthesis of insulin has been achieved, but it is difficult to place the disulfide crosslinks in the proper positions. New approaches mimic nature, which carries out the crosslinking in proinsulin (Fig. 11-9). Not only is synthetic insulin needed because of the large demand for the hormone isolated from animals slaughtered for human use but also alterations in a synthetic molecule might lead to properties desirable for particular groups of patients.[g]

[a] V. Apgar and J. Beck, "Is My Baby Allright?" Trident Press, New York, 1972.
[b] T. H. Maugh, II, *Science* **193,** 220–222 and 252 (1976).
[c] The lower of these values is more typical for mild diabetes because when the glucose concentration exceeds the renal threshold of ~8 mM, the excess is excreted into the urine.
[d] A. White, P. Handler, and E. L. Smith, "Principles of Biochemistry," 5th ed., pp. 1094–1101. McGraw-Hill, New York, 1973.
[e] See pp. 216, 284, and 669, for information on structure, cell surface receptors, and "potentiation" by chromium ions.
[f] H. G. Hers, *Annu. Rev. Biochem.* **45,** 167–189 (1976).
[g] Anonymous, *Chem. & Eng. News* **52** (April 29), 19–22 (1974).

acid peptide hormone, acts primarily on liver cells. Glucagon is secreted by the α cells of the islets of Langerhans in the pancreas, the same tissue that produces insulin. However, glucagon has antagonistic effects to that of insulin and tends to promote an increase in the blood glucose level by stimulating breakdown of liver glycogen. It also stimulates gluconeogenesis, both effects being mediated by cyclic AMP.[46] **Glucocorticoids** (Chapter 12, Section I,3,b) promote gluconeogenesis and the accumulation of glycogen in the liver by mechanisms discussed in Section F,7.

3. The Synthesis and Catabolism of Glycogen

One of the best studied metabolic control systems regulates the formation and breakdown of glycogen in mammalian muscle.[47–49] The system is pictured in Fig. 11-10. The heavy dashed line indicates the conversion of glycogen (left, lower center) to glucose 1-phosphate catalyzed by glycogen phosphorylase (Chapter 7, Section C,5; Chapter 8, Section E,3,e).

The synthetic sequence from glucose-1-P to

UDPG and glycogen is indicated by the heavy solid lines. In the resting muscle glycogen synthetase (labeled E_3 in the figure) is active, but phosphorylase is present as the inactive form, phosphorylase *b* (E_2 in the drawing). Although normally inactive, phosphorylase *b* is activated allosterically by AMP. A sudden burst of muscle action may provide enough AMP to turn on the

phosphorolysis of glycogen. However, more important controlling factors are provided by hormones and by nervous stimulation.

If the epinephrine concentration of the blood increases, this hormone binds to receptors on the cell membrane surface activating the formation of cyclic AMP (Chapter 7, Section E,8). Likewise, in the liver, glucagon receptors bind that hormone

---

BOX 11-D

TRACE ELEMENTS: CHROMIUM

In 1959 it was established that animals deficient in chromium grow poorly and have a reduced life-span. Chromium- deficient animals also show a decreased "glucose tolerance," i.e., glucose injected into the bloodstream is removed only half as fast as normally.[a-c] This is the same kind of response observed in a deficiency of insulin. Fractionation of yeast led to the isolation of a chromium-containing **glucose tolerance factor** which may be a complex of Cr^{3+}, nicotinic acid, and amino acids.[d] The chromium in the glucose tolerance factor is thought to react with insulin and in some way to potentiate its effect.[e,f] In line with this, it is observed that the normal chromium level of serum, ~0.03 mM, decreases sharply when glucose is injected.[g] This suggests that the chromium is actively utilized during sugar metabolism, perhaps in the binding of insulin to cell membrane receptors. A fall in the serum chromium level during acute infections (despite increased insulin levels) suggests that chromium metabolism of humans deserves careful attention.

Concentrations of chromium in tissues of animals are usually less than 2 mM, but tend to be much higher in the caudate nucleus of the brain. High concentrations of Cr^{3+} have also been found in RNA-protein complexes.[h]

An interesting experimental use of chromium is in the formation of Cr(III)-nucleotide complexes, e.g., from ATP.[i] Chromium replaces the magnesium ion ordinarily chelated with the nucleotides to yield extremely stable complexes. Their characteristic light absorption bands and the para-

magnetic properties of the chromium nucleus make these complexes interesting tools for investigation of mechanisms of action of phosphotransferases.

Recently the formation of the "exchange-inert" complexes of either Cr(III) or Co(III) has been introduced as a method of labeling metal-binding sites of proteins and other macromolecules.[j-l] A bound divalent ion, e.g., Mg^{2+} or Zn^{2+} is replaced by Co^{2+} or Cr^{2+}. Then an oxidizing agent, often H_2O_2, is used to convert the bound ion to the trivalent state. This occurs readily only at sites with an octahedral geometry. The trivalent Co or Cr binds tenaciously to the ligands surrounding it. It may be possible to digest away a large part of the macromolecule leaving a small fragment complexed with the metal.

[a] W. Mertz, *Fed. Proc. Fed. Am. Soc. Exp. Biol.* **26**, 186–193 (1967).

[b] W. Mertz and W. E. Cornatzer, "Newer Trace Elements in Nutrition." Dekker, New York, 1971.

[c] E. Frieden, *Sci. Am.* **227**, 52–60 (July 1972).

[d] W. Mertz, *Fed. Proc., Fed. Am. Soc. Exp. Biol.* **33**, 659 (1974).

[e] G. W. Evans, E. E. Roginski, and W. Mertz, *Fed. Proc.* **31**, 264 (Abstr.) (1972).

[f] E. E. Roginski, *Fed. Proc., Fed. Am. Soc. Exp. Biol.* **33**, 659 (1974).

[g] R. S. Pekarek, E. C. Hauer, E. J. Rayfield, R. W. Wannemacher, Jr., and W. R. Beisel, *Fed. Proc., Fed. Am. Soc. Exp. Biol.* **33**, 660 (1974).

[h] B. L. O'Dell and B. J. Campbell, *Compr. Biochem.* **21**, 179–266 (1971).

[i] M. L. DePamphilis and W. W. Cleland, *Biochemistry* **12**, 3714–3724 (1973).

[j] R. A. Anderson and B. L. Vallee, *PNAS* **72**, 394–397 (1975).

[k] J. K. Wright, J. Feldman, and M. Takahashi, *Biochemistry* **15**, 3704–3710 (1976).

[l] S. L. Rose and E. W. Westhead, Abstr., *172nd ACS Natl. Meet. San Francisco, 1976*, BIOL. p. 34.

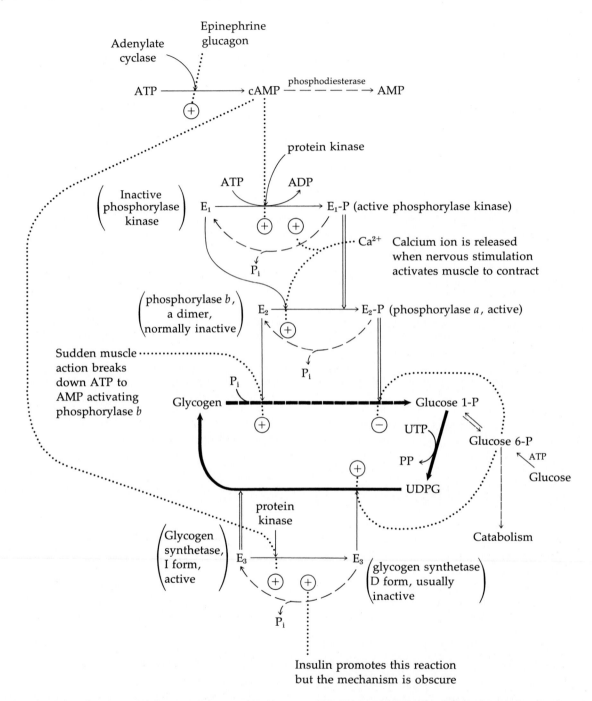

Fig. 11-10 Regulation of the breakdown and the synthesis of glycogen in muscle: (⟶) synthesis, (–→ and ⇢) catabolism, (⟶) reactions of phosphorylation using ATP, (--→) reactions of hy- drolysis catalyzed by phosphatases, and (⇒) effects of active forms of modifiable enzymes.

BOX 11-E

GENETIC DISEASES OF GLYCOGEN METABOLISM

In 1951, B. McArdle described a patient who developed pain and stiffness in muscles after moderate exercise. It was discovered that this individual completely lacked muscle glycogen phosphorylase. Since that time over 20 other persons have been found with the same defect. Glycogen accumulates in muscle tissue in this disease, one of the several types of **glycogen storage disease**. It is perhaps surprising that the disease is not more serious than it is. While severe exercise is damaging, and there are some other problems, steady moderate exercise can be tolerated. Up until the time of McArdle's discovery, it was assumed that glycogen was synthesized by reversal of the phosphorylase reaction. No hint of the UDPG pathway had appeared. It was, therefore, a surprise that glycogen accumulated in the muscles of these patients. With Leloir's discovery of UDPG at about the same time, the answer became apparent.

A number of other rare heritable diseases also lead to accumulation of glycogen for essentially the same reason: Its breakdown through the glycolysis pathway is strongly inhibited. The enzyme deficiencies include those of muscle phosphofructokinase, liver phosphorylase kinase, liver phosphorylase, and liver glucose-6-phosphatase. In the latter case, glycogen accumulates because the liver stores cannot be released to the blood as free glucose. In one of the storage diseases the branching enzyme of glycogen synthesis is lacking and glycogen is formed with unusually long outer branches. In another the "debranching" enzyme is lacking and only the outer branches of glycogen can be removed readily.

The most serious of the storage diseases involve none of the enzymes mentioned above. Pompe's disease is a fatal generalized glycogen storage disease in which a lysosomal α-1,4-glucosidase is lacking. This observation suggests that another completely different pathway of degradation of glycogen to free glucose in the lysosomes is essential.

A few cases of glycogen synthetase deficiency have been reported. Little or no glycogen is stored in muscle or liver and patients must be fed at regular intervals to prevent hypoglycemia.

and stimulate production of cAMP. Cyclic AMP, in turn, activates protein kinases which modify various proteins including a specific **phosphorylase kinase** (E_1 in Fig. 11-10) as well as glycogen synthetase. Phosphorylase kinase in the resting muscle is in an "inactive" form, and phosphorylation by the protein kinase converts it to an active form* that catalyzes phosphorylation (by ATP) of a specific serine residue in phosphorylase b. The modified enzyme, known as **phosphorylase a**, is the *normally active* enzyme responsible for glycogen breakdown. Phosphorylase a shows still another allosteric regulatory feature, *inhibition* by glucose 6-phosphate (Fig. 11-10). Thus, any accumulation of the latter metabolite will serve to turn down the rate of glycogen breakdown.

At the same time that cAMP turns on phosphorylase through the cascade mechanism just described, it activates **protein kinase** to phosphorylate the active form (I form, independent form) of glycogen synthetase. In this instance, the phosphorylated form (known as the D form, dependent form) is inactive in the absence of a specific activator. Thus, at the same time that phosphorolysis of glycogen is initiated, the further synthesis of glycogen is inhibited. The D form of glycogen synthetase is dependent upon allosteric activation by glucose 6-phosphate. Thus, if a rapid buildup of that metabolite occurs, it not only inhibits the phosphorylase reaction but also initi-

---

* Phosphorylase kinase is a very large molecule of MW = 1.3 million containing three kinds of subunits, possibly with the composition $\alpha_4\beta_4\gamma_4$. The cAMP-dependent protein kinase phosphorylates both α and β subunits, phosphorylation of the β subunit apparently being responsible for the activation.[49a,49b] Phosphorylase kinase may also undergo autoactivation involving phosphorylation of sites other than those acted on by protein kinase.[49a]

ates glycogen synthesis, even if all of the glycogen synthetase has been converted to the D form.

Like other kinases, protein kinases and phosphorylase kinase require Mg^{2+} for activity. In addition, phosphorylase kinase in its "inactive" form is allosterically activated by **calcium ions.** Recall that initiation of muscular contraction originates with nerve impulses that stimulate the release of calcium ions from vesicles of the endoplasmic reticulum. Thus, Ca^{2+} released to trigger muscular contraction also promotes phosphorylation of phosphorylase b to phosphorylase a. Now part of the cascade mechanism becomes clear. The "extra" step involving phosphorylase kinase is needed to permit the latter to be specifically influenced by calcium ions in response to nervous stimulation. On the other hand, activation of phosphorylase kinase by phosphorylation with protein kinase permits sensitivity to hormonal stimuli.

It is characteristic of control mechanisms depending upon reversible modification of proteins that special enzymes exist to return the modified proteins to their "resting" states: Cyclic AMP is hydrolyzed by a phosphodiesterase to AMP; all of the phosphorylated proteins generated undergo hydrolysis to remove the phosphate groups under the action of **phosphoprotein phosphatases.**[50] These relaxation reactions are indicated by the dashed lines in Fig. 11-10. Action of the phosphatases is doubtless controlled, too, but we know less about the mechanisms. However, insulin, possibly in an indirect way, promotes rapid conversion of the D form of liver glycogen synthetase to the I form when injected into diabetic rats.[51]

Phosphorylase and glycogen synthetase alone are insufficient to synthesize and degrade glycogen. As previously mentioned, synthesis requires the action of the branching enzyme **amylo-(1,4 → 1,6,-transglycosylase.** Likewise, degradation of glycogen requires "debranching" after the long nonreducing ends of the polysaccharide have been degraded to branch points. This is accomplished by hydrolytic removal of glucose units with **amylo-1,6-glucosidase.** These enzymes, too, must be under some kind of control.

Fig. 11-11 The interlocking pathways of glycolysis, gluconeogenesis, and fatty acid oxidation and synthesis with indications of some control features: (→) reactions of glycolysis and of oxidation via the citric acid cycle. Heavy arrows show the pathway of carbon from glycogen (upper right hand corner) to CO_2. (↦) biosynthetic pathways. Heavy dashed arrows show the gluconeogenic pathway from pyruvate via oxaloacetate and malate.

4. Phosphofructokinase, a Key Regulatory Enzyme

The metabolic interconversions of glucose 1-phosphate, glucose 6-phosphate, and fructose 6-phosphate are thought to be at or near equilibrium within most cells. However, the phosphorylation by ATP of fructose 6-phosphate to fructose 1,6-diphosphate (Fig. 11-11, top center) is usually far from equilibrium. This fact has been established by comparing the mass action ratio [fructose diphosphate] [ADP]/[fructose 6-phosphate] [ATP] measured within tissues with the known equilibrium constant for the reaction. At equilibrium the two should be the same (cf. Section C,3). The experimental techniques for determining the four metabolite concentrations in tissues (needed for evaluation of the mass action ratio) are of interest. The tissues must be frozen very rapidly. This is often done by compressing them between large liquid nitrogen-cooled aluminum clamps.* Tissues can be cooled to $-80°C$ in less than 0.1 s in this manner. The frozen tissue is then powdered, treated with a frozen protein denaturant such as perchloric acid, and analyzed.

For the phosphofructokinase reaction a mass action ratio of 0.03 has been found in heart muscle.[45] This is far lower than the equilibrium constant of over 3000 calculated from the value of $\Delta G'$ (pH 7) $= -20.1$ kJ mol^{-1}. Since this essentially irreversible reaction is so far from equilibrium in tissues, phosphofructokinase can be identified as a pacemaker enzyme in glycolysis (Chapter 6, Section F,1). Consistent with this role is the observa-

* See Newsholme and Start,[45] pp. 30–32, for details.

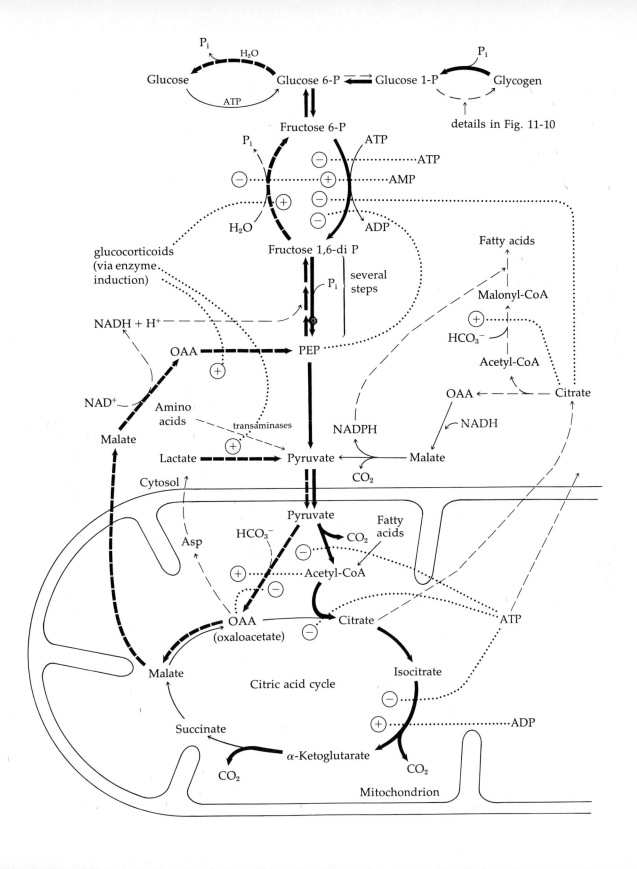

tion that the enzyme is sensitive to a variety of allosteric inhibitors and activators (Fig. 11-11). Phosphofructokinase is inhibited by high ATP concentrations but is activated by AMP. Thus, a crucial factor in regulating this enzyme and many others in both glycolysis and gluconeogenesis is the phosphorylation state of the adenylate system. It appears that AMP "turns on" this first unique step in the glycolysis sequence. The state of the adenylate system also affects later steps in glycolysis and in the citric acid cycle. Thus, a drop in [ATP] releases inhibition of the oxidation of pyruvate and isocitrate. Furthermore, inorganic phosphate is required, both in the initial phosphorolysis of glycogen and in the oxidation of triose phosphates. Thus, rapid utilization of ATP by a cell (e.g., for muscular contraction) leads to both a drop in [ATP] and increases in [AMP] and [P_i]. All of these changes activate glycolysis. However, when muscular activity ceases and the ATP level builds up, there is inhibition at several points (Fig. 11-11).

5. Gluconeogenesis[46,52]

If a large amount of lactate enters the liver it is oxidized to pyruvate which enters the mitochondria. There part of it is oxidized through the tricarboxylic acid cycle. However, if [ATP] is high pyruvate dehydrogenase is blocked (p. 530). If this happens, the amount of pyruvate converted to oxaloacetate and malate (Eq. 9-9) may increase. Malate may leave the mitochondrion to be reoxidized to oxaloacetate which is then converted to PEP and on to glycogen (heavy dashed arrows in Fig. 11-11). When [ATP] is high, phosphofructokinase is blocked, but the special phosphatase (**fructose-1,6-diphosphatase***) that hydrolyzes one phosphate group from fructose diphosphate (Eq. 11-19, step d) is active. If the glucose content of blood is low, the glucose 6-phosphate in the liver is hydrolyzed by another specific phosphatase **glucose-6-phosphatase** (located in the ER) to form free glu-

cose. Otherwise, most of the glucose 6-phosphate is converted to glycogen. Muscle is almost devoid of glucose-6-phosphatase, the export of glucose not being a normal activity of that tissue.

Gluconeogenesis in liver is strongly promoted by glucagon and epinephrine. The effects, mediated by cAMP, may include stimulation of fructose-1,6-diphosphatase and inhibition of phosphofructokinase.[46] An effect on a reaction between pyruvate and PEP is also indicated. The latter may be an indirect effect of stimulation of α-ketoglutarate metabolism.[53]

6. Substrate Cycles

The joint actions of phosphofructokinase and fructose-1,6-diphosphatase (Eq. 11-19, step d, Fig. 11-11) pose a dilemma. Fructose 6-phosphate is being phosphorylated to fructose diphosphate at the same time that the latter is being hydrolyzed back to fructose 6-P. This creates a useless cycle (often called a **futile cycle** or **substrate cycle**) which accomplishes nothing but the cleavage of ATP to ADP and P_i (ATPase activity). Cycles of this type are frequent in metabolism and the fact that they do not ordinarily cause a disastrously rapid loss of ATP is a consequence of the tight control of the metabolic pathways involved. In general, only one of the two enzymes of Eq. 11-19, step d, is fully activated at a time. Depending upon the metabolic state of the cell, degradation may occur with little biosynthesis or biosynthesis with little degradation. Some of the control features are indicated in Fig. 11-11. The ATP and AMP levels are most important, low [AMP] turning on the kinase and turning off the phosphatase. Depending on the species ATP, PEP, or citrate may act as feedback inhibitors of glycolysis. Further control mechanisms for fructose-1,6-diphosphatase may remain to be discovered.

Other substrate cycles may be recognized in the conversion of glucose to glucose 6-phosphate and hydrolysis of the latter back to glucose (Fig. 11-11, upper left-hand corner), the synthesis and breakdown of glycogen (upper right), and the conver-

---

* Among the allosteric inhibitors affecting this enzyme is Zn^{2+}, which binds with a value of $K_i \approx 0.3\ \mu m$.[45a]

sion of PEP to pyruvate and the reconversion of the latter to PEP via oxaloacetate and malate (partially within the mitochondria).

While one might suppose that cells always keep substrate cycling to a bare minimum, experimental measurements on rat livers *in vivo* indicate rates as high as ~3 μmol kg^{-1} s^{-1} for the fructose-1,6-diphosphatase–phosphofructokinase cycle.[54] A surprisingly high rate of cycling has been reported for pyruvate → oxaloacetate → PEP → pyruvate[55]. It has been suggested that by maintaining a low rate of substrate cycling under conditions in which the carbon flux is low (in either the glycolytic or glucogenic direction) the system is more sensitive to allosteric effectors. Theoretical calculations indicate a very large amplification of the response to effectors when a significant rate of cycling is maintained (see Newsholme and Start[45], Chapter 2). In the presence of the appropriate allosteric effectors substrates may flow in one direction or the other with little cycling.

While substrate cycles ordinarily remain under control, they are suspected as a cause of uncontrolled heat production in some pathological conditions (Box 11-F). It also seems likely that substrate cycles are sometimes deliberately activated for the purpose of producing heat (see also Box 10-F). Thus, for a bumblebee to fly the thoracic temperature must reach at least 30°C. On cool days it is thought that the insects use substrate cycling catalyzed by phosphofructokinase and fructose-diphosphatase to warm their flight muscles.[56]

7. The Fasting State

During prolonged fasting glycogen supplies are depleted throughout the body and fats become the principal fuels. Both glucose and pyruvate are in short supply. While the hydrolysis of lipids provides some glycerol (which is oxidized to dihydroxyacetone and phosphorylated) the quantity of glucose precursors formed in this way is limited. (Bear in mind that the animal body cannot recon-

---

BOX 11-F

MALIGNANT HYPERTHERMIA AND STRESS-PRONE PIGS

Very rarely during an operation the temperature of a patient suddenly starts to rise uncontrollably. Even when heroic measures are taken, sudden death may ensue. This **malignant hyperthermia syndrome** is often associated with administration of halogenated anesthetics.[a,b] Although it is possible that some subclinical muscular defect is present, there is often no warning that the patient is abnormally sensitive to anesthetic. Biochemical investigation of the hyperthermia syndrome has been greatly facilitated by the discovery of a similar condition that is prevalent among certain breeds of pigs. Such "stress-prone" pigs are likely to die suddenly of hyperthermia induced by some stress such as shipment to market. The sharp rise in temperature with muscles going into a state of rigor is accompanied by a dramatic lowering of the ATP content of the muscles.

One explanation of hyperthermia for which experimental support exists[b] is that substrate cycling (Section F,6) involving the phosphofructokinase and fructose-diphosphatase enzymes is responsible for the generation of heat and for the sudden hydrolysis of ATP. How an anesthetic would initiate this response remains unexplained, but it could be through action on the cell membrane that would cause derangement of the normal hormonal regulatory systems. Another possibility is that the anesthetic affects the membranes of the mitochondria.[c]

[a] R. A. Gordon, B. A. Britt, and W. Kalow, eds., "International Symposium on Malignant Hyperthermia." Thomas, Springfield, Illinois, 1973.
[b] M. G. Clark, C. H. Williams, W. F. Pfeifer, D. P. Bloxham, P. C. Holland, C. A. Taylor, and H. A. Lardy, *Nature* (London) **245**, 99–101 (1973).
[c] G. Eikelenboom and W. Sybesma, *J. Anim. Sci.* **38**, 504–506 (1974).

vert acetyl-CoA to pyruvate.) Yet a continuing need for both glucose and pyruvate exists. The former is needed for biosynthetic processes, and the latter is a precursor of oxaloacetate, the regenerating substrate of the citric acid cycle. For this reason, during fasting the body readjusts its metabolism. Glucocorticoids (e.g., cortisol, Chapter 12, Section I,3,b) are released from the adrenal glands. Via enzyme induction mechanisms these hormones increase the amounts of a variety of enzymes within the cells of target organs such as the liver. Glucocorticoids also appear to increase the sensitivity of cell responses to cAMP, hence to hormones such as glucagon.[57] It has been suggested that the effect arises because the corticoids are essential for preservation of the normal ionic environment, e.g., of normal concentrations of Ca^{2+}, K^+, or Na^+.

The overall effects of glucocorticoids include an increased release of glucose from the liver (increased activity of glucose-6-phosphatase), an elevated blood glucose and liver glycogen, and a decreased synthesis of mucopolysaccharides. The reincorporation of amino acids released by protein degradation is inhibited and synthesis of enzymes degrading amino acids is enhanced. Among these enzymes are tyrosine and alanine aminotransferases, enzymes that initiate amino acid degradation yielding ultimately the glucogenic precursors fumarate and pyruvate.

Adjustment made by the body during prolonged starvation is a shift from carbohydrate to lipid metabolism. For example, as much as 75% of the glucose need of the brain can be replaced gradually by ketone bodies derived from breakdown of fat (Chapter 9, Section A,7).[44]

8. Ketosis

Sometimes the inability of the animal body to form the glucose precursors, pyruvate or oxaloacetate from acetyl units is a cause of severe metabolic problems. The condition known as ketosis develops when too much acetyl-CoA is produced and its efficient combustion in the citric acid cycle

is blocked by lack of oxaloacetate (Chapter 9, Section B,4). Ketosis develops in diabetes when insulin is lacking (Box 11-C), during starvation, and during fevers. In cows, whose metabolism is much more based on acetate than is ours, spontaneously developing ketosis is a frequent problem. These facts suggest that under some conditions animals are on the border of being unable to muster enough carbohydrate precursors to meet biosynthetic needs and to keep the concentration of the regenerating substrate, oxaloacetate, high enough for efficient operation of the citric acid cycle.

A standard treatment for ketosis in cattle is the administration of a large dose of propionate which is presumably effective because of the ease of its conversion to oxaloacetate via the methylmalonyl-CoA pathway (Chapter 9, Section D,2). It is possible that this pathway was developed by animals as a means of capturing propionyl units, scanty though they may be, for conversion to oxaloacetate and use in biosynthesis. In ruminant animals, the pathway assumes even greater importance. For while we have 5.5 mM glucose in our blood, the cow has only half as much, and a substantial fraction of this glucose is derived (in the liver) from the propionate provided by rumen microorganisms.[58] The need for vitamin B_{12} in the formation of propionate by these organisms accounts for the high requirement for cobalt in the ruminant diet (Box 8-K).

9. Lipogenesis

Consider another metabolic situation. A high carbohydrate meal leads to an elevated blood glucose concentration. The glycogen reserves within cells are filled. The ATP level rises, blocking the citric acid cycle, and citrate is exported from mitochondria (Fig. 11-11). Outside the mitochondria citrate is cleaved by the ATP-requiring citrate lyase (Eq. 7-70) to acetyl-CoA and oxaloacetate. The oxaloacetate can be reduced to malate and the latter oxidized with $NADP^+$ to pyruvate (Eq.

11-13), which can again enter the mitochondrion. In this manner acetyl groups are exported from the mitochondrion as acetyl-CoA. The latter can then be carboxylated, under the activating influence of citrate, to form malonyl-CoA, the precursor to fatty acids. The NADPH formed from oxidation of the malate provides part of the reducing equivalents needed for fatty acid synthesis. Additional NADPH is available from the pentose phosphate pathway. Thus, excess calories in the form of carbohydrate are readily converted to fat within our bodies. While these reactions doubtless occur to some extent in most cells, they are quantitatively most important in the fat cells of adipose tissue.

While not all of the regulatory mechanisms concerned with metabolism of glucose and lipids have been discovered, it is clear that within the cell there is a network of controlling interactions that permit the overall set of metabolic cycles to respond in different ways to different conditions. The response is always such as to meet the energy needs of the cell, and at the same time to ensure restoration of the steady state concentrations of cell constituents.

REFERENCES

1. R. L. Switzer, in "The Enzymes," 3rd ed. (P. D. Boyer, ed.), Vol. 10, pp. 607–629. Academic Press, New York, 1974.

2. D. Soll and P. R. Schimmel, in "The Enzymes," 3rd ed. (P. D. Boyer, ed.), Vol. 10, pp. 489–538. Academic Press, New York, 1974.

3. S. H. Mudd, in "The Enzymes," 3rd ed. (P. D. Boyer, ed.), Vol. 8, pp. 121–154. Academic Press, New York, 1973.

4. H. D. Peck, Jr., in "The Enzymes," 3rd ed. (P. D. Boyer, ed.), Vol. 10, pp. 651–669. Academic Press, New York, 1974.

5. H. B. White, III, O. Mitsuhashi, and K. Bloch, JBC **246**, 4751–4754 (1971).

6. R. W. McGilvery, "Biochemistry, A Functional Approach," pp. 631–632. Saunders, Philadelphia, Pennsylvania, 1970.

6a. F. G. Prendergast, C. M. Veneziale, and N. G. Deering, JBC **250**, 1282–1289 (1975).

7. D. H. Williamson, P. Lund, and H. A. Krebs, BJ **103**, 514–527 (1967).

8. M. Stubbs, R. L. Veech, and H. A. Krebs, BJ **126**, 59–65 (1972).

9. H. A. Krebs and R. L. Veech, Mitochondria: Struct. Funct. Fed. Eur. Biochem. Soc., Meet., 5th, 1968 FEBS Symp., **17**, 101–109 (1969).

10. R. L. Veech, R. Guynn, and D. Veloso, BJ **127**, 387–397 (1972).

11. R. Lundquist and B. M. Olivera, JBC **246**, 1107–1116 (1971).

12. B. B. Buchanan, JBC **244**, 4218–4223 (1969).

13. U. Gehring and D. I. Arnon, JBC **247**, 6963–6969 (1972).

14. R. Y. Stanier, M. Doudoroff, and E. A. Adelberg, "The Microbial World," 3rd ed. Prentice-Hall, Englewood Cliffs, New Jersey, 1970.

15. A. R. Salem, A. J. Hacking, and J. R. Quayle, BJ **136**, 89–96 (1973).

16. H. L. Kornberg and J. G. Morris, BJ **95**, 577–586 (1965).

17. R. D. Dyson, J. M. Cardenas, and R. J. Barsotti, JBC **250**, 3316–3321 (1975).

18. Y. Milner and H. G. Wood, PNAS **69**, 2463–2468 (1972).

19. T. J. Andrews and M. D. Hatch, BJ **114**, 117–125 (1969).

20. H. Weinhouse and M. Benziman, EJB **28**, 83–88 (1972).

21. T. Sugiyama, Biochemistry **12**, 2862–2868 (1973).

22. G. Michaels, Y. Milner, and G. H. Reed, Biochemistry **14**, 3213–3219 (1975).

23. S. J. Wakil and E. M. Barnes, Jr., Compr. Biochem. **18S**, 57–104 (1971).

24. T. W. Orme, J. McIntyre, F. Lynen, L. Kühn, and E. Schweizer, EJB **24**, 407–415 (1972).

25. C. Y. Lin and S. Kumar, JBC **246**, 3284–3290 (1971).

25a. H. M. Miziorko, K. D. Clinkenbeard, W. D. Reed, and M. D. Lane, JBC **250**, 5768–5773 (1975).

26. F. Lipmann, Science **173**, 875–884 (1971).

27. K. Kurahashi, Annu. Rev. Biochem. **43**, 445–459 (1974).

28. L. F. Leloir, Science **172**, 1299–1303 (1971).

28a. F. Fiedler and L. Glaser, JBC **249**, 2684–2689 (1974).

29. R. D. Glass and D. Doyle JBC **247**, 5234–5242 (1972).

30. L. W. Johnson and S. F. Velick, JBC **247**, 4138–4143 (1972).

31. A. L. Goldberg and J. F. Dice, Annu. Rev. Biochem. **43**, 835–869 (1974).

32. I. M. Spector, Nature (London) **249**, 66 (1974).

33. J. R. Russell and D. M. Geller, JBC **250**, 3409–3413 (1975).

34. A. E. Kitabchi, W. C. Duckworth, F. B. Stentz, and S. Yu, C.R.C. Crit. Rev. Biochem. **1**, 59–94 (1972).

35. P. T. Grant and T. L. Coombs, Essays Biochem. **6**, 69–92 (1970).

36. W. Kemmler, J. D. Peterson, and D. F. Steiner, JBC **246**, 6786–6791 (1971).

37. W. K. Paik and S. Kim, Science **174**, 114–119 (1971).

38. M. J. Leibowitz and R. L. Soffer, JBC **246**, 5207–5212 (1971).

38a. F. J. Manasek, in "Current Topics in Developmental Biology" (A. A. Moscona, and A. Monroy, eds.), Vol. 9, pp. 35–102. Academic Press, New York, 1975.

38b. W. T. Butler, H. Birkedal-Hansen, W. F. Beegle, R. E. Taylor, and E. Chung, JBC **250**, 8907–8912 (1975).

38c. R. M. Gryder, M. Lamon, and E. Adams, JBC **250**, 2470–2474 (1975).

39. K. R. Cutroneo, N. A. Guzman, and M. M. Sharawy, *JBC* **249**, 5989–5994 (1974).

40. P. Bornstein, *Annu. Rev. Biochem.* **43**, 567–603 (1974).

40a. J. Rosenbloom, R. Endo, and M. Harsch *JBC* **251**, 2070–2076 (1976).

41. P. H. Byers, E. M. Click, E. Harper, and P. Bornstein, *PNAS* **72**, 3009–3013 (1975).

41a. L. N. Lukens, *JBC* **251**, 3530–3538 (1976).

42. M. L. Tanzer, *Science* **180**, 561–566 (1973).

42a. A. J. Bailey, S. P. Robins, G. Balian, *Nature (London)* **251**, 105–109 (1974).

43. P. F. Davison, *C.R.C. Crit. Rev. Biochem.* **1**, 201–245 (1973).

43a. A. S. Nararyanan and R. C. Page, *JBC* **251**, 1125–1130 (1976).

44. H. A. Krebs, in "Metabolic Regulation and Enzyme Action" (A. Sols and S. Grisolia, eds.), pp. 53–54. Academic Press, New York, 1970.

45. E. A. Newsholme and C. Start, "Regulation in Metabolism." Wiley, New York, 1973.

45a. G. A. Tejwani, F. O. Pedrosa, S. Pontremoli, and B. L. Horecker, *PNAS* **73**, 2692–2695 (1976).

46. M. G. Clark, N. M. Kneer, A. L. Bosch, and H. A. Lardy, *JBC* **249**, 5695–5703 (1974).

47. I. Pastan, *Sci. Am.* **227**, 97–105 (Aug. 1972).

48. E. H. Fischer, A. Pocker, and J. C. Saari, *Essays Biochem.* **6**, 23–68 (1970).

49. J. T. Stull and S. E. Mayer, *JBC* **246**, 5716–5723 (1971).

49a. J. H. Wang, J. T. Stull, T.-S. Huang, and E. G. Krebs, *JBC* **251**, 4521–4527 (1976).

49b. H. G. Hers, *Ann. Rev. Biochem.* **45**, 167–189 (1976).

50. C. Nakai and J. A. Thomas, *BBRC* **52**, 530–536 (1973).

51. W. K. Nichols and N. D. Goldberg, *BBA* **279**, 245–259 (1972).

52. H. A. Lardy, *Diabetes Mellitus, Proc. Nobel Symp., 13th, 1969* pp. 199–214 (1970).

53. M. Ui, J. H. Exton and C. R. Park, *JBC* **248**, 5350–5359 (1973).

54. M. G. Clark, D. P. Bloxham, P. C. Holland, and H. A. Lardy, *JBC* **249**, 279–290 (1974).

55. R. Rognstad and J. Katz, *JBC* **247**, 6047–6054 (1972).

56. M. G. Clark, D. P. Bloxham, P. C. Holland, and H. A. Lardy, *BJ* **134**, 589–597 (1973).

57. J. H. Exton, N. Friedmann, E. H.-A. Wong, J. P. Brineaux, J. D. Corbin, and C. R. Park, *JBC* **247**, 3579–3588 (1972).

58. E. Weigand, J. W. Young, and A. D. McGilliard, *BJ* **126**, 201–209 (1972).

STUDY QUESTIONS

1. (a) If the free energy change $\Delta G'$ (pH 7) for the reaction A → B is +25 kJ/mol at 25°C, what would the ratio of [B]/[A] be at equilibrium? (b) Suppose that the reaction were coupled to the cleavage of ATP as follows:

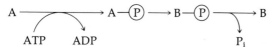

   Suppose further that the group transfer potential ($-\Delta G'$) for the phosphoryl group of A—Ⓟ at 25°C, pH 7 is 12 kJ/mol and that the equilibrium constant for conversion of A—Ⓟ to B—Ⓟ is the same as that for conversion of A to B. Calculate the concentrations of A, B, A—Ⓟ, and B—Ⓟ at equilibrium if the phosphorylation state ratio R_p is $10^4\ M^{-1}$.

2. ^{14}C-Carboxyl labeled palmitic acid is fed to a fasted rat. There is no increase in liver glycogen, but the glucose units of the glycogen contain ^{14}C. (a) Outline, using appropriate equations, the reaction sequence by which the carbon atoms of glucose become labeled. (b) Explain why there is no net synthesis of glycogen from the fatty acid.

3. (a) Write the reactions that most *dietary* tripalmitin will undergo in the body of an adult human in order to be deposited in adipose tissue as tripalmitin. (b) What is the minimum amount of ATP (high energy bonds) normally required to deposit the one mole of dietary tripalmitin in adipose tissue? Count only ATP involved in tripalmitin metabolism and consider the source of glycerol in the adipose tissue.

4. Describe the biochemical effects of injecting (a) insulin, (b) glucagon, (c) epinephrine into a normal animal.

5. Suggest a biosynthetic pathway for formation of the fungal metabolite **agaricic acid**:

$$CH_3-(CH_2)_{15}-CH-COOH$$
$$|$$
$$HOC-COOH$$
$$|$$
$$CH_2-COOH$$

6. The ketone **palmiton** $CH_3(CH_2)_{14}-{}^*CO-(CH_2)_{14}CH_3$ is formed by mycobacteria. The carbon marked by an asterisk was found to be labeled after feeding of [1-^{14}C]palmitic acid. Suggest a biosynthetic pathway.

Metabolite	Abbreviation	Concentrations (μmol/l of cells)	
		Before addition of antimycin	After addition of antimycin
Glucose 6-phosphate	G6P	460	124
Fructose 6-phosphate	F6P	150	30
Fructose 1,6-diphosphate	FDP	8	33
Triose phosphates	TP	18	59
3-Phosphoglycerate	3PGA	45	106
2-Phosphoglycerate	2PGA	26	19
Phosphoenolpyruvate	PEP	46	34
Pyruvate	Pyr	126	315
Lactate	Lac	1125	8750
ATP		2500	1720
ADP		280	855
AMP		36	206

7. Reticulocytes (immature red blood cells) contain mitochondria that are capable of both aerobic and anaerobic oxidation of glucose. In an experiment using these cells, incubated in oxygenated Krebs–Ringer solution with 10 mM glucose, the addition of antimycin A produced the following changes in metabolite concentration after 15 min [From A. K. Ghosh and H. A. Sloviter, *JBC* **248**, 3035–3040 (1973)]. Interpret the observed changes in ATP, ADP, and AMP concentrations (see tabulation above). Express the concentration of each component after addition of antimycin as a percentage of that before addition. Then plot the resulting figures for each compound in the sequence found in glycolysis, i.e., label the X axis as

G6P   F6P   FDP   TP   etc.

and plot the percentage values as ordinates and connect the points. The resulting figure is known as a crossover diagram (the graph crosses the 100% line in one or more places).

8. It has been proposed that the substrate cycle involving phosphofructokinase and fructose diphosphatase is used by bumblebees to warm their flight muscles to 30°C before flight begins. Clark *et al.*[56] found maximal rates of catalytic activity for both enzymes to be about 44 μmol/min/g of fresh tissue. In flying bees glycolysis occurred at a rate of about 20 μmol/min/g of tissue with no substrate cycling. In resting bees at 27°C no cycling was detected, but at 5°C substrate cycling occurred at the rate of 10.4 μmol/min/g while glycolysis had slowed to 5.8 μmol/min/g. If the cycling provides heat to warm the insect, estimate how long it would take to reach 30°C if a cold (5°) bee could carry out cycling at the maximum rate of 40 μmol/min/g and if no heat were lost to the surroundings.

12

Some Specific Pathways of Metabolism of Carbohydrates and Lipids

In the preceding chapter general principles of biosynthesis were considered. The pathways of formation of the major carbohydrate and lipid precursors were described as were the processes of gluconeogenesis, synthesis of glucose 6-phosphate from free glucose, and formation of fatty acids. Also considered were typical polymerization pathways and control features of glycogen synthesis and degradation. In this chapter we will examine a number of specific aspects of metabolism of monosaccharides, oligosaccharides, polysaccharides, lipids, and related materials.

A. INTERCONVERSIONS OF MONOSACCHARIDES

As previously noted (Chapter 11, Section D,1) chemical interconversions are easiest at the level of oxidation of carbohydrates. Thus, a profusion of reactions are known by which one sugar can be changed into another. Most of the transformations take place at the sugar nucleotide level. They are summarized in Fig. 12-1, which indicates how glucose 6-phosphate or fructose 6-phosphate can be converted to many of the other sugars found in living things. Galactose and mannose can also be interconverted with the other sugars. A kinase forms **mannose 6-phosphate** which equilibrates with fructose 6-phosphate. Galactokinase forms **galactose 1-phosphate** from the free sugar. The latter can be converted to UDPgalactose and thence to UDPglucose. By contrast, free fructose, an important constituent (usually as sucrose) of human diets, can also be phosphorylated in the liver by a special **fructokinase** to fructose 1-phosphate. However, there is no mutase able to convert the latter compound to fructose 6-phosphate. Rather there is a special aldolase that cleaves fructose 1-phosphate to dihydroxyacetone phosphate and free glyceraldehyde. Lack of this aldolase leads to the occasionally observed cases of fructose intolerance. Glyceraldehyde formed from fructose can be metabolized by reduction to glycerol followed by phosphorylation of the latter (glycerol kinase) and reoxidation to dihydroxyacetone phosphate.

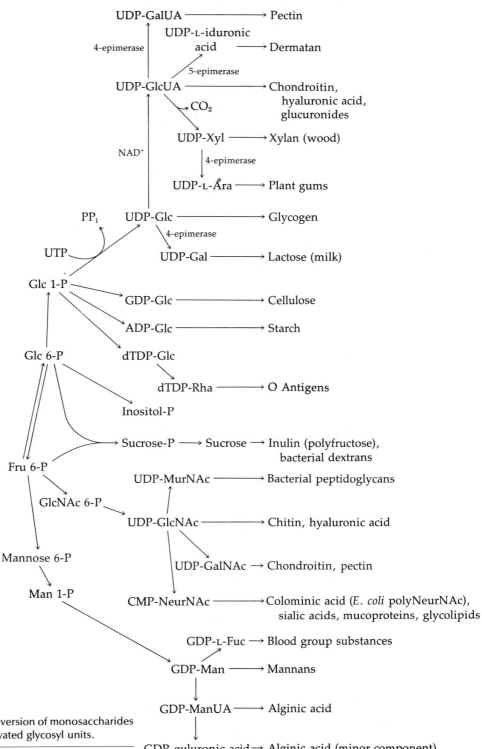

Fig. 12-1 Some routes of interconversion of monosaccharides and of polymerization of the activated glycosyl units.

1. The Metabolism of Galactose

The reactions of galactose are of special interest because of the occurrence of the hereditary disorder **galactosemia.** When this defect is present, the body cannot transform galactose into glucose metabolites but excretes it in the urine. Unfortunately, this is not the only effect. Severe gastrointestinal troubles often appear within a few days or weeks of birth, and death may come quickly from liver damage. Cataracts develop in the eyes and mental retardation occurs in patients that survive. It has been suggested that galactose 1-phosphate may be responsible for the toxic effect in liver, glands, and brain[1]; on the other hand, reduction of galactose to the corresponding sugar alcohol may initiate cataract formation.[2] Fortunately, galactose-free diets can be prepared for young infants and if the disease is diagnosed promptly, no serious damage is done, a happy success for nutritional biochemistry.*

In some galactosemic patients **galactokinase** is absent, but it is more often **galactose-1-phosphate uridyltransferase** (Eq. 12-1) that is missing. This

$$\text{Gal} \xrightarrow{\text{ATP}} \text{Gal 1-P} \qquad\qquad \text{Glc 1-P}$$

Galactokinase

$$\text{Gal-1-P-uridyltransferase}$$
$$a$$
$$\text{UDP-Glc} \qquad\qquad \text{UDP-Gal}$$
$$b$$
$$\text{4-Epimerase}$$

Further metabolism   (12-1)

enzyme transforms galactose 1-phosphate to UDPgalactose by displacing glucose 1-phosphate from UDPglucose. The UDPgalactose is then isomerized with the special NAD^+-dependent 4-epimerase (Eq. 12-1, step b). See also Eq. 8-44 and accompanying discussion. The overall effect of the reactions of Eq. 12-1 is to transform galactose efficiently into glucose 1-phosphate. At the same time the 4-epimerase can operate in the other direction to convert UDPglucose to UDPgalactose when the latter is needed for biosynthesis (Fig. 12-1).

An enzyme present in only small amounts at birth becomes more active in later years. This is an

"activating enzyme" that forms UDPgalactose directly from galactose 1-phosphate and UTP (Eq. 12-2, step a). After isomerization of UDP-Gal to UDP-Glc (Eq. 12-2, step b) the latter can be cleaved (Eq. 12-2, step c) to glucose 1-phosphate.* The ef-

$$\text{Gal 1-P} \xrightarrow[a]{\text{UTP} \quad \text{PP}_i} \text{UDP-Gal} \xrightarrow[b]{\text{epimerase}}$$
$$\text{UMP}$$
$$\text{UDP-Glc} \xrightarrow[c]{} \text{Glc 1-P} \quad (12\text{-}2)$$

fect is to provide an alternate pathway from galactose to glucose 1-phosphate. The development of this enzyme may account for the fact that galactosemic patients tend to become resistant to galactose as they mature.

Before leaving the topic of galactose metabolism, it should be noted that a very common biochemical problem with sugar metabolism is intolerance to the disaccharide lactose.[3] In this case it is a matter of the inability of the intestinal mucosa to make enough **lactase** to hydrolyze the sugar to its monosaccharide components (galactose and glucose). Among most of the peoples of the earth, only infants have a high lactase level, and the use of milk as a food for adults leads to a severe diarrhea. The same is true for most animals. In fact, baby seals and walruses, which drink lactose-free milk, become very ill if fed cow's milk.

2. Interconversions of Sugar Nucleotides

Returning to Fig. 12-1, note that UDPglucose is oxidized in two steps[3a] with NAD^+ to **UDPglucuronic** acid and that the latter is epimerized to **UDPgalacturonic** acid. Likewise, at the bottom of the scheme, **guanosine diphosphate-mannose** is oxidized in a similar fashion to **GDPmannuronic acid** which undergoes 4-epimerization to **GDPguluronic acid.** Looking again at the top of the

---

* While galactosemia is a rare disease, there are perhaps 30 cases per million births.

* It is usually assumed that the cleavage is by phosphorolysis of UDP-Glc with inorganic pyrophosphate and the formation of UTP. However, it seems more likely that a phosphodiesterase or a phosphorylase using P_i catalyzes this reaction.

scheme, note that UDP-D-glucuronic acid may be epimerized at the 5 position to **UDP-L-iduronic acid.** However, the iduronic acid residues in dermatan sulfate arise by inversion at C-5 of D-glucuronic acid residues in the polymer.[4] The mechanism of these reactions, like that of the decarboxylation of UDPglucuronic acid to UDP-xylose (near the top of Fig. 12-1) apparently have not been well investigated. However, the reader should be able to propose mechanisms.

3. Inositol and D-Glucuronic Acid

Related to the monosaccharides is the hexahydroxycyclohexane, *myo*-inositol (Eq. 12-3). This compound, which is apparently present universally within cells,[4a] can be formed from glucose 6-phosphate according to Eq. 12-3. The cyclization is accompanied by a change in configuration at C-5, a fact that should suggest to the student a reaction mechanism. Synthesis by animals seems to be limited and *myo*-inositol is sometimes classified as a vitamin. Mice grow poorly and lose some

Glucose 6-phosphate

myo-Inositol   (12-3)

of their hair if deprived of dietary inositol. Deficient rats develop fatty livers and yeast cells also accumulate neutral lipids if lacking in inositol.[4b] The compound is obviously needed in formation of **phosphoinositides** (inositol-containing phospholipids; Table 2-8).

In plants, inositol is a component of galactinol (Eq. 12-11) which may be a specific precursor of

cell wall polysaccharides. Various phosphate esters of inositol occur in nature. Large amounts of the hexaphosphate (**phytic acid**) are present in grains, usually as the calcium or mixed Ca^{2+}-Mg^{2+} salts **phytin.** The two apical cells of the 28-cell larvae of mesozoa (Fig. 1-10,A) contain enough magnesium phytate in granular form to account for up to half of the weight of the larvae.[4c] Inositol pentaphosphate is an allosteric activator for hemoglobin in birds (Chapter 4, Section E,6).[4d]

In bacteria inositol may be converted to D-glucuronic acid (Fig. 12-2) with the aid of an oxygenase. Free glucuronic acid is also formed by animals and undergoes a number of important metabolic alterations. However, its origin in animal tissues is uncertain. Some possibilities are indicated in Fig. 12-2.

Reduction of glucuronic acid with NADH yields **L-gulonic acid.** Note that this is an aldonic acid that could be formed by oxidation at the aldehyde end of the sugar **gulose.** Because C-6 of the glucuronic acid has become C-1 of gulonic acid, the latter belongs to the L family of sugars. Gulonic acid can readily be converted to a cyclic lactone which in a two-step process involving dehydrogenation and enolization (Fig. 12-2) is converted to **L-ascorbic acid,** vitamin C. This reaction occurs not only in plants, which make vitamin C in large quantities, but in most higher animals. However, the dehydrogenation step is lacking in man, in other primates, and in the guinea pig. One might say that we and the guinea pig have a genetic defect at this point which obliges us to eat relatively large quantities of plant materials to satisfy our bodily needs for ascorbic acid (see Box 10-G). Ascorbic acid is readily oxidized to dehydroascorbic acid which may be hydrolyzed to L-diketogulonic acid. The latter, after decarboxylation and reduction, is converted to L-xylulose, a compound that can also be formed by a standard oxidation and decarboxylation sequence on L-gulonic acid (Fig. 12-2).

At this point we meet an interesting metabolic variation, **idiopathic pentosuria.** Affected individuals cannot reduce xylulose to xylitol and hence excrete large amounts of the pentose into the urine, especially if the diet is rich in glucuronic acid. The "defect" seems to be absolutely harmless. The only problem is that the sugar in the urine can cause the condition to be mistaken for diabetes.

To complete our description of the reactions of Fig. 12-2, note that by reduction of xylulose to xylitol and by oxidation of the latter with NAD$^+$ we obtain D-xylulose. This sugar can be phosphorylated with ATP and enter the standard pentose phosphate pathways. In summary, the overall function of the reactions of Fig. 12-2 is first to provide a catabolic pathway for glucuronic acid, albeit a rather complex one. Second, they provide a route for synthesis of ascorbic acid in most species and (for humans, too) a pathway for catabolizing ascorbic acid.

4. Transformations of Fructose 6-Phosphate

Synthesis of D-**glucosamine 6-phosphate** is accomplished by reaction of fructose 6-phosphate with glutamine (Eq. 12-4):

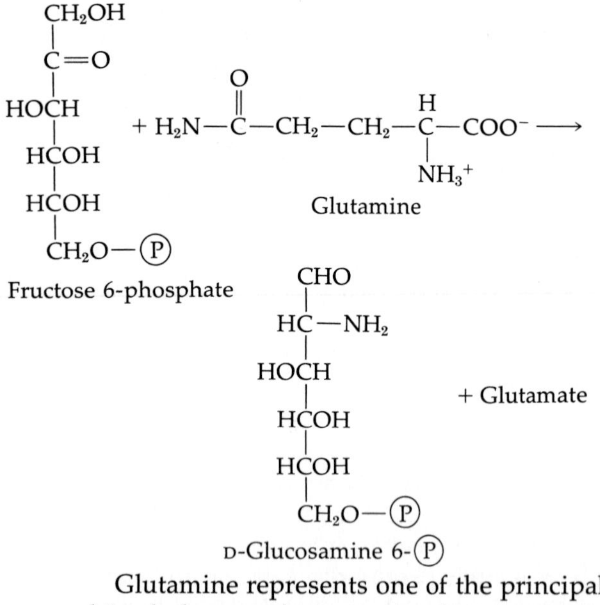

Fructose 6-phosphate

D-Glucosamine 6-(P)

+ Glutamate   (12-4)

Glutamine represents one of the principal combined forms of ammonia that is transported throughout the body (Chapter 14, Section B). At some point in the synthetic reaction of Eq. 12-4,

the amide linkage in glutamine must be hydrolytically cleaved with release of the ammonia. Think of this ammonia as reacting with the carbonyl group of fructose 6-phosphate to form an imine (Schiff base) which then undergoes essentially the same reaction as that catalyzed by sugar isomerases but with the production of D-glucosamine 6-phosphate as the product. An acetyltransferase next adds the N-acetyl group (from acetyl-CoA) to form N-acetylglucosamine 6-phosphate. The latter is isomerized to N-acetylglucosamine 1-phosphate and is converted to UDP-N-acetylglucosamine (UDP-GlcNac) in reactions paralleling those of UDPG synthesis (Eq. 11-24).

One of the products formed from UDP-GlcNac is **UDP-N-acetylmuramic acid.**[5] The initial step in synthesis of this compound is an unusual type of displacement on the α-carbon of PEP by the 3-hydroxyl group of the sugar (Eq. 12-5, step a). In-

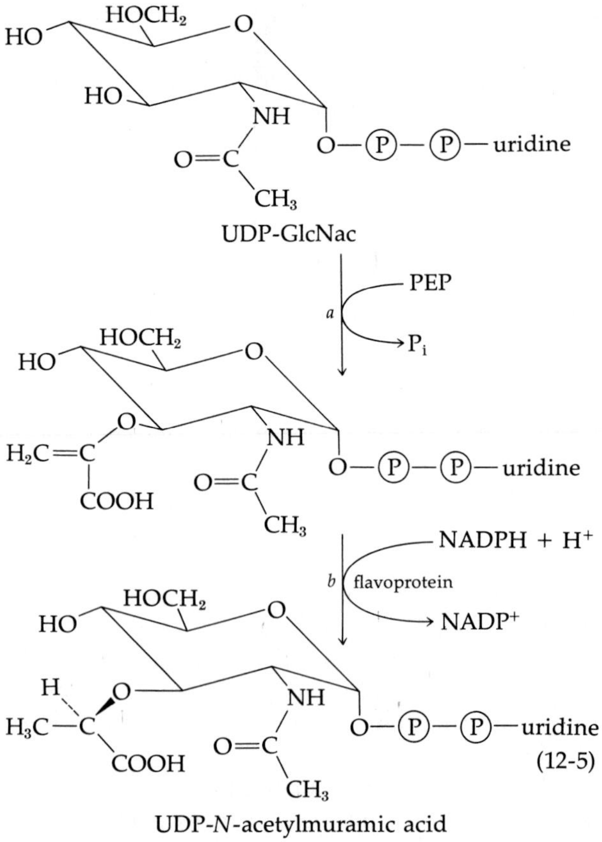

UDP-GlcNac

UDP-N-acetylmuramic acid   (12-5)

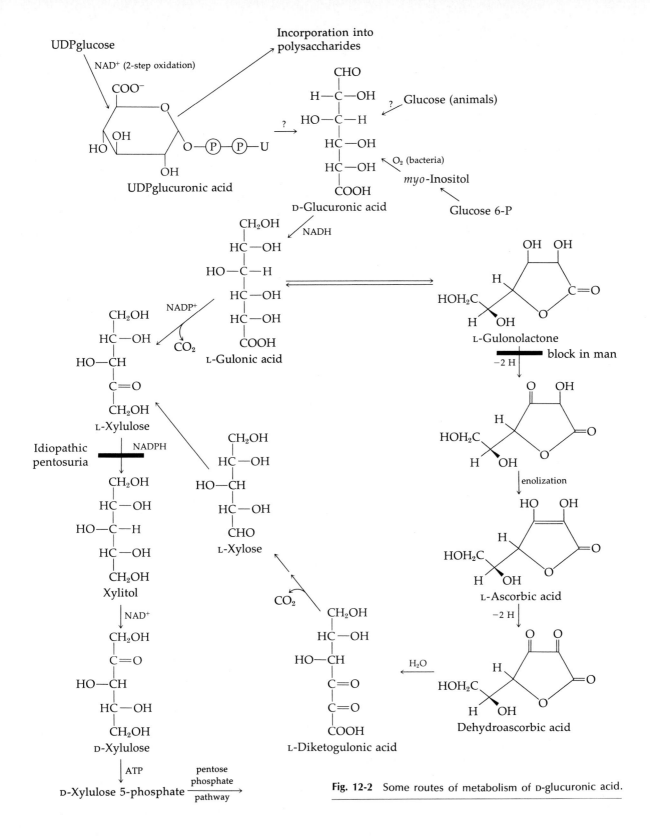

Fig. 12-2 Some routes of metabolism of D-glucuronic acid.

organic phosphate is displaced with formation of an enol pyruvate derivative of UDP-GlcNac. This derivative is then reduced with NADPH (Eq. 12-5, step b). The second sugar nucleotide derived from the same intermediate is **UDP-N-acetylgalactosamine,** formed by another 4-epimerase. Still a third biosynthetic intermediate is **CMP-N-acetylneuraminic acid.** Its formation requires a more complex pathway beginning with epimerization of UDP-GlcNac to UDP-ManNac with concurrent elimination of UDP (Eq. 12-6). The student should be able to write an elimination–addition mechanism by which UDP is split out and water is then added. One wonders whether a transient oxidation of an adjacent hydroxyl to a carbonyl is required for activation. The N-acetylmannosamine is phosphorylated on the

6-hydroxyl (Eq. 12-6, step b). In Eq. 12-6, step c, PEP is split to give P_i; at the same time the enolate anion released condenses with the N-acetylmannosamine phosphate (compare with the PEP carboxylase reaction, Eq. 7-79). The immediate product is N-acetylneuraminic acid 9-phosphate, which is cleaved through phosphatase action (Eq. 12-6, step d) and is activated to the CMP derivative by reaction with CTP (Eq. 12-6, step e). Note that this activated monosaccharide differs from the others in being a derivative of a CMP rather than of CDP. The nucleoside diphosphate derivative of the L-glycero-D-mannoheptose (Fig. 5-11) present in the lipopolysaccharide of gram-negative bacteria is formed from sedoheptulose 7-phosphate in a 4-step process.[5a]

$$(12\text{-}6)$$

5. Synthesis of Deoxy Sugars

Metabolism of sugars often involves dehydration to α,β-unsaturated carbonyl compounds. An example is the formation of 2-keto-3-deoxy derivatives of sugar acids (Eq. 7-59). Sometimes a carbonyl group is created by oxidation of an —OH group, apparently for the sole purpose of promoting dehydration. For example, the biosynthesis of L-rhamnose from D-glucose is a multistep process (Eq. 12-7) that takes place while the

sugars are attached to **deoxythymidine diphosphate**.[5,6] The first step is the introduction of the carbonyl group by oxidation. The dehydration occurs in Eq. 12-7, step b. To complete the sequence, the double bond formed by dehydration is reduced. The complex enzyme that catalyzed Eq. 12-7, steps a–c, contains its own oxidizing agent (NAD^+) which stores the reducing equivalents produced in Eq. 12-7, step a, and uses them to carry out the reduction in Eq. 12-7c. An inversion at C-3 occurs at some point in the sequence. Finally, a separate enzyme is needed for a second reduction (Eq. 12-7, step d).

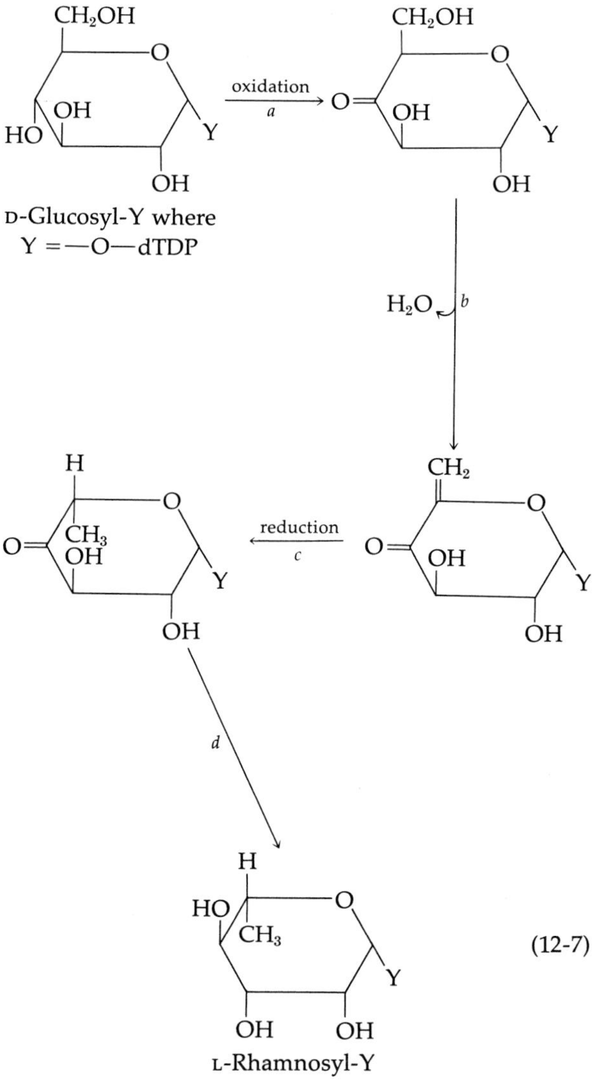

Biosynthesis of dideoxy sugars needed for bacterial surface polysaccharides is considered in Chapter 8, Section E,4.

B. SYNTHESIS AND UTILIZATION OF OLIGOSACCHARIDES

Our most common food sugar **sucrose** is formed in all green plants and nowhere else. Serving in plants principally as a transport sugar, it is made both in the chloroplasts and in the vicinity of other starch stores. Sucrose is extremely water soluble and chemically inert* because the hemiacetal groups of both sugar rings are blocked. However, sucrose is thermodynamically reactive, the glucosyl group having a group transfer potential of 29.3 kJ mol^{-1}. Transport of sugar in the form of a disaccharide provides an advantage to plants in that the disaccharide has a lower osmotic pressure than would the same concentration of sugar transported as a monosaccharide.

Biosynthesis of sucrose[7] utilizes both UDPglucose and fructose 6-phosphate (Eq. 12-8). Metabo-

$$\text{UDP-Glc} \xrightarrow{\qquad} \text{Sucrose 6-P} \xrightarrow{\text{H}_2\text{O}} \text{Sucrose} \quad (12\text{-}8)$$
$$\text{Fru 6-P} \qquad\qquad \text{P}_i$$

lism of sucrose in the animal body begins with the action of sucrase (invertase) which hydrolyzes the disaccharide to fructose and glucose (Eq. 12-9,

D-Glucosyl-Y where
Y = —O—dTDP

L-Rhamnosyl-Y

(12-7)

---

* However, sucrose is extremely sensitive toward hydrolysis catalyzed by acid.

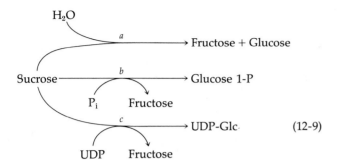

$$(12-9)$$

step *a*). The same enzyme is also found in higher plants and fungi. Cleavage of sucrose by sucrose phosphorylase (Eq. 12-9, step *b*) makes available (to certain bacteria) the activated glucose 1-phosphate which may enter catabolic pathways directly. Cleavage of sucrose for biosynthetic purposes can occur by reaction 12-9, step *c*, which yields UDPglucose in a single step.

A disaccharide with many of the same properties as sucrose is **trehalose** ("mushroom sugar," Chapter 2, Section C,2) which consists of two α-glucose units in $1 \rightarrow 1$ linkage. Trehalose is more inert chemically than sucrose, being remarkably resistant to acid hydrolysis. The biosynthetic pathway (Eq. 12-10) is entirely parallel to that for

UDP-Glc ⟶ Trehalose 6'-P

Glc-6-P

$P_i \leftarrow H_2O$

α,α'-Trehalose
[αGlc(1 → 1)αGlc]   (12-10)

sucrose. Trehalose is found in fungi and also in many insects.[8] Trehalose serves not only as the primary transport sugar in the hemolymph of insects but also acts as an "antifreeze."

Lactose, the characteristic sugar of milk, is formed by transfer of a glycosyl unit from UDP-galactose directly to glucose (Eq. 12-11, step *a*). The similar transfer of a galactosyl unit to N-acetylglucosamine (Eq. 12-11, step *b*) occurs in

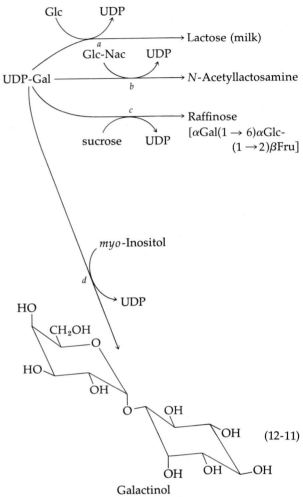

$$(12-11)$$

Galactinol

many animal tissues. A striking example of regulatory modification of an enzyme is seen here. The transferase catalyzing Eq. 12-11, step *b*, becomes, in the presence of **α-lactalbumin, lactose synthetase,** the enzyme catalyzing reaction *a* of Eq. 12-11. Lactalbumin was identified as a milk constituent long before its role as a regulatory protein became known.

Many higher plants contain the trisaccharide **raffinose** and related oligosaccharides. Raffinose arises from UDPgalactose by transfer of a galactosyl unit onto the 6-hydroxyl of the glucose ring of sucrose (Eq. 12-11, step *c*). Another important reaction is the transfer of a galactosyl unit to *myo*-inositol (Eq. 12-11, step *d*). The product **ga-**

lactinol is widespread in its occurrence within the plant kingdom. Galactinol, in turn, can serve as a specific donor of activated galactosyl groups. Thus, many plants contain **stachyose** and higher homologs, all of which are formed by transfer of additional α-D-galactosyl units onto the 6-hydroxyl of the galactose unit of raffinose. These sugars appear to serve as antifreeze agents in the plants.

An important reaction of UDPglucuronic acid is a formation of **glucuronides** (glucosiduronides). These are excretion products found in urine and derived by displacement of UDP from UDP-glucuronic acid by such compounds as phenol and benzoic acid. Phenol is converted to phenyl glucuronide, while benzoic acid (also excreted in part as hippuric acid, Box 9-A) yields an ester by the same type of displacement reaction (Eq. 12-12).

$$\text{Phenol} \xrightarrow{\text{UDP GlcUA}}$$

Phenyl glucuronide

$$(12\text{-}12)$$

Among the interesting sugar derivatives found in nature are numerous antibiotics, often containing amino groups (e.g., see Box 12-A, Fig. 12-10).[9,10]

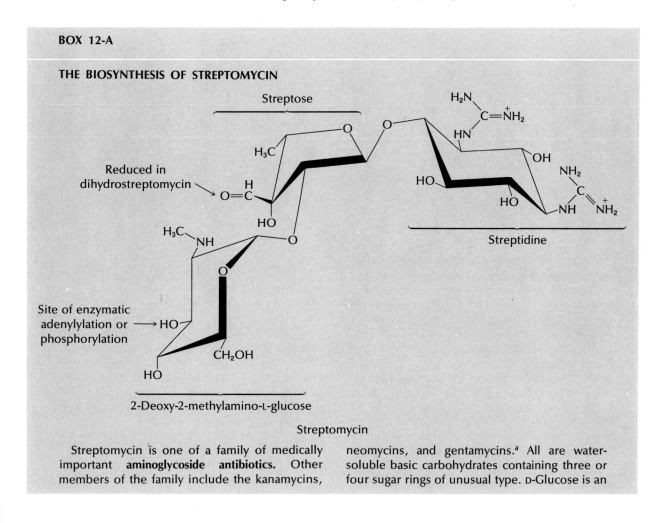

BOX 12-A

THE BIOSYNTHESIS OF STREPTOMYCIN

Streptose

Reduced in dihydrostreptomycin

Streptidine

Site of enzymatic adenylylation or phosphorylation

2-Deoxy-2-methylamino-L-glucose

Streptomycin

Streptomycin is one of a family of medically important **aminoglycoside antibiotics.** Other members of the family include the kanamycins, neomycins, and gentamycins.[a] All are water-soluble basic carbohydrates containing three or four sugar rings of unusual type. D-Glucose is an

Box 12-A (*Continued*)

effective precursor of streptomycin, all three rings being derived from it. While the route of biosynthesis of 2-deoxy-2-methylamino-L-glucose is not clear, the pathways to L-streptose and streptidine, the other two rings, have been characterized.[b,c] The starting material for streptidine synthesis is a nucleoside diphosphate sugar which is an intermediate in the synthesis of L-rhamnose (Eq. 12-7). The carbon-carbon chain undergoes an aldol cleavage between C-2 and C-3 as shown in step *a* of the following equation:

Intermediate
from Eq. 12-7

Streptosyl-Y

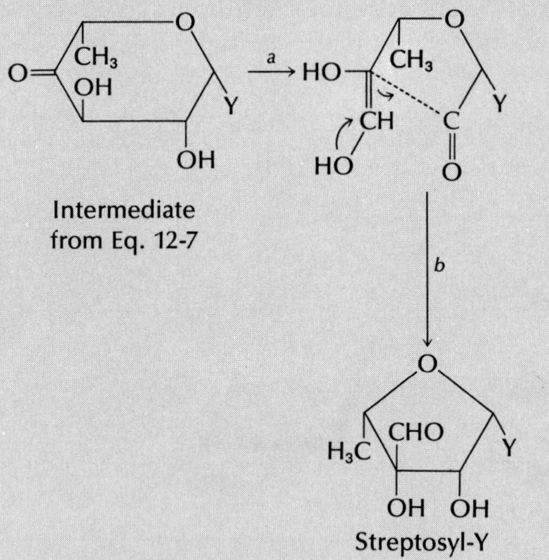

Note that the ring-open product is written as an enediol which is able to recyclize in a second aldol step *b* to form a five-membered ring with a branch at C-3. The L-streptosyl nucleoside diphosphate formed in this way serves as the donor of streptose to streptomycin.

The basic cyclitol streptidine is derived from *myo*-inositol (Eq. 12-3), which in turn is formed from glucose 6-phosphate. The guanidino groups are introduced by oxidation of the appropriate hydroxyl group to a carbonyl group followed by transamination from a specific amino donor. In the first step, illustrated by the following equation, glutamine is the amino donor for the transamination, the keto acid product being α-ketoglutaramic acid (KGam).

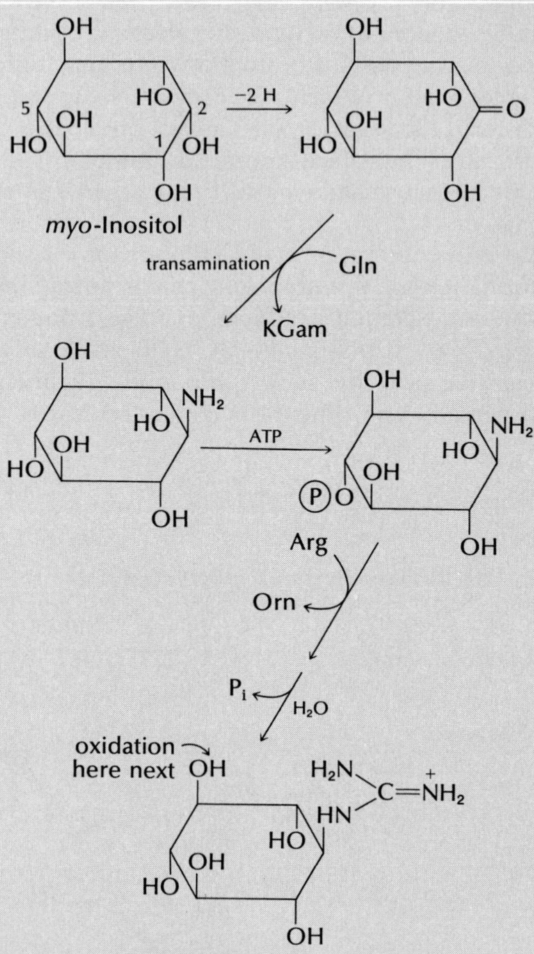

myo-Inositol

transamination   Gln

KGam

ATP

Arg

Orn

P_i H_2O

oxidation
here next

The amino group introduced into the ring now receives an amidine group

$$H_2N-C=\overset{+}{N}H_2$$

transferred from arginine by nucleophilic displacement. However, there is first a phosphorylation at the 5 position and after the amidine transfer has occurred to form the guanidino group, the phosphoryl group is hydrolyzed off by a phosphatase. Here again is a phosphorylation–dephosphorylation sequence (Chapter 11, Section B,3) designed to drive the reaction to completion in the desired direction. The second guanidino group is introduced in an entirely analogous way by oxidation at the 3 position followed

Box 12-A (*Continued*)

by transamination, this time with the amino group being donated by alanine. Again, a phosphorylation is followed by transfer of an amidine group from arginine. However, the final hydrolytic removal of the phosphoryl group (which this time is added at C-6) does not occur until the two other sugar rings have been transferred on from nucleoside diphosphate precursors to form streptomycin phosphate.

Streptomycin is subject to inactivation by enzymes encoded by genetic resistance factors (Chapter 15, Section D,7). Among these are en-

zymes that transfer phosphoryl groups or adenylyl groups onto streptomycin at the site indicated by the arrow in the structure. Thus, dephosphorylation at one site generates the active antibiotic as the final step in the biosynthesis, while phosphorylation at another site inactivates the antibiotic.

[a] R. Benveniste and J. Davies, *Annu. Rev. Biochem.* **42,** 471–506 (1973).
[b] M. Luckner, "Secondary Metabolism in Plants and Animals," pp. 78–80. Academic Press, New York, 1972.
[c] J. B. Walker and M. Skorvaga, *JBC* **248,** 2441–2446 (1973).

C. SYNTHESIS OF POLYSACCHARIDES

The synthesis of one homopolysaccharide, glycogen, has been considered previously (Chapter 11, Section F,3). While glycogen is formed from UDPglucose in animals, it is synthesized in bacteria via ADPglucose. The latter compound is also the glucosyl donor for starch biosynthesis (Fig. 12-1).[10a] The bushlike glycogen and amylopectin molecules grow at the numerous nonreducing ends. The combination of growth and degradation from the same ends provides a means of rapidly storing and utilizing glucose units. A similar synthetic pattern of donation of glycosyl units from UDP derivatives to the nonreducing ends also applies to the oligosaccharides discussed in the preceding section and to many oligosaccharide groups attached to proteins and lipids.

1. Highly Specific Transferases

The problem of synthesizing a complex polysaccharide according to a genetically determined pattern has been discussed briefly (Chapter 11, Section F,3). A specific example is the formation of alternating polysaccharides such as hyaluronic acid and chondroitin sulfates together with their

special terminal units (Chapter 2, Sections C-3 and C-4). A whole series of specific transferases appear to be required. The first one transfers a xylosyl residue from UDPxylose to a serine —OH group in the protein. Then an enzyme, with proper specificity, transfers a galactose from UDPgalactose, joining it in 1,4 linkage. A third enzyme transfers another galactose onto the first one, in 1,3 linkage. Then comes a special transferase for glucuronic acid with a specificity different from that used in creating the main chain. In this way the terminal unit attached to serine is completed. Then two enzymes transfer the alternating units in sequence. In the case of hyaluronic acid one enzyme is specific for UDP-GlcNac and transfers the sugar unit only to the end of a glucuronic acid ring. The second enzyme is specific for UDPglucuronic acid but attaches it only to the end of an acetylglucosamine unit. Formation of chondroitin sulfates requires a different pair of enzymes with different specificities (see structure in Fig. 2-16). Another specific transferase is needed to add the sulfate groups in their proper places.

The addition of sugar units to proteins of the cell membrane and to proteins that are excreted from the cells is promoted by other transferases. An example is the synthesis of the blood group substances. The role of specific glycosyltransferases in determining blood type has already been discussed (Chapter 5, Section C,1).

$$(12\text{-}13)$$

2. Chain Growth by Insertion

A quite different mechanism of polysaccharide chain growth operates in some cases. Thus, **dextran sucrase** of *Leuconostoc* and *Streptococcus* adds glucosyl units at the *reducing* ends of the chains of the polysaccharide dextran (Chapter 2, Section C,3).* Sucrose is the direct donor of the glucosyl groups, and the initial step appears to be the formation of a glucosyl enzyme (the glucosyl group becomes attached to the nucleophilic group Y' in Eq. 12-13, step *a*). As the polysaccharide chain

* If suitable compounds can be found to inhibit this enzyme they might be useful as additives to toothpastes to prevent dental plaque formation (suggestion of John Robyt).

grows, it remains attached to the enzyme. Each new glucose unit added is inserted between the enzyme and the attached polysaccharide chain. This insertion mechanism is understandable if it is assumed that the enzyme has two sites *for binding activated glycosyl groups.* The growing chain is bound at one site and a single activated glucosyl unit is bound at the other (Eq. 12-13). The enzyme catalyzes a transfer reaction by which the growing chain is shifted from some nucleophilic group Y of the enzyme to the free 6-hydroxyl group of the glucosyl unit being inserted.[11]

The foregoing observation could explain a puzzling question about the synthesis of starch. The branched component, amylopectin, presumably grows in much the same way as does glycogen. The only difference is that the outer branches become longer before new branches are formed. The special branching enzyme ("Q enzyme"), like the one acting on glycogen, transfers part of the chain to the —OH group of a glucose unit in an adjacent polysaccharide chain lying parallel to the first. Since amylose and amylopectin are intimately intermixed in the starch granules, how does it happen that the branching enzyme never puts a branch on the straight-chain amylose molecules? One possibility is that the linear amylose chains are oriented in the opposite direction from the amylopectin chains. The nonreducing ends of the amylose molecules may be located toward the center of the starch granule, and growth may occur by an insertion mechanism at the reducing ends. These ends may move out continually as the granule grows.[12] While this is a speculative picture, it serves to emphasize that there are many unresolved questions concerning the synthesis of polysaccharides.

3. Lipid Carriers

A "cluster" of sugar units of specific structure makes up the repeating unit of the "O antigen" of *Salmonella* (Fig. 5-11). Recall that this O antigen is attached to a complex lipopolysaccharide. It appears that both this lipopolysaccharide and the attached O antigen are synthesized inside the bacterial cell by enzymes found in the cytoplasmic membrane.[13,14] Complete lipopolysaccharide units are then translocated from the inner membrane to the outer membrane of the bacteria. The details of synthesis of the "core" of the lipopolysaccharide are only now being learned but the biosynthesis of the O antigen is relatively well understood and is indicated schematically in Fig. 12-3. A striking feature of this biosynthetic pathway is the dependence upon a special polyprenyl alcohol **undecaprenol** (bactoprenol). This C_{55} compound containing two trans double bonds and 9 cis double bonds is one of a family of isoprenoid lipids that occur throughout living things and participate as carriers in biosynthetic processes at

$$H-(-CH_2-\underset{\underset{CH_3}{|}}{C}=CH-CH_2-)_{\overline{10}}CH_2-\underset{\underset{CH_3}{|}}{C}=CH-CH_2-OH$$

Undecaprenol

or in cell membranes. A series of related compounds, the **dolichols,** contain 16–20 prenyl units of which the terminal one, bearing the OH group, is completely saturated.[15] Like undecaprenol,[15a] these compounds are converted to phosphate esters. The phosphate group on a polyprenyl alcohol presumably protrudes into the cytoplasm while the lipid chain provides a firm anchor in the membrane. Of course, there may be more to it than this. The specific geometry of the polyprenyl chain suggests the possibility of a sophisticated structural or mechanical role that is still unknown to us.

During synthesis of the O antigens, undecaprenol phosphate (abbreviated P-lipid in Fig. 12-3) reacts with UDPgalactose with transfer of a phosphogalactosyl unit onto the lipid carrier. Then the oligosaccharide repeating unit of the O antigen is constructed by the consecutive transferring action of three more transferases. For the particular antigen shown in Fig. 12-3 one enzyme transfers a rhamnosyl unit, another a mannosyl unit, and another an abequosyl unit from the appropriate sugar nucleotides. Then the entire growing O antigen chain, which is attached to a second molecule of undecaprenol diphosphate, is transferred onto the end of the newly assembled oligosaccharide unit as indicated in Fig. 12-3. Thus, the oligosaccharide is *inserted* at the *reducing* end of the growing chain. Elongation continues by the transfer of the entire chain onto yet another tetrasaccharide unit. As each oligosaccharide unit is added an undecaprenol diphosphate unit is released and a phosphatase cleaves off the terminal phosphoryl unit to regenerate the original undecaprenol phosphate carrier. Finally, when the O antigen is long enough it is attached to the core lipopolysaccharide.

Another biosynthetic cycle (Fig. 12-4) that depends upon an undecaprenol phosphate is the for-

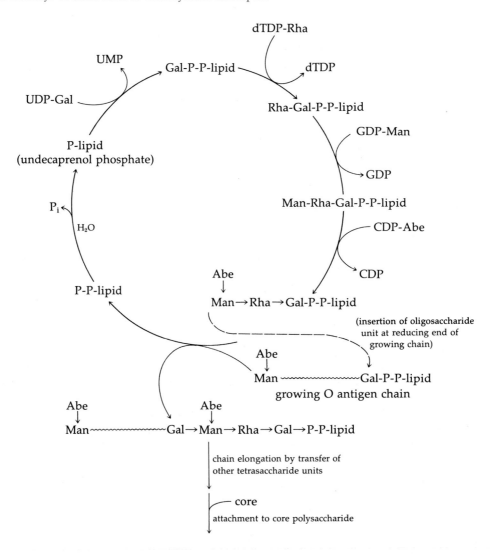

Fig. 12-3 Biosynthesis of an O antigen polysaccharide of *Salmonella typhimurium*.

mation of the peptidoglycan of bacterial cell walls (Fig. 5-9). The first step is synthesis of UDP-*N*-acetylmuramic acid as shown in Eq. 12-5. Then L-alanine is attached to the OH group of the lactyl unit of the muramic acid in a typical ATP-requiring process (Fig. 12-4, step *a*). Next D-glutamic acid, L-lysine, and D-alanyl-D-alanine are joined in sequence, each in an ATP-requiring step. The entire unit assembled in this way is transferred to undecaprenol phosphate (step *e*). An *N*-acetylglucosamine unit is added by action of another transferase (step *f*) and, in an ATP-requiring process, ammonia is (sometimes) added to cap the free α-carboxyl group of the D-glutamyl residue (step *g*). Next (step *h*), in sequence, five, glycyl units are added, each from a molecule of glycyl-tRNA. The completely assembled repeating unit, together with the connecting peptide chain

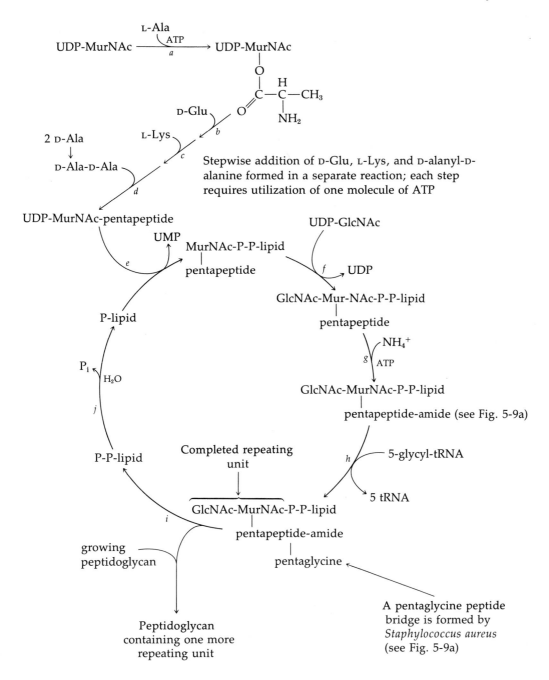

Fig. 12-4 Biosynthesis of the peptidoglycan of *Staphylococcus aureus*. See Fig. 5-9A for details of the peptidoglycan structure.

needed in the cross-linking reaction (see Box 7-D), is then transferred onto the growing chain (step i). Again, as in formation of the *Salmonella* O antigen and dextrans, growth is at the reducing end (insertion). The polyprenyl diphosphate is released and the cycle is completed by the action of a phosphatase (step j).

A somewhat different role for a polyprenyl carrier is observed in the synthesis of the α-1,6-linked yeast **mannan,** a polysaccharide that also contains short branches (1–3 mannose units) attached through α-1,3 and α-1,2 linkages. In this case, it appears that within yeast cells mannosyl units are transferred from GDPmannose to phosphate esters of a mixture of dolichols containing 14–18 prenyl units.[16] The mannosyl units are then transferred from the lipid phosphate carrier onto growing mannan chains. There is now much evidence for a similar mechanism of transfer of mannosyl and *N*-acetylglucosyl groups to glycoproteins in animal tissues.[17-19] An intermediate with the structure $(\alpha\text{-Man})_4\text{-}(1\rightarrow4)\text{-}\beta\text{Man-}(1\rightarrow4)\text{-}$ GlcNAc-$(1\rightarrow4)$-GlcNAc-P-P-dolichol has been proposed.[19] While there is no evidence that polyprenyl alcohols function as carriers in proteoglycan formation (Section C,1) it appears likely that they are often involved in the assembly of various oligosaccharide units (as in Fig. 12-3) of glycoproteins.[19a] The completed oligosaccharides are transferred onto amino acid side chains to form the glycoproteins which are secreted or incorporated into the plasma membrane.

4. Cellulose and Chitin

Despite their obvious importance, little is known of the details of synthesis of the regular β-1,4-linked polymers cellulose and chitin. Cellulose synthetases transfer glucosyl units from GDP-glucose or UDPglucose and chitin synthetases from UDP-*N*-acetylglucosamine. A solubilized form of a fungal chitin synthetase exists as 35–100 nm ellipsoidal granules from which, in the pres-ence of substrate, microfibrils of 12–18 nm diameter arise.[20] Presumably a multienzyme aggregate forms many polymer chains side by side to generate the hydrogen-bonded microfibrils. A similar process has been suggested for cellulose formation.[21,21a]

D. THE INTRACELLULAR BREAKDOWN OF POLYSACCHARIDES AND GLYCOLIPIDS

The attention of biochemists has been drawn to the importance of pathways of degradation of complex polysaccharides through the existence of several inherited metabolic diseases. There are at least seven types of **mucopolysaccharidosis** in which mucopolysaccharides such as hyaluronic acid accumulate to abnormal levels in tissues and may be excreted in the urine. The diseases cause severe skeletal defects; varying degrees of mental retardation; and early death from liver, kidney, or cardiovascular problems. The mucopolysaccharidoses are **lysosomal storage diseases,**[22] which result from a lack of one of the 40-odd known lysosomal hydrolases. As in other lysosomal diseases, undegraded material is stored in intracellular inclusions lined by a single membrane. Various tissues are affected to different degrees and the diseases tend to progress with time.

First described in 1919 by Hurler, mucopolysaccharidosis I (the Hurler syndrome) leads to accumulation of partially degraded dermatan sulfate.[22] Recall (Fig. 2-16) that this protein-bound polysaccharide consists of alternating units of iduronic acid and β-D-*N*-acetylgalacturonic acid-4-sulfate.

$$[\ \alpha\text{-L-IduUA}\ (1\rightarrow3)\ \beta\text{GalNAc-}(1\rightarrow4)\ ]_n$$
$$|$$
$$4\text{-sulfate}$$

A standard procedure in the study of diseases of this type is to culture fibroblasts from a skin biopsy. Cells of a fibroblast culture from patients with the Hurler syndrome accumulate the polysaccharide, but when fibroblasts from a normal person are cultured in the same vessel the defect is "corrected." It has been shown that a protein secreted by the normal fibroblasts is taken up by the defective fibroblasts, permitting them to complete the degradation of the stored polysaccharide. This "Hurler corrective factor" has been identified as an **α-L-iduronidase.** In a similar manner, the missing enzyme in several other disorders has been identified. Thus, in the Hunter syndrome dermatan sulfate and heparan sulfate accumulate. The missing enzyme is a **sulfatase** for sulfated iduronic residues.[23] The Sanfilippo A corrective factor is a heparan sulfatase (probably an *N*-sulfatase), while the Sanfilippo B factor is an *N*-acetyl-α-glucosaminidase. In another disease, β-glucuronidase is lacking. The picture given here is somewhat simplified. For many lysosomal diseases there are mild and severe forms and infantile or juvenile forms to be contrasted with adult forms. Some of the enzymes exist as multiple isozymes. An enzyme can be completely lacking or low in concentration. Perhaps some forms of the diseases result from lack of regulatory proteins.

1. Glycolipids

Glycolipids, like the glycoproteins, are thought to be synthesized in the membranes of the endoplasmic reticulum, perhaps being placed at the same time into the membrane surface facing the cisternae of the endoplasmic reticulum. From there, they are transported into the Golgi apparatus and eventually outward to join the outer surface of the plasma membrane. A major class of glycolipids are the sphingolipids known as cerebrosides and gangliosides. Both are derived from the *N*-acylated sphingosine derivatives known as **ceramides** (Table 2-8). Some of the biosynthetic pathways leading from sphingosine to these substances are indicated in Fig. 12-5. Acyl, glycosyl, and sulfuryl groups are transferred from appro-

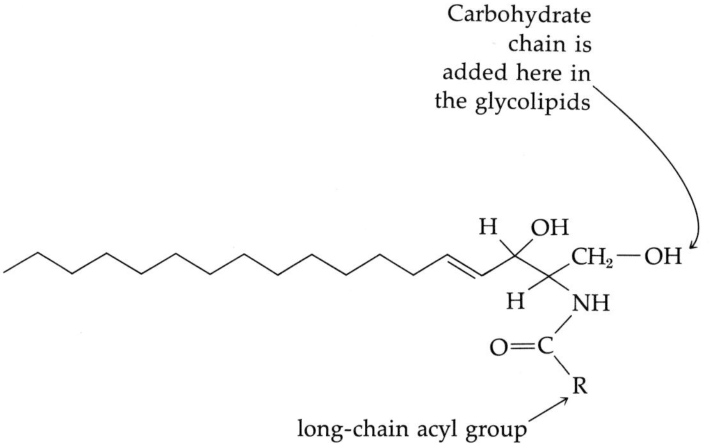

A ceramide

priate derivatives of CoA, CDP, UDP, CMP, and from PAPS. The biosynthesis of a sphingomyelin is also shown in this scheme, but is discussed later in the chapter.

Each of the biosynthetic steps indicated in Fig. 12-5 is catalyzed by a specific transferase. Most of these enzymes are present in membranes and have not been studied in detail. Furthermore, the sequence by which the transferases act may not always be fixed,* and a complete biosynthetic scheme would probably be far more complex than is shown in the figure.

A striking aspect of the metabolism of the glycolipids is the existence of at least ten lysosomal storage diseases (sphingolipidoses)[24–28] whose biochemical basis is indicated in Fig. 12-5 and in

* One alternate sequence is the synthesis of galactosyl ceramide by transfer of galactose to sphingosine followed by acylation:

$$\text{Sphingosine} \xrightarrow{\text{UDP-Gal}}$$

$$O\text{-Galactosylsphingosine (psychosine)} \xrightarrow{\text{acyl-CoA}} \text{Gal-Cer}$$

However, the pathway shown in Fig. 12-5 is probably more important.

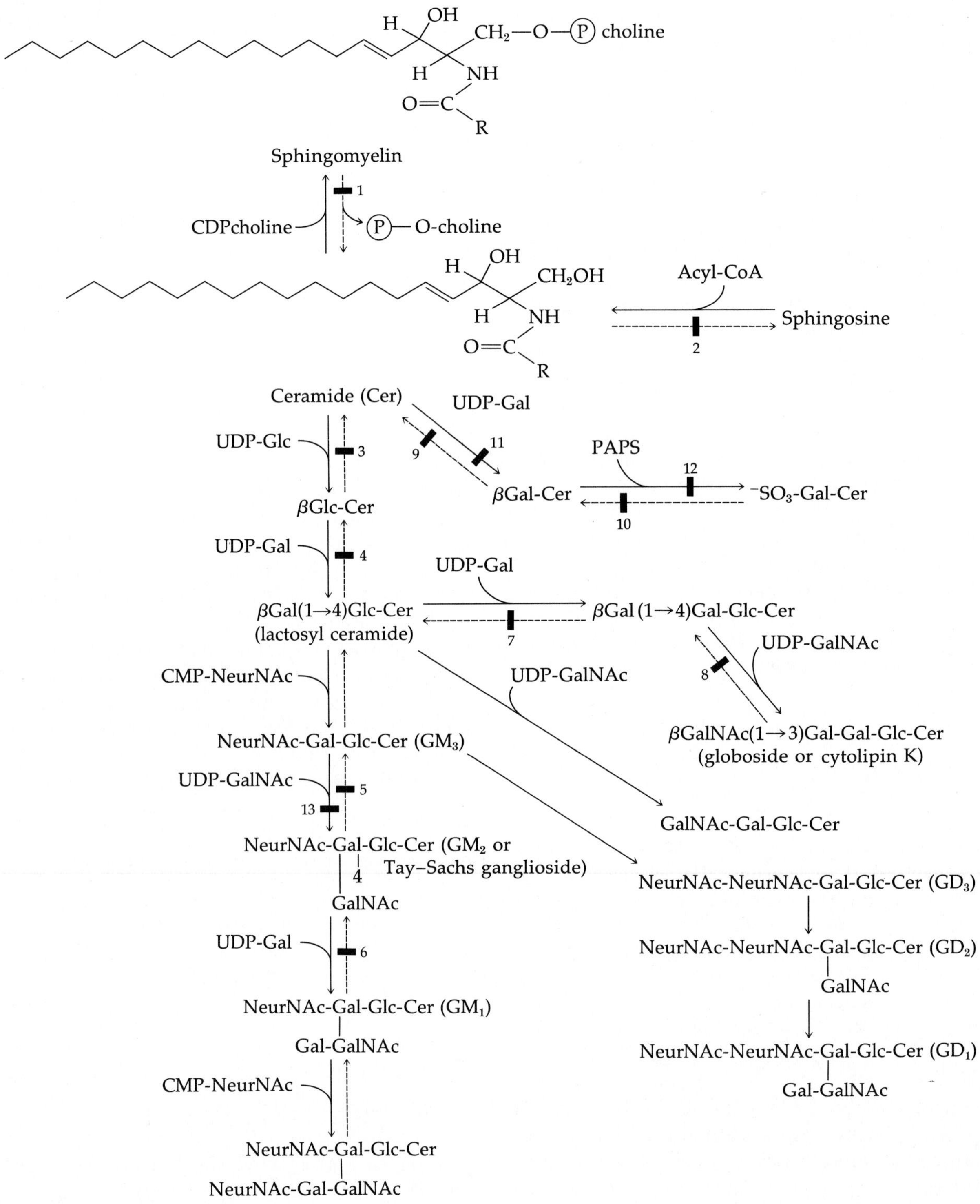

$$H \quad OH$$
$$CH_2-O-\textcircled{P}\,choline$$
$$H \quad NH$$
$$O=C$$
$$R$$

Sphingomyelin

1

CDPcholine $\longrightarrow$ $\textcircled{P}$—O-choline

$$H \quad OH$$
$$CH_2OH \qquad Acyl\text{-}CoA$$
$$H \quad NH$$
$$O=C \qquad \qquad \qquad \qquad Sphingosine$$
$$R \qquad \qquad \qquad \qquad 2$$

Ceramide (Cer)   UDP-Gal

UDP-Glc $\longrightarrow$ 3   9   11   PAPS   12

βGlc-Cer βGal-Cer $\longrightarrow$ $^-SO_3$-Gal-Cer

UDP-Gal $\longrightarrow$ 4   10

βGal$(1\rightarrow4)$Glc-Cer $\longrightarrow$ βGal$(1\rightarrow4)$Gal-Glc-Cer
(lactosyl ceramide)   UDP-Gal   7   UDP-GalNAc   8

CMP-NeurNAc   UDP-GalNAc βGalNAc$(1\rightarrow3)$Gal-Gal-Glc-Cer
(globoside or cytolipin K)

NeurNAc-Gal-Glc-Cer (GM$_3$)

UDP-GalNAc $\longrightarrow$ 5   GalNAc-Gal-Glc-Cer
13

NeurNAc-Gal-Glc-Cer (GM$_2$ or   NeurNAc-NeurNAc-Gal-Glc-Cer (GD$_3$)
Tay–Sachs ganglioside)

4   NeurNAc-NeurNAc-Gal-Glc-Cer (GD$_2$)

GalNAc   GalNAc

UDP-Gal $\longrightarrow$ 6   NeurNAc-NeurNAc-Gal-Glc-Cer (GD$_1$)

NeurNAc-Gal-Glc-Cer (GM$_1$)   Gal-GalNAc

Gal-GalNAc

CMP-NeurNAc

NeurNAc-Gal-Glc-Cer

NeurNAc-Gal-GalNAc

TABLE 12-1
Storage Diseases of Glycolipids (Sphingolipidoses)

No. in Fig. 12-5	Name	Defective enzyme
1.	Niemann–Pick disease	Sphingomyelinase
2.	Farber's disease (lipogranulomatosis)	Ceramidase[a]
3.	Gaucher's disease	β-Glucosidase
4.	Lactosyl ceramidosis	β-Galactosyl hydrolase
5.	Tay–Sachs disease	Hexosaminidase A[b]
6.	Generalized gangliosidosis	β-Galactosidase[c]
7.	Fabry's disease	α-Galactosidase[d,e]
8.	Sandhoff's disease	Hexosaminidases A and B
9.	Krabbe's leukodystrophy	Galactocerebrosidase
10.	Metachromatic leukodystrophy	Sulfatase
13.	Hematoside (GM$_3$) accumulation	GM$_3$-N-acetylgalactosaminyltransferase[f]

[a] M. Sugita, J. T. Dulaney, and H. W. Moser, *Science* **178**, 1100–1102 (1972).

[b] J. F. Tallman and R. O. Brady, *JBC* **247**, 7570–7575 (1972).

[c] The same disorder is found in a strain of Siamese cats; H. J. Baker, Jr., J. R. Lindsey, G. M. McKhann, and D. F. Farrell, *Science* **174**, 838–839 (1971).

[d] E. Beutler and W. Kuhl, *JBC* **247**, 7195–7200 (1972).

[e] J. C. Crawhall and M. Banfalvi, *Science* **177**, 527–528 (1972).

[f] P. H. Fishman, S. R. Max, J. F. Tallman, R. O. Brady, N. K. Maclaren, and M. Cornblath, *Science* **187**, 68–70 (1975).

Table 12-1. The first of these diseases to be investigated was **Gaucher's disease,** a result of an autosomal recessive trait that permits **glucosyl ceramide** to accumulate. The liver and spleen are seriously damaged, the latter becoming enlarged to four or five times normal size in the adult form of the disease. In the more severe juvenile form, mental retardation occurs. It was shown by R. O. Brady that the cerebroside is synthesized at a normal rate in the individuals affected. In 1965, it was shown that a lysosomal hydrolase was missing and that the catabolic pathway indicated by

Fig. 12-5 Biosynthesis and catabolism of glycosphingolipids. The heavy bars indicate metabolic blocks in known metabolic diseases.

dashed arrows in Fig. 12-5 was blocked (block No. 3 in the figure). It has been found that in Fabry's disease, an X-linked gene providing for removal of galactosyl residues from cerebrosides is defective. This leads to accumulation of the **triglycosylceramide** whose degradation is blocked at point 7 in Fig. 12-5.

The best known and commonest of the sphingolipidoses is **Tay–Sachs disease.** Over 500 cases have been reported since it was first described in 1881. A severe disease, it is accompanied by mental deterioration, blindness, paralysis, dementia, and death by the age of three. It is estimated that 30 children a year are born in North America with this condition, and the world figure must be 5–7 times this.[25]

2. Can Lysosomal Diseases Be Treated?

Intense interest now exists in the possibility of enzyme replacement therapy for lysosomal deficiency diseases.[22] That cells can take up enzymes has been demonstrated in tissue culture. It appears that in the process of pinocytosis external membrane is engulfed to form pinocytotic vacuoles which then fuse with lysosomes. The lysosomal hydrolases then degrade the polysaccharides in a process that is apparently essential to the normal function of the cell. Pinocytosis also provides the means for uptake of enzyme from the external medium and the hope of possible therapy by enzyme replacement. A problem arises because injection of enzymes into the bloodstream can lead to allergic responses. An alternate approach that might be feasible when accumulated substances are excreted into the bloodstream is microencapsulation of the enzymes, perhaps in ghosts from the patients' own erythrocytes.[22] In the case of Gaucher's disease and Fabry's disease, it is hoped that treatment of infants and young children may prevent brain damage. Some success has already been reported in treatment of Fabry's disease. However, in Tay–Sachs disease, the primary sites of accumulation of the ganglioside GM$_2$

are the ganglion and glial cells of the brain. Because of the "blood–brain barrier" and the severity of the damage it seems less likely that the disease can be treated successfully.

The approach presently used consists of identifying carriers of highly undesirable genetic traits and offering genetic counseling.* If both parents are carriers, the risk of bearing a child with Tay–Sachs disease is one in four. Thus, in one group of 32 women each of whom had borne a previous child with the disease, the genetic status of the fetus was established by amniocentesis.† As predicted, 8 of the 32 carried the disease. In this case, 7 of the women chose abortion; the diagnosis was made too late in the eighth case and the child was born with the disease.[27]

Other diseases of carbohydrate catabolism are also known. For example, **α-fucosidosis** results from inability to remove fucose residues from the surface polysaccharides.[29–31] We see that there are a large variety of problems in the catabolism of body constituents. On the other hand, fewer cases are on record of deficiencies in biosynthetic pathways. Perhaps such deficiencies are more often absolutely lethal and lead to early spontaneous abortion. However, blockages in the biosynthesis of cerebrosides are known in the special strains of mice known as Jimpy, Quaking, and msd (myelin synthesis deficient).[32–34] The transferases (points 11 and 12 of Fig. 12-5) are not absent but are of low activity. The mice have distinct neurological defects and poor myelination of nerves in the brain. Recently, a human ailment involving impaired conversion of GM_3 to GM_2 (with accumulation of the former; point 13 of Fig. 12-5) has been reported.

---

* However, in many cases tests for identification of heterozygotic carriers are not yet available.

† A sample of the amniotic fluid surrounding the fetus is withdrawn during the sixteenth to eighteenth week of pregnancy. The fluid contains fibroblasts that have become detached from the surface of the fetus. These cells are cultured for 2–3 weeks to provide enough cells for a reliable assay of the appropriate enzymes.

Because of the apparent loss of communication between cells in cancer, the cell surfaces of tumor cells are being examined biochemically. Changes in the ganglioside composition in tumor cells have been attributed to depression of specific glycosyltransferase activities.[35,36]

Ganglioside GM_1 (Fig. 12-5) appears to be the natural receptor for the cholera toxin (Chapter 6, Section F,5).[36a] Binding of the B subunit of this toxic protein to the exposed oligosaccharide chains of several ganglioside molecules presumably initiates a chain of reactions by which the A subunit of the toxin is released to activate adenylate cyclase.[36b]

E. SPECIFIC ASPECTS OF FATTY ACID METABOLISM

Considerable structural variation is encountered in nature within the fatty acids and their immediate derivatives.[36c] Part of this results from the use of more than one "starter piece." Thus, if acetyl-CoA is the starter piece, chain elongation via malonyl-CoA (Fig. 11-2) leads to fatty acids with an even number of carbon atoms. Degradation of the branched chain amino acids valine, isoleucine, and leucine creates a series of branched starter pieces (Table 12-2) whose utilization leads to formation of the iso and anteiso series of fatty acids. These are found in the lipids of tobacco and wool, in the "sound lens" of echo-locating porpoises,[37] and in many other materials.[38] Propionyl-CoA serves as an intermediate for introduction (via methylmalonyl-CoA) of branches at various other parts of a fatty acid chain. Thus, 2R- and 4R-methylhexanoic acids, 2,4,6,8-tetramethyldecanoic acid, and a variety of other branched chain acids are esterified with long-chain alcohols (mainly 1-octadecanol) to form the waxes of the preen glands of ducks and geese.[39] The C_{32} **mycocerosic acid** of the tubercle bacteria is closely related.

2R,4R,6R,8R-Tetramethyloctacosanoic acid (mycocerosic acid)

In the majority of plants, animals, and bacteria fatty acids with an even number of carbon atoms predominate. The fatty acid synthetase complex yields CoA derivatives of primarily the C_{16} and C_{18} straight-chain acids in both animals and plants (Fig. 12-6). However, in plants and animals alike, enzyme systems of the endoplasmic reticulum (microsomes), using malonyl-CoA and NADPH, are able to further elongate the fatty acids. Elongation can also occur in mitochondria by reactions that are essentially the reverse of β oxidation. The only deviation from an exact reversal of β oxidation is the use of NADPH as the reductant for enoyl-CoA reductase. It has been suggested that elongation of fatty acids in the *outer* membrane of mitochondria, followed by transport of the elongated chains into the mitochondria, may constitute another shuttle for transport of reducing equivalents from NADH into mitochondria (Chapter 10, Section E,11).[40]

1. Unsaturated Fatty Acids and Their Modification Products

Fatty acids containing one or more double bonds provide necessary fluidity to cell mem-

TABLE 12-2
Starter Pieces for Biosynthesis of Fatty Acids

Starter piece	Fatty acid products
CH_3—C(=O)—S CoA Acetyl-CoA	Acid with even number of carbon atoms
CH_3CH_2—C(=O)—S CoA Propionyl-CoA	Acid with odd number of carbon atoms
Valine → Isobutyryl-CoA	Iso series (even)
Leucine → Isovaleryl-CoA	Iso series (odd)
Isoleucine →	Anteiso series (odd)

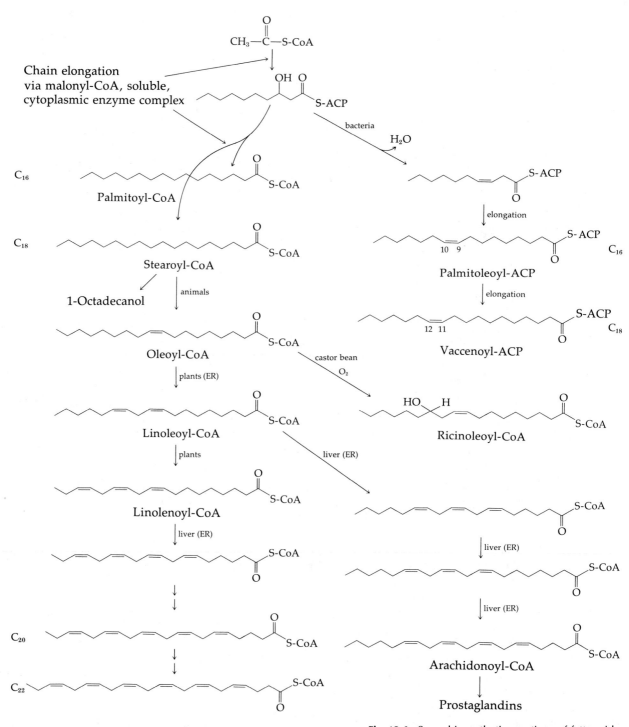

Fig. 12-6 Some biosynthetic reactions of fatty acids.

branes in many organisms and serve as precursors to other components of cells. A significant difference between the introduction of double bonds into fatty acids is observed between animals, protozoa, fungi and certain bacteria, and bacteria such as *E. coli* that can live anaerobically. In the latter, **vaccenic acid** is often the principal unsaturated fatty acid. It is formed by chain elongation after introduction of a cis double bond at the C_{10} stage of synthesis. The bacteria possess a **β-hydroxydecanoyl thioester dehydratase** which catalyzes elimination of a β-hydroxyl group to yield primarily the cis-β,γ rather than the trans-α,β-unsaturated product (Eq. 12-14). The mechanism

trans-α,β-Unsaturated product

cis-β,γ-Unsaturated product

chain elongation

Longer unsaturated fatty acids

where ACP stands for acyl carrier protein (Chapter 8, Section B,3)   (12-14)

may resemble that of enoyl hydratase (Eq. 7-42, reverse), the bracketed trans-α,β-unsaturated intermediate (enzyme bound) being isomerized to the cis-β,γ-unsaturated product through an allylic rearrangement. The product can then be elongated to the C_{16} **palmitoleoyl**-CoA and C_{18} **vaccenoyl**-CoA derivatives (Fig. 12-6, right side). However,

* While both CoA and ACP derivatives are indicated in Fig. 12-6, many steps have been omitted. Often synthesis occurs while the fatty acid chain is attached to an acyl carrier protein but the final product is released as a coenzyme A derivative (Chapter 11, Section D,6).

dehydration of β-hydroxydecanoyl ACP* lies at a branch point in the biosynthetic sequences.[41] The trans-α,β-unsaturated fatty acyl compound lies on the usual route of chain elongation to palmitoyl-CoA (left side, Fig. 12-6).

In higher plants, animals, protozoa, and fungi, saturated fatty acids are acted upon by **desaturases** (Chapter 10, Section G,3) to introduce double bonds, usually of the cis configuration. The introduction of the first double bond, a process occurring in both animals and plants, takes place in the cytosol. The resulting oleoyl coenzyme can be converted to CoA derivatives of **linoleic, linolenic, arachidonic,** and other polyenoic acids by reactions indicated in Fig. 12-6. The desaturation steps required take place in the endoplasmic reticulum of plant cells and require NADPH and light-generated ferredoxin as well as O_2.

The conversion of oleoyl-CoA to linoleoyl-CoA does not occur in animals. As a result of this biosynthetic deficiency polyunsaturated fatty acids such as linoleic, linolenic, and the C_{20} arachidonic acid are necessary in the diet. When these **essential fatty acids** of plant origin* are lacking, animals grow poorly and develop skin lesions, kidney damage, and impaired fertility. Now it is clear that one, though perhaps not the only essential function of these compounds is to serve as precursors to the "local hormones" known as **prostaglandins** (Section E,3).[42] Another role for arachidonic acid has been identified in blood platelets. A lipoxygenase converts it to 12-L-hydroxy-5,8,10,14-eicosatetraenoic acid, a chemotactic factor for neutrophils (see Box 5-G).

The origin of ricinoleic acid, a special constituent of castor beans (Fig. 2-32), is also shown in Fig. 12-6. Some organisms contain **cyclopropane fatty acids** (see Fig. 2-32).[43] The extra carbon of the cyclopropane ring is added from S-adenosylmethionine (SAM) at the site of a double bond in a fatty acyl group of phosphatidylethanolamine present in a membrane (Eq. 12-15).[44-45] The

* Arachidonic acid is not ordinarily found in plants and is formed in animals from linoleic acid as is indicated in Fig. 12-6.

$$-CH=CH- \xrightarrow{SAM} \underset{H}{\overset{CH_3}{\underset{\underset{+}{|}}{-C-CH-}}} \xrightarrow[a]{H^+} -\underset{H}{\overset{CH_2}{\underset{|}{C}}}\diagup\overset{CH_2}{\underset{}{\diagdown}}CH-$$

$$\downarrow^{b} \; H^+$$

$$\underset{\parallel}{\overset{CH_2}{-C}}-CH_2- \xrightarrow{NADPH} \underset{|}{\overset{CH_3}{-CH}}-CH_2-$$

(12-15)

same type of carbonium ion can yield a cyclopropane fatty acid (Eq. 12-15, step *a*) or a methenyl fatty acid (Eq. 12-15, step *b*). The latter can be reduced to a branched fatty acid. This is an alternative way of introducing methyl branches that is used by some bacteria.[44]

Cyclopropane fatty acids are catabolized via β oxidation which is modified as in Eq. 12-16[46]

$$R-\overset{*CH_2}{\underset{}{\underset{\diagup\;\diagdown}{CH-CH}}}-CH_2-\overset{O}{\overset{\parallel}{C}}-S-CoA$$

First two steps
of β oxidation

$$R-\overset{*CH_2}{\underset{\underset{OH}{|}}{\underset{\diagup\;\diagdown}{CH---C}}}-CH_2-\overset{O}{\overset{\parallel}{C}}-S-CoA$$

$$R-CH_2-*CH_2-\underset{\underset{O}{\parallel}}{C}-CH_2-\overset{O}{\overset{\parallel}{C}}-S-CoA$$

Further β
oxidation

(12-16)

when the chain degradation reaches the cyclopropane ring. The ring opening of the cyclopropanol derivative is of a type that occurs readily, even with mild nonenzymatic acid-base catalysis.

Another alteration of unsaturated fatty acids is the formation of acetylenic groups: $(-C\equiv C-)$. This apparently occurs by dehydrogenation of $-CH=CH-$, but the enzymes have been little studied. Examples of naturally occurring acetylenes are **crepenynic acid** (Fig. 2-32), **alloxanthin** (Section H,3), and the following remarkable hydrocarbon from the common cornflower *Centaurea cyanus*[45]:

$$H_3C-C\equiv C-C\equiv C-C\equiv C(CH=CH)_2(CH_2)_4-CH=CH_2$$

2. The Lipids of Cell Surfaces

Special fatty materials are often secreted to form external surfaces of organisms.[38,47] An example already mentioned is the secretion of the preen glands of water fowl. In the goose this material is 90% wax consisting of monoesters of various acids with predominantly **1-octadecanol** as the long-chain fatty alcohol.[38] The latter is formed by reduction of stearoyl-CoA as shown in Fig. 12-6. A large variety of branched fatty acids, both free and combined, are among the many compounds present in the skin lipids of humans. It is thought that the branched fatty acids may play a role in maintaining the ecological balance among microorganisms of the skin. They also impart to each individual a distinct odor or "chemical fingerprint."[47]

Plant surface lipids contain waxes, commonly having C_{10}–C_{30} chains in both acid and alcohol components. Free fatty acids, free alcohols, and alkanes are also present. The alkanes are thought to be formed by elongation of a C_{16} acid to as high as C_{30} followed by direct decarboxylation (Eq. 12-17).[48] While the mechanism of decarboxylation

$$C_{16} \text{ Acyl-CoA} \xrightarrow[\text{elongation}]{} C_{30} \text{ acid}$$

$$\downarrow \searrow CO_2$$

$$C_{29} \text{ alkane}$$

$$\downarrow O_2$$

$$C_{29}\text{-15-ol}$$

$$\downarrow$$

$$C_{29}\text{-15-one} \qquad (12\text{-}17)$$

is not known, the common occurrence of hydrocarbons of an odd chain length suggests the widespread importance of the reaction. Hydroxylation of the resulting alkanes can lead to alcohols and ketones (Eq. 12-17). Hydrocarbon formation can occur in other parts of a plant as well as in the cuticle. Thus, normal **heptane** constitutes up to 98% of the volatile portion of the turpentine of *Pinus jeffreyi*.[38]

Similar chemical reactions occur in insects. Thus, a major hydrocarbon of cockroaches is 6,9-heptacosadiene.[38] Unsaturated hydrocarbons as well as long-chain alcohols and their esters often form the volatile **pheromones** with which insects communicate. Thus, the female pink bollworm attracts a male with a sex pheromone consisting of a mixture of the cis,cis and cis,trans isomers of 7,11-hexadecadienyl acetate[49] and European corn borer males are attracted across the cornfields of Iowa by cis-11-tetradecenyl acetate.[50] Addition of a little of the trans isomer makes the latter sex attractant much more powerful. Since more than one species uses the same attractant, it is possible that the males can distinguish between different ratios of isomers or of mixtures of closely related substances.

3. Prostaglandins

As early as 1930 it was recognized that seminal fluid contains materials that promote contraction of uterine muscles. The active compounds, the prostaglandins, were crystallized in 1960 and were identified shortly thereafter. There is a large family of prostaglandins, as many as 14 closely related compounds being found in human seminal fluid, one of the richest known sources. Prostaglandins are present in this fluid at a total concentration of ~ 1 mM; their action on smooth muscles has been observed at a concentration as low as 10^{-9} M.

The structures of several of the prostaglandins as well as their biosynthesis are indicated in Fig. 12-7. Prostaglandins are usually abbreviated PG with an additional letter and numerical subscript added to indicate the type. For example, the E type are β-hydroxyketones, the F type 1,3-diols, and the A type α,β-unsaturated ketones. Series 2 prostaglandins arise from arachidonic acid, while series 1 and 3 arise from fatty acids containing one less or one more double bond, respectively (Fig. 12-7). Additional forms are known.[51-56]

Biosynthesis of prostaglandins is thought to begin with the release of the 20-carbon polyenoic acid precursor from phospholipids (phospha-tidylinositol or phosphatidylcholine) through the action of phospholipase A. From studies with stereospecifically synthesized ^{3}H-containing fatty acid precursors, it has been established that the next step involves removal of the pro-S proton at C-13 of the fatty acid (step a, Fig. 12-7). The O_2-requiring **cyclooxygenase** resembles lipoxygenase (Eq. 10-48).[54] The product is a peroxy acid, possibly in the form of the peroxide radical shown in Fig. 12-7. This radical (or peroxide anion) undergoes cyclization with synchronous attack by a separate O_2 molecule at C-15 (Eq. 12-7, step b) to give the endoperoxide PGG. Reduction of the latter to an OH group yields PGH which can break down in two ways to give the E and F series of prostaglandins. In the first, the proton at C-9 is eliminated (step d) as indicated by the small arrows by the PGH$_2$ structure of Fig. 12-7. The F prostaglandins are formed by reductive cleavage of the endoperoxide. The A series and other prostaglandins arise by secondary reactions, one of which is shown in Fig. 12-7. In some tissues (e.g., lung and blood platelets) PGH is also transformed into various nonprostaglandin products.[51,55,56] One of these is a labile hemiacetal derivative **thromboxane A** (Fig. 12-7). The latter can be converted further to thromboxane B which contains an —OH group at C-15.

A biochemical characteristic of the prostaglandins is rapid catabolism. The product shown in Fig. 12-7 (lower right) arises by oxidation of the 15-OH to a carbonyl group, permitting reduction of the adjacent trans double bond. Two steps of β oxidation as well as ω oxidation are also required to produce the dicarboxylic acid product shown. The whole story is more complex. A series of products appears and the distribution varies among species. Catabolism of prostaglandins is especially active in the lungs and any prostaglandins entering the bloodstream are removed by a single pass through the lungs. This observation has led to the conclusion that prostaglandins are not hormones in the classical sense. However, on a more local basis it is possible that prostaglandins released by one organ affect another organ or an adjacent tissue. On the other

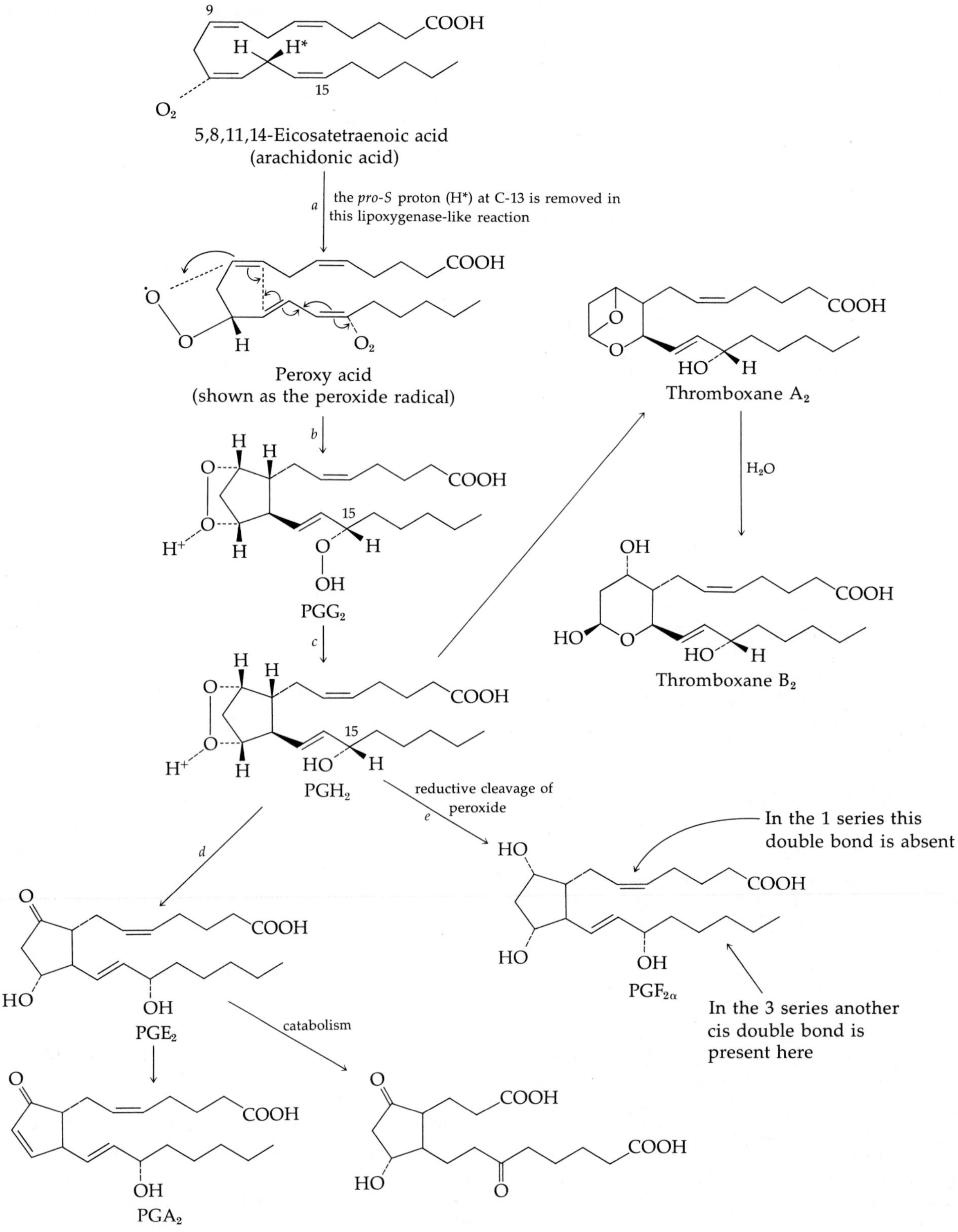

9

H   H*

O_2

5,8,11,14-Eicosatetraenoic acid
(arachidonic acid)

a | the *pro-S* proton (H*) at C-13 is removed in
this lipoxygenase-like reaction

Peroxy acid
(shown as the peroxide radical)

Thromboxane A_2

b

H^+ H

15

OH

PGG_2

H_2O

Thromboxane B_2

c

H^+ H   HO   H

15

PGH_2

reductive cleavage of
peroxide

e

— In the 1 series this
double bond is absent

d

HO

HO   OH

$PGF_{2\alpha}$

HO   OH

PGE_2

catabolism

In the 3 series another
cis double bond is
present here

O

HO

PGA_2

O

HO   O

hand, it may be that the principal action of prostaglandins is within the same cells in which they are formed.

Prostaglandins have the most varied physiological effects but the chemical basis remains elusive. It has often been noted that some prostaglandins such as PGE_1 mimic cAMP in their influence on cells. This has led to the suggestion that prostaglandins may mediate the production of cAMP in response to the binding of a hormone to a cell surface. Perhaps a prostaglandin is the true second messenger which crosses the membrane to cause allosteric activation of adenylate cyclase.[57] Since PGE_2 often appears to have opposite physiological effects from PGE_1, it could function in a different way.

Special interest in the prostaglandins has focused on the processes of inflammation and allergic responses.[58] The medical significance is easy to see. Five million Americans have **rheumatoid arthritis,** an inflammatory disease. Asthma and other allergic diseases are equally important. Our most common medicine is **aspirin,** an

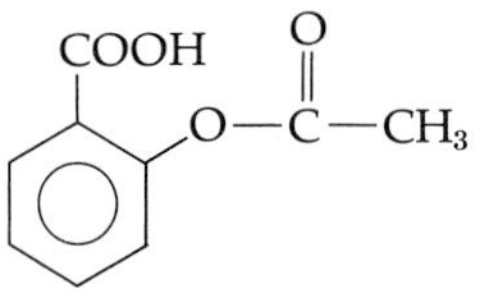

Aspirin (acetylsalicylic acid)

anti-inflammatory drug. Both the inflammatory response and the immune response are normal parts of the defense mechanisms of the body, but both are potentially harmful, and it is their regulation that is probably faulty in rheumatoid arthritis and asthma.

Prostaglandins have been implicated in both the induction of inflammation and in its relief. In inflammation small blood vessels become dilated and fluid and proteins leak into the interstitial spaces to produce the characteristic swelling (edema). Many polymorphonuclear leukocytes (Chapter 1, Section E,2b) migrate into the in-

flamed area, engulfing dead tissue and bacteria. In this process, lysosomes of the leukocytes release phospholipase A, which in turn hydrolyzes phospholipids and releases polyunsaturated fatty acids, the precursors of prostaglandins. It is not known whether prostaglandins are among the compounds initiating the inflammatory response but, if so, their effect might be reinforced by the release of polyunsaturated fatty acids. It is also established that cAMP can suppress inflammation and that PGE_2 seems to have a similar effect. Indeed, while the F prostaglandins induce allergic responses E prostaglandins, when inhaled in small amounts, relieve asthma.

Aspirin inhibits the synthesis of prostaglandins, apparently by acetylation of the cyclooxygenase, and this may represent the mode of action of the drug.[55,59] The possibility of learning how to control inflammation and allergic responses is a hoped-for practical result of future prostaglandin research. Since PGE_1 is a potent **pyrogen** (fever-inducing agent), a relationship to the ability of aspirin to reduce fever is also suggested. Dissolution of bone by release of lysosomal enzymes from bone cells is also stimulated by PGE_1 as well as by the parathyroid hormone (Box 5-E). Aspirin also counteracts this release of enzymes.[60]

Another process in which antagonistic effects of prostaglandins are observed is reproduction. While prostaglandins may be required for conception, remarkably small amounts induce abortion.

F. METABOLISM OF TRIGLYCERIDES, PHOSPHOLIPIDS, AND GLYCOLIPIDS

Reduction of dihydroxyacetone phosphate yields sn-glycerol 3-phosphate, the starting compound for formation of the glycerol-containing lipids (Fig. 12-8, step a). Transfer of two acyl groups from CoA to the hydroxyl groups of this compound (steps b and c) yields a 1,2-diglyceride

Fig. 12-7 Biosynthesis of prostaglandins and some catabolic reactions.

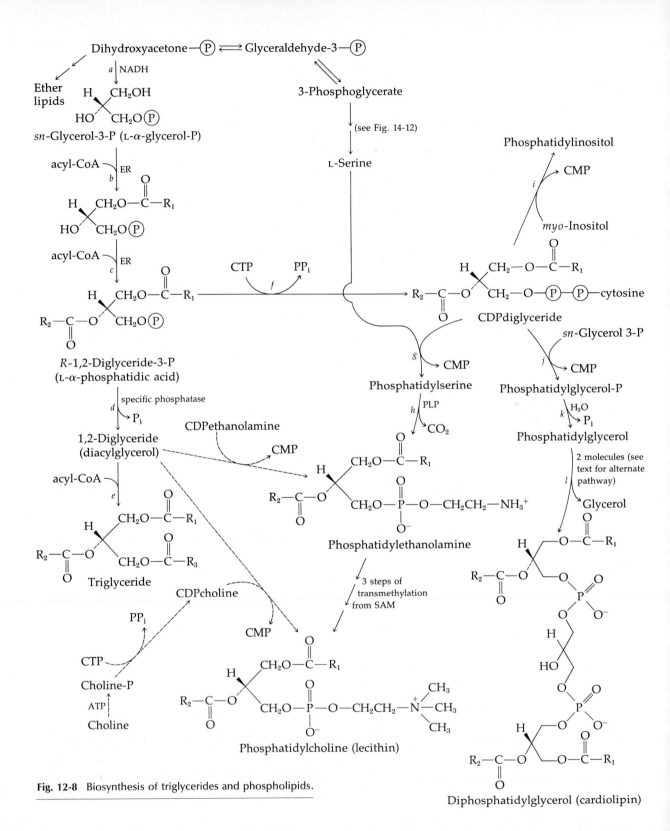

Fig. 12-8 Biosynthesis of triglycerides and phospholipids.

3-phosphate (a phosphatidic acid). An alternative route (that has been studied in liver) is the transfer of one acyl group onto dihydroxyacetone phosphate and reduction prior to addition of the second acyl group (Eq. 12-18).

$$
\begin{array}{c}
O=C\begin{array}{l}CH_2OH \\ \\ CH_2O-\text{(P)}\end{array}\xrightarrow{\;\;\;a\;\;\;}\begin{array}{c}\text{acyl-CoA}\end{array}
\end{array}
$$

$$
O=C\begin{array}{l}CH_2O-\overset{\displaystyle O}{\overset{\|}{C}}-R_1 \\ \\ CH_2O-\text{(P)}\end{array}\xrightarrow{\;\;b\;\;}
$$

$$
\begin{array}{c}
H\;\;\;CH_2-O-\overset{\displaystyle O}{\overset{\|}{C}}-R_1 \\
\diagup\diagdown \\
HO\;\;\;CH_2O-\text{(P)}
\end{array}\quad (12\text{-}18)
$$

Phosphatidic acids lie at a metabolic branch point. On the one hand, the phosphoryl group can be removed by a specific phosphatase (step d) and another acyl group (most often an *unsaturated* acyl group) may be transferred onto the resulting diglyceride (step e) to form a triglyceride. Alternatively, the phosphatidic acid can be converted to a **CDPdiglyceride** (step f) in a reaction resembling that of CTP with sugar phosphates (Eq. 11-24) or with choline phosphate (Eq. 11-26). This is the first step in phospholipid synthesis in bacteria.

The pathway of triglyceride formation shown in Fig. 12-8 predominates in adipose tissues, but in the intestinal epithelium 2-monoglycerides absorbed from digested food are esterified directly to triglycerides by reaction with two molecules of acyl-CoA.[44]

1. Control of Triglyceride Metabolism

A principal regulatory point for lipid synthesis is in the activation of acetyl-CoA carboxylase by citrate (Chapter 8, Section C,2, Fig. 11-1). Beyond that a complex hormonal control is exerted on both biosynthesis and catabolism of triglycerides stored in liver and adipose tissues. Thus, adrenaline and glucagon by stimulating production of cAMP cause activation of lipases that cleave triglycerides, thus mobilizing depot fats. Insulin, on the other hand, promotes lipid storage. It not only increases the activity of the enzymes of lipogenesis from the ATP-dependent citrate cleavage enzyme (Eq. 7-70), but it also inhibits cAMP production, thus blocking lipolysis within cells. On the other hand, the serum **lipoprotein lipase** (also known as "clearing factor") hydrolyzes lipids carried in by the serum lipoproteins, while the latter are in the small capillaries. The released fatty acids can then be taken up by cells and reconverted to lipids.[44]

2. Synthesis of Phospholipids and Glycolipids

Within bacteria most phospholipids are formed following conversion of phosphatidic acids to CDPdiglycerides (Fig. 12-8, step f) which are then able to undergo reaction with a variety of nucleophiles with displacement of CMP. Thus, reaction with L-serine (step g) leads to **phosphatidylserine** and reaction with inositol (step i) to **phosphatidylinositol.** The enzyme catalyzing the formation of phosphatidylserine appears to occur naturally in a form bound to ribosomes.[61,62] In contrast, most of the other enzymes of phospholipid formation are closely associated with or embedded in the cytoplasmic membrane. One of the membrane-bound enzymes catalyzes decarboxylation of phosphatidylserine to **phosphatidylethanolamine** (step h, Fig. 12-8).[63] Phosphatidylcholine is not usually a major component of bacterial lipids, but it can be formed from phosphatidylethanolamine by three steps of transmethylation from S-adenosylmethionine.

In animals an alternative pathway is available for formation of both phosphatidylcholine and phosphatidylethanolamine. This is illustrated by the dashed lines in Fig. 12-8. The free base, either choline or ethanolamine, is phosphorylated with ATP. Choline phosphate formed in this manner is then converted by reaction with CTP (Eq. 11-26) to CDPcholine. The latter transfers a phosphoryl choline group to a 1,2-diglyceride to form the lecithin. In an entirely analogous way, CDPethanolamine is used to form phosphatidylethanolamine.

The formation of phosphatidylserine in animal tissues is accomplished by an exchange reaction (Eq. 12-19). The decarboxylation of phosphatidyl-

$$\text{Phosphatidylethanolamine (or phosphatidylcholine)} \xrightarrow[\text{Serine}]{\text{Ethanolamine (or choline)}} \text{Phosphatidylserine} \xrightarrow{\text{SAM}} \text{Phosphatidylcholine} \tag{12-19}$$

serine back to phosphatidylethanolamine takes place simultaneously, the net effect being a catalytic cycle for decarboxylation of serine to ethanolamine. The latter can react with CTP to initiate synthesis of new phospholipid molecules. Another conversion of significance is the direct methylation of phosphatidylethanolamine to phosphatidylcholine.

Turning again to Fig. 12-8, note that glycerol phosphate can react with CDPdiglyceride in the same manner as does serine to form **phosphatidylglycerol phosphate** (step j). After removal of a phosphate, the resulting phosphatidylglycerol can be converted to **diphosphatidylglycerol** (cardiolipin). One manner in which this is accomplished in some bacteria is indicated by step l of Fig. 12-8. One molecule of glycerol is displaced as two molecules of phosphatidylglycerol are coupled. In mitochondria and perhaps in some bacteria an alternative pathway is followed. CDPdiglyceride transfers the entire phosphatidic acid group to phosphatidylglycerol with displacement of CMP[64]:

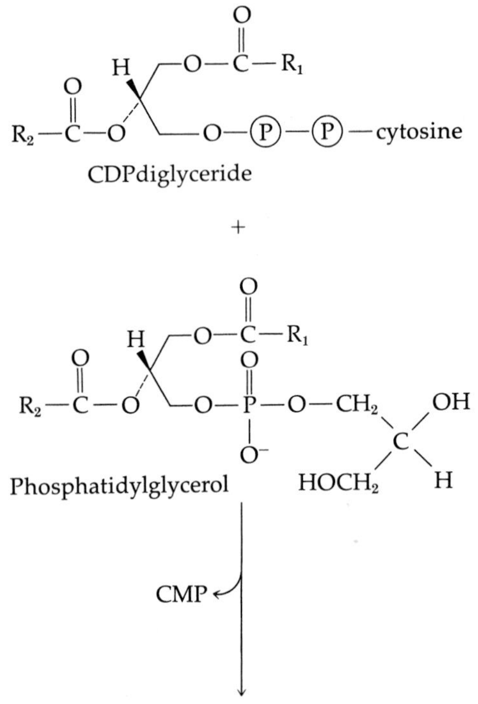

CDPdiglyceride

+

Phosphatidylglycerol

$$\tag{12-20}$$ Diphosphatidylglycerol

Another class of compounds originating from 1,2-diglycerides are the galactolipids of chloroplasts (Fig. 2-25). A second galactosyl ring can be transferred onto the first product to form a **digalactosyldiglyceride.**

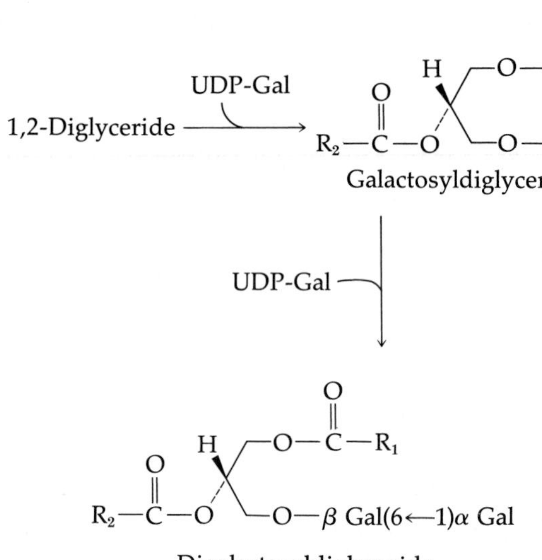

Galactosyldiglyceride

Digalactosyldiglyceride $\tag{12-21}$

3. The Ether-Linked Lipids

The **ether lipids** are closely related to both the triglycerides and phospholipids. They differ by containing in place of one acyl group an *alkyl* (—OR) or *alkenyl* (—O—CH=CH—R) group.[65] Lipids containing the alk-1-enyl group, the **plasmalogens**, were first recognized in 1924 by R. Feulgen and K. Voit who were developing histological staining procedures. They observed that treatment of tissue slices with acid resulted in the liberation of aldehydes. Later investigation showed that the aldehydes were formed by breakdown of the alkenyl lipids.

$$Y-O-\overset{H}{\underset{}{C}}=CH-R + H_2O \xrightarrow{H^+}$$

$$Y-OH + O=\overset{H}{\underset{}{C}}-CH_2-R \quad (12\text{-}22)$$

Over 10% of the lipid in the human central nervous system is plasmalogen and about 1% is alkyl lipid. The proportion in molluscs is especially high, ether lipids constituting up to 35% of the total phospholipid. While ether lipids are usually regarded as animal constituents, small amounts have been identified in plants.

The first step in biosynthesis of ether lipids appears to be the formation of fatty acyl derivatives of dihydroxyacetone phosphate. The acyl group is then apparently displaced, along with the oxygen atom to which it is attached, by an alkoxy group of a long-chain fatty alcohol (Eq. 12-23, step *a*). The oxygen of the alcohol (designated by an asterisk) is retained in the product.[66-68] The reaction is remarkable and differs significantly from displacements discussed in Chapter 7. One of the hydrogen atoms (marked by the dagger, †) exchanges with the medium during the reaction suggesting that enolization takes place.[68] However, it is hard to see how this in itself would aid in the displacement reaction. Perhaps a coenzyme is needed.

Once an alkoxy derivative of dihydroxyacetone is formed the reduction, further acylation, and

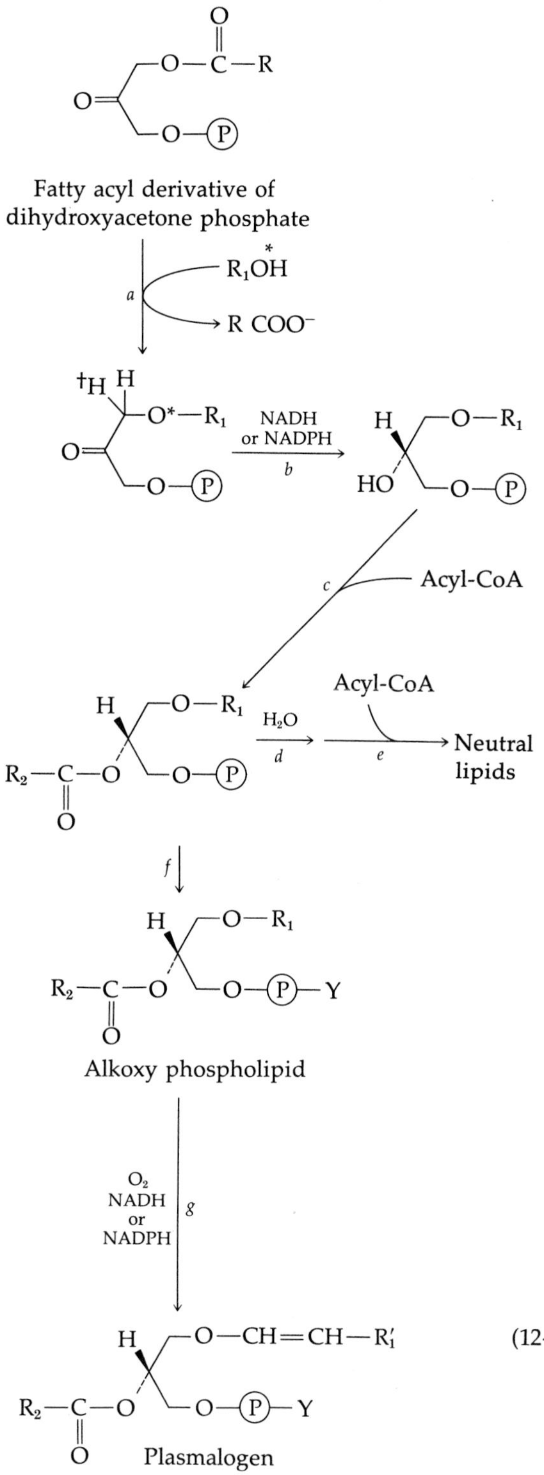

Fatty acyl derivative of dihydroxyacetone phosphate

Alkoxy phospholipid

$$(12\text{-}23)$$

Plasmalogen

conversion to various phospholipids and neutral lipids can occur. The pathways (Eq. 12-23, steps *b–f*) are closely akin to those of Fig. 12-8. The conversion of alkoxy lipids to plasmalogens (alk-1-enyl glycerolipids) occurs by oxidative desaturation (Eq. 12-23, step *g*).[69]

A final comment on ether-linked lipids has to do with extremely "halophilic" bacteria that live only in saturated or near saturated NaCl solutions. Large amounts of an unusual glycerol ether derived from a polyprenol are present in these bacteria.

2,3-Di-*O*-(3′*R*,7′*R*,11′*R*,15-tetramethylhexadecyl)-*sn*-glycerol

4. Sphingolipids

Sphingolipids are phospholipids and glycolipids derived from sphingosine and other "long-chain bases." To date there have been at least 60 substances of this type identified.[70] The molecules vary in chain length from C_{14} to C_{22} and include members of the iso and anteiso series. Up to two double bonds may be present. The compound usually called sphingosine is a C_{18} compound derived from condensation of palmitoyl-CoA with serine (Eq. 12-24). Carbon dioxide is lost from the serine during the condensation reaction (Eq. 12-24, step *a*; Chapter 8, Section E,3,c) and the resulting ketone is reduced with NADPH (Eq. 12-24, step *b*) to form **sphinganine,** a common component of animal sphingolipids. The latter is oxidized by a flavoprotein to sphingosine (Eq. 12-24, step *c*) or is hydroxylated to phytosphingosine (Eq. 12-24, step *d*). The conversion of the sphingosine bases to sphingomyelins and cerebrosides is summarized in Fig. 12-5. Biodegradation of sphingosine is thought to take place by a PLP-mediated chain cleavage to palmitaldehyde.[71]

While pathways of synthesis of complex lipids have been described, we are far from understanding the dynamics of the synthesis and turnover of the membranous structures built from them. Of interest is the fact that phospholipids have been shown to exchange between different membranes. For example, phosphatidylcholine, phosphatidylethanolamine, and phosphatidylino-

(12-24)

sitol exchange between isolated mitochondria and microsomes. The exchange of phosphatidylcholine has been shown to be catalyzed by a specific **exchange protein.**[72]

G. THE POLYKETIDES

In 1907, J. N. Collie proposed that polymers of ketene (CH_2=C=O) might be precursors of such compounds as **orsellinic** acid, a common constituent of lichens. The hypothesis was modernized in 1953 by Birch and Donovan who proposed that several molecules of acetyl-CoA are condensed (Eq. 12-25) but *without the two reduction steps required in biosynthesis of fatty acids* (Fig. 11-2).[73] It is presumed that the condensation occurs via malonyl-CoA and an acyl carrier group of an enzyme.[74] The resulting **β-polyketone** can react in various ways to give the group of compounds identified as polyketides.

$$n \; CH_3-\overset{O}{\overset{\|}{C}}-S\text{-CoA} \longrightarrow H\left(\!-CH_2-\overset{O}{\overset{\|}{C}}\!\right)_{\!n}\!S-\text{enzyme}$$

β-Polyketone

(12-25)

β-Polyketones are often stabilized by ring formation through ester or aldol condensations. Following cyclization the remaining carbonyl groups can be reduced to hydroxyl groups and the latter can be eliminated as water to form benzene or other **aromatic rings.** Figure 12-9 illustrates two ways in which cyclization can occur. One involves a Claisen ester condensation during which the enzyme and its SH group are eliminated. Enolization of the product gives a trihydroxyacetophenone. The second cyclization reaction is the aldol condensation. Following the condensation a molecule of water is eliminated and the product is hydrolyzed and enolized to form orsellinic acid. Another product of fungal metabolism is

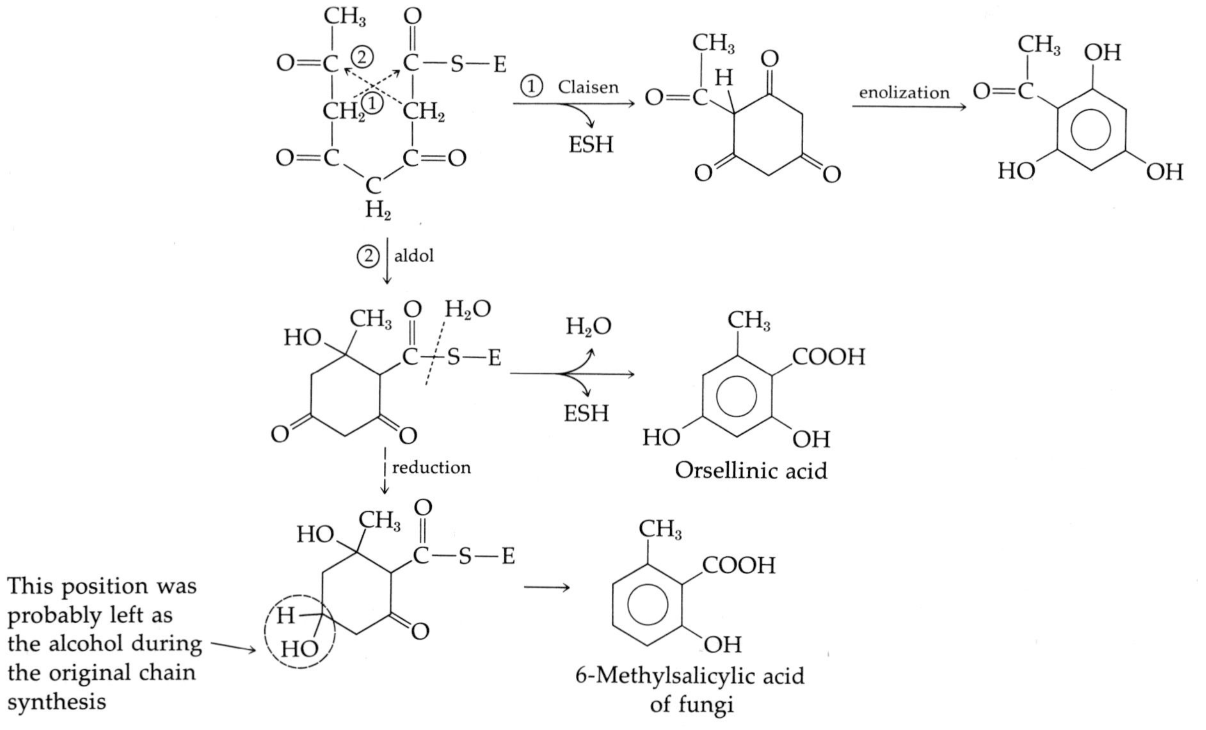

Fig. 12-9 Postulated origin of orsellinic acid and other polyketides.

6-methylsalicylic acid, lacking one OH group of orsellinic acid. The synthesis is easy to rationalize if it is assumed that the carbonyl group at C-5 of the original β-polyketone was reduced to an OH group at some point during the biosynthesis. Elimination of two molecules of water together with enolization of the remaining ring carbonyl gives the product (Fig. 12-9).

By allowing a few variations in the basic polyketone structure, the biosynthesis of a large number of unusual compounds can be explained.[74] Sometimes extra oxygen atoms are added by hydroxylation. Extra methyl groups may be transferred from SAM to form methoxyl groups. Occasionally a methyl group is transferred directly to the carbon chain. Often starter pieces other than acetyl-CoA initiate polyketide synthesis. The branched chain acids derived from valine, leucine, and isoleucine, as well as nicotinic and benzoic acids, can serve. Malonic acid amide, presumably as a CoA derivative, appears to be the starter piece for synthesis of the antibiotic **tetracycline** (Fig. 12-10). The polyketide origins of some other important antibiotics are also indicated in Fig. 12-10.

H. POLYPRENYL (ISOPRENOID) COMPOUNDS

The **terpenes, carotenoids,** and **steroids** all arise in a direct way from the prenyl group of isopentenyl pyrophosphate (Fig. 12-11).[75–78] Biosynthesis of this five-carbon branched unit has been discussed previously (Chapter 11, Section D,10, Fig. 11-8) and is briefly recapitulated in Fig. 12-11. One step in the formation of mevalonic acid, a two-step reduction of 3-hydroxy-3-methylglutaryl-CoA, is highly controlled. In the human, regulation of this reaction in the liver is thought to be a major factor in controlling cholesterol synthesis.[44,79] The enzyme is sensitive to feedback inhibition by cholesterol itself or by metabolites of cholesterol.

Before polyprenyl formation begins, one molecule of isopentenyl pyrophosphate must be isomerized to dimethylallyl pyrophosphate (Fig. 12-11, Eq. 7-84). In this process the hydrogen that was in the 4-*pro-S* position of mevalonic acid (the *pro-R* position of isopentenyl pyrophosphate) is lost. Dimethylallyl pyrophosphate serves as the

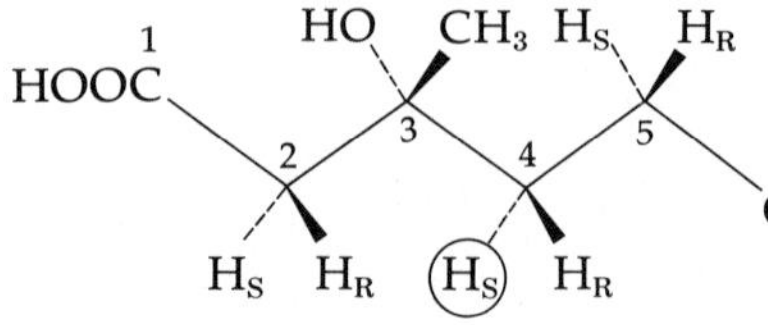

3(*R*)-Mevalonic acid

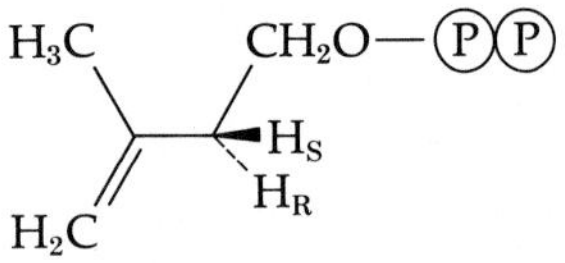

Isopentenyl pyrophosphate

starter piece and additional prenyl units are added, with elimination of pyrophosphate, as indicated in Fig. 12-11. In each of these reactions a hydrogen that was originally the 4-*pro-S* hydrogen of mevalonic acid is lost as a proton.[44,76–78]

Polymerization of prenyl units can continue with the formation of high molecular weight polyprenyl alcohol such as the dolichols (Section C,3) or of the high polymer **rubber.** It is noteworthy that in the latter case almost entirely cis double bonds are formed, whereas most polyprenyl compounds contain mostly trans double bonds. Consistent with this observation is the fact that during the formation of rubber the *pro-R* proton (rather than the *pro-S* proton) of mevalonic acid is lost during the polymerization.

Chain elongation during polymerization of prenyl units can be terminated in one of a number of ways. The pyrophosphate group may be hydrolyzed to a monophosphate or to a free alcohol. Alternatively, two polyprenyl compounds may join "head to head" to form a symmetric dimer. The C_{30} terpene **squalene,** the precursor to cholesterol, arises in this way, as does **phytoene,** precursor of the C_{40} carotenoids.

Fig. 12-10 Some important polyketide antibiotics.

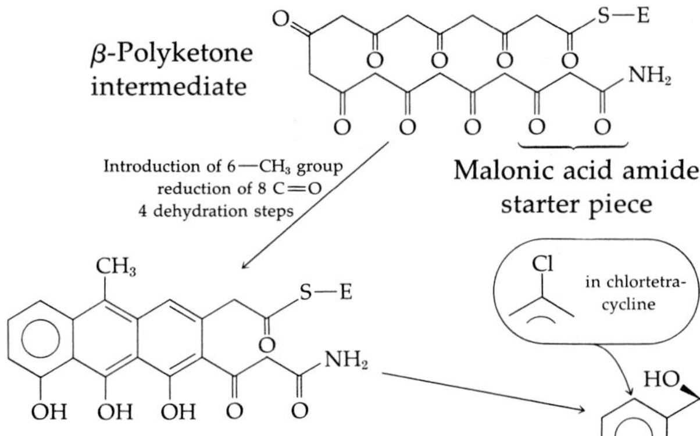
β-Polyketone intermediate
S—E
NH₂
Malonic acid amide starter piece
Introduction of 6—CH₃ group
reduction of 8 C=O
4 dehydration steps
CH₃
S—E
NH₂
OH OH OH O O

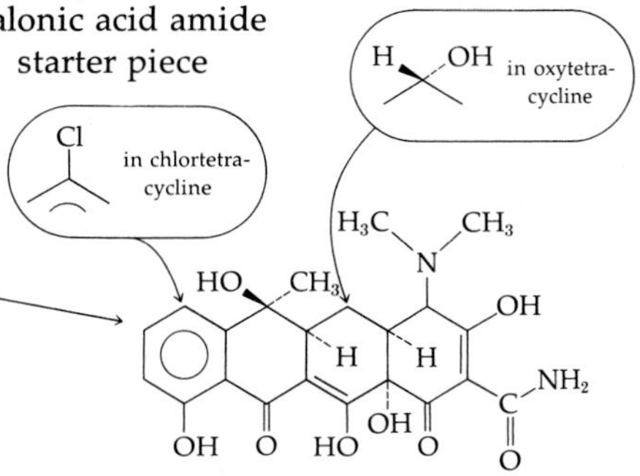
Cl
in chlortetra-cycline
H OH
in oxytetra-cycline
H₃C CH₃
N
HO CH₃
OH
H H
NH₂
C
OH
OH O HO O O

Tetracycline, one of the most widely used broad spectrum antibiotics, from *Streptomyces rimonus*

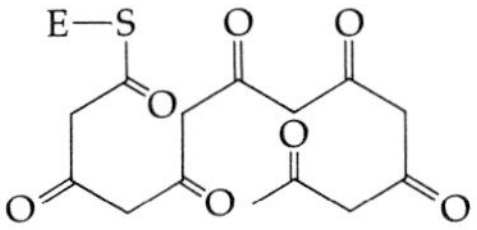
E—S

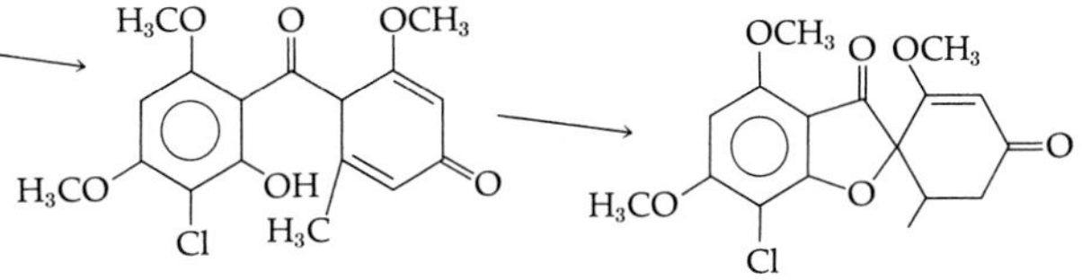
H₃CO O OCH₃
H₃CO OH
Cl H₃C
OCH₃ O OCH₃
H₃CO
Cl

Griseofulvin for fungal infections of the skin, from *Penicillium*

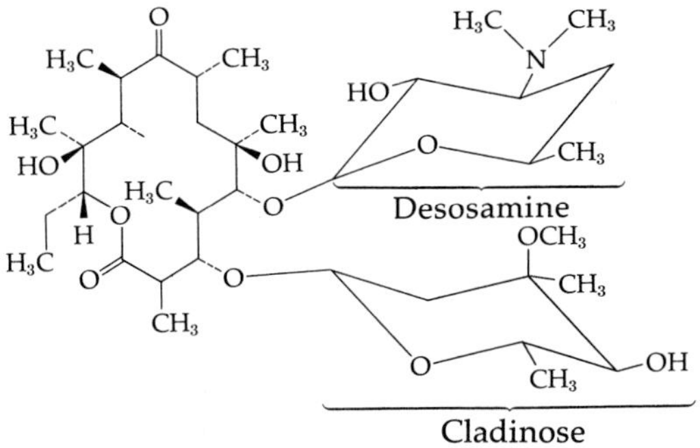
H₃C CH₃
N
HO
CH₃
Desosamine
OCH₃
CH₃
OH
CH₃
Cladinose

Erythromycin, another broad spectrum antibiotic from *Streptomyces erythreus*

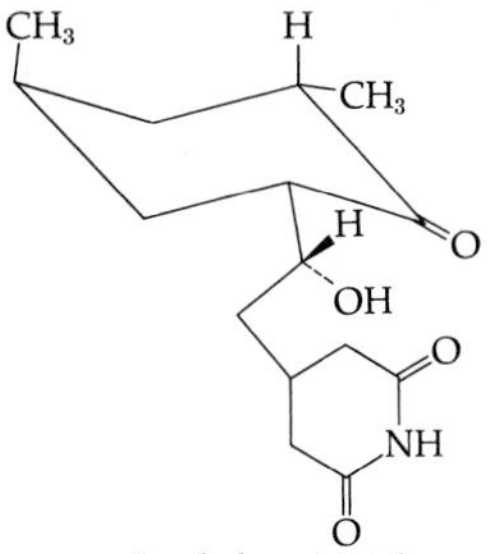
CH₃ H
CH₃
H
OH
NH

L-Cycloheximide from *Streptomyces naraensis*

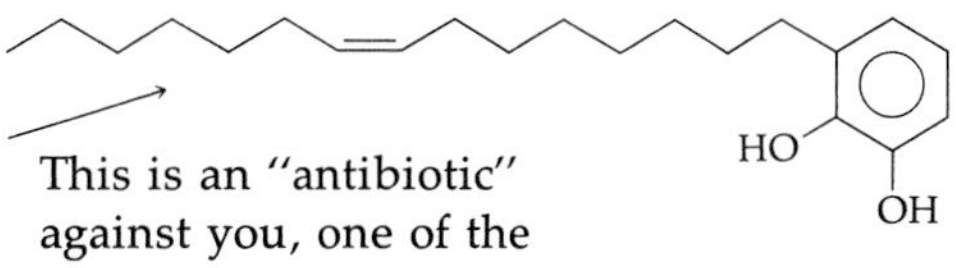
HO
OH

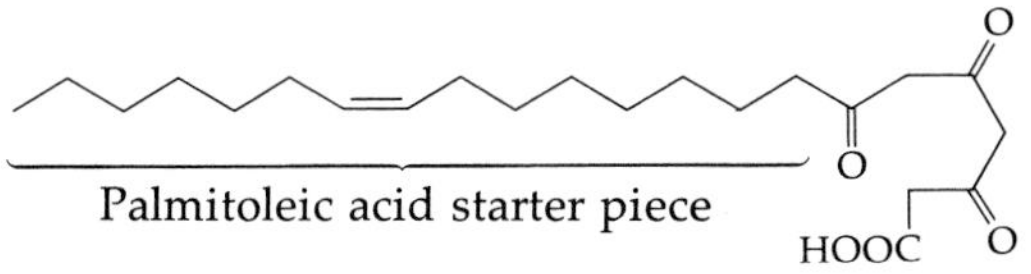
Palmitoleic acid starter piece
HOOC

This is an "antibiotic" against you, one of the components of urushiol of poison ivy

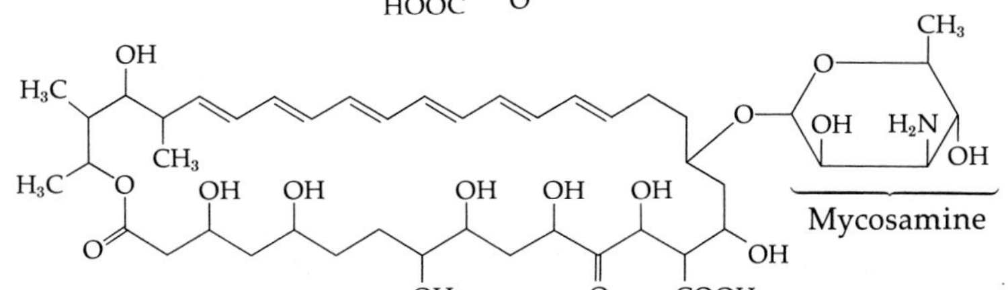
OH
H₃C
CH₃
H₃C
OH OH OH OH OH
OH
OH O COOH
CH₃
OH H₂N
OH
Mycosamine

Amphotericin B, a polyene antibiotic used for deep fungal infections

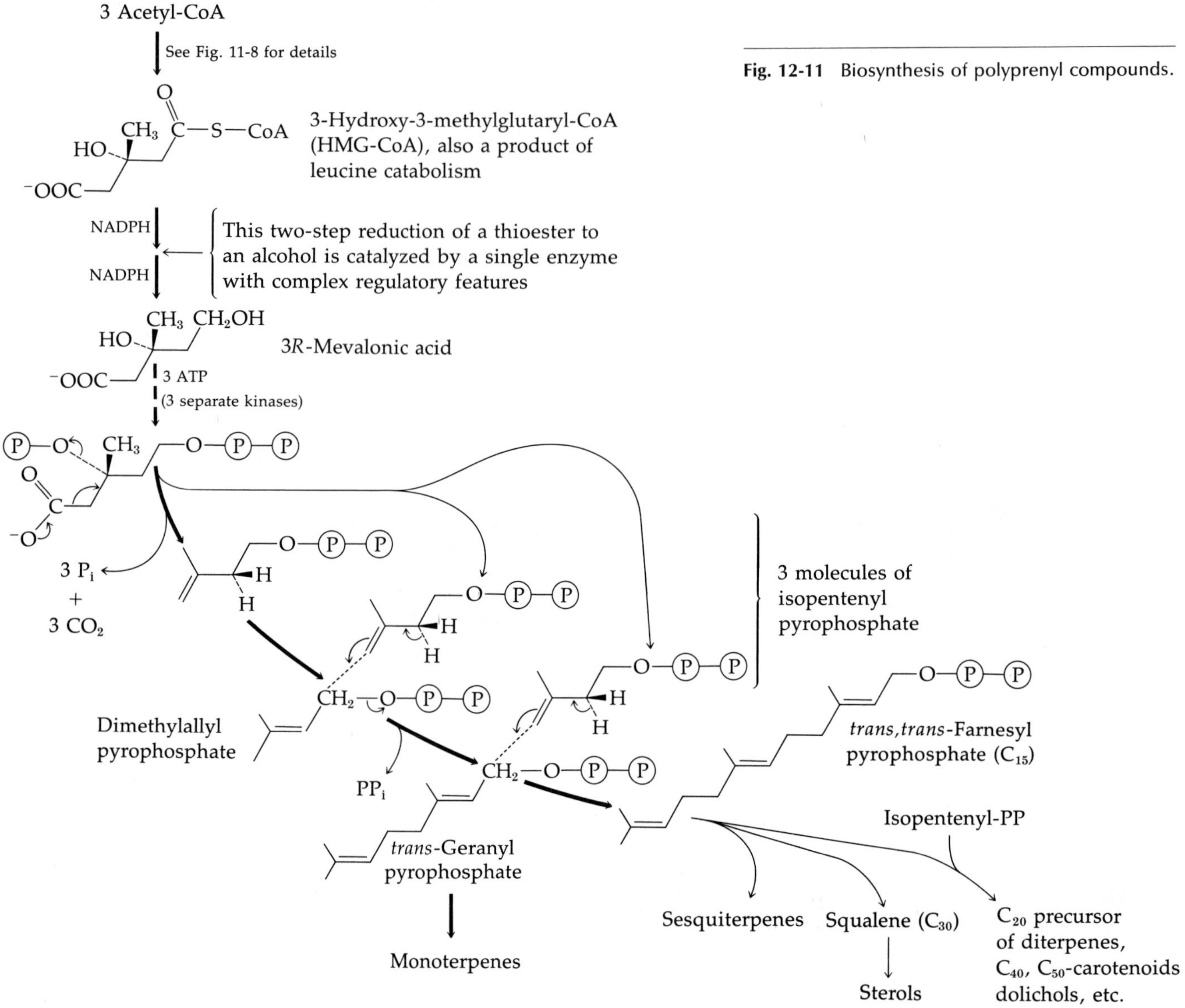

Fig. 12-11 Biosynthesis of polyprenyl compounds.

3 Acetyl-CoA

See Fig. 11-8 for details

3-Hydroxy-3-methylglutaryl-CoA (HMG-CoA), also a product of leucine catabolism

NADPH
NADPH

This two-step reduction of a thioester to an alcohol is catalyzed by a single enzyme with complex regulatory features

3R-Mevalonic acid

3 ATP
(3 separate kinases)

3 P_i
+
3 CO_2

Dimethylallyl pyrophosphate

PP_i

trans-Geranyl pyrophosphate

Monoterpenes

3 molecules of isopentenyl pyrophosphate

trans,trans-Farnesyl pyrophosphate (C_{15})

Isopentenyl-PP

Sesquiterpenes   Squalene (C_{30}) C_{20} precursor of diterpenes, C_{40}, C_{50}-carotenoids dolichols, etc.

Sterols

BOX 12-B

HOW THE FLOWERS MAKE THEIR COLORS

The pigments of flowers arise from an interesting polyketide precursor. Phenylalanine is converted to **trans-cinnamic acid** (Eq. 8-36) and cinnamoyl-CoA. The latter acts as the starter piece for chain elongation via malonyl-CoA (step *a* in the accompanying scheme). The resulting β-polyketone derivative can cyclize in two ways. The aldol condensation (step *b*) leads to **stilbenecarboxylic acid** and to such compounds as pinosylvin of pine trees. The Claisen condensation (step *c*) produces **chalcones, flavonones,** and **flavones.** These in turn can be converted to the yellow **flavonol pigments** and to the red, purple, and blue **anthocyanidins.**[a,b]

At the bottom of the synthetic scheme on the next page the structures and names of three

Box 12-B *(continued)*

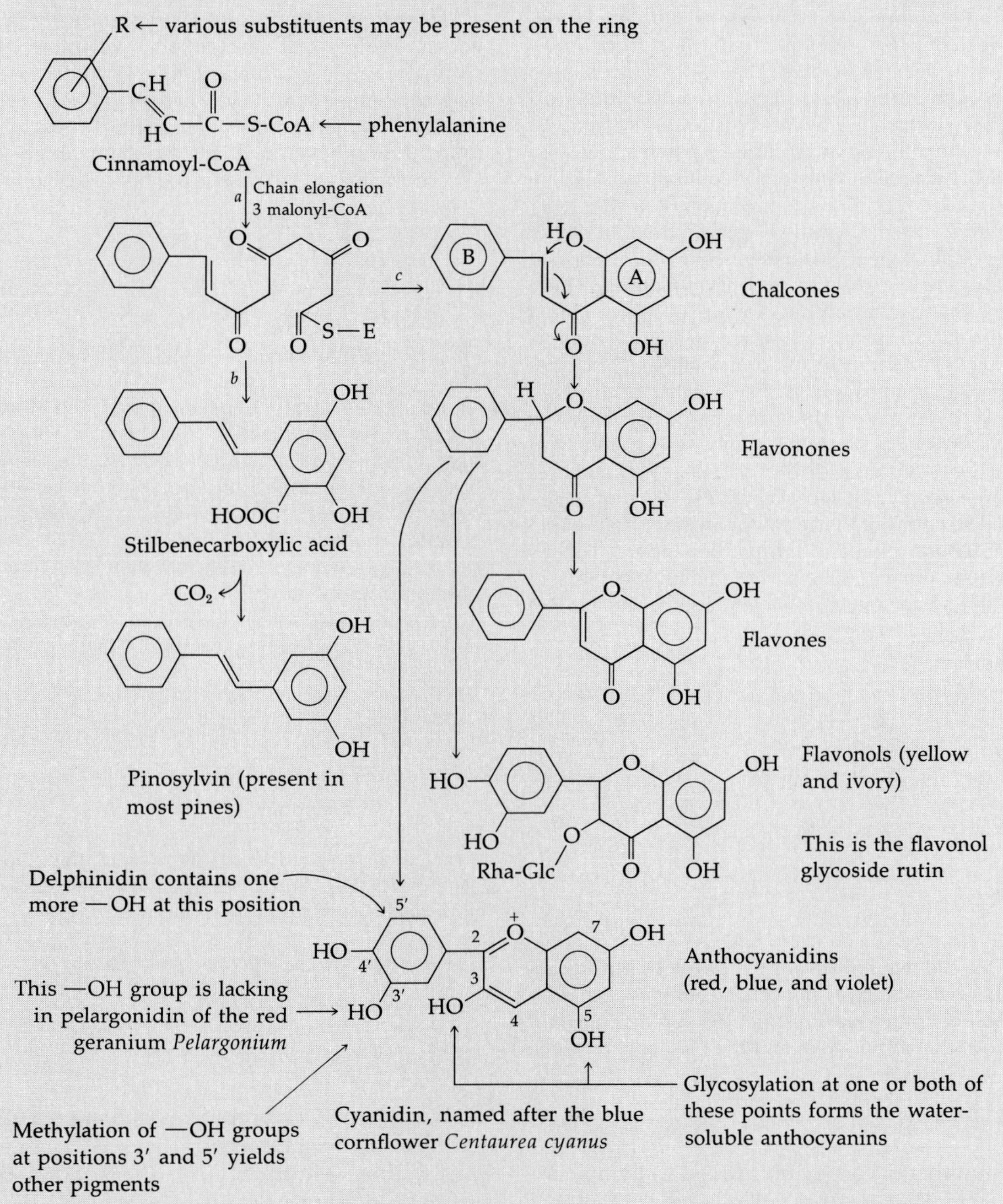
R ← various substituents may be present on the ring
S-CoA ⟵ phenylalanine
Cinnamoyl-CoA
a
Chain elongation
3 malonyl-CoA
S—E
c
B
A
OH
Chalcones
b
OH
Flavonones
HOOC
OH
Stilbenecarboxylic acid
CO_2
OH
Flavones
Pinosylvin (present in most pines)
HO
Flavonols (yellow and ivory)
This is the flavonol glycoside rutin
Rha-Glc
Delphinidin contains one more —OH at this position
5′
4′
3′
2
3
4
5
7
Anthocyanidins (red, blue, and violet)
This —OH group is lacking in pelargonidin of the red geranium *Pelargonium*
Methylation of —OH groups at positions 3′ and 5′ yields other pigments
Cyanidin, named after the blue cornflower *Centaurea cyanus*
Glycosylation at one or both of these points forms the water-soluble anthocyanins

Box 12-B (*continued*)

common anthocyanidins are shown. The names are derived from those of flowers from which they have been isolated. The colors depend upon the number of hydroxyl groups and on the presence or absence of methylation and glycosylation. In addition to the three pigments indicated in the diagram, three other common anthocyanidins are formed by methylation. **Peonidin** is 3′-methylcyanidin. Methylation of delphinidin at position 3′ yields **petunidin**, while methylation at both the 3′ and 5′ positions gives **malvidin.** There are many other anthocyanidins of more limited distribution. Anthocyanidins are very insoluble but they exist in plants principally as glycosides known as **anthocyanins.** The number of different glycosides among the many species of flowering plants is large. Both the 3- and 5-OH groups may be glycosylated with Glc, Gal, Rha, Ara, and by a large variety of oligosaccharides.

The colors of the anthocyanins vary from red to violet and blue and are pH dependent. For example, **cyanin** (diglucosyl cyanidin) is red in acid solution and becomes violet upon dissociation of the 4′-hydroxyl group:

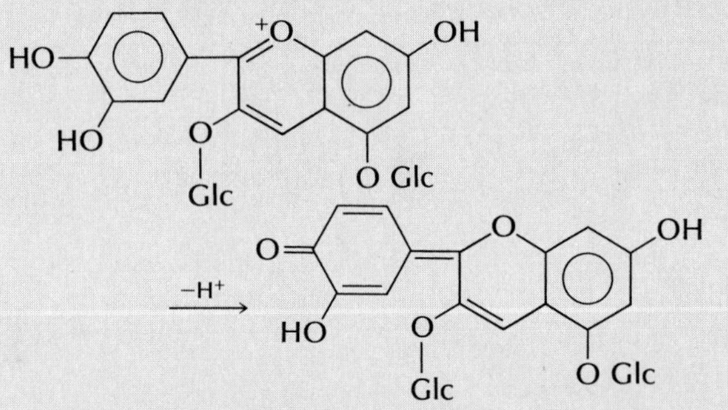

At still higher pH values it becomes blue as additional hydroxyls dissociate. Note that a large number of resonance structures can be drawn for both the anthocyanin and the dissociated forms.

The yellow pigments of flowers are usually flavonols. The most common of all is **rutin,** the 3α-rhamnosyl-D-glucosyl derivative of **quercitin** (see diagram). An extraordinary number of other flavonols, flavones, and related compounds are found throughout the plant kingdom. A well-known example is **phlorhizin,** a dihydrochalcone,

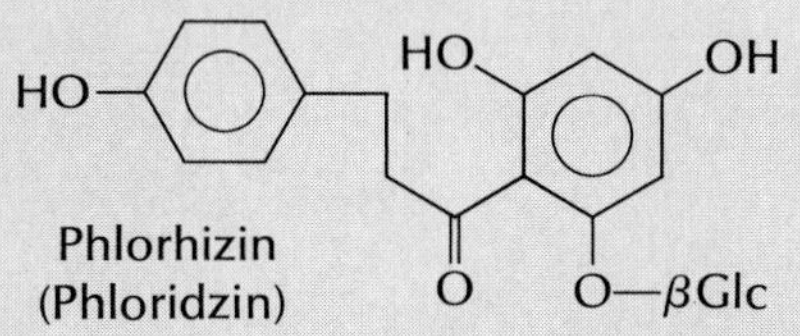

Phlorhizin
(Phloridzin)

found in the root bark of pears, apples, and other plants of the rose family. Phlorhizin is widely used in physiological studies because it specifically blocks resorption of glucose by kidney tubules. As a result, the drug induces a strong glucosuria. The biochemical basis is uncertain but the action on kidney tubules may be related to inhibition of mutarotase.[c]

A flavone glycoside is **hesperidin,** a substance making up 8% of the dry weight of orange peels.

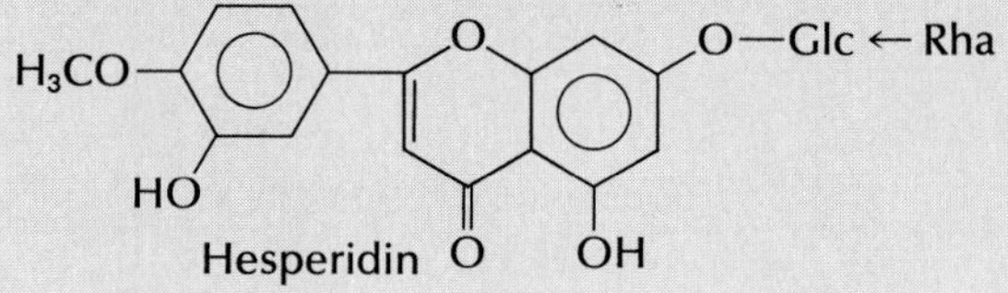

Hesperidin

It has been claimed (but not proved) that this compound, also known as **vitamin P** and **citrus bioflavonoid,** is essential to good health.

[a] S. Clevenger, *Sci. Am.* **210,** 85–92 (June 1964).

[b] J. B. Harborne, "Comparative Biochemistry of Flavonoids." Academic Press, New York, 1967.

[c] A. White, P. Handler, and E. L. Smith, "Principles of Biochemistry," 5th ed., pp. 415–416. McGraw-Hill, New York, 1973.

1. Terpenes

The number of compounds found in plants, animals, and bacteria that arise from isopentenyl pyrophosphate is staggering. Just a few are shown in Figs. 12-12 and 12-13. The compounds of Fig.

12-12 contain 10 carbon atoms and are usually referred to as **monoterpenes.** They occur largely in plants, but a number of them function in arthropods as pheromones. In general, the biosynthetic pathways of plant terpenes have not been worked out in the same detail as have the

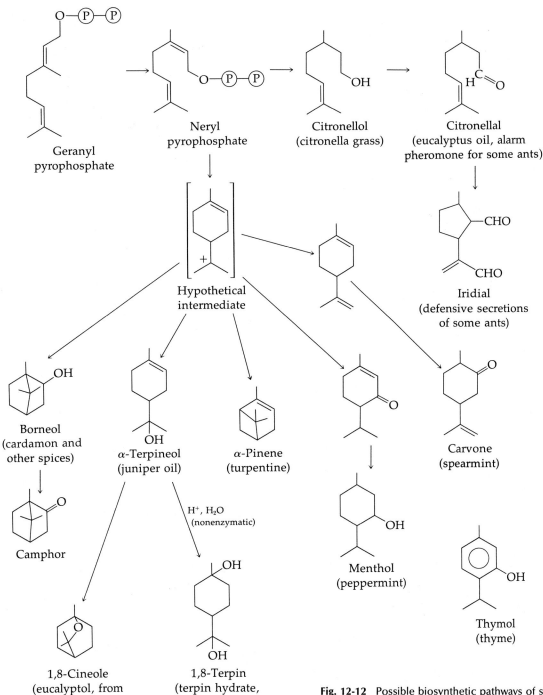

Fig. 12-12 Possible biosynthetic pathways of some monoterpenes and related substances. Some of the natural sources are indicated.

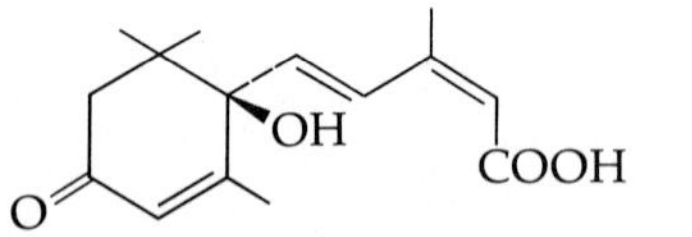

Abscisic acid, an important
hormone of green plants

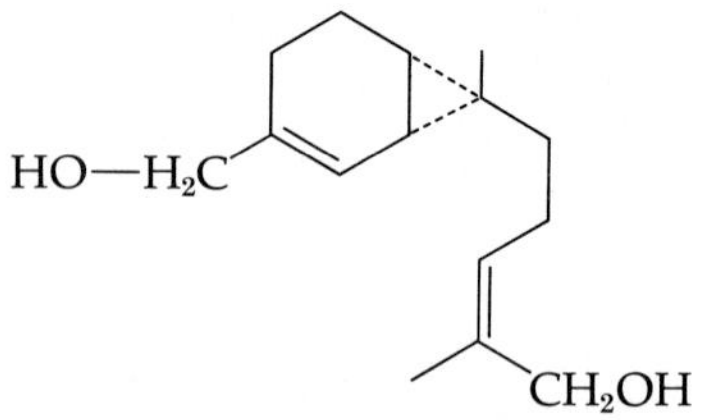

Sirenin, a sperm attractant
produced by female gametes
of the phycomycete *Allomyces*

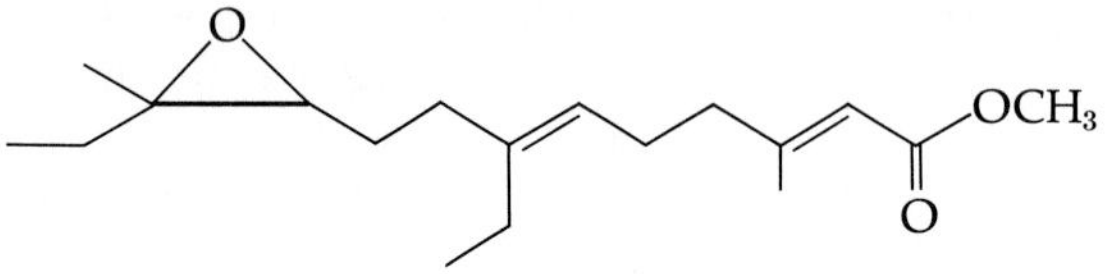

Juvenile hormone, a substance acting
to maintain the larval stage of insects

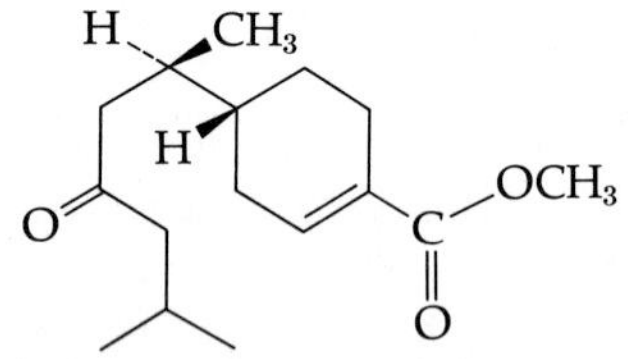

Juvebione, a product of the balsam fir
with juvenile hormone activity

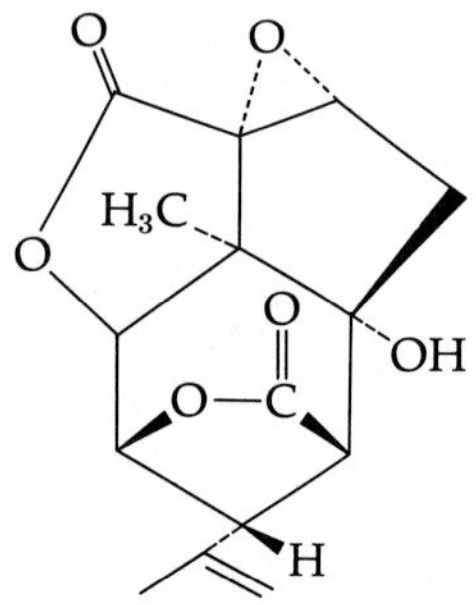

Picrotoxin, a plant toxin used
in southeast Asia to kill fish

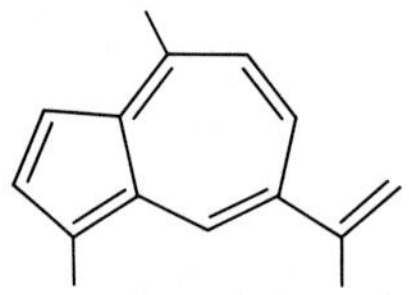

Lactarazulene, dye from
the mushroom *Lactarius
deliciosus*

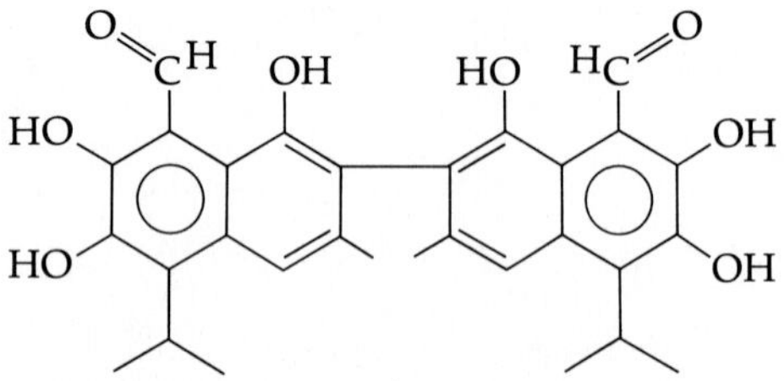

Gossypol, toxic, dimeric sesquiterpene
from cottonseed

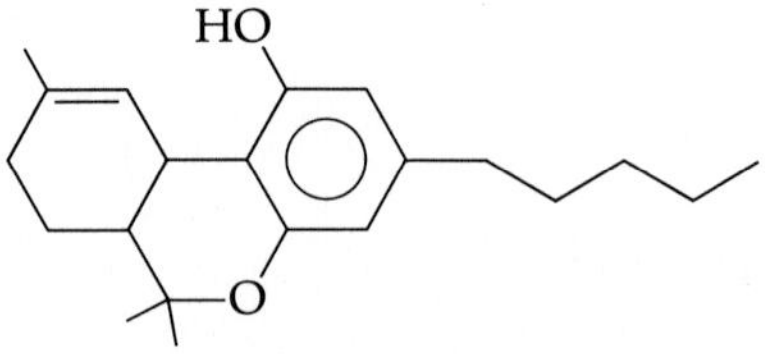

Δ'-3,4-*trans*-Tetrahydrocannabinol
(Δ'-THC) a psychotropic
component of marijuana

Fig. 12-13   More terpenes and related substances.

major metabolic pathways of animal organisms. Purified enzymes have not been isolated. Nevertheless, the "feeding" of radioactively labeled acetate to plants produces characteristic labeling patterns in the terpenes. These have been traced for most of the compounds indicated and are those expected. A given plant usually contains a large number of different terpenes which, in many plants, are concentrated in specialized "oil glands" or resinous duct tissues. Lesser amounts, often as glycosides of terpene alcohols, may be present within cells. Some terpenes occur in truly enormous amounts. Thus, turpentine may contain 64% of α-pinene and juniper oil 65% α-terpineol.[80]

Most of the compounds shown in Fig. 12-13 are derived from the C-15 **farnesyl pyrophosphate**. **Abscisic acid** is one of five known types of plant hormone of general distribution throughout higher plants (Chapter 16, Section A,3). Another class of plant hormones are the **gibberellins,** the first member of which was originally isolated as a product of plants infected with a *Fusarium* fungus. The rice plants grew in an abnormally tall, weak form. Subsequently, the gibberellins, a multimembered class of highly modified C_{20} (and C_{19}) terpenes have been shown to function in all higher plants.

The biosynthesis of gibberellins is complex.[78,81] Equation 12-26 is an abbreviated equation for formation of gibberellin A_1. Step a of Eq. 12-26 is presumably a single enzymatic reaction, an isomerization leading to ring closure. In addition to the movement of electrons required to close the two rings, one proton is removed from a methyl group and a proton is added at the end of the double bond to the left of the first structural formula. Equations 12-26, steps b and c, are complex multistep reactions. Note that in Eq. 12-26, step b, pyrophosphate is eliminated and that the methyl group that becomes a methylene in kaurene undergoes migration. Equation 12-26, step c is even more complex. There are numerous hydroxylation and oxidation steps as well as a ring contraction through which one methylene group (after oxidation) ends up as a carboxyl group in the final product.

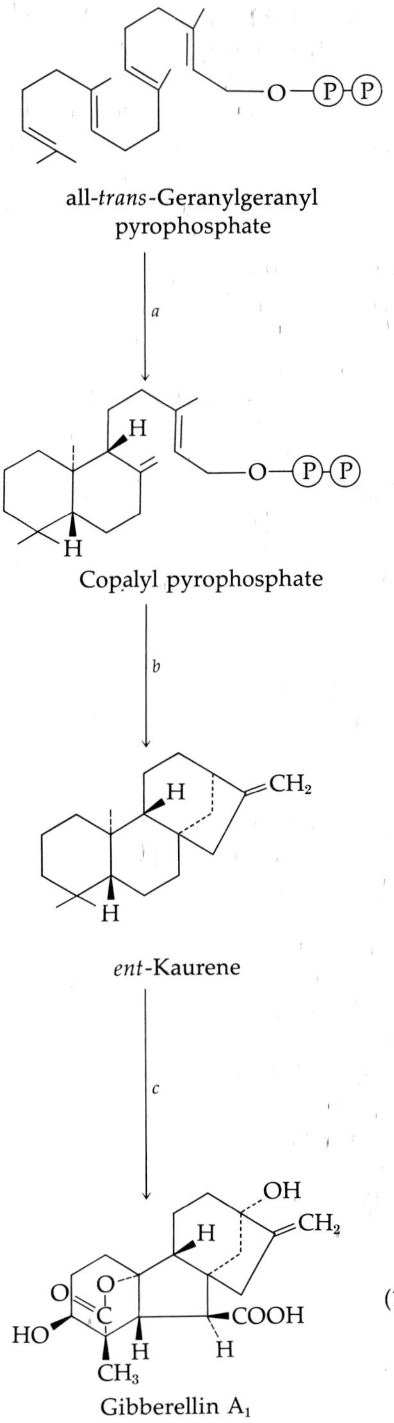

all-*trans*-Geranylgeranyl pyrophosphate

Copalyl pyrophosphate

ent-Kaurene

Gibberellin A_1

(12-26)

The **juvenile hormone** of insects (Fig. 12-13) is also of polyprenyl origin. However, note that two of the methyl groups have been converted to (or replaced by) ethyl groups. The isolation and identification of the structure of the juvenile hormone was a difficult task. After its completion it was a surprise to researchers to discover that a large variety of synthetic compounds, sometimes with only a small amount of apparent structural similarity, also serve as juvenile hormones, keeping insects in the larval stage or preventing insect eggs from hatching. Furthermore, a number of plant products such as **juvebione** (Fig. 12-13), originally isolated from paper, have the same effect. Thus, in nature, products of plant metabolism have a profound effect upon the development of insects that eat the plants. There is great interest in the possible use of juvenile hormone, or of synthetic compounds mimicking its action, as insecticides.

2. Formation of the Symmetric Terpenes, Squalene and Phytoene

Two molecules of the C_{15} farnesyl pyrophosphate can be joined "head to head" to form the C_{30} squalene. Similarly, two C_{20} **geranylgeranyl pyrophosphate** molecules can be joined to form the C_{40} phytoene, a precursor of carotenoid pigments of plants. These are remarkable condensation reactions which apparently differ significantly one from the other.

In the synthesis of squalene, both pyrophosphate groups are eliminated from the precursor molecules and one proton from C-1 of one of the molecules of farnesyl pyrophosphate is lost. The other three C-1 hydrogens are retained. At the same time, one proton (*pro-S*) is introduced from

the B side of NADPH. One suggested [82] mechanism is depicted in Eq. 12-27. A nucleophilic

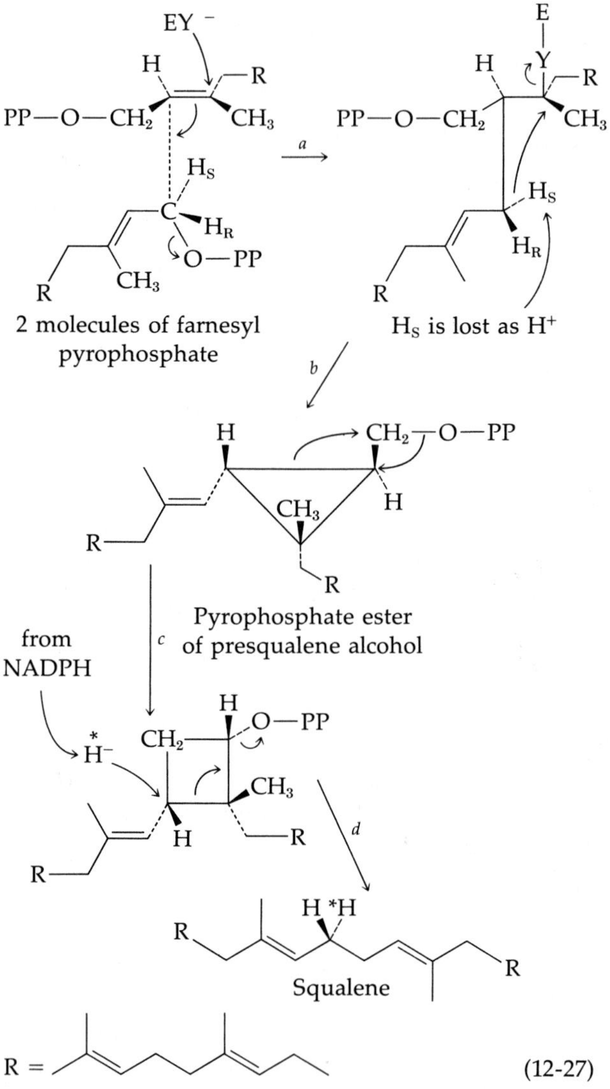

2 molecules of farnesyl pyrophosphate

H_S is lost as H^+

Pyrophosphate ester of presqualene alcohol

from NADPH

Squalene

$$R = $$ (12-27)

group, Y^- attached to an enzyme adds to a double bond adjacent to the pyrophosphate group of one molecule of farnesyl pyrophosphate. This initiates a concerted displacement of the pyrophosphate group of a second molecule of farnesyl pyrophosphate (Eq. 12-27, step a). The *pro-S* hydrogen adjacent to the central double bond in the resulting structure is dissociated as a proton, presumably

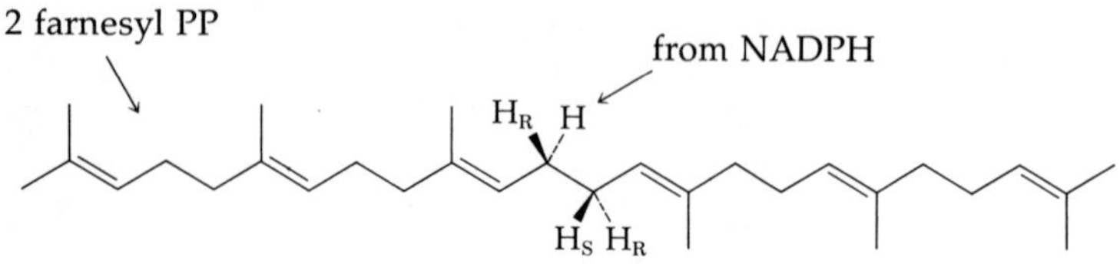

2 farnesyl PP

from NADPH

H_R H

H_S H_R

through action of a basic group in the enzyme. The resulting anion displaces group Y^- with formation of a cyclopropane derivative (Eq. 12-27, step b). This product, **presqualene alcohol pyrophosphate,** was isolated from yeast as the free alcohol by Rilling and associates.[83,83a] The ring could open to the cyclobutane structures shown (Eq. 12-27, step c). The latter could be reduced by NADPH with displacement of pyrophosphate to yield squalene (Eq. 12-27, step d). A partially purified enzyme that catalyzes these reactions has been dissociated from membrane particles of yeast.[84]

An alternative mechanism assumes that the pyrophosphate ester of presqualene alcohol loses pyrophosphate (Eq. 12-28) to form a carbonium ion whose structure is expected to be that of a nonclassical **bicyclobutonium** ion.[85,86] Two

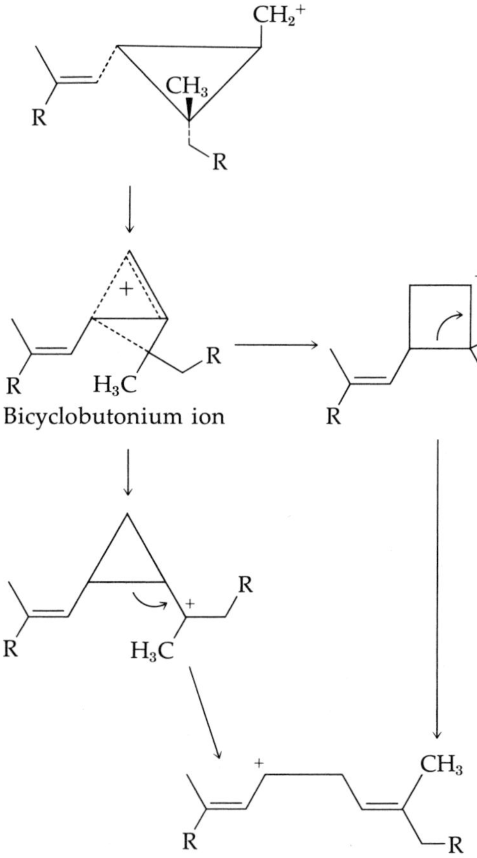

pathways by which such an ion could rearrange are indicated in Eq. 12-28. The final carbonium ion product could be reduced by NADPH to squalene.

The mechanism of formation of phytoene is similar except that no reduction by NADH is involved. It is known that the *pro-5R* hydrogen atoms of mevalonic acid are retained to the center in *cis*-phytoene as indicated in Fig. 12-14. *Trans*-Phytoene is also produced in plants and contains one *pro-S* and one *pro-R* hydrogen of the geranylgeranyl pyrophosphate at the central double bond.[87]

3. Carotenes and Their Derivatives

While it is not established with certainty that formation of carotenes is through phytoene, the process can easily be formulated in that way. Figure 12-14 indicates the pathways of formation of **lycopene,** the red pigment of tomatoes, **β-carotene,** and some substances derived from these compounds by further metabolism. Note that lycopene is an all-trans compound and that desaturation occurs through the trans loss of hydrogen atoms. Desaturation takes place in a stepwise fashion and many intermediate compounds with fewer double bonds are known. Ring closure at the ends of the lycopene molecule to form the carotenes can most readily be formulated (Eq. 12-29) through an acid-catalyzed carbonium ion

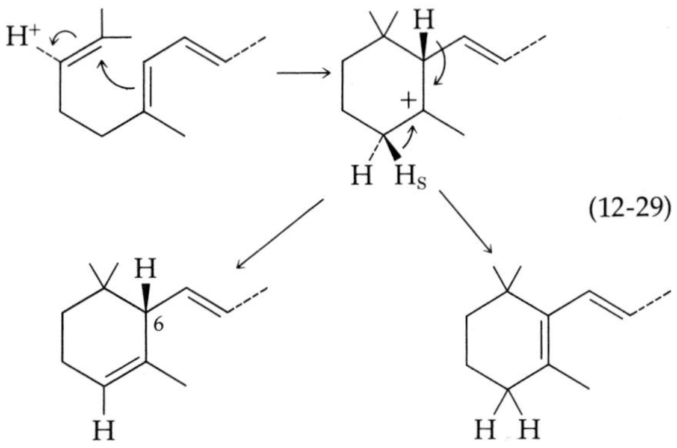

Bicyclobutonium ion

(12-28)

(12-29)

α Ring (6S configuration) β Ring

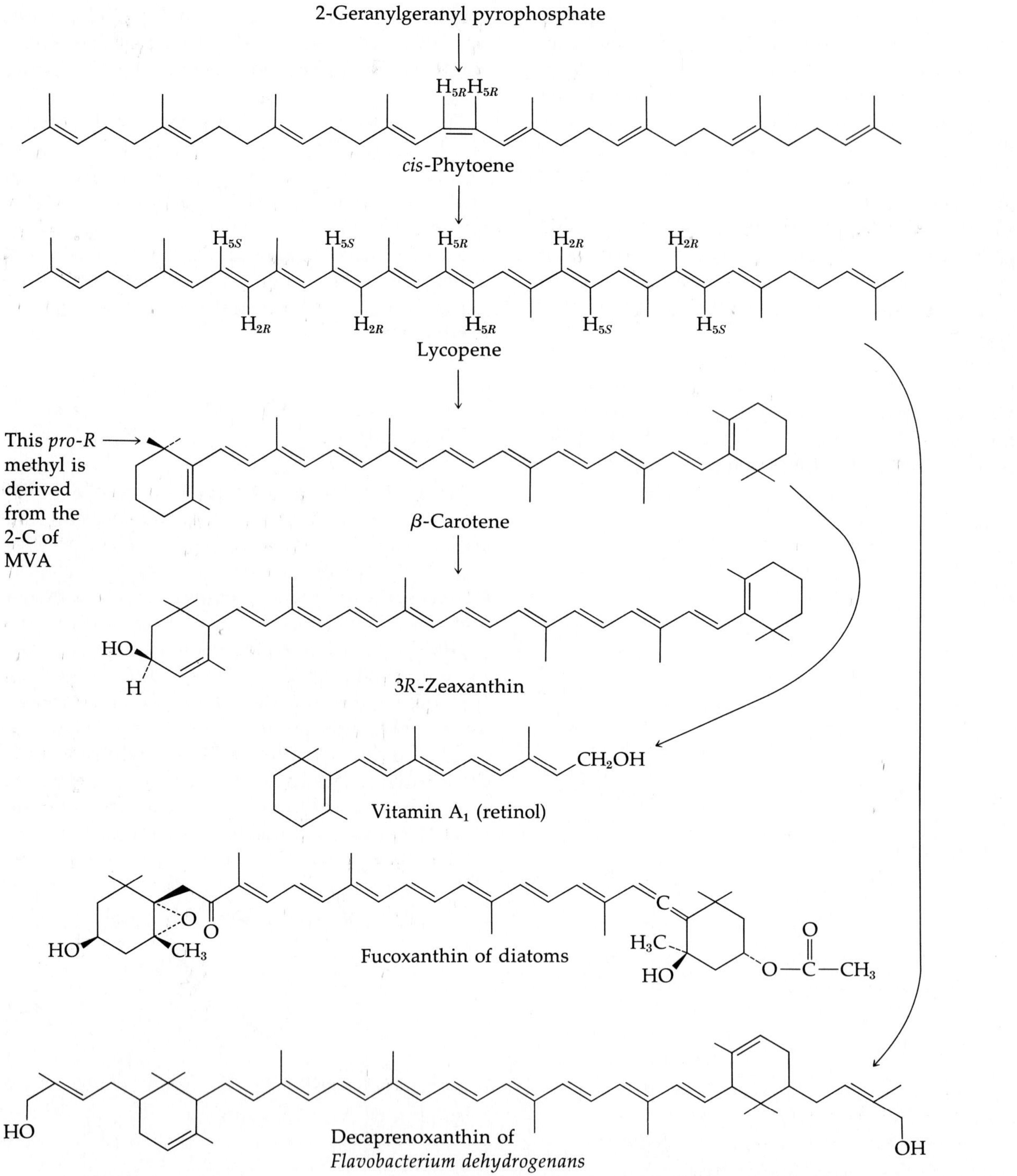

2-Geranylgeranyl pyrophosphate

$H_{5R}H_{5R}$

cis-Phytoene

H_{5S} H_{5S} H_{5R} H_{2R} H_{2R}

H_{2R} H_{2R} H_{5R} H_{5S} H_{5S}

Lycopene

This *pro-R* methyl is derived from the 2-C of MVA

β-Carotene

HO

H

3*R*-Zeaxanthin

CH₂OH

Vitamin A₁ (retinol)

HO   CH₃   Fucoxanthin of diatoms   H₃C   O   O—C—CH₃   HO

HO   Decaprenoxanthin of *Flavobacterium dehydrogenans*   OH

Fig. 12-14 Biosynthesis of some carotenoid pigments.

mechanism. Loss of one or the other of two protons adjacent to the positive charge leads to the β ring of β-carotene or to the α ring of α-carotene.[88]

Carotenes can by hydroxylated and otherwise modified in a number of ways. The structure of only one of the resulting **xanthophylls, zeaxanthin,** is indicated in Fig. 12-14. The structure of the characteristic brown pigment of diatoms, **fucoxanthin,** is also shown. Note that one end of that molecule contains an epoxide, also formed by the action of O_2. The other end contains an **allene** structure, rare in nature. (Even so fucoxanthin may be the most abundant carotenoid of all.)[88] Below is the allene-containing end of the fuco-

xanthin molecule (turned over from that shown in Fig. 12-14). Note that Fig. 12-14 does not indicate the stereochemistry of the allene group correctly. The remainder of the carotenoid chain R_1 protrudes behind the ring as drawn here.

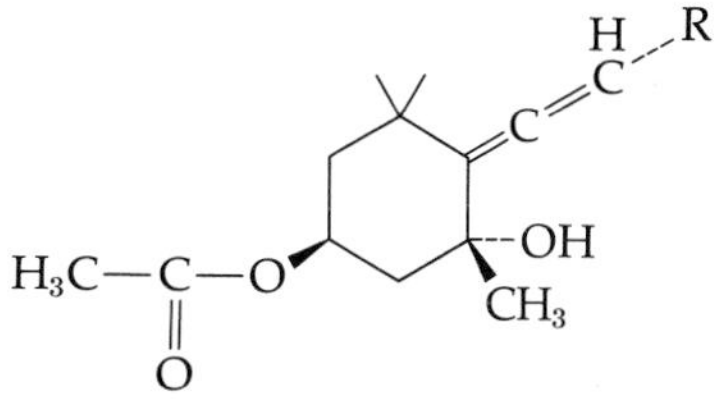

Another algal carotenoid **violaxanthin** contains epoxide groups in the rings at both ends of the

---

BOX 12-C

THE VITAMINS: VITAMIN A

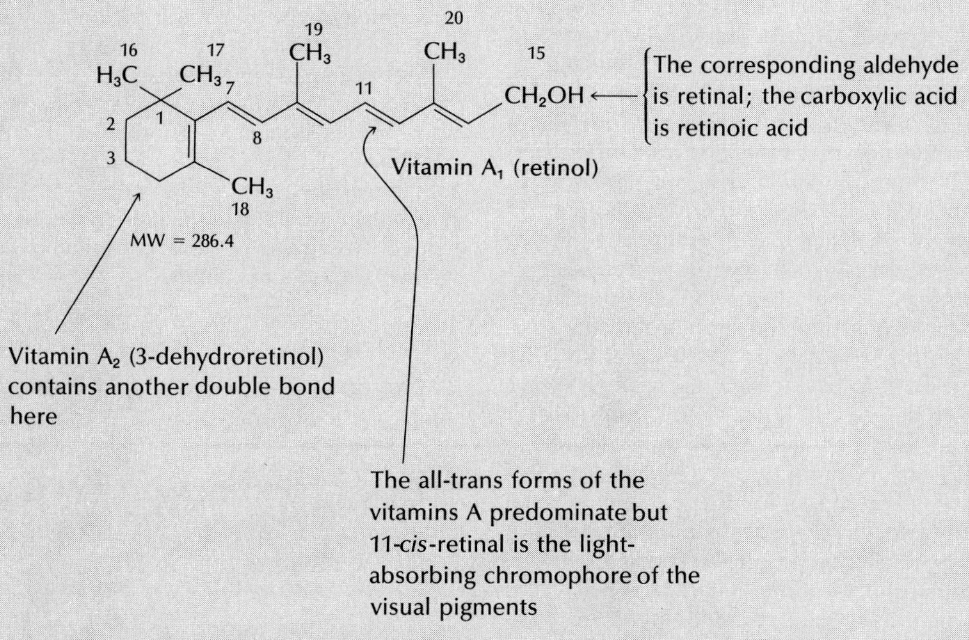

Vitamin A$_1$ (retinol)

The corresponding aldehyde is retinal; the carboxylic acid is retinoic acid

MW = 286.4

Vitamin A$_2$ (3-dehydroretinol) contains another double bond here

The all-trans forms of the vitamins A predominate but 11-*cis*-retinal is the light-absorbing chromophore of the visual pigments

The recognition in the 1920's of vitamin A (Box 8-A) was soon followed by its isolation from fish liver oils.[a] Both vitamin A$_1$ (**retinol**) and vitamin A$_2$ are 20-carbon polyprenyl alcohols. They are formed by cleavage of the 40-carbon β-carotene (Fig. 12-14) or other carotenoids containing one

β-ionone ring. While the carotenes are plant products, vitamin A is produced only in animals. Carotene chains are cleaved in the center by an oxygenase to form vitamin A aldehyde **retinal.**[b] Vitamin A exists in tissues both as the free alcohol and as esters of palmitic and other fatty acids. It is

Box 12-C (*continued*)

one of the few vitamins that can be stored in animals in relatively large quantities. It accumulates in the liver, mainly as retinyl palmitate, in special fat storage cells.[c]

A deficiency of vitamin A leads to a variety of symptoms including dry skin and hair, conjunctivitis of the eyes, retardation of growth, and low resistance to infection. A striking early symptom is **night blindness.** The skin symptoms are particularly noticeable in the internal respiratory passages and alimentary canal lining. About 0.7 mg/day of vitamin A is required by an adult. The content of vitamin A in foods is often expressed in terms of international units: 1.0 mg of retinol equals 3333 I.U.

Vitamin A, as retinal, has a clearly established role in vision (Chapter 13, Section F,2) and it apparently has a specialized function in reproduction. In vitamin A deficiency no sperm cells are formed in males, and fetal resorption occurs in females. Rats deprived of vitamin A but fed **retinoic acid** become blind and sterile but otherwise appear healthy.[d] Evidently it is either the alcohol or the aldehyde that functions in reproduction whereas bone growth and maintenance of mucous secretions only requires retinoic acid.[e]

In vitamin A deficiency, the internal epithelial surface, which is usually rich in special mucous secreting cells and in ciliated cells, develops thick layers of keratinizing squamous cells similar to those on the external surface of the body. The production of fucose-containing glycopeptides is strikingly decreased.[f] It has been shown that addition of retinyl acetate to cell cultures of epidermis leads to an increase in cellular RNA.[g] Addition of insulin and a glucocorticoid enhance the effect. These results suggest that retinoic acid or some metabolite could act as a regulatory molecule to transform undifferentiated epithelial cells into mucous-secreting cells. Another form of vitamin A could function similarly in reproductive tissues. It is plausible to believe that these regulatory functions involve control of the transcription of RNA. Steroid hormones and metabolites of vitamin D may function in a similar way to "turn on" or "turn off" genes specifying particular protein products.

It is not clear what special chemistry of vitamin A is involved in its function. Retinal could form Schiff bases with protein groups as it does in the visual pigments. Redox reactions could occur. Conjugative elimination of water from retinol to form **anhydroretinol** is catalyzed nonenzymatically

Anhydroretinol

by HCl. Anhydroretinol does occur in nature, but it is not thought to have a biological function.

Vitamin A is transported in blood plasma by a special **retinol-binding protein** of MW ∼ 21,000. This protein is normally almost saturated with retinol and is bound to another serum protein **prealbumin.**[h,i] Many tissues also contain retinol-binding proteins in addition to other proteins that bind retinoic acid.[j] Evidence has been presented that zinc is necessary for maintenance of normal amounts of vitamin A in the plasma.[k]

[a] T. Moore, "Vitamin A." Elsevier, Amsterdam, 1957.

[b] J. A. Olson, *Vitam. Horm.* (*N.Y.*) **26**, 1–63 (1968).

[c] K. Kobayashi, Y. Takahashi, and S. Shibasaki, *Nature* (*London*), *New Biol.* **243**, 186–188 (1973).

[d] J. E. Smith, P. O. Milch, Y. Muto, and D. S. Goodman, *BJ* **132**, 821—827 (1973).

[e] G. H. Clamon, M. B. Sporn, J. M. Smith, and V. Saffiotti, *Nature* (*London*) **250**, 64–66 (1974).

[f] L. DeLuca, M. Schumacher, and G. Wolf, *JBC* **245**, 4551–4558 (1970).

[g] M. B. Sporn, N. M. Dunlop, and S. H. Yuspa, *Science* **182**, 722–723 (1973).

[h] A. Vahlquist and P. A. Peterson, *Biochemistry* **11**, 4526–4532 (1972).

[i] Y. Muto, J. E. Smith, P. O. Milch, and D. S. Goodman, *JBC* **247**, 2542–2550 (1972).

[j] D. E. Ong and F. Chytil, *JBC* **250**, 6113–6117 (1975).

[k] J. C. Smith, Jr., E. G. McDaniel, F. F. Fan, and J. A. Halsted, *Science* **181**, 954–955 (1973).

molecule. An isomerase converts violaxanthin (Eq. 12-30) into **neoxanthin,** another compound thought to contain (at one end) the allene structure. Subsequent acetylation yields fucoxanthin.

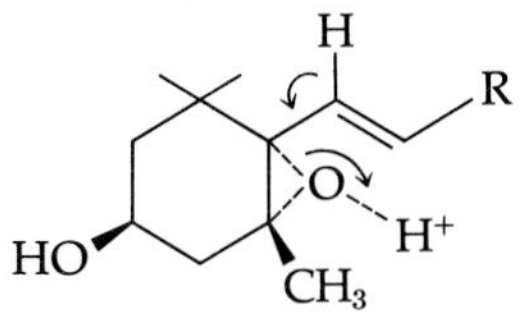

Violaxanthin (ring structure
present at both ends)

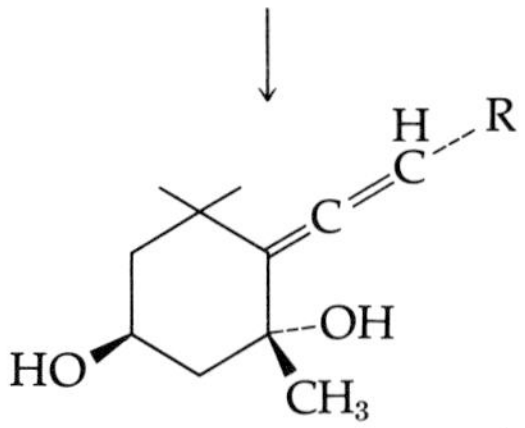

Neoxanthin (this structure
occurs at one end only)   (12-30)

Other algal carotenoids contain acetylenic triple bonds. For example, **alloxanthin** has the following structure at both ends of the symmetric molecule.

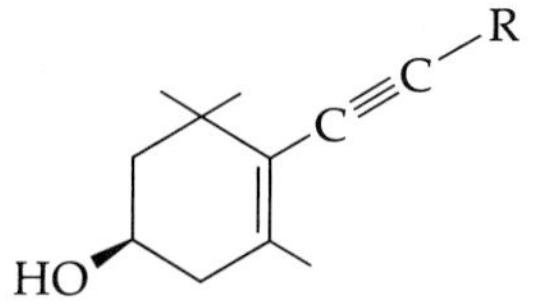

The foregoing descriptions deal with only a few of the many known structural modifications of carotenoids.[44,78,89]

4. Polyprenyl Side Chains

Among other important polyprenyl compounds are the side chains of vitamin K, the ubiquinones, plastoquinones, tocopherols, and the phytyl group of chlorophyll. In all cases, a polyprenyl alcohol in the form of a pyrophosphate ester serves as an alkyl group donor. Introduction of the polyprenyl chain into aromatic groups such as those of the quinones (Fig. 10-8) occurs at a position ortho to a hydroxyl group in the reduced quinone (hydroquinone). The reader should be able to pro-

pose a reasonable mechanism involving elimination of pyrophosphate and participation of the hydroxyl group. However, there is a possibility that decarboxylation of a precursor is associated with the condensation[90] (Chapter 14, Section F,4).

I. STEROID COMPOUNDS

The large class of **steroids** contain a characteristic 4-ring nucleus consisting of three fused six-membered rings and one five-membered ring. **Cholestanol** (dihydrocholesterol) may be taken as a representative steroid alcohol or **sterol.**

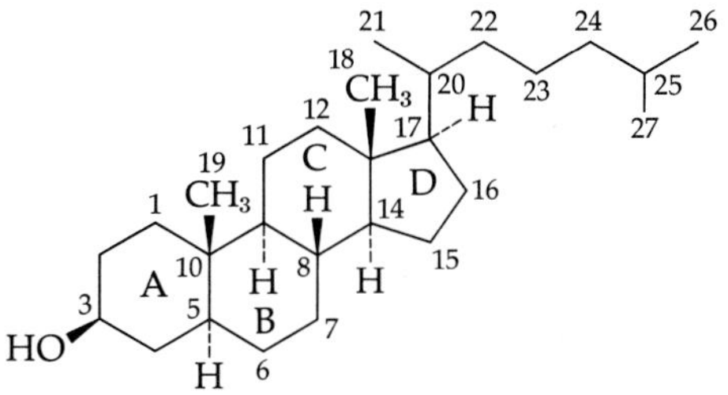

Cholestanol, a minor sterol of the animal body, illustrating the numbering system for steroids. Note that the configuration at C-20 is R

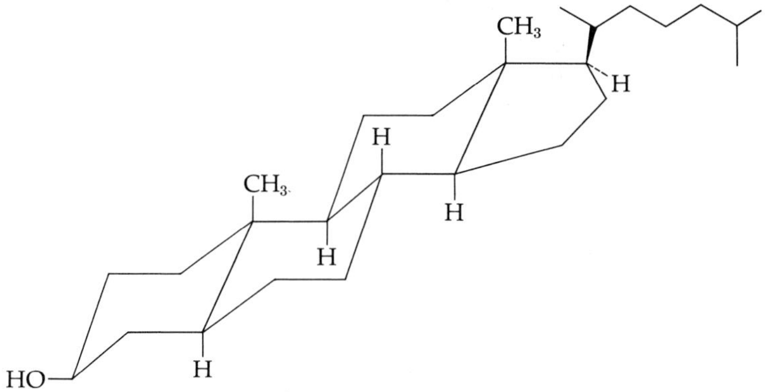

Conformation of cholestanol. Note that all ring junctions are trans

Most sterols, including cholestanol, contain an 8- to 10-carbon side chain at position 17. The polyprenyl origin of the side chain is suggested by the structure. Steroid compounds usually contain an oxygen atom at C-3. This atom is present in an

—OH group in the sterols and frequently in a carbonyl group in other steroids. Most steroids contain two methyl groups attached to the ring system and numbered C-18 and C-19. Note that these groups are axially oriented. In the customary projection formulas, they are to be thought of as extending forward toward the viewer. In the same manner, the equatorially oriented 3-OH group of cholestanol and the side chain at C-17 also project forward toward the viewer in the projection formula.

The "angular" methyl groups (C-18 and C-19) and the 3-OH groups and the side chain of cholestanol are all on the same side of the steroid ring in the projection formula and are all said to have a **β orientation.** Substituents projecting from the opposite side of the ring system are **α oriented.** While the methyl groups (C-18 and C-19) almost always have the β orientation, the 3-OH group is oriented to α in some sterols. Dotted lines are customarily used to connect α-oriented substituents and solid lines are used for β-oriented substituents in structural formulas.

In cholestanol, the ring fusions between rings A and B, B and C, and C and D are all trans; that is, the hydrogen atoms or methyl groups attached to the bridgehead carbon atoms project on opposite sides of the ring system. This permits all three of the six-membered rings to assume relatively unstrained chair conformations. However, the intro-

duction of a double bond alters the shape of the molecule significantly. Thus, cholesterol contains a double bond between C-5 and C-6 (Δ^5) which distorts both the A and B rings from the chair conformation. In some steroid compounds the junction between rings A and B is cis. This greatly alters the overall shape of the steroid from the relatively flat one of cholestanol to a distinctly bent one. An example is β-coprostanol, a product of bacterial action on cholesterol and a compound occurring in large amounts in the feces. In some sterols, notably the estrogenic hormones, ring A is completely aromatic and the methyl group at C-19 is absent.

1. Biosynthesis of Sterols

Most animal steroids arise from cholesterol which in turn is derived from squalene.* The transformations, which take place in most animal tissues, are initiated by a microsomal enzyme system that utilized O_2 and NADPH to form **squalene 2,3-oxide.**[91–94] The reaction can be visualized[77] as proceeding through a carbonium ion created by attack of a proton on the oxygen atom of the epoxide ring (Eq. 12-31). A flow of electrons (either synchronous or stepwise) effects the closure of all four rings, leaving a carbonium ion at the junction of the side chain with ring D (Eq. 12-31, step a). The rearrangement of this carbonium ion to **lanosterol** (Eq. 12-31, step b) is a remarkable reaction that requires the shift of one hydride ion and of two methyl groups, as indicated by the arrows in Eq. 12-31. In addition, a hydrogen at C-9 is lost as a proton. Lanosterol† is the precursor of other sterols in the animal body. However, in plants, which contain little or no cholesterol, **cycloartenol** appears to be the key intermediate in sterol biosynthesis. As indicated in Eq.

---

* Squalene derives its name from that of the dogfish *Squalus* in whose liver it accumulates as a result of blockage in oxidation to cholesterol. It is also a prominent constituent of human skin lipids.

† Lanosterol is a constituent of lanolin, the waxy fat from wool whose principal constituent is cholesterol.

β-Coprostanol

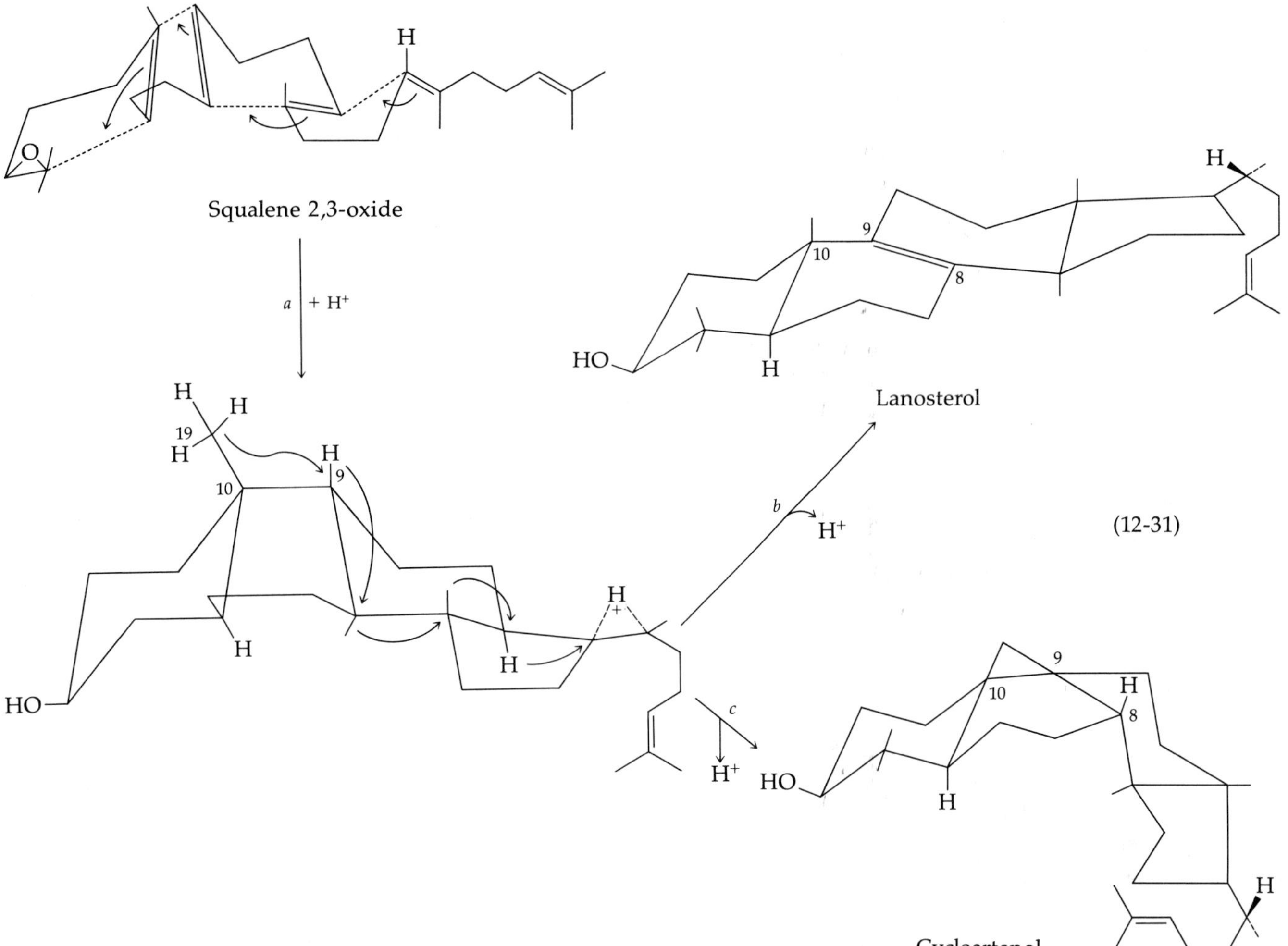

Squalene 2,3-oxide

a | + H^+

Lanosterol

(12-31)

b ↗ H^+

c ↘ H^+ HO

Cycloartenol

12-31, step c, cycloartenol can be formed if the proton at C-9 is shifted (as a hydride ion) to displace the methyl group from C-8. A proton is lost from the adjacent methyl group to close the cyclopropane ring.

The conversion of lanosterol to cholesterol is a complex process that has been estimated to require at least 25 steps. Many of the enzymes involved are bound to membranes of the endoplasmic reticulum.[95] At least one soluble cytoplasmic protein is needed in addition to the enzymes embedded in the membrane. This **sterol carrier protein** may transport the sterol from one enzyme to another during the transformations and may also affect reactivity.[96,97]

It has been established that the removal of the three methyl groups of lanosterol, the migration of the double bond within the β ring, and the saturation of the double bond in the side chain may occur in more than one sequence. Two variations are indicated in Fig. 12-15. Details of the demethylation at the C-D ring junction are uncertain, but the corresponding reactions of the methyl groups at C-4 on the A ring have been clarified.[95] Each of the methyl groups is hydroxylated in turn by a microsomal system similar to the cytochrome P-450 system (Chapter 10, Section G,2,f)[97a] but able to accept electrons from NADH rather than NADPH. Oxidation of the resulting alcohol to a carboxylic

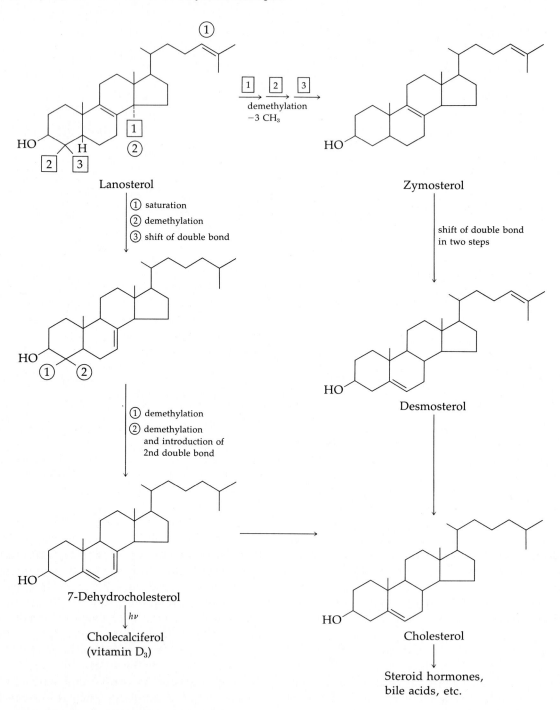

Fig. 12-15 Conversion of lanosterol to cholesterol. The numbers in circles and boxes indicate the sequence in which the reaction steps occur at the indicated positions.

acid followed by oxidation of the 3-OH group to a ketone allows for efficient β-decarboxylation.

Other sterols formed in the animal body are **7-dehydrocholesterol,** prominent in skin and a precursor of vitamin D. Both β-cholestanol and its isomer β-coprostanol are formed by bacteria in the intestinal tract. Many plant sterols contain one or more additional carbon atoms (to those found in cholesterol) in the side chain. These are donated from S-adenosylmethionine. As indicated previously, plant sterols are thought to be formed in most cases through cycloartenol. One of the sterols with an extra carbon is methylenecycloartenol, a substance present in grapefruit peel (and in many plants). The side chain has the following structure.

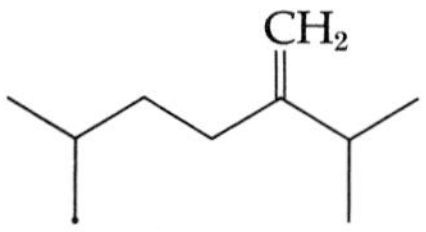

Side chain of
methylenecycloartenol

Campesterol, another plant sterol, has the Δ^5-unsaturated ring of cholesterol but a side chain with one more methyl group.

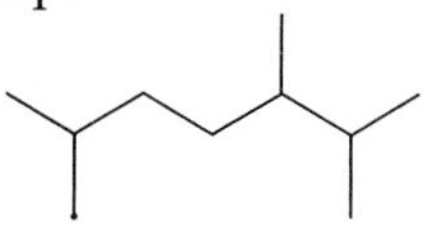

Ergosterol, the most common sterol in fungi, contains the $\Delta^{5,7}$ ring system of 7-dehydrocholesterol as well as an extra double bond in the side chain. Among higher plants, **sitosterol** and **stigmasterol** are the most common sterols. Each contains an extra ethyl group in the side chain. Sitosterol is formed by the methylation (SAM) of ergosterol. For the guinea pig, stigmasterol is a vitamin, the "antistiffness factor" necessary to prevent stiffening of the joints.

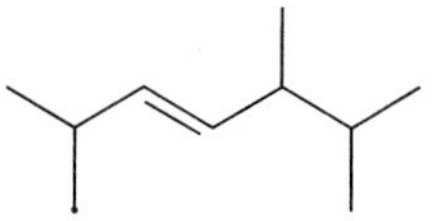

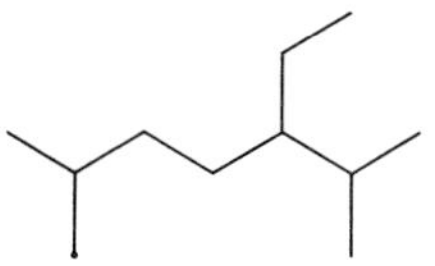

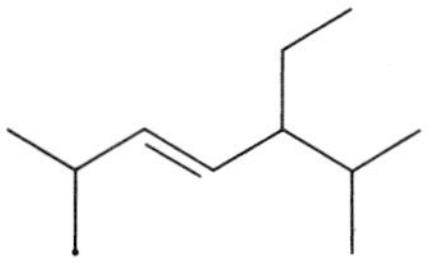

2. Metabolism of Cholesterol

Cholesterol is both absorbed from the intestinal tract in man and synthesized from acetate via squalene, principally in the liver. The quantities produced are substantial. Not only is there a large amount of cholesterol in the brain and other nervous tissues, but about 1.7 g of cholesterol per liter is present in blood plasma, about two-thirds of it being esterified, principally to unsaturated fatty acids. The cholesterol content of blood varies greatly with diet, age, and sex. By age 55 it averages 2.5 g/l and may be considerably higher. Females up to the age of menopause have distinctly lower cholesterol in the blood.

Most serum cholesterol is carried by the low density lipoprotein (LDL, Box 2-A) which delivers the cholesteryl esters directly to cells that need cholesterol. The LDL-cholesterol complex binds to specific LDL receptors on the cell surface and then is taken into the cell by endocytosis.[97b] Lysosomal enzymes attack the lipoprotein and a specific acid lipase cleaves the cholesteryl esters to release free cholesterol. Uptake is regulated by a feedback mechanism which controls the number of LDL receptor molecules synthesized. Cholesterol also inhibits the reduction of 3-hydroxy-3-methylglutaryl-CoA, (Fig. 11-8), an essential step in cholesterol synthesis.[97b,97c] Several abnormalities of cholesterol metabolism are known. In one form of familial **hypercholesterolemia** a mutation in the LDL receptor protein prevents normal binding and uptake of LDL and its cholesteryl esters. In a cholesteryl ester storage disease the lysosomal lipase is lacking. Further studies in this area of metabolism are important because they may shed light on the cause of atherosclerosis, ("hardening of the arteries"), an extremely common human disease.[97d]

Cholesterol is metabolized to a variety of substances including the steroid hormones. However, the quantitatively most important metabolites are the bile acids (Fig. 12-16). These powerful emulsifying agents flow from the liver into the bile duct and the small intestine. Later a large fraction is reabsorbed in the duodenum and is returned to the liver for reuse. Formation of bile acids involves the removal of the double bond of cholesterol, inversion at C-3 to give a 3α-hydroxyl group followed by hydroxylation and β oxidation of the side chain. The bile acids (or their CoA derivatives) are then conjugated with glycine and taurine to form **bile salts,** such as **glycocholic** and **taurocholic acids** (Fig. 12-16).

3. The Steroid Hormones[98,99]

In the animal body three important groups of hormones are formed by the metabolism of cholesterol: the **progestins,** the **sex hormones,** and the **adrenal cortical hormones.** Some features of the major pathways of biosynthesis are outlined in Fig. 12-17. The side chain is shortened to two carbon atoms through hydroxylation and oxidative cleavage. This leaves a two-carbon side chain in the key intermediate **pregnenolone** and in the adrenal cortical hormones. Oxidation of the 3-OH group of pregnenolone to C=O is followed by a shift in the double bond, the ketosteroid isomerase reaction (Eq. 7-56, step b), the product is the α,β-unsaturated ketone **progesterone.**

a. Progestins

Progesterone is the principal hormone of the **corpus luteum,** the endocrine gland developing in the ovarian follicle after release of an ovum. Progesterone is also formed in the adrenals, testes, and placenta. It is metabolized extremely rapidly, largely by reduction to alcohols which may then be conjugated and excreted as glucuronides (Eq. 12-12). Reduction of the double bond within the A ring of progesterone leads to complete loss of

activity, an indication that the α,β-unsaturated ketone group may play an essential role in the action of the hormone. Progesterone has a special role in the maintenance of pregnancy and, together with the estrogenic hormones, it regulates the menstrual cycle.

b. Adrenal Cortical Steroids

Within the adrenal cortex (the outer portion of the adrenal glands) progesterone is converted to two groups of hormones of which **cortisol** and **aldosterone** are the most important representatives. Cortisol, whose production is controlled by the pituitary hormone **corticotropin,** is secreted by the adrenals in amounts of 15–30 mg daily in an adult. The hormone, which is essential to life, circulates in the blood, largely bound to the plasma protein **transcortin.** As mentioned in Chapter 11, Section F,2, cortisol is a glucocorticoid which promotes gluconeogenesis and the accumulation of glycogen in the liver. It inhibits protein synthesis in muscle and other tissues and leads to breakdown of fats to free fatty acids in adipose tissue.

Cortisol and its close relative **cortisone** are probably best known for their anti-inflammatory effect in the body. The mechanism of the action may involve stabilization of lysosomal membranes as

Prednisolone

Dexamethasone

Fig. 12-16 Formation of the bile acids.

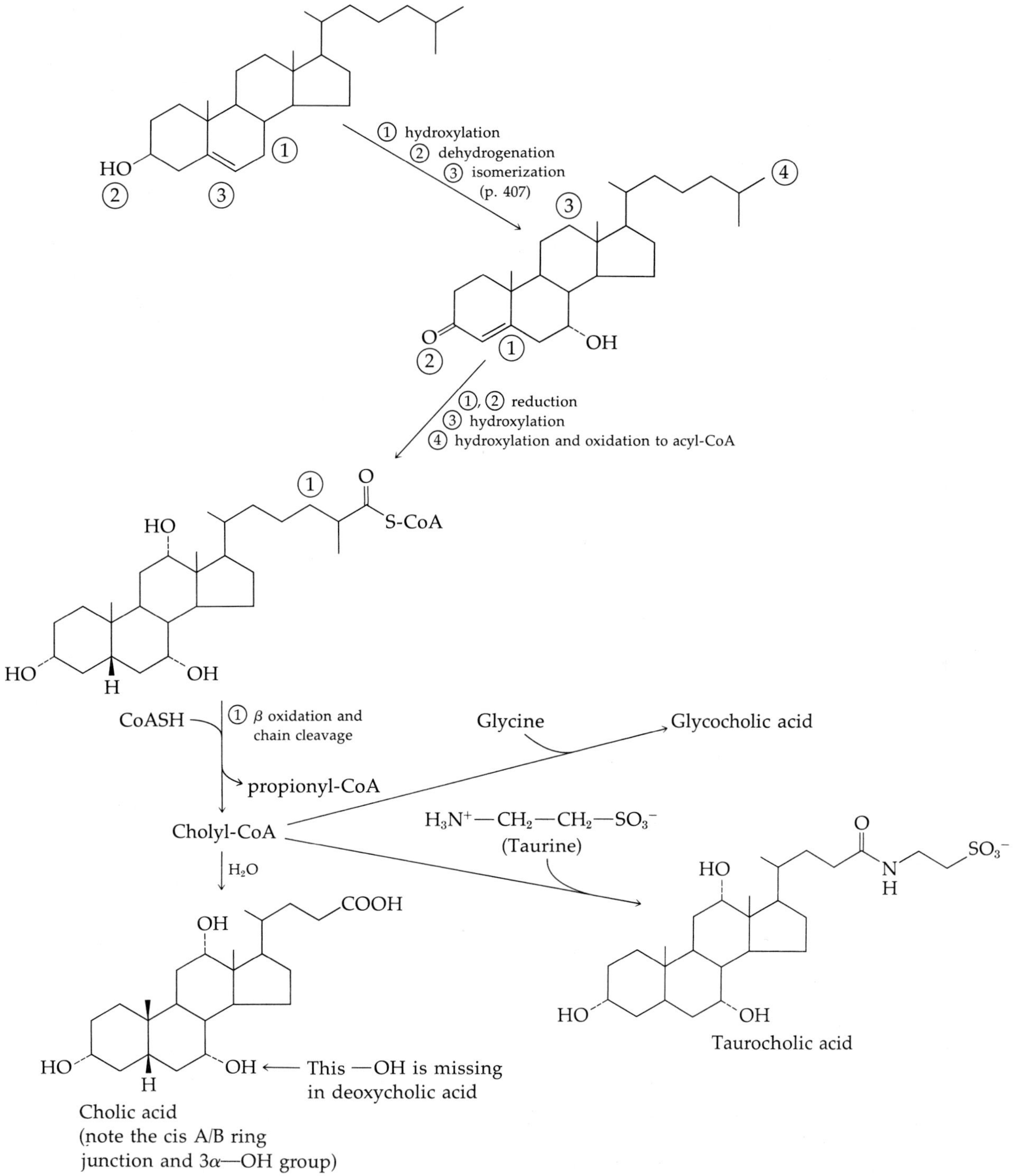
HO
②
③
①
① hydroxylation
② dehydrogenation
③ isomerization
(p. 407)
④
③
O
②
①
OH
①, ② reduction
③ hydroxylation
④ hydroxylation and oxidation to acyl-CoA
①
O
S-CoA
HO
HO
H
OH
CoASH
① β oxidation and
chain cleavage
propionyl-CoA
Glycine
Glycocholic acid
Cholyl-CoA
$H_3N^+—CH_2—CH_2—SO_3^-$
(Taurine)
H_2O
O
SO_3^-
N
H
HO
COOH
OH
HO
OH
Taurocholic acid
HO
H
OH ← This —OH is missing
in deoxycholic acid
Cholic acid
(note the cis A/B ring
junction and 3α—OH group)

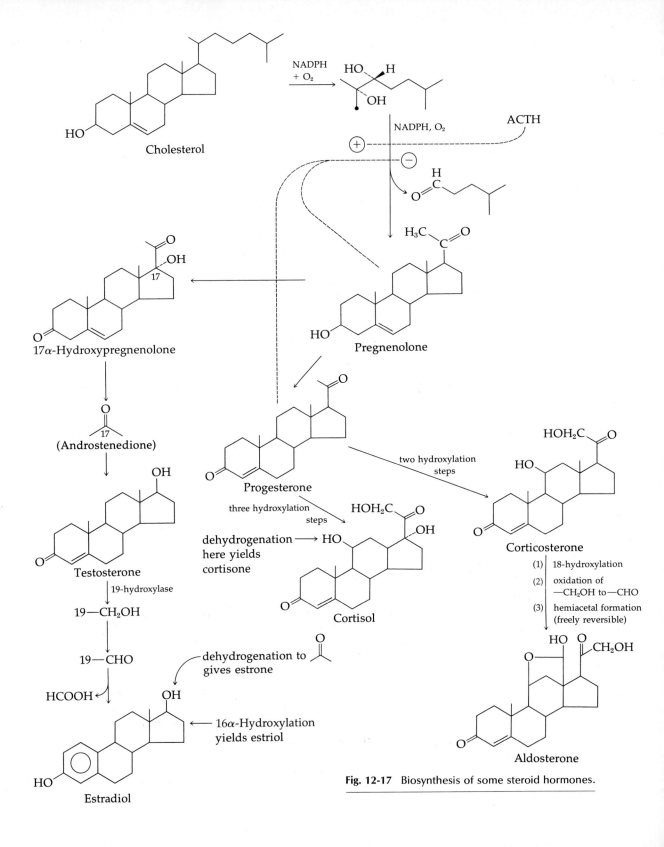

Fig. 12-17 Biosynthesis of some steroid hormones.

well as destruction of lymphocytes. Cortisone and synthetic analogues such as prednisolone and dexamethasone are among the modern "wonder drugs." They are used in controlling acute attacks of arthritis and of serious inflammations of the eyes and other organs. However, prolonged therapy can have serious side effects including wasting of muscles and resorption of bone. The latter effect results from a specific inhibition of calcium absorption from the gastrointestinal tract. In this, glucocorticoids are antagonistic to the action of vitamin D (Box 12-D).

Aldosterone, which is classified as a **mineralocor-**ticoid, is produced under the control of the **rennin-angiotensin** hormone system. This system is stimulated when sodium ion receptors in the kidneys detect an imbalance. Aldosterone promotes the resorption of sodium ions in the kidney tubules and thus regulates water and electrolyte metabolism. An adult on a diet of normal sodium content secretes about 0.1–0.2 mg/day of the hormone. Glucocorticoids also have weak mineralocorticoid activity and it is found that most patients with adrenocortical insufficiency (Addison's disease) can be maintained with glucocorticoids alone if their salt intake is adequate.

BOX 12-D

VITAMIN D

A lack of vitamin D causes rickets, a disease of humans and other animals in which the bones are soft, deformed, and poorly calcified.[a] Rickets was recognized by some persons to result from a dietary deficiency well over a hundred years ago, and the use of cod liver oil to prevent the disease was introduced in about 1870. By 1890 an association of rickets with a lack of sunlight had been made. However, it was not until 1924, when H. Steenbock and A. F. Hess showed that irradiation of certain foods generated protective activity against the disease, that vitamin D (**calciferol**) was recognized as a second lipid-soluble vitamin. Vitamin D is a family of compounds formed by the irradiation of $\Delta^{5,7}$-unsaturated sterols such as ergosterol and 7-dehydrocholesterol.[b] The former yields **ergocalciferol** (vitamin D_2) and the latter **cholecalciferol** (vitamin D_3).

$\Delta^{5,7}$-Sterol   Precalciferol   Calciferol

R =   Vitamin D_2 (ergocalciferol), from irradiation of ergosterol

R =   Vitamin D_3 (cholecalciferol), from irradiation of 7-dehydrocholesterol

Box 12-D (*continued*)

At low temperature the intermediate **precalciferol** can be isolated. Irradiation sets up a photochemical steady state equilibrium between the $\Delta^{5,7}$-sterol and the precalciferol. At higher temperatures, the latter is converted to calciferol. Other products, including toxic ones, are produced in slower photochemical side reactions. Therefore, the irradiation of ergosterol for food fortification must be done with care.

With normal exposure to sunlight enough 7-dehydrocholesterol is converted to cholecalciferol in the skin that no dietary vitamin D is required. This is particularly true for adults. It is usually recommended that children receive ~20 µg (400 I.U.) of ergocalciferol per day in their diet. Larger amounts are undesirable, and at a tenfold higher level vitamin D is seriously toxic.

The principal function of vitamin D is in the control of calcium metabolism. Recently, it has become clear that this is accomplished through the mediation of polar, hydroxylated metabolites. Hydroxylation occurs at three positions, the most polar metabolite yet isolated being **1,24,25-trihydroxycholecalciferol.**[c,d] These metabolites

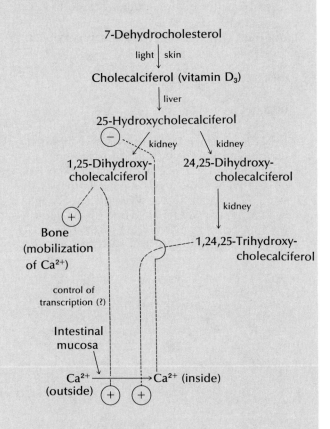

1,24,25-Trihydroxycholecalciferol

may be properly described as steroid hormones and vitamin D itself as a hormone formed in the skin. The established hydroxylation reactions of vitamin D are summarized in the accompanying scheme. It is noteworthy that the first hydroxylation—to 25-hydroxycholecalciferol—occurs in the liver, whereas subsequent hydroxylations

take place in the kidneys. Since it is the di- and trihydroxy derivatives that are essential for control of calcium ion metabolism, human patients with damaged kidneys often suffer severe demineralization of their bones (renal osteodystrophy). An anticipated result of recent research will be the synthetic production of polyhydroxylated vitamin D derivatives for administration to kidney patients.[e]

Both 1,25-dihydroxycholecalciferol and the trihydroxy derivative act to increase calcium ion uptake by the intestinal mucosa. It appears that these compounds, like other steroid hormones, act to regulate gene transcription.[f] For example, it has been shown that calcium-binding proteins, which may participate in transport of calcium ions into the body, increase in concentration in

Box 12-D (*continued*)

response to vitamin D.[c,g,h]

The ratio of 1,25-hydroxycholecalciferol to the 24,25 derivative formed in the body varies with metabolic state. In particular, the presence of calcium ions inhibits the 1-hydroxylase as indicated in the preceding scheme.[j] A direct inhibitory action of calcium ions is possible, as is an indirect action through the parathyroid hormone.[j] Whereas the 1,24,25-trihydroxy-vitamin D has a specific action on intestinal mucosa, the 1,25-dihydroxy derivative also acts on bone cells leading to mobilization of calcium ions. This effect may result in part from stimulation of calcium-activated ATPase of the outer membrane of bone cells.[j] The 1,25-dihydroxy-vitamin D also stimulates reabsorption of inorganic phosphate in the kidney.[k] Vitamin D metabolites also act on muscle and other tissues.

[a] A related disease of cattle is "milk fever."

[b] S. F. Dyke, "The Chemistry of the Vitamins," pp. 271–317. Wiley (Interscience), New York, 1965.

[c] H. F. DeLuca and H. K. Schnoes, *Annu. Rev. Biochem.* **45**, 631–666 (1976).

[d] M. F. Holick, A. Kleiner-Bossaller, H. K. Schnoes, P. M. Kasten, I. T. Boyle, and H. F. DeLuca, *JBC* **248**, 6691–6696 (1973).

[e] M. F. Holick, H. F. DeLuca, P. M. Kasten, and M. B. Korycka, *Science* **180**, 964–966 (1973).

[f] D. A. Procsal, W. H. Okamura, and A. W. Norman, *JBC* **250**, 8382–8388 (1975).

[g] H. F. DeLuca, *Fed. Proc. Fed. Am. Soc. Exp. Biol.* **33**, 2211–2219 (1974).

[h] J. S. Emtage, D. E. M. Lawson, and E. Kodicek, *Nature* (*London*) **246**, 100–101 (1973).

[i] L. Galante, K. W. Colston, I. M. A. Evans, P. G. H. Byfield, E. W. Matthews, and I. MacIntyre, *Nature* (*London*) **244**, 438–440 (1973).

[j] R. G. G. Russel, A. Monod, J.-P. Bonjour, and H. Fleisch, *Nature* (*London*) *New Biol.* **240**, 126–127 (1972).

[k] Y. Tanaka and H. F.DeLuca, *PNAS* **71**, 1040–1044 (1972).

c. Androgens

The principal **androgenic** or male sex hormone is **testosterone,** formed from pregnenolone through removal of the side chain at C-17. About 6–10 mg are produced daily in men, and smaller amounts (~0.4 mg) are synthesized in women. It is of interest that testosterone is a precursor of the female hormones. Testosterone is carried in the blood as a complex with a β-globulin and affects a variety of target tissues including the reproductive organs. A striking effect is the stimulation of the growth of the beard. Another effect of testosterone is to cause premature death of follicles of head hair in genetically susceptible individuals. However, a bald man can usually grow a full beard and follicles of the beard type, when transplanted to the head, remain immune to the action of androgen. No one knows what regulatory differences explain this fact. Baldness might be cured by use of suitable antagonists of the androgenic hormones, but the beard would fall out and sexual interest would be lost. A challenge to the pharmaceutical chemist is to design a drug that will block the action of androgens on head hair follicles only.

Androgens also have a generalized "anabolic" effect causing increased protein synthesis, especially in muscles. They promote bone growth, and the adolescent growth spurt in both males and females is believed to result from androgens. The greater height attained by men results from the higher concentration of androgen than is present in women. Many synthetic steroids have been made in an attempt to find "anabolic hormones" with little or no androgenic activity. The effort has been at least partially successful and the use of anabolic hormones by athletes has become both widespread and controversial.

Norandrolone phenyl propionate, a synthetic drug having a five times higher ratio of anabolic to androgenic activities as does testosterone

d. Estrogens

The principal **estrogenic** or female hormone is estradiol-17β. It is formed by oxidative removal of

C-19 of testosterone followed by aromatization of the A ring.[100] All of the estrogenic hormones have this aromatic ring. The estrogens are formed largely in the ovary and, during pregnancy, in the placenta. Estrogens are also synthesized in the testes. In fact, the estrogen content of the horse testis is the highest of any endocrine organ. Target tissues for estrogens include the mammary glands, the uterus, and many other tissues throughout the body. Estrogens act on the growing ends of the long bones to stop growth. They are responsible for the overall higher fat content of the female body and for the smoother skin of the female as compared to that of the male.

The cooperative action of progesterone and estradiol regulate the menstrual cycle. At the beginning of the cycle the levels of both estrogen and progesterone are low. Estrogen synthesis increases as a result of release of **follicle-stimulating hormone** (FSH) from the anterior pituitary. This hormone stimulates growth of the graafian follicles of the ovary which in turn produce estrogen. At about the midpoint of the cycle, as a result of the action of the pituitary **luteinizing hormone** (LH), an ovum is released and progesterone secretion begins. The latter is essential to maintenance of pregnancy. If a blastocyst is not implanted, hormone production decreases and menstruation occurs.

Administration of estrogens and progestins inhibits FSH and LH secretion from the pituitary (feedback inhibition) and hence ovulation. This effect is the basis for the action of contraceptive pills. A small amount of the synthetic estrogen 17-ethynylestradiol may be taken daily for 10–15 days followed by a combination of estrogen plus a progestin such as ethynodiol diacetate for 10–15 days. Alternatively, a progestin alone may be

17-Ethynylestradiol

Ethynodiol diacetate

ingested over the entire period. Another synthetic compound with estrogenic activity is diethylstilbestrol. Its once widespread use in pro-

Diethylstilbestrol

moting growth of cattle and other animals has been largely discontinued because of demonstration of a carcinogenic action in rats fed large amounts of the compound. Another controversial use is a "morning after pill." The compound prevents implantation of a fertilized ovum.

Mention should also be made of the occurrence of plant flavonoids with estrogenic activity.

4. Other Steroids

The **saponins** are a series of steroid glycosides with detergent properties that are widespread among higher plants. Some are very toxic and among these toxic materials are compounds of extraordinary medical importance. Best known are the steroid glycosides of *Digitalis*, among them **digitonin** (Fig. 12-18). The particular arrangement of sugar units in this molecule imparts a specificity toward heart muscle. The compound is extremely toxic; in small amounts it acts to increase the tone of heart muscle and is widely used in treatment of congestive heart failure. The maintenance dose is only 0.1 mg/day. Another toxic glycoside and heart stimulant is **ouabain** (Fig. 12-18). Of great interest to biochemists is the fact that ouabain is a specific inhibitor of the

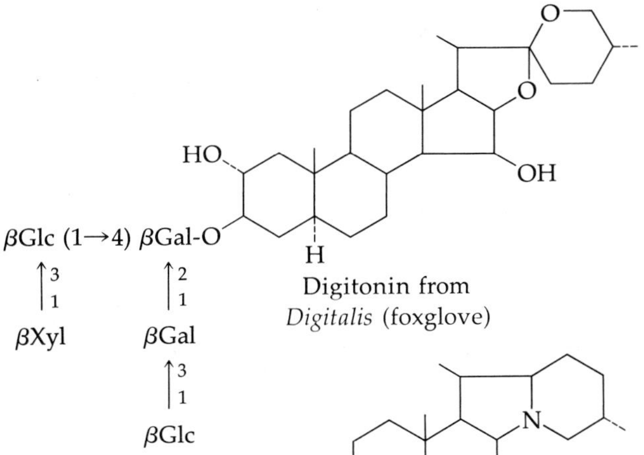

βGlc (1→4) βGal-O

  ↑3   ↑2
  |1 |1

βXyl   βGal

                ↑3
                |1

              βGlc

Digitonin from
Digitalis (foxglove)

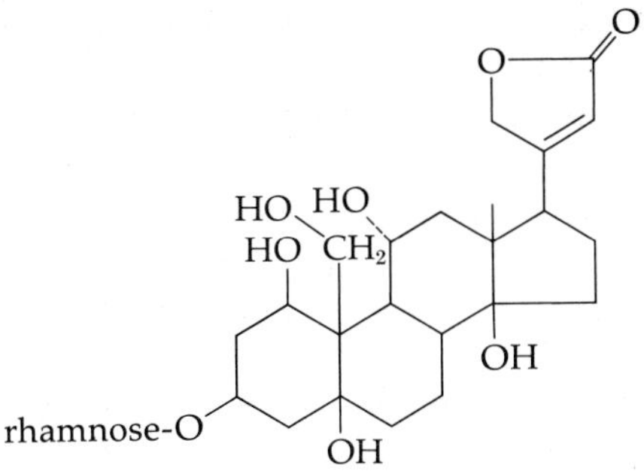

rhamnose-O

Ouabain (*G*-strophanthin)

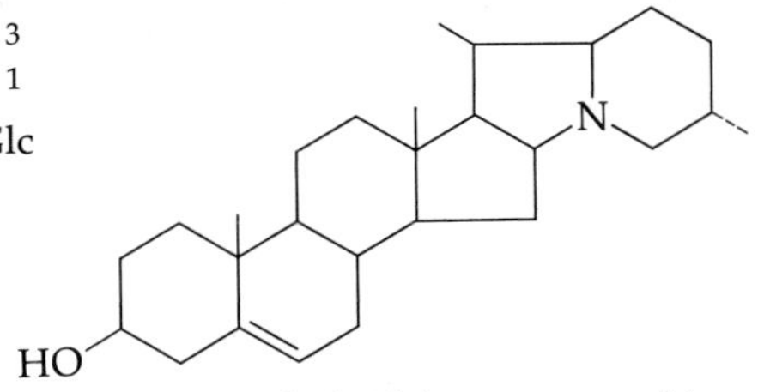

Solanidine, alkaloid from potato skins

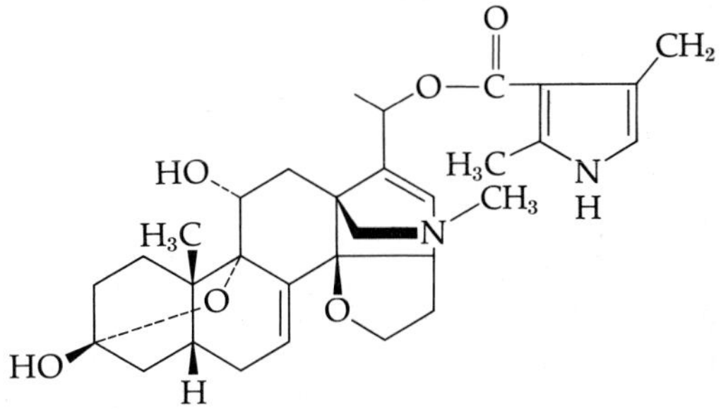

Batrachotoxin, from the Columbian
poison arrow frog
Phyllobates aurotaenia

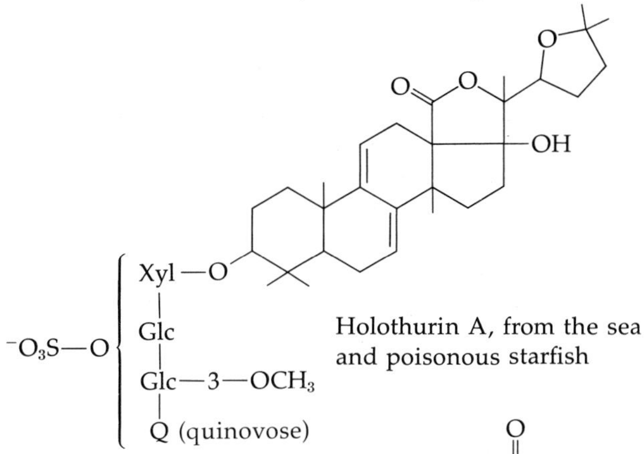

Xyl—O

Glc

Glc—3—OCH$_3$

Q (quinovose)

^-O_3S—O

Holothurin A, from the sea cucumber
and poisonous starfish

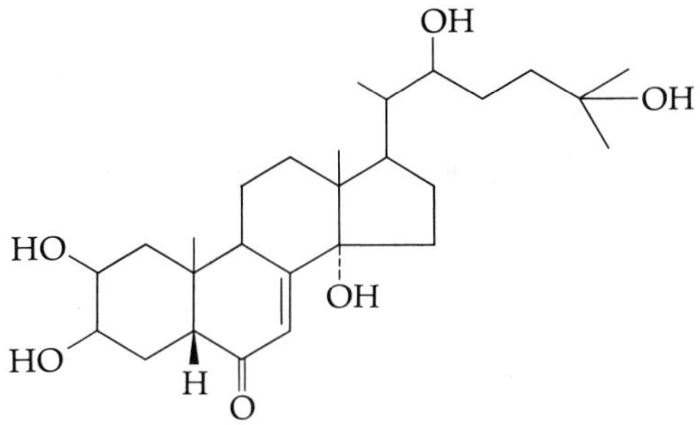

Ecdysone, a molting hormone of insects

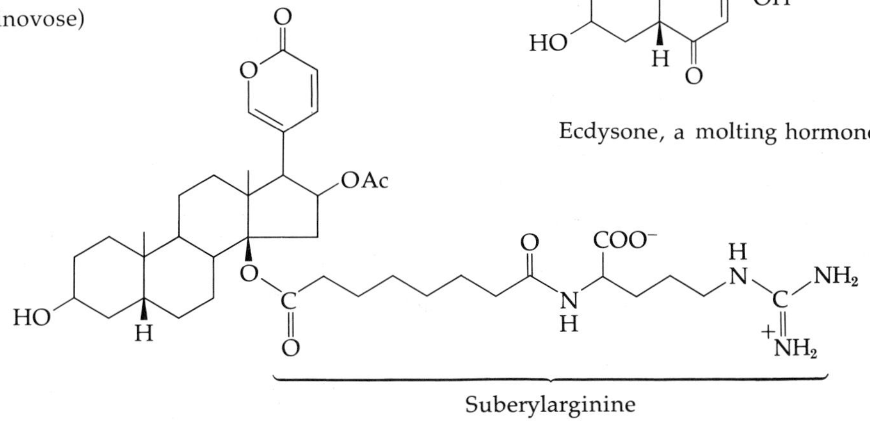

Suberylarginine

Bufotoxin, from the skin glands of toads

Fig. 12-18 More steroids, mostly toxic.

membrane-bound $(Na^+ + K^+)$-ATPase believed to be the "ion pump" that keeps intracellular K^+ concentrations high and Na^+ concentrations low (Chapter 5, Section B,2,c). Similar glycosides account for the extreme toxicity of the leaves of the oleander and the roots of the lily of the valley. A steroid glycoside from red squill is used as a rat poison.

Some animals also contain toxic steroids. **Batrachotoxin** of the Columbian poison arrow frog (Fig. 12-18) is present in amounts of only 50 μg per frog. The toxin acts on nerves to block transmissions to the muscle by increasing the permeability of membranes to sodium ions. It is specifically antagonized by **tetrodotoxin** (Fig. 16-7). Some echinoderms make powerful steroid toxins such as **holothurin A** (Fig. 12-18), a surface active agent that causes irreversible destruction of the excitability of neuromuscular tissues. The common

toad produces steroid toxins in its skin sufficiently powerful to teach a dog to leave toads alone.

In the plant kingdom a number of alkaloids (nitrogenous bases) are derived from steroids. An example is **solanidine** (Fig. 12-18). The compound is present in the skins and sprouts of potatoes making both quite toxic.

Ecdysone, a highly hydroxylated steroid (Fig. 12-18) is a molting hormone for insects. Several molecules with ecdysone activity are known and some of these are produced by certain plants. Although ecdysones are needed by insects for larval molting, they are toxic in excess. Perhaps plants protect themselves from insects by synthesizing these substances.

Among the antibiotics **fusidic acid** is a steroid. It is highly inhibitory to staphylococci but almost noninhibitory to *E. coli*. Note the boat conformation of the B ring.

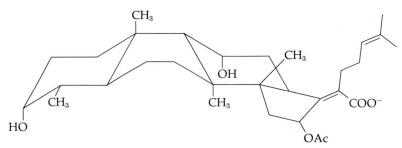

Fusidic acid

REFERENCES

1. R. Mahler, in "Biochemical Disorders in Human Disease" (R. H. S. Thompson and I. D. P. Wootton, eds.), pp. 95–127. Academic Press, New York, 1970.
2. S. D. Varma, I. Mikuni, and J. H. Kinoshita, *Science* **188,** 1215–1216 (1975).
3. N. Kretchmer, *Sci. Am.* **227,** 70–78 (Oct. 1972).
3a. W. P. Ridley, J. P. Houchins, and S. Kirkwood, *JBC* **250,** 8761–8767 (1975).
4. A. Malmström, L.-A. Fransson, M. Höök, and U. Lindahl, *JBC* **250,** 3419–3425 (1975).
4a. F. Eisenberg, Jr. (ed.) *Ann. N. Y. Acad. Sci.* **165,** 513–819 (1969).
4b. E. Hayashi, R. Hasegawa, and T. Tomita, *JBC* **251,** 5759–5769 (1976).
4c. E. A. Lapan, *Exp. Cell Res.* **94,** 277–282 (1975).
4d. L. F. Johnson and M. E. Tate, *Ann. N. Y. Acad. Sci.* **165,** 527–282 (1975).

5. G. W. Wickus, P. A. Rubenstein, A. D. Warth, and J. L. Strominger, *J. Bacteriol,* **113,** 291–294 (1973).
5a. L. Eidels and M. J. Osborn, *JBC* **249,** 5642–5648 (1974).
6. L. Glaser and H. Zarkowsky, in "The Enzymes," 3rd ed. (P. D. Boyer, ed.), Vol. 5, pp. 465–480. Academic Press, New York, 1971.
7. H. Nikaido and W. Z. Hassid, *Adv. Carbohydr. Chem.* **26,** 351–483 (1971).
8. M. Florkin, "A Molecular Approach to Phylogeny." Elsevier, Amsterdam, 1966.
9. S. Hanessian and T. H. Haskell, in "The Carbohydrates," 2nd ed. (W. Pigman and D. Horton, eds.), Vol. 2A, pp. 139–211. Academic Press, New York, 1970.
10. J. F. Snell, "Biosynthesis of Antibiotics," Vol. 1. Academic Press, New York, 1966.
10a. J. Preiss, in "The Enzymes," 3rd ed. (P. D. Boyer, ed.), Vol. 8, pp. 73–119. Academic Press, New York, 1973.

11. J. F. Robyt, B. K. Kimble, and T. W. Walseth, *ABB* **165,** 634–640 (1974).

12. D. French and J. F. Robyt, *Abstr., 166th Natl. Meet., Am. Chem. Soc.* Abstract 65 BIOL (1973).

13. L. Glaser, *Annu. Rev. Biochem.* **42,** 91–112 (1973).

14. V. Braun and K. Hantke, *Annu. Rev. Biochem.* **43,** 89–121 (1974).

15. J. B. Ward and H. R. Perkins, *BJ* **135,** 721–728 (1973).

15a. R. B. Gennis and J. L. Strominger, *JBC* **251,** 1264–1269 (1976).

16. P. Jung and W. Tanner, *EJB* **37,** 1–6 (1973).

17. A. M. Adamany and R. G. Spiro, *JBC* **250,** 2842–2854 (1975).

18. W. J. Lennarz, *Science* **188,** 986–991 (1975).

19. W. W. Chen, W. J. Lennarz, A. L. Tarentino, and F. Maley, *JBC* **250,** 7006–7013 (1975).

19a. C. J. Waechter and W. J. Lennarz, *Annu. Rev. Biochem.* **45,** 95–112 (1976).

20. J. Ruiz-Herrera, V. O. Sing, W. J. Van der Woude, and S. Bartnicki-Garcia, *PNAS* **72,** 2706–2710 (1975).

21. G. G. Leppard, L. C. Sowden, and J. R. Colvin, *Science* **189,** 1094–1095 (1975).

21a. R. M. Brown, Jr. and D. Montezinos, *PNAS* **73,** 143–147 (1976).

21b. A. Dorfman and R. Matalon, *PNAS* **73,** 630–637 (1976).

22. E. F. Neufeld, T. W. Lim, and L. J. Shapiro, *Annu. Rev. Biochem.* **44,** 357–376 (1975).

23. G. Bach, F. Eisenberg, Jr., M. Cantz, and E. F. Neufeld, *PNAS* **70,** 2134–2138 (1973).

24. S. Gatt and Y. Barenholz, *Annu. Rev. Biochem.* **42,** 61–90 (1973).

24a. H. J. Baker, J. A. Mole, J. R. Lindsey, and R. M. Creel, *Fed. Proc.* **35,** 1193–1201 (1976).

25. R. O. Brady, *in* "Current Topics in Biochemistry" (C. B. Anfinsen, R. F. Goldberger, and A. N. Schechter, eds.), pp. 1–48. Academic Press, New York, 1972.

26. J. S. O'Brien, *Fed. Proc., Fed. Am. Soc. Exp. Biol.* **32,** 191–199 (1973).

27. K. O. Raivio and J. E. Seegmiller, *Annu. Rev. Biochem.* **41,** 543–576 (1972).

28. J. B. Stanbury, J. B. Wyngaarden, and D. S. Frederickson, eds., "The Metabolic Basis of Inherited Disease," 3rd ed., Chapters 29–35. McGraw-Hill, New York, 1972.

29. V. Patel, I. Watanabe, and W. Zeman, *Science* **176,** 426–427 (1972).

30. J. A. Alhadeff, A. L. Miller, H. Wenaas, T. Vedvick, and J. S. O'Brien, *JBC* **250,** 7106–7113 (1975).

31. B. M. Turner, N. G. Beratis, V. S. Turner, and K. Hirschhorn, *Nature (London)* **257,** 391–392 (1975).

32. N. Neskovic, J. L. Nussbaum, and P. Mandel, *Brain Res.* **21,** 39–53 (1970).

33. A. Brenkert, R. C. Arora, N. S. Radin, H. Meier, and A. D. MacPike, *Brain Res.* **36,** 195–202 (1972).

34. P. Morell and E. Constantino-Ceccarini, *Lipids* **7,** 266–268 (1972).

35. T. W. Keenan and D. J. Morré, *Science* **182,** 935–937 (1973).

36. J. Tooze, ed., "The Molecular Biology of Tumor Viruses," Chapter 3. Cold Spring Harbor Lab., Cold Spring Harbor, New York, 1973.

36a. P. H. Fishman, J. Moss and M. Vaughn, *JBC* **251,** 4490–4494 (1976).

36b. P. Boquet and A. M. Pappenheimer, Jr. *JBC* **251,** 5770–5778 (1976).

36c. D. E. Kolattukudy, *in* "Recent Advances in the Chemistry and Biochemistry of Plant Lipids" (T. Galliard and E. I. Mercer, eds.), pp. 203–246. Academic Press, New York, 1975.

37. U. Varanasi, H. R. Feldman, and D. C. Malins, *Nature (London)* **255,** 340–343 (1975).

38. P. E. Kolattukudy, *Science* **159,** 498–505 (1968).

39. G. Odham and E. Stenhagen, *Acc. Chem. Res.* **4,** 121–128 (1971).

40. A. F. Whereat, M. W. Orishimo, J. Nelson, and S. J. Phillips, *JBC* **244,** 6498–6506 (1969).

41. S. R. Rosenfeld, G. D'Agnolo, and P. R. Vagelos, *JBC* **248,** 2452–2460 (1973).

42. J. G. Coniglio, *Fed. Proc., Fed. Am. Soc. Exp. Biol.* **31,** 1429 (1972).

42a. S. R. Turner, J. A. Tainer, and W. S. Lynn, *Nature (London)* **257,** 680–681 (1975).

43. G. S. Cox, E. Thomas, H. R. Kaback, and H. Weissbach, *ABB* **158,** 667–676 (1973).

44. N. M. Packter, "Biosynthesis of Acetate-Derived Compounds." Wiley, New York, 1973.

45. F. Bohlmann, T. Burkhardt, and C. Zdero, "Naturally Occurring Acetylenes." Academic Press, New York, 1973.

46. C. L. Tipton and N. M. Al-Shathir, *JBC* **249,** 886–889 (1974).

47. N. Nicolaides, *Science* **186,** 19–26 (1974).

48. P. E. Kolattukudy, J. S. Buckner, and L. Brown, *BBRC* **47,** 1306–1313 (1972).

49. H. E. Hummel, L. K. Gaston, H. H. Shorey, R. S. Kaae, K. J. Byrne, and R. M. Silverstein, *Science* **181,** 873–874 (1973).

50. J. A. Klun, O. L. Chapman, K. C. Mattes, P. W. Wojtkowski, M. Beroza, and P. E. Sonnet, *Science* **181,** 661–662 (1973).

51. B. Samuelsson, E. Granström, K. Green, M. Hamberg, and S. Hammarström, *Annu. Rev. Biochem.* **44,** 669–695 (1975).

52. P. W. Ramwell, ed., "The Prostaglandins." Plenum, New York, 1973.

53. P. Needleman, S. Moncada, S. Bunting, J. V. Vane, M. Hamberg, and B. Samuelsson, *Nature (London)* **261,** 558–560 (1976).

54. M. Hemler, W. E. M. Lands, and W. L. Smith, *JBC* **251,** 5575–5579 (1976).

55. M. Hamberg and B. Samuelsson, *PNAS* **71,** 3400–3404 (1974).

56. M. Hamberg, J. Svensson, and B. Samuelsson, *PNAS* **72**, 2994–2998 (1975).

57. H. S. Kantor, P. Tao, and H. C. Kiefer, *PNAS* **71**, 1317–1321 (1974).

58. A. L. Willis, *Science* **183**, 325–327 (1974).

59. G. J. Roth, N. Stanford, and P. W. Majerus, *PNAS* **72**, 3073–3076 (1975).

60. T. J. Powles, D. M. Easty, G. C. Easty, P. K. Bondy, and A. Munro-Neville, *Nature (London), New Biol.* **245**, 83–84 (1973).

61. W. J. Lennarz, *Acc. Chem. Res.* **5**, 361–367 (1972).

62. C. R. H. Raetz and E. P. Kennedy, *JBC* **247**, 2008–2014 (1972).

63. W. Dowhan, W. T. Wickner, and E. P. Kennedy, *JBC* **249**, 3079–3084 (1974).

64. S. Gatt and Y. Barenholz, *Annu. Rev. Biochem.* **42**, 61–90 (1973).

65. F. Snyder, ed., "Ether Lipids: Chemistry and Biology." Academic Press, New York, 1972.

66. R. L. Wykle, C. Piantadosi, and F. Snyder, *JBC* **247**, 2944–2948 (1972).

67. E. F. LaBelle, Jr. and A. K. Hajra, *JBC* **247**, 5825–5834 (1972).

68. S. J. Friedberg and A. Heifetz, *Biochemistry* **12**, 1100–1106 (1973).

69. F. Paltauf and A. Holasek, *JBC* **248**, 1609–1615 (1973).

70. K.-A. Karlsson, *Lipids* **5**, 878–891 (1970).

71. H. Wiegandt, *Adv. Lipid Res.* **9**, 249–289 (1971).

72. R. A. Demel, K. W. A. Wirtz, H. H. Kamp, W. S. M. Geurts van Kessel, and L. L. M. van Deenen, *Nature (London), New Biol.* **246**, 102–105 (1973).

73. A. J. Birch, *Science* **156**, 202–206 (1967).

74. J. W. Corcoran and F. J. Darby, *in* "Lipid Metabolism" (S. J. Wakil, ed.), pp. 431–479. Academic Press, New York, 1970.

75. T. A. Geissman and D. H. G. Crout, "Organic Chemistry of Secondary Plant Metabolism." Freeman, San Francisco, California, 1969.

75a. E. D. Beytía and J. W. Porter, *Annu. Rev. Biochem.* **45**, 113–142 (1976).

76. M. Luckner, "Secondary Metabolism in Plants and Animals." Academic Press, New York, 1972.

76a. G. Britton, *in* "Chemistry and Biochemistry of Plant Pigments," 2nd ed. (T. W. Goodwin, ed.), Vol. 1, pp. 262–327. Academic Press, New York, 1976.

77. T. W. Goodwin, *BJ* **123**, 293–329 (1971).

78. T. W. Goodwin, ed., "Aspects of Terpenoid Chemistry and Biochemistry." Academic Press, New York, 1971.

79. D. J. McNamara, F. W. Quackenbush, and V. W. Rodwell, *JBC* **247**, 5805–5810 (1972).

80. E. P. Claus, V. E. Tyler, and L. R. Brady, "Pharmacognosy," 6th ed. Lea & Febiger, Philadelphia, Pennsylvania, 1970.

81. R. R. Fall and C. A. West, *JBC* **246**, 6913–6928 (1971).

82. J. Edmond, G. Popják, S. M. Wong, and V. P. Williams, *JBC* **246**, 6254–6271 (1971).

83. W. W. Epstein and H. C. Rilling, *JBC* **245**, 4597–4605 (1970).

83a. F. Muscio, J. P. Carlson, L. Kuehl, and H. C. Rilling, *JBC* **249**, 3746–3749 (1974).

84. I. Schechter and K. Bloch, *JBC* **246**, 7690–7696 (1971).

85. H. C. Rilling, C. D. Poulter, W. W. Epstein, and B. Larsen, *JACS* **93**, 1783–1785 (1971).

86. G. A. Olah, D. P. Kelly, C. L. Jeuell, and R. D. Porter, *JACS* **92**, 2544–2546 (1970).

87. D. E. Gregonis and H. C. Rilling, *Biochemistry* **13**, 1538–1542 (1974).

88. G. Britton, *in* "Aspects of Terpenoid Chemistry and Biochemistry" (T. W. Goodwin, ed.), pp. 255–289. Academic Press, New York, 1971.

89. P. J. Scheuer, "Chemistry of Marine Natural Products." Academic Press, New York, 1973.

90. R. M. Baldwin, C. D. Snyder, and H. Rapoport, *Biochemistry* **13**, 1523–1530 (1974).

91. H-H. Tai and K. Bloch, *JBC* **247**, 3767–3773 (1972).

92. E. E. van Tamelen, *Acc. Chem. Res.* **1**, 111–120 (1968).

93. E. Heftmann, "Steroid Biochemistry." Academic Press, New York, 1969.

94. W. Templeton, "An Introduction to the Chemistry of Terpenoids and Steroids." Butterworths, London, 1969.

95. M. M. Bechtold, C. V. Delwiche, K. Comai, and J. L. Gaylor, *JBC* **247**, 7650–7656 (1972).

96. M. V. Srikantiah, E. Hansbury, E. D. Loughran, and T. J. Scallen, *JBC* **251**, 5496–5504 (1976).

97. M. C. Ritter and M. E. Dempsey, *JBC* **246**, 1536–1547 (1971).

97a. I. C. Gunsalus, T. C. Pederson, and S. G. Sligar, *Annu. Rev. Biochem.* **44**, 377–407 (1975).

97b. M. S. Brown and J. L. Goldstein, *Science* **191**, 150–154 (1976).

97c. A. Fogelman, J. Edmond, J. Seager, and G. Popjak, *JBC* **250**, 2045–2055 (1975).

97d. E. B. Smith, *Adv. Lipid Res.* **12**, 1–49 (1974).

98. M. H. Briggs and J. Brotherton, "Steroid Biochemistry and Pharmacology." Academic Press, New York, 1970.

99. K. W. McKerns, "Steroid Hormones and Metabolism." Appleton, New York, 1969.

100. K. C. Reed and S. Ohna, *JBC* **251**, 1625–1631 (1976).

STUDY QUESTIONS

1. For each of the following processes, name the organ or tissue in the adult human which is most active in carrying out the process.
   a. Cholesterol $\longrightarrow$ vitamin D_3
   b. Cholecalciferol $\longrightarrow$ 25-hydroxycalciferol
   c. 25-Hydroxycholecalciferol $\longrightarrow$
   $\qquad$ 1,25-dihydroxycholecalciferol
   d. Palmitic acid $\longrightarrow$ β-hydroxybutyrate
   e. Secretion of protein
   f. Cholesterol $\longrightarrow$ hydrocortisone
   g. Synthesis of ACTH
   h. Acetoacetate $\longrightarrow$ CO_2
   i. Cholesterol $\longrightarrow$ cholic acid
   j. Creatine $\longrightarrow$ creatine phosphate
   k. $O_2 \longrightarrow H_2O$
   l. Synthesis of glucagon
2. Write a balanced equation comparable to Eq. 11-16 for the synthesis of sucrose by reduction of CO_2.
   a. If the phosphorylation state ratio R_p in a leaf is 100 M^{-1}, the ratio [NADPH]/[NADP$^+$] is 10 and the sucrose concentration is 10 mM, what will be the CO_2 pressure at equilibrium? Compare this with the CO_2 tension in normal air (about 3.3×10^{-4} atm).
   b. How would this answer differ if the phosphorylation state ratio were 1?
   c. Why do you suppose that it is necessary to couple ATP cleavage to reduction of CO_2 in chloroplasts?
3. When β-D-galactose-1-undecaprenol pyrophosphate was treated with 1 N NaOH, undecaprenol phosphate and β-galactose-1,2-cyclic phosphate were obtained as products. Write a mechanism for this reaction. Compare this reaction with the alkaline degradation of RNA.
4. Suggest a detailed pathway for biosynthesis of 1-methylvaleric acid from glucose by *Ascaris lumbricoides*.
5. Suggest two biosynthetic origins for **tiglic acid**

$$CH_3-CH=\underset{\underset{CH_3}{|}}{C}-COOH$$

One pathway is from an amino acid and occurs in plants and animals. Another is used by *Ascaris*.
6. Show the reactions by which an organism synthe-sizes lecithin from its constituents (the products formed by hydrolysis of lecithin).
7. Suggest a detailed biosynthetic pathway for **cycloheximide** (Fig. 12-10).
8. Suggest a biosynthetic pathway from acetate for formation of **coniine,**

the alkaloid in the poison hemlock that killed Socrates.
9. How many isoprene units are contained in the structure of (a) squalene, (b) β-carotene, (c) vitamin A? Starting with acetyl-CoA write out the biosynthetic pathways for these substances.
10. Suggest a pathway from geranylgeranyl pyrophosphate to **pimaradiene**

11. Show a detailed pathway for conversion of leucine into isopentenyl pyrophosphate.
12. Suggest a route of biosynthesis of the fungal metabolite **tenuazonic acid.**

The carbon atoms marked with asterisks are labeled when [1-^{14}C]acetate is fed.
13. Name four groups of steroid hormones and describe their characteristic physiological properties. What are structural features peculiar to each group?
14. Which carbon atoms in cholesterol would you expect to have labeled with ^{14}C if you started with: (a) [1-^{14}C]acetate, (b) [2-^{14}C]acetate, (c) [1-^{14}C]mevalonic acid, (d) [2-^{14}C]mevalonic acid?

13

Light in Biology

The earth is bathed in light from the sun, and from this light comes not only warmth but also energy for all living organisms. Of the 3×10^4 kJ m^{-2} of light energy falling on the earth each day,[1,2] ~ 30 kJ are captured by photosynthesis.[3] High in the stratosphere high energy light reacts with oxygen to create a protective blanket of ozone. Light penetrating the atmosphere allows us to see and provides color to our environment. It controls the flowering of plants and the germination of many kinds of seeds and spores. In the biochemical laboratory light and other forms of electromagnetic radiation of a wide range of energies are used for experimental purposes. X-Rays, ultraviolet light, infrared light, and microwaves all have served in the study of the molecules of which we are made. Light plays a pervasive role in man's life and the interaction of light with biomolecules is of extraordinary importance. This chapter will provide a brief introduction to the subject but will serve principally to point out some sources for further reading.

A. PROPERTIES OF LIGHT

Light is a form of electromagnetic radiation and possesses characteristics of both waves and particles (**photons**). The energy of a photon is usually measured by the frequency (or by the wavelength in a vacuum to which it is inversely related, Table 13-1). In Fig. 13-1 a portion of the electromagnetic spectrum is shown on a logarithmic scale.[4] At the high energy end (off the scale of the figure to the

TABLE 13-1
Some Properties of Light

Velocity of light in a vacuum:	$c = 2.998 \times 10^8$ m s^{-1}
Velocity of light in a medium:	$c' = c/n$ where n = refractive index
Wave number:	$\tilde{\nu} = 1/\lambda$; $\tilde{\nu}$ (in cm^{-1}) = $10^7/\lambda$ (in nm) 1 cm^{-1} = 1 kayser
Frequency:	$\nu = c/\lambda = c\tilde{\nu}$ ν (in hertz) = $2.998 \times 10^{10}\tilde{\nu}$ (cm^{-1}) in a vacuum
Energy of quantum	$E = h\nu = hc\tilde{\nu}$ E (joules) = $1.986 \times 10^{-23} \tilde{\nu}$ (cm^{-1}) E (eV) = $1.240 \times 10^{-4} \tilde{\nu}$ (cm^{-1})
Energy of einstein	$E = Nh\nu = Nhc\tilde{\nu}$ $= 6.023 \times 10^{23} hc\tilde{\nu}$ E (joules) = $11.961 \tilde{\nu}$ (cm^{-1}) E (kcal) = $2.859 \times 10^{-3} \tilde{\nu}$ (cm^{-1})

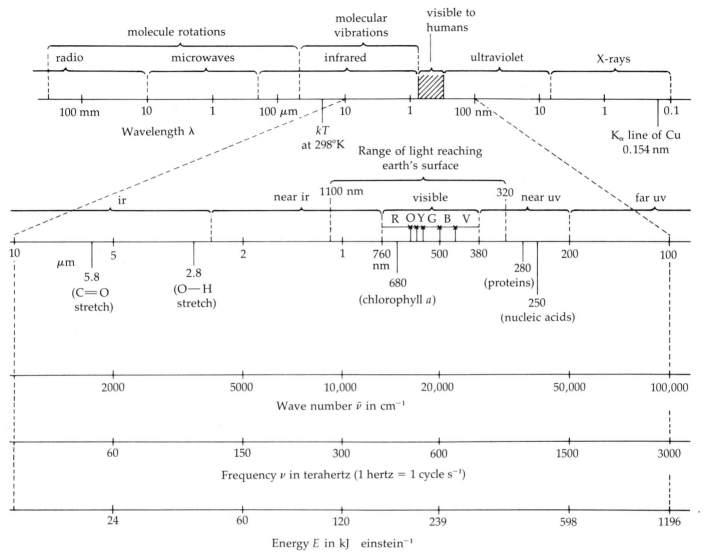

Fig. 13-1 A part of the electromagnetic spectrum. The letters V, B, G, Y, O, R over the visible part of the spectrum refer to the colors of the light. The position marked "K_α line of Cu" is the wavelength of X-rays most widely employed in X-ray diffraction studies of proteins and other organic materials.

right) are cosmic rays and gamma rays, while at the low energy end radio waves extend to wavelengths of many kilometers. The narrow range of wavelengths from about 100 nm to a few micrometers, which is the subject of this chapter, includes the ultraviolet, visible, and near infrared ranges. The second line of Fig. 13-1 shows this region expanded. Note that the range of light reaching the

earth's surface is narrow, largely being confined to wavelengths of 320–1100 nm. The human eye responds to an even more limited range of 380–760 nm, in which all the colors of the rainbow can be found. The aromatic rings of proteins and nucleic acids absorb maximally at 280 and 260 nm, respectively. Even though these wavelengths are largely screened out by the ozone layer of the stratosphere, enough light penetrates to cause many mutations and to damage the skin of the unwary sunbather.

Chemists are increasingly using, as a measure of light energy, **wave numbers** or **frequencies.** The wave number $\tilde{\nu}$ is the reciprocal of wavelength and is customarily given in units of cm^{-1} (**reciprocal centimeters,** sometimes called **kaysers**). In the future the unit μm^{-1} (10,000 cm^{-1}) and mm^{-1} may be preferred. Most of the absorption spectra in this book are plotted against wave number in cm^{-1}. The frequency ν in **hertz** is equal to $c'\tilde{\nu}$, where c' is the velocity of light. (The velocity of light in *a vacuum* is designated c and is equal to 3.00×10^8 m s^{-1}.) The energy of a quantum of light E is equal to $h\nu$, where h is Planck's constant, 6.626×10^{-34} J s^{-1}. From a chemical viewpoint, we are more interested in the energy of one **einstein,** i.e., one "mole" of light (6.023×10^{23} quanta). The energy in kJ per einstein is 11,960 $\tilde{\nu}$ (in cm^{-1}, vacuum). Energy relationships are summarized in Table 13-1. The lower three scales of Fig. 13-1 also show the relationships of ν, $\tilde{\nu}$, and E to wavelength.

A light wave is characterized by oscillating electrical and magnetic fields.[5-7] For propagation of light in the x direction, the electric field vector $\mathbf{E}$, which is customarily plotted in the y direction, is a function of the wavelength λ and the time (Eq. 13-1).

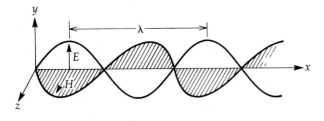

$$E_y = A \sin 2\pi (x/\lambda - \nu t + \phi) \quad (13-1)$$

The magnetic vector $\mathbf{H}$ is at right angles to the electric vector and is given by Eq. 13-2.

$$H_z = (\epsilon/\mu)^{1/2} A \sin 2\pi (x/\lambda - \nu t + \phi) \quad (13-2)$$

The velocity of propagation of light in a medium (Eq. 13-3) depends upon both ϵ, the **dielectric con-**stant of the medium, and μ, the **magnetic permeability.**

$$\text{Velocity in a medium} = c' = c/(\epsilon/\mu)^{1/2} = c/n \quad (13-3)$$

The **refractive index** of a medium relative to a vacuum is given the symbol n. It is the factor by which the velocity of light in a vacuum is diminished in a medium. It is a function of wavelength. For the 589 nm sodium line n is 1.00029 for air and 1.33 for water at 25°C.

The term ϕ in Eqs. 13-1 and 13-2 is a **phase factor.** Most light is called **incoherent** because ϕ varies for the many photons making up the beam. **Coherent light** produced by **lasers** contains photons all with the same phase relationship. If the electric vectors of all the photons in a beam of light are in the same plane (as will be the case for light emerging after passage through certain kinds of crystals), the light is called **plane polarized.** The direction of polarization is that of the electric vector $\mathbf{E}$. **Circularly polarized light,** in which the electric vector rotates and traces out either a left-handed or right-handed helix, can also be generated. A beam of left-handed circularly polarized light, together with a comparable beam polarized in the right-handed direction, is equivalent to a beam of plane polarized light. Conversely, plane polarized light can be resolved mathematically into right- and left-handed circularly polarized components.

B. ABSORPTION OF LIGHT BY MATTER

Absorption of light is fundamental to all aspects of photochemistry and provides the basis for absorption spectroscopy.[5-10] Light absorption is always **quantized,** i.e., it can take place only when the energy $h\nu$ of a quantum is equal to the difference in energy between two energy levels of the absorbing molecule (Eq. 13-4).

$$E_2 - E_1 = h\nu \quad (13-4)$$

An understanding of light absorption therefore rests upon an understanding of the energy levels of molecules. Not only must the difference $E_2 - E_1$ be correct for absorption but also there must always be a change in the dipole moment of the molecule in going from one energy level to another. Only when this is true can the electric field of the light wave interact with the molecule. A further limitation comes from the symmetry properties of the wave functions associated with particular energy levels. Quantum mechanical considerations make it clear that **transitions** between certain energy levels are **allowed,** while others are **forbidden.** While the consideration of such matters is beyond the scope of this book, the student should be aware that the quantum mechanical **selection rules** that express this fact are an important determinant of light absorption.

1. Quantitative Measurement of Light Absorption

An absorption spectrum is a plot of some measure of the intensity of absorption as a function of wavelength or wave number. The **transmittance** of a sample held in a **cell** (or **cuvette**) is the fraction of incident light that is transmitted, i.e., transmittance = I/I_0 where I_0 is the intensity of light entering the sample and I is that emerging. The transmittance is usually defined for a single wavelength, i.e., for **monochromatic** light. The absorbance (or optical density) is defined by Eq. 13-5, which also states the **Beer–Lambert law.**

$$\text{Absorbance} = A = \log_{10}(I_0/I) = \epsilon cl \quad (13\text{-}5)$$

The length (in centimeters) of the light path through the sample is l, c is the concentration in moles per liter, and ϵ is the **molar absorptivity** (molar extinction coefficient) whose units are **liter mol^{-1} cm^{-1}**. The reader can readily derive Eq. 13-5 by assuming that in a thin layer of thickness dx the number of light quanta absorbed is proportional to the number of absorbing molecules in the layer. Integration from $x = 0$ to l gives the Beer–Lambert law. Equation 13-5 generally holds very well for solutions containing single ionic or molecular

forms. However, it is valid only for monochromatic light.

2. The Energy Levels of Molecules

The energy of molecules consists of **kinetic** (translational), **rotational, vibrational,** and **electronic** components. The rotational, vibrational, and electronic energy levels are always quantized. Light quanta of wavelengths 0.2–20 mm (50–0.5 cm^{-1}; frequencies of 1.5×10^{12} to 1.5×10^{10} s^{-1}) with energies of 0.6–0.006 kJ/einstein are sufficient to excite molecules from a given rotational energy level to a higher one. Spectra in this "far infrared" or "microwave" region often consist of a closely spaced series of lines. For example, the rotational spectrum of gaseous HCl is a series of lines at 20.7 cm^{-1} intervals beginning at that wave number and reaching a maximum at about 186 cm^{-1} (54 μm). Note that the energies involved in absorption of such light are far lower than energies of activation for common chemical reactions and lower than the average translational energy of molecules in solution at ordinary temperatures ($\frac{3}{2}k_BT$ or 3.7 kJ/mol at 25°C). However, they are still much higher than energies involved in the nuclear transitions of nmr spectra (Chapter 2, Section H,7). Thus, compare 100 MHz (10^8 s^{-1}) of nmr spectroscopy with the $10^{10} - 10^{12}$ s^{-1} frequencies of microwave spectra.

Vibrational energies range from about 6 to as much as 100 kJ mol^{-1} with corresponding frequencies of ~500–8000 cm^{-1}. The resulting absorption spectra are in the infrared region. Excited electronic energy levels are ~120–1200 kJ mol^{-1}, and the spectral transitions are at 10,000–100,000 cm^{-1} (1000–100 nm wavelengths) in the visible and ultraviolet region.

3. Infrared Spectra

Absorption in the near infrared region is dominated by changes in vibrational energy levels. A typical frequency is that of the "amide A" band at 3300 cm^{-1} (3.0 μm wavelength), approximately 10^{14} s^{-1}. Discussions of infrared spectra usually begin with consideration of the stretching vibrations of a diatomic molecule. The two nuclei of the molecule can be thought of as connected with a spring. The energy of oscillation is approximately that of a harmonic oscillator. The application of quantum theory shows that the discrete energy levels that can be assumed by the oscillator are equally spaced. The difference between each pair of successive energy levels is $h\nu$, where ν is the frequency of light that must be absorbed to raise the energy from one level to the next. In the ground state (unexcited state) the molecule still possesses a "zero-point energy" equal to half the energy needed to induce a transition.

While the harmonic oscillator is a good approximation to the behavior of a molecule in the lower vibrational energy states, marked deviations occur at higher energies. At the lower energy levels the change in the distance between the atomic centers during the course of the vibration amounts to only ±10% or less, but as the energy becomes greater the bond stretches more and the motion becomes **anharmonic.** The energy states of molecules are often in the form of **Morse curves** in which energy is plotted against internuclear distance (Fig. 13-2). As the internuclear distance becomes very short, the energy rises steeply. As the bond is stretched, there comes a point at which addition of more energy will rupture the bond and dissociate a diatomic molecule into atoms and more complex molecules into fragments. Vibrational energy levels can be portrayed as horizontal lines at appropriate heights on the Morse curve (Fig. 13-2).

Because there are many rotational energy levels corresponding to each vibrational energy level, infrared spectra contain absorption bands resulting from simultaneous changes in both the vibrational and rotational energy levels of molecules. Thus, instead of single peaks corresponding to single transitions in vibrational energy, progressions of sharp bands at closely spaced intervals are observed. An example is provided by the band corresponding to the stretching fre-

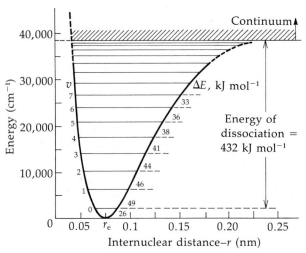

Fig. 13-2 The potential energy of the hydrogen molecule as a function of internuclear distance, and the position of its vibrational energy levels. ΔE values are energy differences between successive levels; v designates vibrational quantum numbers. Adapted from Calvert and Pitts,[5] p. 135.

quency of the H–Cl bond in gaseous HCl at 2886 cm^{-1} (3.46 μm). There is actually no band at this frequency but a series of almost equally spaced bands on either side of the fundamental frequency from ~2600 to ~3100 cm^{-1} at intervals of ~21 cm^{-1}, i.e., that of the rotational frequency seen in the microwave spectrum (Herzberg,[8] p. 55). The effect is to broaden the band as seen in a low resolution spectrum.†

While the infrared spectra of diatomic molecules are relatively easy to interpret, for more complex substances the infrared absorption bands often cannot be associated with an individual chemical bond. Instead, they are associated with the **fundamental vibrations** (normal vibrations) of the *molecule*. Fundamental vibrations are those in which the center of gravity does not change. For a molecule containing n atoms, there are $3n - 6$ such vibrations. While they are sometimes dominated by a vibration of a single bond, they may involve synchronous motion of many atoms. The fundamental vibrations of a molecule are described by

† This is one cause of the broadening of infrared bands in solution, but another cause is interaction with solvent to provide a heterogeneity in environments of the absorbing molecules.

such words as stretching, bending (in-plane and out-of-plane), twisting, and deformation. Rarely are all $3n - 6$ bands seen in an infrared spectrum. One reason is that some of the vibrations are not accompanied by any change in dipole moment, e.g., the symmetric stretching of the linear CO_2 molecule. Other bands are simply too weak to be observed clearly.

Vibrations involving many atoms in a molecule, i.e., **skeletal vibrations,** are often found in the region of 700–1400 cm^{-1} (14–7 μm). Vibrational frequencies that are dominantly those of individual functional groups can often be identified in the range 1000–5000 cm^{-1} (10–2 μm). Examples of the latter are the stretching frequencies of C–H, N–H, and O–H bonds which usually occur approximately in the regions of 2900, 3300, and 3600 cm^{-1}, respectively. Note that the energy (and frequency) of the vibrations increases as the difference in electronegativity between the two atoms increases. When a bond connects two heavier atoms, the frequency is lower, e.g., the C–O frequency in a primary alcohol is ~ 1053 cm^{-1}. For a double bond, frequency increases, e.g., for C=O it is ~ 1700 cm^{-1}. (This C=O stretching frequency usually gives rise to one of the strongest bands observed in infrared spectra.) Hydrogen bonding has a strong and characteristic effect. Thus, the O–H frequency at ~ 3600 cm^{-1} is decreased to 3500 cm^{-1} when hydrogen bonding occurs.

Theory predicts that for a harmonic oscillator only a change from one vibrational energy level to the next higher is allowed, but for anharmonic oscillators, weaker transitions to higher vibrational energy levels can occur. The resulting "overtones" are found at approximate multiples of the frequency of the fundamental. In addition, combination frequencies representing sums and differences of frequencies of individual infrared bands may be seen. While the intensities of these bands are low, their presence at relatively high energies in the near infrared region (4000–12,500 cm^{-1}) means that they may be easier to observe than the fundamental frequencies in the more crowded infrared region.

a. Vibrational Frequencies of Amide Groups

Because of the presence of amide groups in both proteins and in the purine and pyrimidine bases, the infrared absorption bands of the amide group have attracted a great deal of attention. Of the various known amide frequencies (which are described by Fraser and MacRae[11]), three have been studied intensively.

The **amide I** band at ~ 1680 cm^{-1} is associated with an in-plane normal mode of vibration that involves primarily the C=O stretching. The **amide II** band at 1500 cm^{-1} is also from an in-plane mode that involves N–H bending, while the higher frequency **amide A** band at ~ 3450 cm^{-1} involves N–H stretching. The amide A band is shifted to ~ 3300 cm^{-1} when the N–H is hydrogen bonded. An example[12] of an infrared spectrum of a protein, including the amide bands A, I and II, is shown in Fig. 13-3. Note that the band shapes have a complex character. The shape of the amide I band has been found to depend upon the conformation of the peptide chain. An empirical rule is that the amide groups in α helices absorb about 20 cm^{-1} higher than do those in β structures. However, more careful analyses of the normal vibrations of peptide chains have led to conflicting opinions.[11,13–15]

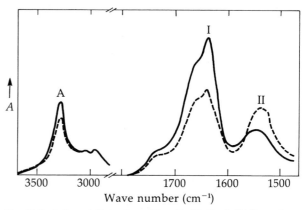

Fig. 13-3 Infrared dichroism of insulin fibrils. Solid line, electric vector parallel to fibril axis; broken line, electric vector perpendicular to fibril axis. Reprinted with permission from M. J. Burke and M. A. Rougvie, *Biochemistry* **11**, 2437 (1972). Copyright by the American Chemical Society.

An important technique that can be used with oriented protein chains is **infrared dichroism.** Absorption spectra are recorded by passing light through the protein in two mutually perpendicular directions, either with the electric vector parallel to the peptide chains or perpendicular to the chains. Such a pair of spectra are shown in Fig. 13-3 for oriented fibrils of insulin. In this instance, it is believed that the insulin molecules assume a β conformation and are stacked in such a way that they extend transverse to the fibril axis (a cross-β structure). Thus, when the electric vector is parallel to the fibril axis it is perpendicular to the peptide chains. Since the amide I band is dominated by a carbonyl stretching motion that is perpendicular to the peptide chains in the β structure, this band is enhanced when the electric vector is also perpendicular to the peptide chains and is diminished when the electric vector is parallel to the

peptide chains (perpendicular to the fibril axis, Fig. 13-3). The same is true of the amide A band which is dominated by an N–H stretch. On the other hand, the dichroism of the amide II band is the opposite because it tends to be dominated by an N–H bending which is in the plane of the peptide group but is longitudinal in direction.

Infrared spectroscopy has also been applied to study of the amide bands of pyrimidines.[16] Figure 13-4A shows the spectrum of 1-methyluracil in H_2O and also in D_2O. Note that the amide II band is totally lacking in D_2O. This illustrates another use of infrared spectroscopy which has been particularly important in the study of proteins. The loss of the amide II band when a proton-containing protein is placed in D_2O provides a technique for following proton exchange from hydrogen-bonded positions in structured regions of proteins.[10] Figure 13-4 also shows the infrared

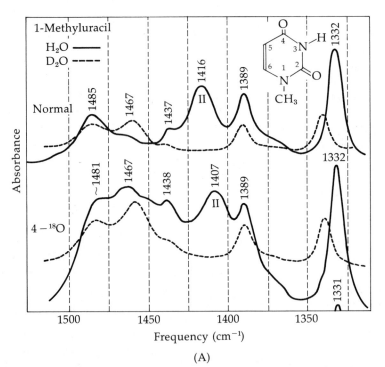
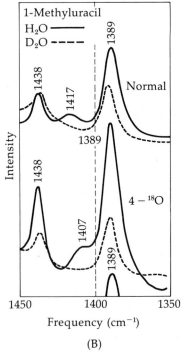

Fig. 13-4 (A) Infrared and (B) Raman spectra of 1-methyluracil in H_2O and D_2O. Spectra for normal 1-methyluracil and for the specific isotopic derivative with ^{18}O in the 4 position are shown. From Miles et al.[16]

spectrum of 1-methyluracil containing ^{18}O in the 4 position. Note the shift of 7 cm^{-1} in the amide II position. This is part of the evidence that the NH bending vibration is extensively coupled to C=O and C=C stretching modes.

b. Raman Spectra

The Raman technique deals with spectra of scattered radiation. In a collision between a photon and a molecule the photon may undergo **elastic collision** in which the photon loses no energy but changes its direction of travel. Such scattering is known as **Rayleigh scattering** and forms the basis for a method of molecular weight determination. Sometimes **inelastic** collisions occur in which both the molecule and the photon undergo changes in energy. Since such changes must be quantized and involve vibrational and rotational levels of the molecule, an analysis of the spectrum of the scattered light (Raman spectrum) can give much the same information as an ordinary infrared spectrum. There is one difference: The selection rules are not the same. Some transitions are "infrared active" and others are "Raman active." Thus, it is very desirable to measure both infrared and Raman spectra on the same sample. Until recently, Raman spectroscopy has been little used because of the very low intensity of the scattered light. However, the use of laser excitation has vastly enhanced the importance of Raman spectra.[16–20] An example of a Raman spectrum of 1-methyluracil is shown in Fig. 13-4B. Note that the intensity of the amide II band is much lower (relative to the amide I band) in the Raman spectrum than in the infrared absorption. Of special interest is **resonance Raman spectroscopy.**[19–21] In this technique a laser beam of a wavelength that is absorbed in an electronic transition is used. The scattered light is often strongly enhanced at frequencies differing from that of the laser by Raman frequencies of groups within the chromophore or of groups in another molecule adjacent to the chromophore. Although there are some experimental difficulties, the technique permits one to examine structural features at a particular site in a macromolecule.

4. Electronic Spectra

Biochemists make extensive use of spectroscopy in the ultraviolet (uv) and visible range. Visible light begins at the red end at ~12,000 cm^{-1} (12 kK, 800 nm) and extends to 25,000 cm^{-1} (400 nm). The ultraviolet range begins at this point and extends upward, the upper limit accessible to laboratory spectrophotometers being ~55,000 cm^{-1} (180 nm). The energies covered in the visible–uv range are from ~140 to ~660 kJ/mol. Note that the latter is greater than the bond energy of all but the strongest double and triple bonds (Table 3-6). Thus, it is understandable that uv light is effective in inducing photochemical reactions. Even the lower energy red light which is used by plants in photosynthesis contains enough energy per einstein to make it feasible to generate ATP, to reduce NADP$^+$, and to carry out other photochemical processes. Although the energies of light absorbed in electronic transitions are large, the geometry of molecules in the excited states is often only slightly altered from that in the ground state. Generally speaking, the amount of vibration is increased and the molecule expands moderately in one or more dimensions. The significance of light absorption in biochemical studies lies partly in the great sensitivity of electronic energy levels of molecules to their immediate environment. The fact that spectrophotometers are precise and very sensitive contributes to the usefulness of electronic spectroscopy. Furthermore, the related methods based on circular dichroism and fluorescence have widespread utility. The fact that proteins, nucleic acids, coenzymes, and many other biochemical substances contain intensely absorbing "chromophores" has made all of these methods popular.

a. Shapes of Absorption Bands

Electronic absorption bands are usually quite broad, the width of the band at half-height commonly being 3000–4000 cm^{-1}. The explanation is found largely in the coupling of electronic excitation to changes in the vibrational and rotational energy levels. Inhomogeneity of environments in the solvent also contributes to band broadening. Shapes of absorption bands are to a large extent determined by the **Franck–Condon principle.** Since the frequency of light absorbed during electronic transitions is $\sim 10^{15}$ to 10^{16} s^{-1}, the absorption of light energy occurs within 10^{-15} to 10^{-16} s (the time equivalent to the passage of one wavelength of light). During this period the vibrational motions of the nuclei are almost insignificant because of the much lower frequencies of vibration. The Franck–Condon principle states that no significant change in the positions of the atomic nuclei of the molecule occurs during the time of the electronic transition. Consider Fig. 13-5, which shows two types of potential energy curves for excited states of molecules.[5] In the first case, the geometry of the molecule is little changed between ground state and excited state. It is important to realize that at room temperature most molecules are in the lowest energy states of at least the most energetic of the various vibrational modes of the molecule ($\frac{3}{2} k_B T \approx 300$ cm^{-1}). Thus, the most probable transitions in the case of Fig. 13-5A occur from the lowest vibrational states of the ground electronic states. The most probable intranuclear distance for a molecule in the ground state is the equilibrium

distance r_e (Fig. 13-2). Since that distance is the same in all of the vibrational levels of the electronically excited state, transitions to any of these states may occur. The transition to the first vibrational level of the excited state is most likely. The result is an absorption spectrum in which the sharp band representing the "0-0 transition" is most intense and in which there are progressively weaker bands corresponding to the 0-1, 0-2, 0-3, etc., transitions (Fig. 13-5A).

A second type of spectrum is illustrated in Fig. 13-5B. In this instance, the molecule has expanded in the excited state and r_e is greater than in the ground state. The Franck–Condon principle suggests that a transition is likely only to those vibrational levels of the excited state in which the internuclear distance is compressed for a significant fraction of the time, approximately to that of r_e in the ground state. Examination of Fig. 13-5B explains why the resulting absorption spectra tend to have weak 0-0 bands and stronger bands corresponding to transitions to higher levels.

For real spectra, especially of polyatomic molecules, the situation is much more complex. For one thing, some molecules in the ground state do occupy higher vibrational levels of the less energetic modes. Therefore, there will be weaker lines, some of which lie on the low energy side of the 0-0 transition. Since in polyatomic molecules there are several normal modes of vibration, there will be other progressions of absorption bands paralleling those shown in Fig. 13-5, and filling in the valleys between them. All of the bands are broadened by rotational coupling and by interactions with solvent.

An example of a molecule giving a spectrum of the type shown in Fig. 13-5B is toluene (methylbenzene). The vapor phase spectrum contains a large number of sharp lines, some of which can be seen in the low resolution spectrum of Fig. 13-6. A number of progressions can be identified.[22] One begins with the intense 0-0 line at 37.48 kK and in which spacing of ~ 930 cm^{-1} between lines corresponds to a vibration causing symmetric expansion of the ring (ring breathing frequency), a frequency that can also be observed in the infrared spectrum. Other progressions beginning at the 0-0

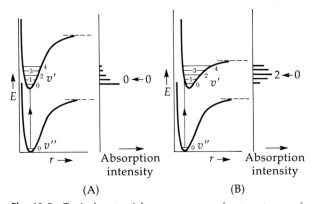

Fig. 13-5 Typical potential energy curves for two types of band spectra: (A) For a transition in which the equilibrium internuclear distances r_e are about equal in the ground and excited states. (B) For a transition in which r_e' (excited state) $> r_e$ (ground state). From Calvert and Pitts,[5] p. 179.

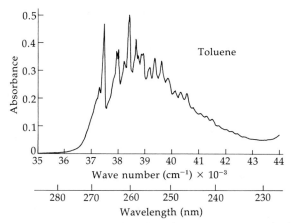

Fig. 13-6 The spectrum of the first electronic transition of toluene vapor at low resolution. Cary 1501 spectrophotometer.

line involve additional modes of vibration with frequencies (in the excited electronic state) of 460, 520, and 1190 cm⁻¹. Other weaker bands are "buried" in the valleys in Fig. 13-6.

When the spectrum of toluene is measured in solution the sharp lines are broadened but there are still indications of vibronic structure. The spectra of phenylalanine and of its derivatives[22] are very similar to that of toluene as is illustrated in Fig. 13-7 (the 0-0 line comes at 37,310 cm⁻¹). The vibronic structure of phenylalanine can be seen readily in the spectra of many proteins (e.g., see Fig. 13-14). The spectrum of tyrosine is also similar (Fig. 13-7) but the 0-0 peak is shifted to a lower energy of ~35,300 cm⁻¹ (in water). Progressions with spacings of 1200 and 800 cm⁻¹ are prominent.[23]

It is often proposed that Gaussian (normal distribution functions) curves be used to describe the shape of the overall envelope of the many vibronic subbands that make up one electronic absorption band. In some instances, e.g., for the copper-containing blue protein of *Pseudomonas* (Fig. 13-8), Gaussian bands are appropriate and permit the resolution of the spectrum into components representing individual electronic transitions. Each transition is described by a **peak position, height** (molar absorptivity), and **width** (as mea-

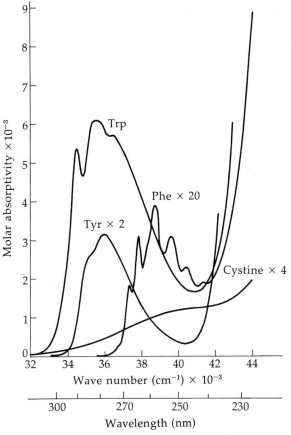

Fig. 13-7 The spectra of the first electronic transitions of the N-acetyl derivatives of the ethyl esters of phenylalanine, tyrosine, and tryptophan together with that of the dimethyl ester of cystine in methanol at 25°C. The spectra for the Tyr, Phe, and cystine derivatives have been multiplied by the factors given on the graph.[41]

sured at the half-height, in cm⁻¹). Most absorption bands of organic compounds are not symmetric but are *skewed* toward the high energy side.† It is best to fit such bands with a skewed function such as the **log normal** distribution curve.[24,25] In ad-

† Absorption spectra plotted as a linear function of wavelength are sometimes fitted with Gaussian curves. While Gaussian curves occasionally give a good fit for such spectra, it is undesirable to measure bandwidths in nanometers. It is wave number that is proportional to energy. Spectral bands tend to have similar widths across the visible ultraviolet range when plotted against wave number but not when plotted against wavelength.

dition to position, height, and width, a fourth parameter provides a measure of **skewness.** Computer-assisted fitting with log normal curves gives precise values for the positions, widths, and intensities. Note that in general the peak position located by such a procedure is somewhat to the high energy side of the 0-0 transition. Bandwidths vary but often lie between 3000 and 4000 cm^{-1}.

Another useful approach to the quantitative description of spectra is to fit the major progressions of vibronic subbands with a series of Gaussian curves.[26,27]

b. Classification of Transitions

The intense 600 nm absorption band of the copper blue protein in Fig. 13-8 is attributed to a *d-d* transition of an electron in the metal ion.[28] The intensity may arise from bonding to an S atom of a methionine residue in the protein.[29] The electronic transitions in most organic molecules are of a different type. Transitions lying at energies $<55,000$ cm^{-1} are classified either as π-π^* or n-π^*. In the π-π^* transitions an electron is moved from a bonding π-molecular orbital to an antibonding (π^*) orbital. Such a transition is present in ethylene at 61,540 cm^{-1} (162.5 nm) with a maximum molar absorptivity ϵ_{max} of $\sim 15,000$ M^{-1} cm^{-1}. The n-π^* transitions result from the raising of an electron in an unshared pair of an oxygen or nitrogen atom into a π^* antibonding orbital. These transitions are invariably weak. For example, acetone in H_2O shows an n-π^* transition at 37,740 cm^{-1} (265 nm). The value of ϵ_{max} is ~ 240 and the width is about 6400 cm^{-1}. A characteristic of n-π^* transitions is a strong shift to lower energies as the compound is moved from water into less polar solvents. Thus, the peak of the acetone band lies at 36,920 cm^{-1} in methanol and at 35,970 cm^{-1} (278 nm) in hexane. Such a solvent shift is often taken as diagnostic of an n-π^* transition, and it is often stated that the π-π^* bands shift in the opposite direction upon change of solvent character. However, the latter is *not true* for many of the polar chromophores found in biochemical substances. Thus, the π-π^* bands of tyrosine also shift to

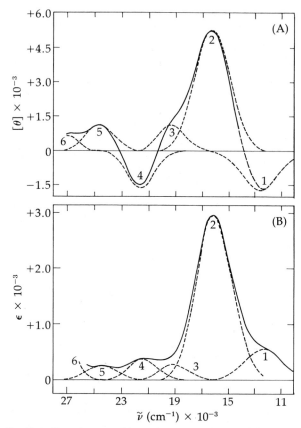

Fig. 13-8 *Pseudomonas* blue protein. Resolution of the visible CD (A) and (B) absorption spectra of the *Pseudomonas* blue protein into a series of overlapping Gaussian bands: (---) the individual Gaussian ellipticity and absorption bands. The numbers 1 to 6 refer to bands of identical position and width in both spectra. (—) (A) CD or (B) absorption envelopes resulting from the sum of the set of overlapping Gaussian bands. Each envelope corresponds within the error of the measurement to the experimental spectra. The *dashed part* of the CD envelope above 700 nm was completed by the curve fitter with the use of a band in the position of *band 1* of the absorption spectrum. From Tang *et al.*[28]

lower energies when the molecule is moved from water into hexane. However, the magnitude of the shift is much less than for the n-π^* band of acetone.

A molecule can have several different excited states of increasing energies. Thus, in benzene and its derivatives there are three easily detectable

π-π^* transitions (see Fig. 13-9). The first is a weak band of $\epsilon = 10^2$ to 10^3. The second is a band at a higher frequency (at 1.35 ± 0.10 times the frequency of the first band) with ϵ_{max} often as high as 10^4. The third band is found at still higher energies with ϵ_{max} reaching 5×10^4. The excited state energy levels represented by these transitions are labeled ^1L_b, ^1L_a, and ^1B_a according to one frequently used designation (that of Platt). Other authors described the levels in terms of the symmetries of the molecular orbitals. Thus, the ground state is $^1A_{1g}$, while the three excited states are $^1B_{2u}$, $^1B_{1u}$, and $^1E_{1u}$. In these symbols the superscript 1 indicates that the excited states are **singlet** in nature; that is, the electrons remain paired in the excited states (absorption of visible and ultraviolet light almost always leads to singlet excited states initially). For more complex ring systems, the number of possible electronic transitions increases. In many cases an attempt is made to relate these transitions back to those of benzene itself.

The intensities of electronic transitions vary greatly. The area ($\mathscr{A}$) under the absorption band when ϵ is plotted against wave number $\tilde{\nu}$ is directly proportional (Eq. 13-6) to a dimensionless quantity called the **oscillator strength** f.

$$f = \frac{2.303 \, m_e c^2}{\pi N e^2} F \mathscr{A} = 4.32 \times 10^{-9} F \mathscr{A} \qquad (13\text{-}6)$$

In this equation m_e and e are the mass and charge of the electron, c is the velocity of light, N is Avogadro's number, and $\mathscr{A}$ is the area in a plot of ϵ vs. $\tilde{\nu}$ in cm^{-1}; F is a dimensionless correction factor that is related to the refractive index of the medium and is near unity for aqueous solutions. If the area is approximated as that of a triangle of height ϵ_{max} and width (at half-height) W, we find that for a typical absorption band of $\epsilon_{max} = 10^4$ and $W = 3000 \, cm^{-1}$, $f = 0.13$.

According to absorption theory, the oscillator strength is related to the probability of a transition and can become approximately 1 only for the strongest electronic transitions. The oscillator strength is rarely this high. For example, it is $\sim 10^{-4}$ for Cu^{2+} and $\sim 2 \times 10^{-3}$ for the toluene absorption band shown in Fig. 13-6. The low intensity of absorption bands of benzene derivatives is related to the fact that these transitions are quantum mechanically forbidden for a completely symmetric molecule. It is only because of coupling with asymmetric vibrations of the ring that the ^1L_b transition of benzene becomes weakly allowed. In the benzene spectrum the 0-0 transition is completely absent, and only those progressions involving uptake of an additional $520 \, cm^{-1}$ of a nonsymmetric vibrational energy are allowed. In the case of toluene and phenylalanine, the asymmetry of the ring introduced by the substituents permits the 0-0 transition to occur and leads to a higher oscillator strength than that observed with benzene.

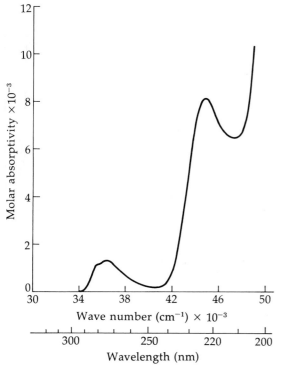

Fig. 13-9 The spectrum of *N*-acetyltyrosine ethyl ester in an aqueous phosphate buffer of pH 6.8. Note the three π-π^* transitions of increasing intensity. The third π-π^* transition of the aromatic ring is $\sim 52{,}000 \, cm^{-1}$ and reaches a molar absorptivity of $\sim 40{,}000$. The n-π^* and π-π^* transitions of the amide group in this compound also contribute to the high energy end of the spectrum.

The ^1L_a transition of benzenoid derivatives is also partially forbidden by selection rules and only the third band begins to approach an oscillator strength of 1.

c. Polarization of Transitions

The intensity of a spectral transition is directly related to the **transition dipole moment** (or simply the transition moment), a vector quantity that depends upon the dipole moments of the ground and excited states. For aromatic ring systems, the transition dipole moments of the π-π^* transitions lie in the plane of the ring. However, both the directions and intensities for different π-π^* transitions within a molecule vary.

The transition moment has a dimension of length (usually given in angstroms) and can be thought of as a measure of the extent of the charge migration during the transition. Light is absorbed best when the direction of polarization (i.e., of the electric vector of the light and of the transition moment coincide). This fact can easily be verified by light absorption measurements on crystals. As with infrared spectra of oriented peptide chains (Fig. 13-3), the electronic spectra of crystals display a distinct dichroism.

In contrast to π-π^* transitions, the n-π^* transitions of heterocyclic compounds and carbonyl-containing rings are often polarized in a direction perpendicular to the plane of the ring.

d. Relationship of Absorption Position and Intensity to Structure

While quantum mechanical calculations permit prediction of the correct number and approximate positions of absorption bands, they are imprecise. For this reason, electronic spectroscopy relies upon a combination of empirical rules and atlases of spectra that can be used for comparison purposes.[30,31] The following comments may help to orient the reader. The position of an absorption band shifts **bathochromically** (to longer wavelength, lower energy) when the number of conjugated double bonds increases. Thus, **butadiene** absorbs at 46,100 cm^{-1} (217 nm) vs. the

61,500 cm^{-1} of ethylene. As the number of double bonds increases further, the bathochromic shifts become progressively smaller (but remain more nearly constant in terms of wavelength than wave number). For **lycopene** (Fig. 12-14) with 11 conjugated double bonds the absorption band is located at 21,300 cm^{-1} and displays distinct vibrational structure (Fig. 13-10). Certain ring molecules such as the porphyrins and chlorophylls have spectra that can be related back to those of the linear polyenes. Note (Fig. 10-2) that the porphyrin α and β bands represent vibrational structure of a single electronic transition, whereas the intense Soret band results from a different transition.

Substituted benzenes almost invariably absorb at lower energies than the parent hydrocarbon. The stronger the electron withdrawing or donating ability of the substituent, the larger the

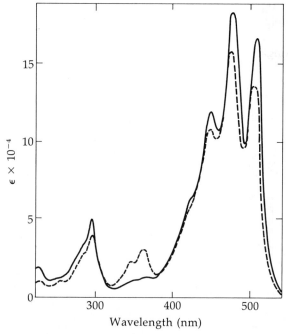

Fig. 13-10 The absorption spectrum of lycopene (plotted vs. wavelength). Note the vibrational structure which has a spacing of ~1200–1500 cm^{-1}. The solid line is for all-*trans*-lycopene while the dashed line is that of the sample after refluxing 45 min in the dark. Note the new peak at ~360 nm which arises from isomers containing some cis double bonds.

bathochromic shift. The magnitude of the shift has been correlated with the Hammett σ constants. Thus, the first absorption band of tyrosine in water is shifted 2600 cm^{-1} toward the red from that of benzene, while that of the dissociated tyrosine anion is shifted 4700 cm^{-1}, very roughly in proportion to the σ_p values of Table 3-9. Especially large shifts are observed when functional groups of opposite types (that is, an electron donating group vs. an electron accepting group) are both present in the same ring. The effects of ortho and meta substituent pairs are closely similar (in contrast to the differing electronic effects of ortho and meta pairs in chemical reactivity). Substituent pairs in para positions yield somewhat different spectral shifts. When there are more than two substituents, the two strongest groups often dominate in determining the character of the spectrum. Useful empirical rules have been developed by Petrushka[32] and Stevenson.[33]

e. Spectra of Nucleic Acids and Proteins

Most proteins have a strong light absorption band at 280 nm (35,700 cm^{-1}) which arises from the aromatic amino acids tryptophan, tyrosine, and phenylalanine.[34] The shape of the band can often be quite closely approximated as the sum of the absorption bands of the aromatic amino acid components taken in proportion to their contents. The absorption bands of simple amide derivatives of phenylalanine, tyrosine, and tryptophan in this region are shown in Figs. 13-7 and 13-9. The low energy band of tryptophan consists of two overlapping transitions ^1L_a and ^1L_b.[26] The ^1L_b transition has well-resolved vibronic bands, whereas those of the ^1L_a transition are more diffuse. Tryptophan derivatives in hydrocarbon solvents show 0-0 bands for both of these transitions at approximately 289.5 nm (34,540 cm^{-1}). However, within proteins, the ^1L_a band may be shifted 3–10 nm (up to 1100 cm^{-1}) toward lower energies. This shift appears to be the result of hydrogen bonding to other groups in the protein. The largest shifts can occur when the NH group of the indole ring is hydrogen bonded to COO$^-$, a ring nitrogen of histi-

dine, or a carbonyl group of amides.[35] When tryptophan is placed in an aqueous medium, the ^1L_b band shifts to higher energies and the ^1L_a band to lower energies compared to those in a hydrocarbon solvent. Note from Fig. 13-7 that the contribution of an equivalent number of tryptophan residues to the light absorption of proteins is much greater than that of either tyrosine or phenylalanine. Thus, in most proteins, the tryptophan absorption dominates.

In addition to the three aromatic amino acids, disulfide bonds absorb in the near ultraviolet region as indicated in Fig. 13-7. Since the absorption characteristics depend upon the dihedral angles in the disulfide bridges, it is difficult to accurately evaluate the contribution of this chromophore to the 280 nm band.

Tyrosine, tryptophan, and phenylalanine all have additional transitions that must be taken into account in the high energy uv region of the spectrum. Quantitatively even more important to the high energy end of protein spectra are the absorption bands of the amide groups which become significant above 45,000 cm^{-1}. These include a weak n-π^* transition at $\sim$47,500 cm^{-1} (210 nm) overlapped by a strong π-π^* transition centered at 52,600 cm^{-1} (190 nm). Histidine also has absorption bands in this region.

As with polypeptides, the light absorption properties of a polynucleotide reflect those of the individual components. The spectra of the purine and pyrimidine bases as ribonucleosides are shown in Figs. 13-11 and 13-12. The number of individual electronic transitions and their origins are not immediately obvious, but many attempts have been made to understand and correlate the spectra.[36] The same can be said for flavins whose spectra contain at least four intense transitions (Fig. 8-16).[37]

Whereas proteins have their low energy absorption band at $\sim$280 nm, the polynucleotides typically have maxima at $\sim$260 nm (38,500 cm^{-1}). A phenomenon of particular importance in the study of nucleic acids is the **hypochromic effect.** Whereas, in a denatured polynucleotide, the absorption is approximately the sum of that of the

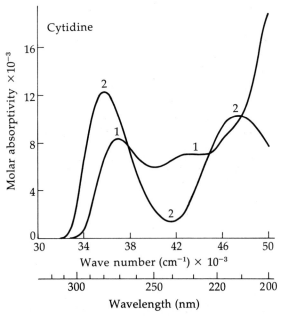

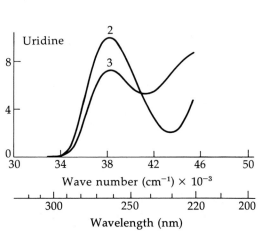

Fig. 13-11   Near ultraviolet absorption spectra of cytidine and uridine. 1. Monoprotonated form of cytidine (for which pK_a = 4.2). 2. Neutral forms (pH ~ 7). 3. Monoanionic form of uridine (for which pK_a = 9.2).

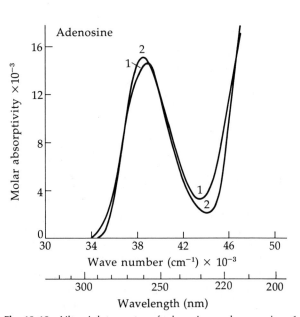

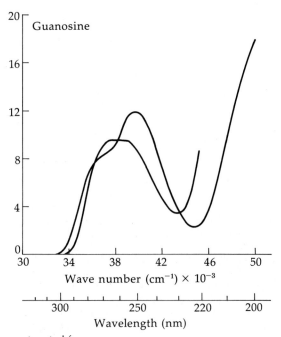

Fig. 13-12   Ultraviolet spectra of adenosine and guanosine. 1. Monoprotonated form of adenosine (pK_a = 3.5). 2. Neutral forms. 3. Monoanion of guanosine (pK_a = 9.2).

individual components, when a double helical structure is formed and the bases are stacked together, there is as much as a 34% depression in the absorbance at 260 nm. This provides the basis for optical measurement of melting curves (Fig. 2-28). The physical basis for the hypochromic effect is found in the interaction between the closely stacked base pairs.[38]

f. Difference Spectroscopy

Changes in light absorbing properties of proteins and nucleic acids are often measured as a function of some quantity such as pH, temperature, ionic environment, or the presence or absence of another interacting molecule. Since the induced changes in the spectrum are small, it is a common practice to measure only the *difference* between two spectra, one "unperturbed" and the other in the presence of some "perturbant." The perturbant might be an additional reagent, an altered solvent (e.g., with added glycerol, D_2O, etc.) a change in pH or temperature. The **difference spectrum** shown in Fig. 13-13B arises from the binding of an inhibitor succinate together with a

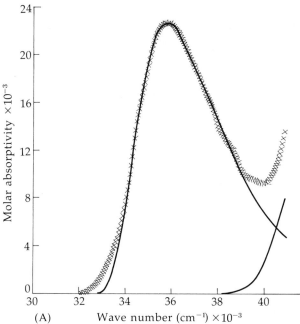

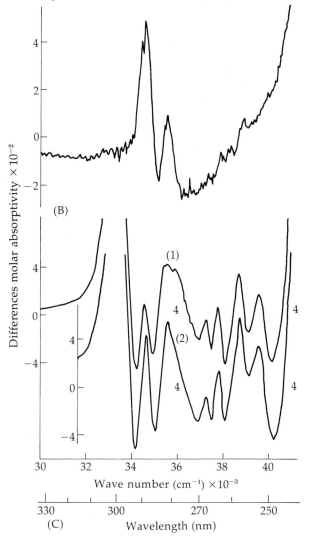

Fig. 13-13 The difference spectrum for the catalytic subunit of aspartate transcarbamoylase in the presence and absence of succinate and carbamoyl phosphate. (A) The spectrum of the unperturbed enzyme (points) fitted with two log normal curves (solid line). (B) The difference between the spectrum of enzyme plus 0.09 *M* succinate and 4.3 m*M* carbamoyl phosphate and that shown in (A) (compare with published difference spectrum for intact aspartate transcarbamoylase.[39]). (C) Curve 1, "fine structure plot" obtained by subtracting the spectrum in A from the smooth curve obtained by summing the two log normal curves. Curve 2, a similar plot for enzyme plus succinate and carbamoyl phosphate. The enzyme was supplied by G. Nagel and H. K. Schachman and the spectra were recorded by I.-Y. Yang.

substrate carbamoyl phosphate, to the catalytic subunit of aspartate carbamoyl transferase (Chapter 4, Section D,8).[39] The difference spectrum appears as a pair of peaks and a valley in the aromatic amino acid region. With proper interpretation (caution!) these difference spectra can be used to infer something about the change in environment of aromatic amino acids in a protein.[40]

Difference spectra are usually recorded by placing the unperturbed spectrum in the *reference* light beam of a spectrophotometer and the perturbed solution in the *sample* beam in carefully matched cuvettes. However, the spectrum shown in Fig. 13-13B was obtained by recording (on punched cards) the two spectra independently and subtracting them with the aid of a computer. The same data have been treated in another way by fitting two log normal curves (Section B,4,a) to the absorption bands and plotting the differences between the mathematically smooth fitted curve and the experimental points taken at close intervals[41] as shown in Fig. 13-13C. The two "fine structure plots" obtained in this way are an alternative way of representing the same data that gave rise to the difference spectrum. The method has the advantage that information about the overall band shape is obtained from the computer-assisted curve fitting process. Thus, the binding of succinate and carbamoyl phosphate caused an almost insignificant shift (of 20 cm^{-1}) in the overall bond position and a very slight broadening. The principal effect is seen to be an enhancement in the vibrational structure at 34,600 cm^{-1} in the 0-0 band of the two tryptophan residues present in the subunit. The cause of this change is not entirely obvious, a weakness of difference spectroscopy.

5. Circular Dichroism and Optical Rotatory Dispersion[42–45]

The circular dichroism of a sample is the difference between the molar absorptivities for left-handed and right-handed polarized light (Eq.

13-7) and is observed only for chiral molecules.

$$\Delta\epsilon = \epsilon_L - \epsilon_R \quad \text{(units are } M^{-1} \text{ cm}^{-1}) \quad (13\text{-}7)$$

The **dichrograph** gives a direct measure of $\Delta\epsilon$. A circular dichroism (CD) spectrum often resembles an absorption spectrum, the peaks coming at the same positions as the peaks in the absorption spectrum of the same sample. However, the CD can be either positive or negative and may be positive for one transition and negative for another (Fig. 13-8). An increasingly common practice is to plot $\Delta\epsilon$ directly as a function of wavelength or wave number. However, much of the literature makes use of the **molar ellipticity** (Eq. 13-8):

$$\text{Molar ellipticity} = [\theta] = 3299 \, \Delta\epsilon \quad (13\text{-}8)$$
$$(\text{units are degrees cm}^2 \text{ decimole}^{-1})$$

The **rotational strength** may also be evaluated (Eq. 13-9):

$$\text{Rotational strength} = \int [(\Delta\epsilon)/\lambda] d\lambda \quad (13\text{-}9)$$

The integration is carried out over the entire absorption band for a given transition.

Circular dichroism is closely related to **optical rotatory dispersion,** the variation of optical rotation with wavelength. Optical rotation depends upon the difference in refractive index $(n_L - n_R)$ between left-handed and right-handed polarized light. Rotation α is measured as an angle, in degrees or radians. Data are customarily reported in terms of **specific rotation,** that of a hypothetical solution containing 1 g/ml in a 1 dm (decimeter) tube. Specific rotation is calculated (Eq. 13-10) from the observed rotation, the concentration c' in g ml^{-1} and the length of the tube l' in decimeters.

$$\text{Specific rotation} = [\alpha] = \alpha_{\text{obs}}/c'l' \quad (13\text{-}10)$$

The **molecular rotation** is defined by Eq. 13-11 in which MW is the molecular weight and c and l are in moles per liter and cm, respectively.

$$\text{Molecular rotation} = [\phi] = 100 \, \alpha_{\text{obs}}/cl$$
$$= [\alpha]\text{MW}/100 \quad (13\text{-}11)$$

It is often multiplied by a factor of $3/(n^2 + 2)$ to correct for a minor effect of the polarizability of the field acting on the molecules. The rotation in radians per centimeter of light path can be related (Eq. 13-12) directly to the wavelength of the light and the refractive indices n_L and n_R.

$$\alpha \text{ (radians/cm)} = [\alpha]c'/1800 = \pi/\lambda[n_L - n_R] \quad (13\text{-}12)$$

In contrast to circular dichroism, optical rotary dispersion (ORD) extends far from absorption bands into spectral regions in which the compound is transparent. As an absorption band is approached, the optical rotation increases in either the positive or negative direction. Then, within the absorption band it drops abruptly through zero and assumes the opposite sign on the other side of the band (the Cotton effect). Although the occurrence of optical rotation in nonabsorbing regions of the spectrum provides an advantage to ORD measurements, the interpretation of ORD spectra is somewhat more complex than that of CD spectra. In principle, the two can be related mathematically one to the other and both are able to give the same kind of chemical information. Because CD and ORD spectra can be measured easily and are sensitively dependent on conformational and environmental changes, they are among the most popular optical tools for biochemists.

An example of a CD spectrum is that of the blue copper protein (Fig. 13-8). The CD in the d-d bands of the copper spectrum arises in part from the fact that within the protein the copper ion is in an asymmetric environment. For a similar reason the aromatic amino acids of proteins often give rise to circular dichroism. In the case of tyrosine, the sign of the CD can be either positive or negative but is the same throughout a given transition. Thus, the CD bands are similar in shape to the absorption bands.[19,46] On the other hand, the behavior of phenylalanine is more complex. The progression of vibronic bands at 930 cm^{-1} intervals above the 0-0 band all have the same sign, and the intensities relative to that of the 0-0 band are similar to those in absorption. However, the vibrations of wave numbers equal to that of the 0-0 transition plus 180 and 520 cm^{-1} sometimes give rise to CD bands of the opposite sign and the relative intensity relationships are variable.[22,46] Thus,

the CD spectrum may be more complex and difficult to interpret than the absorption spectrum for the same chromophore.

It is often observed that the binding of a symmetric chromophore to a protein or nucleic acid induces CD in that chromophore. For example, the bands of enzyme-bound pyridoxal and pyridoxamine phosphates shown in Fig. 8-8 are positively dichroic in CD but the band of the quinonoid intermediate at 20,400 cm^{-1} (490 nm) displays negative CD. This behavior is striking and it seems as if it should be possible to learn something about the environment of the chromophores from the CD. However, it is not yet possible to make simple chemical interpretations.

Dichroism in n-π^* transitions of carbonyl compounds are better understood. A useful series of octant rules make it possible to predict the sign and magnitude of CD to be expected for simple compounds.[47] Theoretical approaches to the CD and ultraviolet absorption of proteins in the high energy ultraviolet region have also been developed. In a regular β structure, in an α helix or in a crystalline array, the transitions of adjacent amide groups may be **coupled,** the excitation energy being delocalized. This **exciton** delocalization leads to a splitting (Davydov splitting) into two transitions of somewhat differing energies and polarized in different directions.[7,44] Thus, the amide absorption band at 52,600 cm^{-1} is split in an α helix into components at ~48,500 and 52,600 cm^{-1}. Furthermore, low energy π-π^* and n-π^* states are close together in energy, a fact that allows mixing of the two states and appearance of rotational strength in the π-π^* band with a sign opposite to that in the n-π^* band (see Bayley[44]). Both the sign and intensity of the CD bands also depend upon conformation. Well-defined differences are observed between α helices, β structure, and random coil conformations. Measurements are being extended into the "vacuum ultraviolet" region—up to 60,000 cm^{-1} in aqueous solutions.[48]

While progress has been made in theoretical predictions, an empirical method of correlation is often favored. For example, Fig. 13-14 shows the

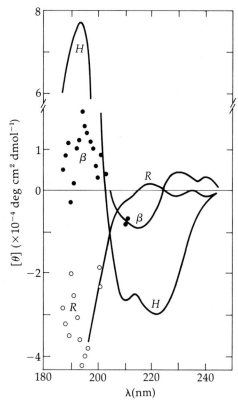

Fig. 13-14 Circular dichroism of the helix (H), β, and unordered (R) form computed from the CD of five proteins. Points are plotted when a smooth curve could not be drawn. Reprinted with permission from Y.–H. Chen *et al., Biochemistry* **13,** 3353 (1974). Copyright by the American Chemical Society.

CD spectra of helices, β structures, and unordered peptide chains computed from measured spectra together with an examination of actual structures deduced from X-ray crystallographic data.[49] Note that the CD curve for the α helix has a deep minimum at 222 nm, whereas the β form has a shallower minimum. The random structure has almost no CD at the same wavelength. The approximate helix content of a protein is often estimated from the depth of the trough at 222 nm in the CD spectrum.

Recent progress has also been made in attempts to predict the optical rotation of molecules from quantitative values for the polarizabilities of individual atoms.[50]

C. FLUORESCENCE AND PHOSPHORESCENCE

An electronically excited molecule is able to lose its excitation energy and return to the ground state in a number of ways. One of these is to reemit a quantum of light as fluorescence.[51-55] The intensity and spectral properties of such fluorescent emission can be measured by illuminating a sample in a cuvette with four clear faces and with the measuring photomultiplier set at right angles to the exciting light beam. In absorption spectrophotometry we measure a difference between the light intensity of the beam entering the sample and that emerging from the sample. In fluorescence spectroscopy we measure the absolute intensity of the light emitted as fluorescence. Even though this intensity is small, the measurement can be made extremely sensitive, far more so than can light absorption. Thus, measurement of fluorescence is an important analytical tool (e.g., see Chapter 2, Section H,5). It is also a useful and sensitive way of monitoring the chemical properties of electronically excited states. Figure 13-15 shows the fluorescence **excitation spectra**[†] (which are very nearly the same as the absorption spectra) for two different molecules, a flavin[56] and tryptophan, together with the fluorescence **emission spectra.** The heights of the latter have been adjusted to the same scale as that of absorption. Note that the fluorescent emission is always at a lower energy than the absorption and that excitation and emission spectra overlap only slightly. Furthermore, the shape of the emission spectrum tends to be an approximate mirror image of that of the absorption. To understand this, refer to the diagram in Fig. 13-16. Recall that absorption usually leads

† The excitation spectrum is an example of an **action spectrum,** a measure of some response to absorbed light. At very low concentrations of pure substances action spectra tend to be identical to absorption spectra. However, since the observed response (fluorescence in this case) is proportional to light *absorbed,* action spectra should be compared to plots of $1 - T$ (T stands for transmittance, Section B,1) vs. wavelength rather than to plots of ϵ vs. λ. The two plots are proportional at low concentrations. For a good discussion of action spectra see Clayton.[51]

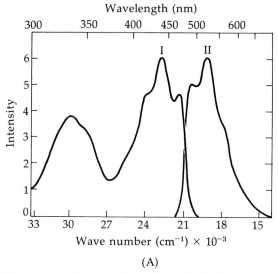

Wavelength (nm)

(A)

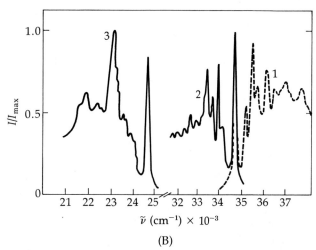

(B)

Fig. 13-15 (A) Corrected emission and excitation spectra of riboflavin tetrabutyrate in *n*-heptane. Concentration, about 0.4 mg l^{-1}. Curve I: excitation spectrum; emission at 525 nm. Curve II: emission spectrum; excitation at 345 nm. From Ko-

taki and Yagi.[56] (B) Indole in cyclohexane, T = 196°C. 1, Fluorescence excitation spectrum; 2, fluorescence spectrum; and 3, phosphorescence spectrum. From Konev.[54]

to a higher vibrational energy state after light absorption than before. However, before much fluorescent emission can occur, the excess vibrational energy is dissipated and the excited molecule finds itself in the lowest vibrational state of the upper electronic state. It is from this state that the bulk of the fluorescent emission takes place. Furthermore, whereas absorption usually occurs from the lowest vibrational state of the ground electronic level, fluorescence can populate many excited vibrational states of the ground electronic state (Fig. 13-16). Thus, as indicated in the figure, the fluorescent emission spectrum consists of a series of vibronic subbands at lower energies than those observed in the absorption spectrum. The

Fig. 13-16 Potential energy diagram for the ground state S_0 and the first excited singlet S_1 and triplet T_1 states of a representative organic molecule in solution. G is a point of intersystem crossing $S_1 \rightarrow T_1$. For convenience in representation, the distances r were chosen $r_{S_0} < r_{S_1} < r_{T_1}$; thus, the spectra are spread out. Actually, in complex, fairly symmetric molecules, $r_{S_0} \sim r_{S_1} < r_{T_1}$ and the 0-0 absorption and fluorescence bands almost coincide, but phosphorescence bands are significantly displaced to the longer wavelengths. From Calvert and Pitts,[5] p. 274.

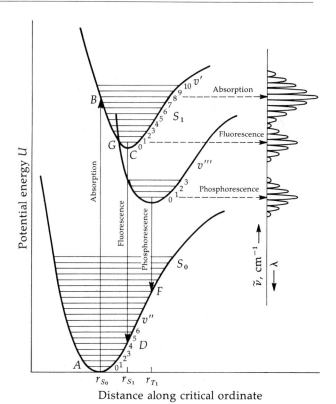

Distance along critical ordinate

two spectra have only the 0-0 transition in common. In fact, as can be seen from Fig. 13-15, in most cases, the two 0-0 transitions do not coincide exactly, the 0-0 band in fluorescence being shifted slightly toward lower energies than in absorption. This shift results from the fact that during or immediately following absorption of a photon, there is some rearrangement of solvent molecules around the absorbing molecule to an energetically more stable arrangement. Just as excess vibrational energy is dissipated in the excited state, so the relaxation of these solvent molecules around the excited chromophore leads to a small shift in energy. Another similar relaxation process occurs in the ground state of a molecule that has just emitted a photon as fluorescence. This also contributes to the shift in position of the 0-0 band in fluorescence (see Parker,[52] p. 13).

Figure 13-16 illustrates another important aspect of the reactivity of excited molecules. They can pass from the excited singlet state to a lower energy **triplet state,** in which two electrons are now unpaired and the molecule assumes something of the properties of a diradical. The triplet state is a very long-lived state and is responsible for much photochemical behavior of molecules. It also gives rise to the delayed light emission known as **phosphorescence,** as illustrated in the figure.

Why are some molecules fluorescent, while others are not? The possibility for fluorescent emission is determined to a considerable extent by the radiative lifetime τ_r, which is related by Eq. 13-13 to the first-order rate constant for exponential decay of the excited state by fluorescence.

$$\tau_r = 1/k_f \qquad (13\text{-}13)$$

The radiative lifetime is a function of the wavelength of the light and of the oscillator strength of the transition. For molecules absorbing in the near uv, the approximation of Eq. 13-14 is often made.

$$1/\tau_r \approx 10^4 \epsilon_{max} \qquad (13\text{-}14)$$

Thus, if $\epsilon = 10,000$, the radiative lifetime (the time in which the fluorescence decays to $1/e$, its initial value) is $\sim 10^{-8}$ s (10 ns). If the absorption is more intense, the lifetime is shorter and if it is less intense, it is longer. There are other modes of deexcitation that compete with fluorescence; therefore, the shorter the radiative lifetime the more likely fluorescence is to be observed.

Among the processes competing with fluorescence are **internal conversion** and the **intersystem crossing,** by which the triplet state is formed, and **photochemical reactions** of the singlet excited state. Internal conversion is the process by which a molecule moves from the lowest vibrational state of the upper electronic level to some high vibrational state of the unexcited electronic level. This is the principal means of depopulating the electronic state and competes directly with fluorescence. Thus, the actual lifetime† of an excited molecule is usually less than τ_r. The fluorescence efficiency can be defined by Eq. 13-15.

$$\text{Fluorescence efficiency } \phi_f = \tau/\tau_r \qquad (13\text{-}15)$$

For a highly fluorescent molecule like riboflavin, ϕ_f may be 0.25 or more.[58]

The rate of deexcitation by nonradiative pathways can be increased by addition of **quenchers.** While quenching of fluorescence occurs by various mechanisms, it most commonly involves collision of the excited chromophore with the quenching molecule. Some substances such as iodide ion are especially effective quenchers. The fluorescence efficiency of a substance in the absence of a quencher can be expressed (Eq. 13-16) in terms of the rate constants for fluorescence (k_f), for nonradiative decay (k_d), and for phosphorescence (k_p):

$$\phi_f = k_f/(k_f + k_d + k_p) \qquad (13\text{-}16)$$

In the presence of a quencher there is an additional rate process for deexcitation. Under these circumstances the ratio of the fluorescence efficiency in the absence of (ϕ_f^0) and the presence of a quencher is given by the **Stern–Volmer equation** (Eq. 13-17):

$$\phi_f^0/\phi_f = 1 + K[Q] = 1 + k_Q\tau_0[Q] \qquad (13\text{-}17)$$

---

† The directly measured lifetime τ for riboflavin 5'-phosphate (ϵ_{max} = 12,200 at 450 nm) at 25°C is $\sim$5 ns.[57]

The constant K is known as the Stern–Volmer quenching constant.

An example of the application of fluorescence quenching to the study of the proteins is the use of quenching by acrylamide to determine whether tryptophan side chains are accessible to solvent or are "buried" in the protein.[59,60]

1. Polarization of Fluorescence

Light emitted from excited molecules immediately after absorption is always partially polarized, whether or not the exciting beam consists of plane polarized light. As time elapses and the molecules become randomly oriented the polarization of luminescent emission disappears. The extent of depolarization of fluorescence can give valuable information about rates of rotation of the macromolecules to which fluorescent chromophores are bound or about the mobility of chromophoric groups within a macromolecule, cell membrane, etc.[52,53,55,61] The rotational rates obtained from polarization measurements are strongly affected by the viscosity of the medium (see Eq. 6-32). The embedding of fluorescent "probes" such as **1-anilinonaphthalene-8-sulfonate** in membranes,

1-Anilinonaphthalene-8-sulfonate

contractile fibers, etc., affords a means of examining changes in mobility that might accompany alterations in physiological conditions. Thus, possible molecular changes occurring in membranes during nerve conduction and in mitochondria during electron transport are being studied.[61,62]

2. Intramolecular Energy Transfer

It has often been observed that electronic excitation of one chromophore may elicit fluorescence from a different chromophore that is located nearby. For example, excitation of a monomolecular layer of dye has been shown to cause fluorescence in a layer of another dye spaced 5 nm away. Excitation of tyrosine residues in proteins can lead to fluorescence from tryptophan, and excitation of tryptophan can cause fluorescence in dyes attached to the surface of a protein or in an embedded coenzyme.[57] Such **resonant transfer of energy** is expected for molecules when the fluorescence spectrum of one overlaps the absorption spectrum of the other. The mechanism is not one of fluorescence emission and absorption but of nonradiative transfer of energy. Resonant transfer of energy has a major biological significance in photosynthesis. Light is a very dilute source of energy. While a molecule of $\epsilon = 3 \times 10^4$ in direct sunlight will absorb about 12 quanta of light per second, a monomolecular layer of chlorophyll would still absorb only about 1% of the total quanta falling on a leaf surface.[63] For this reason many molecules of chlorophyll are located in the numerous thin lamellae of the chloroplasts. However, only a few special chlorophyll molecules are located at the **reaction centers** where the photochemical processes take place. The rest of the chlorophyll molecules absorb light and transfer energy in a stepwise fashion to the reaction center.

Förster[63a] has calculated that the rate of energy transfer k_t should be proportional to the rate of fluorescence k_f, to an orientation factor K^2, to the spectral overlap interval J, to the inverse fourth power of the refractive index and to the inverse sixth power of the distance r separating the two chromophores (Eq. 13-18):

$$k_t \propto k_f K^2 J n^{-4} r^{-6} \tag{13-18}$$

Besides predicting the inverse sixth power dependence of energy transfer, Förster provided a formula for calculating R_0 the distance between chromophores at which 50% efficient singlet–singlet energy transfer takes place. R_0 is commonly of the order of 2.0 nm. Making use of these relationships, Stryer has proposed a method of measuring distances between chromophores. He calibrated the method by constructing a series of molecules containing various lengths of the rigid threefold polyproline helix to which dansyl groups (Chapter 2, Section H,4) were attached at one end and naphthyl groups at the other.[64,65] By exciting the

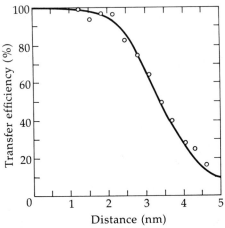

Fig. 13-17 Efficiency of energy transfer as a function of distance between α-naphthyl and dansyl groups at the ends of a polyprolyl "rod" (L-prolyl)$_n$. The observed efficiencies of transfer for $n = 1$ to 12 are shown as points. The solid line corresponds to an r^{-6} distance dependence. From L. Stryer, (Ref. 64) *Science* **162**, 530 (1968). Copyright 1968 by the American Association for the Advancement of Science.

Dansyl   L-Prolyl   Naphthyl

$(CH_3)_2N$

$n = 1–12$

naphthyl groups, which have the higher energy absorption band and are strongly fluorescent, the characteristic lower energy of the dansyl group could be observed if energy transfer took place. Since the fluorescent emission band of the naphthyl group overlaps the absorption band of the dansyl group, efficient transfer was expected. The results of a plot of transfer efficiency against distance is shown in Fig. 13-17. Note that the inverse sixth power dependence is followed quite accurately with a value of $R_0 \approx 3.4$ nm. Having calibrated his "spectroscopic ruler," Stryer turned his attention to biochemical macromolecules. Attaching the same kinds of fluorescent probe to the visual light receptor rhodopsin, Wu and Stryer were able to estimate distances between specific parts of the molecule and to draw some conclusions about the overall shape.

Fluorescence of terbium(III) in a calcium-binding site of thermolysin (Chapter 7, Section D,4) as a result of energy transfer from a cobalt(II) ion in the zinc-binding site has been observed. Application of the Förster equation gave a value of 1.37 nm for the Ca^{2+}–Zn^{2+} distance in this enzyme, identical to that measured by X-ray diffraction.[66]

3. Triplet States

One of the methods of deexcitation of a singlet state is intersystem crossing to the triplet state. Because selection rules forbid transitions between excited triplet state and ground state, the radiative lifetime of the triplet state is long. This accounts in part for the fact that it is often the triplet state that participates in photochemical reactions. The diradical character of the triplet state also makes it unusually reactive. Despite its forbidden character, nonradiative deexcitation of the triplet state is possible and phosphorescence is observed for most molecules only at low temperatures when the solvent is immobilized as a glass.

D. PHOTOCHEMISTRY

Because of their high energy the chemistry of molecules in photoexcited states (both singlet and triplet) is much more varied than that of unexcited molecules.[67,67a] Protons may be dissociated from or taken up by excited molecules as a result of changes in the pK_a values of functional groups. Bonds may be cleaved by dissociation into either ions or radicals. Photoelimination and photoaddition reactions both occur. Molecules may be isomerized, a process of importance in visual receptors. Excited molecules may become strong oxidizing agents able to accept hydrogen atoms or electrons from other molecules. An example is the **photooxidation** of EDTA by riboflavin (which undergoes photoreduction as shown in Fig. 8-15). A biologically more important example is in photosynthesis where excited chlorophyll molecules carry out **photoreduction** of another molecule and are themselves transiently oxidized. One of the most frustrating aspects of investigation of photochemical reactions is that the variety of reactions possible often leads to a superabundance of photochemical products (e.g., see the thin layer chromatogram of cleavage products of riboflavin in Fig. 2-34).

1. Chemical Equilibria in the Excited State

When pyridoxamine with a dipolar ionic ring structure (Eq. 8-27) and an absorption peak at 30,700 cm^{-1} is irradiated, fluorescence emission is observed at 25,000 cm^{-1}. When basic pyridoxamine with an anionic ring structure and an absorption peak at 32,500 cm^{-1} is irradiated, fluorescence is observed at 27,000 cm^{-1}, again shifted ~5500 cm^{-1} from the absorption peak. However, when the same molecule is irradiated in acidic solution, where the absorption peak is at 34,000 cm^{-1}, the luminescent emission at 25,000 cm^{-1} is the same as from the neutral dipolar ionic form and abnormally far shifted (9000 cm^{-1}) from the 34,000 cm^{-1} absorption peak.[68,69] The phenomenon, observed for most phenols, has been explained in terms of a rapid dissociation of a proton from the phenolic group in the photoexcited state. Thus, the excited pyridoxamine cation in acid solution is rapidly converted to a dipolar ion upon excitation. In other words, the phenolic group is more acidic in the excited state than in the ground state.

Study of the variation of fluorescence intensity with pH can give direct information about the pK_a in the excited state. The following indirect procedure (suggested by Förster) is also used to estimate excited state pK_a values for phenols. Let E_1 represent the energy of the 0-0 transition (preferably measured as the mean of the observed 0-0 transition energies in absorption and fluorescent emission spectra); E_2 represents the energy of the 0-0 transition in the dissociated (anionic in the case of a phenol) form, while ΔH and ΔH^* represent the enthalpies of dissociation in the ground and excited states, respectively.

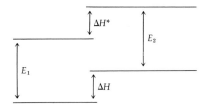

It is immediately evident that Eq. 13-19 holds.

$$E_1 - E_2 = \Delta H - \Delta H^* \qquad (13\text{-}19)$$

Furthermore, if we assume that the changes in entropy for the reaction are the same in the ground and excited state, Eqs. 13-20 and 13-21 follow.

$$\log_{10}(K^*/K) = Nh\ (\Delta\tilde{\nu})/(2.3RT) \qquad (13\text{-}20)$$

or

$$pK^* = pK - (2.1 \times 10^{-3})\Delta\tilde{\nu}\ (cm^{-1}) \quad \text{at } 25°C \quad (13\text{-}21)$$

This equation states that a shift in the spectrum of the basic form by 1000 cm^{-1} to a lower wavenumber compared with the acid form corresponds to a decrease of 2.1 units in pK_a for dissociation of the acid form. Whereas it is generally thought desirable to use both absorption and fluorescent measurements to locate the approximate positions of 0-0 bands, absorption measurements alone are often used and the positions of the band maxima are taken. Thus, for pyridoxamine the shift in absorption maximum from 34,000 cm^{-1} in the protonated form to 30,700 cm^{-1} in the dissociated form suggests that the pK_a of pyridoxamine of 3.4

in the ground state is shifted by 6.9 units to -3.5 in the excited state.†

While phenols and amines are usually more acidic in the singlet excited state than in the ground state, some substances, e.g., aromatic ketones, may become more *basic* in the photoexcited state.

An elegant use of an abnormally large shift in the position of fluorescent emission was made by Johnson *et al.*[70] A 330 nm (30,300 cm^{-1}) absorption band of pyridoxal phosphate in glycogen phosphorylase could be interpreted either as an adduct of some enzyme functional group with the Schiff base of PLP and a lysine side chain (structure A) or

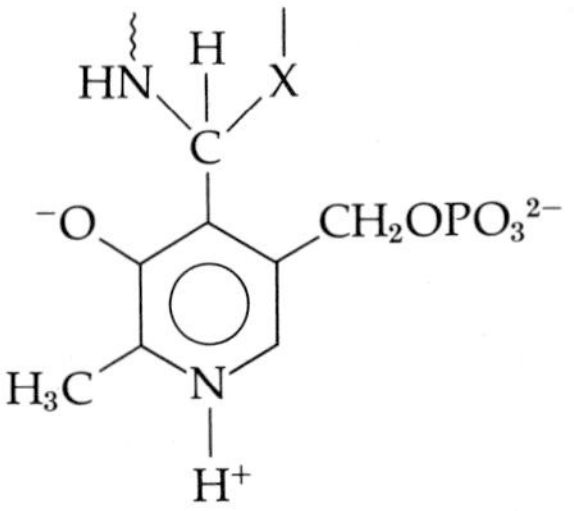

A 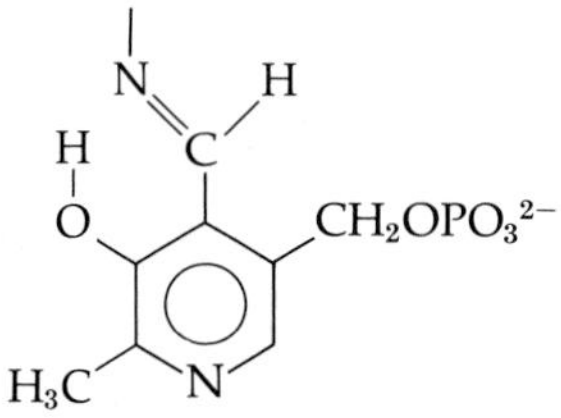   B

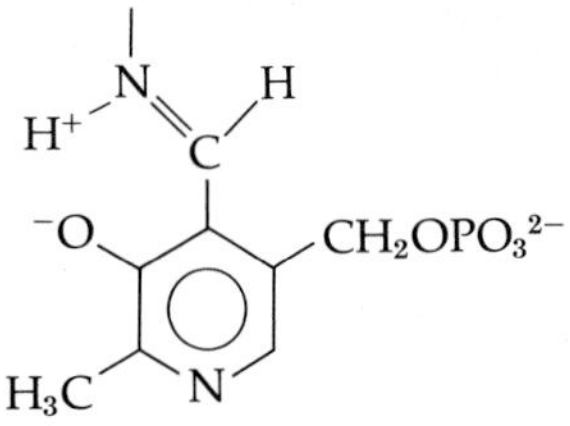

C

as a nonionic ring form of a Schiff base in a hydrophobic environment (structure B). For structure A, the fluorescent emission would be expected at a position similar to that of pyridoxamine. On the other hand, Schiff bases of the type indicated by structure B would be expected[52,70] to undergo a photoinduced proton shift (phototautomerization) to form structure C with an absorption band at 430 nm (23,300 cm^{-1}) and fluorescent emission at a still lower energy. Since the observed fluorescence

was at 530 nm, it was judged that the chromophore does have structure B.

The rate of proton dissociations from the excited states of molecules can now be measured directly by nanosecond fluorimetry.[71]

2. Photoreactions of Nucleic Acid Bases

Photochemical reactions of the purines and pyrimidines assume special significance because of the high molar absorptivities of the nucleic acids present in cells. This means that light is likely to be absorbed by nucleic acids and to induce photoreactions that lead to mutations.[72,73] Both pyrimidines and purines† undergo a variety of photochemical alterations, but only two kinds of reaction of pyrimidines are ordinarily significant. One is photohydration (Eq. 13-22) a reaction observed readily with cytidine.

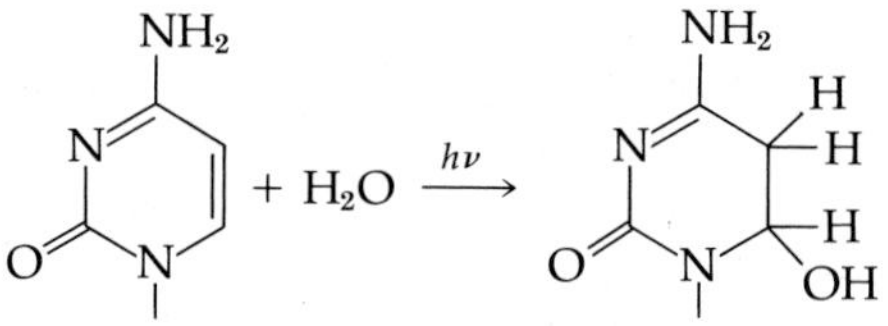

(13-22)

The reaction is the photochemical analogue of the hydration of α,β-unsaturated carboxylic acids (reaction type 2B). Photohydration also occurs with uracil derivatives.

The second and more important reaction is the photodimerization of thymine (Eq. 13-23), a reac-

(13-23)

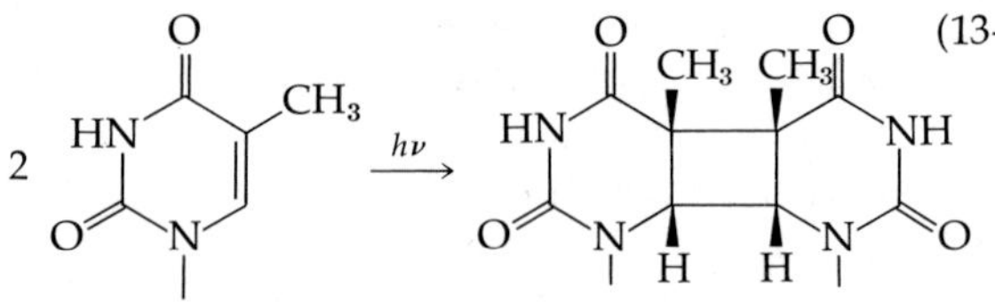

tion also observed with uracil. A variety of stereoisomers of the resulting cyclobutane-linked structure are formed. The one shown in the equation predominates after irradiation of frozen

---

† Bridges *et al.*[69] obtained -4.25 for this calculated pK^* and observed from the pH dependence of fluorescence p$K^* \approx -4.1$.

† Purines are only about one-tenth as sensitive as pyrimidines.

thymine solutions. The significance of photo-dimerization of thymine is that the resulting **cyclobutane dimers** block DNA replication. This accounts for much of the lethal and mutagenic effect of ultraviolet radiation on organisms. The matter is sufficiently important that a special "excision repair" process is used by cells to cut out the thymine dimers (Chapter 15, Section H,2).

A curious observation was made many years ago. Bacteria given a lethal dose of ultraviolet radiation could be saved by irradiating with visible or near ultraviolet light. This **photoreactivation** permitted many of the bacteria to survive. It has now been shown that photoreactivation results from the action of a photoreactivating enzyme (DNA photolyase)[74-76] that absorbs light maximally around 380 nm and carries out a photochemical reversal of Eq. 13-23. The enzyme is present in cells in such small amounts that it has not yet been possible to investigate the mechanism of this intriguing enzymatic process. However, the significance cannot be doubted, for photoreactivation enzymes appear to be found in almost all organisms including animals.

E. PHOTOSYNTHESIS

The photochemical reduction of CO_2 to organic materials is the basic source of energy for the biosphere. Nevertheless, the process is limited to a few genera of photosynthetic bacteria (Table 1-1) including the blue-green algae, the various eukaryotic algae, and the higher green plants. The idea that the photoprocesses in these organisms generate NADPH or reduced ferredoxin plus ATP (Chapter 11, Section D,2) is now thoroughly accepted.[77-79] However, this notion was not always obvious. Consider the overall equation (Eq. 13-24) for formation of glucose by photosynthesis in higher plants:

$$6\ CO_2 + 6\ H_2O \longrightarrow 6\ O_2 + C_6H_{12}O_6 \quad (13\text{-}24a)$$

$$6\ CO_2 + 12\ H_2O^* \longrightarrow 6\ O_2^* + C_6H_{12}O_6 + 6\ H_2O$$
$$(13\text{-}24b)$$

The stoichiometry of Eq. 13-34a suggests that all 12 of the oxygen atoms of the evolved O_2 might have come from CO_2 or that some came from CO_2 and some from H_2O. In fact, it is now believed that water supplies all of the oxygen atoms for formation of O_2, as in Eq. 13-24b. The idea was suggested by C. B. van Niel[79a] in 1933. He pointed out that in bacterial photosynthesis no O_2 is produced and that bacteria must have access to a reducing agent to provide hydrogen for the reduction of CO_2 (Eq. 13-25).

$$H_2A \xrightarrow{h\nu} A + 2[H] \quad (13\text{-}25)$$

In this equation, H_2A might be H_2S (in the purple sulfur bacteria), elemental H_2, isopropanol, etc. From a consideration of these various reactions, van Niel reached the logical conclusion that in the O_2-producing blue-green algae and eukaryotic plants, water serves as the oxidizable substrate in Eq. 13-25. Thus, H_2O is cleaved to form O_2 and to provide hydrogen atoms for reduction. It is interesting that this photochemical cleavage is the only known biological oxidation reaction of H_2O. No oxidizing agents present in living things are powerful enough to dehydrogenate water except for the photochemical reaction centers of the photosynthetic organisms.

1. Two Photosystems

A further clarification of the difference between photosynthetic bacteria and green plants came from experiments of R. Emerson and associates[79b] in 1956. It was known that light of wavelength 650 nm was much more efficient than that of 680 nm. However, Emerson *et al.* showed that a combination of light of 650 nm *plus* that of 680 nm gave a higher rate of photosynthesis than either kind of light alone. This result suggested that there might be two separate photosystems. What is now known as **photosystem I** is excited by far red light (~700 nm), while **photosystem II** depends upon the higher energy red light of ~650 nm. Much evidence supports the idea. For example, Hill had shown many years before[79c] that mild oxidizing agents such as ferricyanide and benzoquinone can serve as substrates for photoproduction of O_2, while H. Gaffron[79d] showed that some green algae could be adapted to photooxidize H_2 to protons (Eq. 13-25) and to use the electrons to reduce NADP. Thus, photosystem I can be disconnected from photosystem II.

The powerful herbicide **dichlorophenyldimethylurea** (DCMU) was found to block electron trans-

3-(3,4-Dichlorophenyl)-1,1-dimethylurea

port between the two photosystems. In the presence of DCMU electrons from such artificial donors as ascorbic acid or an indophenol dye could be passed through photosystem I.

The result of these and other experiments was the development of the series formulation or "Z scheme" of photosynthesis as shown in Fig. 13-18. Passage of an electron through the system requires two quanta of light. Thus, four quanta are required for each NADPH formed and eight quanta for each CO_2 incorporated into carbohydrate. Most authorities now agree that at least eight or nine quanta are, in fact, required to reduce one molecule of CO_2.

Another important experiment of Emerson and Arnold[79e] employed very short flashes of light and measurement of the quantum efficiency of photosynthesis during those flashes. A striking fact was observed. At most, during a single turnover of the photosynthetic apparatus of the leaf, one molecule of O_2 would be released for each 3000 chlorophyll molecules. However, it could be calculated that for each O_2 released only about eight quanta of light had been absorbed. It followed that about 400 chlorophyll molecules were involved in the uptake of one quantum of light. This finding suggested that a large number of chlorophyll molecules act as a single light receiving unit able to feed energy to one **reaction center.** The concept is now fully accepted.

2. Photophosphorylation

Recall from Chapter 11 that both NADPH and ATP are needed for conversion of CO_2 to sugars by means of the Calvin cycle. The stoichiometry, as well as we know it, is given by Eq. 11-16. In addition to the two molecules of NADPH required to reduce one molecule of CO_2, three molecules of ATP are needed. We must ask how these are formed. The Z scheme provides a simple answer: The electron transport system connecting the upper end of photosystem II with the lower end of photosystem I provides enough drop in potential to permit synthesis of ATP by electron transport. It is likely that only one molecule of ATP is formed for each pair of electrons passing through this chain. Since, according to Eq. 11-16, one and a half molecules of ATP are needed per NADPH, some other mechanism must exist for the synthesis of additional ATP. Furthermore, there are doubtless many other processes in chloroplasts that depend upon ATP so that the actual need for photogenerated ATP may be larger than this.

Arnon[79f] demonstrated that additional ATP can be formed in chloroplasts by means of **cyclic photophosphorylation:** Electrons from the top of photosystem I are recycled according to the dashed lines in Fig. 13-18. An electron transport system that may be connected with that of the Z scheme or which may be independent is used to synthesize ATP. In fact, Arnon believed that there are three photosystems in chloroplasts, that photosystem I is involved in cyclic photophosphorylation and that photosystem II is made up of two parts needed for the Z scheme.[80]

What is the relationship of the scheme of Fig. 13-18 to bacterial photosyntheses? An obvious conclusion is that bacteria contain only photosystem I and that photosystem II, which releases O_2, is absent. Experimentally, it is found that the formation of reducing equivalents (reduced ferredoxin or NADPH) by photosynthetic bacteria requires only about half as many light quanta as in green plants where H_2O must be split.

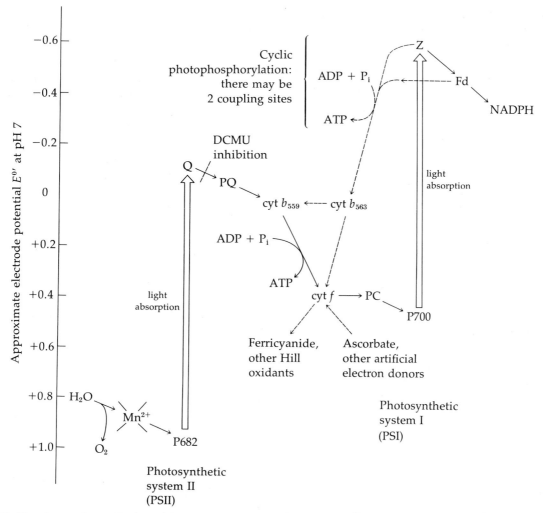

Fig. 13-18 The zigzag scheme (Z scheme) for a two-quantum per electron photoreduction system of chloroplasts. Abbreviations are: P682 and P700, reaction center chlorophylls; Q, quencher of fluorescence of P682 and presumed initial electron acceptor; PQ, plastoquinone; cyt, cytochromes; PC, plastocyanin; Z, initial acceptor of electrons for PSI; Fd, ferredoxin; and DCMU, dichlorophenyldimethylurea. Note that the positions of P682, P700, Q, and Z on the $E^{0'}$ scale are uncertain. The $E^{0'}$ values for P682 and P700 should be for the (chlorophyll/chlorophyll cation radical) pair in the reaction center environment. These may be lower than are shown. Likewise the redox potentials of Q and Z are uncertain and could be higher than is indicated.

3. The Pigments and Their Environments

The chlorophylls (Fig. 13-19) are related in structure to the hemes (Fig. 10-1), but ring IV is not fully dehydrogenated as is the corresponding ring D of the porphyrins. This **chlorin** ring system is further modified in chlorophyll by the addition of a fifth ring (V) containing a ketone group and a methyl ester. Note that ring V has been formed by crosslinking between the propionic acid side chain of ring III and a methine bridge carbon. The parent ring system formed in this way is known as **pheoporphyrin.** The chlorophylls contain appro-

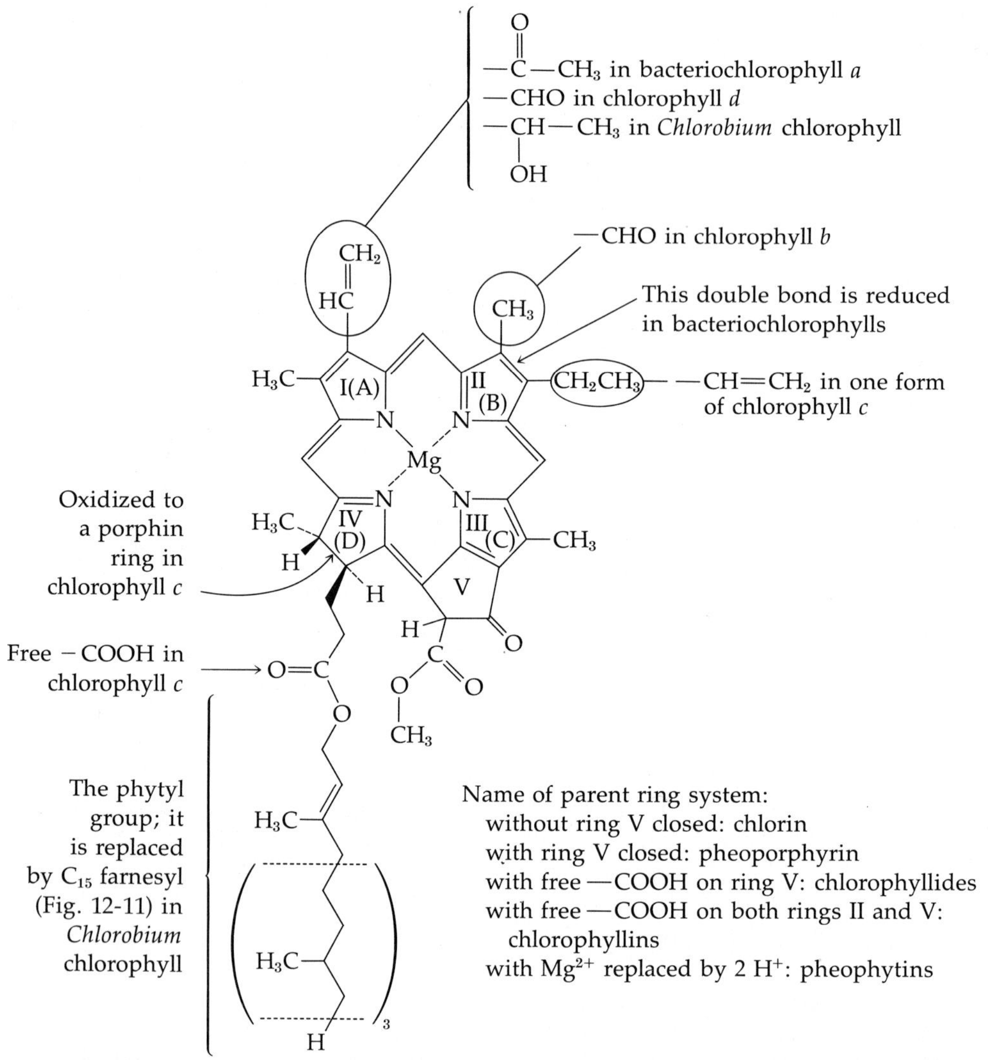

Fig. 13-19 Structures of the chlorophylls.

priate constituents around the periphery that indicate a common origin with the porphyrins. However, one of the carboxyethyl groups is esterified with the long chain phytyl group in most of the chlorophylls. Chlorophyll *a* is the major pigment of chloroplasts and is the centrally important chromophore for photosynthesis in green plants. The other forms of chlorophyll, as well as the carotenoids and certain other pigments, are referred to as **accessory pigments** and are thought to function primarily in "light gathering." Their relative numbers in a photosynthetic unit (see Section 4) of spinach chloroplasts are given in Table 13-2.

In 80% acetone chlorophyll *a* has a sharp absorption band at 663 nm (15,100 cm^{-1}) but within chloroplasts the absorption maximum is shifted toward the red, the majority of the chlorophyll absorbing at 678 nm. Chlorophyll *b* (Fig. 13-19) is also nearly always present in green leaves. The absorption peak in acetone is at 635 nm

TABLE 13-2
Approximate Composition of an Average Photosynthetic Unit in a Spinach Chloroplast[a,b]

Component	Number of molecules
Chlorophyll a	160
Chlorophyll b	70
Carotenoids	48
Plastoquinone A	16
Plastoquinone B	8
Plastoquinone C	4
α-Tocopherol	10
α-Tocopherylquinone	4
Vitamin K_2	4
Phospholipids	116
Sulfolipids	48
Galactosylglycerides	490
Iron	12 atoms
Ferredoxin	5
Cytochrome b_{563}	1
Cytochrome b_{559}	
Cytochrome f	1
Copper	6 atoms
Plastocyanin	1
Manganese	2 atoms
Protein	928,000 daltons

a After Gregory [data of H. K. Luchtenthaler and R. B. Park, *Nature (London)* **198**, 1070 (1963)] and A. White, P. Handler, and E. L. Smith, "Principles of Biochemistry," 5th ed., p. 528. McGraw-Hill, New York, 1973.

b Numbers of molecules assuming 2 Mn^{2+} ions per unit.

(15,800 cm^{-1}). Chlorophyll c found in diatoms, brown algae (Phaeophyta), and dinoflagellates (Fig. 1-7) lacks the phytyl group and is thought to be a mixture of two compounds. Chlorophyll d, found together with chlorophyll a in some Rhodophyta (Chapter 1, Section D,3), has not yet been thoroughly characterized.[77]

Photosynthetic bacteria contain **bacteriochlorophylls** in which ring II is reduced (Fig. 13-19). The absorption band is shifted to the red from that of chlorophyll a to ~770 nm. The principal chlorophyll of green sulfur bacteria *Chlorobium* **chlorophyll** contains hydroxyethyl and farnesyl side chains. Among the chlorophyll derivatives are **pheophytins**, formed by splitting the Mg^{2+} out from chlorophyll by weak acid. **Chlorophyllides** are formed by hydrolysis of the methyl ester group and **chlorophyllins** by removal of both the methyl and phytyl groups.

Since chlorophyll can be readily and completely removed by the use of mild solvent extraction,[81] it might appear that it is simply dissolved in the lipid portion of the membranes. However, the absorption spectrum of chlorophyll in leaves has bands that are shifted to the red by up to 900 cm^{-1} from the position of chlorophyll a in acetone. Most green plants contain at least four major chlorophyll bands at 662 nm (15,120 cm^{-1}), 670 nm (14,940 cm^{-1}), 677 nm (14,770 cm^{-1}), and 683 nm (14,630 cm^{-1}).[82] Minor bands at 14,420 and 14,230 cm^{-1} may also be present (Fig. 13-20). This fact suggests that the chlorophyll within membranes exists in a number of different environments. As a result, the absorption is spread over a broader region leading to more efficient capture of light. The reaction centers are also thought to contain chlorophyll; that for photosynthetic system I (PSI) absorbs at ~700 nm (14,290 cm^{-1}) and that for photosystem II (PSII) at ~682 nm (14,660 cm^{-1}).

In a similar manner, bacteriochlorophyll in *Chromatium* has three absorption bands with peak positions at 800, 850, and 890 nm. The latter includes the reaction center bacteriochlorophyll and is the only form that fluoresces. A water-soluble bacteriochlorophyll-containing protein has been crystallized from the green photosynthetic bacterium *Chlorobium*. The three-dimensional structure, determined by X-ray crystallography,[83] shows that each subunit (of MW = 50,000) in the trimeric molecule contains seven embedded bacteriochlorophyll molecules as illustrated in Fig. 13-20B. Chlorophyll in green plants may also be present as a complex with more hydrophobic proteins. Two chlorophyll-protein complexes, presumably one from photosystem I and one from photosystem II, have been isolated.[84]

Irradiation of chloroplasts leads to easily measurable fluorescence from chlorophyll a, but no fluorescence is observed from chlorophyll b or from other forms of chlorophyll, carotenoids, or

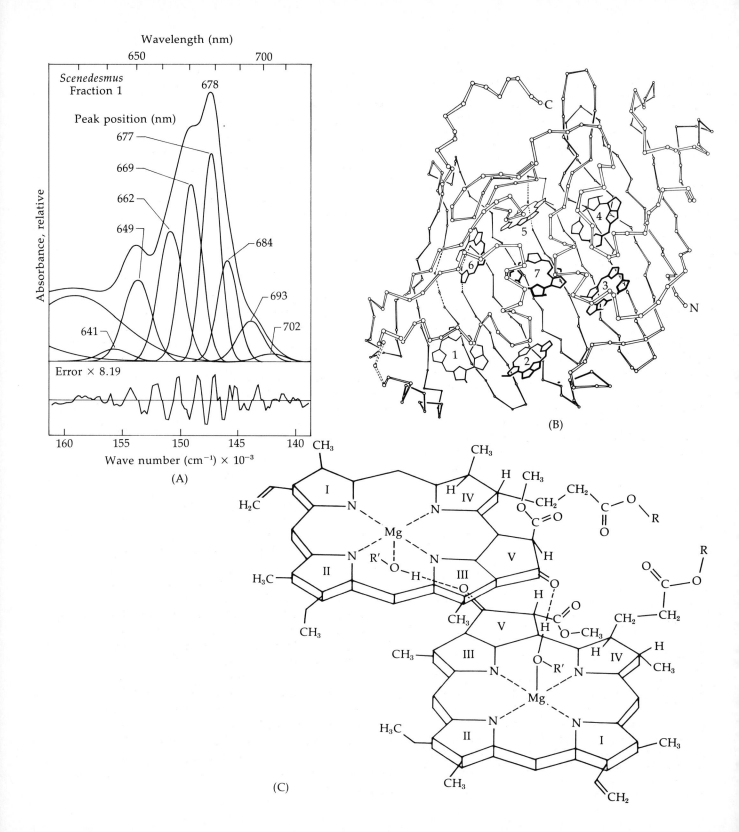

Wavelength (nm)

Scenedesmus
Fraction 1

Peak position (nm)

677

669

662

649

684

693

702

678

641

Error × 8.19

Absorbance, relative

Wave number (cm⁻¹) × 10⁻³

(A)

(B)

(C)

other pigments. It appears that the latter all serve as accessory pigments and that they efficiently transfer their energy to chlorophyll *a* at the reaction centers. As is evident from Fig. 13-21, the accessory pigments generally have higher energy absorption bands than do the reaction centers. Thus, a broad range of wavelengths of light are absorbed by an organism, and energy from all of them is funneled into the reaction centers.

Among the important accessory pigments are the carotenes (Fig. 12-14), the major component in most green plants being β-carotene. Green sulfur

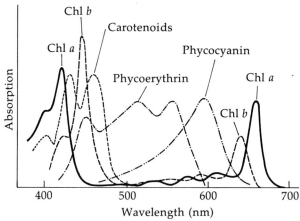

Fig. 13-21 Absorption spectra of chlorophylls and accessory pigments compared. Redrawn from G. and R. Govindjee, *Sci. Am.* **231**, 68–82 (Dec. 1974). Copyright (1974) by Scientific American, Inc. All rights reserved.

bacteria contain γ-carotene in which one end of the molecule has not undergone cyclization and resembles lycopene. Chloroplasts also contain a large variety of oxygenated carotenoids (xanthophylls). Of these, neoxanthin, violaxanthin (Eq. 12-30), and lutein predominate in higher plants and green algae. Lutein resembles zeaxanthin, but the ring at one end of the chain has been isomerized by a shift in double bond position to the accompanying structure:

Euglena and related microorganisms contain much antheraxanthin (Section E,7), while the brown algae and diatoms contain mostly fucoxanthin and zeaxanthin (Fig. 12-14). The purple sulfur bacterium *Rhodospirillum rubrum* synthesizes its own special spirilloxanthin which contains the accompanying structure at both ends of the molecule.

A third class of accessory pigments of more limited distribution are the **open tetrapyrroles**.[85] These are often referred to as "plant bile pigments" because of their relationship to the pigments of animal bile (Fig. 14-14). The **phycocyanins** provide the characteristic color to blue-green bacteria. They are conjugated proteins (biliproteins) containing phycocyanobilin (Fig. 13-22) as a bound pigment. In a similar manner the red **phycoerythrins** of the Rhodophyta contain bound phycoerythrobilin (Fig. 13-22). Although the pigments are sometimes described as linear tetrapyrroles, their true structure is probably helical (Fig. 13-22). The algal biliproteins (phycocyanins and phycoerythrins) appear to be aggregated in special granules that are on the outsides of the photosynthetic membranes. The granules in the blue-green bacteria are known as **phycobilisomes.**

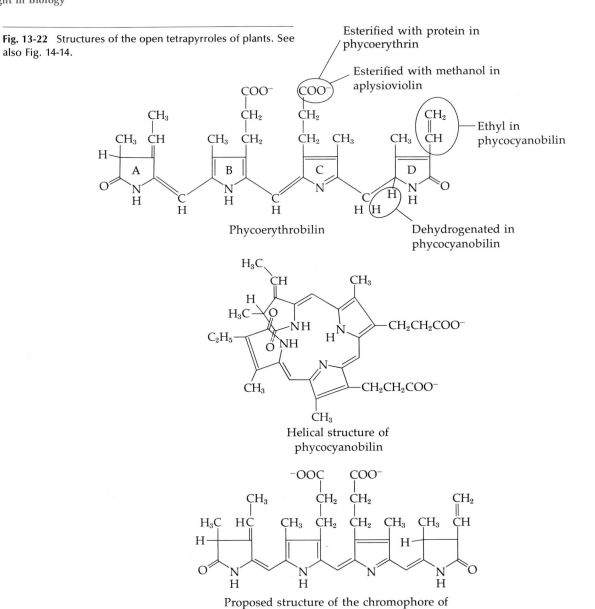

Fig. 13-22 Structures of the open tetrapyrroles of plants. See also Fig. 14-14.

Esterified with protein in phycoerythrin

Esterified with methanol in aplysioviolin

Ethyl in phycocyanobilin

Phycoerythrobilin

Dehydrogenated in phycocyanobilin

Helical structure of phycocyanobilin

Proposed structure of the chromophore of phytochrome (625 nm absorbing form)

4. Chloroplast Structure[86]

Like the other energy-producing organelles (the mitochondria), chloroplasts contain an internal membrane system. Within the colorless **stroma** are stacks of flattened discs known as **grana.** The discs themselves (the **thylakoids**) consist of pairs of closely spaced membranes 9 nm thick (Fig. 13-23) separated by a thin internal space or **loculus** (Fig. 13-23). It has been suggested that the outer layer contains protein subunits of 4 nm diameter, while the inside consists of a lipid layer containing the

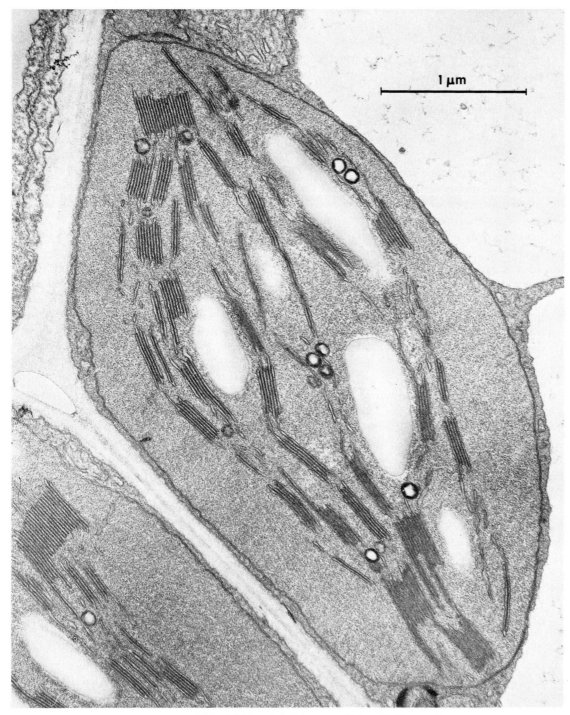

1 μm

Fig. 13-23 Electron micrograph of alfalfa leaf chloroplast. Courtesy of Harry T. Horner, Jr., Iowa State University.

chlorophyll, carotenoids, and certain special lipids. However, the chlorophyll may be present largely as complexes with protein (see Section E,3). The high content of **galactosyl diglyceride, digalactosyl glyceride,** and **sulfolipid** (Fig. 2-32) in chloroplasts has been mentioned previously. The possible function of these lipid components in interacting with the chlorophyll molecules and providing the characteristic differences in environment that are postulated for chlorophyll have yet to be clarified. There is evidence from measurements of dichroism (Gregory,[77] p. 111) that the chlorophyll molecules within the membranes have a definite orientation with respect to the planes of the thylakoids.

There has been controversy about the possible organization of thylakoid membranes into *photosynthetic units* or **quantosomes.**[86, 87] As revealed by freeze etching techniques, the quantosomes are ~20 nm in diameter and ~10 nm thick. However, some investigators find only smaller particles of cubic shape which could be molecules of ribulose diphosphate carboxylase (Chapter 7, Section J,3,g), ~12 nm on a side, and of a coupling factor for ATP synthesis (Section E,6), ~10 nm on a side.

The photosynthetic unit has also been defined chemically by the number of various types of molecules present in a chloroplast membrane for each pair of manganese atoms (Table 13-2). Each unit would presumably contain one reaction center for photosystem I and one for photosystem II.

5. The Reaction Centers and the First Photochemical Process

One of the most intriguing questions about photosynthesis is that of the nature of the first photochemical reaction. We do not yet know whether it is the singlet excited state of a chlorophyll that reacts or whether there is first a conversion to a triplet state. Whichever it may be, it is thought that the excited chlorophyll (Chl*) donates an electron to some acceptor to form a radical $\cdot A^-$, and to leave an oxidized chlorophyll Chl^+ radical (Eq. 13-26).

$$Chl \xrightarrow{h\nu} Chl^* \xrightarrow{A} Chl^+ + A^- \qquad (13\text{-}26)$$

In the scheme of Fig. 13-18, acceptor A would be Q for PSII and Z for PSI. The oxidized chlorophyll (Chl^+) quickly reacts further by receiving an electron from some donor. In the case of PSI plastocyanin is a likely candidate for the immediate donor.

The photooxidation of chlorophyll shown in Eq. 13-26 is accompanied by bleaching in the principal absorption band. However, since there is so much light gathering chlorophyll for each reaction center, the effect is small. The study of the process has been aided greatly by preparation of *isolated photochemical reaction centers* from bacterial chromatophores (Chapter 1, Section A,11). Although the exact composition varies, the properties of the reaction centers from several genera of both purple and green bacteria are similar. They contain three proteins of MW 21,000, 24,000, and 28,000 in a 1:1:1 ratio,[88] together with four molecules of bacteriochlorophyll, two of bacteriopheophytin, a molecule of ubiquinone, an atom of nonheme iron, and (except for carotenoidless mutants) carotenoid. For example, the centers from *Rhodospirillum rubrum* contain one molecule of spirilloxanthin.[89] There is evidence, some from spectral observations suggesting exciton splitting,[90] that the chlorophyll involved in the photoreaction is a dimer.

A chlorophyll dimer has also been proposed for the reaction centers of the chloroplasts of green plants.[91] A possible structure for the dimer is suggested by X-ray diffraction studies of crystalline ethyl chlorophyllide *a*. The molecules pack in a polymeric arrangement with water molecules bridging between adjacent chlorophyll rings, the Mg^{2+} ions lying about 0.04 nm above the plane of the four pyrrole nitrogens on the same side as the coordinated water.[92] A symmetrical dimer involving a pair of chlorophyll molecules hydrated in a similar way [93] or combined with an OH, SH, or NH group[93a] has also been proposed (Fig. 13-20 C).

Picosecond kinetic studies[94] of the bleaching of

bacteriochlorophyll in isolated reaction centers show that the initial photochemical oxidation to form Chl^+ occurs within 10^{-10} s (0.1 ns). In line with this value, the lifetime τ of the fluorescence of the chlorophyll in photosystem I of chloroplasts is estimated as 0.13 ns, a value which can be compared with the natural lifetime τ_0 of 19 ns.[95] The rapid occurrence of the electron transfer from chlorophyll to acceptor accounts for the low value of τ in chloroplasts. The fluorescence lifetime of the photosystem II chlorophyll is about 10 times larger (1.5 ns).[95,†]

Accompanying the bleaching of bacteriochlorophyll is the development of an epr signal at $g = 1.82$, presumably arising from the reduced primary acceptor, possibly a nonheme iron.[96] However, other evidence suggests that a quinone (ubiquinone in *Rhodopseudomonas*) may be the acceptor.[97] A related model reaction is the photoreduction of benzoquinone with bleaching of chlorophyll in alcoholic solution.[98] The primary acceptor in photosystem I of chloroplasts appears to be a special iron—sulfur protein.[99]

What is the chemical nature of Chl^+? It has been suggested that the carbonyl group on ring V may be involved in stabilizing a radical ion formed by ejection of one electron from the adjacent ring III (Eq. 13-27). Note that in the resonance structure to the right the conjugation of the double bonds around the ring has been broken, perhaps accounting for the bleaching of the chromophore. Chlorophyll can be photoreduced under some circumstances and it has been proposed that the initial photochemical act may be transfer of an electron from one chlorophyll molecule to another within a dimer,[100] or, in bacteria, from bacteriochlorophyll to bacteriopheophytin.

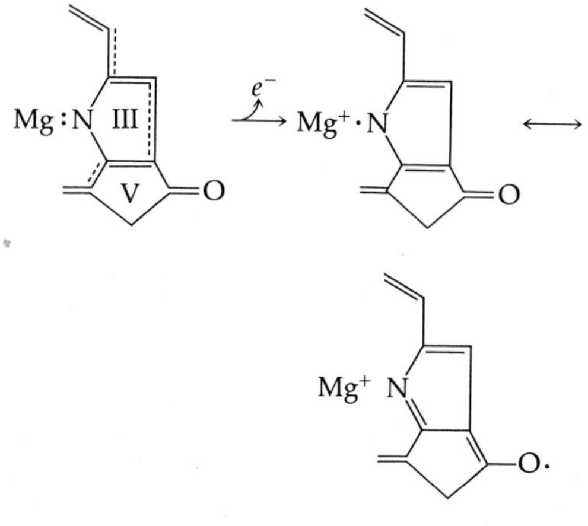

6. The Electron Transport Chains of Chloroplasts[101]

Now let us look more carefully at the processes by which the reduced electron acceptor A^- (of Eq. 13-26) reacts and by which an electron is returned to the oxidized Chl^+. The electrode potential of P700 of the chloroplast PSI (Fig. 13-18) is $+0.43$ V. The absorption of a quantum of 700 nm light would provide enough energy (1.77 eV) to photoreduce a carrier that was 1.77 V more negative, i.e., at a potential of -1.3 V. However, it is unlikely that such a powerful reducing agent is formed. In the first place, it would be impossible to capture light energy with 100% efficiency.† If the photoreaction occurs through a triplet state, the energy loss would necessarily be substantial. Furthermore, no known carriers have such negative redox potentials. Even so, it is clear that we do not know the electrode potential of the hypothetical carrier Z. Most authors put it at ~ -0.55 to -0.6 V, not far from that of ferredoxin, and it has been suggested that Z may be a bound ferredoxin molecule. More recent evidence indicates a special iron—sulfur protein.[99] Whatever it may be, Z is able to reduce ferredoxin rapidly, possibly through as-yet uncertain electron transport proteins. Reduced chloroplast ferredoxin, a one-electron carrier (Chapter 10, Section C), can transfer electrons to $NADP^+$ through a flavoprotein **Fd-NADP reductase** to accomplish the generation of NADPH.

$$(13-27)$$

† Recent experiments[95a] with chloroplasts of *Chlorella* have indicated a true lifetime of excited chlorophyll as 0.6 ns rather than the 0.13 ns cited above.

† It is not commonly appreciated that light carries entropy as well as energy. An important consequence of this fact is that at 700 nm at most 78% of the energy of sunlight could be harnessed for chemical work (R. S. Knox, *Biophys. J.* **9,** 1351, 1969).

While the terminal end of the electron transport chain appears to be well defined in green plants, it is less so in bacteria. It has not been possible to demonstrate photochemical generation of reduced intermediates at the potential of ferredoxin. Rather, it is thought that a quinone (probably ubiquinone) is the electron acceptor.[102] Since the $E^{0'}$ value of the quinone is nearly zero, it must be necessary to use ATP-driven "reverse electron transport" to reduce $NADP^+$. This could be accomplished if half of the photoreduced quinone were reoxidized through an electron transport chain by the oxidized Chl^+ of the reaction centers with accompanying ATP synthesis (cyclic photophosphorylation).

Returning to chloroplasts, the copper-containing protein **plastocyanin** (PC) is a likely candidate for the immediate electron *donor* to P700 in the electron transport chain between PSII and PSI (Fig. 13-18). The essentiality of plastocyanin has been shown by study of copper-deficient *Scenedesmus* (Fig. 1-9). The photoreduction of CO_2 by H_2 is impaired in these cells, but the Hill reaction occurs at a normal rate. At the other end of the chain is Q, the electron acceptor for PSII. It is abbreviated Q for **quencher** because it quenches the fluorescence of chlorophyll P682, the reaction center chlorophyll *a* of PSII. If chloroplasts are irradiated with 650 nm light PSII is activated but PSI is not. Under these conditions Q becomes reduced and the fluorescence of Chl* increases, presumably because the electron acceptor Q is absent. If PSI is activated by addition of far-red light, Q remains more oxidized and fluorescence is quenched.

The exact nature of Q is uncertain, but most investigators believe that it is one of the **plastoquinones**. Plastoquinone A, the predominant form in spinach chloroplasts, has the structure shown in Fig. 10-8 with nine isoprenoid units in the side chain. Spinach chloroplasts also contain at least six other plastoquinones. Plastoquinones C that are hydroxylated in side chain positions are widely distributed. In plastoquinones B these hydroxyl groups are acylated. Many other modifications exist including variations in the number of isoprene units in the side chains.[103, 104] Thus, there may be more than one kind of plastoquinone in the electron transport chain. It has been estimated that there are about five molecules of plastoquinone for each reaction center and that the plastoquinones may serve as a kind of electron buffer between the two photosynthetic systems. According to this view, Q is a small pool of plastoquinone (PQ) associated with the reaction center and separated from the larger pool by a reaction step inhibited by DCMU.

Two carriers usually shown in the chain between PSII and PSI are **cytochrome** b_{559} and **cytochrome** *f*, a *c*-type cytochrome.[104a] While a series of parallel pathways has been suggested,[105] most investigators believe that the sequence is that of Eq. 13-28:

$$Q \longrightarrow PQ \longrightarrow \text{cyt } b_{559} \longrightarrow \text{cyt } f \longrightarrow PC \quad (13\text{-}28)$$

where PC stands for plastocyanin. The sequence $PQ \longrightarrow \text{cyt } b_{559} \longrightarrow \text{cyt } f$ appears to be closely homologous to that in mitochondria[106]: ubiquinone $\longrightarrow$ cyt $b_T \longrightarrow$ cyt c_1. This part of the mitochondrial electron transport chain includes coupling site II for ATP synthesis (Fig. 10-11). As indicated in Fig. 13-18, the synthesis of ATP is believed to be coupled to the corresponding electron transport sequence in chloroplasts.

The similarity of the chloroplast and mitochondrial systems became even more striking when it was found that a **chloroplast coupling factor**, CF_1, analogous to the mitochondrial protein F_1 (Chapter 10, Section E,8), is required for ATP synthesis. Like the mitochondrial coupling factor, CF_1 consists of subunits of five different kinds.[107, 108] Like mitochondria, chloroplasts (when illuminated) also pump protons across their membranes. However, the protons accumulate on the *inside* of the thylakoids, while they are pumped to the *outside* of mitochondria. The coupling factor CF_1 is found on the outside of the thylakoids, facing the stromal matrix, while F_1 lies on the inside of mitochondrial membranes. It seems likely that the same mechanism of ATP formation is used in both chloroplasts and mitochondria.

It is often suggested that the cyclic oxidation and reduction of plastoquinone is involved in the pumping of protons across thylakoid membranes, just as ubiquinone may participate in the corresponding process in mitochondria. In addition, it is thought that PSII may be located on the inside of the thylakoids. The splitting of one water molecule would then leave two protons (one per electron) inside the thylakoids, while the electrons would be "photoejected" through the lipid bilayer to acceptor Q on the outside. The chlorophyll in PSII would likewise be on the inside of the bilayer with acceptor Z (Fig. 13-18) on the outside. Since the conversion of NAD^+ to NADH on the outside generates a proton, the overall reaction would be the pumping of one and a half protons per electron passing through the Z scheme.[107,109] According to the chemiosomotic hypothesis (Chapter 10, Section E,9,c) the pH and membrane potential developed as a result of the proton transport is the source of free energy for synthesis of ATP.

The pathways involved in cyclic photophosphorylation in chloroplasts are not yet established. **Cytochrome b$_{563}$** (cytochrome b_6) has been implicated in the process, but there are uncertainties as to whether electrons then flow to plastoquinones or directly to cytochrome f.

Of great interest is the photophosphorylation of inorganic phosphate to pyrophosphate (PP_i) by chromatophores from *R. rubrum*. There is evidence that the PP_i formed in this way may be used in a variety of energy-requiring reactions in the chromatophores.[110] An example is formation of NADH by reverse electron transport. These observations raise a question about the assumption that PP_i is always immediately hydrolyzed (Chapter 11, Section B,2). It may be that in other organisms as well as in bacteria use can sometimes be made of PP_i as an alternate energy source.

Turning to the electron donor for PSII, there is strong evidence that the 2–4 Mn^{2+} ions found in each reaction center are intimately involved.[111] Thus, the chain is often drawn from H_2O to Mn^{2+} to P682. A specific proposal[112] for the connection between oxidized chlorophyll and a metal-bound hydroxyl ion from water is as follows (Eq. 13-29):

$$\qquad (13\text{-}29)$$

(as in Eq. 13-27)   (OH radical bound to Mn)

Just as the mechanism of reduction of O_2 by the cytochrome oxidase of the electron transport chain of mitochondria appears unusually complex (Chapter 10, Section B,5), so the 4-electron dehydrogenation of two water molecules to give one molecule of O_2 in chloroplasts cannot be pictured in a simple way. It has been established from experiments on oxygen evolution in the presence of repeated short flashes of light that a 4-quantum process at PSII is required.[112a] There must be some way of storing oxidizing equivalents at this end of the chain until enough are present to snap together an oxygen molecule. No doubt the whole thing happens at a metalloenzyme center (see Box 13-A), perhaps between two different ions of Mn^{2+}. Each Mn^{2+} upon acquiring a radical as in Eq. 13-29 could undergo oxidation to Mn^{3+} permitting another light absorption event to generate another bound OH radical. The production of O_2 would be represented as in Eq. 13-30:

$$2[H_2O-Mn(II)]$$

$$\qquad (13\text{-}30)$$

Another aspect of electron transport at this end of the chain is the possibility that there is an energy conservation site for ATP formation.[113] If so,

BOX 13-A

MANGANESE

Tissues usually contain less than one part per million of manganese on a dry weight basis or less than 0.01 mM in fresh tissues (this compares with a total content in animal tissues of the more abundant Mg^{2+} of 10 mM). A somewhat higher content (3.5 ppm) is found in bone. Nevertheless, manganese is nutritionally essential[a,b] and its deficiency leads to well-defined symptoms including ovarian and testicular degeneration, shortening and bowing of legs, and various other skeletal abnormalities (including a "slipped tendon disease" in chicks). The organic matrix of bones and cartilage tends to be developed poorly. The galactosamine content of cartilage is decreased as is the content of hexuronic acids and chondroitin sulfates. Manganese is also essential for plant growth.

Manganese lies in the center of the first transition series of elements. The stable Mn^{2+} (manganous ion) contains five 3d electrons in a high-spin configuration. The less stable Mn^{3+} (manganic ion) appears to be of importance in some enzymes and possibly in the photosyn-

thetic evolution of oxygen (Eq. 13-30).

A surprising number of enzymes specifically require Mn^{2+}. These include galactosyl and N-acetylgalactosaminyltransferases[c] needed for synthesis of mucopolysaccharides (Chapter 12, Section C,1) and lactose synthetase (Eq. 12-11a). Pyruvate carboxylase (Eq. 8-2) contains four atoms of tightly bound Mn^{2+}, one for each biotin molecule present. The manganese ion is essential for the transcarboxylation step in the action of this enzyme while either Mn^{2+} of Mg^{2+} is also needed in the initial step of carboxylation of the biotin.

Manganese is a component of a wine-red superoxide dismutase (Eq. 8-61; see also Box 10-H) from $E.$ $coli$. This enzyme of MW 40,000 contains two atoms of Mn(III). Similar enzymes have been obtained from the mitochondria of chicken livers and of yeast. The yeast enzyme is a tetramer, each subunit of MW = 24,000 containing one atom of bound manganese.[d] A protein known as avimanganin appears to be an inactive form of the chicken enzyme. It is of interest that the cytoplasmic superoxide dismutases from the same sources are zinc-copper enzymes (Box 10-H).[e]

more steps would be required than are indicated in Eq. 13-30 and the process would be even more reminiscent of a cytochrome oxidase system acting in reverse. Although this O$_2$ end of the chain remains very much a mystery, the recent discovery of specific inhibitors is facilitating investigation. Thus, hydroxylamine appears to block oxidation of H$_2$O but to permit electron transport from artificial donors through both PSII and PSI.

Since incorporation of CO$_2$ via the Calvin cycle (Chapter 11, Section D,2) appears to account for the principal route to carbon-containing photosynthetic products, the reducing equivalents must come from cleavage of six water molecules with release of O$_2$. Otherwise Eq. 13-25 could not be balanced. Nevertheless, some experiments suggested that bicarbonate ion is the immediate source of the oxygen atoms for formation of O$_2$.[114] More recent experiments show that ^{18}O from bicarbonate

is not incorporated into O$_2$ but that bicarbonate can stimulate oxygen evolution,[115] perhaps as an allosteric effector.

7. Photoreactions of Carotenoid Pigments

It is a striking fact that there are no naturally occurring green plants that lack carotenoid pigments.[116] Although carotenoidless mutants are used in photosynthesis research, they apparently cannot survive under natural conditions. Carotenoids confer protection to chlorophyll against light-synthesized destruction by molecular oxygen. The mechanism of the protection is not understood.

A second function of carotenoids (already men-

Box 13-A (*Continued*)

The manganese ions in superoxide dismutases are presumed to alternate between the III and II states during catalysis of the reaction of Eq. 8-61. The same may be true for the manganese-containing protein or proteins of chloroplasts (Eq. 13-30). Another Mn^{2+}-containing protein is concanavalin A (Fig. 5-7).

A large number of enzymes that require Mg^{2+} can utilize Mn^{2+} in its place, a fact that has been exploited by chemists in studying the active sites of enzymes.[f] The highly paramagnetic Mn^{2+} is the most favorable ion for epr studies (Box 5-B) and for investigations of paramagnetic relaxation of nmr signals (Chapter 7, Section E,7). Manganese can also replace Zn^{2+} in zinc-dependent enzymes, sometimes causing interesting changes in catalytic properties.[g]

An important function of manganese may be in the regulation of the activity of enzymes. For example, glutamine synthetase (Chapter 14, Section B,2) in one form requires Mg^{2+} for activity but upon adenylylation binds Mn^{2+} tightly.[h] Many nucleases and DNA polymerases show altered specificity when Mn^{2+} substitutes for Mg^{2+}. The significance of these differences *in vivo* is obscure but they should not be overlooked.

A striking accumulation of Mn^{2+} occurs within bacterial spores (Chapter 16, Section C,1). *Bacillus subtilus* has an absolute requirement for Mn^{2+} for initiation of sporulation. During logarithmic growth the bacteria can concentrate Mn^{2+} from 1 μM in the external medium to 0.2 mM internally; during sporulation the concentrations become much higher.[i]

[a] B. L. O'Dell and B. J. Campbell, *Compr. Biochem.* **21**, 191–203 (1971).

[b] R. M. Leach, Jr., *in* "Trace Element Metabolism in Animals-2" (W. G. Hoekstra *et al.*, eds.), pp. 51–59. Univ. Park Press, Baltimore, Maryland, 1974.

[c] A. P. Baker, L. J. Griggs, J. R. Munro, and J. A. Finkelstein, *JBC* **248**, 880–883 (1973).

[d] J. J. Villafranca, F. J. Yost, Jr., and I. Fridovich, *JBC* **249**, 3532–3536 (1974).

[e] S. D. Ravindranath and I. Fridovich, *JBC* **250**, 6107–6112 (1975).

[f] A. S. Mildvan, *Annu. Rev. Biochem.* **43**, 357–399 (1974).

[g] P. H. Haffner, F. Goodsaid-Zalduondo, and J. E. Coleman, *JBC* **249**, 6693–6695 (1974).

[h] J. J. Villafranca and F. C. Wedler, *Biochemistry* **13**, 3286–3291 (1974).

[i] T. F. Deuel and S. Prusiner, *JBC* **249**, 257–264 (1974).

tioned) is that of accessory light-gathering pigments. A third function is suggested by the observation of light-induced reductive deoxygenation of epoxycarotenoids (Eq. 13-31, step *a*)[117,118]:

$$(13\text{-}31)$$

Epoxycarotenoids are found only in photosynthetic O_2 evolving organisms. Their photodeoxygenation and reoxygenation in a nonphoto-chemical process (Eq. 13-31, step *b*) constitutes the **violaxanthin cycle.** Violaxanthin contains the epoxy structure at both ends of the molecule. Photoreduction of one end produces antheroxanthin and of both ends zeaxanthin. These three carotenoids are found in almost all higher green plants and algae. Relatively little is known about either the enzymology or the biological significance of the violaxanthin cycle. It may have a control function. The de-epoxidation (Eq. 13-31, step *a*) appears to be mediated by ascorbic acid and to occur in the loculus (inner space) of the thylakoids. The reaction is favored by the low pH developed during illumination of the chloroplasts. Epoxidation (Eq. 13-31, step *b*) is catalyzed by an "external" monooxygenase located on the stromal side. The result is a transmembrane cycle depending upon movement of the carotenoids through the membrane.[118]

8. Control of Photosynthesis

The key reaction of the Calvin cycle of CO_2 reduction is the carboxylation of ribulose diphosphate (Eq. 7-81). The carboxylase is allosterically activated by fructose 6-phosphate, but is inhibited by fructose 1,6-diphosphate (Fig. 13-24).[119] Thus, the accumulation of fructose 1,6-diphosphate is a signal to turn off the carboxylation process, while the appearance of fructose 6-phosphate in high concentrations serves to turn on the Calvin cycle. From this it is apparent that, like the reactions of gluconeogenesis (Chapter 11, Section F,5), the operation of the Calvin cycle is critically dependent on the highly regulated fructose-1,6-diphosphatase. The latter enzyme, in chloroplasts, is activated by light through the mediation of reduced ferredoxin. The latter, together with a "protein factor," converts (presumably by reduction of some group in the enzyme) inactive fructose diphosphatase to the active form (Fig. 13-24).[119] Several other enzymes of the Calvin cycle are also activated by light[120] and the important glyceraldehyde-3-phosphate dehydrogenase appears to undergo a change in specificity from NADH to NADPH under the influence of light.[121] Another aspect of the regulatory process is the activation of synthesis of ADP-glucose from glucose 1-phosphate by 3-phosphoglycerate, a "feed-ahead" type of regulation (Fig. 13-24). Additional aspects of the control of photosynthesis are discussed by Bassham.[122]

9. Special Adaptations

The first product of incorporation of CO_2 into the Calvin cycle is 3-phosphoglyceric acid (Chapter 11, Section D,2). It was the rapid appearance of radioactivity from $^{14}CO_2$ in phosphoglycerate and other C_3 compounds that permitted Calvin and associates to work out the complex cycle as it is shown in Fig. 11-4. Plants such as the green algae (with which Calvin worked), spinach, and other common crop plants are often known as C_3 **plants.** Another group of plants, mostly of trop-ical origin and capable of extremely fast growth (e.g., sugar cane, corn, and crabgrass), behave differently.[123–127] Radioactive CO_2 is found first in the C_4 compounds oxaloacetate, malate, and aspartate. The C_4 **plants** are characterized by high efficiencies in photosynthesis, the explanation for the rapid growth of crabgrass and the high yield of corn. Maximum rates of CO_2 incorporation for these plants may attain 40–60 mg of CO_2 per square decimeter of leaf surface per hour (~0.3 mmol CO_2 $m^{-1}s^{-1}$ or ~0.10 mol CO_2 per mol of total chlorophyll per second), more than twice that for common crop plants.

a. The Low Photorespiration of C_4 Plants

Like all other organisms, plants respire in the dark, but illumination of C_3 plants markedly increases the rate of oxygen utilization. This light-enhanced respiration (**photorespiration**) may attain 50% of the net rate of photosynthesis. Photorespiration prevents plants from achieving a maximum yield in photosynthesis. For this reason, its understanding and control assume great importance in agriculture. The possibility of breeding plants with lower photorespiration rates or inhibiting glycolate synthesis has been suggested.[126,127]

The rate of photorespiration is difficult to measure. For this reason the literature on the subject often refers instead to the **CO_2 compensation point.**† This is the CO_2 concentration (at a given constant light intensity) at which photosynthetic assimilation and respiration balance. Normal air has a CO_2 content of ~0.03% or 300 ppm. For common C_3 crop plants, the CO_2 compensation point is ~40–60 ppm at 25°C. The C_4 plants are characterized by a much lower CO_2 compensation point, often less than 10 ppm. The significance is great for in strong sunlight, the CO_2 level of air in a field of growing plants drops. Furthermore, as the temperature rises on a hot day, the CO_2 compensation point rises. The result is a serious de-

† **The light compensation point** is the light intensity at which the rate of photosynthetic CO_2 incorporation and that of respiration exactly balance.

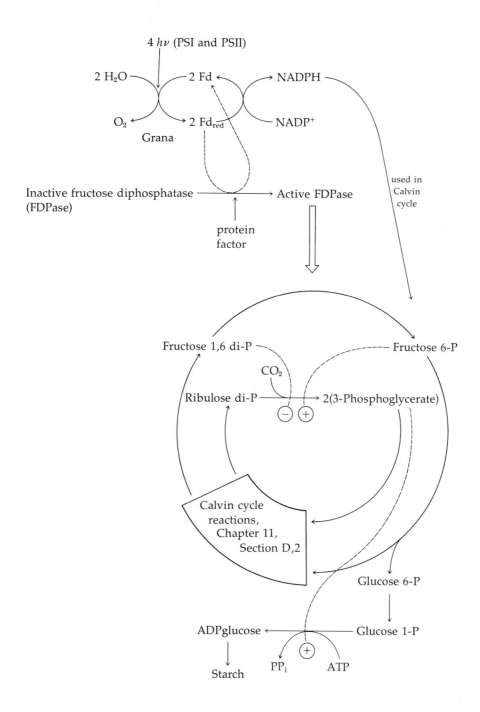

Fig. 13-24 Some control mechanisms for photosynthetic assimilation of carbon dioxide. After Buchanan and Schürmann.[119]

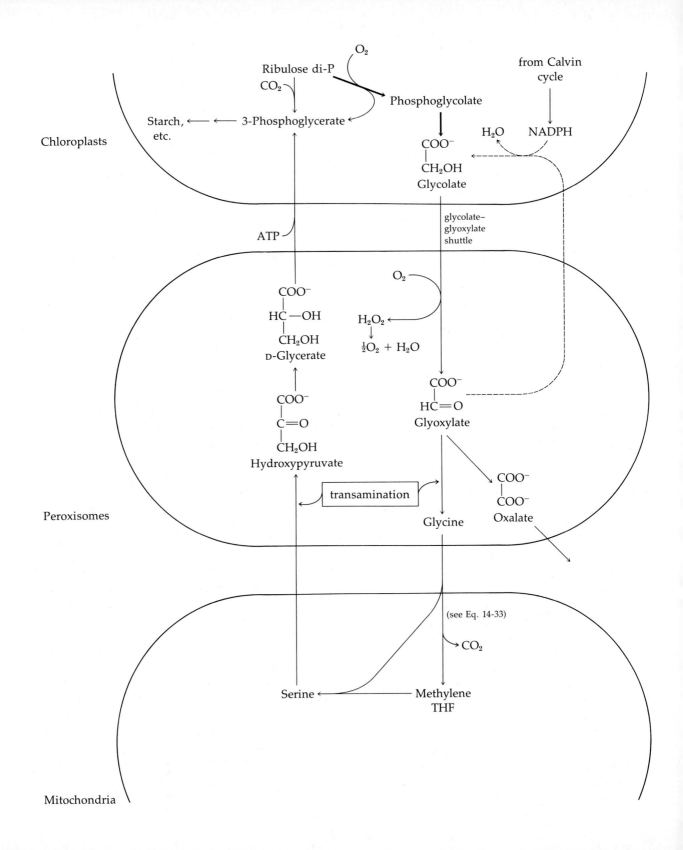

crease in efficiency of photosynthesis for the C_3 plants but not for the C_4 plants.

b. Association of Photorespiration with Metabolism of Glycolic Acid

The 2-carbon **glycolic acid** is formed in large quantities in the chloroplasts of C_3 plants and moves out into the cytosol.[128] One of the sources of glycolate is probably phosphoglycolate whose formation is catalyzed by ribulosediphosphate carboxylase of the chloroplasts (Eq. 7-82). Glycolic acid arises through competition of O_2 for the CO_2 binding site of the enzyme (it is easy to understand why an increase in the O_2 pressure in air increases the CO_2 compensation point for a plant). A second source of glycolate is transketolase which may yield glycolaldehyde as a side product (Eq. 8-15). Glycolaldehyde can be readily oxidized to glycolate. Other sources of glycolate may exist.

Glycolate has a very rapid metabolism, not in the chloroplasts, but in the peroxisomes (microbodies, Chapter 1, Section B,6). There a flavoprotein oxidase converts glycolate to glyoxylate with formation of H_2O_2 (Fig. 13-25).[129] While some of the hydrogen peroxide formed may react nonenzymatically to decarboxylate glyoxylate to formate and CO_2, most is probably destroyed by peroxidases or catalase (strangely, the latter enzyme is lacking in chloroplasts—one reason why oxidation of glycolate must occur in the microbodies). Glyoxylate undergoes transamination to glycine which can be decarboxylated (Fig. 14-32) in the mitochondria. It can also be converted to serine, some of which may return to the peroxisomes to be oxidized to hydroxypyruvic acid and glyceric acid (Fig. 13-25). The latter can be synthesized into glucose. The net result is the stimulation of a large amount of metabolism that ultimately produces CO_2 and apparently accounts for the light-induced respiration of plants. However, our understanding of the matter is still quite incomplete.

Fig. 13-25 Production of glycolate by chloroplasts and some pathways of its metabolism in peroxisomes and in mitochondria. After Tolbert.[129]

c. The C_4 Cycle for Concentration of Carbon Dioxide

In what way do the C_4 plants reduce photorespiration? While the incorporation of CO_2 into oxaloacetate first suggested that there might be an alternative to the Calvin cycle for CO_2 reduction, later studies have indicated that the secret of the C_4 plants lies in a CO_2 concentrating mechanism which enables them to avoid the competition from O_2. All species of C_4 plants have a characteristic internal leaf anatomy in which a single dense layer of dark green cells surrounds the vascular bundles in the leaves. This **bundle sheath** is surrounded by a loosely packed layer of cells, the **mesophyll**. The arrangement is sometimes called the "Kranz anatomy." In C_4 plants there is a compartmentation of chemical reactions between the mesophyll and the bundle sheath cells. The incorporation of CO_2 (actually bicarbonate ion) into oxaloacetate in the mesophyll cells is principally through the action of PEP carboxylase (Fig. 13-26). Oxaloacetate is reduced to malate by light-generated NADPH. Alternatively, it undergoes transamination to aspartate. Both malate and aspartate then diffuse out of the mesophyll cells and into bundle sheath cells. There the malate undergoes oxidative decarboxylation (malic enzyme, Eq. 7-75) to pyruvate (Fig. 13-26). Aspartate also can be converted to oxaloacetate, malate, and pyruvate in the same cells. Note that the effect is to transport CO_2 from the mesophyll cells into the bundle sheath cells along with two reducing equivalents, which appear as NADPH following the action of the malic enzyme. The CO_2, the NADPH, and additional NADPH generated in the chloroplasts of the bundle sheath cells are then used in the Calvin cycle reactions to synthesize 3-phosphoglycerate and other materials. Of the CO_2 used in the bundle sheath cells, it is estimated that 85% comes via the C_4 cycle and only 15% enters by direct diffusion. The advantage to the cell is a higher CO_2 tension, less competition with O_2, and a marked reduction in photorespiration.

The pyruvate produced in the bundle sheath cell is largely returned to the mesophyll cells where it is phosphorylated using **pyruvate-phosphate di-**

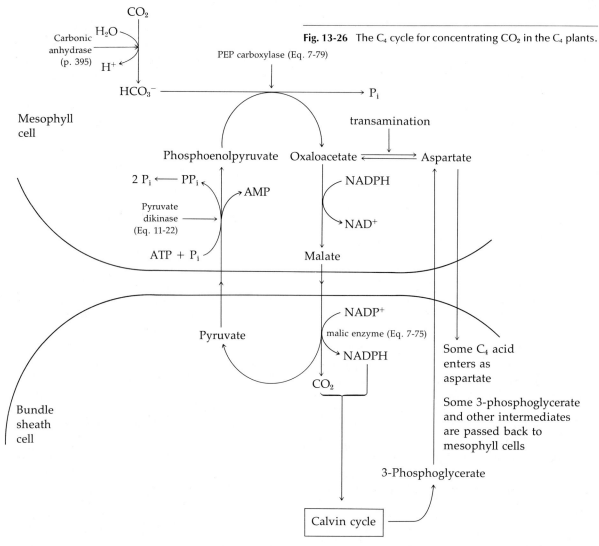

Mesophyll cell

CO₂

Carbonic anhydrase (p. 395)

H₂O

H⁺

HCO₃⁻

PEP carboxylase (Eq. 7-79)

Pᵢ

transamination

Phosphoenolpyruvate   Oxaloacetate ⇌ Aspartate

2 Pᵢ ⟵ PPᵢ

AMP

Pyruvate dikinase (Eq. 11-22)

ATP + Pᵢ

NADPH

NAD⁺

Malate

Pyruvate

NADP⁺

malic enzyme (Eq. 7-75)

NADPH

CO₂

Some C₄ acid enters as aspartate

Some 3-phosphoglycerate and other intermediates are passed back to mesophyll cells

Bundle sheath cell

3-Phosphoglycerate

Calvin cycle

kinase. This unusual enzyme (Eq. 11-22) splits ATP to AMP and PP_i, which is in turn degraded to P_i. In effect, two high energy linkages are needed for each molecule of pyruvate recycled. For this reason, it is thought that cyclic photophosphorylation is probably more important in the chloroplasts of the mesophyll cells than in the bundle sheath cells.

d. Metabolism in the Family Crassulaceae

The crassulacean plants are a large group that includes many ornamental succulents such as

Sedum. These plants have a remarkable metabolism by which they synthesize large amounts of malic and isocitric acids at night. During the day, when photosynthesis occurs, these acids disappear. It is also observed that the stomata (Chapter 1, Section E,4) in the leaves stay closed during the day and open only at night, an adaptation that permits the plants to live with little water. The problem for the plant is how to accumulate carbon dioxide by night and to incorporate it photosynthetically into organic compounds by day. A possible mechanism is shown in Fig. 13-27. On the left side of the figure are reactions by which starch can be broken down at night to phospho-enolpyruvic acid (PEP). While it would also be possible to produce that compound by glycolysis reactions, labeling studies have indicated that reactions of the pentose phosphate pathway are more important.[96] The PEP acts as the CO_2 acceptor to create oxaloacetate which is then reduced to malic acid. A balanced fermentation reaction (Fig. 13-27, Eq. 13-32) can be written by using the NADPH formed in the conversion of

glucose 6-P to ribulose 5-P. During the day, when ATP and NADPH are available in abundance from photoreactions, the conversions on the right side of the figure can take place. The initial step, the release of CO_2 from malic acid by the malic enzyme, is the same as that employed by C_4 plants. In this case, it is used to release the CO_2 stored by night, making it available for incorporation via the Calvin cycle. The remaining pyruvate is reconverted to starch.

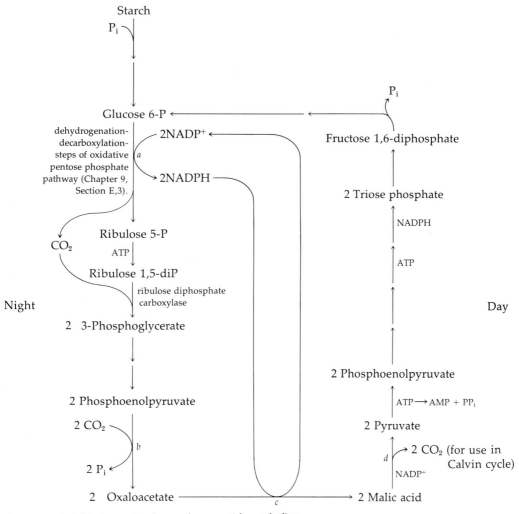

Fig. 13-27 A proposed night–day cycle of crassulacean acid metabolism.

$$C_6H_{10}O_5 \text{ (starch)} + 2\,CO_2 \text{ (g)} \longrightarrow$$
$$2\,C_4H_4O_5{}^{2-} \text{ (malate)} + 4\,H^+ \quad (13\text{-}32)$$
$$\Delta G' \text{ (pH 7)} = -159 \text{ kJ mol}^{-1}$$

It is of interest that many plants store substantial amounts of malate in their cytoplasm and in vacuoles. It apparently serves as a ready reserve for carbohydrate synthesis.

e. Direct Reduction of Formate

The direct reduction of CO_2 to formate by reduced ferredoxin (Eq. 13-33) has been proposed to

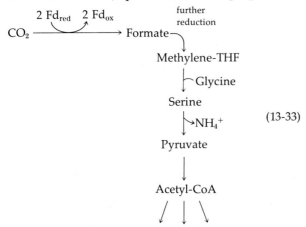

$$(13\text{-}33)$$

account for specific labeling of glutamate in the C_5 position by $^{14}CO_2$ in *Vicia fava*, a type of bean.[130] The label could be introduced as indicated using glycine formed from glycolate. Compare also with the cyclic process for formate incorporation shown in Fig. 11-5.

10. Photosynthetic Formation of Hydrogen

A system consisting of chloroplasts, ferredoxin, and hydrogenase has been used to generate H_2 photosynthetically.[131] It has been suggested that this may be a prototype of a method of solar energy generation for human use. Another photochemical hydrogen-generating system makes use of both the nitrogen-fixing heterocysts and photosynthetic vegetative cells of *Anabaena cylindrica*, a blue-green bacterium.[132] In this instance hydrogen production is accomplished via the nitrogenase system (Eq. 14-5).

Photogeneration of H_2 by bacteria is just one example of a variety of kinds of photometabolism observed among photosynthetic microorganisms.[132a] Another example is the light-induced uptake and conversion of acetate to poly-β-hydroxybutyrate by purple bacteria.

F. VISION

The light receptors of the eye perform a very different function from those of chloroplasts. Visual receptors are designed to initiate a nerve impulse and their primary requirement is a high sensitivity. Indeed, the most sensitive receptors are able to trap essentially every photon that strikes them. This is accomplished by the use of stacked membranes containing a high concentration of an intensely absorbing molecule.[133,133a]

The retina of the human eye contains over 10^8 tightly packed receptor cells of two types. The **rod cells** are extremely sensitive and may respond to as few as five quanta of light. Designed for nighttime vision, the rods give a "black and white picture" and are concentrated around the periphery of the retina. The less sensitive **cone cells,** which are most abundant in the center of the retina, are of three types with different spectral sensitivities. They provide color vision.

1. Structure of the Rod Outer Segment

The rods (Fig. 13-28), the most studied of the retinal receptors, are cells with a very active metabolism. Human rod cells may live and function for a hundred years.[134] A remarkable self-renewal process leads to a casting off of the older membranous discs from the end of the rod[135] and replacement by new discs at the end nearest to the nucleus. The rod outer segment is surrounded by a plasma membrane. Within the membrane, but apparently not attached to it are ~500 stacked discs of ~2 μm diameter and with a repeat dis-

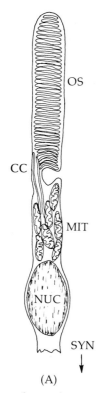

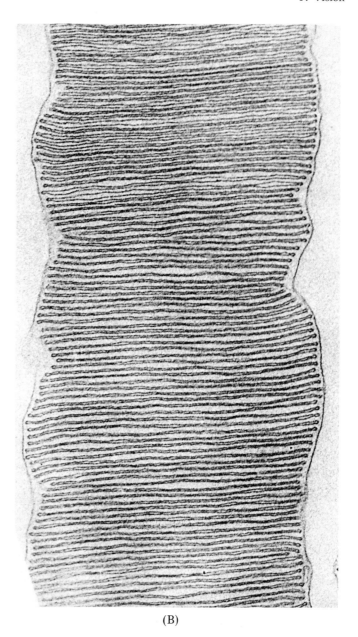

Fig. 13-28 (A) Diagram of a vertebrate rod cell. OS, outer segment; CC, connecting cilium; MIT, densely packed mitochondria; NUC, nucleus; and SYN, synaptic terminal. From Abrahamson and Fager.[133] (B) Electron micrograph of a longitudinal section of the outer segment of a retinal rod of a rat. Courtesy of John E. Dowling.[133b]

(A)

(B)

tance between centers of ~32 nm. Each disc is enclosed by a pair of membranes each ~7 nm thick with a very narrow space between them. From electron micrographs it appears that this space within the discs is sealed off at the edges. A somewhat larger space separates adjacent discs.

The membranes of the rod discs are ~60% protein and 40% lipid (Table 5-1). About 80% of the protein is **rhodopsin** (visual purple), a lipoprotein that is insoluble in water but soluble in detergent solutions. Digitonin is widely used to disperse rhodopsin molecules because it causes no change in the optical properties of the latter. Mammalian rhodopsin has a molecular weight of ~28,000–35,000, and each molecule contains a

single chromophore with an absorption maximum at 500 nm (Fig. 13-29). Since it does constitute the bulk of the membrane protein, it is clear that the rhodopsin molecules are closely packed one against the other. Freeze etching of fractured rods reveals the presence of round particles of ~4–5

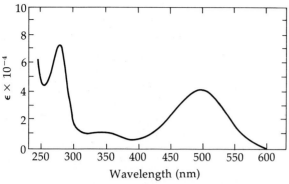

Fig. 13-29 Absorption spectrum of cattle rhodopsin in aqueous dispersion with a nonionic detergent. From H. Shichi et al., *JBC* **244**, 529–536 (1969).

11-*cis*-Retinal

all-*trans*-Retinal   (13-34)

nm diameter, somewhat too large to be individual rhodopsin molecules unless each one is separated from adjacent molecules by lipid. The energy transfer experiments of Wu and Stryer (Section C,2) suggest that rhodopsin is an elongated molecule of maximum dimension 7.2 nm and able to extend completely through the membrane. One suggestion is that there is a large approximately spherical head on the inside of the membrane and a thin tail extending through it.[136] The chromophore, which is attached to or embedded in the rhodopsin, lies with its transition dipole moment parallel to the plane of the discs (i.e., perpendicular to the direction of travel of the incoming photons).

2. The Chemical Nature of the Visual Chromophores[137-139]

Rhodopsin contains bound 11-*cis*-retinal and it is believed that the photochemical event of light absorption induces isomerization (Eq. 13-34) to all-*trans*-retinal. The latter dissociates from the protein leaving the apoprotein **opsin**. Opsin recombines spontaneously with 11-*cis*-retinal to reform rhodopsin.

Other closely related visual pigments are known. The combinations of cone opsins with 11-*cis*-retinal are known as **iodopsins** and usually absorb at somewhat lower energies than does rhodopsin.† In a few freshwater marine species the visual pigments (**porphyropsins**) contain **3-dehydroretinal**. The peak positions of light absorption depend both upon the nature of the bound aldehyde and on the protein, the latter having the larger effect (Table 13-3). Thus, retinal-based pigments absorb in the entire range 467–528 nm (18,900–21,400 cm^{-1}).

When native rhodopsin is treated with sodium borohydride, little reduction is observed, but after the protein is bleached by light reduction becomes rapid and the retinal is incorporated into a secondary amine. This suggests that in rhodopsin the retinal is combined as a Schiff base. Conflicting results have been obtained with respect to the identity of the amino group involved. Some experiments suggest an ε-amino group of the opsin and others an amino group of phosphatidylethanolamine. Recently, a slow reduction of unbleached rhodopsin with cyanoborohydride has given a single product suggesting a Schiff base with lysine as the native form of the photopigment.[133] Studies of model systems indicate that the large bathochromic shift between the spectra

† However, color vision in primates is apparently dependent on three pigments with absorption maxima at 447 nm (blue-violet), 540 nm (green), and 577 nm (yellow): E. F. MacNichol, Jr., *Sci. Am.* **211**, 48–56 (Dec. 1964). Note that one of these absorbs at a higher energy than does rhodopsin.

TABLE 13-3
Positions of Major Absorption Band for Retinals and Visual Pigments

	Free aldehyde	Visual pigment	
Retinal (all-trans)	387 nm 25,800 cm^{-1}		
Retinal (11-cis)	376 nm 26,600 cm^{-1}	Rhodopsin	500 nm 20,000 cm^{-1}
		Iodopsin (cones)	562 nm 17,800 cm^{-1}
3-Dehydro-retinal (11-cis)	393 nm 25,400 cm^{-1}	Porphyropsin	522 nm 19,200 cm^{-1}

of free retinal and the visual pigments requires a protonated Schiff base structure as well as a strong interaction between the polyene chain of the retinal and the protein.

3. The Light-Induced Transformation

When light strikes rhodopsin a sequence of readily detectable spectral changes is observed.[133,140] The relaxation times indicated in Eq. 13-35 are for 20°C.

Rhodopsin (500 nm, 20,000 cm^{-1})

$h\nu \downarrow \tau < 6 \times 10^{-12}$ s

Prelumirhodopsin (bathorhodopsin)
    (545 nm, 18,300 cm^{-1})

$\downarrow \tau \approx 50 \times 10^{-9}$ s

Lumirhodopsin (497 nm, 20,100 cm^{-1})

$\downarrow \tau \approx 50 \times 10^{-6}$ s

Metarhodopsin I (480 nm, 20,800 cm^{-1})

$\downarrow$

Metarhodopsin II (380 nm, 26,300 cm^{-1})

$\downarrow$

Metarhodopsin (?) (465 nm, 21,500 cm^{-1})

$\downarrow$

trans-Retinal (387 nm, 25,800 cm^{-1}) + opsin   (13-35)

The first step, which has been observed using laser flash photolysis and picosecond spectroscopy,[141] occurs within 6×10^{-12} s (6 ps). The absorption maximum of the product **bathorhodopsin** (prelumirhodopsin) is shifted bathochromically. This suggests an increase in conjugation such as might be accomplished through the isomerization of Eq. 13-34. The time is so short that such an isomerization seems unlikely, but rapid formation of a strained all-trans-retinal might be possible.[141a] Perhaps a simple transfer of charge with formation of a carbonium ion (Eq. 13-36) should be considered†:

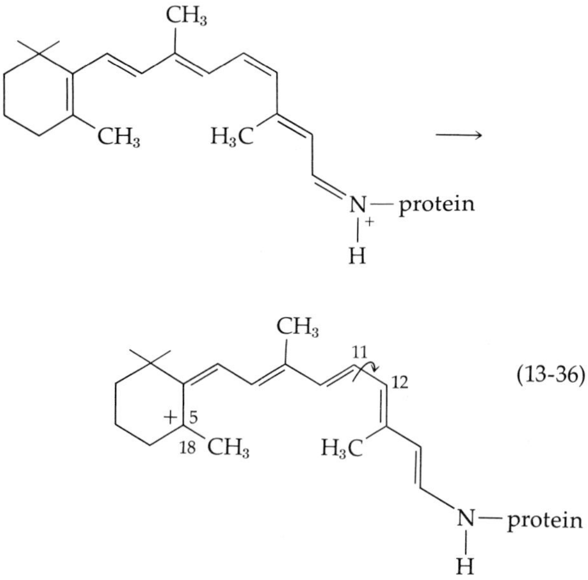

(13-36)

A slower rotation about the single bond could be followed by reversion to the all-trans isomer. Another possibility is that a proton is lost at C-18 (as if from the carbonium ion of Eq. 13-36) to yield a compound with an additional double bond between C-5 and C-18. The resulting structure has been suggested for bathorhodopsin.[141c] Still other possibilities include formation of a diradical (triplet)[142] or of a charge transfer complex with a tryptophan side chain.[143]

† Recently Mathes and Stryer have provided direct evidence for charge migration of the type indicated in Eq. 13-36 in the first singlet excited state of Schiff bases of retinal.[141b]

At present the exact nature of this first photochemical step, as well as that of the subsequent steps which occur in the dark, is not known.[141,144] The reaction sequence of Eq. 13-35 can be stopped at various stages by lowering the temperature. An additional component may be added to the sequence under some conditions. Thus, at 7°K a form called **hypsorhodopsin,** absorbing at 437 nm ($22,900$ cm^{-1}) is the first product observed. A great deal of interest in the transformation **metarhodopsin I** to **metarhodopsin II** exists because this is the slowest step that could trigger a nerve impulse (which must travel the length of the rod to the synapse in about 1 ms). There are also indications that conformational changes may accompany this step. The details of the subsequent steps, which lead to the release of *trans*-retinal, are the subject of controversy. However, they are too slow to be of significance in triggering a nerve impulse.

How can the sequence of transformations indicated by Eq. 13-35 initiate a nerve signal? A simple answer is that the conformational change within the retinal molecule during isomerization from 11-*cis*-retinal to all-*trans*-retinal (Eq. 13-34) could induce a change in conformation of the protein. This could lead to the acquisition of some enzymatic activity by the protein. Thus, metarhodopsin II could be an enzyme that initiates a cascade of chemical transformations culminating in the nerve impulse. However, no experimental support for this theory has been found. Another possibility is that an induced conformational change causes the opening of a pore through the disc membrane permitting some material on the inside to diffuse out. This theory is currently favored with Ca^{2+} the probable substance released. The distances from the disc membranes to the plasma membrane of the rod are such that diffusion could lead to an effect on the plasma membrane (where the nerve signal originates) within the necessary time.

Since rhodopsin constitutes 80% or more of the protein in the disc membranes, it may be rhodopsin itself that functions as a light-controlled "gate."[145] Possibly each rhodopsin molecule has a channel through the center, or along the axis of an oligomeric aggregate. This possibility is supported by the observation that two-thirds of the peptide hydrogens in the disc membranes appear to be hydrogen bonded to solvent water (as judged by hydrogen-tritium exchange rates).[146] This finding is especially remarkable in view of the hydrophobic properties of rhodopsin and the fact that typical globular proteins have only about one-third of rapidly exchangeable peptide group protons. These observations could be explained if many of the peptide group protons of rhodopsin line a water-filled channel, with a light-controlled gate operated by the embedded retinal-Schiff base.[146]

An alternative is that some cooperative behavior permits a signal from one rhodopsin molecule to activate a gate controlled by another protein some distance away. It is even possible that cooperativity is extensive enough that a signal can be physically propagated along the disc membranes to the outer edges and that some chemical signal is released there—close to the plasma membrane.

It is well known that the sensitivity of the retina falls rapidly when the overall level of illumination is increased. This may be a result of phosphorylation of a site on the rhodopsin molecule by an **opsin kinase** which acts specifically on bleached rhodopsin. The phosphorylated rhodopsin is apparently less permeable to Ca^{2+} upon illumination than is normal rhodopsin.[147] In addition, bleaching of rhodopsin causes activation of a phosphodiesterase that is especially effective in hydrolyzing cyclic GMP.[148] Thus, illumination could cause many secondary alterations stemming from a decreased level of cGMP.

4. The Nerve Impulse[134]

The essential consequence of light absorption from the viewpoint of nerve action is an alteration in the membrane potential in the vicinity of the absorbed photon with the resulting propagation of an impulse down the plasma membrane to the synapse by cable conduction (Chapter 5, Section B,3). The type of potential change that is transmitted differs among vertebrates and invertebrates. In the case of mammalian photoreceptors, the rod outer segment is surprisingly permeable to sodium ions so that a large "dark current" of sodium ions flows in through the plasma membrane and is pumped out by the sodium pumps. The effect of visual stimulation is for this permeability to Na$^+$ to be decreased with an increase in polarization of the membrane. Hagins suggested that calcium ions are a logical internal messenger because they can effectively block the sodium pores and cause the observed hyperpolarization.[134] He estimated that at least 10–20 calcium ions, probably more, would have to be released for each photon absorbed. Thus, if light absorption opens channels from the internal space of the rod discs, calcium ions may be released to quickly diffuse to the plasma membrane and block the entrance of sodium ions.

5. Regeneration of Visual Pigments

Since all-*trans*-retinal is released from photobleached pigments, there must be a mechanism of generating 11-*cis*-retinal for the regeneration of these pigments. On the one hand, new 11-*cis*-retinol is continuously brought in from the bloodstream and oxidized to retinal. Thus, the isomerization can occur in other parts of the body. However, there is good evidence that much of the isomerization takes place in the retina. In the case of cephalopods the inner segments of the receptor cells contain a second pigment **retinochrome** that carries out the reverse transformation converting all-*trans*-retinal to 11-*cis*-retinal.[149] There is evidence that in the mammalian retina phosphatidylethanolamine may form a Schiff base with all-*trans*-retinal and promote a light-induced transformation back to the 11-cis form.[150]

6. Bacteriorhodopsin

The salt-loving *Halobacterium halobium* under certain conditions forms a rhodopsinlike protein which it inserts into patches of **purple membrane** in the surface of the cell. These membranes, which may constitute up to 50% of the cell surface, apparently serve as light-operated proton pumps to translocate protons from the inside to the outside of the cells.[151–154] In this manner they may provide energy for a variety of cell functions including ion transport and ATP synthesis. The retinal-containing protein, of MW = 26,000, makes up 75% of the mass of the membrane. It is a globular protein 4.5 nm long. Each peptide chain is coiled into seven closely packed helices extending parallel to the long axes of the molecule and roughly perpendicular to the plane of the membrane. The spaces between the protein molecules are filled by lipid bilayer.[154]

It has been reported that the chromophore, which absorbs at ~560 nm, undergoes transient bleaching with a shift to 415 nm. The bleaching is followed by a rapid return to the initial state with a proton translocation accompanying the process. While it is hard to correlate this behavior with that of rhodopsin of visual receptors, it is possible to imagine how a light-promoted charge shift such as that of Eq. 13-36 could be linked to the release of a proton from a protein surface.

G. OTHER LIGHT RESPONSES

The list of ways in which organisms respond to light is large.[79] **Phototaxis,** the ability of organisms to move toward a source of light or to orient themselves with respect to a source of light, is found in species from bacteria to higher plants. In the latter the chloroplasts assume an orientation that maximizes efficiency of light absorption. Plants grow toward light **(phototropism)** and some organisms avoid light. The greening of plants in light depends upon photochemical reactions as does the tanning of human skin by light. The daily cycles of organisms are often established by light, and the germination and flowering of many plant species are also light-dependent.

1. Phytochrome

In 1951 it was discovered that a flash of red light (maximum activity at 660 nm) during an otherwise dark period promoted a variety of responses in plants.[155,155a] These included flowering, germination of seeds (e.g., those of lettuce), and the expansion of leaves in dark-grown pea seedlings. A most interesting fact was that the effect of the short flash of red light could be *completely reversed* if followed by a flash of *far-red light* (730 nm). This discovery led to the isolation in 1959 of the chromoprotein **phytochrome,** a kind of molecular switch that initiates a whole series of far-reaching effects in plants.[155–157] The phototransformation is completely reversible (Eq. 13-37), and the switch

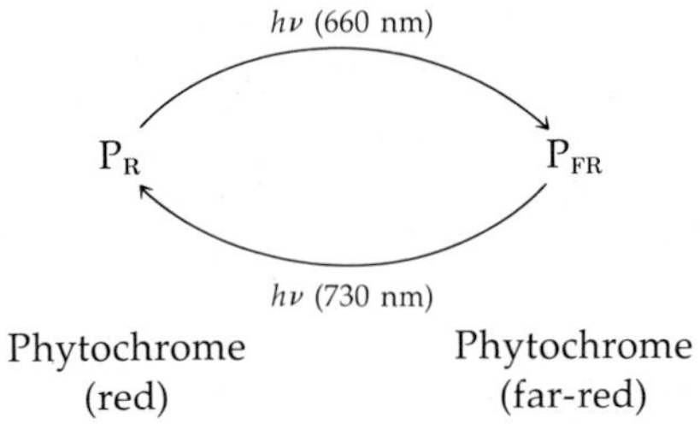

(13-37)

can be thrown in one direction or the other many times in rapid succession by light flashes.

Phytochrome has a molecular weight of ~120,000 but can be degraded by proteolytic enzymes into smaller photoactive fragments. The chromophore has been identified as an open tetrapyrrole related to phycocyanobilin. A proposed structure is shown in Fig. 13-22. However, the exact structure and the conformation of the tetrapyrrole is unknown. If the structure shown is correct for the red-sensitive form of phytochrome, the following transformation (Eq. 13-38) might give rise to the far-red-sensitive chromophore.

The increased conjugation accompanying the

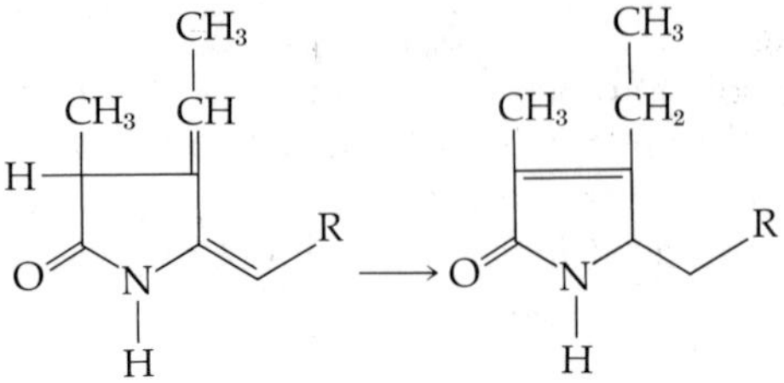

(13-38)

photoinduced isomerization would account for the shift in peak position. However, it has been shown that there are intermediates in the transformations of Eq. 13-37. In the P_R to P_{FR} direction a bleached form P_{BL} is observed and another intermediate (A) is seen in the opposite transformation (Eq. 13-39).

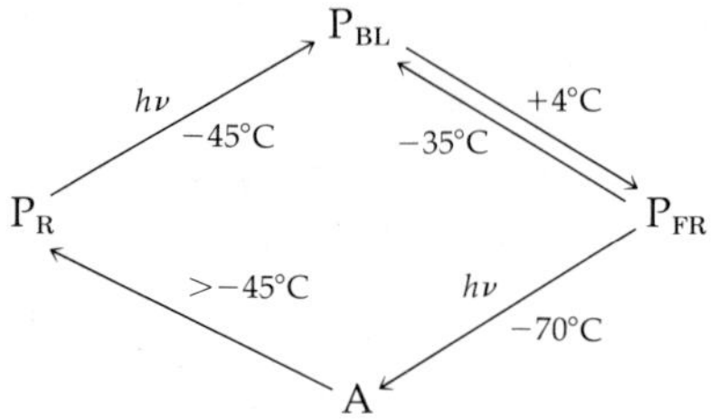

(13-39)

It has been argued that the bleached intermediate can result only if the cis–trans isomerization accompanies the reaction; possible structures have been proposed.[158] Another possibility is that no drastic overall conformational change takes place but that the open tetrapyrrole may undergo photoinduced rotation about a single bond to form P_{BL}.

Regardless of the chemistry behind the changes in the phytochrome chromophore, the principal question (as in the case of rhodopsin) is "How are the biological responses triggered?" As with rhodopsin, we can assume that the photoreaction induces a drastic conformational or chemical alteration in the protein and that this in turn leads to the varied physiological responses. The *slow responses* to phytochrome may result from an effect on gene transcription.

However, one response under phytochrome control is the closing of leaflets of *Mimosa* at the onset of darkness. The response occurs within 5 min, too short a time to be the result of transcriptional control. This and the finding that some phytochrome is tightly bound to membranes have led to the proposal that the primary effect of phytochrome is to alter membrane properties. It is not certain whether it is P_R or P_{FR} that is active in causing a response, but P_{FR} seems to be the most likely candidate for the "active" form. According to a recent suggestion, phytochrome in plastid membranes may mediate the release of gibberelins stored within the plastids.[158a]

2. Blue Light Responses

Numerous biological responses to light of wavelength 400–500 nm are known. These include phototropism in higher plants and the phototaxis of *Euglena*. On the basis of action spectra (p. 762, footnote) both carotenoids and flavins have been proposed as photoreceptors. The arguments have been summarized by Song and Moore who concluded that flavins are the most likely candidates.[159] Photochemical breakdown of flavins to lumichrome within plants has been observed.[160] A flavin-mediated photoreduction of a *b*-type cytochrome has been proposed as the event that initiates blue light responses in the fungus *Phycomyces*.[161] Recent experiments[161a,b] strengthen the view that a flavin acts as the photoreceptor in this organism.

H. BIOLUMINESCENCE

The emission of visible light by living beings is one of the most fascinating of natural phenomena. Luminescent bacteria, glowing toadstools, protozoa that can light up ocean waves, luminous clams, fantastically illuminated worms, and the amazing firefly have all been the objects of the biochemists' curiosity.[162–164] The chemical problem is an interesting one. The firefly's light with a wavelength of ~560 nm (17,900 cm^{-1}) has an energy of 214 kJ/einstein. A natural question is "What kind of chemical reaction can lead to an energy yield that high?" It is far too much energy to be provided by the splitting of ATP. Even the oxidation of NADH by oxygen would barely provide the necessary energy.

A clue comes from the fact that chemiluminescence is very common when O_2 is used as an oxidant in nonenzymatic processes. The slow oxidation of alcohols, aldehydes, and many nitrogen compounds is accompanied by emission of light visible to the eye. Chemiluminescence is especially pronounced in those reactions that are thought to occur by radical mechanisms. The recombination of free radicals provides enough energy to permit the release of visible light.

In view of these facts it is perhaps not so surprising that a great variety of organisms have mastered the ability to channel the energy released in an oxygenation reaction into light emission.

Attempts to extract luminous materials from organisms date from the last century when the French physiologist, R. DuBois, in 1887 prepared both a cold water extract and a hot water extract of luminous clams.[162] He was able to show that the material in the cold water extract, which he named **luciferase,** caused emission of light when a heat-stable material present in hot water extract was added. DuBois called the heat-stable material **luciferin.** These names have been retained and are now used in a general way. Thus, the luciferins are a family of compounds whose structures have been determined for a number of species (Fig. 13-30).†

Firefly luciferin is a carboxylic acid, but it must be activated in an ATP-requiring reaction to give **luciferyl adenylate,** whose structure is shown in the figure. The latter emits light in the presence of O_2 and luciferase. It can be seen that the original carboxyl group becomes CO_2, while the ring becomes oxidized. In addition, the acyl adenylate linkage is broken. In the case of the "sea pansy" *Renilla reniformis* (a coelenterate), the luciferin has quite a different structure.[165] However, the reaction with O_2 to produce CO_2 and an oxidized product causes the light emission, just as in the firefly. In *Renilla* the luciferin is stored as a luciferyl sulfate, possibly having the structure shown. To convert this storage form to the active luciferin the sulfuryl group is transferred onto adenosine 3',5'-diphosphate to form 3'-phosphoadenosine 5'-phosphosulfate, the reverse of Eq. 11-4, step *d*.

Another interesting aspect of *Renilla* bioluminescence is the occurrence of a special green fluorescent protein in the light-emitting cells. When this protein is absent the luminescence is blue with a maximum at 20,500 cm^{-1} (488 nm). In the presence of the protein the emission is in

† The reader should be aware that luciferins have been characterized using minute amounts of material. A number of errors will be found in older literature.

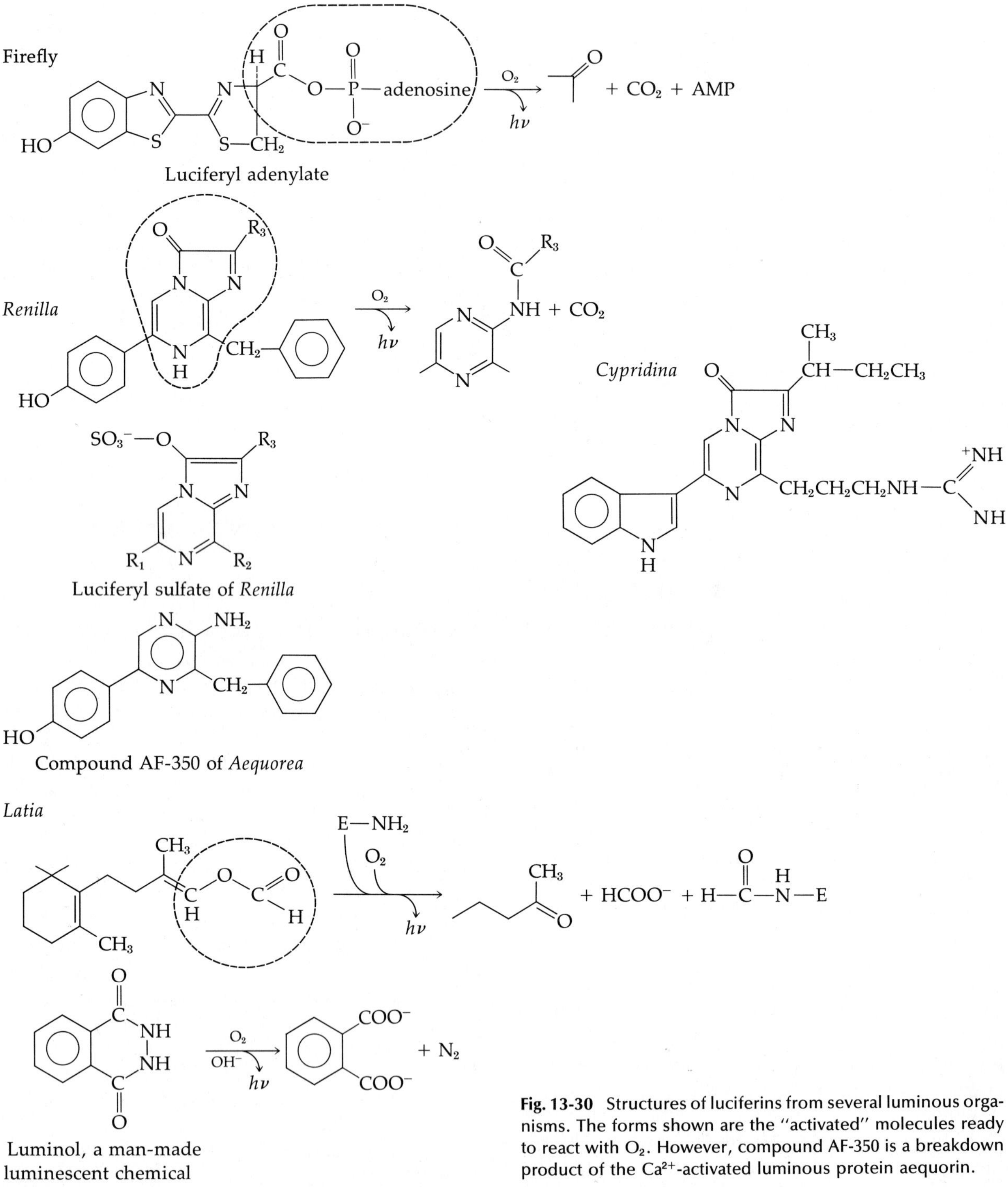

Fig. 13-30 Structures of luciferins from several luminous organisms. The forms shown are the "activated" molecules ready to react with O_2. However, compound AF-350 is a breakdown product of the Ca^{2+}-activated luminous protein aequorin.

a narrow band with a maximum at 19,600 cm^{-1} (509 nm). Apparently, efficient energy transfer between the two chromophores occurs.[164]

The luciferin of the ostracod crustacean *Cypridina* has a structure very close to that from *Renilla*. In *Cypridina* the luciferin and luciferase are produced in separate glands and are secreted into the surrounding water where they mix and produce light. A very different light-producing reaction is used by the limpet *Latia*. The luciferin is an unusual terpene derivative that lacks any chromophore suitable for light emission.[166] Evidently oxidation of this luciferin causes electronic excitation of some other molecule, presumably a "purple protein" which is also needed for luminescence. A complex of luciferin plus the purple protein is believed to react with the luciferase (abbreviated E–NH$_2$ in Fig. 13-30). It is thought that the formyl group is released from its enolic ester linkage in the luciferin. A Schiff base of the resulting aldehyde is thought to form with the enzyme and to react with oxygen (Eq. 13-40):

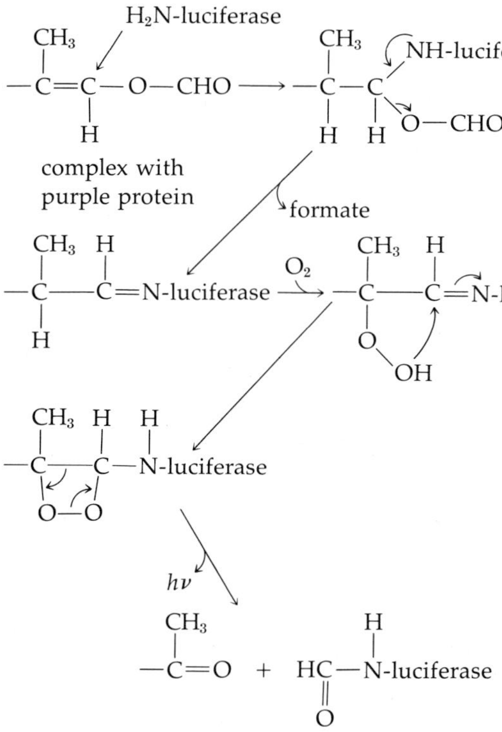

In most mechanisms suggested for luciferase action it is proposed that O$_2$ reacts at the carbon atom that becomes the carbonyl group in the product. In the case of *Renilla* luciferin this can easily be visualized as a result of electron flow from the pyrazine nitrogen (at the bottom of the structure in Fig. 13-30) into the O$_2$. In the case of firefly luciferin, the proton on the carbon could be removed, making use of the electron accepting properties of the adjacent ring system before addition of the O$_2$. The reactions should be compared to those catalyzed by oxygenases, e.g., Eq. 10-50. According to one proposal, the peroxide group formed adds to the carbonyl (to RC$=$N— in the case of *Latia* luciferin) to form a four-membered dioxetane ring (Eqs. 13-40 and 13-41). The latter

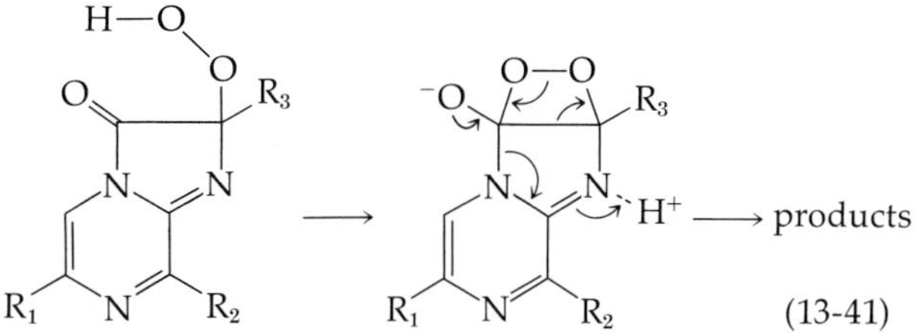

$$(13\text{-}41)$$

opens in a completely concerted process (as indicated by the arrows) to give the products.

This theory was tested using ^{18}O$_2$. In the case of *Cypridina* luciferin, the expected incorporation of one atom of ^{18}O into CO$_2$ was observed, but with firefly and *Renilla* luciferins no ^{18}O entered the CO$_2$. Thus, in these two cases, a somewhat different mechanism may hold. A possibility is that a hydroxyl ion adds to the carbonyl group in the hydroperoxide intermediate. Again a completely concerted breakup to products would be accompanied by light emission. The fact that two molecules of ^{18}O from H$_2$ ^{18}O enter the CO$_2$ supports the idea. Both oxygen atoms of CO$_2$ could become labeled in this way as a result of an exchange reaction (Eq. 13-42) of the adduct with the solvent.

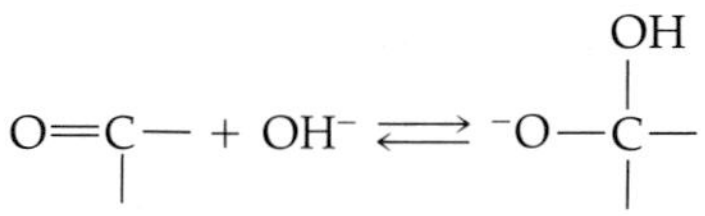

$$O{=}C{-} + OH^- \rightleftharpoons \ ^-O{-}\underset{|}{\overset{|}{C}}{-} \qquad (13\text{-}42)$$

Bioluminescent systems of the jellyfish *Aequorea* and related coelenterates have attracted much

$$(13\text{-}40)$$

interest.[167] *Aequorea* contains a **photoprotein** which emits light when calcium ions are present. Since light emission can be measured with great sensitivity (modern photomultipliers can be used to count light quanta) the protein **aequorin** and related photoproteins are now used as a sensitive indicator of calcium ion concentration. (In a similar way the firefly luciferin–luciferase system, which requires ATP for activation, is widely used in a sensitive assay for ATP.)

To identify the chromophore in aequorin over 4000 kg of jellyfish were used to obtain 125 mg of electrophoretically pure photoprotein.[169] From this 1 mg of a chromophoric substance (AF-350) was isolated and characterized as shown in Fig. 13-30. The close relationship to the *Renilla* and *Cypridina* luciferins is obvious. Presumably, in the intact aequorin the fused imidazole ring is added to AF-350. It is postulated that aequorin and the other photoproteins may contain a stabilized oxygen-containing intermediate such that no additional oxygen is needed to complete the reaction of Eq. 13-41 when Ca^{2+} acts to alter the conformation of the protein.[170] Energy transfer to other fluorescent proteins is also observed in these coelenterates.[171]

Quite different light-producing reactions occur in luminescent bacteria. Reduced riboflavin 5′-phosphate is oxidized by O_2. At the same time a long-chain aldehyde such as palmitaldehyde is required. It appears that oxidation of the aldehyde to a carboxylic acid (Eq. 13-43) (where FH_2 is riboflavin 5′-P) provides an essential part of the energy for the light emission.

$$FH_2 + R\text{—}CHO + O_2 \longrightarrow F + H_2O + R\text{—}COOH$$
$$(13\text{-}43)$$

The luminescent emission spectrum is the same as the fluorescence spectrum of the oxidized flavin ring. Evidence for an enzyme-bound reduced flavin hydroperoxide (as in Eq. 10-50) has been presented. While this hydroperoxide can decompose to flavin and H_2O_2, it can also carry out the oxidation of the aldehyde with emission of light.[172,173] An unidentified new flavin has been reported in some luciferases.[174]

REFERENCES

1. S. L. Miller and H. C. Urey, *Science* **130**, 245–251 (1959).
2. G. M. Woodwell, *Sci. Am.* **223**, 64–74 (Sep 1970).
3. E. I. Rabinowitz, "Photosynthesis and Related Processes" Vols. I and II, Wiley (Interscience), New York, 1945, 1951.
4. G. Wald, *Sci. Am.* **201**, 92–108 (Oct 1959).
5. J. G. Calvert and J. N. Pitts, Jr., "Photochemistry." Wiley, New York, 1966.
6. H. Suzuki, "Electronic Absorption Spectra and Geometry of Organic Molecules." Academic Press, New York, 1967.
7. J. N. Murrell, "The Theory of the Electronic Spectra of Organic Molecules." Wiley, New York, 1963.
8. G. Herzberg, "Molecular Spectra and Molecular Structure," 2nd ed., Vol. I. Van Nostrand-Reinhold, Princeton, New Jersey, 1950.
9. R. Chang, "Basic Principles of Spectroscopy." McGraw-Hill, New York, 1971.
10. F. S. Parker, "Applications of Infrared Spectroscopy in Biochemistry, Biology and Medicine." Plenum, New York, 1971.
11. R. D. B. Fraser and T. P. MacRae, "Conformation in Fibrous Proteins and Related Synthetic Polypeptides," pp. 95–106. Academic Press, New York, 1973.
12. M. J. Burke and M. A. Rougvie, *Biochemistry* **11**, 2435–2439 (1972).
13. T. Miyazawa and E. R. Blout, *JACS* **83**, 712–719 (1961).
14. S. Krimm and Y. Abe, *PNAS* **69**, 2788–2792 (1972).
15. W. H. Moore and S. Krimm, *PNAS* **72**, 4933–4935 (1975).
16. H. T. Miles, T. P. Lewis, E. D. Becker, and J. Frazier, *JBC* **248**, 1115–1117 (1973).
17. P. R. Carey, H. Schneider, and H. J. Bernstein, *BBRC* **47**, 588–595 (1972).
18. N. B. Colthup, L. H. Daly, and S. E. Wiberley, "Introduction to Infrared and Raman Spectroscopy," 2nd ed. Academic Press, New York, 1975.
18a. Y. Nishimura, A. Y. Hirakawa, M. Tsuboi, and S. Nishimura, *Nature* (*London*) **260**, 173–174 (1976).
19. T. G. Spiro, *Acc. Chem. Res.* **7**, 339–344 (1974).
20. T. G. Spiro, *BBA* **415**, 169–189 (1975).
20a. O. Siiman, N. M. Young, and P. R. Carey, *JACS* **98**, 744–748 (1976).
21. R. Mathies, A. R. Oseroff, and L. Stryer, *PNAS* **73**, 1–5 (1976).
22. J. Horwitz, E. H. Strickland, and C. Billups, *JACS* **91**, 184–190 (1969).
23. E. H. Strickland, M. Wilchek, J. Horwitz, and C. Billups, *JBC* **247**, 572–580 (1972).
24. D. B. Siano and D. E. Metzler, *J. Chem. Phys.* **51**, 1856–1861 (1969).

25. D. E. Metzler, C. M. Harris, R. J. Johnson, D. B. Siano, and J. A. Thomson, *Biochemistry* **12**, 5377–5392 (1973).

26. J. Horwitz, E. H. Strickland, and C. Billups, *JACS* **92**, 2119–2129 (1970).

27. J. Horwitz and E. H. Strickland, *JBC* **246**, 3749–3752 (1971).

28. S. W. Tang, J. E. Coleman, and Y. P. Myer, *JBC* **243**, 4286–4297 (1968).

29. T. E. Jones, D. B. Rorabacher, and L. A. Ochrymowycz, *JACS* **97**, 7485–7486 (1975).

30. H. H. Perkampies, I. Sandeman, and C. J. Timmons (eds.), UV Atlas of Organic Compounds,'' Vols. 1–5, Plenum, New York, 1966–1971.

31. L. Lang, ed., "Absorption Spectra in the Ultraviolet and Visible Region." Academic Press, New York, 1961; a serial publication numbering 18 volumes in 1973.

32. J. Petruska, *J. Chem. Phys.* **34**, 1120–1136 (1961).

33. P. E. Stevenson, *J. Mol. Spectrosc.* **15**, 220–256 (1965).

34. J. W. Donovan, *in* "Physical Principles and Techniques of Protein Chemistry" (S. J. Leach, ed.), Part A, pp. 101–170. Academic Press, New York, 1969.

35. E. H. Strickland, C. Billups, and E. Kay, *Biochemistry* **11**, 3657–3662 (1972).

36. W. Hug and I. Tinoco, Jr., *JACS* **95**, 2803–2813 (1973).

37. H. Harders, S. Förster, W. Voelter, and A. Bacher, *Biochemistry* **13**, 3360–3364 (1974).

38. J. N. Murrell, "The Theory of the Electronic Spectra of Organic Molecules," Chapter 7. Wiley, New York, 1963.

39. K. D. Collins and G. R. Stark, *JBC* **246**, 6599–6605 (1971).

40. J. W. Donovan, *in* "Methods in Enzymology" (C. H. W. Hirs and S. N. Timasheff, eds.), Vol. 27, Part D, pp. 497–525. Academic Press, New York, 1973.

41. D. E. Metzler, C. Harris, I-Y. Yang, D. Siano, and J. A. Thomson, *BBRC* **46**, 1588–1597 (1972).

42. J. G. Foss, *J. Chem. Educ.* **40**, 592–597 (1963).

43. K. E. Van Holde, "Physical Biochemistry." Prentice-Hall, Englewood Cliffs, New Jersey, 1971.

44. P. M. Bayley, *Prog. Biophy. Mol. Biol.* **27**, 1–76 (1973).

45. C. A. Bush, *Phys. Tech. Biol. Res., 2nd Ed.* **1A**, 348–408 (1971).

46. E. H. Strickland, *Crit. Rev. Biochem.* **2**, 113–175 (1974).

47. W. Moffitt, R. B. Woodward, A. Moscowitz, W. Klyne, and C. Djerassi, *JACS* **83**, 4013–4018 (1961).

48. W. C. Johnson, Jr. and I. Tinoco, Jr., *JACS* **94**, 4389–4390 (1972).

49. Y.–H. Chen, J. T. Yang, and K. H. Chau, *Biochemistry* **13**, 3350–3359 (1974).

50. J. Applequist, *JACS* **95**, 8255–8262 (1973).

51. R. K. Clayton, "Light and Living Matter," Vol. I. McGraw-Hill, New York, 1970.

51a. R. F. Chen, and H. Edelhoch, eds., "Biochemical Fluorescence: Concepts." Vol. I. Dekker, New York, 1975.

52. C. A. Parker, "Photoluminescence of Solutions." Elsevier, Amsterdam, 1968.

53. G. Weber, *Annu. Rev. Biophys. Bioeng.* **1**, 553–570 (1972).

54. S. V. Konev, "Fluorescence and Phosphorescence of Proteins and Nucleic Acids." Plenum, New York, 1967.

55. R. F. Chen, H. Edelhoch, and R. F. Steiner, *in* "Physical Principles and Techniques of Protein Chemistry" (S. J. Leach, ed.), Part A, pp. 171–244. Academic Press, New York, 1969.

56. A. Kotaki and K. Yagi, *J. Biochem. (Tokyo)* **68**, 509–516 (1970).

57. P. Wahl, J.–C. Auchet, A. J. W. G. Visser, and C. Veeger, *EJB* **50**, 413–418 (1975).

58. J. Koziol and E. Knobloch, *BBA* **102**, 289–300 (1965).

59. D. R. Sellers and C. A. Ghiron, *Photochem. Photobiol.* **18**, 393–402 (1973).

60. M. R. Eftink and C. A. Ghiron, *Biochemistry* **15**, 672–680 (1976).

61. G. K. Radda, *Curr. Top. Bioenerg.* **4**, 81–176 (1971).

62. G. K. Radda, *BJ* **122**, 385–396 (1971).

63. L. N. M. Duysens, *in* "Photobiology of Microorganisms" (P. Halldal, ed.), p. 2. Wiley (Interscience), New York, 1970.

63a. T. Förster, *Ann. Physik.* (Leipzig) **2**, 55–75 (1948).

64. L. Stryer, *Science* **162**, 526–533 (1968).

65. C.–W. Wu and L. Stryer, *PNAS* **69**, 1104–1108 (1972).

66. W. DeW. Horrocks, Jr., B. Holmquist, and B. L. Vallee, *PNAS* **72**, 4764–4768 (1975).

67. N. J. Turro and G. Schuster, *Science* **187**, 303–312 (1975).

67a. L. Salem, *Science* **191**, 822–830 (1976).

68. N. P. Bazhulina, Yu. V. Morozov, M. Ya. Karpeisky, V. I. Ivanov, and A. I. Kuklin, *Biofizika* **11**, 42–47 (1966).

69. J. W. Bridges, D. S. Davies, and R. T. Williams, *BJ* **98**, 451–468 (1966).

69a. E. L. Wehry and L. B. Rogers, *JACS* **87**, 4234–4238 (1965).

70. G. F. Johnson, J.–I. Tu, M. L. S. Bartlett, and D. J. Graves, *JBC* **245**, 5560–5568 (1970).

71. M. R. Loken, J. W. Hayes, J. R. Gohlke, and L. Brand, *Biochemistry* **11**, 4779–4786 (1972).

72. A. J. Varghese, *Photophysiology* **7**, 207–274 (1972).

73. R. B. Setlow, *Prog. Nucleic Acid Res. Mol. Biol.* **8**, 257–295 (1968).

74. S. Minato and H. Werbin, *Biochemistry* **10**, 4503–4508 (1971).

75. B. M. Sutherland, M. J. Chamberlin, and J. C. Sutherland, *JBC* **248**, 4200–4205 (1973).

76. B. M. Sutherland, P. Runge, and J. C. Sutherland, *Biochemistry* **13**, 4710–4715 (1974).

77. R. P. F. Gregory, "Biochemistry of Photosynthesis." Wiley (Interscience), New York, 1971.

78. R. P. Levine, *Sci. Am.* **221**, 58–70 (Dec 1969).

79. R. K. Clayton, "Light and Living Matter," Vol. 2. McGraw-Hill, New York, 1971.

79a. C. B. van Niel, *Cold Spring Harbor Symp. Quant. Biol.* **3**, 138–150 (1935).

79b. R. Emerson, R. Chalmers, and C. Cederstrand, *PNAS* **43**, 133–143 (1957).

79c. R. Hill, *Nature (London)* **139**, 881–882 (1937).

79d. H. Gaffron, in "Plant Physiology" (F. C. Steward, ed.), Vol. IB, pp. 176–180. Academic Press, New York, 1960.

79e. R. Emerson and W. Arnold, *J. Gen. Physiol.* **16,** 191–205 (1932).

79f. D. I. Arnon, H. Y. Tsujimoto, and B. D. McSwain, *Nature (London)* **207,** 1367–1372 (1965).

80. D. I. Arnon, *PNAS* **68,** 2883–2892 (1971).

81. H. H. Strain and J. J. Katz, *Prog. Photosynth. Res., 1st, 1969,* Vol. II. pp. 539–546 (1969).

82. C. S. French and J. S. Brown, *Photosynth. Two Centuries after Its Discovery by Joseph Priestley, Proc. Int. Congr. Photosynth. Res., 2nd, 1971,* pp. 291–306 (1972).

83. R. E. Fenna and B. W. Matthews, *Nature (London)* **258,** 573–577 (1975).

84. J. M. Anderson, *Nature (London)* **253,** 536–537 (1975).

85. H. W. Siegelman, D. J. Chapman, and N. J. Cole, in "Porphyrins and Related Compounds" (T. W. Goodwin, ed.), pp. 107–120. Academic Press, New York, 1969.

85a. R. J. Beuhler, R. C. Pierce, L. Friedman and H. W. Siegelman, *JBC* **251,** 2405–2411 (1976).

86. K. Mühlethaler, in "Structure and Function of Chloroplasts" (M. Gibbs, ed.), pp. 7–34. Springer-Verlag, Berlin and New York, 1971.

87. W. P. Williams, *Nature (London)* **225,** 1214–1217 (1970).

88. M. Y. Okamura, L. A. Steiner, and G. Feher, *Biochemistry* **13,** 1394–1403 (1974).

89. M. van der Rest and G. Gingras, *JBC* **249,** 6446–6453 (1974).

90. K. D. Philipson and K. Sauer, *Biochemistry* **11,** 1880–1885 (1972).

91. S. G. Boxer, G. L. Class, and J. J. Katz, *JACS* **96,** 7058–7066 (1974).

92. C. E. Strouse, *PNAS* **71,** 325–328 (1973).

93. F. K. Fong, *JACS* **97,** 6890–6892 (1975).

93a. L. L. Shipman, T. M. Cotton, J. R. Norris, and J. J. Katz, *PNAS* **73,** 1791–1794 (1976).

94. T. L. Netzel, P. M. Rentzepis, and J. Leigh, *Science* **182,** 238–241 (1973).

95. G. S. Beddard, G. Porter, C. J. Tredwell, and J. Barber, *Nature (London)* **258,** 166–168 (1975).

95a. A. J. Campillo, V. H. Kollman, and S. L. Shapiro, *Science* **193,** 227–229 (1976).

96. P. L. Dutton, J. S. Leigh, Jr., and D. W. Reed, *BBA* **292,** 654–664 (1972).

97. M. Y. Okamura, R. A. Isaacson, and G. Feher, *PNAS* **72,** 3491–3495 (1975).

98. K. Seifert and H. T. Witt, *Prog. Photosynth. Res., 1st, 1969,* Vol. II. pp. 750–756 (1969).

99. M. C. W. Evans, C. K. Sihra, J. R. Bolton, and R. Cammack, *Nature (London)* **256,** 668–670 (1975).

100. J. J. Katz and J. R. Norris, Jr., *Curr. Top. Bioenerg.* **5,** 41–75 (1973).

101. A. Trebst, *Annu. Rev. Plant Physiol.* **25,** 423–458 (1974).

101a. L. N. M. Duysens, in "The Photochemical Apparatus, Its Structure and Function," Brookhaven Symposium in Biology No. 11, pp. 18–19. Brookhaven Natl. Lab., Upton, New York, 1958.

102. W. W. Parson and R. J. Cogdell, *BBA* **416,** 105–149 (1975).

103. R. A. Morton, *Biol. Rev. Cambridge Philos. Soc.* **46,** 47–96 (1971).

104. D. R. Threlfall and G. R. Whistance, in "Aspects of Terpenoid Chemistry and Biochemistry" (T. W. Goodwin, ed.), pp. 372–374. Academic Press, New York, 1971.

104a. R. P. Ambler and R. G. Bartsch, *Nature (London)* **253,** 285–288 (1975).

105. P. M. Wood, *BBA* **357,** 370–379 (1974).

106. R. E. Dickerson and R. Tinkovich, in "The Enzymes," 3rd ed. (P. D. Boyer, ed.), Vol. 17, pp. 397–547. Academic Press, New York, 1975.

107. J. M. Anderson, *BBA* **416,** 191–235 (1975).

108. L. C. Cantley and G. G. Hammes, *Biochemistry* **15,** 1–14 (1976).

109. W. Junge, *Proc. Int. Congr. Photosynth., 3rd, 1974,* Vol. I, pp. 273–286 (1975).

110. D. L. Keister and N. J. Raveed, *JBC* **249,** 6454–6458 (1974).

111. R. Radmer and B. Kok, *Annu. Rev. Biochem.* **44,** 409–433 (1975).

112. V. M. Katyurin, *Photosynth., Two Centuries After Its Discovery by Joseph Priestley, Proc. Int. Congr. Photosynth. Res., 2nd, 1971,* pp. 93–105 (1972).

112a. T. Wydrzynski, N. Zumbulyadis, P. G. Schmidt, H. S. Gutowsky, and Govindjee, *PNAS* **73,** 1196–1198 (1976).

113. G. Hauska, S. Reimer, and A. Trebst, *BBA* **357** 1–13 (1974).

114. H. Metzner, *J. Theor. Biol.* **51,** 201–231 (1975).

115. A. Stemler and R. Radmer, *Science* **190,** 457–458 (1975).

116. T. W. Goodwin, ed., "Aspects of Terpenoid Chemistry and Biochemistry," pp. 346–348. Academic Press, New York, 1971.

117. T. W. Goodwin, in "Phytochemistry" (L. P. Miller, ed.), Vol. I, pp. 136–137. Van Nostrand-Reinhold, Princeton, New Jersey, 1973.

118. D. Siefermann and H. Y. Yamamoto, *ABB* **171,** 70–77 (1975).

119. B. B. Buchanan and P. Schürmann, *Curr. Top. Cell. Regul.* **7,** 1–20 (1973).

120. L. E. Anderson, *Proc. Int. Congr. Photosynth., 3rd, 1974,* Vol. II, pp. 1393–1405 (1975).

121. J. M. O'Brien and R. Powls, *Proc. Int. Congr. Photosynth., 3rd, 1974,* Vol. II, pp. 1431–1440 (1975).

122. J. A. Bassham, in "Phytochemistry" (L. P. Miller, ed.), Vol. I, pp. 66–69. Van Nostrand-Reinhold, Princeton, New Jersey, 1973.

123. I. Zelitch, *PNAS* **70,** 579–584 (1973).

124. I. Zelitch, "Photosynthesis, Photorespiration, and Plant Productivity." Academic Press, New York, 1971.

125. C. C. Black, Jr., *Annu. Rev. Plant Physiol.* **24,** 253–286 (1973).

126. I. Zelitch, *Science* **188,** 626–633 (1975).

127. I. Zelitch, *Annu. Rev. Biochem.* **44**, 123–145 (1975).

128. N. E. Tolbert, *Annu. Rev. Plant Physiol.* **22**, 45 (1971).

129. N. E. Tolbert, *Curr. Top. Cell. Regul.* **7**, 21–50 (1973).

130. S. S. Kent, *JBC* **247**, 7293–7302 (1972).

131. J. R. Benemann, J. A. Berenson, N. O. Kaplan, and M. D. Kamen, *PNAS* **70**, 2317–2320 (1973).

132. J. R. Benemann and N. M. Weare, *Science* **184**, 174–175 (1974).

132a. H. W. Doelle, "Bacterial Metabolism," 2nd ed., pp. 116–135. Academic Press, New York, 1975.

133. E. W. Abrahamson and R. S. Fager, *Curr. Top. Bioenerg.* **5**, 125–200 (1973).

133a. R. W. Rodieck, "The Vertebrate Retina; Principles of Structure and Function." Freeman, San Francisco, California, 1973.

133b. G. Wald and P. K. Brown, *Cold Spring Harbor Symp. Quant. Biol.* **30**, 346, (1965).

134. W. A. Hagins, *Annu. Rev. Biophys. Bioeng.* **1**, 131–158 (1972).

135. R. W. Young, *Sci. Am.* **223**, 81–91 (Oct 1970).

136. B. Honig and T. G. Ebrey, *Annu. Rev. Biophys. Bioeng.* **3**, 151–177 (1974).

137. R. Hubbard and A. Kropf, *Sci. Am.* **216**, 64–76 (June 1967).

138. C. R. Worthington, *Annu. Rev. Biophys. Bioeng.* **3**, 53–80 (1974).

139. H. Langer, ed., "The Biochemistry and Physiology of Visual Pigments." Springer-Verlag, Berlin and New York, 1973.

140. G. E. Busch, M. L. Applebury, A. A. Lamola, and P. M. Rentzepis, *PNAS* **69**, 2802–2806 (1972).

141. M. L. Applebury, D. M. Zuckerman, A. A. Lamola, and T. M. Jovin, *Biochemistry* **13**, 3448–3458 (1974).

141a. A. Warshel, *Nature (London)* **260**, 679–683 (1976).

141b. R. Mathies and L. Stryer, *PNAS* **73**, 2169–2173 (1976).

141c. M. R. Fransen, W. C. M. M. Luyten, J. van Thuijl, L. Lugtenburg, P. A. A. Jansen, P. J. G. M. van Breugel, and F. J. M. Daemen, *Nature (London)* **260**, 726–727 (1976).

142. L. Salem and P. Bruckman, *Nature (London)* **258**, 526–528 (1975).

143. R. Bensasson, E. J. Land, and T. G. Truscott, *Nature (London)* **258**, 768–770 (1975).

144. A. R. Oseroff and R. H. Callender, *Biochemistry* **13**, 4243–4248 (1974).

145. D. S. Papermaster and W. J. Dreyer, *Biochemistry* **13**, 2438–2444 (1974).

146. N. W. Downer and S. W. Englander, *Nature (London)* **254**, 625–627 (1975).

147. M. Weller, N. Virmaux, and P. Mandel, *Nature (London)* **256**, 68–70 (1975).

148. J. J. Keirns, N. Miki, M. W. Bitensky, and M. Keirns, *Biochemistry* **14**, 2760–2766 (1975).

149. K. Azuma, M. Azuma, K. Sakaguchi, and Y. Kito, *Nature (London)* **253**, 206–207 (1975).

150. H. Shichi and R. L. Somers, *JBC* **249**, 6570–6577 (1974).

151. E. Racker and W. Stoeckenius, *JBC* **249**, 662–663 (1974).

152. A. Danon and W. Stoeckenius, *PNAS* **71**, 1234–1238 (1974).

153. R. Henderson, *JMB* **93**, 123–138 (1975).

154. R. Henderson and P. N. T. Unwin, *Nature (London)* **257**, 28–32 (1975).

155. K. Mitrakos and W. Shropshire, Jr., eds., "Phytochrome." Academic Press, New York, 1972.

155a. H. Smith, "Phytochrome and Photomorphogenesis." McGraw-Hill, New York, 1975.

155b. B. F. Erlanger, *Annu. Rev. Biochem.* **45**, 267–283 (1976).

156. W. R. Briggs and H. V. Rice, *Annu. Rev. Plant Physiol.* **23**, 293–324 (1972).

157. W. Shropshire, Jr., *Photophysiology* **7**, 33–72 (1972).

158. M. J. Burke, D. C. Pratt, and A. Moscowitz, *Biochemistry* **11**, 4025–4031 (1972).

158a. A. Evans and H. Smith, *PNAS* **73**, 138–142 (1976).

159. P. S. Song and T. A. Moore, *Photochem. Photobiol.* **19**, 435–441 (1974).

160. G. E. Treadwell and D. E. Metzler, *Plant Physiol.* **49**, 991–993 (1972).

161. K. L. Poff and W. L. Butler, *Nature (London)* **248**, 799–801 (1974).

161a. J. J. Wolken, "Photoprocesses, Photoreceptors and Evolution." Academic Press, New York, 1975.

161b. M. Debruck, A. Katzir, and D. Presti, *PNAS* **73**, 1969–1973 (1976).

162. W. D. McElroy and H. H. Seliger, *Sci. Am.* **207**, 76–89 (Dec. 1962).

163. M. J. Cormier, D. M. Hercules, and J. Lee, eds., "Chemiluminescence and Bioluminescence." Plenum, New York, 1973.

164. M. J. Cormier, J. Lee, and J. E. Wampler, *Annu. Rev. Biochem.* **44**, 255–272 (1975).

165. K. Hori, J. E. Wampler, J. C. Matthews, and M. J. Cormier, *Biochemistry* **12**, 4463–4468 (1973).

166. O. Shimomura and F. H. Johnson, *Biochemistry* **7**, 2574–2580 (1968).

167. W. W. Ward and H. H. Seliger, *Biochemistry* **13**, 1500–1510 (1974).

168. C. C. Ashley and E. B. Ridgway, *in* "Calcium and Cellular Function" (A. W. Cuthbert, ed.), pp. 42–53. St. Martin's Press, New York, 1970.

169. O. Shimomura and F. H. Johnson, *Biochemistry* **11**, 1602–1608 (1972).

170. W. W. Ward and M. J. Cormier, *PNAS* **72**, 2530–2534 (1975).

171. H. Morise, O. Shimomura, F. H. Johnson, and J. Winant, *Biochemistry* **13**, 2656–2662 (1974).

172. J. W. Hastings and C. Balny, *JBC* **250**, 7288–7293 (1975).

173. C. Kemal and T. C. Bruice, *PNAS* **73**, 995–999 (1976).

174. K. Matsuda and T. Nakamura, *J. Biochem. (Tokyo)* **72**, 951–955 (1972).

STUDY QUESTIONS

1. Define the following: photon, a quantum of energy, wave number, an einstein, circularly polarized light, action spectrum, fluorescence, phosphorescence.

2. What is meant by the Franck–Condon principle?

3. Exactly 0.1 ml of a stock solution of adenosine in distilled water was diluted into a neutral phosphate buffer of pH 7.0 and made to a volume of exactly 25 ml. The absorbance at 259 nm was 0.77. The molar absorptivity of adenosine at 259 nm is 1.54×10^4 $M^{-1} cm^{-1}$. What is the concentration of adenosine in the stock solution? What was the transmittance of the sample at 259 nm?

4. A difference spectrum was measured between a protein solution of absorbance 2.0 at 280 nm and a matched solution containing a given concentration of an allosteric modifier. A series of positive and negative bands was seen in the 260–300 nm region of the difference spectrum. When the experiment was repeated with a solution of absorbance 3.0 at 280 nm and 1.5 times the amount of allosteric modifier, the intensity of the peaks and valleys in the difference plot was increased by considerably less than the 1.5-fold expected. Explain.

5. Titration of tyrosyl residues in proteins can be followed spectrophotometrically. (a) Explain how this could be done. (b) At what wavelength or wavelengths would you suggest the measurements be made? (c) Show that for a compound with a single dissociable group and molar absorbtivities ϵ_{HA} for the undissociated form and ϵ_A for the dissociated form, the following equation holds, where ϵ is the observed apparent molar absorptivity at a given pH.

$$pK_a = pH - \log \frac{\epsilon - \epsilon_{HA}}{\epsilon_A - \epsilon}$$

6. Anthracene crystals absorb ultraviolet light at 339 nm. Calculate the height of the activated energy level above the ground state in kcal mol^{-1}, kJ mol^{-1}, and in cm^{-1} corresponding to this absorption.

7. Calculate the energy per einstein in the light at the wavelengths corresponding to maximum absorp-

tion in the two bands of chlorophyll at 430 and 660 nm.

8. Why is it that molecules which fluoresce at room temperatures commonly phosphoresce at very low temperatures, for example, at $-180°C$?

9. Under certain conditions chlorophyll molecules dimerize. How would this dimerization affect the electronic spectrum?

10. What are the possible fates (as discussed in the text) of an absorbed photon? Which is the most applicable to the functioning chloroplast?

11. What is meant by the "light reactions" and "dark reactions" of photosynthesis?

12. What is the ultimate source of electrons needed to reduce NADP during photosynthesis in green plants?

13. Why is the "Emerson enhancement effect" (Section E,1) not observed with photosynthetic bacteria?

14. Name some other differences between photosynthesis in bacteria and that used by blue-green algae or eukaryotic plants.

15. What features do mitochondria and chloroplasts have in common?

16. The CO_2 concentrating mechanism (Fig. 13-26) used by C_4 plants can be described by the following equation:

$$(CO_2 + NADPH)_m + ATP \longrightarrow$$
$$(CO_2 + NADPH)_{bs} + 2 P_i + AMP$$

where the subscripts m and bs refer to mesophyll and bundle sheath, respectively. If $[AMP][P_i]^2/[ATP]$ were equal to 10^{-3} and the concentration of NADPH were the same in the mesophyll and bundle sheath cells, what would the concentration ratio $[CO_2]_{bs}/[CO_2]_m$ be at equilibrium?

17. In what positions would ^{14}C appear in the following molecules after a few seconds of photosynthesis in $^{14}CO_2$ (a) 3-phosphoglycerate, (b) fructose 6-phosphate, (c) serine, (d) oxaloacetate?

18. What is the minimum agricultural land area that could support your personal energy needs by photosynthesis?

14

The Metabolism of Nitrogen-Containing Compounds

Nitrogen, because it is found in so many compounds, has a complex metabolism. The inorganic forms of nitrogen found in our surroundings range from the highly oxidized nitrate ion, in which N has an oxidation state of $+5$, to ammonia, in which the oxidation state is -3. Living cells both reduce and oxidize these inorganic forms. The organic forms of nitrogen are most often derived by incorporation of ammonium ion into amino groups or amide groups. Once it has been incorporated into an organic compound, nitrogen can be transferred into many other carbon compounds. Certain compounds including glutamic acid, aspartic acid, glutamine, asparagine, and carbamoyl phosphate are especially active in these transfer reactions. They constitute a **nitrogen pool** from which nitrogen can be withdrawn and to which it can be returned.

In addition to the pathways for synthesis and degradation of amino acids, nucleotides, and other nitrogenous substances, many organisms have specialized metabolism for incorporation of excess nitrogen into relatively nontoxic excretion products. All of these aspects of nitrogen metabolism will be dealt with in this chapter, but because of the enormous complexity of the subject the treatment is abbreviated. We will look first at the reactions by which organic nitrogen compounds are formed from inorganic compounds, then at the reactions of the nitrogen pool. After that we will examine the specific reactions of synthesis and catabolism of individual nitrogenous compounds.

A. FIXATION OF N_2 AND OTHER REACTIONS OF INORGANIC NITROGEN COMPOUNDS

Most of the nitrogen of the biosphere exists as the very unreactive N_2 which makes up 80% of the molecules of air. The "fixation" of N_2 occurs principally by the action of lightning (which forms oxides of nitrogen and eventually nitrate and nitrite) and by the action of bacteria.[1] Man also contributes a significant share through production of chemical fertilizer. The interconversions of nitrate and nitrite with ammonia and with organic nitrogen compounds are active biological processes. A number of the pertinent reactions have been discussed in Chapter 10. For example, the oxidation of NH_3 to NO_2^- and NO_3^- by bacteria (Chapter 10, Section F,1) and the reduction of NO_3^- to NO_2^- (Eq. 10-32). For many bacteria and higher plants this reduction of nitrate is the first step in an important assimilatory sequence by which the nitrite is reduced all the way to NH_3. However, the chemistry of the reduction of nitrite to NH_3 (Eq. 10-36) is still poorly understood.

1. Reduction of Elemental Nitrogen

One of the most remarkable reactions of nitrogen metabolism is the conversion of dinitrogen (N_2) to ammonia. It has been estimated that in 1974 this biological nitrogen fixation added 17.5×10^{10} kg of nitrogen to the earth (compared with 4×10^{10} kg fixed by chemical reactions).[2] The quantitative significance can be more easily appreciated by the realization that one square meter of land planted to nodulated legumes such as soybeans can fix 10–30 g of nitrogen per year.

Fixation of N_2 by *Clostridium pasteurianum* and a few other species was recognized by Winogradsky[2a] in 1893. Subsequent nutritional studies indicated that both iron and molybdenum were required for the process. Inhibition by CO and N_2O was observed. While ammonia was the suggested product, the possibility remained that more oxidized compounds, such as hydroxylamine were the ones first incorporated into organic substances. When cell-free preparations capable of fixing nitrogen were obtained in 1960 rapid progress became possible.[3-5] An important development came from the discovery that nitrogen-fixing bacteria are invariably able to reduce acetylene to ethylene. The two catalytic abilities appear to go hand in hand. A simple, sensitive **acetylene reduction test** permits easy measurement of the nitrogen-fixing potential of cells.

Application of this test soon led to the discovery that nitrogen fixation is not restricted to a few species, but is a widespread ability of many prokaryotes. Among the organisms most studied are *Azotobacter*, Winogradsky's *Clostridium pasteurianum*, *Klebsiella* (a close relative of *E. coli*), and the symbiotic bacterium of root nodules of legumes *Rhizobium*. *Rhizobium* deserves special

attention, for reduction of N_2 takes place in nodules developed by infected roots.* Within the nodules the bacteria degenerate into "bacteroids" and a special hemoglobin (**leghemoglobin**)[7,8] specified by a plant gene is synthesized. Legumes are not the only plants containing nitrogen-fixing symbionts. Some other angiosperms are hosts to nitrogen-fixing actinomycetes and some gymnosperms contain nitrogen-fixing blue-green algae. Leaf nodules of certain plants infected with *Klebsiella* fix nitrogen. While the nutritional significance is uncertain, nitrogen-fixing strains of *Klebsiella* have also been found in the intestinal tracts of humans in New Guinea.

Of the free-living nitrogen-fixing organisms blue-green bacteria (algae) appear to be of most importance quantitatively. For example, in rice paddy fields, blue-green algae may fix from 2.4 to 10 g of nitrogen per square meter per year.

2. The Nitrogenase Enzyme System

Cell-free nitrogenases have been isolated from a number of organisms. These enzymes all share the property of being readily inactivated by oxygen, a fact that impeded early work. Apparently nitrogen fixation occurs in anaerobic regions of cells. Indeed, it has been suggested that leghemoglobin protects the nitrogen-fixing enzymes in root nodules from oxygen. Leghemoglobin may also function to deliver O_2 by facilitated diffusion to the aerobic mitochondria of the bacteroids at a stable, low pressure.[8]

The nitrogenase system catalyzes the six-electron reduction of N_2 to ammonia (Eq. 14-1):

$$N_2 + 6\,H^+ + 6\,e^- \longrightarrow 2\,NH_3 \qquad (14\text{-}1)$$

It is also able to reduce many other compounds. For example, the reduction of acetylene to ethylene (Eq. 14-2) is a two-electron process:

$$HC\equiv CH + 2\,H^+ + 2\,e^- \longrightarrow H_2C\equiv CH_2 \quad (14\text{-}2)$$

* However, it has now been shown that some free-living rhizobia can also fix nitrogen.[6]

$$N_3^- + 4\,H^+ + 2\,e^- \longrightarrow N_2 + NH_4^+ \qquad (14\text{-}3)$$

$$CN^- + 8\,H^+ + 6\,e^- \longrightarrow CH_4 + NH_4^+ \qquad (14\text{-}4)$$

$$2\,H^+ + 2\,e^- \longrightarrow H_2 \qquad (14\text{-}5)$$

Azide is reduced to N_2 and NH_4^+ in another two-electron reduction (Eq. 14-3). Cyanide ions yield methane and ammonia (Eq. 14-4). Alkyl nitriles as well as N_2O are also reduced, and nitrogenases invariably catalyze reduction of protons to H_2 (Eq. 14-5).

In early experiments it was found that sodium pyruvate was required for fixation of N_2 in cell-free extracts. It was observed that large amounts of CO_2 and H_2 accumulated. Investigation showed that pyruvate was being cleaved by the pyruvate-formate-lyase reaction (Fig. 8-19) and was supplying cells with two important products: ATP and reduced ferredoxin. Pyruvate could be replaced by a mixture of ATP plus Mg^{2+} and Fd_{red}. Furthermore, the nonbiological reductant dithionite ($S_2O_4^{2-}$) could replace the reduced ferredoxin. Since ADP is inhibitory to the nitrogenase system, it was found best to supply ATP from an ATP-generating system such as a mixture of creatine phosphate (Box 10-F), creatine kinase, and a small amount of ADP.

In every case investigated the nitrogenase system can be separated easily into two components. One of these, **azoferredoxin** (azoFd, also known as component II), is an extremely oxygen-sensitive iron–sulfur protein. Azoferredoxin consists of two idential peptide chains, each of MW ~ 30,000. Each dimer contains four iron atoms, 4 S^{2-}, and 12 titratable thiol groups. The other component **molybdoferredoxin** (MoFd, also called component I) contains both iron and molybdenum as well as labile sulfide. This protein contains two kinds of peptide chains of MW ~ 51,000 and 60,000 in a mixed ($\alpha_2\beta_2$) tetramer. Each mixed tetramer contains two molybdenum atoms, ~24 iron atoms, ~24 sulfide ions, and ~30 titratable thiol groups, possibly in the form of three Fe_4S_4 clusters (Chapter 10, Section C). The proteins associate in a ratio of two dimeric azoFd molecules to one MoFd to form the nitrogenase complex (Fig. 14-1).

When azoFd is reduced, an epr signal at $g = 1.94$, typical of iron–sulfur proteins (Chapter 10, Section C), is observed. This signal is altered by interaction with Mg^{2+} and ATP, whereas ATP has no effect on the complex set of epr signals produced upon reduction of MoFd. These are among the observations that led to the concept that azoFd is an electron carrier responsible for reduction of the molybdenum in MoFd. The Mo(IV) formed in this way could then reduce N_2 in a two-electron step with formation of Mo(VI) (Eq. 14-6).

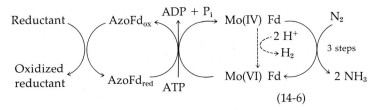

$$(14\text{-}6)$$

According to this picture, three successive 2-electron steps are required to completely reduce N_2 to two molecules of ammonia. Alternatively, two protons can be reduced to H_2 as indicated by the dotted lines in Eq. 14-6. ATP appears to drive the electron flow in a manner analogous to the ATP-driven "reverse electron flow" in the respiratory chain (Chapter 10, Section E,7). This idea is also indicated schematically in Eq. 14-6. It has been shown[9] that ATP binds strongly to azoFd and lowers its $E^{0'}$ (pH 7.5) from -0.29 to -0.40 V. A surprising fact is that most workers have observed a requirement of 4–5 molecules of ATP per two electrons. The reduction of N_2 by reduced ferredoxin (Eq. 14-7) is thermodynamically spontaneous:

$$N_2 + 6\,Fd_{red} + 8\,H^+ \longrightarrow 2\,NH_4^+ + 6\,Fd_{ox}^+ \qquad (14\text{-}7)$$
$$\Delta G'\,(\text{pH 7}) = -89.3 \text{ kJ mol}^{-1}$$
$$\text{or } -29.8 \text{ kJ per 2 electrons}$$

However, N_2 is chemically exceedingly unreactive. Thus, ATP cleavage may have to be coupled to the nitrogenase reduction system in more than one way[10] to overcome a high activation energy. It is also possible that the ATP requirement is lower *in vivo* than in isolated systems.

In some bacteria, e.g., the strictly aerobic *Azotobacter*, NADPH is the electron donor for reduction

BOX 14-A

MOLYBDENUM

Long recognized as an essential element for the growth of plants, molybdenum has never been conclusively demonstrated as a necessary animal nutrient. Nevertheless, it is found in at least three enzymes of the body, as well as in four additional enzymes of bacteria and plants.[a,b] **Aldehyde oxidase, xanthine oxidase** of liver (p. 485), and the related **xanthine dehydrogenases** of certain bacteria contain molybdenum that is essential for catalytic activity. **Sulfite oxidase** of liver (Chapter 14, Section G), the bacterial and plant enzyme **nitrate reductase** (Chapter 10, Section F,2), the bacterial **formate dehydrogenase** (Chapter 9, Section C,3), and **nitrogenase** (this section of the text) complete the list of identified molybdenum-dependent enzymes.

Molybdenum is a metal of the second transition series, one of the few heavy elements known to be essential to life. The most stable oxidation state, Mo(VI) contains a filled $4s$ shell and has four d orbitals available for coordination with anionic ligands. Coordination numbers of 4 and 6 are preferred, but molybdenum can accommodate at least 8 ligands. Most of the complexes are formed from the oxycation MoO_2^{2+}. If two molecules of water are coordinated with this ion, the protons are so acidic that they dissociate completely to give MoO_4^{2-}, the molybdate ion. Other oxidation states vary from Mo(III) to Mo(V). In these lower oxidation states, the tendency for protons to dissociate from coordinated ligands is less, e.g., $Mo(H_2O)_6^{3+}$ does not lose protons even in a very basic medium. Molybdenum tends to form dimeric or polymeric oxygen-bridged ions.

All of the molybdenum-containing enzymes are of MW = 100,000 or greater and often contain two atoms of molybdenum. However, there is no evidence that the metal atoms are associated as a dimer; rather the protein subunits to which they are bound may be dimeric. Nitrate reductase of *E. coli* appears to contain a large subunit of MW = 150,000, a smaller peptide of MW = 55,000, and one atom of Mo and 12 non-heme iron atoms and 12 acid-labile sulfides.[c] Dimers and tetramers also are formed. Evidence has been presented that the molybdenum in all of these enzymes may be present in the form of a low molecular weight cofactor.[d]

The exact mechanism of participation of molybdenum in catalysis remains uncertain. The species Mo(III) and Mo(V) are paramagnetic, but only the signal from Mo(V) is readily detectable in epr spectroscopy. A characteristic 6-line hyperfine structure makes this signal easy to identify. It has been found in xanthine oxidase, nitrate reductase, and sulfite oxidase during action on substrates. However, nitrogenase has yielded only epr signals characteristic of iron. Furthermore, there is no evidence that N_2 reacts directly with the molybdenum atoms of nitrogenase. Nevertheless, it is tempting to speculate that the ability of Mo(VI) to accept three electrons to form Mo(III) may provide the basis for its presence in nitrogenase. Two atoms of Mo(III) could each donate three electrons of the six required according to Eq. 14-10.

Tungsten competes with molybdenum in organisms. Thus, rats receiving 100 ppm of tungsten in their diet form tungsten-containing sulfite oxidase that does not function normally.[e] However, a larger amount of metal-free apoenzyme is produced. Likewise, rats receiving tungsten produce inactive metal-free xanthine oxidase.[f] Evidently tungsten interferes with incorporation of molybdenum into enzymes. A molybdenum storage protein contains most of the molybdenum present in the nitrogen-fixing *Azotobacter*.[g]

[a] F. L. Bowden, *in* "Techniques and Topics in Bioinorganic Chemistry" (C. A. McAuliffe, ed.), pp. 207–267. Macmillan, New York, 1975.

[b] A Symposium on Molybdenum, *J. Less-Common Met.* **36**, 405–533 (1974).

[c] K. Lund and J. A. DeMoss, *JBC* **251**, 2207–2216 (1976).

[d] A. Nason, K. Y. Lee, S.-S. Pan, and R. H. Erickson, *J. Less-Common Met.* **36**, 449–459 (1974).

[e] J. J. Johnson, H. J. Cohen, and K. V. Rajagopalan, *JBC* **249**, 5046–5055 (1974).

[f] J. J. Johnson, W. R. Wand, H. J. Cohen, and K. V. Rajagopalan, *JBC* **249**, 5056–5061 (1974).

[g] W. J. Brill, *Am. Chem. Soc., Cent. Meet., New York, 1976* Abstracts INOR 138 (1976).

of N_2. AzoFd is thought to accept electrons from a chain that includes at least the ordinary bacterial ferredoxin (Fd) and a special one-electron-accepting **azotoflavin**.[11] This flavoprotein, which is somewhat larger in size than the flavodoxins (Chapter 8, Section I,5), appears to play a specific role in N_2 fixation. The proposed electron transport system is indicated schematically in Eq. 14-8:

$$\text{NADPH} \longrightarrow \text{Fd} \longrightarrow \text{azotoflavin} \xrightarrow{\quad S_2O_4^{2-} \quad} \text{AzoFd} \xrightarrow{\text{ATP}}$$

$$\text{MoFd} \longrightarrow N_2 \qquad (14\text{-}8)$$

3. The Mechanism of Reduction

While N_2 is very unreactive, it does form nitrides with metals and complexes with some metal chelates. These complexes are generally of an end-on nature, e.g., $N\equiv N-Fe$. Stiefel suggested that N_2 first forms a complex of this type with an iron atom of the molybdoferredoxin molecule.[12] Then an atom of Mo(IV) can donate two electrons to the N_2 (Eq. 14-9a) to form a complex of N_2 and Mo(VI). Addition of two protons (Eq. 14-9, step b)

$$\text{Mo(IV)} + N\equiv N \xrightarrow{\;a\;} \overset{N}{\underset{N}{\text{Mo}\;\|\;}} \xrightarrow[b]{2H^+} \text{Mo(VI)} + HN\!=\!NH$$

$$\text{diimide} \qquad (14\text{-}9)$$

yields a molecule of **diimide,** which would stay bound at the iron site while the molybdenum underwent another round of reduction. Then the diimide could be reduced to hydrazine and finally to ammonia (Eq. 14-10):

$$N_2 \xrightarrow{2\,e^-} HN=NH \xrightarrow{2\,e^-} H_2N-NH_2 \xrightarrow{2\,e^-} 2\,NH_3$$

$$(14\text{-}10)$$

Stiefel pointed out that Mo(VI) attracts electrons sufficiently strongly that protons bound to surrounding ligands, e.g., H_2O tend to dissociate completely. Thus, the molybdate ion MoO_4^{2-} is not protonated. The same would be true of nitrogenous ligands such as amino groups of a protein that might be coordinated with the bound

molybdenum. On the other hand, the reduced Mo(IV) will tend to be surrounded by protonated ligands. In the initial complex with N_2 the Mo(IV) may be bound by protonated nitrogenous ligands. Concurrently with the electron transport from molybdenum to N_2 these protons could be transferred to the N_2 molecule.

$$\underset{^2H}{\overset{H}{\diagdown}}C=C\underset{^2H}{\overset{H}{\diagup}}$$

The fact that strictly *cis*-dideuteroethylene is formed from acetylene in the presence of ^2H_2O is in accord with this idea.

Likhtenshtein and associates[13] suggested that the N_2 molecule is bound between *two* molybdenum centers and receives electrons from both halves of the dimeric structure (Fig. 14-1).

Because of the tremendous practical significance to agriculture there is intense interest in devising better nonenzymatic processes for fixing nitrogen. Numerous attempts are being made to devise nitrogenase models mimicking the natural biological reaction. For example, a 1:1 mixture of cysteine and sodium molybdate was treated with a reducing agent such as $NaBH_4$. It is presumed that

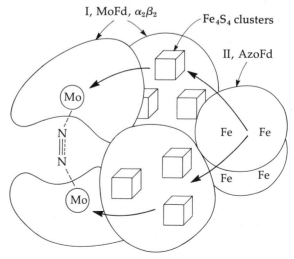

Fig. 14-1 Hypothetical structure of a nitrogenase molecule.

an Mo(IV) complex resulted. This complex was able to coordinate with alkylnitriles which could be reduced by $NaBH_4$ all the way to an alkene and ammonia (Eq. 14-11). Needless to say, the reaction

$$Mo(VI) + CH_3R + NH_3 \qquad (14\text{-}11)$$

must occur in several steps. It is of interest that the rate is significantly increased by the presence of ATP and iron–sulfur cluster compounds and that the reaction is inhibited by CO and N_2.[14,15] Thus, the model system displays a number of characteristics remarkably similar to those of nitrogenases. Another model system that uses a different type of molybdenum complex and an iron–sulfur cluster to reduce N_2 has also been reported.[16] However, the rates of reaction for all of the model reactions are very much slower than those of nitrogenases.

There is also great interest in improving the efficiency of biological nitrogen fixation. Thus, genetic manipulations to "derepress" the nitrogenase genes, making their expression "constitutive" (Chapter 15, Section B,1) can produce bacteria able to fix nitrogen in soil or root nodules more rapidly than do the usual strains. Nitrogenase genes are ordinarily repressed by the accumulation of glutamine within cells as detailed in Section B,2. Nitrogen fixation genes have been found only in prokaryotes. An important advance in agriculture will be made if these genes can be transferred and made to function in green plants (Chapter 15, Section H,4).

B. INCORPORATION OF NH_3 INTO AMINO ACIDS AND PROTEINS

Prior to 1940, amino acids were generally regarded as relatively stable nutrient building blocks. That concept was rapidly abandoned when R. Schoenheimer initiated studies of the metabolism of $^{15}NH_3$ and of ^{15}N-containing amino acids. It was quickly apparent that nitrogen could often be shifted rapidly between one carbon skeleton and another. These results confirmed the proposal put forth earlier by Braunstein (Chapter 8, Section E). Braunstein had pointed out that the C_4 and C_5 amino acids, aspartate and glutamate, which are closely related to the tricarboxylic acid cycle, are able to rapidly exchange their amino groups with those of other amino acids via transamination (Eq. 14-12, steps b and c). Since ammo-

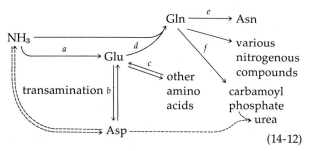

$$(14\text{-}12)$$

nia can be incorporated readily into glutamate (Eq. 14-12, step a, see next section) a general means is available for the synthesis of amino acids.

It soon became apparent that glutamine and asparagine are to be regarded as soluble, nontoxic carriers of additional ammonia in the form of their amide groups. An active synthetase converts glutamate and ammonia to glutamine (Eq. 14-12, step d) and another enzyme transfers the amide nitrogen onto aspartate to form asparagine (Eq. 14-12 step e). The amide nitrogen of glutamine is used in a great variety of biochemical processes, including formation of carbamoyl phosphate (Eq. 14-12, step f, Section C,2), glucosamine (Eq. 12-4), NAD^+ (Section I), purines (Section K,3), CTP (Section K,1), p-aminobenzoate (Section H,3), and histidine (Section J).

Glutamate, glutamine, and aspartate also play central roles in removal of nitrogen from organic compounds.[17] Transamination is reversible and usually represents the first step in catabolism of excess amino acids. Ketoglutarate is the recipient of the nitrogen and the resulting excess glutamate can be deaminated to form ammonia, then glutamine. It can also donate its nitrogen to form aspartate. In the animal body both aspartate and glutamine (via carbamoyl phosphate) are precursors of **urea,** the principal nitrogenous excretion product. These relationships are summarized in Eq. 14-12 and details are provided in later sections.

While reductive amination of glutamate appears to be the major pathway for incorporation of nitrogen into amino groups, other routes may exist. Direct amination of pyruvate and other α-keto acids in reactions analogous to that of glutamate dehydrogenase has been proposed for plants.[17a] A bacterial enzyme catalyzes reversible addition of ammonia to fumarate to form aspartate (Chapter 7, Section H,6,d).

1. Glutamate Dehydrogenase and Glutamate Synthetase

It has usually been assumed that the glutamate dehydrogenase reaction (Fig. 14-2, step a; see also Chapter 8, Section H,4) provides the principal means of reversibly incorporating ammonia into glutamic acid. The reductant can be either NADH or NADPH. In eukaryotic cells the enzyme is found principally in the mitochondria. The action of transaminases, both within and without mitochondria, distributes nitrogen from glutamic acid into most of the other amino acids. Especially active is the aspartate aminotransferase (Chapter 8, Sections E,3 and E,7) which equilibrates aspartate and oxaloacetate with the α-ketoglutarate–glutamate couple.

In $E.\ coli$ and many other bacteria it is **glutamate synthetase** (Fig. 14-2, reaction b) that carries out the reductive amination of α-ketoglutarate. In this reaction glutamine donates the nitrogen. There is every reason to believe that enzymes of this type actually hydrolyze the glutamine to ammonia at the active site of the enzyme. Thus, the formation of a Schiff base and reduction with NADPH may proceed exactly as in reaction a (Chapter 8, Section H,4). The difference is that one of the two glutamate molecules formed in reaction b must be re-

converted by glutamine synthetase (Section B,2) to glutamine with the utilization of one molecule of ATP. Because of this coupling of ATP cleavage to the reaction the equilibrium in reaction b lies far toward the synthesis of glutamate. A very low value of K_m for NH_4^+ is a characteristic of glutamate synthetase.

Glutamate synthetase from $E.\ coli$ is a large protein of MW ~800,000 containing flavin, iron, and S^{2-} in a 1:4:4 ratio.[18] While NADPH is the reductant, some experiments suggest that the bound flavin, after reduction, is the immediate donor of electrons to the Schiff base of α-ketoglutarate and ammonia.[18] It is not clear why the flavin and iron-sulfide prosthetic groups are needed. A possibility is that the iron–sulfide group provides a way of coupling the enzyme to reduced ferredoxin as well as to NADPH.

Recent findings suggest that glutamate synthetase may also provide the major route of incorporation of nitrogen into amino acids of yeasts[19] and higher green plants. In the latter case reduced ferredoxin may be the reducing agent.[20]

2. Glutamine Synthetase

The formation of glutamine from glutamate (Eq. 14-13) also depends upon a coupled cleavage of ATP:

$$Glu \xrightarrow[\text{ATP} \quad \text{ADP} + P_i]{NH_4^+} Gln \qquad (14\text{-}13)$$

The enzyme glutamine synthetase, as isolated from $E.\ coli$, contains 12 identical subunits, each of MW ~50,000. These are arranged in the form of two rings of six subunits each with a center-to-center spacing of 4.5 nm. The units in one layer lie almost directly above those in the next[21] and the center-to-center spacing between the two layers is also 4.5 nm. The enzyme displays extraordinarily complex regulatory properties[22–24] which are summarized in Fig. 14-3. Glutamine synthetase exists in two forms. **Active glutamine synthetase** requires Mg^{2+} in addition to the three substrates gluta-

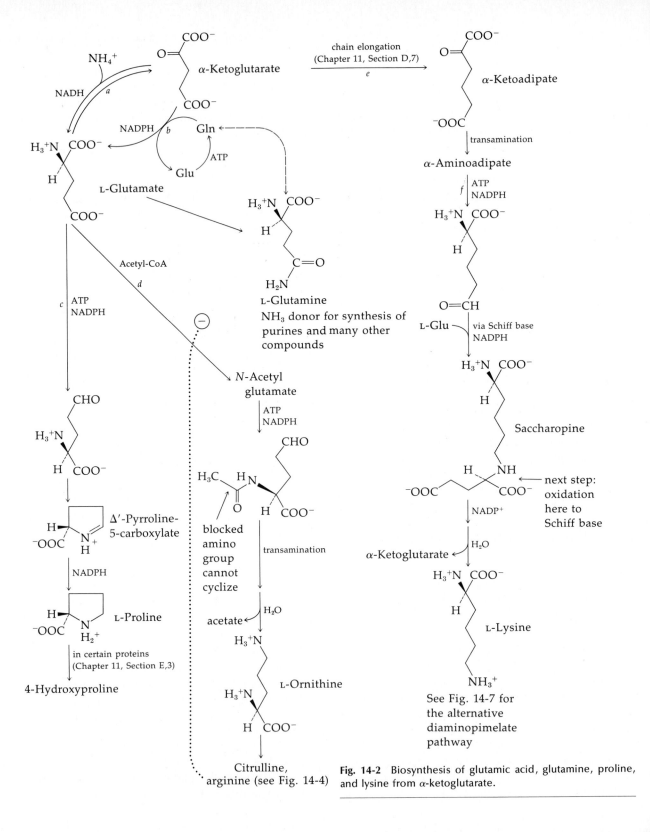

Fig. 14-2 Biosynthesis of glutamic acid, glutamine, proline, and lysine from α-ketoglutarate.

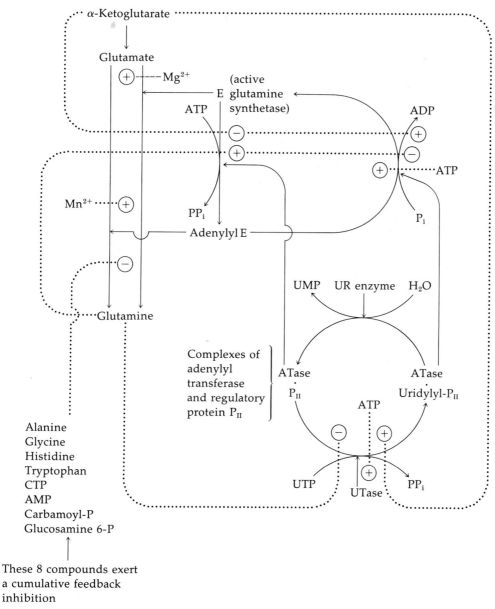

Fig. 14-3 Regulation of glutamine synthetase of *E. coli*: (⊕) activation and (⊖) inhibition.

mate, NH_4^+, and ATP. If the precursor, α-ketoglutarate, is present in excess the enzyme tends to remain in the active form because conversion to a modified form is inhibited. However, when the ketoglutarate concentration falls to a low value and glutamine accumulates, alteration is favored. The modifying enzyme **adenylyltransferase** (ATase) transfers an adenylyl group from ATP to a tyrosine hydroxyl on glutamine synthetase to give an adenylyl enzyme. This modified enzyme requires Mn^{2+} instead of Mg^{2+}. It is far more sensitive to feedback inhibition by a whole series of end products of glutamine metabolism as indicated in the lower left side of Fig. 14-3. All eight of the feedback inhibitors seem to bind to their own specific sites on the enzyme surface exerting a cumulative inhibition.

Relaxation of adenylylated glutamine synthetase to the unmodified form is not catalyzed by a separate hydrolase as in the case of phosphorylase (Fig. 11-10). Rather, removal of the adenylyl group is promoted by a modified form of the adenylyl-transferase that catalyzed adenylylation. It has been found that ATase exists as a complex with a regulatory protein designated P_{II}. The ATase $\cdot$ P_{II} complex catalyzes adenylylation (transfer of an adenylyl group from ATP to the tyrosine of glutamic synthetase). The regulatory subunit P_{II} can be uridylylated, by action of a special **uridylyltransferase** (UTase) and UTP. The complex ATase $\cdot$ P_{II}-uridylyl catalyzes phosphorolytic deadenylylation of glutamine synthetase (with P_i displacing the adenylyl group to form ADP). Finally, removal of the uridylyl group from uridylyl-P_{II} is catalyzed by a fourth enzyme, the "UR enzyme." The cycle of interconversions of P_{II} catalyzed by the UTase and UT enzymes is shown at the lower right side of Fig. 14-3. From the allosteric modification reactions indicated by the dotted lines, it is seen that glutamine not only promotes the adenylylation of glutamine synthetase but also blocks the uridylylation of P_{II} preventing ATase from removing the adenylyl group from the synthetase. Furthermore, it allosterically inhibits the deadenylylation reaction itself. On the other hand, α-ketoglutarate acts in exactly the opposite way at all three sites. A somewhat different regulatory mechanism has been worked out for *B. subtilis*.[25]

Glutamine synthetase has an important function in addition to that of glutamine synthesis. Active glutamine synthetase (but not the adenylylated enzyme) binds to promoter regions of DNA (Chapter 15, Section B,1,b) to activate the transcription of a variety of genes including those for nitrogenase.[25a,b] Thus, a deficiency of glutamine turns on a number of genes involved in nitrogen metabolism. Accumulation of glutamine promotes

modification of the synthetase and loss of this gene activation.

The transfer of nitrogen from glutamine into other substrates has been discussed by Buchanan.[26] A number of antibiotic analogues of glutamine have been useful in studying these processes. Examples are the streptomyces antibiotics **L-azaserine** and 6-diazo-5-oxo-L-norleucine (DON).

$$^-N{=}\overset{+}{N}{=}CH{-}\overset{\overset{O}{\|}}{C}{-}O{-}CH_2{-}\underset{\underset{NH_3{}^+}{|}}{CH}{-}COO^-$$

C is the electrophilic center

O is replaced by CH_2 in DON

L-azaserine

These compounds act as alkylating agents, N_2 being released and a nucleophilic group from the enzyme becoming attached at the carbon atom indicated.

3. Uptake of Amino Acids by Cells

While cells of autotrophic organisms can make all of their own amino acids (by pathways described in succeeding sections), other cells utilize many preformed amino acids. Humans and other higher animals require a number of **essential amino acids** in the diet. Furthermore, cells of a given tissue may take up amino acids made in another tissue.

The active transport systems by which bacteria can take up amino acids have been mentioned briefly (Chapter 5, Section B,2). Another interesting active transport system, the **γ-glutamyl cycle**,[27] functions in mammalian cells. The cycle makes use of the γ-carboxyl group of glutamate, the same carboxyl that carries ammonia in the form of glutamine. In the amino acid transport process glutathione (Box 7-G) supplies the activated γ-glutamyl group. The amino acid to be transported reacts by **transpeptidation** (Eq. 14-14, step *a*), presumably in the cell membrane:

$$\text{(14-14)}$$

The resulting **γ-glutamylamino acid** enters the cytoplasm and releases the free amino acid by an internal displacement by the free amino group (Eq. 14-14, step *b*). Note that in this reaction the natural tendency of the 5-carbon glutamate to undergo cyclization is used to provide the driving force for release of the bound amino acid. The cyclic product 5-oxoproline is then opened in an ATP-requiring reaction (Eq. 14-14, step *c*).[28]

Cysteinylglycine formed in the initial transpeptidation step of Eq. 14-14 is hydrolyzed by a peptidase. Then, glutathione is regenerated in two ATP-dependent steps as indicated.

The significance of the γ-glutamyl cycle is not fully understood. However, the finding of a mentally retarded individual who excretes 25–50 g/day of 5-oxoproline in the urine (possibly because of a defective 5-oxoprolinase) suggests that the pathway is a very active one.[28]

4. Nitrogen Turnover within Cells

The reactions by which amino acids are incorporated into proteins have been considered briefly in Chapter 11, Section E,1 and are discussed further in Chapter 15, Section C. However, the formation of biologically active catalysts, hormones, and structural proteins is often not complete as a peptide chain leaves the ribosome and folds into a preferred conformation. Thus, proteins are very often hydrolyzed at specific sites and may undergo a variety of covalent modifications as discussed in Chapter 11, Section E,2.

While some of the chemical processing of newly formed peptides probably occurs in the cytoplasm,[29] some takes place after "segregation" of proteins that are to be secreted from cells in the cisternae (microsomal cavities) of the endoplasmic reticulum.[30] It is thought that the ribosomes forming these proteins are attached to the cytoplasmic side of the membranes of the ER and that the newly formed peptide chains are extruded through the membrane into the cisternal space. There, a variety of modifying enzymes may act.

One of the surprises coming from the studies of Schoenheimer (Section B, beginning) was the discovery that proteins within cells are in a continuous steady state of synthesis and degradation. Thus, synthetic and hydrolytic pathways together form a metabolic loop (Chapter 11, Section A,1). A generalization is that proteins secreted into extracellular fluids often undergo more rapid turnover than do those that remain within cells. Within cells some proteins are degraded much more rapidly than others, an important aspect of metabolic control. Protein turnover rates tend to be relatively low in plants.

While some proteases degrading intracellular proteins are probably located in the cytoplasm, the **cathepsins,** with acid pH optima, are present in the lysosomes[31,32] and neutral protease has been found in peroxisomes. Cathepsins as well as collagenases and elastases are apparently secreted into intercellular spaces where they participate in degradation of connective tissue. Overactivity of these enzymes may be important in some disease states.[33]

C. SYNTHESIS AND CATABOLISM OF THE GLUTAMIC ACID FAMILY OF COMPOUNDS

The 5-carbon skeleton of glutamic acid gives rise directly to those of proline, ornithine, and arginine. The reactions are outlined in Fig. 14-2. Arginine, in turn, is directly involved in the urea cycle (Fig. 14-4) and is the biosynthetic precursor of the polyamines.

1. Proline and Arginine

The ATP-dependent reduction of the γ-carboxyl group of glutamate by NADPH (reaction c, Fig. 14-2) is of a standard biosynthetic reaction type, the opposite of the oxidation reaction of Fig. 8-14. Like the latter it presumably occurs via an acyl phosphate or acyl adenylate intermediate. Glutamate semialdehyde, the oxidation product, cyclizes spontaneously and is converted to proline by further reduction (Fig. 14-2).

If the amino group of glutamate is blocked by acetylation prior to the reduction to the semialdehyde (Fig. 14-2, step d) cyclization is prevented. The γ-aldehyde group can be transaminated to an

amino group and the acetyl blocking group removed to form **ornithine.*** The latter is converted on to arginine by the reactions of Fig. 14-4. These reactions not only provide pathways for biosynthesis of arginine in all organisms but also provide for the synthesis of urea, the principal nitrogenous end product in mammals and many other organisms. An interesting finding is that *Neurospora* grown in a minimal medium accumulates large amounts of both ornithine and arginine, over 98% of which is sequestered in vesicles within the cytoplasm.[33b]

2. The Urea Cycle

In 1932, H. A. Krebs and K. Z. Henseleit[33c] proposed that in liver slices urea is formed by a cyclic process in which ornithine is converted first to **citrulline** and then to arginine. The hydrolytic cleavage of arginine produces urea and regenerates ornithine (Fig. 14-4, bottom). Subsequent experiments fully confirmed this proposal. Let us trace the entire route of nitrogen removed by the liver from excess amino acids. Transaminases (step a, Fig. 14-4, right center) transfer the nitrogen to α-ketoglutarate to form glutamate. Since urea contains two nitrogen atoms, two molecules of glutamate must donate their amino groups. One of these molecules of glutamate is directly deaminated by glutamate dehydrogenase to form ammonia (step b). This ammonia is combined with bicarbonate (step c) to form carbamoyl phosphate which transfers the carbamoyl group onto ornithine to form citrulline (step d). The second molecule of glutamate transfers its nitrogen by transamination to oxaloacetate (reaction e) to form aspartate. The aspartate molecule is incorporated intact into **argininosuccinate** by reaction with citrulline (reaction f). Undergoing a simple elim-

---

* Ornithine is not usually a constituent of proteins. However, a urate-binding glycoprotein (MW $\simeq$ 67,000) of plasma is reported to contain 43 residues of ornithine. It is postulated that a special arginase is needed to form these residues and that it may be lacking in some cases of gout in which the urate-binding capacity of blood is impaired.[33a]

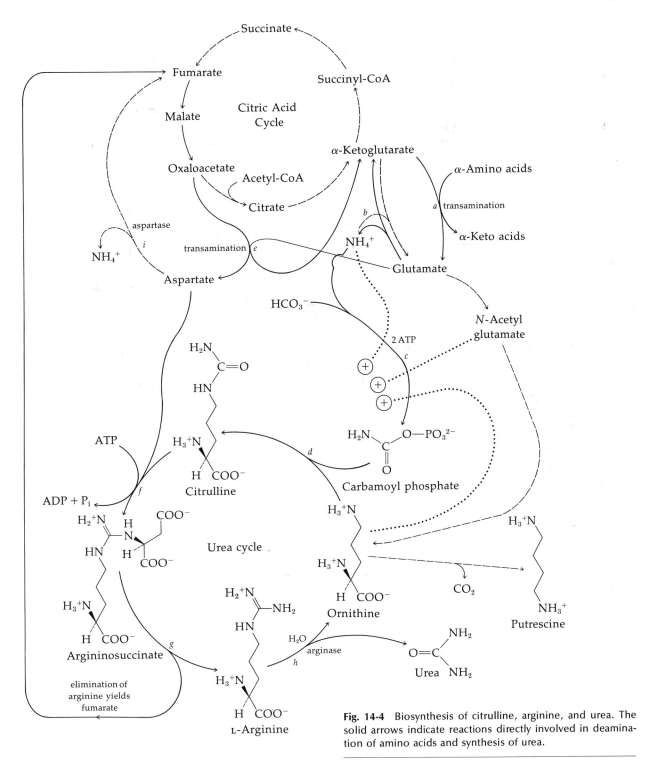

Fig. 14-4 Biosynthesis of citrulline, arginine, and urea. The solid arrows indicate reactions directly involved in deamination of amino acids and synthesis of urea.

ination reaction, the 4-carbon chain of argininosuccinate is converted to fumarate (step g) with arginine appearing as the elimination product. Finally, the hydrolysis of arginine (step h) yields urea and regenerates ornithine.

The first of the individual steps in the urea cycle is the formation of carbamoyl phosphate. Carbon dioxide and ammonia spontaneously equilibrate with carbamic acid (Eq. 14-15):

$$CO_2 + NH_3 \rightleftharpoons H_2N-\overset{\overset{\displaystyle O}{\|}}{C}-O^- + H^+ \quad (14\text{-}15)$$

It was observed in early experiments with bacteria that a kinase is able to convert carbamate into carbamoyl phosphate (Eq. 14-16):

$$NH_2-\overset{\overset{\displaystyle O}{\diagup\!\!\!\!\diagdown}}{C}{\diagdown O^-} + ATP \rightleftharpoons H_2N-\overset{\overset{\displaystyle O}{\diagup\!\!\!\!\diagdown}}{C}{\diagdown O-\textcircled{P}} + ADP \quad (14\text{-}16)$$

However, the equilibrium constant is low (0.04 at pH 9, 10°C). It is now believed that carbamate kinase normally functions in the opposite direction, providing a means of synthesis of ATP for bacteria degrading arginine (Section C,5,d).

Carbamoyl phosphate synthetases harness the cleavage of *two* molecules of ATP to formation of one molecule of carbamoyl phosphate (reaction c, Fig. 14-4).[34] In bacteria such as *E. coli* there is a single synthetase that provides carbamoyl phosphate for biosynthesis of both arginine and the pyrimidines (Fig. 14-29). On the other hand, in fungi and higher animals there are two: Carbamoyl phosphate synthetase I provides substrate for formation of citrulline from ornithine, while carbamoyl phosphate II functions in pyrimidine synthesis. Synthetase I is found in mitochondria and synthetase II in the cytoplasm.

The carbamoyl-phosphate synthetase of *E. coli* contains biotin (Chapter 8, Section C,2). Presumably a carboxybiotin intermediate is used to carboxylate ammonia to an enzyme-bound carbamate which is phosphorylated by ATP. The enzyme contains two subunits of MW $\sim 130,000$ and $\sim 42,000$. A striking fact about the *E. coli* enzyme is that *either* free ammonia or glutamine can act as an amino donor.[26,35] Furthermore, the light subunit has **glutaminase** activity; i.e., it is able to hydrolyze glutamine to ammonia. In fact, it appears that glutamine may be hydrolyzed to free ammonia in the active sites of all those enzymes catalyzing reactions in which glutamine serves as an ammonia donor (see p. 325). The heavy subunit of the synthetase not only contains biotin but also undergoes allosteric modification by a number of effectors. Both ornithine and ammonia are activators (Fig. 14-4), whereas uracil, a pyrimidine end product, exerts feedback inhibition.

Mammalian carbamoyl-phosphate synthetase is not known to contain biotin, but it does function in a multistep mechanism. It is thought that an initially formed carboxyl phosphate (enzyme-bound) reacts with ammonia to form an enzyme-bound carbamate. A powerful allosteric effector for the liver synthetase is **N-acetylglutamate** (Fig. 14-4), a precursor of ornithine.[17]

Continuing around the urea cycle, the equilibrium constant for ornithine transcarbamoylase (reaction d, Fig. 14-4) is very high so that ornithine is completely converted to citrulline. The conversion of citrulline to argininosuccinate and the subsequent breakdown to fumarate and arginine appears remarkably complex.

For the argininosuccinate synthetase reaction it is established that the ureido group of citrulline is activated by ATP (Eq. 14-17, step a):

$$
\begin{array}{ccc}
\underset{\substack{\displaystyle | \\ R}}{\overset{\displaystyle H_2N}{\diagdown}}{C}={}^{18}O & \xrightarrow[a]{ATP\ PP_i} & \underset{\substack{\displaystyle | \\ R}}{\overset{\displaystyle HN}{\diagdown}}{C}-{}^{18}O-\overset{\overset{\displaystyle O}{\|}}{\underset{\underset{\displaystyle O^-}{|}}{P}}-O-Ad & \xrightarrow[b]{AMP} & \underset{\substack{\displaystyle | \\ R}}{\overset{\displaystyle HN}{\diagdown}}{C}-NH-Asp \\
\end{array}
\quad (14\text{-}17)
$$

aspartate

Citrulline   Argininosuccinate

Thus, ^{18}O present in this group is transferred into AMP. A citrulline adenylate intermediate is likely (Eq. 14-17, center). **Argininosuccinase** (reaction g, Fig. 14-4) catalyzes the elimination of arginine with formation of fumarate. It is entirely analogous to the bacterial aspartase that eliminates ammonia from aspartate to form fumarate. Like the latter enzyme (Chapter 7, Section H,6,d), argininosuccinase promotes a trans elimination. The fumarate produced can be reconverted through reactions of the citric acid cycle to oxaloacetate, which can be reaminated to aspartate.

Aspartate is used to introduce amino groups in an entirely similar way in other metabolic sequences such as in the formation of adenylic acid from inosinic acid (Fig. 14-32).

The cleavage of arginine to ornithine converts the biosynthetic route to arginine into a cycle for the synthesis of urea. This cyclic pathway is unique to organisms that excrete nitrogenous wastes as urea, but the biosynthetic path is nearly ubiquitous.

Human adults excrete approximately 20 g of urea nitrogen per day. If this rate decreases ammonia accumulates in the blood to toxic levels. Normally, plasma contains 0.5 mg l^{-1} of ammonia

and only 2–3 times this level is required to produce toxic symptoms. Therefore, it is not surprising that a number of hereditary enzyme deficiencies affecting the urea cycle have been observed. One of the most common (**argininosuccinic aciduria**) is a deficiency of the breakdown of argininosuccinic acid. Both lethal and nonlethal variants of this disease are known. Over 20 cases of the latter have been observed. A common feature of all of the hereditary defects of the urea cycle is an intolerance to high protein intake and mental symptoms. Toxic accumulation of ammonia in blood is often seen in **alcoholic liver cirrhosis** as a result of a decreased capacity of the liver for synthesis of urea.

3. Amidino Transfer and Creatine Synthesis

The terminal amidino group

$$H_2N-\overset{\displaystyle |}{C}=NH_2^+$$

of arginine is transferred intact to a number of other substances in simple displacement reactions. An example is the formation of **guanidinoacetic acid** (Eq. 14-18):

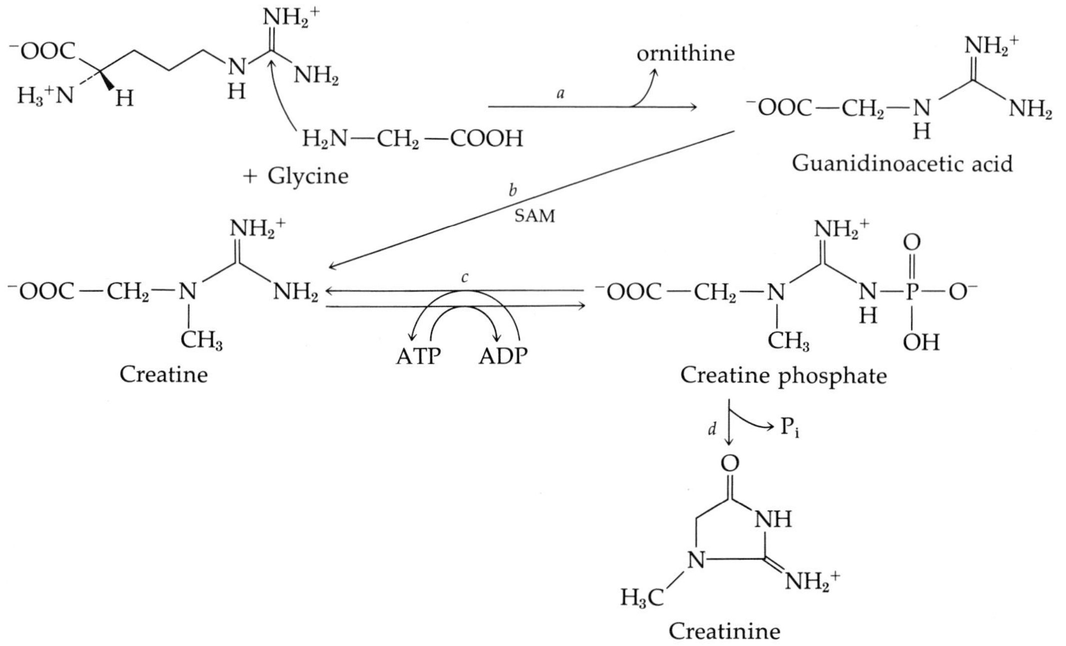

Guanidinoacetic acid

Creatine

Creatine phosphate

Creatinine

(14-18)

Transmethylation from *S*-adenosylmethionine (Eq. 14-18, step *b*) converts guanidinoacetic acid on to **creatine,** a compound of special importance in muscle. Creatine kinase reversibly transfers the phosphoryl group of ATP to creatine to form the *N*-phosphate (Eq. 14-18, step *c*). **Creatine phosphate** serves as an important "energy buffer" for muscular contraction (Box 10-F). Through the reversible action of creatine kinase it is able to rapidly transfer its phosphoryl group back onto ADP as fast as the latter is formed during the hydrolysis of ATP in the contraction process. An end product of creatine phosphate metabolism is the anhydride **creatinine** formed from creatine phosphate as is indicated in Eq. 14-18, step *d* as well as directly from creatine. The urinary creatinine excretion for a given individual is extremely constant from day to day, the amount excreted apparently being directly related to the muscle mass of the person. Another example of the transfer of amidino groups from arginine is found in the synthesis of streptomycin (Box 12-A).

4. The Polyamines[36–40]

A series of related compounds derived in part from arginine, the polyamines are present in all cells in relatively large amounts (frequently in millimolar concentrations). The content of polyamines in cells often bears a stoichiometric relationship to that of RNA. However, the T-even bacteriophage and most bacteria contain polyamines in association with DNA. Polyamines are thought to have many functions. They can substitute to some extent for cellular K^+ and Mg^{2+}, and they may play essential controlling roles in nucleic acid and protein synthesis.[36] A specific role of spermidine in cell division seems likely.[40a] Polyamines may interact with double helical nucleic acids by bridging between strands, the positively charged amino groups interacting with the phosphates of the nucleic acid backbones.[40] In one model (Tsuboi[40b]) the tetramethylene portion of the polyamine lies in the minor groove bridging three base pairs and the trimethylene portions (one in spermidine, two in spermine) bridge adja-

cent phosphate groups in one strand. Polyamines may also stabilize supercoiled or folded DNA.

$$H_3{}^+N \diagdown\diagup\diagdown\diagup\diagdown NH_3{}^+$$
Putrescine (dication of 1,4-diaminobutane)

$$H_3{}^+N \diagdown\diagup\diagdown\diagup\diagdown\diagup NH_3{}^+$$
Cadaverine

$$H_3{}^+N \diagdown\diagup\diagdown\underset{H}{N}\diagdown\diagup\diagdown\diagup NH_3{}^+$$
Spermidine

$$H_3{}^+N \diagdown\diagup\diagdown\underset{H}{N}\diagdown\diagup\diagdown\underset{H}{N}\diagdown\diagup NH_3{}^+$$
Spermine

The 4-carbon putrescine arises most simply by decarboxylation of ornithine. However, it can also be formed by decarboxylation of arginine to agmatine and hydrolysis of the latter compound (Eq. 14-19)[37]:

$$\text{Arginine} \xrightarrow{\text{CO}_2} \text{agmatine} \xrightarrow[\text{H}_2\text{O}]{\text{urea}} \text{putrescine} \quad (14\text{-}19)$$

Putrescine is present in all cells, and all cells are able to convert it on to spermidine. This is accomplished by decarboxylation of *S*-adenosylmethionine (Eq. 14-20, step *a*) and transfer of the propylamine group from the resulting decarboxylation product onto an amino group of putrescine (Eq. 14-20, step *b*)[41]:

$$SAM \xrightarrow[a]{CO_2} \text{adenine}-\text{ribose}-\overset{\underset{\displaystyle CH_3}{|}}{S^+}-CH_2CH_2CH_2NH_3{}^+ \xleftarrow{\text{putrescine}-NH_2}$$

$$\downarrow b$$

$$\text{adenine}-\text{ribose}-S-CH_3 \quad\qquad \searrow \text{spermidine}$$
Methylthioadenosine

$$(14\text{-}20)$$

When *E. coli* cells enter the stationary phase of the growth curve (Chapter 6, Section C), most of the spermidine is converted to glutathionylspermidine (α-glutamylcysteinylglycylspermidine).[42] Acetylation of spermidine also occurs.

The more complex spermine is found only in eukaryotes. An interesting historical note is that Anthony von Leeuwenhoek with one of his first microscopes observed crystals of the phosphate salt of spermine in human semen in 1678. The 5-carbon diamine cadaverine arises from decarboxylation of lysine.

The functions and further metabolism of the polyamines are only now being investigated intensively. Within *E. coli* 1,4-diaminobutane undergoes transamination to α-aminobutyraldehyde which cyclizes (Eq. 14-21). Diamine oxidases of

$$\text{H}_2\text{N CH}_2\text{CH}_2\text{CH}_2 \text{ CHO} \longrightarrow \qquad\qquad (14\text{-}21)$$

$$\Delta^1\text{-Pyrroline}$$

animal tissues oxidize 1,4-diaminobutane with formation of the same products. A copper-containing oxidase of beef serum oxidizes spermidine to a mono-aldehyde and spermine to a dialdehyde.[38] Although the latter are highly toxic materials it is thought that they may play an essential role in regulation of nuclear metabolism.

Oxidative cleavages of spermine to spermidine and of the latter to 1,4-diaminobutane also appear to occur in the animal body and a substantial amount is excreted in the urine.[38]

5. Catabolism of Glutamate and Related Amino Acids

The reversibility of glutamic dehydrogenase means that excess glutamate can be converted back to α-ketoglutarate readily. Ketoglutarate can be degraded to succinyl-CoA and via β oxidation to malate, pyruvate, and acetyl-CoA. The latter can reenter the citric acid cycle and be oxidized to CO_2 (Eq. 14-22):

Glu $\rightarrow \alpha$-ketoglutarate $\rightarrow$ succinyl-CoA $\rightarrow \rightarrow \rightarrow$
   malate $\rightarrow$ pyruvate $\rightarrow$ acetyl-CoA $\rightarrow CO_2$ (14-22)

Closely analogous pathways of degradation apply to many other amino acids. In most cases there is transamination to the corresponding α-keto acid. Then β oxidation and breakdown to such compounds as pyruvate and acetyl-CoA follow.

a. Catabolism Initiated by Decarboxylation

An alternative pathway for glutamate degradation is the γ-aminobutyrate shunt discussed in Chapter 9 (Fig. 9-4). This pathway is initiated by a pyridoxal phosphate-dependent decarboxylation rather than by a deamination or transamination. Since decarboxylases are known for most amino acids, there are usually alternative breakdown pathways initiated in this way. γ-Aminobutyrate is believed to function in the brain as an important neurotransmitter (Chapter 16, Section B,4,b).

b. Fermentation of Glutamate

Special problems face anaerobic bacteria subsisting on amino acids. Their energy needs must be met by a balanced fermentation reaction. Two such fermentations of glutamate are outlined in Fig. 14-5. The first (Eq. 14-23)[43] is the breakdown of glutamate to CO_2, ammonia, acetate$^-$ and butyrate$^-$:

2 Glutamate$^-$ + 2 H_2O + $H^+ \longrightarrow$ 2 CO_2 + 2 NH_4^+
$\qquad\qquad\qquad$ + 2 acetate$^-$ + butyrate$^-$ (14-23)
$\Delta G'$ (pH 7) = -131 kJ

The sequence begins with the γ-aminobutyrate shunt reactions (steps *a* and *b*, Fig. 14-5), but succinic semialdehyde is reduced to γ-hydroxybutyric acid using the NADH generated in the trans-deamination process. With the aid of a CoA-transferase (step *d*), two molecules of the CoA ester of this hydroxy acid are formed at the expense of two molecules of acetyl-CoA. Use is then made of a β,γ elimination of water, analogous to that involved in the formation of vaccenic acid (Eq. 12-14). Isomerization (perhaps by the same enzyme that catalyzes elimination) forms crotonyl-CoA (step *f*). The latter undergoes dismutation, one-half being reduced to butyryl-CoA and one-half being hydrated and oxidized to

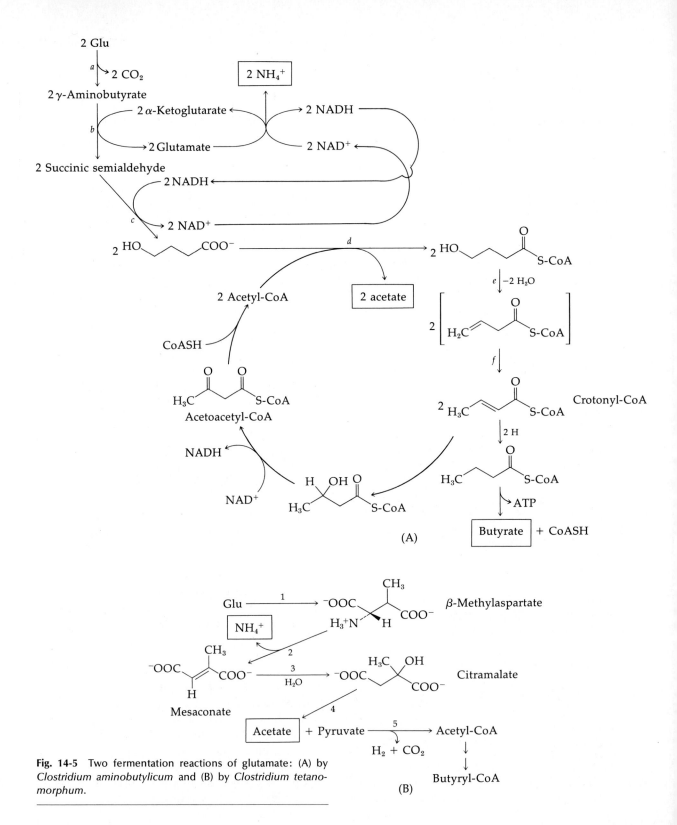

Fig. 14-5 Two fermentation reactions of glutamate: (A) by *Clostridium aminobutylicum* and (B) by *Clostridium tetano-morphum*.

acetoacetyl-CoA in the standard β-oxidation sequence. Acetoacetyl-CoA is cleaved to regenerate the two molecules of acetyl-CoA. The organism can gain one molecule of ATP through cleavage of the butyryl-CoA. Perhaps a second can be gained by oxidative phosphorylation between the NADH produced in the formation of acetoacetyl-CoA and the reduction of crotonyl-CoA to butyryl-CoA. The two processes take place at sufficiently different redox potentials to permit this kind of coupling.

The second glutamate fermentation is initiated by isomerization of glutamate to β-methylaspartate, a reaction catalyzed by a vitamin B_{12} containing mutase (Table 8-6). This rearrangement of structure permits α,β elimination of ammonia, a process not possible in the original glutamate. Hydration of the resulting unsaturated compound to **citramalate** and aldol cleavage yields acetate and pyruvate. Acetate is one of the usual end products of the fermentation. The pyruvate can be cleaved to H_2, CO_2, and acetyl-CoA by the pyruvate-formate-lyase system and cleavage of the acetyl-CoA can provide ATP. Alternatively, two molecules of acetyl-CoA can be coupled and reduced to butyryl-CoA. This reaction would necessitate the use of the reducing power generated in the cleavage of pyruvate to reduce crotonyl-CoA rather than to be released as H_2. Thus, the stoichiometry would be identical to that in the fermentation by *Clostridium aminobutylicum*.

c. Degradation of Proline

One route of catabolism of proline is essentially the reverse of its formation from glutamate. **Proline oxidase** yields Δ^1-pyrroline-5-carboxylate.

Δ^1-Pyrroline-5-carboxylate

The corresponding open chain aldehyde, formed by hydrolysis, is oxidized back to glutamate. Alternatively, degradation can be initiated by oxidation on the other side of the ring nitrogen to form Δ^1-pyrroline-2-carboxylate. The metabolic fate of

Δ^1-Pyrroline-2-carboxylate

this compound is uncertain. Anaerobic bacteria may reduce proline to 5-aminovalerate (Eq. 8-34) and couple this reaction to the oxidative degradation of another amino acid (Stickland reaction).

A corresponding pathway for breakdown of 4-hydroxy-L-proline of collagen yields glyoxylate and pyruvate (Eq. 14-24):

$$\text{pyruvate} + \text{glyoxylate} \qquad (14\text{-}24)$$

Oxidation on the other side of the ring nitrogen of hydroxyproline is utilized by some pseudomonads to convert the amino acid to α-ketoglutarate.

d. Catabolism of Arginine

Arginine is also converted back to glutamate and α-ketoglutarate. The initial step is removal of the guanidino group to form ornithine. This can occur by the action of arginase with formation of urea (Fig. 14-4). The alternative **arginine dihydrolase** pathway is initiated by a special hydrolase that cleaves arginine to citrulline and ammonia. Phosphorolysis of citrulline then yields carbamoyl phosphate. The breakdown of the latter to CO_2 and ammonia (catalyzed by carbamate kinase, Eq. 14-16) can be utilized for generation of ATP by microorganisms that subsist on arginine.

Degradation of L-arginine by *Streptomyces griseus* is initiated by a hydroxylase that causes decarboxylation and conversion of the amino acid into an amide (Eq. 14-25):

$$R-\overset{\overset{\displaystyle H}{|}}{\underset{\underset{\displaystyle NH_3^+}{|}}{C}}-COO^- \xrightarrow[O_2 \quad H_2O]{CO_2} R-\overset{\overset{\displaystyle}{||}}{\underset{\underset{\displaystyle O}{}}{C}}-NH_2 \qquad (14\text{-}25)$$

The reaction is strictly analogous to that catalyzed by lysine oxygenase (Eq. 10-49). The product in the case of arginine is γ-guanidinobutyramide which is further degraded by the hydrolysis of the amide group and cleavage of the guanidino group to form urea and γ-aminobutyrate. *Pseudomonas putida* initiates degradation of arginine by transamination to the corresponding α-keto acid and oxidative decarboxylation with a thiamine diphosphate-requiring enzyme to γ-guanidino-butyraldehyde. Dehydrogenation and hydrolysis lead, again, to γ-aminobutyrate.[44]

D. COMPOUNDS DERIVED FROM ASPARTATE

The 4-carbon aspartate molecule is the starting point for synthesis of pyrimidines and of the amino acids lysine, methionine, isoleucine, and asparagine. The pathways are summarized in Fig. 14-6. Note that there are several branch points. Aspartate can be converted directly to either asparagine, carbamoyl aspartate (the precursor of pyrimidines), or to β-aspartyl phosphate and aspartate semialdehyde. The latter can be converted in one pathway to lysine and in another to homoserine. Homoserine can yield either homocysteine and methionine or threonine. While threonine is one of the end products and a constituent of proteins, it can also be converted further to α-ketobutyrate, a precursor of isoleucine.

Most of the chemistry has been considered already. The reduction of aspartate via β-aspartyl phosphate is a standard one. Conversion to methionine can occur in two ways. In *E. coli*, homoserine is succinylated with succinyl-CoA. The γ-succinyl group is then replaced by the cysteine molecule in a PLP-dependent γ-replacement reaction (Fig. 14-6). The product **cystathionine** (Eq.

8-22) undergoes β elimination to form homocysteine. On the other hand, in *Neurospora* a more direct γ replacement of the hydroxyl of homocysteine by a sulfide ion occurs. Methylation of homocysteine to methionine (Eq. 8-45) has been considered previously, as has the conversion of homoserine to threonine by the PLP-dependent **threonine synthetase** (Chapter 8, Section E,3,d). A standard PLP-requiring β elimination converts threonine to α-**ketobutyrate**, a precursor to isoleucine (Fig. 14-10).

The formation of **asparagine** is similar to that of glutamine. However, the asparagine synthetase of *E. coli*[45] cleaves ATP to AMP and PP_i rather than to ADP. A β-aspartyladenylate intermediate is thought to exist. In higher animals glutamine serves as the ammonia donor for synthesis of asparagine but NH_4^+ can also function.[46]

L-Asparaginase, a bacterial hydrolyzing enzyme is an effective antileukemic drug which, when injected into the blood, acts to deprive fast growing tumor cells of the exogenous asparagine needed for rapid growth.[47] Tissues with a low asparagine synthetase activity are also damaged, limiting the clinical usefulness of asparaginase.

1. Control of Biosynthetic Reactions of Aspartate

In *E. coli* there are three **aspartokinases** catalyzing the conversion of aspartate to β-aspartyl phosphate. All three enzymes catalyze the same reaction but they have very different regulatory properties, as is indicated in Fig. 14-6. Each enzyme is responsive to a different set of end products. The same is true for the two **aspartate semial-dehyde reductases** catalyzing the third step. Note that both repression of transcription and feedback inhibition of the enzymes themselves[48] are involved. It should be noted further that many more regulatory features could be added to the scheme in Fig. 14-6.

Fig. 14-6 Some biosynthetic reactions of aspartic acid: (⊖) feedback inhibition and (⊟) feedback repression.

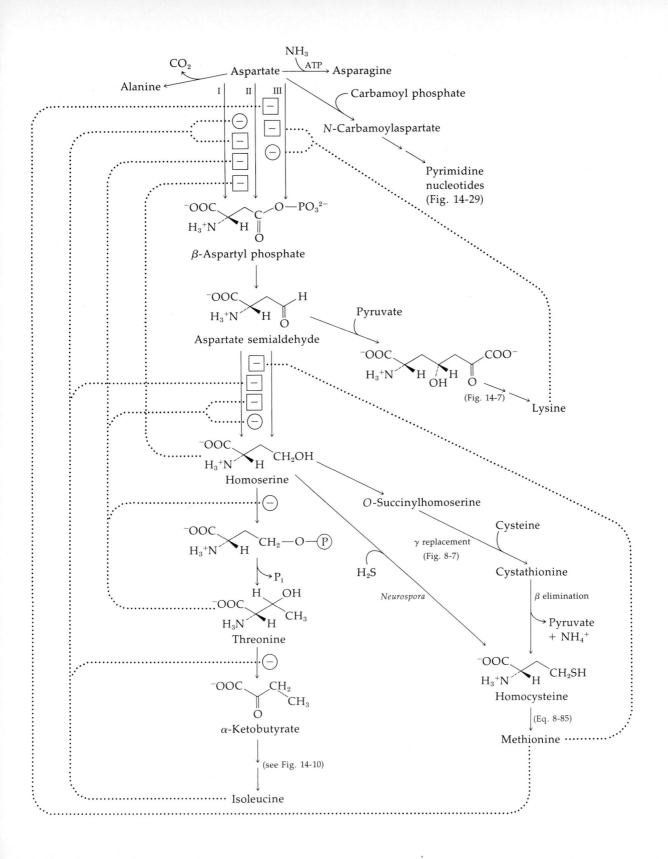

2. Lysine Diaminopimelate, Dipicolinic Acid, and Carnitine

Although lysine cannot be made at all by animals and is a nutritionally essential amino acid, there are two distinct pathways for its formation in other organisms. The **α-aminoadipate pathway** (shown in Fig. 14-2) is limited to a few lower fungi, the higher fungi, and euglenids. The 5-carbon α-ketoglutarate is the starting compound. Bacteria, other lower fungi, and green plants all

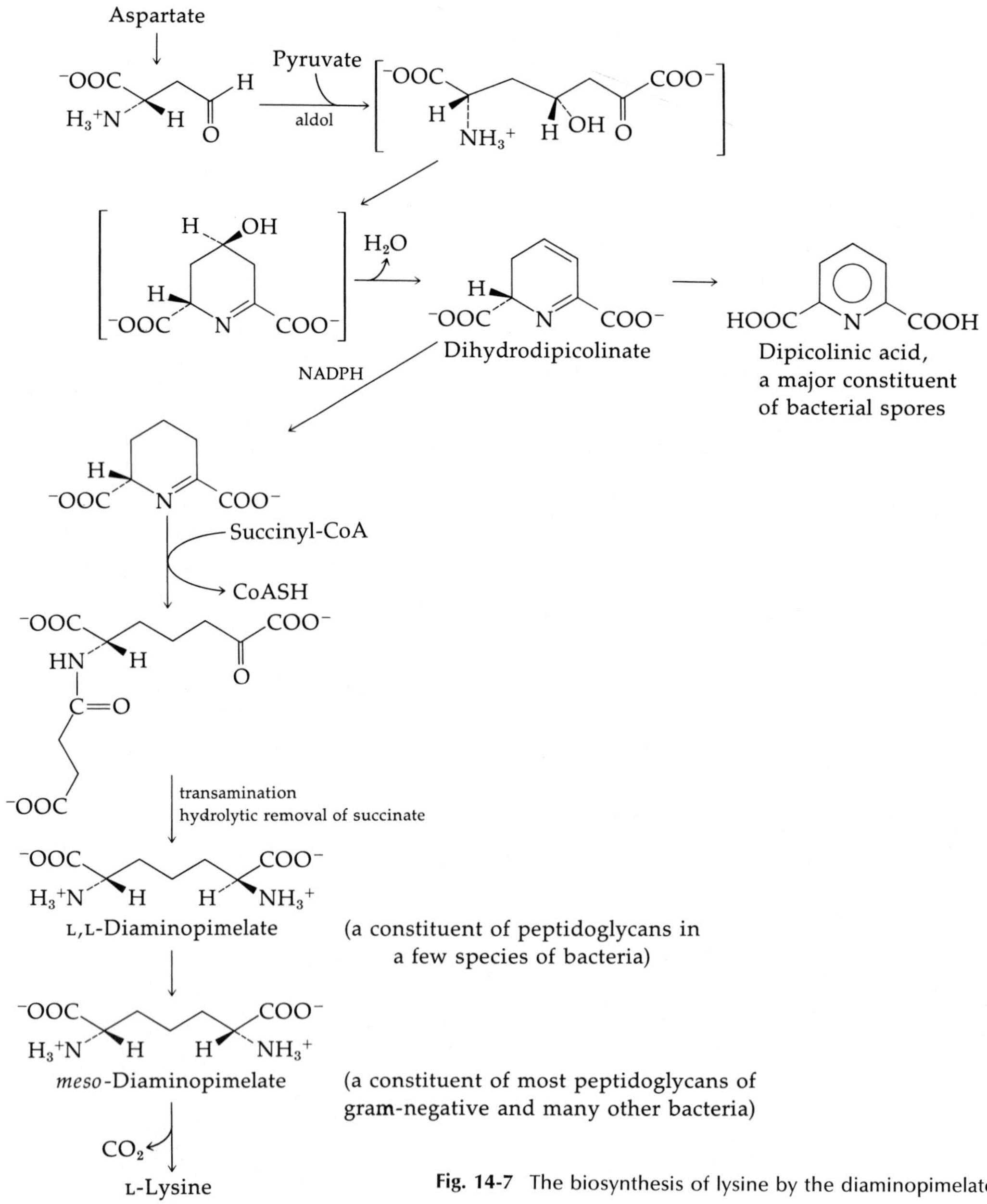

L,L-Diaminopimelate   (a constituent of peptidoglycans in a few species of bacteria)

meso-Diaminopimelate   (a constituent of most peptidoglycans of gram-negative and many other bacteria)

Fig. 14-7 The biosynthesis of lysine by the diaminopimelate pathway.

use the **diaminopimelate** pathway (Fig. 14-7) which originates with the 4-carbon aspartate.

The α-aminoadipate pathway (Fig. 14-2) parallels that of ornithine biosynthesis. α-Ketoglutarate undergoes chain elongation (Fig. 11-7) to α-ketoadipate which is transaminated to α-aminoadipate. This is followed by ATP-dependent reduction to the aldehyde. The final step of transamination is not accomplished in the usual way (with a PLP-dependent enzyme), but through formation of a Schiff base with glutamate and reduction to **saccharopine.** Oxidation now produces the Schiff base of lysine with ketoglutarate.

In the diaminopimelate pathway of lysine synthesis (Fig. 14-7) aspartate is converted to aspartate semialdehyde and a 2-carbon unit is added via aldol condensation with pyruvate. Decarboxylation at the end of the sequence yields lysine. A series of cyclic intermediates exist, but it is noteworthy that the initial product of the aldol condensation (bracketed in Fig. 14-7) is converted to diaminopimelic acid by a simple sequence involving α,β elimination of the hydroxyl group, reduction with NADPH, and transamination. The process is complicated by the natural tendency for ring closure. The required succinylation step (Fig. 14-7) serves to shift the equilibrium back in favor of open chain compounds. This pathway is of special significance to prokaryotic organisms for the reason that **dipicolinic acid** is formed as an important side product and because the diaminopimelic acids are formed as intermediates. The cyclic dipicolinic acid is a major constituent of bacterial spores but is rarely found elsewhere in nature. Both the L,L- and *meso*-diaminopimelic acids are constituents of petidoglycans of bacterial cell walls (Chapter 5, Section D).

Lysine is not only a constituent of proteins but also it can be methylated and degraded to γ-butyrobetaine.[49,50]

H₃C, CH₃ ... (γ-Butyrobetaine structure)

γ-Butyrobetaine

This compound is an intermediate in the biosynthesis of carnitine (Eq. 14-26)[50a]:

$$\text{Lysine} \longrightarrow \epsilon\text{-trimethyllysine} \longrightarrow \longrightarrow \longrightarrow$$

$$\gamma\text{-butyrobetaine} \xrightarrow{\text{(Eq. 10-56)}} \text{carnitine} \quad (14\text{-}26)$$

3. The Catabolism of Lysine

An unusual feature of lysine metabolism is that the α-amino group does not equilibrate with the "nitrogen pool." However, catabolism is initiated by deamination and proceeds by β oxidation. At least six variations of the β-oxidation process have been proposed for lysine degradation. The evolutionary differences concern the manner in which the two amino groups are removed from the carbon skeleton. In the seemingly simplest pathway (A in Fig. 14-8), which is used by *Flavobacterium fuscum*,[51] the ϵ-amino group is removed in a direct (but atypical) transamination. The resulting α-aminoadipate semialdehyde is oxidized to α-aminoadipate. The latter is degraded in *a sequence that is characteristic for the catabolism of amino acids: Transamination is followed by oxidative decarboxylation of the resulting α-keto acid and β oxidation of the coenzyme A derivative.* A decarboxylation step by which the terminal carboxyl group is removed is interposed in the β-oxidation sequence for lysine degradation.

Perhaps the initial transamination in pathway A is chemically difficult, for most organisms use more complex reaction sequences to form α-ketoadipate. In pathway B (which takes place in liver mitochondria and is believed to be the predominant pathway in mammals),[52] the ϵ-amino group is reductively coupled with α-ketoglutarate to form saccharopine. The latter is in turn oxidized on the opposite side of the bridge nitrogen to form glutamic acid and α-aminoadipate semialdehyde. The overall process is the same as direct transamination and just the opposite of that occurring in the aminoadipate pathway of biosynthesis.

Pathway C was formerly believed important in mammals but may be used only for breakdown of D-lysine. This route, which has been established

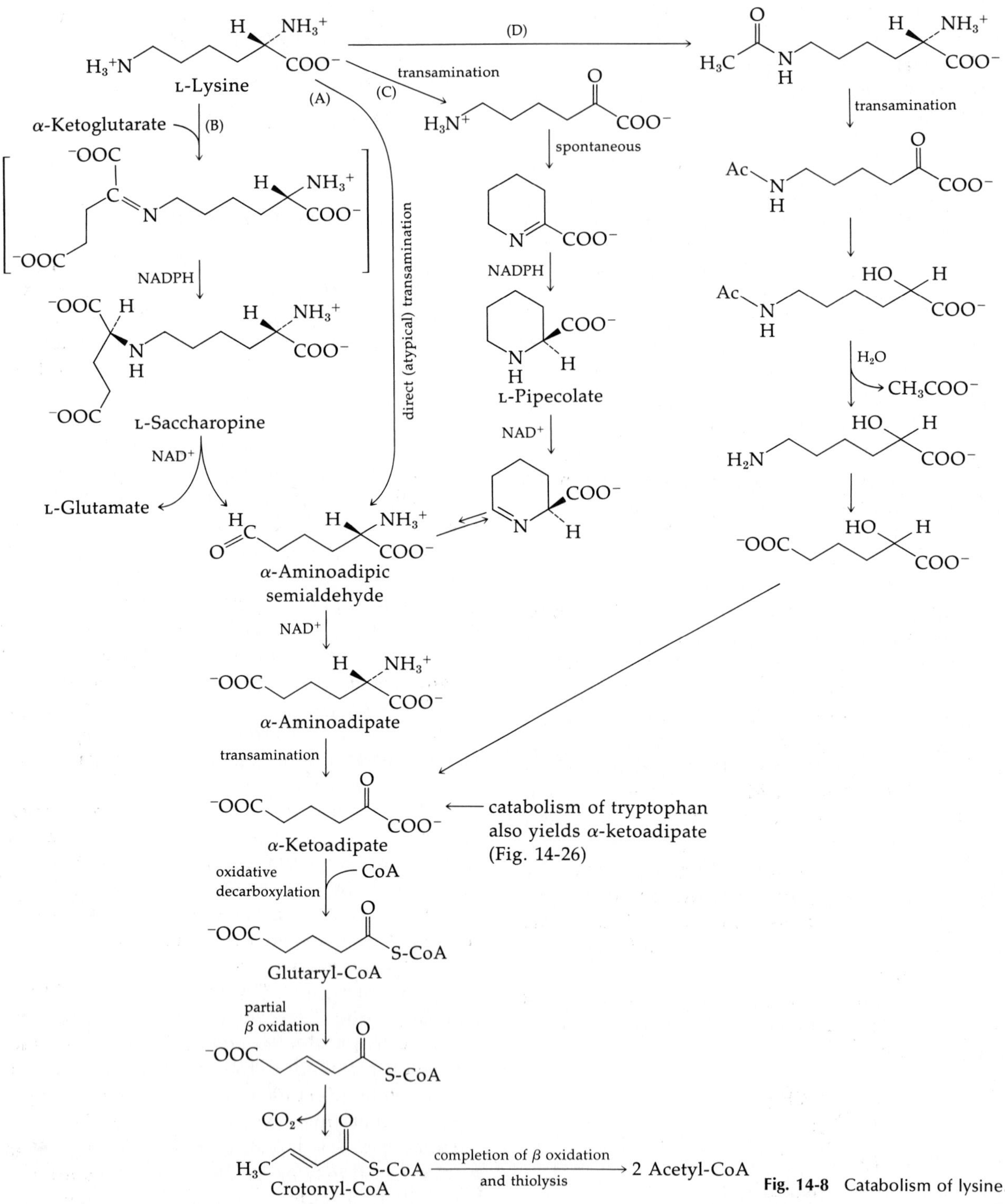

Fig. 14-8 Catabolism of lysine.

for *Pseudomonas putida*,[53] also occurs by a transamination through a reduction–oxidation sequence. This time it is strictly internal, the oxidizing carbonyl group being formed via transamination of the α-amino group of lysine. Pathway D, apparently used by yeasts,[54] avoids cyclic intermediates by acetylation of the ε-amino group prior to transamination. Then the α-keto group is effectively blocked by reduction to an alcohol, the blocking group is removed from the ε-amino group, and that end of the molecule is oxidized in a straightforward way to a carboxyl group. Now the hydroxyl introduced at position 2 is presumably oxidized back to the ketone which again can be converted to give α-ketoadipate. (However, there is some uncertainty about the point at which pathway D rejoins the others.)

Some bacteria, e.g., *Pseudomonas putida*,[53] degrade L-lysine with an oxygenase (Eq. 10-49) to δ-aminovaleramide:

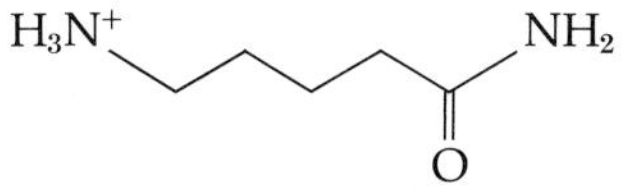

The product is hydrolyzed and oxidized to **glutaryl-CoA,** rejoining the pathways shown in Fig. 14-8. A remarkable and very different approach to lysine breakdown has been developed by bacteria of the genus *Clostridium*,[55] which obtain energy from the fermentation of Eq. 14-27:

L-Lysine + 2 H$_2$O $\longrightarrow$
$\qquad$ butyrate$^-$ + acetate$^-$ + 2 NH$_4^+$ (14-27)

The reaction is coupled to formation of one molecule of ATP from ADP and P$_i$. Two pathways have been worked out. In the first lysine is acted upon by a pyridoxal phosphate-dependent **L-lysine 2,3-aminomutase**[56] (Eq. 14-28, step *a*) to convert it to β-lysine (3,6-diaminohexanoate). The latter is further isomerized (Eq. 14-28, step *b*) by the vitamin B$_{12}$ and PLP-dependent β-lysine mutase.[57] Oxidative deamination to a 3-keto compound (Eq. 14-28, step *c*) permits chain cleavage. The reader can easily propose the remaining reactions of chain cleavage, ATP synthesis, elimination of ammonia,[58] and balancing of the redox steps. An alternative pathway begins with a racemase (Eq. 14-28, step

d) and isomerization of the resulting D-lysine by another B$_{12}$ and PLP-dependent enzyme (Eq. 14-28, step *e*). Oxidative deamination presumably occurs but the mechanism for chain cleavage is not so obvious. It does occur between C-4 and C-5 as indicated by the dotted line in Eq. 14-28.

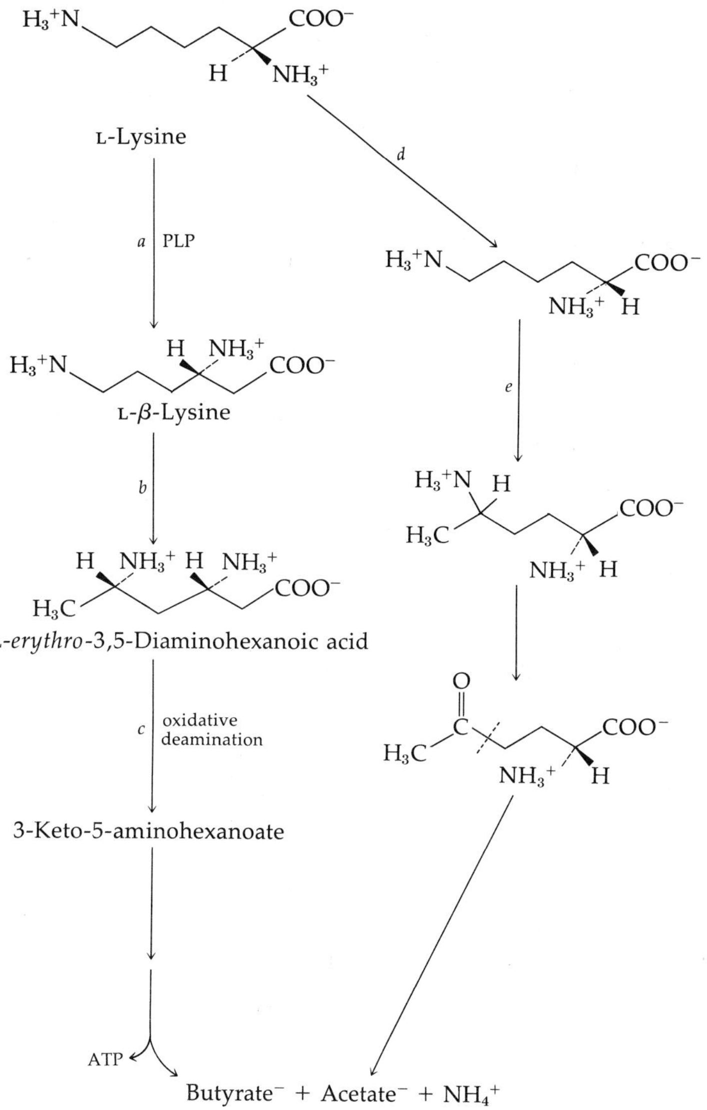

(14-28)

Why are there so many pathways of lysine breakdown? The answer is probably related to the ease of spontaneous formation of cyclic intermediates as occur in the pipecolate pathway (pathway C, Fig. 14-8). These intermediates may have been too stable for efficient metabolism so the indirect pathways were evolved. In the case of the fermentations additional constraints are imposed on the pathways by the need for balanced redox processes with a net free energy decrease.

4. Metabolism of Methionine

Methionine is incorporated into proteins as such and as **N-formylmethionine** at the N-terminal ends of bacterial proteins (steps a and b, Fig. 14-9). Methionine can undergo transamination to the corresponding keto acid (step c) in both animals

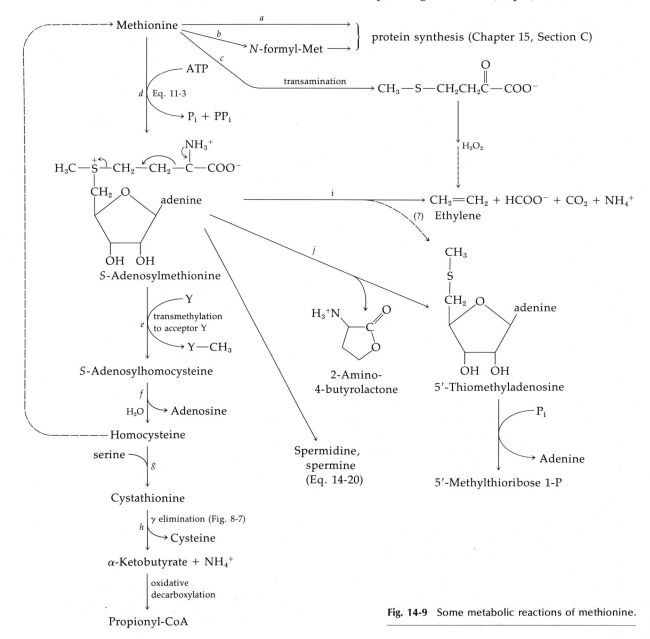

Fig. 14-9 Some metabolic reactions of methionine.

and plants, but the reaction does not appear to be important quantitatively. A major route of metabolism of methionine involves its conversion to **S-adenosylmethionine** (SAM, step *d*, Fig. 14-9). The reaction has been discussed in Chapter 11, Section B,2 and the function of SAM in transmethylation (step *e*) has been considered in Chapter 7, Section C,2. The product of transmethylation, **S-adenosylhomocysteine,** is converted to homocysteine in an unusual hydrolytic reaction by which adenosine is removed (step *f*).* Homocysteine can be reconverted to methionine, as indicated by the dashed line in Fig. 14-9 and in Eq. 8-85. Another important pathway of homocysteine metabolism is conversion to cysteine through steps *g* and *h*, Fig. 14-9. The sequence is discussed in Section G. Another product of this pathway is α-ketobutyrate, which can be oxidatively decarboxylated to propionyl-CoA and further metabolized or which can be converted to leucine (Fig. 14-10).

A very interesting reaction which may proceed via SAM is the formation of **ethylene** in plants.[59,60] Ethylene has been recognized since 1858 as causing a thickening of stems of plants and a depression in the rate of elongation. In 1917 it was established that the compound is produced in fruit and that addition of ethylene hastened the ripening of fruit. Now ethylene is regarded as an established plant hormone having a variety of effects including retardation of mitosis. The synthesis of ethylene is influenced in part by the hormone **auxin** (Chapter 16, Section A,3) and by red light. It is known that the keto acid analogue of methionine can be oxidatively decarboxylated by H_2O_2 to form ethylene and two molecules of CO_2. However, in plants CO_2 is formed from C-1 of methionine but C-2 becomes formate (Fig. 14-9, step *i*). Furthermore, the S—CH$_3$ group is somehow recycled into methionine again. If SAM is the precursor of ethylene, another product would logically be **5'-thiomethyladenosine** as indicated in Fig. 14-9. The latter compound is formed in an-

imals in an important catabolic reaction of SAM (step *j*, Fig. 14-9). The reaction is an internal displacement on the γ-methylene group by the carboxylate group of SAM.

5. Metabolism of Threonine

Excess threonine is degraded principally through the action of L-threonine dehydratase (Eq. 14-29, step *a*), a typical β-elimination reaction.

$$\text{(14-29)}$$

This pyridoxal phosphate-requiring enzyme is produced in high amounts in *E. coli* grown on a medium devoid of glucose and oxygen. Under these circumstances the reaction provides a source of propionyl-CoA which can be converted to propionate with generation of ATP. This **biodegradative threonine dehydratase** (threonine deam-

---

* The enzyme has been shown to contain tightly bound NAD$^+$. It is likely that this causes an oxidation of the 3' position of the adenosine permitting elimination of homocysteine. Addition of water and reduction yields adenosine.[58a]

inase)[61,62] is allosterically activated by AMP, an appropriate behavior for a key enzyme in energy metabolism. A second **biosynthetic threonine dehydratase** is also produced by *E. coli* and is specifically required for production of α-ketobutyrate needed in the biosynthesis of isoleucine.[63] In 1956, H. E. Umbarger[63a] showed that this enzyme is inhibited by isoleucine, the end product of the synthetic pathway. This discovery was instrumental in establishing the concepts of feedback inhibition in metabolic regulation (Chapter 6, Section F,4) and of allostery.

A second catabolic reaction of threonine (Eq. 14-29, step *b*) is cleavage to glycine and acetaldehyde, catalyzed by serine hydroxymethyltransferase (Eq. 8-19). A third and quantitatively more important route is dehydrogenation (Eq. 14-29, step *c*) and decarboxylation to **aminoacetone** (Eq. 14-29, step *d*). Aminoacetone is a urinary excretion product but can also be oxidized (Eq. 14-29, step *e*) to **methylglyoxal** which can be converted to D-lactate through the action of glyoxalase (Chapter 7, Section K). Aminoacetone is also the source of 1-amino-2-propanol for the biosynthesis of vitamin B_{12} (step *f*, Box 8-K). It has been postulated that methylglyoxal is a natural growth regulator that prevents excessive proliferation of animal cells.[63b]

E. ALANINE AND THE BRANCHED CHAIN AMINO ACIDS

As indicated in Fig. 14-10, pyruvate is the starting material for the formation of both L- and D-alanine and of the branched chain amino acids valine, leucine, and isoleucine. The chemistry of the reactions has been discussed in the sections indicated in Fig. 14-10. In *Neurospora,* isoleucine and valine are synthesized in the mitochondria. There is evidence that a complex of the five enzymes needed for synthesis of valine from pyruvate and of isoleucine from threonine and pyruvate functions as a biosynthetic unit.[64]

An additional series of reactions[65] shown in this figure leads to **pantoic acid, pantetheine,** and **coenzyme A.** The initial reactions of the sequence do not occur in the animal body, explaining our need for pantothenic acid as a vitamin. Alanine also gives rise to a precursor of the vitamin biotin after condensation with the 7-carbon dicarboxylic acid unit of pimeloyl-CoA in a reaction analogous to that of Eq. 8-20.[65a,65b]

1. Catabolism

Degradation of amino acids most often begins with transamination to the corresponding α-keto acid and oxidative decarboxylation of the latter (reaction sequence 7B, Fig. 8-19). Alanine, valine, leucine, and isoleucine are all treated this way in the animal body. Alanine gives pyruvate and acetyl-CoA directly, but the others yield CoA derivatives that undergo the β-oxidation schemes shown in Fig. 14-11. There are some variations from the standard β-oxidation sequence for fatty acids (Fig. 9-1); for example, in the case of valine, the sequence proceeds only to the stage of addition of water to form the β-hydroxy derivative. The latter is converted to free 3-hydroxyisobutyrate and β oxidation is then completed by oxidation to methylmalonyl semialdehyde. The latter is decarboxylated to propionate,[66] propionate is converted to propionyl-CoA, and the latter is carboxylated to *S*-methylmalonyl-CoA. Further metabolism of the latter is indicated in Fig. 9-6.

In the degradation of isoleucine, β oxidation proceeds to completion in the normal way with generation of acetyl-CoA and propionyl-CoA. However, in the catabolism of leucine, after the initial dehydrogenation in the β-oxidation sequence, carbon dioxide is added using a biotin enzyme (Chapter 8, Section C). The double bond conjugated with the carbonyl of the thioester makes this carboxylation analogous to a standard β-carboxylation reaction. Why add the extra CO_2? The methyl group in the β position blocks complete β oxidation but an aldol cleavage would be possible to give acetyl-CoA and acetone. However, acetone is not readily metabolized further.

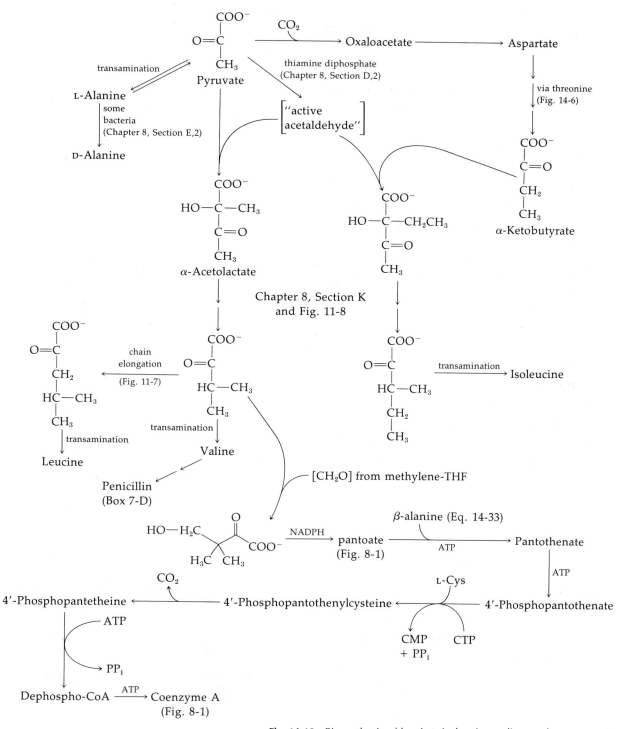

Fig. 14-10 Biosynthesis of leucine, isoleucine, valine, and coenzyme A.

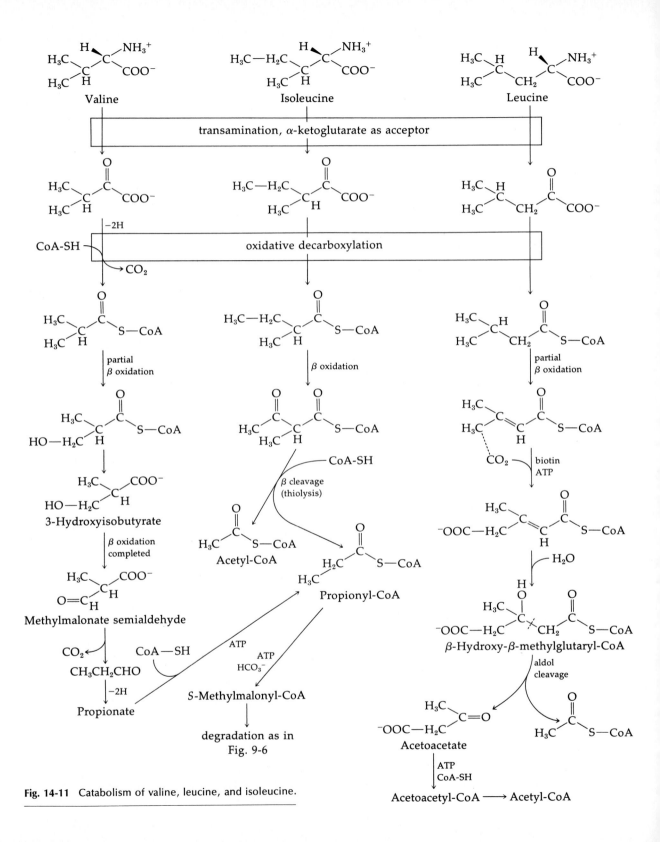

Fig. 14-11 Catabolism of valine, leucine, and isoleucine.

BOX 14-B

THE "MAPLE SYRUP URINE" DISEASE AND JAMAICAN VOMITING SICKNESS

In over 50 cases of a rare autosomal recessive condition (discovered in 1954) the urine and perspiration of the victims has a maple syrup odor.[a] The urine contains high concentrations of the branched chain α-keto acids formed by transamination of valine, leucine, and isoleucine. The odor arises from decomposition products of these acids. The biochemical defect lies in the enzyme catalyzing oxidative decarboxylation of the keto acids, as is indicated in Fig. 14-11.

Maple syrup urine disease (which may affect one person in ~200,000) is fatal in early childhood if untreated. Patients may survive on a low protein (gelatin) diet with supplementation with essential amino acids. However, treatment is difficult and a sudden relapse is apt to prove fatal.

It is of interest that a similar biochemical defect has been reported in a mutant of *Bacillus subtilus*.[b] This bacterium requires branched chain fatty acids in its cell membrane (Chapter 5, Section A,4) and requires branched acyl-CoA derivatives as starter pieces for their synthesis (Chapter 12, Section E). With the oxidative decarboxylation of the necessary keto acids blocked, the mutant is unable to grow unless supplemented with branched chain fatty acids.

A rare defect of catabolism of leucine is **isova-** leric acidemia, a failure in oxidation of isovaleryl-CoA. The symptoms of this disease are also present in the Jamaican vomiting sickness, caused by eating unripe ackee fruit. The fruit contain a toxin **hypoglycin A** with the following structure:[c,d]

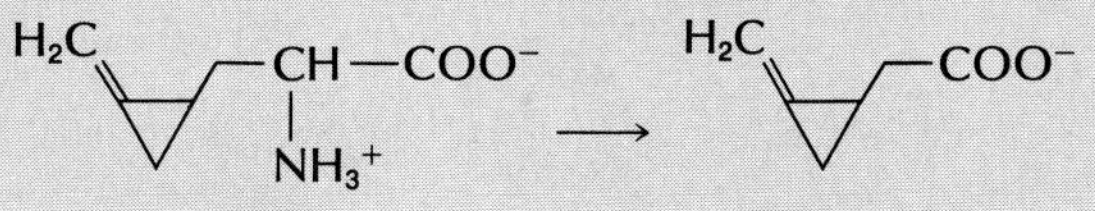

Hypoglycin A   Toxic metabolite

The compound specifically inhibits isovaleryl-CoA dehydrogenase, causing an accumulation of isovaleric acid in the blood. It has been suggested that depression of the central nervous system by isovaleric acid in the blood may be responsible for some of the symptoms. However, it is generally thought that death from the highly fatal Jamaican vomiting sickness comes from the hypoglycemic effect. Blood glucose levels may fall as low as 0.5 mM, one-tenth the normal concentration.

[a] J. Dancis and M. Levits, *in* "The Metabolic Basis of Inherited Disease" (J. B. Stanbury, J. B. Wyngaarden, and D. S. Fredrickson, eds.), 3rd ed., pp. 426–439. McGraw-Hill, New York, 1972.
[b] K. Willecke and A. B. Pardee, *JBC* **246**, 5264–5272, (1971).
[c] K. Tanaka, K. J. Isselbacher, and V. Shih, *Science* **175**, 69–71 (1972).
[d] K. Tanaka, *JBC* **247**, 7465–7478 (1972).

By adding CO_2 the product becomes acetoacetate which can readily be completely metabolized through conversion to acetyl-CoA.

2. Ketogenic and Glucogenic Amino Acids

According to a long used classification amino acids are **ketogenic** if (like leucine) they are converted to acetyl-CoA (or acetyl-CoA and acetoacetate). When fed to a starved animal, ketogenic amino acids cause an increased concentration of acetoacetate and other "ketone bodies" in the blood and urine. On the other hand, **glucogenic** amino acids such as valine, when fed to a starved animal, promote the synthesis of glycogen (in the case of valine via methylmalonyl-CoA, succinate, and oxaloacetate). Examination of Fig. 14-11 shows that isoleucine is both ketogenic and glucogenic, a fact that was known long before the pathway of catabolism was worked out.

F. SERINE AND GLYCINE

Serine originates in a very direct pathway from 3-phosphoglycerate (pathway *a*, Fig. 14-12) that involves dehydrogenation, transamination, and hydrolysis by a phosphatase. The principal route

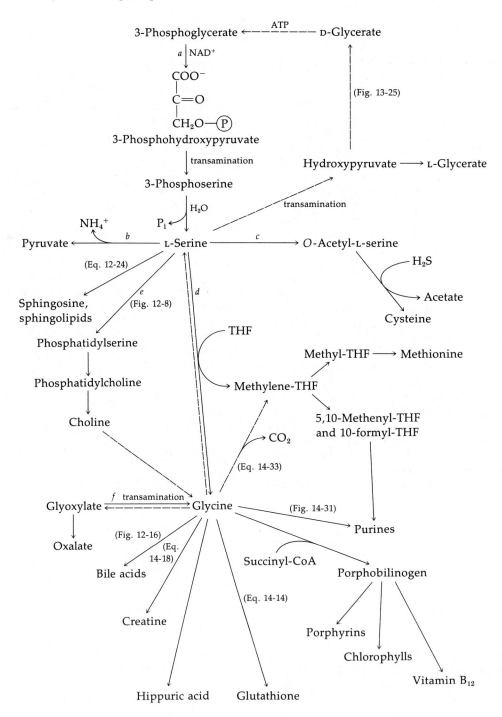

Fig. 14-12 Metabolism of serine and glycine.

of *catabolism* of serine under most circumstances is probably deamination to pyruvate (pathway Fig. 14-12, step *b*), a reaction discussed in Chapter 8, Section E,3. An alternative catabolic pathway is transamination to **hydroxypyruvate** which, as in plants (Fig. 13-25), can be reduced to D-glycerate and back to 3-phosphoglycerate. That this pathway is of importance in the human is suggested by the occurrence of a rare metabolic defect **primary hyperoxaluria** *type II*, also known as **L-glyceric-aciduria.** *,67–69 It is thought that the biochemical defect may be in the lack of reduction of hydroxypyruvate to D-glycerate. When hydroxypyruvate accumulates, lactate dehydrogenase effects its reduction to L-glyceric acid which is excreted in large amounts (0.3–0.6 g/24 h) in the urine. Surprisingly, the defect is accompanied by excessive production of oxalate from glyoxylate. This is apparently an indirect result of the primary defect in utilization of hydroxypyruvate. It has been suggested that oxidation of glyoxylate by NAD^+ is coupled to the reduction of hydroxypyruvate by NADH.[67]

1. Biosynthetic Pathways from Serine

As previously discussed in the sections indicated in Fig. 14-12, L-serine gives rise to many other substances including **sphingosine** and the **phosphatides.** Conversion to *O*-acetyl-L-serine (step *c*, Fig. 14-12) provides for the formation of **cysteine** by a β-replacement reaction.

Serine is also the major source of glycine (step *d*) and of the single-carbon units needed for the synthesis of methyl and formyl groups. The principal route of formation of glycine from serine[70] is that catalyzed by **serine hydroxymethylase** (step *d*, Fig. 14-12), but a lesser portion comes via phosphatidylserine and phosphatidylcholine and free choline (Eq. 14-30). Because the body's capacity to generate methyl groups is limited, **choline** under

many circumstances is a dietary essential and has been classified as a vitamin. However, in the presence of adequate amounts of folic acid and vitamin B_{12}, it is not absolutely essential. Choline can be used directly to reform phosphatidylcholine (Fig. 12-8), but an excess can be dehydrogenated to **betaine** (Eq. 14-30). The latter quaternary nitrogen compound is one of a small number of substances that, like methionine, are able to donate methyl groups to other substances and which are also capable of methylating homocysteine to form methionine. However, the product of transmethylation from betaine, dimethylglycine, is no longer a methylating agent. The two methyl groups are removed oxidatively as formic acid to produce glycine (Eq. 14-30).

$$\text{Phosphatidylcholine} \longrightarrow (CH_3)\overset{+}{N}—CH_2CH_2OH$$
$$\text{Choline}$$
dehydrogenation
$$(CH_3)_3N^+—CH_2COO^-$$
$$\text{Betaine}$$
homocysteine
methionine
$$\text{Dimethylglycine} \xrightarrow{\text{HCOO}^-} \text{Monomethylglycine}$$
$$\text{(sarcosine)}$$
$$\searrow \text{HCOO}^-$$
$$\text{Glycine}$$

$$(14\text{-}30)$$

A third source of glycine is transamination of glyoxylate (step *f*, Fig. 14-12). The equilibrium constant favors glycine strongly in transamination from almost any amino donor.

2. Catabolism of Glycine

While glycine is formed from glyoxylate by transamination, the oxidation of glycine by an amino acid oxidase (Table 8-4) permits excess glycine to be converted back to glyoxylate. That

---

* The hyperoxalurias are very serious diseases characterized by the formation of calcium oxalate crystals in tissues and death from renal failure before the age of 20.

this pathway, too, is quantitatively important in man is suggested by the existence of **type I hyperoxaluria.**[67] It is thought that some normal pathway for utilization of glyoxylate is blocked in this condition leading to its oxidation to oxalate. The nature of the biochemical defect is still uncertain but it may lie in a thiamine-dependent enzyme that condenses glyoxylate with α-ketoglutarate to form 2-hydroxy-3-ketoadipate (Eq. 14-31).

$$^-OOC-CHO + O=\underset{\underset{\underset{CH_2-COO^-}{|}}{\overset{|}{CH_2}}}{C}-COO^- \longrightarrow CO_2 + \underset{\underset{\underset{CH_2-COO^-}{|}}{\overset{|}{CH_2}}}{\overset{|}{\underset{O=C}{\overset{|}{HC-OH}}}}COO^-$$

Glyoxylate

$$(14\text{-}31)$$

While the function of this reaction is uncertain, it is easy to see that the product could undergo further decarboxylation and oxidation to regenerate α-ketoglutarate. In effect, this would provide a cyclic pathway (closely paralleling the dicarboxylic acid cycle, Fig. 9-5) for oxidation of glyoxylate that does not depend upon formation of oxalate. Bear in mind that the demonstrated enzymatic condensation reactions of glyoxylate are numerous and that its metabolism in most organisms is still not well understood.

An alternative route of catabolism is used by *Diplococcus glycinophilus* which is able to grow on glycine as a sole source of energy, of carbon, and of nitrogen.[71] The initial reaction is decarboxylation, with oxidation by NAD$^+$, release of ammonia and transfer of the decarboxylated α-carbon of glycine to tetrahydrofolic acid (THF) to form methylene tetrahydrofolic acid. The C-1 methylene unit of the latter can condense with another molecule of glycine (Eq. 8-19, reverse) to form serine, which in turn can be converted to pyruvate (Eq. 14-32):

$$Gly + THF \xrightarrow[\substack{NAD^+ \;\; NADH \\ CO_2}]{PLP \\ \searrowNH_4^+} Methylene\text{-}THF \overset{}{\underset{\underset{\underset{Pyruvate}{\downarrow}}{Ser}}{\frown}} Gly$$

$$(14\text{-}32)$$

Pyruvate can be oxidized for energy or used for synthesis of cell constituents. The overall pathway appears to be important in the metabolism of both plants and animals[72] as well as bacteria.

The decarboxylation step in Eq. 14-32 requires four proteins, one of which (P1, MW ~125,000) contains two molecules of PLP and presumably reacts with glycine to form a Schiff base. A hypothetical mechanism for further reaction (Eq. 14-33)

$$(14\text{-}33)$$

is based upon proposals of Baginsky and Huennekens.[73] One of the α-hydrogens of glycine is presumably removed in step a to form the quinonoid intermediate which then reacts (step b) with protein P2, a small heat-stable molecule of MW ~10,000 and resembling thioredoxin (Chapter 8,

Section I,2). Following decarboxylation (step c), reduced P2 is released (step d) and is reoxidized by flavoprotein P3 (MW $\sim 120,000$ and containing one molecule of FAD) in step e. Reduced P3 is in turn reoxidized by NAD$^+$.

All that remains of the glycine attached to the PLP at P1 is the CH$_2$ fragment and the amino group. The student can easily write a series of displacement and elimination reactions by which N-5 of tetrahydrofolic acid (Chapter 8, Section K,1) acts as the acceptor of the CH$_2$ unit and by which NH$_3$ is released at the same time that an internal Schiff base of PLP is formed with an ϵ-amino group of the enzyme (step f). The role of P4 is unknown but may involve the transfer of the CH$_2$ unit to tetrahydrofolic acid.

3. Biosynthetic Pathways from Glycine

As indicated in Fig. 14-12, a variety of products can be formed from glycine. Several of the reactions have been discussed previously, as indicated in Fig. 14-12. **Hippuric acid** (Box 9-A), the usual urinary excretion product in the "detoxification" of benzoic acid, is formed via benzoyl-CoA (Eq. 14-34):

$$ (14\text{-}34) $$

Hippuric acid

The formation of porphobilinogen and the various pyrrole pigments derived from it and the synthesis of the purine ring represent two other major routes for glycine metabolism.

4. Porphobilinogen, Porphyrins, and Related Substances

In 1946, D. Shemin and D. Rittenberg[73a] described one of the first successful uses of radio-tracers in the study of metabolism. In this classical paper, it was demonstrated that the atoms of the porphyrin ring in heme have their origins in the simple compounds acetate and glycine. As we now know, acetate is converted to succinyl-CoA in the citric acid cycle. Within the mitochondrial matrix of animal cells succinyl-CoA condenses with glycine to form **δ-aminolevulinic acid** (Eq. 8-20),[73b] which is converted to **porphobilinogen** (Chapter 10, Section B,1), the immediate precursor to the porphyrins. By degradation of ^{14}C-labeled porphyrins formed from labeled acetate and glycine molecules, Shemin and Rittenberg established the labeling pattern for the pyrrole ring that is indicated in Fig. 14-13 for porphobilinogen itself. The solid circles mark those atoms that were found to be derived from methyl carbon atoms of acetate (bear in mind that acetyl groups of acetyl-CoA pass around the citric acid cycle more than once to introduce label from the methyl group of acetate into both the 2 and 3 positions of succinyl-CoA). Those atoms marked with open circles in Fig. 14-13 were found to be derived mainly from the methyl carbon of acetate and in small part from the carboxyl carbon. Atoms marked with asterisks were found derived from glycine, while unmarked carbon atoms came from the carboxyl carbon of acetate.

In plants a different route is used. The intact 5-carbon skeleton of α-ketoglutarate enters δ-aminolevulinate. A possible pathway begins with an internal oxidation–reduction reaction (perhaps after conversion of α-ketoglutarate to a thioester), resembling the glyoxalase I reaction (Chapter 7, Section K) in reverse. The product would be γ,δ-dioxovalerate,

which could be reductively aminated to δ-aminolevulinate.[74]

As indicated in Fig. 14-13, the conversion of two molecules of δ-aminolevulinic acid into porphobilinogen is a multistep reaction initiated by an aldol condensation (step b). It is thought that the enzyme catalyzing this reaction forms a Schiff base with a carbonyl group of one of the substrate

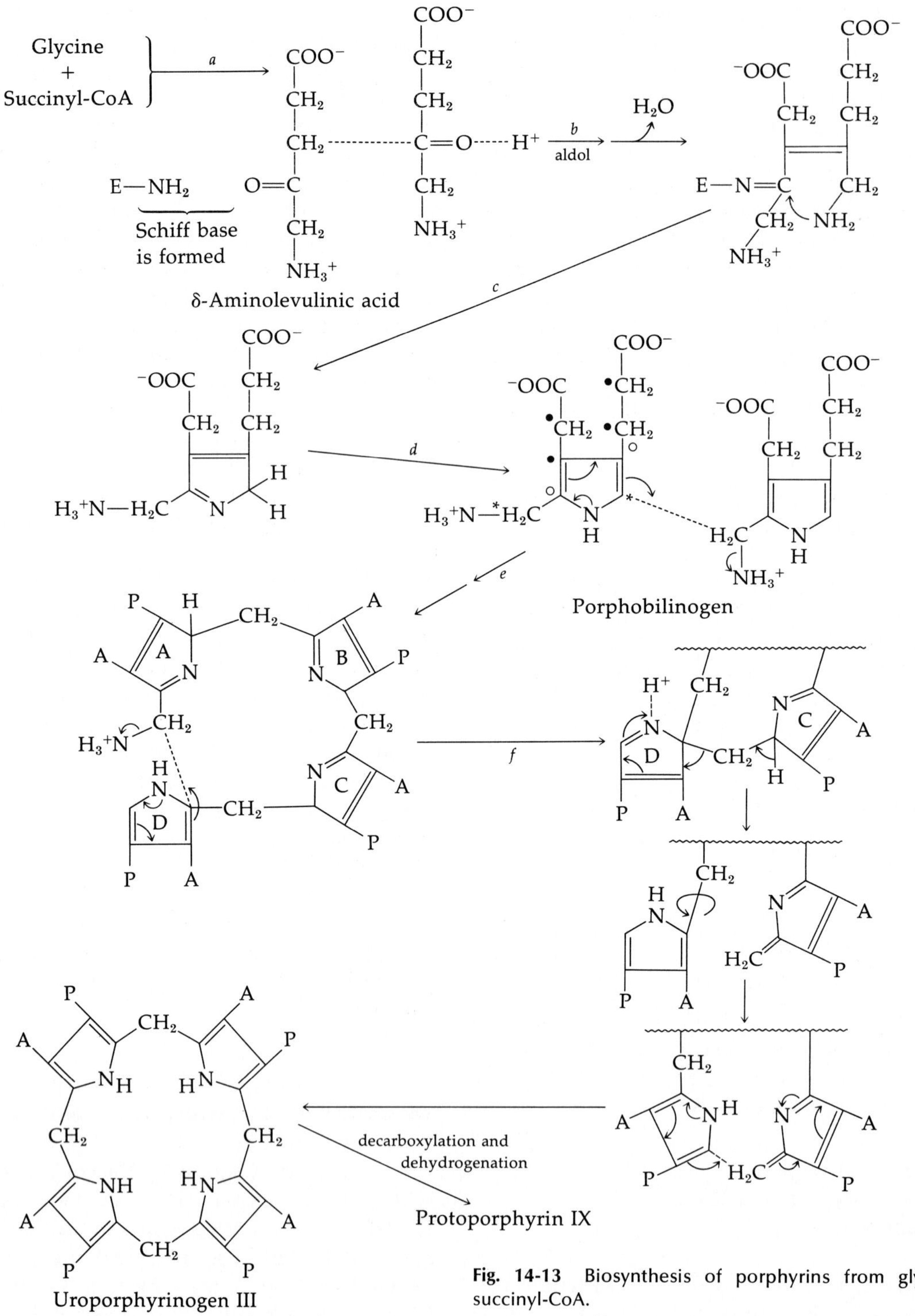

Fig. 14-13 Biosynthesis of porphyrins from glycine and succinyl-CoA.

molecules, as indicated in the figure.[75,76] The aldol condensation is followed by dehydration to form a carbon–carbon double bond and ring closure (step *c*) occurs by a *transimination* sequence analogous to that described on p. 459. Tautomerization step (*d*) is then required to produce porphobilinogen. The condensation of the latter to form porphyrins requires two enzymes, **porphobilinogen deaminase** and **uroporphorinogen III cosynthetase.** The deaminase catalyzes step *e* of Fig. 14-13. Ammonia is eliminated but not necessarily by the direct displacement indicated. It might occur by an electron flow from the adjacent nitrogen in the same pyrrole ring. It is easy to see how the condensation process could be repeated four times to produce the symmetric precursor of **uroporphyrin I** (Fig. 10-1). In the presence of the cosynthetase, a different reaction takes place. Note that the five-membered ring in porphobilinogen has a symmetric arrangement of double bonds. Thus, a condensation reaction can occur at either of the positions α to the ring nitrogen. The sequence of condensation (step *f*, Fig. 14-13) is followed by tautomerization, cleavage, and reformation of the ring with the uroporphyrinogen III pattern of the carboxymethyl and carboxyethyl side chains. A series of decarboxylation and oxidation reactions[77] then leads directly to protoporphyrin IX.

Insertion of a ferrous ion into protoporphyrin requires a special enzyme **protoheme ferro-lyase** (ferrochelatase).[78,79] The enzyme is found firmly bound to the inner membrane of mitochondria of animal cells, chloroplasts of plants, and chromatophores of bacteria. While Fe^{2+} is apparently the only metallic ion ordinarily inserted into a porphyrin, the $[Zn^{2+}]$protoporphyrin chelate accumulates in substantial amounts in yeast and a Cu^{2+} complex is known (Chapter 10, Section B,1).

a. Chlorophyll[80,81]

The first step in the conversion of protoporphyrin IX into chlorophyll may be the insertion of Mg^{2+}. This reaction is catalyzed and does not occur readily spontaneously. Subsequently, the carboxyethyl side chain on ring III is methylated (Eq. 14-35):

$$\text{Proto IX} \xrightarrow{\quad Mg^{2+} \quad} \text{Mg Proto IX} \xrightarrow{\quad SAM \quad} \text{Mg Proto IX}$$
$$\text{(Protoporphyrin IX)} \qquad\qquad\qquad\qquad \text{Methyl ester}$$

$$(14\text{-}35)$$

The remaining steps in formation of chlorophyll include the saturation of the vinyl group on ring IV, the closure of ring V, and addition of the phytyl group (see Fig. 13-19 for the chlorophyll structures). Closure of ring V occurs following β oxidation of the 3-carbon side chain as indicated in Eq. 14-36. This is followed by an oxidative ring closure to **protochlorophyllide *a*.**

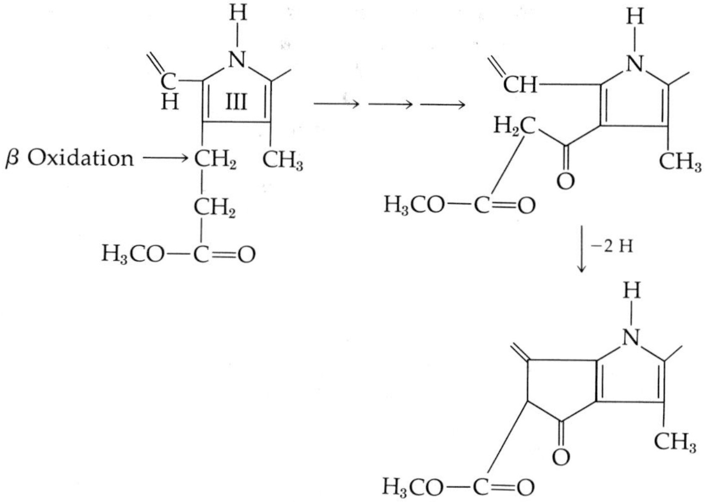

$$(14\text{-}36)$$

The latter is coupled with phytol (probably via phytyl pyrophosphate) to form chlorophyll *a*. It is likely that chlorophyll *b* is derived from chlorophyll *a* and that the bacteriochlorophylls are synthesized from chlorophyllide *a*, the phytyl group being added after reduction of ring IV.

b. Corrins

The formation of vitamin B_{12} and other corrins requires a ring contraction with elimination of the methine bridge between rings A and D of the porphyrins (see Box 8-K). It is natural to assume that

BOX 14-C

METABOLISM OF IRON

Iron is one of the most abundant elements in the earth's crust, being present to the extent of ~4% in a typical soil. Its functions in living cells are numerous and diverse.[a-c] The average overall iron content of both bacteria and fungi is ~1 mmol/kg, but that of animal tissues is usually less. Seventy percent of the 3–5 g of iron present in the human body is located in the red blood corpuscles whose overall iron content is ~20 mM. However, in other tissues the total iron amounts to ~0.3 mM and consists principally of storage forms. The total concentration of all iron-containing *enzymes* amounts to ~0.01 mM. Although these average concentrations are low, iron is concentrated in oxidative enzymes of membranes and may attain much higher concentrations locally. Surprisingly, one group of anaerobic bacteria, the lactic acid bacteria, possess no oxygen-requiring enzymes and appear to be totally devoid of both iron and copper. All other organisms appear to require iron for life.

Within the human body, as well as in other animals, green plants, and fungi, much of the iron occurs in the form of ferritin, a red-brown water soluble protein.[d-f] Ferritin represents a store of Fe(III) in a soluble, nontoxic, readily available form. An unusual protein ferritin contains 17–23% iron as a dense core of hydrated ferric hydroxide 7 nm in diameter and surrounded by a protein coat consisting of 24 subunits in a cubic array much like that in Fig. 8-17. The outer diameter is ~ 12 nm. The molecular weight of apoferritin is 445,000 and that of its subunits 18,500. A completely filled (23% Fe) ferritin molecule contains over 2000 atoms of iron in a largely crystalline lattice. The core is readily visible in the electron microscope, and ferritin is often used in various ways as a labeling reagent in microscopy. Another storage form of iron, **hemosiderin**, seems to consist of molecules of ferritin together with extra iron. Deposits of hemosiderin in the liver can rise to toxic levels if excessive amounts of iron are absorbed.

A major problem for all organisms is posed by the relative insolubility of ferric hydroxide and other compounds from which iron must be extracted by the organism. A consequence is that iron is often taken up in a chelated form and is transferred from one organic ligand (often protein) to another with little or no existence as free Fe^{3+}. Typically the formation constants for chelates of Fe^{2+} lie between those of Mn^{2+} and Co^{2+} (Chapter 4, Section C,8b; Table 4-2). For example, log K_1 = 14.3 for the Fe^{2+} chelate of EDTA. As might be expected, the smaller and more highly charged Fe^{3+} is bonded more strongly (log K_1 = 25.0). A fact of considerable biochemical significance is the preference of Fe^{3+} for oxygen-containing ligands. On the other hand, Fe^{2+} tends to bind preferentially to nitrogen. It is also of significance that Fe^{3+} bound to oxygen ligands tends to exchange readily with other ferric ions in the medium whereas Fe^{3+} bound to nitrogen-containing ligands such as heme exchanges very slowly. This observation may be of importance in consideration of iron-binding transport compounds and enzymes.

If a suitably high iron content is maintained in the external medium (e.g., 50 μM or more for *E. coli*) bacteria and other microorganisms have little problem with uptake of iron. However, when the external iron concentration is low, special compounds (**siderochromes**) are provided to help render the iron soluble.[g] Thus, at iron concentrations of 2 μM or less, *E. coli* and related enterobacteria excrete large amounts of the specific chelating agent enterobactin (Fig. 2-39) into the medium. The very stable complex of this substance with Fe^{3+} is taken up by a specific transport system of the bacteria. Within the cells the enterobactin is cleaved by an esterase and the iron may be reduced to Fe^{2+} to facilitate dissociation. Among other siderochromes are several such as the peptide **ferrochrome** (produced by certain baccilli) that contain hydroxamate groups

$$\begin{array}{ccc} & O & O^- \\ & \parallel & | \\ R- & C-N- & R' \end{array}$$

at the iron-binding centers. Note that oxygen atoms form the bonds to iron in these com-

Box 14-C (*Continued*)

pounds also. Ferrichrome binds Fe^{3+} tenaciously,[h] log K for formation from Fe^{3+} plus the iron-free trihydroxamate ligand being 29.

An average human diet contains ~15 mg of iron of which ~1 mg is absorbed daily. This is ordinarily enough to compensate for the small losses from the body, principally through the bile. The human body appears to have no mechanism for excretion of excessive amounts of iron; the iron content of the body is regulated solely by the rate of uptake. This rate is increased during pregnancy and, in young women, to compensate for iron lost in menstrual bleeding. Excessive amounts of iron can be highly toxic. The mechanism of control of iron absorption is still uncertain but once in the body, Fe^{3+} is bound to **transferrin,** a protein of MW = 80,000 and containing two iron-binding sites.[i] An anion is bound along with each ion of Fe^{3+}. Transferrin of chickens appears to be identical to the egg white iron-binding protein conalbumin. However, the red iron-binding protein of milk, lactoferrin, has a different sequence from transferrin of blood. The iron-binding proteins of body fluids are sometimes given the group name **siderophilins.**

The principal function of transferrin is transport of iron throughout the body, but it also may serve as an iron buffer and it is possible that the inflow of iron through the intestinal mucosa is regulated by the state of saturation of the transferrin in the blood. Reduction of Fe^{3+} to Fe^{2+} is necessary for transfer from transferrin into heme in the young blood cells forming in the bone marrow. Reduction of ferric iron is probably also essential for release from ferritin.[j] The mechanisms are uncertain, but ascorbic acid or glutathione may be the reductants. Ferric iron in hemoglobin (methemoglobin) is reduced by a NADH-dependent enzyme (Box 10-A). On the other hand, Fe^{2+} may sometimes have to be oxidized to Fe^{3+} by the copper-containing ferroxidase (ceruloplasmin, Box 10-H). Once in the body, iron is carefully conserved. Thus, the 9 billion red blood cells destroyed daily yield 20–25 mg of iron which is almost all reused or stored.

[a] J. B. Neilands, ed., "Microbial Iron Metabolism." Academic Press, New York, 1974.

[b] A. Jacobs and M. Worwood, eds., "Iron in Biochemistry and Medicine." Academic Press, New York, 1974.

[c] B. L. O'Dell and B. J. Campbell, *Compr. Biochem.* **21,** 179–265 (1970).

[d] P. M. Harrison and T. G. Hoy, in "Inorganic Biochemistry" (G. L., Eichhorn, ed.), Vol. I, pp. 253–279. Elsevier, Amsterdam, 1973.

[e] R. J. Hoare, P. M. Harrison, and T. G. Hoy, *Nature* (*London*) **255,** 653–654 (1975).

[f] W. H. Massover and J. M. Cowley, *PNAS* **70,** 3847–3851 (1973).

[g] J. B. Neilands, in "Inorganic Biochemistry" (G. L. Eichhorn, ed.), Vol. I, pp. 167–2020. Elsevier, Amsterdam, 1973.

[h] H. Rosenberg and I. G. Young, in "Microbiol Iron Metabolism" (J. B. Neilands, ed.), pp. 67–82. Academic Press, New York, 1974.

[i] P. Aisen, in "Inorganic Biochemistry" (G. L. Eichhorn, ed.), Vol. I, pp. 280–305. Elsevier, Amsterdam, 1973.

[j] I. Cavill, M. Worwood, and A. Jacobs, *Nature* (*London*) **256,** 328–329 (1975).

the methyl group at C-1 of the corrin ring might arise from the same precursor carbon atom as does the methine bridge in porphyrins. (It is easy to visualize a modified condensation reaction by which ring closure at step *f* in Fig. 14-13 occurs by nucleophilic addition to the C=N bond of ring A.) However, ^{13}C-nmr data have ruled out this possibility. When vitamin B_{12} was synthesized in the presence of ^{13}C-methyl-containing methionine and the product was examined, it was found that seven methyl groups contained ^{13}C. All of the "extra" methyl groups around the periphery of the molecule as well as the one at C-1 were labeled.[82] Other experiments have established that uroporphyrinogen III is a precursor of vitamin B_{12}. Therefore, it appears that the ring may first close in a normal way and then reopen between rings A and D with removal of the carbon that forms the methylene bridge.[83] Alternative mechanisms have also been proposed.[83a]

c. The Porphyrias[84–86]

The human body does not use all of the porphobilinogen produced and a small amount is normally excreted in the urine, principally as copro-porphyrins (Chapter 10, Section B,1). In a number of hereditary and acquired conditions blood porphyrin levels are elevated and enhanced urinary excretion (porphyria) is observed. While porphyrias may be mild and almost without symptoms, in other cases the intensely fluorescent free porphyrins are deposited under the skin and cause photosensitivity and ulceration. In extreme cases, in which the excreted porphyrins may color the urine a wine red, patients often have acute neurological attacks and a variety of other symptoms.* In one type of congenital porphyria, uroporphyrin I is excreted in large quantities. The biochemical defect appears to be a deficiency of the cosynthetase that is required for formation of protoporphyrin IX. Another type of porphyria results from overproduction of δ-aminolevulinic acid in the liver. Administration of benzoate or p-aminobenzoate has been suggested as a possible treatment.[87] The idea is to shunt the glycine into production of hippuric acid (Box 9-A) or its p-amino derivative, thus decreasing the rate of porphyrin synthesis.

In some of the mild forms of porphyria ingestion of drugs can precipitate an acute attack. Drugs and other chemicals sometimes cause porphyria by inducing excessive synthesis of δ-aminolevulinate synthetase. Among compounds having this effect are hexochlorobenzene and tetrachlorodibenzodioxin. The latter is one of the most potent inducers of the synthetase known.[88]

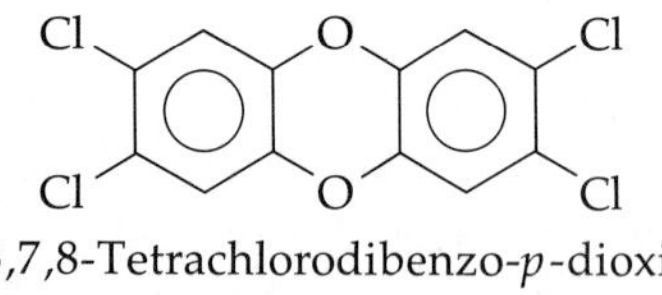

2,3,7,8-Tetrachlorodibenzo-p-dioxin†

---

* A lucid account of the symptoms of which King George III of England is believed to have suffered is given by Macalpine and Hunter.[84]

† The tendency for this dioxin to be present as an impurity in the herbicide 2,4,5-trichlorophenoxyacetic acid (2,4,5-T) has caused concern. The dioxin may be the most toxic small molecule known, the oral LD_{50} for guinea pigs being only 1 μg/kg body weight.[88] It is also a potent teratogenic agent.

d. The Bile Pigments

The enzymatic degradation of heme represents an important metabolic process if only because it releases iron to be reutilized by the body. Some of the pathways are illustrated in Fig. 14-14. It is thought that oxygenation occurs first at the α-methine carbon (between the A and B rings). The hydroxylated product is cleaved with the release of carbon monoxide. The reactions are catalyzed by microsomal hydroxylases.[89–90a] When $^{18}O_2$ is used, it is found that the open tetrapyrrole **biliverdin,** that is formed, contains two atoms of ^{18}O, and that the CO contains one. A large number of open tetrapyrroles can be formed from biliverdin by reduction and oxidation reactions.

Biliverdin, the first product of the ring opening, is reduced to **bilirubin,** which is transported to the liver as a complex with serum albumin. In the liver bilirubin is converted into glucuronides (Eq. 12-12), glycosylation occurring on the propionic acid side chains. A variety of these bilirubin conjugates are excreted into the bile. In the intestine the conjugates are hydrolyzed back to free bilirubin, which is reduced by the action of intestinal bacteria to urobilinogen, stercobilinogen, and *meso*-bilirubinogen. These compounds are colorless but are readily oxidized by oxygen to **urobilin** and **stercobilin.** Some of the urobilin and other bile pigments is reabsorbed into the blood and excreted into the urine where it provides the familiar yellow color.

The yellowing of the skin known as **jaundice** can occur if the heme degradation system is overburdened (e.g., from excessive hemolysis), if the liver fails to conjugate bilirubin, or if there is obstruction of the flow of heme breakdown products into the intestinal tract.

The open tetrapyrroles of algae and the chromophore of phytochrome have been mentioned previously (Fig. 13-22). They are all derived from **phycoerythrobilin,** which is related to biliverdin, as indicated in Fig. 14-14.

Fig. 14-14 The degradation of heme and the formation of open tetrapyrrole pigments.

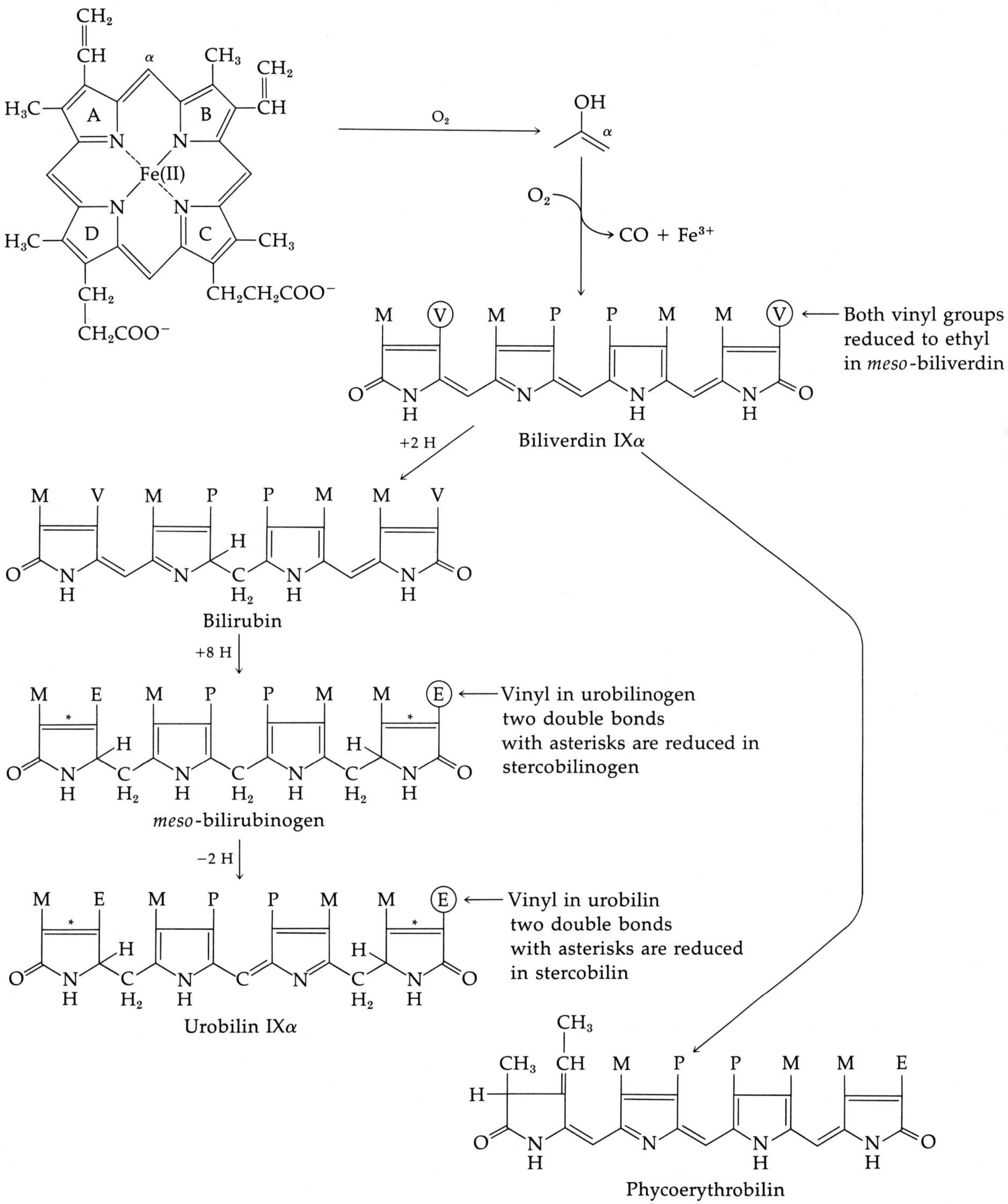
O2
CO + Fe3+
Both vinyl groups reduced to ethyl in meso-biliverdin
Biliverdin IXα
+2 H
Bilirubin
+8 H
Vinyl in urobilinogen two double bonds with asterisks are reduced in stercobilinogen
meso-bilirubinogen
−2 H
Vinyl in urobilin two double bonds with asterisks are reduced in stercobilin
Urobilin IXα
Phycoerythrobilin

G. CYSTEINE AND SULFUR METABOLISM

Cysteine is not only an essential constituent of proteins but also lies on the major route of incorporation of inorganic sulfur into organic compounds. As such, it is involved in a variety of metabolic processes. Autotrophic organisms carry out the stepwise reduction of sulfate to sulfite and sulfide (H_2S). These reduced sulfur compounds are the ones that are incorporated into organic substances. Animals make use of the organic sulfur compounds formed by the autotrophs and have an active oxidative metabolism by which the compounds can be decomposed and the sulfur reoxidized to sulfate.

Several aspects of inorganic sulfur metabolism have been discussed in Chapter 10. Thus, the reduction of sulfate to H_2S by sulfate-reducing bacteria is considered in Chapter 10, Section F,2,b. The initial step in the assimilative pathway of sulfate reduction, used by plants and by E. coli, is also the formation of adenylyl sulfate (step a, Fig. 14-15; see also Eq. 10-38). The sulfate-reducing bacteria are able to reduce adenylyl sulfate directly to sulfite (Eq. 10-38, step b), but the assimilative pathway of reduction in E. coli proceeds through 3'-phospho-5'-adenylyl sulfate (PAPS, a compound whose function as "active sulfate" has been considered in Chapter 11, Section B,3). Reduction of PAPS to sulfite (Fig. 14-15, step d) is accomplished by an NADPH-dependent enzyme.

The reduction of sulfite to sulfide by E. coli is catalyzed by **sulfite reductase,** a large (MW = 670,000) molecule containing four molecules of bound riboflavin phosphate (FMN), four molecules of FAD, 20–21 atoms of iron, 14–15 atoms of labile sulfur, and 3–4 molecules of a special heme.[91] The enzyme uses NADPH as a reductant and passes electrons through its own internal electron transport chain to the special heme which contains a tetrahydroporphyrin of the isobacteriochlorin type (adjacent, reduced pyrrole rings, Fig. 14-16).[92] The protein appears to have the peptide composition of $\alpha_8\beta_4$ where α is a flavoprotein of MW $\sim$54,000 and β is a heme-iron–sulfur protein of MW $\sim$60,000.[93] The same pathway is found in the alga *Chlorella,* but a second route of sulfate reduction is probably more important.[94] Adenylyl sulfate transfers its sulfuryl group to a thiol group of a carrier (Eq. 14-37, step a). The resulting thio-

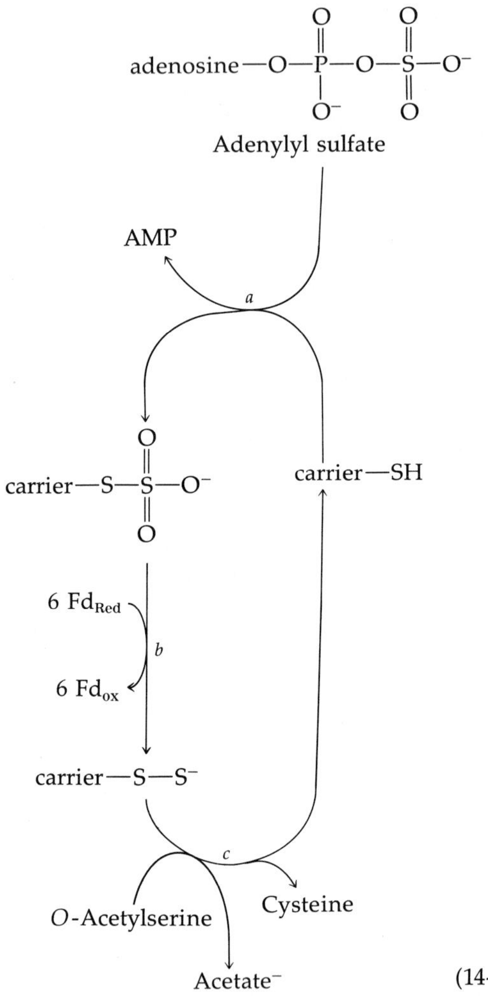

$$\underset{\text{Adenylyl sulfate}}{\text{adenosine}-\text{O}-\overset{\overset{\displaystyle O}{\|}}{\underset{\underset{\displaystyle O^-}{|}}{P}}-\text{O}-\overset{\overset{\displaystyle O}{\|}}{\underset{\underset{\displaystyle O}{\|}}{S}}-\text{O}^-}$$

$$\text{AMP}$$

$$a$$

$$\text{carrier}-\text{SH}$$

$$\underset{}{\text{carrier}-\text{S}-\overset{\overset{\displaystyle O}{\|}}{\underset{\underset{\displaystyle O}{\|}}{S}}-\text{O}^-}$$

$$6\ \text{Fd}_{\text{Red}}$$
$$b$$
$$6\ \text{Fd}_{\text{ox}}$$

$$\text{carrier}-\text{S}-\text{S}^-$$

$$c$$

$$\text{Cysteine}$$

$$O\text{-Acetylserine}$$

$$\text{Acetate}^- \hspace{3cm} (14\text{-}37)$$

sulfonate is reduced by a ferredoxin-dependent reductase. Finally, a sulfide group is transferred from the —S—S⁻ group of the reduced carrier

Fig. 14-15 Pathways of biosynthesis and catabolism of cysteine and some other aspects of sulfur metabolism. Solid arrows are major biosynthetic pathways. The dashed arrows represent more specialized pathways; they also show processes occurring in the animal body to convert methionine to cysteine and to degrade the latter.

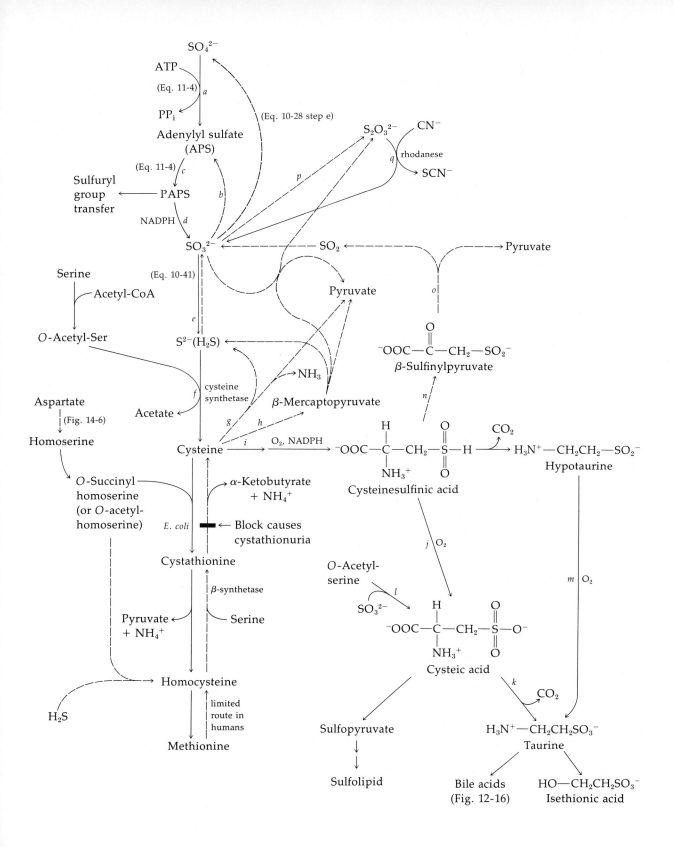

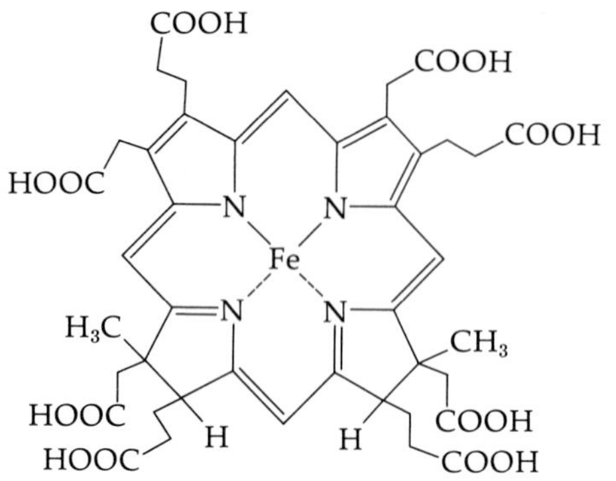

Fig. 14-16 Special heme (siroheme) from sulfite reductase of *E. coli.*

directly into cysteine in a β-substitution reaction analogous to that described in the next paragraph.

Cysteine is formed from sulfide and serine after the latter has been acetylated by transfer of an acetyl group from acetyl-CoA (Fig. 14-15, step *f*). The reaction is a standard pyridoxal phosphate-dependent β replacement (Fig. 8-7) and is catalyzed by **cysteine synthetase.**[94a] A similar enzyme is used by some cells to introduce sulfide ion directly into homocysteine, either via *O*-succinyl homoserine or *O*-acetyl homoserine, as indicated by the dashed arrows in the lower left corner of Fig. 14-15. However, in *E. coli* the principal route of formation of methionine is doubtless through cystathionine, as outlined in Eq. 8-22 and as indicated on the left side of Fig. 14-15 by the solid arrows.

In animals the converse process, the conversion of methionine to cysteine, is important. Animals are unable to incorporate sulfide directly into cysteine, and the latter must be either provided in the diet or formed from dietary methionine. The latter process is limited and cysteine is an essential dietary constituent for infants. The steps involved in formation of cysteine from methionine are indicated by the vertical dashed arrows in Fig. 14-15. The sequence is initiated by conversion of methionine to SAM and on to homocysteine by the sequence shown in Fig. 14-9 (steps *d* to *f*). Then homocysteine reacts with serine under the influence

of **cystathionine β-synthetase** and cystathionine is cleaved to cysteine and α-ketobutyrate by a "γ-cleavage enzyme" found in animal tissues (Fig. 14-9, lower left). The reaction types are shown in Fig. 8-7. The pathway somewhat resembles that of Eq. 8-22 in reverse.

Cysteine can be catabolized or it can enter a variety of biosynthetic pathways. A simple degradation, employed by some bacteria,[95] is a pyridoxal phosphate-dependent α,β elimination to form H_2S, pyruvate, and ammonia (reaction *g*, Fig. 14-15, Fig. 8-6, reaction *b*). Another is transamination (Fig. 14-15, reaction *h*) to β-mercaptopyruvate. The latter can be reductively cleaved to pyruvate and sulfide. An interesting β-replacement reaction (pyridoxal phosphate-dependent) of cysteine leads to β-cyanoalanine, the lathyritic factor (Box 11-B) present in some plants.[96]

A quantitatively important pathway of cysteine catabolism in animals is oxidation to **cysteine sulfinic acid** (Fig. 14-15, reaction *i*), a two-step hydroxylation requiring O_2, NADPH or NADH, and Fe^{2+}. Cysteine sulfinic acid can be further oxidized to **cysteic acid,** which can be decarboxylated to **taurine.** The latter is a component of bile salts (Fig. 12-16) and may be a neurotransmitter (Chapter 16, Section B,4). Taurine may have a special function in retinal photoreceptor cells. It appears to be an essential dietary amino acid for cats.[96a] Taurine can also be reduced to **isethionic acid,** a component of nervous tissue. Cysteic acid can arise in another way from *O*-acetylserine and sulfite (reaction *l*, Fig. 14-15), and taurine can also be formed by decarboxylation of cysteine sulfinic acid to **hypotaurine** and oxidation of the latter (reaction *m*).

An interesting metabolic fate of cysteic acid is conversion to the sulfolipid of chloroplasts (Fig. 2-32, Table 13-2). The probable sequence begins with transamination to β-sulfopyruvate, reduction of the latter (presumably to sulfolactaldehyde), and aldol condensation with dihydroxyacetone phosphate as indicated in Eq. 14-38. Isomerization yields **6-sulfoquinovose,** the sugar present in sulfolipid.[96b]

Cysteic acid $\xrightarrow{\text{transamination}}$

$$\begin{array}{c} \text{COO}^- \\ | \\ \text{C}=\text{O} \\ | \\ \text{CH}_2 \\ | \\ \text{SO}_3^- \end{array}$$

β-Sulfopyruvate

reduction

$$\begin{array}{c} \text{HC}=\text{O} \\ | \\ \text{HC}-\text{OH} \\ | \\ \text{CH}_2\text{SO}_3^- \end{array} \xrightarrow[\text{dihydroxyacetone P}]{\text{aldol}} \longrightarrow$$

Sulfolactaldehyde

$$\begin{array}{c} \text{CHO} \\ | \\ \text{HCOH} \\ | \\ \text{HOCH} \\ | \\ \text{HCOH} \\ | \\ \text{HCOH} \\ | \\ \text{CH}_2\text{SO}_3^- \end{array}$$

6-Sulfoquinovose

(14-38)

Returning to cysteine sulfinic acid, note that it can be transaminated to β-sulfinylpyruvate, a compound that undergoes ready loss of SO_2 in a reaction analogous to the decarboxylation of oxaloacetate (reaction o, Fig. 14-15). This probably represents one of the major routes by which sulfur is removed from organic compounds in the animal body. However, before being excreted, the sulfite must be oxidized to sulfate. The enzyme **sulfite oxidase** is one of an increasing number of proteins found to contain molybdenum (Box 14-A). Sulfite oxidase also contains a b_5-like cytochrome and passes electrons directly to cytochrome c in the mitochondrial electron transport chain. Evidence for the essentiality of the enzyme to humans is provided by a report of a child deficient in sulfite oxidase secreting no urinary sulfate and suffering severe neurological defects.[97]

A reaction that is ordinarily of minor consequence in the animal body but which may be enhanced during deficiency of sulfite oxidase is the oxidative coupling of two molecules of sulfite to thiosulfate (reaction p, Fig. 14-15). Thiosulfate participates in an interesting reaction: An enzyme with the odd name **rhodanese,** found in liver, catalyzes the displacement of sulfite from thiosulfate by cyanide ion (Eq. 14-39). The reaction serves to detoxify the latter.

$$\text{CN}^- \quad {}^-\text{S}-\overset{+}{\underset{\text{O}}{\overset{\text{O}^-}{\text{S}}}}-\text{O}^- \longrightarrow \text{N}\equiv\text{C}-\text{S}^- + \text{SO}_3^- \quad (14\text{-}39)$$

A series of organic hydrodisulfide derivatives such as **thiocysteine** are known and occur in animals in small amounts.

$$\begin{array}{c} {}^-\text{OOC}-\text{CH}-\text{CH}_2-\text{S}-\text{SH} \\ | \\ \text{NH}_3^+ \end{array}$$

Thiocysteine

Thioglutathione and the more oxidized **thiotaurine** are among other compounds of this type. One way in which thiocysteine can arise is through action of a cystathionine cleaving enzyme on cystine. Thiocysteine is eliminated with production of pyruvate and ammonia from the rest of the cystine molecule.

Most of the sulfate generated in the body is excreted unchanged in the urine, but a significant fraction is esterified with oligosaccharides and phenolic compounds. These, of course, are formed by sulfuryl transfer from PAPS (Eq. 11-4). Many readers (perhaps ~40%) will be aware that after eating asparagus a strong odor appears in their urine. These genetic "stinkers" secrete S-methyl thioacrylate,

$$\begin{array}{c} \text{O} \\ \| \\ \text{CH}_2=\text{CH}-\text{C}-\text{SCH}_3 \end{array}$$

and related compounds, but the plant constituent giving rise to them is unknown.[98]

H. METABOLISM OF AROMATIC COMPOUNDS

One means of formation of aromatic rings is through the polyketides (Chapter 12, Section G). However, more important in most autotrophic organisms* is the **shikimic acid pathway** (Fig.

* Animals are unable to synthesize the ring systems of the aromatic amino acids. Phenylalanine and tryptophan are dietary essentials. However, tyrosine can be formed in the animal body by hydroxylation of phenylalanine (Section H,5).

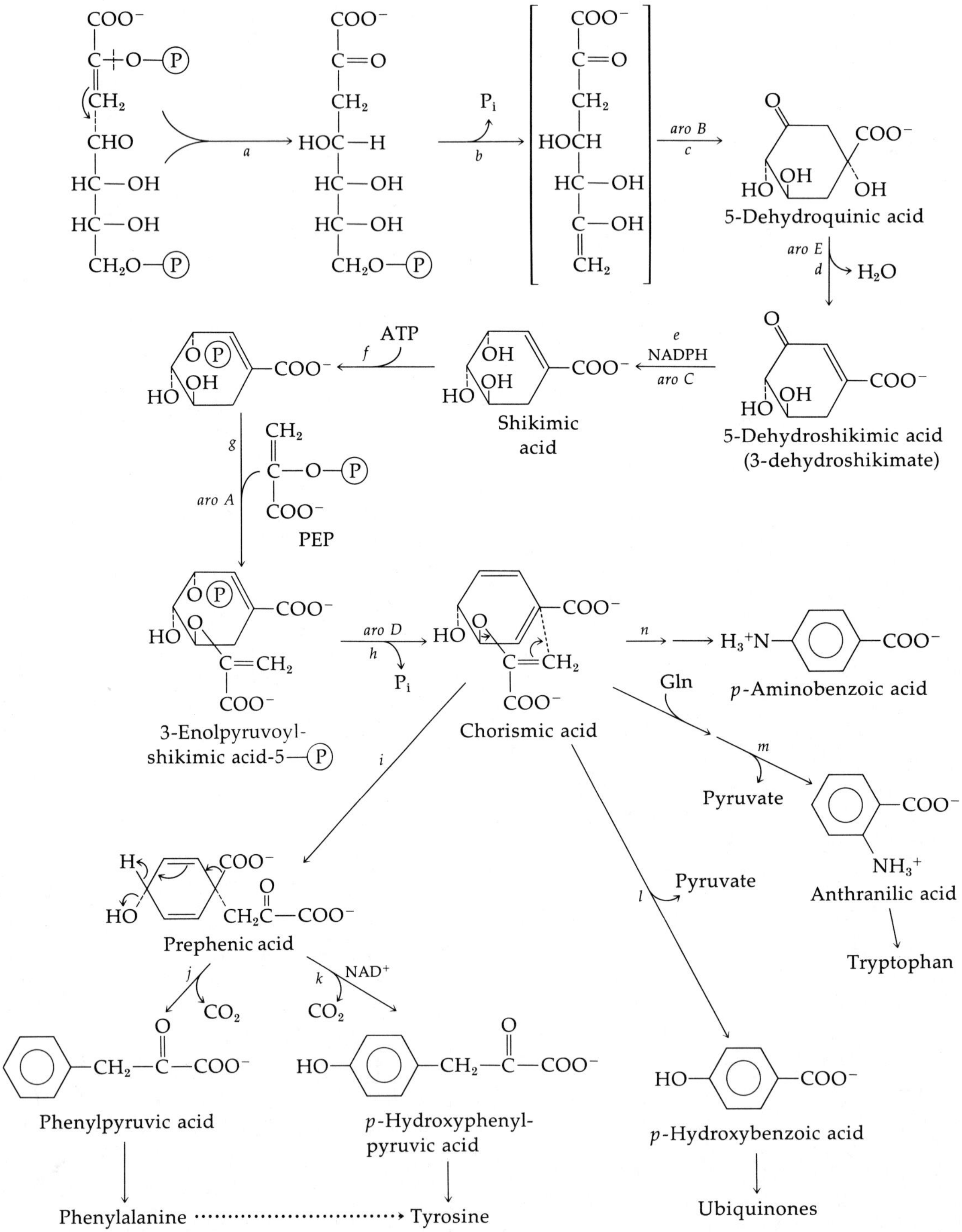
5-Dehydroquinic acid
aro B
aro E
aro C
aro A
aro D
NADPH
ATP
Shikimic acid
5-Dehydroshikimic acid (3-dehydroshikimate)
PEP
3-Enolpyruvoyl-shikimic acid-5—(P)
Chorismic acid
p-Aminobenzoic acid
Gln
Pyruvate
Anthranilic acid
Tryptophan
Prephenic acid
NAD+
Phenylpyruvic acid
p-Hydroxyphenyl-pyruvic acid
p-Hydroxybenzoic acid
Phenylalanine
Tyrosine
Ubiquinones

14-17). This metabolic route was worked out to a large extent through the use of ultraviolet light-induced mutants of *E. coli, Aerobacter aerogenes,* and *Neurospora.* In 1950, using the penicillin enrichment technique (Chapter 15, Section A,1,c), B. D. Davis obtained a series of mutants of *E. coli* that would not grow without the addition of aromatic substances.[98a] A number of the mutants required tyrosine, phenylalanine, tryptophan, *p*-aminobenzoic acid, and a trace of *p*-hydroxybenzoic acid. It was a surprise to find that the requirements for all five compounds could be met by the addition of what was then regarded as a rare plant acid, shikimic acid. Thus, shikimic

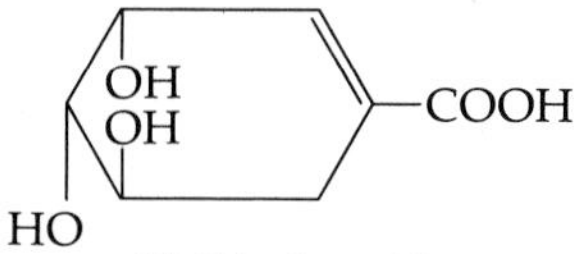

Shikimic acid

acid, which is far from being an aromatic compound, was implicated as an intermediate in the biosynthesis of the three aromatic amino acids and of other essential aromatic substances.[98b]

The mutants that grew in the presence of shikimic acid evidently had the biosynthetic pathway blocked at one or more earlier stages. Among these mutants, certain pairs were found that could not grow alone but that could grow when plated together (the phenomenon is called **syntropism**). Thus, mutant 83-2, which we now know to be blocked in the conversion of 5-dehydroshikimic acid to shikimic acid, accumulated shikimic acid and permitted mutant 83-1 or 83-3 to grow by providing them with a precursor that could be converted onto the end products (Eq. 14-40). Eventually, the entire pathway was traced. The enzymes have been isolated and studied and the locations of the genes in the *E. coli* chromosome have been mapped[99] as shown in Fig. 15-1.

Fig. 14-17 Aromatic biosynthesis by the shikimic acid pathway. The gene symbols for several of the genes coding for the required enzymes are indicated. Their locations on the *E. coli* chromosome map are shown in Fig. 15-1.

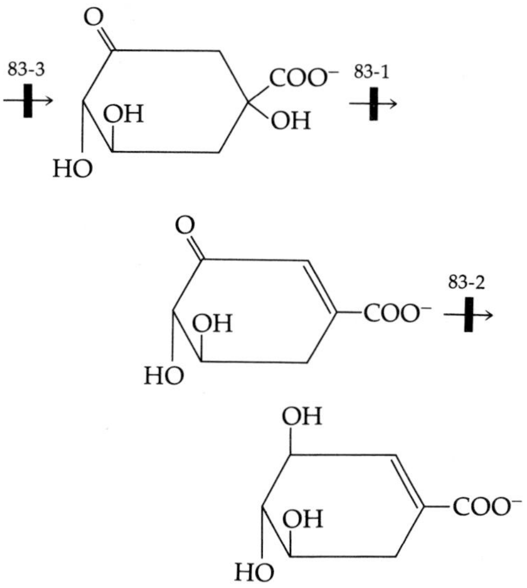

(14-40)

1. Enzymes of the Shikimic Acid Pathway

The six carbons of the benzene ring of the aromatic amino acids are derived from the four carbons of erythrose 4-phosphate and two of the three carbons of phosphoenolpyruvate. The initial step in the pathway (Fig. 14-17, step *a*) is the condensation of erythrose 4-P with PEP. Closely analogous to an aldol condensation, the mechanism poses some mysteries.[100] When PEP containing ^{18}O in the oxygen bridge to the phosphoryl group reacts, the ^{18}O is retained in the eliminated phosphate; biochemical intuition would suggest that it should stay in the carbonyl group of the product. Most bacteria and fungi have three isozymes, each controlled by feedback inhibition by one of the three products tyrosine, phenylalanine, or tryptophan. The product 3-deoxy-2-keto-D-*arabino*-heptulosonic acid-7-phosphate is cyclized to 5-dehydroquinic acid (Fig. 14-17, steps *b* and *c*). The proposed enolic intermediate (bracketed in Fig. 14-17) should condense readily to form the product. It has been proposed that the elimination of phosphate in step *b* may be assisted by a transient oxidation of the hydroxyl group at C-5 to a carbonyl group.[101] The enzyme system requires NAD, lending support to this idea. Both steps b and c are catalyzed by a single enzyme, the product of gene *aro* B.

Step d of Fig. 14-17 is the first of three elimination reactions needed to generate the benzene ring. This dehydration is facilitated by the presence of the carbonyl group. After reduction of the product to shikimic acid (step e) a phosphorylation reaction (step f)[102] sets the stage for a future elimination of P_i. In step g, condensation with PEP provides three carbon atoms that will become the α, β, and carboxyl carbon atoms of phenylalanine and tyrosine. Note that the reaction occurs by displacement of P_i from the α-carbon atom of PEP and resembles a reaction (Eq. 12-5, step a) in the synthesis of N-acetylmuramic acid. When the reaction is carried out in ${}^3H$-containing water, tritium enters the methylene group,[103] suggesting an addition–elimination mechanism (Eq. 14-41):

3-Enoylpyruvylshikimate

After the condensation with PEP, elimination of P_i (Eq. 7-50) yields chorismate (chorismic acid).[104]

2. Chorismic Acid

Chemical properties appropriate to a compound found at a branch point of metabolism are displayed by chorismic acid. Warming the compound yields a mixture of **prephenic acid** (reaction i, Fig. 14-17) and 4-hydroxybenzoic acid (reaction l). Note that the latter reaction is a simple elimination of the enolate anion of pyruvate. As indicated in Fig. 14-17, these reactions correspond to only two of several metabolic reactions of the chorismate ion. The formation of **phenylpyruvate** (steps i and j, Fig. 14-17) is catalyzed by a single protein molecule with two distinctly different enzymatic activities.[105,106] In both reaction steps the enzyme chorismate mutase-prephenate dehydratase promotes reactions that occur spontaneously upon warming in acidic solution. Phenylpyruvate is readily transaminated to phenylalanine to complete the biosynthesis of that amino acid.

A second enzyme chorismate mutase-prephenate dehydrogenase causes the oxidative decarboxylation of prephenate to p-hydroxyphenylpyruvate, which can be converted by transamination to tyrosine.[105] Again, a single protein catalyzes both steps i and k, being in one of its activities an isoenzyme of the protein catalyzing steps i and j. A slightly different pathway for tyrosine formation has been found in blue-green bacteria. Prephenate undergoes transamination to pretyrosine. The latter is then oxidatively

Pretyrosine of
blue-green bacteria

decarboxylated to tyrosine.[107] Chorismate can also be isomerized, hydrolyzed, and dehydrogenated to 2,3-dehydroxybenzoate, a precursor to enterobactin (Fig. 2-44).[107a] The genes (*ent*) for the requisite enzymes are clustered at 14 min on the *E. coli* chromosome map (Fig. 15-1).

3. Anthranilic Acid and Tryptophan

Tryptophan as well as a variety of specialized metabolites arise from anthranilic acid (*o*-aminobenzoic acid) formed by the action of **anthranilate synthetase.** One of the two subunits of this enzyme contains a glutamine-binding site at which ammonia is thought to be generated hydrolytically (Section B,1). A possibility (Fig. 14-18, reaction *a*) is that an amide of chorismic acid is formed first and that the nitrogen is then transferred to the benzene ring, simultaneously with elimination of hydroxyl ion. The resulting bicyclic intermediate would then eliminate pyruvate. Another possibility is that ammonia adds, as indicated in reaction *b* of Fig. 14-18, with elimination of hydroxyl ion in the initial step and elimination of pyruvate in a second step.[108,109]

An isomer of anthranilate is ***p*-aminobenzoate,** a precursor of folic acid. Its synthesis (reaction *c*, Fig. 14-18) has much in common with that of anthranilate but involves replacement of the hydroxyl group by an amino group on the same carbon atom. Evidence for the following intermediate has been obtained[110]:

$$\begin{array}{c}\text{COO}^-\\ \\ \text{H}\\ \text{H}\\ \\ \text{H}\\ \text{OH}\\ \text{NH}_2\end{array}$$

During the conversion of anthranilate to tryptophan, two additional carbon atoms must be incorporated to form the indole ring. These are derived from **phosphoribosyl pyrophosphate (PRPP),** and an important intermediate in the synthesis of both nucleotides and amino acids. PRPP is formed from ribose 5-phosphate by transfer of a *pyrophosphoryl* group from ATP[111]. The —OH group on the anomeric carbon of the ribose phosphate displaces AMP by attack on P_β of ATP (Eq. 14-42, Fig. 7-7). Phosphoribosyl pyrophosphate is the donor of phosphoribosyl groups for biosynthesis of nucleo-

$$\begin{array}{c}\text{ATP}\\ +\\ \text{Ribose-5-P}\end{array} \longrightarrow$$

5-Phosphoribosyl 1-pyrophosphate   (14-42)

tides (Section K). In tryptophan biosynthesis it is converted to an aminoglycoside of anthranilic acid by displacement of the pyrophosphate group by the amino group (Fig. 14-18, step *d*). The aminoglycoside then undergoes an internal redox reaction known as the **Amadori rearrangement.** In this rearrangement C-1 of the sugar becomes reduced, while C-2 is oxidized to a carbonyl group. The product has an open chain. The reaction can be formulated in three steps (*e*, *f*, and *g*) through the hypothetical bracketed intermediates of Fig. 14-18. First, the pentose ring opens with formation of a protonated Schiff base, the kind of reaction believed to account for the sensitivity of nucleotides toward acid hydrolysis (Chapter 2, Section H,2). The same chemistry is probably involved in the acid modification of NADH (Chapter 8, Section H,7,b). Tautomerism of the Schiff base generates a carbon–carbon double bond in an enolic structure. In the third step the enol is tautomerized to the more stable ketone. Decarboxylation and ring closure (reactions *h*, *i*, and *j*) yields **indoleglycerol phosphate.**

A β-replacement reaction catalyzed by the pyridoxal phosphate-dependent **tryptophan synthetase** converts indoleglycerol phosphate and serine to tryptophan. Tryptophan synthetase from *E. coli* consists of two subunits associated as an $\alpha_2\beta_2$ tetramer. The β subunit contains pyridoxal phosphate and presumably generates from serine the Schiff base of aminoacrylic acid, as indicated in Fig. 14-18. This subunit is able to catalyze the addition of free indole to the Schiff base to form tryptophan. The α subunit, on the other hand, is able to catalyze the cleavage (essentially a reverse aldol) of indoleglycerol phosphate to glyceraldehyde 3-phosphate and free indole. Either it gener-

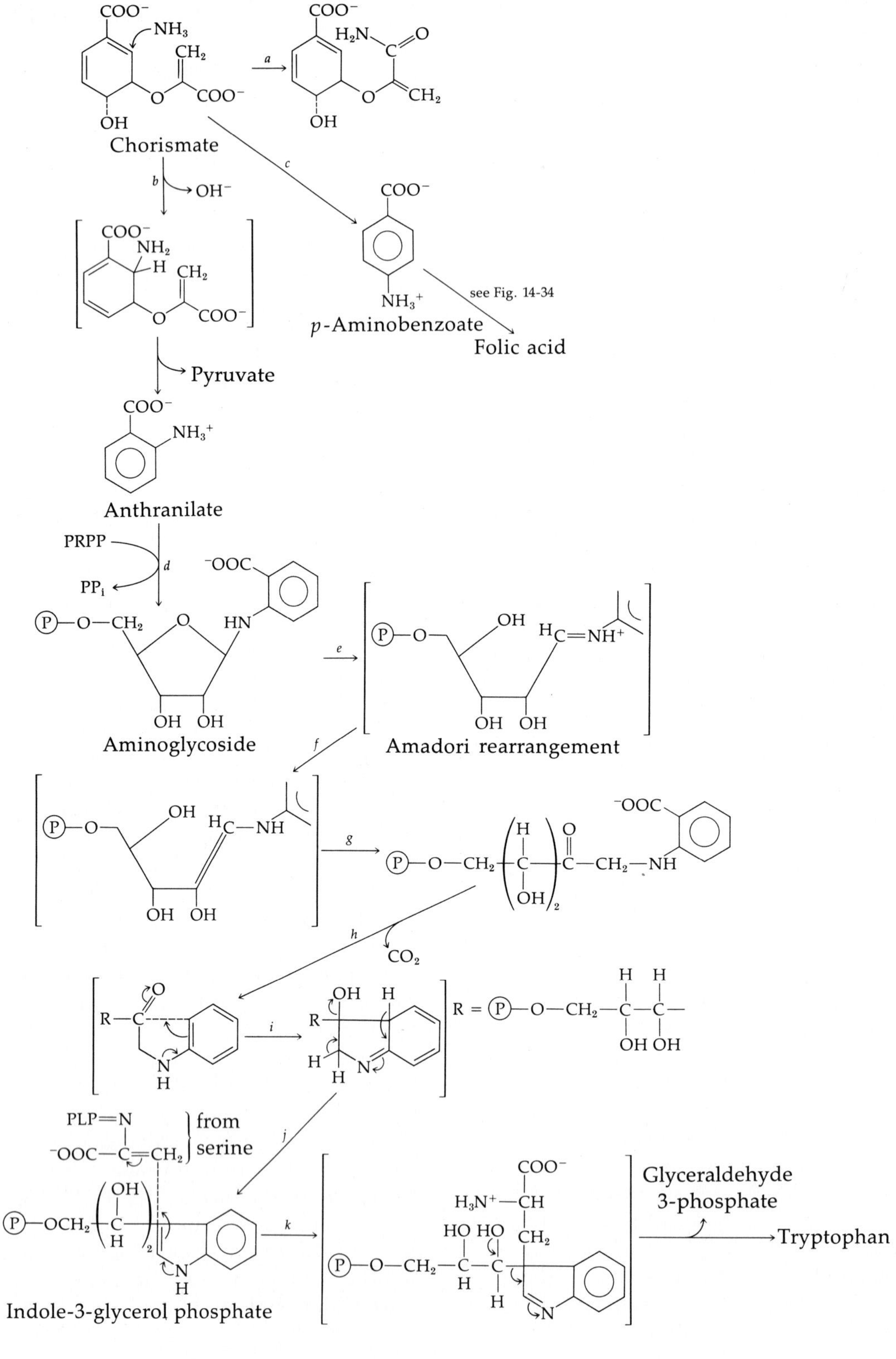

Chorismate

a

H_2N

b → OH^-

c

p-Aminobenzoate

see Fig. 14-34

Folic acid

Pyruvate

Anthranilate

PRPP

PP_i

d

Aminoglycoside

e

Amadori rearrangement

f

g

CO_2

h

$R = \text{(P)}-O-CH_2-\underset{OH}{\underset{|}{C}}-\underset{OH}{\underset{|}{C}}-$

i

PLP=N

$^-OOC-\underset{}{C}=CH_2$ } from serine

j

Indole-3-glycerol phosphate

k

Glyceraldehyde 3-phosphate

→ Tryptophan

←

Fig. 14-18 The biosynthesis of tryptophan from chorismic acid.

ates free indole and provides it to the β subunit in a tight complex, or the complex catalyzes the condensation with the dehydrated Schiff base prior to the aldol cleavage, as is suggested in Fig. 14-18.

While the tryptophan synthetase of *E. coli* is made up of two different subunits, that of *Neurospora* is a single polypeptide chain. It has been proposed that there were originally two separate genes, as in *E. coli,* but that they became fused during the course of evolution. Since this proposal was made, gene fusion has been demonstrated experimentally in *Salmonella* by introduction of two consecutive "frame shift mutations" between two genes of histidine biosynthesis (Chapter 15, Section D). Because of the frame shift, the stop signal for protein synthesis is apparently misread with the result that the organism makes a single long protein corresponding to both genes. Since a number of instances are known in which the two distinctly different catalytic activities are possessed by the same protein, it may be that gene fusion occurs in nature quite commonly.[111a] Examples are the aspartokinase-homoserine dehydrogenase of *E. coli,* which catalyzes both reactions *a* and *c* of Fig. 14-6, and anthranilate synthetase of *Neurospora.* The latter, which has an $\alpha_2\beta_2$ structure, contains one *trifunctional* subunit which catalyzes steps *d* (phosphoribosylanthranilate isomerase) and *e–j* (indoleglycerol-phosphate synthetase) of Fig. 14-18 and provides a glutamine-binding site for the anthranilate synthetase reaction (step *b*).[109] Two other bifunctional enzymes have been mentioned in the discussion of the conversion of chorismate to phenylalanine and tyrosine.

4. Synthesis of the Ubiquinones, Plastoquinones, Tocopherols, and Vitamin K[112–117]

Radioactive carbon of [^{14}C]shikimate is efficiently incorporated into the quinones and toco-pherols that are the topic of this section. These chemically related (Chapter 10, Section D) redox agents also have a related biosynthetic origin, which has been elucidated in greatest detail for ubiquinone as synthesized by bacteria. *p*-Hydroxybenzoate is formed directly from chorismate in bacteria (Fig. 14-17), but in plants it may originate from tyrosine and *p*-coumarate as indicated in Fig. 14-19. The conversion of *p*-hydroxybenzoate on to the ubiquinones is also shown in Fig. 14-19. A polyprenyl group is transferred onto a position ortho to the hydroxyl (see Chapter 12, Section H,4). Then a series of consecutive hydroxylation and transmethylation reactions (from SAM) lead directly to the ubiquinones. Several quinones that can serve as precursors to ubiquinones have been isolated from bacteria. While two of these quinones are shown as intermediates in Fig. 14-19, chemical considerations would suggest that both the methylations and hydroxylations occur on the reduced dihydroxy derivatives. A similar pathway, possibly via the CoA ester of *p*-hydroxybenzoate, is used for synthesis of ubiquinone on the inner membranes of liver mitochondria.[115]

Labeling experiments have shown that the plastoquinones of chloroplasts as well as the tocopherols bear one methyl group (marked with an asterisk in Fig. 14-19) that originates from chorismate. The dihydroxy compound **homogentisate** is believed to be an intermediate.[116] It is a normal catabolite of tyrosine in the animal body (Fig. 14-20). Again, prenylation and methylation from SAM are required to complete the synthesis of the plastoquinones and tocopherols.

The vitamins K and other naphthoquinones probably arise from **O-succinylbenzoate**[117] (Fig. 14-19) whose synthesis from chorismate and α-ketoglutarate may depend upon a thiamine diphosphate-bound intermediate, as indicated in Fig. 14-19. The proposed condensation to form *O*-succinylbenzoate is of a type similar to that proposed for the conversion of chorismate to anthranilate. The remaining reactions of decarboxylation, methylation, and prenylation (Fig. 14-19) resemble those of ubiquinone synthesis.

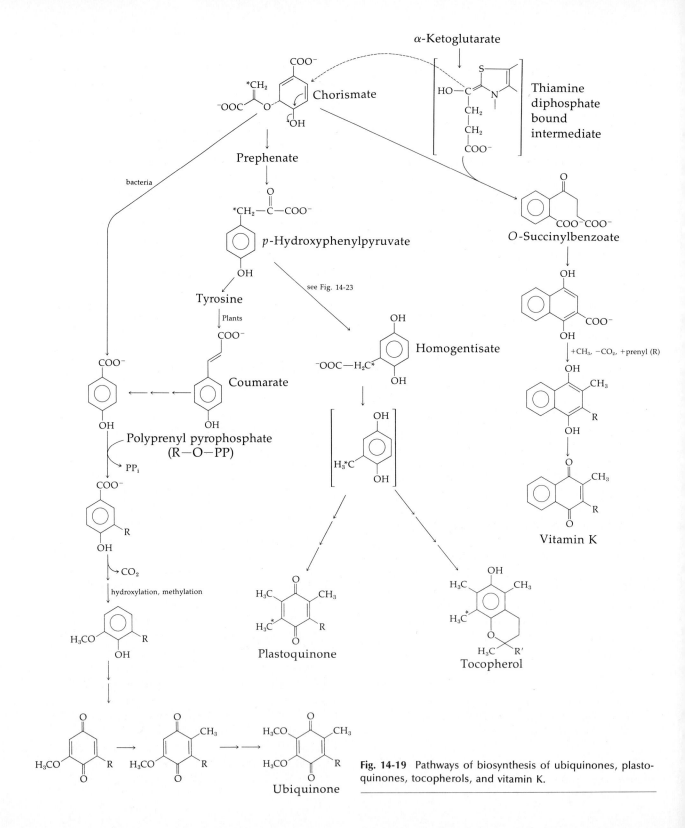

Fig. 14-19 Pathways of biosynthesis of ubiquinones, plasto-quinones, tocopherols, and vitamin K.

5. Metabolism of Phenylalanine and Tyrosine in Animals and Bacteria

Figure 14-20 shows the principal catabolic pathways, as well as a few biosynthetic reactions, of phenylalanine and tyrosine in animals. Transamination to phenylpyruvate (reaction *a*) occurs readily, and the product may be oxidatively decarboxylated to phenylacetate. The latter may be excreted after conjugation with glutamine (recall Knoop's experiments in which phenylacetate was excreted by dogs after conjugation with glycine, Box 9-A). Although it does exist, this degradative pathway for phenylalanine must have a limited function in man, for an excess of phenylalanine is toxic* unless it can be oxidized to tyrosine (reaction *b*, Fig. 14-20).

The pterin-dependent hydroxylation of phenylalanine to tyrosine (Eq. 10-52) has received a great deal of attention, in part because of the occurrence of the metabolic disease **phenylketonuria,**[118] in which this reaction does not take place. Infants born with phenylketonuria appear normal but mental retardation sets in rapidly. However, considerable success has been reported in rearing children with the defect on a low phenylalanine diet, which supplies only enough of the amino acid for essential protein synthesis. Some children have already been raised from infancy to young adulthood on such diets. Much of the brain damage usually encountered has been prevented, and the tolerance to phenylalanine has been found to increase with age.

a. Catabolism of Tyrosine

The major route of degradation of tyrosine in animals begins with transamination (reaction *c*) to **p-hydroxyphenylpyruvate.** The enzyme tyrosine aminotransferase has been studied a great deal because of its induction in the liver in response to the action of glucocorticoid hormones (Chapter 11, Section F,7). The synthesis of the enzyme is also

controlled at the translational level,[119] release of the newly formed protein from liver ribosomes being stimulated by cyclic AMP. Furthermore, the enzyme is subject to posttranscriptional modification including phosphorylation,[120] and it undergoes unusually rapid turnover.[121]

The keto acid *p*-hydroxyphenylpyruvate is decarboxylated by the action of a dioxygenase which has been considered in Chapter 10 (Eq. 10-55). The product **homogentisic acid** is acted on by a second dioxygenase, as indicated in Fig. 14-20, with eventual conversion to fumarate and acetoacetate.

One of the first "inborn errors of metabolism" to be recognized was **alkaptonuria,** a lack of the oxygenase that cleaves the ring of homogenisic acid.[122] The condition is readily recognized by a blackening of the urine upon standing (caused by oxidation of the homogentisate). Alkaptonuria was correctly characterized by Garrod (Box 1-D) in 1909 as a defect in the catabolism of tyrosine.

b. The Thyroid Hormones[123,124]

An important product of tyrosine metabolism in the animal is the thyroid hormone of which the principal and most active forms are **thyroxine** and **triiodothyronine.** The thyroid gland is rich in io-

Triiodothyronine

a fourth iodine is present here in thyroxine

dide ion, which is actively concentrated from the plasma to $\sim 1 \ \mu M$ free I^-. This iodide reacts under the influence of a peroxidase (see Eq. 10-7 and accompanying discussion) to iodinate tyrosyl residues of the high molecular weight (dimeric) **thyroglobulin** (MW $\sim 660,000$).[124a] Several of the tyrosine side chains are iodinated to form **mono-** and **diiodotyrosine** residues (Eq. 14-43). The nature of the coupling reaction by which the aromatic group from one residue of mono- or diiodotyrosine is joined in ether linkage with a second residue is uncertain. However, it is known that the

---

* However, the mechanism by which the excess amino acid causes brain damage is unknown.

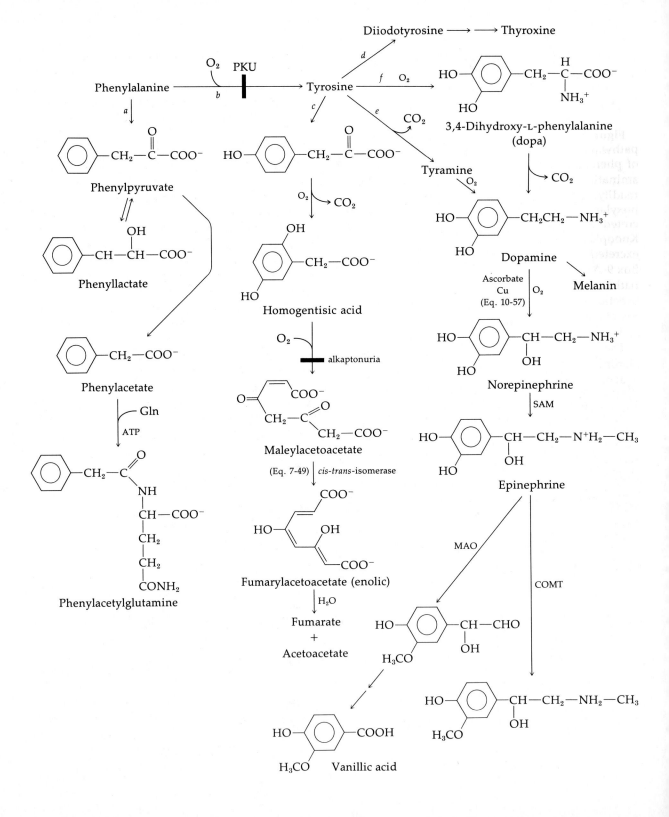

Fig. 14-20 Some routes of metabolism of phenylalanine and tyrosine in animals.

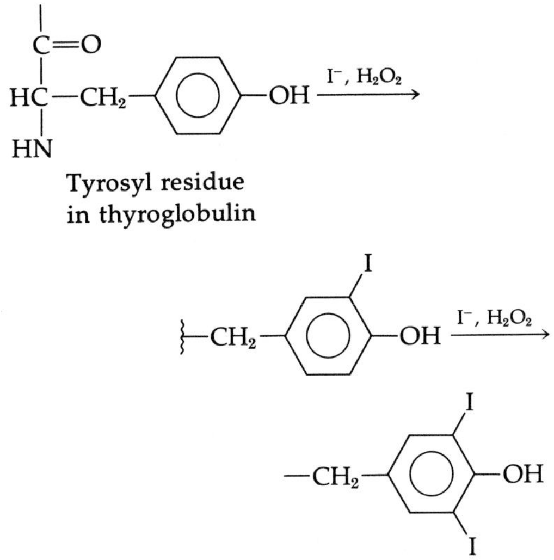

reaction will take place readily in the presence of oxygen and a peroxidase. It is easy to visualize formation of an electron-deficient radical which could undergo β elimination to produce a dehydroalanine residue and an aromatic radical. The latter could couple with a second radical to form triiodothyronine or thyroxine (Eq. 14-44). Alternatively, a pyridoxal phosphate-dependent β elimi-

nation of the radical could be used. A third possibility involves oxidative attack on the keto acids derived from the iodotyrosines.[125]

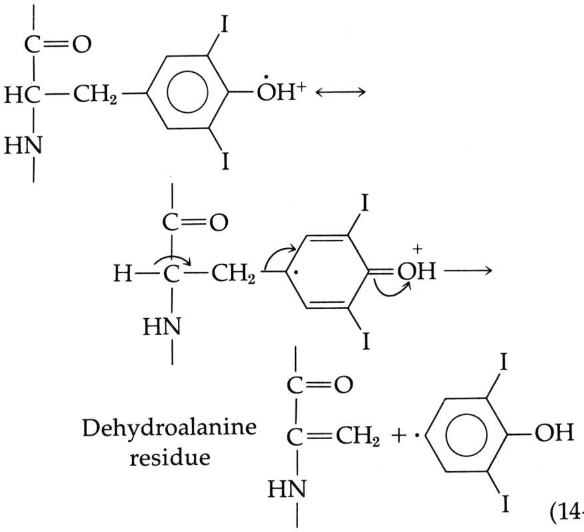

Thyroxine and triiodothyronine are released from thyroglobulin through the action of a series of proteases. Both the protease action and the release of the thyroid hormones into the bloodstream are stimulated by the pituitary thyrotropic hormone (TSH). This thyroid stimulating hormone, like glucagon, probably acts through cAMP-mediated mechanisms. The hormones are carried throughout the body while bound to a special **thyroid-binding globulin** which serves as a carrier. Some of the hormone molecules are carried by other serum proteins as well. Both thyroxine and triiodothyronine have powerful hormonal effects on tissues, but the lag time for a response is shortest for triiodothyronine. Thus, it is thought that thyroxine may undergo loss of one iodine atom to form the more active triiodo form of the hormone within the target cells.

A principal function of thyroxine and triiodothyronine is to stimulate energy metabolism in other tissues. It has long been recognized that a deficiency of thyroid hormone is reflected in an overall lower basal metabolic rate (Chapter 3, Section A,5). However, no totally satisfactory chemical theory of the action of the hormones has appeared. An important observation of H. Lardy and associates is that thyroxine uncouples oxidative phosphorylation (Chapter 10, Section E,5) in isolated mitochondria. When mitochondria from animals receiving extra thyroxine are compared with those from control animals, an increased rate of electron transport is observed. However, little or no change in the P/O ratio is seen. Thus, it appears that *in vivo* the hormone is able to increase the rate of electron transport without decreasing the overall efficiency of ATP synthesis. One idea is that uncoupling occurs at a specific step at one phosphorylation site.[126]

Some investigators believe that uncoupling of phosphorylation is a secondary effect of thyroid hormone and that the primary action on mitochondria is to induce swelling of the organelles. A variety of metabolic consequences would follow.

Evidence has been presented in support of a direct effect of triiodothyronine on gene transcription.[127] Binding of the hormone to a specific nuclear protein has been observed.[127a] Binding to other cellular components also occurs.[127b] An important action of the thyroid hormone is to stimulate the mobilization of fat from the adipose tissue. It has been argued that this response rests on an inhibition by thyroid hormones of the membrane-bound cyclic AMP phosphodiesterase (Eq. 7-25).[128]

A number of thyroid diseases are known. They are usually evident by enlargement of the thyroid gland (**goiter**). The deficiencies may involve transport of iodide into the thyroid, formation of iodinated thyroglobulin, or inefficient coupling to form the iodinated thyronine residues.[124]

c. The Catecholamines

A combination of decarboxylation and hydroxylation of the ring of tyrosine produces derivatives of o-dihydroxybenzene (catechol) which play important roles as neurotransmitters. They are also precursors to the black pigment of skin and hair known as **melanin.** One route of formation of the catecholamines is through decarboxylation of tyrosine into tyramine (reaction e, Fig. 14-20) and subsequent oxidation. However, the quantitatively more important route is hydroxylation by the reduced pterin-dependent tyrosine hydroxylase[129] to 3,4-dihydroxyphenylalanine, better known as **dopa.** This compound has been much in the news because of its effectiveness as a drug in treating Parkinson's disease. The debilitating symptoms of this disease are thought to result from a lack of the decarboxylation product **dopamine** (Fig. 14-20) in certain regions of the brain. Dietary dopa permits the brain more effectively to form dopamine.

Hydroxylation of dopamine by an ascorbic acid and copper-requiring enzyme (Eq. 10-57) produces norepinephrine (noradrenaline). Methylation yields epinephrine (adrenaline) an important hormone. There are two principal catabolic routes for destruction of the catecholamines. They are illustrated for adrenaline in Fig. 14-20. **Monoamine oxidase** (MAO) causes oxidative cleavage with deamination. Subsequent oxidative fission of the side chain together with methylation yields such end products as vanillic acid, which is excreted in the urine. The second catabolic route is immediate O-methylation by **catecholamine O-methyltransferase** (COMT), a very active enzyme in neural tissues. The metabolites are relatively inactive physiologically and may be secreted as such or may undergo further oxidative degradation.

d. The Melanins[130]

Dihydroxyphenylalanine (dopa) darkens rapidly when exposed to oxygen. The process is hastened greatly by tyrosinase (Box 10-H) which also catalyzes reaction f of Fig. 14-20, the oxidation of tyrosine to dopa. Tyrosinase is found in animals only in the organelles known as melanosomes, which are present in the melanin-producing **melanocytes.** A series of enzymatic and nonenzymatic oxidation, decarboxylation, and coupling reactions forms the pigments. The initial steps are indicated in Fig. 14-21. Oxidation of dopa to dopaquinone is followed by an intramolecular addition reaction, together with tautomerization to the indole derivative, leucodopachrome. A second oxidation by tyrosinase is followed by decarboxylation and tautomerization to 5,6-dihydroxyindole. The latter can undergo a third oxidation step to indole-5,6-quinone. Coupling of these last two products as indicated in Fig. 14-21 yields a dimer which is able to continue the addition of dihydroxyindole units with oxidation to form a high polymer. A related series of red polymers, found in red hair and feathers, are formed by addition of cysteine to dopaquinone.[131] Addition is possible at more than one position. The resulting adducts (only one is shown) can undergo oxidative ring closure in the manner indicated.

e. Bacterial Catabolism of Phenylalanine, Tyrosine, and Other Benzenoid Compounds

Bacteria play an essential role in the biosphere by breaking down the many aromatic products of plant metabolism.[132] The latter include **lignin,** a

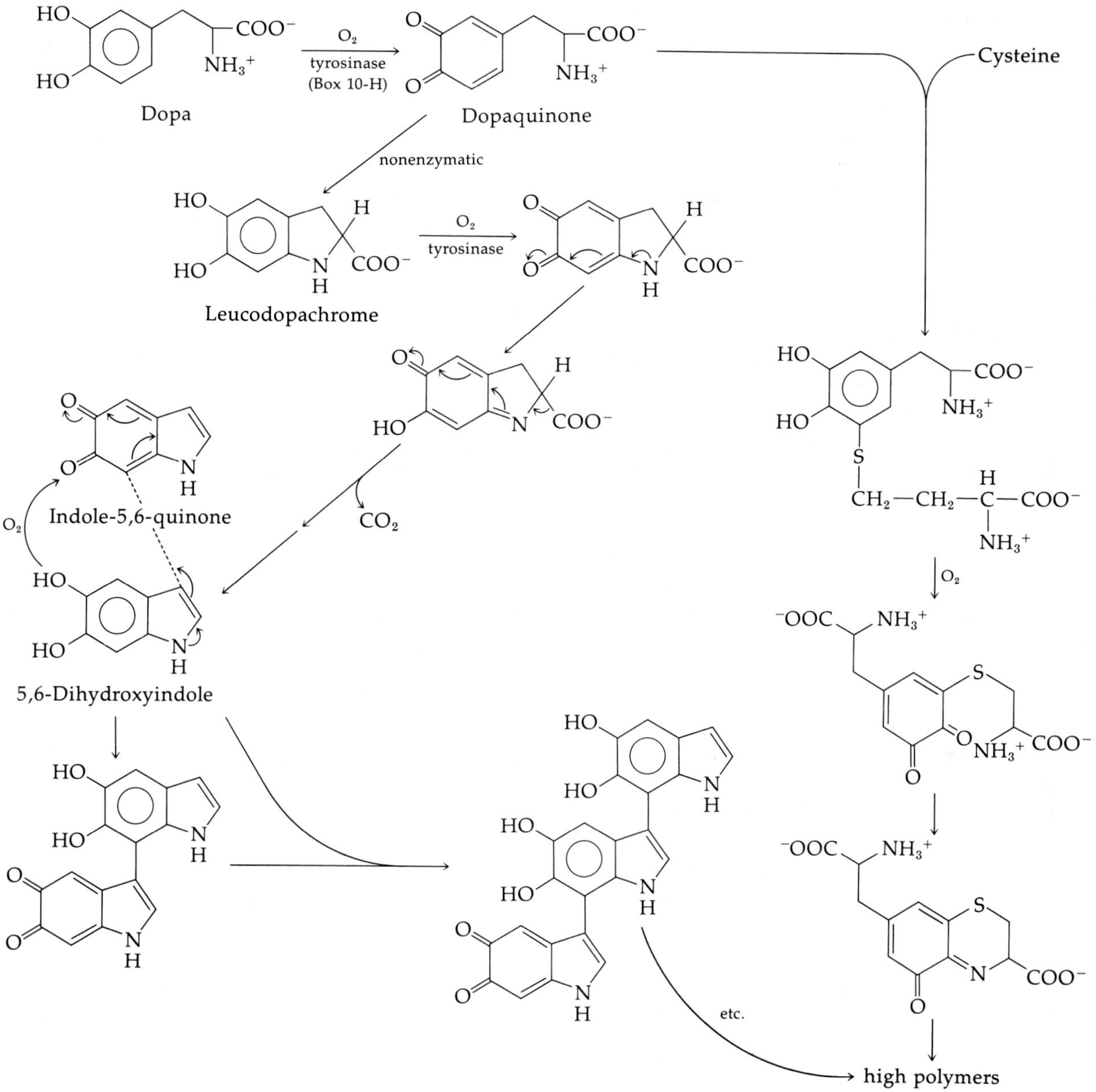

Fig. 14-21 Some postulated pathways for synthesis of black pigment melanin and pigments of reddish hair and feathers.

major constituent of wood, and a plant product second only to cellulose in abundance.

Sometimes elimination reactions are utilized to initiate bacterial degradation of aromatic compounds. For example, certain bacteria release phenol from tyrosine by β elimination. More often

hydroxylation and oxidative degradation of side chains lead to derivatives of benzoic acid or of the various hydroxybenzoic acids.[133] A few examples are shown in Fig. 14-22. Note that in every case a dioxygenase is required for the opening of the benzene ring. The pathways contain interesting

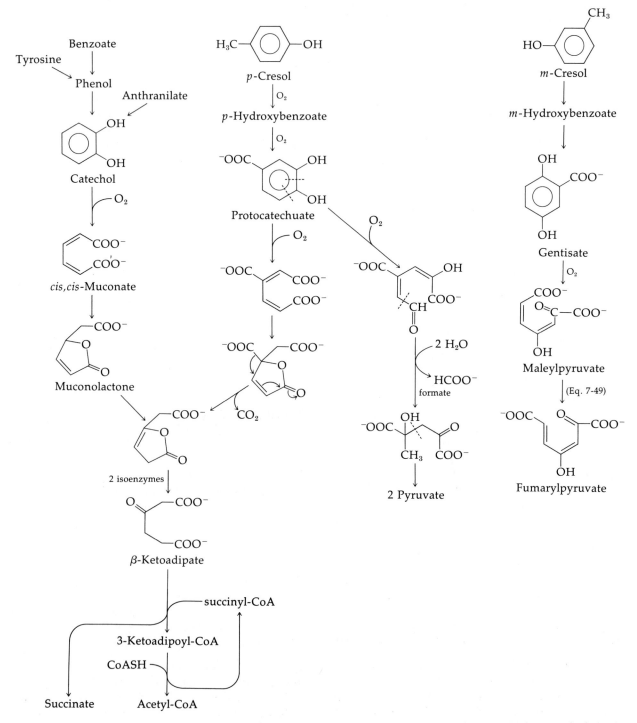

Fig. 14-22 A few pathways of catabolism of aromatic substances by bacteria.

isomerization steps,[134] some of which have been discussed on previous pages.

Another degradative reaction of an aromatic compound deserves mention because of the unusual enzymatic reactions required. This is the breakdown of various forms of vitamin B_6 by bacteria.[136] In one pathway the hydroxymethyl group in the 5 position and the substituent in the 4 position are both oxidized in the early steps to carboxylate groups. Then, as indicated in Eq. 14-45, a de-

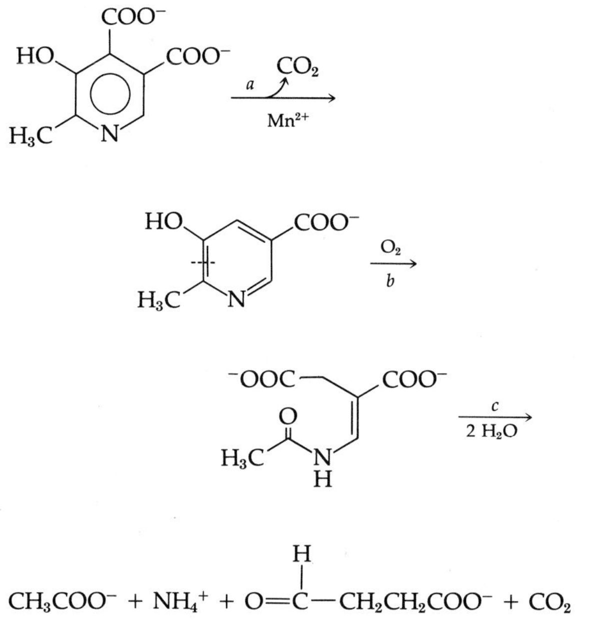

$$CH_3COO^- + NH_4^+ + O{=}\overset{\underset{|}{H}}{C}{-}CH_2CH_2COO^- + CO_2$$

(14-45)

carboxylation is followed by the action of an unusual dioxygenase. The dioxygenase, isolated from a strain of *Pseudomonas*, contains bound FAD, which must be reduced by external NADH. The enzyme, like a typical dioxygenase, introduces two atoms of oxygen into the product. However, it also uses the reduced FAD to reduce the double bond system (either before or after the attack by oxygen). Another enzyme of the same bacterium is remarkable in degrading hydrolytically the product of the oxygenation reaction to four different products without the accumulation of intermediates.

6. The Metabolism of Phenylalanine and Tyrosine in Plants

Some of the pathways of animal and bacterial metabolism of these amino acids also are used in plants. However, more important are the reactions initiated by the **phenylalanine** and **tyrosine ammonia-lyases** discussed in Chapter 8, Section F,5 (Eq. 8-36). Figure 14-23 illustrates the principal pathway by which the two amino acids are converted to *trans*-cinnamate and to mono-, di-, and trihydroxy derivatives. Cinnamoyl-CoA gives rise to the anthocyanins, to other flavonoid pigments, and to the polymeric condensed tannins (Box 12-B). The dihydroxy and trihydroxy methylated products are the starting materials for formation of lignins. In addition, a large series of plant products, many of which impart characteristic fragrance to spices and other plants, are illustrated in this figure. Note, however, that protocatechuate, also a product of bacterial catabolism (Fig. 14-22); can arise in plants by a simple dehydration of 5-dehydroshikimate (Fig. 14-17) followed by enolization. Hydroxylation of protocatechuate leads to gallate, which in the form of esters and other deri-

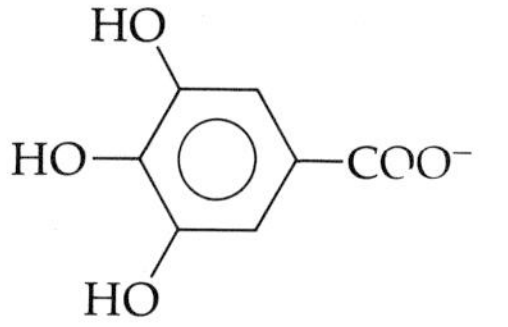

Gallate

vatives constitutes the "hydrolyzable tannins." These materials accumulate in the vacuoles of the plants and are also deposited in the bark, along with the "condensed tannins", polymeric flavonoid compounds (Box 12-B).[135]

Lignin is a very complex material of molecular weight greater than 10,000. It is remarkably stable, being insoluble in hot 70% sulfuric acid. Lignin is often described as a "statistical polymer of oxyphenylpropane units." It arises from oxidative coupling of **coniferyl alcohol** (Fig. 14-23) and related monomers.[137–139] The enzyme responsible for the polymerization may be a peroxidase which

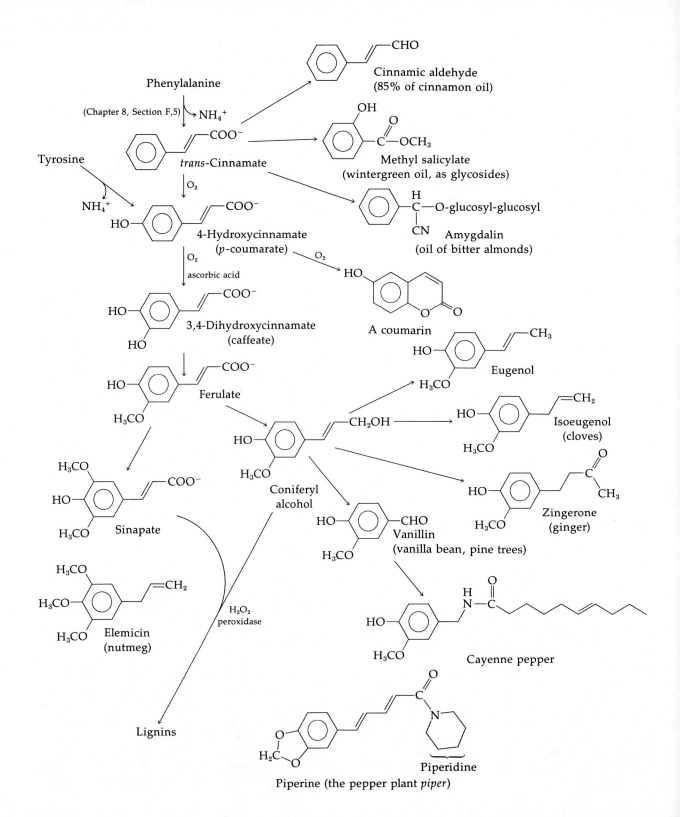

Phenylalanine

(Chapter 8, Section F,5) → NH₄⁺

Cinnamic aldehyde
(85% of cinnamon oil)

trans-Cinnamate

Methyl salicylate
(wintergreen oil, as glycosides)

Tyrosine

NH₄⁺

4-Hydroxycinnamate
(*p*-coumarate)

Amygdalin
(oil of bitter almonds)

3,4-Dihydroxycinnamate
(caffeate)

A coumarin

Eugenol

Ferulate

Isoeugenol
(cloves)

Coniferyl
alcohol

Sinapate

Vanillin
(vanilla bean, pine trees)

Zingerone
(ginger)

Elemicin
(nutmeg)

H₂O₂
peroxidase

Cayenne pepper

Lignins

Piperidine

Piperine (the pepper plant *piper*)

Fig. 14-23 Formation of some plant metabolites from phenylalanine and tyrosine.

catalyzes formation of lignin from the monomeric alcohols and H_2O_2. A radical generated by loss of an electron from a phenolate anion of coniferyl alcohol consists of a number of resonance forms in which the unpaired electron may be present not only on the oxygen but at the positions marked by asterisks in the following structure:

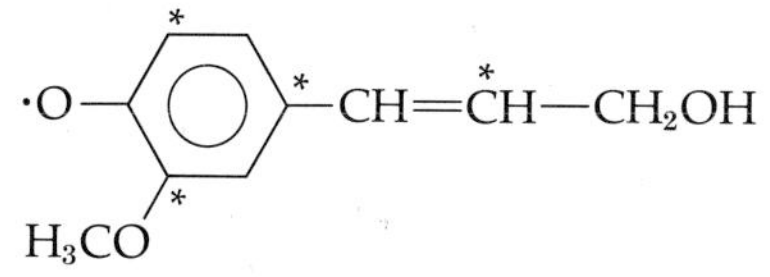

Coupling of such radicals can yield a great variety of products. For example, dimerization gives the stable ether link structure shown in Eq. 14-46. The

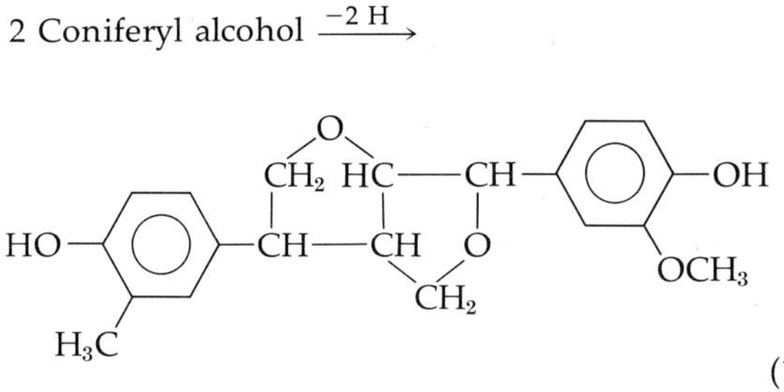

$$2 \text{ Coniferyl alcohol} \xrightarrow{-2\,H}$$

(14-46)

dimer formed still contains hydroxyl groups capable of radical formation and addition to other units. At least ten other types of intermonomer linkage are shown in Fig. 14-24. Lignin represents an enormous potentially valuable industrial source of aromatic raw materials, but little success has been achieved in utilizing it to date.

Oxidative degradation of lignin produces **humic acid,** an important organic constituent of soils.[138]

Alkaloids[140,141]

Over 2500 miscellaneous nitrogen-containing compounds produced by plants have been identified. These **alkaloids** are especially numerous in certain families of plants. Alkaloids are often thought of simply as end products of nitrogen metabolism in plants. However, most plants do not make alkaloids and there may be ecological

reasons why some do. Many alkaloids have potent physiological effects on animals.

Many of the alkaloids are derived directly from aromatic amino acids, as was first recognized by R. Robinson[141a,141b] in 1917. Robinson proposed that alkaloids may arise from **Mannich reactions** of amines and aldehydes. In the Mannich reaction (Eq. 14-47) an amine and an aldehyde (probably

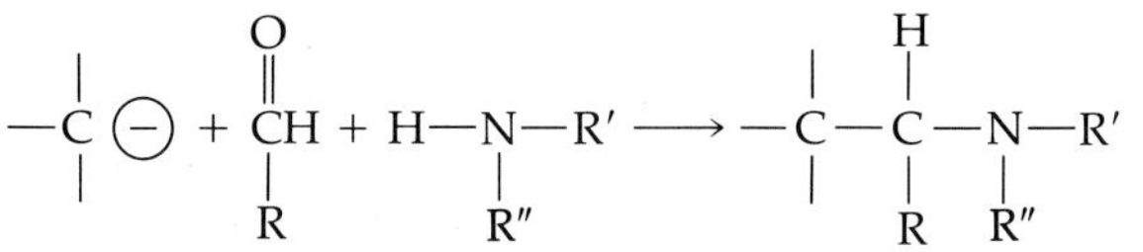

(14-47)

through a Schiff base) react with a nucleophilic carbon such as that of an enolate anion. A variety of amines can be formed by decarboxylation of amino acids and aldehydes can arise by oxidative decarboxylation (via transamination and α decarboxylation) of amino acids. Thus, amino acids can provide both of the major reactants for alkaloid synthesis. Furthermore, nucleophilic centers in the aromatic rings, e.g., in positions para to hydroxyl substituents, are observed to be frequent participants in Mannich condensations in the biosynthetic pathways. An example is shown in Fig. 14-25. Dopa is decarboxylated to dopamine, and is oxidized to 3,4-dihydroxybenzaldehyde. A Mannich reaction (via the Schiff base as shown) leads to ring closure. Oxidation of the ring produces an **isoquinoline** ring, a structural characteristic of a large group of alkaloids. Methylation produces **papaverine,** found in the opium poppy. A related alkaloid **morphine** (Fig. 14-25), at first glance appears dissimilar. However, further inspection suggests a straightforward biosynthetic route for morphine. Tyramine is used in place of dopamine for the same type of condensation that leads to papaverine. The two rings are then oxidatively coupled through one C—C bond and one ether linkage.

While Robinson's ideas on alkaloid biosynthesis were initially speculative, they have been confirmed by isotopic labeling experiments. Nevertheless, questions remain. The postulated aldehydes are not proved intermediates. It is con-

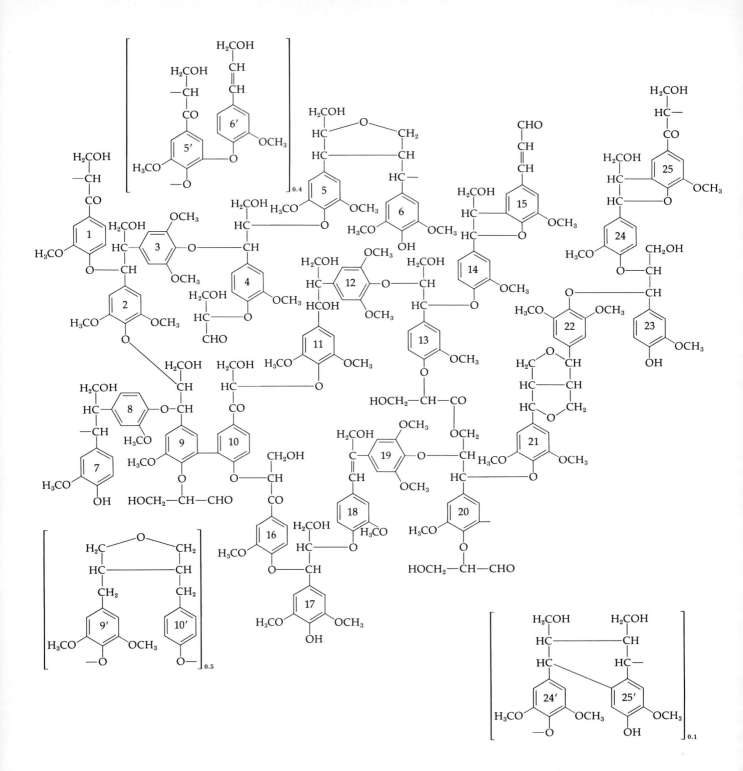

Fig. 14-24 The proposed structure of beech lignin. There are 25 different C_9 units, of which several can, to some extent, be replaced by the three dimeric units in brackets. From Nimz, [139] p. 317.

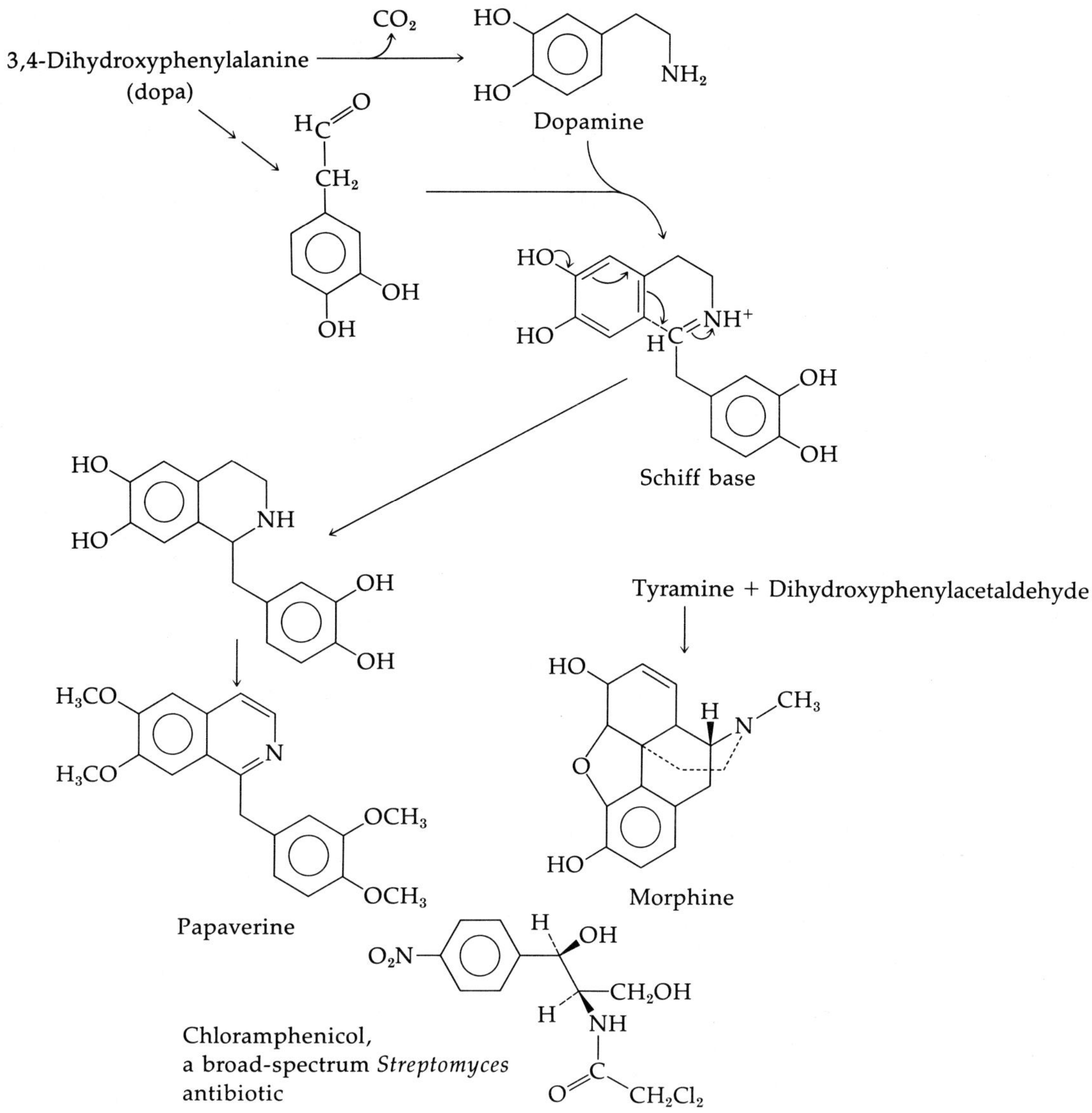

Fig. 14-25 The formation of some alkaloids and other substances from tyrosine metabolites.

ceivable that the Mannich condensation occurs with keto acids prior to decarboxylation.

Another alkaloid that arises from phenylalanine and tyrosine is colchicine (Box 4-A). The six-membered ring originates from phenylalanine, while the seven-membered tropolone ring is formed from tyrosine by ring expansion.

I. THE METABOLISM OF TRYPTOPHAN AND THE SYNTHESIS OF NAD

The biosynthesis of tryptophan, outlined in Fig. 14-18, has been discussed in Section H,3. The catabolism in animals is indicated in Fig. 14-26. One

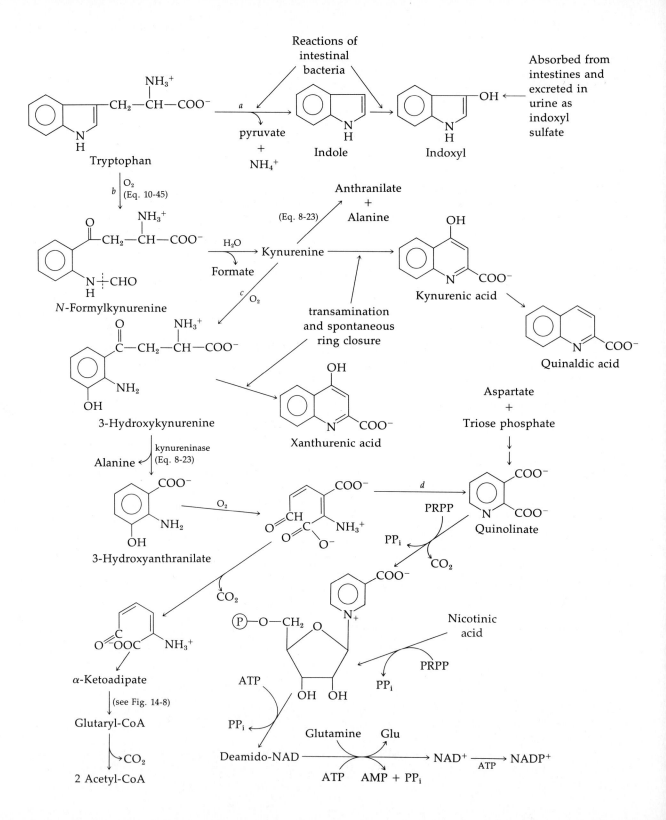

series of reactions (beginning with step *a*) are those of intestinal bacteria. Indole, produced by β elimination, is hydroxylated to **indoxyl.** Some of the latter is absorbed into the bloodstream and is excreted in the urine as indoxyl sulfate. The primary catabolic pathway in animal cells is initiated (step *b*, Fig. 14-26) with **tryptophan 2,3-dioxygenase** (Eq. 10-45). This enzyme has been much studied because of its inducibility in animal tissues and its regulation by hormone action. The enzyme is induced by both glucocorticoids and by tryptophan.[17,142]

The hydrolytic removal of formate from the product of tryptophan dioxygenase action yields **kynurenine,** a compound that is acted upon by a number of enzymes. Kynureninase (Eq. 8-23) cleaves the compound to anthranilate and alanine while transamination leads to the cyclic kynurenic acid. The latter is dehydroxylated in an unusual reaction to quinaldic acid, a prominent urinary excretion product.

Another major pathway of kynurenine metabolism is hydroxylation to 3-hydroxykynurenine (step *c*, Fig. 14-26), which in turn can undergo transamination to the cyclic xanthurenic acid or can be cleaved by kynureninase to 3-hydroxyanthranilate. The latter is opened under the action of a dioxygenase with eventual degradation to glutaryl-CoA, as indicated. An alternative pathway is of considerable nutritional significance in animals. The aldehyde produced by the ring opening reaction can reclose (step *d*) to a pyridine ring in the form of quinolinic acid. The latter, in a reaction that is also accompanied by decarboxylation, is coupled with a phosphoribosyl group of PRPP to form **nicotinate mononucleotide.** Adenylylation produces deamido NAD which is converted to **NAD** by a glutamine-dependent amination of the carboxyl group.

As indicated in Fig. 14-26, free nicotinic acid can also be used to form NAD. Not surprisingly, nicotinic acid, an essential vitamin, is about 60 times more efficient than tryptophan as a source of

NAD. Nevertheless, a high tryptophan diet does partially overcome a deficiency in dietary intake of nicotinic acid. The effectiveness of a diet containing only maize as a source of protein in inducing the deficiency disease pellagra (Box 8-H) is in part a result of the low tryptophan content of maize protein. An alternative pathway for synthesis of quinolinate from aspartate and a triose phosphate is believed to exist in plants and to provide the major route of nicotinic acid synthesis in nature.

Tryptophan is the precursor to a variety of alkaloids and other metabolites. Some of these are indicated in Fig. 14-27. The alkaloid harmine, which is found in several families of plants, can be formed from tryptophan and acetaldehyde (or pyruvate) in the same manner as is indicated for the formation of papaverine in Fig. 14-25.

Tryptophan is hydroxylated to 5-hydroxytryptophan which is decarboxylated to **serotonin,** an important neurotransmitter substance and a component of plants and animals alike. The latter may be methylated and acetylated to **melatonin,** the pineal hormone (Fig. 14-27). Some characteristic plant metabolites such as psilocybine, an hallucinogenic material from the mushroom *Psilocybe aztecorum,* are formed directly from serotonin. The important plant hormone **indole-3-acetate (auxin)** is presumably derived by oxidative decarboxylation of the α-keto acid corresponding to tryptophan. The corresponding aldehyde may be an intermediate. Its reduction product indole-3-ethanol also occurs in plants and is metabolically active.[142a]

For many years the alkaloid gramine from barley was regarded as a curiosity because only one carbon atom separates the nitrogen atom from the indole ring. It is now believed that tryptophan is cleaved in a PLP-dependent reaction analogous to that of serine transhydroxymethylase (Chapter 8, Section E,3c, Fig. 14-27). Other alkaloids arise in a more conventional fashion. Condensation of an isopentenyl group on the indole ring of tryptamine initiates the formation of **lysergic acid.** The indole ring of tryptophan is clearly visible in the structure of **reserpine** of *Rauwolfia,* a compound of

Fig. 14-26 Some catabolic reactions of tryptophan and synthetic reactions leading to NAD and NADP.

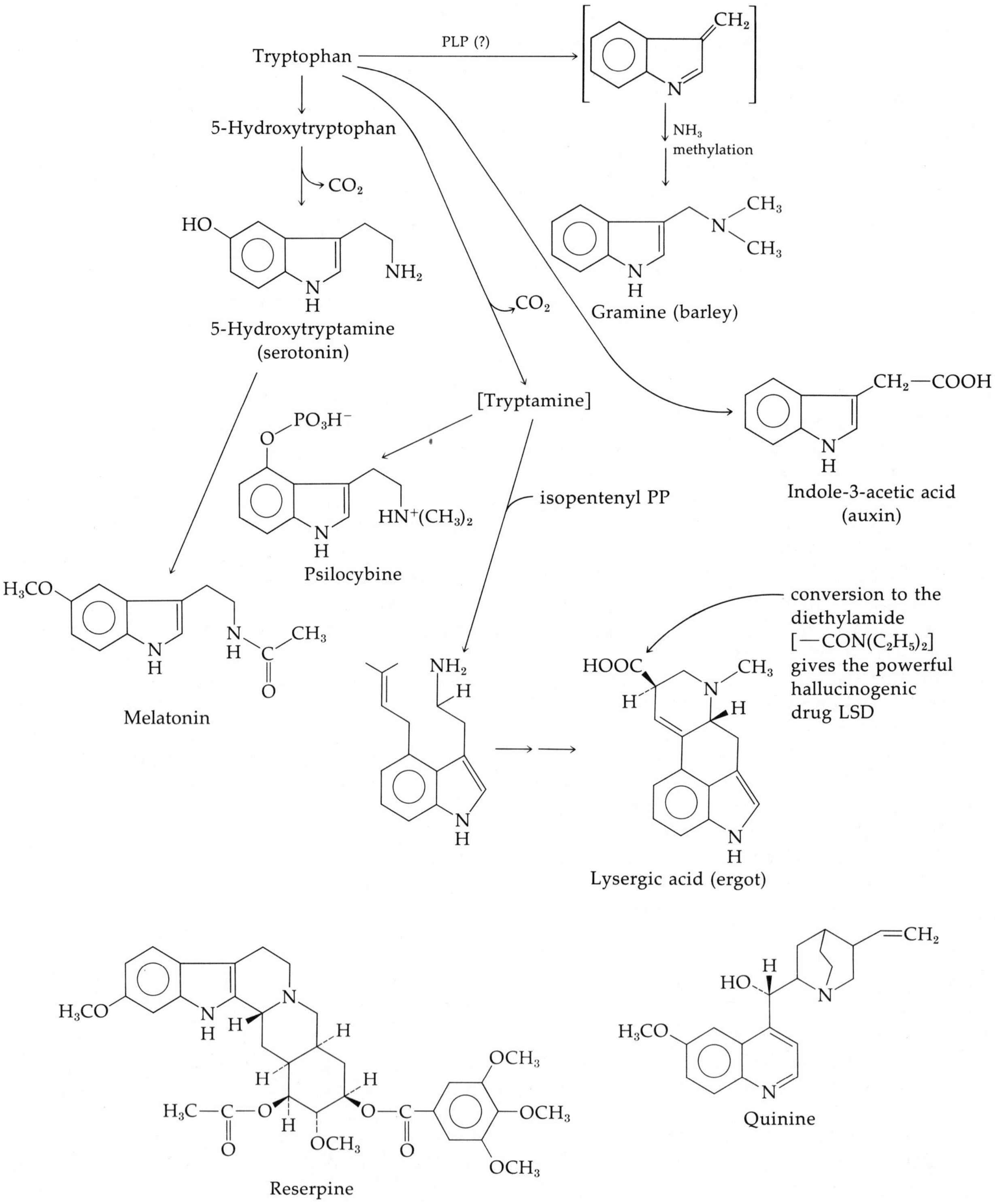

Tryptophan

PLP (?)

5-Hydroxytryptophan

CO_2

HO

NH_2

5-Hydroxytryptamine
(serotonin)

CH_2

NH_3
methylation

CH_3
N
CH_3

H

Gramine (barley)

CO_2

[Tryptamine]

CH_2—COOH

Indole-3-acetic acid
(auxin)

PO_3H^-
O

$HN^+(CH_3)_2$

H

Psilocybine

isopentenyl PP

H_3CO

N
H

CH_3

O

Melatonin

NH_2
H

N
H

HOOC

H

CH_3
N
H

N
H

Lysergic acid (ergot)

conversion to the
diethylamide
$[-CON(C_2H_5)_2]$
gives the powerful
hallucinogenic
drug LSD

H_3CO

N
H H

H

H

H

H_3C—C—O

O

OCH_3

OCH_3

O—C

O

OCH_3

OCH_3

OCH_3

Reserpine

$=CH_2$

H

HO

N

H_3CO

N

Quinine

Fig. 14-27   Structures and some biosynthetic pathways for some indole alkaloids and other metabolites of tryptophan.

great interest to medicine because of its effect in lowering blood pressure and in depleting nervous tissues of serotonin, dopamine, and noradrenaline. Reserpine also contains a benzene ring which is derived from tryptophan by a ring expansion.

J. THE METABOLISM OF HISTIDINE

The last but by no means the least important amino acid to be considered in this chapter is histidine. The biosynthesis of this amino acid, which might be regarded as the "super catalyst" of enzyme active centers, begins with a remarkable reaction of ATP, the "super coenzyme" of cells. The reaction is a displacement by N-1 of the adenine ring on C-1 of PRPP (step a, Fig. 14-28). The resulting product undergoes a ring opening reaction, step b, followed by an Amadori rearrangement (step c). The rearrangement product is cleaved hydrolytically with the release of the ribotide of 5-aminoimidazole-4-carboxamide, an established intermediate in the synthesis of ATP (Fig. 14-31). The other product of the cleavage contains the five carbons of the original ribosyl group of PRPP, together with one nitrogen and one carbon split out from the ATP molecule. Glutamine donates another nitrogen and ring closure (step e) forms the imidazole group which is attached to a glycerol phosphate molecule. The glycerol phosphate end of the molecule undergoes dehydration and ketonization of the resulting enol to a product which can be transaminated and dephosphorylated to histidinol. Oxidation of this alcohol forms histidine.

In all, ten different genes code for the enzymes of histidine biosynthesis in *Salmonella typhimurium*. They are clustered as the **histidine operon,** a consecutive series of genes which are transribed into messenger RNA as a unit.[143] The gene

symbols *His A*, *His B*, etc., are indicated in Fig. 14-28 and their positions on the *E. coli* gene map are indicated in Fig. 15-1. The gene *His B* codes for a complex protein with two different enzymatic activities as shown in Fig. 14-28.

1. Regulation of Amino Acid Metabolism

The presence of an excess of histidine in a bacterial cell brings about repression of synthesis of all of the enzymes of the histidine biosynthetic pathway. This is a result of the arrangement of genes in a single **operon.**[144] Details of the functioning of an operon are considered in Chapter 15, Section B,1. It is important to realize that "coordinate repression" of a series of enzymes organized as an operon is a characteristic of bacteria. The regulation of amino acid metabolism in eukaryotic organisms is probably quite different and is less well understood.[145]

Histidine is also an allosteric inhibitor for the first enzyme of the biosynthetic sequence, i.e., step a of Fig. 14-28. Thus, instantaneous inhibition of the biosynthesis occurs if an excess of histidine accumulates.

Similar patterns of both repression and feedback inhibition exist for many of the pathways of amino acid biosynthesis. Detailed reviews are available.[22,146]

2. Catabolism of Histidine

The first steps of the major degradative pathway for histidine metabolism have already been discussed. Elimination of ammonia, followed by hydration and ring cleavage to **formiminoglutamate,** involves unusual reactions (Eq. 14-48) which have been discussed earlier. Transfer of the formyl group to tetrahydrofolic acid and its further metabolism have also been considered (Chapter 8, Section K,3).

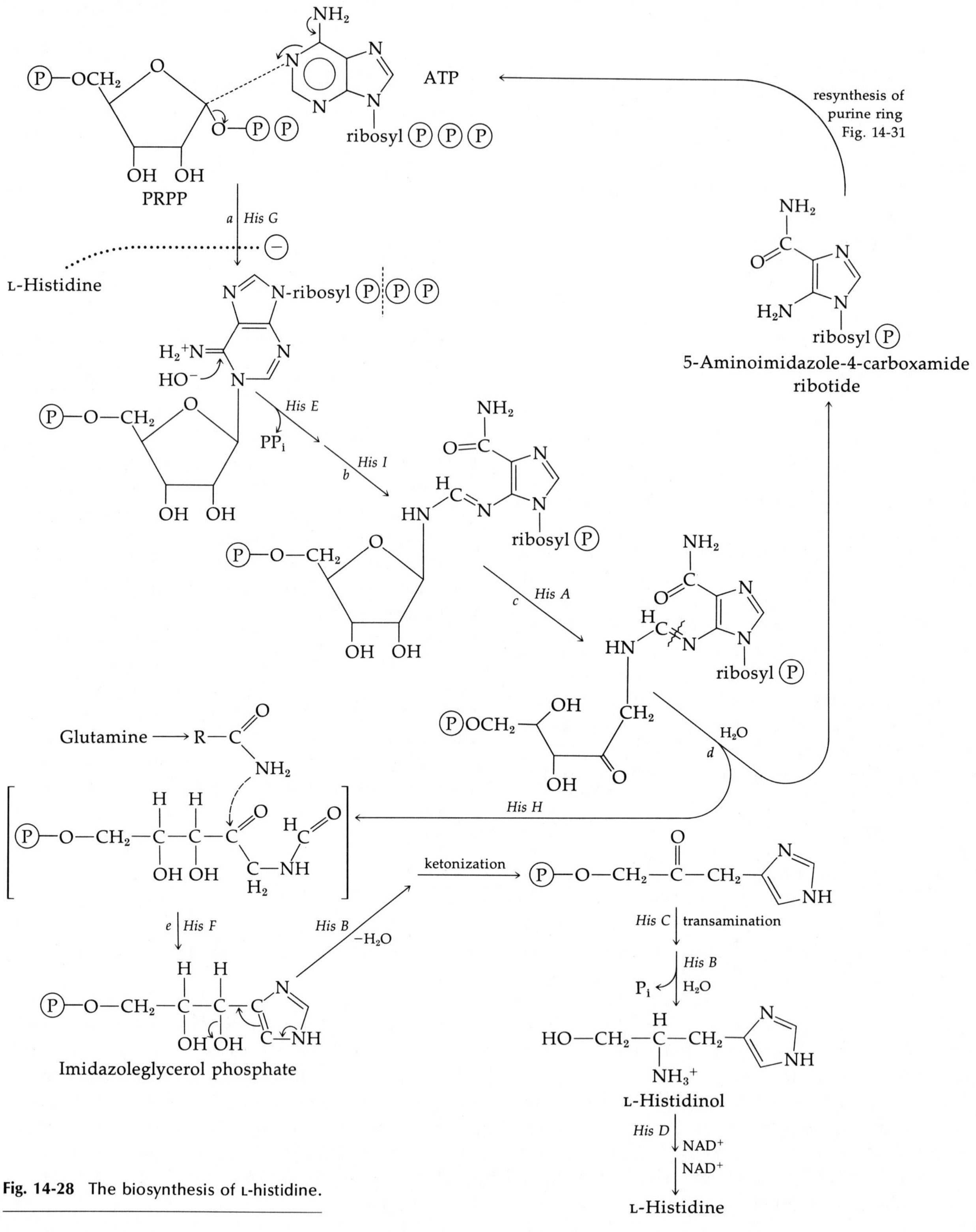

Fig. 14-28 The biosynthesis of L-histidine.

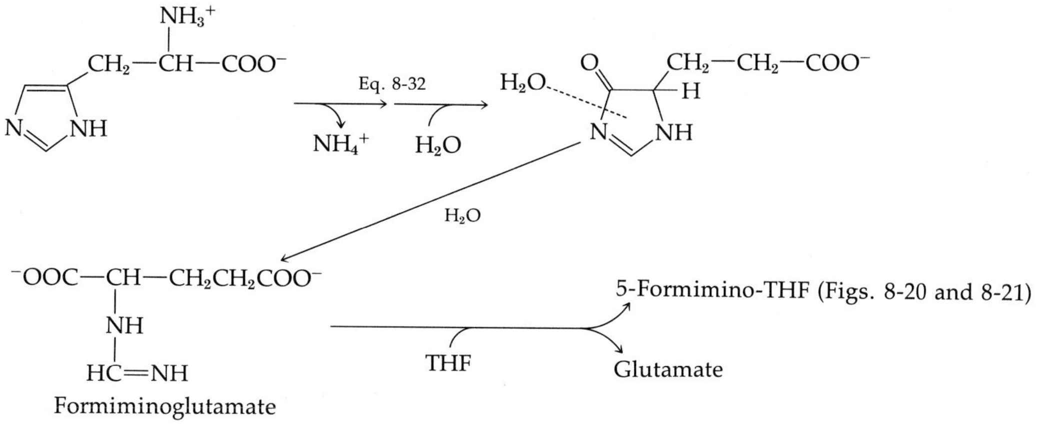

Eq. 8-32
NH4+
H2O
H2O
THF
5-Formimino-THF (Figs. 8-20 and 8-21)
Glutamate
Formiminoglutamate

(14-48)

Other products from histidine include the hormonal substance **histamine** formed by decarboxylation, the oxidation product, imidazole acetic acid, and N^1- and N^3-methylhistidines. Histamine plays a role in allergic responses (Chapter 5, Section C,4) and drugs (antihistamines) that inhibit its release are in widespread use.

K. THE METABOLISM OF PYRIMIDINES AND PURINES

All cells must be able to make pyrimidine and purine bases to be used in synthesis of nucleic acids and coenzymes. In many organisms the pathway of purine formation is greatly enhanced because **uric acid** or a related substance is the major excretory product derived from excess nitrogen. Such is the case for birds and reptiles which excrete uric acid rather than urea and for spiders which excrete guanine.

1. Biosynthesis of Pyrimidines[147,148]

Transfer of a carbamoyl group from carbamoyl phosphate to aspartate (Fig. 14-29, step *a*) produces a substrate able to immediately cyclize by elimination of water to form **dihydroorotate.** The transcarbamoylase is a highly controlled enzyme which is being studied intensively (Chapter 4, Section D,8; Chapter 6, Section B,7). Dihydroorotate is oxidized by a flavoprotein with NAD^+ as the external oxidant. In the next step (Fig. 14-29, step *d*) the product **orotic acid** is combined with a phosphoribosyl group from PRPP (Eq. 14-42) to form the first nucleotide **orotidine 5′-phosphate.**

Orotidine 5′-phosphate undergoes an unusual type of decarboxylation (step *e*, Fig. 14-29) which apparently is not assisted by any coenzyme. It has been suggested that the enzyme stabilizes a dipolar ionic tautomer of the substrate. Decarboxylation (Eq. 14-49) would be assisted by the adja-

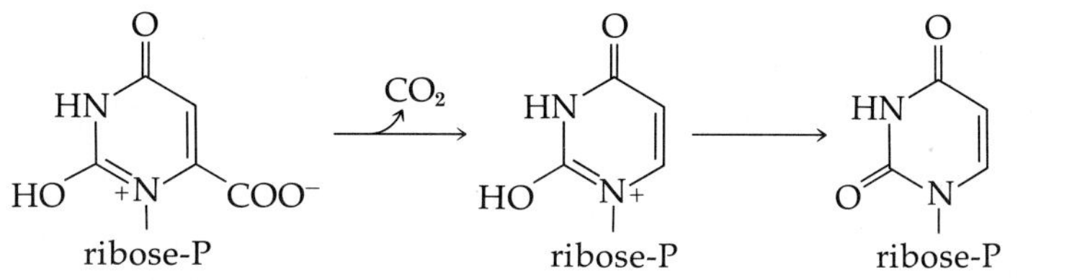

CO2
ribose-P
ribose-P
ribose-P

Orotidine 5′-phosphate (dipolar ionic tautomer)

Uridine 5′-phosphate (UMP)

(14-49)

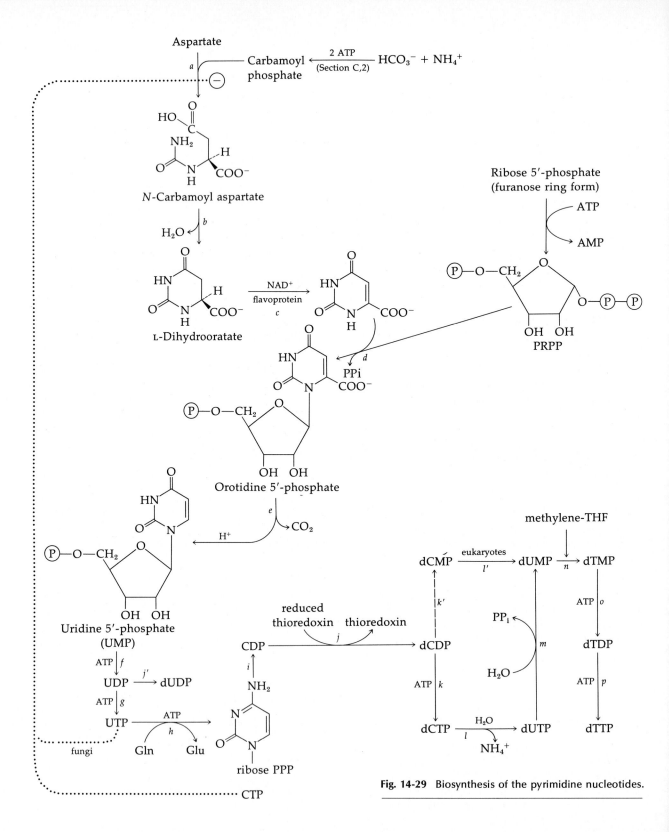

Fig. 14-29 Biosynthesis of the pyrimidine nucleotides.

cent positive charge.[149] Thus, **uridine 5'-phosphate** (UMP) is formed from aspartate in a relatively direct and simple way. Phosphorylation with ATP in two steps produces UDP and UTP.

The **cytosine nucleotides** are formed from UTP, the initial step being amination to CTP (step h, Fig. 14-29). This amination reaction resembles in many respects the conversion of citrulline to arginine, a reaction depending upon ATP and involving transfer of the nitrogen of aspartate (Section C,2). However, in the formation of CTP it is glutamine that serves as the nitrogen donor (NH_4^+ can substitute). CTP can be incorporated into RNA and into such metabolic intermediates as CDP- choline or it can be dephosphorylated to CDP. It is CDP that serves as the principal precursor for the deoxyribonucleotides **dCDP** and **thymidine diphosphate.**

a. Formation of Deoxyribonucleotides

A chain involving NADPH, a flavoprotein, thioredoxin (Chapter 8, Sections I,2 and L,4), and ribonucleotide reductase converts the ribonucleoside diphosphates to the corresponding 2-deoxy forms (step j, Fig. 14-29) as indicated in Eq. 14-50.

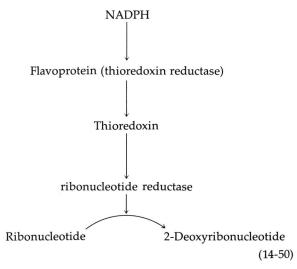

$$\text{(14-50)}$$

Ribonucleotide reductase of *E. coli* is composed of two nonidentical subunits, one of which contains nonheme iron.[150] In a few species of bacteria and in *Euglena* vitamin B_{12} is required and reduction is at the nucleoside *triphosphate* level.[151] The system from *Lactobacillus* has been discussed in Chapter 8, Section L,4. The mammalian ribonucleotide reductase, which may be similar to that of *E. coli*, is regarded as an appropriate target for anticancer drugs. The enzyme is highly regulated by a complex set of feedback mechanisms which apparently insure that DNA precursors are synthesized only in amounts needed for DNA synthesis.[152] Because an excess of one deoxyribonucleotide can inhibit reduction of all ribonucleoside diphosphates it is possible to inhibit DNA synthesis with either deoxyadenosine or with high levels of thymidine, despite the fact that both compounds are DNA precursors.

Phosphorylation of dCDP to dCTP (step k, Fig. 14-29) completes the biosynthesis of the first of the pyrimidine precursors of DNA. The uridine nucleotides arise in two ways. Reduction of UDP yields dUDP (step j', Fig. 14-29). More often deoxycytidine nucleotides are hydrolytically deaminated (reaction l and l', Fig. 14-29). Methylation to thymine nucleotides occurs via dUMP. The latter can be formed by hydrolytic removal of phosphate from dUDP or from the conversions dCDP → dCMP → dUMP (steps k' and l', Fig. 14-29) in eukaryotes. On the other hand, *E. coli* employs a more roundabout pathway: dCDP → dCTP → dUTP → dUMP (steps k, l, and m Fig. 14-29). One of the intermediates is dUTP. It is an interesting fact that DNA polymerases are able to incorporate this compound into polynucleotides. The only reason that this does not happen within cells (thereby forming uracil-containing DNA) is that dUTP is rapidly converted to dUMP by a pyrophosphatase (step m, Fig. 14-29).

Formation of **thymidylic acid,** dTMP, from dUMP (step n, Fig. 14-29) is catalyzed by **thymidylate synthetase.** The reaction (Eq. 14-51) is a transfer of a 1-carbon unit from methylene tetrahydrofolic acid.[153] A slightly different mechanism involving assistance from a neighboring nucleophilic group adding at C-6 has been proposed on the basis of model experiments.[154]

BOX 14-D

THYMIDYLATE SYNTHETASE, A TARGET ENZYME IN CANCER CHEMOTHERAPY[a]

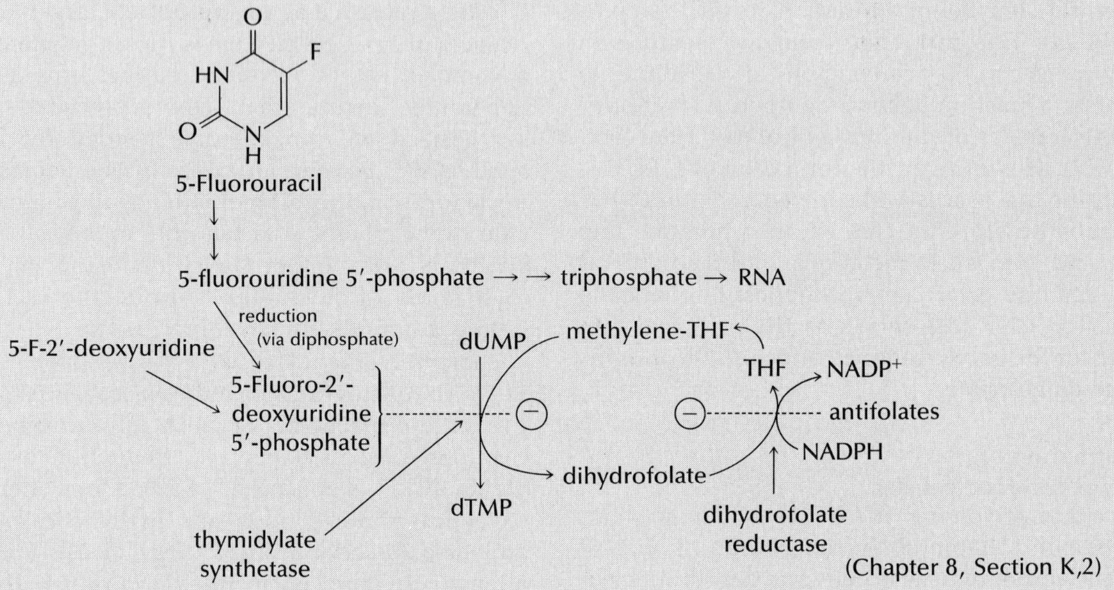

5-Fluorouracil

5-fluorouridine 5'-phosphate $\rightarrow \rightarrow$ triphosphate $\rightarrow$ RNA

reduction (via diphosphate)

5-F-2'-deoxyuridine

5-Fluoro-2'-deoxyuridine 5'-phosphate

dUMP   methylene-THF

THF   NADP⁺

antifolates

NADPH

dihydrofolate

dTMP

thymidylate synthetase

dihydrofolate reductase

(Chapter 8, Section K,2)

If a cell, either of a bacterium or of an animal, is deprived of thymine it can no longer make DNA. However, the synthesis of proteins and of RNA continues. This can be demonstrated experimentally with thymine-requiring mutants. Nevertheless, such cells sooner or later lose their vitality and die. The cause of this **thymineless death** is not certain. Perhaps thymine is needed to repair damage to DNA and if it is not available transcription eventually becomes faulty. Whatever the cause, the phenomenon provides the basis for some of the most effective chemotherapeutic attacks on cancer. Rapidly metabolizing cancer cells are especially vulnerable to thymineless death. Consequently, thymidylate synthetase becomes an important target enzyme for inhibition. One powerful inhibitor is the monophosphate of **5-fluoro-2'-deoxyuridine.** The inhibition was originally discovered when 5-fluorouracil was recognized as a useful cancer chemotherapeutic agent. Fluorouracil has many effects in cells, including

incorporation into RNA,[b] but the inhibition of thymidylate synthetase by the reduction product may be the most useful effect in cancer chemotherapy. In fact, 5-fluoro-2'-deoxyuridine is a much less toxic and more potent drug than 5-fluorouracil. Note that thymidylate synthetase requires methylene tetrahydrofolate as a reductant and that the reduction of dihydrofolate is an important part of the process. As has already been pointed out (Chapter 8, Section K,2), such folic acid analogues as **methotrexate** (N^{10}-methyl-4-amino-4-deoxyfolic acid) are among the most useful anticancer drugs. They act by inhibiting dihydrofolate reductase thereby (among other things) depriving thymidylate synthetase of a necessary substrate.

[a] Title of Section V of M. Friedkin's article on thymidylate synthetase, Adv. Enzymol. **38,** 235–292 (1973).

[b] J. Horowitz, C.-N. Ou, M. Ishaq, J. Ofengand, and J. Bierbaum, JMB **88,** 301–312 (1974).

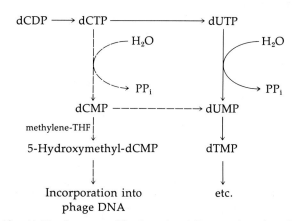

methylene-THF
(see Chapter 8, Section K,3)

dRibose-P
(dUMP)

Attack here by OH⁻ on a
corresponding intermediate
derived from dCMP gives
5-hydroxymethyl-dCMP
without oxidation of THF

dRibose-P

dihydrofolate

thymidylate

dRibose-P

(14-51)

Interesting alterations in nucleotide metabolism occur in cells of *E. coli* infected by T-even bacteriophate. Genes carried by the phage are transcribed and the corresponding proteins are synthesized by the host cell.[155] A number of these viral gene products are enzymes affecting nucleotide metabolism. Three are indicated in Fig. 14-30 by the dashed arrows. One enzyme catalyzes the hydrolytic conversion of dCTP to dCMP and another promotes the synthesis of 5-hydroxymethyl-dCMP. Such virus-specified enzymes may be appropriate target sites for antiviral drugs.

b. Reuse of Bases

Just as orotic acid is converted to a ribonucleotide in step *d*, Fig. 14-29, other free pyrimidine and purine bases can react with PRPP to give monoribonucleotides plus PP$_i$. The reactions constitute a *salvage pathway* by which purine and pyrimidine bases freed by the degradation of nucleic acids can be reused. However, it should be noted that thymine is not reused ordinarily. Nevertheless, to biochemists the ability to introduce radioactive thymine or thymidine into the DNA of an organism is an important experimental tool.

Thymidine is observed to be rapidly phosphorylated by the action of successive kinases to dTTP. Another important reaction of the salvage pathways for pyrimidines is the conversion of cytosine to uracil, the same kind of hydrolytic deamination represented by step *l* in Fig. 14-29.

Fig. 14-30 Some modifications in pathways of nucleotide metabolism in *E. coli* induced by infection with T-even bacteriophage: (→) normal pathways and (⤏) phage induced pathways.

2. Catabolism of Pyrimidine Nucleotides and Nucleosides

Nucleic acids within cells (as well as in the digestive tract) are continually under attack by a variety of *nucleases*. For example, messenger RNA is degraded, often quite rapidly, as an essential part of the control of protein synthesis. Although DNA is very stable, nucleases are called upon to cut out damaged segments of single strands, part of an essential repair process (Chapter 15, Section H,2). Thus, there is an active breakdown of polynucleotides to mononucleotides, which are hydrolyzed to nucleosides by phosphatases. Nucleosides are converted to free bases by the action of **nucleoside phosphorylases** (Eq. 14-52). The further

$$\text{Nucleoside} \xrightarrow{\text{P}_i} \text{free-base} + \text{ribose 1-P}$$
$$\textit{or} \text{ deoxyribose 1-P} \qquad (14\text{-}52)$$

degradation of cytosine is initiated by deamination to uracil, a reaction mentioned in the preceding section. Catabolism of uracil starts with a reduction by NADPH according to Eq. 14-53. The

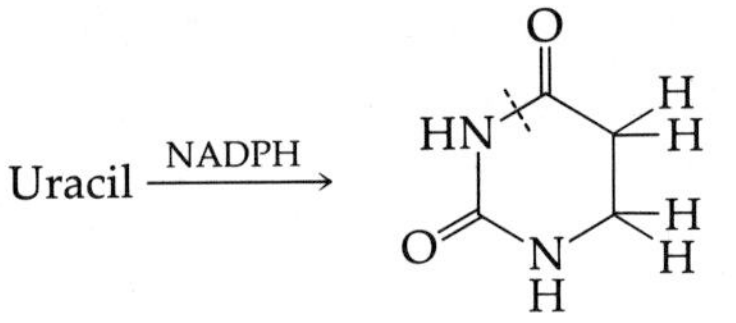

5,6-Dihydrouracil

β-Ureidopropionate

$$NH_4^+ + CO_2 + H_2N^+ - CH_2CH_2 - COO^- \qquad (14\text{-}53)$$
β-Alanine

end product **β-alanine** can be oxidatively degraded to malonic semialdehyde and malonyl-CoA (see Fig. 9-6), but it also serves as a biosynthetic precursor of pantothenic acid and coenzyme A (Fig. 14-10) and of the peptides **carnosine*** (β-ala-

* Do not confuse with *carnitine* (Chapter 9, Section A,6).

nylhistidine), and its N²-methyl derivative **anserine.** These peptides are found in high concentration in muscle. The function is unknown but may be related to the high buffering capacity of histidine derivatives between pH 6 and 7.

Thymine undergoes degradation in a pathway analogous to that of Eq. 14-53, but with the formation of **β-aminoisobutyrate.** The latter can be oxidatively converted to methylmalonate (Eq. 14-54) which can enter the methylmalonyl pathway (Fig. 9-6).

$$(14\text{-}54)$$

3. Biosynthesis of Purines

The first decisive experiments shedding light on the biosynthetic origins of purines were done with pigeons, which form uric acid very actively. Labeling experiments established the complex pattern indicated in the box in the upper left hand corner of Fig. 14-31. Two carbon atoms were found derived from glycine, one from CO_2, and two from formate. One nitrogen was derived from glycine, two from glutamine, and one from aspartate. In the case of adenine, the 6-NH₂ group was also found derived from aspartate.

The detailed biosynthetic pathway, for which enzymes have now been isolated and studied, is indicated in Fig. 14-31. The first "committed step" in purine synthesis is the reaction of PRPP with glutamine to form **phosphoribosylamine** (step *a*). This is another example of a glutamine-dependent amination. Pyrophosphate is displaced by ammonia generated *in situ* from glutamine (see p. 818). The amino group of the intermediate so formed is coupled with glycine in a standard manner (step *b*)

Fig. 14-31 Biosynthesis of purine nucleotides from ribose 5'-phosphate.

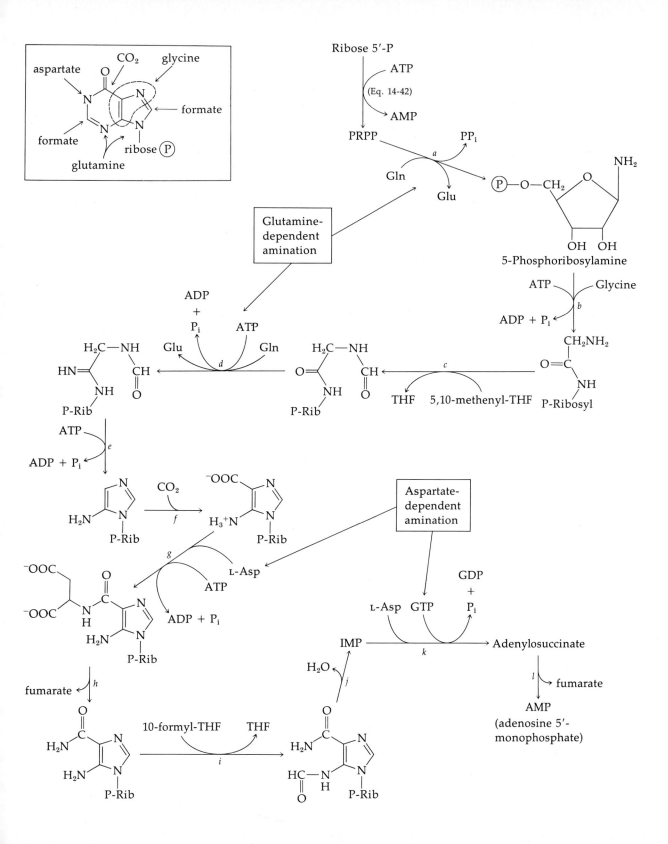

and the resulting product is formylated by 5,10-methenyltetrahydrofolic acid. The latter can be generated from free formate, accounting for the labeling pattern indicated in the box in Fig. 14-31.

In step *d* of Fig. 14-31 a second glutamine-dependent amination takes place, possibly through aminolysis of an intermediate enol phosphate. An ATP-requiring ring closure and tautomerization (step *e*) serve to complete the formation of the imidazole ring. A molecule of CO_2 is now incorporated (step *f*) in what appears to be an unusual type of carboxylation. This is followed by a two-step amination (*g* and *h*) by which nitrogen is transferred from aspartate in a manner strictly comparable to that in urea synthesis in which argininosuccinic acid is an intermediate (Fig. 14-4). As in urea formation, the carbon skeleton of

the aspartate molecule is eliminated as fumarate (step *h*), leaving the nitrogen in the purine precursor. The final carbon atom is added from 10-formyltetrahydrofolic acid (step *i*). It is not clear why in this instance the enzyme uses 10-formyl-THF, while step *c* requires the closely related 5,10-methenyl-THF. Spontaneous ring closure is followed by dehydration to **inosinic acid** (inosine 5′-phosphate, IMP). This initial purine product is then converted to AMP in another two-step aspartate-dependent amination via the intermediate adenylosuccinate.

The steps from IMP to AMP and thence to ADP, ATP, and the corresponding deoxyribonucleotides are recapitulated in Fig. 14-32. In addition, the conversion to **guanosine** and deoxyguanosine nucleotides is indicated. An NAD⁺-dependent

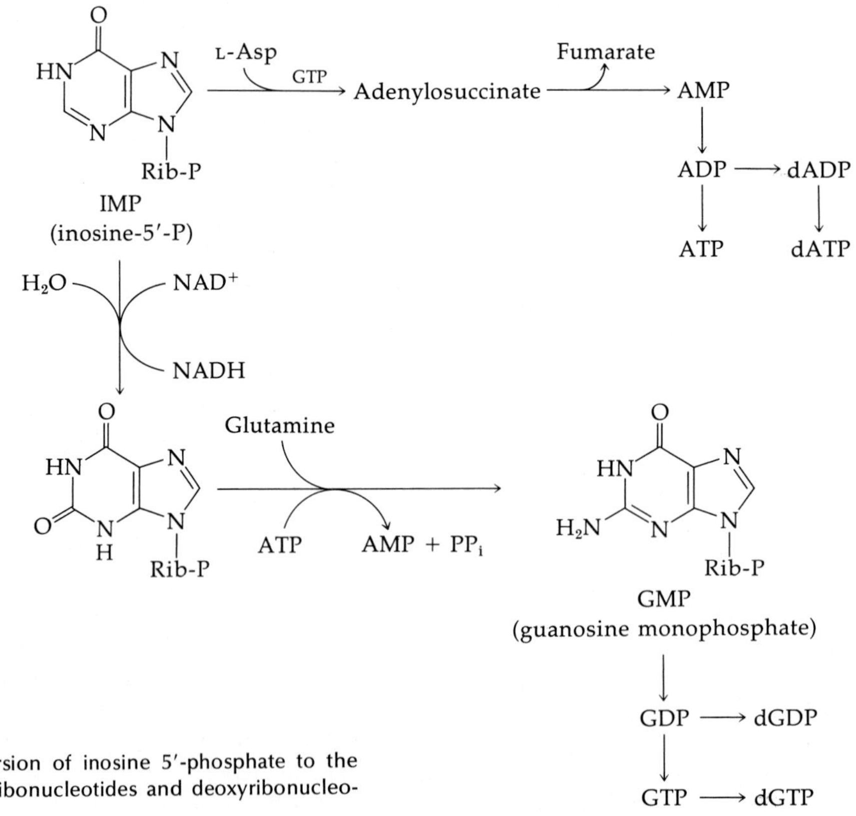

Fig. 14-32 The conversion of inosine 5′-phosphate to the adenine and guanine ribonucleotides and deoxyribonucleotides.

oxidation converts IMP to the corresponding **xanthine** ribotide, which is aminated in a glutamine-dependent process, as indicated.

Synthesis of purines is under complex control. Some of the mechanisms found in bacteria are outlined in Fig. 6-17. Again, in bacteria all of the final end product nucleotides inhibit the initial reaction of step *a* in Fig. 14-31.

4. "Salvage" Pathways

Both the catabolism of nucleic acids and the ingestion of nucleic acids in food provides a supply of preformed purine bases. Just as the free pyrimidines, these purine bases are able to react with PRPP under the influence of **phosphoribosyltransferases** (ribonucleotide pyrophosphorylases). Two such enzymes are known to act on purines. One converts adenine to AMP and also acts upon 5-aminoimidazole-4-carboxamide. The latter com-

5-Aminoimidazole-4-carboxamide

pound played a significant role in early investigations of purine metabolism. The compound was isolated in 1945 from cultures of *E. coli* treated with sulfonamides. The latter, being antagonists of *p*-aminobenzoic acid (Box 6-A), interfered with completion of purine synthesis by depriving the cell of the folic acid derivatives so essential to purine synthesis. Its structure immediately suggested that 5-aminoimidazole-4-carboxamide might be a purine precursor. Later it was shown that it is actually the corresponding ribonucleotide* that lies on the main route of purine synthesis.

* This ribonucleotide is also believed to be a precursor to the pyrimidine ring of thiamine, but the exact pathway is obscure.[155a]

A second phosphoribosyltransferase (hypoxanthine-guanine phosphoribosyltransferase) acts on either hypoxanthine or guanine.

5. Oxidative Metabolism of Purines

As indicated in Fig. 14-33, free adenine released from catabolism of nucleic acids can be deaminated hydrolytically to hypoxanthine. Likewise, guanine can be deaminated to xanthine. Xanthine oxidase, a molybdenum-containing enzyme (Chapter 8, Section I,6), oxidizes hypoxanthine to xanthine and the latter on to uric acid. Another reaction of xanthine occurring in some plants is conversion to the trimethylated derivative **caffeine.**

Whereas uric acid is the end product of purine metabolism in the human, in most species the copper enzyme **urate oxidase** converts uric acid to **allantoin,** the excretory product in most mammals other than primates. Most fish hydrolyze allantoin to **allantoic acid** and some excrete that compound as an end product. However, most fish continue the hydrolysis to form urea and glyoxylate. In some invertebrates the urea may be further hydrolyzed to ammonia.

In organisms that hydrolyze uric acid to urea or ammonia, this pathway is used only for degradation of purines from nucleotides. Excess nitrogen from catabolism of amino acids is either excreted directly as ammonia or is converted to urea by the urea cycle (Section C,2).

6. The Purine Nucleotide Cycle

Muscular work is accompanied by the production of ammonia, the immediate source of which is adenosine 5'-phosphate (AMP).[156,157] This fact has led to the recognition of another apparently useless "substrate cycle" (Chapter 11, Section F,6) that functions by virtue of the presence of a biosynthetic pathway and of a degradative enzyme in

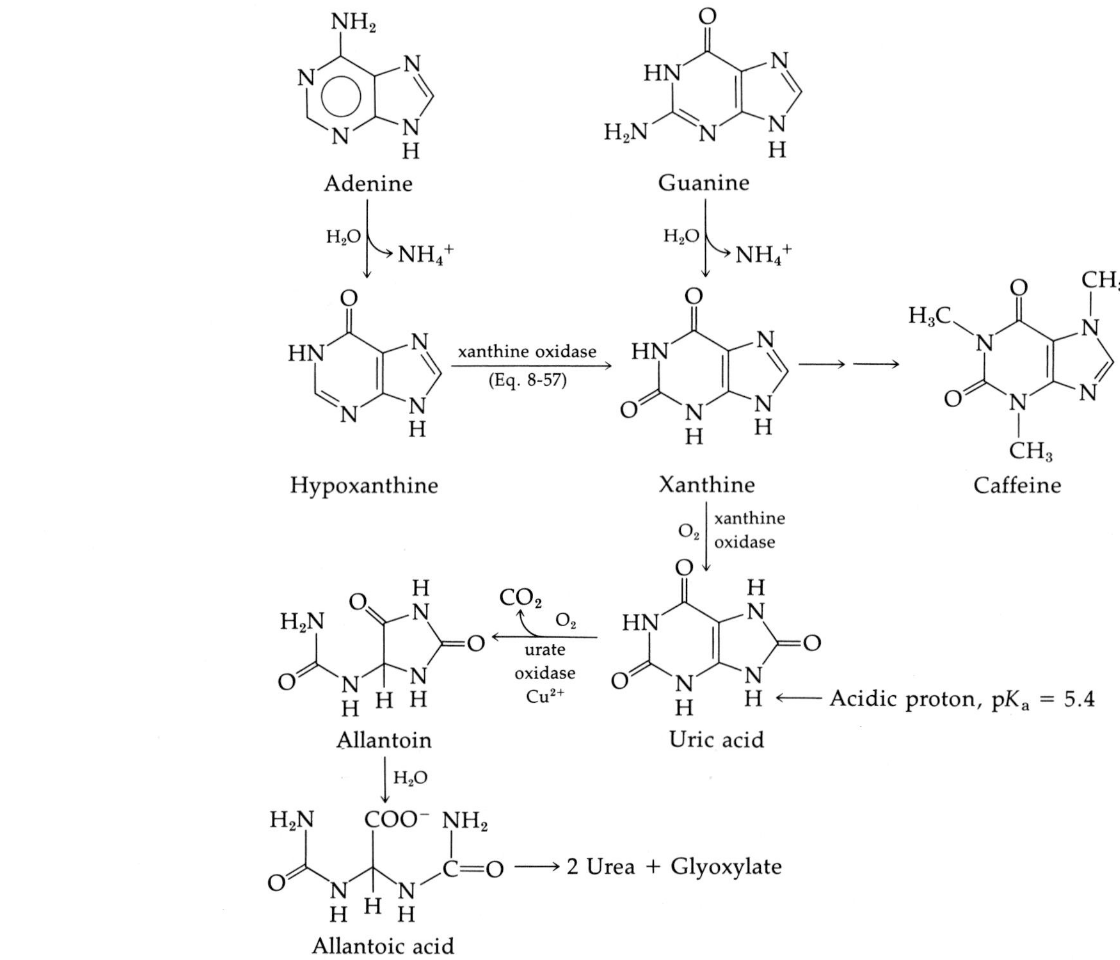

Fig. 14-33 The degradation of purines to uric acid and other excretion products.

the same cells (Eq. 14-55). Whether the cycle has an important function in muscle is still not clear.

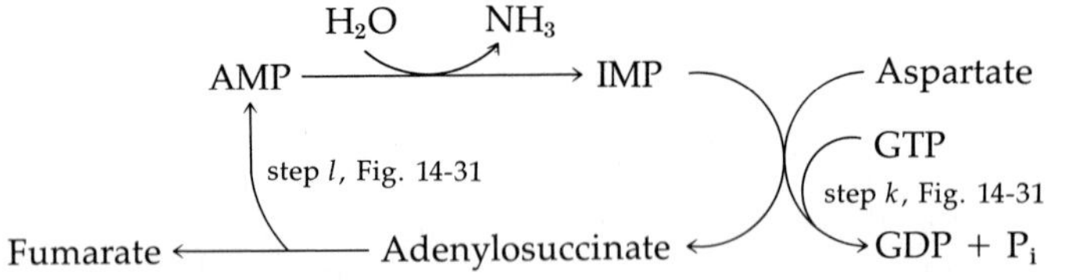

$$\text{AMP} \xrightarrow[\text{step } l, \text{ Fig. 14-31}]{H_2O \quad NH_3} \text{IMP} \longrightarrow \text{Aspartate}$$

Fumarate ← Adenylosuccinate ← GTP, step k, Fig. 14-31, GDP + P_i

(14-55)

A related cycle involves hydrolysis of AMP to adenosine and hydrolytic cleavage of the latter to inosine through the action of **adenosine deaminase.**

The cycle would be completed by further hydrolysis of inosine to hypoxanthine, conversion of the latter to IMP with hypoxanthine-guanine phosphoribosyltransferase and reconversion of IMP to AMP as in Fig. 14-32. Although doubt has been cast on the importance of this cycle,[157a] intense interest is focused on adenosine deaminase for the following reason. Hereditary lack of this enzyme is linked to a severe (usually fatal) immunodeficiency in which the numbers of lymphocytes are inadequate to combat infection.[157b,c] Whether the adenosine deaminase deficiency is the cause or whether there is a deficiency of more than one closely linked gene is uncertain.

Box 14-E

GOUT AND OTHER DEFECTS OF PURINE METABOLISM

An extremely common metabolic derangement with an incidence of ~3 per 1000 persons is **hyperuricemia** or **gout**.[a] As with most metabolic defects, there is a family of diseases ranging from mild to severe. In acute gouty arthritis, a sudden attack occurs, usually in the night, when sodium urate crystals precipitate in one or more joints. In half the cases the victim is awakened by a terrible pain in the big toe. The disease most often strikes adult males. The heredity is apparently complex and not fully understood. The primary biochemical defect in gout is usually an overproduction of uric acid in at least one instance as a result of an overactive PRPP synthetase.[b] In other cases a kidney defect interferes with excretion.

If properly controlled, gout may have few adverse effects. However, a related defect leads to a horrible result.[c] In the **Lesch–Nyhan syndrome** there is a deficiency of the salvage enzyme, hypoxanthine-guanine phosphoribosyltransferase. Instead of being reused, the free purines are oxidized to uric acid, causing symptoms of gout. The disease is characterized by mental deficiency and by an uncontrollable tendency of patients toward self-mutilation of the gums and hands by biting. To date, no drug has been found to ameliorate this problem.

Drugs are available for treatment of gout and the gouty symptoms of the Lesch–Nyhan syndrome. Colchicine (Box 4-A), in a manner which is not understood, alleviates the painful symptoms caused by the deposits of sodium urate in joints and tissues. More important is adequate regulation of the serum urate concentration through dietary means (keeping purine intake low) and through inhibition of xanthine oxidase. One drug, which is very effective, is an isomer of hypoxanthine known as **allopurinol**

Allopurinol

Patients receiving this drug excrete much of their purine as xanthine. Other drugs increase excretion of uric acid, and injection of the uric acid-hydrolyzing enzyme **uricase** has been tested experimentally.

[a] J. B. Wyngaarden and W. N. Kelley, in ''The Metabolic Basis of Inherited Disease'' (J. B. Stanbury, J. B. Wyngaarden, and D. S. Fredrickson, eds.), 3rd ed., pp. 889–968. McGraw-Hill, New York, 1972.

[b] M. A. Becker, P. J. Kostel, and L. J. Meyer, JBC **250**, 6822–6830 (1975).

[c] W. N. Kelley and J. B. Wyngaarden, in ''The Metabolic Basis of Inherited Disease'' (J. B. Stanbury, J. B. Wyngaarden, and D. S. Fredrickson, eds.), 3rd ed., pp. 969–1002. McGraw-Hill, New York, 1972.

L. FOLIC ACID, FLAVINS, AND DIMETHYLBENZIMIDAZOLE

Tracer studies have established that both riboflavin and folic acid originate from a derivative of guanosine. All of the atoms of the purine ring are conserved in the products except for C-8 of the five-membered ring. Figure 14-34 presents hypothetical pathways from guanosine triphosphate, the starting compound in at least some organisms. The first step, which has been verified with cell-free enzyme systems capable of forming pterins, is

the hydrolytic removal of formate. This is followed by an Amadori rearrangement, the product undergoing a simple ring closure between the carbonyl and adjacent amino group.[158–162] The product is **dihydroneopterin triphosphate** (Chapter 8, Section K), designated X in Fig. 14-34. It has the side chain indicated in Eq. 14-56. An aldol cleav-

Neopterin monophosphate

Biopterin

(14-56)

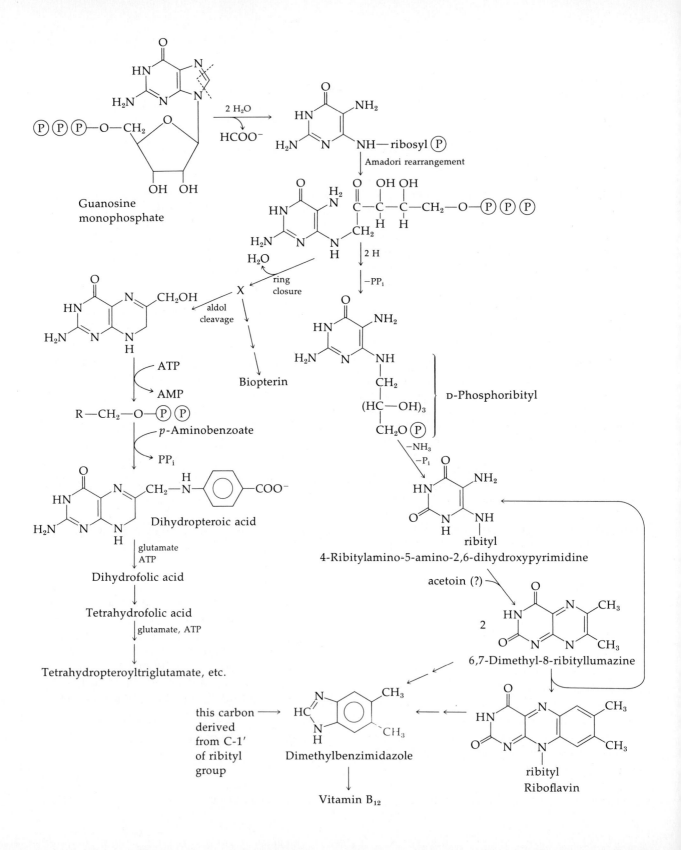

Guanosine
monophosphate

2 H₂O

HCOO⁻

Amadori rearrangement

ring
closure

aldol
cleavage

X

Biopterin

2 H

−PPᵢ

D-Phosphoribityl

−NH₃
−Pᵢ

ATP

AMP

R—CH₂—O—℗ ℗

p-Aminobenzoate

PPᵢ

CH₂—N—⟨⟩—COO⁻

Dihydropteroic acid

4-Ribitylamino-5-amino-2,6-dihydroxypyrimidine

acetoin (?)

glutamate
ATP

Dihydrofolic acid

Tetrahydrofolic acid

glutamate, ATP

6,7-Dimethyl-8-ribityllumazine

2

Tetrahydropteroyltriglutamate, etc.

this carbon →
derived
from C-1′
of ribityl
group

Dimethylbenzimidazole

ribityl
Riboflavin

Vitamin B₁₂

Fig. 14-34 The biosynthesis of folic acid, riboflavin, and the dimethylbenzimidazole group of vitamin B_{12}.

age followed by a series of reactions shown in Fig. 14-34 leads directly to the folate coenzymes. Other pterins may arise by simple modifications. For example, biopterin might be formed from neopterin phosphate by elimination of phosphate, ketonization, and reduction (Eq. 14-56). Both the inversion at C-1' in this equation and the occurrence of **sepiapterin** (Chapter 8, Section K), which contains a 3'-carbonyl group, suggest a transient dehydrogenation to a carbonyl at C-1' prior to elimination of phosphate.

Both the fungus *Eremothecium* (Box 8-H) and mutants of *Saccharomyces* have been used[163–165] to deduce the pathways of riboflavin synthesis outlined in Fig. 14-34. Reduction of the Amadori rearrangement product (whether the product is the triphosphate as shown in Fig. 14-34 or not is uncertain)*, deamination, and dephosphorylation lead to the flavin precursor 4-ribitylamino-5-amino-2,6-dihydroxypyrimidine. Additional carbon atoms to build the benzene ring of riboflavin are supplied in two stages. While it has not been established with certainty, it is believed that pyruvate provides these atoms in the form of acetoin (Eq. 9-28) or biacetyl. Spontaneous combi-

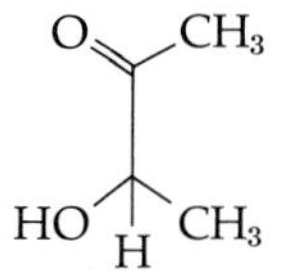

* Recent evidence obtained with a mutant of *Salmonella* suggests that guanosine rather than the triphosphate shown in Fig. 14-34 is the direct precursor to riboflavin.[165a]

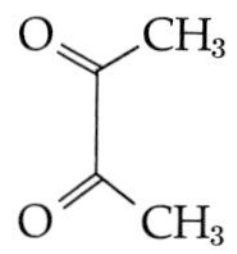

nation of biacetyl with the diamino precursor is known to yield **6,7-dimethyl-8-ribityllumazine.** Completion of the flavin ring requires an additional four carbon atoms which are supplied by a second unit of biacetyl. This is not done directly but by transfer of the biacetyl unit from a second molecule of 6,7-dimethyl-8-ribityllumazine, as indicated in Fig. 14-34. The reaction appears remarkable but is less so when one considers that the bimolecular reaction occurs spontaneously under mild conditions. However, the enzyme-catalyzed process, studied by Plaut, may proceed somewhat differently than the nonenzymatic reaction. Detailed mechanisms have been proposed. The diamino precursor (4-ribitylamino-5-amino-2,6-dihydroxypyrimidine) is regenerated in this process.

Dimethylbenzimidazole may also arise from 6,7-dimethyl-8-ribityllumazine in a process resembling that of riboflavin synthesis but in which the riboflavin formed is hydrolytically degraded to remove the pyrimidine ring and to form the imidazole ring.[166] In fact, the possibility that a separate pool of free riboflavin reacts in this way cannot be excluded.

REFERENCES

1. H. Dalton and L. E. Mortensen, *Bact. Rev.* **36**, 231–260 (1972).
2. R. W. F. Hardy and U. D. Havelka, *Science* **188**, 633–643 (1975).
2a. S. Winogradsky, *C. R. Acad. Sci.* **116**, 1385–1388 (1893).
3. S. L. Streicher and R. C. Valentine, *Annu. Rev. Biochem.* **42**, 279–302 (1973).
3a. W. G. Zumft and L. E. Mortenson, *BBA* **416**, 1–52 (1975).
3b. H. C. Winter and R. H. Burris, *Annu. Rev. Biochem.* **45**, 409–426. (1976).
3c. K. J. Skinner, *Chem. and Eng. News,* Oct. 4, pp. 22–35 1976.
4. R. W. F. Hardy, ed., "Dinitrogen Fixation." Wiley (Inter-

science), New York, 1975.
4a. M. G. Yates, *Trends Biochem. Sci.* pp. 17–20 (Jan. 1976).
5. R. R. Eady and J. R. Postgate, *Nature (London)* **249,** 805–810 (1974).
6. J. Postgate, *Nature (London)*, **256**, 363 (1975).
7. B. K. Vainshtein, E. H. Harutyunyan, I. P. Kuranova, V. V. Borisov, N. I. Sosfenov, A. G. Pavlovsky, A. I. Grebenko, N. V. Konareva, *Nature (London)* **254,** 163–164 (1975).
8. C. A. Appleby, N. A. Nicola, J. G. R. Hurrell, and S. J. Leach, *Biochemistry* **14**, 4444–4450 (1975).
9. G. Zumft, L. E. Mortenson, and G. Palmer, *EJB* **46,** 525–535 (1974).
10. L. E. Bennet, *Prog. Inorg. Chem.* **18**, pp. 1–176 (1973).
11. D. C. Yoch, *BBRC* **49**, 335–342 (1972).
12. E. I. Stiefel, *PNAS* **70**, 988–992 (1973).

13. A. V. Kulikov, L. A. Syrtsora, G. I. Likhtenshtein, and T. N. Pysarskaja, *Mol. Biol.* **9,** 203–212 (1975).

14. G. N. Schrauzer, G. W. Kiefer, K. Tano, and P. A. Doemeny, *JACS* **96,** 641–652 (1974).

15. K. Tano and G. Schrauzer, *JACS* **96,** 5404–5408 (1974).

16. E. E. van Tamelen, J. A. Gladysz, and C. R. Brûlet, *JACS* **96,** 3020–3021 (1974).

17. B. Schepartz, "Regulation of Amino Acid Metabolism in Mammals." Saunders, Philadelphia, Pennsylvania, 1973.

17a. Z. S. Kagan, V. L. Kretovich, and V. A. Polyakov, *Biokhimiya* **31,** 355–364 (1966).

18. R. E. Miller and E. R. Stadtman, *JBC* **247,** 7407–7419 (1972).

19. C. M. Brown, V. J. Burn, and B. Johnson, *Nature (London),New Biol.* **246,** 115–116 (1973).

20. P. J. Lea and B. J. Miflin, *Nature (London)* **251,** 614–616 (1974).

21. A. Ginsburg, *Adv. Protein Chem.* **26,** 1–79 (1972).

22. E. R. Stadtman, *in* "The Enzymes" (P. D. Boyer, ed.), 3rd ed., Vol. 1, pp. 397–459. Academic Press, New York, 1970.

23. R. M. Wohlhueter, E. Ebner, and D. H. Wolf, *JBC* **247,** 4213–4218 (1972).

24. S. P. Adler, D. Purich, and E. R. Stadtman, *JBC* **250,** 6264–6272 (1975).

25. T. F. Deuel and S. Prusiner, *JBC* **249,** 257–264 (1974).

25a. B. Tyler, A. B. Deleo, and B. Magasanik, *PNAS* **71,** 225–229 (1974).

25b. K. T. Shanmugam and R. C. Valentine, *Science* **187,** 919–924 (1975).

26. J. M. Buchanan, *Adv. Enzymol.* **38,** 91–183 (1973).

27. A. Meister, *Science* **180,** 33–39 (1973).

28. P. Van Der Werf, O. W. Griffith, and A. Meister, *JBC* **250,** 6686–6692 (1975).

29. S. S. Rothman, *Science* **190,** 747–753 (1975).

30. G. Palade, *Science* **189,** 347–358 (1975).

31. M. Hayashi, Y. Hiroi, and Y. Natori, *Nature (London) New Biol.* **242,** 163–166 (1973).

32. R. T. Dean, *Nature (London)* **257,** 414–416 (1975).

33. G. S. Lazarus and J. F. Goggins, *Science* **186,** 653–654 (1974).

33a. K. Sletten, I. Aakesson, and J. O. Alvsaker, *Nature (London), New Biol.* **231,** 118–119 (1971).

33b. J. N. Karlin, B. J. Bowman, and R. H. Davis, *JBC* **251,** 3948–3955. (1976).

33c. H. A. Krebs and K. Henseleit, *Hoppe-Seyler's Z. Physiol. Chem.* **210,** 33–66 (1932).

34. S. Ratner, *Adv. Enzymol.* **39,** 1–90 (1973).

35. P. P. Trotta, V. P. Wellner, L. M. Pinkus, and A. Meister, *PNAS* **70,** 2717–2721 (1973).

36. S. S. Cohen, "Introduction to the Polyamines." Prentice-Hall, Englewood Cliffs, New Jersey, 1971.

36a. U. Bachrach, "Functions of Naturally Occurring Polyamines." Academic Press, New York, 1973.

37. W. H. Wu and D. R. Morris, *JBC* **248,** 1687–1695 (1973).

38. C. W. Tabor and H. Tabor, *Annu. Rev. Biochem.* **45,** 285–306 (1976).

39. H. G. Williams-Ashman, J. Jänne, G. L. Coppoc, M. E. Geroch, and A. Schenone, *Adv. Enzyme Regul.* **10,** 225–245 (1972).

40. T. T. Sakai and S. S. Cohen, *Prog. Nucleic Acid Res. Mol. Biol.* **17,** 15–42 (1976).

40a. P. S. Mamont, P. Böhlen, P. D. McCann, P. Bey, F. Schuber, and C. Tardif, *PNAS* **73,** 1628–1630 (1976).

40b. M. Tsuboi, *Bull. Chem. Soc. Japan* **37,** 1514–1522 (1964).

40c. I. Flink and D. E. Pettijohn, *Nature (London)* **253,** 62–63 (1975).

41. W. H. Bowman, C. W. Tabor, and H. Tabor, *JBC* **248,** 2480–2486 (1973).

42. H. Tabor and C. W. Tabor, *JBC* **250,** 2648–2654 (1975).

43. J. K. Hardman and T. C. Stadtman, *JBC* **238,** 2088–2093 (1963).

44. A. S. Vanderbilt, N. S. Gaby, and V. W. Rodwell, *JBC* **250,** 5322–5329 (1975).

45. H. Cedar and J. H. Schwartz, *JBC* **244,** 4112–4121 (1969).

46. B. Horowitz and A. Meister, *JBC* **247,** 6708–6719 (1972).

47. J. C. Wriston, Jr. and T. O. Yellin, *Adv. Enzymol.* **39,** 185–248 (1973).

48. J. D. Funkhouser, A. Abraham, V. A. Smith, and W. G. Smith, *JBC* **249,** 5478–5484 (1974).

49. V. W. Rodwell, *Metab. Pathways, 3rd ed.* **3,** 317–373 (1969).

50. V. Tanphaichitr, D. W. Horne, and H. P. Broquist, *JBC* **246,** 6364–6466 (1971).

50a. C. J. Rebouche and H. D. Broquist, *Fed. Proc., Fed. Am. Soc. Exp. Biol.* **33,** 1545 (1974).

51. K. Soda, H. Misono, and T. Yamamoto, *Biochemistry* **7,** 4102–4109 (1968).

52. J. A. Grove, T. G. Linn, C. J. Willett, and L. M. Henderson, *BBA* **215,** 191–194 (1970).

53. D. L. Miller and V. W. Rodwell, *JBC* **246,** 2758–2764 (1971).

54. M. Rothstein, *ABB* **111,** 467–476 (1965).

55. T. C. Stadtman, *Adv. Enzymol.* **38,** 413–448 (1973).

56. T. P. Chirpich, V. Zappia, R. N. Costilow, and H. A. Barker, *JBC* **245,** 1778–1789 (1970).

57. J. J. Baker, C. van der Drift, and T. C. Stadtman, *Biochemistry* **12,** 1054–1063 (1973).

58. I.-M. Leng and H. A. Barker, *JBC* **249,** 6578–6584 (1974).

58a. J. L. Palmer and R. H. Abeles, *JBC* **251,** 5817–5819 (1976).

59. S. P. Burg, *PNAS* **70,** 591–597 (1973).

60. D. P. Murr and S. F. Yang, *Plant Physiol.* **55,** 79–82 (1975).

61. W. A. Wood, *Curr. Top. Cell. Regul.* **1,** 161–182 (1969).

62. K. W. Rabinowitz, R. A. Niederman, and W. A. Wood, *JBC* **248,** 8207–8215 (1973).

63. D. H. Calhoun, R. H. Rimerman, and G. W. H tfield, *JBC* **248,** 3511–3516 (1973).

63a. H. E. Umbarger, *Science* **123,** 848 (1956).

63b. A. Svent-Györgyi, *Life Sci.* **15**, 863–875 (1974).

64. A. Bergquist, E. A. Eakin, D. K. Murali, and R. P. Wagner, *PNAS* **71**, 4352–4355 (1974).

65. Y. Abiko, *Metab. Pathways, 3rd ed.* **7**, 1–25 (1975).

65a. M. A. Eisenberg, in "Metabolic Pathways," 3rd ed. (D. M. Greenberg, ed.), Vol VII, pp. 27–56. Academic Press, New York, 1975.

65b. G. L. Stoner and M. A. Eisenberg, *JBC* **250**, 4029–4036 (1975).

66. K. Tanaka, I. M. Armitage, H. S. Ramsdell, Y. E. Hsia, S. R. Lipsky, and L. E. Rosenberg, *PNAS* **72**, 3692–3696 (1975).

67. H. E. Williams and L. H. Smith, Jr., *Science* **171**, 390–391 (1971).

68. H. E. Williams and L. H. Smith, Jr., in "The Metabolic Basis of Inherited Disease" (J. B. Stanbury, J. B. Wyngaarden, and D. S. Fredrickson, eds.), 3rd ed., pp. 196–219. McGraw-Hill, New York, 1972.

69. W. L. Nyham, ed., "Heritable Disorders of Amino Acid Metabolism." Wiley, New York, 1974.

70. T. Yoshida and G. Kikuchi, *J. Biochem. (Tokyo)* **72**, 1503–1516 (1972).

71. S. M. Klein and R. D. Sagers, *JBC* **242**, 297–300, 301–305 (1967).

72. Y. Motokawa and G. Kikuchi, *ABB* **164**, 624–633 and 634–640 (1974).

73. M. L. Baginsky and F. M. Huennekens, *BBRC* **23**, 600–605 (1966).

73a. D. Shemin and D. Rittenberg, *JBC* **166**, 621–625 (1946).

73b. M. J. Whiting and S. Granick, *JBC* **251**, 1340–1346 (1976).

74. S. I. Beale, S. P. Gough, and S. Granick, *PNAS* **72**, 2719–2723 (1975).

75. D. Shemin, in "The Enzymes" (P. D. Boyer, ed.), 3rd ed., Vol. 7, pp. 323–356. Academic Press, New York, 1972.

76. T. W. Goodwin, ed., "Porphyrins and Related Compounds," p. 215. Academic Press, New York, 1969.

77. R. Poulson and W. J. Polglase, *JBC* **250**, 1269–1274 (1975).

78. H. Sawada, M. Takeshita, Y. Sugita, and Y. Yoneyama, *BBA* **178**, 145–155 (1969).

79. R. McKay, R. Druyan, G. S. Getz, and M. Rabinowitz, *BJ* **114**, 455–461 (1969).

80. T. W. Goodwin, in "Structure and Function of Chloroplasts" (M. Gibbs, ed.), pp. 215–276. Springer-Verlag, Berlin and New York, 1971.

81. L. Bogorad, in "Chemistry and Biochemistry of Plant Pigments" (T. W. Goodwin, ed.), pp. 64–148. Academic Press, New York, 1976.

82. C. E. Brown, D. Shemin, and J. J. Katz, *JBC* **248**, 8015–8021 (1973).

83. A. I. Scott, C. A. Townsend, K. Okada, and M. Kajiwara, *JACS* **96**, 8054–8069 and 8069–8080 (1974).

83a. A. I. Scott, K. S. Ho, M. Kajiwara, and T. Takahashi, *JACS* **98**, 1589–1591 (1976).

84. I. Macalpine and R. Hunter, *Sci. Am.* **221**, 38–46 (Jul. 1969).

85. J. L. York, "The Porphyrias." Thomas, Springfield, Illinois, 1972.

86. H. S. Marver and R. Schmid, in "The Metabolic Basis of Inherited Disease" (J. B. Stanbury, J. B. Wyngaarden, and D. S. Fredrickson, eds.), 3rd ed., pp. 1087–1140. McGraw-Hill, New York, 1972.

87. W. N. Piper, L. W. Condie, and T. R. Tephly, *ABB* **159**, 671–677 (1973).

88. A. Poland and E. Glover, *Science* **179**, 476–477 (1973).

89. C. O. Heocha, in "Porphyrins and Related Compounds" (T. W. Goodwin, ed.), pp. 91–105. Academic Press, New York, 1969.

90. R. Tenhunen, H. Marver, N. R. Pimstone, W. F. Trager, D. Y. Cooper, and R. Schmid, *Biochemistry* **11**, 1716–1720 (1972).

90a. M. D. Maines and A. Kappas, *PNAS* **71**, 4283–4287 (1974).

91. L. M. Siegel, P. S. Davis, and H. Kamin, *JBC* **249**, 1572–1586 (1974).

92. M. J. Murphy, L. M. Siegel, and H. Kamin, *JBC* **248**, 2801–2814 (1973).

93. L. M. Siegel and P. S. Davis, *JBC* **249**, 1587–1598 (1974).

94. A. Schmidt, W. R. Abrams, and J. A. Schiff, *EJB* **47**, 423–434 (1974).

94a. P. F. Cook and R. T. Wedding, *JBC* **251**, 2023–2029 (1976).

95. J. M. Collins and K. J. Monty, *JBC* **248**, 5943–5949 (1973).

96. T. N. Akopyan, A. E. Braunstein, and E. V. Goryachenkova, *PNAS* **72**, 1617–1621 (1975).

96a. K. C. Hayes, R. E. Carey, and S. Y. Schmidt, *Science* **188**, 949–951 (1975).

96b. T. H. Haines, in "Lipids and Biomembranes of Eukaryotic Microorganisms" (J. A. Erwin, ed.), p. 207. Academic Press, New York, 1973.

97. H. J. Cohen, S. Betcher-Lange, D. L. Kessler, and K. V. Rajagopalan, *JBC* **247**, 7759–7766 (1972).

98. R. H. White, *Science* **189**, 810 (1975).

98a. B. D. Davis, *Experientia* **6**, 41–50 (1950).

98b. E. Haslam "The Shikimate Pathway", Butterworths, London, 1974.

99. E. Gollub, H. Zalkin, and D. B. Sprinson, *JBC* **242**, 5323–5328 (1967).

100. A. B. DeLeo, J. Dayan, and D. B. Sprinson, *JBC* **248**, 2344–2353 (1973).

101. M. Adlersberg and D. B. Sprinson, *Biochemistry* **3**, 1855–1860 (1964).

102. L. Huang, A. L. Montoya, and E. W. Nester, *JBC* **250**, 7675–7681 (1975).

103. W. E. Bondinell, J. Vnek, P. F. Knowles, M. Sprecher, and D. B. Sprinson, *JBC* **246**, 6191–6196 (1971).

104. H. Morell, M. J. Clark, P. F. Knowles, and D. B. Sprinson, *JBC* **242**, 82–90 (1967).

105. B. E. Davidson, E. H. Blackburn, and T. A. A. Dopheide, *JBC* **247**, 4441–4446 (1972).

106. T. A. A. Dopheide, P. Crewther, and B. E. Davidson, *JBC*

247, 4447–4452 (1972).

107. R. A. Jensen and D. L. Pierson, *Nature (London)* **254,** 667–671 (1975).

107a. H. Rosenberg and I. G. Young, *in* "Microbial Iron Metabolism" (J. B. Neilands, ed.), pp. 67–82. Academic Press, New York, 1974.

108. H. Zalkin, *Adv. Enzymol.* **38,** 1–39 (1973).

109. F. M. Hulett and J. A. DeMoss, *JBC* **250,** 6648–6652 (1975).

110. K. H. Hitendorf, H. Gilch, and F. Lingens, *FEBS Lett.* **16,** 95–98 (1971).

111. K. R. Schubert, R. L. Switzer, and E. Shelton, *JBC* **250,** 7492–7500 (1975).

111a. P. Truffa-Bachi and G. N. Cohen, *Annu. Rev. Biochem.* **42,** 113–134 (1973).

112. R. Bentley and I. M. Campbell, *in* "The Chemistry of the Quinonoid Compounds" (S. Patai, ed.), Part 2, pp. 683–736. Wiley, New York, 1974.

113. F. Gibson, *Biochem. Soc. Trans.* **1,** 317–326 (1973).

114. T. W. Goodwin, *in* "Structure and Function of Chloroplasts" (M. Gibbs, ed.), pp. 215–276. Springer-Verlag, Berlin and New York, 1971.

115. B. L. Trumpower, R. M. Houser, and R. E. Olson, *JBC* **249,** 3041–3048 (1974).

116. G. H. N. Towers and P. V. Subba Rao, *Recent Adv. Phytochem.* **4,** 1–43 (1972).

117. R. M. Baldwin, C. D. Snyder, and H. Rapoport, *Biochemistry* **13,** 1523–1530 (1974).

118. W. E. Knox, *in* "The Metabolic Basis of Inherited Disease" (J. B. Stanbury, J. B. Wyngaarden, and D. S. Fredrickson, eds.), 3rd ed., pp. 266–295. McGraw-Hill, New York, 1972.

119. C. Chong-Cheng and I. T. Oliver, *Biochemistry* **11,** 2547–2553 (1972).

120. K.-L. Lee and J. M. Nickol, *JBC* **249,** 6024–6026 (1974).

121. R. W. Johnson and F. T. Kenney, *JBC* **248,** 4528–4531 (1973).

122. B. N. La Du, *in* "The Metabolic Basis of Inherited Disease" J. B. Stanbury, J. B. Wyngaarden, and D. S. Fredrickson, eds.), 3rd., pp. 308–325. McGraw-Hill, New York, 1972.

123. A. White, P. Handler, and E. L. Smith, "Principles of Biochemistry," 5th ed., pp. 1030–1041. McGraw-Hill, New York, 1973.

124. J. B. Stanbury, *in* "The Metabolic Basis of Inherited Disease" (J. B. Stanbury, J. B. Wyngaarden, and D. S. Fredrickson, eds.) 3rd ed., pp. 223–265. McGraw-Hill, New York, 1972.

124a. J. R. Tata, *Nature (London)* **259,** 527–528 (1976).

125. H. J. Cahnmann and K. Funakoshi, *Biochemistry* **9,** 90–98 (1970).

126. H. A. Lardy and G. F. Maley, *Recent Prog. Horm. Res.* **10,** 129–155 (1954).

127. M. I. Surks, D. Koerner, W. Dillman, and J. H. Oppenheimer, *JBC* **248,** 7066–7072 (1973).

127a. B. J. Spindler, K. M. MacLeod, J. Ring, and J. D. Baxter, *JBC* **250,** 4113–4119 (1975).

127b. J. R. Tata, *Nature (London)* **257,** 18–23 (1975).

128. K. J. Armstrong, J. E. Stouffer, R. G. Van Inwegen, W. J. Thompson, and G. A. Robison, *JBC* **249,** 4226–4231 (1974).

129. V. H. Morgenrath, III, M. Boadle-Biber, and R. H. Roth, *PNAS* **71,** 4283–4287 (1974).

130. G. A. Swan, *Fortschr. Chem. Org. Naturst.* **31,** 521–582 (1974).

131. R. H. Thomson, *Angew Chem., Int. Ed. Engl.* **13,** 305–312 (1974).

132. S. Dagley and D. E. Nicholson, "An Introduction to Metabolic Pathways." Wiley, New York, 1970.

133. B. F. Johnson and R. Y. Stanier, *J. Bacteriol.* **107,** 476–485 (1971).

134. R. N. Patel, R. B. Meagher, and L. N. Ornston, *JBC* **249,** 7410–7419 (1974).

135. M. Thomas, S. L. Ranson, and J. A. Richardson, "Plant Physiology," 5th ed. Longmans, Green, New York, 1973.

136. E. E. Snell and B. E. Haskell, *Compr. Biochem.* **21,** 47–71 (1970).

137. W. J. Schubert, *Compr. Biochem.* **20,** 93–230 (1968).

138. C. Steelink, *Recent Adv. Phytochem.* **4,** 239–271 (1972).

139. H. Nimz, *Angew. Chem., Int. Ed. Engl.* **13,** 313–321 (1974).

140. S. W. Pelletier, "Chemistry of the Alkaloids." Van Nostrand-Reinhold, Princeton, New Jersey, 1970.

141. I. D. Spencer, *Compr. Biochem.* **20,** 321–413 (1968).

141a. R. Robinson, *J. Chem. Soc.* **111,** 876–899 (1917).

141b. R. Robinson, "The Structural Relations of Natural Products." Oxford Univ. Press., London, 1955.

142. S. N. Young, M. Oravec, and T. L. Sourkes, *JBC* **249,** 3932–3936 (1974).

142a. H. M. Brown and W. K. Purves, *JBC* **251,** 907–913 (1976).

143. M. Brenner and B. N. Ames, *Metab. Pathways, 3rd ed.* **5,** 349–387 (1971).

144. R. F. Goldberger and J. S. Kovach, *Curr. Top. Cell. Regul.* **5,** 285–308 (1972).

145. W. T. Brashear and S. M. Parsons, *JBC* **250,** 6885–6890 (1975).

146. B. D. Sanwal, M. Kapoor, and H. W. Duckworth, *Curr. Top. Cell. Regul.* **3,** 1-115 (1971).

147. J. N. Davidson, "The Biochemistry of Nucleic Acids," 7th ed., pp. 215–232. Academic Press, New York, 1972.

148. J. F. Henderson and A. R. P. Paterson, "Nucleotide Metabolism." Academic Press, New York, 1973.

149. P. Beak and B. Siegel, *JACS* **98,** 3601–3606 (1976).

150. L. Thelander, B. Larsson, J. Hobbs, and F. Eckstein, *JBC* **251,** 1398–1405 (1976).

151. F. D. Hamilton, *JBC* **249,** 4428–4434 (1974).

152. E. R. Stadtman, *in* "The Enzymes" (P. D. Boyer, ed.), 3rd ed., Vol. 1, pp. 442–443. Academic Press, New York, (1970).

153. M. Friedkin, *Adv. Enzymol.* **38,** 235–292 (1973).

154. D. V. Santi and C. F. Brewer, *JACS* **90,** 6236–6238 (1968).

155. J. F. Koerner, *Annu. Rev. Biochem.* **39,** 291–322 (1970).

155a. I. G. Leder, *Metab. Pathways, 3rd ed.* **7**, 57–85 (1975).

156. J. Lowenstein and K. Tornheim, *Science* **171**, 397–400 (1971).

157. C. J. Coffee and W. A. Kofke, *JBC* **250**, 6653–6658 (1975).

157a. L. W. Brox and J. F. Henderson, *Can. J. Biochem.* **54**, 200–202 (1976).

157b. M. B. Van der Weyden and W. N. Kelley, *JBC* **251**, 5448–5456 (1976).

157c. G. C. Mills, F. C. Schmalstieg, K. B. Trimmer, A. S. Goldman, and R. M. Goldblum, *PNAS* **73**, 2867–2871 (1976).

158. A. W. Burg and G. M. Brown, *JBC* **243**, 2349–2358 (1968).

159. F. Foor and G. M. Brown, *JBC* **250**, 3545–3551 (1975).

160. J. Cone and G. Guroff, *JBC* **246**, 979–985 (1971).

161. T. Shiota, *Compr. Biochem.* **21**, 111–152 (1970).

162. R. L. Blakley, "The Biochemistry of Folic Acid and Related Pteridines." North-Holland Publ., Amsterdam, 1969.

163. A. Bacher and F. Lingens, *JBC* **246**, 7018–7022 (1971).

164. G. W. E. Plaut, *Compr. Biochem.* **21**, 11–45 (1970).

165. G. W. E. Plaut and C. M. Smith, *Annu. Rev. Biochem.* **43**, 899–922 (1974).

165a. B. Mailänder and A. Bacher, *JBC* **251**, 3623–3628 (1976).

166. S.-H. Lu and W. L. Alword, *Biochemistry* **11**, 608–611 (1972).

STUDY QUESTIONS

1. Can you comment (with respect to mechanism) on the following observation? Nitrogenase reduces acetylene to ethylene but ethylene is *not* reduced to ethane. However, cyclopropene is reduced to a mixture of cyclopropane and propene [see C. E. McKenna, M. McKenna and M. T. Higa, *JACS* **98**, 4657–4659 (1976)].

2. There is currently interest in breeding nitrogen-fixing bacteria with an increased activity of hydrogenase. The latter can oxidize H_2 to H^+. Explain why this might lead to greater efficiency in nitrogen fixation in root nodules of legumes [see R. O. D. Dixon, *Nature (London)* **262**, 173 (1976)].

3. Trace the pathway by which some bacteria can convert (a) N_2 into the nitrogen atom of glutamic acid and into glutamine, (b) NO_3^- into the nitrogen atom of alanine, (c) N of glutamic acid into that of porphobilinogen, (d) N of glutamine into that of adenylic acid, (e) N of aspartic acid into that of lysine.

4. What 10 amino acids are essential for the nutrition of the rat? Which amino acids are essential to man?

5. What is meant by "nitrogen balance." Of what utility is this concept in nutritional studies?

6. Propose a pathway for biosynthesis of tropine, a

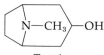

Tropine

component of *Datura* (Jimson weed) alkaloids. Starting materials are glutamic acid (first converted to ornithine) and acetoacetate.

7. Propose a detailed biosynthetic route to fusaric acid, a product of species of fungi of the genus *Fu-*

$$H_3C \quad\quad\quad\quad\quad\quad\quad N \quad COOH$$

Fusaric acid

sarium. Aspartate and acetyl-CoA are the starting materials.

8. Propose a pathway for metabolism of threonine to glycine and acetyl-CoA. This appears to be used by many bacteria in the breakdown of threonine [see S. C. Bell and J. M. Turner, *Biochem. Soc. Trans.* **4**, 497–500 (1976)].

9. While HCN is very toxic to most organisms, many higher plants can use HCN in biosynthetic pathways. (a) Propose a route of conversion of serine and HCN to asparagine and to α,γ-diaminobutyric acid. Propose a pathway of synthesis by fungi of alanine from acetaldehyde, HCN and ammonia (NOTE: The corresponding nonenzymatic reaction is the well-known Strecker synthesis of amino acid).

10. Ketopantoate hydroxymethylase of *E. coli*, the enzyme that catalyzes the first "committed" step in biosynthesis of pantothenic acid (see Fig. 14-10) is not inactivated by sodium borohydride in the presence of excess substrate and is activated by Mg^{2+} How would you classify this enzyme according to the scheme used in Chapter 7? [See S. G. Powers and E. E. Snell, *JBC* **251**, 3786–3793 (1976).]

11. Propose a pathway of biosynthesis of echimidinic

acid, a product of some plants of the family Borraginaceae.

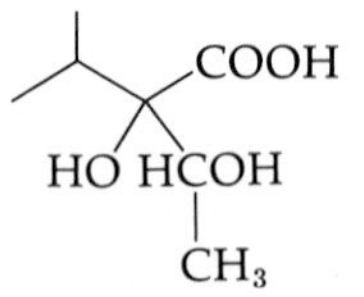

Echimidinic acid

12. Draw the structure of protoporphyrin IX. (a) Mark with an asterisk and a circle, respectively, all of the carbon and nitrogen atoms in this structure which are derived directly from the carboxylate carbon and amino nitrogen atoms of glycine. (b) From what precursor do other atoms arise? (c) What additional precursor molecule donates atoms quite directly to the chlorophyll structure? (d) To vitamin B_{12}?

13. Some plants ("cyanogenic plants"), such as *Prunus amygdalis* and *Sorghum*, utilize cyanide and convert it to cyanogenic glycosides such as amygdalin (Fig. 14-23). In addition to the possible biosynthetic route shown in that figure, it is thought that these glycosides may arise from the corresponding amino acid in the following manner: The amino group is hydroxylated to form an *N*-hydroxyamino acid which can be dehydrogenated to an oxime. The latter can be converted to the nitrile and (following another hydroxylation step) to the glycoside.

Recently a new carbon-nitrogen cycle has been proposed [(R. T. Thatcher and T. L. Weaver, *Science*

192, 1234–1235, 1976)]. Cyanogenic plants, a fungus that converts cyanide to formamide, a pseudomonad and the nitrifying organisms *Nitrosomas* and *Nitrobacter* are all involved. Sketch the proposed cycle and consider the chemistry of the individual reactions.

14. Describe two major pathways of aromatic biosynthesis. By which pathway do you think the following would be synthesized?

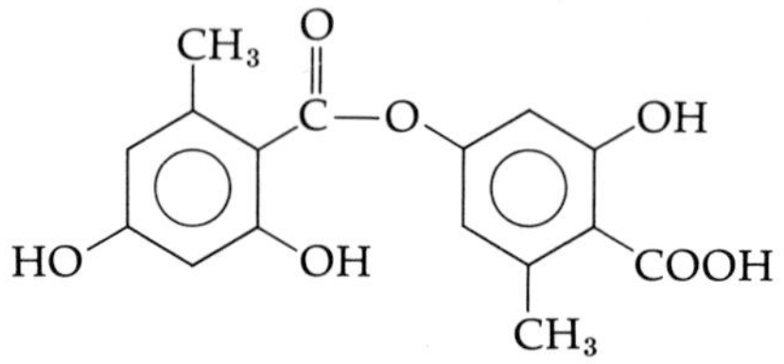

Lecanoric acid, a product
of the lichen *Umbillicaria
pustulata*

15. Gallic acid can arise in plants in the shikimic acid pathway (Section H,6) but it can also be formed in fungi via the polyketide route. Propose a detailed pathway.

16. Propose a three-step pathway for formation of salicylic acid (*o*-hydroxybenzoic acid) from chorismic acid.

17. Propose a detailed mechanism for step *d* of Fig. 14-31, the glutamine- and ATP-dependent amination of formylglycineamide ribotide. The reaction is said to be catalyzed by a single enzyme.

15

Biochemical Genetics and the Synthesis of Nucleic Acids and Proteins

Some of the most exciting biological discoveries of recent years have to do with the unraveling of the genetic code (Chapter 1, Section A,3) and the understanding of the ways in which nucleic acids and proteins are synthesized. The construction of both the nucleotides and the amino acids has been considered already (Chapter 14) as has the basic chemistry of the polymerization processes (Chapter 11, Section E). The present chapter deals with the means by which the polymerization reactions are controlled and by which the correct sequences of nucleotides or amino acids are obtained. The understanding of these matters is a development of genetics as well as of biochemistry; hence, the chapter title.[1-5]

A. HOW THE PRESENT CONCEPTS EMERGED

The subject of this chapter is quite complex and utilizes a specialized vocabulary that is unfamiliar to many students of chemistry. Therefore, it seems appropriate initially to provide a brief survey of the entire field using a historical approach. The section also introduces one method for "map-ping" a bacterial chromosome. In subsequent sections the chemical details of transcription and translation of the genetic messages in the DNA are considered. Finally, after an examination of some genetic methods the chemistry of DNA synthesis and of mutations is discussed.

1. DNA as the Genetic Material

The discovery of deoxyribonucleic acid dates to 1869 when Friedrich Miescher isolated a new chemical substance from white blood cells (of pus) and later from sperm cells. The material became known as nucleic acid. In time it was recognized to occur in both plants and animals, thymus gland and yeast cells being among the best sources. Chemical studies soon made it clear that the nucleic acids isolated from thymus gland and from yeast were different. As we now know, thymus nucleic acid was primarily DNA and yeast nucleic acid primarily RNA. For a while it was suspected that animals contained only DNA and plants only RNA; it was not until the early 1940's that it became clear that both substances were present in all organisms.[5,6]

a. The "Transformation" of Bacteria

In 1928 an important experiment with cells of *Diplococcus pneumoniae* showed that genetic information controlling properties of the capsular polysaccharides (Chapter 5, Section D) could be transferred from one strain of bacteria to another. The experiments showed that some material present in killed cells and cell-free extracts permanently altered the capsular properties of cells exposed to the material. This "transformation" of bacteria remained a mystery for many years. At the time of the experiments, there was no hint of the genetic role of nucleic acids which were generally regarded as strange materials. Furthermore, the covalent structure of nucleic acids was uncertain. A popular idea was that a tetranucleotide served as a repeating unit for some kind of regular polymer. Genes were most commonly thought to be protein in nature.

In 1944, O. T. Avery* and associates showed that purified DNA extracted from pneumococci could carry out transformation.[7,8] The purified transforming principal appeared to contain little protein. It was not inactivated by proteolytic enzymes but was inactivated by deoxyribonuclease.

Thus, the bacterial transformation experiments provided strong evidence that DNA was the genetic material. Other experiments were pointing to the same conclusions. For example, DNA was found localized in the nuclei of eukaryotic cells. The absolute amount per cell was found constant for a given species. Proof that DNA is the genetic material of certain viruses was obtained in 1952 by A. D. Hershey and M. Chase.[8a] They showed that when a bacterial virus infects a cell the viral DNA enters the bacterium but the protein "coat" remains outside. This fact was demonstrated by preparing two types of labeled bacteriophage T2 (Box 4-E). One contained ^{32}P which labeled the DNA and the second contained ^{35}S which had been incorporated into the proteins. Cells of *E. coli* were infected with the la-

beled phage preparations and were then agitated violently in a Waring blender to shear off the phage particles. Over 80% of the ^{35}S was removed from the bacteria by this treatment, but most of the ^{32}P entered the bacteria and could even be recovered in the next generation of progeny bacteriophage.[3]

b. The Double Helix[9]

As newer methods of investigation of the chemical composition of nucleic acids were devised, it was discovered (by E. Chargaff) that the base composition of DNA was extremely variable. However, in all DNA examined the molar ratio of adenine to thymine was nearly 1:1, as was that of cytosine to guanine.[10] This observation provided the basis for the concept of base pairing in the structure of DNA. The final important information was supplied by X-ray diffraction studies of stretched fibers of DNA which showed that molecules of DNA were almost surely helical structures containing more than one chain. The crucial experiments were done by R. Franklin[11] and M. H. F. Wilkins, whose data were used by J. D. Watson and H. F. C. Crick in 1953 in constructing their model of the double helical structure[12,13] (Fig. 2-21). Once established, the structure of DNA itself suggested both the coding properties of the molecule and a natural replication mechanism. It seemed clear that the genetic code had to lie in the sequence of nucleotides. Base pairing provided a mechanism by which the two mutually complementary strands could be separated with biosynthesis of a new complementary strand alongside each one. In this way genes could be replicated precisely. In a similar fashion, RNA could be synthesized alongside a DNA "template" and then be transported out to the cytoplasm.

The presence of RNA in the cytoplasm had been linked to protein synthesis by experiments done in the early 1940's. After the discovery of the double helix, the concept followed quickly that DNA was the master "blueprint" from which secondary blueprints of RNA could be copied. The RNA copies, later identified as **messenger RNA**

---

* As pointed out by Chargaff,[7] Avery was age 67 at the time of this discovery, refuting the popular contention that all important scientific discoveries are made by young people.

(mRNA, Chapter 1, Section A,4), provided the genetic information for determining protein sequence. The flow of information from DNA to RNA to protein could be symbolized as in Eq. 15-1.

$$\text{DNA} \xrightarrow{\text{transcription}} \text{RNA} \xrightarrow{\text{translation}}$$

$$\text{proteins} \longrightarrow \text{control of metabolism} \qquad (15\text{-}1)$$

Proteins, in one way or another, control all of metabolism. This includes the reactions that form the nucleotide precursors of the nucleic acids and that lead to polymerization of the amino acids and nucleotides. Thus, the flow of information from DNA to proteins is only part of a larger loop of metabolic processes. However, DNA is replicated with great accuracy. Thus, genetic information always flows from DNA out into the cell and copies of the master blueprint are passed nearly unchanged from generation to generation. The simple concepts implied by Eq. 15-1 quickly caught the imagination of almost the entire community of scientists and led to a rapid blossoming of the field of biochemical genetics.

c. The Chromosome Map of *E. coli*

Let us now turn our attention to a most important aspect of DNA structure, the nucleotide sequence which carries the genetic information. There are 3.8 million nucleotides in the circular DNA molecule that is the chromosome of *E. coli*. We have only started to learn the details of the sequence at a few points around the chromosome. However, in a less detailed way, we know a great deal. It is well established that the chromosome is a linear array of the individual genes. As of 1972, the locations of 460 genes in the linkage map of *E. coli* had been established (Fig. 15-1, see also Table 15-1).

To understand how the chromosome map of Fig. 15-1 was obtained, it is necessary to consider briefly some of the methods of genetic investigation. (The topic will be pursued further in Section B.) Genetic mapping of bacteria began with the recognition that mutants with specific growth requirements could be found. Ordinary "wild type"

cells of *E. coli* can grow on a minimal medium containing a carbon compound as a source for energy together with inorganic nutrients. Irradiation with ultraviolet light or treatment with mutagenic chemicals produces many mutant cells that fail to grow on such a minimal medium. However, addition of one or more specific compounds, such as an amino acid or vitamin, usually permits growth. Selection of the nutritional **auxotrophs,** as the mutants are called, is most often accomplished by "plating out" large numbers of the irradiated or chemically treated cells on a solid, rich nutrient medium. Colonies (clones) are allowed to develop by multiplication of the individual bacteria. The auxotrophs are selected by **replica plating** to plates containing a minimal medium plus specific supplements.*

A nutritional auxotroph of a bacterium usually has a defective gene specifying a protein needed for the biosynthesis of the required nutrient. Individual genes recognized in this way are named with a genetic symbol. For example, gene *trpA* specifies one of the protein subunits of tryptophan synthetase. Other kinds of mutations, e.g., those affecting motility or other properties of the cells, can also be detected and are given appropriate symbols. A few of these genetic symbols are indicated on the chromosome map in Fig. 15-1.

Bacteria usually reproduce by a simple cell division. The DNA in the chromosome is doubled in quantity and the cell divides, each daughter cell receiving an identical chromosome. However, as was shown by Lederberg and Tatum[13a] in 1946,

---

* In replica plating a sterile velveteen pad is pressed against the nutrient agar plate containing small colonies of bacteria and is then used to "print" replica plates containing minimal medium. The colonies on the initial and replica plates are compared and the colonies of auxotrophs (which do not grow on the minimal medium) are selected. In a second stage, the auxotrophs may be replica plated to minimal medium supplemented with various nutrients (amino acids, purines, pyrimidines, vitamins, etc.). Selection is made easier by pretreatment of the irradiated cells suspended in minimal medium with penicillin (Box 7-D). Penicillin kills the growing cells, but the auxotrophs, which do not grow on the minimal medium, survive. The penicillin is then destroyed by adding penicillinase (Box 7-D) leaving a suspension much enriched in the percentage of auxotrophic mutants.[3]

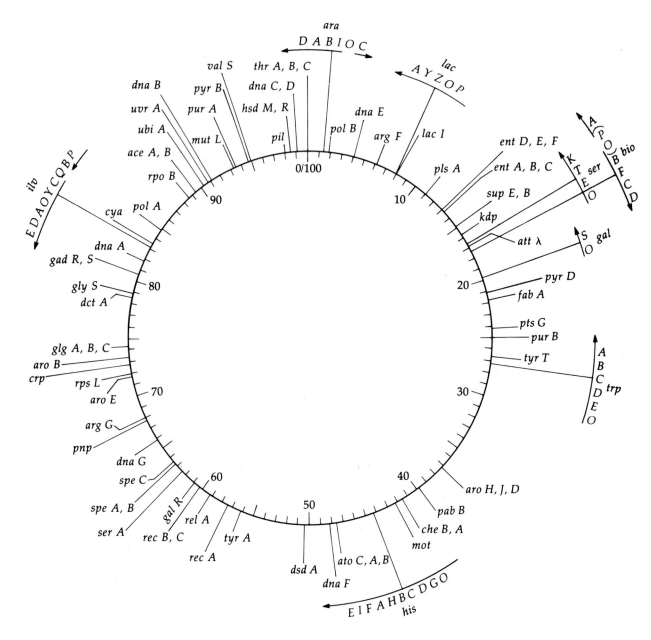

Fig. 15-1 Genetic map of *E. coli*, strain K-12. Data from B. J. Bachmann, K. B. Low, and A. L. Taylor, *Bacteriol. Rev.* **40**, 116–167 (1976). The scale in minutes is based on the results of interrupted conjugation experiments. The *thr* locus is arbitrarily taken as zero. The genetic symbols are defined in Table 15-1. Only a single strand of the DNA molecule is normally transcribed into RNA over any given region. Because of the antiparallel nature of the DNA helix transcription from one strand proceeds clockwise and transcription from the other counterclockwise (the *r* strand is conventionally taken as the one that is transcribed in the clockwise direction when the chromosome is oriented as in the drawing). The directions of transcription of certain operons are indicated by the arrows. A map is available for *Salmonella typhimurium* and was obtained by similar methods. K. E. Sanderson, *Bacteriol. Rev.* **36**, 558–586 (1972).

TABLE 15-1
Some Genes of *E. coli*[a]

Gene symbol	Mnemonic	Map position (min)[a]	Alternate gene symbols; phenotypic trait affected
aceA	Acetate	89	Isocitrate lyase
aceB	Acetate	89	Malate synthetase A
araA	Arabinose	1	L-Arabinose isomerase
araB	Arabinose	1	L-Ribulokinase
araC	Arabinose	1	Regulatory gene
araD	Arabinose	1	L-Ribulose-5-phosphate-4-epimerase
araI	Arabinose	1	Initiator locus
araO	Arabinose	1	Operator locus
argF	Arginine	6	Ornithine carbamoyltransferase
argG	Arginine	68	Argininosuccinic acid synthetase
aroB	Aromatic	73	Dehydroquinate synthetase
aroD	Aromatic	37	Dehydroquinate dehydratase
aroE	Aromatic	71	Dehydroshikimate reductase
aroH	Aromatic	37	DAHP synthetase (tryptophan-repressible isoenzyme)
aroJ	Aromatic	37	Probable operator locus for aroH
atoA	Acetoacetate	48	Coenzyme A transferase
atoB	Acetoacetate	48	Thiolase II
atoC	Acetoacetate	48	Regulatory gene
attλ	Attachment	17	Integration site for prophage λ
bioA	Biotin	17	Group II; 7-oxo-8-aminopelargonic acid (7 KAP) → 7,8-di-aminopelargonic acid (DAPA)
bioB	Biotin	17	Conversion of dethiobiotin to biotin
bioC	Biotin	17	Block prior to pimeloyl-CoA
bioD	Biotin	17	Dethiobiotin synthetase
bioF	Biotin	17	Pimeloyl-CoA → 7 KAP
bioO	Biotin	17	Operator for genes *bio*B through *bio*D
bioP	Biotin	17	Promoter site for genes *bio*B through *bio*D
cheA	Chemotaxis	42	Chemotactic motility
cheB	Chemotaxis	42	Chemotactic motility
crp		73	Cyclic adenosine monophosphate receptor protein
cya		83	Adenylate cyclase
dctA		79	Uptake of C_4-dicarboxylic acids
dnaA	DNA	82	DNA synthesis; initiation defective
dnaB	DNA	91	DNA synthesis
dnaC	DNA	99	*dnaD*; DNA synthesis; initiation defective
dnaE	DNA	4	*polC*, DNA polymerase III and mutator activity
dnaF	DNA	48	*nrdA*; ribonucleoside diphosphate reductase
dnaG	DNA	66	DNA synthesis
dsdA	D-Serine	50	D-Serine deaminase
entA	Enterochelin	13	2,3-Dihydro-2,3-dihydroxybenzoate dehydrogenase
entB	Enterochelin	13	2,3-Dihydro-2,3-dihydroxybenzoate synthetase
entC	Enterochelin	13	Isochorismate synthetase
entD,E,F	Enterochelin	13	Unknown steps in conversion of 2,3-dihydroxybenzoate to enterochelin
fabA		22	β-Hydroxydecanoylthioester dehydratase
gadR		81	Regulatory gene for *gadS*
gadS		81	Glutamic acid decarboxylase

(continued)

TABLE 15-1 (*Continued*)

Gene symbol	Mnemonic	Map position (min)[a]	Alternate gene symbols; phenotypic trait affected
galE	Galactose	17	Uridine diphosphogalactose 4-epimerase
galK	Galactose	17	Galactokinase
galO	Galactose	17	Operator locus
galT	Galactose	17	Galactose 1-phosphate uridyltransferase
galR	Galactose	61	Regulatory gene
glgA	Glycogen	74	Glycogen synthetase
glgB	Glycogen	74	α-1,4-Glucan: α-1,4-glucan 6-glucosyltransferase
glgC	Glycogen	74	Adenosine diphosphate glucose pyrophosphorylase
glyS	Glycine	79	Glycyl-transfer RNA synthetase
hisA	Histidine	44	Isomerase
hisB	Histidine	44	Imidazole glycerol phosphate dehydrase: histidinol phosphatase
hisC	Histidine	44	Imidazole acetol phosphate transaminase
hisD	Histidine	44	Histidinol dehydrogenase
hisE	Histidine	44	Phosphoribosyl-adenosine triphosphate-pyrophosphohydrolase
hisF	Histidine	44	Cyclase
hisG	Histidine	44	Phosphoribosyl-adenosine triphosphate-pyrophosphorylase
hisH	Histidine	44	Amidotransferase
hisI	Histidine	44	Phosphoribosyl-adenosine monophosphate-hydrolase
hisO	Histidine	44	Operator locus
hsdM	Host specificity	98	Host modification activity: DNA methylase M
hsdR	Host specificity	98	Host restriction activity: endonuclease R
ilvA	Isoleucine-valine	83	Threonine deaminase (dehydratase)
ilvB	Isoleucine-valine	83	Acetohydroxy acid synthetase I
ilvC	Isoleucine-valine	83	α-Hydroxy-β-keto acid reductoisomerase
ilvD	Isoleucine-valine	83	Dehydrase
ilvE	Isoleucine-valine	83	Transaminase B
ilvO	Isoleucine-valine	83	Operator locus for genes *ilv*A,D,E
ilvP	Isoleucine-valine	83	Operator locus for gene *ilv*B
ilvQ	Isoleucine-valine	83	Induction recognition site for *ilv*C
ilvY	Isoleucine-valine	83	Positive control element for *ilv*C induction
kdp	K accumulation	16	Defect in potassium ion uptake
lacA	Lactose	8	Thiogalactoside transacetylase
lacI	Lactose	8	Regulator gene
lacO	Lactose	8	Operator locus
lacP	Lactose	8	Promoter locus
lacY	Lactose	8	Galactoside permease (M protein)
lacZ	Lactose	8	β-Galactosidase
mot	Motility	42	Flagellar paralysis
mutL	Mutator	93	Generalized high mutability (AT ⇄ GC)
pabB	*p*-Aminobenzoate	40	Requirement
pil	Pili	98	Presence or absence of pili (fimbriae)
plsA	Phospholipid	11	Glycerol-3-phosphate acyltransferase
pnp		68	Polynucleotide phosphorylase
polA	Polymerase	85	DNA polymerase I
polB	Polymerase	2	DNA polymerase II
ptsG	Phosphotransferase system	24	Catabolite repression

TABLE 15-1 *(Continued)*

Gene symbol	Mnemonic	Map position (min)[a]	Alternate gene symbols; phenotypic trait affected
purA	Purine	93	Adenylosuccinic acid synthetase
purB	Purine	25	Adenylosuccinase
pyrB	Pyrimidine	95	Aspartate carbamoyltransferase
pyrD	Pyrimidine	21	Dihydroorotic acid dehydrogenase
recA	Recombination	58	Ultraviolet sensitivity and competence for genetic recombination
recB	Recombination	60	Ultraviolet sensitivity, genetic recombination; exonuclease V subunit
recC	Recombination	60	Ultraviolet sensitivity, genetic recombination; exonuclease V subunit
relA	Relaxed	59	Regulation of RNA synthesis
rpoB	RNA polymerase	89	RNA polymerase: β subunit (*rif* gene)
rpsL	Ribosomal protein, small	72	Ribosomal protein S12 (*strA* gene, streptomycin resistance)
serA	Serine	62	3-Phosphoglyceric acid dehydrogenase
serO	Serine	20	Operator locus
serS	Serine	20	Seryl transfer RNA synthetase
speA	Spermidine	63	Arginine decarboxylase
speB	Spermidine	63	Agmatine ureohydrolase
speC	Spermidine	63	Ornithine decarboxylase
supB	Suppressor	15	Suppressor of *ochre* mutations
supE	Suppressor	15	Suppressor of *amber* mutations (*su-2*)
thrA	Threonine	0	Aspartokinase I-homoserine dehydrogenase I complex
thrB	Threonine	0	Homoserine kinase
thrC	Threonine	0	Threonine synthetase
trpA	Tryptophan	27	Tryptophan synthetase, A protein
trpB	Tryptophan	27	Tryptophan synthetase, B protein
trpC	Tryptophan	27	N-(5-Phosphoribosyl) anthranilate
trpD	Tryptophan	27	Phosphoribosyl anthranilatetransferase
trpE	Tryptophan	27	Anthranilate synthetase
trpO	Tryptophan	27	Operator locus
tyrA	Tyrosine	56	Chorismate mutase T-prephenate dehydrogenase
tyrT	Tyrosine	27	Tyrosine transfer RNA_1 (*su-3* gene; amber suppressor)
ubiA	Ubiquinone	90	4-Hydroxybenzoate → 3-octaprenyl 4-hydroxybenzoate
uvrA	Ultraviolet	91	Repair of ultraviolet radiation damage to DNA, UV endonuclease
valS	Valine	95	Valyl-transfer RNA synthetase

[a] This list contains 125 of the over 650 genes that have been mapped.[15] Their positions are shown diagrammatically in Fig. 15-1.

sexual reproduction is also possible. While initially there was no direct evidence that bacteria mate, it was found that when cells of two different mutants of strain K-12 of *E. coli* were mixed together and allowed to grow for a few generations, a few individual bacteria regained the ability to grow on a minimal medium. Since each of the two strains had one defective gene, the creation of an individual with neither of the two defects required combining of genetic traits from both strains. Thus, the existence of bacterial conjugation was recognized. In time it was established that true **genetic recombination** can occur during conjugation. That is, genes from the two mating cells can be integrated into a single molecule of bacterial DNA.

d. Bacterial Sex Factors

We know now that some cells of *E. coli* strain K-12 contain a small extra piece of DNA that acts as a "sex factor" (**F agent** or fertility factor). The presence of the F agent in a bacterial cell confers a male character to the individual. Among other things, the F agent contains the genes needed to direct the synthesis of the **F pili** (sex pili). These tiny appendages, 8.5 nm in diameter, grow out quickly during a period of 4–5 min to a length of about 1.1 μm (see also Chapter 1, Section A,6 Fig. 4-7). The end of an F pilus becomes attached to a female cell. It has been suggested by Brinton[14] that DNA is actually transferred through the pilus from the male into the female cell.* However, some workers have seen what appear to be cytoplasmic bridges between cells in close contact and the actual mechanism of DNA transfer remains uncertain.

The importance of the F agent for chromosome mapping lies in the fact that on rare occasions the F agent becomes integrated into the chromosome of the bacterium. Both the F agent and the chromosome have been shown by direct electron microscopy to be circular. The integration process requires the enzymatic cleavage of the DNA of both the chromosome and the F agent and the rejoining of the ends in such a way that a continuous circle is formed (Fig. 15-2). Special enzymes catalyze these reactions which are considered further in Section G. Different F agents can be incorporated into the chromosome at different points around the circle. A strain of bacteria containing the integrated F agent is known as an *Hfr* (high frequency of recombination) strain.

When an Hfr strain conjugates with an F⁻ (female) replication of the entire chromosome commences at some point near the end of the

* However, it is not certain whether the F pilus is a tube with an internal diameter of about 2.5 nm (Fig. 4-7) or a more open structure.

integrated F agent and genes of the bacterial chromosome followed by those of the F factor are transferred into the female. It is thought that only a single strand of DNA (customarily referred to as the "plus" strand) is extruded from the cell, possibly through an F pilus, and into the recipient cell (Fig. 15-2). There the complementary "minus" strand is synthesized to form a complete double-stranded DNA molecule bearing the genes from the Hfr cell. It is only rarely that a copy of the entire chromosome of the donor cell enters the female cell. More often the DNA strand, or perhaps the pilus itself, breaks and only part of the chromosome is transferred.

Partial chromosome transfer from a male cell transforms the F⁻ cell into a partial diploid (**merozygote**) containing double the usual number of many of the genes. Within this partial diploid, mixing of genetic information (**genetic** recombination) between the two chromosomes takes place (Fig. 15-2). The chemical basis for this process, important to all organisms with sexual reproduction, will be discussed in Section G. The end result of the recombination process is that the daughter cells formed by subsequent division contain only single chromosomes with the usual number of genes. However, some genes come from each of the two parental strains. Thus, it may happen that an F⁻ mutant unable to grow on a medium deficient in a certain nutrient receives a gene from the male and is now able to grow on a minimal medium. Even though the number of such **recombinants** is small, they are easily selected from the very large number of mutant bacteria that are mixed together initially.

e. Chromosome Mapping by Interrupted Bacterial Mating

A chromosome map for *E. coli* can be obtained by mixing Hfr and F⁻ cells and allowing conjugation to proceed for a certain length of time. Then the cells are agitated violently, e.g., in a Waring blender. This breaks all of the conjugation bridges and interrupts the mating process. Mating is interrupted at different times, and the recipient bacteria are tested for the presence of genes trans-

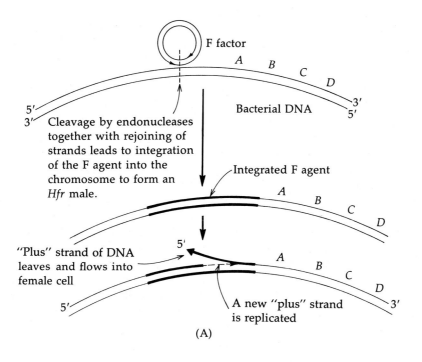

F factor

A
B
C
D

Bacterial DNA

5′
3′ Cleavage by endonucleases
together with rejoining of
strands leads to integration
of the F agent into the
chromosome to form an
Hfr male.

3′
5′

Integrated F agent

A
B
C
D

"Plus" strand of DNA
leaves and flows into
female cell

5′

A
B
C
D

5′

A new "plus" strand
is replicated

3′

(A)

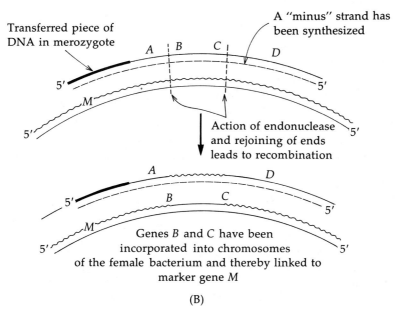

Transferred piece of
DNA in merozygote

A "minus" strand has
been synthesized

A   B   C   D

5′   5′

M

5′   5′

Action of endonuclease
and rejoining of ends
leads to recombination

A   D

5′
B   C
5′

M

5′   5′

Genes *B* and *C* have been
incorporated into chromosomes
of the female bacterium and thereby linked to
marker gene *M*

(B)

Fig. 15-2 Integration of an F agent into a bacterial chromosome and transfer of some bacterial genes into another cell.
(A) Incorporation of the F agent into *E. coli* genome and
transfer of the "plus" strand of DNA out to a female recipient
cell. (B) Genetic recombination between a piece of transferred
DNA and the genome of the recipient cell.

ferred from the donor strain. Using this technique it was found that complete transfer of the chromosome takes ~ 100 min at 37°C and that the *approximate location of any gene on the chromosome can be determined by the length of time required for transfer of that gene into the recipient cell.* It is a little more complex than this. Because complete chromosome transfer is rare, substrains of *E. coli* K-12 with the F agent integrated at different points are used. In each case, those genes lying clockwise* around the circle in Fig. 15-1 immediately beyond the point of integration are transferred quickly and with high frequency.

The map in Fig. 15-1 is based not only on interrupted matings but also on the use of "transduction" by bacteriophage P1.[15] Transduction by phage, discussed in more detail in Section D, permits the transfer of a short fragment of DNA, about 2 min in length, on the *E. coli* map. Joint transduction, i.e., joint incorporation of two genes into the chromosome of the receptor, occurs with a frequency related to the map distance between these two genes. Thus, finer mapping has been done within many segments of the *E. coli* chromosome.

Note that while the map in Fig. 15-1 is calibrated in minutes, this is regarded as a temporary expedient. It will soon be possible to express the linkage map directly as micrometers of DNA length (total length ~ 1100 μm) or in thousands of nucleotide units, sometimes called **kilobases** (kb). The total length is ~ 3800 kb.[†]

2. The Genetic Code

The general nature of the genetic code was suggested by the structure of DNA itself. Both DNA and proteins are linear polymers. Thus, it seemed logical to suppose that the sequence of the bases in DNA coded for the sequence of amino acids. Since there are only four bases in DNA but 20 different amino acids in proteins (at the time of their synthesis) each amino acid must be specified by some combination of more than one base. While 16 *pairs* of bases are possible, this is still too few to specify 20 different amino acids. Therefore, it appeared that at least a *triplet* group of three nucleotides would be required to code for one amino acid.[16] Sixty-four (4^2) such triplet **codons** exist, as is indicated in Tables 15-2 and 15-3.

TABLE 15-2
The Genetic Code[a]

Amino acid	Codons	Total number of codons
Alanine	GCX	4
Arginine	CGX, AGA, AGG	6
Asparagine	AAU, AAC	2
Aspartic acid	GAU, GAC	2
Cysteine	UGU, UGC	2
Glutamic acid	GAA, GAG	2
Glutamine	CAA, CAG	2
Glycine	GGX	4
Histidine	CAU, CAG	2
Isoleucine	AUU, AUC, AUA	3
Leucine	UUA, UUG, CUX	6
Lysine	AAA, AUG	2
Methionine (also initiation codon)	AUG	1
Phenylalanine	UUU, UUC	2
Proline	CCX	4
Serine	UCX, AGU, AGC	6
Threonine	ACX	4
Tryptophan	UGG	1
Tyrosine	UAU, UAC	2
Valine (GUG is sometimes an initiation codon)	GUX	4
Termination	UAA (*ochre*) UAG (*amber*) UGA	3
Total		64

[a] The codons for each amino acid are given in terms of the sequence of bases in messenger RNA. From left to right the sequence is from the 5' end to the 3' end. The symbol X stands for any one of the four RNA bases. Thus each codon symbol containing X represents a group of four codons.

---

* With one type of F factor. Others are integrated in the opposite direction.

[†] This value is somewhat uncertain. Thus the map in Fig. 15-1 is based[15] on a total length of 4100 kb or a molecular weight of 2.7×10^9.

TABLE 15-3
The Sixty-Four Codons of the Genetic Code

5'-OH Terminal base	Middle base				3'-OH Terminal base
	U(T)	C	A	G	
U(T)	Phe	Ser	Tyr	Cys	U(T)
	Phe	Ser	Tyr	Cys	C
	Leu	Ser	Term	Term	A
	Leu	Ser	Term	Trp	G
C	Leu	Pro	His	Arg	U
	Leu	Pro	His	Arg	C
	Leu	Pro	Gln	Arg	A
	Leu	Pro	Gln	Arg	G
A	Ile	Thr	Asn	Ser	U
	Ile	Thr	Asn	Ser	C
	Ile	Thr	Lys	Arg	A
	Met[a]	Thr	Lys	Arg	G
G	Val	Ala	Asp	Gly	U
	Val	Ala	Asp	Gly	C
	Val	Ala	Glu	Gly	A
	Val[a]	Ala	Glu	Gly	G

[a] Initiation codons. The methionine codon AUG is the most common starting point for translation of a genetic message but GUG can also serve. In such cases it codes for methionine rather than valine.

Simplicity argued that the genetic blueprint specifying amino acid sequences in proteins should consist of *consecutive, nonoverlapping triplets*. However, there was initially no proof of this and other possibilities were actively considered, but within a few years genetic experiments (some of which are discussed in Section D) together with chemical experiments considered in the following section provided unequivocal proof for a nonoverlapping code.

Deciphering the Code

Even after the triplet nature of the genetic code became evident, many questions remained. Were all of the 64 possible codons used by the living cell? If so, were they all used to code for amino acids or were some set aside for other purposes? How many codons were used for a single amino acid? Was the code "universal," applying to all organisms, or did different organisms use dif-

ferent codes? How could one decipher the code? Despite the complexity of these questions, they all seem to have been answered definitively.

An important experiment[17] was performed by M. Nirenberg* and H. Matthaei in 1961. Using a typical biochemist's approach, Nirenberg had isolated ribosomes from *E. coli*. He mixed these with crude extracts of soluble materials, also from *E. coli* cells. The extracts included tRNA molecules and amino acid activating enzymes. The 20 amino acids, ATP, and an ATP-generating system (PEP + pyruvate kinase) were added. Nirenberg was able to show that under such conditions protein was synthesized by ribosomes in response to the presence of added RNA. For example, RNA from tobacco mosaic virus (Chapter 4, Section D,2) was very effective in stimulating protein synthesis. The crucial experiment (originally done simply as a "control") was one in which a synthetic polynucleotide consisting solely of uridylic acid units was substituted for mRNA. In effect, this was a synthetic mRNA containing only the codon UUU repeated over and over. To Nirenberg's surprise, the ribosomes read this code and synthesized a peptide containing only phenylalanine. Thus, poly(U) gave polyphenylalanine and *UUU was identified as a codon specifying phenylalanine*. The first nucleotide triplet had been identified! In the same manner CCC was identified as a proline codon and AAA as a lysine codon. Study of mixed copolymers containing two different nucleotides in a random sequence suggested other codon assignments. However, it was a few years later, after H. G. Khorana had supplied the methods for synthesis of oligonucleotides and of regular alternating polymers of known sequence, that the remaining codons were identified.

An important technique was based on the observation that synthetic trinucleotides induced the binding to ribosomes of specific tRNA molecules "charged" with their specific amino acids.[18,19] For example, the trinucleotides UpUpU and ApApA stimulated the binding of ^{14}C-labeled phenyl-

---

* In 1968 Nirenberg and Khorana together with R. Holley, who first determined the sequence of a transfer RNA, were awarded a Nobel Prize.

alanyl-tRNA and lysyl-tRNA, respectively. However, the corresponding dinucleotides had no effect, an observation that not only verified the two codons but also provided direct evidence for the triplet nature of the genetic code. Another powerful approach was the use of artificial RNA polymers, synthesized by combined chemical and enzymatic approaches.[20] For example, the polynucleotide CUCUCUCUCU . . . led to the synthesis by ribosomes of a regular alternating polypeptide of leucine and serine.

Table 15-2 shows the codon assignments, as we now know them, for each of the 20 amino acids. Table 15-3 shows the same 64 codons in a rectangular array. Note that in addition to those codons assigned to specific amino acids, three are designated as **chain termination codons:** UAA, UAG, and UGA. These are frequently referred to as "nonsense" codons. The termination codons UAA and UAG are also known as *ochre* and *amber,** respectively.[21] The codons AUG (methionine) and much less often GUG (valine) have been found to serve as the **initiation codons** in bacterial protein synthesis. Thus, the N-terminal amino acid in most newly synthesized bacterial proteins is methionine, as an *N*-formyl derivative. As explained in Section B, *N*-formylmethionyl-tRNA is specifically bound to initiation sites containing the AUG codon in the mRNA–ribosome complex.

A variety of studies suggest that the genetic code as worked out for *E. coli* may be universal. For example, in the laboratories of Wittman and of Fraenkel-Conrat, RNA extracted from tobacco mosaic virus was treated with nitrous acid, a procedure known to deaminate many cytosine residues to uracil residues. Such treatment could change the codon UCU (serine) to UUU (phenylalanine). Likewise, it could change the codon CCC (proline) to CUC (leucine). When the nitrous acid-treated RNA was used to infect tobacco plants and virus particles were prepared in quantity from the resultant mutant strains, it was found that the amino acid sequence of the virus coat protein had indeed been altered.[22] Many of the alterations were exactly those that would be predicted from Table 15-3. Likewise, the amino acid substitutions in known defects of hemoglobin (Fig. 4-17) could be accounted for in most cases by single base alterations. Thus, hemoglobin S could have arisen by one of the following changes in the seventh codon: GAA (Glu) → GUA (Val) or GAG (Glu) → GUG (Val). Another argument favoring a universal code is based on the observation that mRNA coding for a hemoglobin chain can be translated by ribosomes and tRNA molecules from *E. coli* to accomplish the synthesis of authentic hemoglobin.[23]

3. Replication of DNA

That the DNA content doubles prior to cell division was established by microspectrophotometry. It was clear that both daughter cells must receive one or more identical molecules of DNA. However, it was not clear whether the original double-stranded DNA molecule was copied in such a way that an entirely new double-stranded DNA was formed or whether the two chains of the original molecule separated. In the latter case (called **semiconservative** replication) each of the separated strands would have a new complementary strand synthesized along it to form two identical double-stranded molecules.

The first definitive experimental evidence in favor of this semiconservative replication was reported[24] by M. Meselson and F. W. Stahl in 1958. Cells of *E. coli* were grown on a medium containing isotopically pure $^{15}NH_4^+$ ions as the sole source of nitrogen. After a few generations of growth in this medium, the DNA contained exclusively ^{15}N. Then the cells were abruptly transferred to a medium containing $^{14}NH_4^+$. The cells were allowed to grow and to double and quadruple in number. At various stages DNA was isolated and subjected to ultracentrifugation in a cesium chloride gradient. Small but easily detectable differences in density led to separation of

---

* The originators of these terms explain that "it is often safer to give a new discovery a silly name than a speculatively descriptive one!" The name *amber* was proposed for a class of mutations discovered with the help of a graduate student named Bernstein (the German word for amber).[21]

double-stranded DNA molecules containing only ^{15}N from those containing partly ^{15}N and from those containing only ^{14}N. At the beginning of the experiment only a single kind of DNA containing entirely ^{15}N was present. However, after one generation of growth in the ^{14}N-containing medium, the density of *all* the DNA was such as to indicate a content of one-half ^{15}N and one-half ^{14}N. After a second generation of growth half of the DNA still contained both nitrogen isotopes in equal quantity, whereas half contained only ^{14}N. These are exactly the results to be expected on the basis of "semiconservative" replication.

A few years later a technique of direct autoradiography of DNA using ^3H-labeled thymidine was applied by Cairns to the study of replication.[25] Cells of *E. coli* were grown on a medium containing the radioactive thymidine for various times but typically for 1 h (~2 generations). The cells were then ruptured and the DNA was spread on a thin membrane filter. Autoradiograms were prepared. When the DNA contained [^3H]thymidine the exposed trace in the autoradiogram could be followed around the entire 1.1–1.4 mm circumference of the spread DNA molecule. Furthermore, molecules partially labeled and in the process of replication could be identified. Thus, after a 2-h exposure about half of the bacterial DNA was fully labeled but half contained regions that were labeled only half as heavily. They presumably contained ^3H label in a single strand and therefore represented unreplicated regions. (All of the molecules had undergone one round of replication with tritiated thymidine to yield the lightly labeled molecules; parts of the molecules had not completed the second round.) The more heavily labeled regions were interpreted as fully replicated. The shapes of the "replication figures" suggested that DNA is synthesized in a continu-ous manner starting from one point and continuing around the circular molecule at a constant rate. Although subsequent experiments (considered in Section E) show that replication is more often **bidirectional,** the experiments of Cairns were extremely important because they introduced a technique for direct vizualization of replication *in vivo.*

a. DNA Polymerases

What are the precursors of DNA? Early experiments showed that the nucleoside [^3H]thymidine was efficiently incorporated into DNA. However, for energetic reasons it seemed unlikely that thymidine was an immediate precursor. Evidence favoring the nucleoside triphosphates was provided in 1958 when A. Kornberg identified a **DNA polymerase** from *E. coli.* Kornberg's enzyme, now commonly referred to as **DNA polymerase I,** has been isolated in the amount of 600 mg from 90 kg of bacterial cells[4,26] (over 400 molecules of enzyme per cell). The enzyme displays many of the properties expected of a DNA-synthesizing enzyme. It requires a **template strand** of DNA as well as a shorter **primer strand.** As indicated in Eq. 15-2, the enzyme recognizes the 3' end of the primer strand and binds the proper nucleoside triphosphate to pair with the next base in the template strand. Then it catalyzes the displacement of pyrophosphate, at the same time linking the new nucleotide unit onto the 3' end of the primer strand. Continuing in this way, the enzyme is able to turn a single-stranded template DNA into a double-stranded DNA in which the newly synthesized strand contains, at each point, the base complementary to the one in the template strand.

Although the action of the DNA polymerase I, according to Eq. 15-2, provides a straightforward

$$(15\text{-}2)$$

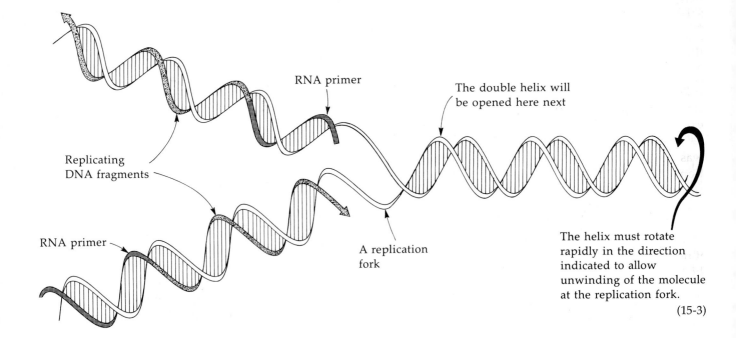

RNA primer

The double helix will
be opened here next

Replicating
DNA fragments

RNA primer

A replication
fork

The helix must rotate
rapidly in the direction
indicated to allow
unwinding of the molecule
at the replication fork.

(15-3)

way to form a complementary strand of DNA, it
does not explain how double-stranded DNA can
be copied. One problem is that the two strands
must be separated and unwound. If unwinding
and replication occur at a single "replication fork"
in the DNA, as indicated by Cairns' experiment,
the entire molecule must spin at a speed of 300
revolutions per second to permit replication of the
E. coli chromosome in 20 min. It also requires that
some kind of a "swivel" (at least, a "nick" in one
chain) be present in the chromosome (Eq. 15-3).

A more serious problem concerns the fact that
the two chains in DNA have opposite orientations. Thus, at the replication fork one of the new
chains must grow by addition of a new nucleotide
at the 3' end, while the other chain must grow at
the 5' end. This observation suggests that there
should be two kinds of DNA polymerase, one specific for polymerization at each of the two kinds of
chain end. Nevertheless, despite intensive search,
the only polymerases (DNA polymerases I, II, and
III) found add new units at the 3' ends.

b. Replication Fragments and DNA Ligase

In 1968 Okazaki reported that during replication
of DNA bacterial cells contain short fragments of
DNA, now known as **replication fragments** (or
Okazaki fragments).[27] A second recent development was the discovery of a new enzyme **DNA
ligase,**[28,29] which is able to join two pieces of DNA
to form a continuous chain. The specific action appears to be that of repairing "nicked" DNA. Such
a DNA strand, as indicated in Eq. 15-4, has a
break in one strand and contains a 3'-hydroxyl
group and a 5' phosphate which must be rejoined.
The DNA ligase from *E. coli* activates the phosphate group in an unusual way by transfer of an
adenylyl group from NAD^+, with displacement of
nicotinamide mononucleotide (Eq. 15-4, step *a*).
The reaction is completed by displacement of
AMP as indicated in Eq. 15-4, step *b*. It is noteworthy that in cells infected by bacteriophage T4,
a special ligase is induced. This enzyme utilizes
ATP rather than NAD^+ as the activating reagent.

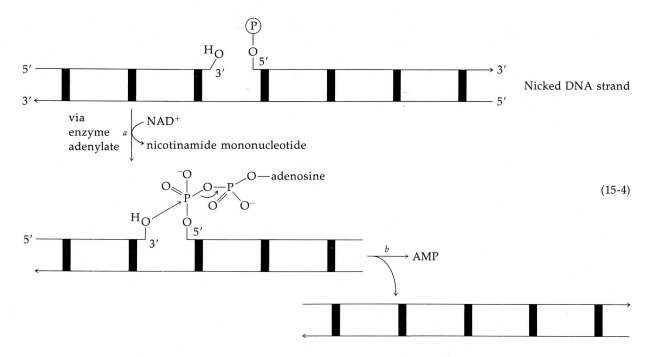

(15-4)

c. The Current Picture of Replication Includes an RNA Primer

Recently, it has been found that a short chain of RNA bound in the form of an RNA–DNA hybrid can serve as the primer for the action of DNA polymerase *in vitro*. This information, together with the appearance of Okazaki fragments and the presence of ligase, has led to a picture of events at a replication fork somewhat as follows: The double-stranded DNA is opened up in a local region, probably with the aid of **unwinding proteins** (see Section E). A short RNA primer, which pairs with the DNA, is synthesized at a special primer region. A DNA polymerase then elongates this RNA chain using deoxyribonucleotide triphosphates to produce the replication fragments. Synthesis occurs alongside both chains and in the directions indicated in Eq. 15-3. The RNA primer ends are then digested away. Gaps in the strand being synthesized are filled by further polymerase action and the nicks are closed by the ligase. According to this mechanism, one strand may be synthesized continuously for its entire length, but the other must be created discontinuously, by the joining of replication fragments. However, in some organisms, both strands may be synthesized discontinuously.

Because much of the newer literature on replication depends heavily on genetic methods, it will be appropriate to defer further consideration of the subject to Section E.

4. Ribonucleic Acids and Proteins

By 1942 it was clear from ultraviolet cytophotometry developed by Caspersson[30] and from cytochemical work of Brachet[31] that RNA had something to do with protein synthesis. Use of radioautography with ^{3}H-containing uridine established that RNA was synthesized in the nucleus of eukaryotic cells and was transported out into the cytoplasm.[32,33]

Ribosomes were discovered by electron microscopists examining the structure of the endoplasmic reticulum of cytoplasm using ultrathin sectioning techniques. Their presence in cells was firmly established by 1956, and the name ribosome was proposed in 1957. Within the next few years isolated ribosomes became the objects of intensive study in the hope of understanding the biosynthesis of proteins. At first it was found difficult to study protein synthesis *in vitro*. No *net* synthesis could be detected. However, Hoagland *et al.*[33a] measured the rate of incorporation of ^{14}C-labeled amino acids into protein. This exceedingly sensitive method permitted measurement of very small amounts of protein synthesis in cell-free preparations from rat liver and paved the way toward studies with ribosomes themselves.

Immediately after the Watson–Crick proposals were made in 1953 it was generally thought that ribosomal RNA (rRNA), which constitutes up to 90% of the total RNA of some cells, carried the genetic message from the nucleus to the cytoplasm. However, by 1960 it seemed unlikely. For one thing, the size and composition of rRNA was very similar for different bacteria, despite large differences in base composition of the DNA (Chapter 2, Section D,8).[34] Furthermore, by this time it was clear that a relatively unstable, short-lived form of RNA must carry the message. Ribosomal RNA, however, was found quite stable.[35]

a. Messenger RNA

Evidence for a labile form of RNA was obtained in 1956 by Volkin and Astrachan[35a] who detected a rapidly labeled RNA in phage-infected bacterial cells. Also very important to the recognition of mRNA were studies of enzyme induction (Chapter 6, Section F,2). It was observed that many bacteria, including *E. coli,* when grown on glucose as the sole source of energy and then suddenly switched to lactose, are unable to utilize the new sugar immediately. However, within a period of 2 min after transfer to lactose the synthesis of new proteins needed for the metabolism of lactose begins. Among these new proteins is a **permease** for lactose and a β-**galactosidase** that cleaves the disaccharide to glucose and galactose. When the lactose is exhausted, the level of the induced enzymes drops almost as quickly. These results

suggested that the RNA that carries the genetic message for synthesis of the new enzymes must be unstable. It must be produced rapidly in response to the presence of the inducing sugar and must disappear rapidly in its absence.

In 1961 F. Jacob and J. Monod postulated a short-lived messenger RNA (mRNA).[36] By this time an abundance of additional evidence supported the proposal. For example, RNA molecules produced after infection of *E. coli* by bacteriophage T4 were found to undergo hybridization (Chapter 2, Section D,10) with denatured DNA of the bacteriophage. Furthermore, this virus-specified mRNA became associated with preexisting bacterial ribosomes and provided the template for synthesis of phage proteins.[37] The experiment provided direct evidence for transcription of mRNA from genes of the viral DNA.

b. Transfer RNA

In 1957 Crick[37a] suggested that special "adapter" molecules might be needed to align amino acids with their codons in the RNA transcript. Crick thought that the adapters might be polynucleotides. At the same time, chemical studies of the RNA of cells were revealing that a low molecular weight RNA made up 15% of the total RNA of *E. coli.* This RNA was recognized in the same year (1957) as constituting the needed adapters when Hoagland demonstrated the enzymatic "activation" of amino acids and their subsequent incorporation into protein. The name **transfer RNA** (tRNA, Fig. 2-24) was proposed.

In recent years an "army" of diligent students of protein synthesis have provided new knowledge of the subject. Rather than attempting to trace the history further, let us now examine some of the details beginning with the transcription of RNA, a process about which a great deal is known.

B. THE TRANSCRIPTION OF RNA MOLECULES

The copying of genetic information from DNA to form mRNA molecules is the initial step in the chain of reactions leading to synthesis of the multitude of proteins needed by cells. As such, it should not be surprising that it is under strict controls.

1. The Operon

The phenomenon of enzyme induction was among those that led Jacob and Monod to the **operon model**[36] for regulation of the transcription of DNA into mRNA. For this work they received the Nobel Prize in 1965. An updated version[38–40] of the Jacob–Monod proposal is shown in Fig. 15-3. The operon is a regulated gene cluster, the one shown in Fig. 15-3 being the *lac* operon of *E. coli*. This operon, found at position 8 min on the genetic map of Fig. 15-1, is probably the most intensively studied group of *E. coli* genes. There are three structural genes coding for the amino acid

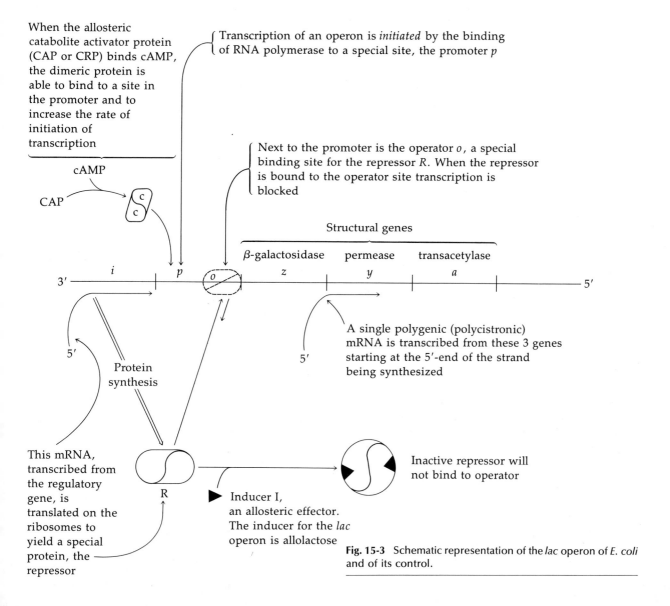

When the allosteric catabolite activator protein (CAP or CRP) binds cAMP, the dimeric protein is able to bind to a site in the promoter and to increase the rate of initiation of transcription

cAMP

CAP

{ Transcription of an operon is *initiated* by the binding of RNA polymerase to a special site, the promoter *p*

{ Next to the promoter is the operator *o*, a special binding site for the repressor *R*. When the repressor is bound to the operator site transcription is blocked

Structural genes

β-galactosidase   permease   transacetylase

3′   *i*   *p*   *o*   *z*   *y*   *a*   5′

5′

Protein synthesis

A single polygenic (polycistronic) mRNA is transcribed from these 3 genes starting at the 5′-end of the strand being synthesized

This mRNA, transcribed from the regulatory gene, is translated on the ribosomes to yield a special protein, the repressor

R

Inducer I, an allosteric effector. The inducer for the *lac* operon is allolactose

Inactive repressor will not bind to operator

Fig. 15-3 Schematic representation of the *lac* operon of *E. coli* and of its control.

sequences of the β-galactosidase (z gene*), permease (y gene), and a transacetylase of uncertain function (a gene). To account for the apparently synchronous control of these three genes, Jacob and Monod proposed that they form a single **transcriptional unit** and that the entire unit is transcribed as a single piece of mRNA.

An operon is distinguished from other transcriptional units in being under the control of a special segment of the DNA molecule located at the beginning (3' end of the template chain) of the operon. The first part of this **control region** is known as the **promoter** (p). The promoter is the site of the initial binding of the RNA polymerase to the DNA, the binding constants for the association being very high. The rates of association and of initiation may be influenced strongly by various special controlling proteins. One, the **cAMP receptor protein** (CAP) is important to the *lac* operon. It also binds in the promoter region (Fig. 15-3).

Immediately adjacent to the promoter is the **operator** (o), a binding site for a **repressor** (R). When the operator is free, transcription is initiated and proceeds through the operator region and on to the genes coding for the three proteins. On the other hand, if the repressor is bound to the operator, transcription is blocked. When the operon model was first proposed the chemical nature of the repressor was unknown. However, in several cases repressors have been identified now as proteins. The well-characterized repressors are all oligomeric proteins able to undergo allosteric alteration. Thus, the *lac* repressor is made up of four identical subunits, each of molecular weight 37,200 and containing 347 amino acid residues. Each subunit has a binding site for the operator and another (allosteric) binding site for an effector (the drawing in Fig. 15-3 is simplified to show only two of the four subunits).

The *lac* operon is ordinarily subject to repression and is activated by the presence of an **inducer.*** Therefore, Jacob and Monod postulated that the free repressor protein binds to the operator. In the presence of an inducer, a conformational change takes place, destroying the affinity of the repressor protein for the operator site. Thus, in the presence of inducer, the operator is not blocked and transcription takes place.[36]

Important to the control of the operon is the **regulatory gene** which codes for the synthesis of the repressor protein. In the case of the *lac* operon, this gene (known as the i gene) is located immediately preceding the *lac* operon (Fig. 15-3). However, for some operons, the regulatory gene is located a considerable distance away. Thus, for the *gal* operon of *E. coli*[41] (coding for synthesis of a number of enzymes concerned with galactose metabolism) located at map position 17 min, the position of the regulatory gene is 61 min.

Regulatory genes are normally transcribed at a very slow but steady rate, presumably because RNA polymerase initiates RNA chains less rapidly at the promoter sites of regulatory genes. Thus, each cell of *E. coli* normally contains only about 10 molecules of the *lac* repressor protein. Because of the importance of repressors in controlling metabolism the regulatory genes are sensitive sites for mutation. For example, a mutation in a regulatory gene may lead to a defective repressor which no longer binds to the operator. In such a case transcription of the operon is uncontrolled and mRNA is produced more rapidly. In such a mutant strain (designated i^- in contrast to the normal i^+ strain) production of the enzyme representing the gene product becomes "constitutive," just as is the formation of the enzymes of the central pathways of metabolism. The latter also appear to be produced regularly in large amounts without control by a repressor. Constitutive **operator mutants** (o^c) are also well known. In these the repressor fails to bind to the operator site in the DNA because of an altered base sequence in the operator region.

---

* The gene symbols used in Table 15-1 and Fig. 15-1 are *lac Z*, *lac Y*, etc. However, the genes are often designated as here by lower case letters.

* While it was long thought that lactose itself was the inducer, it is actually allolactose [β-D-Gal-p-$(1 \rightarrow 6)$D-Glc] that plays this role. In experimental work artificial inducers such as isopropyl-β-D-thiogalactoside are most often used.

a. Nucleotide Sequence of the *lac* Control Region

A remarkable recent accomplishment is the establishment of the nucleotide sequence in the *E. coli* DNA representing the promoter-operator region of the *lac* operon. The sequence includes the end of the *i* gene and the beginning of the *z* gene (Fig. 15-4).[39] Detailed genetic mapping of the region has made it possible to assign the operator and promoter regions with confidence as indi-

cated. Note the series of codons representing the peptide sequence Glu-Ser-Gly-Gln-Stop at the left-hand end. This corresponds to the known C-terminal sequence of the repressor, the *i*-gene product. At the right-hand end the three codons are those of formyl Met-Thr-Met, the known N-terminal sequence of β-glactosidase, the *z* gene product.

A 63 base segment of mRNA transcribed from this region of the *lac* operon has also been se-

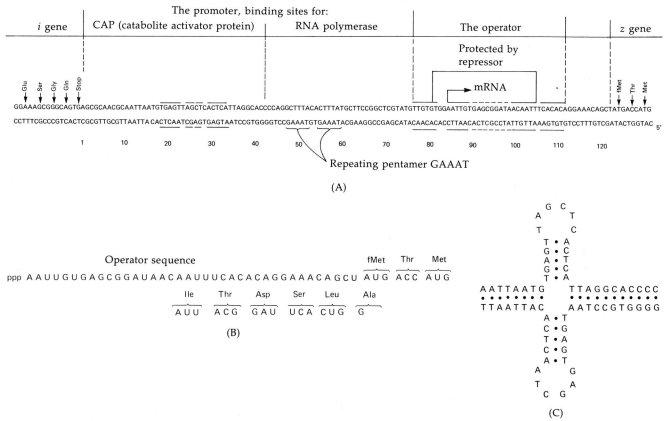

Fig. 15-4 (A) Nucleotide sequence of the *lac* promoter-operator region of the *E. coli* chromosome.[43] The proposed locations of the *i* gene, the promoter, which contains CAP and RNA polymerase binding sites, the operator, and the beginning of the *z* gene (β-galactosidase) are shown. Note the two regions of local 2-fold rotational symmetry which are marked by bars and central dots representing the 2-fold axes of rotation and the repeating pentamer. (B) The sequence of an mRNA molecule initiated in the *lac* promoter-operator region of a mutant strain of *E. coli* with an altered promoter.[46,47] The peptide initiation site is identified by the symbol fMet and the successive amino acids from the known N-terminal sequence of β-galactosidase have been matched with the codons.[46] (C) Cruciform structure that could arise by the looping out of the complementary strands of the DNA in the CAP-binding site.

quenced.[42] The mRNA transcript begins in the operator region* as shown in Fig. 15-4. The initiation codon for the z gene is 39 bases from the end. The fragment sequenced codes for 8 amino acids of the galactosidase.

The operator region was sequenced[43] by digesting the DNA with deoxyribonuclease in the presence of the repressor protein. The bound repressor protected a region of 27 base pairs as indicated in the figure. A striking fact is that the operator is centered on a region of local twofold rotational symmetry (Chapter 2, Section D,11). Thus, as in Figs. 2-30 and 15-4 it is possible for the DNA to loop out into a **cruciform structure.**[44] Such a structure might reasonably be expected to bind more easily to the tetrameric repressor protein than would the linear form.

A change from a linear helical duplex to cruciform structure would require a substantial unwinding of the helix and would mean that negatively superhelical DNA molecules should bind repressor much more tightly than DNA without superhelical turns (Chapter 2, Section D,9). Experimental measurements indicate only a 40°–90° unwinding by the *lac* repressor. Thus, it probably binds to a helical duplex with a relatively small amount of distortion.[45] Nevertheless, the twofold symmetry is probably important in providing tight binding to two subunits of the symmetric tetrameric protein. It is also possible that repressor molecules move along DNA chains in a one-dimensional diffusion process and that the symmetry of the operator site facilitates recognition by a protein moving from either direction.

b. Positive Control

Cyclic AMP (Chapter 7, Section E,8) is an important controlling substance in bacterial cells. An example of a phenomenon that may be mediated by cAMP is **catabolite repression.** This is the inhi-

---

* In fact, the mRNA sequenced came from a strain containing a mutation in the promoter region. It is believed that the initiation site of the mRNA in wild type *E. coli* may lie further to the left.

bition of the transcription of genes for enzymes needed in catabolism of lactose and other energy-yielding substrates when the more efficient energy source glucose is present. Although the mechanism is unknown, the presence of glucose causes a decrease in the level of cAMP.

It is established that cAMP stimulates the initiation of transcription of many operons in bacteria. This is accomplished through the mediation of a special binding protein, commonly abbreviated CAP or CRP (catabolite activator protein or cAMP receptor protein).[46] The CAP–cAMP complex apparently binds to the promoter adjacent to the RNA polymerase site. The exact site of this binding is thought to be marked by a second local center of approximate twofold symmetry[39] in the DNA as is indicated in Fig. 15-4. The CAP molecule is a dimer of two identical subunits (MW ~22,000) of a symmetry appropriate to binding at this site. An important question remaining is "How does the binding of the CAP–cAMP complex increase the rate of initiation of mRNA transcription?" The question will be considered later after a discussion of the properties of RNA polymerases.

Another type of positive control is observed with the arabinose (*ara*) operon (at map position 1 min in the *E. coli* chromosome). In this case, an inducer not only causes detachment of the repressor from the operator site but also converts the repressor to an activator that functions like the CAP–cAMP complex to cause more effective initiation of transcription.

c. Feedback Repression

A simple modification of the operon model as presented in Fig. 15-3 accounts for feedback repression (Chapter 6, Section F,2) by end products of biosynthetic sequences. In such cases, some product, e.g., an amino acid, binds to the aporepressor causing an allosteric modification. In this case, it is not the aporepressor but the **effector–repressor complex** that binds to the operon and serves to turn off transcription.[47] The corepressors for certain amino acid operons, e.g., those for synthesis of histidine and valine, appear to be the aminoacyl-tRNA molecules derived from those amino acids.[48,49] In the case of the tryptophan operon of E. coli, a tryptophan–repressor complex binds at the operator to prevent transcription. However, another mechanism of regulation has also been found.[50] A "leader region" of about 160 nucleotides (of known sequence) precedes the gene transcript in the mRNA. Within the corresponding DNA sequence are two palindromes which are proposed as "attenuator" sites. If these sites are occupied by an appropriate protein, premature termination occurs and the five genes that follow in the operon are not transcribed.[50]

2. RNA Polymerases and Their Regulation

RNA polymerases are the enzymes that carry out transcription of the genetic message from a DNA chain. Since DNA exists in a double-stranded form within cells, we must ask how formation of the single-stranded RNA can be accomplished using a double-stranded template. Part of the answer comes from the fact that purified RNA polymerases can also synthesize RNA from the four ribonucleoside triphosphates using single-stranded DNA as the template. This suggests that the basic mechanism of transcription, like that of DNA replication, involves base pairing. In line with this conclusion is the fact that the single-stranded DNA obtained from bacteriophage ϕX174 (Box 4-C) is converted by RNA polymerases into a double-stranded RNA–DNA hybrid molecule. However, when double-stranded DNA serves as the template free single-stranded RNA is produced. Thus, it would appear that at the site of the polymerase action the double-stranded DNA is momentarily pulled apart into single strands and that one of these is copied by the polymerase.

A striking fact is that within cells only one of the two strands of the DNA double helix is normally transcribed within a given genetic region. This was clearly demonstrated using the double-stranded "replicating" form of ϕX174 DNA. This is a circular molecule of DNA, one strand of which is actively transcribed within cells of E. coli infected with this virus. When the mRNA so transcribed is isolated from the bacteria, it is found that it will not hybridize with the single-stranded ϕX174 DNA, the kind of molecules packaged in the mature virus particles. This DNA strand is usually described as the "plus" strand. Thus, it appears that the transcribed RNA is also a plus strand and therefore must have been transcribed from the "minus" strand of DNA in the replicative form. This conclusion is confirmed by the fact that the isolated RNA readily hybridizes with denatured DNA from the double-stranded replicative form, the latter containing both plus and minus DNA strands.

RNA polymerases are large protein molecules, some of which are made up of subunits.[40,51–53] Curiously, the eukaryotic mitochondrial RNA polymerase may contain only a single subunit of MW = 64,000 and be less complex than that found in bacteria. Bacterial viruses sometimes induce their own RNA polymerases and these, too, may be monomeric. Thus, phage T7 induces an RNA polymerase of MW = 100,000 which appears to contain only a single peptide chain. However, the most studied RNA polymerase is that from E. coli, an oligomeric molecule of MW ~500,000 consisting of four kinds of subunits α, β, β', and σ and containing bound Zn^{2+}. The composition of the oligomer is $\alpha_2\beta\beta'\sigma$. The molecular weights are α, 39,000; σ, 86,000; β, 155,000; and β', 165,000. Of these, σ plays a unique role in initiation of transcription. It is required for the correct selection of promoter sites but is not needed for elongation of an RNA chain once synthesis has started. At least three different classes of RNA polymerases are found in eukaryotic cells. Polymerase I (or A) is found in the nucleolus and presumably transcribes pre-rRNA (Section B,3) genes. Polymerases II (or B) and III (or C) are found in the nucleoplasm proper. The type II enzymes are believed to transcribe most genes while the type III enzymes transcribe genes for the transfer RNA's. Like the bacterial polymerase, eukaryotic RNA polymerases are complex oligomers.[53,53a]

a. Initiation of Transcription

A current model of initiation assumes that RNA polymerase repeatedly associates with and dissociates from DNA at random sites until a promoter region is located. The enzyme is thought to "recognize" the promoter through specific interactions with bases in the major groove of the DNA helix (see Fig. 2-23). It has been calculated that a minimum of about 12 base pairs would be required to construct a unique recognition sequence that would be unlikely to arise by chance in the *E. coli* chromosome.[40]

The initial specific polymerase-promoter complex has been referred to as a **closed complex** because it is thought that the bases in the DNA chain are all still paired. It is postulated that the closed complex is in equilibrium, through a large conformational change, with an **open complex** which is ready to initiate mRNA synthesis.[39,40] In the open complex the hydrogen bonds holding together the base pairs have been broken and the bases of the template chain are available for pairing with incoming ribonucleotide triphosphates.

Turning again to Fig. 15-4A, note the sequence GAAATGTGAAAT (in the lower chain). The sequence does *not* contain a region of local symmetry but it does contain the repeating pentamer GAAAT two times. Lying at the center of the presumed polymerase binding region, it may represent the recognition region of the promoter. The asymmetric structure may be appropriate for the binding of a protein whose function is to move in a specified direction. Furthermore, the region is "AT-rich," meaning that the opening of the helix at this point would be easier than in a GC-rich region (Chapter 2, Section D,6). Thus, it may represent the point of entry of RNA polymerase to form the open complex.[39]

Dickson *et al.* proposed that the CAP–cAMP complex facilitates entry at the adjacent promoter site, possibly by causing destabilization of the GC-rich region between the CAP and polymerase binding sites.[39] Once an open complex is formed it

may be necessary (as is suggested by Fig. 15-4) for the polymerase to migrate a short distance along the DNA chain to the starting site for the mRNA.

An RNA chain is initiated by reaction of either ATP or GTP with a second ribonucleotide triphosphate (Eq. 15-5) to form a dinucleotide still

$$\text{ATP(GTP)} + \text{XTP} \longrightarrow \text{PP}_i + \text{pppPupX} \quad (15\text{-}5)$$

bearing a triphosphate at the 5' end. Further addition of nucleotide units at the 3' end by the same type of reaction leads to rapid transcription at a rate of ~ 50 nucleotides s^{-1} at 25°C (this is about one-thirtieth the rate of replication).

b. Effects of Antibiotics

The antibiotic **rifamycin** appears to interfere with initiation by competing for the binding of the initial purine nucleoside 5'-triphosphate (Box 15-A). The same bacterial RNA polymerase that synthesizes mRNA also transcribes both rRNA and the tRNA's. Thus, the synthesis of all forms of RNA is inhibited by rifamycin. When a population of bacteria is subjected to this antibiotic, certain individuals survive. These rifamycin-resistant mutants are no longer sensitive to the antibiotic. Among the resistant mutants are some that produce an RNA polymerase containing an altered β subunit. Since the mutant polymerases do not bind rifamycin, it is concluded that rifamycin binds to the β subunit and that the rifamycin-resistance gene *rpoB* or *rif* (which maps at 89 min, Fig. 15-1) is the gene for the β subunits of RNA polymerase.

A number of other antibiotics also interfere with transcription. Streptolydigin inhibits elongation of RNA chains as well as initiation. **Actinomycin D** inhibits both DNA polymerases and RNA polymerases, the latter at a concentration of only $10^{-6} M$ (Box 15-B). The eukaryotic RNA polymerases are not inhibited by rifamycin, but RNA polymerases II and III are completely inhibited by the mushroom poison **α-amanitin** (Box 15-C).

c. Nucleotide Selection by Base Pairing

How do cells achieve a high degree of accuracy in the selection of the correct base during the pro-

BOX 15-A

**THE ANTIBIOTICS RIFAMYCIN
AND RIFAMPICIN[a]**

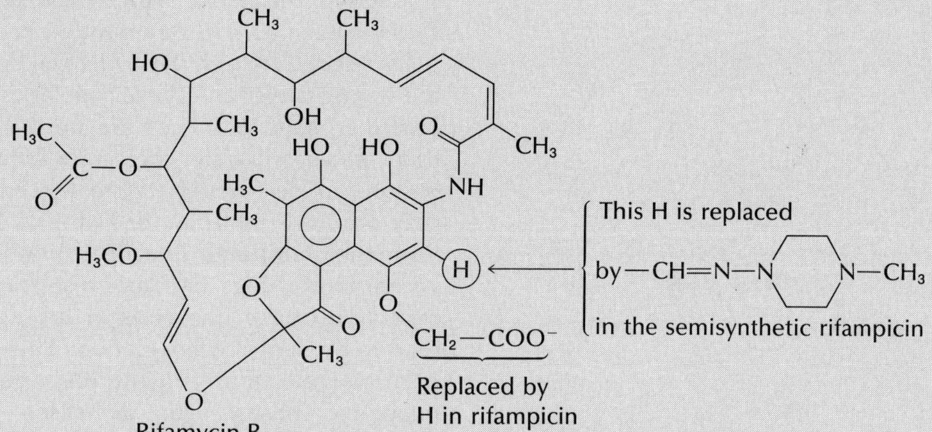

Rifamycin B

This H is replaced
by —CH=N—N   N—CH₃
in the semisynthetic rifampicin

Replaced by
H in rifampicin

Rifamycin, an antibiotic produced by *Strepto-myces mediterranei,* is of medical value because it affects **acid-fast** as well as gram-positive bacteria. The semisynthetic rifampicin has been especially useful in the treatment of tuberculosis. Note the ring at the right in the structural formula. The ether linkage at the bottom is cleaved and the resulting hydroquinone is oxidized to a quinone within the bacteria.

Rifampicin is an extremely effective inhibitor of bacterial RNA polymerase, a 50% inhibition being observed with a $2 \times 10^{-8} M$ concentration of the antibiotic. Rifampicin does not prevent the bind-ing of polymerase to DNA but inhibits initiation of transcription. Mutants of *E. coli* resistant to rifampicin (*rif* gene,) produce RNA polymerase whose β subunit has been altered (sometimes with a change in electrophoretic mobility). The related antibiotic **streptolydigin** also binds to the β subunit of RNA polymerase and blocks elongation. Resistant mutants map very close to *rif* mutants.

[a] I. H. Goldberg and P. A. Friedman, *Annu. Rev. Biochem.* **40,** 775–810 (1971).

cesses of replication, transcription, and precise codon–anticodon matching during protein synthesis? Early workers often attributed the specificity in base pairing entirely to the strength of the two or three hydrogen bonds formed together with the stabilization provided by the adjacent helix. However, the free energy of formation of the base pairs is small (Chapter 2, Section D,6) and the additional free energy of the binding to the end of an existing helix is insufficient to account for the specificity of pairing. Our present knowledge of enzymology suggests that the enzyme itself plays a major role in ensuring correct pairing.

RNA and DNA polymerases are rather large molecules. Thus, the binding site on the enzyme can completely surround the double helix. This being the case, it is easy to imagine a selection process such as that indicated in Fig. 15-5. The figure shows a guanine ring of the template strand of a DNA molecule imagined to be at the point where the complementary strand (either of DNA or of RNA) is growing from the 3′ end. The proper nucleoside triphosphate must be fitted in to form the correct GC base pair before the displacement reaction takes place to link the new nucleotide unit to the growing chain. Let us suppose that the en-

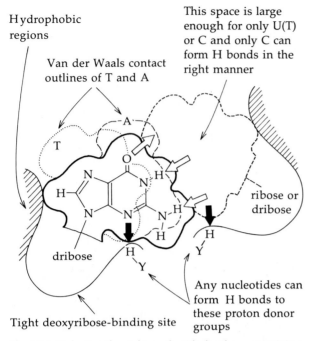

Fig. 15-5 Selecting the right nucleotide for the next unit in a growing RNA or DNA chain. A deoxyguanosine unit of the template chain is shown bound to a hypothetical site of a DNA polymerase.

zyme possesses binding sites for the deoxyribose unit of the template nucleotide and for the sugar unit of the incoming nucleoside triphosphate and that the two binding sites are held at a fixed distance one from the other. As indicated in Fig. 15-5, some group H–Y might also be present at each binding site to hydrogen bond to the nitrogen or oxygen indicated by the heavy black arrows. Note that all four of the bases can form hydrogen bonds of this type in the same position relative to the deoxyribose or ribose rings. We might further imagine a hydrophobic region (indicated by the hatched area in the figure) providing additional stabilization. With such an arrangement, it is easy to see how the correct nucleoside triphosphate can be selected no matter which one of the four bases occupies the binding site on the left side of the figure. (The outlines of the thymine and adenine rings have been drawn in with dotted and dashed lines, respectively.)

The significant thing to note in Fig. 15-5 is that if a purine is present on the left side, as is shown in the drawing, there is room on the right side only for a pyrimidine ring. Thus, A and G are excluded and the choice is only between C and U (or T). However, U will be excluded because the dipoles needed to form the hydrogen bonds point in the wrong direction. These dipolar groups are hydrated in solution. They are unlikely to give up their associated water molecules unless hydrogen bonds can be formed within the base pair. Not only would a molecule of U (or T) be unable to form the stabilizing hydrogen bonds within the vacant site of Fig. 15-5, but also the electrostatic repulsion of the like ends of the dipoles would tend to prevent the association. This would result in a lowered affinity of the RNA polymerase for mispaired bases. This decreased affinity (increased apparent K_m) has been observed experimentally, at least in the case of a DNA polymerase from bacteriophage T4. Mutants of this DNA polymerase are also known. One, a "mutator" polymerase, makes many more pairing mistakes because it does not discriminate as much in the binding process. Likewise, an "antimutator" polymerase makes fewer mistakes than the wild type. It binds proper pairs even more tightly relative to noncomplementary pairs than does the wild type enzyme.[53b] The matter is discussed further by Drake and Baltz.[53c]

Base pairs possess interesting tautomeric properties, and it is attractive to think that advantage may be taken of these to verify that a proper base pair is formed and to send some kind of signal to the active site of a polymerase where the formation of the new phosphodiester linkage takes place. A purely hypothetical way in which this could occur is indicated in Fig. 15-6. Here the group H–Y of Fig. 15-5 is depicted as an imidazole group which is hydrogen bonded to other protein groups able to undergo tautomeric shifts (Chapter 6, Section E,5,f). As indicated by the curved arrows, an electron pair could move from the ring nitrogen on the right side, perhaps under the influence of the approach of some negatively charged group of a protein. If the hydrogen bond

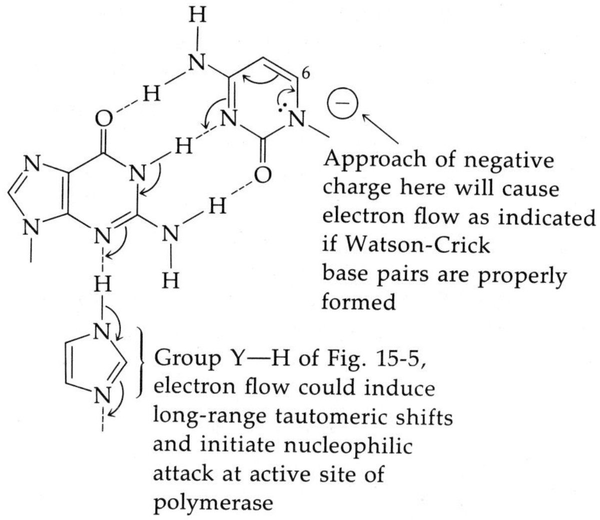

Approach of negative charge here will cause electron flow as indicated if Watson-Crick base pairs are properly formed

Group Y—H of Fig. 15-5, electron flow could induce long-range tautomeric shifts and initiate nucleophilic attack at active site of polymerase

Fig. 15-6 Hypothetical scheme by which an electronic signal might be sent through a base pair to initiate the reaction at the active site of a nucleic acid polymerase.

pairs are formed correctly, the concerted flow of electrons could take place across the base pair and out into group H–Y and beyond through the postulated tautomeric chain. If the base pair were not correctly formed, the signal could not be transmitted (except during an occasional mispairing with a minor tautomer, Chapter 2, Section D,7). Note that another reciprocal electron transfer in the opposite direction to that shown in the figure is also possible through the same base pair. Similar tautomeric shifts are possible for all legitimate base pairs. It is also possible that initiation of a signal of the type shown could occur by the addition of some nucleophile to a purine or pyrimidine ring, e.g., to C-6 of the cytosine ring in Fig. 15-6. In the case of ribosomes, such electronic signals could be passed in turn through each of the base pairs involved in codon–anticodon recognition, and they could also be transmitted through other base pairs formed within loops of ribosomal RNA. While little is usually said about it, there is a possibility that the purine and pyrimidine bases both of nucleic acids and of coenzymes are the sites of active electronic alterations during the course of their function.

Bear in mind that the foregoing ideas are speculations and may be quite incorrect. It has been pointed out that the triphosphate of the cytotoxic nucleoside deazanebularin is incorporated into RNA by RNA polymerase and that it replaces (poorly) *either* adenosine or guanosine.[54] Deazanebularin can form one H bond to U, but it cannot

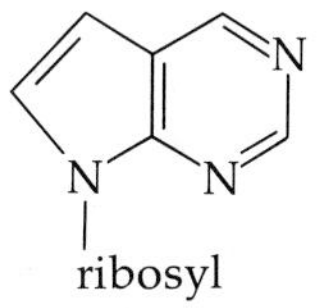

ribosyl
Deazanebularin

form any H bonds with the normal tautomer of C. This has been interpreted to mean that only steric factors operate in base selection and that H bonds are not essential. It is hoped that the reader will examine this argument and develop her/his own thoughts about the mechanism of base selection, one of the most fundamentally important of all biological processes.

d. Termination of Transcription

Encoded in DNA are not only the initiation signals for transcription but also **termination signals.** Termination of RNA chain elongation is poorly understood at present. Some termination signals are sensed by the bacterial RNA polymerase itself. The "reading" of others requires auxiliary proteins. One of these may be the **rho factor** (ρ) which induces termination of RNA chains *in vitro*.[55] The ρ factor of E. coli appears to be a protein of MW = 200,000 with ATPase activity.[55a]

e. Processing of mRNA

The end result of the action of RNA polymerase in bacteria is to produce a series of mRNA molecules of variable length, some corresponding to polycistronic (polygenic, see Section D,3) and some to monocistronic operons. Most of the mRNA molecules produced are unstable with an average lifetime of about 2 min; however, some, such as those produced in bacteria about to un-

BOX 15-B

ACTINOMYCIN D, A TOXIC ANTIBIOTIC

The actinomycins, antibiotics from *Streptomyces*, not only kill bacteria but also are among the most potent antitumor agents known.[a] Because of their extreme toxicity they are of little value in treatment of cancer. Nevertheless, actinomycin is an important tool for biochemists because of its highly specific inhibition of RNA polymerases.

Actinomycin D (actinomycin C1) contains a planar phenoxazone chromophore bearing two carboxyl groups. To each of the latter is linked an identical cyclic peptide made up of L-threonine, D-valine, L-proline, sarcosine (*N*-methylglycine), and L-methylvaline.

Box 15-B (*Continued*)

An ester linkage joins the methylvaline residue of the peptide back to the side chain hydroxyl of threonine. Two cis peptide linkages are present. Ignoring the obvious structural differences on the two sides of the phenoxazine ring, actinomycin possesses approximate twofold symmetry. Biochemical experiments have shown that the antibiotic binds tightly to double-stranded DNA in regions containing guanine.

A 2:1 deoxyguanosine–actinomycin complex has been crystallized, and the structure has been determined by X-ray diffraction.[b–d] The phenoxazine ring lies at the center of the complex, one peptide loop extending above it and the other below it (at the right side of the drawing. The peptide structure can be traced most easily for the upper loop). The twofold symmetry is present in the dideoxyguanosine complex as well as in actinomycin itself. The phenoxazine ring lies between the two flat guanosine rings in van der Waal's contact. Note that the two amino groups of the guanine rings form strong hydrogen bonds with the carbonyl groups of the threonine residues. There are also weaker, nonlinear hydrogen bonds from the N-3 atoms of the guanines to the NH groups of the same threonines. A symmetric pair of hydrogen bonds join the two carbonyl and NH groups of the D-valine residues in the peptide loops.

Model building studies show that a similar complex can be formed with double-stranded DNA. Note that while the two amino groups of the guanine rings in the drawing are hydrogen-bonded to the actinomycin, the other hydrogen atoms of the same amino groups as well as the N-1 hydrogen atoms and the carbonyl groups of the guanine ring are available for hydrogen bonding to form a GC base pair. Thus, the structure above can be modified readily into part of a double-stranded DNA molecule in which the phenoxazone ring of actinomycin is intercalated between two CG pairs (Fig. 2-21). To do this the normal DNA structure has to be unwound by 18° at the point of insertion of the extra ring.

The significance of the crystal structure determination for the actinomycin–deoxyguanosine complex lies first in the graphic demonstration of the reality of intercalation of planar rings into nucleic acid molecules. It also demonstrates how a special reagent possessing approximate twofold symmetry can selectively bind at points in the DNA molecule that also have twofold symmetry.

The research has medical implications. Understanding the mode of action of actinomycin as a toxin, we may be able to modify the molecule chemically to lower its toxicity for normal cells while retaining toxicity to tumor cells.[b] Another possibility, suggested by Sobell, is that a modified actinomycinlike antibiotic might be designed to specifically attack double helical RNA of viruses.

[a] D. Perlman, *in* "Medicinal Chemistry" (A. Burger, ed.), 3rd ed., Part I, pp. 309–316. Wiley (Interscience), New York, 1970.

[b] The structure, as sketched here, is slightly distorted from the computer-generated drawings of Sobell to show the bonding arrangement more clearly. H. M. Sobell, *Prog. Nucleic Acid Res. Mol. Biol.* **13**, 153–190 (1973).

[c] H. M. Sobell, S. C. Jain, T. D. Sakore, and G. Ponticello, *Cold Spring Harbor Symp. Quant. Biol.* **36**, 263–270 (1971).

[d] H. M. Sobell, *Sci. Am.* **231**, 82–91 (Aug 1974).

dergo sporulation, are much longer lived. Messenger RNA may also undergo processing before it reaches the ribosomes. For example, following infection of *E. coli* cells by phage T7, a special ribonuclease cleaves the large (MW = 1.8×10^6) "early RNA" transcript (see Section D,8) from the virus DNA into five defined fragments.[56] Each fragment presumably carries the message for a single viral gene. Processing of messenger RNA in eukaryotic cells (Section B,5) is considerably more complex than in bacteria.

3. Ribosomal RNA

Quantitatively, the most important RNA, making up 90% of that present in cells, is ribosomal RNA. Synthesis of rRNA must be rapid for an *E. coli* cell produces 5–10 new ribosomes per

BOX 15-C

A POWERFUL POISON FROM MUSHROOMS: α-AMANITIN

α-Amanitin

Several deadly species of the genus *Amanita* produce colorless, toxic octapeptides, the **amanitins**.[a] The very toxic α-amanitin contains two residues of glycine, one of L-isoleucine, one of the unusual L-dihydroxyisoleucine, one of L-asparagine, and one of L-hydroxyproline. In the center a modified tryptophan residue has been oxidatively combined with an SH group of a cysteine residue. The LD_{50} for mice is 0.3 mg kg^{-1} (50 g of fresh *Amanita phalloides* may be sufficient to kill a person). Amanitins act slowly and it is impossible to kill mice in less than 15 h, no matter how high the dose. If the dihydroxyisoleucine residue of α-amanitin is replaced with unhydroxylated leucine the resulting compound, known as amanullin, is completely nontoxic.

α-Amanitin completely blocks transcription by eukaryotic RNA polymerases II and III. Polymerase II appears to be the major nuclear RNA polymerase, and its inhibition blocks virtually all protein synthesis by the cell. Note that the amanitin molecule is semisymmetric overall, much as is actinomycin (Box 15-B), with an aromatic group protruding from behind in the center. Perhaps the amanitins also fit to a symmetric site in the polymerase or in the polymerase-DNA complex.

The same mushrooms contain a series of fast-acting toxic heptapeptides, the **phalloidins**, whose structures are similar to those of the amanitins. However, they contain a reduced sulfur atom (—S—) in the cross-bridge. The mode of action of the phalloidins is unknown. The same mushrooms also contain an antidote to the phalloidins, **antamanide**. This cyclic decapeptide, like

$$\text{Pro} \rightarrow \text{Ala} \rightarrow \text{Phe} \rightarrow \text{Phe} \rightarrow \text{Pro}$$
$$\uparrow \qquad\qquad\qquad\qquad \downarrow \quad \text{antamanide}$$
$$\text{Pro} \leftarrow \text{Val} \leftarrow \text{Phe} \leftarrow \text{Phe} \leftarrow \text{Pro}$$

the toxins, is made up entirely of L-amino acids. It apparently competes for the binding site of the phalloidins. Unfortunately, it is of little value in treating cases of mushroom poisoning.

Antamanide is a specific sodium-binding ionophore (Fig. 5-5).

[a] T. Wieland and O. Wieland, *Microb. Toxins* **8**, 249–280 (1972).

second, or 2×10^4 molecules of RNA per generation. Ribosomes contain three pieces of RNA. In bacteria they have the sedimentation constants, molecular weights, and numbers of nucleotides given in the accompanying tabulation.

Sedimentation constant (S)	MW	No. of nucleotides
23	1.1×10^6	3300
16	0.56×10^6	1700
5	41,000	120

All three pieces appear in cells as larger precursor molecules (pre-rRNA) with extra nucleotide sequences at both the 3' and 5' ends. For *E. coli* two precursors of the 5 S RNA have been identified, one containing 150 nucleotides and the other 180.

It is now believed that there are about six rRNA regions in the *E. coli* chromosome. Each region consists of a single transcriptional unit containing one gene each for 16 S, 23 S, and 5 S rRNA in that order. A single transcript (which can, in certain mutant strains, appear as a 30 S molecule) is cut by an endonuclease (RNase III) into the smaller pre-rRNA molecules.[57,58]

A remarkable accomplishment has been the electron microscopy or active regions of chromosomes of growing *E. coli* cells. Chains believed to be 16 S and 23 S RNA have been seen in the process of formation[59,59a-c] (Fig. 15-7).

4. Transfer RNA

The most intensively studied forms of RNA are the small 4 S tRNA molecules of MW ~26,000 and consisting of 75 ± 5 nucleotides (Figs. 2-24 and 15-8). The size and the basic structures seem to be the same in bacteria and eukaryotic cells. The need for "adapters" to carry amino acids to the proper positions along the mRNA template had been predicted prior to the discovery of tRNA. It had been expected that there would be a base sequence constituting an **anticodon** which would fit against the proper codon at some binding site on the protein-synthesizing machinery. This is just

what tRNA molecules do, but their chemistry contained many surprises. The molecules are longer than seems necessary for the formation of adapters and many bases are modified greatly from their original form.[60]

Another surprise about the structure of tRNA molecules is that the anticodons are not all made up of "standard" bases. Thus, hypoxanthine (whose nucleoside is inosine) occurs in some anticodons.

The conventional "cloverleaf" representation of a tRNA molecule is shown in Fig. 15-8A. The actual three-dimensional structure, determined independently by two groups,[61,62] is closely related as is indicated in Fig. 15-8B.

Note the four hydrogen bonded "stems" of the structure in Fig. 15-8. One stem terminates in the **amino acid acceptor end** which carries an activated amino acid generated as in Eq. 11-2. The other three stems terminate in loops which usually contain a large number of modified bases. The **dihydro U loop** contains 5,6-dihydrouridine in various amounts and in varying positions. The **anticodon loop** always contains the anticodon directly opposite the amino acid acceptor end in the cloverleaf drawing. On the 5' side of the anticodon, there is always a U (circled in the drawing) followed by another pyrimidine. Next to the 3' side of the anticodon there is usually a "hypermodified" base. The **TψC loop** contains the specific nucleotide sequence for which the loop is named. While the TψC sequence has been found in all bacterial tRNA's involved in protein synthesis (for which sequences have been determined) it may be replaced by UCG in eukaryotic **initiation tRNA** molecules.[63-65]

The genes for tRNA molecules in both bacteria and mammalian cells are grouped in clusters which are transcribed as large precursor RNA molecules, sometimes containing more than one kind of tRNA.[66-69] At least three different nucleases are needed for the "cutting and trimming" to form the mature tRNA molecules.[57,66,69]

In the case of one tyrosine tRNA* of *E. coli*, the

---

* The minor tyrosine tRNA encoded by supressor gene *supF* (Section D,6).

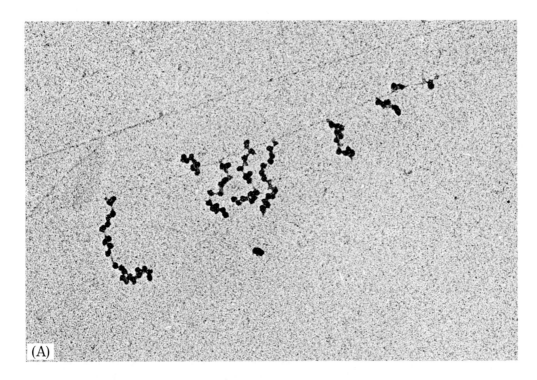

(A)

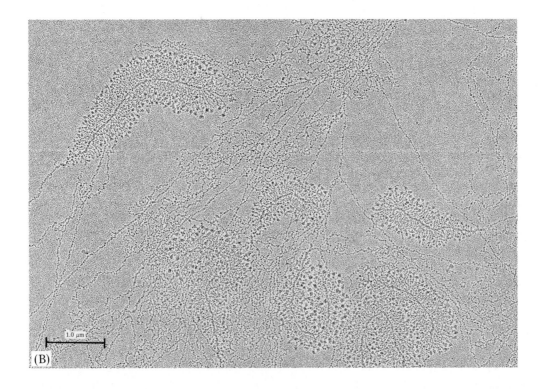

(B)

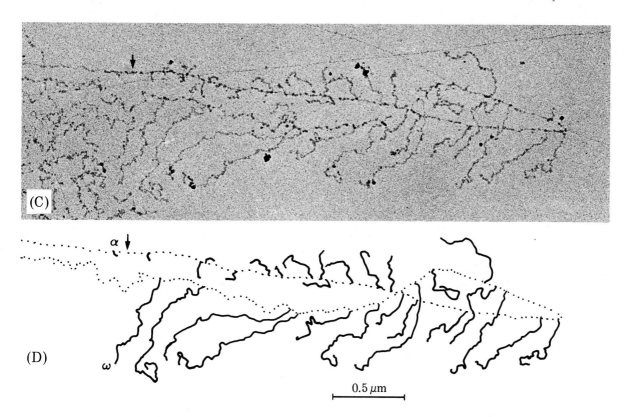

Fig. 15-7   (A) Electron micrograph showing transcription from an unidentified operon in *E. coli*. Note the DNA duplexes (horizontal) and the mRNA chains with ribosomes attached. The mRNA chains are shorter at the right side where transcription begins and larger to the left where transcription has proceeded for a longer time. From O. L. Miller *et al.* *Science* **169**, 392–395. Copyright 1970 by the American Association for the Advancement of Science. (B) Ribosomal RNA genes from an embryo of *Drosophila melanogaster* (the fruit fly) in the process of transcription. A fully activated transcriptional unit is seen in the upper left. The shorter matrices below are not yet mature. Note the densely packed ribonucleoprotein strands containing increasing lengths of the transcribed rRNA. Also note the characteristic granular knobs at the tips of the strands. From McNight and Miller.[59c] (C) Electron micrograph of a nonribosomal transcription unit of an embryo of the milkweed bug *Oncopeltus fasciatus*. (D) Interpretative drawing of the micrograph in (C). Dotted line represents chromatin and the solid lines ribonucleoprotein fibers. Careful measurement showed that the lengths of these 29 fibers increase from the shortest (marked α) to the longest (marked ω) in proportion to the distance along the chromatin strand. The least squares line fitted to the plot of fiber length versus distance intersects the x axis at the point corresponding to the positions of the arrows in (C) and (D). Transcription is thought to begin at about this point and to continue for a distance of $\sim 6.0\ \mu m$ to the point of termination. (The longest fiber in the micrograph is 5.8 μm from the origin.) The length of DNA (B form) in this length of partially folded chromatin is estimated as 9.6 μm corresponding to $\sim 21,000$ base pairs. From Foe *et al.*[59a] Courtesy of Charles D. Laird.

immediate precursor (Fig. 15-9) contains 129 nucleotides, 44 more than in the mature tRNA. Of these extra nucleotides, 41 are on the 5′ side and three on the 3′ end. A special nuclease (RNase P) cleaves only one bond in the precursor, removing the 41 nucleotide fragment from the 5′ end. Another nuclease (RNase PIII) acts to remove the

three-nucleotide piece from the other end.[66] Additional enzymes appear to degrade the fragments removed.[70]

An important chemical achievement is the synthesis by H. G. Khorana and associates of the double-stranded DNA segment coding for this *E. coli* tyrosine tRNA[71] and its precursor.[71a] The work

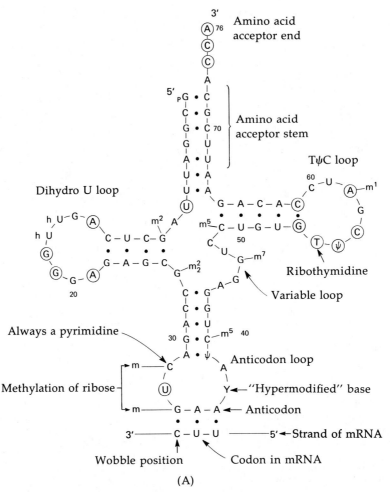

Fig. 15-8 (A) Schematic cloverleaf structure of phenylalanyl transfer RNA of yeast. The dots represent pairs or triplets of hydrogen bonds. Nucleosides common to almost all tRNA molecules are encircled. Other features common to most tRNA molecules are also marked. The manner in which the anticodon may be matched to a codon of mRNA is indicated at the bottom.

has been extended to include the gene termination region which lies beyond the CCA end of tRNA. A DNA sequence of 23 nucleotides beyond this end was determined[72,73] and is shown in Fig. 15-9. Two noteworthy features appear. A local center of twofold rotational symmetry (indicated by vertical bars and a central dot in Fig. 15-9), which may serve as a termination signal, is present. In addition, there are two places in which a short sequence is repeated in the inverse direction, e.g.,

TGAAAGT. Whether or not these represent a new type of genetic signal is uncertain.* The 29-

---

* The word palindrome (Chapter 2, Section D,11) is often used to describe regions of local twofold symmetry in a DNA duplex.[74] Unfortunately, Kornberg[73] (p. 376) and others have used the word for the inversely repeated sequences in a single DNA chain. While the latter may *appear* to read the same in both directions, they will not be read the same by any enzymatic machinery moving along the DNA duplex (nor along a single strand).

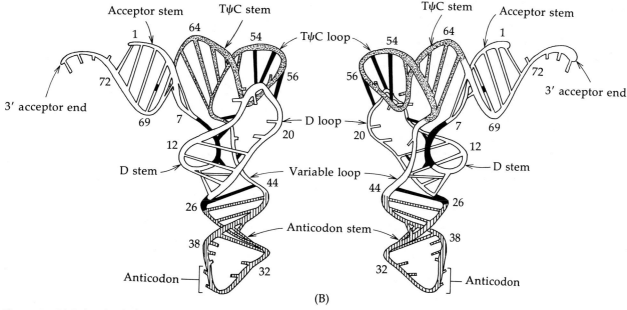

Fig. 15-8 (B) A drawing indicating the three-dimensional structure of yeast phenylalanine tRNA. From Kim *et al.*[61]

nucleotide sequence *preceding* the tyrosine tRNA gene contains the operator, whose sequence is also known.[75]

In addition to the process of "trimming" a precursor by nuclease action, extensive modification of purine and pyrimidine bases is required to generate a mature tRNA.[64] Sixty or more modification reactions are known, the number and the extent of modification depending upon the species. Structures of some of the modified bases are indicated in Fig. 15-10. As indicated by the structure for uridine in that figure, a number of modifications are possible. One of the most common is methylation, which can occur either on the base or on the 2'-hydroxyl of the sugar. Methylation at the 5 position of uridine yields **ribothymidine.** Cytidine can be modified in the same position. Reduction of the 5,6 double bond of uridine gives dihydrouridine. Replacement of the oxygen at position 4 by sulfur gives rise to **4-thiouridine.** Positions in the guanosine structure that can be methylated are also indicated in Fig. 15-10. The symbol m is commonly used to designate methyl-

ation in nucleic acid bases (m^2 indicated dimethylation).

A remarkable transformation is that of uridine into pseudouridine (ψ).

Pseudouridine (ψ)

Pseudouridine is formed by rearrangement of uridine in the original transcript, but the chemistry is unclear. Note that pseudouridine can form a base pair with adenine in the same manner as does uracil. A base designated Y is a highly modified guanine. Two hypermodified adenosines are shown in Fig. 15-10. The N^6-isopentenyl-adenosine is found at the 3' end of the anticodons

Fig. 15-9 Sequence of an *E. coli* tyrosine tRNA precursor drawn in a hypothetical secondary structure. Nucleotides found modified in the mature tRNA are indicated with their modifications. From K. P. Schaeffer, S. Altman, and D. Söll, *PNAS* **70**, 3626–3630 (1973). A partial sequence of the tRNA gene past the CCA end is also shown. Note the region of local 2-fold rotational symmetry (indicated by the bars and the dot) and two inversely repeated segments.

Part of DNA sequence extending past the CCA end of the tRNA gene

Regions containing inversely repeated segments

3′ end of tRNA

5′ end of tRNA

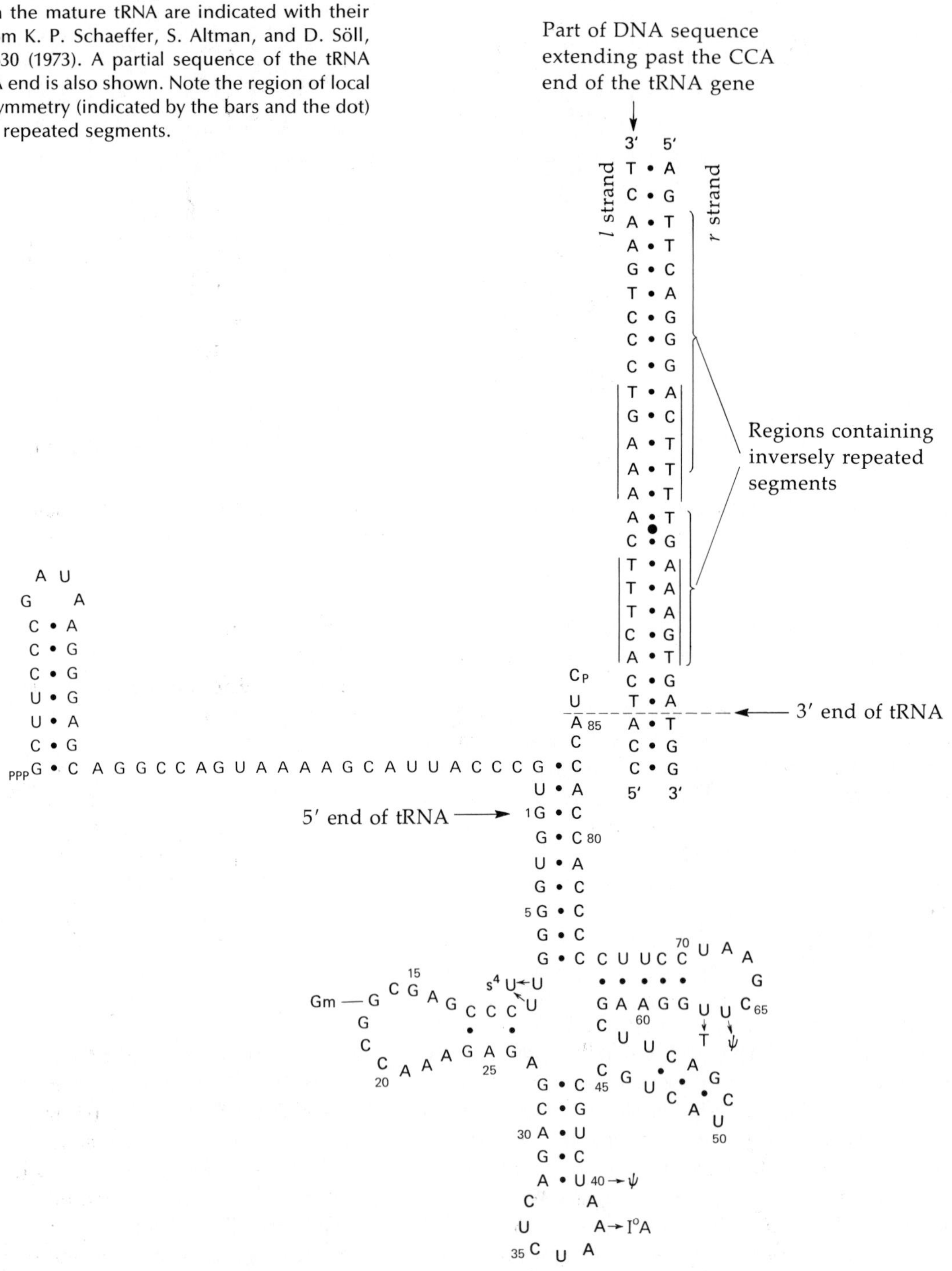

that pair with codons starting with U. It is of interest that this compound is also a plant hormone, a **cytokinin** (Chapter 16, Section A,3).

Another highly modified purine is threonylcarbamoyladenine, which occurs adjacent to the end of anticodons pairing with codons starting with A. The function of these highly modified bases is

uncertain, but they appear to be needed for proper binding to ribosomes. A special enzyme transferring the isopentenyl group from isopentenyl pyrophosphate (Chapter 12, Section H) has been purified.[76]

An interesting feature of tRNA metabolism, which has so far not been explained, concerns the

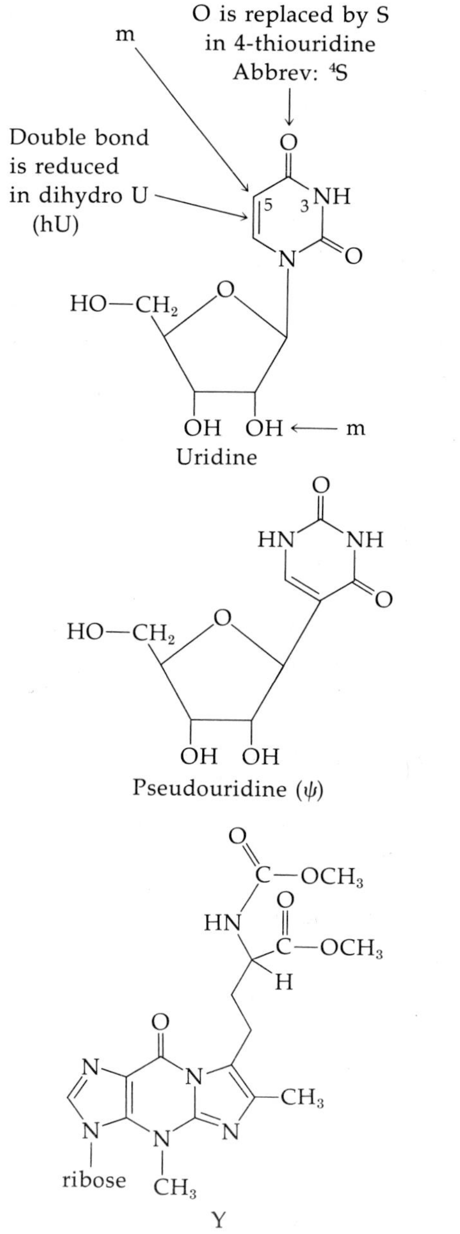

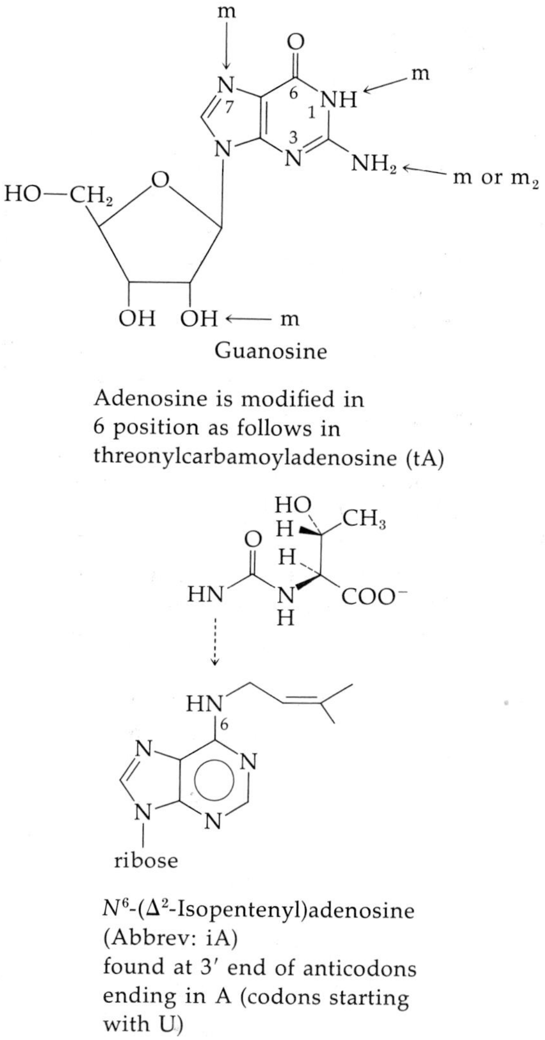

Fig. 15-10   Structures of some nucleosides containing modified bases and found in tRNA molecules. Positions where methylation may occur are designated m.

3'-terminal group of three nucleotides: CCA. This group is invariant among all tRNA molecules and is labile, undergoing active removal and resynthesis. The rate of this "turnover" is sufficient to involve about 20% of the tRNA molecules of a cell per generation, but it is very much slower than the rate of participation of the tRNA molecules in protein synthesis. Thus, it would appear not to be directly linked with peptide bond formation.

5. Transcription in Eukaryotic Cells

In cells with true, membrane-enclosed nuclei, the messenger RNA molecules are relatively long lived. They must move out from the nucleus to the sites of protein synthesis in the cytoplasm. In addition to the obvious need for eukaryotic mRNA to travel further and last longer than that of bacteria, a number of other differences, still not fully understood, are evident. For one thing, it may be that eukaryotic mRNA's are transcribed from single genes and that polygenic operons do not ordinarily function. Furthermore, a new kind of RNA, **heterogeneous nuclear RNA** (HnRNA), constitutes a major fraction of the RNA in the nucleus and is thought to be a precursor of mRNA. Like mRNA, it has a base composition resembling that of DNA. The molecular weights vary from 10^5 to 2×10^7 (1500 to 30,000 nucleotide units). A characteristic of HnRNA is that it turns over rapidly, most of it having a half-life of only about 10 min. However, some may last as long as 20 h. Surprisingly, only about 10% of the HnRNA ever leaves the nucleus, most being degraded without export to the cytoplasm.[77-79]

An unexpected finding provided the clue that HnRNA is truly the precursor of mRNA. Both HnRNA and mRNA of eukaryotes have been shown to contain, at the 3' end, long chains of polyadenylic acid [**poly(A)**]. [An apparent exception to the presence of poly(A) tracts in mRNA exists for the mRNA's coding for the synthesis of histones, the basic nuclear proteins found in eukaryotic cells.] Typically, 200 adenylic acid residues are added to the end of an HnRNA chain, presumably by a special enzyme acting after transcription is completed.[79a] It has been reported that after the HnRNA has been cleaved and the mRNA transported to the cytoplasm, 50–75 adenylic acid units still remain. The adenylic acid units appear to be gradually removed from mRNA, but apparently not completely so. What is the function of the poly(A) "tracts"? The answer is far from clear. One idea is that the poly(A) is somehow needed for transport from the nucleus. Another theory proposed that each time the message of the mRNA is translated by a ribosome one or more nucleotides are removed from the 3' end of the mRNA chain as it leaves the ribosome. The adenylic acid units of the poly(A) tract are "tickets," each one good for one trip through a ribosome. Eventually all of the A units are removed, a signal for the cell to degrade the mRNA. Although poly(A) tracts in mRNA are regarded as characteristic of eukaryotic cells, rapidly degraded poly(A) sequences have also been reported for *E. coli*.[80]

Another surprise was the more recent discovery that the 5' ends of many mRNA molecules of eukaryotic cells as well as of viruses are "capped" by a terminal structure containing 7-methylguanosine from which a proton has dissociated to form a dipolar ion[81-83]:

Note the triphosphate structure connecting the 7-methylguanosine to the mRNA. This may arise in the following way.[82] The 5' end of an RNA transcript initially contains a triphosphate group arising from the fact that a nucleotide triphosphate serves as the primer in initiating transcription. One of these phosphoryl groups may be removed to leave a diphosphate which is guanylated by GTP. The 7-methyl group as well as the 2'-methyl group on the 5'-terminal ribose* of the polynucleotide are then transferred from S-adenosylmethionine by appropriate methylase.

Poliovirus mRNA is unusual in that it remains uncapped in infected cells.[83a]

The Nucleolus

A variety of studies have established conclusively that nucleoli, of which there may be one or several per nucleus, are sites of synthesis of ribosomal RNA. Eukaryotic ribosomes contain four different RNA molecules of the sizes given below.

Sedimentation constant (S)	MW	No. of nucleotides
28	1.7×10^6	5000
18	0.65×10^6	2000
5.8	5×10^4	150
5	4×10^4	120

The 28 S, 18 S, and 5.8 S molecules all originate from a high molecular weight (MW = 4×10^6) 45 S pre-rRNA. This precursor is transcribed in the "core" region of the nucleolus. As the molecules move away into an outer "cortex" cleavage occurs in a number of steps (Eq. 15-6).[57,84,85] Elec-

$$
\begin{array}{ccccc}
& & & & 28\ S \\
& & 32\ S & \nearrow & (1.7 \times 10^6) \\
& & \nearrow (2.1 \times 10^6) \searrow & & 5.8\ S \\
45\ S & \longrightarrow & 41\ S & & (5 \times 10^4) \\
(4.1 \times 10^6) & & (3.1 \times 10^6) \searrow & 20\ S & \\
& & & (0.9 \times 10^6) \searrow & 18\ S \\
& & & & (0.65 \times 10^6)
\end{array}
$$

$$(15\text{-}6)$$

tron microscopy has provided direct confirmation of the relationship of one precursor molecule to another (Fig. 15-11).[85a,b] Note that the 18 S portion of the 45 S RNA lies nearest to the 5'-end just as does the 16 S rRNA in the large transcript of the prokaryotic rRNA genes (Section B,3). There are probably more steps than are shown. Thus, 46 S and 47 S molecules may precede the 45 S RNA.

A dramatic development in the study of the nucleolus has been the direct observation with the electron microscope of portions of unwound cores of nucleoli.[59,86] Fibrils of RNA coated with protein can be seen growing from the DNA strands of the pre-rRNA genes (Fig. 15-7). Approximately 80–100 RNA chains of different length are transcribed concurrently from a single gene. The overall gene length in the electron microscope is 2.3 μm, only a little less than the calculated length for a fully extended DNA molecule (in the B form). However, judging by the lengths of the transcripts formed, the pre-rRNA chains are extensively folded.

As a rule, it appears that organisms of all types usually contain only one gene of a given type per haploid genome. However, there are multiple copies of the genes for ribosomal RNA's. Thus, in *Drosophila* there are 130–190 copies of the 45 S RNA gene.

The gene for eukaryotic ribosomal 5 S RNA is separate from that of 45 S RNA and is not located in the nucleolus. In *Drosophila* about 500 copies of the 5 S RNA gene are located in the right arm of chromosome 2. RNA polymerase III is responsible for the synthesis of 5 S RNA, whose sequence is shown in Fig. 15-12. A characteristic of 5 S RNA is that it can be folded in several ways, and it is not clear which folding pattern or patterns exists in ribosomes.[87]

C. TRANSLATION OF THE GENETIC MESSAGE: PROTEIN SYNTHESIS

The protein synthesizing machinery of *E. coli* is found in the 15,000 ribosomes which account for

* The next adjacent ribose is also sometimes methylated and N^6-methyladenosine may be present near the poly(A) tract at the 3' end.[83]

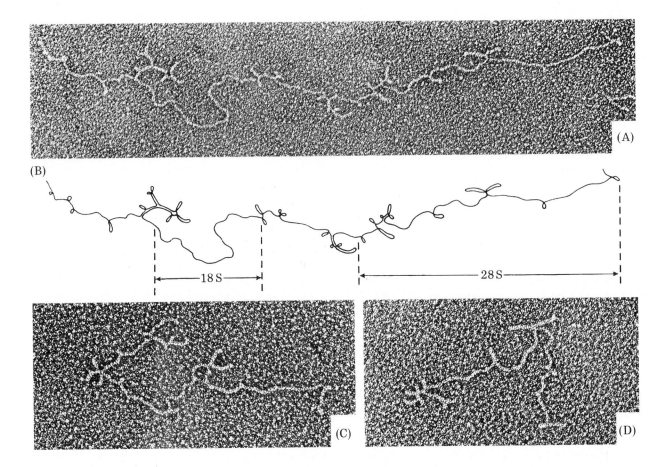

Fig. 15-11 (A) Electron micrograph of the 45 S precursor of rRNA from HeLa cells after spreading from 80% formamide and 4 *M* urea. The molecule is shown in reverse contrast. (B) Tracing of molecule in (A) showing several regions of second- ary structure as hairpin loops. The 28 S and 18 S rRNA regions are indicated. (C) 32 S rRNA. (D) 28 S rRNA. Notice that the same secondary structure can be seen in the 28 S RNA as in its 32 S and 45 S precursors. From Wellauer and Dawid.[85a]

one-fourth of the total mass of the cell. Under- standing these little molecular machines, which appear not much more than blurred dots under the electron microscope, is one of the current major efforts of molecular biology.[88]

1. The Chemical Composition of Ribosomes

Ribosomes of *E. coli* have a mass of ~2.7 × 10⁶ daltons and are made up of ~65% RNA and 35% protein. Ribosomes of eukaryotic organisms are larger, by a factor of ~1.6 in mass (4.3 × 10⁶ daltons). Under some conditions, such as a low [Mg^{2+}], the complete ribosomes (referred to as **70 S ribosomes** when of bacterial origin) dissociate to two subunits of unequal size, referred to as **30 S** and **50 S ribosomal subunits,** respectively. The larger 50 S subunit is about twice the size of the smaller one and contains the 23 S and 5 S RNA molecules (Table 15-4). The small 30 S subunit contains the 16 S RNA, a chain of about 1700 nu- cleotides that, if fully extendĕd, would stretch to a length of over 500 nm. Its complete nucleotide se- quence is known.

TABLE 15-4
Masses of Ribosomes of *E. coli*

Particle	Ribosome ($\times 10^6$ daltons)	RNA ($\times 10^6$ daltons)	Protein (by difference) ($\times 10^6$ daltons)
50 S subunit	~1.8	23 S ~1.1 5 S ~0.04	~0.7
30 S subunit	~0.9	16 S ~0.56	~0.3
70 S ribosome	~2.7	~1.7	~1.0

In addition to the highly folded RNA molecule, the 30 S subunit contains ~21 proteins, each one unique in its amino acid composition and sequence (Table 15-5). Many of these proteins, designated S1, S2, S3, etc., are of relatively low molecular weight. Many are strongly basic. They contain many lysine and arginine residues and are doubtless designed to interact with RNA molecules. However, some neutral and acidic proteins are also present. The 50 S ribosomal subunit contains ~34 proteins, several of which are present as two or more copies. The protein composition of ribosomes is variable and has been difficult to establish with certainty. It does appear that most of the proteins (often called "unit" proteins) are present in a strict 1:1 ratio. Others may be lacking in some of the ribosomes. Likewise extra copies of some subunits may be present in only a fraction of the ribosomes. A number of other proteins are bound to the ribosomes transiently during their function in protein synthesis.

It has been difficult to establish the size and shape of ribosomes exactly, but they are ~22 nm in diameter and perhaps 30 nm in the third dimension. Eukaryotic ribosomes are of the order of 1.17 times larger in linear dimensions and usually contain a substantially greater number of proteins, ~30 in the small subunit and ~40 in the large.[89] However, it has been claimed that the number of essential proteins is the same as in *E. coli* ribosomes.[90] Interestingly, the eukaryotic ribosomal proteins (like the rRNA molecules) are substantially larger than those of bacteria. Mitochondrial ribosomes resemble those of bacteria in some respects but are larger and contain ~66%

protein (compared with 35% for *E. coli* ribosomes).

Various evidence argues for the view that the proteins in ribosomes are compact molecules whose surfaces are mostly accessible to added reagents. The RNA is also largely accessible from

TABLE 15-5
Ribosomal Proteins

Proteins of 30 S ribosomal subunits			Proteins of 50 S ribosomal subunits		
Designation	MW	Binding[a]	Designation	MW	Binding[a]
S1	65,000		L1	22,000	
S2	27,000		L2	28,000	+
S3	28,000		L3	23,000	
S4	25,000	+	L4	28,500	
S5	21,000		L5	17,500	
S6	17,000		L6	21,000	+
S7	26,000	+	L7	15,500	
S8	16,000	+	L8	19,000	
S9	17,500		L9		
S10	17,000		L10	21,000	
S11			L11	19,000	
S12	17,000		L12	15,500	
S13	14,000		L13	20,000	
S14	15,000		L14	18,500	
S15	13,000	+	L15	17,000	
S16	13,000		L16	22,000	+
S17	10,000		L17	15,000	+
S18	12,000		L18	17,000	+
S19	14,000		L19	17,500	+
S20	13,000	+	L20	16,000	+
S21	13,000		L21	14,000	
			L22	17,000	
Sum	405,000		L23	12,500	+
			L24	14,500	+
			L25	12,500	+
			L26	12,500	
			L27	12,000	
			L28	15,000	
			L29	12,000	
			L30	10,000	
			L31		
			L32		
			L33	9,000	
			L34		
			Sum	549,000	

[a] A plus sign indicates direct binding to ribosomal RNA.

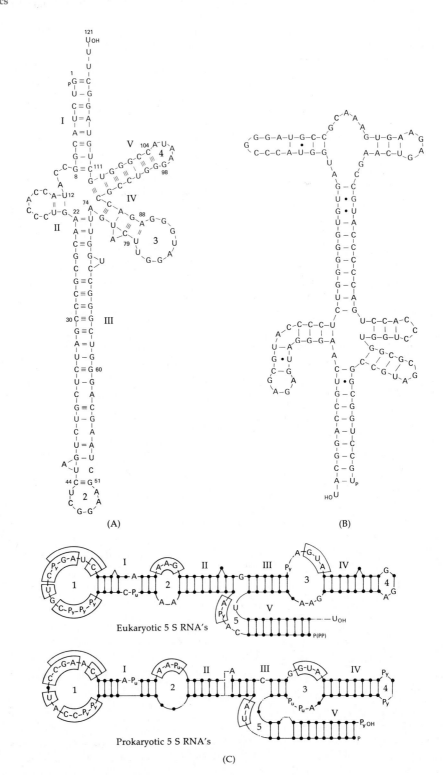

(A)

(B)

Eukaryotic 5 S RNA's

Prokaryotic 5 S RNA's

(C)

the outside and the ribosome contains ~50% of its mass as internal hydration. Thus, the ribosome is a relatively open structure. Much of the RNA (perhaps 60–70%) is folded into base-paired loops as in tRNA. Many experiments are currently being done to try to understand the physical relationship of one ribosomal subunit to another. It has been found that the S4 and S20 proteins bind directly to 16 S RNA near the 5' end. Proteins S8 and S15 bind near the center and S7 near the 3' end of the 16 S RNA. It has been suggested that the S4 protein plays an especially important organizing role.[88] It has now been possible to completely dissociate both the 30 S ribosomal subunits[91] and the 50 S subunits[92] of *E. coli* into individual protein and RNA molecules and to reconstitute them in a functional form. In these reassembly experiments it has been found that the order of addition of the proteins is important. The experiments suggest that in addition to S4, S7, S8, S15, and S20 proteins S9, S13, and S17 bind to 16 S RNA and S16 binds to S4 and S20.[93] Other proteins are not necessary for reconstitution of the structure of the ribosome but are essential for function. These include S3, S10, S12, S14, and S19. Most of the ribosomal proteins have no known enzymatic activity, but they may possess undetected catalytic properties.

Proteins L5, L18, and L25 of the 50 S subunit have been shown to bind specifically to 5 S RNA, a nucleic acid whose sequence is known (Fig. 15-12). Furthermore, the L5·L18·L25·5S RNA complex binds the oligonucleotide TpψpCpGp. This suggests an interaction between the 5 S RNA and the TψC arm of a tRNA molecule bound to the ribosome. In addition, it has been observed that L18 + either L5 or L25 bind 5 S RNA to 23 S RNA. Table 15-4 indicates other proteins of 50 S ribosomal subunits that bind to RNA.

Many experiments are now being performed with cross-linking reagents, for example, bifunctional compounds that covalently bind to two different SH groups or NH$_2$ groups.[93–95] Using such techniques the following cross-linked pairs have been identified[93,95]: S2-S3, S4-S6, S4-S8, S4-S9, S4-S12, S5-S8, S5-S9, S7-S8, S7-S9, S11-S18, S13-S19, and S18-S21. Another approach is to prepare antibodies to specific ribosomal proteins and by electron microscopy, to map the binding sites of the antibodies on the ribosomal subunit surface.[96,97] In this manner, the locations of numerous proteins in both the 30 S and 50 S subunits have been identified as is indicated in Fig. 15-13. In a number of instances more than one distinct antibody binding site was found for a given protein. Thus, pairs of sites on S2, S12, S15, and S18 were 8–19 nm apart, an indication that these proteins assume an elongated or fibrous conformation (Fig.

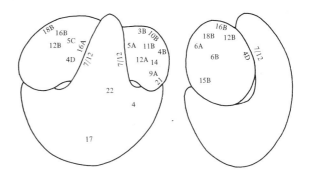

Fig. 15-12 Nucleotide sequence of ribosomal 5 S RNA of human cells (KB carcinoma) with two possible modes of folding [(A) and (B)]. From J. L. Sirlin, "Biology of RNA," p. 67. Academic Press, New York, 1973. (C) A possible generalized base pairing pattern for eukaryotic 5 S RNA's (upper) and prokaryotic 5 S RNA's (lower). Letters in the loops indicate nucleosides common to human KB cell, *X. laevis*, and *T. utilis* 5 S RNA's (upper), and those common to *E. coli*, *P. fluorescens*, and *B. stearothermophilus* 5 S RNA's (lower). The sequences in boxes are commonly found at corresponding positions in the models for both eukaryotic and prokaryotic 5 S RNA's. From K. Nishikawa and S. Takemura, *J. Biochem. (Tokyo)* **76**, 935–947 (1974).

Fig. 15-13 Two views of a three-dimensional representation of a 70 S ribosome from *E. coli*. The numbers indicate locations of binding of antibodies against a particular ribosomal protein (numbered as in Table 15-5). The designation S or L is not given because it is clear from the drawing. When more than one antibody binding site has been located for a protein, these sites are designated A, B, C, and D. Many other sites have been located and can be seen in other views of the ribosome. From Wittman.[97]

2-12) in the 30 S subunit. Protein S4 is also thought to extend for at least 17 nm,[95] accounting for the large number of cross-linked pairs involving S4. It is clear that the ribosome is an extraordinarily complex machine.

2. Protein Synthesis

The formation of polypeptide bonds on ribosomes is usually discussed in terms of three processes: **initiation, elongation,** and **termination.**[98] Protein synthesis begins with an **initiation codon** most commonly that for methionine:AUG. The codon GUG, when properly placed in a mRNA chain, can also serve as an initiation codon. In such cases it codes for methionine rather than for valine. The sequence of bases preceding the initiation codon may also be important for recognition of the "start" signal. This seems likely in view of the fact that both codons AUG and GUG occur at places other than initiation points.

a. Initiation[99]

Peptide chains in bacteria are always initiated with the amino acid **N-formylmethionine.** Thus, the first step in protein synthesis is the alignment of the proper initiation codon correctly on the ribosome and the binding to it of a molecule of tRNA carrying N-formylmethionine.* The process by which this occurs is relatively complex, partly because it is essential for the ribosomes to distinguish the true initiation codon from the many AUG and GUG codons in internal positions in the message. This appears to be accomplished by base pairing between an ACCUCCU sequence next to the 3' end of the 16 S ribosomal RNA and a complementary initiator sequence in the mRNA (Fig. 15-14).[100,101] Ribosomal protein S1 (also known as an *i* factor) seems to be required for this binding and may function by holding the 3' end of the 16 S

RNA in an "open" conformation rather than as a hairpin loop.[102]

In addition to the ribosomal proteins, there are three essential protein **initiation factors: IF-1, IF-2,** and **IF-3.** The latter exists in at least two forms, IF-3α and IF-3β. The role of IF-3 is apparently to bind to and stabilize the mRNA-16 S RNA initiation complex (Fig. 15-15, step *a*). Now a "charged" molecule of formylmethionyl-tRNA (fMet-tRNA) is bound by a second initiation factor IF-2, which also has bound to it a molecule of GTP (step *b*).* This complex is then bound to the 30 S subunit, presumably in such a way that the anticodon of the tRNA pairs with the starting codon of the message (step *c*, Fig. 15-15). The binding of the IF-2 complex is in some way assisted by the third initiation factor IF-1 whose function is not very clear. Now IF-3 leaves the complex (step *d*) and the 50 S ribosomal subunit binds to form the complete ribosome with the expulsion of IF-1 (step *e*).

While a great deal of uncertainty exists about what is really happening in these reactions, the drawing in Fig. 15-15 shows the charged methionyl-tRNA binding to what is known as the "P site" (**peptidyl site**) in the 50 S ribosomal subunit. An earlier proposal, which should probably still be considered, holds that initial binding is at the "A site" (**amino acid site**) and that a **translocation** to the P site then occurred. The reason for this proposal was that hydrolysis of GTP accompanies the last step in the initiation sequence (step *f*, Fig. 15-15) in which protein IF-2 is released as a GDP complex. As explained in the next section, hydrolysis of GTP is required for translocation during peptide chain elongation.

The hydrolysis of GTP during initiation is truly essential, as is shown by the fact that 5'-guanylmethylene diphosphonate (a GTP analogue containing a methylene bridge between the terminal and central phosphorus atoms) can substitute for GTP in the initial binding step. The analogue replaces GTP in all steps up to and including the

---

* This binding has the secondary effect of shifting the equilibrium between 30 S and 50 S subunits and intact 70 S ribosomes in favor of a greater degree of dissociation.[91]

* Some investigators believe that the fMet-tRNA-IF-2 complex binds to the 30 S ribosomal subunit prior to the binding of mRNA.[103]

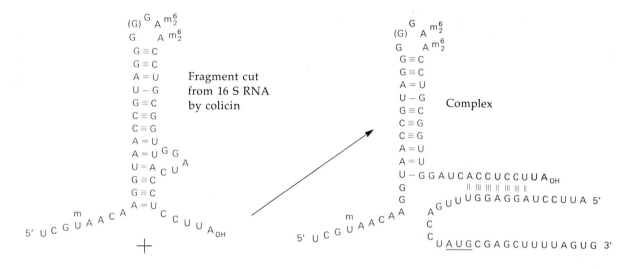

5' A U U C C U A G G A G G U U U G A C C U <u>A U G</u> C G A G C U U U U A G U G 3'

R 17, a protein initiator region

Fig. 15-14 Postulated hydrogen bonding between a fragment of 16 S RNA and the A protein initiator region from R17 phage RNA. The secondary structure drawn for the 16 S RNA fragment is predicted to be stable under physiological conditions. Not shown is an alternative hydrogen bonding scheme of comparable predicted stability which would enlarge the bulge loop to 9 bases and pair the CCUU sequence directly adjacent to the 3' end with the AAGG on the 5' side of the lower portion of the stem. Ragged ends on the messenger fragment are denoted by small capital letters. From Steitz and Jakes.[101]

binding of the 50 S ribosome, but it cannot function in the final step because it cannot undergo hydrolysis. However, the function of GTP hydrolysis is not yet clear. It could provide energy for some major rearrangement of ribosomal components (as suggested above) or it could simply be required for release of the IF-2-GDP complex (i.e., the IF-2-GTP complex could be bound with a high affinity whereas the IF-2-GDP complex could be bound weakly). In fact, these two explanations are not necessarily incompatible.

Some information about spatial arrangements of ribosomal proteins involved in initiation is provided by the fact that antibodies against proteins S19 and S21 block the formation of the initial complex with fMet-tRNA. Furthermore, antibodies against S2, S18, and S20 block the binding of IF-3. Cross-linking experiments show that IF-2 and S19 are close together and that IF-3 is close to S12.

b. Elongation of Peptide Chains[104]

Once the initiating fMet-tRNA is in place in the P site, peptide chain growth can commence. Amino acid residues are added in turn by insertion at the C-terminal end of the growing peptide chain. Elongation occurs in three steps repeated over and over until the entire peptide is formed:

1. Codon-specific binding of a charged tRNA bearing the next amino acid at the A site.
2. Formation of the peptide bond. This step frees the tRNA in the P site and transfers the growing peptide chain onto the tRNA in the A site.
3. "Translocation" of the peptidyl tRNA from the A site to the P site. This step also involves release of the used tRNA from the P site and movement of the mRNA to bring the next codon into place in the A site. The process of translocation re-

quires energy which is provided by the hydrolysis of one molecule of GTP.

The first of these three steps, the binding of an aminoacyl-tRNA to the A site, depends upon a protein transfer factor (**elongation factor T,** or EF-T). The factor is a mixed dimer Ts·Tu in which

Ts is a stable protein of MW ~42,000. Factor Tu is a membrane-associated protein[104a] of MW ~44,000 present in severalfold excess over Ts. Protein Tu forms a complex with GTP as well as with aminoacyl-tRNA molecules (however, it does not bind well to fMet-tRNA). As indicated in Fig.

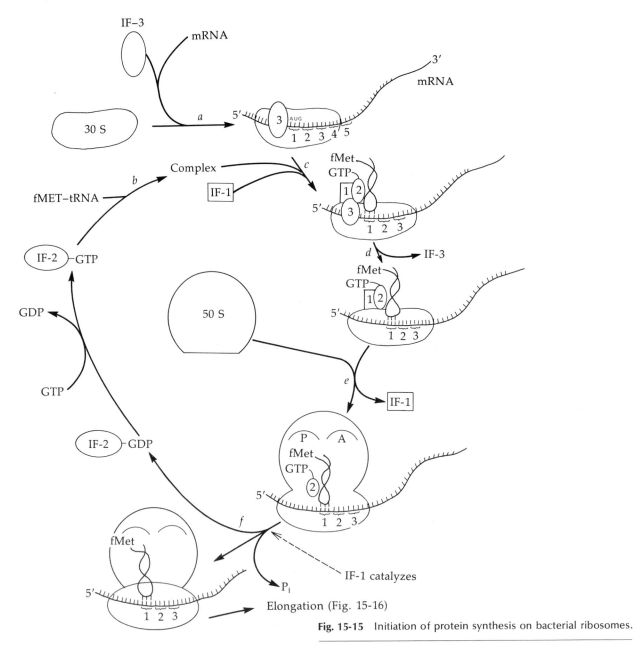

Fig. 15-15 Initiation of protein synthesis on bacterial ribosomes.

15-16, GTP reacts with a Tu · Ts complex with release of Ts and formation of the GTP-Tu complex (step *a*). The latter then combines with aminoacyl-tRNA and with the ribosome (steps *b* and *c*). These reactions all occur at the **peptidyl-transferase center** in the 50 S subunit, a region that includes the proteins L7 and L12.

In the next step (step *d*) the peptide chain is transferred to the amino group of the aminoacyl-tRNA occupying the A site, a simple displacement reaction. Nevertheless, the reaction is more complex than this, for the bound GTP is cleaved with the release of P_i and of a Tu·GDP complex. As indicated in the figure, the latter reacts with Ts to

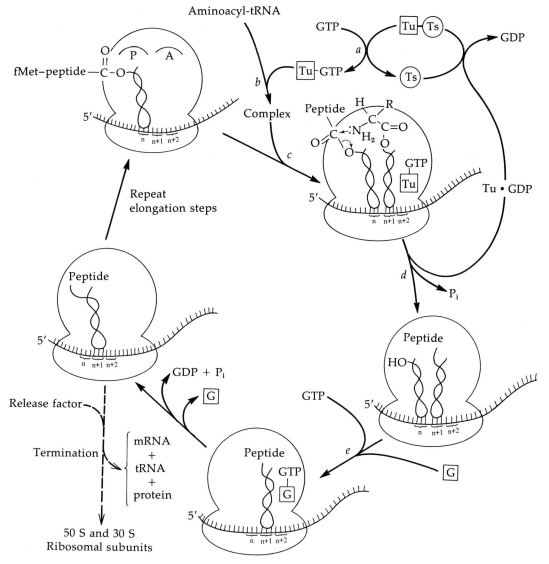

Fig. 15-16 Elongation of the growing peptide chain.

regenerate the Tu·Ts dimer and to give free GDP. Thus, the overall reaction is the cleavage of GTP coupled to the synthesis of a peptide linkage. From the point of view of the chemistry of the reaction, the GTP hydrolysis is unnecessary. However, we do not know how close together the ends of the two adjacent tRNA molecules are. They may be quite far apart. The proteins L7 and L12 are unusual in having high alanine content and a high percentage of α helix. In this respect they resemble myosin of muscle. It has been suggested that they may be part of a "mini-muscle" which uses the free energy of hydrolysis of GTP to move some part of the ribosomal complex, bringing together the amino group and the peptidyl group for the peptidyltransferase reaction.

An interesting chemical question that has been difficult to answer is "Which hydroxyl group, the 2' or the 3', of the terminal adenosine of tRNA carries the activated aminoacyl group or peptidyl group?" Rapid equilibration between the two via an orthoester (tetrahedral adduct) may occur (Eq. 15-7).

(15-7)

Modified yeast phenylalanyl-tRNA terminating in 3'-deoxyadenosine is aminoacylated by its amino acid activating enzyme (aminoacyl-tRNA synthetase, Eq. 11-2) but that terminating in 2'-deoxyadenosine does not react. This suggests the 2' hydroxyl as the initial site of aminoacylation. However, a synthetic 2'-phenylalanyl-tRNA terminating in 3'-deoxyadenosine, when tested as a peptide acceptor in the peptidyltransferase reaction on ribosomes, was found inactive. The isomeric 3'-phenylalanyl-2'-deoxy compound did react.[105] Thus, it appears likely that aminoacylation occurs initially on the 2' hydroxyl and that isomerization (Eq. 15-7) to the 3'-aminoacyl-tRNA precedes the transpeptidation reaction. However, more recent studies make it clear that for both baker's yeast and *E. coli* some tRNA's are initially aminoacylated at the 2'-hydroxyl group, others at the 3' hydroxyl, and still others at either position.[106,107]

The third step in the elongation sequence on ribosomes depends upon another elongation factor EF-G, a molecule displaying "GTPase" activity. There is evidence that factor G also binds at or near L7 and L12 and competes in its location with EF-Tu. Again, the exact function is not clear, but it is known that hydrolysis of GTP is required to effect the translocation reaction and that the used tRNA cannot be released from the P site until EF-G binds.

Mutants of the genes for EF-Tu, EF-Ts, and EF-G have been identified establishing an *in vivo* role for these proteins (however, this has not yet been accomplished for the initiation factors IF-1, IF-2, and IF-3).

c. Termination of Peptide Synthesis[108]

The ribosome faithfully translates the genetic message, adding amino acids to the peptide chain until a stop codon is reached. Then a **termination factor** acts, probably by binding directly to the stop codon on the mRNA. In *E. coli* termination factor RF-1, a protein of MW ~44,000 is thought to recognize UAA or UAG, while RF-2 of MW ~47,000 may recognize UAA or UGA. There are several hundred molecules per cell of these termination factors. In some manner the termination factor not only must recognize the right codon but also must catalyze the hydrolytic removal of the peptidyl chain from the tRNA. As amino acid sequences of large portions of viral mRNA molecules have emerged recently, it has been a surprise to discover that genes coding for the coat proteins of RNA phages (Fig. 15-19) are sometimes terminated by a succession of two stop codons. Thus, there is evidently a safety factor to prevent translation from continuing in case the first stop codon is missed. Note that at the end of the *i* gene of the *lac* operon of *E. coli* (Fig. 15-4) there is a second stop codon in phase with the TGA codon marked in the figure and located five codons further to the right. It is not clear whether doubling of termination codons at the ends of genes is a general phenomenon.

d. Polyribosomes

Under suitable conditions ribosomes isolated from cells are found to sediment together in clusters, often of five or more. These **polyribosomes** (or **polysomes**) can be shown to be held together by chains of RNA. It is now accepted that polyribosomes arise because a single mRNA molecule is being translated by several ribosomes at once. As the 5' terminus of the mRNA emerges from one ribosome it may soon combine with another and initiate translation of a second peptide chain, etc. The length of the mRNA determines how many ribosomes are likely to be associated in a polyribosome.

e. Protein Synthesis in Eukaryotic Cells

Many similarities exist between protein synthesis in eukaryotic cells and in bacteria. In eukaryotic initiation a special methionyl-tRNA (Met-tRNA$_F$) serves as the initiator tRNA. (The subscript refers to the fact that it can be formylated by a bacterial enzyme system. However, within eukaryotic cells this methionyl-tRNA remains without a formyl group.) As in bacteria, another transfer RNA (Met-tRNA$_M$) serves to provide the

methionine for internal positions in proteins. At least three protein initiation factors eIF-1–eIF-3 are required.[109] A significant difference from the prokaryotic system is that the initiator aminoacyl-tRNA is bound first, then mRNA is bound. The 7-methylguanosine-containing cap at the 5' end of eukaryotic mRNA (Section B,5) may be required[109a] for recognition by eIF-3. There are elongation factors[110] EF-1 and EF-2 and a releasing factor, just one, rather than the two identified for bacteria.

f. Pairing of Codon and Anticodon

We now return to the fundamental problem of bringing up the right amino acid at the right time during synthesis of the peptide chain. Everything depends upon correct "recognition" by an anticodon in the tRNA of the complementary codon in the mRNA. A surprise was the discovery of inosine (I) in anticodons of yeast tRNA (but not in most *E. coli* tRNA's). Another unexpected finding was that less than 61 kinds of tRNA exist in a given cell (61 = 64 − 3 stop codons). Consideration of these matters led Crick in 1966 to propose the "wobble hypothesis."[111] According to this proposal the first two bases at the 5' end of the codon (and at the 3' end of the anticodon) must pair in the same ways as do the bases in DNA. However, the third base pair (3' end of the codon and 5' end of the anticodon) is under a less severe steric restriction. That is, there may be some "wobble." Crick proposed the accompanying "rule" for pairing of the third base.

Base in anticodon	Paired base in codon
G	C or U
C	G
A	U
U	A or G
I	C, A, or U

All of the observed deviations from the AU, CG pairing of a Watson–Crick helix could be explained in this way. Thus, an anticodon with G at the 5' end could pair with codons ending in either

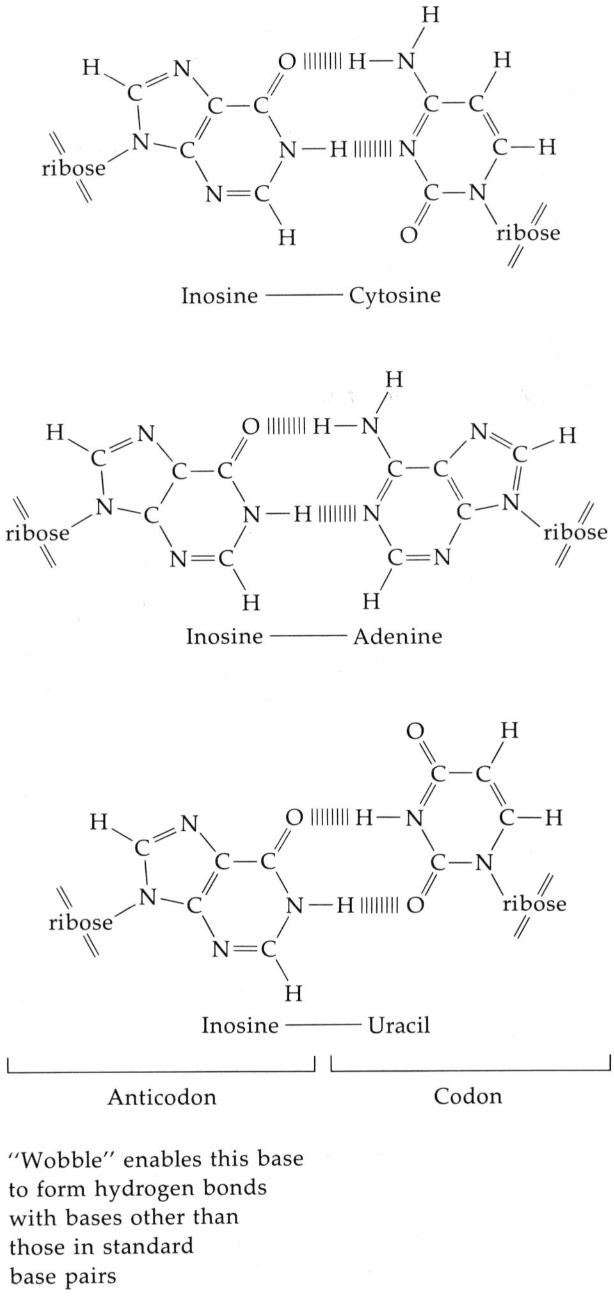

Inosine ——— Cytosine

Inosine ——— Adenine

Inosine ——— Uracil

Anticodon   Codon

"Wobble" enables this base
to form hydrogen bonds
with bases other than
those in standard
base pairs

Fig. 15-17 Pairing of inosine with cytosine (a Watson-Crick pair) and of inosine with adenine and uracil (wobble pairs). From James D. Watson, "Molecular Biology of the Gene," 3rd ed., p. 358. Copyright 1976, 1970, 1965 by W. A. Benjamin, Inc.

C or U. Anticodons ending in C or A would pair strictly. Anticodons ending in U could pair with codons containing either A or G in the 3' position. Anticodons with I in the 5' position could recognize codons with any of the three bases in the third position. Comparison with Table 15-2 makes it immediately clear why less than 61 anticodons are needed. Many codons represent the same amino acid and frequently the nature of the base in the 3' position of the codon is immaterial to the meaning of that codon. Thus, there is an economy in using less than the full array of anticodons. Crick showed that the proposal was chemically feasible if the spatial relationships for the wobble pair were allowed to vary from the usual ones in Watson–Crick base pairs. This is illustrated in Fig. 15-17 for binding of inosine to C (a normal Watson–Crick base pair) and to A and to U. While the word wobble does not necessarily convey the correct meaning, the hypothesis has predicted many things correctly. For example, according to the wobble hypothesis at least three tRNA's are required to recognize the six serine codons, but it is not necessary to have six. In fact, three are found in *E. coli*.

g. Aminoacyl-tRNA Synthetases[112]

Another critical recognition process is required in protein synthesis. That is the selection by an aminoacyl-tRNA synthetase of the correct amino acid and of the transfer of the activated amino acid to the correct tRNA (Eq. 11-2). In bacteria there is one aminoacyl-tRNA synthetase for each of the 20 amino acids. Each synthetase must select a specific amino acid and a correct tRNA for that amino acid. The same enzyme transfers an activated amino acid to all of the **isoacceptor tRNA's** specific for a given amino acid.

Many attempts have been made to learn what part or parts of tRNA molecules are involved in recognition by synthetases. Nucleotide sequences of isoacceptor tRNA's have been compared. Chemically modified and fragmented tRNA molecules have been studied and a number of other approaches have been used. The results suggest that no method of recognition is universal. In some cases, a synthetase does not aminoacylate a chemically modified tRNA if the anticodon structure is incorrect. This is remarkable because the anticodon is ~7.5 nm away from the CCA end of the tRNA in crystals (Fig. 2-24). While the synthetases are large enzymes (MW ≈ 100,000 for most), it is still difficult to understand how a precise recognition of the anticodon can occur unless there is some conformational change in the tRNA that brings it closer to the CCA end. For some tRNA's the anticodon is not involved in recognition. In the case of yeast Phe-tRNA there is evidence that residues in the stem of the dihydrouridine loop and at the upper end of the amino acid acceptor stem are critical.[112]

Rich has pointed out that one side of the folded tRNA molecule has a relatively invariant structure, probably designed to interact with some part of the ribosomal machinery. The other side (inside the angle of the L, Fig. 2-24) is more variable and includes residues of the acceptor stem and the dihydrouridine loop. It may be the site of binding to the aminoacyl-tRNA synthetases. It has also been suggested that one or more conformational changes may accompany aminoacylation and binding to ribosomes.[113] It is possible that the TψCG loop, which is inaccessible in the native tRNA, opens up and interacts with 5 S RNA in the ribosome. The short arm of the L structure may rock at the end of the longer arm carrying the anticodon. A technique that may become important in revealing conformational changes is proton nmr (Chapter 2, Section H,7), which permits direct observation of signals arising from the individual hydrogen bonded protons within the interior of the tRNA molecule.[114]

Selection of the correct amino acid by an aminoacyl-tRNA synthetase is extremely important. Yet it must be difficult to devise an active site that would accurately discriminate between two such similar structures as those of isoleucine and valine. One way in which selection can be made more precise is through a kinetic "proofreading" mechanism similar to that recognized for DNA polymerase I (Section E,4). Indeed, it has been shown that valine, when mistakenly attached to

an isoleucine-specific tRNA undergoes rapid hydrolysis catalyzed by the synthetase.[114a] This much reduces the likelihood of valine being inserted in a wrong position in a protein.

h. Action of Antibiotics on Ribosomes

Many of our most effective antibiotics act by blocking protein synthesis on ribosomes. The usefulness of many of these remarkable drugs in human medicine depends upon the fact that they inhibit protein synthesis by bacterial 70 S ribosomes but do not affect eukaryotic ribosomes. In other instances the selective toxicity of an antibiotic depends upon a much greater permeability of bacterial membranes than those of animal cells.

The list of antibiotics acting on ribosomes is a long one[115,116] that includes compounds which have been of importance in unraveling the mechanism of protein synthesis. The aminoglycoside antibiotics **streptomycin** (Box 12-A), the neomycins, and kanamycin have one structural unit in common but bind to ribosomes in distinctly different ways. Streptomycin has the interesting effect of causing the ribosomes to misread the code. It appears to affect primarily the first base of the codon. Thus, when poly(U) serves as a messenger RNA, the expected polyphenylalanine product contains 40% isoleucine.

When a bacterial population is subjected to the action of antibiotics it is possible to select mutants that are able to grow in the presence of the antibiotic. Thus, streptomycin-resistant mutants of *E. coli* have been obtained (even though they arise at a very low frequency of ~10^{-12}). The gene affected (*rpsL* or *strA*) has been mapped at 72 min.* Subsequently, it has been shown that streptomycin binds to ribosomal protein S12 and that *rpsL* is the gene for this protein. Among streptomycin-resistant bacteria mutants can be selected that have become dependent upon the antibiotic and which will now not grow in its absence. This streptomycin dependence has been shown to re-

---

* Genes for ribosomal protein S7 and for elongation factors EF-G and EF-Tu map at a similar position and appear to be part of the same transcriptional unit.[116a]

sult from modification in ribosomal protein S4. From these experiments it is clear that a single point mutation altering one amino acid is all that is necessary to enormously change the sensitivity of a living organism to a particular toxin, or even to make the organism dependent upon that toxin.

The antibiotic **spectinomycin** binds to protein S5, as indicated by analysis of resistant mutants. The gene *spcA* also maps at 64 min and it appears that there is a ribosomal protein operon in this region of the *E. coli* chromosome. Kasugamycin inhibits the binding of fMet-tRNA (initiation). In this case resistant mutants appear in which it is

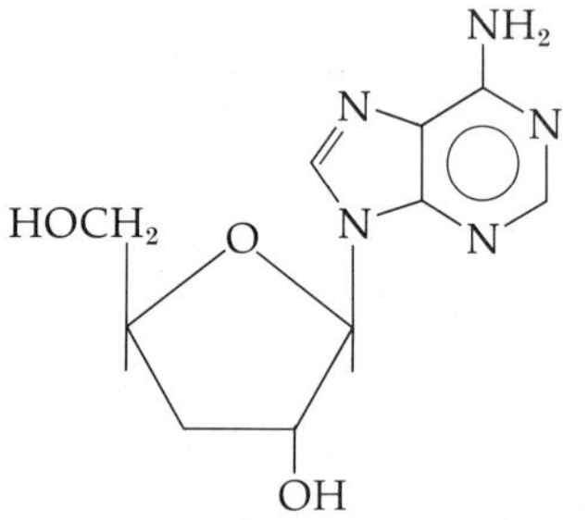

Cordycepin
(3′-deoxyadenosine)
preferentially inhibits
synthesis of ribosomal
and tRNA

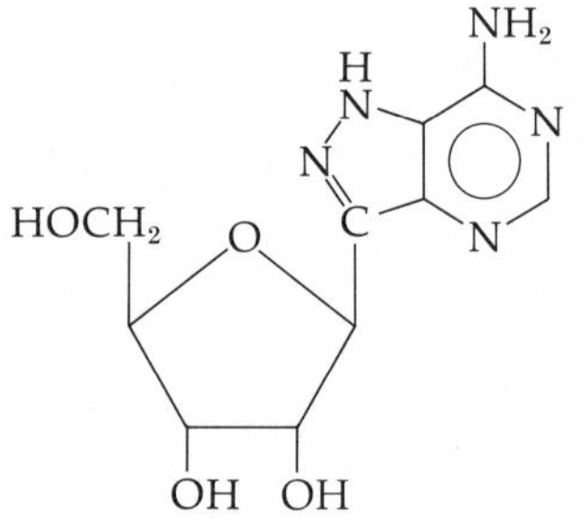

Formycin

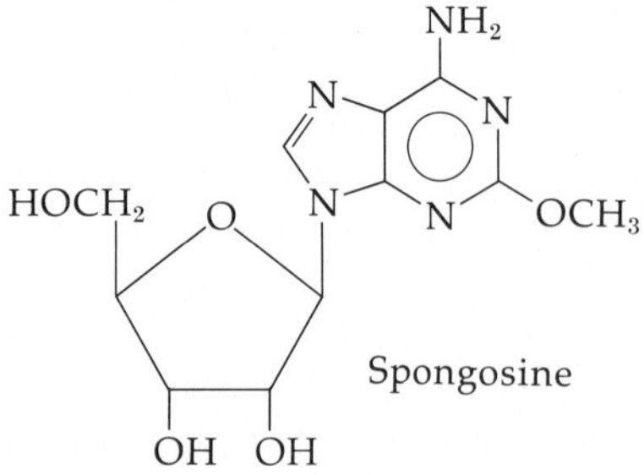

Spongosine

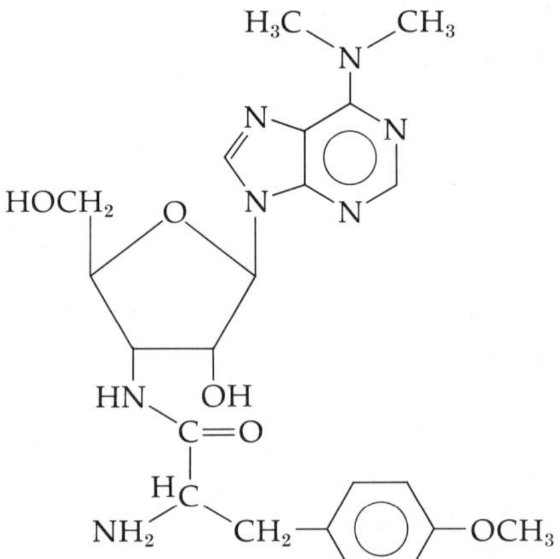

Puromycin—compare structure
with that of an aminoacyl-tRNA

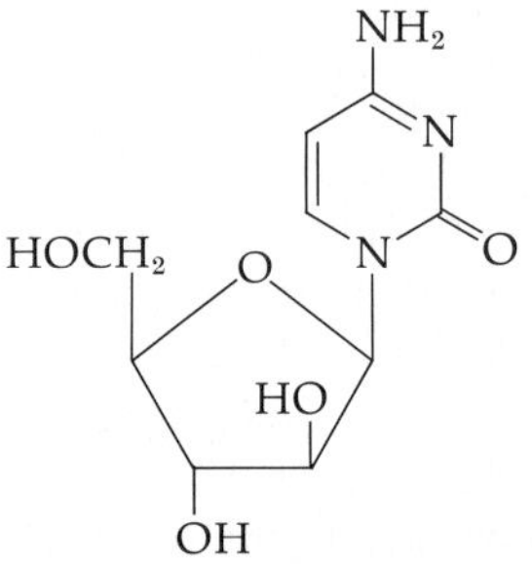

Cytosine arabinoside (*ara*-C)—
the most effective drug for acute
myeloblastic leukemia. Metabolism
yields *ara*-CTP, a potent inhibitor
of DNA synthesis

Fig. 15-18 Structures of some inhibitory analogues of nucleosides occurring in nucleic acids. See R. J. Suhadolnik, "Nucleoside Antibiotics." Wiley-Interscience, New York, 1970 and R. Meyers, V. A. Malathi, R. P. Cox, and, R. Silber, *JBC* **248**, 5909–5913 (1973).

not a protein subunit that has been modified but the 16 S RNA. In resistant strains there is less methylation of this RNA than in normal strains.

A few other antibiotics may be mentioned briefly. The **tetracyclines** (Fig. 12-10) inhibit the binding of aminoacyl-tRNA at the A site in the 30 S ribosomal subunit. Lincomycin, sparsomycin, and chloramphenicol (Fig. 14-25) are inhibitors of peptide bond formation and act on the 50 S subunit. Chloramphenicol also causes an accumulation of the compound ppGpp (Section C,2j). The polypeptide antibiotics thiostrepton, bryamycin, and siomycin interfere with action of the G factor and the Tu factor. Erythromycin (Fig. 12-10), other macrolide antibiotics, cycloheximide (Fig. 12-10), and fusidic acid (Chapter 12, Section I,4) prevent translocation. Fusidic acid also inhibits accumulation of ppGpp. Puromycin (Fig. 15-18) binds to the 50 S subunit and causes premature termination of peptide synthesis. A glance at its structure reveals how it can do this. It resembles in fine detail the structure of the 3' end of a tRNA molecule bearing an aminoacyl group. However, it is no aminoacyl group and once the growing peptide chain has been transferred onto the puromycin, further chain elongation is impossible.

i. Studying the Sequences of the Messages

Only recently has it been possible to obtain nucleotide sequences of parts of mRNA molecules and to study directly the coded instructions contained in them. A convenient source of mRNA molecules carrying a single message are the RNA-containing bacteriophages.[117] The genetic information for these viruses is carried by RNA molecules consisting of only 3500–4500 nucleotides and containing only three genes (Box 4-D). The RNA from phages f2, R17, MS2, and a more distant relative Qβ have been intensively studied. Complete amino acid sequences of some of the proteins specified by the RNA molecules are known, and the corresponding complete nucleotide sequence of one of the virus RNA molecules is known.[118]

To identify the beginning of a gene, a useful technique, developed by Steitz,[119] is to allow isolated ribosomes from *E. coli* to bind to the mRNA to form the initiation complex (Fig. 15-15). In the absence of additional amino acids and tRNA molecules, the initiation complex is stable. When this complex is treated with pancreatic or T$_1$ RNase, most of the RNA is degraded, but one piece is protected by the ribosome to which it is bound. A protected piece of RNA from phage Qβ has the following sequence:

$$5'—\text{AAUUUGAUCAUGGCAAAAUUAGAGAC}—3'$$

fMet Ala Lys Leu Glu Thr

Ribosome-protected 5' end of RNA
segment coding for Qβ coat protein

Note that the protected region includes the initiation codon AUG and that the codon sequence following that corresponds exactly to the known amino acid sequence of the N-terminal end of the virus coat protein. One more thing is of interest about the sequence. The two regions designated by the braces and asterisks are such that they can be paired. This would leave the initiation codon in a hairpin loop in the RNA. This loop feature of the structure of initiation regions in RNA is not universal, but it has been observed in a number of instances.

Further to the left in the foregoing Qβ nucleotide sequence are a group of four nucleotides that could bind to 16 S RNA in the same manner as is depicted in Fig. 15-14 for the initiator region for the A protein in the RNA of phage R17. Similar ribosome-protected initiator sequences have been determined for a large number of viral RNA molecules as well as for several specific mRNA molecules.[101,102]

The complete sequence of 3569 nucleotides is known for the RNA of phage MS2.[118] Parts of the sequence are shown in Fig. 15-19. The 5' end (upper left center) still bears the triphosphate group of the initiating GTP. Following a number of hairpin loops, there is a ribosome-protected

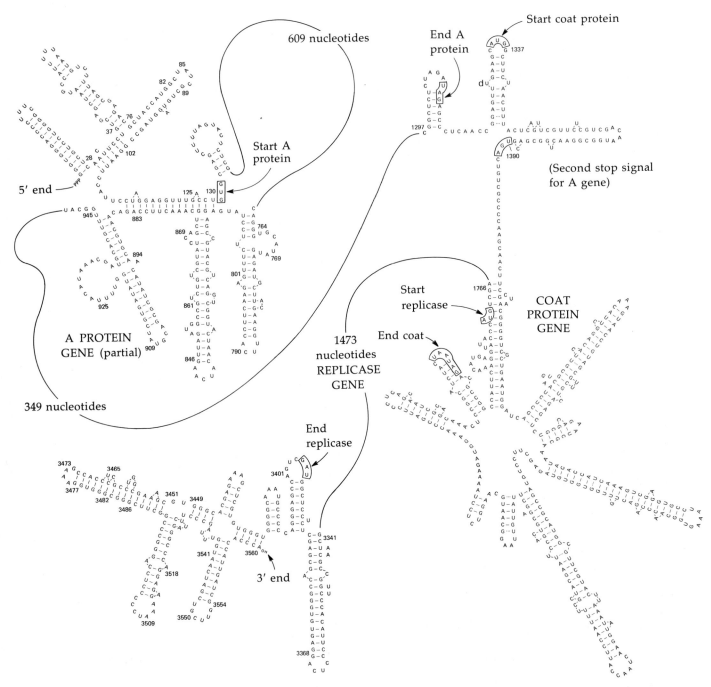

Fig. 15-19 Partial sequence and secondary structure model of RNA of bacteriophage MS2. The initiation and termination codons for each of the three genes (A protein, coat protein, and replicase) are enclosed in boxes as is the second stop signal that is in phase for the A protein gene but out of phase for the coat protein gene. The entire coat protein gene is shown but less than one-third of the entire sequence is given. From W. Fiers and associates.[118,119a,120]

BOX 15-D

REPLICATION OF RNA-CONTAINING BACTERIOPHAGE

The small icosahedral RNA-containing bacteriophage are of interest because of the small number of genes and the possibility of obtaining a detailed understanding of their replication.[a,b] The complete sequence of the 3569-nucleotide MS2 RNA is known (Fig. 15-19). The three genes code for the A protein (maturation protein), the coat protein, and a replicase subunit. One molecule of the A protein is incorporated into the mature virus. It is required for proper encapsulation of the RNA and is essential for binding of the phage to a pilus on the host cell. The approximately 180 molecules of coat protein envelop the RNA molecule. The replicase is required to duplicate the RNA molecules. The virus Qβ is a little more complex, containing RNA of 4.5 kb length and containing a few molecules of a fourth protein A_1 in addition to the maturation protein, which is known as A_2. Protein A_1 is unusual in having, at the N-terminus, the same 130 amino acid sequence as does the coat protein. The coat protein is terminated by the stop codon UGA. However, even wild type *E. coli* contains a small amount of a UGA-specific suppressor tRNA that permits translation to continue on for another ~270 residues before termination occurs at a double stop signal UGAUAA. As a consequence a large amount of coat protein and a much smaller amount of protein A_2 are formed.

The Qβ replicase has been studied intensively. Besides the replicase subunit encoded by the virus, three bacterial proteins are needed to form the complete replicase complex. They are ribosomal protein S1 and elongation factors EF-Tu and EF-Ts. All three are proteins that normally function in translation of mRNA. However, their ability to associate with RNA has been exploited by the phage for a quite different purpose.

Replication of a single-stranded virus must take place in two steps. From the plus strand present in the virus, a complementary minus strand is first formed. Initiation of this step requires another bacterial protein, the host factor HF[b] and GTP. The minus strands formed do not associate with the plus strands. They are apparently released from the replicase in a single-stranded form and presumably fold into highly structured molecules with many hairpin loops (as for the plus strand of MS2 RNA shown in Fig. 15-19). The minus strands are then copied (HF is not needed for this) to make a large number of new plus strands for incorporation into the finished virus particles.

The Qβ replicase is able to synthesize *in vitro* complete complementary strands to either plus or minus viral RNA molecules. However, the system is specific for the viral RNA and will not copy any arbitrary nucleotide sequence. Presumably certain sequences at the 3' end are essential for initiation of replication. During replication in the test tube mistakes are made including premature termination and mispairing of bases. Thus mutation takes place and it is possible to select RNA molecules much smaller than the original viral RNA that will be replicated readily by the QB replicase system. One such fragment contains only 114 nucleotides in a known sequence.[c]

[a] C. Weissman, *FEBS Letters* **40**, S10-S78 (1974).
[b] A. W. Senear and J. A. Steitz, *JBC* **251**, 1902–1912 (1976).
[c] D. R. Mills, F. R. Kramer, C. Dobkin, T. Nishihara and S. Spiegelman, *PNAS* **72**, 4252–4256 (1975).

region[119a] which begins with the initiation codon GUG. Here is direct evidence that GUG as well as AUG is a biologically important initiation codon. Following the initiation codon the nucleotide sequence shown codes exactly for the almost completely established amino acid sequence of the protein. The termination codon UAG is enclosed in a box in the figure. Following this is a short intergenic region which includes one side of a hairpin loop with the initiator codon AUG for the next gene at the end. The nucleotide sequence following this codes exactly for the experimentally established N-terminal end of the coat protein.[120] One other feature of the sequence shown is the UGA termination codon in a box shortly after the beginning of the coat protein gene (at position

1390). This termination signal is out of phase with the initiator codon AUG; hence, it does not represent a termination point for the coat gene. However, it is in phase with the UAG termination codon for the A protein. In the presence of various host *amber* suppressor genes (see Sections D,5 and D,6) the A protein is elongated and terminated at this UGA signal.

The coat gene, containing only 390 nucleotides, is shown in its entirety. The secondary structure proposed resembles a flower.[120] The gene ends with a double stop signal UAAUAG. Following an intergenic sequence of 36 nucleotides the very long replicase gene starts with an AUG codon. It ends at position 3395 leaving an untranslated segment of 174 nucleotides at the 3' end.

Another interesting example of the use of ribosomes to protect a nucleic acid from digestion has been accomplished[121] with the single-stranded DNA of bacteriophage φX174. Ribosomes protected a sequence of nucleotides containing the initiation codon ATG. This codon plus the next seven codons correspond to the known N-terminal amino acid sequence of the gene G spike protein of this bacteriophage.

j. Other Functions of Ribosomes

Not only do ribosomes make proteins but also they appear to participate actively in regulatory mechanisms influencing the entire cell. One intriguing phenomenon is the "stringent response."[122,123] Many amino acid-requiring auxotrophs of *E. coli* and other bacteria, when deprived of an essential amino acid, respond immediately by decreasing their production of ribosomal RNA, ribosomal proteins, purine nucleoside triphosphates, lipids, and other essential materials. However, mutations in the gene *rel* (relaxed) lead to continued production of rRNA even in the absence of an essential amino acid. (The stringent response is "relaxed.") To add to the mystery surrounding this curious behavior, it was found that the **guanosine polyphosphates ppGpp** and **pppGpp** (originally termed MS or "magic spot" compounds) accumulate in stringent (rel^+) strains but not in relaxed (rel^-) strains.

The concentration of ppGpp reaches 1 mM. It is now clear that the guanosine polyphosphates are synthesized on the ribosomes by transfer of a pyrophosphoryl group from ATP (Eq. 15-8):

$$\text{ATP} + \text{GDP (GTP)} \longrightarrow$$
$$\text{ppGpp(pppGpp)} + \text{AMP} \quad (15\text{-}8)$$

A special **stringent factor,** a ribosomal protein of MW ~75,000, in a single polypeptide chain is required.[124] The ribosomes involved must be bound to mRNA and must contain codon-selected uncharged tRNA in the A sites.

Just as mutations in the *rel* gene cancel the regulatory effects by blocking synthesis of the guanosine polyphosphates, so binding of tetracycline and fusidic acid mimic these mutations in permitting synthesis of ribosomal RNA to continue in stringent strains during amino acid starvation. The exact nature of the effect of the ppGpp and pppGpp on synthesis of tRNA and rRNA is still obscure, but an effect on transcriptional regulation through a complex enzyme system is suggested.[125-127]

An unexpected finding was that **phosphatidylserine synthetase** of *E. coli* is tightly bound to ribosomes.[128] This enzyme, which incorporates serine into phospholipids according to step *g* of Fig. 12-8, is responsible for synthesis of the principal membrane lipid of *E. coli*. Localization of this important enzyme on ribosomes may be linked in some way to the joint regulation of the synthesis of proteins and lipids.

D. GENETIC METHODS

It is hard to overemphasize the importance of the methods of genetics in establishing our present knowledge of molecular biology. It is important that biochemists understand these methods. For one thing, the biochemical literature is increasingly filled with the jargon of the geneticists. More important, genetic methods provide a powerful approach to the investigation of many complex biochemical phenomena. Furthermore, in looking to man's future it is essential for us to understand the problems of mutation and the variability of genes.

1. Types of Mutation

Changes in the structure of DNA occur only rarely. For example, the average gene may be duplicated 10^6 times before a single detectable mutation occurs.[128a] Nevertheless, by using bacteria or bacterial viruses, it is possible to screen enormous numbers of individuals for the occurrence of mutations. Thus, if one million virus particles are spread on an agar plate under conditions where mutation in a certain gene can be recognized, on the average one mutant will be found. The most common mutations are **base pair switches** (*point mutations*) that result from incorporation of the wrong base during replication or repair. In these mutations, one base of a triplet codon is replaced by another forming a different codon and causing the substitution of one amino acid by another* in the corresponding protein. Changes involving replacement of one pyrimidine by another (C → T or T → C) or of one purine by another are sometimes called **transition mutations,** whereas if a pyrimidine is replaced by a purine, or vice versa, the mutation is known as a **transversion.** Transition mutations are by far the most common, one possible cause being pairing with a minor tautomer of one of the bases (Chapter 2, Section D,7).† Thus, A could pair with a minor tautomer of C, causing a mutation from T to C. Note that substitution of an incorrect base in one strand will lead, in the next round of replication, to correct pairing again but with an AT pair replaced by GC, or vice versa, in one of the daughter DNA duplex strands.

From the observed rate of appearance of point mutations (one mutation per 10^6 gene duplications), we can estimate that one mutation occurs per 10^9 replications at a single nucleotide site. Point mutants tend to "back mutate," often at almost the same rate as is observed for the forward mutation. That is, one in 10^9 times a mutation of the same nucleotide will take place to return the code to its original form. The phenomenon is easy to understand. For example, if T should be replaced by C because the latter formed a minor tautomer and paired with A, the mutation would appear in progeny duplexes as a GC pair. When this pair was replicated, there would be a finite probability that the C of the parental DNA strand would again assume the minor tautomeric structure and pair with A instead of G, leading to a back mutation.

While the rates of spontaneous mutation are low, they can be greatly increased by mutagenic chemicals (Section H,1) or by irradiation. Thus, it is perfectly practical to measure the rates of both forward and back mutation. When this was done, it was found that certain chemicals, e.g., acridine dyes, induce mutations that undergo reverse mutation at a very much lower frequency than normal. It was eventually shown that these mutations resulted from either **deletions** of one or more nucleotides from the chain or from **insertions** of extra nucleotides. It seems likely that deletion and insertion mutations are a result of errors during genetic recombination and repair at times when the DNA chain is broken.

Mutations involving deletion or insertion of one or a few nucleotides are called **frame-shift mutations.** Think about RNA transcribed from a DNA containing a deletion or insertion. The messenger RNA is read by the protein synthesizing machinery from some starting point. The codons are read three bases at a time and the proper amino acid corresponding to each codon is inserted. However, when the deletion or insertion (present now in the mRNA just as in the DNA) is met, all subsequent codons are misread because the "reading frame" has been shifted forward or backward by one or two nucleotides.* As a result, the protein

---

* A base substitution does not always cause an amino acid replacement because of the "degeneracy" of the code, i.e., the fact that more than one codon specifies a given amino acid.

† There is a possibility that a purine base might pair with another purine if the latter assumed the energetically less favorable syn configuration (Chapter 2, Section D,4).[129a]

* If the reading frame is shifted by three nucleotides a protein lacking one amino acid or containing an extra amino acid residue but otherwise normal will be formed.

synthesized bears little resemblance to that formed by the nonmutant organism and is usually completely nonfunctional. However, a frame shift mutation near the 3' end of a gene, while causing "read through" of the termination codon, may allow synthesis of a functional protein containing an elongated C-terminus. Several hemoglobin variants are thought to have arisen in this way.[129a,b]

Frame-shift mutations are classified (+) or (−) depending upon whether they involve insertion or deletion of a small number of bases. Thus, mutations may be classified as +1, +2, etc. Large deletion and addition mutations are also observed. For example, a large addition will occur if a piece of foreign DNA is incorporated within a gene. Losses or gains of large pieces of chromosomes may also be classified as addition and deletion mutations.

2. Mapping the Chromosome of a Bacteriophage

Much of our knowledge of biochemical genetics stems from studies of the bacteriophage. Intensive work on the "T-even" phage T2, T4, and T6 was begun in 1938 by Max Delbrück and associates. Although these viruses are tiny objects, they turned out to be among the most complex of known viruses (Box 4-E). The genetic information is carried in a single linear DNA molecule which, in the case of T4 contains 2×10^5 base pairs, enough to code for about 200 genes. The positions of over 60 of these genes have been mapped. The method by which this was accomplished is briefly considered below.

When a bacteriophage infects a cell of E. coli, it injects its DNA through the cell wall and into the cytoplasm. Approximately 20 min later the cell bursts and about 100 fully formed replicas of the original virus particle are released. This rapid rate of production of progeny means that it is possible to carry out in a test tube in 20 min a genetic experiment that would require the entire population of the earth if humans were used. The ap-

proach is explained nicely by Seymour Benzer, the man who first mapped the fine structure of a gene.[130] Bacteriophage particles, like bacteria, can be "plated out" on agar plates. The difference is that the agar must contain a uniform suspension of bacteria susceptible to the virus. Wherever a virus particle lies a bacterium is infected. Soon the infection spreads to neighboring bacteria with production of a transparent "plaque" (Fig. 15-20). The number of active virus particles present in a suspension can be determined easily by plating and counting of the plaques.

Mutant bacteriophage can be identified in various ways, but one of the easiest depends upon looking at some character that affects the *appearance of the plaque*. Other easily detected mutant

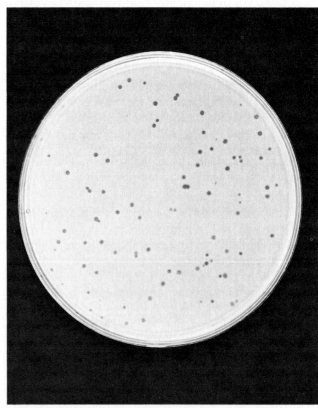

Fig. 15-20 Plaques formed by bacteriophage φ11 growing on *Staphylococcus aureus*. Each transparent (dark) plaque is the result of lysis of bacteria by the progeny of a single bacteriophage particle. Courtesy of Peter Pattee.

traits include alteration in the specificity toward particular strains of the host bacterium. A key discovery that made genetic mapping possible for bacteriophage was that *genetic recombination between two phage particles can take place within a host bacterium.* Recombination can be demonstrated as follows. Two different mutant strains of a bacteriophage are grown in large numbers and are mixed together in excess with a large number of bacteria. From the progeny phage a few are found to contain both mutant traits in the same virus and an equal number are "wild type." Although recombination between mutations that are located close together in the DNA are rare, their frequency still greatly exceeds that of new mutations. Thus, while the type of experiment described above gives no hint as to the nature of the events involved in recombination, it shows unequivocally that recombination occurs.

Study of recombination frequencies between different strains of phage soon revealed that some sites of mutation are **closely linked.** Recombination between these sites occurs rarely. Other sites are very weakly linked and recombination occurs often. This behavior is reminiscent of that established many years earlier for genes of the fruit fly *Drosophila,** corn, and other higher organisms. The basic idea behind chromosome mapping in any organism is the assumption that *recombination frequencies between two mutations are directly proportional to the distance between them on the genetic map.* For the T4 phage, a recombination frequency of 1% is taken as one unit. The total T4 map is 700 units long. The fact that this is greater than 100% means that if genes are located at opposite ends of the chromosome, multiple recombination events can occur between them. In fact, a maximum of 50% crossing over is observed for distant gene pairs and the linearity of map distance and recombination frequency holds only for distances of 10 units or less.[131]

How can recombinant bacteriophage be identified rapidly? Benzer used two strains of *E. coli,* the

B strain and the K strain, as hosts. Mutants in gene *rII* form characteristic plaques on strain B but do not grow on strain K. To determine the recombination frequency between two different *rII* mutants, the viruses are added to a liquid culture of B cells (in which they replicate) and recombination is allowed to occur. Recombination not only permits the emergence of a phage containing *both* mutations but also of a "standard" phage in which both mutations have been eliminated by the recombination process. Since only recombinants of the latter type grow in strain K, it is possible to detect a single recombinant among one billion progeny. Now consider that the total DNA length in phage T4 is 200,000 base pairs (286 base pairs per unit of map length). An 0.01% recombination frequency between two mutations means that the two mutations are only three base pairs apart in the DNA. Thus, Benzer concluded that he could easily observe the expected recombination frequencies that involved mutations even on immediately adjacent bases in the DNA.

To make fine genetic mapping practical, one more technique had to be introduced. A series of bacteriophage containing deletion mutations involving large segments of the *rII* gene were located. Using these, it was possible to quickly establish in which segment of the gene a particular mutation lay. Then, recombination experiments with previously identified mutations in that same general region allowed the mutations to be pinpointed. In this way Benzer was able to identify over 300 sites of mutations within the *rII* region. He concluded that the minimum distance between two mutable sites was completely compatible with the Watson–Crick structure of the gene.

3. The Cistron

How can one tell whether two mutations are in the same gene or in nearby or adjacent genes? The answer can be supplied by a test of **complementation.** If two mutant bacteriophage are altered in different genes, they can often reproduce within a host if the bacterium is infected with both of the

* Recombination by "crossing over" in the chromosomes of *Drosophila* was established by T. H. Morgan and associates in 1911.

mutant phage. Since each one has a good gene for one of the two proteins involved, all of the gene functions are fulfilled. On the other hand, if both mutant phages are defective in the same gene (although at different locations), they cannot complement each other in a coinfection. The experiment is often referred to as a cis-trans comparison. The coinfection with the two different mutants is the trans test. A control, the cis test, uses a recombinant containing both of the mutations in the same DNA and coinfection with a standard phage. Normal replication is expected in this instance.

When the cis-trans test was applied to various mutants in the rII region, it became clear that there are two genes, rIIA and rIIB. This demonstration depended entirely upon complementation in the cis-trans test. The name **cistron** was proposed by Benzer to represent that length of DNA identifiable in this fashion as a genetic unit. For most purposes the terms *gene* and *cistron* are synonymous and both are used freely in the biochemical literature. However, from a genetic viewpoint, cistron is a more precise term.

When mapping of the *rII* region was done, there was no information about the functions of the proteins specified by these two cistrons. Recently, however, both the rIIA and rIIB proteins have been shown to become incorporated in the membranes of phage-infected bacterial cells.[132,133] There they affect the ease of lysis of the infected cells and in that manner cause rII plaques to be larger and to have sharper edges than standard plaques.

4. Establishing the Correspondence of a Genetic Map and an Amino Acid Sequence

Although the studies of the *rII* region of the T4 chromosome established that genetic mapping could be carried to the level of individual nucleotides in the DNA, it was still necessary to prove a linear correspondence between the nucleotide sequence in the DNA and the amino acid sequence in proteins. This was done by C. Yanofsky[134] and associates through study of the enzyme tryptophan synthetase of E. coli and by Sarabhai et al. through a study of the T4 coat protein.

Tryptophan synthetase (p. 853) consists of two subunits A and B (or α and β), the former containing only 268 amino acids. A fine structure map of the A gene was prepared as follows. A large series of mutant bacteria unable to grow in the absence of added tryptophan (tryptophan auxotrophs) were isolated. Genetic crosses were carried out with the aid of a special **transducing bacteriophage** Plkc.[134] Transducing bacteriophage, while multiplying in susceptible bacteria, sometimes incorporate a portion of the bacterial chromosome into their own DNA. Then, when the virus infects other bacteria, some of the genetic information can be transferred through a recombination process into the chromosome of bacteria that survive infection. Use of a series of deletion mutants, as in the *rII* mapping, permitted division of the A gene into a series of segments and observation of recombination frequencies permitted fine structure mapping.

The second part of the proof of colinearity of DNA and protein sequences was the determination of the complete amino acid sequence of tryptophan synthetase and peptide mapping (Chapter 2, Section H,2) of fragments of the mutant enzymes. From the peptide maps it was possible to identify altered peptides and to establish the exact nature of the amino acid substitutions present in a variety of different tryptophan auxotrophs. When this was done, it was found that those mutations that mapped very close together had amino acid substitutions at adjacent or nearly adjacent sites in the peptide chain.

The same problem was approached by Sarabhai and associates[134a] through the nonsense mutations (Section 5) which lead to premature chain termination during protein synthesis. During late stages of the infection of E. coli by phage T4, the major fraction of protein synthesis is that of a single protein of the virus head. Synthesis of protein by infected cells was allowed to proceed in the presence of specific ^{14}C-labeled amino acids. Then cell extracts were digested with trypsin or chymotrypsin, and the head protein peptides were separated by electrophoresis and autoradiograms were prepared. A series of T4 nonsense mutants that mapped within the head protein gene were shown to give rise to incomplete head protein chains. The peptide fragments were of varying lengths. By examining the radioautograms prepared from the enzymatically fragmented peptides, it was possible to arrange the mutants in a sequence based on the length of peptide formed and to show that this was the same as that deduced by genetic mapping.

As we have already seen, the colinearity of codon and amino acid sequences has been verified by the determination of actual nucleotide sequences in RNA and DNA molecules and of the corresponding amino acid sequences (Section C,2,i).

Before the triplet nature of codons had been finally established, Crick and associates made use of frame-shift mutations in a clever way to demonstrate that the genetic code did consist of triplets of nucleotides. Consider what will happen if two strains of bacteria, each containing a frame-shift mutation (e.g., a −1 deletion), are mated. Genetic recombination can occur to yield mutants containing *both* frame-shift mutations. However, it would be difficult to recognize such recombinants because (according to almost any theory of coding) they would still produce completely defective proteins. However, Crick *et al.* were able to introduce a third frame-shift mutation of the same type in the same gene and to observe that the recombinants containing all three deletions (or insertions) were able to synthesize at least partially active proteins. The explanation is simple. Introduction of one or two single nucleotide deletions completely inactivates a gene but deletion of three nucleotides close together within a gene shortens the total message by three nucleotides. The gene will contain only a short region in which the codons are scrambled. Thus, the protein specified will be normal except for a small region where some amino acid substitutions

will be found and where one amino acid will be completely missing. We already know that a relatively small number of amino acids are completely invariant in most proteins. Thus, functional gene products are very often possible if a small region of the gene is modified just as long as the reading frame has not been shifted.

5. Conditionally Lethal Mutations

While the study of nutritional auxotrophs has been very important to the development of biochemistry, it is directed narrowly at one gene or group of genes involved in synthesis of a particular nutrient. It is desirable to have a means of detecting mutations in the whole host of other genes present within cells. The problem is that most mutations are **lethal** and this effect cannot be overcome by adding any nutrient. Earlier genetics studies had shown that lethal mutations are very common in higher organisms. However, since eukaryotic cells have pairs of homologous chromosomes, lethal mutations can be carried in one chromosome and the individual survives. With bacteria and viruses there is only one chromosome, and lethal mutants cannot survive.

Nutritional auxotrophs can be described as **conditionally lethal mutants;** that is, they can survive only if the medium is supplemented with the nutrient whose synthesis depends upon the missing enzyme. Other kinds of conditional lethal mutations have been found that permit study of almost every gene in an organism. One class of such mutations is called **temperature sensitive.**[135,136] Temperature-sensitive (*ts*) mutants grow perfectly well at a low temperature, e.g., 25°C, but do not grow at a higher temperature, e.g., 42°C. Many temperature-sensitive mutations involve amino acid replacement of such a kind that the affected protein is less stable to heat than is the wild type protein. Others involve a loss in protein-synthesizing ability for reasons that may be obscure. Many temperature-sensitive mutations doubtless occur in nature, an example being the gene that controls hair pigment in Siamese cats.[136] The gene (or gene product) is inactivated at body temperatures but is active in the cooler parts of the body, the paws, tail, and nose, with the result that the cat's hair is pigmented only in those regions.

Screening for conditionally lethal temperature-sensitive mutants of bacteriophage T4 permitted isolation of hundreds of different mutants involving sites at random over the entire viral chro-

mosome. Complementation studies permitted assignment of these to individual genes which at first were identified only by number (Fig. 15-21). Later specific functions were associated with many of the genes.[137,138] Thus, the product of gene 42 has been identified as a special enzyme required in the synthesis of hydroxymethyl-dCMP (Chapter 14, Section K,1). Genes 20–24, among others, must code for head proteins because mu-

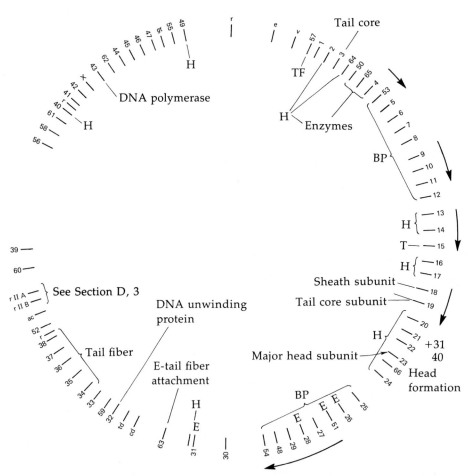

Fig. 15-21 Linkage map for bacteriophage T4. From H. A. Sober, ed., "CRC Handbook of Biochemistry," 1st ed., p. I-25. Chem. Rubber Publ. Co., Cleveland, Ohio, 1968, and F. A. Eiserling and R. C. Dickson, *Annu. Rev. Biochem.* **41**, 467–502 (1972). Abbreviations: H, head; T, tail; TF, tail fiber; BP, base plate; and E, proteins with enzymatic rather than structural functions. The curved arrows indicate some genes that are transcribed together.

tants produce normal tails but no heads. Gene 23 appears to code for the major head subunits, while gene 20 has something to do with "capping" the end of the head. Mutants produce cylindrical "polyheads" in place of the normal heads. Mutants of genes 25–29 have defective base plates and do not form tails while mutants 34–38 lack tail fibers.

A second type of conditionally lethal mutation leads to alteration of an amino acid codon to one of the three **chain termination codons** UAG, UAA, and UGA (Table 15-2).[139,140] These are often called **nonsense mutations** in contrast to **missense mutations** in which one amino acid is replaced by another. A chain termination mutant synthesizes only part of the product of the defective gene, whereupon the peptide chain is released from the ribosome because of the presence of the termination codon. A remarkable aspect of chain termination mutations is that they can be **suppressed** by other mutations in distant parts of the virus or bacterial chromosome. Thus, many otherwise lethal mutations of bacteriophage T4 were discovered by their ability to grow in certain mutant strains of E. coli (which contained **suppressor genes**)[140a] and their inability to grow in the normal B strain. Three different suppressor genes supD, supE, and supF (also known as su1, su2, and su3, respectively),* that were found to suppress mutations to UAG, are commonly known as **amber** suppressor genes (see footnote, p. 902). Other suppressors of this class have since been found. A second class of chain termination mutations was suppressed by genes supB and supC and involve mutation to UAA. These are popularly known as **ochre** mutants. More recently, suppressors for mutation to UGA have also been found.[140b]

Like the temperature-sensitive mutants, amber and ochre mutants can be obtained in almost any genes of a bacterial virus. Chain termination mutants of unessential genes in bacteria can be recognized by transferring the genes by mating or by viral transduction into a strain (sup⁺) containing a desired suppressor gene.

* The symbols supD, supE, and supF are those currently recommended by Bachmann et al.[15]

Conditionally lethal mutants have been of great value in developing our understanding of the genetics of bacterial viruses. They also provide a powerful technique for approaching complex problems of bacterial physiology. For example, how complex is the machinery needed for a bacterium to be able to sense the presence of the foodstuff in the medium and to swim toward it? It has been established that bacteria are "preprogrammed" to sense gradients of concentrations of chemical attractants and to alter their direction of travel in ways that lead them into higher concentrations.[141-143] It would be nice to know how many proteins are needed to sense the attractant, to pass whatever message is required to the flagella (Box 4-B), and to direct the latter either to rotate in a manner that leads to forward motion or in a way that causes random tumbling (Chapter 16, Section B,7).

Even though few clues as to the basic chemistry underlying these phenomena have been obtained, the use of temperature-sensitive mutants and complementation tests can permit us to establish the total number of genes involved as well as their map positions on the E. coli chromosome. This can often be an important step toward a more complete elucidation of a biological phenomenon.

6. The Nature of Suppressor Genes

What is the chemical basis for the suppression of the effect of one mutation by a second mutation at a different point in the chromosome? No single answer can be given. Rarely a mutation is suppressed by another mutation within the same gene. Such a result can be described as **intragenic complementation.** Suppose that a mutation leads to an amino acid replacement which disrupts the structural stability or function of a protein. It is possible that mutation at another site involving a residue which interacts with the first amino acid replaced will change the way in which the two residues interact and will lead to a restoration of a functional protein. For example, if the first amino acid side chain is small and is replaced by mutation with a larger side chain, a second mutation leading to a decrease in the size of another side chain might permit the protein to fold and function properly. An example was found among mutants of tryptophan synthetase.[144] Mutants in which Gly-211 was replaced by Glu or Tyr-175 by Cys both produced inactive enzymes, but the double mutant with both replacements synthesized active tryptophan synthetase. It is thought that most cases of intragenic suppression involve changes in the subunit interactions in oligomeric proteins.

The best known suppressor genes are those that suppress many different chain termination mutations. The explanation of the chemical nature of these genes was discerned, in part, from experiments involving transfer of suppressor gene *supF* (*su3*) into the DNA of a bacteriophage. It was observed that this DNA specifically hybridized with a minor species of **tyrosyl-tRNA (tyrosyl-tRNA$_1$)**. Subsequent investigation showed that su_{III} is a structural gene for a minor tyrosyl tRNA in which the normal GUA anticodon has been replaced with CUA. The latter can pair with the chain termination codon UAG (the *amber* codon) permitting the ribosome to insert tyrosine at the site of chain termination signals introduced in *amber* mutations. It is a little puzzling that a tRNA that prevents chain termination should not interfere with the synthesis of all of the other essential proteins within the bacterium. However, suppression is typically less than 30% efficient. Hence, many protein chains will be terminated normally. If the presence of two chain termination signals is a general feature of genes, it is probable that most protein synthesis in the presence of the suppressor tRNA terminates normally. However, the premature chain termination caused by *amber* mutations will be partially inhibited, permitting the cell to make enough of the missing enzyme to survive. The nucleotide sequence of a further mutated *supF* tRNA and of its longer precursor is shown in Fig. 15-9.

A number of other suppressor genes have also been identified as specific tRNA structural genes.[140] Recently, a frame-shift suppressor mutation has been found in a glycine tRNA gene of *Salmonella typhimurium*.[145] In this tRNA, at the anticodon position there is a nucleotide quadruplet CCCC instead of the usual CCC triplet anticodon. It is the only known tRNA with eight unpaired nucleotides in the anticodon loop instead of the usual seven.

Suppressor genes are not limited to bacteria. Thus, the vermilion eye color mutation of *Drosophila* is suppressed by a mutation in a tryptophanyl tRNA gene.[140] The vermilion mutation leads to a loss of brown eye pigments because of the inactivity of tryptophan oxygenase (Eq. 10-45). It was found that the tryptophan oxygenase from the *vermilion* mutant is inhibited by tRNA$_2$^{Trp}, one of the two tryptophanyl tRNA's. The suppressor mutation alters the tRNA in such a way that the inhibition is relieved.[140]

Suppressor mutations have been used in an impressive way by Miller, Lu, and associates[145a,b] to systematically obtain a series of over 300 mutant variants of the *E. coli lac* repressor protein. The first step was to induce *amber* mutations into the gene at over 80 positions. Then, the mutated genes were transferred into episomes (next section) for cloning. The viruslike episomes were then used to infect five strains of bacteria, each carrying a suppressor mutation that would introduce a different amino acid when the (termination) codon UAG was encountered. From these infected bacteria large quantities of the mutant forms of the *lac* repressor were isolated. It was found that many mutations near the N-terminal end interfere with binding of the repressor to DNA, whereas mutations near the center interfere with binding to the inducer.

7. Plasmids and Episomes

An important tool is provided by the existence of small genetic elements that remain separate from the bacterial chromosome. One group of such factors, the F agents, have previously been considered (Section A,1,d). These small circular DNA molecules are members of a much larger group of such agents known as **plasmids** and **episomes.**[14,146,147] Among them are the **colicinogenic factors** and **drug resistance factors** (R factors). Plasmids replicate independently of the chromosome and may be present as one or several copies for each bacterial chromosome. Episomes are plasmids that are able to become integrated into the bacterial chromosome. Some extrachromosomal elements are episomes in one host and plasmids in another. Plasmids may be infectious (transferable) or noninfectious. In the former case, they contain genes for synthesis of sex pili (Section A,1,d) and are able to transfer their DNA into another cell. The plasmid is defined as a **sex factor** if it is able to integrate with a chromosome and later to come out and transfer other genes with it. We have already seen that F-factor mediated gene transfer has been an important process in the mapping of bacterial chromosomes.

Plasmids and episomes vary in size. The F-1 sex factor is a circular molecule of supercoiled DNA of MW $\sim 62 \times 10^6$. It is large enough to contain about 90 genes and has a length of ~ 30 nm, about $2\frac{1}{2}\%$ that of the *E. coli* chromosome. Colicinogenic factors,[148] which may also be present in *E. coli* in as many as 10–15 copies per bacterial chromosome, are often much smaller with MW ~ 4–5×10^6. Some larger colicinogenic factors are also sex factors. These plasmids carry genes for toxic protein antibiotics known as **colicins** that attack other strains of *E. coli*. The plasmid also carries a gene or genes conferring resistance to the toxins on the host bacterium. Colicin E-3 acts by entering a susceptible bacterial cell and inhibiting protein synthesis by cutting out a small fragment from the 3' end of each of the 16 S RNA molecules of the bacterial ribosomes[149] (see Fig. 15-14).

Much attention has been directed to the drug resistance factors. As antibiotics came into widespread use, an unanticipated problem arose in the rapid development of resistance by the bacteria. The problem was made acute by the indiscriminate use of antibiotics and by the fact that resistance genes are easily transferred from one bacterium to another by the infectious R-factor plasmids.[150–152] Since resistance genes for a variety of different antibiotics may be carried on the same plasmid, "super bacteria," resistant to a large variety of antibiotics, may develop. Hospitals are the most likely places for their development and serious epidemics of drug resistant infections have resulted. The mechanisms of resistance often involve inactivation of the antibiotics. Aminoglycosides such as streptomycin (Box 12-A) and kanamycin are inactivated by enzymes catalyzing phosphorylation or adenylylation of specific hydroxyl groups on the sugar rings. Penicillin (Box 7-D) is inactivated by a penicillinase that hydrolytically cleaves the β-lactam ring. Chloramphenicol (Fig. 14-25 is inactivated by acylation on one or both of the hydroxyl groups.

What is the origin of the drug resistance factors? Why do genes for inactivation of such unusual molecules as the antibiotics exist widely in nature? A possible answer is that the drug resistance genes fulfill some normal biosynthetic roles in nature and that an antibiotic-containing environment leads to selection of mutants of such genes with specific drug-inactivating properties. Nevertheless, it is not entirely clear why drug resistance factors appear so promptly in a population treated with an antibiotic. A partial solution to the resistance problem has been the development of semisynthetic modifications of naturally occurring antibiotics. Since the R factors carry genes for enzymes that modify specific sites on the antibiotic, it is sometimes possible to chemically alter those sites in such a way that they can no longer undergo the enzyme-catalyzed reaction induced by the R factor.

The similarity of infectious plasmids to viruses has often been noted.[153] Thus, filamentous bacteriophages, (Box 4-C and Fig. 4-8)[154] emerge from a bacterial cell by accumulating hydrophobic protein subunits within the membrane and extruding them in the form of thin microtubules of ~ 6 nm in diameter and containing DNA molecules ready to be transferred into other bacteria. The phage absorb to the F pili of male bacteria and the DNA, in a manner that is not understood, enters the cell.[155]

Just as the filamentous bacteriophage carry genes for the protein subunits of their coat, so F sex factors carry genes for synthesis of pilins. Pilins also accumulate within the cell membrane and are extruded to generate F pili. Thus, a close relationship between plasmids and viruses is suggested. One result of DNA transfer from Hfr into F$^-$ bacteria is the introduction of a copy of the F agent into the female bacterium. Since this converts the recipient into a male, Brinton referred to "bacterial sex as a virus disease." There is also a remarkably close similarity between episomes that can be integrated into bacterial chromosomes and the "temperate" bacteriophage considered in the next section.

8. Temperate Bacteriophage; Phage λ

When DNA from a bacteriophage enters a bacterial cell it ordinarily seizes control of the metabolic machinery of the cell almost immediately and directs it entirely toward the production of new virus particles. This leads within a period of about 20 min to the production of one or two hundred progeny viruses and to the lysis and death of the cell. A striking exception to this behavior is provided by temperate phage. After entering the cell, the DNA from a temperate phage may become repressed and integrated with the bacterial genome in the same fashion as can an F factor (Fig. 15-2). In the resulting **prophage** or **lysogenic** state, the repressed phage DNA is replicated as part of the bacterial genome but does no harm to the cell unless some factor "activates" the incorporated genetic material by release of the repression. Replication of the phage and lysis of the bacterium then ensues. Temperate phage may also exist as plasmids (e.g., P1).

The best known temperate phage is **phage lambda** of *E. coli*.[156-158] A tailed virus somewhat similar to the T-even phages (Box 4-E), phage λ has a smaller DNA genome of MW ~31×10^6 (~46,500 base pairs). Within the bacterial cell the ends of the λ DNA may be joined to form a cir-

cular replicative form of the virus. In many of the infected cells (typically about 30%) the λ DNA becomes integrated into the *E. coli* chromosome at the special site, *att* λ, which is located at 17 min on the *E. coli* chromosome map. The incorporated phage DNA now occupies a linear segment amounting to about 1.2% of the total length of the *E. coli* chromosome. It is replicated along with the rest of the chromosome and for the most part goes unnoticed.

Bacteriophage λ is of interest to biochemists for a number of reasons. Perhaps the foremost is that it provides an opportunity to answer basic questions about the control of transcription. Thus, we can ask how it is that most of the genes of the λ prophage can remain unexpressed for many generations but can be turned on with the synthesis of active viruses under some circumstances. The λ genome is small so that we may hope to understand its genetics rather completely. The host, *E. coli* K12, has also been well studied from a genetic viewpoint and contains useful *amber* suppressors that make it easy to detect mutations in the bacteriophage. Furthermore, the integrated prophage can undergo mutations of almost any type, including large deletion mutations, and can still be investigated through complementation studies with other strains of virus. Thus, a whole family of modified **defective λ phage** is available. When the λ prophage is excised from the bacterial chromosome, adjacent bacterial genes are occasionally carried with it. This fact has led to development of a group of λ transducing phage that carry certain bacterial genes and are able to transfer them into bacteria lacking these genes. The possibility that such genes may be transferred into the genome of plants and other higher organisms has created great excitement and controversy. Another development from the study of λ is a method of more accurately mapping gene positions.

a. Transcriptional Controls

The answer to the question of how the λ genes can remain unexpressed has been found in a repressor protein.[159-161] One short operon of the λ prophage is continuously transcribed by the *E.*

coli RNA polymerase. This operon contains genes *cI* and *rex*. They are transcribed from the *l* strands of the prophage DNA as indicated in Fig. 15-22. The protein specified by gene *cI* is the repressor, an oligomeric molecule (predominantly a dimer) with a monomer molecular weight of 27,000. This repressor binds to two operator sites in the prophage DNA. One operator (o_L) is to the left and the other (o_R) to the right of the *cI* gene (Fig. 15-22). From a study of fragments of DNA protected from

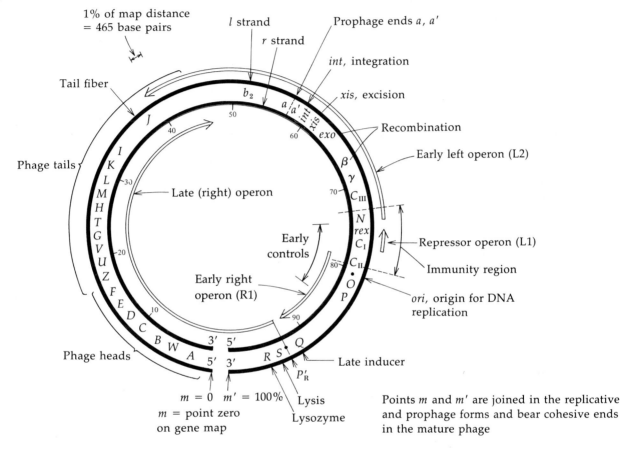

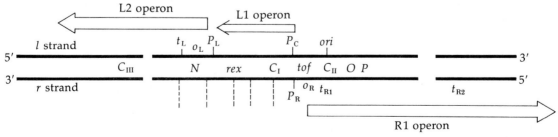

Fig. 15-22 Genetic and physical map of the λ phage genome. After Szybalski.[156] See A. Honigman, S.-L. Hu, R. Chase and W. Szybalski, *Nature (London)* **262**, 112–116 (1976) for a more detailed diagram of the immunity region.

nuclease digestion by the repressor, it has been concluded that each operator has three subsites which are filled (from left to right at o_L and from right to left at o_R) successively by up to six repressor monomers. The nucleotide sequence of the region has been determined, and it has been found that each of the presumed subsites has a twofold axis of approximate rotational symmetry (a nearly palindromic sequence). In each 17-base-pair subsite one half-site contains the sequence TATCACCGC or a clearly related one while the other half-site is somewhat more variable. Presumably a dimeric repressor binds to each subsite with the monomers in quasiequivalent conformations.[159]

By blocking these operators, the repressor prevents synthesis of enzymes that permit **excision** of the λ DNA and replication and transcription of the rest of the genes. The matter is more complex than this since gene products *cIII* and *cII,* from the **early left** and **early right operons,** respectively, appear to stimulate the transcription of *cI* and are needed for establishing the lysogenic state initially. Once established, these genes do not function since they are never transcribed.

It is thought that there are only a few molecules of the λ repressor present in a cell. This is ordinarily sufficient to maintain the prophage state. On the other hand, irradiation of the bacterium with ultraviolet light (apparently acting indirectly through depression of DNA synthesis) leads to an inactivation of repressor and to transcription of the other λ operons.

Transcription of the left operon begins at p_L. The product of the first gene *N* is a protein that permits transcription to continue on past points t_L and t_R.[161] It is an unstable, short-lived molecule of $t_{1/2} \approx 2$ min.[162] Leftward transcription proceeds through genes *exo* and *β* which are involved with recombination and *xis* which is required for excision. When the λ DNA is integrated into the *E. coli* chromosome it is cut at points *aa'* (Fig. 15-22) and is inserted just to the right of the *gal* operon (Fig. 15-1). Prophage transcription can now continue past point a' and into the genes of the bacterium. Translation of the mRNA formed from this early

left operon generates the enzymes needed to free the prophage and to permit reformation of the circular replicative form of the phage DNA. The excision is also made near point a' and it is easy to see how the nearby *gal* genes can sometimes be included in the excised λ genome.

The product of gene *N* also permits rightward transcription through genes *O*, *P*, and *Q*, and at a slower rate on along the rest of the chromosome to point a. Genes *O* and *P* code for proteins that permit the host replication system to initiate formation of new λ DNA molecules. Replication begins at the point *ori* and occurs in both directions as is discussed in Section E. Gene *Q* codes for a protein that greatly increases the transcription of the **late genes** beginning at promoter P_R.

As indicated in Fig. 15-22, the chromosome is customarily divided into four operons, the short one that produces repressor, and the early left, early right, and late operons.* The early operons code largely for replication and recombination enzymes and control proteins. The late operon is concerned with production of proteins needed for assembly of the virus particles and must be transcribed at an even higher rate; hence, the need for the product of gene *Q*. Within the late operon, genes *A* to *F* are involved in packaging of λ DNA and in formation of heads, while genes *z* to *j* are concerned with the production and assembly of tails. Genes *S* and *R* produce proteins that lead to destruction of the host membrane and to lysis of the cell. During the late stages of lytic growth the early genes are largely shut off by another λ repressor (gene *cro*). We can see that even in a virus the control of transcription can be a rather complex process.

b. "Sticky Ends" of DNA Molecules

While the replicative form of phage λ is circular, the DNA in the mature virus is known to be linear. Unlike the linear DNA of T-even phage, that of λ spontaneously forms either circles or linear aggregates when released from the virion.

---

* A somewhat similar organization of the genes is found in T-even phage, in phage T7, and other bacterial viruses.

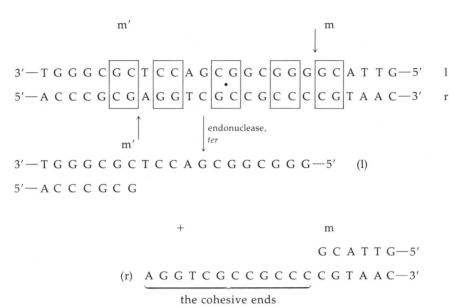

Fig. 15-23 The cohesive ends (sticky ends) of the DNA of phage and their generation from the replicative form by an endonuclease. Note the local two-fold axis of approximate symmetry. (·), an appropriate site for interaction with a symmetric dimeric enzyme. Symmetrically disposed base pairs are enclosed in boxes. The points m m' correspond to those on the gene map of Fig. 15-22.

This suggested that λ DNA contains "sticky" ends that cling together because of specific base pairing. Direct sequence determination has verified the postulate. Figure 15-23 shows the nucleotide sequence at the points m,m' (Fig. 15-22) where the replicative form is known to open, together with the cohesive 5' ends of the l and r strands. The circular form is believed to be opened by the action of an endonuclease (ter function). Not only did the sequence determination verify the assumed nature of the sticky ends, but also it revealed that the two points of hydrolytic cleavage (indicated by arrows in Fig. 15-23) are separated by 12 nucleotide pairs in a region of the DNA that has a striking degree of symmetry.[163,164] Thus, we meet another palindrome marking a special site in a DNA molecule.

c. Heteroduplexes and Physical Mapping of Genes

The availability of phages with large deletions in various parts of the genome has made possible the development of a new method of gene mapping that makes use of direct observation with the electron microscope.[165] DNA is isolated from two different phage strains, for example, from wild type λ and from a mutant phage with a particular gene or genes deleted. The λ DNA can readily be denatured and separated into r strands and l strands. Then, if the l strand of one strain is mixed with the r strand of another strain and annealed, a double-stranded DNA will be formed. However, since there is a deletion in one strain, the homologous region in the normal λ DNA will form a single-stranded loop that can be readily visualized in the electron microscope. Figure 15-24 shows an example of a micrograph of such a **heteroduplex** molecule with a deletion loop and also a "bubble" where a segment of nonhomologous DNA has been substituted in one strand.[165] Since distances can be measured very accurately on the electron micrographs, precise "physical maps" can be obtained in this way. The chromosome map of Fig. 15-22 is a physical map, and as such is a more pre-

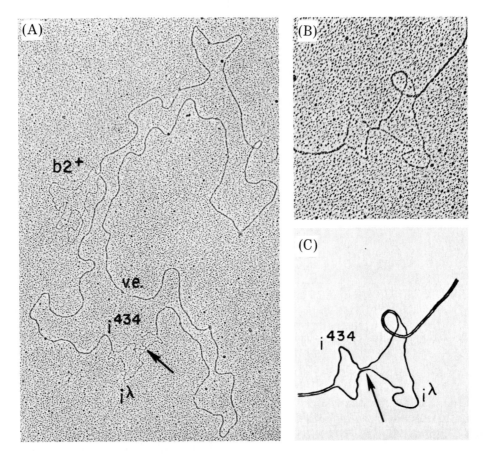

Fig. 15-24 (A) Electron micrograph of a heteroduplex DNA molecule constructed from complementary strands of phages λb2 and λ*imm*434. In λb2 a segment of λ DNA has been deleted producing a deletion loop (labeled b2) and in λ*imm*434 a piece of DNA from phage 434 has been substituted for λ DNA resulting in a "nonhomology bubble" (labeled *i*434/*i*λ). The vegetative (cohesive) ends of the DNA are labeled v.e. (B) Enlargement of the nonhomology bubble. (C) Interpretative drawing of view in (B). Arrow marks a short (20–150 nucleotide) region of apparent homology. From B. C. Westmoreland, *et al. Science* **163,** 1343–1348 (1969). Copyright 1969 by the American Association for the Advancement of Science.

cise representation of distances than is the *E. coli* map of Fig. 15-1. The method has also been applied to colicinogenic factors.[166]

9. The Genetics of Eukaryotic Organisms

Whereas DNA synthesis takes place almost continuously in a rapidly growing bacterium, replication occupies a more limited part of the **cell cycle** of eukaryotic cells.[167] Thus, in a mammalian cell, mitosis proper may typically require about 1 h (Fig. 15-25). It is followed by a "gap" period, usually designated G_1. The length of this period is quite variable and depends upon the nutritional state of the cell and other factors. About 10 h is typical. During the S phase (~9 h) active DNA replication takes place. This is followed by a second gap (G_2) that occupies 4 h in the 24 h cell cycle shown in Fig. 15-25. Bear in mind that the length of the different segments of the cell cycle varies quite widely among different organisms. Also note that it is only in a rapidly growing

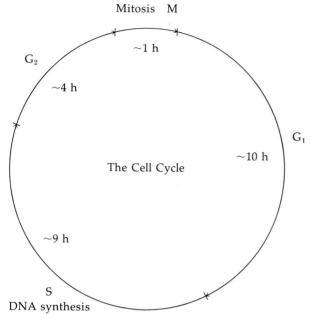

Fig. 15-25 The cell cycle. The times are typical for a mammalian cell but vary greatly among different organisms. The period $G_1 + S + G_2$ is also referred to as **interphase.**

DNA replication during the S phase of the cell cycle. As the folding of the chromosomes occurs (during prophase) the nuclear envelope often completely fragments or dissolves.

An important event, that *precedes* the main stages of mitosis, is the formation of **poles** in the cell. In animal cells, the poles are formed by the **centrioles** which move apart and take up positions at opposite sides of the cell. Each of the centrioles is accompanied by a smaller "daughter" centriole

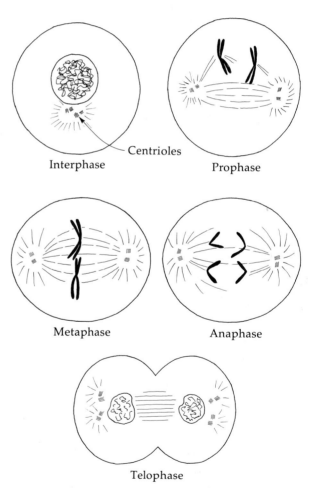

Fig. 15-26 Mitosis. Illustrated for a cell with one homologous pair of chromosomes. After D. Mazia, *Sci. Am.* **205,** 101–120 (Sept. 1961). Copyright by Scientific American, Inc. All rights reserved.

culture that all, or most, cells can follow the same cycle. A characteristic of cells in the adult body is that most of them are inhibited from division most of the time. Thus, the concept of a cell cycle has been challenged.[168]

Now, before considering genetics of higher organisms, let us review briefly some facts about the processes of cell division known as mitosis and meiosis.

a. Mitosis

The distribution of chromosomes to daughter cells of somatic cells undergoing division is accomplished by mitosis (Chapter 1, Section C,3). The successive phases of mitosis are referred to as **prophase, metaphase, anaphase,** and **telophase** (Fig. 15-26). As the chromosomes condense during prophase, it is seen that they actually consist of two separate entities coiled together. These are the identical **chromatids,** each of which is formed from one of the two identical double-stranded DNA molecules (or groups of molecules) formed by the

lying at right angles to the larger parent. In plant cells which lack centrioles a more diffuse pole is formed. As the cell prepares for mitosis, fine microtubular fibers (~ 15 nm diameter) can be seen radiating from the poles. At the end of prophase the microtubules run from one pole to the other to form the **spindle.** They also attach to the chromosomes at the **centromeres.**

At metaphase, the chromosomes are lined up in the center of the cell to form the **metaphase plate.** Now the centromere divides, permitting the sister chromatids to be completely separated. During anaphase, the separated chromatids, now referred to as the **daughter chromosomes,** move to opposite poles as if pulled by contraction of the spindle fibers. However, the mechanism of chromosome motion is not known. Telophase is the final stage in which new nuclear envelopes are formed around each set of daughter chromosomes and the cell either pinches in two or (in plants) new plasma membranes and cell wall are constructed through the center of the cell.

b. Meiosis

The mechanism by which chromosomes are distributed into germ cells (gametes), i.e., during the formation of egg and sperm cells, is known as meiosis (Chapter 1, Section C,3). Formation of gametes involves a halving of the chromosome content of a cell, each gamete receiving only one chromosome of each homologous pair. Genes found in the same chromosome are said to be **linked** because of their tendency to be passed together to the offspring. Genes present in different chromosomes are not linked and their inheritance follows the pattern of **random segregation** established in Mendel's famous studies.

The simple fact that the genetic material is put up in several different packages (chromosomes) is sufficient to provide for considerable mixing of genetic information between different individuals in sexual reproduction. Note, however, that it provides no means for changes *within* the chromosomes. Mixing of genetic information within chromosomes does occur by genetic recombination occurring during **crossing-over,** an aspect of

meiosis with an essential biological role. In the S phase preceding meiosis, DNA is duplicated just as it is prior to mitosis. Now there is sufficient genetic material present to produce four haploid cells and meiosis consists of two consecutive cell divisions (Fig. 15-27). Crossing-over occurs prior to the first of these divisions, at the four strand stage. The two homologous chromosomes of a pair come together to form what is called a **bivalent** or **tetrad** made up of four chromatids. Each chromatid is seen to come into intimate contact with a chromatid in the other homologous chromosome at points known as **chiasmata.** During metaphase of the first meiotic cell division, the homologous chromosomes (each still containing two chromatids) separate. Each chromatid now carries with it some genetic information that was previously found in the other member of the homologous pair and vice versa (Fig. 15-27). Now without further replication of DNA in the second meiotic cell division, the chromatids separate to form haploid cells.

The process of crossing-over provides a means by which genes that are linked on the same chromosome can be separated, providing offspring with mixtures of genetic traits other than those predicted by simple Mendelian theory. The effects of crossing-over were first studied extensively by T. H. Morgan with the fruit fly *Drosophila melanogaster.* The first genetic maps were made by assuming a direct relationship between the frequency of crossing-over and the linear distance between genes in a chromosome. Thus, the same approach to genetic mapping that was used later with *E. coli,* i.e., the measurement of recombination frequencies, was applied much earlier to crossing-over in the chromosomes of *Drosophila.* Extensive genetic maps involving many mutations were obtained for the four chromosomes of this organism. Similar techniques have also been applied to many other organisms including maize, yeast, and the fungi such as *Neurospora crassa* and *Aspergillus nidulans.*

An important advantage in using fungi for genetic studies is that, like prokaryotes, they are haploid during much of their life cycle. Biochem-

Fig. 15-27   Meiosis. Cell division leading to formation of haploid gametes.

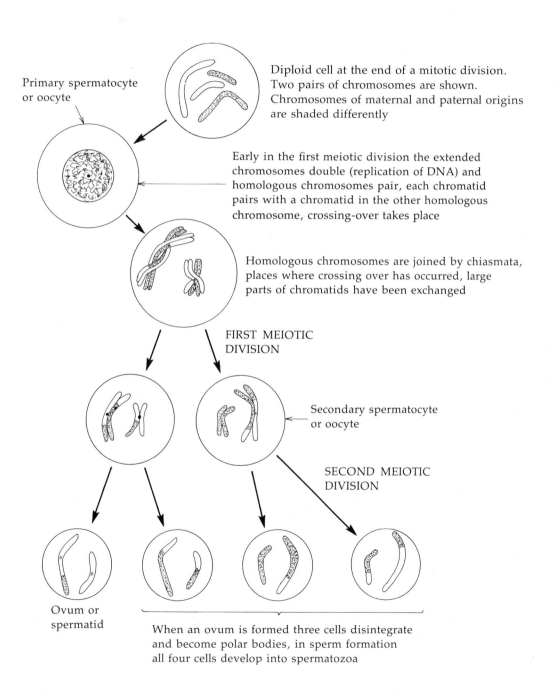

Primary spermatocyte
or oocyte

Diploid cell at the end of a mitotic division.
Two pairs of chromosomes are shown.
Chromosomes of maternal and paternal origins
are shaded differently

Early in the first meiotic division the extended
chromosomes double (replication of DNA) and
homologous chromosomes pair, each chromatid
pairs with a chromatid in the other homologous
chromosome, crossing-over takes place

Homologous chromosomes are joined by chiasmata,
places where crossing over has occurred, large
parts of chromatids have been exchanged

FIRST MEIOTIC
DIVISION

Secondary spermatocyte
or oocyte

SECOND MEIOTIC
DIVISION

Ovum or
spermatid

When an ovum is formed three cells disintegrate
and become polar bodies, in sperm formation
all four cells develop into spermatozoa

ical defects such as the inability to synthesize a particular nutrient can be recognized readily at this stage. At the same time genetic crosses can be made and crossing-over frequencies can be measured and used for genetic mapping. Thus, the study of nutritional auxotrophs of *Neurospora,* initiated by G. W. Beadle and F. L. Tatum in 1940, is usually regarded as the beginning of biochemical genetics. It preceded by several years Lederberg's discovery of recombination in bacteria.

c. Polytene Chromosomes

While most cells of higher organisms are normally diploid, the chromosome number may sometimes be doubled or increased even more. A cell with twice the diploid number of chromosomes is **tetraploid** and with higher multiples of the haploid number it is **polyploid.** Plant breeders have succeeded in producing many tetraploid varieties of flowering plants often with increased size. While most of our body cells are diploid, we too, have polyploid cells. For example, some are always found in the liver. The most spectacular example of an increase in the normal DNA content of cells is provided by the giant **polytene** chromosomes of dipteran (fly) larvae. The DNA of cells in the salivary glands and some other parts of these organisms doubles about 13 times without cell division to give a several thousandfold (i.e., 2^{13}-fold) increase. The supercoiled, duplicated DNA molecules all line up side by side in a much more extended form than in ordinary chromosomes. The total length of the four giant chromosomes of *Drosophila* is $\sim$2nm, compared to 7.5 μm in diploid cells. The giant chromosomes have a banded structure, $\sim$3000 bands being visible along the length of the chromosome. Since it has been possible to correlate visible changes in the appearance of these bands with particular mutations in the DNA, study of polytene chromosomes has provided a second important method of mapping genes in the chromosomes of the fruit fly. The maps produced by the two methods agree well.

d. Mapping Human Chromosomes

Until recently little was known about the location of genes in human chromosomes with the ex-

ception of sex-linked traits (Chapter 1, Section C,4) that could be localized in the X chromosomes. Several recent developments have led to rapid progress and to the systematic mapping of a large number of human genes.[169-171] Most important is the technique of **somatic cell fusion** (Box 15-E). Human lymphocytes are often fused with rodent cells under the influence of inactivated Sendai virus which causes the cells to adhere and then to fuse. From such human–mouse or human–hamster hybrid cells it is possible to obtain strains in which the nuclei have also fused. Although such cell lines can be propagated for many generations, they tend to lose chromosomes, especially those of human origin. By observing loss of particular biochemical traits, e.g., of particular human enzymes (separable from the hamster enzymes by electrophoresis) it is possible to assign a particular gene to a given chromosome. This also requires identification of the chromosomes lost at each stage in the experiment. New staining techniques have made it possible to identify each of the 26 pairs of human chromosomes. Under current development are methods of fine genetic mapping that can be applied to the cultured cells.[171]

e. Cytoplasmic Inheritance

Not all hereditary traits follow the Mendelian patterns expected for chromosomal genes. Some are inherited directly from the maternal cell as if their genes were carried in the cytoplasm rather than the nucleus.[172] An example is provided by *Chlamydomonas* (Fig. 1-9) for which certain mutations to streptomycin resistance are cytoplasmic. Other examples are known in a wide variety of organisms. One cytoplasmic gene leads to "male sterility" in corn. The maternal cytoplasm carries a gene that prevents the formation by the plant of viable, mature pollen grains. Other mutations in nuclear genes that neutralize this and restore fertility have also been found.[173] These mutations have been cleverly exploited in the production of hybrid seed corn. Maternal plants carrying the cytoplasmic gene for male sterility do not have to be "detassled" to prevent self-pollination. If these plants are fertilized with pollen from another

BOX 15-E

HAPLOID PLANTS AND CELL FUSION

Recently described new techniques of propagation of plants from single cells and of plant cell fusion hold promise of revolutionizing plant breeding and also of providing a new method for studying the control of gene expression in plants. For example, it has been possible to induce the haploid nuclei of pollen grains to grow into entire haploid plants.[a] Since a haploid plant cell presumably contains only a single copy of many genes, mutations induced by chemical means or by irradiation can be detected easily. This should facilitate breeding efforts.

In another type of experiment differentiated, chlorophyll-containing mesophyll cells of leaves of tobacco (*Nicotiana*) were treated with cellulose- and pectin-digesting enzymes to remove the cell wall. The resulting protoplasts could be induced to fuse with protoplasts from another species of tobacco to yield nonsexually produced hybrids.[b] While many obstacles lie in the road of practical application, the technique may make possible the rapid development of new varieties of plants.

[a] J. P. Nitsch, *Z. Pflanzenzüecht.* **67**, 3–18 (1972) (in English).
[b] P. S. Carlson, H. H. Smith, and R. D. Dearing, *PNAS* **69**, 2292–2294 (1972).

strain carrying the nuclear restorer gene, the resulting hybrid seeds will produce the self-fertile plants necessary for crop production. Unfortunately, the mutant cytoplasmic male sterility gene also makes the plants highly susceptible to attack by a certain race of the fungus *Helminthosporium maydis*, which causes Southern corn leaf blight. A near disaster in United States agriculture occurred in 1970 when the disease struck large areas of the country. Use of these strains has therefore been largely abandoned.

Where in the cell are cytoplasmic genes found? There are three known locations: the mitochondria, the chloroplasts, and certain other membrane-associated sites.[174,175] An example of the latter is found in "killer" strains of yeast. Cells with the killer trait release a toxin that kills sensitive cells but are themselves immune. The genes are carried in **double-stranded RNA** rather than DNA, but are otherwise somewhat analogous to the colicin factors of enteric bacteria. Similar particles (κ factors) are found in *Paramecium*.[176]

Mitochondrial DNA has been most studied as a carrier of cytoplasmic inheritance.[177–179] In animal cells mitochondrial DNA consists of circular double-stranded molecules with lengths of $\sim 5 \mu m$ (MW $\sim 10^7$, $\sim 15,000$ base pairs) and a coding capacity of about 15 genes. The mitochondrial DNA of *Tetrahymena* is somewhat larger

($\sim 15 \mu m$ in circumference) and that of plants may be even larger. Yeast mitochondrial DNA consists of $25-26 \mu m$ circles. There is good evidence that mitochondrial DNA codes for the special ribosomal RNA of mitochondria. Mitochondrial DNA also codes for several tRNA molecules. In animals 20–25% of the mitochondrial genome may be dedicated to production of rRNA and tRNA molecules.

Striking evidence that some of the remaining genes of the mitochondrial DNA code for proteins necessary to the functioning of mitochondria come from the so-called "petite" mutants of yeast. These mutants, which arise spontaneously, can be produced more readily by treatment of yeast cells with high concentrations of intercalating agents such as **ethidium bromide** (Fig. 2-27). Long treatment with such compounds leads to a virtually complete breakdown of mitochondrial DNA and to the production of mutants that grow to a small size and which may be totally lacking in mitochondria. Since yeast can be grown anaerobically (using fermentation to supply energy), such cells can survive. Even though most of the proteins of mitochondria are synthesized according to instructions in nuclear genes, the ethidium bromide-induced loss of DNA leads to either the complete loss of mitochondria or to the formation of mitochondria deficient in the cytochromes of the electron transport chain. Thus, some essential information must be carried in mitochondrial DNA. Several proteins of the inner mitochondrial membrane appear to have genes within the mitochondria. Three are subunits of the cytochrome oxidase complex, which also contains four subunits made in the cytoplasm.[179] Four hydrophobic protein subunits of the mitochondrial inner membrane, which are synthesized on mitochondrial ribosomes, are associated with the F_1-ATPase complex (Chapter 10, Section E,8).[179]

The DNA of chloroplasts is about 10 times larger than that of animal mitochondria (MW $\sim 1 \times 10^8$)[179a] with a coding capacity of ~ 150 proteins. Ten to 100 copies may be present in each chloroplast.[180] As with mitochondria, the DNA of chloroplasts appears to code for ribosomal and transfer RNA molecules. It is also established that the major subunit (MW $\approx 55,000$) of ribulose-diphosphate carboxylase (Chapter 7, Section J,3,g) is encoded in the chloroplast DNA.[181,182] A smaller subunit (MW $\approx 15,000$) of the same enzyme is encoded in the nucleus. Several hydrophobic membrane proteins of chloroplasts are also synthesized on chloroplast ribosomes using mRNA transcribed from chloroplast DNA. There is evidence that chloroplasts of *Chlamydomonas* contain some genes for proteins of chloroplast ribosomes.[183]

Ethidium bromide also inhibits the replication of chloroplast DNA and causes partial degradation of existing DNA in chloroplasts without interfering with replication of DNA in the nucleus. The effect is similar to that of the same drug on mitochondrial DNA. However, cells of *Chlamydomonas* treated with ethidium bromide are able later to regenerate their chloroplast DNA. This result has been interpreted to mean that there may be one or a few "master copies" of chloroplast DNA in specially protected locations. The result should also be considered in relationship to the observation that although nuclear and organelle DNA molecules replicate at different times in the cell cycle, constant proportions of the organelle and nuclear DNA tend to be maintained. Thus, there must be some kind of control mechanism leading to a coupling of DNA replication in the nucleus, mitochondria, and chloroplasts.[184]

E. REPLICATION OF DNA

Now let us return to a subject that was at one time generally thought to be relatively simple—the synthesis of DNA. Today we know that replication is a complex process that requires the cooperative action of at least ten different gene products and perhaps an association with membrane sites.[185] Furthermore, it has been difficult to devise satisfactory *in vitro* test systems for DNA synthesis. The matter is made more confusing by the fact that some of the enzymes involved in replication may also be required in the processes of genetic recombination, in repair of damaged DNA molecules, and in certain defensive systems of cells.

1. Physical and Topological Problems

The chromosome of *E. coli* contains over 1 mm of DNA folded in a cell that is only 2 μm long. The diploid length of DNA in a 20 μm cell of a human is about 1.5 m. The unwinding of the DNA duplex at replication forks requires a rapid rotation of the chains (Section A,3,a). Although, in chemical terms a rate of unwinding of 3000 bases per second presents no problem, it is hard to understand, even for *E. coli,* how the two copies of the replicating chromosome are separated without becoming entangled. Part of the answer may be found in DNA unwinding proteins (see Section E,5,c) and "DNA-relaxing" or "untwisting" enzymes[185-186] (see also Fig. 2-27). Also of importance is the organization of the chromosome.

Recent experiments have confirmed what has long been suspected—that the *E. coli* chromosome is attached to a number of specific sites on the *membrane*.[187] The DNA helix appears to be highly folded in the form of 12–80 supercoiled loops. It is important to keep this fact in mind when viewing electron micrographs such as those of Fig. 15-26. Before these micrographs were prepared the DNA molecules were opened up into circles or other expanded forms. It seems possible that the sites of replication are fixed in special membrane locations and that as the DNA moves through these replication points, a complex of membrane-bound enzymes acts upon them. Similarly in eukaryotic cells, there is evidence that the DNA binds to the nuclear membrane.[188,188a]

2. The Direction of Replication

Present evidence favors a replication mechanism similar to that shown in Eq. 15-3. As the replication fork opens up, pieces of new DNA are synthesized alongside the parental strands. An important question is "Does replication occur in one direction only or do two forks form at the origin point from which transcription begins and travel in opposite directions around the chromosome?" To answer this question both genetic methods and electron microscopy have been employed.

One technique of establishing the direction of replication in *E. coli* was to insert a λ prophage at the *att* side (Fig. 15-1) and phage Mu-1 DNA at a variety of other sites around the chromosome.[189] Phage Mu-1 is especially useful because it can be integrated at many different sites within well-mapped genes. Integration within a gene inactivates that gene (an addition mutation) and allows the precise localization of the Mu prophage. A series of strains of bacteria were prepared containing both λ and Mu-1 prophage, the latter at various sites. The bacteria were also auxotrophic for certain amino acids. Because of this, replication could be stopped by amino acid starvation. (However, the bacteria usually completed any replication cycle in progress.) When the missing amino acids were added, replication began again starting from the replication origin. Bromouracil,

which enters DNA in place of thymine, was added at the same time. Consequently, the newly synthesized DNA strands were denser than the parent strands. After various times of replication* the newly formed strands were separated by centrifugation in a CsCl gradient (Chapter 2, Section H,1e) and were tested for hybridization with both λ and Mu-1 DNA.

From the observed ratios of Mu-1 DNA to λ DNA for the various strains it was possible to map the progress of replication beginning at an origin near gene *ilv* at 74 min (Fig. 15-1). Replication was found to progress bidirectionally around the chromosome and to terminate between genes *trp* and *his* at ∼25 min.

The use of autoradiographic methods has also confirmed bidirectional replication in *E. coli*. Special strains of amino acid auxotrophs with small nucleoside triphosphate pools were used. The addition of amino acids, after starvation, led to initiation of replication with only a 6 min lag. The cells were labeled with [³H]thymidine and after the replication forks had moved a short distance from the origin of replication the cells were given a pulse of "super-hot" [³H]thymidine. Using autoradiography it was possible to observe clearly bidirectional replication forks[190] (Fig. 15-28).

Replication of DNA in *Drosophila* chromosomes has also been studied in rapidly dividing nuclei by electron microscopy.[191] The replication rate in these nuclei is ∼300,000 bases per second, but it has been estimated that replication forks in animal chromosomes move no faster than approximately 50 bases per second. Thus, we would anticipate at least 6000 forks, or one fork per 10,000 bases. Indeed, this large number of forks was observed.[191] The forks occur in pairs, and careful examination shows many short regions containing single-stranded DNA as if one strand at the fork is replicated more rapidly than the other. The arrange-

---

* The cells do not all begin replication at the same time after addition of amino acids. Therefore, a variety of lengths of newly replicated DNA are present.

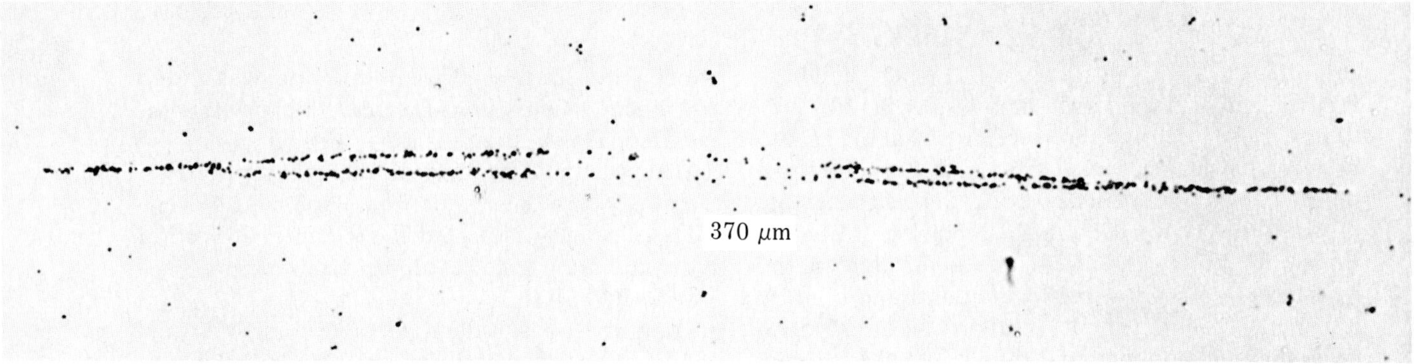

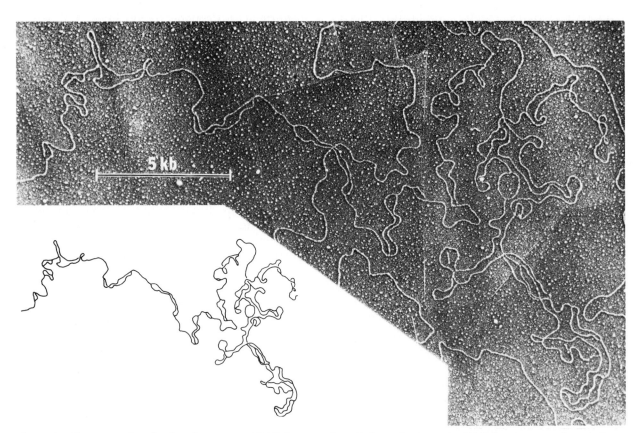

Fig. 15-28 Bidirectional replication of the *E. coli* chromosome. The autoradiographic pattern showing a pair of replication forks was produced by a chromosome that initiated replication with [³H]thymine (5 Ci/mmol) and was subsequently labeled with [³H]thymidine (52 Ci/mmol) for 6 min. The total length of the grain track is 370 μm. From P. L. Kuempel *et al.*, *in* "DNA Synthesis *in Vitro*" (R. Wells and R. Inman, eds.), pp. 463–472. Copyright 1972 University Park Press, Baltimore.

370 μm

5 kb

Fig. 15-29 Fragment of replicating chromosomal DNA from cleavage nuclei of *Drosophila melanogaster*. The DNA, which was spread in the presence of formamide, contains several "eyes" formed where the DNA has been replicated. From Kreigstein and Hogness.[191] See Fig. 2-23B.

ment of the single-stranded regions at the two forks in a pair strongly suggests bidirectional replication (Fig. 15-29). In the case of *Bacillus subtilus,* bidirectional replication also appears to occur, but the forks travel at different rates in the two directions.[192] Replication of DNA of the phages λ and T7 is also bidirectional[193] but that of mouse mitochondrial DNA is apparently unidirectional.[194]

3. Denaturation Maps

Replication in phage and in mouse mitochondria has been studied by direct electron microscopy using **denaturation loops** as markers. These loops are small denatured regions in the DNA that arise under the conditions of spreading for microscopy. Presumably the DNA in the loop regions contains a low content of CG pairs or for some other reason is less stable than the bulk of the DNA. The loops are reminiscent of those arising in heteroduplexes (Section D,8,c) but come from a different cause. There are two such loops in the circular mouse mitochondrial DNA approximately 180° apart and identifiable one from the other. With the aid of these loops as markers, it was possible to trace the direction of replication. The technique is now being applied to a wide variety of investigations.

4. The Enzymes of DNA Synthesis

After the identification of DNA polymerase I (Section A,3,a), it was generally assumed that the principal chain elongation enzyme for synthesis of DNA had been found. However, an *amber* mutant of *E. coli* was discovered that was deficient in DNA polymerase I (gene *polA,* Fig. 15-1) but nevertheless carried out DNA synthesis normally. This finding stimulated an intensive search for new DNA polymerases. Two other enzymes, *DNA polymerases II* (gene *polB*) and *III,* have been found, both in amounts less than 25% of that of DNA polymerase I.[195,196] Both enzymes have prop-

erties somewhat similar to those of DNA polymerase I. However, there are notable differences.

A fact not mentioned in our earlier discussion of DNA polymerase I is that the enzyme not only catalyzes the growth of DNA chains at the 3' end of a primer strand but also catalyzes, at about a 10-fold slower rate, the hydrolytic removal of nucleotides from the 3' end. Furthermore, the same enzyme can catalyze hydrolysis of nucleotides from the 5' end of DNA chains. This latter activity has been shown to reside in a different part of the same protein molecule, as if the DNA polymerase represents the product of two fused genes.[197] DNA polymerases II and III differ from DNA polymerase I in not catalyzing this hydrolysis from the 5' end.

It is proposed that the 3'—5' exonuclease action of DNA polymerase I fulfills a kind of "proofreading function. The polymerase acts at the 3' end of the growing DNA chain. According to the proofreading proposal, before moving on to the next position,* the enzyme verifies that the correct base pair has been formed in the preceding polymerization event. If it has not, the exonuclease action occurs to remove the incorrect nucleotide and to allow the polymerase to add the correct one. Thus, each base pair is checked twice, first before polymerization and then after polymerization. A schematic picture is shown in Fig. 15-30, which also indicates how the 5'–3' exonuclease activity can come into play when the polymerase reaches the end of a gap.

Important information has been provided by the genetic approach to replication.[196,198] Thus, a series of temperature-sensitive mutants of *E. coli* that are unable to carry out DNA synthesis has been obtained. In this manner genes *dnaA, B, C, D, E, F,* and *G* have been identified at various points on the chromosome map. The product of genes A and possibly that of C is essential for initiation of replication but is not needed for elongation. On the other hand, genes *B, D, E,* and *G* are

---

* It is not known whether the polymerase verifies the base pairing before or after shifting to the next site of polymerization. What is clear is that the enzyme removes any 3'-terminal nucleotide that is not properly paired.

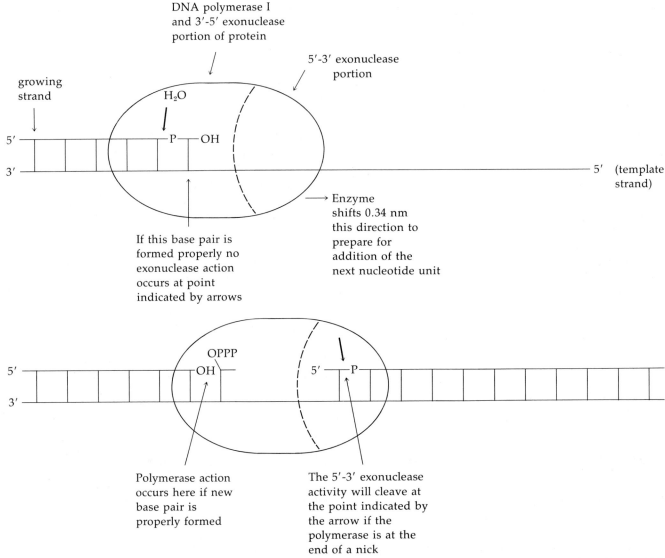

Fig. 15-30 Schematic drawing indicating three enzymatic activities of DNA polymerase I. The top drawing illustrates the 3'-5' exonuclease or "proofreading" activity of the enzyme. (While the drawing indicates that the enzyme shifts one position forward before the next step occurs, the point at which this shift occurs is uncertain.) The lower drawing illustrates both the polymerization reaction and the 5'-3' exonuclease action.

involved in elongation. Genes *C* and *D* map close together at 89 min, and it now appears that they are one gene. Perhaps it is a gene specifying a protein with two functions, one in initiation and one in elongation. The gene *F* product has been identi-

fied as ribonucleotide reductase (Eq. 14-50). This leaves products of genes *B*, *D*, *E*, and *G* that appear to play essential roles in elongation. None of these genes code for DNA polymerase I but gene *dnaE* has been identified as that of DNA polymerase III. Thus, we have strong genetic support for a key function for this polymerase. However, the polymerase by itself is unable to replicate double-stranded DNA and must function in cooperation with other proteins.

The DNA polymerases of eukaryotic cells include three forms, α, β, and γ, which occur in nuclei (the α form is also found in cytoplasm) and a mitochondrial (mt−) enzyme.[199]

Whereas it seems well established that replication in *E. coli* consists of a specific initiation event followed by elongation around the chromosome in two directions, termination of replication has been studied less well. Some experiments indicate that termination in some fashion triggers the synthesis of a specific mRNA and a protein required for cell division.[200] Thus, the cell cycle appears to consist of a series of consecutive events, each one triggering the next.

5. The Replication of Viral DNA

In an attempt to find simpler systems in which to study the synthesis of DNA, many investigators have turned to the small DNA-containing viruses such as ϕX174 and M13. The tailed bacteriophage including phages λ, T7, and T4 as well as the plasmid of colicin E-1 have all received attention. The advantage of these systems is that replication of DNA can more easily be achieved in cell extracts and that the viral or plasmid DNA has already been extensively studied genetically. In many cases replication depends upon the genes of both the host organism and of the virus. For example, mutations in the *E. coli* genes *dnaB, D, E, F,* and *G* lead to a loss in ability to support growth of phage λ as well as loss of ability to reproduce under conditions where the *ts* genes are inactivated. On the other hand, phage λ can still replicate in mutants involving genes *A* and *C*. Many viruses, including the T-even phages, contain genes specifying their own special DNA poly-

merases and other proteins essential for replication.

a. Replicative Forms

It was established sometime ago that the first step in replication of viruses ϕX and M13 is conversion of the single-stranded closed circular DNA of the infecting virus particle into a circular double-stranded replicative form (RF). The double-stranded circle than undergoes several replications to give a number of RF circles which serve as templates for the synthesis of many single strands of viral (+) DNA that are incorporated into the mature viruses. Both steps, the conversion of single-stranded DNA into replicative forms and the duplication of the latter, are under investigation.[201] For both ϕX and M13 the first stage requires the presence of an active polymerase III gene (*dnaE*) in the host bacteria.† Nevertheless, the previously purified polymerase III was found inactive in the test tube system. New purification efforts led to isolation of a new *dnaE* gene product known as **DNA polymerase III***, which is apparently a higher polymer of polymerase III.[202] Another protein, of MW = 77,000, known as **copolymerase III***, is also required. A mixed tetramer of polymerase III and copolymerase III* (polymerase III holoenzyme) is most active.[202] The system also requires spermidine and ATP (the ATP being split to ADP plus P_i during initiation of polymerization) as well as the four deoxyribonucleoside triphosphates.

b. RNA Primers

A second surprise was in store for students of the replication of viruses M13 and ϕX. Formation of the RF DNA requires RNA polymerase. This was one of many pieces of evidence that have led to the concept that a small segment of RNA is required to prime DNA synthesis (Eq. 15-3). Similar observations were made for replication of DNA of the colicin E-1 plasmid. This process is sensitive

† Replication of ϕX also depends upon products of genes *dnaB*, C,D and G, an "unwinding protein" (Section E,5,c) and two other "factors."

to rifampicin, a specific inhibitor of RNA polymerase (Box 15-A). The four ribonucleoside triphosphates as well as the deoxyribonucleoside triphosphates are needed for replication.[203]

Subsequent studies have shown that fragments of RNA are present in the closed, circular plasmid DNA and may be present in the DNA of viruses and of *E. coli*. In the case of phage ϕX, this RNA is synthesized by a special rifampicin-resistant RNA polymerase of MW = 64,000 and encoded by *E. coli* gene *dnaG*. This enzyme may also be the one needed for priming normal DNA synthesis in *E. coli*.[204] A special ribonuclease **RNase H** that specifically degrades the RNA chain from an RNA–DNA hybrid, may play a role in removing the primer RNA. RNase H would leave a gap that could be sealed by the action of a DNA polymerase.[205] However, the question of what enzyme actually removes the primer RNA remains open. RNase H does not remove the RNA completely under *in vitro* conditions. A possibility is that the 5'-exonuclease activity of DNA polymerase I removes primer RNA (Fig. 15-30)[205a].

c. "Unwinding" Proteins

Genetic analysis of replication of the DNA of phage T4 has revealed that at least five genes of the virus are required for formation of virus DNA within the cells of *E. coli*. One of these, gene 43, specifies a T4 DNA polymerase while gene 32 codes for a protein known as the **DNA unwinding protein**.[206] This protein has a greater affinity for single-stranded DNA than for double-stranded DNA. As a consequence, it binds to a length of single-stranded DNA causing unwinding of the double helix and exposure of the purine and pyrimidine bases of the template strand. Genetic studies show that the protein is required both for replication and for genetic recombination. A similar protein induced by phage T7, as well as another protein of the same type from uninfected *E. coli*, have also been isolated.[207,208] Eukaryotic DNA-binding proteins are also known.[208a] The real molecular function of these "unwinding" proteins is still quite obscure. Thus, the present stage of the study of replication is one of identifying the essential proteins and finding ways in which the component parts can be reassembled in a test tube to mimic replication as it occurs in living cells.

d. Replication of Viral Double-Stranded RF Molecules

The specific gene A in phage ϕX is known to be essential for initiation of replication. Recently, it has been shown to code for a protein of MW = 56,000, a specific endonuclease that places a nick in the viral strand of the RF to start the replication process.[209] It is presumably after this nick is formed that a small segment of primer RNA is synthesized. Whereas DNA replication most often occurs in a bidirectional fashion (Section E,2), that of the ϕX replicating form probably occurs unidirectionally by a "rolling circle" mechanism.[210] In this process, as a new viral strand is being synthesized along the complementary (minus) strand as a template, the original viral DNA (plus strand) is displaced (Eq. 15-9) as a single-stranded tail.

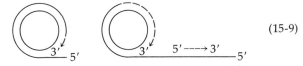

$$(15-9)$$

A complementary strand to the single-stranded tail is then formed, possibly in segments. A complete turn of the circle produces a viral strand twice the normal length. Cleavage by a suitable endonuclease and closure of the complementary strand circle by a ligase completes the replication.

F. RESTRICTION AND MODIFICATION OF DNA

A continuous battle rages between viruses and bacteria with both the invading phage and the host organisms having a variety of offensive and defensive weapons. For example, after invasion of a bacterial cell, many viruses not only turn off the synthesis of host DNA but also digest it with special enzymes (both endonucleases and exonu-cleases) that are encoded in the virus genome.[211–214] Since the host bacteria have their own weapons, the DNA of viruses is frequently modified in such a way that destruction by the bacteria can be avoided.* Thus, enzymes specified by T-even phages convert cytidine monophosphate to 5-hydroxymethyl-CTP and the modified nucleotide is incorporated into the DNA of the viruses.[212] Furthermore, the newly introduced hydroxyl group is glucosylated to varying extents (Chapter 2, Section D,8).

1. Restriction Endonucleases

Bacteria frequently digest and destroy DNA of invading viruses or DNA injected during mating with a bacterium of an incompatible strain. Investigation of this remarkable phenomenon, known as **restriction,** reveals that the DNA of viruses that are able to replicate within a particular host appear to be *marked* in some fashion at specific sites in the molecule. In many cases, the marking appears to consist of the presence of methyl groups. Properly methylated DNA is not degraded, but unmethylated DNA is cleaved by a highly specific endonuclease at the same sites that are normally methylated. Each species of bacteria (and often an individual strain within a species) has its own restriction enzymes. Restriction enzymes are very specific and often cut DNA chains at or near only a few spots that possess particular unique base sequences. About 45 enzymes with different specificities have been isolated.

The restriction endonucleases specified by the chromosome of *E. coli* are large proteins of MW ~300,000–400,000 and composed of at least three kinds of polypeptide chains. They apparently bind at specific sites and cleave the DNA ran-domly nearby. They require ATP, Mg^{2+}, and S-adenosylmethionine and have the unusual property of promoting the hydrolysis of large amounts of ATP.[215] The significance of these properties is still unknown. A second class of restriction enzymes consists of relatively small monomeric or dimeric proteins of MW = 50,000–100,000. The sites of attack, in most instances, are nucleotide sequences with a twofold axis of local symmetry.[216,217] For example, the following sites of cleavage have been identified for two restriction endonucleases encoded by the DNA of R-factor plasmids of *E. coli* and for a restriction enzyme from *Hemophilus influenzae*. In the diagrams ⇕ are sites of cleavage, * are sites of methylation, and · are local twofold axis.

E. coli R factor (*Eco*RI)

$$5'—(T\ or\ A)—G—A\overset{\downarrow}{—}A\overset{*}{—}T—T—C—(A\ or\ T)—3'$$
$$3'—(A\ or\ T)—C—T—T\underset{*}{—}A\overset{\cdot}{—}A—G—(T\ or\ A)—5'$$

E. coli R factor (*Eco*RII)

$$5'—G\overset{\downarrow}{—}C—\overset{*}{C}—A—G—G—C—3'$$
$$3'—C—G—\underset{*}{G}—T\overset{\cdot}{—}C—C—G—5'$$

H. influenzae (*Hin*d III)

$$5'—\overset{*}{A}\overset{\downarrow}{—}A—G—C—T—T—3'$$
$$3'—T—T—C—G\overset{\cdot}{—}A—A—5'$$

It appears that in many cases the restriction enzymes create breaks in each of the two strands in positions symmetrically arranged around the local twofold axis of symmetry. This is what we might expect if a dimeric enzyme binds in the major or minor groove of the double helix, each active site attacking one of the polynucleotide chains.

---

* Curiously, bacteria have evolved in a way that permits phage to grow. A number of mutants (*grov*) block the growth of phage (λ, T3, T4, T7, etc.) but grow normally themselves. That such mutants have not become the wild type suggests that coexistence with phage is very important to bacteria.

2. Modification Methylases

The most common form of modification of DNA is transmethylation from SAM to specific sites on the DNA. These sites appear to be most frequently the 6-amino group of adenine and the C-5 atom of cytosine. In general, there is a methylase to go with each restriction nuclease.

3. Restriction Enzymes in DNA Sequence Analysis

Because they cleave DNA chains at highly specific sequences that may occur only a few times within a long DNA molecule, the restriction endonucleases are being used widely to assist in the determination of DNA sequences.[217] They are employed in much the same way as is trypsin in breaking polypeptide chains into smaller fragments (Chapter 2, Section H,2). A good example is provided by the study of the DNA of the virus SV40.[218,219] This mammalian virus, which is capable of being integrated into the host genome and transforming cells into a tumorigenic state, contains a circular duplex DNA of ~5000 nucleotide pairs. One restriction enzyme of *Hemophilus influenzae* cleaves the SV40 DNA into 11 fragments, whereas an enzyme from *Hemophilus parainfluenzae* produces only four fragments. The enzyme *EcoRI* from *E. coli* produces a single break in the circle at a unique point. Analysis of partial digestion products and overlapping fragments has permitted arrangement of all of the fragments in a circular map. Using pulse-labeling with [³H]thymidine, the origin and directions of bidirectional replication have already been established with reference to the restriction enzyme fragments. Likewise, the directions of early and late transcriptions of mRNA in productively infected cells are known. Sequence studies on the individual fragments have progressed to the point that we expect to know the entire nucleotide sequence of the SV40 DNA molecule soon. Similar approaches are being applied to mitochondrial and chloroplast DNA and to chromosomal DNA segments as well.

G. RECOMBINATION, INTEGRATION, AND EXCISION

Now let us consider briefly the poorly understood chemical phenomena underlying genetic recombination and the integration and excision of virus DNA into and out of the host genome. That recombination is a complex process is indicated by the fact that mutants deficient in recombination ability map at several loci on the *E. coli* chromosome; the genes are designated *recA, B, C, F, G,* and *H*. Mutants at several of these loci are unusually sensitive to ultraviolet light, a fact that stems from their inability to repair ultraviolet damage to DNA (Chapter 13, Section D,2). Thus, it appears that several of the recombination enzymes are also required in repair of ultraviolet damage. However, the specific functions of most of the gene products are not yet fully understood. It is thought that there are two complete systems for **general recombination** in *E. coli*. Phage λ carries genes for another recombination system which functions in addition to the products of λ genes *int* and *xis* (Fig. 15-15). The latter are required for integration and excision of the viral genetic material, processes that also lead to a certain amount of **site-specific recombination** between host and viral genes.

1. Recombination Mechanisms

The most puzzling aspect of recombination is the problem of bringing homologous regions of two different DNA duplexes together. As illustrated schematically in Eq. 15-10, the strand exchange must occur at exactly the same point in each duplex.

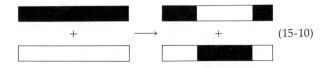

$$+ \longrightarrow + \qquad (15\text{-}10)$$

One of the first mechanisms proposed for genetic recombination was based on the assumption that recombination was associated directly with DNA synthesis. This "copy choice" mechanism assumed that replication occurred along one DNA strand up to some random point at which the polymerase jumped and began to copy from the second of a pair of homologous chromosomes. According to this mechanism, the newly formed DNA molecule would be complementary to parts of one parental DNA duplex and to parts of the second duplex. To test the idea, Meselson and Weigle infected *E. coli* with two strains of λ phage containing ^{13}C- and ^{15}N-labeled DNA, respectively.[220] Recombinant DNA was found to contain some ^{13}C and some ^{15}N, as judged by density gradient centrifugation. Thus, it was clear that DNA from both parents was incorporated into DNA of recombinant progeny, a finding that ruled out the copy choice hypothesis and suggested a **chain-cleavage** mechanism.

If recombination occurs by enzymatic cutting of two homologous duplex DNA molecules (followed by rejoining), how is it possible to avoid inactivation of genes by addition or deletion of genetic material? Recombination cannot depend upon the random action of a nonspecific enzyme with random rejoining. Yet, general recombination can occur at any point and with an essentially constant frequency throughout the DNA chain. It is generally agreed that the explanation of these facts lies in the occurrence of base pairing between homologous regions of strands of the two different DNA duplexes.

One proposed model for recombination is based upon studies with phage λ and T4. The λ gene *exo* (Fig. 15-22) is not essential to replication but is required for general recombination. The product of this gene has been identified as a 5'–3' exonuclease. A way in which this enzyme could function in recombination is indicated in Fig. 15-31. The process would begin with an *endonuclease* which would make single-stranded nicks at random in the DNA duplexes. The special exonuclease would then enlarge these nicks to form gaps. Homologous regions uncovered in this way would tend to pair (Fig. 15-31, step *b*) to give H-shaped structures. A process of **branch migration**

(Fig. 15-31, step *c*) would produce an elongated heteroduplex region as well as a short branch. Electron micrographs of branched DNA molecules of the types shown in Fig. 15-29 have been obtained[221] from replicating phage T4. The branched structures would be acted on by an endonuclease (Fig. 15-31, step *d*) to introduce nicks. Any single-stranded *gaps* could be filled by the action of a DNA polymerase (Fig. 15-31, step *c*) and the nicks could be sealed by polynucleotide ligase.

Objection can be raised to a recombination mechanism that requires production of long lengths of single-stranded DNA, and Holliday has suggested a process that does not necessarily require the formation of such "tails."[222] The recombination process could be initiated at special points on the duplexes, recognizable by a recombination enzyme (Fig. 15-32A). A short amount of unraveling would be followed by strand exchange with the two broken strands being rejoined by a ligase as indicated in Fig. 15-32A. The cross-over points would then migrate up or down the chains as the two helices turned about their own axes. Long regions of heteroduplex DNA could be generated in this way and the process could be terminated at a random distance from the starting point, accounting for the observed uniformity of genetic recombination events. Chain cleavage and rejoining of two of the strands would terminate the process. If these were the same strands broken in the initiation event (cleavage at points *aa'* in Fig. 15-32A), genes lying outside the heteroduplex region would not be recombined but cleavage of the other chains (points *bb'*) would lead to their recombination. Intermediates of the type predicted by the Holliday model have been observed by electron microscopy (Fig. 15-32B).[222a]

It has been pointed out that the kind of cross-stranded structure shown in Fig. 15-32 can be formed with all base pairs in both duplexes intact.[223,224] All that is required is formation of a nick in each of the two polynucleotide chains and a rejoining of the backbones across the close gap between the duplexes. Heteroduplex formation could by propagated up the chains by mutual rotation. This model could also well account for the

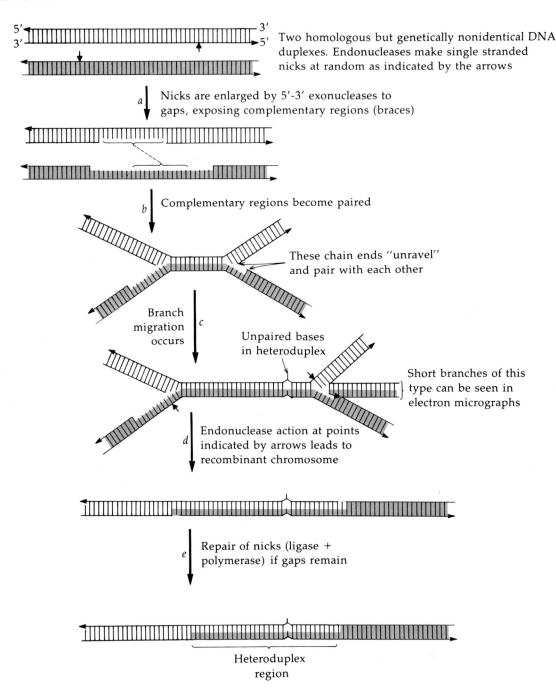

Fig. 15-31 A possible mechanism of recombination involving nicking of homologous DNA duplexes at random, enlargement of the nicks to gaps, and association of complementary regions. A branch migration mechanism then permits formation by gap filling and "resealing" produces a recombinant DNA molecule.

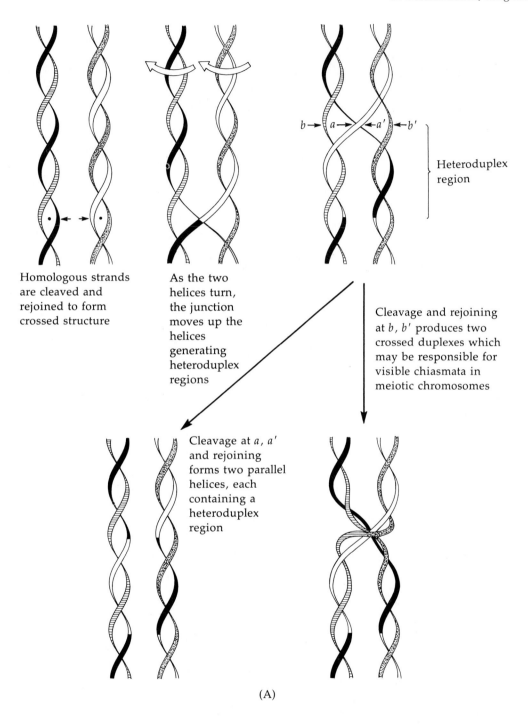

Homologous strands are cleaved and rejoined to form crossed structure

As the two helices turn, the junction moves up the helices generating heteroduplex regions

Heteroduplex region

Cleavage and rejoining at b, b' produces two crossed duplexes which may be responsible for visible chiasmata in meiotic chromosomes

Cleavage at a, a' and rejoining forms two parallel helices, each containing a heteroduplex region

(A)

Fig. 15-32 (A) A recombination mechanism involving single-stranded exchanges. After Holliday.

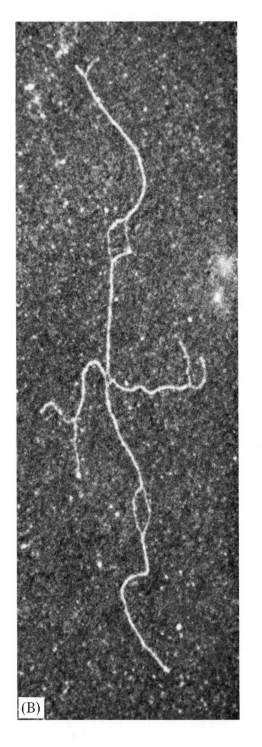

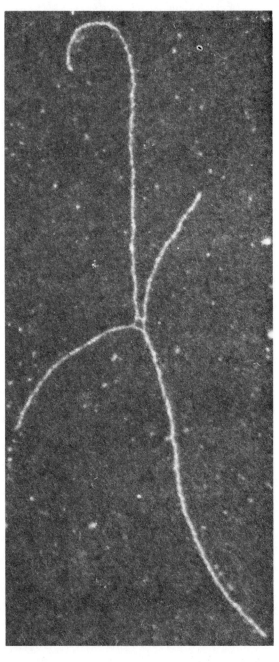

Fig. 15-32 (B) "Chi forms" of DNA from the colicin E1 plasmid. These forms are thought to be derived from recombination intermediates of the Holliday type which appear as "figure eight"-shaped molecules twice the length of the colicin genome. These figure eight forms were cut at a specific site that occurs only once in the genome (twice in the figure eight) by restriction enzyme *Eco* R1 (Section F,1) to give the chi forms. The pairs of long and short arms are believed to represent homologous duplexes. The single-strands in the crossover have pulled apart revealing the strand connections clearly. Such a structure would be expected from the Holliday intermediate [upper right corner of (A)], e.g., if one of the two vertical duplexes were rotated end over end. The DNA at the right was prepared in a high concentration of formamide which introduced denaturation loops and frayed ends in regions rich in AT pairs (see Section E,3). From Potter and Dressler.[222a]

cutting of the two crossed strands at exactly equivalent points to terminate the process.

Several variations on the model of Fig. 15-32 have been considered.[223,225] Proposals by Sobell[226,227] and others[228] visualize the initial nicking process to occur in cruciform structures formed at palindromic sites (see Fig. 15-4).

Circular dimers are observed to function as intermediates in recombination in phage S13 and ϕX.[229] Such double length circles, as well as **catenated** (interlocked) single length circles, could all be produced through recombination mechanisms involving strand cleavages of the type discussed in the foregoing paragraphs.

Our knowledge of recombination is still very incomplete and it is hard to know how to assess some observations. For example, the *recB* and *recC* genes of *E. coli* specify proteins that apparently combine to form an unusual DNA-hydrolyzing enzyme.[230] With a total molecular weight of ~340,000, this enzyme can, among other things, cleave single-stranded DNA in an "endo" fashion. Although it cannot "nick" intact duplex DNA, it could participate in some of the postulated endo- or exonucleolytic cleavage events of recombination. Remarkably, the enzyme requires ATP for its action and, in the test tube, over 20 ATP molecules are hydrolyzed to ADP and P_i for each phosphodiester bond of DNA broken. In the absence of a suitable DNA substrate, no ATP is cleaved. It

seems rather doubtful that the enzyme acts in a totally normal fashion *in vitro*, but its interesting properties and essentiality to recombination make it an important object for further study. A similar enzyme from *Bacillus subtilus* has been reported to bring double-stranded DNA molecules together (as judged by electron microscopy), an effect whose significance is unknown.[231]

Recombination in eukaryotic cells occurs principally during meiosis in a **synaptonemal complex**, a structure lying between the pair of homologous chromatids. It has been suggested that transient "nodules" observed in this complex may be associated with recombination.[232] In any event, it is clear that crossing-over and recombination in eukaryotic cells is a deliberate, complex, and organized process about which little is known at present.

2. Nonreciprocal Recombination

Any acceptable mechanism of recombination must take account of the phenomenon of **gene conversion** or **nonreciprocal recombination**.[220] The phenomenon was first recognized in genetic studies of fungi for which the four haploid miotic products can be examined individually (**tetrad analysis,** Chapter 1, Section D,2). It is sometimes found that instead of the normal Mendelian ratio of 2:2 for the gene distribution in the progeny at a heterozygous locus a ratio of 3:1 is observed. Thus, one of the recombinant chromosomes appears to have been altered to a parental type. A reasonable mechanism by which this can occur arises from the fact that heteroduplex regions contain defects in base pairing. Most often, at the point of a mutation, one strand of the heteroduplex will have a base that does not properly pair with the base in the other strand. Or one strand will have an extra base that will be looped out from the heteroduplex. Since cells contain repair mechanisms that search for defects and carry out a repair process, there is a likelihood that one strand in the heteroduplex region will be altered to restore perfect base pairing, thus causing the observed gene conversion.

3. Unequal Crossing-Over

Recombination depends upon the bringing together of homologous regions of two DNA duplexes (of two chromatids in the case of meiosis). Since repeated sequences occur in DNA (by chance if for no other reason) crossing-over may sometimes occur between locations in the two duplexes that are not exactly equivalent. Such unequal crossing-over has the effect of lengthening one duplex and shortening the other. It may be a very important factor in evolution.[232a]

4. Integration and Excision of DNA

The temperate bacteriophage λ, the F factors and R factors of bacteria can all be integrated into the DNA of the host through a process involving gene cleavage and thus resembling recombination in its chemical nature. However, in the case of λ phage, separate genes *int* and *xis* are required for integration and excision. These are not the same as the enzymes of the *rec* loci of the bacterium or the general recombination gene (*red*) of the phage. Nevertheless, the chemical similarity to the recombination events is striking.

On rare occasions the excision of phage λ leads to the inclusion in the viral genome of a substantial piece of the host chromosome. For instance, strains of λ that contain genes for galactose catabolism (*gal*, Fig. 15-1) and biotin synthesis (*bio*) have been obtained. The genes carried by these transducing phages can be introduced into other strains of bacteria, a fact that has greatly facilitated the mapping of bacterial chromosomes. The map in Fig. 15-1 was obtained with the aid of transducing phage P1 which can be integrated in at least ten places. The frequencies of cotransduction of nearby markers provided many of the recorded map distances.

While phage λ is integrated in the *E. coli* genome at a specific site, the transducing phage

Mu appears able to integrate at any point (Section E,2). However, there is a fixed point in the phage genome at which the circular viral DNA is cleaved. Insertion of the entire genome of one bacterial plasmid into the DNA of another plasmid at a palindromic site has also been observed.[233] In many instances it appears that the same palindromic sequence, often a long one (700–1400 base pairs) is present in both plasmid and bacterial chromosome. For example, the **insertion sequences,** IS1, IS2, and IS3, permit insertion of *E. coli* plasmid DNA into the *gal* and *lac* operons of the *E. coli* chromosome.[233a–c] It has been suggested that the translocation of plasmid DNA may result from the action of enzymes that resemble restriction endonucleases and which are specific for these sites. The same mechanism may underlie site-specific recombination.

It has been known for many years that genes and whole segments of chromosomes in higher organisms can sometimes move from one place to another. In corn "controlling elements" move from one site to another altering gene expression and behaving much as do integratable plasmids in bacteria.[233c] The presence of many long palindromic sequences in eukaryotic DNA could possibly be related to this phenomenon.[235a]

The ease with which foreign DNA is integrated into chromosomes of bacteria is striking. Does the same thing take place within the human body? The answer is probably "yes," but the extent to which our cells are vulnerable to alteration by virus incorporation is not so clear. We do know that tumor-producing (**oncogenic**) viruses can be integrated into the genome of animal cells. Among these are the simplest tumor viruses such as polyoma and SV40 (Box 4-C). After incorporation of the viral DNA into a chromosome of the host, some viral genes continue to be transcribed. Others are kept in an inactive state as in the case of phage λ. A rare effect of the incorporation of viral DNA into the host genome is **transformation** to a tumorlike state. Whether this results from the action of a specific viral gene product, from some alteration in expression of host genes, or through mutation (as by incorporation of phage Mu in *E. coli*) is unknown. It is clear that the surface characteristics of the transformed cells are altered and contact inhibition (Chapter 1, Section E,3,c) is decreased permitting the transformed cells to grow many layers deep. Thus, a principal characteristic of tumors results from the incorporation of viral DNA into a normal cell.[234,235]

5. Reverse Transcriptase

The RNA-containing tumor viruses have created considerable excitement in recent years. Most investigators of biochemical genetics and of the functions of nucleic acids thought that DNA was formed only by replication of other DNA molecules. While transcription of RNA from DNA could occur, the reverse process, formation of DNA from an RNA template, was not regarded as likely. It was therefore a surprise when it was found that many oncogenic RNA viruses, including some that cause animal leukemias, contain an **RNA-dependent DNA polymerase** (reverse transcriptase). The enzyme is found in the mature virus particles. Purified most extensively from the avian myeloblastosis virus, the enzyme consists of two protein subunits of MW = 110,000 and 70,000 and contains two atoms of bound Zn^{2+}. The enzyme requires a short primer as well as an RNA template chain and synthesizes first a DNA–RNA hybrid and later (probably after digestion of the RNA strand by RNase H, Section E,5,b) double-stranded DNA. Thus, infection with these viruses is followed by formation of DNA that can serve as a template for formation of other virus particles. The DNA can also be incorporated into the host DNA where it can cause tumors.[236–240]

The ability of the reverse transcriptase to copy RNA molecules has been of value in preparing DNA chains of the same sequences as unique RNA molecules (e.g., mRNA produced by a highly specialized cell such as an erythrocyte precursor in which only hemoglobin is being formed). The DNA copies (sometimes called cDNA) can then be used in a variety of biochemical studies. Under suitable conditions DNA polymerase I can also accomplish accurate copying of RNA sequences.[241]

H. MUTATIONS, CANCER, AND GENETIC ENGINEERING

One of the most striking characteristics of living things is the high degree of mutability of genes. Harmful mutations take a heavy toll of human life at an early age. The very high incidence of cancer in older persons is thought to be partly a result of the accumulation of somatic mutations. Many mutations may arise as a result of mistakes in the replication of DNA and in repair and recombination processes. The rate of mutation is increased by the presence of mutagenic chemicals, by physical agents such as ultraviolet light, X-rays and heat,[241a] and by the occasional insertion of virus DNA into chromosomes.

1. Mutagenic Chemicals[242–244]

A simple chemical way of mutating CG pairs to AT pairs is treatment with nitrous acid, HNO_2, which deaminates amino groups to hydroxyl groups. Thus, cytosine is converted to uracil, which pairs with A instead of with G. A simple transition mutation (Section D,1) results. Nitrous acid converts adenine to hypoxanthine which (like guanine) tends to pair with C rather than with T. (Guanine may be converted to xanthine but pairing is probably not much changed.) Many other chemical modifications of bases are also mutagenic. Thus, the weakly mutagenic hydroxylamine adds to pyrimidines at the 6 position. Among the most powerful mutagenic compounds are **alkylating agents.** Both those that act by S_N1 and S_N2 mechanisms tend to attack position N-7 of guanine residues preferentially. For some reason the alkylated guanine induces an increased number of pairing mistakes.*

---

* The positive charge placed on the purine ring by methylation on N-7 leads to easy hydrolysis of the N-glycoside linkage and depurination. However, this may be a lethal, rather than a mutagenic event. It is possible that methylation on O-6 of guanine is more important in inducing mutations.[244]

Among the most toxic and potent alkylating agents are the nitrogen and sulfur "mustards," *bis*-(2-chloroethyl)sulfide. Bifunctional com-

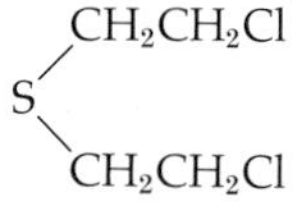

$$CH_2CH_2Cl$$
$$S$$
$$CH_2CH_2Cl$$

Bis(2-chloroethyl)sulfide

pounds of this type are highly toxic and cause a great deal of lethal cross-linking of DNA chains. The monofunctional "half-mustards" are mutagenic but with less acute toxicity. Another potent class of mutagenic alkylating agents are the nitrosamines:

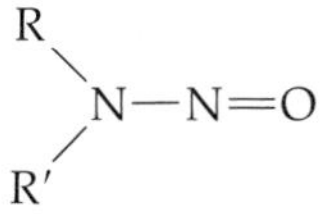

$$R$$
$$N—N=O$$
$$R'$$

Much used in the laboratory is **N-methyl-N-nitro-N-nitrosoguanidine,** one of the most effective

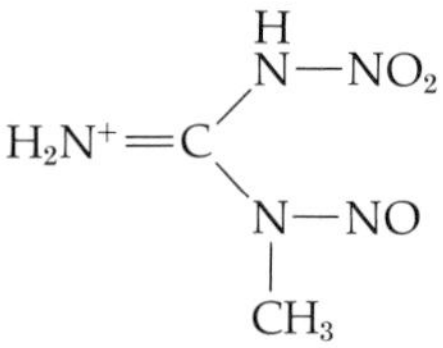

$$H$$
$$N—NO_2$$
$$H_2N^+=C$$
$$N—NO$$
$$CH_3$$

N-Methyl-N'-nitro-N-nitrosoguanidine

of known mutagenic substances. Nitrosamines are among the most carcinogenic agents known and are suspected of being major causative agents of cancer in humans.[245] Any secondary amine will react with nitrous acid to form a nitrosamine (Eq. 15-11). This reaction can occur in the stomach readily and the nitrosamines formed may be absorbed into the system and cause cancer in a variety of places. Since all plants contain some nitrate and some, such as spinach and beets, have large amounts, the possibility of reduction to nitrite and reaction with secondary amines in the stomach according to Eq. 15-11 is a real one. Bacon and

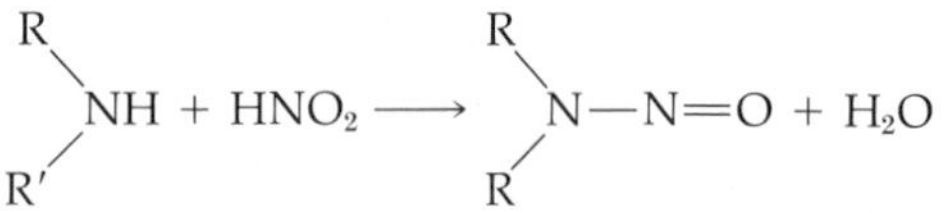

$$R \qquad\qquad R$$
$$NH + HNO_2 \longrightarrow N—N=O + H_2O \qquad (15\text{-}11)$$
$$R' \qquad\qquad R$$

other cured meats contain both nitrites and nitrates. Since many drugs and natural food constituents are secondary amines, there is a possibility that these substances play an important role in inducing cancer in man. Since tertiary amines can also react (with loss of one alkyl group) the significance of the problem is broad.

A second way in which chemicals can induce base substitution mutation is through incorporation into the structure of DNA itself. Thus, 5-bromodeoxyuridine (or bromouracil) can replace thymidine in DNA where it serves as a very efficient mutagenic agent. Less effective agents that presumably act in a similar way are 2-aminopurine and 2,6-diaminopurine.

Less common than base pair exchanges are frame-shift mutations (Section D,1). A characteristic of such mutations is that they do not revert (back mutate) as readily as do base substitution mutations, and reversion is not induced by chemicals known to cause base substitution. However, reversion of frame-shift mutations is induced by acridines and other flat molecules that are known to act as intercalating agents in DNA helices (Chapter 2, Section D,9). The same compounds promote frame-shift mutations. They appear to be especially effective in causing mutations of regions in which long repeated sequences of a single base such as AAAAAAAA occur. Deletion of two base pairs from a "hot spot" (site of frequent mutation) in the *Salmonella* histidine[246] operon of the following structure is induced by 2-nitrosofluorene causing reversion of a (−1)

$$—CGCGCGCG—$$
$$—GCGCGCGC—$$
Base sequence

histidine-requiring mutant. Whereas simple intercalating agents are often not very mutagenic, compounds that are both an intercalating agent and an alkylating agent are especially potent. An example is the following compound, a powerful frame-shift mutagen with an intercalating ring and a half-mustard side chain:

(CH$_2$)$_3$NHCH$_2$CH$_2$Cl

HN

OCH$_3$

Cl

When the CH$_2$Cl group of the side chain is replaced by CH$_2$OH, the compound is 100 times less mutagenic.

2. Repair of Damaged DNA

The retention by a cell of a functioning copy of the genome is essential for survival. Thus, it is not surprising that cells contain a battery of enzymes that apparently move along the DNA duplexes repairing defects and acting to reduce the frequency of mutation. The best known of the repair systems are those that undo the damage caused by ultraviolet radiation. Mutants of *E. coli* with a decreased repair ability and a consequent increased sensitivity to ultraviolet radiation map at several different positions. The genes are *uvrA, B, C, D,* and *F* (identical to *recF*) and *phr*. Mutants of genes *recA, B,* and *C* also show an enhanced sensitivity to ultraviolet light, suggesting that the *recB,C* nuclease and the *recA* product may be required for ultraviolet repair as well as for recombination.

One of the principal effects of ultraviolet radiation on DNA is to produce cyclobutane dimers (Chapter 13, Section D,2) between adjacent pyrimidine rings in a single DNA chain. To repair this damage, the dimer can be excised and replaced with new monomeric units. It has been shown that genes *uvrA* and *B* code for proteins constituting a **uv endonuclease**[247] that causes a nick (incision) on the 5' side of the dimer to expose a 3'-OH group (Fig. 15-33). Attack by this uv endonuclease is presumably followed by 5'-3' exonucleolytic action which cuts out the cyclobutane dimer. Either the 5'-3' exonuclease activity of DNA polymerase I or a special exonuclease may be responsible. The resultant single-stranded gap (Fig. 15-33) can be repaired through the action of a DNA polymerase plus a ligase.

A second mechanism of repair of damage by ultraviolet radiation is photoreactivation by visible or near-ultraviolet light (see Chapter 13, Section D,2). The bacterial gene *phr* codes for the special DNA photolyase enzyme.

Repair systems for ultraviolet damage are not limited to bacteria but seem to be found in all organisms. Of interest is the rare skin condition **xeroderma pigmentosum,** an autosomal recessive, hereditary disease. Homozygous individuals are unusually sensitive to ultraviolet radiation and

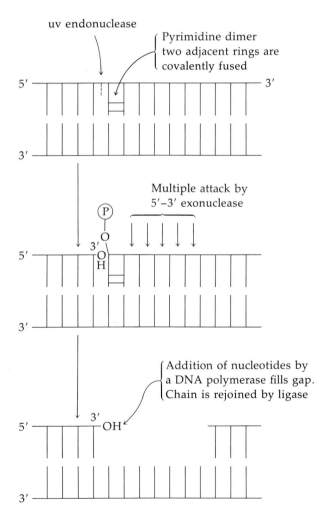

Fig. 15-33 Excision repair of damage by ultraviolet light.

tend to develop a high incidence of multiple carcinomas. The disease appears to have several causes,[248] one being a deficiency of an ultraviolet specific endonuclease. Some xeroderma cells have been shown deficient in photoreactivation ability.[249] A third cause of xeroderma may be the lack of an endonuclease specific for excision of depurinated DNA segments.[249a]

Other human disease syndromes in which DNA repair may be deficient are **progeria,** a condition involving premature aging and many other biochemical abnormalities and **ataxia telangiectasia.** Skin fibroblasts cultured from patients are deficient in the ability to repair **X-ray type damage**[250,250a]—strand breaks in DNA produced by gamma rays emitted by [60]Co. Fibroblasts from individuals with ataxia telangiectasia appear to lack a functional γ-endonuclease, the enzyme thought to initiate excision repair for γ-modified bases. This condition contrasts with that of xeroderma, because cells from patients with the latter disease have a normal capacity for repair of X-ray damage.

Enzymes that recognize mismatched base pairs in hetero-duplexes and correct them[251] and which excise bases modified by the action of carcinogens[252] have also been identified.

3. Mutagens in the Environment

Increasing concern is developing over the exposure of the human population to mutagenic substances. Over 500 new chemicals are introduced industrially each year. Some widely used drugs, e.g., **hycanthone** (Fig. 2-27), are mutagenic. Powerful mutagens are present in some foods.[253] The **aflatoxins,** dangerous carcinogens produced

An aflatoxin

by *Aspergillus flavus,* may be present in infected peanuts and other foodstuffs.

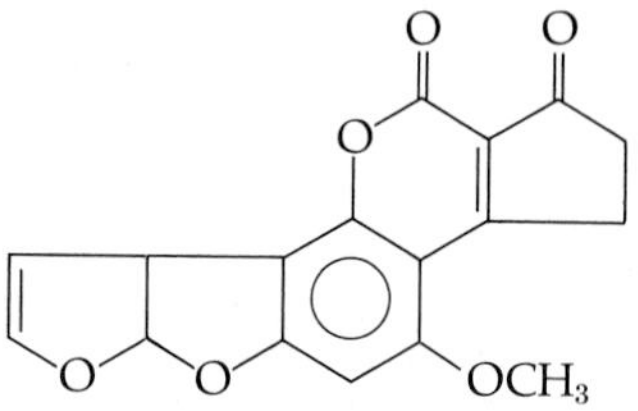

Many compounds that are carcinogenic or mutagenic do not appear unusually reactive chemically. However, these compounds often undergo hydroxylation in the animal body with a resulting conversion to a mutagen. An example is the formation of carcinogenic epoxides of aromatic hydrocarbons (Eq. 15-12).[244,254]

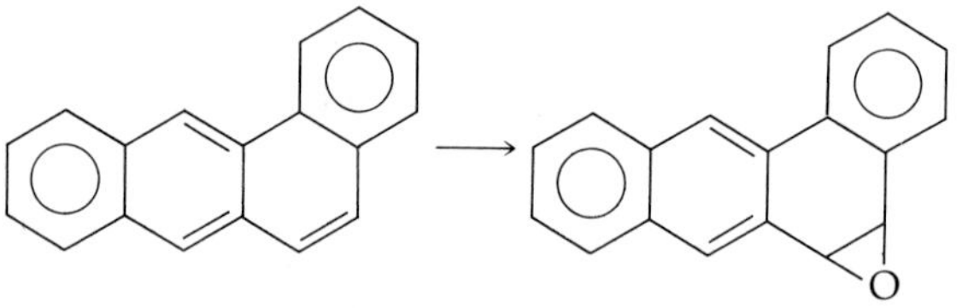

Benz[*a*]anthracene   5,6-Epoxide (carcinogenic)

(15-12)

Another example is conversion of 2-acetylaminofluorene to an *N*-sulfate (Eq. 15-13)[255]:

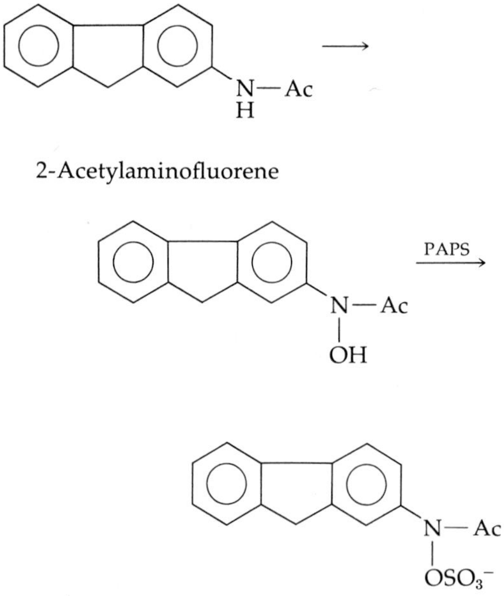

2-Acetylaminofluorene

Carcinogen   (15-13)

How can compounds be recognized as mutagenic? An important technique makes use of special **tester strains** of bacteria developed by Ames and associates. These are *Salmonella* mutants that are unable to synthesize their own histidine, but which can grow when a mutagenic agent produces a back mutation. One of the strains can be mutated by agents causing base exchanges, while the other three, which contain different types of frame-shift mutations, are affected differently by various mutagens. About 10^9 bacteria are spread on a petri plate and a small amount of the mutagenic chemical is introduced in the center of the plate. Where back mutation has occurred, a colony of the bacteria appears. The strains all carry a mutation in the main DNA excision-repair system so that most mutations are not repaired and the test is very sensitive. Addition of a liver homogenate plus an NADPH-generating system allows "activation" of many aromatic chemicals by hydroxylation.[256] Careful vigilance using these tests as well as longer term testing in animals can help us to keep the mutation at a lower level than it would otherwise be. However, the accumulation of deleterious mutations from natural causes and from past exposure to chemicals and to radiation has already created a serious problem.[257,257a]

4. Genetic Surgery

With our present knowledge of the action and the control of genes and with the ability to transfer pieces of DNA from one bacterium to another, attempts are being made to find ways of correcting genetic defects by introducing new genes into human beings. At first such an idea may seem sheer fantasy, but we already know of viruses such as SV40 that can be integrated into the animal genome. Although SV40 is tumorigenic, it should be possible to tailormake SV40-like DNA particles that carry various "normal" genes excised (perhaps with the aid of other viruses) from cells in culture. Alternatively, genes can be borrowed from bacteria or completely synthetic genes can be made and incorporated into transducing viruses.

Workable procedures for the chemical joining of chromosomal fragments have been developed.[258,259] One approach is to use restriction endonucleases to generate DNA fragments with sticky ends. It has already been possible to incorporate eukaryotic genes into bacterial R factors and genes of SV40 into λ phage.[260] Likewise, genes of the *E. coli gal* operon have been introduced, via λ phage, into SV40. An important feature of these methods has been the "molecular cloning" of new combinations of DNA incorporated in a bacterial plasmid,[261] usually one that can replicate within cells of *E. coli.*

While there seems to be little doubt that it will eventually be possible to artificially introduce genes into human cells, we know little about the control of transcription and translation of animal genes. Future investigations will doubtless reveal the nature of such controls and may some day permit successful "genetic surgery." One goal will be to find a means to correct the metabolic defect that causes the atrophy of the insulin-secreting pancreatic β cells in juvenile diabetics. The number of individuals that could benefit from such treatment is enormous (Box 11-C).

Evidence that bacterial genes can be transferred into plant cells via transducing bacteriophage and can be expressed there has been presented.[261–263] There are even reports that bacterial DNA can be incorporated directly (as in bacterial transformation, Section A,1) into plant cells.[264] This possibility raises the prospect of a revolution in methods of plant breeding. There is enormous interest in introducing genes for nitrogen fixation (from bacteria such as *Rhizobium*) into higher plants. As a step in this direction, genes of the nitrogen-fixing operon (*nif*) have already been transferred from *Klebsiella* into *E. coli.*[263–266] Observations of gene transfer from prokaryotes into eukaryotes are still regarded as uncertain. However, a gene required in histidine biosynthesis (that for imidazoleglycerol phosphate dehydratase, Fig. 14-28) has apparently been successfully transferred from yeast into an *E. coli* histidine auxotroph and has been expressed there.[266a]

Some biologists foresee a future in which man will learn to control his own genes and will be able to prevent genetic deterioration resulting from the accumulation of harmful mutations. They find it exciting to think of the possibility that humans will some day in an intelligent way elect to continue their evolution in a desired direction.[267] However, others caution that our present knowledge is such that attempts to eliminate all "bad" genes from the population might be disastrous.[268] They point to the hemoglobin S gene (Box 4-D) and the role it once had in preserving life in a malaria-infested environment and urge that the only selection that man attempt to make at this stage is one of a maximum heterogeneity of genetic type. They emphasize the dangers of allowing genetics to control human lives. In the past "eugenic" doctrines have been used to justify racist laws and (in Nazi Germany) genocide.

Nevertheless, the problem of the accumulation of deleterious mutations in the population cannot be ignored. As more and more defects are recognized, more individuals are deciding not to have children. Because of prenatal diagnosis, some parents are electing to terminate a pregnancy rather than to bear a hopelessly defective infant (Chapter 12, Section D,2). This approach also has its dangers. Even if abortion were universally acceptable, it is frightening to think of a society in which amniocentesis would be mandatory.

There are other and more immediate dangers. In 1974 a Committee on Recombinant DNA Molecules of the United States National Academy of Sciences called for a moratorium on two types of experiments that might be hazardous to the population at large.[269] The Committee pointed out that the use of *E. coli* to clone recombinant molecules may be dangerous because *E. coli* is a normal inhabitant of the human intestine and is also able to exchange genetic information with bacteria pathogenic to man. The two kinds of experiments that the committee believed should be voluntarily halted are those that might lead to accidental introduction of genes for antibiotic resistance or toxin production or which might induce tumors. Great caution is called for in all plans to link fragments of animal DNA to DNA of bacterial plasmids or phage. Regulations are being drafted by various agencies that support biochemical research.[269]

I. THE EUKARYOTIC CHROMOSOME AND ITS CONTROL

The amount of DNA in the genome of the simple eukaryotic slime mold *Dictyostelium* is 11 times that of *E. coli*. The "higher" organism with the smallest amount of DNA is *Drosophila*, whose haploid genome size is 24 times that of *E. coli*. Man has about 600 times the coding capacity of a bacterium (Table 1-3). Part of the problem of understanding the eukaryotic genome is one of dealing with this large amount of DNA. Furthermore, within eukaryotes there tend to be large changes in the transcription of genes with both time and environment. Thus, the control of gene expression must be very complex.

1. Organization

The genome of higher organisms consists of a number of separate chromosomes, each one apparently containing a single DNA duplex. This DNA is intimately associated with other components including approximately 75% protein and 10% RNA (Chapter 1, Section B,2). Until recently little in the way of organization below the level of the chromosomes could be discerned. However, during the prophase of mitosis or meiosis the extended chromosomes sometimes assume a bead-like appearance. The tiny beads of DNA-rich material, known as **chromomeres,** like the bands of the polytene chromosomes of *Drosophila* (Section D,9c), may represent some kind of unit of genetic information. Their existence suggests that the DNA of the chromosome is in some fashion divided into units, possibly analogous to the operons of bacteria.

Some of the bands in a polytene chromosome bulge out as **puffs,** the largest of which are called **Balbiani rings.** In these puffs, the DNA is in a more extended form and is thought to be undergoing active transcription. Since the average band in a polytene chromosome contains a length of DNA consisting of nearly 10^5 base pairs, a very large mRNA transcript would be possible.[270] Indeed, recent studies show that a single Balbiani ring yields a 75 S RNA of mass $15–35 \times 10^6$ daltons.[271] This large RNA molecule appears to be carried intact to the cytoplasm where very active protein synthesis is occurring. Is this mRNA the transcript of a single gene together with a large amount of DNA that does not code for a specific protein, or does it carry the code for a single protein repeated many times? This question has relevance to our general understanding of the organization of the genetic information in eukaryotic chromosomes. At the same time, it must be remembered that polytene chromosomes represent a special modification found in cells that have undergone terminal differentiation and will not reproduce.

a. Lampbrush Chromosomes

A second specialized type of chromosome that has contributed much to our understanding of eukaryotic nuclei is found in the meiotic prophase of oocytes. These "lampbrush chromosomes" have been studied intensively in amphibians such as *Xenopus*. A lampbrush chromosome is in fact a homologous pair of chromosomes, each of which in turn consists of two closely associated chromatids. The chromosomes are highly expanded and about 5% of the DNA is extended in the form of ~4000 perfectly paired loops visible with an electron microscope. Each loop has a length of ~50 nm of DNA or ~150,000 bases. No evidence of any breaks in the DNA is seen, a fact that supports the belief that a single DNA molecule extends from one end of the chromosome to the other through all of the loops.

Like the puffs of polytene chromosomes (which may have a similar structure), lampbrush chromosomes appear to be actively engaged in transcription. It is thought that approximately 3% of the DNA is functional in producing mRNA that is accumulated within the oocyte and functions during early embryonic development.[272] It is logical to suppose that a single loop in a lampbrush chromo-

some, like a single band in a polytene chromosome, represents a transcriptional unit. However, there is a paradox. A band or loop contains enough DNA to code for 30–35 average-sized proteins. Nevertheless, genetic fine structure analysis has failed to locate more than one complementation unit in each band of a *Drosophila* chromosome.[273] This suggests that only 3% of the DNA of *Drosophila* contains structural genes for essential protein. What does the rest do, and why don't mutations in the rest of the DNA harm the organism? The answers to these questions are still unclear.

b. Reiterated Sequences

Chemical investigations are now beginning to give a more detailed picture of the organization of chromosomes. In one kind of experiment DNA is cut into fragments of ~10,000 base pairs and the fragments are denatured by heat. It is found that renaturation of the resulting single-stranded fragments upon cooling takes place in two or more steps. Some material reforms double helices rapidly, whereas other material is slow to renature (Fig. 15-34).[273–275]

The most rapidly renaturing DNA fragments are often found to have a different base composition than the bulk of the DNA. As a result of this, upon centrifugation of the fragmented DNA in a CsCl gradient, the rapidly renaturing fraction tends to separate as a small "satellite" band. This **satellite DNA** is found to consist of highly repetitive short sequences.[276,277] For example, DNA of a satellite band of DNA from the kangaroo rat contains the repeated sequence 5'-GGACACAGCG-3'. This highly reiterated sequence accounts for 11% of the entire DNA of the cell. Satellite DNA is usually associated with regions of the chromosome that do not unravel in telophase as does the bulk of the DNA. The functions of satellite DNA are unknown. It has been suggested that repetitive sequences arise naturally during evolution as a result of unequal crossing-over between sister chromatids during mitosis.[232a] Using restriction endonucleases human DNA has been cut into a

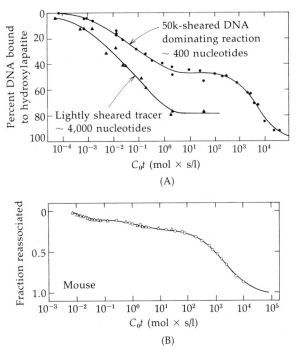

(A)

(B)

Fig. 15-34 Renaturation curves for mammalian DNA. (A) The demonstration of intermixing of repeated and nonrepeated sequences in the calf genome. The upper curves show the reassociation of calf DNA fragments sheared to a length of ~400 nucleotides. Incubation at 60°C in 0.12 M phosphate buffer and hydroxyapatite assay (Chapter 2, Section D,10) under the same conditions. The lower curve shows the reassociation of a small quantity of labeled 4000-nucleotide-long fragments with the majority of 400-nucleotide-long fragments. The DNA was subsequently fractionated into three portions: (1) Fast reassociating ($C_0t \sim 3 \times 10^{-4}$), ~$10^6$ copies per genome; satellite DNA; (2) intermediate fraction ($C_0t \sim 0.01$), 37%, ~66,000 copies; (3) slow ($C_0t \sim 1000$), nonrepetitive DNA, see Fig. 2-29. From R. J. Britten and J. Smith, *Carnegie Inst. Wash. Yearb.* **68**, 378–391 (1970). (B) Reassociation of sheared denatured mouse DNA in 48% formamide and 5 × SSC (SSC refers to a solution 0.15 M in NaCl and 0.015 M in sodium citrate) at 37°C. Reactions were carried out at two concentrations: 50 μg/ml in 1 cm path length glass-stoppered cuvettes (△), and 1 mg/ml in 1 mm path length cells (○). The hypochromicity was monitored at 270 nm, and the fraction of DNA reassociated at a given time was plotted as a function of the initial concentration of DNA (moles of phosphorus/liter) multiplied by the reaction time (seconds). The sample is estimated to consist of 10% fast-renaturing satellite DNA, 15% intermediate, partially redundant DNA, and 75% unique sequence DNA. From B. L. McConaughy and B. J. McCarthy, *Biochem. Genet.* **4**, 425–446 (1970).

series of pieces. By comparing electrophoretic patterns of the DNA fragments of males and females, it was possible to deduce that the male Y chromosome contains a sequence that is repeated in tandem several thousand times.

Other observations suggest that some reiterated sequences are spread more randomly throughout the genome. For example, when fragments of DNA from *Drosophila* are renatured, **rings** of DNA, visible with the electron microscope, are formed.[278] Fragmentation within the repeated sequences would permit ring formation to occur during renaturation. The chromosomes of *Xenopus* may contain ~25% of similar reiterated sequences. Electron microscopy suggests that random reassociation of DNA fragments leads to duplex regions, containing the repeated sequences, with single-stranded "tails." The latter ordinarily do not pair because they represent unique sequences coming from different genes. In *Xenopus* the repetitive segments of DNA contain ~300 nucleotides, and the nonrepetitive or single-copy DNA between the repetitive segments contains ~800 nucleotides.[275]

Much about the organization of DNA can be learned from examination of RNA transcripts. When mRNA from the cytoplasm of *Dictyostelium* is compared with that from the nucleus, substantial differences are found.[279] Nuclear mRNA contains ~1600 nucleotides per chain but that in the cytoplasm has been cut down to ~1300. Both nuclear and cytoplasmic mRNA contain tracts of polyadenylic acid (Section B,5) which can be isolated by enzymatic digestion and analyzed for length. It is found that the nuclear mRNA contains tracts of ~25 adenylic acid units. This poly(A)$_{25}$ is apparently transcribed along with the rest of the genetic message for the DNA itself contains ~15,000 sequences of poly(dT)$_{25}$. This is about the right number to provide one poly(dT)$_{25}$ sequence for each gene. Similar poly (dT) tracts have been found in the DNA of all metazoans. For example, ~6000 sequences are found in the DNA of *Drosophila*, corresponding to the 5000 bands of the polytene chromosomes. The mammalian genome contains ~100,000 poly(dT) sequences.

Turning to the cytoplasmic mRNA, despite the shorter size, more poly(A) is present. In addition to the poly(A)$_{25}$ tracts of *Dictyostelium,* an equal number of longer poly(A) tracts containing ~ 100 adenylic acid units are present. These are apparently added after transcription. A working hypothesis (Eq. 15-14)[280] is that repeated sequences,

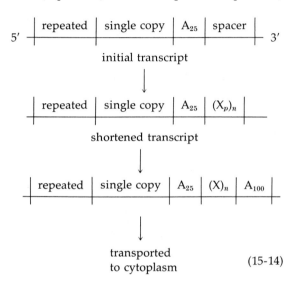

$$(15\text{-}14)$$

single copy regions, poly(A)$_{25}$, and a "spacer region" are present in a sequence for each gene. After transcription the RNA corresponding to the spacer is digested away, leaving an unknown number of nucleotides $(X_p)_n$ at the 3' end of the poly(A)$_{25}$. Then ~ 100 additional units of adenylic acid are added before transport into the cytoplasm.[278] In addition, other "tandem repeats" are thought to occur and possibly to represent "reiterated genes" that are needed in more than one copy to supply adequate transcripts to the cell.

c. Genes for Ribosomal RNA

While it seems probable that most genes are present as only one copy per chromosome, the genes of ribosomal RNA and tRNA are reiterated many times. In the case of *Xenopus,* ~ 450 repeats of the 28 S and 18 S ribosomal RNA genes are present on one chromosome. About 24,000 copies of the 5 S RNA genes are found at the ends of the long arms of most of the chromosomes.[281] The 28 S

and 18 S rRNA genes are transcribed together along with a spacer region between them. Nontranscribed spacer regions lie between the repeats of the gene pairs, as can be seen with the electron microscope in Fig. 15-11. The gene for 5 S RNA is known to contain a high GC content; hence, the regions coding for it should be resistant to heat denaturation. Indeed, a denaturation map of DNA shows easily denatured regions separated by shorter 120 base sequences, apparently of high GC content and presumably coding for the 5 S RNA. The easily denatured (high AT) spacer regions are ~ 630 bases long. Using specific restriction enzymes, much of this high AT region has been cut into repeating units that contain repeats within repeats. A basic 15 unit polynucleotide contains the sequence $A_4CUCA_3CU_3G$ repeated about 30 times.[282]

In *Drosophila* a restriction endonuclease map of the 5 S RNA region of chromosome 2 indicates the presence of two tandomly repeated gene clusters, each containing about 90 genes. While the orientation is unknown it has been suggested that the two clusters might be arranged as a long palindrome and that conversion to the cruciform configuration might occur as a step in mismatch repair. This would have the effect of maintaining homogeneity in the cluster.[282a] The same argument would apply to the many long palindromes present in eukaryotic DNA.[235a]

d. Gene Amplification

Under some circumstances a segment of the genome may be "amplified" by repeated replication of a gene or genes. The best known example is the amplification of the ribosomal RNA genes of amphibian oocytes. In *Xenopus* excess DNA accumulates around the nucleoli and later breaks up to form 1000 or more separate nucleoli. As many as 3000 copies of the rDNA (which forms a distinct satellite band upon centrifugation) may be present. Amplified rDNA represents a convenient material for biochemical investigation. For example, the structural studies mentioned in the preceding paragraph were performed on this type of DNA.

The mechanism of DNA amplification is unknown but is under active study.[283] It has been suggested that the many copies of the rDNA are generated by a rolling circle mechanism similar to that in Eq. 15-9. The purpose of the amplification of rDNA is apparently to produce large numbers of ribosomes to facilitate rapid protein synthesis.

2. The Proteins of the Nucleus

Within bacterial cells the negatively charged phosphate groups of the DNA may be neutralized to a large extent by the positively charged polyamines. However, basic proteins also tend to partially coat the DNA. Within the mature heads of sperm cells of fish, the tightly packaged DNA is neutralized by the **protamines,** special proteins of low molecular weight (MW ~5000) and rich in arginine residues. Similar basic proteins are found in mammalian sperm.[284] However, within somatic cells, the charges on DNA are balanced principally by a heterogeneous group of basic proteins known as the **histones.** There are five classes of histones which range in molecular weight from ~11,000–21,500.[285–287]

H1   (or I or f1)
  Lysine-rich

H2a (or IIb1 or f2a2)
H2b (or IIb2 or f2b)
Moderately lysine-rich

H3   (or III or f3)
H4   (or IV or f2a1)
  Arginine-rich

The arginine-rich histones are noteworthy in possessing a strongly conserved amino acid sequence. Thus, histone H4 from pea seedlings differs from that of the bovine thymus by only two amino acids. On the other hand, the lysine-rich H1 is almost species-specific in its sequence.

Histone H3 of calf thymus contains 135 residues,[288] the first 53 of which contain a net charge of +18. This is probably the portion that binds to DNA. On the other hand, the carboxyl terminal end is hydrophobic and only slightly basic. Interesting clusters of basic amino acids are found at intervals[289] in the sequence of histone H2a. One of the interesting features of histone structure is the presence of a substantial amount of **micromodification** in the form of phosphorylation of serine residues, acetylation, and methylation of lysine residues and methylation of arginine side chains. For example, in histone H3, lysines 14 and 23 are N-acylated, while Lys 9 and Lys 27 are partially ϵ-N-methylated, each site containing partly mono-, partly di-, and partly trimethyl derivatives.

What functions do histones have other than that of neutralizing the negative charge on the DNA? An early idea was that these proteins might serve as gene repressors similar to those in bacteria. However, no evidence to support this idea has been found. Histones do appear to form some kind of a complex with the DNA fibers. The newest electron micrographs show chromatin fibers to exist in a regular repeating structure resembling beads on a string. The beads (or ν bodies) are of 7–10 nm diameter with 2–14 nm of DNA "string" between them (Fig. 15-35).[290–294] The "beads" have a high DNA content and neutron diffraction experiments suggest that within the ν bodies DNA is wound around the outside of a histone, oligomer (Fig. 15-36).[295] Histones H2a, H2b, H3, and H4 are found in nearly equivalent amounts, one molecule of each per 100 base pairs in the DNA. An octamer containing two subunits of each type of histone has been obtained in solution.[296]

Digestion of chromatin by nucleases causes rapid cleavage to fragments containing 205 ± 15 base pairs and slower cleavage to 170 base pair fragments. These results collectively suggest a structure in which 200 base pair segments of DNA are folded around a histone octomer to the extent that the 68 nm extended length is condensed into a 10 nm ν body. A very short length of DNA con-

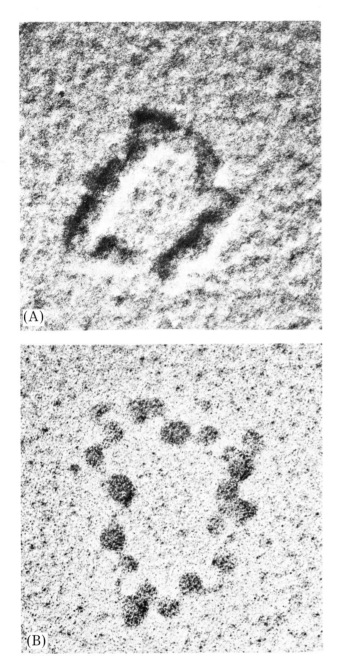

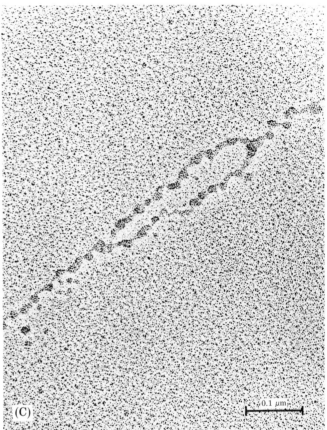

Fig. 15-35 (A) Electron micrograph of ''minichromosome'' formed from virus SV40 growing in monkey cells in culture.[292] In this native form the nucleoprotein fiber is ~11 nm in diameter and ~210 nm in length. (B) Beaded form of minichromosome observed when the ionic strength was lowered. The 21 beads have diameters of ~11 nm and are joined by bridges roughly 2 nm in diameter and 13 nm long. Deproteinization and relaxation of the DNA reveals that the overall length of the DNA present is 7 times the length of the native minichromosome. (C) Electron micrograph of chromatid of a blastoderm-stage embryo of *Drosophila melanogaster* in the process of replication. Note the presence of nucleosomal particles immediately adjacent to the replication forks. Courtesy of Steven L. McKnight and Oscar L. Miller, Jr.

nects adjacent ν bodies. It has been suggested that the DNA, in a typical double helix, may be kinked sharply at 20 base pair intervals as it wraps around the histone core.[297] Each kink would be associated with a 15°–20° unwinding of the helix. Histone H1, present in smaller amounts than the other histones, may have a cross-linking function (Fig. 15-35). Other evidence[296a] suggests that one negative superhelix turn is associated with each ν body. Thus, the number of ν bodies seen in the

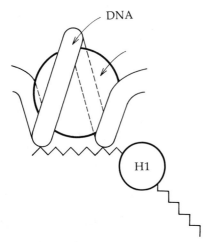

DNA

H1

Fig. 15-36 Schematic representation of a possible model for the chromatin subunit structure. The protein core is a complex of the apolar segment of the four histones indicated, the basic segments of the histones being complexed with DNA on the outside of the unit. Histone H1, possibly on the outside of the chain of globular subunits, may have a cross-linking role, either between subunits in the same chain or between subunits in different chains. The pitch of the DNA, which need not be uniformly coiled, is 5.5 nm with a mean diameter of about 10 nm. From Baldwin et al.[295]

minichromosome of Fig. 15-35 matches the numbers of supercoils in the SV40 DNA (Fig. 2-27). It reacts preferentially with superhelix DNA.[296b] Chemical cross-linking studies suggest that H1 histones are often arranged contiguously with few other histones interspersed and that H2B is next to H2A, H3, and H4 with equal frequency.[298] The same basic structure for chromatin has been found in animals, fungi, and green plants.[299]

The findings described in the preceding paragraphs have led to the current view that histones function primarily to determine DNA folding. However, H1 is sometimes regarded as a general repressor, holding the chromatin tightly folded and preventing transcription. The observation that H1 becomes phosphorylated by a special protein kinase during the initiation step of mitosis is suggestive of another role.[300] Other histones, especially F4, undergo a variety of modification reactions including acetylation and phosphorylation (reversible reactions) and methylation (irreversible).[301] The function of these reactions in controlling events such as transcription and replication is still unknown.

What proteins in addition to histones are found in the nucleus? Polyacrylamide gel electrophoresis has revealed over 450 components in HeLa cell nuclei. Most are present in small amounts ($<10{,}000$ molecules per cell) and are not detectable in cytoplasm.[302] Among the more acidic proteins are a large number of enzymes including the RNA polymerases. There are certainly gene repressors, largely unidentified, hormone binding proteins and many others.[303] While attention has been focused on nuclear proteins a little-studied class of **small nuclear RNA's** (snRNA's) has also been implicated in control of gene expression. These molecules, from 65 to 200 nucleotides long, may stimulate transcription of specific genes by binding to complementary sites in DNA. Thus, information transcribed from one site on a chromosome may influence events at another site on another chromosome.[303a]

3. Poly(ADP-Ribose)

An unusual component of the nuclei of all higher cells is formed by polymerization of ADP-

$$\left(\begin{array}{c} CH_2\!-\!O\!-\!\textcircled{P}\!-\!\textcircled{P}\!-\!O\!-\!CH_2 \\ \end{array} \right)_n$$

OH HO

OH O

NH$_2$

Poly(ADP-ribosyl)

ribosyl groups derived from NAD^+. A special enzyme catalyzes the polymerization, displacing off nicotinamide and creating a glycosidic linkage from the 1-carbon of the ribose from which the nicotinamide was displaced and the 2'-carbon of the ADP portion of the next monomer unit. The function is not known, but poly(ADP-ribosyl) groups are found covalently attached to nuclear proteins, and also to cytoplasmic components[304,305] (see Box 15-F).

BOX 15-F

TOXIC PROTEINS: THE DIPHTHERIA TOXIN[a]

Until a suitable vaccine was developed, an infection by *Corynebacterium diphtheriae* was one of the dread diseases of childhood. Despite the fact that the bacteria caused only superficial membranous lesions in the throat, the patient often died with evident damage to many organs. The cause is a potent heat-labile toxic protein. Oddly, the bacterium produces the toxin only when infected by a temperate bacteriophage[b] carrying the *tox* gene and when the inorganic iron of the surroundings has been largely depleted.

Diphtheria toxin is a protein of MW $\sim 62,000$ with a minimum lethal dose of only 0.16 mg kg^{-1} for the guinea pig. Tests in cell culture show that the toxin blocks incorporation of amino acids into proteins by inactivation of elongation factor EF-2 required for translocation in mammalian ribosomes. The toxin acts as an enzyme which transfers an ADP-ribosyl group from NAD^+ to EF-2.

$$EF\text{-}2 + NAD^+ \xrightarrow{\text{toxin}}$$
$$ADP\text{-ribosyl} —EF\text{-}2 + nicotinamide$$

The modified EF-2 reacts normally with GTP, but the complex so formed is unable to participate in the translocation reaction. A concentration of the toxin in the cytoplasm of only 10^{-8} M is sufficient to promote the fatal reaction.

How does a protein toxin of this type enter a cell? The evidence is that part of the molecule contains a special structure that binds at discrete sites on the cell membrane. Perhaps binding at these sites stimulates pinocytosis. Partial proteolytic digestion occurs before the toxin acts.

What is the origin of the *tox* gene and why is it carried by a virus? Pappenheimer and Gill suggested[a] that the gene was somehow derived from a eukaryotic gene coding for some functional protein. This gene became incorporated into a virus and after a period of evolution came to specify the toxic protein. The fact that cell nuclei contain poly(ADP-ribose) (Section I,3) suggests a possible origin of the *tox* gene. NAD^+ is the substrate for a synthesis of this nuclear polymer, and the synthetase promotes a cleavage of the ribosyl-nicotinamide bond with formation of a new glycosidic linkage between the 1' carbon of the ribose and the 2' hydroxyl of the adenosine in the next monomer unit. Perhaps it was the gene for the synthetase that became modified to yield the diphtheria toxin.

A group of proteins toxic to bacteria, the colicins, have been mentioned in Section D,7. These also appear to bind to special receptors in the outer membrane of bacteria such as *E. coli*. Neilands and associates discovered that the colicin M receptor in *E. coli* is also a receptor for the peptide siderochrome ferrochrome (See Box 14-C) and for bacteriophage T5. It is also the site of binding of the antibiotic albomycin. It is hypothesized that early in evolution bacteria developed iron-chelating molecules and that as these became larger their size exceeded that which would allow diffusion through the outer membrane of the cells. Hence specific transport systems developed. These were later exploited by phage and by colicin-producing strains.[c]

[a] A. M. Pappenheimer, Jr. and D. M. Gill, *Science* **182**, 353–364 (1973).
[b] The toxin of *Clostridium botulinin* also seems to be specified by a phage gene. M. W. Eklund, F. T. Poysky, S. M. Reed, and C. A. Smith, *Science* **172**, 480–482 (1971).
[c] M. Luckey, R. Wayne and J. B. Neilands, *BBRC* **64**, 687–693 (1975).

REFERENCES

1. General references include the following book by Watson and the books cited in references 2–5; J. D. Watson, "Molecular Biology of the Gene," 3rd ed. Benjamin, New York, 1976.
2. B. Lewin, "Gene Expression," 2 vols. Wiley, New York, 1974.
3. W. Hayes, "The Genetics of Bacteria and Their Viruses." Wiley, New York, 1968.
4. A Kornberg, "DNA Synthesis." Freeman, San Francisco, California, 1974.
5. J. N. Davidson, "The Biochemistry of the Nucleic Acids," 7th ed. Academic Press, New York, 1972.
6. A. E. Mirsky, *Sci. Am.* **218**, 78–88 (June 1968).
7. E. Chargaff, *Science* **172**, 637–642 (1971).
8. R. Olby, *Nature* (*London*) **248**, 782–785 (1974).
8a. A. D. Hershey and M. Chase, *J. Gen. Physiol* **36**, 39–56 (1952).
9. J. D. Watson, "The Double Helix." Atheneum, New York, 1968 (Watson's account of the history of the discovery of the DNA structure).
10. E. Chargaff, *Experientia* **6**, 201–209 (1950).
11. A. Sayre, "Rosalind Franklin and DNA." Norton, New York, 1975.
12. J. D. Watson and F. H. C. Crick, *Nature* (*London*) **171**, 737–738 (1953).
13. F. Crick, *Nature* (*London*) **248**, 766–769 (1974).
13a. J. Lederberg and E. L. Tatum, *Nature* (*London*) **158**, 558 (1946).
14. C. C. Brinton, *Crit. Rev. Microbiol.* **1**, 105–160 (1971).
15. B. J. Bachmann, K. B. Low, and A. L. Taylor, *Bacteriol. Rev.* **40**, 116–167 (1976).
16. F. H. C. Crick, *Cold Spring Harbor Symp. Quant. Biol.* **31**, 3–9 (1966), an interesting account of the history of the deciphering of the genetic code.
17. M. W. Nirenberg and J. H. Matthaei, *PNAS* **47**, 1588–1602 (1961).
18. M. Nirenberg and P. Leder, *Science* **145**, 1399–1407 (1964).
19. J. H. Matthaei, H. P. Voigt, G. Heller, R. Neth, G. Schöch, H. Kübler, F. Amelunxen, G. Sander, and A. Parmeggiani, *Cold Spring Harbor Symp. Quant. Biol.* **31**, 25–38 (1966).
20. H. G. Khorana, H. Büchi, H. Ghosh, N. Gupta, T. M. Jacob, H. Kössel, R. Morgan, S. A. Narang, E. Ohtsuka, and R. D. Wells, *Cold Spring Harbor Symp. Quant. Biol.* **31**, 39–49 (1966).
21. R. S. Edgar and R. H. Epstein, *Sci. Am.* **212**, 70–78 (Feb. 1965).
22. H. Fraenkel-Conrat, *Sci. Am.* **211**, 46–54 (Oct. 1964).
23. B. Lewin, "Gene Expression," Vol. 2, p. 258. Wiley, New York, 1974.
24. M. Meselson and F. W. Stahl, *PNAS* **44**, 671–682 (1958).
25. J. Cairns, *JMB* **6**, 208–213 (1963); *Cold Spring Harbor Symp. Quant. Biol.* **28**, 43–46 (1963).
26. A. Kornberg, *Science* **163**, 1410–1418 (1969).
27. K. Sugimoto, T. Okazaki, and R. Okazaki, *PNAS* **60**, 1356–1362 (1968).
28. P. Modrich, Y. Anraku, and I. R. Lehman, *JBC* **248**, 7495–7501 (1973).
29. I. R. Lehman, *Science* **186**, 790–797 (1974).
30. T. Caspersson, *Naturwissenschaften* **29**, 33–43 (1941).
31. J. Brachet, *Arch. Biol.* **53**, 207–257 (1942).
32. D. M. Prescott, *Prog. Nucleic Acid Res. Mol. Biol.* **3**, 33–57 (1964).
33. J. L. Sirlin, "Biology of RNA." Academic Press, New York, 1972.
33a. M. B. Hoagland, E. B. Keller, and P. Zamecnik, *JBC* **218**, 345–358 (1956).
34. A. N. Belozersky and A. S. Spirin, *in* "The Nucleic Acids" (E. Chargaff and J. N. Davidson, eds.), Vol. 3, pp. 147–185. Academic Press, New York, 1960.
35. C. I. Davern and M. Meselson, *JMB* **2**, 153–160 (1960).
35a. E. Volkin and L. Astrachan, *Virology* **2**, 149–161 (1956).
36. F. Jacob and J. Monod, *JMB* **3**, 318–356 (1961).
37. S. Brenner, F. Jacob, and M. Meselson, *Nature* (*London*) **190**, 576–581 (1961).
37a. F. H. C. Crick, *Biochem. Soc. Symp.* **14**, 25–26 (1957).
38. S. Bourgeois, *Curr. Top. Cell. Regul.* **4**, 39–75 (1971).
39. R. C. Dickson, J. Abelson, W. M. Barnes, and W. S. Reznikoff, *Science* **187**, 27–35 (1975).
40. M. J. Chamberlin, *Annu. Rev. Biochem.* **43**, 721–775 (1974).
41. S. Nakanishi, S. Adhya, M. Gottesman, and I. Pastan, *JBC* **248**, 5937–5942 (1973).
42. N. M. Maizels, *PNAS* **70**, 3585–3589 (1973).
43. W. Gilbert and A. Maxam, *PNAS* **70**, 3581–3584 (1973).
44. A. Gierer, *Nature* (*London*) **212**, 1480–1481 (1966).
45. J. C. Wang, M. D. Barkley, and S. Bourgeois, *Nature* (*London*) **251**, 247–249 (1974).
46. R. L. Perlman and I. Pastan, *Curr. Top. Cell. Regul.* **3**, 117–134 (1971).
47. M. A. Savageau, *Nature* (*London*) **258**, 208–214 (1975).
48. R. F. Goldberger and J. S. Kovach, *Curr. Top. Cell. Regul.* **5**, 285–308 (1972).
49. J. A. Lewis and B. N. Ames, *JMB* **66**, 131–142 (1972).
50. K. Bertrand, L. Korn, F. Lee, T. Platt, C. L. Squires, C. Squires, and C. Yanofsky, *Science* **189**, 22–26 (1975).
51. M. J. Chamberlin, *in* "The Enzymes", 3rd ed., (P. D. Boyer, ed.), Vol. 10, pp. 333–374. Academic Press, New York, 1974.
52. P. Chambon, *in* "The Enzymes" (P. D. Boyer, ed.), 3rd ed., Vol. 10, pp. 261–331. Academic Press, New York, 1974.
53. V. E. F. Sklar, L. B. Schwartz, and R. G. Roeder, *PNAS* **72**, 348–352 (1975).
53a. P. Valenzuela, G. L. Hager, F. Weinberg, and W. J. Rutter, *PNAS* **73**, 1024–1028 (1976).
53b. F. D. Gillen and N. G. Nossal, *JBC* **251**, 5225–5232 (1976).
53c. J. W. Drake and R. H. Baltz, *Ann. Rev. Biochem.* **45**, 11–37 (1976).
54. D. C. Ward and E. Reich, *JBC* **247**, 705–719 (1972).
55. J. W. Roberts, *Nature* (*London*) **224**, 1168–1174 (1969).

55a. B. H. Howard and B. de Crombrugghe, *JBC* **251**, 2520–2524 (1976).

56. J. J. Dunn and F. W. Studier, *PNAS* **70**, 3296–3300 (1973).

57. R. P. Perry, *Annu. Rev. Biochem.* **45**, 605–629 (1976).

58. E. Lund, J. E. Dahlberg, L. Lindahl, S. R. Jaskunas, and P. P. Dennis, and M. Nomura, *Cell* **7**, 165–177 (1976).

59. O. L. Miller, Jr., *Sci. Am.* **228**, 34–42 (Mar. 1973).

59a. V. E. Foe, L. E. Wilkinson, and C. D. Laird, *Cell* **9**, 131–146 (1976).

59b. O. L. Miller, B. A. Hamkalo, and C. A. Thomas, Jr., *Science* **169**, 392–395 (1970).

59c. S. L. McNight and O. L. Miller, Jr., *Cell* **8**, 305–319 (1976).

60. R. W. Holley, J. Apgar, G. A. Everett, J. T. Madison, M. Marquisee, S. H. Merrill, J. R. Penswick, and A. Zamir, *Science* **147**, 1462–1465 (1965). This is a report of the first tRNA sequence determined.

61. G. J. Quigley, A. H. J. Wang, N. C. Seeman, F. L. Suddath, A. Rich, J. L. Sussman, and S. H. Kim, *PNAS* **72**, 4866–4870 (1975).

62. J. D. Robertus, J. E. Ladner, J. T. Finch, D. Rhodes, R. S. Brown, B. F. C. Clark, and A. Klug, *Nature (London)* **250**, 546–551 (1974).

63. H. D. Robertson, S. Altman, and J. D. Smith, *JBC* **247**, 5243–5251 (1972).

64. P. W. Piper and B. F. C. Clark, *Nature (London)* **247**, 516–518 (1974).

65. M. Simsek, V. L. RajBhandary, M. Boisnard, and G. Petrissant, *Nature (London)* **247**, 518–520 (1974).

66. E. K. Bikoff, B. F. LaRue, and M. L. Gefter, *JBC* **250**, 6248–6255 (1975).

67. J. G. Seidman and W. H. McClain, *PNAS* **72**, 1491–1495 (1975).

68. V. Daniel, J. I. Grimberg, and M. Zeevi, *Nature (London)* **257**, 193–197 (1975).

68a. C. Ilgen, L. L. Kirk, and J. Carbon, *JBC* **251**, 922–929 (1976).

69. H. Sakano and Y. Shimura, *PNAS* **72**, 3369–3373 (1975).

70. D. Söll, *Science* **173**, 293–299 (1973).

71. H. G. Khorana, K. L. Agarwal, P. Besmer, H. Büchi, M. H. Caruthers, P. J. Cashion, M. Fridkin, E. Jay, K. Kleppe, R. Kleppe, A. Kumar, P. C. Loewen, R. C. Miller, K. Minamoto, A. Panet, U. L. RajBhandary, B. Ramamoorthy, T. Sekiya, T. Takeya, and J. H. van de Sande, *JBC* **251**, 565–570 (1976).

71a. B. Ramamoorthy, R. G. Lees, D. G. Kleid, and H. G. Khorana, *JBC* **251**, 676–694 (1976).

72. P. C. Loewen, T. Sekiya, and H. G. Khorana, *JBC* **249**, 217–226 (1974).

73. A. Kornberg, "DNA synthesis," pp. 364–368. Freeman, San Francisco, California, 1974.

74. D. A. Wilson and C. A. Thomas, Jr., *JMB* **84**, 115–144 (1974).

75. T. Sekiya, R. Contreras, H. Küpper, A. Landy, and H. G. Khorana, *JBC* **251**, 5124–5140 (1976).

76. N. Rosenbaum and M. L. Gefter, *JBC* **247** 5675–5680 (1972).

77. J. E. Darnell, W. R. Jelinek, and G. R. Molloy, *Science* **181**, 1215–1221 (1973).

78. R. A. Weinberg, *Annu. Rev. Biochem.* **42**, 329–354 (1973).

79. G. Brawerman, *Annu. Rev. Biochem.* **43**, 621–642 (1974).

79a. L. A. Haff and E. B. Keller, *JBC* **250**, 1838–1846 (1975).

80. P. R. Srinivasan, M. Ramanarayanan, and E. Rabbani, *PNAS* **72**, 2910–2914 (1975).

81. B. Griffin, *Nature (London)* **255**, 9 (1975).

82. M. J. Ensinger, S. A. Martin, E. Paoletti, and B. Moss, *PNAS* **72**, 2525–2529 (1975).

83. J. M. Adams and S. Cory, *Nature (London)* **255**, 28–33 (1975).

83a. A. Nomoto, Y. F. Lee, and E. Wimmer, *PNAS* **73**, 375–380 (1976).

84. B. E. H. Maden, *Trends Biochem. Sci.* **1**, 196–199 (1976).

85. R. N. Nazar, T. W. Owens, T. O. Sitz, and H. Busch, *JBC* **250**, 2475–2481 (1975).

85a. P. K. Wellauer and I. B. Dawid, *PNAS* **70**, 2827–2831 (1973).

85b. I. B. Dawid and P. K. Wellauer, *Cell* **8**, 443–448 (1976).

86. B. A. Hamkalo and O. L. Miller, Jr., *Annu. Rev. Biochem.* **42**, 379–396 (1973).

87. G. E. Fox and C. R. Woese, *Nature (London)* **256**, 505–507 (1975).

88. R. A. Garrett and H.-G. Wittmann, *Endeavour* **32**, 8–14 (1973).

89. E. Collatz, I. G. Wool, A. Lin, and G. Stöffler, *JBC* **251**, 4666–4672 (1976).

90. E. H. McConkey, *PNAS* **71**, 1379–1383 (1974).

91. M. Nomura, *Science* **179**, 864–873 (1973).

92. K. H. Nierhaus and F. Dohme, *PNAS* **71**, 4713–4717 (1974).

93. W. A. Held, B. Ballou, S. Mizushima, and M. Nomura, *JBC* **249**, 3103–3111 (1974).

94. A correspondent, *Nature (London)* **254**, 555–556 (1975).

95. A. Sommer and R. R. Traut, *JMB* **97**, 471–481 (1975).

96. G. W. Tischendorf, H. Zeichhardt, and G. Stöffler, *PNAS* **72**, 4820–4824 (1975).

97. H.-G. Wittmann, *EJB* **61**, 1–13 (1976).

98. R. Haselkorn and L. B. Rothman-Denes, *Annu. Rev. Biochem.* **42**, 397–438 (1973).

99. S. Ochoa and R. Mazumder, in "The Enzymes," 3rd ed., (P. D. Boyer, ed.), Vol. 10, pp. 1–51. Academic Press, New York, 1974.

100. J. Shine and L. Dalgarno. *Nature (London)* **254**, 34–38 (1975).

101. J. A. Steitz and K. Jakes, *PNAS* **72**, 4734–4738 (1975).

102. A. E. Dahlberg and J. E. Dahlberg, *PNAS* **72**, 2940–2944 (1975).

103. G. Jay and R. Kaempfer, *JBC* **250**, 5742–5748 (1975).

104. J. Lucas-Lenard and L. Beres, in "The Enzymes," 3rd ed., (P. D. Boyer, ed.), Vol. 10, pp. 53–86. Academic Press, New York, 1974.

104a. G. R. Jacobson and J. P. Rosenbusch, *Nature (London)* **261**, 23–26 (1976).

105. S. M. Hecht, J. W. Kozarich, and F. J. Schmidt, *PNAS* **71**,

4317–4321 (1974).

106. M. Sprinzl and F. Cramer, *PNAS* **72,** 3049–3053 (1975).

107. T. H. Fraser and A. Rich, *PNAS* **72,** 3044–3048 (1975).

108. W. P. Tate and C. T. Caskey, *in* "The Enzymes," 3rd ed., (P. D. Boyer, ed.), Vol. 10, pp. 87–118. Academic Press, New York, 1974.

109. H. Weissbach and S. Ochoa, *Ann. Rev. Biochem.* **45,** 191–216 (1976).

109a. D. A. Shafritz, J. A. Weinstein, B. Safer, W. C. Merrick, L. A. Weber, E. D. Hickey, and C. Baglioni, *Nature (London)* **261,** 291–294 (1976).

110. W. C. Merrick, W. M. Kemper, J. A. Kantor, and W. F. Anderson, *JBC* **250,** 2620–2625 (1975).

111. F. H. C. Crick, *JMB* **19,** 548–555 (1966).

112. D. Söll and P. R. Schimmel, *in* "The Enzymes," 3rd ed. (P. D. Boyer, ed.), Vol. 10, pp. 489–538. Academic Press, New York, 1974.

113. A. Stein and D. M. Crothers, *Biochemistry* **15,** 160–168 (1976).

114. B. R. Reid and G. T. Robillard, *Nature (London)* **257,** 287–291 (1975).

114a. J. J. Hopfield, T. Yamane, V. Yue, and S. M. Coutts, *PNAS* **73,** 1164–1168 (1976).

115. L. Gorini, *Sci. Am.* **214,** 102–109 (Apr 1966).

116. S. Pestko, *Annu. Rev. Microbiol.* **25,** 487–562 (1971).

116a. S. R. Jaskunas, L. Lindahl, M. Nomura, and R. R. Burgess, *Nature (London)* **257,** 458–462 (1975).

117. C. Weissmann, M. A. Billeter, H. M. Goodman, J. Hindley, and H. Weber, *Annu. Rev. Biochem.* **42,** 303–328 (1973).

118. W. Fiers, R. Contreras, F. Duerinck, G. H. Haegeman, D. Iserentant, J. Merregaert, W. Min Jou, F. Molemans, A. Raeymaekers, A. Van den Berghe, C. Volckaert, M. Ysebaert, *Nature (London)* **260,** 500–507 (1976).

119. J. A. Steitz, *Nature (London)* **224,** 957–964 (1969).

119a. W. Fiers, R. Contreras, F. Duerinck, G. Haegeman, J. Merregaert, W. Min Jou, A. Raeymakers, G. Volckaert, M. Ysebaert, J. Van de Kerckhove, F. Nolf, and M. Van Montagu, *Nature (London)* **256,** 273–278 (1975).

120. W. Min Jou, G. Haegeman, M. Ysebaert, and W. Fiers, *Nature (London)* **237,** 82–88 (1972).

121. H. D. Robertson, B. G. Barrell, H. L. Weith, and J. E. Donelson, *Nature (London), New Biol.* **241,** 38–40 (1973).

122. S. Kaplan, A. G. Atherly, and A. Barrett, *PNAS* **70,** 689–692 (1973).

123. J. Sy and F. Lipmann, *PNAS* **70,** 306–309 (1973).

124. R. Block and W. A. Haseltine, *JBC* **250,** 1212–1217 (1975).

125. G. Reiness, H.-L. Yang, G. Zubay, and M. Cashel, *PNAS* **72,** 2881–2885 (1975).

126. P. P. Dennis and M. Nomura, *Nature (London)* **255,** 460–465 (1975).

127. M. Aboud and I. Pastan, *JBC* **250,** 2189–2195 (1975).

128. C. R. H. Raetz and E. P. Kennedy, *JBC* **247,** 2008–2014 (1972).

129. F. Vogel and R. Rathenberg, *Adv. Hum. Genet.* **5,** 223–318 (1975).

129a. M. D. Topal and J. R. Fresco, *Nature (London)* **263,** 285–289 (1976).

129b. M. Seid- Akhavan, W. P. Winter, R. K. Abramson, and D. L. Rucknagel, *PNAS* **73,** 882–886 (1976).

130. S. Benzer, *Sci. Am.* **206,** 70–84 (Jan 1962).

131. W. Hayes, "The Genetics of Bacteria and Their Viruses," pp. 52–54. Wiley, New York, 1968.

132. S. B. Weintraub and F. R. Frankel, *JMB* **70,** 589–615 (1972).

133. H. L. Ennis and K. D. Kievitt, *PNAS* **70,** 1468–1472 (1973).

134. C. Yanofsky, *Sci. Am.* **216,** 80–94 (May 1967).

134a. A. S. Sarabhai, A. O. W. Stretton, S. Brenner, and A. Bolle, *Nature (London)* **201,** 13–17 (1964).

135. J. W. Watson, "Molecular Biology of the Gene," 3rd ed. Benjamin, New York, 1976.

136. R. S. Edgar and R. H. Epstein, *Sci. Am.* **212,** 70–78 (Feb 1965).

137. F. A. Eiserling and R. C. Dickson, *Annu. Rev. Biochem.* **41,** 467–502 (1972).

138. W. B. Wood and R. S. Edgar, *Sci. Am.* **217,** 60–74 (Jul 1967).

139. A. Garen, *Science* **160,** 149–159 (1968).

140. U. Z. Littauer and H. Inouye, *Annu. Rev. Biochem.* **42,** 439–470 (1973).

140a. P. E. Hartman and J. R. Roth, *Adv. Genet.* **17,** 1–105 (1973).

140b. See B. Lewin "Gene Expression," Vol. 1, p. 213. Wiley, New York, 1974.

141. N. Tsang, R. Macnab, and D. E. Koshland, Jr., *Science* **181,** 60–63 (1973).

142. H. C. Berg and D. A. Brown, *Nature (London)* **239,** 500–504 (1972).

143. J. S. Parkinson, *Nature (London)* **252,** 317–319 (1974).

144. C. Yanofsky and I. P. Crawford, *in* "The Enzymes," 3rd ed., (P. D. Boyer, ed.), Vol. 7, pp. 1–31. Academic Press, New York, 1972.

145. D. L. Riddle and J. Carbon, *Nature (London), New Biol.* **242,** 230–234 (1973).

145a. G. B. Kolata, *Science* **191,** 373 (1976).

145b. H. Sommer, P. Lu, and J. H. Miller, *JBC* **251,** 3774–3779 (1976).

146. R. C. Clowes, *Bacteriol. Rev.* **36,** 361–405 (1972).

147. G. G. Meynell, "Bacterial Plasmids." MIT Press, Cambridge, Massachusetts, 1973.

148. D. Sherratt, *Nature (London)* **254,** 559–560 (1975).

149. K. Jakes, N. D. Zinder, and T. Boon, *JBC* **249,** 438–444 (1974).

150. R. Benveniste and J. Davies, *Annu. Rev. Biochem.* **42,** 471–506 (1973).

151. J. E. Davies and R. Rownd, *Science* **176,** 758–768 (1972).

152. R. C. Clowes, *Sc. Am.* **228,** 19–27 (Apr 1973).

153. F. Jacob and E. L. Wollman, *Sci. Am.* **204,** 93–107 (Jun 1961).

154. A. Kornberg, "DNA Synthesis," p. 242. Freeman, San Francisco, California, 1974.

155. A. Kornberg, "DNA Synthesis," pp. 246–248. Freeman,

San Francisco, California, 1974.

156. W. Szybalski, *in* "Uptake of Informative Molecules by Living Cells" (L. Ledoux, ed.), pp. 59–82. North-Holland Publ., Amsterdam, 1972.

157. A. D. Hershey, ed., "The Bacteriophage Lambda." Cold Spring Harbor Lab., Cold Spring Harbor, New York, 1971.

157a. H. Echols, *Annu. Rev. Biochem.* **40**, 827–854 (1973).

158. H. Echols, *in* "Genetic Mechanisms of Development" (F. H. Ruddle, ed.), pp. 1–11. Academic Press, New York, 1974.

159. T. Maniatis and M. Ptashne, *Sci. Am.* **234**, 64–76 (Jan 1976).

160. M. Ptashne, K. Backman, M. Z. Humanyun, A. Jeffrey, R. Maurer, B. Meyer, and R. T. Sauer, *Science* **194**, 156–161 (1976).

161. R. P. Dottin and M. L. Pearson, *PNAS* **70**, 1078–1082 (1973).

162. J. Greenblatt, *PNAS* **70**, 421–424 (1973).

163. P. H. Weigel, P. T. Englund, K. Murray, and R. W. Old, *PNAS* **70**, 1151–1155 (1973).

164. K. Murray and N. E. Murray, *Nature (London), New Biol.* **243**, 134–139 (1973).

165. B. C. Westmoreland, W. Szybalski, and H. Ris, *Science* **163**, 1343–1348 (1969).

166. J. Inselburg, *Nature (London), New Biol.* **241**, 234–237 (1973).

167. J. M. Mitchison, "The Biology of the Cell Cycle." Cambridge Univ. Press, New York, 1971.

168. J. A. Smith and L. Martin, *PNAS* **70**, 1263–1267 (1973).

169. F. H. Ruddle, *Adv. Hum. Genet.* **3**, 173–235 (1972).

169a. R. L. Davidson, ed., "Somatic Cell Hybridization." Raven Press, New York, 1974.

170. B. Lewin, "Gene Expression," Vol. 2, pp. 387–416. Wiley, New York, 1974.

171. S. J. Goss and H. Harris, *Nature (London)* **255**, 680–684 (1975).

172. R. Sager, *Sci. Am.* **212**, 71–79 (Jan 1965).

173. D. N. Duvick, *Adv. Genet.* **13**, 1–56 (1965).

174. U. W. Goodenough and R. P. Levine, *Sci. Am.* **223**, 22–29 (Nov 1970).

175. M. H. Vodkin and G. R. Fink, *PNAS* **70**, 1069–1072 (1973).

176. M. M. K. Nass, *Science* **165**, 25–35 (1969).

177. G. Schatz and T. L. Mason, *Annu. Rev. Biochem.* **43**, 51–87 (1974).

177a. A. M. Kroon and C. Saccone, "The Biogenesis of Mitochondria." Academic Press, New York, 1974.

178. J. T. O. Kirk, *Annu. Rev. Biochem.* **40**, 161–196 (1971).

179. G. Schatz and T. L. Mason, *Annu. Rev. Biochem.* **43**, 51–87 (1974).

179a. R. Kolodner and K. K. Tewari, *JBC* **250**, 8840–8847 (1975).

180. V. R. Flechtner and R. Sager, *Nature (London), New Biol.* **241**, 277–279 (1973).

181. L. Bogorad, *Science* **188**, 891–898 (1975).

182. H. Smith, *Nature (London)* **254**, 13 (1975).

183. N. Ohta, R. Sager, and M. Inouye, *JBC* **250**, 3655–3659 (1975).

184. A. Klein and F. Bonhoeffer, *Annu. Rev. Biochem.* **41**, 301–332 (1972).

185. J. C. Wang, *JMB* **55**, 523–533 (1971).

185a. W. Keller, *PNAS* **72**, 2550–2554 (1975).

186. J. J. Champoux and B. L. McConaughy, *Biochemistry* **15**, 4638–4642 (1976).

187. M. L. Gefter, *Annu. Rev. Biochem.* **44**, 45–78 (1975).

188. A. A. Infante, R. Nauta, S. Gilbert, P. Hobart, and W. Firshein, *Nature (London), New Biol.* **242**, 5–8 (1973).

188a. A. A. Infante, W. Firshein, P. Hobart, and L. Murray, *Biochemistry* **15**, 4810–4817 (1976).

189. R. E. Bird, J. Louarn, J. Martuscelli, and L. Caro, *JMB* **70**, 549–566 (1972).

190. D. M. Prescott and P. L. Kuempel, *PNAS* **69**, 2842–2845 (1972).

190a. P. L. Kuempel, D. M. Prescott, and P. Maglothin, *in* "DNA Synthesis *in Vitro*," (R. Wells and R. Inman, eds.), pp. 463–472. University Park Press, Baltimore, 1972.

191. H. J. Kriegstein and D. S. Hogness, *PNAS* **71**, 135–139 (1974).

192. R. G. Wake, *JMB* **77**, 569–575 (1973).

193. A. Kornberg, "DNA Synthesis," pp. 178–182. Freeman, San Francisco, California, 1974.

194. H. Kasamatsu and J. Vinograd, *Nature (London), New Biol.* **241**, 103–105 (1973).

195. T. Kornberg and M. L. Gefter, *JBC* **247**, 5369–5375 (1972).

196. A. Kornberg, "DNA Synthesis," pp. 123–136. Freeman, San Francisco, California, 1974.

197. A. Kornberg, "DNA Synthesis," pp. 67–121. Freeman, San Francisco, California, 1974.

198. Y. Hirota, J. Mordoh, T. Scheffler, and F. Jacob, *Fed. Proc. Fed. Am. Soc. Exp. Biol.* **31**, 1422–1427 (1972).

199. A. Weissbach, D. Baltimore, F. Bollum, R. Gallo, and D. Korn, *Science* **190**, 401–402 (1975).

200. N. C. Jones and W. D. Donachie, *Nature (London), New Biol.* **243**, 100–103 (1973).

201. R. Schekman, A. Weiner, and A. Kornberg, *Science* **186**, 987–993 (1974).

202. W. Wickner and A. Kornberg, *JBC* **249**, 6244–6249 (1974).

203. Y. Sakakibara and J. Tomizawa, *PNAS* **71**, 1403–1407 (1974).

204. J. P. Bouché, K. Zechel, and A. Kornberg, *JBC* **250**, 5995–6001 (1975).

205. I. Berkower, J. Leis, and J. Hurwitz, *JBC* **248**, 5914–5921 (1973).

205a. I. R. Lehman and D. G. Uyemura, *Science* **193**, 963–969 (1976).

206. B. Alberts and L. Frey, *Nature (London)* **227**, 1313–1318 (1970).

207. I. J. Molineux, S. Friedman, and M. L. Gefter, *JBC* **249**, 6090–6098 (1974).

208. J. H. Weiner, L. L. Bertsch, and A. Kornberg, *JBC* **250**, 1972–1980 (1975).

208a. G. Herrick and B. Alberts, *JBC* **251**, 2124–2132 (1976).

209. J. Ikeda, A. Yudelevich and J. Hurwitz, *PNAS* **73**, 2661–2673 (1976).

210. C. H. Schroder and H.-C. Kaerner, *JMB* **71**, 351–362 (1972).

211. F. W. Studier, *Science* **176**, 367–376 (1972).

212. S. E. Luria, *Sci. Am.* **222**, 88–102 (Jan 1970).

213. H. W. Boyer, *Fed. Proc. Fed. Am. Soc. Exp. Biol.* **33**, 1125–1127 (1974).

214. M. Meselson, R. Yuan, and J. Heywood, *Annu. Rev. Biochem.* **41**, 447–468 (1972).

215. K. Horiuchi, G. F. Vovis, and N. D. Zinder, *JBC* **249**, 543–552 (1974).

216. B. Polisky, P. Greene, D. E. Garfin, B. J. McCarthy, H. M. Goodman, and H. W. Boyer, *PNAS* **72**, 3310–3314 (1975).

217. D. Nathans and H. O. Smith, *Annu. Rev. Biochem.* **44**, 273–293 (1975).

218. D. Nathans, S. P. Adler, W. W. Brockman, K. J. Danna, T. N. H. Lee, and G. H. Sack, Jr., *Fed. Proc. Fed. Am. Soc. Exp. Biol.* **33**, 1135–1138 (1974).

219. B. E. Griffin, M. Fried, and A. Cowie, *PNAS* **71**, 2077–2081 (1974).

219a. F. Vedel, F. Quetier and M. Bayen, *Nature (London)* **263**, 440–442 (1976).

220. M. Meselson and J. J. Weigle, *PNAS* **47**, 857–868 (1961).

221. T. R. Broker and I. R. Lehman, *JMB* **60**, 131–149 (1971).

222. R. Holliday, *Genet. Res.* **5**, 282–304 (1964).

222a. H. Potter & D. Dressler, *PNAS* **73**, 3000–3003 (1976).

223. N. Sigal and B. Alberts, *JMB* **71**, 789–793 (1972).

224. M. Meselson, *JMB* **71**, 795–798 (1972).

225. M. S. Meselson and C. M. Radding, *PNAS* **72**, 358–361 (1975).

226. H. M. Sobell, *Adv. Genet.* **17**, 411–490 (1973).

227. H. M. Sobell, *PNAS* **72**, 279–283 (1975).

228. R. E. Wagner, Jr. and M. Radman, *PNAS* **72**, 3619–3622 (1975).

229. J. Doniger, R. C. Warner, and I. Tessma, *Nature (London), New Biol.* **242**, 9–12 (1973).

230. P. J. Goldmark and S. Linn, *JBC* **247**, 1849–1860 (1972).

231. S. Ohi, D. Bastia, and N. Sueoka, *Nature (London)* **248**, 586–588 (1974).

232. A. T. C. Carpenter, *PNAS* **72**, 3186–3189 (1975).

232a. G. P. Smith, *Science* **191**, 528–535 (1976).

233. D. J. Kopecko and S. N. Cohen, *PNAS* **72**, 1373–1377 (1975).

233a. S. N. Cohen and D. J. Kopecko, *Fed. Proc., Fed. Am. Soc. Exp. Biol.* **35**, 2031–2036 (1976).

233b. H. Ohtsubo and E. Ohtsubo, *PNAS* **73**, 2316–2320 (1976).

233c. G. B. Kolata, *Science* **193**, 392–394 (1976).

233d. T. Cavalier-Smith, *Nature (London)* **262**, 255–256 (1976).

234. R. Dulbecco, *Sci. Am.* **216**, 28–37 (Apr 1967).

235. D. Baltimore, *Science* **192**, 632–636 (1976).

235a. R. Dulbecco, *Science* **192**, 437–440 (1976).

236. *Cold Spring Harbor Symp. Quant. Biol.* **39**, (1974). The entire volume is devoted to tumor viruses.

237. E. M. Scolnick, *in* "Current Topics in Biochemistry" (C. B. Anfinsen, R. F. Goldberger, and A. N. Schechter, eds.), Academic Press, New York, pp. 49–64, 1973.

238. H. Temin and D. Baltimore, *Adv. Virus Res.* **17**, 129–186 (1973).

238a. H. M. Temin, *Science* **192**, 1075–1080 (1976).

239. D. S. Auld, H. Kawaguchi, D. M. Livingston, and B. L. Vallee, *PNAS* **71**, 2091–2095 (1974).

240. P. Bentvelzen, *FEBS Symp.* **22**, 1–13 (1972).

241. L. A. Loeb, K. D. Tartof, and E. C. Travaglini, *Nature (London), New Biol.* **242**, 66–69 (1973).

241a. R. H. Baltz, P. M. Bingham, and J. W. Drake, *PNAS* **73**, 1269–1273 (1976).

242. M. S. Legator and W. G. Flamm, *Annu. Rev. Biochem.* **42**, 683–708 (1973).

243. L. Fishbein, H. L. Falk, and W. G. Flamm, "Chemical Mutagens." Academic Press, New York, 1970.

244. C. Heidelberger, *Annu. Rev. Biochem.* **44**, 79–121 (1975).

245. I. A. Wolff and A. E. Wasserman, *Science* **177**, 15–18 (1972).

246. K. Isono and J. Yourno, *PNAS* **71**, 1612–1617 (1974).

247. A. Braun and L. Grossman, *PNAS* **71**, 1838–1842 (1974).

248. A. R. Lehmann, S. Kirk-Bell, C. F. Arlett, M. C. Paterson, P. H. M. Lohman, E. A. de Weerd-Kastelein, and D. Bootsma, *PNAS* **72**, 219–223 (1975).

249. B. M. Sutherland, M. Rice, and E. K. Wagner, *PNAS* **72**, 103–107 (1975).

249a. U. Kuhnlein, E. E. Penhoet, and S. Linn, *PNAS* **73**, 1169–1173 (1976).

250. J. Epstein, J. R. Williams, and J. B. Little, *PNAS* **70**, 977–981 (1973).

250a. M. C. Paterson, P. H. M. Lohman, A. K. Anderson, and L. Fishman, *Nature (London)* **260**, 444–447 (1976).

251. A. Ahmad, W. K. Holloman, and R. Holliday, *Nature (London)* **258**, 54–56 (1975).

252. R. P. P. Fuchs, *Nature (London)* **257**, 151–152 (1975).

253. J. A. Miller and E. C. Miller, *Fed. Proc. Fed. Am. Soc. Exp. Biol.* **35**, 1316–1321 (1976).

254. B. N. Ames, P. Sims, and P. L. Grover, *Science* **176**, 47–49 (1972).

255. E. C. Miller and J. A. Miller, *in* "The Molecular Biology of Cancer" (H. Busch, ed.), p. 386. Academic Press, New York, 1974.

256. J. McCann, E. Choi, E. Yamasaki, and B. N. Ames, *PNAS* **72**, 5135–5139 (1975).

257. H. J. Muller, *Sci. Am.* **193**, 58–68 (Nov 1955).

257a. J. McCann and B. N. Ames, *PNAS* **73**, 950–954 (1976).

258. A. Kornberg, "DNA Synthesis," pp. 368–373. Freeman, San Francisco, California, 1974.

259. J. F. Morrow, S. N. Cohen, A. C. Y. Chang, H. W. Boyer, H. M. Goodman, and R. B. Helling, *PNAS* **71**, 1743–1747 (1974).

260. V. Hershfield, H. W. Boyer, C. Yanofsky, M. A. Lovett, and D. R. Helinski, *PNAS* **71**, 3455–3459 (1974).

261. R. Higuchi, G. V. Paddock, R. Wall and W. Salser, *PNAS*

73, 3146–3150 (1976).

262. P. S. Carlson, *PNAS* **70,** 598–602 (1973).

263. C. H. Doy, P. M. Gresshoff, and B. Rolfe; *in* "The Biochemistry of Gene Expression in Higher Organisms" (J. K. Pollack and J. W. Lee, eds.), pp. 38–55. Reidel Publ., Dordrecht, Netherlands, 1972.

264. L. Ledoux, R. Huart, and M. Jacobs, *Nature (London)* **249,** 17–21 (1974).

265. R. A. Dixon and J. R. Postgate, *Nature (London)* **237,** 102–103 (1972).

266. K. T. Shanmugam and R. C. Valentine, *Science* **187,** 919–924 (1975).

266a. K. Struhl, J. R. Cameron, and R. W. Davis, *PNAS* **73,** 1471–1475 (1976).

267. J. Bonner, *Plant Sci. Bull.* **11,** 1–7 (1965).

268. C. Wills, *Sci. Am.* **222,** 98–107 (Mar 1970).

269. C. Norman, *Nature (London)* **258,** 561–564 (1975).

269a. N. Wade, *Science* **194,** 303–306 (1976).

270. W. Beermann and U. Clever, *Sci. Am.* **210,** 50–58 (Apr 1964).

271. B. Danehalt and H. Hosick, *PNAS* **70,** 442–446 (1973).

272. J. L. Sirlin, "Biology of RNA," pp. 162–164. Academic Press, New York, 1972.

273. W. J. Peacock, *in* "The Biochemistry of Gene Expression in Higher Organisms" (J. K. Pollack and J. W. Lee, eds.), pp. 3–20. Reidel Publ., Dordrecht, Netherlands, 1972.

274. R. J. Britten and D. E. Kohne, *Sci. Am.* **222,** 24–31 (Apr 1970).

275. E. H. Davidson, B. R. Hough, C. S. Amenson, and R. J. Britten, *JMB* **77,** 1–23 (1973).

276. A. Kornberg, "DNA Synthesis," pp. 23–25. Freeman, San Francisco, California, 1974.

277. G. P. Smith, *Science* **191,** 528–535 (1976).

277a. H. Cooke, *Nature (London)* **262,** 182–186 (1976).

278. R. E. Pyeritz and C. A. Thomas, Jr., *JMB* **77,** 57–73 (1973).

279. A. Jacobsen, R. A. Firtel, and H. F. Lodish, *PNAS* **71,** 1607–1611 (1974).

280. H. F. Lodish, A. Jacobsen, R. A. Firtel, T. Alton, and J. Tuchman, *PNAS* **71,** 5103–5108 (1974).

281. P. K. Wellauer, R. H. Reeder, D. Carroll, D. D. Brown, A. Deutch, T. Higashinakagawa, and I. B. Dawid, *PNAS* **71,** 2823–2827 (1974).

282. D. Carroll and D. D. Brown, *Cell* **7,** 467–475 (1976).

282a. J. D. Procunier and K. D. Tartof, *Nature (London)* **263,** 255–257 (1976).

283. A. Bird, E. Rogers, and M. Birnstiel, *Nature (London), New Biol.* **242,** 226–230 (1973).

283a. P. K. Wellauer, R. H. Reeder, I. B. Dawid and D. D. Brown, *JMB* **105,** 487–505 (1976).

284. W. S. Kistler, M. E. Geroch, and H. G. Williams-Ashman, *JBC* **248,** 4532–4543 (1973).

285. S. C. R. Elgin and H. Weintraub, *Annu. Rev. Biochem.* **44,** 725–774 (1975).

286. R. D. Kornberg and J. O. Thomas, *Science* **184,** 865–868 (1974).

287. M. O. J. Olson, W. C. Starbuck, and H. Busch, *in* "The Molecular Biology of Cancer" (H. Busch, ed.), pp. 309–353. Academic Press, New York, 1974.

288. R. J. DeLange, J. A. Hooper, and E. L. Smith, *JBC* **248,** 3261–3274 (1973).

289. L. C. Yeoman, M. O. J. Olson, N. Sugano, J. J. Jordan, C. W. Taylor, W. C. Starbuck, and H. Busch, *JBC* **247,** 6018–6023 (1972).

290. R. D. Kornberg and J. O. Thomas, *Science* **184,** 865–871 (1974).

291. D. E. Olins and A. L. Olins, *J. Cell Biol.* **53,** 715–736 (1972).

292. J. D. Griffith, *Science* **187,** 1202–1203 (1975).

293. J. P. Langmore and J. C. Wooley, *PNAS* **72,** 2691–2695 (1975).

294. G. B. Kolata, *Science* **188,** 1097–1099 (1975).

295. J. P. Baldwin, P. G. Boseley, E. M. Bradbury, and K. Ibel, *Nature (London)* **253,** 245–249 (1975).

296. J. O. Thomas and R. D. Kornberg, *PNAS* **72,** 2626–2630 (1975).

296a. W. Keller, *PNAS* **72,** 4876–4880 (1975).

296b. T. Vogel and M. F. Singer, *JBC* **251,** 2334–2338 (1976).

297. F. H. C. Crick and A. Klug, *Nature (London)* **255,** 530–533 (1975).

298. R. Chalkley and C. Hunter, *PNAS* **72,** 1304–1308 (1975).

299. J. D. McGhee and J. D. Engel, *Nature (London)* **254,** 449–450 (1975).

300. E. M. Bradbury, R. J. Inglis, H. R. Matthews, and T. A. Langan, *Nature (London)* **249,** 553–556 (1974).

301. A. Ruiz-Carrillo, L. J. Waugh, and V. G. Allfrey, *Science* **190,** 117–127 (1975).

302. J. L. Peterson and E. H. McConkey, *JBC* **251,** 548–554 (1976).

303. G. S. Stein, J. S. Stein; and L. J. Kleinsmith, *Sci. Am.* **232,** 46–57 (Feb 1975).

303a. L. Goldstein, *Nature (London)* **261,** 519–521 (1976).

304. I. L. Goldknopf, C. W. Taylor, R. M. Baum, L. C. Yeoman, M. O. J. Olson, A. W. Prestayko, and H. Busch, *JBC* **250,** 7182–7187 (1975).

305. O. Hayashi, *Trends Biochem. Sci.* **1,** 9–10 (1976).

STUDY QUESTIONS

1. What is the nature of the evidence that hereditary characteristics are actually encoded in the base sequence of DNA?

2. Describe the Watson-Crick model for DNA structure. What biological and chemical facts does it explain adequately?

3. Describe the experimental proof that the strands in double-stranded DNA are antiparallel. Discuss the specificity of any enzymes used for this proof.

4. If all the DNA molecules in your body were joined into a single double-stranded helix of the Watson–Crick type, how long would it be?

5. Describe three types of RNA that have been implicated in the biosynthesis of proteins. Describe what is known about each in terms of chemistry, physical properties, base composition, and biosynthesis.

6. The lac repressor protein contains only two tryptophan residues at positions 190 and 209. When the inducer isopropyl-β-D-thiogalactoside binds to the protein the fluorescence emission maximum undergoes a small (~ 3 nm) but distinct shift to longer wavelength. Modified inducers have been produced by introduction of nonsense mutations and suppression by suitable suppressor genes [see Section D,5 and H. Summer, P. Lu, and J. H. Miller, JBC **251,** 3774–3779 (1976)]. When Trp 190 was replaced by Tyr the shift in fluorescence still occurred when the inducer was bound, but when Trp 209 was replaced with Tyr no shift was seen. Comment on these results. Can you suggest two possible explanations? How could you hope to distinguish between them?

7. A bacterial ribosome of diameter 23 nm is 35% protein. If we assume that 35% of the volume is available to hold the proteins, approximately how many protein molecules of average MW = 17,300 could be present in a ribosome? (Assume closest packing, see also Table 15-5.)

8. Transfer RNA molecules interact specifically with several enzymes in the course of their synthesis and function. Write the equations for *three* reactions in which an enzyme recognizes tRNA.

9. Name three functions of certain aminoacyl-tRNA molecules other than their participation in protein biosynthesis.

10. A glycine-specific tRNA of *E. coli* contains the codon GCC. It may be paired either with the codon GGU or GGC on the ribosomes. What is the generally accepted explanation of the ability of one tRNA to recognize several different codons?

11. A portion of a messenger RNA molecule was isolated and its base sequence was determined to be

UGAAGCAUGGCUUCUAACUUU
↑

What would be the effect on the peptide coded for by this messenger RNA of a mutation which resulted in a change of the U indicated by the arrow to a C?

12. Many mRNA molecules have a long sequence at the 5' end of the molecule that is not translated by the ribosomal protein synthesizing system. Describe an experiment by which you might find the beginning of a message in a mRNA molecule. Assume that you have a pure preparation containing only one kind of mRNA.

13. 7-Methylguanosine 5'-monophosphate inhibits protein synthesis in a cell-free system prepared from reticulocytes [D. A. Shafritz *et al., Nature (London)* **261,** 291–294 (1976)]. Most messenger RNA molecules in eukaryotic cells contain a 5'-terminal 7-methylguanosine. However, chemical removal of this group from the mRNA of vesicular stomatitis virus did not prevent translation in the cell-free reticulocyte system [J. K. Rose and H. F. Lodish, *Nature (London)* **262,** 32–37 (1976)]. Comment on the significance of these observations.

14. Define and describe the significance of each of the following in studies of biochemical genetics.

Nutritional auxotrophs   Complementation
Replica plating   Suppressor genes

15. Contrast the following, as used in biochemical genetics.

Transformation   Operon
Transduction   Codon
Cistron

16. Describe the following forms of DNA (or bodies which contain DNA).

Plasmid   R factor
Episome   F factor
Colicinogenic factor Lysogenic phage

17. How may a foreign piece of DNA be inserted into a plasmid? Why is the insertion of eukaryotic DNA into a plasmid of ethical, as well as of biochemical significance? What is one important biochemical problem that may be solved in the future by the use of this technique?

18. How may the location of foreign DNA incorporated in a chromosome be determined by heteroduplex analysis?

19. (a) What are the differences between the chromosomes of *E. coli* and of a mammal? How may these differences be detected? (b) Define euchromatin, heterochromatin, satellite DNA, repetitious DNA, and histones. Indicate some of the possible functions which have been proposed for each of these.

20. Define each of the following types of mutant.
    Base pair switch   Temperature sensitive
    Frame-shift   Nonsense

21. Describe the chemical basis for the action of each of the following mutagenic chemicals or agents.
    Hydroxylamine   Methyl iodide
                        (weakly mutagenic)
    5-bromouracil   Ultraviolet light
    9-aminoacridine   Bacteriophage Mu

22. What special property would you predict for a tyrosyl-tRNA whose anticodon is GψA if it undergoes mutation to a tRNA with anticodon CψA?

23. When mitochondrial DNA isolated from normal maize was digested with a restriction endonuclease and subjected to agarose gel electrophoresis about 50 bands were observed. When the mitochondrial DNA of maize carrying the male sterility trait mentioned in Section D,9,e was examined in the same way distinct differences in the pattern were seen. Several bands present in the "fingerprint" of normal mitochondrial DNA were absent [C. S. Levings, III and D. R. Pring, *Science* **193,** 158–160 (1976)]. Comment on this observation and also on the fact that when crosses between normal maize and that carrying the male sterility trait were crossed the mitochondrial DNA pattern was always that of the female parent.

24. How would you account for the fact that a hinny has a gentler disposition than does a mule?

16

Growth, Differentiation, and Chemical Communication between Cells

The regulation of complex multicellular organisms depends heavily upon chemical messages sent between cells. The secretion of hormones into the circulatory system is one major type of communication. Less well understood is the chemical transfer of information through communicating cell junctions (Chapter 1, Section E,3,b). The phenomenon is best established for nerve cells, and **neurochemistry** has become one of the major branches of biochemistry. Embryonic development and differentiation of tissues also depend upon communication between cells. However, growing cells also appear to be regulated internally through **developmental programs** encoded in the DNA. This chapter deals briefly with these matters and also with communication between organisms, i.e., with the biochemistry of ecological relationships.

A. THE HORMONES[1]

Most hormones appear to act by one of two types of mechanism. In the first, a hormone binds to a receptor on a cell membrane. Thus, glucagon, epinephrine, and ACTH bind to cell surfaces and promote the synthesis of cyclic AMP (Chapter 5, Section C,5), which in turn promotes chemical modifications of proteins. It is possible that synthesis of prostaglandins (Chapter 12, Section E,3) is stimulated in a similar way. The second mechanism depends upon binding of a hormone to a cytoplasmic receptor with control of the transcription of RNA as the end result. The steroid hormones, thyroxine and the **growth hormone** (somatotropin) are among the compounds that appear to function in this manner. The steroid hormone receptors, which are found initially in the cytoplasm, bind incoming steroid molecules tenaciously.[2] After an "activation" step, the complex enters the nucleus where it binds at many sites in the chromatin, apparently through the mediation of certain nonhistone proteins.[3] Although the chemical details are still obscure, the net result is an increase in initiation of transcription of selected genes in the hormone-responsive cells.[3a]

A generalization is that hormone action is usually part of a feedback loop. Thus, insulin promotes the uptake of glucose by tissues. However, a decrease in the glucose concentration of the blood leads, through feedback inhibition, to a decreased rate of secretion of insulin from the pancreas. Similar regulatory loops can be traced for most hormones. Sometimes they involve several stages and involve sensing devices in the central nervous system. In such cases neural impulses stimulate the **hypothalamus** of the brain (Section B,2) to release **neurohormones** that travel to the anterior lobe of the pituitary gland. The pituitary in turn releases hormones such as corticotropin (adrenocorticotropic hormone, ACTH) which stimulate the adrenal cortex to release its hormones. The latter, among other effects, exerts feedback inhibition upon the hypothalamus to decrease the secretion of ACTH by the pituitary. Using ^{3}H-labeled steroid hormones it has been possible to locate specific brain cells sensitive to a given hormone by autoradiography.[3b]

A characteristic of hormonal effects is that they are seldom unique and that they are often balanced by counteracting effects of other hormones. For example, both glucagon and epinephrine promote the release of glucose from liver glycogen into the bloodstream. The glucocorticoids stimulate the rate of production of glucose from other body constituents (Chapter 11, Section F,7), Growth hormone tends to increase glucose levels by inhibiting utilization of sugar by tissues. On the other hand, insulin acts to promote an uptake of glucose by tissues and a more efficient utilization while the thyroid hormone increases the overall rate of metabolism of cells and also tends to promote a decrease in blood glucose.

Gordon Tompkins proposed a generalized model for hormone-mediated intercellular communication.[4] He suggested that a small number of intracellular "symbols," molecules such as cAMP or, in prokaryotic cells, ppGpp, serve to control particular "domains" of metabolic processes within cells. While only the cyclic nucleotides are firmly established as symbols, in eukaryotic cells metal ions such as Ca^{2+}, Na^+, and K^+ are also likely candidates. In bacteria ppGpp serves as a symbol of nitrogen or amino acid deficiency. In cells, ranging from those of bacteria to animals, cAMP is a symbol for carbon-source starvation. Thus, in *E. coli* cAMP levels increase during

carbon-source starvation and stimulate the initiation of transcription of many bacterial operons (Chapter 15, Section B,1,b). In *Dictyostelium* (Chapter 1, Section D,1) cAMP is released by cells when substrate depletion occurs. In this instance

---

BOX 16-A

RADIOIMMUNOASSAY

Among the important techniques that have permitted rapid progress in studies of hormone action is the use of specific antibodies formed against hormones or hormone–protein conjugates. Using radioactive hormones, the radioimmunoassay methods permit measurement of as little as a femtomole of hormone (i.e., the amount present in 1 ml of a 10^{-12} M solution).[a] Methods are available for virtually every pure hormone.[b]

In a typical assay, various amounts of a sample containing an unknown quantity of hormone, e.g., insulin, are placed in a series of tubes. Additional tubes containing known amounts of the hormone are also prepared. Then a standard quantity of radiolabeled hormone (often iodinated with a γ emitter such as ^{125}I) is added to each tube together with a standard quantity of the specific antibody to the hormone. The solution is then incubated for minutes or hours to obtain equilibrium between hormone (the antigen) and antibody–hormone complex. The antibody–hormone complex is then separated, e.g., by gel filtration or ammonium sulfate precipitation, and the radioactivity of the complex is measured. In the tubes containing higher concentrations of hormone, the labeled hormone has been diluted more, and the amount bound to antibody is less than in tubes with lower concentrations of hormone. The tubes of known concentrations are used to construct a standard curve from which the unknown concentrations can be read.

[a] B. Brooker, W. L. Terasake, and M. G. Price, *Science* **194**, 270–276 (1976).
[b] B. M. Jaffe and H. R. Behrmann, eds., "Methods of Hormone Radioimmunoassay." Academic Press, New York, 1974.

the cyclic nucleotide acts as a hormone transmitting a signal to other cells.

However, whereas cAMP may be used by lower organisms as a hormone, its metabolic lability makes it unsuitable for higher animals. Thus, in our bodies the hormones glucagon and epinephrine carry a message to cell surfaces where binding to receptors stimulates cAMP production. This, in turn, leads to mobilization of metabolic stores such as those of glycogen and triglycerides, just as if these cells had also been subjected to acute starvation. According to Tompkins, hormones are produced by "sensor" cells in direct contact with environmental signals and travel to and activate more sequestered "responder" cells. The picture can be generalized further by realizing that neurotransmitters are largely derivatives of amino acids. Tompkins suggested that these amino acids may have originally served as intracellular symbols representing changes in environmental amino acid concentration but were later utilized in short-range intercellular communication within the nervous system.

1. The Vertebrate Hormones[1,5]

The principal established vertebrate hormones are listed in Table 16-1. Also given are references to other parts of the text where particular hormones are discussed. Note that the hormones can be divided into three groups on the basis of chemical structure: (1) peptides and proteins, (2) derivatives of the aromatic amino acids, and (3) steroids and the prostaglandins.

The hormones of the pituitary (**hypophysis**) de-

TABLE 16-1
Vertebrate Hormones

Type, name, and source of hormone	Principal site of action	Chapter, Section
A. Peptides and proteins		
1. Pituitary gland (hypophysis)		
a. Adenohypophysis (anterior portion)		
Somatotropin (growth hormone, GH)	All tissues	
Corticotropin (ACTH)	Adrenal cortex, adipose tissue	Fig. 2-2; Chapter 6, Section F,5
Thyrotropin (thyroid-stimulating hormone, TSH)	Thyroid	Chapter 14, Section H,5b
	Adipose tissue	
Follitropin (FSH)	Ovary, testis	Chapter 12, Section I,3d
Lutropin (luteinizing hormone, ICSH or LH)	Ovary, testis	Chapter 12, Section I,3d
Prolactin (mammatropin)	Mammary gland	
Lipotropin		Chapter 16, Section A,1
b. Neurohypophysis (posterior portion)	Uterus, mammary gland	Fig. 2-2
Oxytocin		
Vasopressin (antidiuretic hormone)	Kidney, arteries	Fig. 2-2
c. Pars intermedia		
Melanotropins (MSH)	Melanophores	
2. Pancreas		
Insulin	All cells	Chapter 4, Section D,7; Chapter 5, Section C,5; Chapter 6, Section F,5; Chapter 12, Section F,1

TABLE 16-1   (*continued*)

Type, name, and source of hormone	Principal site of action	Chapter, Section
Glucagon	Liver, adipose tissue	Chapter 6, Section F,5; Chapter 11, Section F,5; Chapter 12, Section F,1
3. Ovary (corpus luteum)		
Relaxin	Pelvic ligaments	
4. Thyroid		
Calcitonin (thyrocacitonin)	Bones, kidney	Box 5-E
5. Parathyroid		
Parathyrin (parathyroid hormone)	Bones, kidney	Box 5-E; Chapter 12, Section E,3 and Box 12-D; Chapter 16, Section A,1
6. Kidney		
Erythropoietin	Bone marrow	
Renin	Adrenal cortex	
7. Alimentary tract		
Gastrin	Stomach	
Enterogastrone	Stomach	
Cholecystokinin	Gall bladder	
Secretin	Pancreas	
Pancreozymin	Pancreas	
B. Amino acid derivatives		
1. Thyroid		
Thyroxine and triiodothyronine	Most cells	Chapter 14, Section H,5b
2. Adrenal medulla		
Epinephrine, norepinephrine (adrenaline, noradrenaline)	Most cells	Chapter 6, Section F,6; Chapter 12, Section F,1
3. Pineal gland		
Melatonin	Melanophores	Chapter 14, Section I, Fig. 14-27
4. Nerves and other cells		
Serotonin (5-hydroxytryptamine)	Arterioles, central nervous system	
C. Steroids and prostaglandins		
1. Testes		
Testosterone	Most cells	Chapter 12, Section I,3,c
2. Ovaries		
Estrogen (estradiol-17β)	Most cells	Chapter 12, Section I,3d
3. Corpus luteum		
Progesterone	Uterus, mammary glands	Chapter 12, Section I,3a
4. Adrenal cortex		
Corticosterone, cortisol	Most cells	Chapter 11, Section F,2 and F,7; Chapter 12, Section I,3b
Aldosterone	Kidney	Chapter 12, Section I,3b
5. Various tissues		
Prostaglandins	Smooth muscle	Chapter 12, Section E,3

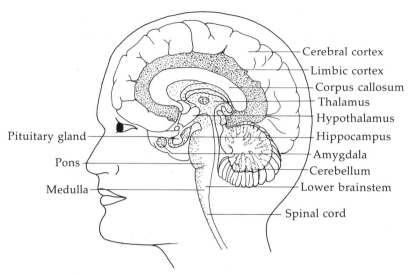

Fig. 16-1   Median sagittal section of the human brain. From Maya Pines, *Saturday Review* Aug. 9, p. 14 (1975).

serve special comment. Connected to the brain by a stalk (Fig. 16-1) the pituitary gland releases at least ten peptide or protein hormones that regulate the activity of other **endocrine** (hormone-producing) glands in distant parts of the body.

The pituitary is composed of several distinct parts. As previously mentioned, the anterior lobe (**adenohypophysis**) releases its hormones in response to at least nine neurohormones known as **releasing factors.**[6,7] Minute quantities of the releasing factors are secreted by the hypothalamus. It is only recently that the structures of some of the releasing factors have been determined. As indicated in Fig. 2-2, several are modified simple peptides. The **melanocyte-stimulating hormone** (melanotropin) of the intermediate portion of the pituitary is also released in response to a hypothalamic factor **melanoliberin.**

The hypothalamus also synthesizes a series of **release-inhibiting factors.**[8] One of these, **somatostatin,** inhibits release of somatotropin, thus counteracting the effect of **somatoliberin.** Somatostatin has attracted a great deal of interest for another reason. It acts not only on the pituitary but also on the pancreas where it inhibits the release of both insulin and glucagon. The result is a lowering of blood glucose which suggests a new approach to the treatment of diabetes (Box 11-C).

The size of the peptide chains in pituitary hormones varies markedly. Some are average size proteins. For example, human growth hormone has a molecular weight of 21,500 and is highly specific. Growth hormones from other sources cannot substitute. The thyroid-stimulating hormone (**thyrotropin,** TSH) is a glycoprotein of MW = 28,000. In contrast, the hormones of the **neurohypophysis** (posterior lobe of the pituitary), **vasopressin** and **oxytocin** are simple peptides containing nine amino acid residues (eight if cystine is counted as a single amino acid, Fig. 2-2). As the name implies, the neurohypophysis consists of neural tissue whose secretions are directly controlled by the central nervous system. Vasopressin is a major regulator of blood volume and pressure and its secretion is influenced by stress. Oxytocin acts on smooth muscles of the uterus during childbirth and triggers the release of milk from the mammary glands. The latter response is partially

$$
\begin{array}{l}
\mathrm{H_3^+N-Ala-Gly-Cys-Lys-Asn-Phe-Phe} \\
\qquad\qquad\quad\ |\qquad\qquad\qquad\qquad\quad \searrow \mathrm{Trp} \\
\qquad\qquad\quad \mathrm{S}\qquad\qquad\qquad\qquad\qquad\quad\ | \\
\qquad\qquad\quad \mathrm{S}\qquad\qquad\qquad\qquad\qquad\quad \mathrm{Lys} \\
\qquad\quad ^-\mathrm{OOC-Cys-Ser-Thr-Phe-Thr} \nearrow
\end{array}
$$

Somatostatin

controlled by the suckling of the infant, which induces the nervous system to release oxytocin into the bloodstream.

An interesting relationship is observed among the pituitary hormones. Several of them contain a *common heptapeptide unit* which is marked in the following structure:

Ac·Ser·Tyr·Ser·Met·Glu·His·Phe·Arg·Tyr·Gly·Lys·Pro·ValNH$_2$

Heptapeptide core

Structure of α-melanotropin from pig, beef, and horse

Not only the heptapeptide but also the entire amino acid sequence of α-melanotropin is found in corticotropin (Fig. 2-2), but this is followed by an additional 29 amino acids at the C-terminal end. The same heptapeptide is also found in the **lipotropins,** also secreted by the pituitary. The lipotropins contain 46 amino acids preceding the heptapeptide and another 5–37 at the C-terminal end. The presence of the common heptapeptide suggests an evolutionary relationship for this group of hormones. Furthermore, corticotropin may be a precursor of α-melanotropin and of other biologically active peptides.[9] Indeed, it appears that the processing of protein and peptide hormones by post-transcriptional cleavage by proteases and through other modifications is a very general phenomenon[10,11] (Chapter 11, Section E,2).

Another example is provided by the 84-residue **parathyrin** (parathyroid hormone). In the secretion granules the peptide is present as a 90-residue prohormone containing six extra residues at the N-termini. The primary biosynthetic product **preproparathyrin** appears to contain an additional 25 residues at the N-terminal end.[11a,b] Thus, modification involves at least two steps which have been dubbed pre-processing and pro-processing. A similar situation may hold for insulin, a **preproinsulin** being cleaved to proinsulin (Fig. 11-9). The second step in the processing often involves a protease with trypsinlike properties.

It is proposed that some prohormones are converted to active forms by **aminolysis** rather than hydrolysis of the peptide chain. Cleavage by displacement with NH$_3$ would yield the C-terminal amide group so frequently found in small peptide hormones. Vasopressin, oxytocin, α-melanotropin, the pituitary releasing factors, and others may have arisen in this manner.

It has also been suggested that covalent modification (e.g., by phosphorylation) could be a feature of control mechanisms affecting storage of prohormones and duration of action of hormones.[11]

Among the gastrointestinal hormones listed in Table 16-1 **gastrin** and **secretin** are relatively small polypeptides containing 17 and 27 amino acids, respectively.[12] The kidney hormone **renin** is of interest for it acts as a specific protease to cleave a decapeptide **proangiotensin** from a serum α$_2$-globulin.[1] A second enzyme[13] removes two more amino acids from the C-terminal end to form **angiotensin,** the most potent hypertensive com-

Asp·Arg·Val·Tyr·Ile·His·Pro·Phe

Angiotensin

pound known. Angiotensin stimulates the smooth muscles of blood vessels. It reduces blood flow through the kidneys and decreases the excretion of fluid and salts. The hormone also increases secretion of aldosterone by the adrenal cortex with a resultant increase in sodium ion reabsorption.

2. Insect Hormones

The neuroendrocrine systems of invertebrates have many features in common with those of mammals. For example, in insects, a series of neurohormones are formed.[14] One of these, the **activation hormone,** controls the secretion of the corpora allata, paired glands that synthesize the **juvenile hormone** (Fig. 12-13) in insect larvae.[15] While the structure of the juvenile hormone varies somewhat with species, it is usually an isoprenoid ester. A specific binding protein provides the hormone with protection from degradative enzymes. However, in the tobacco hornworm, an esterase, able to hydrolyze the protein-bound juvenile hormone, is produced at the start of pupal differentiation.[16]

A steroid hormone **ecdysone** (Fig. 12-18) is secreted by the insect's prothoracic gland. Ecdysone, also known as the molting hormone, is required for the periodic replacement of the exoskeleton of the larvae. It induces molting in crayfish and other arthropods and appears to be needed by such members of lower phyla as schistosomes and nematodes. In addition to α-ecdysone, hydroxylated derivatives, 20-hydroxyecdysone, 26-hydroxyecdysone, and 20,26-dihydroxyecdysone have been identified in insects.[17] It has been suggested that different ecdysones may function at different stages of insect development.

Ecdysone has been shown to stimulate the synthesis of RNA in tissues. Visual demonstration of the effect has been provided by its action on polytene chromosomes of fly larvae. Fifteen minutes after the application of ecdysone, a puff (Chapter 15, Section I,1) is induced in one band of the chromosome; a second puff forms at a later time while a preexisting puff diminishes. Thus, like steroid hormones in mammals, ecdysone appears to have a direct controlling effect on transcription.

3. Plant Hormones[18–22]

Plants also possess a kind of circulatory system by which fluids are transported from the roots upward in the xylem and downward from the leaves through the phloem. A variety of compounds are carried between cells in this manner, while others are transported across cell membranes and against concentration gradients by active transport processes. A number of compounds that move between cells in either of these two manners have been classified as hormones, and the number is bound to increase. At present there are five compounds or groups of compounds accepted as plant hormones: **auxins** (Chapter 14, Section I), **gibberellins** (Chapter 5, Section E; Chapter 12, Section H,1), **cytokinins** (Chapter 15, Section B,4), **abscisic acid** (Fig. 12-13), and **ethylene** (Chapter 14, Section D,4).

Plant hormones tend to have multiple and overlapping functions. This makes it difficult to discuss their functions briefly. Indeed, at the molecular level little is yet known. Most studied are the auxins, of which the principal member is indole-3-acetic acid (Fig. 14-27). This compound has been implicated as a controlling agent for cell division and cell elongation. In this capacity, auxin influences a great variety of plant processes. Produced principally by growing shoots, auxin diffuses down the stem and inhibits the growth of lateral buds. However, the hormone stimulates the growth of stems, thus establishing the apical dominance of the tip of a plant. Other hormones also have an influence. Auxin is well established as the controlling agent in phototropism, the tendency of a plant to bend toward the sun. Thanks to the development of a sensitive test (the bending of the coleoptile of *Avena sativa*, the common oat), it is possible to measure readily 3 pmol of auxin. Using this assay, it was shown that auxin is transported laterally away from the illuminated side of plants, thus causing the darker side to elongate more rapidly.

The molecular mechanism of action of auxin is not known, but it is possible that like many other hormones it acts to increase the rate of transcription of RNA.

The gibberellins are active in helping to determine the *form* of plants. They are synthesized in mature leaves and are transported downward. Because of their very active effect in stimulating RNA synthesis in dwarf varieties of vegetables, it has been suggested that gibberellins serve as gene activators to promote RNA synthesis. A possible function in the **geotropic** response of plant roots is suggested by the presence of higher concentrations in the upper half of horizontal roots than in the lower half.[23] On the other hand, auxin has long been known to have a higher concentration on the lower side of the root, and it has been assumed to inhibit elongation (in contrast to stimulation in stems).

The cytokinins are a family of isopentenyladenosine derivatives (Fig. 15-10) which may be further hydroxylated or substituted in the 2 position by a methylthio group. The role of cytokinins may be at the level of gene transcription or of translation. Their hormonal effect in plants appears to be quite independent of the function in tRNA. The most striking effect of cytokinins in solution is on differentiation of plant cells (Section C,3).

Abscisic acid appears to block the growth promoting effects of hormones such as the gibberellins and cytokinins. It is sometimes regarded as a **general gene repressor** which prepares plants for dormancy. Synthesis of abscisic acids occurs in response to the short day–long night pattern of the fall.

The effect of ethylene is harder to pinpoint. It not only hastens the ripening of fruit but also tends to promote senescence in all parts of plants.

Other important plant regulating compounds include the vitamins, thiamine, pyridoxine, and nicotinic acid, which are synthesized in the leaves and transported downward to the roots. Since they promote growth of roots, they are sometimes referred to as **root growth hormones.** However, they are more commonly regarded as nutrients universally needed by cells. There is good evidence for a special **flowering hormone,** and recently there has been interest in the effect of synthetic plant "bioregulators" including derivatives of chalcones (Box 12-B) and compounds such as diethyloctylamine.[24] Another important aspect of plant growth regulation is **photomorphogenesis,** the effect of light in regulating plant development.

B. NEUROCHEMISTRY[25−32]

The nervous system, a network of neurons in active communication, reaches its ultimate development in the human brain. While many invertebrates (e.g., leeches, crayfish, insects, and snails) have brains containing no more than 10^4 to 10^5 neurons,[33,34] the human brain contains $\sim 10^{11}$. Each of these neurons interconnects through **synapses** with hundreds or thousands of other neurons. The number of connections is estimated to be as many as 60,000 with each Purkinje cell of the human cerebellum.

In addition to neurons, the brain contains 5–10 times as many **glial** cells of several types (the neuroglia occupy 40% of the volume of brain and spinal cord in the human). Some glial cells seem to bridge the space between neurons and bloodcarrying capillaries. Others synthesize myelin. Some are very irregular in shape.

1. Properties of Neurons

While neurons have many shapes and forms, a common pattern is evident. At one end of the elongated cell (Fig. 16-2) is a series of **dendrites,** thin fibers often less than 1 μm in diameter. The ends of the dendrites form synapses with other neurons and act as receivers of incoming messages. Additional messages come into synapses on the **cell body,** while the **axon** serves as the output end of the cell. The axon, a long fiber of diameter 1–20 μm, is also branched. As a consequence, the nervous system contains both highly branching and highly converging pathways.

The ends of the fine nerve fibers are thickened to form the **synaptic knobs** which make synaptic contacts with dendrites on cell bodies of other neurons. In most instances the arrival of a nerve signal at the **presynaptic** end of a neuron causes the release of a transmitter substance (or neurohormone). The transmitter passes across the 10–50 nm (typically 20 nm) **synaptic cleft** between the two cells and causes depolarization of the **postsynaptic** membrane of the next neuron (Fig. 16-3).[35] The resultant postsynaptic potential is propagated to the cell body and axon and, under appropriate circumstances, may initiate an **action potential** (Chapter 5, Section B,3) in the axon.

A characteristic of neurons is the *all-or-none response* or "firing." An action potential passes down the axon only if sufficient depolarization exists. In general, a stimulus must reach a neuron through *more than one synapse* before the neuron will fire. Furthermore, neurons are often *inhibitory,* releasing transmitters that counter the excitatory synapses and tend to prevent firing.[36] It is thought that inhibition is important in damping

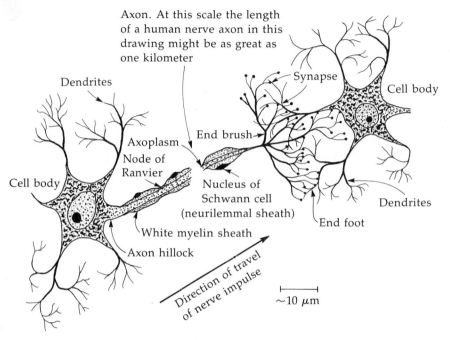

Axon. At this scale the length
of a human nerve axon in this
drawing might be as great as
one kilometer

Dendrites

Synapse

Cell body

Axoplasm

End brush

Node of
Ranvier

Nucleus of
Schwann cell
(neurilemmal sheath)

Cell body

Dendrites

White myelin sheath

End foot

Axon hillock

Direction of travel
of nerve impulse

~10 μm

Fig. 16-2 Schematic drawing of a neuron (after Brand and Westfall,[30] p. 1192).

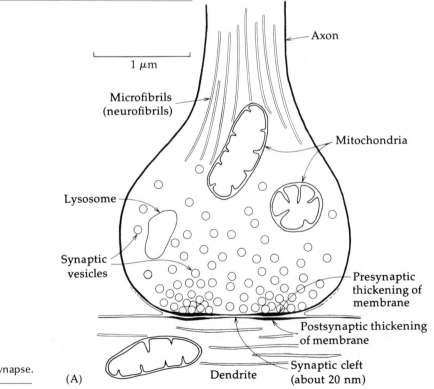

Axon

1 μm

Microfibrils
(neurofibrils)

Mitochondria

Lysosome

Synaptic
vesicles

Presynaptic
thickening of
membrane

Postsynaptic thickening
of membrane

Fig. 16-3 (A) Schematic drawing of a synapse.

(A)

Dendrite

Synaptic cleft
(about 20 nm)

Fig. 16-3 (B) Electron micrograph showing the synaptic junctions in the basal part pedicle) of a retinal cone cell of a monkey.[35a] Each pedicle contains synaptic contacts with ~12 triads, each made up of processes from a bipolar cell (center that carries the principle output signal and from two horizontal cells that also synapse with other cones. A ribbon structure within the pedicle is characteristic of these synapses. Note that numerous synaptic vesicles in the pedicle, some arranged around the ribbon, the synaptic clefts and characteristic thickening of the membranes surrounding the cleft (below the ribbons). Micrograph courtesy of John Dowling. See also Michael.[35b]

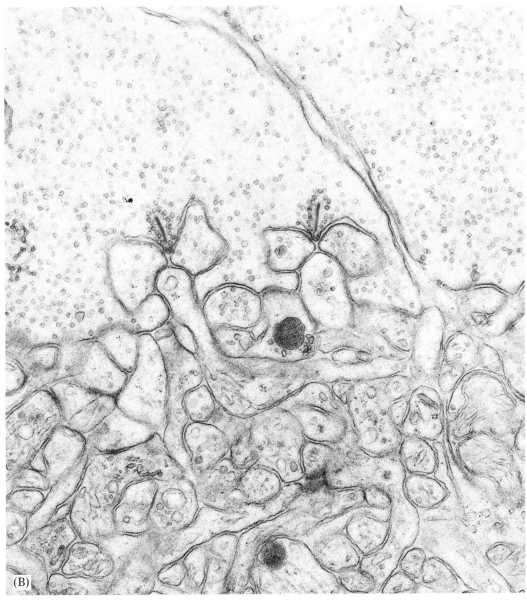

(B)

out small excitations, thus sharpening the response of the nervous system toward strong stimuli.

Another general property that is of basic importance to the operation of the brain is that neurons fire at longer or shorter intervals depending upon the strength and duration of the stimulus. The stronger the stimulus to a given neuron, the more rapid the train of "spikes" that passes down the axon. Thus, the brain functions to a large extent in decoding trains of impulses. The frequency of the impulses from neurons varies from a few per second to a maximum of about 200 s^{-1} in most nerves (up to 1600 s^{-1} in the Renshaw cells of the spinal cord). The maximum frequency is dictated by the refractory period of ~1 ms (Chapter 5, Section B,3).

While the concepts of neuronal function outlined in the preceding paragraphs have been accepted for many years, recent discoveries require that they be modified somewhat. Dendrites seem to be able to transmit information as well as to receive it. Furthermore, while information is certainly transmitted long distances by "spike" action potentials, shorter neurons and dendrites may communicate extensively by exchange of chemicals through low resistance gap junctions **(electrotonic junctions),** (Chapter 1, Section E,3). Small changes in membrane potential may be transmitted through these junctions and alter the behavior of adjacent neurons. Chemical transmitters may not always have an electrical effect on postsynaptic neurons but may influence metabolism or gene transcription.[36a]

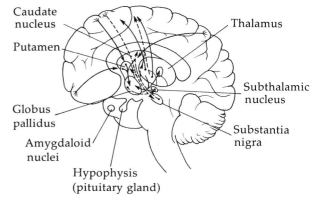

Fig. 16-4 Diagram illustrating some of the major interconnections of the "extrapyramidal system" of the brain. Arrows indicate major direction of projections. The nigrostriatal (substantia nigra to striatum) and related neuronal pathways are indicated with dashed lines. After Noback and Demarest,[26a] pp. 182 and 183.

2. Organization of the Brain*

The two cerebral hemispheres make up the largest part of the **cerebrum,** the upper portion of the brain. The deeply folded outermost layer, the **cerebral cortex,** consists of **gray matter,** a mass of cell bodies and fine, unmyelinated nerve fibers. Beneath this lies a layer of **white matter** made up of myelin-covered axons connecting the cerebral cortex with other parts of the brain. The two cerebral hemispheres are connected by the **corpus callosum,** a band of $\sim 2 \times 10^8$ nerve fibers. Remarkably, these fibers can be completely severed with a relatively minimal disruption of the nervous system.[†] Deeper in the cerebrum lie the **basal ganglia** which include the caudate, lenticular, and amygdaloid nuclei. The lenticular nuclei are further divided into putamen (an outer portion) and the globus pallidus. The putamen and caudate nuclei together are known as the **striatum** (Fig. 16-4). The lower lying subthalamic nuclei and sub-

stantia nigra are sometimes also included in the basal ganglia.

The outer parts of the cerebrum, including the basal ganglia, make up the telencephalon. Deep in the center of the brain is the diencephalon consisting of the **thalamus** (actually two thalami), **hypothalamus,** and **hypophysis** and other attached regions. A major structure at the back of the brain is the **cerebellum.** Like the cerebrum, its cortex is highly folded. The 30 billion neurons of the cerebellum are organized in a highly regular fashion.[37] The interconnections of the seven types of neurons present in this part of the brain have been worked out in fine detail.

The basal part of the brain or **brain stem** consists of the medulla oblongata and the pons. While the bulk of the tissue consists of myelinated nerve tracts passing into the spinal cord, synaptic regions such as the olivary nucleus are also present.

---

* The anatomy of the brain is quite complex, and only a few terms will be defined here.

† The corpus callosum has sometimes been severed to control almost incessant epileptic seizures that cannot be prevented by drugs. The "split-brain" patients suffer relatively little disability as long as both eyes function normally. Studies of these patients have provided remarkable insights into the differing functions of the two hemispheres of the cerebrum.[26]

3. Neuronal Pathways and Systems

Consider a message originating with a nerve receptor in the skin or in another sense organ. A nerve signal passes via a **sensory neuron (afferent fiber)** upward toward the brain. It may pass through two or more synapses (often through one in the spinal cord and one in the thalamus) finally reaching a spot in the sensory region of the cerebral cortex. From there the signal in modified form spreads through the **interneurons** of virtually the entire cortex. In each synapse, as well as in the cortex, the impulse excites inhibitory fibers that dampen impulses flowing through adjacent fibers. Likewise, if a given impulse is not strong enough, it will itself be inhibited before reaching the cortex. Among the most important sensory neurons are those from the seven million cone cells and 100 million rod cells of the eye. The nerve signals pass out of the retina by way of a million axons from retinal ganglion cells reaching, among other parts of the brain, the **visual cortex** (Fig. 16-5).

The neuronal events that occur within the cerebral cortex are extraordinarily complex and little understood. In what way the brain is able to initiate voluntary movement of muscles is totally obscure. However, it is established that the signals that travel out of the brain down the **efferent fibers** to the muscles arise from large **motor neurons** of the **motor cortex,** a region that extends in a band across the brain and adjacent to the sensory cortex (Fig. 16-5). The axons of these cells form the **pyramidal tract** that carries impulses downward to synapses in the spinal cord and from there to the **neuromuscular junctions.** At these latter specialized synapses acetycholine is released, carrying the signal to the muscle fibers themselves. Passing over the cell surface and into the T tubules (Chapter 4, Section F,1, Fig. 4-22A), the wave of depolarization initiates the release of calcium and muscular contraction.

At the same time that the motor neurons send signals to the muscles, branches travel into other parts of the brain including the olivary nuclei which send neurons into the cerebellum. The cerebellum acts as a kind of computer needed for fine tuning of the impulses to the muscles. Injury to the cerebellum leads to difficulty in finely coordinated motions. The output from the cerebellum,

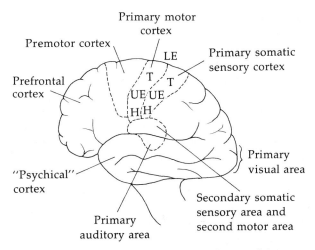

Fig. 16-5 The location of several functional areas of the cerebral cortex. The representation of body parts on the primary motor and somatic sensory cortices include the head (H), upper extremity (UE), trunk (T), and lower extremity (LE). After Noback and Demarest,[26a] p. 193.

via the Purkinje cells, is entirely inhibitory. The Purkinje cells synapse in the cerebellar nuclei with neurons that lead back to the cerebral cortex, into the thalamus, and down the spinal cord. The pathway to the cortex completes an inhibitory feedback loop, of which there are many in the nervous system.

In addition to the **somatic motor system** that operates the voluntary (striated) muscles via the pyramidal tract, there is the **autonomic system** which controls the involuntary (smooth) muscles, glands, heartbeat, blood pressure, and body temperature. This system has its origins in both the cerebral cortex and hypothalamus. It is subdivided into two systems, the **sympathetic** and **parasympathetic** systems which are anatomically distinct. The sympathetic system is geared to the fight and fright reactions. Its **postganglionic fibers** (those below the ganglia in the spinal cord) liberate norephinephrine (noradrenaline) and include the adrenal medulla which consists of specialized neurons, the **chromaffin** cells. The parasympathetic system has to do more with homeostasis and maintenance of body systems. Biochemically it is characterized by the release of acetylcholine as a transmitter substance.

The hypothalamus, a 4 g portion of the brain, receives a great deal of biochemical attention because of its function in the autonomic nervous system, in homeostasis and endocrine secretion. Its liberation of neurohormones that stimulate the hypophysis has already been considered (Section A). The hypothalamus is also involved in the regulation of the body temperature, water balance, and possibly of glucose concentration.

Two other systems of importance in the brain are the **reticular system** and the **limbic system.** The former is the mediator of the sleep–wake cycle and is responsible for characteristic waves in the electroencephalogram. The limbic system is the mediator of **affect** or mood and of instincts. It is anatomically complex with centers in the amygdala, other subcortical nuclei, and in the limbic lobe of the cortex. The limbic cortex forms a ring lying largely within the longitudinal fissure between the two hemispheres. It includes the olfactory cortex, the **hippocampus,** and other evolutionarily older regions of the cerebral cortex. Within the limbic lobe are the **pleasure centers.** When electrodes are implanted in these regions animals will repeatedly push levers that are designed to electrically stimulate these centers. There are also **punishing centers** whose stimulation causes animals to avoid further stimulation.

4. Neurotransmitters

Studies of the neuromuscular junctions of the autonomic nervous system as early as 1904 led to the suggestion that epinephrine might be released at the nerve endings. Although the true transmitter was later shown to be norepinephrine,* the concept of chemical communication in synapses was formulated. By 1921 it was shown that acetylcholine is released at nerve endings of the parasympathetic system, and it later became clear that

---

* Epinephrine does serve as a transmitter at the neuromuscular junction in amphibians, but in mammals its function appears to be primarily that of a hormone.

motor nerve endings of the somatic system also release acetylcholine.*

a. Cholinergic Synapses

With the advent of electron microscopy, the fine structure of the synaptic contacts became evident. The synaptic knobs were found to contain vesicles of ~30–80 nm diameter which were later shown by chemical analysis and staining procedures to contain the neurotransmitters (Fig. 16-3). In the case of the acetylcholine-releasing synapses (**cholinergic synapses**) each 80 nm vesicle contains ~40,000 molecules of acetylcholine,[40] the concentration in the vesicle being of the order of 0.5 M. To show that the acetylcholine released at a synapse stimulates the postsynaptic membrane to initiate an impulse, the technique of **electrophoretic injection (microiontophoresis)** was developed.[41] By using ultramicrocapillaries a small pulse of current, e.g., 3×10^8 amp for 1 ms, can be used to inject electrically a compound directly into a synaptic cleft. The results may be observed with separate recording electrodes, one of which is inserted into an axon or a muscle fiber. By this means it was shown that amounts of acetylcholine comparable to those released at the neuromuscular junction do cause muscles to contract.

How does the release of neurotransmitter occur? That the release is "quantal," i.e., involving the entire content of a vesicle, was established from the observation of **miniature end-plate potentials.** These are fluctuations in the postsynaptic potential observed under conditions of weak stimulation of the presynaptic neuron. They reflect the random release of neurotransmitter from individual vesicles.[42] Normally, a strong impulse will release on the order of 100–200 quanta of transmitter, enough to initiate an action potential in the postsynaptic neuron. What are the chemical events that trigger the release of transmitter? It appears that depolarization of the membrane at the synaptic ending permits a rapid inflow of calcium

---

* Although the role of the acetylcholine as a synaptic transmitter is generally accepted, Nachmansohn argued against it and pointed out that formation, storage, release, and hydrolysis of acetylcholine are also needed for propagation of action potentials (see Section B,8).[38,39]

ions.[43,44] The transient increase in intracellular [Ca^{2+}] triggers the release of the contents of vesicles by causing fusion of the vesicle membranes with the plasma membrane. About four calcium ions are needed to release one vesicle. Although the significance is uncertain, the vesicles have a lattice-like coat consisting of a unique protein clathrin[44a] of MW 180,000.

After release of the transmitter the synaptic vesicles reform. That this must occur at a rapid rate follows from the fact that there are only enough vesicles in the average synaptic ending to provide for ~2000–5000 impulses. This is enough for only a few minutes of action under strong stimulus. Thus, the synaptic endings of axons must have a very strong synthetic mechanism for transmitters as well as a storage mechanism. In addition, neurons frequently reabsorb the transmitter from the synaptic cleft and store it for reuse.

What does a neurotransmitter do at the postsynaptic membrane? In the case of acetylcholine the principal action appears to be one of depolarizing the membrane and of opening the sodium and potassium pores. The membrane alteration may be the same as that triggered by an action potential (Chapter 5, Section B,3) at the leading edge of a nerve impulse. Acetylcholine must bind to a special receptor and the binding must somehow open the Na$^+$ pores. Recently, a large protein that appears to have the properties of the acetylcholine receptor has been isolated from electrical organs of electric eels.[45] With a minimum molecular weight of ~330,000, it may be a trimer of units of MW ≈110,000 which, in turn, are composed of 2–4 peptides of MW = 34,000–54,000. How the receptor functions is still quite unclear (Chapter 5, Section C,5).

After a pulse of transmitter is released it must be removed or inactivated quickly to prepare the synapse for arrival of a new nerve impulse. This is accomplished in two ways in cholinergic synapses: the first is via hydrolytic destruction by acetylcholinesterase (Box 7-B). This esterase is present in the synaptic membrane itself. The second mechanism is transport of acetylcholine into the neuron

for reuse, an energy-dependent active transport process.

Among the acetylcholine-releasing neurons are the motor neurons that form synapses at neuromuscular junctions, the preganglionic neurons of the entire autonomic system, and the postganglionic neurons of the parasympathetic system. A large number of other cholinergic synaptic regions within the brain have also been discovered.

Important in the study of neurotransmitters is the identification of specific agonists, which mimic the action of a transmitter; and of antagonists, which block the action of the transmitter. Two groups of compounds influence cholinergic neurons, leading to the classification of these neurons either as muscarinic (activated by muscarine, Fig. 16-6) or nicotinic (stimulated by nicotine).[46] The muscarinic receptors, which are found in many autonomic neurons, are specifically inhibited by atropine and decamethonium (Fig. 16-6). The nicotinic synapses occur in ganglia and skeletal muscle. They are inhibited by curare and its active ingredient, D-tubocurarine (Fig. 16-6) and by the protein snake venom α-bungarotoxin (Fig. 16-7). This toxin has been used to titrate the number of acetylcholine receptors in the motor end plate of the rat diaphragm. About 4×10^7 receptors per end plate (or 13,000/μm^2) were found.[47] A large number of other snake venoms have closely homologous structures.[48,49] The three-dimensional structure of one of them has been determined.[49a]

It has been suggested that muscarinic receptors are designed to bind acetylcholine in a folded conformation similar to that of muscarine (Fig. 16-6).[49b] Nicotinic receptors may bind the fully extended conformer. However, the similarity to nicotine is not as clear (Fig. 16-6) and other explanations have been suggested.

A serious disease, myasthenia gravis (believed to be an autoimmune disease, Section C,7), is associated with a decrease in the number of functional postsynaptic receptors.[50,51] The resulting extreme muscular weakness can be fatal. An interesting treatment consists of the administration of physostigmine (Fig. 16-6), diisopropylphosphofluori-

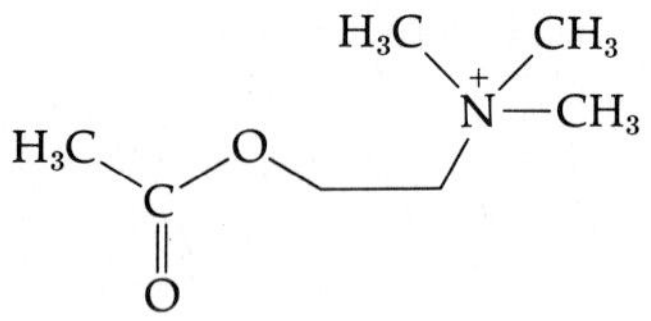

Acetylcholine in a skewed conformation

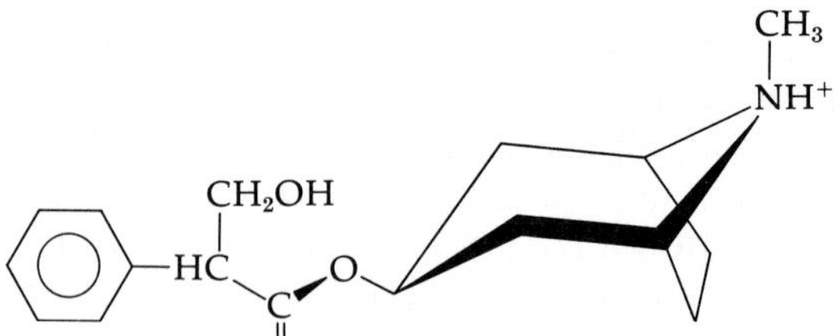

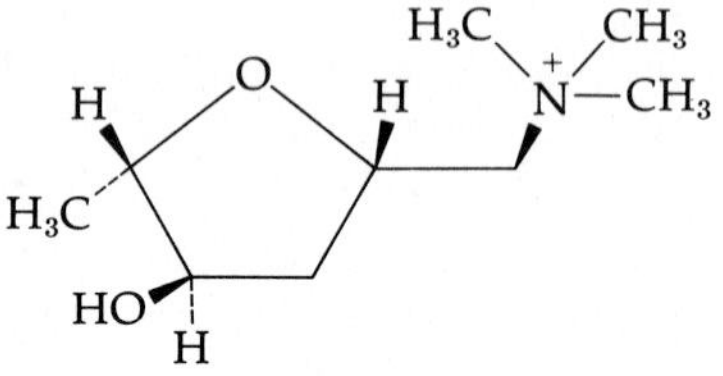

L(+)-Muscarine, a cholinergic agonist from the mushroom *Amanita muscaria*

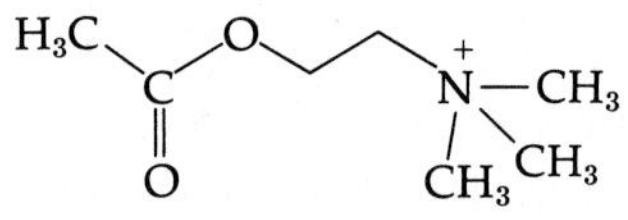

Atropine

Acetylcholine in extended conformation

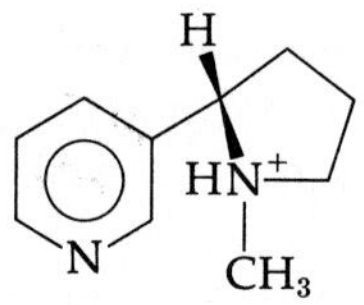

Nicotine (protonated)

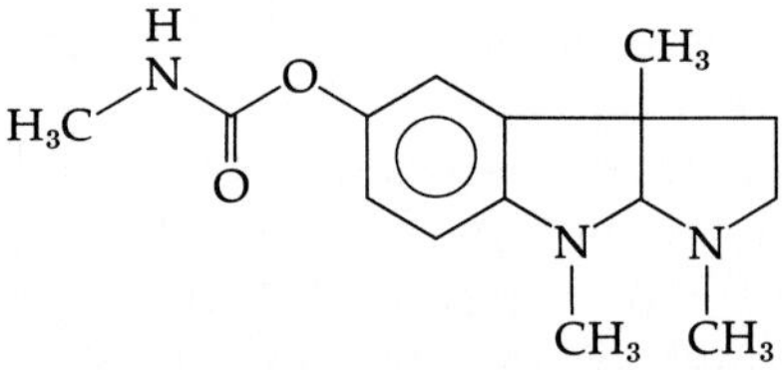

Physostigmine, an acetylcholinesterase inhibitor widely used in treatment of glaucoma. Compare with other carbamate esters (Box 7-B)

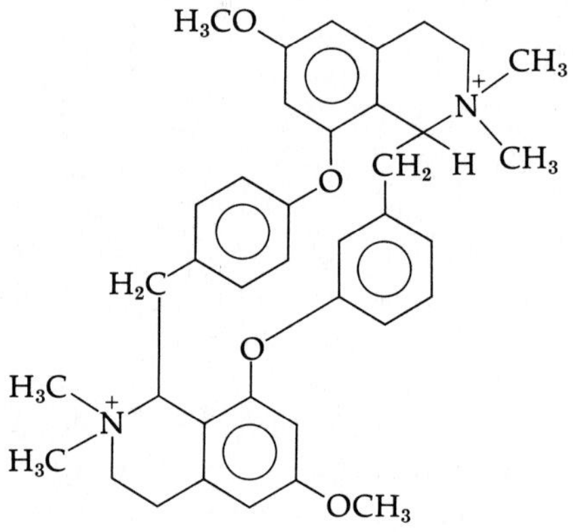

D-Tubocurarine, the principal ingredient of South American arrow poisons. Blocks cholinergic receptors in skeletal muscle

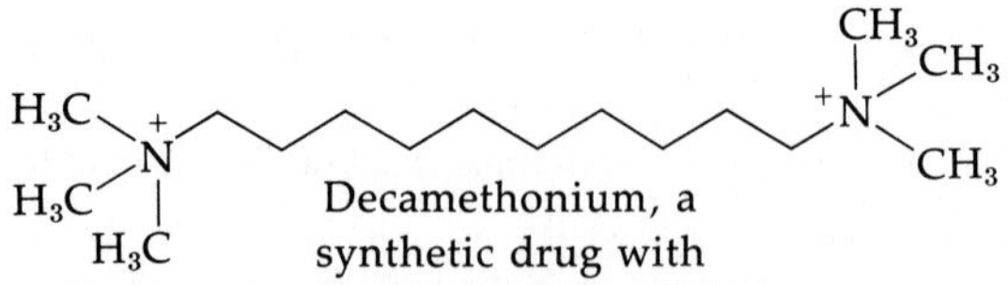

Decamethonium, a synthetic drug with potent curarelike activity

α-Bungarotoxin and related proteins of snake venoms also block nicotinic receptors in skeletal muscle

Fig. 16-6. Some inhibitors of cholinergic synapses.

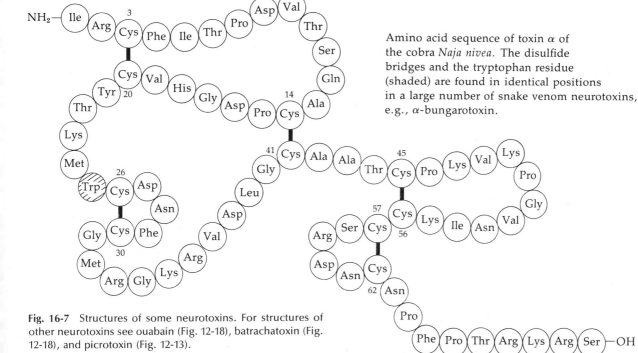

Tetrodotoxin of the puffer fish

Bicuculline, an antagonist of γ-aminobutyrate

Saxitoxin of *Gonyaulax*, a dinoflagellate whose toxin accumulates in shellfish. As little as 0.3 mg of toxin can be fatal

Strychnine

Amino acid sequence of toxin α of the cobra *Naja nivea*. The disulfide bridges and the tryptophan residue (shaded) are found in identical positions in a large number of snake venom neurotoxins, e.g., α-bungarotoxin.

Fig. 16-7 Structures of some neurotoxins. For structures of other neurotoxins see ouabain (Fig. 12-18), batrachatoxin (Fig. 12-18), and picrotoxin (Fig. 12-13).

date (Chapter 7, Section D,1), or other acetylcholinesterase inhibitors (Box 7-B). These very toxic compounds, when administered in controlled amounts, permit accumulation of higher acetylcholine concentration with a resultant activation of muscular contraction.*

In addition to antagonists that are presumed to act directly on receptors, inhibitors may influence several other steps. For example, **botulinum toxin,** one of the most poisonous substances known, inhibits the release of the vesicles of acetylcholine, as do some snake venoms.[52,53] **Tetrodotoxin** (Fig. 16-7), of the puffer fish, blocks the sodium pores in postsynaptic membrane, again preventing nerve transmission.[54] **Saxitoxin** (Fig. 16-7)[55] has a similar action. **Batrachotoxin** (Fig. 12-18)[56] causes an *increased* sodium permeability in muscle membranes. Tetanus toxin, a large protein, blocks transmission in the central nervous system and in muscle, perhaps by interfering with calcium transport.[57]

b. Other Neurotransmitters

Acetylcholine is regarded as an established neurotransmitter which meets five important criteria: (1) a synthetic mechanism exists within the presynaptic neuron (acetylcholine is formed by transfer of an acetyl group from acetyl-CoA under the influence of a special acetyltransferase); (2) a mechanism of storage (in vesicles) is evident; (3) the transmitter is released in proportion to the strength of the stimulus (frequency of firing); (4) a postsynaptic action of the transmitter can be demonstrated directly by microiontophoresis; and (5) an efficient means for inactivation of the transmitter is present. The same five criteria must be met by other compounds if they are to be considered as transmitters.

At present, in addition to acetylcholine, norepinephrine, epinephrine (in amphibians), and γ-aminobutyrate (GABA) are regarded as established transmitters. A large number of probable ("putative" or "candidate") transmitters are

* The same compounds are widely used in the treatment of glaucoma.

also known. Of these dopamine, 5-hydroxytryptamine (serotonin), glutamate, and glycine are well on their way to being accepted as transmitters. Other compounds such as aspartate, taurine, and a number of peptides including the hypothalamic pituitary releasing factors are under consideration.[58] It is possible that the list of accepted neurotransmitters will grow rapidly. It appears that a single neuron ordinarily releases a single transmitter. However, there are now some doubts about this assumption.

Glutamate is one of the most important excitatory transmitters in the central nervous system (CNS) of invertebrates and may also be important in humans. Aspartate may also be a transmitter. Both γ-aminobutyric acid and glycine are thought to be major inhibitory transmitters. Whereas excitatory transmitters lead to depolarization of the postsynaptic membrane, inhibitory transmitters cause **hyperpolarization,** apparently by increasing the conductance of K^+ and Cl^-. The result is that it is more difficult to excite the postsynaptic membrane in the presence of than in the absence of these materials.

c. Adrenergic Synapses:
The Catecholamines[30,59,60]

The three closely related tyrosine metabolites, dopamine, norepinephrine, and epinephrine, known collectively as catecholamines, are important products of neuronal metabolism; dopamine and norepinephrine serve as neurotransmitters. In many invertebrates **octopamine,**[61] synthesized via tyramine (Fig. 16-8), is also important. Note the precursor–product relationship between dopamine, norepinephrine, and epinephrine. The synthetic pathways to these neurotransmitters involve decarboxylation and hydroxylation, reaction types that are important in formation of other transmitters as well. The most important process for terminating the action of released catecholamine transmitters is reuptake by the neurons. A high affinity uptake system transports the catecholamine molecules back into the neurons and then into the storage vesicles. The uptake is specifically blocked by the drug **reserpine** (Fig. 14-27).

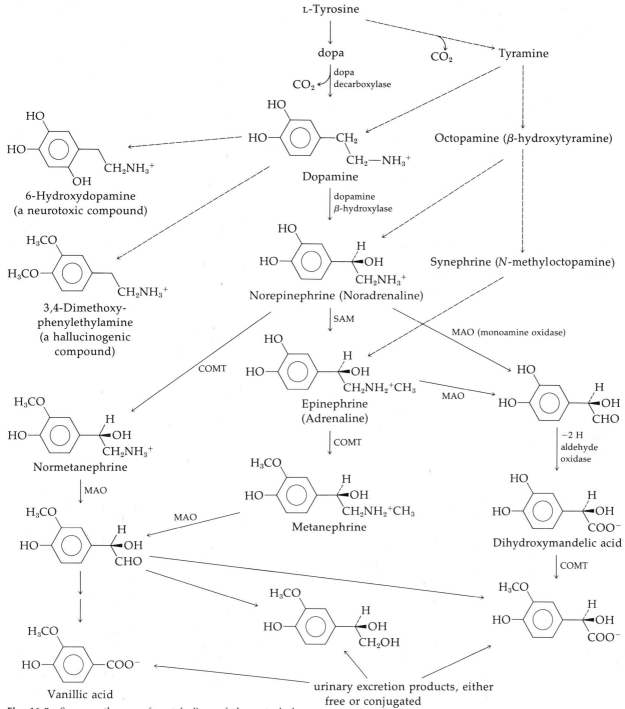

Fig. 16-8 Some pathways of metabolism of the catecholamines. See also Fig. 14-2.

Much of the transmitter is catabolized by two enzymes. One is monoamine oxidase (MAO, Chapter 8, Section I,3), an enzyme present within the mitochondria of neurons (as well as in other cells in all parts of the body). A second catabolic enzyme is catechol-O-methyltransferase (COMT). This enzyme is found in postsynaptic membranes, in liver and kidney and other tissues. It apparently provides the principal means of inactivating circulating catecholamines.

Both epinephrine and norepinephrine stimulate smooth muscles throughout the body and have a hypertensive effect. A comparison of these two compounds and of effects of various analogues has led to the classification of postsynaptic receptors into two classes α and β. The α receptors usually provoke an excitatory response. However, in intestinal smooth muscles they are inhibitory. Epinephrine is usually more active at α receptors than is norepinephrine. A specific antagonist is **dibenane** (Fig. 16-9). The β receptors usually induce muscular relaxation but cause myocardial stimulation. Norepinephrine is usually more active than epinephrine. In most cases the β receptors of the postsynaptic membrane respond to the neurotransmitter by causing a hyperpolarization of the cell membrane and inhibition of nerve impulses. A specific antagonist is **propranolol** (Fig. 16-9).

While it has been easy to establish the role of the catecholamines as transmitters in the sympathetic nervous system and in the peripheral ganglia, the function in the central nervous system is only now being clarified. Catecholamines are present in varying quantities throughout the brain, and fluorescent histochemical techniques[62] have made it possible to visualize both dopamine and norepinephrine-containing neurons as a result of green fluorescence produced by reaction with formaldehyde or glyoxylate.[63] (The reactions are presumably analogous to those in Fig. 14-25.) Another method for tracing dopamine receptors in the central nervous system is through labeling with specific antibodies to dopamine-β-hydroxylase (Eq. 10-57), the enzyme that converts dopamine to norepinephrine.

It is now clear that there are catecholamine neurons running throughout the brain, including the cortical and cerebellar regions. Very large dopamine-containing neurons have also been found in the brains of gastropod molluscs, and the responses of the individual neurons of this type are being studied.[64]

Within the rat brain fluorescent histochemical techniques have demonstrated the presence of a prominent series of dopamine neurons running from the substantia nigra to the caudate nuclei and putamen of the striatum. Neurons of this **nigrostriatal** pathway (Fig. 16-4) degenerate in Parkinson's disease, a condition accompanied by severe tremors and rigidity. The significance of dopamine in the brain was dramatically illustrated when it was found that the precursor amino acid L-dopa is a "miracle drug" for the treatment of many cases of Parkinson's disease. By making dopamine available more readily* in the basal ganglia of the brain there is apparently a compensation for the deficiency resulting from the neuronal degeneration.

d. Cyclic Nucleotides

The fact that adrenaline activates the adenylate cyclase in membranes of muscle and other cells is well established (Chapter 6, Section F,5). There is good evidence that norepinephrine, upon combination with β receptors, and dopamine (upon activating certain receptors) do the same.[64a] The resultant increase in cAMP levels causes phosphorylation of some membrane proteins, a slow hyperpolarization and a decreased response to excitatory stimuli.[65,66] An example is found in sympathetic ganglia. Impulses enter from the brain on cholinergic fibers and cause a rapid depolarization of postganglionic neurons via the nicotinic cholinergic receptors. However, the incoming neurons also synapse with small dopamine-containing interneurons that synapse with the same postganglionic neurons that are excited by acetylcholine. Thus, after a time delay synaptic transmission is inhibited. The matter is

* Dopamine and the other catecholamines do not cross the blood-brain barrier but dopa does.

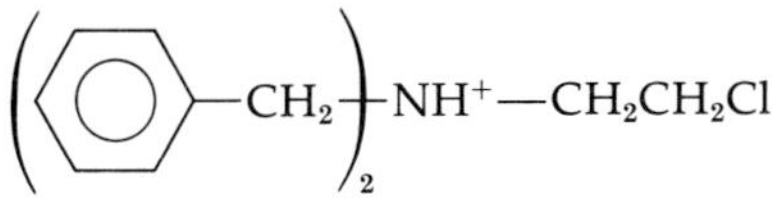

Dibenamine, an antagonist of adrenergic α receptors

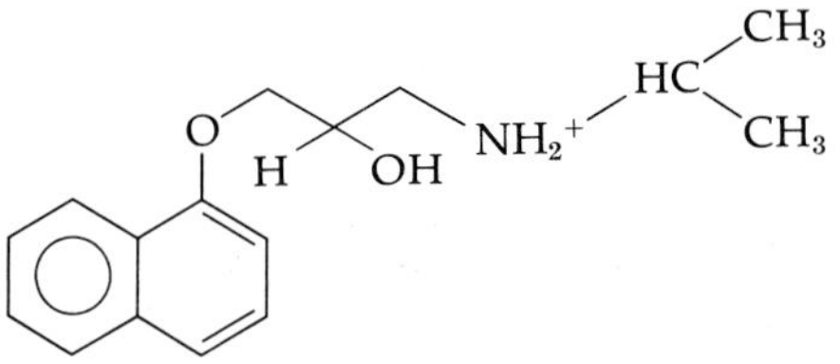

Propranolol, an antagonist of adrenergic β receptors

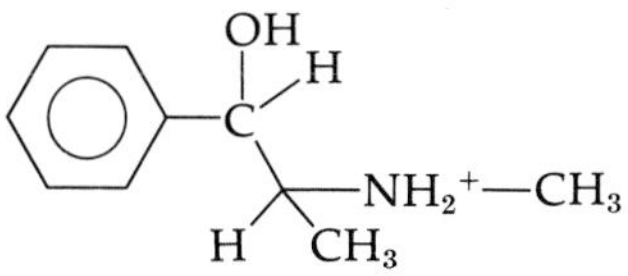

Ephedrine, a drug of low toxicity used for treating asthma

N-Methylation yields methedrine (the S isomer)

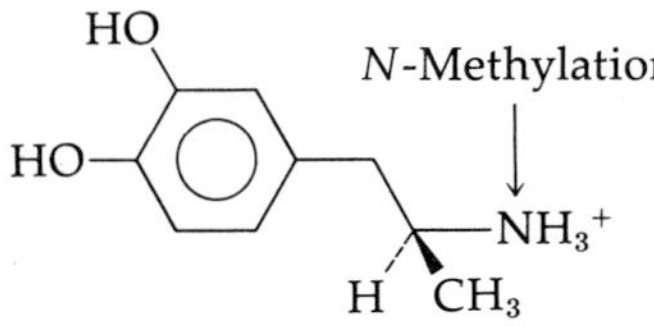

Amphetamine (benzedrine) the S(+) form (dexedrine) is most active

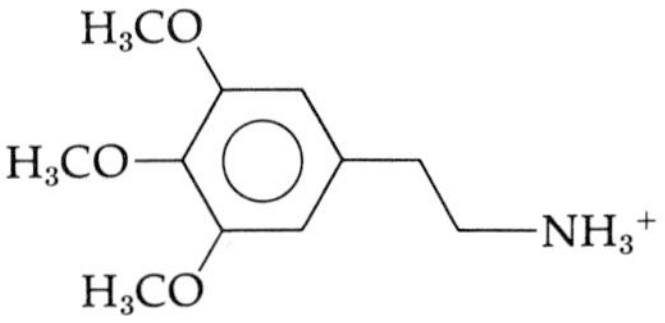

Mescaline

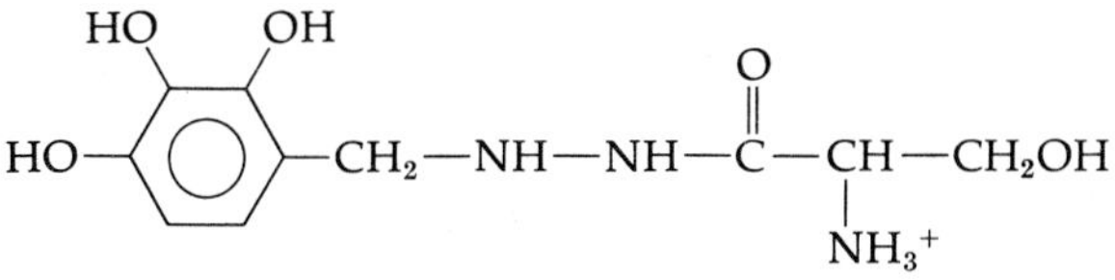

Two dopa decarboxylase inhibitors used as drugs against hypertension

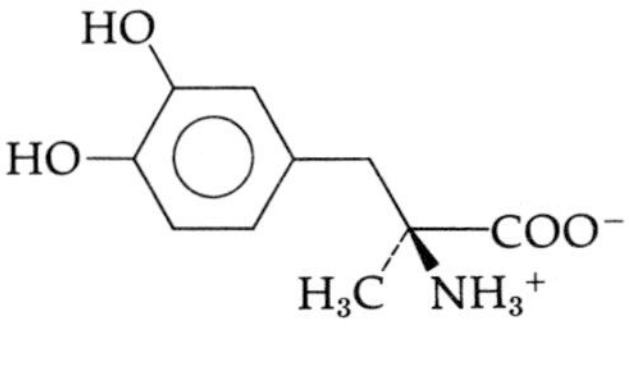

α-Methyldopa

Fig. 16-9 Some agonists and antagonists of adrenergic synapses (shown as cations in most cases).

made more complex by the fact that activation of the postganglionic muscarinic acetylcholine receptors stimulates guanylate cyclase to form cyclic GMP. This causes a slow depolarization, the electric effects of the two cyclic nucleotides being opposite, but not necessarily of the same magnitude.[65-67] The significance of these alterations which "modulate" cholinergic transmission is not yet clear.

Released cyclic nucleotides may also cause longer lasting changes in neurons. For example, stimulation of the chromaffin cells of the adrenal medulla by acetylcholine liberated at synapses causes induction of increased tyrosine 3-monooxygenase activity needed for catecholamine formation. This is presumed to result from action of a cytoplasmic protein kinase which enters the nucleus and acts on lysine-rich histone H1.[67a]

e. Serotonin (5-Hydroxytryptamine)

Another proposed neurotransmitter, the indole-alkyl amine serotonin, is found in all mammalian brains and in invertebrates as well. Its distribution is limited, serotonin-containing neurons being found in the raphe nuclei of the brainstem. They ascend into the brain and down the spinal cord.[68] Serotonin-containing neurons have been traced within brains of snails using ^{3}H-labeled serotonin.[69] Studies with these simpler brains have revealed both inhibitory and excitatory responses to these neurons.

The biosynthesis of serotonin is via tryptophan to 5-hydroxytryptophan with decarboxylation of the latter (Fig. 14-27). Within the **pineal** body of the brain, serotonin is acetylated to N-acetylserotonin which is methylated to **melatonin,** the pineal hormone (Fig. 14-27). A specific inhibitor of serotonin synthesis is p-chlorophenylalanine, and studies with this and other inhibitors suggest that serotonin is required for sleep.[69a]

The serotonin content of the brain is influenced by the diet, being higher after a meal rich in carbohydrates. On this basis it has been suggested that serotonin may serve as a chemical message sent from one set of neurons to the rest of the brain, reporting on the nature of dietary intake.[68,69a] Another suggested function is based on the observation that the inhibition of pituitary secretion by corticosteroids is less effective in animals in which brain serotonin has been depleted.[70]

f. Glutamate, γ-Aminobutyrate, and Glycine

The concentrations of glutamate and of its decarboxylation product γ-aminobutyrate are high in all regions of the brain (Chapter 9, Section C,1). The two compounds are generated sequentially in the γ-aminobutyrate shunt, a pathway that accounts for a quantitatively significant part of the total metabolism of the brain.

The role of glutamate as an excitatory transmitter has been well established in the neuromuscular junction of arthropods.[71] Likewise, the function of GABA as an inhibitory transmitter has been shown unequivocally in special inhibitory neurons present in the peripheral nervous system of arthropods. However, the role of these compounds in vertebrates is in doubt. It is often argued that substances present in such high concentrations must have some quantitatively more important metabolic role. However, the γ-aminobutyrate shunt appears to account for less than 10% of the total oxidative metabolism of brain.[72] γ-Aminobutyrate is scarcely present in other tissues, and its concentration varies at least threefold in different parts of the brain as might be expected for a neurotransmitter.

Although it is a major constituent of all animal tissues, the concentration of glutamate is much higher in brain than in other tissues and it is higher in neurons than in glia. Microiontophoretic application of glutamate to the brain cortex leads to very strong excitatory responses, and glutamate is considered a likely candidate for the major excitatory transmitter in the central nervous system. (However, it should be noted that by microiontophoresis both aspartate and cysteic acid are also potent exciters, while their decarboxylation products, β-alanine and taurine, are inhibitors.)

Other evidence favoring glutamate and GABA as transmitters is the specific and rapid uptake of both compounds by glial cells.[73,74] It has been proposed that glutamate may be converted to glutamine in glial cells and then transferred back into neurons. Specific antagonists for GABA are the alkaloid convulsants **bicuculline** (Fig. 16-7)[75] and **picrotoxin** (Fig. 12-13). One cause of convulsions may be a deficiency of GABA in the brain, and convulsions are one of the most striking symptoms of a severe vitamin B_6 deficiency. Convulsive agents such as 1,1-dimethylhydrazine are thought to act by interfering with pyridoxal phosphate-dependent enzymes (Box 8-G). A primary target appears to be glutamate decarboxylase, and depressed GABA concentrations have been found in brains of animals deficient in pyridoxine and those treated with hydrazines. However, convulsions induced by hydroxylamine are reported to be accompanied by an increased GABA concentration.

The hereditary disease, Huntington's chorea (which has an incidence of 4–7 per 100,000 persons and which affects principally persons of age over 40) may result from a specific deficiency of GABA in basal ganglia.[76]

A second inhibitory transmitter thought to be important in the human brain is **glycine.** Glycine has a concentration of 3–5 mM in the spinal cord and in the medulla of the brain but is low in the cerebral cortex. **Strychnine** (Fig. 16-7) appears to be a specific antagonist of glycine receptors in the spinal synapses. Evidence has been presented that tetanus toxin acts to inhibit the release of glycine from neurons.[77,78]

The compounds **diazepam** and **chlordiazepoxide** (Fig. 16-10) are antianxiety drugs and muscle relaxants and are said to be the two most frequently prescribed drugs in the United States. Since they are able to displace ^{3}H-labeled strychnine from brainstem and spinal cord receptors in the rat, it is suggested that they may mimic glycine and thereby enhance postsynaptic inhibition.[79]

g. Peptide Transmitters

A peptide known as **substance P,** isolated from the brainstem in 1931, has been characterized[80] as

Arg-Pro-Lys-Pro-Gln-Gln-Phe-Phe-Gly-Leu-Met-NH$_2$

Evidence has been obtained for a transmitter or modulator function in synapses. Substance P appears to be present in certain small sensory neurons in both central and peripheral nervous systems.[81]

Recently, considerable interest has been aroused by reports that peptide pituitary hormones such as ACTH, melanocyte-stimulating hormone, and vasopressin as well as the hypothalamic neurohormones may have effects on learning and behavior.[82,83] Transmitter roles are postulated for some of these peptides. Also of interest are the morphinelike peptides, described in Section 6, for which a transmitter role seems possible.

5. Mental Illness and Drugs[83–87]

While many metabolic defects affect only a small number of individuals, emotional illness is a major health problem that at one time or another afflicts a large fraction of the population. Among the most baffling of mental illnesses are the **schizophrenias,** a group of diseases involving thought disorder, disturbance of the affect and withdrawal from interactions with other people. Hallucinations and paranoid feelings are common.[83]

A revolution in the treatment of the schizophrenias and related disorders, as well as in our thinking about mental illnesses, took place following the synthesis, in 1950, of the antipsychotic drug **chlorpromazine** (Fig. 16-10). At about the same time the effect of the *Rauwolfia* alkaloid reserpine (Fig. 14-27) in calming mentally disturbed persons was rediscovered. (The Indian plant *Rauwolfia* had been used for centuries in Hindu medicine for the same purpose.) Prior to

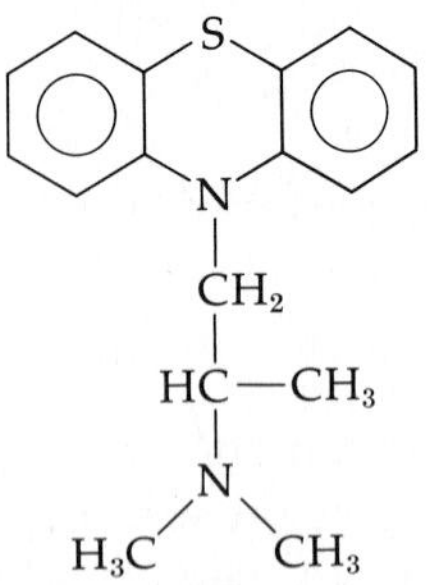

Promethazine,
an "antihistamine"

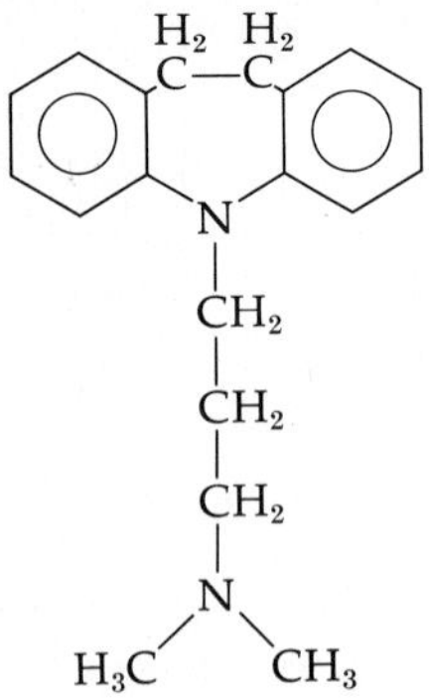

Imipramine, an
"antidepressant" drug

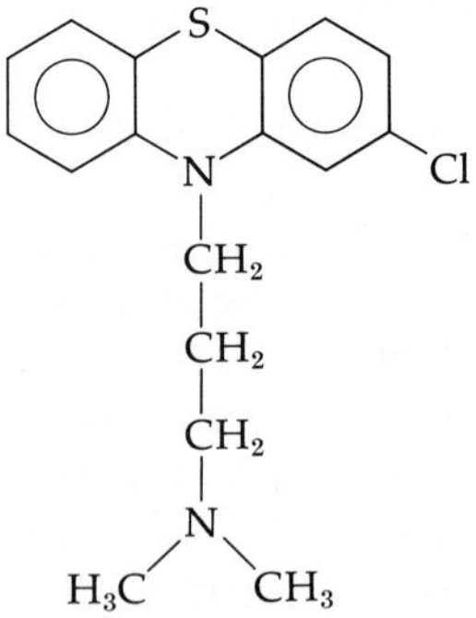

Chlorpromazine,
an "antipsychotic" drug

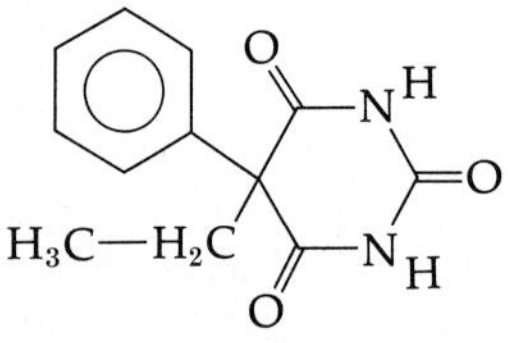

Phenobarbital, a
CNS depressant

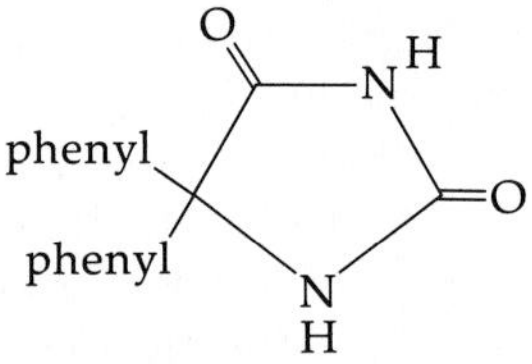

Diphenylhydantoin, an
anticonvulsant used
in treatment of epilepsy

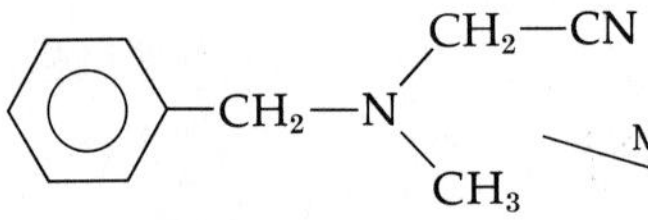

Pargyline,
an antidepressant
monamine oxidase
inhibitor

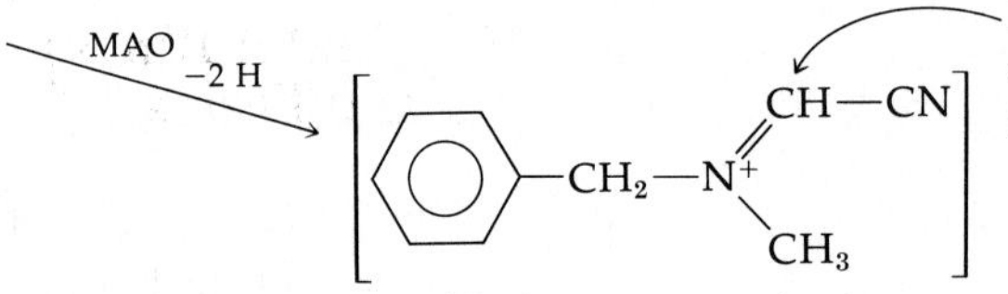

$\xrightarrow[\text{-2 H}]{\text{MAO}}$

Nucleophilic addition of an
enzyme group results in
irreversible inhibition

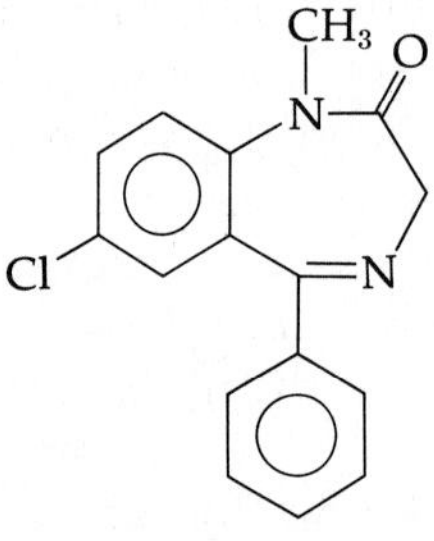

Diazepam
(Valium)

Antianxiety
drugs

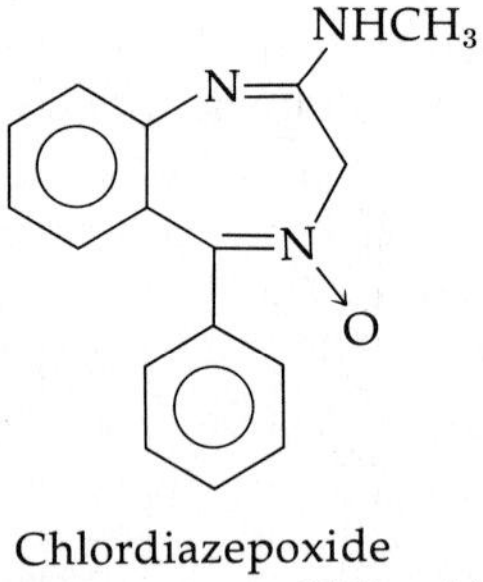

Chlordiazepoxide
(Librium, as HCl salt)

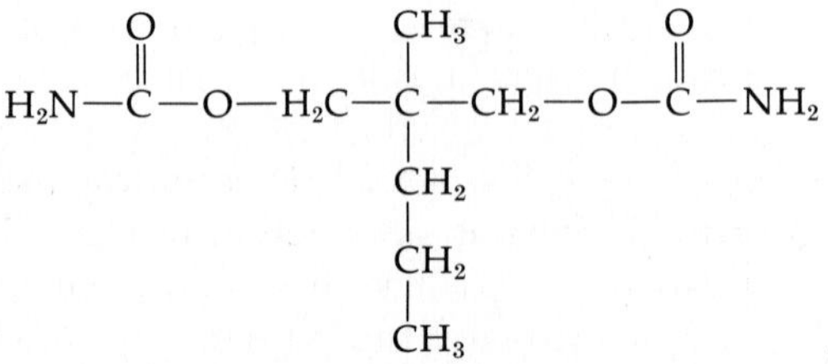

Meprobamate
(Miltown)

Fig. 16-10 Some drugs used in treatment of mental disorders.

the synthesis of chlorpromazine tricyclic pheno-thiazines such as promethazine (Fig. 16-10) had been found to have powerful **antihistamine** activities and to be useful in treatment of allergic disorders. It was the search for better antihistamine drugs that led to the synthesis of chlorpromazine.[88] While the latter was of no value as an antihistamine, its powerful antipsychotic effect was soon recognized. It has been estimated that as many as 250 million people throughout the world may have been treated with chlorpromazine and related drugs in the 20 years following its discovery.

What does chlorpromazine do? A possible clue comes from the fact that it sometimes induces serious "extrapyramidal" side effects, including tremors and other symptoms of Parkinson's disease. This finding suggested that chlorpromazine may block dopamine receptors in the corpus striatum, thereby precipitating a functional deficiency of dopamine.[89] If so, it seemed reasonable to propose that *schizophrenia may result from an overactivity of dopamine neurons,* perhaps including some of the same neurons that are hypoactive in Parkinson's disease. Supporting this view is the observation that amphetamines (Fig. 16-9) tend to worsen the symptoms of schizophrenia and in very high doses may induce striking schizophrenia-like symptoms in normal individuals. There is reason to believe that amphetamines substitute for dopamine.

Amphetamines are known to induce stereotyped compulsive behavior in both humans and in laboratory animals. This fact provides the basis for an interesting method of measuring the action of drugs on amphetamine-sensitive centers of the brain. A lesion in the nigrostriatal bundle on one side of the brain is made by injection of a neurotoxic compound such as 6-hydroxydopamine. This causes degeneration of dopamine-containing neurons on one side of the brain. When rats that have been injured in this way are given amphetamines, a compulsive rotational behavior follows. Administration of chlorpromazine and several other antipsychotic drugs neutralizes this behavior and in direct proportion to the efficacy in clin-ical use. This observation also tends to support the theory that schizophrenia involves overactivity of dopamine neurons. However, some antipsychotic drugs do not significantly alter the rat rotational behavior.[90]

Involvement of brain cholinergic neurons in the action of chlorpromazine has also been suggested.[91] In this respect it is of interest that blockage of muscarinic acetylcholine receptors in the brain by belladonna alkaloids such as atropine (Fig. 16-6) has often been used in treatment of Parkinson's disease. Apparently antagonizing acetylcholine action is to some extent functionally equivalent to increasing dopamine concentrations.

If schizophrenia results from an elevated dopamine content of the brain, the fault might lie with either an oversupply or a reduced rate of metabolism of dopamine. The possibilities of reduced activity of monoamine oxidase or of dopamine β-hydroxylase have both been suggested, but it has been difficult to check these ideas.

Numerous theories of mental illness have embodied proposals that some toxic metabolite is produced in abnormal quantities. An example is 6-**hydroxydopamine** (Fig. 16-8). This neurotoxic compound is known to damage dopamine-containing neurons.[92] Overactive methylation of catecholamines has also been suggested as a cause of mental disorders.[93] In support of this idea, 3,4-**dimethoxyphenylethylamine** (Fig. 16-8) has been reported to be present in urine during acute schizophrenic attacks. Nevertheless, the variability observed in urine samples from mental patients and normal individuals is high and no definite conclusion can yet be reached. Methylation of indole alkylamines has also been suggested on the basis of the observation that administration of tryptophan and methionine to schizophrenic patients exacerbates their illness.[94] N-Methylation of serotonin yields **bufotenin** (N-methylserotonin) and **N-dimethylserotonin,** known hallucinogenic agents. Enzymatic synthesis of the latter by human brain and other tissues has been demonstrated.[94,95]

Among the interesting theories is that of endogenous alkaloid formation. Since the initial products of oxidation by COMT are aldehydes, it is conceivable that they might condense with amines to form Schiff bases and alkaloids, as in Fig. 14-25. This "plant chemistry," if it occurred in the brain, could have a potent effect. Indeed, incubation of tryptamine derivatives with 5-methyltetrahydrofolic acid and an enzyme preparation from brain gives closely related substances known as **tryptolines.**

this carbon came from 5-methylTHF, possibly after oxidation to the level of formaldehyde

5-Hydroxytryptoline

Depression[96] is one of the most common mental problems. The **biogenic amine hypothesis** states that depression results from the depletion of neurotransmitters in the areas of the brain involved in sleep, arousal, appetite, sex drive, and psychomotor activity. An excess of transmitters is proposed to give rise to the manic phase of the manic-depressive cycle that is sometimes observed. In support of this hypothesis is the observation that administration of reserpine precipitates depression, sometimes of a serious sort, in 15–20% of hypertensive patients receiving the drug. Similar side effects are observed with the dopa decarboxylase inhibitor **α-methyldopa** (Fig. 16-9). Since L-tryptophan shows some antidepressant activity in man, but L-dopa does not, it has been suggested that the indole-alkylamines, rather than the catecholamines, may be involved.

The strongest support for the biogenic amine theory is the observation that inhibitors of monoamine oxidase have a powerful antidepressant effect. An example is **pargyline** (Fig. 16-10), which forms a covalent adduct with the flavin of MAO.[96a] Although effective, this drug is somewhat dangerous. Because their monoamine oxidase activity is

so low, patients taking pargyline have been killed by ingesting compounds such as tyramine, which occurs in cheese. Less easy to understand but clinically very important are the tricyclic antidepressants such as **imipramine** (Fig. 16-10). Note the close similarity to chlorpromazine but the greater flexibility of the central ring.[97] An important advance in treating manic-depressive illness is the administration of lithium salts which tends to be very effective. However, the chemical basis for its effects is unknown.[98] It is also noteworthy that Mg^{2+} and Mn^{2+} are powerful CNS depressants and can cause general anesthesia.

Many other aspects of brain chemistry are under study in an attempt to understand mental illnesses. For example, the possible significance of zinc deficiency, copper toxicity, and other metal imbalances in the nervous system is the subject of much current investigation.[99]

6. Addictive and Psychotropic Drugs

The depressive drugs, including morphine and other narcotics (Fig. 16-11), barbiturates (Fig. 16-10), and alcohol have a strong potential for addiction in susceptible individuals. The phenomenon is most striking in the case of the opiates. Addiction leads to physical dependence, a situation in which painful withdrawal symptoms occur in the absence of the drug. At the same time a striking tolerance to the drug is developed. The addicted individual can survive what would otherwise be a fatal dose without ill effect. Indeed, aside from the pathological hunger for the drug, the addict can function normally in almost every respect.[100]

That there are specific morphine receptors in the central nervous system was suggested by the striking specificity for molecules of a particular shape (Fig. 16-11) and by the observed cross tolerance toward other narcotics observed in animals addicted to morphine. More recently, direct binding of highly radioactive opiate has permitted localization of the receptors.[101] Most narcotics are

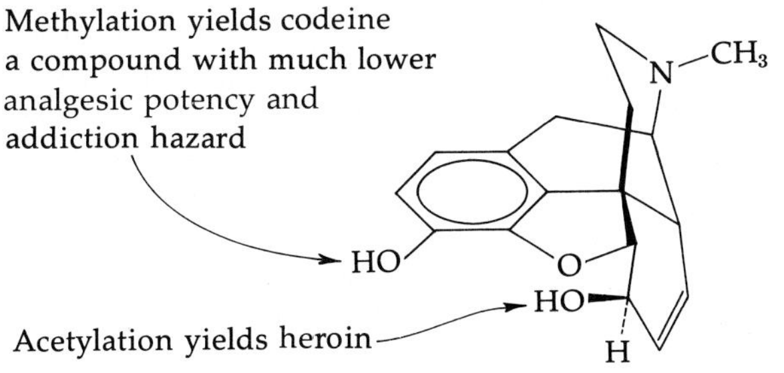
Methylation yields codeine
a compound with much lower
analgesic potency and
addiction hazard
HO
HO
O
N
CH3
H
Acetylation yields heroin

Morphine (See also Fig. 14-25)

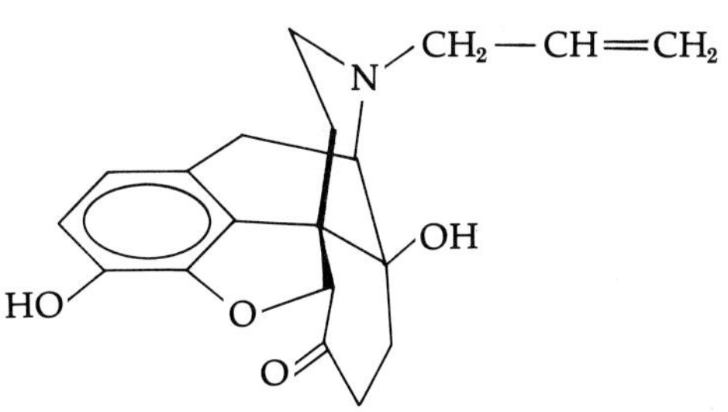
N
CH2 — CH=CH2
OH
HO
O
O

Naloxone

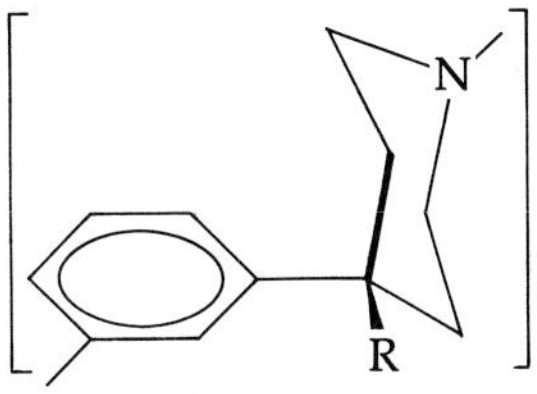
N
R

Structure common to many narcotic drugs

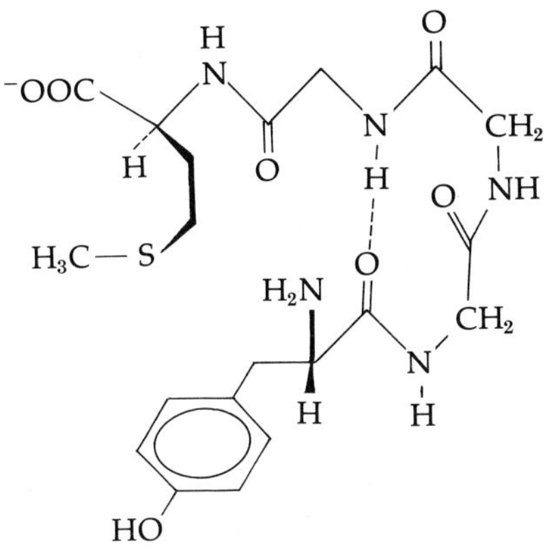
-OOC
H
N
H
O
N
H
O
CH2
NH
H3C — S
O
H2N
O
CH2
N
H
H
HO

Tyr-Gly-Gly-Phe-Met

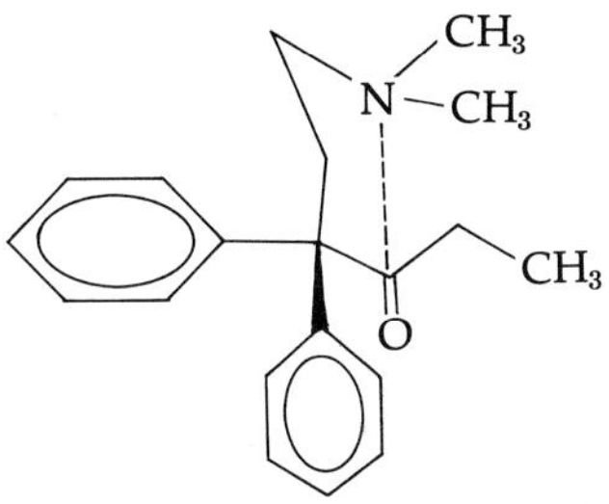
CH3
N
CH3
CH3
O

Methadone

Fig. 16-11 The structures of morphine and some analogues including a peptide isolated from brain and possessing opiate-like activity. Also shown is a structure common to many narcotic drugs.

polycyclic in nature and share the grouping indicated in Fig. 16-11. However, the flexible molecule **methadone** also binds to the same receptors.[102] Specific antagonists that block the euphoric effects of opiates are also known, among the most effective antagonists is **naloxone** (Fig. 16-11).

What is the natural function of opiate receptors? It is logical to suppose that they are designed to bind some neurotransmitter or modulator. Recently, the following two pentapeptides (known as **enkephalins**), as well as longer peptides isolated from brain tissue, have been shown to have potent opiate agonist activity[103]:

Tyr·Gly·Gly·Phe·Met	Tyr·Gly·Gly·Phe·Leu
Methionine enkephalin	Leucine enkephalin

While the test system is based on opiate receptors in nerves of the guinea pig intestine rather than the brain, the activity in the test is believed to closely parallel that in the brain.[103a] The obvious tentative conclusion is that opiate drugs mimic the normal action of one or more brain peptides.[103b] It is of special interest that a fragment from the C terminus of lipotropin (Section 2,A) contains the methionine enkephalin at its N-terminal end.[103c, d]

An important pharmacological problem is related to the fact that opiates are the most powerful **analgesic agents** known and that their efficiency in diminishing pain is directly related to their addiction potential. To date, it has been impossible to design a nonaddictive analgesic drug of the potency of morphine.

Theories of addiction usually postulate that some compensatory change in the receptor-agonist system results from the occupation of the receptor sites by the drug. In the case of narcotics, specific receptors have been identified in the central nervous system and also in tumor-derived cells in culture. One proposal is based on studies of opiate receptors in tumor-derived cells in culture. It is postulated that morphine acts on neurons like an inhibitory hormone, lowering the internal level of cAMP.[104] The neuron then makes compensatory changes, increasing the number or activity of adenylate cyclase molecules to restore the internal cAMP level. This leads to dependence upon morphine because in its absence the cAMP level rises too high. The increased number of adenylate cyclase molecules and associated receptors also accounts for the tolerance observed.

Despite much study, little is known about the chemistry of addiction to alcohol.[105,106] As with morphine addiction, a distinct tolerance is observed and a lack of ethanol produces withdrawal symptoms. The principal route of metabolism of ethanol (both ingested and the small amount of endogenous alcohol) is believed to be oxidation in the liver to the chemically reactive acetaldehyde.* The latter is further oxidized to acetate. Many theories of alcoholism assume that addiction (and possibly also the euphoric feeling experienced by some drinkers) results from abnormal metabolism of ethanol in the brain. For example, it has been assumed that acetaldehyde might form alkaloids by reaction with neurotransmitters, as has been proposed for some forms of mental illness (Fig. 14-25). However, it has been clearly shown that there is no cross reactivity between morphine and alcohol in addicted mice[107] and it is currently believed that acetaldehyde is probably not the addictive agent.[108] Alternative pathways of metabolism of alcohol are found in the smooth endoplasmic reticulum of the liver.[109] This provides another possible route to abnormal metabolites.

Interesting experiments with mice and rats have established a genetic propensity toward addiction to alcohol. Some strains of animals shun alcohol and become addicted only if force-fed for prolonged periods. Other strains accept the alcohol more readily and become addicted quickly. That a similar situation holds for humans is quite possible.

Of the psychotropic or mind-changing drugs, the hallucinogenic compounds are a source of special fascination to many people. The presence of the indole ring in the powerful hallucinogen **lysergic acid diethylamide** (LSD, Fig. 14-27) suggests that this compound might mimic the action of serotonin. However, some experiments suggest antagonism of dopamine receptors in the striatum.[110] A common site of action for a large variety of hallucinogens has been suggested.[111]

Despite the intense interest in understanding the effects of **marijuana,** very little is known at the molecular level. It is hard to see similarities in molecular geometry between tetrahydrocannabinol (Fig. 12-13) and known neurotransmitters. A single site of action has been postulated for tetrahydrocannabinol and **thujone,** a psychotropic plant product present in the liquor absinthe.[112]

Another important family of drugs that affect the nervous system are the anesthetics.[113] Although some are molecules of moderate size, e.g., **barbiturate** derivatives, others are very simple molecules such as **diethyl ether** or **halothane** ($CF_3CHClBr$). The latter is today the most widely used inhalation anesthetic. Many theories of the action of anesthetics have been suggested. It has been most often proposed that the effectiveness is related to solubility in lipids, but it is very difficult to pinpoint the site of action in nerves. A recent theory proposes that anesthetics break hydrogen bonds.[114] A principal effect of anesthetics is a reduction in sodium ion conductance by nerve membranes.[114a]

---

* It is of interest that administration of D-fructose increases the rate of oxidation of ethanol.[105]

7. Odor and Taste

A question that we have not considered is how sensory neurons are activated. The functioning of taste and smell receptors has been especially intriguing to biochemists. It is obvious that compounds have different tastes and odors, but it is not easy to understand the relationship between these qualities and chemical structures.

Remarkably enough, it seems that bacteria possess something akin to the ability to taste. Bacteria are attracted toward certain compounds that they can metabolize, a process known as **chemotaxis.**[115–117] Thus, *E. coli* will swim toward higher concentrations of L-serine (but not of D-serine) or of D-ribose. Other compounds, such as phenol, are repellent. By what mechanism can a miniscule prokaryotic cell sense the direction of a concentration gradient? It appears quite certain that the plasma membrane contains receptors whose response is in some fashion linked to control of the flagella that provide the motion. Since the dimensions of a bacterium are so small, it is hard to believe that they can sense the difference in concentration between one end and the other end of the cell. Individual bacteria (*E. coli* or *Salmonella*) swim in straight lines, periodically "tumbling" and then swimming in a new random direction (Box 4-B). The chemotatic response apparently results from the fact that a bacterium swims for a relatively long time without tumbling when it senses that the concentration of the attractant is increasing with *time.* When it swims in the opposite direction and the concentration of attractant decreases, it tumbles sooner. A plausible theory is that as the membrane receptors become increasingly occupied with the attractant molecule, the rate of formation v_f of some compound X, within the membrane or within the bacterium, is increased (Eq. 16-1).

$$\xrightarrow{v_f} [X] \xrightarrow{v_d} \tag{16-1}$$

When $[X]$ rises higher than a certain threshold level, the tumbling behavior is induced. At the same time, X is destroyed at a velocity of v_d. Subsequently, a readjustment of v_f and v_d occurs such that the concentration of X falls to its normal steady state level.

Many examples of chemotaxis are known among the lower invertebrates such as *Euglena.* Among the Cnidaria, interesting chemically controlled feeding behavior is observed. For example, chemoreceptors in *Hydra* sense glutathione that flows from the broken tissue of their prey. Other related organisms respond to proline. In the sea anemone *Anthopleura* asparagine induces the bending of the tentacles while glutathione induces swallowing.[118] Many other examples could be cited, and it is hard to believe that these phenomena are not essentially the same as those involved in smelling and tasting in the human.

Turning to the human senses, it has been difficult to relate taste and odor to physical or chemical qualities or to molecular shape.[119,120] The matter is quite complicated and it has been suggested that there may be 20–30 primary odors, perhaps reflecting 20–30 different types of receptor proteins. It seems likely that the fundamental physical basis for taste and odor will be found in the binding of the molecules being sensed to proteins, in the induction of conformational changes leading to depolarization of a portion of the nerve membrane, and in initiation of an action potential just as in synaptic transmission. An interesting observation is that some peptides have unusually sweet tastes and that there are **chemostimulatory proteins.** Two of the latter which taste sweet have been discovered. A third protein, which occurs in a tropical fruit, modifies taste so that the taste of acids is changed from sour to sweet after the tongue has been treated with the protein.[121] Exposure of the tongue to artichokes often makes water taste sweet.[122] Thus, it appears that the response of taste receptors can be temporarily altered by binding of other substances, perhaps at adjacent sites on the sensitive membrane.

8. Neuronal Metabolism and the Chemistry of Thinking

Neurons are characterized by an unusually active metabolism, a substantial part of which can be attributed to the sodium ion pumps in the membranes and to the maintenance of the excitable state. The chemistry associated with the transmission of a nerve impulse along an axon has been described in Chapter 5, Section B,3. The sequential opening of Na^+ channels then K^+ channels seems well established. Less clear is whether any special enzymatic processes underlie the changes in ion conductance needed for propagation of the action potential. Nachmansohn has pointed out that acetylcholinesterase is present in high concentration throughout neuronal membranes, not just in the synapses.[38,39] He proposed that permeability to sodium is increased by the cooperative binding of several molecules of acetylcholine to a membrane receptor that is either the sodium gate itself or that controls the sodium gate. Acetylcholine would be released from a storage site on the membrane as a result of membrane depolarization. A change in electrical field in the membrane could easily cause a conformational change in a protein that would release acetylcholine. The sodium ion permeability would then return to normal within a short time as a result of the activity of acetylcholinesterase. This sequence of events would be essentially the same as those postulated for cholinergic synapses except that the acetylcholine would be stored in a protein-bound form rather than in vesicles. It has been proposed that the concentration of calcium ion controls the K^+ channels. A field-sensitive calcium-binding protein would release Ca^{2+} which would activate the K^+ gates with the observed time delay, relative to opening of the Na^+ gates, arising from differing rate constants for the two processes.[123] Hydrolysis of ATP is proposed as an energy source for closing the K^+ gates again. Various other mechanisms of conduction have been suggested.[124] Some assume no special metabolism other than the pumping of Na^+ ions.

Another factor, peculiar to neurons, doubtless contributes to their rapid metabolism. The nucleus and most of the ribosomes are found in the cell body. However, many proteins are needed in high concentrations in the axons and synaptic endings. Among these are enzymes catalyzing synthesis and catabolism of neurotransmitters and membrane proteins. If an axon is cut, the separated synaptic endings soon atrophy, an observation that long ago suggested that essential materials might flow from the cell body. It has now been experimentally established that many materials do move at the rate of 1–10 mm/day from the cell body down the axon. More remarkable, is the more recent discovery of **fast axonal transport.** Proteins and other materials move at rates of up to 0.4 m/day. This transport is specifically blocked by compounds such as vinblastine (Box 4-A) and batrachotoxin (Fig. 12-18). It has been suggested that an ATP-hydrolyzing protein chemically related to the myosin heads functions together with fine filaments and microtubules to provide a kind of miniature railway for moving materials along the microfilaments.[125,126] Transport is sometimes in the opposite direction, i.e., from the synaptic endings to the cell body. This **retrograde axonal transport** could be of importance in altering neuronal properties in response to electrical activity at synaptic endings.[127]

There is some evidence that brain cells transcribe an unusually large fraction of the genome.[128,129] Thus, ~20% of the DNA of human brain was found to hybridize with mRNA formed by brain cells. In other tissues, only about half this amount of DNA appears to be transcribed. The amount of transcription is higher in the human than in the mouse.[128] A related observation that seems surprising is the absence of common electrophoretic variants of enzymes in the brain.[129] The significance of these observations is not yet clear.

What is known about the chemistry underlying thinking and the generation of the stream of consciousness within the brain? The passage of nerve impulses into the brain strongly affects the signals that are sent out through the motor neurons. However, it is also clear that the brain has its own endogenous electrical rhythms that do not depend upon input through sensory neurons. In the simpler invertebrates, the source of these rhythms appears to reside in special **pacemaker neurons.** These are neurons that spontaneously fire at regular intervals. Their cell membranes apparently undergo a cyclic series of changes in ionic permeabilities sufficient to initiate action potentials. Three types of pacemaker output from molluscan neurons[130] are illustrated in Fig. 16-12. It seems likely that similar phenomena may underlie brain function in humans. Rhythms from endogenous pacemakers may combine with pulses from sensory neurons to evoke conscious thought. Looking at more primitive organisms, it is of interest to compare the spontaneous output of pacemaker neurons with the periodic release of cAMP by cells of *Dictyostelium* (Chapter 6, Section F,5). The two phenomena may be closely related.

Especially intriguing are questions of the chemical basis of memory. If thinking consists of the passage of some pattern of electrical waves throughout the network of neurons in the cortex, where is the memory trace or **engram** stored, and what is its form? Experiments have shown that there is a **temporary memory** with a relatively low storage capacity and a **long-term memory.** Stored information can be transferred from the temporary to permanent forms. It is most commonly believed that information in the temporary memory is stored in reverberating circuits of the cortex and dies out rapidly. Temporary memory can be totally lost from a blow to the head. The permanent memory, by contrast, is retained for such a long time that we must seek some stable change in the chemistry or even in the physical connections between neurons.

In view of the all-or-none character of nerve impulses, it seems reasonable to suppose that the synapses are the sites of any alteration that leads to memory. When individual synapses are studied, it is found that **facilitation** and **habituation** occur. Facilitation refers to the fact that a second impulse will often be transmitted through a synapse more effectively than the first, while habituation refers to a decreased response to repeated stimuli. Memory could consist of facilitation and habituation phenomena at specific synapses. Recall that excitation of muscarinic receptors at cholinergic synapses stimulates the release of cyclic GMP within cells while excitation by catechola-

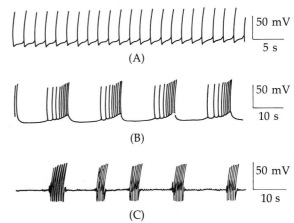

Fig. 16-12 Intracellular recordings from isolated neurons of the mollusc *Aplysia:* (A) beating pacemaker, (B) bursting pacemaker, and (C) oscillating pacemaker. From Chen *et al.*[130]

mines stimulates formation of cyclic AMP. The cyclic nucleotides in turn regulate numerous intracellular enzymes. Not only are the properties of the synaptic membranes altered, but also gene transcription and many other cellular processes may be affected. The activity of RNA polymerase II is increased by a cAMP-dependent phosphorylation as is that of ornithine decarboxylase, needed for formation of polyamines (Chapter 14, Section C,4). Thus, passage of an impulse through a synapse may have a long-term influence on the properties of that synapse.

Many experimental results confirm the chemical basis of memory. For example, learning is facilitated by administration to animals of small doses of strychnine.[131] Other drugs such as puromycin (Fig. 15-18) interfere with learning.[29,132] Learning processes in animals are associated with increased synthesis of mRNA and of proteins in neurons. It is significant that the synthesis of polypeptides and nucleic acids takes place in the cell body rather than at the ends of axons or dendrites. Cell bodies are usually covered with synaptic knobs, and it may be that stimulation on the membrane surface of the cell bodies induces synthesis of macromolecules.

An alternative to the idea of habituation and facilitation providing the chemical basis for learning is **molecular coding.** Thus, it has been reported that a specific 15-amino acid peptide isolated from rats trained to avoid the dark carries behavioral information. When this peptide is injected into brains of untrained rats, they also avoid the dark.[133] This is only one of many reports of transfer of some learned behavior through chemical substances extracted from the brain. However, in view of the complexity of the brain and our knowledge of neuronal behavior, these ideas are hard to accept. Nevertheless, they cannot be dismissed lightly. We know of peptide hormones and releasing factors formed by neurons (Section A,1). The possibility that long-term memory is associated with transcription in specific neurons of specific amino acid sequences must be considered.

Learning takes place in many parts of the brain. Because the organization of the cerebellum is so well understood, it is a logical organ in which to consider the learning process. Chemically based theories of cerebellum memory have been proposed.[134]

C. DIFFERENTIATION AND DEVELOPMENTAL BIOLOGY[135−139]

One of the most fascinating of all biological phenomena is the development of an animal from a fertilized egg. From the early embryonic cells, which appear to be very much alike, there arise during the course of a very few cell divisions differentiated organs and tissues such as liver, brain, kidney, muscle, skin, and red blood cells. The biochemical properties of differentiated cells are often highly specialized. Red blood cells make hemoglobin, while muscle cells make large amounts of myosin and actin. The endocrine cells of the pancreas make insulin or glucagon, while the exocrine cells form the digestive enzymes that are secreted into the intestinal tract. In general, it is estimated that in a given tissue no more than about 10% of the total genes are transcribed at any one time (the

exception may be brain, see Section B,8). Chemical analysis makes it clear that the specialized cells contain a full complement of DNA, yet 90% of the genes are turned off.

While the precise chemical basis for differentiation remains largely unknown, enough has been learned to make it clear that chemical signals reaching a cell from the external environment and from adjacent cells play a very important role. These signals interact with internal genetically coded **developmental programs** to determine the developmental pattern followed by a given cell. The precision with which this can be accomplished is well illustrated by the rotifers and annelid worms (Fig. 1-10), which tend to have an almost invariant total number of cells in a given species. Thus, the nematode *Oxyuris equi* contains exactly 251 nerve cells, one excretory cell, 18 midgut cells, and 64 muscle cells.[140]

1. Physiological Modulation and Alternative Developmental Programs

Before considering differentiation in complex multicellular organisms it is instructive to examine unicellular and colonial forms. Under suitable conditions bacteria and eukaryotic cells alike follow a uniform "cell growth and division cycle" (Fig. 15-25) that gives rise to exponential growth (Eq. 6-60). However, this pattern can easily be altered by a change in external conditions. Thus, substrate depletion not only decreases growth rate but alters gene transcription. This occurs in *E. coli* as a result of a rise in internal cAMP concentration. The presence of an alternative energy source such as lactose can induce specific changes in gene transcription (Chapter 15, Section B,1). Many other examples can be cited. All may be described as "physiological modulation."

More striking is the fact that environmental signals can trigger a cell to "switch" to an "alternative developmental program" by which enough new genes are activated as to essentially rebuild the cell into a new form. An example is spore formation by certain bacteria (Chapter 1, Section

A,8), a process that occurs when external conditions become unfavorable for vegetative growth.

Many chemical changes occur during sporulation.[141-143] Synthesis of rRNA is completely turned off at the outset, new classes of mRNA molecules are transcribed and several new proteins are synthesized. Most striking is the production of large amounts of dipicolinic acid (Fig. 14-7), a process that requires the appearance of at least one new enzyme. In addition, as the spores are formed the bacteria take up large amounts of Ca^{2+} and substantial concentrations of Mn^{2+} and other metal ions. In many bacteria 3-L-sulfolactic acid is also formed.

$$^-OOC - \overset{H}{\underset{CH_2SO_3^-}{C}} - OH$$

3-L-Sulfolactate

These components account for the following percentages of the total dry weight of spores of *Bacillus subtilis:* dipicolinic acid, 10%; sulfolactic acid, 3–6%; Ca^{2+}, 3%; and Mn^{2+}, 0.3%. Bacterial spores are remarkably resistant to heat and can survive boiling water for prolonged periods. It is often suggested that the dipicolinic acid and other ions protect the proteins from denaturation. However, the heat resistance may arise from the maintenance of the core of the spore in a highly dehydrated state.[144] When conditions become appropriate for growth again the spore germinates and the bacterium again follows the cell growth and division program.

More complex alternative developmental programs are followed by colonial forms of bacteria such as the myxobacteria, but the chemical signals are as yet not clear.[145] However, in the eukaryotic cellular slime molds such as *Dictyostelium* (Chapter 6, Section F,5), which follow a similar developmental program, bursts of cAMP signal substrate depletion.* The cAMP is sensed by other cells which then change their synthetic patterns in a way that leads to differentiation and to

* However, sensitivity to cAMP may arise as a secondary response to excretion of a macromolecular differentiation-stimulating factor.[144a]

formation of a fruiting body.[135,136,146] Some of the cells begin to produce cellulose as well as a mucopolysaccharide. Trehalose is formed and is stored in the spores. New enzymes have to be formed to synthesize these materials.

In the phycomycete water mold *Blastocladiella*, HCO_3^-, an even simpler chemical messenger than cyclic AMP, determines whether a thin-walled sporangium or a thick-walled heat resistant sporangium will appear. On the other hand, chemical signaling with macromolecules, perhaps proteins, can also occur among the simpler eukaryotes. Thus, an alternative developmental pattern for some strains of *Dictyostelium* is formation of macrocysts between cells of two different mating types. A diffusible inducing factor of MW ~12,000 appears to be released by cells of one strain.[147]

Because of their relative simplicity, some plant tissues are attractive subjects for the study of differentiation. The cambium layer of stems (Fig. 1-12) continuously differentiates to form phloem on the outside of the cambium and xylem on the inside. At the same time, cambium cells are retained. Thus, at each cell division, one daughter cell becomes a differentiated cell, while another remains the less differentiated cambium. This pattern of continuous differentiation from a **stem line** with constant properties is commonly found in animals as well as in plants. In the differentiation of cambium, it appears that chemical signals obtained from the surrounding cells on either the inside or the outside of the cambium layer determine the nature of the differentiated cell. It is known that sucrose, auxin, and cytokinins are all involved.

Ordinarily asexual propagation of plants occurs by virtue of the ability of embryonic meristematic tissue (Chapter 1, Section E,4) to differentiate into roots and shoots. On the other hand, if isolated phloem cells or other more differentiated cells are cultured, the result is often the formation of a **callus,** a dedifferentiated mass of cells somewhat reminiscent of embryonic cells. Under the proper conditions, e.g., in a coconut milk culture and in the presence of a proper auxin to cytokinin ratio, some carrot root phloem cells have been induced to revert to embryonic cells and to develop into intact plants.[136] This experiment is important because it provides definitive proof that the differentiated carrot phloem cells contained a complete genome for the plant. However, it is significant that the experiment cannot be done easily with most plants and that dedifferentiation is not always automatic. Nevertheless, it occurs in enough cases to establish the **totipotency** of the nucleus of differentiated cells.

The foregoing and many other experiments establish that differentiation is in part a response of cells to chemical signals from adjacent cells and from the external medium.

2. Development of Animal Embryos

In considering the differentiation of cells in an embryo it is necessary to realize that within the nearly spherical ovum (egg cell) there is a strong polarity. The nucleus lies nearer to one "end," referred to by the old name **animal pole,** than to the other. The opposite end, which in many eggs is rich in yolk granules, is the **vegetal pole.** In eggs of amphibians, the animal pole is highly pigmented but the vegetal pole is less so and on one side, below the equator, there is a gray crescent. The sperm cell in some animals enters the egg opposite this gray crescent (Fig. 16-13). The gray crescent marks the future back (dorsal) side of the organism and the opposite part of the cell, the future ventral side. There is good evidence that the cytoplasm of the mature ovum contains an unequal distribution of many materials with a well-developed bilateral symmetry. That this distribution is important is seen from the fact that centrifugation of eggs prior to fertilization often leads to formation of abnormal embryos because of displacement of preformed ribosomes and other materials. Fertilization of the egg is itself a complex process from a biochemical viewpoint. Often the final stages of the last mitotic division of the egg are delayed until after penetration of the sperm. This event somehow leads to "activation" of the egg. In lower organisms, this activation can often be carried out by chemical or physical treatment in the absence of a sperm cell, with formation of **parthenogenic** offspring.

The fertilized (activated) ovum rapidly undergoes several mitotic divisions, known as **cleavage,** during which no overall growth occurs. The number of cells increases and the DNA replicates at each division, but the overall size of the resulting cell cluster is the same as that of the original ovum (Fig. 16-13). A stage is soon reached in which a single layer of cells (called **blastomeres** at this stage) surrounds an internal cavity forming a **blastula.** In the sea urchin, the blastula is made up of a single layer of cells, but in many other organisms such as the frog there are two or more layers. In mammals a solid cell mass (**morula**) forms first and is later transformed to a **blastocyst,** a hollow ball with an internal cavity.

In many invertebrates and in amphibians, the next stage in embryo formation is the invagination of the blastula at the vegetal pole to form a **gastrula.** At this stage, the embryo has distinct ectoderm and endoderm cell layers. The cavity formed in the gastrulation process and connecting to the outside is referred to as the **archenteron** and is the forerunner of the gastrointestinal tract or enteron. Gastrula formation is more complex in the frog embryo, and in the human embryo a still more complex and somewhat different pattern is seen.

In all but the most primitive of animals a third layer of **mesosomal** cells is formed between the endoderm and ectoderm. These three **germ layers** differentiate further as follows. The ectoderm yields the skin and nervous system; the mesoderm the skeleton, muscles, connective tissues, and circulatory system; and the endoderm the digestive tract, lungs, and other internal organs and germ cells.

A striking demonstration of the totipotency of differentiated cells of amphibian embryos has been provided by Gurdon.[148,149] Nuclei from cells of intestinal epithelia and other tissues have been substituted, by transplantation techniques, for the

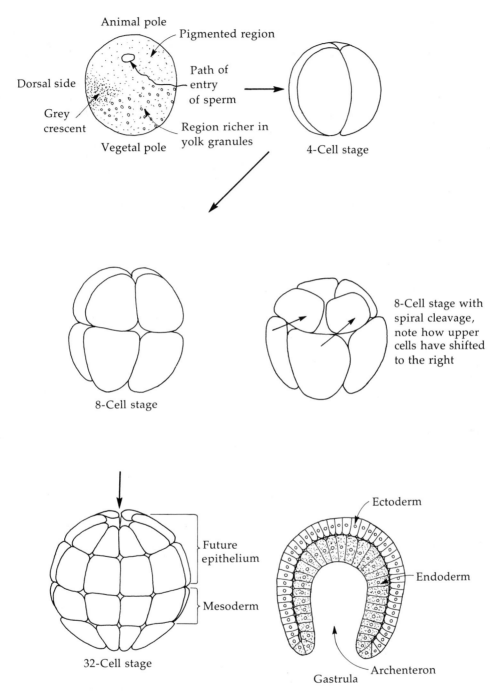

Fig. 16-13 Cleavage and gastrula formation. Fertilization and cleavage of an amphibian egg is illustrated at the left. Spiral cleavage, usual among molluscs, is shown in the center right. The gastrula on the lower right might be from an echinoderm or some lower invertebrate.

nuclei of egg cells. The resulting eggs in some cases grew into adult toads. Thus, the full genetic information of the toad was present in the differentiated cells. However, it has not been possible to accomplish this result with nuclei of neurons. Thus, the possibility of **irreversible differentiation** in some cells is not excluded.

3. Hormonal Control of Growth

Organ development occurs largely by infoldings of cells from the endoderm and ectoderm. Ample evidence suggests that these infoldings are induced by chemical substances secreted by cells of an adjacent germ layer. Thus, ectodermal cells form the **neural plate,** the prospective brain, and spinal chord in response to induction by mesodermal cells lying beneath the neural plate area. The mammary glands also arise from interactions of mesodermal and ectodermal cells, while the formation of the pancreas, liver, and lungs depends upon interactions of groups of cells from endoderm and mesoderm. Since induction can occur through thin (20 μm) Millipore filters without any cell–cell contact, specific chemical agents are believed responsible.[150]

It now seems likely that a large number of molecules of various types serve as **local hormones** or **autocoids** involved in communication between cells. These may induce growth or differentiation. They may serve as chemotactic agents. Some doubtless inhibit growth. Only a few of the known examples will be mentioned in the following paragraphs.

A striking example of chemical induction of a developmental process is observed with embryonic sensory and sympathetic neurons. The **nerve growth factor,**[151–153] a small protein containing 118 amino acids and three disulfide bridges, stimulates profuse outgrowths of neurites from the embryonic neurons (Fig. 16-14). Resembling proinsulin in size and general structure, nerve growth factor can be regarded as a true hormone. It is formed by many tissues, but the submaxillary

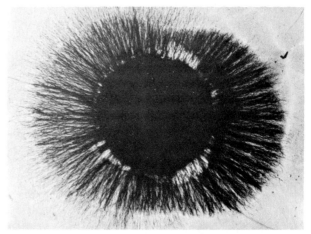

Fig. 16-14. Effect of 1 ng of nerve growth factor in promoting the production of neurites in a chick embryonic sensory ganglion. From W. A. Frazier, C. E. Oblendorf, L. F. Boyd, L. Aloe, E. M. Johnson, J. A. Ferrendelli, and R. A. Bradshaw, *PNAS* **70,** 2448–2452 (1973).

(salivary) glands of male mice have proved the richest source. Like the pancreas, these glands contain both exocrine and endocrine tissues. Fibroblasts also secrete nerve growth hormone, and the compound has been found in malignant sarcomas. It has been suggested that within the central nervous system glial cells may produce the hormone. Nerve growth factor appears to act by altering membrane properties in a way that encourages the rapid outgrowth of the long slender neurites. Another protein of higher molecular weight acting in a somewhat similar manner on embryonic brain cells has been reported.[154]

A very interesting developmental question arises from examination of the organization of the brain and observation of the way in which the interconnections between neurons develop in the embryo.[155] The entire brain is "prewired" before it starts to function, and in the process axons must grow from one region of the brain into another, often traveling extraordinary distances. What is even more remarkable is that the connections are made in a very regular way. For example, the image formed on the retina of the eye is transmitted without scrambling to specific regions of the brain in which the image is projected. The interconnections between the retina and the brain arise by growth of neurons which behave as if each one contained a specific address in the cerebrum.[155] We cannot even guess at this point what chemically encoded information is required to permit this to take place.

Similar to nerve growth factor and also formed by the submaxillary glands of male mice is the **epidermal growth factor,** a 53-residue polypeptide.[156,157] A human form has been identified[157] and may be identical to **urogastrone.**[158] The latter is an inhibitor of the development of peptic ulcers and is found in relatively large amounts in the urine of pregnant women (who tend not to develop ulcers).

Just as neuronal processes move and synapse with specific target sites so entire cells often move during the process of development. **Cell migration** is thought to be guided by chemotaxis (Section B,7). The process has been much studied in the *Hydra* (Fig. 1-10), a simple organism containing only 10 cell types. One of these is an embryonic reserve of mesodermal cells. From this stem line, among other things, nematocytes (stinging cells) are formed and move up the body of the *Hydra* and take up residence in the tentacles.[159,160]

The role of chemotaxis in the defense system of the human body has previously been mentioned (Box 5-G). A somewhat controversial postulate is that every cell produces a tissue-specific local hormone known as **chalone** that inhibits the mitotic activity of other cells of the same tissue. These hormones are thought to be essential in the control of cellular division and in the prevention of malignant growth. Several chalones have been identified as proteins and peptides of varying molecular weights.[161,162]

In summary, local hormones can produce a variety of effects on nearby cells including stimulation or inhibition of growth, cell migration, and cell differentiation. While specific peptides or proteins play this role in many cases, simple molecules such as small peptides, histamine, sero-

tonin, and even bicarbonate ion may also serve.[163] Local hormones must pass through cell membranes in moving from one cell to another. While this often occurs via passage out into the surrounding fluids, there is sometimes a more direct route through the **gap junctions** (Chapter 1, Section E,3,a). These junctions contain specialized channels that permit exchange of many small molecules between the cytoplasms of adjacent cells, but which apparently do not allow for exchange of proteins.[164,165,166a] Synapses represent another form of highly directed communications pathway.

4. Cell–Cell Recognition

Another process of cell migration is involved in the unscrambling of mixed cells[166] as described in Chapter 1, Section E,3,c. Like cells often associate and this tendency is essential to the formation of tissues. By what mechanism do cells recognize their counterparts in a mixture of different cells? It has been possible to isolate **tissue-specific adhesive molecules** which cause a particular type of embryonic cells to associate.[167–170] One isolated factor causes embryonic cerebral cells to associate while another factor causes retinal cells to aggregate. The adhesive molecules or aggregation factors appear to be glycoproteins, and the differences in the oligosaccharides borne by those molecules appear to provide the specificity.[169] Specific inhibiting molecules that block aggregation have also been solubilized from plasma membrane surfaces.[171]

An interesting theory postulates an association of an oligosaccharide chain of a glycoprotein attached to one cell with a specific glycosyltransferase of another cell.[172,173] The specific interaction could hold cells together, but addition of another glycosyl unit to the oligosaccharide by the transferase would alter the surface properties of the cell carrying the glycoprotein (Fig. 16-15). This, in turn, could cause disaggregation of the cells. Glycosyltransferases as well as acceptor molecules can be found on the outer surfaces of cell membranes.[174,175] The enzymes, sometimes referred to as **ectoenzymes,** may have been present originally

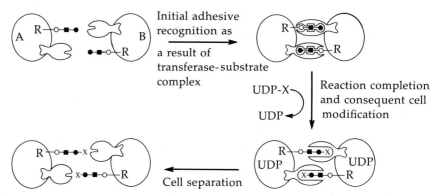

Fig. 16-15 Schematic representation of the possible role of surface glycosyltransferases and their substrates in intercellular adhesive recognition. Cells A and B have both substrates (R-containing groups) and enzymes with active sites exposed. The theory allows for surface modification when the reaction is completed; this could provide a possible explanation for some contact phenomena such as contact inhibition and induction. From Roth et al.,[168] p. 536.

within the cell as components of the glycosylating systems of the endoplasmic reticulum and were then transported out into the plasma membrane.[175] There is intense interest in these systems because of the possibilities that alteration of their properties is a factor in loss of "contact inhibition" in cancer (Chapter 1, Section E,3,c).

Edelman[175a] suggested that cell-cell recognition may be related to the problem of recognition by the immune system (Section 7 and Chapter 5, Section C,4). A cascade of proteolytic cleavage reactions, similar to that in the complement system (Box 5-G) as well as the cytoskeletal network (Chapter 5, Section C,4) may participate.

5. Developmental Programs in Multicellular Organisms

We have seen that cells are continuously receiving chemical signals from adjacent cells and fluids and that they may respond by releasing chemical substances and by altering their surface structures in various ways. However, we might question whether the 200 cell types of the vertebrate body could all arise as a result of such interactions. The fact that even bacterial cells may switch to alternative developmental programs suggests that it may be possible. For lower animals

at some time in the development of an ovum DNA synthesis is switched off and the cell accumulates vast amounts of RNA that is utilized in embryonic development. During the early stages of embryo growth the development seems to be directed largely by the polarity and gradients of all constituents set up in the ovum. That is, the nuclei are responding to external stimuli to provide the initial polarity in the embryo.

In the earliest stages differentiation is readily reversible. However, it later becomes difficult or impossible to convert a differentiated cell into one resembling an embryonic cell. The experiments of Gurdon (Section C,2) show that nuclei of differentiated cells often (if not always) retain all of their genetic material. However, this observation is not in conflict with a large body of experimental data that suggest that early in development cells in different locations in the embryo begin to follow different internal genetic programs. Cells seem to become committed to a certain pathway of differentiation, and in some cases a kind of "developmental clock" seems to be set in motion and to completely determine future differentiation.

One fact that supports this idea is the presence of stem cells retaining some embryonic character and producing with each cell division a new stem cell plus a differentiated cell. It is difficult to explain this behavior solely as a response to chemical signals from the environment. Some observations suggest that animal cells seem to have a limited potential for doubling.[176,177] For example, normal human diploid embryonic fibroblasts have been observed to grow in culture and double their number approximately 50 ± 10 times. Regardless of cultural conditions, the cells die after this number of doublings. Cells taken from older humans underwent a smaller number of doublings before dying, as did those taken from shorter lived animals such as the mouse (14–28 doublings).[177] From these experiments we might conclude that there is an internal program by which cells are scheduled to die. However, other observations are contradictory,[178] and the concept that there is an upper limit to the number of doublings possible for differentiated cells remains an uncertain hypothesis.

A third piece of evidence favoring the existence of internal developmental programs comes from careful embryological studies. For example, in the development of the chick embryo limb bud, the very tip (the last 20 cell diameters of length) contains cells that differentiate into the various elements of the limb in a remarkably autonomous manner. If this **progress zone** from one limb bud is grafted onto the end of another limb bud, the bones and cartilaginous elements of the limb are repeated.[179]

Current developmental theories usually assume that some genetic programs exist and that development depends upon a combination of responses to hormones and inducers together with the effect of the internal genetic developmental programs.[179] At this point, we can only begin to guess the nature of the internal programs. However, clever schemes have been proposed by which developmental clocks count the number of cell divisions and at the appropriate time turn off one set of genes and turn on another.[180] Concrete suggestions about the chemistry of such developmental clocks have been made. Thus, it has been pointed out that whereas we usually regard DNA as very stable, it is readily mutated by chemical action. There may be special enzymes that deliberately modify DNA at certain sites. Indeed, we know that DNA molecules contain a certain number of extra methyl groups which may mark selected sites (Chapter 2, Section D,8). Another possibility is that amino group-containing bases may be deaminated at selected sites, e.g., within palindromic sequences.

The following reaction sequence could take place. An enzyme could deaminate the adenine in an AT pair to inosine. Upon replication and cell division, one daughter cell would receive an unaltered DNA molecule. The other would now contain in place of the AT base pair an IC pair. Following a second replication, a GC pair would be formed. Thus, an AT to GC mutation would have occurred at a specific site in the DNA of some of the daughter cells. Such a simple change, occurring in response to a special enzyme formed at a certain stage of development, could alter the expression of genes in some cells. It is also possible that another enzyme might revert the modified base pair to its original form. Thus, deamination of the cytosine followed by replication would yield an AU pair, which, after a second replication, would be reconverted to the original AT. If the specific palindromic sites involved were suitably designed and reiterated, the site of action of a modification enzyme could progressively move in two directions down the chromosomes. This could lead to the turning off of specific genes after a given number of cell replications (see Holliday and Pugh[180] for details).

Figure 16-16 shows the manner in which a hypothetical enzyme E1 could modify a site by methylating a base in one of the two strands of a palindromic sequence. The enzyme, which would have a rather unusual specificity, would also methylate a second site in the complementary strand but outside the palindromic region. Replication would form one unaltered DNA molecule but a second that was a substrate for enzyme E2. The latter would methylate the other half of the palindrome and all progeny DNA. The presence of both E1 and E2 in cells would lead to the continuous differentiation of the modified cells from unmodified ones, the situation found in stem line

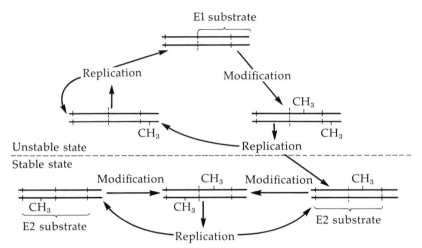

Fig. 16-16 The segregation of methylated DNA from an un-methylated precursor. The first modification enzyme E1 methylates one-half of a palindromic sequence and an adjacent sequence in the complementary strand. Replication provides one substrate for the second enzyme E2, which methylates the other half of the palindrome and all subsequent progeny molecules. In the presence of E1 and E2, unmodified or partially modified cells continually give rise to stable modified ones. If E1 is inactivated or disappears, stable modified and unmodified cells are formed. From R. Holliday and J. E. Pugh, *Science* **187**, 227 (1975). Copyright 1975 by the American Association for the Advancement of Science.

cells undergoing differentiation. A somewhat different scheme (again making use of an enzyme of rather unusual specificity) would provide a mechanism for counting cell divisions similar to that postulated for AT—GC transitions.[180]

What of the totipotency of nuclei of differentiated cells? There is abundant evidence that the cytoplasm of the ovum contains factors that turn off transcription of specialized gene products. There appears to be some kind of *"resetting"* of the developmental clocks that led to differentiation. As long as no DNA is actually lost from the genome, it is possible that enzymatic processes convert the modified DNA molecules back to unmodified ones. In the case of the methylated DNA molecules, it is significant that if E1 and E2 (Fig. 16-16) were absent in the cytoplasm of the ovum, no further methylation would occur during the cleavage divisions. At the stage of gastrulation, when control by developmental clocks might begin, most of the cells would not contain methylated bases in their DNA.

6. Alterations in Amount of DNA

There are many cases known in which differentiation leads to a temporary or permanent change in the genome. Thus, "amplification" of rRNA genes (Chapter 15, Section I,1,d) of oocytes leads to a temporary increase in the total DNA content of the cell. Some highly specialized cells, e.g., the Purkinje cells of the cerebellum and many cells of Diptera larvae (Chapter 15, Section D,9,c), are polyploid. Such cells generally represent a terminal stage of differentiation and do not divide. Polyploid cells tend to contain their full complement of genes (of which most are not expressed) in each copy of their DNA.

On the other hand, some cells undergo irreversible differentiation in which part of the genome is lost. The extreme case is that of the human red blood cell from which the entire nucleus is expelled. In other cases individual chromosomes are destroyed by a cell. In still other cases a chromosome or part of a chromosome is permanently inactivated and may remain in the cell as compactly folded **heterochromatin.** This term refers to densely staining regions of the nucleus. Some heterochromatin contains highly reiterated sequences (Chapter 15, Section I,1,b), but groups of inactivated genes may also be found in heterochromatic regions. A very interesting case is the total inactivation of one of the two X chromosomes in cells of female mammals.[181] The entire chromosome appears as heterochromatin. The inactivation occurs early in embryo development and is random with respect to the two X chromosomes. In some cells the maternal chromosome, in others the paternal chromosome, becomes inactive. However, upon further cell divisions the same chromosome in each clone remains inactive. As a consequence the female body is a mosaic, with respect to genes in the X chromosome for which the individual is heterozygous.

The mechanism by which inactivation or selective destruction of chromosomes occurs is unknown, but it has been suggested that the basic chemistry is related to that of bacterial modification and restriction (Chapter 15, Section F).[182] A methylating system could mark one chromosome and leave the other for attack by restriction endonucleases elaborated subsequently. Alternatively, some other enzyme could initiate conversion to the heterochromatic state.

While some genes are selectively inactivated or are turned off and on sequentially, others may be irreversibly lost from some differentiated cells. Genetic recombination events appear to occur in some cells and chromosomes during mitosis. Crossing over between sister chromatids has been demonstrated. However, this does not alter the genetics of the progeny cells if equal amounts of genetic material are exchanged. On the other hand, if two or more similar base sequences occur in tandem in a DNA molecule, it is possible for unequal crossing over (Chapter 15, Section G,3) to occur with the loss of genetic material from a chromosome of one of the progeny cells. In fact, this could be a deliberately programmed route of dif-

ferentiation for some cells.

Alternatively, loss of genes from a chromosome could occur through a **looping out excision** mechanism.[183] Like the excision of the prophage λ (Chapter 15, Section D,8) from the chromosome of *E. coli,* this loss of genes would presumably occur at specific sites in the DNA. One process in which a permanent loss of genetic material has been postulated is differentiation of multipotential stem cells that form blood cells. The multipotential cells first give rise to three other lines of stem cells, the **myeloid,** the **erythroid,** and the **lymphoid,** which differentiate further as indicated in Eq. 16-2.

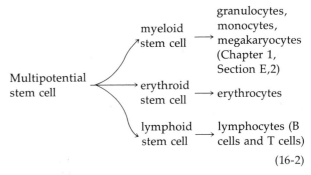

$$(16\text{-}2)$$

The erythroid stem cells are precursors to the hemoglobin-containing erythrocytes. Recall (Chapter 4, Section E,7) that mammalian hemoglobins contain two α chains and two other chains, either β, γ, δ, or ϵ. Adult hemoglobin is mainly $\alpha_2\beta_2$ but contains small amounts of $\alpha_2\delta_2$. In early embryos the hemoglobin is $\alpha_2\epsilon_2$, but later in embryonic life the two fetal hemoglobin chains $^G\gamma$ and $^A\gamma$ replace the ϵ chains. Genetic evidence indicates that the ϵ-, γ-, β-, and δ-globin genes are all closely linked.[188] A possible explanation for the exclusive occurrence of one hemoglobin type in a given erythrocyte is that a single promotor exists for this series of genes. If there is a terminator signal at the end of each gene, only the gene immediately adjacent to the promoter would be transcribed. Excision of that gene at some point in development would permit transcription of the next gene, etc., and would account for the progressive changes in gene expression in red blood cells. Another feature of differentiation of erythro-

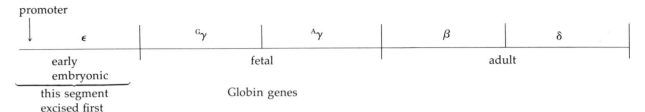

this segment excised first

Globin genes

cytes is the sensitivity to the hormone **erythropoietin,** a glycoprotein hormone formed in the kidney.[184-186] The action of this hormone on a differentiated stem cell is required to initiate massive hemoglobin synthesis and the terminal differentiation of the erythrocyte.[186a]

7. The Immune Response

The best established example of a permanent change in the organization of the DNA of specialized cells comes from studies of the differentiation of antibody-forming lymphoid cells of the spleen, lymph nodes, bone marrow, and other parts of the body. A variety of genetic evidence supports the view that the variable (V) regions of immunoglobulin chains (Box 5-F) are encoded by different genes than are the constant (C) regions of the same chains. This unusual situation probably reflects the necessity of constructing many thousands of different antibodies. Genetic studies indicate that antibody genes are inherited in classical Mendelian fashion with multiple non-

identical V-region and C-region genes clustered in what is sometimes called a **translocon.**[187]

The number of genes in the clusters is unknown. To form antibodies to as many as 2 million antigenic determinants, it has been estimated that from 100 to several thousand V_H, V_κ, and V_λ genes would be required.[188-190] However, some experimental evidence suggests that the number of "germ line" genes for immunoglobulins is much less than this. It is possible that some as yet unrecognized process "scrambles" parts of the V genes to generate a diversity of different sequences in antibody genes of lymphocytes. Be that as it may, the hypervariable regions of the immunoglobulin genes in lymphoid cells appear to be subjected to an unusually high rate of mutation (somatic mutation).[190a,b] Furthermore, final assembly of a gene for a specific immunoglobulin requires that the DNA be modified by unequal crossing over, looping out excision, or translocation of a segment of the DNA. In the antibody-forming cell only one kind of immunoglobulin is formed, an indication that only one of the homologous chromosomes is actively transcribed.

$V_{\kappa I}$ $V_{\kappa II}$ $V_{\kappa III}$ . . . C_κ light chains

$V_{\lambda I}$ $V_{\lambda II}$ $V_{\lambda III}$ $V_{\lambda IV}$ $V_{\lambda V}$ . . . $C_{\lambda 1}$ $C_{\lambda 2}$ $C_{\lambda 3}$

V_{HI} V_{HII} V_{HIII} . . . $C_{\mu 1}$ $C_{\mu 2}$ $C_{\gamma 2}$ $C_{\gamma 3}$ $C_{\gamma 1}$ $C_{\gamma 4}$ $C_{\alpha 1}$ $C_{\alpha 2}$ C_δ C_ϵ heavy chains

Arrangement of human antibody genes in clusters[187]

Recently, an mRNA molecule coding for an immunoglobulin light chain has been studied and shown to carry information for both the variable and constant regions of the chain.[191] Genetic evidence indicates that the joining of V and C regions probably occurs in the DNA prior to transcription. The pathway of differentiation of antibody-producing cells is complex and may be intimately tied to the complexity of the immune response itself.[192,193] The B cells and the T cells (Chapter 5, Section C,4), both of which are sometimes referred to as **small lymphocytes,** are derived from common stem cells. The B cells develop in birds in a special organ, the bursa of Fabricius, and in other locations. In mammals they probably arise principally in the bone marrow. The T cells develop in the thymus where they come under the influence of the hormone **thymosin**[194,195] which alters the course of development in an unknown way.

The immune response is initiated by binding of an antigen to the B lymphocyte receptors and presumably also to receptors on the T cells, the nature of which is still not certain. The binding of an antigen may trigger the B cell to divide and, after a number of cell divisions, to develop into **plasma cells,** the actively secreting cells present in a variety of lymphatic tissues. In addition, the B cells give rise to **memory cells,** long-lived lymphocytes that can be triggered into rapid proliferation many years later if the same antigen is encountered.

The complexity in the immune response arises in part from the fact that other cells, notably the T cells and macrophages alter the response of B cells to an antigen. Without triggering by an antigen, most lymphocytes are inhibited from further cell division. T cells, of which there are at least three kinds, may stimulate cell division upon antigen binding or they may further suppress division. Suppression probably occurs when the immune system recognizes that the antigen contains an antigenic determinant also present on the cell surfaces of the organism. It is clearly important for the immune system to distinguish between "self" and "foreign" antigens. Just as the nervous system is largely inhibited at all times and only oc-

casionally transmits a message, so it appears that the immune system is largely inhibited and only occasionally permits a clone of plasma cells to develop. Part of the inhibition may consist of synthesis of antibodies against other antibodies, namely, those serving as receptors on B cell surfaces.

A puzzling and important feature of the immune system is the occurrence of autoimmune diseases in which the body makes antibodies against its own cells. **Rheumatoid arthritis** appears to be such a disease in which the serum and joint fluids contain abnormal complexes of IgG with unknown antigens. In the severe autoimmune disease **systemic lupus erythematosus** the immune system often forms antibodies against the victim's own DNA. The antibodies then attack any tissue, e.g., the red blood cells. Although the cells of the immune system are ordinarily kept apart from those of the nervous system by means of the blood-brain barrier, allergic encephalomyelitis in which antibodies attack the myelin sheath can easily be induced in mice (p. 264). Another form of immune complex disease is **amyloidosis,** a deposition of an extracellular protein-carbohydrate complex.[196] An important observation is that the amounts of autoantibodies and of amyloid deposits increase with age. Immune complex disease is suspected of being a major cause of aging. Another finding of great medical significance is that primary **glomerulonephritis,** a major kidney disease, may be caused by a cross-reaction between the membrane of streptococci and the glomerular basement membranes of the kidney.

A major function of the immune system is thought to be destruction of cancer cells. Again, recognition of foreign antigens, in this case altered cell surface carbohydrates or proteins (Chapter 5, Section C), elicits an antibody response with destruction of the offending cells. That this process works imperfectly may explain why the incidence of cancer increases with age and also why the concentration of autoantibodies increases. The further study of these complex matters will doubtless be important both to medicine and to the unraveling of mechanisms of cell–cell recognition and differentiation.

D. ECOLOGICAL MATTERS
(Author's Personal Postscript)

The final section of this chapter on communication should deal with interactions among different species. As humans, beset by problems arising from our inability to communicate with other humans, we may feel that the ecological relationships are relatively unimportant. However, any careful look at what can be regarded as an extension of metabolic cycles into the biosphere should convince the student of the importance of this aspect of biochemistry. Recall that the original development of eukaryotic creatures may have started with a symbiotic relationship between two prokaryotes. Symbiosis between algae and non-photosynthetic organisms may have led to development of higher plants.

Associations between species are still important today. For example, the bacteria in the protozoa of the digestive tract of ruminant animals are essential to production of meat. Our own bodies play host to various species of bacteria, fungi, and other organisms with whom we have to try to maintain friendly relations. We depend upon antibiotics produced by bacteria or by fungi to fight our bacterial infections. More important is our dependence on plants to provide both essential nutrients and oxygen. Our environment has been created in large part by other living forms that coexist with ecological checks and balances. It is clear that a great growth of knowledge will occur in the area of chemical ecology. This will include not only the effects of one group of organisms on another but also the effects of human activities on plants and animals of all degrees of complexity. The consequences of environmental pollution, of depletion of atmospheric ozone or other alterations that affect the radiant energy reaching us, and the possible effects of the availability to humans of excessive amounts of energy must all be considered. Just as a steady state within cells is often essential to the life of organisms, maintenance of a steady state in the chemical cycles of the biosphere may be a necessity.

It seems certain that in the future biochemists will be called upon to play an increasing role in medicine, agriculture, and industry. As such, they must be prepared to help in the making of important decisions that may affect the future of life on earth. Biochemical approaches will undoubtedly be required to cope with many important problems. The mutation problem (Chapter 15, Section H,3) and the consequences of pollution with a growing number of synthetic chemicals need careful attention. There are ethical as well as scientific questions. For example, it may be impossible for humans to avoid genetic deterioration without either effective selection, which does not exist at the present time, or "genetic surgery." As we develop the ability to alter genetic traits, there is no reason in principle why human beings cannot evolve into healthier, more intelligent, and more thoughtful people than they are at present. However, we must realize that our knowledge is very incomplete and that our pathway is filled with pitfalls.

Despite attempts to ignore it, we cannot avoid facing the war problem. The possibility of virtually total destruction of the more complex forms of life by genetic damage from radiation is real. That we have lived with nuclear weapons as long as we have is encouraging but continuing threats to use them as a last resort may possibly bring eventual catastrophe. Perhaps biochemists, who understand the technical problems of mutation and other radiation damage, have a special obligation to point out the hazard to others.

Just as threatening as radiation are the possibilities of biochemical warfare, e.g., the use of artificial viruses. Biological weapons have been little used because of their lack of discrimination between friend and foe. However, our increasing knowledge of molecular biology could make possible insidious attacks on a population of unvaccinated persons. Since biochemical work does not require elaborate facilities, the development of biological weapons could be carried on by a small group in a clandestine manner.

Why worry? Since biochemistry is unable to ascribe any purpose to life, shouldn't we scientists stick to science? Science is amoral, isn't it? And besides, won't society do just what it wants to regardless of our opinions? Questions like these will always be with us, but most of the best scientists in the world seem to act with a great deal of responsibility. Not only do they want the pleasure and excitement of discovery and recognition for their work, but also they want a world for their children and grandchildren. They tend to feel compassion for other human beings. A very large number of them will give as a principal motivation for becoming biochemists the desire to contribute to the understanding of living things for the purpose of improving health, medical care, nutrition, etc. Most of them would not like to see the evolution of human beings ended through a disaster with nuclear weapons or by irreversible pollution of land and sea.

Returning to the problem of biological weapons, at a conference in Berkeley in 1971, Joshua Lederberg, discoverer of genetic recombination in bacteria, talked about these matters. Lederberg asked if fairness and objectivity are possible outside the laboratory. He thought so. He pointed out that the nations of the world have agreed to stop production of biological weapons and that genuine steps have been taken to decrease some of

the hazards facing us. Nevertheless, there is always some resistance. Some will insist on inspection for violation of agreements. But how could one inspect thoroughly enough? Lederberg suggested that the only possible form of control is now evolving. It must come from scientists themselves who must step out of their roles as "pure" scientists and accept the responsibility of preventing foolish uses of new biological discoveries. It may seem impossible that there could be a scientific community which could be counted on always to act in a responsible way, but it may be the only way that human beings can survive for long on this planet. Lederberg believes it possible (and so do I).

If this book has helped to bring to the reader some awareness of the knowledge and power of molecular biology, I hope that these final words may lead the reader to heed the advice of Professor Lederberg. I sincerely hope that all the young people now studying biochemistry and modern biology will commit themselves to using the fantastic new knowledge available to us for the betterment of mankind and proceeding with caution and responsibility as they move into positions of influence in the scientific community.

REFERENCES

1. E. Frieden and H. Lipner, "Biochemical Endocrinology of the Vertebrates." Prentice-Hall, Englewood Cliffs, New Jersey, 1971.
2. S. S. Simons, Jr., H. M. Martinez, R. L. Garcea, J. D. Baxter, and G. M. Tomkins, *JBC* **251,** 334–343 (1976).
3. B. W. O'Malley and W. T. Schrader, *Sci. Am.* **234,** 32–43 (Feb 1976).
3a. R. J. Schwartz, R. W. Kuhn, R. E. Buller, W. T. Schrader, and B. W. O'Malley, *JBC* **251,** 5166–5177 (1976).
3b. B. S. McEwen, *Sci. Am.* **235,** 48–58 (Jul 1976).
4. G. M. Tomkins, *Science* **189,** 760–763 (1975).
5. H. S. Tager and D. F. Steiner, *Annu. Rev. Biochem.* **43,** 509–538 (1974).
6. R. Guillemin and R. Burgus, *Sci. Am.* **227,** 24–33 (Nov 1972).
7. A. V. Schally, A. Arimura, and A. J. Kastin, *Science* **179,** 341–350 (1973).
8. T. H. Maugh, II, *Science* **188,** 920–923 (1975).
9. A. P. Scott, J. G. Ratcliffe, L. H. Rees, J. Landon, H. P. J. Bennett, P. J. Lowry, and C. McMartin, *Nature (London), New Biol.* **244,** 65–67 (1973).
10. D. F. Steiner, W. Kemmler, H. S. Tager, and J. D. Peterson, *Fed. Proc., Fed. Am. Soc. Exp. Biol.* **33,** 2105–2115 (1974).
11. D. G. Smyth, *Nature (London)* **257,** 89–90 (1975).
11a. J. E. Habener, J. T. Potts, Jr., and A. Rich, *JBC* **251,** 3893–3899 (1976).
12. M. I. Grossman, *Fed. Proc., Fed. Am. Soc. Exp. Biol.* **27,** 1312–1313 (1968).
13. M. Das and R. L. Soffer, *JBC* **250,** 6762–6768 (1975).
14. K. Sláma, M. Romaňuk, and F. Šorm, "Insect Hormones and Bioanalogues." Springer-Verlag, Berlin and New York, 1974.
15. C. M. Williams, *Sci. Am.* **198,** 67–74 (Feb 1958).
16. L. L. Sanburg, K. J. Kramer, F. J. Kezdy, J. H. Law, and H. Oberlander, *Nature (London)* **253,** 266–267 (1975).
17. J. N. Kaplanis, W. E. Robbins, M. J. Thompson, and S. R. Dutky, *Science* **180,** 307–308 (1973).
18. N. T. Spratt, Jr., "Developmental Biology." Wadsworth, Belmont, California, 1971.

19. L. J. Audus, "Plant Growth Substances," Vol. I. Leonard Hill, London, 1972.
20. J. G. Torrey, *Science* **181,** 1075–1076 (1973).
21. D. W. Krogmann, "The Biochemistry of Green Plants." Prentice-Hall, Inc., Englewood Cliffs, New Jersey, 1973.
22. M. Thomas, S. L. Ransom, and J. A. Richardson, "Plant Physiology," 5th ed. Longmans, Green, New York, 1973.
22a. T. J. V. Higgins, J. A. Zwar, J. V. Jacobsen, *Nature (London)* **260,** 166–169 (1976).
23. H. M. M. El-Antably and P. Larsen, *Nature (London)* **250,** 76–77 (1974).
24. T. H. Maugh, II, *Science* **184,** 655 (1974).
25. R. W. Albers, G. J. Siegel, R. Katzman, and B. W. Agronoff, "Basic Neurochemistry." Little, Brown, Boston, Massachusetts, 1972.
26. J. C. Eccles, "The Understanding of the Brain." McGraw-Hill, New York, 1973.
26a. C. R. Noback and R. J. Demarest, "The Nervous System; Introduction and Review." McGraw-Hill, New York, 1972.
27. A. Rosenfeld and K. W. Klivington, *Saturday Rev.* Aug. 9, pp. 13–15 (1975).
28. F. O. Schmitt, ed., "The Neurosciences: Second Study Program." Rockefeller Univ. Press, New York, 1970.
29. R. Fried, *J. Chem. Educ.* **45,** 322–335 (1968).
30. E. D. Brand and T. C. Westfall, *in* "Medicinal Chemistry" (A. Burger, ed.), 3rd ed., Part II, pp. 1190–1234. Wiley (Interscience), New York, 1970.
31. J. R. Cooper, F. E. Bloom, and R. H. Roth, "The Biochemical Basis of Neuropharmacology," 2nd ed. Oxford Univ. Press, London and New York, 1974.
32. H. McIlwain and H. S. Bachelard, "Biochemistry and the Central Nervous System," 4th ed. Livingston, Edinburgh, 1971.
33. E. R. Kandel, *Sci. Am.* **223,** 57–70 (July 1970).
34. J. G. Nicholls and D. Van Essen, *Sci. Am.* **230,** 38–48 (Jan 1974).
35. J. Eccles, *Sci. Am.* **212,** 56–66 (Jan 1965).
35a. J. E. Dowling, *Science* **147,** 57–59 (1965).
35b. C. R. Michael, *Sci. Am.,* **220** 105–114 (May 1969).
36. V. J. Wilson, *Sci. Am.* **214,** 102–110 (May 1966).
36a. F. O. Schmitt, P. Dev, and B. H. Smith, *Science* **193,** 114–120 (1976).
37. R. R. Llinas, *Sci. Am.* **232,** 56–71 (Jan 1975).
38. D. Nachmansohn and E. Neumann, "Chemical and Molecular Basis of Nerve Activity." Academic Press, New York, 1975.
39. D. Nachmansohn, *PNAS* **73,** 82–85 (1976).
40. Z. W. Hall, *Annu. Rev. Biochem.* **41,** 925–952 (1972).
41. K. Krnjevic, *Methods Neurochem.* **1,** 129–172 (1971).
42. B. Katz, *Science* **173,** 123–126 (1971).
43. R. P. Rubin, "Calcium and the Secretory Process." Plenum, New York, 1974.
44. R. Llinás and C. Nicholson, *PNAS* **72,** 187–190 (1975).
44a. B. M. F. Pearse, *PNAS* **73,** 1255–1259 (1976).

45. S. J. Edelstein, W. B. Beyer, A. T. Eldefrawi, and M. E. Eldefrawi, *JBC* **250,** 6101–6106 (1975).
46. J. E. Gearien, *in* "Medicinal Chemistry" (A. Burger, ed.), 3rd ed., Part II, pp. 1296–1313. Wiley (Interscience), New York, 1970.
47. D. M. Fambrough and H. C. Hartzell, *Science* **176,** 189–191 (1972).
48. D. P. Botes, *JBC* **246,** 7383–7391 (1971).
49. A. T. Tu, *Annu. Rev. Biochem.* **42,** 235–258 (1973).
49a. B. W. Low, H. S. Preston, A. Sato, L. S. Rosen, J. E. Searl, A. D. Rudko, and J. S. Richardson, *PNAS* **73,** 2991–2994 (1976).
49b. C. Chothia, R. W. Baker, and P. Pauling, *JMB* **105,** 517–526 (1976).
50. V. Lennon, *Nature (London)* **258,** 11–12 (1975).
51. S. Satyamurti, D. B. Drachman, and F. Slone, *Science* **187,** 955–957 (1975).
52. J. Halpert and D. Eaker, *JBC* **250,** 6990–6997 (1975).
53. P. N. Strong, J. Goerke, S. G. Oberg, and R. B. Kelly, *PNAS* **73,** 178–182 (1976).
54. F. A. Fuhrman, *Sci. Am.* **217,** 60–71 (Aug 1967).
55. E. J. Schantz, V. E. Ghazarossian, H. K. Schnoes, F. M. Strong, J. P. Springer, J. O. Pezzanite, and J. Clardy, *JACS* **97,** 1238–1239 (1975).
56. I. L. Karle, *PNAS* **69,** 2932–2936 (1972).
57. J. P. Robinson, J. B. Picklesimer, and D. Puett, *JBC* **250,** 7435–7442 (1975).
58. L. L. Iversen, *Nature (London)* **252,** 630 (1974).
59. U. S. von Euler, *Science* **173,** 202–206 (1971).
60. D. J. Triggle, *in* "Medicinal Chemistry" (A. Burger, ed.), 3rd ed., Part II, pp. 1235–1295. Wiley (Interscience), New York, 1970.
61. J. M. Saavedra, M. J. Brownstein, D. O. Carpenter, and J. Axelrod, *Science* **185,** 364–365 (1974).
62. L. L. Iversen, *Nature (London)* **250,** 700–701 (1974).
63. S. B. Kater and C. Nicholson, "Intracellular Staining in Neurobiology." Springer-Verlag. Berlin and New York, 1973.
64. M. S. Berry and G. A. Cottrell, *Nature (London), New Biol.* **242,** 250–253 (1973).
64a. D. Schubert, H. Tarikas, M. La Corbiere, *Science* **192,** 471–472 (1976).
65. J. B. Kebabian, F. E. Bloom, A. L. Steiner, and P. Greengard, *Science* **190,** 157–159 (1975).
66. L. L. Iversen, *Science* **188,** 1084–1089 (1975).
67. T. W. Stone, D. A. Taylor, and F. E. Bloom, *Science* **187,** 845–846 (1975).
67a. E. Costa, A. Kurosawa, and A. Guidotti, *PNAS* **73,** 1058–1062 (1976).
68. J. D. Fernstrom and R. J. Wurtman, *Sci. Am.* **230,** 84–91 (Feb 1974).
69. V. W. Pentreath and G. A. Cottrell, *Nature (London)* **250,** 655–658 (1974).
69a. G. B. Kolata, *Science* **192,** 41–42 (1976).
70. P. A. Berger, J. D. Barchas, and J. Vernikos-Danellis, *Na-*

ture (London) **248,** 424–426 (1974).

71. J. L. Johnson, *Brain Res.* **37,** 1–19 (1972).
72. P. A. Srere, *Adv. Enzyme Regul.* **9,** 221–233 (1971).
73. A. M. Benjamin and J. H. Quastel, *BJ* **128,** 631–646 (1972).
74. F. A. Henn, M. N. Goldstein, and A. Hamberger, *Nature (London)* **249,** 663–664 (1974).
75. R. D. Gilardi, *Nature (London) New Biol.* **245,** 86–88 (1973).
76. T. J. Crow, *Nature (London)* **252,** 634–635 (1974).
77. R. H. Osborne and H. F. Bradford, *Nature (London), New Biol.* **244,** 157–158 (1973).
78. K. Dismukes, *Nature (London)* **252,** 442–443 (1974).
79. A. B. Young, S. R. Zukin, and S. H. Snyder, *PNAS* **71,** 2246–2250 (1974).
80. M. M. Chang, S. E. Leeman, and H. D. Niall, *Nature (London), New Biol.* **232,** 86–87 (1971).
81. T. Hökfelt, J. O. Kellerth, G. Nilsson, and B. Pernow, *Science* **190,** 889–890 (1975).
82. J. L. Marx, *Science* **190,** 367–370 and 544–545 (1975).
83. S. H. Snyder, S. P. Banerjee, H. I. Yamamura, and D. Greenberg, *Science* **184,** 1243–1253 (1974).
84. G. E. Crane, *Science* **181,** 124–128 (1973).
85. H. Weil-Malherbe, "The Biochemistry of Functional and Experimental Psychoses." Thomas, Springfield, Illinois, 1971.
86. J. Mendels, ed., "Biological Psychiatry." Wiley, New York, 1973.
87. H. E. Himwich, ed., "Biochemistry, Schizophrenias and Affective Illnesses." Williams & Wilkins, Baltimore, Maryland, 1970.
88. C. L. Zirkle and C. Kaiser, *in* "Medicinal Chemistry" (A. Burger, ed.), 3rd ed., Part II, pp. 1410–1469. Wiley (Interscience), New York, 1970.
89. P. Seeman and T. Lee, *Science* **188,** 1217–1219 (1975).
90. T. J. Crow and C. Gillbe, *Nature (London), New Biol.* **245,** 27–28 (1973).
91. M. Trabucchi, D. Cheney, G. Racagni, and E. Costa, *Nature (London)* **249,** 664–666 (1974).
92. L. Stein and C. D. Wise, *Science* **171,** 1032–1036 (1971).
93. A. J. Friedhoff, *in* "Biological Psychiatry" (J. Mendels, ed.), pp. 113–129. Wiley, New York, 1973.
94. R. J. Wyatt, E. Erdelyi, J. R. Do Amaral, G. R. Elliott, J. Renson, and J. D. Barchas, *Science* **187,** 853–855 (1975).
95. J. M. Saavedra and J. Axelrod, *Science* **175,** 1365–1366 (1972).
96. H. S. Akiskal and W. T. McKinney, Jr., *Science* **182,** 20–29 (1973).
96a. A. L. Maycock, R. H. Abeles, J. I. Salach, and T. P. Singer, *Biochemistry* **15,** 114–125 (1976).
97. M. L. Post, O. Kennard, and A. S. Horn, *Nature (London)* **252,** 493–495 (1974).
98. D. S. Segal, M. Callaghan, and A. J. Mandell, *Nature (London)* **254,** 58–59 (1975).
99. C. C. Pfeiffer, ed., "Neurobiology of the Trace Metals Zinc and Copper." Academic Press, New York, 1972.
100. V. P. Dole, *Annu. Rev. Biochem.* **39,** 821–840 (1970).

101. S. H. Snyder, *Nature (London)* **257,** 185–189 (1975).
102. H. B. Bürgi, J. D. Dunitz, and E. Shefter, *Nature (London), New Biol.* **244,** 186–188 (1973).
103. J. Hughes, T. W. Smith, H. W. Kosterlitz, L. A. Fothergill, B. A. Morgan, and H. R. Morris, *Nature (London)* **258,** 577–579 (1975).
103a. J. Belluzzi, N. Grant, V. Garsky, D. Sarantakis, C. D. Wise, and L. Stein, *Nature (London)* **260,** 625–626 (1976).
103b. A. S. Horn and J. R. Rodgers, *Nature (London)* **260,** 795–797 (1976).
103c. C. Hao Li and D. Chung, *PNAS* **73,** 1145–1148 (1976).
103d. A. F. Bradbury, D. G. Smyth, C. R. Snell, N. J. M. Birdsall, and E. C. Hulme, *Nature (London)* **260,** 793–795 (1976).
104. S. K. Sharma, W. A. Klee, and M. Nirenberg, *PNAS* **72,** 3092–3096 (1975).
105. J. H. Mendelson, *in* "Biological Psychiatry" (J. Mendels, ed.), pp. 443–468. Wiley, New York, 1973.
106. H. Kalant, *Fed. Proc., Fed. Am. Soc. Exp. Biol.* **34,** 1930–1941 (1975).
107. A. Goldstein and B. A. Judson, *Science* **172,** 290–292 (1971).
108. R. A. Deitrich and V. G. Erwin, *Fed. Proc., Fed. Am. Soc. Exp. Biol.* **34,** 1962–1968 (1975).
109. E. Rubin and C. S. Lieber, *Science* **172,** 1097–1102 (1971).
110. K. von Hungen, S. Roberts, and D. F. Hill, *Nature (London)* **252,** 588–589 (1974).
111. W. Keup, ed., "Origin and Mechanism of Hallucinations." Plenum, New York, 1970.
112. J. Del Castillo, M. Anderson, and G. M. Rubottom, *Nature (London)* **253,** 365–366 (1975).
113. A. R. Patel, *in* "Medicinal Chemistry" (A. Burgher, ed.), 3rd ed., Part II, pp. 1314–1326. Wiley (Interscience), New York, 1973.
114. T. D. Paolo and C. Sandorfy, *Nature (London)* **252,** 471–472 (1974).
114a. A. G. Lee, *Nature (London)* **262,** 545–548 (1976).
115. J. L. Spudich and D. E. Koshland, Jr., *PNAS* **72,** 710–713 (1975).
115a. J. Adler, *Sci. Am.* **234,** 40–47 (Apr 1976).
116. N. Tsang, R. Macnab, and D. E. Koshland, Jr., *Science* **181,** 60–63 (1973).
117. H. C. Berg and D. A. Brown, *Nature (London)* **239,** 500–504 (1972).
118. K. J. Lindstedt, *Science* **173,** 333–334 (1971).
119. C. Pfaffmann, ed., "Olfaction and Taste." Rockefeller Univ. Press, New York, 1969.
120. S. S. Schiffman, *Science* **185,** 112–117 (1974).
121. R. H. Cagan, *Science* **181,** 32–35 (1973).
122. L. M. Bartoshuk, C.-H. Lee, and R. Scarpellino, *Science* **178,** 988–990 (1972).
123. D. M. Dubois and E. Schoffeniels, *PNAS* **71,** 2858–2862 (1974); **72,** 1749–1752 (1975).
124. L. G. Abood, *in* "Basic Neurochemistry" (R. W. Albers *et al.* eds.), pp. 41–65. Little, Brown, Boston, 1972.

125. S. Ochs, *Science* **176**, 252–260 (1972).

126. S. Ochs and R. Worth, *Science* **187**, 1087–1089 (1975).

127. J. H. LaVail and M. M. LaVail, *Science* **176**, 1416–1417 (1972).

128. L. Grouse, G. S. Omenn, and B. J. McCarthy, *J. Neurochem.* **20**, 1063–1073 (1973).

129. R. Caplan, S. C.-Y. Cheung, and G. S. Omenn, *J. Neurochem.* **22**, 517–520 (1974).

130. C. F. Chen, R. von Baumgarten, and R. Takeda, *Nature (London), New Biol.* **233**, 27–29 (1971).

131. H. P. Alpern and J. C. Crabbe, *Science* **177**, 722–724 (1972).

132. B. W. Agranoff, *Sci. Am.* **216**, 115–122 (Jun 1967).

133. G. Ungar, *Naturwissenschaften* **59**, 85–91 (1972).

134. P. F. C. Gilbert, *Brain Res.* **70**, 1–18 (1974).

135. M. Sussman, "Developmental Biology." Prentice-Hall, Englewood Cliffs, New Jersey, 1973.

136. C. A. Pasternak, "Biochemistry of Differentiation." Wiley (Interscience), New York, 1970.

137. J. Brachet, "Introduction to Molecular Embryology." Springer-Verlag, Berlin and New York, 1974.

138. O. A. Schjeide and J. de Vellis, eds., "Cell Differentiation." Van Nostrand-Reinhold, Princeton, New Jersey, 1970.

139. W. J. Rutter, R. L. Pictet, and P. W. Morris, *Annu. Rev. Biochem.* **42**, 601–646 (1973).

140. M. Sussman, "Developmental Biology," p. 149. Prentice-Hall, Englewood Cliffs, New Jersey, 1973.

141. R. Losdick, *in* "Control of Transcription" (B. B. Biswas *et al.*, eds.), pp. 15–19. Plenum, New York, 1974.

142. R. H. Doi and T. J. Leighton, *Spores* **5**, 225–232 (1972).

143. D. L. Nelson, J. A. Spudich, P. P. M. Bonsen, L. L. Bertsch, and A. Kornberg, *Spores* **4**, 59–71 (1969).

144. G. W. Gould and G. J. Dring, *Nature (London)* **258**, 402–405 (1975).

144a. C. Klein and M. Darmon, *PNAS* **73**, 1250–1254 (1976).

145. J. W. Wireman and M. Dworkin, *Science* **189**, 516–523 (1975).

146. J. T. Bonner, *Ann. Rev. Microbiol.* **25**, 75–92 (1971).

147. D. H. O'Day and K. E. Lewis, *Nature (London)* **254**, 431–432 (1975).

148. J. B. Gurdon, *Sci. Am.* **219**, 24–35 (Dec 1968).

149. J. B. Gurdon, "The Control of Gene Expression in Animal Development." Harvard Univ. Press, Cambridge, Massachusetts, 1974.

150. N. K. Wessells and W. J. Rutter, *Sci. Am.* **220**, 36–44 (March 1969).

151. W. A. Frazier, R. H. Angeletti, and R. A. Bradshaw, *Science* **176**, 482–488 (1972).

152. J. L. Marx, *Science* **185**, 930–932 (1974).

153. R. Levi-Montalcini, *Science* **187**, 113 (1975).

154. R. Lim and K. Mitsunobu, *Science* **185**, 63–66 (1974).

155. M. Jacobson and R. K. Hunt, *Sci. Am.* **228**, 26–35 (Feb 1973).

156. C. R. Savage, Jr., T. Inagami, and S. Cohen, *JBC* **247**, 7612–7621 (1972).

157. S. Cohen and G. Carpenter, *PNAS* **72**, 1317–1321 (1975).

158. H. Gregory, *Nature (London)* **257**, 325–327 (1975).

159. A. Gierer, *Sci. Am.* **231**, 44–54 (Dec 1974).

160. R. L. Herlands and H. R. Bode, *Nature (London)* **248**, 387–390 (1974).

161. W. S. Bullough, *in* "Humoral Control of Growth and Differentiation" (J. LoBue and A. S. Gordon, eds.), Vol. 1, pp. 3–21. Academic Press, New York, 1973.

162. R. W. Holley, *Nature (London)* **258**, 487–490 (1975).

163. R. G. Mitchell and J. F. Porter, *in* "The Biochemistry of Development" (P. Benson, ed.), pp. 204–223. Lippincott, Philadelphia, Pennsylvania, 1971.

164. J. D. Pitts, *Nature (London)* **255**, 371–372 (1975).

165. R. P. Cox, "Cell Communication." Wiley, New York, 1974.

165a. E. Lawrence and M. Robertson, *Nature (London)* **261**, 99–100 (1976).

166. G. Van de Vyver, *Curr. Top. Dev. Biol.* **10**, 123–140 (1975).

167. R. E. Hausman and A. A. Moscona, *PNAS* **72**, 916–920 (1975).

168. S. Roth, *Q. Rev. Biol.* **48**, 541–563 (1973).

169. J. Balsamo and J. Lilien, *Biochemistry* **14**, 167–171 (1975).

169a. K. M. Yamada, S. S. Yamada, and I. Pastan, *PNAS* **73**, 1217–1221 (1976).

170. G. B. Kolata, *Science* **188**, 718–719 (1975).

171. R. Merrell, D. I. Gottlieb, and L. Glaser, *JBC* **250**, 5655–5659 (1975).

172. S. Roseman, *Chem. Phys. Lipids* **5**, 270–297 (1970).

173. S. Roth, E. J. McGuire, and S. Roseman, *J. Cell Biol.* **51**, 536–547 (1971).

174. R. J. McLean and H. B. Bosmann, *PNAS* **72**, 310–313 (1975).

175. C. W. Porter and R. J. Bernacki, *Nature (London)* **256**, 648–650 (1975).

175a. G. M. Edelman, *Science* **192**, 218–226 (1976).

176. L. Hayflick, *Sci. Am.* **218**, 32–37 (Mar 1968).

177. L. Hayflick. *Fed. Proc., Fed. Am. Soc. Exp. Biol.* **34**, 9–13 (1975).

178. R. R. Kohn, *Science* **188**, 203–204 (1975).

179. L. Wolpert and J. H. Lewis, *Fed. Proc., Fed. Am. Soc. Exp. Biol.* **34**, 14–20 (1975).

180. R. Holliday and J. E. Pugh, *Science* **187**, 226–232 (1975).

181. H. S. Chandra and S. W. Brown, *Nature (London)* **253**, 165–168 (1975).

182. R. Sager and R. Kitchin, *Science* **189**, 426–433 (1975).

183. D. Kabat, *Science* **175**, 134–140 (1972).

184. P. A. Marks and R. A. Rifkind, *Science* **175**, 955–961 (1972).

185. A. S. Gordon, E. D. Zanjani, A. S. Gidari, and R. A. Kuna, *in* "Humoral Control of Growth and Differentiation" (J. LoBue and A. S. Gordon, eds.), Vol. 1, pp. 25–49. Academic Press, New York, 1973.

186. C. Peschle and M. Condorelli, *Science* **190**, 910–912 (1975).

186a. P. R. Harrison, *Nature (London)* **262**, 353–356 (1976).

187. G. M. Edelman, *in* "The Biochemistry of Gene Expression in Higher Organisms" (J. K. Pollak and J. W. Lee, eds.), pp. 555–573. Reidel Publ., Dordrecht, Netherlands, 1972.

188. M. Potter, *Fed. Proc., Fed. Am. Soc. Exp. Biol.* **34**, 21–23 (1975).

189. A. R. Williamson, E. Premkumar, and M. Shoyab, *Fed. Proc., Fed. Am. Soc. Exp. Biol.* **34**, 28–32 (1975).

190. F. F. Richards, W. H. Konigsberg, R. W. Rosenstein, and J. M. Varga, *Science* **187**, 130–137 (1975).

190a. L. Hood, *Fed. Proc., Fed. Am. Soc. Exp. Biol.* **35**, 2158–2167 (1976).

190b. T. H. Rabbitts, *Trends Biochem. Sci.,* **1**, 86–88 (1976).

191. C. Milstein, G. G. Brownlee, E. M. Cartwright, J. M. Jarvis, and N. J. Proudfoot, *Nature (London)* **252**, 354–359 (1974).

192. H. N. Eisen, "Immunology." Harper, New York, 1974.

193. N. K. Jerne, *Sci. Am.* **229**, 52–60 (July 1973).

194. B. Mandi and T. Glant, *Nature (London), New Biol.* **246**, 25 (1973).

195. T. D. Luckey, ed. "Thymic Hormones." Univ. Park Press, Baltimore, Maryland, 1973.

196. N. Eriksen, L. H. Erecsson, N. Pearsall, D. Lagunoff, and E. P. Benditt, *PNAS* **73**, 964–967 (1976).

STUDY QUESTIONS

1. A corticoid hormone–receptor complex formed at low temperature and ionic strength does not bind to chromatin until it is "activated." The activation reaction is first order and yields a monomolecular product. The reaction, which proceeds more rapidly at higher temperatures, reaches an equilibrium in which ~60% of the molecules of complex are activated. The following values were observed for the process [M. Atger and E. Milgrom, *JBC* **251**, 4758–4762 (1976)].

$$\Delta G^{\ddagger} = 89 \text{ kJ mol}^{-1}$$
$$\Delta H^{\ddagger} = 131 \text{ kJ mol}^{-1}$$
$$\Delta S^{\ddagger} = 142 \text{ J/}^{\circ}\text{C}$$

To what type of reaction would you attribute the activation?

2. Many peptides secreted as hormones appear to be derived from larger peptides by proteolysis, sometimes in two or more steps. What possible advantages could this arrangement offer to an organism?

3. An insoluble phosphoprotein of rat brain acquires radioactivity from P_i more rapidly during sleep than during wakefulness. The phosphoprotein was purified and was found to be a glucose-6-phosphatase [J. M. Anchors and M. L. Karnovsky, *JBC* **250**, 6408–6416 (1976)]. Offer a possible explanation for these observations. Can you devise a chemical theory to explain sleep?

4. A single behavioral characteristic of an organism could be altered by mutation in a gene for any of the following proteins.

Adenylate cyclase   cAMP receptor
Phosphodiesterase   protein
Prostaglandin-synthesizing   A protease acting
    enzymes   on lipotropin
Monoamine oxidase

Explain the molecular basis for this common effect.

5. Certain *N*-formyl peptides serve as attractants for phagocytes [S. Aswanikumar, E. Schiffman, B. A. Corcoran, and S. M. Wahl, *PNAS* **73**, 2439–2442 (1976)]. What could be the biological significance of this fact? It has been suggested that peptidase activity of phagocytes may also be important in chemotaxis. Explain.

Appendix

The Construction of Molecular Models

Although excellent commercial molecular models are available, they are relatively expensive, a fact that tends to limit their use. However, very good models can be made, and the student can learn much by constructing them. Here are a few suggestions.

A. BALL-AND-STICK MODELS

Small balls representing atoms are connected by sticks which serve as bonds. Balls of Styrofoam, e.g., of diameter 2.54 cm, are very suitable and are inexpensive (2.54 cm balls may be obtained in quantities for little more than 1 cent apiece).* Chenille wire, available from hobby shops, is a superior material for the connecting "sticks" if the latter are short, e.g., with a scale of 2 cm = 0.1 nm. Chenille wire holds without glue and is generally superior to toothpicks or wooden dowels. However, you will need a good pair of wire cutters.

* One supplier of Styrofoam balls in carton quantities to schools ($75 minimum order) is The Frostee-Foam Co., 181 Ada Avenue, Antioch, Illinois 60002. The Edmund Scientific Co., 555 Edscorp Bldg., Barrington, New Jersey 08007 also markets a large selection of balls.

The center-to-center bond lengths can be obtained from Table 2-1 (p. 49). Bond angles (see p. 47) are most easily marked on the surface of the balls by using a cardboard template like the following (one for 120° and another for 109.5° serve most needs): The cardboard is cut to fit against the

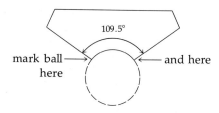

ball with the two edges pointing to the center. If three 109.5° templates are fastened at the edges in a triangular arrangement, 3 points on a tetrahedral carbon atom may be located at once. It is convenient to insert toothpicks to mark these points temporarily.

Atoms may be painted. A commonly used scheme is hydrogen, white; carbon, black; oxygen, red; nitrogen, blue; sulfur, yellow. With Styrofoam take care to obtain enamels that will not melt the balls. Spray enamels (now free of fluorocarbon propellents) are convenient. Spray a group of balls in a carton, shaking to turn frequently.

B. SPACE-FILLING MODELS

The radii of the atoms in these models approach the van der Waals radii (Table 2-1, p. 49). As a practical matter, models cannot be constructed with radii quite this large because the packing is too tight to permit easy assembly. A convenient compromise which still gives a good impression of the actual shapes of the molecules is to use 3.8-cm Styrofoam balls to represent the C, H, O and N atoms and 5.1 cm balls for P and S. If 0.5

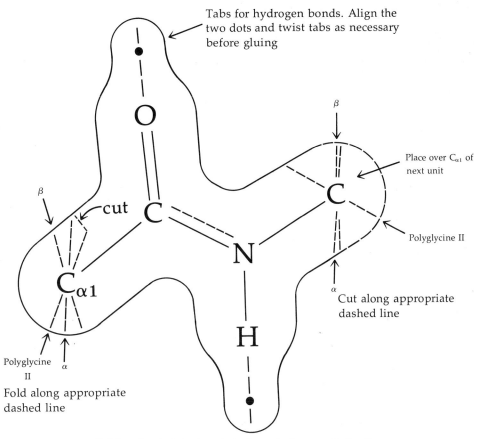

Tabs for hydrogen bonds. Align the two dots and twist tabs as necessary before gluing

Place over $C_{\alpha 1}$ of next unit

Polyglycine II

Cut along appropriate dashed line

Fold along appropriate dashed line

Fold and cut at designated places to form units for the α helix, antiparallel β pleated sheet or polyglycine II. For α helix, all units, as drawn, will face the outside of the helix. For polyglycine II and β structure, alternate units must be flipped over before gluing

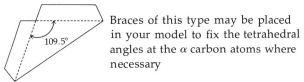

Braces of this type may be placed in your model to fix the tetrahedral angles at the α carbon atoms where necessary

Fig. A-1 Cutout guide for construction of polypeptide models from cardboard. Scale 2 cm = 0.1 nm.

cm is cut off from the C, O, and N balls in the direction of each bond, the remaining 1.4 cm will represent the approximately 0.07 nm covalent radius at a scale of 2 cm = 0.1 nm (1 Å). From the hydrogen balls, it is necessary to cut 1.3 cm in order to give the proper 0.03 nm covalent radius. The 1.9 cm radius of the balls is equivalent to 0.095 nm, about $\frac{2}{3}$ of the actual van der Waals radii for C, N, and O and 0.8 the radius of hydrogen. Another good scale is 1.25 cm = 0.1 nm. This is used in the commercially available "CPK models" developed by the American Society of Biological Chemists, Inc. Homemade models can be combined with commercial ones if this scale is used. Balls of 2.54, 3.8, and 3.18 cm diameter, respectively, can serve for H, C, and O and N.

For space-filling models the balls must be cut rather carefully to give the correct covalent radii. This can be done simply as follows: The *centers* of the faces to be cut are first marked by sticking toothpicks into the balls. If there are tetrahedral angles, these are established using a cardboard guide as explained above. The faces are cut by placing the ball in a hole in a metal sheet, the diameter of which is just right to permit the desired amount to stick out the other side of the sheet. The protruding part of the ball is then cut off with a *very sharp* butcher knife. The holes in the metal (e.g., aluminum) sheet must be cut with precision, but the template, once prepared, will last a long time.

Large, flat chemical groups, such as the base pairs in DNA, may be sculptured from Styrofoam sheets.

C. CARDBOARD MODELS

Many biological "macromolecules" such as proteins and nucleic acids involve so many atoms that construction from individual balls becomes both tedious and expensive. Cardboard (or metal or plastic) cutouts provide a solution. For example, a cutout based on the dimensions in Fig. 2-3 (p. 63) can be used to form planar peptide units (Fig. A-1). Calculating the angle to make the

fold to give proper ϕ and ψ angles (see Fig. 2-4 and Table 2-3) is a tricky geometrical problem. (It can be solved easily by actual geometrical construction using pieces of cardboard.) Very nice helical structures can be made in this way (some examples are shown by Karlson[1]). Styrofoam side chains can be glued to the α-carbons if desired. Along the same lines, a helix laid out on the surface of a cardboard cylinder can be decorated with styrofoam side chains.

Flat base pairs from cardboard (or preferably from a tougher material) can be strung on a wire representing the axis with pieces of rubber or plastic tubing serving as spacers to form DNA or RNA helices. Folded paper cutouts or ball-and-stick models can be used to construct the sugar phosphate backbones along the edges. (For decorative purposes these may be formed by using two colors of crepe paper ribbon.) The correct geometry of the base pairs is given by Arnott and Hukins.[2]

REFERENCES

1. P. Karlson, "Introduction to Modern Biochemistry," 3rd Ed., pp. 43–55. Academic Press, N. Y., 1967. (Note: a different type of model is shown in the fourth edition).
2. S. Arnott and D. W. L. Hukins, *JMB* **81**, 93–105 (1975).

Index

Page numbers set in **boldface** refer to major discussions.
The symbol *s* after a page number refers to a chemical structure.